MOLECULAR DETECTION OF HUMAN PARASITIC PATHOGENS

MOLECULAR DETECTION OF HUMAN PARASITIC PATHOGENS

EDITED BY
DONGYOU LIU

CRC Press
Taylor & Francis Group
Boca Raton London New York

CRC Press is an imprint of the
Taylor & Francis Group, an **informa** business

CRC Press
Taylor & Francis Group
6000 Broken Sound Parkway NW, Suite 300
Boca Raton, FL 33487-2742

Printed in the United States of America on acid-free paper
Version Date: 20120309

International Standard Book Number: 978-1-4398-1242-6 (Hardback)

Library of Congress Cataloging-in-Publication Data

Molecular detection of human parasitic pathogens / edited by Dongyou Liu.
p. ; cm.
Includes bibliographical references and index.
Summary: "A book on the molecular detection of major human viral, bacterial, fungal, and parasitic pathogens, this text provides coverage of the organisms with an emphasis on the sample preparation and molecular detection procedures. Chapters consist of concise reviews of pathogens in respect to taxonomy, biology, epidemiology, and pathogenesis; a description of a clinical sample collection preparation procedures; a summary of molecular detection methods available; and a discussion on the challenges, limitations, and advantages for current methods. Written for practitioners and students, this textbook is a guide for future development of improved molecular methods for pathogens. Offers an in-depth review of state-of-the-art molecular detection and identification methods for human parasitic pathogens. Provides comprehensive coverage of important human parasitic pathogens. Provides an authoritative summary of biology, epidemiology, and pathogenesis of major human parasitic pathogens. Includes chapter entries written by experts in respective fields of human parasitic pathogen collection"--Provided by publisher.
ISBN 978-1-4398-1242-6 (hardback : alk. paper)
I. Liu, Dongyou.
[DNLM: 1. Parasitic Diseases--microbiology. 2. Molecular Diagnostic Techniques--methods. 3. Parasites--genetics. 4. Parasites--isolation & purification. 5. Parasitic Diseases--diagnosis. WC 695]

616.9'6--dc23 2012009684

Visit the Taylor & Francis Web site at
http://www.taylorandfrancis.com

and the CRC Press Web site at
http://www.crcpress.com

This volume is dedicated to a group of international parasitologists whose willingness to share their in-depth knowledge and technical expertise has made an all-inclusive coverage of major human parasitic pathogens possible.

Contents

SECTION II Ciliophora and Apicomplex

SECTION III Platyhelminthes: Cestoda

SECTION IV Platyhelminthes: Trematoda

SECTION VI Arthropoda: Arachnida

SECTION VII Arthropoda: Insecta

SECTION VIII Arthropoda: Pentastomida

Preface

Parasites are a group of eukaryotic organisms that may be free living or form a symbiotic or parasitic relationship with their hosts. Consisting of over 800,000 recognized species, parasites may be unicellular (protozoa) or multicellular (helminths and arthropods). In the current six kingdom systems for biological organisms (i.e., Bacteria, Protozoa, Fungi, Chromista, Plantae, and Animalia), protozoa constitute an independent kingdom, while helminths and arthropods make up parts of the kingdom Animalia.

The association of parasites with human populations had occurred long before the emergence of civilization. The identification of hookworm ova and *Paragonimus* ova in human coprolites dated c. 5000 BC and c. 2500 BC in Brazil and Chile, respectively, reinforces the longstanding relationship between parasitic pathogens and human races. In spite of our unrelenting efforts against parasitic diseases, parasites continue to cause significant human morbidity and mortality. For example, *Plasmodium*, the malaria parasite, inflicts >250 million people worldwide and causes >1 million deaths per annum; soil-transmitted helminths such as roundworm (*Ascaris*), whipworm (*Trichuris*), and hookworm (*Ancylostoma* and *Necator*) affect >2 billion people globally, resulting in >200,000 yearly deaths. Other important human parasitic pathogens include *Schistosoma*, *filariae*, *Trypanosoma cruzi*, and *Leishmania*, which together cause miseries to hundreds of millions of people worldwide.

In view of their diverse range and variety, it is critical that parasites are correctly identified so that effective treatment and control of parasitic diseases can be implemented. Traditionally, laboratory identification of parasites has relied upon various phenotypic procedures that detect their morphological, biological, and immunological features. Because these procedures tend to be time consuming and require specialized skills, molecular methods based on nucleic acid amplification technologies have been increasingly utilized for rapid, sensitive, and specific characterization of parasites. As a consequence, a large number of original molecular protocols and subsequent modifications have been reported. This has created a dilemma to laboratory personnel who are not directly involved in the development of original or modified protocols to know which are most appropriate to adopt for accurate identification of parasites of interest.

The purpose of the current book is to address this issue, with international scientists in respective parasite research and diagnosis providing expert summaries on current diagnostic approaches for major human parasitic pathogens. Each chapter consists of a brief review on the classification, epidemiology, clinical features, and diagnosis of an important parasite genus or group; an outline of clinical sample collection and preparation procedures; a selection of representative stepwise molecular protocols; and a discussion on further research requirements relating to improved diagnosis. This book constitutes a reliable and convenient reference on molecular detection and identification of major human parasitic pathogens; an indispensable tool for upcoming and experienced medical, veterinary, and industrial laboratory scientists engaged in parasite characterization; and an essential textbook for undergraduate and graduate students majoring in parasitology.

A comprehensive book such as this is clearly beyond an individual's capacity. I am fortunate and honored to have a large panel of international parasitologists as chapter contributors. Their detailed knowledge and technical insights have greatly enriched this book. Additionally, the professionalism and dedication of executive editor Barbara Norwitz and senior project coordinator Jill Jurgensen at CRC Press have enhanced its presentation. Finally, the compilation of this all-encompassing volume would have been impossible without the understanding and support of my family, Liling Ma, Brenda, and Cathy.

Editor

Dongyou Liu, PhD, undertook his veterinary science education at Hunan Agricultural University, China, and pursued his postgraduate training in parasitology at the University of Melbourne, Australia, where he worked toward improved immunological diagnosis of human hydatid disease.

Over the last two decades, he has crisscrossed between research and clinical laboratories in Australia and the United States of America, with focuses on molecular characterization and virulence determination of microbial pathogens such as ovine footrot bacterium (*Dichelobacter nodosus*), dermatophyte fungi (*Trichophyton*, *Microsporum*, and *Epidermophyton*), and listeriae (*Listeria* species).

He is the first author of over 50 original research and review articles in various international journals and the editor of several recent biomedical books published by Taylor & Francis CRC Press: *Handbook of* Listeria monocytogenes (2008), *Handbook of Nucleic Acid Purification* (2009), *Molecular Detection of Foodborne Pathogens* (2009), *Molecular Detection of Human Viral Pathogens* (2010), *Molecular Detection of Human Bacterial Pathogens* (2011), and *Molecular Detection of Human Fungal Pathogens* (2011).

Contributors

Emily R. Adams
Koninklijk Instituut voor de Tropen Biomedical Research
Amsterdam, the Netherlands

Ross H. Andrews
Department of Parasitology and Liver Fluke and Cholagiocarcinoma Research Center
Khon Kaen University
Khon Kaen, Thailand

Supatra Areekit
Department of Biochemistry
Srinakharinwirot University
Bangkok, Thailand

Kjetil Åsbakk
Norwegian School of Veterinary Science
Section of Arctic Veterinary Medicine
Tromsø, Norway

Frank W. Austin
Department of Pathobiology and Population Medicine
Mississippi State University
Mississippi State, Mississippi

Anna Borecka
Institute of Parasitology
Polish Academy of Sciences
Warsaw, Poland

Jean-Louis Bourges
Department of Ophthalmology
Université Paris
Paris, France

Dwight D. Bowman
College of Veterinary Medicine
Cornell University
Ithaca, New York

Walter M. Boyce
Department of Pathology, Microbiology, and Immunology
University of California, Davis
Davis, California

S.V. Brant
Department of Biology
University of New Mexico
Albuquerque, New Mexico

Marina Busi
Department of Public Health Science
Sapienza University of Rome
Rome, Italy

Monica Caffara
Department of Veterinary Public Health and Animal Pathology
University of Bologna
Bologna, Italy

R. Callejón
Department of Microbiology and Parasitology
University of Seville
Seville, Spain

Serena Cavallero
Department of Public Health Science
Sapienza University of Rome
Rome, Italy

Isaac Cervantes-Sandoval
Department of Infectomic and Molecular Pathogenesis
National Polytechnic Institute
Mexico City, Mexico

Jong-Yil Chai
Department of Parasitology and Tropical Medicine
Seoul National University College of Medicine

and

Institute of Endemic Disease
Seoul National University Medical Research Center
Seoul, Korea

Kosum Chansiri
Department of Biochemistry
Srinakharinwirot University
Bangkok, Thailand

Jianshun Chen
Zhejiang Fisheries Technical Extension Center and Zhejiang Aquatic Disease Prevention and Quarantine Center
Zhejiang, China

K. Chiers
Department of Pathology, Bacteriology, and Avian Diseases
Ghent University
Merelbeke, Belgium

Muriel Cornet
Department of Microbiology
Université Paris
Paris, France

Graça Costa
Centre for Macaronesian Studies
University of Madeira
Funchal, Portugal

Paul R. Crosbie
Department of Biology
California State University
Fresno, California

Tania Crucitti
STD/HIV Research and Intervention Unit
Institute of Tropical Medicine
Antwerp, Belgium

C. Cutillas
Department of Microbiology and Parasitology
University of Seville
Seville, Spain

Stefano D'Amelio
Department of Public Health Science
Sapienza University of Rome
Rome, Italy

Paron Dekumyoy
Department of Helminthology
Mahidol University
Bangkok, Thailand

Luís Pérez del Villar
Tropical Medicine Research Center
University of Salamanca
Salamanca, Spain

M. de Rojas
Department of Microbiology and Parasitology
University of Seville
Seville, Spain

R. Ducatelle
Department of Pathology, Bacteriology, and Avian Diseases
Ghent University
Merelbeke, Belgium

Jean Dupouy-Camet
Laboratoire de Parasitologie-Mycologie
Université Paris Descartes
Paris, France

Ron Dzikowski
Department of Microbiology and Molecular Genetics
The Hebrew University–Hadassah Medical School
Jerusalem, Israel

John Ellis
School of Medical and Molecular Biosciences
University of Technology
Sydney, New South Wales, Australia

Weihuan Fang
Institute of Preventive Veterinary Medicine
Zhejiang University
Zhejiang, China

Sarra Farjallah
Institute of Applied Sciences and Technology of Gabes
Rue Amor Ben El Khatab
Gabes, Tunisia

Sanford H. Feldman
Center for Comparative Medicine
University of Virginia
Charlottesville, Virginia

Yaoyu Feng
School of Resource and Environmental Engineering
East China University of Science and Technology
Shanghai, China

Maria Letizia Fioravanti
Department of Veterinary Public Health and Animal Pathology
University of Bologna
Bologna, Italy

Bernard Fried
Department of Biology
Lafayette College
Easton, Pennsylvania

Lorella Garagnani
Dipartimento ad Attività Integrata di Laboratori Anatomia Patologica e Medicina Legale
Azienda Ospedaliero–Universitaria di Modena
Modena, Italy

P. Geldhof
Department of Virology, Parasitology, and Immunology
Ghent University
Merelbeke, Belgium

Elisabetta Giuffra
Parco Tecnologico Padano
Polo Universitario
Lodi, Italy

Andrea Gustinelli
Department of Veterinary Public Health and Animal Pathology
University of Bologna
Bologna, Italy

Aaron Harmon
Department of Biology and Microbiology
South Dakota State University
Brookings, South Dakota

Michael B. Hildreth
Department of Biology and Microbiology
South Dakota State University
Brookings, South Dakota

P. Horák
Department of Parasitology
Charles University in Prague
Prague, Czech Republic

K. Houf
Department of Veterinary Public Health and Food Safety
Ghent University
Merelbeke, Belgium

Sun Huh
Department of Parasitology
Hallym University
Chuncheon, Korea

Sofia Ingrosso
Department of Public Health Science
Sapienza University of Rome
Rome, Italy

Yuji Ishikawa
Department of Parasitology
Asahikawa Medical College
Hokkaido, Japan

Akira Ito
Department of Parasitology
Asahikawa Medical College
Hokkaido, Japan

Somchai Jongwutiwes
Department of Parasitology
Chulalongkorn University
Bangkok, Thailand

Jenny Knapp
Department of Parasitology
Asahikawa Medical College
Hokkaido, Japan

L. Kolářová
Institute of Immunology and Microbiology
Charles University in Prague
Prague, Czech Republic

Xiangang Kong
Harbin Veterinary Research Institute
Chinese Academy Agricultural Sciences
Heilongjiang, China

Thewarach Laha
Department of Parasitology and Liver Fluke and Cholangiocarcinoma Research Center
Khon Kaen University
Khon Kaen, Thailand

Giuseppe La Rosa
Department of Infectious, Parasitic, and Immunomediated Diseases
Istituto Superiore di Sanità
Rome, Italy

Michael G. Levy
Department of Population Health and Pathobiology
North Carolina State University
Raleigh, North Carolina

Janice L. Liotta
Department of Microbiology and Immunology
Cornell University
Ithaca, New York

Dongyou Liu
Royal College of Pathologists of Australasia Quality Assurance Programs
Surry Hills, New South Wales, Australia

Silvia S. Longoni
Department of Parasitology
University of Granada
Granada, Spain

Vincenzo Lorusso
Department of Veterinary Public Health
University of Bari
Bari, Italy

Xuedong Lu
Futian Hospital
Guandong Medical College
Guangdong, China

Clotilde Marín
Department of Parasitology
University of Granada
Granada, Spain

Simonetta Mattiucci
Department of Public Health Sciences
Sapienza University of Rome
Rome, Italy

Heinz Mehlhorn
Department of Parasitology
Heinrich Heine University
North Rhine–Westphalia, Germany

B.R. Mirdha
Department of Microbiology
All India Institute of Medical Sciences
New Delhi, India

Ramiro Morales-Hojas
Instituto de Biologia Molecular e Celular
Universidade do Porto
Porto, Portugal

Boris Müller
Department of Parasitology
Heinrich Heine University
North Rhine–Westphalia, Germany

Antonio Muro
Tropical Medicine Research Center
University of Salamanca
Salamanca, Spain

Minoru Nakao
Department of Parasitology
Asahikawa Medical College
Hokkaido, Japan

Kazuhiro Nakaya
Department of Parasitology
Asahikawa Medical College
Hokkaido, Japan

Giuseppe Nascetti
Department of Ecology and Sustainable Economic Development
Tuscia University
Viterbo, Italy

Fousseyni S. Touré Ndouo
Medical Parasitology Unit
International Centre for Medical Research of Franceville
Franceville, Gabon

Munehiro Okamoto
Primate Research Institute
Kyoto University
Kyoto, Japan

Domenico Otranto
Department of Veterinary Public Health
University of Bari
Bari, Italy

Kerry A. Padgett
California Department of Public Health
Vector Borne Disease Section
Richmond, California

Michela Paoletti
Department of Public Health Sciences
Sapienza University of Rome
Rome, Italy

Raffaele Peduzzi
Istituto cantonale di microbiologia
Bellinzona, Switzerland

Jesús M. Pérez
Departamento de Biología Animal, Biología Vegetal y Ecología
Universidad de Jaén
Jaén, Spain

John H. Porter
Department of Environmental Sciences
University of Virginia
Charlottesville, Virginia

Edoardo Pozio
Department of Infectious, Parasitic, and Immunomediated Diseases
Instituto Superiore di Sanità
Rome, Italy

P.K. Prasad
Department of Zoology
North-Eastern Hill University
Meghalaya, India

Kristin Wear Prestrud
Norwegian School of Veterinary Science
Section of Arctic Veterinary Medicine
Tromsø, Norway

Chaturong Putaporntip
Department of Parasitology
Chulalongkorn University
Bangkok, Thailand

Yvonne Qvarnstrom
Department of Health and Human Services
Centers for Disease Control and Prevention
Atlanta, Georgia

Achariya Rangsiruji
Department of Biology
Srinakharinwirot University
Bangkok, Thailand

Shabeena Rehman
Department of Biosciences
Jamia Millia Islamia
New Delhi, India

Francesco Rivasi
Dipartimento ad Attività Integrata di Laboratori Anatomia Patologica e Medicina Legale
Azienda Ospedaliero–Universitaria di Modena
Modena, Italy

Luca Rossi
Dipartimento di Produzione Animali, Epidemiologia ed Ecologia
Università degli Studi di Torino
Torino, Italy

B. Roy
Department of Zoology
North-Eastern Hill University
Meghalaya, India

Una M. Ryan
School of Veterinary and Biomedical Science
Murdoch University
Murdoch, Western Australia, Australia

Yasuhito Sako
Department of Parasitology
Asahikawa Medical College
Hokkaido, Japan

Manuel Sánchez-Moreno
Department of Parasitology
University of Granada
Granada, Spain

F.M. Schets
National Institute for Public Health and Environment Centre for Infectious Disease Control Netherlands
Bilthoven, the Netherlands

Jürgen Schmidt
Department of Parasitology
Heinrich Heine University
North Rhine–Westphalia, Germany

Jan M. Schwenkenbecher
Institute of Biological and Environmental Sciences
University of Aberdeen
Aberdeen, Scotland, United Kingdom

Jesús Serrano-Luna
Department of Cell Biology
National Polytechnic Institute
Mexico City, Mexico

Takhellambam Shantikumar Singh
Department of Microbiology
Sikkim Manipal Institute of Medical Sciences
Sikkim, India

Mineko Shibayama
Department of Infectomic and Molecular Pathogenesis
National Polytechnic Institute
Mexico City, Mexico

Paiboon Sithithaworn
Department of Parasitology and Liver Fluke and Cholangiocarcinoma Research Center
Khon Kaen University
Khon Kaen, Thailand

Damien Stark
Division of Microbiology
SydPath St. Vincent's Hospital
Darlinghurst, New South Wales, Australia

Sacha Stelzer-Braid
Virology Research Laboratory
Prince of Wales Hospital
Randwick, New South Wales, Australia

Hiromu Sugiyama
Department of Parasitology
National Institute of Infectious Diseases
Tokyo, Japan

Veena Tandon
Department of Zoology
North-Eastern Hill University
Meghalaya, India

Yi-Wei Tang
Clinical Microbiology Service
Memorial Sloan-Kettering Cancer Center
New York, New York

Urusa Thaenkham
Department of Helminthology
Mahidol University
Bangkok, Thailand

Rafael Toledo
Departamento de Biología Celular y Parasitología
Universidad de Valencia
Burjassot-Valencia, Spain

Donato Traversa
Department of Comparative Biomedical Sciences
University of Teramo
Teramo, Italy

Víctor Tsutsumi
Department of Infectomic and Molecular Pathogenesis
National Polytechnic Institute
Mexico City, Mexico

L. Vangeel
Department of Pathology, Bacteriology, and Avian Diseases
Ghent University
Merelbeke, Belgium

J. Vercruysse
Department of Virology, Parasitology, and Immunology
Ghent University
Merelbeke, Belgium

Jaco J. Verweij
Department of Parasitology
Leiden University Medical Center
Leiden, the Netherlands

Govinda S. Visvesvara
Department of Health and Human Services
Centers for Disease Control and Prevention, Public Health Service
Atlanta, Georgia

Jitra Waikagul
Department of Helminthology
Mahidol University
Bangkok, Thailand

Stephen C. Webb
Cawthron Institute
Nelson, New Zealand

Barbara Wicht
Istituto cantonale di microbiologia
Bellinzona, Switzerland

Changyou Xia
Harbin Veterinary Research Institute
Chinese Academy of Agricultural Sciences
Heilongjiang, China

Lihua Xiao
Division of Foodborne and Waterborne Disease Prevention
Centers for Disease Control and Prevention
Atlanta, Georgia

Tetsuya Yanagida
Department of Parasitology
Asahikawa Medical College
Hokkaido, Japan

Rongchang Yang
School of Veterinary and Biomedical Science
Murdoch University
Murdoch, Western Australia, Australia

Hélène Yera
Laboratory of Parasitology–Mycology
Université Paris Descartes
Paris, France

Chong Yin
College of Veterinary Medicine
Hunan Agricultural University
Hunan, China

Tai-Soon Yong
Department of Environmental Medical Biology
Yonsei University College of Medicine
Seoul, Korea

Tippayarat Yoonuan
Department of Helminthology
Mahidol University
Bangkok, Thailand

Hisao Yoshikawa
Department of Biological Sciences
Nara Women's University
Nara, Japan

Jae-Ran Yu
Konkuk University School of Medicine
Seoul, Korea

Xing-Quan Zhu
Lanzhou Veterinary Research Institute
Chinese Academy of Agricultural Sciences
Gansu, China

1 Introductory Remarks

Dongyou Liu

CONTENTS

1.1 PREAMBLE

Parasites (from Greek parasitos, para—beside, by; sitos—wheat; meaning one who eats at the table of another) are eukaryotic organisms that may be free living, or form symbiotic or parasitic relationship with their hosts. Eukaryotes are characterized by having a well-defined chromosome in a nuclear membrane (which is packed with histones to form linear chromatin), possessing various independent membrane-bounded cytoplasmic organelles, and undergoing endocystosis and exocytosis. This is in clear contrast with prokaryotes (e.g., bacteria) that lack nuclear membrane, contain few independent membrane-bounded cytoplasmic organelles, and do not undergo endocytosis and exocytosis.

Consisting of over 800,000 identified species, parasites come under the following three categories: protozoa, helminths, and arthropods. In the current six kingdom systems for biological organisms (i.e., Bacteria, Protozoa, Fungi, Chromista, Plantae, and Animalia), protozoa constitute an independent kingdom of their own, while helminths and arthropods constitute parts of the kingdom Animalia [1,2].

Based on their interactions with their hosts, parasites are often referred to as endoparasites (which live inside the host, e.g., parasitic worms) and ectoparasites (which live on the surface of the host, e.g., mites). Endoparasites may inhabit spaces in the host's body (intercellular form) or cells in the host's body (intracellular form). Intracellular parasites often rely on a third organism (generally known as carrier or vector) for transmission to the host. In addition, some parasites spend their entire life on the host (e.g., *Enterobius*), while others may undergo freeliving and parasitic stages in their life cycle (e.g., *Strongyloides*).

The study of parasitic protozoa, helminths, and insects, their hosts, and the host–parasite relationship is commonly known as parasitology. The discipline of parasitology examines the morphology, life cycle, pathogenesis and identification of parasites, and draws on techniques from fields such as cell biology, ecology, biochemistry, immunology, molecular biology, bioinformatics, genetics, and evolution. Although fungi represent another category of eukaryotes with the potential to lead a parasitic life in plant, animal, and human hosts, they are discussed in a distinct branch of science (mycology). The kingdom Chromista comprises unicellular plant-like organisms, to which a previous protozoan taxon *Blastocystis* belongs.

1.2 CLASSIFICATION, MORPHOLOGY, AND BIOLOGY

Parasites infecting humans commonly belong to three categories: protozoa, helminths, and arthropods.

1.2.1 Protozoa

Protozoa (singular protozoon) are small, unicellular, phagotrophic, nonphotosynthetic, eukaryotic organisms without cell walls. Previously, protozoa (meaning "first animals," with Protista and Protoctista, being alternative names) have been classified in the kingdom Animalia, and more recently form an independent, basal eukaryotic kingdom within the six kingdom system (Bacteria, Protozoa, Chromista, Plantae, Fungi, and Animalia). As protozoa can be of both purely auto- and heterotrophic, they can fit into a plant [auto (photo) trophic], fungal (heterotrophic), or animal (heterotrophic) context. Additionally, some protozoa (e.g., dinoflagellates) demonstrate an intermediate trophic capability (thus called mixotrophic). Other previous protozoan taxa such as microsporidia and pneumocystis have been transferred to the kingdom Fungi (which covers eukaryotic heterotrophic organisms lacking plastids but possessing cell walls containing chitin and glycans). Moreover, there is another previous protozoan taxon *Blastocystis* that has been redesignated in the kingdom Chromista (covering unicellular, photosynthetic, filamentous, or colonial organisms ("algae"); some with secondary loss of plastids). Being the only parasitic member of the kingdom Chromista, Blastocystis will be covered in Section I, Sarcomastigophora of the book for convenience in spite of its changed taxonomical classification.

Measuring from 10 to 200 μm in size (with the parasitic forms tending toward the lower end of this range), protozoa demonstrate considerable variations in their locomotory structures (pseudopodia, flagella, cilia, or no organelle of locomotion), which have underlined the traditional classification system for parasitic organisms within this group (i.e., phyla Sarcomastigophora, Apicomplexa, and Ciliophora) (Table 1.1).

The phylum Sarcomastigophora is further separated into subphyla Mastigophora and Sarcodina. While the subphylum Mastigophora (flagellates) possesses one or more flagella as locomotive organ and reproduces by binary fission as well as sexual reproduction in some species, the subphylum Sarcodina (amoebae) moves by pseudopodia (although flagella may be present in the reproductive stages) and reproduces by binary fission. The phylum Apicomplexa (sporozoans) has no special locomotive organs, reproduces by multiple fission and gamogony and forms spores at one stage of their life. The phylum Ciliophora (ciliates) moves by cilia, has two nuclei (one macronucleus and one micronucleus), and reproduces by transverse binary fission, conjugation, autogamy, and cytogamy.

Structurally, many protozoa possess a pellicle that is composed of an outer plasma membrane and an inner cytoskeleton, and that plays a pivotal role in the maintenance of cell shape. The plasma membrane further separates into a thin outer ectoplasm and an inner endoplasm; the underlying cytoskeleton may include additional membranes, microtubules, microfilaments, or plates of cellulose or protein. Within the cytoplasm exist nucleus, nuclear membrane, chromosomes, endoplasmic reticulum, mitochondria, golgi body, ribosomes, and various specialized structures (e.g., vacuoles that may be stomach-like for food digestion, or contractible for eliminating excess water).

Protozoa have the ability to reproduce sexually and asexually, with each individual protozoan being both male and female. Cell division in protozoa usually involves a closed spindle (constructed inside the intact nuclear envelope), in contrast to the open spindle of multicellular animals. In particular, parasitic protozoa tend to have short generation times and high rates of reproduction.

Protozoa are both diverse and abundant. With >200,000 named species, and present in soil, freshwater, and marine environments, protozoa make one of the largest biomasses on earth. As heterotrophic organisms, protozoa prey on unicellular or filamentous algae, bacteria and microfungi, and also form an important food source for microinvertebrates, contributing to the transfer of bacterial and algal production to successive trophic levels. Moreover, protozoa may be symbiotic or parasitic, living attached to or inside other organisms. In fact, of the >200,000 named species, nearly 10,000 protozoa are parasitic in invertebrates and in almost every species of vertebrate.

Protozoa may enter human hosts via the nasal passage (e.g., *Naegleria fowleri*), inoculation (e.g., malaria transmitted by mosquitoes), ingestion of cysts (e.g., amoebic dysentery), fecal contamination of wounds (e.g., Chagas disease transmitted by assassination bugs), intrauterine infection (e.g., malaria), kissing (e.g., *Trichomonas tenax*), or sexual intercourse (*Trichomonas vaginalis*). Examples of significant protozoan diseases in humans include malaria (*Plasmodium*), sleeping sickness (African trypanosomes), Chagas disease (*Trypanasoma cruzi*), and leishmaniasis (*Leishamania*).

1.2.2 Helminths

The term helminths (singular helminth) commonly refers to organisms belonging to the phyla Platyhelminthes (platyhelminths or flatworms) and Nematoda (nematodes or roundworms) (Table 1.1).

The phylum Platyhelminthes (platyhelminths or flatworms) is divided into four classes: Cestoda (tapeworms), Trematoda (flukes), Monogenea, and Turbellaria. The class Cestoda (tapeworms) is further separated into two subclasses, the Cestodaria (which lacks a scolex, is monozoic with a set of reproductive organ, and parasitic in fishes and turtles) and the Eucestoda (the majority of which are segmented and polyzoic, and are parasitic to humans). The class Trematoda (flukes) comprises two subgroups, the Digenea

TABLE 1.1
Classification and Characteristics of Major Human Parasitic Pathogens

Kingdom	Phylum	Subphylum/ Class	Brief Description	Notable Human Pathogens
Protozoa	Sarcomastigophora	Sarcodina (amoebae)	Obligate and facultative amoebae; move by pseudopodia; cytoplasmic streaming assists movement. Asexual reproduction occurs by fission of the cell	*Acantamoeba* *Balamuthia* *Entamoeba* *Naegleria*
		Mastigophora (flagellates)	Intestinal, tissue, and blood-dwelling flagellates; move by flagella; reproduction in these protozoa generally occurs by fission, although sexual reproduction is observed in some species	*Dientamoeba* *Giardia* *Leishmania* *Trichomonas* *Trypanosoma*
	Apicomplexa (sporozoans)		Sporozoa; unicellular organisms possessing at some stage an apical complex composed of polar rings, rhoptries, micronemes, and typically a conoid; elaborate life cycles involving a sexual process; all parasites; no organelle of locomotion; forming spores at one stage in their life cycle	*Babesia* *Cryptosporidium* *Isospora* *Plasmodium* *Sarcocystis* *Toxoplasma*
	Ciliophora (ciliates)		Unicellular organisms possessing many cilia for locomotion and complex oral ciliature for feeding; having somatic, polyploid macronuclei and generative, diploid micronuclei; conjugation may be used for sexual reproduction, and binary fission also occurs	*Balantidium*
Animalia	Nematoda	Adenophorea Secrenentea	Round worms; appear round in cross section; have body cavities, a straight alimentary canal and an anus; have separate sexes (dioecious reproductive system)	*Ascaris* *Ancylostoma* *Enterobius* *Necator* *Strongyloides* *Trichuris*
	Platyhelminthes	Cestoda	Tape worms; have a head (scolex) with sucking organs, a segmented body, but no alimentary canal; each body segment is hermaphrodite; adult tapeworms live in the intestine of the definitive host	*Diphyllobothrium* *Echinococcus* *Taenia*
		Trematoda	Flukes; a nonsegmented, usually leaf-shaped body, with two suckers but no distinct head; have an alimentary canal (but no anus) and are usually hermaphrodite; however, Schistosomes are thread-like, and have separate sexes	*Fasciolopsis* *Opisthorchis* *Schistosoma*
Animalia	Arthropoda	Arachnida	Two segments (cephalothorax and abdomen); eight legs; one pair of chelicerae; no attennae Consisting of orders Astigmata (mites) and Metastigmata (ticks)	*Argas* *Ixodes* *Sarcoptes*
		Insecta	Three body segments (head, thorax, and abdomen); six legs; one pair of attennae Consisting of medically relevant orders Anoplura (lice), Diptera (flies), and Siphunculata (fleas)	*Pediculus* *Phthirus* *Tunga*
		Pentastomida	Tongue worms	*Armillifer* *Linguatula*

and the Aspidogastrea, with only the former occurring in humans and other mammals. The class Monogenea covers mainly ectoparasites of fishes; and the class Turbellaria includes a variety of free-living organisms in aquatic and terrestrial environments.

The platyhelminths or flatworms in the phylum Platyhelminthes have a bilaterally symmetrical, dorsoventrally flattened body with three well-defined germ layers (ectoderm, endoderm, and mesoderm), a definite head end, but without a coelom (a fluid-filled cavity between the outer body wall and the gut). Many are hermaphroditic (with both male and female sex organs), undergo self-fertilization, and have an indirect lifecycle.

The adult tapeworms in the subclass Eucestoda have an elongate, segmented, flat body, which is covered by a syncytial outer layer formed by the tegumental cells (known as tegument). The anterior end of the body contains a holdfast organ called a scolex, which may have a rostellum, suckers, bothria, bothridia, tentacles, hooks, and/or spines for attachment to the gut of the host. The scolex is followed by a short unsegmented portion (called the neck) and the remainder of the body (or strobila) is composed of a number of segments (or proglottids). Each segment (proglottid) harbors one or more sets of the male and female reproductive organs. There is a lack of body cavity and alimentary canal.

Of the two orders within the subclass Eucestoda, the order Pseudophyllidea is characterized by its lack of suckers and hooks in the scolex, and its use of a pair of bothria (dorsal and ventral longitudinal grooves on the scolex) for attachment; and the order Cyclophyllidea is characterized by presence of a ring of four suckers and sometimes one or more rings of apical hooks on the scolex. The presence or absence of scolex, the shape of the segments (proglottids) and the arrangement and form of the reproductive system(s) within the segments (e.g., the position of the genital pore, the size of the cirrus sac, the shape of the ovary, uterus, and vitellarium or yolk gland) are useful features for differentiation of cestodes. The medically important members in the order Pseudophyllidea are found in the family Diphyllobothriidae, while those in the order Cyclophyllidea are found in the families Anoplocephalidae, Davaineidae, Dipylidiidae, Hymenolepididae, Mesocestoididae, and Taeniidae.

During their lifecycle, cestodes generally involve a final host, and one or two intermediate hosts; the only cestode species capable of autoinfection without the use of intermediate hosts is *Hymenolepis*. Adult cestodes live in the intestine of vertebrates, and the eggs or gravid segments containing eggs are discharged with the feces. The egg of the cyclophyllideans is eaten by a terrestrial invertebrate (e.g., an arthropod) or a vertebrate, in which it hatches and releases a hexacanth (six-hooked) larva (called onchosphere). The egg of the pseudophyllideans hatches in water and releases a ciliated, motile hexacanth larva (called coracidium), which is eaten by an aquatic arthropod (e.g., a copepod). The hexacanth penetrates the gut wall of intermediate host and develops into a procercoid in the body cavity. The procercoid may further develop, either in the same host or in a second intermediate host if the first host is eaten, into a resting, encysted, stage that is referred to as cysticercus, cysticercoid, plerocercoid, tetrathyridium, or hydatid. The final host acquires the worm by feeding on the intermediate host harboring the encysted stage. In rare instances, asexual multiplication of larval heads (protoscoleces) can take place in vertebrates which act as an intermediate host.

The presence of adult tapeworm in humans may result in diarrhea, loss of appetite, and illthrift. On the other hand, the presence of encysted larval tapeworms in the tissues (e.g., the brain, liver, and lungs) may have serious consequences. For example, about 50,000 people die annually of cysticercosis, an infection with *Taenia solium* larvae.

The adult worms of the digenetic trematodes in the class Trematoda (flukes) are usually dorsoventrally flattened, with some having long and narrow, leaf-shaped, or thick fleshy bodies. Adult trematodes typically have an anterior oral sucker surrounding the mouth, and a ventral sucker (sometimes termed the acetabulum) on the ventral surface. Most digeneans have a forked digestive system (consisting of oral sucker, the pharynx, esophagus, and blind-ending caeca) that opens at the mouth as an anus is absent; and a single set of reproductive organs. The tegument of digeneans is syncitial (which consists of a mass of protoplasm containing many nuclei but not divided into cells) and is involved in the direct nutrient uptake. The digenetic trematodes are normally hermaphroditic (with both male and female reproductive systems), occasionally partly or entirely dioecious (e.g., schistosomes have separate sexes). The female reproductive system comprises an ovary, oviduct, sometimes a seminal receptacle (for storing sperm from the copulatory partner), ootype (where the egg is formed) surrounded by Mehlis' gland, yolk gland (vitellarium) which opens into the ootype or near it, and the uterus (in which eggs mature), which sometimes forms a muscular terminal called metraterm, and which opens through a common gonopore together with the male system. A narrow duct, Laurer's canal, links the female system near the ootype to the dorsal surface. The male reproductive system comprises of one testis or several testes, and sperm ducts which unite and widen terminally to form a seminal vesicle in a cirrus pouch or sac.

The lifecycle of digenetic trematodes includes an adult stage in the final host, and five larval stages (miracidium, sporocyst, redia, cercaria, and metacercaria) in one, two, or more intermediate hosts. The eggs of the digenea are oval and operculate and are normally passed in the faces of the final host. Under suitable conditions of moisture, warmth, light, and salinity, the egg hatches to release a ciliated larva, miracidium. The miracidium enters a first intermediate host (a snail), and forms a sporocyst after losing ciliated coat. Within the sporocyst the germinal cells multiply and produce either daughter sporocysts (which lacks a pharynx) or rediae (which has a pharynx or a sac-like gut), and the sporocyst or the redia then produces the cercaria. After leaving the snail host through an opening, or the tegument or being expelled, the cercaria encysts on or in a second intermediate host (typically an invertebrate) or on vegetation, and the encysted cercaria matures to produce the infective stage, the metacercaria. The metacercaria contained in the second intermediate host is ingested by a definitive host and develops into the adult worm in the gut, liver, bile ducts, lungs, or blood system. However, the metacercaria is absent in the family Schistosomatidae and its cercaria actively penetrates the skin of the definitive host.

The medically important trematodes are found in the orders Strigeida (family Schistosomatidae), Echinostomida (families Echinostomatidae and Fasciolidae), and Plagiorchiida (families Dicrocoeliidea, Heterophyidae, Opsithorchiidae, Paragonimidae, and Plagiorchiidae). Features useful for distinguishing trematode groups include the presence or absence of spines on the tegument, the position and number of suckers, the form and arrangement of the reproductive organs, and the shape of the gut and excretory vesicle. For species level discrimination, details such as sucker ratio and egg size are informative.

Besides taking in nutrients through their body surface, digeneans also feed by browsing. The feeding process and the presence of spines on the trematode body can cause damage and irritation. The toxins produced by digeneans can elicit inflammation, tissue reactions, fibrosis, obstruction, and so on.

The phylum Nematoda is divided into two classes, Adenophorea and Secernentea, which reflect the presence and absence of small sensory structures (phasmids) on the tail and the nature of the excretory system. In addition, in the Adenophora the first-stage larva is infective to the definitive (final) host, whereas in the Secernentea it is the third-stage larva. A majority of animal parasites are found in the Secernentea.

The phylum Nematoda is characterized by (i) a bilateral symmetrical and triploblastic (or triblastic) body; (ii) a body wall consisting of an outer layer of cuticle and an inner layer of longitudinal muscles with scattered nuclei but without cell boundaries; (iii) a pseudocoel body cavity between the body wall and the digestive tract, that is not lined by the mesodermal (or coelomic) epithelium; (iv) a complete digestive system with a mouth at the anterior end, a muscular esophagus and an intestine leading to anus; (v) a lack of the circulatory and respiratory system; (vi) a circumventric nerve ring with longitudinal nerves in the nervous system; (vii) a pair of longitudinal excretory canals with excretory pores in the excretory system; (viii) separate sexes, with males and females differing in size and certain morphological features (sexual dimorphism); (ix) eggs with plug at either end. The nature of the anterior regions of the alimentary canal (e.g., the esophagus), the form of the head (presence, number of lips, teeth, etc.), the form of the male tail such as the arrangement of caudal papillae (sensory structures used during copulation), and the length and shape of the spicule(s) (sclerotized copulatory aids) are important features for nematode identificaiton. In the majority of cases, males carry more taxonomically useful information than females, with the females being often unidentifiable at the specific level in the absence of the males.

Nematodes undergo five developmental stages (four larval and one adult), which are separated by a molt of the cuticle, with the first one or two molts occurring within the egg. The life cycle of nematodes may be direct or indirect. Direct life cycle involves the ingestion of eggs or larvae with food or direct penetration of larvae through the skin. Indirect life cycle utilizes invertebrate or vertebrate (or other invertebrate) as intermediate or paratenic hosts. The larvae normally reside in the tissues of intermediate hosts, while the adults occur in the alimentary canal of vertebrates, and occasionally in other parts of the body following larval migration.

With around 12,000 described species, nematodes (roundworms) are ubiquitous in freshwater, marine, and terrestrial environments, and occur in both invertebrate and vertebrate as parasites. The definitive host may acquire the infection by (i) ingestion or penetration of an active, nonparasitic third-stage larva; (ii) ingestion of a passive infective egg containing an infective larva; (iii) ingestion of an intermediate host containing the infective larva. Once inside the definitive host, nematode larvae migrate through the body and develop in adult worms after the third and fourth molts (ecdyses) in their normal habitat.

Heavily armed with teeth or other sclerotized mouthparts, nematodes such as hookworms browse on the gut wall, causing considerable damage. Migrating adults in the tissues and larvae (termed as larva migrants) can also cause serious problems in sensitive regions such as the brain, liver, or eyes. The medically important families in the class Adenophorea include Trichinellidae and Trichuridae; those in the class Secernentea include Ancylostomatidae, Anisakidae, Ascarididae, Dracunculidae, Onchocercidae, Gnathostomatidae, Angiostrongylidae, Oxyuridae, Strongyloididae, Thelaziidae, and Trichostrongylidae.

Apart from phyla Platyhelmithes and Nematoda, there is another phylum of parasites Acanthocephalans ("thorny-headed worms") that commonly occur in animals, and rarely in humans (causing intestinal perforation and peritonitis). As dioecious worms with a body cavity, Acanthocephalans have lost their digestive system and reduced their muscular, nervous, and excretory systems. The body of Acanthocephalans consists of a large trunk and a retractable anterior proboscis armed with a regular array of hooks used for attachment to the gut wall. The number of rows of hooks and the number of hooks per row in the proboscis are of diagnostic importance. The body wall is a syncytium with large, fragmented nuclei, along with a complex outer tegument and an inner layer of muscles, as well a lacunar system of interconnecting canals that may serve as a circulatory system. The males harbor two testes and a series of accessory glands (called cement glands). The females have ovarian balls, and the body cavity of mature females is full of eggs.

During the lifecycle of Acanthocephalans, female adults in the intestine produce eggs that are voided with the feces. After ingestion by a suitable arthropod, the egg releases a larva armed with hooks (called an acanthor), which then enters the hemocoel and develops into an acanthella with the internal organs of the adult beginning to emerge. When the proboscis develops, it invaginates into the body and the acanthella becomes encysted and is known as a cystacanth. The final host acquires the infection either by feeding directly on an arthropod harboring the cystacanth larva or a vertebrate paratenic host harboring a juvenile form (a reencysted cystacanth). Inside the intestine of a suitable final host, the parasite attaches to the gut wall and matures into adult.

The proboscis with its armature of hooks penetrates the gut wall, causing some local inflammation and abdominal pain. Some juvenile parasites may perforate the intestine, causing peritonitis. Due to their rare occurrences in humans, Acanthocephalans are not discussed in details in the book.

1.2.3 Arthropods

Arthropods are invertebrate animals that are characterized by their bilaterally symmetrical, segmented bodies (with each segment containing a pair of appendages), jointed limbs or appendages (specialized for feeding, locomotion, sensing), and exoskeleton or cuticle (containing a noncellular material made of α-chitin, a derivative of glucose, which is secreted by the epidermis).

The cuticle is composed of three main layers: the epicuticle (a thin outer waxy coat that moisture-proofs the other layers and provides protection), the exocuticle (which consists of chitin and chemically hardened proteins), and the endocuticle (which consists of chitin and unhardened proteins). Together, the exocuticle and endocuticle are known as the procuticle. While each body segment and limb section is encased in hardened cuticle, the joints between body segments and between limb sections are covered by flexible cuticle. The cuticle can have setae (bristles) which grow from specialized cells in the epidermis, and which function as appendages.

Other notable features of arthropods include a coelom (or hemocoel), which is a membrane-lined cavity between the gut and the body wall that runs most of the length of the body and through which blood flows, and that accommodates the internal organs (e.g., the reproductive, excretory, and circulatory systems, as well as ladder-like nervous system). The head is formed by fusion of varying numbers of segments, and the brain is formed by fusion of the ganglia of these segments that encircle the esophagus. The heart of arthropods is a muscular tube that situates under the back and runs for most of the length of the hemocoel. By contracting in ripples that run from rear to front, the heart pushes blood forward. Along the heart are a series of paired ostia, nonreturn valves that allow blood to enter the heart but prevent it from leaving before it reaches the front. Arthropods rely on various combinations of compound eyes and pigment-pit ocelli for vision. Arthropods reproduce by internal fertilization via either an appendage or the ground (rather than by direct injection).

Taxonomically, arthropods are classified in the phylum Arthropoda (from Greek árthron, "joint"; podós "foot"; together "jointed feet"). The phylum Arthropoda is further divided into five classes: Arachnida (or Chelicerata, e.g., ticks, mites, spiders, and scorpions), Chilopoda (e.g., centipedes), Diplopoda (e.g., millipedes), Crustacea (e.g., crabs, shrimp, lobsters, shrimp, water fleas, copepods, barnacles, sowbugs, and woodlice), and Insecta (or Hexapoda, e.g., flies, fleas, wasps, and bees). A majority of medically important arthropods are found in the classes Arachnida and Insecta.

Class Arachnida have two distinct body regions or tagmata (cephlothorax and abdomen), two pairs of mouthparts (the first pair being chelicerae, which are three-segmented and pincher-like; the second pair being six-segmented pedipalps), and four pairs of legs (most with seven segments; compared to insects, an extra segment located between the third segment (the femur) and the fourth segment (the tibia) is called the patella), but no antennae. Class Arachnida can be further separated into the orders Araneae (spiders, consisting of >2500 described species), Acari or Acarina (mites and ticks, consisting of >30,000 described species), and Scorpiones (scorpions, with >40 species in the United States).

Class Insecta have three distinct body regions or tagmata (head, thorax, and abdomen), three pairs of legs (most with six segments), and one pair of antennae. Many insects have either one or two pairs of wings. The head contains mouthparts (which are used for biting, sucking, stinging, licking), antennae (which are used as tactile organs or as olfactory organs), and eyes (which may be compound eyes or simple eyes called "ocelli" or "little eyes"). The thorax is further divided into three segments (pro-thorax, meso-thorax, and meta-thorax), with each segment containing one pair of legs (with each leg being made up of coxa, trochanter, femur, tibia, and tarsus). The abdomen has the gonopore (genital opening) at the posterior end.

As the exoskeleton is unstretchable and restrictive to growth, arthropods undergo molting to shed (replace) the old exoskeleton after growing a new one. On the basis of their differences in the type of metamorphosis (showing variable external changes after each molt), class Insecta can be distinguished into apterygota, exopterygota, and endopterygota. The apterygota have no metamorphosis (i.e., except for the size, all larval stages resemble the adults, which are wingless), the exopterygota undergo a simple metamorphosis (i.e., there is a gradual change in the external appearance from egg, the nymphal stages to an adult), and the endopterygota has a complete metamorphosis (i.e., there are significant morphological changes from egg, larvae, pupa to adult).

Of the 20 or so orders within the class Insecta, the following are most notable: Collembola (springtails), Ephemeroptera (mayflies), Odonata (dragonflies), Orthoptera (grasshoppers and crickets), Blattaria (cockroaches), Mantodea (mantids), Phasmida (walkingsticks), Isoptera (termites), Anoplura (sucking lice), Heteroptera (cicadas, hoppers, and aphids), Coleoptera (beetles), Lepidoptera (butterflies and moths), Diptera (flies), Siphonaptera (fleas), and Hymenoptera (ants, wasps, and bees). To date, around 1,000,000 insect species have been described.

1.3 PHENOTYPIC CHARACTERIZATION

Phenotypic techniques for the identification of parasitic protozoa, helminths, and arthropods include macroscopic and microscopic assessments (morphometrics) of parasitic materials in clinical and histological specimens, biochemical and immunological characterization, and *in vitro* and *in vivo* cultivation, and so on.

1.3.1 Macroscopic and Microscopic Assessment

Macroscopic examination is useful for detection of proglottids of tapeworms and sometimes adult ascarid worms in stools. Direct saline preparations may be employed for identification of trophozoite stages of protozoa. Use of the Kato–Katz method is helpful for quantification of the number of eggs present in stools [3]. Use of modified acid-fast preparations or UV illumination for the examination of auramine-stained preparations facilitates easy identification of *Cryptosporidium* and *Isospora*.

To improve the sensitivity of diagnosing protozoal cysts and helminth eggs, stool samples may be concentrated using the formol–ether method and its variants, which also helps remove fecal debris [4,5]. Flotation, sedimentation,

McMaster method, and Baermann method are other common techniques for coprological examination of intestinal parasites such as *Entamoeba* and *Giardia*, with a minimum of three samples of stool being necessary to make a diagnosis [6,7]. Similarly, concentration techniques are essential for identification of parasites in blood. For example, the thick film technique utilizes the hypotonic Field's stain to lyse red cells, allowing a large number of red cells to be scanned quickly for the presence of the characteristic ring-shaped merozoites of malaria parasites. Buffy-coat concentration permits the diagnosis of *Leishmania donovani* and African trypanosomes, while centrifugation of blood through a DEAE column also works well for African trypanosomes. In addition, filtration of a measured volume of blood through a micropore membrane is useful to concentrate microfilariae. Subsequent staining and microscopic examination enables species identification of microfilariae [8].

Rectal biopsy samples obtained at sigmoidoscopy or proctoscopy is valuable in the diagnosis of schistosomiasis, especially in travelers who have few eggs present in stools. For some parasites, biopsy represents the only practical means of making the diagnosis (e.g., cutaneous leishmaniasis or onchocerciasis). Additionally, string test may be used to diagnose giardiasis, strongyloidiasis, and cryptosporidiosis. In this method, a length of nylon "string" with a weighted end is swallowed and travels to the upper part of the jejunum (bile staining of the string indicates that the string has reached the jejunum). After a period of time, the string is pulled up and the fluid expressed is examined.

As morphometric analysis relies on having material in good condition, specimens may be fixed by using 10% formalin, 70–80% alcohol, Berland's fluid (19 parts glacial acetic acid to 1 part pure formalin), or boiling water (especially in the case of cestodes). Apart from large tapeworms and acanthocephlans, specimens should not be fixed under the pressure (e.g., coverslip or glass slide). After fixation for about a minute, the specimens are stored in 80% alcohol.

Nematodes and acanthocephalans are often examined as temporary mounts between a glass slide and a coverslip in a clearing agent (e.g., beechwood creosote, lactophenol, or glycerine). Glycerine is only useful for smaller worms (of <15 mm in length). Trematodes and cestodes are mounted on slides permanently, followed by staining in a carmine-based stain (e.g., Mayer's paracarmine for 1–20 min, depending on size), destaining in acid alcohol (until the worm is a pale pink color), dehydrating, clearing, and mounting (in Canada balsam or other mountant).

1.3.2 Biochemical and Immunological Tests

Biochemical tests (e.g., liver function tests) may be employed for preliminary diagnosis of blood-dwelling parasites causing malaria, theileriosis, babesiosis, trypanosomiasis, and filariasis, and so on. Many patients suffering from malaria and filariasis often show elevated levels of bilirubin, alanine aminotransferase (ALAT), aspartate aminotransferase (ASAT), and alkaline phosphatase in serum samples. However, as a general rule, changes detected by the biochemical tests are non-specific, and further confirmation is invariably required to achieve a correct diagnosis.

Immunological tests [e.g., complement fixation test (CFT), immunodiffusion (ID), indirect hemagglutination (IHA), indirect immunofluorescent antibody test (IFA), enzyme-linked immunosorbent assay (ELISA), radioimmunoassay (RIA), and Western blot] demonstrate a higher specificity than biochemical tests for diagnosis of parasitic infections. For instance, application of fluorescent antibodies permits rapid and specific diagnosis of *P. falciparum* and *P. vivax*. In some cases, immunological tests facilitate differentiation between recent and latent infections, identification of carrier status, and verification of parasite elimination after therapies. Nonetheless, given the frequent antigen sharing among various closely related parasites, the results of immunological tests need to be interpreted with accompanied clinical manifestations and parasitological findings.

1.3.3 *In Vitro* and *In Vivo* Cultivation

Although culture methods are infrequently used in parasitology, for diagnosis of certain parasites (e.g., Strongyloides and hookworms), an improved yield is obtainable by culturing stools, especially when the numbers of larvae are low [9]. In addition, *in vitro* cultivation of schistosome eggs yields the miracidia for easy identification. Typically, fresh stool or urine specimens are placed in 10 volumes of unchlorinated water (e.g., spring water) in an Erlenmeyer flask, which is covered. The sediment is then examined after 24 h for the presence of miracidia. Similarly, blood and buffy coat may be cultured for Leishmania on Schneider's insect medium for two weeks to visualize promastigotes.

For sensitive diagnosis of *Trypanosoma cruzi* in blood, uninfected triatomid bugs are permitted to feed on a suspected patient for ~30 min. The bugs are kept and sacrificed monthly for 3 months. Microscopic examination of the dissected guts helps reveal the presence of trypanosomes.

1.4 GENOTYPIC CHARACTERIZATION

Since phenotypic characterization of parasitic protozoa, helminths, and arthropods is not only labor intensive, but also requires specialized skills, molecular techniques have been increasingly utilized for rapid and reliable identification of parasites.

1.4.1 Target Genes

In general, while introns and noncoding regions (e.g., the internal transcribed spacer or ITS of ribosomal RNA (rRNA) gene) exhibit a high mutational rate (with a high level of interspecific sequence variation and a low level of intraspecific difference), the coding genes with specific translational products (e.g., the mitochondrial DNA or mtDNA) demonstrate a relatively low mutational rate (with a low level of interspecific sequence variation). For differentiation of protozoa, trematoda, cestoda, nematodes, and arthropods at species level, a target gene or region with a low intraspecific variability is preferred to a gene with a high degree of variation. On the other hand, for comparison of two larval populations of the same species (e.g., *Hypoderma lineatum*) originating from two geographic areas, a target gene with a high level of intraspecific variation is preferred (e.g., tandemly repeated sequences or microsatellites of ITS, which are units of repetition between one and five nucleotides) [10].

Ribosome is an essential cellular organelle that is involved in protein synthesis in all living organisms. As the key component of the ribosome, rRNA molecules comprises two complex folded subunits of differing sizes (small and large), whose main functions are to provide a mechanism for decoding messenger RNA (mRNA) into amino acids (at the center of the small ribosomal subunit) and to interact with transfer RNA (tRNA) during translation by providing peptidyltransferase activity (large subunit). The two rRNA subunits in eukaryotes have sedimentation coefficient values of 40S (Svedberg units) and 60S. The small rRNA subunit (40S) in eukaryotes contains a single RNA species (i.e., 18S rRNA) whereas the large rRNA subunit (60S) in eukaryotes comprises three RNA species (5S, 5.8S, and 25–28S rRNA) [11].

The eukaryotic rRNA genes exist as a family of multiple-copy genes that are arranged in a head-to-toe manner, with each copy consisting of 18S RNA, ITS 1 (internal transcribed spacer 1), 5.8S RNA, ITS 2, 28S RNA, IGS 1 (intergenic spacer), 5S, and IGS 2. Having been homogenized through concerted evolution, the tandemly repeated copies of rRNA genes contain highly similar nucleotide sequence, and thus are often treated as a single-locus gene. Along with ITS, the 18S, 5.8S, and 25–28S rRNAs are transcribed as a 35–40S precursor, with all spacers being later spliced out of the transcript. A nontranscribed or IGS region exists between the copies of the 18S, 5.8S, and 25–28S rRNA repeats, serving to separate the repeats from one another on the chromosome. A 5S RNA gene takes a position within the IGS region and is transcribed in the opposite direction.

In general the 25–28S RNA gene is conserved across widely divergent taxa, and only the first 600–900 bases contain three divergent domains (D1, D2, D3) that are useful for phylogenetic study of eukaryotic organisms such as parasitic worms and arthropods. The 18S rRNA (small subunit (SSU)) gene also includes alternating regions of sequence conservation and heterogeneity. The conserved regions are often targeted for phylogenetic analysis of higher taxonomic orders (e.g., phylum, family, and genus), while the regions of sequence diversity are valuable for characterization of isolates to the genus or species level (with isolates showing sequence identity >97% in the 18S rRNA gene being considered as the identical species). Compared to the rRNA genes, the rRNA ITS 1 and ITS 2 regions tend to be more variable, offering valuable targets for species-specific identification. The identity of the testing organism is defined by its ITS sequence similarity (%) to the type strain or control isolate: species, ≥99%; genus, 93–99%; and inconclusive, ≤93%. The rRNA IGS regions also experience more genetic drift and are not as highly conserved. This makes the rRNA IGS regions another potential target for identification.

mtDNA consists of two rRNA genes, 13 protein coding genes (including cytochrome *c* oxidase subunit 1 (*cox1*), cytochrome *b* (*cob*), and NADH dehydrogenase subunit 1 (*nad1*)) and 22 tRNA genes, which are highly conserved within vertebrates and insects, with the exception of tRNAs. Because of its conserved sequences and its multiple copies, mtDNA is more readily detected than nuclear DNA markers and thus is widely used for taxonomic, population, and evolutionary investigations in eukaryotes including mammals, parasites, and arthropods [12]. However, the existence of mitochondrial pseudogenes integrated into a nuclear genome may sometimes impact negatively on the reliability of polymerase chain reaction (PCR)-based mitochondrial studies [13].

Being the terminal catalyst in the respiratory mitochondrial chain, the cytochrome *c* oxidase subunit I (*cox*1) of the mitochondrial genome is particularly suitable as a molecular marker for the taxonomic differentiation and evolutionary studies of insects because of its large size and the presence of highly conserved and variable regions. The COI amino acid sequence comprises 12 transmembrane helices (M1–M12), six external loops (E1–E6), five internal loops (I1–I5), one carboxyl (COOH), and one amino (NH_2) terminal. The nicotinamide dehydrogenase subunit 4 (*nad*4) gene is also useful for studying genetic variation both within and among populations of key parasitic helminthes.

1.4.2 Template Amplification and Product Detection

1.4.2.1 Template Amplification

The early molecular procedures are nonamplification based and suffer from poor sensitivity and time-consumingness. With the development of PCR and other nucleic acid amplification technologies (such as ligase chain reaction (LCR), nucleic acid sequence-based amplification (NASBA), transcription-mediated amplification (TMA), strand displacement amplification (SDA), rolling circle amplification (RCA), cycling probe technology (CPT), branched DNA (bDNA), and loop-mediated isothermal amplification (LAMP)), it has become possible to rapidly and specifically detect a single copy of nucleic acid in a few hours.

Given their efficiency, simplicity, robustness, and versatility, PCR and its derivatives have been widely adopted in both research and clinical laboratories for identification and determination of parasitic protozoa, helminths, and arthropods. The amplification formats have also evolved from single-round PCR to nested PCR, multiplex PCR, reverse transcription-PCR (RT-PCR), real-time PCR, and quantitative PCR (Q-PCR), and so on, leading to enhanced assay sensitivity (nested PCR) and versatility (multiplex PCR detecting multiple genes and/or organisms; and RT-PCR detecting RNA instead of DNA) as well as accurate quantitation (Q-PCR) and instant result availability (real-time PCR) [14].

In particular, PCR-restriction fragment-length polymorphism (PCR-RFLP) has been widely applied for the identification of closely related taxa of insects (e.g., Calliphoridae and Sarcophagidae families). Single-strand conformation polymorphism analysis (SSCP) is a PCR-based mutation scanning technique that is based on the different electrophoretic mobility of ssDNA in a nondenaturing gel. The SSCP analysis is useful for assessing genetic distance and identifying haplotypes existing in insect populations from different geographic areas belonging to the same species (e.g., flies or mosquitoes) or to identify morphologically indistinguishable species (e.g., mosquitoes). Random amplified polymorphic DNA (RAPD) analysis relies on the use of a single arbitrary primer under low annealing temperature conditions to amplify DNA sequences and screen genetic variation across a genome.

1.4.2.2 Product Detection

In its conventional (standard) form, PCR products are separated by agarose gel electrophoresis with or without modification (e.g., enzymatic digestion), stained with a DNA-binding dye (e.g., ethidium bromide or gel star), and visualized under UV light. For PCR products <100 bp or for distinction of products with minor size differences, polyacrylamide gel electrophoresis, and its derivatives (e.g., single-strand conformational polymorphism analysis (SSCP), denaturing gradient gel electrophoresis (DGGE), and temperature-gradient gel electrophoresis (TGGE)) may be employed.

To further improve the sensitivity of PCR product detection, enzymatic signal amplification (e.g., ELISA and flow cytometry), DNA microarray (also known as DNA chip, gene or genome chip, or gene array), and line probe assay (LiPA) (which uses a nitrocellulose strip instead of a glass, silicon, plastic chip as in the case of DNA array) have been developed. Recent advances in instrument automation and fluorescent dye chemistry facilitate instant monitoring of PCR amplicons (real-time PCR) without additional manual handling. Real-time PCR may be based on a double-stranded DNA intercalating dye (e.g., SYBR Green), or specifically designed probes that target a region of amplicon and incorporate a fluorescent dye. Examples of these probes include hydrolysis dual-labeled probes (TaqMan®), hybridization probes (LightCycler), molecular beacons, peptide nucleic acid (PNA) probes, TaqMan minor groove binding (MGB™) probes, locked nucleic acid (LNA®) primers and probes, and scorpions™. DNA sequencing analysis provides a most accurate way to ascertain the identity of PCR amplicons. The classical chain termination sequencing method (or Sanger method) utilizes primers or dideoxynucleotides that are labeled with radioactive isotope or fluorescent tag, and the sequencing products are detected by exposure to x-rays or UV-light. More recently, the 454 Genome Sequencer (Roche), the Genome Analyzer II (Illumina), and the SOLiD3 System (Applied Biosystems) (collectively referred to as "next-generation sequencing" sequencing or NGS technologies) have been adopted for high-throughput sequencing analysis of PCR products. The nucleotide sequences of the PCR amplicon are then compared with those stored at reference databases such as GenBank, and the phylogenetic relationships of related parasitic protozoa, helminths, and arthropods are displayed in the form of trees, constructed with distance matrix methods (resulting in phenograms), and maximum-parsimony methods (resulting in cladograms).

1.4.3 Data Analysis

Pairwise nucleotide comparison of test sample with publicly available gene databases (such as GenBank) by BLASTn analysis provides a straightforward means to identify the presence or absence of a particular sequence motif or even a single nucleotide and to ascertain the relatedness of organisms of interest. For evolutionary studies, multiple sequence alignment and the true homology between variable sites and portions of sequences must be addressed. One or more alignment approaches may be employed: (i) on the basis of secondary structure and functional domains; (ii) using one of a range of specialist alignment programs with various weighting options and gap penalties (e.g., ClustalX); (iii) by eye, often in relation to previously aligned sequences. Three main categories of phylogenetic analysis with molecular data include distance methods (e.g., phenetic methods), cladistics/parsimony, and maximum-likelihood analysis. Bootstrap analysis is the most commonly employed statistical method to provide support for the evolutionary relationships presented [15].

1.5 RESULTS INTERPRETATION, STANDARDIZATION, AND QUALITY CONTROL AND ASSURANCE

1.5.1 Key Performance Characteristics

The performance of a diagnostic assay is often evaluated by using several key parameters, including detection limit, sensitivity, specificity, accuracy, intra-assay precision, interassay precision, and linearity (as in the case of a quantitative assay). Detection limit (or limit of detection) is the lowest concentration or quantity of bacteria that can be detected by a given assay. Sensitivity is the percentage of samples containing bacteria of interest that are identified by the assay as positive for the bacteria. Specificity is the percentage of samples without bacteria of interest that are identified by the assay as negative for the bacteria. Accuracy (or trueness) is the degree of conformity of an assay's measurements to the actual (true) value. It is often estimated by analyses of reference materials or comparisons of results with those obtained by a reference method. The closer an assay's measurements to the accepted value, the more accurate the assay is. Precision is the degree of mutual agreement among a series of assay's individual measurements, values, or results. Usually characterized in terms of the standard deviation of the measurements, precision can be stratified into (i) repeatability—the variation arising using the same instrument and operator in a single run (i.e., intra-assay precision) or repeating during a short time period; and (ii) reproducibility—the variation arising using the same measurement process among different instruments and operators from one run to another (i.e., inter-assay precision) or over longer time periods. Linearity refers to the tendency of measurements by a quantitative assay to form a straight line when plotted on a graph. Data from linearity experiments may be subjected to linear regression analysis with an ideal regression coefficient of 1. In the case of a nonlinear curve, other objective, statistically valid methods may be utilized.

1.5.2 Results Interpretation

A positive result by a molecular assay for a given pathogen normally confirms the etiologic relationship if the clinical syndrome is compatible with the pathogen identified. Considering the sensitive nature of the amplified methods such as PCR, it is important to rule out the possibility of a false-positive result. Occasionally, false-positive results may originate from the low diagnostic specificity of the assay, in which primers bind to irrelevant sequences and occasionally a homologous sequence that is shared among related or unknown bacteria. More often, false-positive results in the molecular testing come from contamination, which may arise during manual handling of the samples in the testing laboratory either at the pre or postextraction (while setting up the PCR) stages. This risk is heightened when a high copy-number polynucleotide (or plasmid) is used as a quantification standard and distributed around the laboratory, contaminating reaction source. Additionally, contamination may be attributable to samples that are referred from other laboratories, which do not utilize manipulation techniques that are mandatory for the molecular testing. These may include the use of unplugged pipette tips, infrequent changing of gloves and using pipette for long periods without decontamination. Another cause of contamination is by amplification products from previous tests. Contamination may also occur by leakage from tubes or microtiter plates with lids not tightly closed or by breakage of glass capillaries leading to spillage of the amplification mixture. Besides the adoption of stringent laboratory practice, the risk of contamination with PCR products may be reduced by replacing nucleotide dTTP with dUTP in PCR, and implementing a digestion step with Uracil-DNA-glycosylase (UNG) to remove previous PCR products containing dUTP prior to each amplification reaction. Furthermore, inclusion of multiple negative controls, such as no-template controls (NTC) and no-amplification controls (NAC) may help identify the likely source of contamination and prevent false-positive results. Moreover, microbial DNA may come with PCR reagents.

Similarly, a negative result by a molecular assay for a given pathogen normally indicates the absence of the pathogen. However, it is equally important to rule out the possibility of false-negative results. One possible cause is due to the low sensitivity of the assay employed. Alternatively, insufficient amount of bacteria may be present in the sample (due to sample degradation or prior antibiotic treatment). Another may be due to the impurity of the processed sample. Enzymes (e.g., DNA polymerase, reverse transcriptase) used in PCR and RT-PCR are impeded by components in blood and feces (e.g., heme, hemoglobulin, lactoferrin, immunoglobulin G, leukocyte DNA, polysaccharides, and urea), in foods (e.g., phenolics, glycogen, calcium ions, fat, and other organic substances), in environmental specimens (e.g., phenolics, humic acids, and heavy metals); and in added anticoagulants (e.g., EDTA and heparin) as well as nucleic acid purification reagents (e.g., detergents, lysozyme, NaOH, alcohol, EDTA, EGTA, phenol, and high salt concentrations) [16]. Any impurities and contaminations present in the samples after nucleic acid isolation may contribute to false-negative results. A useful way to determine the effective of nucleic acid purification procedure for removing inhibitory substances is to spike samples with well-defined DNA or RNA prior to and after sample preparation (as process and amplification internal controls). In light of the high sensitivity of PCR, the occurrence of false-negative results is probably a truly underestimated problem.

1.5.3 Standardization and Validation

As molecular tests such as PCR and sequencing offer improved sensitivity, specificity, accuracy, precision, and result availability for identification and diagnosis of parasites, they have been increasingly adopted and applied in routine diagnostic laboratories. Considering the possibility of false-positive and false-negative results that may occur in these highly sensitive tests, it is essential to properly standardize and validate them prior to their adoption, and to put in place appropriate quality control measures to ensure their consistent performance.

Standardization of molecular tests addresses the need for standardized reagents and common units, contamination control mechanisms, inhibition control mechanisms, clinically relevant dynamic ranges and internal controls, and so on. Validation helps to verify the sensitivity, specificity, accuracy, repeatability (intra-assay precision), reproducibility (inter-assay precision), detection limit, and linearity (if quantitative) of molecular tests.

Before validating a method, it is important to have all instruments calibrated and maintained throughout the testing process. The validation process may involve a series of steps including: (i) testing of dilution series of positive samples (or plasmid construct) to determine the limits of detection of the assay and their linearity over concentrations to be measured in quantitative test (using minimal number of reference calibrators such as previously tested patient samples or pooled sera); (ii) evaluating the sensitivity and specificity of the assay, along with the extent of cross-reactivity with other genomic material; (iii) establishing the day-to-day variation of the assay's performance; (iv) assuring the quality of assembled assays using quality control procedures that monitor the performance of reagent batches; and (v) aligning the in-house primer and probe sequences with a genome sequence databank to avoid extended specificity testing.

1.5.4 Quality Control and Assurance

1.5.4.1 Quality Control

Quality control strategies for nucleic acid-based tests include (i) designation of a "clean" area for reaction setup (e.g., room under negative air pressure; positive-displacement pipettes; aerosol-block pipette tips; UV-equipped PCR cabinet); (ii) use of personal protective equipment (PPE) (e.g., disposable gloves and lab coats to prevent introduction of contaminating DNA or nucleases); (iii) use of uracil-*N*-glycosylase (UNG) in real-time PCR (to eliminate cross-over amplicon contamination); (iv) use of a "hot-start" method (to minimize false priming events by withholding a crucial reaction component until appropriate temperature is reached); (v) Use of external positive and negative controls (to monitor reaction performance and contamination) and homologous or heterologous internal controls (to monitor presence of inhibitors).

A variety of test controls may be considered for diagnostic PCR. These include (i) internal amplification control (IAC) (negative sample spiked with sufficient pathogen and processed throughout the entire protocol); (ii) processing positive control (PPC) (negative sample spiked with sufficient closely related, but nontarget, strain processed throughout the entire protocol); (iii) reagent control (blank) (containing all reagents, but no nucleic acid apart from the primers); (iv) premises control (tube containing the master mixture left open in the PCR setup room to detect possible contaminating DNA in the environment (carried out at regular intervals as part of the quality assurance program); (v) standard (three to four samples containing 10-fold dilution series of known number of target DNA copies in a range).

1.5.4.2 Quality Assurance

One way to assess preparedness of the diagnostic laboratories is through the conduct of an external quality assurance (EQA) program providing characterized specimens containing pathogens of interest. The design of a quality assurance program has the following components: (i) internal quality control (IQC) materials are distributed every month and comprising three pools of clinical samples of known pathogen status (typically one negative, one positive containing 1 log10 over the lower limit of detection of the assay, and one low positive containing up to 1 log10 of the lower limit of detection of the assay). These are incorporated in test runs on a weekly basis. The purpose of IQC is to provide samples of known status for repeated testing in parallel with clinical samples to ensure reproducibility of the test system in an individual laboratory. (ii) EQA distributions of panels of five unknown samples distributed quarterly. Results are returned to the QA laboratory for assessment. EQA compares the performance of different testing sites using specimens of known, but undisclosed, content. (iii) Aliquots of all samples sent from the reference laboratory are posted back to Site A for repeat testing to check for integrity of the pools and for transport problems. (iv) A final element of the pilot program involves Sites B, C, and D sending an aliquot of every 50th sample to Site A to check for reproducibility. (v) A detailed record of distributions is kept to provide an audit trail.

1.6 CONCLUSION

Parasites are eukaryotic organisms that are separated into three categories: protozoa, helminths, and arthropods. Although a majority of the 800,000 or so parasite species are free-living or form symbiotic relationship with their hosts, causing no obvious adverse effects, a small proportion have the capacity to induce clinical symptoms of various severity in affected individuals. Due to the fact that many parasite species share considerable morphological and biological similarities, and that clinical features resulting from the infection with parasitic protozoa, helminths, and arthropods are often nonspecific, it is important that laboratory techniques are applied for their correct identification in order to facilitate the implementation of effective control and prevention measures.

While phenotypic procedures such as macroscopic and microscopic examination, biochemical and immunological assays, and *in vitro/in vivo* culture provide a useful means for diagnosis of parasitic infections, they suffer from the drawbacks of being slow, variable at times, costly, and technically demanding. The recent development of molecular methodologies offers an opportunity to improve the sensitivity, specificity, and speed of parasite detection and identification. In particular, PCR amplification of ribosomal RNA and other gene regions followed by sequencing analysis represents a rapid and accurate approach for identification and differentiation of parasitic pathogens.

There is no doubt that with an increasing adoption and utilization of molecular methods in and clinical laboratories, reliable identification and detection of parasitic protozoa, helminths, and arthropods directly from patient materials is no longer an impossible dream. Furthermore, since molecular techniques enable precise epidemiological tracking and phylogenetic analysis of parasitic pathogens involved in disease epidemics, they will play an even more prominent role in the future control and prevention campaigns against parasitic infections in human populations.

REFERENCES

1. http://www.uniprot.org/; last accessed on 20 May 2011.
2. http://www.data.gbif.org/; Global Biodiversity Information Facility data portal, published by Field Museum of Natural History, Museum of Vertebrate Zoology, University of Washington Burke Museum, and University of Turku; last accessed on 20 May 2011.
3. Katz, N., Chaves, A., and Pellegrino, J. A simple device for quantitative stool thick-smear technique in *Schistosomiasis mansoni*. *Rev. Inst. Med. Trop. Sao Paulo* 14, 397, 1972.
4. Allen, A.V.H., and Ridley, D.S. Further observations on the formol-ether concentration technique for fecal parasites. *J. Clin. Pathol.* 23, 545, 1970.
5. Ramsay, A. et al. A field evaluation of the formol detergent method for concentrating fecal parasites. *J. Trop. Med. Hyg.* 94, 210, 1991.

6. Bell, D.R. Diagnosis of parasitic diseases by filtration. *Ann. Soc. Belg. Med. Trop.* 55, 489, 1975.
7. Peters, P.A. et al. Rapid, accurate quantification of schistosome eggs via nuclepore filters. *J. Parasitol.* 62, 154, 1976.
8. Scheiber, P., Braun-Munzinger, R.A., and Southgate, B.A. A new technique for the determination of microfilariael densities in onchocerciasis. *Bull. World Health Organ.* 53, 130, 1976.
9. Arakaki, T.J. et al. Efficacy of agar-plate culture in detection of *Strongyloides stercoralis* infection. *Parasitology* 76, 425, 1990.
10. Barker, G.C. Microsatellite DNA: A tool for population genetic analysis. *Trans. R. Soc. Trop. Med. Hyg.* 96, S21, 2002.
11. Hillis, D.M., and Dixon, M.T. Ribosomal DNA: Molecular evolution and phylogenetic inference. *Quart. Rev. Biol.* 66, 411, 1991.
12. Meusemann, K. et al. A phylogenomic approach to resolve the arthropod tree of life. *Mol. Biol. Evol.* 27, 2541, 2010.
13. Jex, A.R., Littlewood, D.T., and Gasser, R.B. Toward next-generation sequencing of mitochondrial genomes—Focus on parasitic worms of animals and biotechnological implications. *Biotechnol. Adv.* 28, 151, 2010.
14. Bretagne, S., and Costa, J.-M. Towards a nucleic acid-based diagnosis in clinical parasitology and mycology. *Clin. Chim. Acta* 363, 221, 2006.
15. Liu, L. et al. Estimating species trees using multiple-allele DNA sequence data. *Evolution* 62, 2080, 2008.
16. Wilson, I.G. Inhibition and facilitation of nucleic acid amplification. *Appl. Environ. Microbiol.* 63, 3741, 1997.

Section I

Sarcomastigophora

2 Acanthamoeba

Hélène Yera, Muriel Cornet, and Jean-Louis Bourges

CONTENTS

2.1 INTRODUCTION

2.1.1 Classification, Morphology, and Biology

Acanthamoeba spp. are found as free-living amoebae in the aquatic environment (bottled water, tap water, chlorinated swimming pools, jacuzzis, seawater, sewage, power plant, etc.), soil and air, and can be isolated from rhino pharynx and stools of healthy humans [1]. Recently, the International Society of Protozoologists adopted a system based on modern morphological approaches, biochemical pathways, and molecular phylogenetic analysis [2]. According to this scheme, *Acanthamoeba* is classified under Super Group Amoebozoa: Acanthamoebidae.

During its life cycle, *Acanthamoeba* undergoes two stages: a vegetative trophozoite and a resistant cyst. Spiny surface projections, called acanthopodia, are presented on the trophozoites and distinguish *Acanthamoeba* from other free-living amoebae that infected human beings (*Balamuthia mandrillaris*, *Naegleria fowleri*, and *Sappinia pedata*) [3]. During the trophozoite stage, *Acanthamoeba* actively feed on bacteria, algae, yeasts, or small organic particles. This stage is maintained at neutral pH, appropriate temperature (i.e., 30°C) and osmolarity (50–80 mOsmol). Harsh conditions (i.e., lack of nutrients, extreme variation in environmental pH, temperatures, and osmolarity) induce the transformation of trophozoites into cyst stage. The cysts possess a double wall and pores called ostioles, which are capable of monitoring environmental changes [4]. The cysts are resistant to several biocides, chlorination and antibiotics, and survive low temperature (0–2°C) [3]. On return to favorable growth conditions, the trophozoite emerges from the cyst by dislodging the operculum which covers the pore and by reverting to a trophic form [5].

Acanthamoeba spp. are nonopportunistic or opportunistic pathogens in humans and are responsible for corneal infections in immunocompetent individuals and for brain or disseminate infections in immunocompromised individuals. The host immune status, as well as the material characteristics of contact lens, plays a part in pathogenesis [6], which is also dependent on strain virulence. Pathogenic isolates exhibit higher binding to host cells and increased cytotoxicity [7]. Extracellular proteases are major *Acanthamoeba* virulence factors, being involved in host tissue breakdown [8,9].

2.1.2 Clinical Features and Pathogenesis

Acanthamoeba keratitis (AK) is a sight-threatening infection. The greatest populations at risk are soft contact lens wearers associated with poor contact lens hygiene [10] or ineffective lens disinfection [6] and they have been mainly described in industrialized countries. In developing regions, data concern mostly noncontact lens wearers and subjects with corneal injuries with plants or soil [11,12]. AK cases could also be observed after dirty water projection or dust in eye [11,13].

Contact lenses serve as vectors for transmitting trophozoites to the corneal epithelium and abrasion or mild trauma to the corneal surface enhances the level of trophozoite binding and is required for the establishment of corneal infection [8]. Trophozoites adhere to mannosylated glycoproteins upregulated on the surfaces of wounded corneal epithelium and secrete a mannose-induced cytolytic protein [8,14]. They breech Bowman's membrane, enter the collagenous stroma and produce several proteases, including a collagenase and a plasminogen activator, that contribute to the dissolution of the corneal stroma. Then, trophozoites cluster around corneal nerves, producing radial keratoneuritis and severe pain [8].

At the onset of the infection, the clinical diagnosis is difficult to assess due to the lack of distinct pathological signs and due to the similarity with herpetic keratitis. Most patients complain of photophobia, pain, tearing, and blush. When the clinical signs become typical (keratitis with a stromal infiltrate) and severe, *Acanthamoebae* evolves deep in the cornea. AK develops in corneal ulcer that tending to heal with scar tissue, opacification of the cornea and decreased visual acuity, then in corneal perforation with iris prolapse, pus in the anterior chamber (hypopyon), or panophthalmitis. Patients can undergo penetrating keratoplasty "à chaud" to prevent corneal perforation. AK usually reaches one eye, but it can be occasionally bilateral. Topical treatment requires a combination of eye drops containing biguanides and diamidines (i.e., 0.02% polyhexamethylene biguanide with hexamidine or 0.1% propamidine and 12,500 UI/mL colimycin) and can be efficient if administered at disease onset (i.e., each hour for 3 days, then each two hours the day and each four hours in the night for 8 days, then five to four times a day for a number of weeks according to the clinical evolution and the microbiological results). In severe cases presenting a corneal ulceration and stromal infiltrate, the treatment should be prolonged to avoid recurrences and the corneal scars may require keratoplasty to recover visual acuity. But, the ocular toxicity of the antiamoebicidal agents may compromise the effectiveness of the treatment [15]. Thus, an early AK diagnosis significantly improves the prognosis if appropriate therapy is administrated [16].

Acanthamoeba infections out of the cornea have been rarely described [15]. One case of *Acanthamoeba* chorioretinitis was reported after penetrating keratoplasties in a patient presenting recurrent keratitis [17], one case of *Acanthamoeba* endophthalmitis in a patient with acquired immunodeficiency syndrome (AIDS), presenting also a disseminated infection [18], one case of scleral invasion, and one case of keratitis, uveitis, and meningoencephalitis. *Acanthamoeba* causes granulomatous amoebic encephalitis (GAE), primarily in patients with HIV/AIDS or suffering from chronic diseases like diabetes or have undergone organ transplantation or are otherwise weakened with no recent history of exposure to recreational freshwater [1,3,19,20]. A few cases have been documented from individuals with no obvious signs of immunodepression [1]. *Acanthamoeba* may disseminate and be responsible for infections of the skin, the lung, the sinus, and the nasal passages, especially in patients with HIV/AIDS [3,4,19,20]. Infections of the skin but without dissemination to the central nervous system (CNS) have been observed [1].

Because the prodromal period may last several weeks or months, the precise portal of entry is not clearly determined. Trophozoites and/or cysts may enter the body by a nasopharyngeal route or by breaks in the skin, resulting in hematogenous dissemination to the lung and brain. Inhalation of cysts may lead to the same dissemination [1]. Trophozoites and cysts are often observed in the brain around blood vessels and within necrotic tissue [19]. Serine proteases produced by pathogenic isolates may increase the blood–brain barrier permeability [21] and degraded extracellular matrix, fibrinogen, IgG, IgA, albumin, and hemoglobin contributing to brain tissue damage [22]. In immunocompromised patients, the chronic inflammatory reaction is minimal and without multinucleated giant cells [1,19].

The infections are perhaps either misdiagnosed or undiagnosed due to their nonspecific clinical signs. At the onset, GAE is slow and develops as a chronic disease for several weeks to months. The clinical manifestations include headache, stiff neck, confusion, nausea, vomiting, fever, lethargy, cerebellar ataxia, visual disturbance, hemiparesis, and coma [1,3]. GAE can mimic meningeal tuberculosis, brain tumors, viral encephalitis, neurocysticercosis, cerebral toxoplasmosis, and acute disseminated encephalomyelitis [20]. The prognosis of these infections is poor. The therapeutic strategies require a combination of agents with a reduced successful clinical outcome if treatment is administrated late [1].

2.1.3 Diagnosis

2.1.3.1 Conventional Techniques

Conventional laboratory diagnosis of AK has been dependent on direct examination of deep corneal scrapings and culture.

2.1.3.1.1 Direct Examination

Corneal samples can be smeared onto glass slides and stained with Gram, May–Grünwald–Giemsa pH 7.4, Giemsa, lactophenol cotton blue, or calcofluor-white for microscopic examination. The *Acanthamoeba* trophozoites are 15–50 μm in length and have a single nucleus with a prominent central nucleolus. The cysts are 10–25 μm long and have a double wall; most of the species at the points where the two layers join are pore closed by an operculum, frequently giving rise to a polygonal shape.

2.1.3.1.2 Culture

The material obtained from the corneal samples is discharged into a small portion of 2% agar gel and cultivated into flasks containing 5 mL of 0.9% sodium chloride and a killed *Escherichia coli* suspension (0.5 McFarland). The flasks are incubated at 30°C for 20 days and observed daily for detection of amoebae by inverted microscopy. *Acanthamoeba* trophozoites show slow movements by extending hyaline pseudopodia, and may contain one or more contractile vacuoles. The cysts are also characterized by their double wall (Figure 2.1).

These conventional methods lack sensitivity. About 60% of AK cases are misdiagnosed using direct detection methods [23–25] and positive cultures may be obtained only after prolonged incubation, particularly if antibiotics with potential amoebicidal activity have been used previously [23].

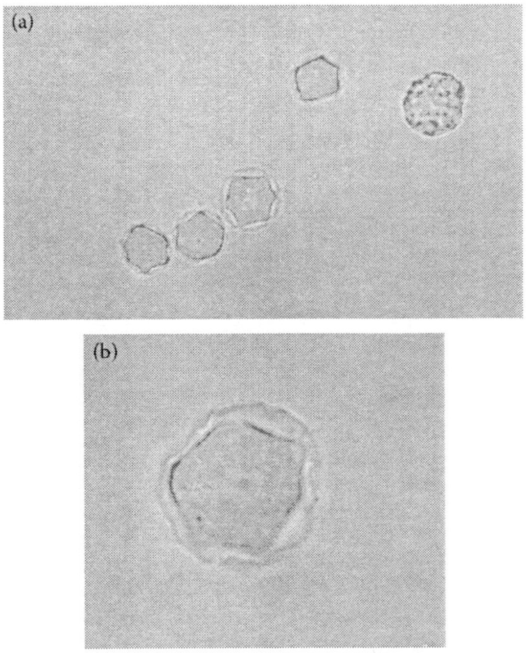

FIGURE 2.1 *Acanthamoeba* trophozoite and cysts in culture 200× magnification (a) and *Acanthamoeba* cyst in culture 400× magnification (b) (M. Cornet).

Conventional laboratory diagnosis of other *Acanthamoeba* infections can be accessed by histologic examination, serology, and culture.

2.1.3.1.3 Histologic Examination

Detection of *Acanthamoeba* is often confirmed postmortem with histopathologic examination. Trophozoites and cysts can be observed in tissue sections after hematoxylin–eosin, periodic acid–Schiff, or trichrome staining. The amoebae are distinguishable from host cells by their prominent central nucleolus and their location in the perivascular areas in brain tissue [1]. *Acanthamoeba* spp. infecting the CNS are not readily found in the cerebrospinal fluid (CSF), although a few cases reported their isolation from the CSF [1]. Immunostaining with fluorescent antibodies is useful to identify *Acanthamoeba* trophozoites from tissue sections [26].

2.1.3.1.4 Testing for Acanthamoeba Antibodies

Detection of circulating antibodies by indirect immunofluorescence can be utilized to screen acanthamoebiasis in patients presenting clinical, laboratory, and neuroimagings data suggestive of amoebic encephalitis [27]. Low antibody titers are found in >50% of asymtomatic individuals [28]. However, high titers (1:128 or more) or significant increase in titers have been reported in GAE [29,30]. Nevertheless, antibody testing seems not to reflect the real picture of infection in immunocompromised patients [30].

2.1.3.1.5 Culture

Disrupted tissues or CSF are inoculated on nonnutrient agar plates covered with killed *E. coli* and incubated at 30°C or 37°C [30–32] or on sea-salt agar plates seeded with a *Phyllobacterium* species [33]. The plates are observed daily for the detection of amoebae by inverted microscopy.

2.1.3.2 Molecular Diagnosis

2.1.3.2.1 Standard PCR

Several studies have confirmed the usefulness of standard polymerase chain reaction (PCR) in corneal samples for the early and sensitive diagnosis of AK and positive PCR results could be obtained in samples from patients previously treated with antibiotics presenting amoebicidal activity [23,34,35].

Acanthamoeba standard PCR targeting regions of the *Acanthamoeba* nuclear small-subunit rRNA (*Rns*) gene [12,23,24,34–37] or the *Acanthamoeba* nuclear large-subunit rRNA gene [37] have been reported. On *Rns* gene, the following primer pairs have been used: P1GP.2379for–2632rev primers [23], ACARNA.1383for–1655rev primers [23,24,35], Nelson primers [24,34,35], and JDP (JDP1–JDP2) primers [12,35–37]. Nelson primers were designed to produce an amplimer nearly identical to that of the P1GP primers [24]. The combination of primer pairs increased the sensitivity of PCR up to 94% [23,24,35,37]. A semi-nested *Acanthamoeba* PCR was developed with JDP1 and JDP2 as the outer primers in a first PCR then with A1 as the inner forward primer in a second PCR [38]. The semi-nested PCR showed better sensitivity than the simplex PCR with JDP1 and JDP2 (100% vs. 27%) and 100% specificity. A duplex PCR assay was developed with primers targeting the *Acanthamoeba* nuclear large-subunit rRNA gene and JDP primers [37]. The assay based on the two targets provides more reliable detection than the assay performed only with JDP primers and based on the *Rns* gene.

Recently, PCR has been employed as an adjunct in the diagnosis of noncorneal acanthamoebiasis [20,39–43]. The assays can be performed in CSF, bronchoalveolar lavages, brain tissues, and others tissues. Positive PCR results can also be obtained in formalin-fixed paraffin-embedded samples, but this testing scheme shows limited sensitivity and reproducibility [42].

Standard PCR targeting the *Acanthamoeba* mitochondrial small-subunit rRNA (*rns*) gene with the Aca16Sf1010–Aca16Sr1180 primers [20,39,42] or the *Acanthamoeba Rns* gene with the JDP primers have been described [40,41,43].

2.1.3.2.2 Real-Time PCR

Recently, *Acanthamoeba* real-time PCR assays have been reported. First, a TaqMan® PCR assay targeting the *Rns* gene was designed for the quantification of *Acanthamoeba* cells from cultures [44]. Second, a TaqMan PCR assay targeting the *Acanthamoeba Rns* gene has been designed in a triplex assay developed for the simultaneous detection of *Acanthamoeba* spp., *Balamuthia mandrillaris*, and *Naegleria fowleri* in cultures and clinical samples [45]. This assay recognizes isolates belonging to genotypes T1, T4, T7, and T10 and was more broadly reactive than the first. Third, TaqMan PCR assays targeting the *Rns* gene has been adapted from the two first assays by adding locked nucleic acids to probe and were used on ocular samples [46]. Finally, a fast real-time TaqMan PCR assay targeting the *Acanthamoeba rns* gene was devised for detecting strains available from the American Type Culture Collection (ATCC) related to genotypes T1, T2, T4, T5, T7 to T9 and was tested on ocular isolates [47].

2.1.3.2.3 *Molecular Typing of Acanthamoeba Strains*

The genus *Acanthamoeba* can sometimes be differentiated by its morphology into three morphological groups [48]. Classification based on this criterion is not reliable due to alternating cyst appearance and species-level identification requires more complex methods [49].

The molecular classification of *Acanthamoeba* isolates is based on amplification and sequencing of the nearly complete *Rns* genes [50,51]. Seventeen genotypes, designated T1–T17, have been described [50–54]. Parts of the *Rns* gene contain sufficient sequence differences that are phylogenetically relevant. Such sequence corresponds to the variable region DF3 of the amplimer ASA.S1, obtained with the JDP primer pair, which discriminates most of the *Acanthamoeba* genotypes [36]. Sequence analysis of complete *Rns* genes or of the DF3 region is useful for the characterization of the *Acanthamoeba* genotypes in clinical or environmental samples. Around 90% of *Acanthamoeba* isolates associated with AK in different areas belong to genotype T4 [35–37,55–59]. Other closely related genotypes (T3, T11) and more distantly related genotypes (T2, T6, T5, T15) have been identified in AK [35,36,56,58,60,61]. The genotype T4 is also predominant in non-AK infections and in environmental sources [56,58]. The genotype T4 is predominant in GAE, but isolates from genotypes T1, T2, T5, T7, T10, and T12 have also been described [4,42,43,56,62].

The analysis of the complete *Acanthamoeba* mitochondrial small-subunit rRNA (*rns*) genes confirmed the identification of the *Acanthamoeba* genotypes based on *Rns* genes [63]. Recently, examination of the *Acanthamoeba* first internal transcribed spacer (ITS1) demonstrates that ITS1 is a more variable region than the *Rns* gene and that the ITS1 clusters correlate well with the *Rns* clusters [64]. In fact, the variability in ITS1 is 10-fold higher than that in the *Rns* gene, and therefore it can be used to discriminate between different genotypes or even within genotype T4.

2.2 METHODS

We provide here techniques to detect and identify *Acanthamoeba* in clinical samples by using standard or real-time PCR assays and sequencing the DF3 region of the *Rns* gene. JDP1 and JDP2 currently represent the best primers set to genus-specific detection and identification of all *Acanthamoeba* genotypes. Some real-time PCR assays are promising but their specificity remains to be evaluated before being routinely used. However, as real-time PCR is easier, faster to perform, and avoids amplicons contaminations, they are therefore also included and compared in this section.

2.2.1 Sample Preparation

2.2.1.1 Sample Collection

In patients with clinical signs suggesting AK, the identification (and if possible semiquantification) of *Acanthamoeba* in the corneal structures should be conducted. Samples consist of either corneal scrapings or corneal biopsies. Scrapings are performed with sterile sponges, stainless-steel vaccinostyles, or surgical blades in the periphery of the lesion, deeply in the cornea. Vaccinostyles or surgical blades may give a better quality of corneal scrapings than sponges. Samples are collected in sterile microtubes and conserved at −20°C before processing in the laboratory. Fluorescein solution or oxybuprocain anesthetic, used routinely by ophthalmologists for corneal examination and sampling, inhibit even diluted 20 times the capacity of PCR assays to detect *Acanthamoeba* DNA [65]. To minimize the risks of misdiagnosis, the eye surface should be rinsed intensively with 0.9% sodium chloride solution before sampling. Polyhexamethylene biguanide, used for AK treatment, inhibits, even diluted 640 times, the capacity of PCR assays to detect *Acanthamoeba* DNA [46]. The DNA extraction step can reduce but not eliminate the inhibitory effects of the drug. To avoid misdiagnosis, sampling should be performed before the beginning of specific treatment.

2.2.1.2 DNA Extraction

Corneal samples or 10–25 mg of tissues previously cut up or disrupted are discharged in 200 μL sterile phosphate-buffered saline (pH 7.2). Biological fluids (i.e., CSF) are pellet in a 200 μL volume.

DNA is extracted manually by using the commercial kit: QIAamp® DNA Mini kit (Qiagen, France) with the tissue protocol. Briefly, the sample is mixed with 180 μL of buffer ATL and 20 μL of proteinase K, and incubated for one night at 56°C. Then 200 μL of buffer AL is added and the sample is incubated for 10 min at 70°C. Next, 200 μL of ethanol 100% is added. The mix is applied in QIAamp column and centrifuged for 1 min at 6000 × *g* (8000 rpm) to bind DNA to the silica membrane inside the column. The DNA is washed first with 500 μL of buffer AW1, which is eliminated by centrifugation 1 min at 6000 × *g* (8000 rpm) then with 500 μL of buffer AW2, which is eliminated by centrifugation 5 min at 17,900 *g* (13,000 rpm). A last centrifugation is performed for 1 min at 17,900 × *g* (13,000 rpm) to eliminate residual buffer. The DNA is eluted by adding 100 μL of sterile distilled water in the column, incubating 5 min and centrifuging for 1 min at 6000 × *g* (8000 rpm).

DNA from corneal samples can also be extracted automatically by the MagNa Pure® Compact (Roche Diagnostics, France) automate using the MagNA Pure Nucleic Acid isolation kit (Roche Diagnostics) after 10 min proteinase K pretreatment [47,66]. Briefly, the samples are mixed with 200 μL of ATL buffer and 40 μL of proteinase K, and incubated at 56°C for 10 min. The mix is applied in the automate system for DNA isolation based on magnetic-bead technology with the MagNA Pure Nucleic Acid isolation kit. DNA is eluted in 100 μL of sterile distilled water.

2.2.2 Detection Procedures

2.2.2.1 Standard PCR Detection

DNA extracts can be examined with three different simplex PCR assays targeting the *Acanthamoeba Rns* gene with the Nelson (forward: 5′-GTTTGAGGCAATA ACAGGT-3′ and reverse: 5′-GAATTCCTCGTTGAAGAT-3′) [24], ACARNA (ACARNA.1383 for: 5′-TCCCCTAGCAGC TTGTG-3′ and ACARNA.1655rev: 5′-GTTAAGGTCTCG TTCGTTA-3′) [23], and JDP (JDP1:5′-GGCCCAGATCG TTTACCGTGAA-3′ and JDP2:5′-TCTCACAAGCTGCTA GGGGAGTCA-3′) primers [55].

2.2.2.1.1 Procedure

1. Prepare PCR mixture (50 μL) containing 5 μL of GeneAmp 10x PCR Gold buffer (Applied Biosystem, France), 5 μL of 25 mM of $MgCl_2$, 8 nmol of each DNTP (dATP, dCTP, dGTP), 16 nmol of dUTP, 10 pmol of each primer set (i.e., Nelson, ACARNA, or JDP), 0.4 U of Uracil-DNA Glycosylase heat-labile (Roche Diagnostics, France), 3 U of AmpliTaq Gold DNA polymerase (Applied Biosystems), sterile distilled water to a volume of 40 μL, and 10 μL of DNA (neat or diluted 10-fold). Include a tube with 40 μL of PCR mixture and 10 μL of sterile water as a negative control.
2. Subject the PCR tubes to the following cycling program: one cycle of 94°C for 5 min; 44 cycles of 94°C for 30 s, 58°C for 1 min, and 72°C for 40 s; one cycle of 72°C for 15 min.
3. Separate PCR products by 2% agarose gel containing ethidium bromide and visualized on ultraviolet lamp.

Note: DNA extracts can also be tested with a triplex PCR assay using the three primer sets in a 25 μL reaction volume containing 2.5 μL of incubation buffer T. Pol (Q-Biogene, MP Biomedicals), 5 μL of 25 mM of $MgCl_2$, 4 nmol of each DNTP (dATP, dCTP, dGTP), 8 nmol of dUTP, 5 pmol of each primer, 0.2 U of Uracil-DNA Glycosylase heat-labile (Roche Diagnostics), 1 U of Taq DNA Pol (Q-Biogene, MP Biomedicals), sterile distilled water to 20 μL, and 5 μL of DNA (neat or diluted 10-fold). A tube with 20 μL of PCR mixture and 5 μL of sterile water is included as a negative control. The tubes are subjected to one cycle of 94°C for 10 min; 44 cycles of 94°C for 30 s, 58.5°C for 1 min 20 s, and 72°C for 1 min 20 s; one cycle of 72°C for 10 min. Then, 20 μL of PCR products is electrophoresed in 2% agarose gel containing ethidium bromide and visualized on an ultraviolet lamp.

Positive samples give single 229, 272, and 423–551 bp fragments using the Nelson, ACARNA, and JDP primer pairs, respectively. We have detected one to five cysts from *Acanthameba* suspensions with either simplex or triplex PCR assays according to the assay (Figure 2.2).

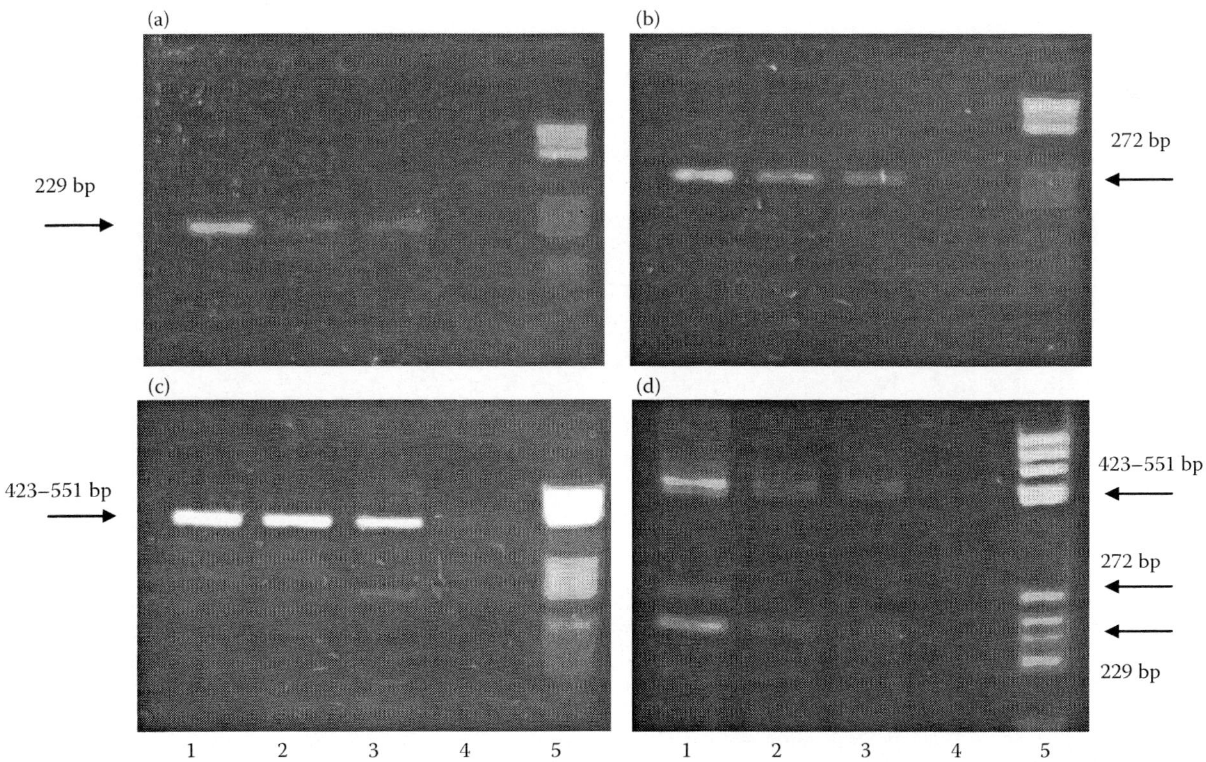

FIGURE 2.2 PCR detection of *Acanthamoeba*. Simplex PCR assays with the Nelson primers (a), the ACARNA primers (b), the JDP primers (c). Triplex PCR assay with the three primer pairs (d). In a–d, lanes 1–4, a treatment with proteinase K overnight when applying QIAamp DNA Mini kit tissue extraction protocol (Qiagen) (with lane 1 containing 50 cysts; lane 2, 5 cysts; lane 3, 1 cyst; and lane 4, negative control); lane 5, molecular weight marker. bp, base pairs.

2.2.2.2 Real-Time TaqMan PCR Detection

DNA extracts can be examined using real-time TaqMan PCR assays targeting the *Acanthamoeba Rns* gene with the Acant primers-probe set (AcantF900:5′-CCCAGATCGTTTACCG TGAA-3′, AcantR1100:5′-TAAATATTAATGCCCCCAAC TATCC-3′, and probe AcantP1000:5′-(FAM)-CTGCCACC GAATACATTAGCATGG-(TAMRA)-3′) [45] or the Ac primers-probe set (AcF1:5′- CGACCAGCGATTAGGAGACG-3′, AcR1:5′-CCGACGCCAAGGACGAC-5′, and probe AcP1:5′-(VIC)-TGAATACAAAACACCACCATCGGCGC-(TAMRA)-3′) [44].

They can be examined also using the TaqMan PCR assay targeting the *Acanthamoeba rns* gene which employs forward primer: 5′-GCAGTCGCGGTAATACGGA-3′, reverse primer: 5′-ACCACCTACGCACCCTTTACA-3′, and probe: 5′-(FAM)-AGTGTTATTCGCATTGACTGGGTGTAA-(TAMRA)-3′ [47].

We have adapted the assays for reactions in ABI Prism 7000 sequence detector system (Applied Biosystems).

2.2.2.2.1 Procedure

1. Prepare real-time PCR mixture (25 μL) containing 12.5 μL of TaqMan Universal Master Mix containing Uracil-DNA Glycosylase (Applied Biosystems), 12.5 or 20 pmol of each primer set, 12.5 or 5 pmol of TaqMan probe, sterile distilled water to 20 μL, and 5 μL of DNA (neat or diluted 10-fold) using the *rns* primers–probe set or the Acant and Ac primers–probe ones. Include a tube containing 20 μL of PCR mixture and 5 μL of distilled water as a negative control.
2. Subject the reaction mixture to one cycle of 50°C for 2 min; one cycle of 95°C for 10 min; 45 cycles of 95°C for 15 s, and 60°C for 1 min; using an ABI Prism 7000 sequence detector system (Applied Biosystems).
3. Detect and analyze data using the ABI Prism 7000 system software (Applied Biosystems).

Note: Positive samples result in an increase in fluorescence during the amplification. The kinetic of fluorescence takes the form of a sigmoidal curve when plotted as cycle number versus fluorescence emission. The point at which fluorescence surpasses the background noise, called the threshold cycle, is proportional to starting template concentration.

Our extraction protocol enabled us to detect 0.1 cyst per reaction using the *rns* primers–probe set, 0.5 cyst using the Acant primers–probe set or the Ac primers–probe one.

In order to assess the detection rate of PCR assays, the real-time and the standard PCR assays have been tested with *Acanthamoeba* DNA extracts from 17 cultures of ocular samples and one corneal scraping from patients with clinical signs or risk factors suggesting AK. The standard PCR assays using the Nelson or the JDP1–JDP2 primers and the real-time PCR assay targeting the *Acanthamoeba rns* gene gave the best detection rate in the *Acanthamoeba* genotypes studied (genotypes T3, T5, T6, T11, and T14). The DNA extracts from two *Hartmannella* spp. cultures were also tested. Only ACARNA primers led to false-positive results with *Hartmannella* spp. DNA (Table 2.1).

2.2.2.3 Genotyping

The DF3 sequences of the amplimer ASA.S1, obtained with the JDP primer pair, can be used to identify most *Acanthamoeba* genotypes [36]. The amplimers are purified using a PCR purification kit (e.g., QIAquick PCR Purification kit, Qiagen) and are sequenced in both directions using an automated DNA sequencer. In 30 μL of positive PCR reaction is mixed with five volumes of buffer PBI and applied to the QIAquick column. DNA binds to the silica membrane after centrifugation for 1 min at 17,900 × *g* (13,000 rpm). DNA is then washed with 750 μL of buffer PE, which is eliminated by centrifugation 1 min at 17,900 *g* (13,000 rpm). A last centrifugation is performed for 1 min at 17,900 × *g* (13,000 rpm) to eliminate residual buffer. The DNA is eluted by adding 50 μL sterile distilled water in a column, incubating for 5 min and centrifuging for 1 min at 17,900 × *g* (13,000 rpm). An amount of 60 ng of purified amplimer is mixed with 4 pmol of JDP1 primer or JDP2 primer and is sequenced in an automatic sequencing platform.

The amplimers producing multiple sequencing ladders are cloned into a vector (e.g., pGEM-T Easy, Promega) and amplified in competent cells (JM109 High Efficiency Competent cells, Promega). Preparations of plasmid are performed using a miniprep kit (e.g., QIAprep Spin Miniprep kit, Qiagen). Six clones of each amplimer are sequenced on both strands.

Amplimers obtained with the ACARNA primer pair are also sequenced when the sample is negative with the JDP primer pair. An amount of 30 ng of purified amplimer is mixed with 4 pmol of ACARNA.1383for or ACARNA.1655rev and is sequenced in an automatic sequencing platform.

The sequences are aligned and compared with GenBank reference sequences using BLAST searches.

2.3 CONCLUSIONS AND FUTURE PERSPECTIVES

Early and sensitive diagnosis and treatment on time may prevent the poor outcome of *Acanthamoeba* infections. The PCR assays described for *Acanthamoeba* diagnosis mainly target *Acanthamoeba Rns* gene [23,24,34–37,45,46]. The JDP primers are genus-specific and allow sequencing of the amplimers for further identification of most *Acanthamoeba* genotypes [36]. We and others showed that the combination of several primer pairs increased the sensitivity of PCR [23,24,35].

False-negative results are possible despite the combination of several primer pairs [35]. In fact, the quality of corneal scrapings may lead to reduce number of *Acanthamoeba* in the samples. The cellular riches of samples could be

TABLE 2.1
Molecular Detection of *Acanthamoeba* Isolated from Clinical Specimens

Isolate	Sample	Real-Time PCR Ac/Acant /*rns*[a]	PCR Nelson /ACARNA /JDP[b]	*Acanthamoeba* Genotype[c] (GenBank Accession Number)
BOUJ	CL	+ / + / +	+ / + / +	T3
CHAM	CL	+ / + / +	+ / + / +	T4
PETI	CS	+ / + / +	+ / + / +	T4
TROL	CL	+ / + / +	+ / + / +	T4
BARS	CLC	+ / + / +	+ / + / +	T4
SPAE	CS	+ / + / +	+ / + / +	T4
PEIG	CS	+ / + / +	+ / + / +	T4
TORO	CL	+ / + / +	+ / + / +	T4
DIPL	CL	+ / + / +	+ / + / +	T4
BLON	CLC	+ / + / +	+ / + / +	T4
S32[d]	CS	+ / + / +	+ / + / +	T4 + T4 (EU146068 + EU146069)
S11[d]	CL + CLC	+ / + / +	+ / + / +	T4 (DQ087306)
S36[d]	CLC	+ / + / +	+ / + / +	T4 (EU146073)
S35[d]	CLC	− / − / +	+ /–/ +	T5 (EU146072)
S7[d]	CL + CLC	+ / + / +	+ / + / +	T6 (DQ087289)
143GUIL[e]	CS[f]	+ / + / +	+ / + / +	T11 (DQ087292)
AFRI	CS	− / + / +	+ / + / +	T11
ARO	CLC	− / + / +	nd/nd/ +	T14
SKAL	CL + CLC	− / − / −	− / + / −	*Hartmannella* sp.
S8[d]	CLC	− / − / −	− / + / −	*Hartmannella* sp. (DQ087330)

Note: CL, contact lens; CLC, contact lens case solution; CS, corneal scraping.
[a] Ac, primers and probe designed by Rivière, D. et al. *J. Microbiol. Methods* 64, 78, 2006; Acant, primers and probe designed by Qvarnstrom, Y. et al. *J. Clin. Microbiol.* 44, 3589, 2006; *rns*, primers and probe designed by Goldschmidt, P. et al. *Br. J. Ophthalmol.* 93, 1086, 2009.
[b] Nelson, Nelson primers; ACARNA, ACARNA.1383for-1655rev primers; JDP, JDP1- JDP2 primers.
[c] Sequence analysis using BLAST searches.
[d] Isolates shared with reference Yera, H. et al. *Br. J. Ophthalmol.* 92, 1139, 2008.
[e] Isolate shared with reference Yera, H. et al. *Eur. J. Clin. Microbiol. Inf. Dis.* 26, 221, 2007.
[f] DNA was extracted directly from corneal scraping.

assessing by the amplification and the quantification of a human gene from the DNA extracted. PCR inhibitors, including topical solutions used for corneal examination or anti-amoebicidal treatment, could also cause false-negative results [46,65]. Therefore, sampling should be performed before the beginning of specific treatment and the eye surface should be rinsed intensively with 0.9% sodium chloride solution before sampling. An internal control could be added in assays performed on clinical samples. The use of commercial kits (QIAamp DNA Mini kit, Qiagen, MagNA Pure Nucleic Acid isolation kit, Roche Diagnostics) partially improves the sensitivity of the PCR assay by eliminating plus or minus the inhibitors present in the proteolytic mixtures during the DNA extraction [46,66].

False-positive results were observed with *Hartmannella* isolates when using the ACARNA primer pair here and elsewhere [59]. ACARNA and P1GP primers were predicted to produce amplimer with amoebae from the genera *Hartmannella* and *Balamuthia* [36]. A false-positive result was described by using the Nelson primers on one corneal scraping from a patient without AK; sequence analysis of the amplimer revealed DNA from Vexilliferidae, another free-living amoeba [35].

Real-time PCR are sensitive tools for the detection and quantification of microorganisms, allowing faster analysis. The risk of amplicon dispersion and cross-contamination is minimized. TaqMan PCR assays have recently been described for the detection and quantification of *Acanthamoeba* from cultures [44] or artificial and clinical samples [45–47]. We confirmed that the Acant primers–probe set [45] is able to detect a larger spectrum of genotypes than the Ac one [44] when testing isolates from genotypes T1, T3, T4, T6, T7, T10, T11, and T14. However, these two assays were unable to detect one isolate related to genotype T5 and obtained from a patient's contact lens case solution [59]. The genotype T5 has mainly been described in the environment [58] but one isolate has been observed in a keratitis case, [60] another in a disseminated infection [32], and a last one in a GAE [62]. Among the real-time PCR assays, this targeting the *Acanthamoeba rns* gene [47] gave the best detection rate in the *Acanthamoeba* genotypes studied here. The specificity of the PCR stays to be determined.

The pretreatment with proteinase K before DNA extraction procedures and the combination of primer pairs increase the sensitivity of PCR. The conditions of sampling are also primordial to avoid misdiagnosis.

Other method could be combined with PCR to increase the sensitivity of *Acanthamoeba* infections diagnosis. Authors used PCR to confirm confocal microscopy diagnosis of AK [15] in contact lens wearers [24].

AK and other *Acanthamoeba* infections can be misdiagnosed particularly at the onset of the disease due to a lack of specific clinical signs [23,24,34]. Direct examination presented low sensitivity and requires well-trained microscopists. Culture lacks sensitivity in particular if the patients have been treated by antibiotics with potential amoebicidal activity. The usefulness of PCR has been demonstrated for the confirmation of AK and is promising for the diagnosis of GAE or other *Acanthamoeba* infections. Because the genome of *Acanthamoeba* is highly variable between the isolates, the combination of primer sets is recommended to increase the detection rates of PCR.

Finally, a better and wider knowledge of the acanthamoebiasis will lead to an early diagnosis and treatment on time, improving the prognosis.

ACKNOWLEDGMENT

The authors are grateful to Professor Marie-Laure Dardé (Laboratory of Parasitology-Mycology, Faculté de Médecine, CHU, Limoges, France) for her steady interest and for her substantial contributions involving the triplex PCR assay.

REFERENCES

1. Visvesvara, G.S., Moura, H., and Schuster, F.L. Pathogenic and opportunistic free-living amoebae: *Acanthamoeba* spp., *Balamuthia mandrillaris*, *Naegleria fowleri*, and *Sappinia diploidea*. *FEMS Immunol. Med. Microbiol.* 50, 1, 2007.
2. Adl, S.M. et al. The new higher level classification of eukaryotes with emphasis on the taxonomy of protists. *J. Eukaryot. Microbiol.* 52, 399, 2005.
3. Marciano-Cabral, F., and Cabral, G. *Acanthamoeba* spp. as agents of disease in humans. *Clin. Microbiol. Rev.* 16, 273, 2003.
4. Khan, N.A. *Acanthamoeba*: Biology and increasing importance in human health. *FEMS Microbiol. Rev.* 30, 564, 2006.
5. Page, F.C. Re-definition of the genus *Acanthamoeba* with descriptions of three species. *J. Protozool.* 14, 709, 1967.
6. Illingworth, C.D., and Cook, S.D. *Acanthamoeba* keratitis. *Surv. Ophthalmol.* 42, 493, 1998.
7. Alsam, S. et al. *Acanthamoeba* interactions with human brain microvascular endothelial cells. *Microb. Pathog.* 35, 235, 2003.
8. Clarke, D.W., and Niederkorn, J.Y. The pathophysiology of *Acanthamoeba* keratitis. *Trends Parasitol.* 22, 175, 2006.
9. Kim, W.T. et al. Comparison of specific activity and cytopathic effects of purified 33 kDa serine proteinase from *Acanthamoeba* strains with different degree of virulence. *Korean J. Parasitol.* 44, 321, 2006.
10. Radford, C.F. et al. Risk factors for *Acanthamoeba* keratitis in contact lens users: A case–control study. *Br. Med. J.* 310, 1567, 1995.
11. Sharma, S., Garg, P., and Rao, G.N. Patient characteristics, diagnosis, and treatment of non-contact lens related *Acanthamoeba* keratitis. *Br. J. Ophthalmol.* 84, 1103, 2000.
12. Pasricha, G. et al. Use of 18S rRNA gene-based PCR assay for diagnosis of *Acanthamoeba* keratitis in non-contact lens wearers in India. *J. Clin. Microbiol.* 41, 3206, 2003.
13. Syam, P.P., Narendran, R., and van der Hoek, J. Persistent *Acanthamoeba* keratitis in a non-contact lens wearer following exposure to bird seed dust. *Br. J. Ophthalmol.* 89, 388, 2005.
14. Hurt, M. et al. Pathogenic *Acanthamoeba* spp. secrete a mannose-induced cytolytic protein that correlates with the ability to cause disease. *Infect. Immun.* 71, 6243, 2003.
15. Dart, J.K., Saw, V.P., and Kilvington, S. *Acanthamoeba* keratitis: Diagnosis and treatment update, 2009. *Am. J. Ophthalmol.* 148, 487, 2009.
16. Cardine, S. et al. Clinical management and prognosis in *Acanthamoeba* keratitis: A retrospective study of 25 cases. *J. Fr. Ophtalmol.* 25, 1007, 2002.
17. Moshari, A. et al. Chorioretinitis after keratitis caused by *Acanthamoeba*: Case report and review of the literature. *Ophthalmology* 108, 2232, 2001.
18. Heffler, K.F. et al. *Acanthamoeba* endophthalmitis in acquired immunodeficiency syndrome. *Am. J. Ophthalmol.* 122, 584, 1996.
19. Martinez, A.J., and Visvesvara, G.S. Free-living, amphizoic and opportunistic amebas. *Brain Pathol.* 7, 583, 1997.
20. Yagi, S., Schuster, F.L., and Visvesvara, G.S. Demonstration of *Balamuthia* and *Acanthamoeba* mitochondrial DNA in sectioned archival brain and other tissues by polymerase chain reaction. *Parasitol. Res.* 102, 491, 2008.
21. Alsam, S. et al. Extracellular proteases of *Acanthamoeba castellanii* (encephalitis isolate belonging to T1 genotype) contribute to increased permeability in an *in vitro* model of the human blood–brain barrier. *J. Infect.* 51, 150, 2005.
22. Kong, H.H., Kim, T.H., and Chung, D.I. Purification and characterization of a secretory serine proteinase of *Acanthamoeba healyi* isolated from GAE. *J. Parasitol.* 86, 12, 2000.
23. Lehmann, O.J. et al. Polymerase chain reaction analysis of corneal epithelial and tear samples in the diagnosis of *Acanthamoeba* keratitis. *Invest. Ophthalmol. Vis. Sci.* 39, 1261, 1998.
24. Mathers, W.D. et al. Confirmation of confocal microscopy diagnosis of *Acanthamoeba* keratitis using polymerase chain reaction analysis. *Arch. Ophthalmol.* 118, 178, 2000.
25. Stothard, D.R. et al. Fluorescent oligonucleotide probes for clinical and environmental detection of *Acanthamoeba* and the T4 18S rRNA gene sequence type. *J. Clin. Microbiol.* 37, 2687, 1999.
26. Epstein, R.J. et al. Rapid diagnosis of *Acanthamoeba* keratitis from corneal scrapings using indirect fluorescent antibody staining. *Arch. Ophthalmol.* 104, 1318, 1986.
27. Schuster, F.L. et al. Detection of antibodies against free-living amoebae *Balamuthia mandrillaris* and *Acanthamoeba* species in a population of patients with encephalitis. *Clin. Infect. Dis.* 42, 1260, 1986.
28. Chappell, C.L. et al. Standardized method of measuring *Acanthamoeba* antibodies in sera from healthy human subjects. *Clin. Diagn. Lab. Immunol.* 8, 724, 2001.

29. Grunnet, M.L., Cannon, G.H., and Kushner, J.P. Fulminant amebic meningoencephalitis due to *Acanthamoeba*. *Neurology* 31, 174, 1981.
30. Bloch, K.C. and Schuster, F.L. Inability to make a premortem diagnosis of *Acanthamoeba* species infection in a patient with fatal granulomatous amebic encephalitis. *J. Clin. Microbiol.* 43, 3003, 2005.
31. McKellar, M.S. et al. Fatal granulomatous *Acanthamoeba* encephalitis mimicking a stroke, diagnosed by correlation of results of sequential magnetic resonance imaging, biopsy, *in vitro* culture, immunofluorescence analysis, and molecular analysis. *J. Clin. Microbiol.* 44, 4265, 2006.
32. Barete, S. et al. Fatal disseminated *Acanthamoeba lenticulata* infection in a heart transplant patient. *Emerg. Infect. Dis.* 13, 736, 2007.
33. Petry, F. et al. Early diagnosis of *Acanthamoeba* infection during routine cytological examination of cerebrospinal fluid. *J. Clin. Microbiol.* 44, 1903, 2006.
34. Zamfir, O. et al. Diagnostic par PCR des kératites à *Acanthamoeba* spp. *J. Fr. Ophthalmol.* 29, 1034, 2006.
35. Yera, H. et al. Comparison of PCR, microscopic examination and culture in the early diagnosis and characterization of *Acanthamoeba* isolates from ocular infections. *Eur. J. Clin. Microbiol. Inf. Dis.* 26, 221, 2007.
36. Schroeder, J.M. et al. Use of subgenic 18S ribosomal DNA PCR and sequencing for genus and genotype identification of *Acanthamoeba* from humans with keratitis and from sewage sludge. *J. Clin. Microbiol.* 39, 1903, 2001.
37. Sharma, S. et al. *Acanthamoeba* keratitis in non-contact lens wearers in India: DNA typing-based validation and a simple detection assay. *Arch. Ophthalmol.* 122, 1430, 2004.
38. Dhivya, S. et al. Comparison of a novel semi-nested polymerase chain reaction (PCR) with a uniplex PCR for the detection of *Acanthamoeba* genome in corneal scrapings. *Parasitol. Res.* 100, 1303, 2007.
39. Shirwadkar, C.G. et al. *Acanthamoeba* encephalitis in patient with systemic lupus, India. *Emerg. Infect. Dis.* 12, 84, 2006.
40. MacLean, R.C. et al. Identification of *Acanthamoeba* sp. in paraffin-embedded CNS tissue from an HIV+ individual by PCR. *Diagn. Microbiol. Infect. Dis.*, 57, 289, 2007.
41. Meersseman, W. et al. Rapidly fatal *Acanthamoeba* encephalitis and treatment of cryoglobulinemia. *Emerg. Infect. Dis.* 13, 469, 2007.
42. Yagi, S., Schuster, F.L., and Bloch, K. Demonstration of presence of *Acanthamoeba* mitochondrial DNA in brain tissue and cerebrospinal fluid by PCR in samples from a patient who died of granulomatous amebic encephalitis. *J. Clin. Microbiol.* 45, 2090, 2007.
43. Walochnik, J. et al. Granulomatous amoebic encephalitis caused by *Acanthamoeba* amoebae of genotype T2 in a human immunodeficiency virus-negative patient. *J. Clin. Microbiol.* 46, 338, 2008.
44. Rivière, D. et al. Development of a real-time PCR assay for quantification of *Acanthamoeba* trophozoites and cysts. *J. Microbiol. Methods.* 64, 78, 2006.
45. Qvarnstrom, Y. et al. Multiplex real-time PCR assay for the simultaneous detection of *Acanthamoeba* spp., *Balamuthia mandrillaris*, and *Naegleria fowleri*. *J. Clin. Microbiol.* 44, 3589, 2006.
46. Thompson, P.P. et al. Validation of real-time PCR for laboratory diagnosis of *Acanthamoeba* keratitis. *J. Clin. Microbiol.* 46, 3232, 2008.
47. Goldschmidt, P. et al. New tool for the simultaneous detection of 10 different genotypes of *Acanthamoeba* available from the American Type Culture Collection. *Br. J. Ophthalmol.* 93, 1086, 2009.
48. Pussard, M., and Pons, R. Morphologie de la paroi kystique et taxonomie du genre *Acanthamoeba* (Protozoa, Amoebida). *Protistologica* 13, 557, 1977.
49. Visvesvara, G.S. Classification of *Acanthamoeba*. *Rev. Infect. Dis.* 13(Suppl 5), S369, 1991.
50. Gast, R.J. et al. Subgenus systematics of *Acanthamoeba*: Four nuclear 18S rDNA sequence types. *J. Eukaryot. Microbiol.* 43, 498, 1996.
51. Stothard, D.R. et al. The evolutionary history of the genus *Acanthamoeba* and the identification of eight new 18S rRNA gene sequence types. *J. Eukaryot. Microbiol.* 45, 45, 1998.
52. Hewett, M.K. et al. Identification of a new *Acanthamoeba jacobsi* Sawyer, Nerad and Visvesvara (1992) (Lobosea: Acanthamoebidae). *Acta Protozool.* 42, 325, 2003.
53. Horn, M. et al. Novel bacterial endosymbionts of *Acanthamoeba* spp. related to the *Paramecium caudatum* symbiont *Caedibacter caryophilus*. *Environ. Microbiol.* 1, 357, 1999.
54. Gast, R.J., Development of an *Acanthamoeba*-specific reverse dot-blot and the discovery of a New ribotype. *J. Eukaryot. Microbiol.* 48, 609, 2001.
55. Booton, G.C. et al. 18S ribosomal DNA typing and tracking of *Acanthamoeba* species isolates from corneal scrape specimens, contact lenses, lens cases, and home water supplies of *Acanthamoeba* keratitis patients in Hong Kong. *J. Clin. Microbiol.* 40, 1621, 2002.
56. Fuerst, P.A. et al. Genotypic identification of no-keratitis infections caused by the opportunistically pathogenic ameba genus *Acanthamoeba*. *J. Eukaryot. Microbiol.* 50 (Suppl), 512, 2003.
57. Zhang, Y. et al. Identification of 18S ribosomal DNA genotype of *Acanthamoeba* from patients with keratitis in North China. *Invest. Ophthalmol. Vis. Sci.* 45, 1904, 2004.
58. Booton, G.C. et al. Identification and distribution of *Acanthamoeba* species genotypes associated with nonkeratitis infections. *J. Clin. Microbiol.* 43, 1689, 2005.
59. Yera, H. et al. The genotypic characterization of *Acanthamoeba* isolates from human ocular samples. *Br. J. Ophthalmol.* 92, 1139, 2008.
60. Spanakos, G. et al. Genotyping of pathogenic *Acanthamoebae* isolated from clinical samples in Greece—Report of a clinical isolate presenting T5 genotype. *Parasitol. Int.* 55, 147, 2006.
61. Di Cave, D. et al. *Acanthamoeba* T4 and T15 genotypes associated with keratitis infections in Italy. *Eur. J. Clin. Microbiol. Inf. Dis.* 28, 607, 2009.
62. Lackner, P. et al. Acute granulomatous *Acanthamoeba* encephalitis in an immunocompetent patient. *Neurocrit. Care* 12, 91, 2010.
63. Ledee, D.R. et al. Advantages of using mitochondrial 16S rDNA sequences to classify clinical isolates of *Acanthamoeba*, *Invest. Ophthalmol. Vis. Sci.* 44, 1142, 2003.
64. Köhsler, M. et al. ITS1 sequence variabilities correlate with 18S rDNA sequence types in the genus *Acanthamoeba* (Protozoa: Amoebozoa). *Parasitol. Res.* 98, 86, 2006.
65. Goldschmidt, P. et al. Effects of topical anaesthetics and fluorescein on the real-time PCR used for the diagnosis of Herpesviruses and *Acanthamoeba* keratitis. *Br. J. Ophthalmol.* 90, 1354, 2006.
66. Goldschmidt, P. et al. Resistance of *Acanthamoeba* to classic DNA extraction methods used for the diagnosis of corneal infections. *Br. J. Ophthalmol.* 92, 112, 2008.

3 Balamuthia

Govinda S. Visvesvara and Yvonne Qvarnstrom

CONTENTS

3.1 INTRODUCTION

Until 1986, members of only two genera of free-living amoebae were considered as agents of central nervous system (CNS) disease of humans: *Acanthamoeba* (several species) and *Naegleria (Naegleria fowleri)*. However, there were several cases of meningoencephalitis where culture of the amoeba could not be established and amoebae that were seen in the tissues could not be specifically identified by immunohistochemical tests. These unknown amoebae were considered as unidentified species of *Vahlkampfia* or *Acanthamoeba* because of the presence of cysts in the tissues (*N. fowleri* does not produce cysts in brain tissue) [1,2]. But in 1986 an amoeba was isolated from the brain tissue of a pregnant mandrill baboon that died in the San Diego Zoological Park after developing signs of meningoencephalitis. It was initially identified as leptomyxid amoeba (CDC:V039) [3] because of similarity of the cyst stage with that of *Leptomyxa*, a soil amoeba. Further studies including electron microscopy revealed that this amoeba was substantially different from the amoebae belonging to the family Leptomyxidae and therefore, it was described as a new taxon and designated as *Balamuthia mandrillaris* [4]. The genus was named after the late Professor William Balamuth, an eminent protozoologist and a Professor of Zoology at the University of California, Berkeley, CA. The species designation denotes the source of the index isolate, the brain of a mandrill baboon. An antiserum to this amoeba (CDC:V039) produced in rabbits was found to be reactive only with *Balamuthia* and not with a number of different free-living and parasitic amoebae including *Acanthamoeba*, *Vahlkampfia*, *Hartmannella*, *N. fowleri*, *Willaertia*, and *Entamoeba histolytica* [3,4]. Subsequently, this anti-*B. mandrillaris* serum was used on tissue sections that were previously negative in the immunofluorescence assay. The amoebae present in those tissue sections were found to be reactive with this new anti-serum, indicating that those cases of encephalitis were all caused by *B. mandrillaris* [2–7].

It is now well established that small free-living amoebae belonging to the genera *Acanthamoeba*, *Balamuthia*, and *Naegleria* cause infections in humans and other animals

[1–8]. Several species of *Acanthamoeba* (*A. castellanii, A. culbertsoni, A. divionensis, A. healyi, A. lenticulata, A. rhysodes,* and *A. polyphaga*) and the only known species of *Balamuthia, B. mandrillaris*, cause a chronic CNS disease, granulomatous amoebic encephalitis (GAE). *N. fowleri*, the only pathogenic species among more than 40 species included in the genus *Naegleria*, causes an acute and fulminating CNS disease known as primary amoebic meningoencephalitis (PAM) [6–8]. PAM occurs primarily in healthy children and young adults with a history of recent swimming, diving, and engaging in other water sports in freshwater lakes and ponds. GAE caused by *Acanthamoeba* occurs mainly in immunosuppressed, alcoholic, and otherwise debilitated individuals, whereas GAE caused by *Balamuthia* occurs in both immunocompromised and immunocompetent persons [3–10]. Both *Acanthamoeba* and *Balamuthia* cause, in addition to GAE, disseminated infections of other organs including the skin, kidneys, adrenals, spleen, and lungs, whereas *N. fowleri* is confined to the CNS [1–10]. *Acanthamoeba* also causes sight-threatening infection of the human eye, *Acanthamoeba* keratitis [1,2,6–8]. Recently, another free-living amoeba—*Sappinia pedata* (previously identified as *S. diploidea*)—has been identified as causing encephalitis in an otherwise healthy young man [11–13], which underscores the importance of the small free-living amoebae as agents of CNS infections in humans.

3.1.1 Classification, Morphology, and Biology

3.1.1.1 Classification

When *Balamuthia* was initially isolated it was considered as a leptomyxid amoeba because of similarities of its life cycle, including an actively feeding trophic stage like those of *Leptomyxa* and *Gephyramoeba* and a thick-walled cyst stage like that of *Leptomyxa*. However, it could not be classified as belonging to either of the genera *Leptomyxa* or *Gephyramoeba* because of fundamental differences in other morphologic as well as physiologic and antigenic characteristics. Therefore, it was designated as *B. mandrillaris*, nov. gen., nov. sp [4]. Amaral-Zettler et al. [14] compared sequences of several leptomyxid amoebae and showed *Balamuthia* to be distinct from the leptomyxid amoebae. It was instead included in the family Acanthamoebidae, based on similarities in the 18S ribosomal RNA (rRNA) genes between *B. mandrillaris* and *Acanthamoeba*, the other opportunistic free-living amoeba that causes GAE in humans and other animals [15]. Further, Booton et al. [16] performed phylogenetic analysis on rRNA genes from seven *B. mandrillaris* isolates and the resulting phylogenetic trees confirm that *B. mandrillaris* and *Acanthamoeba* are closely related but belong to separate genera. These and other studies also indicate that all isolates, whether clinical or environmental, belong to a single species, *B. mandrillaris* [17–19]. The free-living amoebae, according to the classic taxonomic system, were classified under the subphylum Sarcodina and super class Rhizopodea [20]. Recently, however, the International Society of Protistologists abandoned the older hierarchical system of classification consisting of the traditional "Kingdom," "Phylum," "Class," "subclass," and so on and replaced it with a new classification system that emphasizes modern morphologic approaches, biochemical pathways and molecular phylogenetics. According to this new schema the Eukaryotes have been classified into six clusters or "Super Groups," namely Amoebozoa, Opisthokonta, Rhizaria, Archaeplastida, Chromalveolata, and Excavata. *Acanthamoeba* and *Balamuthia* are included under Super Group Amoebozoa: (Acanthamoebidae), along with *Sappinia* (Flabellinea: Thecamoebidae), and *N. fowleri* under Super Group Excavata: (Heterolobosia: Vahlkampfiidae) [21].

3.1.1.2 Morphology

B. mandrillaris has two stages, the trophozoite and the cyst, in its life cycle (Figure 3.1). The trophozoite is uninucleate, although binucleate forms are occasionally seen. The binucleate amoebae may have completed karyokinesis but not cytokinesis. The nucleus measures 5 μm and contains a large, centrally placed, dense nucleolus; occasionally, however, amoebae with two or three nucleolar bodies have been seen, especially in infected tissues. The trophic form is highly irregular and sometimes highly reticulate. The trophozoite measures from 12 to 60 μm with a mean of 30 μm [4]. These dimensions were based upon actively feeding and growing cells although frequent variation in size was noticed, especially in older cultures wherein all the tissue culture cells have been consumed and the amoebae appear to have shrunk. These shrunken amoebae upon inoculation into fresh monolayers regain their normal size. The trophic

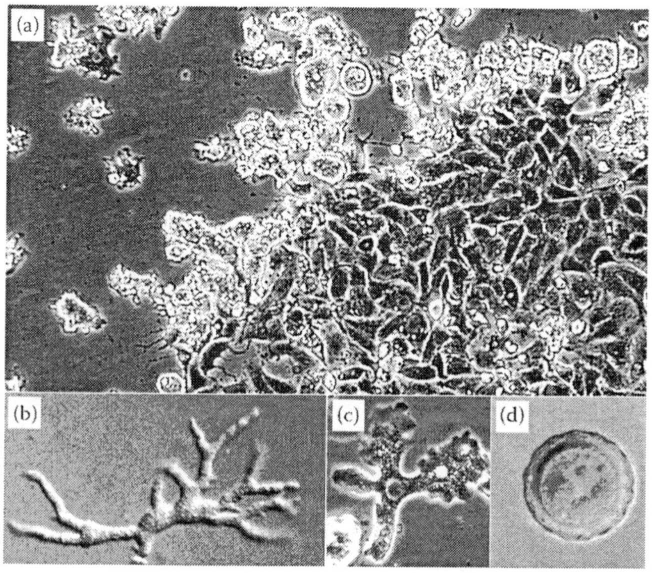

FIGURE 3.1 *B. mandrillaris* trophozoites and cyst. (a) *B. mandrillaris* trophozoites growing on monkey kidney cell culture. Note the amoebae have cleared a portion of the monkey kidney monolayer, ×300. (b) and (c) Trophozoite with different morphology. (d) A round mature cyst.

forms exhibit two different patterns of locomotion in tissue culture vessels when observed using an inverted microscope. In one type, when the amoeba is actively feeding on the cell monolayer, it forms a broad pseudopodium and moves like a slug. When most of the cell monolayer had been destroyed the amoeba forms finger-like pseudopodia and exhibit spider-like walking movement and move across the floor of the tissue culture flask [3,4].

At the ultrastructural level, the cytoplasm can be distinguished into an inner dense endoplasm containing numerous vacuoles and mitochondria and a peripheral ectoplasm free of these cytoplasmic inclusions. The nucleus is round and vesicular and is the major feature seen in sectioned amoebae. The cytoplasm contains numerous mitochondria with tubular cristae. Vesicles of endoplasmic reticulum are also found in relatively large numbers. A distinct golgi complex is not observed. Numerous food vacuoles of varying sizes containing electron-dense debris of broken down tissue culture cells is seen (Figure 3.2a). Occasionally mitotic stages are also seen. The nucleolus is present during the early stages of mitosis, and the nuclear envelope remains intact. A microtubule organizing center is seen near the nucleus and microtubules radiates from this electron-dense region. As mitosis continues, the nuclear envelope breaks down and the nucleolus disappears. Centrioles appear to be absent from the amoebae [3,4].

Cysts are typically round although oval shapes are also seen. Cysts are uninucleate, and range in size from 12 to 30 μm with a mean of 15 μm. As with trophic amoebae, variation in size of the cysts is frequently noticed in older cultures. Cysts, when examined with an optical light microscope, appear to be double walled, the outer wall being wavy and the inner wall round. The cyst wall has no pores and at excystation, the wall breaks down and the amoeba emerges and resumes its trophic existence. In mature cysts a layer of refractile granules is seen at the cytoplasmic periphery below the inner cyst wall. When examined under the transmission electron microscope the cyst wall have three layers: an outer thin and irregular ectocyst which gives a rough appearance to the mature cyst, an inner thick electron-dense endocyst, and a middle amorphous fibrillar mesocyst (Figure 3.2b) [3,4]. The mature cysts are hardy and resistant to environmental pressures including drying and freezing [22]. This is probably due partly to the presence of complex carbohydrates and cellulose in the cyst wall [23].

3.1.1.3 Cultivation

Balamuthia can be grown in the laboratory on a number of different mammalian cell cultures including monkey kidney, human lung fibroblasts, and human brain microvascular endothelial cells (HBMEC). *Balamuthia* feeds on the cell monolayer and consumes the monolayer within a week [3,4,24]. Kiderlen et al. [25] have demonstrated that murine mastocytoma cells, that are used as feeder cells for culturing *Balamuthia*, swell and rupture probably because of contact with the amoebae. If the amoebae were separated from target cells by a semipermeable membrane destruction of the cells was prevented. In an elegant study, Dunnebacke [26] has shown that *Balamuthia* trophozoite growing in cell culture contacts the target cell by sending a pseudopodium and penetrating the target cell, causing visible damage to the target cell upon withdrawal of the pseudopodium. Further, according to Dunnebacke the amoeba enters and moves around the mammalian cell and remains quiescent within the cell for hours, a feature that has been described for another nonpathogenic free-living amoeba, *Paravahlkampfia francinae*, recently [27]. *Balamuthia* also produces amoebastomes or feeding cups and ingests bits and pieces of cells and thus causes destruction of tissue culture monolayers [28].

Balamuthia can be grown in a complex chemical medium supplemented with 10% fetal bovine serum fetal calf serum, vitamins, nonessential amino acids, liver extract, yeast RNA, and hemin [29,30]. *Balamuthia* multiplies in such medium but has a long generation time of ~20–30 h when grown at the optimal temperature of 37°C and yield ~1.5 to 3×10^5 amoebae/mL. As the culture ages the trophozoites differentiates

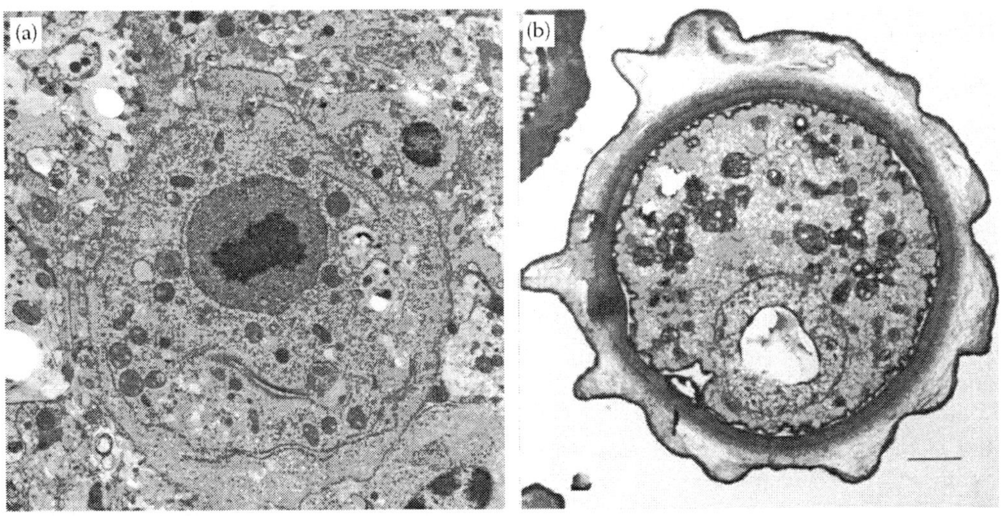

FIGURE 3.2 Electron micrograph of *B. mandrillaris* trophozoite (a) and cyst (b). Magnification ×7500.

into cysts in both tissue culture flasks and axenic medium. In cultures that are several weeks old the amoebae decrease in size ranging from 6 to 10 μm. However, the amoebae regain their normal size when inoculated into fresh tissue culture monolayers or axenic medium. According to Kiderlen et al. [25] *Balamuthia* amoebae, like *Acanthamoeba* and *N. fowleri*, lose their virulence potential after prolonged *in vitro* culture, whereas *Balamuthia* maintained on tissue culture monolayers retain their virulence properties.

Balamuthia cannot grow on agar plates coated with bacteria although it can survive as trophozoites for more than 10 days in the presence of bacteria. In contrast, when cultured without bacteria the amoebae became rounded up and inactive within 24 h [30]. Matin et al. [31] showed that *Balamuthia* trophozoites ingest fluorescently labeled, heat-killed bacteria, but do not multiply in the presence of bacteria. Thus, amoebae *in vitro* may be able to utilize nutrients that diffuse outward from bacteria. It is clear that *B. mandrillaris* lives in the environment; Dunnebacke et al. [19] and Schuster et al. [18] isolated *Balamuthia* amoebae from sealed agar plates that were inoculated with soil and kept at 17°C for several months. They found small numbers of *Balamuthia* amoebae on the agar surface of these plates, *B. mandrillaris* has also been isolated from airborne dust [32]. Although it is not known what are its nutritional requirements it is believed that it feeds on small amoebae like *Acanthamoeba* and *Naegleria* present in the environment [18,19,31]. For example, *B. mandrillaris* when incubated with *Acanthamoeba* trophozoites caused partial damage to some acanthamoebae but the intact *Acanthamoeba* trophozoites differentiated into cysts. However, *B. mandrillaris* amoebae were unable to consume *Acanthamoeba* cysts [31].

3.1.1.4 Cryopreservation

In our laboratory, we have successfully cryopreserved eight isolates of *Balamuthia*. Five isolates were from human brain tissue originating from Georgia, Nevada, California, New York, and Australia, and three isolates were from animals; a mandrill baboon, a horse, and a gorilla. These isolates have been cryopreserved immediately after being established in cell culture. The ideal method is to scrape or trypsinize the infected cell culture and mix it with an equal volume of the growth medium containing 15% dimethyl sulfoxide. The mixture is then aliquoted in 1.0 mL amounts into plastic cryovials. The vials are placed in a controlled-rate cooling apparatus such as the Linde cell freezer and cooled at the rate of 1°C/min to a temperature of −60°C, at which point the vials are placed in the vapor phase of a liquid nitrogen freezer as described previously [33]. Linde cell freezer is very expensive and may not be available in many laboratories. Therefore, in our laboratory, we use a relatively simple and inexpensive apparatus, the Cell Freezer (NalgeneCryo 1°C Freezing Container, Catalog no. 5100-0001; Nalgene Nunc International, Rochester, NY). It consists of two components, a cylindrical plastic shell with a screw-cap closure and a molded plastic insert containing 18 wells (tube or vial holders) that accept 1 mL cryovials. Approximately 50 mL of either isopropanol or ethylene glycol is introduced into the plastic shell, and the cryovials containing the isolates to be preserved are placed in the molded plastic insert. The Cell Freezer is then placed in a −70°C freezer overnight (~18 h). The vials are then removed and stored in the vapor phase of a liquid nitrogen freezer. The frozen isolates, now referred to as stabilates, can be stored indefinitely in liquid nitrogen vapor (approximately −150°C). Viable cultures are established from frozen stabilates by immediately transferring the vials from the freezer to a 37°C water bath and inoculating the thawed material into E6 or human lung fibroblast (HLF) cell cultures. When cryopreserved *B. mandrillaris* amoebae were inoculated into E-6 cell cultures, they caused characteristic cytological changes and destroyed the cell cultures within 3–4 days. The organisms that were recovered from such cultures were infective to mice and caused their death within 1 week. We have used this method to cryopreserve not only *B. mandrillaris* but also *Acanthamoeba* spp., *Naegleria* spp., *Entamoeba histolytica*, *E. invadens*, *Giardia intestinalis*, *Trichomonas vaginalis*, as well as microsporidia such as *Anncalia algerae*, *Encephalitozoon* spp., and *Vittaforma corneae*.

3.1.1.5 *Balamuthia* as Host for Other Microorganisms

It is well known that *Acanthamoeba* spp. act as hosts for a number of pathogenic microorganisms [34] and recently has it been shown that *Balamuthia* may also act as hosts for pathogenic *Legionella* [35]. Further, according to Michel et al. [36] the index strain (CDC:V039) of *B. mandrillaris* supported the intracellular growth of a Chlamydia-like bacterium, *Simkania negevensis*, and allowed it to produce and periodically shed the infectious stage of the bacterium—the elementary bodies.

3.1.2 Epidemiology

Balamuthiasis is a chronic and subacute disease that develops over time spanning from 2 weeks to 2 years. *B. mandrillaris* is also known to cause skin infection, similar to that caused by *Acanthamoeba*. Cases of *Balamuthia* have been recorded from all over the world [37,38–88]. Although a large number of cases have been described from the United States [37,38,39–60] cases have also been described from many countries in South America including Argentina [61,62], Brazil [63,64], Chile [65,66], Mexico [67–69], Peru [70–74], and Venezuela [75,76]. Cases have also been recorded from Africa [77,78], Australia [79,80], Europe [81–84], Japan [85], India [86], and Thailand [87,88]. The African case described by Bhagwandeen et al. [77] as a case of hartmannellid amoebic encephalitis in 1975 was actually due to *Balamuthia* because the amoebae measured 30–80 μm and failed to react with the anti-*Acanthamoeba* sera in the immunofluorescence test [77]. The authors speculate that the portal of entry was probably the superficial ulcer in the paraumbilical region of the Zambian patient. The possible *Acanthamoeba* case described by Rutherfoord in a patient in Cape Town, South Africa, was also caused by *Balamuthia* because of the tripartite cyst wall figured in that publication [78].

Since *Balamuthia* has been isolated from soil and recently from dust in air it is possible that playing with dirt, gardening, or inhalation of cysts in soil carried on the wind may be the source of infections [18,19,32]. For example, one of the early cases of *Balamuthia* GAE suffered a farm accident during which a limb was severed and came into contact with farm dirt [42]. One of the few patients who had recovered from balamuthiasis probably contracted the infection via soil when he was pricked by a rose bush thorn while digging in his backyard [52]. Similarly, another patient who also recovered, a 60 year-old woman, had a history of gardening in composted soil [53]. An Australian case, a young girl, probably inhaled soil dust and developed nasal infection that subsequently spread to CNS, leading to fulminant infection and death [80]. Several other cases also appear to have had contact with soil, such as playing with soil in a flower pot, or inhaling dust while riding motor bicycle over dusty desert terrain [51,57,60]. Another interesting feature is that Hispanic Americans, representing a large proportion of workforce dealing with agriculture business and thus having exposure to soil and dirt, share a disproportionately high burden of *Balamuthia* GAE cases [60,89]. Recently, it has been shown that Hispanics are less likely to develop antibodies to *Acanthamoeba* [90]. It is quite possible that they are also less likely to develop antibodies to *Balamuthia* and therefore be genetically predisposed to develop *Balamuthia* GAE.

3.1.3 Clinical Features

Infections caused by *Balamuthia* are rare and the symptoms are inconspicuous. The clinical course is insidious and may last weeks to several months. The patients may be afebrile or develop low-grade fever and complain of headache, nausea, nuchal rigidity, lethargy, and vomiting and hence the index of suspicion among clinicians is poor. Neurologic symptoms include papilledema, meningeal inflammation, aphasia, ataxia, photophobia, myalgia, weight loss, cranial nerve palsies (mainly 3rd and 6th cranial nerves), personality changes, hemiparesis, tonic-clonic seizures, and coma. These symptoms are not unique to balamuthiasis and may mimic other types of encephalitis thus contributing to the difficulty in diagnosis [2,5,57].

Balamuthiasis occurs in both immunosuppressed and immunocompetent persons. Many of the immunocompetent cases, however, have been in young children or elderly persons suggesting that there may be subtle immunological deficiencies involved [1–8]. Many patients with *Balamuthia* GAE have developed skin lesions on the face, limbs, and chest [42,55,60,61,66,80,82,87,88], ulcer in the sinus cavities and rhinitis [41,49], otitis media [43,44,48,54,80,86–89], and hydrocephalus [45,48,51]. Additionally, the infection can disseminate into lungs [51] and kidney and adrenal and uterus [41,64]. Peruvian patients develop skin lesions before developing CNS infection [69–74]. In the Peruvian patients the skin lesions appear initially as single painless plaques a few millimeters thick and one to several centimeters wide. Some may develop multiple lesions. The lesions are commonly seen in the center of the face but occasionally they may occur on the trunk, and hands, and feet [74]. The time between the appearance of skin lesions and the onset of neurological symptoms may range from 1 month to about 2 years. However, CNS involvement may not always occur. For example, in a male Peruvian patient skin biopsy from a facial ulcer showed the presence of cysts and the patient appeared to have spontaneously recovered without treatment. A follow-up examination 8 years later revealed that the patient's skin lesion had healed and he appeared to be well and without any symptoms [74]. In these Peruvian cases differential diagnoses included leishmaniasis, sporotrichosis, lupus vulgaris, and Wegener's granulomatosis [74]. A plaque appearing on the face or trunk should alert the clinician to the possibility of cutaneous balamuthiasis. Additionally a persistent or chronic sore that does not heal with general or topical application of antimicrobials should raise concern [80]. Confirmation is made by a biopsy of the affected area with examination of the sections for amoebae with vesicular nuclei.

3.1.4 Diagnosis

3.1.4.1 Conventional Techniques

Cerebrospinal fluid (CSF) samples are usually negative for bacteria. CSF, however, reveals lymphocytic pleocytosis with counts of >1000 cells/mm^3 (normal 0–5 cells/mm^3), protein from normal (15–45 mg/dL) to over 1000 mg/dL), and normal or low glucose. Intracranial pressure may be elevated and thus may lead to herniation of the uncal and cerebellar tonsils and death [2]. *Balamuthia*, except in one case [24], has not been isolated or identified in the CSF either by conventional staining such as hematoxylin & eosin (H&E) or by immunostaining although *B. mandrillaris* DNA has been detected in the CSF [91]. Neuroimaging such as magnetic resonance imaging (MRI) or computerized tomography (CT) have been helpful in the diagnosis of balamuthiasis in the CNS [92,93]. However, these scans may show the presence of single or multiple heterogeneous, hypodense, nonenhancing, "space-occupying lesions," which may mimic a brain abscess, brain tumor, or intracerebral hematoma, and, hence, may not specifically identify balamuthiasis. Disease may present as ring-enhancing lesions [92] or as a space-occupying mass. Single lesions may be seen initially but may develop as multifocal lesions as the disease progresses [93]. The lesions are most frequently hypodense, though hyperdense and isodense types are seen in a minority of cases. Most CT scans show enhancement with contrast. In some cases *Balamuthia* GAE has been erroneously diagnosed as either neurotuberculosis or neurocysticercosis resulting in delay in the initiation of appropriate antimicrobial therapy. Differential diagnoses include neurotuberculosis, neurocysticercosis, bacterial abscess, viral or bacterial encephalitis, acute disseminated encephalomyelitis, leptomeningitis, and brain tumor. Diagnosis of balamuthiasis based on clinical presentation alone is elusive and requires confirmatory evidence from histopathology or immunopathology.

MRI images obtained from two patients who survived the infections have shown changes in cranial lesions following recovery. Ring-enhancing lesions were seen in early MRIs of a 5-year-old child. The lesions came to be enclosed by calcified walls which, in turn, were surrounded, by edematous areas [52,92]. With time, the lesions regressed in size and the edematous areas resolved. The child showed no gross neurologic sequelae. A similar regression of lesions was seen in a 64-year-old man after several years [52] although this patient had significant neurologic deficits.

Several cases have been diagnosed premortem by brain or skin biopsies; hence, they are important diagnostic procedures. As a result of premortem diagnosis a few patients have survived the infection because of treatment with a combination of antimicrobials. However, most cases of balamuthiasis have been identified at autopsy. Autopsies are less likely to be performed because of the cost or reluctance of family members to agree to the procedure based on religious beliefs. There have been several cases in California that, were it not for a proactive program to identify causal agents of encephalitis (the California Encephalitis Project), would have been classified as "encephalitis of unknown origin" [59]. In Africa, the Indian subcontinent and Southeast Asia where the HIV/AIDS epidemic is largely uncontrolled, *Balamuthia* is probable one of the many opportunistic pathogens that infect HIV/AIDS patients but, for lack of diagnostic facilities, are not identified and certainly do not appear in the literature [5].

3.1.4.2 Molecular Techniques

Specific polymerase chain reaction (PCR) assays for *B. mandrillaris* have been developed with all available DNA sequences as targets: the small subunit rRNA genes and the gene for RNase P RNA. As previously mentioned, both the nuclear small subunit rRNA (18S rRNA) and the mitochondrial small subunit rRNA (16S rRNA) genes were originally analyzed to clarify the phylogeny of *B. mandrillaris*. All 18S rRNA genes from *B. mandrillaris* examined so far have been identical [16], but a low degree of sequence variation (0–1.8%) was found in the 16S rRNA gene [17]. Isolates with identical 18S rRNA genes could be distinguished based on their 16S rRNA sequences, suggesting that the latter gene can be used as a genotyping tool.

Conventional (gel-based) PCR assays have focused on the mitochondrial 16S rRNA gene (Figure 3.3). A primer pair that amplifies 1075 base pairs can be used both to detect the amoebae and to genotype them in samples containing sufficient amoebae [17–19,94,95]. It can be combined with primers specific for the 16S rRNA gene in *Acanthamoeba* for multiplex detection of both amoebae [94]. For samples containing too little amoebic DNA for this PCR to work, a modified version that amplifies a smaller fragment can be used [91].

For faster analysis, real-time PCR can be performed. A triplex assay is available that simultaneously detects the three pathogenic free-living amoebae *Acanthamoeba*, *Balamuthia*, and *N. fowleri* [96]. This assay uses species-specific TaqMan probes to detect free-living amoebae in clinical and environmental samples (Figure 3.4). It amplifies a 171 base-pair fragment of the 18S rRNA gene from *B. mandrillaris* and can detect as little as one amoeba in a sample. An alternative approach is a recently published real-time PCR targeting the RNase P gene [97]. It showed sensitivity equal to that of the triplex assay, but the RNase P assay needs further evaluation as the primer design was based on the sequence of just one strain and it was validated with very few samples in the initial study. The authors suggest using this assay to combine it with the *B. mandrillaris* part of the triplex assay for increased reliability, since targeting different genetic regions minimizes the risk for potential false results due to unforeseen sequence variability.

3.1.5 Pathology and Pathogenesis

3.1.5.1 Pathology

Histopathology may reveal a granulomatous process, though this may be absent in immunocompromised individuals. Areas of the brain more likely to be affected are the cerebral and cerebellar hemispheres, the pons, the brain

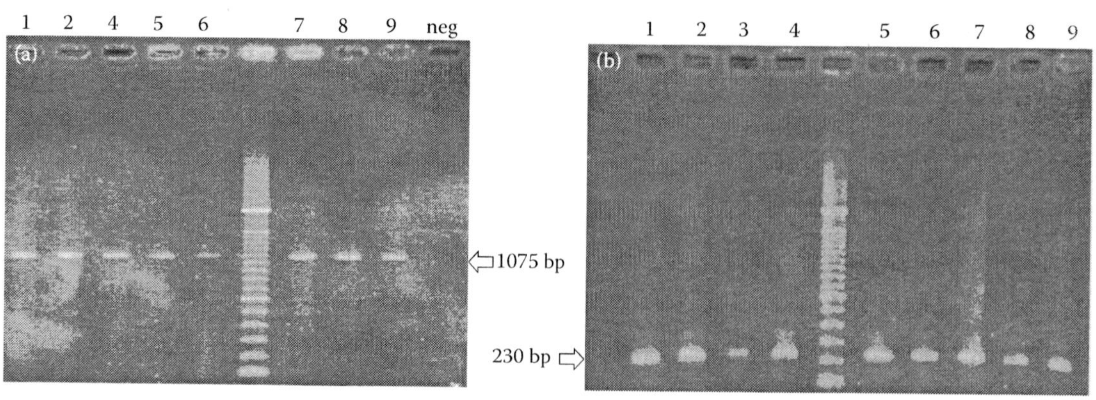

FIGURE 3.3 Agarose gel electrophoresis image of PCR products using primers specific for the *B. mandrillaris* mitochondrial 16S rRNA. (a) the large fragment 16S rRNA PCR; (b) the small fragment 16S rRNA PCR. 1, brain tissue from gorilla; 2, brain tissue from orangutan; 3, CSF from human patient A; 4, brain biopsy from human patient A; 5, brain tissue from human patient B; 6, brain tissue from human patient C; 7, cultured amoebae isolate CDC:V039; 8, cultured amoebae isolate CDC:V451; and 9, cultured amoebae isolate CDC:V416.

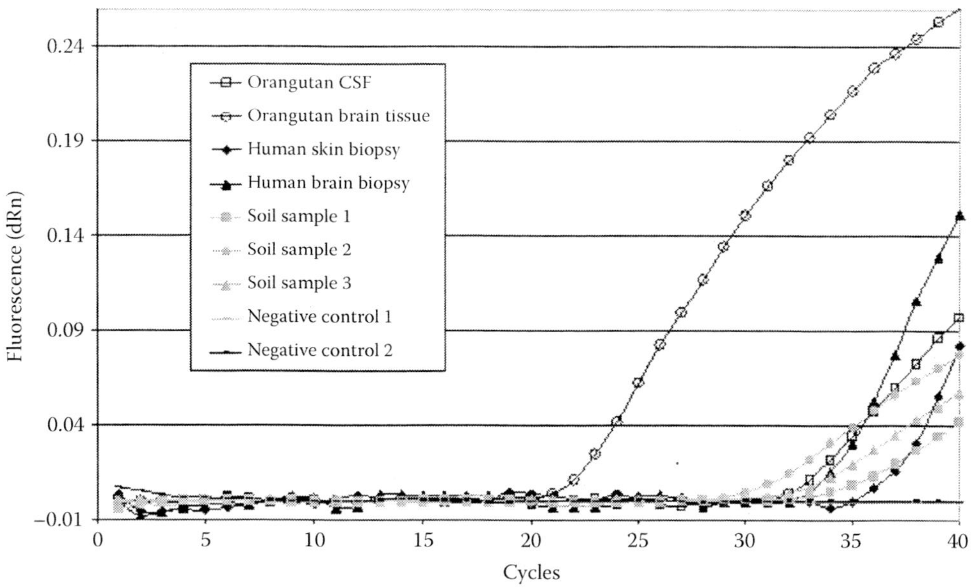

FIGURE 3.4 Amplification plots from real-time PCR detection of *B. mandrillaris* in animal, human, and environmental samples (figure is a compilation of several individual experiments).

stem, and areas of the gray–white junction. Amoebae are found in many areas of the brain, causing hemorrhagic necrosis in the midbrain, thalamus, brainstem, and cerebellum (Figure 3.5). Specific identification of *B. mandrillaris* normally requires immunohistological procedures

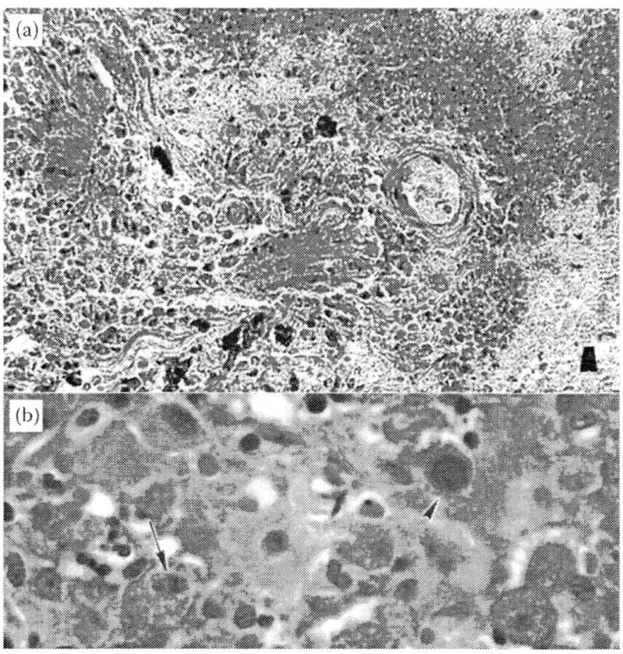

FIGURE 3.5 CNS section with *B. mandrillaris* amoebae amidst brain parenchyma. (a) CNS section showing the extensive inflammation and necrosis of the brain tissue with *B. mandrillaris* amoebae interspersed within the brain tissue, H&E, ×100. (b) Trophozoites and a cyst (at arrow head) in the CNS section at higher magnification, H&E, ×1000. Note the presence of multiple nucleoli (at arrows) in the nucleus of the trophozoites.

such as immunofluorescence or immunoperoxidase or electron microscopy since both trophozoites and cysts of *Acanthamoeba* spp. and *B. mandrillaris* have similar morphology in stained (H&E or periodic acid Schiff) tissue sections at light microscopy. Rarely, however, it may be possible to identify *Balamuthia* in H&E-stained sections by searching for amoebae with multiple nucleoli in the nucleus (Figure 3.5b). Only *Balamuthia* amoebae possess two or more nucleoli. In the absence of this feature *Balamuthia* can be differentiated from *Acanthamoeba* by immunohistochemical analysis of the tissue sections using rabbit anti-*Acanthamoeba* or anti-*B. mandrillaris* sera [1–3,98].

Light and electron microscopic examination of the CNS sections reveals that the CNS parenchyma was infiltrated by a prominent subacute and chronic inflammatory exudate associated with tissue necrosis. Numerous amoebic trophozoites and occasionally cysts were found interspersed in the affected brain tissue. Amoebae were frequently found in clusters around blood vessels. Occasionally two or three nucleolar elements were seen within the nucleus of several amoebic trophozoites. In other areas of the brain, a diffuse and chronic inflammatory infiltrate with a few multinucleated giant cells and diffuse lymphocytic and plasma cells was seen. Additionally, foci of polymorphonuclear leukocytes were seen scattered within the affected areas. Activated microglia (mesenchymal microglia cells with numerous cytoplasmic organelles) or macrophage-like cells were also seen surrounding the amoebae in the brain tissue. Astrocytosis was also observed in the less necrotic areas. In general, mononuclear cells were the predominant cell type found in the host tissue. Amoebae in the tissue sections reacted intensely with the rabbit anti-*B. mandrillaris* serum, but no reaction was seen when the sections were treated with *anti-Acanthamoeba* or *anti-Naegleria* sera [1–3].

3.1.5.2 Pathogenesis

Balamuthia invades the CNS most likely at the sites of the blood–brain barrier [9,10]. The precise mechanisms by which *B. mandrillaris* crosses the blood–brain barrier are unknown. The blood–brain barrier prevents penetration of the CNS by pathogens and is controlled by HBMEC. Based on *in vitro* study using *Balamuthia* and HBMEC, Khan's group at the Birkbeck University in England [99–104], postulated that *Balamuthia* disrupts the blood–brain barrier following adhesion of the amoebae and release of proteolytic enzymes. This, in turn, leads to the injury of blood–brain barrier as a result of the host inflammatory responses. Recent studies have shown that a galactose-binding protein expressed on the surface membranes of *B. mandrillaris* binds to galactose residues of the plasma membrane glycoconjugates of cultured HBMEC [9,10,103,104]. Another study by Khan's group has shown that *Balamuthia* stimulates cytokine production in the host [101]. The presence of the amoebae induces release of the cytokine interleukin-6 (IL-6), after stimulating the cells to produce phosphatidylinositol kinase. The release of IL-6 causes alterations in the blood–brain barrier facilitating movement of leukocytes (and perhaps *Balamuthia*) across the barrier. Other recent studies have shown that *Balamuthia* trophozoites induce HBMEC to release a proinflammatory cytokine, IL-6, which is known to play a role in initiating early inflammatory response [100].

According to recent reports many uninfected individuals possess antibodies against *B. mandrillaris* [3]. Matin et al. [103] have shown that normal human sera inhibit the growth of *B. mandrillaris* in culture and may possess amoebicidal activity. The serum also inhibits the ability of the amoeba to bind to HBMEC and neutralizes the proteases [99] that may contribute to the destruction of the cells. If the amoebae are able to break through into the brain, they will multiply and destroy the brain rapidly [99–103]. In another study it has been shown that protein synthesis, cytoskeletal rearrangements, and possibly ergosterol biosynthesis are important pathways in *B. mandrillaris* encystment and that inhibiting these pathways blocked amoeba-mediated cytopathogenicity in cultured primary HBMEC [104].

It is possible that *Balamuthia* secretes enzymes that are toxic to host cells. Matin et al. [99] have shown that *Balamuthia* produces metalloprotease suggesting its ability to weaken or destroy the extracellular matrix (ECM) that holds brain cells together. The metalloproteases attack collagen, elastin, and plasminogen—the components of the ECM—and thus facilitate the penetration of *Balamuthia* amoebae into the brain parenchyma. Additionally, other enzymes such as lipase, phospholipase A, and lysophospholipase A may also be involved in the destruction of the brain parenchyma [105].

According to a study by Rocha-Azevedo [28] the index isolate of *B. mandrillaris* interacts with the ECM proteins such as collagen-1, fibronectin, and laminin-1, which are usually present in host connective tissue. Further, these authors have shown by scanning electron microscopy that the binding to ECM is associated with changes in the shape of the trophic amoebae and production of surface projections or food cups reminiscent of those seen with *N. fowleri* and *Acanthamoeba* spp.

Balamuthia amoebae may also produce a number of enzymes that function both intra- and extracellularly. For example, these amoebae possess enzymes such as esterase and malate dehydrogenase that are commonly seen in other free-living amoebae including *Acanthamoeba* and *Naegleria*. Such enzymatic activity can be visualized by reacting the electrophoretically separated proteins with appropriate substrates and staining agents. Although differences in the banding patterns have been seen between different isolates it is not clear whether these enzymes have any activity with reference to pathogenicity or virulence of the isolates [106].

3.1.6 *Balamuthia* Infecting Other Animals Besides Humans

Balamuthia GAE has also been reported in a variety of animals including gorillas [107,108], baboons [1,108], gibbons [108], orangutan [109], monkeys [108], horses [110,111], sheep [112], and dogs [91,113]. One of the early cases described by Anderson et al. [107] as a possible *Acanthamoeba* infection in a low land gorilla was actually caused by *Balamuthia* [1–3,108]. The CNS pathology in the infected animals is similar to that which occurs in humans including foci of necrosis, hemorrhage, edema, and inflammation. Although neutrophils were predominant, there were variable numbers of lymphocytes, eosinophils, reactive astrocytes, and vacuolated macrophages were seen. Additionally, prominent mononuclear to mixed inflammatory infiltrates and edema in the leptomeninges and Virchow–Robin spaces were seen. There was also necrotizing vasculitis characterized by mural necrosis, fibrin deposition, and mixed inflammatory infiltrates [108]. Extra CNS infections consisted of nodular masses resembling malignant neoplasms, especially in the gorillas, and these masses were disseminated throughout the abdominal cavity. Also, smaller masses, randomly distributed throughout the liver, kidneys, pancreas, and hilar lung regions [108]. Numerous amoebic trophozoites measuring 15–30 μm and few cysts were seen in the CSF and other organs [108]. Light and electron microscopic examination of the CNS sections of the baboon revealed that the CNS parenchyma was infiltrated by a prominent subacute and chronic inflammatory exudate associated with tissue necrosis. Numerous amoebic trophozoites and occasionally cysts were found interspersed in the affected brain tissue. Amoebae were frequently found in clusters around blood vessels.

Janitschke et al. [114] using severe combined immunodeficient mice developed an animal model to study *Balamuthia* GAE. Mice that were inoculated with the amoebae either intranasally or intraperitoneally developed signs of meningoencephalitis such as ruffled fur, aimless wandering, partial paralysis, coma, and finally death within 8–10 days. Amoebae were either isolated in mammalian cell culture from the brains as well as the lungs of dead mice or were demonstrated in tissue sections using conventional light microscopic examination and immunofluorescence techniques. The CNS lesions in the mice, whether infected intranasally or intraperitoneally, were similar to those of the index case in the baboon, but less pronounced. Intranasal inoculation of mice with suspensions of 1000 trophozoites and cysts caused death in 70% of the animals within a period of 2–4 weeks. Amoebae were found in the brain parenchyma of dead mice without any evidence of inflammatory response. Kiderlen and Laube [115] showed that in intranasally infected immunodeficient mice, the amoebae adhered to the nasal epithelium, migrated along the olfactory nerves, traversed the cribriform plate, and invaded the brain, which is somewhat similar to the invasion pathway described for *N. fowleri*. In another exciting study, Kiderlen et al. [116] demonstrated that *Balamuthia* amoebae migrate to CNS even after mice were infected gastrointestinally. They introduced ~10,000 trophozoites and cysts by gavage into the stomach of both immunocompetent and immunocompromised mice and found amoebae and their antigens in the CNS of 20% of the immunocompetent and 40% of immunodeficient mice. Evidence of *Balamuthia* antigen and amoebae was detected by immunofluorescent staining in the CNS, as well as stomach lining, intestinal tract, lungs, liver, and gall bladder of the dead mice. Amoebic antigen but not intact amoebae were found in the fecal pellets. Animal-to-animal transmission did not occur when infected and noninfected mice were housed in the same cage; however, *Balamuthia* antigen appeared in the intestinal tract of the latter group of animals.

3.1.7 Immune Response

Antibodies of the IgG and IgM classes to *Balamuthia*, with titers ranging from 1:64 to 1:256, have been demonstrated in the sera from healthy people in South Australia using flow cytometry techniques [117]. Schuster et al. [118–120] used immunofluorescent antibody (IFA) to test serum samples collected from patients hospitalized with encephalitis in California and found that about 3% of 225 serum samples had titers ranging from 1:128 to 1:256 for anti-*B. mandrillaris* antibody. These Californian patients with high titered antibody were subsequently confirmed as being infected with *B. mandrillaris* by specific immunostaining of biopsy sections using rabbit anti-*B. mandrillaris* serum and by PCR. Further, two of the patients with a titer of 1:128 in the acute phase survived the infection after treatment and the antibody titers had dropped to 1:64 in the convalescent serum [52]. However, about 10% of the patients who did not have balamuthiasis had titers of 1:32 or greater, suggesting a humoral immune response to these amoebae or, possibly, cross-reactivity with other antigens [120]. Since low levels of anti-*Balamuthia* antibodies are present in the general population, serology is not recommended as a diagnostic procedure. However, serum antibody titers to *Balamuthia* should be measured since a high antibody titer may indicate active infection with *Balamuthia*.

3.1.8 Therapy

Most cases of balamuthiasis have been detected at autopsy. Since these cases had no clear-cut clinical profile they had been treated empirically with a number of antibacterial, antifungal, and antiviral agents and steroids with little or no effect upon the course of the infection. In many of these cases treatment with steroids may have aggravated the infection by suppressing the inflammatory response. A few cases, however, that were clearly identified premortem as due to *Balamuthia* have been treated successfully with a cocktail of drugs. For example, two patients from California, a 60-year-old man, and a 6-year-old girl, and a 70-year-old woman from New York survived this infection after treatment with a combination of pentamidine isothionate, flucytosine (5-fluorocytosine), fluconazole, and a macrolide antibiotic (azithromycin or clarithromycin), sulfadiazine and a phenothiazine (thioridazine or trifluoroperazine) [52,53]. These drugs were used because they have been found to have considerable activity against several isolates of *Balamuthia in vitro* [121,122]. During the course of treatment spanning several years, some drugs had to be suspended intermittently or permanently because of side effects: hyperglycemia and elevated creatinine levels caused by pentamidine; elevated liver enzyme levels caused by fluconazole; rigidity or seizures caused by trifluoroperazine; and elevated creatinine levels and concern about interstitial nephritis caused by azithromycin [52]. Recently, a wide variety of drugs have been tested *in vitro* against *Balamuthia* and miltefosine, an alkyl phosphocholine, appear to have activity against *Balamuthia* and can be considered for inclusion in the treatment of balamuthiasis [123]. In addition to the above, a few Peruvian patients with cutaneous lesions caused by *Balamuthia* survived the infection after surgical intervention and prolonged therapy with albendazole and itraconazole, whereas one patient recovered without any treatment [74]. According to Bravo et al. [74], surgical excision of a large lesion in the chest wall of one patient may have reduced the parasite load, thus aiding the recovery process. Optimal therapy for balamuthiasis is not clear. Given the problems with the diagnosis of infection and the lack of effective antimicrobial agents, the prognosis for patients is poor.

3.2 METHODS

3.2.1 Sample Preparation

PCR detection of *B. mandrillaris* has been accomplished from CSF, unfixed brain tissue, paraffin-embedded formalin-fixed brain tissue, and lung and kidney tissue from human and animal origin [59,60,84,91,95,124]. *B. mandrillaris* does not require any special lysis method to release its DNA, so commonly used DNA extraction techniques can be used. Methods used successfully in the literature include proteinase K digestion, treatment with lysis buffers containing guanidinium thiocyanate acid or UNSET (Tris-ethylenediamine tetraacetic acid buffer with urea, NaCl, and SDS), and commercially available genomic DNA extraction kits from QIAGEN and BioRad [24,91,95–97].

The most easily accessible clinical specimen from patients suspected of free-living amoebic infections is CSF. *Balamuthia* DNA has been detected by PCR [95] and it is believed that its presence in the CSF may be a result of necrosis of brain tissue and/or lysis of amoebae. It is currently not known as to what factors influence the probability of finding amoebic DNA in CSF. To increase the likelihood of detecting *B. mandrillaris* in the CSF from an infected patient it is recommended to spin down as much of the liquid as possible to collect any cellular material [95]. After removal of most of the supernatant, DNA extraction can then be performed on the pellet and a small volume of liquid.

Tissue sections on microscopy slides can be used for PCR detection, although some preparations may result in poor yield or quality of the DNA due to the formalin fixation of the tissue. Formalin is known to cause problems for obtaining good-quality DNA so the shorter time the tissue is fixed in formalin the better are the chances of successful PCR amplification [125]. The age of the final paraffin-embedded preparation does not seem to be important; successful PCR detection has been reported from an archived tissue slide more than 20 years old [124].

3.2.2 Detection Procedure

3.2.2.1 Conventional Techniques

In our laboratory we routinely use indirect immunofluorescence (IIF) test using anti-*Balamuthia* serum raised in rabbits in conjunction with goat-anti-human or anti-rabbit fluorescein isothiocyanate conjugate (FITC) [1–3]. This is usually performed on biopsied or necropsied brain or other types of tissues (e.g., kidney, lung, etc.). After exposure of deparaffinized tissue sections to rabbit anti-*Balamuthia* serum, goat anti-rabbit FITC is applied for specific detection of *Balamuthia* amoebae which will exhibit bright green fluorescence. Typically, clusters of amoebae will be found in the perivascular areas of blood vessels. Transmission electron microscopy can also be performed on glutaraldehyde-fixed and plastic-embedded (Epon) specimens or by deparaffinizing formalin-fixed autopsy tissues followed by fixation in Karnovsky's fluid and embedment in Epon. Ultrastructure demonstrates the characteristic features of trophozoites (one to several nucleolar masses) and cysts (triple wall) and thus is helpful in the identification of the specific amoeba. Cysts may or may not be present in the H&E-stained brain and skin sections. Depending on the part of the brain biopsied, few amoebae may be present and may require an intensive search of the sectioned brain or skin. Additionally, amoebae may be mistaken for macrophages, gitter cells, and so on and may pass unrecognized.

Although *in vitro* culture may not be a useful diagnostic procedure because of the time involved in the development of culture it should be attempted and clinical specimens should be processed as soon as possible. Many isolates of *Balamuthia* have been established in culture by inoculating fresh or frozen tissue specimens. It is possible that trophozoites may be destroyed by freezing but cysts, if present, may survive freezing. Because *Balamuthia* will not grow on bacteria-coated agar plates, biopsy specimens should be inoculated onto monolayers of mammalian cell cultures such as monkey kidney, human lung fibroblasts, or HBMEC [3–5,24,25,108].

For the serum antibody testing, the indirect IFA staining, whole *Balamuthia* amoebae fixed and dried on slides are treated with dilutions of patient serum (1:2–1:4096) to determine antibody titer [3–5]. This is followed by incubation with goat anti-human FITC. In a positive reaction the amoebae will produce an apple green fluorescence. Most commonly, serum titers for *Balamuthia* antibodies range from none to about 1:32 [3,117–120,126]. Although titer alone may be misleading, a titer of 1:64 or higher is indicative of an active immune response to the presence of *Balamuthia*. No cross-reactivity between *Balamuthia* and *Acanthamoeba* antibodies, which are usually tested for simultaneously, has been detected [3,117–120,126].

Elevated titers, if observed, may be a result of cross-reactivity with an unknown antigen. Therefore, other techniques such as PCR of CSF or examination of H&E-stained slides should be considered [120]. Caution should be exercised in those cases when serum samples show low antibody titers because the patient is immunosuppressed due to HIV/AIDS or high-dose corticosteroid therapy. In such cases brain sections can be positive with either few or numerous amoebae depending on the area biopsied.

The technique for isolation from brain tissue or CSF is to add macerated brain tissue or CSF to the tissue culture [3–5,24,25]. Amoebae may become apparent within a few days or it may take several weeks for the amoebae to appear. The amoebae will feed on the monolayer, initially producing virus-like plaques, and soon thereafter consuming virtually all the tissue culture cells. All isolates in culture, clinical as well as environmental, will grow at 37°C. They will grow at lower temperatures but at a slower rate. The ability of the amoebae to tolerate temperatures above 37°C has not been critically tested.

3.2.2.2 Molecular Techniques

Table 3.1 lists the primers and probes currently available in the literature for species-specific detection of *B. mandrillaris* by PCR. PCR can be performed to confirm the identity on cultured amoebae after isolation from clinical or environmental material. More importantly, it can be applied directly to clinical specimens for a fast and specific diagnosis. The PCR amplifying the larger fragment of the mitochondrial 16S rRNA gene was validated with brain tissue and CSF samples from suspected GAE patients and was found to be highly sensitive and specific [95]. However, because of a large amplicon size it may not work with samples with low amounts or poor quality of the amoebic DNA. For this reason a modified version of this assay was developed, featuring a new reverse primer that resulted in a 230 base pair amplicon [91]. This assay was successful in amplifying *B. mandrillaris* DNA in archived tissue section slides as well as more recent preparations of brain tissue from a human GAE case and brain and

TABLE 3.1
PCR Primers and Probes for Specific Detection of *Balamuthia mandrillaris*

PCR Assay	Oligonucleotide Name	Sequence (5′–3′)	Amplicon Size (Base Pairs)	Detection Mode	Reference
1	5′Balspec16S	CGCATGTATGAAGAAGACCA	1075	Gel electrophoresis	18
	3′Balspec16S	TTACCTATATAATTGTCGATACCA			
2	5′Balspec16S	CGCATGTATGAAGAAGACCA	230	Gel electrophoresis	95
	Bal16Sr610	CCCCTTTTTAACTCTAGTCATATAGT			
3	BalaF1451	TAA CCT GCT AAA TAG TCA TGC CAA T	171	Fluorescence (real-time PCR)	96
	BalaR1621	CAA ACT TCC CTC GGC TAA TCA			
	BalaP1582	(R)-AGT ACT TCT ACC AAT CCA ACC GCC A-(Q)			
4	RNaseP/FW	GGC AGG TTC CGA GGA GAC A	82	Fluorescence (real-time PCR)	97
	RNaseP/RV	GTG GCC TTG TGT ATT GAA CTT AAC ATT			
	RNaseP/probe	(R)-TG GAA CCA TAC CTT GGG TGA CAC GAT G-(Q)			

kidney tissue from an infected dog [84,91,124]. This assay is also being used within the California Encephalitis Project, which aims at identifying the cause of unexplained encephalitis. So far they have detected 10 cases of *B. mandrillaris* GAE using PCR [60].

In our laboratory we routinely use real-time PCR targeting the 18S rRNA gene on all specimens received for free-living amoeba testing [96]. One great advantage of this assay is that it simultaneously detects and identifies all free-living amoebae commonly infecting humans. It is a fast assay with little hands-on time and a turn-around time of just a few hours. To date, we have confirmed 18 cases of human *B. mandrillaris* encephalitis using this real-time PCR, in addition to a few infections in animals. Furthermore, we have successfully used this assay to detect *B. mandrillaris* in soil samples associated with a fatal GAE case in Arizona.

3.3 CONCLUSION AND FUTURE PERSPECTIVE

Balamuthia mandrillaris is an important cause of central nervous system and disseminated diseases in both immunocompromised and immunocompetent persons. *B. mandrillaris* is now increasingly recognized as infectious agent of persons receiving organ transplantation since three clusters of transplant-transmitted cases of *Balamuthia* GAE have been identified. Although both a real-time PCR and immunofluorescence test are available for quick diagnoses of this terrible infection, recognition of this disease by the care givers is poor. Therefore, it is believed that many cases have gone undiagnosed.

Although *Balamuthia* GAE is a disease with low incidence, it has a huge impact because of the high mortality rate (>95%). Current drug treatments are not effective: the drug cocktail including pentamidine, 5 fluorocytosine, sulfadiazine, fluconazole, and azithromycin/clarithromycin currently used in the treatment regimen of few surviving patients has huge disadvantage because of toxicity. Additionally, many other patients treated similarly have succumbed to this disease. Therefore, potential new drugs with no or low toxicity is urgently needed. Currently, we do not know the various factors that are necessary for *B. mandrillaris* to cause infection in humans and other animals. Additionally, we do not know the infectious dose, the virulence of the infecting strain, and potential predisposing host factors that allow for infection, and whether infectious strains of *B. mandrillaris* can be transmitted from animals to humans. It is hoped that analysis of the whole-genome sequence of *B. mandrillaris*, when available, would greatly facilitate studies to (a) investigate exact mechanisms of *pathogenesis*, (b) help identify potential targets for drug discovery, (c) potential molecular markers to be used for improved noninvasive diagnostic methods and environmental detection. Together, these will improve our understanding of the pathogenesis of this deadly amoeba of increasing importance.

REFERENCES

1. Visvesvara, G.S., and Stehr-Green, J.K. Epidemiology of free-living ameba infections. *J. Protozool.* 37, 359, 1990.
2. Martinez, A.J., and Visvesvara, G.S. Free-living, amphizoic and opportunistic amebas. *Brain Pathol.* 7, 583, 1997.
3. Visvesvara, G. S. et al. Leptomyxid ameba, a new agent of amebic meningoencephalitis in humans and animals. *J. Clin. Microbiol.* 28, 2750, 1990.
4. Visvesvara, G. S., Schuster, F. L., and Martinez, A. J. *Balamuthia mandrillaris*, N. G., N. Sp., agent of amebic meningoencephalitis in humans and other animals. *J. Eukaryot. Microbiol.* 40, 504, 1993.
5. Schuster, F.L., and Visvesvara, G.S. *Balamuthia mandrillaris*. In: Khan, N.A. (ed.), *Emerging Protozoan Pathogens*. Taylor & Francis, London, pp. 71–118, 2008.
6. Visvesvara, G.S., Moura, H., and Schuster, F.L. Pathogenic and opportunistic free-living amoebae: *Acanthamoeba* spp., *Balamuthia mandrillaris*, *Naegleria fowleri*, and *Sappinia diploidea*. *FEMS Immunol. Micribiol.* 50, 1, 2007.
7. Schuster, F.L., and Visvesvara, G.S. Free-living amoebae as opportunistic and non-opportunistic pathogens of humans and animals. *Int. J. Parasitol.* 34, 1001, 2004.
8. Martinez, A.J., and Visvesvara, G.S. *Balamuthia mandrillaris* infection (Editorial). *J. Med. Microbiol.* 50, 205, 2001.
9. Siddiqui, R., and Khan, N.A. *Balamuthia* amoebic encephalitis: An emerging disease with fatal consequences. *Microb. Pathogen.* 44, 89, 2008.

10. Matin, A. et al. Increasing importance of *Balamuthia mandrillaris*. *Clin. Microbiol. Rev*. 21, 435, 2008.
11. Gelman, B.B. et al. Amoebic encephalitis due to *Sappinia diploidea*. *JAMA* 285, 2450, 2001.
12. Gelman, B.B. et al. Neuropathological and ultrastructural features of amebic encephalitis caused by *Sappinia diploidea*. *J. Neuropathol. Exp. Neurol*. 62, 990, 2003.
13. Qvarnstrom, Y. et al. Molecular confirmation of *Sappinia pedata* as causative agent of amebic encephalitis. *J. Infect. Dis*. 199, 1139, 2009.
14. Amaral-Zettler, L.A. et al. A molecular reassessment of the leptomyxid amoebae. *Protist* 151, 275, 2000.
15. Stothard, D.R. et al. The evolutionary history of the genus *Acanthamoeba* and the identification of eight new 18S rRNA gene sequence types. *J. Eukaryot. Microbiol*. 45, 45, 1998.
16. Booton, G.C. et al. Genotyping of *Balamuthia mandrillaris* based on nuclear 18S and mitochondrial 16S rRNA genes. *Am. J. Trop. Med. Hyg*. 68, 65, 2003.
17. Booton, G.C. et al. *Balamuthia mandrillaris*: Identification of clinical and environmental isolates using genus-specific PCR. *J. Eukaryot. Microbiol*. 50 (Suppl), 508–509, 2003.
18. Schuster, F.L. et al. Environmental isolation of *Balamuthia mandrillaris* associated with a case of amebic encephalitis. *J. Clin. Microbiol*. 41, 3175, 2003.
19. Dunnebacke, T.H. et al. *Balamuthia mandrillaris* from soil samples. *Microbiology* 150, 2837, 2004.
20. Page, F.C. *A New Key to Freshwater and Soil Gymnamoebae*. Freshwater Biological Association, Ambleside, UK, 1988.
21. Adl, S.M. et al. The new higher level classification of eukaryotes with emphasis on the taxonomy of protists. *J. Eukaryot. Microbiol*. 52, 399, 2005.
22. Siddiqui, R., Ortega-Rivas, A., and Khan, N.A. *Balamuthia mandrillaris* resistance to hostile conditions. *J. Med. Microbiol*. 57, 428, 2008.
23. Siddiqui, R., Khan, N.A., and Jarroll, E.L. The cyst wall carbohydrate composition of *Balamuthia mandrillaris*. *Parasitol. Res*. 104, 1439, 2009.
24. Jayasekera, S. et al. Post-mortem culture of *Balamuthia mandrillaris* from the brain and cerebrospinal fluid of a case of granulomatous amoebic meningoencephalitis, using human brain microvascular endothelial cells. *J. Med. Microbiol*. 53, 1007, 2004.
25. Kidederlen, A.F. et al. Cytopathogenicity of *Balamuthia mandrillaris*, an opportunistic causative agent of granulomatous amoebic encephalitis. *J. Eukrayot. Microbiol*. 53, 456, 2006.
26. Dunnebacke, T.H. The amoeba *Balamuthia mandrillaris* feeds by entering into mammalian cells in culture. *J. Eukaryot. Microbiol*. 54, 451, 2007.
27. Visvesvara, G.S. et al. *Paravahlkampfia francinae* n. sp. masquearding as an agent of primary amebic meningoencephalitis. *J. Eukaryot. Microbiol*. 56, 357, 2009.
28. Rocha-Azevedo, B. et al. The interaction between the amoeba *Balamuthia mandrillaris* and extracellular glycoproteins in vitro. *Parasitology* 134, 51, 2007.
29. Schuster, F.L., and Visvesvara, G.S. Axenic growth and drug sensitivity studies of *Balamuthia mandrillaris*, an agent of amebic meningoencephalitis in humans and other animals. *J. Clin. Microbiol*. 34, 385, 1996.
30. Schuster, F.L. Cultivation of pathogenic and opportunistic free-living amebas. *Clin. Microbiol. Rev*. 15, 342, 2002.
31. Matin, A. et al. Evaluation of prokaryotic and eukaryotic cells as food source for *Balamuthia mandrillaris*. *Arch. Microbiol*. 186, 261, 2006.
32. Niyyati, M. et al. Isolation of *Balamuthia mandrillaris* from urban dust, free of known infectious involvement. *Parasitol. Res*. 106, 279, 2009.
33. Visvesvara, G.S. *In vitro* cultivation of microsporidia of clinical importance. *Clin Microbiol. Rev*. 15, 401, 2002.
34. Greub, G., and Raoult, D. Microorganisms resistant to free-living amoebae. *Clin. Microbiol. Rev*. 17, 413, 2004.
35. Shadrach, W.S. et al. *Balamuthia mandrillaris*, a free-living ameba and opportunistic agent of encephalitis, is a potential host for *Legionella pneumophila* bacteria. *Appl. Environ. Microbiol*. 71, 2244, 2005.
36. Michel, R. et al. Free-living amoebae serve as a host for the Chlamydia-like bacterium *Simkania negevensis*. *Acta Protozool*. 44, 113–121, 2005.
37. Hoffmann, E.O. et al. A case of primary amebic meningoencephalitis. Light and electron microscopy and immunohistologic studies. *Am. J. Trop. Med. Hyg*. 27, 29, 1978.
38. Cox, E.C. Amebic meningoencephalitis caused by *Acanthamoeba* species in a four month old child. *J. S. C. Med. Assoc*. 76, 459, 1980.
39. Matson, D.O. et al. *Acanthamoeba* meningoencephalitis masquerading as neurocysticercosis. *Pediatr. Infect. Dis*. 7, 121, 1988.
40. Kidney D.D., and Kim, S.H. CNS infections with free-living amebas: Neuroimaging findings. *Am. J. Radiol*. 171, 808, 1988.
41. Anzil, A.P. et al. Amebic meningoencephalitis in a patient with AIDS caused by a newly recognized opportunistic pathogen. *Arch. Pathol*. 115, 21, 1991.
42. Gordon, S.M. et al. Culture isolation of *Acanthamoeba* spp. and leptomyxid amebas in AIDS and non-AIDS patients with amebic meningoencephalitis. *Clin. Infect. Dis*. 15, 1024, 1992.
43. Popek, E.J., and Neafie, R.C. Cae 4. Granulomatous meningoencephalitis due to leptomyxid ameba. *Pediatr. Pathol*. 12, 871, 1992.
44. Griesemer, D.A. et al. Amebic meningoencephalitis caused by *Balamuthia mandrillaris*. *Pediatr. Neurol*. 10, 249, 1994.
45. Rowen, J.L. et al. *Balamuthia mandrillaris*: A newly recognized agent for amebic meningoencephalitis. *Pediatr. J. Infect. Dis*. 14, 705, 1995.
46. Lowichik, A. et al. Leptomyxid amebic meningoencephalitis mimicking brain stem glioma. *Am. J. Neuroradiol*. 16 (Suppl), 926, 1995.
47. Denney, C.F. et al. Amebic meningoencephalitis caused by *Balamuthia mandrillaris*: Case report and review. *Clin. Infect. Dis*. 25, 1354, 1997.
48. Duke, B.J. et al. *Balamuthia mandrillaris* meningoencephalitis presenting with acute hydrocephalus. *Pediatr. Neurol*. 26, 107, 1997.
49. Deol, I. et al. Encephalitis due to a free-living amoeba (*Balamuthia mandrillaris*): Case report with literature review. *Surg. Neurol*. 53, 611, 2000.
50. Katz, J.D. et al. A case of *Balamuthia mandrillaris* meningoencephalitis. *Arch. Neurol*. 57, 1210, 2000.
51. Bakardjiev, A. et al. Amebic encephalitis caused by *Balamuthia mandrillaris*: Report of four cases. *Pediatr. Infect. Dis*. 22, 447, 2003.
52. Deetz, T.R. et al. Successful treatment of *Balamuthia* amoebic encephalitis: Presentation of two cases. *Clin. Infect. Dis*. 37, 1304, 2003.

53. Jung, S. et al. *Balamuthia mandrillaris* meningoencephalitis in an immunocompetent patient. *Arch. Pathol. Lab. Med.* 128, 466, 2004.
54. Rahimian, J., Kleinman, G., and Parta, M. Amebic encephalitis caused by Balamuthia. *Infect. Med.* 22, 382, 2005.
55. Zagardo, M.T., Castellani, R.J., Zoarski, G.H., and Bauserman, S.C. Granulomatous amebic encephalitis caused by leptomyxid amebae in an HIV-infected patient. *Am. J. Neuroradiol.* 18, 903, 1997.
56. Pritzker, A.S. et al. Fatal granulomatous amebic encephalitis caused by *Balamuthia mandrillaris* presenting as a skin lesion. *J. Am. Acad. Dermatol.* 50, S38–S41, 2004.
57. CDC, Balamuthia amebic encephalitis—California. 1999–20 *Mort. Morb. Wkly Rep.* 57, 768, 2008.
58. Kansagra, A.P. et al. *Balamuthia mandrillaris* meningoencephalitis in an immunocompromised patient. *J. Neurosurg.* 111, 301, 2009.
59. Glaser, C.A. et al. Beyond viruses: Clinical profiles and etiologies associated with encephalitis. *Clin. Infect. Dis.* 43, 1565, 2006.
60. Schuster, F.L. et al. Under the radar: *Balamuthia* amebic encephalitis. *Clin. Infect. Dis.* 48, 879, 2009.
61. Taratuto, A.L. Leptomyxid amoeba encephalitis: Report of the first case in Argentina. *Trans. R. Soc. Trop. Med. Hyg.* 85, 77, 1991.
62. Galarza, M. et al. Pediatric granulomatous cerebral amebiasis: A delayed diagnosis. *Pediatr. Neurol.* 26, 153, 2002.
63. Chemelli, L. et al. Granulomatous amoebic encephalitis due to leptomyxid amoebae: Report of the first Brazilian case. *Trans. Roy. Soc. Trop. Med. Hyg.* 86, 635, 1992.
64. Silva-Vergara, M.L. et al. Disseminated *Balamuthia mandrillaris* amoeba infection in an AIDS patient from Brazil. *Am. J. Trop. Med. Hyg.* 77, 1096, 2007.
65. Oddó, D., Ciani, S., and Vial, P. Encefalitis amebiana granulomatosa por *Balamuthia mandrillaris*. Primer caso diagnostico en Chile. *Rev. Chil. Infect.* 23, 232, 2006.
66. Cuevas, M. et al. Meningoencefalitis granulomatosa por *Balamuthia mandrillaris*: Reporte de un caso y revisión de la literature. *Rev. Chil. Infect.* 23, 237, 2006.
67. Riestra-Castaneda, J.M. et al. Granulomatous amebic encephalitis due to *Balamuthia mandrillaris* (Leptomyxiidae): Report of four cases from Mexico. *Am. J. Trop. Med. Hyg.* 56, 603, 1997.
68. Uribe-Uribe, N.O. et al. Granulomatous amebic encephalitis by *Balamuthia mandrillaris. Patologia* 39, 141, 2001.
69. Candanedo-Gonzalez, F., and Gamboa-Domínguez, A. Acute amebic thyroiditis by *Balamuthia mandrillaris. Patologia* 38, 31–34, 2000.
70. Recavarren-Arce, S. et al. Amoeba angeitic lesions of the central nervous system in *Balamuthia mandrillaris* amoebiasis. *Human Pathol.* 30, 269, 1999.
71. Ballona, R., and Aquije, M. Compromiso cutáneo en encefalitis branulomatosa amebiana fatal causada por *Balamuthia mandrillaris. Folia Dermatol.* 14, 28, 2003.
72. Seas, C.R., and Bravo, F.P. Encefalitis amebiana granulomatosa por *Balamuthia mandrillaris*: un enfermedad fatal reconocida cada vez más frecuentemente en América Latina. *(Editorial) Rev. Chil. Infect.* 23, 197, 2006.
73. Valverde, J., Arrese, J.E., and Pierard, G.E. Granulomatous cutaneous centrofacial and meningocerebral amebisis. *Am. J. Dermatol.* 7, 267, 2006.
74. Bravo, F.G. et al. Cutaneous manifestations of infection by free-living amebas. In: Tyring, S.K., Lupi, O., and Hengge, U.R. (eds), *Tropical Dermatology*. Elsevier Inc., Philadelphia, pp. 49–55, 2006.
75. Martinez, J. et al. Granulomatous amebic encephalitis: A review and report of a spontaneous case from Venezuela. *Acta Neuropathol.* 87, 430, 1994.
76. Rodríguez, O. et al. Infección del sistema nervioso central por amebas de vida libre: Comunicación de tres nuevos casos venezolanos. *Rev. Neurol.* 26, 1005, 1998.
77. Bhgwandeen, S.B. et al. A case of hartmannellid amebic meningoencephalitis in Zambia. *Am. J. Clin. Pathol.* 63, 483, 1975.
78. Rutherfoord, G.S. Amoebic meningo-encephalitis due to a free-living amoeba. A case report. *S. Afr. Med. J.* 69, 52, 1986.
79. Carter, R.F. et al. A fatal case of meningoencephalitis due to a free-living amoeba of uncertain identity—Probably *Acanthamoeba* sp. *Pathology* 13, 51, 1981.
80. Reed, R.P. et al. Fatal granulomatous amoebic encephalitis caused by *Balamuthia mandrillaris. Med. J. Aust.* 167, 82, 1997.
81. Kodet, R. et al. Amebic encephalitis caused by *Balamuthia mandrillaris* in a Czech child: Description of the first case from Europe. *Pathol. Res. Pract.* 194, 423, 1998.
82. White, J.M.L. et al. Granulomatous amoebic encephalitis. *Lancet* 364, 20, 2004.
83. Bodi, I. et al. Fatal granulomatous amoebic meningoencephalitis due to *Balamuthia mandrillaris. Pathol. Res. Practice.* 204, 925, 2008.
84. Tavares, M. et al. Diagnosis of first case of Balamuthia amebic encephalitis in Portugal by immunofluorescence and PCR. *J. Clin. Microbiol.* 44, 2660, 2006.
85. Shirabe, T., Monobe, Y., and Visvesvara, G.S. An autopsy case of amebic meningoencephalitis. The first Japanese case caused by *Balamuthia mandrillaris. Neuropathology* 22, 213, 2002.
86. Prasad, K. et al. Fatal subacute necrotizing brainstem encephalitis in a young man due to a rare parasitic (*Balamuthia*) infection. *Pract. Neurol.* 8, 122, 2008.
87. Sangruchi, T. et al. Spontaneous granulomatous amebic encephalitis. Report of 4 cases from Thailand. *SouthEast Asian J. Trop. Med. Public Health* 25, 309, 1994.
88. Intalpaporn, P. et al. *Balamuthia mandrillaris* meningoencephalitis: The first case in Southeast Asia. *Am. J. Trop. Med. Hyg.* 70, 666, 2004.
89. Schuster, F.L. et al. *Balamuthia* amebic encephalitis risk, Hispanic Americans. *Emerg. Infect. Dis.* 10, 1510, 2004.
90. Chappell, C.I. et al. Standardized method of measuring *Acanthamoeba* antibodies in sera from healthy subjects. *Clin. Diagn. Lab. Immunol.* 8, 724, 2001.
91. Foreman, O. et al. Disseminated infection with *Balamuthia mandrillaris* in a dog. *Vet. Pathol.* 41, 506, 2004.
92. Healy, J.F. *Balamuthia* amebic encephalitis: Radiographic and pathologic findings. *AJNR* 23, 486, 2002.
93. Schumacher, D.J., Tien, R.D., and Lane, K. Neuroimaging findings in rare amebic infections of the central nervous system. *Am. J. Neuroradiol.* 16, 930, 1995.
94. Booton, G.C. et al. Identification of *Balamuthia mandrillaris* by PCR assay using the mitochondrial 16S rRNA gene as a target. *J. Clin. Microbiol.* 41, 453, 2003.
95. Yagi, S. et al. Detection of *Balamuthia* 16S rRNA gene DNA in clinical specimens by PCR. *J. Clin. Microbiol.* 43, 3192, 2005.
96. Qvarnstrom, Y. et al. Multiplex real-time pCR assay for simultaneous detection of *Acanthamoeba* spp., *Balamuthia mandrillaris* and *Naegleria fowleri. J. Clin. Microbiol.* 44, 3589, 2006.
97. Kiderlen, A.F., Radam, E., and Lewin, A. Detection of *Balamuthia mandrillaris* DNA by real-time PCR targeting the RNase P gene. *BMC Microbiol.* 8, 210, 2008.

98. Guarner, J. et al. Histopathologic spectrum and immunohistochemical diagnosis of amebic meningoencephalitis. *Modern Pathol.*, 20, 1230, 2007.
99. Matin, A. et al. *Balamuthia mandrillaris* exhibits metalloprotease activities. *FEMS Immunol. Med. Microbiol.* 47, 83, 2006.
100. Matin, A., and Khan, N.A. Demonstration and partial characterization of ecto-ATPase in *Balamuthia mandrillaris* and its possible role in the host–cell interactions. *Lett. Appl. Microbiol.* 47, 348, 2008.
101. Jayasekera, S. et al. *Balamuthia mandrillaris* stimulates interleukin-6 release in primary brain microvascular endothelial cells via a phosphatidyliositol 3-kinase-dependent pathway. *Microbes Infect.* 7, 1345, 2005.
102. Matin, A. et al. *Balamuthia mandrillaris* interactions with human brain microvascular endothelial cells in vitro. *J. Med. Microbiol.* 56, 1110, 2007.
103. Matin, A. et al. Effects of human serum on *Balamuthia mandrillaris* interaction with human brain microvascular endothelial cells. *J. Med. Microbiol.* 56, 30, 2007.
104. Siddiqui, R. et al. Effect of antimicrobial compounds on *Balamuthia mandrillaris* encystment and human brain microvascular endothelial cell cytopathogenicity. *Antimicrob. Agents Chemother.* 51, 4471, 2007.
105. Shadrach, W.S. et al. The pathogenic amoeba *Balamuthia mandrillaris* possesses cell-associated phospholipase A, lysophospholipase A, and lipase activities. *J. Med. Microbiol.* 293 (Suppl. 38), 100, 2004.
106. Visvesvara, G.S., Moura, H., and Schuster, F.L. *Balamuthia mandrillaris*: *In vitro* cultivation and isoenzyme electrophoresis reveal differences in certain strains. In: Billot-Bonef, S., Cabanes, P.A., Marciano–Cabral, F., Pernin, P., and Pringuez, E. (eds), *Proceedings of the IXth International Meeting on the Biology and Pathogenicity of Free-Living Amoebae.* John Libbey Eurotext, Montrouge, France, pp. 166–171, 2001.
107. Anderson, M.P. et al. Pneumonia and meningoencephalitis due to amoeba in a lowland gorilla. *J. Zool. Anim. Med.* 17, 87, 1986.
108. Rideout, B.A. et al. Fatal infections with *Balamuthia mandrillaris* (a free-living amoeba) in gorillas and other old world primates. *Vet. Pathol.* 34, 15, 1997.
109. Canfield, P.J. et al. Amoebic meningoencephalitis in an orang utan (*Pongo pygmaeus*) caused by *Balamuthia mandrillaris. Austral. Vet. Pathol.* 75, 97, 1997.
110. Kinde, H. et al. Amebic meningoencephalitis caused by *Balamuthia mandrillaris* (leptomyxid ameba) in a horse. *J. Vet. Diagn. Invest.* 10, 378, 1998.
111. Kinde, H. et al. Infections caused by pathogenic free-living amebas (*Balamuthia mandrillaris* and *Acanthamoeba* sp.) in horses. *J. Vet. Diagn. Invest.* 19, 317, 2007.
112. Fuentealba, I.C. et al. Amebic meningoencephalitis in a sheep. *JAVMA*, 200, 363–365, 1992.
113. Finnin, P.J. et al. Multifocal *Balamuthia mandrillaris* infection in a dog in Australia. *Parasitol. Res.* 100, 423, 2006.
114. Janitschke, K. et al. Animal model for *Balamuthia mandrillaris* CNS infection: Contrast and comparison in immunodeficient and immunocompetent mice: A murine model of "granulomatous" amoebic encephalitis. *J. Neuropathol. Exp. Neurol.* 55, 815, 1996.
115. Kidederlen, A.F., and Laube, U. *Balamuthia mandrillaris*, an opportunistic agent of encephalitis, infects the brain via the olfactory nerve pathway. *Parasitol. Res.* 94, 49, 2004.
116. Kiderlen, A.F. et al. Oral infection of immunocompetent and immunodeficient mice with *Balamuthia mandrillaris* amoebae. *Parasitol. Res.* 100, 775, 2007.
117. Huang, Z.H., Ferrante, A., and Carter, R.F. Serum antibodies to *Balamuthia mandrillaris*, a free-living amoeba recently demonstrated to cause granulomatous amoebic encephalitis. *J. Infect. Dis.* 179, 1305, 1999.
118. Schuster, F.L. et al. Detection of antibodies against free-living amoebae *Balamuthia mandrillaris* and *Acanthamoeba* species in a population of patients with encephalitis. *Clin. Infect. Dis.* 42, 1260, 2006.
119. Schuster, F.L. et al. *Balamuthia mandrillaris*, agent of amebic encephalitis: Detection of serum antibodies and antigenic similarity of isolates by enzyme immunoassay. *J. Eukaryot. Microbiol.* 55, 313, 2008.
120. Schuster, F.L. et al. Survey of sera from encephalitis patients for *Balamuthia mandrillaris* antibody. *J. Eukaryot. Microbiol.* IWOP: 10S, 2001.
121. Schuster, F.L., and Visvesvara, G.S. Efficacy of novel antimicrobials against clinical isolates of opportunistic amebas. *J. Eukaryot. Microbiol.* 45, 612, 1998.
122. Schuster, F.L., and Visvesvara, G.S. Opportunistic amoebae: Challenges in prophylaxis and treatment. *Drug Resist. Updates* 7, 41, 2004.
123. Schuster, F.L., Guglielmo, B.J., and Visvesvara, G.S. In-vitro activity of miltefosine and voriconazole on clinical isolates of free-living amoebas: *Balamuthia mandrillaris*, *Acanthamoeba* spp., and *Naegleria fowleri. J. Eukaryot. Microbiol.* 53, 121, 2006.
124. Yagi, S., Schuster, F.L., and Visvesvara, G.S. Demonstration of *Balamuthia* and *Acanthamoeba* mitochondrial DNA in sectioned archival brain and other tissues by the polymerase chain reaction. *Parasitol. Res.* 102, 211, 2008.
125. Greer, C.E., Lund, J.K., and Manos, M.M. PCR amplification from paraffin-embedded tissues; recommendation on fixatives for long-term storage and prospective studies. *PCR Methods Appl.* 1, 46, 1991.
126. Kiderlen, A.F., Radam, E., and Tata, P.S. Assessment of *Balamuthia mandrillaris*-specific serum antibody concentrations by flow cytometry. *Parasitol. Res.* 104, 663, 2009.

4 Blastocystis

Hisao Yoshikawa

CONTENTS

4.1 INTRODUCTION

Ever since the first description in 1912 by Brumpt [1] as harmless yeast, *Blastocystis hominis* is a common intestinal protozoan parasite in the current textbooks of parasitology and microbiology. Despite being reported 100 years ago, the nature of *Blastocystis* is not fairly elucidated. Although a detailed review on the history and basic biology of *Blastocystis* has been made by Zierdt [2], several descriptions have not yet been demonstrated and are still confusing as mentioned in many reviews [3–7]. Based on suitable diagnostic methodologies and expertise in laboratory diagnosis, *B. hominis* is found to be one of the most frequently encountered protozoan parasites in human stool samples [5,8–11]. Morphologically *B. hominis*-like organisms have been widely found in various animals, including nonhuman primates and domestic, wild, pets, and zoo mammals or birds, and occasionally reptiles, amphibians, and insects [12–25].

In the last 10 years, a large number of molecular studies have revealed that *Blastocystis* isolates from humans and animals appear to be extensively polymorphic in genetics [26–34]. Interestingly, dendrograms inferred from the sequence of the small subunit ribosomal RNA (SSU rRNA) genes show a mixed population of *Blastocystis* isolates from both humans and animals within each clade [28,33,35–37], suggesting a zoonotic potential of the parasite. A consensus terminology of *Blastocystis* sp. is recently proposed for the isolates from mammalian and avian animal hosts including humans and it classifies all isolates into nine subtypes (STs) based on the phylogeny inferred with SSU rRNA gene [38]. The reason due to which *Blastocystis* sp. rather than *B. hominis* was proposed is because some reptilian and amphibian species seem to fall within the range of variation by the mammalian and avian clades [36,37]. If the reptilian or amphibian isolates are distinct species, all of the nine clades (nine STs) cannot be *B. hominis*. However, a most recent phylogenetic study inferred with the homologous gene including more wide range of isolates from vertebrate to invertebrate hosts has apparently shown that most isolates from poikilothermal animal hosts (i.e., reptiles, amphibians, and insects) are positioned into the basal clades among the genus of *Blastocystis*, suggesting that the isolates from poikilothermal animal hosts are different from homoiothermal animal hosts (i.e., mammals and birds) [24]. Moreover, recent findings of human infections with *Blastocystis* sp. STs 5 and 8 and 9 [39–41] clearly show that humans are specific hosts capable of being infected by any *Blastocystis* STs. Interestingly, molecular analysis on human *Blastocystis* isolates in various countries indicated that the majority of human isolates are typical genotypes of ST3, suggesting that "*B. hominis*" should be used for the isolates of ST3. Therefore, it seems more suitable to use *Blastocystis* sp. when the genetic characterization is not clarified, because various STs are found among isolates from humans and animals.

4.1.1 Morphology, Biology, and Classification

Morphology of *Blastocystis* isolates from humans and animals have been elucidated by both light and electron microscopy using the samples from fresh fecal materials and axenic or xenic culture. *Blastocystis* is an anaerobic eukaryotic unicellular organism possessing multinuclei, mitochondria-like organelles, Golgi apparatus, and a large central vacuole that contains various substances, such as carbohydrates, lipids, and basic proteins [42–44]. Interestingly, pinocytosis via crathrin-based endocytosis on the surface of the organism clearly shows that nutrients are accumulated into the central vacuole via small vesicles in the cytoplasm [45]. These evidences indicate a storage function of the central vacuole. There is a great variation in cell size and the features of polymorphic morphology are observed not only within a single sample, but also among isolates [3,4]. Therefore, it is difficult to assign for a standard feature for diagnosis of clinical samples.

There are several morphologically distinct forms, including vacuolar, granular, amoeboid, and cyst forms. These forms except for cysts are easily observed in fecal samples or *in vitro* culture under the light microscopy. Although some other forms, multivacuolar and avacuolar forms, were reported by several authors, these forms were small in size and only characterized by electron microscopy [4,46]. In fresh fecal samples, the vacuolar and granular forms are relatively small in size, 5–50 μm, while they reach up to 200 μm or more *in vitro* culture.

The amoeboid forms are generally more frequently observed in the samples of *in vitro* culture than in fecal materials [46–48], while predominance excretion of the amoeboid form in symptomatic patients has recently reported [49,50]. However, irregular shape is more likely artifact in the processing for transmission electron microscopy [7] and disruption of this anaerobic parasite with exposing air is also concerned. Irregular shape of the amoeboid form is difficult to be distinguished from the amoebic forms of other intestinal protozoa, but locomotion of the amoeboid form of *Blastocystis* has not been recognized.

The cyst form is mainly seen in the stool samples and rare *in vitro* culture [13,51–54]. The cyst form is relatively small in size, 3–5 μm in diameter; hence it is difficult to identify it in the fecal samples by routine diagnostic examinations. Several cyst-concentration techniques to detect cysts from fresh fecal samples have been reported [55,56].

The vacuolar form is characterized by a large central vacuole (also known as the central body form), occupying 80–90% of the cell's volume. Cytoplasm containing several nuclei, mitochondria-like organelle and Golgi apparatus spreads into a thin peripheral rim. These features are more clearly observed when using Nomarski or phase-contrast microscopy. The central vacuole is empty under light microscopy, whereas the fine-to-flocculent materials in the vacuole can be observed under electron microscopy. The organisms from *in vitro* culture show the great variation in the density of the materials containing the central vacuole from fine to large granules with variations in size and distribution by transmission or freeze-fracture electron microscopy [48,57–59]. Granular form resembles the vacuolar form except that they possess granules in the central vacuole. The terminology of vacuolar (or central body form) and granular forms were originally named for identifying the organisms under the light microscopy [60]. No clear boundary between vacuolar and granular forms is observed by electron microscopy, indicating that these forms are variable depending on their physiological state.

The life cycle of *Blastocystis* is still not clear. The most conflicting mode is the reproductive stages in the proposed life cycle [2–4,6,7,60–64]. Reproductive stages, such as schizogony [2,64], plasmotomy (budding) [65], endodyogeny [66], and suc-like pouches [67] are not accepted generally for the true modes of *Blastocystis*. A life cycle including thin- and thick-walled cysts via the schizogony process is also postulated [63]. The thin-walled cysts are autoinfectious, whereas the thick-walled cysts are responsible for external transmission, but there is no corroborating evidence. So far, the only accepted mode of reproduction is binary fission, which is frequently observed in cultured samples [3,7]. Recently, Tan [7] proposed a most acceptable life cycle that includes vacuolar, granular, amoeboid and cyst forms, and binary fission reproductive form.

B. hominis was originally reported as yeast by Brumpt [1], subsequently proposed as a sporozoan and then reclassified as a sarcodine by Zierdt [2]. Molecular studies based on the sequence of SSU rRNA gene revealed that the genus *Blastocystis* belongs to a member of the heterogeneous group of stramenopiles [68], which was confirmed by the recent molecular phylogenetic study of cytosolic-type heat-shock protein 70 [69]. Stramenopiles are a diverse heterogeneous group of unicellular and multicellular algae and protists, including heterotropic and photosynthetic representatives [70]. This diverse group includes brown algae, diatoms, oomycetes, and several other lineages. The closest relative of *Blastocystis* in the stramenopiles is *Proteromonas lacertae*, a commensal flagellate in the hindgut of lizards and amphibian [71]. However, unlike other stramenopiles, no flagellar and tubular hairs have ever been observed in *Blastocystis*.

4.1.2 Epidemiology

In general, prevalence of *Blastocystis* infection is higher in developing countries than in developed countries [4,6,7]. The major reason of the higher prevalence in developing countries is the poor sanitary infrastructure and/or low level of care to environmental hygiene in developing countries. High prevalence was observed in developing countries, even using the routine diagnostic technique in epidemiological survey, for example, 22.9–27.2% in Argentina [72,73], 33.3% in Egypt [74], 40.9% in Brazil [75], 52.3% in Malaysia [10], 53.8% in Zambia [76], and 54.5% in Iran [77], and also in developed countries, for example, 22.9% in the United States [8]. Low prevalence was observed in developed countries even using the sensitive method of *in vitro* culture, for example, 3.3% in Singapore [78] and 2.5% in Japan [41].

The prevalence of *Blastocystis* in animals is greatly varied among different countries or among different regions in the same country. Different prevalent rates have been reported in various animals, for example, 74.4% in Australia [17] and 100% in Japan [22] in domestic hens, 58% in domestic ostriches in Europe [20], 80.0% in the pheasants, and 56.3%

in the ducks in the zoo of Japan [12]. In mammals, domestic pigs were 46.8% in Spain [79], 60% in Great Britain [80], 95.1% in Japan [12], and 93% in Czechoslovakia [19]. Domestic cattle are 1.8% in Spain [81] and 70.9% in Japan [12]. The prevalence in zoo primates is 2.1% in Malaysia [82], 57.7–85.3% in Japan [12,22], 66.6% in Spain [14], and 16.0% in the wild primates in Ethiopia [18]. In pet macaques, the prevalence was 43.2% in Indonesia [16]. High prevalence rates with 70.8% in domestic dogs and 67.3% in cats were also reported in Australia [15].

4.1.2.1 Zoonotic Potential

Since the first report of a zoonotic *Blastocystis* isolate from a chicken [32], zoonotic potential of *Blastocystis* isolates from various vertebrate hosts have been studied based on the sequencing of the SSU rRNA gene [28,33,35–37,83–87] or various PCR-based methodologies [26,27,29,34,40,88,89]. One problem that we are facing is how to assign species name for these *Blastocystis* isolates from humans and animals [36,37]. The accumulation of the recent molecular data indicates that the majority of human isolates were subtype 3 (ST3) in most countries [41,78,83,90–97]. Although a few showed that the major genotype of human isolates was ST1 rather than ST3 [85,98], in the studies nearly a half of the isolates was unidentified ST [85]. Conversely, only a few isolates from animals without primates are found to belong to ST3 [27,89,99]. Interestingly, all these animal ST3 isolates are found to be from firm or pet animals, which are potentially in close contact with humans. These evidences suggest that only a genetically distinct genotype (ST3) of *Blastocystis* can easily infect humans, whereas other genotypes are sporadic or more suitable to animal hosts.

Several circumstance evidences also support the zoonotic feature of *Blastocystis*. High prevalence rate among food handlers, animal handlers, and abattoir workers is observed [100–102], indicating that certain occupations, such as close contact with animals is at high risk of infection with zoonotic *Blastocystis*. Moreover, molecular epidemiological studies in the local survey areas also support the transmission from animals to humans [87,103]. Infectivity of various zoonotic *Blastocystis* STs from human isolates was tested experimentally in rats and chickens [104]. There was variability in the infectivity of the isolates in relation to rodent ST4 and avian ST6. Interestingly, both human ST3 isolates could not infect rats and chickens, whereas avian ST7 isolates could only infect chickens. These results also strongly support that ST3 is the human origin genotype.

4.1.2.2 The Mode of Transmission

The actual mode of transmission of *Blastocystis* between humans is not conclusively demonstrated. Molecular approach on human *Blastocystis* infections in a small closed-community might be the best model for elucidating the transmission mode of human-to-human of the parasite. This is evidenced by the study of genotype analysis of the *Blastocystis* isolates from two closed communities, facilities A and B. The investigation showed that when a certain genotype of *Blastocystis* had spread widely among patients and staff members in a facility A, the patients of another facility B were also infected by the same genotype after two patients of facility A moved to facility B [105]. Another local epidemiological study of children in Kathmandu, Nepal showed the possible transmission from animals to humans of the specific genotype [103]. A high risk of *Blastocystis* infection among homosexual males is also reported [106]. These facts support the direct or indirect transmission of *Blastocystis* between individuals via fecal–oral route similar to other intestinal protozoan, such as *Entamoeba* spp., *Giardia intestinalis*, and *Cryptosporidium* spp. Indeed, in the survey of human *Blastocystis* infection, mixed infection with *Blastocystis* and other common intestinal protozoa, which are known to be transmitted by the fecal–oral route, is frequently observed [8,10,72,74,75,96,97,107].

Mode of transmission and the form of the parasite to contribute infection are recently well characterized by using animal models of rats and chickens. A series of oral inoculation of cultured organisms or cyst form revealed that only cyst form was an infective stage [108]. Fecal–oral transmission was also demonstrated in both rats and chickens from an infected one to an uninfected one when both animals had been housed in the same cage [108,109]. Administration of ten cysts into three groups of ten rats showed the different infectivity from 10% to 100% [108], suggesting that contaminated water and food with a few number of *Blastocystis* cysts can establish an infection. A finding of viable cysts in Scottish and Malaysian sewage samples also support the possible transmission by drinking water [110]. On taking account of this issue, several molecular epidemiological studies indicate that consumption of unboiled or raw water plants are infective sources of *Blastocystis* [92,98,111–113]. Therefore, only the cyst form sustains transmission of *Blastocystis* infection via direct or indirect fecal–oral route.

4.1.3 Clinical Features and Pathogenesis

Blastocystis is commonly found in healthy individuals, as well as in patients with gastrointestinal disorders; hence its pathogenic potential remains an enigma. There are many conflicting reports or debates on the pathogenicity of this parasite [9,77,114–127]. Recently, a possible link to irritable bowel syndrome (IBS) with *Blastocystis* infection is of concern [128–131], while no correlation with *Blastocystis* infection was also reported [132].

The most common clinical symptoms among patients infected with *Blastocystis* are diarrhea, abdominal discomfort, fatigue, constipation, and flatulence, which are similar to the patients infected with pathogenic bacteria or virus, as well as other pathogenic intestinal parasites. Therefore, it is necessary to eliminate other factors causing gastrointestinal symptom when the pathogenic potential of *Blastocystis* infection is suspected symptomatically. Although most of the reports had excluded the patients infected with the known pathogens, a series of examinations for laboratory diagnosis are limited.

Accumulated molecular epidemiological studies, by comparing the genotypes between symptomatic and asymptomatic cases, suggested the correlation between clinical symptoms and a certain ST of *Blastocystis* [133–137]. There are several reports suggesting a possible correlation between ST1 and symptomatology [133–135] while other data proved otherwise [136]. Conversely, some studies reported no association between ST and symptomatology [90,91,94,97,138–140]. In the present, there is not enough evidence to prove that *Blastocystis* is nonpathogenic, while there are more accumulating data to suggest the pathogenic potential of this parasite. This situation is quite similar to the finding of pathogenic *Entamoeba histolytica* and nonpathogenic *Entamoeba disper*, which are morphologically indistinguishable.

4.1.4 Diagnosis

Diagnosis of *Blastocystis* in humans and animals has been accomplished from histological staining of fecal samples to *in vitro* culture, and recently progresses to sensitive assays for detecting and genotyping by using PCR-amplification with specific target genes. Simple and sensitive methods are expected for detecting the parasite in stool samples of hosts. Considerable labor and expertise in microscopy are formerly required for the identification of this parasite when only small numbers of the organisms exist in the sample or only cyst form is excreted. Comparative studies indicate that the short-term *in vitro* culture assay is superior to the methods of direct smear or permanent smear, and the formalin ethyl acetate concentration technique, which is poor in sensitivity for detecting *Blastocystis* [96,139,141–143].

The vacuolar and granular forms are the most indisputable organisms for the diagnosis of *Blastocystis* under the light microscopy. However, the morphology of these organisms is not usually kept in the stained specimen, because they degenerate when exposed to oxygen. In addition, low contrast of these round-shaped organisms makes it difficult to be distinguished from lipid droplets or yeast, especially in its small size or less number. Immunofluorescent assay (IFA) is an alternative method for identifying the organisms, as well as small cysts in the samples of *in vitro* culture [53]. An IFA kit for detection of *Blastocystis* organisms in stool samples is now commercially available (Antibodies Inc., USA, 13-310, FITC-labeled polyclonal antibody), which is currently applied for laboratory tests for the diagnosis of *Blastocystis*. However, antigenic differences among isolates or STs may limit its usefulness in detecting all nine STs. Therefore, the short-term *in vitro* culture assay is now considered to be a gold standard technique for the clinical diagnosis of human *Blastocystis* in stool samples.

Progression has been made in immunological assays for detecting serum antibody or secreted IgA in stool samples against *Blastocystis* infection. The significant levels of serum IgA, IgG, or IgG2 antibodies to *Blastocystis* antigens were recognized in the patients with *Blastocystis* infection, using the enzyme-linked immunosorbent assay (ELISA) [144–146], and an ELISA kit for the detection of *Blastocystis* cells in human stool samples has been developed and commercially available (Savyon® Diagnostics Ltd. Israel, the data sheet shows 94.7% and 96.1% sensitivities in two tests). IgA response was detected in all fecal samples and in 83.3% serum samples of symptomatic patients [145]. High level of IgG antibodies were detected in the IBS patients irrespective of the presence or absence of *Blastocystis* in the stools [144], suggesting the possible infection of *Blastocystis* in the past or undetectable of the current infection due to poor detection methods used. When detected with IFA, 70% of the sera obtained from asymptomatic patients with *Blastocystis* infection, were positive in a dilution of 1:60, whereas dilution more than 1:60 was not detected [147]. Another study showed that 20% of the control sera were positive fluorescence in 1:50 and 97% of the patient sera were positive reactions in a series of dilution from 1:100 to 1:800 [148]. Some different results were also reported, for example, no serum IgG antibody response was assessed by the immunoblot analysis to the protein components of *Blastocystis* [149], and there were no serum IgA response among symptomatic patients [146].

However, since there are wide antigenic differences among *Blastocystis* isolates from humans [150–152] or even among isolates of the same ST [153], immunological assays using the antigen from a certain ST of *Blastocystis* are not suitable for the evaluation of results. Therefore, more ideal techniques are required for testing the performance of serological assay in corresponding to the current nine STs.

Several molecular techniques have been developed for diagnosis by PCR targeting DNA extracted from fresh fecal materials or stool cultured samples. There have been two pairs of primer (Blf/Blr[154] and bl 1400ForC/bl 1710RevC[155]) reported (Table 4.1). Unfortunately, both pairs of primer showed strong cross-amplification with common intestinal protozoan (Table 4.2) that are known to be frequently mixed-infection with *Blastocystis*, which may cause false positive. Therefore, it is expected to develop highly sensitive and specific primers for detecting *Blastocystis* so as to reduce the risk of erroneous diagnosis.

4.1.4.1 Conventional Techniques

Fresh stool specimens are necessary for the laboratory examination, because preservation, for example, using 10% buffered formalin, sodium acetate–acetic acid–formalin (SAF), or 2.5% aqueous potassium dichromate that is used for preservation of *Cryptosporidium* oocysts cannot preserve *Blastocystis* organisms. Actually, microscopic diagnosis of SAF-fixed stool samples for *Blastocystis* results in poor diagnosis among all five European reference laboratories [156]. As for routine laboratory examinations, direct smear of stool samples in saline or Lugol's iodine solution, or permanent stained smear by using trichrome or Giemsa may detect parasites if there are many *Blastocystis* in stool samples. It must be kept in mind that *Blastocystis* is an anaerobic organism and be quickly damaged when exposed to air, resulting in the change of its spherical morphology to irregular shape or corrupted, even covered the samples with a coverslip. Therefore,

TABLE 4.1
List of Primers Used for Identification of *Blastocystis* and Sequencing or RFLP Analyses to Classify *Blastocystis* STs

Primer	Product Size (bp) (Position)[a]	Sequence of Forward (F) and Reverse (R) Primers (5′–3′)	Purpose	Reference
Blf/Blr	450 bp (144–593)	F: TAA CCG TAG TAA TTC TAG GGC R: AAC GTT AAT ATA CGC TAT TGG	Diagnosis	154
RD5/RD3	1779 bp (−4+5)	F: GGA AGC TTA TCT GGT TGA TCC TGC CAG TA R: GGG ATC CTG ATC CTT CCG CAG GTT CAC CTA C	RFLP	29
F1/R1	1099 bp (445–1543)	F: GGA GGT AGT GAC AAT AAA TC R: CGT TCA TGA TGA ACA ATT AC	RFLP	138
F1/RB	1112 bp (445–1556)	F: GGA GGT AGT GAC AAT AAA TC R: ACT AGG AAT TCC TCG TTC ATG	RFLP	78
SR1F/SR1R	1770 bp (1–1770)	F: GCT TAT CTG GTT GAT CCT GCC AGT R: TG ATC CTT CCG CAG GTT CAC CTA	RFLP	105
F1/BHCRSeq3	555 bp (445–999)	F: GGA GGT AGT GAC AAT AAA TC R: TAA GAC TAC GAG GGT ATC TA	Sequencing	96
RD5/BhRDr	607 bp (5–611)	F: ATC TGG TTG ATC CTG CCA GT R: GAC CTT TTT AAC TGC AAC AAC G	Sequencing	83
bl 1400ForC/bl 1710RevC	317 bp (1345–1661)	F: GGA ATC CTC TTA GAG GGA CAC TAT ACA T R: TTA CTA AAA TCC AAA GTG TTC ATC GGA C	Diagnosis/ sequencing	155

[a] The numbers indicate the corresponding position of the SSU rRNA gene sequence of *B. hominis* Nand II (Accession number in GenBank U51151).

TABLE 4.2
PCR Amplification of Several Intestinal Protozoan Parasites with Primers Practically Reported for Diagnosis of *Blastocystis* and RFLP or Sequencing for Classification of *Blastocystis* STs

Primer[a]	Gl	Em	Eh	Ed	Cm	Cp	Bh	Band Size (bp)
Blf/Blr	+++	+++	++	+++	++	++	++++	450
RD5/RD3	—	+	++	++++	—	+	++++	1800
F1/R1[b]	—	—	—	—	—	—	++++	1100
SR1F/SR1R	—	—	—	+	—	+	++++	1800
F1/RB	—	—	—	—	—	—	++++	1100
F1/BHCRSeq3	+++	+++	++	++	+++	+++	++++	550
RD5/BhRDr	—	+	+	+	+	+	++++	610
bl 1400ForC/bl 1710RevC	+	+	+	+	+	+	++	310

Note: Gl, *Giardia intestinalis* (Portland 1); Em, *Entamoeba moskovuski* (Laredo); Eh, *E. histolytica* (HM1:IMSS); Ed, *E disper* (AS16-IR); Cm, *Cryptosporidium muris* (RN66); Cp, *Cryptosporidium parvum* (HNJ-1); Bh, *Blastocystis hominis* (Nand II).
Except for *C. muris* and *C. parvum*, all other parasite DNAs were extracted from axenic cultures, whereas DNAs of *Cryptosporidium* were extracted from oocysts purified from fecal materials of the infected mice.
The grade + to ++ ++ indicates intensity of the amplicons observed ethidium bromide staining.

[a] These primers were designed based on the sequence of SSU rRNA gene of *Blastocystis*.

[b] Inability of amplification with some genetically different sequences had recently reported [78].

the adequate amount of stool samples is necessary for the diagnostic examination and the surface area of the stool samples must be avoided for use.

Number in *Blastocystis* stool samples is quite variable. Therefore, multiple stool samples should be tested before a negative diagnostic interpretation is reported. A short-term *in vitro* culture assay is "gold standard," because it is cost-effective and does not need any specific equipment, and increasing the number of the organisms does not require expertise of laboratory diagnostician. A comparative study also shows that *in vitro* culture assay has same sensitivity as the method of a combination of direct extraction of DNA from stool samples, followed by PCR assay [96]. Even applying the most sensitive assay, multiple examinations are necessary in order to increase the detective sensitivity. As indicated by Abe and Yoshikawa [157], the positive rate was only 16.7% when using a single fecal examination of the *in vitro* culture method, which was much lower than 53.0% positive rate when four consecutive examinations by the same method were employed in the same group. These results also care a risk of less advantage for recovering *Blastocystis* in a single sampling from patients when using commercial kits for direct extraction of DNA from stool samples, because only a small amount of the samples (usually 200–300 mg) can be applied in most of the kits.

A shortcoming regarding *in vitro* culture assay should be considered. In culture, *Blastocystis* organisms rapidly proliferate, reach a plateau, and then rapidly degenerate. Therefore, consecutive check of the culture samples is required (normally 2–5 days). Several culture media have been well studied and are now commonly used for detection of *Blastocystis* in the fecal samples of patients or animals, for example, monophasic media, Jones' medium [25,40,41,78, 84–86,96,107,111,138,141–143], and Ringer's solution containing 10% horse serum and 0.05% asparagine (RSHA) medium [41,87,90–93,103–105,108,109,139]. As diphasic media, Locke-Egg (LE) medium [22,30,31,80,105,137,147] modified from the original report by Boeck and Drbohlav's medium [158], Robinson's medium [83,99], and agar slant medium composed of RSHA medium [12,23,26–28,89,106] were also employed to culture. The details of these culture media are described in a recent review [159] or described in Section 4.2.2. Despite no comparative studies on detection efficiency among these media, there was no difference of the detection efficiency from human fecal samples between Jones' and RSHA media (H. Yoshikawa, unpublished data). The agar slant medium based on RSHA can support parasite growth of amphibian [23] and cockroach isolates (H. Yoshikawa, unpublished data), while the Jones' medium was reported as inability to grow *Blastocystis* isolates from some animal fecal samples [86]. Although the short-term *in vitro* culture assay required at least 2 days for positive diagnosis, there are several advantages in the RSHA medium rather than other media or molecular diagnosis. First, this medium is composed of the simplest constituents among any medium reported and can be stored for more than 1 month in a refrigerator if the medium is sterilized. Second, it is cost-effective and does not require any specific techniques and equipment for clinical laboratory. Third, increased number of organisms in cultures can easily diagnose the parasite by microscopy and exclude other organisms that were not supported in this culture medium. Fourth, propagated organisms yield a large amount DNA and can keep them at -80°C or in liquid nitrogen for further study in which viable organisms are required. However, the liquid medium does not support repetitive subcultures, it must replace the diphasic agar-slant medium composed of RSHA without agar.

4.1.4.2 Molecular Techniques

Although many molecular techniques to identify and classify different STs of *Blastocystis* have been developed, a standard methodology is not yet established. The techniques, commonly employed for identifying STs of *Blastocystis*, are PCR with restriction-fragment-length polymorphism of SSU rRNA (PCR–RFLP) [26,27,29,34,79,88, 98,137,138,140,160], sequencing the gene of SSU rRNA [33,83,94,96–99,107,114], and PCR with ST-specific sequenced-tagged site (STS) primers [26,27,34,40,88–93, 103,105,133–136,139,161]. Each of these methods has its advantage or disadvantage when compared with the other two. The most widely applied method is the PCR–RFLP analysis with the combined restrictions of three restriction enzymes, such as *Alu* I, *Hinf* I, and *Rsa* I, which allows identifying the 7ST, as first reported by Clark [29]. However, there are some limitations for this method in identifying STs, for example, too many variations of band profiles [34,88,137,160], same band profiles among different STs [34], which may lead to confusing in identification. Furthermore, a variety of primers applied for amplification of partial or entire sequence of the SSU rRNA gene for application of RFLP (see Table 4.1) also raise confusion. Mixed infection with different STs leads to a complicated RFLP profile. Similarly, various primer combinations for PCR or nested PCR for direct or cloning sequencing are also applied to ascertain subtypes. This technique must concern mixed infections with other protozoa due to the cross-amplification with other common intestinal protozoa (Table 4.2), or with other STs of *Blastocystis* results in mixed peaks in sequencing chromatograms. Recently, insensitive results by using a nested PCR protocol based on RD3/RD5 and F1/R1 primers were also reported [84]. Conversely, PCR employing with ST-specific STS primers can detect directly different STs in agarose gel even in mixed infections with other STs, while this technique requires a series of PCR reactions for each sample and is only applicable for STs1–7 [88,105,161] (Table 4.3). In the mixed infection with different STs, a risk of outgrow of certain ST in stool culture and subsequently only a single ST infection can be detected by PCR, is recently reported [40,86]. In contrast, many culture positive samples (13.9–45.8%) show negative PCR with DNA extracted directly from fresh fecal samples is also reported [85,97,107]. If these PCR-negative results are caused by unsuitable primers due to sequence variation of the SSU rRNA gene among the genus *Blastocystis*, various primers

TABLE 4.3
Primer Pairs of STS Primers for Identification of *Blastocystis* STs

Subtype	Primer	Product Size (bp)	Sequence of Forward (F) and Reverse (R) Primers (5′–3′)	GenBank Accession No.
1	SB83	351	F: GAAGGACTCTCTGACGATGA R: GTCCAAATGAAAGGCAGC	AF166086
1-variant	SB82	462	F: TCTTGCTTCATCGGAGTC R: CCTTCTCGCAGTTCTTTATC	AF166085
2	SB340	704	F: TGTTCTTGTGTCTTCTCAGCTC R: TTCTTTCACACTCCCGTCAT	AY048752
3	SB227	526	F: TAGGATTTGGTGTTTGGAGA R: TTAGAAGTGAAGGAGATGGAAG	AF166088
	SB228	473	F: GACTCCAGAAACTCGCAGAC R: TCTTGTTTCCCCAGTTATCC	AF166089
	SB229	631	F: CACTGTGTCGTCATTGTTTTG R: AGGGCTGCATAATAGAGTGG	AF166090
4	SB337	487	F: GTCTTTCCCTGTCTATTCTGCA R: AATTCGGTCTGCTTCTTCTG	AY048750
5	SB336	317	F: GTGGGTAGAGGAAGGAAAACA R: AGAACAAGTCGATGAAGTGAGAT	AY048751
6	SB332	338	F: GCATCCAGACTACTATCAACATT R: CCATTTTCAGACAACCACTTA	AF166091
7	SB155	650	F: ATCAGCCTACAATCTCCTC R: ATCGCCACTTCTCCAAT	AF166087

Sources: Yoshikawa, H. et al. *J.Parasitol.* 89, 585, 2003; Yoshikawa, H. et al. *J. Clin. Microbiol.* 38, 1324, 2000; Yoshikawa, H. et al. *Mol. Cell. Probes* 12, 153, 1998.

developed for identification STs are likely limited for application. Although there are a few negative samples with stool cultures that are positive with PCR [96,114], *in vitro* culture assay is a highly cost-effective diagnostic assay and a few negative results due to parasite degeneration may be reduced by repetitive sampling. Thus, diagnosticians or researchers must account for the above situations and select or combine several methods for their purpose.

Recently, a novel ST, ST10, was proposed, based on the analysis of partial sequence (about 600 bp) of the SSU rRNA gene [99]. Further study is required to assess the ST10 with sequence and PCR–RFLP analysis of the full-length gene of the SSU rRNA, because phylogenetic tree inferred from the partial sequence does not fit to the dendrograph inferred from the entire sequence of the SSU rRNA gene sometimes.

Some new approaches have been made to type *Blastocystis* recently. A high-throughput pyrosequencing technique was reported to identify ST-specific nucleotide polymorphism in the SSU rRNA gene [95], which could identify 94% identical STs with dideoxy sequencing. However, further confirmation is needed for the usefulness of this approach in subtyping, because this technique could not identify mixed infection and not all STs have been tested with the technique. Menounos et al. tried PCR-single-strand conformation polymorphism (SSCP) for the subtyping and detection of *Blastocystis* [162]. However, there is a limitation for this method in subtyping and detecting *Blastocystis*, because the DNA polymorphism within same STs [28] makes it difficult to obtain identical band profiles, due to its high sensitivity that can detect one mutation of sequence. Although real-time PCR assay is a promising method in subtyping and detecting *Blastocystis*, a higher amount of samples needed for detection (more than 760 *Blastocystis* organisms in one reaction) may limit its application [163].

So far, many PCR primers have been developed for detection or identification *Blastocystis*, as shown in Table 4.1. Unfortunately, most of these primers also amplify other common intestinal protozoan parasites at different magnitudes (Table 4.2). For example, F1/R1 primer set [138], which is frequently employed for RFLP analysis cannot amplify all kinds of *Blastocystis* isolates, as pointed out by Wong et al. [78]. A modified primer set has been developed by Wong et al. [78] recently. This primer set could amplify a broad range of *Blastocystis* isolates, including reptilian and amphibian isolates, but showed no amplification with any common protozoa tested (Table 4.2). Therefore, this primer set can be recommended for developing the most practical strategy for diagnosis of *Blastocystis* at the molecular level. In the next section, the most practical strategy for diagnosis of *Blastocystis* in DNA samples and a combination of RFLP and STS–PCR method for identification of *Blastocystis* STs are described. These methodologies can identify *Blastocystis* sp. STs 1–7, whereas the remaining STs 8 and 9, and novel ST could be suspected, because identification of both the STs is still based on sequencing of the SSU rRNA gene [38].

4.2 METHODS

4.2.1 Sample Preparation

There are two distinct methodologies to recover DNA of *Blastocystis*, one is directly to extrude DNA from fresh stool samples by using commercially available kits and the other is to recover it from *in vitro* cultures. When *Blastocystis*-positive samples were assessed *in vitro* culture assay, propagated organisms can be used for DNA extraction, or used for subculture of the isolate. DNA extracted with any kinds of commercially available kits should be kept at −20°C until used for PCR.

The well-achieved direct DNA extraction from fecal samples is QIAamp DNA Stool Mini Kit (Qiagen). Recently, however, a comparative study among several commercial kits showed that ZR Fecal DNA MiniPrep™ (Zymo Research Corp.) was superior than others [164]. To increase sensitivity of detection, the volume of elution buffer of the final step must be modified to 50 μL from the original volume.

4.2.2 Detection Procedures

4.2.2.1 *In Vitro* Culture Assay for Diagnosis

1. Prepare culture tubes by filling up 6–8 mL RSHA medium composed of Ringer's solution with 10% horse serum and 0.05% asparagine in an appropriate 8–12 mL disposable plastic tube. (10× stock Ringer's solution contains 86.0 g NaCl, 3.3 g $CaCl_2$, 3.0 g KCl, and 5.0 g asparagine; adjust to pH 7.0–7.4 with 1 N NaOH, make up the volume to 1.0 L with distilled water and then autoclave for 20 min at 121°C. 10× stock solution without horse serum can keep in refrigerator for >1 year after autoclaving).
2. Warm up the culture tubes at 25–37°C before use.
3. Take approximately 1 g stool samples by means of a wooden stick and mix well in the medium.
4. Incubate the culture tubes at 37°C.
5. Check the cultures from 2 to 5 days after culturing.
6. Suck up the surface portion of the sediment with pipette, drop on the slide, mixed well by pipetting, and cover with a coverslip. If the drop is excess, take up a part of the drop appropriately, and put on a coverslip.
7. Observe with conventional light microscopy with 10× objective. If small round *Blastocystis*-like organisms are found, observe with 40× objective for easy identification. Check the samples everyday until 5 days after cultures, even if no organisms are found, then negative results can be reported.

4.2.2.2 Molecular Subtyping with PCR–RFLP Followed by STS–PCR

The protocol for subtyping with PCR–RFLP followed by STS–PCR is shown in Table 4.4.

TABLE 4.4
Protocol for Classification and/or Identification of *Blastocystis* STs

Step I. Classification into four groups by RFLP profiles digested with *Spe* I enzyme and 1.1 kb products amplified primer pair F1/RB[a]

Group	Number of Bands	Band Profile with *Spe* I Enzyme	Subtype
A	2 or 3	230, 880 or 230, 430–450 or 230, 410, 470	1, 2
B	2	500–510, 600	3, 4, 8
C	2	470–480, 650	5, 7
D	1	1,100	6, 9

Step II. Identification of ST by PCR with STS primers

Group	Primer[b]	Subtype Identified by Size of Amplicon
A	SB83, SB82, SB340	351 bp (ST1), 462 bp (variant of ST1), 704 bp (ST2),
B	SB227, SB228, SB229, SB337	526 bp (ST3), 473 bp (ST3), 631 bp (ST3), 487 bp (ST4), No amplicon (ST8)[c]
C	SB336, SB155	317 bp (ST5), 650 bp (ST7)
D	SB332	338 bp (ST6), No amplicon (ST9)[c]

[a] Adapted from Wong et al. (2008).
[b] Multiplex PCR is not recommended.
[c] Identification of ST requires sequencing of the SSU rRNA gene.

Step I: Diagnosis of *Blastocystis* infection in DNA samples is based on positive or negative amplification within PCR with F1/RB primer pairs (Table 4.1).

1. Mix 1 μL of the sample DNA in a PCR tube with 9 μL PCR mixture containing the F1/RB primer pair. If the sample is already confirmed *Blastocystis* DNA, such as originated from cultured organisms, the total volume should be changed to 15 μL for convenience of Step II.
2. Set the PCR tubes thermal-cycler at the following condition; the first denature at 94°C for 5 min, followed by 35 cycles (replaced from the original 30 cycles) of 94°C for 1 min, 52°C for 1 min (replaced from the original 49°C), 72°C for 1.5 min, then final extension at 72°C for 10 min [78].
3. Load 1 μL of the amplicons in 1.5% agarose gel with size marker, such as 100 bp ladder and achieve electrophoresis.
4. Stain the gel with ethidium bromide and check presence or absence of the 1.1 kb target size and intensity of the band by ultraviolet irradiator (Lane 6 or 9, in Figure 4.1). If nonspecific extra band(s) with the 1.1 kb band is observed, the DNA sample should dilute appropriately with TE buffer.

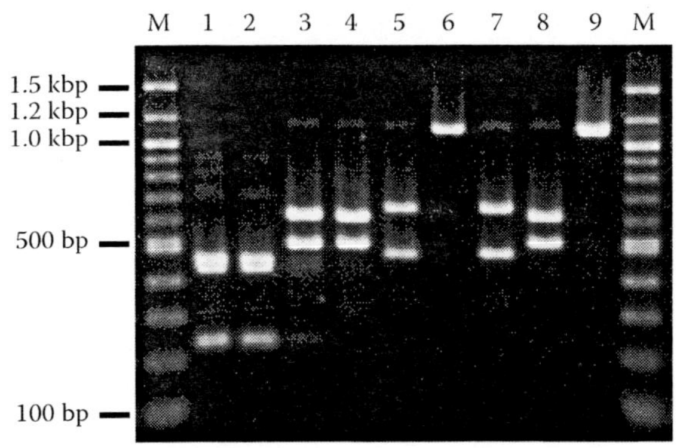

FIGURE 4.1 RFLP profiles of 1.1 kb products of SSU rRNA gene digested with *Spe* I. All nine STs of *Blastocystis* are divided into four groups based on the band profiles as listed in Table 4.4. Lanes 1–9 correspond to STs 1–9, respectively. M is the molecular marker of 100 bp ladder.

Step II: Classification of *Blastocystis* into four groups by RFLP analysis with *Spe* I enzyme.

1. Concentrate the remaining 14 μL of 1.1 kb amplicons into 3–4 times by using a commercially available concentration kit (e.g., Montage PCR, Millipore) for PCR product and finally dissolve it into 3–4 μL distilled water.
2. Incubate the concentrated PCR products with *Spe* I enzyme at 37°C for 2 h under the manufacturer's protocol. Usually the following recipe will be applicable [88]. Volume of the DNA should be adjusted by the intensity of 1.1 kb bands appeared in Step I.

DNA	2–4 μL
10 × buffer	1 μL
BSA	0.1 μL
Spe I (10 U/μL)	0.25 μL
Distilled water	6.65–4.65 μL

3. Load the samples in 2.0–2.5% agarose gel with size marker, such as 100 bp ladder and achieve electrophoresis.
4. Stain the gel with ethidium bromide, check the band profiles, and all samples should be classified into four groups as shown in Table 4.4. If the band profiles are not fitted to any group, the sample might be a new ST.

Step III: Identification of ST by STS–PCR among the four groups.

1. Each group classified by the RFLP analysis can identify one of the STs 1–7 by STS–PCR as shown in Table 4.4.
2. Mix 1 μL of the DNA in a PCR tube with 9 μL PCR reaction mixture containing one of the STS primer pair under the manufacturer's protocol. Multiprex PCR is not recommended.
3. Set the PCR tubes thermal-cycler at the following condition; the first denature at 94°C for 3 min, followed by 30 cycles of 94°C for 30 s, 57°C for 30 s, 72°C for 1 min, then final extension at 72°C for 7 min [161].
4. Load the amplicons in 1.5% agarose gel with size marker, such as 100 bp ladder and achieve electrophoresis.
5. Stain the gel with ethidium bromide and identify ST by the presence or absence of target size of the amplicons listed in Table 4.4. If no band was observed with SB332 for ST6, SB227, SB228, and/or SB229 for ST3, and SB337 for ST4, possible explanation of the sample is ST8 or ST9. However, currently no specific primers are available for these STs8 and 9. For identification of STs8 and 9, sequencing of the entire SSU rRNA gene is required.

Note: Molecular subtyping with PCR–RFLP with *Spe* I enzyme followed by STS–PCR may fit to the large scale of epidemiological survey on *Blastocystis* STs. However, a relatively small number of the samples should directly identify ST with a series of STS–PCR.

4.3 CONCLUSIONS AND FUTURE PERSPECTIVES

The genus *Blastocystis* is probably one of the most indistinguishable eukaryotic unicellular organisms present in the stool samples obtained from vertebrate to invertebrate animals. Recently, there have been increased evidences of the zoonotic and pathogenic potentials of *Blastocystis* isolates from humans and animals. However, it should be investigated to what extent such potentially zoonotic STs are correlated with disease, and clearly, more studies should aim at clarifying issues related to transmission routes and host specificity.

Molecular diagnostic tool may be the most powerful methodology for the detection of the parasite or the typing of different *Blastocystis* STs, and to elucidate transmission route and/or origin of the infectious parasite. Currently, the available molecular data covering the broad range of the genus *Blastocystis* are only from SSU rRNA gene. However, a remarkable heterogeneous sequence among the *Blastocystis* isolates, even among the same genotype poses a problem for the practical design of universal primers for detection of *Blastocystis* and/or classification of STs of *Blastocystis* sp. Therefore, currently published primers for the molecular tool and further newly developed primers must carefully evaluate its performance or cross-amplification within PCR to other common intestinal parasitic protozoa, which are frequently mixed-infected with *Blastocystis*. In addition, further investigation of *Blastocystis* isolates from humans and animals may find novel genotype(s) and full-length sequence of the SSU rRNA gene should be deposited in GenBank for the reference of it.

PCR amplification with DNA directly extracted from stool samples require careful evaluation of the results, because it cannot rule out false positive with the contamination by other organisms DNA or possible misreading of sequence. Currently, no practical methodology is available to ascertain *Blastocystis* STs 8 and 9 without sequencing the SSU rRNA gene. Therefore, cloning of *Blastocystis* cells may be necessary for the identification of STs of mixed infections with different STs or ST8 or ST9. It is also expected to develop simpler and more applicable molecular tools for diagnostic laboratory.

REFERENCES

1. Brumpt, E. *Blastocystis hominis* n sp. et formes voisines. *Bull. Soc. Pathol. Exot.* 5, 725, 1912.
2. Zierdt, C.H. *Blastocystis hominis*—past and future. *Clin. Microbiol. Rev.* 4, 61, 1991.
3. Boreham, P.F.L., and Stenzel, D.J. *Blastocystis* in humans and animals: Morphology, biology, and epizootiology. *Adv. Parasitol.* 32, 1, 1993.
4. Stenzel, D.J., and Boreham, P.F.L. *Blastocystis hominis* revisited. *Clin. Microbiol. Rev.* 9, 563, 1996.
5. Tan, K.S.W., Singh, M., and Yap, E.H. Recent advances in *Blastocystis hominis* research: Hot spots in terra incognita. *Int. J. Parasitol.* 32, 789, 2002.
6. Tan, K.S.W. *Blastocystis* in humans and animals: New insights using modern methodologies. *Vet. Parasitol.* 126, 121, 2004.
7. Tan, K.S.W. New insights on classification, identification, and clinical relevance of *Blastocystis* spp. *Clin. Microbiol. Rev.* 21, 639, 2008.
8. Amin, O.M. Seasonal prevalence of intestinal parasites in the United States during 2000. *Am. J. Trop. Med. Hyg.* 66, 799, 2002.
9. Cirioni, O. et al. Prevalence and clinical relevance of *Blastocystis hominis* in diverse patient cohorts. *Eur. J. Epidemiol.* 15, 389, 1999.
10. Noor Azian, M.Y. et al. Prevalence of intestinal protozoa in an aborigin community in Pahang, Malaysia. *Trop. Biomed.* 24, 55, 2007.
11. Windsor, J.J. et al. Incidence of *Blastocystis hominis* in faecal samples submitted for routine microbiological analysis. *Br. J. Biomed. Sci.* 59, 154, 2002.
12. Abe, N. et al. A survey of *Blastocystis* sp. in livestock, pets, and zoo animals in Japan. *Vet. Parasitol.* 106, 203, 2002.
13. Chen, X.Q. et al. A survey of *Blastocystis* sp. in rodents. *Lab. Anim. Sci.* 47, 91, 1997.
14. Cordón, G.P. et al. Intestinal parasitism in the animals of the zoological garden "Peña Escrita" (Almuñecar, Spain). *Vet. Parasitol.* 156, 302, 2008.
15. Duda, A., Stenzel, D.J., and Boreham, P.F. Detection of *Blastocystis* sp. in domestic dogs and cats. *Vet. Parasitol.* 76, 9, 1998.
16. Jones-Engel, L. et al. Prevalence of enteric parasites in pet macaques in Sulawesi, Indonesia. *Am. J. Primatol.* 62, 71, 2004.
17. Lee, M.G., and Stenzel, D.J. A survey of *Blastocystis* in domestic chickens. *Parasitol. Res.* 85, 109, 1999.
18. Legesse, M., and Erko, B. Zoonotic intestinal parasites in *Papio anubis* (baboon) and *Cercopithecus aethiops* (vervet) from four localities in Ethiopia. *Acta Trop.* 90, 231, 2004.
19. Pakandl, M. Occurrence of *Blastocystis* sp. in pigs. *Folia Parasitol.* 38, 297, 1991.
20. Ponce Gordo, F. et al. Parasites from farmed ostriches (*Struthio camelus*) and rheas (*Rhea americana*) in Europe. *Vet. Parasitol.* 107, 137, 2002.
21. Teow, W.L. et al. A survey of *Blastocystis* in reptiles. *Parasitol. Res.* 78, 453, 1992.
22. Yamada, M. et al. Light microscopical study of *Blastocystis* spp. in monkeys and fowls. *Parasitol. Res.* 73, 527, 1987.
23. Yoshikawa, H. et al. A survey of *Blastocystis* infection in anuran and urodele amphibians. *Vet. Parasitol.* 122, 91, 2004.
24. Yoshikawa, H. et al. Ultrastructural and phylogenetic studies on *Blastocystis* isolates from cockroaches. *J. Eukaryot. Microbiol.* 54, 33, 2007.
25. Zaman, V. et al. Isolation of *Blastocystis* from the cockroach (Dictyoptera: Blattidae). *Parasitol. Res.* 79, 73, 1993.
26. Abe, N., Wu, Z., and Yoshikawa, H. Molecular characterization of *Blastocystis* isolates from birds by PCR with diagnostic primers and restriction fragment length polymorphism analysis of the small subunit ribosomal RNA gene. *Parasitol. Res.* 89, 393, 2003.
27. Abe, N., Wu, Z., and Yoshikawa, H. Zoonotic genotypes of *Blastocystis hominis* detected in cattle and pigs by PCR with diagnostic primers and restriction fragment length polymorphism analysis of the small subunit ribosomal RNA gene. *Parasitol. Res.* 90, 124, 2003.
28. Arisue, N., Hashimoto, T., and Yoshikawa, H. Sequence heterogeneity of the small subunit ribosomal RNA genes among *Blastocystis* isolates. *Parasitology* 126, 1, 2003.
29. Clark, C.G. Extensive genetic diversity in *Blastocystis hominis*. *Mol. Biochem. Parasitol.* 87, 79, 1997.
30. Hoevers, J. et al. Restriction-fragment-length polymorphism analysis of small-subunit rRNA genes of *Blastocystis hominis* isolates from geographically diverse human hosts. *Parasitol. Res.* 86, 57, 2000.
31. Snowden, K. et al. Restriction-fragment-length polymorphism analysis of small-subunit rRNA genes of *Blastocystis* isolates from animal hosts. *Parasitol. Res.* 86, 62, 2000.
32. Yoshikawa, H. et al. DNA polymorphism revealed by arbitrary primers polymerase chain reaction among *Blastocystis* strains isolated from humans, a chicken, and a reptile. *J. Eukaryot. Microbiol.* 43, 127, 1996.
33. Abe, N. Molecular and phylogenetic analysis of *Blastocystis* isolates from various hosts. *Vet. Parasitol.* 120, 235, 2004.
34. Abe, N., Wu, Z., and Yoshikawa, H. Molecular characterization of *Blastocystis* isolates from primates. *Vet. Parasitol.* 113, 321, 2003.
35. Noël, C. et al. Phylogenetic analysis of *Blastocystis* isolates from different hosts based on the comparison of small-subunit rRNA gene sequences. *Mol. Biochem. Parasitol.* 126, 119, 2003.
36. Noël, C. et al. Molecular phylogenies of *Blastocystis* isolates from different hosts: Implications for genetic diversity, identification of species, and zoonosis. *J. Clin. Microbiol.* 43, 348, 2005.
37. Yoshikawa, H. et al. Problems in speciation in the genus *Blastocystis*. *Trends Parasitol.* 20, 251, 2004.
38. Stensvold, C.R. et al. Terminology for *Blastocystis* subtypes—a consensus. *Trends Parasitol.* 23, 93, 2007.
39. Stensvold, C.R. et al. Symptomatic infection with *Blastocystis* sp. subtype 8 successfully treated with trimethoprim-sulfamethoxazole. *Ann. Trop. Med. Parasitol.* 102, 271, 2008.
40. Yan, Y. et al. *Blastocystis* sp. subtype 5: A possibly zoonotic genotype. *Parasitol. Res.* 101, 1527, 2007.
41. Yoshikawa, H. et al. Polymerase chain reaction-based genotype classification among human *Blastocystis hominis* populations isolated from different countries. *Parasitol. Res.* 92, 22, 2004.
42. Yoshikawa, H., and Oishi, K. Ultrastructural localization of basic proteins of *Blastocystis hominis*. *Protoplasma* 200, 31, 1997.

43. Yoshikawa, H., Kuwayama, N., and Enose, Y. Histochemical detection of carbohydrates of *Blastocystis hominis*. *J. Eukaryot. Microbiol.* 42, 70, 1995.
44. Yoshikawa, H., Satoh, J., and Enose, Y. Light and electron microscopic localization of lipids in *Blastocystis hominis*. *J. Electron Microsc.* 44, 100, 1996.
45. Stenzel, D.J., Dunn, L.A., and Boreham, P.F.L. Endocytosis in cultures of *Blastocystis hominis*. *Int. J. Parasitol.* 19, 787, 1989.
46. Stenzel, D.J., Boreham, P.F.L., and McDougall, R. Ultrastructure of *Blastocystis hominis* in human stool samples. *Int. J. Parasitol.* 21, 807, 1991.
47. Chen, X.Q. et al. Description of a *Blastocystis* species from *Rattus norvegicus*. *Parasitol. Res.* 83, 313, 1997.
48. Dunn, L.A., Boreham, P.F.L., and Stenzel, D.J. Ultrastructural variation of *Blastocystis hominis* stocks in culture. *Int. J. Parasitol.* 19, 43, 1989.
49. Tan, T.C., and Suresh, K.G. Predominance of amoeboid forms of *Blastocystis hominis* in isolates from symptomatic patients. *Parasitol. Res.* 98, 189, 2006.
50. Tan, T.C., and Suresh, K.G. Amoeboid form of *Blastocystis hominis*—a detailed ultrastructural insight. *Parasitol. Res.* 99, 737, 2006.
51. Moe, K.T. et al. Observations on the ultrastructure and viability of the cystic stage of *Blastocystis hominis* from human feces. *Parasitol. Res.* 82, 439, 1996.
52. Stenzel, D.J., and Boreham, P.F. A cyst-like stage of *Blastocystis hominis*. *Int. J. Parasitol.* 21, 613, 1991.
53. Yoshikawa, H. et al. Freeze-fracture and cytochemical studies on the *in vitro* cyst form of reptilian *Blastocystis pythoni*. *J. Eukaryot. Microbiol.* 50, 70, 2003.
54. Zaman, V., Howe, J., and Ng, M. Ultrastructure of *Blastocystis hominis* cysts. *Parasitol. Res.* 81, 465, 1995.
55. Zaman, V. The diagnosis of *Blastocystis hominis* cysts in human faeces. *J. Infect.* 33, 15, 1996.
56. Zaman, V., and Khan, K.Z. A concentration technique for obtaining viable cysts of *Blastocystis hominis* from faeces. *J. Pak. Med. Assoc.* 44, 220, 1994.
57. Pakandl, M. *Blastocystis* sp. from pigs: Ultrastructural changes occurring during polyxenic cultivation in Iscove's modified Dulbecco's medium. *Parasitol. Res.* 85, 743, 1999.
58. Yoshikawa, H., and Hayakawa, A. Morphological changes in the central vacuole of *Blastocystis hominis* during *in vitro* culture. *Protoplasma* 194, 63, 1996.
59. Yoshikawa, H., Yamada, M., and Yoshida, Y. Freeze-fracture study of *Blastocystis hominis*. *J. Protozool.* 35, 522, 1988.
60. Zierdt, C.H. Studies of *Blastocystis hominis*. *J. Protozool.* 20, 114, 1973.
61. Burghelea, B., and Radulescu, S. Ultrastructural evidence for a possible differentiation way in the life-cycle of *Blastocystis hominis*, Roum. *Arch. Microbiol. Immunol.* 50, 231, 1991.
62. Govind, S.K., Khairul, A.A., and smith, H.V. Multiple reproductive processes in *Blastocystis*. *Trends Parasitol.* 18, 528, 2002.
63. Singh, M. et al. Elucidation of the life cycle of the intestinal protozoan *Blastocystis hominis*. *Parasitol. Res.* 81, 446, 1995.
64. Tan, K.S.W., and Stenzel, D.J. Multiple reproductive processes in *Blastocystis*: Proceed with caution. *Trends Parasitol.* 19, 290, 2003.
65. Tan, T.C., and Suresh, K.G. Evidence of plasmotomy in *Blastocystis hominis*. *Parasitol. Res.* 101, 1521, 2007.
66. Zhang, X. et al. Morphology and reproductive mode of *Blastocystis hominis* in diarrhea and *in vitro*. *Parasitol. Res.* 101, 43, 2007.
67. Suresh, K. et al. Sac-like pouches in *Blastocystis* from the house lizard *Cosymbotus platyurus*. *Parasitol. Res.* 83, 523, 1997.
68. Silberman, J.D. et al. Human parasite finds taxonomic home. *Nature* 380, 398, 1996.
69. Arisue, N. et al. Phylogenetic position of *Blastocystis hominis* and of stramenopiles inferred from multiple molecular sequence data. *J. Eukaryot. Microbiol.* 49, 42, 2002.
70. Patterson, D.J. Stramenopiles: Chromophytes from a protistan perspective. In: Green, J.C., Leadbeater, B.S.C., and Diver, W.L. (ed.), *The Chromophyte Algae: Problems and Perspectives.* Oxford, Clarendon Press, pp. 357–379, 1989.
71. Leipe, D.D. et al. 16S-like rDNA sequences from *Developayella elegans*, *Labyrinthuloides haliotidis*, and *Proteromonas lacertae* confirm that the stramenopiles are a primarily heterotrophic group. *Europ. J. Protistol.* 32, 449, 1996.
72. Basualdo, J.A. et al. Intestinal parasitoses and environmental factors in a rural population of Argentina 2002–2003. *Rev. Inst. Med. Trop. S. Paulo* 49, 251, 2007.
73. Minvielle, M.C. et al. Epidemiological survey of *Giardia* spp. and *Blastocystis hominis* in an Argentinian rural community. *Korean J. Parasitol.* 42, 121, 2004.
74. Rayan, H.Z., Ismail, O.A., and El Gayar, E.K. Prevalence and clinical feature of *Dientamoeba fragilis* infections in patients suspected to have intestinal parasitic infection. *J. Egypt. Soc. Parasitol.* 37, 599, 2007.
75. Aguiar, J.I.A. et al. Intestinal protozoa and helminths among Terena Indians in the State of Mato Grosso do Sul: High prevalence of *Blastocystis hominis*. *Rev. Soc. Bras. Med. Trop.* 40, 631, 2007.
76. Graczyk, T.K. et al. The association of *Blastocystis hominis* and *Endolimax nana* with diarrheal stools in Zambian schoolage children. *Parasitol. Res.* 98, 38, 2005.
77. Arani, A.S. et al. Prevalence of intestinal parasites in a population in south of Teheran, Iran. *Rev. Inst. Med. Trop. S. Paulo* 50, 145, 2008.
78. Wong, K.H. et al. Predominance of subtype 3 among *Blastocystis* isolates from a major hospital in Singapore. *Parasitol. Res.* 102, 663, 2008.
79. Navarro, C. et al. High prevalence of *Blastocystis* sp. in pigs reared under intensive growing systems: frequency of ribotypes and associated risk factors. *Vet. Parasitol.* 153, 347, 2008.
80. Burden, D.J., Anger, H.S., and Hammet, N.C. *Blastocystis* sp. infection in pigs. *Vet. Microbiol.* 3, 227, 1978/1979.
81. Quílez, J. et al. Occurrence of *Blastocystis* sp. in cattle in Aragón, northeastern Spain. *Parasitol. Res.* 81, 703, 1995.
82. Lim, Y.A.L. et al. Intestinal parasites in various animals at a zoo in Malaysia. *Vet. Parasitol.* 157, 154, 2008.
83. Scicluna, S.M., Tawari, B., and Clark, C.G. DNA barcoding of *Blastocystis Protist* 157, 77, 2006.
84. Stensvold, C.R. et al. *Blastocystis*: unravelling potential risk factors and clinical significance of a common but neglected parasite. *Epidemiol. Infect.* 137, 1655, 2009.
85. Thathaisong, U. et al. *Blastocystis* isolates from a pig and a horse are closely related to *Blastocystis hominis*. *J. Clin. Microbiol.* 41, 967, 2003.
86. Parkar, U. et al. Direct characterization of *Blastocystis* from faeces by PCR and evidence of zoonotic potential. *Parasitology* 134, 359, 2007.
87. Rivera, W.L. Phylogenetic analysis of *Blastocystis* isolates from animal and human hosts in the Philippines. *Vet. Parasitol.* 156, 178, 2008.

88. Yoshikawa, H. et al. Molecular comparative studies among *Blastocystis* isolates obtained from humans and animals. *J. Parasitol.* 89, 585, 2003.
89. Yoshikawa, H., Abe, N., and Wu, Z. PCR-based identification of zoonotic isolates of *Blastocystis* from mammals and birds. *Microbiology* 150, 1147, 2004.
90. Dogruman-Al, F. et al. A possible link between subtype 2 and asymptomatic infections of *Blastocystis hominis*. *Parasitol. Res.* 103, 685, 2008.
91. Dogruman-Al, F. et al. *Blastocystis* subtypes in irritable bowel syndrome and inflammatory bowel diseases in Ankara, Turkey. *Mem. Inst. Oswald Cruz* 104, 724, 2009.
92. Li, L.H. et al. Molecular epidemiology of human *Blastocystis* in a village in Yunnan province, China. *Parasitol. Int.* 56, 281, 2007.
93. Li, L.H. et al. Cross-sectional surveys and subtype classification of human *Blastocystis* isolates from four epidemiological settings in China. *Parasitol. Res.* 102, 83, 2007.
94. Özyurt, M. et al. Molecular epidemiology of *Blastocystis* infections in Turkey. *Parasitol. Int.* 57, 300, 2008.
95. Stensvold, C.R. et al. *Blastocystis*: Subtyping isolates using pyrosequencing™ technology. *Exp. Parasitol.* 116, 111, 2007.
96. Stensvold, C.R. et al. Detecting *Blastocystis* using parasitologic and DNA-based methods: A comparative study. *Diagn. Microbiol. Infect. Dis.* 59, 303, 2007.
97. Souppart, L. et al. Molecular epidemiology of human *Blastocystis* isolates in France. *Parasitol. Res.* 105, 413, 2009.
98. Leelayoova, S. et al. Drinking water: A possible source of *Blastocystis* spp. subtype 1 infection in schoolchildren of a rural community in central Thailand. *Am. J. Trop. Med. Hyg.* 79, 401, 2008.
99. Stensvold, C.R. et al. Subtype distribution of *Blastocystis* isolates from synanthropic and zoo animals and identification of a new subtype. *Int. J. Parasitol.* 39, 473, 2009.
100. Khan, Z.A., and Alkhalife, I.S. Prevalence of *Blastocystis hominis* among "healthy" food handlers in Dammam, Soudi Arabia. *J. Egypt. Soc. Parasitol.* 35, 395, 2005.
101. Sadek, Y. et al. Intestinal parasites among food-handlers in Qualyobia Governorate, with reference to the pathogenic parasite *Blastocystis hominis*. *J. Egypt. Soc. Parasitol.* 27, 471, 1997.
102. Salim, H.R. et al. *Blastocystis* in animal handlers. *Parasitol. Res.* 85, 1032, 1999.
103. Yoshikawa, H. et al. Molecular characterization of *Blastocystis* isolates from children and rhesus monkeys in Kathmandu, Nepal. *Vet. Parasitol.* 160, 295, 2009.
104. Iguchi, A. et al. Infectivity of different genotypes of human *Blastocystis hominis* isolates in chickens and rats. *Parasitol. Int.* 56, 107, 2007.
105. Yoshikawa, H. et al. Genomic analysis of *Blastocystis hominis* strains isolated from two long-term health care facilities. *J. Clin. Microbiol.* 38, 1324, 2000.
106. Stark, D. et al. Prevalence of enteric protozoa in human immunodeficiency virus (HIV)-positive and HIV-negative men who have sex with men from Sydney, Australia. *Am. J. Trop. Med. Hyg.* 76, 549, 2007.
107. Rene, B.A. et al. Subtype analysis of *Blastocystis* isolates from *Blastocystis* cyst excreting patients. *Am. J. Trop. Med. Hyg.* 80, 588, 2009.
108. Yoshikawa, H. et al. Fecal–oral transmission of the cyst form of *Blastocystis hominis* in rats. *Parasitol. Res.* 94, 391, 2004.
109. Tanizaki, A. et al. Infectivity of *Blastocystis* isolates from chickens, quails and geese in chickens. *Parasitol. Res.* 96, 57, 2005.
110. Suresh, K., Smith, H.V., and Tan, T.C. Viable *Blastocystis* cysts in Scottish and Malaysian sewage samples. *Appl. Environ. Microbiol.* 71, 5619, 2005.
111. Leelayoova, S. et al. Evidence of waterborne transmission of *Blastocystis hominis*. *Am. J. Trop. Med. Hyg.* 70, 658, 2004.
112. Nimri, L., and Batchoun, R. Intestinal colonization of symptomatic and asymptomatic schoolchildren with *Blastocystis hominis*. *J. Clin. Microbiol.* 32, 2865, 1994.
113. Taamasri, P. et al. Transmission of intestinal blastocystosis related to the quality of drinking water. *Southeast Asian J. Trop. Med. Public Health* 31, 112, 2000.
114. Stensvold, C.R. et al. Pursuing the clinical significance of *Blastocystis*—diagnostic limitations. *Trends Parasitol.* 25, 23, 2009.
115. Carrascosa, M., Martínez, J., and Pérez-Castrillon, J.L. Hemorrhagic proctosigmoiditis and *Blastocystis hominis* infection. *Ann. Intern. Med.* 124, 278, 1996.
116. Chen, T.L. et al. Clinical characteristics and endoscopic findings associated with *Blastocystis hominis* in healthy adults. *Am. J. Trop. Med. Hyg.* 69, 213, 2003.
117. Kaya, S. et al. Pathogenicity of *Blastocystis hominis*, a clinical reevaluation. *Turkiye Parazitol. Derg.* 31, 184, 2007.
118. Kuo, H-Y. et al. Clinical significance of *Blastocystis hominis*: Experience from a medical center in northern Taiwan. *J. Microbiol. Immunol. Infect.* 41, 222, 2008.
119. Leder, K. et al. No correlation between clinical symptoms and *Blastocystis hominis* in immunocompetent individuals. *J. Gastroenterol. Hepatol.* 20, 1390, 2005.
120. Sun, T. et al. Questionable clinical significance of *Blastocystis hominis* infection. *Am. J. Gastroenterol.* 84, 1543, 1989.
121. Zaki, M. et al. Clinical report of *Blastocystis hominis* infection in children. *J. Trop. Med. Hyg.* 94, 118, 1991.
122. Qadri, S.M.H., Al-Okaili, G.A., and Al-Dayel, F. Clinical significance of *Blastocystis hominis*. *J. Clin. Microbiol.*, 27, 2407, 1989.
123. Doyle, P.W. et al. Epidemiology and pathogenicity of *Blastocystis hominis*. *J. Clin. Microbiol.* 28, 116, 1990.
124. Markell, E.K., and Udkow, M.P. Association of *Blastocystis hominis* with human disease? *J. Clin. Microbiol.* 28, 1085, 1990.
125. Rosenblatt, J.E. *Blastocystis hominis*. *J. Clin. Microbiol.* 28, 2379, 1990.
126. Zierdt, C.H. Pathogenicity of *Blastocystis hominis*. *J. Clin. Microbiol.* 29, 662, 1991.
127. Lee, M.J. Pathogenicity of *Blastocystis hominis*. *J. Clin. Microbiol.* 29, 2089, 1991.
128. Boorom, K.F. et al. Oh my aching gut: Irritable bowel syndrome, *Blastocystis*, and asymptomatic infection. *Parasite Vector* 1, 40, 2008.
129. Giacometti, A. et al. Irritable bowel syndrome in patients with *Blastocystis hominis* infection. *Eur. J. Clin. Microbiol. Infect. Dis.* 18, 436, 1999.
130. Stark, D. et al. Irritable bowel syndrome: A review on the role of intestinal protozoa and the importance of their detection and diagnosis. *Int. J. Parasitol.* 37, 11, 2007.
131. Yakoob, J. et al. Irritable bowel syndrome in search of an etiology: Role of *Blastocystis hominis*. *Am. J. Trop. Med. Hyg.* 70, 383, 2004.
132. Tungtrongchitr, A. et al. *Blastocystis hominis* infection in irritable bowel syndrome patients. *Southeast Asian J. Trop. Med. Pub. Health* 35, 705, 2004.
133. Eroglu, F. et al. Identification of *Blastocystis hominis* isolates from asymptomatic and symptomatic patients by PCR. *Parasitol. Res.* 105, 1589, 2009.

134. Hussein, E.M. et al. Pathophysiological variability of different genotypes of human *Blastocystis hominis* Egyptian isolates in experimentally infected rats. *Parasitol. Res.* 102, 853, 2008.
135. Yan, Y. et al. Genetic variability of *Blastocystis hominis* isolates in China. *Parasitol. Res.* 99, 597, 2006.
136. Tan, T.C., Suresh, K.G., and Smith, H.V. Phenotypic and genotypic characterization of *Blastocystis hominis* isolates implicates subtype 3 as a subtype with pathogenic potential. *Parasitol. Res.* 104, 85, 2008.
137. Kaneda, Y. et al. Ribodemes of *Blastocystis hominis* isolates in Japan. *Am. J. Trop. Med Hyg.* 65, 393, 2001.
138. Böhm-Gloning, B., Knobloch, J., and Walderich, B. Five subgroups of *Blastocystis hominis* from symptomatic and asymptomatic patients revealed by restriction site analysis of PCR-amplified 16S-like rDNA. *Trop. Med. Int. Health* 2, 771, 1997.
139. Dogruman-Al, F. et al. PCR-based subtyping of *Blastocystis* isolates from symtomatic and asymptomatic individuals in a major hospital in Ankara, Turkey. *Parasitol. Res.* 106, 263, 2009.
140. Motazedian, H., Ghasemi, H., and Sadjjadi, S.M. Genomic diversity of *Blastocystis hominis* from patients in southern Iran. *Ann. Trop. Med. Parasitol.* 102, 85, 2008.
141. Leelayoova, S. et al. *In-vitro* cultivation: A sensitive method for detecting *Blastocystis hominis*. *Ann. Trop. Med. Parasitol.* 96, 803, 2002.
142. Suresh, K., and Smith, H. Comparison of methods for detecting *Blastocystis hominis*. *Eur. J. Clin. Microbiol. Infect. Dis.* 23, 509, 2004.
143. Termmathurapoj, S. et al. The usefulness of short-term *in vitro* cultivation for the detection and molecular study of *Blastocystis hominis* in stool specimens. *Parasitol. Res.* 93, 445, 2004.
144. Hussain, R. et al. Significantly increased IgG2 subclass antibody levels to *Blastocystis hominis* in patients with irritable bowel syndrome. *Am. J. Trop. Med. Hyg.* 56, 301, 1997.
145. Mahmoud, M.S.E., and Saleh, W.A. Secretory and humoral antibody responses to *Blastocystis hominis* in symptomatic and asymptomatic human infections. *J. Egypt. Soc. Parasitol.* 33, 13, 2003.
146. Zierdt, C.H., Zierdt, W.S., and Nagy, B. Enzyme-linked immunosorbent assay for detection of serum antibody to *Blastocystis hominis* in symptomatic infections. *J. Parasitol.* 81, 127, 1995.
147. Kaneda, Y. et al. Serologic response to *Blastocystis hominis* infection in asymptomatic individuals. *Tokai J. Exp. Clin. Med.* 25, 51, 2000.
148. Garavelli, P.L. et al. Serum antibody detected by fluorescent antibody test in patients with symptomatic *Blastocystis hominis* infection. *Recenti Prog. Med.* 86, 398, 1995.
149. Chen, J. et al. Lack of serum immune response to *Blastocystis hominis Lancet* 330, 1021, 1987.
150. Kukoschke, K.G. and Müller, H.E., SDS–PAGE and immunological analysis of different axenic *Blastocystis hominis* strains. *J. Med. Microbiol.* 35, 35, 1991.
151. Lanuza, M.D. et al. Soluble-protein and antigenic heterogeneity in axenic *Blastocystis hominis* isolates: Pathogenic implications. *Parasitol. Res.* 85, 93, 1999.
152. Müller, H.E. Four serologically different groups within the species *Blastocystis hominis*. *Zbl. Bakt.* 280, 403, 1994.
153. Yoshikawa, H. et al. Antigenic differences between *Blastocystis hominis* and *Blastocystis* sp. revealed by polyclonal and monoclonal antibodies. *J. Protozool. Res.* 5, 118, 1995.
154. Grabensteiner, E., and Hess, M. PCR for the identification and differentiation of *Histomonas meleagridis*, *Tetratrichomonas gallinarum* and *Blastocystis* spp. *Vet. Parasitol.* 142, 223, 2006.
155. Stensvold, R. et al. Detection of *Blastocystis hominis* in unpreserved stool specimens by using polymerase chain reaction. *J. Parasitol.* 92, 1081, 2006.
156. Utzinger, J. et al. Microscopic diagnosis of sodium acetate–acetic acid–formalin-fixed stool samples for helminths and intestinal protozoa: A comparison among European reference laboratories. *Clin. Microbiol. Infect.* 16, 267, 2010.
157. Abe, N., and Yoshikawa, H. Stool examination and diagnosis of recently attracted intestinal parasitic protozoa *Blastocystis hominis*. *Rinsho Kensa* 41, 938 (1997) (in Japanese).
158. Boeck, W.C., and Drbohlav, J. The cultivation of *Entamoeba histolytica*. *Am. J. Hyg.* 5, 371, 1925.
159. Clark, C.G., and Diamond, L.S, Methods for cultivation of luminal parasitic protists of clinical importance. *Clin. Microbiol. Rev.* 15, 329, 2002.
160. Rivera, W.L., and Tan, M.A. Molecular characterization of *Blastocystis* isolates in the Philippines by riboprinting. *Parasitol. Res.* 96, 253, 2005.
161. Yoshikawa, H. et al. Genomic polymorphism among *Blastocystis hominis* strains and development of subtype-specific diagnostic primers. *Mol. Cell. Probes* 12, 153, 1998.
162. Menounos, P.G. et al. Direct detection of *Blastocystis* sp. in human faecal samples and subtype assignment using single strand conformational polymorphism and sequencing. *Mol. Cell. Probes* 22, 24, 2008.
163. Jones II, M.S. et al. Detection of *Blastocystis* from stool samples using real-time PCR. *Parasitol. Res.* 103, 551, 2008.
164. Yoshikawa et al. Evaluation of DNA extraction kits for molecular diagnosis of human *Blastocystis* subtypes from fecal samples. *Rarasitol. Res.* 109, 1045, 2011.

5 Dientamoeba

Damien Stark

CONTENTS

5.1 INTRODUCTION

5.1.1 Classification, Morphology, and Biology

5.1.1.1 Classification

Dientamoeba fragilis is a pathogenic human intestinal protozoan parasite with a worldwide distribution. Although it was discovered in 1909 by the microbiologist Wenyon, *D. fragilis* was described in scientific literature by M. Jepps and C. Dobell in 1918 [1].

Jepps and Dobell [1] described *D. fragilis* as an amoeba measuring 8–10 μm in diameter with a binucleate structure. They observed the fragile nature of the organism as it degenerated rapidly outside the human body—hence the name *D. fragilis*. Initially considered to be an amoeba, it was placed in the family Endamoebidae. However, its exact phylogenic position has been a matter of ongoing contention and still remains unclear [2].

Although *D. fragilis* was initially considered to be an amoeboid organism Dobell soon challenged the validity of this scientific assumption. After careful observation he concluded that the nuclear apparatus and the mode of nuclear division of *D. fragilis* was flagellate like and that it was structurally similar to the amoeboflagellate *Histomonas meleagridis* (a common pathogen of many galliform and some anseriform birds and the causal agent of a type of entero-hepatitis termed as "blackhead"). Having reviewed all the available scientific evidence Dobell postulated that *Dientamoeba* represented a stage in the life cycle of a flagellate which somewhere in its evolutionary development had permanently lost its flagella [3].

Analysis of antigenic relationships using qualitative gel diffusion methods and immunoelectrophoresis techniques indicated that *Trichomonas*, *Histomonas*, and *Dientamoeba* shared some structurally identical or closely related antigens [4–6]. The antigenic similarities between *Dientamoeba* and *Histomonas* were greater than those between *Dientamoeba* and *Trichomonas*.

The advent of the electron microscopy provided further evidence that *D. fragilis* was closely related to flagellates [7,8]. In 1980, *Dientamoeba* was reclassified in the order *Trichomonadida* along with *Histomonas*, *Monocercomonas*, and *Trichomonas*. However, it was not until 1996 when molecular techniques were used as an alternative to traditional phenotypic markers to determine the taxonomic position of *Dientamoeba*. The phylogenic position of *H. meleagridis* was determined by analysis of the SSU rRNA (small-subunit ribosomal RNA) gene. Analysis of the sequence data showed significant similarity to other parabasalids sequences and phylogenic analysis determined a close relationship between *H. meleagridis* and *D. fragilis* [2,9,10].

Currently, the evidence supports that *D. fragilis* is a trichomonad that has permanently lost flagella and kinetosomes from all stages of its life cycle. Through microscopy, antigenic analysis and molecular characterization of the SSU rRNA gene of *D. fragilis* has been shown to be closely related to the trichomonad *H. meleagridis*. However, its exact phylogenic position still remains to be completely resolved. It has taken over 60 years and the advent of molecular techniques for Dobell's hypothesis that *D. fragilis* was not an amoeba but a flagellate that had permanently lost its flagella to be finally strongly supported by the application of more modern technologies.

5.1.1.2 Taxonomical Status of *D. fragilis*

Kingdom	Protista
SubKingdom	Protozoa
Phylum	Sarcomastigophora
SubPhylum	Mastigophora
Class	Zoomastigophora
Order	Trichomonadida
Family	Monocercomonadidae
Genus	*Dientamoeba*
Species	*D. fragilis* [1]

5.1.1.3 Morphology

D. fragilis is a pleomorphic trophozoite (Figure 5.1) ranging in size from 4 μm to 20 μm with most trophozoites typically in the range of 5–15 μm [11]. *D. fragilis* may not have a cyst stage. Although some have claimed to have seen *D. fragilis* cysts in permanently stained smears, these findings have since been dismissed as they were not substantiated by other parasitologists [12].

The trophozoites of *D. fragilis* are typically binucleate with only 30–40% uninucleate [3]. Dobell [3] described multinucleated trophozoites with up to four nuclei, but usually only one or two are found. Nuclear pleomorphism is common and the size of the nucleus varies in relation to the rest of the cell. In stained smears the nuclear membrane is delicate and does not possess any peripheral chromatin. The karyosome contains 4–8 chromatin granules which often appear as chromatin packets [11].

The cytoplasm of *D. fragilis* appears granular and may be vacuolated and contain food inclusions, ingested bacteria, and yeasts. It has been postulated that extensive vacuolation is indicative of degenerative condition and is often seen in stained smears made from older fecal specimens [13].

In freshly passed stools and from culture media the organisms may be motile by pseudopodia. The trophozoites become less motile as the stool approaches room temperature and no motility is seen if the specimen has been refrigerated.

5.1.1.4 Transmission

The mode of transmission of *D. fragilis* is yet to be determined. Unlike many other intestinal protozoa *D. fragilis* apparently has no cyst stage. The lack of a cyst stage makes it unlikely that the organism can survive in the environment for any length of time. Estimates of the survival time of trophozoites in stool specimen vary from 6 to 48 h [14]. Given the above data it must be assumed that transmission must occur directly. Transmission via contaminated food and water would be unlikely as the organism does not survive outside the human body for long periods of time.

Dobell postulated that, because the amoeboflagellate *H. meleagridis* and *D. fragilis* are similar that they both might have similar modes of transmission. *H. meleagridis*, a common pathogen of domestic fowl, is transmitted from bird to bird via the eggs of the nematode *Heterakis gallinae*. Several studies have examined the role of nematode ova as vectors of transmission. Some researchers have reported high rates of coinfection between *D. fragilis* and *Enterobius vermicularis* giving rise to the hypothesis that pin worm may have a role in the transmission of *Dientamoeba* [15–18]. However, all attempts to isolate *D. fragilis* from *E. vermicularis* eggs have been unsuccessful [19]. Other studies have found no correlation between the two organisms [20–22]. This would argue against the theory that *E. vermicularis* plays a significant role in the transmission of *D. fragilis*. It is

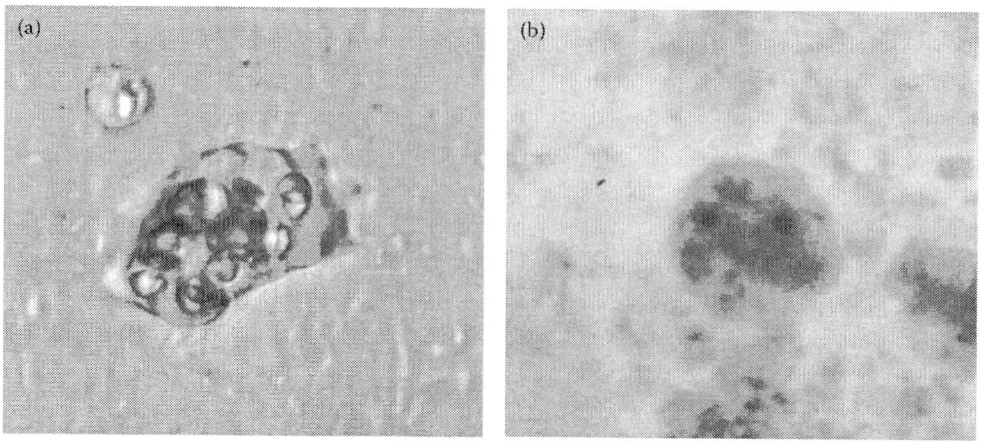

FIGURE 5.1 Photomicrograph of (a) *D. fragilis* (~17 μm) grown in xenic culture media showing irregularly protruding pseudopodia, note the ingested rice starch; (b) an iron hematoxylin-stained *D. fragilis* trophozoite (~12 μm).

yet to be proven what role helminth ova play in the transmission of *D. fragilis*.

Numerous studies have shown that a large portion of patients with *D. fragilis* infection often have other enteric protozoa present, all of which are transmitted via the fecal oral route. One study found that 54% of patients with *Dientamoeba* had other parasites or enteropathogens present [23], while other studies detected that 40% of patients with dientamoebiasis were coinfected with other parasites [22]. A large study investigating 1497 *D. fragilis* infected patients found coinfection with *E. vermicularis* in only 5% of patients, but coinfection rates of *Blastocystis hominis* and *Endolimax nana* were 40 and 24%, respectively. These above findings would support the hypothesis of direct transmission of *D. fragilis* via the fecal oral route [24].

Clearly, further study is required to elucidate the exact nature of the mode of transmission of *D. fragilis*. Due to the fragile nature of this organism and the fact that no cyst stages have been demonstrated it would be advantageous for *D. fragilis* to be transmitted by a vector. Several studies have found a higher anticipated coincidence of *D. fragilis* and *E. vermicularis* infections. Transmission by means of fecally contaminated food or water seems unlikely, as there are no cyst stages and the fact that the parasite cannot survive for any length of time in the environment. However, direct transmission could occur and is seen in other intestinal parasites that do not have cyst stages such as *Pentatrichomonas hominis*, which does infect humans. Higher rates of *D. fragilis* infection are also seen where personal hygiene is poor. These observations would also support direct transmission.

5.1.1.5 Genetic Diversity

Only a few studies have addressed the issue of genetic diversity of the organism. Using polymerase chain reaction (PCR)-restriction fragment-length polymorphism (RFLP) analysis of ribosomal genes of 12 *D. fragilis* isolates researchers found that organisms currently being reported as *D. fragilis* represent at least two significantly different genetic entities [25]. They estimated a sequence divergence of 2% between the two SSU rRNA genotypes of *D. fragilis*; this was later supported by research conducted in the Netherlands by sequencing a 558 bp region of the SSU rDNA from *D. fragilis* isolates and comparing this with available sequence data [26]. Windsor et al. [27] undertook PCR-RFLP analysis of the SSU rRNA gene of 33 *D. fragilis* isolates from the United Kingdom. All 33 isolates gave the same RFLP patterns indicating that *D. fragilis* displays very little variation in its SSU rDNA among UK strains.

Sequence data generated by Stark et al. [22] from the entire SSU rDNA region of Australian isolates of *D. fragilis* showed a sequence divergence of 4% between the Australian genotypes and the *D. fragilis* ATCC strain Bi/PA (ATCC 30948). All the Australian strains sequenced were identical supporting the notion that *D. fragilis* is clonal. The Australian isolates were found to be similar to those found in a recent study in the Netherlands and do not contain the polymorphic *Dde* I restriction site (CTTAG) at position 644 found in the *D. fragilis* strain Bi/PA. RFLP analysis was performed on 50 Australian samples to determine the genotypes present in the Australian population and the extent of genetic diversity. All *D. fragilis* isolates showed no variation and corresponded to genotype A. A recent retrospective study in Belgium reported genotyping of *D. fragilis* isolates by PCR-RFLP of the SSU rRNA gene. Only genotype 1 (Genotype A) was detected in *D. fragilis*-infected patients [28].

Hussein et al. [29] concluded that genetic variation based on the SSU rRNA using RFLP is not useful for typing *D. fragilis* isolates. By using high-resolution melting curve (HRM) analysis following RT-PCR they detected four subtypes. The exact extent of this genetic diversity is unknown and further study is required to resolve this issue.

5.1.2 Clinical Features and Pathogenesis

5.1.2.1 Clinical Features

D. fragilis has been reported throughout the world. The prevalence of *D. fragilis* varies widely between 0.4% and 53% [14]. Some of the major studies are summarized in Table 5.1.

In contrast to the many pathogenic protozoa which have a high prevalence in developing regions of the world, high prevalence rates of *D. fragilis* have been reported from countries normally regarded as having high standards of health: 4.5% in Italy [30], 5.2% in Australia, 6.3% in Belgium, 9.4% in the United States [31], 11.7% in Denmark [32], and 16.9% in the British Isles [33]. Several reports have also identified *D. fragilis* as the most common pathogenic protozoan found in stool when appropriate diagnostic methods are used [34,35].

Initially Jepps and Dobell [1] concluded that this parasite was nonpathogenic; however, researchers questioned the pathogenicity of *D. fragilis* almost immediately. Numerous studies have highlighted the pathogenic potential of this organism [11,12,14].

A study over a 6-month period detected *D. fragilis* in 4.1% of 857 stool samples submitted to the Department of Microbiology and Immunology at the Sultan Qaboos University Hospital in Oman. *D. fragilis* was the most common enteropathogen encountered. Of the patients with *D. fragilis* infection 83% had abdominal pain and 50% had diarrhea,

TABLE 5.1
Worldwide Prevalence of *D. fragilis*

Country	Number of Samples Studied	Positive (%)	Reference
Oman	857	5.1	[23]
Turkey	400	8.8	[36]
Australia	6750	0.9	[22]
Belgium	1270	6.3	[28]
Denmark	103	11.7	[32]
Italy	3139	3.7	[37]
British Isles	543 and 421	14.6 and 16.9	[33]

the duration of which varied from 1 month to 2 years [12]. In a more recent study in Turkey *D. fragilis* was as prevalent and pathogenic as *Giardia intestinalis*. In stool samples from 400 patients *D. fragilis* was found in 8.8% of cases and *G. intestinalis* in 8.5%. The most common symptoms were abdominal pain and diarrhea for both infections [36]. In 2002–2003, a survey was carried out on enteroparasites in fecal samples of outpatients attending a day care center in the town of Perugia (Central Italy). Of 1989 enrolled people *D. fragilis* was found to be more prevalent than *Giardia duodenalis* [38]. Vandenberg et al. [28] found *D. fragilis* (6.3%) and *Giardia lamblia* (7.1%) were the most common pathogenic protozoa isolated from 448 patients studied. Symptoms most frequently encountered with *D. fragilis* and *G. lamblia* infection were abdominal pain (69.2% and 72.4%, respectively) and diarrhea (61.5% and 79.3%, respectively). However, patients with *D. fragilis* infections suffered significantly less frequently from nausea and/or vomiting, anorexia and weight loss. After treatment, all *D. fragilis*- and *G. lamblia*-infected patients presenting a negative follow-up stool test also reported a complete resolution of their symptoms.

A recent prospective study was conducted over a 30-month period, in which fecal specimens from 6750 patients were submitted to the Department of Microbiology at St. Vincent's Hospital Sydney, Australia. Trophozoites of *D. fragilis* were detected in 60 (0.9%) patients by permanent staining and confirmation was performed by PCR. Gastrointestinal symptoms were present in all patients, with diarrhea and abdominal pain the most common symptoms. Thirty-two percent of patients presented with chronic symptoms. Previous reports have highlighted the propensity of the parasite to cause prolonged infection [22,35] with chronic infections reported in the literature to last as long as 2 years.

Many studies have shown that the elimination of *D. fragilis* with antimicrobial agents usually relieves clinical symptoms [11,12]. As such, the treatment of symptomatic patients with *D. fragilis* infections is warranted. The most common antimicrobials used to treat *D. fragilis* include iodoquinol (diidohydroxyquin) [39,40], metronidazole [41,42], tetracycline [43], paramomycin [44], newer nitromidazole derivatives such as secnidazole and ornidazole [36,45] or combination therapy [46]. However, there is currently no consensus regarding best practice in the treatment of dientamoebiasis, and no large-scale randomized control trials have been undertaken to evaluate treatment options.

5.1.2.2 Pathology

Unfortunately there has been insufficient scientific research on the pathogenesis of *D. fragilis* infections. Published information regarding the mechanisms of the pathology caused by *D. fragilis* infection is minimal.

Burrows et al. [15] were the first to report pathological findings in four appendices infected with *D. fragilis*. They found several histopathological changes. Marked fibrosis of the appendix was evident and it was postulated that *D. fragilis* elaborates a low-grade irritation that induces an inflammatory response which results in fibrosis.

Kean and Malloch [20] infected the cecum of rats with *D. fragilis*. Edema of the mucosa became evident, and the trophozoites attached to the cecal mucosa and caused damage to the cells. However, no invasion of cells or ulceration was demonstrated.

D. fragilis has been shown by a number of researchers to ingest red blood cells. Dobell was the first to report this finding [3]. In later years, two more parasitologists demonstrated the erythrophagocytic capacity of *D. fragilis* [15,16].

A lack of a suitable animal model has hampered the study of the clinical manifestations of *D. fragilis* infection. Macaque monkeys, chickens, kittens, and rats have been trialed as animal models [3,20] with little or no success. Many questions regarding the pathogenesis remain unanswered. More research is required to ascertain the exact pathological processes that result from infection by *D. fragilis*.

5.1.3 Diagnosis

5.1.3.1 Conventional Techniques

It has been more than 90 years since *D. fragilis* was first described. Definitive diagnosis is still based on permanent stains of fixed fecal smears. In wet preparations *D. fragilis* appears as a nonspecific rounded mass and the nuclear structure cannot be seen in either saline or iodine preparations [12]. As trophozoites degenerate rapidly prompt fixation of the specimen is necessary [11]. Successful diagnosis of *D. fragilis* is closely associated with the use of permanent stains of fecal smears.

Many different stains and fixatives have been used successfully with *D. fragilis.*, including polyvinyl alcohol, sodium-acetate-acetic acid-formalin, modified Schaudinn's fixative and merthiolate-iodine-formalin [11]. A wide variety of permanent stains have been used to detect *D. fragilis* the most common being iron hematoxylin and trichrome stain [12]. As permanent stains are time consuming and require a highly trained microscopist to read the stains, another diagnostic method would be helpful. Care must be taken when reading the slides as *D. fragilis* may be difficult to distinguish from nonpathogenic protozoa such as *E. nana*.

As with other enteric protozoan infections the collection of multiple stool specimens is essential to aid in diagnosis. Shedding of *D. fragilis* is typically intermittent and the degree of shedding can be highly variable within a day [26,47]. Hiatt et al. [48] compared the sensitivity of examining one stool specimen to that of three specimens. Using conventional permanent staining it was found that the additional stool examinations increased the percentage of positive results by 31.1% for *D. fragilis*. These data suggest that even in symptomatic patients the examination of a single stool specimen could cause a large number of *D. fragilis* infections to be undetected.

Parasite culture techniques have been successfully used for the detection of *D. fragilis*. As with fixatives and stains a large number of culture systems have been used, including: Boeck and Drbohlav's, Robinson' medium, Dobell and Laidlow's medium, Cleavland–Collier's medium, Balamuth's medium, and TYGM-9 [11]. In 1929, Dobell and Svensson were the first to grow "mono-protist" *D. fragilis* cultures by using a diphasic medium devised by Dobell, an inspissated horse serum slant overlaid with diluted egg whites in Ringer's solution and supplemented with rice. He also reported that it grew best at 41°C, a temperature much higher than expected for a human intestinal protist [3].

Several studies have shown that culture techniques are more sensitive than permanent stains. Sawangjaroen et al. [49] found culture to be more sensitive than microscopy in diagnosing *D. fragilis*. A more recent study showed that culturing for *D. fragilis* using a modified Robinson medium dramatically increased the detection rate for *D. fragilis* [50].

However, the cultivation of luminal parasitic protists is technically difficult, time consuming, and often unrewarding [51]. As such, these techniques are usually restricted to specialist parasitology laboratories, and are not offered by routine laboratories. *D. fragilis* is regarded difficult to establish in long-term culture, but can often be grown for a few subcultures before dying out [51]. One drawback of culture systems is that specimens need to be inoculated promptly after collection as reduced temperatures adversely affect *D. fragilis* and trophozoites degenerate rapidly. Specimens cannot be stored refrigerated as this greatly reduces recovery rates [49].

5.1.3.2 Molecular Techniques

Molecular biology techniques such as PCR offer a highly sensitive and specific alternative to traditional diagnostic approaches such as microscopy. Regions of ribosomal DNA are ideal targets for PCR development as they exhibit interspecific variability yet are highly conserved. rRNA genes are among the most ubiquitous and conserved DNA sequences in nature. For this reason they have been extensively used for phylogenic analysis and as diagnostic probes. PCR techniques have been used for the diagnosis of a wide variety of parasites from clinical samples.

Peek et al. [26] was the first to report the development of a conventional PCR by amplification of the SSU rRNA gene to detect *D. fragilis*; however, the sensitivity and specificity of the assay was not determined in this study. In this study the detection limit of PCR was the equivalent of approximately 0.1 *D. fragilis* trophozoites per sample (Table 5.2).

A more recent conventional PCR and real-time PCR assay based on the SSU rRNA gene of *D. fragilis* for the specific detection of *D. fragilis* DNA in fresh unpreserved stool samples was developed [52,53]. To determine the sensitivity of both conventional and real-time PCR, the entire SSU rRNA gene was cloned and a known number of copies were amplified. The detection limit was 100 plasmid copies or an equivalent of approximately one *D. fragilis* trophozoite for conventional PCR. The detection limit for real-time PCR was determined at one plasmid copy (a crossing point of 27.87) of the SSU rRNA gene which is equivalent to approximately 0.01 *D. fragilis* trophozoite. This shows that the real-time PCR was 100 times more sensitive than the conventional PCR. When the conventional and RT-PCR assays were compared the conventional PCR was shown to have a sensitivity of 88.9% and a specificity of 100%, while the real-time PCR was shown to be 100% sensitive and specific [11] (Table 5.3).

Real-time PCR offers several advantages over conventional PCR for a diagnostic laboratory. Amplification and detection is in an integrated system and the technology obviates the need for gel electrophoresis. Coupled with the sealed reaction capillaries the risk of contamination from previously amplified products is greatly reduced. The TaqMan probes provide constant reaction monitoring and rapid cycling times enable high specimen throughput and greatly reduce the time to diagnosis. However, the major advantage of real-time PCR compared to conventional PCR is its increased sensitivity. It has been previously demonstrated that amplification of large fragments, especially using DNA extracted from feces, limits the sensitivity of diagnostic PCR assays [54].

5.2 METHODS

5.2.1 Sample Preparation

Fecal samples are considered the most complex specimens for direct PCR testing due to the presence of PCR inhibitors such as heme, bilirubins, bile salts, and complex carbohydrates which are often coextracted with the target DNA. Optimization of the fecal DNA extraction procedure is critical.

The transport and preservation conditions, of fecal samples, need careful consideration. Transportation of fecal samples containing *D. fragilis* at ambient temperature may result in the degeneration of parasite DNA. Therefore, the sensitivity of DNA assays using unpreserved samples is time dependent. Specimens to be examined by PCR may be preserved by refrigeration or freezing. Many laboratories commonly use fixatives to preserve parasite morphology for microscopic examination and for safe handling of specimens. However, preservation in fixative may decrease the sensitivity of PCR. Sodium acetate acetic acid formalin (SAF) has been found to be inhibitory to both conventional and RT-PCR for the detection of *D. fragilis* [52]. Consequently, if testing a fresh sample is not feasible then freezing a fresh specimen at −20°C before extraction of DNA may be the best strategy. Freezing does not affect the sensitivity of the molecular assays and allows for storage that is not time dependent.

Simple and effective methods for the isolation of DNA from feces have been developed. Commercially available DNA extraction kits are recommended. They are easy to use, minimize extraction time, and DNA can be extracted directly from the feces without the need to culture the parasite. The QIAamp stool kit has become the most widely accepted method for DNA extraction from fecal material.

The following method is adapted from Verweij et al. [54] for parasite DNA isolation from feces using QIAamp spin columns (QIAamp tissue kit).

1. Suspend 0.1 g of feces in 200 μL of ethanol.
2. Centrifuge for 1 min full speed and wash twice with 1 mL of phosphate-buffered saline.

TABLE 5.2
Conventional PCR Assays for Detection of *D. fragilis*

Gene Target	Primer	Sequence (5′ to 3′)	Amplicon Size	Cycling Profile	Primer Concentration	Detection Method	Reference
Conventional PCR SSU rRNA	DF1 DF2	CTC ATA ATC TAC TTG GAA CCA ATT3 TTA TAG TTT CTC TTA TTA GCC CC	662 bp	40 cycles of 94°C for 60 s, 52°C for 60 s, 72°C for 1 min	0.5 μM	1.0% agarose gel stained with ethidium bromide	Peek et al. (2004)
Conventional PCR SSU rRNA	DF400 DF1250	TATCGGAGGTGGTAATGACC3 CATCTTCCTCCTGCTTAGACG	887 bp	94°C for 3 min, 30 cycles of 94°C for 1 min, 57°C for 1.5 min, and 72°C for 2 min	0.5 μM	1.0% agarose gel stained with ethidium bromide	Stark et al. (2005)

TABLE 5.3
Real-Time PCR Assays for Detection of *D. fragilis*

Gene Target	Primer/Probe	Sequence (5′ to 3′)	Amplicon Size	Cycling Profile	Primer/Probe Concentration	Detection Method	Reference
Taqman SSU rRNA	DF3	GTTGAATACGTCCCTGCCCTTT	77 bp	95°C for 10 min, followed by 30 cycles of 95°C, 58°C for 10 s, 72°C for 30 s or 95°C for 10 min, followed by 30 cycles of 95°C for 15 s, 60°C for 60 s	0.25 μM of each primer/0.2 μM of each probe	Fluorescence at the end of each annealing plateau	Stark et al. (2000)
	DF4	TGATCCAATGATTTACACCGAGTC					
	DF-TMP	FAM-CACACCGCCCGTCGCTCCTACCG-TAMRA					
MGB TaqMan SSU rRNA	Df-124F	CAACGGATGTCTTGGCTCTTA	98 bp	95°C for 15 min, followed by 50 cycles of 95°C for 15 s, 60°C for 60 s	3 pmol of each primer/5 pmol of probe	Fluorescence during the annealing/elongation step	Verweij et al. (2007)
	Df-221R	TGCATTCAAAGATCGAACTTATCAC					
	Df-172revT	FAM-CAATTCTAGCCGCTTAT-MGB-Dark Quencher					

3. Resuspend the pellet into 200 μL of phosphate buffered saline containing 2% polyvinylpolypyrolidone (PVPP*) in a microfuge tube.
4. Heat feces-suspension for 10 min at 100°C.
5. Add 180 μL of tissue lysis buffer (ATL) and 20 μL of proteinase K from the QIAamp kit to the sample and mix by vortexing.
6. Incubate for 2 h (or overnight) at 55°C in a heating block, after 1 h vortex briefly.
7. Add 400 μL of AL buffer to 400 μL of sample, mix thoroughly by vortexing.
8. Incubate for 70°C for 10 min.
9. Centrifuge for 10 s full speed and transfer the supernatant to a microfuge tube containing 400 μL of ethanol (96–100%), and mix thoroughly by vortexing.
10. Place a spin column in a 2 mL collection tube. Add 600 μL of the mixture from step 7 to the spin column without moistening the rim, close the cap and centrifuge at 10,000 rpm for 1 min.
11. Place spin column in a clean 2 mL collection tube a repeat with the rest of the mixture from step 7.
12. Place the spin column in a clean 2 mL collection tube.
13. Add 500 μL of AW1 buffer (from kit) to the spin column and centrifuge at 10,000 rpm for 1 min.
14. Empty the collection tube and add 500 μL of AW2 (from kit) buffer and centrifuge for 3 min at full speed.
15. Place the spin column in a clean 1.5 mL microfuge tube and elute DNA by adding 200 μL of AE buffer (from kit) and incubate for 1 min at room temperature.
16. Centrifuge at 8000 rpm for 1 min. The eluted DNA is now ready for use in PCR.

5.2.2 Detection Procedures

5.2.2.1 Conventional PCR

The following PCR assay was developed by Stark et al. This assay employs forward primer (5′-TATCGGAGG TGGTAATGACC-3′) and reverse primer (5′-CATCTTCCT CCTGCTTAGACG-3′) derived from the SSU (18S) rRNA gene for amplification of an 880 bp fragment from *D. fragilis* only [52].

5.2.2.1.1 Procedure

1. Prepare PCR mixture (25 μL) containing ~1.5 U Taq DNA polymerase, 10 mM Tris-Hcl, 50 mM KCl, 1.5 mM $MgCl_2$, 200 mM dNTPs, 0.5 μM of each primer, 2.0 μL purified DNA, and distilled water (up to 25 μL).
2. Perform PCR amplification in a thermocycler with the following cycling program: 94°C for 3 min, 30 cycles of 94°C for 1 min, 57°C for 1.5 min, and 72°C for 2 min.

* PVPP (Sigma P6755. PVPP does not dissolve, it stays in a suspension and so regular stirring is needed when in use.)

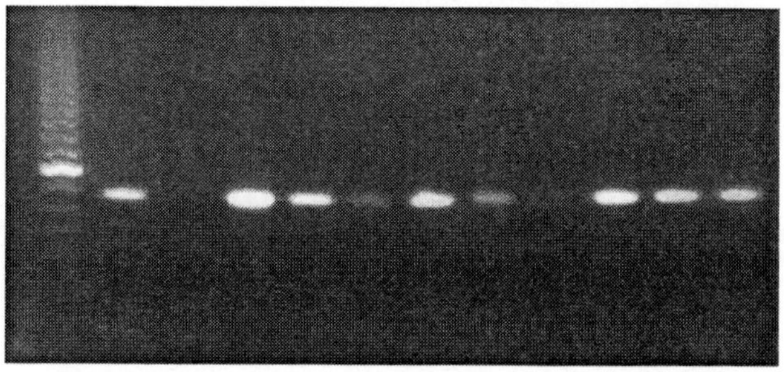

FIGURE 5.2 *D. fragilis* conventional PCR as described by Stark et al. [52]. Lane 1 molecular ladder, lane 2 positive control, lane 3 negative control, lanes 4–8 *D. fragilis* positive clinical samples, lane 9 negative *D. fragilis* clinical sample, lanes 10–12 *D. fragilis* positive clinical samples.

3. Separate the PCR product on 1% agarose with 0.5 μg/mL ethidium bromide. Visualize and photograph the product (Figure 5.2).

5.2.2.2 Real-Time PCR

The following PCR assay that was developed by Veweiji et al. for specific detection of *D. fragilis*. This assay employs forward primer (5′-CAACGGATGTCTTGGCTCTTA-3′) and reverse primer (5′-TGCATTCAAAGATCGAACTTATCAC-3′) and MGB Taqman probe (FAM-5′-CAATTCTAGCCGCTTAT-3′-MGB-Dark Quencher) derived from the SSU (18S) rRNA gene for amplification of a 98 bp fragment [54].

5.2.2.2.1 Procedure

1. Prepare PCR mixture (50 μL) containing 1× PCR buffer (HotstarTaq master mix, Qiagen), 5 mM $MgCl_2$, 200 μM dNTPs, 3 pmol of each primer, 7.5 pmol of MGB probe, 10 μL purified DNA, and distilled water (up to 50 μL).
2. Conduct PCR amplification/detection in a real-time detection system with the following cycling program: 95°C for 15 min, followed by 50 cycles of 95°C for 15 s, 60°C for 60 s. Fluorescence of FAM measured during the annealing elongation step (Figure 5.3).

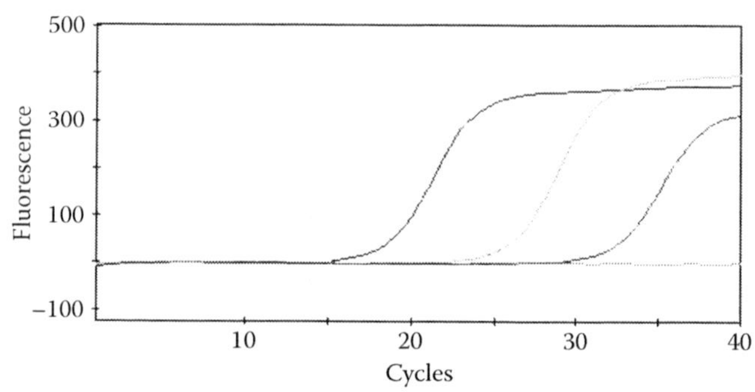

FIGURE 5.3 *D. fragilis* RT-PCR analysis as described by Stark et al. [11] using Smartcycler.

5.3 CONCLUSIONS AND FUTURE PERSPECTIVES

Numerous clinical and epidemiological studies have substantiated *D. fragilis* as a significant enteropathogen. It is therefore inexcusable that so little research has been undertaken on this organism.

Clifford Dobell, one of the parasitologists to first describe *D. fragilis*, wrote the following about this organism in 1940.

> To the protozoologist—if not the physician—*D. fragilis* is now, perhaps, the most interesting of all the intestinal amoebae of Man: for we know less about it than about any of the others, and its life history and activities are still mysterious .... Ever since I first saw this curious organism in 1917, I have been intrigued by its peculiarities, and have taken every opportunity of studying it further: yet after more than 20 years' work and cogitation, I am still baffled ...

Over 70 years have passed since the distinguished parasitologist wrote these remarks and unfortunately little has changed.

Of all the pathogenic protozoan parasites that infect humans *D. fragilis* is the least understood. Its life cycle and mode of transmission are poorly defined. The pathogenic mechanisms are unknown. No animal models or axenic culture systems have been developed for the study of this organism. The diagnostic tests available are limited when compared with those available for the diagnosis of other protozoa. There are no commercially available immunofluorescent antibody tests for microscopy, enzyme immunoassays, or immunochromatographic assays for the detection of *D. fragilis*. No commercially available PCR assay, either real-time or conventional, is available for the diagnosis of *D. fragilis* infections. Currently no multiplex assays incorporating *D. fragilis* as a target pathogen with other enteric pathogens has been developed nor is available.

More research is needed into the epidemiology, clinical syndromes, and pathology of dientamoebiasis. The life cycle of *D. fragilis* still needs to be determined. Molecular epidemiology studies also should be undertaken as there is already evidence that two genetic stains of *D. fragilis* exist. Animal models would provide a greater understanding of various aspects of not only the life cycle of *D. fragilis* but also of its pathogenicity. Such research will provide a better understanding of the epidemiology, pathogenesis, diagnosis, and treatment of *D. fragilis* infections. More diagnostic modalities are also needed for the detection of *Dientamoeba* to help laboratories diagnose this enteropathogen.

REFERENCES

1. Jepps, M.W., and Dobell, C. *Dientamoeba fragilis* n.g., n. sp., new intestinal amoeba from man. *Parasitology* 10, 352–367, 1918.
2. Delgado-Viscogliosi, P. et al. Molecular phylogeny of parabasalids based on small subunit rRNA sequences, with emphasis on the Trichomonadinae subfamily. *J. Eukaryot. Microbiol.* 47, 70–75, 2000.
3. Dobell, C. Researches on the intestinal protozoa of monkeys and man. X. The life history of *Dientamoeba fragilis*: Observations, experiments, and speculations. *Parasitology* 32, 417–461, 1940.
4. Dwyer, D.M. Analysis of the antigenic relationships among *Trichomonas*, *Histomonas*, *Dientamoeba*, and *Entamoeba*. II. Gel diffusion methods. *J. Protozool* 19, 326–332, 1972.
5. Dwyer, D.M. Analysis of the antigenic relationships among *Trichomonas*, *Histomonas*, *Dientamoeba*, and *Entamoeba*. I. Quantitative fluorescent antibody methods. *J. Protozool* 19, 316–325, 1972.
6. Dwyer, D.M. Analysis of the antigenic relationships among *Trichomonas*, *Histomonas*, *Dientamoeba* and *Entamoeba*. 3. Immunoelectrophoresis technics. *J. Protozool* 21, 139–145, 1974.
7. Camp, R.R., Mattern, C.F. and Honigberg, B.M. Study of *Dientamoèba fragilis* Jepps & Dobell. I. Electron microscopic observations of the binucleate stages. II. Taxonomic position and revision of the genus. *J. Protozool* 21, 69–82, 1974.
8. Bird, R.G., Sargeaunt, P., and Upton, C.P. Uni- and binucleate trophozoites of *Dientamoeba fragilis*. *Trans. R. Soc. Trop. Med. Hyg.* 64, 18, 1970.
9. Silberman, J.D., Clark, C.G., and Sogin, M.L. *Dientamoeba fragilis* shares a recent common evolutionary history with the trichomonads. *Mol. Biochem. Parasitol.* 76, 311–314, 1996.
10. Gerbod, D. et al. Phylogenetic position of the trichomonad parasite of turkeys, *Histomonas meleagridis* (Smith) Tyzzer, inferred from small subunit rRNA sequence. *J. Eukaryot. Microbiol.* 48, 498–504, 2001.
11. Stark, D.J., Beebe, N., Marriott, D., Ellis, J.T., and Harkness, J. Dientamoebiasis: Clinical importance and recent advances. *Trends Parasitol.* 22, 92–96, 2006.
12. Windsor, J.J., and Johnson, E.H. *Dientamoeba fragilis*: The unflagellated human flagellate. *Br. J. Biomed. Sci.* 56, 293–306, 1999.
13. Wenrich, D.H. Studies on *Dientamoeba fragilis* (protozoa). IV. Further observations, with an outline of present-day knowledge of this species. *J. Parasitol.* 30, 322–337, 1944.
14. Johnson, E.H., Windsor, J.J., and Clark, C.G. Emerging from obscurity: Biological, clinical, and diagnostic aspects of *Dientamoeba fragilis*. *Clin. Microbiol. Rev.* 17, 553–570, table of contents, 2004.
15. Burrows, R.B., Swerdlow, M.A., Frost, J.K., and Leeper, C.K. Pathology of *Dientamoeba fragilis* infections of the appendix. *Am. J. Trop. Med. Hyg.* 3, 1033–1039, 1954.
16. Swerdlow, M.A., and Burrows, R.B. *Dientamoeba fragilis*, an intestinal pathogen. *J. Am. Med. Assoc.* 158, 176–178, 1955.
17. Burrows, R.B., and Swerdlow, M.A. *Enterobius vermicularis* as a probable vector of *Dientamoeba fragilis*. *Am. J. Trop. Med. Hyg.* 5, 258–265, 1956.
18. Girginkardesler, N., Kurt, O., Kilimcioglu, A.A., and Ok, U.Z. Transmission of *Dientamoeba fragilis*: Evaluation of the role of *Enterobius vermicularis*. *Parasitol. Int.* 57, 72–75, 2008.
19. Yang, J., and Scholten, T. *Dientamoeba fragilis*: A review with notes on its epidemiology, pathogenicity, mode of transmission, and diagnosis. *Am. J. Trop. Med. Hyg.* 26, 16–22, 1977.
20. Kean, B.H., and Malloch, C.L. The neglected ameba: *Dientamoeba fragilis*. A report of 100 "pure" infections. *Am. J. Dig. Dis.* 11, 735–746, 1966.
21. Cuffari, C., Oligny, L., and Seidman, E.G. *Dientamoeba fragilis* masquerading as allergic colitis. *J. Pediatr. Gastroenterol. Nutr.* 26, 16–20, 1998.

22. Stark, D., Beebe, N., Marriott, D., Ellis, J., and Harkness, J. Prospective study of the prevalence, genotyping, and clinical relevance of *Dientamoeba fragilis* infections in an Australian population. *J. Clin. Microbiol.* 43, 2718–2723, 2005.
23. Windsor, J.J., Rafay, A.M., Shenoy, A.K., and Johnson, E.H. Incidence of *Dientamoeba fragilis* in faecal samples submitted for routine microbiological analysis. *Br. J. Biomed. Sci.* 55, 172–175, 1998.
24. Ayadi, A., and Bahri, I. *Dientamoeba fragilis*: Pathogenic flagellate?. *Bull. Soc. Pathol. Exot.* 92, 299–301, 1999.
25. Johnson, J.A., and Clark, C.G. Cryptic genetic diversity in *Dientamoeba fragilis*. *J. Clin. Microbiol.* 38, 4653–4654, 2000.
26. Peek, R., Reedeker, F.R., and van Gool, T. Direct amplification and genotyping of *Dientamoeba fragilis* from human stool specimens. *J. Clin. Microbiol.* 42, 631–635, 2004.
27. Windsor, J.J., Clark, C.G., and Macfarlane, L. Molecular typing of *Dientamoeba fragilis*. *Br. J. Biomed. Sci.* 61, 153–155, 2004.
28. Vandenberg, O. et al. Clinical and microbiological features of dientamoebiasis in patients suspected of suffering from a parasitic gastrointestinal illness: A comparison of *Dientamoeba fragilis* and *Giardia lamblia* infections. *Int. J. Infect. Dis.* 10, 255–261, 2006.
29. Hussein, E.M., Al-Mohammed, H.I., and Hussein, A.M. Genetic diversity of *Dientamoeba fragilis* isolates of irritable bowel syndrome patients by high-resolution melting-curve (HRM) analysis. *Parasitol. Res.* 105, 1053–1060, 2009.
30. Crotti, D., and D'Annibale, M.L. Role of *Dientamoeba fragilis* in human bowel infections. *Infez. Med.* 15, 30–39, 2007.
31. Spencer, M.J., Garcia, L.S., and Chapin, M.R. *Dientamoeba fragilis*. An intestinal pathogen in children? *Am. J. Dis. Child.* 133, 390–393, 1979.
32. Stensvold, C.R., Arendrup, M.C., Molbak, K., and Nielsen, H.V. The prevalence of *Dientamoeba fragilis* in patients with suspected enteroparasitic disease in a metropolitan area in Denmark. *Clin. Microbiol. Infect.* 13, 839–842, 2007.
33. Schuster, H., and Jackson, R.S. Prevalence of *Dientamoeba fragilis* among patients consulting complementary medicine practitioners in the British Isles. *J. Clin. Pathol.* 62, 182–184, 2009.
34. Rayan, H.Z., Ismail, O.A., and El Gayar, E.K. Prevalence and clinical features of *Dientamoeba fragilis* infections in patients suspected to have intestinal parasitic infection. *J. Egypt. Soc. Parasitol.* 37, 599–608, 2007.
35. Crotti, D., and D'Annibale, M.L. Intestinal infections caused by *Dientamoeba fragilis* and *Giardia duodenalis* in our experience. *Recenti Prog. Med.* 98, 361–366, 2007.
36. Girginkardesler, N., Coskun, S., Cuneyt Balcioglu, I., Ertan, P., and Ok, U.Z. *Dientamoeba fragilis*, a neglected cause of diarrhea, successfully treated with secnidazole. *Clin. Microbiol. Infect.* 9, 110–113, 2003.
37. Crotti, D., and D'Annibale, M.L. Human intestinal parasitosis: Role of *Dientamoeba fragilis* in human infections. *Ann. Ig.* 19, 27–34, 2007.
38. Crotti, D. et al. *Dientamoeba fragilis* is more prevalent than *Giardia duodenalis* in children and adults attending a day care centre in Central Italy. *Parasite* 12, 165–170, 2005.
39. Butler, W.P. *Dientamoeba fragilis*. An unusual intestinal pathogen. *Dig. Dis. Sci.* 41, 1811–1813, 1996.
40. Shein, R., and Gelb, A. Colitis due to *Dientamoeba fragilis*. *Am. J. Gastroenterol.* 78, 634–636, 1983.
41. Preiss, U., Ockert, G., Bromme, S., and Otto, A. *Dientamoeba fragilis* infection, a cause of gastrointestinal symptoms in childhood. *Klin. Padiatr.* 202, 120–123, 1990.
42. Norberg, A., Nord, C.E., and Evengard, B. *Dientamoeba fragilis*—A protozoal infection which may cause severe bowel distress. *Clin. Microbiol. Infect.* 9, 65–68, 2003.
43. Dardick, K.R. Tetracycline treatment of *Dientamoeba fragilis*. *Conn. Med.* 47, 69–70, 1983.
44. Vandenberg, O., Souayah, H., Mouchet, F., Dediste, A., and van Gool, T. Treatment of *Dientamoeba fragilis* infection with paromomycin. *Pediatr. Infect. Dis. J.* 26, 88–90, 2007.
45. Kurt, O., Girginkardesler, N., Balcioglu, I.C., Ozbilgin, A., and Ok, U.Z. A comparison of metronidazole and single-dose ornidazole for the treatment of dientamoebiasis. *Clin. Microbiol. Infect.* 14, 601–604, 2008.
46. Oxner, R.B., Paltridge, G.P., Chapman, B.A., Cook, H.B., and Sheppard, P.F. *Dientamoeba fragilis*: A bowel pathogen? *N. Z. Med. J.* 100, 64–65, 1987.
47. van Gool, T., Weijts, R., Lommerse, E., and Mank, T.G. Triple Faeces test: An effective tool for detection of intestinal parasites in routine clinical practice. *Eur. J. Clin. Microbiol. Infect. Dis.* 22, 284–290, 2003.
48. Hiatt, R.A., Markell, E.K., and Ng, E. How many stool examinations are necessary to detect pathogenic intestinal protozoa? *Am. J. Trop. Med. Hyg.* 53, 36–39, 1995.
49. Sawangjaroen, N., Luke, R., and Prociv, P. Diagnosis by faecal culture of *Dientamoeba fragilis* infections in Australian patients with diarrhoea. *Trans. R. Soc. Trop. Med. Hyg.* 87, 163–165, 1993.
50. Windsor, J.J., Macfarlane, L., Hughes-Thapa, G., Jones, S.K., and Whiteside, T.M. Detection of *Dientamoeba fragilis* by culture. *Br. J. Biomed. Sci.* 60, 79–83, 2003.
51. Clark, C.G., and Diamond, L.S. Methods for cultivation of luminal parasitic protists of clinical importance. *Clin. Microbiol. Rev.* 15, 329–341, 2002.
52. Stark, D., Beebe, N., Marriott, D., Ellis, J., and Harkness, J. Detection of *Dientamoeba fragilis* in fresh stool specimens using PCR. *Int. J. Parasitol.* 35, 57–62, 2005.
53. Stark, D., Beebe, N., Marriott, D., Ellis, J., and Harkness, J. Evaluation of three diagnostic methods, including real-time PCR, for detection of Dientamoeba fragilis in stool specimens. *J. Clin. Microbiol.* 44, 232–235, 2006.
54. Verweij, J.J. et al. Real-time PCR for the detection of *Dientamoeba fragilis* in fecal samples. *Mol. Cell. Probes* 21, 400–404, 2007.

6 Entamoeba

Damien Stark and John Ellis

CONTENTS

6.1 INTRODUCTION

Entamoeba histolytica is an intestinal protozoan parasite and the causative agent of human amoebiasis. It remains a significant cause of morbidity and mortality worldwide. It is estimated that one-tenth of the world's population is infected with *E. histolytica* and this infection results in up to 100,000 deaths each year [1]. *E. histolytica* has a worldwide distribution but is more commonly found in developing countries, and is more prevalent in the tropics and subtropics [2].

6.1.1 History, Classification, and Morphology

E. histolytica (phylum Sarcomastigophora, family Entamoebidae) is defined as a unicellular eukaryote whose life cycle contains two main stages: trophozoites (diameter 20–40 μm) with a single nucleus and cysts (diameter 10–16 μm) with four nuclei when mature; the immature cyst has a single nucleus often with a small central karyosome [3].

The trophozoite form of *E. histolytica* was first described in 1875 by Fedor Losch, from dysenteric stools and colonic ulcerations from a Russian farmer with a fatal case of dysentery [4]. Although Losch was able to induce dysentery in a dog that was inoculated rectally with the patients stools, he failed to recognize the causal relationship between the organism and disease. It was not until 1891 that clinical and pathological evidence for the association of *E. histolytica* with dysentery and liver abscesses was described by Councilman and LaFleur [5]. In 1903, Quincke and Roos described the cyst form of the organism and in 1903, Schaudin named it *Entamoeba histolytica* and differentiated it from other *Entamoeba* species [6]. Leonard Rogers designated emetine as the first effective treatment for amoebiasis in 1912 [6]. In 1913, Walker and Sellards conducted controlled experiments with *E. histolytica* on prisoners in the Philippines [7]. They found that transmission occurred via cysts, not trophozoites, asymptomatic carriers acted as reservoirs and were responsible for transmission, there were differences between individuals' disease risk and there were differences in organism strain virulence [7]. In 1925, Brumpt formulated the theory of the existence of two distinct but morphologically identical species, namely, the pathogenic invasive form *E. histolytica* and the nonpathogenic, noninvasive form *E. dispar* [6]. His hypothesis was dismissed at the time and gained little support until isoenzyme typing differentiated pathogenic and nonpathogenic strains [8]. In 1993, Clark and Diamond, on

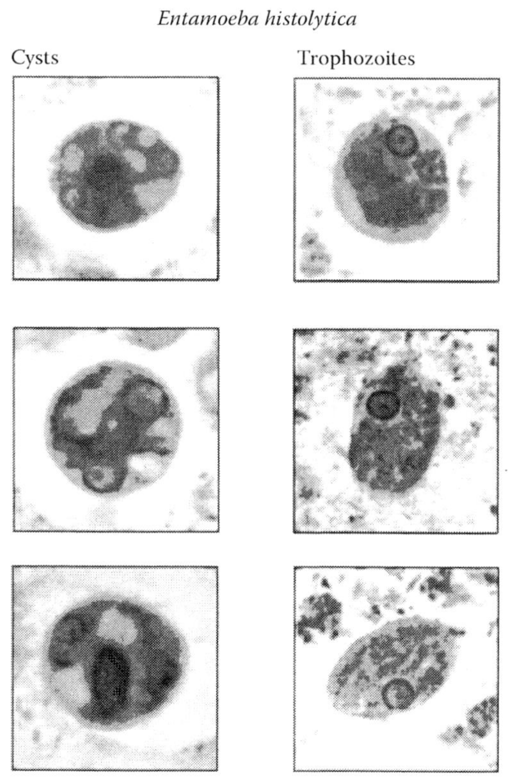

FIGURE 6.1 Cysts and trophozoites of *Entamoeba histolytica* permanently stained using a modified iron-haematoxylin stain at 1000× magnification.

the basis of molecular, immunological and biochemical analysis redescribed the protozoa, retaining the name *E. histolytica* for the pathogenic invasive parasite, and renaming the noninvasive parasite *E. dispar* [9,10].

The morphological characteristics of *E. histolytica* can be seen in Figure 6.1. *E. histolytica* trophozoites can vary in size from 10 to 50 μm, with most generally measuring between 15 and 20 μm. The trophozoites are unnucleated and may often exhibit pseudopoding motility which is often described as sporadic, rapid, and unidirectional. The cytoplasm of the trophozoite is finely granular and may contain ingested material. The nuclear structure consists of a centrally located, small, compact karyosome with evenly arranged peripheral chromatin present that may have a beaded appearance [11]. Cysts of the organism range in size from 10 to 20 μm and may contain one, two, or four nuclei. Inclusions may be present in the cysts such as chromatoidal bars and a glycogen mass [11]. The nuclear structure is similar to what is found in trophozoites. The morphology is often difficult to see in unstained preparations. When the organism is examined on a permanent stained smear the morphological characteristics are readily seen.

6.1.2 *Entamoeba* Species That Infect Humans

There are many described species in the Genus *Entamoeba* several of which may infect humans, including those that reside in the intestinal lumen: *E. histolytica*, *E. dispar*, *E. moshkovskii*, *E. coli*, and *E. hartmanni*. There have also been reports of *E. polecki* and *E. chattoni* infections in humans though they are normally associated with pigs and nonhuman primates, respectively [12].

E. histolytica is the only species pathogenic to humans; all the others are considered nonpathogenic [13]. *E. histolytica* is morphologically identical to *E. dispar* and *E. moshkovskii*. Microscopy is unable to differentiate these three indistinguishable species [13]. Until recently, *E. moshkovskii* was considered to be a mainly a free-living amoeba, with environmental isolates recovered from sewage, fresh brackish water, and salt water sediments [14]. Reports on *E. moshkovskii* in human specimens have recently emerged from around the world [13,15–18].

The existence of these morphologically indistinguishable species of *Entamoeba* highlights the need to differentiate the three species to determine clinically if the patient is infected with the pathogenic *E. histolytica* or the nonpathogenic *E. dispar*/*E. moshkovskii*. In a clinical setting a patient infected with the nonpathogenic *E. dispar*/*E. moshkovskii* could be treated unnecessarily with antiamoebic chemotherapy.

6.1.3 Life Cycle and Epidemiology

The life cycle is seen in Figure 6.2. The life cycle was first described by Dobell [19] and unlike many protozoan parasites have a simple life cycle, comprising trophozoites, precyst, cyst, metacyst, and metacystic trophozoites stages. The cyst

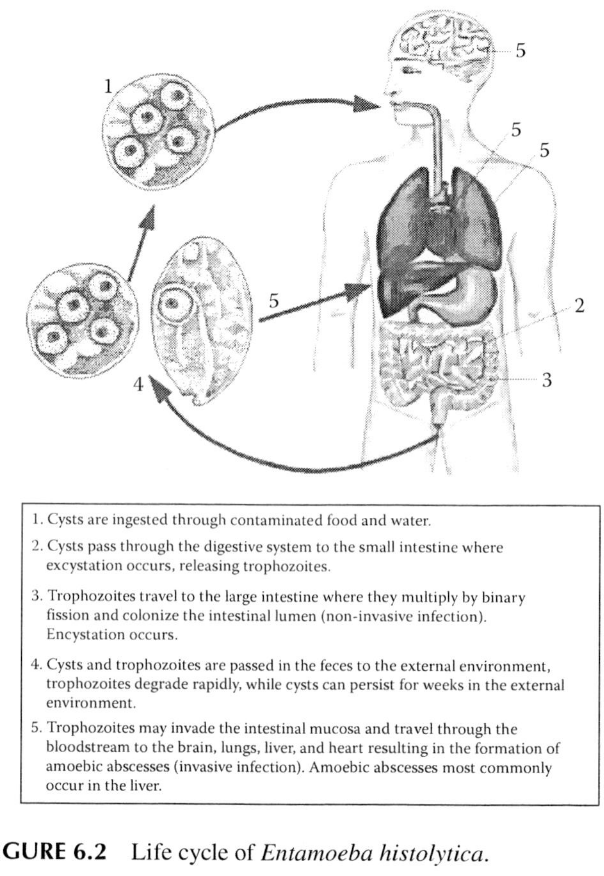

FIGURE 6.2 Life cycle of *Entamoeba histolytica*.

form is the infective stage for humans and can survive in contaminated food and water [6]. Ingestion of food and drink contaminated with *E. histolytica* cysts from human feces and direct fecal–oral contact are the most common means of infection. Infection begins with the ingestion of infective cysts. After cysts are ingested they survive the acid of the stomach and travel to the small intestine. Within the terminal ileum or colon, where the pH becomes neutral or slightly alkaline, the encysted organisms develop (during excystation) into four separate metacystic trophozoites which then mature into normal trophozoites that establish infections in the large intestine. Cyst formation occurs in the intestinal tract, and infective cysts are passed in the feces. Humans, and some nonhuman primates, are the only natural hosts [1]. *E. histolytica* infections occur worldwide but are more prevalent in the tropics [6]. The prevalence of infection varies from area to area as does the severity of disease from patient to patient. It has been suggested that 12% of the world's population is infected with this organism [20]. The epidemiology of *E. histolytica* remains uncertain as most of the existing data were obtained using methods incapable of distinguishing the three morphologically identical species. Most studies that have investigated the prevalence of *E. histolytica* and *E. dispar* have not considered the presence of *E. moshkovskii*. The worldwide prevalence of *E. histolytica* and *E. dispar* as separate species is not well defined and the prevalence of *E. moshkovskii* is unknown. *E. dispar* appears to be much more common than *E. histolytica* with most of the 500 million people infected with *E. histolytica*/*E. dispar*/*E. moshkovskii* carrying *E. dispar* [21]. Most morbidity and mortality due to amoebiasis occur in developing regions such as Central and South America, Africa, and the Indian subcontinent [22]. In developed countries most infections are due to *E. dispar*. High-risk groups for *E. histolytica* infection include immigrants from or travelers to areas of endemicity, men who have sex with men and institutionalized populations [18,23].

6.1.4 Pathogenesis, Clinical Presentation, and Treatment

6.1.4.1 Pathogenesis

E. histolytica is unique among the intestinal amoebae as it has the ability to invade host tissue. The factors that control the invasiveness of *E. histolytica* are not completely understood. The main key to the pathogenesis of this organism is its ability to directly lyse host cells and cause tissue destruction. The mechanisms of pathogenicity include adherence, extracellular cytolysis, and phagocytosis [24–26]. Host cell leukocytes also play a major role in pathogenesis along with numerous other parasite factors [6,27].

Disease begins with the adherence of *E. histolytica* to colonic epithelial cells. A number of *E. histolytica* adherence receptors have been identified, but the galactose-specific lectin receptor is thought to be responsible for mediating the attachment to colonic mucus and colonic epithelial cells [28]. Human colonic epithelial cells and mucus contain a large number of *N*-acetylgalactosamine residues [29]. Mammalian cells without N-terminal galactose or *N*-acetylgalactosamine residues are resistant to adherence by amoebic trophozoites and to killing by *E. histolytica*, consistent with a physiological role for the lectin in adhesion [30].

The cytolytic capabilities of *E. histolytica* have been demonstrated to be caused by amoebapores (a family of small peptides capable of forming pores in lipid bilayers). The pore-forming proteins are transferred from the trophozoites to the target cell, causing disruption of the transmembrane gradient and subsequent lysis [31]. Cysteine proteinases are also important virulence factors and are directly implicated in invasion and inflammation of the gut. These proteinases are major proteolytic enzymes and can degrade cellular attachment and matrix proteins such as collagen, laminin, and fibronectin. There is a direct correlation between the amount of proteinase activity and the pathogenicity of *E. histolytica* [32].

The host's cellular and humoral immune responses to *E. histolytica* infection also play a role in the pathogenesis of the organism. As invasive strains of *E. histolytica* have been shown to be resistant to complement-mediated lysis, cell-mediated immunity represents the major human host defense [6]. Intestinal epithelial cells produce interleukin-1B, interleukin-8, and cyclo-oxygenase (COX)-2 in response to *E. histolytica* [33]. These mediators attract neutrophils and macrophages. This inflammation can contribute to tissue damage early in disease.

6.1.4.2 Clinical Presentation

Clinical syndromes associated with amoebiasis vary widely with the host and the organism both playing a role. Syndromes range from asymptomatic infection to disseminated fatal disease.

6.1.4.2.1 Asymptomatic Infection

Most human *E. histolytica* infections are asymptomatic. It is estimated that 90% of those infected will not go on to develop the disease [34]. In such individuals the organism acts as a harmless commensal. Many asymptomatic individuals never become symptomatic and usually only excrete cysts in the stool for a short period, and often clear the infection without any signs of the disease [35]. This pattern is found in patients infected with pathogenic or nonpathogenic strains [10]. However, asymptomatic colonization with *E. histolytica*, if left untreated, can lead to disease and it is estimated that 4–10% of asymptomatic individuals may develop gastrointestinal or invasive disease [36].

6.1.4.2.2 Symptomatic Infection

6.1.4.2.2.1 Gastrointestinal Approximately 10% of individuals infected with *E. histolytica* will develop clinical symptoms, with intestinal or extraintestinal involvement [34]. Patients with noninvasive disease may exhibit nonspecific gastrointestinal symptoms including abdominal pain and diarrhea [6]. The symptoms associated with invasive gastrointestinal amoebiasis are usually also nonspecific. Patients with amoebic colitis present with bloody diarrhea with abdominal pain and tenderness [37]. The onset is usually gradual, over weeks, with abdominal pain, diarrhea, dysentery, and weight loss depending on severity of disease. When amoebic colitis is diagnosed it is usually associated with ulceration of the colon. Perforation of the colon may occur in over 20% of patients with acute amoebic [38]. Perforations occur most commonly in the cecum and may result in peritonitis [39]. Occasionally, patients may develop a fulminant amoebic colitis, with profuse bloody diarrhea, fever, leucocytosis, dehydration and abdominal pain, often with peritoneal signs, and extensive involvement of the colon [40]. Complications of intestinal amoebiasis include toxic megacolon, amoeboma [41], cutaneous amoebiasis [42], and rectovaginal fistulae [43].

6.1.4.2.2.2 Extraintestinal Extraintestinal disease is the result of previous intestinal infection. Dissemination of *E. histolytica* infection to extraintestinal sites most frequently involves the liver and to a lesser extent the lungs and rarely pericardium, brain, and skin. Invasive amoebiasis is associated with significant morbidity and mortality. Amoebic liver abscesses (ALA) is caused by the hematogenous spread of the invasive trophozoites from the colon, which reaches the liver via the portal vein [6]. This explains the frequent occurrence of abscesses in the right hepatic lobe, which receives most of the blood draining from the cecum and ascending colon [44]. ALA can occur concurrently with colitis, often there is intestinal infection and stool microscopy is usually negative for trophozoites, but more often they have no gastrointestinal symptoms. Onset may be acute or subacute. Symptoms of upper right abdominal pain, fever, and substantial hepatic tenderness are commonly seen [34].

Hematogeneous spread to other sites including the brain, lung, pericardium, and other sites is possible, with *E. histolytica* having been reported in almost every soft organ and tissue of the body. Amoebiasis of the lungs may develop when hepatic abscesses rupture through the diaphragm, along with originating as a secondary site from the original intestinal infection. Rupture of ALA into the peritoneum or pericardium may occur [34]. Amoebic brain abscesses are rare and usually arise from hepatic or pulmonary involvement. Sudden onset of symptoms and rapid progression to death has been reported in a majority of patients with amoebic brain abscesses [45]. Other ectopic sites include the urinary tract (kidney, ureters, urinary bladder, urethra), genital involvement (penis, clitoris, rectovaginal fistulae, perianal region) and cutaneous lesions have all been described as case reports [46–48]. Amoebiasis cutis involves the skin and usually develops as a perianal extension of amoebic colitis or on the abdominal wall due to rupture of intestinal or hepatic lesions. It may also occur as a venereal infection [49].

6.1.4.3 Treatment

Treatment of amoebiasis is divided into two categories; treatment with luminal amoebicides and treatment with tissue amoebicides [34]. Lumen agents used in the treatment of amoebic infections include Diloxanide, Paromomycin, and Iodoquinol. These agents act on the organism in the intestinal lumen and are not effective against organisms in tissue [6]. Tissue amoebicides include the nitroimidazole derivatives (metronidazole, tinidazole, ornidazole), chloroquine or dehydroemetine. These agents are effective in the treatment of invasive amoebiasis but are ineffective in the treatment of the organisms in the intestinal lumen [6].

Asymptomatic carriers of *E. histolytica* should be treated with a luminal agent to not only minimize the spread of disease but to also prevent the risk of developing invasive disease [50]. In patients with invasive disease a tissue amoebicide should be used in conjunction with a luminal agent to eradicate the organism [6]. The management of ALA is unique and can be cured without drainage, and even with one dose of metronidazole [51]. The role of surgery is generally limited to patients with complications of invasive disease, as often cure can be achieved with medical therapy alone [52]. All treated patients should be followed up clinically using appropriate laboratory tests.

6.1.5 Conventional and Molecular Diagnosis

6.1.5.1 Microscopy

Numerous diagnostic techniques have been used for the diagnosis of amoebic infection. Traditional diagnosis of amoebiasis relies on conventional microscopy of wet preparations, concentrations, and permanently stained smears for the identification of *E. histolytica* in feces. Identification is based on the morphological characteristics of the organism. Although microscopic diagnosis of amoebic infection is relatively quick and cheap, considerable expertise is required by the microscopist to differentiate the different amoeba. However, the sensitivity and specificity of microscopy are poor (<60%) and can be confounded by false-positive results due to misidentification [13,53]. Also the trophozoites of *E. histolytica* degenerate rapidly in unfixed fecal samples and as such specimens should be preserved with a fixative to prevent morphological degradation [54]. Due to the intermittent shedding of the parasite in fecal samples, for optimal microscopic detection of the organism, a minimum of three stool samples taken over no more than 10 days is recommended. Microscopy performed on aspirated material from ALA, are often negative for trophozoites, thus limiting the usefulness of microscopy as a diagnostic test for patients with ALA. The existence of *E. dispar* and *E. moshkovskii* in humans, however, has made the definitive diagnosis of *E. histolytica* by microscopy impossible as the three species cannot be differentiated microscopically [13].

6.1.5.2 Culture

Various culture techniques have been successfully used for the diagnosis of amoebiasis. Culture media include xenic (diphassic and monophasic) and axenic systems. Xenic cultures are defined as the growth of parasite in the presence of an unidentified flora [55]. Axenic cultures involve the cultivation of the parasites in the absence of any other metabolizing cells [55]. In 1925, Boeck and Drbohlav were the first to successfully culture *E. histolytica* from clinical samples, using a xenic diphasic egg slope slant medium, with modifications of this media that is still used today [55]. Other xenic culture systems developed for the cultivation of *E. histolytica* include; Balamuth medium [56], Jones's medium [57], TYSGM-9 [58], Locke-egg [55], and Robinson's medium [59]. Axenic cultivation of *E. histolytica* was first described by Diamond in 1961 [60]. Modifications to this medium were later developed and used in different research laboratories. Currently, TYI-S-33 and YS-S medium are the most commonly used axenic media for the cultivation of *E. histolytica* [55].

Various clinical specimens can be used for the culturing of *E. histolytica* including fecal samples, biopsy specimens, and liver abscess aspirates. The success rate for the cultivation of *E. histolytica* from clinical samples is approximately 50% [55]. Culture usually is performed only in research laboratories because of the fact that the procedures are technically difficult, labor intensive, and often unrewarding. Overgrowth of the medium is a major problem and cultures have a significant false-negative rate. As such, culture is not recommended for use in routine diagnostic laboratories.

6.1.5.3 Isoenzyme Analysis

Isoenzyme analysis can differentiate *Entamoeba* species on the basis of zymodeme patterns [8]. A zymodeme is defined as a group of amoeba strains that share the same electrophoretic pattern and mobility for several enzymes (malic enzyme, hexokinase, glucose phosphate isomerase, phosphoglucomutase isoenzyme) [61]. Isoenzyme analysis can differentiate *E. histolytica* from *E. dispar* and was considered the gold standard for diagnosing amoebic infection prior to the development of DNA-based techniques. The technique does have a number of disadvantages. The test is difficult to perform, complex, costly, is time consuming, and relies on the establishment of amoeba in culture. The process is not always successful, and cultivation of the amoeba may lead to selection bias. Zymodeme analysis is now restricted mainly to research laboratories and in the diagnostic lab it has been superseded by polymerase chain reaction-based methods.

6.1.5.4 Antibody Detection

A number of different serological assays are commercially available for the detection of *E. histolytica* antibodies. Indirect hemagglutination, latex agglutination, complement fixation, indirect immunofluorescence assay, and enzyme-linked immunosorbent assays (ELISA) have all been developed [13]. Serological testing may be helpful from a diagnostic perspective in certain situations. In patients suspected of having extraintestinal disease, serological tests are warranted; however, testing in patients with intestinal disease is normally not recommended.

The serological tests are specific but have varying sensitivity depending on the presence or absence of invasive disease and the type of invasive disease. The sensitivity for serology is about 100% in patients with ALA [62] and 82% with invasive intestinal disease [13]. Antibody testing to diagnose carriage or noninvasive disease is unhelpful, as the sensitivity is low at around 8% [20].

Serum IgG antibodies persist for years and antibody titers can remain high for years after successful therapy and/or eradication of the organism. This limits the usefulness of the test in areas where infection is endemic and people have been exposed to *E. histolytica*, due to the inability of serological tests to distinguish past from current infection [13]. However, serological tests may be helpful from a diagnostic perspective in industrialized countries, where infections and exposure to *E. histolytica* is uncommon.

6.1.5.5 Antigen Detection

Stool antigen assays are reported to outperform microscopy and to be as sensitive (80–85%) and specific (99%) as culture with isoenzyme analysis for the detection of *E. histolytica* [63]. They are also relatively cheap, and easy to perform. Several kits are commercially available for the detection of *E. histolytica* antigens in fecal samples. These tests can either determine the presence of the *E. histolytica/E. dispar* group as being distinct from the rest of the *Entamoeba* species or they can actually differentiate *E. histolytica* from *E. dispar*. Most of these tests are in enzyme immunoassay (EIA) formats. While most have reported excellent sensitivity and specificity there are a number of conflicting reports with researchers finding lower than expected sensitivities/specificities of the assays [13].

The triage parasite panel (TTP) is an immunochromatographic assay for the simultaneous detection of *E. histolytica/E. dispar*, *Giardia lamblia*, and *Cryptosporidium parvum*. Garcia et al. reported that the sensitivity and specificity of the TTP were 96% and 99.1%, respectively [64]. However, other studies have shown that the sensitivity of the assay was low (68.3%) when compared with the ProSpecT EIA test [65] and when compared with PCR the TTP performed poorly [66].

Antigen-based ELISA kits that are specific for *E. histolytica* use monoclonal antibodies against the Gal/GalNAc-specific lectin (*E. histolytica* II, TechLab; *Entamoeba* CELISA, Cellabs,; ProspecT EIA, Remel Inc) or monoclonal antibodies against the serine-rich antigen of *E. histolytica* (Optimin S kit, Merlin Diagnostika). The TechLab antigen kit has been used over several years in different laboratories for the detection of *E. histolytica* in both endemic and nonendemic regions of the world. Results obtained using the TechLab antigen kit are conflicting with studies conducted in *E. histolytica* high endemic countries reporting high sensitivities between 95% and 100% [67,68]. However, a recent study conducted in a high *E. histolytica* endemic region of northern Ecuador found that the Techlab *E. histolytica* II test performed poorly with a reported sensitivity of 14.3% and a specificity of 98.4% when compared with isoenzyme analysis [69]. Stark et al. found that the CELISA PATH kit showed 28% sensitivity and 100% specificity when PCR was used as a reference standard. In the same study the TechLab ELISA kit failed to identify any of the *E. histolytica* samples, which were positive by PCR [70]. In low endemic settings the

TechLab ELISA was documented to have a lower sensitivity than that of microscopy [71]. Similar results were obtained when the TechLab ELISA was compared with real-time PCR (RT-PCR) as a reference test [72].

The ProspecT EIA (Remel Inc.) is a microplate EIA that while detecting *E. histolytica* and *E. dispar* has the inability to distinguish the two. When compared with microscopy the sensitivity of this assay was reported between 73.5% and 78% while the specificity was 97.7 and 99% [73,74]. When the assay was compared with culture and isoenzyme analysis the sensitivity and specificity were lower at 54.5% and 94%, respectively [69].

Quantitative assessments have shown that EIA kits are 100–1000 times less sensitive than PCR for detecting *E. histolytica* [70,75]. These studies, however, used cell lysates of *E. histolytica* to calculate the analytical sensitivity of the ELISA's, and it is not clear if the ELISA's have comparable limits of detection of trophozoites, cysts, and cell lysates.

Another disadvantage of antigen detection tests is that the antigens detected are denatured by fixation of the stool specimen, limiting the detection to fresh or frozen specimens [13].

6.1.5.6 Molecular Diagnosis

The PCR method for the detection of *E. histolytica* is a powerful tool and offers excellent sensitivity and specificity. DNA-based tests for the diagnosis of amoebiasis have already become the gold standard by which other diagnostic techniques (microscopy, antibody, and antigen detection) are measured. It should be noted, however, that PCR is susceptible to cross-contamination and false-negative results due to inhibitors of DNA polymerase in stool samples.

6.1.5.6.1 Conventional PCR

To date, a large number of conventional PCR methods have been described for the detection of *E. histolytica* targeting many different genes (Table 6.1). The initial studies used PCR to amplify the gene which encodes the 125-kDa surface antigen [76] and this was further adapted by others to distinguish between the different *Entamoeba* species using restriction digestion [77]. These initial experiments were performed on laboratory-maintained control isolates of *Entamoeba* species. Primers targeting highly repetitive sequences present in *E. histolytica* were developed for PCR and used on DNA extracted from cultured trophozoites from feces; these primers amplify *E. histolytica* and *E. dispar* [78,79].

The most widely used assays for *E. histolytica* PCR target the 18S rDNA. These targets are present on multicopy, extrachromosomal plasmids in the amoeba, making the 18S rDNA more easily detected and thus more sensitive than single copy gene targets [80]. The consistent genetic diversity found in the 18S rDNA allows for the differentiation between *E. histolytica* and the closely related *E. dispar* and *E. moshkovskii*. PCR methods targeting this loci have proved to be highly sensitive and specific for detecting *Entamoeba* DNA [14,15,18,81–85]. The 18S rDNA targets can be successfully amplified from either DNA extracted from laboratory cultured trophozoites or DNA extracted directly from feces. Both single round and nested PCR have been developed using 18S rDNA. A single-round multiplex PCR assay was developed by Hamzah et al. for the detection and differential diagnosis of *E. histolytica*, *E. dispar*, and *E. moshkovskii* in humans [86]. The PCR targeted the 18S rDNA by utilizing a conserved forward primer and reverse primers that are species specific to each of the three *Entamoeba* species. The assay was able to detect as little as 10 pg of DNA while exhibiting no cross-reactivity with other bacterial or protozoan parasites.

Other targets include primers that amplify other regions of the extrachromosomal circular DNA [87,88]. The 29-kDa/30-kDa antigen gene was also used to distinguish between *E. histolytica* and *E. dispar* [89]. This method was used with both DNA extracted from culture [90] and direct fecal samples fixed in formalin [91,92]. Protein encoding genes that have demonstrated polymorphisms in the coding region have also been utilized as targets for PCR. The serine-rich protein (SREPH) gene and the chitinase gene have both been used [93,94]. In addition, a novel PCR assay based on the hemolysin gene HLY6 was developed for the detection of *E. histolytica* DNA from both fecal and ALA samples and was shown to have excellent sensitivity and specificity [95].

Nested PCR has also been developed in an attempt to increase the sensitivity of detection of *E. histolytica*. When utilizing a nested multiplex PCR for the simultaneous detection and differentiation of *E. histolytica* and *E. dispar* the sensitivity and specificity were 94% and 100%, respectively [96]. A more recent study developed a nested multiplex PCR targeting the 18S rDNA for the differential detection of *E. histolytica*, *E. dispar*, and *E. moshkovskii* simultaneously in stool samples [97]. This PCR demonstrated a sensitivity of 94% and a specificity of 100% for the demonstration of *E. histolytica*, *E. dispar*, and *E. moshkovskii* DNA in stool samples.

6.1.5.6.2 Amoebic Liver Abscesses

An important and serious complication of intestinal infection with *E. histolytica* is the involvement of the liver (hepatic amoebiasis) [34]. Hepatic amoebiasis is usually diagnosed by the clinical picture (pain in the right upper quadrant and fever), ultrasound examination, and positive serology. However, none of these tests are definitive and the picture overlaps with pyogenic liver abscess caused by bacteria. Direct demonstration of *E. histolytica* by conventional microscopy and *in vitro* culture in pus obtained from ALA are often unsuccessful [13]. PCR has been shown to be more reliable and a better alternative diagnostic test for ALA. A comparative study was done to verify the sensitivity of 10 pairs of primers specific for detecting *E. histolytica* in stools [98]. Of the 10 pairs of previously published primers tested, two pairs of primers (PI + P2 and P11 + P12) both targeting the extrachromosomal circular DNA of *E. histolytica* were found to give 100% sensitivity. Based on these results, the authors recommended that PCR assay can be successfully used to confirm the diagnosis of ALA with the primers identified [98]. Previously, Tachibana et al. used a PCR that employed primers specific for the gene encoding

TABLE 6.1
Conventional PCR Assays for *E. histolytica* Detection

Assay Gene Target	Primer/ Probe	Sequence (5′ to 3′)	Amplicon Size	Cycling Profile	Primer Concentration	Detection Method	Reference
Conventional PCR SSU rRNA	PSP5 PSP3	GGCCAATTCATTCAATGAATTGAG CTCAGATCTAGAAACAATGCTTCTC	877 bp	95°C for 5 min, 45 cycles of 95°C for 15 s, 65°C for 15 s, 72°C for 1 min, with a final extension at 72°C for 10 min	0.4 μM	2% agarose gel stained with ethidium bromide	[10]
Conventional PCR SSU rRNA	EntaF EhR	ATGCACGAGAGCGAAAGCAT GATCTAGAAACAATGCTTCTCT	166 bp	94°C for 3 min, 30 cycles of 90°C for 1 min, 58°C for 1 min, and 72°C for 1 min, with a final extension at 72°C for 7 min.	0.1 μM	1.5% agarose gel stained with ethidium bromide	[86]
Conventional PCR SSU rRNA	Eh5 Eh3	GTAACTTACTTAACCGGTAAAACATG TCTCTTCGTAACAAAGATCTAGACTC	880 bp	94°C for 3 min, 40 cycles of 90°C for 1 min, 60°C for 1 min, 72°C for 2 min, with a final extension at 72°C for 10 min	18 pmol	1% agarose gel stained with ethidium bromide	[81]
Conventional PCR Tandem repeats in extrachromosomal circular rDNA	EhP1 EhP2	CGATTTTCCCAGTTAGAAATTA CAAAATGGTCGTCGTCTAGGC	132 bp	30 cycles of 92°C for 1 min, 55°C for 1 min, and 70°C for 90 s	0.4 μM	10% acrylamide gel stained with ethidium bromide	[96]
Nested PCR SSU rRNA	E-1 E-2	TAAGATGCACGAGAGCGAAA GTACAAAGGGCAGGGACGTA	439 bp	1st round—96°C for 2 min, 30 cycles of 92°C for 60 1 min, 56°C for 1 min, 72°C for 90 s, with a final extension at 72°C for 7 min	0.3 μM	1.8% agarose gel stained with ethidium bromide	[97]
	EH-1 EH-2	AAGCATTGTTTCTAGATCTGAG AAGAGGTCTAACCGAAATTAG		2nd round—96°C for 2 min, 30 cycles of 92°C for 1 min, 48°C for 1 min, 72°C for 90 s, with a final extension at 72°C for 7 min			
Nested PCR SSU rRNA	E-1 E-2	TTTGTATTAGTACAAA GTA[A/G]TATTGATATACT	900 bp	1st round —96°C for 2 min, 50 cycles of 92°C for 1 min, 43°C for 1 min, and 72°C for 90 s	0.4 μM	1.3% agarose gel stained with ethidium bromide	[121]
	Eh-1 Eh-2	AATGGCCAATTCATTCAATG TTTAGAAACAATGCTTCTCT		2nd round—96°C for 2 min, 50 cycles of 92°C for 1 min, 62°C for 1 min, and 72°C for 90 s			

the 30 kDa molecule of *E. histolytica* [99]. Liver abscess fluids (19 samples), from 14 patients with a presumptive amoebic liver abscess, were examined microscopically and by the PCR method. Only two of the 19 samples were positive microscopically, whereas all 19 samples tested positive by PCR [99]. A more recent study using another set of primers found the sensitivity and specificity of PCR for detecting ALA to be 83% and 100%, respectively. The overall sensitivity of PCR was higher when compared with serological tests [100].

6.1.5.6.3 Real-Time PCR

A number of distinct RT-PCR protocols have been published for the identification and detection of *E. histolytica* (Table 6.2). RT-PCR offers several advantages over conventional PCR including elimination of post-PCR analysis reducing the risk of amplicon contamination, and faster cycling times resulting in reduced assay turnaround times. RT-PCR also allows for the quantitation of organisms. These benefits do come at a cost with RT-PCR significantly being more expensive than conventional PCR.

Like conventional PCR assays the majority of the RT-PCR assays have targeted the rDNA. An SYBR green assay that did not utilize any specific probes was described by Qvarnstrom et al. [101]. Blessmann et al. reported high sensitivity and specificity utilizing an RT-PCR assay performed with the LightCycler system using fluorescein-labeled detection probes and primers amplifying a 310-bp fragment from the rDNA [102]. The assay was able to detect as little as 0.1 parasite per g of feces. Two RT-PCR assays were developed using TaqMan probes, [103] one targeting the single subunit (SSU) rDNA [104] and the other targeting the episomal repeats [105]. An RT-PCR assay using a molecular beacon targeting the 18S rDNA for use on fecal and ALA specimen has been described [106].

An evaluation study conducted by Qvarnstrom et al., compared three *E. histolytica* RT-PCR techniques (LightCycler assay targeting the 18S rDNA, a TaqMan assay targeting episomal repeats, and a TaqMan assay targeting the 18S rDNA) and an SYBR green RT-PCR method [101]. The limits of detection and efficiency of each RT-PCR assay were determined and major differences in detection limits and assay performance were observed among the evaluated tests. Two of the assays could not reliably distinguish *E. histolytica* from *E. dispar*: the LightCycler assay and the TaqMan assay targeting episomal repeats. The LightCycler assay was found to be prone to false positives and clearly illustrates a lack of specificity of the primers. The TaqMan assay targeting episomal repeats also produced false results. Other publications have reported peculiar results concerning detection of *E. dispar* in conventional PCRs targeting these episomal-repeat sequences questioning the usefulness of this target for molecular testing [83,107]. In conclusion, this study identified the TaqMan targeting the 18S rRNA gene as a superior RT-PCR assay for specific and quantitative diagnosis of amoebiasis. The SYBR Green assay also performed well and offered a good alternative to the TaqMan assay [101].

6.2 METHODS

6.2.1 Reagents and Equipment

- Thermocycler
- RT-PCR system
- DNA template that contains the region of the DNA fragment to be amplified
- One or more primers, which are complementary to the DNA regions at the 5′ and 3′ ends of the DNA region that is to be amplified
- DNA polymerase (e.g., Taq polymerase or another DNA polymerase with a temperature optimum at around 70°C), used to synthesize a DNA copy of the region to be amplified
- Deoxynucleotide triphosphates (dNTPs), from which the DNA polymerase builds the new DNA
- Buffer solution, which provides a suitable chemical environment for optimum activity and stability of the DNA polymerase

Divalent cation, magnesium or manganese ions; generally Mg^{2+} is used, but Mn^{2+} can be utilized for PCR-mediated DNA mutagenesis, as higher Mn^{2+} concentration increases the error rate during DNA synthesis.

- Monovalent cation potassium ions
- RT-PCR probes
- PCR grade sterile H_2O
- Pipettes
- Aerosol barrier filter tips

6.2.2 Sample Collection and Preparation

Fecal samples are considered the most complex specimens for direct PCR testing because of the presence of PCR inhibitors such as heme, bilirubin, bile salts, and complex carbohydrates that are often co-extracted along with the target DNA [13]. Therefore, optimization of the fecal DNA extraction procedure is critical.

The transport and preservation conditions, of fecal samples, needs to be taken into consideration. A recent study showed that transportation of fecal samples containing *E. histolytica* at ambient temperature may result in the degeneration of parasite DNA, especially for the labile trophozoite stage [108]. Therefore, the sensitivity of DNA assays using unpreserved samples is time dependent. Specimens may be preserved by refrigeration or freezing. Many laboratories commonly use fixatives to preserve parasite morphology and for safe handling of specimens. Samples preserved in fixative may decrease the sensitivity of PCR with time [81]. Several researchers have shown that formalin-fixed samples can be used for DNA extraction [109]. No reduction in the ability to perform PCR amplifications of *E. histolytica* DNA fixed in 1–10% formalin was noted for 7 days [110]. However, other researchers have shown formalin to interfere with the amplification [81]. Consequently, if testing of fresh sample is not an option the freezing of a fresh specimen at −20°C before extraction of DNA may be the best strategy as this does not affect the sensitivity of the molecular assays and allows for storage that is not time dependent [15].

Recent, simple and effective methods for the isolation of DNA from feces have been developed. Commercially available DNA extraction kits are recommended due to the ease and success as they minimize extraction time and the DNA can be extracted directly from the feces without the need to culture the parasite. The following kits were used for the direct extraction of DNA from fecal samples: XTRAX

TABLE 6.2
RT-PCR Assays for *E. histolytica* Detection

Assay Gene Target	Primer/Probe	Sequence (5′ to 3′)	Amplicon Size	Cycling Profile	Primer/Probe Concentration	Detection Method	Reference
SYBR Green RT-PCR SSU rRNA	PSP5 PSP3	GGCCAATTCATTCAATGAATTGAG CTCAGATCTAGAAACAATGCTTCTC	877 bp	95°C for 15 min, followed by 50 cycles of 95°C for 15 s, 60°C for 1 min, 72°C for 1.5 min, 80°C for 30 s + melting curve	0.1 μM of each primer	Fluorescence at the end of the 80°C incubation plateau and during the melting curve	[101]
LightCycler SSU rRNA	Eh-S26C Eh-Ed-AS25 Eh-Ed-24-R Eh-Ed-25-D	GTACAAAATGGCCAATTCATTCAACG GAATTGATTTTACTCAACTCTAGAG LC640-TCGAACCCCAATTCCTCGTTATCCp GCCATCTGTAAAGCTCCCTCTCCGA-FAM	307 bp	95°C for 15 min, followed by 50 cycles of 95°C for 10 s, 58°C for 10 s, 72°C for 20 s	0.5 μM of each primer/0.1 μM of each probe	Fluorescence at the end of each annealing plateau	[102]
TaqMan 1 SSU rRNA	Ehd-239F Ehd-88R *Histolytica*-96T	ATTGTCGTGGCATCCTAACTCA GCGGACGGCTCATTATAACA FAM-UCAUUGAAUGAAUUGGCCAUUU-BH	231 bp	95°C for 3 min, followed by 40 cycles of 95°C for 15 s, 60°C for 30 s, 72°C for 30 s	0.5 μM of each primer/0.1 μM of probe	Fluorescence at the end of each extension plateau	[105]
TaqMan 2 Episomal repeats	*Histolytica*-50F *Histolytica*-132R *Histolytica*-78T	CATTAAAAATGGTGAGGTTCTTAGGAA TGGTCGTCGTCTAGGCAAAATATT FAM-TTGACCAATTTACACCGTTGATTTTCGGA-BH	83 bp	95°C for 3 min, followed by 40 cycles of 95°C for 15 s, 60°C for 30 s, 72°C for 30 s	0.5 μM of each primer/0.1 μM of probe	Fluorescence at the end of each extension plateau	[12]
MGB-TaqMan probe SSU rRNA	Ehd-239F Ehd-88R *Histolytica*-96T	CATTAAAAATGGTGAGGTTCTTAGGAA TGGTCGTCGTCTAGGCAAAATATT FAM-TTGACCAATTTACACCGTTGATTTTCGGA-BH	172 bp	95°C for 15 min followed by 40 cycles of 95°C for 15 s, 60°C for 30 s, and 72°C for 30 s	6.25 pmol of each primer/1.75 pmol of probe	Fluorescence at the end of each extension plateau	[104]
Molecular beacon	Ehf Ehr Molecular beacon probe	AACAGTAATAGTTTCTTTGGTTAGTAAAA CTTAGAATGTCATTTCTCAATTCAT Texas Red-GCGAGC-ATTAGTACAAAATGGCCA ATTCATTCA-GCTCGC-dR Elle	134 bp	45 cycles of 95°C for 15 s, 55°C for 30 s, and 72°C for 15 s. The ramping of the machine was 3.3°C/s in every step	25 pmol of each primers/ 6.25 pmol of probe	Fluorescence at the end of each extension plateau	[106]

DNA extraction kit [111] (Gull Laboratories, Salt Lake City, UT), Extract Master Faecal DNA extraction kit 9Epicentre Biotechnologies, WI) and the Genomic DNA Prep Plus kit [112]. Of the methods for DNA extraction Qiagen kits have been used the most widely with the QIAamp stool kit now predominating. QIAamp tissue kit spin columns (QIAGEN) have also been used for the extraction and purification of DNA from fecal samples [83]. Various modifications including optimizing the duration and temperature of proteinase K digestion, adding additional wash steps, and treatment of the sample with 2% polyvinylpolypyrolidone (PVPP) have all been used [104,106]. The QIAamp stool kit has proven to be the most widely accepted method for DNA extraction providing successful and reliable recovery of DNA from fecal material [15,16,18,70]. Aspirated pus from ALA can also be tested by PCR and has shown excellent sensitivity and specificity [98,99].

The following method is adapted from Verweij et al. [104,113] for parasite DNA isolation from feces using QIAamp spin columns (QIAamp DNA minikit cat# 51306).

1. Suspend 0.1 g of feces in 200 μL of phosphate-buffered saline (PBS) with 2% PVPP* in a microfuge tube.
2. Heat feces-suspension for 10 min.
3. Add 180 μL of tissue lysis buffer (ATL) and 20 μL of proteinase K from the QIAamp kit to the sample and mix by vortexing.
4. Incubate for 2 h (or overnight) at 55°C in a heating block, after 1 h vortex briefly.
5. Add 400 μL of AL buffer to 400 μL of sample, mix thoroughly by vortexing.
6. Incubate at 70°C for 10 min.
7. Centrifuge for 10 s at full speed >14,000 rpm and transfer the supernatant to a microfuge tube containing 400 μL of ethanol (96–100%), and mix thoroughly by vortexing.
8. Place a spin column in a 2 mL collection tube. Add 600 μL of the mixture from step 7 to the spin column without moistening the rim, close the cap and centrifuge at 10,000 rpm for 1 min.
9. Place spin column in a clean 2 mL collection tube a repeat with the rest of the mixture from step 7.
10. Place the spin column in a clean 2 mL collection tube.
11. Add 500 μL of AW1 buffer (from kit) to the spin column and centrifuge at 10,000 rpm for 1 min.
12. Empty the collection tube and add 500 μL of AW2 (from kit) buffer and centrifuge for 3 min at full speed.
13. Place the spin column in a clean 1.5 mL microfuge tube and elute DNA by adding 200 μL of AE buffer (from kit) and incubate for 1 min at room temperature.
14. Centrifuge at 8000 rpm for 1 min. The eluted DNA is now ready for use in PCR.

* PVPP (Sigma P6755) does not dissolve and it stays in a suspension, so regular stirring is needed when in use.

6.2.3 Detection Procedure

6.2.3.1 Principle

We present here a PCR assay that was developed by Troll et al. for specific detection of pathogenic *E. histolytica* [81]. This assay employs forward primer Eh5 (5′-GTAACTTACTTAACCGGTAAAACATG-3′) and reverse primer Eh3 (5′-TCTCTTCGTAACAAAGATCTAGACTC-3′) derived from the SSU (18S) rRNA gene for amplification of an 880 bp fragment from *E. histolytica* only.

6.2.3.2 Procedure

1. Prepare PCR mixture (50 μL) containing 1× PCR buffer, 3 mM $MgCl_2$, 200 μM dNTPs, 18 pmol of each primer (Eh5 and Eh3), 1.0 μL purified DNA, and distilled water (up to 50 μL). Include a tube containing the PCR mixture but not template DNA as negative control.
2. Conduct PCR amplification in a thermocycler with the following cycling program: 1 cycle of 94°C for 1 min; 40 cycles of 94°C for 1 min, 60°C for 1 min, and 72°C for 2 min; 1 cycle of 72°C for 10 min.
3. Separate the PCR product on 1% agarose with 0.5 μg/mL ethidium bromide. Visualize and photograph the product.

Note: This method has been shown recently to be useful for identification of *E. histolytica* from stool samples [15].

6.3 CONCLUSIONS AND FURTHER PERSPECTIVES

It is now a little over 10 years since an expert committee on amoebiasis suggested that there was a need to improve and advance the methods for diagnosis of *E. histolytica*. They acknowledged the difficulty of differentiating among this species and *E. dispar*; and this has been a major driving force behind research efforts, made more complex by the recent recognition of *E. moshkovskii* in human stool. It is therefore a rewarding experience to reflect on the achievements of the past 10 years, and to recognize the advances that ELISA and PCR have contributed to this discipline.

For the diagnostic laboratory, any diagnostic test used for the diagnosis of amoebiasis needs to meet specific criteria in addition to the age-old issue of cost. Determination of specificity and sensitivity of the test are paramount and the usefulness of the test within different settings raises considerations that may not be at first obvious. In developed countries such as Australia where infections with *E. histolytica* are not common, inclusion of gut parasites is often not the first choice in a differential diagnosis. Nevertheless, increased awareness of amoebiasis within the medical profession and the availability and provision of well-characterized diagnostic tests by hospital laboratories will prevent amoebiasis becoming a major public health problem [114].

Microscopy, despite all is limitations and especially the need for experienced, well-trained microscopists, remains a relatively sensitive way to detect *Entamoeba* species in human stool. The need for provision of confirmatory tests, based on either ELISA or PCR, reduces the cost effectiveness of providing microscopy alone as a diagnosis. Fecal antigen detection tests represent an attractive follow-up test because they are relatively quick to perform, easy to use and because they are often configured to the EIA format, they are not considered "unusual" to implement into the laboratory environment. In contrast technology such as RT-PCR, though probably the most effective diagnostic tool available at this moment in tie, is still surrounded by a little "voodoo" and because of the cost involved (experienced staff, reagents, machines) is still unlikely to become the mainstay for diagnosis.

It is worth continuing the discussion on how to integrate technology into the diagnostic laboratory. Universities and colleges of science and technology have a roll to play in this through their educational programs and courses on professional development. Although molecular biology has been well integrated into many, if not most, educational courses in the area of biomedical sciences, and graduates emerging from these courses are familiar with the terminology and jargon associated with molecular techniques, there is still resistance (cost) to introduce molecular technologies as routine diagnostic tests. Companies selling reagents and equipment must shoulder some of the blame here, since the prohibitive cost associated with technology such as RT-PCR, means the services are simply not provided. In this modern world where diagnosis can aid in the prevention of life-threatening diseases, the failure to provide diagnostic services is almost unethical.

How do we resolve this situation? As always, technology may provide a solution to this dilemma [115]. The detection of many pathogens in a single assay is an achievable goal in this day and age, and some simple tests are already commercially available. One can imagine, however, that chip-based technology may ultimately provide a means to detect all pathogens currently assayed for by diagnostic laboratories in addition to *Entamoeba* species [116,117]. The added attraction in this approach is that it is also possible to include tests that monitor pathogenesis and disease occurrence at the same time. Once again arrays for numerous cellular pathways are commonly used by research laboratories, and their incorporation into tests for pathogens, should provide a powerful approach for diagnosis of pathogen and disease at the same time.

What role is there for the immunoassay in a world based on molecular diagnosis? Nanotechnology may also provide a future for the immunoassay and the development of biosensors as immunosensors may eventually come forth with useful tools [118]. Indeed advances in the detection of bacteria using immunosensors have recently been reviewed [119] and the data suggest that this technology may yet evolve into an efficient high-throughput device for use in diagnostic labs [120]. Ironically, parasites that infect humans do not yet appear to be included in immunosensor prototypes at this moment in time [121].

Finally, whatever the future holds for diagnosis of parasitic infections such as those caused by *E. histolytica*, it must be remembered that such tests require applicability in developing countries, especially those where amoebiasis is endemic. This raises further challenges, not only of cost, but also of how diagnostic approaches can be integrated into management of public health. Such issues, unlike advances in technology, are frequently difficult for Governments to manage at a local level.

REFERENCES

1. Haque, R. et al. Amoebiasis. *N. Engl. J. Med.* 348, 1565, 2003.
2. Petri, W.A., Jr. et al. Estimating the impact of amoebiasis on health. *Parasitol. Today* 16, 320, 2000.
3. Anonymous. WHO News and activities, *Entamoeba* taxonomy. *Bull. WHO* 75, 291, 1997.
4. Losch, F.A. Massive development of amoebas in the large intestine. *Am. J. Trop. Med. Hyg.* 24, 383, 1875.
5. Councilman, W.T., and LaFleur, H.A. Amoebic dysentery. *Johns Hopkins Hosp. Rep.* 2, 395, 1891.
6. Brukner, D.A. Amoebiasis. *Clin. Microbiol. Rev.* 5, 356, 1992.
7. Walker, E.L., and Sellards, A.W. Experimental entamoebic dysentery. *Philipp. J. Sci.* 8, 253. 1913.
8. Sargeaunt, P.G., Williams J.E., and Grene, J.D. The differentiation of invasive and non-invasive *Entamoeba histolytica* by isoenzyme electrophoresis. *Trans. R. Soc. Trop. Med. Hyg.* 72, 519, 1978.
9. Clark, C.G., and Diamond, L.S. *Entamoeba histolytica*: An explanation for the reported conversion of "nonpathogenic" amoebae to the pathogenic form. *Exp. Parasitol.* 77, 456, 1993.
10. Diamond, L.S., and Clark, C.G. A redescription of *Entamoeba histolytica* Schaudin (1903) (amended Walker, 1911) separating it from *Entamoeba dispar* Brumpt, 1925. *J. Eukaryotic. Microbiol.* 40, 340, 1993.
11. Tanyuksel, M., and Petri, W.A., Jr. Laboratory diagnosis of Amoebiasis. *Clin. Microbiol. Rev.* 16, 713, 2003.
12. Verweij, J.J. et al. *Entamoeba histolytica* infections in captive primates. *Parasitol. Res.* 90, 100, 2003.
13. Fotedar, R. et al. Laboratory diagnostic techniques for *Entamoeba* species. *Clin. Microbiol. Rev.* 20, 511, 2007.
14. Clark, C.G., and Diamond, L.S. Intraspecific variation and phylogenetic relationships in the genus *Entamoeba* as revealed by riboprinting. *J. Eukaryot. Microbiol.* 44, 142, 1997.
15. Fotedar, R. et al. Detection and identification of *Entamoeba histolytica*, *Entamoeba dispar* and *Entamoeba moshkovskii* in stool samples from an Australian population. *J. Clin. Microbiol.* 45, 1035, 2007.
16. Fotedar, R. et al. *Entamoeba moshkovskii* infections in Sydney, Australia. *Eur. J. Clin. Microbiol. Infect. Dis.* 27, 133, 2008.
17. Ali, I.K. et al. *Entamoeba moshkovskii* infections in children, Bangladesh. *Emerg. Infect. Dis.* 9, 580, 2003.
18. Stark, D. et al. Prevalence of enteric protozoa in HIV-positive and HIV-negative men who have sex with men from Sydney, Australia. *Am. J. Trop. Med. Hyg*, 76, 549, 2007.
19. Dobell, C. Research on the intestinal protozoa of monkeys and man. *Parasitology* 20, 357, 1928.

20. World Health Organization. World Health Organization/Pan American Health Organization/UNESCO report of a consultation of experts on amoebiasis. *Wkly. Epidemiol. Rec.* 72, 97, 1997.
21. Hutson, C.D., and Petri, W.A. Jr. Amoebiasis: Clinical implications of the recognition of *Entamoeba dispar. Curr. Infect. Dis. Rep.* 1, 441, 1999.
22. Walsh, J.A. Problems in recognition and diagnosis of amoebiasis: Estimation of the global magnitude of morbidity and mortality. *Rev. Infect. Dis.* 8, 228, 1986.
23. Petri, W.A. Jr. Recent advances in amoebiasis. *Crit. Rev. Clin. Lab. Sci.* 33, 1, 1996.
24. Leroy, A.G. et al. Contact-dependent transfer of the galactose-specific lectin of *Entamoeba histolytica* to the lateral surface of enterocytes in culture. *Infect. Immun.* 63, 4253, 1995.
25. Stanley, S.L. et al. Role of the *Entamoeba histolytica* cysteine proteinase in amoebic liver abscess formation in severe combined immunodeficient mice. *Infect. Immun.* 63, 1587, 1995.
26. Vazquez, J. et al. Characterization of adhesion plates induced by the interaction of *Entamoeba histolytica* trophozoites with fibronectin. *Cell Motil. Cytoskeleton* 32, 37, 1995.
27. Weiss, S.J. Tissue destruction by neutrophils. *N. Engl. J. Med.* 320, 365, 1988.
28. Mann, B.J. Structure and function of the *Entamoeba histolytica* Gal/GalNAc lectin. *Int. Rev. Cytol* 216, 59, 2002.
29. Chadee, K. et al. Rat and human colonic mucins bind to and inhibit adherence lectin of *Entamoeba histolytica. J. Clin. Invest.* 80, 1245, 1991.
30. Ravdin, J.I. et al. Characterization of cell surface carbohydrate receptors for *Entamoeba histolytica* adherence lectin. *Infect. Immun.* 57, 2179, 1989.
31. Leippe, M. Amoebapores. *Parsitol. Today* 13, 178, 1997.
32. Que, X., and Reed, S.L. Cysteine proteinases and the pathogenesis of amoebiasis. *Clin. Microbiol. Rev.* 13, 196, 2000.
33. Stenson, W.F. et al. Amoebic infection in the human colon induces cyclooxgenase-2. *Infect. Immun.* 69, 3382, 2001.
34. Stanley, S.L. Jr. Amoebiasis. *Lancet* 361, 1025, 2003.
35. Gathiram, V., and Jackson, T.F.H.G. A longitudinal study of asymptomatic carriers of pathogenic zymodemes of *Entamoeba histolytica. S. Afr. Med. J.* 72, 669, 1987.
36. Haque, R. et al. Amoebiasis and mucosal IgA antibody against the *Entamoeba histolytica* adherence lectin in Bangladeshi children. *J. Infect. Dis.* 183, 1787, 2001.
37. Lewis, E.A., and Antia, A.U. Amoebic colitis: Review of 295 cases. *Trans. R. Soc. Trop. Med. Hyg.* 63, 633, 1969.
38. Brandt, H., and Perez-Tamayo, R. Pathology of human amoebiais. *Hum. Pathol.* 1, 351, 1970.
39. Kapoor, O.P., Nathwani, B.N., and Joshi, V.R. Amoebic peritonitis: A study of 73 cases. *J. Trop. Med. Hyg.* 75, 11, 1972.
40. Takahashi, T. et al. Fulminant amoebic colitis: Analysis of 55 cases. *Dis. Colon Rectum* 40, 1362, 1997.
41. Adams, E.B., and MacLeod, I.N. Invasive amoebiasis. I. Amoebic dysentery and its complications. *Medicine* 56, 315, 1977.
42. Mhlanga, B.R. et al. Amoebiasis complicating carcinomas—A diagnostic dilemma. *Am. J. Trop. Med. Hyg.* 46, 759, 1992.
43. Lysy, J. et al. Crohn's colitis complicated by superimposed invasive amoebic colitis. *Am. J. Gastroenterol.* 86, 1063, 1991.
44. Rustgi, A.K., and Richter, J.M. Pyogenic and amoebic liver abscess. *Med. Clin. N. Am.* 73, 487, 1989.
45. Orbison, J.A. et al. Amoebic brain abscess: Review of the literature and report of five additional cases. *Medicine* 30, 247, 1951.
46. Kirsc, D., and Diaz-Rivera, R.S. Perinephric abscess: A previously unrecognized complication of amoebiasis. *Am. J. Med. Sci.* 206, 373, 1943.
47. Heinz, K.P. Amoebic infection of the female genital tract: A report of three cases. *S. Afr. Med. J.* 47, 1795, 1973.
48. O'Leary, R.K., and Posen, J. Amoebiasis of the penis. *S. Afr. Med. J.* 65, 113, 1984.
49. Ramdial, P.K. et al. Amoebiasis cutis revisited. *J. Cutan. Pathol.* 34, 620, 2007.
50. Pehrson, P.O., and Bengtsson, E. A long-term follow up study of amoebiasis treated with metronidazole. *Scand. J. Infect. Dis.* 16, 195, 1984.
51. Lasserre, R. et al. Single-day drug treatment of amoebic liver abscess. *Am. J. Trop. Med. Hyg.* 32, 723, 1983.
52. Akgun, Y., Tacyildiz, I.H., and Celik, Y. Amoebic liver abscess: Changing trends over 20 years. *World J. Surg.* 23, 102, 1999.
53. Pillai, D.R., and Kain, K.C. Immunochromatographic strip-based detection of *Entamoeba histolytica-E. dispar* and *Giardia lamblia* coproantigen. *J. Clin. Microbiol.* 37, 3017, 1999.
54. Proctor, E.M. Laboratory diagnosis of amoebiasis. *Clin. Lab. Med.* 11, 829, 1991.
55. Clark, C.G., and Diamond, L.S. Methods for cultivation of luminal parasitic protists of clinical importance. *Clin. Microbiol. Rev.* 15, 329, 2002.
56. Balamuth, W. Improved egg yolk medium for the cultivation of *Entamoeba histolytica* and other intestinal protozoa. *Am. J. Clin. Pathol.* 16, 380, 1946.
57. Jones, W.R. The experimental infection of rats with *Entamoeba histolytica*; with a method for evaluating the anti-amoebic properties of new compounds. *Ann. Trop. Med. Parasitol.* 40, 130, 1946.
58. Diamond, L.S. A new liquid medium for xenic cultivation of *Entamoeba histolytica* and other lumen dwelling protozoa. *J. Parasitol.* 68, 958, 1982.
59. Robinson, G.L. The laboratory diagnosis of human parasitic amoebae. *Trans. R. Soc. Trop. Med. Hyg.* 62, 285, 1968.
60. Diamond, L.S. Axenic cultivation of *Entamoeba histolytica. Science* 134, 336, 1961.
61. Sargeaunt, P.G. et al. The reliability of *Entamoeba histolytica* zymodemes in clinical laboratory diagnosis. *Arch. Invest. Med.* 18, 69, 1987.
62. Zengzhu, G. et al. Analysis by enzyme-linked immunosorbent assay and PCR of human liver abscess aspirates from patients in China for *Entamoeba histolytica. J. Clin. Microbiol.* 37, 3034, 1999.
63. Haque, R. et al. Rapid diagnosis of *Entamoeba* infection by using *Entamoeba* and *Entamoeba histolytica* stool antigen detection kits. *J. Clin. Microbiol.* 33, 2558, 1995.
64. Garcia, L.S., Shimizu, R.Y., and Bernard C.N. Detection of *Giardia lamblia, Entamoeba histolytical Entamoeba dispar*, and *Cryptosporidium parvum* antigens in human fecal specimens using the Triage parasite panel enzyme immunoassay. *J. Clin. Microbiol.* 38, 3337, 2000.
65. Pillai, D.R. et al. *Entamoeba histolytica* and *Entamoeba dispar:* Epidemiology and comparison of diagnostic methods in a setting of nonendemicity. *Clin. Infect. Dis.* 29, 1315, 1999.
66. Levia, B. et al. Overdiagnosis of *Entamoeba histolytica* and *Entamoeba dispar* in Nicaragua: A microscopic, triage parasite panel and PCR study. *Arch. Med. Res.* 37, 529, 2006.
67. Haque, R. et al. *Entamoeba histolytica* and *Entamoeba dispar* infection in children in Bangladesh. *J. Infect. Dis.* 175, 734, 1997.
68. Solaymani-Mohammadi, S. et al. Comparison of a stool antigen detection kit and PCR for diagnosis of *Entamoeba*

histolytica and *Entamoeba dispar* infections in asymptomatic cyst passers in Iran. *J. Clin. Microbiol.* 44, 2258, 2006.
69. Gatti, S. et al. Amoebic infections due to the *Entamoeba histolytica-Entamoeba dispar* complex: A study of the incidence in a remote rural area of Ecuador. *Am. J. Trop. Med. Hyg.* 67, 123, 2002.
70. Stark, D. et al. Comparison of stool antigen detection kits with PCR for the diagnosis of amoebiasis. *J. Clin. Microbiol.* 46, 1678, 2008.
71. Gonin, P., and Trudel, L. Detection and differentiation of *Entamoeba histolytica* and *Entamoeba dispar* isolates in clinical samples by PCR and enzyme-linked immunosorbent assay. *J. Clin. Microbiol.* 41, 237, 2003.
72. Visser, L.G. et al. Diagnostic methods for differentiation of *Entamoeba histolytica* and *Entamoeba dispar* in carriers: Performance and clinical implications in a non-endemic setting. *Int. J. Med. Microbiol.* 296, 397, 2006.
73. Jelinek, T. et al. Evaluation of an antigen-capture enzyme immunoassay for detection of *Entamoeba histolytica* in stool samples. *Eur. J. Clin. Microbiol. Infect. Dis.* 15, 752, 1996.
74. Ong, S.J. et al. Use of the ProSpecT microplate enzyme immunoassay for the detection of pathogenic and non-pathogenic *Entamoeba histolytica* in faecal specimens. *Trans. R. Soc. Trop. Med. Hyg.* 90, 248, 1996.
75. Mirelman, D., Nuehamowitz, Y., and Stolarsky, T. Comparison of use of enzyme-linked immunosorbent assay-based kits and PCR amplification of rRNA genes for simultaneous detection of *Entamoeba histolytica* and *E. dispar. J. Clin. Microbiol.* 35, 2405, 1997.
76. Edman, U. et al. Characterisation of an immuno-dominant variable surface antigen from pathogenic and non-pathogenic *Entamoeba histolytica. J. Exp. Med.* 172, 879, 1990.
77. Tannich, E., and Burchard, G.D. Differentiation of pathogenic from non-pathogenic *Entamoeba histolytica* by restriction fragment analysis of a single gene amplified in vitro. *J. Clin. Microbiol.* 29, 250, 1991.
78. Garfinkel, L. et al. DNA probes specific for *Entamoeba histolytica* possessing pathogenic and nonpathogenic zymodemes. *Infect. Immun.* 57, 926, 1989.
79. Romero, J.L. et al. Use of polymerase chain reaction and non-radioactive DNA probes to diagnose *Entamoeba histolytica* infection. *Arch. Med. Res.* 23, 277, 1992.
80. Bhattacharya, S. et al. Circular DNA of *Entamoeba histolytica* encodes ribosomal RNA. *J. Protozool.* 36, 455, 1989.
81. Troll, H., Marti, H., and Weiss, N. Simple differential detection of *Entamoeba histolytica* and *Entamoeba dispar* in fresh stool specimens by sodium acetate-acetic acid-formalin concentration and PCR. *J. Clin. Microbiol.* 35, 1701, 1997.
82. Ramos, F. et al. Prevalence of *Entamoeba histolytica* and *Entamoeba dispar* in a highly endemic rural population. *Arch. Med. Res.* 31, 34, 2000.
83. Verweij, J.J. et al. Differentiation of *Entamoeba histolytica* and *Entamoeba dispar* cysts using polymerase chain reaction on DNA isolated from faeces with spin columns. *Eur. J. Clin. Microbiol. Infect. Dis.* 19, 358, 2000.
84. Heckendorn, F. et al. Species-specific field testing of *Entamoeba* spp. in an area of high endemicity. *Trans. Roy. Soc. Trop. Med. Hyg.* 96, 521, 2002.
85. Ramos, F. et al. *Entamoeba histolytica* and *Entamoeba dispar*: Prevalence infection in a rural Mexican community. *Exp. Parasitol.* 110, 327, 2005.
86. Hamzah, Z. et al. Differential detection of *Entamoeba histolytica, Entamoeba dispar*, and *Entamoeba moshkovskii* by a single-round PCR assay. *J. Clin. Microbiol.* 44, 3196, 2006.
87. Acuna-Soto, R. et al. Application of the polymerase chain reaction to the epidemiology of pathogenic and nonpathogenic *Entamoeba histolytica. Am. J. Trop. Med. Hyg.* 48, 58, 1993.
88. Britten, D. et al. An improved colorimetric PCR-based method for the detection and differentiation of *Entamoeba histolytica* and *Entamoeba dispar. J. Clin. Microbiol.* 35, 1108, 1997.
89. Tachibana, H., et al. Distinguishing pathogenic isolates of *Entamoeba histolytica* by polymerase chain reaction. *J. Infect. Dis.* 164, 825, 1991.
90. Tachibana, H. et al. Asymptomatic cyst passers of *Entamoeba histolytica* but not *Entamoeba dispar* in institutions for the mentally retarded in Japan. *Parasitol. Int.* 49, 31, 2000.
91. Rivera, W.L. et al. Differentiation of *Entamoeba histolytica* and *E. dispar* DNA from cysts present in stool specimens by polymerase chain reaction: Its field application in the Philippines. *Parasitol. Res.* 82, 585, 1996.
92. Rivera, W.L., Santos S.R., and Kanbara, H. Prevalence and genetic diversity of *Entamoeba histolytica* in an institution for the mentally retarded in the Philippines. *Parasitol. Res.* 98, 106, 2006.
93. Ayeh-Kumi, P.F. et al. *Entamoeba histolytica*: Genetic diversity of clinical isolates from Bangladesh as demonstrated by polymorphisms in the serine-rich gene. *Exp. Parasitol.* 99, 80, 2001.
94. Freitas, M.A. et al. A single step duplex PCR to distinguish *Entamoeba histolytica* from *Entamoeba dispar. Parasitology* 128, 625, 2004.
95. Zindrou, S. Specific detection of *Entamoeba histolytica* DNA by hemolysin gene targeted PCR. *Acta Trop.* 78, 117, 2001.
96. Nunez, Y.O. et al. Multiplex polymerase chain reaction amplification and differentiation of *Entamoeba histolytica* and *Entamoeba dispar* DNA from stool samples. *Am. J. Trop. Med. Hyg.* 64, 293, 2001.
97. Khairnar, K., and Parija, S.C. A novel nested multiplex polymerase chain reaction (PCR) assay for differential detection of *Entamoeba histolytica, E. moshkovskii* and *E. dispar* DNA in stool samples. *BMC Microbiol.* 7, 47, 2007.
98. Zaman, S. et al. Direct amplification of *Entamoeba histolytica* DNA from amoebic liver abscess pus using polymerase chain reaction. *Parasitol. Res.* 86, 724, 2000.
99. Tachibana, H. et al. Detection of pathogenic *Entamoeba histolytica* DNA in liver abscess fluid by polymerase chain reaction. *Int. J. Parasitol.* 22, 1193, 1992.
100. Khan, U. et al. Detection of *Entamoeba histolytica* using polymerase chain reaction in pus samples from amoebic liver abscess. *Indian J. Gastroenterol.* 25, 55, 2006.
101. Qvarnstrom, Y. et al. Comparison of real-time PCR protocols for differential laboratory diagnosis of amoebiasis. *J. Clin. Microbiol.* 43, 5491, 2005.
102. Blessman, J. et al. Real-time PCR for detection and differentiation *of Entamoeba histolytica* and *Entamoeba dispar* in fecal samples. *J. Clin. Microbiol.* 40, 4413, 2002.
103. Kebede, A. et al. The use of real-time PCR to identify *Entamoeba histolytica* and *E. dispar* infections in prisoners and primary-school children in Ethiopia. *Ann. Trop. Med. Parasitol.* 98, 43, 2004.
104. Verweij, J.J. et al. Simultaneous detection of *Entamoeba histolytica, Giardia lamblia*, and *Cryptosporidium parvum* in faecal samples by using multiplex real-time PCR. *J. Clin. Microbiol.* 42, 1220, 2004.
105. Verweij, J.J. et al. Short communication: Prevalence of *Entamoeba histolytica* and *Entamoeba dispar* in northern Ghana. *Trop. Med. Int. Health.* 8, 1153, 2003.

106. Roy, S. et al. Real-time PCR assay for the diagnosis of *Entamoeba histolytica* infection. *J. Clin. Microbiol.* 43, 2168, 2005.
107. Furrows, S.J., Moody, A.H., and Chiodini, P.L. Comparison of PCR and antigen detection methods for diagnosis of *Entamoeba histolytica* infection. *J. Clin. Pathol.* 57, 1264, 2004.
108. Lebbad, M., and Svard, S.G. PCR differentiation of *Entamoeba histolytica* and *Entamoeba dispar* from patients with amoeba infection initially diagnosed by microscopy. *Scand. J. Infect. Dis.* 37, 680, 2005.
109. Katzwinkel-Wladarsch, S., Loscher, T., and Rinder, H. Direct amplification and differentiation of pathogenic and nonpathogenic *Entamoeba histolytica* DNA from stool specimens. *Am. J. Trop. Med. Hyg.* 51, 115, 1994.
110. Ramos, F. et al. The effect of formalin fixation on the polymerase chain reaction characterisation of *Entamoeba histolytica. Trans. R. Soc. Trop. Med. Hyg.* 93, 335, 1999.
111. Evangelopoulos, A. et al. A nested, multiplex, PCR assay for the simultaneous detection and differentiation of *Entamoeba histolytica* and *Entamoeba dispar* in faeces. *Ann. Trop. Med. Parasitol.* 94, 233, 2000.
112. Myjak, P., Kur, J., and Pietkiewicz, H. Usefulness of new DNA extraction procedure for PCR technique in species identification of *Entamoeba* isolates. *Wiad. Parazytol.* 43, 163, 1997.
113. Verweij, J.J., Polderman, A.M., and Clark, C.G. Genetic variation among human isolates of uninucleated cyst-producing *Entamoeba* species. *J. Clin. Microbiol.* 39, 1644, 2001.
114. van Hal, S.J. et al. Amoebiasis: Current status in Australia. *Med. J. Aust.* 186, 412, 2007.
115. Paul, J., Srivastava, S., and Bhattacharya, S. Molecular methods for diagnosis of *Entamoeba histolytica* in a clinical setting: An overview. *Exp. Parasitol.* 116, 35, 2007.
116. Wang, Z., Vora, G.J., and Stenger, D.A. Detection and genotyping of *Entamoeba histolytica, Entamoeba dispar, Giardia lamblia*, and *Cryptosporidium parvum* by oligonucleotide microarray. *J. Clin. Microbiol.* 42, 3262, 2004.
117. Palacios, G. et al. Panmicrobial oligonucleotide array for diagnosis of infectious diseases. *Emerg. Infect. Dis.* 13, 73, 2007.
118. Cruz, H.J., Rosa, C.C., and Oliva, A.G. Immunosensors for diagnostic applications. *Parasitol. Res.* 88, S4, 2002.
119. Ricci, F. et al. A review on novel developments and applications of immunosensors in food analysis. *Anal. Chim. Acta* 605, 111, 2007.
120. Johnson-White, B., Lin, B., and Ligler, F.S. Combination of immunosensor detection with viability testing and confirmation using the polymerase chain reaction and culture. *Anal. Chem.* 79, 140, 2007.
121. Haque, R. et al. Comparison of PCR, isoenzyme analysis, and antigen detection for diagnosis of *Entamoeba histolytica* infection. *J. Clin. Microbiol.* 36, 449, 1995.

7 Giardia

Yaoyu Feng and Lihua Xiao

CONTENTS

7.1 INTRODUCTION

Giardia is the most common intestinal parasite of humans in developed countries and about 200 million people in Asia, Africa, and Latin America have symptomatic infections with about 50,000 cases reported each year [1]. This flagellated protozoan causes a generally self-limited clinical illness (i.e., giardiasis) characterized by diarrhea, abdominal cramps, bloating, weight loss, and malabsorption; asymptomatic infection also occurs frequently [2,3]. *Giardia* is also one of the most common enteric parasites of domestic animals, including livestock, dogs, and cats [4,5]. It is also a common parasite of wildlife [6].

7.1.1 Classification, Biology, and Epidemiology

7.1.1.1 Classification

Giardia species are established based on morphology and host specificity. Six species are generally recognized, including *Giardia agilis* in amphibians, *Giardia ardeae* and *Giardia psittaci* in birds, *Giardia muris* in rodents, *Giardia microti* in muskrats and voles, and *Giardia duodenalis* in mammals. Humans are believed to be infected by a single species, variably termed as *G. lamblia*, *G. intestinalis*, or *G. duodenalis*. For consistency, *G. duodenalis* is used in this chapter. Within this species the current trend has been to identify a complex of assemblages based on genetic differences and host specificity. These assemblages are identified based on sequence analysis of conserved genetic loci [7]. Currently, there are seven well-defined assemblages of *G. duodenalis*, designated A through G. Assemblages A and B have the broadest host specificity, having been found in humans and other mammals, including dogs, cats, livestock, and wildlife [7]. Assemblage A consists of two major subgroups, AI and AII, and a few other new subgroups that have been identified recently. In contrast, there is no clear subgrouping in assemblage B [2,8–12]. Assemblages C and D have been found largely in dogs [13–15], assemblage E in cloven-hoofed mammals [16], assemblage F in cats, and assemblage G in rodents [8].

7.1.1.2 Biology

Giardia has a direct life cycle and the infective stage of the parasite, the cyst, is encysted when released in the feces and is immediately infectious [17]. Cysts remain infectious for months in cool, damp areas and rapidly accumulate in environments. They can survive in water for considerable periods. Following exposure to gastric acid, gastric and pancreatic enzymes, cysts excyst in the duodenum and release the trophozoites. The latter attach to the surface of the intestinal microvilli beneath the mucus layer via a ventral adhesive disc. Trophozoites undergo repeated mitotic division.

Environmentally resistant cysts form from trophozoites in response to stimuli of intestinal conditions such as the presence of bile salts. They pass through the intestine in the feces and contaminate the environment. Infection of new hosts is established by ingestion of cysts.

Giardia infection is transmitted by the fecal–oral route and results from the ingestion of cysts in contaminated food and water, or contact with infected persons (person-to-person transmission), animals (zoonotic transmission), or contaminated fomites. Some characteristics of *Giardia* markedly influence the epidemiology of the infection: (i) the infective dose is low (1–10 cysts); (ii) cysts are immediately infectious when excreted in feces, and can be transmitted by direct person-to-person contact; (iii) cysts are remarkably stable and can survive for weeks to months in the environment; and (iv) environmental dispersal can lead to the contamination of drinking water, food, and fomites. Direct and indirect transmission between infected hosts and susceptible individuals is favored by high population densities (e.g., in daycare or nursing homes, or during lambing or calving), close contact, or contamination of food or water. In industrialized nations, human giardiasis is commonly transmitted through drinking and recreational water, and there have been numerous outbreaks of giardiasis following the ingestion of contaminated water [18,19]. The contributors to environmental contamination include infected humans, livestock, and feral hosts as well as transport hosts. Cysts in feces can contaminate water directly or indirectly, and the disposal of human and animal wastes remains a significant public health issue in many parts of the world. In developing countries, anthroponotic transmission via direct person-to-person contact is probably the dominant transmission mode [20]. Therefore, control strategies in poor countries must give greater emphasis to policies geared toward reducing person-to-person transmission for the prevention of giardiasis, along with improvement in environmental and sanitary conditions, public education, source water protection, and drinking water treatment.

7.1.1.3 Epidemiology

Since giardiasis is a notifiable disease in many developed countries and disease surveillance, prevention, education, and control are better in these countries, consequently, the prevalence of giardiasis in developed countries is normally lower than in developing countries. In the United Kingdom, the prevalence of *Giardia* was 1.3% in preschool children [21]. The prevalence of *Giardia* in humans in Italy ranged between 0.9% and 6.2% [22]. In the Nordic countries of Denmark, Finland, Norway, and Sweden, the prevalence of giardiasis was estimated to be 2.97% in the general population and 5.81% in the symptomatic population [23]. The prevalence of *G. duodenalis* infections in Portugal was 3.7% [24]. In other European countries, prevalence rates were mostly between 2% and 7% [24].

In contrast, the prevalence of *Giardia* is generally higher in developing countries. Prevalence rates of 5.6–44.0% were reported in Albania [25]. Likewise, prevalence rates of 18–50% were seen in Mexico [26–29], 16.7–26.9% in Brazil [30–34], 23.0–29.0% in Argentina [35], 10.6–23.8% in Peru [36,37], and 15.2% in Colombia [38]. In western Nepal, the highest prevalence of *G. duodenalis* reached 73.4% in pre-school and school children [39]. In Iran, the prevalence rate was 11.5% [40]. In north Lebanon, the prevalence of *Giardia* was 15.4% in 17,126 patients examined between 1997 and 2001 [41]. The percentage of people excreting *Giardia* cysts was 34.6% in Egypt [42]. In North Africa, giardiasis was seen in 34.2% of children in Sahrawi [43] and 11.7% of adults living in rural areas in Morocco [44], whereas in west Africa, *G. duodenalis* was found in 14.8% of samples in Guinea-Bissau [45] and 29% of samples in Sierra Leone [46]. In South Africa, the prevalence of giardiasis was 17.3% [47].

Giardiasis affects persons of all ages, but the infection rate is significantly higher in children than in adults. Generally, a bimodal infection pattern is seen for *Giardia* infection in most developed countries. The populations affected mostly include children of 0–9 years and adults of 20–44 years of age [48–51]. The higher giardiasis incidence in adults of sexual active and child-bearing age is possibly a result of transmission of giardiasis from children to their caregivers (e.g., daycare staff, family members, and other household contacts), more outdoor activities of the adults (such as traveling and swimming), and unsafe sexual practice (such as anal–oral sex) [2,51,52]. Differences in climate, socioeconomic conditions, and cultural practices possibly contribute to the different age distribution of human giardiasis in different countries. In developing countries, the prevalence of *G. duodenalis* is high in children less than 15 years old and decreases with age thereafter until the age of 40, after which the prevalence is kept stable, with a small rebound in old men [17,39,53–55].

A marked seasonality in the onset of illness occurs in early summer through early fall in United States, Canada, the United Kingdom, and New Zealand [48–50,56]. This is due to the increased use of rivers, lakes, swimming pools and water parks, or travel during the summer vacation period. In studies conducted in Albania and China, however, the prevalence of giardiasis doubled in spring/summer compared to autumn/winter [25]. The prevalence of giardiasis was much higher in the rainy season in Thailand and Tanzania [57,58]. The seasonal pattern of giardiasis in developing countries is probably a reflection of the better environmental survival of *Giardia* cysts at low temperatures and in wet seasons.

Risk factors for *Giardia* infection include (i) travel to disease-endemic areas, (ii) attending daycare facilities, (iii) close contact with infected persons and/or animals, (iv) ingesting or swallowing contaminated water, (v) eating contaminated food, (vi) men who have sex with men, and (vii) socioeconomic conditions (size of household, age of the family members, education levels of adults, family income, employment status, etc.) [50]. Some factors, such as contact with ill persons, consumption of contaminated drinking water or food, and poor socioeconomic status are especially important in developing countries.

Molecular diagnostic tools have recently been used in characterization of the transmission of giardiasis, especially the role of zoonotic transmission in giardiasis epidemiology. Only *G. duodenalis* assemblages A and B have so far been definitively associated with human infections. As both assemblages A and B also infect animals, zoonotic transmission has long been suspected to play a role in epidemiology of human giardiasis. Although assemblage A has been considered the most important *G. duodenalis* genotype involved in zoonotic giardiasis, subtyping data obtained so far do not support widespread occurrence of zoonotic transmission. The two most common subtypes of assemblage A, A-I and A-II, differ significantly in host preference. Humans are mostly infected with A-II, in contrast, animals are mostly infected with A-I, although there are exceptions to this observation [59].

7.1.2 Clinical Features and Pathogenesis

Giardia can cause asymptomatic infection, acute self-limiting diarrhea, or chronic diarrhea and malabsorption depending on host susceptibility and pathogen genotypic differences [60,61]. Human giardiasis presents diverse clinical manifestations, ranging from asymptomatic, to acute, intermittent, or chronic nonbloody diarrhea. Other symptoms include abdominal cramping, nausea, vomiting, steatorrhea, anorexia, and weight loss [17]. In endemic areas, asymptomatic infections are common in both children and adults. Patients who have giardiasis often have symptoms for more than 7–10 days. Patients who develop persistent or chronic diarrhea, defined as diarrhea for more than 14 days, may have profound malaise, lassitude, occasional headache, weight loss, and diffused abdominal and epigastric discomfort that is exacerbated by eating. Stools from patients who have chronic diarrhea often are greasy and foul smelling or frothy, yellowish, and frequently occur in small volumes. Periods of diarrhea may alternate with periods of constipation or normal bowel habits over a period of months until spontaneous resolution or an antimicrobial therapy is given.

The pathogenesis of *Giardia* is postulated to involve direct damage to the intestinal brush border and mucosa by trophozoites, alteration of bile content or duodenal flora, induction of a host immune response that results in the secretion of fluid, and apoptosis in the small intestinal epithelial cells [17]. Disruption of the intestinal epithelial cell brush border may explain the lactose intolerance that develops commonly. *In vitro* studies showed that *Giardia* disrupts tight junctional zona occludens, increases permeability, and induces apoptosis in small intestinal epithelial cells [62]. Increased permeability of the small intestinal epithelial cells occurs by way of a myosin light chain kinase-dependent phosphorylation of F-actin [63,64]. However, mucosal invasion occurs rarely and no *Giardia* enterotoxin has been found.

Giardia is capable of undergoing antigenic variation in its cysteine-rich surface proteins through a palmitoylated process during human infection and during the encystation and the excystation cycle [65,66]. Antigenic variation protects the parasite from the activity of intestinal proteases, provides oxygen stability, and possibly allows adaptation to different hosts [17]. The clinical significance of *Giardia* antigenic variation is not known.

7.1.3 Diagnosis of *G. Duodenalis*

7.1.3.1 Conventional Diagnosis

Traditionally, the diagnosis of *Giardia* infection has depended on microscopic examination of stool. This requires an experienced microscopist and a specialized laboratory. Stool samples should be examined fresh or placed immediately in a preservative. Motile trophozoites may be seen in a saline wet mount of fresh liquid stool from a patient who has acute diarrheal illness. Cysts can be identified after iodine staining or after preservation in 10% buffered formalin or polyvinyl alcohol and subsequent trichrome or iron hematoxylin staining. Repeated samplings may be necessary. The yield of cysts may be increased with formalin-ethyl acetate sedimentation or zinc sulfate flotation concentration techniques [67,68]. Nevertheless, even under ideal conditions, organisms are identified in a single stool sample in only half of specimens [17].

Giardia antigen detections by enzymatic immunoassays and immunochromatography, and direct fluorescence antibody (DFA) detection of cysts are becoming common diagnostic tests for giardiasis in the United States. Commercial antigenic detection tests are similar in cost to microscopy with better reproducibility and more rapid return of results [69,70]. Antigen tests have a sensitivity of 85–98% and a specificity of 90–100% compared with microscopy [69–73]. A commonly used assay, the Merifluor DFA (Meridian Bioscience, Inc.), has a reported sensitivity and specificity of 96–100% [70,71].

Other tests that may be used for the diagnosis of *Giardia* include the string test or EnteroTest (HDC Corporation, Milpitas, California), duodenal aspiration, duodenal biopsy, and serology. A string test that yields bile-stained mucus from the duodenum can be examined by wet mount for trophozoites. Although more invasive, these tests offer the advantage of examining for trophozoites and other enteric pathogens.

The above methods do not differentiate *Giardia* species and the assemblages of *G. duodenalis*. Therefore, molecular typing tools are now used to provide insight into the species and assemblages that occur in clinical specimens and water and food samples. Molecular diagnostic tools also have the advantage in its potential in assessing the contamination source and human-infective potential of *Giardia* cysts in samples. Molecular methods usually have higher detection sensitivities than microscopy, which makes them applicable in detecting low numbers of *Giardia* cysts in water and food samples.

7.1.4 Molecular Characterization

Molecular biology has provided powerful new tools for characterizing *Giardia*, and the analysis of previously unrecognized genetic differences within this genus has revolutionized our understanding of both the taxonomy and epidemiology of human giardiasis. Although some simple polymerase chain reaction (PCR) assays have been used in the detection of *Giardia* in clinical and environmental samples, more recent molecular tools have been developed to differentiate *Giardia* at the species/assemblage and subtype levels. These tools are widely used in the identification and *G. duodenalis* genotypes and subtypes in clinical specimens [8,13,74–79]. Variations between the methods lie in the gene(s) utilized (such as the small subunit rRNA—SSU rRNA, glutamate dehydrogenase—GDH, triose phosphate isomerase—TPI, elongation factor 1 alpha—EF1a, β-giardin, and variant surface protein—VSP), regions of the gene targeted, the specificity of the assay (*Giardia* specific, *G. duodenalis* specific, or assemblage specific), and downstream procedures (restriction fragment-length polymorphisms analysis or DNA sequencing of PCR products). The usage of these loci in genotyping and subtyping of *G. duodenalis* and their sequence characteristics have been reviewed recently [13]. Some published molecular targets for species, genotype, and subtype differentiation of *Giardia* in animal and human specimens and water samples are listed in Table 7.1. Among all the genes studied, SSU rRNA, GDH, TPI, and β-giardin are used more frequently.

Since different *Giardia* loci differ in total, synonymous, and nonsynonymous substitution rates, their applications vary from species differentiation, genotyping (assemblage-differentiation), to subtyping. For example, substitution rates for the partial SSU rRNA, β-giardin, GDH, and TPI genes were reported to be 0.01, 0.03, 0.06, and 0.12 substitutions per nucleotide, respectively [12]. Thus, the conserved SSU rRNA gene is traditionally used for species and assemblage differentiation (mostly genotyping) [13,76,106–107] whereas the most variable locus, TPI, is frequently used for subtyping clinical specimens [85,108] and the GDH locus, with a substitution rate midway between them, has a broad application spectrum [8,78]. Although β-giardin has a low substitution rate, it has demonstrated prominent subgrouping within some common *G. duodenalis* assemblages, and therefore, has also been used for determining and differentiating the subtypes within assemblages A, B, and E [12]. However, β-giardin is yet to demonstrate consistent subtyping with respect to the other loci.

7.1.4.1 Genotyping Tools

The genotyping primers for SSU rRNA gene locus of *Giardia* from Weiss et al. [74], van Keulen et al. [95], or Hopkins et al. [13] amplify only the first fifth or the last tenth of the gene (1400 bp in total), and the products produced by these primer sets are too small to differentiate assemblages. Notably assemblage F (of cats) is identical to assemblage A until base 499 [12]. This has led to some inaccurate results. In the study by Berrilli et al. [109], cat isolates were genotyped as assemblage A using the 292 bp-product primers of Hopkins et al. [13] and in a study by Fayer et al. [110], cat isolates were genotyped as assemblage F using the same primers. This discrepancy may be due to an unsubstantiated substitution at the 5′ end of the SSU rRNA gene and the limited sequences available in GenBank [12]. In contrast, other loci were comparatively good for genotyping usage. The coverage of the genes also varied. The sequence of the GDH gene covered at least the first 40–60% of the gene, and the TPI and β-giardin sequences covered 60% of the length of the genes by most primers. Therefore, these lengths were sufficient for genotyping samples to the assemblage and possibly subtype levels.

7.1.4.2 Subtyping Tools

Subgrouping within the zoonotic assemblages A and B is of interest as there may be a link between specific subgroups and their hosts, with particular attention to those that may infect both humans and animals. It is therefore important to establish the specific subgroups, based on their intra-genotypic substitution patterns, and the continuity of these subgroups across genetic loci. The A-I and A-II subgroups are quite robust [13]. In the SSU rRNA gene, they are differentiated at the 3′ end by a single substitution [74]. At the GDH locus, there are numerous A-I/A-II-specific polymorphisms and the majority of isolates can be easily differentiated [12]. The TPI gene has only two A-I/A-II-specific polymorphisms in the region examined by most primers, and there are other mutations defining other groups [12]. It is noteworthy that all these loci have produced consistent distinction between the two close subgroups. The conserved β-giardin alignment in contrast had only one substitution clearly differentiating the A-I/A-II subgroups and several other substitutions producing additional groups [12]. The significance of the additional TPI and β-giardin subgroups within the assemblage A and the intergenic linkage of them are not yet clear. In fact these two genes have consistently demonstrated numerous subgroups within all assemblages. The significance of this is difficult to determine, because the samples that had been characterized at other loci did not always segregate into the same subgroups defined by the β-giardin locus [12].

The BIII and BIV subgroups originally described by allozyme electrophoretic studies [111] are not reproducible in DNA sequence analyses [13]. There are insufficient lengths of targets characterized for reference isolates for some of the genes (such as SSU rRNA), insufficient reference isolates characterized at some loci (such as β-giardin), lack of clusters that are well supported by bootstrap analysis, inconsistent grouping at the different loci, and the presence of subgroups other than BIII and BIV [86].

It is also interesting to note that with increasing number of samples characterized subgroups have been identified in assemblage E. Assemblage E, like assemblages A and B, has a broad host range (various hoofed livestock), therefore likely may contain host-specific subgroups. The subgrouping is again most obvious at the β-giardin locus. The nucleotide sequences do not yet appear to segregate in the host-specific manner as the allozyme data had demonstrated [9], but the number of representative sequences from the different livestock hosts is still small.

It is not possible to speculate on the subgroups formation in other assemblages (C, D, F, and G) as there are too few data available. For assemblage D, there is one apparently

TABLE 7.1
List of the Gene, Primer, Assay Type, and Main Use of Some Published *G. duodenalis* Genotyping Tools

Gene	Primer (5′ to 3′)	Size (bp)	Specificity	Assay Type	Usage	Sample Type in Original Study	Sample Type in Other Studies	Reference
TPI	TPI-F1: CCAGAATGTGTACCTAGAGGGG TPI-R1: TACTCTCCGAGCTCCTTCTGC TPI-F2: ATGCCTGCTCGTCGCCCCTTC TPI-R2: CACTGGCCAAGCTTCTCGCAG	812 683	Unknown	Nested PCR, RFLP, sequencing	Genotyping	Human, animal (beaver, cat, guinea pig, dog)		[77]
TPI	TPI-F3: GGAGACCGACGAGCAAAGC TPI-R3: CTTGCCAAGCGCCTCAA	148	Assemblage A specific	PCR	Genotyping	Human		[80]
TPI	TPI-F4: AATAGCAGCACARAACGTGTATCTG TPI-R4: CCCATGTCCAGCAGCATCT	81	Assemblage B specific	PCR	Genotyping	Human		[80]
TPI	TPI-F5: AAATIATGCCTGCTCGTCG TPI-R5: CAAACCTTITCCGCAAACC TPI-F6: CCCTTCATCGGIGGTAACTT TPI-R6: GTGGCCACCACICCCGTGCC	605 532	Genus-specific	Nested PCR, sequencing	Genotyping and subtyping	Human, animal (dog, cattle, muskrat, beaver, rat, rabbit)	Wastewater, cattle, lamb	[11,81–83]
TPI	TPI-F7: ATYAAGAGCCACGTRGCGKC TPI-R7: CCATGATTCTRCGYCTTTCAG	253	Unknown	PCR, sequencing	Genotyping and subtyping	Human, dog		[84]
TPI	TPI-F8: CGAGACAAGTGTTGAGATGC TPI-R8: GGTCAAGAGCTTACAACACG TPI-F9: CCAAGAAGGCTAAGCGTGC TPI-R9: GCCACATGCCTATGTACGGG	576 452	Assemblage A specific	Nested PCR, RFLP, sequencing	Genotyping and subtyping	Human	Wastewater	[85,86]
TPI	TPI-F10: GTTGCTCCCTCCTTTGTGC TPI-R10: GGCCTTGCGTTCATCCAGG TPI-F11: GCACAGAACGTGTATCTGG TPI-R11: CTCTGCTCATTGGTCTCGC	210 141	Assemblage B specific	Nested PCR, sequencing	Genotyping and subtyping	Human	Wastewater	[85,86]
TPI	TPI-F12: ATCGGYGGTAAYTTYAARTG TPI-R12: CACTGGCCAAGYTTYTCRCA TPI-F13: CCCTTCATCGGYGGTAAC TPI-R13: CCCGTGCCRATRGACCACAC TPI-R14: ACRTGGACYTCCTCYGCYTGCTC	520	Unknown	Nested PCR, sequencing	Genotyping and subtyping	Human, animal (cat, dog, pig, rat)		[87]
GDH	GDH-F1: TCCACCCCTCTGTCAACCTTTC GDH-R1: AATGTCGCCAGCAGGAACG	220	Genus-specific	PCR, sequencing. This set of primer could combine with GDH-F3 and GDH-R3 in nested PCR	Genotyping	Dog	Dog, cat, calf, monkey	[88,89]
GDH	GDH-F2: GAGAGGATCCTTGARCCNGAGCGCG GDH-R2: ACCTTCTAGAANCCNGCDATGTTNGCGCC	1170	Could amplify most *Giardia* species and some bacteria	PCR, RFLP, sequencing	Genotyping and subtyping	Human, animal (guinea pig, slow loris, rat)		[78]
GDH	GDH-F3: ATCTTCGAGAGGATGCTTGAG GDH-R3: AGTACGCGACGCTGGGATACT	778	Genus-specific	PCR, RFLP sequencing	Genotyping and subtyping	Human	Ferret, cattle	[90–92]

continued

TABLE 7.1 (continued)
List of the Gene, Primer, Assay Type, and Main Use of Some Published *G. duodenalis* Genotyping Tools

Gene	Primer (5′ to 3′)	Size (bp)	Specificity	Assay Type	Usage	Sample Type in Original Study	Sample Type in Other Studies	Reference
GDH	GDH-F3: ATCTTCGAGAGGATGCTTGAG GDH-R4: TGTCCTTGCACATCTCCTCCA	528	Genus-specific	PCR, RFLP, sequencing	Genotyping and subtyping	Animal (dog, sheep, goat, roe deer)		[93]
GDH	GDH-F4: CGAGCGCGAGCCGAAGTATATCC GDH-R5: GATGTTYGCRCCCATCTGRTAGTTC	1190	Genus-specific	PCR, sequencing	Genotyping and subtyping	Human, animal (dog, cat, cattle)		[94]
GDH	GDH-F5: TCAACGTYAAYCGYGGYTTCCGT GDH-R6: GTTRTCCTTGCACATCTCC GDH-F6: CAGTACAACTCYGCTCTCGG GDH-R6:	432	Genus-specific	Semi-nested PCR, RFLP	Genotyping and subtyping	Human, animal (cat, dog, cattle)	Wastewater	[10,86]
SSU rRNA	SSU-F1: TCCGGTCGATCCTGCCGGA SSU-R1:GCTCTCCGGAGTCGAAC	298	Genus-specific	PCR, RFLP, sequencing	Genotyping	Human, wastewater, animal (dog, cat, deer mouse)		[76,95]
SSU rRNL	SSU-F2: CATCCGGTCGATCCTGCC SSU-R2: AGTCGAACCCTGATTCTCCGCCCAGG	292	Genus-specific	PCR, sequencing	Genotyping	Human, dog	Ferret, cattle, cat	[13,90,96–98]
SSU rRNA	SSU-F3: AAGTGTGGTGCAGACGGACTC SSU-R3: CTGCTGCCGTCCTTGGATGT	497	Unknown	Nested PCR (combine with primers SSU-F2 and SSU-R2), sequencing	Genotyping	Cattle		[99]
SSU rRNA	SSU-F4: GACGCTCTCCCCAAGGAC SSU-R4: CTGCGTCACGCTGCTCG	130	Unknown	Nested PCR (combine with primers SSU-F2 and SSU-R2), sequencing	Genotyping and subtyping	Human, dog	Cat	[84,96]
SSU rRNA	SSU-F5 : GCGCACCAGGAATGTCTTGT SSU-R5 : TCACCTACGGATACCTTGTT	183	Amplify most protozoa parasites	PCR, sequencing	Genotyping and subtyping	Human, animal (beaver, cat, guinea pig, muskrat, dog, sheep, mouse, blue heron)		[74]
SSU rRNA	SSU-F6: TCCGGTYGATTCTGCC SSU-R6: CTGGAATTACCGCGGCTGCT	480	Unknown	PCR, sequencing	Genotyping and subtyping	Human, animal (cat, dog, pig, rat)		[87]
β-Giardin	Giardin-F1: AAGTGCGTCAACGAGCAGCT Giardin-R1: TTAGTGCTTTGTGACCATCGA	171	Genus-specific	PCR, hybridization	Genotyping	Cattle, human	Water	[100,101]
β-Giardin	Giardin-F2: CATAACGACGCCATCGCGGCTCTCAGGAA Giardin-R2: TTTGTGAGCGCTTCTGTCGTGGCAGCGCTA	218	*G. duodenalis*-specific	PCR, hybridization	Genotyping	Cattle, human		[101]

β-Giardin	Giardin-F3: AAGCCCGACGACCTCACCCGCAGTGC Giardin-R3: GAGGCCGCCCTGGATCTTCGAGACGAC Giardin-F4: CATAACGACGCCATCGCGGCTCTCAGGAA Giardin-R3:	753 384	Unknown	Semi-nested PCR, RFLP, sequencing	Genotyping and subtyping	Human	Animal (dog, cat), wastewater	[79,86,102]
β-Giardin	Giardin-F5: GAACGAACGAGATCGAGGTCCG Giardin-R4: CTCGACGAGCTTCGTGTT	511	Unknown	Nested PCR (combine with Giardin-F3 and Giardin-R3), RFLP sequencing	Genotyping and subtyping	Human, animal (dog, cat, calf)		[103]
Ef1-α	EF1-α-F1: GCCGAGGAGTTCGACTACATC EF1-α-R1: GACGCCSGAGATCTTGTAGAC	191	Unknown	PCR, sequencing	Genotyping and subtyping	Human, dog		[84]
Ef1-α	EF1-α-F2: GCTCSTTCAAGTACGCGTGG EF1-α-R2: GCATCTCGACGGATTCSACC	650	Unknown	PCR, sequencing	Genotyping and subtyping	Human, animal (cat, dog, pig, rat)		[87]
VSP	VSP-F1: GTTGGATCCGCCACCCAGCAGGTCG-GCGGCCTAATG VSP-R1 GTATCC(T/C)GGAGCTCAC(G/A) CCTTACATGTTGTAGCTGCTCCA	1908	Gene tsa 417-specific	PCR, RFLP, sequencing	Genotyping and subtyping	Human, animal (sheep, dog, calves)		[104]
VSP	VSP-F2: CGTATCGATGAGCTGAGAAGGATTTTAATG VSP-R1:	1755	Gene tsp 11-specific	PCR, RFLP sequencing	Genotyping and subtyping	Human, animal (sheep, dog, calves)		[104]
VSP	VSP-F3: ACGACGGTACTAAGGGCACGTG VSP-R1:	520	Gene tsa 417 and tsp 411-specific	PCR, RFLP, sequencing	Genotyping	Human		[75]
VSP	VSP-F4: TTACTCTAGAATAAGCTGAGGGAAACTTCAATG VSP-R2: ACGCCTCTAGACYRCANABGAACCACCARCA where B = T, C or G, R = A or G, Y = C or T, and N = A, T, C or G	1619	Gene vsp 1267-specific	PCR, RFLP, sequencing	Genotyping and subtyping	Human, animal (sheep, dog, calves)		[104]
VSP	VSP-F5: TTGCTCGAGCAACTGGAAATGAAGGTGGGTG VSP-R2:	1290	Gene vsp 1267-specific	PCR, RFLP sequencing	Genotyping and subtyping	Human, animal (sheep, dog, calves)		[104]
rRNA	rDNA-F1: CGATAGCAGGTCTGTGATGC rDNA-R1: GCTACTGATATGCTTAAGTTC	615	Genus-specific	PCR, sequencing	Genotyping	Cattle		[105]
Unknown	A-F1: CTGCAGGGGCAAGGCGTAGAT A-R1: CCACCGTGCCAGTCTTCTGGG	800	Assemblage A-specific	PCR, RFLP, sequencing	Subtyping	Human	Animal	[91,93]
Unknown	B-F1: CTGCAGTAACACTGGCAAG B-F2: CTGCAGAGTTCTCCGCAGCG	1500	Assemblage B-specific	PCR, RFLP sequencing	Subtyping	Human	Animal	[91,93]

Abbreviations: EF-1α, elongation factor 1α; GDH, glutamate dehydrogenase; RFLP, restriction fragment-length polymorphism; TPI, triose phosphate isomerase; VSP, variant surface protein.

host-specific polymorphism in the β-giardin alignment where samples have included dogs and a coyote described by Santin et al. [112], but for assemblage G, there is no host-specific polymorphism in the GDH alignment of sequences from rats and a mouse [13].

7.1.4.3 Multilocus Tools

The quantity and quality of current genotyping tools have limited the understanding of the molecular epidemiology of human giardiasis, and direct characterizations of human and nonhuman-derived cysts have mostly been based on analysis of individual genetic loci [7]. The vast majority of studies have relied on the analysis of the SSU rRNA, β-giardin, GDH, EF-1α, TPI, the GLORF-C4 (C4) genes, and recently, the intergenomic rRNA spacer region [113]. In early studies, there was a strong bias toward the use of SSU rRNA because of its multicopy nature and high degree of sequence homology. As a consequence, the reliability of different genetic loci in the assignment of isolates to specific *G. duodenalis* assemblages was not assessed. However, some genes, such as SSU rRNA gene and EF-1α genes provided poor genetic resolution of the *G. duodenalis* isolates recovered from humans and dogs within various assemblages, and analysis of the TPI and some other genes provided better genetic resolution, stressing the importance of using multiple loci when inferring genotypes of *Giardia* [84].

In order to provide most robust information and to determine the intra-assemblage level of genetic variation at different loci, Caccio et al. [114] suggested a multilocus genotype (MLG) model for the appropriate nomenclature of subassemblage based on the β-giardin, GDH, and TPI genes. However, recent studies based on a multilocus approach have shown that a number of isolates, of both human and animal origin, cannot be unequivocally assigned at the assemblage level, because the genotype designations were not consistent among loci [84,114,115]. In one study in which sequence analyses of four genes (SSU rRNA, β-giardin, GDH, and TPI) of human and animal isolates were performed, results obtained showed that multilocus genotypes can be readily defined for *G. duodenalis* isolates of assemblage A but not for assemblage B. For assemblage A isolates, congruent genotyping results were obtained at the four genetic loci investigated. On the other hand, there were variable levels of intra-isolate sequence heterogeneity in assemblage B isolates, which prevented the unambiguous identification of MLGs. Furthermore, in some human isolates and one nonhuman primate isolate, it was difficult to assign the isolate to an assemblage because the genotyping result obtained at one locus was not consistent with those obtained at the other three loci [114]. Similar results were reported in a few previous studies of both human and animal isolates, using the same genetic targets (with the exception of β-giardin) [10,84,116]. This inconsistency in genotyping results was frequently reported in isolates from dogs where, depending on the markers used, isolates were typed as either host-adapted, nonzoonotic assemblages C and D or assemblage B, which has zoonotic potential [10,84]. This has very important implications as different conclusions may be inferred, depending on which marker is used and how genotyping data are interpreted. These findings also raise concerns about the interpretation of genotyping data based on single markers. As a consequence, 15% isolates among 2400 *Giardia* sequences (including those from GenBank) in the ZOOnotic Protozoa NETwork (ZOOPNET) have inconsistent typing between two markers among the markers of SSU-rRNA, β-giardin, GDH, and TPI genes, and this inconsistence was predominantly observed in humans and dogs [117].

Grouping the samples into different MLGs is further complicated by the fact that many isolates exhibit double peaks at specific positions in otherwise clean sequences. In one recent study, double peaks in chromatograms were seen in assemblage B, C, D, and E isolates but were never observed in assemblage A, F, and G isolates, which makes multilocus genotyping a highly discriminatory and useful tool in the determination of zoonotic subgroups within assemblage A [118].

7.2 METHODS

The following *Giardia* genotyping and subtyping techniques are used in the Molecular Epidemiology Laboratory at the Division of Foodborne and Waterborne Disease Prevention, Centers for Disease Control and Prevention. For fecal specimens, DNA is extracted using either the QIAamp® DNA Stool Kit (QIAGEN) [119] or the FastDNA® SPIN Kit for Soil Samples (Q-BIOgene) [120]. The latter requires the use of a FastPrep instrument, which may not be available in some laboratories.

7.2.1 Sample Preparation

7.2.1.1 Collection and Preservation

Store at least 2 g of fecal specimens containing *Giardia* cysts unpreserved at 4°C, or in 2.5% potassium dichromate solution at 4°C. If specimens are stored frozen at –20°C, proceed to DNA extraction without the washing step.

7.2.1.2 DNA Extraction from Fecal Specimen

i. Extraction of DNA using the QIAamp DNA Stool Kit
   1. Transfer ~200 μL of fecal specimens to a 2.0 mL microfuge tube.
   2. Wash at least twice with distilled water by centrifugation at 10,000×*g* for 10 min. By-pass the washing step if the fecal specimen is stored frozen.
   3. Add 66.6 μL of 1 M KOH, and 18.6 μL of 1 M Dithiothreitol (DTT) to the pellet, and mix thoroughly by stirring with a pipette tip.
   4. Incubate at 65°C for 15 min.
   5. Neutralize the alkaline with 8.6 μL of 25% HCl and 160 μL of 2 M Tris-HCl pH 8.3. Vortex the tube.
   6. Add 250 μL of phenol: chloroform: isoamyl alcohol (25:24:1) solution to the tube and vortex.
   7. Centrifuge at 500× *g* for 5 min.

8. Transfer the supernatant to a 2.0 mL microfuge tube.
9. Add 1 mL of ASL buffer from the QIAamp DNA Stool Kit (QIAGEN) to the supernatant. Incubate the mixture at 80°C for 5 min.
10. Add one InhibitEX tablet to each specimen and vortex immediately for 1 min or until the tablet is completely dissolved.
11. Follow the remaining procedures specified in the QIAamp DNA Stool Kit. Elute the DNA using 200 μL of the AE Elution Buffer, and store the extraction at –20°C.

ii. Extraction of DNA using the FastDNA SPIN Soil Kit

1. Transfer ~200 μL of fecal specimens to a 2.0 mL microfuge tube.
2. Wash at least twice with distilled water by centrifugation at 10,000×*g* for 10 min. By-pass the washing step if the fecal specimen is stored frozen.
3. Aspirate the supernatant. Use the pellet immediately for DNA extraction.
4. Resuspend the washed water or fecal pellet in 978 μL of sodium phosphate buffer from the FastDNA SPIN Soil Kit (Q-BIOgene).
5. Transfer the suspension to a Lysing Matrix E Tube.
6. Add 122 μL of MT buffer to the Lysing Matrix E Tube.
7. Secure the tube in FastPrep® Instrument (Q-BIOgene) and process for 30 s at the speed setting 5.5.
8. Centrifuge the Lysing Matrix E Tube at 10,000×*g* for 30 s.
9. Transfer supernatant to a clean tube. Add 250 μL of PPS reagent from the kit and mix by shaking the tube by hand 10 times.
10. Follow the remaining procedures specified in the FastDNA SPIN Soil Kit. Elute the DNA using 100 μL of DES (DNase/pyrogen-free water), and store the extraction at –20°C.

7.2.2 Detection Procedures

In the protocol described below, nested PCR is used in analysis of the *Giardia* TPI and GDH genes, using primers listed below. *G. duodenalis* genotypes and subtypes are determined by sequence analysis of the secondary PCR products. Fecal specimens are generally analyzed twice with 2 μL of extracted DNA. Positive (DNA of genotypes or subtypes unexpected in specimens under analysis) and negative (reagent-grade water) are used in each PCR run. The size of the primary PCR product is 605 bp for TPI and 603 bp for GDH, whereas the size of the secondary PCR product is 530 bp for TPI and 394 bp for GDH. For both primary and secondary PCR, the annealing temperature is 50°C for TPI and 55°C for GDH.

Primary TPI PCR primers:

1. Forward (TPI-F5 in Table 7.1): 5′-AAATIATGCC TGCTCGTCG-3′
2. Reverse (TPI-R5 in Table 7.1): 5'-CAAACCTTIT CCGCAAACC-3′

Secondary TPI PCR primers:

1. Forward (TPI-F6 in Table 7.1): 5′-CCCTTCATCGGI GGTAACTT-3′
2. Reverse (TPI-R6 in Table 7.1): 5′-GTGGCCACCA CICCCGTGCC-3'

Primary GDH PCR primers:

1. Forward: 5′-TTC CGT GTC CAG TAC AAC TC-3′
2. Reverse: 5′-GCC AGC TTC TCC TCG TTG AA-3′

Secondary GDH primers:

1. Forward: 5′-CGC TTC CAC CCC TCT GTC AAT-3′
2. Reverse: 5′-TGT TGT CCT TGC ACA TCT C-3′

i. Primary PCR

1. For each PCR reaction, prepare the following master mixture:

10× Perkin-Elmer PCR buffer	10 μL
dNTP (1.25 mM)	16 μL
Primary forward primer (40 ng/μL or 4 μM)	2.5 μL
Primary reverse primer (40 ng/μL or 4 μM)	2.5 μL
$MgCl_2$ (25 mM)	6 μL
Nonacetylated bovine serum albumin (10 mg/mL)	4 μL
Distilled water	56.5 μL
Taq DNA polymerase	0.5 μL
Total	98 μL

2. Add 98 μL of the master mixture to each PCR tube.
3. Add 2 μL of DNA suspension to each tube.
4. Run the following PCR program: 94°C for 3 min; 35 cycles of 94°C for 45 s, 50°C (TPI) or 55°C (GDH) for 45 s and 72°C for 1 min; 72°C for 7 min; 4°C soaking.

ii. Secondary PCR

1. For each PCR reaction, prepare the following master mixture:

10× Perkin-Elmer PCR buffer	10 μL
dNTP (1.25 mM)	16 μL
Secondary forward primer (40 ng/μL or 4 μM)	5 μL
Secondary reverse primer (40 ng/μL or 4 μM)	5 μL
$MgCl_2$ (25 mM)	6 μL
Distilled water	55.5 μL
Taq DNA polymerase	0.5 μL
Total	98 μL

2. Add 98 μL of the master mixture to each PCR tube.
3. Add 2 μL of the primary PCR reaction to each tube.
4. Run the following PCR program: 94 °C for 3 min; 35 cycles of 94°C for 45 s, 50°C (TPI) or 55°C (GDH) for 45 s and 72°C for 1 min; 72 °C for 7 min; 4°C soaking

iii. Detection of secondary PCR products

iv. Run electrophoresis on 1.5% agarose gel with 15 μL of the secondary PCR product, using the 100-bp ladder as the control.

v. DNA sequencing.

1. Clean the positive secondary PCR products of the expected size using the Montage-PCR kit (Millipore). The cleaned PCR products can be stored at −20°C.
2. The cleaned PCR products are sequenced using the forward and reverse primers used in the secondary PCR.
3. For each sequencing reaction, prepare the following using components of the BigDye® Terminator V3.1 Cycle Sequencing Kit (Applied Biosystems).

a.	BigDye Terminator Buffer	4 μL
b.	BigDye Terminator	1 μL
c.	Primer (40 ng/μL)	2 μL
d.	Reagent water	9.5 μL

4. Add 19.5 μL of the master mixture to a PCR tube.
5. Add 0.5 μL cleaned PCR products to each tube.
6. Run the following program in a PCR machine: 25 cycles of 96°C for 10 s, 50°C for 5 s, and 60°C for 4 min; 1 cycle of 4°C soaking.
7. Clean the 20 μL of sequencing reaction products using the Centrisep Spin Column (Princeton Separations).
8. Dry the cleaned sequencing products in a vacuum centrifuge. The cleaned products can be stored in a freezer.
9. Add 15 μL of BigDye Formamide to the dehydrated sequencing products.
10. Transfer the DNA suspension to the MicroAmp Optical 96-well reaction plate (Applied Biosystems).
11. Cover the plate with a 96-well Plate Septa (Applied Biosystems).
12. Heat the plate in a PCR machine at 94°C for 5 min, and cool the plate at −20°C for at least 3 min.
13. Load the reaction product onto a 3130 AB Prism Genetic Analyzer for sequencing using a proper program.

vi. Sequence analysis

1. Read the electropherograms generated by the sequencer using the ChromasPro software (www.technelysium.com.au/ChromasPro.html) or other software.
2. Aligned the TPI or GDH sequences generated with each other and reference sequences using the software ClustalX (ftp://ftp-igbmc.u-strasbg.fr/pub/ClustalX/).
3. Check the sequence alignment for sequencing accuracy using the software BioEdit (www.mbio.ncsu.edu/BioEdit/bioedit.html).
4. Recheck the electropherograms for any sequence uncertainty.
5. Determine *G. duodenalis* genotypes and subtypes based on sequence identity to reference sequences.

7.3 CONCLUSIONS AND FUTURE PERSPECTIVE

Molecular diagnostic tools have been used to provide insights into the taxonomy, host specificity, host range, and transmission routes of *Giardia* and the epidemiology of human disease [121]. These tools are essential for understanding the public health importance of different routes of transmission. Specific aspects of *Giardia* biology—including their low infectious doses, prolonged survival of cysts in moist environment, and host specificity and host adaptation—have particular pertinence in determining transmission routes. To maximize the use of molecular epidemiological tools, they must be applicable to both clinical (human and nonhuman) and environmental (including food) samples, preferably collected from the same community or area in surveillance or well-designed case–control or longitudinal cohort studies, for both genotype and subtype identifications.

An ongoing issue in *Giardia* research is the question of zoonotic transmission [5,7,122]. Data accumulated so far have shown only a limited role of zoonotic transmission in giardiasis epidemiology, especially infections caused by the assemblage A. Nevertheless, the number of human and animal studies that have used genotyping and subtyping tools is still small, which does not allow the generation of firm conclusions. This is especially true for the assemblage B, whose prevalence in animals is not well known and population structure is not clear. Clearly, infections with assemblage B are common in some animals that have long been considered sources of human infections, such as beavers and muskrats [12]. Whether there is any host adaptation within the assemblage B remains to be answered. Likewise, answers to questions on the significance of other subtypes identified at the TPI, GDH, and β-gardin loci within the assemblage A and their linkage across genetic loci will have important implications in assessing disease burdens attributable to zoonotic transmission.

Based on a clonal population structure, some authors consider the use of a single, highly polymorphic marker suitable for epidemiological investigations, with a resolution comparable to multilocus sequence typing [11]. However, clonality in *G. duodenalis* was challenged by both the heterozygocity at some genetic locus such as TPI and the observed genetic recombination. The single genetic locus approach therefore is probably dated. Investigations of the transmission of routes and risk factors and tracking infection sources source and cross-species transmission require more discriminatory fingerprinting techniques, which can identify individual isolates or clonal lineages. Nevertheless, the high occurrence of mixed infections, allelic heterozygocity, and occurrence of genetic recombination makes the development of high-resolution subtyping tools more difficult and the comparison with the available sequence data intricate. The described MLG approach using a standardized procedure for identifying *Giardia* subassemblages is very important, as it would facilitate comparison of data from different studies, which for the time being is hampered by the use of different methodologies. Hopefully, this goal will be achieved in future by thorough testing of a large number of reference strains and isolates from various sources using the same loci and the same primer sets.

ACKNOWLEDGMENTS

We acknowledge the financial support from the National Natural Science Foundation of China (Nos. 30771881 and 30928019). The findings and conclusions in this report are those of the authors and do not necessarily represent the views of the Centers for Disease Control and Prevention.

REFERENCES

1. Yason, J.A., and Rivera, W.L. Genotyping of *Giardia* duodenalis isolates among residents of slum area in Manila, Philippines. *Parasitol. Res.* 101, 681, 2007.
2. Thompson, R.C. *Giardia*sis as a re-emerging infectious disease and its zoonotic potential. *Int. J. Parasitol.* 30, 1259, 2000.
3. Hellard, M.E. et al. Prevalence of enteric pathogens among community based asymptomatic individuals. *J. Gastroenterol. Hepatol.* 15, 290, 2000.
4. Thompson, R.C. The zoonotic significance and molecular epidemiology of *Giardia* and giardiasis. *Vet. Parasitol.* 126, 15, 2004.
5. Thompson, R.C., and Monis, P.T. Variation in *Giardia*: Implications for taxonomy and epidemiology. *Adv. Parasitol.* 58, 69, 2004.
6. Appelbee, A.J., Thompson, R.C., and Olson, M.E. *Giardia* and *Cryptosporidium* in mammalian wildlife–current status and future needs. *Trends Parasitol.* 21, 370, 2005.
7. Caccio, S.M. et al. Unravelling *Cryptosporidium* and *Giardia* epidemiology. *Trends Parasitol.* 21, 430, 2005.
8. Monis, P.T. et al. Molecular systematics of the parasitic protozoan *Giardia intestinalis*. *Mol. Biol. Evol.* 16, 1135, 1999.
9. Monis, P.T. et al. Genetic diversity within the morphological species *Giardia intestinalis* and its relationship to host origin. *Infect. Genet. Evol.* 3, 29, 2003.
10. Read, C.M., Monis, P.T., and Thompson, R.C. Discrimination of all genotypes of *Giardia duodenalis* at the glutamate dehydrogenase locus using PCR-RFLP. *Infect. Genet. Evol.* 4, 125, 2004.
11. Sulaiman, I.M. et al. Triosephosphate isomerase gene characterization and potential zoonotic transmission of *Giardia duodenalis*. *Emerg. Infect. Dis.* 9, 1444, 2003.
12. Wielinga, C.M., and Thompson, R.C. Comparative evaluation of *Giardia duodenalis* sequence data. *Parasitology* 134, 1795, 2007.
13. Hopkins, R.M. et al. Ribosomal RNA sequencing reveals differences between the genotypes of *Giardia* isolates recovered from humans and dogs living in the same locality. *J. Parasitol.* 83, 44, 1997.
14. Monis, P.T., and Andrews, R.H. Molecular epidemiology: Assumptions and limitations of commonly applied methods. *Int. J. Parasitol.* 28, 981, 1998.
15. Leonhard, S. et al. The molecular characterisation of *Giardia* from dogs in southern Germany. *Vet. Parasitol.* 150, 33, 2007.
16. Ey, P.L. et al. Genetic analysis of *Giardia* from hoofed farm animals reveals artiodactyl-specific and potentially zoonotic genotypes. *J. Eukaryot. Microbiol.* 44, 626, 1997.
17. Huang, D.B., and White, A.C. An updated review on *Cryptosporidium* and *Giardia*. *Gastroenterol. Clin. North Am.* 35, 291, 2006.
18. Karanis, P., Kourenti, C., and Smith, H. Waterborne transmission of protozoan parasites: A worldwide review of outbreaks and lessons learnt. *J. Water Health* 5, 1, 2007.
19. Savioli, L., Smith, H., and Thompson, A. *Giardia* and *Cryptosporidium* join the "Neglected Diseases Initiative." *Trends Parasitol.* 22, 203, 2006.
20. Ferrer, S.R. et al. A hierarchical model for studying risk factors for childhood diarrhoea: A case–control study in a middle-income country. *Int. J. Epidemiol.* 37, 805, 2008.
21. Davies, A.P. et al. Asymptomatic carriage of protozoan parasites in children in day care centers in the United Kingdom. *Pediatr. Infect. Dis. J.*, 28, 838, 2009.
22. Giangaspero, A., Berrilli, F., and Brandonisio, O. *Giardia* and *Cryptosporidium* and public health: The epidemiological scenario from the Italian perspective. *Parasitol. Res.* 101, 1169, 2007.
23. Horman, A. et al. Meta-analysis in assessment of the prevalence and annual incidence of *Giardia* spp. and *Cryptosporidium* spp. infections in humans in the Nordic countries. *Int. J. Parasitol.* 34, 1337, 2004.
24. Almeida, A.A. et al. Genotype analysis of *Giardia* isolated from asymptomatic children in northern Portugal. *J. Eukaryot. Microbiol.* 53, S177, 2006.
25. Berrilli, F. et al. Prevalence and genotyping of human isolates of *Giardia duodenalis* from Albania. *Parasitol. Int.* 55, 295, 2006.
26. Morales-Espinoza, E.M. et al. Intestinal parasites in children, in highly deprived areas in the border region of Chiapas, Mexico. *Salud Publica Mex.* 45, 379, 2003.
27. Diaz, E. et al. Epidemiology and control of intestinal parasites with nitazoxanide in children in Mexico. *Am. J. Trop. Med. Hyg.* 68, 384, 2003.
28. Paniagua, G.L. et al. Two or more enteropathogens are associated with diarrhoea in Mexican children. *Ann. Clin. Microbiol. Antimicrob.* 6, 17, 2007.
29. Sanchez-Vega, J.T. et al. Cryptosporidiosis and other intestinal protozoan infections in children less than one year of age in Mexico City. *Am. J. Trop. Med. Hyg.* 75, 1095, 2006.
30. Silva, R.R. et al. Association between nutritional status, environmental and socio-economic factors and *Giardia lamblia* infections among children aged 6–71 months in Brazil. *Trans. R. Soc. Trop. Med. Hyg.* 103, 512, 2009.
31. Maia, M.M. et al. Intestinal parasitic infection and associated risk factors, among children presenting at outpatient clinics in Manaus, Amazonas state, Brazil. *Ann. Trop. Med. Parasitol.* 103, 583, 2009.

32. Korkes, F. et al. Relationship between intestinal parasitic infection in children and soil contamination in an Urban slum. *J. Trop. Pediatr.*, 2008.
33. de Carvalho, T.B., de Carvalho, L.R., and Mascarini, L.M. Occurrence of enteroparasites in day care centers in Botucatu (Sao Paulo State, Brazil) with emphasis on *Cryptosporidium* sp., *Giardia duodenalis* and *Enterobius vermicularis*. *Rev. Inst. Med. Trop. Sao Paulo*, 48, 269, 2006.
34. Teixeira, J.C., Heller, L., and Barreto, M.L. *Giardia duodenalis* infection: Risk factors for children living in sub-standard settlements in Brazil. *Cad Saude Publica*, 23, 1489, 2007.
35. Guignard, S. et al. Prevalence of enteroparasites in a residence for children in the Cordoba Province, Argentina. *Eur. J. Epidemiol.* 16, 287, 2000.
36. Perez Cordon, G. et al. Prevalence of enteroparasites and genotyping of *Giardia lamblia* in Peruvian children. *Parasitol. Res.* 103, 459, 2008.
37. Cabrera, M., Verastegui, M., and Cabrera, R. Prevalence of entero-parasitosis in one Andean community in the Province of Victor Fajardo, Ayacucho, Peru. *Rev. Gastroenterol. Peru*, 25, 150, 2005.
38. Chaves Mdel, P. et al. *Giardia duodenalis* prevalence and associated risk factors in preschool and school-age children of rural Colombia. *Biomedica*, 27, 345, 2007.
39. Dib, H.H., Lu, S.Q., and Wen, S.F. Prevalence of *Giardia lamblia* with or without diarrhea in South East, South East Asia and the Far East. *Parasitol. Res.* 103, 239, 2008.
40. Nematian, J., Gholamrezanezhad, A., and Nematian, E. *Giardia*sis and other intestinal parasitic infections in relation to anthropometric indicators of malnutrition: A large, population-based survey of schoolchildren in Tehran. *Ann. Trop. Med. Parasitol.* 102, 209, 2008.
41. Hamze, M. et al. Prevalence of infection by intestinal parasites in north Lebanon: 1997–2001. *East. Mediterr. Health J.* 10, 343, 2004.
42. Foronda, P. et al. Identification of genotypes of *Giardia intestinalis* of human isolates in Egypt. *Parasitol. Res.*, 103, 1177, 2008.
43. Lalle, M. et al. High genetic polymorphism among *Giardia duodenalis* isolates from Sahrawi children. *Trans. R. Soc. Trop. Med. Hyg.*, 103, 834, 2009.
44. El Kettani, S., Azzouzi, E.M., and Maata, A. Prevalence of *Giardia intestinalis* in a farming population using sewage water in agriculture, Settat, Morocco. *Med. Mal. Infect.* 36, 322, 2006.
45. Perch, M. et al. Seven years' experience with *Cryptosporidium parvum* in Guinea-Bissau, West Africa. *Ann. Trop. Paediatr.* 21, 313, 2001.
46. Gbakima, A.A. et al. Intestinal protozoa and intestinal helminthic infections in displacement camps in Sierra Leone. *Afr. J. Med. Sci.* 36, 1, 2007.
47. Adams, V.J. et al. Paradoxical helminthiasis and giardiasis in Cape Town, South Africa: Epidemiology and control. *Afr. Health Sci.* 5, 276, 2005.
48. Pollock, K.G. et al. *Giardia* surveillance in Scotland, 1988–2003. *Eur. J. Clin. Microbiol. Infect. Dis.* 24, 571, 2005.
49. Hoque, E. et al. A descriptive epidemiology of giardiasis in New Zealand and gaps in surveillance data. *N Z Med. J.* 117, U1149, 2004.
50. Yoder, J.S., and Beach, M.J. *Giardia*sis surveillance–United States, 2003–2005. *MMWR Surveill. Summ.* 56, 11, 2007.
51. Hlavsa, M.C., Watson, J.C., and Beach, M.J. *Giardia*sis surveillance–United States, 1998–2002. *MMWR Surveill. Summ.* 54, 9, 2005.
52. Furness, B.W., Beach, M.J., and Roberts, J.M. *Giardia*sis surveillance–United States, 1992–1997. *MMWR CDC Surveill. Summ.* 49, 1, 2000.
53. Saksirisampant, W. et al. Intestinal parasitic infestations among children in an orphanage in Pathum Thani province. *J. Med. Assoc. Thai.* 86, S263, 2003.
54. Mohammed Mahdy, A.K. et al. *Giardia intestinalis* genotypes: Risk factors and correlation with clinical symptoms. *Acta Trop.* 112, 67, 2009.
55. Sharma, B.K. et al. Prevalence of intestinal parasitic infestation in schoolchildren in the northeastern part of Kathmandu Valley, Nepal. *Southeast Asian J. Trop. Med. Public Health* 35, 501, 2004.
56. Odoi, A. et al. Geographical and temporal distribution of human giardiasis in Ontario, Canada. *Int. J. Health Geogr.* 2, 5, 2003.
57. Wongstitwilairoong, B. et al. Intestinal parasitic infections among pre-school children in Sangkhlaburi, Thailand. *Am. J. Trop. Med. Hyg.* 76, 345, 2007.
58. Vargas, M. et al. Etiology of diarrhea in children less than five years of age in Ifakara, Tanzania. *Am. J. Trop. Med. Hyg.* 70, 536, 2004.
59. Xiao, L., and Fayer, R. Molecular characterization of species and genotypes of *Cryptosporidium* and *Giardia* and assessment of zoonotic transmission. *Int. J. Parasitol.* 38, 1239, 2008.
60. Haque, R. et al. *Giardia* assemblage A infection and diarrhea in Bangladesh. *J. Infect. Dis.* 192, 2171, 2005.
61. Eligio-Garcia, L., Cortes-Campos, A., and Jimenez-Cardoso, E. Genotype of *Giardia intestinalis* isolates from children and dogs and its relationship to host origin. *Parasitol. Res.* 97, 1, 2005.
62. Chin, A.C. et al. Strain-dependent induction of enterocyte apoptosis by *Giardia lamblia* disrupts epithelial barrier function in a caspase-3-dependent manner. *Infect. Immun.* 70, 3673, 2002.
63. Scott, K.G. et al. Intestinal infection with *Giardia* spp. reduces epithelial barrier function in a myosin light chain kinase-dependent fashion. *Gastroenterology* 123, 1179, 2002.
64. Buret, A.G. et al. *Giardia lamblia* disrupts tight junctional ZO-1 and increases permeability in non-transformed human small intestinal epithelial monolayers: Effects of epidermal growth factor. *Parasitology*, 125, 11, 2002.
65. Nash, T.E. Surface antigenic variation in *Giardia lamblia*. *Mol. Microbiol.*, 45, 585, 2002.
66. Touz, M.C., Conrad, J.T., and Nash, T.E. A novel palmitoyl acyl transferase controls surface protein palmitoylation and cytotoxicity in *Giardia lamblia*. *Mol. Microbiol.* 58, 999, 2005.
67. Hiatt, R.A., Markell, E.K., and Ng, E. How many stool examinations are necessary to detect pathogenic intestinal protozoa? *Am. J. Trop. Med. Hyg.* 53, 36, 1995.
68. Mank, T.G. et al. Sensitivity of microscopy versus enzyme immunoassay in the laboratory diagnosis of giardiasis. *Eur. J. Clin. Microbiol. Infect. Dis.* 16, 615, 1997.
69. Aldeen, W.E. et al. Comparison of nine commercially available enzyme-linked immunosorbent assays for detection of *Giardia lamblia* in fecal specimens. *J. Clin. Microbiol.* 36, 1338, 1998.
70. Garcia, L.S., and Shimizu, R.Y. Evaluation of nine immunoassay kits (enzyme immunoassay and direct fluorescence) for detection of *Giardia lamblia* and *Cryptosporidium parvum* in human fecal specimens. *J. Clin. Microbiol.* 35, 1526, 1997.
71. Zimmerman, S.K., and Needham, C.A. Comparison of conventional stool concentration and preserved-smear methods with Merifluor *Cryptosporidium/Giardia* direct

immunofluorescence assay and ProSpecT *Giardia* EZ microplate assay for detection of *Giardia lamblia*. *J. Clin. Microbiol.* 33, 1942, 1995.
72. Chan, R. et al. Evaluation of a combination rapid immunoassay for detection of *Giardia* and *Cryptosporidium* antigens. *J. Clin. Microbiol.* 38, 393, 2000.
73. Aziz, H. et al. A comparison study of different methods used in the detection of *Giardia lamblia*. *Clin. Lab. Sci.* 14, 150, 2001.
74. Weiss, J.B., van Keulen, H., and Nash, T.E. Classification of subgroups of *Giardia lamblia* based upon ribosomal RNA gene sequence using the polymerase chain reaction. *Mol. Biochem. Parasitol.* 54, 73, 1992.
75. Ey, P.L. et al. *Giardia intestinalis*: Detection of major genotypes by restriction analysis of gene amplification products. *Int. J. Parasitol.* 23, 591, 1993.
76. van Keulen, H. et al. A three nucleotide signature sequence in small subunit rRNA divides human *Giardia* in two different genotypes. *J. Eukaryot. Microbiol.* 42, 392, 1995.
77. Baruch, A.C., Isaac-Renton, J., and Adam, R.D. The molecular epidemiology of *Giardia lamblia*: A sequence-based approach. *J. Infect. Dis.* 174, 233, 1996.
78. Monis, P.T. et al. Molecular genetic analysis of *Giardia intestinalis* isolates at the glutamate dehydrogenase locus. *Parasitology*, 112, 1, 1996.
79. Caccio, S.M., De Giacomo, M., and Pozio, E. Sequence analysis of the beta-giardin gene and development of a polymerase chain reaction-restriction fragment length polymorphism assay to genotype *Giardia duodenalis* cysts from human faecal samples. *Int. J. Parasitol.* 32, 1023, 2002.
80. Bertrand, I., Albertini, L., and Schwartzbrod, J. Comparison of two target genes for detection and genotyping of *Giardia lamblia* in human feces by PCR and PCR-restriction fragment length polymorphism. *J. Clin. Microbiol.* 43, 5940, 2005.
81. Aloisio, F. et al. Severe weight loss in lambs infected with *Giardia duodenalis* assemblage B. *Vet. Parasitol.* 142, 154, 2006.
82. Sulaiman, I.M. et al. Distribution of *Giardia duodenalis* genotypes and subgenotypes in raw urban wastewater in Milwaukee, Wisconsin. *Appl. Environ. Microbiol.* 70, 3776, 2004.
83. Feng, Y. et al. High intragenotypic diversity of *Giardia duodenalis* in dairy cattle on three farms. *Parasitol. Res.* 103, 87, 2008.
84. Traub, R.J. et al. Epidemiological and molecular evidence supports the zoonotic transmission of *Giardia* among humans and dogs living in the same community. *Parasitology*, 128, 253, 2004.
85. Amar, C.F. et al. Sensitive PCR-restriction fragment length polymorphism assay for detection and genotyping of *Giardia duodenalis* in human feces. *J. Clin. Microbiol.* 40, 446, 2002.
86. Robertson, L.J. et al. Application of genotyping during an extensive outbreak of waterborne giardiasis in Bergen, Norway, during autumn and winter 2004. *Appl. Environ. Microbiol.* 72, 2212, 2006.
87. Monis, P.T. The importance of systematics in parasitological research. *Int. J. Parasitol.* 29, 381, 1999.
88. Abe, N., Kimata, I., and Iseki, M. Identification of genotypes of *Giardia intestinalis* isolates from dogs in Japan by direct sequencing of the PCR amplified glutamate dehydrogenase gene. *J. Vet. Med. Sci.* 65, 29, 2003.
89. Itagaki, T. et al. Genotyping of *Giardia intestinalis* from domestic and wild animals in Japan using glutamete dehydrogenase gene sequencing. *Vet. Parasitol.* 133, 283, 2005.
90. Abe, N., Read, C., Thompson, R.C., and Iseki, M. Zoonotic genotype of *Giardia intestinalis* detected in a ferret. *J. Parasitol.* 91, 179, 2005.
91. Homan, W.L. et al. Characterization of *Giardia duodenalis* by polymerase-chain-reaction fingerprinting. *Parasitol. Res.* 84, 707, 1998.
92. Matsubayashi, M., Kimata, I., and Abe, N. Identification of genotypes of *Giardia intestinalis* isolates from a human and calf in Japan. *J. Vet. Med. Sci.* 67, 337, 2005.
93. van der Giessen, J.W. et al. Genotyping of *Giardia* in Dutch patients and animals: A phylogenetic analysis of human and animal isolates. *Int. J. Parasitol.* 36, 849, 2006.
94. Souza, S.L. et al. Molecular identification of *Giardia duodenalis* isolates from humans, dogs, cats and cattle from the state of Sao Paulo, Brazil, by sequence analysis of fragments of glutamate dehydrogenase (gdh) coding gene. *Vet. Parasitol.* 149, 258, 2007.
95. van Keulen, H. et al. Presence of human *Giardia* in domestic, farm and wild animals, and environmental samples suggests a zoonotic potential for giardiasis. *Vet. Parasitol.* 108, 97, 2002.
96. Santin, M. et al. *Cryptosporidium*, *Giardia* and *Enterocytozoon bieneusi* in cats from Bogota (Colombia) and genotyping of isolates. *Vet. Parasitol.* 141, 334, 2006.
97. O'Handley, R.M. et al. Prevalence and genotypic characterisation of *Giardia* in dairy calves from Western Australia and Western Canada. *Vet. Parasitol.* 90, 193, 2000.
98. Trout, J.M., Santin, M., and Fayer, R. Prevalence of *Giardia duodenalis* genotypes in adult dairy cows. *Vet. Parasitol.* 147, 205, 2007.
99. Appelbee, A.J. et al. Prevalence and genotyping of *Giardia duodenalis* from beef calves in Alberta, Canada. *Vet. Parasitol.* 112, 289, 2003.
100. Mahbubani, M.H. et al. Detection of *Giardia* in environmental waters by immuno-PCR amplification methods. *Curr. Microbiol.* 36, 107, 1998.
101. Mahbubani, M.H. et al. Differentiation of *Giardia duodenalis* from other *Giardia* spp. by using polymerase chain reaction and gene probes. *J. Clin. Microbiol.* 30, 74, 1992.
102. Volotao, A.C. et al. Genotyping of *Giardia duodenalis* from human and animal samples from Brazil using beta-giardin gene: A phylogenetic analysis. *Acta Trop.* 102, 10, 2007.
103. Lalle, M. et al. Genetic heterogeneity at the beta-giardin locus among human and animal isolates of *Giardia duodenalis* and identification of potentially zoonotic subgenotypes. *Int. J. Parasitol.* 35, 207, 2005.
104. Ey, P.L. et al. Comparison of genetic groups determined by molecular and immunological analyses of *Giardia* isolated from animals and humans in Switzerland and Australia. *Parasitol. Res.* 82, 52, 1996.
105. Hunt, C.L., Ionas, G., and Brown, T.J. Prevalence and strain differentiation of *Giardia intestinalis* in calves in the Manawatu and Waikato regions of North Island, New Zealand. *Vet. Parasitol.* 91, 7, 2000.
106. Sogin, M.L. et al. Phylogenetic meaning of the kingdom concept: An unusual ribosomal RNA from *Giardia lamblia*. *Science* 243, 75, 1989.
107. van Keulen, H. et al. Nucleotide sequence of the 5.8S and large subunit rRNA genes and the internal transcribed spacer and part of the external spacer from *Giardia ardeae*. *Nucleic Acids Res.* 19, 6050, 1991.
108. Lu, S.Q., Baruch, A.C., and Adam, R.D. Molecular comparison of *Giardia lamblia* isolates. *Int. J. Parasitol.* 28, 1341, 1998.
109. Berrilli, F. et al. Genotype characterisation of *Giardia duodenalis* isolates from domestic and farm animals by SSU-rRNA gene sequencing. *Vet. Parasitol.* 122, 193, 2004.

110. Fayer, R. et al. Detection of *Cryptosporidium felis* and *Giardia duodenalis* assemblage F in a cat colony. *Vet. Parasitol.* 140, 44, 2006.
111. Andrews, R.H. et al. *Giardia intestinalis*: Electrophoretic evidence for a species complex. *Int. J. Parasitol.* 19, 183, 1989.
112. Santin, M. et al. First report of *Giardia* in coyotes (*Canis latrans*). *J. Eukaryot. Microbiol.* 50, S709, 2003.
113. Caccio, S.M., and Ryan, U. Molecular epidemiology of giardiasis. *Mol. Biochem. Parasitol.* 160, 75, 2008.
114. Caccio, S.M. et al. Multilocus genotyping of *Giardia duodenalis* reveals striking differences between assemblages A and B. *Int. J. Parasitol.*, 2008.
115. Gelanew, T. et al. Molecular characterization of human isolates of *Giardia duodenalis* from Ethiopia. *Acta Trop.* 102, 92, 2007.
116. Geurden, T. et al. Multilocus genotyping of *Cryptosporidium* and *Giardia* in non-outbreak related cases of diarrhoea in human patients in Belgium. *Parasitology*, 136, 1161, 2009.
117. Caccio, S.M., and Sprong, H. *Giardia duodenalis*: Genetic recombination and its implications for taxonomy and molecular epidemiology. *Exp. Parasitol.*, 124, 107, 2009.
118. Lebbad, M. et al. From mouse to moose: Multilocus genotyping of *Giardia* isolates from various animal species. *Vet. Parasitol.*, 168, 231, 2009.
119. Xiao, L. et al. Host adaptation and host-parasite co-evolution in *Cryptosporidium*: Implications for taxonomy and public health. *Int. J. Parasitol.* 32, 1773, 2002.
120. Jiang, J. et al. Development of procedures for direct extraction of *Cryptosporidium* DNA from water concentrates and for relief of PCR inhibitors. *Appl. Environ. Microbiol.* 71, 1135, 2005.
121. Smith, H.V. et al. Tools for investigating the environmental transmission of *Cryptosporidium* and *Giardia* infections in humans. *Trends Parasitol.* 22, 160, 2006.
122. Hunter, P.R., and Thompson, R.C. The zoonotic transmission of *Giardia* and *Cryptosporidium*. *Int. J. Parasitol.* 35, 1181, 2005.

8 Leishmania

Clotilde Marín, Silvia S. Longoni, and Manuel Sánchez-Moreno

CONTENTS

8.1 INTRODUCTION

Leishmaniasis is a group of infectious diseases caused by protozoan parasites of the genus *Leishmania*, of which there are more than 20 species that affect humans. Several clinical syndromes are subsumed under the term leishmaniasis—visceral, cutaneous, and mucocutaneous leishmaniasis—which result from the replication of the parasite within macrophages in the mononuclear phagocyte system, dermis, and nasopharyngeal tissue, respectively. These parasitoses are widely distributed in all continents except in Antarctica, representing a grave problem for world health, as mobility and mortality are rising [1].

Knowledge of these New World diseases dates to pre-Columbian times, as shown in the sculpture of the Mochica (330 BC–500 AD) and Chimú (1000–1400 AD) cultures. The first clinical descriptions of the diseases appeared during the Spanish colonization in texts of the period (Fernando de Oviedo, 1535, Pedro Pizarro, 1571, and Fernando de Santillán, 1572), which mentioned the risk suffered by farmers who returned from the Andes with skin ulcers, known as the valley sickness or Andean disease. In the eighteenth century, Cosme Bueno and Hipólito Ruiz described for the first time the role of the insects in transmitting the disease. These authors mentioned a corrosive blister that was difficult to cure, called "uta," which appeared on the face [2]. Alexander Russel (1756) masterfully described the clinical form of cutaneous leishmaniasis of the striking scars of a Turkish patient, calling it the "Aleppo boil." Throughout history, leishmaniasis has received different names: Jericho button, Delhi boil, salek, and uta [3].

In 1901, the disease began to be known as leishmaniasis when William Boog Leishman identified ovoid organisms extracted from the spleen of an Irish soldier who had died from a fever called "dum dum" (name of a town near Calcutta, considered until then unhealthy). These organisms were initially described as Trypanosomes, but later Charles Donovan (1903) confirmed the discoveries of Leishman and identified the pathogen as a separate organism and called it *Leishmania donovani*.

Currently, leishmaniasis is endemic to 98 countries where there are an estimated 12 million people infected and 350 million at risk, with some 2 million new cases recorded each year. This disease constitutes a grave public-health problem which levies heavy psychological, sociocultural, and economic costs. Nevertheless, these data are not accurate, because in 33 countries the announcement of the disease is obligatory and therefore most of the cases are not registered. In recent years, leishmaniasis has been considered an emerging group of diseases, given the increasing number of patients, due to climatic change, population migration, resistance of mosquitoes to insecticides, the difficulty of eliminating infection in animals acting as reservoirs, and coinfection with the HIV.

An estimated 250,000 people in the state of Bihar, India have contracted visceral leishmaniasis since 1993 while in southern Sudan, from 1998 to 2003, around 100,000 persons have died (59,000 in 2001 alone) [4,5]. In temperate countries, visceral leishmaniasis is located in the Mediterranean basin (Spain, Italy, Greece, France, Croatia, Serbia, Montenegro, Portugal, Turkey, and Macedonia), but it is rare to find a person with the disease, and the few cases registered involve children [4,5]. Rather, it is a typical disease in developing countries, with rural transmission in areas far from the urban suburbs, where there are poor sanitary and socioeconomic conditions, where there is no diagnosis of the disease, either, for lack of medical attention. In India, for example, 88% of the persons who contract the disease earn less than $2 per day. The World Health Organization (WHO) considers leishmaniasis one of the most important tropical diseases, together with malaria, schistosomiasis, filariasis, trypanosomiasis, and leprosy, and therefore it is included in the WHO/TDR program supporting the search for new prevention, diagnosis, and treatment techniques.

8.1.1 Classification, Morphology, Biology, and Epidemiology

8.1.1.1 Classification

Leishmania classification was at first based on etiological criteria such as the vector, geographic distribution, tropism, antigen properties, and clinical manifestations, for example, *Leishmania (Viannia) guyanensis*, isolated in Guyana; *Leishmania (V.) peruviana*, isolated in Peru; or *Leishmania (L.) infantum*, isolated in a child from Tunisia. This was because morphologically it is impossible to distinguish the different species, but biochemical and molecular analyses demonstrated that the pathological and geographical criteria are often inadequate and that other criteria such as the polymorphism patterns reflected by the DNA markers of the kinetoplast (DNAk), proteins or antigens, are being used to classify the genus *Leishmania*. New methods for detection, isolation, and genetic identification have resulted in a greater number of species described. Currently, 30 species are known and about 20 of these are pathogenic for humans.

Leishmania belongs to the subkingdom of Protozoa, the order Kinetoplastida, the family Trypanosomatidae, and the genus *Leishmania*, which in turn is divided into two subgenera: *Leishmania* and *Viannia*. In the Old World, only the subgenus *Leishmania* appears, and includes four complexes: complex donovani, which includes the species *L. (L.) donovani*, *L. (L.) infantum*, and *Leishmania (L.) archibaldi;* the tropical complex, with *Leishmania (L.) tropica*, and *Leishmania. (L.) killicki*; the major complex; and the aethiopica complex, the latter two with a representative species each: *Leishmania (L.) major* and *Leishmania (L.) aethiopica*, respectively. Meanwhile, in the New World, two subgenera are found: the subgenus *Viannia*, with four complexes: the Brazilian complex, which contains the species *Leishmania (V.) braziliensis*, *L. (V.) peruviana*; the guayanian complex, with the species. *(V.) panamensis*, *Leishmania (V.) guayanensis*, and *Leishmania (V.) shawi*; the naiffi complex, with only *Leishmania (V.) naiffi*; and the laisoni complex, with only the species *Leishmania (V.) lansoni*. Meanwhile, the subgenus *Leishmania* in the Americas is represented by two complexes: the Mexican complex, represented by *Leishmania (L.) mexicana*, *Leishmania (L.) amazonensis*, *Leishmania (L.) garnhami*, *Leishmania (L.) venezuelensis*, *Leishmania (L.) forattinii*, and *Leishmania (L.) pifanoi*; and the donovani complex. *(L.) chagasi*.

Furthermore, there are other species that have not yet been assigned to any complex: *Leishmania (V.) colombiensis*, *Leishmania (V.) equatoriensis*, and *Leishmania (L.) aristidesi*, *Leishmania (L.) enrietti*, *Leishmania (L.) deanei*, and *Leishmania (L.) hertigi* [6,7].

8.1.1.2 Morphology and Biology

Leishmania is a heteroxene parasite that, during its life cycle, presents two morphologies adapted to different environments in which it must survive [8]. The free extracellular form is the promastigote, which develops in the digestive tract of the phlebotome and is the infective form for the vertebrate host. It has an elongated form of about 15–20 μm in length and has a free flagellum in its anterior part which can measure twice the body length of the parasite. Electron microscope views reveal a central nucleus, ribosomes, endoplasmic reticulum, Golgi body, a single mitochondrion, and a kinetoplast, which appears as an electrodense body in the anterior zone [9,10]. The intracellular form, aflagellated, is the amastigote, found in the parasitized tissues of the vertebrate hosts, generally inside the cells of the mononuclear system (macrophages). The amastigotes are usually surrounded by a vacuole (parasitophore vacuole), which is formed after the fusion of the lysosome and the phagosome once the parasite has been internalized [11]. These are ovoid forms measuring 2–5 μm in diameter, with a central nucleus and an elongated kinetoplast, with a flagellar sack and a vestigial flagellum [11,12] (Figure 8.1).

The vector of this parasites is the sandfly, a small Diptera of the family *Psycosidae*, subfamily *Phlebotominae*, which measures some 2–3.5 mm long and is yellow/brown in color, has lanceolate wings, and it entire body is covered with tiny hairs. There are more than 1000 species belonging to the family *Phlebotominae* throughout the world, with an estimated 70 vectors. In the Old World, the genus *Phlebotomus* appears, while in the New World there are 150 species in the genus *Lutzomyia*. Only female phlebotomes are haematophagous, since they need blood to complete their reproductive cycle [13]. During the development inside the sandfly, the parasite adopts various morphologies that are intermediate between the flagellate and aflagellate form. When the sandfly ingests blood of an infected individual, the development of the parasite begins within the digestive apparatus. During an initial period of 6–9 days (depending on the species), *Leishmania* differentiates in its different stages (flagellate–aflagellate–flagellate) until changing into metacyclic promastigotes. The transformation process of the amastigotes into metacyclic promastigotes is known as metacyclogenesis [14]. Approximately from the seventh day after ingestion, the first infective forms

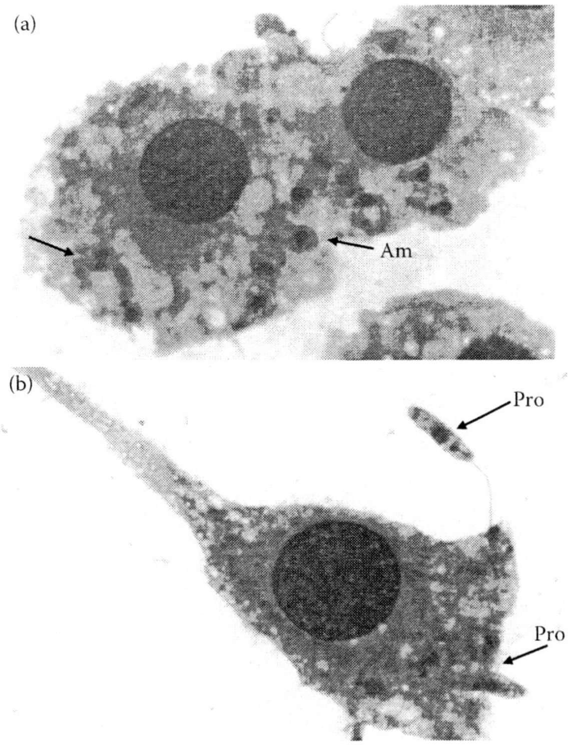

FIGURE 8.1 Macrophages infected by *L. infantum*: (a) Intracellular amastigotes (Am), (b) promastigote (Pro).

appear, and the metacyclic promastigotes are released from the epithelium of the insect after undergoing a change in the sugars on their surface [15,16]. The metacyclic promastigotes are transmitted to the mammal host through the bite of the phlebotome vector. Once in the interior of the vertebrate, the parasite has the capacity to evade the defence barriers of the host and even benefit from them. The promastigotes manage to resist lysis due to the great number of long lipophosphoglycan (LPG) molecules in its membrane (up to a million copies per cell). In addition, these molecules enable adhesion to the CR3 receptors of the macrophage membranes, facilitating phagocytosis of the promastigote [17]. The circulating parasitized monocytes are ingested by another sandfly, in the intestine of which the amastigotes are released and resume the form of promastigotes. After several days, they attain infective capacity, now near the proboscis, remaining there, and ready to be inoculated into a vertebrate host and thus completing the cycle.

At times, an infection occurs without the active participation of the vector. That is, the visceral form can be transmitted directly by a blood transfusion or organ transplant, whereas the cutaneous form can be transmitted by needles or syringes containing infective material [13].

Leishmaniasis is a typically anthropozoonotic disease. Most types of the leishmaniasis are rural or peri-urban zoonosis, except for the only form of urban anthroponotic leishmaniasis caused by *L. (L.) tropica* present only in the New World.

8.1.1.3 Epidemiology

Leishmaniasis both in its cutaneous as well as visceral form has been reported in Europe, Asia, Africa, and America. More than 90% of the cases of cutaneous leishmaniasis take place in Iran, Afghanistan, Syria, Saudi Arabia, Brazil, and Peru. More than 90% of the cases of visceral leishmaniasis occur in Bangladesh, Brazil, India, and Sudan. In former years, the cases of leishmaniasis remained constant and stable, but currently, the risk patterns to contract the disease have changed [13]. The factors that increase the incidence of the disease are caused primarily by humans themselves, as for example, migration, deforestation, urbanization, agricultural development, and changes in the susceptibility of the hosts to infection such as malnutrition and especially the increase in immune-depressed humans, mainly HIV. To date, the greatest prevalence of *Leishmania*–HIV coinfections occurs in the Mediterranean area, with more than 2000 cases, of which 90% were in Spain, France, Italy, and Portugal. In any case, it is considered that in southern Europe the coinfection of visceral leishmaniasis–HIV is not as high as in other countries of the world due to good highly active antiretroviral therapy (HAART) [13,18].

There are different eipidemiological cycles: (1) a primitive or silvatic cycle (infection in the human is accidental and the transmission occurs in the wild focal point), for example, *L. (V.) braziliensis*; (2) a secondary or peridomestic cycle (the reservoir is a peridomestic or domestic animal, and the parasite is transmitted to the humans by anthropophilic mosquitoes), for example, *L. (L.) infantum*; and (3) tertiary cycle, strictly anthroponotic, in which the animal reservoir has disappeared (or still has not been identified) and the mosquito vectors are totally anthroponotic, for example, *L. (L.) donovani*. *Leishmania* spp. are parasites with great success, capable of infecting a great variety of orders of mammals which are potential reservoirs: rodents, canids, marsupials, primitive ungulates, and primates [19]. Humans are possible hosts, but in most cases are accidental hosts, especially patients that are tourists and more often military and workers who enter the wild habitat of the parasite. The only definitive host known to date is the dog, in the case of infection by *L. infantum*, and all cases of infection in dogs by other *Leishmania* species are considered accidental.

8.1.2 Clinical Features and Pathogenesis

The disease is manifested in at least four main clinical forms: cutaneous, diffuse cutaneous, mucocutaneous, and visceral. All the forms always begin with the bite of a sandfly that inoculates promastigotes, which involve only monocytes for their multiplication [20].

The cutaneous disease (typical of the Old World) is commonly ulcerous lesions, although the pathology is usually characteristic of each *Leishmania* species and normally disappears spontaneously. The lesion usually appears at the site of the mosquito bite and at times, though rarely, satellite lesions are visible around the main sore. First, a papule forms by the inflammatory reaction with hyperplasia of the epithelium and necrosis of the dermis, provoking ulceration of the sore in six months. The ulcers usually have the aspect of a crater, rounded, painless, with definite borders, and hyperemic. These ulcers usually become infected with bacteria present in the environment, but when there are no coinfections, a clear exudate (not purulent) appears. The immunological response of the host determines the growth of the ulcer and the time of spontaneous cure (6 months to 3 years) [20]. An extreme case occurs if the cutaneous infection is complicated by multiple lesions that cannot scar, producing stigmas for life, which are called leishmaniasis relapses; these are very difficult to treat, long lasting, and destructive in the sense of disfiguring the patient.

Diffuse cutaneous leishmaniasis (also called Uta in the New World) affects immunodepressed individuals and its importance is due to lesions that never disappear spontaneously and relapses tend to occur after treatment. Thus, this is considered a special problem of public health. The lesions do not present well-defined borders but rather blend with normal skin. Isolated or grouped nodules, macules, papules, infiltrated plaques, and in some cases verrugous lesions similar to leprosy appear.

The mucocutaneous form (also known as Espundia in the New World) causes extensive destruction in the oro-nasal cavities and pharynx with horrendous disfigurement and mutilation of the face, causing deep lifelong anguish for the patient. The mucosal lesions are usually secondary due to the migration of parasites of the primary lesion (cutaneous form) in the zone of the mucosas. The period of incubation in the muco-nasal region is not very clear, though most cases usually develop 5 years after the spontaneous curing of the primary lesion, but mucocutaneous lesions have been reported after 20 or 30 years of infection.

Visceral leishmaniasis (Mediterranean, Chinese, American, Indian, and African form, known as kala-azar) is the most serious form, since it can compromise the life of the patient if not treated. It is characterized by undulating fever, weight loss, splenomegalia, and or anemia. After recuperation, patients can develop a chronic cutaneous form called post-kala-azar dermal leishmaniasis (PKDL), which generally requires long and expensive treatment. PKDL is normally a consequence of kala-azar treatment, although some cases have been related without the history of kala-azar. PKDL normally appears two years before the complete cure of the visceral infection and begins with the aspect of spots on the skin similar to freckles. It appears that the different clinical forms are closely related to the immune response of the host and the nature of the pathogenic species in question. How *Leishmania* spp. causes disease in humans and why the clinical symptoms are so variable continue to be an enigma [20].

8.1.3 Diagnosis

For a confident diagnosis of leishmaniasis, it is first necessary to consider the epidemiological antecedents, as for example, the origin of the patient, previous residences, duration of stays in or visits to endemic areas of leishmaniasis, presence of prior cutaneous lesions that could have been catalogued as leishmaniasis or not, and so on [21]. After considering all these aspects, the clinical diagnosis can be examined to help define the type of leishmaniasis. Finally, as definitive confirmation, a laboratory analysis is performed.

It is crucial to make a careful differential diagnosis with other nosological entities that can produce similar lesions. The place of residence of the patient should be taken into account as well as the pathologies most frequent in that zone that could be confused with leishmaniasis. The differential diagnosis for visceral leishmaniasis includes, among other diseases: malaria, tropical splenomegaly syndrome, schistosomiasis, cirrhosis with portal hypertension, African trypanosomiasis, milliary tuberculosis, brucellosis, typhoid fever, bacterial endocarditis, histoplasmosis, malnutrition, lymphoma, and leukemia. Also, many skin diseases are frequently misdiagnosed as early lesions of cutaneous leishmaniasis. Some of the diseases that should be distinguished from cutaneous leishmaniasis are tropical ulcers from other causes: impetigo, infected insect bites, leprosy, lupus vulgaris, yaws, blastomycosis, and skin cancer [22]. The primary lesions of mucocutaneous leishmaniasis are usually confused with paracoccidioidomycosis, polymorphic reticulosis, Wegener's granulomatosis, lymphoma, histoplasmosis, yaws, tuberculosis, nasopharyngeal carcinoma, and other lesions [22,23].

8.1.3.1 Conventional and Serological Techniques

The definitive diagnosis of leishmaniasis requires the demonstration of the parasite, which can be observed in the amastigote form, in samples from lesions, and/or in its promastigote form when isolated from cultures. The presence of the parasite can be demonstrated through field smears, culturing, histopathology, and/or inoculation in animals with a biopsy sample or tissue aspirated from the spleen or marrow. Sensitivity is great when the spleen is aspirated, but extremely difficult for the problem of hemorrhages. Occasionally, the amastigotes have been demonstrated in a liver biopsy, aspirated from lymph nodules and buffy-coat smears, particularly in cases of *Leishmania*–HIV coinfection [22]. However, this diagnosis is often difficult to establish, especially in very chronic lesions. Many other methods for demonstrating parasites such as histochemical and immunohistochemical examination of smears have been described [24]. Culture-based diagnosis of mucocutaneous leishmaniasis has very low sensitivity because the organisms are often scarce. The individual sensitivities of the methods for patients and Montenegro-positive healthy controls were: histopathology 14% and 16%; impression smear 19% and 21%; dermal scraping 22% and 26%; aspirate-culture 58% and 64%; aspirate-hamster 38% and 41%; biopsy-culture 50% and 55%; and biopsy-hamster 52% and 57%, respectively. The sensitivity is slightly better for younger lesions than for lesions older than 6 months [25].

The main problem with conventional smears or culture techniques is low sensitivity, particularly in detecting the hidden and subclinical infections. These techniques are also cumbersome and time consuming and not suitable for the field. The serological diagnosis is based on the presence of specific humoral response, as in cases of visceral leishmaniasis or cell-mediated immune response, as in cases of cutaneous and mucocutaneous leishmaniasis [26]. A wide range of serological methods varying in sensitivity and specificity are available for the diagnosis of visceral leishmaniasis. However, this type of test is rarely used to diagnose cutaneous or mucocutaneous leishmaniasis, given that the number of antibodies circulating against the parasite is low and therefore its sensitivity is variable [27]. The specificity can also vary, especially in areas where other parasites coexist and can give a crossed reaction, especially *Trypanosoma cruzi*.

These serological methods can be grouped into nonspecific and specific tests: the nonspecific tests have been used in a routine way, but their positive results are not definitive [22]. Indirect hemagglutination (IHA), counter-current immunoelectrophoresis (CCIEP), and immunodiffusion (ID) are some examples of these nonspecific tests. In addition, they are slow, with low sensitivity and specificity, and therefore they are not commonly used [25]. Notable specific tests include those mentioned below.

8.1.3.1.1 Montenegro Reaction Test or Leishmanin Skin Test (MST or LST)

Delayed hypersensitivity is characteristic of the cutaneous forms of human leishmaniasis and can be measured by this type of test. Crossed reaction with Chagas disease is unknown, although such reaction has been found in cases of glandular tuberculosis and lepromatous leprosy. Generally, this test has been used as an indicator of the prevalence of cutaneous and mucocutaneous leishmaniasis in humans and in animal populations, the main advantages being their simplicity of use and their high sensitivity and specificity [28]. This type of test, although has had some success, is not strongly recommended for visceral leishmaniasis [29,30].

8.1.3.1.2 Indirect Fluorescent Antibody Test

Indirect fluorescence (IFAT) is another of the most sensitive methods available at present. It is based on the detection of antibodies, the presence of which has been demonstrated in the early phase of infection and which are practically undetectable at 6 or 9 months after the patient is cured. If the antibodies persist at a low titering, it is a good indication of a possible relapse of the disease. Crossed reaction often takes place with sera from Chagas patients, although this problem can usually be overcome by using amastigotes instead of promastigotes as the antigen [31]. Although this test is more sensitive (96%) and more specific (98%) than ELISA with soluble antigen, it is slower and it is not applicable in field studies [31].

8.1.3.1.3 Direct Agglutination Test

This is a highly sensitive test, not expensive, and is very simple to conduct, ideal both for field studies and the laboratory. A complete parasite extract can be used, stained promastigotes both in suspension as well as lyophilized. The lyophilized forms are very stable at the ambient temperature, facilitating the use of direct agglutination test (DAT) in field studies. However, the main disadvantage of DAT is that the incubation time is too long (18 h) and needs the use of a series of dilutions and sera [32]. Recently, the researchers who figured out this test developed a fast agglutination-screening test (FAST) for quickly detecting (<3 h) anti-*Leishmania* antibodies in serum and blood samples collected on filter paper. In addition, FAST uses a single-serum dilution. This test has been evaluated for the diagnosis of cutaneous leishmaniasis using promastigotes of *L. major* and has proved to be highly specific and sensitive (90.5%) [33].

8.1.3.1.4 Enzyme-Linked Immunosorbent Assay

Enzyme-linked immunosorbent assay (ELISA) is an extremely valuable, too, in the serodiagnosis of leishmaniasis. The technique is useful for analysis in the laboratory as well as for field applications. However, the sensitivity and specificity of ELISA depend on the antigen used; that is, the use of the total antigens presents limited specificity (65%) due to the presence of epitopes shared with other parasites, for example, *T. cruzi* and *Plasmodium* sp. [34]. Recently, diverse recombinant antigens have been developed and tested, such as gene B protein (GBP) from *L. donovani* [35], ORFF from *L. infantum* [36], K9, K26 [37], ribosomal-acid proteins [38], heat-shock proteins [39]; and rgp63 and K39 from *L. donovani* [40,41], among others. Of all of these, the antigen K39 has shown great sensitivity and a predictive value of the manifestations of cutaneous leishmaniasis. In recent years, in our laboratory, we have characterized a new antigen that can be of high interest not only for the visceral forms of the disease but also for the cutaneous and mucocutaneous forms. This is superoxide dismutase (SOD), which is excreted by *Leishmania* species and which has been demonstrated to be highly immunogenic [42,43]. This enzyme, SOD excreted (SODe) can be used to diagnose canine leishmaniasis as well as cutaneous and mucocutaneous leishmaniasis with better results than conventional serological methods.

Our data agree with previously published findings which show that the sensitivity and specificity of diagnostic methods depend on the type, source, and purity of antigen employed, as some of the leishmanial antigens have common cross-reactive epitopes shared with other microorganisms, particularly *Trypanosoma*, *Mycobacteria*, *Plasmodium*, and *Schistosoma* [44]. To define the protein determinants of the SODe which could discriminate between leishmaniasis and Chagas disease, our analysis enabled us to confirm that the SODeby *Leishmania* species is recognized only by leishmaniasis sera and not by the sera from the Chagas patients. The absence of cross-reactivity with Chagas disease sera is important, since the endemic areas of the two diseases often overlap [45].

8.1.3.1.5 Immunoblotting

Most works on this technique have been made on visceral leishmaniasis, which reaches high levels of anti-*Leishmania* antibodies [46,47] and on cutaneous leishmaniasis. In the latter clinical form, low levels of antibodies occur and therefore the immunoblot is a technique that needs to be greatly improved for its complexity and reported contradictions

[48]. However, some authors have demonstrated that the immunoblot is a valid technique when using soluble antigens of 27 and/or 30 kDa for the diagnosis of cutaneous leishmaniasis [49]. In the same way, the use of the excreted SOD (SODe-Li, SODe-Lp, and SODe-Lb) as the antigen fraction in the immunoblot improves its sensitivity and specificity for diagnosing both canine as well cutaneous and mucocutaneous leishmaniasis [42,43].

In addition, it has been demonstrated that with specific antigens and by immunoblotting, crossed reactions can be avoided between species and even between members of the same family [45]. However, the technique is not easy to use and is limited to only laboratory research [49].

8.1.3.1.6 Antigen Detection

The detection of antigens is an ideal method to diagnose an infection. The antigen levels are correlated with the parasite load. These methods should be an alternative to the antibody-detection methods, particularly in immunocompromised patients, where the antibody response is very poor. However, the detection of antibodies in the sera of patients is complicated by the presence of high antibody levels, circulating immunological complexes, serum amyloid, rheumatoid factor, and autoantibodies, all of which can immunologically mask important antigen determinations or competitively inhibit the bonds of free antigens. Although some works have been published in this line, the antigen-detection method is not yet available [50]. Recently, a latex agglutination test (KAtex) has been developed to detect *Leishmania* antigens in the urine of patients with visceral leishmaniasis [51]. The results from the test, using samples collected from patients from different places indicate that the results were good regardless of the origin of the sample. The test has 100% specificity and a sensitivity of 68–100%. Although the test can be applied to detect asymptomatic cases of visceral leishmaniasis, this has not been confirmed. There are no antigen-detection methods applicable in a common way for cutaneous and mucocutaneous leishmaniasis. The KAtex test is simple to use and economical, and its two main advantages are the rapid diagnosis, especially when the patient travels long distances for the test, and the simplicity, in that no electric appliance is needed. The test discriminates between active disease and subclinical and latent infections [52].

8.1.3.2 Molecular Techniques

Although different molecular methods have been successfully tested for diagnosing leishmaniasis (e.g., pulsed-field gel electrophoresis and multilocus enzyme electrophoresis), polymerase chain reaction (PCR)-based assays currently constitute the main proposal of molecular diagnosis for researchers and health professionals. PCR has proved more sensitive and specific, although it is limited to hospitals and research laboratories. The amplification of the genes by PCR has several advantages compared with traditional techniques. First, it is highly sensitive, quick, and capable of action over a broad range of clinical specimens [53]. Also, the detection or identification of the causative agent is possible directly from the clinical specimens. Unlike the ELISA-based antibody-detection methods, no host–species-specific reagents are required and the same reagents can be used for specimens from humans, dogs, or any other animal host [54]. Various gene targets and nucleic acids can be used in PCR. The important gene targets are 18S-rRNA, small subunit rRNA (SSU rRNA), a repetitive genomic sequence of DNA, the miniexon (spliced ladder) gene repeat, the β-tubulin gene region, gp63 gene locus, internal transcribed spacer (ITS) regions; microsatellite DNAs such as maxi- and minicircles of kinetoplast DNA [55,56].

Different PCR formats are available and in general terms can be classified as low, medium, and high technology. The proposal of medium technology is probably the one most commonly used and includes the conventional PCR test, in which the amplified PCR products are resolved by electrophoresis, generally after the breakage of the restriction enzyme PCR–restriction fragment length polymorphism (RFLP), and visualized afterward with ethidium bromide dye [57]. This proposal does not require very sophisticated laboratory equipment and is only a question of time. The high-technology proposals are methods in which the PCR products are analyzed during their application (the so-called real-time PCR) after they are stained with SYBR-green I dye or hybridization with fluorogenic probes (e.g., TaqMan or fluorescent resonance energy transfer (FRET)) [58]. These techniques have the notable advantage of detecting the fluorescence in closed tubes, thereby avoiding grave problems of contamination. Although the technique is rapid, the equipment is very expensive and not all laboratories have it.

The low-technology proposals refer to simple PCR and can be used in any minimally equipped laboratory. It can be done simply in two stages: amplification and detection of the PCR products. Loop-mediated isothermal amplification (LAMP) represents a promising possibility, since it requires only a bath and visualization using SYBR-green I dye; if there is amplified product, the coloration changes to green and if not, the color remains orange. This method has been tested for *Trypanosoma brucei* and appears to be 100-fold more sensitive than conventional PCR [59], but up to now it has not been tested on leishmaniasis diagnosis. PCR–ELISA, also called reverse hybridization, is very simple. This method is based on the capture of the amplified products by PCR using specific probes immobilized on microtiter plates and colorimetric visualization. However, this technique requires relatively sophisticated equipment. More recently, other methods have been developed, such as oligochromatography–PCR (OC–PCR), which represent a promising alternative and appear to be very sensitive for diagnosing sleeping sickness and even *Leishmania* [60].

In practice, several biological and/or clinical questions can be answered by the methods based on nucleic acids. One of these, PCR, provides high sensitivity and specificity (around 100%) in *Leishmania* detection, regardless of the species or genus. This application is necessary for the differential diagnosis before beginning therapy, and it has been demonstrated that PCR performance is better than microscopic examination or parasite culturing, particularly in samples of low levels of parasites, for example, in patients with mucocutaneous leishmaniasis [61] or in noninvasive samples, such as blood or conjunctive tissue [62]. The contribution of PCR also appears to be important to diagnose leishmaniasis in patients coinfected with HIV [58]. Parasite detection by PCR for the confirmation of recovery has been observed fully in the case of visceral leishmaniasis, but it should be further investigated in the case of patients with cutaneous leishmaniasis, since it has been found that 80% of these patients remain PCR positive even 8 years after being clinically cured [63].

Another question is that the parasites in host tissues can be quantified by PCR, which is of paramount importance for the monitoring of a disease in the case of leishmanicide therapy [64].

Several studies have reported that PCR assay could detect parasitemia a few weeks before the appearance of any clinical signs or symptoms. Recently, Martin-Sanchez et al. [65], using PCR-ELISA found 24% asymptomatic individuals carrying *Leishmania* kinetoplast DNA (kDNA) in their blood. It was pointed out that these individuals could be potential sources of transmitting the infections in the community. These authors also found a good correlation between the antibody titers, skin-test positives, and PCR positives. The PCR methods could also be useful where transfusion-transmitted kala-azar is a potential threat in multiply transfused patients in whom serology has a limited role [66].

Chronic cutaneous leishmaniasis patients are the greatest diagnostic challenge and are easily misdiagnosed by clinical criteria because they are often atypical. These patients often have low or no *Leishmania* antibodies, and thus serological tests are not accurate. In such cases, PCR has proved to be the most useful tool for diagnosis.

The sensitivity of PCR in this disease is reportedly 100%. Oliveira et al. [67], using PCR, confirmed the diagnosis in all 50 dogs studied. In mucocutaneous leishmaniasis, PCR was detected parasites in 17 of 24 (71%) patients, whereas by conventional techniques detected only four (17%) patients [68].

8.2 METHODS

8.2.1 Sample Preparation

8.2.1.1 Serological Techniques

8.2.1.1.1 The Parasite Suspension

For MST or LST and IFAT, the sample contains the promastigote forms of any *Leishmania* sp., which are harvested from the culture in the stationary growth phase, washed three times with sterile phosphate-buffered saline (PBS), and centrifuged at 1000 g for 10 min. In the case of LST, the precipitate is resuspended with a phenolized buffered solution (5 g NaCl, 2.75 g $NaHCO_3$, 4.0 g phenol in one liter of distilled and sterile water) to make a concentrated stock solution (5×10^6–4×10^8 promastigotes/mL). For this test, this solution must be diluted to 1×10^6 promastigotes/mL. These promastigote solutions can be stored at 4°C for at least 12 months. The phenol can be changed to tiomersal to prevent solution contamination. For IFAT, the promastigote forms harvested from the culture are resuspended at a concentration of 10^7 promastigotes/mL in PBS. An amount of 10 μL of these suspensions is added to the multispot slides. After drying at 37°C for 2 h or at room temperature until complete evaporation of the buffer and adhesion of the cells to the surface, the slides are fixed with acetone at room temperature and afterward can be stored at 4°C until further use. Similarly, DAT uses promastigotes in suspension or lyophilized.

Recently, the parasite suspension (5×10^6 parasites/mL in 2% formalin) obtained by centrifugation at 7000 rpm for 10 min (cells precipitated and washed three times in PBS at pH 7.2) is used to detect IgG specific for *Leishmania* promastigotes by applying inflow cytometry [69].

8.2.1.1.2 Soluble and Membrane Fractions

This antigen fraction is a parasite homogenate or the cell-protein pool: crude soluble antigen (CSA), soluble fraction (*Leishmania major* antigen (LMS), soluble *Leishmania* antigen (SLA), or soluble nonmembranous *Leishmania donovani* (SLD)), or membranes (membrane proteins antigens (MBA)). There are different extraction protocols (see Table 8.1) but the essential procedure consists of harvesting the parasites from the culture (amastigotes and/or promastigotes) which are resuspended in different lysis buffers and disrupted by freeze–thawing or by sonic disintegration. The resulting suspension is centrifuged to separate the soluble fraction (supernatant) from the membrane fraction (precipitate).

8.2.1.1.3 Specific Purified Proteins

A step forward in molecular diagnosis is the further purification processes from cell lysates of the soluble or membrane fraction (Table 8.1). These fractions are submitted to SDS/PAGE separation and electroelution, sequential precipitation with ammonium sulfate, and, in the last step, chromatography (ionic exchange, filtration, affinity, HPLC, FPLC), or even matrix-assisted laser desorption/ionisation-time of flight (MALDI-TOF)/TOF mass spectrometry. That is, from the soluble fraction, proteins are purified, such as 68–97.4 kDa heat shock and SOD, or the P32/P33 protein, from membrane fraction. Meanwhile, glucose phosphate isomerase (GPI)-anchored membrane proteins are obtained from the cellular lysate using the partition Triton X-114 method (Table 8.1).

8.2.1.1.4 Exoantigens

This involves protein excreted/secreted by the parasite into the culture media: soluble exogenous antigens (SEAgs) or excreted/secreted (ES) molecules. The different protocols have high cellular density of promastigotes (2×10^7–5×10^9 parasites/mL) when cultured *in vitro* in serum-free or in protein-free media: 3 days at 26°C in XOM [95], in MTL for 24 h [43] or in RPMI for 6 h [96]. Afterward, the spent medium is harvested by centrifugation twice at $9000 \times g$ for 30 min, or at $3500 \times g$ for 15 min, after which the supernatant is recentrifuged and the resulting clear supernatant is further filtered through 0.45 μm pore-size membranes to collect the secreted exoantigens [95]. Alternatively, the supernatants are concentrated 10-fold with Centricon YM-10, 20-fold

TABLE 8.1
General Procedures for Antigenic Fractions of *Leishmania*

Antigen Fraction	Procedure	Buffer	Reference
Cell lysates	1. Wash promastigotes twice with PBS at 2000 rpm for 15 min. 2. Resuspending in 1–2 mL of lysis buffer. 3. Lysis by sonication briefly (10–15 s) and snap frozen in liquid nitrogen.	**Lysis Buffer**: 20 mM Tris–HCl, 40 mM NaCl, pH 7.4, and proteolytic inhibitor (10 mM sodium ethylenediaminetetracetate, 2 mM ioacetamine, 2 mM phenylmethylsulfonyl fluoride, and 1 mM leupeptin)	[70]
Crude soluble antigen (CSA)	1. Parasites washed 3 times in PBS at 2000 rpm for 10 min. 2. Lysis by 3 cycle of sonication at 1 min/40 W. 3. Centrifugation at 18,500 rpm for 90 min at 4°C. Pellet discarded. 4. Supernatant dialyzed through PBS for 36 h. 5. Filtration of the supernatant-dialyzed in disposable sterile 22 μm filters. Further purification is carried out by gel filtration chromatography over a column of Sephadex G200.	PBS pH 7.2	[71,72]
Homogenate fraction (H)	1. Parasites washed 2 times in PBS at 2500 rpm for 10 min 2. Cellular pellet is dissolved in STE buffer and disrupted by 3 cycles of sonic disintegration (30 s each to 60 V). 3. Centrifugation at 2500 rpm for 10 min, 4°C, and the supernatant was recovered and desalted in a Sephadex G-25 column, balanced with phosphate buffer.	**STE buffer:** 0.25 M sucrose, 25 mM Tris–HCl, 1 mM EDTA pH 7.8 **Phosphate buffer**: 20 mM potassium phosphate buffer pH 7.8, with1 mM EDTA	[43]
Parasite extract	1. Parasites washed 3 times in PBS and resuspended in PBS with protease inhibitor cocktail. 2. Lysis by 3 cycles of freezing and thawing and centrifuged at 13,800× *g* for 10 min at 4°C. 3. Pellet extraction with Triton X-100 (1% in PBS) with antiprotease cocktail and recentrifuged at 13,800× *g* for 10 min at 4°C to collect the Triton X-100 fractions.	**Protease inhibitor cocktail**: 2 mM AEBSF hydrochloride, 1.6 μM apronitin, 100 μM bestatin, 30 μM E-64, 40 μM leupeptin hemisulfate, and 20 μM pepstatin A.	[73]
Soluble Antigen (SA)	1. Parasites washed three times in PBS at 1500× *g* for 15 min at 8°C. 2. Lysis by 5 cycles of frozen in liquid nitrogen and thawed in a 37°C water bath, ultrasonication (twice for 2 min). 3. Centrifugation at 15,000× *g* for 60 min. Pellet discarded. 4. Supernatant dialyzed at 4°C overnight against PBS.		[74]
Leishmania major antigen (LMS)	1. Parasites washed 3 times in PBS at 800× *g* for 15 min at 4°C. 2. Lysis by resuspension in sterile distilled water and there rapid freeze–thaw cycles (liquid nitrogen, 25°C). 3. Centrifugation at 100,000× *g* for 1 h at 4°C. Further purification must made by R. Barbosa de Deus, C.A. Pereira Taveres, M.L. dos Mares Guia, patent application number PI-9700830-3, 8 December 1998, Brazilian Patent Office.	PBS pH 7.2	[75]
Soluble *Leishmania* Antigen (SLA) *91 method modified*	1. Parasites washed 4 times in PBS and resuspended in lysis buffer and incubated on ice for 10 min 2. Lysis by sonication at 4°C with two 20 s blasts. 3. Centrifugation at 27,000× *g* for 20 min, collected at supermatant and recentrifuged at 100,000× *g* for 4 h (SLA fraction) 4. Dialysis of SLA against lysis buffer and then filtered through 0.22 μm. 5. Fractionation by FPLC (Mono Q column) with a linear NaCl gradient (0–1 M) at a flow rate of 0.75 mL for min. 6. Fractions dialyzed against PBS and concentrated using Centricon 3 (MW-Co 3000).	**Lysis Buffer**: 100 mM Tris–HCl, 1 mM EDTA (pH 8) with 50 μg/mL leupeptin, 50 μg/mL antipapain, 50 μg/mL aprotinin, and 1.6 mM PMSF.	[76]
	1. Parasites washed 3 times in PBS. 2. Resuspension in lysis buffer and disrupted. 3. Centrifugation at 13,000× *g* for 15 min	**Lysis Buffer**: 6 mM CHAPS in 50 mM Tris–HCl buffer containing 150 mM NaCl pH 7.5	[77]

TABLE 8.1 (continued)
General Procedures for Antigenic Fractions of *Leishmania*

Antigen Fraction	Procedure	Buffer	Reference
	1. Cell pellets washed and resuspended in lysis buffer (LBi). 2. Incubation on ice for 20 min and lysis with a Douce homogenizer and 4 cycles of sonication of 10 s each. 3. Centrifugation at 1500× *g* for 10 min at 4°C. 4. The pellet is submitted to lysis and centrifugation steps. 5. The two supernatants (S_{1500}) pooled, are centrifuged at 10,000× *g* for 30 min at 4°C to generate a supernatant (soluble antigens) and a pellet fraction (MBA).	**Lysis Buffer LBi**: 10 mM Tris–HCl, 2 mM EDTA (pH 8) with 1 mM PMSF, 2 mM EGTA, 2 μg/mL pepstatin, 2.5 mM N-ethylmaleimide, and 0.127 IU/mL aprotinin.	[78]
	1. Parasites washed 4 times in PBS. 2. Resupension in lysis buffer and sonicated on ice by 6 times 10 s pulses. 3. The sonicated cell lysate is frozen and thawed 6 times using liquid nitrogen and left at 4°C for 1 h. 4. Centrifugation at 2500× *g* for 25 min at 4°C. 5. Collect the supernatant, dialyze against sterile PBS and filter through 0.22 μm. Further purification by affinity chromatography (CNBr-activated sepharose 4B) with Mab 7F9.	**Lysis Buffer**: 100 mM Tris–HCl, 1 mM EDTA (pH 8.4)	[79]
	1. Parasites washed 4 times in PBS. 2. Resupension in lysis buffer and sonicated at 4°C with 2 times 30 s blasts. 3. Centrifugation at 27,000× *g* for 30 min at 4°C. 4. Collect the supernatant and centrifuged at 100,000× *g* for 1 h. 5. Collect the supernatant as SLA and follow with the pellet to obtain MBA.	**Lysis Buffer**: PBS (pH 7.7) with 0.1 mM EDTA, 0.5 mM DTT, and 1 mM p-APMSF..	[80]
	1. Parasites washed 3 times in PBS at 5300× *g* for 10 min at 4°C. 2. Resuspension in lysis buffer and disrupted by freeze (dry, ice/ethanol)–thaw (water bath at 60°C) cycles with sonication. 3. Centrifugation at 10,700× *g* for 10 min at 4°C and collect the supernatant as SLA.	**Lysis Buffer**: PBS (pH 7.2) with 1 mM iodoacetamine, 1 mM PMSF, and phenanthroline.	[81]
	1. Parasites washed 3 times in Hanks' saline (pH 7.4) at 3000× *g* for 10 min at 4°C, then precipitate resuspension in Con-A buffer or Jaca buffer. 2. Cellular lysis by six cycles of freezing and thawing and one cycle of ultrasonication. 3. Centrifugation at 20,000× *g* for 60 min at 4°C and collect the supernatant as SLA. Further purification by affinity chromatography on concanavalin-A (Con-A)-Sepharose 4B or jacalin (Jaca)-agarose columns.	**Con-A buffer**: 10 mM Tris–HCl pH 4.4, 0.5 M NaCl, 1 mM $CaCl_2$, and 1 mM $MnCl_2$ **Jaca buffer**: 0.1 M Na_2HPO_4 and 0.1 M NaH_2PO_4, pH 8. Both containing 0.2% *n*-octyl-β-D-glucopyranoside and protease inhibitors: 1.6 mM PMS and 1 mM benzamidine.	[82]
Soluble nonmembranous *Leishmania donovani* proteins (SLD) [75] and [83] *methods modified*	1. Parasites washed 4 times in PBS. 2. Resupension in PBS containing protease inhibitors cocktail and sonicated. 3. Centrifugation at 40,000× *g* for 30 min and collect the supernatant (SLD). Further purification by precipitation with ammonium sulphate or TCA and wash 3 times with acetone. Then, pellet solubilised in rehydratation buffer and recentrifuged. This sample is separated by IEF followed by 2-D SDS/PAGE and the peptides purified by electro elution or gel digestion plus C18 reverse-phase minicolumn (ZipTip C18) and finally, MALDI-TOF/TOF Mass Spectrometer.	**Rehydratation buffer**: 7 M urea, 4% w/v CHAPS, 100 mM DTT, 0.5% v/v biolyte (3–10), and 40 mM Tris.	[84,85]
Membrane proteins Antigens (MBA)	1. Parasites washed once in PBS at 2000 rpm for 15 min. 2. Resuspension in lysis buffer and disruption in a Parr cell disruption bomb (10 min incubation at 1500 psi). 3. Centrifugation at 39,000× *g* for 30 min and collect the pellet as MBA. Purification of the membrane proteins MBA is dissolved in resuspension buffer, and incubated 1h at 4°C. Then, centrifuged at 39,000× *g* for 30 min and the supernatant primary purified by affinity resin: wash by centrifugation –1500× *g* for 5 min twice with 0.05%Tween-20 in 30 mM Tris–HC1, pH 8.2; twice with 0.5% DOC in the same Tris buffer. And loaded into a mini-column for elution with elution buffer into a microfuge tube containing 100 ~1 of 2 M Tris-base or solid glycine. Further purification is made by SE-HPLC and dialyzed extensively against double-distilled water.	**Lysis buffer**: 20 mM Tris–HC1, 40 mM NaC1 pH 7.4, with 10 mM sodium ethylenediaminetetracetate, 2 mM iodoacetamide, 2 mM phenylmethylsulfonyl fluoride, and 1 mM leupeptin. **Resuspension buffer**: 30 mM Tris–HC1 pH 8.2 containing 0.5% (w/v) sodium deoxycholate (DOC) and proteolytic inhibitors **Elution buffer:** 50 mM diethylamine, pH 11.5 containing 0.01% DOC	[70,86]

continued

TABLE 8.1 (continued)
General Procedures for Antigenic Fractions of *Leishmania*

Antigen Fraction	Procedure	Buffer	Reference
	1. Parasites washed in PBS, cellular pellet dissolved in resuspension buffer and kept on ice for 20 min. Add PMSF to a final concentration of 1 mM. 2. Disruption in a Dounce homogenizer and centrifuged at 10,000× *g* for 30 min. Further purification by ultracentrifugation at 33,000× *g* for 90 min at 4°C, on a 7–17% discontinuous sucrose gradient.	**Resuspension buffer**: 10 mM Tris–HCl-2 mM EDTA (pH 8)	[87]
	1. Obtained the $P_{10,000}$ as 103, it is resuspended in 5 mL of Tris-buffered saline and named as crude MBAs. 2. Centrifuged the crude MBAs at 2000× *g* for 10 min at 4°C. Collect the pellet and resuspended in 1 mL of 10 mM Tris–HCl (pH 11)–2 mM EDTA–30% sucrose and incubated at 4°C under agitation for 1 h. 3. Lysis by one freeze–thaw cycle and ultracentrifuge at 9000× *g* for 4 h at 4°C. The supernatant is the peripheral MBAs and the integral MBAs (IMAs, $P_{90,000}$) were sedimented in the pellet.	**Tris-buffered saline**: 10 mM Tris–HCl (pH 7.4), 150 mM NaCl, supplemented with protease inhibitors.	[88]
	1. The residual pellet ($P_{10,000}$) obtained according to 105 is dissolved in wash buffer and precipitated by centrifugation at 100,000× *g* for 1 h ($P_{100,000}$ = MBA). Further purification is achieved by MBA nonreducing SDS/PAGE separation, electrotransfer to PVDF membranes. The antigenic bands of interest are excised, placed in a tube containing 0.2–0.5 mL of a detergent solution (2% SDS, 1% Triton X-100, 0.1% DTT in 50 mM Tris–HCl, pH 9.0), and eluted according to Simpson et al. [115]. The eluent is filtered on a YM-10 membrane filter and then precipitated by cold acetone. Also, RP–HPLC analyses could be conducted on the eluent.	**Wash Buffer**: 0.25 M sucrose containing 0.15 M NaCl and 5 mM Tris–HCl, pH 8.	[80]
	Specific Purified Proteins		
68–97.4 kDa heat-shock	1. Promastigotes harvested by centrifugation at 5000× *g* for 20 min at 10°C. 2. The pellet was washed in PBS and then suspended in 300 μL of solubilization buffer and heated in a boiling water bath for 3 min to be electrophoresed in 10% acrylamide gels. Further purification by Gelatin affinity chromatography, or SDS/PAGE followed by electroelution.	**Solubilization buffer:** 0.1 M Tris Cl, pH 7/, 2% $NaDodSO_4$/10% (vol/vol) glycerol, 3% 2-mercaptoethanol, and 0.1 mM PMSF.	[89–91]
Superoxide dismutase (SOD)	1. The cellular pellet is washed twice and resuspended in ice-cold STE buffer. 2. The resuspended pellet is disrupted by three cycles of sonic disintegration (30 s each at 60 V). 3. The sonicated homogenate is centrifuged at 1500× *g* for 10 min at 4°C, and the pellet was washed three times with STE buffer for a total supernatant fraction of 9 mL. This fraction is recentrifuged (2500× *g* for 10 min at 4°C). Collect the supernatant and solid ammonium sulphate added to precipitate at between 35 and 85% salt concentration. Afterward is centrifuged (9000× *g* for 20 min at 4°C) and redissolved in 2.5 mL of 20 mM potassium phosphate buffer (pH 7.8) containing 1 mM EDTA and dialyzed on a Sephadex G-25 column, bringing it to a final volume of 2.5 mL.	**STE buffer:** (0.25 M sucrose, 25 mM Tris–HCl, 1 mM EDTA, pH 7.8	[42]
Membrane proteins P32 and P33 protein	(A)-By electroelution from the gel after MBA SDSPAGE: Two vertical strips are excised from both sides of the gel and stained with Coomassie blue to localize the P32 antigen; then, horizontal strip containing the relevant antigen is cut out from the gel and placed into a dialysis bag filled with 2 mL of PBS. Electroelution was performed at 120 V for 1 h at 4°C, and the eluted polypeptides were extensively dialyzed against PBS. (B)-By HPLC, the $P_{90,000}$ pellet [86] is resuspended in 1 mL of PBS containing one of the following detergents: 1% Triton X-100 (TX-100), 1% digitonin, 1% CHAPS, 1% sodium cholate, 1% OG, or 0.5% LDAO. And incubated for 1 h at 4°C under vigorous agitation and centrifuged at 90,000× *g* for 1 h at 4°C. The supernatants are submitted to electrophoresis and the two prominent bands (P32 and P33) are purified on a C18-DEAE high-pressure liquid chromatography (HPLC) column by using linear gradients of 2–35% and 2–45% acetonitrile in 0.1% trifluoroacetic acid over 40 min, respectively. The purified peptides were further characterized by (i) analytical C18 reversed-phase HPLC, (ii) amino acid analysis after acidolysis, and (iii) mass determination by matrix-assisted laser desorption ionization–time of flight mass spectrometry.		[87,88]

TABLE 8.1 (continued)
General Procedures for Antigenic Fractions of *Leishmania*

Antigen Fraction	Procedure	Buffer	Reference
GIPI-anchored membrane proteins	By partition Triton X-114 method described by [92] and modified by [93]: 1. The collected pellet is resuspended in buffer 1, with shaking during 1 h at 0°C, and spun at 8000 g for 10 min at 0°C. 2. 2-The pellet is collected and washed with buffer 1 and kept at −20°C. 3. 3-Collect the supernatant which is kept at −20°C during 24 h and then slowly unfrozen at room temperature and later placed during 12 min at 32°C to obtain the first partition (P1). P1 is then spun at 3000× *g* for 3 min, obtaining two phases. The aqueous phase containing hydrophilic proteins (HP) was separated and kept at −20°C. The detergent phase was re-suspended in 3 vol buffer 1 kept at 0°C for 10 min. The P1 process is repeated and the aqueous phase containing the rest of HP is discharged. The detergent phase was then mixed with 3 vol buffer 1, kept at 0°C for 10 min and spun at 18,000× *g* for 10 min at 0°C. The supernatant is collected and submitted to a second partition process (P2). The proteins contained at the detergent phase are precipitated with 3 vol of cold acetone and spun at 3000× *g* for 3 min, to obtain the GPI-anchored proteins.	**Buffer 1**: 10 mM Tris–HCl, 150 mM NaCl, 2% Triton X-114 pH 7.4 buffer containing a pool of protease inhibitors	[94]
78 kDa protein	From MBA fraction by a monoclonal antibody tagged CNBr sepharose CL-4B column is purified the 78 kDa protein.		[94]

Note: CHAPS: 3-[(3-cholamidopropyl)dimethylamino]-1-propanesulfonate; LDAO: lauryldimethylamine oxide; OG: octyl_D-glucopyranoside; PMSF: phenylmethylsulfonyl fluoride; RP-HPLC: Reversed-phase high-performance liquid chromatography; SE-HPLC: Size-exclusion high-performance liquid chromatography.

with 6-mL Vivaspin tube concentrator, and 200-fold using Centricon Plus 80 [96].

The SODe is obtained by collecting the supernatant after being centrifuged at 1500× *g* for 10 min and being passed through a filter of 0.45 μm pore size. The filtrate is subjected to two ice-cold consecutive additions of ammonium sulfate between 35% and 85% (wt/vol) salt concentration and is then centrifuged (9000× *g* for 20 min at 4°C) to obtain a precipitate, which is finally redissolved in 2.5 mL of Buffer 2 and desalted in a Sephadex G-25 column (PD 10; Pharmacia, Barcelona, Spain), previously balanced with Buffer 2, raising it to a final volume of 2.5 mL (fraction SODe). For further purification, this fraction is transferred to a QAE Sephadex A-50 column balanced with 20 mM potassium phosphate buffer (pH 7.8) containing 1 mM ethylenediaminetetraacetic acid (EDTA). The column is thoroughly washed with the same buffer, and the absorbed protein is eluted with a linear gradient of KCl (0–0.6 M). The eluted fractions exhibiting SOD activity (SODe) are pooled and concentrated by ultrafiltration in Centriprep-10 tubes to a final concentration of 2 μg/μL [43].

8.2.1.2 Molecular Techniques

8.2.1.2.1 Recombinant Proteins

The recombinant DNA-derived *Leishmania* proteins have a potential use in diagnostics (immunodetection, ELISA, or radioimmunoassay (RIA)) [97]. Most *Leishmania* recombinant proteins (rP) that have been identified are expressed in both host-specific amastigote and vector-specific promastigote stages [98].

The technique consists of five steps: (1) obtaining the gene, (2) cloning in the vector, (3) transformation in *E. coli*, (4) selection of the positive clone, and (5) expression/production and rP purification. The first step can be achieved easily if the sequence of interest is known, simply by PCR with the specific oligonucleotide primers [99,100]. However, if the sequence is unknown, it is necessary to clone the entire genome (genomic library) or construct a cDNA library (using "reverse genetics") and, by the gene encoding, screening the protein of interest with leishmaniasis-serum patients [101–103], and then sequencing. The second step is cloning in a vector, which is a current method in research labs [104]. The most commonly used vectors are: pBluescript, pET (15b and 17b), pGEX, pUC18, pMAL-CR1, pQE-30, and Lambda ZAPII, coupled with their *E. coli* competent cells to carry out the rP expression and production. Finally, to purify the selected rP, the simplest method is to electrophorese the purified material (bacteria harvested and lysed) by SDS/PAGE and afterward to purify the rP band by electroelution. In several rP, such as rK39, the purification is achieved from a 25% to 40% ammonium sulfate fraction of bacterial lysate by preparative isoelectric focusing (IEF) with the Rotofor IEF cell and 1% 3/10 ampholytes (pH range 3.5–9.5; Bio-Rad) in the presence of 8 M urea and 10 mM dithiothreitol; then peak fractions are concentrated by ammonium sulfate precipitation and dialyzed against tris buffered saline (TBS) [101]. In other cases, if the rP is synthesized in insoluble inclusion bodies in *E. coli*, it could be largely separated from host proteins by multiple low-speed centrifugation [99].

In most cases, researchers use affinity chromatography by gravity-flow chromatography in Ni-NTA agarose when the rP has been produced with a 6X-tag in 5′ [98], CNBr-activated sepharose [105], on amylose affinity chromatography for rP produced as maltose-binding protein (MBP) by the vector pMAL-c derived [106] or glutathione-Sepharose gel for purification of rP expressed with glutathione *S*-transferase (GST) [102].

8.2.1.2.2 DNA Extraction (Genomic and Kinetoplast)

Genomic DNA (gDNA) is routinely extracted by phenol–chloroform protocol and ethanol precipitation [104]. The method used to extract the kDNA is by following the procedure described by Gonçalves et al. [107].

8.2.2 Detection Procedures

8.2.2.1 Serological Detection of the Parasite Suspension

8.2.2.1.1 MST or LST Detection

1. 0.1 mL of antigen is injected intradermally on the volar surface of the forearm (after being cleaned with ethanol 70%).
2. On the opposed forearm the buffered solution must be injected intradermally as a control test.
3. The induration diameter is measured 48–72 h later in two perpendicular directions using the ball-point pen method [108].

The induration perimeter can be marked with a ball-point pen, and afterward the impression can be taken by applying a strip of adhesive band, which with the ink mark provides a permanent record of each patient (instead of an adhesive band, a piece of paper moistened with alcohol can be used). Following the international consensus definition, the LST result is considered positive when the mean of the two measurements is ≥5 mm [28,108,109].

8.2.2.1.2 Indirect Fluorescence

1. Promastigote forms harvested from the culture are resuspended at a concentration of 10^7 promastigotes/mL in PBS.
2. Cells are settled on the coverwell for immunofluorescence until evaporation of the buffer.
3. The cells are fixed with acetone at room temperature.
4. The cells are immunofluorescence stained.
5. The cells are incubated with the serum for 30 min at room temperature. Then, the coverwell is washed three times in PBS, and incubated 30 min at room temperature with the secondary antibody (anti-IgG labeled with FITC).
6. The coverwell is washed and the coverslip put in place with PBS-buffered glycerine (9 vol. of glycerine, 1 vol. of PBS) for examination under a fluorescence microscope.

8.2.2.2 Serological Detection of Specific Antigens

The specific antigens (purified peptides/proteins and recombinants proteins) are detected by immunodetection (Western blot, dot blot), ELISA, or RIA.

8.2.2.2.1 Dot Blot

1. The membrane (nitrocellulose, nylon or polyvinylidene fluoride [PVDF]) is cut into individual circles or a grid is drawn by pencil to indicate the region to be blotted.
2. The membrane is equilibrated in PBS.
3. From 0.5 μL to 2 μL of the specific antigen (at the chosen dilution and concentration) is loaded as a spot on the membrane at the center of the grid. The area that the solution penetrates (usually 3–4 mm diameter) is minimized by applying it slowly.
4. The membrane is allowed to dry.
5. The nonspecific sites are blocked by soaking in 5% BSA in TBS-T (20 mM Tris–HCl 150 mM NaCl, pH 7.5, plus 0.3% Tween 20) for 30 min at room temperature.
6. TBS-0.1% Tween 20 is used to wash three times (3 × 5 min).
7. Incubation is carried out for 30 min at room temperature with the serum of the suspected patients (primary antibody) in a chosen dilution made with TBS-0.1% Tween 20.
8. Step 6 is repeated.
9. A second incubation is carried out with secondary labeled antibody, normally peroxidase or alkaline phosphatase and combined with the IgG (anti-IgG conjugated with peroxidase at a dilution 1:800) for 30 min at room temperature.
10. Step 6 is repeated.
11. The reaction is revealed on adding the substrate of the enzyme used (for peroxidase: 0.5 mg/mL 3,3-diamino-benzidine in 0.1 M Tris–HCl, pH 7.4 and 1:5000 H_2O_2).

The appearance of the signal indicates the positive samples.

8.2.2.2.2 Western Blot and ELISA

The Western blot and ELISA indirect test follows the same principle (secondary Ag–Ac–Ac reaction). For the Western blot the antigen is transferred from an electrophoresed polyacrylamide gel to the equilibrated membrane. Meanwhile, for ELISA indirect the antigen is fixed to a polystyrene microtiter plate (96 wells) and the reaction is determined quantitatively by reading the absorbance at 405 nm in a microplate reader (Metertech Σ 960). Mean and standard deviations (SD) of the optical densities of the negative control sera is used to calculate the cut-off value (mean + 3 × SD).

8.2.2.2.3 Radioimmuno Assy

The rP is used to generate specific antibodies that are afterward used to analyze tissue samples in search of intracellular forms of *Leishmania* by RIA. The technique is based on the permeabilization of the cells from tissue in order to facilitate the entry of the Ac into the intracellular space and its bonding to the parasite proteins. Afterward, secondary Ac marked with colorimetric enzymes, colloidal gold, or radioactivity is used for identification. The general protocol was developed by Stocker and Heusser [110].

1. The cells are harvested and washed three times in PBS and finally resuspended at a concentration of 2.0×10^7 cell/mL.
2. The cells are fixed to flexible polyvinyl chloride microtiter plates with U or V bottoms by the glutaraldehyde method.
3. Fifty μL of cell suspension is added per well and the plates are centrifuged at 50–100 g for 5 min. Without disturbing the cell layer, the plates are then gently immersed in a 1 liter glass beaker containing freshly prepared 0.25% glutaraldehyde solution in PBS at 4°C. The immersion is performed carefully to prevent disturbance of the cell layer or trapping air in the wells.
4. After 5 min the plate is removed and glutaraldehyde solution discarded by gently flicking the plate over a sink. Free glutaraldehyde is washed away by immersing the plate in three successive beakers containing PBS, flicking between each immersion.
5. Then, the plate is dipped into a beaker containing PBS with 1% bovine serum albumin, 1% normal rabbit serum, and 0.2% sodium azide (RIA buffer) and incubated for 1 h at room temperature to saturate protein-binding sites on the plastic. After this, the plate is either used immediately or stored at 4°C without changing the buffer.
6. The RIA buffer is flicked out and 50 μL of dilutions of antisera or supernatant media containing monoclonal antibodies are added. This reaction is carried out at room temperature for 1 h.
7. Afterward, the plate is washed as above by successively flicking and immersing in PBS.
8. Twenty-five microliters RIA buffer containing $2–4 \times 10^4$ counts per min (cpm) [^{125}I]RaMIg is added to each well and the plate is kept at room temperature for another 1 h.
9. Unbound radioactive antibody is removed by washing as described above, except that the bulk of the radioactivity is first removed. Three drops of PBS are added to each well and the solution is sucked away or absorbed by paper towels backed by aluminum foil.
10. After four wash cycles the trays are allowed to dry and then cut with a heated wire. During cutting, the trays are supported by a rigid, polystyrene microtiter plate with its border trimmed to hold the assay tray.
11. This subsequently acts as a holder for the individual cut wells, which are then transferred to tubes for counting the radioactivity.

8.2.2.3 DNA Detection

PCR is a sensitive and rapid method to detect *Leishmania* gDNA or kDNA [111]. PCR primers have become available and target either kDNA or rRNA gene, ITSs, mini-exon genes, and specific gene sequences (e.g., glycoproteins, heat-shock proteins, and cysteine proteinases); primers can be genus, subgenus, or species specific.

8.3 CONCLUSIONS AND PERSPECTIVES

Leishmaniasis is caused by a kinetoplastid protozoan parasite transmitted from one host to another through the bites of female sandfly, or occasionally through nonvector routes including blood transfusion, congenital, sexual, laboratory acquired, or person-to-person contact. The disease may manifest itself in three main clinical forms, of which the visceral form known as kala-azar is most severe and may be fatal if not treated. The causative species of the parasite is specifically identified but in immunosuppressed patients the manifestation may be atypical. The diagnosis is made by the demonstration of the amastigote stage of the parasite in the bone marrow, spleen, liver, or lymph node aspirates or by using the histopathological and smear preparations. Rarely can the parasite be demonstrated in peripheral blood. The parasite is culture isolated *in vitro* in enriched media or in experimental animals for taxonomic and other biological studies on the isolates. However, with the advances in the noninvasive serological and molecular methods, direct methods are rarely used in clinical practice except for research purposes. Serological methods are sensitive, specific, and cost effective, while molecular tools are extremely sensitive and useful in molecular studies of epidemiology and diagnosis [112–114].

Despite the great number of techniques and protocols currently available, a number of problems persist in diagnosis, especially for professionals who would like to have a simple, reliable test to diagnose the disease. Among health professionals and researchers who work in tropical and subtropical areas, there is great demand for a quick test, given that a quick, sensitive, and specific diagnostic test could be not only of great usefulness for the application of protocol, but also of great help in field samples and eradication campaigns, as the results could be known immediately and could be taken as *in situ* control measures.

In the near future, leishmaniasis diagnosis could increase as a consequence of the development of tests, both serological as well as PCR based, and these could be used not only to evaluate the state of the infection in a patient but also to know the patient's immunological state. This would augment sensitivity and specificity values in the diagnosis while helping to improve the characterization of the host/parasite relationship.

Unquestionably, PCR-based methods are the ones used today in the most advanced laboratories. The sensitivity and specificity compensate for the relatively high costs involved, especially in these diseases in which the conventional methods are impractical or imprecise. There are sufficient arguments to believe that in the near future the molecular detection methods (PCR) will replace conventional methods. What is almost certain is that all methods to be developed will be based on chemical properties of the double helix of DNA.

However, PCR-based protocols urgently need standardization and optimization. Recommendations include the use of extraction controls, internal controls, a *Leishmania* standard control, replicate assays, and participation in an external quality-control programme. Compared with other diagnostic techniques available, the molecular approaches remain expensive and require technological expertise, so that efforts should be made to make PCR platforms more user friendly and cost effective, especially in remote areas where leishmaniasis is endemic.

REFERENCES

1. WHO. http://www.who.int/Leishmaniasis/Leishmaniasis_maps/en/index.html. Accessed September 2009.
2. Sánchez-Saldaña, L. et al. Leishmaniasis. *Dermatol. Peruana.* 14, 82–98, 2004.
3. Oumeish, O.Y. Cutaneous Leishmaniasis: A historical perspective. *Clin. Dermatol.* 17, 249, 54, 1999.
4. Catania, S. *Leishmania viscerale* in età pediatrica. Simposio. *Controllo della Leishmaniosi: Veterinari e Medici a confronto.* Domenica, Roma, pp. 8–9, 2003.
5. Davies, C.R. Effetto di una banda protettiva a base di deltametrina sull' incidenza di leishmanisi viscerale zoonotica nei bambini. Simposio. *Controllo della Leishmaniosi: Veterinari e Medici a confronto.* Domenica, Roma, pp. 13–14, 2003.
6. Farrel, J.P. New World Leishmaniasis: The ecology of leishmaniasis and the diversity of leishmanial species in Central and South America, In: Black, S.J., and Seed J.R. (series eds), *World Class Parasite: Vol. 4. Leishmania.* Kluwer Academic Publishers, Springer-Verlag, New York, 2002.
7. Campino, L.M. Canine reservoirs and Leishmaniasis: Epidemiology and disease. In: Black, S.J., and Seed J.R. (series eds), *World Class Parasite: Vol. 4. Leishmania.* Kluwer Academic Publishers, Springer-Verlag, New York, 2002.
8. Trager, W. The development of *Leishmania donovani in vitro* at 37°C, effects of the kind of serum. *J. Exp. Med.* 97, 177, 1953.
9. Lofgren, R. The structure of *Leishmania tropica* as revealed by phase and electron microscopy. *J. Bacteriol.* 60, 617, 1950.
10. Crowther, S., Fulton, J.D., and Joyner, L.P. The metabolism of *Leishmania donovani* in culture. *Biochem. J.* 56, 182, 1954.
11. Chang, K.P., and Dwyer, D.M. *Leishmania donovani.* Hamster macrophage interactions in vitro: Cell entry, intracellular survival, and multiplication of amastigotos. *J. Exp. Med.* 147, 515, 1978.
12. Magill, A.J. Leishmaniasis, In: Strickland, G.T. (ed.), *Hunter's Tropical Medicine and Emerging and Infectious Diseases.* Saunders WB Company, Philadelphia, PA, 2000.
13. Murray, H.W. et al. Advances in leishmaniasis. *Lancet* 366, 1561, 2005.
14. Kamhawi, S. Phlebotomine sand flies and *Leishmania* parasites: Friends or foes. *Trends Parasitol.* 22, 439, 2006.
15. Cunningham, A.C. Parasitic adaptive mechanisms in infection by *Leishmania. Exp. Mol. Pathol.* 72, 132, 2002.
16. Chang, K.P., Chaudhuri, G., and Fong, D. Molecular determinants of *Leishmania* virulence. *Annu. Rev. Microbiol.* 44, 499, 1990.
17. Cruz, I. et al. *Leishmania*/HIV co-infections in the second decade. *Indian J. Med. Res.* 123, 357, 2006.
18. Laison, R., and Saw, J.J. Evolution, classification and geographical distribution. In: Peters, W., and Killick-Kendrick, R. (eds.), *The Lesihmaniases in Biology and Medicine.* Academic Press, New York, 1987.
19. Mehlhorn, H. (ed.), Morphology. In *Parasitology in Focus: Facts and Trends*, 1st edn. Springer Verlag, Berlin, Heidelberg, New York, London, Paris, Tokyo, 1988.
20. Cáceres, O.R., and Montoya, Y.P. Diseño y evaluación de tres oligonucleótidos para la detección de *Leishmania* por PCR. *Rev. Peru Med. Exp. Salud Publica* 19, 109, 2002.
21. Leishmaniasis. Módulos Técnico. Serie Documentos Monográficos nº 8. Oficina General de Epidemiología (OGE) e Instituto Nacional de Salud (INS), Lima. 2000.
22. Herwaldt, B.L. Leishmaniasis. *Lancet* 354, 1191, 1999.
23. El-Hassan, A.M., and Zijlstra, E.E. Leishmaniasis in Sudan. Mucosal leishmaniasis. *Trans. R. Soc. Trop. Med. Hyg.* 95, 519, 2001.
24. Singh, S., Mohapatra, D.P., and Sivakumar, R. Successful replacement of foetal calf serum with human urine for *in vitro* culture of *Leishmania donovani. J. Commun. Dis.* 32, 289, 2000.
25. Rosbotham, J.L. et al. Imported mucocutaneous leishmaniasis. *Clin. Exp. Dermatol.* 21, 288, 1996.
26. Sing, R., and Silvakumar, R. Recent advances in the diagnosis of leishmaniasis. *J. Postgrad. Med.* 49, 55, 2003.
27. Reithinger, R., and Dujardin, J.C. Molecular diagnosis of leishmaniasis: Current status and future applications. *J. Clin. Microbiol.* 45, 21, 2007.
28. Weigle, K.A. et al. Leishmanin skin test standardization and evaluation of safety, dose, storage, longevity of reaction and sensitization. *Am. J. Trop. Med. Hyg.* 44, 260, 1991.
29. Singh, S. New developments in diagnosis of leishmaniasis. *Indian J. Med. Res.* 123, 311, 2006.
30. Bensoussan, E. et al. Comparison of PCR assays for diagnosis of cutaneous leishmaniasis, *J. Clin. Microbiol.* 44, 1435, 2006.
31. Sassi, A. et al. Leishmanin skin test lymphoproliferative responses and cytokine production after symptomatic or asymptomatic *Leishmania major* infection in Tunisia. *Clin. Exp. Immunol.* 116, 127, 1999.
32. Schallig, H.D. et al. Development and application of "simple" diagnostic tools for visceral leishmaniasis. *Med. Microbiol. Immunol.* 190, 69, 2001.
33. Schoone, G.J. et al. A fast agglutination-screening test (FAST) for the detection of anti-*Leishmania* antibodies. *Trans. R. Soc. Trop. Med. Hyg.* 95, 400, 2001.
34. Montoya, I. et al. Diagnostico molecular para la leishmaniasis. *Rev. Perú. Med. Exp. Salud Publica* 14, 25, 1997.
35. Jensen, A.T. et al. Serodiagnosis of *Leishmania donovani* infections: Assessment of enzyme-linked immunosorbent assays using recombinant *L. donovani* gene B protein (GBP) and a peptide sequence of *L. donovani* GBP. *Trans. R. Soc. Trop. Med. Hyg.* 93, 157, 1999.
36. Raj, V.S. et al. Serodiagnosis of leishmaniasis with recombinant ORFF antigen. *Am. J. Trop. Med. Hyg.* 61, 482, 1999.
37. Bhatia, A. et al. Cloning, characterization and serological evaluation of K9 and K26: Two related hydrophilic antigens of *Leishmania chagasi. Mol. Biochem. Parasitol.* 102, 249, 1999.
38. Padilla, C.R., and Montoya, Y.P. Caracterización e inmunoreactividad de la proteína acídica ribosomal P2β de *L. (V.) braziliensis. Rev. Per. Med. Exp. Salud Publica.* 20, 92, 2003.
39. Amorin, A.G. et al. Identification of the C-terminal region of 70 kDa heat shock protein from *Leishmania (Viannia) braziliensis* a target for the humoral immune response. *Cell Stress Chaperones* 1, 177, 1996.
40. Singh, S. et al. Diagnostic and prognostic value of K39 recombinant antigen in Indian leishmaniasis *J. Parasitol.* 81, 1000, 1995.
41. Singh, S., Kumasi, V., and Singh, N. Predicting kala-azar disease manifestations in asymptomatic patients with latent

Leishmania donovani infection by detection of antibody against recombinant K39 antigen. *Clin. Diagn. Lab. Immunol.* 9, 568, 2002.
42. Marín, C. et al. *Leishmania infantum*: Use of excreted superoxide dismutase in ELISA and Western blotting for serologic diagnostic in naturally infected dogs. *Parasitol. Res.* 101, 801, 2007.
43. Marín, C. et al. Enzyme-linked immunosorbent assay for superoxide dismutase-excreted antigen in diagnosis of sylvatic and andean cutaneous leishmaniasis of Peru. *Am. J. Trop. Med. Hyg.* 80, 55, 2009.
44. McCord, J.M., and Fridovich, I., Superoxide dismutase: An enzymatic function in erythrocuprein (hemocupresin). *J. Biol. Chem.* 244, 6049, 1969.
45. Marín, C., and Sánchez-Moreno, M., The trypanosomiasis diagnosis: A review on the role of the superoxide dismutase as molecular marker. *Curr. Res. Microbiol.* I, 1, 2009.
46. Marty, P.A. et al. Detection by Westem blot of four antigens characterizing acute clinical leishmaniasis due to *Leishmania infantum*. *Trans. R. Soc. Trop. Med. Hyg.* 89, 690, 1995.
47. Mary, C.D. et al. Westem blot analysis of antibodies to *Leishmania infantum* antigens: Potential of the 14-kD and 16-kD antigens for diagnosis and epidemiologic purposes. *Am. J. Trop. Med. Hyg.* 47, 764, 1992.
48. Montoya, Y. et al. Recombinant antigens for specific and sensitive serodiagnosis of Latin American tegumentary leishmaniasis. *Trans. R. Soc. Trop. Med. Hyg.* 91, 674, 1997.
49. Brito, M.E. et al. Identification of potentially diagnostic *Leishmania braziliensis* antigens in human cutaneous leishmaniasis by immunoblot análysis. *Clin. Diagn. Lab. Immunol.* 7, 318, 2000.
50. Senaldi, G. et al. Serological diagnosis of visceral leishmaniasis by a dot-enzyme immunoassay for the detection of a *Leishmania donovani*-related circulating antigen. *J. Immunol. Methods.* 193, 9, 2001.
51. Attar, Z.J. et al. Latex agglutination test for the detection of urinary antigens in visceral leishmaniasjs. *Acta Trop.* 78, 11, 2001.
52. Gavgani, A.S.M., KhademVatan, S., and Ghazanchael, A. KAtex antigen-detection test as a diagnosis tool for latent visceral leishmaniasis cases. *African J. Biotechnol.* 7, 852, 2008.
53. Tavares, C.A.P., Fernandez, A.P., and Melo, M.N. Molecular diagnosis of leishmaniasis. *Expert Rev. Mol. Diagn.* 3, 657, 2003.
54. Schallig, H.D., and Oskam, L. Molecular biological applications in the diagnosis and control of leishmaniasis and parasite identification. *Trop. Med. Int. Health.* 7, 641, 2002.
55. Gangneux, J.P. et al. Prospective value of PCR amplification and sequencing for diagnosis and typing of Old World Leishmania infections in an area of nonendemicity. *J. Clin. Microbiol.* 41, 1419, 2003.
56. da Silva, E.S. et al. Diagnosis of human visceral leishmaniasis by PCR using blood samples spotted on filter paper. *Genet. Mol. Res.* 3, 251, 2004.
57. Singh, S., Dey, A., and Sivakumar, R. Applications of molecular methods for *Leishmania* control. *Exp. Rev. Mol. Diagn.* 5, 251, 2005.
58. De Doncker, S. et al. A new PCR-ELISA for diagnosis of visceral leishmaniasis in blood of HIV-negative subjects. *Trans. R. Soc. Trop. Med. Hyg.* 99, 25, 2005.
59. Kuboki, N. et al. Loop-mediated isothermal amplification for detection of African trypanosomes. *J. Clin. Microbiol.* 41, 5517, 2003.
60. Deborggraeve, S. et al. Molecular dipstick test for diagnosis of sleeping sickness. *J. Clin. Microbiol.* 44, 2884, 2006.
61. Garcia, A.L. et al. American tegumentary leishmaniasis: Antigen-gene polymorphism, taxonomy and clinical pleomorphism. *Infect. Genet. Evol.* 5, 109, 2005.
62. Strauss-Ayali, D. et al. Polymerase chain reaction using non invasively obtained samples, for the detection of *Leishmania infantum* DNA in dogs. *J. Infect. Dis.* 189, 1720, 2004.
63. Maurya, R. et al. Evaluation of PCR for diagnosis of Indian kala-azar and assessment of cure. *J. Clin. Microbiol.* 43, 3038, 2005.
64. Mary, C. et al. Quantification of *Leishmania infantum* DNA by a real-time PCR assay with high sensitivity. *J. Clin. Microbiol.* 42, 524, 2004.
65. Martin-Sanchez, J. et al. Detection of *Leishmania infantum* kinetoplast DNA in peripheral blood from asymptomatic individuals at risk for parenterally transmitted infections: Relationship between polymerase chain reaction results and other *Leishmania* infection markers. *Am. J. Trop. Med. Hyg.* 70, 545, 2004.
66. Otero, A.C. et al. Occurrence of *Leishmania donovani* DNA in donated blood from seroreactive Brazilian blood donors. *Am. J. Trop. Med. Hyg.* 62, 128, 2000.
67. de Oliveira, C.I. et al. Clinical utility of polymerase chain reaction-based detection of *Leishmania* in the diagnosis of American cutaneous leishmaniasis. *Clin. Infect. Dis.* 37, 49, 2003.
68. Mimori, T. et al. Usefulness of sampling with cotton swab for PCR-diagnosis of cutaneous leishmaniasis in the New World. *Acta Trop.* 81, 197 (2002)
69. Santiago, M.A. et al. Differentiation between canine cutaneous and visceral leishmaniasis by the detection of immunoglobulin G specific for *Leishmania (Viannia) braziliensis* and *Leishmania (Leishmania) chagasi* antigens using flow cytometry. *Vet. Parasitol.* 154, 341, 2008.
70. Jaffe, L., and Zalis, M., Purification of two *Leishmania donovani* membrane proteins recognized by sera from patients with visceral leishmaniasis. *Mol. Biochem. Parasitol.* 27, 53, 1988.
71. Rosário, E.Y. et al. Evaluation of enzyme-linked immunosorbent assay using crude *Leishmania* and recombinant antigens as a diagnostic marker for canine visceral leishmaniasis. *Mem Inst Oswaldo Cruz* 100, 197, 2005.
72. Arora, S.K., and Sehgal, S. Identification of major antigens of *Leishmania donovani* using kala-azar sera. *Med. Microbiol. Immunol.* 178, 81, 1989.
73. Tonui, W. K. et al. Immunization with *Leishmania major* exogenous antigens protects susceptible BALB/c mice against challenge infection with *L. major. Infect. Immun.* 72, 5654, 2004.
74. Mettler, M. et al. Evaluation of enzyme-linked immunosorbent assays, an immunofluorescent-antibody test, and two rapid tests (immunochromatographic-dipstick and gel tests) for serological diagnosis of symptomatic and asymptomatic *Leishmania* infections in dogs. *J. Clin. Microbiol.* 43, 5515, 2005.
75. Barbosa-de-Deus, R. et al. *Leishmania major*-like antigen for specific and sensitive serodiagnosis of human and canine visceral leishmaniasis. *Clin. Diagn. Lab. Immunol.* 9, 1361, 2002.
76. Rafati, S. et al. Biochemical analysis and immunogenicity of *Leishmania major* amastigote fractions in cutaneous leishmaniasis. *Clin. Exp. Immunol.* 110, 203, 1997.

77. Atta, A.M. et al. Antileishmanial IgG and IgE antibodies recognize predominantly carbohydrate epitopes of glycosylated antigens in visceral leishmaniasis. *Mem. Inst. Oswaldo Cruz* 99, 525, 2004.
78. Kamoun-Essghaier, S. et al. Proteomic approach for characterization of immunodominant membrane-associated 30- to 36-kilodalton fraction antigens of *Leishmania infantum* promastigotes, reacting with sera from Mediterranean visceral leishmaniasis patients. *Clin. Diagn. Lab. Immunol.* 12, 310, 2005.
79. Kabiri A.R. et al. Leishmanin skin test in guinea pig with a single purified protein of *Leishmania major. Exp. Parasitol.* 111, 239–243, 2005.
80. Zeinali, M., Ardestani, S.K., and Kariminia, A. Purification and biochemical characterization of two novel antigens from *Leishmania major* promastigotes. *Korean J. Parasitol.* 45, 287, 2007.
81. Ribeiro, F.C. et al. Use of ELISA employing *Leishmania (Viannia) braziliensis* and *Leishmania (Leishmania) chagasi* antigens for the detection of IgG and IgG1 and IgG2 subclasses in the diagnosis of American tegumentary leishmaniasis in dogs. *Vet. Parasitol.* 148, 200, 2007.
82. Gomes-Silva et al. Serological reactivity of different antigenic preparations of *Leishmania (Leishmania) amazonensis* and the *Leishmania braziliensis* complex. *Rev. Soc. Bras. Med. Trop.* 41, 135, 2008.
83. Choudhury, A. et al. Enzyme linked immunosorbant assay in diagnosis of kala-azar in Bhadohi (Varanasi) India. *Trans. R. Soc. Trop. Med. Hyg.* 84, 363, 1990.
84. Garg, R. et al. *Leishmania donovani*: Identification of stimulatory soluble antigenic proteins using cured human and hamster lymphocytes for their prophylactic potential against visceral leishmaniasis. *Vaccine* 24, 2900, 2006.
85. Gupta, S.K. et al. Proteomic approach for identification and characterization of novel immunostimulatory proteins from soluble antigens of *Leishmania donovani* promastigotes. *Proteomics* 7, 816, 2007.
86. Jaffe, C.L., and McMahon-Pratt, D., Monoclonal antibodies specific for *Leishmania tropica* I. Characterization of antigens associated with stage- and species-specific determinants. *J. Immunol.* 131, 1987, 1983.
87. Tebourski, F. et al. Identification of an immunodominant 32-kilodalton membrane protein of *Leishmania donovani infantum* promastigotes suitable for specific diagnosis of Mediterranean visceral leishmaniasis *J. Clin. Microbiol.* 32, 2474, 1994.
88. Kamoun-Essghaier, S. et al. Proteomic approach for characterization of immunodominant membrane-associated 30 to 36 kDa fraction antigens of *Leishmania infantum* promastigotes reacting with sera from Mediterranean visceral leishmaniasis patients. *Clin. Diagn. Lab. Immunol.* 12, 310, 2005.
89. Lawrence, F., and Robert-Gero, M. Induction of heat shock and stress proteins in promastigotes of three *Leishmania* species. *PNAS USA* 82, 4414, 1985.
90. Nandan, D. et al. A rapid, single-step purification method for immunogenic members of the hsp 70 family: Validation and application. *J. Immunol. Methods* 176, 255, 1994.
91. Jeronimo, S.M. et al. Identification of *Leishmania* chagasi antigens recognized by human lymphocytes *J. Infect. Dis.* 172, 1055, 1995.
92. Ko, Y.G., and Thompson, G.A. Purification of glycosylphosphatidylinositol-anchored proteins by modified Triton X-114 partitioning and preparative gel electrophoresis. *Anal. Biochem.* 224, 166, 1995.
93. Añez-Rojas, N. et al. Isolation, purification and characterization of GPI-anchored membrane proteins from *Trypanosoma rangeli* and *Trypanosoma cruzi. Acta Trop.* 97, 140, 2006.
94. Mukherjee, M., Bhattacharyya, A., and Duttagupta, S. Serodiagnostic and immunoprophylactic potential of a 78 kDa protein of *Leishmania donovani* of Indian origin. *Med Sci Monit.* 8, 117, 2002.
95. Rajasekariah, G-H.R. et al. A novel exo-antigen-based ELISA for the detection of canine leishmaniasis. *Am. J. Trop. Med. Hyg.* 78, 616, 2008.
96. Chenik, M. et al. Approaches for the identification of potential excreted/secreted proteins of *Leishmania major* parasites. *Parasitology* 132, 493, 2006.
97. Kubar, J., and Fragaki, K. Recombinant DNA-derived *Leishmania* proteins: From the laboratory to the field. *Lancet Infect. Dis.* 5, 107, 2005.
98. Kubar, J., and Fragaki, K. *Leishmania* proteins derived from recombinant DNA: Current status and next steps. *Trends Parasitol.* 22, 111, 2006.
99. Rafati, A.H. et al. Identification of *Leishmania major* cysteine proteinases as targets of the immune response in humans. *Mol. Biochem. Parasitol.* 113, 15, 2001.
100. Pinheiro, P.H. et al. A recombinant cysteine proteinase from *Leishmania (Leishmania) chagasi* as an antigen for delayed-type hypersensitivity assays and serodiagnosis of canine visceral leishmaniasis. *Vet. Parasitol.* 26(162), 32, 2009.
101. Burns, J.M. et al. Molecular characterization of a kinesin-related antigen of *Leishmania chagasi* that detects specific antibody in African and American visceral leishmaniasis. *Proc. Natl. Acad. Sci., USA* 90, 775, 1993.
102. Fragaki, K. et al. A novel *Leishmania infantum* nuclear phosphoprotein Lepp12 which stimulates IL1-beta synthesis in THP-1 transfectants. *BMC Microbiol.* 30, 3, 2003.
103. Suffia, B. et al. A novel *Leishmania infantum* recombinant antigen which elicits interleukin 10 production by peripheral blood mononuclear cells of patients with visceral leishmaniasis. *Infect. Immun.* 68, 630, 2000.
104. Sambrook, J., Fritsch, E.F., and Maniatis, T. *Molecular Cloning. A Laboratory Manual*, 2nd edn. Cold Spring Harbor Laboratory Press, 1989.
105. Handman, B., Button, L.L., and McMaster RW, *Leishmania major*: Production of recombinant gp63, its antigenicity and immunogenicity in mice. *Exp. Parasitol.* 70, 427, 1990.
106. Soto, M. et al. Multicomponent chimeric antigen for serodiagnosis of canine visceral leishmaniasis. *J. Clin. Microbiol.* 36, 58, 1998.
107. Gonçalves, A.M., Nehme, N.S., and Morel, C.M. Trypanosomatid characterization by schizodeme analysis, In: Morel, C.M. (ed.), *Genes and Antigens of Parasites (A Laboratory Manual)*. Fundaçao Oswaldo Cruz, Rio de Janeiro, RJ, Brazil, 1984.
108. Sokal, J.E. Measurement of delayed skin-test responses. *N. Engl. J. Med.* 293, 501, 1975.
109. Gramiccia, M. et al. Leishmaniasis in Sardinia. 5. Leishmanin reaction in the human population of a focus of low endemicity of canine leishmaniasis. *Trans. R. Soc. Trop. Med. Hyg.* 84, 371, 1990.
110. Stocker, J.W., and Heusser, C.H. Methods for binding cells to plastic: Application to a solid-phase radioimmunoassay for cell-surface antigens. *J. Immunol. Methods* 26, 87, 1979.
111. Bañuls, A.L., Hide, M., and Prugnolle, F. *Leishmania* and the Leishmaniases: A parasite genetic update and advances in taxonomy, epidemiology and pathogenicity in humans. *Adv. Parasitol.* 64, 1, 2007.

112. Pampiglione, S. et al. Studies in Mediterranean leishmaniasis. 3. The leishmanin skin test in kala-azar. *Trans. R. Soc. Trop. Med. Hyg*. 69, 60–68, 1975.
113. Scott, P. et al. Vaccination against cutaneous leishmaniasis in murine model II. Immunological properties of protective and non protective subfractions of a soluble promastigotes extract. *J. Immunol.* 139, 3118–3125, 1987.
114. Rojas, A. et al. Isolation, purification, characterization and antigenic evaluation of GPI-anchored membrane proteins from *Leishmania (Viannia) braziliensis. Acta Trop*. 105, 139–144, 2008.
115. Simpson R.J. et al. Peptide mapping and internal sequencing of proteins electroblotted from two-dimentional gels onto polyvinylidene difluoride membranes. A chromatographic procedure for separating proteins from detergents. *J Chromatogr.* 476, 345–361, 1989.

9 Naegleria

Mineko Shibayama, Jesús Serrano-Luna, Isaac Cervantes-Sandoval, and Víctor Tsutsumi

CONTENTS

9.1 INTRODUCTION

Naegleria fowleri, *Acanthamoeba* spp., *Balamuthia mandrillaris*, and *Sappinia* spp. are free-living amoebae. However, under certain conditions, these amoebae can behave as parasites of humans and other mammals and are known as amphizoic organisms. These free-living amoebae are distributed worldwide in nature and can be isolated from samples of soil and water [1]. *N. fowleri* is an amoeboflagellate that normally grows in tropical or subtropical climates with a temperature about 37°C. This amoeba has three morphological stages in its life cycle: the trophozoite, the biflagellate, and the cyst. The parasite causes an infection of the central nervous system (CNS) known as primary amoebic meningoencephalitis (PAM), an acute, fulminant rapidly fatal disease that occurs mainly in apparently healthy children or young people with a recent history of swimming [2]. Therefore, infection is acquired by exposure to water in ponds, pools, or lakes contaminated with *N. fowleri* [1].

9.1.1 CLASSIFICATION

Amoebae of the genus *Naegleria* belong to the family Vahlkamfiidae. The classification of the different species of *Naegleria* has been based on the morphological characteristics of the cyst, tolerance to temperature, immunological criteria, and their pathogenic capacity and enzymatic patterns. However, this classification system is highly subjective, because it has been shown that these features vary depending on the conditions and the period during which strains have been maintained in axenic cultures. Classification of *Naegleria* species using molecular criteria was introduced by De Jonckheere [3,4], who analyzed the sequence of internal transcribed spacers 1 and 2 (ITS1 and ITS2), and the sequence of the 5.8S subunit of ribosomal DNA (rDNA). In that study, the author analyzed nearly 80 different isolates of diverse sources from different continents. Based on the sequence and size of individual fragments of the small subunit rDNA (SSU rDNA), he concluded that different molecular species can be identified within the genus *Naegleria*. In this new classification system, De Jonckheere proposed the existence of 11 new species of *Naegleria* [4]. More recently, De Jonckheere [5] used the sequence of the large subunit of RNA polymerase II (RPB1) to construct a parallel classification system in order to demonstrate the reproducibility of the classification based on the SSU rDNA sequence. In this study he found that the RPB1 sequence confirmed the classification obtained with the SSU rDNA. Furthermore, he found that there are two types of *N. fowleri* based on the sequence of this gene. Currently, nearly 40 different species of the *Naegleria* genus have been listed in the National Center for Biotechnology Information taxonomy database. Although this database is not an authoritative source of nomenclature or classification, it shows the high diversity of species within a single genus, of which only *N. fowleri* is considered to be pathogenic in humans. Other species such as *Naegleria australiensis*, *Naegleria italica*, and *Naegleria philippiniensis* are pathogenic in animal models, but not in humans. *Naegleria lovaniensis* has attracted attention recently because it is the first nonpathogenic thermophilic species described for this genus, and it maintains a close evolutionary relationship with *N. fowleri*. It has been proposed therefore that *N. lovaniensis* is the nonpathogenic counterpart of *N. fowleri* and that it will be useful for comparative studies of pathogenic mechanisms. Although the identification of species by the above system remains artificial in asexual protists, the nomenclature of species is needed to enable international exchange of information on the diversity and distribution of these amoebae [5].

9.1.2 Morphology

The trophozoites of *N. fowleri* measure from 15 to 25 μm in diameter. In culture medium the cells are very active and change their shapes constantly (Figure 9.1a and 9.1b). The amoeba presents cytoplasmic projections, or pseudopodia, which are related to locomotion. In addition, it presents other cytoplasmic projections called food cups (Figure 9.1c) that allow it to phagocytose bacteria, yeast, and cellular debris. At the ultrastructural level, the trophozoites of *N. fowleri* show the typical features of eukaryotic cells [6], except for the centrioles, which have not been identified [7]. The trophozoites have a cytoplasmic membrane of about 10 nm thickness. The cytoplasm presents a large number of free ribosomes that are associated with the membranes that form the rough endoplasmic reticulum. It also contains smooth endoplasmic reticulum. Other membranes associated with a large number of vesicles and organized similar to Golgi apparatus have been identified. In addition, abundant vacuoles of different sizes, either empty or containing different types of materials, have been observed. The mitochondria have a characteristic curved "dumb-bell" shape. The dumb-bell forms average 1 μm in length and 0.25 μm in breadth (Figure 9.1d). The outer membrane is separated by a space of 20 nm from the inner binding membrane [8]. Lysosomes have also been identified by histochemical staining with acid phosphatases [9]. The presence of contractile vacuoles has also been reported in several species of *Naegleria* [6]. The nucleus of *Naegleria* is prominent with a conspicuous nucleolus (Figure 9.1d). The nuclear envelope is typical for eukaryonts, having a double membrane and a large number of pores, with the outer nuclear membrane associated with ribosomes.

When trophozoites of *N. fowleri* are grown in buffers free of nutrients they may differentiate to a transitional flagellate stage that cannot divide or feed. This flagellate differentiation involves a change in the cell shape from the pleomorphic trophozoite to a pear-shape with a pair of flagella at the distal end. During the first minutes of differentiation, no ultrastructural changes can be observed, but after 2 h, the number of cytoplasmic vacuoles decreases considerably. This is caused by the ejection of the entire contents of the vacuoles to the surrounding environment, accompanied by the appearance of the flagella in the region of the basal bodies [10]. The development of the flagellar apparatus occurs before the change of cell shape, and some authors believe that there is a transition phase between the amoeba and the flagellate.

The morphological differences in the cysts of *Naegleria* spp. have been used as screening criteria for distinguishing between species. However, these criteria are not used currently, because the morphological features change constantly in relation to growing conditions. The mature cyst of *N. fowleri* is usually spherical, smooth, double walled, and refractile, and measures about 10 μm in diameter [11]. The wall is composed mainly of cellulose and mucopolysaccharides [12]. This cell

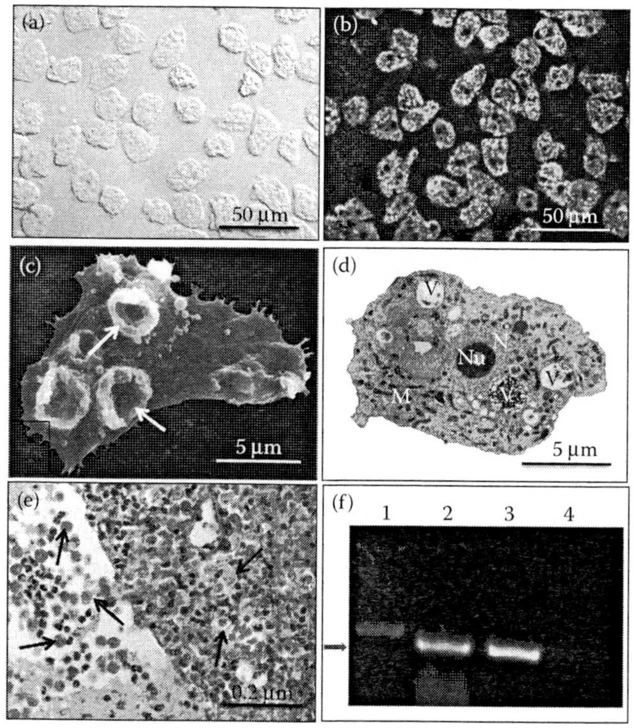

FIGURE 9.1 (a–d). *N. fowleri* trophozites observed with different morphological techniques: (a) Differential interference contrast (DIC. Nomarski optics). Multiple pleomorphic cells with irregular profiles and granular surfaces are seen. (b) Immunofluoresce of trophozoites using anti-*N. fowleri*. (c) Scanning electron microscopy. A trophozoite shows a flat irregular profile and several large food-cups (arrows) on the surface. (d) Transmission electron microscopy. Trophozoite with a large nucleus (N) and conspicuous nucleoli (Nu) is seen at the cell center. Cytoplasm contains multiple small, dense, and elongated mitochondria (M) and empty or cell debris containing vacuoles (V). (e) Central nervous system damage produced by *N. fowleri*. Parenchymal necrosis and hemorrhage areas associated with inflammatory infiltrates are seen associated with multiple trophozoites (arrows). (f) PCR with specific primers for *N. fowleri*. Lane 1: MR, lambda/Hind III; lane 2 *N. fowleri* (ATCC 30808); lane 3 *N. fowleri* strain (Mexicali, Mexican origin); lane 4 negative control (without DNA). Expected 1500 bp product is indicated (arrow).

wall contains pores that are sealed by a thin mucoid layer. During the early stages of cyst formation, the mitochondria become elongated and the cisterns of the endoplasmic reticulum are widened. The material of the cyst wall is synthesized and packaged by rough endoplasmic reticulum. Once the cyst is mature, the nucleus, and its nucleolus are less pronounced than in the trophozoite. In addition, vacuoles containing ribonucleoproteins, which are considered to be autophagosomes, have been observed. During the excystment of *Naegleria*, the cyst wall remains intact and the trophozoites emerge through the pores [2].

9.1.3 Biology, Isolation, and Culture Conditions

The trophozoite is the vegetative form and is the stage in which food uptake and cell division take place. The division of the trophozoites of *N. fowleri* is achieved by a mechanism known as promitosis, which is characterized by an intact nuclear membrane and nucleolus during karyokinesis [13]. When the trophozoites are incubated in nonnutrient solutions, such as isotonic saline or phosphate buffer, they suffer a morphological and physiological differentiation toward the biflagellate stage [14]. This change reflects a condition in which the trophozoites are found in water bodies exhausted of their source of nutrients, or with modifications in temperature, salinity, or pH that are unsuitable for their growth. It is believed that such challenges stimulate the dispersal of *N. fowleri* amoebae to reach new habitats where the environmental conditions are more favorable for their reproduction.

The biflagellate *N. fowleri* is unable to eat or divide, although the presence of flagellated amoebae capable of division has been reported in other species of *Naegleria* [15], as well as in other genera of amoeboflagellates of the Vahlkamfiidae family. On the other hand, the presence of at least three species of *Naegleria* that do not exhibit flagellates during their life cycle has also been reported [16], and it has even been noticed that the trophozoites of *N. fowleri*, when maintained for long periods in axenic cultures, lose their ability to differentiate, and that this property cannot be recovered by any known technique [17].

The cyst is the resistant stage that provides protection against adverse conditions such as desiccation, lack of nutrients, accumulation of waste products, lack of oxygen, changes in pH, high salinity, and even overpopulation. The transformation to a cyst in *Naegleria* spp. begins with the expulsion of water via contractile vacuoles, the reorganization of its cytoplasm and the synthesis of a double cell wall. The cyst returns to its vegetative form when the environmental conditions favor its growth. It is known that the cysts and trophozoites are the infective stages in PAM, although it has been demonstrated experimentally that the flagellate is able to infect mice [18]. However, the flagellate and cyst forms probably become trophozoites before penetrating the tissues of the host, because neither of these two forms has ever been observed in tissues of experimental animals or in human autopsy cases.

Naegleria spp. have also been isolated from some animals, such as amphibians and the intestines of fish. Soil is the usual habitat of the free-living amoebae, because it provides a source of oxygen, a suitable temperature, and bacteria as a food source.

The vertical distribution of potentially pathogenic amoebae in freshwater lakes has been correlated with physical, chemical, and biological features. Kyle and Noblet [19] noted that there is a significant increase in the number of amoebae in water layers where large amounts of algae and other eubacteria can serve as a food source. In the same way, *N. fowleri* has been isolated from strata that contain large amounts of exogenous iron and manganese [19]. Apparently, *N. fowleri* can grow in environments with a wide range of pH, so that pH does not seem to be a limiting factor for its distribution [19]. Pathogenic amoebae have been isolated from fresh water at 10°C and soil at 8°C; however, most strains of *N. fowleri* have been isolated from water with high temperatures, either natural or artificially heated. The temperature is in fact of paramount importance for the growth of *N. fowleri* populations, and these amoebae can survive at high temperatures, unlike other free-living microorganisms. Griffin [20] proposed a hypothesis that explains the success of this species in such habitats. The alteration of the normal environment by human actions (e.g., thermal pollution, chlorination of water) results in the disappearance of the protozoan normal fauna. This eliminates the competition and leaves empty and optimal the habitats of *N. fowleri*, which can colonize these habitats because they can withstand temperatures up to 45°C. Moreover, the amoeboflagellate stage of *N. fowleri* has the advantage of being a mobile phase that allows dispersal. In studies of population ecology, it has been observed that the numbers of this amoeba increase in the warmer months of the year, and can approach 5000 amoebae/L of water. A study in the northern region of Mexico during the hot months of the year reported a peak abundance of amoebae of 1724 amoebae/L and a minimum of 16 amoebae/L [21]. The number of *N. fowleri* in different water bodies has a direct relationship with the risk of acquiring the infection, and the French government has deemed as unsuitable for aquatic activities the water bodies in which the number of amoebae is approximately 100/L [22]. In Western Australia, an even lower limit has been implemented in the water park "Ski"; here, the allowable limit is 5 amoebae/L [23].

Soil amoebae can behave as vectors of other microorganisms. One of the microorganisms most often associated with *N. fowleri* is the bacterium *Legionella pneumophila*. *Naegleria* spp. usually destroy *L. pneumophila* within the first 24 h after phagocytosis. However, some *L. pneumophila* can survive and grow inside the amoebae, and Newsome et al. [24] suggested that *L. pneumophila* maintains its virulence through its association with *Naegleria* spp.

9.1.4 Pathogenesis

Once *N. fowleri* enters the nasal cavity of patients, the amoebae invade the olfactory neuroepithelium and migrate through the olfactory nerves, crossing the cribiform plate to reach the brain. The cribiform plate is more porous in children than in adults, which could be a possible explanation for the higher incidence of disease in children. The olfactory bulbs are the first areas of the brain to be invaded because of their proximity to the route of entry of the parasites; however, at later stages of infection the parasites can be located at the base of the brain and in the cerebellum.

The autopsy usually reveals inflammation of the brain with severe tissue damage throughout the area of invasion, with ulceration of the nasopharyngeal mucosa and necrosis of the olfactory nerves. Furthermore, there is hemorrhagic necrosis in the region of the olfactory bulbs. The histological analysis of brain sections shows a large number of trophozoites throughout the cerebral parenchyma, and particularly in perivascular areas.

It has been demonstrated experimentally that the trophozoites adhere to and penetrate the olfactory neuroepithelium but that they are unable to invade the respiratory epithelium and other tissues. Similarly, it has been shown that the invasion does not occur by the hematogenous route [25]. The trophozoites invade the nerves of the submucosal plexus by migrating through the olfactory nerves crossing the cribiform plate, to arrive at the olfactory bulbs of the CNS [26]. Later in the course of experimental infection, a significant inflammatory infiltrate develops with similar characteristics to those seen in postmortem analysis of human patients [8,27]. The use of animal models to study the pathogenesis of PAM has shown three main characteristics of the infection. (1) No inflammation is observed during the migration of trophozoites from the olfactory epithelium to the olfactory bulbs, excluding that observed initially in the nasal cavity, prior to penetration of the epithelial tissue by amoebae. (2) Damage to the brain tissue is not apparent during the early stages of invasion, and lytic behavior of the amoebae seems to be low *in vivo*. (3) There is an intense inflammatory infiltrate in the final stages of the infection and this is always associated with large areas of necrosis with hemorrhage, suggesting that inflammation plays an important role in the pathogenesis of PAM.

A detailed analysis of the morphological changes at later stages of infection of the brain parenchyma has shown the presence of trophozoites in the olfactory bulbs at 72 and 96 h postinoculation. Their number increases considerably, and the inflammatory foci surrounding the trophozoites can be visualized clearly until 102 h. These foci consist mainly of eosinophils and neutrophils. In the later stages of infection, tissue damage is characterized by several areas of necrosis with hemorrhagic zones and the presence of cell debris [27]. These lesions are associated with the presence of inflammatory cells, which show signs of cell lysis. These data suggest that inflammation contributes significantly to the tissue damage. To test this hypothesis, the role of the inflammatory response in the pathogenesis of PAM has been evaluated using $CD38^{-/-}$ knock-out mice that have a deficiency in the chemotaxis of neutrophils and macrophages in response to various stimuli. The results showed that the inflammation appeared later and was less intense when compared with the inflammatory response that occurs in mice of the parental strain. These data are directly related to survival. Based on these results, the authors suggested that the inflammatory response and cell lysis mediated by polymorphonuclear cells contribute significantly to the pathogenesis of PAM [27]. The delayed onset of inflammation can be attributed to the privileged immune condition of the CNS [28]. It has been reported that inflammation of the brain after an injection of lipopolysaccharide occurs after three days [29]. Moreover, the intracranial administration of pro-inflammatory cytokines, such as TNF-α, IL-1β, or IL-8, induces little or no exudate or infiltration of cells in the CNS parenchyma [30]. This is due to a strict control mechanism that regulates the inflammatory process in the brain [28]. *N. fowleri* is capable of inducing the production of pro-inflammatory signals in the brain [31]; however, it is possible that these signals are insufficient to initiate the inflammatory process. Inflammation begins only when the number of amoebae in the brain parenchyma is very high, and this leads to a concomitant increase in pro-inflammatory signals that disrupt the blood–brain barrier and result in an intense inflammatory exudate [27].

9.1.5 Epidemiology

The genetic heterogeneity of *N. fowleri* strains has been documented by restriction fragment-length polymorphism (RFLP) analysis of whole cell DNA [32,33]. Examination of a combination of isoenzyme profiles and RFLP analysis of geographically diverse *N. fowleri* isolates led to the recognition of different groups of strains from Australia and Europe, and the presence of both genotypic groups in the United States [3]. The populations of *N. fowleri* in the United States have been postulated to be the source of two groups that spread globally eastward and westward to Europe and to Australia and New Zealand, respectively. In another study involving RFLP analysis, the genotypic heterogeneity among strains of *N. fowleri* was confirmed, leading to the recognition of three groups: European, Oceanic, and American [34]. In addition, biochemical assays such as isoenzyme analysis have been developed for specific identification of *N. fowleri* amoebae cultured from the cerebrospinal fluid (CSF) and brain specimens from patients, as well as from environmental sources (water and soil) [35,36].

As mentioned above, most cases of PAM have occurred subsequent to recreational water activities in water bodies where the parasite is present [37]. Owing to the wide distribution of these amoebae, any warm water body that is not treated properly carries a risk of infection. As stated earlier, the parasite enters through the nostrils and travels through the olfactory neuroepithelium to reach the CNS [8,26]. The disease usually presents in young, previously healthy children with a recent history of swimming, and it therefore contrasts with the infections caused by amoebae of the genus *Acanthamoeba* and *Balamuthia mandrillaris*, which occur in immunocompromised individuals. An additional possible explanation for the increased incidence of PAM in children or in young people is that these people stay longer and are more active in the water activities [38]. The same explanation is given for the fact that the incidence of PAM is higher in boys than in girls. Recently, McKee et al. [39] studied the history of a case of PAM in one child of a total group of 15 children. This study concluded that the only child infected with *N. fowleri* had spent more than 2 h in the water, swimming and diving, and was more active than the rest of the group. In an effort to explain the contradiction between the low incidence of PAM and the wide distribution of this parasite, Cabanes et al. [22] constructed a mathematical model to calculate the risk of infection with *N. fowleri*. They found a maximum risk of 8.5×10^{-8} in water containing 10 trophozoites of *N. fowleri*/L; this represents a risk of approximately 10^{-6} for 1000 trophozoites/L.

So far, over 200 cases of PAM have been reported throughout the world, most of them from North America [38]. The Centers for Disease Control (CDC) in Atlanta, Georgia, reports four to six cases annually in the United States. The first case of PAM was detected in Mexico in 1978 in a 16-year-old boy who had been swimming in an irrigation canal in the valley of Mexicali, Baja California. This case was published in 1984 and since then, more than 30 cases of PAM have been reported in Mexico, from which over 80% are from Baja California State [40].

9.1.6 Clinical Feature

The presence of PAM should be suspected in young adults and children with acute neurological symptoms who have a history of recent activities in fresh water (lakes, rivers, swimming pools, or hot springs). Symptoms occur from one day to two weeks after exposure. Given that there are no distinctive clinical features that differentiate PAM from acute pyogenic or bacterial meningitis, it is very important that physicians obtain information on the clinical history of the patient related to contact with fresh water during the past week. The earliest symptoms appear suddenly and are characterized by bifrontal or bitemporal headache, high fever, and stiff neck, followed by anorexia, vomiting, and irritability. Nuchal rigidity usually occurs, with positive Kerning and Brudzinski signs. Photophobia occurs late in the clinical course, followed by neurological abnormalities, including diplopia, lethargy, altered mental status, seizures, and coma. Cranial nerve palsies (of the third, fourth, and sixth cranial nerves) may indicate brain edema. Intracranial pressure is usually raised to 600 mm H_2O. Death typically occurs three to seven days after the onset of symptoms. Autopsy findings usually show hemorrhagic necrosis of the olfactory bulbs and cerebral cortex [41–43].

9.1.7 Laboratory Diagnosis

9.1.7.1 Imaging Methods

Imaging features of the brain lesions in studies of amoebic meningoencephalitis that use computed tomography (CT) and magnetic resonance imaging (MRI) are nonspecific. A spectrum of findings has been seen in the form of multifocal parenchymal lesions, pseudotumoral lesions, meningeal exudates, hemorrhagic infarcts, and necrosis in the brain. In one case of PAM reported by Kidney and Kim [44], both CT and MRI revealed a pattern of brain edema and hydrocephalus with rapid progression of the disease.

9.1.7.2 Microscopy

Given that PAM is a rare disease but is almost always fatal, successful differential diagnosis and an appropriate choice of therapy depend on precise and rapid laboratory identification of *N. fowleri*. The disease is often misdiagnosed because no distinctive differences in diagnostic findings exist to distinguish PAM from viral or bacterial meningoencephalitis.

Laboratory examination of the CSF of patients with PAM is of importance to make a timely and accurate diagnosis. The CSF may vary from grayish to yellowish-white in color, and it sometimes becomes tinged red with a few erythrocytes (250 mm^{-3}) in the early stages of the disease. However, as the disease progresses, the red blood cell number increases to as high as 24,600 mm^{-3}. The leukocytes, predominantly comprised of polymorphonuclear cells, also vary from 300 to 26,000 cells mm^{-3} [41]. Bacteria are usually not present in the CSF. The CSF pressure is generally high (300–600 mm H_2O). The protein concentration varies from 100 mg per 100 mL to 1000 mg per 100 mL, and the glucose concentration may be 10 mg per 100 mL or lower [45,46]. In *Naegleria* infections, the diagnosis can be made by microscopic examination of CSF. A wet mount may detect motile trophozoites, and Giemsa or Wright-stained smears will show trophozoites with typical morphology. On microscopy, *N. fowleri* is recognized by the amoeboid form of the trophozoite. Their size is between 10 and 35 μm. The cytoplasm is granular, contains large dense karyosomes and lacks peripheral chromatin. A Gram stain is not useful in the identification of *N. fowleri*.

The amoebic flagellation response of trophic amoebae is often employed as a confirmatory test for recognition of environmental *Naegleria* spp. or to verify the identity of clinical amoebal isolates as *N. fowleri*. Some isolates, however, flagellate poorly under laboratory conditions, and in the absence of flagella the amoebae could be misidentified. In these cases, immunofluorescence microscopy using specific antibodies, enzyme-linked immunosorbent assay (ELISA) assays or polymerase chain reaction (PCR) is necessary to confirm the presence of *N. fowleri* [47].

9.1.7.3 Antibody Detection Methods for *N. fowleri*

Because most patients affected by PAM die soon after infection there is insufficient time to produce a detectable immune response. Previous attempts to detect an antibody response to *N. fowleri* using immunofluorescence assays have therefore been largely unsuccessful [38]. In studies conducted at the CDC using immunoblot assays on the serum of patients who had recovered from PAM, IgM was the main class of antibody generated [43]. On the other hand, polyclonal antibodies produced in rabbits, or monoclonal antibodies, can be used to identify amoebae in the CSF.

A commercially available ELISA (Indicia, Oulin, France) based on a specific monoclonal antibody (5D12) that recognizes a glycosylated epitope on *N. fowleri* can be used to diagnose the infection [48].

9.1.7.4 Molecular Techniques

Modern techniques using highly specific and sensitive PCR and real-time PCR assays have been developed for the detection of *N. fowleri* in clinical and in environmental samples [49–52]. Although these techniques are available in research laboratories, they are not available in most clinical laboratories. A PCR assay has been developed for the detection of *N. fowleri* using primers for the complete ribosomal internal transcribed spacer (ITS) region. This PCR allows the discrimination of *Naegleria* species, and a species-specific assay permits the detection of *N. fowleri* [49]. In addition, a PCR assay that detects *N. fowleri* in fresh brain tissue has been reported [53]. Qvarnstrom et al. [51] have also developed a fast and sensitive multiplex real-time PCR assay based on the use of probes targeting the partial or full-length nuclear small subunit ribosomal gene (18 sRNA gene) for simultaneous detection of *Naegleria*, *Balamuthia*, and *Acanthamoeba*.

9.2 METHODS

9.2.1 Sample Preparation

9.2.1.1 Samples for Microscopy

For clinical diagnosis using CSF, amoebae can be cultured in plates coated with bacteria or by placing into axenic growth medium. The trophozoites can be fixed with 2% paraformaldehyde and stained with Giemsa for light microscopy analysis [54]. Trophozoites of *N. fowleri* can also be identified in CSF by specific indirect immunofluorescence antibody essays using a polyclonal or monoclonal antibody bound to a fluorescein-conjugated antibody (Figure 9.1b) [55,56].

In autopsy cases, brain specimens are fixed with 10% buffered formalin for histological examination. Amoebae can be observed in tissues following H&E staining or by immunolabeling (Figure 9.1e) [57].

9.2.1.2 Samples for PCR

The early applications of PCR methodology were for the specific amplification of *N. fowleri* DNA in order to identify the amoeba in environmental samples [48,58]. The method was sensitive enough to detect a single amoeba, and was also able to detect cysts [48,58]. More recently, a species-specific probe was designed from cDNA for use with nested PCR assays [4,50,59]. The technique does not require cultivation of, or DNA extraction from, the amoebae, and is sensitive enough to detect 5 pg (five trophic amoebae) of *N. fowleri* (Figure 9.1f).

Sequencing of the 5.8S rRNA gene and the internal transcribed spacers 1 and 2 (ITS1 and ITS2) of *N. fowleri* has shown that this genomic region can be used to differentiate strains of *N. fowleri* [49,59,60]. Two species-specific primers that are complementary to variable end-regions of ITS1 and ITS2 have been designed, making it possible to distinguish *N. fowleri* from other *Naegleria* spp. [49]. In addition, based on the sequencing of the ITS of clinical isolates, it was shown that two strains of *N. fowleri*, isolated from two patients affected by PAM who visited the same hot spring in California but at different times, belonged to the same genotype [59]. A recently developed real-time multiplex PCR assay identifies DNA in the CSF and brain tissue samples obtained from PAM patients. This test identifies three genotypes known to be present in the United States. The time taken to identify *N. fowleri* from the arrival of the specimen at the laboratory is about 5 h and this can be shortened to 2 h if the specimen is CSF. It is important that the diagnosis is obtained in the shortest time possible because of the acute progression of the disease and the requirement for rapid treatment. A new, sensitive, rapid, and discriminating technique that uses a single primer set and the DNA intercalating dye SYT09 for real-time PCR and melting-curve analysis of the 5.8S rRNA gene flanking the ITS has been developed that distinguishes several *Naegleria* species in an environmental sample [61]. The new Matrix-Assisted Laser Desorption/Ionization-Time of Flight-Mass Spectrometry (MALDI-TOF-MS) technique [62] can identify *N. fowleri* specifically.

9.2.2 Detection Procedures

9.2.2.1 Microscopic Examination

The trophozoite or invasive form of *N. fowleri* can be observed by direct light microscopy (Figure 9.1a). An alternative and corroborative procedure is the so-called "enflagellation," which consists in placing the trophozoites in an isotonic saline solution to transform to a flagellate form within 2–4 h at 35.5°C. This flagellate form can also be observed by light microscopy.

When antibodies are used for immunofluorescence, the positive reaction can be achieved in CSF samples from *N. fowleri*-infected patients. The sample is first washed with an isotonic buffer and then incubated with polyclonal or monoclonal antibodies directed against *N. fowleri*; after washing again, a secondary antibody labeled with fluorescein isothiocyanate (FITC) is used and the sample is examined finally under a fluorescence microscope. Strong FITC-positive cells must be observed in the sample to confirm the diagnosis (Figure 9.1b).

Immunohistochemical assay has been used to confirm the presence of *N. fowleri* trophozoites in brain tissue from autopsy cases. The technique is described by Guarner et al. [63] as follows: tissue sections of 3 μm thick are deparaffinized and rehydrated through graded alcohols and distilled water. Tissue sections are washed in Tris-saline-Tween 20 buffer (pH 7.5) and placed in a DAKO Autostainer (Dako, Carpinteria, CA, USA). Digestion is performed by soaking the slides in 0.1 mg/mL proteinase K (Roche, Indianapolis, IN, USA) in 0.6 Tris-saline-Tween/0.1% $CaCl_2$ (buffer), followed by a blocking step with 20% normal sheep serum in Tris-saline-Tween. The tissue sections are incubated for 60 min with polyclonal rabbit antibodies against *N. fowleri*. The slides are washed, and a porcine anti-rabbit biotinylated link antibody, alkaline phosphatase-labeled streptavidin, and naphthol fast red chromogenic substrate are applied sequentially. The sections are then counterstained in Mayer's hematoxylin, mounted with a coverslip, and analyzed under a standard light microscope.

9.2.2.2 Molecular Identification

9.2.2.2.1 cDNA Library

Previous to the PCR technique, a cDNA library is necessary to obtain. This method has been reported in detail by Réveiller et al. [50] An expression cDNA library for *N. fowleri* is obtained from amoebic RNA. The RNA is purified using the guanidinium isothiocyanate method [64] and polyA + mRNA is selected using an oligo-dT cellulose column [65]. cDNA is synthesized using superscript reverse transcriptase and the BRL Superscript plasmid system (Bethesda Research Laboratory, Gaitherburg, MD, USA). The double-stranded DNA is constructed with *Not*I and *Sal*I adapters to enable directional cloning. Portions of cDNA greater than 500 bp long are ligated into a predigested *Not*I and *Sal*I cut pSPORT1 plasmid vector. Vector ligand cDNA is introduced to *Escherichia coli* ELECTROMAX DH10B cells by electroporation. Blue/white screening is performed on Luria/Bertani medium containing ampicillin, 5-bromo 4-chloro-3-indolyl-β-D-galactoside, and isopropylthiol-β-D-galactoside (IPTG). White colonies are selected. The cDNA expression library is induced by growing *E. coli* in the presence of IPTG on nitrocellulose filters. The library is then immunoscreened with rabbit polyclonal antiserum against *N. fowleri* and antiperoxidase-conjugated goat anti-rabbit IgG antiserum. Antibody reactivity is visualized using a chloro-naphthol/H_2O_2 substrate. Colonies positive for the expression of *N. fowleri* proteins are purified and the clones are sequenced using the Sequenase version 2.0 sequencing kit (United States Biochemical, Cleveland, OH, USA).

9.2.2.2.2 Restriction Digestion of gDNA and Southern Blot Analysis

To determine whether the cDNA hybridize with the genomic DNA (gDNA) of several *Naegleria* spp., a Southern blot can be performed. The gDNA of each species of *Naegleria* is isolated using the Easy-DNA kit (Invitrogen, Carlsbad, CA, USA). The gDNA (10 μg) is digested with *Eco*RI (30 U) for 4 h at 37°C. The digested and undigested samples of gDNA (10 μg /lane) are separated by electrophoresis through a 0.8% agarose gel and transferred by capillary action on to a supported nitrocellulose membrane (MSI, Westborough, MA, USA) with 29SSC as transfer buffer. The blot is hybridized overnight with a ^{32}P-labeled (dATP, 3000 Ci/mM; ICN, Costa Mesa, CA, USA) cDNA insert in hybridization buffer (50% formamide, 5 × SSC, 5 × Denhardt's solution, 250 μg/mL yeast tRNA, 50 mM sodium phosphate pH 6.5, 0.1% SDS) at 42°C. The blot is washed thoroughly three times with 2 × SSC containing 0.1% SDS (20 min each at 21–23°C), twice with 0.1 × SSC containing 0.1% SDS (30 min each at 62°C), and exposed to Kodak XAR film for three days.

9.2.2.2.3 Restriction Digestion of the cDNA Insert

This technique was reported by Réveiller et al. [50] and is described as follows: plasmid DNA can be prepared from *E. coli* clone Mp2C15 using a Wisard Plus Miniprep Kit (Promega, Madison, WI, USA) and co-digested (10 μg) with *Sal*I and *Xba*I (30 U) overnight at 37°C. The digestion products are loaded on to a 1% agarose gel and electrophoresed. The two restriction fragments of the cDNA insert are purified from the gel using the QIAquick Gel Extraction kit (Qiagen, Valencia, CA, USA). The samples are radiolabeled with α^{32}-P-dATP as described above and used as two different probes for Southern blot analysis.

9.2.2.2.4 Nested PCR

The PCR assay published by Réveiller et al. [50] is a nested PCR, of which the main feature is to amplify a fragment of gDNA from *N. fowleri* and a second fragment within the first one, thus increasing the sensitivity and specificity of the test. The construction of the primers was based on the sequence data from the clone Mp2C15. The forward primer Mp2C15 (5′-TCTAGAGATCCAACCAATGG-3′ (positions 435–454)) and the reverse primer Mp2C15 (5′ATTCTATTCACTCCACAATCC-3′ (positions 582–601)) amplify a 166-bp fragment of the Mp2C15.P sequence. The PCR is performed in a final volume of 50 μL consisting of Taq DNA polymerase buffer (10 mM Tris-HCl (pH 8.3), 50 mM KCl, 2.5 mM $MgCl_2$), 0.2 mM of deoxynucleoside triphosphate (dNTP), 0.6 μM of primer, 2.5 U of Taq polymerase, and the gDNA. The temperature program is 5 min at 95°C for one cycle and 1 min at 95°C, 1 min at 65°C, and 2 min at 72°C for 35 cycles. For the negative control PCR-grade water lacking the DNA template is used. The positive control consists of 10 ng of plasmid DNA template. The products are analyzed on a 1.5% agarose gel stained with ethidium bromide.

To increase the sensitivity of the test, the first PCR amplification with Mp2C15 forward and Mp2C15 reverse are followed by an additional PCR step. The primers used for this reaction are located within Mp2C15 forward and Mp2C15 reverse, making this a nested PCR. The forward primer in Mp2C15 is (5′-GTACATTGTTTTTATTAATTTCC-3′ (positions 458–480)) and the reverse primer in Mp2C15 is (5′GTCTTTGTGAAAACATCACC-3′ (positions 549–548)), which amplify a 110-bp fragment of the Mp2C15.P sequence. The second (nested) PCR is also performed in a 50 μL volume consisting of Taq DNA polymerase buffer, 0.2 mM of dNTP, 0.5 μM of each primer, 2.5 U of Taq polymerase and 2 μL of the first PCR product. The standard temperature program is 1 min at 95°C, 1 min at 55°C, and 1 min at 72°C for 35 cycles. The negative control is PCR-grade water lacking a DNA template. The positive control is a 1 μL volume of the first PCR product from the positive control sample, diluted 50 times in PCR-grade water.

9.2.2.2.5 Quantitative PCR (qPCR)

Real-time PCR melting-curve analysis allows differentiation of several free-living species of amoeba [66]. When qPCR is used, distinguishing features can be found for *N. fowleri* and other *Naegleria* species and even for other free-living amoebae, such as *Hartmanella vermiformis* and *Willaertia magna*. The species-specificity of the amplicons is confirmed using agarose gel electrophoresis. The amplification efficiency is above 90%, indicating the quantitative potential of the assay. The amplification reaction is achieved in a final mixture volume of 30 μL containing PCR buffer, 0.25 μM of each primer, 0.125 μM of dNTPs, 0.25 mM of dUTP, 0.625 U of Taq DNA polymerase, 2 mM of $MgCl_2$, and 5 μL of template. The temperature-cycling conditions for the PCR are started with 2 min at 50°C and 10 min at 95°C, followed by 50 cycles of 30 s at 95°C, 1 min at 57°C, and 1 min at 60°C. The PCR products are run electrophoretically on 1.5% agarose gels stained with ethidium bromide to check the amplicon size.

9.2.2.2.6 Southern Blot Analysis of Nested PCR Products

The PCR products can be confirmed to be derived from *N. fowleri* by internal probe hybridization. A measure of 5 ng of gDNA extracted from various isolates of *Naegleria* is amplified by nested PCR. A volume of 2 μL of PCR products diluted 1:50 in distilled water is separated by electrophoresis through a 1.5% agarose gel and transferred by capillary action on to a supported nitrocellulose membrane with transfer buffer, then Southern blot is performed. The blot is hybridized overnight with an internal probe ^{32}P-labeled in hybridization buffer (6 × SSC, 1 × Denharedt's solution, 100 μg/mL yeast tRNA, and 0.05% sodium pyrophosphate) at 48°C. The internal probe, termed as Mp2C15.Sprobel (5′-GACAGCCTTCAACAAGTGATCTT-3′, positions 493–515)), which is located between inner primers Mp2C15.for-in and Mp2C15.rev-in, and was designed based on the sequence data of Mp2C15.P, is 5′end-radiolabeled with [γ^{32}-P]dATP (7000 Ci/mM; ICN, Costa Mesa, CA, USA), using the forward T4 kinase reaction (RTS T4 Kinase Labeling System, Bethesda Research Laboratory, Gaithersburg, MD, USA). The blot is washed five times with 6 × SSC (15 min each at room temperature) and once (30 min at 65°C) with 0.05% sodium pyrophosphate, and is exposed to Kodak XAR film for 3 h [50].

9.3 CONCLUSIONS AND FUTURE PERSPECTIVES

The ample distribution of free-living amoebae in nature, in soil as well as in different bodies of water, could lead one to suppose in principle that a large number of encephalopathies caused by *N. fowleri* should be reported. However, these infections are considered to be relatively rare. However, there has been an increase in cases of PAM in recent years. The increase in knowledge of the disease, which includes its clinical features, the physiopathological mechanisms of the parasitic infection, and the biology of the etiological agent, have all contributed, at least partially, to more accurate and opportune diagnosis of the infection. However, in rural areas of developing countries accessibility to modern diagnostic procedures and, even more pertinently, to relatively simple laboratory analyses, are lacking. Added to the particular characteristics of the infection and the absence of adequate medical services, this greatly increases the possibility that cases of PAM are either misdiagnosed, treated inappropriately, or simply ignored.

The introduction of highly sophisticated methodologies of molecular biology has improved our knowledge of *N. fowleri*, especially in basic areas related to signaling and transduction pathways, intrinsic metabolic mechanisms, genetics and the taxonomic aspects of the parasite. However, up to the present, all these data have contributed only little to implementing better, and especially rapid, opportune, and sensitive diagnostic tests for PAM. Some promising techniques that are being improved and implemented in clinical laboratories are the rapid ELISA, the fluorescent assays using antibodies to detect amoebae in CSF and the use of PCR probes, among others. However, as mentioned, use of these technologies in developing countries is greatly limited.

In addition to the requirement for rapid diagnosis of PAM, the specific treatment of this severe and commonly fatal infection constitutes an additional unsolved problem. The low efficacy of the available anti-amoebic treatment has contributed to a high rate of mortality in PAM, and pharmacological research into novel and more specific drugs is needed urgently.

Prevention seems to constitute one of the best ways to lower the morbidity and mortality caused by this parasite. Although it is difficult to prevent exposure, owing to the ubiquity of the organisms, there are different measures that have been carried out in areas considered as endemic. Posting warning signs near soil, lakes, irrigation canals, or other bodies of water suspected to be contaminated is an important step for preventing PAM. Strict parasitological analysis of the water used in swimming pools in recreational areas constitutes a control measure used already in many countries. Another necessary measure is the introduction of educational programs by the health authorities directed at primary care physicians in endemic areas, which will allow them to be aware of and to detect rapidly patients with suspected *N. fowleri* infection.

In conclusion, although many aspects of the parasite have been investigated, including both basic science and clinical features, there are still several unsolved practical problems, mainly related to obtaining a rapid and efficient diagnosis and the specific pharmacological treatment of PAM. Research currently being carried out by different groups around the world will hopefully contribute to a better understanding of the physiopathology of the disease and, in consequence, to an improvement in the management and treatment of this important free-living amoebal infection of the CNS.

REFERENCES

1. Rodríguez-Zaragoza, S. Ecology of free-living amoebae. *Crit. Rev. Microbiol.* 20, 225, 1994.
2. Marciano-Cabral, F. Biology of *Naegleria* spp. *Microbiol. Rev.* 52, 114, 1988.
3. De Jonckheere, J.F. Geographic origin and spread of pathogenic *Naegleria fowleri* deduced from restriction enzyme patterns of repeated DNA. *BioSystems* 21, 269, 1988.

4. De Jonckheere, J.F. Molecular definition and ubiquity of species in the genus *Naegleria. Protist* 155, 89, 2004.
5. De Jonckheere, J.F. Key note lecture: Molecular taxonomy of the genus *Naegleria* and other vahlkampfiids. *XIth International Meeting on the Biology and Pathogenicity of Free-Living Amoebae Proceedings*, Ceské Budejovice, Czech Republic, 2005.
6. Schuster, F. An electron microscope study of the amoebo-flagellate, *Naegleria gruberi* (Schardinger) I. The ameboid and flagellate stage. *J. Protozool.* 10, 297, 1963.
7. Fulton, C., and Dingle, A.D. Basal bodies but not centrioles, in *Naegleria. J. Cell Biol.* 51, 826, 1971.
8. Carter, R.F. Description of a *Naegleria* sp. isolated from two cases of primary amoebic meningoencephalitis, and of the experimental pathological changes induced by it. *J. Pathol.* 100, 217, 1970.
9. Feldman, M.R. *Naegleria fowleri*. Fine structural localization of acid phosphatase and heme proteins. *Exp. Parasitol.* 41, 290, 1977.
10. Patterson, M. et al. Ultrastructure of *Naegleria fowleri* enflagellation. *J. Bacteriol.* 147, 217, 1981.
11. Marshall, M.M. et al. Waterborne protozoan pathogens. *Clin. Microbiol. Rev.* 10, 67, 1997.
12. Werth, J.M., and Kahn A.J. Isolation and preliminary chemical analysis of the cyst wall of the amoeba-flagellate *Naegleria gruberi. J. Bacteriol.* 94, 1272, 1967.
13. Fulton, C., and Guerrini, A.M. Mitotic synchrony in *Naegleria* ameba. *Exp. Cell Res.* 56, 194, 1969.
14. Willmer, E.N. Factors which influence the acquisition of flagella by the amoeba. *Naegleria gruberi. J. Exp. Biol.* 33, 583, 1956.
15. De Jonckheere, J.F., and Brown, S. SSU rDNA analysis reveals the existence of another *Naegleria* sp. with dividing flagellates: *N. robinsoni* sp. nov. *Eur. J. Protisol.* 35, 264, 1999.
16. De Jonckheere, J.F. et al. The amoeba-to-flagellate transformation test is not reliable for the diagnosis of the genus *Naegleria*. Description of three new *Naegleria* spp. *Protist.* 152, 115, 2001.
17. John, D.T., Cole, T.B., and John, R.A. Flagella number among *Naegleria* flagellates. *Folia Parasitol.* 38, 289, 1991.
18. Singh, B.N., and Das, S.R. Intranasal infection of mice with flagellate stage of *Naegleria aerobia* and its bearing on the epidemiology of human meningo-encephalitis. *Curr. Sci.* 41, 625, 1972.
19. Kyle, D.E., and Noblet, A.D. Vertical distribution of potentially pathogenic free-living amoebae in freshwater lakes. *J. Protozool.* 32, 99, 1985.
20. Griffin, J.L. The pathogenic amoeboflagellate *Naegleria fowleri* environmental isolations, competitors, ecologic interactions, and the flagellate-empty hypothesis. *J. Protozool.* 30, 403, 1983.
21. Lares-Villa F. A quantitative study of thermophilic *Naegleria* during summer season, in natural waters used for recreational purposes in Sonora, México. *XIth International Meeting on the Biology and Pathogenicity of Free-Living Amoebae Proceedings*, Ceské Budejovice, Czech Republic, 2005.
22. Cabanes, P.A. et al. Assessing the risk of primary amoebic meningoencephalitis from swimming in the presence of environmental *Naegleria fowleri. Appl. Environ. Microbiol.* 67, 2927, 2001.
23. Lugg, R., and Psaila-Savona, P. A legal standard for thermophilic *Naegleria* in freshwater body in Western Australia. *IXth International Meeting on the Biology and Pathogenicity of Free-Living Amoebae Proceedings*, Paris, France, p. 291, 2001.
24. Newsome, A.L. et al. Interaction between *Naegleria fowleri* and *Legionella pneumophilia. Infect. Immun.* 50, 449, 1985.
25. Jarolim, K.L., McCosh, J.K., and Howard, M.J. The role of blood vessels and lungs in the dissemination of *Naegleria fowleri* following intranasal inoculation in mice. *Folia Parasitol.* 49, 183, 2002.
26. Jarolim, K.L. et al. A light microscopy study of the migration of *N. fowleri* from the nasal submucosa to the central nervous system during the early stage of primary amebic meningoencephalitis in mice. *J. Parasitol.* 86, 50, 2000.
27. Cervantes-Sandoval, I. et al. Characterization of brain inflammation during primary amoebic meningoencephalitis. *Parasitol. Int.* 57, 307, 2008.
28. Galea, I., Bechmann, I., and Perry, V.H. What is immune privilege (not)? *Trends. Immunol.* 28, 12, 2007.
29. Andersson, P.B., Perry, V.H., and Gordon, S. The acute inflammatory response to lipopolysaccharide in CNS parenchyma differs from that in other body tissues. *Neuroscience*, 48, 169, 1992.
30. Andersson, P.B., Perry, V.H., and Gordon, S. Intracerebral injection of proinflammatory cytokines or leukocyte chemotaxins induces minimal myelomonocytic cell recruitment to the parenchyma of the central nervous system. *J. Exp. Med.* 176, 255, 1992.
31. Oh, Y.H. et al. Cytopathic changes and pro-inflammatory cytokines induced by *Naegleria fowleri* trophozoites in rat microglial cells and protective effects of an anti-Nfa1 antibody. *Parasite. Immunol.* 27, 453, 2006.
32. De Jonckheere, J.F., Characterization of *Naegleria* species by restriction endonuclease digestion of whole-cell DNA. *Mol. Biochem. Parasitol.* 24, 55, 1987.
33. McLaughlin, G.L., Brandt, F.H., and Visvesvara, G.S. Restriction fragment length polymorphisms of the DNA of the selected *Naegleria* and *Acanthamoeba* amoebae. *J. Clin. Microbiol.* 26, 1655, 1988.
34. Kilvington, S., and Beeching, J. Detection of a novel restriction fragment length polymorphism type in *Naegleria fowleri* (free-living amoeba) isolates from electricity power stations in England and France. *Eur. J. Protisol.* 33, 186, 1997.
35. De Jonckheere, J.F. Isoenzyme patterns of pathogenic and non-pathogenic *Naegleria* spp using agarose gel isoelectrofocusing. *Ann. Microbiol.* 133A, 319, 1982.
36. Moss, D.M. et al. High- resolution polyacrylamide gradient gel electrophoresis (PGGE) of isoenzymes from *Naegleria* species. *J. Protozool.* 38, 26, 1988.
37. John, D.T., and Nussbaum, S.L. *Naegleria fowleri* infection acquired by mice through swimming in amebae-cantaminated water. *J. Parasitol.* 69, 871, 1983.
38. Schuster, F.L., and Visvesvara G.S. Free-living amoeba as opportunistic and non-opportunistic pathogens of humans and animals. *Int. J. Parasitol.* 34, 1001, 2004.
39. Mckee, T. et al. Primary amoebic meningoencephalitis-Georgia, 2002. *MMWR* 52, 962, 2003.
40. Lares-Villa, F. Free-living amoebae infections in Mexico. *IXth International Meeting on the Biology and Pathogenicity of Free-Living Amoebae Proceedings*, Paris, France, p. 13, 2001.
41. Hannisch, W., and Hallagan, L.F. Primary amebic meningoencephalitis: A review of the clinical literature. *Wild. Environ. Med.* 8, 211, 1997.
42. Schumacher, D.J., Tien, R.D., and Lane, K. Neuroimaging findings in rare amebic infections of the central nervous system. *Am. J. Neuroradiol.* 16, 930, 1993.
43. Visvesvara, G.S., Moura, H., and Schuster F.L. Pathogenic opportunistic free-living amoebae: *Acantamoeba* spp.,

Balamuthia mandrillaris, *Naegleria fowleri*, and *Sappinia diploidea*. FEMS *Immunol. Med. Microbiol.* 50, 1, 2007.
44. Kidney, D.D., and Kim, S.H. CNS infections with free-living amebas: Neuroimaging findings. *Am. J. Roentgenol.* 17, 809, 1998.
45. Martinez, A.J. *Free-Living Amebas: Natural History, Prevention, Diagnosis, Pathology and Treatment of Disease*, CRC Press, Boca Raton, FL, 156pp, 1985.
46. Visvesvara, G.S., and Maguire, J.H. Pathogenic and opportunistic free-living amebas: *Acanthamoeba* spp., *Balamuthia mandrillaris*, *Naegleria fowleri*, and *Sappinia diploidea*. In Guerrant, R.L., Walker, D.H., and Weller, P.F. (eds), *Tropical Infectious Diseases. Principles, Pathogens, & Practice*, Vol. 2, Churchill Livingstone, Philadelphia, PA, pp. 1114–11125, 2006.
47. Behets, J. et al. Detection of *Naegleria* spp. and *Naegleria fowleri*: A comparison of flagellation tests, ELISA and PCR. *Water. Sci. Technol.* 47, 117, 2003.
48. Sparagano, O. et al. Use of monoclonal antibodies to distinguish pathogenic *Naegleria fowleri* (cysts, trophozoites, or flagellate forms) from other *Naegleria* species. *J. Clin. Microbiol.* 31, 2758, 1993.
49. Pelandakis, M., Serre, S., and Pernin, P. Analysis of the 5.8S rRNA gene and the internal transcribed spacers in *Naegleria* spp. and in *N. fowleri*. *J. Eukaryot, Microbiol.* 47, 116, 2000.
50. Réveiller, F., Cabanes, P.A., and Marciano-Cabral, F. Development of a nested PCR assay to detect the pathogenic free-living amoeba *Naegleria fowleri*. *Parasitol. Res.* 88, 443, 2002.
51. Qvarnstrom, Y. et al. Multiplex real-time PCR assay for simultaneous detection of *Acanthamoeba* spp., *Balamuthia mandrillaris*, and *Naegleria fowleri*. *J. Clin. Microbiol.* 44, 3589, 2006.
52. Behets, J. et al. A duplex real-time PCR assay for the quantitative detection of *Naegleria fowleri* in water samples. *Water. Res.* 41, 118, 2007.
53. Schild, M. et al. PCR-based diagnosis of *Naegleria* sp. Infection in formalin-fixed and paraffin embedded brain sections. *J. Clin. Microbiol.* 45, 564, 2007.
54. Cervantes-Sandoval, I. et al. Characterization of *Naegleria fowleri* strains isolated from human cases of primary amoebic meningoencephalitis in Mexico. *Rev. Invest. Clin.* 59, 342, 2007.
55. Marciano-Cabral, F., and Cabral, G. The immune response to *Naegleria fowleri* amebae and pathogenesis of infection. *FEMS Immunol. Med. Microbiol.* 51, 116, 2007.
56. Martínez, A.J., and Visvesvara, G.S. Laboratory diagnosis of pathogen free-living amoebas: *Naegleria*, *Acanthamoeba*, and *Leptomyxid*. *Clin. Lab. Med.* 11, 861, 1991.
57. Da Rocha-Azevedo, B., Tanowitz, H.B., and Marciano-Cabral, F. Diagnosis of infections caused by pathogenic free-living amoebae. *Interdiscip. Perspect. Infect. Dis.* 2009, 251406, 2009.
58. Kilvington, S., and Beeching, J. Identification and epidemiological typing of *Naegleria fowleri* with DNA probes. *Appl. Environ. Microbiol.* 61, 2071, 1995.
59. Zhou, L. et al. Genetic variations in the internal transcribed spacer and mitochondrial small subunit rRNA gene of *Naegleria* spp. *J. Eukaryot. Microbiol.* 50, 522, 2003.
60. De Jonckheere, J.F. Sequence variation in the ribosomal internal transcribed spacer, including 5.8S, of *Naegleria* spp. *Protist* 149, 221, 1998.
61. Robinson, B.S., Monis, P.T., and Dobson, P.J. Rapid, sensitive, and discriminating identification of *Naegleria* spp. by real-time PCR and melting-curve analysis. *Appl. Environ. Microbiol.* 72, 5857, 2006.
62. Moura, H. et al. Detection of protein fingerprints of emerging and opportunistic pathogens by MALDI-TOF and proteome database search. *Am. J. Mass Spectrom.* 14, S615, 2003.
63. Guarner, J. et al. Histopathologic spectrum and immunohistochemical diagnosis of amebic meningoencephalitis. *Mod. Pathol.* 20, 1230, 2007.
64. Chirgwin, J.M. et al. Isolation of biologically active ribonucleic acid from sources enriched in ribonuclease. *Biochemistry* 18, 5294, 1979.
65. Aviv, H., and Leder, P. Purification and properties of biologically active globin messenger RNA by chromatography on oligothymidylic acid-cellulose. *Proc. Natl Acad. Sci. USA*, 69, 1408, 1972.
66. Behets, J. et al. Quantitative detection and differentiation of free-living amoeba species using SYBR Green-based real time PCR melting curve analysis. *Curr. Microbiol.* 53, 506, 2006.

10 Pentatrichomonas

Tania Crucitti

CONTENTS

10.1 INTRODUCTION

10.1.1 Classification, Morphology, and Biology

Pentatrichomonas hominis belongs to the order of Trichomonadida, which is a group of flagellated protists of the phylum Parabasala [1]. One of the major subfamilies is the Trichomonadinae, including the genus *Trichomitus*, *Trichomonas*, *Tetratrichomonas*, and *Pentatrichomonas*. These species have four to six flagella, which are free or attached to an undulating membrane. The organisms possess a rigid microtubular axostyle, which runs through the body and appears to project from its posterior end. They also have a pelta, a costa, and filamenta, and lamellae associated with the basal bodies. They are facultative anaerobic protists lacking mitochondria and peroxisomes; instead, they contain an unusual organelle involved in carbohydrate production that is called a hydrogenosome. The lifecycle of trichomonads is not yet fully understood, but generally the flagellates reproduce by mitotic division and longitudinal binary fission. True cyst forms are not formed by species found in humans.

More than 100 separate species of the Trichomonadinae are distinguished; most of them are commensal organisms of birds and mammals. Four of the species are found in humans: *Trichomonas vaginalis*, *Trichomonas tenax*, *Dientamoeba fragilis*, and *P. hominis*. *T. vaginalis* is found in the urogenital tract; *T. tenax* is found in the oral cavity; *D. fragilis*; and *P. hominis* are both found in the intestinal tract.

P. hominis was observed by the French medical scientist Casimir Joseph Davaine in 1860. The nomenclature and morphology have been subject for discussion over the years. It was suggested by Wenrich in 1947 that the parasite, which Davaine called *Trichomonas hominis*, should be placed under the genus *Pentatrichomonas*, based on the characteristic of the five anterior flagella and being so different from *T. vaginalis* [2]. *P. hominis* is the name used today, and the names *Trichomonas intestinalis* and *T. hominis* are used as synonyms.

P. hominis is pear- or teardrop-shaped, 6–15 μm long and 4–6.5 μm wide. It has five anterior flagella, with the fifth anterior flagellum originating quite apart from the four typical flagella. The movements of the fifth flagellum resemble more the whipping movements of the recurrent flagellum. The recurrent flagellum emerges from the cell dorsal to the anterior flagella, a short distance posterior to the periflagellar canal. The undulating membrane extends for the length of its body, along with the axostyle, which forms posteriorly a pointed spine [3]. The costa is about the same length as the undulating membrane. The nucleus with small karyosome is situated close to the origin of the flagella. In contrast to *T. vaginalis* and *T. tenax*, *P. hominis* does not form pseudopods commonly, but autotomy has been observed.

P. hominis is found in humans, and morphologically identical trophozoites are found in a number of animal species including monkeys, cats, dogs, cattle, and rodents. Experimental cross-infections support the conclusion that they are also physiologically identical [2].

Very little is known about the parasite and its biochemical characteristics. Biochemical studies on *P. hominis* are scarce, and only a few researchers studied the carbohydrate utilization and the enzyme activities of superoxide dismutase and cysteine protease [4–6]. The purine metabolism has been investigated and the purine salvage pathway shows not only some similarities but also differences with the salvage pathways of *T. vaginalis* [7].

The 5.8S rRNA gene and internal transcribed spacer (ITS) regions of trichomonads were sequenced for taxonomy purposes. The study showed that *P. hominis* is more closely related to the genus *Trichomonas* than to *Tritrichomonas*. The ITS regions flanking the 5.8S rRNA gene are very short and the authors question if this protozoon diverged early in the eukaryotic evolution [8].

The organism has a worldwide distribution, but could be more prevalent in tropical and subtropical regions. A summary of 72 publications addressing the prevalence rates of *P. hominis*, covering the period of 1912–1985, is presented in the book by Honigberg. The listed publications differ in geographical location, study population, and method of stool examination. In general, the frequency of *P. hominis* infection is low, and varies from 0.1% to 30.9% [9]. Table 10.1 presents more recent publications including *P. hominis* detection rates, and the same variations in frequencies are noted, ranging from 0.01% to 39.6% [10–28].

In general, the parasite was detected more frequently among children with diarrhea or among children having abdominal complaints. A surveillance of diarrheal diseases in children in China found a large number of *P. hominis*, mainly during the autumn and winter seasons, suggesting a seasonal variability [14]. Interestingly is the study performed in Nepal by Sherchand et al., where they found *P. hominis* in only an adult population [19]. In overall, general hygienic standards, such as the method of disposal of excrements, purity of drinking water, and crowding, are associated with infection by this organism. This also explains why the prevalence of the parasite is higher in young children who play in the soil and place everything in their mouth.

10.1.2 Clinical Features and Pathogenesis

P. hominis is regarded as a harmless commensal when found in the human gastrointestinal tract. In many instances the infection is asymptomatic, but it may cause diarrhea and has been associated with abdominal discomfort [14,17]. The mechanisms of how the parasite causes diarrhea are not known. Mancilla-Ramirez et al. reported the case of a female newborn infant with protracted diarrhea. The diarrhea persisted for >30 days until furazolidone treatment was initiated, after which they identified *T. hominis* trophozoites [29]. In 1988, Chunge et al. reported 45 cases of long standing (1–3 months) afebrile diarrhea in patients of 9–22 years old. The stools were examined by direct wet mount microscopy, and all showed heavy loads of *P. hominis*. The authors observed that the stools of the infected individuals did not contain blood and only very little mucus. They therefore suggested that the mechanism of diarrhea is most probably irritation of the intestinal mucosa. The patients were treated with Tinidazol until parasitological cure was achieved [30].

Like *T. vaginalis* and *T. tenax*, *P. hominis* has no true cyst stage, and infection with the parasite occurs through the ingestion of the *P. hominis* trophozoites. Pseudocysts, which are rounded-up immotile forms with internalized flagella are described *in vitro*, but were never observed in fresh material.

Food, drinking water, fingers, or eating devices can be contaminated or soiled with infected feces. The trophozoites can survive under various environmental conditions in feces, water, and insects. The organisms can live up to 8 days in feces and up to 7 days in fecal samples on moist garden soil; the survival time is shorter on sandy soil [31]. The parasite remains viable in tap water, distilled water, and spring water for several days. Synanthropic flies, particularly the common house fly (*Musca domestica*), have been identified as vectors of protozoan parasites such as *P. hominis*. The flagellates can be carried on with feces-contaminated legs of the insect, or are excreted by the insect after feeding on contaminated food [32]. The trophozoites are resistant to saliva and gastric juices [33], and when Dobbel infected himself with a *P. hominis* strain isolated from macaques, the infection persisted for 4½ years without causing symptoms [34].

P. hominis has occasionally been identified outside the gastrointestinal tract. A fatal case of systemic lupus erythematosus in a woman harboring *P. hominis* in both stool and an exudative pleural effusion has been reported. The identification of *P. hominis* was confirmed by DNA sequencing [35].

Masur et al. describes a case of meningitis in a 69-year-old man due to a mix of microbial organisms and *Trichomonas* sp. The motile trichomonads were seen in the patients' cerebrospinal fluid. Identification of the *Trichomonas* sp. was not done, but the authors assumed that the source was the gastrointestinal tract, and in that case, the trichomonad could have been *P. hominis* [36]. *Trichomonas* sp. were also found in a subhepatic abscess of a 51-year-old man. Microscopic examination of the pus revealed flagellates but again no exact identification of the trichomonads was performed [37].

P. hominis has been detected using PCR assays in self-collected vaginal specimens. A total of 3077 young women living in the rural area of Mwanza region, Tanzania, were asked to collect vaginal specimen for the detection of *T. vaginalis* and *P. hominis* using DNA amplification assays. A total of 840 (27.3%) specimens tested positive for *T. vaginalis* and two other specimens were positive for *P. hominis* [38]. Albeit that inoculation studies performed in the period of 1930–1940, inoculating *P. hominis* and *T. tenax* in the vagina, concluded that both trichomonads could not survive in the vagina [39,40]. The *P. hominis* found in the vagina of the two young women in Mwanza, could have been introduced in the vagina by cross contamination from the rectum; however, this could not

Table 10.1
Frequency of *P. hominis*

Year of Publication	Location	Study Population	Detection Rate	Remarks	Diagnostic Method	Reference
1985	Kampuchea	1407 adults, consulting hospital	20.1% in age group 3–5 years		Native preparation and the flotation method of Faust with subsequent staining with Lugol solution	Giboda [290]
1988	Bolivia	381 healthy population	0.8%		Direct examination	Cancrini [291]
1989	India	1020 healthy population	0.9%		(a) Direct smear in physiological saline and D'Antoni's iodine, (b) Zinc sulfate concentration method and (c) by culture in modified Boek and Drbolhav medium	Subbannayya [292]
1989	Jordan	22,970 patients	1.5%		Direct examination	Ali-Shtayeh [294]
1990	China	Children with diarrhea <5 years	3.7%		Culture	Yang [214]
1991	Kenya	Children from community, 10–28 months; 2258 stools	7.3%	10 months of follow-up	Direct examination	Chunge [295]
1992	Nigeria	300 villagers	1.0%		Light microscope after formal-ether centrifugation	Okafor [296]
1995	Zaire	328 children <5 years	5.5%		Direct examination	Henry [297]
1996	Nepal	Healthy children and adults	$N = 34$		Direct examination	Shercland [298]
2003	Thailand	106 children 10–82 months	39.6%		Simple smear, formalin-ether concentration, Boeck and Drbohlav's Locke-Egg-Serum medium culture and modified acid-fast and modified trichrome staining techniques	Saksirisampant [299]
2004	Ghana	162 hospitalized infants, 0–5 years	1.9%		By standard microbiological methods	Addy [300]
2005	Turkey	5057	0.01%		Direct examination and flotation methods	Degerh [301]
2006	Turkey	3679 patients with diarrhea	1.8%		Native-Lugol method	Culha [302]
2006	Turkey	1838 children from primary school	0.1%		Direct mounting methods	Celik [303]
2006	Egypt	3180 patients attending Mansoura University Hospitals' Clinics	4.2%		Direct wet smear, and concentration methods	El Shazly [304]

be confirmed. The specimens were self-administered vaginal swabs and contamination of the swab with fecal material could not be excluded. In addition it is not known whether the presence of *P. hominis* in the vagina is of clinical importance. The possible presence of *P. hominis* in the vagina could not be confirmed in a recent study performed in Ndola, Zambia. The study included adolescents, pregnant women, and commercial sex workers and looked for *P. hominis* DNA using the amplification method described under methods. The study participants collected specimens from the vagina, the mouth, and the rectum. *P. hominis* was found in the rectal specimens of the adolescent girls (1/136) and pregnant women (2/226), but not of the commercial sex workers. The specimens collected from the mouth and the vagina did not harbor *P. hominis* DNA [41].

P. hominis has been associated with chronic diarrhea in cats and dogs. Using gene sequence analysis and PCR *P. hominis* was identified in young dogs with diarrhea and in cats, although it has to be noted that the cats were also coinfected with *Tritrichomonas foetus*.

Zoonotic transmission has never been studied as such, but because of the close contact between humans and cats or dogs, human exposure may be a relatively frequent event [42–44].

Very little is known on the immune response to *P. hominis*. A study by Shaio and Chen showed that *P. hominis* was killed by normal human serum through classical complement pathway activation. The complement system was activated by IgM but not by IgG. On the contrary, they found that high concentration of IgG inhibited the IgM binding to the parasite and blocked the killing activity. The authors also hypothesized that the presence of the parasite specific IgM antibodies could be due to a previous unapparent infection or that they are naturally occurring antibodies resulting from stimulation by other organisms within the environment. It has to be said that the authors used for their experiments serum from 24 healthy donors, who had no recent history of diarrhea and no *P. hominis* trophozoites were found in six consecutive stool examinations [45].

Treatment, if indicated, is effective with metronidazole or tinidazole. It is orally administered as a 600–900 mg daily dose for 3–4 days or as a single 2 g dose. Furazolidone was also successful in the treatment of the female infant with persistent diarrhea [29].

10.1.3 Diagnosis

10.1.3.1 Conventional Techniques

Usually, the diagnosis is made by the detection of the trophozoites in wet mount preparations of stool specimens in saline. The organisms are easily recognized in fresh preparations through their jerky movements and rapid beats of flagella and undulating membrane.

Fixed (Schaudinn's fluid or sodium acetate acetic acid formalin) fecal smearsstained with iron hematoxylin, trichroom, or May–Grünwald–Giemsa can be microscopically observed. The iron hematoxylin stains the costa and nucleus well; the trichrome staining is superior and stains flagella, undulating membrane, and axostyle.

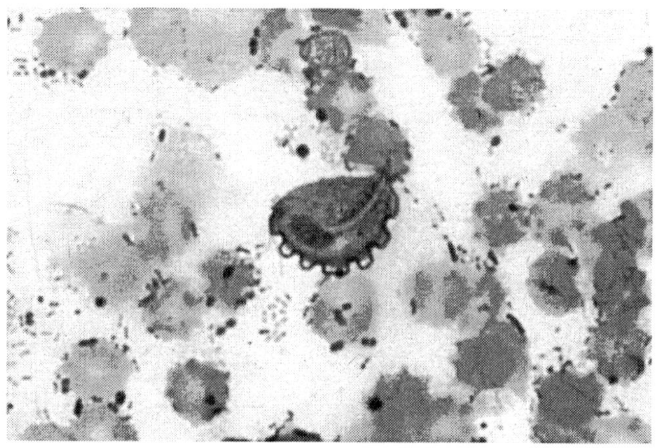

FIGURE 10.1 *P. hominis* in a stool sample containing blood. Photograph courtesy of Marc Lontie.

Figure 10.1 is a photograph of *P. hominis* in a stool sample containing blood. The Giemsa-stained smear shows very clearly the long undulating membrane and the axostyle.

Specific culture of *P. hominis* is not a standard practice; however, it is used in veterinary medicine for identification of the parasites in feces of domestic animals.

Culture is especially helpful in case the parasite concentration in the sample is low. It is also recommended to examine more than one sample.

Because *P. hominis* is an obligate parasite and lacks the ability to synthesize many molecules such as purines, pyrimidines, and many lipids [7], the culture media have to include all essential macromolecules, vitamins, and minerals. Axenic culture is possible [46] and *P. hominis* will grow in some of the culture media used for culture of *T. vaginalis* such as InPouch TV medium (Remel, Lenexa, USA) and Diamond medium. Best culture results are obtained using Modified Trypticase Yeast Maltose (TYM) medium with pH adjusted to 6.0–7.0 and 0.2–0.5 mL of heat-inactivated horse or lamb serum added before use to each culture tube of 9 mL. The inoculated culture tubes are incubated at 35–37°C in upright position, in preferably anaerobic atmosphere. The parasites will also grow under aerobic atmosphere, but will remain mainly at the bottom of the culture tubes. *P. hominis* grows in InPouch TF and InPouch TF-feline. Inpouch TF and InPouch TF-feline media are used in veterinary medicine to culture *T. foetus* from bovine genital samples or from fecal specimens from cats, respectively. *T. foetus* is a sexually transmitted infection which causes bovine trichomoniasis with early embryonic death and abortion in impregnated cows. In cats it can cause chronic diarrhea.

Cryopreservation of the *P. hominis* cultures in liquid nitrogen is identical to the *T. vaginalis* cryopreservation, and is performed using TYM with 5% dimethyl sulfoxide (DMSO).

10.1.3.2 Molecular Techniques

The identification of trichomonads may not always be conclusive: *P. hominis* was initially reported to have been isolated from four kittens with chronic diarrhea. The diagnosis

```
1 368 TTTCGGAGGAGGTAATGACCAGTT-TCA-TGTG-AAGCTTATGCTTCTGTGA
2 388 ........................-C..-.CGA-TGC.A.TC.G.AT....G
3 387 ........................-C..-.TG.-TGC..TT..G.A.....G
4 392 .A.......T..............A.A.A.AA.G..T-..-.C...T.-.-.

1 816 TTCCTAC-TGTAAACGATGCCGACAGAGGTTT-GTCATTTT--ATAATGGCA
2 838 .......-CT................GA....-....C..G--T..G.....
3 834 .......-CT................GA....-....C..G--T........
4 898 .......C.-...............-..RTTTT..TT..Y.TT....AA...

1 1130 CAATTATATACCTCGCATTGGCTTC-AGTGTTGAGG-TT--CTAATTGGGA
2 1152 .......T...T...TC.AA.TCCT-TAGA.GC.A.-..--..........
3 1148 .......T...T.T..CGAA.TCCT-TCGG.TA.A.-..--..........
4 1215 ......A...TA.AAT.A.TTT..TT.AAA..A.TTA..TT..........
```

FIGURE 10.2 Parts of alignment in SSU rRNA gene sequences of trichomonads found in human beings: (1) *P. hominis* (GenBank accession number: Af124609); (2) *T. tenax* (ac: D49495 and ac: U37711.1); (3) *T. vaginalis* (ac: U17510); (4) *D. fragilis* (ac: U37461). The sequence of *T. tenax* (ac: D49495) included in this alignment is identical to the sequence of *T. tenax* (ac: E37711). The underline indicates the position of the primers Th4, Th3, and Th5, respectively. Dots indicate nucleotides identical to those of *P. hominis*, dashes are gaps inserted to optimize alignment. Numbering is based on the data registered in GenBank. The symbol R represents simultaneous A + G and Y represents simultaneous C + T.

was made by observation of trophozoites in a smear of the fecal sample stained with new methylene blue stain [47]. A later publication concluded that these flagellates had been misdiagnosed and that rRNA gene sequencing showed that they were *T. foetus* [48].

Amplification assays have been developed; a universal amplification assay described by Felleisen et al. targets the ITS regions and the 5.8S rRNA gene flanking it. Although the amplification assay was developed for taxonomy purposes, it can also be used for the detection of *P. hominis*. The assay in itself is not specific and amplifies all trichomonadid protozoa. The amplicon of *P. hominis* is the shortest one, 339 bp, while the amplicons of both *T. vaginalis* and *T. tenax* have the same size of 368 bp [8].

A more specific and sensitive PCR amplification of *P. hominis* 18S rRNA, allowing immediate identification of the amplicon without sequencing, has shown promising results [38,43,44]. The primer pairs Th3–Th5 and Th4–Th5 target both a sequence of the Small SubUnit (SSU) rRNA gene of *P. hominis*. Figure 10.2 shows parts of alignment in SSU rRNA gene sequences of trichomonads found in human beings and the sequences of the primers [38].

The primer pair Th3–Th5 amplifies a sequence of 339 bp and Th4–Th5 a sequence of 785 bp. Both primer pairs showed to be highly specific for the detection and identification of *P. hominis*. The following organisms were not amplified by the primer pairs; *T. vaginalis*, *D. fragilis*, *Trichomonas gallinae*, *Tritrichomonas suis*, *T. tenax*, *T. foetus*, *Tetratrichomonas gallinarum*, *Escherichia coli*, *Enterococcus faecalis*, *Neisseria gonorrhoeae*, *Staphylococcus aureus*, *Gardnerella vaginalis*, *Candida albicans*, and *Giardia lamblia*.

The analytical sensitivity of the primer pairs was assessed by testing a serial dilution of a known concentration of *P. hominis* trophozoites in TE buffer (10 mM Tris-HCl pH 8.0, 1 mM EDTA). The concentration of the trophozoites was determined by counting the trophozoites in a Nuebauer counting chamber. The analytical sensitivity of both primer pairs was equivalent to 1 *P. hominis* trophozoite per 10 µL TE buffer.

10.2 METHODS

All pipetting is performed using filtered tips or positive displacement pipettes. The specimen preparation is performed in a class II biological safety cabinet.

Three physically separate areas are used for performing the amplification assays. Specimens are prepared in premixture rooms which are free of amplified DNA, the reagent mixtures are prepared in DNA-free rooms and amplification is performed in a specific amplification area.

In overall the assays should be carried out according to standard quality assurance guidelines for molecular diagnostics.

10.2.1 Sample Preparation

Fixed specimens are not suitable for amplification assays, as the fixative fluid may interfere with the amplification reaction and may inhibit the activity of the polymerase.

Clinical stool specimens are processed by dipping a synthetic sterile swab into the stool after homogenization of the sample, or by preparing aliquots of 1 g of stool. The stool specimens can then be stored at –20°C for 5 days before DNA extraction is performed. Stool specimens should be extracted as soon as possible as the endonucleases produced by the organisms of the intestinal flora can disintegrate the *P. hominis* DNA.

Other biological unfixed specimens are kept frozen at –20°C until extraction is performed.

10.2.1.1 Specimen Elution

1. Thaw the swabs at room temperature for 30 min.
2. Transfer 1500 µL of 1:10 diluted phosphate-buffered saline (PBS) pH 7.4 (1 part of PBS is mixed with 9 parts of saline) into the swab container.
3. Mix the solution and the tip of the swab by gently vortexing for 5–10 s.
4. Express the liquid against the side of the swab container and discard the swab.
5. Mix again by gently vortexing for 5–10 s.

6. Distribute the well-mixed specimen elutes in aliquots of 250 μL into polypropylene, sterile, and conical screw cap tubes.
7. Store the aliquots for a longer period at −20°C.

10.2.1.2 Specimen Extraction

DNA extraction is performed using commercial extraction methods. Although in-house extraction methods can be also used, the validation of these is more cumbersome. Stool specimens very often contain PCR inhibitors. The inhibitors can be eliminated during the extraction procedure if the method includes a filtration step or specific DNA capture.

A suitable extraction method for stool specimens is the QIAamp DNA Stool MiniKit, Qiagen. For other biological specimens the QIAamp DNA Mini Kit, Qiagen, can be used.

10.2.1.3 Method Using the QIAamp DNA Stool MiniKit

1. Centrifuge the aliquot of 250 μL at 12,500 rpm for 10 min. After centrifugation, discard the supernatant.
2. Add 1.4 mL of AST buffer to the pellet, vortex to resuspend the pellet; the addition of 1.4 mL is done in two steps of 700 μL.
3. Place the tubes for 5 min in a heating bloc at 70°C.
4. Vortex for 15 s and centrifuge the tubes for 1 min at 13,000 rpm.
5. Transfer 1.2 mL of the supernatant in a new polypropylene, sterile, and conical screw cap tube.
6. Add 1 InhibitEX tablet and vortex immediately until the tablet is completely dissolved.
7. Incubate for 1 min at room temperature; the tablet will adsorb the inhibitors.
8. Centrifuge for 3 min at 13,000 rpm.
9. Transfer 15 μL Proteinase K in a new 1.5 mL polypropylene, sterile, and conical screw cap tube.
10. Add 200 μL supernatant to the tube containing the Proteinase K.
11. Add 200 μL AL buffer and vortex for 15 s.
12. Place the tubes for 10 min in a heating bloc at 70°C.
13. Add 200 μL ethanol (96–100%) and vortex vigorously.
14. Transfer the total volume of elute on the QIAmp spin column.
15. Centrifuge the column for 1 min at 13,000 rpm.
16. Transfer the column in a new polypropylene, 2 mL collection tube.
17. Open the column and add 500 μL AW1 buffer, centrifuge for 1 min at 13,000 rpm.
18. Place the column in a new polypropylene, 2 mL collection tube.
19. Open the column and add 500 μL AW2 buffer and centrifuge for 3 min at 13,000 rpm.
20. Place the column in a new, sterile screw cap 1.5 mL tube.
21. Open the column and add 200 μL AE buffer on the membrane.
22. Incubate for 1 min at room temperature and centrifuge for 1 min at 13,000 rpm and elute the DNA.
23. Take the column out of the 1.5 mL tube and discard.
24. Close the 1.5 mL tube and store the extract in the refrigerator at 2–8°C, for 24 h or at −20°C for long-term storage.

10.2.2 Detection Procedures

10.2.2.1 Reagent Mixture

Prepare the reagent mixture in a DNA-free room using filtered tips.

Per PCR reaction the mixture consists of:

GeneAmp 10× PCR Gold Buffer	5 μL
dNTPs (dATP, dCTP, dGTP, dUTP)	280 μM
Primers[a] (each)	0.15 μM
$MgCl_2$	2 mM
AmpliTaq Gold® DNA Polymerase	2 U
Water	q.s. to 40 μL

[a] Following primer pairs can be used Th3-Th5 or Th4-Th5.

The following primer pairs can be used Th3–Th5 or Th4–Th5. Mix well by inverting the mixture.

The AmpliTaq Gold is purchased in a kit format, which includes Gold® 10× buffer, $MgCl_2$ solution, and AmpliTaq Gold DNA polymerase (Applied Biosystems, Roche Molecular Systems). The Gold PCR buffer consists of 15 mM Tris-HCl pH 8.3 and 50 mM KCl.

10.2.2.2 Controls

Add to each amplification run at least one negative and one positive control.

Negative control: water.

Positive control: DNA extract of a culture of *P. hominis* reference strain.

10.2.2.3 Amplification

1. Prepare a tray with the number of PCR reaction tubes needed, including two tubes for the negative and positive run controls.
2. Aliquot 40 μL of reagent mixture in each of the PCR reaction tubes using filtered tips.
3. Close the PCR reactions tubes very tightly and bring it over to the premix room.
4. Work in a laminar air flow cabinet and use filtered tips, wear over sleeves and powder-free gloves in the cabinet.
5. Transfer 10 μL of DNA extract in each of the PCR reaction tubes, open and close the tubes one per one.
6. Add the controls after all biological specimens have been transferred.
7. Amplify in a thermocycler (Perkin Elmer 9600) starting with an initial denaturation step of 5 min at 95°C and followed by a total of 35 cycles of 1 min at 94°C, 1 min at 64°C, and 1 min at 72°C. The 35th cycle is followed by a final extension of 7 min at 72°C.

10.2.2.4 Gel Electrophoresis

Upon completion of the DNA amplification, 15 μL of each amplified specimen is analyzed by electrophoresis in a 2% agarose gel in Tris-acetate-EDTA buffer (TAE) pH 8.5.

The gel is prepared by boiling 1 g of agarose gel powder in 50 mL TAE buffer pH 8.5. Pour the gel in a gel chamber, and place the comb into it. Let it solidify for 20–40 min.

Fill the electrophoresis tank with 500 mL TAE buffer, and place the gel and gel chamber in the tank and add 500 mL of distilled water to cover the gel completely.

Take the comb out of the gel and load the gel with for each amplified product a mixture of 15 μL of amplified product and 3 μL of loading buffer. The loading buffer consists of 4 g sucrose, 150 μL 1% bromophenol blue and 20 mL distilled water. Include at both sides of the gel a 100 bp commercial weight marker.

Close the electrophoresis tank with a lid and start the electrophoresis for 15 min.

Once the electrophoresis is completed take the gel out of the tank and stain it in an ethidium bromide solution (1 μg/L) for 20 min.

After staining photograph the gel under short UV light.

Assess the sizes of the amplified product by comparison with the commercial weight.

10.2.2.5 Analysis of Results

The primer set Th3–Th5 generates amplicons of *P. hominis* of 339 bp long, the amplicons obtained with the Th4–Th5 primers are 785 bp long.

10.2.2.6 Quality Control

To assess PCR inhibition in the stool specimens and the adequacy of DNA extraction, a PCR assay for the detection of *E. coli* can be performed. A PCR for the human β2 microglobulin gene or other housekeeping gene can be carried out on the other biological specimens.

10.3 CONCLUSION AND FUTURE PERSPECTIVE

P. hominis was observed for the first time in 1860. Almost 150 years later, the parasite remains understudied. The biology, the pathogenicity, and the epidemiology of the parasite are not fully known or understood, and still a lot of the parasite's characteristics have to be revealed.

Why was this parasite ignored over all these decades? This is most probably due to the very little evidence of virulence of *P. hominis*. Although in certain cases the parasite is associated with diarrhea, in the majority of the cases the infection is asymptomatic. But it also has to be said that in the case of diarrhea it is hard to confirm that *P. hominis* solely is the cause of the symptoms. There are many causes of diarrhea and multiple infections are rather frequent.

P. hominis is found to be associated with diarrhea in children in especially tropical and subtropical regions were hygiene conditions are poor and were contact with the parasite is more frequent.

Knowing that *P. hominis* can be killed by normal human serum of adults, we could speculate that *P. hominis* causes less frequently diarrhea in adults because of acquired immunity against the parasite. This hypothesis has also been proposed by the researchers in veterinary medicine. *P. hominis* has also been associated with diarrhea in kittens and young dogs but not in adult animals, the researchers hypothesized that adult animals might have had acquired immunity against the parasite. If indeed adults acquire immunity against *P. hominis*, then the parasite may be an important cause of opportunistic infections in immunocompromised and immunodeficient hosts.

P. hominis might be virulent if found outside the gastrointestinal tract, and this has been demonstrated by the few case reports found in the literature. On the other hand, *P. hominis* DNA was detected in the vagina of young adults, but no symptoms were reported or observed. The detection and identification of *P. hominis* was performed using PCR; however, it could not be confirmed that the DNA was extracted from viable organisms. The clinical importance of finding *P. hominis* in the vagina remains unclear.

P. hominis is a parasite found in humans but also in animals living close with humans such as pets and cattle. The risk of zoonotic infection with this organism maybe low, but it has never been studied.

Until recently, diagnosis of *P. hominis* was made microscopically, even after culture the trophozoites are observed with the use of a light microscope. Trichomonads are microscopically difficult to distinguish, they have very similar morphologies and the number of flagella, which is one of their morphological characteristics, is not always easy to determine because of the motility of the flagella and the parasite itself. Misdiagnosis could have occurred in the past due to the lack of the specificity of microscopy. Microscopy is also known to be less sensitive compared to culture or amplification assays, by consequence the epidemiological data available today may be underestimated.

With the introduction of molecular amplification assays the diagnosis of *P. hominis* will become more accurate and the epidemiologists will be able to provide more reliable prevalence data.

However, the scientific community should be made aware of the limited amount of information available on *P. hominis*. Researchers should be encouraged to study this neglected parasite with the overall aim to enrich our knowledge about *P. hominis*.

REFERENCES

1. Cavalier-Smith, T. A revised six-kingdom system of life. *Biol. Rev.* 73, 203, 1998.
2. Wenrich, D.H. Morphology of the intestinal trichomonad flagellates in man and of similar forms in monkeys, cats, dogs and rats. *J. Morphol.* 74, 189, 1947.
3. Honigberg, B.M., Mattern, C.F., and Daniel, W.A. Structure of *Pentatrichomonas hominis* (Davaine) as revealed by electron microscopy. *J. Protozool.* 15, 419, 1968.
4. Solomon, J.M. Studies on the physiology of *Trichomonas hominis*. *J. Parasitol.* 43, 39, 1957.
5. Nadler, S.A., and Honigberg, B.M. Genetic differentiation and biochemical polymorphism among trichomonads. *J. Parasitol.* 74, 797, 1988.
6. Lockwood, B.M. et al. The use of a highly sensitive electrophoretic method to compare the proteinases of trichomonads. *Mol. Biochem. Parasitol.* 24, 89, 1987.

7. Tang, P., and Hsin-Sheng L. *Pentatrichomonas hominis*: Purine salvage pathway. *Comp. Biochem. Physiol.* 90B, 419, 1988.
8. Felleisen, R.S. Comparative sequence analysis of 5.8S rRNA genes and internal transcribed spacer (ITS) regions of trichomonadid protozoa. *Parasitology* 115, 111, 1997.
9. Honigberg, B.M. Trichomonads found outside the urogenital tract of humans. In: Honigberg, B.M. (ed.), *Trichomonads Parasitic in Humans*, Chapter 19. Springer-Verlag, New York, 1990.
10. Giboda, M. Intestinal parasites in Kampuchea, Takeo province. *J. Hyg. Epidemiol. Microbiol. Immunol.* 29, 377, 1985.
11. Cancrini, G. et al. Intestinal parasites in the Camiri, Gutierrez and Boyuibe areas, Santa Cruz Department, Bolivia. *Parasitologia* 30, 263, 1988.
12. Subbannayya, K. et al. *Entamoeba histolytica* and other parasitic infections in south Kanara district, Karnataka. *J. Commun. Dis.* 21, 207, 1989.
13. Ali-Shtayeh, M.S. et al. Prevalence and seasonal fluctuations of intestinal parasitic infections in the Nablus area, West Bank of Jordan. *Ann. Trop. Med. Parasitol.* 83, 67, 1989.
14. Yang, C.-R. et al. Diarrhoea surveillance in children aged under 5 years in a rural area of Hebei province, China. *J. Diarrhoeal Dis. Res.* 8, 155, 1990.
15. Chunge, R.N. et al. Longitudinal study of young children in Kenya: Intestinal parasitic infection with special reference to *Giardia lamblia*, its prevalence, incidence and duration, and its association with diarrhoea and with other parasites. *Acta Trop.* 50, 39, 1991.
16. Okafor, C.N., and Azubike, C.N. Studies in intestinal parasitic disease agents in stools of people in a rural area of Nigeria. *West. Afr. J. Med.* 11, 106, 1992.
17. Ogunsanya, T.I., Rotimi, V.O., and Adenuga, A. A study of the aetiological agents of childhood diarrhoea in Lagos, Nigeria. *J. Med. Microbiol.* 40, 10, 1994.
18. Henry, M.C. et al. Community survey of diarrhoea in children under 5 years in Kinshasa, Zaire. *Ann. Soc. Belg. Med. Trop.* 75, 105, 1995.
19. Sherchand, J.B., Larsson, S., and Shrestha, M.P. Intestinal parasites in children and adults with and without abdominal discomfort from the Kathmandu area of Nepal. *Trop. Gastroenterol.* 17, 15, 1996.
20. Saksirisampant, W. et al. Intestinal parasitic infestations among children in an orphanage in Pathum Thani province, *J. Med. Assoc. Thai.* 86, S263, 2003.
21. Addy, P.A., Antepim, G., and Frimpong, E.H. Prevalence of pathogenic *Escherichia coli* and parasites in infants with diarrhoea in Kumasi, Ghana, *East. Afr. Med. J.* 81, 353, 2004.
22. Degerli, S., Ozcelik, S., and Celiksoz, A. The distribution of intestinal parasites in patients presenting at the Parasitology Laboratory of the Cumhuriyet University. *Turkiye Parazitol. Derg.* 29, 116, 2005.
23. Culha, G. The distribution of patients with intestinal parasites presenting at the Parasitology Laboratory of the Mustafa Kemal University Medical Faculty. *Turkiye Parazitol. Derg.* 30, 302, 2006.
24. Celik, T. et al. Incidence of intestinal parasites among primary school children in Malatya. *Turkiye Parazitol. Derg.* 30, 35, 2006.
25. El Shazly, A.M. et al. Intestinal parasites in Dakahlia governorate, with different techniques in diagnosing protozoa. *J. Egypt. Soc. Parasitol.* 36, 1023, 2006.
26. Alaofé, H. et al. Intestinal parasitic infections in adolescent girls from two boarding schools in southern Benin. *Trans. R. Soc. Trop. Med. Hyg.* 103, 452, 2008.
27. Haghighi, A. et al. Frequency of enteric protozoan parasites among patients with gastrointestinal complaints in medical centers of Zahedan, Iran. *Trans. R. Soc. Trop. Med. Hyg.* 103, 452, 2009.
28. Nasiri, V. et al. Intestinal parasitic infections among inhabitants of Karaj city, Tehran Province, Iran in 2006–2008. *Korean J. Parasitol.* 47, 265, 2009.
29. Mancilla-Ramirez, J., and Gonzalez-Yunes, R. Diarrhea associated with *Trichomonas hominis* in a newborn infant. *Bol. Med. Hosp. Infant. Mex.* 46, 623, 1989.
30. Chunge, C.N. et al. Pathogenicity of *Trichomonas hominis* in Kenya. *East. Afr. Med. J.* 65, 532, 1988.
31. Hegner, R. Differential reactions of species and strains of trichomonad flagellates to changes of the environment. *Am. J. Hyg.* 15, 513, 1932.
32. Khan, A.R., and Huq, F. Disease agents carried by flies in Dacca city. *Bangladesh Med. Res. Counc. Bull.* 4, 86, 1978.
33. Caruso, G. Human trichomoniasis. *Acta Med. Ital. Mal. Infett. Parassit.* 15, 231, 1960.
34. Dobbel, C. Research on the intestinal protozoa of monkeys and man. VI Experiments with the trichomonads of man and macaques. *Parasitology* 24, 531, 1934.
35. Jongwutiwes, S., Silachamroon, U., and Putaporntip, C. *Pentatrichomonas hominis* in empyema thoracis, *Trans. R. Soc. Trop. Med. Hyg.* 94, 185, 2000.
36. Masur, H., Hook, E. III, and Armstrong, D.A. *Trichomonas* species in a mixed microbial meningitis. *JAMA* 236, 1978, 1976.
37. Jakobsen, E.B., Friis-Moller, A., and Friis, J. *Trichomonas* species in a subhepatic abscess. *Eur. J. Clin. Microbiol.* 6, 296 (1987).
38. Crucitti, T. et al. Detection of *Pentatrichomonas hominis* DNA in biological specimens by PCR. *Lett. Appl. Microbiol.* 38, 510, 2004.
39. Stabler, R.M., Feo, L.G., and Rakoff, A.E. Implantation of intestinal trichomonads (*T. hominis*) into the human vagina. *Am. J. Trop. Med. Hyg.* 34, 114, 1941.
40. Stabler, R.M., and Feo, L.G. Survival time of intravaginally implanted *Trichomonas hominis*. *Am. J. Trop. Med. Hyg.* 22, 633, 1942.
41. Crucitti, T. et al. *Trichomonas vaginalis* is highly prevalent in adolescent girls, pregnant women, and commercial sex workers in Ndola, Zambia. *Sex. Transm. Dis.* 37, 223, 2010.
42. Gooikin, J.L. et al. Molecular characterization of Trichomonads from feces of dogs with diarrhea. *J. Parasitol.* 91, 939, 2005.
43. Gookin, J.L., Stauffer, S.H., and Levy, M.G. Identification of *Pentatrichomonas hominis* in feline fecal samples by polymerase chain reaction assay. *Vet. Parasitol.* 145, 11, 2007.
44. Gookin, J.L. et al. Optimization of a species-specific polymerase chain reaction assay for identification of *Pentatrichomonas hominis* in canine fecal specimens. *Am. J. Vet. Res.* 68, 783, 2007.
45. Shaio, M.-F., and Chen, J.-R. Immunoglobulin M-dependent classical complement pathway activation in killing of *Pentatrichomonas hominis*. *Infect. Immun.* 57, 902, 1989.
46. Lo, H.S., and Shaio, M. A method for the isolation of *Trichomonas hominis* in axenic culture. *Ann. Trop. Med. Parasitol.* 75, 567, 1981.
47. Romatowski, J. *Pentatrichomonas hominis* infection in four kittens. *JAVMA* 8, 1270, 2000.
48. Levy, M.G. et al. *Tritrichomonas foetus* and not *Pentatrichomonas hominis* is the etiologic agent of feline trichomonal diarrheae. *J. Parasitol.* 89, 99, 2003.

11 Trichomonas

Dongyou Liu, Xiangang Kong, and Changyou Xia

CONTENTS

11.1 INTRODUCTION

The genus *Trichomonas* covers several flagellate protozoan species that are infective to humans and animals. In particular, *Trichomonas vaginalis* parasitizes the human vagina, prostate gland, and urethra, resulting in trichomoniasis that is considered the most common nonviral sexually transmitted disease (STD), affecting more than 180 million people annually worldwide. Besides causing vaginitis, cervicitis, and urethritis in women, *T. vaginalis* infection is implicated in premature membrane rupture and delivery in pregnant women, and in the delivery of low-birth-weight infants. In addition, *T. vaginalis* is responsible for prostatitis, urethritis, and epididymitis in man. Due to its relatively mild and often asymptomatic nature, trichomoniasis has not been regarded as a notifiable disease. However, recent data indicate that *Trichomonas* plays a role as cofactor in the transmission of human immunodeficiency virus (HIV), as well as a risk factor in the development of cervical cancer.

11.1.1 Classification, Morphology, Life Cycle, and Epidemiology

11.1.1.1 Classification

The genus *Trichomonas* consists of a group of flagellate protozoa in the family Trichomonadidae, order Trichomonadida, class Parabasalia, phylum Zoomastigina. Being one of the seven genera in the family Trichomonadidae (i.e., *Cochlosoma*, *Pentatrichomonas*, *Pentatrichomonoides*, *Pseudotrypanosoma*, *Tetratrichomonas*, *Trichomitopsis*, and *Trichomonas*), the genus *Trichomonas* currently contains five recognized species (i.e., *Trichomonas canistomae*, *Trichomonas equibuccalis*, *Trichomonas gallinae*, *Trichomonas tenax*, and *T. vaginalis*), in addition to five unassigned species (i.e., *Trichomonas* sp. 24K, *Trichomonas* sp. 7895-C1, *Trichomonas* sp. duck_salpingitis, *Trichomonas* sp. RWG-2007-1, and *Trichomonas* sp. RWG-2007-2) [1].

While many species in the genera *Tritrichomonas* (e.g., *Tritrichomonas foetus*) and *Tetratrichomonas* (e.g., *Tetratrichomonas gallinarum*) are pathogenic for various animals (sheep, pigs, cats, birds, etc.), three *Trichomonas* species (i.e., *T. vaginalis*, *T. tenax*, and *Trichomonas hominis*) have been shown to infect humans. Of these, *T. vaginalis* is most common and important; this parasite colonizes the vagina of women and the urethra (sometimes prostate) of men, causing vaginitis and cervicitis in women, urethritis in both sexes, and pulmonary infection in newborn babies (in the form of respiratory difficulty). *T. tenax* is found in oral gingival and tracheobronchial sites (e.g., the oral cavity and dental plaque), and has been implicated in debilitating or pulmonary diseases (carcinoma, abscess, bronchiectasis) in individuals with poor dental hygiene, malnutrition, and alcoholism. *T. hominis* (also known as *Pentratrichomonas hominis*) is located in the intestinal tract and represents an occasional cause of necrotizing pulmonary abscess and pleural effusions (after reaching the respiratory apparatus by aspiration or through a bronchoenteral fistula [2].

11.1.1.2 Morphology

The genus *Trichomonas* is characterized by possessing four free anterior flagella, an undulating membrane incorporating a fifth lateral flagellum, a costa at the base of the undulating membrane, and an axostyle extending through the cell [2].

T. vaginalis is typically pyriform in culture, and may appear amoeboid in vaginal tissues. Similar to other trichomonads, *T. vaginalis* has only the trophozoite stage. The trophozoites measure 23–39 μm in length (including an 8–13 μm body length and 8–15 μm flagella length) and 5–8 μm in width. The four anterior flagella [9(2) + 2 arrangement] are 8–15 μm in length, the undulating membrane harboring a fifth flagellum extends about 2/3 of the distance to the posterior end of the body, and the costa originates in the kinetosomal complex at the anterior of the parasite. The axostyle (3–14 μm in length) is slightly curved and runs through the cell from the anterior end to the posterior end. Internal organelles consist of a prominent nucleus, a number of vacuoles and round- to oval-shaped hydrogenosomes (formerly called siderophil granules), which are present as paraxostylar and paracostal chromatic granules by light microscopy and as osmiophilic granules by electron microscopy [3]. The hydrogenosomes serve as mitochondria, participate in energy production and drug activation, and produce molecular hydrogen as the end product of metabolism. However, hydrogenosomes differ from mitochondria in that hydrogenosomes lack cytochromes, mitochondrial respiratory chain enzymes, and DNA.

11.1.1.3 Life Cycle and Epidemiology

T. vaginalis has a simple life cycle, with only the trophozoite stage and without a cyst form. Transmitted through sexual contact, the trophozoite does not survive well in external environment. *T. vaginalis* divides by binary fission (which is favored by low acidity of pH 3.5–4.5) and generates a population in the lumen and on the mucosal surfaces of the vagina and the urethra (sometimes prostate) of humans [4].

T. vaginalis has a worldwide distribution, and humans are its only natural host. Affecting an estimated 180 million people per annum, *T. vaginalis* has been found in 5–70% of sexually active women of all age groups (especially those between the ages of 16 and 35) and in 1–68% of men with nongonococcal urethritis [5]. Being commonly associated with other STDs (which have a higher prevalence among adolescents and young adults), *T. vaginalis* infections represent a marker of high-risk sexual behavior. Trichomoniasis has an incubation period of 4–28 days [6].

11.1.2 Clinical Features and Pathogenesis

11.1.2.1 Clinical Features

First described by Donne in 1836 and long regarded as a harmless commensal of the human vagina, *T. vaginalis* has been later incriminated as a cause of vaginitis and cervicitis, in women, and urethritis in both sexes [7–9].

Although approximately 50–80% of women and a majority of men infected with *T. vaginalis* are asymptomatic, the parasite is known to cause nonspecific clinical signs, ranging from vaginal discharge, pruritus (itching), erythema, vaginal wall edema, leucorrhea, foul odor, low abdominal discomfort, dysuria, vulvar pain, and dyspareunia in women, and a self-limiting mild urethritis in men. [2].

As the most common manifestation associated with vaginal trichomoniasis, vaginal discharge is frequently profuse and frothy, along with burning, itching, or chafing. The vaginal mucosa sometimes appears diffusely hyperemic, with bright-red punctate lesions. Colpitis macularis (strawberry cervix) initially develops as a strong irritable erythema and progresses into a group of small punctate hemorrhagic spots of a "strawberry appearance" on the vagina close to the cervical mucosa. This specific clinical sign for vaginal trichomoniasis is detected only by colposcopy [10,11].

Furthermore, trichomoniasis has been linked to increased risk of HIV transmission and pregnancy complications (e.g., premature rupture of placental membranes, premature labor and low infant birth weight) [12–14].

11.1.2.2 Pathogenesis

Adhesion of *T. vaginalis* parasites to target cells (e.g., vaginal epithelial cells) represents a prerequisite for the establishment of infection, cytotoxicity toward mammalian cells and thus pathogenesis of trichomoniasis [7,15,16]. Adherence is a multifactor process, involving microtubules, microfilaments, four adhesins, and cysteine proteinase. Cysteine proteinases and cell-detaching factors with trypsin-like activity degrade proteins (e.g., laminin and vitronectin) and other components of the extracellular matrix, leading to the release of host cells from tissue and the contact-dependent damage to the epithelium of the infected organ. Cytotoxic molecules with perforin-like activity mediate cytotoxicity through damage of the target cell plasma membrane [17].

The adherence process also induces the recruitment of neutrophils, which produce interleukin-8 and initiate host inflammatory responses that aggravate the symptoms and signs of *T. vaginalis* infection [18].

In addition, *T. vaginalis* employs a variety of mechanisms to evade and subvert the host's immune system, including its generation of surface antigen variations to disguise the parasite, and its degradation of human secretory IgA, and serum IgA and IgG with cysteine proteinases [2].

11.1.3 Diagnosis and Treatment

11.1.3.1 Diagnosis

Traditionally, diagnosis of *T. vaginalis* infections has relied on microscopic examination (wet mount or Papanicolaou smear) of Giemsa-stained clinical or cultured specimens for the presence of pathogenic flagellates with characteristic active motility [19–26]. Growth of *T. vaginalis* on Diamond's medium or pouch medium for 3–5 days enhances microscopic identification of the parasite [27–36]. It is important to distinguish *T. vaginalis* trophozoites from those of the related flagellate *T. hominis* (i.e., *P. hominis*).

Given the morphological similarity among *Trichomonas* species and other flagellates, the relatively low sensitivity of microscopy (requiring more than 10^3/mL of live protozoa) [37,38], and the inability of some *T. vaginalis* isolates to grow in culture [39], serological and molecular techniques have been developed for improved diagnosis of *T. vaginalis* infections [40].

Among the serological methods reported for trichomoniasis diagnosis, enzyme-linked immunosorbent assay, indirect fluorescent antibody test and latex agglutination are useful for detection of serum anti-*T. vaginalis* IgG and IgM antibodies in human patients [41–51].

Focusing on the nucleic acids of the parasite, molecular procedures such as polymerase chain reaction (PCR) provide a rapid, sensitive, and specific approach for detection of *T. vaginalis* [52–71]. *T. vaginalis* genes that have been targeted by specific PCR assays include the ferredoxin gene [72,73], the β-tubulin gene [74,75], a highly repeated 2-kb DNA sequence [76,77], the 18S ribosomal RNA (rRNA) gene [78], and the adhesion protein gene [79]. Several primer sets (e.g., TVK3/TVK7, TVA5/TVA6, BTUB9/BTUB2, IP1/IP2, TV1/TV2) have been reported for conventional PCR detection of *T. vaginalis* [77,80,81]. Evolving from the original labor-intensive protocols based on separation of specific PCR product by agarose gel electrophoresis, advances in instrument automation and probe chemistry enable the development of real-time PCR with instant result availability [66,67,73,82–85]. Real-time PCR has added advantage over conventional PCR in that it does not require post-PCR processing of amplicons, thereby eliminating the possibility of amplicon cross-contamination.

11.1.3.2 Treatment

Metronidazole and other 5-nitroimidazoles (e.g., tinidazole, ornidazole, and secnidazole) are effective against *T. vaginalis*. The adverse effects of metronidazole include myopia, neuralgia, and allergic dermatitis as well as potential carcinogenicity at high doses for protracted periods of time. With a lower incidence of adverse effects than metronidazole, tinidazole is useful for treatment of metronidazole-resistant *T. vaginalis* infections. Sodium nitrite, sodium nitroprusside, and Roussin's black salt offer another potential means of treating metronidazole-sensitive and resistant *T. vaginalis* isolates. Application of vinegar douche, personal hygiene, and condoms are valuable for prevention of *T. vaginalis* transmission.

11.2 METHODS

11.2.1 Sample Preparation

Swabs for direct microscopy and culture are transported in Transwabs with charcoal medium. Swabs for PCR are collected in 2-SP transport medium (0.2 M sucrose–0.02 M phosphate, Roche). First void urine is also useful for *T. vaginalis* detection [86].

On arrival in the laboratory the samples in charcoal medium are immediately tested by direct wet mount microscopy. When *T. vaginalis* is not detected by wet mount microscopy, the sample is also inoculated on a modification of Diamond's modified trypticase/yeast extract/maltose (TYM) medium supplemented with bovine fetal serum (10%), penicillin G (2000 U/mL) and streptomycin sulfate (2 mg/mL), and amphotericin B (25 μg/mL) at 37°C medium. The culture is further examined by wet mount microscopy after 3 and 5 days. Specimens are considered positive by active motility of flagellates in direct wet-mount microscopy or in wet-mount microscopy after culture [86].

For extraction of *T. vaginalis* genomic DNA, 3 mL of urine sample is centrifuged and the pellet mixed with 200 μL of DNA lysis buffer (320 mM saccharose, 10 mM Tris, 5 mM $MgCl_2$, 1% SDS). After incubation at 37–40°C, the lysis is mixed with 200 μL saturated phenol and centrifuged. A volume of 200 μL of chloroform is added to the supernatant and centrifuged. Then 1/10 of the supernatant is added to natrium acetate, and twice the amount is added to absolute ethanol, incubated at −20°C for 60 min. After incubation, each sample is centrifuged and the supernatant is discarded, washed with 100 μL of 70% ethanol, air dried, and dissolved in 30 μL sterile water [87].

Alternatively, 0.5 mL urine may be diluted with 0.5 mL of Urine Wash Buffer (Roche Specimen Preparation Kit), concentrated by centrifugation, reconstituted in 250 μL lysis buffer and diluted with 250 μL Specimen Diluent. In addition, 100 μL urine may be processed with the MagnaPure LC total NA kit (Roche diagnostics), and eluted in 100 μL [86]. Furthermore, *T. vaginalis* DNA may be extracted from 140 μL aliquots of cervico-vaginal lavage (CVL) fluid using the QIAamp DNA mini-kit (Qiagen), and from 500 μL aliquots of urine using a viral RNA kit (Qiagen) [64].

11.2.2 Detection Procedures

11.2.2.1 Conventional PCR Detection

Kengne et al. [77] developed a conventional PCR assay with primers TVK3 (5′AT TGT CGA ACA TTG GTC TTA CCC TC-3′) and TVK7 (5′-TCT GTG CCG TCT TCA AGT ATG C-3′) for amplification of a 261-bp product in a *T. vaginalis*-specific repeat DNA fragment.

11.2.2.1.1 Procedure

1. PCR mixture (50 μL) is composed of 1 μL of deoxynucleoside triphosphates (0.2 mM each dATP, dCTP, dGTP, and dTTP), 0.75 μL of dUTP (0.3 mM; Applied Biosystems), 6 μL of $MgCl_2$ (1.5 mM), 0.2 μM each primer, 1 U of uracil DNA glycosylase (Roche Diagnostics), 2.5 U of AmpliTaq Gold polymerase (Applied Biosystems), 10 μL of 10× PCR buffer (Applied Biosystems), and 20 μL of template DNA.
2. Amplification is conducted at 20°C for 10 min and 95°C for 5 min, followed by 45 cycles of 90°C

for 1 min, 60°C for 30 s, and 72°C for 1 min. The final extension consisted of one cycle of 72°C for 7 min.
3. The PCR products are separated on 1.5% agarose gel, stained with ethidium bromide and visualized under UV light.

11.2.2.2 Real-Time PCR Detection

11.2.2.2.1 Protocol of López-Villaseñor et al. [86]

López-Villaseñor et al. [86] reported two different *T. vaginalis*-specific TaqMan primer/probe sets for real-time PCR detection of *T. vaginalis*. One primer/probe set targets part of the 2 kb repeated sequence (L23861) and the other primer/probe set detects a sequence of the β-tubulin protein (L05468) (Table 11.1).

11.2.2.2.1.1 Procedure

1. For both assays, PCR mixture (25 μL) is prepared with a real-time PCR kit (Applied Biosystems) and contains 5 μL of extracted sample.
2. Amplification and detection are conducted on ABI Prism 7700, with the following temperature profile: 2 min at 50°C (uracil-DNA glycosylase (UDG) treatment), 10 min at 95°C (UDG inactivation), and 40 cycles of 15 s at 95°C, 1 min at 60°C.
3. Sample (including the positive control) is considered to be positive if the cycle threshold (C_t) of the specific reaction is <40, while the negative control sample gives no detectable signal after 40 cycles.
4. For each clinical sample, an additional real-time PCR spiked with *T. vaginalis* DNA is carried out as a control for the presence of inhibitors. The spiked samples usually give a C_t of about 32. A sample is considered to inhibit the PCR when the C_t of the spiked sample is >35.

11.2.2.2.2 Protocol of Simpson et al. [88]

Simpson et al. [88] described two real-time PCR assays using fluorescence resonance energy transfer (FRET) hybridization probes for the detection of *T. vaginalis*. A pair of FRET hybridization probes consists of two probes, donor and sensor, each labeled with a different fluorophore (fluorescein, LC Red 640 or 705). When the probes are bound to target DNA, the donor fluorophore excites the sensor, which yields an emission fluorescence. The FRET probes and target DNA hybrids are distinguishable at a characteristic temperature (T_m) that is determined using melting-curve analysis.

The first PCR with primers BTUB3f and BTUB_bkmt targets a 195 bp sequence of the *T. vaginalis* β-tubulin gene. The FRET hybridization probes for the β-tubulin gene are BTUB FL and BTUB [67,89]. The second PCR with primers TV16Sf-2 and TV16Sr-2 amplifies a 323 bp product of the 18S rRNA gene of *T. vaginalis*. The FRET hybridization probes for the 18S rRNA gene PCR consist of TV16_FL and TV16_LC (Table 11.2).

11.2.2.2.2.1 Procedure

1. The β-tubulin gene LightCycler PCR mix (20 μL) contains 1× LC FastStart DNA Master hybridization probe buffer (Roche Diagnostics), 4 mM $MgCl_2$, 0.5 μM of each primer, 0.2 μM BTUB FL and 0.4 μM BTUB LC, and 2 μL template DNA.
2. Cycling conditions are as follows: 95°C for 10 min followed by 50 amplification cycles of 95°C for 10 s, 55°C for 10 s, and 72°C for 10 s. Fluorescence is detected throughout the annealing phase of each amplification cycle.
3. After amplification is complete, melting-curve analysis is performed at 95°C for 0 s, 40°C for 60 s followed by a gradual increase in temperature (transition rate of 0.1°C s^{-1}) to 80°C with continual fluorescence detection. A final cooling step is then performed at 40°C for 30 s. All steps are performed at a temperature transition rate of 20°C s^{-1}. Data are

TABLE 11.1
Primers and Probes for Real-Time PCR Detection of *T. vaginalis*

Target Gene	Primers/Probe	DNA-Sequence (5′–3′)
2 kb repeated sequence	L23861 forward primer	AAG ATG GGT GTT TTA AGC TAG ATA AGG T
	L23861 reverse primer	CGT CTT CAA GTA TGC CCC AGT AC
	L23861 probe	FAM-CCG AAG TTC ATG TCC TCT CCA AGC GT-TAMRA
β-Tubulin protein	L05468 forward primer	CAC TCA AGC TCA CAA CAC CAA CA
	L05468 reverse primer	TGC CGG ACA TAA CCA TGG A
	L05468 probe	FAM-ACG GCG ATC TTA AC-MGBNFQ

TABLE 11.2
Primers and Probes for *T. vaginalis* Real-Time PCR

Target Gene	Primers/Probe	DNA-Sequence (5′–3′)
β-Tubulin	BTUB3f	TCCAAAGGTTTCCGATACAGT
	BTUB_bkmt	GTTGTGCCGGACATAATCATG
	BTUB FL	CCGTACACTCAAGCTCACAACACCAAC-FL
	BTUB	LC Red 640-CGGCGATCTTAACCACCTTGTTTCC
18S rRNA	TV16Sf-2	TGAATCAACACGGGGAAAC
	TV16Sr-2	ACCCTCTAAGGCTCGCAGT
	TV16_FL	CAGGTGGAAGAGGGTAGCAATAACA-FL
	TV16_LC	LC Red 640-TCCGTGATGCCCTTTAGATGCTCTG-p

analyzed in the F2/F1 channel using the LightCycler software (version 3.5).

4. The 18S rRNA gene LightCycler PCR mix (20 μL) contains 1× LC FastStart DNA Master hybridization probe buffer, 4 mM $MgCl_2$, 0.5 μM of each primer and 0.2 μM of both FRET probes, and 2 μL template DNA. The cycling conditions and fluorescence acquisition are the same as the β-tubulin gene FRET probe PCR, except that the number of amplification cycles is reduced to 45 and the extension phase of each amplification is performed for 13 s.

11.2.2.3 PCR-Restriction Fragment-Length Polymorphism (RFLP) Analysis

Simões-Barbosa et al. [90] described a PCR-RFLP procedure for genetic analysis of *T. vaginalis* isolates, with primers (IGSf2 and IGSr2) derived from the intergenic spacer (IGS) of the rRNA. The IGS corresponding to the region between the 28S rRNA and 18S rRNA in eukaryotes is highly polymorphic, and the primer pair IGSf2 (5′-CAC GTG ATG TTG GAC CGA TA-3′) and IGSr2 (5′-CAA TCG CAT GTA TTA GCA CCA-3′) amplifies a 2 kb fragment from the IGS region. After digestion of the amplicon with restriction enzyme *TSP*509I, the genetic relatedness of *T. vaginalis* isolates is ascertained.

11.2.2.3.1 Procedure

1. PCR mix (50 μL) is made up of 200 μM each of the four deoxynucleoside triphosphates, 0.5 μM each primer, 2.5 mM $MgCl_2$, 0.25 U of *Taq* DNA polymerase (Invitrogen), and 50 ng of DNA.
2. Amplification is performed with a predenaturation at 94°C for 1 min 30 s, 40 cycles of 94°C for 1 min, 59°C for 1 min, and 72°C for 1 min 30 s, followed by a final postextension at 72°C for 5 min.
3. After verification of the PCR products (3 μL) in a 1% (w/v) agarose gel, 3 μL of each product is digested with restriction enzyme *TSP*509I (New England Biolabs).
4. The digested PCR products are loaded along with 100 bp ladder (Invitrogen) in a 3% (w/v) agarose gel and electrophoresed at 100 V for 2–3 h. The products are visualized after ethidium bromide staining.

11.3 CONCLUSION

Within the flagellate protozoan genus *Trichomonas*, *T. vaginalis* is responsible for causing human trichomoniasis, a most common STD in the world that affects >180 million people annually. The clinical spectrum of *T. vaginalis* infection ranges from vaginitis, cervicitis, urethritis to asymptomatic carrier state. In addition, the disease appears to increase risk of HIV transmission and pregnancy complications such as premature rupture of placental membranes, premature labor, and low infant birth weight.

While metronidazole and other 5-nitroimidazoles (e.g., metronidazole, tinidazole, ornidazole, and secnidazole) provide reliable and inexpensive treatments for trichomoniasis, correct identification of *T. vaginalis* plays an essential role in the effective treatment and control of the parasite. Although microscopic examination and culture as well as serological methods are useful for diagnosis of trichomoniasis, the recent development and application of molecular techniques, especially PCR, have contributed to the enhanced detection and discrimination of *T. vaginalis* in terms of sensitivity, specificity, and speed.

REFERENCES

1. http://www.uniprot.org/; last accessed on 20 May 2011.
2. Schwebke, J.R., and Burgess, D., Trichomoniasis. *Clin. Microbiol. Rev.* 17, 794, 2004.
3. Nielsen, M., and Nielsen, R., Electron microscopy of *Trichomonas vaginalis* Donne: Interaction with vaginal epithelium in human trichomoniasis. *Acta Pathol. Microbiol. Scand. Sect. B* 83, 305, 1975.
4. Moldwin, R.M., Sexually transmitted protozoal infection: *Trichomonas vaginalis*, *Entamoeba histolytica* and *Giardia lamblia*. *Urol. Clin. North Am.* 19, 93, 1992.
5. Van der Pol, B., *Trichomonas vaginalis* infection: The most prevalent nonviral sexually transmitted infection receives the least public health attention. *Clin. Infect. Dis.* 44, 23, 2007.
6. Bowden, F.J., and Garnett, G.P., *Trichomonas vaginalis* epidemiology: Parameterising and analysing a model of treatment interventions. *Sex. Transm. Infect.* 76, 248, 2000.
7. Krieger, J.N., Ravdin, J.I., and Rein, M.F., Contact-dependent cytopathogenic mechanism of *Trichomonas vaginalis*. *Infect. Immun.* 50, 778, 1985.
8. Imandel, K., Aflatoni, M., and Behjatnia, Y., Clinical manifestations of female trichomoniasis and comparison of direct microscopy and culture media in its diagnosis. *Bull. Soc. Pathol. Exot. Filiales* 78, 360, 1985.
9. Petrin, D. et al., Clinical and microbiological aspects of *Trichomonas vaginalis*. *Clin. Microbiol. Rev.* 11, 300, 1998.
10. Spence, M.R. et al., The clinical and laboratory diagnosis of *Trichomonas vaginalis* infection. *Sex. Transm. Dis.* 7, 168, 1980.
11. Wolner-Hanssen, P. et al., Clinical manifestations of vaginal trichomoniasis. *JAMA* 261, 571, 1989.
12. Soper, D.E., Bump, R.C., and Hurt, W.G., Bacterial vaginosis and trichomonas vaginitis are risk factors for cuff cellulitis after abdominal hysterectomy. *Am. J. Obstet. Gynecol.* 163, 1016, 1990.
13. Cotch, M.F. et al., *Trichomonas vaginalis* associated with low birth weight and preterm delivery. *Sex. Transm. Dis.* 24, 361, 1997.
14. Shafir, S.C., Sorvillo, F.J., and Smith, L., Current issues and considerations regarding trichomoniasis and human immunodeficiency virus in African-Americans. *Clin. Microbiol. Rev.* 22, 37, 2009.
15. Arroyo, R., Engbring, J., and Alderete, J., Molecular basis of host epithelial cell recognition by *Trichomonas vaginalis*. *Mol. Microbiol.* 6, 853, 1992.
16. Alderete, J.F., Lehker, M.W., and Arroyo, R., The mechanisms and molecules involved in cytoadherence and pathogenesis of *Trichomonas vaginalis*. *Parasitol. Today* 11, 70, 1995.

17. Fiori, P.L., Rappelli, P., and Addis, M.F., The flagellated parasite *Trichomonas vaginalis:* New insights into cytopathogenicity mechanisms. *Microb. Pathog.* 1, 149, 1999.
18. Graves, A., and Gardner, W.A., Pathogenicity of *Trichomonas vaginalis. Clin. Obstet. Gynecol.* 36, 145, 1993.
19. de Carli, G.A. et al., Laboratory diagnosis of *Trichomonas vaginalis* among women attending a venereal diseases control division. First report. *Rev. Latinoam. Microbiol.* 29, 301, 1987.
20. Fouts, A.C., and Kraus, S.J., *Trichomonas vaginalis*: Reevaluation of its clinical presentation and laboratory diagnosis. *J. Infect. Dis.* 141, 137, 1993.
21. Weinberger, M.W., and Harger, J.H., Accuracy of the Papanicolaou smear in the diagnosis of asymptomatic infection with *Trichomonas vaginalis. Obstet. Gynecol.* 82, 425, 1993.
22. Schwebke, J.R., Morgan, S.C., and Pinson, G.B., Validity of self-obtained vaginal specimens for diagnosis of trichomoniasis. *J. Clin. Microbiol.* 35, 1618, 1997.
23. Schwebke, J., Venglarik, M., and Morgan, S., Delayed versus immediate bedside inoculation of culture media for diagnosis of vaginal trichomonosis. *J. Clin. Microbiol.* 37, 2369, 1999.
24. Tabrizi, S.N. et al., A self-administered technique for the detection of sexually transmitted diseases in remote communities. *J. Infect. Dis.* 176, 289, 1997.
25. Wiese, W. et al., A meta-analysis of the Papanicolaou smear and wet mount for the diagnosis of vaginal trichomoniasis. *Am. J. Med.* 108, 301, 2000.
26. Kissinger, P.J. et al., Vaginal swabs versus lavage for detection of *Trichomonas vaginalis* and bacterial vaginosis among HIV-positive women. *Sex. Transm. Dis.* 32, 227, 2005.
27. Garber, G.E. et al., Cell culture compared with broth for detection of *Trichomonas vaginalis. J. Clin. Microbiol.* 25, 1275, 1987.
28. Krieger, J.N. et al., Diagnosis of trichomoniasis. Comparison of conventional wet-mount examination with cytologic studies, cultures, and monoclonal antibody staining of direct specimens. *JAMA* 259, 1223, 1988.
29. Gelbart, S.M. et al., Comparison of Diamond's medium modified and Kupferberg medium for detection of *Trichomonas vaginalis. J. Clin. Microbiol.* 27, 1095, 1989.
30. Schmid, G.P. et al., Evaluation of six media for the growth of *Trichomonas vaginalis* from vaginal secretions. *J. Clin. Microbiol.* 27, 1230, 1989.
31. Boeke, A.J., Dekker, J.H., and Peerbooms, P.G., A comparison of yield from cervix versus vagina for culturing *Candida albicans* and *Trichomonas vaginalis. Genitourin. Med.* 69, 41, 1993.
32. Draper, D. et al., Detection of *Trichomonas vaginalis* in pregnant women with the InPouch TV culture system. *J. Clin. Microbiol.* 31, 1016, 1993.
33. Borchardt, K.A. et al., A comparison of the sensitivity of the InPouch TV, Diamond's and Trichosel media for detection of *Trichomonas vaginalis. Genitourin. Med.* 73, 297, 1997.
34. Levi, M.H. et al., Comparison of the InPouch TV culture system and Diamond's modified medium for detection of *Trichomonas vaginalis. J. Clin. Microbiol.* 35, 3308, 1997.
35. Ohlemeyer, C.L. et al., Diagnosis of *Trichomonas vaginalis* in adolescent females: InPouch TV culture versus wet-mount microscopy. *J. Adolesc. Health* 22, 205, 1998.
36. Gilbert, R. et al., Cytopathogenic effect of *Trichomonas vaginalis* on human vaginal epithelial cells cultured *in vitro. Infect. Immun.* 68, 4200, 2000.
37. Thomason, J.L. et al., Comparison of four methods to detect *Trichomonas vaginalis. J. Clin. Microbiol.* 26, 1869, 1988.
38. Rompalo, A.M. et al., Evaluation of use of a single intravaginal swab to detect multiple sexually transmitted infections in active-duty military women. *Clin. Infect. Dis.* 33, 1455, 2001.
39. Beal, C. et al., The plastic envelope method, a simplified technique for culture diagnosis of trichomoniasis. *J. Clin. Microbiol.* 30, 2265, 1992.
40. Kurth, A. et al., Performance of a new, rapid assay for detection of *Trichomonas vaginalis. J. Clin. Microbiol.* 42, 2940, 2004.
41. Mason, P.R., Serodiagnosis of *Trichomonas vaginalis* infection by the indirect fluorescent antibody test. *J. Clin. Pathol.* 32, 1211, 1979.
42. Alderete, J.F., Enzyme linked immunosorbent assay for detecting antibody to *Trichomonas vaginalis*: Use of whole cells and aqueous extract as antigen. *Br. J. Vener. Dis.* 60, 164, 1984.
43. Smith, R.F., Detection of *Trichomonas vaginalis* in vaginal specimens by direct immunofluorescence assay. *J. Clin. Microbiol.* 24, 1107, 1986.
44. Watt, R.M. et al., Rapid assay for immunological detection of *Trichomonas vaginalis. J. Clin. Microbiol.* 24, 551, 1986.
45. Yule, A. et al., Detection of *Trichomonas vaginalis* antigen in women by enzyme immunoassay. *J. Clin. Pathol.* 40, 566, 1987.
46. Carney, J.A. et al., New rapid latex agglutination test for diagnosing *Trichomonas vaginalis* infection. *J. Clin. Pathol.* 41, 806, 1988.
47. Lisi, P.J. et al., Monoclonal-antibody-based enzyme-linked immunosorbent assay for *Trichomonas vaginalis. J. Clin. Microbiol.* 26, 1684, 1988.
48. Bickley, L.S. et al., Comparison of direct fluorescent antibody, acridine orange, wet mount, and culture for detection of *Trichomonas vaginalis* in women attending a public sexually transmitted diseases clinic. *Sex. Transm. Dis.* 16, 127, 1989.
49. Gombosová, A., and Valent, M., Dot-immunobinding assay with monoclonal antibody for detection of *Trichomonas vaginalis* in clinical specimens. *Genitourin. Med.* 66, 447, 1990.
50. Romia, S.A., and Othman, T.A., Detection of antitrichomonal antibodies in sera and cervical secretions in trichomoniasis. *J. Egypt Soc. Parasitol.* 21, 373, 1991.
51. Sharma, P. et al., A comparison of wet mount, culture and enzyme linked immunosorbent assay for the diagnosis of trichomoniasis in women. *Trop. Geogr. Med.* 43, 257, 1991.
52. DeMeo, L.R. et al., Evaluation of a deoxyribonucleic acid probe for the detection of *Trichomonas vaginalis* in vaginal secretions. *Am. J. Obstet. Gynecol.* 174, 1339, 1996.
53. Heine, R.P. et al., Polymerase chain reaction analysis of distal vaginal specimens: A less invasive strategy for detection of *Trichomonas vaginalis. Clin. Infect. Dis.* 24, 985, 1997.
54. Lin, P.R., Shaio, M.F., and Liu, J.Y., One-tube, nested-PCR assay for the detection of *Trichomonas vaginalis* in vaginal discharges. *Ann. Trop. Med. Parasitol.* 91, 61, 1997.
55. Shaio, M.F., Lin, P.R., and Liu, J.Y., Colorimetric one-tube nested PCR for detection of *Trichomonas vaginalis* in vaginal discharge. *J. Clin. Microbiol.* 35, 132, 1997.
56. Ryu, J. et al., Diagnosis of trichomoniasis by polymerase chain reaction. *Yonsei Med. J.* 40, 56, 1999.
57. van der Schee, C. et al., Improved diagnosis of *Trichomonas vaginalis* infection by PCR using vaginal swabs and urine specimens compared to diagnosis by wet mount microscopy, culture, and fluorescent staining. *J. Clin. Microbiol.* 37, 4127, 1999.

58. Lawing, L., Hedges, S., and Schwebke, J., Detection of trichomonosis in vaginal and urine specimens from women by culture and PCR. *J. Clin. Microbiol.* 38, 3585, 2000.
59. Patel, S.R. et al., Systematic review of diagnostic tests for vaginal trichomoniasis. *Infect. Dis. Obstet. Gynecol.* 8, 248, 2000.
60. Knox, J. et al., Evaluation of self-collected samples in contrast to practitioner- collected samples for detection of *Chlamydia trachomatis*, *Neisseria gonorrhoeae*, and *Trichomonas vaginalis* by polymerase chain reaction among women living in remote areas. *Sex. Transm. Dis.* 29, 647, 2002.
61. Schwebke, J., and Lawing, L., Improved detection by DNA amplification of *Trichomonas vaginalis* in males. *J. Clin. Microbiol.* 40, 3681, 2002.
62. Wendel, K.A. et al., *Trichomonas vaginalis* polymerase chain reaction compared with standard diagnostic and therapeutic protocols for detection and treatment of vaginal trichomoniasis. *Clin. Infect. Dis.* 35, 576, 2002.
63. Kaydos-Daniels, S. et al., Validation of a urine-based PCR-enzyme-linked immunosorbent assay for use in clinical reseach settings to detect *Trichomonas vaginalis* in men. *J. Clin. Microbiol.* 41, 318, 2003.
64. Pillay, A. et al., Comparison of a TaqMan-based real-time polymerase chain reaction with conventional tests for the detection of *Trichomonas vaginalis*. *Sex. Transm. Infect.* 83, 126, 2007.
65. Rojas, L., Fraga, J., and Sariego, I., Genetic variability between *Trichomonas vaginalis* isolates and correlation with clinical presentation. *Infect. Genet. Evol.* 4, 53, 2004.
66. Smith, K.S. et al., Comparison of conventional testing to polymerase chain reaction in detection of *Trichomonas vaginalis* in indigenous women living in remote areas. *Int. J. STD AIDS* 16, 811, 2005.
67. Hardick, J. et al., Comparison between the Gen-Probe transcription-mediated amplification *Trichomonas vaginalis* research assay and real-time PCR for *Trichomonas vaginalis* detection using a Roche LightCycler instrument with female self-obtained vaginal swab samples and male urine samples. *J. Clin. Microbiol.* 44, 4197, 2006.
68. Radonjic, I.V. et al., Diagnosis of *Trichomonas vaginalis* infection: The sensitivities and specificities of microscopy, culture and PCR assay. *Eur. J. Obst. Gynecol. Reprod. Biol.* 126, 116, 2006.
69. Upcroft, J.A. et al., Genotyping *Trichomonas vaginalis*. *Int. J. Parasitol.* 36, 821, 2006.
70. Munson, E. et al., *Trichomonas vaginalis* transcription-mediated amplification-based analyte-specific reagent and alternative target testing of primary clinical vaginal saline suspensions. *Diagn. Microbiol. Infect. Dis.* 68, 66, 2010.
71. Nye, M.B., Schwebke, J.R., and Body, B.A., Comparison of APTIMA *Trichomonas vaginalis* transcription-mediated amplification to wet mount microscopy, culture, and polymerase chain reaction for diagnosis of trichomoniasis in men and women. *Am. J. Obst. Gynecol.* 200, 188.e1, 2009.
72. Riley, D.E. et al., Development of a polymerase chain reaction based diagnosis of *Trichomonas vaginalis*. *J. Clin. Microbiol.* 30, 465, 1992.
73. Jordan, J.A., Lowery, D., and Trucco, M., Taqman-based detection of *Trichomonas vaginalis* DNA from female genital specimens. *J. Clin. Microbiol.* 39, 3819, 2001.
74. Katiyar, S.K., and Edlind, T.D., β-Tubulin genes of *Trichomonas vaginalis*. *Mol. Biochem. Parasitol.* 64, 33, 1994.
75. Madico, G. et al., Diagnosis of *Trichomonas vaginalis* infection by PCR using vaginal swab samples. *J. Clin. Microbiol.* 36, 3205, 1998.
76. Paces, J., Urbankova, V., and Urbanek, P., Cloning and characterization of a repetitive DNA sequence specific for *Trichomonas vaginalis*. *Mol. Biochem. Parasitol.* 54, 247, 1992.
77. Kengne, P. et al., *Trichomonas vaginalis* repeated DNA target for highly sensitive and specific polymerase chain reaction diagnosis. *Cell Mol. Biol.* 40, 819, 1994.
78. Mayta, H. et al., 18S ribosomal DNA-based PCR for diagnosis of *Trichomonas vaginalis*. *J. Clin. Microbiol.* 38, 2683, 2000.
79. Alderete, J.F. et al., Cloning and molecular characterization involved in *Trichomonas vaginalis* cytoadherence. *Mol. Microbiol.* 17, 69, 1995.
80. Jeremias, J. et al., Detection of *Trichomonas vaginalis* using the polymerase chain reaction in pregnant and non-pregnant women. *Infect. Dis. Obstet. Gynecol.* 2, 16, 1994.
81. Crucitti, T. et al., Comparison of culture and different PCR assays for detection of *Trichomonas vaginalis* in self collected vaginal swab specimens. *Sex. Transm. Infect.* 79, 393, 2003.
82. Caliendo, A.M. et al., Real-time PCR improves detection of *Trichomonas vaginalis* infection compared with culture using self-collected vaginal swabs. *Infect. Dis. Obstet. Gynecol.* 13, 145, 2005.
83. Schirm, J. et al., *Trichomonas vaginalis* detection using real-time TaqMan PCR. *J. Microbiol. Methods* 68, 243, 2007.
84. Depuydt, C.E. et al., Epidemiology of *Trichomonas vaginalis* and human papilloma virus infection detected by real-time PCR in flanders. *Gynecol. Obstet. Invest.* 70, 273, 2010.
85. Diaz, N. et al., Rapid detection of coinfections by *Trichomonas vaginalis*, *Mycoplasma hominis*, and *Ureaplasma urealyticum* by a new multiplex polymerase chain reaction. *Diagn. Microbiol. Infect. Dis.* 67, 30, 2010.
86. López-Villaseñor, I. et al., *Trichomonas vaginalis* ribosomal DNA: Analysis of the intergenic region and mapping of the transcription start point. *Mol. Biochem. Parasitol.* 137, 175, 2004.
87. Vatanshenassan, M. et al. *Trichomonas vaginalis*: Investigation of a novel diagnostic method in urine samples using cysteine proteinase 4 gene and PCR technique. *Exp. Parasitol.* 126, 187, 2010.
88. Simpson, P. et al., Real-time PCRs for detection of *Trichomonas vaginalis* beta-tubulin and 18S rRNA genes in female genital specimens. *J. Med. Microbiol.* 56, 772, 2007.
89. Hardick, J. et al., Use of the Roche LightCycler Instrument in a real-time PCR for *Trichomonas vaginalis* in urine samples from females and males. *J. Clin. Microbiol.* 41, 5619, 2003.
90. Simões-Barbosa, A. et al., *Trichomonas vaginalis*: Intrastrain polymorphisms within the ribosomal intergenic spacer do not correlate with clinical presentation. *Exp. Parasitol.* 110, 108, 2005.

12 Trypanosoma

Emily R. Adams

CONTENTS

12.1 INTRODUCTION

12.1.1 Classification, Morphology, and Biology

Trypanosomes are single-celled, eukaryotic parasites ubiquitous in all vertebrate and invertebrate classes. The trypanosomes are members of the genus *Trypanosoma* (order Kinetoplastida). The Kinetoplastida are characterized by a DNA containing organelle, the kinetoplast, which is unique to this group (Figure 12.1). Trypanosomes are obligate parasites with digenic (two-host) lifestyles. They are transmitted cyclically from insect vector, commonly tsetse flies, leeches, triatomine bugs, and ticks, to their vertebrate host.

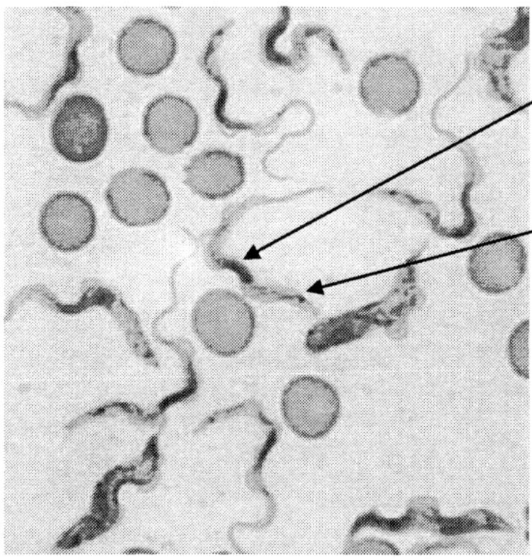

FIGURE 12.1 Field of trypanosomes. The nucleus contains the genomic DNA and the kinetoplast contains the mitochondrial DNA.

Some species of trypanosome are able to cause disease in humans and in livestock resulting in a severe social burden and major economic loss in sub-Saharan Africa, South America, and some parts of Asia. In sub-Saharan Africa two subspecies of *Trypanosoma brucei* causes human African trypanosomiasis (HAT) or sleeping sickness, and in America *T. cruzi* causes American trypanosomiasis (AT) or Chagas disease. Both diseases are included on the WHO list of most neglected diseases and are part of the priority medicines program due to their disproportionate effect on the poor (http://whqlibdoc.who.int/hq/2004/WHO_EDM_PAR_2004.7.pdf).

HAT is caused by the parasites *T. brucei gambiense* and *T. brucei rhodesiense* and in West and East Africa, respectively. *T. b. gambiense* is responsible for the majority of infections (~90%) and the main host is considered to be man, whilst *T. b. rhodesiense* is considered a zoonotic disease with humans as an accidental host.[1] The parasites are transmitted through the infected bite of a tsetse fly (*Glossina* sp.).[2] *T. b. gambiense* causes chronic disease which can persist for many months and years; conversely *T. b. rhodesiense* infection results in acute disease which causes death within weeks to months. Both diseases can be fatal if left untreated.

Glossina species are found exclusively in the tsetse belt of sub-Saharan Africa; spatial segregation of the tsetse flies has led to discrete populations of HAT patients in approximately 300 foci within 36 countries. Incidence of HAT was severely reduced in the 1960s; however, a resurgence was witnessed due to the breakdown of control programs and networks of treatment. The rising prevalence in the late twentieth century was classed as a category I by WHO/TDR, this situation has now been curbed and the numbers of patients have reduced; sustainable surveillance and control will be the primary means of further reducing the number affected by this disease (http://whqlibdoc.who.int/hq/2007/WHO_CDS_NTD_IDM_2007.6_eng.pdf). Currently, around 60 million people are still at risk from contracting HAT, and the number reporting for treatment every year is around 10,000;[3] patients are mainly seen in Angola, Democratic Republic of the Congo, and Sudan.

AT or Chagas disease is caused by the trypanosome *T. cruzi*, and is transmitted via bugs belonging to the subfamily Triatominae in the family Reduviidae.[1] Parasites enter the blood stream of a human when an infected triatomine bug bites the skin followed by defecation; parasites then pass through the irritated region through fecal contact or through the eyes when they are rubbed or scratched. Infection can also occur transplacentally or through organ and blood transfusion. In addition, oral transmission, through eating contaminated food, has also become an important route of infection in some areas.[4] Recent investigations estimate around 8–14 million people are infected with Chagas disease[5] this is less than previous calculations due to control measures, for example, the Southern Cone Initiative which started in 1991 and introduced measures including screening and vector elimination. However, it is still estimated that only 1% of infected individuals receive any treatment for Chagas due to the few adequate treatments available, expense and the lack of accessible health care programs. The endemic regions cover Central and South America, worst-affected countries include Bolivia, Peru, and Guatemala. Increasing numbers of patients are now seen outside of the endemic areas due to the movement of people to America and other parts of the world. Recently the Food and Drug Administration (FDA) in the United States have issued guidance to screen blood due to contamination of their blood banks.[6] The South American trypanosome *T. rangeli* can also infect humans but is presumed nonpathogenic; this occurrence makes diagnosis difficult.

In addition to HAT and AT, there are a wide range of trypanosome species that cause disease in livestock inflicting widespread economic losses for endemic countries. Estimated losses from livestock production from animal African trypanosomiasis (AAT) or Nagana, are estimated at approximately US 1.0–1.2 billion, and losses in terms of potential agricultural Gross Domestic Product (GDP) maybe as much as US 4.75 billion (FAO 2005). Animal hosts of *T. brucei* subspecies can provide the major reservoir to human disease in endemic areas of Africa especially *T. b. Rhodesiense*.[7,8] Furthermore, human infections with presumed nonpathogenic trypanosomes have been reported from Ghana,[9] India,[10] and Thailand[11] thus indicating the importance of zoonotic trypanosome species to humans.

12.1.2 Clinical Features and Pathogenesis

12.1.2.1 African Trypanosomiasis

HAT is distinguished by two stages of disease, I and II. (i) Stage I: Infection occurs through the bite of an infected tsetse fly, trypanosomes pass into the blood stream though the saliva of the fly and a chancre or swelling may develop at this site around 5–15 days post-infection. Parasites then multiply in the bloodstream and tissues and travel primarily to the lymph nodes but also the spleen, liver, heart, and endocrine system.[12] At this stage, the parasites avoid the host immune system with a technique called antigenic variation. A glycoprotein coat surrounding the trypanosome protects against complement-mediated lysis; once an antibody response has been raised to this glycoprotein the majority of the parasites will die; however, a small percentage are able to switch to a new, antigenically distinct coat. This population then grows in number till a similar response is raised against this glycoprotein. Trypanosomes have around 1000 different genes encoding the variant surface glycoproteins (VSGs); sequential expression of these allows parasite survival.[1] The symptoms of this stage of disease are nonspecific, including fever, headaches, lethargy, and anemia. (ii) Stage II: After a few weeks or months the parasites cross the blood–brain barrier and invade the central nervous system; generally *T. b. rhodesiense* cross this barrier earlier.[1] Symptoms for this second stage of disease include severe headaches, disturbance in circadian rhythms with nightly insomnia and urges to sleep during the day, weight loss, muscle twitches and impotence; a wide range of psychological problems are also evident, with anxiety, irritability, hallucinations and delirium. Progressive disease without intervention can lead to eventual coma and death.

12.1.2.2 American Trypanosomiasis

In AT, like HAT, an acute stage of disease is frequently observed; inflammation at the site of infection can lead to local swelling, this is especially true if infection has occurred in the eye where swelling of the eyelids leads to a condition known as Romana's sign. Generic symptoms can also be observed including fever and general malaise; this early stage of disease is often not diagnosed unless active screening programs are present.[13] Death occurs infrequently in patients with acute infection; however, patients infected orally have a different pathogenesis and death can commonly be witnessed in this early stage. If the acute stage remains untreated then the patient will enter the chronic stage of disease which is asymptomatic in nature. The *T. cruzi* parasites invade host cells where they multiply as intracellular amastigotes by binary fission and avoid the immune system. The parasites transform from promastigotes into trypomastigote form and burst out of the host cell where they reinfect other cells. Approximately 70–85% of infected patients continue in this state for the rest of their lives.[1] However, 10–30 years after initial infection approximately 15–30% of patients develop serious clinical manifestations typified by severe organ pathologies primarily in the cardiac and intestinal systems;[14,15] this is most common in male individuals between 20 and 50 years old.[14] Arrhythmia, thromboses, palpitations, shortness of breath, chest pain, heart failure, and sudden death are frequent outcomes in this stage of disease. Pathogenesis of the intestine is observed with prominent geographical separation; megaviscera producing megacolon leading to constipation and occasionally torsion of the intestine causing severe problems. Recent studies estimate around 14,000 people die of Chagas disease every year.[5] Unfortunately, due to the silent nature of this disease and the poverty of the endemic areas, accurate estimates of disease prevalence are not available. Contamination of blood banks and emigration of the endemic population have led to increasing infection in nonendemic areas of the world notably in the United States and Spain. The year 2009 was the hundredth year since the discovery of the *T. cruzi* parasite by Carlos Chagas, and campaigns to highlight this disease to promote research into diagnostics and drug therapies were held.[5]

12.1.3 Diagnosis

12.1.3.1 Conventional Diagnosis

The conventional, gold-standard diagnostic for HAT remains microscopy. In East Africa, infections with *T. b. rhodesiense* are characterized by high parasitemia which is often sufficient for direct diagnosis by microscopic observation of blood. However, for *T. b. gambiense* infections, which are typified by cyclical parasitemia, microscopy alone may not be sufficient for diagnosis.[16] Only around 10,000 parasites/mL can be detected by simple wet blood film microscopy; therefore, methods have been developed for the concentrations of parasites. Only with these improved concentration methods can microscopy detect as low as 50 parasites/mL.[17] Microhematocrit centrifugation involves taking a capillary tube containing anticoagulant, and blood from a finger prick; by high-speed centrifugation the trypanosomes collect with the white blood cells and buffy coat which can be observed under a microscope.[18,19] The quantification buffy coat (QBC) technique relies on similar principles with the addition of ethylenediaminetetraacetic acid (EDTA) and acridine orange in the capillary tubes; the acridine orange stains the DNA and kinetoplast of the trypanosome, this can be visualized with UV light. This technique is as sensitive as the microhematocrit, yet has the advantage of distinguishing trypanosomes from malarial parasites, common in HAT patients.[16] Other conventional diagnostics include fluorescent microscopy which also uses acridine orange to stain trypanosomes, and Mini-anion-exchange centrifugation technique (mAECT) which separates trypanosomes based upon charge; eluted parasites are concentrated by low-speed centrifugation followed by direct microscopy.[20] This later technique is the most sensitive, and recent developments in the manufacture of the collection tube, viewing chamber, reagents, and quality control have led to improved sensitivity and robustness.[17]

Serological tests have also been developed; the card agglutination test (CATT) detects anti-*T. brucei gambiense* antibodies in the blood and serum, and high sensitivity can be achieved. This assay was developed in the late 1970s and can be used as a screening tool as it is quick, cheap, and practical with only basic equipment requirements (refrigerator and rotator).[21,22] Positive samples show agglutination on the card; low and high titers are easy to read, however, when titer is between 1:4 and 1:8 the result is ambiguous and the patient must be followed up.[16,22] Positive test results must be confirmed with microscopic observation of parasites from the blood lymph before treatment can be initiated. Other forms of the CATT test have been developed such as micro-CATT and latex-CATT; however, the same sensitivity cannot be achieved with these tests.[22]

The staging of HAT is of critical importance as stage I treatments cannot cross the blood-brain barrier (BBB), and stage II treatments can be extremely toxic causing 5–10% mortality in patients, therefore it is not desirable to treat patients unless necessary. According to the WHO (1998), parasites in the central spinal fluid (CSF), a white blood cell (WBC) count of >5 cells/μL and increased protein content (>370 mg/L) is considered confirmation of stage II disease. However, these values are not universally accepted. Often a higher value of white blood cells, 20 cells/μL, is used. Kennedy et al. suggest that a gradient of early CNS invasion occurs where parasites have passed the BBB but have not yet invaded the brain parenchyma, and that these patients may be treatable with stage I drugs and hence a higher WBC count should be used.[12] Microscopy of the CSF by lumbar puncture is examined to determine if parasites have crossed the BBB; this stage is carried out for all HAT patients in order to initiate correct treatment for both stage I and II patients. It should always be kept in mind that taking a lumbar puncture is a painful procedure for the patient and may introduce infection into the new wound. Diagnostics that are able to avoid this invasive and painful process are highly desirable.

Elevated protein concentrations are also recommended by the WHO to stage HAT patients; however, the levels used are highly controversial and this method is seldom used.[16] Another marker of CNS invasion is the elevation of IgM levels in the CSF.[23,24] Latex agglutination cards have been developed to monitor this invasion[24] and have been shown to predict treatment failure in combination with WBC, IL-10 levels, trypanosome antibodies, and protein concentration.[25]

Conventional diagnosis for AT also relies upon microscopy and serological testing. In the acute phase of disease, patients may be diagnosed by Romana's sign, swelling round the eye, or by direct microscopy which is possible due to high parasitemia. However, patients with chronic Chagas disease are typified by low parasitemia; therefore, detection must be made indirectly by the detection of antibodies which are present at differing titers.[13] Serological diagnosis is usually performed with enzyme-linked immunosorbent assay (ELISA), indirect hemagglutination assay (IHA), and the indirect immunofluorescence assay (IFA). Due to problems with specificity the WHO recommends the use of two antibody detection tests that use two different antigens; if the tests are not in agreement then a third test should be performed. ELISA is a highly specific and sensitive test and the ability to process many samples; however, a cold chain is required for the reagents and a skilled technician must be present to pipette and read the results. IFA is a sensitive test which uses fluorescence microscopy to visualize the results, which gives both qualitative and quantitative results. However, a cold chain and an expensive fluorescent microscope are required for this test. IHA is not as sensitive as ELISA or IFA and has a low reproducibility of results; again the reagents must be kept in a cold chain and the readings are more subjective than that of the other tests.

Rapid diagnostic tests (RDTs) are coming to the forefront of Chagas diagnostics; these tests are cheaper, are easy to use with dipstick formatting, and are much quicker than the previous range of serological tests. An analysis of the STATPAK serological test showed good sensitivity (93–98%) and specificity of 97% for less than 2$.[26] However, there is still a call for better quality RDTs, which are effective for both acute and chronic stage, affordable, and have quality control for batch-to-batch testing in order to ensure ongoing reliability.[13]

Traditional xenodiagnosis is still used in some Latin American countries for the diagnosis of Chagas disease. Four boxes of 10 nymphs each are placed on the skin of a patient for 30 min. After 30–90 days the intestine of the nymphs is dissected and examined for the presence of parasites using a light microscope. However, this method is extremely time consuming and has low sensitivity;[27] in addition no distinction can be made between *T. cruzi* and the nonpathogenic but human infective *T. rangeli*.

12.1.3.2 Molecular Diagnosis

Molecular techniques may have potential to be used for the diagnosis of HAT and AT due to their high sensitivity and specificity; although thorough validation of their diagnostic accuracy must take place before implementation. Here promising developments for the molecular diagnosis of the trypanosomiasis are described.

12.1.3.2.1 Polymerase Chain Reaction

The PCR technique was first developed in the 1980s as a tool for the amplification of millions of copies of specifically targeted DNA by an *in vitro* enzymatic reaction.[28] The incorporation of hot-start polymerases, proof-reading enzymes, nested PCR strategies have all helped to improve the sensitivity, specificity, ease of use, and popularity of PCR diagnostics.

12.1.3.2.2 Species-Specific Approach

The application of PCR-based techniques has greatly aided molecular-based diagnosis for the *Trypanosoma*. The first primer sets for the specific amplification of the trypanosomes typically targeted noncoding, repetitive DNA elements in the satellite DNA repeats which form the bulk of the mini-chromosomes.[29] Since then there has been a rapid increase in sequence data publically available through cell biology, taxonomy, and genome projects; for example, a single copy gene, the serum resistance-associated gene (SRA) was discovered that enables *T. b. rhodesiense* to survive in humans;[30] this led to the development of a specific PCR test for this gene.[31] The differentiation of this trypanosome from the morphologically indistinguishable *T. b. brucei* is important as they share many of the same hosts, yet *T. b. brucei* cannot infect humans, and discrimination from *T. b. gambiense* is also crucial as the campaign continues to keep the two human pathogens geographically separate. In addition, the *T. gambiense*-specific glycoprotein (TgsGP) is only present in *T. b. gambiense*; consequently primers designed here can differentiate this species from the *Trypanozoon*.[32] Both TgsGP and SRA are single copy genes and, therefore, the sensitivity of these assays is lower than that of a multicopy gene. Recently, Picozzi et al. (2008) developed a multiplex PCR to differentiate *T. b. rhodesiense*; this PCR contains two sets of specific primers, one for the amplification of the SRA gene, and one which targets a single copy gene to indicate whether the DNA is of sufficient quality and quantity to amplify single copy genes.[33]

A systematic review which assessed the diagnostic accuracy of molecular amplification tests for the diagnosis of HAT showed that very few articles have been published that are of a high enough quality and good design as to evaluate this aim.[34] Only 14 articles evaluated the accuracy of PCR to diagnose HAT patients as compared with the gold standard of microscopy. Summary sensitivity for PCR on blood was 99.0% (95% CI 92.8–99.9) and the specificity was 97.7% (95% CI 93.0–99.3) (See Figure 12.2).

Differences in study design and readout method did not significantly change estimates although use of satellite DNA as a target significantly lowered specificity. Sensitivity and specificity of PCR on CSF for staging varied from 87.6% to 100%, and 55.6% to 82.9%, respectively although the scarcity of these articles is a cause for concern and the value of PCR positivity in the CSF is not yet known. In this study, PCR seemed to have sufficient accuracy to replace microscopy where facilities allow, although that conclusion was

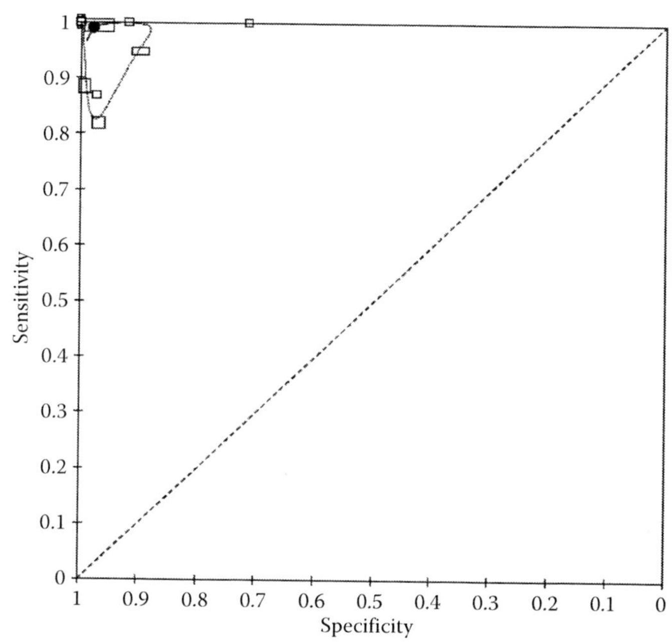

FIGURE 12.2 Raw ROC plot. Summary ROC plot for PCR and PCR-OC. Shows summary estimate (black dot), summary curve and confidence ellipse around the summary estimate. Width of the symbols reflects the number of nondiseased patients; height of the symbol reflects the number of diseased patients. (Adapted from Mugasa, C.M. et al. *PLoS. Negl. Trop. Dis.*, 6, e1438, 2012.)

based on multiple reference standards and a patient population that was not always representative. It must be concluded that future studies should include patients for which PCR may become the test of choice and consider well-designed diagnostic accuracy studies to provide extra evidence on the value of PCR in practice.[34]

The *Trypanosoma* are members of the order Kinetoplastida, these protists have a unique DNA-containing organelle called the kinetoplast which is also a successful source of species-specific primers for PCR. Minicircle DNA, held within the kinetoplast, is a repetitive target with high copy number, thereby enhancing the sensitivty of molecular tests as there are more areas for the primers to bind; this is therefore a common target. Several PCRs designed to amplify the kinetoplast DNA have been compared for the sensitive and reliable detection *T. cruzi* infection. The minicircle DNA guides RNA for editing mitochondrial DNA transcripts; there is a high level of sequence variation within the minicircles which has been used to differentiate *T. cruzi* and the human infective but nonpathogenic trypanosome *T. rangeli*.[35] In addition, most PCRs are not able to distinguish between the sublineages of the *T. cruzi* complex, and in fact report varying sensitivities of PCRs between the lineages.[36] Although the treatment protocol for all lineages remains the same, the differentiation of the sublineages can give important information regarding the severity of disease to be expected and the organs that may be targeted. A sensitive PCR that is able to make this discrimination is highly desirable.

Studies comparing the results between serological testing and PCR testing of patients on the whole show good agreement, for example, Gil et al. (2007) tested patients with ELISA, IFAT, and PCR based on the histone H2A/SIRE gene.[37] They found the sensitivity to be 88% and the specificity to be 92% concluding that PCR is a good diagnostic tool to be used in a complementary way to serological testing.

12.1.3.2.3 Generic Approaches

Since there are many species of trypanosome that are able to infect animals and humans a species-specific approach can be time consuming and expensive. To overcome these limitations, a range of methods have been developed that use generic primers in conserved areas of the genome to amplify sections of DNA from any, or specific groups of, trypanosome species. For example, the internal transcribed spacer (ITS) regions of the ribosomal RNA locus have a high copy number, vary both in length and sequences between species and primers can be designed in the conservative regions of the 18S, 28S, or 5.8S. These properties have been exploited for species identification in the *Trypanosoma*. Desquesnes et al.[38] designed primers to amplify the ITS-1 regions with generic primers in the 18S and 5.8S regions; sufficient interspecies size variation in this region enabled the differentiation of the major tsetse-transmitted trypanosomes. Redesign of these original primers and the addition of a nested PCR-step increased the specificity and sensitivity.[39,40] Unfortunately, members of the *Trypanozoon* cannot be distinguished with this method due to the acknowledged close relationship of the *T. brucei* subspecies with each other. This method has been used to identify a trypanosome infection in a sick child in Thailand;[11] upon further analysis this infection was found to be an atypical human infection of a *T. lewisi*-like trypanosome, a group of trypanosomes more commonly associated with rats. This method is also useful to screen livestock; by identification of the species of trypanosomes that the herd carries the disease risk to humans can then be identified; that is, if *T. b. gambiense* or *T. b. rhodesiense* is present, then the risk of transmission to humans, via tsetse flies, is high. Therefore, control strategies can be targeted more efficiently.

Length variation in the 24S rRNA gene has been used to differentiate the three parasites commonly associated with triatomine bugs, namely: *T. cruzi, T. rangeli*, and *Blastocrithidia triatoma*.[41] This generic PCR procedure provides an accurate and fast way of differentiating pathogenic trypanosomes from other trypanosomatids in field surveys of disease vectors and laboratory triatomines used from xenodiagnosis.

A new identification technique based upon generic PCR was developed by Hamilton et al. (2008) called fluorescent fragment-length bar-coding (FFLB).[42] This technique uses four sets of generic primers, with one of each pair labeled with a fluorescent tag; these amplify regions of the 18S and 28S rRNA with known interspecies size variation. The size of each product is determined accurately on an automated DNA sequencer allowing sizing to within 1 bp, far more accurate than conventional gel electrophoresis. Using the four primer sets, a barcode specific for each species is generated, thereby allowing species identification. The great advantage of this technique is to recognize new species by a unique barcode even in mixed infections. This has led to the discovery of a new trypanosome, closely related to the human-infective *T. brucei*.[42] However, due to the need for specialized staff and equipment, and due to the cost of reagents, this technique is limited to research purposes only.[2]

12.1.3.2.4 Standardization

The standardization of PCR for molecular analysis is crucial for accurate and reliable diagnosis. The design of primers for PCR and optimization is a relatively easy process; consequently, many laboratories have their own "in-house" PCR leading to a huge number of possible protocols with various reported sensitivities and specificities. Recently, attempts have been made to standardize the entire procedure from DNA extraction to diagnostic readout for detection of both *T. brucei* subspecies and *T. cruzi*.[43,44] Principally a DNA extraction was carried out on blood samples of patients, followed by PCR amplification with inclusion of an internal control and simple read out using oligochromatographic dipsticks. In the future it is hoped that more standardized kits, with ongoing quality control, will be produced so that PCR diagnostics for the *Trypanosoma* can be more reliable, accurate, and consistent for the benefit of patients.

As shown, PCR is readily applicable to field collected material; the tests are highly specific and sensitive often with detection to 0.1 pg of DNA, roughly equivalent to one trypanosome. However, a cold chain for enzymes to be transported, a continual supply of electricity in order to complete the PCR reaction and specialized equipment are absolute requirements. Therefore, these techniques cannot be used at the point-of-care and can only be completed in a well-equipped laboratory with trained staff. Improvements in the standardization of PCR, cost of equipment, and an increasing skill base in the technique have enabled PCR to be at the forefront of molecular diagnostics for the *Trypanosoma*; however, further improvements must be made before the benefits of this highly sensitive and reliable technique can be unanimously achieved. In addition, the low number of well-designed diagnostic accuracy studies for molecular diagnostics for HAT, as observed by Mugasa et al.[34] shows that further validation studies must take place before these techniques can be implemented.

12.1.3.2.5 PCR-Restriction Fragment-Length Polymorphism

The PCR technique can also be used in combination with restriction fragment-length polymorphism (RFLP) analysis to distinguish species from one another. This technique involves PCR amplification using generic primers followed by digestion of the PCR products with restriction enzymes; the product is cut in specific places producing a number of smaller fragments; the fragments are run on a gel by electrophoresis and a banding pattern unique for each species may be achieved. This technique is particularly important for members of the *T. cruzi* sublineages in order to identify the correct lineage and, therefore, possible clinical manifestations.[45–48] Common regions for PCR amplification followed by RFLP include the variable regions of the 18S rDNA,[45,49] ITS regions,[47] and kinetoplast DNA (kDNA);[48] these regions have been chosen due to their high copy number and intraspecies sequence variation.

This technique is able to distinguish between the major members of the *Trypanosoma*. However, the enzymatic digestion step is time-consuming and is prone to contamination as the PCR products are further manipulated. Since there are PCR methods available that can diagnose without the need for enzymatic digestion, the benefits of using RFLP for the diagnosis of human infections, are limited. This method is of greater use for the identification of trypanosome species in epidemiological studies especially for livestock where mixed infections are frequent.[49]

12.1.3.2.6 Real-Time PCR

Conventional PCRs have proved to be highly sensitive and specific for the diagnosis of the trypanosomiases; however, the detection methods are often cumbersome, for example, gel electrophoresis and, importantly, it is not possible to accurately quantify parasitemia. Therefore, a range of quantitative PCRs, or real-time PCRs, which overcome these problems, have been developed. Fluorescent real-time monitoring of amplicon accumulation allows quantification of parasitemia by determining the threshold cycle (C_T) at which amplification exceeds the base level fluorescence thereby monitoring exponential amplification of the DNA product. The C_T of the unknown samples is compared against set of known parasitemias and the sample can be quantified.

Becker et al. (2004) developed a real-time assay based on a 177 bp repetitive sequence for the detection of *T. brucei* DNA in human blood samples.[50] This reported detection limit was similar to conventional PCR of parasites in blood samples; they were able to show that real-time PCR could be used for human diagnosis with the advantage of quantification, single tube format and no post-amplification handling.

Assays have also been developed for the specific amplification of *T. cruzi*.[51–54] Most recently, an assay incorporating a commercial kit for samples processing, a single tube reaction, a correction of the parasitic load, according to variation in the number of target sequences between the different lineages of *T. cruzi*, and an internal control has been developed.[54] This technique has been successfully used to monitor clinical reactivation after chemotherapy treatment and heart transplantation.

The use of quantitative techniques which can monitor clinical reactivation are of great significance in clinical trials where early assessment of efficacy or failure is mandatory.

12.1.3.2.7 Sequencing

DNA sequences are increasingly used for species detection and identification. By direct comparison of DNA from previous studies on publically held databases such as GENBANK, exact species identification can be made. A large data set for sequences encoding the rRNA genes exists for members of the *Trypanosoma*; the most variable section of which is the 18S rRNA (D7–D8), commonly used for species identification. In addition, over 100 full-length 18S rRNA (~2000 bp) sequences are available for comparison from a variety of hosts and geographical areas. Another large data set has been created for the glycosomal glyceraldehyde phosphate dehydrogenase (*gGAPDH*) gene from over 70 different trypanosomes; this set provides confirmation of the 18S rRNA sequence and they are often used in parallel for phylogenetic analysis.[42,55] The *gGAPDH* gene is shorter than the full-length 18S rRNA, at approximately 800 bp, yet the two genes have similar phylogenetic resolution, therefore, the *gGAPDH* gene is quicker and negates the need for end linking.[56] More recently other genes have been sequenced and small-to-medium size databases have been formed. For example, the cathepsin L-like enzymes, which are cysteine proteases, playing a vital role in metabolism, infectivity, cell differentiation, and pathogenicity have been targeted for drug treatments, vaccines, and diagnostics;[57] these genes have also been sequenced for a range of tsetse-transmitted trypanosomes and can now be used for identification within epidemiological and population genetics studies.[58] Studies have been conducted using the *cytochrome B* gene (*Cyt b*); this kinetoplast (mitochondrial) gene is encoded on part of the maxicircle. It has been used extensively for analysis of within-species diversity of *T. cruzi*[59] where differentiation of the sublineages of *T. cruzi* can be observed. The trypanosomal internal transcribed spacer 1 (ITS1) has also been used for species identification and phylogenetic placement; this region varies in length between 200 and 800 bp, depending on species. It evolves approximately 6 times faster than the D7–D8 18S rDNA region and is frequently used in combination with the 18S rRNA for accurate identification.[60]

Sequencing can be an extremely accurate and reliable system for the diagnosis and identification of trypanosomes; however, if the sequences being compared are not identical then identification is difficult and there is no consensus on the degree of similarity necessary to identify a known species.[42,29] In these cases, precise phylogenetic placement must then be completed in order to accurately identify the species of trypanosome present. For routine diagnosis and species identification, sequence analysis is time consuming, expensive, and can only be completed with highly specialized equipment and staff. The development of high-throughput technologies such as 454 sequencing[61] may enable sequencing to become a more attractive option in the future.

12.1.3.2.8 Nucleic Acid Sequence-Based Amplification

Nucleic acid sequence-based amplification (NASBA) is an isothermal reaction that is based upon the amplification of single-stranded RNA; it is a sensitive and specific technique with a run time of 90 min, significantly shorter than PCR.[62] NASBA amplifies RNA during a single temperature step (41°C); as there is no repeated heating and cooling this reaction can be carried out in a water bath without the need for a thermocycler, this reduces the cost of equipment associated with PCR-based techniques. Three enzymes are involved in the reaction: avian myeloblastosis virus reverse transcriptase, RNase H, and T7-dependant RNA polymerase. Mugasa et al.[63] developed the technique for the amplification of *T. brucei* subspecies based on the 18S rRNA gene and was able to achieve amplification of 10 parasites/mL in spiked blood.[63] This technique was subsequently adapted for use in combination with oligochromatography (Section 12.2.2) and evaluated in a ring trial where it is performed comparably with PCR-OC.[64] This technique is promising for further diagnostic use due to its isothermal nature, the high sensitivities and specificities achieved, and a standardized easy-read-out system in. In addition, due to the amplification of RNA only, NASBA can be used to detect active infection in patients as RNA rapidly degrades after parasite death. This can, therefore, be used to monitor treatment outcome and to diagnose active infection. However, further developments to reduce cost and eliminate the cold-chain needed for the transport of enzymes are required before this assay can be implemented. Since NASBA targets RNA, which is more liable to degradation than DNA, addition care must be taken during extraction of RNA and subsequent specimen storage.

12.1.3.2.9 Loop-Mediated Isothermal Reaction (LAMP)

A recent advance in diagnostics has been the development of loop-mediated isothermal reaction (LAMP).[65] This nucleic acid amplification technique utilizes only one enzyme (*Bst* DNA polymerase) and is able to amplify large amounts of DNA within 30–60 min by the intricate design of primers and auto-strand displacement DNA synthesis. The reaction has several advantages; first, the reaction takes place between 60°C and 65°C, and can be completed without the use of a thermo-cycler. Second, the specificity of the reaction is high due to the design of six primers. Third, the product can be visualized directly utilizing simple detection methods such as visualization with fluorescent dyes under UV light. Therefore, LAMP has potential to emerge as a powerful tool for diagnostics. This technology was first implemented for the direct amplification of trypanosomes by Kuboki (2003), who designed primers targeting amplification of the paraflagella rod protein A (*PFR A*); this test was able to identify samples from the subgenus *Trypanozoon* and *T. congolense*.[66] In 2007 Thekisoe et al. designed primers based upon the 18S rRNA for the generic amplification of all trypanosomes including both *T. brucei* and *T. cruzi*.[67] More recently the technology has been implemented for the specific amplification of *T. brucei rhodesiense* by targeting the *serum resistance-associated (SRA)* gene.[68,69] In addition, recent publications have shown the use of LAMP without a cold-chain for the transport of enzymes.[70] This highly sensitive technique has so far proved accurate, fast, cost-effective and the results are easy to detect without the need for specialized equipment. These assays have strong diagnostic potential in remote clinical settings, but must be fully validated before implementation.

12.1.3.2.10 Nonamplification Techniques

Promising alternatives to the PCR and isothermal-based methods are amplification-independent techniques to detect parasite DNA or RNA. In the 1980s amplification-independent identification for the trypanosomes was made with species-specific DNA probes.[71] Probes designed to highly repetitive sequences such as satellite DNAs were targeted which offered high sensitivity and specificity. Crude tissue preparations were taken and radioactively labeled species-specific DNA probes were hybridized to the preparation and radioactive labels were detected by exposure to autoradiography film. This technique was easy to prepare and store in the field with a sensitivity of 100 parasites/mL. However, extensive laboratory work to process results and the use of radioactive chemicals restricted its use and the method was superseded by PCR methods in the 1990s.[71] Recently, techniques working on similar principles but focusing on easy and safe readout systems with higher sensitivity have been developed. The design of nonradioactive branched DNA-based technique by Harris et al. (1996) showed how probes could hybridize to their targets; here, visualization of the parasite could be made directly with chemiluminescence.[72] This process was developed into a more accessible technique by using Peptide Nucleic Acid (PNA) combined with fluorescence *in situ* hybridization (FISH).[73] Briefly, fluorescein-labeled PNA probes specific for the African trypanosomes were developed in the 18S rRNA; these hybridize with complementary nucleic acids, and can be analyzed with simple fluorescence microscopy. The detection limit for this technique was relatively low at approximately 500 trypanosomes/mL; however, with the inclusion of a cytospin step the limit can be improved to approximately 5 trypanosomes/mL; sensitivity close to that of the amplification techniques.[73] More recently, a further advance was made for both *T. brucei* and *T. cruzi* by using RNA aptamers, which directly target whole organisms.[74,75] Aptamers, small serum-stable RNAs, bound with biotin, hybridize with whole organisms. The surface coat RNAs are expressed in large numbers on the parasite and can be visualized by tagging the bound biotin with fluorescently labeled streptavidin and visualized with a LED microscope. The sensitivity on clinical samples is not yet known, but it is hoped that the sensitivity will be beyond that of previous techniques due to the high copy number of the RNA.

This class of detection techniques has the potential to be implemented into first-line diagnostics as there is no nucleic-acid amplification step and the techniques are therefore less complex.[76] Further development of these protocols will reveal the true specificities and sensitivities with clinical samples and, their future use as diagnostics.

12.2 METHODS

12.2.1 Sample Preparation

PCR, sequencing, and NASBA techniques require a nucleic acid extraction prior to amplification. This can be completed in a variety of ways, most commonly with a standardized kit, for example, Qiagen (Hilden), or with a simple procedure such as the Boom extraction.[77] Blood, CSF, serum, and so on can be stored in buffers provided by the Qiagen kit or in L6 solution for Boom extraction. Nucleic acid extraction is a bottleneck to implementation to first-line diagnostics, as the protocols involve the use of centrifuges, pipettes, heat-blocks, and a number of reagents; moreover, the procedure takes some time, requires trained staff, laboratory space, and eventual storage of samples at −20°C. FTA® cards have been utilized as a simple way of storing blood samples in the field; cards are left at room temperature for storage and extraction of DNA can be achieved with relative ease in the laboratory (www.whatman.com).

The LAMP reaction shows high tolerance to biological substances which often inhibit PCR reactions.[78] In recent studies standard nucleic acid preparations have been compared with heat-treated buffy coat preparations, boiled blood, and CSF preparations; here buffy-coat preparations were the most sensitive, this maybe due to the concentration of parasites by centrifugation; however, these protocols need to be further developed, evaluated, and validated.

Sample preparation for the amplification independent reactions is also in the developmental stages due to the preliminary nature of these techniques. Generally, samples of live cultured parasites and blood samples from patients were denatured, lysed, neutralized and incubated with labeled probes at set temperatures and time (55°C for 30 min for the nonradioactive branching technique; 45°C or 55°C for 90 min for the PNA FISH technique).[72,73] For the RNA aptamers technique live parasites were incubated with a RNA aptamer with a modified biotin side group. For detection the bound biotin group was tagged with fluorescently labeled streptavidin which could be visualized with an LED microscope.[74]

12.2.2 Detection Procedures

See Table 12.1 for details of primer sequences, amplification conditions, and target genes of all major PCR-based, LAMP, and NASBA techniques.

PCR and generic PCR products are run on standard gel electrophoresis; infection is recognized by a specific-sized band on the gel. For generic PCR, for example, the ITS region, multiple species can be identified by the unique size of the product; mixed infections are recognized by two distinct bands of different sizes. Whilst these detection methods are easy to operate, they require electricity and specific equipment with UV light. Attempts have been made to standardize and simplify the detection procedure; to this end several authors have used oligochromatographic dipstick technology to detect the products amplified during PCR and NASBA (see Figure 12.3). The technique works on a principle of gold-conjugated probes. Probes capture PCR and NASBA products which have been labeled with biotin, on a dipstick. PCR and NASBA products from the internal control and target DNA or RNA hybridize with their respective probes to produce a positive test line. Deborggraeve et al.[43,44] extracted DNA with a QIAamp DNA mini kit (Qiagen), followed by PCR with primers targeting the conserved 18S rRNA gene of the *Trypanozoon* and satellite DNA of *T. cruzi* respectively; the reverse primers were biotinylated and an internal control was included. After PCR an oligochromatographic dipstick was used to detect the products of the PCR. Mugasa et al.[63] applied this technology in combination with NASBA amplification; the RNA products of the NASBA reaction were run on specifically designed dipstick.[63] This detection method is quick and simple to use and importantly allows standardization of PCR and NASBA diagnostics by the inclusion of internal standards and quality control (Figure 12.3).

For FFLB, one primer pair in each set was labeled with a fluorescent tag. A volume of 1 μL of PCR product was pooled from each reaction, totalling 4 μL; this mixture was combined with a size standard, buffers and run through an automated sequencer to accurately size the PCR products. Once products were sized a barcode unique to each species was generated; for new sample collections the barcodes are compared to reference isolates, any unknown barcodes are potentially new or previously unrecognized species and should be taken forward for further sequencing analysis.

Real-time PCR does not require any further manipulation after completion of the reaction; this ensures that contamination risk for this technique is minimal. There are two types of real time (RT)-PCR reaction (i) sybr green is included in the initial reaction, when bound to double-stranded (ds) DNA the sybr green fluoresces and this is detected with an excitation beam in the RT-PCR machine; for each cycle twice the amount of dsDNA is created and twice as much sybr green

TABLE 12.1
Details of Target Genes, Primer Sequences, Amplification Conditions, and Detection Formats for the Molecular Diagnostics for the *Trypanosoma*

Molecular Diagnostic	Target Species	Target Gene	Primer Name	Primer Sequences (5′–3′)	Amplification Conditions	Detection Format	Reference
PCR-species specific	HAT	TgsGP	TgsGP-sense	GCTGCTGTGTTCGGAGAGC	63°C	Electrophoresis	Radwanska et al. 32
			TgsGP-antisense	GCCATCGTGCTTGCCGCTC	"		
	HAT	SRA	SRA 651	GAAGAGCCCGTCAAGAAGGTTTG	63°C	Electrophoresis	Picozzi et al. 33
			SRA 652	TTTTGAGCCTTCCACAAGCTTGGG	"		
		VSG + SRA	SRA 02	GGAGCCAAAACCAGTGGGCACATC	"		
			SRA 03	AAGTAGCGCTGTCCTGTAGACGCTTC	"		
		PLC	PLC 657	CGCTTTGTTGAGGAGCTGCAAGCA	"		
			PLC 658	TGCCACCGCAAAGTCGTTATTTCG	"		
	HAT	18S rRNA	18S-f	CGCCAAGCTAATACATGAACCAA	60°C	OC	Deborggraeve et al. 43
			18S-r	ATTAAAGTAAGTAAGCGACCTGC	"		
	AT	Tcz1/Tcz2	Tcz1	CGAGCTCTTGCCCACACGGGTGCT	57°C	Electrophoresis	
			Tcz2	CCTCCAAGCAGCGGATAGTTCAGG	"		
	AT	k-minicircle	S35	AAATAATGTACGGGKGAGATGCATGA	60°C	Electrophoresis	Vallejo et al. 35
			S36	GGTTCGATTGGGGTTGGTGTAATATA	"		
	AT	18S rRNA	Tc-Sat-F	CACTCTCTGTCAATGTCTGTTTGCGTG	65°C	OC	Deborggraeve et al. 43
			Tc-Sat-R	GAAATTCCTCCAAGCAGCGGATA	"		
PCR-generic	Tryp	ITS	TRYP 1	AAG CCA AGT CAT CCA TCG	54°C	Electrophoresis	Adams et al. 40
			TRYP 2	TAG AGG AAG CAA AAG	"		
			TRYP 3	TGC AAT TAT TGG TCG CGC	"		
			TRYP 4	CTT TGC TGC GTT CTT	"		
FFLB	Tryp	18S and 28S rRNA	18S-1f	ACCGWTTCGGCTTTTGTTGG	62°C	Capillary sequencing	Hamilton et al. 42
			18S-2r[a]	CGGTCTAAGAATTTCACCTC	"		
			18S-3f	GACCRTTGTAGTCCACACTG	"		
			18S-4r[a]	CCCCCTGAGACTGTAACCTC	"		
			28S-1f[a]	GAAAGAGAGTGACATAGAAC	"		
			28S-2r	TGTTTCAAGACGGGTGGGGGC	"		
			28S-2f[a]	CCCCCACCCGTCTTGAAACA	"		
			28S-3r	GGGTCCAAACAGGCACACTC	"		

continued

TABLE 12.1 (continued)
Details of Target Genes, Primer Sequences, Amplification Conditions, and Detection Formats for the Molecular Diagnostics for the *Trypanosoma*

Molecular Diagnostic	Target Species	Target Gene	Primer Name	Primer Sequences (5′–3′)	Amplification Conditions	Detection Format	Reference
RFLP	Tryp	ITS	ITS-1	GCTGTAGGTGAACCTGCAGCAGCTGGATCATT	Touch-down	Electrophoresis	Luna-Marin et al. 47
			ITS-2	GCGGGTAGTCCTGCCAAACACTCAGGTCTG	"		
	Tryp	18S rRNA	18STnF2	CAACGATG-ACACCCATGAATTGGGGA	58°C	Electrophoresis	Delespaux et al. 49
			18STnR3	TGCGCGACCAATAATTGCAATAC	"		
			18STnR2	GTGTCTTGTTCTCACTGACATTGTAGTG	Nested 58°C		
Quantitative RT-PCR	HAT	18S rRNA	Tb177F	AACAATGCGCAGTTAACGCTAT	66°C		Becker et al. 50
			Tb177R	ACATTAAACACTAAAGAACAGCGTTG	"		
	AT	Satellite DNA	Sat Fw	GCAGTCGGCKGATCGTTTTCG	65°C		Duffy et al. 54
			Sat Rv	TTCAGRGTTGTTTGGTGTCCAGTG	"		
NASBA	HAT	18S rRNA	TrypnasF7	GATGCAAGGTCGCATATGAG	41°C	OC	Mugasa et al. 63 and 64
			Trypnas6T7r	AATTCTAATACGACTCACTATAGGGAGAAGGCTCGGACTCTTGTTCTC	"		
LAMP	HAT	SRA	SRA-F3	GCGGAAGCAAGAATGACC	62°C	Turbidity or fluorescence	Njiru et al. 68 and 69
			SRA-B3	TCTTACCTTGTGACGCCTG	"		
			SRA-FIP	GGACTGCGTTGAGTACGCATCCGCAAGCACAGACCACAGC	"		
			SRA-BIP	CGCTCTTCAAGTCTTGCGCCCTTCTGAGATGTGCCCACTG	"		
			SRA-LF	CGCGGCATAAAGCGCTGAG	"		
			SRA-LB	GCAGCGACCAACGGAGCC	"		

Note: Where multiple primer combinations exist, the most popular and/or most recent have been shown. References to all protocols should be studied before attempting each test. HAT = human African trypanosomiasis, AT = American trypanosomiasis, Tryp = *Trypanosoma*, OC = oligochromatography.

[a] Primers containing fluorescence labels.

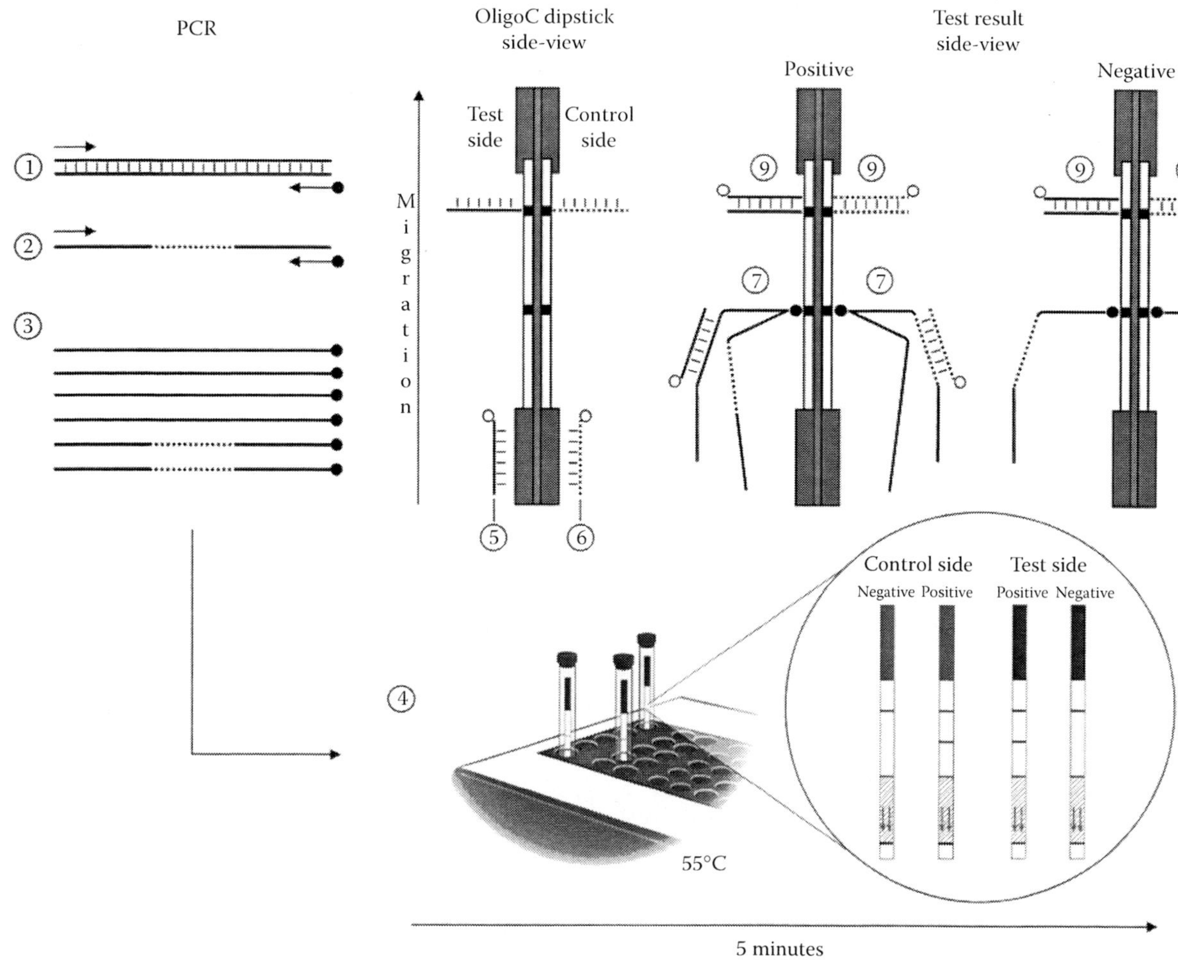

FIGURE 12.3 Schematic overview of the *T. cruzi* OligoC-TesT. Polymerase chain reaction (PCR) amplification of a sample containing *T. cruzi* DNA (1) is performed using a PCR mix containing single-stranded internal control (IC) template (2). The IC sequence is the same as the target sequence within the *T. cruzi* satellite DNA but with a specific internal sequence (dotted line). Both templates are amplified with the same primers of which the reverse primer is biotinylated (*N*). When the PCR and subsequent denaturation is completed, the PCR product solution contains single-stranded *T. cruzi* and IC DNA (3), and is mixed with an equal volume of migration buffer preheated at 55uC. The *T. cruzi* Oligo-Strip is dipped into the mixture and test results are read after 5 minutes migration at 55uC (4). During migration, the solution takes up the gold-labelled (#) detection probes. The *T. cruzi* detection probes (5) at the test side and the IC detection probes (6) at the control side hybridize with their respective amplicons. The biotinylated PCR products accumulate on the neutralite avidin lines at both sides of the dipstick and the *T. cruzi* and IC amplicons are visualized by their respective detection probes (7). In case of a negative sample, only the IC amplicon is present and visualized at the control side of the dipstick (8). The excess of detection probes migrate further and hybridise on the complementary probes coated at both sides of the dipstick as a control for migration (9). doi:10.1371/journal.pntd.0000450.g002. (Adapted from Deborggraeve, S. et al. *PLoS. Negl. Trop. Dis.* 3, e450, 2009.)

is bound and fluorescent. (ii) Unlabeled primers amplify a double-dyed probe which hybridizes to the DNA; the probe is labeled on one end with a quencher and on the other end with a fluorescent tag; the quencher prevents fluorescence until amplification takes places when fluorescence is detected with the excitation beam. This later method is more reliable as sybr-green is also able to bind to primer dimer giving false-positive results.

The LAMP reaction amplifies large amounts of DNA within a short time; DNA can be detected with the use of a turbidimeter, this machine detects white turbidity caused by the existence of magnesium pyrophosphate in the by-products of amplification (http://loopamp.eiken.co.jp/e/products/la100/index.html). For diagnostic purposes, easy to use fluorescence detection procedures have been established. Eiken's own Fluorescent Detection Reagent (FDR) can be added to the reaction to give a fluorescent green color under UV light; the distinct contrast between positive and negative is used to identify positive samples.

The amplification-independent reactions generally use fluorescence and chemiluminescence for the detection of positive samples; however, identifying positive infections requires skilled microscopy. Both FISH and the RNA aptamer technique are able to utilize the Zeiss Primo Star iLED (www.zeiss.com) which has been recently developed for straightforward use in developing countries; one switch allows the cross over from bright field to LED fluorescence and the battery power option enables its use without continuous electricity.

12.3 CONCLUSIONS AND FUTURE PERSPECTIVES

This chapter describes the available molecular techniques for the diagnosis and identification of the *Trypanosoma*. These methods can have a higher sensitivity and specificity than conventional methods that are regularly applied for diagnostics. They have repeatedly been successfully used for epidemiological studies and have given us great insight in the status, spread of disease and identified previously unknown species. However, none of the current molecular techniques can be implemented for routine diagnostics at the primary health care level where they are most at need. Reasons include the cost of reagents, equipment and the highly trained staff that apply these techniques; other causes are the general requirement for continual electricity to run the reactions and a cold chain to prevent degradation of the reagents and most importantly they have not been validated by well-designed diagnostic accuracy studies. The WHO highlights the requirements of safe diagnostics that are sensitive, specific, simple to perform and interpret, do not require expensive equipment, are thermo-stable for prolonged storage and, importantly, cost-effective. Unless all of these requirements are met then the molecular diagnostics highlighted here will not be integrated into health systems in the resource poor settings where this spectrum of disease occurs most frequently.

A few of these techniques show potential for diagnostics and meet many of the WHO requirements; for example, LAMP is a relatively low-cost diagnostic which does not need a cold chain, requires only low-technology equipment and the results can be easily interpreted. In addition, PCR-OC and NASBA-OC have been standardized and, with laboratory strengthening, have the potential to be used at the district hospital level. It is imperative that these molecular diagnostics undergo rigorous and independent evaluation in clinical trials (phases I–III). They must also address the ongoing need for quality control, emphasized by the WHO, in order that they can be relied upon and trusted by the public for whom they are intended.

Further challenges in this area are the development of diagnostics which are able to test for successful treatment outcomes and treatment failures; these tests will be essential for patients who may relapse and require ongoing treatment and, looking into the future, for elimination programs. For sleeping sickness there remains a desperate need for a better stage two diagnostic tools that avoids the use of lumbar punctures and indicates if stage II drug regimes are necessary. For Chagas disease, a diagnostic for monitoring chronic patients is crucial so treatment can be given as soon as reactivation occurs.

Other challenges remain concerning the implementation of these diagnostics at the point-of-care so that as many patients as possible can have access to the high sensitivity and specificity that they offer. Laboratory strengthening efforts are ongoing as are projects to stabilize enzymes to limit the need for cold chains. PCRs which are completed on a chip, the so-called "lab on a chip" are being developed; these reactions are fully standardized and are able to purify DNA and complete a PCR reaction with minimal staff involvement http://www.path.org/dxcenter/index.php. Currently, these reactions are prohibitively expensive; in the future it is hoped that the cost will be reduced.

Increasingly, nonamplification DNA methods are being developed; these do not require the use of an enzymatic reaction or a DNA purification step and therefore bypass many of the bottlenecks of the more widely used molecular techniques. The challenge for this set of diagnostics is to reach the same analytical sensitivity as the amplification methods.

The molecular methods have great potential for diagnostics and identification of the *Trypanosoma*; implementation would greatly improve treatment outcome for the thousands of patients affected each year. Improving the tests so they are easy to use, require low technology equipment, minimal electricity, and training will aid implementation. Cost, standardization, and quality control must also be at the forefront of all manufacturers' minds in order that patients benefit from such technologies.

ACKNOWLEDGMENT

I would like to thank Dr. Stijn Deborggraeve, Institute of Tropical Medicine, Antwerp, Belgium for providing Figure 12.3 and for reviewing the chapter.

REFERENCES

1. Barrett, M.P. et al. The trypanosomiases. *Lancet* 362, 1469–1480, 2003.
2. Adams, E.R. et al. New molecular tools for the identification of trypanosome species. *Future. Microbiol.* 3, 167–176, 2008.
3. Simarro, P.P. et al. The Atlas of human African trypanosomiasis: A contribution to global mapping of neglected tropical diseases. *Int. J. Health Geogr.* 9, 57, 2010.
4. Coura, J.R. et al. Emerging Chagas disease in Amazonian Brazil. *Trends Parasitol.* 18, 171–176, 2002.
5. Ribeiro, I. et al. New, improved treatments for Chagas disease: From the r&d pipeline to the patients. *PLoS. Negl. Trop. Dis.* 3, e484, 2009.
6. FDA/CDER. Guidance for industry: Use of serological tests to reduce the risk of transmission of *Trypanosoma cruzi* infection in whole blood and blood components for transfusion and human cells, tissues, and cellular and tissuebased products. FDA/CDER, Washington D.C., 2010. http://www.fda.gov/BiologicsBloodVaccines/GuidanceComplianceRegulatoryInformation/Guidances/default.htm or http://www.regulations.gov.
7. Fevre, E.M. et al. The origins of a new *Trypanosoma brucei rhodesiense* sleeping sickness outbreak in eastern Uganda. *Lancet* 358, 625–628, 2001.
8. Zoller, T. et al. Analysis of risk factors for *T. brucei rhodesiense* sleeping sickness within villages in south-east Uganda, *BMC. Infect. Dis.* 8, 88, 2008.

9. Deborggraeve, S. et al. Molecular analysis of archived blood slides reveals an atypical human *Trypanosoma* infection. *Diagn. Microbiol. Infect. Dis.* 61, 428–433, 2008.
10. Joshi, P.P. et al. Human trypanosomiasis caused by *Trypanosoma evansi* in India: The first case report. *Am. J. Trop. Med. Hyg.* 73, 491–495, 2005.
11. Sarataphan, N. et al. Diagnosis of a *Trypanosoma lewisi*-like (Herpetosoma) infection in a sick infant from Thailand. *J. Med. Microbiol.* 56, 1118–1121, 2007.
12. Kennedy, P.G. Diagnostic and neuropathogenesis issues in human African trypanosomiasis. *Int. J. Parasitol.* 36, 505–512, 2006.
13. Yun, O. et al. Feasibility, drug safety, and effectiveness of etiological treatment programs for Chagas disease in Honduras, Guatemala, and Bolivia: 10-year experience of medecins sans frontieres. *PLoS Negl. Trop. Dis.* 3, e488, 2009.
14. Prata, A. Clinical and epidemiological aspects of Chagas disease. *Lancet Infect. Dis.* 1, 92–100, 2001.
15. Marin-Neto, J.A. et al. Pathogenesis of chronic Chagas heart disease. *Circulation* 115, 1109–1123, 2007.
16. Chappuis, F. et al. Options for field diagnosis of human African trypanosomiasis. *Clin. Microbiol. Rev.* 18, 133–146, 2005.
17. Buscher, P. et al. Improved models of mini anion exchange centrifugation technique (mAECT) and modified single centrifugation (MSC) for sleeping sickness diagnosis and staging. *PLoS. Negl. Trop. Dis.* 3, e471, 2009.
18. Woo, P.T. The haematocrit centrifuge technique for the diagnosis of African trypanosomiasis. *Acta Trop.* 27, 384–386, 1970.
19. Woo, P.T. Evaluation of the haematocrit centrifuge and other techniques for the field diagnosis of human trypanosomiasis and filariasis. *Acta Trop.* 28, 298–303, 1971.
20. Zillmann, U. et al. Improved performance of the anion-exchange centrifugation technique for studies with human infective African trypanosomes. *Acta Trop.* 62, 183–187, 1996.
21. Magnus, E. et al. A card-agglutination test with stained trypanosomes (C.A.T.T.) for the serological diagnosis of *T. b. gambiense* trypanosomiasis. *Ann. Soc. Belg. Med. Trop.* 58, 169–176, 1978.
22. Truc, P. et al. Evaluation of the micro-CATT, CATT/*Trypanosoma brucei gambiense*, and LATEX/T b gambiense methods for serodiagnosis and surveillance of human African trypanosomiasis in West and Central Africa. *Bull. World Health Organ* 80, 882–886, 2002.
23. Greenwood, B.M. et al. Cerebrospinal-fluid IgM in patients with sleeping-sickness. *Lancet*, 2, 525–527, 1973.
24. Lejon, V. et al. Neuro-inflammatory risk factors for treatment failure in "early second stage" sleeping sickness patients treated with pentamidine. *J. Neuroimmunol.* 144, 132–138, 2003.
25. Lejon, V. et al. Novel markers for treatment outcome in late-stage *Trypanosoma brucei gambiense* trypanosomiasis. *Clin. Infect. Dis.* 47, 15–22, 2008.
26. Chippaux, J.P. et al. Sensitivity and specificity of Chagas Stat-Pak test in Bolivia. *Trop. Med. Int. Health*, 14, 732–735, 2009.
27. Ferreira, A.W. et al. Laboratory diagnosis of Chagas' heart disease. *Sao Paulo Med. J.* 113, 767–771, 1995.
28. Mullis, K.B. et al. Specific synthesis of DNA *in vitro* via a polymerase-catalyzed chain reaction. *Methods Enzymol.* 155, 335–350, 1987.
29. Gibson, W. Resolution of the species problem in African trypanosomes. *Int. J. Parasitol.* 37, 829–838, 2007.
30. Xong, H.V. et al. A VSG expression site-associated gene confers resistance to human serum in *Trypanosoma rhodesiense. Cell*, 95, 839–846, 1998.
31. Welburn, S.C. et al. Identification of human-infective trypanosomes in animal reservoir of sleeping sickness in Uganda by means of serum-resistance-associated (SRA) gene. *Lancet* 358, 2017–2019, 2001.
32. Radwanska, M. et al. Novel primer sequences for polymerase chain reaction-based detection of *Trypanosoma brucei gambiense. Am. J. Trop. Med. Hyg.* 67, 289–295, 2002.
33. Picozzi, K. et al. A multiplex PCR that discriminates between *Trypanosoma brucei brucei* and zoonotic *T. b. rhodesiense*, *Exp. Parasitol.* 118, 41–46, 2008.
34. Mugasa, C.M. et al. Diagnostic accuracy of molecular amplification tests for human African trypanosomiasis—Systematic review. *PLoS. Negl. Trop. Dis.* 6, e1438, 2012.
35. Vallejo, G.A. et al. Species specific detection of *Trypanosoma cruzi* and *Trypanosoma rangeli* in vector and mammalian hosts by polymerase chain reaction amplification of kinetoplast minicircle DNA. *Acta Trop.* 72, 203–212, 1999.
36. Velazquez, M. et al. *Trypanosoma cruzi*: An analysis of the minicircle hypervariable regions diversity and its influence on strain typing. *Exp. Parasitol.* 120, 235–241, 2008.
37. Gil, J. et al. Comparison of a PCR test based on the histone H2A/SIRE genes with classical serological tests for the diagnosis of chronic Chagas disease in Colombian patients. *Biomedica.* 27(Suppl 1), 83–91, 2007.
38. Desquesnes, M. et al. Detection and identification of Trypanosoma of African livestock through a single PCR based on internal transcribed spacer 1 of rDNA. *Int. J. Parasitol.* 31, 610–614, 2001.
39. Cox, A. et al. A PCR based assay for detection and differentiation of African trypanosome species in blood. *Exp. Parasitol.* 111, 24–29, 2005.
40. Adams, E.R. et al. Trypanosome identification in wild tsetse populations in Tanzania using generic primers to amplify the ribosomal RNA ITS-1 region. *Acta Trop.* 100, 103–109, 2006.
41. Schijman, A.G. et al. Differential detection of *Blastocrithidia triatomae* and *Trypanosoma cruzi* by amplification of 24 s alpha ribosomal RNA genes in faeces of sylvatic triatomine species from rural northwestern Argentina. *Acta Trop.* 99, 50–54, 2006.
42. Hamilton, P.B. et al. A novel, high-throughput technique for species identification reveals a new species of tsetse-transmitted trypanosome related to the *Trypanosoma brucei* subgenus. *Trypanozoon. Infect. Genet. Evol.* 8, 26–33, 2008.
43. Deborggraeve, S. et al. Molecular dipstick test for diagnosis of sleeping sickness. *J. Clin. Microbiol.* 44, 2884–2889, 2006.
44. Deborggraeve, S. et al. T. cruzi OligoC-TesT: a Simplified and standardized polymerase chain reaction format for diagnosis of Chagas disease. *PLoS. Negl. Trop. Dis.* 3, e450, 2009.
45. Stothard, J.R. et al. On the molecular taxonomy of *Trypanosoma cruzi* using riboprinting. *Parasitology* 117(Pt 3), 243–247, 1998.
46. Bastrenta, B. et al. Restriction fragment length polymorphism of 195 bp repeated satellite DNA of *Trypanosoma cruzi* supports the existence of two phylogenetic groups. *Mem. Inst. Oswaldo Cruz* 94, 323–328, 1999.
47. Luna-Marin, K.P. et al. ITS-RFLP- and RAPD-based genetic variability of *Trypanosoma cruzi* I, human and vector strains in Santander, Colombia. *Parasitol. Res.* 105, 519–528, 2009.
48. Campos, R.F. et al. Comparative analysis by polymerase chain reaction amplified minicircles of kinetoplast DNA of a stable strain of *Trypanosoma cruzi* from Sao Felipe, Bahia, its clones and subclones: Possibility of predominance of a principal clone in this area. *Mem. Inst. Oswaldo Cruz* 94, 23–29, 1999.

49. Delespaux, V. et al. PCR-RFLP using Ssu-rDNA amplification: Applicability for the diagnosis of mixed infections with different trypanosome species in cattle. *Vet. Parasitol.* 117, 185–193, 2003.
50. Becker, S. et al. Real-time PCR for detection of *Trypanosoma brucei* in human blood samples. *Diagn. Microbiol. Infect. Dis.* 50, 193–199, 2004.
51. Cummings, K.L. et al. Rapid quantitation of *Trypanosoma cruzi* in host tissue by real-time PCR. *Mol. Biochem. Parasitol.* 129, 53–59, 2003.
52. Freitas, J.M. et al. Real time PCR strategy for the identification of major lineages of *Trypanosoma cruzi* directly in chronically infected human tissues. *Int. J. Parasitol.* 35, 411–417, 2005.
53. Piron, M. et al. Development of a real-time PCR assay for *Trypanosoma cruzi* detection in blood samples. *Acta Trop.* 103, 195–200, 2007.
54. Duffy, T. et al. Accurate real-time PCR strategy for monitoring bloodstream parasitic loads in Chagas disease patients. *PLoS. Negl. Trop. Dis.* 3, e419, 2009.
55. Hamilton, P.B. et al. Phylogenetic analysis reveals the presence of the *Trypanosoma cruzi* clade in African terrestrial mammals. *Infect. Genet. Evol.* 9, 81–86, 2009.
56. Hamilton, P.B. et al. Trypanosomes are monophyletic: Evidence from genes for glyceraldehyde phosphate dehydrogenase and small subunit ribosomal RNA. *Int. J. Parasitol.* 34, 1393–1404, 2004.
57. Sajid, M. et al. Cysteine proteases of parasitic organisms. *Mol. Biochem. Parasitol.* 120, 1–21, 2002.
58. Cortez, A.P. et al. Cathepsin L-like genes of *Trypanosoma vivax* from Africa and South America—Characterization, relationships and diagnostic implications. *Mol. Cell Probes.* 23, 44–51, 2009.
59. Marcili, A. et al. Comparative phylogeography of *Trypanosoma cruzi* TCIIc: New hosts, association with terrestrial ecotopes, and spatial clustering. *Infect. Genet. Evol.* 9, 1265–1274, 2009.
60. Da Silva, F.M. et al. Phylogeny, taxonomy and grouping of *Trypanosoma rangeli* isolates from man, triatomines and sylvatic mammals from widespread geographical origin based on SSU and ITS ribosomal sequences. *Parasitology* 129, 549–561, 2004.
61. Metzker, M.L. Emerging technologies in DNA sequencing. *Genome Res.* 15, 1767–1776, 2005.
62. Deiman, B. et al. Characteristics and applications of nucleic acid sequence-based amplification (NASBA). *Mol. Biotechnol.* 20, 163–179, 2002.
63. Mugasa, C.M. et al. Detection of *Trypanosoma brucei* parasites in blood samples using real-time nucleic acid sequence-based amplification. *Diagn. Microbiol. Infect. Dis.* 61, 440–445, 2008.
64. Mugasa, C.M. et al. Nucleic acid sequence-based amplification with oligochromatography for detection of *Trypanosoma brucei* in clinical samples. *J. Clin. Microbiol.* 47, 630–635, 2009.
65. Notomi, T. et al. Loop-mediated isothermal amplification of DNA. *Nucleic Acids Res.* 28, E63, 2000.
66. Kuboki, N. et al. Loop-mediated isothermal amplification for detection of African trypanosomes. *J. Clin. Microbiol.* 41, 5517–5524, 2003.
67. Thekisoe, O.M. et al. Species-specific loop-mediated isothermal amplification (LAMP) for diagnosis of trypanosomosis. *Acta Trop.* 102, 182–189, 2007.
68. Njiru, Z.K. et al. Loop-mediated isothermal amplification (LAMP) method for rapid detection of *Trypanosoma brucei rhodesiense. PLoS. Negl. Trop. Dis.* 2, e147, 2008.
69. Njiru, Z.K. et al. African trypanosomiasis: Sensitive and rapid detection of the sub-genus *Trypanozoon* by loop-mediated isothermal amplification (LAMP) of parasite DNA. *Int. J. Parasitol.* 38, 589–599, 2008.
70. Thekisoe, O.M. et al. Stability of loop-mediated isothermal amplification (LAMP) reagents and its amplification efficiency on crude trypanosome DNA templates. *J. Vet. Med. Sci.* 71, 471–475, 2009.
71. Gibson, W. Species-specific probes for the identification of the African tsetse-transmitted trypanosomes. *Parasitology* 136, 1501–1507, 2009.
72. Harris, E. et al. Detection of *Trypanosoma brucei spp.* in human blood by a nonradioactive branched DNA-based technique. *J. Clin. Microbiol.* 34, 2401–2407, 1996.
73. Radwanska, M. et al. Direct detection and identification of African trypanosomes by fluorescence *in situ* hybridization with peptide nucleic acid probes. *J. Clin. Microbiol.* 40, 4295–4297, 2002.
74. Lorger, M. et al. Targeting the variable surface of African trypanosomes with variant surface glycoprotein-specific, serum-stable RNA aptamers. *Eukaryot. Cell* 2, 84–94, 2003.
75. Ulrich, H. et al. *In vitro* selection of RNA aptamers that bind to cell adhesion receptors of *Trypanosoma cruzi* and inhibit cell invasion. *J. Biol. Chem.* 277, 20756–20762, 2002.
76. Deborggraeve, S. et al. Molecular diagnostics for sleeping sickness: What is the benefit for the patient? *Lancet Infect. Dis.* 10, 433–439, 2010.
77. Boom, R. et al. Rapid and simple method for purification of nucleic acids. *J. Clin. Microbiol.* 28, 495–503, 1990.
78. Kaneko, H. et al. Tolerance of loop-mediated isothermal amplification to a culture medium and biological substances. *J. Biochem. Biophys. Methods* 70, 499–501, 2007.

Section II

Ciliophora and Apicomplex

13 Babesia

Dongyou Liu and Xuedong Lu

CONTENTS

13.1 INTRODUCTION

Babesia is a tick-borne, intraerythrocytic protozoan that causes a hemolytic disease known as babesiosis (whose former or alternative names include piroplasmosis, Texas cattle fever, redwater fever, tick fever, and Nantucket fever), with clinical symptoms resembling those of malaria. Recognized as a disease of veterinary importance, babesiosis occurs commonly in cattle but also in horses, sheep, pigs, and dogs. In fact, *Babesia* spp. represent the second most common blood-borne parasites of mammals after trypanosomes. In recent years, babesiosis has attracted increasing attention as a zoonotic infection in human [1–3].

Babesia was first described in cattle by Victor Babes in 1888 [4], and the role of tick in its transmission was identified 5 years later [5]. The first case of human babesiosis was observed in a splenectomized patient in 1957 [6], and cases of babesiosis in nonsplenectomized patients were documented since 1969. As an accidental dead-end host of *Babesia*, humans do not usually acquire infections from each other through tick bites, although the parasite may be transmitted via blood transfusion. Of over 100 *Babesia* species identified to date, *Babesia microti* and *Babesia divergens* are the most common human pathogens, with *B. microti* being predominant in the Americas and *B. divergens* in Europe. Human infections with other species of *Babesia* (e.g., *Babesia bovis*, *Babesia canis*, *Babesia duncani*, and *Babesia venatorum*) have also been documented occasionally [7].

13.1.1 Classification, Life Cycle, and Epidemiology

13.1.1.1 Classification

On the basis of their exclusive invasion of erythrocytes, multiplication by budding rather than schizogony, and lack of hemozoin, members of the genus *Babesia* (formerly *Piroplasma*, *Babesiella*, and *Microbabesia*) are classified taxonomically in the family Babesiidae, order Piroplasmida, class Aconoidasida, and phylum Apicomplexa [3]. Of over 100 recognized species within the genus *Babesia*, the following are most notable: *Babesia bennetti*, *Babesia bicornis*, *Babesia bigemina*, *Babesia bovis*, *Babesia caballi*, *Babesia canis*, *Babesia capreoli*, *Babesia conradae*, *Babesia crassa*, *Babesia divergens*, *Babesia duncani*, *Babesia equi*, *Babesia felis*, *Babesia gibsoni*, *Babesia hylomysci*, *Babesia kiwiensis*, *Babesia leo*, *Babesia major*, *Babesia microti*, *Babesia motasi*, *Babesia muratovi*, *Babesia occultans*, *Babesia odocoilei*, *Babesia orientalis*, *Babesia ovate*, *Babesia ovis*, *Babesia poelea*, *Babesia rodhaini*, *Babesia rossi*, and *Babesia vesperuginis* [8].

In accordance with their morphology, host/vector specificity, and susceptibility to drugs, *Babesia* organisms are often grouped into small babesias (with trophozoites of 1.0–2.5 μm; including *B. gibsoni*, *B. microti*, and *B. rodhaini*) and large babesias (with trophozoites of 2.5–5.0 μm; including *B. bovis*, *B. caballi*, and *B. canis*). Sequence analysis of nuclear small subunit-ribosomal RNA (18S rRNA) gene revealed that the large and small babesias fall into two phylogenetic clusters, with the small babesias being more closely related to *Theileria* spp. than the large babesias [1].

Further molecular investigations suggested division of the piroplasms into five distinct clades: (i) *B. microti* group containing *B. rodhaini*, *B. felis*, *B. leo*, *B. microti*, and a *B. microti*-like canine isolate; (ii) Western US *Theileria*-like group, containing *B. conradae*; (iii) *Theileria*-group, containing all *Theileria* spp. from bovines; (iv) a first group of "true" *Babesia* spp. (sensu stricto) including *B. canis* and *B. gibsoni* from canines together with *B. divergens* and *Babesia odocoilei*; and (v) a second *Babesia* spp. sensu stricto group composed mainly of *Babesia* spp. from ungulates: *B. caballi*, *B. bigemina*, *B. ovis*, *B. bovis*, and other *Babesia* spp. from cattle.

B. microti, long considered as a single species, is now regarded as a genetically diverse species complex that is only distantly related to *Babesia* spp. sensu stricto (*B. bigemina*, *B. bovis*, and *B. divergens*). Within the *B. microti* species complex, three clades are identified. Clade 1 contains mostly rodent parasites and also the majority of strains thought to be zoonotic, Clade 2 contains carnivore parasites, and Clade 3 contains rodent parasites that are probably not zoonotic. A *Babesia* of mice and other small rodents, *B. microti* was thought to occur mainly in Americas [9]. However, human babesiosis due to *B. microti* was also reported recently from Europe, and *B. microti*-like species were identified in China, India, Egypt, South Africa, Brazil, and Mexico as the course of human infections [10–14].

B. divergens (formerly *Piroplasma divergens*) is one of the two *B.* species (the other being *B. major*) infecting cattle in Northern Europe, with *B. divergens* being more common [15]. The natural reservoir of *B. divergens* is cattle, and about 40 cases of *B. divergens* infection in Europe have been described, especially the countries with extensive cattle industry, such as France, Ireland, and Great Britain. Sporadic cases have also been reported from Sweden, Switzerland, Spain, Portugal, and Croatia (index case). *B. divergens* poses risks to farmers or people vacationing in rural areas. Recently, several cases due to *B. divergens*-like organisms were reported in Missouri, Kentucky, and Washington State in the United States [16,17]. The Missouri isolate (MO1) and the Kentucky isolate (KY) are identical to each other and to piroplasms found in eastern cottontail rabbits on Nantucket Island, but differ from the Washington State isolate and from the *B. divergens* organisms of European cattle by a few base pairs in their 18S rRNA gene. The likely vector for transmission of The Missouri isolate (MO1) may be *Ixodes dentatus* [3].

B. duncani, previously designated WA1 *Babesia*, displays only subtle morphological differences from *B. microti*. Together with isolates from dogs and wildlife, *B. duncani* forms a separate phylogenetic group, and is indistinguishable from two other human isolates, CA6 and WA2-clone 1, but clearly distinguishable from the dog isolate *B. conradae* [18]. *B. duncani* is responsible for human infections on the west coast of the United States through transfusion of blood products and tick bites, with clinical manifestations resembling those of *B. microti* infection. The reservoir hosts and vectors of *B. duncani* remain unknown at present.

Babesia venatorum, originally referred to as EU1, occurs in deer in Europe, with possible transmission by *Ixodes ricinus*. Recent phylogenetic studies further uncovered the existence of *B. venatorum* EU1-3 in Europe and *B. divergens*-like organisms in the United States [19–21]. Along with *B. odocoilei* that infects white-tail deer in the United States, *B. venatorum* forms a sister group to *B. divergens*. To date, a few nonfatal cases of infection with *B. venatorum* were documented in men of over 50 years of age with splenectomy in Austria, Italy, and Germany [20,22].

13.1.1.2 Life Cycle

Babesia spp. are maintained in mammals (e.g., rodents, voles, deer, and cattle), which act as reservoirs, and ticks (of the family *Ixodidae*), which are the definitive host [23–26]. In general, *Babesia* spp. undergo merogony, gamogony, and sporogony in their life cycle. Typically, an *Ixodes* tick takes up the gametocytes during a blood meal from a *Babesia*-infected mammal. During subsequent gametogenesis, the gametes develop into "Strahlenkörper," which fuse to form zygotes, and further develop into kinetes in the gut of the tick. Gaining access to the hemolymph of the tick, kinetes replicate and invade various organs. Interestingly, members of the *Babesia* spp. sensu stricto groups can infect the ovaries and be transmitted transovarially via the eggs, so that all stages (larvae, nymphs, and adults females) are potentially infective. On the other hand, members of the *B. microti*-like groups are only transmitted from one stage to the next (transstadially), so that larvae are rarely infected. In the salivary glands, kinetes initiate sporogony, forming multinucleated sporoblasts, and sporozoites, which are then introduced into another mammal (or human, which represents an incidental and dead-end host) upon inoculation at the bite of an infected tick. After entry into erythrocytes, *Babesia* sporozoites develop into trophozoites. After undergoing asexual reproduction (binary fission), trophozoites turn into merozoites with a unique tetrad structure (also referred to as a Maltese-cross form), which upon Geimsa staining of a thin-blood smear, serves as a distinguishing feature of *Babesia* from *Plasmodium falciparum*, a protozoan of similar morphology that causes malaria. Besides further evolving into sporozoites, some merozoites/trophozoites mature into gametocytes, the infectious parasitic bodies, which spread throughout the blood, with the capacity to initiate infection in the tick vector [3].

13.1.1.3 Epidemiology

B. microti organisms are capable of infecting a variety of rodent, insectivore, and carnivore hosts, such as the white-footed mouse (*Peromuscus leucopus*), microtus voles (*Microtus* spp.), and the white-tailed deer (*Odocoileus virginianus*). The human-biting hard tick *Ixodes scapularis* (commonly known as a deer tick) plays a vital role in the transmission of *B. microti* to humans [27]. Other vectors of *B. microti* complex include *Ixodes trianguliceps*, *I. ricinus*, *Ixodes ovatus*, *Ixodes spinipalpis*, *Ixodes angustus*, and *Ixodes muris* [28,29]. European strains of *B. microti* are maintained in rodents and field vole (*Microtus agrestis*) and their transmission to humans is facilitated by two vectors *I. trianguliceps* and *I. ricinus*. In Japan, *I. ovatus* is implicated in the transmission of *B. microti*, with microtine rodents being the likely reservoir hosts [3].

Possibly due to the expansion of the deer population (which are the primary host for adult ticks), heightened awareness of the disease by local physicians, encroachment of humans on wildlife habitat, and increasing travel of immunosuppressed individuals to endemic areas, there has been a dramatic increase in the number of reported cases of *B. microti* infections over the last two decades throughout the world [30]. In the northern hemisphere, the prevalence of babesiosis is most documented during the months of May–September, when there is high tick activity in endemic regions. Risk factors include splenectomy, advanced age, and depressed cellular immunity. In addition, cases of babesiosis due to packed red blood cell (PRBC) transfusions and organ transplantation, as well as transplacental transmission have also been documented [24,31].

B. divergens is the main agent of bovine babesiosis (also known as redwater fever) in Europe, which is transmitted by the tick *I. ricinus*. In areas where *I. ricinus* is absent (e.g., the Canary Islands), *Ixodes ventalloi* may act as a vector. Increasingly, clinical cases of *B. divergens* infection have been reported in splenectomized humans. The parasite can be transmitted experimentally by needle injection from infected cattle to splenectomized mouflon, rat, red, roe, fallow deer, and intact reindeer. Although these animals (except for the reindeer) display clinical signs of only very mild infections with low parasitemias, their blood is infectious to susceptible cattle. The Mongolian gerbil (*Meriones unguiculatus*), either intact or splenectomized, is the only laboratory animal that is fully susceptible to *B. divergens*. With high inocula ($\geq 10^7$ infected erythrocytes), the gerbils develop fever, anemia, jaundice, anorexia, hemoglobinuria, and splenomegaly, leading to death within a few days. Some gerbils may recover from infections, developing sterile rather than residual immunity as in the bovine host [32]. *B. divergens* can be maintained in the long-term cultures of bovine, human, rat, and sheep erythrocytes.

13.1.2 Clinical Features and Pathogenesis

13.1.2.1 Clinical Features

Following the description of the first case of human babesiosis involving a splenectomized farmer in 1957, several hundred cases of *Babesia* infections have been reported in the United States, Europe, and other parts of the world [21,25,33–36]. The main clinical symptoms of human babesiosis include intravascular hemolysis with hemoglobinuria and jaundice, persistent high fever, shaking chills, sweating, headache, myalgia, and lumbar and abdominal pain. Splenectomy represents the major predisposing factor, with babesiosis in patients with intact spleens being usually less severe [37–43].

Clinical signs of *B. microti* infection usually appear 1–9 weeks after tick bites. These may range from asymptomatic infection, mild-to-moderate viral-like illness, and severe disease with a fulminant course, leading to occasional death or a persistent relapsing illness [44].

Asymptomatic infection accounts for a third of the cases, which may persist for months or years following resolution of symptomatic babesiosis. Patients with asymptomatic infection may transmit the infection through blood donation. Most cases of *B. microti* infection are a mild-to-moderate viral-like illness. Typically beginning with a gradual onset of malaise and fatigue, the patients often display intermittent fever, chills, sweats, headache, arthralgia, myalgia, anorexia, cough, and nausea. Other less common manifestations include gastrointestinal symptoms, weight loss, conjunctival injection, and emotional lability. Physical examination often reveals fever, together with pallor, mild splenomegaly, or hepatomegaly. The illness may last for a week to months, and sometimes more than a year. Severe cases of babesiosis usually occur in patients with suppressed immune functions, such as those with HIV, malignancy, splenectomy, and older age. Physical examination shows jaundice, retinal infarcts, or ecchymoses, and petechiae, with most common complications being acute respiratory failure, congestive heart failure, DIC, liver and renal failures, splenic rupture, and with mortality rates of 5–21%. The surviving patients may experience a prolonged, relapsing course of illness, sometimes lasting more than 1 year [3].

B. duncani induces symptoms that are similar to those of *B. microti* infection, consisting of fever, chills, headache, sweats, myalgia, nausea, vomiting, diarrhea, dark urine, and fatigue. Of the nine reported cases of *B. duncani*, parasitemia ranged from 1% to 54% [18].

Signs of *B. divergens* infection begin 1–3 weeks after tick transmission and consist of high fever (40–41°C), headache, shaking chills, intense sweating, myalgia, abdominal pain, vomiting, and diarrhea, with severe intravascular hemolysis, leading to hemoglobinemia, hemoglobinuria, and jaundice. Other common complications include renal failure, pulmonary edema, ecchymoses, petechiae, congestive heart failure, and coma, with death or prolonged convalescence occurring in more than a third of patients.

Of the three reported cases of *B. venatorum* infection, all had a history of Hodgkin's disease and splenectomy, with fever, dark urine, fatigue, chills, headache, confusion, jaundice, sweats, and shortness of breath being the main symptoms.

13.1.2.2 Pathogenesis

Babesia parasitism alters membrane structures of erythrocytes, resulting in reduced survival and enhanced erythrocyte removal upon spleen passage in animals, contributing to anemia, hyperbilirubinuria, hemoglobinuria, kidney failure, adult respiratory distress syndrome (ARDS), and central nervous system impairment.

Both cellular and humoral immunity may play a role in controlling babesiosis. Infection with *Babesia* parasite stimulates an overwhelming production of pro-inflammatory cytokines, such as TNF-α and interferon-γ, which may be partially responsible for causing elevated transaminases, alkaline phosphatases, unconjugated bilirubin, and lactic dehydrogenase in patients suffering from babesiosis [45]. While depletion of host macrophages and natural killer cells increases susceptibility to infection, the development of antibodies to *Babesia* within the infected host is associated with clearing of the parasite and recovery [46].

13.1.3 Laboratory Diagnosis

As human babesiosis often presents relatively nonspecific clinical signs (e.g., fever, fatigue, chills, sweats, headache, myalgia, anorexia, cough, arthralgia, nausea, emotional lability, depression, hyperesthesia, sore throat, abdominal pain, conjunctival injection, photophobia, and weight loss, along with splenomegaly and hepatomegaly), laboratory testing is required for the correct diagnosis. The commonly used laboratory methods for the specific diagnosis of *Babesia* include microscopic observation of the organism in Giemsa-stained thin-blood smears, detection of babesial antibody in acute-and convalescent-phase sera, isolation of the organism after cultivation on animal blood or following the injection of patient blood into laboratory animals, and amplification and sequencing analysis of babesial nucleic acids.

13.1.3.1 Microscopy

In Giemsa-stained thin-blood smears, *B. microti* and other *Babesia* spp. appear as intra-erythrocyte ring (annular) forms containing nonstaining vacuoles in the center with light-blue cytoplasm and a darker-stained nucleus around the periphery, or pyriform (pear-shaped) inclusions occurring either singly or in pairs assembled at their pointed extremities, or filamentous or amorphous shapes at or just after very high levels of parasitemia. Because *Plasmodium* spp. causing malaria may also show intraerythrocytic rings, it is essential to differentiate *Babesia* from *Plasmodium*. The key distinguishing features for *Babesia* spp. include their varying shapes and sizes, their tendency to contain vacuoles, and their lack of pigment production (although early parasitic stages of malaria parasites may also lack pigments). In comparison with *Plasmodium* spp., *Babesia* bloodstream stages do not divide simultaneously, permitting observation of all divisional stages in a single blood smear from a parasitized host [47,48].

Among *Babesia* spp., trophozoites in a tetrad formation (Maltese cross) within an erythrocyte are indicative of small *Babesia* spp., such as *B. microti* and *B. duncani*, with *B. duncani* displaying more tetrad forms than *B. microti*, *B. divergens* and the related organisms tend to display paired piriforms in a subcentral, central, or peripheral position within human RBCs. Furthermore, *B. microti*-infected erythrocytes often show several parasites appearing as rings, sometimes with incomplete closures, with basket-shaped merozoites measuring 1.5–2.5 μm in size. While *B. microti* parasites are monomorphic (typically of tetrad forms) in human blood smears, *B. divergens*, *B. capreoli*, and *B. duncani* parasites appear pleomorphic with accolé forms, and paired pyriform ring forms, with merozoites measuring 1–3 μm in size. In addition, *B. divergens* and *B. venatorum* typically show paired forms at an obtuse or diverging angle usually in the center of human erythrocytes. Despite being valuable for diagnostic purposes, microscopic examination of *Babesia* parasites in blood smears without additional molecular confirmation is not reliable for the species identification.

13.1.3.2 Serology

Serological techniques are useful for confirming *Babesia* infection after observation of ring-like structures in blood smears [49]. Indirect fluorescent antibody (IFA) test provides a sensitive and specific approach for detecting anti-babesial antibody in infected patients, assessing serum prevalence in asymptomatic individuals, and screening for the disease in blood donations. While the presence of IgM antibody is indicative of recent infection, an increase in IgG titers signifies active or recent infection with *Babesia* [46]. Usually, microscopic observation of IFA slides starts with a magnification of 400×, and the fluorescence pattern is verified with a magnification of 1000× with oil to exclude rare nonspecific fluorescence of babesial antigen, which mainly involves the apex of the parasite. However, occasional false IFA-positive results are noted in *Plasmodium* spp. and *Toxoplasma gondii* infections, and in lupus erythematosus or rheumatoid arthritis.

Enzyme-linked immunosorbent assay (ELISA) is another common serological technique used for *Babesia* detection. ELISA has the advantage of being less subjective in result interpretation and more amenable to automation than IFA. Nonetheless, because the specific antibodies are generally undetectable until at least 1 week after the onset of illness when infection may have proven fatal, serological tests may be of limited value for the early diagnosis of *Babesiai* infection. Furthermore, cross-reactions among some *Babesia* spp. are observed in ELISA [3].

13.1.3.3 *In Vitro* Blood Culture and *In Vivo* Models

Babesia spp. may be cultivated directly from blood by *in vitro* culture using autologous erythrocyte preparations. The parasites (after confirmation by blood smears) can be adapted to culture in blood after several passages. In addition, injection of patient blood into golden hamsters, gerbils, jirds, or mice via the intravenous or intraperitoneal route offers a useful way to detect and confirm *Babesia* parasitemia. While hamsters are highly susceptible to *B. microti*, and *B. duncani*, jirds easily succumb to *B. divergens* but are relatively resistant to *B. venatorum*. Typically, 1.0 mL ETDA whole blood is inoculated into the peritoneum of the laboratory animal, and blood smears from the animal often become positive within 2–4 weeks. However, besides their high cost and the necessity to obtain prior institutional approval, *in vivo* assays involving laboratory animals are relatively slow (requiring about 2–4 weeks after inoculation to give a positive smear) and time-consuming (as blood smears need to be monitored daily).

13.1.3.4 Nucleic Acid Detection

Polymerase chain reaction (PCR) provides a highly sensitive and specific test for detection of *Babesia* DNA in blood and identification of the *Babesia* spp. [46,50–52]. PCR is particularly useful when parasites are not observed in blood smears but symptoms and history are suggestive of babesiosis. Under normal circumstances, *Babesia* parasitemia seldom exceeds 5% in normal hosts but may reach as high as 85% in asplenic individuals.

Often, universal primers are employed to amplify a portion or the whole *Babesia* 18S ribosomal RNA (rRNA) gene (about 1700 bp) followed by the sequence analysis [24,53]. This will not only allow identification of *Babesia* spp., but also facilitate subsequent development of specific primers for individual *Babesia* spp. on the basis of unique 18S rRNA signature sequence. As PCR has the analytical sensitivity of approximately three merozoites and the potential to generate a result in a day, it represents a valuable adjunct to conventional methods for diagnosing human babesiosis, for monitoring of ongoing chemotherapy, and for enhancing epidemiological surveillance for this emerging pathogen [54–56].

Exploiting the genetic variability in *Babesia* β-tubulin gene, PCR assays have been developed for the detection and identification of nine of the most common pathogenic bovine, equine, and rodent piroplasms, including *B. divergens*. Following specific nested OCR amplification of the parasite DNA, the parasite species is confirmed by PCR-restriction fragment-length polymorphism analysis [57,58].

To ensure the authencity of the *Babesia* PCR results, the 12S rDNA of ticks or the glyceraldehyde-3-phosphate dehydrogenase (GAPDH) gene may be used as housekeeping genes to control for high quality of DNA extraction from ticks or human blood, respectively [59].

13.1.4 Treatment and Prevention

Historically, oral or intravenous Clindamycin with oral quinine (or intravenous quinidine) have been used as the standard treatment for symptomatic individuals with severe *Babesia* infections [60]. From the year 2000, oral Atovaquone with oral azithromycin are increasingly applied to avoid quinine side effects and to improve disease outcome [61,62]. Another useful drug for treating *Babesia* infections is oral doxycycline [63]. In addition, blood exchange transfusions may be performed to reduce the parasitic load in the individual with severe disease [64].

The most effective measures to prevent *Babesia* infections are avoidance of tick exposure by not visiting endemic areas (especially during high tick season between May and September), remaining covered with light clothing, searching for ticks after being outdoors and removing discovered ticks from the skin. Another preventative measure is applying diethyltoluamide (DEET), a common and effective bug repellent against ticks and other insects.

13.2 METHODS

13.2.1 Sample Preparation

13.2.1.1 Microscopy

Thin-smeared blood films are stained with Giemsa-stain and examined under microscope to for *Babesia* organisms in erythrocytes.

13.2.1.2 Cultures

Blood collected into Alsever's solution is centrifuged to pellet the cells, and the resulting cell pellet is washed by centrifugation. Packed washed erythrocytes (0.2 mL) are dispensed into 0.8 mL complete culture medium (HL-1 with 20% bovine serum) in a well of a 24-well culture plate. In addition, cultures containing 0.1 mL packed erythrocytes from the principals and 0.1 mL washed packed human erythrocytes (Rockland Immunochemicals, Gilbertsville, PA) in 0.8 mL HL-1 medium with 20% human serum (Rockland Immunochemicals) are prepared. The cultures are incubated as above. Each day, 0.8 mL supernatant medium is removed from the cultures and replaced with the fresh appropriate medium. Giemsa-stained erythrocyte smears are made daily from the cultures and examined microscopically at 1000× under oil immersion to screen for the appearance of *Babesia* organisms. The cultures are maintained until parasites are detected or until 2 weeks after initiation [65].

13.2.1.3 Parasite Isolation

Heparinized blood samples from the patient are washed with phosphate-buffered saline (PBS; pH 7.2) by repeated centrifugation, and aliquots of RBCs are inoculated into hu-RBC-SCID mice or splenectomized hamsters for parasite isolation. To prepare hu-RBC-SCID mice, the peripheral RBCs in splenectomized NOD/shi-*scid* mice are replaced with human type O RBCs. The rate of replacement with human RBCs in hu-RBC-SCID mice is monitored by flow cytometry with biotin-labeled anti-human RBC mouse immunoglobulin G (IgG) Fab fragment and phycoerythrin-labeled streptavidin (Life Technologies, Rockville, MD) and is maintained at over 90% by repeatedly transfusing 0.5 mL of a packed cell volume of human RBCs (~6 × 10^9) into the mice at 2- to 4-day intervals, together with administration of anti-mouse RBC monoclonal rat antibody clone 2E11 and anti-erythropoietin rabbit serum. The isolates are further propagated by subpassage into new animals [66].

13.2.1.4 Extraction of *Babesia* DNA from Ticks

Total DNA is extracted from each tick by using a DNeasyTissue kit (Qiagen). Briefly, each sample is minced with a disposable sterile scalpel within a microtube. After digestion with proteinase K (20 μg/mL) samples are applied to the columns for absorption and washing of DNA. Finally, DNA is eluted in 200 μL of buffer available from the kit and stored at 4°C [59]. Alternatively, ticks are thawed, immersed in 70% ethanol for 10 min, and air-dried on microscope slides. Adult ticks and nymphs are cut lengthwise with sterile scalpel blades, and larvae are processed whole. Total DNA is isolated with a DNeasy blood and tissue kit (Qiagen) and eluted in 40 μL AE buffer (Qiagen) [26].

13.2.1.5 Extraction of *Babesia* DNA from Human Blood Specimens

Parasite DNA is isolated from blood specimen with a QIAmp DNA Blood Mini kit (Qiagen) [59].

13.2.2 Detection Procedures

13.2.2.1 Nested PCR Identification of *B. microti*

Wei et al. [11] in 2001 reported the use of two nested PCR assays targeting the 18S rRNA gene for specific identification of *B. microti* in blood samples (Table 13.1).

13.2.2.1.1 Procedure

1. Extract *Babesia* DNA from 500 μL of the heparinized whole-blood specimens with a DNA Extractor WB kit (Wako Pure Chemical Industries, Osaka, Japan).
2. Prepare the first-round PCR mixture (50 μL) containing 50 pmol (each) of the primer set of Bab1A and Bab4A (or Bab5 and Bab8), 10 mM Tris pH 8.3, 50 mM KCl, 2.5 mM $MgCl_2$, 200 mM (each) deoxyribonucleoside triphosphates, 1 U of AmpliTaq Gold DNA polymerase (Applied Biosystems, Foster City, CA), and 1/10 of each extracted specimen.
3. Perform amplification in a thermal cycler with 35 cycles of 94°C for 1 min, 55°C for 1 min, and 72°C for 1 min.
4. Prepare the second round PCR mixture (50 μL) as above, but with 50 pmol (each) of the primer set of Bab2A and Bab3A (or Bab6 and Bab7), and 1 μL of the first-round PCR product; and carry out amplification using the same cycling program.
5. Separate the PCR products from both round on 2% agarose gel in 1× TBE, stain with ethidium bromide and visualize the specific band under ultraviolet (UV) light.

Note: The first-round PCR using primers Bab1A and Bab4A yields a 238-bp product; the second-round PCR using primers Bab2A and Bab3A gives a 154-bp product [54]. The identity of *B. microti* can be further verified by a separate nested PCR using primers Bab5 and Bab8 followed by primers Bab6 and Bab7 [11].

13.2.2.2 Nested PCR Identication of *B. divergens*

Holman et al. [65] reported a nested PCR for specific identification of *B. divergens* and the Nantucket *Babesia* isolate (NR831) with primers derived from the 18S rRNA gene (Table 13.2).

13.2.2.2.1 Procedure

1. Purify genomic DNA from 200 μL packed washed erythrocytes by a standard phenol/chloroform extraction protocol; resuspend the ethanol precipitated DNA pellets in 20 μL TE (10 mM Tris–HCl, pH 8.0, and 1 mM ethylene diamine tetracetic acid (EDTA)).
2. Prepare the primary PCR mixture (25 μL) with Advantage 2 PCR Kit (BD Biosciences, Palo Alto, CA) containing 1 μL of DNA, 1 pmol each forward primer A and reverse primer B.
3. Carry out amplification with an initial denaturation step at 96°C for 3 min followed by 35 cycles of 94°C for 10 s, 60°C for 10 s, and 72°C for 2 min, with a final extension at 72°C for 10 min in a PCR thermocycler.

TABLE 13.1
Primers for Nested PCR Identification of *B. microti*

PCR	Primer	Sequence (5′–3′)	Reference
1st round	Bab1A	GTCTTAGTATAAGCTTTTATACAGCG	Persing et al. [54]
	Bab4A	GATAGGTCAGAAACTTGAATGATACATCG	
2nd round	Bab2A	CAGTTATAGTTTATTTGATGTTCGTTTTAC	Persing et al. [54]
	Bab3A	CGGCAAAGCCATGCGATTCGCTAAT	
1st round	Bab5	AATTACCCAATCCTGACACAGG	Wei et al. [11]
	Bab8	TTTCGCAGTAGTTCGTCTTTAACA	
2nd round	Bab6	GACACAGGGAGGTAGTGACAAGA	Wei et al. [11]
	Bab7	CCCAACTGCTCCTATTAACCATTAC	

TABLE 13.2
Primers for Nested PCR Identification of *B. divergens*

PCR	Primer	Sequence (5′–3′)	Product (bp)
First round	Forward A	AACCTGGTGATCCTGCCAGT	
	Reverse B	GATCCTTCTGCAGGTTCACCTAC	
Second round	srBdvF	GGTGTTAATATTGACTAATGTCGAGATTGC	1026
	srBdvR	CCAAGCCGACGAATCGGAAAGGCC	

4. Dilute the primary PCR products 1:20 in sterile water and use 1 μL as template in each reaction mixture as above in the nested PCR, but with 1 pmol each primer srBdvF and srBdvR and the initial denaturation reduced to 1 min.
5. Separate the products by electrophoresis through a 1% agarose gel, stain with ethidium bromide, and view by UV transillumination.

Note: A specific product of 1026 bp is observed from DNA of *B. divergens* and the Nantucket *Babesia* isolate.

13.2.2.3 Sequencing Analysis of *Babesia* spp.

Blaschitz et al. [26] employed 18S rRNA gene primers for PCR amplification and sequencing analysis of a conserved 620-bp fragment from *Babesia* spp. (Table 13.2). For species differentiation within the genus *Babesia*, a second primer pair (HSP70for/HSP70rev) is designed to amplify a 354-bp fragment within the more variable heat-shock protein 70 (HSP70) gene. In addition, two more primer pairs that amplify highly variable ribosomal internal transcribed spacer region 1 (ITS1) and ITS2 of *Babesia* (ITS1for/ITS1rev and ITS2for/ITS2 rev) are designed for the 18S rRNA gene, the 5.8S rRNA gene, and the 28S rRNA gene. ITS1for is the reverse complement of Babrev except that it has one additional nucleotide at the 5′ end; ITS1rev is the reverse complement of ITS2for (Table 13.3).

13.2.2.3.1 Procedure

1. Prepare the PCR mixture (50 μL) containing 3 μL of total DNA, amplify using the cycling program outlined in Table 13.2.
2. Separate PCR products in 3% agarose gel in Tris–acetate–EDTA buffer, stain with ethidium bromide and visualize under UV light.
3. Purify positive PCR products using a PCR DNA and gel band purification kit (GE Healthcare Bio-Sciences AB) and subject to cycle sequencing using an ABI PRISM Big Dye kit (AB, Langen, Germany) and a standard sequencing program (30 cycles of 96°C for 10 s, 50°C for 5 s, and 60°C for 4 min).
4. Align sequences, using the GeneDoc sequence editor computer program, analyze the data using SPSS 14.0 software. Calculate correlations using Pearson and Spearman coefficients.

13.3 CONCLUSION

The genus *Babesia* consists of a large group of protozoa that circulate between mammals (especially rodents) and *Ixodes* ticks. As zoonotic intraerythrocytic pathogens transmitted through tick bites, *Babesia* spp. cause a hemolytic disease known as babesiosis (or piroplasmosis, Texas cattle fever, redwater fever, tick fever, and Nantucket fever) in cattle, horses, sheep, pigs, and dogs, as well as humans. Being an accidental dead-end host of *Babesia*, humans do not usually acquire infections from each other via tick bites, although transmission of the parasites is possible via blood transfusion.

Of over 100 recognized species in the genus, *B. microti*, *Babesia divergens B. bovis*, *B. canis*, *B. duncani*, and *B. venatorum*, are associated with human disease. While *B. microti* commonly occurs in the United States, *B. divergens* is predominant in Europe. However, human babesiosis due to *B. microti* has been described recently in Europe, and human

TABLE 13.3
Primers for PCR Amplification and Sequencing Analysis of *Babesia* spp.

PCR	Primer	Sequence (5′–3′)	Condition
18S rRNA	Babfor	GACTAGGGATTGGAGGTC	35 cycles of 95°C for 1 min, 53°C for 1.50 min, and 72°C for 1.50 min
	Babrev	GAATAATTCACCGGATCACTC	
ITS1	ITS1for	CGAGTGATCCGGTGAATTATTC	40 cycles of 95°C for 1 min, 52°C for 2 min, and 72°C for 3 min
	ITS1rev	CCTTCATCGTTGTGTGAGCC	
ITS2	ITS2for	GGCTCACACAACGATGAAGG	40 cycles of 95°C for 1 min, 54°C for 2 min, and 72°C for 3 min
	ITS2rev	CTCGCCGTTACTAAGGGAATC	
HSP70	HSP70for	GCTATTGGTATTGACTTGGG	40 cycles of 95°C for 1 min, 54°C for 2 min, and 72°C for 3 min
	HSP70rev	CCTTCATCTTGATAAGGACC	

infecting *B. microti*-like species are identified in Asia, Africa, and South America. Furthermore, *B. divergens*-like organisms are described as cause of human infections in Missouri, Kentucky, and Washington State in the United States.

Clinically, human babesiosis may appear as asymptomatic infection, mild-to-moderate illness, or fulminant disease resulting in death. In most healthy adults, babesiosis tends to present nonspecific symptoms, such as fever, fatigue, chills, sweats, headache, myalgia, anorexia, cough, arthralgia, nausea, emotional lability, depression, hyperesthesia, sore throat, abdominal pain, conjunctival injection, photophobia, and weight loss, along with splenomegaly and hepatomegaly. In individuals with splenectomy and compromised immune functions (due to advanced age, malignancy, HIV, or immunosuppressive therapy), a more severe form of the disease may develop. Although asymptomatic carriers do not suffer any obvious illness, they pose a blood safety risk when donating blood.

Because babesiosis resembles malaria and other blood-borne infections clinically, there is a need to apply laboratory procedures for correct diagnosis. Traditional methods for diagnosing babesiosis are based on microscopic identification of *Babesia* organisms in Giemsa-stained thin-blood smears, detection of specific antibody to *Babesia* in acute- and convalescent-phase sera, isolation of the organisms on animal blood or in experimental animals, and amplification and sequencing analysis of babesial nucleic acids. As *Babesia* and malaria-causing *Plasmodium* spp. often show similar ring forms in erythrocytes under microscope, additional assays (e.g., IFAT, *in vitro* culture, and *in vivo* models) are required for confirmation of babesiosis. Owing to the time-consuming and variable features of serological tests, *in vitro* culture, and *in vivo* models, molecular techniques, such as PCR and sequencing have increasingly been applied for speedy diagnosis in suspected cases, and precise determination of zoonotic *Babesia* spp.

REFERENCES

1. Homer, M.J. et al., Babesiosis. *Clin. Microbiol. Rev.*, 13, 451, 2000.
2. Vannier, E., Gewurz, B.E., and Krause, P.J., Human Babesiosis. *Infect. Dis. Clin. North Am.*, 22, 469, 2008.
3. Vannier, E., and Krause, P.J., Update on Babesiosis. *Interdiscip Perspect Infect Dis.*, 2009, 984568, 2009.
4. Babes, V., Sur l'hemoglobinurie bacterienne du boeuf. *C. R. Acad. Sci.*, 107, 692, 1888.
5. M'Fadyean, J., and Stockman, S., A new species of piroplasm found in the blood of British cattle. *J. Comp. Pathol.*, 24, 340, 1911.
6. Skrabalo, Z., and Deanovic, Z., Piroplasmosis in man; report of a case. *Doc. Med. Geograph. Trop.*, 9, 11, 1957.
7. Hunfeld, K.-P., Hildebrandt, A., and Gray, J.S., Babesiosis: Recent insights into an ancient disease. *Int. J. Parasitol.*, 38, 1219, 2008.
8. Allsopp, M.T., and Allsopp, B.A., Molecular sequence evidence for the reclassification of some *Babesia* species. *Ann. N. Y. Acad. Sci.*, 1081, 509, 2006.
9. Kogut, S.J. et al., *Babesia microti*, upstate New York. *Emerg. Infect. Dis.*, 11, 476, 2005.
10. Shih, C.-M. et al., Human babesiosis in Taiwan: Asymptomatic infection with a *Babesia microti*-like organism in a Taiwanese woman. *J. Clin. Microbiol.*, 35, 450, 1997.
11. Wei, Q. et al., Human babesiosis in Japan: Isolation of *Babesia microti*-like parasites from an asymptomatic transfusion donor and from a rodent from an area where babesiosis is endemic. *J. Clin. Microbiol.*, 39, 2178, 2001.
12. Rios, L., Alvarez, G., and Blair, S., Serological and parasitological study and report of the first case of human babesiosis in Colombia. *Rev. Soc. Bras. Med. Trop.*, 36, 493, 2003.
13. Marathe, A. et al., Human babesiosis—a case report. *Indian J. Med. Microbiol.*, 23, 267, 2005.
14. Hildebrandt, A. et al., First confirmed autochthonous case of human *Babesia microti* infection in Europe. *Eur. J. Clin. Microbiol. Infect. Dis.*, 26, 595, 2007.
15. Zintl, A. et al., *Babesia divergens*, a bovine blood parasite of veterinary and zoonotic importance. *Clin. Microbiol. Rev.*, 16, 622, 2003.
16. Herwaldt, B.L. et al., A fatal case of babesiosis in Missouri: Identification of another piroplasm that infects humans. *Ann. Intern. Med.*, 124, 643, 1996.
17. Herwaldt, B.L. et al., *Babesia divergens*-like Infection, Washington State. *Emerg. Infect. Dis.*, 10, 622, 2004.
18. Conrad, P.A. et al., Description of *Babesia duncani* n.sp. (Apicomplexa: Babesiidae) from humans and its differentiation from other piroplasms. *Int. J. Parasitol.*, 36, 779, 2006.
19. Herwaldt, B.L. et al., Endemic babesiosis in another eastern state: New Jersey. *Emerg. Infect. Dis.*, 9, 184, 2003.
20. Haselbarth, K. et al., First case of human babesiosis in Germany—clinical presentation and molecular characterisation of the pathogen. *Int. J. Med. Microbiol.*, 297, 197, 2007.
21. Krause, P.J. et al., Persistent and relapsing babesiosis in immunocompromised patients. *Clin. Infect. Dis.*, 46, 370, 2008.
22. Herwaldt, B.L. et al., Molecular characterization of a non-*Babesia divergens* organism causing zoonotic babesiosis in Europe. *Emerg. Infect. Dis.*, 9, 942, 2003.
23. Armstrong, P.M. et al., Diversity of *Babesia* infecting deer ticks (*Ixodes dammini*). *Am. J. Trop. Med. Hyg.*, 58, 739, 1998.
24. Kjemtrup, A.M., and Conrad, P.A., Human babesiosis: An emerging tick-borne disease. *Int. J. Parasitol.*, 30, 1323, 2000.
25. Holman, P.J., Phylogenetic and biologic evidence that *Babesia divergens* is not endemic in the United States. *Ann. New York Acad. Sci.*, 1081, 518, 2006.
26. Blaschitz, M. et al., *Babesia* species occurring in Austrian *Ixodes ricinus* ticks. *Appl. Env. Microbiol.*, 74, 4841, 2008.
27. Rodgers, S.E., and Mather, T.N., Human *Babesia microti* incidence and *Ixodes scapularis* distribution, Rhode Island, 1998–2004. *Emerg. Infect. Dis.*, 13, 633, 2007.
28. Cieniuch, S., Stańczak, J., and Ruczaj, A., The first detection of *Babesia* EU1 and *Babesia canis canis* in *Ixodes ricinus* ticks (Acari, Ixodidae) collected in urban and rural areas in northern Poland. *Pol. J. Microbiol.*, 58, 231, 2009.
29. Reye, A.L. et al., Prevalence and seasonality of tick-borne pathogens in questing *Ixodes ricinus* ticks from Luxembourg. *Appl. Env. Microbiol.*, 76, 2923, 2010.
30. Eskow, E.S. et al., Southern extension of the range of human babesiosis in the Eastern United States. *J. Clin. Microbiol.*, 37, 2051, 1999.
31. Gubernot, D.M. et al., Babesia infection through blood transfusions: Reports received by the US Food and Drug Administration, 1997–2007. *Clin. Infect. Dis.*, 48, 25, 2009.
32. Gray, J. et al., Transmission studies of *Babesia microti* in *Ixodes ricinus* ticks and gerbils. *J. Clin. Microbiol.*, 40, 1259, 2002.

33. Ortiz, J.M., and Eagle, R.C., Jr., Ocular findings in human babesiosis (Nantucket fever). *Am. J. Ophthalmol.*, 93, 307, 1982.
34. Foppa, I.M. et al., Entomologic and serologic evidence of zoonotic transmission of *Babesia microti*, eastern Switzerland. *Emerg. Infect. Dis.*, 8, 722, 2002.
35. Okabayashi, T. et al., Detection of *Babesia microti*-like parasite in filter paper-absorbed blood of wild rodents. *J. Vet. Med. Sci.*, 64, 145, 2002.
36. Kim, J.-Y. et al., First case of human babesiosis in Korea: Detection and characterization of a novel type of *Babesia* sp. (KO1) similar to ovine Babesia. *J. Clin. Microbiol.*, 45, 2084, 2007.
37. Benach, J.L., and Habicht, G.S., Clinical characteristics of human babesiosis. *J. Infect. Dis.*, 144, 481, 1981.
38. Bush, J.B. et al., Human babesiosis—a preliminary report of 2 suspected cases in South Africa. *South Afr. Med. J.*, 78, 699, 1990.
39. Gorenflot, A. et al., Human babesiosis. *Ann. Trop. Med. Parasitol.*, 92, 489, 1998.
40. Beattie, J.F., Michelson, M.L., and Holman, P.J., Acute babesiosis caused by *Babesia divergens* in a resident of Kentucky. *N. Engl. J. Med.*, 347, 697, 2002.
41. Fox, L.M. et al., Neonatal babesiosis: Case report and review of the literature. *Pediatr. Infect. Dis. J.*, 25, 169, 2006.
42. Krause, P.J. et al., Diagnosis of babesiosis: Evaluation of a serologic test for the detection of *Babesia microti* antibody. *J. Infect. Dis.*, 169, 923, 1994.
43. Berman, K.H. et al., Fatal case of babesiosis in postliver transplant patient. *Transplantation*, 87, 452, 2009.
44. Froberg, M.K. et al., Case report: Spontaneous splenic rupture during acute parasitemia of *Babesia microti*. *Ann. Clin. Lab. Sci.*, 38, 390, 2008.
45. Hemmer, R.M., Ferrick, D.A., and Conrad, P.A., Role of T cells and cytokines in fatal and resolving experimental babesiosis: Protection in TNFRp55 – /– mice infected with the human *Babesia* WA1 parasite. *J. Parasitol.*, 86, 736, 2000.
46. Krause, P.J. et al., Comparison of PCR with blood smear and inoculation of small animals for diagnosis of *Babesia microti* parasitemia. *J. Clin. Microbiol.*, 34, 2791, 1996.
47. Healy, G.R., and Ruebush, T.K., II Morphology of *Babesia microti* in human blood smears. *Am. J. Clin. Pathol.*, 73, 107, 1980.
48. Sun, T., Tenenbaum, M.J., and Greenspan, J., Morphologic and clinical observations in human infection with *Babesia microti*. *J. Infect. Dis.*, 148, 239, 1983.
49. Ryan, R. et al., Diagnosis of babesiosis using an immunoblot serologic test. *Clin. Diagn. Lab. Immunol.*, 8, 1177, 2001.
50. Gubbels, J.M. et al., Simultaneous detection of bovine *Theileria* and *Babesia* species using reverse line blot hybridization. *J. Clin. Microbiol.*, 37, 1782, 1999.
51. Skotarczak, B., and Cichocka, A., Isolation and amplification by polymerase chain reaction DNA of *Babesia microti* and *Babesia divergens* in ticks in Poland. *Ann. Agric. Environ. Med.*, 8, 187, 2001.
52. Pieniazek, N., Sawczuk, M., and Skotarczak, B., Molecular identification of *Babesia* parasites isolated from *Ixodes ricinus* ticks collected in northwestern Poland. *J. Parasitol.*, 92, 32, 2006.
53. Holman, P.J. et al., Ribosomal RNA analysis of *Babesia odocoilei* isolates from farmed reindeer (*Rangifer tarandus tarandus*) and elk (*Cervus elaphus canadensis*) in Wisconsin. *Parasitol. Res.*, 91, 378, 2003.
54. Persing, D.H. et al., Detection of *Babesia microti* by polymerase chain reaction. *J. Clin. Microbiol.*, 30, 2097, 1992.
56. Centeno-Lima, S. et al., A fatal case of human babesiosis in Portugal: Molecular and phylogenetic analysis. *Trop. Med. Int. Health*, 8, 760, 2003.
57. Caccio, S. et al., The beta-tubulin gene of *Babesia* and *Theileria* parasites is an informative marker for species discrimination. *Int. J. Parasitol.*, 30, 1181, 2000.
58. Zamoto, A. et al., Epizootiologic survey for *Babesia microti* among small wild mammals in northeastern Eurasia and a geographic diversity in the beta-tubulin gene sequences. *J. Vet. Med. Sci.*, 66, 785, 2004.
59. Casati, S. et al., Presence of potentially pathogenic *Babesia* sp. for human in *Ixodes ricinus* in Switzerland. *Ann. Agric. Environ. Med.*, 13, 65, 2006.
60. Denes, E. et al., Management of *Babesia divergens* babesiosis without a complete course of quinine treatment. *Eur. J. Clin. Microbiol. Infect. Dis.*, 18, 672, 1999.
61. Krause, P.J. et al., Atovaquone and azithromycin for the treatment of babesiosis. *N. Engl. J. Med.*, 343, 1454, 2000.
62. Raju, M. et al., Atovaquone and azithromycin treatment for babesiosis in an infant. *Pediatr. Infect. Dis. J.*, 26, 181, 2007.
63. Weiss, L.M., Wittner, M., and Tanowitz, H.B., The treatment of babesiosis. *N. Engl. J. Med.*, 344, 773, 2001.
64. Vyas, J.M., Telford, S.R., and Robbins, G.K., Treatment of refractory *Babesia microti* infection with atovaquone-proguanil in an HIV-infected patient: Case report. *Clin. Infect. Dis.*, 45, 1588, 2007.
65. Holman, P.J. et al., Comparative infectivity of *Babesia divergens* and a zoonotic *Babesia divergens*-like parasite in cattle. *Am. J. Trop. Med. Hyg.*, 73, 865, 2005.
66. Saito-Ito, A. et al., Transfusion-acquired, autochthonous human babesiosis in Japan: Isolation of *Babesia microti*-like parasites with hu-RBC-SCID mice. *J. Clin. Microbiol.*, 38, 4511, 2000.

14 Balantidium

Dongyou Liu

CONTENTS

14.1 INTRODUCTION

Protozoa (proto, meaning "first"; zoa, meaning "animals" in Greek) are small unicellular eukaryotic organisms (usually of 10–52 μm in size) with the ability to move. Consisting of >92,000 known species, protozoa are divided into four phyla: Sarcomastigophora, Apicomplexa, Ciliophora, and Myxozoa. A previous protozoan phylum Microspora has now been redesignated as a separate phylum (Microsporidia) in the kingdom Fungi. The phylum Sarcomastigophora is further separated into two subphyla Mastigophora and Sarcodina; the subphylum Mastigophora harbor one or more flagella as locomotive organ and reproduce by binary fission as well as sexual reproduction in some species, while the subphylum Sarcodina move by pseudopodia (although flagella may be present in the reproductive stages) and reproduce by binary fission. The phylum Apicomplexa possesses no special locomotive organs, reproduce by multiple fission and gamogony, and form spores at one stage of their life. The phylum Ciliophora move by cilia, have two nuclei (one macronucleus and one micronucleus), and reproduce by transverse binary fission, conjugation, autogamy, and cytogamy. The phylum Myxozoa contains amoeboid germinal elements (sporoplasms) [1].

Although most protozoa are free-living organisms that prey upon unicellular or filamentous algae, bacteria, and microfungi, and serve as an important food source for microinvertebrates, some are known to lead a parasitic life in human and animal hosts. To date, virtually all human pathogenic protozoa belong to the phyla Sarcomastigophora (e.g., *Giardia*, *Leishmania*, *Trichomonas*, and *Trypanosoma* in the subphylum Mastigophora; *Acanthamoeba*, *Babesia*, *Dientamoeba*, *Entamoeba*, and *Naegleria* in the subphylum Sarcodina), Apicomplexa (e.g., *Cryptosporidium*, *Isospora*, *Plasmodium*, *Sarcocystis*, and *Toxoplasma*), and Ciliophora (e.g., *Balantidium*).

Balantidium is a genus of ciliated binucleate protozoa that possesses macronuclei and micronuclei (features of Ciliophora); rows of monokinetid somatic cilia (arising from single rows of subsurface kinetosomes and their associated fibrils); a slit-like, depressed anteroventral oral cavity (a feature of Litostomea) with dense ciliation (hairy mouths; features of Trichostomatia); a vestibular groove (of less than one-half the body length) leading into the oral apparatus (features of Vestibulifera and Balantidiidae) [2]. As the largest protozoan and the only ciliate known to parasitize humans and nonhuman primates within the genus, *Balantidium coli* is widely distributed in the environment. Pigs serve its main reservoir host, while rodents and nonhuman primates may function as its alternative reservoir hosts. Humans acquire the infection through ingestion of contaminated food or water, leading to balantidiasis, whose clinical manifestations range from asymptomatic, persistent diarrhea, dysentery (bloody and mucus-containing diarrhea), colitis, abdominal pain, and weight loss, especially in debilitated persons.

14.1.1 Classification, Morphology, and Life Cycle

14.1.1.1 Classification

The genus *Balantidium* (balantidion, meaning "little bag" in Greek) is classified in the family Balantidiidae, order Vestibuliferida, class Litostomatea, Phylum Ciliophora. The family Balantidiidae is characterized by its lack of concrement vacuoles, and by having an oral cavity in the anterior half of the body. As the sole genus of the family Balantidiidae, *Balantidium* encompasses about 50 species, the most notable of which are *B. coli* in pigs, *Balantidium entozoon* in frogs, *Balantidium caviae* in guinea pigs, and *Balantidium jocularum* in herbivorous surgeonfish [3–6].

B. coli (formerly *Paramecium coli*) is infective to pigs and humans. Previously, *Balantidium* parasites infecting these two hosts have been proposed to form separate species (i.e., smaller *Balantidium suis* in pigs and larger *B. coli* in humans). However, this separation has not been widely accepted, and the parasites from both hosts continue to be regarded as a single species.

B. entozoon (formerly *Bursaria entozoon*) is the type species of the genus *Balantidium*, which occurs mainly in frogs. The small subunit ribosomal RNA (rRNA) gene sequence in *B. entozoon* has been shown to differ from that of *B. coli* by 5% [2].

14.1.1.2 Morphology

Members of the genus *Balantidium* undergo two distinct stages (forms) in their life cycle: a dormant (nonreproductive) cyst stage and a proliferative (reproductive) trophozoite (tropho, meaning "to nourish" in Greek) stage. The cyst (of 40–60 μm in diameter, with some up to 200 μm in length) is spherical in shape, resistant to environmental stress (e.g., extreme temperatures, harmful chemicals, absence of nutrients, water, or oxygen, due to its thickened wall of 400–500 nm in thickness) and infective (transmissive); the trophozoite (of 30–300 μm in length and 25–120 μm in width) is oval in shape, actively motile (in a rotary motion similar to a thrown football) and invasive. The conversion of a cyst to a trophozoite is known as excystation, whereas the transformation of a trophozoite to a cyst is known as encystation.

Under light microscope (with interference or phase-contrast capability), *Balantidium* organisms are ovoid and slightly flattened on one side, with a mouth (oral apparatus) at the tapering anterior end, a cytopyge (anus) at the rounded posterior end, a large, kidney-shaped macronucleus (of $10 \times 20\ \mu m^2$ in size, which is situated near the middle of the body and enclosed in a nuclear membrane) and a small, spherical micronucleus (of $3 \times 5\ \mu m^2$ in size, which lies in close proximity to and often hidden by the macronucleus) in the cytoplasm. However, only the macronucleus is visible in the cyst stage. The mouth appears as a funnel-shaped depression, consisting of a peristome (a rigid structure with striations), cytopharynx, and cytosome, with the food passing from the peristome to the elongated cytopharynx and then to the cytosome. The parasite has an excretory opening at the posterior end known as the cytopyge, which is circular and much smaller than the peristome. The whole trophozoite is covered with cilia (the organs of locomotion), with the peristomal cilia (used for propelling food into the cytopharynx) being larger than the somatic (body) cilia (1.55 μm). On the other hand, the cyst is only partially covered with cilia. Living trophozoites and cysts appear yellowish or greenish in color, and their ciliary movement is easily observed.

Examined by scanning electron microscopy, *Balantidium* organisms show a specialized "Villeneuve–Brachon's" (V–B) field of somatic cilia to the right of the vestibule (or in a dextroral location). The anterior–posterior length of the V-B field is about 11 μm. The V-B field consists of one row of relatively short cilia of uniform length flanked on each side by 2–3 rows, or more, of very short cilia (some of which are clavate). The cilia are embedded in the pellicle in longitudinal rows known as kineties (with each row containing about 20 cilia, and total number of rows ranging from 36 to 106). The kinetids have typical litostome structure (including kinetodesmal fibrils, postciliary microtubules, transverse (T-1 and T-2) microtubules, and nematodesmal microtubules). Hydrogenosome-like structures (relict mitochondria) are present between and around the somatic kinetosomes and within the inter-kinetal ridges. Other cytoplasmic components consist of endoplasmic reticulum, ribosomes, mucocysts, two contractile vacuoles (for maintaining the proper osmotic pressure in the cell) and numerous vacuoles filled with food particles [2,7].

14.1.1.3 Biology

B. coli infection begins when a host ingests cysts via contaminated water or food. The ingested cysts are capable of surviving in the acidic environment of the stomach and the basic environment of the small intestine, where encystation takes place, with cysts becoming trophozoites. Trophozoites move with their cilia (in a rotary-type motion), colonize the lumen of the large intestine (cecum and colon), and feed on the intestinal flora and nutrients. Some trophozoites invade the mucosa of the large intestine, and multiply by asexual transverse binary fission or sexual conjugation (syngamy), with some returning to the lumen. Some trophozoites gain entry to the lymphoid tissue and travel to the rest of the body.

During asexual transverse binary fission, the mother cell elongates and forms a transverse structure through the middle. The elongated mother cell constricts in the middle and finally separates into two asymmetrical daughter individuals: an anterior (proter) and a posterior (opisthe) cell. While the proter retains the oral apparatus and develops a new excretory pore, the opisthe generates a new mouth. Two successive fissions (meiosis) are followed by the formation of conjugants via either equal or unequal divisions. The two conjugants attach at the oral apparatus and exchange micronuclear products of meiosis for a few moments, and then detach. There is no increase in the number of conjugants during sexual conjugation.

In the lumen, trophozoites may disintegrate or undergo encystation (transforming into cyst), which is triggered by dehydration of the intestinal contents usually in the distal large intestine (i.e., the colon and rectum) as well as outside of the host in feces. Trophozoites and cysts are discharged with the feces and released into the environment for later infection of a new host. Interestingly, *B. coli* cysts are not produced in cultures, nor in cases of acute dysentery (which may harbor only trophozoites).

Balantidiasis is a zoonotic disease caused by *B. coli*, which is regarded as a commensal organism of the large intestine of pigs [8]. The parasite has been also reported in camels, horses, cattle, rodents, nonhuman primates, and humans [9–14]. Despite causing no symptoms in pigs, pig-to-pig transmission is very common, especially in piggeries with poor hygienic conditions and poor water systems. Infected pigs shed vast volumes of *B. coli* cysts in their feces, contaminating water sources, and, without water sanitation, cysts are consumed with the contaminated water. Humans in close contact with pigs are vulnerable to infection through the fecal–oral route [15].

B. coli has a cosmopolitan distribution, with a relatively high incidence in South and Central America, the Philippines, Iran, Central Asia, Papua New Guinea, Easter Timor, and some Pacific islands, where over 50% of pigs and 6–29% of humans may be infected [16,17]. *B. coli* infection is most likely to occur in people with malnutrition due to their low stomach acidity or people with compromised immune systems. Thus, it appears that the major factors leading to human balantidiasis are (i) close contact with pigs (including visitation to endemic areas), (ii) lack of appropriate waste disposal system to prevent swine and human excrement from contaminating drinking water wells, and streams, and food, and (iii) subtropical and tropical climatic conditions with warmth and humidity which favor survival of cysts.

14.1.2 Clinical Features, Pathogenesis, and Treatment

14.1.2.1 Clinical Features

As an uncommon human disease of tropical and subtropical regions with poor sanitary standards, balantidiasis due to *B. coli* induces a range of mild-to-severe clinical presentations, from asymptomatic, chronic, to fulminating infection [18–24].

1. Asymptomatic infection. Most people infected with *B. coli* are asymptomatic, namely, the infected individuals shed cysts or trophozoites in their feces, but are free of any other symptoms. These individuals are carriers of disease and serve as reservoirs of infection in the community.
2. Chronic infection. Individuals with this type of infection have mucoid and rarely bloody diarrhea (with alternating periods of constipation), cramping, halitosis, nausea, foul breath, colitis (inflammation of the colon), and abdominal pain (secondary to trophozoite invasion of the large intestine) [25–27]. Persistent diarrhea can lead to high fluid loss and dehydration.
3. Fulminating acute infection. This form of balantidiasis appears suddenly and with great intensity. Explosive diarrhea may occur as often as every 20 min. Dysentery (diarrhea with passage of blood or mucus, mimicking acute amoebic dysentery) may lead to severe dehydration, perforation of the colon, fever, pain and tenderness in the cecal region (mimicking appendicitis), peritonitis, pleuritis, ulcer, liver abscess, pneumonia, weight loss, tenesmus, renal failure, shock, and death, especially in immunocompromised and malnourished individuals [28–31]. Left untreated, it has a case fatality rate of 30%, and some remaining cases can become chronic balantidiasis.

The gross pathologic characteristics of balantidiasis include multiple ulcers with necrotic bases and undermined edges. Microscopically, *B. coli* appears in clusters in the submucosa or at the bases of crypts, and its macronucleus stains deeply with hematoxylin and eosin. Many lymphocytic cells plus some plasma cells are present.

Sharma and Harding [32] reported a case of *B. coli* causing a thick-walled right upper lobe cavity in a 42-year-old organic farmer who had contact with aerosolized pig manure and traveled to the South American countries of Paraguay and Argentina. With a history of insulin-dependent diabetes mellitus and previous diarrheal illnesses from intestinal protozoa, the patient presented with two episodes of hemoptysis (with up 50–200 mL of bright-red blood). A 2 cm cavitary right upper lobe lesion with thick walls and fibrotic strands extending from the wall measuring 3–4 mm were noted by computerized tomography scan. Bronchoalveolar lavage (BAL) fluid examination revealed trophozoites of *B. coli* in large numbers on wet preparation and iron–hematoxylin stain. The trophozoites (of 50–70 μm in length and 40–50 μm in breadth) were ovoid in shape, and greenish–yellowish in color, with two nuclei and several contractile vacuoles. Covered by longitudinal cilia, the trophozoites two nuclei and several contractile vacuoles moved in a spiral motion. The cysts (of 45–65 μm in diameter) also appeared greenish–yellow in color. Treatment with doxycycline hyclate 100 mg/daily resulted in marked improvement.

In addition, Anargyrou et al. [33] described a case of pulmonary balantidiasis in a 59-year-old woman suffering from chronic lymphocytic leukemia. Direct microscopic examination of BAL revealed ciliated trophozoites of *B. coli*. Administration of metronidazole led to rapid improvement of condition after 24–48 h. More recently, Koopowitz et al. [34] documented a case of *B. coli*-induced pulmonary hemorrhage with iron deficiency in a 20-year-old, immune-competent man. Examination of bronchial biopsy and lavage showed numerous trophozoites of *B. coli* with a background of acute inflammatory cells. The patient was treated with oxytetracyline and metronidazole and was discharged within 4 days.

In a separate study, Yazar et al. [35] presented a case of *B. coli*-associated diarrhea and abdominal pain in a 47-year-old female patient with non-Hodgkin's lymphoma from Turkey. The patient had watery, bad smelling, bloody diarrhea (10 times/day), and abdominal pain. Upon examination of the stool samples by native, lugol, and flotation methods, moving trophozoites of 60 μm long and 35 μm broad with cilia were observed and diagnosed as *B. coli*. After treatment with metronidazole 750 mg three times daily for five days, the symptoms disappeared rapidly and completely resolved.

14.1.2.2 Pathogenesis

Upon ingestion by a host, *B. coli* cysts withstand acidic condition in the stomach and alkaline condition in the small intestine, and undergo excystation to become trophozoites. Trophozoites use their cilia to move around in the lumen of intestine, invade the mucosa of the large intestine (cecum and colon) and cause flask-shaped lesions in the submucosa (where they form clusters called nides or nests) and ulcerations. Trophozoites' invasion of the mucosa is assisted by their secretion of hyaluronidase and other proteolytic enzymes. In particular, hyaluronidase digests hyaluronic acid, a component of the "glue" holding mucosal epithelial cells together in intestinal tissue, facilitating penetration of the mucosa. Other intestinal bacteria (e.g., *Salmonella*) then enter the host through the hemorrhage and perforation induced by *B. coli*, and cause secondary infections and ulcer. Malnutrition, alcoholism, or a compromised immune function may reduce host's resistance to *B. coli*, contributing to infection and disease.

14.1.2.3 Treatment

B. coli is sensitive to a number of drugs, including metronidazole, tetracycline, iodoquinol, nitazoxanide, secnidazole, tinidazole, trimethoprim, sulfamethoxazole, sulfaguanidine, and sulfadimidine. Currently, tetracycline, metronidazole, and iodoquinol are the drugs of choice for treatment of balantidiasis in humans. Tetracycline is often prescribed at 500 mg, 4 times a day for 10 days; metronidazole (Flagyl) at 750 mg, 3 times a day for 5–7 days; and iodoquinol at 650 mg, 3 times daily for 20 days. Because most cases of *B. coli* infection are asymptomatic, it is important to treat asymptomatic individuals as well to halt further transmission of the disease.

As it is transmitted by a fecal–oral route, Balantidiasis is preventable by improving hygiene practices, water sanitation, and proper disposal of fecal material. More specifically, pigs, the main animal reservoir of *B. coli* cysts, should be fenced and provided with a separate water source from humans. In the absence of water cleaning facilities, humans should seek water from sources far from pig habitation and human toilets, and drink only boiled water. Human waste and animal waste should be disposed properly. Personal hygiene such as hand washing should be maintained. In the laboratory setting, precautions should be taken in handling formed porcine or human fecal matter that might contain *B. coli* cysts. Sodium hypochlorite (1%) is an effective disinfectant for *B. coli* cysts which may survive drying on bench tops, and instruments.

14.1.3 Diagnosis

Balantidiasis is a relatively uncommon infection, producing nonspecific symptoms such as diarrhea, dysentery, and abdominal pain. Patients with a history of working closely with pigs or other livestock, recent travel to a region with poor water sanitation, or contact with infected persons may be suggestive of balantidiasis.

Microscopic examination of stool (wet mount) for trophozoites in acutely infected patients and cysts in patients with chronic infections is critical for making a diagnosis of balantidiasis, which may resemble other conditions such as amoebic dysentery (amoebiasis).

To search for cysts in formed stools, a portion is dissolved in phosphate-buffered saline (PBS) or fixative (10% phosphate-buffered formalin or polyvinyl alcohol) and filtered through gauze or a sieve to remove large pieces of debris. The resulting fluid is then examined with a phase-contrast microscope for internal structures of unstained living or fixed ciliates (e.g., the oral apparatus, uniform somatic ciliation, and the macronucleus). Stains such as Lugol's iodine (1:5 to 1:100 dilutions) and hematoxylin–eosin may be employed. However, heavily stained cysts may be occasionally mistaken for helminth ova. Methods for concentration of parasites from stool samples, including sedimentation and floatation, may also be utilized [36]. As *B. coli* is passed periodically, stool samples are collected over several days and examined immediately (or preserved) to make a definitive diagnosis. *Balantidium* cysts are larger (40–60 μm) than amoeba cysts (~25 μm in diameter) and are binucleate (macro- and micronuclei), and at times spinning trophozoites are observed within the cyst wall due to ciliary activity. On the other hand, trophic *Entamoeba* move on the slide surface by means of an anterior ectoplasmic pseudopod.

Examination of biopsy specimens from the rectum and the sigmoid colon (taken with the help of a sigmoidoscope) or of scrapings of an ulcer for trophozoites is also informative. Use of the hematoxylin–eosin or Diff-Quik stain is valuable for evaluating the extent of damage to the wall due to as *B. coli* or for distinguishing ciliocytophthoria (the degenerative fragments of epithelial cells).

Gastric mucin media (incorporating calf serum and rice powder) are useful for *Balantidium* growth and maintenance for as long as 30 months. Biphasic media in tubes containing an agar, inspissated egg yolk, or serum butt overlaid with nutrient medium as well as antibiotics (e.g., penicillin–streptomycin) permit growth of balantidia (at the bottom of the tube) under microaerophilic or anaerobic conditions. Diamond's trypticase-yeast extract-serum-gastric mucin (TYSGM) medium (containing trypticase (casein digest), yeast extract, serum, and porcine gastric mucin, with starch powder added to tubes at the time of inoculation) also supports the xenic growth of balantidia. Currently, there is a lack of an animal model for study of the pathogenicity of the ciliate [37].

As the identification of *Balantidium* species based on morphological characteristics is often difficult and sometimes insufficiently accurate, serological procedures such as indirect fluorescence antibody assay may be used to enhance microscopic detection. More recently, molecular techniques targeting *Balantidium* rRNA gene, internal transcribed spacers 1 and 2 (ITS 1 and ITS 2), and other gene regions have been exploited for their identification and diagnosis [2,38–43].

14.2 METHODS

14.2.1 Sample Preparation

14.2.1.1 Sample Collection

Fecal samples are collected and fresh wet mounts are prepared to detect trophozoites of *B. coli* using a light microscope (at ×100 and ×400 magnifications). *B. coli* can be cultivated in a liquid medium of Ringer's solution (consisting of NaCl, KCl, $CaCl_2$, and $NaHCO_3$) supplemented with 10% heat-inactivated horse serum and a sprinkle of heat-sterilized rice starch at room temperature [43].

For scanning electron microscopy, motile trophozoites of *B. coli* are collected by centrifugation at 2000 × *g* for 5 min. The pellets are fixed with 2.5% glutaraldehyde, washed three times with PBS, secondarily fixed with 1% OsO_4, dehydrated in a series of ethanol solutions, substituted with isoamyl acetate and then critical-point-dried using CO_2. Samples are coated with gold and observed in Hitachi S-510 scanning electron microscope. For transmission electron microscopy, *B. coli* cells fixed with 2.5% glutaraldehyde and postfixed in 1% OsO_4 are washed with PBS, dehydrated in a graded series of acetone solutions, and gradually impregnated in Epon resin with heat polymerization. Semithin survey sections are cut with glass knives stained with toluidine blue and are used to orientate sections. Ultra-thin sections mounted on uncoated copper grids are stained with uranyl acetate and lead citrate and viewed in a JEOL JEM TEM 1010 electron microscope [43].

14.2.1.2 DNA Extraction

B. coli cells from culture tubes are pooled and genomic DNA is extracted using the Chelex method [43]. Alternatively, cells are lysed in a solution containing 10 mM Tris–HCl pH 8.0, 100 mM EDTA pH 8.0, 1% SDS, and proteinase K at a concentration of 50 μg/mL. The lysate is extracted 3 times with phenol/chloroform followed by precipitation with two volumes of 100% ethanol in 0.4 M LiCl. After 1 h of incubation at −20°C, the lysate is centrifuged for 20 min at 16,000 × *g*, and the pellet is washed twice with 1 mL of 70% ethanol and air dried for 10 min. The DNA is further purified with a second precipitation step, by incubating the sample with 1 vol of 25% polyethylene glycol in 2.5 mM NaCl, and the pellet is suspended in 10 mM Tris–HCl pH 7.5 before use as a template for the polymerase chain reaction (PCR) amplification [2].

14.2.2 Detection Procedures

Nilles-Bije and Rivera [43] utilized primers (Euk A: 5′-AACCTGGTTGATCCTGCCAGT-3′ and Euk B: 5′-TGATCCTTCTGCAGGTTCACCTAC-3′) [44] from the small subunit (16S) rRNA gene for PCR identification and phylogenetic analysis of *B. coli* isolates.

14.2.2.1 Procedure

1. PCR mixture is prepared using Go Taq® Hot Start Polymerase Kit (Promega) and primers Euk A and Euk B.
2. The cycling parameters include an initial denaturation step at 94°C for 130 s; 35 cycles of 94°C for 30 s, 50°C for 45 s, and 72°C for 130 s; and a final extension step at 72°C for 7 min.
3. The resulting amplicons (1543 bp) are separated by gel electrophoresis on a 1.5% agarose gel and visualized with ethidium bromide stain.
4. The amplicons are excised from the gel and purified using magnetic beads. The purified amplicons are sequenced using the same primers with ABI PRISM Big Dye® Terminator v1.1 Cycle Sequencing Kit (Applied Biosystems) on an automated sequencer (ABI PRISM 3100, Applied Biosystems).
5. Sequences are uploaded into the Basic Local Alignment Search Tool program to search for the most similar reference sequences, and are aligned and removed of ambiguous nucleotide positions along with reference sequences. The trees are constructed using the model-based neighbor-joining (NJ) as well as the nonmodel-based maximum parsimony (MP), with *Dasytricha ruminantium* as an outgroup. The program PAUP* v. 4.0b10 is used for NJ and MP analyses with 1000 bootstrap replicates. Clusters in the phylogenetic trees are considered valid if these have bootstrap support values of >50% from NJ and MP analyses. The trees were viewed using the version 1.6.6 of TreeView.

14.3 CONCLUSION

B. coli is a commensal organism in the large intestine of pigs and other mammals. Although *B. coli* appears harmless to pigs, its infective cysts are discharged with the pig feces, contaminating water sources. Subsequent ingestion of *B. coli* cysts by humans facilitates excystation and development of invasive trophozoites, which penetrate the linings of the mucosa and submucosa of the damaged intestine, enter the lymphoid tissue and travel to other parts of the body. A range of clinical manifestations are associated with *B. coli* infection (balantidiasis), including diarrhea, dysentery, cramping, halitosis, nausea, colitis, and abdominal pain, and so on, especially in malnourished and immunosuppressed individuals. To facilitate effective antibiotic treatment, accurate identification of *B. coli* cysts and trophozoites in stool is essential. Considering the difficulty in determining *Balantidium* species on morphological criteria alone, molecular analysis of rRNA and other gene sequences has played an increasingly important role in the diagnosis and management of balantidiasis.

REFERENCES

1. Adl, S.M. et al., The new higher level classification of eukaryotes with emphasis on the taxonomy of protists. *J. Eukaryot. Microbiol.*, 52, 399, 2005.
2. Grim, J.N., and Buonanno, F., A re-description of the ciliate genus and type species, *Balantidium entozoon*. *Eur. J. Protistol.*, 45, 174, 2009.
3. Grim, J.N., Description of somatic kineties and vestibular organization of *Balantidium jocularum*, sp. n., possible taxonomic implications for the class Litostomatea and the genus *Balantidium*. *Acta Protozool.*, 32, 37, 1993.
4. Grim, J.N., Clements, K.D., and Byfield, T., New species of *Balantidium* and *Paracichlidotherus* (Ciliophora) inhabiting the intestines of four surgeonfish species from the Tuvalu Islands, Pacific Ocean. *J. Eukaryot. Microbiol.*, 49, 146, 2002.
5. Li, M. et al., First report of two *Balantidium* species from the Chinese giant salamander, *Andrias davidianus*: *Balantidiuim sinensis* Nie 1935 and *Balantidium andianusis* n. sp. *Parasitol. Res.*, 102, 605, 2008.
6. Schuster, F.L., and Ramirez-Avila, L., Current world status of *Balantidium coli*. *Clin. Microbiol. Rev.*, 21, 626, 2008.
7. Skotarczak, B., Ultrastructural and cytochemical identification of peroxisomes in *Balantidium coli*, Ciliophora. *Folia Biol.*, 45, 117, 1997.
8. Hindsbo, O. et al., Age-dependent occurrence of the intestinal ciliate *Balantidium coli* in pigs at a Danish research farm. *Acta Vet. Scand.*, 41, 79, 2000.
9. Yang, Y. et al., Diarrhoea in piglets and monkeys experimentally infected with *Balantidium coli* isolated from human faeces. *J. Trop. Med. Hyg.*, 98, 69, 1995.
10. Nakauchi, K., The prevalence of *Balantidium coli* infection in fifty-six mammalian species. *J. Vet. Med. Sci.*, 61, 63, 1999.
11. Cox, F.E., Human balantidiasis in Iran: Are camels reservoir hosts? *Trends Parasitol.*, 21, 553, 2005.
12. Headley, S.A., Kummala, E., and Sukura, A., *Balantidium coli*-infection in a Finnish horse. *Vet. Parasitol.*, 158, 129, 2008.
13. Bilal, C.Q. et al., Prevalence and chemotherapy of *Balantidium coli* in cattle in the River Ravi region, Lahore (Pakistan). *Vet. Parasitol.*, 163, 15, 2009.
14. Pomajbíková, K. et al., Discrepancies in the occurrence of *Balantidium coli* between wild and captive African great apes. *J. Parasitol.*, 96, 1139, 2010.
15. Karanis, P., Kourenti, C., and Smith, H., Waterborne transmission of protozoan parasites: A worldwide review of outbreaks and lessons learnt. *J. Water Health*, 5, 1, 2007.
16. Esteban, J.-G. et al., Balantidiasis in Aymara children from the northern Bolivian altiplano. *Am. J. Trop. Med. Hyg.*, 59, 922, 1998.
17. Maleky, F., Case report of *Balantidium coli* in human from south of Tehran, Iran. *Indian J. Med. Sci.*, 52, 201, 1998.
18. Baskerville, L., Ahmed, Y., and Ramchand, S., *Balantidium* colitis. Report of a case. *Am. J. Dig. Dis.*, 15, 727, 1970.
19. Francesco, R., and Giannotti, T., *Balantidium coli* in cervicovaginal cytology. A case report. *Pathologica*, 75, 439, 1983.
20. Currie, A.R., Human balantidiasis. A case report. *S. Afr. J. Surg.*, 28, 23, 1990.
21. Dodd, L.G., *Balantidium coli* infestation as a cause of acute appendicitis. *J. Infect. Dis.*, 163, 1392, 1991.
22. Cermeño, J.R. et al., *Balantidium coli* in a HIV-infected patient with chronic diarrhea. *AIDS*, 17, 941, 2003.
23. Barsoum, R.S., Parasitic infections in organ transplantation. *Exp. Clin. Transplant.*, 2, 258, 2004.
24. Agapov, M., Balantidiasis in a patient with suspected Crohn's disease. *Endoscopy*, 38, 655, 2006.
25. Ladas, S.D. et al., Invasive balantidiasis presented as chronic colitis and lung involvement. *Dig. Dis. Sci.*, 34, 1621, 1989.
26. Oberhuber, G., Karpitschka, G., and Stolte, M., *Balantidium coli*: A rare cause of colonic ulcer. *Eur. J. Gastroenterol. Hepatol.*, 5, 755, 1993.
27. Gonzalez de Canales Simon, P. et al., Colonic balantidiasis. *Gastroenterol. Hepatol.*, 23, 129, 2000.
28. Dorfman, S., Rangel, O., and Bravo, L.G., Balantidiasis: Report of a fatal case with appendicular and pulmonary involvement. *Trans. R. Soc. Trop. Med. Hyg.*, 78, 833, 1984.
29. Clyti, E. et al., A case of coinfection by *Balantidium coli* and HIV in French Guiana. *Bull. Soc. Pathol. Exot.*, 91, 309, 1998.
30. Vasilakopolou, A. et al., *Balantidium coli* pneumonia in an immunocompromised patient. *Scand. J. Infect. Dis.*, 35, 144, 2003.
31. Ferry, T. et al., Severe peritonitis due to *Balantidium coli* acquired in France. *Eur. J. Clin. Microbiol. Infect. Dis.*, 23, 393, 2004.
32. Sharma, S., and Harding, G., Necrotizing lung infection caused by the protozoan *Balantidium coli*. *Can. J. Infect. Dis.*, 14, 163, 2003.
33. Anargyrou, K. et al., Pulmonary *Balantidium coli* infection in a leukemic patient. *Am. J. Hematol.*, 73, 180, 2003.
34. Koopowitz, A. et al., *Balantidium coli*-induced pulmonary haemorrhage with iron deficiency. *S. Afr. Med. J.*, 100, 534, 2010.
35. Yazar, S. et al., Dysentery caused by *Balantidium coli* in a patient with non-Hodgkin's lymphoma from Turkey. *World J. Gastroenterol.*, 10, 458, 2004.
36. Maino, A. et al., Urinary balantidiasis: Diagnosis at a glance by urine sediment examination. *J. Nephrol.*, 23, 732, 2010.
37. Clark, C.G., and Diamond, L.S., Methods for cultivation of luminal parasitic protists of clinical importance. *Clin. Microbiol. Rev.*, 15, 329, 2002.
38. Lynn, D.H., and Sogin, M.L., Assessment of phylogenetic relationships among ciliated protists using partial ribosomal RNA sequences derived from reverse transcripts. *Biosystems*, 21, 249, 1988.
39. Lynn, D.H., Morphology or molecules: How do we identify the major lineages of ciliates (Phylum Ciliophora)? *Eur. J. Protistol.*, 39, 356, 2003.
40. Strüder-Kypke, M.C. et al., Molecular phylogeny of litosome ciliates (Ciliophora, Litostomatea) with emphasis on free-living haptorian genera. *Protist*, 157, 261, 2006.
41. Strüder-Kypke, M.C., Kornilova, O.A., and Lynn, D.H., Phylogeny of trichostome ciliates (Ciliophora, Litostomatea) endosymbiontic in the Yakut horse (*Equus caballus*). *Eur. J. Protistol.*, 43, 319, 2007.
42. Gao, S. et al., Phylogeny of six genera of the subclass Haptoria (Ciliophora, Litostomatea) inferred from sequences of the gene coding for small subunit ribosomal RNA. *J. Eukaryot. Microbiol.*, 55, 562, 2008.
43. Nilles-Bije, M.L., and Rivera, W.L., Ultrastructural and molecular characterization of *Balantidium coli* isolated in the Philippines. *Parasitol. Res.*, 106, 387, 2010.
44. Medlin, L. et al., The characterization of enzymatically amplified eukaryotic 16S-like rRNA-coding regions. *Gene*, 71, 491, 1988.

15 Cryptosporidium

Una M. Ryan and Rongchang Yang

CONTENTS

15.1 INTRODUCTION

Cryptosporidium is an important enteric parasite and has been identified as the cause of numerous waterborne, foodborne, and day-care outbreaks of diarrheal disease worldwide [1]. The oocyst is the environmentally stable stage and is able to survive and penetrate routine wastewater treatment and is resistant to inactivation by commonly used drinking water disinfectants [2]. Currently, cryptosporidiosis represents the major public health concern of water utilities in developed nations [3] and has been responsible for 50.8% (165/325) of all water-associated outbreaks of parasitic protozoan disease reported in North American and Europe [4].

15.1.1 Classification, Morphology, and Biology

Although originally classified as a coccidian parasite [5], *Cryptosporidium* demonstrates several peculiarities that separate it from any other coccidian. These include

1. The location of *Cryptosporidium* within the host cell where the endogenous developmental stages are confined to the apical surfaces of epithelial cells (intracellular but extracytoplasmic).
2. The attachment of the parasite to the host cell where a multimembranous attachment or feeder organelle is formed at the base of the parasitophorous vacuole to facilitate the uptake of nutrients from the host cell.
3. The presence of two morphofunctional types of oocyst, thick walled and thin walled, with the latter responsible for the initiation of the autoinfective cycle in the infected host.
4. The small size of the oocyst (7.4×5.6 μm for *C. muris* and 5.0×4.5 μm for *C. parvum*) which lacks morphological structures such as sporocyst, micropyle, and polar granules.
5. The insensitivity of *Cryptosporidium* to all anticoccidial agents tested so far from *Cryptosporidium* species [6].

Molecular and biological evidence suggests that *Cryptosporidium* is more closely related to gregarine parasites [6–10]. More recently, phylogenetic comparisons between the gregarine, *Ascogregarina taiwanensis* and *Cryptosporidium*, inferred from a concatenated protein sequence comprised of 10,750 amino acid positions, as well as the large subunit rRNA genes, robustly supported the phylogenetic affinity of *Ascogregarina* with *Cryptosporidium* at the base of the apicomplexan clade [11].

Due to the small size of the oocysts (4–6 μm) and the lack of morphological differences between most species, genetic characterization has become an essential element in defining new species and validating established species. Currently, 24 species of *Cryptosporidium* are regarded as valid (Table 15.1) and over 40 genotypes are recognized, with new genotypes continually being identified [12–15]. Five *Cryptosporidium* species/genotypes are commonly identified in humans: *C. hominis*, *C. parvum*, *C. meleagridis*, *C. felis*, and *C. canis*, with *C. hominis* and *C. parvum* responsible for the majority of infections [15], although, in some endemic areas *C. meleagridis* infection rate is as high as *C. parvum* [16].

Upon ingestion by a suitable host, the sporulated oocysts, which contain four sporozoites, excyst and undergo a complex monoxenous life cycle where all stages of development

TABLE 15.1
Currently Recognized Species of *Cryptosporidium*

Species	Major Host	Minor Host	Site of Infection
C. hominis	Humans	Dugongs, sheep, cattle	Small intestine
C. parvum	Humans, livestock		Small intestine
C. meleagridis	Birds, humans		Small intestine
C. tyzzeri	Mice		Small intestine
C. cuniculus	Rabbits	Humans	Small intestine
C. fayeri	Red Kangaroo		Unknown
C. marcopodum	Gray Kangaroo		Unknown
C. bovis	Cattle	Sheep	Small intestine
C. ubiquitum	Deer, sheep, wildlife	Humans	Small intestine
C. ryanae	Cattle		Small intestine
C. xiaoi	Sheep		Small intestine
C. suis	Pigs	Humans	Small intestine
C. felis	Cats	Humans	Small intestine
C. canis	Dogs	Humans	Small intestine
C. wrairi	Guinea pigs		Small intestine
C. baileyi	Poultry	Wide variety of bird species	Bursa
C. galli	Finches, chickens	Capercailles, pine grosbeaks and others	Proventriculus
C. muris	Rodents		Stomach
C. andersoni	Cattle	Sheep	Abomasum
C. fragile	Frogs		Stomach
C. serpentis	Snakes, lizards		Stomach
C. varani	Lizards	Snakes	Stomach and small intestine
C. molnari	Fish		Stomach
C. scophthalmi	Fish		Intestine

(asexual and sexual) occur within one host. Following fertilization, oocysts may differentiate into thin-walled oocysts, which are involved in autoinfection, or thick-walled oocysts which exit the host via the feces, permitting further transmission via contaminated food/water/fomites and direct contact with infected animals/humans or contaminated surfaces. Thick-walled oocysts are environmentally robust and are highly resistant to environmental stresses, including chlorine treatment of community water supplies. Oocysts of some species such as *C. parvum* can remain infectious in cool, wet conditions for 6 months or longer [17].

15.1.2 Clinical Features and Pathogenesis

In immunocompetent persons, *Cryptosporidium* species infect the intestinal and respiratory epithelial cells resulting in self-limiting diarrhea or pulmonary cryptosporidiosis, with intestinal infection most common. *Cryptosporidium* can involve either children or adults but is a major cause of diarrhea in children (<5 years old) in both developed and developing countries [15,18,19]. There is wide variability in the clinical presentation of infection and patients may be asymptomatic or experience acute or chronic diarrhea depending on their age and immune status [12,15,19]. Immunocompromised individuals can experience severe chronic diarrhea and respiratory cryptosporidiosis that frequently leads to wasting and sometimes death [20]. Infection rarely resolves spontaneously in these individuals. Studies also indicate that immunocompromised individuals are more susceptible to infection from a wider range of *Cryptosporidium* species and genotypes [21,22].

Currently, nitazoxanide (NTZ) is approved for treatment of cryptosporidiosis in children and immunocompetent adults in the United States; however, NTZ is not effective without an appropriate immune response and is therefore ineffective for the treatment of immunocompromised individuals [23].

The attachment of a sporozoite to the apical surface of epithelial cells leads to loss of the surface epithelium, causing changes in the villous architecture, with villous atrophy, blunting and crypt cell hyperplasia, and mononuclear cell infiltration in the lamina propia [24]. Osmotic, inflammatory, and secretory aspects of diarrhea have all been investigated for cryptosporidiosis; yet, the causative mechanisms have not been fully elucidated. They are, however, multifactorial, consisting of the effect of the parasite and its products on the epithelial layer and the immunological and inflammatory responses of the host, leading to impaired intestinal absorption and enhanced secretion [24]. The nature and intensity of the diarrhea suggests enterotoxin involvement but none has been demonstrated [24,25].

It has been shown that the apoptotic response of infected intestinal epithelial cells is actively suppressed by *C. parvum* via upregulation of a host antiapoptotic protein (survivin), favoring parasite infection [26]. Another study, which examined Se-deficient and Se-adequate mouse models, reported that a nutritional selenium (Se) deficiency can lead to a higher susceptibility of the host to *Cryptosporidium* infection and decreased immune responses [27]. Recently, the role played by human mannose-binding lectin (MBL) (a protein that binds to microbial surfaces and promotes opsonophagocytosis) in a large prospective cohort of Bangladeshi preschool children was investigated [28]. MBL is known to bind to *Cryptosporidium* sporozoites and analysis of *Cryptosporidium* infections revealed that serum MBL deficiency and polymorphisms in the MBL gene (MBL2) were strongly associated with *Cryptosporidium* infections, particularly recurrent infection [28].

15.1.3 Diagnosis

Detection of *Cryptosporidium* involves a variety of methods including microscopy, immunofluorescent antibody (IFA) staining, and other immunological detection methods such as enzyme-linked immunosorbent assay (ELISAs) and DNA-based detection methods.

15.1.3.1 Conventional Techniques

The method of detection of *Cryptosporidium* oocysts in environmental, water, food, and fecal samples that still dominates in most diagnostic laboratories is conventional microscopy and diagnosis by recognition of the morphological features of oocysts. However, morphological characters for identifying *Cryptosporidium* are few, and identification based on light microscopy alone is unreliable and relatively time consuming [29]. When using conventional microscopy, differential staining techniques are usually required due to the fact that oocysts are similar in size and shape to yeasts, fecal components, and other debris [30]. A safranin–methylene blue stain has been described [31], which stains oocysts red and yeasts and other fecal debris blue. Acid-fast stains such as Kinyoun Ziehl–Neelsen and dimethyl sulfoxide (DMSO)-carbol fuchsin [32–34], stain oocysts red with the background counterstained and are commonly used in many laboratories. Differential staining techniques, however, are time consuming and vary in their sensitivity and specificity [31,35,36]. In addition, unless samples have been fixed overnight in formalin, oocysts may frequently take up the counterstain. Negative staining techniques using nigrosin [34], light green and merbromide [37], and malachite green [38], which stain background yeasts and bacteria but not the oocysts, have also been developed. Negative staining methods are faster but are considered by some to be less sensitive than conventional staining techniques [39,40]. In general, detection methods that rely on conventional microscopy lack sensitivity, are time consuming, and require an experienced microscopist due to the resemblance of oocysts to background fecal debris, which can make identification difficult.

Immunological detection methods offer superior specificity [41], and IFA staining has been widely used for the detection of *Cryptosporidium* in fecal smears, water, and food [42–44]. Both polyclonal and monoclonal antibodies (MAbs) have been raised against surface-exposed epitopes on purified oocysts and fluorescently labeled antibody preparations are readily available commercially. A limitation of polyclonal immunodetection methods is that they suffer from cross-reaction with other organisms [6,45].

MAbs offer increased specificity and both direct and indirect detection methods are available. Commercially available fluorescein isothiocyanate (FITC)-MAbs are used routinely for detecting and enumerating oocysts in environmental samples [44,46]. There are no commercially available antibody preparations specific for epitopes on human pathogen *Cryptosporidium* oocysts. Different MAbs used commercially recognize different sets of surface epitopes [47], and most, if not all, commercially available MAbs have been raised against epitopes on a limited number of *C. parvum* isolates [44]. Therefore, if the FITC-MAb supplied in the kit does not bind to a particular isolate due to antigenic variability of the oocyst wall epitopes, false negatives can occur. Oocysts of different species and genotypes of *Cryptosporidium* can also express a range of oocyst wall epitopes and differences in staining intensity has been observed in purified *C. hominis* and *C. parvum* oocyst isolates using three commercially available FITC-MAb preparations [44]. False negatives can also occur due to the fact that the target epitopes can be easily stripped from the organism by mild environmental perturbations [48]. In general, IgG MAbs appear to have increased avidity to oocyst surface antigens compared with IgM, allowing more efficient oocyst concentration, resulting in increased recoveries using immuno-magnetic separation (IMS) from water samples with high turbidity [49]. IgG1 antibodies have been shown to achieve a better diagnostic specificity compared with IgG3 or IgM for the detection of *Cryptosporidium* oocysts in environmental samples [50].

ELISAs and enzyme immunoassays (EIAs) for *Cryptosporidium* are also available and have the advantage of reducing assay times and being amenable to automation. An immunochromatographic (dipstick) assay is also available. However, the sensitivity of all immunological-based methods is relatively low. For example, between 1×10^4 and 5×10^4 oocysts per gram of unconcentrated feces are necessary to obtain 100% detection efficacy using either conventional differential staining or commercially available FITC-MAbs methods [51]. The sensitivity of ELISAs has been reported to be as low as 3×10^5 to 10^6 oocysts per mL [52]. Biosensor chips that detect and quantitate *C. parvum* in real time via anti-*C. parvum* IgM binding have also been developed [53,54].

A major limitation of all these techniques, however, is their inability to speciate due to morphological similarity between many species. For example, oocysts of *C. parvum*, *C. hominis*, *C. meleagridis*, and *C. bovis* have no apparent internal features that are unique among them and they overlap in size, ranging from 4.2–4.8 to 4.8–6.0 μm [14,55].

15.1.3.2 Molecular Techniques

Since the development of the polymerase chain reaction (PCR), PCR-based techniques have permitted specific and sensitive detection of *Cryptosporidium* spp. for clinical diagnosis and environmental monitoring [15,44]. The advantages of PCR-based detection of *Cryptosporidium* in clinical and environmental samples are its high sensitivity, specificity, and rapidity enabling large numbers of samples to be analyzed at a time; ease of use; relatively low cost; and ability to genotype, therefore, eliminating false positives due to cross-contamination and allowing the detection of the range of pathogens present [56,57]. The sensitivity and specificity of PCR-based detection technique allow for the detection of a single oocyst in clinical and environmental samples [58,59].

There is no standard locus recommended for detecting *Cryptosporidium*, and a wide variety of loci have been utilized (Table 15.2), but PCR-RFLP or sequence analysis of the 18S rRNA gene is the most appropriate method as, due to its conserved nature, it results in amplification of all *Cryptosporidium* species and genotypes. A nested PCR-RFLP 18S assay has been developed which is widely used and can detect most species and genotypes of *Cryptosporidium* [60,61]. The primary PCR product produced by the external primers is 1325 bp long and the internal product is 826 bp. The Xiao et al. primers are specific for *Cryptosporidium* but as they amplify a relatively large PCR product, samples that contain very low numbers of oocysts, or partially degraded DNA, will not always be amplified. As the numbers of oocysts extracted from water samples is typically low, primers which amplify a smaller 18S rRNA

TABLE 15.2
Some of the Loci Used to Detect *Cryptosporidium*

Locus	Product Size (bp)	Type of PCR	Reference
Acetyl-CoA	347	Nested PCR	[83]
Actin	1095	Nested PCR	[84]
Beta-tubulin	~592	PCR	[85]
COWP	553	PCR-RFLP	[86]
COWP	1033	Nested PCR	[87]
DHFR	408	Nested PCR-RFLP	[88]
DHFR		Nested PCR	[89]
DHFR	357 (*C. hominis*), 190 (*C. parvum*)	Species-specific PCR	[90]
HSP-70	1950	Nested PCR	[91]
HSP-90	~835–844	Nested PCR-RFLP	[92]
ITS1 of rRNA	513 (*C. hominis*), 519 (*C. parvum*)	Species-specific PCR	[93]
Poly-T	318	PCR-RFLP	[94]
TRAP-C1	1200	PCR-RFLP	[95]
18S rRNA	556	PCR-RFLP	[96]
18S rRNA	445	PCR	[97]
18S rRNA	1750	PCR-RFLP	[98]
18S rRNA	~825	Nested PCR-RFLP	[60]
18S rRNA	590–593	Nested PCR-RFLP	[59]
18S rRNA	597	Nested PCR	[62]
Unknown	1353	PCR-RFLP	[99]
Unknown	*C. hominis* (411), 311 (*C. parvum*)	Species-specific PCR	[100]

PCR product offer more sensitivity [62]. PCR tools based on antigen genes, housekeeping genes such as the heat shock protein 70 (HSP70), or the *Cryptosporidium* oocyst wall protein (COWP) gene can sometimes fail to detect the DNA of some of the human pathogenic species, including *C. felis*, *C. canis*, and *C. muris*, and are therefore more effectively used as secondary tools for detecting common species such as *C. hominis* and *C. parvum* [63].

Real-time quantitative PCR (RT-qPCR) has overcome several limitations of the conventional PCR and allows for the quantitation of the number of starting copies of a specific template in the sample. In RT qPCR, the amplified product is detected using fluorescent dyes. These fluorescent dyes are linked to oligonucleotide probes and/or primers, which bind specifically to or are incorporated into, the amplified PCR product. Changes in the fluorescence intensities are monitored during the PCR reaction (in "real time") during the exponential phase. The threshold cycle (or Ct) when the product is first detected is compared with a standard curve (based on Ct value versus known concentrations of DNA standards) to estimate the amount of material in the test sample. In addition, the whole process is faster than regular PCR as no electrophoresis is required, reducing the possibility for cross-contamination and false-positive results [64]. Quantitative real-time PCR detection methods have been developed to quantitate the numbers of *Cryptosporidium* oocysts present in fecal and water samples [57,65,66]. qPCR can also be coupled with high-resolution melt (HRM) curve analysis to differentiate different species of *Cryptosporidium*. HRM curve analysis involves the gradual dissociation of relatively AT-rich regions ("lower" melting domains) within the amplicon prior to the ultimate melting. Single-nucleotide polymorphisms or heteroduplexes can result in slight but measurable changes in the HRM profile by affecting amplicon dissociation [67,68].

The development of subgenotyping tools has facilitated the tracking of transmission in outbreak situations. The term "subgenotype" refers to genetic differences *within* *Cryptosporidium* species/genotypes (i.e., variation within *C. hominis* or *C. parvum*, etc.) [15,69]. Subgenotype analysis is more informative than genotyping because it can elucidate genetic variation on a finer scale and is therefore more accurate to track contamination sources in outbreak investigations or contamination of food. Subgenotyping is also important in assessing the zoonotic potential of the parasite, as it has been shown that some subtypes of *C. parvum* only circulate within the human population [15].

Currently, there are several subgenotyping tools developed for the genetic characterization of *Cryptosporidium*, including microsatellite analysis, sequence analysis of the GP60 gene, the HSP70 gene, a double-stranded (ds) RNA virus that is present in most *Cryptosporidium* species, a 47-kDa protein (CP47), a mucin-like protein (Mucin1), a serine repeat antigen (MSC6-7), a 56-kDa transmembrane protein (CP56), and a T-rich gene fragment (Chrom3T) [15,69,70].

Microsatellites are short sequence repeats that constitute a rich source of polymorphisms, which has subsequently led them to be used extensively for subgenotyping/fingerprinting [71,72]. Microsatellites located in noncoding regions are more likely to accumulate mutations than protein-coding sequences, and therefore have been chosen for subgenotyping studies [73]. The GP60 gene, also known as Cpgp15/45, exhibits a high degree of sequence polymorphism, far greater than any other coding *Cryptosporidium* locus examined to date [74,75]. It is a single-copy gene encoding a precursor protein that is proteolytically cleaved during the intracellular stages of the parasite to yield mature cell surface glycoproteins, known as gp45 and gp15. Both glycoprotein products are expressed on the apical and surface regions of the invasive stages of the *Cryptosporidium* life cycle and have been implicated in the attachment to and invasion of host cells [74,76,77]. There are a number of significant and epidemiologically important advantages of using sequence analysis of the GP60 gene, including (i) the high degree of sequence polymorphism, (ii) the full comparability of data across laboratories, (iii) the availability of data from several waterborne and food-borne outbreaks, and (iv) the biological relevance [74,77]. An international interlaboratory comparative trial revealed that GP60 subgenotyping is a more reliable method than microsatellite analysis as a fingerprinting tool for *Cryptosporidium* [78].

Within *C. hominis*, *C. parvum*, and *C. meleagridis*, several GP60 subtype families have been identified: 7 subtype families in *C. hominis* (Ia–Ig), 2 zoonotic (IIa, IId), and 10 nonzoonotic (IIb, IIc, IIe–IIl) subtype families in *C. parvum*, and 6 subtype families in *C. meleagridis* (IIIa–IIIf) [15,75]. There are also further "subtypes" within each subgenotype, which mostly differ from each other in the number of TCA repeats, and each subgenotype may contain more than one subtype [15,75].

15.2 METHODS

15.2.1 Sample Preparation

Due to the sensitivity of the nested PCR methods employed for the detection of *Cryptosporidium*, total DNA is usually extracted from *Cryptosporidium*-positive fecal samples without prior purification of the oocysts. Various methods are available for purification of oocysts from the feces, however, including discontinuous density sucrose gradient centrifugation, isopycnic Percoll centrifugation, cesium chloride gradient centrifugation, or ficoll gradient centrifugation [41,79,80].

For DNA extraction, various methods have been compared and the QIAamp DNA stool minikit (Qiagen) and the FastDNA® SPIN kit (BIO 101) give comparable results. The addition of 400 ng/μL of nonacetylated BSA (Sigma) to the PCR results in significant reduction of PCR inhibitors. For DNA extraction using the QIAamp DNA stool minikit, the following modification should be used. Extract DNA from 100 mg of feces by adding 180 μL of the ATL buffer from the QIAamp DNA minikit, transfer the suspension into a 1.5-mL Eppendorf tube and subject to five cycles of freeze–thaw (–70°C for 30 min and 56°C for 30 min). Continue with the extraction using the manufacturer-recommended procedures from this point. If using the FastDNA SPIN kit for extraction from feces, dilute 100 mg of feces to 0.5 mL with H_2O and transfer into a 2-mL tube containing the lysing matrix E from the FastDNA SPIN kit. Add 978 μL of sodium phosphate buffer and 122 μL of MT buffer from the kit and vortex in a FastPrep instrument (BIO 101) for 30 s at the speed setting 5.5. Process samples further in accordance with the manufacturer-recommended procedures.

15.2.2 Detection Procedures

15.2.2.1 Nested PCR Targeting a 825-bp 18S rDNA Fragment

15.2.2.1.1 Principle

A two-step nested PCR protocol was described by Xiao et al. [60] to amplify an 826-bp region of the 18S rDNA gene.

15.2.2.1.1.1 Procedure

1. Primary PCR: A PCR product of about 1325 bp is amplified using primers 5′-TTCTAGAGCTAATAC ATGCG-3′ and 5′-CCCATTTCCTTCGAAACAG GA-3′. PCR components: water, 10 μL of 10× PCR buffer, 6 mM $MgCl_2$, 200 μM (each) deoxynucleoside triphosphate (dNTP), 100 nM (each) primer, 2.5 U of *Taq* polymerase, and 0.25 to 1 μL of DNA template in a total 100 μL reaction mixture. Amplification conditions: a total of 35 cycles, each consisting of 94°C for 45 s, 55°C for 45 s, and 72°C for 1 min, with an initial hot start at 94°C for 3 min and a final extension at 72°C for 7 min.
2. Secondary PCR: A secondary PCR product of 825–864 bp (depending on the species/genotype) is then amplified from 2 μL of the primary PCR with primers 5′-GGAAGGGTTGTATTTATTAGATAA AG-3′ and 5′-AAGGAGTAAGGAACAACCTC CA-3′. The PCR and cycling conditions are identical to primary PCR, except that 3 mM $MgCl_2$ is used.
3. RFLP analysis: 20 μL of the secondary PCR product is digested in a total of 50 μL of reaction mixture, consisting of 20 U of *Ssp* I or *Vsp* I and 5 μL of respective restriction buffer at 37°C for 1 h, under conditions recommended by the supplier. The digested products should then be fractionated on 2.0% agarose gels.

Note: Table 15.3 lists the predicted RFLP fragment sizes for some species and genotypes of *Cryptosporidium*. *Vsp* I digestion is required to differentiate between *C. parvum* and *C. hominis* parasites. Because *C. andersoni* and *C. muris* had identical *Ssp* I and *Vsp* I restriction patterns, they can be differentiated by digesting secondary PCR products with 20 U of *Dde* I at 37°C for 1 h under conditions recommended by the supplier [61].

15.2.2.2 Nested PCR Targeting a 587-bp 18S rDNA Fragment

15.2.2.2.1 Principle

A two-step nested PCR protocol was described by Ryan et al. [62] to amplify a 587-bp fragment of the 18S rDNA gene.

15.2.2.2.1.1 Procedure

1. Primary PCR: a PCR product of 763 bp is amplified using the forward primer 18SiCF2 (5′-GAC ATA TCA TTC AAG TTT CTG ACC-3′) (bp position 292) and the reverse primer 18SiCR2 (5′-CTG AAG GAG TAA GGA ACA ACC-3′) (bp position 1007). PCR components: water, 200 μM each of dNTP, 1 × PCR buffer, 1.5 mM $MgCl_2$, 0.5 U of *Taq* DNA polymerase and 12.5 pmol of forward and reverse primers in a total of 25 μL reaction. PCR amplification: 45 cycles (94°C for 30 s, 58°C for 30 s, 72°C for 30 s) with an initial hot start (94°C for 5 min) and a final extension (72°C for 10 min).
2. Secondary PCR: a fragment of ~587 bp is amplified using 1 μL of primary PCR product and the nested forward 18SiCF1 (5′-CCT ATC AGC TTT

TABLE 15.3
Predicted RFLP Fragment Sizes Following Digestion with *Ssp* I and/or *Vsp* I for Various *Cryptosporidium* Species and Genotypes

Species	Source	PCR Fragment (bp)	*Ssp* I Digestion (bp)	*Vsp* I Digestion (bp)
C. andersoni	Cattle, camel, hyrax	833	385, 448	102, 731
C. serpentis	Snake	831	14, 33, 370, 414	102, 729
C. baileyi	Chicken	826	254, 572	102, 104, 620
C. felis	Cat	864	15, 33, 390, 426	102, 104, 182, 476
C. meleagridis	Turkey	833	11, 11, 108, 254, 449	102, 104, 171, 456
C. wrairi	Guinea pig	834	11, 11, 109, 254, 449	102, 104, 628
C. varanii	Desert monitor	834	19, 33, 109, 255, 418	102, 104, 628
C. hominis	Human	837	11, 12, 111, 254, 449	70, 102, 104, 561
C. parvum Type A	Bovine	834	11, 12, 108, 254, 449	102, 104, 628
C. parvum Type B	Bovine	831	9, 119, 254, 449	102, 104, 625
Mouse genotype I	Mouse	838	11, 12, 112, 254, 449	102, 104, 175, 457
C. canis	Dog	829	20, 33, 105, 254, 417	94, 102, 633
Ferret genotype	Ferret	837	11, 12, 111, 254, 449	102, 104, 174, 457
C. suis	Pig	838	9, 11, 365, 453	102, 104, 632
C. fayeri	Kangaroo, koala	837	33, 109, 254, 441	102, 104, 631

Source: Xiao et al. *Appl. Environ. Microbiol.* 65, 1578, 1999.

AGA CGG TAG G-3′) (bp position 289) and nested reverse 18SiCR1 (5′-TCT AAG AAT TTC ACC TCT GAC TG-3′) (bp position 851) primers. The PCR conditions for the secondary PCR are identical to the primary PCR. Sequence analysis is then used to identify to the species/genotype level.

15.2.2.3 Nested PCR Targeting the GP60 Locus

15.2.2.3.1 Principle

A two-step nested PCR was reported by Strong et al. [74] and Peng et al. [81] to amplify the GP60 locus.

15.2.2.3.1.1 Procedure

1. Primary PCR: a PCR product of ≈995 bp is amplified using the forward primer, StrongGP60 F1 (5′-ATG AGA TTG TCG CTC ATT ATC-3′), and the reverse primer, StrongGP60 R1 (5′-TTA CAA CAC GAA TAA GGC TGC-3′) [74] PCR reaction components: 0.01–1.0 ng of DNA should be amplified in a total volume of 25 μL reaction mixture consisting of water, 0.5 U of *Tth* polymerase, 10× DNA reaction buffer, 1.5 mM $MgCl_2$, 0.2 mM of each dNTP, and 12.5 pmol of forward and reverse primers. Amplification conditions: preliminary cycles of 94°C for 2 min, 56°C for 1 min, and 72°C for 2 min, followed by 45 PCR cycles of 94°C for 30 s, 56°C for 20 s, 72 C for 1 min, and a final extension of 72°C for 7 min.
2. Secondary PCR: The primers CDCGP60 F2 (5′-TCC GCT GTA TTC TCA AGC C-3′) and CDCGP60 R2 (5′-GCA GAG GAA CCA GCA TC-3′) are used to amplify a fragment of ≈832 bp [81]. The secondary PCR is set up as for the primary PCR. The secondary PCR cycling conditions are identical to that of the primary reaction, with the exception that the annealing temperature is reduced from 56°C to 54°C. Sequence analysis is then used to identify to the subtype level.

15.3 CONCLUSION AND FUTURE PERSPECTIVES

Cryptosporidium has emerged as an important gastrointestinal parasite of humans and animals, with a global distribution. It currently represents the major public health concern of water utilities in developed nations. It is also considered a potential bioterrorist agent (www3.niaid.nih.gov/topics/BiodefenseRelated/Biodefense/research/CatA.htm), because the oocysts produced by *Cryptosporidium* are extremely hardy, easily spread via water, resistant to chlorine, and difficult to inactivate or remove from water intended for consumption without the use of filtration [82].

A variety of molecular tools have been developed to detect and differentiate *Cryptosporidium* at the species/genotype and subtype levels [15]. These tools are now increasingly being used in characterizing the transmission of *Cryptosporidium* spp. in humans and tracking zoonotic transmission from animals. Results of these molecular epidemiologic studies have led to improved detection of *Cryptosporidium* as well as a better understanding of infection sources in humans and the public health importance of *Cryptosporidium* species/genotypes in various animals.

REFERENCES

1. Fayer, R. et al. Examination of naturally exposed bottlenose dolphins (*Tursiops truncatus*) for microsporidia, *Cryptosporidium*, and *Giardia*. *J. Parasitol.* 94, 143, 2008.
2. Rose, J.B. et al. Survey of potable water supplies for *Cryptosporidium* and *Giardia*. *Environ. Sci. Technol.* 25, 1393, 1991.
3. Fayer, R. et al. Epidemiology of *Cryptosporidium*: Transmission, detection and identification. *Int. J. Parasitol.* 30, 1305, 2000.
4. Karanis, P. et al. Waterborne transmission of protozoan parasites: A worldwide review of outbreaks and lessons learnt. *J. Water Health* 5, 1, 2007.
5. Levine, N.D. Taxonomy and review of the coccidian genus *Cryptosporidium* (protozoa, *apicomplexa*). *J. Protozool.* 31, 94, 1984.
6. Hijjawi, N.S. et al. Successful *in vitro* cultivation of *Cryptosporidium andersoni*: Evidence for the existence of novel extracellular stages in the life cycle and implications for the classification of *Cryptosporidium*. *Int. J. Parasitol.* 32, 1719, 2002.
7. Carreno, R.A. et al. *Cryptosporidium* is more closely related to the gregarines than to coccidia as shown by phylogenetic analysis of apicomplexan parasites inferred using small-subunit ribosomal RNA gene sequences. *Parasitol. Res.* 85, 899, 1999.
8. Hijjawi, N.S. et al. Complete development of *Cryptosporidium parvum* in host cell-free culture. *Int. J. Parasitol.* 34, 769, 2004.
9. Leander, B.S. et al. Molecular phylogeny and surface morphology of *Colpodella edax* (Alveolata): Insights into the phagotrophic ancestry of apicomplexans. *J. Eukaryot. Microbiol.* 50, 334, 2003.
10. Rosales, M.J. et al. Extracellular like-gregarine stages of *Cryptosporidium parvum*. *Acta Trop.* 95, 74, 2005.
11. Templeton, T.J. et al. Genome sequence survey for Ascogregarina taiwanensis supports evolutionary affiliation, but metabolic diversity between a gregarine and *Cryptosporidium*. *Mol. Biol. Evol.* 27, 235, 2010.
12. Fayer, R. et al. *Cryptosporidium ryanae* n. sp. (*Apicomplexa: Cryptosporidiidae*) in cattle (*Bos taurus*). *Vet. Parasitol.* 156, 191, 2008.
13. Fayer, R. and Santin, M. *Cryptosporidium xiaoi* n. sp. (*Apicomplexa: Cryptosporidiidae*) in sheep (*Ovis aries*). *Vet. Parasitol.* 164, 192, 2009.
14. Fayer, R. Taxonomy and species delimitation in *Cryptosporidium*. *Exp. Parasitol.* 124, 90, 2010.
15. Xiao, L. Molecular epidemiology of cryptosporidiosis: An update. *Exp. Parasitol.* 124, 80, 2010.
16. Cama, V.A. et al. *Cryptosporidium* species and genotypes in HIV-positive patients in Lima, Peru. *J. Eukaryot. Microbiol.* 50, S531, 2003.
17. Xiao, L., and Fayer, R. Molecular characterisation of species and genotypes of *Cryptosporidium* and *Giardia* and assessment of zoonotic transmission. *Int. J. Parasitol.* 38, 1239, 2008.
18. Xiao, L. et al. Genetic diversity of *Cryptosporidium* spp. in captive reptiles. *Appl. Environ. Microbiol.* 70, 891, 2004.
19. Warren, C.A., and Guerrant, R.L. Clinical disease and pathology. In: Fayer, R., and Xiao, L. (eds), *Cryptosporidium and Cryptosporidiosis*. CRC Press, Boca Raton, FL, p. 235, 2008.
20. Xiao, L. and Feng, Y. Zoonotic cryptosporidiosis. *FEMS Immunol. Med. Microbiol.* 52, 309, 2008.
21. Morgan, U. et al. Molecular characterization of *Cryptosporidium* isolates obtained from human immunodeficiency virus-infected individuals living in Switzerland, Kenya, and the United States. *J. Clin. Microbiol.* 38, 1180, 2000.
22. Guyot, K. et al. Molecular characterization of *Cryptosporidium* isolates obtained from humans in France. *J. Clin. Microbiol.* 39, 3472, 2001.
23. Gargala, G. Drug treatment and novel drug target against *Cryptosporidium*. *Parasite* 15, 275, 2008.
24. Farthing, M.J. Clinical aspects of human cryptosporidiosis. Contrib. *Microbiol. Immunol.* 6, 50, 2000.
25. Okhuysen, P.C. and Chappell, C.L. *Cryptosporidium* virulence determinants—Are we there yet? *Int. J. Parasitol.* 32, 517, 2002.
26. Liu, J. et al. Inhibition of apoptosis in *Cryptosporidium parvum*-infected intestinal epithelial cells is dependent on survivin. *Infect. Immun.* 76, 3784, 2008.
27. Wang, C. et al. Induced susceptibility of host is associated with an impaired antioxidant system following infection with *Cryptosporidium parvum* in Se-deficient mice. *PLoS One* 4, 4628, 2009.
28. Carmolli, M. et al. Deficient serum mannose-binding lectin levels and MBL2 polymorphisms increase the risk of single and recurrent *Cryptosporidium* infections in young children. *J. Infect. Dis.* 200, 1540, 2009.
29. Fall, A. et al. Morphology is not a reliable tool for delineating species within *Cryptosporidium*. *J. Parasitol.* 89, 399, 2003.
30. O'Donoghue, P.J. *Cryptosporidium* and cryptosporidiosis in man and animals. *Int. J. Parasitol.* 25, 139, 1995.
31. Baxby, D. et al. The development and performance of a simple, sensitive method for the detection of *Cryptosporidium* oocysts in faeces. *J. Hyg. (Lond)*, 93, 317, 1984.
32. Ma, P. and Soave, R. Three-step stool examination for cryptosporidiosis in 10 homosexual men with protracted watery diarrhoea. *J. Infect. Dis.* 147, 824, 1983.
33. Henricksen, S.A. and Pohlenz, J.F.L. Staining of *cryptospordia* by a modified Ziehl-Neelsen technique. *Acta Vet. Scand.* 22, 594, 1981.
34. Pohjola, S. et al. Dimethylsulphoxide-Ziehl-Neelsen staining technique for the detection of cryptosporidial oocysts. *Vet. Rec.* 115, 442, 1984.
35. Smith, H.V. et al. An analysis of staining methods for the detection of *Cryptosporidium* spp. oocysts in water-related samples. *Parasitology*, 99, 323, 1989.
36. Moodley, D. et al. *Cryptosporidium* infections in children in Durban. Seasonal variation, age distribution and disease status. *S. Afr. Med. J.* 79, 295, 1991.
37. Chichino, G. et al. New rapid staining methods of *Cryptosporidium* oocysts in stools. *J. Protozool.* 38, 212, 1991.
38. Elliot, A. et al. Improved staining method for detecting *Cryptosporidium* oocysts in stools using malachite green. *J. Gen. Appl. Microbiol.* 45, 139, 1999.
39. Casemore, D.P. et al. Laboratory diagnosis of cryptosporidiosis. *J. Clin. Pathol.* 38, 1337, 1985.
40. Garcia, L.S. et al. Techniques for the recovery and identification of *Cryptosporidium* oocysts from stool specimens. *J. Clin. Microbiol.* 18, 185, 1983.
41. Arrowood, M.J. and Sterling, C.R. Comparison of conventional staining methods and monoclonal antibody-based methods for *Cryptosporidium* oocyst detection. *J. Clin. Microbiol.* 27, 1490, 1989.
42. Geurden, T. et al. Prevalence and molecular characterisation of *Cryptosporidium* and *Giardia* in lambs and goat kids in Belgium. *Vet. Parasitol.* 155, 142, 2008.
43. Ortega, Y.R. et al. Efficacy of gaseous chlorine dioxide as a sanitizer against *Cryptosporidium parvum*, *Cyclospora cayetanensis*, and *Encephalitozoon intestinalis* on produce. *J. Food Prot.* 71, 2410, 2008.

44. Smith, A.L. Diagnostics. In: Fayer, R. and Xiao, L. (eds), *Cryptosporidium and Cryptosporidiosis*. CRC Press, Boca Raton, FL, p. 173, 2008.
45. Wang, Z. et al. Detection and genotyping of *Entamoeba histolytica, Entamoeba dispar, Giardia lamblia* and *Cryptosporidium parvum* by oligonucleotide microarray. *J. Clin. Microbiol.* 42, 3262, 2004.
46. Smith, H.V. et al. A microscopic system with a dual band filter for the simultaneous enumeration of *Cryptosporidium parvum* oocysts and sporozoites. *Water Res.* 37, 2525, 2003.
47. Smith, H.V. and Ronald, A. *Cryptosporidium*: The analytical challenge. In: Smith, H.V. and Thompson, K. (eds), *Cryptosporidium: The Analytical Challenge*. The Royal Society of Chemistry, Cambridge, p. 1, 2001.
48. Millar, C.B. et al. *Cryptosporidium* in foodstuffs—An emerging aetiological route of human foodborne illness. *Trends Food Sci. Technol.* 13, 168, 2002.
49. Paton, C.A. et al. Immunomagnetisable separation for the recovery of *Cryptosporidium* sp. oocysts. In Clark, S.A. et al. (eds), *Rapid Detection Assays for Food and Water.* The Royal Society of Chemistry, Cambridge, U.K., p. 38, 2001.
50. Ferrari, B.C. et al. Comparison of *Cryptosporidium*-specific and *Giardia*-specific monoclonal antibodies for monitoring water samples. *Water Res.* 33, 1611, 1999.
51. Weber, R. et al. Threshold of detection of *Cryptosporidium* oocysts in human stool specimens: Evidence for low sensitivity of current diagnostic methods. *J. Clin. Microbiol.* 29, 1323, 1991.
52. Webster, K.A. et al. Detection of *Cryptosporidium parvum* oocysts in faeces: Comparison of conventional coproscopical methods and the polymerase chain reaction. *Vet. Parasitol.* 61, 5, 1996.
53. Kang, J.M. et al. Identification and characterization of a mitochondrial iron-superoxide dismutase of *Cryptosporidium parvum*. *Parasitol. Res.* 103, 787, 2008.
54. Campbell, G.A. and Mutharasan, R. Near real-time detection of *Cryptosporidium parvum* oocyst by IgM-functionalized piezoelectric-excited millimeter-sized cantilever biosensor. *Biosens. Bioelectron.* 23, 1039, 2008.
55. Fayer, R. et al. *Cryptosporidium bovis* n. sp. (*Apicomplexa*: *Cryptosporidiidae*) in cattle (*Bos taurus*). *J. Parasitol.* 91, 624, 2005.
56. Amar, C.F.L. et al. Blinded application of microscopy, bacteriological culture, immunoassays and PCR to detect gastrointestinal pathogens from faecal samples of patients with community-acquired diarrhoea. *Eur. J. Clin. Microbiol. Infect. Dis.* 23, 529, 2004.
57. Amar, C.F. et al. Detection and identification by real time PCR/RFLP analyses of *Cryptosporidium* species from human faeces. *Lett. Appl. Microbiol.* 38, 217, 2004.
58. Sulaiman, I.M. et al. Evaluation of *Cryptosporidium parvum* genotyping techniques. *Appl. Environ. Microbiol.* 65, 4431, 1999.
59. Sturbaum, G.D. et al. Species-specific, nested PCR-restriction fragment length polymorphism detection of single *Cryptosporidium parvum* oocysts. *Appl. Environ. Microbiol.* 67, 2665, 2001.
60. Xiao, L. et al. Phylogenetic analysis of *Cryptosporidium* parasites based on the small-subunit rRNA gene locus. *Appl. Environ. Microbiol.* 65, 1578, 1999.
61. Xiao, L. et al. Molecular characterization of *Cryptosporidium* oocysts in samples of raw surface water and wastewater. *Appl. Environ. Microbiol.* 67, 1097, 2001.
62. Ryan, U. et al. Identification of novel *Cryptosporidium* genotypes from the Czech Republic. *Appl. Environ. Microbiol.* 69, 4302, 2003.
63. Jiang, J. and Xiao, L. An evaluation of molecular diagnostic tools for the detection and differentiation of human-pathogenic *Cryptosporidium* spp. *J. Eukaryot. Microbiol.* 50(Suppl), 542, 2003.
64. Loeb, K.R. et al. High-throughput quantitative analysis of hepatitis B virus DNA in serum using the TaqMan fluorogenic detection system. *Hepatology* 32, 626, 2000.
65. Higgins, J.A. et al. Real-time PCR for the detection of *Cryptosporidium parvum*. *J. Microbiol. Methods* 47, 323, 2001.
66. Garces-Sanchez, G. et al. Evaluation of two methods for quantification of hsp70 mRNA from the waterborne pathogen *Cryptosporidium parvum* by reverse transcription real-time PCR in environmental samples. *Water Res.* 43, 2669, 2009.
67. Monis, P.T. et al. Emerging technologies for the detection and genetic characterization of protozoan parasites. *Trends Parasitol.* 21, 340, 2005.
68. Rasmussen, J.P. et al. Use of DNA melting simulation software for *in silico* diagnostic assay design: Targeting regions with complex melting curves and confirmation by real-time PCR using intercalating dyes. *BMC Bioinformat.* 8, 107, 2007.
69. Xiao, L. et al. *Cryptosporidium* taxonomy: Recent advances and implications for public health. *Clin. Microbiol. Rev.* 17, 72, 2004.
70. Gatei, W. et al. Multilocus sequence typing and genetic structure of *Cryptosporidium hominis* from children in Kolkata, India. *Infect. Genet. Evol.* 7, 197, 2007.
71. Caccio, S. et al. Large sequence variation at two microsatellite loci among zoonotic (genotype C) isolates of *Cryptosporidium parvum*. *Int. J. Parasitol.* 31, 1082, 2001.
72. Widmer, G. et al. Genotyping of *Cryptosporidium parvum* with microsatellite markers. *Methods Mol. Biol.* 268, 177, 2004.
73. Feng, X. et al. Extensive polymorphism in *Cryptosporidium parvum* identified by multilocus microsatellite analysis. *Appl. Environ. Microbiol.* 66, 3344, 2000.
74. Strong, W.B. et al. Cloning and sequence analysis of a highly polymorphic *Cryptosporidium parvum* gene encoding a 60-kilodalton glycoprotein and characterization of its 15- and 45-kilodalton zoite surface antigen products. *Infect. Immun.* 68, 4117, 2000.
75. Plutzer, J. and Karanis, P. Genetic polymorphism in *Cryptosporidium* species: An update. *Vet. Parasitol.* 165, 187, 2009.
76. Leav, B.A. et al. Analysis of sequence diversity at the highly polymorphic Cpgp40/15 locus among *Cryptosporidium* isolates from human immunodeficiency virus-infected children in South Africa. *Infect. Immun.* 70, 3881, 2002.
77. Alves, M. et al. Subgenotype analysis of *Cryptosporidium* isolates from humans, cattle, and zoo ruminants in Portugal. *J. Clin. Microbiol.* 41, 2744, 2003.
78. Chalmers, R.M. et al. Direct comparison of selected methods for genetic categorisation of *Cryptosporidium parvum* and *Cryptosporidium hominis* species. *Int. J. Parasitol.* 35, 397, 2005.
79. Arrowood, M.J. and Donaldson, K. Improved purification methods for calf-derived *Cryptosporidium parvum* oocysts using discontinuous sucrose and cesium chloride gradients. *J. Eukaryot. Microbiol.* 43, 89S, 1996.
80. Meloni, B.P. and Thompson, R.C. Simplified methods for obtaining purified oocysts from mice and for growing *Cryptosporidium parvum in vitro*. *J. Parasitol.* 82, 757, 1996.
81. Peng, M.M. et al. Molecular epidemiology of cryptosporidiosis in children in Malawi. *J. Eukaryot. Microbiol.* 50, S557, 2003.
82. Striepen, B. and Kissinger, J.C. Genomics meets transgenics in search of the elusive *Cryptosporidium* drug target. *Trends Parasitol.* 20, 355, 2004.

83. Foo, C. et al. Novel *Cryptosporidium* genotype in wild Australian mice (*Mus domesticus*). *Appl. Environ. Microbiol.* 73, 7693, 2007.
84. Sulaiman, I.M., Lal, A.A., and Xiao, L. Molecular phylogeny and evolutionary relationships of *Cryptosporidium* parasites at the actin locus. *J. Parasitol.* 88, 388, 2002.
85. Cacciò, S. et al. Genetic polymorphism at the beta-tubulin locus among human and animal isolates of *Cryptosporidium parvum*. *FEMS Microbiol. Lett.* 170, 173, 1999. Erratum in: *FEMS Microbiol. Lett.* 173, 273, 1999.
86. Spano, F. et al. PCR-RFLP analysis of the *Cryptosporidium* oocyst wall protein (COWP) gene discriminates between *C. wrairi* and *C. parvum*, and between *C. parvum* isolates of human and animal origin. *FEMS Microbiol. Lett.* 150, 209, 1997.
87. Xiao, L. et al. Sequence differences in the diagnostic target region of the oocyst wall protein gene of *Cryptosporidium* parasites. *Appl. Environ. Microbiol.* 66, 5499, 2000.
88. Gibbons, C.L., and Awad-El-Kariem, F.M. Nested PCR for the detection of *Cryptosporidium parvum*. *Parasitol. Today* 15, 345, 1999.
89. Morgan, U.M. et al. Phylogenetic relationships among isolates of *Cryptosporidium*: Evidence for several new species. *J. Parasitol.* 85, 1126, 1999.
90. Giles, M. et al. A multiplex allele specific polymerase chain reaction (MAS-PCR) on the dihydrofolate reductase gene for the detection of *Cryptosporidium parvum* genotypes 1 and 2. *Parasitology* 125, 35, 2002.
91. Sulaiman, I.M. et al. Phylogenetic relationships of *Cryptosporidium* parasites based on the 70-kilodalton heat shock protein (HSP70) gene. *Appl. Environ. Microbiol.* 66, 2385, 2000.
92. Feng, Y. et al. 90-kilodalton heat shock protein, Hsp90, as a target for genotyping *Cryptosporidium* spp. known to infect humans. *Eukaryot. Cell* 8, 478, 2009.
93. Carraway, M., Tzipori, S., and Widmer, G. Identification of genetic heterogeneity in the *Cryptosporidium parvum* ribosomal repeat. *Appl. Environ. Microbiol.* 62, 712, 1996.
94. Carraway, M., Tzipori, S., and Widmer, G. A new restriction fragment length polymorphism from *Cryptosporidium parvum* identifies genetically heterogeneous parasite populations and genotypic changes following transmission from bovine to human hosts. *Infect. Immun.* 65, 3958, 1997.
95. Spano, F. et al. *Cryptosporidium parvum*: PCR-RFLP analysis of the TRAP-C1 (thrombospondin-related adhesive protein of *Cryptosporidium*-1) gene discriminates between two alleles differentially associated with parasite isolates of animal and human origin. *Exp. Parasitol.* 90, 195, 1998.
96. Awad-el-Kariem, F.M., Warhurst, D.C., and McDonald, V. Detection and species identification of *Cryptosporidium* oocysts using a system based on PCR and endonuclease restriction. *Parasitology* 109, 19, 1994.
97. Johnson, D.W. et al. Development of a PCR protocol for sensitive detection of *Cryptosporidium* oocysts in water samples. *Appl. Environ. Microbiol.* 61, 3849, 1995.
98. Leng, X., Mosier, D.A., and Oberst, R.D. Differentiation of *Cryptosporidium parvum*, *C. muris*, and *C. baileyi* by PCR-RFLP analysis of the 18S rRNA gene. *Vet Parasitol.* 62, 1–7, 1996.
99. Bonnin, A. et al. Genotyping human and bovine isolates of *Cryptosporidium parvum* by polymerase chain reaction-restriction fragment length polymorphism analysis of a repetitive DNA sequence. *FEMS Microbiol. Lett.* 137, 207, 1996.
100. Morgan, U.M. et al. Differentiation between human and animal isolates of *Cryptosporidium parvum* using rDNA sequencing and direct PCR analysis. *J. Parasitol.* 83, 825, 1997.

16 Cyclospora

Dongyou Liu

CONTENTS

16.1 INTRODUCTION

16.1.1 Classification and Biology

The genus *Cyclospora* encompasses a group of protozoan species belonging to the family Eimeriidae, order Eucoccidiorida, subclass Coccidiasina, phylum Apicomplexa. Being coccidian organisms, the members of the genus *Cyclospora* resemble those of the genera *Cryptosporidium*, *Isospora*, and *Sarcocystis* morphologically and biologically. However, *Cyclospora* oocysts harbor two sporocysts, each containing two sporozoites, in comparison with *Cryptosporidium* (whose oocysts have no sporocyst but have four sporozoites), *Isospora*, and *Sarcocystis* (whose oocysts all have two sporocysts, each containing four sporozoites). In addition, *Cyclospora* oocysts (measuring between 8–36 μm × 7–22 μm in diameter, and *Cyclospora cayetanensis* oocysts measuring 8 μm × 10 μm) are larger than those of *Cryptosporidium* (5.2 μm × 4.9 μm), but smaller than those of *Isospora belli* (30 μm × 12 μm in diameter) and *Sarcocystis* (30 μm × 15 μm) [1].

Although the first *Cyclospora* organism was described in 1870 from the intestines of the mole *Talpa europaea*, the genus *Cyclospora* was not established until 1881 referred to as *C. glomericola* observed in the millipede [2]. For a long time, *Cyclospora* species have been considered as parasites of invertebrates, and attracted little attention. However, following the reports of undescribed coccidian diarrhea in Papua New Guinea in the 1970s, traveler's diarrhea in Haiti, Nepal, and Peru in the 1980s, and chronic diarrhea in acquired immunodeficiency syndrome (AIDS) patients in the 1990s the importance of *Cyclospora* as a human pathogen has been recognized [3]. With the early epidemiological and taxonomic work at the Universidad Peruana Cayetano Heredia in Lima, Peru, the *Cyclospora* species that is infective to humans is named *Cyclospora cayetanensis* [4]. Among the 18 *Cyclospora* species that are identified to date, most are found in insects, reptiles, and mammals. *C. cayetanensis* is the only one capable of causing cyclosporiasis in humans (Table 16.1) [1,5,6]. Morphologically, the *C. cayetanensis* oocysts are 8–10 μm in size and spherical with a characteristic blue or green outer ring fluorescence when examined under ultraviolet (UV) light, whereas oocysts from other species within the *Cyclospora* genera are larger and in different shapes.

Like other coccidian pathogens, *C. cayetanensis* is an obligate intracellular parasite that completes its life cycle within the host's epithelial cells and gastrointestinal (GI) tract where it undergoes both asexual (merogony or schizogony) and sexual (gametogony) reproduction. Sporulated oocysts of *C. cayetanensis* are ingested by the human host through contaminated food or water. In the small bowl (typically the jejunum) the oocysts will excyst with the assistance of bile, trypsin, and other proteolytical factors in the GI tract. Crescent-shaped sporozoites (1.2 × 9.0 μm in size) are then released. After gaining entry into epithelial cells of the small intestine, sporozoites undergo asexual reproduction (merogony) in parasitophorous vacuoles within epithelial cells to generate large numbers of merozoites. Apart from having the capacity to infect a new host cell and initiate

TABLE 16.1
***Cyclospora* Species Identified to Date**

Species	Main Host	Year of First Description
C. glomericola	Insect	1881
C. caryolytica	Insect	1902
C. viperae	Reptile	1923
C. babaulti	Reptile	1924
C. scinci	Reptile	1924
C. tropidonoti	Reptile	1924
C. zamenis	Reptile	1924
C. niniae	Reptile	1965
C. talpae	Insect	1968
C. megacephali	Mammal/insect	1988
C. ashtabulensis	Mammal/insect	1989
C. parascalopi	Mammal	1989
C. angimurinensis	Mammal/rodent	1990
C. cayetanensis	Primate (Human)	1994
C. colobi	Primate	1999
C. cercopitheci	Primate	1999
C. papionis	Primate	1999
C. schneideri	Reptile	2005

Sources: Shields, J.M., and Olson, B.H. *Cyclospora cayetanensis*: *Int. J. Parasitol.* 33, 371, 2003; Lainson, R. The genus *Cyclospora* (Apicomplexa: Eimeriidae), with a description of *Cyclospora schneideri* n. sp. in the snake *Anilius scytale scytale* (Aniliidae) from Amazonian Brazil—A review. *Mem. Inst. Oswaldo Cruz* 100, 103, 2005.

another cycle of merogony, the merozoites can also proceed to sexual production in parasitophorous vacuoles via gametogony, by which merozoites turn into both male macrogamonts (which subsequently produce many microgametes) and female macrogamonts. The fusion of male microgamete and female macrogamont results in the formation of a zygote that then matures into an oocyst. After disruption of the host cell, the oocyst is passed with the stool. Unlike the oocyst of *Cryptosporidium*, which exits the host as fully sporulated oocysts and is immediately infective, the oocyst of *Cyclospora* passed in the stool is unsporulated and not infectious. The *Cyclospora* oocysts can survive in water at 4°C for 2 months and at 37°C for 7 days [7,8]. To become infectious, the *Cyclospora* oocyst needs to go through a sporulation process (with its content, i.e., sporont, dividing into two sporocysts, each containing two banana-shaped sporozoites), which takes 7–13 days at temperatures between 22 and 32°C. However, sporulation of the *Cyclospora* oocyst can be induced with 2.5% potassium dichromate, which is commonly used to induce sporulation in other coccidians. Alternatively, physical rupture of the *Cyclospora* oocyst wall and exposure to 0.5% trypsin and 1.5% sodium taurocholate in phosphate-buffered saline (PBS) may also speed up the sporulation process. Even though the *Isospora belli* oocyst is also shed in unsporulated form in human stool, it differs from the *Cyclospora* oocyst in size and shape as well as in the number of sporocysts (2) and sporozoites (4) after sporulation.

16.1.2 Clinical Features and Epidemiology

Clinically, *C. cayetanensis* infections in humans often display GI symptoms such as loose or watery diarrhea, nausea, vomiting, and abdominal cramps accompanied by loss of appetite and unintentional weight loss [9,10]. In addition; human cyclosporiasis may present some nonspecific clinical signs such as fever, chills, muscle aches, joint aches, generalized body aches, headache, or fatigue [11–13]. Although a case of *C. cayetanensis* infection is usually defined as diarrhea with three or more loose stools per day, asymptomatic infections or infection without watery diarrhea are known to occur. The onset of symptoms in endemic countries usually begins ~5–8 days (with a median incubation period of 7.5 days and a range of 1–23 days) after infection and the symptoms may persist for a month to six weeks. The severity and length of the illness varies significantly in relation to the age and immunological condition of the host [14]. The potential complications associated with cyclosporiasis include the Guillain–Barré syndrome, the reactive arthritis syndrome (formerly the Reiter syndrome), and acalculous cholecystitis. Trimethoprim–sulfamethoxazole provides an effective treatment for cyclosporiasis, and without therapeutic intervention, the disease may assume a remitting–relapsing course.

C. cayetanensis is distributed in various parts of the world including countries of Central and South America, Asia, Africa, Europe, and Australia as it can persist on food, in water, and in the environment [1,15–25]. The parasite has been isolated from sewage water, plants, and animal feces (e.g., chicken, ducks, dogs, rats, and mice) [26–33], although the role of animals in the transmission of human cyclopsoriasis is unknown. In fact, as Eberhard et al. [34] failed to find morphologically compatible oocysts in chicken, ducks, turkeys, pigeons, pigs, cattle, horses, goats, dogs, cats, and guinea-pigs, which lived in or near houses with human infection, the authors concluded that humans are the only host for *C. cayetanensis*, and they acquire the infection by way of human stool-contaminated food or water [12]. Because *C. cayetanensis* is unsporulated and thus noninfectious when discharged in stool, it is incapable of direct fecal–oral infection. Food items associated with the disease outbreaks include raspberries, blackberries, fresh basil, fresh baby lettuce leaves, potato salad, and so on. For example, human cyclosporiasis outbreaks that caused acute diarrhea in the United States and Canada between 1996 and 1998 were traced to the imported, unwashed raspberries [3,35]. Other human cyclosporiasis outbreaks in the United States and Germany were linked to the consumption of fresh vegetables in salads and side-dishes [3,36–38]. Given that coccidial oocysts are resistant to chlorine and most other chemical disinfectants, inadequately filtered drinking water as well as untreated or poorly treated water used for irrigation of fruits and vegetables may act as a vehicle of transmission for the parasite. Additionally, swimming pools were also found to harbor oocysts of *C. cayetanensis*. Cyclosporiasis appears to be a seasonal disease with nearly all (97.8%) of cases of cyclosporiasis occurring during the spring and summer months (April–July).

Being an obligate intracellular pathogen, *C. cayetanensis* resides in epithelial cells of human intestines where it reproduces both asexually (via merogony) and sexually (via gametogony). The parasite is known to generate an electron-dense phospholipid membrane/myelin-like material in the intestine that appears to increase with length of the disease [39]. It also induces other inflammatory changes in the infected areas of the upper small intestine (particularly the jejunum) and other parts of the body that are attributable to diarrhea and clinical symptoms associated with cyclosporiasis in humans.

16.1.3 Diagnosis

16.1.3.1 Phenotypic Techniques

Considering that *C. cayetanensis* typically produces cycles of diarrhea with anorexia, malaise, nausea, and cramping as well as a low-grade fever, which resemble those caused by *Cryptosporidium*, Microsporidia, and *Isospora belli* [40], it is necessary to apply laboratory diagnostic techniques to differentiate *C. cayetanensis* from other parasitic and bacterial pathogens, to trace the source of an outbreak and to monitor vehicles of transmission [41].

In general, identification of protozoa involves concentration, purification, detection, and viability assessment. For diagnosis of cyclosporosis in symptomatic or asymptomatic individuals, it can be achieved on the basis of the following criteria: (i) observation of *C. cayetanensis* oocysts in stool by microscopic examination, which are of characteristic size and shape (8–10 μm, as measured with a microscope eyepiece graticule, and spherical); (ii) recovery of oocysts in intestinal fluid or biopsy specimens; (iii) observation of two sporozoites, each containing two sporocysts after *in vitro* oocyst sporulation; or (iv) detection of *C. cayetanensis* DNA by nucleic acid amplification. While the first three criteria are dependent on the application of phenotypic procedures that are often time-consuming and imprecise, the fourth criterion relies on the measurement of genetic features that are intrinsically stable.

Microscopic examination provides a straightforward way to diagnose *C. cayetanensis* oocyst, although its efficiency can be variable since *C. cayetanensis* is relatively small, located intracellularly, and incapable of uptaking various stains. Therefore, phase contrast microscopy is often used to check for the spherical oocysts characteristic of *C. cayetanensis*, modified acid-fast staining is performed to confirm acid-fast oocysts, and autofluorescence with UV lights is applied to identify oocysts (as *C. cayetanensis* oocysts autofluoresce dark blue under a 365 nm dichromatic filter and autofluoresce mint green under a 450–490 nm dichromatic filter). Additionally, the use of safranin can aid in the visualization of *C. cayetanensis* oocyst [42]. For rapid screening of large numbers of samples, the flow cytometry method can be employed [43]. Frequently, centrifugation of a sample of feces in a sucrose solution (also known as sucrose or Sheather's flotation technique) prior to mounting on a slide increases the sensitivity of microscopic detection of *C. cayetanensis* (to ~84%).

The fact that oocysts sometimes give the appearance in wet-mount slides of nonrefractile spheres under a light microscope and demonstrate variability in staining with a modified Ziehl–Nielsen stain and acid-fast stain [44] often complicates the direct microscopic observation of the *C. cayetanensis* oocyst in stool. In particular, it is known that the takeup of the acid-fast stain by *C. cayetanensis* varies from nonstaining to full staining, in spite of the value of this stain for diagnosis of *C. cayetanensis* in stool specimens. This can lead to its misidentification. Attention should also be paid to the confusion caused by the copresence of *Cryptosporidium parvum* in stool. Besides being smaller than *C. cayetanensis*, *C. parvum* appears red using modified acid-fast staining and does not autofluorescence under UV light while *C. cayetanensis* does. This is especially so when dealing with environmental samples where presence of other substances tends to create additional difficulty for microscopic detection.

To enhance its detection, it is useful to apply filtration, centrifugation, and flotation techniques to recover *C. cayetanensis* oocysts from food products and environmental specimens in which a relatively low number of organisms occur in contrast with stool containing about 10^2–10^4 oocysts/g. Depending on the washing and detection procedures utilized, the rates of recovery from leafy vegetables and herbs ranged from 12% to 14% [45,46], and those from raspberries were higher [45,47,48]. The next step in the identification of *C. cayetanensis* involves *in vitro* sporulation, which is normally performed by subjecting the recovered oocysts to 2 weeks incubation in 2.5% potassium dichromate solution [29,49]. This not only helps ascertain the characteristic features of *C. cayetanensis* oocysts (i.e., two sporocysts, each containing two sporozoites), but also offers a means of determining the viability of organisms, which is vital for the assessment of its risk associated with potentially contaminated water [50].

Although conventional techniques based on histochemical staining and microscopy are still largely in use, they do not always provide a conclusive determination of *C. cayetanensis* species identity. While filtration, centrifugation, and flotation techniques are useful for recovery of *C. cayetanensis* oocysts from clinical, food, and environmental samples, they tend to be labor-intensive and show varied efficiency in dealing with different sample types. *In vitro* sporulation is undoubtedly valuable for the confirmation of *C. cayetanensis* identity, but the length of time it takes to complete compromises its use in epidemic investigations where speedy identification is desired.

16.1.3.2 Molecular Techniques

Molecular techniques such as polymerase chain reaction (PCR) provide a more sensitive, specific, and rapid alternative to conventional, phenotype-based methods for diagnosis of *C. cayetanensis* [51–55]. Shortly after the description of

C. cayetanensis as a human pathogen by Ortega et al. [4] a nested PCR assay with primers (F1E-R2B, F3E-R4B) targeting 18S ribosomal DNA gene sequences of eukaryote and *C. cayetanensis*, respectively, was devised by Relman et al. [56] for *C. cayetanensis* identification. This method has been used successfully for sensitive detection of *C. cayetanensis* oocysts in human stool [47,48,57]. To eliminate the cross-reaction of these primers with some *Eimeria* species (showing the identical size of a 294 bp amplicon), a restriction enzyme *Mnl*I digestion step was incorporated into the assay (i.e., PCR–restriction-fragment-length polymorphism (PCR-RFLP)) that helped differentiate *C. cayetanensis* from *Eimeria* species [47]. This modification extended the applicability of the assay to foodstuffs [37,47] and environmental waters [58]. Following the identification of three new *Cyclospora* species: *C. cercopitheci*, *C. colobi*, and *C. papionis* from nonhuman primates in 1999 (see Table 16.1) [59,60], it is apparent that the 18S rRNA regions from these species demonstrate a high similarity to that of *C. cayetanensis*, and the use of *Mnl*I digestion in the PCR does not permit discrimination among the species. This led to the development of another nested PCR assay by Shields and Olson [52], which employs three new primers from 18S rRNA sequences from *Cyclospora* and *Eimeria* species: AO1-AR1 and AI2-AR1. After digestion with the restriction enzyme *Alu*I, *C. cayetanensis* can be distinguished from *Eimeria* and other *Cyclospora* species in water and soil samples without microscopic confirmation.

In addition, focusing on the 18S rRNA gene of *C. cayetanensis* as well as those of nonhuman primate-derived *Cyclospora* spp., Chu et al. [61] described a multiplex PCR for differentiation of primate-derived *Cyclospora* and *Eimeria* from human *C. cayetanensis*. Whereas the primer pair CC719–CRP999 amplifies a 298-bp fragment from the 18S rRNA gene of *C. cayetanensis*, the primer pair PDCL661–CRP999 produces a 361-bp band from the 18S rRNA gene of nonhuman primate-derived *Cyclospora* spp. and the primer pair ESSP841–CRP999 forms a 174-bp amplicon with the 18S rRNA gene of *Eimeria* spp. Without the need to digest the amplicons by restriction enzyme, this multiplex PCR enables rapid discrimination of *Cyclospora* and *Eimeria* in an efficient and precise manner. Further, Orlandi et al. [62] took advantage of minute differences at the 18S SSU-rRNA gene amount *Cyclospora* spp. and designed a series of genus- and species-specific forward primers that distinguish sites of limited sequence heterogeneity in the same gene regions for differentiation between *C. cayetanensis*, nonhuman primate species of *Cyclospora*, and *Eimeria* species. Hussein [63] also utilized a multiplex PCR for the detection of *Cyclospora* organisms in children suffering from diarrhea. Moreover, Varma et al. [43] reported a real-time TaqMan assay for *C. cayetanensis* detection that also targets the 18S rRNA gene sequences. Using the forward primer (CcayF1) and reverse primer (CcayR1) together with the probe (CcayP1), the authors showed that the assay was highly specific for *C. cayetanensis* as non-*C. cayetanensis* control organisms were invariably negative. Hussein et al. [64] described an RT-PCR for the detection of *C. cayetanensis* in stool with a species-specific primer set and dual-fluorescent labeled *C. cayetanensis* hybridization probe which originated from unique regions of 18S rRNA gene sequence. All symptomatic and asymptomatic patients were detected.

In addition to 18S (small subunit) rRNA regions, the internal transcribed spacer region 1 (ITS1) of *C. cayetanensis* has been used for PCR identification of *C. cayetanensis* [65,66]. The ITS gene-based PCR assay by Adam et al. [65] was useful for the identification of different genotypes of *C. cayetanensis*, while that by Olivier et al. [66] facilitated differentiation of *C. cayetanensis* from *C. papionis*. However, considering enormous variation among *C. cayetanensis* oocysts in the same stool sample, this latter PCR assay may have limited value in the determination of its geographic source [66].

Even though PCR technology provides a highly effective way to generate multiple copies of gene target with purified nucleic acids for subsequent detection and analysis, it is relatively inefficient in dealing with clinical, food, and environmental samples. This is due to the fact polymerase enzymes used in the PCR process are susceptible to inhibition by substances that are commonly found in unprocessed samples. Notable polymerase-interfering compounds include heme, hemoglobulin, lactoferrin, immunoglobulin G, leukocyte DNA, and polysaccharides in blood; phenolics, glycogen, calcium ions, fat, and other organic substances in foods; phenolics, humic acids, and heavy metals in environmental samples. To reduce and eliminate inhibitory materials from these samples, it is critical that efficient sample preparation methods are utilized. Additionally, since food, water, and other environmental samples often contain low numbers of *C. cayetanensis*, inclusion of an enrichment protocol may be necessary to increase organism number and thus assay sensitivity. However, this extra step creates delay in test turnover.

Toward this goal, Orlandi and Lampel [48] described an extraction-free, FTA filter-based protocol to prepare *C. cayetanensis* DNA templates for PCR identification of *C. cayetanensis* oocysts. Originally designed as a matrix for blood storage, FTA filter (Whatman Inc., Newton, MA) incorporates denaturants, chelating agents, and free-radical traps that lyse *C. cayetanensis* on contact and sequesters DNA within the matrix while other materials remain unbound. The adsorbed *C. cayetanensis* can then be used for DNA preparation. Adopting this procedure in the PCR, the authors detected 3–10 oocysts directly applied to filters and 10–30 oocysts in complex mixed samples (e.g., 100 g of fresh raspberries) (which amounted to as few as 0.3 oocyst per g of raspberries). Robertson et al. [46] utilized lectin wheat germ agglutinin-coated paramagnetic beads to improve recovery and purification of *C. cayetanensis* oocysts. With an affinity for β-linked poly-*N*-acetyl hexosamine-containing polysaccharides, which constitute a component of *C. cayetanensis* oocyst walls, lectin aids the isolation of the *C. cayetanensis* oocyst from other substances. This approach helped us to clarify and concentrate the final sample volume, and make microscopic examination easier. Lalonde and Gajadhar [67] described a reliable DNA extraction for *C. cayetanensis* oocysts in water for PCR detection, with a limit of between 1 and 10 oocysts.

16.2 METHODS

16.2.1 Sample Preparation

16.2.1.1 Sample Processing

16.2.1.1.1 Feces

Upon collection, fecal specimens are examined by differential interference contrast microscopy for the presence of *Cyclospora* oocysts of 8–10 μm in size, and oocysts can be further identified by UV epifluorescence microscopy. *Cyclospora* oocysts are then concentrated by using a discontinuous sucrose density purification method, or by using a 10% formalin–ethyl acetate concentration in conjunction with Fecal Parasite Concentrators (Evergreen Scientific, Los Angeles, CA). For the latter method, fecal specimens (100–200 μL of packed debris) are washed twice with 1 mL of distilled water and pelleted by centrifugation. The washed debris was then suspended in 1 mL of distilled water and extracted with 0.25 mL of ethyl acetate, vortexed for 20–30 s, and centrifuged. After removing the upper, organic phase, any debris at the interface, and the lower aqueous phase, the sedimented debris is then washed twice with 1.0 mL of distilled water, suspended in 1 mL of distilled water and passed through a glass wool column.

16.2.1.1.2 Foods

Fresh fruits and vegetables (100 g) are washed in 100 mL of distilled water in a container, and agitated for 30 min at room temperature [37,40]. The wash liquid is decanted and centrifuged at 1870 × *g* for 20 min at 4°C. The resulting sediment is then suspended in 5–10 mL of distilled water and applied to a Poly-Prep chromatography column (Bio-Rad, Hercules, CA) containing tightly packed glass wool presoaked with distilled water. The column is washed once with 5 mL of distilled water, and the eluent is centrifuged at 20,000 × *g* for 15 min at 4°C. The resulting pellet is then suspended in 50–100 μL of distilled water.

16.2.1.1.3 Water

Being in large volume, water samples are filtered using cartridge, hollow-fiber ultra-filters, or capsule filters to capture oocysts. An alternative method of collection and concentration is flocculation that is not affected by turbidity. Namely, calcium chloride and sodium bicarbonate are added to a water sample and the pH is raised to 10.0 with sodium hydroxide. Acting as a coagulant, calcium carbonate increases the density and diameter of any particle in the water, which help its settling to the bottom of the container. The water is aspirated and the calcium carbonate is dissolved with a small volume of a mild acid (10% sulfamic acid) [68]. Resultant pellets are added with digestion buffer for genomic DNA extraction and/or 2.5% potassium dichromate for microscopic analysis.

16.2.1.2 DNA Isolation

i. *Phenol–chloroform extraction*. The pellet is digested with proteinase K overnight, and centrifuged at 14,000 × *g* for 1 min to remove debris. Phenol–chloroform is then used to remove proteins. The resultant aqueous layer is precipitated with ethanol–NH_4 acetate and washed with cold 70% ethanol. The purified DNA is dissolved in TE buffer.
ii. *FastDNA Spin Prep Kit*. da Silva et al. [69] used the FastDNA Spin Prep Kit for *Cryptosporidium* and Microsporidia genera in stool samples by the addition of polyvinylpyrrolidone (0.5% w/v) during the protein precipitation step, which helped in removal of additional PCR inhibitors. Briefly, in 100 μL of packed pellet, 300 μL of digestion buffer, and 50 mg of glass beads are added. After agitation in a bead beater, the crude DNA lysate is briefly centrifuged (~6000 × *g* for 30 s), the supernatant is removed, and more buffer and detergent are added to the pellet. This is repeated four times, and the final centrifugation is at 14,000 × *g* for 5 min. The crude lysates are pooled, and crude DNA is precipitated overnight in 5 M ammonium acetate (0.1 × volume) and ethanol (2.2 volumes) at −20°C. The crude DNA samples are then centrifuged (20 min at 14,000 × *g*), and the resultant pellet is washed with 70% cold ethanol, air dried, and then resuspended in 500 μL of phosphate buffer for further purification.
iii. *FTA filter purification*. After spotting 25–50 μL of packed pellet (or 10–20 μL of stool) onto FTA filter and subsequent drying on a 56°C heating block, 6 mm FTA disks are punched out using an individual hole-punch, and placed in a 1.5-mL microcentrifuge tube. FTA disks are washed twice with 0.5 mL of FTA purification buffer (Life Technologies, Gaithersburg, MD) for 2 min, twice with 0.5 mL of 10 mM Tris (pH 8.0) containing 0.1 mM ethylenediaminetetraacetic acid (EDTA) for 2 min and again air-dried on a 56°C heating block. The washed filters are then used directly as the source of template in PCR [48].

16.2.2 Detection Procedures

16.2.2.1 PCR-RFLP

16.2.2.1.1 PCR with F1E-R2B, F3E-R4B Primers, and RFLP with MnlI

A nested PCR protocol based on 18S rRNA gene was first developed by Relman et al. [56] and later modified by Orlandi's group [48,62]. This method has been widely applied for routine detection of *C. cayetanensis*. The first-round PCR utilizes primers F1E (5′-TACCCAATGAAAACAGTTT-3′) and R2B (5′-CAGGAGAAGCCAAGGTAGG-3′), leading to a product of 636 bp; and the second-round PCR uses primers F3E (5′-CCTTCCGCGCTTCGCTGCGT-3′) and R4B (5′-CGTCTTCAAACCCCCTACTG-3′), leading to a product of 294 bp from both

Cyclospora spp. and *Eimeria* spp. Digestion of the second-round PCR product with the restriction enzyme *Mnl*I results in the formation of 140, 106, and 48 bp fragments for *Cyclospora* and 126, 106, and 62 bp fragments for *Eimeria* [61].

16.2.2.1.1.1 Procedure

1. Prepare the first-round PCR mixture (100 μL) containing Hot-StarTaq Master Mix (Qiagen), containing 2.0 mM $MgCl_2$, and 200 μM of each dNTP, 0.2 μM of primers F1E and R2B, and 1–5 μL of purified DNA.
2. Conduct the first-round PCR amplification with 1 cycle of 95°C for 15 min (for the activation of the HotStarTaq DNA polymerase); 35 cycles of 94°C for 30 s, 53°C for 30 s, and 72°C for 90 s; 1 cycle of 72°C for 10 min followed by soaking at 4°C. Run 10 μL of the PCR product on 1% agarose gel to verify the successful amplification of the 636 bp fragment.
3. Prepare the second-round PCR mixture (50 μL) with HotStarTaq Master Mix, 2.0 mM $MgCl_2$, 200 μM of each dNTP, and 0.2 μM each of primers F3E and R4B, and 1–5 μL of the first-round PCR product as the template.
4. Conduct the secondary nested PCR amplification with 1 cycle of 95°C for 15 min (for the activation of the HotStarTaq DNA polymerase); 35 cycles of 94°C for 30 s, 60°C for 30 s, and 72°C for 90 s; 1 cycle of 72°C for 10 min followed by soaking at 4°C.
5. Load 10 μL of PCR amplicon on a 1.5% agarose gel and stain with 0.2 μg/mL of ethidium bromide. Visualize the specific amplicon of 294 bp under UV transillumination.
6. Add 2.5 μL of 10 × reaction buffer and 2 U *Mnl*I (New England Biolabs) to 20 μL of the secondary PCR amplicon (total volume 25 μL), and incubate for 2 h at 37°C.
7. Separate the digests on a 5% Nusieve 3:1 agarose gel (Biowittaker Molecular Applications) containing 0.2 μg/mL of ethidium bromide and visualize under UV transillumination.

Note: Although this routinely used nested PCR-RFLP protocol is capable of detecting and distinguishing *Cyclospora* from *Eimeria* spp., it is unable to differentiate between *C. cayetanensis* and nonhuman primate-derived *Cyclospora* spp.

16.2.2.1.2 PCR with AO1-AR1, AI2-AR1 Primers, and RFLP with AluI

While the PCR-RFLP protocol with F1E-R2B, F3E-R4B primers, and RFLP with *Mnl*I [56] is capable of identifying *C. cayetanensis* oocysts in stool samples, it requires microscopic confirmation to produce reliable results for environmental samples, due to the fact that *C. cayetanensis*, *C. cercopitheci*, *C. colobi*, and *C. papionis* all show a high degree of homology in the inner (CYCF3E and CYCR4B) PCR products that are not distinguished with *Mnl*I digestion. Therefore, Shields and Olson [52] designed three new primers from 18S rRNA sequences from *Cyclospora* and *Eimeria* species: AO1 (5′-ATAACGAACGAGACCTTAGCCT-3′), AR1 (5′-5′-AAGGATGCAAAAGTCGTAACAC-3′), AI2 (5′-CAGGTCTGGGTAATCTTTTGAG-3). The first-round PCR is performed with AO1-AR1, and the second-round PCR with AI2-AR1. This is followed by RFLP with the restriction enzyme *Alu*I. The method allows differentiation of *C. cayetanensis* from *Eimeria* and other *Cyclospora* species in water and soil samples without microscopic confirmation.

16.2.2.1.2.1 Procedure

1. Prepare first-round PCR mixture (50 μL) containing 50 mM KCl, 10 mM Tris-HCl pH 8.3, 200 μg of nonfat dry milk, 200 μM of deoxynucleoside triphosphate, 2.0 mM $MgCl_2$, 0.4 μM each of AO1 and AR1 primers, 1 U of AmpliTaq polymerase, and 5 μL of template.
2. Conduct the first-round PCR amplification for 1 cycle of 94°C for 5 min; 45 cycles of 94°C for 30 s, 59.5°C for 30 s, and 72°C for 90 s; and 1 cycle of 72°C for 9 min.
3. Prepare the second-round PCR mixture (50 μL) containing 50 mM KCl, 10 mM Tris-HCl pH 8.3, 200 μM of deoxynucleoside triphosphate, 2.0 mM $MgCl_2$, 0.4 μM each of AI2 and AR1 primers, 1 U of AmpliTaq polymerase, and 2.5 μL of the first-round PCR product. Conduct the second-round PCR amplification as above.
4. Load 10 μL of PCR product on 1.75% agarose gels supplemented with 0.5 μg of ethidium bromide per mL at constant current in Tris-borate-EDTA buffer. Include three molecular size markers: 25- and 50-bp ladders (Promega) and DNA weight marker VIII (a mixture of pUCBM21 DNA cleaved with *Hpa*II and pUCBM21 DNA cleaved with *Dra*I and *Hind*III; Boehringer Mannheim). Visualize the product by UV transillumination.
5. Digest the positive PCR products from the second reaction (10 μL) with 1 U of *Alu*I (New England Biolabs) at 37°C for 2 h. Separate the fragments, visualize, and photograph as described above.

Note: The expected second-round PCR product from *Cyclospora* species using AI2 and AR1 primers is 251 bp, which shows four bands at 98, 88, and 50 bp and a faint band at 15 bp for *C. cayetanensis* after *Alu*I digestion. However, the second-round PCR product becomes 153–159 bp and/or 94–98 bp for *Eimeria* and *C. papionis*, respectively and three bands for *Isospora robini* and the other two *Cyclospora* species after *Alu*I digestion.

16.2.2.2 Multiplex PCR

A multiplex PCR was described by Chu et al. [61] to distinguish the presence of primate-derived *Cyclospora* and *Eimeria* from human *C. cayetanensis*. Using the first-round PCR product of 636 bp (which is generated with primers F1E: 5′-TACCCAATGAAAACAGTTT-3′ and R2B:

5′-CAGGAGAAGCCAAGGTAGG-3′) as template, the multiplex PCR is conducted with common reverse primer CRP999 (5′-CGTCTTCAAACCCCCTACTGTCG-3′), species-specific forward primers CC719 (5′-GTAGCCTTCCGCGCTTCG-3′); PDCL661 (5′-CTGTCGTGGTCATCTGT.CCGC-3′), and ESSP841 (5′-GTTCTATTTTGTTGGTTTCTAGGACCA-3′). While the primer pair CC719–CRP999 generates a 298-bp amplicon from the 18S rRNA gene of *C. cayetanensis*, the primer pair PDCL661–CRP999 generates a 361-bp amplicon from the 18S rRNA gene of nonhuman primate-derived *Cyclospora* spp., and primer pair ESSP841–CRP999 generates a 174-bp amplicon from the 18S rRNA gene of *Eimeria* spp.

16.2.2.2.1 Procedure

1. Prepare multiplex PCR mixture (50 μL) containing HotStarTaq Master Mixture, 2 mM $MgCl_2$, 0.2 μM of CRP999, CC719, PDCL661, and ESSP841 primers and 1–5 μL of PCR product generated with F1E, and R2B (as shown in Section 16.2.2.1).
2. Perform PCR amplification with 1 cycle of 95°C for 15 min and 25 cycles of 94°C for 15 s and 66°C for 15 s.
3. Separate the PCR products on a 1.5% agarose gel with 0.2 μg/m of ethidium bromide and visualize on a UV transilluminator.

Note: The primer pair CC719–CRP999 generates an amplicon of 298 bp only with *C. cayetanensis*; the primer pair PDCL661–CRP999 generates an amplicon of 361 bp with other *Cyclospora* spp (e.g., *C. cercopitheci*); and the primer pair ESSP841–CRP999 generates an amplicon of 174 bp with *Eimeria* spp (e.g., *E. tenella*) [61,62].

16.2.2.3 Real-Time PCR

A real-time *C. cayetanensis*-specific TaqMan assay targeting 18S rRNA gene sequences was developed by Varma et al. [43] The forward primer (CcayF1) is located at nucleotide positions of 1671–1692 (5′-TGAACTCATTGGACTGACCA GC-3′), the reverse primer (CcayR1) at 1753–1731 (5′-ACTT TTGCATCCTTTAGAGGGCT-3′), and the probe (CcayP1) at 1698–1721 (5′-FAM-TTCGCGGAGCTGGTCGGAAAGT TG-TAMRA-3′), based on the *C. cayetanensis* sequence in GenBank accession no. AF111183. While the forward and reverse primers are synthesized unlabeled, the probe is synthesized with dual labels having a reporter fluorescent dye [6-carboxyfluorescein (FAM)] on the 5′-end and a quencher fluorescent dye [6-carboxytetramethylrhodamine (TAMRA)] on the 3′-end. The 3′-end of the probe is also phosphorylated to prevent probe extension.

16.2.2.3.1 Procedure

1. Prepare real-time PCR mixture (25 μL) in 0.5-mL MicroAmp optical PCR tubes (Applied Biosystems) containing 5 μL of extracted *C. cayetanensis* DNA (diluted as described), 2.4 μL of a 450 nM solution of the dual-labeled *C. cayetanensis*-specific hybridization probe (CcayP1), 2.4 μL of a 5.2 μM solution of the forward primer (CcayF1), 2.4 μL of a 5.2 μM solution of the reverse primer (CcayR1), 0.9 μL of a 2 mg/mL bovine serum albumin Fraction V (Sigma-Aldrich), and 11.9 μL of the 2 × TaqMan Universal PCR Master Mix (AmpliTaq Gold DNA polymerase, Amperase UNG, dNTPs, passive reference 6-carboxy-X-rhodamine (ROX), $MgCl_2$, and buffer components in amounts undisclosed by the manufacturer; Applied Biosystems)). Insert the optical tubes into a 96-well tray/retainer set and seal. Centrifuge the tray for 5 min at 500 × *g* before inserting into an ABI-Prism 7700 Sequence Detector (Applied Biosystems). Five replicates are prepared for each sample.
2. Conduct PCR amplification for 1 cycle of 50°C for 2 min; 1 cycle of 95°C for 10 min; 40 cycles of 95°C for 0.25 min and 60°C for 1 min.

Note: The forward primer shows a high degree of specificity for the *C. cayetanensis* ribosomal DNA gene; while the reverse primer is broadly specific for a variety of protists, the hybridization probe is also specific to several *Cyclospora* spp. The combination of forward primer and probe specificity results in a highly specific assay and with a C_T value of at least 40, no positive results are obtained with non-*C. cayetanensis* control organisms tested [43].

16.3 CONCLUSION

The genus *Cyclospora* consists of at least 18 known species of protozoa sharing morphological, biological, and molecular similarities with *Cryptosporidium*, *Isospora*, and *Sarcocystis* [6]. Being obligate intracellular parasites, 14 of the 18 *Cyclospora* species are found in insects and reptiles, three (*C. colobi*, *C. cercopitheci*, and *C. papionis*) are infective to nonhuman primates, while one (*C. cayetanensis*) causes cyclosporiasis in humans after gaining entry into the host via contaminated food and water (Table 16.1). Considering the nonspecificity of clinical symptoms (e.g., diarrhea and general illnesses) caused by *C. cayetanensis* as well as other pathogens (e.g., *Cryptosporidium*, *Isospora*, *Sarcocystis*, bacterial, and viral infections), correct identification of *C. cayetanensis* is essential for selecting appropriate chemotherapy and effective control measures [70–72].

Conventional techniques for the identification of *C. cayetanensis* are dependent on the detection of the morphological characteristics of this parasite. Specifically, the oocysts (8 × 10 μm) of *C. cayetanensis* are larger than those of *Cryptosporidium* (5.2 × 4.9 μm), but smaller than those of *Isospora belli* (30 × 12 μm^2 in diameter) and *Sarcocystis* (30 × 15 μm^2). In addition, *C. cayetanensis* oocysts comprise two sporocysts, each containing two sporozoites, *Cryptosporidium* oocysts have no sporocyst but four sporozoites, *Isospora* and *Sarcocystis* oocysts all possess two sporocysts, each containing four sporozoites. Therefore, through a combination of microscopic examination and *in vitro* sporulation, it is possible to differentiate *Cyclospora* from *Cryptosporidium*, *Isospora*, and *Sarcocystis*. The use of immunological assays such as ELISA adds another dimension to the phenotypic characterization of *Cyclospora* spp. However, besides being time-consuming and lacking precision, these phenotypic techniques do not allow precise discrimination among *Cyclospora* species.

The development and application of nucleic acid amplification technologies such as PCR during the past decades have facilitated improved laboratory identification and diagnosis of *C. cayetanensis*. By employing primers derived from 18S rRNA genes, ITS, and other gene regions in various PCR formats (e.g., standard PCR, PCR-RFLP, multiplex PCR, real-time PCR, and reverse-transcription PCR), *C. cayetanensis* can be rapidly, precisely, and sensitively distinguished from closely related *C. colobi*, *C. cercopitheci*, and *C. papionis* as well as *Eimeria* species. These "state-of-the-art" molecular detection procedures offer a valuable tool in the laboratory diagnosis of *C. cayetanensis* and epidemiological investigations of human cyclosporiasis outbreaks, contributing to improved control and prevention of this important foodborne disease [73,74]. It is envisaged that future optimization of prior sample processing (e.g., filtration, centrifugation, and concentration) and nucleic acid isolation protocols will enhance the utility of molecular assays for direct detection of low numbers of *C. cayetanensis* oocysts that exist in clinical, environmental, water, and food samples. In addition, there is scope for further improvement in the applicability of the molecular detection techniques in field situations or in places where access to expensive equipment is restricted and where rapid and precise diagnostic techniques for *C. cayetanensis* are most needed.

Despite our past efforts, human *C. cayetanensis* infection remains a significant health issue throughout the world. Indeed, there is indication that *C. cayetanensis* is far from being under control as new cases of cyclosporiasis are described with regularity in the literature [75–81]. Therefore, besides further refinement in the molecular diagnostic procedures, there is a pressing need to improve our understanding of the epidemiology and pathogenesis of *C. cayetanensis* organisms. Given the current lack of a reliable and effective animal model for *C. cayetanensis* infection, concerted research on this line will yield new insights into the pathogenic mechanisms of cyclosporiasis [82].

REFERENCES

1. Eberhard, M.L. and Arrowood, M.J., *Cyclospora* spp., *Curr. Opin. Infect. Dis.*, 15, 519, 2002.
2. Curry, A., and Smith, H.V., Emerging pathogens: *Isospora*, *Cyclospora* and microsporidia, *Parasitology*, 117, S143, 1998.
3. Herwaldt, B.L., *Cyclospora cayetanensis*: A review, focusing on the outbreaks of cyclosporiasis in the 1990s, *Clin. Infect. Dis.*, 31, 1040, 2000.
4. Ortega, Y.R., Gilman, R.H., and Sterling, C.R., A new coccidian parasite (Apicomplexa: Eimeriidae) from humans, *J. Parasitol.*, 80, 625, 1994.
5. Shields, J.M., and Olson, B.H., *Cyclospora cayetanensis*: A review of an emerging parasitic coccidian, *Int. J. Parasitol.*, 33, 371, 2003.
6. Lainson, R., The genus *Cyclospora* (Apicomplexa: Eimeriidae), with a description of *Cyclospora schneideri* n.sp. in the snake *Anilius scytale scytale* (Aniliidae) from Amazonian Brazil—A review, *Mem. Inst. Oswaldo Cruz*, 100, 103, 2005.
7. Smith, H.V. et al., Sporulation of *Cyclospora* sp. oocysts, *Appl. Environ. Microbiol.*, 63, 1631, 1997.
8. Ortega, Y.R., Sterling, C.R., and Gilman, R.H., *Cyclospora cayetanensis*, *Adv. Parasitol.*, 40, 399, 1998.
9. Chacin-Bonilla, L. et al., *Cyclospora cayetanensis* infections among diarrheal patients from Venezuela, *Am. J. Trop. Med. Hyg.*, 65, 351, 2001.
10. Chacin-Bonilla, L. et al., Prevalence and pathogenic role of *Cyclospora cayetanensis* in a Venezuelan community, *Am. J. Trop. Med. Hyg.*, 68, 304, 2003.
11. Fleming, C.A. et al., A foodborne outbreak of *Cyclospora cayetanensis* at a wedding: Clinical features and risk factors for illness, *Arch. Intern. Med.*, 158, 1121, 1998.
12. Mead, P.S., Food-related illness and death in the United States, *Emerg. Infect. Dis.*, 5, 607, 1999.
13. Crist, A. et al., Outbreak of cyclosporiasis associated with snow peas—Pennsylvania, 2004, *MMWR*, 53, 876, 2004.
14. Gascon, J. et al., Cyclosporiasis: A clinical and epidemiological study in travellers with imported *Cyclospora cayetanensis* infection, *Med. Clin. (Barc.)*, 116, 461, 2001.
15. Ashford, R.W., Occurrence of an undescribed coccidian in man in Papua New Guinea, *Ann. Trop. Med. Parasitol.*, 73, 497, 1979.
16. Nunez, F.A. et al., The first report in Cuba of human intestinal infection by *Cyclospora cayetanensis*, Ortega, 1993, *Rev. Cubana Med. Trop.*, 47, 211, 1995.
17. Clarke, S.C., and McIntyre, M., The incidence of *Cyclospora cayetanensis* in stool samples submitted to a district general hospital, *Epidemiol. Infect.*, 117, 189, 1996.
18. Madico, G. et al., Epidemiology and treatment of *Cyclospora cayetanensis* infection in Peruvian children, *Clin. Infect. Dis.*, 24, 977, 1997.
19. Graczyk, T.K., Ortega, Y.R., and Conn, D.B., Recovery of waterborne oocysts of *Cyclospora cayetanensis* by Asian freshwater clams (*Corbicula fluminea*), *Am. J. Trop. Med. Hyg.*, 59, 928, 1998.
20. Bern, C. et al., Epidemiologic studies of *Cyclospora cayetanensis* in Guatemala, *Emerg. Infect. Dis.*, 5, 766, 1999.
21. Fryauff, D.J. et al., *Cyclospora cayetanensis* among expatriate and indigenous populations of West Java, Indonesia, *Emerg. Infect. Dis.*, 5, 585, 1999.
22. Sherchand, J.B. and Cross, J.H., Emerging pathogen *Cyclospora cayetanensis* infection in Nepal, *Southeast Asian J. Trop. Med. Public. Health*, 32, S143, 2001.
23. Wang, K.X. et al., *Cyclospora cayetanensis* in Anhui, China, *World J. Gastroenterol.*, 8, 1144, 2002.
24. Lopez, A.S. et al., Outbreak of cyclosporiasis associated with basil in Missouri in 1999, *Clin. Infect. Dis.*, 32, 1010, 2001.
25. Nimri, L.F., *Cyclospora cayetanensis* and other intestinal parasites associated with diarrhea in a rural area of Jordan, *Int. Microbiol.*, 6, 131, 2003.

26. Zerpa, R., Uchima, N., and Huicho, L., Cyclospora cayetanensis associated with watery diarrhoea in Peruvian patients, *J. Trop. Med. Hyg.*, 98, 325, 1995.
27. Marshall, M.M. et al., Waterborne protozoan pathogens, *Clin. Microbiol. Rev.*, 10, 67, 1997.
28. Wright, M.S., and Collins, P.A., Waterborne transmission of *Cryptosporidium*, *Cyclospora* and *Giardia*, *Clin. Lab. Sci.*, 10, 287, 1997.
29. Sherchand, J.B. et al., Study of *Cyclospora cayetanensis* in health care facilities, sewage water and green leafy vegetables in Nepal, *Southeast Asian J. Trop. Med. Public Health*, 30, 58, 1999.
30. Sterling, C.R., and Ortega, Y.R., *Cyclospora*: An enigma worth unraveling, *Emerg. Infect. Dis.*, 5, 48, 1999.
31. Mota, P., Rauch, C.A., and Edberg, S.C., Microsporidia and *Cyclospora*: Epidemiology and assessment of risk from the environment, *Crit. Rev. Microbiol.*, 26, 69, 2000.
32. Dowd, S.E. et al., Confirmed detection of *Cyclospora cayetanesis, Encephalitozoon intestinalis* and *Cryptosporidium parvum* in water used for drinking, *J. Water Health*, 1, 117, 2003.
33. Mansfield, L.S., and Gajadhar, A.A., *Cyclospora cayetanensis*, a food- and waterborne coccidian parasite, *Vet. Parasitol.*, 126, 73, 2004.
34. Eberhard, M.L. et al., *Cyclospora cayetanensis* infections in Haiti: A common occurrence in the absence of watery diarrhea, *Am. J. Trop. Med. Hyg.*, 60, 584, 1999.
35. Chalmers, R.M., Nichols, G., and Rooney, R., Foodborne outbreaks of cyclosporiasis have arisen in North America. Is the United Kingdom at risk? *Commun. Dis. Public Health*, 3, 50, 2000.
36. Ho, A.Y. et al., Outbreak of cyclosporiasis associated with imported raspberries, Philadelphia, Pennsylvania, 2000, *Emerg. Infect. Dis.*, 8, 783, 2000.
37. Lopez, F.A. et al., Molecular characterization of *Cyclospora*-like organisms from baboons, *J. Infect. Dis.*, 179, 670, 1999.
38. Döller, P.C. et al., Cyclosporiasis outbreak in Germany associated with the consumption of salad, *Emerg. Infect. Dis.*, 8, 992, 2002.
39. Connor, B.A., Reidy, J., and Soave, R., Cyclosporiasis: Clinical and histopathologic correlates, *Clin. Infect. Dis.*, 28, 1216, 1999.
40. Ortega, Y.R. et al., Isolation of *Cryptosporidium parvum* and *Cyclospora cayetanensis* from vegetables collected in markets of an endemic region in Peru, *Am. J. Trop. Med. Hyg.*, 57, 683, 1997.
41. Eberhard, M.L, Pieniazek, N.J., and Arrowood, M.J., Laboratory diagnosis of *Cyclospora* infections, *Arch. Pathol. Lab. Med.*, 121, 792, 1997.
42. Visvesvara, G.S. et al., Uniform staining of *Cyclospora* oocysts in fecal smears by a modified safranin technique with microwave heating, *J. Clin. Microbiol.*, 35, 730, 1997.
43. Varma, M. et al., Detection of *Cyclospora cayetanensis* using a quantitative real-time PCR assay, *J. Microbiol. Methods*, 53, 27, 2003.
44. Abou el Naga, I.F., Studies on a newly emerging protozoal pathogen: *Cyclospora cayetanensis*, *J. Egypt. Soc. Parasitol.*, 29, 575, 1999.
45. Ortega, Y.R. et al., Pathologic and clinical findings in patients with cyclosporiasis and a description of intracellular parasite life-cycle stages, *J. Infect. Dis.*, 176, 1584, 1997.
46. Robertson, L.J., Gjerde, B., and Campbell, A.T., Isolation of *Cyclospora* oocysts from fruits and vegetables using lectin-coated paramagnetic beads, *J. Food Prot.*, 63, 1410, 2000.
47. Jinneman, K.C. et al., Template preparation for PCR and RFLP of amplification products for the detection and identification of *Cyclospora* sp. and *Eimeria* spp. oocysts directly from raspberries, *J. Food Prot.*, 61, 1497, 1998.
48. Orlandi, P.A., and Lampel, K.A., Extraction-free, filter-based template preparation for rapid and sensitive PCR detection of pathogenic parasitic protozoa, *J. Clin. Microbiol.*, 38, 2271, 2000.
49. Sturbaum, G.D. et al., Species-specific, nested PCR-restriction fragment length polymorphism detection of single *Cryptosporidium parvum* oocysts, *Appl. Environ. Microbiol.*, 67, 2665, 2001.
50. Dalton, C. et al., Viability of *Giardia intestinalis* cysts and viability and sporulation state of *Cyclospora cayetanensis* oocysts determined by electrorotation, *Appl. Environ. Microbiol.*, 67, 586, 2001.
51. Quintero-Betancourt, W., Peele, E.R., and Rose, J.B., *Cryptosporidium parvum* and *Cyclospora cayetanensis*: A review of laboratory methods for detection of these waterborne parasites, *J. Microbiol. Methods*, 49, 209, 2002.
52. Shields, J.M., and Olson, B.H., PCR-restriction fragment length polymorphism method for detection of *Cyclospora cayetanensis* in environmental waters without microscopic confirmation, *Appl. Environ. Microbiol.*, 69, 4662, 2003.
53. Steele, M., Unger, S., and Odumeru, J., Sensitivity of PCR detection of *Cyclospora cayetanensis* in raspberries, basil, and mesclun lettuce, *J. Microbiol. Methods*, 54, 277, 2003.
54. Veiweij, J.J. et al., Detection of *Cyclospora cayetanensis* in travellers returning from the tropics and subtropics using microscopy and real-time PCR, *Int. J. Med. Microbiol.*, 293, 199, 2003.
55. Mundaca, C.C. et al., Use of PCR to improve diagnostic yield in an outbreak of cyclosporiasis in Lima, Peru, *Trans. R. Soc. Trop. Med. Hyg.*, 102, 712, 2008.
56. Relman, D.A. et al., Molecular phylogenetic analysis of *Cyclospora*, the human intestinal pathogen, suggests that it is closely related to *Eimeria* species, *J. Infect. Dis.*, 173, 440, 1996.
57. Jinneman, K.C. et al., An oligonucleotide-ligation assay for the differentiation between *Cyclospora* and *Eimeria* spp. polymerase chain reaction amplification products. *J. Food Prot.*, 62, 682, 1999.
58. Sturbaum, G.D. et al., Detection of *Cyclospora cayetanensis* in wastewater, *Appl. Environ. Microbiol.*, 64, 2284, 1998.
59. Eberhard, M.L. et al., Morphologic and molecular characterization of new *Cyclospora* species from Ethiopian monkeys: *C. cercopitheci* sp. n., *C. colobi* sp. n., and *C. papionis* sp. n, *Energ. Inf. Dis.*, 5, 651, 1999.
60. Lopez, A.S. et al., Epidemiology of *Cyclospora cayetanensis* and other intestinal parasites in a community in Haiti, *J. Clin. Microbiol.*, 41, 2047, 2003.
61. Chu, D.M. et al., Detection of *Cyclospora cayetanensis* in animal fecal isolates from Nepal using an FTA filter-base polymerase chain reaction method, *Am. J. Trop. Med. Hyg.*, 71, 373, 2004.
62. Orlandi, P.A. et al., Targeting single-nucleotide polymorphisms in the 18S rRNA gene to differentiate *Cyclospora* species from *Eimeria* species by multiplex PCR, *Appl. Environ. Microbiol.*, 69, 4806, 2003.
63. Hussein, E.M., Molecular identification of *Cyclospora* spp. using multiplex PCR from diarrheic children compared to others conventional methods, *J. Egypt Soc. Parasitol.*, 37, 585, 2007.

64. Hussein, E.M. et al., Real-time PCR and flow cytometry in detection of *Cyclospora* oocysts in fecal samples of symptomatic and asymptomatic pediatrics patients, *J. Egypt Soc. Parasitol.*, 37, 151, 2007.
65. Adam, R.D. et al., Intervening transcribed spacer region 1 variability in *Cyclospora cayetanensis*, *J. Clin. Microbiol.*, 38, 2339, 2000.
66. Olivier, C. et al., Sequence variability in the first internal transcribed spacer region within and among *Cyclospora* species is consistent with polyparasitism, *Int. J. Parasitol.*, 31, 1475, 2001.
67. Lalonde, L.F., and Gajadhar, A.A., Highly sensitive and specific PCR assay for reliable detection of *Cyclospora cayetanensis* oocysts, *Appl. Environ. Microbiol.*, 74, 4354, 2008.
68. Vesey, G. et al., A new method for the concentration of *Cryptosporidium* oocysts from water, *J. Appl. Bacteriol.*, 75, 82, 1993.
69. da Silva, A.J. et al., Fast and reliable extraction of protozoan parasite DNA from fecal specimens, *Mol. Diagn.*, 4, 57, 1999.
70. Ortega, Y.R., and Sanchez, R., Update on Cyclospora cayetanensis, a food-borne and waterborne parasite, *Clin. Microbiol. Rev.*, 23, 218, 2010.
71. Chacín-Bonilla, L., Epidemiology of Cyclospora cayetanensis: A review focusing in endemic areas, *Acta Trop.*, 115, 181, 2010.
72. Helmy, M.M., *Cyclospora cayetanensis*: A review, focusing on some of the remaining questions about cyclosporiasis, *Infect. Disord. Drug Targets*, 10, 368, 2010.
73. Riner, D.K. et al., Intragenomic sequence variation of the ITS-1 region within a single flow-cytometry-counted Cyclospora cayetanensis oocysts, *J. Parasitol.*, 96, 914, 2010.
74. Taniuchi, M. et al., Multiplex polymerase chain reaction method to detect Cyclospora, Cystoisospora, and Microsporidia in stool samples, *Diagn. Microbiol. Infect. Dis.*, 71, 386, 2011.
75. Masucci, L. et al., First Italian case of cyclosporiasis in an immunocompetent woman: Local acquired infection, *New Microbiol.*, 31, 281, 2008.
76. Naito, T. et al., *Cyclospora* infection in an immunocompetent patient in Japan, *Jpn. J. Infect. Dis.*, 62, 57, 2009.
77. Ozdamar, M., Hakko, E., and Turkoglu, S., High occurrence of cyclosporiasis in Istanbul, Turkey, during a dry and warm summer. *Parasit. Vectors*, 3, 39, 2010.
78. Insulander, M. et al., A foodborne outbreak of *Cyclospora* infection in Stockholm, Sweden, *Foodborne Pathog. Dis.*, 7, 1585, 2010.
79. Hall, R.L., Jones, J.L., and Herwaldt, B.L., Surveillance for laboratory-confirmed sporadic cases of cyclosporiasis—United States, 1997-2008, *MMWR Surveil.l Summ.*, 60, 1, 2011.
80. Iqbal, J. et al., *Cyclospora cayetanensis*: First report of imported and autochthonous infections in Kuwait, *J. Infect. Dev. Ctries.*, 5, 383, 2011.
81. Milord, F. et al., *Cyclospora cayetanensis*: A description of clinical aspects of an outbreak in Quebec, Canada. *Epidemiol. Infect.*, 140, 626, 2012.
82. Eberhard, M.L. et al., Attempts to establish experimental *Cyclospora cayetanensis* infection in laboratory animals, *J. Parasitol.*, 86, 577, 2000.

17 Isospora

Chaturong Putaporntip and Somchai Jongwutiwes

CONTENTS

17.1 INTRODUCTION

Several members of protozoa belonging to the phylum Apicomplexa are human parasites; they include the genera *Plasmodium* and *Babesia* infecting mainly erythrocytes, *Toxoplasma* and *Sarcocystis* inhabiting various tissues, and enteric coccidia in the genera *Isospora*, *Cyclospora*, and *Cryptosporidium* mainly involving intestinal epithelial cells. All enteric coccidian protozoa cause indistinguishable gastrointestinal symptoms; therefore, a laboratory investigation is crucial to make definite diagnosis and, as a consequence, a proper management. *Isospora belli* is the only species pathogenic to humans, causing isosporiasis. *I. belli* is a coccidian protozoa that parasitizes epithelium cells of human small intestines, causing isosporiasis. Historically, the description of characteristic oocysts of *I. belli* in stools was initially observed among military personnel stationed in Egypt, the Middle East, and eastern Mediterranean countries during 1914 and 1921 [1]. Nowadays, it is well recognized that isosporiasis has a cosmopolitan distribution. Although oocysts of *I. natalensis* were reportedly found in human fecal samples, several studies have failed to reaffirm the presence of these oocysts as a causative agent of diarrheal illness in humans since its first report in 1953, raising the possibility that it may not be a valid pathogenic species in humans [2]. Most of the enteric coccidian parasites of man are considered to be opportunistic pathogens except the genus *Sarcocystis*. Thus, the prevalence of these coccidian infections seems to increase as the number of immunocompromised patients increases, especially after the acquired immunodeficiency syndrome (AIDS) becomes pandemic. In general, the severity of illness caused by *I. belli* seems to be more aggressive in patients with compromised immunity than those with normal immune status. However, chronic emaciating infections have been reported in immunocompetent patients, suggesting that host immunity per se may not determine the severity of isosporiasis [3,4]. Meanwhile, difference in intrinsic virulence of this organism has not yet been explored.

17.1.1 Classification and Phylogeny of *Isospora*

The members of protozoa belonging to the phylum Apicomplexa are intracellular organisms that share common structural features by possessing apical complexes at certain stages of their life cycles. Apical complexes are subcellular organelles consisting of polar ring, rhoptries, micronemes, conoid, and subpellicular microtubules that can be identified under a transmission electron microscope. Several lines of evidences indicate that these organelles play a crucial role in host cell invasion and thus exist only at the invasive stages of these protozoa. Parasitic protozoa in the phylum Apicomplexa inhabit tissues or circulation of the host and most of them, including the genus *Isospora*, belong to the class Sporozoea. Taxonomic position based on the biology and characteristics of protozoa in the subclass Coccidia has placed the genus *Isospora* in suborder Eimeriina and family Eimeriidae whose members include genera *Eimeria* and *Cyclospora*. Meanwhile, phylogenetic inference from the small subunit ribosomal RNA sequences of several members of apicomplexan protozoa has placed members of mammal host species of *Isospora*

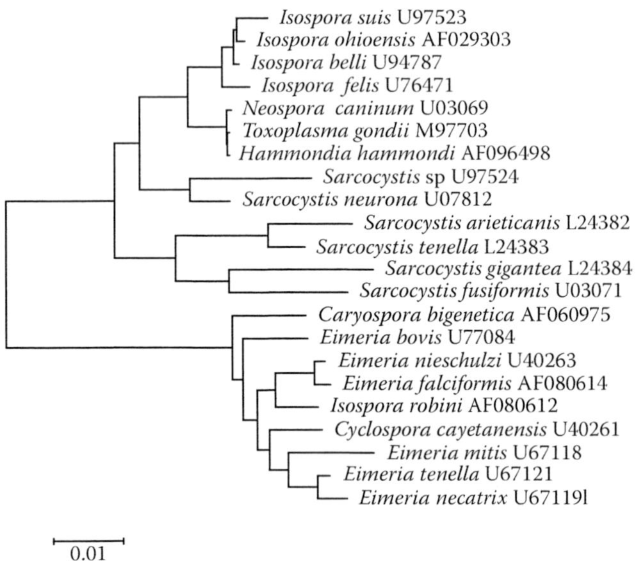

FIGURE 17.1 Phylogeny of *Isospora* and other apicomplexan protozoa inferred from the neighbor-joining method based on the small subunit ribosomal RNA sequences. The evolutionary distances were computed using the maximum composite likelihood method and are in units of the number of base substitutions per site as implemented in the MEGA version 4.0 program [12].

within tissue-cyst-forming coccidia that belong to the family Sarcocystidae including genera *Toxoplasma*, *Neospora*, *Hammondia*, and *Sarcocystis* as a monophyletic group. On the other hand, an avian host species *Isospora robini* is clustered within another monophyletic group with *Caryospora*, *Eimeria*, and *Cyclospora*, all of which are nontissue-cyst-forming coccidia in the family Eimeriidae (Figure 17.1) [5–7]. The presence of tissue cyst in the genus *Isospora* favors phylogenetic rather than phenetic classification [8–10]. Misconceptions on the identification and classification of *I. belli* over seven decades ago led to a description of *I. hominis* which *de facto* belongs to the genus *Sarcocystis*. Other species of *Isospora* are parasitic in wild and domestic animals, for example, *I. arctopitheci* in nonhuman primates, *I. canis* and *I. ohioensis* in dogs, *I. rivolta* and *I. felis* in cats, and *I. suis* in pigs [11]. Over 300 species of *Isospora* have been proposed and found to infect a wide range of animals that includes amphibians, reptiles, birds, and mammals. However, some of these species may require further validation.

17.1.2 Biology, Morphology, and Life Cycle

Isospora is an intracellular parasite undergoing asexual multiplication and sexual reproduction followed by sporogonic development. The life cycle of *I. belli* completes within the human host without known animal reservoirs. Humans get infection by ingestion of food or water contaminated with oocysts (Figure 17.2). The infective stage of *I. belli*

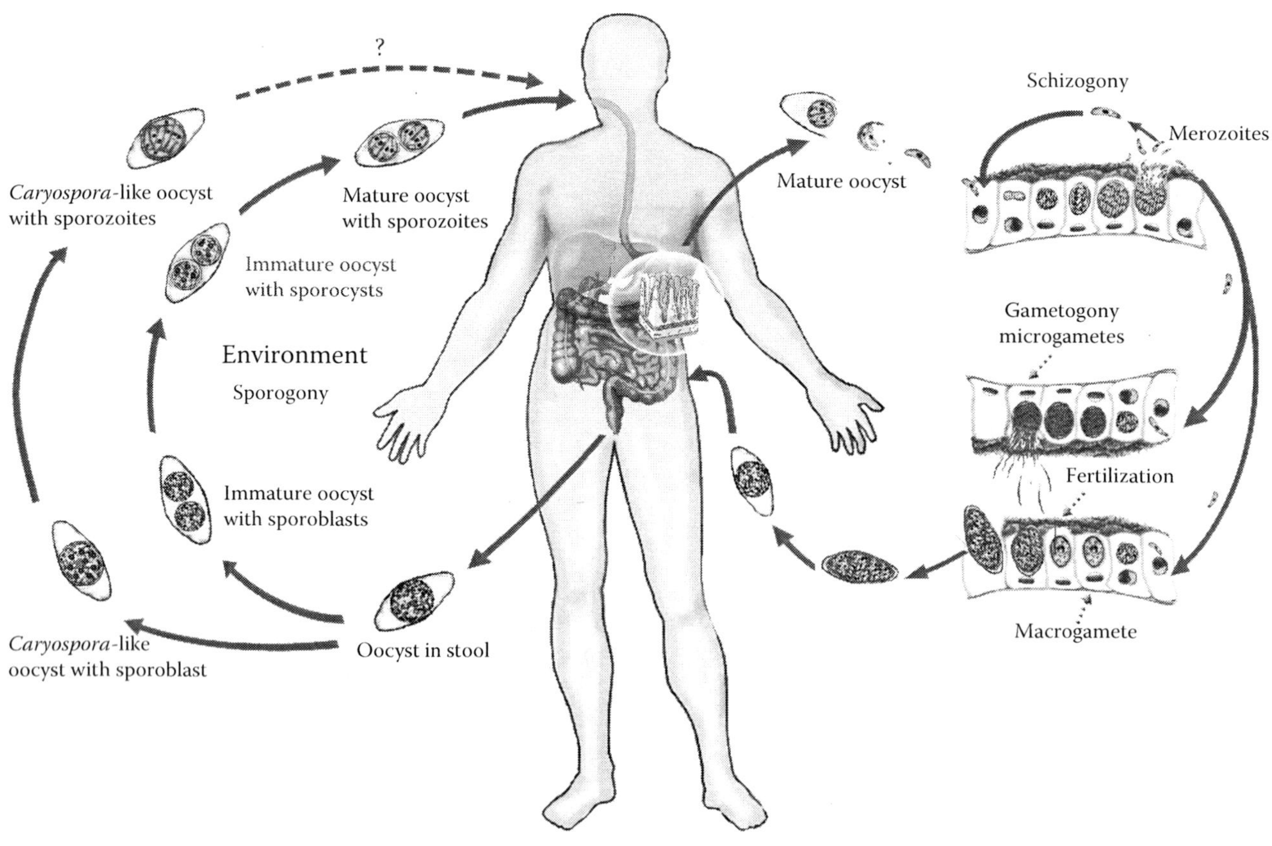

FIGURE 17.2 Life cycle of *Isospora belli* depicting schizogony, gametogony, and sporogony. (Illustration by Urassaya Pattanawong.)

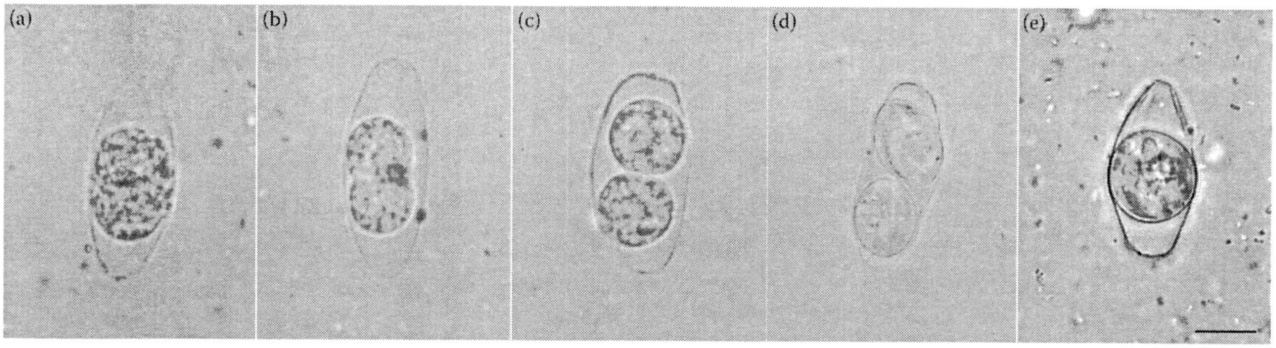

FIGURE 17.3 Oocysts of *Isospora belli*. (a) Oocyst from freshly passed stool containing one sporocyst. (b), (c) Developing oocysts with two sporoblasts. (d) Mature oocyst with two sporocysts, each with four sporozoites. (e) *Caryospora*-like oocyst with eight sporozoites in one sporocyst. Scale = 10 μm.

is characterized by an oval or rugby-shaped oocyst with a wide range of size variation in terms of the maximum length and width, both within and between isolates. The average length measured from more than 700 oocysts is 28.3 μm (range = 17–37). A remarkable narrowing with a neck-like appearance may be observed at one end of some oocysts, while most oocysts are nearly symmetric at both ends. The maximum width of oocysts averages 13.5 μm (range = 8–21). The mean shape index or the ratio of length to width of oocysts is 2.1 (range = 1.3–3.3) [6]. The oocyst wall is bilayer, thin, smooth, and transparent for all stages of development. The inner wall is membranous and the outer wall is rigid and relatively impermeable to external fluids [13]. When freshly passed in feces, the oocyst remains immature, containing a spherical or slightly elongated oval-shaped sporoblast that develops within another rigid cyst wall called the sporocyst. Occasionally, some freshly passed oocysts in stool may contain two sporocysts, each of which possesses one sporoblast. The sporocyst of *I. belli* lacks a proteinaceous plug, known as a Stieda body. The dimension of the sporocyst is 12–14 by 7–9 μm.

Sporogony, the process of producing infective sporozoites, of *Isospora* usually occurs outside the host (Figure 17.3). The maturation period of oocysts usually completes within 24 h to 10 days [6]. The mature oocyst possesses two sporocysts, each with four crescent-shaped or banana-shaped sporozoites clustering together with a clump of granular mass or residual bodies surrounded by the sporocyst wall. Detailed investigation of *ex vivo* development has shown that the percentage of oocyst maturation varies between isolates. *In vitro* studies reveal that only 27–30% of oocysts excreted from stools of patients become fully sporulated at ambient temperature. Importantly, despite the frequently identified characteristic mature oocyst that contains two sporocysts, about 5% of the oocysts that undergo the maturation process are *Caryospora*-like, characterized by eight sporozoites in a single sporocyst [6,14]. The formation of *Caryospora*-like oocysts requires 5–14 days after excretion in feces [6]. Both types of mature oocysts contain a number of granular residual bodies clumping together. Environmental factors such as moisture, temperature, and oxygen composition in the atmosphere have been suggested to influence the maturation of *Isospora* oocysts. It is noteworthy that incubation of *I. rivolta* oocysts at 50°C for 5 min enhances the development of *Caryospora*-like oocysts and they are infective to mice and cats [15].

Sporozoites are released from the oocyst and become free in the lumen of the small intestine, especially around distal duodenum and jejunum. Thereafter, sporozoites invade small intestinal epithelial cells where asexual multiplication ensues. All endogenous stages of *Isospora* reside in parasitophorous vacuoles in the cytoplasm of host cells. Sporozoites of some mammalian *Isospora* species, for example, *I. suis* and *I. felis*, initially undergo repeated endodyogeny generating type I merozoites, in which two daughter cells are generated from a mother cell so that these daughter cells are surrounded by the mother cell membrane [16,17]. Repeated endodyogeny probably occurs and generates a number of merozoites (Figure 17.4). Subsequently, type II merozoites are generated from type I merozoites by schizogonic development in which repeated nuclear division within a single cytoplasmic mass occurs. Cytoplasmic division of multinucleate schizont produces a number of merozoites. Meanwhile, schizogonic development occurs in *I. belli* but asexual multiplication by endodyogeny has not been identified. Fully developed merozoites disintegrate host cells, invade other epithelial cells, and undergo further asexual multiplication cycles.

Extraintestinal tissue cysts of *I. belli* have been found in sections of lymph nodes, spleen, and liver at autopsy of AIDS patients, suggesting that these stages may represent hypnozoites or dormant forms that could be responsible for relapse of symptoms [9,10,18,19]. Ultrastructural study has revealed that the extraintestinal tissue cyst is surrounded by a parasitophorous vacuole membrane and characterized by oval or ellipsoidal body measuring about 11 × 8 μm. A conspicuous cyst wall is observed with a thickness varying from 0.7 to 4 μm, enclosing a crescent-shaped zoite. Occasionally, the surface of zoites possesses grooves or small projections. Each zoite contains a single rather than multiple nuclei. The anterior part of zoites possesses polysaccharide-like and lipid-like granules and apical organelles consisting of rhoptry, microneme, and dense bodies. Although the origin of this developmental process remains elusive, abundant numbers of tissue cysts found at autopsy imply that they probably

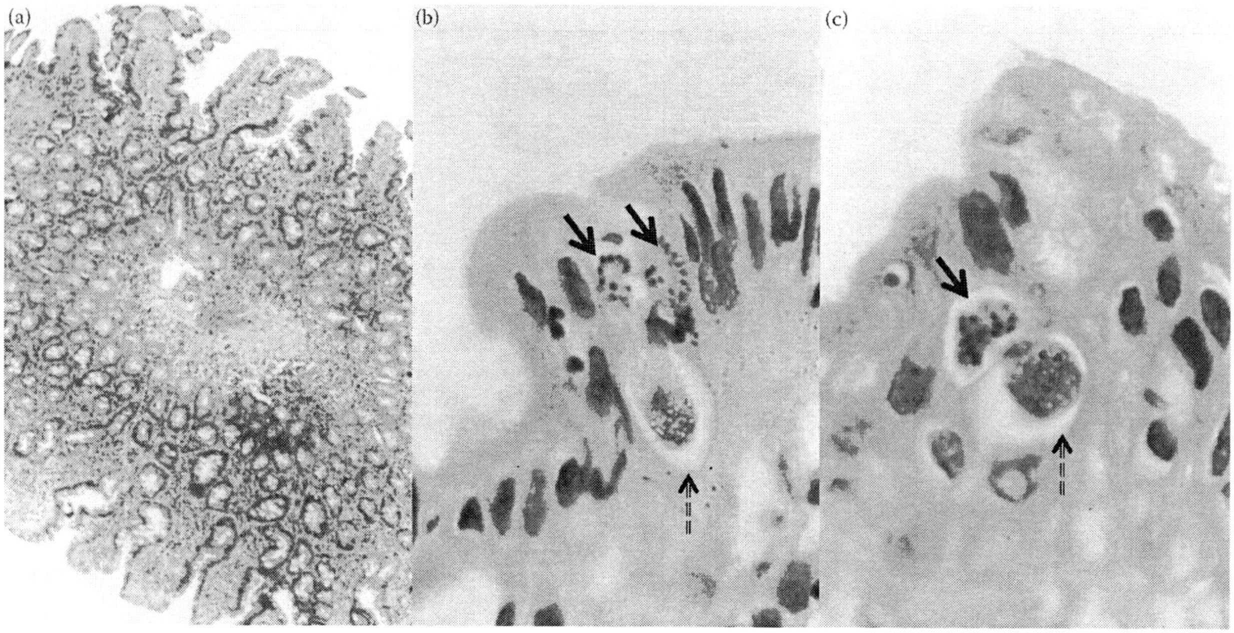

FIGURE 17.4 Upper intestinal biopsy showing (a) shortening of villi, and (b), (c) endogenous stages of *Isospora belli* in epithelium cells. Single and broken arrows indicate schizonts and oocysts, respectively.

originated from merozoites, rather than sporozoites from the oocyst stage [20]. It is noteworthy that extraintestinal stages have been found in feline and canine *Isospora* species [8].

Sexual reproduction or gametogony usually occurs after the initiation of asexual cycle for about 1 week. Female gamont develops intracellularly without nuclear division and becomes macrogamete while male gamont undergoes multiple nuclear and cytoplasmic divisions, resulting in the production of a number of microgametes. Microgametes disrupt host cells and enter other epithelial cells where macrogametes are located. After fertilization, zygote is formed within the epithelial cell and sequentially develops into an oocyst. Intraepithelial oocysts containing granular bodies are excreted in feces upon rupture of host cells. The oocysts are resistant to the environment, including common disinfectants, and remain viable in cool and moist conditions for months [11].

17.1.3 Epidemiology

Transmission of *I. belli* is solely anthroponotic because humans are the only known natural hosts. Before 1935, the prevalence of human isosporiasis seems to be low as only 200 cases were reported [21]. However, in certain circumstances when living condition is overcrowded or sanitation is below the standard level, cases of *I. belli* infections could be encountered. The majority of infected cases were identified mainly among military or related personnel during 1914–1921 [1]. Until 1961, after awareness of the infection along with the competency of microscopic detection, more than 800 cases were detected in various localities in South America [21]. Before the pandemic of human immunodeficiency virus (HIV) infections, sporadic or endemic isosporiasis has been reported from several countries both in tropical and in temperate zones. The infection rates of *I. belli* among immunocompetent individuals are usually less than 2%, while a remarkably higher rate was observed among those with underlying immunodeficiency. The prevalence of isosporiasis in AIDS patients in Zaire, Haiti, Venezuela, Brazil, and Thailand is 19%, 15%, 14%, 9.9%, and 7.6%, respectively [6,22–25]. Nevertheless, endemicity, public sanitation, preexisting anticoccidial drug treatment and other factors may also affect such prevalence. An 8 year surveillance of intestinal infections among AIDS patients in Los Angeles County revealed the presence of *I. belli* in 1% of pathogen-positive stool samples [26]. Interestingly, most of the *I. belli*-positive samples were from foreign-born patients, especially those from Central America. Travel-related infections may be a contributing factor.

17.1.4 Pathology and Clinical Aspects

Infection with *I. belli* is usually confined to mucosal epithelial cells of upper small intestine. Endogenous stages vary in size and structure depending on their growth and development. Both schizogonic and sporogonic developmental cycles may persist continuously for an indefinite period if untreated or probably due to the failure of the gut immune response to control the progression of infection. No obvious pathological change is observed in the early phase of infection except for some leucocyte infiltration in mucosa and lamina propria, most of which are plasma cells, lymphocytes, and eosinophils [9,27]. In chronic isosporiasis, pathological features include shortening or blunting of villi or even atrophy, hypertrophied crypts and lamina propria infiltrated with eosinophils, lymphocytes, and neutrophils [4,28]. In some cases, histopathology of isosporiasis may masquerade eosinophilic enteritis [4]. However, administration of corticosteroid cannot alleviate the symptoms or can even worsen its clinical course. Postmortem examinations reveal that some isosporiasis cases who had underlying AIDS exhibit extraintestinal involvement of lymphoid tissues, causing enlargement of mesenteric lymph nodes, liver, and spleen [9,10,18,19].

A spectrum of clinical manifestations of isosporiasis has been noted. Although both immunocompetent and immunocompromised patients are susceptible to *I. belli* infections, higher prevalence of isosporiasis afflicts the latter more frequently than the former. Incubation period takes about 1 week after ingestion of infective oocysts [29,30]. Asymptomatic infection with spontaneous clearance of oocysts is rarely observed. Symptomatic infections with *I. belli* in immunocompetent hosts present with watery diarrhea, abdominal discomfort, colicky abdominal pain, low-grade fever, malaise, anorexia, nausea, vomiting, headache, and dehydration. Experimental infections in human volunteers show that symptoms precede patency of oocysts in stool for a few days. The duration of symptoms continues for about 1 day while oocysts are discharged in stool for more than 1 month [29]. Diarrheal episode in some patients can be severe with up to 20 stools per day [31]. Diarrheal symptoms usually become chronic if left undiagnosed. Chronic infection may turn into malnutrition, steatorrhea, and cachexia. The excretion of oocysts may last for months or years in chronic isosporiasis. A few reports have shown chronic diarrheal symptoms lasting for over a decade in immunocompetent patients [3,4,28].

Isosporiasis in immunosuppressed patients, such as those who receive immunosuppressive drugs and prolonged corticosteroid treatment, and more commonly in AIDS patients, is generally more severe than those with normal immune status in terms of duration of symptoms and volume of stools. The diarrheal symptoms caused by *I. belli* seem to be a secretory process that may lead to massive fluid loss, resulting in dehydration and electrolyte imbalance. Fever is frequently observed in dehydrated cases. Coinfections with other enteric pathogens can be detected, especially in AIDS patients [6,23,24,32].

Recurrent symptomatic isosporiasis occurs in both immunocompetent and immunodeficient individuals. In a study of isosporiasis in Haitian AIDS patients, relapse occurs in almost half of the cases [23,32]. Extraintestinal isosporiasis identified in lymphoid tissues at postmortem in AIDS patients has suggested that these stages could reinvade mucosal epithelial cells and are responsible for clinical relapse [9,10,18,19]. Although relapse is believed to be caused by reactivation of these dormant forms, its pathogenesis remains to be elucidated. An unusual case of endometrial isosporiasis has been reported in an otherwise immunologically uncompromised host presenting with granulomatous endometritis. Numerous intracytoplasmic cysts resembling *I. belli* oocysts, but not unicellular zoite, were found in the endometrial layer of the uterus [33].

Laboratory tests frequently show eosinophilia in the majority of infected cases but a remarkable alteration of other hematological profiles is not observed. Patients with normal immune status tend to have a higher level of eosinophils than those with suppressed immune system [6]. Charcot–Leyden crystals in stools of isosporiasis patients are frequently observed, indicating active proliferation and degradation of eosinophils in gut tissues. Because endogenous stages of *I. belli* neither cause apparent ulceration of intestinal epithelial cells nor induce intense inflammatory response, a remarkable number of erythrocytes or leucocytes is not found in stool. Identification of characteristic oocysts in stool allows definite diagnosis in routine laboratory practice. The efficiency of detection depends on various factors such as the microscopist's experience, shedding pattern of oocysts that may be fluctuating during the course of infection as well as the choice of diagnostic techniques.

In general, diarrheal symptoms of isosporiasis respond well to treatment with trimethoprim (160 mg) and sulfamethoxazole (800 mg) given orally four times a day for 10 days [23,32]. However, almost half of the patients may require long-term prophylactic treatment to prevent relapse. Several antirelapse regimens have been reported which include trimethoprim plus sulfamethoxazole, and pyrimethamine alone or in combination with sulfadoxine. Because enteric coccidia respond to anti-infective agents differently, definite diagnosis is crucial for proper treatment of isosporiasis. Some differential features of coccidian protozoa that shed their oocysts in host stool samples are listed in Table 17.1 [34].

17.1.5 Conventional Diagnosis of *I. belli*

Conventional laboratory diagnosis of isosporiasis relies on the identification of characteristic oocysts in stool or intestinal content. Occasionally, oocysts of *I. belli* can be overlooked because of its transparency when the adjustment of the microscope is not appropriate. Although endogenous stages can be observed from tissue biopsy samples, they need to be differentiated from other coccidians, especially those of *Cyclospora cayetanensis*. Alternatively, molecular techniques can be applied for the detection of *I. belli* DNA from oocysts in stool or intestinal fluid as well as other endogenous stages in tissue samples.

17.1.5.1 Sample Collection and Preparation

17.1.5.1.1 Stool Sample

Collection of stool samples containing or suspected to contain oocysts of *I. belli* essentially follows standard guidelines for parasitological diagnosis. Fresh stool samples should be collected without contamination of urine or other materials and kept in a clean container with a lid cover to prevent flies or insects that may lay eggs on them. Stool samples should be carefully handled because they may contain various kinds of infectious pathogens, ranging from viruses to parasites. Use universal precautions and wear gloves when handling stool specimens. Although immediate examination of stool specimen is not a crucial issue as those for amoebic trophozoites, it is recommended to examine within 24 h after passage. If the examination of stool samples is not possible within one

TABLE 17.1
Comparative Characteristics of *Isospora belli, Cyclospora cayetanensis, Cryptosporidium* sp., and *Sarcocystis hominis*

Feature	*I. belli*	*C. cayetanensis*	*Cryptosporidium* spp.	*S. hominis*
First report in humans	1914	1977–1978	1974	1925
Location in enterocytes of small intestine	Intracytoplasmic in apical supranuclear region	Intracytoplasmic in apical supranuclear region	Intracellular, extracytoplasmic at luminal surface	Lamina propria
Zoonotic potential	No	Unknown	Yes	Yes
Complete life cycle within humans	Yes	Yes	Yes	No
Average size of oocyst	13.5 μm × 28.3 μm	8–10 μm	4.5–6 μm	13.5 μm × 19.5 μm (sporocyst: 9.3 μm × 14.7 μm)
Acid-fast staining of oocysts	Acid fast	Variably acid fast	Acid fast	Not acid fast (due to thick sporocyst wall)
Number of sporocysts in mature oocyst	2 (1 for *Caryospora*-like oocyst)	2	No sporocyst	2
Number of sporozoites per sporulated oocyst	8	4	4	8
Sporulation outside the host	Yes	Yes	No	No
Autofluorescence of the oocyst wall	Yes	Yes	Yes	Unknown. Oocyst usually disrupted when passed in stool.
Number of oocysts in stools	Low	Low to moderate	Low to very high	Low
Infectivity of oocysts in freshly passed stool	No	No	Yes	Yes (Infective to cattle, not human)
Resistance of oocysts to chlorine	Yes	Yes	Yes	Unknown

Source: Herwaldt, B.L. *Cyclospora cayetanensis*: A review, focusing on the outbreaks of cyclosporiasis in the 1990s. *Clin. Infect. Dis.* 31, 1040 (2000).

day, the stool should be kept at less than 10°C, but not frozen, in order to minimize bacterial overgrowth and decay of fecal materials. In case of handling a large number of samples that cannot be processed within a few days, especially during field survey or excessive workload in the laboratory, adding appropriate preservation is mandatory.

Preservation of stool samples not only maintains the structure of adult worms, ova, trophozoites, and cysts/oocysts, but also halts the development of almost all parasite stages. A number of stool fixatives are available, some of which are suitable for specific purposes. Of these, formalin solution in the concentration of 5% or 10% is most widely used and meets several objectives for subsequent parasitological examination. It has been recommended that 5% formalin solution is appropriate for preserving protozoan cysts, coccidian oocysts, and microsporidian spores, while 10% solution is for helminthes, larvae, and eggs. Formalin solution can be diluted with water, phosphate buffer saline, or normal saline solution. Preparation of 10% formalin solution is by diluting 1 volume of the commercially available formalin with 9 volumes of appropriate diluents. In fact, the actual concentration of the solution is only 3.7–4% because the commercial stock of formalin is supplied as 37–40% solution. Alternatively, other preservatives that are used infrequently can also be used for fixing stool containing *I. belli* oocysts, such as merthiolate–iodine–formalin (MIF) and polyvinyl alcohol (PVA). Usually equal amounts of preservative and stool sample are sufficient to retain the structure of cysts and ova. However, it is important to mix stool and preservative suspension thoroughly by an applicator stick before storage [35].

Although coccidian oocysts can be preserved in 5% or 10% formalin solution for microscopy-based diagnostic purposes, it is not suitable for subsequent biological study or molecular analysis of coccidian parasites. For keeping the viability of *I. belli* oocysts, the stool sample should be preserved in 2–3% (weight/volume) potassium dichromate in water as used for the preservation of viable oocysts of *Cryptosporidium* spp. and *C. cayetanensis* [14]. When molecular analysis of *I. belli* oocysts is required, it is recommended to preserve the stool sample in ethanol [6]. This is by adding 3–4 volumes of absolute ethanol to 1 volume of stool sample and mixing thoroughly to ensure that fecal materials are evenly distributed. Ethanol-preserved stool samples can be stored at ambient temperature for subsequent DNA extraction despite the rise of temperature to 40°C in tropical countries [36]. Although ethanol is not classified as a parasitological fixative, preservation of coccidian oocysts in ethanol also gives good results for subsequent acid-fast stain. Structures of *I.belli* and *Cryptosporidium* oocysts kept in 80% ethanol at ambient temperature for more than 2 years are also well preserved for acid-fast stains [6,36].

17.1.5.1.2 Duodenal Content

Like *Giardia duodenalis* and *Strongyloides stercoralis*, aspiration of duodenal content for direct examination of *I. belli* can be an adjunctive source other than stool samples. Duodenal content should be submitted to the parasitology laboratory without preservative, and examination using the wet smear method should be done soon to visualize motility of parasites such as falling leaf pattern of *G. duodenalis* trophozoites and rapid sine-curve movement of *S. stercoralis* larvae. Usually, nonmotile parasite stages are not vulnerable to time interval between sample collection and microscopic examination. If the sample cannot be examined within a few hours after collection, keeping at less than 10°C is recommended. Preservation of the content with 5–10% formalin solution is adequate for parasitological detection but not for molecular analysis. For molecular diagnosis, centrifugation of the sample prior to preservation in ethanol can reduce the amount of ethanol required. Oocysts of *I. belli* can be seen by direct smear of duodenal content or the acid-fast staining procedure. Mucus in the intestinal tract attracts a number of oocysts so that the sampling of content containing mucus material may help in increasing the recovery rate. In case of voluminous content, examination of sediment after centrifugation at 500 × *g* for 10 min should be done. If the volume of content is small and precludes several preparative diagnostic methods, some authors suggest processing the sample for staining procedure or molecular diagnosis, instead of preparing for direct wet smear examination. Duodenal content can be obtained from various procedures such as the duodenal capsule technique or duodenal aspirate during endoscopic examination of upper gastrointestinal mucosa [37].

17.1.5.1.3 Histological Sample

Biopsy material from upper small intestine during endoscopic examination provides a useful diagnostic sample for *I. belli* as well as other pathogens that parasitize these regions of the gastrointestinal tract such as *G. duodenalis*, *Cryptosporidium* spp., and *C. cayetanensis*. Biopsy material should never be allowed to dry but should be fixed with 10% formalin solution for histopathological study. It is possible to diagnose these parasites in hematoxylin–eosin-stained sections under light microscopy. However, the sensitivity of the test may be variable because the sampling of the region used for biopsy may or may not involve the affected areas. The procedure for hematoxylin–eosin stain as well as preparation of biopsy sections is essentially the same as the standard pathological method. Various stages of *I. belli* can be identified in the cytoplasm of enterocytes lining the upper small intestine, while *Cryptosporidium* spp. are seen as tiny deeply stained bodies at the tips of the mucosal epithelial cells [27,38,39]. Biopsy materials are also good sample sources for molecular detection provided that they should be kept frozen at less than −20°C without preservatives.

17.1.5.2 Conventional Diagnostic Techniques

17.1.5.2.1 Stool Examination

17.1.5.2.1.1 Direct Wet Smear The principle is to sample a small amount of stool or other specimens to search for parasites, ova, trophozoites, cysts, and oocysts. Although the characteristic movement of amoebic trophozoites can be observed within 30 min after stool passage, such immediate attention is not imperative for diagnosis of *I. belli* oocysts. The procedure is as follows:

1. Add one drop of normal saline solution at the center of a clean glass slide (25 mm × 75 mm).
2. Use an applicator stick to sample a small amount of formed or soft stool, about 2 mg, and apply it to the center of the slide.
3. Mix and smear thoroughly within a circle of diameter less than 1 cm.
4. Cover with a clean coverslip (22 mm × 22 mm).
5. Search for *I. belli* oocysts under 10× objective lens by scanning systematically so that all areas of the slide are examined. Confirmation of parasite objects should be done under 40× objective lens.

In case of liquid stool, a drop of normal saline may or may not be required for diluting the sample, depending on the amount of fecal debris. Usually, oocysts of *I. belli* can be best confirmed under 40× objective lens and the microscope diaphragm should be reduced so that a sharp contrast occurs. The transparency of oocysts may preclude identification under light microscopy if contrast of the microscope field is not properly set. Some microscopists may prefer adding a weak iodine solution to the edge of the coverslip. Alternatively, a new slide may be prepared by adding one drop of iodine solution instead of normal saline.

17.1.5.2.1.2 Concentration by Sedimentation Like most helminth eggs and protozoan cysts, oocysts of *I. belli* can be concentrated by formalin–ether or formalin–ethyl acetate sedimentation [40–42].

1. Use about 2–4 g or half a teaspoon of stool specimen and transfer into 10 mL of 5% or 10% formalin solution in a wide-mouth container. Use an applicator stick to mix the sample thoroughly. Allow the sample to be fixed for about 10 min or longer. If the sample has already been preserved in formalin solution, proceed to the next step.
2. Strain the suspension through wet gauze into a conical 15 mL centrifuge tube. If the specimen contains a lot of mucus, it is not recommended to strain the suspension through wet gauze because the mucus that usually traps a lot of coccidian oocysts may stick to, but not pass through, the gauze. Therefore, the mucus portion should be taken for the examination procedure directly.

3. Add 0.85% sodium chloride solution or 5–10% formalin solution to the suspension so that the total volume of fecal suspension from step 2 is 10–12 mL. Centrifuge at 500 × *g* for 10 min. After decanting out the supernatant, the remaining volume of sediment should be 0.5 or 1 mL. Optionally, this step can be repeated to ensure elimination of excess fecal debris.
4. Add 6–8 mL of 5–10% formalin solution. Total volume of fecal suspension is now about half the volume of the tube.
5. Add 3–5 mL of ethyl acetate or diethyl ether. Use a stopper or rubber plug to close the tube. Shake the tube vigorously for at least 30 s while pressing the stopper or rubber plug with some fingers to counter the effect of vapor pressure in the tube. Leave the tube at room temperature for about 30 s. Carefully and slowly remove the stopper or rubber plug.
6. Centrifuge at 500 × *g* for 10 min.
7. Gently taking the tube from the centrifuge without disturbing the four layers that should occur in the suspension, comprising in order from the top to the bottom of the tube: a layer of ethyl acetate (or diethyl ether), a plug of fecal debris, a layer of formalin, and a small amount of sediment at the bottom.
8. Remove the plug of debris by ringing the plug with an applicator stick while decanting all the supernatant part. Usually a few drops of fluid remain attached to the inner side of the tube that will make a small volume of fluid to the sediment. If no fluid remains or the sediment sticks tightly to the bottom of the conical tube, a few drops of saline solution or 5–10% formalin solution can be added.
9. Apply one or a few drops of the sediment fluid to a clean glass slide and cover with a 22 mm × 22 mm coverslip. Optionally, one drop of weak iodine solution may be added prior to covering.
10. Examine systematically for the presence of *I. belli* oocysts as well as other parasites.

17.1.5.2.1.3 Concentration by Flotation Besides the sedimentation method that can be used to concentrate oocysts of *I. belli* (and other parasite ova and protozoan cysts), flotation can be an alternative method. The procedure for flotation concentration was originally described by Faust and colleagues [43]. The modified procedure of Garcia is as follows [42]:

1. Use about 2–4 g or half a teaspoon of stool specimen and transfer it into 10 mL of 5% or 10% formalin solution in a wide-mouth container. Use an applicator stick to mix the sample thoroughly. Allow the sample to be fixed in about 10 min or longer. If the sample has already been preserved in formalin solution, proceed to the next step.
2. Strain the suspension through wet gauze into a round-bottom 15 mL tube. If the specimen contains a lot of mucus, it is not recommended to strain the suspension through wet gauze because the mucus that usually traps a lot of coccidian oocysts may stick to, but not pass through, the gauze. Therefore, the mucus portion should be taken for the examination procedure directly.
3. Add 0.85% sodium chloride solution or 5–10% formalin solution to the suspension so that the total volume of fecal suspension from step 2 is 10–12 mL. Centrifuge at 500 × *g* for 10 min. After decanting out the supernatant, the remaining volume of sediment should be 0.5 or 1 mL. Optionally, this step can be repeated to ensure elimination of excess fecal debris.
4. Resuspend the sediment with 1 or 2 mL of zinc sulfate (33% aqueous solution, specific gravity 1.18). Add additional zinc sulfate solution to fill the tube within 2–3 mm of the rim.
5. Centrifuge at 500 × *g* for 2 min. Allow the centrifuge to stop by its own inertia without deceleration or breaking.
6. Gently take the tube from the centrifuge without disturbing the solution in the tube. Alternatively, some authors recommend not to take the tube out of the centrifuge, but to proceed to the next step while the tube is still in the centrifuge machine.
7. Because the floated materials are on the surface of the solution, harvesting the concentrated samples should be done with care. A Pasteur pipette should be placed exactly at, not below, the surface. Alternatively, a clean wire loop can be used just to touch the surface without dipping into it.
8. Apply the fluid to a clean glass slide and cover with a 22 mm × 22 mm coverslip. Optionally, one drop of weak iodine solution may be added prior to covering.
9. Examine systematically for the presence of *I. belli* oocysts as well as other parasites.

17.1.5.2.1.4 Acid-Fast Stain Conventional microscopy without any staining procedure can detect oocysts of *I. belli* and has been used in routine diagnostic laboratory. The acid-fast staining method, intentionally deployed for detection of *Cryptosporidium* spp. and *Cyclospora* oocysts, has also been found to be useful for the diagnosis of *I. belli* oocysts [6]. A number of acid-fast staining methods are available. Specimen can be versatile, ranging from fresh stool, sediment from formalin–ethyl acetate sedimentation, ethanol-preserved stool sample, duodenal fluid, or other body fluid. One of these is a modified Kinyoun acid-fast stain [44].

1. Make a smear on the clean slide by applying one or two drops of liquid specimen or 2 mg of formed stool. If stool contains mucus, it is recommended to sample the mucus portion because oocysts tend to concentrate in it. Smear should not be too thick.

A proper thickness of smear is the one that should be able to see partly through the background. Allow the smear to dry at ambient temperature.
2. Dip the smear into or flood the smear with absolute methanol for 1 min.
3. Stain by dipping the smear into or flooding the slide with Kinyoun's carbol fuchsin solution. Allow staining for 1 min.
4. Rinse the stained slide excessively with tap water.
5. Dip or flood the stained slide with 10% sulfuric acid for 1 or more min until no more color runs from the slide as a decolorization step.
6. Rinse the stained slide excessively with tap water.
7. Counterstain the slide with methylene blue or brilliant green stain for 1 min.
8. Rinse the stained slide thoroughly with tap water.
9. After the slide is completely dry, examine the stained slide with objectives 10× or 40×. If other coccidian oocysts are to be searched for, use objective 100×.

17.1.5.2.1.5 Autofluorescence Oocysts of *I. belli*, *C. cayetanensis*, and *Cryptosporidium* spp. become autofluorescent bright green under violet excitation (405 nm), bluish violet autofluorescent under ultraviolet excitation (365 nm), and green autofluorescent under blue-violet light excitation (436 nm) [45]. Faint reddish fluorescence is also observed under green light (546 nm). Characteristic size, shape, and structure of *I. belli* oocysts under autofluorescence microscopy differ from those of *C. cayetanensis* and *Cryptosporidium* spp., and thus the method is useful for diagnosis. Autofluorescence is proven to be more sensitive than bright field microscopy in detecting *Isospora* oocysts [46,47].

1. Prepare stool sample on a clean glass slide following the procedure described for the direct wet smear method. Alternatively, 50–100 μL of unstained stool sediment from the concentration method can be used.
2. Examine systematically using 10× and 40× objective lens under a fluorescent microscope using ultraviolet excitation at 365 nm and/or violet excitation at 405 nm.

17.1.5.2.2 Histopathology

Procedures for the identification of *I. belli* in tissue samples essentially follow the standard histopathological method. Biopsy tissue should be fixed with 10% formalin solution, embedded in paraffin or equivalent materials, and stained with hematoxylin–eosin. The endogenous stages can be seen in the cytoplasmic mass of tissues or at the lamina propria. Usually, the endogenous stages of *I. belli* are found in the duodenum and upper jejunum although involvement of biliary ducts and large intestine has been reported [9,10,18]. Stool examination often gives more sensitive results than histopathology because the sampling areas of biopsy are limited and may not include the affected regions. However, a large study involving 118 adult AIDS patients who suffered from chronic diarrhea identified *I. belli* from duodenal biopsy samples in two patients who had negative results in stool examinations [18]. Duodenal mucosal architecture of patients with chronic isosporiasis usually presents with blunting of villi and hypertrophied crypts. Endogenous stages of *I. belli* in intestinal epitheliums consist of schizogony, gametogony, and early sporogony [27]. Inflammatory responses in upper intestinal tissues of both isosporiasis and cyclosporiasis are similar, showing mixed infiltration of polymorphonuclear cells, plasma cells, lymphocytes, and eosinophils although the eosinophilic infiltration seems to be more prominent in isosporiasis [27,48]. Meanwhile, all endogenous stages of *Cryptosporidium* spp. are located at the intracellular rim at the luminal surface of mucosal epithelium and the diameters of all stages do not exceed 6 μm, making them easily differentiable from *I. belli* [38,39]. Extraintestinal stages of *I. belli*, existing predominantly in mesenteric lymph nodes, liver, and spleen of some AIDS patients, are mostly crescent shaped, centrally located, and contain only one zoite in the tissue cyst. The presence of longitudinal grooves or projections of the zoite surface may cause artifactual findings that 2–3 zoites are in tissue cysts. Neither multinucleate nor sexual stages exist in tissue cyst of isosporiasis [20].

17.1.6 Molecular Diagnosis of *I. belli*

Laboratory diagnosis of *I. belli* can be performed by searching for characteristic oocysts in stool or other clinical samples under light microscopy. However, the accuracy of the method largely depends on experienced microscopists. In certain circumstances, endogenous stages of *I. belli* in tissue samples require differentiation from other intestinal coccidians such as *C. cayetanensis* and *Sarcocystis hominis*. Because limited genetic data of *I. belli* are available and the microscopy-based method has been used for routine laboratory diagnosis, little progress has been made for the polymerase chain reaction (PCR)-based method to detect *I. belli*. Amplification of the small subunit ribosomal RNA gene of *I. belli* by nested PCR has been successfully applied to duodenal biopsy, and bile and stool samples without cross-reactivity to other protozoa [6,49]. Although there is no detailed comparative study of the diagnostic value of PCR and microscopy in detecting *I. belli*, nested PCR targeting the small subunit ribosomal RNA gene of *I. suis* in piglet feces is superior to microscopy [50]. Diagnosis of *I. belli* infection by PCR involves the isolation of *I. belli* DNA from clinical specimens that can be stool, duodenal fluid, tissue sample, or other potential sources of infections. The source of *I. belli* DNA is usually from the oocyst stage in stool, while the tissue sample contains any stages of endogenous development.

17.2 METHODS

17.2.1 Sample Preparation

Freshly collected stool specimen, frozen stool sample, or stool preserved in 80% ethanol can be used for extraction of

I. belli DNA. Ethanol in preserved stool samples should be removed before disruption of oocysts. Duodenal fluid or tissue samples can be kept at −20°C and ethanol should not be used for the preservation of these samples.

17.2.1.1 Disruption of *I. belli* Oocysts from Stool Samples

Because the oocyst wall of coccidian protozoa is solid and resistant to diverse environmental conditions, it is recommended that the oocyst wall should be ruptured so that the internal content of either sporoblasts or sporozoites are exposed to DNA extraction solution. Examples of simple procedures to disrupt the oocyst wall of *I. belli* are mechanical disruption and the freeze–thaw method.

17.2.1.1.1 Mechanical Disruption of I. belli *Oocysts*

1. Transfer approximately 200 mg of the formed stool sample or 0.5 mL of liquid stool into a 1.5 mL tube with cap. Add 0.5 mL of sterile water. Mix well until stool becomes homogeneously dispersed by vortex.
2. Optionally, add 0.3 mL of diethyl ether. Cap the tube tightly with cap lock and shake vigorously for 3–5 min.
3. Centrifuge at 5000 × *g* for 5 min. Discard supernatant.
4. Add 0.5 mL of TE buffer (10 mM Tris.HCl, 1 mM EDTA, pH 8.0) to the remaining pellet.
5. Mix well using vortex. If the pellet keeps sticking to the bottom of the tube, try mixing vigorously by pipetting.
6. Add 200–300 mg of glass beads, diameter 425–600 μm (Cat. No. G9268, Sigma), to the suspension. Shake vigorously for 20 min using a vortex mixer. If possible, set at maximum speed.
7. Spin down briefly.
8. Transfer all suspension to a new microcentrifuge tube. Centrifuge at 13,000 × *g* for 5 min. Discard supernatant and save the pellet for DNA extraction step.

17.2.1.1.2 Freeze–Thaw Method

1. Transfer approximately 200 mg of the formed stool sample or 0.5 mL of liquid stool into a 1.5 mL microcentrifuge tube with cap. Add 0.5 mL of sterile water. Mix well until stool becomes homogeneously dispersed by vortex.
2. Optionally, add 0.3 mL of diethyl ether. Cap the tube tightly with cap lock and shake vigorously for 3–5 min.
3. Centrifuge at 5000 × *g* for 5 min. Discard supernatant.
4. Add 0.5 mL of TE buffer to the pellet. (If a commercial DNA extraction kit will be used in the subsequent DNA extraction process, use 0.5 mL of buffer reagent supplied with the kit such as ASL buffer from the QIAamp DNA Stool Mini kit). Mix well until the suspension looks homogenous.
5. Transfer the suspension to a 2 mL cryopreservation tube that is resistant to ultralow temperature (such as Nunc CryoTube™).
6. Prepare liquid nitrogen in a low-temperature-resistant container. Use enough volume to submerge the cryopreservation tube.
7. Set the water bath at 95–100°C.
8. Use forceps or a metal holder to submerge the cryopreservation tube into liquid nitrogen for 3 min.
9. Immediately heat the cryopreservation tube in boiled water for 3 min.
10. Repeat steps 7 and 8 for 5–7 cycles.
11. Centrifuge at 13,000 × *g* for 5 min. Discard supernatant and save the pellet for DNA extraction step. (Omit this step when using buffer supplied in commercial kits.)

17.2.1.2 DNA Extraction

17.2.1.2.1 In-House DNA Extraction Method

This is a standard protocol using proteinase K digestion with phenol–chloroform extraction.

1. Add 440 μL of TE buffer containing 10 mM Tris-HCl, 1 mM EDTA pH.8.0, 50 μL of 10% SDS and 10 μL of proteinase K (20 mg/mL) to the pellet prepared from either mechanical disruption or freeze–thaw method as described. Mix well by inverting the tube repeatedly.
2. Incubate at 50°C overnight or for at least 4 h.
3. Add equal volume of buffer-saturated phenol: chloroform:isoamyl alcohol (25:24:1).
4. Close the tube tightly and shake vigorously for 5–10 min.
5. Centrifuge at 5000 × *g* for 5 min at room temperature. Two layers appear, the upper part contains DNA and the lower is solvent (phenol/chloroform/isoamyl alcohol).
6. Gently pipette the upper part of the solution to a new tube without disturbing the lower layer and the interface where the whitish layer may appear.
7. Repeat steps 3–6 until no visible whitish layer appears.
8. Add equal volume of chloroform.
9. Shake vigorously for 5–10 min.
10. Centrifuge at 5000 × *g* for 5 min.
11. Save the upper layer of the solution to the new tube.
12. Add 2.5–3 volumes of ice-cold absolute ethanol and 1/10 volume of 3.5 M sodium acetate, pH 5.4. Mix well and keep at −80°C for 1 h.
13. Centrifuge at 13,000 × *g* for 15 min at 4°C. Allow to stop without using the brake.
14. Remove the solution gently without disturbing the DNA pellet at the bottom of the tube by using a pipette.
15. Gently rinse with 0.5 mL of ice-cold 80% ethanol.

16. Centrifuge at 13,000 × *g* for 3 min at 4°C. Allow to stop without using the brake.
17. Remove the solution gently without disturbing the DNA pellet at the bottom of the tube by using a pipette.
18. Dry the sample in desiccators or at room temperature.
19. Resuspend in 10–20 μL of TE buffer.

A remarkable drawback of the conventional method described above is that a significant amount of DNA may be lost, especially when a small amount of the sample is used. The remnant of enzymatic inhibitors from stool may significantly hinder PCR amplification.

17.2.1.2.2 DNA Extraction from Stool Samples Using a Commercial Kit

The QIAamp DNA Stool Mini kit (Qiagen) provides a convenient way to extract DNA from *I. belli* and other biological organisms [51].

1. Add 800 μL of ASL buffer to the suspension prepared from mechanical disruption or add 300 μL of ASL buffer to the sample prepared from the freeze–thaw method. Total volume is now about 800 μL.
2. Add one InhibitEX tablet to each sample and vortex immediately and continuously for 2 min or until the tablet is completely suspended. Incubate the suspension for 2 min at room temperature.
3. Centrifuge the sample at 20,000 × *g* for 3 min to pellet the inhibitors bound to InhibitEX.
4. Pipette all of supernatant into a new 1.5 mL tube and discard the pellet. Centrifuge the sample at 20,000 × *g* for 3 min.
5. Pipette 15 μL of proteinase K into a new 1.5 mL tube.
6. Pipette all of supernatant from step 4 into a tube of step 5 (Note: if the volume of supernatant is less than 200 μL, add more ASL buffer to make 200 μL in the final volume).
7. Add 200 μL of AL buffer supplied in the kit and mix by vortexing for 20 s.
8. Incubate at 75°C for 20 min. Spin down the sample before the next step.
9. Add 200 μL of ethanol (96–100%) to the sample, and mix by vortexing. Spin down the sample before the next step.
10. Label the lid of a new QIAamp spin column placed in a 2 mL collection tube. Carefully apply the complete lysate from step 9 to the QIAamp spin column without moistening the rim. Close the cap and centrifuge at 20,000 × *g* for 1 min. Place the QIAamp spin column in a new 2 mL collection tube, and discard the tube containing the filtrate.
11. Carefully open the QIAamp spin column and add 500 μL of AW1 buffer. Centrifuge at 20,000 × *g* for 1 min. Place the QIAamp spin column in a new 2 mL collection tube, and discard the collection tube containing the filtrate.
12. Carefully open the QIAamp spin column and add 500 μL of AW2 buffer. Centrifuge at 20,000 × *g* for 3 min. Discard the collection tube containing the filtrate.
13. Place the QIAamp spin column in a new 1.5 mL microcentrifuge tube. Centrifuge at 20,000 × *g* for 1 min. Discard the collection tube containing the filtrate.
14. Transfer the QIAamp spin column into a new labeled 1.5 mL microcentrifuge tube and pipette 30–50 μL of AE buffer directly onto the QIAamp membrane. Incubate at room temperature for 5 min and then centrifuge at 20,000 × *g* for 1 min to elute DNA.

17.2.1.2.3 DNA Extraction from Tissue Samples Using a Commercial Kit

Although tissue stages of *I. belli* differ remarkably from *Cryptosporidium* because of the difference in cellular locations of these organisms, sometimes certain endogenous stages of *I. belli*, *C. cayetanensis*, and *S. hominis* may appear almost indistinguishable, leading to difficulty in making a definite diagnosis from tissue biopsy materials based on morphological features per se. Meanwhile, formalin-fixed, paraffin-embedded tissue samples are usually available in most laboratory services of hospitals. These archived samples could potentially provide a wealth of information in retrospective molecular studies of *I. belli* infections provided that they have been appropriately kept and preserved (usually in alcohol or formalin). Several commercial kits offer adequate DNA isolation from these samples for PCR detection. One of these, the DNeasy Blood & Tissue kit (Qiagen), is a convenient process in obtaining DNA from intestinal biopsy materials or from formalin-fixed, paraffin-embedded tissue samples [52]. The length of DNA purified from fixed tissues is usually less than 650 bp; therefore, the target PCR products should be much shorter to obtain acceptable sensitivity.

1. Place a section of paraffin-embedded tissue of not more than 25 mg in a 2 mL microcentrifuge tube.
2. Add 1200 μL of xylene. Vortex vigorously.
3. Centrifuge at 20,000 × g for 5 min at room temperature (15–25°C).
4. Remove supernatant by pipetting. Do not remove any of the pellets.
5. Add 1200 μL of 96–100% ethanol to the pellet to remove residual xylene, and mix gently.
6. Centrifuge at 20,000 × g for 5 min at room temperature.
7. Carefully remove the ethanol by pipetting. Do not remove any of the pellets.
8. Repeat steps 5–7 once.
9. Incubate the open microcentrifuge tube at 37°C for 10–15 min until ethanol has evaporated.

10. Resuspend the tissue pellet in 180 mL of ATL buffer supplied in the kit. Proceed to step 12.
11. For a fresh tissue sample of 10–25 mg, cut it up into small pieces, and place it in a 1.5 mL microcentrifuge tube. Alternatively, tissue can be ground in liquid nitrogen or using a homogenizer.
12. Add 180 μL of ATL buffer.
13. Add 20 μL of proteinase K. Mix by vortexing and incubate at 56°C until tissue is completely lysed. Vortex occasionally during incubation to disperse the sample. The incubation time can be 1–3 h or overnight to ensure complete lysis of tissue.
14. Vortex for 15 s. Add 200 mL of AL buffer (supplied in the kit) to the sample, and mix thoroughly by vortexing.
15. Add 200 μL of 96–100% ethanol immediately and mix again thoroughly by vortexing.
16. Pipette the mixture from step 6, including any precipitate into the DNeasy Mini spin column placed in a 2 mL collection tube. Centrifuge at ≥6000 × *g* for 1 min. Discard the flow-through and collection tube.
17. Place the DNeasy Mini spin column in a new 2 mL collection tube. Add 500 μL of AW1 buffer. Centrifuge at ≥6000 × *g* for 1 min. Discard the flow-through and collection tube.
18. Place the DNeasy Mini spin column in a new 2 mL collection tube. Add 500 μL of AW2 buffer. Centrifuge at 20,000 × *g* for 3 min to dry the DNeasy membrane. Discard the flow-through and collection tube.
19. Place the DNeasy Mini spin column in a clean 1.5 mL microcentrifuge tube, and pipette 100–200 μL of AE buffer onto the DNeasy membrane. Centrifuge at ≥6000 × *g* for 1 min to elute. For maximum DNA yield, repeat elution once as described in step 10.

17.2.2 Detection Procedure

We present below a nested PCR targeting the small subunit ribosomal RNA gene for the detection of *I. belli* [6,46]. Because of the extreme sensitivity of the nested PCR, it is important to work under clean conditions and use clean equipments to minimize cross-contamination and PCR inhibition. Wear clean powder-free gloves to handle all the preparative processes.

1. In a 0.2 μL tube designed to fit a PCR thermal cycler that is precooled on ice, mix together the following:

Template DNA	5.0 μL
Forward PCR primer 1 (0.1 μM)	0.5 μL
Reverse PCR primer 2 (0.1 μM)	0.5 μL
10× PCR buffer (supplied in a PCR kit)	5.0 μL
dNTP mixture (2.5 mM each)	4.0 μL
Taq DNA polymerase (5 U/μL)	0.2 μL
Sterile water (DNase and RNase free)	34.8 μL

For diagnostic PCR, use positive and negative control DNA in separate reaction tubes. Positive control can be genomic or recombinant DNA of *I. belli*. Forward and reverse PCR primers 1 and 2 are IBITS1F0 (5′-ATGGACGGTGAGGAAGCGGT-3′) and IBITS2R0 (5′-CCACCGTATCTCAAAGATGT-3′) whose sequences are derived from internal transcribed spacers 1 and 2, respectively, of *I. belli* (Figure 17.5). The amplified product contains 684 base pairs.

2. Place the tube in the preheated PCR thermal cycler and start the thermal cycling profile as follows: 1 cycle of 94°C for 1 min; 40 cycles of 94°C for 30 s, 60°C for 30 s, and 72°C for 1 min; 1 cycle of 72°C for 5 min.
3. Use 1 μL of the primary PCR sample as template DNA for nested PCR. The composition of PCR mix is as follows:

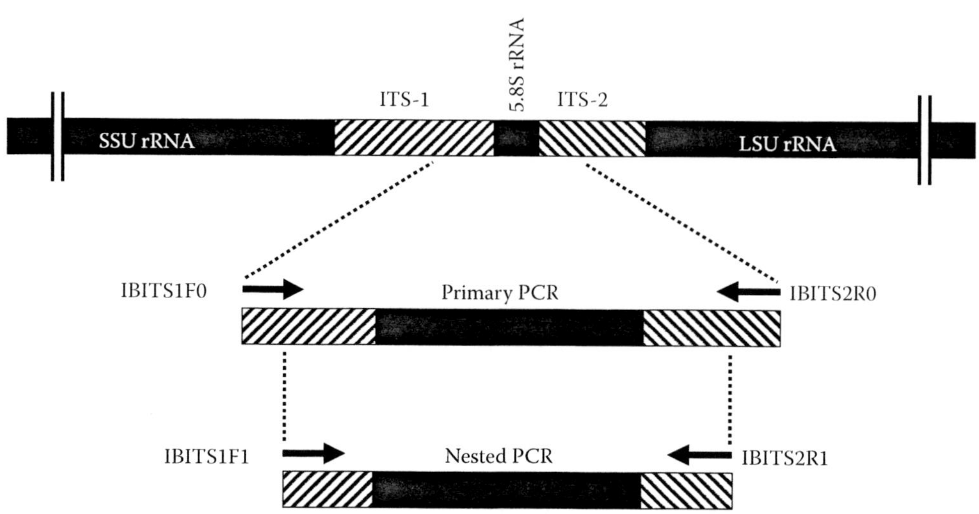

FIGURE 17.5 Schematic representation of the RNA cistron of *Isospora belli*. Locations of forward and reverse PCR primers are indicated by directional arrows.

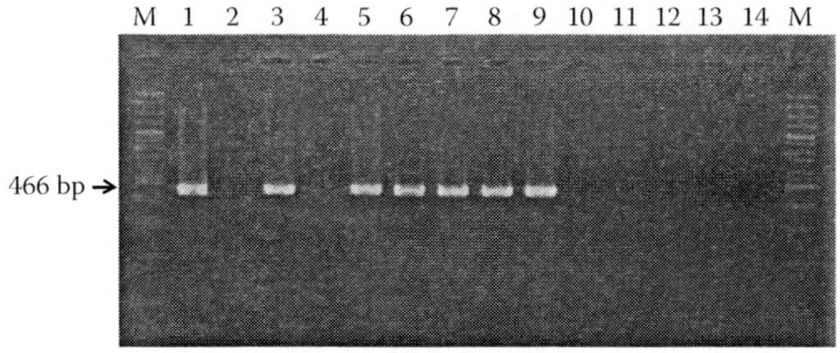

FIGURE 17.6 Agarose gel electrophoresis of nested PCR amplification using primers derived from internal transcribed spacer 1 (ITS-1) and ITS-2. DNA was extracted by using a commercial purification kit (Qiagen). Lanes M, molecular markers of 100-bp DNA ladder; lane 1, positive control; lanes 2 and 4, negative control; lane 3, sample with one oocyst; lanes 5 and 6, samples with 5 oocysts; lanes 7 and 8, samples with 20 oocysts; lane 9, stool sample from a patient with *Isospora belli* oocysts in stool; lanes 10–14, *Giardia duodenalis*, *Entamoeba histolytica*, *Cyclospora cayetanensis*, *Cryptosporidium hominis*, and human DNA, respectively.

Primary PCR product	1.0 μL
Forward PCR primer 3 (0.1 μM)	0.5 μL
Reverse PCR primer 4 (0.1 μM)	0.5 μL
10× PCR buffer (supplied in a PCR kit)	5.0 μL
dNTP mixture (2.5 mM each)	4.0 μL
Taq DNA polymerase (5 U/μL)	0.2 μL
Sterile water (DNase and RNase free)	38.8 μL

Forward and reverse PCR primers 3 and 4 are IBITS1F1 (5′-ATTCTCTCCCACTTGGTGGAA-3′) and IBITS2R1 (5′-ACACTGCCACACGCGTATTC-3′) whose sequences are derived from the internal transcribed spacers 1 and 2, respectively, of *I. belli* (Figure 17.5).

4. Place the tube in the preheated PCR thermal cycler and start the thermal cycling profile as follows: 1 cycle of 94°C for 1 min; 30 cycles of 94°C for 30 s, 60°C for 30 s, and 72°C for 40 s; 1 cycle of 72°C for 5 min.
5. Use 5 μL of the nested PCR sample to check for the presence of amplified DNA products by agarose minigel electrophoresis. The expected PCR product contains 466 base pairs.

The primers derived from internal transcribed spacers 1 and 2 sequences under the PCR conditions as described above give a sensitive and specific result capable of detecting a single oocyst of *I. belli*. A minimum of five oocysts can be reproducibly amplified by nested PCR and these primers do not cross-hybridize with other enteric protozoa as well as human DNA (Figure 17.6).

17.3 CONCLUSION AND FUTURE PERSPECTIVES

It has been almost a century since *I. belli* was recognized as a human pathogen. Since then the prevalence of human isosporiasis seems to be low in the general population compared with other human enteric pathogenic protozoa. Not until the pandemic of HIV infections that *I. belli* has been listed as an important opportunistic parasite causing severe and prolonged debilitating diarrheal illness. However, lack of awareness of this enteric coccidian as a causative agent of diarrhea in immunocompetent hosts as well as the lack of diagnostic concern of this pathogen could account for underestimation of the prevalence.

Like most other enteric coccidians, the developmental cycle of *I. belli* consists of both asexual and sexual reproduction in a single host. Freshly excreted oocysts from infected human stool are not readily infective but require some time in the environment to reach maturation, producing infective eight crescent-shaped sporozoites per oocyst. The oocysts are environmentally resistant and can retain viability for months at low temperature. Invasion of upper small intestinal enterocytes by sporozoites initiates the asexual developmental cycle where schizogony ensues. To date, endodyogenic development preceding schizogony has not been identified in *I. belli* as seen in some other mammalian *Isospora* species. However, extraintestinal unicellular zoites, found mostly in lymphoid tissues of some AIDS patients, are suggestive of the dormant stage that could be responsible for relapse in almost half of the symptomatic isosporiasis cases. The high rate of relapse in *I. belli* infection suggests the presence of the drug-resistant cryptic stage that is not responsive to treatment. Whether the cryptic stage is an extraintestinal dormant stage and occurs in all *I. belli*-infected patients requires further investigation.

Isosporiasis is more prevalent in immunocompromised patients than in immunocompetent individuals and the severity of illness seems to be more in the former. However, chronic emaciating infections have reportedly occurred in patients without compromised immunity, raising the possibility that other factors, in addition to host immunity, such as parasite strain difference could also contribute to disease severity. The lack of appropriate animal models hinders one from conducting experimental studies to elucidate the pathogenesis of *I. belli*. Meanwhile, genomic analysis as well as whole-genome sequence determination of *I. belli* is imperative to shed light on this neglected coccidian.

Although conventional stool examinations by direct smear and concentration methods are routinely used to detect *I. belli* oocysts, the transparency of oocysts can mask correct diagnosis. Staining by acid-fast techniques aiming at diagnosing *Cryptosporidium* and *Cyclospora* is also useful in detecting oocysts of *Isospora*, but the sensitivity is almost the same as that for conventional examination by an expert microscopist. Alternative diagnosis by the examination of wet smear of stool samples by fluorescence microscopy can be used to detect autofluoresced *I. belli* oocysts. Recent advances in laboratory diagnosis of isosporiasis include the development of the PCR-based method targeting the ribosomal RNA cistron. Theoretically and practically, the molecular method is far superior to other diagnostic techniques. Nevertheless, the presence of PCR inhibitors in stool can drastically lower its sensitivity during the DNA extraction process. Addition of recombinant

I. belli DNA as a positive control in tested samples can be useful in ruling out the presence of such interference.

The recent administration of highly active antiretroviral therapy has remarkably reduced the prevalence of several opportunistic infections, including isosporiasis. Nevertheless, *I. belli* can circulate in both immunocompetent and immunocompromised hosts and has a wide geographic range. The clinical significance of *I. belli* infection will certainly remain and be in differential diagnosis of diarrheal illness. In this regard, sensitive and specific diagnostic tools are essential.

Epidemiological surveys have shown a range of prevalence levels of isosporiasis. However, most studies revealed comparatively low prevalence. Because human is the only natural host of *I. belli* and the number of oocysts excreted from infected individuals is usually low, asymptomatic infected persons could serve and maintain *I. belli* to circulate within the population, while the oocysts might effectively survive in harsh environments to continue its infectiousness. Undoubtedly, further research on its biological aspects should form a basis for successful prevention, control, and eradication of this opportunistic parasite.

REFERENCES

1. Wenyon, C.M. Coccidiosis of cats and dogs and the status of the *Isospora* of man. *Ann. Trop. Med. Parasitol.* 17, 231, 1923.
2. Elsdon-Dew, R. *Isospora natalensis* (sp. Nov.) in man. *J. Trop. Med. Hyg.* 56, 149, 1953.
3. Ravenel, J.M., Suggs, J.L., and Legerton, C.W. Human coccidiosis. Recurrent diarrhea of 26 years duration due to *Isospora belli*: A case report. *J. S. Carolina Med. Assoc.* 72, 217, 1976.
4. Jongwutiwes, S., Sampatanukul, P., and Putaporntip, C. Recurrent isosporiasis over a decade in an immunocompetent host successfully treated with pyrimethamine. *Scand. J. Infect. Dis.* 34, 859, 2002.
5. Franzen, C. et al. Taxonomic position of the intestinal protozoan parasite *Isospora belli* as based on ribosomal RNA sequences. *Parasitol. Res.* 86, 669, 2000.
6. Morrison. D.A. et al. The current status of the small subunit rRNA phylogeny of the coccidia (Sporozoa). *Int. J. Parasitol.* 34, 501, 2004.
7. Jongwutiwes, S. et al. Morphologic and molecular characterization of *Isospora belli* oocysts from patients in Thailand. *Am. J. Trop. Med. Hyg.* 77, 107, 2007.
8. Dubey, J.P., and Frenkel, J.K. Extra-intestinal stages of *Isospora felis* and *I. rivolta* (Protozoa: Eimeriidae) in cats. *J. Protozool.* 19, 89, 1972.
9. Restrepo, C., Macher, A.M., and Radany, E.H. Disseminated extraintestinal isosporiasis in a patient with acquired immune deficiency syndrome. *Am. J. Clin. Pathol.* 87, 536, 1987.
10. Michiels, J.F. et al. Intestinal and extraintestinal *Isospora belli* infection in an AIDS patient. A second case report. *Pathol. Res. Pract.* 190, 1089, 1994.
11. Lindsay, D.S., Dubey, J.P., and Blagburn, B.L. Biology of *Isospora* spp. from humans, nonhuman primates, and domestic animals. *Clin. Microbiol. Rev.* 10, 19, 1997.
12. Tamura, K. et al. MEGA4: Molecular evolutionary genetics analysis (MEGA) software version 4.0. *Mol. Biol. Evol.* 24, 1596, 2007.
13. Beaver, P.C., Jung, R.C., and Cupp, E.W. Coccidia, Microsporidia and *Pneumocystis*. In: Beaver, P.C., Jung, R.C., and Cupp, E.W. (eds), *Clinical Parasitology*, 9th edn. Lea and Febiger, Philadelphia, PA, pp. 149–173, 1984.
14. Zaman, V. Observations on human *Isospora*. *Trans. R. Soc. Trop. Med. Hyg.* 62, 556, 1968.
15. Matsui, T. et al. Infectivity and sporogony of *Caryospora*-type oocyst of *Isospora rivolta* obtained by heating. *Parasitol. Res.*, 79, 599, 1993.
16. Ferguson, D.J.P. et al. Ultrastructural observations on multiplication of *Cystoisospora* (*Isospora*) *felis* by endodyogeny. *Z. Parasitenkd.* 63, 289, 1980.
17. Matuschka, F.R. Ultrastructural evidence of endodyogeny in *Isospora suis* from pigs. *Z. Parasitenkd.* 67, 27, 1982.
18. Velásquez, J.N. et al. Isosporosis and unizoite tissue cysts in patients with acquired immunodeficiency syndrome. *Hum. Pathol.* 32, 500, 2001.
19. Frenkel, J.K. et al. *Isospora belli* infection: Observation of unicellular cysts in mesenteric lymphoid tissues of a Brazilian patient with AIDS and animal inoculation. *J. Eukaryot. Microbiol.* 50(Suppl), 682, 2003.
20. Lindsay, D.S. et al. Examination of extraintestinal tissue cysts of *Isospora belli*. *J. Parasitol.* 83, 620, 1997.
21. Faust, E.C. et al. Human isosporosis in the Western hemisphere. *Am. J. Trop. Med. Hyg.* 10, 343, 1961.
22. Henry, M.C. et al. Parasitological observations of chronic diarrhoea in suspected AIDS adult patients in Kinshasa (Zaire). *Trans. R. Soc. Trop. Med. Hyg.* 80, 309, 1986.
23. DeHovitz, J.A. et al. Clinical manifestations and therapy of *Isospora belli* infection in patients with the acquired immunodeficiency syndrome. *N. Engl. J. Med.* 315, 87, 1986.
24. Certad, G. et al. Isosporiasis in Venezuelan adults infected with human immunodeficiency virus: Clinical characterization. *Am. J. Trop. Med. Hyg.* 69, 217, 2003.
25. Lainson, R. and da Silva, B.A. Intestinal parasites of some diarrhoeic HIV-seropositive individuals in North Brazil, with particular reference to *Isospora belli* Wenyon, 1923 and *Dientamoeba fragilis* Jepps & Dobell, 1918. *Mem. Inst. Oswaldo. Cruz.* 94, 611, 1999.
26. Sorvillo, F.J. et al. Epidemiology of isosporiasis among persons with acquired immunodeficiency syndrome in Los Angeles County. *Am. J. Trop. Med. Hyg.* 53, 656, 1995.
27. Brandborg, L.L., Goldberg, S.B., and Breidenbach, W.C. Human coccidiosis: A possible cause of malabsorption. *N. Engl. J. Med.* 283, 1306, 1970.
28. Trier, J.S. et al. Chronic intestinal coccidiosis in man: Intestinal morphology and response to treatment. *Gastroenterol.* 66, 923, 1974.
29. Matsubayashi, H. and Nozawa, T. Experimental infection of *Isospora hominis* in man. *Am. J. Trop. Med.* 28, 633, 1948.
30. McCracken, A.W. Natural and laboratory-acquired infection by *Isospora belli*. *South. Med. J.* 65, 800, 1972.
31. Syrkis, I. et al. A case of severe human coccidiosis in Israel. *Israel. J. Med. Sci.*, 1, 373, 1975.
32. Pape, J.W., Verdier, R.I., and Johnson, W.D. Jr. Treatment and prophylaxis of *Isospora belli* infection in patients with the acquired immunodeficiency syndrome. *N. Engl. J. Med.*, 320, 1044, 1989.
33. de Otazu, R.D. et al. Endometrial coccidiosis. *J. Clin. Pathol.*, 57, 1104, 2004.
34. Herwaldt, B.L. *Cyclospora cayetanensis*: A review, focusing on the outbreaks of cyclosporiasis in the 1990s. *Clin. Infect. Dis.* 31, 1040, 2000.

35. Garcia, L.S. Collection, preservation, and shipment of fecal specimens. In: Garcia, L.S. (ed), *Diagnostic Medical Parasitology*, 4th edn. ASM Press, Washington, DC, pp. 723–740, 2001.
36. Jongwutiwes, S. et al. Simple method for long-term copro-preservation of *Cryptosporidium* oocysts for morphometric and molecular analysis. *Trop. Med. Int. Health.* 7, 257, 2002.
37. Garcia, L.S. Macroscopic and microscopic examination of fecal specimens. In: Garcia, L.S. (ed), *Diagnostic Medical Parasitology*, 4th edn. ASM Press, Washington, DC, pp. 741–785, 2001.
38. Nime, F.A. et al. Acute enterocolitis in a human being infected with the protozoan *Cryptosporidium. Gastroenterol.* 70, 592, 1976.
39. Meisel, J.L. et al. Overwhelming watery diarrhea associated with a *Cryptosporidium* in an immunosuppressed patient. *Gastroenterol.* 70, 1156, 1976.
40. Ritchie, L. An ether sedimentation technique for routine stool examinations. *Bull. U.S. Army Med. Dept.* 8, 326, 1948.
41. Young, K.H. et al. Ethyl acetate as a substitute for diethyl ether in the formalin-ether sedimentation technique. *J. Clin. Microbiol.*, 10, 852, 1979.
42. Garcia, L.S. Sputum, aspirates, and biopsy material. In: Garcia, L.S. (ed), *Diagnostic Medical Parasitology*, 4th edn. ASM Press, Washington, DC, pp. 809–828, 2001.
43. Faust, E.C. et al. A critical study of clinical laboratory technics for the diagnosis of protozoan cysts and helminth eggs in feces. *Am. J. Trop. Med.* 18, 169, 1938.
44. Ma, P., and Soave, R. Three-step stool examination for cryptosporidiosis in 10 homosexual men with protracted watery diarrhea. *J. Infect. Dis.* 147, 824, 1983.
45. Varea, M. et al. Fuchsin fluorescence and autofluorescence in *Cryptosporidium*, *Isospora* and *Cyclospora* oocysts. *Int. J. Parasitol.* 28, 1881, 1998.
46. Daugschies, A. et al. Autofluorescence microscopy for the detection of nematode eggs and protozoa, in particular *Isospora suis*, in swine feces. *Parasitol. Res.* 87, 409, 2001.
47. Bialek, R. et al. Comparison of autofluorescence and iodine staining for detection of *Isospora belli* in feces. *Am. J. Trop. Med. Hyg.* 67, 304, 2002.
48. Sun, T. et al. Light and electron microscopic identification of *Cyclospora* species in the small intestine. Evidence of the presence of asexual life cycle in human host. *Am. J. Clin. Pathol.* 105, 216, 1996.
49. Müller, A. et al. Detection of *Isospora belli* by polymerase chain reaction using primers based on small-subunit ribosomal RNA sequences. *Eur. J. Clin. Microbiol. Infect. Dis.* 19, 631, 2000.
50. Joachim, A. et al. Detection of *Isospora suis* (Biester and Murray 1934) in piglet faeces: Comparison of microscopy and PCR. *J. Vet. Med. B.* 51, 140, 2004.
51. QIAamp DNA Stool Mini Kit Handbook. For DNA purification from stool samples, August 2001.
52. QIAGEN DNeasy® Blood & Tissue Handbook. For purification of total DNA from animal tissue, July 2006.

18 Plasmodium

Dongyou Liu and Yi-Wei Tang

CONTENTS

18.1 INTRODUCTION

The genus *Plasmodium* covers a diverse group of protozoan parasites whose life cycle involves a vertebrate host (e.g., human, monkey, rodent, bird, and reptile) and a mosquito vector. As the causal agents for malaria, a disease that has been recognized since time immemorial and that was thought to result from exposure to bad air or gas from swamps (mal air ia or miasma), *Plasmodium* was only identified in 1880, and the role of mosquito in its transmission was elucidated in 1897. Of >200 species within the genus, five (i.e., *Plasmodium falciparum*, *Plasmodium vivax*, *Plasmodium ovale*, *Plasmodium malariae*, and *Plasmodium knowlesi*) are capable of inducing clinical disease of varying severity in humans, affecting more than 300 million people and resulting in approximately 1.5 million deaths each year.

18.1.1 Classification and Life Cycle

18.1.1.1 Classification

The genus *Plasmodium* is classified in the family Plasmodiide, order Hemosporidia, phylum Apicomplexa. In addition to *Plasmodium*, the family *Plasmodiide* includes the genera of *Bioccala*, *Billbraya*, *Dionisia*, *Hepatocystis*, *Mesnilium*, *Nycteria*, *Polychromophilus*, *Rayella*, and *Saurocytozoon*. The genus *Plasmodium* can be further distinguished into the subgenera of *Asiamoeba*, *Bennettinia* (type species *Plasmodium juxtanucleare*), *Carinamoeba* (type species *Plasmodium minasense*), *Fallisia*, *Garnia*, *Giovannolaia* (type species *Plasmodium circumflexum*), *Hemamoeba* (type species *Plasmodium relictum*), *Huffia* (type species *Plasmodium elongatum*), *Lacertaemoba*, *Laverania* (type species *P. falciparum*), *Novyella* (type species *Plasmodium vaughani*), *Ophidiella* (type species *Plasmodium weyoni*), *Plasmodium*, *Paraplasmodium*, *Sauramoeba* (type species *Plasmodium agame*), and *Vinckeia*. To date, more than 200 species have been described within the various subgenera of the genus *Plasmodium*. Although the full taxonomic name of a *Plasmodium* species comprises the subgenus, the latter is often omitted for simplicity.

Depending on the type of vertebrate hosts they infect, members of the genus *Plasmodium* are often separated into four categories: (i) *Plasmodium* spp. infecting humans and other primates (including *P. falciparum* in the subgenus *Laverania*; *P. malariae* in the subgenus *Plasmodium*; *P. bubalis* in the subgenus *Vinckeia*, etc); (ii) *Plasmodium* spp. infecting mammals other than primates (e.g., rodent-infecting *P. bergei*, *P. chabaudi*, and *P. yoelli*); (iii) *Plasmodium* spp. infecting birds (including members of the subgenera *Giovannolaia*, *Hemamoeba*, *Huffia*, *Novyella*, and *Bennettinia*); and (iv) *Plasmodium* spp. infecting reptiles (e.g., lizard-infecting *P. mexicanum* and snake-infecting *P. wenyoni*).

Traditional classification of the *Plasmodium* subgenera and species is based largely on their morphological and biological features. For example, the mature gametocytes of subgenus *Haemamoeba* are round or oval, while those of the subgenera *Giovannolaia*, *Huffia*, and *Novyella* are elongated. Furthermore, the schizonts of *Giovannolaia* and *Huffia* are large with plentiful cytoplasm (and are often larger than the host cell nucleus and frequently displace it), while those of *Novyella* are small with scanty cytoplasm (or only slightly larger than the host nucleus). In addition, the schizonts of *Carinamoeba* tend to produce less than eight merozoites, whereas those of the subgenus *Sauramoeba* tend to produce more than eight merozoites.

More recently, the genetic characteristics of *Plasmodium* organisms have been exploited for their improved differentiation [1–6]. In contrast to other eukaryotes that have one distinct small subunit (SSU) rRNA (18S rRNA) molecule, *Plasmodium* species possess two or three such molecules (referred to as types A, S, and O) within the genome. While type A is expressed in the asexual stages and type S is expressed in the sexual stages, type O (only known to occur in *P. vivax*) is only expressed in the oocyte. Interestingly, the Asian simian *Plasmodium* species (e.g., *P. coatneyi*, *P. cynomolgi*, *P. fragile*, *P. inui*, *P. fieldi*, *P. hylobati*, and *P. simiovale*) have a single S-type-like gene and several A-type-like genes. Sequence analyses of 18S and 28S (large subunit) rRNA genes and other genetic regions allow distinction of various clades within the *Plasmodium* genus. With the elucidation of several *Plasmodium* genomes (e.g., *P. falciparum*, *P. knowlesi*, *P. vivax*, and *P. yoelli*), it is clear that *Plasmodium* species harbor genomes of about 25 megabasepairs (Mb) organized into have 14 chromosomes, the lengths of which vary from 500 kilobases to 3.5 megabases. Further comparative examination of their genome sequences will undoubtedly reveal new insights into the taxonomical relationships among the species and subgenera of the *Plasmodium* genus.

Among the five species of *Plasmodium* (*P. falciparum*, *P. vivax*, *P. malariae*, *P. ovale*, and *P. knowlesi)* that cause naturally acquired malaria in humans, *P. falciparum* is the most widespread and dangerous. *P. falciparum* is associated with a rapid severe onset of symptoms and is responsible for most malaria-associated deaths (amounting to more than 1,000,000 deaths each year), especially in nonimmune subjects. *P. vivax* is also widely distributed in the tropics and subtropics, and *P. malariae* and *P. ovale* are considerably less widespread and are responsible for only a small proportion of clinical cases [7,8]. Previously regarded as a cause of disease in primates and occasionally misidentified as *P. malariae*, *P. knowlesi* is a common cause of malaria in Malaysia as well as southern Thailand, Myanmar, the Philippines, Singapore, and Vietnam [9–18].

18.1.1.2 Life Cycle

Members of the *Plasmodium* genus have a life cycle that alternates between a vertebrate host and a mosquito vector and that undergo several distinct developmental stages including merogony, gamogony, and sporogony. For a short period of time during its development in the mosquito vector, the parasite has a diploid genome. Throughout the majority of its life cycle in the invertebrate host, it has a haploid genome [7,8].

In general, a mosquito sucks up microgametocytes and macrogametocytes into its gut from a *Plasmodium*-infected invertebrate host during blood meal. In the mosquito gut, microgametes are formed by three mitotic divisions within the microgametocyte. Then the plasmalemmas of male and female gametes fuse to allow the nucleus of the microgamete entering the female cytoplasm to become elongated, motile zygotes called ookinetes. The ookinetes penetrate and escape the midgut and then embed themselves onto the exterior of the gut membrane of mosquito, developing into oocysts. After undergoing sporogony, many tiny elongated sporozoites (12 μm × 1 μm in size, with a complex pellicle for motility) are produced within an oocyst. Upon rupture of oocyst, the liberated sporozoites escape into the hemocoele and penetrate salivary gland cells of mosquito.

Sporozoites are injected to either the blood or the lymphatic system of the invertebrate host from the saliva (containing anticoagulant) of a biting female mosquito. Along with blood circulation, sporozoites migrate to the liver and invade hepatocytes, where *P. falciparum* and *P. malariae* sporozoites trigger immediate schizogony while *P. ovale* and *P. vivax* sporozoites may either trigger immediate schizogony or have a delayed trigger, resulting in latent or dormant hypnozoites. Upon release from the liver, *Plasmodium* merozoites invade the erythrocytes. Within the erythrocyte, merozoites grow first to a ring-shaped form and then to a larger trophozoite form. The parasite feeds by ingesting hemoglobin and other materials from red blood cells and serum, damaging the erythrocytes during the process. After shizogony, the parasite divides several times to produce new merozoites, which leave the red blood cells and travel within the bloodstream to invade new red blood cells. Some merozoites differentiate into male or female sexual forms (microgametocytes and macrogametocytes) in the bone marrow, which move to the bloodstream ready to be taken up by female mosquito in the next blood meal.

18.1.2 Epidemiology

Apicomplexa are characterized by a set of specialized organelles (together called the apical complex) that are located at one end of the parasite, exist in some stages of the parasite's life cycle and enable penetration of the host cells. Possibly derived from dinoflagellates—a large group of photosynthetic protists, Apicomplexa may have evolved by gaining the ability to invade the intestinal cells and losing their photosynthetic ability. This is supported by the findings that many species within the Apicomplexa still possess plastids (the organelle in which photosynthesis takes place in photosynthetic eukaryotes), or plastid genes within their genomes.

The genera *Plasmodium*, *Hepatocystis*, and *Haemoproteus* of the family Plasmodium may have come from one or more *Leucocytozoon* species that infect white blood cells (leukocytes) and liver and spleen cells and are transmitted by black flies. In turn, the genus *Leucocytozoon* may have originated from a parasite that spreads by the orofecal route and infects the intestinal wall, acquiring the ability to infect blood cells and to survive in mosquitoes, and losing the previous orofecal route.

A reptile (specifically a squamate) was likely the first vertebrate host of *Plasmodium*, followed by birds, with mammals being the most recent group of vertebrates infected. The plasmodia causing malaria have evolved biologically and phylogenetically in two ways: (i) *P. vivax*, *P. malariae*, and *P. ovale* may have either coevolved with human mankind, or encountered human species during ancient times; on the other hand, *P. falciparum* has been transmitted to humans by monkeys in a more recent period [19–23].

Of >3500 recognized mosquito species, about 60 of the 360 *Anopheles* species (e.g., *A. gambiae*, *A. atroparvus*, *A. dirus*, *A. freeborni*, *A. albimanus*, *A. quadrimaculatus*, *A. stephensi*, *A. maculatus*, *A. subpictus*, and *A. farauti*) are inolved in the transmission of *Plasmodium* causing malaria in humans. Species of the mosquito genera *Aedes*, *Culex*, *Culiseta*, *Mansonia*, and *Theobaldia* can transmit malaria but not to humans, with *Culex* species carrying malaria to birds. It appears that resistant mosquito species produce substances (e.g., trypsin) toxic to *Plasmodium* within their cells, possibly inhibiting ookinete development. In addition, sporogeny within the mosquito are influenced by environmental temperature as Anopheline mosquitoes are poikilotherms. Notably, only female mosquitoes are involved in the disease transmission as the males feed only on plant juices, not blood.

Currently, malaria is endemic in over 90 countries, resulting in approximately 300 million clinical cases and 1.5 million deaths each year [24]. *P. falciparum* alone is responsible for about 200 million clinical cases and about 1 million deaths per annum, mostly in tropical regions and especially in sub-Saharan Africa. *P. vivax* infects 100 million people per year in both temperate and tropical areas. *P. malariae* is widespread throughout sub-Saharan Africa, southeast Asia, Indonesia, the islands of the western Pacific, and the Amazon Basin of South America as well as Europe and southern parts of the United States. In areas of endemicity in Africa, infections of *P. malariae* may coexist with *P. falciparum* infections [7,8]. While *P. falciparum* is capable of infecting erythrocytes of any age, *P. vivax* attacks mainly reticulocytes, which are erythrocytes aged 36 h or less after differentiation from erythroblasts and that still retain traces of a nucleus.

In endemic regions, children under the age of five are particularly vulnerable; pregnant women are also at risk since malaria accounts for a substantial number of miscarriages and low-birth-weight babies. Population movement further contributes to the spread of drug-resistant *Plasmodium*. Severe anemia and organ complications, especially in the brain, cause death or long-term disability. The overall cost of malaria including health care, treatment, lost production, etc. is estimated to be more than $US1.8 billion for tropical Africa. Ongoing control programs such as adequate surveillance, accurate diagnosis, and prompt treatment remain essential in endemic areas.

18.1.3 Clinical Features and Pathogenesis

18.1.3.1 Clinical Features

Human malaria results from infection with *P. falciparum*, *P. vivax*, *P. ovale*, *P. malariae*, and *P. knowlesi*, leading to significant morbidity and mortality in tropical and subtropical regions, with common clinical signs ranging from intense chills, fever, sweating, headaches, nausea, fatigue, abdominal pain, joint pain, breathlessness, vomiting, cough, anemia, renal failure, tachypnea, pyrexia, and tachycardia [7,8,15,25,26].

Upon injection into either the blood or the lymphatic system from the saliva of a biting female mosquito, *Plasmodium* sporozoites migrate to and replicate in the liver. After approximately 15 days, the parasites return to the blood and penetrate the red cells, where they multiply again, breaking down the red cells and inducing bouts of fever and chill (with duration of 5–32 h occurring on an every-third-day pattern for *P. malariae*) as well as anemia in the infected individual. The infected red cells obstruct the blood vessels in the brain as in the case of cerebral malaria and damage other vital organs, leading to the death of the patient. Depending on the *Plasmodium* species involved, the time until the first day that the parasite is detected on a thick blood film, namely the prepatent period, ranges from 16 to 59 days. Often the maximum parasite counts for *P. malariae* are lower than those for *P. falciparum* or *P. vivax*, because *P. malariae* produces a smaller number of merozoites per erythrocytic cycle, requires an extended developmental cycle (72 h) versus the 48-h cycle of *P. vivax* and *P. falciparum*, prefers to develop in older erythrocytes and allows for earlier buildup of immunity by the human host. The blood stage parasites may persist for long periods, sometimes for the life of the human host.

18.1.3.2 Pathogenesis

During its life cycle, *Plasmodium* produces three distinct invasive forms: sporozoite, merozoite, and ookinete. Under favorable conditions asexual reproduction is superior to sexual reproduction. In times of stress or within a new host, sexual reproduction is generally superior as this produces a shuffling of genes better adapted to the new environment. The disease pathology is due to the asexual forms that may spread via blood transfusion, but are not transmitted by mosquitoes.

Merozoites enter erythrocytes rapidly (within approximately 20 s) through several distinct steps: (i) a random collision between the merozoite and the erythrocyte leads to reversible interactions between proteins on the merozoite surface and the host erythrocyte. The merozoite surface protein-1 (MSP-1), which is uniformly distributed over the merozoite surface, is possibly involved in the binding with erythrocyte [27]. In addition, the circumsporozoite protein may contribute to interaction between sporozoites and heparin sulfate proteoglycans of hepatocytes. (ii) After binding to the erythrocyte, the apical end of the parasite is reoriented and juxtaposed to the erythrocyte membrane, with the help of apical membrane antigen-1 (AMA-1), which is a transmembrane protein localized at the apical end of the merozoite and binds erythrocytes. (iii) Specialized secretory organelles (e.g., micronemes, rhoptries, and dense granules) at the apical end expel their contents (consisting of subtilisin-like proteases, erythrocyte-binding antigen, Duffy-binding protein, sporozoite surface protein-2 or thrombospondin-related adhesive protein, circumsporozoite- and TRAP-related protein, etc) as the parasite invades, resulting in the formation of a junction between the parasite and host cell. (iv) Localized clearing of the erythrocyte submembrane cytoskeleton and formation of the incipient parasitophorous vacuole, and movement of the merozoite through the ring-shaped tight junction formed by the receptor/ligand complex with the assistance of myosin motors associated with the transmembrane parasite ligands moving along actin filaments within the parasite. (v) Closure of the parasitophorous vacuole and erythrocyte membrane.

During the intraerythrocytic stage, the parasite modifies the erythrocyte membrane to render it more permeable to small-molecular-weight metabolites, presumably satisfying the needs of an actively growing parasite, and to facilitate cytoadherence of *Plasmodium*-infected erythrocytes to endothelial cells, thus sequestering the mature parasites in capillaries and post-capillary venules. This sequestration leads to microcirculatory alterations and metabolic dysfunctions, which may account for many of the manifestations of malaria. In addition, the parasite alters the host erythrocyte membrane by inducing the formation of electron-dense protrusions, or knobs, with the help of the knob-associated histidine-rich protein and erythrocyte membrane protein-2 (*Pf*EMP2), also called MESA. The knobs may also play a role in the sequestration of infected erythrocytes. Other erythrocyte membrane-associated proteins (e.g., *Pf*EMP1 and *Pf*EMP3 and cytoadherence ligands (a modified band-3 called pfalhesin, sequestrin, rifins, and clag9)) may be also associated with the interaction of the parasite and erythrocyte.

Host receptors reacting with *Plasmodium* include CD36 (an 88 kDa integral membrane protein found on monocytes, platelets and endothelial cells), intracellular adhesion molecule-1 (ICAM1, a member of the immunoglobulin superfamily localized in the brain with possible functions in cell–cell adhesion) and chondroitin sulfate A (CSA, which is implicated in the cytoadherence of *Plasmodium* within the placenta).

18.2 LABORATORY DIAGNOSIS

Malaria is curable with appropriate antimalarial drugs (e.g., chloroquine or sulfadoxine/pyrimethamine and artemisinin-based combinations in the case of anti-chloroquine resistance). Accurate diagnosis of *Plasmodium* infections is therefore vital for effective treatment and control of the disease. Conventional diagnosis of malaria relies on clinical observations (e.g., fever, shivering, joint pain, and headache) and microscopic examination of the blood. Considering that malaria resembles other parasitic (e.g., babesiosis) and viral diseases clinically and that *Plasmodium* may sometimes produce similar ring forms to *Babesia* in erythrocytes, a variety of serological and molecular techniques have been employed for highly sensitive and specific detection in recent years [28–31].

18.2.1 Microscopy

As merogony occurs both in erythrocytes and in other tissues, microscopic observation of *Plasmodium* merozoites, schizonts or gametocytes in Giemsa-stained peripheral blood films offers a convenient and rapid means of diagnosing malaria [32–34]. Under light microscopy, merozoites tend to display a signet-ring structure within erythrocytes as the formation of a large vacuole forces the parasite's nucleus to one pole; schizonts appear as round to oval inclusions containing deeply stained merozoites; gametocytes are halter-shaped with confined pigment granules [35–39]. However, due to the fact that microscopic examination is subjective and technically demanding, a definitive diagnosis of malaria and precise identification of *Plasmodium* species may require additional confirmation with other techniques, especially nucleic acid tests [39].

18.2.2 Antigen and Antibody Tests

As an adjunct to microscopy for the diagnosis of malaria, immunochromatographic antigen detection, enzyme-linked immunosorbent assays (ELISA) and other serological procedures have been developed for the verification of current disease or past exposure to malaria. By capturing *Plasmodium* soluble proteins (e.g., histidine-rich protein (HRP) and lactate dehydrogenase (LDH)) from a drop of blood with capture antibodies embedded on a nitrocellulose strip, immunochromatographic antigen detection permits colometric visualization of a specific antigen–antibody complex directly from the membrane. This technique is rapid, stable at temperatures up to 40°C, easy to use, cost-effective and thus readily applicable in field setting [40,41]. However, immunochromatographic antigen detection has been limited to the identification of *P. falciparum* and *P. vivax* infections, but not *P. malariae* and *P. ovale* infections. In addition, this technique is relatively inefficient in detecting very-low-density infections. For this reason, ELISA provides a more sensitive alternative for serological diagnosis of malaria [42,43]. Nonetheless, commonly occurring cross-reactions among various *Plasmodium* spp. have restricted the use of ELISA to preliminary screening, and the application of nucleic acid tests is essential for final confirmation.

18.2.3 Nucleic Acid Tests

Besides being more objective and less technically demanding, nucleic acid tests such as polymerase chain reaction (PCR) demonstrate much higher sensitivity, specificity, and speed than microscopy and serology for rapid diagnosis of malaria [44–70].

A diversity of *Plasmodium* gene regions have been targeted for specific detection of malaria. These include SSU 18S ribosomal RNA (rRNA), mitochondrial genes (*cytB* and *cox1*), plastid gene (*tufA*), nuclear gene (*ldh*), and dihydrofolate reductase-thymidylate synthase gene (dhfr-ts) [32,52,55,67,71–75]. The amplification approaches range from single-round, nested, to real-time PCR, while detection formats vary from gel electrophoresis, ELISA to automated real-time detection [55,58,73,76–79].

Several nested PCRs using genus- and species-specific primers from the 18S rRNA gene have been described for diagnostic identification of *Plasmodium* species infecting humans [16,47,52,73,80–82]. Although extremely sensitive, nested PCR protocols require manual handling between the first and second PCR rounds, creating potential risk of PCR carry-over contamination. The combination of single-round PCR with either microtiter plate hybridization or colorimetric detection (e.g., PCR-ELISA) greatly reduces the cross-contamination risk without sacrificing sensitivity [78,83]. Following recent advances in new probe chemistry and instrument automation, multiplex real-time PCR has been developed for sensitive detection of all human-infecting *Plasmodium* species, significantly reducing the cost per test, providing instant results after amplification, and eliminating the risk of cross-contamination at the same time [84–90]. Furthermore, quantitative real-time PCR has proven valuable for highly concordant, accurate, and reliable diagnosis of human malaria and for rapid differentiation of drug-sensitive *Plasmodium* strains [67,91,92].

Apart from PCR-based procedures, loop-mediated isothermal amplification (LAMP) has shown promise for fast and low-cost diagnosis of malaria. Without relying on temperature changes to mediate amplification, LAMP negates the need to use an expensive thermocycler, which is not readily available in impoverished areas where malaria is endemic. Furthermore, LAMP has the convenience of visual detection of the resulting products, permitting rapid test turnaround without costly probe chemistry, which is a presequite for real-time PCR [93].

18.3 METHODS

18.3.1 Sample Preparation

18.3.1.1 Microscopy

Thick blood films are used for microscopic detection of *Plasmodium* species and thin blood films for species identification. Briefly, thick films are prepared by adding one drop of blood (collected in heparinated Vacutubes) onto a glass slide, which is dried at room temperature and fixed with absolute methanol at 37°C for 10 min. Slides are then immersed in a 2.5% Giemsa stain in phosphate-buffered saline (PBS, pH 7.2) for 25 min, washed with water, dried, and cover-slipped before microscopic investigation. Thin blood films are prepared as above, except that thin blood films are fixed in methanol for 5 min and stained using a 10% Giemsa stain in PBS. All thick and thin blood films are examined under a light microscope at a magnification of ×1000 with immersion oil [39].

Plasmodium parasitemia of each patient is calculated from thick blood films by counting the number of parasitized red blood cells per white blood cells in over 100 fields and calculated using the actual white blood cell count of each patient. The percentage of each of the parasite developmental stages is determined based on the number of early trophozoites, late and mature trophozoites, schizonts, and gametocytes in thin blood films [39].

18.3.1.2 Extraction of *Plasmodium* DNA

Plasmodium DNA is extracted from blood by using the Chelex method. Briefly, in an Eppendorf tube, 1 mL of autoclaved distilled water is added to 3 mm of dried blood spot or 3 μL of liquid blood or 3 mm^3 of clotted blood and incubated for 30 min at room temperature, mixed by inverting several times. The sample is centrifuged for 3 min at 14,000 rpm, and the supernatant is removed and made up to 200 μL with 5% Chelex. The sample is incubated for 30 min at 56°C, vortexed for 10 s, heated in a boiling water bath for 10 min and vortexed again for 10 s before being centrifuged for 3 min at high speed. A measure of 3 μL supernatant is used in PCR, and the rest stored at 4°C. For reuse after storage, the remaining sample is vortexed, and centrifuged again before supernatant is taken for PCR [56].

Alternatively, *Plasmodium* DNA is prepared by phenol/chloroform extraction. A volume of 4 mL of blood is centrifuged at 1600 rpm for 5 min, and the cell pellet is resuspended in 400 μL of lysis solution (100 mM Tris-HCl, 10 mM ethylenediaminetetraacetic acid, pH 8.0, 100 mM NaCl, 1% sodium dodecyl sulfate, 1 mg/mL proteinase K; and 100 mg/mL ribonuclease). The suspension is incubated for 2 h at 50°C; after the addition of an equal volume of phenol/chloroform and mixing, it is centrifuged for 5 min at 14,000 rpm. The supernatant is transferred to another Eppendorf tube and the extraction procedure was repeated using chloroform alone. One-tenth of the volume of 3 M potassium acetate is added, followed by two volumes of ethanol, and the mixture is kept at −20°C for at least 1 h and then centrifuged for 15 min at 14,000 rpm. The supernatant is discarded and the pellet is dried and dissolved in 25 μL of water. A volume of 1 μL of this preparation is used as template for PCR [56]. Another way to isolate *Plasmodium* DNA from blood is to treat 100 μL of blood with 0.1 M NaOH for 3 min at 100°C. The supernatant is collected and treated with 250 μL of lysis solution (0.1 M Tris-HCl, Triton X-100, 1 M NaCl, 10 × SDS, and 0.5 M EDTA) and 20 mg/mL proteinase K for 1 h at 37°C. It is then extracted twice with phenol/chloroform (1:1). DNA is precipitated with ethanol, resuspended in 100 μL of distilled water and stored at −20°C.

For blood films on microscope slides, DNA is extracted by moistening the blood film with one drop of Tris-EDTA (TE) buffer, pH 8, and scraping the film of blood into a microcentrifuge tube containing 100 μL TE buffer. A volume of 10 μL of 10 mg/mL proteinase K and 100 μL of lysis buffer (5 mM EDTA, 0.5% sodium dodecyl sulfate, 200 mM NaCl, and 100 mM Tris-Cl, pH 8) are added to the tube and incubated in a thermomixer at 56°C with shaking at 900 rpm for 10 min. An equal volume of phenol–chloroform isoamyl alcohol is then added to each sample, followed by vigorous mixing for 15 s and centrifugation for 2 min at 14,000 rpm. After transferring the aqueous phase into a new microcentrifuge tube, the organic phase is re-extracted by adding 100 μL of TE buffer, with further mixing and centrifugation. The aqueous phase from the second extraction is pooled with that from the first and the DNA is ethanol precipitated. The air-dried DNA pellet is dissolved in 50 μL of TE buffer [53].

18.3.2 Detection Procedures

18.3.2.1 Single Round PCR for the Identification of Four Human-Infecting *Plasmodium* Species

Padley et al. [59] reported a single-round, multiplex PCR for the detection of *P. falciparum*, *P. vivax*, *P. malariae*, or *P. ovale* in human blood, with primers from the conserved 18S SSU RNA genes of the *Plasmodium* parasites (Table 18.1)

18.3.2.1.1 Procedure

1. The PCR mixture (25 μL) is made up of 3.0 mm $MgCl_2$, 400 μM of each dNTP, 2.5 U HotStartTaq DNA polymerase (HotStarTaq Master Mix Kit, Qiagen), 15 pmol of the reverse primer, 15 pmol of the *P. falciparum* forward primer, 17.5 pmol each of the *P. ovale* and *P. malariae* forward primers, 7.5 pmol of the *P. Vivax* forward primer and 5 μL of extracted DNA.

TABLE 18.1
Primers for Single-Round PCR Detection of Four Human-Infecting *Plasmodium* spp.

Primer	Sequence (5′–3′)	Expected Product (bp)	Optimal Concentration (pmol)
Plasmodium Reverse	GTA TCT GAT CGT CTT CAC TCCC		15
P. falciparum Forward	AAC AGA CGG GTA GTC ATG ATT GAG	276	15
P. vivax Forward	CGG CTT GGA AGT CCT TGT	300	7.5
P. ovale Forward	CTG TTC TTT GCA TTC CTT ATGC	375	17.5
P. malariae Forward	CGT TAAGAA TAA ACG CCA AGCG	412	17.5

2. Amplification is performed with an initial, 15-min incubation at 95°C (to activate the HotStart polymerase) followed by 43 cycles of 95°C for 45 s and 60°C for 90 s, and then a final step at 72°C for 5 min.
3. The amplified products are separated on a 2% agarose gel and stained with ethidium bromide. The lengths of the expected products are 276 bp for *P. falciparum*, 300 bp for *P. vivax*, 375 bp for *P. ovale*, and 412 bp for *P. malariae*.

18.3.2.2 Nested PCR for the Identification of Four Human-Infecting *Plasmodium* Species

Rubio et al. [56] described a simplified protocol for the identification of four *Plasmodium* species by semi-nested multiplex polymerase chain reaction (SnM-PCR) in human blood samples. The SnM-PCR targets the SSU nuclear ribosomal ribonucleic acid (RNA) gene (rRNA), a multicopy gene containing both highly conserved domains and domains characteristic of human-infecting *Plasmodium* species. In the first-round PCR, a universal reverse primer (UNR) is combined with two forward primers (PLF and HUF) specific for *Plasmodium* and mammals, respectively (Table 18.2). In the second-round PCR, a *Plasmodium*-specific forward primer (PLF) and four species-specific reverse primers (VIR, OVR, FAR, and MAR) for *P. vivax*, *P. ovale*, *P. falciparum*, and *P. malariae*, respectively, are used (Table 18.2).

18.3.2.2.1 Procedure

1. The first-round PCR mixture (50 μL) is composed of 20 mM Tris-HCl (pH 8.4). A total of 50 mM KCl, 15 mM $MgCl_2$, 0.05% W-1 (BRL detergent), 200 μM of each nucleoside triphosphate, primers UNR, PLF, and HUF, 1 U of Taq DNA polymerase (GibcoBRL), 39 μL of distilled water, and 1 μL of template DNA.

TABLE 18.2
Primers for Nested PCR Detection of Four Human-Infecting *Plasmodium* spp.

Primer	Sequence (5′–3′)	Specificity	Expected Product (bp)	Optimal Concentration (pmol)
UNR	GACGGTATCTGATCGTCTTC	Universal[a]		25
PLF	AGTGTGTATCAATCGAGTTTC	*Plasmodium*	783–821[c]	25
HUF	GAGCCGCCTGGATACCGC	Mammals[b]	231	1.5
VIR	AGGACTTCCAAGCCGAAGC	*P. vivax*	499	2.5
OVR	GCATAAGGAATGCAAAGAACAG	*P. ovale*	436	6.25
FAR	AGTTCCCCTAGAATAGTTACA	*P. falciparum*	395	15
MAR	GCCCTCCAATTGCCTTCTG	*P. malariae*	269	3.12

[a] Including humans.
[b] Position 1229 on the published sequence X03205 for human 18s rRNA.
[c] Size depending upon species: *P. malariae* = 821 bp, *P. falciparum* = 787 bp, *P. vivax* = 783 bp, and *P. ovale* = 794 bp.

2. The first-round PCR amplification is performed with an initial 94°C for 5 min followed by 40 cycles of 94°C for 60 s, 55°C for 60 s, and 72°C for 90 s.
3. The second-round PCR mixture (25 μL) is made up of 20 mM Tris-HCl (pH 8.4), 50 mM KCl, 15 mM $MgCl_2$, 0.05% W-1 (BRL detergent), 200 μM of each nucleoside triphosphate, primers PLF, VIR, OVR, FAR, and MAR, 1 U of Taq DNA polymerase (GibcoBRL), distilled water, and 2 μL of first round PCR product (diluted in distilled water to 1 mL).
4. The second-round PCR amplification is carried out with an initial 94°C for 5 min followed by 35 cycles of 94°C for 30 s, 58°C for 30 s, and 72°C for 60 s.
5. The resulting products from the second-round PCR are separated by 2% agarose gel electrophoresis and specific amplicons are visualized with ethidium bromide stain.

Note: This SnM-PCR is more sensitive and more specific than the standard microscopic examination, and works well with samples collected as dried blood spots on filter paper and heparinized blood [69].

18.3.2.3 Real-Time PCR for the Identification of Five Human-Infecting *Plasmodium* Species Using SYBR Green

Oddoux et al. [17] reported a real-time PCR for the identification and differentiation of five human-infecting *Plasmodium* species, *P. falciparum*, *P. vivax*, *P. ovale*, *P. malariae*, and *P. knowlesi*. The assay targeting the A-type 18S rRNA gene of *Plasmodium* spp. is performed on a LightCycler (Roche Diagnostic) using fluorescent SYBR Green I dye.

18.3.2.3.1 Procedure

1. The PCR mixture (20 μL) contains 5 μL of DNA extracted from blood samples, 5 μL of LightCycler FastStart DNA Master SYBR Green I Buffer (Roche Diagnostics), 2.4 μL of $MgCl_2$ (4 mM), and 0.6 μL of each primer (initial concentration 10 μM and final concentration 0.3 μM) (Table 18.3).
2. The cycling program includes an initial denaturation step of 95°C for 8 min, 40 cycles of denaturation at 95°C for 10 s, annealing at 58°C for 10 s, and extension at 72°C for 30 s.
3. The melting curve is measured after 2 s denaturation at 95°C followed by annealing (20 s at 50°C) and a slow increase in temperature (0.2°C/s until 98°C).
4. The results are analyzed with the threshold cycle (C_t) corresponding to the cycle at which the fluorescence signal becomes significantly different from the baseline signal and with the post-amplification melting curve analysis that is related to the sequence of the PCR products.

Note: Amplifications of each of the five *Plasmodium* species with their specific primers result in specific melting temperatures: 78°C for *P. falciparum*, 79°C for *P. ovale*, 80°C for *P. vivax*, 74°C for *P. malariae*, and 86°C for *P. knowlesi*. For each primer set, nonspecific amplification is excluded by the lack of amplification and melting curves with DNA of the four other species as with noninfected blood samples.

TABLE 18.3
Primers for Real-Time PCR Detection of Five Human-Infecting *Plasmodium* spp.

Primer	Sequence (5′–3′)	Specificity
PSP1AS	CTACTCCTATTAATCGTAACT	
POS1	GCATTCCTTATCCAAAATGTG	*P. ovale*
PFS1	CATTTAAACTGGTTTGGGAAAAC	*P. falciparum*
PM1	ATAACATAGTTGTACGTTAAGAATAACCGC	*P. malariae*
PM2	AAAATTCCCATGCATAAAAAATTATACAAA	
PV1	CGCTTCTAGCTTAATCCACATAACTGATAC	*P. vivax*
PV2	CTTCCAAGCCGAAGCAAAGAAAGTCCTTA	
PKe'F	TCTTTTCTCTCCGGAGATTAGAACTC	*P. knowlesi*
PKg'R	CAAATCCACTATATTCAATTCTACA	

18.3.2.4 Real-Time PCR for the Identification of Four Human-Infecting *Plasmodium* Species Using Minor-Groove Binding (MGB) and TaqMan Probes

Taylor et al. [94] utilized a real-time PCR for the detection of four human-infecting *Plasmodium* species, with primers and probes from the block 9 region of the gene encoding the SSU (18S) of plasmodium rRNA and LDH gene. Primers and probe targeting human GAPDH are incorporated as an internal control (Table 18.4).

Real-time PCR is conducted with 25-μL mixtures consisting of 2 μL of DNA, 12.5 μL of 2× TaqMan Universal PCR Master Mix (Applied Biosystems), forward and reverse primers at 1000 nM, and VIC-labeled MGB probe at 200 nM. The cycling conditions for the Applied Biosystems 7300 system were 50°C for 2 min, 95°C for 10 min, and 40 cycles of 95°C for 15 s and 60°C for 1 min.

TABLE 18.4
Primers and Probes for Real-Time PCR Identification of Five Human-Infecting *Plasmodium* Species

Primer	Sequence (5′–3′)
Pan-species forward	GTT AAG GGA GTG AAG ACG ATC AGA TA
Pan-species reverse	AAC CCA AAG ACT TTG ATT TCT CAT AAG
P. falciparum 18S rRNA forward	ATT GCT TTT GAG AGG TTT TGT TAC TTT
P. falciparum 18S rRNA forward	GCT GTA GTA TTC AAA CAC AAT GAA CTC AA
P. ovale 18S rRNA forward	CCG ACT AGG TTT TGG ATG AAA GAT TTT T
P. ovale 18S rRNA reverse	CAA CCC AAA GAC TTT GAT TTC TCA TAA
P. malariae 18S rRNA forward	AGT TAA GGG AGT GAA GAC GAT CAG A
P. malariae 18S rRNA reverse	CAA CCC AAA GAC TTT GAT TTC TCA TAA
P. falciparum LDH forward	ACG ATT TGG CTG GAG CAG AT
P. falciparum LDH reverse	TCT CTA TTC CAT TCT TTG TCA CTC TTT C
Human GAPDH forward	CCT CCC GCT TCG CTC TCT
Human GAPDH reverse	GCT GGC GAC GCA AAA GA
MGB Probe	
Pan-species	VIC-TCG TAA TCT TAA CCA TAA AC
P. falciparum	FAM-CAT AAC AGA CGG GTA GTC AT
P. ovale	VIC-CGA AAG GAA TTT TCT TAT T
P. malariae	FAM-ATG AGT GTT TCT TTT AGA TAG C
Human GAPDH	VIC-CCT CCT GTT CGA CAG TCA GCC GC
TaqMan Probe	
P. falciparum LDH	FAM-GTA ATA GTA ACA GCT GGA TTT ACC AAG GCC CCA-TAMRA

18.4 CONCLUSION

Members of the genus *Plasmodium*, that is, *P. falciparum*, *P. vivax*, *P. ovale*, *P. malariae*, and *P. knowlesi*, are responsible for human malaria, an important tropical parasitic disease with significant morbidity and mortality. Approximately 300 million clinical cases and ~1 million deaths due to malaria are reported each year. The life cycle of *Plasmodium* parasites is complex, involving multiple tissues and different stages (sexual and asexual) inside mosquitoes and invertebrates including humans.

Because human malaria tends to induce a variety of noncharacteristic clinical symptoms (e.g., intense chills, fever, sweating, headaches, nausea, fatigue, abdominal pain, joint pain, breathlessness, vomiting, cough, anemia, renal failure, tachypnea, pyrexia, and tachycardia), it is vital to correctly diagnose the disease in order to implement effective treatment and prevention strategies. Although malaria and its causal pathogens are identifiable by using conventional laboratory procedures such as microscopic examination of Giemsa-stained blood smears and serological tests, the variable and time-consuming nature of these techniques necessitates the development of improved diagnostic methods.

The recent advances in nucleic acid detection technologies, especially PCR, offer promise for improving the laboratory diagnosis of human malaria. Not surprisingly, a range of PCR-based techniques have been designed and applied for the detection and diagnosis of *Plasmodium* infections. In particular, the development of multiplex real-time PCR targeting 18S rRNA and other gene regions has facilitated simultaneous detection of five human-infecting *Plasmodium* species, enabling identification of mixed infections and subclinical/asymptomatic infections and leading to enhanced success in the control of malaria.

REFERENCES

1. Liu, Q. et al., Sequence variation in the small-subunit rRNA gene of *Plasmodium malariae* and prevalence of isolates with the variant sequence in Sichuan, China. *J. Clin. Microbiol.*, 36, 3378, 1998.
2. Perkins, S.L., and Schall, J.J., A molecular phylogeny of malarial parasites recovered from cytochrome b gene sequences. *J. Parasitol.*, 88, 972, 2002.
3. Tachibana, M. et al., Two types of *Plasmodium ovale* defined by SSU rRNA have distinct sequences for ookinete surface proteins. *Mol. Biochem. Parasitol.*, 122, 223, 2002.
4. Leclerc, M.C. et al., Evolutionary relationships between 15 *Plasmodium* species from new and old world primates (including humans): An 18S rDNA cladistic analysis. *Parasitology*, 129, 677, 2004.
5. Martinsen, E.S. et al., Morphologically defined subgenera of *Plasmodium* from avian hosts: Test of monophyly by phylogenetic analysis of two mitochondrial genes. *Parasitology*, 134, 483, 2007.
6. Nishimoto, Y. et al., Evolution and phylogeny of the heterogeneous cytosolic SSU rRNA genes in the genus *Plasmodium*. *Mol. Phylogenet. Evol.*, 47, 45, 2008.

7. Collins, W.E., and Jeffery, G.M., *Plasmodium ovale*: Parasite and disease. *Clin. Microbiol. Rev.*, 18, 570, 2005.
8. Collins, W.E., and Jeffery, G.M., *Plasmodium malariae*: Parasite and disease. *Clin. Microbiol. Rev.*, 20, 579, 2007.
9. Win, T.T. et al., Wide distribution of *Plasmodium ovale* in Myanmar. *Trop. Med. Int. Health*, 7, 231, 2002.
10. Jongwutiwes, S. et al., Naturally acquired *Plasmodium knowlesi* malaria in human, Thailand. *Emerg. Infect. Dis.*, 10, 2211, 2004.
11. Singh, B. et al., A large focus of naturally acquired *Plasmodium knowlesi* infections in human beings. *Lancet*, 363, 1017, 2004.
12. Vythilingam, I. et al., Natural transmission of *Plasmodium knowlesi* to humans by Anopheles latens in Sarawak, Malaysia. *Trans. R. Soc. Trop. Med. Hyg.*, 100, 1087, 2006.
13. Luchavez, J. et al., Human infections with *Plasmodium knowlesi*, the Philippines. *Emerg. Infect. Dis.*, 14, 811, 2008.
14. White, N.J., *Plasmodium knowlesi*: The fifth human malaria parasite. *Clin. Infect. Dis.*, 46, 172, 2008.
15. Daneshvar, C. et al., Clinical and laboratory features of human *Plasmodium knowlesi* infection. *Clin. Infect. Dis.*, 49, 852, 2009.
16. Van den Eede, P. et al., Human *Plasmodium knowlesi* infections in young children in central Vietnam. *Malar. J.*, 8, 249, 2009.
17. Oddoux, O. et al., Identification of the five human *Plasmodium* species including *P. knowlesi* by real-time polymerase chain reaction. *Eur. J. Clin. Microbiol. Infect. Dis.*, 30, 597, 2011.
18. Sulistyaningsih, E. et al., Diagnostic difficulties with *Plasmodium knowlesi* infection in humans. *Emerg. Infect. Dis.*, 16, 1033, 2010.
19. Escalante, A.A. et al., The evolution of primate malaria parasites based on the gene encoding cytochrome *b* from the linear mitochrondrial genome. *Proc. Natl. Acad. Sci. USA*, 95, 8124, 1988.
20. Greenwood, B., The molecular epidemiology of malaria. *Trop. Med. Int. Health*, 7, 1012, 2002.
21. Krief, S. et al., On the diversity of malaria parasites in African apes and the origin of *Plasmodium falciparum* from bonobos. *PloS Pathog.*, 6, e1000765, 2010.
22. Liu, W. et al., Origin of the human malaria parasite *Plasmodium falciparum* in gorillas. *Nature*, 23, 420, 2010.
23. Ricklefs, R.E., and Outlaw, D.C., A molecular clock for malaria parasites. *Science*, 329, 226, 2010.
24. Zhou, M. et al., High prevalence of *Plasmodium malariae* and *Plasmodium ovale* in malaria patients along the Thai-Myanmar border, as revealed by acridine staining and PCR-based diagnoses. *Top. Med. Int. Health*, 3, 304, 1998.
25. Muehlenbachs, A. et al., An unusual presentation of placental malaria: A single persisting nidus of sequestered parasites. *Hum. Pathol.*, 38, 520, 2007.
26. Bruneel, F. et al., Severe imported falciparum malaria: A cohort study in 400 critically ill adults. *PLoS One*, 5, e13236, 2010.
27. Sawai, H. et al., Lineage-specific positive selection at the merozoite surface protein 1 (msp1) locus of *Plasmodium vivax* and related simian malaria parasites. *Evol. Biol.*, 10, 52, 2010.
28. Shiff, C.J., Premij, Z., and Minjas, J.N., The rapid ParaSight™-F test. A new diagnostic tool for *Plasmodium falciparum* infection. *Trans. R. Soc. Trop. Med. Hyg.*, 87, 29, 1993.
29. Warhurst, D.C., and Williams, J.E., Laboratory diagnosis of malaria. *J. Clin. Pathol.*, 49, 533, 1996.
30. Moody, A., Rapid diagnostic tests for malaria parasites. *Clin. Microbiol. Rev.*, 15, 66, 2002.
31. Rakotonirina, H. et al., Accuracy and reliability of malaria diagnostic techniques for guiding febrile outpatient treatment in malaria-endemic countries. *Am. J. Trop. Med. Hyg.*, 78, 217, 2008.
32. Kawamoto, F., and Billingsley, P.F., Rapid diagnosis of malaria by fluorescence microscopy. *Parasitol. Today*, 8, 69, 1992.
33. Nkrumah, B. et al., Comparison of the novel Partec rapid malaria test to the conventional Giemsa stain and the gold standard real-time PCR. *J. Clin. Microbiol.*, 48, 2925, 2010.
34. Rantala, A.M. et al., Comparison of real-time PCR and microscopy for malaria parasite detection in Malawian pregnant women. *Malar. J.*, 9, 269, 2010.
35. Matsumoto, Y., Matsuda, S., and Yoshida, Y., Ultrastructure of erythrocytic stages of *Plasmodium ovale* in humans. *Am. J. Trop. Med. Hyg.*, 35, 689, 1986.
36. Atkinson, C.T. et al., Ultrastructure of the erythrocytic stages of *Plasmodium malariae. J. Protozool.*, 34, 267, 1987.
37. Millet, P. et al., *In vitro* cultivation of exoerythrocytic stages of the human malaria parasite *Plasmodium malariae. Am. J. Trop. Med. Hyg.*, 38, 470, 1988.
38. McKenzie, F.E., Collins, W.E., and Jeffery, G.M., *Plasmodium malariae* blood-stage dynamics. *J. Parasitol.*, 87, 626, 2001.
39. Lee, K.-S. et al., Morphological features and differential counts of *Plasmodium knowlesi* parasites in naturally acquired human infections. *Malar. J.*, 8, 73, 2009.
40. Ndao, M. et al., Comparison of blood smear, antigen detection, and nested-PCR methods for screening refugees from regions where malaria is endemic after a malaria outbreak in Quebec, Canada. *J. Clin. Microbiol.*, 42, 2694, 2004.
41. Khairnar, K. et al., Multiplex real-time quantitative PCR, microscopy and rapid diagnostic immuno-chromatographic tests for the detection of *Plasmodium* spp: Performance, limit of detection analysis and quality assurance. *Malar. J.*, 8, 284, 2009.
42. Beier, M.S. et al., Identification of malaria species by ELISA in sporozoite and oocyst infected *Anopheles* from western Kenya. *Am. J. Trop. Med. Hyg.*, 39, 323, 1988.
43. Collins, F.H. et al., Monoclonal antibody-based enzyme-linked immunosorbent assay (ELISA) for detection of *Plasmodium malariae* sporozoites in mosquitoes. *Am. J. Trop. Med. Hyg.*, 38, 283, 1988.
44. Relf, W.A. et al., Diagnosis of *Plasmodium vivax* malaria using a specific deoxyribonucleic acid probe. *Trans. R. Soc. Trop. Med. Hyg.*, 84, 630, 1990.
45. McLaughlin, G.L. et al., Evaluation of non-isotopic DNA assay kit for diagnosing *Plasmodium falciparum* malaria in Madagascar. *Am. J. Trop. Med. Hyg.*, 48, 211, 1993.
46. Contamin, H. et al., PCR typing of field isolates of *Plasmodium falciparum. J. Clin. Microbiol.*, 33, 944, 1995.
47. Oliveira, D.A. et al., Polymerase chain reaction and a liquid-phase nonisotopic hybridization for species-specific and sensitive detection of malaria infection. *Am. J. Trop. Med. Hyg.*, 52, 139, 1995.
48. Stoffels, S.A.W., Docters van Leeuwen, W.M., and Post, R.J., Detection of *Plasmodium*-infectious mosquitoes by polymerase chain reaction without dissection of the salivary glands. *Med. Vet. Entomol.*, 9, 433, 1995.
49. Vu, T.T. et al., Screening donor blood for malaria by polymerase chain reaction. *Trans. R. Soc. Trop. Med. Hyg.*, 89, 44, 1995.
50. Kawamoto, F. et al., Sequence variation in the 18S rRNA gene, a target for PCR-based malaria diagnosis, in *Plasmodium ovale* from southern Vietnam. *Microbiology*, 34, 2287, 1996.

51. Singh, B. et al., Detection of malaria in Malaysia by nested polymerase chain reaction amplification of dried blood spots on filter papers. *Trans. R. Soc. Trop. Med. Hyg.*, 90, 519, 1996.
52. Singh, B. et al., A genus- and species-specific nested polymerase chain reaction malaria detection assay for epidemiologic studies. *Am. J. Trop. Med. Hyg.*, 60, 687, 1999.
53. Cox-Singh, J. et al., Increased sensitivity of malaria detection by nested polymerase chain reaction using simple sampling and DNA extraction. *Int. J. Parasitol.*, 27, 1575, 1997.
54. Tahar, R., Ringwald, P., and Basco, L.K., Diagnosis of *Plasmodium malariae* infection by the polymerase chain reaction. *Trans. R. Soc. Trop. Med. Hyg.*, 91, 410, 1997.
55. Rubio, J.M. et al., Usefulness of seminested multiplex PCR in surveillance of imported malaria in Spain. *J. Clin. Microbiol.*, 37, 3260, 1999.
56. Rubio, J.M. et al., Alternative polymerase chain reaction method to identify *Plasmodium* species in human blood samples: The semi-nested multiplex malaria PCR (SnM-PCR). *Trans. R. Soc. Trop. Med. Hyg.*, S199, 96, 2002.
57. Manca, N. et al., Search for malaria parasites by PCR and Southern blot in patients with imported malaria in Italy. *New Microbiol.*, 23, 339, 2000.
58. Perandin, F. et al., Usefulness of genus-specific PCR and Southern blot species-specific hybridization for the detection of imported malaria cases in Italy. *New Microbiol.*, 24, 69, 2001.
59. Padley, D. et al., Use of a rapid, single-round, mutiplex PCR to detect malarial parasites and identify the species present. *Ann. Trop. Med. Parasitol.*, 97, 131, 2003.
60. Berry, A. et al., Contribution of PCR-based methods to diagnosis and management of imported malaria. *Med. Trop.* (Mars), 65, 176, 2005.
61. McNamara, D.T. et al., Diagnosing infection levels of four human malaria parasite species by a polymerase chain reaction/ligase detection reaction fluorescent microsphere-based assay. *Am. J. Trop. Med. Hyg.*, 74, 413, 2006.
62. Boonma, P. et al., Comparison of three molecular methods for the detection and speciation of *Plasmodium vivax* and *Plasmodium falciparum*. *Malar. J.*, 6, 124, 2007.
63. Gama, B.E. et al., Real-time PCR versus conventional PCR for malaria parasite detection in low-grade parasitemia. *Exp. Parasitol.*, 116, 427, 2007.
64. Mens, P.F. et al., Molecular diagnosis of malaria in the field: development of a novel 1-step nucleic acid lateral flow immunoassay for the detection of all 4 human *Plasmodium* spp. and its evaluation in Mbita, Kenya. *Diagn. Microbiol. Infect. Dis.*, 61, 421, 2008.
65. Safeukui, I. et al., Evaluation of FRET real-time PCR assay for rapid detection and differentiation of Plasmodium species in returning travellers and migrants. *Malar. J.*, 7, 70, 2008.
66. Veron, V., Simon, S., and Carme, B., Multiplex real-time PCR detection of *P. falciparum*, *P. vivax* and *P. malariae* in human blood samples. *Exp. Parasitol.*, 121, 346, 2009.
67. Bourgeois, N. et al., Comparison of three real-time PCR methods with blood smears and rapid diagnostic test in *Plasmodium* sp. infection. *Clin. Microbiol. Infect.*, 16, 1305, 2010.
68. Cnops, L. et al., Giemsa-stained thick blood films as a source of DNA for Plasmodium species-specific real-time PCR. *Malar. J.*, 9, 370, 2010.
69. Mixson-Hayden, T., Lucchi, N.W., and Udhayakumar, V., Evaluation of three PCR-based diagnostic assays for detecting mixed Plasmodium infection. *BMC Res. Notes*, 3, 88, 2010.
70. Ta, T.T. et al., First case of detection of *Plasmodium knowlesi* in Spain by real time PCR in a traveller from Southeast Asia. *Malar. J.*, 9, 219, 2010.
71. Lal, A.A. et al., Ribosomal RNA-based diagnosis of *Plasmodium falciparum* malaria. *Mol. Biochem. Parasitol.*, 36, 67, 1989.
72. Goman, M., Mons, B., and Scaife, J., The complete sequence of a *Plasmodium malariae* SSU rRNA gene and its comparison to other plasmodial SSU rRNA genes. *Mol. Biochem. Parasitol.*, 45, 281, 1991.
73. Snounou, G. et al., Identification of the four human malaria parasite species in field samples by the polymerase chain reaction and detection of a high prevalence of mixed infections. *Mol. Biochem. Parasitol.*, 58, 283, 1993.
74. Waters, A.P., The ribosomal RNA genes of *Plasmodium*. *Adv. Parasitol.*, 34, 33, 1994.
75. Tanomsing, N. et al., Accurate and sensitive detection of *Plasmodium* species in humans by use of the dihydrofolate reductase-thymidylate synthase linker region. *J. Clin. Microbiol.*, 48, 3735, 2010.
76. Arai, M. et al., Detection of *Plasmodium falciparum* in human blood by a nested polymerase chain reaction. *Am. J. Trop. Med. Hyg.*, 51, 617, 1994.
77. Patsoula, E. et al., A single-step, PCR-based method for the detection and differentiation of *Plasmodium vivax* and *P. falciparum*. *Ann. Trop. Med. Parasitol.*, 97, 15, 2003.
78. Whiley, D.M. et al., Detection and differentiation of *Plasmodium* species by polymerase chain reaction and colorimetric detection in blood samples of patients with suspected malaria. *Diagn. Microbiol. Infect. Dis.*, 49, 25, 2004.
79. Rosanas-Urgell, A. et al., Comparison of diagnostic methods for the detection and quantification of the four sympatric *Plasmodium* species in field samples from Papua New Guinea. *Malar. J.*, 9, 361, 2010.
80. Imwong, M. et al., Spurious amplification of *Plasmodium vivax* small subunit RNA gene using the primers currently used to detect *P. knowlesi*. *J. Clin. Microbiol.*, 47, 4173, 2009.
81. Cnops, L., Jacobs, J., and Van Esbroeck, M., Validation of a four-primer real-time PCR as a diagnostic tool for single and mixed *Plasmodium* infections. *Clin. Microbiol. Infect.*, 17, 1101, 2011.
82. Genc, A., Eroglu, F., and Koltas, I.S., Detection of *Plasmodium vivax* by nested PCR and real-time PCR. *Korean J. Parasitol.*, 48, 99, 2010.
83. Kimura, M. et al., Species-specific PCR detection of malaria parasites by microtiter plate hybridization: clinical study with malaria patients. *J. Clin. Microbiol.*, 33, 2342, 1995.
84. de Monbrison, F. et al., Simultaneous identification of the four human *Plasmodium* species and quantification of *Plasmodium* DNA load in human blood by real-time polymerase chain reaction. *Trans. R. Soc. Trop. Med. Hyg.*, 97, 387, 2003.
85. Rougemont, M. et al., Detection of four *Plasmodium* species in blood from humans by 18S rRNA gene subunit-based and species-specific real-time PCR assays. *J. Clin. Microbiol.*, 42, 5636, 2004.
86. Mangold, K.A. et al., Real-time PCR for detection and identification of *Plasmodium* spp. *J. Clin. Microbiol.*, 43, 2435, 2005.
87. Babady, N.E. et al., Detection of *Plasmodium knowlesi* by real-time polymerase chain reaction. *Am. J. Trop. Med. Hyg.*, 81, 516, 2009.

88. Shokoples, S.E. et al., Multiplexed real-time PCR assay for discrimination of *Plasmodium* species with improved sensitivity for mixed infections. *J. Clin. Microbiol.*, 47, 975, 2009.
89. Divis, P.C. et al., A TaqMan real-time PCR assay for the detection and quantitation of *Plasmodium knowlesi*. *Malar. J.*, 9, 344, 2010.
90. Dormond, L. et al., Multiplex real-time PCR for the diagnosis of malaria: Correlation with microscopy. *Clin. Microbiol. Infect.*, 17, 469, 2011.
91. Witney, A.A. et al., Determining liver stage parasite burden by real time quantitative PCR as a method for evaluating pre-erythrocytic malaria vaccine efficacy. *Mol. Biochem. Parasitol.*, 118, 233, 2001.
92. Nwakanma, D.C. et al., Quantitative detection of *Plasmodium falciparum* DNA in saliva, blood, and urine. *J. Infect. Dis.*, 199, 1567, 2009.
93. Lucchi, N.W. et al., Real-time fluorescence loop mediated isothermal amplification for the diagnosis of malaria. *PLoS One*, 5, e13733, 2010.
94. Taylor, S.M. et al., High-throughput pooling and real-time PCR-based strategy for malaria detection. *J. Clin. Microbiol.*, 48, 512, 2010.

19 Sarcocystis

L. Vangeel, K. Houf, P. Geldhof, J. Vercruysse, R. Ducatelle, and K. Chiers

CONTENTS

19.1 INTRODUCTION

19.1.1 Classification

Sarcocystis (*S.*), a genus of tissue cyst forming coccidia, belongs to the family of Sarcocystidae, phylum of Apicomplexa, and subkingdom of Protozoa [1]. Related genera include *Besnoitia*, *Cystoisospora*, *Frenkelia*, *Hammondia*, *Neospora*, and *Toxoplasma*. *Sarcocystis* is the largest genus within this family. Over 200 *Sarcocystis* species have been characterized, and more continue to be described in mammals (especially sheep, pigs, and cattle), reptiles, and birds [2].

Sarcocystis tissue cysts were described for the first time by Friedrich Miescher in 1843 when they were discovered in skeletal tissue of domestic house mice. Once Lankester introduced the genus name *Sarcocystis* (sarco = muscle) in 1882, several species were described based on the host species. Confusion occurred, since the life cycle was not known yet and validation of the identity of the species was not possible. In 1972, pioneering research of Rommel and Heydorn revealed the two host (heteroxenous) life cycle of *Sarcocystis* [3]. Heydorn et al. showed also that both intermediate and definitive host can be infected by more than one species [4]. Since the original descriptions were inadequate and it was impossible to confirm that the originally named species were identical to those wherein the life cycle was named, Heydorn et al. proposed new names (Table 19.1) combining the name of the intermediate and definitive hosts [4]. Nevertheless, Levine stated that the old names were valid following the international code of zoological nomenclature [5]. These two opposite views caused considerable confusion.

TABLE 19.1
***Sarcocystis* Species in Cattle and Pigs with Their Synonyms**

Sarcocystis Species	Synonym[a]	Intermediate Host	Definitive Host
S. hominis	*S. bovihominis*	Cattle	Humans (primates)
S. hirsuta	*S. bovifelis*	Cattle	Felids
S. cruzi	*S. bovicanis*	Cattle	Canids
S. suihominis	—	Pigs	Humans (primates)
S. porcifelis	—	Pigs	Felids
S. miescheriana	*S. suicanis*	Pigs	Canids

[a] Adapted from Heydron, A.O. et al. 1975. *Z. Parasitenkd.* 48, 73.

19.1.2 Life Cycle and Morphology

Sarcocystis species are intracellular protozoan parasites with a heteroxenous (obligatory two-host) life cycle, alternating between hosts that serve as each other's predator and prey (definitive—intermediate host) [6].

Various *Sarcocystis* species can infect humans [7]. Man has been shown to serve as definitive host as well as intermediate host, but not for the same species. People act as definitive host for *S. hominis* and *S. suihominis*, with cattle and pigs as intermediate hosts, respectively, causing intestinal sarcocystosis (Table 19.1). Humans sometimes serve as intermediate hosts for hitherto unknown species, causing muscular sarcocystosis. The species involved in this cycle has not yet been determined. The name *Sarcocystis lindemanni* was once proposed, but it was not clearly described, and evidence of multiple morphologically different cysts suggests that several species probably are involved. So the name is no longer used and the life cycle including definitive hosts for these *Sarcocystis* species need to be elucidated [7].

The life cycle of *S. hominis* and *S. suihominis* is presented in Figure 19.1. Humans get infected by ingesting raw or undercooked beef or pork, containing mature sarcocysts. After ingestion, the sarcocyst wall is mechanically disrupted or digested in the stomach and intestine and the enclosed bradyzoites are liberated [8]. They penetrate goblet cells on the surface lining of the small intestine. Each intracellular bradyzoite develops in a male or female stage (micro- and macrogamonts) moving into the lamina propria. Fusion of a macrogamont and a microgamont results in a zygote which develops into an oocyst. Spherical oocysts pass into the intestinal lumen and contain two sporocysts, each with four sporozoites and a granular residual body (*Isospora* type of oocyst). As the oocyst wall is very thin, oval sporocysts are frequently liberated before excretion and individual sporocysts are excreted with the feces. The sporocysts are immediately infective, after being shed in the environment. This is a specific characteristic of the genus *Sarcocystis*, because other members of the family Sarcocystidae shed unsporulated oocysts [9]. The prepatent period (the interval between infection by a parasitic organism and the first diagnostic stage of the organism detectable in that host) for *S. hominis* is 9–10 days, and 11–13 days for *S. suihominis*. The patent period (the period of time in the course of a parasitic disease during which the parasitic organisms can be detected in the body) for both *S. hominis* and *S. suihominis* is more than six weeks [9]. The approximate size of the sporocyst is 15×9 μm for *S. hominis* and 10.5×13.5 μm for *S. suihominis*. Sporocysts cannot be distinguished from one another based on their morphology [6,9].

The appropriate intermediate host (cattle or pigs) becomes infected by ingestion of sporocysts in fecal contaminated feed and/or water. In the small intestine the sporocyst wall ruptures, releasing the sporozoites. They enter the intestinal wall and migrate to the endothelial lining of arteries, particularly those of the intestines and mesenteric lymph nodes. The sporozoites multiply in close association with the endothelium, resulting in first-generation schizonts containing merozoites (*S. suihominis* 5–6 DPI (days postinfection); *S. hominis* unknown). The schizogony in *Sarcocystis* species happens by endopolygeny

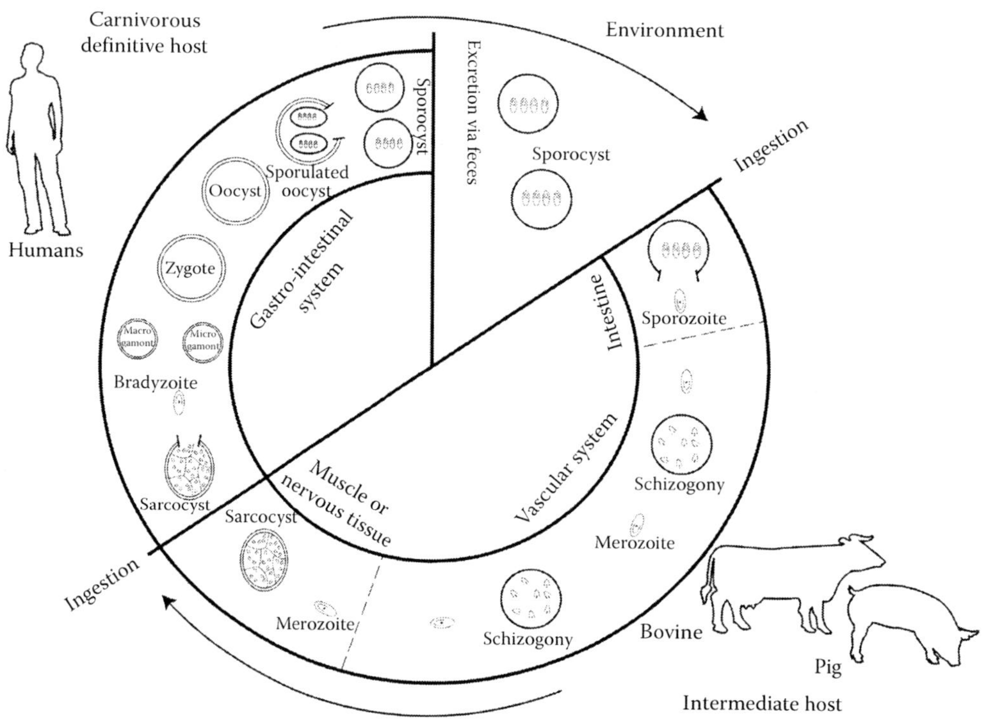

FIGURE 19.1 General life cycle of *S. hominis* and *S. suihominis*. The asexual stages with establishment of tissue cysts (sarcocysts) occur in cattle (*S. hominis*) and pigs (*S. suihominis*), the intermediate hosts. The sexual stages, with the formation of oocysts, take place in the definitive hosts, which are primates (humans, rhesus monkeys, baboons, and possibly chimpanzees).

(a process of asexual reproduction, in which many daughter cells are produced inside a mother cell). Merozoites released from the schizonts enter the bloodstream and develop into second-generation schizonts (*S. suihominis* 12–17 DPI; *S. hominis* unknown) in the endothelium of capillaries of virtually every organ, predominantly in renal glomeruli [10,11]. Both first- and second-generation schizonts are located within the host cytoplasm and not surrounded by a parasitophorous vacuole. Merozoites emerge from the schizonts and circulate extracellularly or may enter mononuclear cells, where they divide [10]. In the life cycle of *S. suihominis* no third-generation schizont is formed, for *S. hominis* no information is available. Merozoites from the terminal schizont generation leave the blood stream and invade muscle cells (skeletal or cardiac) [7]. Merozoites entering muscle cells round up to form metrocytes and initiate sarcocyst formation. A wall develops around the metrocytes, forming an immature sarcocyst. Consequently, the sarcocyst is always located within a parasitophorous vacuole in the host cell. Multiplication by endodyogeny (a process of asexual reproduction, in which two daughter cells are produced inside a mother cell) in the tissue cyst leads to the accumulation of numerous metrocytes (globular mother cells) and the sarcocyst increases in size. In the center of the cyst, the small, rounded, noninfectious metrocytes differentiate in crescent- or banana-shaped bodies, called bradyzoites, which are infective for the definitive host. The bradyzoites measure from 3–5 μm up to 18 μm depending on the species. A cyst is mature when it contains more bradyzoites than metrocytes [9]. Maturation varies for each species and takes minimum two months (62–98 days for *S. hominis*) [12]. In old cysts, bradyzoites that are located centrally degenerate and are replaced by a hollow. Sarcocysts can persist for months or years in the intermediate host. Mature sarcocysts of each species differ in size (from 0.1 mm to 30 mm) and shape. Some always remain microscopic, whereas others (can) become macroscopically visible (e.g., *S. suihominis*). Whether *S. hominis* produces macroscopic cysts has not been determined [13,14]. Sarcocysts develop distinct cyst walls that vary in thickness and organization of the villar protrusions [15,16]. Internally the sarcocyst is compartmented by septae, which is a specific characteristic of *Sarcocystis* species [9].

19.1.3 Epidemiology

Epidemiological data on intestinal as well as muscular sarcocystosis are scarce, especially on the prevalence of *Sarcocystis* in humans. Reasons for this are, for example, the difficulties of diagnosis and lack of (typical) symptoms both in definitive and intermediate hosts.

19.1.3.1 Intestinal Sarcocystosis in Humans

19.1.3.1.1 Prevalence in Humans

There are no reliable data on the prevalence of *Sarcocystis* in humans. The major reasons for this are: (i) the majority of infections are asymptomatic, (ii) intestinal sarcocystosis is characterized by nonspecific transient intestinal complaints [17,18], and (iii) symptoms, when present, are occurring during the prepatent period. This can imply that most cases of natural infections will not be reported. Moreover, a distinction between *S. hominis* and the more pathogenic *S. suihominis* cannot be made [6,9].

In a German study 13.64% of 403 patients with intestinal complaints were infected with intestinal parasites of whom 2% were shedding *Sarcocystis* sporocysts, but a distinction between *S. hominis* and *S. suihominis* was not made [19]. In a Thai study 8.0% of 1124 rural residents were infected with *Sarcocystis* species [20].

19.1.3.1.2 Prevalence in Cattle and Pigs

A high percentage (up to 100%) of cattle worldwide are infected with sarcocysts [18,21–28]. However, cattle can harbor three species (Table 19.1), namely *Sarcocystis cruzi* with canids as definitive host, *Sarcocystis hirsuta* with felids as definitive host and *Sarcocystis hominis* with primates as definitive hosts [4]. Mixed infections are very common (78.2%) [21] but little is known about the proportion of each species in cattle, due to difficulties in correct identification [16]. Surveys indicate that *S. cruzi* is the most prevalent species [7,21,29]. The prevalence of *S. hominis* shows large variations, ranging from 0% to 97% [18,28,29].

Also in pigs three species can occur (Table 19.1), namely *S. miescheriana*, *S. porcifelis*, and *S. suihominis* with canids, felids, and primates, respectively, as definitive hosts. No representative data are available on the current status of individual *Sarcocystis* species infections in domestic pigs. *Sarcocystis* species were found in 8–36% of slaughtered pigs in Germany and Austria [30,31], with a higher prevalence in slaughter sows than in fattening pigs. Reports of infections in slaughter sows in Japan and the USA, show that 16–18% of the sows had intramuscular sarcocysts [32,33].

19.1.3.1.3 Transmission

There is a striking discrepancy between the apparent low prevalence in humans and the high prevalence in cattle. This suggests a successful way for cattle to be infected, and less for humans. Further research is needed on the prevalence of *S. hominis* in humans and on other possible routes of transmission to cattle, to obtain a better insight into the epidemiological cycle of this protozoon.

Humans are infected by ingesting raw or undercooked beef or pork, containing mature sarcocysts. In some countries, where the frequency of consumption of raw or undercooked meat is relatively high, humans should get easily infected [8]. At a temperature of 2–4°C bovine sarcocysts remain infective in meat for 3 weeks [34]. Although data are not available for *S. hominis*, Saleque et al. [35] reported that sarcocysts of *S. miescheriana* in pork became noninfective after heating at 60°C for 20 min, at 70°C for 15 min or at 100°C for 5 min or freezing for 2 days at –4°C or 1 day at –20°C.

Due to the lack of schizogony in the definitive host, humans acquire little or no immunity, and each meal of infected meat can thus initiate a new round of sporocyst production [8].

As a rule, the only way cattle can be infected, is by ingesting sporocysts via fecal contaminated feed and/or water. The fact that *Sarcocystis* oocysts, unlike those of many other species of coccidia, are passed in feces of the definitive host in the infective form, frees them from dependence on weather conditions for maturation and infectivity [6]. Sporocysts can survive in feces of the definitive host for 2 years. Low temperatures or freezing does not influence their infectivity. However, *in vitro* investigations showed that sporocysts are sensitive to dryness, direct sunlight, and strong fluctuations in temperature [36]. Research on the viability of sporocysts of *Sarcocystis neurona* showed that heat treatment is the most effective means of killing. Treatment with bleach (10%, 20%, and 100%), 2% chlorhexidine, 1% betadine, 5% *o*-benzyl-*p*-chlorophenol, 12.56% phenol, 6% benzyl ammonium chloride, and 10% formalin was not effective in killing sporocysts. However, treatment with undiluted ammonium hydroxide (29.5% ammonia) for 1 hour killed sporocysts [37].

19.1.3.2 Muscular Sarcocystosis in Humans

Most cases of muscular sarcocystosis have been reported from South-East Asia and other regions where monkeys are commonly found [38]. Subclinical infections (incidental findings) are quite common among people from South-East Asia, as an autopsy study reported a prevalence of 21% [39,40]. In contrast, clinical cases are not common. Fewer than 100 cases are reported involving skeletal muscle and the myocardium [38,41,42]. These cases of myositis and vasculitis have been reported mostly associated with a history of traveling to the Far East [41].

Humans can acquire muscular sarcocystosis after ingestion of sporocysts of unknown *Sarcocystis* species. Presumably, they acquire this infection incidentally. Probably it is caused by a species normally infecting other animals, such as primates. Carnivorous or omnivorous predators living in close contact with monkeys then may pass fecal sporocysts, which are also infective for humans [43]. The most likely source of sporocysts is contaminated drinking water or using this water for washing or irrigating food [42].

19.1.4 Clinical Features and Pathogenesis

Little is known about the mechanisms that cause pathogenicity of *Sarcocystis* species in the definitive as well as in the intermediate hosts.

19.1.4.1 Intestinal Sarcocystosis in Humans

The clinical course after infection with *S. hominis* in humans is usually mild and in many cases even asymptomatic [42]. It is characterized by transient nonspecific gastrointestinal complaints, like nausea, stomach ache, abdominal pain and distension, and (watery) diarrhea [18,42]. The symptoms appear 3–6 h after eating infected beef and last about 36 h [42]. The severity of the symptoms is correlated with the number of ingested sarcocysts. In some cases abdominal discomfort, vomiting, diarrhea, dizziness, fatigue, anemia, and segmental necrotizing enteritis has been reported [44,45].

Infection with *S. suihominis* can cause bloat, nausea, loss of appetite, stomach ache, vomiting, diarrhea, difficulty in breathing, and rapid pulse, 6–48 h after ingestion of tissue cysts [17]. Symptoms are also infection dose related, but reactions can vary between individuals [42].

19.1.4.2 Muscular Sarcocystosis in Humans

When humans serve as intermediate host, most sarcocysts are located in skeletal muscle and myocard and to a lesser extent in the larynx, pharynx, and upper esophagus [46]. Most cases are incidental findings, in these patients no inflammatory response or clinical symptoms are observed [40]. In clinical cases, symptoms have been associated with disintegration of sarcocysts or with an immune-mediated response to the presence of immature forms of *Sarcocystis* in early infection [41,47]. The spectrum of illness ranges from acute self-limiting infections to chronic, moderately severe disease. Common clinical features include painful muscle swellings, which occur episodically, lasting days to weeks, episodic weakness, muscle tenderness and fasciculations. Other symptoms may include fever, rash, bronchospasm, persistent myalgia, subcutaneous nodules, and dermatomyositis [41,42,47].

Systemic sarcocystosis with dissemination to the liver and muscle tissue is also reported as an opportunistic infection in an AIDS patient [48].

19.1.5 Diagnosis

The main tests used for the diagnosis of human sarcocystosis in many diagnostic laboratories are still based on the morphology of the different development stages of *Sarcocystis* species (oocysts, sporocysts, and sarcocysts containing bradyzoites). The conventional light microscopical techniques can be used only to demonstrate the presence or absence of *Sarcocystis* species.

Therefore, direct diagnostic tests based on the detection/identification of parasite-specific molecules by specific DNA probes or PCR-based techniques could provide a valuable alternative in the diagnosis of *Sarcocystis* infection.

19.1.5.1 Conventional Techniques

Diagnosis of intestinal sarcocystosis requires detection of sporocysts in the stool. Sporocysts can be observed by bright-field microscopy in a fecal flotation wet mount just beneath the coverslip. As an alternative, the traditional flotation technique based on concentration methods (specific gravity 1.15 or more) with saturated sodium chloride (NaCl), sucrose or zinc sulphate ($ZnSO_4$) solutions can be used [7]. The concentration step is necessary, because the number of sporocysts in the feces is often low. The latter technique is more commonly used than the former. The determination of numbers of oocysts/sporocysts can be done by the McMaster technique [49]. In addition, a modified Kato thick smear technique, commonly used for the diagnosis of helminthiasis, has also been tested successfully for the diagnosis of *Sarcocystis* sporocysts [20].

Morphological species identification/differentiation is not possible for the oocyst/sporocyst stages of this parasite. Sporulated oocysts of both species (*S. hominis* and

S. suihominis) are spherical, generally colorless, thin-walled (<1 µm) and contain two elongated sporocysts. The colorless wall is delicate and often ruptures. An oocyst residuum and micropyle are absent. Similarly, the morphology of sporocysts has no taxonomic value. Except for minor variations in size, the structure is the same [9]. Each sporocyst contains four elongated sporozoites and a compact or dispersed granular residual body, but a Stieda body is absent. Because sporozoites are curved in the sporocyst, all four are often not seen in a single plane of focus. Each sporozoite has a central to terminal nucleus and several cytoplasmic granules, but there is no refractile body. Sporocysts of *S. hominis* average 9×15 µm and those of *S. suihominis* average 10.5×13.5 µm [7].

An indirect fluorescent antibody test (IFAT) is described to detect circulating antibodies against *S. hominis* in sera in man [50]. However, Heckeroth and Tenter assumed that data on *Sarcocystis* infections in definitive hosts have been complicated by the fact that these hosts do not develop parasite-specific antibodies [49].

A likely diagnosis of muscular sarcocystosis can be made if a patient has a history of travel in endemic area and blood analysis reveals an eosinophilia, elevated creatinine kinase, and aldolase [41,42,47]. Detection of antibodies in serum by Western blot analysis has been described to be of diagnostic value [41]. However, this technique cannot be performed on a routine basis. Therefore, diagnosis of muscular sarcocystosis is mainly performed by histological examination of muscle biopsies stained with hematoxylin and eosin (H&E), and PAS reaction. Cysts containing banana-shaped bradyzoites can be observed in the cytoplasm of striated muscle cells, the cyst width can range from 30 to 137 µm [51]. As mentioned above, the species involved in muscular sarcocystosis in humans have not yet been identified. As for yet identified *Sarcocystis* species in other intermediate hosts, transmission electron microscopy of the structure of the villar protrusions of the cyst wall is the only morphological technique for more definitive identification [7,15,16]. Even so, this technique is not employed for routine diagnosis but could be of importance in further research of human muscular *Sarcocystis* species. However, some researchers question the description of *Sarcocystis* species based on sarcocyst morphology alone, since this feature may vary depending on the developmental stage, location of the cyst, and the degenerative stage of the parasitized cell [52].

19.1.5.2 Molecular Techniques

To date, there are some molecular based methods that have been developed, though primarily in a research-based context.

Tests targeting rRNA, for example, ribosomal hybridization assay [53], have been developed for some *Sarcocystis* species. These methods have been suggested to be more sensitive and accurate than assays targeting the rRNA genes, as there is up to 50 times more RNA than DNA in a cell, of which 90–95% is rRNA [54]. Since RNA is not stable and easily degraded by enzymatic digestion, these assays are not suitable for the diagnosis of clinical or epidemiological samples [49]. Moreover, standardization of RNA hybridization assays is difficult and the necessity of using radioactively labeled oligonucleotide probes restricts the use to specialized laboratories [55,56].

Since the development of the randomly amplified polymorphic DNA polymerase chain reaction (RAPD-PCR) fingerprinting technology, this method has been used in intermediate hosts to distinguish different *Sarcocystis* species [57,58]. An advantage of this technique is that short regions of genomic DNA can be amplified without prior knowledge of gene sequences. DNA sequence analysis of these PCR products can then be used to design species specific primers. However, this method has not been developed as a diagnostic test. The reason is that RAPD is laden by technical drawbacks, through which standardization of the assay between laboratories is difficult [54].

In recent years the small subunit rRNA gene has been used extensively to differentiate between apicomplexans and other eukaryotic species. Due to its abundance in the genome, the sensitivity of the tests increases. Moreover, its double feature of hypervariable regions interspersed within highly conserved DNA sequences makes it a useful target for developing molecular-based diagnostics [59–62]. The rapidly growing database of 18S rDNA sequences generated by phylogenetic studies of *Sarcocystis* species provides a rich source of sequences that can be exploited to develop specific PCRs [61].

In the last decade, a number of PCR-based techniques (e.g., genus or species specific PCR, nested PCR, PCR-RFLP (restriction-fragment-length polymorphism)) have been established, especially for the detection and identification of tissue cysts in livestock [28,56,63–65]. Although DNA templates are stable targets for PCR amplification and diagnostic samples can be stored at ambient temperature for some time, each of the range of different sources of parasite material (feces, tissue) involves different problems for PCR-based diagnosis, associated with inhibitors and sensitivity [63]. Therefore, PCR assays that have been shown to be specific and sensitive on tissue in livestock, probably cannot be used as such for the diagnosis of *Sarcocystis* species in fecal samples.

19.2 METHODS

In general, molecular formats for a routine diagnosis of muscular sarcocystosis in humans have not been developed. For intestinal sarcocystosis, Xiang et al. have evaluated a genetic approach for the detection and identification of *S. hominis* sporocysts based on a seminested PCR-RFLP assay (described below) [45].

19.2.1 Intestinal Sarcocystosis in Humans

19.2.1.1 Sample Preparation

As described earlier, it may be necessary to take several stool samples over the course of a few days before sporocysts can be detected/identified due to intermittent shedding by the definitive host.

19.2.1.2 Sporocyst Isolation

The following protocol shows the different steps of sporocyst isolation as described by Xiang et al. [45].

1. 1–2 g of the fecal sample is used for obtaining sporocysts by fecal flotation in 15 mL of a saturated $ZnSO_4$ solution (specific gravity 1.15).
2. Transfer the meniscus of the flotation to a microscope slide using an inoculating loop for sporocyst and oocyst exam.
3. Adsorb to a piece of filter paper measuring 2–8 × 1–5 mm (Chinese patent pending no. 200710065686.2 or commercial blank filter paper).
4. Store in an Eppendorf with 500 µL distilled water at −20°C until use.

It should be noted that high-density solutions based on formalin-ethyl acetate cannot be used if PCR needs to be performed, as these products induce cross-linking of DNA to proteins.

19.2.1.3 Oocyst/Sporocyst Lysis

In the study of Xiang et al., sporocysts are submitted to DNA extraction without lysis of the wall [45]. Because sporocysts are resistant [37], we believe that lysis of the sporocyst wall should be achieved as described for related *Sarcocystis* species and related protozoa prior to DNA extraction [66]. A successful technique for lysis of *Sarcocystis neurona* sporocyst walls subjects sporocysts to six cycles of freezing in liquid nitrogen followed by thawing [67]. Other methods described in the literature are based on bead disruption and *in vitro* excystation techniques [66,68]. Incubating sporocysts overnight at 37°C and 5% atmospheric carbon dioxide in horse bile containing 2% trypsin has been shown to be a successful method for excystation of *Sarcocystis neurona* sporocysts [69].

19.2.1.4 DNA Extraction

DNA extraction can be performed using a traditional phenol/chloroform procedure or by using commercial kits for genomic DNA purification [28].

19.2.1.5 Detection Procedure

A step-wise protocol for molecular identification of *S. hominis* and *S. suihominis* in human stool is presented, as reported by Yang et al. and Xiang et al. [45,64,70,71]. This method is based on the amplification of a fragment from 18S rDNA gene using a semi-nested PCR combined with RFLP. Due to the low number of sporocysts excreted by the definitive host, Xiang et al. consider that semi-nested PCR is necessary to increase the sensitivity of the assay [45]. The primers described for this assay target relatively conserved portions of 18S rDNA. Therefore, successful amplification indicates the presence of a *Sarcocystis* species, but does not discriminate between different species. When used in conjunction with RFLP or sequencing, however, species identification is possible [64].

19.2.1.5.1 Procedure

1. Prepare the PCR mixture (25 µL) for the primary PCR [45,70,71]. It contains 1 µL of DNA (about 100 ng/µL) template and a PCR-mix consisting of PCR buffer (20 mM Tris–HCl, pH 8.3, 60 mM KCl), 2 mM $MgCl_2$, 100 µM dNTPs, 0.75 U Taq polymerase, and 400 pM of each primer (primers *2L* 5′- GGA TAA ACC GTG GTA ATT CTA TG -3′ and *2H* 5′-ACC TGT TAT TGC CTC AAA CTTC-3′) [70] and ultrapure water to a final volume of 25 µL.
2. Carry out the PCR amplification on a thermocycler using the following cycling program: 95°C for 2 min, followed by 40 cycles of 94°C for 40 s, 50–57°C for 30 s or 1 min, 72°C for 1 min, followed by one cycle of 72°C for 6 min.
3. For the secondary amplification prepare the PCR mixture [70,71] (50 µL): a 2 µL aliquot of the primary PCR product and a PCR mix consisting of PCR buffer (20 mM Tris–HCl, pH 8.3, 60 mM KCl), 2 mM $MgCl_2$, 200 µM dNTPs, and 1.25 U Taq polymerase, 10 pM of each primer (primer *2L* and primer *3H* 5′- GGC AAA TGC TTT CGC AGT AG - 3′), and ultrapure water to a final volume of 50 µL. Appropriate control samples should be included.
4. Carry out the PCR amplification on a thermocycler using the following cycling program: 95°C for 2 min, followed by 40 cycles of 94°C for 40 s, 50–57°C for 30 s or 1 min, 72°C for 1 min, followed by one cycle of 72°C for 6 min.
5. Purify the semi-nested PCR products. Digestion of 18S rRNA gene PCR products with restriction endonucleases (*Hinf I, Bcl I*) [64]: prepare a 20 µL reaction, containing 5–8 µL of PCR product, 5–10 U of restriction enzyme and the appropriate restriction enzyme buffer. Incubate the reactions for 4–12 h at 37 or 65°C, according to the manufacturer's instructions. Terminate the reactions by adding the loading buffer. Analyze the digestions by 2% agarose gel electrophoresis at 3 V/cm. Run also PCR markers and uncut PCR products as controls and as a means of sizing fragments. Stain the gel with ethidium bromide and visualize the fragments using a UV transilluminator (Table 19.2).

19.2.2 Muscular Sarcocystosis in Humans

In literature no data are available for the detection or identification of human muscular sarcocystosis by molecular means. However, molecular identification of muscular cysts in livestock has already been performed [28,63,64].

Since muscle samples of humans probably will not contain many sarcocysts, an obstacle in the detection/identification of sarcocysts by molecular techniques could be the isolation of *Sarcocystis* DNA. Therefore, we believe that sample preparation is an important step in the diagnosis. The information hereafter is based on techniques for sample

TABLE 19.2
Electromorph Patterns for Taxa/Restriction Digest Combinations

Enzyme	Taxon	Size (bp) before Digestion	Number of Restriction Sites	Size (bp) of Fragments
Hinf I	*S. suihominis*	850	5	250, 250, 150, 70, 65, 65
	S. hominis	850	4	450, 140, 120, 70, 70
Bcl I	*S. suihominis*	850	0	850
	S. hominis	850	1	700, 150

Source: Yang, Z.Q. et al. 2002. *Exp. Parasitol.* 102, 212 for *S. suihominis* and *S. hominis*.

preparation in livestock, which may be applicable to human muscle samples. It should, however, be noted that these are not yet validated for this purpose.

The isolation of parasitic stages (sarcocysts or bradyzoites) in postmortem samples could be achieved by digestion in trypsin or pepsin/HCl solution [72,73] or by a sedimentation technique [28]. The digestion technique is a frequently used, practical, and sensitive method for *Sarcocystis* detection in livestock [22]. After digestion of the sample the muscle fibers and sarcocyst are lysed. The resulting bradyzoites can then be purified on a Percoll gradient [74]. The sedimentation technique on the other hand does not result in lysis of sarcocysts, which allows a stereomicroscopical screening of the sarcocysts. This step should be followed by lysis of the sarcocyst wall with a pepsin/HCl solution before DNA extraction [28]. If only small amounts of muscle tissue are submitted, the sedimentation technique will probably not be sensitive enough.

In case of an ante mortem biopsy, the amount of sample will be the limiting factor in the choice of the technique used for sample preparation. If a nonfixated biopsy is sent in for examination, the recovery of bradyzoites can be performed by incubation of the sample in a trypsin or pepsin/HCl solution [72,73]. However, a more specific method for isolation of sarcocysts from muscle biopsies is the use of laser capture microdissection (unpublished data) [75]. An advantage of this technique is that the sample may be fixated. Moreover, this technique can be applied after conventional histopathological examination, which allows the researcher to examine all pathological alterations.

Since samples will not contain many sarcocysts, a sensitive and efficient method for genomic DNA extraction is probably required.

In brief, the procedure for laser capture microdissection based on De Preter et al. [76], Emmert-Buck et al. [77], and Bonner et al. [78] is as follows:

1. Cut 5 μm serial sections from paraffin-embedded formalin-fixed tissues.
2. Perform H&E staining manually and dehydrate [76].
3. Microdissect the sarcocysts using a PixCell II laser capture microscope equipped with an infrared diode laser (Arcturus Engineering, Santa Clara, CA, USA) [77,78].
4. Capture the sarcocysts individually on CapSure™ Macro LCM Caps (Arcturus Bioscience, Mountain View, CA, USA).
5. Extract DNA from the laser-captured sarcocysts using the Pico pure DNA extraction kit™ (Arcturus Bioscience).

For the detection procedure itself, the primers used in the report of Yang et al. [70] are quite universal primers targeting different *Sarcocystis* species. If this PCR assay is performed in combination with sequence analysis, a diagnosis of muscular *Sarcocystis* infection in humans can probably be made. However, this technique is not yet validated for this purpose. In case of formalin-fixed samples, PCR amplification of fragments of more than 150–200 bp will not be possible due to cross-linking of the DNA.

19.3 CONCLUSIONS AND FUTURE PERSPECTIVES

While it is clear that data about molecular methods for detection and identification of muscular sarcocystosis in livestock have surged forward recently, feasible tests for routine diagnosis are not yet available. This implies that still many questions about the epidemiology of *S. hominis* and *S. suihominis* remain unanswered.

It is also evident that much more is yet to be learned about detection, identification, and epidemiology of the parasite stages in people. Until now, data of human intestinal sarcocystosis are based on results of conventional tests which have a low sensitivity and do not allow species identification. Only recently a molecular technique has been tested on human stool samples, but it is not applied on a routine basis yet. The lack of sensitive diagnostic assays may imply that intestinal sarcocystosis is still an underestimated problem in people.

The difficulties that appear with the conventional techniques for diagnosis of intestinal sarcocystosis (e.g., low sensitivity and no species identification possible) are present as well in the diagnosis of human muscular sarcocystosis. Moreover, molecular tests for the detection and identification of sarcocysts present in human muscle are not yet established.

The improvement and development of molecular diagnostic methods for routine diagnosis, particularly those utilizing PCR, will contribute greatly to the identification and control of these pathogens, by increasing the speed of diagnosis, specificity and sensitivity, reproducibility, and ease of interpretation. Detailed knowledge on the prevalence and transmission dynamics of infections with *Sarcocystis* species is after all central to the development of effective measures for prevention.

ACKNOWLEDGMENTS

Lieve Vangeel received a PhD grant from the Institute for the Promotion of Innovation by Science and Technology in Flanders (I.W.T. Vlaanderen). Peter Geldhof is a postdoctoral fellow of the Fund for Scientific Research Flanders (FWO—Belgium).

REFERENCES

1. Levine, N.D. 1986. The taxonomy of *Sarcocystis* (Protozoa, Apicomplexa) species. *J. Parasitol.* 72, 372.
2. Frenkel, J.K., and Smith, D.D. 2003. Determination of the genera of cyst-forming coccidia. *Parasitol. Res.* 91, 384.
3. Rommel, M., and Heydorn, A.O. 1972. Contributions to the life cycle of Sarcosporidia. 3. *Isospora hominis* (Railliet and Lucet, 1891) Wenyon, 1923, the sporocyst of the Sarcosporidia of cattle and swine. *Berl. Munch. Tierarztl. Wochenschr.* 85, 143.
4. Heydorn, A.O. et al. 1975. Proposal for a new nomenclature of the Sarcosporidia. *Z. Parasitenkd.* 48, 73.
5. Levine, N.D. 1977. Nomenclature of *Sarcocystis* in the ox and sheep and of fecal Coccidia of the dog and cat. *J. Parasitol.* 63, 36.
6. Dubey, J.P. 1976. Review of *Sarcocystis* of domestic animals and of other Coccidia of cats and dogs. *J. Am. Vet. Med. Assoc.* 169, 1061.
7. Dubey, J.P., Speer, C.A., and Fayer, R. 1989. *Sarcocystosis of Animals and Man*, Chapters 1 and 10. CRC Press, Boca Raton, USA.
8. Dubey, J.P., and Lindsay, D.S. 2006. Neosporosis, toxoplasmosis, and sarcocystosis in ruminants. *Vet. Clin. North Am. Food Anim. Pract.* 22, 645.
9. Mehlhorn, H., and Heydorn, A.O. 1978. The Sarcosporidia (Protozoa, Sporozoa): Life cycle and fine structure. *Adv. Parasitol.* 16, 43.
10. Dubey, J.P., and Fayer, R. 1983. Zoonosis in practice—sarcocystosis. *Br. Vet. J.* 139, 371.
11. Mehlhorn, H. 1988. *Parasitology in Focus*. Springer Verlag Berlin Heidelberg, Germany, 32.
12. Heydorn, A.O., Mehlhorn, H., and Gestrich, R. 1975. Light and electron microscope studies on cysts of *Sarcocystis fusiformis* in the muscles of calves infected experimentally with oocysts and sporocysts of *Isospora hominis* Railliet et Lucet, 1891. 2. Fine structure of metrocytes and merozoites. *Zentralbl. Bakteriol. Orig. A.* 232, 373.
13. Bottner, A., Charleston, W.A.G., and Hopcroft, D. 1987. The structure and identity of macroscopically visible *Sarcocystis* cysts in cattle. *Vet. Parasitol.* 24, 35.
14. Dubey, J.P., Fayer, R., and Speer, C.A. 1988. Experimental *Sarcocystis hominis* infection in cattle—Lesions and ultrastructure of sarcocysts. *J. Parasitol.* 74, 875.
15. Dubey, J.P., Speer, C.A., and Charleston, W.A.G. 1989. Ultrastructural differentiation between sarcocysts of *Sarcocystis hirsuta* and *Sarcocystis hominis*. *Vet. Parasitol.* 34, 153.
16. Odening, K. et al. 1995. On the morphological diagnostics and host specificity of the *Sarcocystis* species of some domesticated and wild bovini (cattle, banteng and bison). *Appl. Parasitol.* 36, 161.
17. Heydorn, A.O. 1977. Sarkosporidien enfiziertes fleisch als mogliche krankheitsurache fur den menschen. *Arch. Lebenmittelhyg.* 28, 27.
18. Pena, H.F., de, J., Ogassawara, S., and Sinhorini, I.L. 2001. Occurrence of cattle *Sarcocystis* species in raw kibbe from Arabian food establishments in the city of São Paulo, Brazil, and experimental transmission to humans. *J. Parasitol.* 87, 1459.
19. Gauert, B., Jungmann, R., and Hiepe, T. 1983. Beziehungen zwischen intestinalstöringen und parasitären darminfektionen des menschen unter besonderer berücksichtigung von *Sarcocystis*-befall. *Dt. Gesundh.-Wesen.* 38, 62.
20. Tungtrongchitr, A. et al. 2007. The potential usefulness of the modified Kato thick smear technique in the detection of intestinal sarcocystosis during field surveys. *Southeast Asian J. Trop. Med. Public Health* 38, 232.
21. Bottner, A. et al. 1987. The prevalence and identity of *Sarcocystis* in beef cattle in New Zealand. *Vet. Parasitol.* 24, 157.
22. van Knapen, F., Bouwman, D., and Greve, E. 1987. Study on the incidence of *Sarcocystis* spp in Dutch cattle using various methods. *Tijdschr. Diergeneeskd.* 112, 1095.
23. Vercruysse, J., Fransen, J., and Vangoubergen, M. 1989. The prevalence and identity of *Sarcocystis* cysts in cattle in Belgium. *J. Vet. Med. B Infect. Dis. Vet. Public Health* 36, 148.
24. Fortier, G. et al. 1993. Prevalence of muscular bovine sarcocystosis in the Calvados (Normandy-France). *Rec. Med. Vet. Ec. Alfort.* 169, 779.
25. Woldemeskel, M., and Gebreab, F. 1996. Prevalence of sarcocysts in livestock of Northwest Ethiopia. *J. Vet. Med. B Infect. Dis. Vet. Public Health* 43, 55.
26. Latif, B.M.A. et al. 1999. Prevalence of *Sarcocystis* spp in meat-producing animals in Iraq. *Vet. Parasitol.* 84, 85.
27. De Bosschere, H., and Ducatelle, R. 2001. Inverse correlation between myositis eosinophilica and number of *Sarcocystis* cystozoits in heart of cattle. *Vlaams Diergen. Tijdschr.* 70, 118.
28. Vangeel, L. et al. 2007. Molecular-based identification of *Sarcocystis hominis* in Belgian minced beef. *J. Food Prot.* 70, 1523.
29. Pritt, B. et al. 2008. Detection of *Sarcocystis* parasites in retail beef: A regional survey combining histological and genetic detection methods. *J. Food Prot.* 71, 2144.
30. Hinaidy, H.K., and Supperer, R. 1979. Incidence of Sarcosporidia of pigs in Austria. *Wien. Tierarztl. Monatsschr.* 66, 281.
31. Damriyasa, I.M. et al. 2004. Cross-sectional survey in pig breeding farms in Hesse, Germany: Seroprevalence and risk factors of infections with *Toxoplasma gondii*, *Sarcocystis* spp and *Neospora caninum* in sows. *Vet. Parasitol.* 126, 271.
32. Omata, Y. et al. 1993. Survey of *Sarcocystis* spp infection in slaughtered pigs in East Hokkaido, Japan. *J. Protozool. Res.* 3, 29.
33. Dubey, J.P., and Powell, E.C. 1994. Prevalence of *Sarcocystis* in sows from Iowa. *Vet. Parasitol.* 52, 151.
34. Stephan, R., Tholen, R., and Meier, D. 1998. Greenish discolorations of beef muscle tissue and their assessment according to meat hygiene regulations. A case report. *Tierarztl. Prax. Ausg. Grobtiere Nutztiere.* 26, 31.
35. Saleque, A., Juyal, P.D., and Bhatia, B.B. 1990. Effect of temperature on the infectivity of *Sarcocystis miescheriana* cysts in pork. *Vet. Parasitol.* 36, 343.
36. Savini, G., Robertson, I.D., and Dunsmore, J.D. 1996. Viability of the sporocysts of *Sarcocystis cruzi* after exposure to different temperatures and relative humidities. *Vet. Parasitol.* 67, 153.
37. Dubey, J.P. et al. 2002. Effects of high temperature and disinfectants on the viability of *Sarcocystis neurona* sporocysts. *J. Parasitol.* 88, 1252.

38. Beaver, P.C., Gadgil, K., and Morera, P. 1979. *Sarcocystis* in man: A review and report of five cases. *Am. J. Trop. Med. Hyg.* 28, 819.
39. Wong, K.T., and Pathmanathan, R. 1992. High prevalence of human skeletal-muscle sarcocystosis in South-East Asia. *Trans. R. Soc. Trop. Med. Hyg.* 86, 631.
40. Mehrotra, R. et al. 1996. Diagnosis of human *Sarcocystis* infection from biopsies of the skeletal muscle. *Pathology* 28, 281.
41. Arness, M.K. et al. 1999. An outbreak of acute eosinophilic myositis attributed to human *Sarcocystis* parasitism. *Am. J. Trop. Med. Hyg.* 61, 548.
42. Fayer, R. 2004. *Sarcocystis* spp in human infections. *Clin. Microbiol. Rev.* 17, 894.
43. Pamphlett, R., and Odonoghue, P. 1990. *Sarcocystis* infection of human muscle. *Aust. N. Z. J. Med.* 20, 705.
44. Bunyaratvej, S., and Unpunyo, P. 1992. Combined *Sarcocystis* and gram-positive bacterial infections. A possible cause of segmental enterocolitis in Thailand. *J. Med. Assoc. Thai.* 75, 38.
45. Xiang, Z. et al. 2009. Non-invasive methods for identifying oocysts of *Sarcocystis* spp from definitive hosts. *Parasitol. Int.* 58, 293
46. Lele, V.R., Dhopavkar, P.V., and Kher, A. 1986. *Sarcocystis* infection in man (a case report). *Indian J. Pathol. Microbiol.* 29, 87.
47. Crum-Cianflone, N.F. 2008. Bacterial, fungal, parasitic, and viral myositis. *Clin. Microbiol. Rev.* 21, 473.
48. Velasquez, J.N. et al. 2008. Systemic sarcocystosis in a patient with acquired immune deficiency syndrome. *Hum. Pathol.* 39, 1263.
49. Heckeroth, A.R., and Tenter, A.M. 2007. *Sarcocystosis, Protozoal Abortion in Farm Ruminants: Guidelines for Diagnosis and Control.* Oxfordshire, UK, 172.
50. Tadros, W., Laarman, J.J., and van den Eijk, A.A. 1974. The demonstration of antibodies to *Sarcocystis fusiformis* antigen in sera of *Isospora hominis* carriers, using the indirect fluorescence technique. *Z. Parasitenkd.* 43, 221.
51. Wong, K.T. et al. 1994. Light microscopic and three-dimensional morphology of the human muscular sarcocyst. *Parasitol. Res.* 80, 138.
52. Tenter, A.M. 1995. Current Research on *Sarcocystis* species of domestic-animals. *Int. J. Parasitol.* 25, 1311.
53. Gajadhar, A.A., Marquardt, W.C., and Blair, C.D. 1992. Development of a model ribosomal RNA hybridization assay for the detection of *Sarcocystis* and other coccidia. *Can. J. Vet. Res.* 56, 208.
54. Zarlenga, D.S., and Higgins, J. 2001. PCR as a diagnostic and quantitative technique in veterinary parasitology. *Vet. Parasitol.* 101, 215.
55. MacPherson, J.M., and Gajadhar, A.A. 1993. Ribosomal RNA sequences for the specific detection of *Toxoplasma gondii* by hybridization assay. *Mol. Cell. Probes* 7, 97.
56. Heckeroth, A.R., and Tenter, A.M. 1999. Development and validation of species-specific nested PCRs for diagnosis of acute sarcocystiosis in sheep. *Int. J. Parasitol.* 29, 1331.
57. Macpherson, J.M., and Gajadhar, A.A. 1994. Specific amplification of *Sarcocystis cruzi* DNA using a randomly primed polymerase chain reaction assay. *Vet. Parasitol.* 55, 267.
58. Guclu, F., Aldem, O.S., and Guler, L. 2004. Differential identification of cattle *Sarcocystis* spp by random amplified polymorphic DNA polymerase chain reaction (RAPD-PCR). *Rev. Med. Vet.* 155, 440.
59. Neefs, J.M. et al. 1991. Compilation of small ribosomal-subunit RNA sequences. *Nucleic Acids Res.* 19, 1987.
60. Tenter, A.M., and Johnson, A.M. 1997. Phylogeny of the tissue cyst-forming coccidia. *Adv. Parasitol.* 39, 69.
61. Fischer, S., and Odening, K. 1998. Characterization of bovine *Sarcocystis* species by analysis of their 18S ribosomal DNA sequences. *J. Parasitol.* 84, 50.
62. Prichard, R., and Tait, A. 2001. The role of molecular biology in veterinary parasitology. In: *18th International Conference of the World-Association-for-the-Advancement-Veterinary-Parasitology*. Elsevier Science Bv, Stresa, Italy, 169.
63. Tenter, A.M., Luton, K., and Johnson, A.M. 1994. Species specific identification of *Sarcocystis* and *Toxoplasma* by PCR amplification of small subunit ribosomal RNA gene fragments. *Appl. Parasitol.* 35, 173.
64. Yang, Z.Q. et al. 2002. Characterization of *Sarcocystis* species in dómestic animals using a PCR-RFLP analysis of variation in the 18S rRNA gene: A cost-effective and simple technique for routine species identification. *Exp. Parasitol.* 102, 212.
65. Gonzalez, L.M. et al. 2006. Differential molecular identification of *Taeniid* spp and *Sarcocystis* spp cysts isolated from infected pigs and cattle. *Vet. Parasitol.* 142, 95.
66. da Silva, A.J. et al. 1999. Fast and reliable extraction of protozoan parasite DNA from fecal specimens. *Mol. Diagn.* 4, 57.
67. Elsheikha, H.M. et al. 2003. Purification of *Sarcocystis neurona* sporocysts from opossum (*Didelphis virginiana*) using potassium bromide discontinuous density gradient centrifugation. *Parasitol. Res.* 90, 104.
68. Horn, K., Ono, K., and Heydorn, A.O. 1991. *In vitro* excystation and cryopreservation of ovine and caprine *Sarcocystis* species. *J. Protozool. Res.* 1, 13.
69. Tanhauser, S.M. et al. 1999. Multiple DNA markers differentiate *Sarcocystis neurona* and *Sarcocystis falcatula*. *J. Parasitol.* 85, 221.
70. Yang, Z.Q. et al. 2001. Analysis of the 18S rRNA genes of *Sarcocystis* species suggests that the morphologically similar organisms from cattle and water buffalo should be considered the same species. *Mol. Biochem. Parasitol.* 115, 283.
71. Yang, Z.Q. et al. 2001. Identification of *Sarcocystis hominis*-like (Protozoa: Sarcocystidae) cyst in water buffalo (*Bubalus bubalis*) based on 18S rRNA gene sequences. *J. Parasitol.* 87, 934.
72. Seneviratna, P., Edward, A.G., and Degiusti, D.L. 1975. Frequency of *Sarcocystis* spp in Detroit, Metropolitan Area, Michigan. *Am. J. Vet. Res.* 36, 337.
73. Boch, J., Laupheimer, K.E., and Erber, M. 1978. Sarcosporidia species in slaughter cattle in Southern Germany. *Berl. Munch. Tierarztl. Wochenschr.* 91, 426.
74. Gajadhar, A.A., and Marquardt, W.C. 1992. Ultrastructural and transmission evidence of *Sarcocystis cruzi* associated with eosinophilic myositis in cattle. *Can. J. Vet. Res.* 56, 41.
75. Vangeel, L. et al. unpublished data.
76. De Preter, K. et al. 2003. Application of laser capture microdissection in genetic analysis of neuroblastoma and neuroblastoma precursor cells. *Cancer Lett.* 197, 53.
77. Emmert-Buck, M.R. et al. 1996. Laser capture microdissection. *Science*. 274, 998.
78. Bonner, R.F. et al. 1997. Laser capture microdissection: Molecular analysis of tissue. *Science*. 278, 1481.

20 Toxoplasma

Kjetil Åsbakk and Kristin Wear Prestrud

CONTENTS

20.1 INTRODUCTION

The obligate intracellular protozoan parasite *Toxoplasma gondii* has a cosmopolitan distribution and the capacity to infect all warm-blooded animals, including humans and birds [1]. The parasite was discovered in 1908 in a colony of a desert rodent, the gundi (*Ctenodactylus gundi*), maintained in the Pasteur Institute in Tunis [2], before it was known to cause disease in humans. Infection with *T. gondii* is now known from a vast number of animal species including food animals and marine mammals [3]. In humans, most infections are asymptomatic and clinical toxoplasmosis affects only a minority of individuals. Symptoms are usually mild including manifestations such as lymphadenopathy, fever, and malaise [4]. Infection with *T. gondii* is mainly acquired by ingestion of uncooked or undercooked meat containing tissue cysts or food and drink contaminated with oocysts, the environmentally resistant infectious stage of the parasite. The parasite also infects transplacentally in humans and other species, and congenital toxoplasmosis is a very serious disease, often resulting in abortions or fetal damage or death in humans and other species [4]. In sheep and goat, toxoplasmosis is a significant cause of abortion [5,6]. Since the mid-1980s, *T. gondii* is also recognized as a complicating factor for immunocompromised individuals, particularly AIDS patients.

20.1.1 Classification, Life Cycle, and Morphology

20.1.1.1 Classification

T. gondii is member of Apicomplexa, the phylum comprising protozoa possessing the apical complex, a combination of structures distinguishable only by the electron microscope [1,7]. Apicomplexa is divided into three classes. One is the class Conoidasidae, containing the subclass Coccidiasina, where *T. gondii* is member under order Eucoccidiorida, suborder Eimeriorina, family Sarcocystidae [1]. Apicomplexa classification has been changing rapidly as new information becomes available. Like the other members of the Sarcocystidae family, *T. gondii* has a heteroxenous (involving more than one host) life cycle with both intestinal and tissue stages (Figure 20.1).

20.1.1.2 Life Cycle and Morphology

20.1.1.2.1 Oocysts

Oocysts (10–13 × 9–11 μm of size) result from sexual reproduction of the parasite in a definitive host. Animals of the family Felidae only, both domestic and wild cats, are known as definitive hosts [8]. In cats, the extraintestinal (asexual) and the enteroepithelial (sexual) phase can occur simultaneously [9], while in hosts other than cats (intermediate hosts)

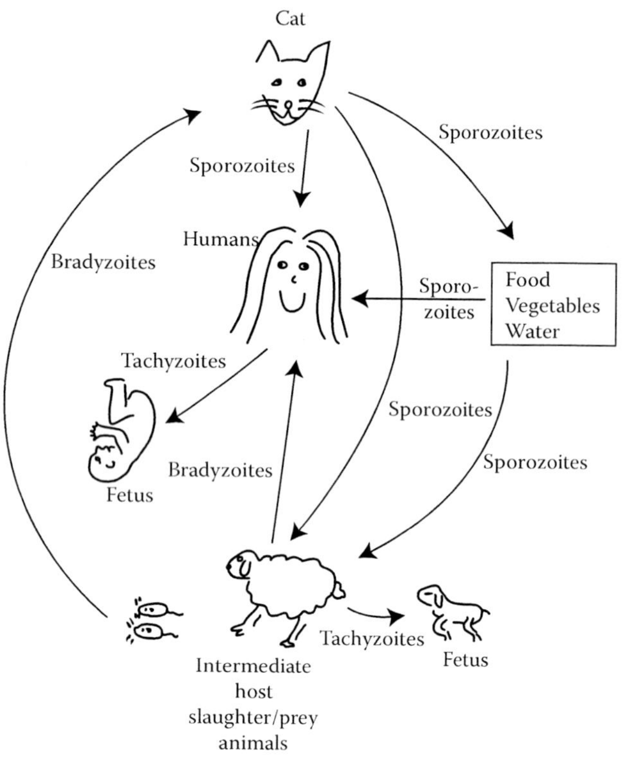

FIGURE 20.1 The classical life cycle and known transmission routes of *T. gondii*. (Reprinted from Weiss, L.M., and Kim, K. (eds), *Toxoplasma gondii. The Model Apicomplexan—Perspectives and Methods*, Dubey, J.P. The history and life cycle of *Toxoplasma gondii*, pp. 1–17, Copyright 2007, with permission from Elsevier.)

only the extraintestinal phase occurs. Intermediate hosts can be considered paratenic (transport) hosts since the sexual life cycle is never completed in those hosts [9].

Oocysts are shed with the feces of the definitive host and are noninfective in freshly passed feces. Sporulation, the process by which oocysts become infective, requires oxygen and occurs outside the host in 1–5 days depending on temperature, humidity, and aeration [1]. Sporulated oocysts contain two ellipsoidal sporocysts each with four sporozoites (2 by 6–8 μm) [10]. Sporozoites are released from the sporulated oocyst by stomach secretions into the lumen of the host's small intestine. The wall of the oocyst is extremely physically and chemically resistant. Oocysts can survive for years in soil [11], withstand periods of freezing at –20°C and temperatures as high as 50°C, and survive for months and even years at 4°C [12–14], sporulate and remain viable in seawater for several months [15]. Even unsporulated oocysts can be viable after months at 4°C [14]. Oocysts are resistant to numerous disinfectants [8].

20.1.1.2.2 Extraintestinal Phase

When an intermediate host is infected by ingestion of sporulated oocysts, sporozoites released into the lumen of the small intestine penetrate the intestinal lining and establish an infection [16]. Once inside a cell, enclosed in a parasitophorous vacuole protected from host defense mechanisms, sporozoites proceed through asexual reproduction cycles by endodyogeny (formation of two membrane-bound daughter cells while still in the mother cell) to produce tachyzoites. Tachyzoites are rapidly dividing stages and motile, often crescent shaped, about 7 by 12 μm of size [10]. They invade host cells by an active penetration process which leads to the formation of the parasitivorous vacuole (PV). The PV is formed as a hybrid between host cell membrane and parasite proteins. The environment inside the PV is well suited for parasite multiplication and protects the parasites against the host immune system [10]. The tachyzoites multiply within the host cell to make it disintegrate and release new tachyzoites which then invade neighboring cells or disseminate via the vascular system to all organs of the body where they proliferate by endodyogeny in various cells to result in cell rupture and tissue destruction [6]. In mice, tachyzoites are initially present in large numbers in the lungs and spleen, and by day 6–10 after infection they have invaded all organs including the brain [16]. In heavy acute infections, tachyzoites can be found free in blood and peritoneal exudate [1]. In the early (acute) phase, the large number of tachyzoites produced elicits a strong immune response which suppresses tachyzoite growth [17,18], and tachyzoites are transformed into bradyzoites (7 by 1.5 μm) that multiply slowly in cells to produce tissue cysts. Tissue cysts are accumulations of large numbers of bradyzoites within host cells, each accumulation surrounded by a tough wall. Tissue cysts grow and remain intracellular (size 5–70 μm, containing a few to several hundred bradyzoites) and are present most frequently not only in the central nervous system and the eye, but also in striated muscle including the heart [16]. Tissue cysts, which represent the quiescent stage of the parasite and are probably harmless as long as they are intact, can persist indefinitely for the life of the host. Viable cysts within muscle (meat) are a significant source of human infection. Disseminated or local infections can develop from rupturing tissue cysts in the immunocompromised individual [10,19]. Stage conversion between tachyzoites and bradyzoites is reversible, and conversion back to actively proliferating and tissue-destructive tachyzoites is a problem in immunocompromised hosts [16,20–22]. In the brains of immunocompetent chronically infected mice, a very small percentage of tissue cysts rupture at any time during chronic infections, resulting in rapid cell-mediated immune response destroying the bradyzoites before they can replicate or undergo stage conversion to tachyzoites [17]. In the immunocompetent host, if immunity to the parasite has vanished, it is restored by the cyst-rupture and bradyzoite release [9].

Bradyzoites, oocysts, and tachyzoites are infectious for all hosts, but the infective dose is much higher for tachyzoites than for the other stages [8]. When a host (intermediate or definitive) is infected by ingestion of tissue cysts, bradyzoites released into the lumen of the small intestine penetrate the intestinal lining and establish an infection basically following the same route as after infection with oocysts [16].

The tachyzoites have the ability to cross the placenta and infect the developing fetus in many species [5,23]. Evidence indicate that serial vertical transmission in some host species [24] may lead to clonal expansion of the parasite in that species and sustain the parasite in the population over generations [25].

20.1.1.2.3 Enteroepithelial Phase

The enteroepithelial phase, resulting in cats shedding oocysts, is initiated when a cat eats oocysts, tissue cysts, and occassionally also tachyzoites [26]. After an unknown number of asexual cycles in cat small intestine tissues, some tachyzoites enter new epithelial cells and start the enteroepithelial phase where they go through a series of morphologically distinct life stages [27]. This process results in production of microgametocytes (males, 2–4% of gametocytes are males, each producing about 15–30 male gametes, microgametes) or macrogametocytes (females, each producing one female gamete, macrogamete) [1,9,16,28,29]. After ingestion of tissue cysts by the cat, the sexual cycle starts 2 days later, and within 3–15 days gametocytes appear in epithelial cells throughout the small intestine and ileum [8]. It is unknown what causes the conversion from asexual to sexual development and what signals a tachyzoite to develop into a male or female gametocyte [16]. Fertilized oocysts break out of the cell into the lumen of the feline host intestine and are passed out with the feces [8,16].

While nearly all cats shed oocysts after ingesting tissue cysts, <50% shed oocysts after ingesting tachyzoites or oocysts [8]. Domestic cats under laboratory conditions shed millions of oocysts after eating one *T. gondii*-infected mouse [27]. The time between initial infection and oocyst shedding vary according to the stage of *T. gondii* ingested; 3–10 days after ingesting tissue cysts and 18 days or longer after ingesting oocysts, respectively, while it may vary after ingesting tachyzoites [30]. Oocysts were demonstrated to be shed with cat feces for <3 weeks, mostly for 1 week [31]. Cats acquire good immunity to reshedding of oocysts under laboratory conditions, but 4 out of 9 cats studied reshed oocysts after feeding with tissue cysts 6 years after primary infection [32]. It is not known as to how often cats shed oocysts under normal conditions [8]. A single oocyst is sufficient to establish infection in mice, and oocysts are highly infective also for other intermediate hosts, while high doses are required to infect cats [33].

20.1.2 Biology and Epidemiology

T. gondii is cosmopolitan in distribution. In the majority of animal species, infection does not cause clinical illness, while in some species it causes life-threatening disease or manifests itself as a disease of pregnancy resulting in abortion, stillbirth, or birth of weak offsprings, for example, in sheep and goat [34]. In humans the parasite causes clinical manifestations ranging from abortions and congenital infection to encephalitis [4] and eye diseases [35]. The most likely sources of infection to humans are ingestion of raw or lightly cooked meat containing live tissue cysts or ingestion of food and water contaminated with sporulated oocysts. Handling of cat litter box filler, garden work, and children's sand pits are also well-known links of exposure to oocysts derived from cat feces [34]. Flies and cockroaches have been implicated as carriers of oocysts between cat feces and food [9].

Tissue cysts are rendered nonviable by freezing at −13°C [36,37]. Some of the salt treatments of meat also kill tissue cysts [38,39]. Tachyzoites are fragile and highly susceptible to salt concentration [40]. Observations of symptoms and severity of disease from oocyst-associated outbreaks of toxoplasmosis in humans [41,42] suggested that oocysts are more pathogenic than bradyzoites or tachyzoites irrespective of the dose [30]. Differentiation of strains of *T. gondii* is based on pathogenicity in mice. While most are avirulent and result in chronic asymptomatic infections, a few are highly virulent and produce fatal acute toxoplasmosis [43].

Numerous genetic markers for differentiation of *T. gondii* strains have been developed [44], and population structure analysis using such markers have shown that most isolates from humans and domestic species in North America, Europe, and Africa are of three clonal lineages (I, II, III), with type II as the predominant form identified in human disease [45,46]. Epidemiological screenings have shown higher levels of genetic variation than previously recognized [47,48].

T. gondii has also been reported in marine mammals such as sea otter and beluga whale [49–52], bearded and ringed seal [53], and polar bear [54,55]. Findings of fatal toxoplasmosis in arctic fox [56] and other animals of habitats with minimal natural exposure to *T. gondii* through the ages, suggest that such animals are particularly vulnerable to the parasite [34]. Toxoplasmosis is recognized as a disease affecting many people and causing congenital infections also among Inuits of northern Canada and other northern people dependent on marine mammals for food, providing a link between the infection and the marine environment [57,58]. For sea otters of a central California coastal system, an increased risk of *T. gondii* infection for otters preying on small marine snails was demonstrated [59]. The existence of a yet unknown life cycle for *T. gondii* in the arctic, independent of cats, was proposed by Measures and coworkers [51]. Fish and other cold-blooded animals are not considered hosts for *T. gondii* [6]. It is not known if sporozoites excyst after ingestion by cold-blooded animals [30].

T. gondii tissue cysts are common in meat for human consumption, such as from sheep, pigs, chickens, and game [60,61]. Pigs and other animals are not tested for *T. gondii* infection at slaughter in any country [39,62]. Currently used molecular and histological methods are insensitive for the detection of *T. gondii* in meat because of the often low concentration of the parasite in meat destined for human consumption (one tissue cyst per 25 g or more) [39]. Although cattle can be successfully infected with *T. gondii* oocysts, cattle are considered a poor host for the parasite, perhaps due to innate resistance [63].

20.1.3 Clinical Features and Pathogenesis

Development of clinical disease and level of pathology is affected by factors such as virulence of the strain involved and age and level of acquired immunity of the host (older hosts more resistant) [9]. In immunocompetent individuals, acute acquired *T. gondii* infections most often appears as asymptomatic diseases, and only a minority of individuals show signs or symptoms of the infection, which may vary from short unspecific illness to severe symptoms [4,41]. The disease may clinically be classified as acute, subacute, chronic, or congenital. In the acute phase, the tachyzoites proliferate in many tissues and tend to rapidly kill host cells. Symptoms may be swollen and painful lymph nodes, fever, headache, muscle pain, anemia, malaise and fatigue, pneumonia, hepatitis, myocarditis, and polymyositis. The most serious lesions often result when tachyzoites proliferate in host cells of the retina and the brain [4,35]. Normally the duration of the acute stage is limited by the build-up of the host's immune response which suppresses tachyzoite proliferation, accompanied by formation of tissue cysts which can remain intact for years without producing clinical symptoms.

Protective immunity to *T. gondii* is mainly achieved through cellular immunity, and is dependent on both CD4+ and CD8+ T lymphocytes and interferon-gamma (IFN-γ) [64,65]. Fatal toxoplasmosis is proposed to be caused by overstimulation of mechanisms that are normally protective. Overproduction of Th1-cytokines like IL-18 may lead to massive tissue damage, for example, liver necrosis and lymphoid degeneration, and may be lethal [66]. In addition, humoral immunity (antibodies) develops during the first weeks of infection [6].

Congenital infection occurs when seronegative pregnant women acquire *T. gondii* infection. Women who are seropositive before conception do not, as a rule, transmit the parasite to the fetus [4]. The trimester in which the infection was acquired is important for the clinical manifestations of congenital toxoplasmosis; infection shortly before or around the time of conception usually do not result in transmission to the fetus, while infection acquired in the first, second, and third trimester results in congenital infection in 10–25%, 30–50%, and 60–70% of fetuses, respectively [67]. The fetus has only a rudimentary immune system in early pregnancy. Congenital toxoplasmosis is most severe and commonly affects the central nervous system if the fetus is infected before week 26 of gestation, while fetuses infected in the third trimester may be asymptomatic at birth and develop symptoms related to the infection later in life [68]. *T. gondii* infection is asymptomatic also in most pregnant women and the risk to the fetus does not correlate with the symptoms in the mother [19]. Congenital infection incidence may be reduced by treatment of the mother with antiparasitic drugs [4], and treatment of children with congenital infection can alter the course of the disease [19]. In immunocompromised women, such as those infected with HIV or treated with corticosteroids, reactivation of latent *T. gondii* infections during pregnancy resulting in congenital infection occurs [69].

Soon after the start of the HIV epidemic in the early 1980s, encephalitis caused by *T. gondii* was found as an important cause of death in HIV patients [4]. Infections with *T. gondii* are usually fatal also in other immunocompromised hosts such as patients with hematological malignancies involving T-cell defects (e.g., Hodgkin's disease), in recipients of organ transplants and in patients receiving other forms of immunosuppressive therapy (corticosteroids, cytotoxic drugs), unless the infection is recognized and treated [19]. Toxoplasmosis in the immunocompromised host is most often the consequence of reactivation of disease acquired before immunosuppression occurred, following rupture of tissue cysts in the central nervous system [4,19].

20.1.4 Diagnosis

20.1.4.1 Conventional Techniques

In general, clinical symptoms of *T. gondii* infection are nonspecific and unreliable for diagnosis [70]. Serological tests and bioassays in mice and/or cats, or a combination of the two approaches, are usually used for the conventional diagnosis of *T. gondii* infection [6]. Each of the different *T. gondii* life cycle stages has stage-specific expression of antigens, with the most immunodominant one being the tachyzoite-specific surface antigen SAG1, for which there is a lack of known cross-reactivity to antigens from other microorganisms [4]. Other antigens identified include the SAG2 and SAG3 tachyzoite-specific surface antigens and the SAG4 and BAG1 bradyzoite-specific antigens, along with the dense granule antigens (GRAs; GRA1, GRA6, GRA7) and the microneme antigens (MICs) [4].

The dye test (DT) [71] is the longest established diagnostic test for *T. gondii*-specific antibodies. It is based on antibody-mediated killing of live *T. gondii* parasites in the presence of complement. The DT is sensitive and considered the most specific test for humans. It gives false or erratic results with cattle sera [63], and the requirement for the live parasites is a disadvantage of the DT. The immunofluorescence assays, introduced in the 1960s, are specific but with a lower sensitivity compared to the DT. Enzyme-linked immunoassay techniques (enzyme-linked immunosorbent assay (ELISA)) for the detection of *T. gondii*-specific antibodies became available in the 1970s and 1980s. Immunosorbent agglutination assays (ISAGAs, introduced in the 1980s), based on whole-cell formalin-fixed *T. gondii*, are considered highly sensitive and specific for *T. gondii*-specific IgM and IgA antibodies. Also immunoblot using single antigens have been used. A direct agglutination test was described by Fulton [72]. In the 1980s, the modified agglutination test (MAT, IgG-induced agglutination of formalin-fixed tachyzoites after destruction of IgM) was established [19]. The MAT has been used extensively for the detection of *T. gondii*-infection in animals. The ELISA, which can handle large numbers of samples, does not rely on human interpretation for the results. Toxoplasmosis in immunocompetent humans is in most cases established by serological tests, but molecular (i.e., polymerase chain reaction, PCR) tests have proven useful for diagnosis of the infection *in utero* as well as in immunocompromised hosts [4].

Bioassaying in laboratory animals, normally mice or cats [73], is very sensitive and generally considered the "gold standard" for detection of infective stages of *T. gondii*. Bioassaying can detect low parasite burdens in sample material since very few bradyzoites are needed to induce *T. gondii* infection in laboratory animals. Bioassaying is therefore also used as a valuable tool for isolation of live *T. gondii* [74]. Subpassage of the original sample material in mice or cats strongly increases sensitivity of *T. gondii* diagnostics [75]. PCR may subsequently be performed on laboratory animal material with high parasite numbers, for example, mouse brains. Disadvantages of bioassaying are that it is expensive and time consuming, and that it involves animal experiments.

Tachyzoites are relatively easily grown in cultures of commercially available cells, such as HeLa [76]. Cell cultures may be infected with tachyzoites from acutely infected mice, harvested, for example, from peritoneal fluid after intraperitoneal inoculation, or from lung exudate [6]. Tissue cysts formation may occur spontaneously in cell cultures, but may also be induced by application of stressors such as high or low pH, high temperature (40°C) or certain chemical compounds to the tachyzoite culture [77].

20.1.4.2 Molecular Techniques

Detection of *T. gondii* DNA from a single tachyzoite by PCR amplification was reported by Burg et al. [78], targeting the B1 repetitive sequence. Now PCR is the molecular method relied on for specific detection or analysis of *T. gondii* DNA [70]. Frequently used target regions, aside from the B1 sequence, are the P30 (SAG1) gene [79] and 18S ribosomal RNA (rRNA) [80]. The copy number of the target sequence (B1: 35 copies; P30: 1 copy; rRNA: 110 repeat units) is important for the sensitivity of the PCR. It is a prerequisite for the use of PCR that the tissue examined contains sufficient amount of the parasite (tissue cysts, tachyzoites) for target DNA to be successfully isolated.

Molecular methods focus either on high sensitivity for the detection of *T. gondii* DNA, or on multilocus genotyping of single copy genes for high-resolution identification of each *T. gondii* isolate [81]. For the detection of *T. gondii* in human and animal tissue samples, the 300 copy 529-bp element [82] currently appears as a preferred marker in nested PCR (nPCR) since targeting of it is 10–100 times more sensitive than the B1 marker [83]. PCR performed directly on sample material from naturally infected individuals may, however, have low sensitivity for the detection of *T. gondii* compared to bioassay protocols, as shown for a PCR targeting the 529-bp-PCR in muscle tissues from pigs [84] and a PCR targeting the SAG1 gene in fox tissue [85].

In some situations diagnosis needs to be rapid and accurate to start treatment (e.g., AIDS patients, pregnant patients, fetal infection), or low-level parasite burdens need to be quantified to monitor the progression of infections under drug treatment. Real-time PCR, which is superior to nPCR with respect to detection threshold [86] and also quantifies the *T. gondii* burden, is extremely useful and is suggested to replace standard bioassay or indirect serology tests for sensitive and rapid detection of clinically relevant parasite burdens [87].

Earlier genotyping studies followed a single marker PCR protocol [88]. As the information on genetic variation in *T. gondii* increased and exotic strains were discovered, the single marker PCR protocol proved insufficient in discriminating between other strains than the traditional three clonal types. Genotyping protocols with increasing resolution have been developed during the last years, aiming at exploring the epidemiology of *T. gondii*, for example, transmission patterns. Two studies on multiplex multilocus PCR were published in 2005, one involving sequencing [89], the other involving nPCR and restriction fragment polymorphism analysis (RFLP) in four genetic markers: SAG2, SAG3, GRA6, and BTUB [90]. Genotyping is now conducted using high-resolution multimarker protocols; one frequently used is a high-resolution PCR-RFLP method involving 10 genetic markers [91], where all 10 markers are preamplified by multiplex PCR using external primers and the resulting PCR products are used to amplify each individual marker by nPCR [81]. This method was applied to reveal the genetic diversity among *T. gondii* isolates from South America [92]. It was also successfully applied to genotype *T. gondii* directly from brain samples from sea otters [93] and arctic foxes [94].

20.2 METHODS

There are numerous papers from the recent years reporting PCR-based detection and diagnosis of *T. gondii* in humans and animals, and over 200 genetic markers that can be used for genotyping of *T. gondii* [81]. There is no consensus on a standard set of DNA targets for PCR-based diagnosis and genotyping of *T. gondii*, and each of the reported methods have their advantages and limitations. Su et al. [70] proposed an integrated approach for the detection and identification of *T. gondii* for further development aiming at generating maximum information for epidemiological, population, and phylogenetic studies of *T. gondii* as a key pathogen.

20.2.1 Sample Preparation

20.2.1.1 Preparation of Blood Samples for Serology

Whole blood is allowed to clot and serum is collected after centrifugation (1000 × g, 10 min) and stored at −20°C until analysis.

20.2.1.2 Preparation of Inoculation Material for Bioassay

20.2.1.2.1 Bioassay in Mice (after Dubey [95])

1. Trim the sample material using a scalpel blade, razor blade, or sharp scissors.
2. Homogenize 50 g material in a blender at low speed, add saline (0.9% NaCl) and blend at top speed. Rinse the blender with 125 mL saline and add this to the sample homogenate.
3. Pour the homogenate into a 1000 mL plastic jar, prewarm at 37°C.
4. Prepare acid pepsin solution; 2.6 g pepsin, 5.0 g NaCl, 7.0 mL HCl and add distilled water to 500 mL, pH ~1.1–1.2.
5. Mix 250 mL prewarmed (37°C) acid pepsin solution with the prewarmed sample homogenate.

6. Incubate at 37°C in shaking water bath for 60 min.
7. Filter the homogenate through two layers of gauze, centrifuge 250 mL of filtered homogenate and centrifuge at 1200 × g for 10 min.
8. Pour off the supernatant, suspend in 20 mL of phosphate-buffered saline (PBS, pH 7.2) and transfer to a 50 mL centrifuge tube with a conical bottom. Neutralize the homogenate with 1.2% sodium bicarbonate pH 8.3, using neutral red as a pH indicator until the color changes to orange. Mix and centrifuge at 1200 × g for 10 min.
9. Pour off the supernatant and mix with 5–10 mL of saline containing 1000 IU penicillin and 100 µg streptocillin per mL.
10. Inject 1 mL homogenate per mouse subcutaneously over the back.

20.2.1.2.2 Bioassay in Cats (after Dubey [32])

1. Trim sample material for excess tissues, for example, skin, hair, and so on, and homogenize in a blender.
2. Material of specific interest, for example, brain tissue, may be processed separately and given perorally to the cats using a syringe without needle.
3. Collect fecal samples from the cats starting 3 days after feeding.

20.2.1.3 Preparation of DNA from Tissues

20.2.1.3.1 Materials

Procedure using QIAamp DNA Mini Kit (QIAamp Mini Spin Columns, 2 mL collection tubes, Buffer AL, Buffer ATL, Buffer AW1, Buffer AW2, Buffer AE (buffers included in kit without being further defined), and Proteinase K).

20.2.1.3.2 Procedure

After QIAamp DNA Mini and Blood Mini Handbook, for purification of total DNA. One milligram of tissue will yield ~0.2–1.2 µg of DNA:

1. Cut up to 25 mg of tissue (up to 10 mg of spleen) into small pieces. Place in a 1.5 mL microcentrifuge tube.
2. Add 180 µL Buffer ATL.
3. Add 20 µL Proteinase K, mix by vortexing, and incubate at 56°C until the tissue is completely lysed. Vortex occassionally during incubation to disperse the sample.
4. Briefly centrifuge the tube to remove drops from the inside of the lid.
5. Add 200 µL Buffer AL to the sample, mix by pulse-vortexing for 15 s, and incubate at 70°C for 10 min. Briefly centrifuge the tube to remove drops from inside the lid.
6. Add 200 µL ethanol (96–100%) to the sample, and mix by pulse-vortexing for 15 s. After mixing, briefly centrifuge the tube to remove drops from inside the lid.
7. Carefully apply the mixture (including the precipitate) to the 2 mL QIAamp Mini Spin Column without wetting the rim. Centrifuge at 6000 × g for 1 min. Place the column in a clean 2 mL collection tube, and discard the tube containing the filtrate.
8. Add 500 µL Buffer AW1 (again without wetting the rim), centrifuge at 6000 × g for 1 min. Place the column in a clean 2 mL collection tube, and discard the tube containing the filtrate.
9. Carefully add 500 µL Buffer AW2, centrifuge at 6000 × g for 1 min. Place the column in a clean 2 mL collection tube, and discard the tube containing the filtrate. Centrifuge at full speed (10,000 × g) for 3 min.
10. Place the QIAamp Mini Spin Column in a new 2 mL collection tube and discard the old collection tube with the filtrate. Centrifuge at full speed for 1 min.
11. Place the QIAamp Mini Spin Column in a clean 1.5 mL microcentrifuge tube, and discard the collection tube containing the filtrate. Add 200 µL Buffer AE or distilled water and incubate at room temperature for 1 min. Then centrifuge at 6000 × g for 1 min.
12. DNA is now ready for the detection of *T. gondii* by nPCR or real-time PCR, or for multilocus nPCR-RFLP genotyping. For later analysis, DNA samples are stored at −20°C.

20.2.2 Detection Procedures

20.2.2.1 Detection of Antibodies against *T. gondii* by MAT/Direct Agglutination Test (DAT)

1. Prepare the reagents needed for the analysis as described by the manufacturer (ToxoScreen DA, BioMerieux), that is, PBS pH 7.2, *Toxoplasma* antigen, 2-Mercaptoethanol, positive and negative control serum.
2. Dilute the test serum samples. For analysis at 1:40, an initial dilution of 1:20 is required (the samples are diluted twofold during later steps of the analysis). Other dilution(s) may be used.
3. Transfer 25 µL of each diluted sample to wells of 96-well round-bottom plate, include positive and negative control serum in every plate.
4. To each well with sample, add 25 µL of 0.2 M 2-mercaptoethanol (destruction of IgM).
5. Add 50 µL of antigen suspension to each well, mix well.
6. Cover with selfadhesive sheet. Leave for 5–18 h at room temperature away from vibrations.
7. *Reading*: Positive reaction—agglutination of the antigen in a mat covering about half of the well base. Negative reaction—sedimentation of the antigen in a button or ring.

20.2.2.2 Bioassay in Mice and Cats

20.2.2.2.1 Mice

Diagnosis of *T. gondii* in mouse brains is done by microscopy of brain smears [6]. A small piece (~1 mm^3) of brain tissue is squeezed between a microscopic slide and a coverslip until the margins of the squash reaches the margins of the coverslip, and the smear is examined for the presence of tissue cysts by low-magnification light microscopy.

20.2.2.2.2 Cats

Oocyst shedding in cats is verified by sucrose floatation and microscopy of a drop from the meniscus as described by Dubey [33].

20.2.2.3 Nested PCR

This section describes the nPCR detection method for the more-than-300-fold-repeated 529-bp element of *T. gondii* [82,83]. For protocols for *T. gondii* detection by real-time PCR and *T. gondii* genotyping by multilocus nPCR-RFLP, see Su and Dubey [81], and Su et al. [70].

20.2.2.3.1 Materials

- FastStart Taq DNA polymerase kit (Roche Applied Science).
- External primers (flanking a 450 bp region within the *T. gondii* specific 529 bp repeat element:
  - Forward, Tox-8 (5′-CCC AGC TGC GTC TGT CGG GAT-3′) [82,83].
  - Reverse, Tox-5 (5′-CTG CAG ACA CAG TGC ATC TGG ATT-3′) [81–83].
- Internal primers (flanking a conserved 162 bp region within the above 450 bp region):
  - Forward, Tox-9 (5′-AGG AGA GAT ATC AGG ACT GTA G-3′) [83].
  - Reverse, Tox-11 (5′-GCG TCG TCT CGT CTA GAT CG-3′) [81,83].
- 2.5 mM dNTPs mix (dATP, dTTP, dCTP, dGTP, each 2.5 mM).
- Extracted DNA samples.

20.2.2.3.2 Procedure (after Su and Dubey [81])

1. PCR reagents except for DNA polymerase are thawed at room temperature. Tubes are tapped for mixing and are spun briefly.
2. Assemble the first PCR mix:

H_2O	26.2 μL
10 × PCR buffer (without Mg^{2+})	5.0 μL
2.5 mM dNTPs mix	4.0 μL
25 mM $MgCl_2$	4.0 μL
External forward and reverse primer (each 50 μM)	0.3 μL of each
FastStart Taq DNA polymerase (5 U/μL)	0.2 μL
Total	40.0 μL

3. Aliquot 40 μL of PCR mix to PCR plate (0.2 mL wells). Add 10 μL of extracted DNA sample. Spin briefly.
4. Thermocycling program: 95°C for 5 min, then 30 cycles of (1) 94°C for 30 s, (2) 55°C for 1 min, (3) 72°C for 2 min. Soak at 4°C.
5. Assemble nPCR mix:

H_2O	31.2 μL
10 × PCR buffer (without Mg^{2+})	5.0 μL
2.5 mM dNTPs mix	4.0 μL
25 mM $MgCl_2$	4.0 μL
Internal (nested) forward and reverse primer (each 50 μM)	0.3 μL of each
FastStart Taq DNA polymerase (5 U/μL)	0.2 μL
Total	45.0 μL

6. Aliquot 45 μL of PCR mix to PCR plate. Add 5 μL of first PCR products. Spin briefly.
7. Thermocycling program: 95°C for 5 min, then 35 cycles of (1) 94°C for 30 s, (2) 55°C for 1 min, (3) 72°C for 1 min. Soak at 4°C.
8. To check nPCR products, take 5 μL of sample, mix with 2 μL of 5 × loading buffer and load to 2% agarose gel (containing 0.3 μg/mL ethidium bromide). Run gel in 1 – TBE buffer at 120 V for about 1 h. Visualize DNA bands under UV light.

20.3 CONCLUSION AND FUTURE PERSPECTIVES

Knowledge regarding the biology of *T. gondii* has been greatly expanded through the last few decades. Still, further studies on epidemiology, population structure, and transmission patterns are needed for this ubiquitous key pathogen and model organism. Future studies on *T. gondii* genotyping should focus mainly on two aspects. First, to conduct large-scale epidemiological studies of *T. gondii* isolates from human and animals to determine if (i) animals serve as the reservoir to transfer the parasite to humans, and (ii) human disease presentation is associated with specific *T. gondii* genotypes. Second, to develop high-resolution serotyping methods to study *T. gondii* isolates in humans and animals. Most *T. gondii* infections in humans and animals are chronic without symptoms. Hence, it is difficult to isolate the parasite from the majority of hosts currently infected. Serotyping has been proved to be a promising approach to overcome this obstacle [96–98]. However, at present, the resolution of serotyping is limited to differentiating type II from nontype II strains, and therefore it has limited use in typing serum samples from South America. A high-resolution, multilocus serotyping method is strongly needed to facilitate epidemiology and population studies in *T. gondii*, and could bring our knowledge of this complex parasite a huge leap forward.

ACKNOWLEDGMENT

We thank Dr. Chunlei Su for valuable comments on the manuscript.

REFERENCES

1. Schmidt, G.D., and Roberts, L.S. *Foundations of Parasitology*, 7th edn. McGraw-Hill, Boston, USA, 2005.
2. Nicolle, C., and Manceaux, L. Sur une infection à corps de Leishman (ou organisms voisins) du gondi. *C. R. Seances Acad. Sci.* 147, 763–766, 1908.
3. Dubey, J.P. The history and life cycle of *Toxoplasma gondii*. In: Weiss, L.M., and Kim, K. (eds), *Toxoplasma gondii. The Model Apicomplexan—Perspectives and Methods*. Elsevier Ltd., Academic Press, London, UK, pp. 1–17, 2007
4. Petersen, E., and Liesenfeld, O. Clinical disease and diagnostics. In: Weiss, L.M., and Kim, K. (eds), *Toxoplasma gondii. The Model Apicomplexan—Perspectives and Methods*. Elsevier Ltd., Academic Press, London, UK, pp. 81–100, 2007.
5. Buxton, D. et al. *Toxoplasma gondii* and ovine toxoplasmosis: New aspects of an old story. *Vet. Parasitol.* 149, 25–28, 2007.
6. Dubey, J.P., and Beattie, C.P. *Toxoplasmosis of Animal and Man*. CRC Press, Boca Raton, FL, USA, pp. 1–120, 1988.
7. Levine, N.D. Taxonomy of the sporozoa. *J. Parasitol.* 56 (sect. II, part 1), 208–209, 1963.
8. Jones, J.L., and Dubey, J.P. Waterborne toxoplasmosis—recent developments. *Exp. Parasitol.* 124, S10–S25, 2010.
9. Bogitsh, B.J., Carter, C.E., and Oeltmann, T.N. *Human Parasitology*, 3rd edn. Elsevier, Academic Press, London, UK, 2005.
10. Dubey, J.P., Lindsay, D.S., and Speer, C.A. Structures of *Toxoplasma gondii* tachyzoites, bradyzoites and sporozoites, and biology and development of tissue cysts. *Clin. Microbiol. Rev.* 11, 267–299, 1998.
11. Frenkel, J.K., Ruiz, A., and Chinchilla, M. Soil survival of *Toxoplasma* oocysts in Kansas and Costa Rica. *Am. J. Trop. Med. Hyg.* 24, 439–443, 1975.
12. Frenkel, J.K., and Dubey, J.P. Effects of freezing on the viability of *Toxoplasma* oocysts. *J. Parasitol.* 59, 587–588, 1973.
13. Dubey, J.P. *Toxoplasma gondii* survival under defined temperatures. *J. Parasitol.* 84, 862–865, 1998a.
14. Lindsay, D.S., Blagburn, B.L., and Dubey, J.P. Survival of nonsporulated *Toxoplasma gondii* oocysts under refrigerator conditions. *Vet. Parasitol.* 103, 309–313, 2002.
15. Lindsay, D.S. et al. Sporulation and survival of *Toxoplasma gondii* oocysts in seawater. *J. Eukaryot. Microbiol.* 50, S687–S688, 2003.
16. Ferguson, D.J.P., and Dubremetz, J.F. The ultrastructure of *Toxoplasma gondii*. In: Weiss, L.M., and Kim, K. (eds), *Toxoplasma gondii. The Model Apicomplexan—Perspectives and Methods*. Elsevier Ltd., Academic Press, London, UK, pp. 19–48, 2007.
17. Ferguson, D.J., Hutchison, W.M., and Pettersen, E. Tissue cyst rupture in mice chronically infected with *Toxoplasma gondii*. An immunocytochemical and ultrastructural study. *Parasitol. Res.* 75, 599–603, 1989.
18. Weiss, L.M., and Kim, K. Bradyzoite development. In: Weiss, L.M., and Kim, K. (eds.), *Toxoplasma gondii. The Model Apicomplexan—Perspectives and Methods*. Elsevier Ltd., Academic Press, London, UK, pp. 133–152, 2007.
19. Weiss, L.M., and Dubey, J.P. Toxoplasmosis: A history of clinical observations. *Int. J. Parasitol.* 39, S895–S901, 2009.
20. Luft, B.J., and Remington, J.S. Toxoplasmic encephalitis in AIDA. *Clin. Infect. Dis.* 15, 211–222, 1992.
21. Soete, M. et al. *Toxoplasma gondii*: Kinetics of bradyzoite–tachyzoite interconversion *in vitro*. *Exp. Parasitol.* 76, 259–264, 1993.
22. Odaert, H. et al. Stage conversion of *Toxoplasma gondii* in mouse brain during infection and immunodepression. *Parasitol. Res.* 82, 28–31, 1996.
23. Schaap, D. et al. Vaccination against toxoplasmosis: Current status and future prospects. In: Weiss, L.M., and Kim, K. (eds), *Toxoplasma gondii. The Model Apicomplexan—Perspectives and Methods*. Elsevier Ltd., Academic Press, London, UK, pp. 721–752, 2007.
24. Morley, E.K. et al. Significant familial differences in the frequency of abortion and *Toxoplasma gondii* infection within a flock of Charollais sheep. *Parasitology* 131, 181–185, 2005.
25. Duncanson, P. et al. High levels of congenital transmission of *Toxoplasma gondii* in a commercial sheep flock. *Int. J. Parasitol.* 31, 1699–1703, 2001.
26. Frenkel, J.K., Dubey, J.P., and Miller, N.L. *Toxoplasma gondii* in cats: Fecal stages identified as coccidian oocysts. *Science* 167, 893–896, 1970.
27. Dubey, J.P., and Frenkel, J.K. Cyst-induced toxoplasmosis in cats. *J. Protozool.* 19, 155–177, 1972.
28. Speer, C.A. et al. Time lapse video microscopy and ultrastructure of penetrating sporozoites, types 1 and 2 parasitophorous vacuoles, and the transformation of sporozoites to tachyzoites of the VEG strain of *Toxoplasma gondii*. *J. Parasitol.* 83, 565–574, 1997.
29. Speer, C.A., and Dubey, J.P. Ultrastructural differentiation of *Toxoplasma gondii* schizonts (types B to E) and gamonts in the intestines of cats fed bradyzoites. *Int. J. Parasitol.* 35, 193–206, 2004.
30. Dubey, J.P. History of discovery of the life cycle of *Toxoplasma gondii*. *Int. J. Parasitol.* 39, 877–882, 2009a.
31. Dubey, J.P. Oocyst shedding by cats fed isolated bradyzoites and comparison of bradyzoites of the VEG strain *Toxoplasma gondii* to cats and mice. *J. Parasitol.* 87, 215–219, 2001.
32. Dubey, J.P. Duration of immunity to shedding of *Toxoplasma gondii* oocysts by cats. *J. Parasitol.* 81, 410–415, 1995.
33. Dubey, J.P. Comparative infectivity of oocysts and bradyzoites of *Toxoplasma gondii* for intermediate (mice) and definitive (cats) hosts. *Vet. Parasitol.* 140, 69–75, 2006.
34. OIE, 2009. Toxoplasmosis. In: Manual of Diagnostic Tests and Vaccines for Terrestrial Animals 2009. OIE, Terrestrial Manual 2008. Updated: August 14, 2009. http://www.oie.int/eng/normes/mmanual/a_summry.htm.
35. Latkany, P. Ocular disease due to *Toxoplasma gondii*. In: Weiss, L.M., and Kim, K. (eds), *Toxoplasma gondii. The Model Apicomplexan—Perspectives and Methods*. Elsevier Ltd., Academic Press, London, UK, pp. 101–131, 2007.
36. Dubey, J.P. Effect of freezing on the infectivity of *Toxoplasma* cysts to cats. *J. Am. Vet. Med. Assoc.* 165, 534–536, 1974.
37. Kotula, A.W. et al. Effect of freezing on infectivity of *Toxoplasma gondii* tissue cysts in pork. *J. Food Prot.* 54, 687–690, 1991.
38. Hill, D.E. et al. Effect of commonly used enhancement solutions on the viability of *Toxoplasma gondii* tissue cysts in pork loin. *J. Food Protect.* 67, 2230–2233, 2004.
39. Dubey, J.P. Toxoplasmosis in pigs—The last 20 years. *Vet. Parasitol.* 164, 89–103, 2009b.
40. Dubey, J.P. Re-examination of resistance of *Toxoplasma gondii* tachyzoites and bradyzoites to pepsin and trypsin digestion. *Parasitology* 116, 43–50, 1998b.
41. Bowie, W.R. et al. Outbreak of toxoplasmosis associated with municipial drinking water. The BC *Toxoplasma* Investigation Team. *Lancet* 350, 173–177, 1997.

42. de Moura, L. et al. Waterborne outbreak of toxoplasmosis, Brazil, from field to gene. *Emerg. Infect. Dis.* 12, 326–329, 2006.
43. Howe, D.K., Summers, B.C., and Sibley, L.D. Acute virulence in mice is associated with markers on chromosome VIII in *Toxoplasma gondii. Infect. Immun.* 64, 5193–5198, 1996.
44. Dardé, M.L., Ajzenberg, D., and Smith, J. Population structure and epidemiology of *Toxoplasma gondii*. In: Weiss, L.M., and Kim, K. (eds), *Toxoplasma gondii. The Model Apicomplexan—Perspectives and Methods*. Elsevier Ltd., Academic Press, London, UK, pp. 49–80, 2007.
45. Howe, D.K., and Sibley, L.D. *Toxoplasma gondii* comprises three clonal lineages: Correlation of parasite genotype with human disease. *J. Infect. Dis.* 172, 1561–1566, 1995.
46. Velmurugan, G.V., Dubey, J.P., and Su, C. Genotyping studies of *Toxoplasma gondii* isolates from Africa revealed that the archetypal clonal lineages predominate as in North America and Europe. *Vet. Parasitol.* 155, 314–318, 2008.
47. Ajzenberg, D. et al. Genetic diversity, clonality and sexuality in *Toxoplasma gondii. Int. J. Parasitol.* 34, 1185–1196, 2004.
48. Dubey, J.P., and Su, C. Population biology of *Toxoplasma gondii*: What's out and where did they come from. *Mem. Inst. Oswaldo Cruz.* 104, 190–195, 2009.
49. Mikaelian, I. et al. Toxoplasmosis in beluga whales (*Delphinapterus leucas*) from the St. Lawrence estuary: Two case reports and a serological survey. *J. Comp. Path.* 122, 73–76, 2000.
50. Dubey, J.P. et al. *Toxoplasma gondii, Neospora caninum, Sarcocystis neurona*, and *Sarcocystis canis*-like infections in marine mammals. *Vet. Parasitol.* 116, 275–296, 2003.
51. Measures, L.N. et al. Seroprevalence of *Toxoplasma gondii* in Canadian pinnipeds. *J. Wildl. Dis.*, 40, 294–300, 2004.
52. Prestrud, K.W. et al. Serosurvey for *Toxoplasma gondii* in arctic foxes and possible sources of infection in the high Arctic of Svalbard. *Vet. Parasitol.* 150, 6–12, 2007.
53. Jensen, S.-K. et al. The prevalence of *Toxoplasma gondii* in polar bears and their marine mammal prey; evidence for a marine transmission pathway? *Polar Biol.*, 33(5), 599–606, 2009.
54. Rah, H. et al. Serosurvey of selected zoonotic agents in polar bears (*Ursus maritimus*). *Vet. Rec.* 156, 7–13, 2005.
55. Oksanen, A. et al. Prevalence of antibodies against *Toxoplasma gondii* in polar bears (*Ursus maritimus*) from Svalbard and East Greenland. *J. Parasitol.* 95, 89–94, 2009.
56. Sørensen, K.K. et al. Acute toxoplasmosis in three wild arctic foxes (*Alopex lagopus*) from Svalbard; one with co-infections of *Salmonella Enteritidis* PT1 and *Yersinia pseudotuberculosis* serotype 2b. *Res. Vet. Sci.* 78, 161–167, 2005.
57. McDonald, J.C. et al. An outbreak of toxoplasmosis in pregnant women in northern Quebec. *J. Infect. Dis.* 161, 769–774, 1990.
58. Messier, V. et al. Seroprevalence of *Toxoplasma gondii* among Nunavik Inuit (Canada). *Zoonoses and Public Health* 56, 188–197, 2009.
59. Johnson, C.K. et al. Prey choice and habitat use drive sea otter pathogen exposure in a resource-limited coastal system. *PNAS* 106, 2242–2247, 2009.
60. Vikøren, T. et al. Prevalence of *Toxoplasma gondii* antibodies in wild red deer, roe deer, moose, and reindeer from Norway. *Vet. Parasitol.* 120, 159–169, 2004.
61. Tenter, A.M., Heckeroth, A.R., and Weiss, L.M. *Toxoplasma gondii*: From animals to humans. *Int. J. Parasitol.* 30, 1217–1258, 2000.
62. Dorny, P., Praet, N., Deckers, N., and Gabriel, S. Emerging food-borne parasites. *Vet. Parasitol.* 163, 196–206, 2009.
63. Dubey, J.P., and Jones, J.L. *Toxoplasma gondii* infection in humans and animals in the United States. *Int. J. Parasitol.* 38, 1257–1278, 2008.
64. Araujo, F.G. Depletion of L3T4 (CD4+) T lymphocytes prevents development of resistance to *Toxoplasma gondii* in mice. *Infect. Immun.* 59, 1614–1619, 1991.
65. Gazzinelli, R.T. et al. Synergistic role of CD4+ and CD8+ T lymphocytes in IFN-γ production and protective immunity induced by an attenuated *Toxoplasma gondii* vaccine. *J. Immunol.* 146, 286–292, 1991.
66. Mordue, D.G. et al. Acute toxoplasmosis leads to lethal overproduction of Th1 cytokines. *J. Immunol.* 167, 4574–4584, 2001.
67. Dunn, D. et al. Mother to child transmission of toxoplasmosis: Risk estimates for clinical counselling. *Lancet* 353, 1829–1833, 1999.
68. Remington, J.S, McLead, R., and Thulliez, P. Toxoplasmosis. In: Remington, J.S., and Klein, J. (eds), *Infectious Diseases of the Fetus and Newborn Infant*. Philadelphia Publishing, WB Saunders, pp. 205–346, 2001.
69. Mitchell, C.D. et al. Congenital toxoplasmosis occurring in infants perinatally infected with human immunodeficiency virus 1. *Pediatr. Infect. Dis. J.* 9, 512–518, 1990.
70. Su, C. et al. Moving towards an integrated approach to molecular detection and identification of *Toxoplasma gondii. Parasitology* 136, 1–11, 2009.
71. Sabin, A.B., and Feldman, H.A. Dyes as microchemical indicators of a new immunity phenomenon affecting a protozoon parasite (*Toxoplasma*). *Science* 108, 660–663, 1948.
72. Fulton, J.D. Studies on the agglutination of *Toxoplasma gondii. Trans. R. Soc. Trop. Med. Hyg.* 59, 694–704, 1965.
73. Dubey, J.P., Thulliez, P., and Powell, E.C. *Toxoplasma gondii* in Iowa sows: Comparison of antibody titers to isolation of *T. gondii* by bioassays in mice and cats. *J. Parasitol.* 81, 48–53, 1995.
74. Dubey, J.P. et al. Genetic and biologic characteristics of *Toxoplasma gondii* infections in free-range chickens from Austria. *Vet. Parasitol.* 133, 299–306, 2005.
75. Hill, D.E. et al. Comparison of detection methods for *Toxoplasma gondii* in naturally and experimentally infected swine. *Vet. Parasitol.* 141, 9–17, 2006.
76. Chatterton, J.M.W. et al. *Toxoplasma gondii in vitro* culture for experimentation. *J. Microbiol. Methods* 51, 331–335, 2002.
77. Bohne, W., Heesemann, J., and Gross, U. Reduced replication of *Toxoplasma gondii* is necessary for induction of bradyzoite-specific antigens: A possible role for nitric oxide in triggering stage conversion. *Infect. Immun.* 62, 1761–1767, 1994.
78. Burg, J.L. et al. Direct and sensitive detection of a pathogenic protozoan, *Toxoplasma gondii*, by polymerase chain reaction. *J. Clin. Microbiol.* 27, 1787–1792, 1989.
79. Savva, D. et al. Polymerase chain reaction for detection of *Toxoplasma gondii. J. Med. Microbiol.* 32, 25–31, 1990.
80. Ellis, J.T. Polymerase chain reaction approaches for the detection of *Neospora caninum* and *Toxoplasma gondii. Int. J. Parasitol.* 28, 1053–1060, 1998.
81. Su, C., and Dubey, J.P. Toxoplasma. In: Liu, D. (ed.), *Molecular Detection of Foodborne Pathogens*. CRC Press, Boca Raton, FL, USA, pp. 741–753, 2009.
82. Homan, W.L. et al. Identification of a 200- to 300-fold repetitive 529 bp DNA fragmentin *Toxoplasma gondii*, and its use for diagnostic and quantitative PCR. *Int. J. Parasitol.* 30, 69–75, 2000.
83. Reischl, U. et al. Comparison of two DNA targets for the diagnosis of Toxoplasmosis by real-time PCR using fluorescence resonance energy transfer hybridization probes. *BMC Infect. Dis.* 3, 7, 2003.

84. Garcia, J.L. et al. *Toxoplasma gondii*: Detection by mouse bioassay, histopathology, and polymerase chain reaction in tissues from experimentally infected pigs. *Exp. Parasitol.* 113, 267–271, 2006.
85. Smith, G.C. et al. Prevalence of zoonotic important parasites in the red fox (*Vulpes vulpes*) in Great Britain. *Vet. Parasitol.* 118, 133–142, 2003.
86. Contini, C. et al. Evaluation of a real-time PCR-based assay using the lightcycler system for detection of *Toxoplasma gondii* bradyzoite genes in blood specimens from patients with toxoplasmic retinochoroiditis. *Int. J. Parasitol.* 35, 275–283, 2005.
87. Jauregui, L.H. et al. Development of a real-time PCR assay for detection of *Toxoplasma gondii* in pig and mouse tissues. *J. Clin. Micriobiol.* 39, 2065–2071, 2001.
88. Howe, D.K. et al. Determination of genotypes of *Toxoplasma gondii* strains isolated from patients with toxoplasmosis. *J. Clin. Immunol.* 35, 1411–1414, 1997.
89. Ajzenberg, D., Dumètre, A., and Dardé, M.-L. Multiplex PCR for typing strains of *Toxoplasma gondii*. *J. Clin. Microbiol.* 43, 1940–1943, 2005.
90. Khan, A. et al. Genotyping of *Toxoplasma gondii* strains from immunocompromised patients reveals high prevalence of Type I strains. *J. Clin. Microbiol.* 43, 5881–5887, 2005.
91. Su, C., Zhang, X., and Dubey, J.P. Genotyping of *Toxoplasma gondii* by multilocus PCR-RFLP markers: A high resolution and simple method for identification of parasites. *Int. J. Parasitol.* 36, 841–848, 2006.
92. Dubey, J.P. et al. Prevalence of *Toxoplasma gondii* in dogs from Colombia, South America and genetic characterization of *T. gondii* isolates. *Vet. Parasitol.*, 145, 45–50, 2007.
93. Sundar, N. et al. Genetic diversity among sea otter isolates of *Toxoplasma gondii*. *Vet. Parasitol.* 151, 125–132, 2008.
94. Prestrud, K.W. et al. Direct high-resolution genotyping of *Toxoplasma gondii* in arctic foxes (*Vulpes lagopus*) in the remote arctic Svalbard archipelago reveals widespread clonal Type II lineage. *Vet. Parasitol.* 158, 121–128, 2008.
95. Dubey, J.P. Refinement of pepsin digestion method for isolation of *Toxoplasma gondii* from infected tissues. *Vet. Parasitol.* 81, 75–77 (1998c).
96. Kong, J.T. et al. Serotyping of *Toxoplasma gondii* infections in humans using synthetic peptides. *J. Infect. Dis.* 187, 1484–1495, 2003.
97. Nowakowska, D. et al. Genotyping of *Toxoplasma gondii* by multiplex PCR and peptide-based serological testing of samples from infants in Poland diagnosed with congenital toxoplasmosis. *J. Clin. Microbiol.* 44, 1382–1389, 2006.
98. Peyron, F. et al. Serotyping of *Toxoplasma gondii* in chronically infected pregnant women: Predominance of type II in Europe and types I and III in Colombia (South America). *Microb. Infect.* 8, 2333–2340, 2006.

Section III

Platyhelminthes: Cestoda

21 Diphyllobothrium

Barbara Wicht, Raffaele Peduzzi, Hélène Yera, and Jean Dupouy-Camet

CONTENTS

21.1 INTRODUCTION

Tapeworms of the genus *Diphyllobothrium* (Cobbold, 1858) are widely distributed all around the world, with some being agents of human diphyllobothriosis, one of the most important fish-borne zoonoses caused by a cestode parasite. The intermediate hosts include both freshwater and marine fish, mainly anadromous species. Piscivorous birds and mammals (including man) are definitive hosts and contract these parasites by eating raw or poorly cooked fish containing plerocercoid larvae. *Diphyllobothrium* cestodes are commonly present in the wildlife among fish, mammal, and bird hosts that make an important reservoir [1–3]. In the early 1970s the worldwide prevalence of human diphyllobothriosis was estimated at 9 million cases [4]. More recent data indicate that 20 million people are infected worldwide [1,3], but no further estimation on global prevalence has been done. The parasitosis is frequently reported from Japan, South Korea, the Baltic countries, Scandinavia, Western and Eastern Russia, and North America (Pacific Northwest). Recent surveys indicate that human diphyllobothriosis is emerging or reemerging in some European countries, particularly in Italian- and French-speaking sub-Alpine regions [5,6], although in South America and other countries sporadic cases have been reported so far [7–9]. A recent review by Scholz et al. [6] gives more precise data on the geographical distribution of this parasitosis.

21.1.1 Classification, Morphology, and Biology

The genus *Diphyllobothrium* belongs to family Diphyllobothriidae, order Diphyllobothriidea (formerly Pseudophyllidea), class Cestoda, and phylum Platyhelminthes [10]. To date, about 50 species have been described but only 14 have been reported as human pathogens, the most frequent being *Diphyllobothrium latum* (cosmopolitan), *Diphyllobothrium nihonkaiense* (Japan, South Korea, Eastern Russia, and probably North America), *Diphyllobothrium dendriticum* (northern part of the Northern Hemisphere), and *Diphyllobothrium pacificum* (Argentina, Brazil, Chile, Peru) [11–14]. However, some species are only quoted in review papers or book chapters but are not mentioned in bibliographic databases (Table 21.1). Infections with *Diplogonoporus balaenopterae* [syn. *D. grandis* (Blanchard, 1894)] have been signaled in Japan and recent molecular studies revealed a close relationship with the cetacean tapeworm *Diphyllobothrium stemmacephalum* [15], which in turn is probably a synonym of *Diphyllobothrium yonagoense* [6]. Other species (*Diphyllobothrium alascense*, *Diphyllobothrium cameroni*, *Diphyllobothrium cordatum*, *Diphyllobothrium dalliae*, *Diphyllobothrium hians*, *Diphyllobothrium orcini*, *Diphyllobothrium ursi*, *Diphyllobothrium scoticum*, and *Diphyllobothrium lanceolatum*) have also been described in humans, but a few details are available in the scientific literature [6].

The life cycle of *D. latum*, probably the most frequent species in humans, is complex and involves several hosts [16]. Released in freshwater, the eggs mature within 8–12 days at a water temperature of 16–20°C, and yield coracidium larvae that are ingested by zooplanktonic crustaceans. About 40 copepod species (Copepoda: Diaptomidae and Cyclopidae) are likely to serve as first intermediate hosts. The coracidium develops into a procercoid larva within the general cavity of the copepod. When a carnivore fish ingests planktonic crustaceans, the procercoid develops into a plerocercoid larva, a

TABLE 21.1
Species of *Diphyllobothrium* Reported from Humans and Availability of Sequences in Genbank (Status February 2011)

Species	Fish Hosts	Other Hosts	Distribution	18S	ITS	Cox1	Cob	NADH3
D. alascense (Rausch & Williamson, 1958)	Burbot, Smelt	Dog	Alaska	0	0	0	0	0
D. cameroni[a] (Rausch, 1969)	Marine fishes	Monk seal	Japan	0	0	0	0	0
D. cordatum (Cobbold, 1869)	Marine fishes	Seals, Walruses	North Atlantic	0	+	0	0	0
D. dalliae (Rausch, 1956)	Alaska blackfish, Dolly varden	Gulls, Dog	Alaska	0	0	0	0	0
D. dendriticum (Nitzsch, 1824)	Salmonids, Coregonids	Fish eating birds	Circumpolar	+	+	+	+	+
D. hians[a] (Diesing, 1850)	Marine fishes	Arctic seals	North Atlantic	0	0	0	0	0
D. lanceolatum (Krabbe, 1865)	*Coregonus*	Fur seals	North Pacific	0	+	0	0	0
D. latum (L., 1758)	Pike, Burbot, Perch, Arctic char, other Percids	Terrestrial mammals	Cosmopolitan	+	+	+	+	+
D. nihonkaiense[b] (Yamane, 1986)	Pacific salmons	Brown bear	Eastern Eurasia, Japan	+	+	+	+	+
D. pacificum (Nybelin, 1931)	Marine fishes	Sea lions, Eared seals	South America	+	+	+	+	0
D. ursi (Rausch, 1956)	Red salmon	Bears	Alaska, British Columbia	0	0	0	0	0
Diplogonoporus grandis[c] (Blanchard, 1889)	Anchovy, Sardine	Cetaceans	Circumpolar, Japan	+	+	+	0	+

[a] No mention of this species in PubMed bibliographic database.
[b] *D. nihonkaiense* and *D. klebanovskii* (Muratov & Psokhov, 1988) are synonymous according to Arizono et al. [47].
[c] This could be synonymous with *D. stemmacephalum* according to Arizono et al. [15].
Data summarized from Dick et al. [2], Chai et al. [1], and Müller [3].

few millimeters long. It migrates into the host musculature or in the viscera, where it can remain inactive for several years, but can re-encyst many times in other predatory fishes (paratenic hosts). In Europe, the species susceptible to harbor *D. latum* plerocercoids are perch (*Perca fluviatilis*), pike (*Esox lucius*), Arctic charr (*Salvelinus alpinus*), and burbot (*Lota lota*). Coregonidae and probably Salmonidae of *Salmo* genus do not host *D. latum* larvae, but they can host *D. dendriticum*. Pacific salmons of the genus *Oncorhynchus* are the usual hosts of *D. nihonkaiense*, *D. pacificum*, and *D. balaenopterae* are hosted by marine fish. Man and other ichthyophagous animals become infected by ingesting raw infected fish. The plerocercoid larva can grow between 5 and 20 cm a day and develops into an adult that yields its first eggs 2–6 weeks after the infestation.

21.1.2 Clinical Features and Pathogenesis

Diphyllobothrium tapeworms are the longest human parasites known (mostly 2–15 m long) and can live for several years. The symptoms of human diphyllobothriosis, although limited, are polymorphous: manifestations may include abdominal discomfort (abdominal pain, diarrhea), weight loss, asthenia, and vertigo. All these symptoms are vague and are quite common among the general population. A study performed 50 years ago during an extensive field investigation in Finnish Karelia compared 295 parasitized persons to 1127 controls [16]. Statistical analysis showed that the worm carriers had a very significant higher frequency of fatigue, weakness, diarrhea, and numbness of extremities. A sensation of hunger and "craving for salt" was also associated with worm infection. Some carriers experience very little discomfort and are just alerted when proglottids are passed in the feces. The parasite can also be found accidentally when performing a colonoscopy. Megaloblastic anemia due to vitamin B_{12} deficiency has been described in case of prolonged infestation in malnourished Finnish population after World War II [16], but it is only rarely reported nowadays. *Diphyllobothrium* parasites are sensitive to praziquantel (25 mg/kg/day in one dose is effective against most human pathogen species) and to niclosamide (2 g on an empty stomach in two doses an hour apart) [6].

21.1.3 Diagnosis

21.1.3.1 Conventional Techniques

All members of the genus *Diphyllobothrium* are morphologically very similar and are sometimes distinguished by hosts and geographical origin. Human infective species can be identified by skillful parasitologists using morpho-anatomical criteria [17,18], which indeed are relative as they vary with the age, the degree of development, and according to physiological modifications [19]. Diagnosis is possible on the whole adult parasite, rarely obtained, showing a medium-sized to large tapeworm with strobila usually segmented. Genital apertures are on the ventral face and the scolex is unarmed with a pair of dorsal and longitudinal grooves,

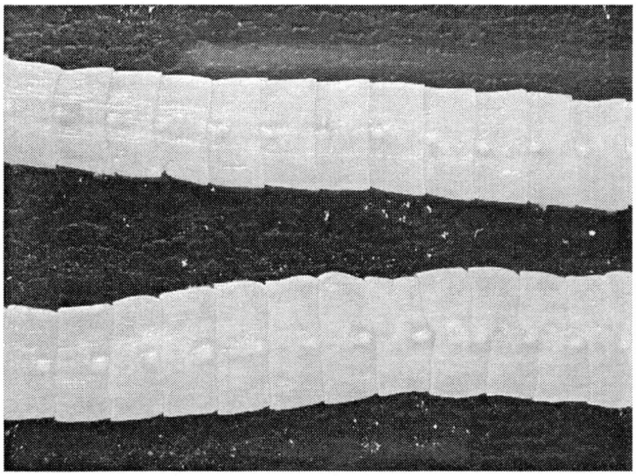

FIGURE 21.1 Proglottids of *Diphyllobothrium latum* obtained from a patient contaminated after eating raw fish from Lake Geneva. Note the ventral opening of the genital pore (Picture courtesy: J. Dupouy-Camet).

named bothria. Segments (proglottids) are wider than long with genital pores opening medioventrally (Figure 21.1). Each segment measures 2–7 mm × 10–12 mm. Eggs are oval and operculated. Those of *D. latum* are grayish brown with a 50–85 μm × 40–60 μm average size [20], while those of *D. nihonkaiense* are smaller (53–59 μm × 35–45 μm; [14,21,22]). However, sizes vary considerably among individual worms and could decrease with the intensity of infection [17]. Eggs can be mistaken with those of *Paragonimus* spp. or *Fasciola hepatica*, whose sizes are larger [3].

The plerocercoid larvae, obtained from fish, are ranging from a few millimeters to a few centimeters long and their identification has always been the job of an expert. They appear as white elongated masses in the skeletal muscle of certain fish families (Figure 21.2). They can be confounded by nonexperts with acanthocephalans (easily distinguishable by the presence of an evertable proboscis armed with spines) or with cestodes of the genus *Proteocephalus*. Andersen and Gibson (1989) published a key to identify three species of larval *Diphyllobothrium* occurring in European and North American freshwater fishes [18]. This key is based on features of gross-morphology visible under a light or a scanning electron microscope, on body length and site within the host, and features visible in histological sections.

FIGURE 21.2 Plerocercoid of *Diphyllobothrium latum* in a perch fillet (*Perca fluviatilis*) from Lake Maggiore, Switzerland (Picture courtesy: B. Wicht).

21.1.3.2 Molecular Techniques

Biochemical techniques (isoenzymatic assay or immunoelectrophoresis) have been used as alternatives to the traditional tools for species identification, but the development of molecular biology resulted in a better knowledge of the *Diphyllobothrium* genus [23–25]. For example, Matsuura et al. [24] used restriction fragment-length polymorphisms of ribosomal DNA (rDNA) to discriminate between *D. latum* and *D. nihonkaiense*, as the profiles generated with restriction enzymes *Sma*I, *Hinf*I, and *Hha*I provided valuable species-specific markers. Since 1998, interesting results in the study of the *Diphyllobothrium* genus have been obtained by sequence analysis of 18S ribosomal RNA (18S rRNA), cytochrome *c* oxidase subunit 1 (cox1 or COI), and NADH dehydrogenase subunit 3 (NADH3) genes, and of the partial internal transcribed spacer 2 (ITS2) region [26–30]. By using nucleotide sequence analysis of partial cox1 gene of mitochondrial DNA (mtDNA) from *D. nihonkaiense* plerocercoids, Ando et al. [29] confirmed the first identification of *D. nihonkaiense* from *Oncorhynchus masou ishikawae* fish in Japan. Upon examination of ITS2 sequences of the family Diphyllobothriidae, Logan et al. [30] elucidated many phylogenetic insights among the genera *Ligula*, *Digramma*, *Diphyllobothrium*, and *Schistocephalus*, including establishment of *Schistocephalus* as the most basal taxon of the Diphyllobothriidae, and ligulids (previously family Ligulidae) as being part of the Diphyllobothriidae.

Recently, we determined the value of 18S rRNA, internal transcribed spacer 1 (ITS1), cox1 and cob (cytochrome *b*) sequences to differentiate among *Diphyllobothrium* isolates. Sequences from samples (plerocercoids, adults or eggs) of *D. latum*, *D. nihonkaiense*, *D. ditremum*, *D. dendriticum*, *D. stemmacephalum*, and *D. pacificum* were examined. Cox1 and cob sequences analyses were clearly more discriminative than those of the ITS1 and 18S rRNA, representing therefore a useful tool for identifying specimens [31,32]. The 18S rRNA gene has already been described as a very conserved genomic target, and analysis of its nucleotide sequence provides a wealth of information about the phylogenetic relationships of biological organisms concerned [33]. This is due to the fact that, because this gene is present in multiple copies, it contains regions of relative conservation as well as diversity, and demonstrates patterns of concerted evolution that occurs among repeated copies. The sequencing of ITS regions and partial cox1 gene of *D. pacificum* indicated that this species is clearly different from other members of the genus (e.g., *D. cordatum*, *D. dendriticum*, and *D. lanceolatum*) and that ITS could not discriminate between *D. ditremum* and

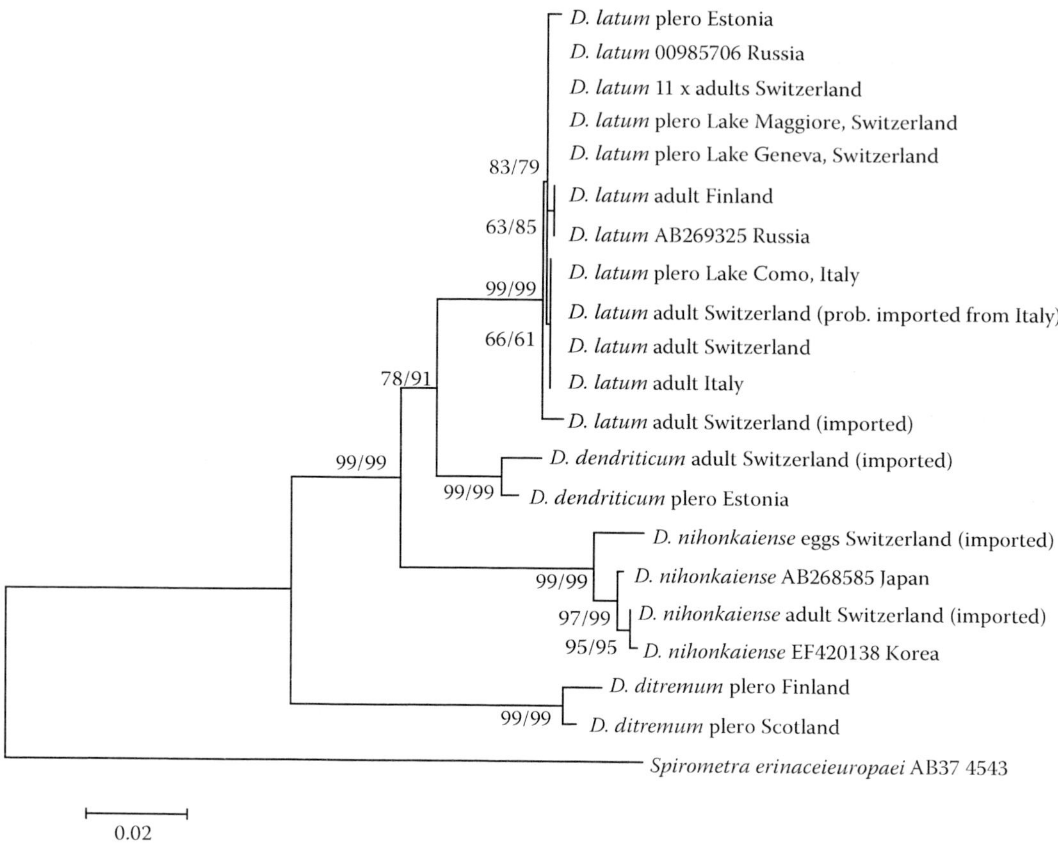

FIGURE 21.3 Neighbor-Joining tree based on 1086 bp fragment of complete cob gene of *Diphyllobothrium* species (Kimura-2 parameters, bootstrap values ME/NJ shown on nodes). (Modified from Wicht, B. et al. *Parasitol. Intern.* 59, 35, 2010.)

D. dendriticum [14,31,34]. At present, cox1 and cob genes seem to provide the most discriminative sequences among the different species (Figure 21.3), suggesting the existence of a link between the haplotypes of *D. latum* and their geographic origin [32]. NADH3 is also promising to discriminate the different species, because sequences of *D. latum*, *D. nihonkaiense*, and *D. dendriticum* are more divergent than cox1 sequences [35]. MtDNA targets are known to be more polymorphic than the nuclear ones, such as 18S rRNA and ITS [36]. Therefore, mtDNA can also be targeted for high-resolution analysis of the evolutionary process among *Diphyllobothrium* tapeworms.

21.2 METHODS

We provide here a technique to identify *Diphyllobothrium* samples (larval stages obtained from fish, eggs, or proglottids obtained from patients or mammals) by amplification and sequencing of mtDNA genes cox1, cob, and NADH3, which currently represent the best targets to identify such parasites to the species level. However, as these genes have not been sequenced yet for some species, nuclear targets such as 18S rRNA or ITS could still be useful for diagnostic purposes; they are therefore also included in this section.

21.2.1 Sample Preparation

21.2.1.1 Sample Collection

Fish fillets are examined for plerocercoids by cutting the flesh in 2–3 mm-thick slices. Segments of adult worms are discharged with the feces. Eggs are concentrated from stools by usual diphasic methods. Larvae and proglottids are washed with 0.9% sodium chloride. Eggs are submitted to three cycles of washing with 0.9% sodium chloride and centrifugation 5 min at 200 × *g*. One larva, one proglottid, or the pellet of feces concentration is used for molecular analysis. All samples must be stored in pure ethanol. The use of fixatives containing formalin, as well as denaturated alcohol must be avoided, because they could irrevocably damage DNA.

21.2.1.2 DNA Extraction

Proglottids can be pretreated by three cycles of freezing at −80°C and thawing at 60°C before DNA extraction, but this pretreatment is not compulsory. DNA from the different samples is extracted by using QIAamp® DNA Mini kit (QIAGEN), following the Tissue Protocol according to the manufacturer's instructions. Twenty microliter of proteinase K are mixed with the sample and 180 μL of buffer ATL, and incubated at 56°C until the tissue is completely lysed.

A total of 200 μL of buffer AL are then added and the sample is incubated for 10 min at 70°C. A total of 200 μL of 100% ethanol are added and DNA is extracted with the use of the QIAamp column and centrifugation for 1 min at 6000 × *g*. DNA is washed with 500 μL of buffer AW1 (eliminated by centrifugation for 1 min at 6000 × *g*), then with 500 μL of buffer AW2 (eliminated by centrifugation for 3 min at 20,000 × *g*). A last centrifugation is performed for 1 min at 20,000 × *g* to eliminate residual buffer. DNA is eluted by adding 100 μL sterile distilled water in the column, which is incubated for 5 min and centrifuged for 1 min at 6000 × *g*.

21.2.2 Detection Procedure

Partial 18S rRNA gene is amplified using the primers 81 and 83, designed for molecular phylogeny of Eucestoda [26]. ITS1 and ITS2 regions including 5.8S rRNA gene are amplified using the primers BD1 and BD2, originally designed for flukes analysis [37]). Partial cox1 gene is amplified with the primers JB3 and JB4.5, designed to analyze *Echinoccocus* spp. [38], or JB6 and JB5R [31]. Complete cob gene is amplified with the primers F2DnihCob and R2DnihCob, designed for the intra-specific characterization of *Diphyllobothrium* species [32]. The mitochondrial tRNA-Pro, tRNA-Ile, tRNA-Lys, NADH3 genes are amplified using primers designed to identify *Diphyllobothrium* spp. [35]. All the primers are described in Table 21.2.

21.2.2.1 Procedure

1. Prepare polymerase chain reaction (PCR) in a 0.5 mL tube by mixing 2–50 ng DNA, 5 μL high fidelity platinum DNA polymerase 10x buffer (Invitrogen), 1.5 or 4 μL $MgSO_4$ 50 mM, 10 nmol of each dNTP (dATP, dTTP, dCTP, dGTP), 10–40 pmol of each primer, 1–3 U high fidelity platinum DNA polymerase (Invitrogen), and sterile distilled water to a final volume of 50 μL (see Table 21.3). A sample containing 10 μL of sterile water instead of DNA is included as negative control.
2. The amplification cycles are listed in Table 21.3, according to the target sequence.
3. Analyze 10 μL of amplicon in 1% agarose gel containing 0.5 μg/mL ethidium bromide and visualize bands with a UV illuminator.
4. Purify the remaining PCR product using a PCR purification kit, for example, QIAquick PCR Purification kit (QIAGEN). Namely, mix the positive PCR amplicon with five volumes of buffer PBI; extract DNA with QIAquick column; centrifuge for 1 min at 17,900 × *g*; wash DNA with 750 μL of buffer PE, which is eliminated by two centrifugations of 1 min at 17,900 × *g*; elute DNA by adding 50 μL sterile distilled water in the column, 5 min incubation, and centrifugation for 1 min at 17,900 × *g*.
5. Perform the sequencing reaction with the BigDye® Terminator Cycle Sequencing Reaction kit (Applied Biosystems). According to the length of the sequence, 10 ng per 100 bp of purified amplicon are mixed with 3 μL Ready Reaction Premix 2.5×, 1.5 μL Big Dye Sequencing Buffer 5×, 2.4 μL of the corresponding forward or reverse primer (1 μM), and sterile distilled water to a final volume of 15 μL. The thermal cycle consists in 96°C/1 min and 25 cycles of 96°C/10 s, 50°C/5 s, 60°C/4 min. Purify the products of the sequencing reaction, that is, by ethanol precipitation or by plate and spin columns filtration (Applied Biosystems. ABI Prism® BigDye® Terminator v. 3.0 Ready Reaction Cycle Sequencing Kit Protocol). Load 8 μL of each sample with 12 μL Hi-Di™ Formamide (Applied Biosystems) in an automated sequencer (ABI PRISM® Genetic Analyzers are recommended if the BigDye® Terminator Cycle Sequencing Reaction kit is used).

TABLE 21.2
Primers for *Diphyllobothrium* Species Identification

>Target	DNA	Primers
18S rRNA	Nuclear	83: 5′-GATACCGTCCTAGTTCTGACCA-3′ 81: 5′-TTCACCTACGGAAACCTTGTTACG-3′
ITS1 and ITS2	Nuclear	BD1: 5′-GTCGTAACAAGGTTTCCGTA-3′ BD2: 5′-TATGCTTAARTTCAGCGGGT-3′
Cox1	Mitochondrial	JB3: 5′-TTTTTTGGGCATCCTGAGGTTTAT-3′ JB4.5: 5′-TAAAGAAAGAACATAATGAAAATG-3′ JB6: 5′-GATAGTAAGGGTGTTGA-3′ JB5R: 5′-CAAGTATCRTGCAAAATATTATCAAG-3′
Cob	Mitochondrial	F2DnihCob: 5′-GTTTTACTGATAGGTTATTTAAAC-3′ R2DnihCob: 5′-CAATTTAAAAAACGAGTTAAAGAT-3′
NADH3	Mitochondrial	DnND3Fo: 5′-CGTTAAGATGAATGAG-3′ DnND3Re: 5′-GACCTTACATCACTTAGTG-3′

TABLE 21.3
Magnesium and Primers Concentrations, DNA Polymerase Units, Amplification Cycles, and Length of Amplicons for *Diphyllobothrium* Species Identification

Primers	$MgSO_4$ 50 mM	Primer Concentration	DNA Polymerase	Cycling Program	Amplicon (bp)
83 & 81	1.5 μL	10 pmol	1 U	1 cycle of 94°C/2 min; 35 cycles of 94°C/30 s, 55°C/40 s, 68°C/1 min 30 s 1 cycle of 68°C/10 min	1000
BD1 & BD2	1.5 μL	10 pmol	1 U	1 cycle of 94°C/5 min 30 cycles of 94°C/1 min, 55°C/40 s, 68°C/2 min 1 cycle of 68°C/10 min	1100
JB3 & JB4.5	4 μL	40 pmol	3 U	1 cycle of 94°C/7 min	440
JB6 & JB5R	1.5 μL	10 pmol	1 U	40 cycles of 94°C/30 s, 50°C/40 s, 68°C/1 min 1 cycle of 68°C/10 min	650
F2DnihCob & R2DnihCob	1.5 μL	30 pmol	2.5 U	1 cycle of 94°C/1 min 35 cycles of 94°C/15 s, 56°C/30 s, 68°C/1 min 1 cycle of 68°C/4min	1107
DnND3Fo & DnND3Re	4 μL	40 pmol	3 U	1 cycle of 94°C/7 min 38 cycles of 94°C/30 s, 55°C/40 s, 68°C/1 min 40 s 1 cycle of 68°C/10 min	708

6. Align the resulting sequences with Clustal® W, make minor corrections by using ChromasPro v. 1.42 (Technelysium Pty Ltd, Australia) and BioEdit v. 5.0 [39], and BLAST the sequences into DDJB/EMBL/GenBank databases, where sequences of *Diphyllobothrium* are available for most species (Table 21.1).

21.3 CONCLUSIONS AND FUTURE PERSPECTIVES

The methods described above have been very useful for the molecular typing of parasitic specimens obtained from fish [29,40] or human stools [14,21,29,35,41], and will certainly allow a more precise classification of the genus. The growing popularity of raw fish specialities has probably favored the increase of diphyllobothriosis in sub-alpine regions and elsewhere, and imported Pacific salmons or travels in endemic countries have led to the occurrence of cases of allocthonous *D. nihonkaiense* and *D. dendriticum* in France, Switzerland, Finland, and New Zealand [6,21,35,41–44]. In the light of these considerations, infections with allocthonous species of *Diphyllobothrium* are probably more widespread than so far estimated. In addition, these parasites show sometimes a high colonization potential, as was the case in South America for introduced *D. latum* and *D. dendriticum* [45]. Although human-infecting species can be easily treated with the same anthelminthic (praziquantel), their correct diagnosis by laboratories and physicians is of great importance from an epidemiological perspective and must be improved. The genetic identification of broad tapeworms and the analysis of anamnestic data of the patients can contribute to clarify the distribution of such parasites and help in detecting the origin of the infections. It is therefore of importance, owing to the high morphological similarity of *Diphyllobothrium* species, to precisely identify all isolates.

Mitochondrial sequences are known to be more polymorphic than those of 18S rRNA and ITS. The cox1 gene is at present the most appropriate target to identify and classify human pathogen *Diphyllobothrium*, because its sequences are enough divergent and are available in databases for most species. Moreover, the recent determination of the complete mtDNA genomes of *D. latum* and *D. nihonkaiense* provides additional tools to study intra-specific variations [46]. Both NADH3 and cob genes are promising to discriminate the different taxa, because the sequences among and within *D. latum*, *D. nihonkaiense*, and *D. dendriticum* are more divergent than those of cox1, probably even according to the geographic origin of the samples [32]. In the future it would be interesting to further investigate the existence of such link, which could be useful for epidemiological studies aimed at tracing the fish source of human infections.

In our own experience, amplification of the partial cox1 gene was obtained with the JB3–JB4.5 primers from all samples except for *D. stemmacephalum*. For the latter species, DNA amplification is possible when the JB5–JB6 primers are used. In addition, recent studies by Arizono et al. (2008) showed that this species could be synonymous with *D. balaenopterae* [15]. *D. dendriticum* and *D. ditremum* could be differentiated only by using mtDNA sequences [31]. Both species are sharing the same geographic area (circumpolar), as well as the same intermediate (coregonids) and definitive hosts (piscivorous birds and mammals). However, only *D. dendriticum* is a human pathogen.

In 2006, we identified the first *D. nihonkaiense* infection in France, acquired by the consumption of a Pacific salmon imported from Canada [35]. This cestode was described in 1986, when Yamane et al. [22] showed morphological and biochemical differences between Finnish and Japanese isolates of *D. latum*; the latter was renamed *D. nihonkaiense*, and has been recently synonymized with *Diphyllobothrium klebanovskii* [47]. This parasite is now recognized as a frequent cause of diphyllobothriosis in Japan, contracted by the consumption of raw Pacific salmon (*Oncorhynchus keta* or *O. masou*), and along the Pacific coast of Russia (Khabarovsk Territory, Sakhalin, and Kamchatka regions), in *O. keta* and *O. gorbuscha* muscular tissue. Typing *D. ursi* isolates would be very interesting to be sure that this species is not *D. nihonkaiense*. Diphyllobothriosis has also been described among *Oncorhynchus* consumers along the western coast of the United States [48], and a significant upsurge was observed in the late 1970s (17 cases in 1979, 64 in 1980). *D. ursi*, usually infecting bears, has been reported among Alaskan residents following the consumption of *Oncorhynchus* spp. [49]. The migratory routes of salmons originated from Asia and America overlap in the north Pacific, and we suggested that the reported absence of *D. nihonkaiense* along the Pacific coast of North America might be in fact due to its misidentification. Such hypothesis seems to be confirmed by the recent record of this cestode in a patient infected in Canada [14].

The taxonomy in the genus *Diphyllobothrium* is evolving very rapidly, thanks to the new techniques provided by molecular biology. Sequencing is certainly a precise approach to type specimens of *Diphyllobothrium*, but it is only feasible with species whose related gene sequences are available in the databases. On the other hand, collaborations with skilful morphologists are still necessary.

REFERENCES

1. Chai, J.Y., Murrell, K.D., and Lymbery, A.J. Fish-borne parasitic zoonoses: Status and issues. *Int. J. Parasitol.* 35, 1233, 2005.
2. Dick, T.A. Diphyllobothriasis: The *Diphyllobothrium latum* human infection conundrum and reconciliation with a worldwide zoonosis. In: Murrell, K.D. and Fried, B. (eds), *Food-Borne Parasitic Zoonoses: Fish and Plant-Borne Parasites*. Springer, New York, pp. 151–84, 2007.
3. Muller, R. *Worms and Human Disease*, 2nd edn. CABI, Wallingford, UK, pp. 63–105, 2002.
4. WHO. Parasitic Zoonoses. Report of a WHO expert committee with the participation of FAO. *WHO Techn. Rep. Ser.* 637, 1, 1979.
5. Dupouy-Camet, J., and Peduzzi, R. Current situation of human diphyllobothriasis in Europe. *Euro Surveill.* 9, 5, 2004.
6. Scholz, T. et al. Update on the human broad tapeworm (genus *Diphyllobothrium*), including clinical relevance. *Clin. Microbiol. Rev.* 22, 146, 2009.
7. Semenas, L., Kreiter A., and Urbanski J. New cases of human diphyllobothriosis in Patagonia, Argentine. *Rev. Saúde Pública* 35, 214, 2001.
8. Sagua, H. et al. Nuevos casos de infección humana por *Diphyllobothrium pacificum* (Nybelin, 1931) Margolis, 1956 en Chile y su probable relación con el fenómeno de El Niño, 1975–2000. *Bol. Chil. Parasitol.* 56, 22, 2001.
9. Sampaio, J.L. et al. Diphyllobothriasis, Brazil. *Emerg. Infect. Dis.* 11, 1598, 2005.
10. Brabec, J., Kuchta R., and Scholz T. Paraphyly of the Pseudophyllidea (Platyhelminthes: Cestoda): Circumscription of monophyletic clades based on phylogenetic analysis of ribosomal RNA. *Int. J. Parasitol.* 36, 1535, 2006.
11. Ashford, R.W., and Crewe, W. *The Parasites of Homo Sapiens: An Annotated Checklist of the Protozoa, Helminths, and Arthropods for which We are Home*. CRC Press, London, 2003.
12. Delyamure, S.L., Skryabin A.S., and Serdiukov A.M. Diphyllobothriata-flatworm parasites of man, mammals and birds. In: *Principles of Cestodology*. Vol. 9. Nauka, Moscow, Russia, 1985. [In Russian].
13. Kamo, H. *Guide to Identification of Diphyllobothriid Cestodes*. Tokyo, Japan, 1999.
14. Wicht B. et al. First record of human infection with the tapeworm *Diphyllobothrium nihonkaiense* in North America. *Am. J. Trop. Med. Hyg.* 78, 235, 2008.
15. Arizono, N. et al. Diplogonoporiasis in Japan: Genetic analyses of five clinical isolates, *Parasitol. Int.* 57, 212, 2008.
16. von Bonsdorff, B. *Diphyllobothriasis in Man*. Academic Press, London, 1977.
17. Andersen, K., and Halvorsen, O. Egg size and form as taxonomic criteria in *Diphyllobothrium* (Cestoda, Pseudophyllidea), *Parasitology*. 76, 229, 1978.
18. Andersen, K., and Gibson, D. A key to three species of larval *Diphyllobothrium* Cobbold, 1858 (Cestoda: Pseudophyllidea) occurring in European and North American freshwater fishes, *Syst. Parasitol.* 13, 3, 1989.
19. Dick, T.A., and Poole, B.C. Identification of *Diphyllobothrium dendriticum* and *Diphyllobothrium latum* from freshwater fishes of central Canada. *Can. J. Zool.* 63, 196, 1985.
20. Garcia, L.S., and Bruckner, D.A. *Diagnostic Medical Parasitology*. American Society for Microbiology, Washington, DC, 1993.
21. Wicht, B., de Marval, F., and Peduzzi, R. *Diphyllobothrium nihonkaiense* (Yamane et al., 1986) in Switzerland: First molecular evidence and case reports. *Parasitol. Int.* 56, 195, 2007.
22. Yamane, Y. et al. *Diphyllobothrium nihonkaiense* sp. nov. (Cestoda: Diphyllobothriidae): Revised identification of Japanese broad tapeworm. *Shimane J. Med. Sci.* 10, 29, 1986.
23. Fukumoto, S. et al. Distinction between *Diphyllobothrium nihonkaiense* and *Diphyllobothrium latum* by immunoelectrophoresis. *Jpn. J. Parasitol.* 37, 91, 1988.
24. Matsuura, T., Bylund, G., and Sugane, K. Comparison of restriction fragment length polymorphisms of ribosomal DNA between *Diphyllobothrium nihonkaiense* and *D. latum*. *J. Helminthol.* 66, 261, 1992.
25. De Vos, T., and Dick, T.A. Differentiation between *Diphyllobothrium dendriticum* and *Diphyllobothrium latum* using isoenzymes, restriction profiles and ribosomal gene probes. *Sys. Parasitol.* 13, 161, 1989.
26. Mariaux, J. A molecular phylogeny of the eucestoda. *J. Parasitol.* 84, 114, 1998.
27. Olson, P.D., and Caira, J.N. Evolution of the major lineages of tapeworms (Platyhelminthes: Cestoda) inferred from 18S ribosomal DNA and elongation factor-1. *J. Parasitol.* 85, 1134, 1999.
28. Isobe, A. et al. Molecular phylogeny of *Diphyllobothrium nihonkaiense* (Yamane et al., 1986) and other diphyllobothiid tapeworms based on mitochondrial cytochrome c oxidase subunit I gene sequence. *Parasitol. Int.* 47 (suppl 1), 138, 1998.

29. Ando, K. et al. Five cases of *Diphyllobothrium nihonkaiense* infection with discovery of plerocercoids from an infective source, *Oncorhynchus masou* ishikawae. *J. Parasitol.* 87, 96, 2001.
30. Logan, F.J. et al. The phylogeny of diphyllobothriid tapeworms (Cestoda: Pseudophyllidea) based on ITS-2 rDNA sequences. *Parasitol. Res.* 94, 10, 2004.
31. Yera, H., Nicoulaud, J., and Dupouy-Camet, J. Use of nuclear and mitochondrial DNA PCR and sequencing for molecular identification of *Diphyllobothrium* isolates potentially infective for humans. *Parasite* 15, 402, 2008.
32. Wicht, B. et al. Inter- and intra-specific characterization of tapeworms of the genus *Diphyllobothrium* (Cestoda: Diphyllobothriidea) from Switzerland, using nuclear and mitochondrial DNA targets. *Parasitol. Int.* 59, 35, 2010.
33. Hillis, D.M., and Dixon, M.T. Ribosomal DNA: Molecular evolution and phylogenetic inference. *Q. Rev. Biol.* 66, 411, 1991.
34. Skerikova, A. et al. Is the human-infecting *Diphyllobothrium pacificum* a valid species or just a South American population of the holarctic fish broad tapeworm, *D. latum*? *Am. J. Trop. Med. Hyg.* 75, 307, 2006.
35. Yera, H. et al. Putative *Diphyllobothrium nihonkaiense* acquired from a Pacific salmon (*Oncorhynchus keta*) eaten in France: Genomic identification and case report. *Parasitol. Int.* 55, 45, 2006.
36. Brown, W.M., George, Jr. M., and Wilson, A.C. Rapid evolution of animal mitochondria DNA. *Proc. Natl Acad. Sci. USA* 76, 1967, 1979.
37. Luton, K., Walker, D., and Blair, D. Comparisons of ribosomal internal transcribed spacers from two congeneric species of flukes (Platyhelminthes: Trematoda: Digenea). *Mol. Biochem. Parasitol.* 56, 323, 1992.
38. Bowles, J., Blair, D., and McManus, D.P. Genetic variants within the genus *Echinococcus* identified by mitochondrial DNA sequencing. *Mol. Biochem. Parasitol.* 54, 165, 1992.
39. Hall, T. *BioEdit. 5.0. Distributed by the Author.* Department of Microbiology, North Carolina State University, Raleigh, NC, 2001.
40. Nicoulaud, J., Yera, H., and Dupouy-Camet, J. Prevalence of Diphyllobothrium latum, L., 1758 infestation in *Perca fluviatilis* from Lake Leman.*Parasite* 12, 362, 2005 (in French).
41. Wicht, B. et al. Imported diphyllobothriasis in Switzerland: Molecular evidence of *Diphyllobothrium dendriticum* (Nitsch, 1824). *Parasitol. Res.* 102, 201, 2008.
42. Shimizu, H. et al. *Diphyllobothriasis nihonkaiense*: Possibly acquired in Switzerland from imported Pacific salmon. *Intern. Med.* 47, 1359, 2008.
43. Paugam A. et al. *Diphyllobothrium nihonkaiense* infection: A new risk in relation with the consumption of salmon. *Presse. Med.* 38, 675, 2009.
44. Yamasaki, H., and Kuramochi, T. A case of *Diphyllobothrium nihonkaiense* infection possibly linked to salmon consumption in New Zealand. *Parasitol. Res.* 105, 583, 2009.
45. Torres, P. et al. Introduced and native fishes as infection foci of *Diphyllobothrium* spp. in humans and dogs from two localities at Lake Panguipulli in Southern Chile. *Comp. Parasitol.* 71, 111, 2004.
46. Nakao, M. et al. Mitochondrial genomes of the human broad tapeworms *Diphyllobothrium latum* and *D. nihonkaiense* (Cestoda: Diphyllobothriidae). *Parasitol. Res.* 101, 233, 2007.
47. Arizono, N. et al. Mitochondrial DNA divergence in populations of the tapeworm *Diphyllobothrium nihonkaiense* and its phylogenetic relationship with *Diphyllobothrium klebanovskii. Parasitol. Intern.* 58, 22, 2008.
48. Ruttenber, A.J. et al. Diphyllobothriasis associated with salmon consumption in Pacific Coast states. *Am. J. Trop. Med. Hyg.* 33, 455, 1984.
49. Rausch, R.L., and Hilliard D.K. Studies on the helminth fauna of Alaska. XLIX. The occurrence of *Diphyllobothrium latum* (Linnaeus, 1758) (Cestoda: Diphyllobothriidae) in Alaska, with notes on other species. *Can. J. Zool.* 48, 1201, 1970.

22 Dipylidium

Dongyou Liu

CONTENTS

22.1 INTRODUCTION

Cestodes (or tapeworms) are a group of helminthic parasites, which are separated into two major groups: cyclophyllideans and pseudophyllideans. Examples of human pathogenic cyclophyllideans include *Echinococcus granulosus* and *Echinococcus multilocularis* (hydatid tapeworm, echinococcosis), *Taenia saginata* (beef tapeworm), *Taenia asiatica*, and *Taenia solium* (pork tapeworm) (taeniasis/cdysticercosis), *Hymenolepis nana* and *Hymenolepis diminuta* (dwarf tapeworm, hymenolepiasis); and examples of human pathogenic pseudophyllideans consist of *Diphyllobothrium latum* (fish tapeworm, diphyllobothriasis), *Spirometra erinaceieuropaei* (sparganosis), *Diphyllobothrium mansonoides* (sparganosis), and *Dipylidium caninum* (dog tapeworm).

D. caninum adult worms are found in the intestine of carnivorous hosts, such as dogs and cats, as well as humans. The larvae are present in the dog flea and cat flea. Humans (especially young children) acquire the infection through accidental ingestion of fleas containing infective larvae (cysticercoids). Although human dipylidiasis is rarely fatal, it can cause a number of intestinal symptoms (e.g., diarrhea and restlessness) in human patients. Thus, *D. caninum* represents a serious concern for pet owners and parents.

22.1.1 Classification, Morphology, and Life Cycle

22.1.1.1 Classification

The genus *Dipylidium* is classified in the family Dipylidiidae, order Cyclophyllidea, class Cestoda, and phylum Platyhelminthes. The genus comprises a single species *D. caninum*.

D. caninum (also known as dog tapeworm, cat tapeworm, cucumber tapeworm, or double-pore tapeworm) is a parasite of dogs, and sometimes cats and humans. The role of *D. caninum* in human illness has been known since the beginning days of the Linnaean classification system in the 1750s.

22.1.1.2 Morphology

D. caninum adult is a tapeworm of about 15–80 cm in length. The body of the adult worm consists of the head (or scolex), the neck, and a segmented section (called the strobilus). The scolex is rhomboid in shape with an apical rostellum and four rows of hooks (characteristic of Cyclophyllidea cestode), and possesses sensory organs (to detect tactile stimulation), which are attached to longitudinal nerves extending down the body. The segmented section comprises multiple segments (which may be in different stages of development: immature, mature, or gravid, with gravid proglottids containing eggs of 20–40 μm in diameter located at the rear end). Each segment (of 0.5–1.0 cm long and 0.1–0.2 cm thick, with ivory appearance resembling cucumber seed when fresh, or rice grain when dried) consists of two proglottids (a proglottid represents one set of male and female reproductive organs) and two genital pores located laterally (the so-called double-pored tapeworm). Segments are mobile when freshly out of the hosts (in fecal material).

The gravid proglottids can contain 5–30 spherical eggs (called egg clusters, egg balls, or egg packets) each. The eggs (oval and 20–40 μm in diameter, and each containing an oncosphere) hatch in fleas (the intermediate host), so that oncosphere becomes cysticercoid (so called because the scolex is enclosed in a fluid-filled cyst at this stage).

22.1.1.3 Biology

D. caninum life cycle involves a definitive host (dog or cat) and an intermediate host (flea, such as *Ctenocephalides canis* and *Ctenocephalides felis*).

The adult worm uses its hooked scolex to attach to the small intestine of the definitive host. Due to its lack of digestive tract (similar to other cestodes), *D. caninum* feeds by absorption of the partially digested food in the intestine through its body covering (or tegument). Segments containing gravid proglottids detach singly or in chains continually and are excreted in the host feces. Through contracting and expanding lengthwise, proglottids may migrate from the anus to the exterior of the host, independent of stool, and release the egg packets within the proglottids. The egg packets contained in the segments are then ingested by the larva of fleas (the intermediate host), where the eggs hatch to release oncospheres, which uses its hooks to transport itself from the gut of the flea into the hemocoel. In the hemocoel, larva continues to develop into cysticercoid. The surface temperature of the mammals (31–36°C), and not the fleas' blood meals, contributes to the cestode's development into the cysticercoid. When the flea is eaten by the definitive host (a vertebrate), the larva grows into sexually mature adults in a month. The lifespan of adult worms is less than 1 year. Since it is monecious, or hermaphroditic (with the ability to self-fertilize), only one worm is necessary for reproduction. In addition, due to its ability to continue to generate segments, this worm has high reproductive potentials (producing millions of eggs in its lifetime).

Besides a variety of canids, felids, and wild carnivores, humans can also act as a definitive host for *D. caninum*. The most common intermediate hosts for this parasite are *C. felis* (dog flea) and *C. felis* (cat flea). Other intermediate hosts include *Pulex irritans* (human flea) and *Trichodectes canis* (dog louse). The parasite can be highly pathogenic for the intermediate host.

Human infections occur most commonly in children and infants, and children playing with canine or feline pets acquire the fleas containing juvenile worms via mouth to mouth (the habit of kissing/licking pets), or hand-to-mouth contact. Infections in humans are usually limited to one or a few worms [1].

D. caninum has a worldwide distribution, and human infections have been reported in Europe, China, Japan, the Philippines, India, Australia, Africa, Argentina, Chile, Guatemala, and the United States [2–4].

22.1.2 Clinical Features and Treatment

22.1.2.1 Clinical Features

D. caninum, the common intestinal tapeworm of dogs and cats, is infective to humans after accidental ingestion of infected fleas. *D. caninum* infection (dipylidiasis) typically occurs in infants (6 months of age or younger) and small children in contact with dogs and cats, with an incubation period of approximately 20 days [5–14]. Rarely, dipylidiasis is also found in adults [15]. While human dipylidiasis is often asymptomatic, when present, the clinical symptoms may include loss of appetite, indigestion, abdominal pain, diarrhea, irritability, poor weight gain, and anal pruritis, with proglottids discharged in feces. *D. caninum* infection may also coexist with other parasitic infections (e.g., pinworm or threadworm *Enterobius vermicularis*), inducing additional clinical symptoms [16].

Molina et al. [17] documented a case of *D. caninum* infection involving a 6-month-old infant girl. Having passed small, white, rice-like bodies in stool over a 3-month period, the baby was otherwise asymptomatic. The parasites were ivory white, rice grain-like, and measured 0.5–1.0 cm long and 0.1–0.2 cm thick. Light microscopic examination of hematoxylin and eosin-stained 4-μm-thick tissue sections of the parasite revealed gravid proglottids with a body wall and internal compartmentalization and eggs in packets. The body wall (of 25–50 μm in thickness) was composed of a tegument, tegumental cells, muscle fibers, and spherical to oval laminated calcareous corpuscles of 10–15 μm. The egg packets contained 8–15 eggs surrounded by a thin membrane. The eggs appeared round to oval and measured 30–40 μm in diameter, each containing an oncosphere. The proglottids and eggs were thus identified as those of *D. caninum*.

In a recent report, Ramana et al. [18] described a case of *D. caninum* infection in a 9-year-old girl. The child had a 3-day history of nausea, vomiting, recurrent abdominal pain, and diarrhea as well as persistent low-grade fever. Macroscopic assessment of stool revealed white to creamish specks suggestive of worms. A wet mount showed the presence of egg cluster. Adult worm with characteristic cucumber seed-shaped segments was observed. Based on the morphology of eggs together with the presence of egg balls or egg clusters and the adult worm with double-pore segments and cucumber seed shape of the proglottids, the parasite was identified as *D. caninum*. The patient was successfully treated with praziquantel.

22.1.2.2 Treatment

Praziquantel (at one dose of 5–10 mg/kg body weight) is very effective against *D. caninum* infection [19]. Niclosamide is an alternative, and can be administered as a single dose 2.0 g for adults, 1.5 g for children weighing over 34 kg, and 1.0 g for children weighing between 11 and 34 kg. However, niclosamide may cause abdominal cramps, diarrhea, and nausea and vomiting in a small number of patients [20].

Prevention of *D. caninum* infection is by controlling the fleas and de-worming of pet dogs and cats. Praziquantel and Niclosamide are available as injection for pets.

22.1.3 Laboratory Diagnosis

Diagnosis of *D. caninum* infection is made by finding characteristic cucumber seed-shaped or rice grain-like proglottids (with double genital pores) in stool specimens and microscopic observation of the egg packets (with each harboring an oncosphere) in the gravid uterus of the unstained proglottids (segments) compressed into glass slides. Sometimes, examination of more than one stool specimen is needed to find the proglottids or eggs. The specimen may be preserved in 70% alcohol or 10% neutral buffered formalin before submission to the laboratory.

D. caninum proglottids in tissue sections show a body wall with tegument, tegumental cells, and smooth muscle fibers; a parenchyma composed of a loose stroma, fluid-filled spaces, mesenchymal fibers, and smooth muscle fibers; absence of a digestive tract; calcareous corpuscles in the body wall or parenchyma; inner medullary zone containing reproductive structures; or a uterine cavity with eggs. Larval cestodes (e.g., cysticerci) in the tissues do not contain reproductive structures or eggs.

Dried proglottids may also be misdiagnosed as maggots or fly larvae. *D. caninum* segments are ribbon-like, and move by contracting and expanding lengthwise, while proglottids from threadworm (*E. vermicularis*) are round with one tapering end, and move in a serpentine wriggling fashion. *D. caninum* eggs are differentiated from the eggs (31–43 μm) of *T. solium*, *T. saginata*, and *Echinococcus granulosus*, which have thick, radially striated shells and do not occur in packets. *D. latum* eggs are much larger (70 × 45 μm) and are operculate.

The recent development and application of PCR and DNA sequencing techniques have facilitated the improved identification of cestodes and other Platyhelminthes. Among the most commonly targeted genes, the nuclear 28S rRNA gene repeat unit harbors different regions (D1–D3) that evolve at varying rates, and has been used extensively in the estimation of the relationships existing within and among the Platyhelminthes. Mitochondrial cytochrome-c oxidase subunit I (mCOI), the most highly conserved among three genes coding for cytochrome oxidase, has been proven valuable for discrimination of closely related organisms [21]. PCR amplification of mitochondrial DNA (mtDNA) gene followed by DNA sequencing or restriction-fragment-length polymorphism (RFLP) analysis, has allowed fast and accurate identification of tapeworm eggs, including *Echinococcus multilocularis*, *E. granulosus*, *Taenia*, *Mesocestoides*, *Dipylidium*, and *Diphyllobothrium* [22].

22.2 METHODS

22.2.1 Sample Preparation

D. caninum eggs are lysed in an alkaline solution, and parasite DNA is purified using 4 M guanidine isothiocyanate buffer and silica (SiO2) particles (glass milk). Glass milk is prepared by resuspending 6 g SiO2 (Sigma S5631) in 50 ml deionized water. The tube is mixed and centrifuged at 1500 × *g* for 1 min, and the pellet is washed with deionized water until supernatant is clear. After a last centrifugation, 45 mL of the liquid is removed and the sediment is resuspended in 5 mL deionized water, autoclaved for 20 min at 120°C, and frozen as 200 μL aliquots at –20°C until use. Taeniid eggs stored at –80°C are thawed at room temperature and centrifuged at 2500 × *g*. About 50 μL of the egg-containing suspension are mixed with 25 μL 1 M potassium hydroxide and 7 μL 1 M dithiothreitol and incubated for 10 min at 65°C for lysing the eggs. Alkaline lysis is followed by a neutralization by adding of 60 μL 2 M Tris–HCl (pH 8.3) and 5 μL 10 M HCl. The lysed sample is mixed with 300 μL guanidinium thiocyanate (GTC) extraction buffer (4 M guanidine isothiocyanate, 40 mM Tris–HCl (pH 6.4), 16 mM ethylene diamine tetraacetic acid (EDTA; pH 8), 1% Tween-20) and incubated for 10 min at 65°C. After adding of 350 μL of absolute ethanol and 5 μL of silica suspension, the mixture is incubated for 5 min at room temperature with gentle shaking to bind the DNA to SiO_2. Afterward, the silica with bound DNA is sedimented by centrifugation at 13,000 × *g* for 1 min and washed once with GTC washing buffer (4 M guanidine isothiocyanate, 40 mM Tris–HCl pH 6.4, and 16 mM EDTA pH 8) and then twice with 70% ethanol. Following the last wash, supernatants are removed without disturbing the pellet and are air-dried. The DNA is eluted with 30 μL TE buffer and separated from glass milk by centrifugation as described above. The DNA is stored at –20°C [23].

Genomic DNA is extracted from the whole adult worms. Each specimen is fixed using 100% ethanol at room temperature or frozen at −80°C. The ethanol in the tissue samples is replaced with 1 M Tris–EDTA buffer (pH 8.0). The frozen worms are lyophilized and lysed with lysis buffer containing 1% SDS, proteinase K (500 μg/mL), and RNase at 37°C for 2–3 h or overnight. The DNA is then extracted in phenol/chloroform, or by using commercial DNA purification kit such as QIAamp DNA kit (Qiagen) [21].

22.2.2 Detection Procedures

Lee et al. [21] targeted the partial 28S ribosomal RNA (rRNA) D1 region and the partial mitochondrial cytochrome-c oxidase subunit 1 (mCOI) DNA sequences for identification and phylogenetic analysis of members of Platyhelminthes. The 28S rRNA D1 region was amplified with primers JB10 (5′-GATTACCCGCTGAACTTAAGCATA-3’) and JB9 (5′-GCTGCATTCACAAACACCCCGACTC-3’); and the mCO1 DNA was amplified with primers JB3 (5’-TTTTTTGGGCATCCTGAGGTTTAT-3′) and JB4.5 (5’-TAAAGAAAGAACATAATGAAAATG-3′) (3021) [24].

22.2.2.1 Procedure

1. PCR mixture (50 μL) is composed of 200 μM of each dNTP, 10 pmoles of each primer (JB10/JB9 or JB3/JB4.5), and 0.5 U of Ex Taq enzyme (TaKaRa Ex Taq Kit, TAKARA), and 50–100 ng DNA template.
2. After incubation at 95°C for 2 min, amplification is conducted with 40 cycles of denaturation at 95°C for 20 s, annealing at 55°C (JB10/JB9) or 48°C (JB3/JB4.5) for 30 s, and extension at 72°C for 30 s, followed by a final extension at 72°C for 6 min.
3. The PCR products (about 280–300 bp for 28S rRNA D1 gene and about 360–480 bp for mCOI gene) are purified and sequenced using PCR primers and ABI Prism Dye Terminator Cycle Sequencing Core Kit (Perkin Elmer) on an automated DNA sequencer (Applied Biosystems model 373A, Perkin Elmer).
4. Sequence analysis is conducted using the nucleotide BLAST program in the NCBI database. The nucleotide sequences are compared using the Multalign program for nucleotide differences and gaps. The resulting aligned output is manually adjusted in order

to improve the homology statements. The alignment gaps are treated as missing data. Sequences corresponding to the PCR amplification primers are excluded prior to multiple sequence alignment and phylogenetic analysis, as primer incorporation during amplification masks potential mismatches (substitutions) that might occur within the priming sites.

5. Phylogenetic analyses are conducted using the PAUP. Each molecular data partition is analyzed by both maximum parsimony (MP) and neighbor-joining (NJ) analysis. The analyses are run using a heuristic search strategy (100 search replicates) with tree bisection–reconnection branch swapping, equally weighted characters and gaps treated as missing data. Bootstrap analysis is performed using a fast heuristic search strategy (1000 replicates where ≥50%) for the 28S rRNA D1 and mCOI data. A NJ analysis is performed using the Kimura 2-parameter model for distance correction; corrected for multiple substitutions.

Note: The 28S rRNA D1 region sequence appears to be more conserved than the mCOI gene sequence. Therefore, the 28S rRNA D1 region is particularly useful for interspecies comparison and the mCOI gene sequence is valuable for intraspecies comparison.

22.3 CONCLUSION

D. caninum (commonly known as dog tapeworm, double-pored tapeworm, or cucumber tapeworm) is a parasitic cestode of dogs and cats. Its life cycle involves dog and cat fleas (*Ctenocephalides canis* and *Ctenocephalides catis*) as intermediate hosts, which acquire parasite by ingesting eggs present in the animal feces. Humans represent an accidental host for the parasite, and become infected after ingestion of fleas containing infective cysticercoids or contact with saliva of pet animals. Infants and crawling children are particularly prone to dipylidiasis due to their tendency to play with pet dog and cats. Although *D. caninum* infections in humans are often asymptomatic, they may be responsible for causing loss of appetite, indigestion, abdominal pain, diarrhea, irritability, poor weight gain, and anal pruritis. As an easily treatable and preventable disease, correct diagnosis plays a crucial role in the implementation of appropriate control strategies. Considering the time-consumingness and technical demands of conventional diagnostic methods for identification of *D. caninum* parasite, molecular approach offers a reliable way to improve the detection and epidemiological tracking of dipylidiasis in humans and animals.

REFERENCES

1. Neafie, R.C., and Marty, A.M., Unusual infections in humans. *Clin. Microbiol. Rev.*, 6, 34, 1993.
2. Turner, J.A., Human dipylidiasis (dog tapeworm infection) in the United States. *J. Pediatr.*, 61, 763, 1962.
3. Shane, S.M. et al., A case of *Dipylidium caninum* in Baton Rouge, Louisiana. *Int. J. Zoonoses*, 13, 59, 1986.
4. Dai, R.S. et al., Severe infection of adult dogs with helminths in Hunan Province, China poses significant public health concerns. *Vet. Parasitol.*, 160, 348, 2009.
5. Currier, R.W. 2nd, Kinzer, G.M., and DeShields, E., *Dipylidium caninum* infection in a 14-month-old child. *South Med. J.*, 66, 1060, 1973.
6. Reddy, S.B., Infestation of a five-month-old infant with *Dipylidium caninum*. *Del. Med. J.*, 54, 455, 1982.
7. Hamrick, H.J. et al., Two cases of dipylidiasis (dog tapeworm infection) in children: update of an old problem. *Pediatrics*, 72, 114, 1983.
8. Wijesundera, M.D., and Ranaweera, R.L., Case reports of *Dipylidium caninum*, a pet associated infection. *Ceylon Med. J.*, 34, 27, 1989.
9. Chappell, C.L., Enos, J.P., and Penn, H.M., *Dipylidium caninum*, an underrecognized infection in infants and children. *Pediatr. Infect. Dis. J.*, 9, 745, 1990.
10. Raitiere, C.R., Dog tapeworm (*Dipylidium caninum*) infestation in a 6-month-old infant. *J. Family Pract.*, 34, 101, 1992.
11. Reid, C.J., Perry, F.M., and Evans, N., *Dipylidium caninum* in an infant. *Eur. J. Pediatr.*, 151, 502, 1992.
12. Ferraris, S. et al., *Dipylidium caninum* in an infant. *Eur. J. Pediatr.*, 152, 702, 1993.
13. Gadre, D.V., Kumar, A., and Mathur, M., Infection by *Dipylidium caninum* through pet cats. *Indian J. Pediatr.*, 60, 151, 1993.
14. Stiles, C.W., A case of infection with the double-pored tapeworm (*Dipylidium caninum*) in an American child. *Am. Med.*, 5, 65, 1993.
15. Jackson, D. et al., Dipylidiasis in a 57-year-old woman. *Med. J. Aust.*, 2, 740, 1977.
16. Samkari, A. et al., *Dipylidium caninum* mimicking recurrent enterobius vermicularis (pinworm) infection. *Clin. Pediatr.* (Phila), 47, 397, 2008.
17. Molina, C.P., Ogburn, J., and Adegboyega, P., Infection by *Dipylidium caninum* in an infant. *Arch. Pathol. Lab. Med.*, 127, e157, 2003.
18. Ramana, K.V. et al., Human dipylidiasis: A case report of *Dipylidium caninum* infection in teaching hospital at Karimnagar. *Online J. Health Allied Sci.*, 10, 28, 2011.
19. Wijesundera, M.D., The use of praziquantel in human infection with *Dipylidium*. *Trans. R. Soc. Trop. Med. Hyg.*, 83, 383, 1989.
20. Jones, W.E., Niclosamide as a treatment for *Hymenolepis diminuta* and *Dipylidium caninum* infection in man. *Am. J. Trop. Med. Hyg.*, 28, 300, 1979.
21. Lee, S.U., Chun, H.C., and Huh, S., Molecular phylogeny of parasitic Platyhelminthes based on sequences of partial 28S rDNA D1 and mitochondrial cytochrome c oxidase subunit I. *Korean J. Parasitol.*, 45, 181, 2007.
22. Trachsel, D., Deplazes, P., and Mathis, A., Identification of taeniid eggs in the faeces from carnivores based on multiplex PCR using targets in mitochondrial DNA. *Parasitology*, 134, 911, 2007.
23. Dyachenko, V. et al., Cost-effective method of DNA extraction from taeniid eggs. *Parasitol. Res.*, 102, 811, 2008.
24. Bowles, J., and McManus, D.P., Genetic characterization of the Asian Taenia, a newly described taeniid cestode of humans. *Am. J. Trop. Med. Hyg.*, 50, 33, 1994.

23 Echinococcus and Echinococcosis

Akira Ito, Tetsuya Yanagida, Yasuhito Sako, Minoru Nakao, Kazuhiro Nakaya, Jenny Knapp, and Yuji Ishikawa

CONTENTS

23.1 INTRODUCTION

Among zoonotic cestodiases, both echinococcosis and cysticercosis (see Chapter 27 in this book) are major human diseases spread across worldwide. The former is maintained by the tiny tapeworms (of 1–10 mm long) of the genus *Echinococcus*, which inhabit and mature in the small intestine of carnivorous definitive hosts belonging to the families Canidae, Felidae, and Hyaenidae [1–3] and produce eggs infective to herbivorous or omnivorous intermediate hosts including humans and contaminate the environment [4,5]. Humans may become the intermediate [6] or accidental host of *Echinococcus* spp and there is no human-to-human infection except for liver transplantation from patients harboring metacestodes of *Echinococcus multilocularis*. However, biology of *Echinococcus* spp is complicated, since (i) it involves four species, *Echinococcus granulosus* sensu lato (s.l.), *E. multilocularis*, *Echinococcus vogeli*, and *Echinococcus oligarthrus* [7], (ii) the pathology and the life cycle differs crucially among these four species, and (iii) recent reevaluation of *E. granulosus* s.l., which causes cystic echinococcosis (CE), has strongly suggested that it is divided into five independent species, *E. granulosus* sensu stricto (s.s), *Echinococcus equinus*, *Echinococcus ortleppi*, *Echinococcus canadensis*, and *Echinococcus felidis* (Table 23.1) [8–10,15], (iv) furthermore, a new species, *Echinococcus shiquicus*, closely related to *E. multilocularis* has been found in Tibetan plateau, China [8,12,15–17], and (v) polycystic echinococcosis (PE) caused by *E. vogeli* is not so rare in Central and South America. More than 100 cases have been reported mainly from Amazon rain forest, Brazil [18]. So far, there is no human case of *E. shiquicus* and *E. felidis* [19], although these two species are very closely related to *E. multilocularis* [12] and *E. granulosus* s.s. [10], which are the most serious causative agents for alveolar echinococcosis (AE) and CE worldwide, respectively [15]. Here, we review the most recent advances in taxonomy, pathology, diagnosis, and epidemiology of *Echinococcus* species, with emphasis on their molecular identification.

23.1.1 Taxonomy, Biology, and Epidemiology

23.1.1.1 Taxonomy

Table 23.1 summarizes the most recent taxonomic concept [15]. The genus *Echinococcus* is one of the main groups in the family Taeniidae, order Cyclophyllidea, subclass Eucestoda, class Cestoda, phylum Platyhelminthes.

The history on the taxonomical studies of *Echinococcus* and *Alveococcus* is briefly summarized by Nakao et al. [15].

TABLE 23.1
Brief Taxonomic Summary of the Genus *Echinococcus*

Species (Genotype)	Human Infections	Main Hosts for Adult	Larva	Distribution[a]	New Definition	Reference
E. granulosus (G1)	Common	Dog	Sheep	Cosmopolitan	*E. granulosus s.s.*	[8]
E. granulosus (G4)	Unknown	Dog	Horse	PA	*E. equinus*	[9]
E. granulosus (G5)	Rare	Dog	Cattle	PA, ET, NT	*E. ortleppi*	[9]
E. granulosus (G6,G7)	Rare	Dog	Camel, Pig	PA, ET, NT	*E. canadensis*	[8]
E. granulosus (G8,G10)	Rare	Wolf	Cervid	PA, NA	*E. canadenisis*	[8]
E. granulosus felidis	Unknown	Lion	Warthog?	ET	*E. felidis*	[10]
E. multilocularis	Common	Fox	Rodents	PA, NA		[11]
E. shiquicus	Unknown	Fox	Pika	PA (Tibetan plateau)		[12]
E. oligarthrus	Rare	Wild felids	Rodents	NT		[13]
E. vogeli	Not rare	Bush dog	Rodents	NT		[14]

Source: Nakao, M. et al. *Infect. Genet. Evol.* 10, 444, 2010 [15].
Note: ET, Ethiopian; NA, Nearctic; NT, Neotropical; OR, Oriental; PA, Palearctic.
[a] The areas are shown as zoogeographic regions.

All *Echinococcus* species need herbivorous and omnivorous intermediate hosts including humans and carnivorous definitive hosts belonging to the families Canidae, Felidae, and Hyaenidae (Table 23.1). Figure 23.1 summarizes the molecular phylogenetic tree of the genus *Echinococcus*. It clearly indicates that (i) both *E. multilocularis* and *E. shiquicus*, and *E. granulosus* s.s. and *E. felidis* are sister species, (ii) all definitive hosts for the genus *Echinococcus* except for *E. felidis* and *E. oligarthrus* which require felids, are canids, (iii) metacestode of *E. multilocularis* is exclusively multivesicular by exogenous budding, whereas *E. granulosus* s.l. is unilocular with daughter cysts inside by endogenous budding, and (iv) *E. shiquicus* and *E. vogeli* are basically unilocular but without daughter cysts and may become polycystic by exogenous budding and called PE.

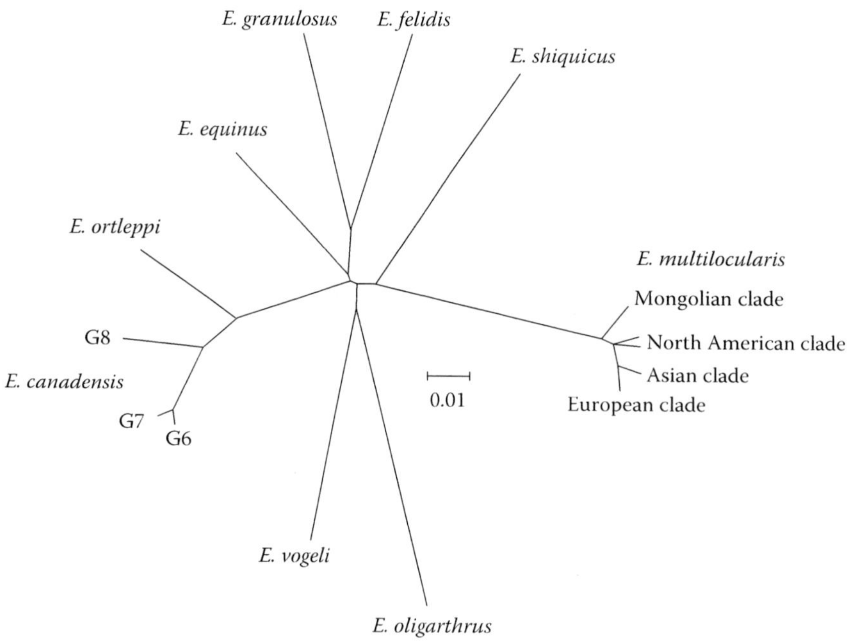

FIGURE 23.1 An unrooted phylogenetic tree of *Echinococcus* inferred from mitochondrial *cob* and *cox1* genes by the neighbor-joining method. The taxa used cover all known species (nine species) in the genus. Scale bar (divergence of 0.01) is shown. The genotypes of *E. canadensis* (G6, G7, and G8) and the geographic clades of *E. multilocularis* are included. (Adapted from Nakao, M. et al. *Parasitology* 134, 713, 2007 [8]; Hüttner, M. et al. *Int. J. Parasitol.* 38, 861, 2008 [10]; Nakao, M. et al. *Parasitol. Int.* 58, 384, 2009 [20]; Ito, A. et al. *Am. J. Trop. Med. Hyg.* 82, 266, 2010 [21].)

E. granulosus s.l. has been well known to cause CE in animals and humans worldwide. However, as discussed by Nakao et al. [8], it is considered to have been over-simplified as one species *E. granulosus* s.l. [22]. Although all adult worms isolated from dogs from different areas of the world appear to be very similar and impossible to differentiate morphologically, the pathology in the intermediate host animals may be highly variable to produce sterile or fertile cysts and to infect humans or not. Therefore, genotypes G1–G10 have been introduced for differentiation of *E. granulosus* s.l., which causes unilocular cyst in the domestic and wild animals, into four independent species, *E. granulosus* s.s. (G1–G3), *E. equinus* (G4), *E. ortleppi* (G5) and *E. canadensis* (G6–G10) [8,17,23]. Another isolate, *E. granulosus felidis* [24] from African lions is now regarded as an independent species, *E. felidis* [10]. Another recent topic is the description of a new species, *E. shiquicus* [8,10,12,15,17,20,25]. Such reevaluation is based on molecular approaches using not only mitochondrial but also nuclear DNA, which have proven instrumental in the determination of sibling or cryptic species of *Echinococcus*.

23.1.1.2 Biology

Identification of *Echinococcus* species is often based on the morphology of adult worms and larval stages. The morphology of adult worms differs notably among the four classic species, *E. oligarthrus*, *E. vogeli*, *E. granulosus*, and *E. multilocularis* [4,14]. However, it is impossible to identify the species using eggs released from adult worms from genera of *Echinococcus* and *Taenia* of the same family Taeniidae, since the morphology of the egg is very similar among all these species in this family. In contrast, the morphology of the larval stages is crucially different among the four classic species. However, *Echinococcus* larval specimens are not readily available for morphological or pathological identification except for surgically removed larval stages from patients or animals.

23.1.1.2.1 Egg

Eggs of the genera *Echinococcus* and *Taenia* released from mature adult worms are infective to suitable intermediate host animals including humans upon ingestion, since infective embryos (oncospheres) have already developed in the gravid proglottid(s). Oncosphere is highly immunogenic and the host animals become immune to reinfection after exposure. This forms the basis of molecular approach for vaccine development against this parasite [26–29].

23.1.1.2.2 Metacestode Causing CE, AE, PE, or Unicystic Echinococcosis

As mentioned above, metacestodes of *E. granulosus* s.l. are unilocular in herbivores and omnivores including humans, and called CE. As *E. granulosus* is now divided into five independent species, it is essential to know which species are infective to human population. *E. granulosus* s.s., previously described as *E. granulosus* (G1), is the main causative agent to humans worldwide (Table 23.1) [8,10]. However, there are still poor data on other species, *E. canadensis*, previously called G6–G10 and the lion strain (*E. granulosus felidis*), in relation to their infectivity to humans, although rare human cases due to *E. canadensis* G5, G6, G7, G8, and G10 have been reported. The fully developed metacestode of *E. granulosus* s.s. is typically unilocular, subspherical in shape, fluid-filled, and exhibits the least complex structure among the four species. The cyst consists of an inner germinal layer supported externally by an acellular laminated layer. Asexual proliferation of the germinal layer and brood capsule formation takes place entirely endogeneously [30,31].

Both *E. oligarthrus* and *E. vogeli* are now called neotropical echinococcosis (NE) [18]. Information on the metacestode of *E. oligarthrus* is scarce but implies unicystic and may be called unicystic echinococcosis (UE). There are only three human cases reported from Brazil, Venezuela, and Surinam. In contrast, the metacestode of *E. vogeli* is expected to show a budding pattern between those of *E. granulosus* and *E. multilocularis*, and called PE [18,32]. The newly described *E. shiquicus* [12] is also unilocular but often become polycystic. Figure 23.2 shows the

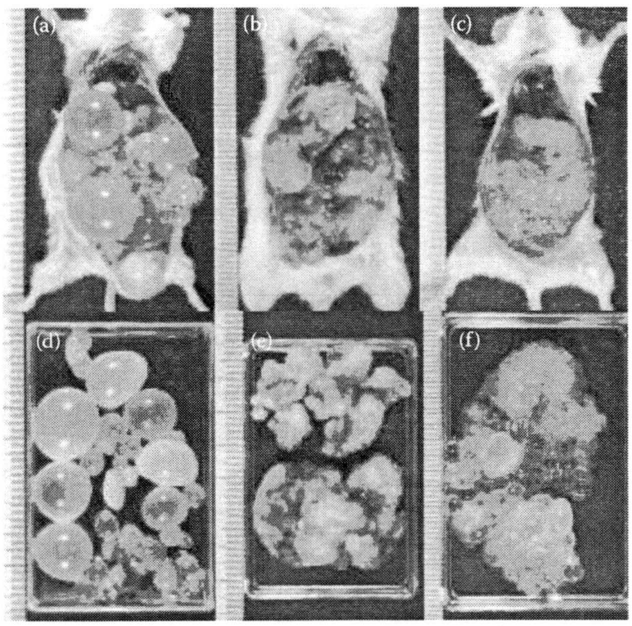

FIGURE 23.2 Secondary metacestodes of *E. granulosus* s.s. (G1) developed in BALB/c mouse (a and d), and *E. multilocularis* (b and e) and *E. shiquicus* (c and f) in severe combined immunodeficiency (*scid*) mice. Vesicles of *E. granulosus* are free in the abdomen and some are ~20 mm in diameter. Protoscoleces have been produced in these vesicles (a and d). Vesicles of *E. multilocularis* infiltrate into the liver parenchyma. Small vesicles multiply by exogenous budding into the inflammatory fibrous tissue (b and e). Vesicles of *E. shiquicus* are ~8 mm in diameter and occur in the abdominal cavity of *scid* mouse (c and f). Multivesicular development is, simultaneously, observed on the liver surface diaphragm side. They appear not to infiltrate into the liver parenchyma or other organs and differ from those of *E. multilocularis*. (Adapted from Nakaya, K. et al. *Parasitol. Int.* 56, S91, 2006 [31].)

metacestodes of *E. multilocularis* and *E. shiquicus* developed in immunodeficient mice [31]. The speed of the development of secondary metacestodes in immunodeficient mice is faster in *E. multilocularis* than in *E. shiquicus*. In contrast, that of *E. multilocularis* is exclusively multilocular by exogenous buddings and called AE. It is highly pathogenic and often misdiagnosed as hepatocellular carcinoma.

The most unique characteristic of the genus *Echinococcus* is asexual reproduction in metacestode stage. The metacestode has a potentially unlimited sequential generative capacity. *E. granulosus* metacestode produces daughter cysts inside the mother cyst by endogenous budding, whereas *E. multilocularis* metacestode generates a huge number of alveolar small vesicles by exogenous budding. Both *E. vogeli* and *E. shiquicus* appear to show budding patterns between those of *E. granulosus* s.l. and *E. multilocularis*, although this concept may require reevaluation on the basis of in vitro culture of metacestodes.

23.1.1.2.3 Adult Worm

Adult worms of *Echinococcus* and *Taenia* are conceived to be hermaphroditic and expected to self-fertilize. However, recent studies indicate that it is not always true [33–35]. The size of *Echinococcus* species is highly variable. The smallest species is *E. shiquicus* distributing in Tibetan plateau, China [12,25], much smaller than *E. multilocularis*. Except for the size, *E. shiquicus* adult worm appears to be similar to *E. multilocularis* in morphology. However, the metacestode stage is crucially different (Figure 23.2). Based on the genetic distances of *E. felidis* and *E. shiquicus* from all known *Echinococcus* species, there is no doubt that *E. felidis* and *E. shiquicus* are independent species. Most recent molecular studies have shown additional evidences to support their crucial differences from *E. granulosus* s.s. and *E. multilocularis*, respectively [10,16,17,25]. Because various *Echinococcus* species are indistinguishable from each other morphologically, introduction of molecular tools for their differentiation is essential.

23.1.1.3 Epidemiology

Echinococcosis is one of the most lethal helminthic infections worldwide [36–41]. One of the biggest foci on echinococcosis is in China [36,42–44]. Both *E. oligarthrus* and *E. vogeli* have been found from Central and Latin America and are often called NE [15,18], whereas *E. multilocularis* has exclusively been found from the Northern hemisphere. In contrast, *E. granulosus* (*E. granulosus* s.l.), which has been conceived to cause CE worldwide, shows global distribution with high variation in morphology and genetic background, and is now divided into five independent species. The major causative species of CE is now confirmed to be *E. granulosus* s.s. (G1). However, it is still not clear if other species, *E. shiquicus* and *E. felidis* are risky to humans [15]. As *E. felidis* is the sister species of *E. granulosus* s.s., we have to keep in mind that it may infect humans more than others.

E. multilocularis and *E. shiquicus* occur in the Northern hemisphere, whereas *E. vogeli* and *E. oligarthrus* occur in the Southern hemisphere. In contrast, *E. granulosus* s.l. is cosmopolitan. Although both *E. vogeli* and *E. oligarthrus* are expected to emerge in South America, there is no doubt that *E. granulosus* s.s. was introduced to South America from other areas by human's migration with domestic animals, mainly sheep. Most recent work by Nakao et al. [25] revealed that *E. granulosus* s.s. (G1) distributing in both China and Peru keeps the major haplotype (Eg01). *E. granulosus* s.s. (G1), the most risky to humans, may be distributed worldwide from a very small population delivered by human migration. Further evaluation of the similar approaches will be highly informative for considering, where *E. granulosus* s.s. emerged at first. Also, it will give us some information on the origin of other species in *E. granulosus* s.l. Further phylogeographical studies using isolates from Africa or the Middle East—India or Europe will be important to understand the history or evolution of this parasite.

E. multilocularis is divided into three clades and one independent isolate: North American, Asian, European clades, and Inner Mongolian isolate [20]. However, additional samples from Mongolia have revealed that there is another Mongolian clade including the Inner Mongolian isolate (Figures 23.1 and 23.3) [21]. Tang et al. [45] found a unique *E. multilocularis* from corsac fox in Inner Mongolia, China, and later described it as a new independent species, *Echinococcus russicensis* [46]. However, it is an intraspecific variation of *E. multilocularis* and placed in the Mongolian clade [20,21]. It will be interesting to obtain additional specimens of the Mongolian clade for assessing the full extent of its distribution. Mitochondrial DNA (mtDNA) analysis indicated that the intraspecific divergence within *E. multilocularis* occurred not so long time ago compared with interspecific divergence between *E. multilocularis* and its sister species, *E. shiquicus* [20,25]. Phylogeographic analysis of mtDNA suggested a possible recent bottleneck effect in the Tibetan population of *E. multilocularis* while the ancestor of *E. shiquicus* emerged in Tibetan plateau with longer history [25]. However, there is another hypothesis that both *E. multilocularis* and *E. shiquicus* emerged at the similar time, but the former has been selected through the environmental drastic changes in lower altitude, whereas the latter has no such selection in the Tibetan plateau where the parasite adapted to the Tibetan fox instead of the red fox. There is no molecular phylogenetic study on *E. vogeli* or *E. oligarthrus*.

Epidemiological studies based on such molecular reconfirmation will be more important to obtain better information on the causative agent [20,25,33,47–54].

23.1.2 Clinical Features and Pathogenesis

23.1.2.1 Clinical Features

Figure 23.4 illustrates typical AE, CE, and PE [55] which are easily diagnosed by the characteristic imaging figures. However, in many cases with various atypical imaging findings similar to either echinococcosis or other diseases, diagnosis is far from straightforward, especially for less experienced clinicians. Among these three echinococcosis, AE is the most difficult to diagnose by imaging picture only (see Figure 23.5) [56–58].

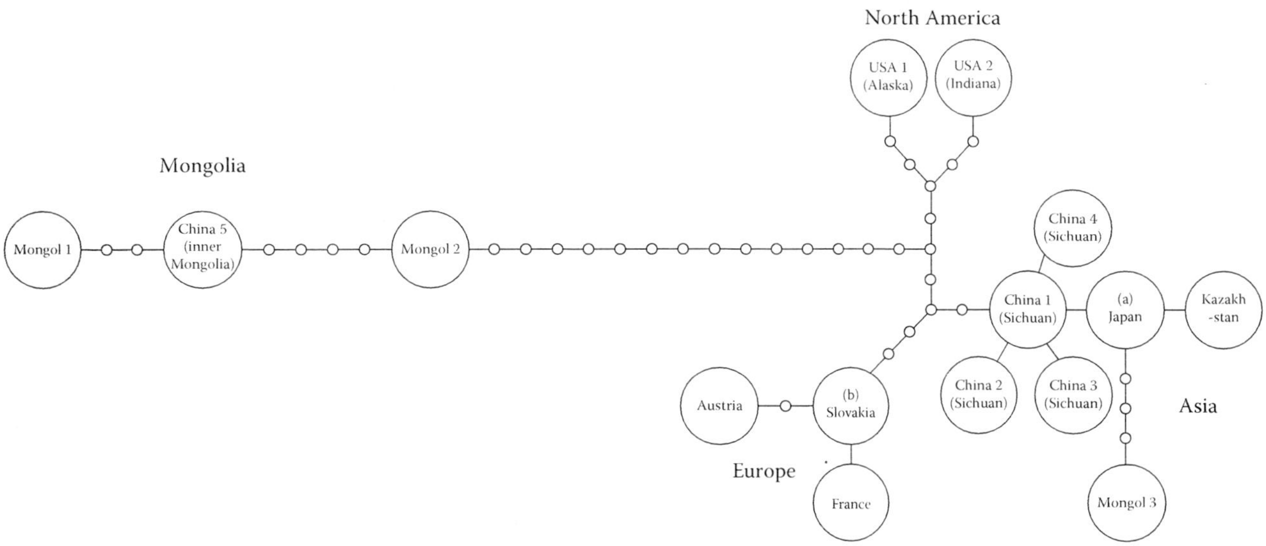

FIGURE 23.3 Haplotype network of *E. multilocularis* based on partial (1543 bp) *cox1* sequences. (Adapted from Ito, A. et al. *Am. J. Trop. Med. Hyg.* 82, 266, 2010 [21].) One isolate from Inner Mongolia was distantly related to all other clades of *E. multilocularis* (*Parasitol. Int.* 58, 384, 2009 [20]). However, of the three AE samples from Mongolia, two of them were closely related to Inner Mongolian haplotype (*Am. J. Trop. Med. Hyg.* 82, 266, 2010 [21]). As this type may distribute widely in Mongolia, it is called a Mongolian clade instead of Inner Mongolian isolate. (a) This haplotype was also obtained from Kazakhstan and Alaska (St. Lawrence Island). (b) This haplotype was also obtained from France, Germany, and Belgium. (Adapted from Nakao, M. et al. *Parasitol. Int.* 58, 384, 2009 [20].)

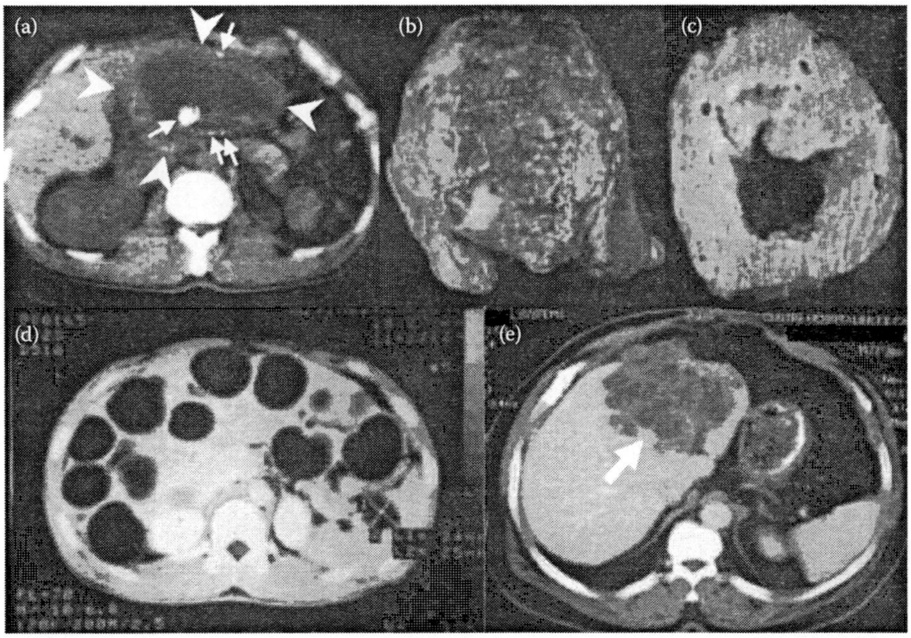

FIGURE 23.4 CT image of hepatic AE (a), macroscopic hepatic AE cyst appearance (b), cut surface of AE cyst(c), CT images of hepatic CE (d), and PE (e). (a) A large low-density area with indistinct margin contains central lower-density area of necrosis (arrowheads) and pathogenomonic calcifications (arrows). (b) Resected AE lesion occupying almost whole right lobe of the liver looks like a malignant tumor and has white or yellowish-white color. (c) Cut surface of the AE lesion shows spongy yellowish appearance with a central necrotic zone. (d) Disseminated CE, multiple, round, homogenous, well-defined low-density areas are individually demonstrated. (e) Multiple closed small round low-density areas form a mass of well-demarcated low-density area. Such CE (d) and PE (e) with typical imaging findings appear to be easily differentiated from each other, AE (a) or other diseases (Adapted from Knapp, J. et al. *Emerg. Infect. Dis.* 15, 2029, 2009 [55]).

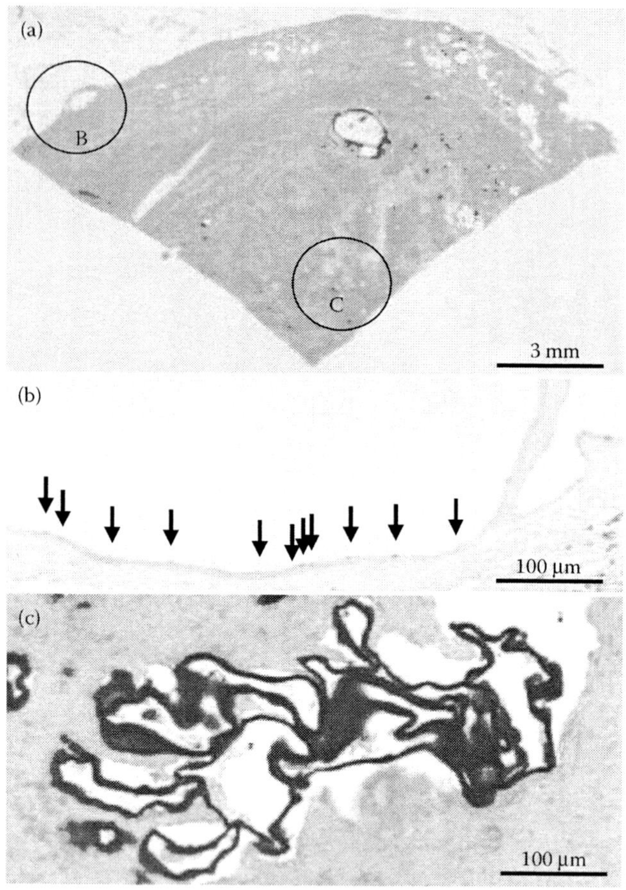

FIGURE 23.5 Histopathology of AE. Active lesions with germinal cells (b) are common at the surface lesion B in (a), whereas necrotic lesion (c) without germinal cells or germinal layer are common in inner or central lesion C in (a). Surface or outer lesions with many germinal layers and germinal cells expand into vesicular lumen. (Adapted from Ito, A. et al. *Am. J. Trop. Med. Hyg.* 82, 266, 2010 [21].)

23.1.2.2 Pathology of *Echinococcus* spp Infection in Humans

Echinococcus species are highly pathogenic to mammalian intermediate hosts. Table 23.1 summarizes the present concept of the species of the genus *Echinococcus*. Among the nine species, *E. multilocularis* is unique, with exogenous budding of the metacestode to develop AE (Figure 23.2b and e) [31]. Although we usually expect that the definitive host is prone to infection with adult worms exclusively, it is not always correct. Recently, a small number of dogs in Europe has been confirmed to harbor metacestode of *E. multilocularis* [59,60]. Such rare cases in dogs may suggest foxes and even cats may also become accidental intermediate host as well as the definitive host.

23.1.3 Diagnosis

23.1.3.1 Serology

Historically, crude antigens have been used in compliment fixation test, indirect hemagglutination test, and enzyme-linked immunosorbent assay (ELISA) for diagnosis of human echinococcosis. However, crude antigens were not suitable for differentiation of specific and nonspecific antigen–antibody responses. For this reason, purified or semipurified specific antigenic components identified by immunoblots and other procedures have been applied for diagnostic application [61–63]. However, there is no difference in the detection of AE using purified or semipurified antigens in both ELISA and immunoblot [64–67]. It was the same for CE [67] and cysticercosis [68,69] when we developed purified antigens for diagnosis. Until 1990, there were many challenges to use crude antigens for serological screenings of CE and AE. ELISA using crude antigens caused nonspecific or false-positive results especially for AE, almost all research groups used them in order to identify and purify diagnostic antigen candidates. Echinococcosis in the intermediate or accidental host is one of the typical chronic diseases. It strongly suggests that the parasite has a very unique strategy for adaptation to mimic or adopt host components in order to escape from the host responses. Antigen B, either from *E. multilocularis* [70] or *E. granulosus* [71–73] has been found to be an immunomodulative factor inducing Th 2 cell responses. The mechanism of the metacestode's development and survival in the mammalian host [74–79] shows crucial difference from the invasive stage, the oncosphere of high immunogenicity [26–28,78].

23.1.3.1.1 Cystic Echinococcosis

In contrast, crude antigens (hydatid cyst fluid) of *E. granulousus* have often been used for screening of CE, since CE cyst is isolated from the host tissue by connected tissue, and one of the major components is Antigen B [61,63,64,80–84]. Therefore, even crude cyst fluid may be useful as diagnostic antigens for screening and even for identification of cases. There are some devices as to the improvement of the quality of cyst fluid from different host animals for diagnosis in humans [85]. Although many candidate antigens are potentially useful for the detection of CE, only few antigens including Antigen B are confirmed to be reliable. Recent studies on Antigen B have revealed that it consisted of 8 kDa subunits [86] and five 8 kDa subunit genes have been cloned [87–91] The expression of these five genes is stage specific and Antigen B8/1 and B8/2 are highly useful for the detection of human CE [91]. As Antigen B is the major component of the cyst fluid, it is expected to become positive in PE and AE when the volume of cyst fluid in PE and AE becomes more than the threshold which may be sufficient to produce antibodies against Antigen B in CE with single cyst.

23.1.3.1.2 Alveolar Echinococcosis

The most reliable serology for the detection of AE are both $Em2^{plus}$-ELISA, and Em18 serology, either ELISA or immunoblot (Gottstein and Ito, WHO guideline in prep) [5]. Most recent blind test to compare the sensitivity of these tools including EM10-, $Em2^{plus}$-, and Em18-ELISA revealed that Em18-ELISA was the only tool that could monitor

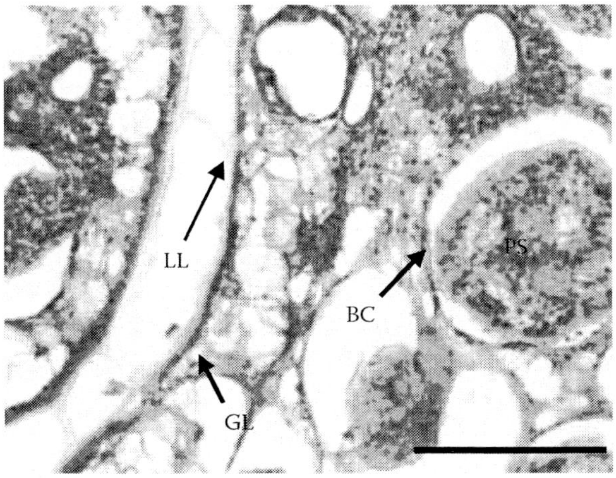

FIGURE 23.6 Immunohistochemical detection of ELP in *E. multilocularis* metacestodes. Parasite tissues were isolated and cryosections were produced. The sections were probed with anti-ELP monoclonal antibody. The following structures are indicated: PS, protoscolex; GL, germinal layer; BC, brood capsule; LL, laminated layer. Scale bar = 100 μm.

the progression of active AE cases [92,93], and IgG4 is the main IgG subclass [94]. As ezrin-like-protein (ELP) is expressed in the metacestode stage other than the laminated layer (Figure 23.6) [5,95], Em18 serology is highly reliable for the detection of active AE with germinal layer. Another additional candidate might be the 14-3-3 protein [74] and in vitro-culture component of 22 kDa [96]. These appear to be specific to echinococcosis, either AE or CE, and either active or inactive. However, as they are components of metacestode, they might not be useful for the detection of the early stage of AE or CE. The data that the component of 22 kDa is expected to be useful by immunoblot but not by ELISA mean that it is not purified yet as discussed above. Em18 serology is essential for differentiation of active and inactive AE and highly useful for monitoring the progression of AE [92–94,97,98].

Further identification of antigenic components in the oncosphere, the invasive stage of the parasite into human tissues, which can share the components of metacestode might be important for earlier detection of echinococcosis, especially in cases with no visible lesion.

23.1.3.1.3 AE, PE, and CE

The antigens such as Em18 and Antigen B are not species specific but genus specific. Nonetheless, these antigens are the most useful antigens for AE and CE, respectively. Antibody responses to Em18, the major component of the germinal layer, may be based on the number of fertile cysts due to exogenous budding, whereas those to Antigen B may be based on the amount of cyst fluid [99]. AE cases with more volume of cyst fluid may show antibody responses to Antigen B as well as those to Em18. Based on such working hypothesis, it is easy for us to expect antibody responses in PE which may show antibody responses between AE and CE (Figure 23.7) [100]. Antibody responses to Em18 in

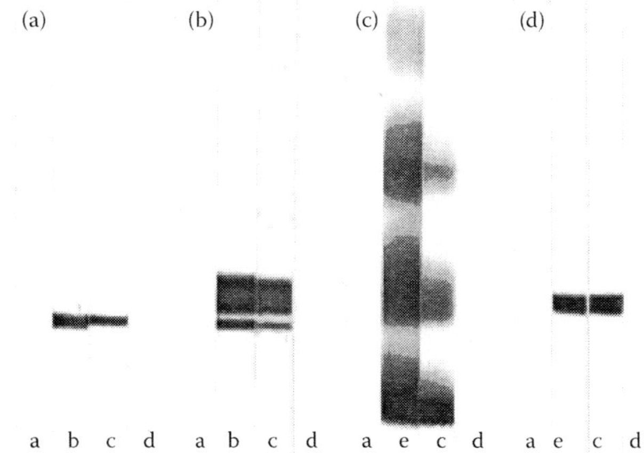

FIGURE 23.7 Immunoblot figures of one PE case using purified native Em18, recombinant Em18, crude antigen B of *E. granulosus* s.s. and recombinant Antigen B8/1. (a) native Em18 affinity purified with monoclonal antibody to ELP (see Figure 23.6), (b) recombinant Em18 for *E. multilocularis*, (c) crude antigen B (from cyst fluid), and (d) recombinant Antigen B8/1 for *E. granulosus*. a and d: negative control, b: positive AE, d: PE case, e: positive CE. (Adapted from Knapp, J. et al. *Emerg. Infect. Dis.* 15, 2029, 2009 [55].)

a recent PE case (see Figure 23.4e) appear to be similar to AE and simultaneously those to Antigen B8/1 appear to be similar to CE. More samples of PE should be analyzed. Figure 23.8 illustrates antibody responses of two AE cases from Mongolia with different genotypes [21] showing strong responses to Em18 but different responses to Antigen B8/1. Metacestode of *E. multilocularis* from Inner Mongolia, now recognized as Mongolian genotype, usually produces more amount of cyst fluid than other *E. multilocularis* from other areas and was expected to be independent species [45,46].

Almost all active AE cases are easily detectable by ELISA or immunoblot against Em18, either native or recombinant. However, the sensitivity and specificity have been greatly improved when recombinant Em18 was applied for serology

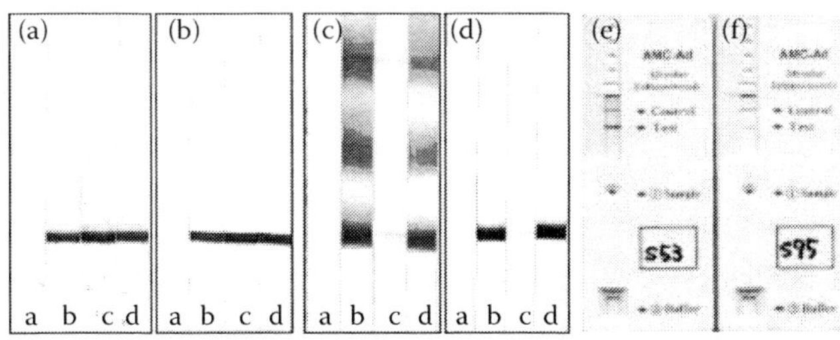

FIGURE 23.8 Immunoblots and IC kits of AE cases with different haplotypes from Mongolia. (a) native Em18 purified by monoclonal antibody to ELP, (b) recombinant Em18, (c) native Antigen B in cyst fluid, (d) recombinant Eg Antigen B8/1. (e and f) rapid IC kits for the detection of AE. a: negative control, b: positive control for AE, c: AE case 4 of Asian clade, d: AE case of Mongolian clade. (Modified from Ito, A. et al. *Am. J. Trop. Med. Hyg.* 82, 266, 2010 [21].)

[65–67,92–94,101]. Although it is a degenerated product of Em10 (=ELP) [102], its amino acid sequence showed the lowest homology with ezrin–radexin–moesin protein [5,66]. Therefore, it is highly expected that it will be much better than any other candidate of the same ELP. This has been confirmed under blind test [92]. Similar results indicating the complete correlation between the new type of imaging, positron emission tomography (PET) and Em18-ELISA, have been found by a French group [93]. Furthermore, a rapid kit using immunochromatography (IC) to detect antibodies to RecEm18 has been developed [103] and quantitative IC is now applicable for monitoring of progression of AE (Sako et al. in prep.) and CE (Sako et al. unpublished).

The current approach for serological diagnosis is to screen and identify patients at once using highly specific and sensitive serology. There is no rational to support the use of ELISA with crude antigens for screening of echinococcosis, and the addition of immunoblot for the detection of specific banding patterns for the identification of patients [104]. As many components of the native antigens are often combined with lipids and carbohydrates, and hydrophobic ligand binding proteins are shared at least in part among all species of platyhelminths [62], it is important to use recombinant proteins as diagnostic markers if such candidate protein markers are available. Native antigens may have higher background of noise shared with other parasites or even some components of the host itself. The immunoblot banding patterns using native antigens to differentiate AE and CE is reasonably useful for screening. It is mainly based on the detection of Em18, Em16, Antigen B (8 kDa) [64,65,84,104–108], but the reliability of the test is much lower than Em2^{plus}-ELISA or Em18-ELISA [101] and the quality checks of the crude antigens and the transblotted membranes are other unstable problems. Therefore, recombinant antigens are much better than the native antigens for preparation of the antigens of high quality and standardize the tools.

Since differences in amino acid sequences of all *Echinococcus* candidate antigens are rather small [95], these antigens are not species specific but genus specific [10,12,20,65,66,91,95]. Therefore, the differential serodiagnosis of AE and CE or PE is based on the crucial difference in pathological features of these three different types of echinococcosis. Serology using Em18 and Antigen B is highly useful for differentiation of the majority of AE and CE in hospitals as differential diagnostic tool and endemic areas for mass screening [64–67, 92–94,101,104]. Based on such serological evidence, we strongly recommend imaging analysis for confirmation, if necessary.

23.1.3.2 Imaging of AE, CE, and PE

Figure 23.4 illustrates typical AE, CE, and PE which are easily diagnosed by such imaging pictures. However, there are many cases showing atypical echinococcosis or other diseases showing some similarity in imaging which inexperienced clinicians might not diagnose correctly.

23.1.3.2.1 Alveolar Echinococcosis

Pathological stages of AE have been launched by WHO and reviewed by Kern et al. [109]. The stage is based on three main factors: (i) location and extension of the primary (original) parasitic lesion within the liver, (ii) involvement of neighboring organs (whether or not the larva has spread to the nearby tissue, including lymph nodes), and (iii) the presence or absence of metastasis (whether or not the larva has spread to distant areas of the body). So, it is called the WHO-PNM system (P = parasite mass in the liver, N = involvement of neighboring organs, M = metastasis) including Stage I, II, IIIa, IIIb, and IV. Ultrasonography, computed tomography (CT), magnetic resonance imaging and PET have been applied for getting morphological evidence of AE lesions [110]. Nonetheless, these pictures are not always specific to AE. AE is often difficult to be differentiated from other cystic liver diseases or liver tumor by imaging diagnosis. An infiltrative AE lesion usually grows larger outward to form a central necrotic cavity. In our previous study, histological difference between different sites of a large hepatic AE lesion was observed under microscopy (Figures 23.4c and 23.5) and biological activity of the surface lesion was confirmed in inoculation experiments using samples from surgically resected specimens [111]. Confusing AE imaging feature may be based on the specific pathologic process. Therefore, to detect tumor markers or AE-specific antibodies including Em18 and Antigen B8/1 is essential for differential diagnosis of these diseases.

23.1.3.2.2 Cystic Echinococcosis

Pathologic stages of CE (CL, CE1, CE2, CE3, CE4, and CE5) have well been reviewed [56]. Among the five stages, CL and CE1 may include other diseases. Serology for the detection of CE cases is also essential as supporting information provided that we have no pathological specimens [57].

23.1.3.3 Combination of Serology and Imaging

Now, rapid IC tests for both AE [103] and CE (Sako et al., unpublished) have been developed and commercially available (see Figure 23.8e and f; Adtec Co. Ltd., Oita, Japan). If such tools for rapid tests become popular, it may become the first choice for screening and identification of patients and imaging analysis will become the second for morphological confirmation near future.

23.1.3.4 Molecular Techniques

In order to identify *Echinococcus* species and genotypes, various molecular approaches have been developed (sequencing of DNA, PCR-RFLP [16,112,113], PCR-SSCP [114], microsatellite analysis) [33,48–52,115]. However, simple diagnostic methods available for all *Echinococcus* spp have not been developed because there are too many species and genotypes to be easily differentiated. Currently, molecular identification of *Echinococcus* spp is mostly based on sequencing of mitochondrial and nuclear DNA genes after PCR amplification. Obtained sequences can be readily compared with previously deposited sequences in public databases (GenBank, EMBL, and DDBJ). As complete mitochondrial genome sequences have been published for most of the species and genotypes [8,116,117], mtDNA genes such as cytochrome *c* oxidase subunit 1 (*cox1*), cytochrome *b* (*cob*), and NADH dehydrogenase subunit 1 (*nad1*) can be good targets for molecular identification. Although several nuclear genes are also available for identifying *Echinococcus* species and genotypes as well as for molecular phylogeny [17], mtDNA genes have advantages in the detection of the parasites at most of the situations. MtDNA is a multicopy genome and thus more detectable than nuclear DNA markers, especially from the samples containing a small amount of parasite DNA and the histopathological specimens fixed with formalin. Several reports demonstrated that small fragments of mtDNA could be amplified even from paraffin-embedded histopathological specimens [21,118,119]. Moreover, sequence data of the nuclear markers have not yet been accumulated from a wide range of host species and locality on each parasite species. Therefore, mtDNA genes seem to be the best targets for the diagnosis in clinical cases.

23.2 METHODS

The currently available molecular primers and probes are summarized in Table 23.2. Although many more primers have been designed and published, only some of them are described in the table. They are chosen because they seem to be applicable for most *Echinococcus* species. As PCR-RFLP is relatively more rapid and cheaper compared with DNA sequencing, it has been tentatively used at a local scale, where the number of target species was limited [16,113,122]. On the other hand, sequencing of DNA is applicable for all species and genotypes, when the universal primers are used. Thus, PCR amplification and subsequent sequencing of nuclear and/or mtDNA genes are currently applied in the differentiation of *Echinococcus* specimens obtained from both human patients and animal hosts. For the identification of species, a partial sequence of *cox1* gene is basically enough, although phylogenetic or phylogeographic analyses including closely related genotypes might need longer sequences from several genes. Here, we summarize the protocol of molecular identification of *Echinococcus* species and genotypes using universal mtDNA primers (Tables 23.2 and 23.3).

TABLE 23.2
Molecular Probes for Identification of *Echinococcus* Species

	Target DNA	Primers	Differential Diagnosis	Reference
PCR-SSCP	*nad1* (525 bp) *cox1* (460 bp)	JB11: AGATTCGTAAGGGGCCTAATA JB12: ACCACTAACTAATTCACTTTC JB3: TTTTTTGGGCATCCTGAGGTTTAT JB4.5: TAAAGAAAGAACATAATGAAAATG	*E. granulosus* (G1/G2), *E. canadensis* (G6/G7)	[114]
PCR-RFLP	ITS2 (600–620 bp)	NC6 > : ATCGACATCTTGAACGCACATTGC < NC2: TTAGTTTCTTTTCCTCCGCT	*E. granulosus*, *E. equinus* (G4), *E. canadensis* (G6), *E. multilocularis*	[113]
	rrn L (570 bp)	Ech-LSU/F: GGTTTATTTGCCTTTTGCATCATGC Ech-LSU/R: ATCACGTCAAACCATTCAAACAAGC	*E. granulosus* (G1), *E. multilocularis*, *E. shiquicus*	[16]
	18S-ITS1-5.8S (1 kb)	BD1: GTCGTAACAAGGTTTCCGTA 4S: TCTAGATGCGTTCGAA(G/A)TGTCGATG	*E. granulosus* (G1), *E. equinus* (G4), *E. ortleppi* (G5), *E. canadensis* (G6/G7), *E. multilocularis*, *E. oligarthrus*	[112]
PCR/sequencing	*cox1* (466 bp)	CO1F: TTTTTTGGGCATCCTGAGGTTTAT CO1R: TAAAGAAAGAACATAATGAAAATG	*Taenia*, *Echinococcus*	[120]
	cox1 (880 bp)	F/COI: TTGAATTTGCCACGTTTGAATGC R/COI: GAACCTAACGACATAACATAATGA	*Echinococcus*, *Taenia*	[121]
	nad1 (471 bp)	JB11: AGATTCGTAAGGGGCCTAATA JB12: ACCACTAACTAATTCACTTTC	*Taenia*, *Echinococcus*	[122]
	cob complete (>1068 bp)	Forward: GTGGTATTGACATTTTGTAGATTA Reverse: AATTTCCAGAGTAGAAATCACCAT	*Echinococcus*	[10]
	cox1 complete (>1608 bp)	Forward: TTACTGCTAATAATTTTGTGTCAT Reverse: GCATGATGCAAAAGGCAAATAAAC	*Echinococcus*	[10]
	ef1a (656 bp)	Forward: AATGGATGCAGTTGATTACAGTGA Reverse: GCAGGGTTAGTCTCCTTGACCTGA	*E. granulosus* (G1), *E. canadensis* (G6/G7) *E. multilocularis*, *E. shiquicus*	[123]

TABLE 23.3
Primers Used for the Amplification of *cox1* Gene Fragments of *E. multilocularis*

Primers	Sequence (5′–3′)	Positions
Em cox1 F1	GACTTTCTCTTTGGTTGGTGTAAG	tRNA-ser
Em cox1 F11	TGTGATTAGGTAGTTGGGTATTGA	11–34
Em cox1 F100	TTAAGTTTTAGTTTGTTGATTCGTG	100–124
Em cox1 F9	TGACTAATCATGGTATAATAATGATC	182–207
Em cox1 F400	TCTTCTTCATATTTTTCTAGGAGTAG	400–425
Em cox1 F500	GTACTTTGTATAGTGTTTTTATGACT	500–525
Em cox1 F4	GTTTTGGCTGCTGCTATTACTATGC	601–625
Em cox1 F2	GTTTATGTTTTGATTCTGCCTGGATT	727–752
Em cox1 F5	TGGGTTGGATGTGAAGACGGCGGT	885–908
Em cox1 F1000	AAGAGTGATCCTATTTTGTGGTGGGT	1000–1025
Em cox1 F1100	TTTTACACGATACTTGATTTGTGGTG	1100–1125
Em cox1 F6	ATTACTGGTTTGAGGTTGAATAAGT	1201–1225
Em cox1 F10	CCTATGCATTATTTTGGTTTATGTG	1282–1306
Em cox1 F7	TCTTTTATATCTGCGTTTAGTGGGTG	1375–1400
Em cox1 R1	AACCTAAACAACCAACTTCACAG	*rrnL*
Em cox1 R11	CTAACCAACAGCAAATACATAATTAC	1608–1583
Em cox1 R10	TACAGGACTCATTAAATAATCCACTA	1500–1475
Em cox1 R1400	CACCCACTAAACGCAGATATAAAAG	1400–1376
Em cox1 R9	CACACACGACGAGGCAACCCACATA	1325–1301
Em cox1 R1200	CAACGGTCACCATCAAATAAACATAA	1200–1175
Em cox1 R4	ATGAGCCACCACAAATCAAGTATCG	1131–1107
Em cox1 R8	TATTTACACTAGAATTAAGCAACAT	1000–976
Em cox1 R7	AAACAACAAACCATAAAACCCAAAC	825–801
Em cox1 R2	CCAAAAAACCAAAACATATGCTGAA	716–692
Em cox1 R600	AGGCAACGTCACTAACAATAAAATA	600–576
Em cox1 R400	ACAATGGAGGATAAAAAGTCCAACCA	400–375
Em cox1 R12	GGCAAATTCAAATCAGACAAACCACC	293–268
Em cox1 R200	ATTATACCATGATTAGTCACCAAAAA	200–175

Source: Ito, A. et al. *Am. J. Trop. Med. Hyg.* 82, 266, 2010 [21].
Note: tRNA-ser, tRNA Gene for Serine; rrnL, Large Subunit Ribosomal RNA.

23.2.1 Sample Preparation

Echinococcus specimens are collected from the infected tissues of patients and animals (larval stages), or from intestinal contents and stools of canid/felid definitive hosts (adult worms). Samples should be preserved in ethanol (>70%) or frozen for molecular examination. Formalin-fixed materials are also useful for molecular work. However, the amplification of long fragment DNA (>500 bp) is not always possible because of the degradation of DNA.

For DNA extraction, many kinds of commercial kits are available. To increase the amount and quality of yielded DNA, appropriate extraction procedures should be chosen, especially for stool samples and formalin-fixed specimens. Specialized DNA extraction kits are recommended for such problematic samples. Extracted DNA should be kept frozen when it is dissolved in water, or buffers (TE buffer, etc.) that can protect DNA from degradation.

23.2.2 Detection Procedures

Cox1 has been intensively analyzed since it was firstly demonstrated as a useful marker for identifying *Echinococcus* species and genotypes [120]. To amplify a partial *cox1* gene (~880 bp), universal primers F/COI (TTG AAT TTG CCA CGT TTG AAT GC) and R/COI (GAA CCT AAC GAC ATA ACA TAA TGA) were designed [121]. The primer set is applicable to all examined species (*E. granulosus*, *E. equinus*, *E. ortleppi*, *E. canadensis* (G6-8), *E. multilocularis*, *E. shiquicus*, *E. vogeli*, *E. oligarthrus*) (Figure 23.9). The amplified region is variable enough to discriminate all species and contains the target of the previously published primer set which has been used in several studies [120,124–127]. Table 23.3 shows the primers for formalin-fixed paraffin-embedded histopathological specimens of *E. multilocularis* (modified from Ito et al. [21]). Combination of the neighboring primers can yield short (100 ~ 300 bp) fragments of DNA.

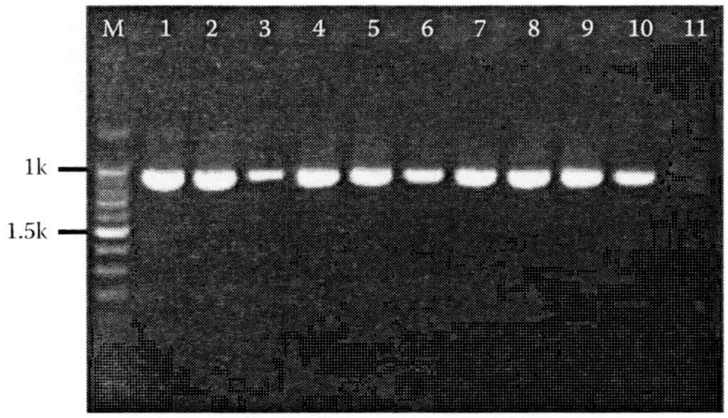

FIGURE 23.9 Agarose gel electrophoresis of PCR amplification using *cox1* universal primers F/COI and R/COI. Lane M, 100 bp DNA ladder; lane 1, *E. granulosus* (G1); lane 2, *E. equinus*; lane 3, *E. ortleppi*; lane 4, *E. canadensis* (G6); lane 5, *E. canadensis* (G7); lane 6, *E. felidis*; lane 7, *E. multilocularis*; lane 8, *E. shiquicus*; lane 9, *E. oligarthrus*; lane 10, *E. vogeli*; lane 11, negative control.

23.2.2.1 Procedure

1. Prepare a master mix as follows. This procedure uses *Ex Taq* (TaKaRa) for PCR amplification.

Reagent	Volume	Final Concentration
10 × PCR buffer	5.0 µL	1×
dNTP (2.5 mM each)	4.0 µL	200 µM each
Primers (10 µM)	2.5 µL	0.5 µM
Taq polymerase (5 U/µL)	0.2 µL	1 U
Template DNA	<500 ng	
Distilled water	Variable	
Total volume	50 µL	

   Reaction volume can be decreased to 10 µL, although the amount of resultant PCR product will also be decreased.
2. Amplify the DNA fragment of the target gene. PCR protocol consists of 30 cycles of denaturation (94°C for 30 s), annealing (52°C for 45 s), and extension (72°C for 60 s).
3. Electrophorese the PCR products in a 1.5% agarose gel (containing 0.5 µg/mL ethidium bromide) in TAE buffer. Then, visualize the DNA band (~880 bp) using a UV transilluminator.
4. Purify the PCR products using any commercial kit or method suitable for subsequent direct sequencing.
5. Perform cycle sequencing. The example of reaction mixture using BigDye Terminator v3.1 (or v1.1) Cycle Sequencing Kit (Applied Biosystems) is shown below.

Reagent	Quantity
Terminator ready reaction mix	8.0 µL
Primer	3.2 pmol
Template	5–20 ng
Distilled water	Variable
Total volume	20 µL

   Reaction volume can be decreased to 10 µL. Cycle sequencing protocol consists of predenaturation (94°C for 5 min), 25 cycles of denaturation (96°C for 10 s), annealing (50°C for 5 s), and extension (60°C for 4 min).
6. Purify the cycle sequencing products as recommended by the manufacturer's instruction. Entire removal of unincorporated dye terminators is important to obtain clear signals.
7. Electrophorese the purified sequencing products using DNA sequencers such as ABI PRISM 3500 Genetic Analyzer.
8. Conduct sequence homology search in public DNA databases using BLAST.

As mentioned in 23.1.1.3, molecular tools are essential for identification of the species, genotypes, and haplotypes of *Echinococcus* spp. A combination of mitochondrial and nuclear DNA analysis will give us better information on the taxonomy [15]. Furthermore, this approach on *E. granulosus* s.l., especially on *E. equinus* (G4), *E. ortleppi* (G5), *E. canadensis* (G6-G10) might give us the evidence of hybrids as shown in *Taenia saginata* and *Taenia asiatica* [35,128,129] or the history of the establishment of this monophyletic group (G4–G10) (Figure 23.1). As CE has been spreading worldwide through migration of humans and domestic animals, it is time to reevaluate the species which pose risk to human infections and their geographic distribution using molecular tools. Although haplotype network studies of CE so far reported suggest G01 as the major type from China and Peru, there might be other haplotypes from other areas including Russia, Africa, Europe, and India to Middle East or even from Latin America other than Peru.

In contrast, AE is rather simple compared with CE. As people are traveling world scale and may get accidental infection with eggs of *E. multilocularis*, we may trace back the patients where they were exposed to eggs, since *E. multilocularis* has been divided into four clades and based on the molecular information, we can estimate where the patients were exposed to eggs [21,130,131].

23.3 PERSPECTIVES

There is a need for obtaining new evidence and information as to how the parasite has adapted to humans to cause a chronic disease [76–79]. The genetic background of exogenous and endogenous budding, host switching from canid to felid or *vice versa* may become clear in the future. Also, new challenges for the development of new drugs for better chemotherapy and better understanding of the molecular basis of vaccine candidates and additional new serological tools may be expected based on such pathologic and pathogenic mechanisms including the host–parasite relationship with more information of molecular approaches [75,76].

In northern hemisphere, we have to get additional data, especially from Russia. More information on AE and CE from Russia and the neighboring countries which consisted of the former Soviet Union will be essential for us to understand the origin of these parasites. Also, in the southern hemisphere, we have to get those on CE and PE and UE. *Echinococcus* spp from wild animals will be another option for real understanding of the genera *Echinococcus* and *Taenia* [8,15,17,121,132,133].

Based on molecular tools already developed and under development, we can obtain better resolution on reevaluation of the morphological taxonomy of this genus *Echinococcus* and more understanding of with biology, developmental biology, pathobiology, and coevolution of this highly pathogenic genus *Echinococcus* with additional molecular data.

Therefore, our living environment contaminated with taeniid eggs is highly risky and we have to establish the monitoring system of contamination of the environment and/or control of definitive host animals infected with adult worms. Simultaneously, we have to establish highly reliable diagnostic tools for echinococcosis for early curative treatment.

ACKNOWLEDGMENTS

The authors' group work has been supported financially by the Special Coordination Funds for Promoting Science and Technology (2010–2012), Translational Research Fund (2007–2011) from the Ministry of Education, Japan (MEXT), Grants-in-Aid for Scientific Research (21256003, 22590376), Asia-Africa Scientific Platform Fund (2006–2011), Japan-China Medical Cooperation Program (2009–2010), Japan-France Research Cooperation Program (2010–2011) from the Japan Society for the Promotion of Science.

REFERENCES

1. Nelson, G.S., Pester, F.R., and Rickman, R. The significance of wild animals in the transmission of cestodes of medical importance in Kenya. *Trans. R. Soc. Trop. Med. Hyg.* 59, 507, 1965.
2. Rausch, R.L., and D'Alessandro, A. The epidemiology of echinococcosis caused by *Echinococcus oligarthrus* and *E. vogeli* in the neotropics. In: Craig, P.S., and Pawlowski, Z. (eds), *Cestode Zoonoses: Echinococcosis an Emergent and Global Problem.* IOS Press, Amsterdam, The Netherlands, pp. 107–113, 2002.
3. Eckert, J., and Deplazes, P. Biological, epidemiological, and clinical aspects of echinococcosis, a zoonosis of increasing concern. *Clin. Microbiol. Rev.* 17, 107, 2004.
4. Eckert, J. et al. *WHO/OIE Manual on Echinococcosis in Humans and Animals: A Public Health Problem of Global Concern.* World Organisation for Animal Health, Paris, France, 265pp., 2001.
5. Ito, A., Nakao, M., and Sako, Y. Echinococcosis: Serological detection of patients and molecular identification of patients and molecular identification of parasites. *Future Microbiol.* 2, 439, 2007.
6. Macpherson, C.N.L. An active intermediate host role for man in the life cycle of *Echinococcus granulosus* in Turkana, Kenya. *Am. J. Trop. Med. Hyg.* 32, 397, 1983.
7. Thompson, R.C.A., and McManus, D.P. Aetiology: Parasites and life-cycles. In: *WHO/OIE Manual on Echinococcosis in Humans and Animals: A Public Health Problem of Global Concern.* World Organisation for Animal Health, Paris, France, p. 1, 2001.
8. Nakao, M. et al. A molecular phylogeny of the genus *Echinococcus* inferred from complete mitochondrial genomes. *Parasitology* 134, 713, 2007.
9. Thompson, R.C.A., and McManus, D.P. Towards a taxonomic revision of the genus *Echinococcus*. *Trends Parasitol.* 18, 452, 2002.
10. Hüttner, M. et al. Genetic characterization and phylogenetic position of *Echinococcus felidis* Ortlepp, 1937 (Cestoda: Taeniidae) from the African lion. *Int. J. Parasitol.* 38, 861, 2008.
11. Vogel, H. Üeber den *Echinococcus multilocularis* Süddeutschlands. I. Das Bandwurmstadium von Stämmen menschlicher und tierischer Herkunft. *Z. Tropenmed. Parasitol.* 8, 404, 1957.
12. Xiao, N. et al. *Echinococcus shiquicus* n. sp., a taeniid cestode from Tibetan fox and plateau pika in China. *Int. J. Parasitol.* 35, 693, 2005.
13. Lühe, M. Cystotänien südamerikanischer Feliden. *Zoologische Jahrbücher, Abteitlung für allgemeine Zool. Physiol. Tiere* 12 Suppl, 787, 1910.
14. Rausch, R.L., and Bernstein, J.J. *Echinococcus vogeli* sp. n. (Cestoda: Taeniidae) from the bush dog, *Speothos venaticus* (Lund). *Z. Tropenmed. Parasitol.* 23, 25, 1972.
15. Nakao, M. et al. State-of-the-art *Echinococcus* and *Taenia*: Phylogenetic taxonomy of human-pathogenic tapeworms and its application to molecular diagnosis. *Infect. Genet. Evol.* 10, 444, 2010.
16. Xiao, N. et al. Dual infection of animal hosts with different *Echinococcus* species in the eastern Qinghai-Tibet plateau region of China. *Am. J. Trop. Med. Hyg.* 75, 292, 2006.
17. Saarma, U. et al. A novel phylogeny for the genus *Echinococcus*, based on nuclear data, challenges relationships based on mitochondrial evidence. *Parasitology* 136, 317, 2009.
18. D'Allessandro, A., and Rausch, R.L. New aspects of neotropical polycystic (*Echinococcus vogeli*) and unicystic (*Echinococcus oligarthrus*) echinococcosis. *Clin. Microbiol. Rev.* 21, 380, 2008.
19. Hüttner, M. et al. A survey of *Echinococcus* species in wild carnivores and livestock in East Africa. *Int. J. Parasitol.* 39, 1269, 2009.
20. Nakao, M. et al. Geographic pattern of genetic variation in the fox tapeworm *Echinococcus multilocularis*. *Parasitol. Int.* 58, 384, 2009.
21. Ito, A. et al. Histopathological, serological and molecular confirmation of alveolar echinococcosis cases in Mongolia. *Am. J. Trop. Med. Hyg.* 82, 266, 2010.
22. Rausch, R.L. Life cycle pattern and geographic distribution of *Echinococcus* species. In: Thompson, R.C.A. and Lymbery, A.J. (eds), *Echinococcus and Hydatid Disease.* CAB International, Oxon, UK, p. 89, 1995.
23. Romig, T., Dinkel, A., and Mackenstedt, U. The present situation of echinococcosis in Europe. *Parasitol. Int.* 55, S187, 2006.
24. Verster, A. Review of *Echinococcus* species in South Africa. *Onderstepoort J. Vet. Res.* 32, 7, 1965.
25. Nakao, M. et al. Genetic polymorphisms of *Echinococcus* tapeworms in China as determined by mitochondrial and nuclear DNA sequences. *Int. J. Parasitol.* 40, 379, 2010.
26. Johnson, K.S. et al. Vaccination against ovine cysticercosis using a defined recombinant antigen. *Nature* 338, 585, 1989.
27. Lightowlers, M.W. et al. Molecular and genetic characterization of the host-protective oncosphere antigens of taeniid cestode parasites. *Int. J. Parasitol.* 33, 1207, 2003.

28. Lightowlers, M.W. Cestode vaccines: Origins, current status and future prospects. *Parasitology* 133, S27, 2006.
29. Jabbar, A. et al. The ultrastructure of taeniid cestode oncospheres and localization of host-protective antigens. *Parasitology* 137, 521, 2010.
30. Thompson, R.C.A. Biology and systematic of *Echinococcus*. In: Thompson, R.C.A., and Lymbery, A.J. (eds), *Echinococcus and Hydatid Disease*, Chapter 1. CAB International, Oxon, p. 1, 1995.
31. Nakaya, K. et al. Usefulness of severe combined immunodeficiency (scid) and inbred mice for studies of cysticercosis and echinococcosis. *Parasitol. Int.* 56, S91, 2006.
32. Rausch, R.L., D'Alessandro, D., and Rausch, V.R. Characteristics of the larval *Echinococcus vogeli* Rausch and Bernstein, 1972 in the natural intermediate host, the paca, *Cuniculus paca* L. (Rodentia: Dasyproctidae). *Am. J. Trop. Med. Hyg.* 30, 1043, 1981.
33. Nakao, M., Sako, Y., and Ito, A. Isolation of polymorphic microsatellite loci from the tapeworm *Echinococcus multilocularis*. *Infect. Genet. Evol.* 3, 159, 2003.
34. Badaraco, J.L. et al. Using mitochondrial and nuclear markers to evaluate the degree of genetic cohesion among *Echinococcus* populations. *Exp. Parasitol.*, 119, 453, 2008.
35. Okamoto, M. et al. Evidence of hybridization between *Taenia saginata* and *Taenia asiatica*. *Parasitol. Int.* 59, 70, 2010.
36. Craig, P.S. et al. A large focus of alveolar echinococcosis in central China. *Lancet* 340, 826, 1992.
37. Craig, P.S. et al. Human echinococcosis: A neglected disease? *Trop. Med. Health* 35, 283, 2007.
38. Schantz, P.M. et al. Epidemiology and control of hydatid disease. In: Thompson, R.C.A., and Lymbery, A.J. (eds), *Echinococcus and Hydatid Disease*. CAB International, Oxon, 233pp., 1995.
39. Gemmell, M.A. et al. Epidemiology. In: *WHO/OIE Manual on Echinococcosis in Humans and Animals:A Public Health Problem of Global Concern*. World Organisation for Animal Health, Paris, France, 143pp., 2001.
40. Craig, P.S., and McManus, D.P. Echinococcosis: Transmission biology and epidemiology. *Parasitology* 127, S1, 2003.
41. Gurbadam, A. et al. Mongolian and Japanese joint conference on "Echinococcosis: Diagnosis, treatment and prevention in Mongolia" June 4, 2009. *Parasit. Vectors* 3, 8, 2010.
42. Li, T. et al. Echinococcosis in Tibetan populations, western Sichuan province, China. *Emerg. Infect. Dis.* 11, 1866, 2005.
43. Schantz, P.M. et al. Echinococcosis on the Tibetan plateau: Prevalence and risk factors for cystic and alveolar echinococcosis in Tibetan populations in Qinghai province, China. *Parasitology* 127, S109, 2003.
44. Ito, A. et al. Control of echinococcosis and cysticercosis: A public health challenge to international cooperation in China. *Acta Trop.*, 86, 3, 2003.
45. Tang, C. et al. Alveolar *Echinococcus* species from *Vulpes corsac* in Hulunbeier, Inner Mongolia, China, and differential development of metacestodes in experimental rodents. *J. Parasitol.* 92, 719, 2006.
46. Tang, C. et al. Studies on the alveolar *Echinococcus* species in northward Daxingan mountains, Inner Mongolia, China. III. *Echinococcus russicensis* sp. nov. *Chin. J. Zoonoses* 23, 957, 2007.
47. Bretagne, S. et al. *Echinococcus multilocularis*: Microsatellite polymorphism in U1 snRNA genes. *Exp. Parasitol.* 82, 324, 1996.
48. Bart, J.M. et al. Genotyping of human cystic echinococcosis in Xinjiang, PR China. *Parasitology* 133, 571, 2006.
49. Bart, J.M. et al. EmsB, a tandem repeated multi-loci microsatellite, new tool to investigate the genetic diversity of *Echinococcus multilocularis*. *Infect. Genet. Evol.* 6, 390, 2006.
50. Knapp, J. et al. Assessment of use of microsatellite polymorphism analysis for improving spatial distribution tracking of *Echinococcus multilocularis*. *J. Clin. Microbiol.* 45, 2943, 2007.
51. Knapp, J. et al. Genetic diversity of *Echinococcus multilocularis* on a local scale. *Infect. Genet. Evol.* 8, 367, 2008.
52. Knapp, J. et al. Genetic diversity of the cestode *Echinococcus multilocularis* in red foxes at a continental scale in Europe. *PLoS Negl. Trop. Dis.* 3, e452, 2009.
53. Knapp, J. et al. The genomic *Echinococcus* microsatellite EmsB sequences: From a molecular marker to the epidemiological tool. *Parasitology* 137, 439, 2010.
54. Casulli, A. et al. Multi-locus microsatellite analysis supports the hypothesis of an autochthonous focus of *Echinococcus multilocularis* in northern Italy. *Int. J. Parasitol.* 39, 837, 2009.
55. Knapp, J. et al. *Echinococcus vogeli* infection in a hunter, French Guiana. *Emerg. Infect. Dis.* 15, 2029, 2009.
56. Brunetti, E. et al. Expert consensus for the diagnosis and treatment of cystic and alveolar echinococcosis in humans. *Acta Trop.* 114, 1, 2010.
57. Li, T. et al. Specific IgG responses to recombinant antigen B and Em18 in cystic and alveolar echinococcosis in China. *Clin. Vaccine Immunol.* 17, 470, 2010.
58. Li, T. et al. Widespread co-endemicity of human cystic and alveolar echinococcosis on the eastern Tibetan plateau, northwest Sichuan/southeast Qinghai, China. *Acta Trop.* 113, 248, 2010.
59. Scharf, G. et al. Radiographic, ultrasonographic, and computed tomographic appearance of alveolar echinococcosis in dogs. *Vet. Radiol. Ultrasound* 45, 411, 2004.
60. Staebler, S. et al. Serological diagnosis of canine alveolar echinococcosis. *Vet. Parasitol.* 141, 243, 2006.
61. Lightowlers, M.W., and Gottstein, B. Echinococcosis/hydatidosis: Antigens, immunological and molecular diagnosis. In: Thompson, R.C.A., and Lymbery, A.J. (eds), *Echinococcus and Hydatid Disease*. CAB International, Oxon, UK, 355pp., 1995.
62. Ito, A. Serologic and molecular diagnosis of zoonotic larval cestode infections. *Parasitol. Int.* 51, 221, 2002.
63. Ito, A., and Craig, P.S. Immunodiagnostic and molecular approaches for the detection of taeniid cestode infections. *Trends Parasitol.* 19, 377, 2003.
64. Ito, A. et al. Differential serodiagnosis for cystic and alveolar echinococcosis using fractions of *Echinococcus granulosus* cyst fluid (antigen B) and *E. multilocularis* protosoclex (Em18). *Am. J. Trop. Med. Hyg.* 60, 188, 1999.
65. Ito, A. et al. Evaluation of an enzyme-linked immunosorbent assay (ELISA) with affinity-purified Em18 and an ELISA with recombinant Em18 for differential diagnosis of alveolar echinococcosis: Results of a blind test. *J. Clin. Microbiol.* 40, 4161, 2002.
66. Sako, Y. et al. Alveolar echinococcosis: Characterization of diagnostic antigen Em18 and serological evaluation of recombinant Em18. *J. Clin. Microbiol.* 40, 2760, 2002.
67. Xiao, N. et al. Evaluation of use of recombinant Em18 and affinity-purified Em18 for serological differentiation of alveolar echinococcosis from cystic echinococcosis and other parasitic infections. *J. Clin. Microbiol.* 41, 3351, 2003.

68. Ito, A. et al. Novel antigens for neurocysticercosis: Simple method for preparation and evaluation for serodiagnosis. *Am. J. Trop. Med. Hyg.* 59, 291, 1998.
69. Sako, Y. et al. Molecular characterization ad diagnostic value of *Taenia solium* low-molecular-weight antigen genes. *J. Clin. Microbiol.* 38, 4439, 2000.
70. Graichen, D.A.S. et al. Expression and diversity of *Echinococcus multilocularis AgB* genes in secondary infected mice: Evaluating the influence of T-cell immune selection on antigenic variation. *Gene.* 392, 98, 2007.
71. Riganò, R. et al. *Echinococcus granulosus* antigen B impairs human dendritic cell differentiation and polarizes immature dendritic cell maturation towards a Th2 cell response. *Infect. Immun.* 75, 1667, 2007.
72. Siracusano, A. et al. Immunomodulatory mechanisms during *Echinococcus granulosus* infection. *Exp. Parasitol.* 119, 483, 2008.
73. Siracusano, A. et al. Molecular cross-talk in host-parasite relationships: The intriguing immunomodulatory role of *Echinococcus* antigen B in cystic echinococcosis. *Int. J. Parasitol.* 38, 1371, 2008.
74. Siles-Lucas, M., and Gottstein, B. The 14-3-3 protein: A key molecule also in parasites. *Trends Parasitol.* 19, 575, 2003.
75. Brehm, K. et al. The molecular mechanisms of larval cestode development: First steps into an unknown world. *Parasitol. Int.* 55, S15, 2006.
76. Brehm, K. *Echinococcus multilocularis* as an experimental model in stem cell research and molecular host-parasite interaction. *Parasitology* 237, 537, 2010.
77. Gottstein, B. et al. Molecular survival strategies of *Echinococcus multilocularis* in the murine host. *Parasitol. Int.* 55, S45, 2006.
78. Haag, K.L. et al. The EG95 antigen of *Echinococcus* spp. contains positively selected amino acids, which may influence host specificity and vaccine efficacy. *PLoS One* 4, e5362, 2009.
79. Mejri, N, Hemphill, A., and Gottstein, B. Triggering and modulation of the host-parasite interplay by *Echinococcus multilocularis*: A review. *Parasitology* 137, 557, 2010.
80. Oriol, R. et al. Purification of lipoprotein antigens of *Echinococcus granulosus* from sheep hydatid fluid. *Am. J. Trop. Med. Hyg.* 20, 569, 1971.
81. Lightowlers, M.W. et al. Subunit composition and specificity of major cyst fluid antigens of *Echinococcus granulosus. Mol. Biochem. Parasitol.* 37, 171, 1989.
82. Siracusano, A. et al. Detection of antibodies against *Echinococcus granulosus* major antigens and their subunits by immunoblotting. *Trans. R. Soc. Trop. Med. Hyg.* 85, 239, 1991.
83. Ioppolo, S. et al. Immunological responses to antigen B from *Echinococcus granulosus* cyst fluid in hydatid patients. *Parasitol. Immunol.* 18, 571, 1996.
84. Ito, A. et al. Evaluation of Em18-, Em16-, antigen B-western blots, Em2^{plus}-ELISA and four other tests for differential serodiagnosis of alveolar and cystic echinococcosis patients in Poland. *Parasitol. Int.* 47, 95, 1998.
85. Mamuti, W. et al. Usefulness of hydatid cyst fluid of *Echinococcus granulosus* developed in mice with secondary infection for serodiagnosis of cystic echinococcosis in humans. *Clin. Diagn. Lab. Immunol.* 9, 573, 2002.
86. Leggatt, G.R., Yang, W., and McManus, D.P. Serological evaluation of the 12 kDa subunit of antigen B in *Echinococcus granulosus* cyst fluid by immunoblot analysis. *Trans. R. Soc. Trop. Med. Hyg.* 86, 1, 1992.
87. Frosch, P. et al. Sequence heterogeneity of the echinococcal antigen B. *Mol. Biochem. Parasitol.* 64, 171, 1994.
88. Fernández, V. et al. Molecular characterization of a novel 8 kDa subunit of *Echinococcus granulosus* antigen B. *Mol. Biochem. Parasitol.* 77, 247, 1996.
89. Chamale, G. et al. *Echinococcus granulosus* antigen B is encoded by a gene family. *Mol. Biochem. Parasitol.* 116, 233, 2001.
90. Arend, A.C. et al. The *Echinococcus granulosus* antigen B shows a high degree of genetic variability. *Exp. Parasitol.* 108, 76, 2004.
91. Mamuti, W. et al. *Echinococcus multilocularis*: Developmental stage-specific expression of antigen B 8-kDa-subunits. *Exp. Parasitol.* 113, 75, 2006.
92. Tappe, D. et al. Close relationship between clinical regression and specific serology in the follow-up of patients with alveolar echinococcosis in different clinical stages. *Am. J. Trop. Med. Hyg.* 80, 792, 2009.
93. Bresson-Hadni, S. et al. Should possible recurrence forbid liver transplantation in patients with end-stage alveolar echinococcosis? A 20-yr follow up. *Liver Transpl.* 17, 855, 2011.
94. Tappe, D. et al. Immunoglobulin G subclass responses to recombinant Em18 in the follow-up of patients with alveolar echinococcosis in different clinical stages. *Clin. Vaccine Immunol.* 17, 944, 2010.
95. Mamuti, W. et al. Molecular cloning, expression, and serological evaluation of an 8-kilodalton subunit of antigen B from *Echinococcus multilocularis. J. Clin. Microbiol.* 42, 1082, 2004.
96. Müller, N. et al. Improved serodiagnosis of alveolar echinococcosis of humans using an in vitro-produced *Echinococcus multilocularis* antigen. *Parasitology* 134, 879, 2007.
97. Fujimoto, N. et al. Usefulness of recombinant Em18-ELISA to evaluate efficacy of treatment in patients with alveolar echinococcosis. *J. Gastroenterol.* 40, 426, 2005.
98. Ishikawa, Y. et al. Serological monitoring of progression of alveolar echinococcosis with multi-organ involvement using recombinant Em18. *J. Clin. Microbiol.* 47, 3191, 2009.
99. Wen, H. et al. Immunoblot evaluation of IgG and IgG subclass antibody responses for immunodiagnosis of human alveolar hydatidosis. *Trans. R. Soc. Trop. Med. Hyg.* 89, 485, 1995.
100. Gottstein, B., D'Alessandro, A., and Rausch, R.L. Immunodiagnosis of polycystic hydatid disease/polycystic echinococcosis due to *Echinococcus vogeli. Am. J. Trop. Med. Hyg.* 53, 558, 1995.
101. Bart, J.M. et al. Comparison of several commercial serologic kits and Em18 serology for detection of human alveolar echinococcosis. *Diagn. Microbiol. Infect. Dis.* 59, 93, 2007.
102. Brehm, K. et al. Characterization of the genomic locus expressing the ERM-like protein of *Echinococcus multilocularis. Mol. Biochem. Parasitol.* 100, 147, 1999.
103. Sako, Y. et al. Development of an immunochromatographic test to detect antibodies against recombinant Em18 for diagnosis of alveolar echinococcosis. *J. Clin. Microbiol.* 47, 252, 2009.
104. Liance, M. et al. Immunodiagnosis of *Echinococcus* infections: Confirmatory testing and species differentiation by a new commercial Western blot. *J. Clin. Microbiol.* 38, 3718, 2000.
105. Ito, A., Schantz, P.M., and Wilson, J.F. Em18, a new serodiagnostic marker for differentiation of active and inactive cases of alveolar hydatid disease. *Am. J. Trop. Med. Hyg.* 52, 41, 1995.

106. Ito, A. et al. Immunodiagnosis of alveolar echinococcosis by enzyme-linked immunosorbent assay using a partially purified Em18/16 enriched fraction. *Clin. Diagn. Lab. Immunol.* 4, 57, 1997.
107. Ito, A. et al. Antibody responses against Em18 and Em16 serological markers in alveolar and cystic echinococcosis patients from northwest China. *Jpn. J. Med. Sci. Biol.* 50, 19, 1997.
108. Ito, A. et al. An imported case of cystic echinococcosis in Japan diagnosed by imaging and serology with confirmation of *Echinococcus granulosus*-specific DNA sequences. *Am. J. Trop. Med. Hyg.* 58, 790, 1998.
109. Kern, P. et al. WHO classification of alveolar echinococcosis: Principles and application. *Parasitol. Int.* 55, S283, 2006.
110. Bresson-Hadni, S. et al. Imaging and non-surgical interventional treatment in human alveolar echinococcosis. *Parasitol. Int.* 55, S267, 2006.
111. Ishikawa, Y. et al. *In vivo* evaluation of the viability of human hepatic alveolar echinococcal tissue from patients. *Clin. Parasitol.* 1, 157, 1990. [In Japanese].
112. Bowles, J., and McManus, D.P. Rapid discrimination of *Echinococcus* species and strains using a polymerase chain reaction-based RFLP method. *Mol. Biochem. Parasitol.* 57, 231, 1993.
113. Gasser, R.B. and Chilton, N.B. Characterisation of taeniid cestode species by PCR-RFLP of ITS2 ribosomal DNA. *Acta Trop.* 59, 31, 1995.
114. Zhang, L. et al. Screening for different genotypes of *Echinococcus granulosus* within China and Argentina by single-strand conformation polymorphism (SSCP) analysis. *Trans. R. Soc. Trop. Med. Hyg.* 93, 329, 1999.
115. Bartholomei-Santos, M.L. et al. Isolation and characterization of microsatellites from the tapeworm *Echinococcus granulosus*. *Parasitology* 126, 599, 2003.
116. Le, T.H. et al. Complete mitochondrial genomes confirm the distinctiveness of the horse-dog and sheep-dog strains of *Echinococcus granulosus*. *Parasitology* 124, 97, 2002.
117. Nakao, M. et al. The complete mitochondrial DNA sequences of the cestode *Echinococcus multilocularis* (Cyclophyllidea: Taeniidae). *Mitochondrion* 1, 497, 2002.
118. Yamasaki, H. et al. Mitochondrial DNA diagnosis for cestode zoonoses: Application to formalin-fixed paraffin-embedded tissue specimens. *Southeast Asian J. Trop. Med. Public Health* 38(Suppl. 1), 166, 2007.
119. Yamasaki, H. et al. Genetic analysis of *Echinococcus multilocularis* originating from a patient with alveolar echinococcosis occurring in Minnesota in 1977. *Am. J. Trop. Med. Hyg.* 79, 245, 2008.
120. Bowles, J. et al. Genetic variants within the genus *Echinococcus* identified by mitochondrial DNA sequencing. *Mol. Biochem. Parasitol.* 54, 165, 1992.
121. Nakao, M. et al. Mitochondrial genetic code in cestodes. *Mol. Biochem. Parasitol.* 111, 415, 2000.
122. Bowles, J., and McManus, D.P. NADH dehydrogenase 1 gene sequences compared for species and strains of the genus *Echinococcus*. *Int. J. Parasitol.* 23, 969, 1993.
123. Moro, P.L. et al. Molecular identification of *Echinococcus* isolates from Peru. *Parasitol. Int.* 58, 184, 2009.
124. Bowles, J., Blair, D., and McManus, D.P. Molecular genetic characterization of the cervid strain ("northern form") of *Echinococcus granulosus*. *Parasitology* 109, 215, 1994.
125. Kamenetzky, L. et al. Several strains of *Echinococcus granulosus* infect livestock and humans in Argentina. *Infect. Genet. Evol.* 2, 129, 2002.
126. Lavikainen, A. et al. Molecular genetic characterization of Fennoscandian cerrvid strain, a new genotypic group (G10) of *Echinococcus granulosus*. *Parasitology* 127, 207, 2003.
127. Moks, E. et al. First report of *Echinococcus granulosus* G8 in Eurasia and a reappraisal of the phylogenetic relationships of 'genotypes' G5-G10. *Parasitology* 135, 647, 2008.
128. Nkouawa, A. et al. Loop-mediated isothermal amplification method for differentiation and rapid detection of *Taenia* species. *J. Clin. Microbiol.* 47, 168, 2009.
129. Nkouawa, A. et al. Evaluation of loop-mediated isothermal amplification method using fecal specimens for differential detection of Taenia species. *J. Clin. Microbiol.* 48, 3350, 2010.
130. Yanagida, T. et al. Neurocysticercosis: Assessing where the infection was acquired from. *J. Travel Med.* 17, 206, 2010.
131. Jongwietiwes, U. et al. Isolated intradural-extramedullary spinal cysticercosis: A case report. *J. Travel Med.* 18, 284, 2011.
132. Lavikainen, A. et al. Molecular characterization of *Echinococcus* isolates of cervid origin from Finland and Sweden. *Parasitology* 133, 565, 2006.
133. Lavikainen, A. et al. Mitochondrial DNA data reveal cryptic species within *Taenia krabbei*. *Parasitol. Int.* 59, 290, 2010.

24 Hymenolepis

B.R. Mirdha and Shabeena Rehman

CONTENTS

24.1 INTRODUCTION

The class Cestoda consists of endoparasitic organisms which are usually segmented and have tape-like morphology. Cestoda species that infect humans include members of the order Cyclophyllidae and members from the order Pseudophyllidea of the family Diphyllobothridae. Cyclophyllidean cestoda includes Anoplocephalidae, Davaineidae, Dilepididae, Taeniidae, Hymenolepididae, and Mesocestoidae [1]. Adult tapeworm is transparent and consists of head, and body made up of short segments called *proglottids*. The head or the scolex is armed with hooks, "suckers," or both. There are different types of suckers. Some types of "suckers" are called *bothria* or *bothridia. Bothria* are usually slit-like grooves as seen in *Diphyllobothrium* species. Another type of sucker-like organ is the cup-like acetabulum. Scolex of cestodes has sensory organs, and nerves which are attached longitudinally extend down the body. These nerve fibers are attached to various parts/organs and the cestode is able to detect tactile stimulation [2]. Below the scolex is the neck and the body or strobila with segments whose width is more than their length. The genital pores are unilateral, or on either side of the segment. The short segments are called *proglottids*. The *proglottids* can be classified as immature, mature, and gravid. Each mature *proglottid* encompasses a complete set of male and female reproductive organs and the genital opening (vaginal canal) opens to the outside. Self-fertilization, involving the sexual structures of a single *proglottid*, or cross-fertilization between *proglottids* of the same worm may occur. Some *proglottids* in mature strobila may become sexually undifferentiated due to the crowding effect [3]. Crowding effect possibly prevents parts of the strobila from coming into contact with the mucosa of the host from which they receive necessary nutrition and protein-synthesizing enzyme. When a *proglottid* becomes old and unable to absorb any more nutrition, it is released and is passed through the host's digestive tract. A gravid *proglottid* contains fertilized eggs, which are sometimes expelled with the feces. However, most of the time, the egg settles in the microvilli of the small intestine, hatch, and the larvae can develop to sexual maturity without ever leaving the host. The life cycle of *Hymenolepis* can be briefly and broadly described as follows: (a) eggs are ingested by definitive hosts, (b) eggs hatch in the duodenum releasing oncospheres and lie in the lymph channels of the villi, (c) oncospheres develop into a *cysticercoid*, which has a tail and a well-formed scolex, and it attaches to the small intestine and matures into an adult, (d) gravid *proglottids* then release and pass out through feces along with eggs or eggs can hatch and infect the original host and start the cycle again within the gut itself, and finally, (e) eggs are excreted to the environment that are ingested by insects or rodents [4] for further transmission. It has been shown that *cysticercoid* burden reaches a plateau depending on the host's nutritional status. An interesting feature of *Hymenolepis* tapeworms is that they undergo a diurnal migration within the gut which is associated with the feeding patterns of the host.

24.1.1 Classification

24.1.1.1 Family Hymenolepididae

Family Hymenolepididae consists of more than 800 species, but the most disease-causing *Hymenolepis* species are *Hymenolepis nana*, *Hymenolepis diminuta*, *Hymenolepis microstoma*, and *Drepanidotaenia lanceolata*. The tapeworm *H. nana* was first described as *Taenia nana* by Von Siebold in 1852 and a morphologically similar parasite from a rodent host was described by Stiles in 1906 and was named *H. nana* var. *fraterna*. Both these species are morphologically similar and Grassi in 1887 suggested that both strains are identical. But Brumpt in 1949 and Yamagut in 1959 considered them as separate species. Issues to resolve the controversy over their status as a single or dual species and host specificity had existed and *H. nana* and *H. fraterna* were known as *Rodentolepis nana* and *Rodentolepis fraterna*, respectively, for some time by taxonomists. The cestode *Rodentolepis* (=*Hymenolepis*) *microstoma* is known to infect the definitive hosts which include mice, rats, and hamsters [5]. The terms *Rodenotolepis* and *Hymenolepis* are synonymous. The other important species that rarely cause human infection is *H. diminuta*. The sylvatic cycles of both *H. nana* and *H. microstoma* involves infection in murine rodents [6].

24.1.1.1.1 Hymenolepis nana (Nanos = Dwarf)

It is called the "dwarf tapeworm" and found worldwide in both temperate and tropical regions. The prevalence is very common in tropics [7]. It is most prevalent in conditions of poor sanitation where people live in a crowded environment. In temperate countries, it has been seen more in institutionalized children [8].

H. nana is a common parasite in man as well as in rats and mice and is cosmopolitan in distribution. As its name implies (nanos = dwarf), it is a small species, seldom exceeding 40 mm in length and 1 mm in width. The adult dwarf tapeworm measures 25–45 mm in length and 1 mm in width. The rostellum of *H. nana* is retractile and possesses a single ring of 20–30 hooks. The scolex has four suckers, or a tetrad and a short rostellum with hooks. The neck is long and slender, and the width of the segments is more than their length. Genital pores are unilateral, and each mature segment contains three testes. After apolysis, gravid segments disintegrate, releasing eggs. The eggs of *H. nana* are round or slightly oval at about 40–60 μm in length and 30–50 μm in breadth with 4–8 polar filaments spread out between the inner and outer membranes. The oncosphere is covered with a thin, hyaline, outer membrane and an inner, thick membrane with polar thickenings that bear several filaments. The heavy embryophores that give taeniaid eggs their characteristic striated appearance are lacking in *H. nana* [9].

An intermediate host is not mandatory for the development of *H. nana*. *H. nana* can maintain its life cycle with only one host or can also go through the normal two-host cycle. Often, the nonmandatory nature of requirement of an intermediate host for sustenance of life cycle is the reason for considering *H. nana* as the most common tapeworm infection throughout the world [10]. *Cysticercoids* of *H. nana* can develop at higher temperatures than those of the other species of *hymenolepidids*. Direct contaminative infection by eggs is probably the most common route in human cases, but accidental ingestion of an infected grain/flour beetle or flea cannot be ruled out. The direct infectiousness of the eggs frees the parasite from its former dependence upon an insect intermediate host, making the infection to spread rapidly and also allows person-to-person spread possible. The short life span and rapid course of development also facilitate the spread and ready availability of this worm. A single tapeworm may lay thousands of eggs a day and 1–2 million eggs during its life time. Eggs of *H. nana* are immediately infective when passed with the stool and cannot survive more than 10 days in the external environment. Eggs which reach the outside may be ingested by grain beetles or fleas in which the oncospheres develop into tailed *cysticercoids*, but often these eggs are swallowed directly by man or mouse. The eggs hatch in the stomach or small intestine. The liberated oncosphere penetrates the host epithelium of small intestine and migrates to lamina propria and finally metamorphoses into the *cysticercoid* stage in the tissue within 96 h after infection. Upon rupture of the villus, the *cysticercoids* return to the intestinal lumen, evaginate their scoleces, attach to the intestinal mucosa, and develop into adults that reside in the ileal portion of the small intestine, producing gravid proglottids within a period of 1–2 weeks. The adult worm lives for 4–6 weeks and releases eggs inside the lumen. Depending on the species, cestode larva can localize in a variety of organs, but they are targeted to a specific site.

Autoinfection may occur if the eggs inside the gut hatch to release oncospheres which invade the villi and develop into *cysticercoids*, which then enter the gut lumen, attach to the mucosa, and grow into adults. The dwarf tapeworm like all other tapeworms lacks a digestive system and feeds by absorption on nutrients in the intestinal lumen. They have nonspecific carbohydrate requirements and it seems likely that they will absorb whatever is being passed through the intestine at that time. When it becomes an adult, it will attach to the intestinal walls with its suckers and toothed rostellum and have its segments reaching out into the intestinal space to absorb food [11]. *H. nana* of human origin introduced into specific pathogen-free hamsters, four mouse strains, and two rat strains failed to establish infections, meaning thereby that specific strains are predilected to cause infection in certain particular type of host/s [12].

24.1.1.1.2 Hymenolepis diminuta

H. diminuta (rat tapeworm) is a tapeworm species of rats and mice. *Cysticercoid* larvae of *H. diminuta* are harbored by arthropods like beetles, fleas, and caterpillars. The life cycle requires an intermediate host in which the *cysticercoid* develops. Many kinds of insects serve as intermediate host, for example, grain/flour beetles (*Tribolium confusum*), earwigs, fleas, and cockroaches. *H. diminuta* eggs are passed in the rat's feces and intermediate arthropod hosts. Upon ingestion of these eggs, the eggs develop into the larva stage in infected individuals. Larvae are capable of evading the hemocytic defense reactions within the intermediate host and avoid encapsulations [13] and successfully establish transmission in their final definitive host. Once within the intermediate host, *H. diminuta* oncosphere develops into larvae. The pattern of larval development of the tapeworms, *H. diminuta* and *H. nana*, has been described by Voge and Heyneman [14]. There are five stages in the growth of the *cysticercoid*: (1) a hexacanth stage develops in the hemocoel; (2) an enlarged spherical larva containing an eccentric cavity is formed; (3) the larva gets elongated with the beginning of a forebody, midbody, and hindbody; (4) this emerging/young *cysticercoid* gets partially differentiated to scolex, and (5) finally a mature scolex at the anterior end and a tail at the posterior end develops. It is also interesting to note that positioning of the worm within its definitive host is determined by a peripheral sensory system. *H. diminuta* is the most studied of all tapeworms and has been the choice parasite for experimental studies and is easily maintained in laboratory settings. *H. diminuta* is a common rat and mouse tapeworm and only an occasional parasite of man [15]. Humans are usually accidental host of this parasite by ingestion of infected intermediate hosts. Transmission of the disease to humans may occur from eating contaminated cereals and due to unusual eating habits [16], and infection due to *H. diminuta* is sporadic. *H. diminuta* is about 45 cm in length, thus being considerably larger than *H. nana*. Eggs of *H. diminuta* are morphologically similar to those of *H. nana* but lack the polar filaments.

24.1.2 Clinical Manifestation and Pathogenesis

24.1.2.1 Disease Pathogenesis

H. nana is the most common cestode parasite of humans in the world and approximately more than 75 million per year are probably infected with *H. nana* [17]. Overall understanding of the pathogenesis of *Hymenolepis* infection has gradually become clear from various studies conducted by several investigators over a long period of time [18]. *H. nana* lodges itself in the intestines and absorbs nutrients from the intestinal lumen. In human adults, the infection due to *H. nana* is relatively less of a health problem, but in small children, it can be dangerous. Usually, it is the larva of this tapeworm that causes disease in children [19]. The larva usually burrows into the walls of the intestine and if the worm burden is heavy in the infected child, severe damage may be inflicted. This is done by absorbing the nutrition from the gut. *H. nana* usually will not cause deaths unless in extreme circumstances and usually in young children or in people who have weakened immune systems [20]. In some parts of the world where the disease is endemic, individuals found to be heavily infected are often attributed to internal autoinfection [21]. Humans become infected with *H. nana* when they ingest infective eggs, most commonly by direct fecal–oral exposure. The eggs pass into the ileum and hatch and the larval form penetrates the lamina propria. Within 3–4 days, the larvae mature into the preadult *cysticercoid*, which then enters the gut lumen to attach to the mucosae of the villus. Mature adults, measuring 35–45 mm in length and comprising 150–200 proglottids, result within about 3 weeks time. Self-mating between adjacent proglottids generates hundreds of eggs, some of which penetrate intestinal villi and some get excreted out in feces. Occasionally, rodents may ingest the eggs excreted in the feces and serve as incidental hosts and reservoir for the spread of infection.

Human infection with *H. diminuta* results from accidental ingestion of intermediate insects hosts (immature fleas, flour/grain beetles, cockroaches) that carry the parasite in their body cavities. Infective eggs are ingested by insects and hatch in their guts. After hatching, they invade the body cavity and become *cysticercoid*, which are infectious for humans. After the insects are consumed and digested, the larvae are released in the small intestine and mature within 25 days into 50-cm adults. When the adult tapeworm begins to pass eggs, insect hosts can become infected again. Most infections produce no symptoms. Myoelectric alterations study has shown that in infected animals, specific products prevent expulsion of the tapeworm from the small intestine and aid its migration and absorption of nutrients. Such parasite-induced changes and modulation of absorption probably establish colitis in *Hymenolepis*-infected animals [22]. Regarding modulation of absorption, it has been observed in a case–control study that *H. nana*-infected patients showed significant higher percentage ($P = 0.04$) of altered intestinal permeability denoting intestinal leakage, lower levels of plasma Vitamin B12 ($P = 0.01$), and folate ($P < 0.04$) against parasite-free controls. The percentage of stunting and wasting were observed more among *H. nana*-infected patients [23].

24.1.2.1.1 Clinical Manifestations of Hymenolepiasis

H. nana lodges itself in the intestines and absorbs nutrients from the intestinal lumen and do not lead to fatal outcome unless in extreme circumstances such as people suffering from either congenital or acquired weakened immune systems and/or in young children. Initial symptoms are due to allergic responses or systematic toxemia caused by excretory–secretory products of the tapeworm. A common laboratory finding is increased eosinophil counts [24]. Light infections are usually asymptomatic. Infection with more than 2000 worms can cause abdominal pain, diarrhea, loss of appetite, restlessness, irritability, and anal pruritus. Rare symptoms include vomiting, nausea, bloody diarrhea, hives, extremity pain, headache, dizziness, and behavioral disturbances. Occasionally, epileptic seizures occur in infected children. Morbidity is uncommon, death may occur when parasite burden is extremely high. No racial or gender predisposition is known for hymenolepiasis. However, because of the increased likelihood of exposure to human feces, younger children have the highest risk of hymenolepiasis. Symptoms of infection are usually subclinical, if any, but toxic reactions such as nervous disorders, sleeplessness, diarrhea, and intestinal pain may occur. Prolonged infection with *H. nana* more than a period of five decades may contribute to weight loss and generalized weakness [25].

24.1.2.1.2 Pediatric Hymenolepiasis

Infection is most common in children aged between 4 and 10 years, in dry, warm regions of Africa, South America, and Asia. *H. nana* infection affects millions of people, primarily children, worldwide. Estimated rates of infection in various regions range from 0.1% to 58%.

H. nana was the most common helminth detected in stool examination in a survey in Thailand. Prevalence of 13% enteric infection due to *H. nana* has been found in 2083 children. Most of these children were from orphanages and from primary schools with a male preponderance. The unique findings are that most of the children were asymptomatic [26]. Study from Africa (Zimbabwe) in primary school children documented an infection rate up to 24% [27]. In a study conducted in Mexico, more than 200,000 stool specimens submitted to the state laboratories in 1987 for ova and parasite analysis showed that 0.4% were positive for *H. nana* [28]. As most infections do not produce symptoms, the true incidence might be higher. Regions with high reported infection rates include Sicily (46%), Argentina (34% of school children), and southern areas of the former Soviet Union (26%). In contrast, only 0.1% of stools examined at a children's hospital in Calgary were positive for *H. nana*. Most cases with associated neurologic symptoms have been reported from the former Soviet Union [29,30]. In our laboratory, an infection rate of 11% was observed from 5000 samples studied with few cases of familial clustering (unpublished data). It is also interesting to note that recrudescence of *H. nana* infection is common in children.

24.1.2.1.3 Concurrent and Extra Intestinal Infection due to Hymenolepis nana

The occurrence of cysticercoids of *H. nana* in extraintestinal sites has been reported [31] and the larvae have been recovered from mesenteric lymph nodes of white rats. It has been shown that oncospheres of *H. nana* injected subcutaneously into mice developed to cysticercoids at that site of inoculation. Similarly, an apparently related species *H. microstoma* has also been able to develop in mice from eggs to adult worms, especially if eggs were kept for 24 h at 4°C before they were fed to mice.

Recurrent cutaneous papular eruption in a 50-year-old man associated with *H. nana* has been reported with no other apparent etiological agents. It is suggested that persistent antigenic stimulation due to autoinfection cycle of *H. nana* may be the cause of the papular eruption [32]. Phlyctenular eye disease has also been associated with *H. nana* infection [33]. *H. nana* infection has also been observed to have some modulatory role in the immunology of tuberculosis [34]. Similarly, it has been observed that immune responses against hymenolepiasis infection are affected by infection with *Trypanosoma cruzi* and *Trypanosoma burcei* causing delayed expulsion of parasite [35]. Both antagonistic and agonistic effects have been observed in concurrent infections with some of the other helminthic infection. Fatal tissue invasion due to larval *Hymenolepis* parasite has been reported [36].

24.1.3 Immunology

The mechanism of parasite survival in the immunized host has been the most interesting aspect of immunobiology of the genus *Hymenolepis*. It has been well analyzed in the natural intermediated hosts for these cestodes, that is, mice and rats. In *H. nana*/mouse system, it has been shown that even a single oncosphere can lead to disease [37].

24.1.3.1 *In Vivo* Models for Studies on *Hymenolepis* Immunology

To elucidate immunological aspects, infection with *Hymenolepis* spp. has been studied in rodent models. In primary infection, experimental animal models acquire immunity against *Hymenolepis* and studied rapid elimination of worms [38]. Similar experiment in athymic nude mice failed to give similar response, suggesting that the cellular response is thymus dependent [39]. It has also been reinforced with the fact that immunosuppressive agents abrogate the acquired resistance [40]. Cross-immunity among different species of *Hymenolepis* do occur and this cross-immunity is protective and is dependent on the burden of parasites [41]. However, difference in susceptibility depends on the type of animal model as well as on the variations in parasite isolates. So, "strain" variation among *Hymenolepis* isolate may produce immune response in different ways. The role of humoral response in primary and secondary infection is not clearly defined. However, passive immunity against infection produced increased production of specific immunoglobulins in resistant hosts [42] especially against a 32-kDa component of oncosphere.

24.1.3.2 Immunobiology of *Hymenolepis nana* in Mouse System

Hymenolepis spp. require beetles and rodents for completion of their life cycle. However, *H. nana* is unique, since it can complete its whole life cycle in the small intestine of a single mouse host. Oncospheres differentiate into metacestodes (cysticercoids) in the mouse intestinal tissue within 4 days or in beetles approximately within 2 weeks at 20°C.

Metacestodes developed either in mice or in beetles develop into mature adult worms in the mouse intestinal lumen within 10 days. *H. nana* adapted to mice does not develop into adult worms in rats [43].

Host–parasite relationship has revealed that (i) an oral infection with even a single oncosphere derived from ingestion of a few eggs evokes complete immunity to reinfection in mice, (ii) all developmental stages of this parasite are recognized as antigens by the infected host, (iii) mice infected with a single oncosphere become completely immune to reinfection with eggs and with metacestodes (the immune response to egg challenge is called "early response" whereas that to metacestode challenge is called "late response"), (iv) autoinfection usually do not occur in most strains of immunocompetent mice, and (v) adult worms are expelled by mechanisms other than the early and late responses against reinfection (worm expulsion response) [44].

24.1.3.3 Immunobiology of *Hymenolepis nana* in Rat System

Rats treated with antilymphocyte serum or congenitally athymic nude rats show basically similar resistance to the lumen phase of *H. nana*. However, when rats were orally inoculated with eggs or cysticercoids (beetle, mouse, or rat derived) and treated with corticosteroid during the expected lumen phase of this parasite, that is, either from day 4 of egg inoculation or from day 0 of the eggs and cysticercoids developed into mature adult worms. Cessation of treatment with corticosteroid induces rapid expulsion of the growing adult worms from the rats. The exact mechanism of the induction of adult cestodes in such unnatural host animals treated with corticosteroid is not well defined. Corticosteroid may possibly affect the rat's physiological condition, thereby making the host suitable for the development of *H. nana* into adult worms, but it might also affect the worm directly [45].

Existence of anti-*Hymenolepis* T-memory cells in passively transferred models established the importance of cellular immunity in the controls of infection [46]. *Hymenolepis*-specific murine T-cell clones have shown to produce IFN-gamma and IL2 that confer passive immunity [47]. In a recent study, it has been observed that serum of humans infected with *H. nana* shows a significant difference in IFN-gamma ($P < 0.001$) and IL3 ($P < 0.05$) besides increase in IL12 and IL5 [48]. The production of inflammatory cytokine is indicative and involves in host protective immunity [49].

24.1.3.4 Role of Eosinophils

It has been proposed that *H. nana* oncospheres are selective chemoattractants for eosinophils and is capable of inducing adhesion molecule CD116, CD18, and CD4 that are responsible for expression and generations of eosinophils. Eosinophils are predominantly attracted toward the cysticercoid. Preferential eosinophil migration phenomenon is a lymphocyte-independent mechanism and is believed that oncosphere-derived molecules with molecular mass of 5.5–9.6 kDa and 30–40 kDa may facilitate eosinophil infiltration/migration [50]. *Hymenolepis* infection has been shown by the increased emigration of eosinophils toward larva within the gut [51]. A relationship between this type of response and expulsion of adult *H. nana* in mice was proposed and established [52].

24.1.3.5 Immunobiology of Other *Hymenolepis* spp. Infection

H. microstoma and *H. diminuta* require insects and rodents (mice and rats) for completion of their life cycles, respectively. These cestode species are intestinal parasite in rodents and have been studied to be good models for mucosal immunity. In the *H. microstoma*/mouse system, noninfected mice kept with infected mice showed antibody response against egg antigens. Immunological memory following primary infection with *H. diminuta* is long lasting [44,53].

24.1.4 Diagnosis

Hymenolepis infection is more common in children compared to adults. Symptoms include abdominal pain, diarrhea, anorexia, and headache. Most of the infections are often asymptomatic and often go unnoticed and an early diagnosis is important in order to start treatment.

24.1.4.1 Conventional Techniques

24.1.4.1.1 Direct Detection of Hymenolepis Infection

24.1.4.1.1.1 Macroscopic Examination of Feces Adult worm of *H. nana* tapeworm ranges from 7 to 45 mm in length and that of *H. diminuta* up to 20–90 cm in length with more than hundreds of proglottids. Proglottids normally do not get disintegrated from the strobila and when they are broken off from the strobila, they can be passed in feces. Each proglottid is roughly three to four times wider than the length and very rarely found in fecal sample [54]. Invariably, whenever there is excretion of proglottids, the eggs are also found in microscopic examination of stool samples.

24.1.4.1.1.2 Microscopic Methods Definitive diagnosis of hymenolepiasis depends on the demonstration of eggs in stool specimens. Collection of three consecutive fecal samples for three consecutive days is recommended. Examination of clinical samples using various concentration techniques such as zinc sulfate floatation technique [55] and formol-ether sedimentation techniques [56] increase the likelihood of detection of low-burden infection. *H. nana* eggs are usually spherical or ovoid with a thin hyaline shell and measure 30–45 μm in diameter. The oncosphere with its three pairs of hooklets lies in the center of the eggs and is separated from the outer shell by considerable space. The oncosphere has an internal membrane with polar thickening from which arise 4–8 filaments extending into the space between it and the colorless hyaline shell. Eggs of *H. nana* must be distinguished from those of *H. diminuta*. Eggs of *H. diminuta* have a moderate thick shell and the oncosphere lacks polar thickening and polar filaments.

24.1.4.1.1.3 Biopsy Studies The degree and character of histopathological changes depend on the intensity of infection. Invasive larval cestodiasis has been diagnosed with distinct morphologic findings [57]. During an outbreak of spontaneous *H. nana* infection in Mongolian gerbils, it was observed that larval form caused extensive inflammation in lamina propria and in surrounding tissues. Large numbers of adult worms were also present in the lumen of the gut [58]. Changes in experimental rats due to *H. diminuta* showed alterations in intestinal wall such as fusion of villi, damage to the epithelial layers with flattened cells with pyknotic nuclei. With severe infections, *H. diminuta* infections showed changes in extraintestinal viscera such as dilatations of sinusoids in liver, alterations in *zona fascicularis* of adrenal glands [58].

24.1.4.1.2 Indirect Detection of Hymenolepis Infection

24.1.4.1.2.1 Eosinophilia *H. nana* infection is characterized by intense tissue eosinophilia around the larva in the intestinal villi [59]. However, there is no peripheral blood eosinophilia. It has been shown that the oncosphere of *H. nana* is a selective chemoattractant for eosinophils and also helps in the expression of adhesion molecule of eosinophils [50].

24.1.4.1.2.2 Enzyme-Linked Immunosorbent Assay Enzyme-linked immunosorbent assay (ELISA) detects the presence of an antibody or an antigen in a sample. In ELISA, an unknown amount of antigen (Ag) or antibody (Ab) is coated to the surface of a microtiter plate, and then a specific Ab or Ag is added over the surface so that it can bind to the coated molecule. This antibody is linked to an enzyme, and in the final step a substance is added so that the enzyme can convert to some detectable signal.

Serum antibody detection of *H. nana* was first studied by Ito et al. by ELISA and immunoprecipitation technique using sodium deoxycholate (DOC)-solubilized antigen prepared from embryonated eggs [60]. ELISA for detection of antibody in serum for *H. nana* infection has been performed using crude antigenic extract prepared from scolex and neck regions of adult worms [61]. Castillo et al. developed an ELISA using soluble portion of a homogenate of adult *H. nana* worm harvested from intestine of infected mice. This technique was developed to measure serum IgG antibodies in 65 *H. nana*-infected patients and 30 noninfected patients in Peru. The technique had a sensitivity of 79% and a specificity of 83%, respectively [62]. Nearly all individuals infected with *H. nana* had antibodies against the parasite and posttreatment antibodies of *H. nana* disappeared by 3 months. It was found to be useful in evaluating drug therapy for hymenolepiasis. When the ELISA was tested with sera from patients with hydatidosis and cysticercosis, fairly high rate of cross-reactivity was observed. The notable finding of the study was that the level of anti-*H. nana* IgG antibodies had correlated well with lower value after elimination of parasite and was higher in sustained infection even after treatment. So, ELISA was useful in monitoring drug therapy and in situation where microscopic findings are negative either due to episodic shedding of parasite or due to very low burden of parasite which may not be detected even after examination of three consecutive stool samples.

24.1.4.1.2.3 Western Blotting Western blotting (WB) is a technique which detects and localizes protein/s of interest from a mixture of a great number of proteins. It provides information about the size of the protein and level of protein expression. Western blotting has been of immense use in the diagnosis of helminthic infections as it greatly reduces cross reactions. Montenegro et al. [63] studied the crude antigens of *H. nana*. The WB assay revealed 49- and 66-kDa proteins. Gonec [64] also looked for specific proteins from sera of mice naturally infected with *H. nana*. Crude antigens were separated by SDS-PAGE and proteins were detected. A 24-kDa band in sera of *H. nana*-infected mice was identified. Identification of more specific and non-cross-reactive proteins will be useful for specific serodiagnosis of *H. nana* infections.

24.1.4.1.2.4 Indirect Fluorescent Antibody Test Indirect fluorescent antibody test (IFAT) detects antibodies present in clinical material to specific antigenic material using fluorescent microscopy. Expression of surface antigen of *H. nana* during differentiation and maturation of parasite appears in the following succession: oncosphere, cysticercoid, adult scolex, and adult strobila, and they differed critically from each other. The specific antibodies of IgG, IgM, and IgA isotypes to all these antigens were detected by IFAT by Ito et al. [65].

24.1.4.1.2.5 Double Diffusion (Ouchterlony Test) This technique uses a gel plate/slide that is cut to form a series of holes and the desired microbial antigen is placed in the center. Patient's serum/sera to be tested are placed in one or more of the outer wells. The slide or plate is left for 48 h to allow the antigens and antibodies to react and migrate in a radial fashion to outer wells. When the antigen and the antibody meet, a complex is formed with a visible precipitin line. Sera from animals following various periods of infection with *H. nana* and *H. diminuta* and sera from animals challenged artificially with homogenates were examined by double diffusion in agar gel [66]. Immunogenic reactivity of affinity-purified tegumental antigens was evaluated using double diffusion [67]. On further analysis of antigens by PAGE and SDS-PAGE, the affinity-purified antigen showed four antigenic components specific of *H. nana*.

24.1.4.1.2.6 Indirect Hemagglutination Test Indirect hemagglutination (IHA) is a frequently used test for the diagnosis of amebiasis, hydatidosis, and toxoplasmosis. In IHA, red blood cells are coated with certain substances like tannic acid, which helps to adsorb soluble antigen onto their surfaces by changing the surface charge of the red blood cells and to some extent increases the surface area of the red blood cells. The red blood cells then agglutinate in the presence of antiserum specific for the adsorbed antigen. Antibodies against soluble extracted antigens of *H. nana* were detected in infected mice sera using IHA test [68].

24.1.4.1.2.7 Passive Cutaneous Anaphylaxis Test Passive cutaneous anaphylaxis (PCA) test in the rat skin was carried out by Mota and Wong [66]. The antibody detected by PCA was of IgE isotype.

24.1.4.1.2.8 Complement Fixation Test Larsh [69] and Coleman et al. [70] observed complement fixing antibodies against soluble extracted antigens of *H. nana*.

24.1.4.2 Molecular Techniques

Limitations of microscopic examination due to sporadic shedding of *H. nana* in stool sample, the criticality in histopathological diagnosis, and cross-reactivity of serological results emphasize the importance of molecular diagnostic approaches. Molecular techniques improve the diagnosis, genetic characterization of etiological agents, and epidemiological tracking of disease.

Several regions of ribosomal DNA of the organism have been used for the detection and differentiation of various species of both human and animal infection due to *Hymenolepis* using molecular methods. Polymerase chain reaction (PCR) is the most commonly used molecular assay. PCR results have been further subjected to molecular typing analysis. Various target regions, 18S ribosomal DNA, mitochondrial DNA (mtDNA), and internal transcribed spacer 1 (ITS1) have been employed to achieve the molecular identification of parasites. Sequencing and phylogenetic analysis have been carried out to study the genetic diversity and to elucidate source of transmission for *Hymenolepis* [71].

24.1.4.2.1 Molecular Characterization and Phylogenetic Analysis of Hymenolepis spp

Three regions of ribosomal DNA, the small subunit 18S, ITS1, and a mitochondrial gene, the cytochrome C oxidase subunit 1 (CO1) gene has been studied. The 18S gene was relatively less discriminatory, when this target was used for genetic characterization of *H. nana* and *H. diminuta*. A large number of human isolates of *H. nana* have been characterized using ITS1. The rationale for using ITS1 is that ITS1 contains signals for the processing of the rRNA transcripts and is moderately conserved. PCR restriction fragment-length polymorphism (PCR-RFLP) results of ITS1 showed highly variable sequence. But when the ITS1 was cloned and sequenced, the sequence variations upon phylogenetic analysis were able to differentiate *H. nana* isolates of rodents and humans. Further sequence analysis of CO1 gene showed a genetic divergence of nearly 5% between *H. nana* isolates of mice and humans. This finding gave the breakthrough information about *H. nana* and suggested that *H. nana* is a complex species. This convincing information obtained upon analysis of mitochondrial gene could be because of the fact that genetic recombination is usually not seen in ITS1 and this gene is inherited as a unit. In addition, mitochondrial DNA contains no introns and noncoding regions. Thus, characterizing phylogenetic relationship between related species using this target is likely to elucidate better diversity.

24.1.4.2.2 Study of Hymenolepis Parasite Population

Studies of *Hymenolepis* spp. on a molecular level dealing with polymorphism, genetic distance, intra- and interspecific difference and variability, and molecular phylogenetics have been studied. Study of genetic characterization or variation in a parasite population is of great importance to understand the epidemiology of a disease. Diverse population within a species may exist which essentially differ in their host specificity, virulence factors, sensitivity to therapeutic agents, disease severity, and outcome. It also helps one to understand the transmission dynamics. Macnish et al. [71] studied the genetic variability by sequencing a 444-bp fragment of the mitochondrial cytochrome c oxidase 1 gene in nine isolates of *H. nana* from rodents and six isolates from humans and could differentiate distinct species of rodent *H. nana* from human isolates of *H. nana*. It was concluded that the *Hymenolepis* species are cryptic and are complex, that is, though they morphologically appear similar they are distinct at the genetic level and are genetically diverse. Okamoto et al. studied CO1 gene and ITS2 gene and observed nucleotide difference (in percentage) between *H. diminuta* and *H. nana* to be 22.9% and 18.2%, respectively, by both genes [72].

24.1.4.2.3 Polymerase Chain Reaction Assay-Based Approaches

The development of a useful molecular tool based on a PCR-RFLP to detect the presence of this parasite both in rodent and in human fecal samples has been studied. The development of this diagnostic technique has enabled the study of *H. microstoma* in human populations. Primers that were designed were to amplify a 646-bp product in *H. nana* and a 748-bp product in *H. diminuta* across the internal transcribed spacer 1 and a 635-bp product for *H. microstoma*. PCR-RFLP study was performed. The RFLP profiles were consistent with the predicted fragments. This provided a potentially useful diagnostic test for the future identification of *Hymenolepis* infections in humans. However, molecular approaches used for *Hymenolepis* infection have essentially provided more of epidemiological or population structure of *Hymenolepis* species than the detection of parasite directly from clinical specimens. Genomic diversity of 16 *H. nana* isolates from mice and rats has been studied using random amplified polymorphic DNA-PCR (RAPD-PCR), but no distinct difference could be obtained from both rats and mouse isolates [73].

24.1.4.2.4 Molecular Detection and Identification of Hymenolepis from Clinical Samples

Definitive diagnosis of human hymenolepiasis relies on the direct demonstration of *Hymenolepis* eggs or very rarely the segment of worms. Furthermore, hymenolepiasis is often asymptomatic to warrant necessary investigations. All these factors necessitate the use of molecular techniques for the detection and identification of *Hymenolepis* spp. in stool samples as well as in extraintestinal tissues. The detection of extraintestinal *Hymenolepis* infection, though very uncommon have been diagnosed by histopathological examination. Extension of application of the PCR technique in such tissue may be of immense help and may be a potential supporting technique for the diagnosis of hymenolepiasis.

24.2 METHODS

24.2.1 Sample Preparation

Appropriate sample collection from different clinical sites and its proper preparation are essential components of diagnosis of a disease. It is more important in the molecular detection and identification of *Hymenolepis* spp. infection using the PCR technique to achieve the diagnosis and to minimize contamination.

24.2.1.1 Collection of Parasite Material

Hymenolepis eggs are inoculated into experimental animals (BALB/c mice, Wistar rats) to obtain adult worms. The mice are sacrificed and adult worms are dissected from the small intestine approximately 2 weeks postinoculation. The worms are washed repeatedly in appropriate buffer saline and stored at −80°C until DNA extraction. A small amount of fecal matter is collected for extraction of DNA.

24.2.1.2 Purification of DNA from Human Feces

A small amount of fecal material was collected and suspended in 250 μL of 10% Chelex 100 (Bio-Rad) in Tris-EDTA buffer. This mixture was boiled for 7 min and then centrifuged for 5 min. The supernatant thus obtained was de-proteinized by mixing with a deproteinizer agent (*ProCipitatae*). The mixture was further centrifuged for 5 min after being kept for 5 min at room temperature. This supernatant was then eluted in 50 μL of Tris-EDTA buffer. About 2.5 μL of template DNA was used for PCR [74,75].

24.2.1.3 Purification of DNA from Adult Worms of *Hymenolepis*

Approximately 25 mg of worm tissue was homogenized with appropriate buffer and was treated with proteinase K and incubated for 2 h at 56°C. After incubation, proteinase K was deactivated. The sample was centrifuged, the supernatant was discarded, and the pellet was collected. The pellet was washed in buffer and dried carefully. DNA was eluted in 300 μL of Tris-EDTA buffer by incubating at 72°C for 5 min and was stored at −20°C.

24.2.2 Detection Procedures

24.2.2.1 Detection Procedure Using ITS1

DNA extracted from samples were used for PCR amplification as described by Macnish et al. [71]. DNA sequences of *H. nana* and *H. diminuta* spanning 3′ end of 18S rRNA gene, ITS1, were used to identify regions conserved between these two species. Primers for these regions designated F3 (5′-GCGGAAGGGATACTTACACGTTC-3′) and R3 (5′-GCTCGACTCTTCATCGATCCACG-3′) were designed. This amplifies a product of approximately 646 bp. DNA was amplified in 67 mM Tris-HCl (pH 8.8), 16 mM $MgCl_2$, 0.5 U *Tth*, 200 μm of each dNTP, and 12.5 pmol of each primer. The cycling conditions comprise of heating at 94°C for 2 min, 63°C for 2 min, 72°C for 1 min, followed by 50 cycles of 94°C for 20 s, 63°C for 20 s, 72°C for 45 s, and a final cycle of 72°C for 7 min. Amplification products were cloned and sequenced. The primers F3 and R3 and the protocols used for the amplification of the ITS1, purification, and cloning of the PCR product and sequence analysis were conducted.

24.2.2.2 Detection Procedure Using Mitochondrial CO1 Gene

A 444 bp segment of mitochondrial CO1 was amplified using forward primer (5′-TGGTTTTTTGTGCATCCTGAGGTTTA-3′) and reverse primer (5′-AGAAAGAACGTAATGAAAATGAGCAAC-3′) [72]. PCR amplification was carried out in 25 μL volume. The cycling conditions consist of heating at 94°C for 50 s followed by 50 cycles at 94°C for 50 s, 55°C for 90 s, 72°C for 90 s, and a final cycle of 72°C for 7 min. Amplified product was cloned and sequenced [72].

24.3 TREATMENT OF HYMENOLEPIASIS

Therapeutic agents like benziimidazote, niclosamide, and praziquintal have been used for *Hymenolepis* infection. Parasite biochemical pathways are usually different from the human host and the various mechanisms of antiparasitic actions have been observed that include (i) inhibition of microtubules inhibiting glucose uptake, (ii) tubulin polymerization inhibition, (iii) depolarizing neuromuscular blockade, (iv) cholinesterase inhibition, (v) increased cell membrane permeability, resulting in intracellular calcium loss, (vi) vacuolization of the tegument, and lastly (vii) increased cell membrane permeability to chloride ions via chloride channels alteration [76]. The activity of benziimidazole derivates (albendazole, mebendazole) have been studied extensively. Benziimidazole drugs are primarily effective against oncosphere and at a higher dose against larvae stage. Beniziimidazoles were reported to act through interaction with parasite tubulin with subsequent inhibition of microtubule polymerization. It also inhibits glucose uptake by worm and diminishes levels of cellular ATP in the parasite [77].

Praziquantel is a bacteriocidal agent, structurally unrelated to other anthelminthics, that kills both adult worms and larvae when given in a single dose. Adult and pediatric dose is 25 mg/kg body weight orally as a single dose. It is advisable to take the medicine with food and should not be chewed. It may rarely cause hypersensitivity. It is the drug of choice as it acts very rapidly against *H. nana* and 80% of the dose is excreted as metabolites in urine within 24 h. *In vitro*, praziquantel causes tegument impairment and vacuolization, which are confined to the neck region of the parasite [78,79]. Niclosamide inhibits parasite glucose uptake [80] and inhibits mitochondrial oxidative phosphorylation and glucose uptake in the parasite. The treatment course is 7 days because it does not reach the cysticercoids in the lamina propria. A single dose of 2 g, followed by 1 g four times daily for 6 days, is effective in adults. This tablet should be chewed 2 h before meal in contrast to praziquantel. In the pediatric group, the treatment is started with 1–1.5 g followed by 500–1000 g for 6 days for the age group above 11 and 34 years, respectively. Gastrointestinal disturbances, drowsiness, dizziness, headache, and rash are associated side effects.

Nitazoxanide elicits antiprotozoal activity by interfering with pyruvate-ferredoxin oxidoreductase (PFOR) enzyme-dependent electron transfer reaction, which is essential for anaerobic energy metabolism.

Preventing fecal contamination of food and water in institutions and crowded areas is of primary importance. General sanitation and rodent and insect control (especially, control of fleas and grain insects) are also essential for the prevention of *H. nana* infection.

24.4 CONCLUSION AND FUTURE PERSPECTIVES

Metazoan parasites may be responsible for previously unrecognized invasive or metastatic disease in humans, especially in immunocompromised hosts. Some of *Hymenolepis* parasites may be zoonotic agents with unsuspected animal reservoirs. Anecdotal evidence from various studies suggest that domestic rodent species such as *Musca domesticus* which inhabit human dwellings and *Mus spretus* which live in habitats away from human dwellings are known hosts of both *H. nana* and *H. microstoma* in some parts of the world. These rodent species may represent a potential reservoir and their role in the transmission in remote communities must therefore be considered.

Distinction between pathogenic *Hymenolepis* species could not be made solely on the basis of the microscopic examination, and it was the molecular analysis that provided conclusive residence of identity. It is clear that the advent of molecular biology has provided a sensitive tool for both the detection and genetic characterization of *H. nana* and *H. diminuta* species that cause human infection. Detection of *H. microstoma* in humans highlights the growing importance of the use of molecular techniques in both the detection and characterization of parasite species in human population. Currently, detailed information regarding species/strains variations among different *Hymenolepis* species on the basis of genetic differences has not been completely resolved in available literature and warrants further studies from various geographical areas worldwide to generate meaningful conclusion. From a public health perspective, a better understanding of the transmission dynamics and comprehensive epidemiological picture is the need of the hour.

REFERENCES

1. Coombs, I. Helminthic species recovered from humans. In: Crompton, D.W.T. and Savioli, L. (eds), *Handbook of Helminthiasis for Public Health*, CRC Press, Taylor & Francis Group, Boca Raton, FL, pp. 12–24, 2006.
2. Wittner, M., and Tanowitz, H.B. Other cestode infections. In: Guerrant, R.L., Walker, D.H., Weller, P.F. (eds.), *Tropical Infectious Diseases: Principles, Pathogens, and Practice*. 1st edn., Churchill Livingston, New York, pp. 1026–1027, 1999.
3. Ghazal, A.M., and Avery, R.A. Population dynamics of *Hymenolepis nana* in mice: Fecundity and the crowding effect. *Parasitology*. 69, 403, 1974.
4. Schantz, P.M. Tapeworms (cestodiasis). *Gastroenterol. Clin. North Am.* 25, 637, 1996.
5. Ito, A. Basic and applied immunology in cestode infections from *Hymenolepis* to *Taenia* and *Echinococcus*. *Int. J. Parasitol.* 27, 1203, 1997.
6. Andreasssen, J. Immunity to adult cestodes: Basic knowledge and vaccination problems: A review. *Parasitologia* 33, 45, 1991.
7. Ito, A., Nakao, M., and Wandra, T. Human teaniasis and cysticercosis in Asia. *Lancet* 362, 1918, 2003.
8. Mehraj, V. et al. Prevalence and factors associated with intestinal parasitic infection among children in an urban slum of Karachi. *PLoS One* 3, e3680, 2008.
9. Richards, F.O. Jr., *Diphyllobothrium*, *Dipylidium*, and *Hymenolepis* species. In: Long, S.S., Pickering, L.K., and Prober, C.G. (eds.), *Principles and Practice of Pediatric Infectious Diseases*, Churchill Livingstone, New York, p. 1351, 2003.
10. Beaver, P.C., Jung, R.C., and Cupp, E.W. (eds), In: *Clinical Parasitology*, 9th ed., Lea & Febiger, Philadelphia, PA, pp. v365–367, 1984.
11. Bogitsh, B.J., Carter, C.E., and Oeltmann, T.N. (eds), In: *Human Parasitology*, 3rd ed., Elsevier Academic Press, London, pp. 274–278, 2005.
12. Andreassen, J., and Birklund, D. *In vivo* effects of immune mouse serum on three species of *Hymenolepis* tapeworms. *Bull. Scand. Soc. Parasitol.* 3, 26, 1993.
13. Hopkins, C.A., and Allen, L.M. *Hymenolepis diminuta*: The role of tail in determining the position of the worm in the intestine of the rat. *Parasitology* 79, 401, 1979.
14. Voge, M., and Heyneman, D., Effect of high temperature on the larval development of *Hymenolepis nana* and *Hymenolepis diminuta* (Cestoda: Cyclophyllidea). *J. Parasitol.* 44, 249, 1958.
15. Katz, M., Despommier, D.D., and Gwadz, R.W., Tapeworms of minor medical importance. *Parasit. Dis.* 16, 88, 1989.
16. Chu, G.S., Palmieri, J.R., and Sullivan, J.T. Beetle eating: A Malaysian folk medical practice and its public health implication. *Trop. Geog. Med.* 29, 422, 1977.
17. Crompton, D.W.T. How much human helminthiasis is there in the world? *J. Parasitol.* 85, 397, 1999.
18. Conchedda, M. et al. Immune response to the cestode *Hymenolepis nana*: Cytokine production during infection with eggs or cysts. *Int. J. Parasitol.* 27, 32, 1997.
19. Mirdha, B.R., and Samantaray, J.C. *Hymenolepis nana*: A common cause of paediatric diarrhoea in urban slum dwellers in India. *J. Trop. Pediatr.* 48, 331, 2002.
20. Lucas, S.B. et al. Abnormal development of *Hymenolepis nana* larvae in immunocompromised mice. *J. Helminthol.* 54, 75, 1980.
21. Sirivichayakul, C. et al. *Hymenolepis nana* infection in Thai children. *J. Med Assoc. Thai.* 83, 1035, 2000.
22. Dwinell, M.B., Bass P., and Oaks, J.A. *Hymenolepis diminuta* fractions but not previous tapeworm infection stimulate intestinal myoelectric alteration *in vivo* in the rat. *J. Parasitol.* 84, 673, 1998.

23. Mohammed, M.A., and Hepazi, M.A. Intestinal permeability in *Hymenolepis nana* as reflected by non-invasive lactulose/mannitol dual permeability test and its impact on nutritional parameters of patients. *J. Egypt. Soc. Parasitol.* 37, 877, 2007.
24. Lloyd, S. Other cestode infections: Hymenolepiosis, diphyllobothriosis, coenurosis, and other adult and larval cestodes. In: Palmer, S.R., Soulsby, L., Simpson, D.I.H. (eds.), *Zoonoses*, Oxford Medical Press, Oxford, London, UK, pp. 651–663, 1998.
25. Pawlowski, P.W. Cestodes: Teaniasis, dyphyllobothriasis, hymenolepiasis, and others. In: Warren, K.S., and Mahmoud, A.A.F. (eds.), *Tropical and Geographical Medicine*, McGraw-Hill Book Company, New York, pp. 471–486, 1984.
26. Dworkin, M.S., and Wan, P.C. Helminth infection in HIV-infected persons in the United States, 1990–1999. *Clin. Infect. Dis.* 30, 241, 2000.
27. Weiis, M.E., and Raszaka, W.V. Jr., Cestode infection in children. *Adv. Pediatr. Infect. Dis.* 12, 109, 1996.
28. Manson, P.R., and Petterson, B.A. Epidemiology of *Hymenolepis nana* infections in primary school children in urban and rural communities in Zimbabwe. *J. Parasitol.* 80, 245, 1994.
29. Quihui-Cota, L. et al. Prevalence and intensity of intestinal parasitic infections in relation to nutritional status in Mexican school children. *Trans. R. Soc. Trop. Med. Hyg.* 98, 653, 2004.
30. Maggi, P. et al. *Hymenolepis nana* parasites in adopted children. *Clin. Infect. Dis.* 41, 571, 2005.
31. Black, R.E., Lopez de, R., and Brown, K.H. Incidence and etiology of infantile diarrhea and major route of transmission in Hausecar, Peru. *Am. J. Epidemiol.* 129, 785, 1989.
32. Gama-Eddin, F.M., Aboul-Atta, A.M., and Hassounah, O.A. Extra-intestinal *Hymenolepis nana* cysticercoidiasis in asthamatic and filarised Egyptian patients. *J. Egypt. Soc. Parasitol.* 16, 517, 1986.
33. Vito, D.L., Cinza, R., and Giuseppe, A. Skin eruption associated with *Hymenolepis nana* infection. *Int. J. Dermatol.* 43, 357, 2004.
34. Al-Hussani, M.K. et al. Phlyctenular eye disease is association with *Hymenolepis nana* in Egypt. *Br. J. Ophthalmol.* 63, 627, 1979.
35. Olds, R.J. The affect of the tapeworm *Hymenolepis nana* on immunity to tuberculosis in mice. *J. Hyg.* 67, 233, 1969.
36. Fagbemi, B.O., and Christensen, N.O. Delayed expulsion of *Hymenolepis diminuta* in *Trgpanosome brucei* infected mice. *Zeit, Parasit.* 70, 663, 1984.
37. Lucas, S.B. et al. Aberrant form of *Hymenolepis nana*: Possible opportunistic infection in immunocompromised patients. *Lancet* 2, 1372–1373, 1979.
38. Alghali, S.T., Hagan, P., and Robinson, M. *Hymenolepis citelli* (Cestoda) and *Nematospiroides dubins* (Nemchi) inter specific interaction in mice. *Exp. Parasitol.* 60, 364, 1985.
39. Ito, A., and Yamamoto, Y. Mechanism of the parasite's survival in the immunized host as well as the mechanism of immunity to reinfection. Primary and secondary infections with *Hymenolepis nana* eggs. 1. Changes in haematological values. *Jap. J. Parasitol.* 31, 203, 1976.
40. Isaak, D.D., Jacobson, R.H., and Reed, N.D. Thymus dependence of tapeworm *Hymenolepis diminuta* elimination from mice. *Infect. Immun.* 12, 1478, 1975.
41. Isaak, D.D., Jacobson, R.H., and Reed, N.D. The course of *Hymenolepis nana* infection in thymus deficient mice. *Int. Arch. Allergy Appl. Immunol.* 55, 504, 1977.
42. Alghali, S.T.O., and Grenics, R.K. Immunity to tapeworms: Intra specific cross-protective interactions between *Hymenolepis citelli, H. diminuta* and *H. microstoma* in mice. *Parasitology* 92, 665, 1986.
43. Ito, A. et al. *Hymenolepis nana*: Immunity against oncosphere challenge in mice previously given viable or non-viable oncosphere of *H. nana, H. diminuta, H. microstoma* and *Taenia taeniaforms*. *Int. J. Parasitol.* 21, 241, 1991.
44. Ito, A., Onitake, K., and Andreasen, J. Lumen phase cross immunity between *Hymenolepis microstoma* and *H. nana* in mice. *Int. J. Parasitol.* 18, 1019, 1988.
45. Ito, A. *Hymenolepis nana*: Survival in the immunized mouse. *Exp. Parasitol.* 49, 248, 1980.
46. Van der Vorst et al. Influence of an *Hymenolepis diminuta* infection on IgI and IgA antibodies to mouse intestinal eosinophus. *Int. Arch. Allergy Appl. Immunol.* 87, 281, 1988.
47. Ito, A. Basic and applied immunology in cestode infections from *Hymenolepis* to *Taenia* and *Echinococcus*. *Int. J. Parasitol.* 27, 1203, 1997.
48. Palmas, C., Wakelin, D., and Gabriele, F. Transfer of immunity against *Hymenolepis nana* in mice with lymphoid cells or serum from infected donors. *Parasitology* 89, 287, 1984.
49. Asano, K. et al. *Hymenolepis nana*: Passive transfer of mouse immunity by T-cell subset of phenotype Lyt-1. *Exp. Parasitol.* 61, 373, 1986.
50. Ajami, A., and Raflee, A. Cytokine production in *Hymenolepis* infection. *Iran. J. Immunol.* 4, 236, 2007.
51. Niwa, A., and Miyazato, T. Enhance of intestinal eosinophilia during *H. nana* infection in mice. *J. Helminthol.* 70, 33, 1996.
52. Ito, A. et al. Analysis of antibody responses to *Hymenolepis nana* infection in mice by the enzyme-linked immunosorbent assay and immunoprecipitator. *Parasite Immunol.* 16, 265, 1988.
53. Bortoletti, G., Gabriele, F., and Palmas, C. Mechanisms of protective immunity in *Hymenolepis nana*/mouse model. *Parasitologia* 34, 17, 1992.
54. Hopkins, C.A. Immunological memory in mice to adult *Hymenolepis diminuta* (Cestoda). *J. Parasitol.* 68, 32, 1982.
55. Craig, P., and Ito, A. Intestinal cestodes. *Curr. Opin. Infect. Dis.* 20, 524, 2007.
56. Voge, M. Concentrated method for helminthic ova and larvae. In: MacInnis, A.J., and Voge, M. (eds), *Experiment and Techniques in Parasitology*, W.H. Freeman, San Francisco, pp. 132–134, 1986.
57. Turant, A.L. Comparison of formalin–ethyl ether sedimentation, formalin–ethyl acetate sedimentation, and zinc sulfate floatation technique for detection of intestinal parasites. *J. Clin. Microbiol.* 13, 882, 1981.
58. Lussier, G., and Loew, F.M. Natural *Hymenolepis nana* infection in Mongolian gerbils (*Meriones unguiculatus*). *Can. Vet. J.* 11, 5, 1970.
59. Movvsesyan, S.O. et al. Experimental hymenolepiasis of rats: Preliminary data on histological changes of visceral organs. *Acta Parasitol.* 53, 193, 2008.
60. Shimoda, K., Furakawa, T., and Miyazato, T. Clinicopathological changes of mice following primary and secondary infection with *Hymenolepis nana* eggs. 1. Changes in haematological values. *Jap. J. Parasitol.* 31, 203, 1982.
61. Ito, A. et al. Analysis of antibody response to *Hymenolepis nana* infection in mice by enzyme-linked immunosorbent assay and immunoprecipitation. *Parasitol. Immunol.* 10, 265, 1988.
62. Gomez-Priego, A., Godinez-Hana, A.L., and Gulterrez-Quiiraz, M. Detection of serum antibodies in human *Hymenolepis* –infection by enzyme immunoassay. *Trans. R. Soc. Trop. Med. Hyg.* 85, 645, 1991.
63. Castillo, R.M. et al. Effect of treatment on serum antibody to *Hymenolepis nana* detected by enzyme-linked immunosorbent assay. *J. Clin. Microbiol.* 29, 413, 1991.

64. Montenegro, T. et al. The diagnostic importance of species specific and cross reactive components of *Taenia solium, Echinococcus granulosus* and *Hymenolepis nana*, *Rev. Inst. Med. Trop. Sao Paulo*. 36, 327, 1994.
65. Gonenc, B. Analysis of the crude antigen of *Hymenolepis nana* from mice by SDS-PAGE and the determination of specific antigens in protein structure by Western blotting. *Turk. J. Vet. Animal Sci*. 26, 1067, 2002.
66. Mota, I. and Wong, D. Homologus and heterologus passive cutaneous anaphylaxis activity of mouse antisera during the course of immunization. *Life Sci*. 8, 813, 1969.
67. Ito, A. IgG and IgE antibodies to *Hymenolepis nana* detected in infected mouse sera by gel diffusion and passive cutaneous anaphylaxis tests. *J. Parasitol*. 70, 170, 1984.
68. Krawczuk, S., Rode, W., and Machnicka, B. Purification and immunologic reactivity of *Hymenolepis diminuta* surface antigens. *Parasitol. Res*. 76, 707, 1990.
69. Larssh, J.E. Serological studies on the mouse strain of the dwarf tapeworm *Hymenolepis nana* var. *fraterna*. *Am. J. Hyg*. 37, 289, 1943.
70. Coleman, R.M., Carty, J.M., and Graziadei, W.D. Immunogenicity and phylogenetic relationship of tapeworm antigens produced by *Hymenolepis nana* and *Hymenolepis diminuta*. *Immunology*, 15, 297, 1968.
71. Macnish, M.G. et al. A molecular phylogeny of nuclear and mitochondrial sequences in *Hymenolepis nana* (Cestoda) supports the existence of a cryptic species. *Parasitology* 125, 567, 2002.
72. Okamoto, M. et al. Phylogenetic relationships of three hymenolepidid species inferred from nuclear ribosomal and mitochondrial DNA sequences. *Parasitology* 115, 661, 1997.
73. Mohammadzadeh, T. et al. Study on the genomic diversity of *Hymenolepis nana* between rat and mouse isolates by RAPD-PCR. *Iran. J. Vet. Res*. 18, 16, 2007.
74. Walsh, P.S., Metzger, D.A., and Higuchi, R. Chelex 100 as a medium for simple extraction of DNA for PCR based typing from forensic material. *BioTechniques* 10, 506, 1991.
75. Paxinos, E. et al. A non-invasive method for distinguishing among canid species: Amplification and enzyme restriction of DNA from dung. *Mol. Ecol*. 6, 483, 1997.
76. Heukelbach, J. et al. Selective mass treatment with ivermectin to control intestinal helminthiases and parasitic skin diseases in a severely affected population. *Bull. W.H.O*. 82, 563, 2004.
77. McCrckon, R.O., Lipkomitz, K.B., and Dronen, N.O. Efficacy of albendazole and mebendazole against *Hymenolepis microstoma* and *Hymenolepis diminuta*. *Parasitol. Res*. 78, 108, 1992.
78. Gupta, S., and Katiyar, J.C. Comparative activity of anticestode drugs praziquantel, niclosamide and compared 77-6 against *Hymenolepis nana*. *J. Helminthol*. 7, 31, 1983.
79. Andrews, P., Thomas, H., and Weber, H. The in-vitro uptake of 14C–Praziquantel by cestodes, trematodes and a nematode. *J. Parasitol*. 66, 920, 1980.
80. Chero, J.C. et al. *Hymenolepis nana* infection: Symptoms and response to nitazoxanide in field conditions. *Trans. R. Soc. Trop. Med. Hyg*. 101, 203, 2007.

25 Mesocestoides

Kerry A. Padgett, Paul R. Crosbie, and Walter M. Boyce

CONTENTS

25.1 INTRODUCTION

Mesocestoides spp. tapeworms are common parasites of wildlife and companion animals; human cases are rare. Although adult *Mesocestoides* spp. tapeworms are easily identified to the genus level using morphological characteristics, molecular diagnostic techniques provide better taxonomic resolution at the species level. In particular, the immature tapeworm stages are notoriously difficult to identify morphologically, and molecular methods are essential in the diagnoses of cases of larval cestodiasis in dogs, and potentially in other animals.

25.1.1 Biology and Morphology

25.1.1.1 Epidemiology and Distribution

Adult *Mesocestoides* tapeworms are found in birds and placental mammals worldwide, with the exception of Australia. While rarely reported, humans can also serve as hosts to adult *Mesocestoides* tapeworms. Most human cases are associated with ingestion of uncooked or undercooked reptiles or poultry. To date, there have been at least 30 documented human cases of mesocestoidiasis worldwide, with the largest number in Japan (14), followed by the United States (10), Korea (2), China (2), Ruanda-Urundi (1), and Greenland (1) [1–10] (CDC-DPDx, unpublished data). In eastern Asia, reported human cases of mesocestoidiasis are typically males over 30 years of age, with a history of eating raw poultry viscera or uncooked snake blood or viscera for medicinal purposes. In contrast, all cases of mesocestoidiasis in the United States and the single case in Ruanda-Urundi were in children, most under the age of 2 years. While the African child likely acquired infection after ingesting raw partridge liver, the source for the other children is unclear and may involve accidental contamination of food or ingestion of infective material as part of mouthing behavior common to toddlers. Human cases of mesocestoidiasis in eastern Asia were associated with *Mesocestoides lineatus*, which occurs in Europe and Asia, while cases in the United States, Africa, and Greenland were identified as *M. variabilis*, *M. lineatus*, or simply *Mesocestoides* spp.

There have been no confirmed human metacestode (larval tapeworm) *Mesocestoides* cases. Several human metacestode cases were suspected to involve *Mesocestoides* tetrathyridia [11–13], but metacestodes recovered from patients have never been definitively identified as *Mesocestoides*. Other pseudophyllidean or cyclophyllidean tapeworms are the likely parasites in these cases, although nonhuman primates can be infected with *Mesocestoides* metacestodes [14–16].

The geographic distribution of documented pathogenic larval cestodiasis cases, caused by proliferative *Mesocestoides* metacestodes in nonhuman hosts, includes the western United States, western Canada, Germany, Italy, Spain, and Turkey [17–24]. The majority of *Mesocestoides* larval cestodiasis cases in domestic animals have been reported from dogs [17,18,21,23–25], but there have been reports in cats [20,22], ferrets (Boyce, unpublished results), and primates [14–16].

25.1.1.2 Classification and Gross Morphology

Mesocestoides Vaillant, 1863 is the only genus in the subfamily Mesocestoidinae Lühe, 1894. (Phylum: Platyhelminthes, Class: Cestoda, Order: Cyclophyllidea, Family: Mesocestoididae). The other subfamily, Mesogyninae Tschertkowa and Kosupko 1977 also has a single genus, *Mesogyna*, Voge, 1952; adult tapeworms of the single species in this genus, *Mesogyna hepatica*, have been reported only from the bile ducts and gall bladder of kit foxes (*Vulpes macrotis*) in California.

Mesocestoides tapeworms differ from other tapeworms in the order Cyclophyllidea to such an extent that it has been proposed that the family Mesocestoididae be placed in its own order [26–29]. Adult Cyclophyllidean tapeworms have an anterior attachment region called a scolex with a heavy muscular wall and four cup-like suckers (acetabula) evenly spaced around the scolex. Although adult *Mesocestoides* have these four suckers, the rostellum, a structure at the tip of the scolex common to many Cyclophyllidean tapeworms, is absent, and the proglottids are unique in that the genital pore is mid-ventral rather than lateral; the pore is for fertilization, not dissemination of eggs. Furthermore, unlike any other tapeworm, and thus diagnostic, *Mesocestoides* have a unique thick-walled paruterine organ within which the eggs or oncospheres develop (Figure 25.1). But perhaps the most unusual feature of Mesocestoidid tapeworms is that some species or strains are capable of asexual replication of the larval stages (metacestodes), which may have suckers (tetrathyridia) or lack suckers (acephalic larva) [25,30,31]. Lastly, unlike other Cyclophyllidean tapeworms that are two-host parasites, *Mesocestoides* is assumed to require three hosts to complete its life cycle.

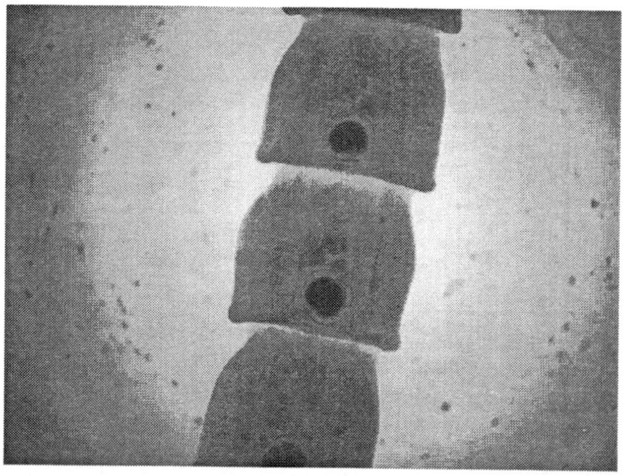

FIGURE 25.1 Mature *Mesocestoides* proglottids; note the diagnostic paruterine organ in the center of proglottids. (*Mesocestoides* tapeworm courtesy of Claude G. Alexander; photograph by Heike Quinn, California Department of Public Health, Microbial Disease Laboratory.)

At the species level, the taxonomy of *Mesocestoides* needs further clarification as many species names are likely synonyms and it is also likely that there are many undescribed species. Identification and species description of *Mesocestoides* tapeworms are difficult as the key morphological characters are not well established [32] and there is a high degree of host-induced nonspecific morphological variation [33]. Furthermore, the larval or metacestode stage may sometimes lack any distinguishing morphological characteristics such as suckers and only the presence of calcareous corpuscles confirms their cestode identity [17,18]. While at least 27 species have been described worldwide [34], it is unclear as to how many of these are true species. In Europe and Asia, there are records of at least 15 species (*M. alaudae, M. angustatus, M. bergini, M. canislagopodis, M. erschovi, M. imbutiformis, M. kirbyi, M. leptothylacus, M. latus, M. lineatus, M. litteratus, M. mesorchis, M. petrowi, M. perlatus*, and *M. zacharovae*). There are at least 14 described species in North America (*M. bassarisci, M. caninum, M. corti, M. insignis, M. jonesi, M. kirbyi, M. latus, M. lineatus, M. manteri, M. rara, M. perlatus, M. petrowi, M. variabilis*, and *M. vogae*), and 8 recorded species in Africa (*M. ambiguus, M. charadrii, M. dissimilis, M. elongatus, M. lineatus, M. litteratus, M. longistriatus*, and *M. perlatus*). Only one species has been recorded from South America, *M. michaelseni*, from a crab-eating fox (*Cerdocyon thous*) in Brazil. There are no keys to differentiate among species and identification is typically made by subjective and descriptive characteristics of adult tapeworms or by geographic location [35]. Furthermore, formal taxonomic assignment of species is difficult as the range of morphometric measurements as well as geographic distribution overlap among nominal species; therefore, other methods such as molecular characterization are needed [36]. Many of these species are listed as *species inquirenda*, such as *M. corti*, which is now generally accepted to be *M. vogae* [36,37]; others are suggested synonyms such as *M. manteri* and *M. variabilis* which may both be *M. vogae* [35]. Moreover, recent molecular analyses of tapeworms removed from wildlife and domestic animals suggest potential additional species [23,36]. As additional sequences from *Mesocestoides* spp. sampled from diverse geographic locations and hosts become available, *Mesocestoides* spp. may be most reliably identified to the species level with a combination of morphological and molecular methods (see below for diagnostic methods). To date, only *M. vogae, M. litteratus*, and *M. lineatus* can reliably be identified to species using morphological and molecular characters [36,38–40]. There are a few *Mesocestoides* spp. that are apparent species based on molecular and morphological characteristics but it is unclear as to whether they are new species or synonyms [23,36].

All species-level taxonomic descriptions to date have been made based on morphology of adult *Mesocestoides*. Tetrathyridia can only be reliably identified to the genus *Mesocestoides*. Molecular methods are the recommended technique to refine identification of immature *Mesocestoides* (i.e., tetrathyridia, acephalic larva, hexacanth embryo) to species or clade.

25.1.1.3 Morphology

Adult *Mesocestoides* have a scolex (219–587 μm) with four unarmed suckers (100–278 μm) and lack a rostellum. The strobila are thin, between 1 and 2 mm wide, and are not typically longer than 25 cm. Mature proglottids near the middle of strobila are wider than long and gravid proglottids near the end are typically longer than wide, and are similar, in size and shape, to a sesame seed. The size and length of strobila vary dramatically within species, typically shorter and thinner if the worms were part of a heavy infection [32]. There is a single bilobed ovary and numerous testes either pre- or postovarian in each proglottid. The vitellarium is postovarian and doubled. The primary diagnostic features are the paruterine organ (90–547 μm) in the center of the proglottid, and the mid-ventral genital pore. In a gravid proglottid, the paruterine organ contains hundreds of motile circular or oval oncospheres (20–40 μm) also known as hexacanth larvae. Each oncosphere is armed with three pairs of hooks, and unlike pseudophyllidean tapeworm infections, oncospheres are not generally found in fecal samples by standard diagnostic techniques, such as flotation, but rather by detecting gravid intact proglottids released in feces.

The typical metacestode forms of *Mesocestoides*, called tetrathyridia, are small white irregular-shaped immature cestodes with four suckers, often invaginated, ranging in size from <1 cm to 7 cm [23]. Often, they are widest at the anterior and narrowest at the posterior end. The metacestode may also be recovered with no suckers and with highly variable morphology; this form is often associated with canine peritoneal larval infections. These "acephalic" metacestodes can range in size from tiny 1-mm flakes to larger 1–2-cm irregularly shaped metacestodes, sometimes with fluid-filled sacs. The presumptive first-stage procercoid or cysticercoid has never been recovered and its morphology as well as its host range is unknown.

25.1.1.4 Life History

Despite over 50 years of dedicated research, life cycles of *Mesocestoides* tapeworms remain undefined. Sexually mature *Mesocestoides* are intestinal parasites of birds and mammals, chiefly canines. Immature tapeworms (metacestodes) are found in the body cavities of a variety of tetrapod vertebrates, including amphibians, reptiles, birds, and mammals [26] (Figure 25.2). In addition to being hosts of adult *Mesocestoides*, domestic dogs and cats can function as intermediate hosts and harbor metacestodes [17,18,21,23]. These metacestode infections in dogs can be serious, often resulting in death [17,18]. As discussed, humans are periodically infected with *Mesocestoides* adult tapeworms, presumably after ingesting uncooked amphibians, reptiles, or poultry [10].

While most cyclophyllidean tapeworms (e.g., *Taenia* spp., *Dipylidium caninum*, or *Hymenolepis* spp.) require only two hosts to complete development, *Mesocestoides* spp. apparently require three hosts, that is, a vertebrate definitive host, a vertebrate secondary intermediate host, and a presumed arthropod first intermediate host [33,41,42]. Carnivores, principally wild canines, are the definitive hosts, developing adult tapeworm infections after ingesting tetrathyridia or acephalic larvae within prey. Adult tapeworms are principally found in the small intestine of canines, but are also found in the colon.

Mature *Mesocestoides* tapeworms shed gravid proglottids in feces; dog and cats begin shedding proglottids 2–3 weeks after ingesting tetrathyridia and foxes shed in less than 2 weeks [32,42]; gravid proglottids are then shed intermittently. Proglottids are of the size and shape of sesame seeds and are usually white or cream colored and are generally found dotting the exterior of feces; few proglottids are found within the feces [18]. Fresh *Mesocestoides* proglottids are motile and move in irregular patterns without dispensing oncospheres [32,43–45].

It is clear that vertebrates do not develop metacestode or adult tapeworm infections after ingesting *Mesocestoides* proglottids that contain viable motile oncospheres [41–43]. A first intermediate host is needed to initiate the life cycle of these tapeworms and despite many studies [41,43,45,46] the identity of this host or hosts is still unknown. It is likely that, similar to some other cyclophyllidean tapeworms (e.g., *Moniezia* spp. and *Raillietiella* spp.), *Mesocestoides* spp. require an arthropod first intermediate host. As the oncospheres are retained in the paruterine organ and are not released into the environment, it is likely the first host ingests all or part of the proglottid, including the paruterine organ. The procercoid or cysticercoid presumably develops in this first host. Arthropods that have been suggested as potential hosts include oribatid mites, dung beetles, and ants [43,45,46]. Ingestion of the cysticercoid or procercoid within the first intermediate host by vertebrates likely results in metacestode (tetrathyridial) infection.

Second intermediate hosts of *Mesocestoides* include amphibians [47], reptiles [48,49], small rodents [42], and birds [40], with tetrathyridia most commonly found within the peritoneal or pleural cavities but also sometimes embedded in liver, mesentery, peritoneum, heart muscle, or subcutaneous tissues [43]. Tetrathyridia have also been found embedded in cutaneous cysts [40]. Periodically, dogs and cats can also be infected with extraintestinal tetrathyridia but are considered dead-end hosts in the life cycle. There are no reports of wild or domestic carnivores with concurrent infections of extraintestinal *Mesocestoides* tetrathyridia and adult tapeworms.

As previously mentioned, in addition to the tetrathyridial form, *Mesocestoides* metacestodes that lack suckers are called "acephalic" metacestodes [17,18]. Other than the appearance of calcareous corpuscles, there are no diagnostic features that identify these acephalic metacestodes as cestodes, or as *Mesocestoides* metacestodes specifically. These acephalic forms are commonly recovered from the peritoneal cavity of domestic dogs in western North America but have also been reported from dogs in Italy and Germany [19,24]. Other cestode larvae that may be potentially confused with *Mesocestoides* spp. are the plerocercoids of the pseudophyllideans *Spirometra* spp. Prior to the development of molecular diagnostic methods it was impossible to confirm suspected cases of acephalic asexually proliferative cestodes in peritoneal cavities as *Mesocestoides* [50,51].

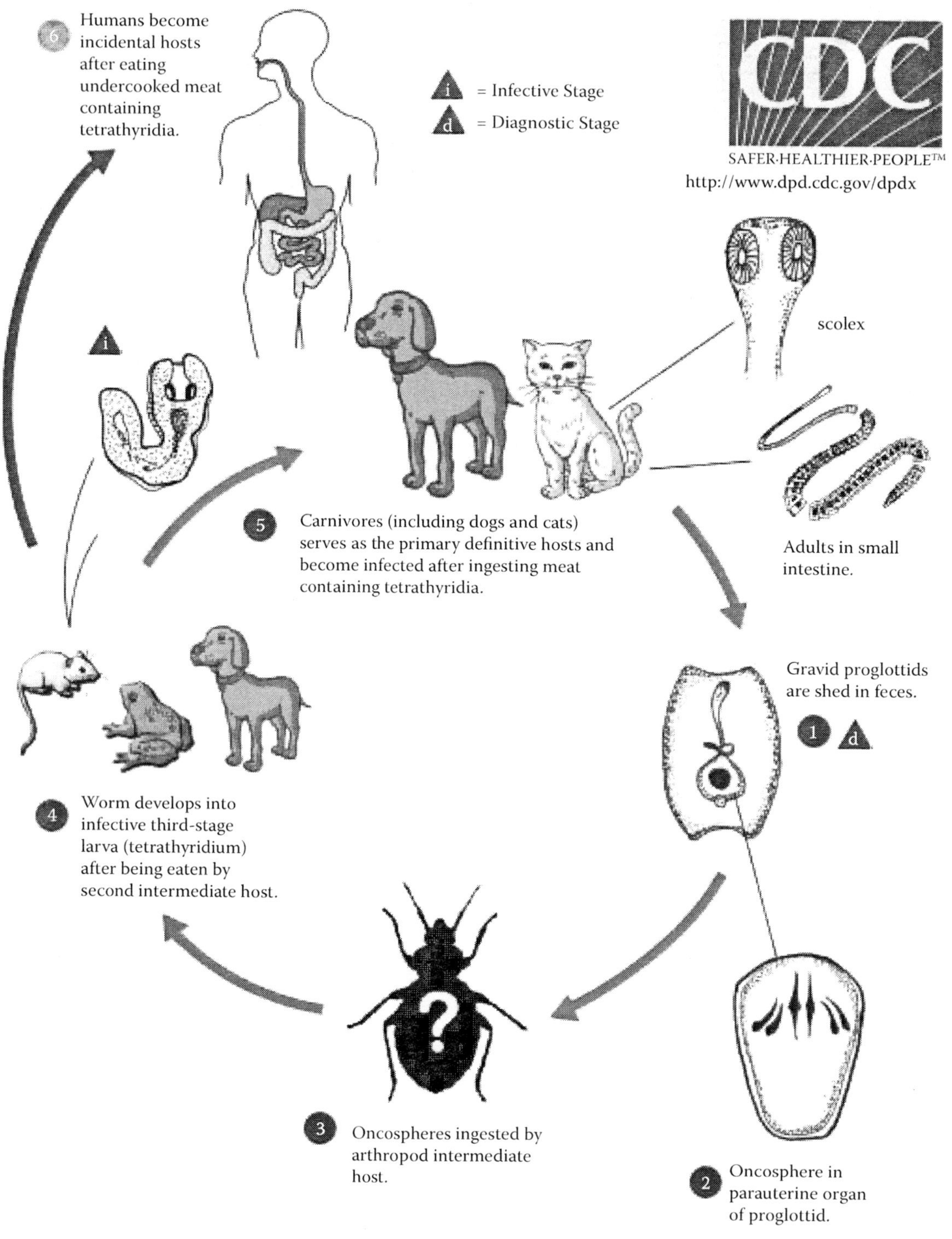

FIGURE 25.2 *Mesocestoides* spp. require a three-host life cycle to complete their development. The definitive hosts are primarily carnivores, including canids, felids, and mustelids. Motile gravid proglottids are shed in feces (1–2). The first intermediate host is presumed to be an arthropod (3), and becomes infected after eating proglottids or oncospheres. Several arthropods have been looked at as potential first intermediate hosts for *Mesocestoides* spp, including ants and oribatid mites; however, none of these species has been definitively demonstrated to serve as an intermediate host. In the first intermediate host, the oncosphere develops into a second-stage larva. When the first intermediate host is eaten by a second intermediate host, including small mammals, birds, reptiles, and amphibians (4), the second-stage larva develops into an infective, third-stage larva (tetrathyridium). Domestic carnivores, such as dogs and cats, which usually serve as the definitive host, may also serve as dead-end intermediate hosts upon ingestion of infected first intermediate hosts. The definitive host ultimately becomes infected after eating meat contaminated with tetrathyridia (5). Upon ingestion, the cestode settles in the small intestine where it matures. Gravid proglottids can be seen in the stool as early as two weeks. Humans are not usual definitive hosts, but can serve as such after eating undercooked meat containing tetrathyridia (6). (Life cycle of *Mesocestoides* spp. courtesy of the CDC-DPDx.)

As a strategy for increased transmission, asexual replication is a relatively common trait of the metacestodes of other cyclophyllidean tapeworms [52]. The mechanism for asexual replication of adult or tetrathyridial forms is by longitudinal fission of the scolex [31]. While asexual proliferation may not be common for all *Mesocestoides* species [37,53], it is well established that tetrathyridia of *M. vogae* can asexually replicate in mice, snakes, and dogs [31,54]. Furthermore, adult *M. vogae* can also asexually replicate within the intestinal tract [31,42]. Acephalic metacestodes retrieved from dogs with larval cestodiasis are commonly proliferative, likely by budding, but the mechanism is still unclear [25]. *Mesocestoides lineatus* acephalic metacestodes associated with cases of canine larval cestodiasis in Italy and Germany appear to be asexually proliferative based on clinical symptoms [19,24]. The acephalic metacestodes of another variant, currently known as "Clade A," also have a propensity to replicate asexually [25,36]. This is the most common form associated with canine peritoneal larval cestodiasis (CPLC) in western North America. Interestingly, the tetrathyridial form of this Clade A variant more commonly found in wildlife, is not asexually proliferative [42].

25.1.2 Clinical Features and Pathogenesis

25.1.2.1 Human Pathogenesis

Human infection with adult *Mesocestoides* spp. tapeworms is relatively rare, with only 30 cases documented in the literature. While diarrhea and gastrointestinal distress and anemia have been reported [1,10], the majority of patients with confirmed infections were asymptomatic and initially sought care because of the presence of proglottids in their stool. There are no confirmed cases of extraintestinal larval infection in humans.

25.1.2.2 Veterinary Pathogenesis

Veterinary infections with adult *Mesocestoides* spp. tapeworms are widespread, especially in wild carnivores, but typically asymptomatic. Clinical disease caused by *Mesocestoides*, while relatively rare, is most commonly seen in dogs with larval, extraintestinal infections (Figure 25.3). Peritoneal larval cestodiasis has been diagnosed in dogs and cats in western Europe and western Asia, but the majority of cases have been reported from western North America [18,19,21–24], and the laboratory of one of the authors (Boyce) has become the leading center for research on this emerging disease that kills >25% of clinically affected dogs [17,18,25,36,42,45]. The route and risk factors for infection are unknown, diagnosis is challenging (it requires exploratory surgery or attaining a peritoneal aspirate with a large bore needle), and clinically affected animals have a guarded prognosis [17,55].

The agents responsible for CPLC are immature tapeworms in the genus *Mesocestoides* that multiply asexually by division or budding within the peritoneal cavity of the dog. Dogs with larval cestodiasis often present with ascites, and severely affected dogs may have liters of fluid and

FIGURE 25.3 Peritoneal cavity of dog with canine peritoneal larval cestodiasis. White flakes are *Mesocestoides* acephalic metacestodes.

parasites in their peritoneal cavity. A history of long-term steroid treatment seems to be a common factor in many of the acute cases, and laboratory studies with rodents suggest that steroids and sex (males are more susceptible) influence parasite multiplication, perhaps through immunomodulation [56,57]. Dogs with larval cestodiasis do not harbor concurrent *Mesocestoides* adult tapeworm infections.

25.1.2.3 Veterinary Treatment

The most effective approach appears to be a combination of surgery and anthelmintic treatment. Regardless of the treatment approach, the prognosis is guarded, especially for dogs presenting with rapidly developing signs or severe disease.

In cases where large quantities (sometimes liters) of fluid and parasites are present, it is highly recommended to physically remove as much of this material as possible prior to treatment.

Peritoneal lavage is useful and effective when parasites and fluid are found free in the peritoneal cavity. If parasites are sequestered in "pockets" or cysts, it may be possible to drain the cysts or surgically remove them. Occasionally, tetrathyridia invade surrounding organs and tissues, and even find their way into the scrotum. Clinical improvement is often noticed immediately after peritoneal lavage, especially if a large volume of fluid is removed. However, infected dogs have varying degrees of peritonitis, and some dogs may develop severe fibrinous peritonitis during and following anthelmintic treatment.

The current suggested treatment is with fenbendazole (Panacur®) at 100 mg/kg per os, twice daily for 28 consecutive days. Anthelmintics used to treat adult tapeworms such as praziquantel (Droncit®) are not effective against *Mesocestoides* larvae in the peritoneal cavity. Recent work on drug uptake in helminth parasites (tapeworms, flukes, nematodes) has shown that passive uptake across the tegument is the primary way anthelmintics reach the target receptors within parasites [58]. Fenbendazole is poorly absorbed across the gastrointestinal tract making it difficult to achieve high concentrations at the tegmental surface. Unfortunately, intraperitoneal administration of fenbendazole and praziquantel do not appear to increase efficacy (Boyce, unpublished).

It may be necessary to retreat animals if clinical disease reoccurs in the months or years following initial treatment. Reoccurrence may be more common in dogs initially harboring the greatest number of tapeworm larvae, since these animals have a demonstrated inability to keep the parasite numbers in check. We also suspect that fenbendazole may fail to completely eliminate all the parasites. Some dogs may need life-long daily treatment with fenbendazole to keep the infection under control.

Special attention should be paid to the animal's condition during treatment. Fenbendazole has a high margin of safety, but there may be adverse reactions (i.e., azotemia) due to the high dose of the drug, or due to the interaction of the drug, parasites, and the host immune response. Dogs undergoing treatment do not need to be hospitalized, but the owner should watch the animal closely for any central nervous system signs, vomiting, or changes in appetite, behavior, and so on. Fenbendazole should be discontinued immediately if adverse reactions are noted. Some dogs will tolerate reinitiation of treatment after a break of several days.

Ultrasound and peritoneal aspirates may be used to monitor treatment efficacy and the potential reoccurrence of parasites. Posttreatment follow-up is especially useful for evaluating and modifying treatment protocols.

25.1.3 Diagnostic Techniques

25.1.3.1 Conventional Diagnostic Techniques

Adult tapeworm infections are traditionally diagnosed by detection of either eggs or proglottids in fecal samples. Pseudophyllidean tapeworms (e.g., *Diphyllobothrium latum*) are usually diagnosed after detection of thin-walled operculate eggs in fecal sediments, although the shedding of entire sections of gravid proglottids is not uncommon. Cyclophyllidean tapeworms are usually detected either by observation of the thick-walled eggs with enclosed hexacanth embryos in fecal flotations, or by gross observation of entire proglottids (e.g., *Dipylidium caninum*) on the exterior of fecal masses. Identification to genus is often possible based on the morphology of the egg, and presumptive species identification may often be made based on the identity of the host, as many tapeworm species have high degrees of host specificity.

Unlike other Cyclophyllidean tapeworms, adult *Mesocestoides* tapeworm infections cannot be diagnosed during routine ova and parasite examination using fecal flotation. Hexacanth embryos are retained in the paruterine organ and are not released in the feces nor are they released in the environment during movement of the proglottid. Diagnosis of *Mesocestoides* sp. adult tapeworm infections is based on finding gravid proglottids with characteristic paruterine organ on the exterior surface of stool samples.

More precise identification of the cestodes may be achieved by fixation and staining of proglottids and scolices; in this case, observation of the paruterine organ and the median genital aperture would provide a definitive identification to genus for *Mesocestoides*. Specimen preparation after collection of proglottids from fecal samples or entire worms at necropsy typically involves relaxation in tap water, and fixation with an ethanol-based fixative prior to staining. The stain of choice for tapeworms is an ethanol-based solution of carmine. In a progressive staining procedure, after fixation, specimens are passed through a graded series of alcohols before immersion in a solution of carmine in 70% ethanol. When correct differentiation of internal structures is achieved (a matter of experience), the specimen is washed in ethanol prior to mounting in Canada balsam [34]. Larval stages (including tetrathyridia and the acephalic forms of *Mesocestoides*) may be prepared in a similar manner.

25.1.3.2 Molecular Diagnostic Techniques

The development of molecular techniques has greatly increased our ability to accurately identify and characterize parasites. The high virulence and amorphous nature of the acephalic larvae of *Mesocestoides* made development of more sensitive and precise identification techniques necessary and valuable.

As yet, there are no biochemical or immunological techniques for identification and characterization of *Mesocestoides* spp. Instead, a technique was developed based on restriction enzyme digest of PCR-amplified DNA fragments, to not only identify *Mesocestoides*, but also to differentiate between that genus and other common tapeworm genera, particularly those that may be found in dogs, for example, *Taenia* and *Dipylidium* [17]. The basis of the technique is the recognition of cestode genus-specific restriction fragment length polymorphisms (RFLP) in a ≈650-bp fragment of the second internally transcribed spacer (ITS-2), a noncoding portion of the nuclear ribosomal gene array. Three RFLP patterns (genotypes) unique for the genus *Mesocestoides* have been described from western North America and each of these is distinct from RFLP patterns shown in all of the other tapeworm genera analyzed [17,25]. The original demonstration of the technique included two *M. corti* larval isolates, two *Mesocestoides* sp. proglottids from coyotes (*Canis latrans*), one *Mesocestoides* sp. proglottid from a dog, eight other tapeworm genera [17], and later-included two adult tapeworm samples from dogs [25].

Using this technique, to identify a sample as belonging to *Mesocestoides* requires extraction of genomic DNA, PCR amplification of the ITS-2 fragment using specific primers, and then digestion of the product of this reaction separately by each of the four restriction endonucleases (*Alu*I, *Cfo*I, *Msp*I, and *Rsa*I). Banding patterns are visualized on agarose gels. This technique definitively identifies a sample, regardless of morphology or amount of available cestode tissue, as *Mesocestoides* [17].

Greater resolution of the species identities and phylogenetic relationships of *Mesocestoides* specimens has been achieved by comparison of sequenced DNA fragments. So far, the evolutionary relationships within *Mesocestoides* have largely been resolved by phylogenetic analyses of sequence data from a portion of the nuclear 18S rDNA gene coupled with sequences from the portion of ITS-2 used in the RFLP analysis described above [17,25,36,42]. These studies were able to identify several distinct clades within the genus. Although species designations were not made, these clades were designated as Clade A, which is possibly *M. kirbyi* Chandler 1944 or *M. manteri* Chandler 1943; Clade B, which is likely synonymous with *M. vogae*; and Clade C, which is possibly a new species, although similar to *M. litteratus* [18,23,36,40]. Additionally, a possible new species of *Mesocestoides* was identified from the Canary Islands, using ITS-2 sequence data, that is the sister taxon (based on maximum parsimony analysis) of Clade C above; this new species also causes CPLC [23].

A 314-bp fragment of mitochondrial 12S rDNA has also been shown to be useful in distinguishing *Mesocestoides* from other cyclophyllidean tapeworms as well as among putative species [36,39]. Comparative sequence analysis among both *Mesocestoides* isolates and *Taenia* species using 12S rDNA agreed with data from the ITS-2 gene region and supported species delimitation [36]. Nevertheless, as for all molecular systematics, it is advisable to include standard morphological descriptions when feasible. For example, analysis of 12S rDNA for the two species *M. lineatus* and *M. leptothylacus* (synonym *M. litteratus*) from red foxes (*Vulpes vulpes*) in Germany had only slight genetic differences but are clearly distinguished morphologically [39]. Whether this may be a case of phenotypic plasticity is unknown.

The only other gene that has been sequenced in molecular studies of *Mesocestoides* is a 366-bp portion of the mitochondrial CO1 gene [22]. Sequences were obtained from 10 individual cestodes recovered from a single infected cat; all were identical, and differed from *M. corti* (*vogae*) at 11.99% of positions. These sequences were not used in any phylogenetic analyses [22].

25.2 METHODS

25.2.1 Sample Preparation

Human, veterinary, and wildlife samples include proglottids or tapeworm strobila removed from fecal samples. Veterinary or wildlife samples may also include metacestodes (acephalic metacestodes or tetrathyridia) removed from the body cavity of vertebrates. Samples can be preserved in 70% ethanol or in frozen sterile saline solution. Genomic DNA can be extracted from 30 mg of cestode tissue using commercial extraction kits (DNeasy Tissue Kit, Qiagen); concentration should be confirmed by the use of a spectrophotometer.

25.2.2 Detection Procedures

Although several genetic fragments can be targeted for molecular detection and characterization of *Mesocestoides* spp. in diagnostic samples, the second internally transcribed spacer region (ITS-2) in the nuclear region has been most widely used for veterinary samples using the protocol described below. Alternatively, the 12S mitochondrial gene fragment, with highly variable noncoding regions, also may be useful in molecular characterization of *Mesocestoides* [39].

1. Amplify a 650-bp portion of DNA from the second internally transcribed spacer (ITS-2) in the nuclear genome using the following cestode-specific primers: NC-6 (5′-ATCGACATCTTGAACGCACATTGC-3′) and NC-2 (5′-TTAGTTTCTTTTCCTCCGCT-3′). Reaction volume is 100 μL, comprised of 5 μL of each primer (10 mM), 2 μL of 10 mM dNTP mix, 10 μL of 10 PCR buffer with $MgCl_2$, 0.4 μL (5 U/μL) Taq polymerase, 1–5 μL of template DNA (optimized to ~400 ng), and a sufficient amount of sterile water to achieve the 100 μL final volume [17,25].
2. Perform each reaction in thin-walled microtubes in a thermocycler. Reaction conditions are as follows: two cycles of 94°C (30 s), one cycle of 65°C (30 s), and one cycle of 72°C (30 s); then 23 cycles of 94°C (30 s), one cycle of 55°C (30 s), and one cycle of 72°C (30 s); hold at 72°C for 4 min, and final storage at 4°C.
3. The product of this reaction can be evaluated, and separated to Clades A, B, and C by RFLP patterns using each of four restriction endonucleases (*Alu*I, *Cfo*I, *Msp*I, and *Rsa*I) at a concentration of 4 U/μL at 37°C for 1 h [17]. Digested material may be electrophoresed on 3% agarose gels containing 0.1 μg/mL of ethidium bromide at 100 mA for 35 min in 1× TBE buffer (45 mM Trizma base, 45 mM boric acid, and 1.0 mM EDTA; pH 8.0).
4. Alternatively, PCR reaction product can be sequenced for comparative sequence analysis. Purify the PCR reaction product using a commercial kit (Qiaquick PCR Purification Kit, Qiagen). Sequence both strands using NC-6 and NC-2 primers. Assemble and edit complementary sequences using the appropriate program (e.g., Sequencher), remembering to remove terminal PCR primer sequences. Conduct comparative sequence analyses with published sequences in GenBank to determine the closest species match.

25.3 CONCLUSION

Mesocestoides infections with tetrathyridia and acephalic larval forms can cause fatal larval cestodiasis in domestic animals, principally dogs. As the life cycle of this tapeworm is incompletely described, it is unknown as to how domestic animals acquire these larval infections. Infections of larval cestodiasis, primarily involving acephalic larvae, have been reported from the western United States as well as in Spain, Italy, Germany, and Turkey. Molecular methods, principally PCR and sequence analysis of the ITS-2 (18S rDNA), 12S rDNA, and CO1 gene regions [22,25,36,39] have proven to be essential in identifying *Mesocestoides* as the causative agent in larval cestodiasis as the lack of distinguishing morphological characteristics of acephalic larvae precludes identification based on morphology.

While adult *Mesocestoides* tapeworms are rarely reported from humans, infections may be underreported or misidentified. As adult *Mesocestoides* intermittently shed proglottids, not eggs, *Mesocestoides* adult tapeworm infections will not be detected during routine ova and parasite screening procedures such as fecal flotation. All reported human *Mesocestoides* cases have been detected when whole tapeworms or proglottids, with the characteristic paruterine organ, were recovered from stool samples. Risk factors for *Mesocestoides* adult infection in humans include ingestion of uncooked reptiles, amphibians, and poultry. Furthermore, in the United States, it appears that children under the age of 2 years are most commonly infected.

Due to the high phenotypic plasticity of *Mesocestoides*, it is notoriously difficult to assign species to *Mesocestoides* associated with human, wildlife, or domestic animal infection. Human cases of *Mesocestoides* in Asia have been identified as *Mesocestoides lineatus* while in other regions such as the United States, cases were identified as *M. variabilis*, *M. lineatus*, or *Mesocestoides* spp. In future cases, molecular characteristics may assist in identification to species and perhaps may clarify factors involved in infection, such as potential intermediate host species ingested or geographic area exposed. As morphological characteristics are unreliable, molecular characteristics can assist in delineating species. It is likely that there are some undescribed species of *Mesocestoides* and conversely, it is also likely that there are many *Mesocestoides* species that are synonyms. As more studies include molecular characteristics of *Mesocestoides* among isolates obtained in diverse regions, this may help to clarify the taxonomy of this group and clearly define the species of *Mesocestoides* that cause infections in people as well as in domestic dogs and perhaps shed light on the enigmatic life cycle of this parasite.

REFERENCES

1. Chandler, A.C. First record of a case of human infection with tapeworms of the genus *Mesocestoides*. *Am. J. Trop. Med.* 22, 493, 1942.
2. Gleason, N.N., and Healy, G.R. Report of a case of *Mesocestoides* (Cestoda) in a child in Missouri. *J. Parasitol.* 53, 83, 1967.
3. Gleason, N.N., Kornblum, R., and Walzer, P. *Mesocestoides* (Cestoda) in a child in New Jersey treated with Niclosamide (Yomesan®). *Am. J. Trop. Med. Hyg.* 22, 757, 1973.
4. Gutierrez, Y., Buchino, J.J., and Schubert, W.K. *Mesocestoides* (Cestoda) infection in children in the United States. *J. Pediatr.* 93, 245, 1978.
5. Hutchison, W.F., and Martin, J.B. *Mesocestoides* (Cestoda) in a child in Mississippi treated with paromomycin sulfate (Humatin®). *Am. J. Trop. Med. Hyg.* 29, 478, 1980.
6. Ohtomo, H. et al. Therapeutic effect of paramomycin sulfate on the 13th case of *Mesocestoides lineatus* infection found in Japan. *Jpn. J. Antibiot.* 36, 632, 1983.
7. Min, D.Y. Cestodes infections in Korea. *Kor. J. Parasitol.* 28, 123, 1990.
8. Eom, K.S., Kim, S.H., and Rim, H.J. Second case of human infection with *Mesocestoides lineatus* in Korea. *Kor. J. Parasitol.* 30, 147, 1992.
9. Schlutz, L.J. et al. *Mesocestoides* (Cestoda) infection in a California child. *Pediatric Infect. Dis. J.* 11, 332, 1992.
10. Fuentes, J.V., Galan-Puchades, M.T., and Malone, J.B. A new case report of human *Mesocestoides* infection in the United States. *Am. J. Trop. Med. Hyg.* 68, 566, 2003.
11. Marty, A.M., and Neafie, R.C. Hymenolepiasis and miscellaneous cyclophyllidiases. In: Meyers, W.M. (ed.), *Pathology of Infectious Diseases*. Chapter 12. Armed Forces Institute of Pathology, American Registry of Pathology, Washington, 2000.
12. Garin, Y.J.F. et al. Case report: Human brain abscess due to a tetra-acetabulate plerocercoid metacestode (Cyclophyllidea). *Am. J. Trop. Med. Hyg.* 72, 513, 2005.
13. Ihalamulla, R.L., Gunatilaka, H., and Fernado, S.D. Human sparganosis in Sri Lanka: A case report and review of the cases. *J. Parasit. Dis.* 32, 30, 2008.
14. Reid, W.A., and Reardon M.J. *Mesocestoides* in the baboon and its development in laboratory animals. *J. Med. Primatol.* 5, 345, 1976.
15. Hubbard, G.B. et al. *Mesocestoides* infection in captive olive baboons (*Papio cynocephalus anubis*). *Lab. Anim. Sci.* 43, 625, 1993.
16. Fincham, J.E. et al. Pleural *Mesocestoides* and cardiac shock in an obese vervet monkey (*Cercopithecus aethiops*). *Vet. Pathol.* 32, 330, 1995.
17. Crosbie, P.R. et al. Diagnostic procedures and treatment of eleven dogs with peritoneal infections caused by *Mesocestoides* spp. *J. Vet. Med. Assoc.* 213, 1578,1998.
18. Crosbie, P.R., Padgett, K.A., and Boyce, W.M. *Mesocestoides* spp. tapeworm infections in dogs in California, *Calif. Vet.* 54, 15, 2000.
19. Bonfanti, U. et al. Clinical, cytological and molecular evidence of *Mesocestoides* sp. infection in a dog from Italy. *J. Vet. Med.* 51, 435, 2004.
20. Haziroglu, R. et al. Peritoneal tetrathyridosis in a Siamese cat—A case report. *Vet. Archiv.* 75, 543, 2005.
21. Venco, L. et al. Ultrasonographic features of peritoneal cestodiasis caused by *Mesocestoides* sp. in a dog and in a cat. *Vet. Radiolol. Ultrasound* 46, 417, 2005.
22. Eleni, C. et al. Proliferative peritoneal and pleural cestoiasis in a cat caused by metacestodes of *Mesocestoides* sp. Anatomohistopathological findings and genetic identification. *Parasite* 14, 71, 2007.
23. Foronda, P. et al. First larval record of *Mesocestoides* in Carnivora of Tenerife (Canary Islands). *J. Parasitol.* 93, 138, 2007.
24. Wirtherle, N. et al. First case of canine peritoneal larval cestodosis caused by *Mesocestoides lineatus* in Germany, *Parasitol. Int.* 56, 317, 2007.
25. Crosbie, P.R. et al. Molecular systematics of *Mesocestoides* spp. (Cestoda: Mesocestoididae) from domestic dogs (*Canis familiaris*) and coyotes (*Canis latrans*), *J. Parasitol.* 86, 350, 2000.
26. Wardle, R.A., McLeod, J.A., and Radinovsky, S. *Advances in the Zoology of Tapeworms 1950–1970*. University of Minnesota Press, Minneapolis, p. 274, 1974.
27. Brooks, D.R., Hoberg, E.P., and Weekes, P.J. Preliminary phylogenetic systematic analysis of the major lineages of the Eucestoda (Platyhelminthes: Cercomeria). *Proc. Biol. Soc. Wash.* 104, 651, 1991.
28. Mariaux, J., A molecular phylogeny of the Eucestoda. *J. Parasitol.* 84, 114, 1998.
29. Hoberg, E.P. Jones, A., and Bray, R.A. Phylogenetic analysis among the families of the Cyclophyllidea (Eucestoda) based on comparative morphology, with new hypotheses for co-evolution in vertebrates. *System. Parasitol.* 42, 51, 1999.

30. Specht, D., and Voge, M. Asexual multiplication of *Mesocestoides* tetrathyridia in laboratory animals. *J. Parasitol.* 52, 268, 1965.
31. Eckert, J., Von Brand, T., and Voge, M. Asexual multiplication of *Mesocestoides corti* (Cestoda) in the intestine of dogs and skunks. *J. Parasitol.* 55, 241, 1969.
32. Loos-Frank, B. Shedding of gravid proglottids and destrobilation in experimental infections of foxes with *Mesocestoides leptothylacus* Loos-Frank, 1980 (Cestoda). *J. Helminthol.* 61, 213, 1987.
33. Rausch, R.L. Family Mesocestoididae Fuhrmann, 1907. In: Khalil, L.F., Jones, A., and Bray, R.A. (eds), *Keys to the Cestode Parasites of Vertebrates*. CAB International, Wallingford, 1994.
34. Schmidt, G.D., *Handbook of Tapeworm Identification*. CRC Press, Boca Raton, FL, 675p, 1986.
35. Voge, M. North American cestodes of the genus *Mesocestoides*. *J. Calif. Pub. Zool.* 59, 125, 1955.
36. Padgett, K.A. et al. Systematics of *Mesocestoides* (Cestoda: Mesocestoididae): Evaluation of molecular and morphological variation among isolates. *J. Parasitol.* 91, 1435, 2005.
37. Etges, F.R. The proliferative tetrathyridium of *Mesocestoides vogae* sp. N. (Cestoda). *J. Helminthol. Soc. Wash.* 58, 181, 1991.
38. Gubányi, A., and Eszterbauer, E. Morphological investigation of *Mesocestoides* (Cestoda, Mesocestoididae) species parasitizing *Vulpes vulpes* in Hungary. *Miscnea Zool. Hung.* 12, 11, 1998.
39. Von Nickisch-Rosenegk, M., Lucius, R., and Loos-Frank, B. Contributions to the phylogeny of the Cyclophyllidea (Cestoda) inferred from mitochondrial 12S rDNA. *J. Mol. Evol.* 48, 586, 1999.
40. Literák, I. et al. *Mesocesotides litteratus* (Batsch, 1786) (Cestoda: Cyclophyllidea: Mesocestoididae) from the red fox: Morphological and 18S rDNA characterization of European isolates. *Helminthologia* 43, 191, 2006.
41. Loos-Frank, B. One or two intermediate hosts in the life cycle of *Mesocestoides* (Cyclophyllidea, Mesocestoididae)? *Parasitol. Res.* 77, 726, 1991.
42. Padgett, K.A., and Boyce, W.M. Life-history studies on two molecular strains of *Mesocestoides* (Cestoda: Mesocestoididae): Identification of sylvatic hosts and infectivity of immature life stages. *J. Parasitol.* 90, 108, 2004.
43. Webster, J.D. Fragmentary studies on the life history of the cestode *Mesocestoides* latus. *J. Parasitol.* 35, 83, 1949.
44. Sharpilo, V.P., and Kornyushin, V.V. Locomotory activity of mature cestode proglottids in external environment and egg dissemination pattern they display. *Vestnik Zoologii.* 4, 24, 1995.
45. Padgett, K.A., and Boyce, W.M. Ants as first intermediate hosts of *Mesocestoides* on San Miguel Island, USA. *J. Helminthol.* 79, 67, 2005.
46. Soldatowa, A.P. A contribution to the study of the development cycle in the cestode *Mesocestoides lineatus* (Goeze, 1782) parasitic in carnivorous mammals. *Doklady Akademii Nauk SSSR.* 45, 310, 1944.
47. McAllister, C.T., and Conn, D.B. Occurrence of tetrathyridia of *Mesocestoides* sp. (Cestoidea: Cyclophyllidea) in North American anurans (Amphibia). *J. Wildl. Dis.* 26, 540,1990.
48. Voge, M. New host records for *Mesocestoides* (Cestoda: Cyclophyllidea) in California. *Am. Midland. Nat.* 49, 249, 1953.
49. McAllister, C.T. et al. A new host and locality record for *Mesocestoides* sp. tetrathyridia (Cestoidea: Cyclophyllidea), with a summary of the genus from snakes of the world. *J. Parasitol.* 77, 329, 1991.
50. Voge, M., and Berntzen, A.K. Asexual multiplication of larval tapeworms as the cause of fatal parasitic ascites in dogs. *J. Parasitol.* 49, 983, 1963.
51. Williams, J.F., Lindsay, M., and Engelkirk, P. Peritoneal cestodiasis in a dog. *JAVMA.* 186, 1103, 1985.
52. Roberts, L.R. and Janovy, J. Jr. *Foundations of Parasitology*, 8th ed., Chapter 20. McGraw Hill, New York, 2005.
53. Conn, D.B. The rarity of asexual reproduction among *Mesocestoides* tetrathyridia (Cestoda). *J. Parasitol.* 76, 453, 1990.
54. Widmer, E.A., and Specht, H.D. Isolation of asexually proliferative tetrathyridia (*Mesocestoides* sp.) from the southern Pacific rattlesnake (*Crotalus viridis helleri*) with additional data from two previous isolates from the Great Basin fence lizard (*Sceloporus occidentalis longipes*). *J. Parasitol.* 78, 921, 1992.
55. Caruso, K.J. et al. Cytologic diagnosis of peritoneal cestodiasis in dogs caused by *Mesocestodies* sp. *Vet. Clin. Pathol.* 32, 50, 2003.
56. Hermánek J. Nonspecific immunomodulation influences resistance of mice to experimental infection with *Mesocestoides corti* and *Ascaris sum*. *J. Helminthol.* 65, 121,1991.
57. Poulin, R. The evolution of life history strategies in parasitic animals. *Adv. Parasitol.* 37, 107, 1996.
58. Alvarez, L.I., Mottier, M.L., and Lanusse, C.E. Drug transfer into target helminth parasites. *Trends Parasitol.* 23, 97, 2007.

26 Spirometra

Dongyou Liu, Jianshun Chen, and Weihuan Fang

CONTENTS

26.1 INTRODUCTION

The genus *Spirometra* consists of several diphyllobothroid cestode species whose adult worms inhabit the small intestine of canines and felines and whose larvae (procercoid and plerocercoid or sparganum) parasitize copepods (freshwater crustaceans), frogs, snakes, chicken, and rodents. Humans acquire the infection (sparganosis) through ingestion of copepods containing procercoids in drinking water, or consumption of raw or partially cooked flesh of fish, frogs, snakes, and chickens containing plerocercoids, as well as exposure of open wounds or sore eyes to poultices made of infected flesh of frogs or snakes (or to water containing procerciod-infested cyclops). Once inside, the plerocercoid larvae migrate to various sites of human body, including subcutaneous tissues, lymph ganglia, the conjunctiva, the viscera, and the cerebrum, forming movable painful nodules, and obstructing the normal functions of vital organs.

Despite its first description by Patrick Manson in China in 1882, sparganosis remains under-recognized and an emerging disease, and continues to be responsible for significant health problems in eastern and southeastern Asia, in addition to causing sporadic cases in the United States, Africa, South America, and Australia.

26.1.1 Classification, Morphology, Life Cycle, and Epidemiology

26.1.1.1 Classification

The genus *Spirometra* is classified in the family Diphyllobothriidae, order Pseudophyllidea, subclass Eucestoda, class Cestoda, phylum Platyhelminthes. The family Diphyllobothriidae (obsolete synonym: Ligulidae) consists of eight genera: *Digramma*, *Diphyllobothrium*, *Diplogonoporus*, *Duthiersia*, *Ligula*, *Probothriocephalus*, *Schistocephalus*, and *Spirometra*. In turn, the genus *Spirometra* (synonym: *Sparganum*) is separated into eight species: *Spirometra decipiens*, *Spirometra erinaceieuropaei* (synonym: *Spirometra erinacei*), *Spirometra folium*, *Spirometra houghtoni*, *Spirometra mansoni* (synonym: *Sparganum mansoni*), *Spirometra mansonoides*, *Spirometra proliferum* (synonym: *Sparganum proliferum*), and *Spirometra ranarum* [1].

While *S. erinaceieuropaei* (*S. erinacei*) and *S. mansoni* are common species affecting humans in Asia, *S. mansonoides* is reported mainly in the southeastern United States. *S. houghtoni* has been identified in China, and *S. proliferum* (synonym: *S. proliferum*) is described in Japan, the United States, Venezuela, and Paraguay [2].

Interestingly, the generic name *Sparganum*, originally created by Diesing in 1854, has been used to refer to the larvae (plerocercoids) of *Spirometra* spp, because this stage was thought to represent a separate genus. In fact, the first reported human case of sparganosis in China in 1882 was attributed by Manson to *S. mansoni* (currently known as *S. mansoni*). Because *S. proliferum* adult stage is yet to be identified and only its larval stage is known, this species is sometimes referred to as *S. proliferum* by some authors. At one stage, *S. proliferum* was thought to be an abnormally differentiated *S. erinaceieuropaei* (*S. erinacei*) because of their morphological similarity. However, comparative analyses of mitochondrial NADH dehydrogenase subunit 3 (ND3), mitochondrial cytochrome *c* oxidase subunit I (COI) and succinate dehydrogenase iron–sulfur protein subunit (*sdhB*) gene sequences confirmed that *S. proliferum* is phylogenetically related to, but distinct from, *S. erinaceieuropaei* [3,4].

26.1.1.2 Morphology

Spirometra adult worms measure 60–110 cm in length, 0.5–0.8 cm in width. Based on experimental infection data, *S. erinaceieuropaei* (*S. erinacei*) adult worm is of 110 cm in length, 0.8 cm in width. The scolex (measuring 1.88 × 0.46 mm) possesses two lateral bothria. Mature proglottids are of 0.94–2.33 mm in length and 6.03–9.99 mm in width. Besides harboring numerous eggs and uterine tubules (forming 3–5 coils), gravid proglottids show genital pore (which opens on the ventral midline near the anterior margin), vagina pore (which opens like a slit just distal to the genital pore), and uterine pore (which is 420 μm distal to the vaginal pore at the end of uterine coils). In addition, Mehlis' gland and dumbbell-shaped ovary are connected to the spiral uterus. On the sagittal section of a gravid proglottid, the cirrus is located within a sac with seminal vesicle. The transverse section of a gravid proglottid shows uterine tubules at the midline with the testes and vitellaria in the lateral field. Small round lobular testes are situated at the center, but vitelline follicles are at the margin under the ventral and dorsal tegument [5].

S. erinaceieuropaei (*S. erinacei*) eggs (65 μm × 35 μm) are slightly asymmetric (ellipsoidal), harbor a germ cell and yolk, and have greenish shell, with a prominent operculum at a sharp pole [5]. The coracidium (emerging from eggs after 8–14 days at 29°C) measures 43.8 × 36.9 μm; the procercoids (of 56–142 μm in length and 20–72 μm in width) after 5 days in the cyclops have minute spines at the anterior end, calcium corpuscles in the body parenchyme and the cercomer at the posterior end; the plerocercoids after 15 days in tadpole are pear-shaped and shorter than 1 mm in the length and are infective to mice [5]. *S. erinaceieuropaei* spargana dissected from wild frog tissues are flat, white worms of 2–115 mm in length and 1–2 mm in width [6]. The growth of *S. erinaceieuropaei* (*S. erinacei*) plerocercoids to egg-producing adults takes 15–18 days in dogs or cats [5].

S. mansoni (formerly *S. mansoni*) adult worms measure 60–100 cm in length and 0.5–0.6 cm in width. The scolex is elongated and spoon-shaped, with two sucking grooves (instead of suckers and hooks); the neck is thin and unsegmented; and the body is composed of about 1000 proglottids. The mature proglottids are of rectangular shape (broader than long), but almost square at the distal end. Each mature proglottid consists of male and female reproductive systems. While the male reproductive system consists of several hundred testes, the female reproductive system is made up of a bilobed ovary, and a coiled uterus, with three openings (vas deferens, vagina, and uterus situated to each other) appearing in the middle of each segment on its ventral surface. Through uterine pores, eggs are discharged by gravid segments. Eggs (of about 60 μm × 35 μm in size) are ovoid, with an inconspicuous operculum and a thin shell. *S. mansoni* eggs with an operculum may be confused with those of trematodes such as *Paragonimus westermani*. *S. mansoni* eggs hatch to become coracidium (the first-stage larva, of 43 × 36 μm in size), then proceroid (the second-stage larva), and plerocercoid (the third-stage larva).

Spirometra plerocercoids (sparganum larvae) recovered from human hosts are whitish, wrinkled, and ribbon-shaped, ranging from 3 to 30 cm in length and 1 to 2 mm in width. The sparganum is a solid-bodied larva with an unsegmented strobila, with bundles of longitudinal muscle fibers scattered throughout the mesenchyme, and a thick tegument [7]. It lacks a bladder, and shows two longitudinal grooves (bothrids) instead of suckers at the anterior end. The absence of a scolex or protoscolex in *Spirometra* sparganum larvae differentiates it from *Taenia solium*. *Spirometra* (*Sparganum*) *proliferum* spargana (of 0.5–38 mm in length and 0.3–3 mm in width) vary from filiform, to round, dentritic, and large branching forms, and appear yellowish to whitish in color, with a notable absence of scolex. While lacking lateral symmetry, *S. proliferum* larvae tend display distinct movements of extension, flexion, elongation, retraction, and torsion. Histologically, *S. proliferum* larvae show two well-defined areas, that is, the tegument and the parenchyma, similar to other cestodes [8].

Among the *Spirometra* spp., *S. mansoni* plerocercoids are larger and more delicate than *S. mansonoides* plerocercoids. Additionally, the uterine shape of the mature proglottid appears piled in *S. erinaceieuropaei* (*S. erinacei*) and C-shaped in *S. mansonoides*.

26.1.1.3 Life Cycle

Spirometra adult worms live in the small intestine of the definitive host (e.g., dog, cat, raccoon, or other mammal) for up to 9 years, and produce many unembryonated eggs that are discharged in feces. The unembryonated eggs hatch in fresh water to become coracidia, which are taken up by the first intermediate host, copepods (freshwater crustaceans), in which the coracidia develop into procercoid larvae (first-larval stage). When the copepods are consumed by the second intermediate host (e.g., fish, reptiles, or amphibians), the procercoid larvae penetrate the intestinal tract, and become plerocercoid larvae (sparganum larvae) and then migrate to the subcutaneous tissues and muscles. When the second intermediate host is eaten by a definitive host predator (e.g., dog and cat), the cycle begins again [9].

As an accidental second intermediate, or paratenic host in the *Spirometra* life cycle, humans acquire the infection by ingestion of the first or second intermediate hosts containing procercoid or plerocercoid larvae in water or meat. Alternatively, exposure of open wounds, lesions, or the eyes to raw poultices (dressings) made of the second intermediate hosts for medicinal or ritualistic reasons will also lead to infection in humans. The plerocercoid larvae (spargana) are released in the intestine, lose the strobila, penetrate the intestinal wall, and migrate to the subcutaneous tissues and other parts (e.g., brain, eye, abdominal cavity, spinal cord) of human body and grow back to their full length (without proliferation). However, *S. proliferum* larvae proliferate throughout the subcutaneous tissues (with several spargana in a single site). Spargana can live up to 20 years in human host.

26.1.1.4 Epidemiology

Sparganosis is a parasitic zoonosis due to invasion of the spargana, the plerocercoid larvae of the genus *Spirometra* [10]. The definitive hosts of *Spirometra* include dog, cat, raccoon, or other mammal (although the definitive host for *Spirometra (Sparganum) proliferum* remains unknown); the first intermediate host is copepods (freshwater crustaceans of the genus *Cyclops*), which represent an important food source for many animals living in and around the water, and are accidentally swallowed by many others; the second intermediate hosts are frogs, fish, snakes, pigs, and mice, with humans being an accidental secondary intermediate or paratenic host [11–16].

Human sparganosis is endemic in Asian countries including China, South Korea, Japan, Vietnam, Thailand, and India. In eastern and southeastern Asia, raw or insufficiently cooked meat of frogs and snakes are considered as delicacy and consumed regularly. In addition, poultices (dressings) made of frog or snake flesh are thought to have healing power and are often placed on open wounds for treatment of skin ulcers (and even hypothyroidism) or on eyes to treat inflammation. Furthermore, in some areas, water containing procercoid-infected copepods is consumed due to a lack of clean drinking water. During an 80-year period (1927–2007), >1000 cases of human sparganosis have been described in 25 Chinese provinces [17]. In Thailand, 52 cases of human sparganosis were reported between 1943 and 2010 [18]. A moderate number of sparganosis cases have been documented in the United States, South America, Africa, Greece, and Australia [6,17]. Overall, human sparganosis in Asia tends to occur after ingestion of infected raw snake, frog, or pig, while the most common cause of infection in other parts of the world is drinking contaminated water.

26.1.2 Clinical Features, Pathogenesis, and Treatment

26.1.2.1 Clinical Features

Once inside human body, plerocercoid larvae migrate to subcutaneous tissue and other locations, where they typically develop into a nodular mass (sparganum or cyst), inducing nonspecific discomforts, seizures, hemiparesis, and headache as well as eosinophilia [19]. The characteristic host responses in human patients with sparganosis include an inflammatory cell infiltration in the first week after the inoculation of the larvae into the soft tissue, tunnel-like structures (an elongated tract-like cavity through which a larva has passed) 2 weeks later, and a fibroblast proliferation between 4 weeks and 6 months postinfection [20]. When the larva migrates and then dies, the tunnels with inflammatory foci are replaced by granulation tissue.

Clinical manifestations of human sparganosis vary with the final location of spargana in the host. Spargana may invade subcutaneous tissue, muscle, breast, orbit, pleural cavity, lungs, urogenital, and abdominal viscera as well as the central nervous system (CNS), resulting in subcutaneous sparganosis, visceral sparganosis, ocular sparganosis cerebral sparganosis and proliferative sparganosis, and so on [19,21–25]. According to a review of 34 sparganosis cases reported in Thailand between 1943 and 2004, the most common clinical presentation is ocular infection (17) (possibly reflecting the local practice of applying fresh frog muscle as a poultice to relieve sore eyes), which is followed by subcutaneous (10), CNS (5), auricular (1), pulmonary (1), intraosseous (1), and intraperitoneal infections (1), with a combined case of ocular and auricular infection and another combined case of subcutaneous and CNS infection [26].

1. Subcutaneous sparganosis occurs after the larvae migrate to subcutaneous connective tissues and superficial muscles (e.g., the abdominal wall, scrotum, lower extremities, and chest wall). The lesion (sparganum) under the skin appears as a palpable, rubber-like, irregular nodule, or lump (1–2 cm in diameter), resembling a lipoma, fibroma, or sebaceous cyst that is slow-growing, itchy, inflamed, painful, and moveable (appearing and disappearing over a period of time). These nodules may persist for months or years without any symptoms then suddenly become painful [21,27]. Occasionally, elephantitis may appear in the location associated with the lymph channels. In genital sparganosis, subcutaneous nodules form in the groin, labia, scrotum, the epididymis, ureter, and urinary bladder, and may be tumor-like [28–31].

Yun et al. [32] documented a case of subcutaneous sparganosis due to *S. erinaceieuropaei* in a 59-year-old farmer. With a history of eating raw snake and frog during his childhood, the patient presented with a 2-cm-sized painless movable hard mass in the left scrotum and complained of slow urine stream and hesitancy. The ultrasound scan revealed a 2.2-cm-sized well-defined round heteroechogenic nodule between the left epididymis and the spermatic cord. At operation, multiple spargana were found within the mass. In a serological test, patient's serum reacted positively to anti-sparganum IgG antibody. Plerocercoids of *S. erinaceieuropaei* was confirmed morphologically and also by PCR-sequencing analysis of the cytochrome *c* oxidase subunit 1 (*cox1*) target fragment (353 bp in length corresponding to the positions 769–1121 bp of the *cox1* gene). Additionally, Lee et al. [33] reported a case of sparganosis in a 60-year-old woman with pain of the right ankle, aggravated on dorsiflexion. The patient had noticed a movable nodular mass at the medial side of the right knee 20 years ago. Upon physical examination, an oval, rubber-like, nontender mass (of 3 cm in diameter) was noted to be just proximal to the lateral malleolus of the right foot. Another mass at the anterior part of the right ankle caused discomfort on dorsiflexion. The plain radiograph of the right leg revealed multiple calcifications extending from the medial part of the knee to the calf. During the surgery, a white-shiny, synovium-like piece of tissue popped out through the incision, which measured 18 cm in length and 0.5 cm in width, wriggling after removal from the patient, and was identified as a sparganum. The patient was a farmer, and drank untreated mineral water from a local mountain. On the third postoperative day, the patient was discharged.

2. Visceral sparganosis relates to the migration and growth of the larvae in the intestine, the breast, abdominal cavity, pleura (lung), and pericardium (heart), resulting in damage to or malfunction of the particular organ or tissue. Sparganosis can cause intestinal obstruction or peritonitis in the case of intestinal perforation. If the inner ear is involved, vertigo, or deafness may result [33–50].

Iwatani et al. [45] described a case of *S. mansoni*-associated lung sparganosis related in a 57-year-old man. The patient had a solitary nodule in the middle lobe of the right lung. After a right middle lobectomy, the lesion was diagnosed as sparganosis by histological and serological examinations. In a recent study, Lee et al. [48] described a case of sparganosis in a 53-year-old woman with a recurrent pericardiopleural effusion and peripheral eosinophilia. The patient had taken medications for diabetes, hypertension, and hypothyroidism for 10 years and had eaten >30 raw frogs in an effort to control her thyroid disease 3 months earlier. The patient presented prickly chest pain (which was triggered mainly by deep inspiration, and unrelated to exercise), dry cough, arthralgias, generalized malaise, a chilling sensation, and dyspnea. Echocardiography demonstrated a pericardial effusion of maximal depth 12 mm. After three consecutive doses of praziquantel (75 mg/kg/day), the amount of pericardiopleural effusion had decreased markedly. The patient has not complained of any symptoms in over 4 years of follow-up.

3. Ocular sparganosis is associated with the larva migration to the conjunctiva and the orbit. Located in the posterior pole, the larva induces periorbital edema, exophthalmos (protrusion of the eyeball), and lagophthalmia, which in turn cause eye pain, epiphora (excessive lacrimation or watering of the eye), ptosis (drooping of the upper eyelid), irritation, orbital cellulitis, and marked swelling of the eyelids, and cornea ulcer. If untreated, ocular sparganosis can lead to blindness [22,49–54].

Mentz et al. [55] reported a case of ocular sparganosis due to *S. mansonoides* a female patient. The patient had three inflammatory, movable nodules in the periocular region, including subpalpebral space, and complained pain in the right eye and frontal region. Two nodules became stationary in the inferior temporal bulbar conjunctiva, accompanied by edema, and red spots on the lower eyelid; and the third one had moved to the nasal bulbar conjunctiva. During surgery, an inert, whitish structure was observed inside each granuloma and a flat parasite (of 0.3×5.5 cm) was recovered, and subsequently identified as plerocercoid larva (sparganum), belonging to the genus *Spirometra*. After the larval removal, the inflammation receded and her clinical conditions improved.

4. Cerebral sparganosis involves the growth of larvae in the cerebral hemispheres, especially the frontoparietal lobes, in some cases extends to the cerebellum. The disease may appear as a cerebral hemorrhage. When spargana settle in the brain or spine, various neurological symptoms may appear, including fatigue, motor weakness, confusion, headache, seizure, memory loss, coma, fever, paresthesias, hemiparesis, brain abscesses, and abnormal skin sensations (e.g., numbness or tingling) [56–60].

Ou et al. [61] reported a case of cerebral sparganosis in an 18-year-old female, who presented with dizziness, headache, decreasing memory, reaction dullness, and seizure. Lumbar puncture revealed high pressure in the cranial cavity, with a rapid dropping of cerebrospinal fluid (CSF). The blood and CSF samples were strongly positive in enzyme-linked immunosorbent assay (ELISA) for antisparganum antibody. T1-weighted imaging (T1WI) revealed an irregular flake low signal focus in the white matter of left parietal lobe with vague edge. T2-weighted imaging (T2WI) showed focus with high signals. The patient ate poorly cooked frog flesh 5 months ago, and was diagnosed with cerebral sparganosis. After taking albendazole (300 mg twice a day) pulsed sodium valproate therapy (200 mg twice a day) for 20 days, the patient was discharged with no signs or symptoms with no neurological relapse during a one-year follow-up.

5. Proliferative sparganosis is caused by *Spirometra (Sparganum) proliferum* plerocercoids (spargana), which invade the subcutaneous region and other tissues of the human body including bone, and undergo continuous branching and budding, producing vast numbers of progeny plerocercoids in a single site, and leading to fatal human sparganosis. It often begins with an open, tumor-like subcutaneous nodule (due to ulceration or scarification) in the thigh, shoulder, or neck, and spreads to other parts of the

skin, the muscles, and the internal organs (e.g., the lungs, abdomen, and brain) over 5–25 years, usually with a grim outcome [62,63].

Meric et al. [64] documented a case of proliferative sparganosis due to *S. proliferum* in a 45-year-old, HIV-positive man. The patient had eaten snake flesh several times 25 years ago and undergone surgery for cutaneous and scrotal lesions 4 years previously. On admission, he presented with abdominal pain, a high fever, and colonic ileus. Computed tomography (CT) scan showed a large intra-abdominal mass. During surgery, purulent necrotic tissues associated with a pseudo-neoplastic lesion of the omentum were removed. Histological examination at low magnification of the omentum, scrotal, cutaneous, and liver tissues revealed numerous cystic structures containing parasites (of 30 μm–1 mm in diameter) surrounded by inflammatory tissues. At higher magnification, layered tegument with a tegumental syncytium, a basal lamina, tegumental muscle cells and vesicles were observed. These features confirmed the diagnosis of a disseminated infection with *S. proliferum*. The patient recovered after surgery.

The long-term survival of spargana inside the host possibly reflects the ability of the parasite to modulate host's immune responses. For instance, an excreted–secreted (ES) cathepsin-S-like protease of *S. mansoni* plerocercoid is capable of cleaving IgG into Fab and Fc fragments [65,66]. In addition, *Spirometra* ES products are shown to downregulate the transcription of inducible nitric oxide synthase, chemokines of KC and JE, tumor necrosis factor-α, and IL-1β. Furthermore, a glycoprotein from crude ES products of spargana exerts a negative impact on the expression of osteoclastogenesis and proinflammatory cytokine genes [20].

26.1.2.2 Treatment

Surgical removal remains the most effective treatment for human sparganosis. However, surgical treatment of the pericardiopleural effusion associated with sparganosis without a pleural or pericardial nodule or lump is impossible. Praziquantel has been shown to be effective for treating pleural sparganosis, and eosinophilic pleuritis due to sparganosis. Nonetheless, praziquantel has no effect on adult worms in the CNS.

Prevention of human sparganosis should center on the provision of clean drinking water (free of Cyclops or larvae) and avoiding drinking unfiltered, untreated, or nonboiled water from ponds and ditches (which may contain infected copepods) in endemic areas. In addition, education on the danger of ingesting raw flesh of the intermediate hosts, (e.g., snakes and frogs), or using them as poultices, is also important. Furthermore, feeding dogs and cats with cooked meat only and treating infected dogs and cats provide another useful approach for the control of sparganosis in humans.

26.1.3 Diagnosis

Preoperative diagnosis of sparganosis includes identification of exposure history and a painful, migratory, subcutaneous nodule. Subcutaneous sparganosis often shows as a single nodule, whereas other cestode infections (e.g., cysticercosis, *Hymenolepis nana*, *Mesocestoides* sp., and *Taenia taeniaeformis*) typically have multiple nodules. Cerebral sparganosis induces characteristic lesions (column-shaped or fusiform-shaped tunnel, multiple conglomerated ring-shaped enhancements, or multifocal low-density lesions along white matter bundles) in the brain as detected by magnetic resonance imaging (MRI). On CT scan, informative signs of cerebral sparganosis affecting the CNS are hypodensity, ventricular dilation, irregular, or nodular-enhancing lesions, and small, punctuate areas of calcification [67]. The differential diagnosis of cerebral sparganosis consists of brain tumors and other inflammatory granulomas. In addition, visceral spargana lesions are sometimes mistaken for tuberculosis lesions as examined by CT and MRI scans.

In any case, surgical removal of the worms and identification of the parasite in a tissue specimen as well as detection of eosinophilia provide a reliable approach for confirmation for sparganosis. Given that the spargana belonging to various *Spirometra* species are virtually indistinguishable, it is difficult to determine their species identity on morphological criteria alone. One useful, but time-consuming, way to identify *Spirometra* to the species level is to infect an appropriate definitive host with a living sparganum and identify the adult after maturation.

In cases of sparganosis presenting pericardial or pleural effusions without a pleural nodule, biopsy, or surgery may be impractical. Use of ELISA and other serological techniques for specific IgG antibodies to 31 and 36 kDa proteins and other sparganum antigens may be helpful [2,68–71]. However, serological cross-reactivity between *Spirometra spargana* and *Cysticercus cellulosae* is occasionally observed. Similarly, ELISA is useful for the detection of parasitic antigens in CSF or serum from patients suspected of cerebral sparganosis. Often, a combination of clinical history, ELISA, and either MRI or CT scans are sufficient to make a sparganosis diagnosis [67].

Recent development and application of molecular techniques have facilitated rapid, sensitive and precise identification, and phylogenetic characterization of *Spirometra* species. The most commonly targeted gene regions include small (18S) and large (28S) subunit ribosomal RNA (rRNA), ribosomal internal transcribed spacers 1 and 2 (ITS 1 and ITS 2), COI, mitochondrial ND3, and *sdhB* genes [3,72–74].

Because of their relatively conserved nature, analysis of the nucleotide sequences of small subunit (18S) and large subunit (28S) rRNA genes offers pertinent insights on the taxonomy and phylogenetic relationship of various eukaryotic organisms. Compared to the 28S rRNA gene, the nucleotide conservation in 18S rRNA gene is even more remarkable. This makes the 18S rRNA gene especially suitable for estimating phylogenetic relatedness among eukaryotic organisms including *Spirometra* species. Forming a catalytic center of cytochrome *c* oxidase, COI subunit is also highly conserved among organisms, rendering it another often targeted gene for phylogenetic studies.

After comparison of the nucleotide sequence of the 28S rRNA, mitochondrial COI, and ribosomal ITS 1 fragments, Lee et al. [75] showed that *S. erinaceieuropaei* (*S. erinacei*) and *S. mansonoides* share a recent common ancestry. Similarly, analysis of the *cox1* and other gene sequences forms a genetic basis for rapid and specific determination of *S. erinaceieuropaei* and *Spirometra* species.

26.2 METHODS

26.2.1 Sample Preparation

26.2.1.1 Sample Collection

Spargana removed from skin or muscle is put in a Petri dish containing physiological saline and identified microscopically according to morphological features and site of predilection of the genus *Spirometra*. The spargana are then fixed in 70% ethanol and stored at −20°C until extraction of genomic DNA.

26.2.1.2 DNA Isolation

Genome DNA is extracted from individual sparganum by SDS/proteinase K treatment, column-purified (Wizard® SV Genomic DNA Purification System, Promega) and eluted into 60 μL H_2O according to the manufacturer's recommendations.

26.2.2 Detection Procedures

26.2.2.1 Sequence Analysis of Partial 18S rRNA Gene

Liu et al. [72] utilized a pair of primers (forward: 5′-ATGACAGGGATCGGGGCT TGGAA-3′; and reverse: 5′-TAATGATCCT TCCGCAGGTTCACCTAC-3′) to amplify a 226 bp fragment from the 3′-terminal region of *S. erinaceieuropaei* 18S rRNA gene (which is 2182 bp in length) for identification and phylogenetic analysis. After determination of the nucleotide sequences by the dideoxynucleotide chain termination method using the *Taq* DyeDeoxy Terminator Cycle Sequencing Kit (Applied Biosystems), the identity of *S. erinaceieuropaei* is revealed by sequence alignments.

26.2.2.2 Sequence Analysis of Partial 28S rRNA Gene

26.2.2.2.1 Protocol of Lee et al. [75]

Lee et al. [75] described the application of primers JB10 and JB10 to amplify the partial nucleotide sequence in the 28S rRNA gene for identification of *Spirometra* spp. The forward primer JB10 (5′-GATTACCCGCTGAACTTAAGCTA-3′) corresponds to nucleotides 21–45 of *Schistosoma mansoni* 28S rRNA gene, while the reverse primer JB9 (5′-GCTGCATTCACAAACACCCCGACTC-3′) correlates to nucleotides 278–302 of *S. mansoni* 28S rRNA gene. The amplification cycles consist of 40 cycles of 95°C for 20 s, 55°C for 30 s, and 72°C for 30 s. The PCR product is analyzed by sequencing technique and the resulting nucleotide sequences are compared with those available at GenBank.

26.2.2.2.2 Protocol of Brabec et al. [73]

Brabec et al. [73] utilized primers LSU5 (5′-TAGGTCGAC CCGCTGAAYTTAAGCA-3′) and LSUD6–3 (5′-GGAACC CTTCTCCACTTCAGTC-3′) to amplify the D1–D3 region (divergent domains) of 28S rRNA gene for phylogenetic analysis of Diphyllobothriidea including *Spirometra* and *Diphyllobothrium*. PCR is conducted in 25 μL volume with the following cycling conditions: five cycles of 96°C for 1 min, 44°C for 1 min, 72°C for 2 min, followed by 25 cycles with annealing temperature at 48°C. Amplified products of expected size are cut out of the gel, purified using the JETQUICK Gel Extraction Spin Kit (GENOMED), and cloned into the pCR® 2.1-TOPO® Cloning Vector using the TOPO TA Cloning® Kit (Invitrogen). Gene fragments are sequenced using CEQ™ Dye Terminator Cycle Sequencing Kit (Beckman Coulter), ethanol-precipitated and run on a CEQ™ 8000 Beckman Coulter automated sequencer. The resulting sequences are assembled and edited with Seqman (DNASTAR).

26.2.2.3 Sequence Analysis of Partial *Cox1* Gene

26.2.2.3.1 Protocol of Miyadera et al. [4]

Miyadera et al. [4] targeted subunit I of mitochondrial cytochrome *c* oxidase (COI) with primers p-1F (5′-TGGTTT TTTGTGCATCCTGAGGTTTA-3′) and p-1R, (5′-AGAA AGAACGTAATGAAAATGAGCAAC-3′) for differentiation between *Spirometra (Sparganum) proliferum* and *S. erinaceieuropaei*. The primer set p-1F and p-1R is derived from the highly conserved region of COI gene of *Dugesia japonica*.

26.2.2.3.1.1 Procedure

1. PCR mixture (50 μL) is composed of 5 μL of 10× *Taq* polymerase buffer, 200 μM of dNTP, 2 μM of $MgCl_2$, 2.5 U of *Taq* DNA polymerase (Boehringer Mannheim), 0.2 μM each of p-1F and p-1R primers, and template DNA.
2. The mixture is subjected to 30 amplification cycles of 95°C for 30 s, 50°C for 30 s, and 72°C for 1 min.
3. The amplicon is sequenced by using each of primers p-1F and p-1R in separate reactions; and the nucleotide sequences are aligned with related sequences available at GenBank.

Note: S. proliferum demonstrates a 99% sequence identity in the COI region covered by primers p-1F and p-1R between different isolates of the same species. However, *S. proliferum* only shares 89% identity with *S. erinaceieuropaei* in this region. Thus, *S. proliferum* is phylogenetically distinct species from *S. erinaceieuropaei*, despite the fact that these two species are morphologically similar.

26.2.2.3.2 Protocol of Liu et al. [76]

Liu et al. [76] utilized primers JB3 (5′-TTTTTTGGGCAT CCTGAGGTTTAT-3′) and JB4.5 (5′-TAAAGAAAGAAC ATAATGAAAATG-3′) targeting a 446bp portion of the mitochondrial *cox1* gene sequences for identification and phylogenetic analysis of *S. erinaceieuropaei* and other *Spirometra* species.

26.2.2.3.2.1 Procedure

1. PCR mixture (25 μL) is made up of 4 mM of $MgCl_2$, 5 μM of each primer, 2.5 μL 10× r*Taq* buffer, 0.2 mM of each dNTPs, 1.25 U of r*Taq* DNA polymerase (TAKARA), and 1 μL of DNA sample.
2. The cycling parameters consists of an initial denaturation at 94°C for 5 min, then 35 cycles of 94°C for 30 s (denaturation), 55°C for 30 s (annealing), 72°C for 30 s (extension); followed by a final extension at 72°C for 10 min.
3. After examination of each amplicon (5 μL) on 1% agarose gel, amplicons are purified and sequenced using the PCR primers above (individually).
4. The partial *cox1* sequences are aligned using the computer program Clustal X 1.81. Pairwise comparisons are made of the level of sequence differences (D) using the formula $D = 1 - (M/L)$, where M is the number of alignment positions at which the two sequences have a base in common, and L is the total number of alignment positions over which the two sequences are compared. Then, *cox1* sequences of representative samples are used for phylogenetic analyses.
5. The partial *cox1* sequences are aligned with reference sequences in the GenBank™ using Clustal X 1.81, with manual adjustment. Three methods, namely neighbor-joining (NJ), maximum likelihood (ML), and maximum parsimony (MP), are used for phylogenetic reconstructions. NJ and MP analyses are carried out using PAUP 4.0 Beta 10 program, and ML analysis is performed using PUZZLE 4.1 under the default setting. The consensus tree is obtained after bootstrap analysis, with 1000 replications, with values above 50% reported. To study the genetic relationships among diphyllobothroid tapeworms, other members of the Diphyllobothriidae are considered, with *T. solium* as the outgroup. Phylograms are drawn using the Tree View program version 1.65.

Note: The 436 bp partial *cox1* sequence offers a suitable genetic marker for the accurate identification of *S. erinaceieuropaei* and other diphyllobothroid tapeworms [4,74,77].

26.3 CONCLUSION

Spirometra is a genus of diphyllobothroid cestodes that are responsible for a rare, but sometimes deadly, parasitic disease known as sparganosis. The adult worms of *Spirometra* species live in the small intestine of canines and felines, and their larvae (procercoid and plerocercoid or sparganum) are found in copepods (freshwater crustaceans), frogs, snakes, chicken, and rodents. Humans represent an accidental, dead-end host and usually acquire the infection through one of three routes: (i) ingestion of copepods containing procercoids in untreated water; (ii) consumption of raw or improperly cooked flesh of fish, frogs, snakes, and chickens containing plerocercoids; (iii) exposure of open wounds or sore eyes to poultices (dressings) made of infected flesh of frogs or snakes (or to water containing procerciod-infested cyclops) [78]. Within the human body, plerocercoid larvae penetrate the intestinal wall, and migrate to other locations such as subcutaneous tissues, intermuscular fascia, the brain, the eyes, the abdomen, and pleural cavity, producing itchy, inflamed, and painful nodules or lumps and interfering with the normal functions of vital organs. Although human sparganosis has been documented in many parts of the world, it shows a high prevalence in eastern and southeastern Asia, where clean drinking water is not always accessible, raw flesh of frogs and snakes is consumed regularly, and flesh of snakes is still used as dressings for open wound and eye sores.

Clinical diagnosis of human sparganosis relies on the identification of exposure history and observation of a painful, migratory, subcutaneous nodule. Cerebral sparganosis may be diagnosed with the help of CT and MRI scans. Surgical removal of spargana and identification of the parasite in tissue specimens offer additional evidence of *Spirometra* infections. However, due to the morphological similarity of spargana among various species, precise determination of *Spirometra* identities is not possible on morphological criteria alone [79]. The availability of this information is critical to the epidemiological investigation and prevention of human sparganosis. While infection of appropriate definitive host with a living sparganum allows accurate identification after maturation of adult, this procedure is unduly time consuming and costly. Application of serological tests such as ELISA is helpful in assisting the speciation of *Spirometra* spargana, the occurrence of cross-reactivity between *Spirometra* spargana and *C. cellulosae* complicates the diagnosis [80]. For these reasons, molecular techniques based on PCR amplification and sequence analysis of conserved genes in *Spirometra* represent a much faster and more specific way to identify and phylogenetically evaluate these parasites, contributing to a more effective control and prevention of human sparganosis.

REFERENCES

1. http://www.uniprot.org/; last accessed on 20 May 2011.
2. Cui, J. et al., Sparganosis, Henan province, central China. *Emerg. Infect. Dis.*, 17, 146, 2011.
3. Kokaze, A. et al., Phylogenetic identification of *Sparganum proliferum* as a pseudophyllidean cestode. *Parasitol. Int.*, 46, 271, 1997.
4. Miyadera, H. et al., Phylogenetic identification of *Sparganum proliferum* as a pseudophyllidean cestode by the sequence analyses on mitochondrial COI and nuclear *sdhB* genes. *Parasitol. Int.*, 50, 93, 2001.
5. Lee, S.H. et al., Experimental life history of *Spirometra erinacei*. *Kisaengchunghak Chapchi*, 28, 161, 1990.
6. Li, M.W. et al., Enzootic sparganosis in Guangdong, China. *Emerg. Infect. Dis.*, 15, 1317, 2009.

7. Andersen, K.L., Description of musculature differences in spargana of *Spirometra* (Cestoda; Pseudophyllidea) and *Tetrathyridia* of Mesocestoides (Cestoda; Cyclophyllidea) and their value in identification. *J. Helminthol.*, 57, 331, 1983.
8. Alarcón de Noya, B., Torres, J.R., and Noya, O., Maintenance of *Sparganum proliferum* in vitro and in experimental animals. *Int. J. Parasitol.*, 22, 835, 1992.
9. Mueller, J.F., The biology of *Spirometra*. *J. Parasitol.*, 60, 3, 1974.
10. Fukushima, T., and Yamane, Y., How does the sparganosis occur? *Parasitol. Today*, 15, 124, 1999.
11. Gray, M.L. et al., Sparganosis in feral hogs (*Sus scrofa*) from Florida. *J. Am. Vet. Med. Assoc.*, 215, 204, 1999.
12. Ooi, H.K. et al., Survey of *Spirometra erinaceieuropaei* in frogs in Taiwan and its experimental infection in cats. *J. Helminthol.*, 74, 173, 2000.
13. Nithiuthai, S. et al., Waterborne zoonotic helminthiases. *Vet. Parasitol.*, 126, 167, 2004.
14. Zhou, P. et al., Food-borne parasitic zoonoses in China: Perspective for control. *Trends Parasitol.*, 24, 190, 2008.
15. Berger, L. et al., Severe sparganosis in Australian tree frogs. *J. Wildl. Dis.*, 45, 921, 2009.
16. Dai, R.S. et al., Severe infection of adult dogs with helminths in Hunan Province, China poses significant public health concerns. *Vet. Parasitol.*, 160, 348, 2009.
17. Li, M.W. et al., Sparganosis in mainland China. *Int. J. Infect. Dis.*, 15, e154, 2011.
18. Anantaphruti, M.T., Nawa, Y., and Vanvanitchaim, Y., Human sparganosis in Thailand: An overview. *Acta Trop.*, 118, 171, 2011.
19. Norman, S.H., and Kreutner, A., Jr., Sparganosis: Clinical and pathologic observations in ten cases. *South Med. J.*, 73, 297, 1980.
20. Kina, Y. et al., A glycoprotein from *Spirometra erinaceieuropaei* plerocercoids suppresses osteoclastogenesis and proinflammatory cytokine gene expression. *Int. J. Parasitol.*, 35, 1399, 2005.
21. Cho, J.H. et al., Subcutaneous and musculoskeletal sparganosis: Imaging characteristics and pathologic correlation. *Skeletal Radiol.*, 29, 402, 2000.
22. Botterel, F., and Bourée, P., Ocular sparganosis: A case report. *J. Travel Med.*, 10, 245, 2003.
23. Pampiglione, S., Fioravanti, M.L., and Rivasi, F., Human sparganosis in Italy. Case report and review of the European cases. *APMIS*, 111, 349, 2003.
24. Koo, J.H. et al., A case of sparganosis mimicking a varicose vein. *Korean J. Parasitol.*, 44, 91, 2006.
25. Koo, M. et al., Cases and literature review of breast sparganosis. *World J. Surg.*, 35, 573, 2011.
26. Wiwanitkit, V., A review of human sparganosis in Thailand. *Int. J. Infect. Dis.*, 9, 312, 2005.
27. Griffin, M.P., Tompkins, K.J., and Ryan, M.T., Cutaneous sparganosis. *Am. J. Dermatopathol.*, 18, 70, 1996.
28. Uh, H.S. et al., Sparganosis infesting in the penis; a case report with review of Korean literatures. *Korean J. Urol.*, 38, 873, 1997.
29. Sakamoto, T. et al., Testicular sparganosis in a child from Uruguay. *Acta Trop.*, 88, 83, 2003.
30. Lim, D., Kim, C.S., and Kim, S.I., Sparganosis presenting as spermatic cord hydrocele in six-year-old boy. *Urology*, 70, 1223.e1, 2007.
31. Sarukawa, S. et al., Case of subcutaneous sparganosis: Use of imaging in definitive preoperative diagnosis. *J. Dermatol.*, 34, 654, 2007.
32. Yun, S.J. et al., A case of vesical and scrotal sparganosis presenting as a scrotal mass. *Korean J. Parasitol.*, 48, 57, 2010.
33. Lee, K.J., Myung, N.H., and Park, H.W., A case of sparganosis in the leg. *Korean J. Parasitol.*, 48, 309, 2010.
34. Koh, T.W. et al., Empyema thoracis associated with sparganosis: A case report. *Korean J. Thorac. Cardiovasc. Surg.*, 21, 761, 1988.
35. Shimada, M. et al., Two cases of sparganosis having onsets of pleural effusion. *Jpn. J. Parasitol.*, 38, 145, 1989.
36. Oh, S.J., Chi, J.G., and Lee, S.E., Eosinophilic cystitis caused by vesical sparganosis: A case report. *J. Urol.*, 149, 581, 1993.
37. Tanaka, S. et al., A case report of pleural sparganosis. *Parasitol. Int.*, 46, 73, 1997.
38. Ishii, H. et al., A rare case of eosinophilic pleuritis due to sparganosis. *Intern. Med.*, 40, 783, 2001.
39. Kim, J.R., and Lee, J.M., A case of intramuscular sparganosis in the sartorius muscle. *J. Korean Med. Sci.*, 16, 378, 2001.
40. Sim, S. et al., A case of breast sparganosis. *Korean J. Parasitol.*, 40, 187, 2002.
41. Kim, H.J. et al., A case of sparganosis in the submandibular triangle. *J. Clin. Otolaryngol. Head Neck Surg.*, 16, 330, 2005.
42. Kim, D.H. et al., Pleural sparganosis: A case report. *Korean J. Thorac. Cardiovasc. Surg.*, 39, 502, 2006.
43. Lee, J.H. et al., Abdominal sparganosis presenting as an abscess with fistulous communication to the bowel. *Am. J. Roentgenol.*, 185, 1084, 2005.
44. Yang, K.M. et al., Sparganosis combined with inguinal hernia. *Korean J. Urol.*, 46, 1366, 2005.
45. Iwatani, K. et al., *Sparganum mansoni* parasitic infection in the lung showing a nodule. *Pathol. Int.*, 56, 674, 2006.
46. Park, J.H. et al., A surgically confirmed case of breast sparganosis showing characteristic mammography and ultrasonography findings. *Korean J. Parasitol.*, 44, 151, 2006.
47. Hong, S.J. et al., Breast and scrotal sparganosis: Sonographic findings and pathologic correlation. *J. Ultrasound Med.*, 29, 1627, 2010.
48. Lee, J.H. et al., A case of sparganosis that presented as a recurrent pericardial effusion. *Korean Circ. J.*, 41, 38, 2011.
49. Sen, D.K. et al., Cestode larva (Sparganum) in the anterior chamber of the eye. *Trop. Geogr. Med.*, 41, 270,
50. Yoon, K.C. et al., Eyelid sparganosis. *Am. J. Ophthalmol.*, 138, 873, 2004.
51. Subudhi, B.N. et al., Ocular sparganosis. *J. Indian Med. Assoc.*, 104, 529, 2006.
52. Kubota, T., and Itoh, M., Sparganosis associated with orbital myositis. *Jpn. J. Ophthalmol.*, 51, 311, 2007.
53. Yang, J.W., Lee, J.H., and Kang, M.S., A case of ocular sparganosis in Korea. *Korean J. Ophthalmol.*, 21, 48, 2007.
54. Otranto, D., and Eberhard, M.L., Zoonotic helminths affecting the human eye. *Parasit. Vectors*, 4, 41, 2011.
55. Mentz, M.B. et al., Human ocular sparganosis in southern Brazil. *Rev. Inst. Med. Trop. Sao Paulo*, 53, 51, 2011.
56. Holodniy, M. et al., Cerebral sparganosis: Case report and review. *Rev. Infect. Dis.*, 13, 155, 1991.
57. Jeong, S.C. et al., Cerebral sparganosis with intracerebral hemorrhage: A case report. *Neurology*, 50, 503, 1998.
58. Cummings, T.J. et al., Parasitic lesion of the insula suggesting cerebral sparganosis: Case report. *Neuroradiology*, 42, 206, 2000.
59. Kwon, J.H., and Kim, J.S., Sparganosis presenting as a conus medullaris lesion; case report and literature review of the spinal sparganosis. *Arch. Neurol.*, 61, 1126, 2004.

60. Rengarajan, S., Nanjegowda, N., and Bhat, D., Cerebral sparganosis: A diagnostic challenge. *Br. J. Neurosurg.*, 22, 784, 2008.
61. Ou, Q., Li, S.J., and Cheng, X.J., Cerebral sparganosis: A case report. *Biosci. Trends*, 4, 145, 2010.
62. Moulinier, R. et al., Human proliferative sparganosis in Venezuela: Report of a case. *Am. J. Trop. Med. Hyg.*, 31, 358, 1982.
63. Nakamura, T. et al., Human proliferative sparganosis. *Am. J. Clin. Pathol.*, 94, 224, 1990.
64. Meric, R. et al., Disseminated infection caused by *Sparganum proliferum* in an AIDS patient. *Histopathology*, 56, 824, 2010.
65. Kong, Y. et al., Cleavage of immunoglobulin G by excretory-secretory cathepsin S-like protease of *Spirometra mansoni* plerocercoid. *Parasitology*, 109, 611, 1994.
66. Kong, Y. et al., Differential expression of the 27 kDa cathepsin L-like cysteine protease in developmental stages of *Spirometra erinacei*. *Korean J. Parasitol.*, 38, 195, 2000.
67. Song, T., Wang, W.S., and Mai, W.W., CT and MR characteristics of cerebral sparganosis. *Am. J. Neuroradiol.*, 28, 1700, 2007.
68. Morakote, N., and Kong, Y., Antigen specificity of 36 and 31 kDa proteins of *Spirometra erinacei* plerocercoid in tissue invading nematodiasis. *Korean J. Parasitol.*, 31, 169, 1993.
69. Yeo, I.S., Yong, T.S., and Im, K., Serodiagnosis of human sparganosis by a monoclonal antibody-based competition ELISA. *Yonsei Med. J.*, 35, 43, 1994.
70. Chung, Y.B. et al., IgG antibody responses in early experimental sparganosis and IgG subclass responses in human sparganosis. *Korean J. Parasitol.*, 38, 145, 2000.
71. Rahman, M., Lee, E.G. and Bae, Y.A., Two-dimensional immunoblot analysis of antigenic proteins of *Spirometra* plerocercoid recognized by human patient sera. *Parasitol. Int.*, 60, 139, 2011.
72. Liu, D.-W., Kato, H., and Sugane, K., The nucleotide sequence and predicted secondary structure of small subunit (18S) ribosomal RNA from *Spirometra erinaceieuropaei*. *Gene*, 184, 221, 1997.
73. Brabec, J., Kuchta, R., and Scholz, T., Paraphyly of the Pseudophyllidea (Platyhelminthes: Cestoda): Circumscription of monophyletic clades basedon phylogenetic analysis of ribosomal RNA. *Int. J. Parasitol.*, 36, 1535, 2006.
74. Okamoto, M. et al., Intraspecific variation of *Spirometra erinaceieuropaei* and phylogenetic relationship between *Spirometra* and *Diphyllobothrium* inferred from mitochondrial CO1 gene sequences. *Parasitol. Int.*, 56, 235, 2007.
75. Lee, S.U., Huh, S., and Phares, C.K., Genetic comparison between *Spirometra erinacei* and *S. mansonoides* using PCR-RFLP analysis. *Korean J. Parasitol.*, 35, 277, 1997.
76. Liu, W. et al., Survey of *Spirometra erinaceieuropaei* spargana infection in the frog *Rana nigromaculata* of the Hunan Province of China. *Vet. Parasitol.*, 173, 152, 2010.
77. Zhu, X.Q. et al., Single-strand conformation polymorphism-based analysis reveals genetic variation within *Spirometra erinacei* (Cestoda: Pseudophyllidea) from Australia. *Mol. Cell. Probes*, 16, 159, 2002.
78. Kim, D.G. et al., Cerebral sparganosis: Clinical manifestations, treatment, and outcome. *J. Neurosurg.*, 85, 1066, 1996.
79. Noya, O. et al., *Sparganum proliferum*: An overview of its structure and ultrastructure. *Int. J. Parasitol.*, 22, 631, 1992.
80. Cui, J. et al., Serodiagnosis of experimental sparganum infections of mice and human sparganosis by ELISA using ES antigens of *Spirometra mansoni* spargana. *Parasitol. Res.*, 108, 1551, 2011.

27 Taenia

Munehiro Okamoto and Akira Ito

CONTENTS

27.1 INTRODUCTION

27.1.1 CLASSIFICATION, MORPHOLOGY, AND BIOLOGY

Taenia Linnaeus 1758 is a genus of tapeworm belonging to the family Taeniidae. The family Taeniidae is a medically important group of tapeworms constituted of the two genera *Echinococcus* Rudolphi 1801 and *Taenia*. The classification of tapeworms belonging to the family Taeniidae has been controversial because of the paucity of adult phenotypic characters and the great plasticity of larvae developed in various intermediate hosts.

Verster [1] surveyed 70 *Taenia* spp., and showed 32 species to be valid. She divided the valid species into two groups, based on the arrangement of the genital ducts. The ducts of group I pass between lateral excretory canals, but those of group II pass the canals ventrally. Group I included human *Taenia* and other species infecting modern canines and felines with the exception of *Taenia taeniaeformis*. Group II consisted of *T. taeniaeformis* and other species infecting the primitive carnivores belonging to the families Viverridae and Mustelidae (e.g., *Taenia martis*, *Taenia mustelae*, and *Taenia parva*). Formerly, several genera had been described within the family Taeniidae. The genera *Multiceps* Goeze 1782, *Hydatigera* Lamarck 1861, and *Tetratirotaenia* Abuladse 1964 were characterized with their larval morphology, whereas the genera *Taeniarhynchus* Weinland 1858, *Monordotaenia* Little 1967, and Fossor Honess 1937 were differentiated from *Taenia* only by the variations of adult rostellar hooks. All of these genera were treated as *Taenia* in the Verster's monograph. Afterwards, Loos-Frank [2] compiled the results of Verster [1] with new findings, and some authors have described several valid species [3–5]. Currently, 42 species have been recognized within this genus, if we approve *Taenia asiatica* as a valid species [3].

Two mammalian hosts showing predator–prey relationships are necessary to maintain their life cycle. Namely, adults parasitic in the intestine of carnivorous mammals and man, and herbivorous mammals serve as intermediate hosts for bladder larvae. Adult worms of *Taenia* are usually large tapeworms, whereas those of *Echinococcus* are very small. The length of mature segments usually exceeds the width. The rostellum in general is present and is armed with a double row of small and large hooks, but some species (*Taenia saginata* and *T. asiatica*) lack hooks. Neck is present. The genital pores are single and irregularly alternating. The testes are numerous and the female genital glands lie in the posterior part of the segment. The uterus has the shape of a longitudinal trunk, from which arise lateral branches [6,7]. There are many eggs in the uterus of the gravid segments. The egg has a capsule which is usually lost in the fecal eggs. Inside the capsule, the inner envelope develops into the emblyophore which is made of blocks and gives its characteristically brown, radiated appearance. Within these, the oncospheral membrane surrounds the hexacanth embryo or oncosphere. The intermediate hosts orally ingest the eggs, and hatched oncosphere invades various tissues to develop into metacestodes, fluid-filled bladder larvae. The metacestode is a cyticercus, a strobilocercus, or a coenurus [7].

Although the accidental invasion of taeniid larvae into human internal tissues were reported from several species including, *Taenia multiceps*, *Taenia serialis*, *Taenia crassiceps*, and *T. taeniaeformis*, it is certain that the majority of

Taenia species causing cysticercosis in humans is *Taenia solium*. On the other hand, only three species use humans as a definitive host. In other words they are the so-called beef tapeworm, *T. saginata*, pork tapeworm, *T. solium* and the third species, *T. asiatica*. The larval stages of these *Taenia* species have been identified as *Cysticercus cellulosae*, *Cysticercus bovis*, and *Cysticercus viscerotropica*, respectively [8]. Humans may harbor adult worms after consuming raw or under-cooked pork, beef, or viscera of swine, respectively, contaminated with metacestodes of these species. These larvae develop into adult tapeworm(s) in the intestine of humans, the only definitive hosts.

As a matter of course, the most important intermediate host of *T. saginata* is cattle worldwide, but buffaloes, giraffes, llamas, and reindeer may serve as the intermediate hosts. On the other hand, that of *T. solium* is mainly pigs. However, dogs [9,10] and humans are also highly susceptible hosts when these hosts have chances to ingest viable eggs orally. Therefore, humans are the only host in which *T. solium* develops into metacestodes from eggs as the intermediate host, and adult tapeworms from metacestodes as the definitive host. There is no confirmed evidence of human cysticercosis due to metacestodes of *T. saginata*. From the viewpoint of public health, *T. solium* is very important, since humans may be infected with metacestodes after ingestion of eggs causing cysticercosis, especially neurocysticercosis (NCC) [11].

The life cycle of these human tapeworms is thoroughly based on the food (meat) consumption by human. *T. saginata* is rather cosmopolitan, because people who eat beef are the majority of human populations except for Hindus in India and in Nepal. The Hindus in Bali, Indonesia are different from the Hindus in India or Nepal and have been eating beef as well as the favorite dish pork [12]. *T. solium* is also cosmopolitan but it basically does not occur in Jewish or Moslem societies. However, outbreaks of cysticercosis of *T. solium* have been reported in Jewish society in the United States [13], in Moslem society in Saudi Arabia [14], and in Kuwait [11,15]. In these cases, patients were infected from eggs which were released from non-Jewish or Moslem tapeworm carriers.

Taeniasis, which is caused by adult tapeworms, is definitely a food-borne parasitic disease. Therefore, if we consume parasite-free beef or pork or well-cooked meat, we cannot be infected with these *Taenia*. Theoretically, it is rather simple to cut-off the life cycle: not to eat uncooked or undercooked meat contaminated with metacestodes infective stage to human. In fact these tapeworm infections, especially *T. solium* in pigs, are nowadays, rather uncommon in developed countries, after pigs have been grown under industrial environment where direct contact of pigs to human feces contaminated with eggs of this parasite is blocked. In developing countries where sanitation is poor, however, pigs and dogs and other domestic animals are roaming with free access to human feces. So, dogs as well as well-known pigs may be full of metacestodes in developing countries where *T. solium* is endemic.

T. saginata has slightly different background situations. In developed countries, meat inspection has in general well been established for the production of safe beef. Cattle are grown in the pasture where strangers cannot access. Under such conditions, however, there are at least sporadic cases of cattle contaminated with metacestodes of *T. saginata* in developed countries like United States or Canada. In contrast, cattle in developing countries are grown in the open field that may easily be contaminated by tapeworm carriers [16]. The wave of globalization in human activities including enormous increase in worldwide trading of foods and livestock, the number of travelers, immigrants, emigrants, and economic or political refugees with destruction of natural environment causes a new disaster age including risks of emergent and reemergent diseases everywhere [16].

Furthermore, a third species, *T. asiatica* has been described from people in Asia and the Pacific [3]. Since *T. saginata* and *T. asiatica* are morphologically very similar, it is impossible for us to differentiate *T. asiatica* from *T. saginata* morphologically [17]. On the other hand, the two taxa clearly differ in biological features including host specificity and organotropism [3,18]. The tapeworm carriers are mainly the local people who have customs to eat uncooked or undercooked viscera of pigs, mainly the liver. There has long been a debate as to the specific status of the cestode *T. asiatica*, with some people regarding it as a distinct species and some preferring to recognize it as a strain of *T. saginata*. The balance of current opinion seems to be that *T. asiatica* is a distinct species. In this regard, however, hybrid-derived offspring between *T. saginata* and *T. asiatica* has been found in a sympatric endemic area as described later.

27.1.2 Clinical Features and Pathogenesis

Taeniasis, infection with adult worms, is not such a serious disease for human, as same as for other carnivorous mammals. It may cause a variety of nonspecific abdominal symptoms such as diarrhea, constipation, epigastric pain, and nausea, but in some cases does not display any symptoms and is therefore asymptomatic. Since gravid segments, especially in the case of *T. saginata* and *T. asiatica*, are capable of fairly energetic locomotion, they may actively come out through the anus by themselves. It was reported that the segments were expelled at a constant rate of 6–8 segments in 24 h [6].

Larval infections in pig and cattle, cysticercosis in animal intermediate hosts, are usually asymptomatic except for extremely heavy infections. However, cysticerci do cause huge amount of economic loss in the livestock by condemnation of contaminated individuals. As mentioned above, metacestodes of several taeniid species can infect humans, but cysticercosis in humans is exclusively caused by cysticercus/cysticerci of *T. solium*. *T. solium* is one of the most important cestodes for human health. Cysticerci can develop in many organs in humans. Same as in pigs, the main intermediate host of *T. solium*, organotropy of cycticerci to muscle in humans is fairly high. Invasion of muscle by cysticerci can sometimes cause myositis, eosinophilia, and muscular pseudohypertrophy, which initiate with muscle swelling. In most cases, it is asymptomatic since the cysticerci die and become calcified. Subcutaneous cysticercosis (SCC) also commonly occurs mainly on the trunk and extremities. Subcutaneous nodules are sometimes painful. In some cases, ophthalmic cysticercosis is found. Depending on the location, it may cause visual difficulties.

No matter how you look at it, however, the most important clinical symptom of human cysticercosis is NCC. In humans, cysticerci may develop in brain as well as in muscle and subcutaneous tissues. Symptoms of NCC can occur months to years after infection, usually when the cysts are in the process of dying. They depend upon where and how many cysticerci (often called lesions) are found in the brain. Epileptic seizures and headaches are the most common symptoms. Cysticercosis may be the most common cause of symptomatic epilepsy worldwide. However, confusion, difficulty with balance, and swelling of the brain may also occur. Death can occur suddenly with heavy infections.

It was revealed that the two genotypes of *T. solium* are endemic worldwide: the Asian type, and the American/African type [19–21]. It has been conceived that the clinical manifestation of cysticercosis in Asia and the Pacific often includes NCC and SCC [9], whereas it mainly includes NCC without SCC in America and Africa with some rare exceptions. In this regard, however, it is not clear as to whether this is due to variations in parasite strain or to those in the host.

A proliferating acephalous form of cestode larvae, the so-called "racemose," is sometimes found in human cerebral tissue. The sequence comparison of parasite DNA clearly reveals that the racemose is an aberrant larva of *T. solium* [22,23], but the mechanism of racemose form is not totally understood.

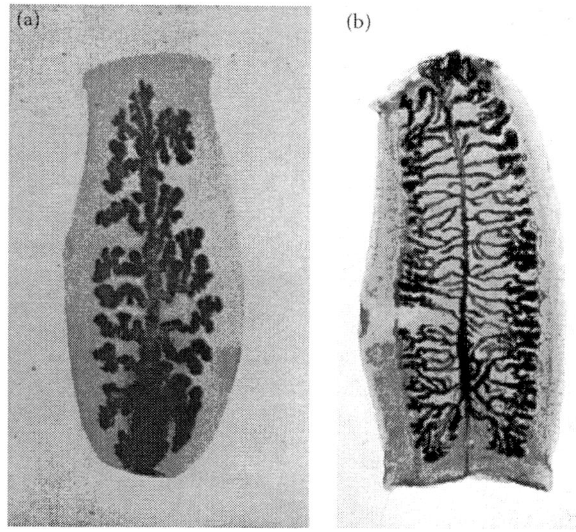

FIGURE 27.1 Gravid proglottids of human *Taenia*. (a) *T. solium* (b) *T. saginata*.

27.1.3 Diagnosis

27.1.3.1 Conventional Techniques

In the case of human taeniasis, eggs and proglottids are passed into the human feces because adult worms parasitize the small intestine. Therefore, we can use conventional stool examination techniques, such as the Kato–Katz technique [24], but sensitivities of conventional techniques are not so high. A centrifugation procedure with sugar solution of high-specific gravity may improve sensitivity [25]. However, the inspection for eggs by the microscopic method is not perfect because the specific gravity of the taeniid egg is relatively high. Furthermore, identification of species is impossible even if the eggs are found in feces. Since eggs of Taeniid cestode including *Echinococcus* are very similar to one another, we cannot distinguish them morphologically.

If gravid proglottids could be obtained, we can identify them morphologically. After a proglottids uterus is stained with India ink, the number of visible uterine branches can help identify the species: unlike the *T. saginata* or *T. asiatica* uteri, *T. solium* uteri have only 5–10 uterine branches on each side (Figure 27.1). However, we cannot distinguish between *T. saginata* and *T. asiatica* by morphology of gravid proglottids. The scolex is generally the most distinctive part of an adult tapeworm. Since the scolex of *T. solium* has armed rostellum, we can easily distinguish it from *T. saginata* or *T. asiatica*, which has nonarmed rostellum. In this regard, however, it is often hard to obtain the intact scolex from patients who have administered anthelminthics, such as praziquantel, which is the most effective drug against cestodes and trematodes. Although original description about *T. asiatica* described morphological difference between *T. asiatica* and *T. saginata*, we cannot actually distinguish *T. asiatica* from *T. saginata* morphologically.

Surprisingly, the questionnaire for patients is sometimes more useful than the stool examination. As mentioned above, since gravid segments are capable of fairly energetic locomotion, they may actively come out through the anus by themselves. The patients must surely become aware of such unpleasant feelings. In our experience, the questionnaire has been the most effective method to find taeniasis patients with *T. saginata* or *T. asiatica*. In the case of *T. solium*, however, the questionnaire may not be always effective because gravid segments of *T. solium* do not move actively. In order to identify species, morphologies of segments expelled must be examined.

27.1.3.2 Molecular Techniques

As described in the preceding paragraph, although eggs of human *Taenia* cannot be distinguished by the conventional method of using a microscope, morphology of gravid proglottids is useful for identification between *T. solium* and *T. saginata*. Molecular identification of taeniid species was encouraged when the third human taeniid species, Asian *Taenia*, was reported [3] because we could not distinguish *T. asiatica* from *T. saginata* morphologically. In addition, it was revealed that the two genotypes of *T. solium* are endemic worldwide: the Asian type, and the American/African type. These genotypes may be related to pathogenicity to humans. In the cases of *T. saginata* and *T. asiatica*, there is also a certain degree of intraspecific genetic variation but their biological features are still uncertain [26]. Of course, there is no way to distinguish their intraspecific variations each other morphologically. However, if we use molecular tools we can distinguish not only species but also intraspecific variants. In addition, identification of species using molecular tools is hard to be influenced by the skill of the experimenter. Therefore, molecular tools should be used positively whenever they are available.

27.1.3.2.1 Preservation of Taeniid Samples

The specimen of the helminthic worm has been stored in 10% formalin traditionally. Adult worms and cysticerci of taeniid cestodes have also not been exceptions. However, preservation in formalin should not be recommended, because it causes not only fragmentation of DNA but also difficulty in dissolving the worms [27]. On the other hand, preservation in 100% ethanol or at −80°C is convenient for DNA, but it is not necessarily suitable for morphological examination. Especially, preservation in 100% ethanol may destroy the structure of the worms. Judging from above, preservation in 70% ethanol is most practical. However, DNA is gradually broken down even if the worm is stored in 70% ethanol. Then, preservation in 70% ethanol at −20°C may be recommended.

27.1.3.2.2 DNA Extraction

Prior to molecular detections, DNA should be extracted from taeniid cestodes with various conditions. Numerous nucleic acid extraction methods have been reported and many commercial kits are also available now. DNA extraction methods, which are commonly used for helminthic worms were reviewed by Okamoto and Ito [28].

If proglottids or cysticerci of human *Taenia* are obtained, we can use a wide variety of extraction techniques. Since helminthic worms are essentially made of soft tissue, the conventional in-house isolation technique from soft tissues may be used. The most commonly applied conventional methods for DNA extraction from isolated worms include phenol/chloroform extraction and cetyltrimethylammonium bromide (CTAB) precipitation. Although there are some disadvantages in phenol/chloroform extraction including the use of toxic organic solvent and sample loss during successive extraction, intact, and large-sized DNA can be obtained. Azuma et al. [29] isolated intact DNA from taeniid cestode, *T. taeniaeformis*, by phenol/chloroform extraction. Isolation of DNA from taeniid cestodes by CTAB precipitation has been reported [30,31]. This technique is simple, rapid, reproducible, and yields undegraded DNA. This technique has also proved effective for mitochondrial DNA (mtDNA) as well as chromosomal DNA from cestodes [30].

DNA has also been extracted from isolated worms by using commercial kits. As DNA is often used as a template for polymerase chain reaction (PCR) diagnosis of parasitic infections, intact DNA is not always needed. Important points are that purified DNA is free of contaminants and enzyme inhibitors. Many commercial DNA extraction kits have been launched; some of them are suitable for this purpose. DNA extraction methods by kits can be generally classified into two types, affinity or nonaffinity methods. Nonaffinity methods, such as GNOME, Easy-DNA Kit, AquaPure, are cheaper methods and sometimes used for helminthic worms [32–34]. However, extraction kits using affinity methods are overwhelmingly employed in recent studies. Most of them adopt the silica method. In these, a series of products by QIAGEN such as DNeasy Blood and Tissue Kit seems to be most common. Several protocols are usually designed for each kit, but the protocol for animal tissue is suitable in many cases.

When DNA is extracted from stool samples of taeniasis patients, main targets are eggs and parasite debris of tegmental turnover. Recently, commercial DNA extraction kits, which can isolate DNA from stool samples, are available. The QIAamp DNA Stool Mini Kit (QIAGEN) is one of them. This kit provides fast and easy purification of total DNA from fresh or frozen stool samples. To ensure removal of PCR-interfering substances, this kit contains InhibitEX Tablets, a unique reagent provided in a convenient tablet form. InhibitEX Tablets efficiently adsorb these substances early in the purification process so that they can easily be removed by a quick centrifugation step. In addition, the kit contains buffer ASL, which is specially developed to remove inhibitory substances from stool samples. This kit is commonly used for DNA diagnosis of parasitic disease from fecal samples [35–37].

In addition to the problem of PCR-interfering substances, the eggshells of taeniid eggs are sometimes strong and hard to destroy by proteinase K digestion. Furthermore, since eggs of *T. solium* are dangerous for humans, stool samples are sometimes fixed by ethanol or boiling. When DNA is extracted from such samples, it is better to disrupt eggs prior to extraction. Several disruption procedures for nematode eggs have been tested including sonication, bead beating, boiling, microwaving, proteinase K/SDS digestion, freezing, and various combinations of the above [38]. In the case of taeniid eggs, previous shaking with glass beads gave best results [39].

27.1.3.2.3 Molecular Detection

Various molecular approaches for the identification of taeniid species have been developed including Southern blot hybridization, PCR-RFLP, PCR-SSCP, RAPD-PCR, sequencing of DNA, multiplex PCR and loop-mediated isothermal amplification (LAMP) method, and so on. Molecular techniques for the identification of species or genotypes of human *Taenia* are overviewed in the previous article [16]. Molecular techniques for the identification of species or genotypes of human *Taenia* are summarized in Table 27.1. Southern blot hybridization using specific DNA probes was utilized to distinguish between *T. solium* and *T. saginata* [40,41,43]. The assay was highly sensitive and specific, but the isotope-labeled probes were impractical in actual use. PCR-RFLP and PCR-SSCP were subsequently developed to diagnose both human taeniases, but those methods were relatively time consuming.

TABLE 27.1
Molecular Probes for Identification of Human Taeniid Species

Hybridization	Target DNA	Probe	Differential Diagnosis	Reference
Southern, dot	Unknown	pTS10	*T. solium* specific	[40]
	Unknown	Total genomic DNA	*T. solium*, *T. saginata*	[41]
	Noncording DNA fragments	HDP1, HDP2	*T. solium*, *T. saginata*	[42]
	Unknown: cox1	pTsol-9, pTsag-16	*T. solium*, *T. saginata*	[43]
PCR				
Method	**Target DNA**	**Primers**	**Differential Diagnosis**	**Reference**
PCR-RFLP	HDP2 (170 bp)	PTs7S35F2: CTTCTCAATTCTAGTCGCTGTGGT PTs7S35R1: GGACGAAGAATGGAGTTGAAGGT	*T. solium*, *T. sagniata*	[44]
	ITS1 (760 bp)	BD1: GTCGTAACAAGGTTTCCGTA 4S: TCTAGATGCGTTCGAAG/ATGTCGATG	*T. solium*, *T. sagniata*	
	ITS2 (Tsag: 650 bp) (Tsol: 600 bp)	NC6: ATCGACATCTTGAACGCACATTGC NC2: TTAGTTTCTTTTCCTCCGCT	*T. solium*, *T. sagniata*	
	cox1(521)	forward primer: TGGGTTAGATGTTAAGACGGC reverse primer: AAAACACCCACTAAAAGCAGA	*T. solium*, *T. sagniata*	[45]
PCR-SSCP	ND1 (530 bp)	JB11: AGATTCGTAAGGGGCCTAATA JB12: ACCACTAACTAATTCACTTTC	*Taenia*	[46]
	cox1 (444 bp)	JB3: TTTTTTGGGCATCCTGAGGTTTAT JB4.5: TAAAGAAAGAACATAATGAAAATG		
RAPD	Unknown	Primers from Operon Technologies	Variation of *T. solium*	[47]
	Unknown	Primers from Operon Technologies	*T. solium*, *T. sagniata*	[48]
SCAR	SCAR marker A03 (864 bp)	A03F: AGTCAGCCACCTGTACACATG A03R: AGTCAGCCACGCAAAGGGTAG	*T. solium* specific	[48][a]
	SCAR marker AD14 (463 bp)	AD14F: GAACGAGGGTGAATCACATAGC AD14R: GAACGAGGGTCGATGTATAAAT	*T. solium* specific	
	SCAR marker AD16 (502 bp)	AD16F: AACGGGCGTCCTAGGAGATGGC AD16R: AACGGGCGTCAGTAGTGAAC	*T. solium* specific	
	SCAR marker D05 (545 bp)	D05F: TGAGCGGACACCGAAAGCTATTGC D05R: TGAGCGGACAAGTAGTGAAGTG	*T. solium* specific	
	SCAR marker S07 (484 bp)	S07F: TCCGATGCTGAGCATGGGTAG S07R: TCCGATGCTGGTCAAAGTGGAAAG	*T. solium* specific	
	SCAR marker G14 (377 bp)	G14F: GGATGAGACCGTACTAGCATG G14R: GGATGAGACCCTTGTAACAAATG	*T. saginata* specific	
	SCAR marker K02 (531 bp)	K02F: GTCTCCGCAACTGTAATGTATC K02R: GTCTCCGCAATAATAATCTTTC	*T. saginata* specific	
BESS T-base	*cox1* (469 bp) *cob* (253bo)	P1: ATATTTACTTTAGATCATAAGCG P2: TAAAATTAATAGAACTAAAAAT P3: TTATGAGATTGTCAAAAGATTCTT P4: TATAGATGTCAAAACAGTAGCAGCCC	*T. saginata*, *T. asiatica* two genotypes of *T. solium*	[49]
Specifc PCR	HDP1 (ladder)	PTs4F1: GCAGTGTGCTGAAGATGAATA PTs4R1: GAATTTGGCTCTCACTGAATG	*T. saginata* specific	[50]
Multiplex PCR	HDP2 (Tsag:170 bp, 600 bp) (Tsol: 170 bp)	PTs7S35F1: CAGTGGCATAGCAGAGGAGGAA PTs7S35F2: CTTCTCAATTCTAGTCGCTGTGGT PTs7S35R1: GGACGAAGAATGGAGTTGAAGGT	*T. solium*, *T. sagniata*	[50]
	cox1 (Tsag: 827 bp) (Tasi: 269 bp) (TsolAm/Af: 720 bp) (TsolAsia: 984 bp)	TsagMpF: TTGATTCCTTCGATGGCTTTTCTTTTG TasiMpF: ACGGTTGGATTAGATGTTAAGACTA TsolAmMpF: GGTAGATTTTTTAATGTTTTCTTTA TsolAsiMpF: TTGTTATAAATTTTTGATTACTAAC TaeniaMpR: GACATAACATAATGAAAATG	*T. saginata*, *T. asiatica* two genotypes of *T. solium*	[35][a]

continued

TABLE 27.1 (continued)
Molecular Probes for Identification of Human Taeniid Species

Method	Target DNA	Primers	Differential Diagnosis	Reference
	*t*RNA(Val)- ND2	Ta4978F: GGGTTTAAGTTATAAATGTGATGT	*T. saginata*, *T. asiatica*,	[51]
	(Tasi: 706 bp)	Ts5058F: ACTACATTTGGTTTGTTTTTGTAG	*T. solium*	
	(Tsag: 629 bp)	Tso7421F: CTAGGCCACTTAGTAGTTTAGTTA		
	(Tsol: 474 bp)	Rev7915: CATAAAACACTCAAACCTTATAGA		
PCR/sequencing	28SrDNA D1 (281 bp)	JB10: GATTACCCGCTGAACTTAAGCATAT	Universal	[52]
		JB9: CCTGCATCACAAACACCCCGACTC		
	cox1 (444 bp)	JB3: TTTTTGGGCATCCTGAGGTTTAT	Universal	
		JP4.5: TAAAGAAAGAACATAATGAAAATG		
	12S rDNA (311 bp)	60.for.:	*Taenia*, *Echinococcus*	[53]
		TTAAGATATATGTGGTACAGGATTAGATACCC		
		375.rev.: AACCGAGGGTGACGGGCGGTGTGTACC		
	cox1 (444 bp)	pr-a: TGGTTTTTTGTGCATCCTGAGGTTTA	Universal	[54]
		pr-b: AGAAAGAACGTAATGAAAATGAGCAAC		
	cox1 (875 bp)	F/CO1: TTGAATTTGCCACGTTTGAATGC	*Taenia*, *Echinococcus*	[55]
		R/CO1: GAACCTAACGACATAACATAATGA		
	ND4,ATP6,ND2 (1628 bp)	F/ND4: TGGAGTTAGATGGTAAGCGTTGAT		
		R/ND2: CAGGAAACTTCATAACAACACTTA	*Taenia*	
	cox1 (complete >1620 bp)	Cox1/F: GTTATGTTAGACTAGATGTTTTCA	*T. saginata*, *T. asiatica*	[56]
		Cox1/R: TCCACTAAGCATAATGCAAAAGGC	two genotypes of	
		Cytb/F: ATAAACTGATAGATTGTGGTTC	*T. solium*	
	cob (complete >1068 bp)	Cytb/R: CATATGACTGTCTAATGAAGAAAA		
LAMP method	*cox1*	Several primer sets were designed for each gene.	*T. saginata*, *T. asiatica*,	[57]
	clp	See original paper.	*T. solium*	

Source: Okamoto, M., and Ito, A. In: *Handbook of Nucleic Acid Purification*, Chapter 13. Taylor & Francis Group, London, pp. 271–291, 2009.
[a] Each primer was not named in the original paper.

On the contrary, PCR using species-specific primers provides rapid and sensitive. Based on the richness of thymine-base in cestode mtDNA, a base excision sequence scanning thymine-base method clearly differentiated not only from *T. saginata* from *T. solium* but also from *T. saginata* from *T. asiatica* [49]. The differentiation method of three human *Taenia* and genotypes of *T. solium* was further improved to a multiplex PCR method [51], which is applicable to the detection of *Taenia*-specific DNA in feces (copro-DNA). Although the multiplex PCR is a simple and rapid method but it is very sensitive and reliable.

Due to further evaluation of polymorphism of mtDNA of human taniid cestodes, more detailed molecular probes to differentiate intraspecific variations might become important. Then, if it is possible, we better analyze the long DNA sequence from each specimen in order to obtain additional information to track back where the patients were exposed to eggs or cysticerci using the retrospective specimens. For this purpose, mitochondrial cytochrome *c* oxidase subunit 1 gene (*cox*1) seems to be suitable. The length of *cox*1 in taeniid cestodes is relatively long (about 1620 bp), and the primer set which can amplify complete sequence of *cox*1 has been designed [57]. In the case of formalin-fixed samples or histopathological specimens, it may be difficult to amplify more than 1-kbp DNA fragments, because of the fragmentation of DNA [27]. In that case, we can select another primer set which amplifies a shorter fragment such as P1/P2 [27].

The majority of *Taenia* species causing cysticercosis in humans is known to be *T. solium*. However, there are many other zoonotic taeniid cestodes species including *T. taeniaeforimis*, *T. pisiformis*, *T. hydatigena*, *T. serialis*, *T. multiceps*, and *Echinococcus* spp. [27]. A lot of the *cox*1 sequence from not only taeniid cestodes but also various organisms have been deposited in the database. When we use universal primer sets such as JB3/JP4.5 or pr-a/pr-b, it may allow us to determine the causative organisms. Further studies to establish additional molecular probes for the identification of other zoonotic cestodes species may be necessary when we face such unexpected species.

In our fieldwork in Thailand, we failed in applying the multiplex PCR methods under nonstable electric power supplies. The multiplex PCR method is very simple, but requires a thermal controller, which controls temperature precisely. For field use, LAMP method without using the PCR machine was recently developed to classify the human taeniid tapeworms [58]. The LAMP method does not require the thermal controller and positive reactions of the LAMP can be confirmed by naked eyes without electrophoresis. The sensitivity of the LAMP method seems to be nearly equal or higher than that of the multiplex PCR method. A field trial of the LAMP method in endemic area is expected.

27.2 METHODS

27.2.1 DNA Extraction from Eggs

For molecular diagnosis of human taeniasis, the DNA extraction step from eggs is essential. Eggs of taeniid worms usually exist in feces. If possible, isolation of eggs by a suitable technique such as a modified Wisconsin sucrose flotation method should be employed before DNA extraction. In many cases, however, DNA is extracted from eggs together with feces. There are large amounts of PCR-interfering substances in feces. Some commercial DNA extraction kits can remove PCR-interfering substances from stool samples. The QIAamp DNA Stool Mini Kit (QIAGEN) is most commonly used for DNA diagnosis of parasitic disease from fecal samples. Protocols for isolation of DNA from stool for pathogen are as follows:

27.2.1.1 Procedure

1. Weigh 180–220 mg stool in a 2 mL microcentrifuge tube and place the tube on ice.
2. Add 1.4 mL buffer ASL to each stool sample. Vortex continuously for 1 min.
3. Heat the suspension for 5 min at 70°C.
4. Vortex for 15 s and centrifuge sample at full speed for 1 min to pellet stool particles.
5. Pipet 1.2 mL of the supernatant into a new 2 mL microcentrifuge tube and discard the pellet.
6. Add 1 InhibitEX Tablet to each sample and vortex immediately and continuously for 1 min or until the tablet is completely suspended. Incubate suspension for 1 min at room temperature to allow inhibitors to adsorb to the InhibitEX matrix.
7. Centrifuge sample at full speed for 3 min to pellet inhibitors bound to InhibitEX matrix.
8. Pipet all the supernatant into a new 1.5 mL microcentrifuge tube and discard the pellet. Centrifuge the sample at full speed for 3 min.
9. Pipet 15 μL proteinase K into a new 1.5 mL microcentrifuge tube.
10. Pipet 200 μL supernatant from step 8 into the 1.5 mL microcentrifuge tube containing proteinase K.
11. Add 200 μL buffer AL and vortex for 15 s.
12. Incubate at 70°C for 10 min.
13. Add 200 μL of ethanol (96–100%) to the lysate, and mix by vortexing.
14. Label the lid of a new QIAamp spin column placed in a 2 mL collection tube. Carefully apply the complete lysate from step 13 to the QIAamp spin column without moistening the rim. Close the cap and centrifuge at full speed for 1 min. Place the QIAamp spin column in a new 2 mL collection tube, and discard the tube containing the filtrate. Close each spin column in order to avoid aerosol formation during centrifugation.
15. Carefully open the QIAamp spin column and add 500 μL Buffer AW1. Close the cap and centrifuge at full speed for 1 min. Place the QIAamp spin column in a new 2 mL collection tube, and discard the collection tube containing the filtrate.
16. Carefully open the QIAamp spin column and add 500 μL Buffer AW2. Close the cap and centrifuge at full speed for 3 min. Discard the collection tube containing the filtrate.
17. Place the QIAamp spin column in a new 2 mL collection tube and discard the old collection tube with the filtrate. Centrifuge at full speed for 1 min.
18. Transfer the QIAamp spin column into a new, labeled 1.5 mL microcentrifuge tube. Carefully open the QIAamp spin column and pipet 200 μL Buffer AE directly onto the QIAamp membrane. Close the cap and incubate for 1 min at room temperature, then centrifuge at full speed for 1 min to elute DNA.

27.2.2 Detection Procedures

27.2.2.1 Multiplex PCR

Compared to PCR-RFLP and PCR-SSCP analysis, multiplex PCR is a simpler and more rapid method. Up to now, two methods have been reported by a Japanese group and a Korean group [35,51]. Former one reported by Yamasaki et al. [35] is easy to differentiate between not only the species of three human *Taenia* but also two genotypes of *T. solium*. The procedure utilizes the following specific primer set from mitochondrial cytochrome *c* oxidase subunit 1 gene (*cox*1):

TsagMpF: TTGATTCCTTCGATGGCTTTTCTTTTG
TasiMpF: ACGGTTGGATTAGATGTTAAGACTA
TsolAmMpF: GGTAGATTTTTTAATGTTTTCTTTA
TsolAsiMpF: TTGTTATAAATTTTTGATTACTAAC
TaeniaMpR: GACATAACATAATGAAAATG

27.2.2.1.1 Procedure

1. Prepare a master mix as follows:

Component	Volume/Reaction	Final Concentration
10 × PCR buffer[a]	2.5 μL	1×
dNTP Mixture[b]	2.0 μL	200 μM of each dNTP
Primer TsagMpF (5 μM)	1.25 μL	0.25 μM
Primer TasiMpF (5 μM)	1.25 μL	0.25 μM
Primer TsolAmMpF (5 μM)	1.25 μL	0.25 μM
Primer TsolAsiMpF (5 μM)	1.25 μL	0.25 μM
Primer TaeniaMpR (5 μM)	1.25 μL	0.25 μM
Taq polymerase	0.1 μL	1 unit/reaction
Template DNA	<1 μg	
Deionized Water	Variable	
Total volume	Volume 25 μL	

[a] Contains 15 mM $MgCl_2$.
[b] 2.5 mM of each.

In general, it is assumed that the concentration and proportion of forward-to-reverse primers are important in the multiplex PCR. The optimal ratio of these primers recommended by Yamasaki et al. [35] was 1:2:4 for TsagMpF and TsolAmMpF (0.08 μM), TasiMpF and TsolAsiMpF (0.16 μM), and TaeniaMpR (0.32 μM), respectively. In our laboratory, however, the above master mix produces clear PCR products. A better reaction composition should be examined in each laboratory. The master mix contains all the components needed for PCR except for the template DNA. Prepare a volume of master mix 10% greater than that required for the total number of PCR assays to be performed. A negative control (without template DNA) and positive controls (*T. saginata*, *T. asiatica*, and two genotypes of *T. solium*) should always be included.

2. Mix the master mix thoroughly and dispense appropriate volumes into PCR tubes.
3. Add template DNA (<1 μg/reaction) to the individual tubes containing the master mix.
4. Proceed directly to the next step, when using a thermal controller with a heated lid. Otherwise, overlay each reaction with mineral oil.
5. Program the thermal controller. In the original paper by Yamasaki and others [41], the multiplex PCR protocols consisted of 35 cycles of denaturation (30 s at 94°C), annealing (30 s at 60°C), and extension (90 s at 72°C), plus one cycle of 5 min at 72°C. In our experiments, however, very weak or no band was obtained at this PCR protocol. On the other hand, nonspecific bands that gave unreliable results were observed at lower annealing temperature. In our case, when protocols consisted of predenaturation (5 min at 94°C), 35 cycles of denaturation (30 s at 94°C), annealing (30 s at 57°C), and extension (70 s at 72°C), plus one cycle of 7 min at 72°C, the most preferable result was obtained. PCR program should be adjusted to the thermal controller used.
6. Place the PCR tubes in the thermal controller and start the cycling program. After amplification, samples can be stored overnight at 2–8°C or at –20°C for longer storage.
7. Loaded the PCR reaction product onto an agarose gel (1%) with prior addition of gel-loading buffer.
8. Close the lid of the gel tank and attach the electrical leads so that the DNA will migrate towards the anode. Run the gel until the dye (bromophenol blue and/or xylene cyanol) has migrated the appropriate distance through the gel.
9. Turn off the electric current. If ethidium bromide is present in the gel and electrophoresis buffer, examine the gel by ultraviolet light and photograph. Otherwise, stain the gel with ethidium bromide (0.5 μg/L) then take a photograph.

27.2.2.2 Sequencing of the PCR Product

Although most of cycticercosis in humans are caused by *T. solium*, the accidental invasion of other taeniid larvae into human internal tissues was rarely found. In these cases, sequencing of the PCR product is very useful. Sequence data from the individual worm are also very informative for epidemiological or retrospective studies.

In the case of human taeniid cestodes, mitochondrial cytochrome *c* oxidase subunit 1 gene (*cox*1) seems to be suitable. The length of *cox*1 in taeniid cestodes is relatively long (about 1620 bp), and the primer set which can amplify complete sequence of *cox*1 has been designed (Cox1/F, Cox1/R) [56]. When it is expected that worms collected from human may be taeniid species and that large molecule of DNA can be extracted from these worms, this primer set is suitable for the PCR amplification. Sequences of these primers are as follows:

Cox1/F: GTTATGTTAGACTAGATGTTTTCA
Cox1/R: TCCACTAAGCATAATGCAAAAGGC

In the case of formalin-fixed samples or histopathological specimens, it may be difficult to amplify more than 1-kbp DNA fragments, because of the fragmentation of DNA. Species of cysticerci recovered from human or domestic animals is not always human *Taenia* species. In these cases, Cox1/F/Cox1/R may not function, and universal primer sets which amplify shorter fragments may be useful. The universal primer set, pr-a/pr-b, can amplify 444 bp fragment of *cox*1 [54]. In our experience, this primer set is applicable to all species of cestodes and trematodes examined, and also amplifies the *cox*1 fragment of some nematode species. Sequences of pr-a/pr-b are as follows:

pr-a: TGGTTTTTTGTGCATCCTGAGGTTTA
pr-b: AGAAAGAACGTAATGAAAATGAGCAAC

27.2.2.2.1 Procedure

1. Amplification of the DNA fragment of the target gene: Prepare a master mix as follows:

Component	Volume/Reaction	Final Concentration
10 × PCR buffer[a]	2.5 μL	1×
dNTP mixture[b]	2.0 μL	200 μM of each dNTP
Forward primer (5 μM)	1.25 μL	0.25 μM
Reverse primer (5 μM)	1.25 μL	0.25 μM
Taq polymerase	0.1 μL	1 unit/reaction
Template DNA	<1 μg	
Deionized Water	Variable	
Total volume	25 μL	

[a] Contains 15 mM $MgCl_2$.
[b] 2.5 mM of each.

PCR protocols for Cox1/F/Cox1/R consisted of 30 cycles of denaturation (30 s at 94°C), annealing (30 s at 56°C), and extension (90 s at 72°C), plus one cycle of 7 min at 72°C. PCR protocols for pr-a/pr-b consisted of predenaturation (5 min at 94°C), 30 cycles of denaturation (50 s at 94°C), annealing (45 s at 42°C), and extension (90 s at 72°C), plus one cycle of 7 min at 72°C.

2. Preparing the templates: For optimal results, purify the PCR product before sequencing. In general, any

method that removes dNTPs and primers such as MinElute PCR Purificaticn Kit and Centricon-100 columns should work.

3. Performing cycle sequencing: Cycle sequencing provides the most reproducible results for sequencing PCR templates. Although PCR fragments can be difficult to denature with traditional sequencing methods, cycle sequencing provides several chances to denature, and extend the template, ensuring adequate signal in the sequencing reaction. In the following, the example using BigDye Terminator v3.1 Cycle Sequencing Kit (Applied Biosystems) is shown.

   The reaction mixtures:

Reagent	Quantity
Terminator ready reaction mix	8.0 μL
Template	10–40 ng for Cox1/F/Cox1/R
	3–10 ng for pr-a/pr-b
Primer	3.2 pmol
Deionized water	Variable
Total	Volume 20 μL

   Place the tubes in a thermal cycler and set to the correct volume. Cycle sequencing protocols consisted of predenaturation (5 min at 94°C), 35 cycles of denaturation (10 s at 96°C), annealing (5 s at 50°C), and extension (4 min at 60°C).

4. Purifying extension products: The best results are obtained when unincorporated dye terminators are completely removed prior to electrophoresis. Excess dye terminators in sequencing reactions obscure data in the early part of the sequence and can interfere with base calling. Spin column such as Centri-Sep can be used for purification.
5. Electrophoresis: Purified extension products will be electrophoresed by the sequencing instrument such as ABI PRISM 3100 Genetic Analyzer.

27.3 CONCLUSIONS AND FUTURE PERSPECTIVES

Morphologically, *Taenia* species and intraspecific variants are not distinguishable. Application of molecular techniques is essential for the identification of taeniid cestodes. Given that there may be unknown variants and PCR primers commonly used in diagnosis of human *Taenia* may not anneal these variants, we should use the microscopic examination together at present.

Recently, the molecular techniques have also been utilized in ecological or population genetic studies. In these studies, microsatellite markers, which consist of short tandemly repeated DNA, are commonly applied. In the case of *Echinococcus*, the single locus or multiloci microsatellite markers have been isolated [58–60]. In fact, the multiloci microsatellite marker was used to classify *E. multilocularis* individuals into genotypes [61] However, microsatellite markers have not been isolated from human *Taenia* in spite of their usefulness [62]. Isolation of the microsatellite marker from human *Taenia* is expected as early as possible.

For molecular diagnosis of human *Taenia*, the most serious problem may be existence of hybrid-derived offspring between *T. saginata* and *T. asiatica*. Recently, two hybrid-derived worms between *T. sginata* and *T. asiatica* were found in Knachanaburi Province, Thailand where the two species are sympatrically endemic [26,63]. In these worms, phylogeny of mtDNA is not always consistent with those of nuclear genes. This finding is evidence of past hybridization between the two taxa. In this chapter, we regard *T. saginata* and *T. asiatica* as the distinct species. However, the existence of hybrid-derived worms contradicts our concept. Under the biological species concept (BSC), distinct species are not expected to exchange genes. It is therefore reasonable to consider that *T. asiatica* is the same species as *T. saginata* according to the BSC.

Is it appropriate for us to assume that *T. asiatica* is equated to *T. saginata*? There are obvious biological differences between the two taxa, including morphology of the cysticerci, host specificity, and organ specificity. In particular, specificity for the intermediate host is very important for medical and veterinary science, because it holds the key to the potential source of infection of the patient. Although *T. asiatica* is still able to hybridize with *T. saginata*, each taxon has its own biological identity. Biomedical researchers need more-pragmatic approaches that can be understood by nonspecialists. At the very least, medical or veterinary researchers should not equate *T. asiatica* with *T. saginata*.

In this way it is important to distinguish *T. asiatica* from *T. saginata*. However, existence of hybridization makes it impossible to identify them by the molecular technique. When eggs of taeniid cestodes are detected in human feces, the determination of the species is usually dependent on the specific PCR or sequencing analysis of the mtDNA. And then, the intermediate host, which would have been the source of infection, is extrapolated from species identified retrospectively. Since the mitochondrial gene does not define host specificity, the occurrence of hybridization between *T. saginata* and *T. asiatica* indicates that the intermediate host extrapolated by analysis of the mitochondrial gene is not always correct. In addition, the intermediate host cannot be extrapolated from analyses of some nuclear genes alone. Unless the gene that regulates host specificity can be identified, the intermediate host of hybrid-derived offspring should not be deduced based on DNA analysis in sympatric endemic areas. The inability to accurately identify intermediate hosts continues to be a significant problem in the epidemiological study. The biological features of hybrid-derived offspring, including host specificity and genetic structure, should become clear as soon as possible.

ACKNOWLEDGMENTS

The projects summarized in this review had financial support from the Japan Society for the Promotion of Science (JSPS) (14256001, 17256002) and from the Ministry of Education,

Japan (Kagaku-Gijutsu Shinko Choseihi 2003–2005) and JSPS-Asia/Africa Science Platform Fund (2006–2008) to A. Ito, from JSPS (18790289) to Y. Sako, and from JSPS (16500277, 18406008) to M. Okamoto.

REFERENCES

1. Verster, A. A taxonomic revision of the genus *Taenia* Linnaeus, 1758s. str. *Onderstepoort J. Vet. Res.* 36, 3–58, 1969.
2. Loos-Frank, B. An up-date of Verster's, 1969."Taxonomic revision of the genus Taenia Linnaeus" (Cestoda) in table format. *Syst. Parasitol.* 45, 155–183, 2000.
3. Eom, K.S., and Rim, H.J. Morphologic descriptions of *Taenia asiatica* sp. *Korean J. Parasitol.* 31, 1–6, 1993.
4. Rausch, R.L. *Taenia pencei* sp. from the Ringtail, *Bassariscus astutus* (Carnivora: Procyonidae), in Texas, U.S.A. *Comp. Parasitol.* 70, 1–10, 2003.
5. Hoberg, E.P., Phylogeny of *Taenia*: Species definitions and origins of human parasites. *Parasitol. Int.* 55(Suppl.), S23–S30, 2006.
6. Abuladze, K.I. Taeniata of animals and man, and diseases caused by them. Akademiya Nauk USSR, Moscow, pp 59–67, 1964. (translated by Israel's Program for Scientific Translation, Jerusalem, 1970.)
7. Soulsby, E.J.L. *Helminths, Arthropods and Protozoa of Domesticated Animals*. Lea & Febiger, Philadelphia, 1982.
8. Eom, K.S. What is Asian *Taenia*? *Parasitol. Int.* 55 (Suppl.), S137–S141, 2006.
9. Ito, A., Nakao, M., and Wandra, T. Human taeniasis and cysticercosis in Asia. *Lancet* 362, 1918–1920, 2003.
10. Ito, A. et al. Dogs as alternative intermediate hosts of *Taenia solium* in Papua (Irian Jaya), Indonesia confirmed by highly specific ELISA and immunoblot using native and recombinant antigens and mitochondrial DNA analysis. *J. Helminthol.* 76, 311–314, 2002.
11. Ito, A. et al. Neurocysticercosis: Clinical manifestation, neuroimaging, serology and molecular confirmation of histopathologic specimens. *Southeast Asian J. Trop. Med. Public Health* 37(Suppl 3), 74–81, 2006.
12. Wandra, T. et al. Current situation of taeniasis and cysticercosis in Indonesia. *Trop. Med. Health* 35, 323–328, 2007.
13. Schantz, P.M. et al. Neurocysticercosis in an orthodox Jewish community in New York City, *N. Engl. J. Med.* 327, 692–695, 1992.
14. Al Shahrani, D. et al. First case of neurocysticercosis in Saudi Arabia. *J. Trop. Pediatr.* 49, 58–60, 2003.
15. Hira, P.R. et al. Cysticercosis: Imported and autochthonous infections in Kuwait. *Trans. R. Soc. Trop. Med. Hyg.* 98, 233–239, 2004.
16. Ito, A. et al. Chapter 62. *Taenia*, In: Liu, D. (ed), *Molecular Detection of Foodborne Pathogens*, Taylor & Francis Group, London, pp. 827–838, 2009.
17. Okamoto, M. et al. Asian *Taenia*: Species or subspecies? *Southeast Asian J. Trop. Med. Public Health*, 38 (Suppl 1), 125–130, 2007.
18. Fan, P.C. Taiwan *Taenia* and taeniasis, *Parasitol. Today*. 4, 86–88, 1988.
19. Okamoto, M. et al. Molecular variation of *Taenia solium* in the world, *Southeast Asian J. Trop. Med. Public Health*, 32(Suppl 2), 90–93, 2001.
20. Nakao, M. et al. A phylogenetic hypothesis for the distribution of two genotypes of the pig tapeworm *Taenia solium* worldwide. *Parasitology* 124, 657–662, 2002.
21. Ito, A. et al. Chapter 5. Mitochondrial DNA of *Taenia solium*: From basic to applied science. In: Singh, G., and Prabhakar, S. (eds), *Taenia solium Cysticercosis*, CAB International, 47–55, 2002.
22. Chung, J.Y. et al. Molecular determination of the origin of acephalic cysticercus. *Parasitology* 130, 239–246, 2005.
23. Hinojosa-Juarez, A.C. et al. Genetic similarity between cysticerci of *Taenia solium* isolated from human brain and from pigs. *Infect. Genet. Evol.* 8, 653–656, 2008.
24. Anantaphruti, M.T. Molecular and serological survey on taeniasis and cysticercosis in Kanchanaburi Province, Thailand. *Parasitol. Int.*, 59, 326, 2010.
25. Dryden, M.W., Payne, P.A., and Smith, V. Accurate diagnosis of *Giardia* spp and proper fecal examination procedures. *Vet. Therapeutics* 7, 4–14, 2006.
26. Okamoto, M. et al. Evidence of hybridization between *Taenia saginata* and *Taenia asiatica*. *Parasitol. Int.* 59, 70–74, 2010.
27. Yamasaki, H. et al. Mitochondrial DNA diagnosis for taeniasis and cysticercosis. *Parasitol. Int.* 55, S81, 2006.
28. Okamoto, M., and Ito, A. Purification of nucleic acids from helminths. In: Liu, D. (ed), *Handbook of Nucleic Acid Purification*, Chapter 13, Taylor & Francis Group, London, pp. 271–291, 2009.
29. Azuma, H. et al. Intraspecific variation of *Taenia taeniaeformis* as determined by various criteria. *Parasitol. Res.* 81, 103, 1995.
30. Yap, K.W., and Thompson, R.C.A. CTAB precipitation of Cestode DNA. *Parasitol. Today*. 3, 220, 1987.
31. Yap, K.W. et al. *Taenia hydatigena*: Isolation of mitochondrial DNA, molecular cloning, and physical mitochondrial genome mapping. *Exp. Parasitol.* 63, 288, 1987.
32. Okamoto, M. et al. Phylogenetic relationships of three hymenolepidid species inferred from nuclear ribosomal and mitochondrial DNA sequences, *Parasitology* 115, 661, 1997.
33. Copeland, C.S. et al. Identification of the Boudicca and Sinbad retrotransposons in the genome of the human blood fluke *Schistosoma haematobium*. *Mem. Inst. Oswaldo Cruz* 101, Online. 2006.
34. Matoba, Y. et al. Detection of a taeniid species *Taenia taeniaeformis* from a feral raccoon *Procyon lotor* and its epidemiological significance. *Mamm. Study*. 28, 157, 2003.
35. Yamasaki, H. et al. DNA differential diagnosis of taeniasis and cysticercosis by multiplex PCR. *J. Clin. Microbiol.* 42, 548, 2004.
36. Gobert, G.N. et al. Copro-PCR based detection of Schistosoma eggs using mitochondrial DNA markers. *Mol. Cell. Probes* 19, 250, 2005.
37. Li, T. et al. Taeniasis/cysticercosis in a Tibetan population in Sichuan Province, China. *Acta Trop.* 100, 223, 2006.
38. Harmon, A.F. et al. Improved methods for isolating DNA from *Ostertagia ostertagi* eggs in cattle feces. *Vet. Parasitol.* 135, 297, 2006.
39. Nunes, C.M. et al. Fecal specimens preparation methods for PCR diagnosis of human taeniosis. *Rev. Inst. Med. Trop., Sao Paulo* 48, 45, 2006.
40. Rishi, A.K., and McManus, D.P. DNA probes which unambiguously distinguish *Taenia solium* from *T. saginata*, *Lancet* 2(8570), 1275–1276, 1987.
41. Flisser, A. et al. Specific detection of *Taenia saginata* eggs by DNA hybridization. *Lancet* 2(8625), 1429–1430, 1988.
42. Harrison, L.J., Delgado, J., and Parkhouse, R.M. Differential diagnosis of *Taenia saginata* and *Taenia solium* with DNA probes, *Parasitology* 100, 459–461, 1990.

43. Chapman, A. et al. Isolation and characterization of species-specific DNA probes from *Taenia solium* and *Taenia saginata* and their use in an egg detection assay. *J. Clin. Microbiol.* 33, 1283–1288, 1995.
44. González, L.M. et al. PCR tools for the differential diagnosis of *Taenia saginata* and *Taenia solium* taeniasis/cysticercosis from different geographical locations. *Diagn. Microbiol. Infect. Dis.* 42, 243–249, 2002.
45. Nunes, C.M. et al. *Taenia saginata*: Differential diagnosis of human taeniasis by polymerase chain reaction-restriction fragment length polymorphism assay. *Exp. Parasitol.* 110, 412–415, 2005.
46. Gasser, R.B., Zhu, X., and Woods, W. Genotyping *Taenia* tapeworms by single-strand conformation polymorphism of mitochondrial DNA. *Electrophoresis* 20, 2834–2837, 1999.
47. Vega, R. et al. Population genetic structure of *Taenia solium* from Madagascar and Mexico: Implications for clinical profile diversity and immunological technology. *Int. J. Parasitol.* 33, 1479–1485, 2003.
48. Dias, A.K. et al. *Taenia solium* and *Taenia saginata*: Identification of sequence characterized amplified region (SCAR) markers. *Exp. Parasitol.* 117, 9–12, 2007.
49. Yamasaki, H. et al. DNA differential diagnosis of human taeniid cestodes by base excision sequence scanning thymine-base reader analysis with mitochondrial genes. *J. Clin. Microbiol.* 40, 3818–3821, 2002.
50. González, L.M. et al. Differential diagnosis of *Taenia saginata* and *Taenia solium* infection by PCR. *J. Clin. Microbiol.* 38, 737–744, 2000.
51. Jeon, H.K. et al. Differential diagnosis of *Taenia asiatica* using multiplex PCR. *Exp. Parasitol.* 121, 151–156, 2009.
52. Bowles, J., and McManus, D.P. Genetic characterization of the Asian *Taenia*, a newly described taeniid cestodes of humans. *Am. J. Trop. Med. Hyg.* 50, 33–44, 1994.
53. von Nickisch-Rosenegk, M., Silva-Gonzalez, R., and Lucius, R. Modification of universal 12S rDNA primers for specific amplification of contaminated *Taenia* spp. (Cestoda) gDNA enabling phylogenetic studies. *Parasitol. Res.* 85, 819–825, 1999.
54. Okamoto, M. et al. Phylogenetic relationships within *Taenia taeniaeformis* variants and other taeniid cestodes inferred from the nucleotide sequence of the cytochrome *c* oxidase subunit I gene. *Parasitol. Res.* 81, 451–458, 1995.
55. Nakao, M. et al. Mitochondrial genetic code in cestodes. *Mol. Biochem. Parasitol.* 111, 415–424, 2000.
56. Nakao, M. et al. A phylogenetic hypothesis for the distribution of two genotypes of the pig tapeworm *Taenia solium* worldwide. *Parasitology* 124, 657–662, 2002.
57. Nkouawa, A. et al. Loop-mediated isothermal amplification method for differentiation and rapid detection of *Taenia* species. *J. Clin. Microbiol.* 47, 168–174, 2009.
58. Nakao, M., Sako, Y., and Ito, A., Isolation of polymorphic microsatellite loci from the tapeworm *Echinococcus multilocularis*. *Infect. Genet. Evol.* 3, 159–163, 2003.
59. Bartholomei-Santos, M.L. et al. Isolation and characterization of microsatellites from the tapeworm *Echinococcus granulosus*. *Parasitology* 126, 599–605, 2003.
60. Bart, J.M. et al. EmsB, a tandem repeated multiloci microsatellite, new tool to investigate the genetic diversity of *Echinococcus multilocularis*. *Infect. Genet. Evol.* 6, 390–400, 2006.
61. Knapp, J. et al. Genetic diversity of the cestode *Echinococcus multilocularis* in red foxes at a continental scale in Europe. *PLoS Negl. Trop. Dis.* 3, e452, 2009.
62. Campbell, G. et al. Genetic variation in *Taenia solium*. *Parasitol. Int.* 55, S121–S126, 2006.
63. Anantaphruti, M.T. et al. Sympatric occurrence of *Taenia solium*, *T. saginata*, and *T. asiatica*, Thailand. *Emerg. Infect. Dis.* 13, 1413–1416, 2007.

Section IV

Platyhelminthes: Trematoda

28 Clonorchis

Heinz Mehlhorn, Boris Müller, and Jürgen Schmidt

CONTENTS

28.1 INTRODUCTION

The genus *Clonorchis* contains a liver fluke species *Clonorchis sinensis* that utilizes fish as intermediate host, and dogs and cats as final (definitive) hosts. Humans become infected with this parasite through consumption of raw or undercooked fish—even in salted preparations, with adult worms (flukes) developing in the liver (liver fluke) [1–3]. Human clonorchiosis is prevalent in East Asian countries such as China, Korea, and Vietnam, where raw or undercooked fish is eaten regularly. Although many infected people are asymptomatic, some with heavy *C. sinensis* infections may show nonspecific symptoms such as anorexia, nausea, vomiting, loose stool, and abdominal or epigastric discomfort. Left untreated, the infection may lead to cholangiocarcinoma.

28.1.1 Classification and Morphology

The genus *Clonorchis* is classified in the family Opisthorchiidae, suborder Opisthorchiata, order Opisthorchiida, subclass Digenea, class Trematoda (sucking worms), phylum Platyhelminthes (flat worms). The name Digenea comes from Greek *digenes* = two sexes and refers to the fact that most of the Digenea are hermaphroditic organisms that combine several generations within their complicated life cycle. Besides *Clonorchis*, the other genera in the family Opisthorchiidae are *Amphimerus*, *Metorchis*, *Opisthorchis*, and *Pseudamphistomum*.

The genus name *Clonorchis* (Greek = *klon*, Latin—*clonus*, Greek—*opisthen* = in the rear, *orchis* = testis, testicle) describes that the two deeply lobulated testes are situated behind the female system at the posterior region of the worm (Figure 28.1). Currently, the genus *Clonorchis* consists of a single species *C. sinensis*—the so-called Chinese liver fluke of humans. Awareness of *C. sinensis* is old, since its eggs were already found in a corpse from the West Han Dynasty in China >2300 years ago. However, the first valid observation of the adult worm was recorded in 1875 by McConnell [4] at the autopsy of a Chinese carpenter in Calcutta, India. The American scientist Cobbold [1] used in 1875 the genus name *Distoma* (= two mouth openings) referring to the two suckers. The German zoologist Looss is considered to have described, in 1907, the first full and valid description of the *C. sinensis* life cycle [5], although the Japanese scientist Ijiima discovered, in 1887, that *C. sinensis* also occurs in some nonhuman hosts that live around humans and thus he established the concept of reservoir hosts, which makes it difficult to eliminate this parasite from humans [6]. Other Japanese—Kobayashi [7] and Muto [8]—contributed to the confirmation of important stages in the life cycle, showing that fresh water fish are the second

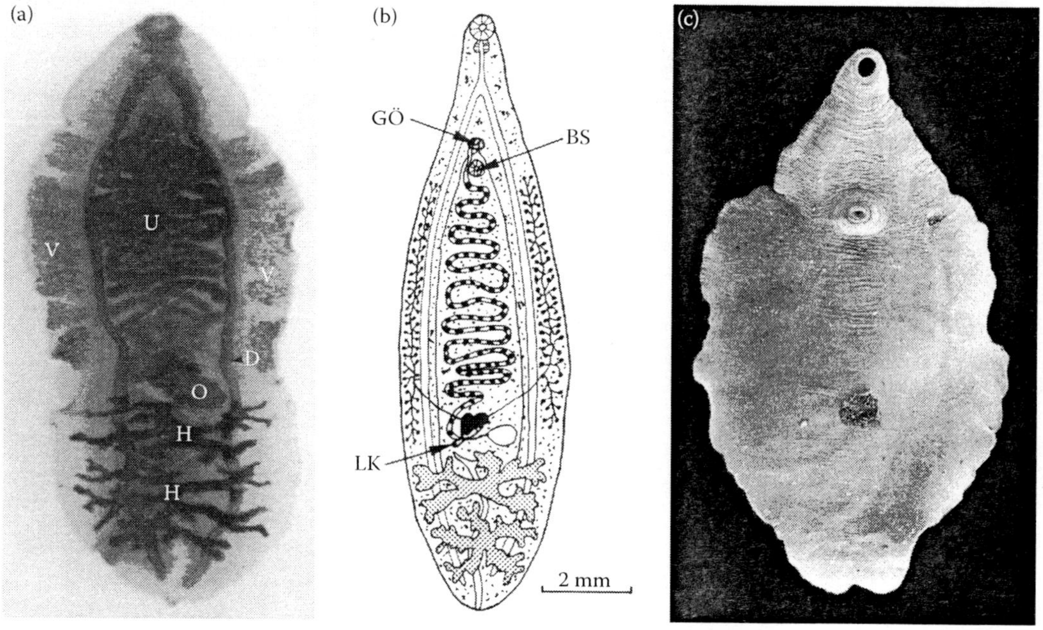

FIGURE 28.1 Adult worms of *Clonorchis sinensis*. (a) Light micrograph of a carmin-red stained *C. sinensis* worm (D = intestine, H = testis, O = ovary, U = uterus filled with eggs, V = vitellarium). (b) Diagrammatic representation of an adult stage of *C. sinensis*. (c) Scanning electron micrograph of an adult, contracted *C. sinensis*, showing the smooth surface and the two suckers, which in former times led to the genus description as "*Distoma*" = two mouths.

intermediate host, while several water snails (e.g., members of the genera *Bulimus, Bithynia*) act as first intermediate hosts, within which the fish-infecting cercariae are produced [7,8]. Recently, the genus name *Clonorchis* is often considered to be synonymous of *Opisthorchis* (Greek—*opisthos* = behind, *orchis*—testis), since two other species *Opisthorchis felineus* and *Opisthorchis viverrini* (see Chapter 38 of this book) are so closely related to *C. sinensis* that two different genus names (with identical meaning) are not justified.

C. (syn. *Opisthorchis*) *sinensis* is found focally in many countries of East Asia (Thailand, Laos, Cambodia, Korea, China, Taiwan, Hong Kong, Macao, and Vietnam) with a decreasing tendency, although a true mass evaluation of its epidemiology had not been undertaken [9,10]. Thus the number of infected humans is only based on estimations, which range between 9 and 30 million humans including thousands of Chinese immigrants in the United States. However, many more millions live under the risk of an infection, since people in the endemic regions love to eat it in many variations of raw or undercooked fish, which harbor metacercariae. When the potentially infected fish are transported to the large cities (e.g., from South China to Hong Kong), they pose health risk for a much larger population base.

Adult *C. sinensis* worms show some characteristics that allow a clear and convincing species diagnosis:

1. *Size*. They are 10–28 mm in length and reach 3–7 mm in width with a thickness of about 1 mm. These variations are due to the fact that living worms may stretch and contract considerably by the activity of their strong layers of circular and longitudinal muscles.
2. *Shape*. In the state of rest the adult worms appear slightly lanceolate or leaf-like (Figures 28.1 and 28.2) with a pointed/tapered anterior and a rounded posterior end.
3. *Appearance*. Living worms appear whitish translucent, so that the internal organs can be seen from outside without means of coloration.
4. *Characteristics*. Two suckers are present at the ventral side of the worms and are used to attach at the walls of the bile ducts. The anterior sucker surrounds the mouth and, thus, is called the oral sucker, while the second one has no connection to any internal organ. This ventral sucker is situated just at a short distance behind the two genital pores at the end of the first third of the worm's body.
5. *Surface*. The tegumental surface of the adults is not provided with hooks, which are found in other trematodes such as *Fasciola hepatica*, some *Schistosoma* species in *Metagonimus yokogawai* or in its own juvenile stages.
6. *Sex*. The adult worms are hermaphrodites with two lobulated testes at the posterior end, situated behind each other. The one in front has four main lobes and the one behind has five, which are further subdivided (Figure 28.1). From the testes the sperms are transported via one rather long vas efferens to a short common vas deferens which opens into the male sexual porus. At the end of the male porus, a short penis-like cirrus can become protruded and introduced into the own or—in the case of a copulation—into a foreign female pore. The female system consists of several compartments: (i) a single ovarium (ovary, germarium); (ii) a single ootype (egg formation chamber) with an opening to the worm's surface via the

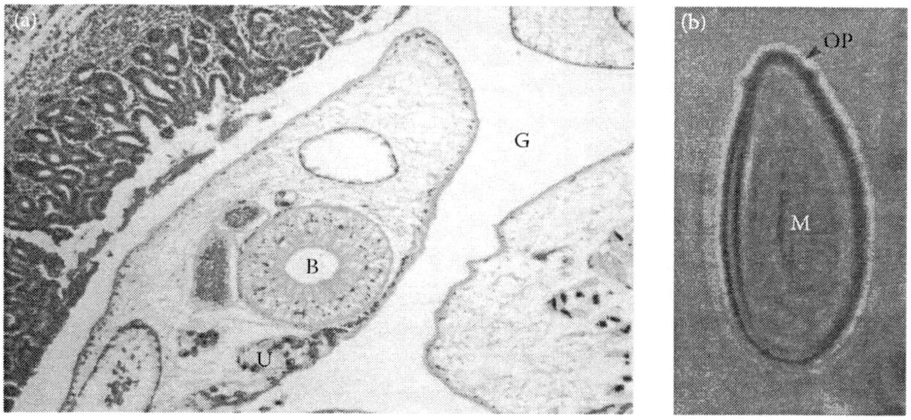

FIGURE 28.2 Light micrographs of *Clonorchis sinensis*. (a) Light micrograph of a section through *Clonorchis* worms in a human bile duct (B = ventral sucker, G = bile duct, U = uterus). (b) Light micrograph of a typical *C. sinensis* egg showing the typical apical operculum (OP) and the developing miracidium (M).

so-called Laurer's channel; (iii) two vitellaria—which are glands situated at each of the two lateral sides of the worm. Both vitellaria produce supplementary cells, which are transported via a common channel into the ootype; (iv) a single receptaculum seminis, which is used as storage sack for sperms that arrive via the convoluted uterus, which has its starting point in the ootype and which transports the fertilized eggs by means of rhythmic constrictions to the female opening at the sexual porus; (v) several single-cell glands (Mehlis' glands) which surround the ootype and help with their products to fortify the egg shell which derives from excretions of the vitellarian cells.

7. *Eggs*. The oocyte—produced in the ovarium—is fertilized inside the ootype, the lumen of which has the shape of the later egg. The zygote becomes surrounded still inside the ootype by up to 20 vitellarian cells which excrete the confluencing egg shell. This combination of zygote and helping vitellarian cells is mostly called "egg," although its organization represents a cocoon. During the transport of the eggs/cocoons to the female sexual opening inside the convoluted uterus, its wall becomes considerably hardened and finally appears yellowish brown with a length of 28–35 μm and a width of 12–19 μm (Figure 28.2). They are characterized by an operculum (lid), at its anterior pole, giving rise to a collar-like swelling. At the posterior end there is a knob-like swelling (dot). Inside the egg/cocoon the larval development starts already while it is still inside the uterus. The development of the larva—a motile miracidium—is finished when the egg has been excreted by the worm. The eggs leave the final host via the bile fluid and feces and must reach a convenient site (i.e., water). In contrast to other trematodes the miracidium of *C. sinensis* does not leave the egg shell in water, but waits until it has been engorged by a water snail. In the intestine of the latter, the operculum is blown up, the miracidium creeps out and enters the midgut gland of the snail (hepatopancreas) to continue there its further development.
8. *Intestine*. The two trunks of the digestive system unite in a short esophagus just before the mouth and end blindly with no rectum or anus at their posterior ends. Thus all remnants of digested food have to become discharged via the mouth.
9. *Excretion system*. Watery excretion is done via the so-called solenocytes = cyrtocytes = terminal cells of the protonephridial system. Their excretory channels unite in an enlarged channel, which ends in a large bladder at the posterior end of the worm. From there the excretory fluids are released to the outside.
10. *Nervous system*. The nervous system is composed of an esophageal commissure with ladder-like pairs of nerve trunks running in each plane of the body until the posterior end of the worm. The nerves are provided with terminal sensory processes which are distributed along the surface of the integument (= neodermis).
11. *Body cavity*. There is no true body cavity in the interior of the worm below the layers of the syncytial integument and the annular and longitudinal muscles. This space is filled by interconnected parenchymal cells which fulfill transportation functions.

28.1.2 Biology

The adult flukes—up to 20,000 specimens have been found within one patient—live within the bile ducts and feed on epithelia (Figure 28.2). Fully embryonated eggs are laid by the worms, pass from the common bile ducts into the small intestine and become excreted with the feces. From early animal experiments [11], it is concluded that each *Clonorchis* worm produces about 1000–4000 eggs/day. This rather limited production of eggs is probably the reason why infections with small numbers of adult *Clonorchis* will hardly be found by fecal examination techniques. If the eggs/cocoons of *Clonorchis* had been passed into water, water snails of various genera (*Bithynia*, *Bulimus*, *Semisulcospira*,

Parafossarulus, etc.) ingest such eggs. Inside their intestine the included larva hatches from the operculated egg and enters the midgut—gland of the snail, where further development is initiated. The ciliated miracidium grows into a sporocyst, within which rediae are produced. The latter produce a second generation of rediae, which finally give rise to motile cercariae. These stages consist of an anterior young worm with the typical anlagen of the later organs and a long tail, which is up to four times longer than the young worm. This tail may propel the cercaria in water after it emerges from the snail. When getting close to a second intermediate host—one of more than 100 species of cyprinid or salmonid fish—the cercaria is attracted by skin odor [12] and bores through the epidermis after discharging the tail. Having reached the dermis, the muscles or the subcutaneous tissues, the small juvenile worm stages transform into metacercariae. Each of them excretes a smooth cover which protects the included worm from host cell attacks. Such metacercariae are spherical in shape with diameters of about 0.3 mm, appear whitish and thus are hardly seen with the naked eye in the whitish muscle tissue or when situated below the fish's scales. As soon as these second intermediate hosts are eaten by humans, dogs, or cats (and other possible predator such as swine when fed with food wastes), the young immature worms hatch from the metacercarian cover inside the small intestine and crawl up the bile duct and establish themselves there, growing to adulthood within two weeks. Generally, *C. sinensis* locates within the small intrahepatic bile ducts. In heavy infections, adult worms may also occur in the extrahepatic bile ducts and in the gall bladder [13]. A tissue migration (as in young *F. hepatica* worm) was never seen. The adults of *C. sinensis* may copulate or produce their eggs after self-fertilization, thus completing the life cycle.

The life span of the adult worms may reach up to 25 years as it was claimed for schistosomes [14,15]. However, in all cases with long-lasting chlonorchiasis—Asians living far away from their native country—it could not be definitively excluded that they had become reinfected, for example, by eating salted raw fish. Thus, summarizing the life cycle of *C. sinensis* inside man, the following data have been recorded in the literature: (i) an incubation period of about 14 days in heavy infections, otherwise without symptoms; (ii) a prepatent period of at least 14 days; in low-grade infections, however, eggs are often found much later; (iii) patency of up to 25 years.

28.1.3 Pathogenesis and Control

28.1.3.1 Pathogenesis

The intensity of pathogenic effects of clonorchiasis depends on the number of adults and preadults at their final sites of location: the bile ducts, which, however, influence the function of the whole gastrointestinal system. The basis of the damages seen in the bile ducts and gall bladder are inflammatory changes induced by the attachment of the two suckers at the wall of the parasitized ducts and their feeding behavior of engorging epithelial cells.

The changes increase in number and in severity with the number of feeding adult worms and include desquamation of the biliary epithelium and formation of crypts. These effects result in a marked proliferation of the mucosa, a thickening of the walls (metaplasia) of the bile ducts and in a considerable dilation of the latter. Consequently, there may be severe damage to the liver, which becomes enlarged in cases of acute disease. Such damages may lead to cirrhosis of the liver, oedema, and formation of ascites. Cancer of the liver and of the ducts were also seen, including cholangiocarcinoma and hepatocarcinoma [2,10,14–22]. In some cases, even pancreatitis has been claimed to occur as a consequence of an etiological relation to severe infections with *C. sinensis* [23].

The above-described pathogenic effects of an infection with *C. sinensis* worms must not introduce well-defined clinical symptoms. If the number of worms is rather low, for example, in old chronic infections, only mild, nonspecific gastrointestinal symptoms occur, if at all. On the other hand, severe infections are mostly associated with intensive right upper quadrant pain and tenderness. In any case an infection with *C. sinensis* must never be neglected—especially with respect to the long-time effects of a duct wall cancer, which has a very poor prognosis, since most of the patients died within 2 years after the diagnosis. The clinical symptoms of clonorchiasis, thus, should be grouped as follows:

1. First acute phase: this phase starts just 1 week after infection and may last for long, showing the following symptoms (singly or in combination): anorexia, feeling of stoutness, diarrhea, fevers, rigors and chills, colic-like abdominal pain, slight icterus, enlargement of liver, weight loss, leucocytosis combined with eosinophilia.
2. Chronic phase: with increasing persistence of the disease, a subsequent chronic phase may show the following symptoms: gall bladder enlargement, edema, meteorism, cachexia, cholangitis, chronic diarrhea, increasing hepatomegalia, retardation of growth and sexual maturity, and neuropsychiatric symptoms.
3. Complications: as complications the following events may occur: ascending cholangitis, intermittent severe cholestasis, pancreatitis, liver cirrhosis, formation of gall bladder stones (hepatolithiasis), obstructive jaundice, ascites, portal hypertension, secondary bacterial infections, night-blindness due to lack of vitamin A, cholangionergic liver carcinoma gastrointestinal bleedings, death due to multiorgan failure.

28.1.3.2 Prophylaxis

With respect to the possibility of severe symptoms of disease as follow-up of an infection with *C. sinensis*, measurements of prophylaxis must be undertaken. At first it is needed to create awareness in the population on the endangered custom to eat raw or undercooked fish. The second need is to cook or boil any fish—at least those from fresh or brackish water, since the latter might have been contaminated by worm-egg-containing human or animal feces and thus are a source of infection of fish with these trematodes. Another possibility of prophylaxis is shock frosting of freshly caught fish. This procedure can be used in industrial plants and would avoid food-borne infections, even if later the fish are eaten raw. Finally, the implementation of a feces collecting system in villages would exclude the import of possibly worm egg-contaminated feces into fish ponds.

28.1.3.3 Therapy

Fortunately, the chemotherapy of clonorchiasis and related worm diseases is rather uncomplicated due to the discovery of the compound praziquantel, which is the drug of choice against trematodes and cestodes since the year 1979. This drug is generally well tolerated and has a very-high efficacy. Thus there are up to now no proven cases of resistance. Praziquantel enters the blood soon after its oral uptake and thus it reaches the tissues within a few minutes [24]. The reported side effects (malaise, headache, dizziness) remain rather mild. Even the occasionally occurring symptoms such as sedation, abdominal discomfort, slight fever, sweating, nausea, eosinophilia, or fatigue are not very important as are not the rarely found pruritus or rash. It was shown that 3 × 25 mg praziquantel per kg bodyweight on each of two successive days or 20 mg praziquantel per kg body weight for three days are sufficient for a complete cure in cases of clonorchiasis and opisthorchiasis [24–27].

28.1.4 Diagnosis of Food-Borne Clonorchiasis

Diagnosis of clonorchiasis has relied for long exclusively on the demonstration of the very small and tiny eggs of this worm by methods of stool examination such as merthiolate-iodine-formaldehyde concentration or sodiumacetate-acetic acid formaline. These ubiquitous methods are not very specific; however, they give a rather good general survey on many parasites that may occur in human or animal feces. In cases of mild infections with a low number of *Clonorchis* worms, the number of the daily excreted eggs is also low so that they may be easily overlooked. Furthermore, there is obviously no possibility of definite discrimination between *Clonorchis* and the eggs of the species *O. felineus* and *O. viverrini*. The same is true when the floatation method (zinc chloride/sodium chloride) is used [2]. Adult worms can also be detected in the bile ducts by ultrasonification or by radiographical techniques or by endoscopic retrograde cholangiopancreatography.

The different serological methods of ELISA, when first established in 1992 by Feldheim and Knobloch [28], turned out to be very helpful, but soon cross-reactions with the antigens of other genera of trematodes (e.g., *Fasciola*, *Paragonimus*, and *Schistosoma*) were seen [26]. Thus without the use of molecular biology methods, which, however, are costly and need more time, a clear diagnosis of *C. sinensis* at the species level remains always somewhat doubtful.

A similar situation is given with respect to the metacercariae inside the intermediate host fish. When practising the regular meat inspection of fish fillets, the metacercariae are very easily overlooked due to their small size of often <1 mm, by their whitish color in whitish fillets, by their small number or by their often hidden position below the scales. Furthermore, genus and species determination remains rather doubtful using this time-wasting method, even if the digestion technique (used in trichellinosis) is added. Thus, the use of molecular biological methods is required in order to detect an infection with *C. sinensis*.

28.2 METHODS

In this chapter, we present procedures for molecular detection of *C. sinensis* rediae in snails, metacercariae in fish, and eggs in human, or animal fecal samples, along with relevant sample collection and preparation techniques.

28.2.1 Detection of *C. sinensis* Rediae in Snails

The following detection method for *C. sinensis* sporocysts and rediae in snails has the advantage that only a very few steps have to be carried out and that only inexpensive chemicals are required.

28.2.1.1 Sample Collection and Preparation

Each snail is collected in a 2 mL Eppendorf tube, and the shell is crushed by using clean sticks. An amount of 1.5 mL of 25 mM potassium hydroxide is added to the tube and the snail's soft body is lysed by incubation at about 95°C for 15 min. After a brief centrifugation to sediment the shell particles, about 10 µL of a saturated aqueous solution, cresol red, is added, and the supernatant is neutralized by the addition of 1 N HCl. The indicator dye appears violet at alkaline pH and becomes yellow when an appropriate pH ≤7.5 is reached. Of this sample, 1 µL is added to a polymerase chain reaction (PCR) assay volume of 50 µL. The muscular food of the snail may not be entirely lysed during the short incubation. However, the rediae reside in the hepatopancreas and are easily freed from this soft tissue. The rediae are completely lysed by alkaline treatment, and the parasite DNA is efficiently liberated. It has been amply evaluated [29] that it is not necessary to apply methods for further purification of the DNA, for example, by using commercial kits.

28.2.1.2 Polymerase Chain Reaction

A suitable primer pair for the specific detection of *C. sinensis* has been designed from the internal transcribed spacer 2 (ITS2) of the ribosomal DNA after examination of the sequences of several related opisthorchiid species [29]. The primer pair consists of CS1 (5′-CGAGGGTCGGCTTATAAAC-3′) as a forward primer and CS2 (5′-GGAAAGTTAAGCACCGACC-3′) as a reverse primer. These primers yield a single DNA fragment of a typical product size of 315 bp. The utility of this PCR for the detection of rediae in snails or metacercariae in fishes was previously assessed with adult *C. sinensis* (stored in 70% ethanol, originating from Korea: Professor Dr Rim) as source material, with DNA extracted from the liver flukes with the commercially available QIAamp DNA mini kit (Qiagen, Hilden, Germany). The specificity of the PCR primers was examined with DNA from the closely related liver flukes *O. viverrini*, *O. felineus*, *Pseudoamphistomum truncatum*, and *Metorchis xanthosomus*. In addition, DNA from the Digenea *Echinostoma caproni*, *F. hepatica*, *Schistosoma mansoni*, *Holostephanus dubenini*, and *Paracoenogonimus ovatus* were tested as the DNA from intermediate host snails *Bithynia fuchsiana* and *Melanoides tuberculata* and the cyprinid fishes *Pseudorasbora parva* and *Carassius auratus auratus*, respectively [29].

The PCR assay is carried out with a final volume of 50 μL with 1 × PCR buffer containing Tris-HCl, KCl, $(NH_4)_2SO_4$, and 1.5 mM $MgCl_2$, pH 8.7 (Qiagen), 1.25 U Taq DNA polymerase (Qiagen), 200 μM of each dNTP, 0.4 μM of each CS primer, and 1 μL of snail lysate (template DNA). PCR amplification is performed in a PTC-0150 MiniCycler (MJ Research, Watertown, MA, USA) with the following cycling conditions: 1 cycle of 94°C for 3 min; 40 cycles of 94°C for 1 min, 62°C for 1 min, and 72°C for 1 min; and 1 cycle of 72°C for 10 min. After the completion of the cycles, the amplified PCR products of 315 bp in size are separated in agarose gel and visualized with ethidium bromide stain.

Note: PCR has been successfully performed with snail lysate from *B. fuchsiana* infected with *C. sinensis* rediae, as the substances inhibiting the enzymatic reaction are eliminated in the lysate preparation. This was confirmed by adding 1 μL of lysate from uninfected snails to the PCR assay, together with serially diluted genomic DNA from *C. sinensis*. The PCR assay based on the primer pair CS1/CS2 has a detection limit of 10^{-13} g DNA. Clearly, the sample processing technique has successfully removed interfering components from the snail tissue as an identical level of sensitivity is achieved with parasite DNA tested alone or in the presence of snail lysate (Figure 28.3). Using a preparation volume of 1.5 mL, the PCR is capable of detecting 1.5×10^{-10} g DNA of rediae from an infected snail. With this level of sensitivity, practically any infected snail will be recognized as positive by the PCR assay. Thus, the PCR procedure provides a reliable means of identifying *C. sinensis*-infected snails.

28.2.2 Detection of Metacercariae in Fish

Traditionally, metacercariae isolated from fish are detected by light microscopy. Given that all mentioned digenean metacercariae form ovoid cysts and have thick-walled, ovoid excretory vesicles filled with dark corpuscles, it is difficult to differentiate unambiguously metacercariae of *C. sinensis* from those of related human pathogenic species of the genera *Opisthorchis* and *Heterophyes* and also from other less known and less pathogenic digenean species. To determine the specimens reliably, it may be necessary to infect susceptible mammals with the metacercariae to obtain adult worms. This is unduly cumbersome and time consuming.

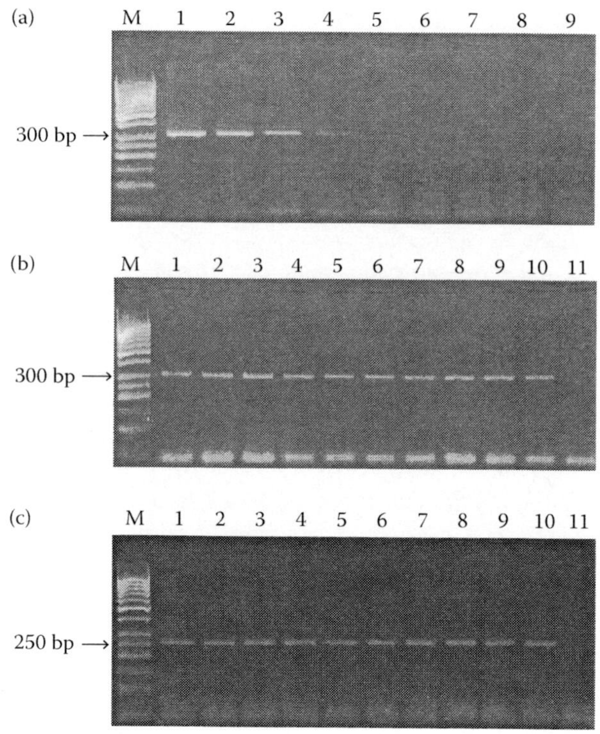

FIGURE 28.3 (a) Sensitivity of the CS1/CS2 primers with purified *C. sinensis* DNA. Lane M, DNA size marker; lane 1, 10^{-10} g; lane 2, 10^{-11} g; lane 3, 10^{-12} g; lane 4, 10^{-13} g; lane 5, 10^{-14} g; lane 6, 10^{-15} g; lane 7, 10^{-16} g; lane 8, 10^{-17} g of *C. sinensis* DNA; lane 9, negative control. (b) Detection of *C. sinensis* metacercariae from fish with the CS1/CS2 primers. One metacercaria was artificially inoculated into each 3 g fish meat prior to processing the samples. Lane M, DNA size marker; lanes 1–10, samples from infected fish; lane 11, negative control containing no metacercaria. (c) Detection of *C. sinensis* eggs in human stool with OP1/OP2 primers. One egg was inoculated into each 0.1 g stool sample before processing. Lane M, 50-bp ladder; lanes 1–10, stool samples; lane 11 negative control containing no egg.

The following method allows detection of *C. sinensis* metacercariae in fish without microscopical inspection [29].

28.2.2.1 Sample Preparation

Fish filets (2–3 g) are taken from *P. parva* fish, finally minced and then transferred to a 50 mL Falcon tube containing artificial gastric juice consisting of 0.9% pepsin, 0.7% HCL, and 1% Triton X-100. The use of detergent Triton X-100 here improves the dissolution of the fish muscles and does not destroy the metacercarial cysts. This mixture is incubated at 38–40°C and occasionally shaken by hand. Digestion is complete within 45–60 min. Then, the samples are briefly centrifuged to sediment the metacercariae. The sediment is washed twice by removing the supernatant and filling it up with water. Subsequently, the supernatant is removed but left a residual volume of 1–1.5 mL. Disruption of the metacercariae and liberation of the DNA are achieved by homogenization with a tissue homogenizer (Omni TH220, Omni International, Marietta, GA, USA) at 10,000 rpm for 1 min. After a brief centrifugation, the supernatant is boiled in a water bath for 5 min to inactivate the pepsin and, subsequently, is neutralized by the addition of 1 N potassium hydroxide. Neutralization is carried out with the aid of cresol red as the indicator dye. Cresol red appears yellow at acid pH and becomes violet when an appropriate pH of around ≥7.5 is reached. Two microliter of this sample is added to 50 μL PCR assay volume. This preparation procedure for metacercariae from fishes is rapid and simple as it does not involve further extraction of DNA by phenol–chloroform or even more complex purification methods. Inhibitors contained in fish samples are effectively removed by simple washing steps, whereby they are diluted to a value that enables the PCR to reliably work.

28.2.2.2 Polymerase Chain Reaction

This is conducted with the CS primers as outlined in Section 28.2.1. Detection of *C. sinensis* rediae in snails, with the positive specimens showing an expected single DNA fragment of 315 bp.

Note: Reports on molecular methods to detect opisthorchiid liver flukes are scarce. To our knowledge for *C. sinensis*, only one study with PCR has been published [30], in which a mitochondrial-based multiplex PCR for the identification of *C. sinensis* and *O. viverrini* was described, with a sensitivity of 7.8×10^{-10} g of genomic parasite DNA. The CS primer pair developed for *C. sinensis* by Müller et al. [31], however, is much more sensitive with a detection limit of 1×10^{-13} g DNA. By evaluating with multiple samples, the reliability of the DNA preparation procedure for metacercariae from fish is validated. To do this, fresh fish fillet from 10 uninfested *P. parva*, bred parasite-free in the laboratory [29], were inoculated with one metacercariae of *C. sinensis* (Figure 28.3). Examination of DNA from closely related opisthorchiid liver fluke species, other Digenea and various intermediate hosts mentioned above showed no cross-reactions with the primers specific for *C. sinensis*.

These results indicate that the PCR-based method is a suitable tool for the detection of *C. sinensis* in infected fish, with sensitivity approaching only one metacercariae per fish sample. In contrast to conventional techniques, PCR-based methods have a higher potential for the differentiation of digenean species, which is essential for epidemiological studies. Thus, application of the PCR enables unambiguous identification of *C. sinensis*, which can be difficult to achieve by light microscopy. This technique offers a better way of tracking the relevant pathways of transmission by identifying host species and infection reservoirs. With respect to the expanding worldwide trade of fresh water fish from aquacultures, it is highly desirable to have a means for control inspections for fish in order to avoid importation of fish being infected with *C. sinensis* metacercariae [32].

28.2.3 Detection of *C. sinensis* Eggs in Human or Animal Fecal Samples

The opisthorchiids live in the bile ducts and the gall bladder (Figure 28.2). All species cause similar symptoms, such as adenomatous hyperplasia of the biliary epithelium or obstruction of the biliary system. Chronic infections may also introduce liver cirrhosis, cholangiocarcinoma, or liver carcinoma [9,33]. It appears that species of the genera *Metorchis* and *Pseudamphistomum* occasionally infect humans and cause the same problems as the better-known liver flukes [34]. Since all species of the opisthorchiids cause the same health problems, it makes sense to use a single test system for their diagnosis. However, for appropriate treatment of patients, for example, with praziquantel [24,26], there is no need to determine the exact species. The development of PCR techniques permits identification of flukes of the family Opisthorchiidae from human and animal fecal specimens, and discrimination of all opisthorchiids from other species of Digenea, especially from the members of the related family Heterophyidae or other so-called minute flukes that colonize in the intestine of humans.

28.2.3.1 Cleanup Procedure

It is well known that feces contain various substances that strongly inhibit amplification by PCR [35]. The conventional methods for extraction of DNA of parasite eggs from stool samples are complex and require the use of expensive commercial kits to remove inhibitory substances. Many of these DNA preparation procedures are based on those for the diagnosis of bacteria and viruses, and do not take into account the fact that DNA of the parasite is enclosed by the egg shell. Not surprisingly, the DNA so obtained does not always give reproducible PCR amplification as the DNA polymerase inhibitors are not completely eliminated from fecal material. The present technique can be carried out rapidly with inexpensive chemicals, and even small amounts of worm eggs in feces can be detected reliably by PCR. As reported by Müller et al. [31] 100 mg of fecal samples are collected in a 2 mL Eppendorf tube and incubated with an aqueous solution consisting of 2% Triton X-100 and 25 mM potassium hydroxide for 10 min at room temperature on a shaker. After centrifugation at 10,000 rpm in an Eppendorf centrifuge for 2 min, the supernatant is removed, and the sediment is washed two or three times with 2% Triton X-100 and 25 mM KOH solution until the supernatant appears clear. Then, the supernatant is discarded and 0.8 mL of sucrose/potassium chloride solution is added to the sediment containing the eggs. The sucrose/KCl solution consists of 82.5 g sucrose, 20 g KCl, and distilled water added up to 100 mL. This solution has a density of 1.39 g/mL. During centrifugation at 14,000 rpm for 5 min, the eggs are distributed within the sucrose/KCl solution and do not float up to the surface of the solution. Thus, they are separated from the sedimenting fecal particles.

The entire supernatant containing the eggs is transferred into a new 2 mL tube, which is then filled up with water and centrifuged at 14,000 rpm for 5 min. By adding water, the density of the solution is lowered so that the eggs became sedimented. The supernatant is removed and the sediment is washed twice with water. After the second washing, the supernatant is removed, leaving a remnant of 0.3 mL. The disruption of the egg shell and release of the parasitic DNA are achieved by homogenization of the sample with a tissue homogenizer (Omni TH220, Omni International, USA) at 10,000 rpm for 2 min. Alternatively, the eggs are opened by autoclaving [36].

However, the autoclaving approach is found to be less efficient since a considerable percentage (often 50%) of the eggs remain closed. After a brief centrifugation, the supernatant is boiled in a water bath for 5 min to inactivate hydrolytic enzymes of the feces. Five microliters of this sample, free of inhibitory substances, is added to the PCR tube in a final volume of 50 μL.

28.2.3.2 Polymerase Chain Reaction

The species of the family Opisthorchiidae share extreme homologies of the ITS2 sequence (Figure 28.4), which is clearly different from that of other Digenea. Thus, this sequence is exploited for the development of specific primers OP1/OP2 for the opisthorchiid species *C. sinensis*, *O. viverrini*, *O. felineus*, *P. truncatum*, and *M. xanthosomus*. Indeed, *C. sinensis* sequences derived from specimens from Korea (AF217094) and from China (AF217099) demonstrate identical compositions. Sequences from *O. viverrini* (AF408147) are available in GenBank. Data for *O. viverrrini* that had been reported for isolates from different areas in northeast Thailand [37] are also identical with one another and with those shown in GenBank. Nucleotide sequences of the ITS2 region of *O. felineus*, *P. truncatum*, and *M. xanthosomus* have not been reported before and are determined (Sequence Laboratories Göttingen GmbH, Germany) (Figure 28.4). Although DNA from *F. hepatica* generates an amplification product, this product is not in the range of the specific products of the opisthorchiid flukes. The sensitivity of the primers is determined in assays with diluted genomic DNA of *C. sinensis* and a minimal amount of 10^{-15} g DNA is achievable.

The primer pair consists of the forward primer OP1 (5′-CGAGGGTCGGCTTATAAAC-3′) and the reverse primer OP2 (5′-AGCCTCAACCAAAGACAAAG-3′), yielding a specific 250 bp product from the family Opisthorchiidae. The PCR is carried out in a final volume of 50 μL, consisting of 1 × PCR buffer (Tris-HCl, KCl, $(NH_4)_2SO_4$, 1.5 mM $MgCl_2$; pH 8.7), 1.25 U Taq DNA polymerase (Qiagen), 200 μM of each dNTP, 0.4 μM of each primer (OP1 and OP2), and 5 μL DNA. PCR amplification is conducted in a PTC-0150 Minicycler (MJ Research, USA) with the following cycling conditions: 1 cycle of 94°C for 3 min; 40 cycles of 94°C for 1 min, 62°C for 1 min, and 72°C for 1 min, and 1 cycle of 72°C for 10 min. The amplified products of 250 bp in size are separated by agarose gels and stained with ethidium bromide.

Note: The cleaning up procedure and efficacy of the PCR have been evaluated with an array of stool samples to confirm their reliability and extremely high sensitivity. DNA from adult *C. sinensis* originating from Korea was used for the establishment of the PCR assay. The DNA extraction of the flukes is also carried out with the commercially available QIAamp DNA Mini Kit (Qiagen, Hilden, Germany). The PCR yielded amplification products of ~250 bp for all tested opisthorchiid flukes, that is, *C. sinensis*, *O. viverrini*, *O. felineus*, *P. truncatum*, and *M. xanthosomus*. To evaluate possible cross-reactions with DNA from other Digeneans, *E. caproni*, *F. hepatica*, *S. mansoni*, *H. dubenini*, and *P. ovatus* are analyzed. DNA from several intermediate hosts, the snails *Bithynia tentaculata*, *Bithynia leachi*, *B. fuchsiana*, *Bithynia funiculata*, and *M. tuberculata* and the cyprinid fishes *P. parva*, *Rutilus rutilus*, and *C. auratus*, are also tested. DNA from other Digenea and from intermediate hosts does not cross-react with the OP1 and OP2 primer pair.

Additionally, the PCR using OP1/OP2 primers is run with each of 10 0.1-g stool samples of uninfested humans experimentally spiked with only one *C. sinensis* egg (Figure 28.3). All samples give a strong signal in PCR. For convenience, the tests can be carried out using Eppendorf tubes containing 0.1 g feces dotted with the lowest number, that is, one egg. Hence, the proven sensitivity of the method can be calculated as 10 eggs per g (EPG) feces. However, if diagnosis is carried out with 1 g stool samples in larger test tubes, the sensitivity may be estimated to be under 10 EPG.

The sample preparation method has originally been developed using feces of cats infected with *C. sinensis*. It is later found to work comparably well using either human or cat feces. The purification method, however, is less suitable for feces of rodents containing large amounts of undigested plant materials.

The technique can also be used for species- or strain-specific diagnosis of other parasites, such as eggs of nematodes or cestodes, oocysts of coccidian species (Apicomplexa), cysts of amoebae and of *Giardia* species, or of other protozoan species occurring in feces.

Diagnosis on conventional light microscopy is difficult in patients with low-grade infections, who shed less than about 200 EPG feces. Methods like Stoll's egg count and formalin–ether techniques easily fail in the case of low infection rates [13,36]. The sensitivity of light microscopy diagnosis for *O. viverrini* eggs in stool samples was examined in comparison with the numbers of worms recovered from the liver of autopsied humans [13]. According to this study, only in 30% of patients harboring one to nine worms (22% of all infected persons) did the stool diagnosis give a positive result. In the case of burdens with 10–19 worms (11% of infected humans), the result was positive only in 77% with the Stoll's and in 61.5% with the formalin-ether technique.

The eggs of opisthorchiids are peculiar because in flotation media, low numbers of eggs are hardly detectable. Harnnoi et al., compared the flotation of *O. viverrini* eggs in saturated sodium nitrate solution (specific gravity 1.4) to the results when using the formalin-ether sedimentation. Attempts to float eggs quantitatively failed; thus, the authors revealed that the flotation technique was not as efficient as the sedimentation. The specific gravity of *O. viverrini* eggs was found to range from 1.27 to 1.30 g/mL [38]. Although the specific gravity of *O. viverrini* eggs is lower than that of the salt solution (specific gravity 1.4), the eggs do not float. An additional flotation technique is the zinc chloride/sodium chloride method [2]. The specific gravity of this solution is 1.3 and failed, like the attempt with saturated sodium nitrate solution (specific gravity 1.4), to float the eggs quantitatively. Thus, it appears that flotation techniques are not very suitable for the detection of opisthorchiid eggs in feces in the case of low infection rates.

The egg production of a single *C. sinensi* worm has been determined to be 100 EPG of feces [11]. Song et al. [26] reported in 1986 in their study on the mean EPG of *C. sinensis* patients

```
                                                              →ITS2
C.sinensis        TTGCGGCCATGGGTTTGCCTGTGGCCACGCCTGTCCGAGGGTCGGCTTATAAACTATCAC
O.viverrini       TTGCGGCCATGGGTTTGCCTGTGGCCACGCCTGTCCGAGGGTCGGCTTATAAACTATCAC
O.felineus        TTGCGGCCATGGGTTTGCCTGTGGCCACGCCTGTCCGAGGGTCGGCTTATAAACTATCAC
M.xanthosomus     TTGCGGCCATGGGTTTGCCTGTGGCCACGCCTGTCCGAGGGTCGGCTTATAAACTATCAC
P.truncatum       TTGCGGCCATGGGTTTGCCTGTGGCCACGCCTGTCCGAGGGTCGGCTTATAAACTATCAC
                  ************************************************************

C.sinensis        GACGCCCAAAAAGTCGTGGCTTGGGTCTTGCCAGCTGGCATGATTTCCCCACACAATTGT
O.viverrini       GACGCCCAAAAAGTCGTGGCTTGGGTCTTGCCAGCTGGCATGATTTCCCCGCGCAATTG-
O.felineus        GACGCCCAAAAAGTCGTGGCTTGGGTCTTGCCAGCTGGCATGATTTCCCCACGCATTTG-
M.xanthosomus     GACGCCCAAAAAGTCGTGGCTTGGGTCTTGCCAGCTGGCATGATTTCCCCGCACGTTTGT
P.truncatum       GACGCCCAAAAAGTCGTGGCTTGGGTCTTGCCAGCCGGCATGATTTCCCCGCACGTTTG-
                  *********************************** ************** * * ****

C.sinensis        GTGTATGTGTGTGGGGTGCCGGATCTATGGCTTTTCCCCAATGTGCCGGACGCAACCATG
O.viverrini       ---------TGTGGGGTGCCGGATCTATGGCTTTTCCCCAATGTGCCGGACGCAACCATG
O.felineus        ---------TGTGGGGTGCCGGATCTATGGCTTTTCCCCAATGTGCCGGACGCAACCATG
M.xanthosomus     ----------GTGGGGTGCCGGATCTGTGGCTTTTCCCCAATGTGCCGGACGCAACCATG
P.truncatum       ---------TGTGGGGTGCCGGATCTGTGGCTTTTCCCCAATGTGCCGGACGCAACCATG
                            **************** *********************************

C.sinensis        TCTGGGCTGACTGCCTAGATGAGGGGGTGGCGGCGGAGTCGTGGCTCAATTGTTGTTATT
O.viverrini       TCTGGGCTGACTGCCTAGATGAGGGGGTGGCGGCGGAGTCGTGGCTCAATTGTTGTTATT
O.felineus        TCTGGGCTGACTGCCTGGATGAGGGGGTGGCGGCGGAGTCGTGGCTCAATTGTTGTTGTT
M.xanthosomus     TCTGGGCTGACTGCCTAGATGAGGGGGTGGCGGCGGAGTCGTGGCTCAATTGTTCTTATT
P.truncatum       TCTGGGCTGACTGCCTAGATGAGGGGGTGGCGGCGGAGTCGTGGCTCAATTGTTATTGTT
                  **************** ************************************* ** **

C.sinensis        ---GTTGTGAATGTGCGCGCTCCGTTGTTGGTCCTTTGTCTTTGGTTGAGGCTTCAGTAT
O.viverrini       GTTGTTGTGAATGCGCGCGCTCCGTTGTTGTTCCTTTGTCTTTGGTTGAGGCTCCAGTAG
O.felineus        ATTGTTGTGAATGTGCGCGCTCCGTTGTTGGTCCTTTGTCTTTGGTTGAGGCTCCAGTGG
M.xanthosomus     GTTGT-GT--ATGTGCGCGCTCCGTTGTTGTTCCTTTGTCTTTGGTTGAAGCTCCAGTAG
P.truncatum       ------GTGGATGTGCGCGCTCCGTTGTTGCTCCTTTGTCTTTGGTTGAGGCTCCAGTAG
                        **  *** **************** ****************** ********

                                                                     ITS2←
C.sinensis        TGGCAATGCATTCGATGCAAATCTGTTTTGCACCGGTCGGTGCTTAAC---TTTCCTGAC
O.viverrini       TGGCAATGCATTCGATGCAAATCGGTTTTGCACT--TTGGTGCTTAACAACTTTCCTGAC
O.felineus        TGGCAATGCATTCGATGCAAATCTGTTTTGCACT--TCGGTGCTTAAC---TTTCCTGAC
M.xanthosomus     TGGCAATGCATTCGATGCAAATCTGTTTTGCACT--TCCGTGCTTAAC---TTTCCTGAC
P.truncatum       TGGCAATGCATTCGATGCAAATCTGTTTTGCACT--TCCGTGCTTA-C---TTTCCT-AC
                  *********************************   *  ******* *   ****** **

C.sinensis        CTCGGATCAGA
O.viverrini       CTCGGATCAGA
O.felineus        CTCGGATCAGA
M.xanthosomus     CTCGGATCAGA
P.truncatum       CATGGATCAGA
                  *  ********
```

FIGURE 28.4 Nucleotide sequences of the ITS2 region of opisthorchiid flukes. Nucleotide sequences of *O. felineus*, *P. truncatum*, and *M. xanthosomus* were determined. The asterisks show those base pairs that are identical in all five species. The sequences of *C. sinensis* (AF217094/AF217099) and *O. viverrini* (AF408147) are registered in GenBank.

in various areas in Korea that the lowest single value found was 100 EPG. Haswell-Elkins et al. [10] and Sithithaworn et al. [13] found in 1991 that one *O. viverrini* worm releases about 50–200 EPG host feces. With the present sample preparation technique and PCR, one egg per 100 mg sample gave a strong signal. Calculating for a 1 g sample, the detection limit would be 10 EPG, and an infection with only a single worm is reliably detectable. Thus, the PCR-based method, along with the reported cleanup procedure reported herein, is more sensitive and reliable than the normal light microscopy methods.

Eggs of closely related opisthorchiid and heterophyid species morphologically appear very similar. Differentiation of the eggs of *C. sinensis*, *O. viverrini*, or *O. felineus* by light microscopy requires considerable skill. The small eggs of these species are ovoid in form, similar in shape, and possess an operculum with a prominent shoulder (Figure 28.2). The present cleanup procedure for stool samples and the application of the PCR system OP1/OP2 is a robust and highly sensitive method of detecting infections with opisthorchiid liver flukes. Thus, the reported techniques represent a suitable diagnostic tool.

By employing the cleanup procedure reported here for stool samples together with a PCR using species-specific primers, it is possible to rapidly and reliably determine the species identity of opisthorchiids, which is important for epidemiological studies. For example, the use of the primer pair CS1/CS2 is recommended for species-specific detection of *C. sinensis* [29,31].

28.3 CONCLUSIONS AND FUTURE PERSPECTIVES

C. sinensis is parasitic trematode whose eggs readily enter suitable snails (as the primary intermediate host) to develop into rediae, which in turn infect fish (as the secondary intermediate host) and develop into metacercariae. When infected fish harboring metacercariae are consumed raw or undercooked by humans and animals (as final hosts), the metacercariae mature into adult worms whose eggs are excreted with the feces. Given that fish make up a significant source of dietary proteins for many countries, clonorchiasis has been an important food-borne disease in regions such as East Asia where raw fish are often consumed.

Traditionally, laboratory diagnosis of clonorchiasis is dependent on the identification of *C. sinensis* eggs in the fecal samples and other stages (rediae, cercariae, and metacercariae) in the intermediate hosts such as snails and fish. Since *C. sinensis* adult worms produce a limited number of eggs per worm, human or animal hosts harboring relatively few worms are not easily diagnosed with fecal examination techniques. The fact that *C. sinensis* rediae and metacercariae are small and have similar morphological features as those from other pathogenic species of *Opisthorchis* and *Heterophyes* genera makes microscopic differentiation difficult. Given that clonorchiasis can become a severe disease in humans and animals, it is vital that a diagnosis even of a low worm burden is achieved as early as possible. The application of biochemical and serological techniques has enhanced the laboratory identification of *C. sinensis* to a large extent, but it only became possible to rapidly and precisely distinguish *C. sinensis* eggs, rediae, and metacercariae with the recent development of PCR-based nucleic acid detection assays. These molecular techniques are also valuable for detecting small amounts of *C. sinensis* eggs that may be easily overlooked in feces with standard microscopical techniques.

In particular, the use of OP1 and OP2 primers derived from the ITS2 of the family Opisthorchiidae facilitates the separation of eggs of opisthorchiid species from those of heterophyid species. Coupled with an efficient cleanup procedure, the PCR using the OP1/OP2 primers offers a quick and sensitive screening system for infections with opisthorchiid liver flukes. Furthermore, the application of another primer pair (CS1/CS2) from the ITS2 of the ribosomal DNA of *C. sinensis* enables species-specific identification of *C. sinensis* rediae and metacercariae [29,31]. There is no doubt that future optimization of a real-time PCR detection platform will further streamline the laboratory diagnosis of *C. sinensis* and contribute to the improved control and prevention of food-borne clonorchiasis in humans worldwide.

Utilizing the current state of technology, diagnosis in human patients with opisthorchiids should be carried out best with the opisthorchiid-specific primer pair OP1/OP2. Assays with these primers are very robust and sensitive, since only 10^{-15} g DNA can be detected. It is no problem to diagnose even the lightest infections (with one self-fertilizing worm) in humans; if there is an egg in a sample, the assay will detect it. The general use of this primer pair should be promoted. The three opisthorchiid species *O. viverrini*, *O. sinenesis*, and *O. felineus* are astonishingly closely related. Pathology and treatment are identical. Hence, it is NOT necessary to determine the exact species in medical diagnosis since treatment is identical.

For scientific purposes, the (*C.*) *O. sinensis*-specific primer pair CS1/CS2 can be utilized, which enables a highly sensitive detection with fecal material also. The detection limit with CS1/CS2 is 10^{-13} g parasite DNA in a single assay. This level of sensitivity easily surpasses the performance of other primers reported in the literature using a nested PCR format. For practical and economic reasons, the nested PCR should be avoided whenever possible. Similarly, despite its convenience and speed, real-time PCR is much too expensive at the moment and, thus, is unsuitable as a widely applied routine diagnostic tool, especially in developing countries [39].

REFERENCES

1. Despommier, D.D., Gwadz, R., and Hotez, P.J. *Parasitic Diseases*. Springer Verlag, Heidelberg, 1995.
2. Mehlhorn, H. et al. *Diagnostik und Therapie der Parasitosen des Menschen*. 2. Auflage, Gustav Fischer Verlag Stuttgart, Jena, New York, pp. 6–7, 1995.
3. Mehlhorn, H. *Encyclopedia of Parasitology*, Vols. 1 and 2. Springer Heidelberg, New York, 2008.

4. McConnell, J.F.P. Anatomy and pathological relations of a new species of liver fluke. *Lancet* 2, 271, 1875.
5. Loos, A. *Clonorchis* (1907). Cited in: Yamaguti, S. (ed.), *Synopsis of Digenetic Trematodes of Vertebrates*. Keigaku, Tokyo, 1971.
6. Ijiima, I. Notes on *Distoma endemicum* Baelz. *J. Coll. Sci. Imperial Univ. Jpn.* 1, 47, 1887.
7. Kobayashi, H. 1914. Fish as intermediate hosts in clonorchiasis. Cited in: Despommier, D.D., Gwadz, R., and Hotez, P.J., *Parasitic Diseases*. Springer Verlag, Heidelberg, 1995.
8. Muto, S. On the primary intermediate host of *Clonorchis sinensis*. *Chuo Igakakai Zasshi* 25, 49, 1918.
9. Song, I.C., Lee, J.S., and Rim, H.J. Epidemiological studies on the distribution of *Clonorchis sinensis* infection in Korea (in Korean). *Korean Univ. Med. J.* 20, 165, 1983.
10. Haswell-Elkins, M.R. Lung and liver flukes. In: Cox, F.E.G. et al. (eds), *Parasitology, Topley and Wilson's Microbiology and Microbial Infections*, 10th edn., Hoddes Arnold, ASM Press, Washington, 2005.
11. Wykoff, D.E. Studies on *Clonorchis sinensis* IV: Production of eggs in experimentally infected rabbits. *J. Parasitol.* 45, 91, 1959.
12. Haas, W., Granzer, M., and Brockelman, C.R. *Opisthorchis viverrini*: Finding and recognition of the fish host by the cercariae. *Exp. Parasitol.* 71, 422, 1990.
13. Sithithaworn, P. et al. Relationship between faecal egg count and worm burden of *Opisthorchis viverrini* in human autopsy cases. *Parasitology* 102, 277, 1991.
14. Piekarski, G. *Medizinische Parasitologie in Tafeln*. Springer Verlag, Heidelberg, 1987.
15. Chai, J.Y., Murrel, K.D., and Lymbery, A.J. Fish-borne parasitic zoonoses: Status and issues. *Int. J. Parasitol.* 35, 1233, 2005.
16. Kim, Y.H. Carcinoma of the gall bladder associated with clonorchiasis: Clinicopathologic and CT evaluation. *Abdom. Imaging* 28, 83, 2003.
17. Leung, J.W. et al. Hepatic clonorchiasis—A study by endoscopic retrograde cholangiopancreatography. *Gastrointest. Endosc.* 35, 226, 1989.
18. Leung, J.W., and Yu, A.S. Hepatolithiasis and biliary parasites. *Bull. Clin. Gastroenterol.* 11, 681, 1997.
19. Chen, M.G., Lu, Y., and Hua, X.J. Progress in assessment of morbidity due to *Clonorchus sinensis* infection: A review of recent literature. *Trop. Dis. Bull.* 91, 7, 1994.
20. Bunnag, D., and Harinasuta, T. Opisthorchiasis, clonorchiasis and paragonimiasis. In: McGraw, R.P., and Mclvor, D. (eds), *Tropical and Geographic Medicine*. McGraw-Hill, New York, 1984.
21. MacFadzean, A.J.S., and Yeung, R.T.T. Acute pancreatitis due to *Clonorchis sinensis*. *Trans. R. Soc. Trop. Med. Hyg.* 60, 466, 1966.
22. Hou, P.C. The relationship between primary carcinoma of the liver and infestation with *Clonorchis sinensis*. *J. Pathol. Bacteriol.* 72, 239, 1965.
23. Komiya, Y. Clonorchis and clonorchiasis. *Adv. Parasitol.* 4, 53, 1966.
24. Mehlhorn, H. et al. Ultrastructural investigations on the effects of praziquantel on human trematodes from Asia (*Clonorchis sinensis, Metagonimus yokogawai, Opisthorchis viverrini, Paragonimus westermani, Schistosoma japonicum*). *Drug Res.* 33, 91, 1983.
25. Rim, H.J. et al. Clinical evaluation of the therapeutic efficacy of praziquantel (Embay 8440) against *Clonorchis sinensis* infection in man. *Ann. Trop. Med. Parasitol.* 75, 27, 1981.
26. Rim, H.J. The current pathology and chemotherapy of clonorchiasis. *Korean J. Parasitol.* 24 (Suppl. 3), 27, 1986.
27. Hänel, H., and Raether, W. Chemotherapy. In: Mehlhorn, H. (ed), *Encyclopedia of Parasitology*, Vols. 1 and 2. Springer, Heidelberg, New York, 2008.
28. Feldheim, W., and Knobloch, J. Serodiagnosis of *Opisthorchis viverrini* infestation by an enzyme immuno-assay. *Tropenmed. Parasitol.* 33, 8, 1982.
29. Müller, B., Schmidt, J., and Mehlhorn, H. Sensitive and species—Specific detection of *Clonorchis sinensis* by PCR in infected snails and fishes. *Parasitol. Res.* 100, 911, 2007.
30. Le, T.H. et al. *Clonorchis sinensis* and *Opisthorchis viverrini*: Development of a mitochondrial-based, multiplex PCR for their identification and discrimination. *Exp. Parasitol.* 112, 109, 2006.
31. Müller, B., Schmidt, J., and Mehlhorn, H. PCR diagnosis of infections with different species of Opisthorchiidae using a rapid clean-up procedure for stool samples and specific primers. *Parasitol. Res.* 100, 905, 2007.
32. WHO, *Manual of Basic Techniques*. Geneva, Switzerland, 1995.
33. Schwartz, D.A. Review: Helminths in the induction of cancer: *Opisthorchis viverrini, Clonorchis sinensis* and cholangiocarcinoma. *Trop. Geogr. Med.* 32, 95, 1980.
34. Schuster, R., and Leberegelbefall, In: Aspöck H., Amoeben, Bandwürmer etc. und parasitäre Erkrankungen des Menschen in Mitteleuropa. *Denisia* 6, 291, 2002.
35. Wilde, J., Eiden, J., and Yoken, R. Removal of inhibitory substances from human fecal specimens for detection of group a rotaviruses by transcriptase and polymerase chain reactions. *J. Clin. Microbiol.* 28, 1300, 1990.
36. Wonggratanacheewin, S. et al. Detection of *Opisthorchis viverrini* in human stool specimens by PCR. *J. Clin. Microbiol.* 40, 3879, 2002.
37. Ando, K. et al. Nucleotide sequence of mitochondrial CO I and ribosomal ITS II genes of *Opisthorchis viverrini* in northeast Thailand. *Southeast Asian J. Trop. Med. Public Health* 32 (Suppl 2), 17, 2001.
38. Harnnoi, T. et al., Specific gravity of *Opisthorichis viverrini* egg. *J. Helminthol.* 72, 359, 1998.
39. Rahman, S.M. et al. Early detection and estimation of infection burden by real-time PCR in rats experimentally infected with *Clonorchis sinensis*. *Parasitol. Res.* 110, 72–78, 2011.

29 Dicrocoelium

Donato Traversa, Vincenzo Lorusso, and Domenico Otranto

CONTENTS

29.1 INTRODUCTION

Dicrocoelium spp. (Greek δικροσ = bifid, κοιλια = gut), a trematode commonly known as "lancet fluke" or "small liver fluke" (as opposed to the "liver fluke" *Fasciola* spp.), is responsible for dicrocoeliosis, a globally diffused parasitic infection. This parasite lives in the bile ducts and gall bladder of domestic and wild ruminants (i.e., sheep, goats, cattle, buffaloes, roe deer, and camels) and occasionally affects rabbits, pigs, dogs, horses, and humans. *Dicrocoelium* cycles in two intermediate hosts, represented by snails and ants.

The main causative agents of dicrocoeliosis are *Dicrocoelium dendriticum* [1] which is reported in Europe, Asia, northern Africa, and North America, and *Dicrocoelium hospes* Looss, 1907 found in Africa [2]. A third species, *Dicrocoelium chinensis* [3] Tang and Tang, 1978, first described as *Dicrocoelium orientalis* [3], was found in several species of cervids in the former Soviet Union [3–5] and sika deer in Japan [5]. In Europe, *D. chinensis* was found in mouflon and roe deer in Austria [6] and in mouflon in northern Italy [7].

While *D. dendriticum* and *D. hospes* exhibit low host specificity, *D. chinensis* has been described only from sika deer, roe deer, and mouflon [8]. Since *D. dendriticum* is the most diffused and investigated species of the genus, the information presented in this chapter refer to *D. dendriticum* unless specifically indicated.

29.1.1 Classification and Morphology

29.1.1.1 Classification

For taxonomic studies of flatworms, ribosomal (rDNA) genes have been widely used, albeit molecular investigations of the order Plagiorchiida, in which *Dicrocoelium* is ranked, are few [9]. For further details, see Section 29.1.4. Currently, *Dicrocoelium* belongs to the family Dicrocoeliidae, in which the most important species are *D. dendriticum*, *D. hospes*, and *D. chinensis*.

29.1.1.2 Morphology

The adult stages of *D. dendriticum* (8–14 × 2–3 mm) have a semitransparent and pied lancet-like body (synonym *Dicrocoelium lanceolatum*) with a black uterus and white vitellaria visible to the naked eye. The eggs are small (35–45 × 22–30 μm), oval, dark brown, and operculate, and typically contain a miracidium [8]. The adult fluke *D. hospes* and *D. chinensis* differ from those of *D. dendriticum* for their greater width and length, and the morphological features of testes and uterus orientation [8,10]. Overall size, level of maximum body width with respect to the middle of the fluke body, testes orientation, oral and ventral sucker diameters, and vitelline glands are significant morphological characters useful to differentiate between *D. chinensis* and *D. dendriticum* (Figure 29.1) [6,8,10–12].

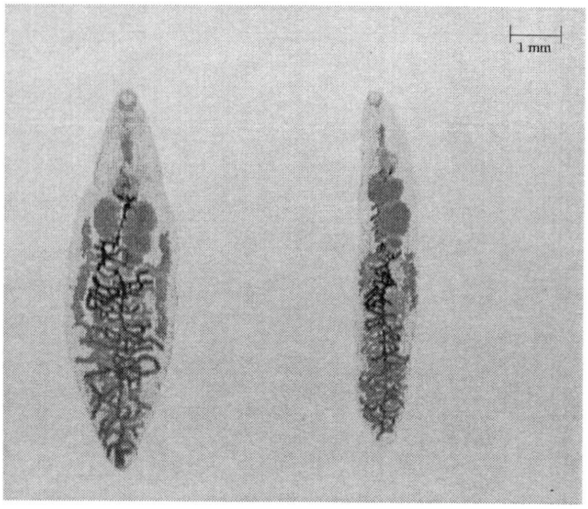

FIGURE 29.1 Adult specimens of *Dicrocoelium chinensis* and *Dicrocoelium dendriticum*. (Reproduced from Otranto, D. et al. *Acta. Trop.*, 104, 91, 2007. With permission.)

29.1.2 Biology and Epidemiology

29.1.2.1 Biology

Due to its complexity, the life cycle of *Dicrocoelium* spp. was not fully understood until more than a century after the first morphological description of the parasite [1,13]. Indeed, *D. dendriticum* develops into two intermediate hosts and its biological cycle (Figure 29.2) takes about 6 months. Several species of land mollusks and ants represent its primary and secondary intermediate hosts, respectively, while ruminants are its definitive host. For many years, snails were considered to be the only intermediate hosts and such assumption was presented by some authors [14,15]. Since the role of *Formica* spp. as the second intermediate host was demonstrated [13], the biological role of ants has been repeatedly confirmed [16]. In the first intermediate hosts the larval stages evolve in about 3–4 months, from the miracidium, which passes into the snails with the eggs to sporocysts of first and second generations. The sporocysts are an asexual multiplication state, which originate numerous cercariae, which migrate from the hepato-pancreas to the pulmonary cavities of the snail when they are completely mature and may use their tail, enzymatic products, and stylets inside their oral sucker (reviewed in [17]). The cercariae are then extruded from the snails in clusters of thousand, enveloped in a mucilaginous substance, commonly known as "slime-balls," which is ingested by ants (i.e., *Formica fusca*, *Formica pratensis*, *Formica rufibarbis*) and become metacercariae within 60 days. One or two cercariae may encyst in the ant's subesophageal ganglion, thus inducing a cataleptic cramp when the temperature is below 15°C. Such a phenomenon blocks the ant on the tips of herbage and grass, making it more readily ingestible by grazing ruminants.

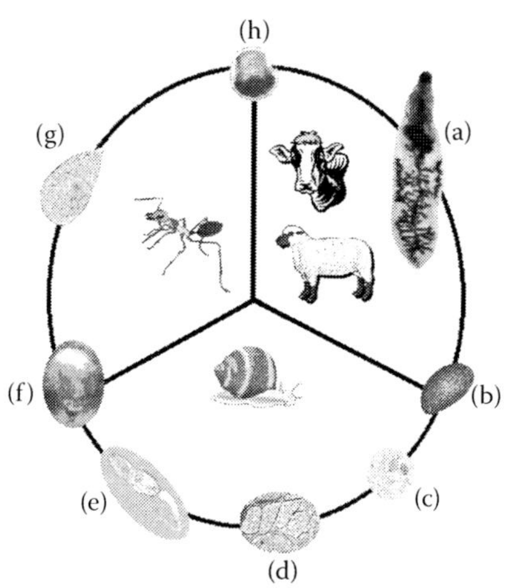

FIGURE 29.2 Biological cycle of *Dicrocoelium dendriticum*. Adults of *D. dendriticum* (a); eggs (b) containing a miracidium (c); sporocysts (d); mature cercariae (e); mucilaginous slime-balls (f); and metacercariae (g–h). (Reproduced from Otranto, D. and Traversa, D. *Trends Parasitol.* 19, 12–15, 2003. With permission.)

The cycle continues when the infected ants are ingested by a competent definitive host, in whose intestine the metacercariae excyst from the insects due to the action of duodenal enzymes and move up to the small and larger bile ducts and to the gall bladder via the choledocus (i.e., without any parenchymal migration into the liver). Thereby they reach sexual maturity and reproduce by hermaphroditism or cross-fertilization. After a prepatent period of about 2 months, the eggs are shed into the environment with the feces [18].

29.1.2.2 Epidemiology

Dicrocoeliosis is a cosmopolitan disease, mainly present in lowland or mountain pastures, which provide adequate conditions for the survival and development of the invertebrate hosts [19]. For instance, small ruminant dicrocoeliosis by *D. dendriticum* is common in sheep flocks with a prevalence of up to 100% in many European and Asian countries (reviewed by [20]) and the infection caused by *D. hospes* has also been reported in western areas of Africa with high prevalence rates [21,22]. *D. dendriticum* also affects species of wild ruminants, such as camelids (e.g., lamas and alpacas) in South America [23,24], and yaks and buffaloes in India [25,26]. In contrast to the knowledge on the ecology and biology of *D. dendriticum*, information for *D. chinensis* is scant [27,28].

The epidemiological factors influencing the disease are multiple, local environment and ecological characteristics (e.g., presence and the phenology of the intermediate hosts and the presence of domestic and/or wild ruminants) being the most important (reviewed in [17]). The occurrence of *D. dendriticum* is mainly related to calcareous or alkaline soils where the presence of favorable biotopes for the reproduction and survival of snails and ants is assessed [29]. In the environment, not only domestic and wild ruminants but also rabbits and hares are responsible for the dispersal of eggs and contamination of pastures [30]. Eggs are highly resistant in the external environment, as they overwinter and remain infective for up to 20 months on pastures. It has been indeed demonstrated that under natural conditions that egg survival is dependent on the season (i.e., 85% in winter and minimal values in summer) and that there is no relationship between infectivity of the eggs and their age [31]. Additionally, the dynamics of egg elimination in ovine feces is seasonal, with a winter peak of 90.5% positive animals and an average number of 323.4 ± 18.5 eggs per gram (EPG) [20]. These data fit with the life cycle of the fluke in Iberian peninsula, where young mollusks become infected at the beginning of spring and the definitive hosts become infected with ants from the end of spring to the beginning of fall [32].

The cosmopolitan diffusion of *Dicrocoelium*, and the related risk of infection in animals and humans as well, is also favored by the large number of species of intermediate hosts that may be involved in its life cycle. More than 90 snail species have indeed been found to act as competent vectors, and several of them (e.g., *Cochlicopa lubrica*) are diffused worldwide. Other species are present only in well-defined areas where they contribute to the distribution of the parasite, like *Helicella corderoi* in Spain, *Zebrina hohenackeri* in the Caucasus, *Helicella obvia* in Germany, and *Cernuella virgata* in Italy [33].

The efficiency of different mollusk species as vectors of *Dicrocoelium* has been investigated under laboratory conditions. For instance, the development of the fluke cercariae found in *Cernuella cespitum arigonis* 110–149 days after the experimental infection depends not only on the conditions in which the infected snails are kept (as high temperatures help larval development) but also on the mollusk species itself [34,35].

In Germany, the infection rate for *D. dendriticum* in *H. obvia* was evaluated over a period of four grazing seasons together with the biology of the snails [36]. Comparison between the infection rate and age of snails indicated that young specimens in the first year were less involved in the epidemiology of dicrocoeliosis (infection rate = 3.77%) than older mollusks (infection rate = 34.77%) [36]. Therefore, snails with a medium shell are likely to be more susceptible to *D. dendriticum* for their active metabolism and good nutritional conditions for the development of sporocysts [36,37]. Conversely, the decrease in the infection rate during the summer of the second year may be explained by the death of heavily infected snails in which the sporocysts cause disruption of the hepatopancreas and then impairment of reproductive activity and shortening of life span [38]. The negative effects of *D. dendriticum* sporocysts on the mollusks were confirmed by experimental infection of *C. cespitum arigonis* and *C. virgata*, in which histopathological alterations (e.g., hemocytic infiltration and cell plasma vacuolization) were observed 3 months postinfection [39]. On the basis of the data presented above, it has been suggested that first- and second-generation sporocysts develop in 2 years and that their maturation follows the snail's life cycle [36,38].

With regard to the second intermediate host, the vectorial efficiency is mainly related to the species and size of the ants, which may influence the size and the number of infective metacercariae. In fact, the larger the ant specimens, the greater the number of metacercariae present in the insect [40]. The paralysis and disorientation mechanisms caused by the encysted cercariae in the ants ("plant-topping") are pivotal to favor the infection of grazing ruminants. They seem to derive from an ancient behavior of some hymenopteran ancestors of ants, which sleep in the same way as the infected ants [41]. In Mediterranean areas, infected ants are usually observed between April and November and in tetania at 7.5–27°C. The highest shedding of eggs by sheep and cattle occurs during winter [42].

Among the different animal species that may act as definitive hosts, sheep are more susceptible to *D. dendriticum* than goats [26] while the relationship existing between animal age and susceptibility to the parasite has not yet been fully elucidated. However, some data suggest a higher mean EPG output (367.6 ± 50.9) in lambs than in adults [20] and in cattle aged up to 2 years than in older animals, with a slight gradual decrease with increasing of host age [43]. On the contrary, other studies showed a rise in the number of infected cattle, from 24.2% in calves younger than 18 months to more than 70% in animals older than 6 years [44]. Other factors, like stress-inducing conditions (e.g., animal transportation and confinement or overstocking) may increase the number and shedding of *Dicrocoelium* eggs as a likely consequence of immunedepression [44].

There is also evidence that the sex of cattle may be related to the incidence of dicrocoeliosis by both *D. dendriticum* [19] and *D. hospes* [22]. In particular, the postmortem examination of more than 2000 livers from cattle with a natural infection by *D. dendriticum* revealed a prevalence of dicrocoeliosis of 62.5% in females and 9.8% in males. This discrepancy may be explained by the typical cattle management, in that dairy heifers and cows graze for several seasons thus acquiring overlapping infections, while the steers and oxen may spend a considerable period of their lives in fattening units [19]. The movement of animals from lowland to mountain pastures may considerably influence the seasonality of this infection, considering that ruminants become infected by the ants in the pastures and then bring the parasites back to the valleys during the winter. As a matter of fact, some species of intermediate hosts like the snails *H. obvia* and *C. lubrica* and the ants *F. pratensis* and *Formica cunicularia* are present in mountain pastures up to altitudes of 1800–2600 m above sea level in European countries (e.g., Switzerland, Spain, and Austria) and in Asia (e.g., India) [26,45]. The migratory period seems to predispose animals to infection not only for the presence of intermediate hosts but also for the high stress induced by the transhumance on pasture-grazing nomadic sheep and goats [17].

29.1.3 Clinical Features and Pathogenesis

The pathogenic effects of *Dicrocoelium* are difficult to be estimated. In experimentally infected hamsters, pathological changes were characterized by bile duct proliferation and enlargement of the surface area with infiltration by collagen, lymphocytes, macrophages, and eosinophils of portal tracts and interlobular septa which ultimately lead to hepatic atrophy [46]. In the mean time, inefficient scavenging of reactive oxygen leading to oxidative liver damage and an increase in alanine transaminase (ALT) and aspartate transaminase (AST) activities were demonstrated by the decrease in superoxide dismutase (SOD) activity in both the hepatocyte cytosol and mitochondria [47]. Knowledge on the pathogenic effects of *Dicrocoelium* in natural infections is difficult to assess mainly due to the presence of concomitant gastrointestinal and pulmonary nematode and distomatid infections, which are characterized by similar symptoms to those of dicrocoeliosis. Nonetheless, some studies revealed that burdens up to 4000 *D. dendriticum* specimens do not cause significant blood or plasma protein loss in sheep and those alterations in those parameters registered during coinfection with *Fasciola hepatica* [48] and *Haemonchus contortus* [49] may be due to the greater pathogenicity of the latter two helminths.

In general, the most important hepatic lesions are due to the buccal stilets of the flukes, which irritate the bile ducts surfaces, thus inducing proliferation of the epithelial cells and changes in the structure of septal bile ducts [50,51].

Compared to fasciolosis, dicroceliosis consists of mild symptoms in affected animals; however, it causes severe economic losses, in terms of milk and meat productions in many countries. The disease can rarely be fatal [50] and it is difficult to be clinically determined due to the high aspecificity of symptoms, even in severe infections. Hence, major lesions are found only at the necroscopic examination of the liver [52]. Animals infected by *Dicrocoelium* show anemia, edema, emaciation, and, in advanced stages, cirrhosis, scarring of the liver surface, and marked distension of bile ducts. In severe infections, a large number of parasites are detectable inside the bile ducts and gall bladder; the liver is swollen, with thickened ducts, cholangitis, whitish spots on the surface, marked scarring, and cirrhosis [26].

Several reports on the increase in anti-*F. hepatica* antibodies in the bile and intestinal mucosa of infected animals [53] represented a foundation for investigations aiming to evaluate the role of somatic antigens of *D. dendriticum* in eliciting the local (bile) antibody responses in dicrocoeliosis-infected patients. Highly specific IgG and IgA isotype antibody responses against soluble antigens were detected in the bile of naturally infected cattle with higher IgAs, and the specificity of the bile antibody response against *D. dendriticum* was also assessed [54].

Although the antibody levels are not related to the host's immune protection to *Dicrocoelium*, their detection is important for early immunodiagnosis.

The infection by *Dicrocoelium* is occasionally zoonotic, even without exhibiting symptoms. Both *D. dendriticum* and *D. hospes* are indeed able to establish in the bile ducts of humans. However, cases of pseudoparasitism due to the ingestion of raw or undercooked liver are still difficult to be discriminated from primary infection. The infection usually occurs by the ingestion of the second intermediate host present on grass, herbs, drinking water, raw vegetables, or fruits [55]. Cases of human dicrocoeliosis have been mostly reported in middle-eastern Asia and Europe where *D. dendriticum* is endemic in livestock with high prevalence rates. For instance, in the last few decades, cases of human dicrocoeliosis have been repeatedly reported in Switzerland and Saudi Arabia with different infection rates [56–58]. The vast majority of the infections caused by *D. dendriticum* generate only mild symptoms, like biliary colic and general digestive distress (e.g., bloating and diarrhea). Other signs are constipation, chronic diarrhea, vomiting, and abdominal pain. In severe cases, both bile ducts and the biliary epithelium are enlarged due to the expanding fibrous tissue around the ducts, leading to hepatomegaly, intense eosinophilia, cholangitis, biliary obstruction, chronic cirrhosis, and urticaria [57,59–62].

With regard to treatment and control, although benzimidazolic compounds are efficacious and show a broad antiparasitic spectrum, their activity against liver flukes may be low, therefore requiring a high dose rate with more than one treatment. The administration of 1, two or three times a year according to farm management practices and the parasitological status of animals, has been proved to be effective in controlling dicrocoeliosis in sheep flocks in several areas. Such a treatment protocol should always be accompanied by copromicroscopic examination and evaluation of animal production, grazing, and animal feeding.

29.1.4 Diagnosis

29.1.4.1 Conventional Techniques

Given the typical subclinical appearance, animal dicrocoeliosis oftentimes remains undiagnosed, thus making the retrieval of the adult flukes in the liver at necropsy or the detection of fecal eggs at microscopical examination the sole possible means of diagnosis in dead and live animals, respectively [19].

Copromicroscopy is the most commonly used approach for the diagnosis of dicrocoeliosis by revealing the presence of typical eggs, which are 40 × 25 mm in size, thick-walled, yellowish-brown, and contain a miracidium. Unlike the results obtained for the diagnosis of other flukes infections (e.g., fasciolosis and paramphistomosis), sedimentation techniques are not efficient when compared with floatation techniques (specific gravity: 1.30–1.45).

A laboratory study evaluated a sedimentation technique and three different high-density solutions, namely $ZnSO_4$ (specific gravity (sp.g.) 1.3–1.45), K_2CO_3 (sp.g. 1.45), and HgI_2/KI (sp.g. 1.44), for the estimation of fecal egg counts (FEC) in sheep feces [63]. The HgI_2/KI solution showed the highest percentage rate of egg recovery from the feces (91.2 ± 9.4%) without significant influence of floatation time on the egg count while sedimentation technique proved to have a very low rate of recovery of *Dicrocoelium* eggs (42.2 ± 1.5%), compared to the floatation method [63]. In lambs experimentally infected with different doses of infective metacercariae (i.e., 1000 and 3000), the prepatent period showed to range from 49 to 79 days postinfection ($\bar{X} = 59$), thus indicating that a negative copromicroscopic examination in this period does not necessarily imply a parasitological negativity [64].

Alternative diagnostic methods to the copromicroscopy and *postmortem* inspections of the liver are represented by several immunological techniques, such as immunofluorescence, passive hemoagglutination test, complement fixation, and enzyme-linked immunosorbent assay (ELISA) [26,65–68]. The somatic (So) and excretory secretory products (ESP) of *Dicrocoelium* flukes have been evaluated for their usefulness in ELISA tests, and their chemical characteristics have been examined for both bile [54] and serum antibody responsiveness [70]. Although proteins and surface molecules of *Dicrocoelium* elicit an intense antibody response, using ESP or purified antigens might affect the specificity and sensitivity of serological tests [54]. Slightly less relative antibody responsiveness was detected by other authors using So rather than ESP antigens [69].

Anti-*D. dendriticum* antibody titers were studied by ELISA and compared to the FEC in lambs with natural dicrocoeliosis. Such studies demonstrated that colostral humoral immunity occurs 24 h after birth and disappears completely within 2–4 months [70]. Following the first natural infection, 4–8 weeks before the appearance of eggs in the feces, high anti-*D. dendriticum* antibody titers may be detected by ELISA showing that this is an effective test for an early immune diagnosis of dicrocoeliosis [71].

Analogously, in sheep experimentally infected with 1000 and 3000 metacercariae, the IgG response may be detected by ELISA 19 and 23 days earlier than eggs shedding, respectively, and no difference was found in the antibody titers in relation to the infective dose [69]. The long persistence of anti-*D. dendriticum* antibody titers was important for seroepidemiological studies, indicating the usefulness of the ELISA technique for an early diagnosis of dicrocoeliosis in lambs (antibodies appear 1 month postinfection) also allowing a prompt treatment of positive animals.

Dicrocoelium dendriticum So and ESP antigens were analyzed by SDS-PAGE and a specific protein, weighing about 130 kDa, showed strong immunoreactivity at the immunoblot analysis. This protein showed to be suitable for the diagnosis of dicrocoeliosis by Western blot using pooled and individual sera samples from experimentally infected sheep [72]. Another recent ELISA test based on ESP antigen of *D. dendriticum* proved to be able to detect specific IgG in experimental infected sheep by 1 month after the infection and 1 month earlier than the egg shedding in the feces. This study demonstrated the highest utility of immunological assays compared to copromicroscopy and the applicability of the ELISA as a test in field epidemiological surveys [73].

The diagnosis of human infections relies on the detection and identification of *Dicrocoelium* eggs in the feces of infected patients. However, the presence of eggs in stool samples from human beings may be a result of ingestion of raw or undercooked infected animal liver and should therefore be considered as a false-positive result or pseudoparasitism.

29.1.4.2 Molecular Techniques

Molecular approaches developed for diagnosing infections by *Dicrocoelium* spp. are scant. So far, molecular and biochemical researches on the population structure of *Dicrocoelium* spp. both within and among host individuals have indeed been restricted only to *D. dendriticum* [74,75]. A study carried out on the isoenzymatic profiles of different *D. dendriticum* isolates from cattle, sheep, and goats revealed little phenotypic variation among flukes from a single host individual or from a single host species collected from the same geographic area [74]. Nonetheless, a genetic analysis based on randomly amplified polymorphic DNA showed a high intrapopulation variability among specimens collected from sheep in a relatively small geographic area [75].

A recent study has been carried out to molecularly differentiate *D. dendriticum* from *D. chinensis* by the analysis of several parasite specimens from different hosts and geographical localities [10]. Individual isolates were molecularly characterized by sequencing regions of the ribosomal DNA (i.e., partial 18S rDNA and ITS-2 including the 5.8S and 28S flanking regions). This study showed the existence of interspecific differences between *D. dendriticum* and *D. chinensis* of 0.14% and 3.8% in 18S rRNA and ITS-2 sequences, respectively. Interspecific divergence of the ITS-2 between *D. dendriticum* and *D. chinensis* (3.8%) was about three times greater than the maximum value of the intraspecific rate (0–1.3%), demonstrating the utility of ITS-2 as marker for species circumscription, which is also supported by analyses carried out on ITS-1 and ITS-2 within several groups of digeneans [9,76–78]. The study by Otranto and colleagues [10] demonstrated the separate status of *D. chinensis* opening new venues for the study of the etiology, biology, and diagnosis of the infection of this poorly known parasitic species. In addition, the use of molecular markers to identify *D. dendriticum* and *D. chinensis* is of particular importance in enabling the identification of the species of snails and ants that may act as competent intermediate hosts (as demonstrated for *D. dendriticum*) [79]. Nonetheless, more studies are needed to pursue the simultaneous molecular identification of *D. dendriticum*, *D. hospes*, and *D. chinensis* both for the innovative diagnosis in definitive hosts and for the accurate detection in intermediate hosts.

In another research, the rDNA 28S region and the ITS-2 plus flanking 5.8S and 28S sequences of *D. dendriticum* from sheep and cattle from Italy and *D. hospes* from a *Bos indicus* from Senegal were amplified by PCR and sequenced. While no intraspecific variation was observed between the 28S of all the *D. dendriticum* specimens studied and the 28S sequence present in GenBank, intraspecific variation was observed in the 28S rDNA of *D. hospes* compared to the sequence present in GenBank. Intra- and interpopulation variations were observed for ITS-2 of *D. dendriticum*, whereas 100% identity was observed in all the ITS-2 sequences of *D. hospes*. Additionally, the ITS-2 of *D. dendriticum* and *D. hospes* differed intraspecifically in 33 positions. The rate of sequence variability (8.2–8.5%) between *D. dendriticum* and *D. hospes* ITS-2 demonstrated the utility of this sequence to differentiate the two species [80]. A phylogenetic analysis carried out via Bayesian inference using ITS-2 and ribosomal 28S sequences of *D. chinensis* and *D. dendriticum* and including also other closely related 20 Digenean species belonging to

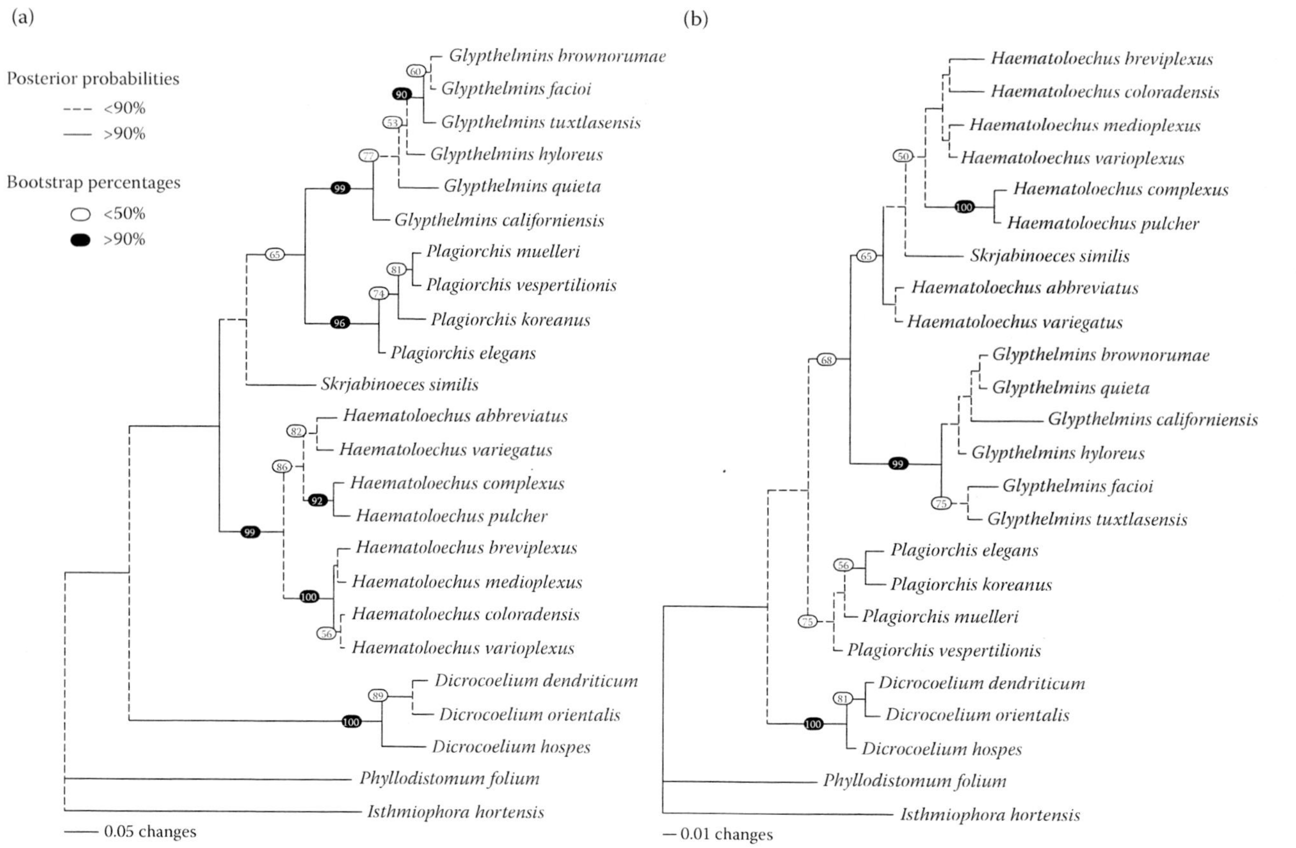

FIGURE 29.3 Phylogenetic analysis by Bayesian inference using ITS-2 (a) and ribosomal 28S sequences (b) of Digenean species belongig to Xiphidiata within the Plagiorchiida Order. (Reproduced from Otranto, D. et al. *Acta. Trop.*, 104, 91, 2007. With permission.)

Xiphidiata within the Plagiorchiida showed a strong concordance in differentiating Dicrocoeliidae, Gorgoderidae, and Plagiorchiidae (Figure 29.3) in agreement with their current classical taxonomical position [10].

29.2 METHODS

29.2.1 Sample Preparation

29.2.1.1 Morphological Identification

The specimens are stained with 10% carmine and stored in 70% ethanol.

29.2.1.2 Molecular Examination

The flukes are preserved in 70% or absolute ethanol before DNA extraction and subsequent analysis.

29.2.2 Detection Procedures

29.2.2.1 Morphological Identification

Morphological features of diagnostic relevance from adult specimens are the overall size, the level of maximum body width with respect to the middle of the fluke body, the testes orientation, the oral and ventral sucker diameters, and the vitelline glands (see [8,10]).

29.2.2.2 PCR Amplification

Genomic DNA can be isolated using a commercial acid nucleic extraction kit (e.g., QIAmp tissue kit, QIAgen, Germany) after washing three times in cold distilled water to remove traces of the preservative as described in Reference 10.

Partial 18S rDNA (~1400 bp) can be amplified in fragments using primers Dd18SF1 and Dd18SR1, and Dd18SF2–Dd18SR2. The ITS-2, plus 5.8S and 28S flanking regions (~600 bp), can be amplified using Dd58SF1 and Dd28SR1. All primers (Table 29.1) reported were based on previously characterized 18S (GenBank accession no. Y11236) and 5.8S–28S (GenBank accession nos. DQ379986 and AF151939) sequences of *D. dendriticum*. PCR amplification described here showed to be useful in amplifying *D. chinensis* DNA also.

PCR is carried out as follows usually in a total of 50 μL mixture. Genomic DNA (~4–5 μL) is added to PCR mix (~45–46 μL) containing 10 mM Tris-HCl pH 8.3 and 50 mM KCl, 2.5 mM $MgCl_2$, 250 μM of each dNTPs, 50 pM of each primer, and 1.25 U of DNA polymerase. PCR is performed with the cycling protocol: 94°C for 10 min; 40 cycles of 98°C for 30 s; 53°C or 55°C for 60 s (for ITS-2 and 18S, respectively); 72°C for 90 s; and 72°C for 7 min. Amplicons are purified and sequenced directly in automated sequencers. Sequences are determined in both directions using the same primers as used for PCR [10].

TABLE 29.1
List of PCR and Sequencing Primers for *Dicrocoelium* spp.

Site	Direction	Name	Sequence (5′–3′)	Reference
18S	Forward	Dd18SF1	GATAACGGGTAACGGGGAAT	[10]
18S	Reverse	Dd18SR1	AACCTCTGACTTTCGCTCCA	[10]
18S	Forward	Dd18SF2	TGGAGCGAAAGTCAGAGGTT	[10]
18S	Reverse	Dd18SR2	GTACAAAGGGCAGGGACGTA	[10]
5.8S	Forward	Dd58SF1	ATATTGCGGCCATGGGTTAG	[10]
28S	Reverse	Dd28SR1	ACAAACAACCCGACTCCAAG	[10]

29.3 CONCLUSIONS AND FUTURE PERSPECTIVES

The limited number of reports on *D. hospes* infection and the scant data on the epidemiology, ecology, and biology of *D. chinensis* are probably the main reasons for which *D. dendriticum* is considered the most important and the best-known species within the genus. The assessment of the potential risk of human infection and the actual pathogenic impact on other host species rather than domestic ruminants still remain to be clearly addressed. Under a practical veterinary standpoint, the more frequently occurring polyparasitism may mask the pathological effects and economic losses related to the presence of liver flukes in livestock. As mentioned above, the diagnosis of dicrocoeliosis by using conventional methodologies may be troublesome. Furthermore, only few effective drugs are available on the market for the control and treatment of the parasite. The development of immunological methods (i.e., ELISA) for an accurate diagnosis of dicrocoeliosis would be of relevance to overcome constraints inherent to the copromicroscopic examinations (e.g., the long prepatent period leading to false-negative results). In fact, immunological assays may provide a timely diagnosis of dicrocoeliosis, and it consequently should lead to early treatments. A better understanding of this disease would also enable the planning of effective control measures of the parasite in the environment, preventing both human and animal infections.

Importantly, more molecular studies are needed for *Dicrocoelium* spp. in order to dispose of ready-to-use diagnostic tools for innovative investigations in both definitive and intermediate hosts. For instance, the use of molecular markers to detect *D. dendriticum*, *D. hospes*, and *D. chinensis* at the same time would indeed be of central importance in enabling the identification of the species of snails and ants that may act as competent intermediate hosts in different countries, as partially demonstrated for *D. dendriticum* [80]. Finally, such comprehensive molecular assays would allow researchers to undertake investigations on the pathogenic role of these flukes in both human beings and animals, overcoming constraints related to the use of conventional diagnostic techniques.

REFERENCES

1. Rudolphi, C.A. *Entozoorum Synopsis cui Accedunt Mantissa Duplex et Incides Loclupetissimi*. Sumtibus Augusti Rucker, Berolini, 1819.
2. Soulsby, E.J.L. *Helminths, Arthropods and Protozoa of Domesticated Animals*. 6th ed., Bailliére, London, pp. 24–26, 1968.
3. Sudarikov, V.E., and Ryjikov, K.M. On the helminth fauna of ungulates of the Baikal region, Trudy Gel'mintol. *Lab. Akad. Nauk SSSR*. 5, 53–58, 1951 (in Russian).
4. Oshmarin, P.G., Parasitic worms of mammals and birds in the Primorje region. *Izd. Akad. Nauk SSSR*. Moskva (in Russian), 1963.
5. Gubanov, N.M., Helminth fauna of mammals of economic importance in Yakutia. *Izd. Nauka*, Moskva (in Russian), 1964.
6. Taira, K. et al. Morphometry on lancet flukes found in Japanese sika deer (*Cervus nippon centralis*) captured in Iwate Prefecture, Japan. *J. Vet. Med. Sci.* 68, 375, 2006.
7. Poglayen, G. et al. Endoparassiti del Muflone (*Ovis ammon musimon*): Indagine in provincia di Belluno. *Suppl. Ric. Biol. Selvaggina*. 24, 73–77, 1996 (in Italian).
8. Euzéby, J. *Les maladies vermineuses des animaux et leur incidence sur la pathologie humaine, Maladies dues aux plathelminthes* Tome 2. Vigot et Fréres, Paris. pp. 619–670, 1971.
9. Tkach, V.V., Pawlowski, J., and Mariaux, J. Phylogenetic analysis of the suborder Plagiorchiata (Platyhelminthes, Digenea) based on partial lsrDNA sequences. *Int. J. Parasitol.* 30, 83, 2000.
10. Otranto, D. et al. Morphological and molecular differentiation between *Dicrocoelium dendriticum* (Rudolphi, 1819) and *Dicrocoelium chinensis* (Sudarikov and Ryjikov, 1951) Tang and Tang, 1978 (Platyhelminthes: Digenea). *Acta Trop.* 104, 91, 2007.
11. Yamaguti, S. *Synopsis of Digenetic Trematodes of Vertebrates*. Keigaku Publishing Ltd, Tokyo, 1971.
12. Hinaidy, H.K. *Dicrocoelium suppereri* nomen novum (syn. *D. orientalis* Sudarikov et Ryjikov 1951), a new trematode for the parasite fauna of Austria, *Zentralbl. Veterinarmed.* B30, 576, 1983 (in German).
13. Krull, W.M., and Mapes, C.R. Studies on the biology of *Dicrocoelium dendriticum* (Rudolphi, 1819) Looss, 1899 (Trematoda Dicrocoeliidae), including its relation to the intermediate host *Cionella lubrica* (Müller) IV, Infection experiments involving definitive hosts. *Parasitol. Res.* 42, 277, 1952.
14. Mattes, O. Der Entwicklungsgang des lanzettegels *Dicrocoelium lanceatum*. *Z. Parasitkde*. 8, 371, 1936.
15. Neuhaus, W. Untersuchungen uber bau und entwicklung der lanzettegel-cercarie (*Cercaris vitrina*) und klarstellung des infections-vorganges beim endwirt. *Z. Parasitenkd.* 8, 431, 1936.
16. Hohorst, W., and Lammler, A. Experimentelle Dicrocoeliose-Studien, Z. *Tropenmed. Parassit.* 13, 377, 1962.
17. Otranto, D., and Traversa, D. A review of dicrocoeliosis of ruminants including recent advances in the diagnosis and treatment. *Vet. Parasitol.* 107, 317, 2002.
18. Krull, W.M. The migratory route of the metacercarie of *Dicrocoelium dendriticum* (Rudolphi, 1819) Looss, 1899 in the definitive host: Dicrocoeliidae. *Cornell Vet.* 48, 17, 1958.
19. Ducommun, D., and Pfister, K. Prevalence and distribution of *Dicrocoelium dendriticum* and *Fasciola hepatica* infections in cattle in Switzerland. *Parasitol. Res.* 77, 364, 1991.
20. Manga-Gonzàlez, M.Y., Gonzàlez-Lanza, C., and Del-Pozo, P. Dynamics of the elimination of *Dicrocoelium dendriticum*

(Trematoda, Digenea) eggs in the faeces of lambs and ewes in porma basin (Leòn, Nw Spain). *Ann. Parasitol. Hum. Comp.* 66, 57, 1991.
21. Malek, A.E. Occurrence of *Dicrocoelium hospes* in Mali and Senegal, West Africa. *J. Helminthol.* 54, 45, 1980.
22. Asanji, M.F., and Williams, M.O. The effect of sex on seasonal variation on single and double infection of cattle in Sierra Leone by *Dicrocoelium hospes* and *Fasciola gigantica. Vet. Parasitol.* 15, 247, 1984.
23. Wenker, C. et al. Dikrozoliose bei Neuweltkameliden. *Tieraztl. Prax.* 26, 355, 1998.
24. Gunsser, I., Hanichen, T., and Maierl, J. Leberegelbefall bei Neuweltkameliden. Liver fluke in South American camelids. *Tierarztl. Prax.* 27, 187, 1999.
25. Somvanshi, R., and Kaul, G.L. Dicrocoeliasis in Pashima goats. *Ind. Vet. Med. J.* 13, 42, 1989.
26. Jithendran, K.P., and Bhat, T.K. Prevalence of dicrocoeliosis in sheep and goats in Himachal Pradesh, India. *Vet. Parasitol.* 61, 265, 1996.
27. Tang, C. et al. Studies on the biology and epizootiology of *Dicrocoelium chinensis* in the eastern Inner Mongol Autonomous region. *Acta Zool. Sinica.* 29, 340 (1983) (in Chinese).
28. Tang, C. et al. Investigations on the biology of a few *Dicrocoelium* species of sheep on Qinghai Plateau. *Acta Zool. Sinica.* 31, 254 (1985) (in Chinese).
29. Chartier, C., and Reche, B. Gastrointestinal helminths and lungworms of French dairy goats: Prevalence and geographical distribution in poitou-charentes. *Vet. Res. Comm.* 16, 327, 1992.
30. Boray, J.C. Flukes of domestic animals In: Gafaar, S.M., Howard, W.E., and Marsh, R.E. (eds), *Parasites, Pests and Predators.* Elsevier, Amsterdam, pp. 179–218, 1985.
31. Alunda, J.M., and Rojo-Vàzquez's, F.A. Survival and infectivity of *Dicrocoelium dendriticum* eggs under field conditions in NW Spain. *Vet. Parasitol.* 13, 245, 1983.
32. Manga-González, M.Y. Some aspects of the biology and helminthofauna of *Helicella (Helicella) itala* (Linnaeus, 1758) (Mollusca) Natural infection by Dicrocoeliidae (Trematoda). *Rev. Ibér Parasitol. Extraordinario,* 131, 1987.
33. Fasanella, A., Lia, R., and Giangaspero, A. *Cernuella virgata* (Mollusca: Gasteropoda: Pulmonata) hôte intermédaire de *Dicrocoelium dendriticum* (Rudholphi, 1819) dans la region de Puglia, Italie. *Parasite.* 2, 331, 1995.
34. Manga-Gonzàlez, M.Y. et al. Studies on experimental infection by *Dicrocoelium dendriticum* (Trematoda) of *Cernuella* (Xeromagna) *cespitum arigonis* (Mollusca) specimens in a natural environment. In: Guerra, A., Rolan, E., and Rocha, F. (eds), *Abst. 12th Int. Malacol. Congr.* Vigo, 1995.
35. Gonzàlez-Lanza, C. et al. Larval development of *Dicrocoelium dendriticum* in *Cernuella* (Xeromagna) *cespitum arigonis* under controlled laboratory conditions. *J. Helminthol.* 71, 311, 1997.
36. Schuster, R. Infection patterns in the first intermediate host of *Dicrocoelium dendriticum. Vet. Parasitol.* 47, 235, 1993.
37. Alunda, J.M., and Rojo-Vàzquez, F.A. Survival and infectivity of *Dicrocoelium dendriticum* eggs under field conditions in NW Spain. *Vet. Parasitol.* 13, 245, 1983.
38. Schuster, R. Zur beeinflussung von *Helicella obvia* durch *Dicrocoelium*- Parthenitae. *Angew. Parasitol.* 33, 61, 1992.
39. Gomez, B.J. et al. Alteraciones histopatologicas producidas por *Dicrocoelium dendriticum* (Trematoda) en dos especies de *Cernuella* (Mollusca) sacrificadas a los tres meses post-infestacion *Iberus.* 14, 189, 1996.
40. Schuster, R., Factors influencing the metacercarial intensity in ants and the size of *Dicrocoelium dendriticum* metacercarial cysts. *J. Helminthol.* 65, 275, 1991.
41. Wikler, W. Evolution-oriented ethology, kin selection, and altruistic parasites. *Z. Tierpsyschol.* 42, 206, 1976.
42. Manga-González, M.Y., and González-Lanza, C. Field and experimental studies on *Dicrocoelium dendriticum* and dicrocoeliasis in northern Spain. *J. Helminthol.* 79, 291, 2005.
43. González-Lanza, C. et al. Coprological study of the *Dicrocoelium dendriticum* (Digenea) egg elimination by cattle in highland areas in León Province, Northwest Spain. *Parasitol. Res.* 79, 488, 1993.
44. Sotiraki, S.T. et al. The effect of transportation and confinement stress on egg production by *Dicrocoelium dendriticum* in sheep. *J. Helminthol.* 73, 337, 1999.
45. Eckert, J., and Hertzberg, H. Parasite control in transhumant situations. *Vet. Parasitol.* 54, 103, 1994.
46. Sanchez-Campos, S. et al. Morphologic and biochemical changes caused by experimentally induced dicroceliosis in hamsters (*Mesocricetus auratus*). *Comp. Med.* 50, 147, 2000.
47. Sanchez-Campos, S., Tunon Gonzalez, P., and Gonzalez-Gallego, J. Oxidative stress and changes in liver antioxidant enzymes induced by experimental dicrocoeliosis in hamsters. *Parasitol. Res.* 85, 468, 1999.
48. Holmes, P.H. et al. The anaemia in fasciolosis studies with G-labelled red cells. *J. Comp. Pathol.* 78, 415, 1968.
49. Dargie, J.D., and Allonby, E.W. Pathophysiology of single and challenge infections of *Haemonchus contortus* in Merino sheep: Study on the red cell kinetics and the "self-cure" phenomenon. *Int. J. Parasitol.* 5, 147, 1975.
50. Wolff, K., Hauser, B., and Wild, P. Dicrocoeliose des schafes: Untersuschungen zur pathogenese und zur regeneration der laber nach therapie. *Berl. Münch Tierärztl. Wschr.* 97, 378, 1984.
51. Camara, L., Pfister, K., and Aeschlimann, A. Analyse Histopathologique de foie de bovin infestè par *Dicrocoelium dendriticum. Vet. Res.* 27, 87, 1996.
52. Theodoridis, Y. et al. Pathophysiological studies on *Dicrocoelium dendriticum* infection in sheep. *Vet. Parasitol.* 39, 61, 1991.
53. Wedrychowicz, H. et al. Local antibody responses in the bile and faeces of sheep infected with *Fasciola hepatica. Res. Vet. Sci.* 37, 44, 1984.
54. Wedrychowicz, H. et al. Somatic antigens of adult *Dicrocoelium dendriticum* recognised by bile antibodies of naturally infected cattle. *Vet. Parasitol.* 56, 47, 1995.
55. Haridy, F.M. et al. A preliminary study on dicrocoeliasis in Egypt, with a general review. *J. Egypt Soc. Parasitol.* 33, 85, 2003.
56. Stahel, E. Spurious infections by the small liver fluke (*Dicrocoelium dendriticum*) in Switzerland 1976–1980 Stahel E. *Schweiz Med Wochenschr.* 111, 1159, 1981.
57. el-Shiekh Mohamed, A.R., and Mummery, V. Human dicrocoeliasis. Report on 208 cases from Saudi Arabia. *Trop. Geogr. Med.* 42, 1, 1990.
58. Helmy, M.M., and Al-Mathal, E.M. Human infection with *Dicrocoelium dendriticum* in Riyadh district (Saudi Arabia). *J. Egypt Soc. Parasitol.* 33, 139, 2003.
59. Ondriska, F. et al. A rare case of human autochthonous dicrocoeliasis in Czechoslovakia. *Bratisl. Lek Listy* 90, 467, 1989.
60. Jelinek, T., Rack, J., and Adusu, E. Human infection with *Dicrocoelium dendriticum, Deut. Med. Wochenschr.* 129, 2538, 2004.

61. Karadag, B. et al. An unusual case of biliary obstruction caused by *Dicrocoelium dentriticum*. *Scand J Infect Dis*. 37, 385, 2005.
62. Sing, A., Tybus, K., and Fackler, I. Acute urticaria associated with *Dicrocoelium dendriticum* infestation. *Indian J. Med. Microbiol.* 26, 97, 2008.
63. Rehbein, S., Kokott, S., and Lindner, T. Evaluation of techniques for the enumeration of *Dicrocoelium* eggs in sheep faeces, Zentralbl. *Veterinarmed A*. 46, 133, 1999.
64. Campo, R., Manga-Gonzàlez, M.Y., and Gonzàlez-Lanza, C. Relationship between egg output and parasitic burden in lambs experimentally infected with different doses of *Dicrocoelium dendriticum* (Digenea). *Vet. Parasitol.*, 87, 139, 2000.
65. Schröder, L., and Geyer, E. Experimentelle Dikrozöliose: Bildung humoraler Antikörper beim Goldhamster (*Mesocricetus auratus*). *Behring Inst. Mitt.* 60, 24, 1976.
66. Calamel, M. Application de la technique d'immunofluorescence indirecte a l'étude épidemiologique de la dicrocoeliose. *Rec. Mèd. Vèt.* 153, 343, 1977.
67. Piergili Fioretti, D. et al. Il metodo ELISA per la ricerca di anticorpi in ovini parassitati da *Dicrocoelium dendriticum*. *Rivista di Parassitologia*. 3,289, 1980.
68. Savitskii, I.G., and Ponomareva, A.M. Diagnosis of ovine dicrocoeliasis by the ELISA. *Veterinariya*. 1, 43 (1984) (in Russian).
69. Gonzàlez-Lanza, C. et al. IgG antibody response to ES or somatic antigens of *Dicrocoelium dendriticum* (Trematoda) in experimentally infected sheep. *Parasitol. Res.* 86, 472, 2000.
70. Baldelli, B. et al. Dicroceliosi ovina: Studio degli anticorpi verso *Dicrocoelium dendriticum* in agnelli nati da pecore spontaneamente infette. *Rivista di Parassitologia*. 1, 23, 1981.
71. Ambrosi, M. et al. Dicroceliosi ovina: Insorgenza e decorso della infezione da *Dicrocoelium dendriticum* studiati con metodi parassitologici e sierologici (ELISA) in quattro gruppi di ovini di traccia. *Rivista di Parassitologia*. 3, 299, 1980.
72. Revilla-Nuín, B. et al. Partial characterization and isolation of 130 kDa antigenic protein of *Dicrocoelium dendriticum* adults. *Vet. Parasitol.* 134, 229, 2005.
73. Broglia, A. et al. Experimental ELISA for diagnosis of ovine dicrocoeliosis and application in a field survey. *Parasitol. Res.* 104, 949, 2009.
74. Campo, R. et al. Characterization of adult *Dicrocoelium dendriticum* by isoelectric focusing. *J. Helminthol.* 72, 109, 1998.
75. Sandoval, H. et al. Preliminary study on genetic variability of *Dicrocoelium dendriticum* determined by random amplified polymorphic DNA. *Parasitol. Int.* 48, 21, 1999.
76. Morgan, J.A., and Blair, D. Nuclear rDNA ITS sequence variation in the trematode genus *Echinostoma*: An aid to establishing relationships within the 37-collar-spine group. *Parasitology*. 111, 609, 1995.
77. Jousson, O. et al. Use of the ITS rDNA for elucidation of some life-cycles of Mesometridae (Trematoda, Digenea). *Int. J. Parasitol.* 28, 1403, 1998.
78. Casey, S.P. et al. Use of ITS rDNA for discrimination of European green- and brown-banded sporocysts within the genus *Leucochloridium* Carus, 1835 (Digenea: Leucochloriidae). *Syst. Parasitol.* 56, 163, 2003.
79. Heussler, V. et al. A DNA probe for the detection of *Dicrocoelium dendriticum* in ants of *Formica* spp. and *Lasius* spp. *Parasitol. Res.* 84, 505, 1998.
80. Maurelli, M.P. et al. Characterization of the 28S and the second internal transcribed spacer of ribosomal DNA of *Dicrocoelium dendriticum* and *Dicrocoelium hospes*. *Parasitol. Res.* 101, 1251, 2007.
81. Otranto, D., and Traversa, D. Dicrocoeliosis of ruminants: A little known fluke disease. *Trends Parasitol.* 19, 12–15, 2003.

30 Echinostomes

Rafael Toledo and Bernard Fried

CONTENTS

30.1 INTRODUCTION

Under the term echinostomes are included those digeneans belonging to the family Echinostomatidae. Echinostomes are a rather heterogeneous group of cosmopolitan hermaphroditic digeneans that inhabit, as adults, the intestine of a great spectrum of vertebrate hosts, such as birds, mammals and, occasionally, reptiles and fishes. They can also parasitize humans, causing the food-borne infection called echinostomiasis. The definitive hosts become infected after ingestion of the second intermediate host harboring the encysted metacercariae. The stage of the echinostomes in the definitive host involves a number of processes such as metacercarial excystment, establishment, habitat location, adult development, mating, and release of eggs. In this chapter, we will review the main features of these trematodes with an emphasis on the molecular methods for the diagnosis of the echinostomiasis.

30.1.1 Classification and Morphology

30.1.1.1 Classification

The family Echinostomatidae constitutes a group of digeneans characterized by great confusion regarding their systematic classification, particularly within the genus *Echinostoma*. This is attributable to a number of factors, including misidentified species or species that have been insufficiently described, as well as the existence of substantial interspecific homogeneity of the morphological characteristics of the adult stages [1,2]. These facts, together with the broad range of final vertebrate hosts and the wide geographical distribution, have resulted in the composition of the family revealed so far (91 nominal genera) [3]. Historically, attempts at pursuing or exploring the systematic relationships within the Echinostomatidae have resulted in an important number of monographs and reviews [1,4]. In general, the Echinostomatidae has been viewed as a monophyletic taxon. However, the morphology and/or the diversity of the criteria adopted by different authors have led to its division into an impressive number of taxa (i.e., a total of 21 nominal subfamilies). The following characteristics traditionally used for identification are considered significant: the degree of development and the morphology of the collar of spines; the number, shape, arrangement, and relative size of the collar spines; the morphology of the male terminal genitalia (position of the cirrus-sac, structure of the internal seminal vesicle, development of the pars prostatica, and size and arrangement of the cirrus); the position of the ovary and testes; the nature of the tegumental armament; the presence of an uroproct; size and shape of the body; and the length of the forebody, uterine, and posttesticular fields [3]. In addition, biological characteristics, including cercarial morphology and chaetotaxy, have also been used for identification [5–7].

A relatively small number of molecular-based studies on the systematics of echinostomes have been carried out to date, and most of them are focused on the genus

Echinostoma. Sequence data from mitochondrial DNA (mtDNA)—nicotinamide adenine dinucleotide dehydrogenase subunit 1 (ND1) and cytochrome *c* oxidase subunit 1 (CO1)—and nuclear ribosomal DNA (rDNA)-internal transcribed spacers 1 and 2 (ITS1 and ITS2) have been used as markers to characterize species of *Echinostoma* [8–14]. Molecular studies in general other than on *Echinostoma* are scant. The ITS sequences of *Echinoparyphium*, *Pseudoechinoparyphium*, *Neoacanthoparyphium*, and *Hypoderaeum* were studied [15]. Moreover, the ND1 and ITS sequences of several species of *Echinostoma*, *Echinoparyphium*, *Hypoderaeum*, and *Isthmiophora* have provided useful information for the construction of phylogenetic trees [14].

30.1.1.2 Morphology

Members of the family Echinostomatidae are morphologically characterized by the presence of a head collar with collar spines around the oral sucker (Figure 30.1a and b). The number and arrangement of collar spines are an important feature for taxonomic purposes. Considerable variation exists in size of echinostomes depending upon species, fixation procedures, definitive host, and crowding effect. Echinostomes are referred to as "small" if up to 5 mm in length, "medium" if 5–10 mm in length, and "large" if longer than 10 mm [3,16]. The spines of the cephalic collar may be arranged in one or two circles (Figure 30.1b) and the number of spines is constant within the species. The collar contains numerous sensory papillae and it is incomplete in the ventral aspect (Figure 30.1b) [16]. The tegument contains scale-like spines on both dorsal and ventral surfaces, although the number and size of the spines are reduced in the posterior half of the body. The oral and ventral suckers are close to each other. The two testes, usually in tandem, are posterior to the ovary. The uterus is intercecal and normally preovarian. The vitellarium is follicular, in two lateral fields, usually in the hindbody but may extend into the forebody [3].

30.1.2 Biology and Epidemiology

30.1.2.1 Biology

Echinostome adults are cosmopolitan hermaphroditic digeneans that live in the intestine and bile ducts of numerous vertebrate hosts, particularly aquatic or semi-aquatic birds and mammals, including humans. In the wild, the life cycle of an echinostome using intermediate and definite hosts associated with fresh water is maintained when a definitive host releases eggs into ponds, streams, and lakes. The fertilized

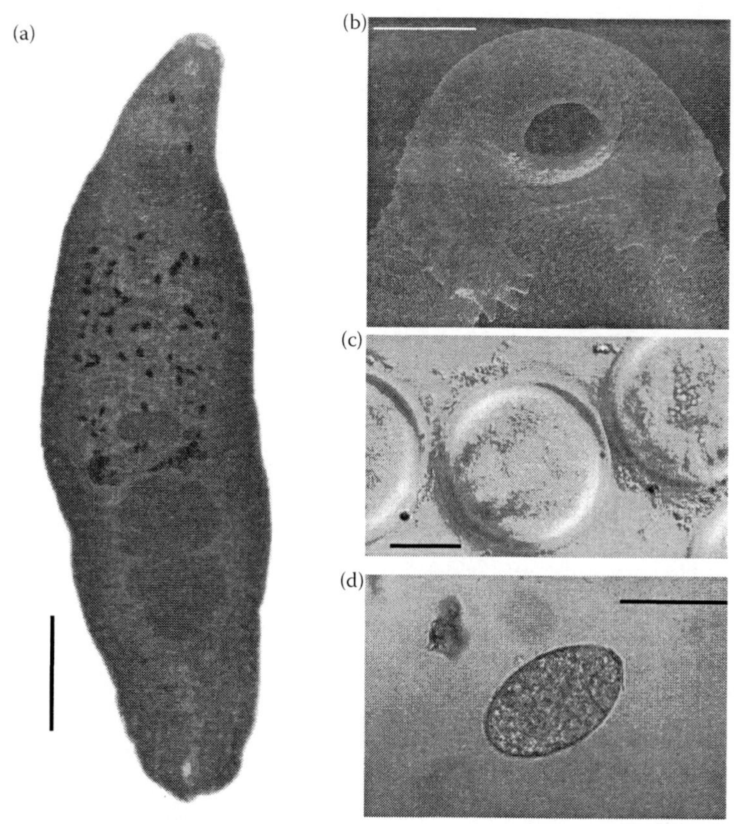

FIGURE 30.1 (a) Specimen of adult Echinostoma caproni collected from hamster experimentally infected; the worm was stained in Grenacher's Borax Carmine (scale bar: 2 mm); (b) Low-magnification scanning electron micrographs of the forebody of an adult worm of *Echinostoma caproni* showing the spined collar (scale bar: 100 μm); (c) Metacercariae of Echinostoma friedi collected from specimens of *Biomphalaria glabrata* experimentally infected hamsters (scale bar: 60 μm); and (d) Egg of Euparyphium albuferensis in feces from experimentally infected hamsters (scale bar: 50 μm).

eggs are undeveloped when laid and take about 2–3 weeks at 22°C to reach the fully developed miracidial stage. Miracidia hatch from eggs actively locate the first intermediate snail host in response to host signals and emitted products [17,18]. Although it was suggested that each species of echinostome infects only one or a few closely related snail species [19,20], recent studies have shown that the spectrum of first intermediate host species for each echinostome species may be broader than previously expected [4,6,21]. Several species of planorbids, lymnaeids, and bulinids have been recorded as first intermediate hosts of echinostomes. Miracidia usually enter the head foot region of the snail, shed their ciliated plates, and transform into sporocysts in the heart. Shortly after larval entry, the primary germinal cells of the sporocyst start to develop into mother rediae [22]. Mother rediae initially reside within the ventricle and aorta of the snail, although when these sites become filled, these rediae begin to colonize the ovotestis of the snail [22]. Mother rediae reproduce asexually and produce daughter rediae which develop in the digestive gland–ovotesis complex.

Cercariae begin to emerge from infected snails 4–6 weeks postinfection. Echinostome cercariae show a low degree of host specificity and several species of snails, frogs, tapdoles, and fishes may serve as second intermediate hosts. The free-living cercariae respond to environmental cues and locate their second intermediate host by chemical signals emitted from the host [23–27]. Cercariae enter the cloacal opening of a tapdole or the excretory pore of a snail and encyst mainly in the tadpole kidney and kidney/pericadial cavity of the snails [28].

Definitive hosts become infected after ingestion of the second intermediate host harboring encysted metacercariae (Figure 30.1c). Echinostomes do not undergo tissue migration in the definitive host. Following infection of the definitive host, the metacercariae excyst in the duodenum and the juvenile parasites migrate to the small intestine where they attach to the mucosa by the ventral sucker [29]. Echinostomes select an optimal microhabitat in the intestine, although the mechanisms and the host cues involved remain unknown. Numerous vertebrates, including humans [30,31], can serve as definitive hosts for echinostomes. However, host preference for some echinostomes has been documented and the course of the infection of an echinostome infection may vary largely in relation to the definitive host species [32–36]. The release of eggs from the definitive host starts at 10–16 days postinfection. Although the release of eggs is continuous from the first day of the patent period, significant variations over the course of the infections in egg output have been documented [37,38].

30.1.2.2 Epidemiology

Echinostomes are commonly found in birds (waterfowl) and mammals, and their distribution is ubiquitous. In general, their specificity toward the definitive host is low and an echinostome species is able to infect several species of vertebrate hosts. The distribution of echinostomes is mainly dependent upon the presence of molluscan first intermediate hosts since these parasites are more restrictive in their specificity toward this host.

Current incidence of human echinostomiasis is difficult to determine with any accuracy because of the unavailability of epidemiological surveys and most of the data rely on historic surveys and occasional case reports. Distribution of human echinostomiasis is strongly determined by dietary habits. Humans become infected when they eat raw or inadequately cooked food, especially fish, snakes, amphibians, clams, and snails containing encysted echinostome metacercariae [39]. Moreover, it has been postulated that humans can also be infected by drinking untreated water containing echinostome cercariae, which could become encysted when exposed to the human gastric juice [40]. Infections are thus most prevalent in areas where traditional cultural practices encourage ingestion of raw or undercooked fish, frogs, snakes, or snails and bivalves or drinking tainted water. In addition, transmission is limited to areas where both intermediate hosts live together in the same habitat and suitable climatic factors are present. Hence, distribution is highly focused, although occasional cases can also occur. Characteristic of several of the food-borne diseases, changing infection patterns have been observed in recent years. Population growth, pollution, poverty, lack of improved sanitation due to declining socioeconomic conditions, and man-made modifications have contributed to increased infection rates in several areas. In contrast, social and economic advances, combined with health-education campaigns and mass drug administration, have reduced prevalences in other settings [41].

Although echinostomiasis occurs worldwide, most human infections are reported from foci in East and Southeast Asia. Echinostomiasis is relatively rare, yet the foci of transmission remain endemic owing to the local dietary preferences. Most of these endemic foci are localized in China, India, Indonesia, Korea, Malaysia, Philippines, Russia, Taiwan, and Thailand [42,43]. Moreover, occasional cases have also been reported in other countries (see Table 30.1).

The number and identity of the echinostome species causing human echinostomiasis are uncertain in relation to the absence of systematic surveys and occasional case reports. Moreover, the problematical taxonomy of the group complicates further the specific diagnosis of the worms found in humans [43]. Haaseb and Eveland [42] listed a total of 21 species infecting humans belonging to eight genera of Echinostomatidae (*Artyfechinostomum*, *Echinochasmus*, *Echinoparyphium*, *Echinostoma*, *Episthmium*, *Himasthla*, *Hypoderaeum*, and *Paryphostomum*). Chai [43] listed 20 species belonging to nine genera (*Acanthoparyphium*, *Artyfechinostomum*, *Echinochasmus*, *Echinoparyphium*, *Echinostoma*, *Episthmium*, *Himasthla*, *Hypoderaeum*, and *Isthmiophora*) that have been found parasitizing humans. Table 30.1 summarizes the species of Echinostomatidae infecting humans and their geographical distribution.

TABLE 30.1
Geographical Distribution of Species of *Echinostomatidae* Infecting Humans

Continent	Species	Country
Africa	*Echinoschasmus liliputanus*	Egypt
	Echinochasmus perfoliatus	Egypt
	Echinoparyphium recurvatum	Egypt
	Echinostoma revolutum	Egypt
America	*Himasthla muehlensi*	Colombia, USA
	Echinostoma echinatum	Brazil
	Isthmiophora melis	USA
Asia	*Acanthoparyphium tyonense*	Japan, Korea
	Artyfechinostomum malayanum	China, India, Indonesia, Malaysia, Philippines, Singapore, Thailand
	Artyfechinostomum mehrai	India
	Artyfechinostomum oraoni	India
	Echinoschasmus angustitestis	China
	Echinoschasmus fujianensis	China
	Echinoschasmus japonicus	China, Korea, Japan
	Echinoschasmus jiufoensis	China
	Echinoschasmus liliputanus	China, Palestina, Syria
	Echinochasmus perfoliatus	China, Japan, Taiwan
	Echinoparyphium recurvatum	Indonesia, Taiwan
	Echinostoma cinetorchis	China, Japan, Korea, Taiwan
	Echinostoma echinatum	Indonesia, Japan, Thailand
	Echinostoma hortense	China, Japan, Korea
	Echinostoma ilocanum	China, India, Indonesia, Malaysia, Philippines, Thailand
	Echinostoma japonicum	Japan, Korea
	Echinostoma macrorchis	Indonesia, Japan, Korea, Taiwan
	Echinostoma revolutum	Thailand
	Episthmium caninum	India, Thailand
	Hypoderaeum conoideum	Thailand
	Isthmiophora melis	China, Taiwan
	Paryphostomum sufrartyfex	Thailand
Europe	*Echinochasmus perfoliatus*	Denmark, Hungary, Italy, Romania, Russia
	Echinostoma echinatum	All Europe
	Echinostoma paraulum	Russia
	Echinostoma revolutum	All Europe
	Hypoderaeum conoideum	All Europe
	Isthmiophora melis	All Europe
Oceania	*Echinostoma revolutum*	Australia

30.1.3 Clinical Features and Pathogenesis

Major clinical symptoms due to echinostome infection may include abdominal pain, diarrhea, easy fatigue, and loss of body weight [31,44,45]. The symptoms in echinostomiasis seem to be more severe than those observed in other intestinal trematode infections [43]. Human morbidity is due to the prolonged latent phase, symptomatic presentations, and similarity of symptoms with other intestinal helminth infections [30,45]. Clinical symptoms depend on the parasite load [31]. Heavy infections are associated with eosinophilia, abdominal pain, watery diarrhea, anemia, edema, and anorexia [46]. Pathological damage includes catharral inflammation, erosion, and even ulceration [47,48].

The clinical signs in echinostomiasis are poorly known. Peripheral blood eosinophilia is commonly observed. In *E. hortense* infections, the level of peripheral blood eosinophilia was dependent on the worm burden [43,49]; 11–24% in patients with more than 100 worms; 4–21% among those with 51–100 worms; and 2–14 among those with less than 50 worms [50]. Several studies on *E. hortense* have shown that patients suffered epigastric and abdominal pain, acid belching, weight loss discomfort, anorexia, headache, nausea, and vomiting [47,51–53]. Patients infected with *E. ilocanum* experienced intestinal colic and diarrhea [43]. It is interesting to note that among 18 American tourists to Kenya who were echinostome-egg positive to fecal examination, 10 had moderately severe abdominal cramps and loose or watery stools [54].

The intestinal histopathology in echinostome infections was studied using laboratory rodents in a few echinostome species, namely *E. revolutum* [55], *E. hortense* [44,49], *E. caproni* [56,57], and *E. trivolvis* [58]. The pathology of echinostomes represents a complex set of reactions. The complexity is dependent on a wide variety of factors including characteristics of the echinostome species, the nature of the host species and the intensity of the infection [45]. Adult echinostomes attach to the intestinal mucosa and may provoke an inflammatory response. Pathological changes are often seen at worm attachment sites. Such areas may show marked dilation, erosion of the villi and lymphocytic infiltration [57,59–62]. Furthermore, increases in the goblet cells, mucosal neutrophils, mononuclear inflammatory cells in the mesentery, and crypt hyperplasia have been observed [57,63]. In the hyperplastic crypts, the mitotic rates were increased over that seen in normal crypts. An increase of collagen fibers and fibroblasts was also observed in the epithelium of the atrophic villi [55,58]. Cellular infiltration of lymphocytes, eosinophils, and plasma cells was observed in lamina propria and submucosa [57,64].

Of particular interest are the endoscopic findings in human *E. hortense* infections. Several adult worms were attached to an ulcerated mucosal layer in the distal part of the stomach [47]. The lesion was accompanied by a stage IIc or stage III early gastric cancer and multiple ulcerations and bleeding in the stomach and duodenum. Ulceration and bleeding appeared to be caused by the worms [47]. Other factors observed by endoscopy are mucosal erosions, ulcerative lesions and signs of chronic gastritis [48,51–53].

Praziquantel is the drug of choice for echinostomiasis. Human infections can be treated with a single oral dose of 10 mg/kg [43]. Niclosamide 150 mg/kg for one or two days is less likely to be effective than praziquantel. Mebendazole and albendazole have also been shown to have an effect against echinostomiasis. Historically, thymol, tetrachlorethylene, bithionol, dichlorophen, and niclofolan have been used but these drugs are variably toxic, relatively ineffective, and now difficult to obtain [42,43].

30.1.4 Diagnosis

30.1.4.1 Conventional Techniques

Human echinostomiasis infections may remain unapparent or become symptomatic depending on the worm burden. The symptoms, even when present, are often unspecific, making clinical diagnosis difficult, particularly in nonendemic areas. Local physicians with the knowledge of endemicity of these infections and greater familiarity with the present signs and symptoms may, however, make clinical diagnosis [42].

Laboratory diagnosis of echinostomiasis is based on demonstration of eggs in feces. The yellow to yellow-brown eggs are thin shelled with an operculum, which may be difficult to see, and a slight thickening of the shell at the abopercular end (Figure 30.1d). They are unembryonated when passed in feces. The size of human-infecting echinostome eggs is in the range 66–145 μm × 43–90 μm [43]. However, determination of echinostome species, and even from other trematodes, on the basis of the egg morphology is difficult, and recovery and identification of adult worms are strongly recommended.

Occasionally, human echinostomiasis has been revealed by gastroduodenal endoscopy performed in relation to severe epigastric symptoms and ulcerative lesions in the stomach and duodenum [47,48,51–53]. In Japan, four cases of *E. hortense* infections were reported based on worm recoveries by upper endoscopy [65].

Immunological methods may also be useful in the diagnosis and monitoring of echinostome infections [45]. Several techniques have been developed based on: (i) antibody detection; (ii) seroantigen detection; and (iii) coproantigen detection. Although the use of these methods has provided interesting results, the information is limited to experimental infections since the potential cross-reactivity with other helminths has not been evaluated.

Using indirect ELISA, anti-*E. caproni*, anti-*E. friedi*, anti-*E. hortense*, and anti-*E. trivolvis* antibodies have been demonstrated [37,66–72]. Positive levels of IgG anti-*E. caproni* and anti-*E. trivolvis* antibodies to glycocalyx membrane antigen were identified in mice by eight and 10 days postinfection, respectively [68,69]. However, a high degree of cross-reactivity between both species was noted. Positive levels of IgG anti-*E. caproni* were also detected in mice using excretory–secretory antigens [71]. Moreover, it has been shown that the earliest response against *E. caproni* is of the IgM class [73].

Toledo et al. [74,75] showed that *E. caproni* antigens were present in sera of infected rodents as assessed by capture ELISA. High levels of circulating antigens were detected from 1 to 2 weeks postinfection. Additionally, Toledo et al. [74,75] designed an antibody-based capture ELISA for the detection of *E. caproni* antigens in the feces of experimentally infected rodents. Coproantigens were positive from one day postinfection and, thereafter, the values were correlated with the worm burden and remained positive until the rejection of the infection.

Metabolic profiling has recently been applied to the diagnosis of *E. caproni* [76]. When stool, urine, or plasma obtained from infected mice was compared with noninfected control animals, an altered metabolic profile was evident from one day postinfection and beyond. Plasma choline, acetate, formate and lactate, and urine creatine decreased when plasma glucose was increased. Furthermore, thin layer chromatography (TLC) showed significant differences in neutral lipid profiles in feces and urine between *E. caproni*-infected and noninfected mice [77]. Recently, Vasta et al. [78] using TLC showed differences in the amino acid profiles of mice infected with *E. caproni* versus controls (uninfected mice).

30.1.4.2 Molecular Techniques

There are still many avenues available for molecular research on echinostomes. Relatively few molecular-based studies have been carried out to date, and the application of molecular methods for the diagnosis of echinostomes is very limited. Most of the molecular studies have focused on species identification and systematic and/or phylogenetic studies. Early molecular studies involved counting chromosomes and comparing isoenzyme patterns [79]. Other early studies used methods such as isoelectric focusing and enzyme electrophoresis to compare species [80,81]. Recently, multilocus enzyme electrophoresis has allowed differentiating between *E. revolutum* and *Hypoderaeum conoideum* [82]. DNA sequencing has been mainly used over the last few years for taxonomic and phylogenetic studies which have allowed for the characterization of a number of echinostome species that were difficult to compare using traditional methods.

DNA sequencing has proved a useful tool for distinguishing among 37 collar-spined species of the genus *Echinostoma*. Morgan and Blair [8] sequenced 1000 nucleotide bases of noncoding rDNA (ITS) from six echinostome species of this group and they differed from each other by 2.2% on average. The use of mitochondrial DNA genes showed a greater sequence divergence [9,10]. Sorensen et al. [12] detected high levels of intraspecific variation among ITS sequences of *E. revolutum* (0.9%) and *E. caproni* (0.6%). Kostadinova et al. [14] analyzed the phylogenetic relationships between several genera of Echinostomatidae.

Random amplification of polymorphic DNA (RAPD-PCR) analysis has also been used to differentiate species of *Echinostoma*. There are PCR primers for the differentiation of *E. caproni* and *E. trivolvis* using genomic DNA from adults [83] and metacercariae [84]. *E. caproni* and *E. paraensei* can also be differentiated by RAPD-PCR analysis [85] and PCR-restriction fragment length polymorphism (RFLP) is also available [12]. However, these methods have been exclusively employed on laboratory samples. In this context, Hertel et al. [86] designed PCR primers against a 192 bp tandem repeated sequence of *E. caproni* (EcSau3A). This allowed for the detection of less than 10 fg of *E. caproni*

DNA, two miracidia of this parasite in 50 mg of snail tissue, and one cercaria in 50 mg of plankton with high specificity. Methods of detecting DNA from echinostomes in human samples (i.e., feces) remain to be developed.

30.2 METHODS

Below are a number of molecular biology protocols that have been successfully used on echinostomes. They are by no means inclusive but provide enough information to extract, amplify, and sequence DNA from echinostomes.

30.2.1 Sample Preparation

30.2.1.1 Genomic DNA Extraction from Parasite Samples

Genomic DNA can be obtained from any of the life-cycle stages of echinostomes. Methods of obtaining these stages from biological samples have been described previously [28].

Samples may be fresh, frozen, or stored in dimethyl sulfoxide (DMSO) or ethanol. Specimens are then homogenized in a 1.5 mL microcentrifuge tube containing 500 μL of extraction buffer, SDS to a final concentration of 1%, and 200 μg of proteinase K. Complete digestion of the mixture varies from 30 min to 2 h depending on the size of the sample [11].

Lysis is followed by a phenol/chloroform extraction to remove protein. An equal amount of phenol:chloroform:isoamyl alcohol is added and the mixture is gently inverted and then centrifuged at 9000 rpm for 5 min. The upper aqueous layer is transferred to a new microcentrifuge tube. This step is repeated until the aqueous layer becomes clear. Then an equal volume of chloroform is added to the aqueous solution, inverted to mix and then centrifuged at 9000 rpm for 5 min. The aqueous solution is transferred to a new microcentrifuge tube. To precipitate the DNA, one-tenth volume of 3 M sodium acetate, pH 5, and 2.5 volumes of ethanol are added to the microcentrifuge tube and then the solution is gently mixed and left at −20°C overnight. The DNA is pelleted by centrifuging for 15 min at 14,000 rpm. After removing the supernatant, the pellet is washed with 500 μL of 70% ethanol, respun for 5 min at 14,000 rpm, decanted, and then the pellet is left to air dry before resuspending in water [11].

30.2.1.2 Genomic DNA of the Parasite from Snail Tissues

This technique is applied to detect echinostomes in the intermediate hosts [86].

DNA is extracted using the whole snail including the shells. Previously crushed snails were incubated for 4 h at 60°C in a buffer containing 2% hexadecyltrimethylammonium bromide (CTAB), 1.4 M NaCl, 0.2% dithiotreitol (DTT), 20 mM EDTA, 100 mM Tris–HCl, pH 8.0, and 10 U/mL proteinase K. After extraction with chloroform:isoamyl alcohol (24:1) the DNA was precipitated with ethanol. All samples were treated with 10 U/mL RNase A at 37°C for 15 min and stored at −20°C in TE (10 mM Tris–HCl, pH 7.5; 1 mM EDTA) [86].

30.2.1.3 Genomic DNA of the Parasite from Plankton

This technique can be used to detect echinostome cercariae in field samples. The DNA of plankton samples can be extracted as described by Hertel et al. [86] The substrate can be crushed in liquid nitrogen and digested for 4 h at 65°C in 1% sodium dodecyl sulfate (SDS), 50 U/mL proteinase K, 100 mM NaCl, 50 mM Tris-HCl, pH 7.5, and 50 mM EDTA, followed by DNA extraction with phenol and ethanol precipitation.

30.2.2 Detection Procedures

Herein, we describe a technique for PCR amplifications that has been used on echinostomes. Until PCR reactions have been optimized, an initial reaction volume of 20 μL is recommended to minimize the potential loss of DNA. To a thin-walled PCR tube, 10–100 ng of DNA template is added to 10 pmol of each primer, 10× Taq buffer, 0.8 mM dNTP, 4 mM $MgCl_2$, and 0.25 U of Taq DNA polymerase. This mix is thermocycled through 30 cycles [11]. The cycling conditions recommended by Morgan and Blair [11] are displayed in Table 30.2, although other alternatives can be used [12,86]. A number of successful PCR primers have been used to amplify and sequence echinostome DNA. The sequences and characteristics of these primers are given in Table 30.3.

TABLE 30.2
PCR Conditions for Specific Amplification of Echinostome DNA

Cycle	Step	Temperature (°C)	Time (s)	Times through Cycle
1	1	96	90	1
	2	50	60	
	3	72	90	
2	1	95	15	4
	2	50	15	
	3	72	90	
3	1	95	5	25
	2	57	5	
	3	72	90	
4	1	72	300	1

Source: Adapted from Morgan, J.A.T., and Blair, D., *Echinostomes as Experimental Models for Biological Research*. Chapter 13. Kluwer Academic Publishers, Dordrecht, 2000.

30.3 CONCLUSIONS AND FUTURE PERSPECTIVES

This chapter introduces the reader to the echinostomes, with an emphasis on the molecular diagnosis of this family of trematodes. Echinostomes are cosmopolitan intestinal parasitic flatworms that may invade humans. Interest in echinostomes in parasitology has come from a variety of topics such

TABLE 30.3
List of PCR, RAPD-PCR, and Sequencing Primers for Echinostomes

Site	Direction	Identity	Sequence (5′-3′)	Reference
18S	Forward	BD1	GTCGTAACAAGGTTTCCGTA	[8]
28S	Reverse	BD2	TATGCTTAAATTCAGCGGGT	[8]
5.8S	Forward	3S	GGTACCGGTGGAATCACGTGGCTAGTG	[8]
5.8S	Reverse	Kim1	GTGTTACCGCGGCG/TGCTG	[8]
ND1	Forward	JB11	AGATTCGTAAGGGGCCTAATA	[9,10]
ND1	Reverse	JB12	ACCACTAACTAATTCACTTTC	[9,10]
CO1	Forward	JB3	TTTTTTGGGCATCCTGAGGTTTAT	[9,10]
CO1	Reverse	JB13	TCATGAAAACACCTTAATACC	[9,10]
18S	Forward	ITS-5	GGAAGTAAAAGTCGTAACAAG	[12]
18S	Reverse	ITS-4	TCCTCCGCTTAGTGATATGC	[12]
ITS1	Forward	ITS-300	TCACTGTTCAAGTGGTTG	[12]
ITS2	Forward	ITS-620	GTCGGCVTTAAACTATCAC	[12]
[a]	[a]	P 1	GCAAGTAGCT	[83]
[a]	[a]	P 2	TCAGCATGCA	[83]
[a]	[a]	P 3	CGGCCCCTGT	[83]
[a]	[a]	P 4	CGGCCATGT	[83]
[a]	[a]	P 5	CGGTCACTGT	[83]
[a]	[a]	SB 45	GTGAAATGTATGCTAATGA	[84]
[a]	[a]	SB 45	GTTGCGTTGCTATGTGGT	[84]
[a]	[a]	SB 46	ACTTTAGCGTTGTGAGACTTAC	[84]
[a]	[a]	SB 46	TAAGGCACACAGGGGTAG	[84]
[a]	[a]	3301	TCGTAGCCAA	[85]
[a]	[a]	3303	TCACGATGCA	[85]
Sau3A	Forward	ECE10	TACACCTCACCTCGTTTTCG	[86]
Sau3A	Reverse	ECE10R	GCATGTGCCACGATGTTTAG	[86]

[a] Used for RAPD-PCR.

as systematic, helminth biology, or their use as experimental models; moreover, echinostomiasis constitutes an important food-borne disease of public health, mainly in Asia and Far East. In spite of this fact, the incidence of human echinostomiasis is poorly known and can be considered a neglected disease. This is probably due to a variety of reasons: (i) unspecificity of the symptoms; (ii) unavailability of systematic surveys; (iii) difficulty entailed in the diagnosis; and (iv) absence of immunological and molecular methods for diagnosis. The diagnosis of echinostomiasis is commonly done by recovery of eggs in feces, although specific diagnosis is difficult because of the morphological similarity of eggs. In this context, the development of PCR-based methods for the diagnosis of human and animal echinostomiasis may be helpful. Although PCR diagnosis is unlikely to become a tool for routine diagnosis in resource-constricted settings, it can be useful to know the current status of echinostomiasis worldwide. In the past, PCR-based methods have been developed to differentiate several echinostome species and to detect larval stages in snail and field samples. However, PCR tools for direct application to human samples (i.e., feces) are not available. Future approaches to human echinostomiasis should consider the development of molecular techniques that may serve to gain further insight into the epidemiology and control of this food-borne disease.

REFERENCES

1. Kostadinova, A., and Gibson, D.I. The systematics of the echinostomes. In: Fried, B., and Gracyzk, T.K. (eds), *Echinostomes as Experimental Models for Biological Research*. Chapter 2. Kluwer Academic Publishers, Dordrecht, 2000.
2. Esteban, J.G., and Muñoz-Antolí, C. Echinostomes: Systematics and life cycles. In: Fried, B., and Toledo, R. (eds), *The Biology of Echinostomes: From the Molecule to the Community*. Chapter 1. Springer, New York, 2009.
3. Kostadinova, A. Family Echinostomatidae Looss, 1899. In: Jones, A., Bray, R.A., and Gibson, D.I. (eds), *Keys to the Trematoda*. Chapter 3. CABI Publishing and The Natural History Museum, London, Vol. 2, 2005.
4. Kostadinova, A. et al. Re-validation of *Echinostoma miyagawai* Ishii, 1932 (Digenea: Echinostomatidae) on the basis of the experimental completion of its cycle. *Syst. Parasitol.* 45, 35–44, 2000.
5. Toledo, R. et al. Cercarial chaetotaxy of *Euparyphium albuferensis* Esteban et al., 1997 (Trematoda: Echinostomatidae), with a review of some genera of the Echinostomatinae. *Syst. Parasitol.* 39, 35, 1998.

6. Toledo, R., Muñoz-Antolí, C., and Esteban, J.G. The life-cycle of *Echinostoma friedi* n. sp. (Trematoda: Echinostomatidae) in Spain and a discussion on the relationships within the "*revolutum*" group based on cercarial chaetotaxy. *Syst. Parasitol.* 45, 199, 2000.
7. Kostadinova, A., Cercarial chaetotaxy of *Echinostoma miyagawai* Ishii, 1932 (Digenea: Echinostomatidae), with a review of the sensory patterns in the "*revolutum*" group. *Syst. Parasitol.* 44, 201, 1999.
8. Morgan, J.A.T., and Blair, D. Nuclear rDNA ITS sequence variation in the trematode genus *Echinostoma*: An aid to establishing relationships within the 37-collar-spine group. *Parasitology* 111, 609, 1995.
9. Morgan, J.A.T., and Blair, D. Mitochondrial ND1 gene sequence used to identify echinostome isolates from Australia and New Zealand. *Int. J. Parasitol.* 28, 493, 1998.
10. Morgan, J.A.T., and Blair, D. Relative merits of nuclear ribosomal internal transcribed spacers and mitochondrial CO1 and ND1 genes for distinguishing among *Echinostoma* species (Trematoda). *Parasitology*. 116, 289, 1998.
11. Morgan, J.A.T., and Blair, D. Molecular biology of echinostomes. In: Fried, B., and Gracyzk, T.K., (eds), *Echinostomes as Experimental Models for Biological Research*. Chapter 13. Kluwer Academic Publishers, Dordrecht, 2000.
12. Sorensen, R.E., Curtis, J., and Minchella, D.J. Intraspecific variation in the rDNA ITS loci of 37-collar-spined echinostomes from North America: Implications for sequence-based diagnoses and phylogenetics. *J. Parasitol.* 84, 992, 1998.
13. Maldonado, A. Jr. et al. Description of the adult worms of a new Brazilian isolate of *Echinostoma paraensei* (Platyhelminthes: Digenea) from its natural vertebrate host *Nectomys squamipes* by light and scanning electron microscopy and molecular analysis. *Parasitol. Res.* 87, 840, 2001.
14. Kostadinova, A. et al. Phylogenetic relationships of *Echinostoma* Rudolphi, 1809 (Digenea: Echinostomatidae) and related genera re-assessed via DNA and morphological analyses. *Syst. Parasitol.* 54, 159, 2003.
15. Grabda-Kazubska, B. et al. A phylogenetic analysis of trematodes of the genus *Echinoparyphium* and related genera based on sequencing of internal transcribed spacer region of rDNA. *Acta Parasitol.* 43, 116, 1998.
16. Huffman, J.E., and Fried, B. *Echinostoma* and echinostomiasis. *Adv. Parasitol.* 29, 215, 1990.
17. Haas, W. Snail-host finding by miracidia and cercariae: Chemical host cues. *Parasitol. Today* 11, 468, 1995.
18. Haberl, B. Host-finding in *Echinostoma caproni*: Miracidia and cercariae use different signals to identify the same snail species. *Parasitology* 120, 470, 2000.
19. Kanev, I. Life-cycle, delimitation and redescription of *Echinostoma revolutum* (Froelich, 1982) (Trematoda: Echinostomatidae). *Syst. Parasitol.* 28, 125, 1994.
20. Kanev, I. Redescription of *Echinostoma trivolvis* (Cort, 1914) (Trematoda: Echinostomatidae) with a discussion on its identity. *Syst. Parasitol.* 32, 61, 1995.
21. Muñoz-Antolí, C. et al. Infectivity of *Echinostoma friedi* miracidia to different snail species under experimental conditions. *J. Helminthol.* 80, 323, 2006.
22. Ataev, G.L. et al. Migration and development of mother sporocysts of *Echinostoma caproni* (Digenea: Echinostomatidae). *J. Parasitol.* 83, 444, 1997.
23. Haas, W. Physiological analyses of host-finding behaviour in trematode cercariae: Adaptations for transmission success. *Parasitology*. 109, S15, 1994.
24. Haas, W. et al. Finding and recognition of the snail intermediate hosts by 3 species of echinostome cercariae. *Parasitology* 110,133, 1995.
25. Körner, M., and Haas, W. Chemo-orientation of echinostome cercariae towards their snail-hosts: Amino acids signal a low host-specificity. *Int. J. Parasitol.* 28, 511, 1998.
26. Körner, M., and Haas, W. Chemo-orientation of echinostome cercariae towards their snail-hosts: The stimulating structure of amino acids and other attractans. *Int. J. Parasitol.* 28, 517, 1998.
27. Muñoz-Antolí, C. et al. Interactions related to non-host snails in the host-finding process of *Euparyphium albuferensis* and *Echinostoma friedi* (Trematoda: Echinostomatidae) miracidia. *Parasitol. Res.* 91, 353, 2003.
28. Toledo, R., Muñoz-Antolí, C., and Fried, B. The use of echinostomes to study host–parasite relationships between larval trematodes and invertebrate and cold-blooded vertebrate hosts. *Parasitol. Res.* 100, 1177, 2007.
29. Fried, B., and Huffman, J.E. The biology of the intestinal trematode *Echinostoma caproni. Adv. Parasitol.* 38, 311, 1996.
30. Graczyk, T.K., and Fried, B. Echinostomiasis: A common but forgotten food-borne disease. *Am. J. Trop. Med. Hyg.* 58, 501, 1998.
31. Fried, B., Graczyk, T.K., and Tamang, L. Food-borne intestinal trematodiases in humans. *Parasitol. Res.* 93, 159, 2004.
32. Hosier, D.W., and Fried, B. Infectivity, growth, and distribution of *Echinostoma revolutum* (Trematoda) in Swiss webster and ICR mouse. *Proc. Helminthol. Soc. Washington* 53, 173, 1986.
33. Odaibo, A.B., Christensen, N.Ø., and Ukoli, F.M.A. Establishment, survival and fecundity in *Echinostoma caproni* (Trematoda) infections in NMRI mice. *Proc. Helminthol. Soc. Washington* 55, 265, 1988.
34. Odaibo, A.B., Christensen, N.Ø., and Ukoli, F.M.A. Further studies on the population regulation in *Echinostoma caproni* infections in NMRI mice. *Proc. Helminthol. Soc. Washington* 56, 192, 1989.
35. Christensen, N.Ø., Simonsen, P.E., Odaibo, A.B., and Mahler, H. Establishment, survival and fecundity of *Echinostoma caproni* (Trematoda) infections in hamsters and birds. *J. Helminthol. Soc. Washington* 57, 104, 1990.
36. Toledo, R. et al. The comparative development of *Echinostoma caproni* (Trematoda: Echinostomatidae) adults in experimentally infected hamsters and rats. *Parasitol. Res.* 93, 439, 2004.
37. Toledo, R. et al. An experimental study of the reproductive success of *Echinostoma friedi* (Trematoda: Echinostomatidae) in the golden hamster. *Parasitology* 126, 433, 2003.
38. Toledo, R. et al. A quantitative approach to the experimental transmission success of *Echinostoma friedi* (Trematoda: Echinostomatidae) in rats. *J. Parasitol.* 92, 16, 2006.
39. Carney, W.P. Echinostomiasis —A snail-borne intestinal disease. *Southeast Asian J. Trop. Med. Public Health* 22, 206, 1991.
40. Xiao, X. et al. *In vivo* and *in vitro* encystment of *Echinochasmus liliputanus* cercariae and biological activity of the metacercariae. *J. Parasitol.* 91, 492, 2005.
41. Keiser, J., and Utzinger, J. Chemotherapy for major food-borne trematodes: A review. *Expert Opin. Pharmacother.* 5, 1711, 2004.
42. Haaseb, M.A., and Eveland, L.K., Human echinostomiasis: Mechanisms of pathogenesis and host resistance. In: Fried, B., and Gracyzk, T.K. (eds), *Echinostomes as Experimental Models for Biological Research*. Chapter 4. Kluwer Academic Publishers, Dordrecht, 2000.

43. Chai, J.Y. Echinostomes in humans. In: Fried, B., and Toledo, R. (eds), *The Biology of Echinostomes: From the Molecule to the Community*. Chapter 7. Springer, New York, 2009.
44. Chai, J.Y., and Lee, S.H. Food-borne intestinal trematode infections in the Republic of Korea. *Parasitol. Int.* 51, 129, 2002.
45. Toledo, R., Esteban, J.G., and Fried, B. Immunology and pathology of intestinal trematodes in their definitive hosts. *Adv. Parasitol.* 63, 285, 2006.
46. Chattopadyay, U.K. et al. A case of echinostomiasis in tribal community in Bengal. *Ann. Trop. Med. Parasitol.* 84, 193, 1990.
47. Chai, J.Y. et al. A case of echinostomiasis with ulcerative lesions in the duodenum. *Korean J. Parasitol.* 32, 201, 1994.
48. Cho, C.M. et al. A human case of *Echinostoma hortense* (Trematoda: Echinostomatidae) infection diagnosed by gastroduodenal endoscopy in Korea. *Korean J. Parasitol.* 41, 117, 2003.
49. Lee, S.H. et al. Chronological observation of intestinal lesions of rats experimentally infected with *Echinostoma hortense. Korean J. Parasitol.* 28, 45, 1990. (In Korean).
50. Lee, S.K. et al. An epidemiological survey of *Echinostoma hortense* infection in Chongsong-gun, Kyongbuk Province. *Korean J. Parasitol.* 26, 199, 1988.
51. Lee, S.J., and Hong, S.J. Gastric echinostomiasis diagnosed by endoscopy. *Gastrointest. Endosc.* 55, 440, 2002.
52. Chang, Y.D. et al. A human infection of *Echinostoma hortense* in duodenal bulb diagnosed by endoscopy. *Korean J. Parasitol.* 43, 57, 2005.
53. Park, C.J., and Kim, J. A human case of *Echinostoma hortense* infection diagnosed by endoscopy in area of southwestern Korea. *Korean J. Med.* 71, 229, 2006. (In Korean).
54. Poland, G.A., Navin, T.R., and Sarosi, G.A. Outbreak of parasitic gastroenteritis among travelers returning from Africa. *Arch. Intern. Med.* 145, 2220, 1985.
55. Bindseil, E., and Christensen, N.Ø. Thymus-independent crypt hyperplasia and villous atrophy in the small intestine of mice infected with the trematode *Echinostoma revolutum. Parasitology* 88, 431, 1984.
56. Kim, S., and Fried, B. Pathological effects of *Echinostoma caproni* (Trematoda) in the domestic chick. *J. Helminthol.* 63, 227, 1989.
57. Toledo, R. et al. *Echinostoma caproni*: Intestinal pathology in the golden hamster, a highly compatible host, and the Wistar rat, a less compatible host. *Exp. Parasitol.* 112, 164, 2006.
58. Fujino, T., and Fried, B. Expulsion of *Echinostoma trivolvis* (Cort, 1914) Kanev, 1985 and retention of *E. caproni* Richard, 1964 (Trematoda, Echinostomatidae) in C3H mice: Pathological, ultrastructural, and cytochemical effects on the host intestine. *Parasitol. Res.* 79, 286, 1993.
59. Huffman, J.E., Iglesias, D., and Fried, B. *Echinostoma revolutum*: Pathology of extraintestinal infection in the golden hamster. *Int. J. Parasitol.* 18, 873, 1988.
60. Mabus, J., Huffman, J.E., and Fried, B. Humoral and cellular response to infection with *Echinostoma revolutum* in the golden hamster, *Mesocricetus auratus. J. Helminthol.*. 62, 127, 1988.
61. Isaacson, A.C., Huffman, J.E., and Fried, B. Infectivity, growth, development and pathology of *Echinostoma caproni* (Trematoda) in the golden hamster. *Int. J. Parasitol.* 19, 943, 1989.
62. Fried, B., Huffman, J.E., and Weiss, P.M. Single and multiple worm infections of *Echinostoma caproni* (Trematoda) in the golden hamster. *J. Helminthol.* 64, 75, 1990.
63. Fujino, T., and Fried, B. The expulsion of *Echinostoma trivolvis* from C3H mice: Differences in glycoconjugates in mouse versus hamster small intestinal mucosa during infection. *J. Helminthol.* 70, 115, 1996.
64. Weinstein, M.S., and Fried, B. The expulsion of *Echinostoma trivolvis* and retention of *Echinostoma caproni* in the ICR mouse: Pathological effects. *Int. J. Parasitol.* 21, 255, 1991.
65. Hamamoto, T. et al. A case of *Echinostoma hortense* infection diagnosed by the upper gastrointestinal endoscopy. *Japan J. Gastroenterol. Soc.* 94, 487, 1997. (in Japanese).
66. Simonsen, P.E., Estambale, B.B., and Agger, M. Antibodies in the serum of golden hamsters experimentally infected with the intestinal trematode *Echinostoma caproni. J. Helminthol.* 65, 239, 1991.
67. Agger, M.K., Simonsen, P.E., and Vennervald, B.J. The antibody response in serum, intestinal wall and intestinal lumen of NMRI mice infected with *Echinostoma caproni. J. Helminthol.* 67, 169, 1993.
68. Graczyk, T.K., and Fried, B. ELISA method for detecting anti-*Echinostoma caproni* (Trematoda: Echinostomatidae) immunoglobulins in experimentally infected ICR mice. *J. Parasitol.* 80, 544, 1994.
69. Graczyk, T.K., and Fried, B. An enzyme-linked immunosorbent assay for detecting anti-*Echinostoma trivolvis* (Trematoda) IgG in experimentally infected ICR mice. Cross-reactivity with *E. caproni. Parasitol. Res.* 81, 710, 1995.
70. Toledo, R. et al. Immunological characterization of somatic and excretory/secretory antigens of *Echinostoma caproni* (Trematoda: Echinostomatidae) in experimentally infected rats. *Comp. Parasitol.* 71, 42, 2004.
71. Toledo, R. et al. Kinetics of antibodies and antigens in serum of mice experimentally infected with *Echinostoma caproni* (Trematoda: Echinostomatidae). *J. Parasitol.* 91, 978, 2005.
72 Cho, Y.K. et al. Differential immune profiles following experimental *Echinostoma hortense infection* in BALB/c and C3H/HeN mice. *Parasitol. Res.* 100, 1053, 2007.
73. Sotillo, J. et al. *Echinostoma caproni*: Kinetics of IgM, IgA and IgG subclasses in the serum and intestine of experimentally infected rats and mice. *Exp. Parasitol.* 116, 390, 2007.
74. Toledo, R. et al., Development of an antibody-based capture enzyme-linked immunosorbent assay for detecting *Echinostoma caproni* (Trematoda) in experimentally infected rats: Kinetics of coproantigen excretion. *J. Parasitol.* 89, 1227, 2003.
75. Toledo, R. et al. Kinetics of *Echinostoma caproni* (Trematoda: Echinostomatidae) antigens in feces and serum of experimentally infected hamsters and rats. *J. Parasitol.* 90, 752, 2004.
76. Saric, J. et al. Metabolic profiling of an *Echinostoma caproni* infection in the mouse for biomarker discovery. *PLoS Negl. Trop. Dis.* 2, e254, 2008.
77. Vasta, J.D., Fried, B., and Sherma, J. High performance thin layer chromatographic analysis of neutral lipids in the urine of BALB/c mice infected with *Echinostoma caproni. Parasitol. Res.* 102, 625, 2008.
78. Vasta, J.D., Fried, B., and Sherma, J. Determination and quantification of amino acids in the urine of BALB/c mice infected with *Echinostoma caproni* by high-performance thin-layer chromatography-densitometry. *J. Liq. Chrom. Relat. Tech.* 32, 1210, 2009.
79. Voltz, A. et al. Isoenzyme analysis of *Echinostoma liei*: Comparison and hybridization with other African species. *Exp. Parasitol.* 66, 13, 1988.
80. Vasilev, I. et al. Comparative electrophoretic studies of certain species of the genus *Echinostoma* with 37-collar spines. *Khelminthologia* 6, 31, 1978.

81. Ross, G.C., Fried, B., and Southgate, V.R. *Echinostoma revolutum* and *E. liei*: Observations on enzymes and pigments. *J. Nat. Hist.* 23, 981, 1989.
82. Saijuntha, W., Sithithaworn, P., and Andrews, R.H. Genetic differentiation of *Echinostoma revolutum* and *Hypodereaum conoideum* from domestic ducks in Thailand by multilocus enzyme electrophoresis. *J. Helminthol.* 28, 1, 2009.
83. Fujino, T., Takahashi, Y., and Fried, B. A comparison of *Echinostoma trivolvis* and *E. caproni* using random amplified polymorphic DNA analysis. *J. Helminthol.* 69, 263, 1995.
84. Fujino, T. et al. Specific primers for the detection of genomic DNA of *Echinostoma trivolvis* and *E. caproni* (Trematoda: Echinostomatidae). *Mol. Cell. Probes* 11, 77, 1997.
85. Petrie, J.L., Burg, E.F., 3rd, and Cain, G.D. Molecular characterization of *Echinostoma caproni* and *E. paraensei* by random amplification of polymorphic DNA (RAPD) analysis. *J. Parasitol.* 82, 360, 1996.
86. Hertel, L.A., Stricker, S.A., and Loker, E.S. Calcium dynamics of hemocytes of the gastropod *Biomphalaria glabrata*: Effects of digenetic trematodes and selected bioactive compounds. *Invert. Biol.* 119, 27, 2000.

31 Fasciola

Dongyou Liu and Xing-Quan Zhu

CONTENTS

31.1 INTRODUCTION

Flukes (trematodes) are a large group of flattened, rhomboidal-shaped worms that parasitize various mollusks and vertebrates. Taxonomically, flukes are classified in the class Trematoda covering 18,000–24,000 species, which is further divided into two subclasses: the Aspidogastrea (comprising about 100 species of obligate parasites of mollusks as well as turtles and fish) and the Digenea (comprising the remaining species of obligate parasites of both mollusks and vertebrates, but rarely cartilaginous fish). The main biological difference between the subclass Aspidogastrea and the subclass Digenea is that Aspidogastrea larvae do not multiply, while Digenea larvae do.

Depending on their preferences for vertebrate host tissues, Digeneans are conveniently separated into two groups: tissue flukes and blood flukes. Tissue flukes infect the bile ducts, lungs, or other biological tissues, and are exemplified by the lung fluke *Paragonimus westermani* and the liver flukes *Clonorchis sinensis* and *Fasciola hepatica*. Blood flukes spend some time in the blood during their life cycle, and are represented by species of the genus *Schistosoma*.

Fasciola is a genus of trematodes that comprises species of economical importance. Besides causing liver condemnation, higher mortality rates, and reduced production of meat, milk, and wool in sheep and cattle industry, *Fasciola* spp. are responsible for infecting millions of people worldwide. The ability to correctly and speedily identify *Fasciola* spp. in both vertebrate (definitive) and snail (intermediate) hosts is therefore vital for the control and prevention of *Fasciola* infection in humans and animals.

31.1.1 Classification, Morphology, Life Cycle, and Epidemiology

31.1.1.1 Classification

The genus *Fasciola* is classified in the family Fasciolidae, superfamily Echinostomatoidea, suborder Echinostomata, order Echinostomida, subclass Digenea, class Trematoda, phylum Platyhelminthes. Apart from *Fasciola*, the family Fasciolidae also consists of the genera of *Fascioloides*, *Fasciolopsis*, *Parafasciolopsis*, and *Protofasciola*. Currently, the genus *Fasciola* includes three recognized species (*Fasciola hepatica*, *Fasciola gigantic*, and *Fasciola jacksoni*) and >40 unassigned species.

First described as sheep liver fluke in 1379, *F. hepatica* (also known as the common liver fluke) is a parasitic flatworm that occurs in the liver of domestic livestock and other mammals, including humans, leading to a disease entity known as fascioliasis (fasciolosis, distomatosis or liver rot). *F. hepatica* has a worldwide distribution and is responsible for causing liver condemnation, higher mortality rates, and reduced production of meat, milk, and wool in sheep and cattle in temperate zones.

F. gigantica is a parasitic flatworm that resides in the liver parenchyma and bile ducts of sheep, cattle, and most other domesticated species and humans. This species is prevalent in tropical regions, such as southern Asia, Southeast Asia, and Africa where sheep and cattle are intensively reared. *F. gigantica* overlaps with *F. hepatica* in many African and Asian countries, although *F. gigantica* demonstrates distinct ecological, pathological, and epidemiological features (such as requirement of different snail species as intermediate host) in comparison with *F. hepatica* [1–4].

In addition, recent examination of the first and/or second internal transcribed spacers (ITS-1 and ITS-2) of ribosomal RNA (rRNA) as genetic markers indicated the possible existence of an "intermediate form" of *Fasciola* between *F. hepatica* and *F. gigantica* from Japan, Korea, China, Vietnam, and Iran [5–7]. Further comparison of mitochondrial gene sequences showed that Japanese *Fasciola* sp. is more closely related to *F. gigantica* than to *F. hepatica* [8–11].

F. jacksoni (obsolete synonyms: *Cladocoelium elephantis*, *Distomum elephantis*, *Distomum jacksoni*, *Distomum jacksonii*, and *Fasciolopsis jacksoni*) inhabits the bile ducts and duodenum of elephants. This species has been identified in elephant populations in India and Burma as well as captive elephants in Malaysia, and is not known to infect humans.

31.1.1.2 Morphology

F. hepatica adult worm measures 2–3 cm in length and 1.3 cm in width, and appears leaf-shaped and gray-brown, with the anterior end being broader than the posterior end. The anterior end is conical and marked off by distinct shoulders from the body, with an oral sucker at the end of the cone and a ventral sucker at the base of the cone (for attachment to the lining of the biliary ducts). The tegument is covered with backwardly projecting spines. Being hermaphroditic, each adult worm possesses both ovaries and testes which are highly branched and allow for individual flukes to produce eggs independently. *F. hepatica* eggs are oval, operculate, yellow, and large (around 150 μm × 90 μm). *F. hepatica* juvenile (at the time of entry into the liver) is 1.0–2.0 mm in length and lancet like.

F. gigantica adult worm measures 4–7.5 cm in length and is larger than *F. hepatica* worm. With a leaf-like appearance, the conical anterior end of *F. gigantica* is very short, so that the shoulder characteristic is barely perceptible in *F. gigantica* (in contrast to *F. hepatica*). *F. gigantica* eggs measure 190 μm × 100 μm, and are slightly larger than those of *F. hepatica*.

F. jacksoni adult worm is 1.2–1.4 cm long, 0.9–1.25 mm wide with a central thickness of 0.15–0.2 cm.

31.1.1.3 Life Cycle

Fasciola spp. undergo seven stages of development in their life cycle, including egg, miracidium, sporocyst, redia, cercaria, metacercaria, and adult, and require mammals as definitive host and snail as intermediate host.

F. hepatica adult worms live in the bile ducts of the mammalian liver, and begin to lay eggs 2–4 months after initial infection. The eggs produced by adult flukes pass down the bile ducts through the gastrointestinal tract and are excreted in the host's feces. In water at temperatures above 10°C, the eggs become embryonated and release ciliated miracidiae (singular, miracidium) within 2 weeks. Within 24 h of hatching, the miracidium burrows into a suitable *Lymnaea* snail host, in which the miracidium loses its cilia and develops into a sac of germinating cells called a sporocyst. Each of these germinating cells within sporocyst develops into a redia (plural, rediae). The rediae burst through the sporocyst, migrate in the hepato-pancreas of the snail, and develop into cercariae (singular, cercaria). The whole process turns a single miracidium into 600–4000 cercariae. Released out of the snail, the cercariae swim through the water using their new tails, and attach to aquatic vegetation (e.g., watercress) in the water and encyst, forming metacercariae (singular, metacercaria), which are infective to mammalian hosts. When the vegetation on which the metacercariae are encysted is consumed by a mammalian host, the metacercariae excyst in the small intestine, pierce through the intestinal wall, migrate across the peritoneal cavity, and enter the liver parenchyma in 6–8 weeks, inducing the symptoms of acute infection. The young flukes enter and settle in the bile ducts, mature into adults in a month, and begin to produce eggs (up to 25,000 eggs per day per fluke). In humans, maturation from metacercariae into adult flukes takes ~3–4 months.

The life cycle of *F. gigantica* is similar to that of *F. hepatica*, the main differences being in the time scale of the cycle and in the habitat requirement. The prepatent period is 10–14 weeks for *F. gigantic* and about 8–12 weeks for *F. hepatica*. In addition, the intermediate hosts for *F. gigantica* are snails that live in deeper water and are close to being true aquatic snails in their behavior, whereas those for *F. hepatica* are amphibious snails that live close to the edge of slow-moving or stagnant water.

31.1.1.4 Epidemiology

F. hepatica is found on every continent with highest prevalence in temperate areas where there is extensive sheep and cattle raising. Fascioliasis is a major parasitic disease of livestock with over 700 million production animals being at risk of infection and worldwide economic losses estimated at >US$3.2 billion per year.

As accidental hosts for *Fasciola* spp., human fascioliasis is common in areas where dietary practices include the consumption of raw aquatic vegetables (e.g., watercress, alfalfa, and water lettuce) or raw liver of an infected animal. Approximately 2.4 million people in more than 60 countries are estimated to be infected and the number of people at risk is more than 180 million globally. In places such as Portugal, the Nile delta, northern Iran, parts of China, and the Andean highlands of Ecuador, Bolivia, and Peru, fascioliasis is a serious public health concern [12,13]. In fact, fascioliasis has been considered as an important plant- and water-borne parasitic zoonosis [14–18].

The intermediate hosts of *Fasciola* spp. are freshwater snails of family Lymnaeidae. Members of the family

Lymnaeidae with the capacity for serving as naturally or experimentally intermediate hosts for *F. hepatica* include *Austropeplea tomentosa*, *Austropeplea ollula*, *Austropeplea viridis*, *Fossaria bulimoides*, *Galba truncatula*, *Lymnaea cousini*, *Lymnaea cubensis*, *Lymnaea diaphana*, *Lymnaea humilis*, *Lymnaea neotropica*, *Lymnaea occulta*, *Lymnaea stagnalis*, *Lymnaea viatrix*, *Omphiscola glabra*, *Pseudosuccinea columella*, *Radix auricularia*, *Radix lagotis*, *Radix natalensis*, *Radix peregra*, *Radix rubiginosa*, *Stagnicola caperata*, *Stagnicola fuscus*, *Stagnicola palustris*, and *Stagnicola turricula* [19–21].

The most important intermediate host for *F. gigantica* is *R. auricularia*. While *Lymnaea rufescens* and *Lymnaea acuminata* are important host snails in the Indian subcontinent, *R. rubiginosa* and *R. natalensis* are common snail hosts in Malaysia and in Africa, respectively. In addition, the Australian *Lymnaea tomentosa* (host of *F. hepatica*) was shown to be receptive to miracidia of *F. gigantica* from East Africa, Malaysia, and Indonesia. Human infections with *F. gigantica* have been reported less frequently than those with *F. hepatica*, largely in Asia, Africa, and Hawaii [22,23].

31.1.2 Clinical Features, Pathogenesis, and Treatment

31.1.2.1 Clinical Features

Clinical manifestations of human *F. hepatica* infections may appear from a few days to several years after infection [24,25]. However, in about 50% of cases, human fascioliasis is asymptomatic.

In individuals with clinical symptoms, fascioliasis shows the following patterns:

1. Acute fascioliasis. The acute phase of fascioliasis occurs after ingestion of a large number of metacercariae (1000–5000) at once, and symptoms appear 4–7 days after ingestion, ranging from fever (40–42°C), tender hepatomegaly, splenomegaly, abdominal pain, vomiting, loss of appetite, flatulence, diarrhea, urticaria (hives), jaundice, ascites to anemia. These symptoms are likely caused by the migration of the *F. hepatica* larvae throughout the liver parenchyma, and may persist for 6–8 weeks until the larvae mature and settle in the bile ducts [26].
2. Chronic fascioliasis. The chronic infection of fascioliasis results from ingestion of a smaller number of metacercariae (200–1000) and may also evolve from acute fascioliasis. Chronic fascioliasis is more common in humans and may present symptoms ranging from biliary cholic, right upper-quadrant abdominal tenderness, hepatomegaly, fatty food intolerance, nausea, pruritus, jaundice, cardiac disorder, pancreatitis, cholangitis, and cholecystitis, as well as severe anemia in children, with portal cirrhosis and death as potential outcomes. The bile ducts may appear while following development of fibrosis, a condition known as pipestem liver. The clinical symptoms of chronic fascioliasis are attributable to the biliary obstruction and inflammation caused by the presence of the large adult worms and their metabolic waste in the bile ducts.
3. Halzoun. This occurs after an individual consumes *F. hepatica*-infected raw liver. The young adult worms attach to the pharyngeal mucosa and cause pain, edema, and bleeding as well as respiration difficulty. The adult worms live in the biliary ducts, and cause symptoms for up to 10 years.
4. Ectopic fascioliasis. Instead of settling in the liver and biliary ducts, *F. hepatica* accidentally moves in the peritoneal cavity, intestinal wall, lungs, subcutaneous tissue, lymph nodes, eye, and other locations, causing ectopic fascioliasis [27,28].

In blood, anemia, hypoalbuminemia, and eosinophilia are observable in all types of fasciolosis. Liver enzyme activities such as glutamate dehydrogenase, gamma-glutamyl transferase (GGT), and lactate dehydrogenase (LDH) are elevated in chronic fasciolosis from 12 to 15 weeks after ingestion of metacercariae.

Sezgin et al. [29] described two cases of human fascioliasis due to *F. hepatica* in Turkey. The first case concerned a 51-year-old female with a sudden onset of nausea and upper abdominal pain. Physical examination revealed severe tenderness in the epigastrium with hypoactive bowel sound and fever (38°C). Laboratory testing showed elevated serum levels of amylase, lipase, alanine aminotransferase (ALT), aspartate aminotransferase (AST), and LDH. Abdominal ultrasonography showed bile duct dilation and hyperechoic materials filling the common bile duct (CBD) and diffuse enlargement of the pancreas. Endoscopic sphincterotomy recovered multiple *F. hepatica* flukes. The patient was treated successfully with a single oral dose of triclabendazole (10 mg/kg). The second case related to a 70-year-old male with nausea, upper abdominal pain, and jaundice for 3 days and a history of watercress consumption. Physical examination revealed severe tenderness in the epigastrium with jaundice. Laboratory tests showed elevated serum levels of amylase, lipase, ALT, and AST. Abdominal ultrasonography revealed bile duct dilation and gallbladder wall thickening (5.2 mm). Cholangiogram depicted dilatation and numerous motile curvilinear filling defect images in the CBD and irregularities at the bile duct wall. Endoscopic sphincterotomy resulted in the extraction of multiple *F. hepatica* flukes. Treatment with triclabendazole as a single oral dose of 10 mg/kg led to complete recovery.

Aslan et al. [30] reported a case of *F. hepatica* infection in a 58-year-old female patient with fatigue, malaise, loss of appetite, and abdominal pain. Laboratory tests showed high blood levels of AST, ALT, GGT, alkaline phosphatase, total bilirubin, and eosinophil (25.1%). Endoscopic retrograde cholangiopancreatography revealed dilatation in the intrahepatic bile ducts and CBD. After a standard sphincterotomy, two living *F. hepatica* were extracted. The extraction of *F. hepatica* flukes and antihelmintic therapy led to full recovery.

Goral et al. [22] documented a case of *F. gigantica*-induced biliary obstruction and cholestasis in a 37-year-old Turkish woman presenting with icterus, ascites, pain in right upper abdominal region, and weight loss. Biochemical tests revealed an increase in leukocytes, AST, ALT, bilirubin levels, and eosinophilia (26.4%). CT scans uncovered lobulated contours in multiple views and some cystic lesions. After contrast agent injection, mild peripheral enhancement of the lesions were seen in the majority (abscess-phlegmon was considered). She had yellow-color ascites and splenomegaly. Serological assay by a standard enzyme-linked immunosorbent assay (ELISA) protocol yielded a titer of 1:20,480 (reference: <1160). Seven live fluke were removed. On the basis of its morphology and shape, the fluke was diagnosed as *F. gigantica*. After a single dose of triclabendazole (3 tablets, oral), clinical symptoms disappeared, and the patient remained completely healthy.

The clinical signs of *F. jacksoni* infection in elephants include anorexia, weight loss, pale mucous membranes, colic, diarrhea or constipation, depraved appetite (eating of earth), malodorous feces, dehydration, jaundice, frequent yawning, thirst, depression, debility, submandibular edema, ventral abdominal edema, hypoproteinemia, anemia, and death. Some adult elephants may show no obvious clinical signs.

31.1.2.2 Pathogenesis

The development of fascioliasis in definitive hosts can be separated into two phases: the parenchymal (migratory) phase and the biliary phase.

The parenchymal phase begins with the excysted juvenile flukes piercing the intestinal wall, and then the flukes migrate within the abdominal cavity and move to the liver or other organs. Although *F. hepatica* has a predilection for the tissues of the liver, the parasite may accidentally enter other locations of flukes such as the lungs, diaphragm, intestinal wall, kidneys, and subcutaneous tissue (the so-called ectopic fascioliasis). The migration of flukes causes mechanical damages to host tissues and elicit inflammation around migratory tracks of flukes [31].

The biliary phase begins after the parasites enter the biliary ducts of the liver, where flukes mature, feed on blood, and produce eggs. Due to the tissue damage caused by the parasite, hypertrophy of biliary ducts occurs, resulting in obstruction of the lumen.

31.1.2.3 Treatment

Triclabendazole (Egaten) at 1–2 oral dose of 10 mg/kg body weight in 24 h represents the drug of choice for treating human fascioliasis [32]. Additionally, nitazoxanide and Bithionol (30 mg/kg body weight) may be used for human fascioliasis treatment in Mexico. Mirazid, an Egyptian drug made from myrrh, has been investigated as an oral treatment of trematode-induced ailments including fascioliasis. Along with pharmaceutical therapy, surgery may be necessary in extreme cases to clear the biliary tract [33].

For treatment of animal fascioliasis, the following five categories of drugs can be used: (i) halogenated phenols such as bithionol (Bitin), hexachlorophene (Bilevon), nitroxynil (Trodax); (ii) salicylanilides such as closantel (Flukiver, Supaverm), rafoxanide (Flukanide, Ranizole); (iii) benzimidazoles such as triclabendazole (Fasinex), albendazole (Vermitan, Valbazen), mebendazol (Telmin), luxabendazole (Fluxacur); (iv) sulfonamides such as clorsulon (Ivomec Plus); and (v) phenoxyalkanes such as diamphenetide (Coriban).

Triclabendazole (Fasinex) is an antihelmintic agent with a high efficacy against adult as well as juvenile flukes. However, long-term use in animals has led to the development of *F. hepatica* resistance to triclabendazole.

Prevention of human fascioliasis should focus on issues concerning the presence of a substantial reservoir, the presence of the intermediate host (*Lymnaea* snails), the opportunity for water source contamination by human and nonhuman hosts and dietary practices that include the consumption of raw, untreated aquatic vegetation. Therefore, it is important to keep domestic animal herds separate from the growing sites of aquatic vegetation, which will help decrease both human infection and the animal reservoir. Washing vegetables in either 6% vinegar or potassium permanganate for 5–10 min will destroy the encysted metacercariae in case raw vegetables are consumed. Application of molluscicides is also helpful in decreasing the population of *Lymnaea* snails [34].

31.1.3 Laboratory Diagnosis

Human fascioliasis tends to induce nonspecific symptoms; it needs to be differentiated from other diseases. For example, cholangitis, cholecystitis, and cholelithiasis due to fascioliasis may be caused also by other pathogens. For this reason, laboratory techniques have been utilized for the identification of *Fasciola* parasites and diagnosis of fascioliasis [33].

In human fascioliasis, radiology, ultrasound, computer tomography, and magnetic resonance may be employed to visualize the adult flukes in the bile ducts, and to reveal the burrow tracts made by the worms and dilation of the bile ducts. Biochemical and hematological examinations of human sera offer tangible evidence (eosinophilia, elevation of liver enzymes) of fascioliasis. Direct observation of *F. hepatica* eggs in a stool sample, duodenal aspirate or biliary aspirate provides a definitive diagnosis [35]. However, due to the fact that the flukes do not begin to produce eggs until 4 months after infection, a negative result does not rule out early stages of fascioliasis. Similarly, in the case of ectopic infections, eggs are not present in the stool. On the other hand, in case of a recent ingestion of infected livers containing eggs, a false positive result (pseudofasciolosis) may be obtained. Furthermore, the close morphological similarity amongst *F. hepatica*, *F. gigantica*, and *Fasciolopsis buski* eggs makes their differentiation impossible on the basis of microscopic examination.

Application of serological tests such as ELISA and Western blot helps overcome the difficulty associated with the absence of *Fasciola* eggs in the clinical specimens [36–43]. The serological techniques have the advantage of being applicable during all stages of the disease, especially during the invasive or acute phase, and in situations where coprological techniques are inappropriate. Serological techniques have also proven useful in monitoring post-treatment evolution. With a sensitivity of 95%, the FAST-ELISA can detect specific *F. hepatica* antibodies as early as 2 weeks after infection, and verify effective cure as antibody levels return to normal in 6–12 months after the cessation of infection. Nevertheless, the usefulness of serological techniques is limited by their low sensitivity and specificity.

Molecular methods offer a rapid, sensitive and precise approach for the identification and phylogenetic analyses of *Fasciola* spp. and diagnosis of human and animal fascioliasis [44–60]. A number of gene regions have been targeted for identification and phylogenetic investigations of *Fasciola* spp. These include the ITS-1 and ITS-2 of nuclear rRNA and mitochondrial DNA (mtDNA) [61–71]. Because of the inability of conventional techniques to differentiate between *F. hepatica* and *F. gigantica* in humans, application of molecular procedure is essential for their diagnosis in areas with overlapping *Fasciola* infections [72–74].

Toward this end, a simple and rapid polymerase chain reaction (PCR)-restriction fragment length polymorphism (PCR-RFLP) assay, using the common restriction enzymes *Ava*II and *Dra*II, has been described for discrimination between *F. hepatica* and *F. gigantica*. Targeting a 618-bp-long sequence of the 28S rRNA, this assay provides unambiguous results and has the potential to be used for both diagnosis of individual cases and epidemiological surveys of humans and animals in endemic regions of sympatry in Africa and Asia [75]. A similar PCR-RFLP assay, targeting the ITS-2 of rRNA and using restriction endonucleases *Hsp92* II and *Rca* I, has permitted differentiation between *F. hepatica*, *F. gigantica* and the "intermediate" *Fasciola* in China [63]. In addition, a PCR-linked single-strand conformation polymorphism (PCR-SSCP) approach, targeting the ITS-1 of rRNA as a genetic marker, also allowed the accurate identification and differentiation between *F. hepatica*, *F. gigantica*, and the "intermediate form" of *Fasciola* in China [76].

Apart from the detection of *Fasciola* species in clinical specimens, molecular procedures have been utilized for improved identification of the flukes in intermediate molluscan hosts [77,78]. The availability of highly sensitive and specific molecular methods for detecting infected snails is critical in the epizootiological studies on *F. gigantica* and *F. hepatica*, enabling the implementation of highly effective strategies for the control of fascioliasis [79]. Indeed, several PCR assays have been developed to detect *Fasciola* sp. infections in the snail intermediate hosts, utilizing various genetic markers, such as cytochrome *c* oxidase subunit 1 gene (*cox*1), ITS, and noncoding repetitive DNA fragment [80–82].

Using the forward primer 5′-ATTCACCCATTTCTGT TAGTCC-3′ and the reverse primer 5′-ACTAGGCTTAAA GGCGTCC-3′ in standard PCR followed by sequence alignment, Krämer and Schnieder [83] demonstrated the sequence heterogeneity within a 120 bp region of the pFh5 gene in *F. hepatica* and *F. gigantica*. Similarly, Mostafa et al. [62] also employed this primer pair for sensitive and specific detection of *F. gigantica* in *Lymnaea natalensis*. In addition, Velusamy et al. [84] also utilized this primer pair for identification of *F. gigantica* infection in the snail intermediate host *Lymnaea auricularia*. The specific primers amplified a *F. gigantica*-specific 124 bp noncoding repetitive DNA fragment from infected *L. auricularia* snails. Caron et al. [80] examined *G. truncatula*, *Radix labiata*, and *Radix balthica* experimentally infected with *F. hepatica* by PCR and showed that *R. labiata* can act as an incidental intermediate host for *F. hepatica*. Cucher et al. [85] developed a PCR assay targeting the cytochrome *c* oxidase subunit 1 gene. Using the forward primer 5′-TATGTTTTGATTTTACCCGGG-3′ and the reverse primer 5′-ATGAGCAACCACAAACCATGT-3′, these authors successfully amplified a 405 bp fragment from *F. hepatica*-infected *P. columella* and *L. viatrix*. Magalhães et al. [78] designed a multiplex PCR assay for detection of *F. hepatica* DNA in *L. viatrix* snails which were formalin-fixed and paraffin-embedded. Specifically, the assay included a pair of primers targeting mtDNA for recognition of *F. hepatica* and another pair targeting the ITS region of the trematode and the molluscan rRNA, to function as an internal control

Recent advances in the molecular detection of fascioliasis include the development and use of real-time PCR for identification of *Fasciola* from intermediate snail host. Schweizer et al. [86] exploited the 5′–3′ exonuclease activity of the Taq DNA polymerase to cleave fluorescent dye-labeled TaqMan probe during PCR for detection of *G. truncatula* snail naturally infected with *F. hepatica*. In addition, real-time PCR platforms have also been established for simultaneous detection and identification of *F. hepatica* and *F. gigantica* [74].

31.2 METHODS

31.2.1 Sample Preparation

Adult trematodes of *Fasciola* are collected from the liver of infected animal hosts at necropsy. Individual flukes were washed extensively in physiological saline, identified morphologically to genus and/or species according to existing keys and descriptions [1–3,5], and then fixed in 70% ethanol until extraction of genomic DNA.

Total genomic DNA can be extracted from a portion of individual specimens by a method of sodium dodecyl sulfate/proteinase K (Merck) treatment, then column-purified using Wizard™ DNA Clean-Up System (Promega) and eluted into 60 μL H_2O by following the manufacturer's recommendations. DNA was also isolated from liver from cattle, buffalo, or goat using the same method as for trematode samples. DNA samples were stored at −20°C until use.

Fasciola DNA in formalin-fixed and paraffin-embedded *Lymnaea* tissues can be extracted by a technique reported by Magalhães et al. [78]. This technique includes deparaffinizing the slides with xylol, enzymatic digestion with proteinase K, purification with phenol or chloroform and precipitation with isopropanol.

31.2.2 Detection Procedures

31.2.2.1 Real-Time PCR Identification of *F. hepatica* and *F. gigantica*

Alasaad et al. [74] described a real-time PCR assay for identification of *F. hepatica* and *F. gigantic*. The assay relies on the use of genus-specific primers from *Fasciola* ITS-2 gene: SSCPFa*F*: 5′-TTGGTACTCAGTTGTCAGTGTG-3′ and SSCPFa*R*: 5′-AGCATCAGACACATGACCAAG-3′ for amplification of a 140 bp product followed by detection with species-specific TaqMan probes for *F. hepatica* (ProFh: 5′-ACCAGGCACGTTCCGTCACTGTCACTTT-3′) and *F. gigantica* (ProFg: 5′-ACCAGGCACGTTCCGTTACTG TTACTTTGTC-3′). Both TaqMan probes are labeled with a BHQ quencher dye at their 3′-end, but at the 5′-end ProFh is labeled with an FAM reporter dye and ProFg with an HEX reporter dye.

31.2.2.1.1 Procedure

1. RT-PCR mixture (20 μL) is composed of 0.3 μM of each primer (SSCPFaF and SSCPFaR), 0.1 μM of each probe (ProFh and ProFg), 1× Master Mix (Kapa Probe Fast qPCR Kit), 1 μL of DNA solution, and nuclease-free water.
2. Amplification is conducted on an Mx-3005P cycler (Stratagene) with the following cycling conditions: an initial 95°C for 2 min and 45 cycles of 95°C for 3 s and 60°C for 30 s.
3. Data are analyzed with the software package MxPro v4.00 (Stratagene).

31.2.2.2 PCR-RFLP Identification of *F. hepatica* and *F. gigantica*

Huang et al. [63] reported a PCR-linked restriction fragment-length polymorphism (PCR-RFLP) for the characterization and differentiation of *Fasciola* spp. The assay uses the primers 3S (5′-GGTGGATCACTGGGCTCGTG-3′) and BD2 (5′-TATGCTTAAATTCAGCGGGT-3′) for amplification of the ITS-2 rRNA from *Fasciola* spp. Subsequent digestion of the amplicon using restriction enzyme *Hsp92*II or *Rca*I permits differentiation of *F. hepatica*, *F. gigantica*, and the "intermediate form" of *Fasciola*.

31.2.2.2.1 Procedure

1. PCR mixture (100 μL) is made up of 1 × PCR buffer (10 mM Tris-HCl, pH 8.4, 50 mM KCl, 2 mM $MgCl_2$), 200 μM each of dNTP, 50 pmol of each primer (3S and BD2), 2 U *Taq* polymerase (Promega), 1 μL purified *Fasciola* genomic DNA (~10–20 ng) and distilled water.
2. Amplification is performed in a thermocycler (Biometra or other brand) under the following conditions: 1 cycle of 95°C for 5 min; 30 cycles of 95°C for 1 min, 55°C for 1 min, and 72°C for 1 min; 1 cycle of 72°C for 10 min.
3. An aliquot (5 μL) of each amplicon is examined on 1.2% agarose-TBE (65 mM Tris-HCl, 22.5 mM boric acid and 1.25 mM EDTA, pH 9.0) gels. A DNA size marker (Takara) is included in a spare lane for estimation of the sizes of the ITS-2 amplicon.
4. The gel is with ethidium bromide (0.5 μg/mL) and photographed using a gel documentation system.
5. The ITS-2 PCR amplicon is purified using spin columns (Wizard PCR-Preps DNA Purification System, Promega).
6. The purified *Fasciola* ITS-2 PCR products (5–15 μL, depending on DNA concentration) are digested with 10 U (1 μL) of restriction enzyme *Hsp92*II or *Rca*I in 20 μL for 4 h or overnight at 37°C.
7. The resulting restriction fragments are separated on 2.5% agarose gel. A DNA size marker (Takara) is also included in a spare lane. The gel is stained and photographed as above.

Note: The ITS-2 amplicon from *Fasciola* genomic DNA is ~550 bp in length. When the purified ITS-2 amplicon is digested with *Hsp*92 II, three bands of ~90, 110, and 230 bp are produced for *F. hepatica* samples; four bands of ~70, 90, 100, and 160 bp are produced for *F. gigantica* samples; and five bands of ~70, 90, 100, 160, and 230 bp are produced for samples representing the "intermediate form" of *Fasciola*, allowing the identification and differentiation of *F. hepatica*, *F. gigantica* and the "intermediate form" of *Fasciola*.

When the restriction endonuclease *Rca*I is used, the purified ITS-2 PCR amplicon from *F. hepatica* samples remains undigested; that from *F. gigantica* samples yields two bands of ~170 and 380 bp; and that from the "intermediate form" of *Fasciola* displays three bands of ~170, 380, and 550 bp, respectively, also allowing the identification and differentiation of *F. hepatica*, *F. gigantica*, and the "intermediate form" of *Fasciola*.

This PCR-RFLP assay offers a valuable tool for the molecular identification and for studying the ecology and population genetic structures of *Fasciola* spp. and has been employed for the unequivocal delineation of the *Fasciola* spp. from Mainland China.

31.3 CONCLUSION

Fascioliasis caused by the liver flukes of the genus *Fasciola* (*F. hepatica* and *F. gigantica*) is a significant food-borne parasitic zoonosis in a number of countries, with millions of people being affected worldwide. Humans become infected with *Fasciola* spp. after ingestion of raw vegetation containing metacercariae (or raw sheep liver harboring the parasite). The metacercariae move down to the small intestine, penetrate the intestinal wall and peritoneal cavity to the liver, and mature into adults in the biliary ducts of the liver. The migration of *Fasciola* metacercariae causes tissue damages and the secretion of metabolic products by *Fasciola* worms elicits inflammatory responses in the host. As a consequence, a range of nonspecific clinical symptoms are present, including fever, abdominal pain, respiratory disturbances, and skin rashes. Conventional diagnosis of human fascioliasis is dependent on the detection of *Fasciola* eggs or specific antibodies in clinical samples. Given the poor sensitivity and specificity of these phenotypic methods, the use of molecular procedures is crucial for improved diagnostic outcome. PCR-based approaches (e.g., PCR-RFLP, PCR-SSCP, species-specific PCR, and real-time PCR assays) have not only enabled the accurate identification of *Fasciola* spp., but also facilitated the efficient epizootiological investigation of these parasites in snail hosts [87,88]. There is no doubt that further application of molecular procedures will contribute to the improved efficiency in the control and prevention of fascioliasis in humans and animals.

REFERENCES

1. Mas-Coma, S., and Bargues, M.D. Human liver flukes: A review. *Res. Rev. Parasitol.* 57, 145, 1997.
2. Mas-Coma, S., Esteban, J.G., and Bargues, M.D. Epidemiology of human fascioliasis: A review and proposed new classification. *Bull. WHO* 77, 340, 1999.
3. Mas-Coma, S., Bargues, M.D., and Valero, M.A. Fascioliasis and other plant-borne trematode zoonoses. *Int. J. Parasitol.* 35, 1255, 2005.
4. Ali, H. et al. Genetic characterisation of *Fasciola* samples from different host species and geographical localities revealed the existence of *F. hepatica* and *F. gigantica* in Niger. *Parasitol. Res.* 102, 1021, 2008.
5. Ashrafi, K. et al. Phenotypic analysis of adults of *Fasciola hepatica*, *Fasciola gigantica* and intermediate forms from the endemic region of Gilan, Iran. *Parasitol. Int.* 55, 249, 2006.
6. Lin, R.Q. et al. Sequence analysis of the first internal transcribed spacer of rDNA supports the existence of the intermediate *Fasciola* between *F. hepatica* and *F. gigantica* in mainland China. *Parasitol. Res.* 101, 813, 2007.
7. Le, T.H. et al. Human fascioliasis and the presence of hybrid/introgressed forms of *Fasciola hepatica* and *Fasciola gigantica* in Vietnam. *Int. J. Parasitol.* 38, 725, 2008.
8. Hashimoto, K.T. et al. Mitochondrial DNA and nuclear DNA indicate that the Japanese *Fasciola* species is *F. gigantica*. *Parasitol. Res.* 83, 220, 1997.
9. Itagaki, T. et al. Taxonomic status of the Japanese triploid forms of *Fasciola*: Comparison of mitochondrial ND1 and COI sequences with *F. hepatica* and *F. gigantica*. *J. Parasitol.* 84, 445–448, 1998.
10. Itagaki, T. et al. Genetic characterization of parthenogenic *Fasciola* sp. in Japan on the basis of the sequences of ribosomal and mitochondrial DNA. *Parasitology* 131, 679, 2005.
11. Itagaki, T. et al. Molecular characterization of parthenogenic *Fasciola* sp. in Korea on the basis of DNA sequences of ribosomal ITS1 and mitochondrial NDI gene. *J. Vet. Med. Sci.* 67, 1115, 2005.
12. Prociv, P., Walker, J.C., and Whitby, M. Human ectopic fascioliasis in Australia: First case report. *Med. J. Aust.* 156, 349, 1992.
13. Walker, S.M. et al. The distribution of *Fasciola hepatica* and *Fasciola gigantica* within southern Tanzania—Constraints associated with the intermediate host. *Parasitology* 135, 495, 2008.
14. Esteban, J.G., Bargues, M.D., and Mas-Coma, S. Geographical distribution, diagnosis and treatment of human fascioliasis: A review. *Res. Rev. Parasitol.* 58, 13, 1998.
15. Haseeb, A.N. et al. A review on fascioliasis in Egypt. *J. Egypt. Soc. Parasitol.* 32, 317, 2002.
16. Nithiuthai, S. et al. Waterborne zoonotic helminthiases. *Vet. Parasitol.* 126, 67, 2004.
17. Zhou, P. et al. Food-borne parasitic zoonoses in China: Perspective for control. *Trends Parasitol.* 24, 190, 2008.
18. Keiser, J., and Utzinger, J. Food-borne trematodiases. *Clin. Microbiol. Rev.* 22, 466, 2009.
19. Gutiérrez, A. et al. *Fasciola hepatica*: Identification of molecular markers for resistant and susceptible *Pseudosuccinea columella* snail hosts. *Exp. Parasitol.* 105, 211, 2003.
20. Bargues, M.D., and Mas-Coma, S. Reviewing lymnaeid vectors of fasciolosis by ribosomal DNA sequence analyses. *J. Helminthol.* 79, 257, 2005.
21. Kozak, M., and Wedrychowicz, H. The performance of a PCR assay for field studies on the prevalence of *Fasciola hepatica* infection in *Galba truncatula* intermediate host snails. *Vet. Parasitol.* 168, 25, 2010.
22. Goral, V. et al. A case of biliary fascioliasis by *Fasciola gigantica* in Turkey. *Korean J. Parasitol.* 49, 65, 2011.
23. Goral, V. et al. Biliary *Fasciola gigantica* case report from Turkey. *Southeast Asian J. Trop. Med. Public Health* 42, 509, 2011.
24. Arjona, R. et al. Fascioliasis in developed countries: A review of classic and aberrant forms of the diseases. *Medicine* (Baltimore). 74, 13, 1995.
25. Almendras-Jaramillo, M. et al. Hepatic fascioliasis in children: Uncommon clinical manifestations. *Arch. Gastroenterol.* 34, 241, 1997.
26. Chen, M.G., and Mott, K.E. Progress in assessment of morbidity due to *Fasciola hepatica* infection: A review of recent literature. *Trop. Dis. Bull.* 87, R1, 1990.
27. Dalimi, A., and Jabarvand, M. *Fasciola hepatica* in the human eye. *Trans. R. Soc. Trop. Med. Hyg.* 99, 798, 2005.
28. Zhou, L. et al. Multiple brain hemorrhages and hematomas associated with ectopic fascioliasis in brain and eye. *Surg. Neurol.* 69, 516, 2008.
29. Sezgın, O. et al. *Fasciola hepatica*-induced acute pancreatitis: Report of two cases and review of the literature. *Turk. J. Gastroenterol.* 21, 183, 2010.
30. Aslan, F. et al. A rare cause of cholangitis: *Fasciola hepatica*. *Turk. J. Gastroenterol.* 21, 329, 2010.
31. Acosta-Ferreira, W., Vercelli-Retta, J., and Falconi, L.M. *Fasciola hepatica* human infection. Histopathological study of sixteen cases. *Virchows. Arch. A. Pathol. Anat. Histopathol.* 383, 319, 1979.
32. Ishii, Y., Nakamura-Uchiyama, F., and Nawa, Y. A praziquantel-ineffective fascioliasis case successfully treated with triclabendazole. *Parasitol. Int.* 51, 205, 2002.
33. Carrada-Bravo, T. Fascioliasis: Diagnosis, epidemiology and treatment. *Rev. Gastroenterol. Mex.* 68, 135, 2003.
34. Kumar, P., Singh, V.K., and Singh, D.K. Combination of molluscicides with attractant carbohydrates and amino acids in bait formulation against the snail *Lymnaea acuminata*. *Eur. Rev. Med. Pharmacol. Sci.* 15, 550, 2011.
35. Kabaalioglu, A. et al. US-guided gallbladder aspiration: A new diagnostic method for biliary fascioliasis. *Eur. Radiol.* 9, 880, 1999.
36. Moustafa, N.E., Hegab, M.H., and Hassan, M.M. Role of ELISA in early detection of *Fasciola* copro-antigens in experimentally infected animals. *J. Egypt. Soc. Parasitol.* 28, 379, 1998.
37. O'Neill, S.M. et al. Immunodiagnosis of *Fasciola hepatica* (Fascioliasis) in a human population in the Bolivian Altiplano

using purified cathepsin-l-cysteine proteinase. *Am. J. Trop. Med. Hyg.* 58, 417, 1998.
38. Maleewong, W. et al. *Fasciola gigantica*-specific antigens: Purification by a continuous-elution method and its evaluation for the diagnosis of human fascioliasis. *Am. J. Trop. Med. Hyg.* 61, 648, 1999.
39. Intapan, P.M. Development of rapid agglutination test using *Fasciola gigantica* specific antigen for serodiagnosis of human fascioliasis. *Southeast Asian J. Trop. Med. Public Health* 35, S313, 2004.
40. Tantrawatpan, C. et al. Serodiagnosis of human fascioliasis by a cystatin capture enzyme-linked immunosorbent assay with recombinant *Fasciola gigantica* cathepsin L antigen. *Am. J. Trop. Med. Hyg.* 72, 82, 2005.
41. Wongkham, C. et al. Evaluation of immunoglobulin G subclass antibodies against recombinant *Fasciola gigantica* cathepsin L1 in an enzyme-linked immunosorbent assay for serodiagnosis of human fasciolosis. *Clin. Diagn. Lab. Immunol.* 12, 1152, 2005.
42. Demerdash, Z.A. et al. Diagnostic efficacy of monoclonal antibody based sandwich enzyme linked immunosorbent assay (ELISA) for detection of *Fasciola gigantica* excretory/secretory antigens in both serum and stool. *Parasit. Vectors* 4, 176, 2011.
43. Rahimi, M.T. et al. Evaluation of Fast-ELISA versus standard-ELISA to diagnose human fasciolosis. *Arch. Iran. Med.* 14, 18, 2011.
44. Luton, K., Walker, D., and Blair, D. Comparisons of ribosomal internal transcribed spacers from two congeneric species of flukes (Platyhelminthes: Trematoda: Digenea). *Mol. Biochem. Parasitol.* 56, 323, 1992.
45. Shubkin, C.D. et al. A nucleic acid-based test for the detection of *Fasciola hepatica. J. Parasitol.* 78, 817, 1992.
46. Heussler, V. et al. DNA probes for the detection of *Fasciola hepatica* in snails. *Mol. Cell. Probes* 7, 261, 1993.
47. Rognlie, M.C., Dimke, K.L., and Knapp, S.E. Detection of *Fasciola hepatica* in infected intermediate hosts using RT-PCR. *J. Parasitol.* 80, 748, 1994.
48. Rognlie, M.C. et al. Seasonal transmission of *Fasciola hepatica* in Montana, USA, with detection of infected intermediate hosts using a DNA-based assay. *Vet. Parasitol.* 65, 297, 1996.
49. Kaplan, R.M. et al. A repetitive DNA probe for the sensitive detection of *Fasciola hepatica* infected snails. *Int. J. Parasitol.* 25, 601, 1995.
50. Kaplan, R.M. et al. The prevalence of *Fasciola hepatica* in its snail intermediate host determined by DNA probe assay. *Int. J. Parasitol.* 27, 1585, 1997.
51. Agatsuma, T. et al. Molecular evidence of natural hybridization between *Fasciola hepatica* and *F. gigantica. Parasitol. Int.* 49, 231, 2000.
52. Ramadan, N.I., and Saber, L.M. Detection of genetic variability in nonhuman isolates of *Fasciola hepatica* and *Fasciola gigantica* by the RAPD-PCR technique. *J. Egypt. Soc. Parasitol.* 34, 679, 2004.
53. Abdel Tawab, A.H., el-Garhy, M.F., and Heckmann, R.A. Molecular polymorphisms associated with Egyptian, European, Chinese and North American *Fasciola hepatica. J. Egypt. Soc. Parasitol.* 36, 389, 2006.
54. Le, T.H. et al., Molecular confirmation that *Fasciola gigantica* can undertake aberrant migrations in human hosts. *J. Clin. Microbiol.* 45, 648, 2007.
55. McGarry, J.W. et al. PCR-based differentiation of *Fasciola* species (Trematoda: Fasciolidae), using primers based on RAPD-derived sequences. *Ann. Trop. Med. Parasitol.* 101, 415, 2007.
56. Li, Q.Y. et al. Sequence-related amplified polymorphism, an effective molecular approach for studying genetic variation in *Fasciola* spp. of human and animal health significance. *Electrophoresis* 30, 403, 2009.
57. Ai, L. et al. Specific PCR-based assays for the identification of *Fasciola* species: Their development, evaluation and potential usefulness in prevalence surveys. *Ann. Trop. Med. Parasitol.* 104, 65, 2010.
58. Ai, L. et al. Rapid identification and differentiation of *Fasciola hepatica* and *Fasciola gigantica* by a loop-mediated isothermal amplification (LAMP) assay. *Vet. Parasitol.* 174, 228, 2010.
59. Dar, Y. et al. Microsatellite analysis of *Fasciola* spp. in Egypt. *Parasitol. Res.* 109, 1741, 2011.
60. Prasad, P.K. et al. PCR-based molecular characterization and in silico analysis of food-borne trematode parasites *Paragonimus westermani, Fasciolopsis buski* and *Fasciola gigantica* from Northeast India using ITS2 rDNA. *Bioinformation* 6, 64, 2011.
61. Adlard, R.D. et al. Comparison of the second internal transcribed spacer (ribosomal DNA) from populations and species of Fasciolidae (Digenea). *Int. J. Parasitol.* 23, 423, 1993.
62. Mostafa, O.M., Taha, H.A., and Ramadan, G. Diagnosis of *Fasciola gigantica* in snail using the polymerase chain reaction (PCR) assay. *J. Egypt. Soc. Parasitol.* 33, 733, 2003.
63. Huang, W.Y. et al. Characterisation of *Fasciola* species from mainland China by ITS-2 ribosomal DNA sequence. *Vet. Parasitol.* 120, 75, 2004.
64. Alasaad, S. et al. Characterization of *Fasciola* samples from different host species and geographical localities in Spain by sequences of internal transcribed spacers of rDNA. *Parasitol. Res.* 101, 1245, 2007.
65. Capuano, F. et al., A comparison of five methods for DNA isolation from liver and rumen flukes to perform ITS-2+ amplification. *Parasitologia* 49, 27, 2007.
66. Ai, L. et al. Genetic diversity and relatedness of *Fasciola* spp. isolates from different hosts and geographic regions revealed by analysis of mitochondrial DNA sequences. *Vet. Parasitol.* 181, 329, 2011.
67. Amor, N. et al. Molecular characterization of *Fasciola* spp. from the endemic area of northern Iran based on nuclear ribosomal DNA sequences. *Exp. Parasitol.* 128, 196, 2011.
68. Amor, N. et al. Molecular characterization of *Fasciola gigantica* from Mauritania based on mitochondrial and nuclear ribosomal DNA sequences. *Exp. Parasitol.* 129, 127, 2011.
69. Ichikawa, M., and Itagaki, T. Discrimination of the ITS1 types of *Fasciola* spp. based on a PCR-RFLP method. *Parasitol. Res.* 106, 757, 2010.
70. Ichikawa, M. et al. Characterization of *Fasciola* spp. in Myanmar on the basis of spermatogenesis status and nuclear and mitochondrial DNA markers. *Parasitol. Int.* 60, 474, 2011.
71. Choe, S.E. et al. Genetic analysis of *Fasciola* isolates from cattle in Korea based on second internal transcribed spacer (ITS-2) sequence of nuclear ribosomal DNA. *Parasitol. Res.* 109, 833, 2011.
72. Marcilla, A., Bargues, M.D., and Mas-Coma, S. A PCR-RFLP assay for the distinction between *Fasciola hepatica* and *F. gigantica. Mol. Cell. Probes* 16, 327, 2002.
73. Rokni, M.B. et al. Identification and differentiation of *Fasciola hepatica* and *Fasciola gigantica* using a simple PCR-restriction enzyme method. *Exp. Parasitol.* 124, 209, 2010.

74. Alasaad, S. et al. A TaqMan real-time PCR-based assay for the identification of *Fasciola* spp. *Vet. Parasitol.* 179, 266, 2011.
75. Itagaki, T. et al. Characterization of genetic divergence among species within the genus *Fasciola* by PCR-SSCP. *Jpn. J. Parasitol.* 44, 244, 1995.
76. Alasaad, S. et al. A fluorescence-based polymerase chain reaction-linked single-strand conformation polymorphism (F-PCR-SSCP) assay for the identification of *Fasciola* spp. *Parasitol. Res.* 108, 1513, 2011.
77. Magalhães, K.G., Passos, L.K., and Carvalho Odos, S. Detection of *Lymnaea columella* infection by *Fasciola hepatica* through multiplex-PCR. *Mem. Inst. Oswaldo Cruz* 99, 421, 2004.
78. Magalhães, K.G. et al. Isolation and detection of *Fasciola hepatica* DNA in *Lymnaea viatrix* from formalin-fixed and paraffin-embedded tissues through multiplex-PCR. *Vet. Parasitol.* 152, 333, 2008.
79 El-Gozamy, B.R., and Shoukry, N.M. Identification of Egyptian *Fasciola* species by PCR and restriction endonucleases digestion of the nuclear small subunit ribosomal RNA gene. *J. Egypt. Soc. Parasitol.* 39, 429, 2009.
80. Caron, Y., Lasri, S. and Losson, B. *Fasciola hepatica*: An assessment on the vectorial capacity of *Radix labiata* and *R. balthica* commonly found in Belgium. *Vet. Parasitol.* 149, 95, 2007.
81. Caron, Y., Rondelaud, D., and Losson, B. The detection and quantification of a digenean infection in the snail host with special emphasis on Fasciola sp. *Parasitol. Res.* 103, 735, 2008.
82. Caron, Y. et al. An optimized DNA extraction and multiplex PCR for the detection of *Fasciola* sp. in lymnaeid snails. *Vet. Parasitol.* 178, 93, 2011.
83. Krämer, F., and Schnieder, T. Sequence heterogeneity in a repetitive DNA element of *Fasciola. Int. J. Parasitol.* 28, 1923, 1998.
84. Velusamy, R., Singh, B.P., and Raina, O.K. Detection of *Fasciola gigantica* infection in snails by polymerase chain reaction. *Vet. Parasitol.* 120, 85, 2004.
85. Cucher, M.A. et al. PCR diagnosis of *Fasciola hepatica* in field-collected *Lymnaea columella* and *Lymnaea viatrix* snails. *Vet. Parasitol.* 137, 74, 2006.
86. Schweizer, G. et al. Prevalence of *Fasciola hepatica* in the intermediate host *Lymnaea truncatula* detected by real time TaqMan PCR in populations from 70 Swiss farms with cattle husbandry. *Vet. Parasitol.* 150, 164, 2007.
87. Itagaki, T., and Tsutsumi, K. Triploid form of *Fasciola* in Japan: Genetic relationships between *Fasciola hepatica* and *Fasciola gigantica* determined by ITS-2 sequence of nuclear rDNA. *Int. J. Parasitol.* 28, 777, 1998.
88. Králová-Hromadová, I. et al. Sequence analysis of ribosomal and mitochondrial genes of the giant liver fluke *Fascioloides magna* (Trematoda: Fasciolidae): Intraspecific variation and differentiation from *Fasciola hepatica. J. Parasitol.* 94, 58, 2008.

32 Fasciolopsis

Veena Tandon, B. Roy, and P.K. Prasad

CONTENTS

32.1 INTRODUCTION

The giant intestinal fluke, *Fasciolopsis buski* (Lankester, 1857) Looss, 1899 (Trematoda: Fasciolidae), is widely distributed in South and Southeast Asia including India and neighboring countries [1]. The parasite infects both humans and pigs and is the causative agent of the disease, commonly known as fasciolopsosis or fasciolopsiasis. The infection of *F. buski* occurs by ingestion of raw aquatic vegetation or food plants that are contaminated with the infective encysted larvae, the metacercariae. The infection mainly prevails among the socioeconomically underdeveloped and impoverished areas of tropics and subtropics where high rainfall and moderate water temperature offer suitable environmental conditions for the propagation of snail intermediate hosts that are an essential link in the life cycle of the parasite [2]. It was in the year 1843 that George Busk, an English surgeon, observed for the first time several flukes in the duodenum of an Indian sailor from eastern India during an autopsy. Lankester [3] named these worms *Distoma buskii* after its discoverer. However, objecting to it, Busk suggested the name *Distoma crassum*. It emerged later that the name was already occupied by another parasite described in 1836! The parasite remained latent in the literature till 1874 when it was again observed in a missionary couple residing in China; the adult flukes passed out in their stool were identified by Busk to be the same flukes as the ones he had first seen decades ago! In the following years, more than once similar-looking worms were reported from Chinese subjects but their identification was always not free from controversy. Finally, Looss [4] proposed the genus *Fasciolopsis* (*Fasciola* (Latin) = fillet; opsis (Greek) = resembling)) for *Distoma buski* Lankester, 1857, a name that continues to be in use until now.

32.1.1 Classification and Biology

32.1.1.1 Systematic Position

Phylum: Platyhelminthes
Class: Trematoda
Subclass: Digenea
Order: Prosostomata
Family: Fasciolidae
Genus: *Fasciolopsis*
Species: *Fasciolopsis buski*

GenBank Taxonomy ID: 27845—("Lineage (full)—cellular organisms; Eukaryota; Fungi/Metazoa group; Metazoa;Eumetazoa;Bilateria;Acoelomata;Platyhelminthes;

Trematoda; Digenea; Echinostomida; Echinostomata; Echinostomatoidea; Fasciolidae; *Fasciolopsis*").

32.1.1.2 Habitat and Morphology

The adult *F. buski* lives in the duodenum and jejunum of man, but in heavy infections may be found in the stomach and most part of the intestine [5]. The fasciolids are also of particular interest with respect to the evolution of body form and habitat among their adult stages. The genus *Fasciolopsis* is the next most basal member taxon of the trematode family Fasciolidae after the type genus *Fasciola*. The two most basal species in the phylogeny, *Protofasciola robusta* and *F. buski*, both colonize the small intestine of their respective hosts. While *F. buski* inhabits the suid and primate (human) hosts, the next species retrieved in the phylogeny, *Parafasciolopsis fasciolaemorpha*, inhabits the duodenum and the bile ducts of its cervid hosts [6]. All the three species characteristically have unbranched caeca and a ventral sucker that is larger than the oral sucker. All other fasciolid species have extensively branched intestinal caeca and relatively small ventral suckers, and live in the biliary tracts (*Fasciola hepatica*, *Fasciola gigantica*, *Fasciola jacksoni*, *Fasciola nyanzae*, and *Tenuifasciola tragelaphi*) or in the liver parenchyma and bile passages (*Fascioloides magna*). The adoption of branched gonads, a characteristic of the subfamily Fasciolinae, is already highly manifested in the basal taxon *F. buski* that harbors the intestinal tract [7].

F. buski is among the largest digeneans. The body of the live parasite is thick, fleshy, pinkish, or dark brown in color, broadly leaf-shaped, ovate, elliptical with rounded posterior end, generally 17–75 mm in length, 8–20 mm in width, and 0.5–3.5 mm in thickness [8]. There is no cephalic cone or shoulders unlike the species of the genus *Fasciola*. The ventral sucker is large and muscular, lies close to the oral sucker and is 1.22–2.48 mm in diameter. The prepharynx is very short and the pharynx measures 0.36–0.72 mm in diameter. The esophagus is short. The intestinal caeca are long, lateral, sinuous but unbranched and extend to the subterminal region of the body. The gonads are highly branched structures; the female organs are found in the anterior half, while the testes mainly occupy the posterior half. The cirrus sac is sinuous. The ovary is branched; the Mehlis' gland is median and postovarian and the uterus transversely coiled and confined to the preovarian region. The eggs are brownish yellow in color, operculate, ovoid in shape, and 11.3–17.1 μm × 80–98 μm in size (Figure 32.1).

The ventral surface of the fluke is provided with densely packed, posteriorly directed broadly rounded scales (Figure 32.2). Little deviation from this normal surface topography and anatomy of the fluke may be observed in specimens originating in different geographical regions [9].

32.1.1.3 Biology and Life Cycle

F. buski follows the typical trematode life pattern, involving an intermediate host, that is, a fresh water planorbid snail and a mammalian definitive host, that is, man or pig (Figure 32.3). About 25,000 eggs escape out of the mammalian host everyday through excreta and develop further in water. Different stages of larval development occur inside the lymph spaces of planorbid snails and finally encyst in stable aquatic vegetation [10,11]. Species of the genera *Segmentina*, *Hippeutis*, and *Gyraulus* act as snail hosts; *Segmentina hemisphaerula* (syn. *Segmentina coenosus*, *Segmentina nitidella*, *Segmentina calathus*, *Segmentina largeillierti*), *Hippeutis cantori* (syn. *Hippeutis smackeri*), *Hippeutis umbilicalis*, and *Gyraulus convexiusculus* are the snail hosts in China, Vietnam, and Taiwan [12–15]. In India, the snail species reported are *Segmentina trochoideus*, *S. hemisphaerula*, and *H. umbilicalis* in Assam besides *Indoplanorbis exustus* and *Helicorbis coenosus* in Uttar Pradesh [16–19]. *S. hemisphaerula*, *S. trochoideus*, and *Gyraulus chinensis* are snail hosts in Thailand, *H. umbilicalis* in Laos, and *S. trochoideus* and *H. umbilicalis* in Bangladesh [10,15,20–22].

Eggs are unembryonated when shed in the feces and must reach fresh water to continue the cycle. In the course of 2–12 weeks (with a mean of 22 days), the first larval stage, the miracidium, develops inside the egg at an optimum

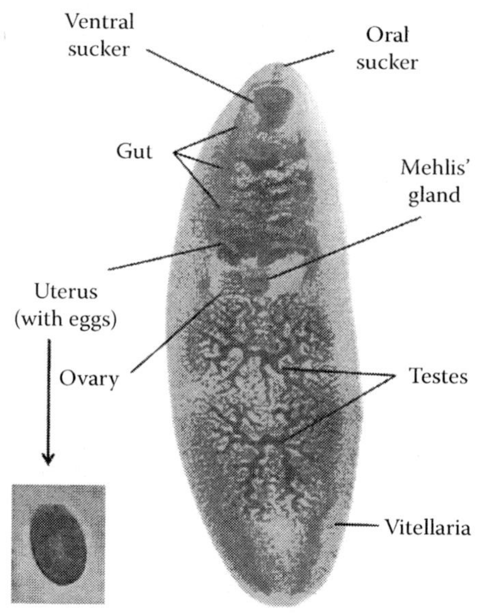

FIGURE 32.1 *Fasciolopsis buski* adult fluke; egg (inset).

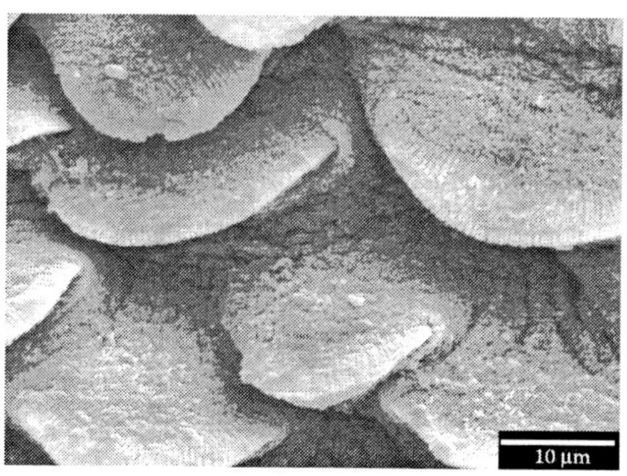

FIGURE 32.2 Scale-like protrusions covering the ventral surface of the fluke.

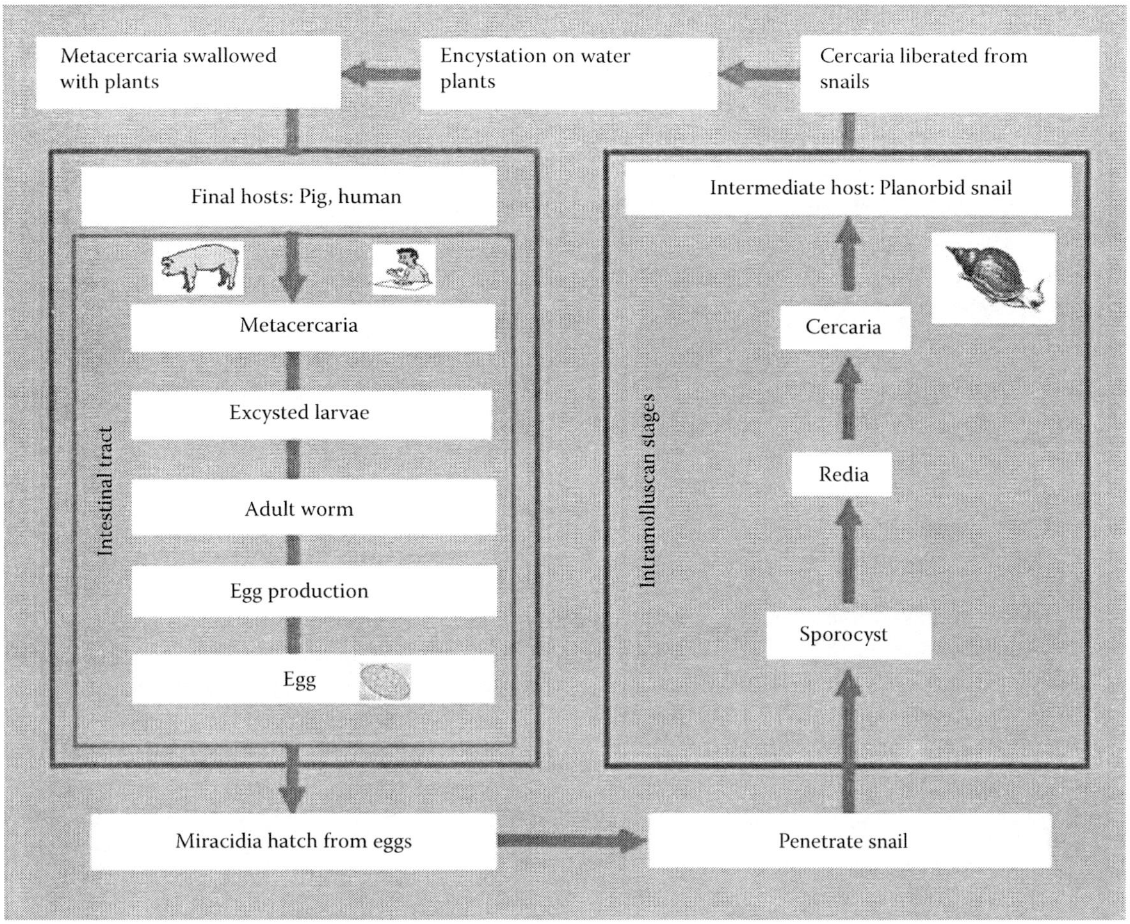

FIGURE 32.3 Life cycle of *Fasciolopsis buski.*

water temperature of 22–25°C and a 6.5–7.2 pH range [23]. On hatching, the miracidia escapes the operculum and finds its way into a suitable snail host. Inside the lymph spaces of the snail, the miracidium changes into a sporocyst. The latter produces the first and second generations of redia, the third larval stage. Finally, the daughter rediae produce tailed larvae, the cercariae which are typically the distome leptocercous type. Cercariae emerge from the rediae after 25–30 days, but they do require a maturation period in snail tissues before they are released from the molluscan host after an average of 49 days although the incubation periods may vary in accordance with the ambient temperature, for example, 46–59 days at 22–24°C and 85–100 days at 18–22°C [17,23,24]. The emergence of cercariae is dependent on light, with great variation in daily emergence patterns [19]. Cercariae swim in water until encystment occurs on a substratum, mainly aquatic plants and debris. The cercarial survival period in water varies from 64 to 72 days [10,25]. Infection of the definitive host takes place through the ingestion of encysted metacercaria along with infested vegetation or water. Important to transmission is the consumption of various aquatic plants such as water caltrop (*Trapa natans* in China, *Trapa bispinosa* in Taiwan, and *Trapa bicornis* in Bangladesh and Thailand), water chestnut (*Eliocharis tuberosa*), water hyacinth (*Eichhornia* sp.), water bamboo (*Zizania* sp.), water lotus or water lily (*Nymphaea* sp.), watercress, gankola (*Otelia* sp.), and water morning glory (*Ipomoea aquatica*). Up to 200 metacercarial cysts may be found on the surface of one water caltrop, but the usual number is about 15–20. Cercariae may also encyst on the water surface; reportedly 10–13% of the patients and 35–40% of the infected pigs had possibly contracted infection by drinking metacercariae-contaminated water [26]. On entering the duodenum, the infective metacercaria excysts and liberates a juvenile worm, which develops into the adult fluke in the course of 3 months time; the adult soon begins egg production and its life span seldom exceeds 6 months [23,27,28]. The eggs pass out in the feces into fresh water, where they hatch into miracidia that resume the cycle of further development through intramolluscan stages.

32.1.1.4 Epidemiology

More than 10 million people in southern Asia may have *F. buski* flukes residing in their intestines. Ingestion of infected raw aquatic food stuffs and other aquatic vegetation (*Salvinia*, *Valisneria*, *Eichornia* spp) contaminated with encysted metacercaria through peeling of these stuffs with the teeth facilitates the entry for the infective form [11,29,30].

In endemic zones, pigs, dogs, and rabbits serve as a reservoir of infection [30,31]. Fresh aquatic green fodder and contaminated water used to raise pigs are the main infection sources for farm animals [26,32]. Different infection rates in pig have been reported: 52% in Taiwan [12]; 10% in China [13]; and 30% in India [16]. As many as 152 flukes were found in a single host's intestine, indicating the poor status of the animal's health. Pig is the only important reservoir host, although it harbors a few flukes (usually only 3–12) and produces fewer eggs per adult than humans [33].

There is still disagreement as to whether there are only one or several strains of *F. buski*, more or less adapted to either humans or animals. The prevalence pattern among these hosts shows variations; in some places the parasite is common in humans but absent in pigs; in other places the situation is reverse, that is, high prevalence in pigs but very rare infection in humans; and in still other areas, the rates of occurrence of infection among humans and pigs are almost equal. Host-related variations related to infectiousness have also been recorded. Mice, rats, monkeys, and dogs were experimentally refractory, and guinea-pigs only partially susceptible, but rabbits were susceptible, as were squirrel monkeys [34,35]. *F. buski* has been rarely reported in rhesus monkeys, and there are no reports incriminating monkeys in endemic areas [36].

In pigs, the prevalence is seasonal, with a peak in June–September, declining thereafter to a low level during winter and early spring, and it is absent during January and February, at least in northeastern India [2].

Cultivation of aquatic, metacercariae-carrying plants for consumption on a large scale and the pollution of the areas in which they are grown with human and animal (mainly pig) excreta are important spreading factors [20,35,37]. Prevalence among people living near water caltrop plantations was much higher than in villages distant from plantations [12,38]. Social and economic factors, such as poverty, malnutrition, a free-food market associated with lack of food inspection, poor sanitation, and other infections of worm origin play a role in aggravating the disease [39–42]. Incidence of infection among human subjects of the same place may vary and the difference can be attributed to factors such as socio-cultural practices, standard of health, and mode of living [43].

Fasciolopsiasis is known in many Asian countries [42]. Animal reservoirs and the use of untreated human feces for fertilizer make it difficult to prevent human infections. In other parts of the world, the giant intestinal fluke does not occur, either because the climate does not support the survival of its transmission or infective stages or because the potential intermediate hosts are absent.

The *F. buski* infection is widely distributed in India, Pakistan, Bangladesh, Burma, Thailand, Malaysia, Indonesia, Taiwan, China, and Vietnam [30,31,43–45]. In China, infections have been reported from many provinces in both humans and pigs; while prevalence is as high as up to 85% in some areas, in other areas it varies from <1–5% [26,46,47]. Recent nationwide surveys suggest a decrease in fasciolopsiasis [40,48]. In endemic areas of China, mass chemotherapeutic practice failed to prevent the infection, particularly among children, because of the widespread practice of eating raw vegetables and ignorance about preventive measures for the disease [49–51]. In Taiwan, the most highly endemic area is in southern region, where infection rates ranged between 48% and 61%, a prevalence of 25% was found in schoolchildren [44,47,52–54].

In India, morbidity due to fasciolopsiasis has been recorded since 1843 when during a postmortem examination Busk for the first time recovered *F. buski* from the duodenum of a sailor from eastern India. Subsequently, the parasite was reported from different provinces of India, namely Assam, Bengal, Bihar, Orissa, Maharashtra, Madhya Pradesh, and Uttar Pradesh, where poverty, and hygienic socio-cultural practices and food habits are associated with the existence of open types of large-scale piggeries [17,30,55–59]. Infection in pigs, with a prevalence of 29% was reported in Mumbai and human foci of infection were also recorded in Maharashtra [60,61]. In Bangladesh, surveys showed 8.6–39.2% infection in two villages [39]; and prevalence in schoolchildren reached 50% in an endemic focus [10].

In Vietnam, early studies reported infection in both Asians and Europeans, and both humans and pigs were found positive [62,63]. In Thailand, the main endemic area is the central part of the country, with an estimate of 100,000 persons infected among 500,000, whereas a lower prevalence (7–13%) was reported in other provinces and northeastern Thailand [64–66]. Human reports are also known from Laos [67,68]. In Cambodia, prevalences of 0.04% in humans and 5% in pigs were reported in Phnom-Penh [67]. The *F. buski* infection has also been reported from Borneo and Sumatra, Indonesia, where a 20–27.0% prevalence was reported in residents of Kalimantan Province [33,69–71]. The parasite has also been detected in Burma, the Philippines and Singapore, and Malaysia [33,67,72]. Reports of infections in Korean and Japanese subjects indicate that the parasite is contracted outside of Korea and Japan [33]. Reports in the United States, Cuba, Venezuela, Guatemala, Israel, and Australia may be associated with immigrants from the Far East or attributed to possible misidentification of fecal eggs [73–75].

The disease is underreported and most prevalent in remote rural places and semiurban areas [46,54,62,65,76,77]. It is mainly so in schoolchildren in which the number of worms per child can exceed 800 [10,26]. The prevalence varied from 8% among adults to 15% in 5–14-year-old children. The severity of infection increased in the 10–14-year age group and decreased in older groups. A higher prevalence among females (16%) than males (11%) was reported in coprological surveys in Thailand [38]. Similar gender and age relationships were found in Thailand, Taiwan, and Bangladesh [12,20,77], in all these regions the prevalence detected in children was reported to be 10%, 25%, and 50%, respectively [10,44,65]. In comparison to Taiwan, the prevalence in mainland China was 57% [26].

32.1.2 Clinical Features and Pathogenesis

Fasciolopsiasis appears to be associated with a few or no disease manifestations in individuals examined in field environments. Light infections are usually asymptomatic except for diarrhea, alternating with periods of constipation and abdominal pain. They may include anemia with eosinophilia, headache, dizziness, stomach ache, gastric pain, loose stools, malnutrition, protuberant abdomen, and abdominal distention [10,78]. Moderate and heavy infections are associated with malnutrition, severe epigastric and abdominal pain, diarrhea or bowel obstruction, poor appetite, mild abdominal colic, vomiting, fever, nausea marked eosinophilia, and leukocytosis, and a significant lowering in serum vitamin B12 content [71,77,79]. Generalized toxic and allergic symptoms, usually in the form of edema, particularly of the face, abdominal wall, and lower extremities, may appear in heavy infections [32,80]. Flukes cause extensive intestinal and duodenal erosions, ulceration, hemorrhage, abscesses, and catarrhal inflammation. Pathological changes may be traumatic, obstructive, and toxic, especially in heavy infections. Feces are profuse, yellow-brown, and contain pieces of undigested food due to malabsorption [79]. Infection in other animals generally shows subclinical symptoms but a heavy parasite load can mimic the human symptoms. Young and adult *F. buski* remain attached to the duodenum and jejunum with the help of their ventral sucker and cause mechanical injury to the mucous membrane of the intestinal wall, leading to local inflammation with hypersecretion of mucus, hemorrhage and ulceration. Abscesses sometimes develop and worms may be passed in the stool.

Morbidity in endemic areas is high, and the disease can be fatal, depending on the worm burden [26,52,65]. However, human morbidity and mortality are difficult to assess for lack of quantitative data exclusive to *F. buski* infections [5]. Mortality has been reported in heavily infected children owing to profound intoxication [38,80,81]. However, little, if any, evidence has been found that this parasite is harmful to humans if present in small numbers [79,82].

32.1.3 Diagnosis

32.1.3.1 Conventional Techniques

Diagnosis of *F. buski* can be made by coprological examination for detection of the fluke eggs in the sample or occasionally by examination of expelled adult fluke vomited or passed in feces [10,26,33,83]. Demonstrating the eggs of the parasite in the excreta (using direct smear or salt floatation methods) provides the definitive diagnosis in all cases of fluke infection. The best quantitative method so as to assess the intensity of infection is the Kato thick smear method, in which a 50 mg feces sample is placed on a slide covered with a plastic coverslip soaked in glycerol and is allowed to clear for 24 h.

32.1.3.2 Molecular Techniques

Eggs of most fluke parasites appear similar in contour and/or size and diagnosis based on morphological characters alone at species level may not be free from ambiguity. However, recent advances in molecular biology, in particular the amplification of specific DNA regions via the polymerase chain reaction (PCR) and improved sequencing techniques, have been employed to resolve taxonomic issues related to various helminth parasites by comparing their DNA, utilizing genetic markers in nuclear ribosomal DNA (rDNA) and mitochondrial DNA in particular [84]. The rDNA cluster, which codes for structural components of ribosomes, is particularly useful for genetic studies because it is highly repeated and contains variable regions flanked by more conserved regions [85]. It enhances PCR amplification because many templates are available for initial priming and allows primers to be designed to anneal the known conserved regions to amplify across unknown variable regions. In the context of metazoan parasites, the rDNA internal transcribed spacers 1 and 2 (ITS1 and ITS2), which occur between the 18S, 5.8S, and 28S coding regions, have proven useful for diagnostic purposes at the level of species [86–91]. ITS2 sequences have been shown to be a sensitive marker at the species level in trematodes, whereas the sequences in the ITS1 might be less conserved than those in the ITS2 [92].

In the context of the fasciolid flukes, nucleotide sequences of the ITS regions of the small subunit rDNA gene have been used in delineating genera and species in the Fasciolidae [93]. Also, mitochondrial genomes (e.g., NADH dehydrogenase subunit 1 (Nad1) and cytochrome c oxidase subunit (Co1)) have been found useful in investigating the phylogenetic relationships among trematodes at many levels [94]. Obtaining the 18S rDNA sequence from an intestinal fluke vomited by a Vietnamese child was recently used to verify the diagnosis [83]. The ITS sequences of *F. buski* have also been determined and have proven to be useful molecular markers for species identification and differentiation at the level of geographical isolates [95,96].

32.2 METHODS

32.2.1 Sample Preparation

32.2.1.1 Parasite Material and DNA Isolation

Live adult *F. buski* can be obtained from the intestine of freshly slaughtered pig, at local abattoirs. The mature adult flukes yield eggs when squeezed between two glass slides. For the purpose of DNA extraction, adult flukes collected from different host animals are processed singly; eggs recovered from each specimen are also processed separately. The flukes are first immersed in digestion extraction buffer (containing 1% sodium dodecyl sulfate, 25 mg proteinase K) at 37°C overnight. DNA is then extracted from lysed individual worms by the standard ethanol precipitation technique [97]. DNA from eggs can also be extracted on FTA cards using Whatman's FTA Purification Reagent (Flowchart 1).

Collect cell pellet
↓ Add 200 μl 1 × TNE
Agitate and keep in ice for 1/2 h
↓
Add 10% SDS (10μl) and agitate
↓ Add 4 μl proteinase K
Mix and incubate at 37°C overnight
↓
Add more proteinase K (3 μl)
↓ Keep for 2 h at 37°C
Add 1/4th vol. 5M NaCl
↓ Keep in ice for 1/2 h
Cold centrifuge
↓ 10,000 rpm for 10–12 min
Collect the supernatant in a fresh tube
↓
To pellet add 500 μl of chilled 70% ethanol
↓ Centrifuge at 10,000 rpm for 10 min
Decant supernatant and keep the pellet to air dry for 20 min
↓
Add autoclaved double distilled water or TE buffer (40–60 μl) and keep overnight at 37°C, finally store at 4°C

Apply metacercariae/eggs on FTA cards
↓
Allow to dry for 1 h at room temperature
↓
Punch out 2–3 sample discs of 1.2 mm size from the desired spot using coring device (assuming a 25 μl reaction)
↓
Place discs in PCR amplification tube and wash with FTA Purification Reagent + TE buffer
↓
Allow discs to dry at room temperature for ~1 h to make it ready for PCR

32.2.2 Detection Procedures

32.2.2.1 DNA Amplification, Sequencing, and Its Analysis

The rDNA region spanning the ITS regions is amplified from DNA obtained from flukes by PCR. The universal trematode primers based on conserved ITS sequences of *Schistosoma* species as detailed below have been successfully used following the procedure of Bowles et al. [98]:

1. *ITS2 region*
   3S (forward)—5′-GGTACCGGTGGATCACTCGG CTCGTG-3′
   A28 (reverse)—5′-GGGATCCTGGTTAGTTTCT TTTCCTCCGC-3′
2. *ITS1 region*
   BD1 (forward)—5′-GTCGTAACAAGGTTTCC GTA-3′
   4S (reverse)—5′-TCTAGATGCGTTCGAA(G/A) TGTCGATG-3′

The PCR amplification is performed following the standard protocol as provided by White [99], with minor modifications in 100 mM Tris-HCl, pH 9.0, 500 mM KCl, 1.5 mM $MgCl_2$, and 0.2 mM deoxynucleotide triphosphate each of dATP, dGTP, dCTP, and dTTP, 0.25 mM of each primer, and 2.5 units of Taq polymerase. DNA is preheated at 94°C for 5 min and added to each PCR reaction. The PCR cocktail (final reaction volume 25 μL) is amplified with the following conditions: 26 cycles of denaturing at 94°C for 30 s, annealing at 55°C for 38 s and extension at 72°C for 42 s followed by a final extension at 72°C for 10 min. The resultant PCR products are separated by electrophoresis through 1.6% (w/v) agarose gels in Tris-acetate-EDTA buffer, stained with ethidium bromide, transilluminated under ultraviolet light, and then photographed.

For DNA sequencing, the PCR products are purified and sequenced in both directions using PCR primers on an automated sequencer. The sequences thus obtained are put to further analysis by using various Bioinformatics tools including the homology-based basic local alignment search tool (BLAST; http://www.ncbi.nlm.nih.gov/blast) and multiple sequence alignment by Clustal W (http://www.ebi.ac.uk/clustalw) for each query DNA sequence (Clustal W aligns sequences in a two-step process. Initially the sequence undergoes pair-wise alignment and for each pair of sequences it introduces gaps in an attempt to maximize the number of matching characters. A positive score is assigned for every match and Clustal W seeks the alignment that maximizes the score. The scores of the pair-wise alignments are employed to create a guide tree that is not a real phylogenetic tree and in turn guides for the progressive alignment in the second stage using a different set of penalties (i.e., mismatches) for creating the final alignment).

For molecular phylogenetic analysis, phylogenetic tree-building methods presume particular evolutionary models. Therefore, while interpreting the results obtained, different tree building models should be considered to entertain possible explanations. Several such models are available for construction of phylogenetic trees, which include character-based methods, namely, maximum parsimony (MP) and maximum likelihood and distance methods, namely the neighbor-joining (NJ), unweighted pair group method using arithmetic mean and minimum evolution. The distance methods are used so as to augment MP because they are less likely to give errors when trees contain long branches [100]. All these analyses can be done using MEGA software [101].

32.2.2.2 Phylogenetic Analysis

The various bioinformatics tools used for the purpose include Clustal W multiple alignment and MP and NJ methods for construction of phylogenetic trees using 1000 bootstrap replicates in MEGA for branch support [101,102] (Bootstrap is a statistical parameter employed for evaluating the robustness of the phylogenetic tree. Several pseudo datasets are random sampled from the original dataset and subsequently pseudo trees are generated. The latter are compared with the original dataset and a bootstrap value is assigned in percent scores. Groupings with 70% or more bootstrap values are reliable. The evolutionary distances are computed using the maximum composite likelihood method and are in units of the number of base substitutions per site. The MP tree is obtained using the close-neighbor-interchange algorithms and all positions containing gaps and missing data are eliminated from the dataset (complete deletion option). Parsimony is based on the assumption that the most likely tree is the one that attains the least possible number of changes to explain the data in the alignment and hence operates by selecting the tree that minimizes the number of evolutionary steps.)

32.2.2.3 Bayesian Phylogenetic Analysis

DNA sequences are aligned using ClustalX 2.0.7 (a standalone graphical user interface tool). The interleaved NEXUS file thus generated is edited manually in order for it to be recognized by MrBayes V3.1.2 program. Phylogenetic analysis is carried out using the Bayesian approach with combined datasets and default parameters, wherein each data partition is allowed to have a different evolution rate. MrBayes is executed in a command-line interface and uses the metropolis-coupled Monte Carlo Markov chain method, which allows a search for sampling trees from the distribution of posterior probabilities. The program starts with generation of a tree, evaluates according to the model specified and records the tree with its likelihood in a file. In several subsequent generations, the program calculates the consensus with the branch lengths to generate a cladogram with the posterior probabilities for each split and a phylogram with mean branch lengths; these are generated and subsequently read by the tree drawing program, Tree view V1.6.6 [103].

32.2.2.4 Motif Identification, Testing, and Validation

The ITS sequence motifs are identified from aligned sequences of the dataset for the species using PRATT software (http://www.ebi.ac.uk/Tools/pratt/index.html). The minimum percentage of sequences to match (C%) parameter is adjusted to report pattern matching at 100% of the sequence input. The motifs are expressed using the DNA alphabet (A, T, C, G) in PROSITE language [104]. The validation of the motifs is performed for each species using a "PATTERN MATCHING" web application (http://genoweb.univ-rennes1.fr/Serveur-GPO/). ITS sequence motifs are subjected to BLAST algorithm against the nonredundant GenBank database of the NCBI. The BLAST output is then analyzed to find only the exact or perfect pair-wise matches showing significantly high scores in terms of percent identity and low expect (E) values for each species.

32.2.2.5 Predicted ITS2 RNA Secondary Structures and Analyses

Secondary structures of ITS2 sequences are reconstructed by aligning the sequences using Bioedit. The acquired structures with restrictions and constrains are submitted in MFOLD. RNA is folded at a fixed temperature of 37°C, and the structure chosen from different output files is the desired six-helicoidal ring or the one with the highest negative free energy if various similar structures are obtained. The potential folding patterns of the ITS2 transcripts are cross compared, to discover the single secondary structure common to them all.

32.2.2.6 Sample Analysis

So far, information regarding the nucleotide sequences available in the GenBank of fasciolid flukes is limited to a few geographical isolates of respective species (Table 32.1). The nucleotide sequences of *F. buski* Indian isolate were obtained for ITS and were compared with sequences of *F. buski* Vietnam isolate and other trematode species obtained from GenBank. The fragments of amplified DNA were estimated to be 480–550 bp long. Phylogenetic trees were obtained by comparing the sequences of *F. buski* and available ITS2 sequences for other digenean trematodes including fasciolid, paragonimid, and echinostomatid flukes, all of which are prevalent potential zoonoses in tropical regions, Asia in particular. Phylogenetic analyses and evolutionary history were inferred using the distance method

TABLE 32.1
Fasciolid Species along with Their Hosts, Geographical Origins, and the GenBank Accession Number for Each Sequenced DNA Region

			GenBank Accession Numbers					
			rDNA			mtDNA		
Species	**Host**	**Location**	**28S**	**ITS1**	**ITS2**	**18S**	**Nad1**	**Co1**
F. buski	Domestic pig	Meghalaya, India	–	DQ351843 DQ351844	DQ351841 DQ351842	–	–	EF027094
F. buski	Domestic pig	Hanoi, Vietnam	EU025870	EF612477	EF612489	AY311386	EF612501	–
Parafasciolopsis fasciolamorpha	European bison	Bialowieza, Poland	EU025869	EF612476	EF612488	–	EF612500	–
Parafasciolopsis robusta	African forest elephant	Africa	EU025868	EF612478	EF612490	–	EF612502	–
F. magna	White-tailed deer	Minnesota, USA	EU025872	EF612475	EF612487	–	EF612499	–
F. gigantica	Domestic cow	Meghalaya, India		EF027103	EF027104	–	–	–
F. gigantica	Domestic cow	Nirju, Kenya	EU025873	EF612472	EF612484	–	EF612496	–
F. hepatica	Domestic water buffalo	Alexandria, Egypt	EU025874	EF612467	EF612479	–	EF612491	–
F. hepatica	Domestic sheep	Tehran, Iran		EF612469	EF612481	–	EF612493	–
F. jacksoni	Asian elephant	Sri Lanka	EU025871	EF612473	EF612485	–	EF612497	–

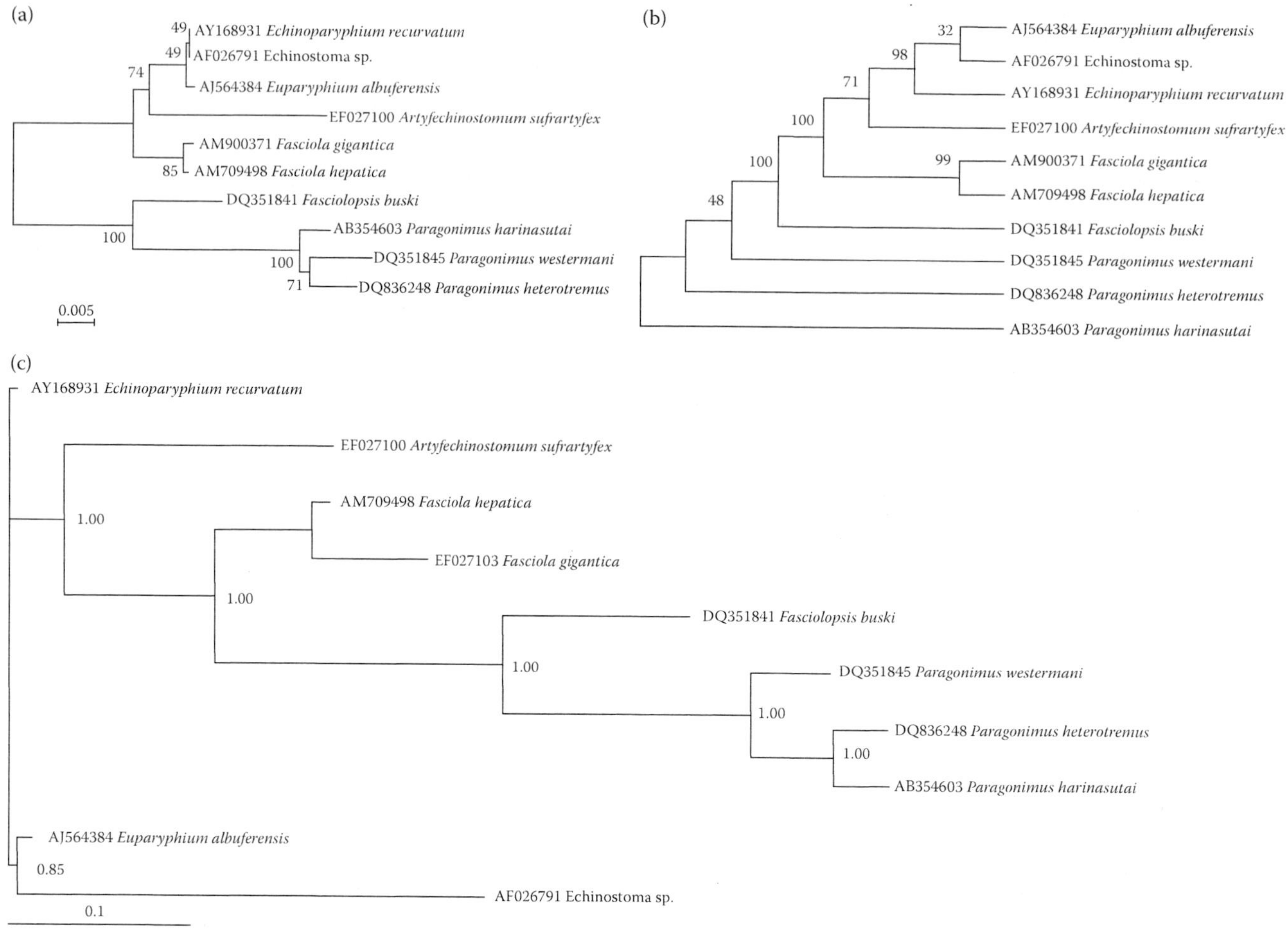

FIGURE 32.4 Phylogenetic trees showing relationships among *Fasciolopsis buski* and other digenean flukes: (a) N–J tree, (b) MP tree, and (c) Bayesian phylogeny.

N–J and character state method MP (Figure 32.4) [95]. The N–J tree showed that the *F. buski* sequence is close to family Paragonimidae, whereas the MP tree showed that the other members of Fasciolidae were the closest relative. Bootstrapping of the sequences revealed significant support (100%) among clades containing members of both Fasciolidae and Paragonimidae. The bootstrap consensus tree inferred from 1000 replicates is taken to represent the evolutionary history of the taxa analyzed. The percentages of replicate trees in which the associated taxa clustered together in the bootstrap test are shown next to the branches. Bayesian analysis of the alignment retained the same topology and supported the same branches as the secondary structures (Figure 32.4).

Predicted RNA secondary structures were reconstructed from the unique sequences with highest negative free energy of *F. buski* Meghalaya, India (DQ351845) and *F. buski* Hanoi, Vietnam (EF612489) to provide the basic information for phylogenetic analysis (Figure 32.5). The secondary structural features of ITS2 regions as shown in the figures were analyzed based on conserved stems and loops.

32.3 CONCLUSIONS AND FUTURE PERSPECTIVES

The taxonomy of *F. buski* has been based mainly on morphological data complemented with ecological, cytological, and pathological results as well as clinical manifestations. Traditional diagnostic techniques in parasitology are now complemented by a variety of molecular tools to help in resolving the taxonomic issues associated with describing new species or strains on the basis of phenotypic characteristics [105–107]. There are scanty data on genotypic diversity of fasciolid flukes, which makes it difficult to perform fine-scale phylogeographic analysis of populations and does not allow the origin of regional populations to be unambiguously determined. With regard to *F. buski*, only two geographical isolates—one from India and the other from Vietnam—have been worked out so far for molecular characterization using PCR-based techniques utilizing the rDNA and mtDNA markers. While information pertaining to ITS (1, 2) sequences is available for both the isolates, that regarding 28S, 18S, and Nad1 is limited to *F. buski* Vietnam isolate.

dG = −169.28 *Fasciolopsis buski* [Meghalaya, India] dG = −121.07 *Fasciolopsis buski* [Hanoi, Vietnam]

FIGURE 32.5 RNA secondary structures.

The fasciolid digeneans, believed to have evolved as parasites of Proboscidea in Africa to begin with, seem to have radiated to a wider range of herbivorous mammals across Europe and Asia. With this adaptive radiation indicating a host shift, a trend toward habitat shift (from intestine to liver and biliary passages) and diverticulation of intestinal caeca (extending to farthest extremities of the body) is also exhibited in the higher taxa of the family. However, in spite of this divergence, the fasciolids, likely all, share a similar pattern of life cycle with the infective metacercarial stage reaching the mammalian definitive host via an ingestive mode and an oral route. More geographical isolates of *F. buski*, particularly from the endemic regions, need to be characterized with the use of newer molecular markers, so as to determine the genetic variability, if any, and reveal genotypic diversity among the various populations of the parasite.

Control can be achieved by pharmacological treatment of people and pigs, preventing reinfection, and using modern pig farming practices [46]. Individual prevention is possible simply by avoiding the consumption of raw, water-derived food. However, it is extremely difficult to enforce this in view of age-old traditions. Such measures demand fundamental changes in the eating habits, customs, and economic conditions of the people, which cannot be brought about in a short frame of time [5]. A general awareness needs to be generated among the affected peoples, particularly in the endemic areas, stressing the importance of thoroughly cooking aquatic plants and boiling water where treated water is not available [26,37,46]. Such educational programs may facilitate this change. In the past, fasciolopsiasis was highly endemic in Taiwan, but the number of human cases decreased in the 1980s, thanks to aggressive educational programs [52,53]. Unfortunately, this tendency was not maintained, because many cultures still enjoy eating raw food [26]. Where it was thought to be controlled, as in the state of Uttar Pradesh, India, and where no cases were detected since the 1990s, there are reports of recent reemerging infections [108,109]. Fasciolopsiasis can be controlled along with other food-borne parasitoses [110]. However, despite control programs, this infection remains a public health problem in endemic areas.

REFERENCES

1. Roy, B., and Tandon, V. *Fasciolopsis buski*. In: Miliotis, M.D., and Bier, J.W. (eds), *International Handbook of Foodborne Pathogens*. Marcel & Decker Inc, Basel, p. 563, 2003.
2. Roy, B., and Tandon, V. Seasonal prevalence of some zoonotic trematode infections in cattle and pigs in the north-east montane zone in India. *Vet. Parasitol.* 41, 69, 1992.
3. Lankester, E.R. Animal parasites belonging to the group Entozoa, In: Kuchenmeister, F. (ed.), *On Animal and Vegetable Parasites of the Human Body. A Manual of their Natural History, Diagnosis and Treatment*, Vol. 1. The Sydenham Society, London, p. 452, 1857.
4. Looss, A. Weitere Beiträge zur kenntnis der Trematoden-Fauna Aegyptens zugleich versuch einer naturlichen Gleiderung des Genus Distomum Retzius. *Zoologische Jahrbücher. Abteilung für Systematik, Geograpbie und Biologie der Tiere.* 12, 521, 1899.
5. Graczyk, T.K., Gilman, R.H., and Fried, B. Fasciolopsiasis: Is it a controllable food-borne disease? *Parasitol. Res.* 87, 80, 2001.
6. Lachowicz, J. Development of *Parafasciolopsis fasciolaemorpha* Ejsmont, 1932 from eggs derived from sheep. *Acta Parasit.* 33, 169, 1988.

7. Jones, A. Family Fasciolidae Railliet, 1895. In: Jones, A., Bray, R.A., and Gibson, D. I. (eds), *Keys to the Trematoda*, Chapter 6. CABI Publishing and the Natural History Museum, Cambridge, pp. 79–85, 2005.
8. Kumar, V. The digenetic trematodes, *Fasciolopsis buski, Gastrodiscoides hominis* and *Artyfechinostomum malayanum*, as zoonotic infections in South Asian countries. *Ann. Soc. Belg. Med. Trop.* 60, 331, 1980.
9. Roy, B., and Tandon, V. Morphological and microtopographical strain variations among *Fasciolopsis buski* originating from different geographical areas. *Acta Parasitol.* 38, 72, 1993.
10. Gilman, R.H. et al. Endemic focus of *Fasciolopsis buski* in Bangladesh. *Am. J. Trop. Med. Hyg.* 31, 796, 1982.
11. Smyth, J.D. *Animal Parasitology, Cambridge Low Price Editions*, Cambridge University Press, Great Britain, p. 212, 1996.
12. Hsieh, H.C. Studies on the epidemiology of *Fasciolopsis buski* in South Taiwan. *Formos. Sci.* 14, 95, 1960.
13. Hsu, P.J. A survey of fasciolopsiasis of pigs in Kwantung province. *Acta Vet. Zootech. Sin.* 7, 143, 1964.
14. Wang, P. et al. Studies on the life-history of *Fasciolopsis buski* (Lankester, 1857) with consideration of its seasonal infection in pigs. *Acta Zool. Sinica.* 23, 88, 1977.
15. Nguen Tkhi Le. A study of the biology of the trematode *Fasciolopsis buski* (Lankester, 1857) in Vietnam. *Med. Parazitol. (Moskva)* 47, 81, 1978.
16. Tripathi, J.C., Srivastava, H.D., and Dutt, S.C. A note on experimental infection of *Helicorbis coenosus* and pig with *Fasciolopsis buski. Indian J. Anim. Sci.* 43, 647, 1973.
17. Buckley, J.J.C. Observations on *Gastrodiscoides hominis* and *Fasciolopsis buski* in Assam. *J. Helminthol.* 17, 1, 1939.
18. Malviya, H.C., and Varma, T.K. On the life history of *Fasciolopsis buski*. Ninth National Congress of Parasitology, Ujjain, January 29–31, p. 16, 1989.
19. Preet, S., and Prakash, S. Cercarial emergence in *Fasciolopsis buski* (Lankester). *J. Parasit. Dis.* 25, 108, 2001.
20. Manning, G.S., and Ratanarat, C. *Fasciolopsis buski* (Lankester, 1857) in Thailand. *Am. J. Trop. Med. Hyg.* 19, 613, 1970.
21. Ditrich, O. et al. Larval stages of medically important flukes (Trematoda) from Vientiane province, Laos, Part II. Cercariae. *Ann. Parasitol. Hum. Comp.* 67, 75, 1992.
22. Graczyk, T.K. et al. Development of *Fasciolopsis buski* (Trematoda: Fasciolidae) in *Hippeutis umbilicalis* and *Segmentina trochoideus* (Gastropoda: Pulmonata) *Parasitol. Res.* 86, 324, 2000.
23. Barlow, C. The life cycle of human intestinal fluke *Fasciolopsis buski* (Lankester). *Am. J. Hyg. Monoger. Ser.* 4, 1, 1925.
24. Lo, C.T. Life history of the snail, *Segmentina hemisphaerula* (Benson), and its experimental infection with *Fasciolopsis buski* (Lankester). *J. Parasitol.* 53, 735, 1967.
25. Nguyen Van Tho. Evolution cycle of *Fasciolopsis buski* in the mollusc-intermediate host and in nature. *Khoa Hoc Ky Thuat Thu Y (Vet. Sci. Tech.).* 9, 43, 2002.
26. Weng, Y.L. et al. Studies on ecology of *Fasciolopsis buski* and control strategy of fasciolopsiasis. *Zhongguo Ji Sheng Chong Xue Yu Ji Sheng Chong Bing Za Zhi.* 7, 108, 1989.
27. Nakagawa, K. On the life cycle of *Fasciolopsis buski* (Lankester). *Kitasato Arch. Exp. Med.* 4, 159, 1921.
28. Nakagawa, K. The development of *Fasciolopsis buski* Lankester. *J. Parasitol.* 8, 161, 1922.
29. Arambulo, P.V., and Moran, N., Food-transmitted parasitic zoonoses—socio-cultural and technological determinants. *Int. J. Zoon.* 7, 135, 1980.
30. Chandra, S.S. Epidemiology of *Fasciolopsis buski* in Uttar Pradesh. *Indian J. Med. Res.* 78, 55, 1984.
31. Jones, M.E.H., Hubbert, W.T., and Hagstad, H.V. *Zoonoses—Recognition, Control and Prevention*, 1st edn. Iowa State University Press, Ames, p. 245, 1995.
32. D'Souza, P.E., Jagannath, M.S., and Prasanna, K.M. A note on the occurrence of *Fasciolopsis buski* in pigs, *Cheiron* 30, 178, 2001.
33. Rim, H.J. Fasciolopsiasis. In: Hillyer, G.V., and Hopla, C.E. (eds), *Handbook Series in Zoonoses*, Vol. 3. CRC Press, Boca Raton, pp. 89–97, 1982.
34. Malviya, H.C. The susceptibility of mammals to *Fasciolopsis buski. J. Helminthol.* 59, 19, 1985.
35. Kuntz, R.E., and Lo, C.T. Preliminary studies on *Fasciolopsis buski* (Lankester, 1857) (giant Asian intestinal fluke) in the United States. *Trans. Am. Microsc. Soc.* 86, 163, 1967.
36. Hartman, H.A. The intestinal fluke (*Fasciolopsis buski*) in a monkey. *Am. J. Vet. Res.* 22, 1123, 1961.
37. Cross, J.H. Changing patterns of some trematode infections in Asia. *Arzneimittelforschung* 34, 1224, 1984.
38. Sadun, E.H., and Maiphoom, C. Studies on the epidemiology of the human intestinal fluke, *Fasciolopsis buski* (Lankester) in Central Thailand. *Am. J. Trop. Med. Hyg.* 2, 1070, 1953.
39. Muttalib, M.A., and Islam, N., *Fasciolopsis buski* in Bangladesh—A pilot study. *J. Trop. Med. Hyg.* 78, 135, 1975.
40. Yu, S. et al. Report of the first nationwide survey of the distribution of human parasites in China. 1. Regional distribution of parasitic species. *Zhongguo Ji Sheng Chong Xue Yu Ji Sheng Chong Bing Za Zhi.* 12, 241, 1994.
41. World Health Organization, Foodborne trematode infections. *Bull. World Health Organ.*, 73, 397, 1995.
42. World Health Organization, Control of foodborne trematode infections. *World Health Organ. Tech. Rep. Ser.* 849, 1, 1995.
43. Jaroonvesama, N. et al. Prevalence of *Fasciolopsis buski* and other parasitic infections in residents of three villages in Sena district, Ayudhaya Province, Thailand. *J. Med. Assoc. Thai.* 63, 493, 1980.
44. Shyu, L.Y., Lee, H.H., and Chen, E.R. A preliminary study on epidemiology of fasciolopsiasis in Tainan Hsien, south Taiwan. *Chin. J. Microbiol. Immunol.* 17, 118, 1984.
45. Thi, L.N., and Sey, O. Infectivity of metacercariae of *Fasciolopsis buski* (Lankester, 1857) and their development in definitive host. *Parasitol. Hung.* 18, 15, 1985.
46. Gai, L. et al. Relationships between the rate of parasitic infection and the knowledge of prevention. *Chin. J. Parasitol. Parasit. Dis.* 13, 269, 1995.
47. Lo, C.T., and Lee, K.M. Intestinal parasites among Southeast Asian labours in Taiwan during 1993–1994. *Zhonghua Yi Xue Za Zhi (Taipei)* 57, 401, 1996.
48. Xu., L. et al. Characteristics and recent trends in endemicity of human parasitic diseases in China. *Zhongguo Ji Sheng Chong Xue Yu Ji Sheng Chong Bing Za Zhi.* 13, 214, 1995.
49. Li, G. et al. Relationship between the rate of parasitic infection and the knowledge of prevention. *Chin. J. Parasitol. Parasit. Dis.* 13, 269, 1995.
50. Longqi, X. et al. Characteristics and recent trends in endemicity of human parasitic diseases in China. *Chin. J. Parasitol. Parasit. Dis.* 13, 214, 1995.
51. Hotez, P.J. et al. Emerging and reemerging helminthiasis and the public health of China. *Emerg. Infect. Dis.* 3, 303, 1997.
52. Lee, H.H. Studies on epidemiology and treatment of fasciolopsiasis in southern Taiwan. *Kaohsiung J. Med. Sci.* 2, 21, 1986.

53. Lee, H.H., Shyu, L.Y., and Chen, E.R. Experimental control study of fasciolopsiasis in Taiwan. *Kaohsiung J. Med. Sci.* 5, 335, 1989.
54. Lee, H.H. *Fasciolopsis buski* infection among children of Liuying Primary School in Tainan Hsien south Taiwan. *Chin. J. Microbiol.* 5, 110, 1972.
55. Misra, S.C. et al. Incidence of gastrointestinal parasites in pigs with evaluation of an effective technique for the recovery of their ova or cysts in the faeces. *Ind. Vet. J.* 49, 140, 1972.
56. Malviya, H.C., and Varma, T.K. Incidence of *Fasciolopsis buski* infection in pigs and snails. *Ninth National Congress of Parasitology*, Ujjain, January 29–31, p. 15, 1989.
57. Roy, B., and Tandon, V. Trematodiasis in North East India: A study on the spectrum of digenetic trematodes among pigs, buffaloes, cattle, goats and sheep. *Indian J. Anim. Health* 31, 5, 1992.
58. Mohan, A., Mirdha, B.R., and Vishaka, M. Fascioliasis in a pregnant woman. *Indian J. Gastroenterol.* 15, 155, 1996.
59. Mahanta, J. Trematode infection of public health importance in Assam. *Second Global Meet on Parasitic Diseases*, Indian Society for Parasitology, Hyderabad, August 18–22, p. 72, 1997.
60. Shah, A., Gadgil, R.K., and Manohar, K.D. Fasciolopsiasis in Bombay. A preliminary communication. *Indian J. Med. Sci.* 20, 805, 1966.
61. Manjarumkar, P.V., and Shah, P.M. Epidemiological study of *Fasciolopsis buski* in Palghar Taluk. *Indian J. Public Health.* 16, 3, 1972.
62. Harinasuta, T., Bunnag, D., and Radomyos, P. Efficacy of praziquantel on fasciolopsiasis. *Arzneimittelforschung* 34, 1214, 1984.
63. Yoshihara, S. et al. Helminths and helminthiosis of pigs in the Mekong Delta, Vietnam with special reference to ascariosis and *Fasciolopsis buski* infection. *Jpn. Agric. Res. Q.* 33, 193, 1999.
64. Manning, G.S., Brockelman, W.Y., and Yiyant, V., An analysis of the prevalence of *Fasciolopsis buski* in central Thailand using catalytic models. *Am. J. Epidemiol.* 93, 354, 1971.
65. Bunnag, D., Radomyos, P., and Harinasuta, T. Field trial on the treatment of fasciolopsiasis with praziquantel. *Southeast Asian J. Trop. Med. Public Health* 14, 216, 1983.
66. Wiwanitkit, V., Suwansaksri, J., and Chaiyakhun, Y. High prevalence of *Fasciolopsis buski* in an endemic area of liver fluke infection in Thailand. *Med. Gen. Med.* 4, 6, 2002.
67. Waikagul, J. Intestinal fluke infections in Southeast Asia. *Southeast Asian J. Trop. Med. Public Health* 22, 158, 1991.
68. Giboda, M. et al. Current status of food-borne parasitic zoonoses in Laos. *Southeast Asian J. Trop. Med. Public Health* 22, 56, 1991.
69. Hadidjaja, P. et al. First autochthonous case of *F. buski* infection in Indonesia. *Am. J. Trop. Med. Hyg.* 31, 1065, 1982.
70. Margono, S.S. Important human helminthiasis in Indonesia. In: Crompton, D.W.T., Montresor, A., Nesheim, M.C., and Savioli, L. (eds), *Controlling Disease due to Helminth Infections.* World Health Organization, Geneva, p. 3, 2003.
71. Handoyo, I. et al. A survey of fasciolopsiasis in Sei Papuyu village of Babirik subdistrict, Hulu Sungei Utara Regency, South Kalimantan Province. *Trop. Biomed.* 3, 113, 1986.
72. Shekhar, K.C. Epidemiological assessment of parasitic zooonoses in Malaysia. *Southeast Asian J. Trop. Med. Public Health* 22, 337, 1991.
73. Basnuevo, J.G., and Seuc-Chiu, A. *Fasciolopsis buski* in Cuba; two cases treated with chloroquine, gentian violet, hexylresorcinol and tetrachloroethylene. *Rev. Cubana. Med. Trop. Parasitol.* 6, 91, 1950.
74. Greenberg, Z. et al. Prevalence of intestinal parasites among Thais in Israel. *Harefuah* 126, 507, 1994.
75. Schubert, S., and Granz, W. Problems in the differential diagnosis of intestinal diseases in patients who have traveled in the tropics. *Z. Gesamte. Med.* 36, 230, 1981.
76. Shah, P.M. et al. *Fasciolopsis buski* infestation in children. *Indian Pediatr.* 10, 721, 1973.
77. Rahman, K.M., Idris, M., and Khan, A.K.A. A study of fasciolopsiasis in Bangladesh. *J. Trop. Med. Hyg.* 84, 81, 1981.
78. Chandra, S.S. A field study on the clinical aspects of *Fasciolopsis buski* infections in Uttar Pradesh. *Med. J. Armed Forces India* 32, 181, 1976.
79. Jaroonvesama, N., Charoenlarp, K., and Areekul, S., Intestinal absorption studies in *Fasciolopsis buski* infection. *Southeast Asian J. Trop. Med. Public Health* 17, 582, 1986.
80. Viranuvatti, V., Stitinimankarn, T., and Tansurat, P. A fatal case of infection with *Fasciolopsis buski* in Thailand. *Ann. Trop. Med. Parasitol.* 47, 132, 1953.
81. Yu, S.H., 'and Mott, K.E., Epidemiology and morbidity of food-borne intestinal trematode infections. *Trop. Dis. Bull.* 91, R125, 1994.
82. Plaut, A.G. Kampanart-Sanyakorn, C., and Mannig, G.S. A clinical study of *Fasciolopsis buski* infection in Thailand. *Trans. R. Soc. Trop. Med. Hyg.* 63, 470, 1969.
83. Le, T.H. et al. Case report: Unusual presentation of *Fasciolopsis buski* in a Vietnamese child. *Trans. R. Soc. Trop. Med. Hyg.* 98, 193, 2004.
84. Blair, D. et al. Geographical genetic structure within the human lung fluke, *Paragonimus westermani*, detected from DNA sequences. *Parasitology* 115, 411, 1997.
85. Hillis, D.M., and Dixon, M.T. Ribosomal DNA: Molecular evolution and phylogenetic inference. *Q. Rev. Biol.* 66, 411, 1991.
86. Morgan, J.A.T., and Blair, D. Nuclear rDNA ITS sequence variation in the trematode genus *Echinostoma:* An aid to establishing relationships within the 37-collar spine group. *Parasitology* 111, 609, 1995.
87. Leon-Regagnon, V., Brooks, D.R., and Perez-Ponce, de Leon, G., Differentiation of Mexican species of Haematoloechus Looss, 1899 (Digenea: Plagiorchiformes): Molecular and morphological evidence. *J. Parasitol.* 85, 935, 1999.
88. Tkach, V.V., Pawlowski, J., and Sharpilo, V.P., Molecular and morphological differentiation between species of the *Plagiorchis vespertilionis* group (Digenea, Plagiorchiidae) occurring in European bats, with a re-description of *P. vespertilionis* group (Muller, 1780). *Syst. Parasitol.* 47, 9, 2000.
89. Kostadinova, A. et al. Phylogenetic relationships of *Echinostoma rudolphi*, 1809 (Digenea: Echinostomatidae) and related genera re-assessed via DNA and morphological analyses. *Syst. Parasitol.* 54, 159, 2003.
90. Scholz, T. et al. A taxonomic study of species of *Bothriocephalus* Rudolphi, 1808 (Cestoda: Pseudophyllidea) from eels in Japan: Morphological and molecular evidence for the occurrence of *B. claviceps* (Goeze, 1782) and confirmation of the validity of *B. japonicus* Yamaguti, 1934. *Syst. Parasitol.* 57, 87, 2004.
91. Nolan, M.J., and Cribb, T.H. The use and implications of ribosomal DNA sequencing for the discrimination of digenean species. *Adv. Parasitol.* 60, 102, 2005.
92. Luton, K., Walker, D., and Blair, D. Comparisons of ribosomal internal transcribed spacers from two congeneric species of flukes (Platyhelminths: Trematoda: Digenea). *Mol. Biochem. Parasitol.* 56, 323, 1992.
93. Adlard, R.D. et al. Comparison of the second internal transcribed spacer (ribosomal DNA) from populations and species of Fasciolidae (Digenea). *Int. J. Parasitol.* 23, 423, 1993.

94. Le, T.H. et al. Phylogenies inferred from mitochondrial gene orders—A cautionary tale from the parasitic flatworms. *Mol. Biol. Evol.* 17, 1123, 2000.
95. Prasad, P.K. et al. PCR-based determination of internal transcribed spacer (ITS) regions of ribosomal DNA of giant intestinal fluke. *Fasciolopsis buski* (Lankester, 1857) Looss, 1899. *Parasitol. Res.* 101, 1581, 2007.
96. Prasad, P.K. et al. Molecular identification of Indian liver fluke, *Fasciola* (Trematoda: Fasciolidae) based on the ribosomal internal transcribed spacer regions. *Parasitol. Res.* 103, 1247, 2008.
97. Sambrooke, J., Fritsch, E.F., and Maniatis, T. *Molecular Cloning: A Laboratory Manual*, 2nd edn. Cold Spring Harbor Laboratory Press, Cold Spring Harbor, NY, 1989.
98. Bowles, J., Blair, D., and McManus, D.P. A molecular phylogeny of the human schistosomes. *Mol. Phylogenet. Evol.* 4, 103, 1995.
99. White, B.A. *PCR Protocols, Current Methods and Applications*, Vol. 15. Humana, Totowa, NJ, 1993.
100. Blair, D., and Barker, S.C., Affinities of the Gylianchenidae: Utility of the 18S rRNA gene for phylogenetic inference in the Digenea (platyhelminthes). *Int. J. Parsitol.* 23, 527, 1993.
101. Tamura, K. et al. MEGA4: Molecular Evolutionary Genetics Analysis (MEGA) software version 4.0. *Mol. Biol. Evol.* 24, 1596, 2007.
102. Hillis, D.M., and Bull, J.J. An empirical test of bootstrapping as a method for assessing confidence in phylogenetic analysis. *Syst. Biol.* 42, 182, 1993.
103. Huelsenbeck, J.P., and Ronquist, F. Bayesian inference of phylogenetic trees. *Bioinformatics.* 17, 754, 2001.
104. Bailey, T.L., and Elkan, C. Fitting a mixture model by expectation maximization to discover motifs in biopolymers. In: *Proceedings of the Second International Conference on Intelligent Systems for Molecular Biology.* AAAI Press, Menlo Park, CA, p. 28, 1994.
105. Thompson, R.C.A. et al. Advances in the diagnosis and systematics of parasites of veterinary importance: New and exciting prospects. In: Molecular systematics and diagnosis. Gasser, R.B. and Zarlenga, D.S. (eds), *Vet. Parasitol.* 125, 69,, 2004.
106. Blair, D. et al. A molecular perspective on the genera *Paragonimus* Braun, *Euparagonimus* Chen and *Pagumogonimus* Chen. *J. Helminthol.* 74, 295, 1999.
107. Iwagami, M. et al. Molecular phylogeographic studies on *Paragonimus westermani* in Asia. *J. Helminthol.* 74, 315, 2000.
108. Bhatti, H.S. et al. Fasciolopsiasis, a re-emerging infection in Azamgarh (Uttar Pradesh). *Indian J. Pathol. Microbiol.* 43, 73, 2000.
109. Muralidhar, S. et al. Fasciolopsiasis, a persisting problem in eastern U.P.: A case report. *Indian J. Pathol. Microbiol.* 43, 69, 2000.
110. Mott, K.E. et al. New geographical approaches to control of some parasitic zoonoses. *Bull. World Health Org.* 73, 247, 1995.

33 Gastrodiscoides

Dongyou Liu

CONTENTS

33.1 INTRODUCTION

The subclass Digenea in the phylum platyhelminthes consists of over 6000 parasitic flatworms (trematodes or flukes) that reside in the digestive tract and other organs of animal hosts. A number of Digenean species have been implicated in human diseases, including *Schistosoma* spp., *Fasciolopsis buski*, *Fasciola hepatica*, *Heterophyes heterophyes*, *Metagonimus yokogawaii*, *Paragonimus westermani*, *Clonorchis sinensis*, *Opisthorchis viverrini*, and *Gastrodiscoides hominis*. Compared with other human pathogenic trematodes, *G. hominis* causes relatively mild clinical diseases such as abdominal pain and diarrhea. Nonetheless, it is important to be able to differentiate gastrodiscoidiasis from intestinal disorders due to other microorganisms. This will not only aid the selection of appropriate treatments measures, but also contribute to the effective control and prevention of gastrodiscoidiasis in humans.

33.1.1 Classification, Morphology, and Biology

33.1.1.1 Classification

The genus *Gastrodiscoides* is classified in the family Paramphistomidae, suborder Paramphistomata, order Echinostomida, subclass Digenea, class Trematoda, phylum Platyhelminthes. The family Paramphistomidae is divided into seven genera: *Calicophoron*, *Gastrodiscoides*, *Gigantocotyle*, *Indosolenorchis*, *Orthocoelium*, *Paramphistomum*, and *Stichorchis*; and the genus *Gastrodiscoides* comprises of a single species *G.* hominis (synonym: *Gastrodiscus hominis*).

First brought to attention in 1939, *G. hominis* (also known as colonic fluke) is a natural parasite of the colon of the pig, and is occasionally involved in human infection (found in the human cecum).

33.1.1.2 Morphology

G. hominis adult worms (6.41–8.04 mm in length and 4.29–5.42 mm in width) are thick, fleshy, pyramidal in shape, and bright pink in color (when alive). Its body can be divided into anterior conical and posterior discoidal regions. The anterior region is short, narrow, and conical-cylindrical (ending with a rounded tip), and bears a small, globular, oral sucker, and a prominent genital cone. The posterior region (up to 8.0 mm in width) is large, discoidal, and ventrally excavated, bears a ventral sucker (concave disc) and contains an oval-shaped ovary located under the 2-lobed testes (the reproductive organs). The testes are situated below the gut caecae and are in front of the vitellaria that surround the intestinal caecae [1].

G. hominis worms recovered from pigs tend to be smaller than those from humans, and have characteristic genital papilla/cone and shape and arrangements of testes. For this reason, some authors regard the parasite recovered from pigs as a subspecies/variety of *G. hominis*.

Similar to other Digenean flukes, the oral sucker in the anterior portion of *G. hominis* has two functions: (i) to hold the parasite to its host and (ii) to assist in feeding. The ventral sucker in the posterior portion has only one function, that is, to attach to its host.

Electron microscope examination of *G. hominis* adult worm reveals that the tegument is folded into concentrically arranged furrows and ridges bearing numerous tightly packed tubercules, and extends into the oral cavity. An area of specialized tegument is present on the ventral surface, anterior to the disc region. Ciliated and nonciliated sensory papillae are present, particularly around the oral opening [2,3].

G. hominis eggs are ovoid (rhomboidal), measure 127–160 × 62–75 μm (average 146 × 66 μm), and have a pale greenish-gray color (in fresh feces) and a thickened aboper-cular end which rarely shows a spine-like elongation.

33.1.1.3 Biology

G. hominis has a complicated life cycle involving at least one intermediate host (an aquatic snail) and a primary (definitive) host (pig), and both asexual and sexual reproduction [4].

G. hominis adult lives in the lumen of the cecum and colon by attaching to the intestinal mucosa and lays eggs that are discharged in host feces. The eggs are ingested by the aquatic snail *Helicorbis coenosus* (synonym: *S. coenosus*), and hatch for 9–14 days at a temperature of 24–33°C to release the miracidia. In the digestive gland, the miracidia develop from the sporocyst stage to one or more generations of rediae. The rediae (148–747 × 45–140 μm in size) are sausage-shaped, lack collar and locomotor organs, but have salient birthpore located near the level of the saccular intestine. The redia transforms into the cercaria (with a keeled tail) in about 20 days. The cercaria (of 403–865 × 192–310 μm) has an ovoid to pyriform body with a narrower anterior end, and possesses a pair of dark-pigmented eye-spots. The snail sheds 1–34 (average 6.8) cercariae per day, intermittently for 6–40 days, with a total of 7–238 (average 74.7) cercariae shed per snail, and lives for 10–147 days after the first shedding day. The cercariae emerge in morning hours, swim for 1–24 h or longer, attach to vegetation (caltrop), and encsyt as metacercaria. The metacercarial cyst (of 201–227 μm, average 216 μm in diameter) is hemispherical and brownish. The metacercaria can remain attached to vegetation (where night soil is used). After ingestion of the contaminated vegetation by a final host, the metacercariae release juvenile flukes in the small intestine, which migrate to the cecum and colon and mature into adults. After self-fertilization, the adult worm lays hundreds of eggs for passage through the feces [4].

The primary (definitive) host of *G. hominis* is pigs (*Sus scrofa domestica* as well as wild pigs), and the intermediate host is *H. coenosus*, an aquatic snail that abounds in the water reservoirs around pigsties [5]. The number of *G. hominis* flukes in an individual pig can be very high (up to 1886, average 227, worms) [6]. Besides pigs, *G. hominis* can also infect monkeys, rats, fish, and other fish-eating animals as well as humans. Indeed, the parasite has been identified from the Napu mouse deer (*Tragulus napu*) in Malaysia, the field rat (*Rattus brevicaudatus*) in Java, rhesus monkeys (*Macaca mulatta*) in India, and American muskrat (*Ondatra zibethica*) introduced in the Volga Delta, Russia.

Humans are considered an accidental host for *G. hominis*. This parasite is usually transmitted to humans through the ingestion of vegetation (aquatic plant caltrop) contaminated by snail. It can also be transmitted after eating raw or undercooked fish, crustaceans (crayfish), squid, molluscs, or amphibians (frogs, tadpoles). These animals may act as a second intermediate host for *G. hominis*, and acquire the parasite after eating infected snails, with the cercariae developing into the metacercaria. *G. hominis* infection in humans can be asymptomatic or cause serious health problems and even mortality.

G. hominis is endemic in India (Assam, Bengal, Bihar, Uttar Pradesh, Madhya Pradesh, and Orissa), Pakistan, Burma, Thailand, Vietnam, the Philippines, China, Kazakstan, and the Volga Delta in Russia [7]. It has also been found in Zambia and Nigeria as well as in Indian immigrants in Guyana. In children in Kamrup district, Assam, eastern India, the prevalence is as high as 41% [8].

33.1.2 Clinical Features and Treatment

33.1.2.1 Clinical Features

While human gastrodiscoidiasis can be asymptomatic, it has been associated with intestinal disorders including diarrhea, abdominal pain, colic, and an increase in mucous production in a number of individuals. In severe cases, tissue reactions can occur in the heart or mesenteric lymphatics, and deaths may result among untreated patients, especially children. The intestinal symptoms caused by *G. hominis* infection are attributable to the inflammation of the mucosa of the cecum and ascending colon [9].

Being a large parasitic fluke of pig and human, *G. hominis* is endemic in the Indian state of Assam (as well as other locations), the Philippines and Southeast Asia, where many cases of human gastrodiscoidiasis have been reported. Sporadical cases of human *G. hominis* infection have been also described elsewhere. Dada-Adegbola et al. [10] documentad a case of gastrodiscoidiasis in a 7-year-old child in Nigeria. The patient appeared malnourished and anemic and was found to harbor both *G. hominis* and *Ascaris lumbricoides* worms. After anthelmintic treatment, the patient recovered well, with a much better state of health.

In pigs, *G. hominis* induces characteristic gross lesions, such as mucosal desquamation, infiltration with eosinophils, lymphocytes and plasma cells, and infiltrated submucosa, leading to a subacute inflammation of the cecum and mucoid diarrhea. Histopathological changes include marked desquamation of epithelial lining, hypersecretion of mucus, necrosis of the mucous glands, massive infiltration of eosinophils in the lamina propria (between the crypts of Lieberkuhn), the thickened muscularis mucosa (due to infiltration of eosinophils, lymphocytes, plasma cells, and macrophages), hyperemic blood vessels, thickened tunica intima, and infiltration by eosinophils and lymphocytes [6].

33.1.2.2 Treatment

G. hominis infection can be treated with tetrachlorothyline (0.1 mg/kg on an empty stomach) or prazinquantal (three doses in one day at 25 mg/kg), which is less toxic and eliminates most of the parasite. Other drugs include thymol, carbon tetrachloride, and mebendazole (a single dose of 500 mg) [10]. Application of a soapsud enema (or soap water) can be effective in removing the worms, as it works to flush the flukes from the colon and thus removes the parasite entirely.

33.1.3 Laboratory Diagnosis

Because clinical manifestations of human gastrodiscoidiasis are generally nonspecific, identification of *G. hominis* adult worms and eggs in feces represents a useful approach for the diagnosis. When passed spontaneously or expelled following treatment with anthelmintics or soap water enemas, the thick, fleshy, pyramidal, bright-pink body of *G. hominis* adult worm is unmistakable. The eggs of *G. hominis* are easily distinguishable from those of *F. buski* (another parasitic fluke of human intestine), as the former are ovoid (rhomboidal) and pale greenish grey while the latter are oval and yellowish brown. Nonetheless, species-specific identification of *G. hominis* eggs based on microscopic examination is technically challenging.

In recent years, the application of DNA-based methods provides an alternative option for the detection and identification of *G. hominis* parasite. The common gene targets for molecular discrimination of *G. hominis* and related trematodes (e.g., *P. westermani*, *F. buski*, *F. hepatica*, *C. sinensis*, and *O. viverrini*) include the highly conserved 18S and 28S ribosomal RNA (rRNA) genes as well as ribosomal noncoding DNA internal transcribed spacer regions (ITS1 and ITS2) [11–13]. Polymerase chain reaction amplification of ITS1 and ITS2 followed by sequencing analysis facilitates accurate resolution of the taxonomy and phylogenetic relationship of *G. hominis* isolates [14].

33.2 METHODS

33.2.1 Sample Preparation

For the preparation of DNA from *G. hominis* adult flukes, individual worms are first digested in extraction buffer (containing 1% SDS, 25 mg proteinase K) at 37°C overnight. DNA is then extracted from lysed individual worms by the standard phenol/chloroform technique.

For the preparation of DNA from *G. hominis* eggs, eggs are removed from the mature adult fluke by squeezing between two slides and placed on the Flinders Technology Associates (FTA) cards (Whatman Biosciences). The cards are dried for 1 h, punched, and two sample discs taken from each card are added into a tube. Sample discs are washed with Whatman's FTA purification reagent and TE buffer, allowed to dry for 1 h at room temperature, and used for PCR analysis.

33.2.2 Detection Procedures

Goswami et al. [14] utilized primers BD1 (forward: 5′-GTCGTAACAAGGTTTCCGTA-3′) and 4S (reverse: 5′-TCTAGATGCGTTCGAA (G/A) TGTCGATG-3′) for amplification of a 841 base pair (bp) fragment in the rRNA region spanning the ITS; and primers 3S (forward: 5′-GGTACCGGTGGATCACTCGGCTCGTG-3′) and A28 (reverse:5′-GGGATCCTGGTTAGTTTCTTTTCCTCCGC-3′) for amplification of a 494 bp fragment in the rRNA region spanning the ITS of *G. hominis* DNA.

33.2.2.1 Procedure

1. For amplification of ITS1 region, the PCR mixture (25 μL) containing buffer, dNTPs, primers BD1 and 4S, Taq DNA polymerase, and *G. hominis* DNA is subjected to initial denaturation at 94°C for 5 min; 35 cycles of 94°C for 60 s, 54°C for 50 s, and 72°C for 80 s; and a final extension at 72°C for 7 min.
2. For amplification of ITS2 region, the PCR mixture (25 μL) containing buffer, dNTPs, primers 3S and A28, Taq DNA polymerase, and *G. hominis* DNA is subjected to an initial denaturation at 94°C for 5 min; 26 cycles of 94°C for 30 s, 55°C for 38 s, and 72°C for 42 s; and a final extension at 72°C for 10 min.
3. The resultant PCR products are separated by 1.5% agarose gel electrophoresis in TAE buffer, stained with ethidium bromide, transilluminated under ultraviolet light, and then photographed. A DNA molecular marker (100 bp ladder) is run along with the samples.
4. PCR products are purified using the Gene Quick PCR purification kit, sequenced in both directions using PCR primers on an automated sequencer.
5. The resulting ITS1 and ITS2 sequences are compared using nucleotide BLAST (blastn) with default settings, and phylogenetic prediction is done by CLUSTALW for each query DNA sequence. ITS sequences are entered in the MEGA 4.0 package for the construction of the phylogenetic trees using maximum parsimony and distance methods, namely the Neighbor-Joining, UPGMA, and Minimum Evolution.

Note: *G. hominis* shows close resemblance with members of Paramphistomidae and Fasciolidae in the ITS1 sequence and with members of families Paramphistomidae and Echinostomatidae in the ITS2 sequence.

33.3 CONCLUSION

The genus *Gastrodiscoides* consists of a single amphistomid fluke species (*G. hominis*) that commonly occurs in the colon of pigs, and is occasionally implicated in human disease. This parasite requires aquatic snails (as intermediate host), vegetation (caltrop), or fish/aquatic animals (as second intermediate host), and pigs (or other animals including humans) as final host for its life cycle. Humans become infected with the parasite after ingestion of contaminated vegetation or raw fish and aquatic products. The clinical symptoms of human gastrodiscoidiasis range from abdominal pain, diarrhea, colic, and an increase in mucous production. Traditional approaches for diagnosis of human gastrodiscoidiasis rely on microscopic identification of *G. hominis* worms and/or eggs in stool samples. Given the close morphological similarity among related human pathogenic trematodes, molecular techniques based on PCR amplification provide a valuable tool for rapid, sensitive, and specific determination of *G. hominis* parasite.

REFERENCES

1. Jones, A., A comparative morphological study of the subfamily *Gastrodiscinae Monticelli*, 1892 (Paramphistomidae) with comments on some members of related subfamilies. *J. Nat. Hist.*, 20, 863, 1986.
2. Brennan, G.P., Hanna, R.E., and Nizami, W.A., Ultrastructural and histochemical observations on the tegument of *Gastrodiscoides hominis* (Paramphistoma: Digenea). *Int. J. Parasitol.*, 21, 897, 1991.
3. Tandon, V., and Maitra, S.C., Surface morphology of *Gastrodiscoides hominis* (Lewis & McConnell, 1876) Leiper, 1913 (Trematoda: Digenea) as revealed by scanning electron microscopy. *J. Helminthol.*, 57, 339, 1983.
4. Dutt, S.C., and Srivastava, H.D., The life history of *Gastrodiscoides hominis* (Lewis and McConnel, 1876) Leiper, 1913, the amphistome parasite of man and pig. *J. Helminthol.*, 46, 35, 1972.
5. Dutt, S.C., and Srivastava, H.D., The intermediate host and the cercaria of *Gastrodiscoides hominis* (Trematoda: Gastrodiscidae) preliminary report. *J. Helminthol.*, 40, 45, 1966.
6. Shrivastav, H.O., and Shah, H.L., On *Gastrodiscoides hominis* (Lewis and McConnell, 1876) Leiper, 1913 from pigs (*Sus scrofa domestica*) in Madhya Pradesh, its pathology and public health importance. *Indian J. Pathol. Bacteriol.*, 13, 68, 1970.
7. Chaisiri, K. et al., Human-dominated habitats and helminth parasitism in Southeast Asian murids. *Parasitol. Res.*, 107, 931, 2010.
8. Kumar, V., The digenetic trematodes, *Fasciolopsis buski, Gastrodiscoides hominis* and *Artyfechinostomum malayanum*, as zoonotic infections in South Asian countries. *Ann. Soc. Belg. Med. Trop.*, 60, 331, 1980.
9. Murty, C.V., and Reddy, C.R., A case report of *Gastrodiscoides hominis* infestation. *Indian J. Pathol. Microbiol.*, 23, 303, 1980.
10. Dada-Adegbola, H.O. et al., *Gastrodiscoides hominis* infection in a Nigerian-case report. *West Afr. J. Med.*, 23, 185, 2004.
11. Barker, S.C. et al., Utility of the D1 domain of nuclear 28S rRNA for phylogenetic inference in the Digenea. *Syst. Parasitol.*, 26, 181, 1993.
12. Capuano, F. et al., A comparison of five methods for DNA isolation from liver and rumen flukes to perform ITS-2+ amplification. *Parassitologia*, 49, 27, 2007.
13. Lotfy, W.M. et al., A molecular approach for identification of paramphistomes from Africa and Asia. *Vet. Parasitol.*, 174, 234, 2010.
14. Goswami, L.M. et al., Molecular characterization of *Gastrodiscoides hominis* (Platyhelminthes: Trematoda: Digenea) inferred from ITS rDNA sequence analysis. *Parasitol. Res.*, 104, 1485, 2009.

34 Haplorchis

Urusa Thaenkham and Jitra Waikagul

CONTENTS

34.1 INTRODUCTION

Several fish-borne trematodes in the genus *Haplorchis* parasitize the small intestines of birds, reptiles, and mammals, including humans. They can coinfect with *Opisthorchis* species—small-liver-flukes—in Greater Mekong Subregion (GMS) countries. Mixed infections are difficult to diagnose, specifically because the eggs of *Haplorchis* and *Opisthorchis* are morphologically similar. The ability to discriminate between haplorchiasis and opisthorchiasis is crucial for early warning of cholangiocarcinoma (CCA) caused by small-liver-fluke infections.

34.1.1 Classification, Morphology, and Biology

34.1.1.1 Classification of *Haplorchis* spp

Haplorchis is a genus of heterophyid intestinal flukes (HIFs) classified in the class Trematoda, subclass Digenea, order Opisthorchiida, family Heterophyidae, subfamily Haplorchinae, with *H. pumilio* (Looss, 1896) as type. The genus comprises 10 species—*H. taichui*, *H. parataichui*, *H. pumilio*, *H. parapumilio*, *H. yokogawai*, *H. sprenti*, *H. vanissimus*, *H. paravanissimus*, *H. wellsi*, and *H. popelkae* [1–3].

34.1.1.2 Morphology of *Haplorchis*

The identification of *Haplorchis* is based on the diagnostic characteristics described by Pearson [1]. In general, the bodies of adult worms in this genus are pyriform. They are ~400 × 150 μm in size [1,2] except that *H. popelkae* is about 1117.4 × 200 μm [3]. The esophagus is longer than the prepharynx. The ceca extend between the mid-ovary and beyond the posterior border of the testis. The ventrogenital sac is median and postbifurcal, and is composed of a large ventral chamber and a small dorsal out-pocket. The ventral sucker is modified by reducing the cavity with a groove mid-dorsally or a notch on the left side, opposite the genital pore. The ventral sucker is divided into a nucleate base with parenchymatous medulla and embedded in the body and armed with sclerites that vary in number, shape, and size by species. It is usually a solidly muscular apex projecting into the ventrogenital sac. The gonotyl is absent. There is only a single testis with one or two vasa efferentia. The thin-walled seminal vesicle is bi- or tripartite. The ejaculatory duct presents with no prostatic bulb and the prostatic glands are free. The ovary is submedian on the left side, closer to the ventral surface than the testis. The seminal receptacle is posterio-dorsal to the ovary on the right side and overlaps between the ovary and testis. Laurer's canal opens dorsally to the left side of the midline. The uterus comprises three descending loops. The vitelline follicle is a single group, primarily dorsal of the gonads, while some follicles are situated ventrally among the coils of the uterus. The ejaculatory duct and the uterus are united to form a short genital sinus. The excretory bladder is sac-like or broadly V-shaped. It extends to the posterior border of the testis. The excretory system is composed of a bilateral system of flame cells that formulate as $2[(n + n + n) + (n + n + n)]$, or $2[(n + n) + (n + n)]$ [1,2]. The morphological characters of *Haplorchis* metacercariae are usually similar to their adult worms, except that the reproductive organs are not yet fully developed. The metacercariae are normally elliptical and vary in size (190 × 160 μm on average). Their excretory bladders are O-shaped, and occupy a large portion of the posterior body [4,5].

34.1.1.3 Life Cycle

Only three of the 10 species in the genus *Haplorchis* have been reported to infect humans—*H. taichui*, *H. pumilio*, and *H. yokogawai* (Figure 34.1) [6–11]. The life cycles of these species require two intermediate hosts. Generally, aquatic snails in the genus *Melanoides* and *Thiara* are the first intermediate hosts, and freshwater fish in the family Cyprinidae are the second intermediate hosts [7]. Within the snails, ingested eggs containing miracidia hatch in the intestine and develop into sporocysts in the rectum, by penetrating the perirectal tissue. The germinal cells within the sporocysts multiply and produce rediae. Within the rediae, cercariae are developed and migrate to the intrahepatic lymph space of the snails, to develop to maturity. The mature cercariae swim freely to attract fish by detecting fish movement. Fortunate cercariae penetrate the fish bodies at the muscle, head, and fins, by discarding their tails, and become metacercariae by encysting their bodies. Definitive hosts can become infected with metacercariae by eating raw or uncooked infected fish. The definitive hosts of *H. taichui*, *H. pumilio*, and *H. yokogawai* are generally cats, dogs, and birds. Humans are also the final host of these three species. After infection, metacercariae move into the intestines and hatch out. The excysted metacercariae penetrate the intestinal surface and become mature worms within 3–8 days, depending on species. The eggs can be examined by the stool examination technique under a light microscope. The eggshells of *Haplorchis*, however, are similar to flukes in the family Opisthorchiidae, such as *Clonorchis sinensis* and *Opisthorchis viverrini*, which are normally distributed within the same areas as *H. taichui*, *H. pumilio*, and *H. yokogawai*. Coinfections of *O. viverrini* and *Haplorchis* spp occur frequently in endemic areas, along the flood plains of the lower Mekong River, except in northern Vietnam, where coinfections of *C. sinensis* and HIFs have been investigated [8]. Incorrect identification of parasites by stool examination can sometimes give a

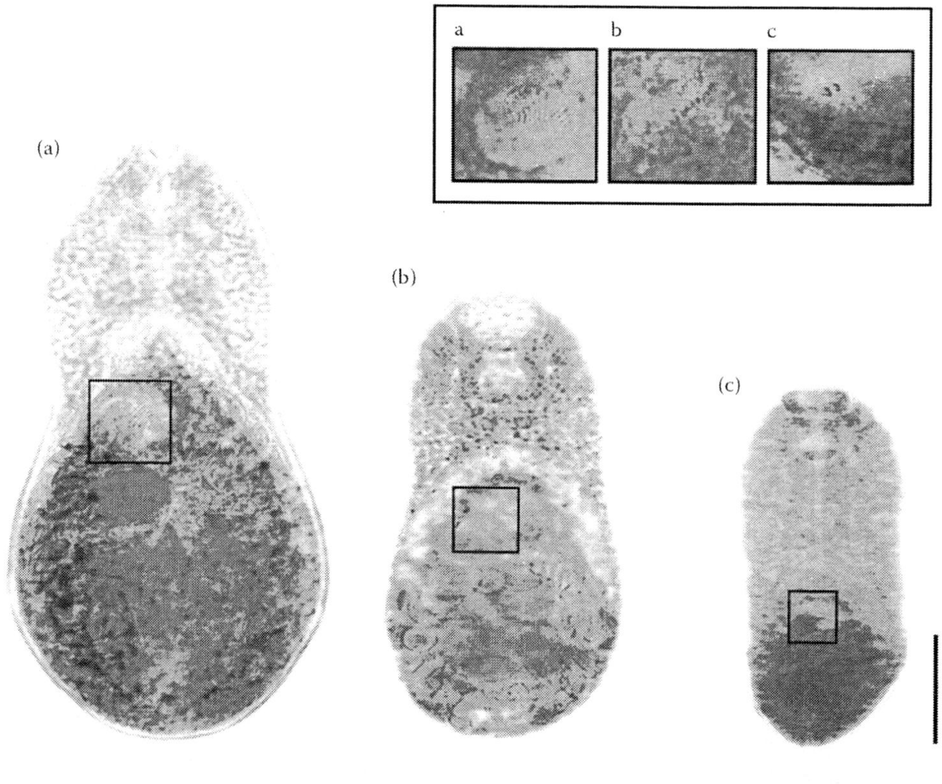

FIGURE 34.1 Adult *Haplorchis* reported to infect humans: *H. taichui* (a), *H. pumilio* (b), and *H. yokogawai* (c). The modified ventral sucker of each species are shown in the rectangular box (a, b, and c, respectively).

false picture of the prevalence of both opisthorchiid liver flukes (OLFs) and *Haplorchis* spp infecting humans.

34.1.1.4 Epidemiology of *Haplorchis*

The epidemiological characteristics of *H. taichui*, *H. pumilio*, and *H. yokogawai* (Figure 34.1) are very similar to those of the OLFs. They can be transmitted to the definitive hosts by ingestion of the metacercariae in infected fish. In endemic areas, the presence of specific hosts of zoonotic flukes can maintain the life cycles of the parasites. A very low level of snail infection (≈0.9%) is sufficient for the release of large numbers of cercariae, so maintaining endemicity [12]. The prevalence of human infections with *H. taichui*, *H. pumilio*, and *H. yokogawai* is related to the habit of consuming inadequately cooked or uncooked fish. These HIFs, which can infect humans, are mainly distributed throughout Southeast Asia, especially Lao PDR, Vietnam, and Thailand [8,9,13–15]. There has, however, been no report of minute intestinal fluke (MIF) in Cambodia [16]. *H. taichui* and *H. yokogawai* have also been reported in Taiwan, China, India, and Egypt [7,10,17–19]. Since many species of freshwater fish, especially those in the family Cyprinidae, are infected with HIFs, which are also the second intermediate hosts of *O. viverrini*, humans who eat raw fish are at risk of infection by both groups of parasites. A very popular dish among both Thai and Lao people is "Koi pla," which contains raw diced fish with garlic, fish sauce, lemon juice, chilli, roasted ground rice, and some local vegetables [20]. Coinfections with HIFs and *O. viverrini* have most frequently been studied among inhabitants of Thailand and Lao PDR [7,9,14,15].

In northern Thailand, examination of cyprinoid fish has revealed *H. taichui* and *H. pumilio* metacercariae at rates of about 380 times those of *O. viverrini* metacercariae [21]. In northern Thailand, mixed infections of *H. taichui* and *O. viverrini* were confirmed by stool examination; the median ratio of *O. viverrini* to *H. taichui* eggs was 2.29 for the 68 specimens examined [22]. In 2005, Kumchoo et al. investigated *H. taichui* metacercariae in 15 species of cyprinoid fish from two districts of Chiang Mai Province. They found high rates of infection with this parasite (~87%; mean intensity = 175) [23]. Heavy *H. taichui* infections have been reported in northern Thailand among freshwater fish in Chiang Mai water reservoirs [13]. Radomyos et al. [24] reported a low prevalence of *Haplorchis* spp in 16 provinces of northeastern Thailand [24]. Human infections with *Haplorchis* spp were around 13 times lower than for *O. viverrini*. In eastern Thailand, coinfections with *Haplorchis* spp and *O. viverrini* metacercariae were also reported among freshwater fish from Sakaew and Prachinburi provinces [25].

In Lao PDR, HIFs were investigated in Vientiane, Savannakhet, and Saravane provinces, in 2005–2009. Stool samples from Lao residents were examined and often showed *H. taichui* coinfection with *O. viverrini* [7,9,14,15]. In 2005, Chai et al. reported that infection rates for *O. viverrini* and other small intestinal flukes differed in intensity by locality in the Mekong River basin. More HIFs and/or lecithodendriid flukes were recovered from Saravane residents than Vientiane [15]. In 2007, Chai et al. investigated the prevalence of liver and intestinal flukes among 981 residents in four Mekong riverside villages in Savannakhet Province, by the Kato–Katz (KK) thick smear technique; the most common species were *O. viverrini* and *H. taichui*, followed by *Prosthodendrium molenkampi*, echinostomes, *H. pumilio*, *Phaneropsolus bonnie*, and *H. yokogawai* [14]. In 2009, Sayasone et al. reported the stool-examination results of 232 residents in Vientiane Province, using the KK method. Coinfection of *O. viverrini* and MIFs constituted 81.4%, and the common MIF species was *H. taichui* [9]. Freshwater fish from Vientiane Municipality and Savannakhet Province were examined by muscle-compression and artificial-digestion methods; *O. viverrini*–*H. taichui* coinfection was common in many species of cyprinoid fish [26].

In Vietnam, De et al. [27] reported the prevalence of foodborne trematode infections from 12 provinces of Vietnam, in 1994–2000. About 20,000 villagers living in the Red River Delta region were infected with *C. sinensis* in the coastal delta region, mountainous, and highland areas, with *O. viverrini* infection in urban coastal areas [27]. Human infections with fish-borne zoonotic intestinal trematodes were reported from a fecal survey of 615 persons in northern Vietnam by Dung et al. in 2007. The prevalence of trematode eggs was 64.5%. After treatment with praziquantel, *C. sinensis*, *H. pumilio*, and *H. taichui* were the major parasitic species recovered [8]. In an Giang Province, southern Vietnam, the prevalence of zoonotic trematodes was investigated in a metacercarial survey in both cultured and wild fish, in June 2005–March 2006. The results showed that *H. pumilio*, *Procerovum* sp., and *O. viverrini* were distributed in this area. However, *O. viverrini* metacercariae were only found in wild-caught snakehead fish (*Channa striata*), together with *H. pumilio* and *Procerovum* sp. [28]. In 2008, Chi et al. reported the prevalence of zoonotic trematodes among cultivated fish in Northern Vietnam. They investigated the overall prevalence among fish ranging from fingerling to market size. Fish cultured in nurseries had higher infection rates than adult fish, ranging from 7.4% to 62.8%. The highest infection rates were for *H. pumilio*, followed by *H. taichui*, *H. yokogawai*, *Centrocestus formosanus*, *Stellantchasmus falcatus*, and *Echinochasmus japonicus*. However, a cross-sectional survey among the villagers in the same area found a very low rate of infection. These results suggested that reservoir hosts (such as dogs, cats, and pigs) were very important for maintaining the life cycles of the parasites [29].

Outside the GMS, infections with *H. pumilio*, *H. taichui*, and *C. sinensis* were investigated in freshwater fish in China and Taiwan; the prevalence of *Haplorchis* spp was higher than that of *C. sinensis* [17,18]. In the southern Philippines, haplorchiasis taichui infections were reported among humans on Mindanao Island, with a prevalence of 36%. In Egypt, Elshazly et al. [30] recovered *H. yokogawai* metacercariae from freshwater fish in Dakahlia Governorate.

The prevalence of *Haplorchis* spp in many endemic areas, especially in the GMS, shows that these kinds of parasites can effectively maintain their life cycles in freshwater, as long as they have the specific intermediate host, the reservoir host, and the human who eats undercooked fish. Humans in endemic areas can be heavily infected by this group of parasites, which can cause serious symptoms [11].

34.1.2 Clinical Features and Pathogenesis

At least three species in the genus *Haplorchis*—*H. taichui*, *H. pumilio*, and *H. yokogawai* (Figure 34.1)—have reportedly infected humans. In general, symptoms are usually mild or asymptomatic [11,31,32]. However, infection remains a concern because of occasional reports of severe pathological tissue changes [10]. In areas endemic for *Haplorchis* spp, people who habitually eat improperly cooked fish can become infected and then suffer from abdominal discomfort, acute abdominal pain, or appendicitis [31,33]. There is a high risk that patients can become chronic, with heavy infections. Chronic heavy infections with *Haplorchis* spp can cause lesions and leakage in the small intestine. The small parasites or their eggs can distribute into the blood circulation and even reach the heart, spinal cord, and brain, causing death [34–36].

In 2005, Sukontason et al. reported three serious human cases of heavy *H. taichui* infection from Lampang Province, Thailand. Microscopic examination revealed mucosal ulceration, mucosal and submucosal hemorrhage, fusion and shortening of the villi, chronic inflammation, and fibrosis of the submucosa. Damage to the small intestine was caused by the oral sucker. The pathological features caused by this parasite were similar to those caused by other minute flukes, such as *Metagonimus* spp, where infections lead to gastrointestinal complaints [11,37].

Haplorchiasis does not have a great public health impact; infections with *Haplorchis* spp may manifest as diarrhea or even be asymptomatic. Reports have focused on *Haplorchis* infection because parasites in this genus are normally involved in coinfection with other parasites. Hotez et al. [38] noted a higher risk of morbidity in coinfections with multiple helminths, even at low intensity [31,38]. The issue of *Haplorchis* spp coinfection with other parasites is especially emphasized in mixed OLF–*Haplorchis* spp infections in the GMS [9,14,26,28]. The OLFs are very important parasites, which can cause hepatobiliary symptoms; chronic infections can develop into CCA [7]. Therefore, diagnoses of OLFs and *Haplorchis* spp have been developed for using in endemic areas to increase the impact of parasite control programs in the GMS.

34.1.3 Diagnosis

34.1.3.1 Conventional Techniques

34.1.3.1.1 Stool Examination

To discriminate between OLFs and HIFs, stool examination has recently been the gold standard. Distinguishing these two groups of parasites is problematic, because their eggs are similar in size and shape, so that an experienced microscopist is required for accurate identification [39,40].

Tesana et al. [41] investigated the eggshell morphology of some OLFs and MIF species in Thailand, that is, *O. viverrini*, *H. taichui*, *H. pumilio*, *P. bonnie*, and *P. molenkampi*. Under light microcopy, the eggshells of *O. viverrini* were different from *H. taichui*, *H. pumilio*, *P. bonnie*, and *P. molenkampi*; they have a rough shell, while the other species have smooth ones. *O. viverrini*, *H. taichui*, and *H. pumilio* are similar in having prominent shoulders. *Haplorchis* spp are the same in shape, but differ in size. The eggshells of *H. pumilio* are larger than *H. taichui*, *P. bonnie*, and *P. molenkampi*, and the eggs are distinctly different from *O. viverrini* and *Haplorchis* spp, having no prominent shoulder, but only an indistinct one. Under scanning electron microscopy, *O. viverrini* eggs appear to have a musk-melon skin and a long knob. *H. taichui* eggshells have a curly, thread-like ridge, and knob, while *H. pumilio* eggs have a stretched ridge resembling Chinese bitter mormodica [41]. Ditrich et al. [39] noted that egg shape was not a suitable characteristic to use in species identification. However, egg shape and eggshell surface structure can be used to distinguish some medically important species of OLF and HIF, such as *O. viverrini*, *C. sinensis*, *H. taichui*, *H. pumilio*, *H. yokogawai*, *S. falcatus*, and *Metagonimus* sp. [39].

34.1.3.1.2 Fish Examination

OLFs and HIFs can also be differentially diagnosed using metacercarial morphology. Scholz et al. [5] revealed the morphological differences of *H. taichui*, *H. pumilio*, *Haplorchoides mehrai*, and *O. viverrini*, from cyprinoid fish in Lao PDR. The morphological characters were studied under a dissecting microscope at low magnification; and the metacercarial characters of those four parasites were described, as follows: (i) *H. taichui* and *H. pumilio* larvae are surrounded by a thin-walled and oval cyst. They move actively and the bodies are folded into C- or S-shape. The excretory bladder is relatively small in size; (ii) *O. viverrini* larvae are very active, with rapid movement inside the oval cyst. The larval body is C-shaped, and has a large excretory bladder; (iii) *H. mehrai* larvae move slowly, and the body is C- or S-shaped, varying in size and shape of the thick cyst wall [5].

Basic knowledge of the morphological characteristics of metacercariae can be used in routine fish examination, to discriminate and separate each species under a stereomicroscope. This is very useful in studying species' prevalence in endemic areas.

34.1.3.1.3 Immunodiagnosis

Human small-fluke infections were differentiated serologically using *H. taichui* and *O. viverrini* antigens obtained from the adult worms of each species, by division into cytoplasmic and membranous antigens. The antigens were tested with 642 sera from Lao people in Vientiane Province by enzyme-linked immunosorbent assay (ELISA) and electro-immuno-transfer blot (EITB) methods. The ELISA results showed serological reactions with cytoplasmic antigen to be more sensitive and more specific than with membranous antigen. However, cross-reactions between *H. taichui* and *O. viverrini* occurred frequently with ELISA. Therefore, the highly positive sera were examined by EITB technique wit the cytoplasmic antigens of both species, to discriminate between them by protein fraction pattern. The results showed that *H. taichui* and *O. viverrini* shared several cytoplasmic antigens fractions; homologous sera reactions were also investigated with *O. viverrini* antigen in the 70 kDa area and *H. taichui* antigen in the 10 kDa area. EITB was performed with 122 serum samples; 25% had specific antibody against *O. viverrini* antigen, with 63% against *H. taichui*. The remainder had antibodies against both species' antigens. The immunoblot patterns of *H. taichui* and *O. viverrini* appeared useful in specifying human infections with small-flukes [42]. The problematic results for the immunodiagnosis of *H. taichui* and *O. viverrini* indicate a need for improvement for detecting haplorchiasis in serum samples from endemic areas of *Haplorchis* spp and OLF coinfection. Molecular techniques provide an alternative for different types of samples collected from endemic areas, such as fecal and infected-fish samples.

34.1.3.2 Molecular Techniques

Several studies have developed effective methods for discriminating and/or detecting *Haplorchis* spp in samples also containing OLFs. The major aim of each study was to differentiate the genetic material of each species in coinfections.

Thaenkham et al. [43] designed experiments to discriminate *O. viverrini* from *H. taichui* using the mitochondrial cytochrome *c* oxidase subunit I (COI) gene. The species-specific primers for each species were designed in the same region as the partial sequences of the mitochondrial COI genes of *O. viverrini* and *H. taichui* deposited in the GenBank databases (accession nos AY055380 and EF055885, respectively). The primers were degenerated in some positions, to amplify both species simultaneously, but the primer designed for each species could amplify at different annealing temperatures. They were called COI-OV-Hap (F) and (R) primers. These primers were tested with adult *O. viverrini*, *H. taichui*, and *C. sinensis* worms to find the best annealing temperature. The metacercariae of *O. viverrini*, *H. taichui*, *H. pumilio*, *H. yokogawai*, and the eggs of *O. viverrini* and *H. taichui* were also amplified to examine the specificity and sensitivity of the primers.

The species-specific amplification was examined using 10 ng of *H. taichui*, *H. pumilio*, *H. yokogawai*, *O. viverrini*, and *C. sinensis* at different annealing temperatures (between 50°C and 56°C) using the COI-OV-Hap(F) and (R) primers. The polymerase chain reaction (PCR) amplicons obtained were about 370 bp. The results revealed that the optimal annealing temperature for *O. viverrini* and *C. sinensis* was 50°C, while that for *H. taichui* was 56°C. The primers could not amplify any specific PCR product from *H. pumilio* or *H. yokogawai*.

The sensitivity of detection at optimal annealing temperatures was analyzed, to assess specificity and sensitivity when detecting *H. taichui* and *O. viverrini* in different developmental stages (egg and metacercariae), and serially diluted genomic DNA extracted from the adult worms (between 10 and 10^{-4} ng). At 56°C, PCR amplicons were obtained from 10 to 10^{-4} ng *H. taichui* genomic DNA, a single metacercaria, and an egg. Although the genomic *H. taichui* DNA was also amplified at 50°C, the PCR sensitivity was lower than that at 56°C. At this temperature, the PCR products were obtained between 10 and 10^{-2} ng; the single egg could not be amplified. At 50°C, the optimal annealing temperature for *O. viverrini*, the PCR amplicons were obtained from 10^{-4} to 10 ng genomic DNA from adult worms, and the experiment was also carried out with a metacercaria and a single egg.

The PCR amplicons from *H. taichui*, *O. viverrini*, and *C. sinensis* were about 370 bp; since there was no significant size difference in the PCR products, discriminating species at 50°C might be problematic. The experiment was designed to use the *Alu*I restriction enzyme, which digested the PCR amplicons in different site(s). The PCR-restriction fragment length polymorphism (RFLP) profile obtained two fragments, of 200 and 169 bp, from *O. viverrini*, three fragments of 119, 52, and 209 bp from *C. sinensis*, and two fragments of 119 and 253 bp from *H. taichui*. PCR-RFLP was performed with PCR amplicons of each species and also with artificial mixed infections of *O. viverrini* and *H. taichui*, *C. sinensis* and *H. taichui*, *O. viverrini* and *C. sinensis*, and all three species. The results revealed different band patterns. Mixed infections of *O. viverrini* and *H. taichui*, and *O. viverrini*, *C. sinensis*, and *H. taichui*, indicated different DNA concentrations at band sizes of 119 and 200 bp.

The study results suggested that the species-specific temperature of *H. taichui*, 56°C, might be useful in detecting small amounts of genomic DNA, while the PCR-RFLP profile might be useful in diagnosing mixed infections of *O. viverrini* and *H. taichui* in endemic areas, and in detecting metacercariae of *O. viverrini*, *C. sinensis*, and *H. taichui* in epidemiological surveys of infected fish [43].

Sato et al. [44] used the internal transcribed spacers (ITS1 and ITS2) of ribosomal DNA as the DNA marker to discriminate among *O. viverrini*, *C. sinensis*, *H. pumilio*, and *H. taichui*, and then used PCR conditions to detect a species' genetic material in fecal samples.

The study aimed to identify and differentiate species and to use the DNA marker for preliminary analysis using fecal samples diagnosed by modified thick smear. The PCR primers used were designed manually from the sequences of *H. taichui* and *H. pumilio* (GenBank accession nos AY245705 and AY245706, respectively). They were called ITS1-F&R and ITS2-F&R for the ITS1 and ITS2 regions, respectively. The species discrimination results revealed that the ITS1 region from each species studied had differently sized PCR amplicons, while only *H. taichui* had a differently sized ITS2 region from the others. The ITS1 region appeared a useful marker for detecting eggs of *O. viverrini*, *C. sinensis*, *H. pumilio*, and *H. taichui* in fecal samples. However, both ITS1 and ITS2 were used as the DNA marker for copro-PCR. Before using the fecal samples for species detection by PCR, 1 g from each of 21 stool samples was examined for small trematode eggs. Sato et al. [44] compared stool extraction methods, between the conventional method using cetyl trimethyl ammonium bromide (CTAB) to remove PCR inhibitor, and a commercial stool-extraction kit (Qiagen Stool kit, Qiagen). The sensitivities of the PCR results of the two methods were not significantly different. It was concluded that it was not necessary to use CTAB to improve PCR sensitivity.

The PCR-amplified products revealed the difference in sensitivity between the ITS1 and ITS2 regions. Under the optimized PCR conditions, the sensitivity of the PCR amplicons from the ITS1 target (76.2%) was lower than from the ITS2 target (95.2%). The limitation of this study was the presence of false positive and negative results, using either ITS1 or ITS2 as the PCR target. Kappa testing showed a moderate association ($k = 0.75$) between the PCR ITS1-region assay and the modified thick-smear method results. The PCR amplicons from the ITS1 region appeared less sensitive than the thick-smear method. The high sensitivity of the PCR method over the thick smear approach was revealed when using the ITS2 region as the PCR target, with a high association ($k = 0.95$). Sato et al. [44] suggested that the ITS2 region might be a useful diagnostic tool, instead of the modified thick-smear method. In that study, both ITS regions were recommended, because the ITS1 region was used to discriminate and screen small-liver-flukes and the ITS2 region for intestinal flukes [44].

Sato et al. [46] used these two regions of ribosomal DNA with a large number of fecal samples ($n = 125$). The ITS-PCRs successfully detected opisthorchiasis cases but apparently failed to detect haplorchiasis taichui, when compared with the actual worm expelled [46]. They suggested that daily egg production of adult worms of *H. taichui* is low per day (82 eggs/worm/day) [45], and many immature worms were expelled before laying eggs. It remains necessary to understand the life cycle and reproductivity of *H. taichui* to improve molecular diagnosis [46].

Traub et al. [47] developed a new PCR-based approach to diagnosing "*Opisthorchis*-like eggs" in feces from the Sanamchaikaet District community in central Thailand, using the ITS2 region of ribosomal DNA as a detection target. This is because the ITS2 region has successfully discriminated species from many digenean families, including the opisthorchiidae and heterophyidae [44]. The new PCR test aimed to evaluate both analytical and diagnostic sensitivity and specificity against the conventional parasitological methods (direct fecal smear (DFS), KK technique, and formalin-ether concentration technique (FECT)) in endemic areas for liver-fluke infection in central Thailand. The results of microscopy were assumed to be the "gold standard" for determining the sensitivity and specificity of the PCR assay.

The PCR primers, named "RTFlukeFa" and "RTFlukeRa," were designed from the ITS2 region of *C. sinensis*, *O. viverrini*, *O. felineus*, *H. taichui*, *H. pumilio*, and *Centrocestus* sp. (GenBank accession nos EF688144, EF688143, AY584735, DQ513403, DQ513405, AY245705, AY245706, AY245699). The ITS2 regions of *O. viverrini*, *C. sinensis*, and *H. taichui* were amplified and yielded specific PCR products of 375, 381, and 529 bp, respectively. The lowest amounts of DNA that could be amplified from individual adults of these species were 0.6, 0.8, and 3 pg, respectively. Titration experiments included separation of the artificially mixed DNA of *H. taichui* with *O. viverrini* and *H. taichui* with *C. sinensis*, in ratios of 1:1, 1:2, 1:3, and 2:1. PCR performed with an artificial mixture of *O. viverrini* and *H. taichui* could be successful, but at a ratio of <1:1. The PCR amplicons of *O. viverrini* and *C. sinensis* could not be differentiated by size. PCR-RFLP was designed using the *Acu*I to discriminate between the products of the two small-liver-fluke species. The experiments used fecal samples screened positive for "*Opisthorchis*-like eggs" by conventional parasitological methods. Compared with microscopy, the sensitivity and specificity of PCR were 71.0% and 76.7%, respectively. This study found no *H. taichui* infection in the Sanamchaikaet District community. It was possible that PCR failed to amplify the fecal sample with mixed infections of *H. taichui* and *O. viverrini*. The experiment showed good analytical sensitivity for detecting a single infection with *H. taichui* and mixed infections of *H. taichui* and *C. sinensis*. Moreover, this study primarily investigated *C. sinensis* infection in the Thai community [47].

Lovis et al. [48] developed a PCR-based diagnostic method for *O. viverrini* and *H. taichui* infection using fecal samples from a Lao community. Since microscopy sometimes yields low sensitivity and specificity, this study aimed to develop PCR tests for detecting and discriminating between *O. viverrini* and *H. taichui* parasite infections. Mitochondrial DNA from the COI gene was used as the PCR target. The PCR method was used as the "gold standard" and validated by stool samples from purged individuals. The PCR results were compared with conventional parasitological methods (KK and FECT) to evaluate the sensitivity and specificity of fecal diagnosis.

A nested PCR (nPCR) protocol was developed to optimize the sensitivity of each species-specific primer set, using the outer (primary PCR; pPCR) and inner primer sets (nPCR) designed from the COI gene of *H. taichui* (GenBank accession no. EF055885), and from the mitochondrial DNA sequence of *O. viverrini* (GenBank accession no. DQ11955). DNA samples were extracted for testing from stool samples using a commercial stool extraction kit (Qiagen, Basel, Switzerland). The positive controls for the study were genomic DNA extracted from adult *O. viverrini* worms from infected hamsters and naturally infected humans in Saravane Province, Lao PDR. The sensitivity and specificity of the primer sets were tested with serial dilutions of *O. viverrini* and *H. taichui* DNA, from 150 to 1.50×10^{-2} ng/μL and from 132 to 1.32×10^{-2} ng/μL, respectively. From each dilution, 5 μL were used to evaluate the sensitivity of the PCR primers. The specificities of the primer sets were assessed with total genomic DNA of human, *H. taichui* or *O. viverrini*, *Schistosoma japonicum*, *Necator americanus*, *Ancylostoma duodenale*, *Strongyloides stercoralis*, *Trichuris trichiura*, *Ascaris lumbricoides*, and *Strongyloides ratti*. The results showed both the sensitivity and specificity of the primer sets for both species. The nPCR yielded *O. viverrini* products at 508 and 226 bp, while the PCR amplicons for *H. taichui* were 245 and 186 bp. The specificity of the primer sets for these two parasites was measured by no cross-reaction of the PCR amplifications. The PCR tests were validated using frozen stool samples collected from 80 Lao adults from purging studies. PCR inhibitor testing showed that one sample with persistent inhibition was removed. Seventy-nine PCR-uninhibited stool samples were measured for PCR validation at 95% confidence interval. The PCR sensitivities of *O. viverrini* and *H. taichui* were 93.7% and 73.3%, respectively; the PCR specificity for *O. viverrini* was undefined, but 63.2% for *H. taichui*.

The PCR test was used to determine the prevalence of *O. viverrini* and/or *H. taichui* infections in 114 stool samples from a cross-sectional community-based survey in Khong District, Lao PDR. PCR inhibition was tested with the samples and six were removed from the study. Copro-PCR diagnosis was used to detect *O. viverrini* and *H. taichui* infections in 108 stool samples. This approach was then used to compare PCR with light-microscopy results (KK and FECT methods). Using PCR as the gold standard, the results showed the PCR-based method to yield better sensitivity and specificity than both KK and FECT methods. Lovis et al. [48] demonstrated the usefulness of the PCR approach as a sensitive diagnostic tool, particularly for low-intensity *O. viverrini* and *H. taichui* infections [48].

Wongsawad et al. [49] developed a high-annealing-temperature random amplified polymorphic DNA (HAT-RAPD) PCR for detecting *H. taichui* infection. Eighteen arbitrary primers (Operon Technologies, USA) were used to generate fingerprints in different polymorphic DNA profiles; the polymorphic patterns were compared with 13 other parasites at an annealing temperature of 48°C. The *H. taichui*-specific HAT-RAPD fragment was investigated at 256 bp. The *H. taichui* fragment was sequenced to determine genetic information, and specific primers for this parasite, named "Hap-t F" and "Hap-t R," were designed. Primer specificity was examined by amplification with sample DNA from all 13 parasites, varying the total amount of $MgCl_2$ and optimizing the annealing temperature. The sensitivity of the PCR reaction was also tested by amplification with serial dilutions of *H. taichui* DNA template (1×10^{-8}–1×10^{-15} g). The specificity of the primer to *H. taichui*, without cross-reaction with other parasites, was investigated. PCR detection sensitivity was minimal with the 0.1 pg DNA template. Wongsawad et al. [49] suggested that the method illustrated a major benefit of using the Sequence Characterized Amplifying Region marker technique to generate a diagnostic tool for rapidly detecting *H. taichui* in snails, fish, and fecal samples, in endemic areas [49].

34.2 METHODS

Molecular diagnostic methods were developed to detect and/or discriminate *Haplorchis* from other OLF species. Samples could be obtained from various sources, such as feces and intermediate hosts.

34.2.1 Sample Preparation

34.2.1.1 Stool Extraction

The stool extraction methods used for releasing genetic material from "*Opisthorchis*-like eggs" differed somewhat [44,46–49]. Fecal samples could be either feces fixed in fixative or fresh samples. The PCR inhibitor into feces has to be concerned before starting extraction and PCR.

Samples for each experiment can be prepared from 0.5 mL feces in fixative or 180–240 mg of fresh stool sample. Feces in fixative must be washed at least twice in distilled water (DW) and frozen at −80°C. Without a lysis buffer, frozen feces are crushed three times with a glass bar to break the eggs and release genetic material. Fresh stool samples are suspended in 1.4 mL lysis buffer, and the *Opisthorchis*-like eggs are subjected to five freeze–thaw cycles using liquid nitrogen, to open the opercula. Where liquid nitrogen is unavailable, the miracidia inside the eggs can be removed by heating the samples at 95°C for 4 min and freezing in dry ice for 8 min twice. The suspensions are then incubated at 95°C for 10 min. After the miracidia are released into the suspension, stool particles are separated out by centrifuging at 14,000 rpm for 1 min. A 1.2 mL portion of the supernatant is incubated with a commercial anti-PCR inhibitor tablet and DNA is extracted from the stool samples, according to the manufacturer's instructions. DNA is eluted by 50 μL of commercial elution buffer, for use in PCR.

Sato et al. [46] developed a stool-extraction technique that may be easy to use in the field and less expensive than previous methods. Each fresh fecal sample was washed in 12 mL tap water containing 2% liquid dish-washing detergent. The sample was washed repeatedly until the supernatant became clear, when the supernatant was removed; 8 mL of 70% ethanol was added to the sediment and the sample was then kept at room temperature until use. Before DNA extraction, the sediments were washed with DW and resuspended in 3 mL DW. Half of the suspension was transferred to a 1.5 mL microtube and then the *O. viverrini*-like eggs were disrupted by microwaving for 60 s. Without using a commercial anti-PCR inhibitor tablet, DNA was extracted using a commercial genomic DNA extraction kit [46].

34.2.1.2 Preparation of Genetic Material without Extraction

1. *From an egg*
   a. An egg is separated from feces or an adult worm and washed in sterile DW for 10 min.
   b. The slightly wet single egg is broken using the sharp corner of a coverslip on a sterile glass slide under a stereomicroscope.
   c. The removed genetic material is washed with 5 μL sterile double-DW and then placed in 0.5 mL sterile Eppendorf.
   d. The genetic content of the egg is boiled at 95°C for 10 min with 5 μL mineral oil and then spun down for use directly in PCR.

2. *From a metacercaria*
   a. A metacercaria is separated from an infected fish and washed in DW twice.
   b. The slightly wet metacercaria is torn using the sharp corner of a sterile cover slide on a sterile glass slide under a stereomicroscope.
   c. The genetic material obtained is washed in 5 μL sterile double-DW.
   d. The genetic content of the metacercaria is boiled at 95°C for 10 min with 5 μL mineral oil and then spun down for use directly in PCR.

34.2.1.3 Extraction of DNA from Metacercariae and Adult Worms

The method normally used for extracting genomic DNA from both metacercariae and adult worms is according to the manufacturer's instructions for the tissue-extraction kit. However, a conventional approach, the phenol/isoamyl alcohol/chloroform method, can be used.

1. *Phenol/isoamyl alcohol/chloroform method*
   a. The metacercariae or adult worms are digested with lysis buffer containing 25 mM Tris buffer (pH 7.5), 0.5% sodium dodecyl sulfate, 100 mM sodium chloride, 10 mM EDTA, and 0.1% proteinase K, at 56°C in a water bath.
   b. The genetic content is extracted by phenol/isoamyl alcohol/chloroform, according to the method described by Maniatis et al. [50].
   c. The DNA is then precipitated with absolute ethanol and 0.25 M sodium acetate (pH 5.2).
   d. The DNA is resuspended with 200 μL of DW.

34.2.2 Detection Procedures

34.2.2.1 Conventional PCR Method

The primers for the DNA targets, or DNA markers, are designed at the conserved regions of the HIF and OLF species. Generally, the ITS regions of the ribosomal DNA and the COI gene of mitochondrial DNA are used, because the DNA sequences are in GenBank. The PCR reaction can be conducted in a volume of 20–50 μL. The PCR mixture normally contains 200 μM of dNTP, 1.5–2.0 mM $MgCl_2$, 20–40 pmol of each primer, 1× *Taq* polymerase buffer, 1.0–1.5 U *Taq* polymerase, and various amounts of DNA template. The PCR samples are amplified in a PCR machine using 30–35 PCR cycles. A cycle is started with initial denaturation at 94°C for 3–5 min, followed by cycles of denaturation at 94°C for 30–45 s, the optimal annealing temperature for 30–60 s, and 72°C for 1.5–2.0 min, with a final extension at 72°C for 5–10 min.

34.2.2.2 PCR-RFLP

PCR-RFLP is designed to distinguish OLFs and HILs in mixed or coinfections. The criterion for using the restriction enzyme is that the cutting site(s) of the enzyme on the PCR products should yield clearly different sizes of digested PCR amplicons.

In the PCR-RFLP method, the DNA sequences of the PCR products of each species are first aligned by a multiple-alignment program, such as CLUSTAL W [51]. The aligned sequences then recognize the cutting site of the enzyme using the restriction profile program in BioEdit version 7.0 [52], Nebcutter version 2.0 (New England Biolabs), and so on. A suitable restriction enzyme is selected to digest the PCR amplicons.

The PCR products are cleaned before digestion with restriction enzyme. The digestion profile is 20–30 μL reaction volume, to which units of restriction enzyme are added, depending on DNA concentration and digestion time. The reaction mixture is incubated at 37°C for 1–3 h(s). The digested PCR-restriction fragment(s) are then loaded and run on 2.0% agarose gel using a 100 bp ladder marker for approximating the digested PCR fragment size(s). The agarose gel is stained with ethidium bromide and destained before visualizing on the UV transilluminator.

34.2.2.3 Nested PCR

nPCR is used to optimize the sensitivity of the PCR products by using the specific primers for pPCR and nPCR. The primers can be designed from the available DNA sequence data in GenBank. The pPCR primers are used in the first PCR amplification cycles with a small amount of DNA template. The products from the pPCR primers are then used as template for the nPCR primers. The nPCR product should be designed to yield different sizes of PCR product per species.

The species-identification PCR was performed separately, using the species-specific primer set. PCR reactions are conducted in a final volume of 50 μL, normally containing 200 μM dNTP, 1.5 mM $MgCl_2$, 25 pmol of each primer, 1× *Taq* polymerase buffer, 0.5 U *Taq* polymerase, and various amounts of DNA template. For nPCR, 2 μL of primary product is used as DNA template. PCR amplifications are carried out in a PCR machine, starting with initial denaturation at 94°C for 5 min, then denaturation at 94°C for 30 s, optimal annealing temperature for 1 min, extension at 72°C for 1 min, and final extension at 72°C for 5–10 min. The nPCR products are run on 1.5% agarose gel, stained with ethidium bromide, and visualized by UV transilluminator to detect the differentiated PCR fragments.

34.2.2.4 High-Annealing Temperature Random Amplified Polymorphic DNAPCR

The HAT-RAPD PCR uses arbitrary 10-mer primers at high temperature in the PCR primer-annealing step. To obtain a species-specific band, many primers are used in amplification with many other species, including closely related species and nonrelated species. Around 50 ng/μL of genomic DNA per species is used as the PCR-RAPD DNA template. The PCR process used is as follows: initial denaturation at 95°C for 5 min, 30 cycles of 95°C for 45 s, optimal annealing temperature for 45 s, 72°C for 1 min, and final extension at 72°C for 7–10 min. The unique band amplified from the species of interest is sliced from the agarose gel after separating the polymorphic band by gel electrophoresis. The separated band

TABLE 34.1
Summary of Molecular Detection Techniques for Heterophyid and Opisthorchiid Flukes

Detection Methods	Species Studied	PCR Target	Source(s) of DNA Template	Specificity/ Sensitivity	Utility
Conventional PCR and PCR-RFLP [42]	*Ht*, *Hp*, *Hy*, *Ov*, and *Cs*	Partial COI mtDNA	Egg, MC, and AD of HIFs and OLFs	High/medium	Discriminating mixed infection
Conventional PCR [44, 46]	*Ht*, *Hp*, *Ov*, and *Cs*	ITS regions of rDNA	Eggs, MC, AD, and fecal samples	High/medium	Revealing the utility of ITS1 and ITS2 for different purposes
Conventional PCR and PCR-RFLP [47]	*Ht*, *Ov*, and *Cs*	ITS2 region of rDNA	AD and fecal samples	High/medium	Detecting and discriminating the species in low mixed infections in fecal samples
Nested-PCR [48]	*Ht* and *Ov*	Partial COI mtDNA	Fecal samples from purged individuals	High/high	Sensitive diagnostic tool for low-intensity *Ht* and *Ov* infections
HAT-RAPD, and conventional PCR [49]	*Ht* and 13 other parasites	The *Ht* specific HAT-RAPD fragment	AD	High/high	Primer developed may be useful for detecting low *Ht* infections in endemic areas

Note: MC and AD refer to metacercaria and adult worm, respectively.

is then purified, cloned with the plasmid vector, and transformed into bacterial cells, to multiply the number of DNA fragments. After sequencing, PCR primers of the species of interest are designed to yield high sensitivity and specificity by the conventional PCR technique at the optimal annealing temperature of the newly designed primers. The specificity of the PCR primers is tested with cross-species' amplification.

34.3 CONCLUSIONS AND FUTURE PERSPECTIVES

Mixed OLF–HIF infections are still significant public-health concerns. Diagnostic methods have been designed to detect and discriminate between these groups of parasites. However, parasitological techniques are limited by misidentification and immunological techniques by cross-reactions with other parasites. Since 2007, molecular detection methods, using PCR-based techniques, have been developed to achieve the high levels of detection sensitivity and specificity (Table 34.1). The PCR targets generally used are held in GenBank, such as the ITS regions of ribosomal DNA (ITS1 and ITS2 regions) and the mitochondrial COI gene. The ultimate aim of developing this method is field application with fecal samples in endemic areas. The PCR-based method should be developed for use as the "gold standard" for detection, because it is easier to develop a specific PCR than other detection approaches. However, limitations of PCR amplification from fecal samples include the inhibitor in the feces, and the sensitivity of the tiny amount of DNA normally present in fecal samples. The sensitivity of PCR detection with *H. taichui* was also probably limited by the small numbers of eggs in feces, which may depend on the maturity of the adult worms in the human's intestine [46,48]. Investigation of the biology of *H. taichui*, to enable the use of fecal samples at the time that eggs are laid, and the development or improvement of stool-extraction methods, to enable removal of anti-PCR inhibitor at low cost, will increase the usefulness of PCR-based techniques in routine diagnostic work.

REFERENCES

1. Pearson, J.C. A revision of the subfamily Haplorchinae Looss, 1899 (Trematoda: Heterophyidae) 1. The *Haplorchis* group. *Parasitology* 54, 601, 1964.
2. Pearson, J.C., and Ow-Yang, C.K. New species of *Haplorchis* from Southeast Asia, together with keys to the *Haplorchis*-group of heterophyid trematodes of the region. *Southeast Asian J. Trop. Med. Public Health* 13, 35, 1982.
3. Snyder, S.D., and Tkach, V.V. *Haplorchis popelkae* n. sp. (Digenea: Heterophyidae) from short-necked turtles (Chelidae) in northern Australia. *J. Parasitol.* 95, 204, 2009.
4. Sohn, W.M. et al. Fish-borne trematode metacercariae in freshwater fish from Guangxi Zhuang Autonomous region, China. *Korean J. Parasitol.* 47, 249, 2009.
5. Scholz, T., Ditrich, O., and Giboda, M. Differential diagnosis of Opisthorchiid and Heterophyid metacercariae (Trematoda) infecting fresh of cyprinid fish from Nam Ngum Dam Lake In Laos. *Southeast Asian J. Trop. Med. Public Health* 22, 171, 1991.
6. Waikagul, J. Intestinal fluke infections in Southeast Asia, Food-borne parasitic zoonosis. *Southeast Asian J. Trop. Med. Public Health* 22, 158, 1991.
7. Chai J.Y., Murrell K.D., and Lymbery, A.J. Fish-borne parasitic zoonoses: Status and issues. *Int. J. Parasitol.* 35, 1233, 2005.
8. Dung D.T. et al. Fish-borne zoonotic intestinal Trematodes, Vietnam. *Emerg. Infect. Dis.* 13, 1828, 2007.
9. Sayasone, S. et al. Diversity of human intestinal helminthiasis in Lao PDR. *Trans. R. Soc. Trop. Med. Hyg.* 103, 247, 2009.
10. Belizario, V.Y. et al. A focus of human infection by *Haplorchis taichui* (Trematoda: Heterophyidae) in the southern Philippines. *J. Parasitol.* 90, 1165, 2004.

11. Sukontason, K. et al. Evidence of *Haplorchis taichui* infection as pathogenic parasite: Three case reports. *Scand. J. Infect. Dis.* 37, 388, 2005.
12. Giboda, M. et al. Current status of food-borne parasitic zoonoses in Laos. *Southeast Asian J. Trop. Med. Public Health* 22, S56, 1991.
13. Nithikathkul, C., and Wongsawad, C. Prevalence of *Haplorchis taichui* and *Haplorchoides* sp. metacercariae in freshwater fish from water reservoirs, Chiang Mai, Thailand. *Korean J. Parasitol.* 46, 109, 2008.
14. Chai, J.Y. et al. High prevalence of liver and intestinal fluke infections among residents of Savannaket province in Laos. *Korean J. Parasitol.* 45, 213, 2007.
15. Chai, J.Y. et al. Mixed infections with *Opisthorchis viverrini* and intestinal flukes in residents of Vientiane Municipality and Saravane Province in Laos. *J. Helminthol.* 79, 283, 2005.
16. Touch, S. et al. Discovery of *Opisthorchis viverrini* metacercariae in freshwater fish in southern Cambodia. *Acta Trop.* 111, 108, 2009.
17. Ooi H.K. et al. Metacercariae in fishes of Sun Moon lake which is an endemic area for *Clonorchis sinensis* in Taiwan. *Southeast Asian J. Trop. Med. Public Health* 28, 222, 1997.
18. Wang, J.J. et al. *Haplorchis* infections in intermediate hosts from a clonorchiasis endemic area in Meinung, Taiwan, Republic of China. *J. Helminthol.* 76, 185, 2002.
19. Elsheikha H.M., and Elshazly A.M. Host-dependent variations in the seasonal prevalence and intensity of heterophyid encysted metacercariae (Digenea: Heterophyidae) in brackish water fish in Egypt. *Vet. Parasitol.* 153, 65, 2008.
20. Rim, H.J. Opisthorchiasis. In: Steele J.H. (ed.), *CRC Handbook Series in Zoonoses, Section C: Parasitic Zoonoses, Vol. III (Trematode Zoonoses)*. CRC Press, Boca Raton, FL, USA, p. 109, 1982.
21. Sukontason, K. et al. Prevalence of trematode metacercariae in cyprinoid fish of Ban Pao district, Chiang Mai Province, northern Thailand. *Southeast Asian J. Trop. Med. Public Health* 30, 365, 1999.
22. Sukontason, K.L. et al. Prevalence of *Opisthorchis viverrini* infection among villagers harboring Opisthorchis-like eggs. *Southeast Asian J. Trop. Med. Public Health* 32, S23, 2001.
23. Kumchoo, K. et al. High prevalence of *Haplorchis taichui* metacercariae in cyprinoid fish from Chiang Mai Province, Thailand. *Southeast Asian J. Trop. Med. Public Health* 36, 451, 2005.
24. Radomyos, P., Radomyos, B., and Tungtrongchitr, A. Multi-infection with helminthes in adults from northeast Thailand as determined by post-treatment fecal examination of adult worms. *Trop. Med. Parasitol.* 45, 133, 1994.
25. Waikagul, J. *Opisthorchis viverrini* metacercaria in Thai freshwater fish. *Southeast Asian J. Trop. Med. Public Health* 29, 324, 1998.
26. Rim, H.J. et al. Fish-borne trematode metacercariae detected in freshwater fish from Vientiane Municipality and Savannakhet Province, Lao PDR. *Korean J. Parasitol.* 46, 253, 2008.
27. De, N.V. et al. The food-borne trematode zoonoses of Vietnam. *Southeast Asian J. Trop. Med. Public Health* 34, 12, 2003.
28. Thu, N.D. et al. Survey for Zoonotic liver and intestinal trematode metacercariae in cultured and wild fish in an Giang Province, Vietnam. *Korean J. Parasitol.* 45, 45, 2007.
29. Chi, T.T. et al. Prevalence of Zoonotic trematodes in fish from a Vietnamese fish-farming community. *J. Parasitol.* 94, 423, 2008.
30. Elshazly, A.M. et al. Natural infection of fresh and brackish water fish with heterophyid encysted metacercariae in Dakahlia Governorate, Egypt. *J. Egypt Soc. Parasitol.* 37, 999, 2007.
31. Keiser, J., and Utznger, J. Food-borne trematodiases. *Clin. Microbiol. Rev.* 22, 466, 2009.
32. Chi, J.K. et al. Intestinal pathology in human metagonimiasis with ultrastructural observations of parasites, *J. Korean Med. Sci.* 3, 171, 1988.
33. Tantachumrun, T., and Kilks, M. Heterophyid infection in human ileum: Report of three cases. *Southeast Asian J. Trop. Med. Public Health* 9, 128, 1978.
34. Africa, C.M., de Leon, W.E.D., and Garcia, E.Y. Intestinal heterophyidiasis with cardiac involvement: A contribution to etiology of heart failure. *J. Philippine Island Med. Assoc.* 15, 358, 1935.
35. Africa, C.M., de Leon, W.E.D., and Garcia, E.Y. Visceral complications in intestinal heterophyidiasis of human. *Acta Med. Philippines Monogr. Ser.* 1, 1, 1940.
36. Fried, B., Graczyk, T.K., and Tamang, L. Food-borne intestinal trematodiasis in human. *Parasiol. Res.* 93, 159, 2004.
37. Chai, Y.Y., and Lee, S.H. Intestinal trematodes of humans in Korea: *Metagonimus*, heterophyids and echinostomes. *Korean J. Parasitol.* 28, 103, 1990.
38. Hotez, P.J. et al. Helminth infections: The great neglected tropical diseases. *J. Clin. Investig.* 118, 1311, 2008.
39. Ditrich, O. et al. Comparative morphology of eggs of the Haplorchiinae (Trematoda: Heterophyidae) and some other medically important heterophyid and opisthorchiid flukes. *Folia Parasitol.* 39, 123, 1992.
40. Radomyos, B. et al. Opisthorchiasis and intestinal fluke infections in northern Thailand. *Southeast Asian J. Trop. Med. Public Health* 29, 123, 1998.
41. Tesana, S. et al. Eggshell morphology of the small eggs of human trematodes in Thailand. *Southeast Asian J. Trop. Med. Public Health* 22, 631, 1991.
42. Ditrich, O. et al. Serological differentiation of human small fluke infections using *Opisthorchis viverrini* and *Haplorchis taichui* antigens. *Southeast Asian J. Trop. Med. Public Health* 22 (Suppl), 174, 1991.
43. Thaenkham, U. et al. Discrimination of *Opisthorchis viverrini* from *Haplorchis taichui* using COI sequence marker. *Acta Trop.* 103, 26, 2007.
44. Sato, M. et al. Discrimination of *O. viverrini, C. sinensis, H. pumilio* and *H. taichui* using nuclear DNA-based PCR targeting ribosomal DNA ITS regions. *Acta Trop.* 109, 81, 2009.
45. Sato, M. et al. Egg laying capacity of *H. taichui* (Digenea: Heterophyidae) in humans. *Korean J. Parasitol.* 47, 3, 2009.
46. Sato, M. et al. Copro-DNA diagnosis of *Opisthorchis viverrini* and *Haplorchis taichui* infections in an endemic area of Laos. *Southeast Asian J. Trop. Med. Public Health* 41, 28, 2010.
47. Traub, R.J. et al. A new PCR-based approach indicates the range of *Clonorchis sinensis* now extend to central Thailand. *PLos. Negl. Trop. Dis.* 3, e367, 2009.
48. Lovis, L. et al. PCR diagnosis of *Opisthorchis viverrini* and *Haplorchis taichui* infections in a Lao community in an area of endemicity and comparison of diagnostic methods for parasitological field surveys. *J. Clin. Microbiol.* 47, 1517, 2009.
49. Wongsawad, C. et al. *Haplorchis taichui*, Witenberg, 1930: Development of a HAT-RAPD marker for the detection of minute intestinal fluke infection. *Exp. Parasitol.* 123 158, 2009.
50. Maniatis, T. et al. *Molecular Cloning*, 1st edn. Cold Spring Harbor Laboratory, New York, 458pp., 1982.
51. Thompson, J.D. et al. The clustal X windows interface: Flexible strategies for multiple sequence alignment aided by quality analysis tools. *Nucleic Acids Res.* 25, 4876, 1997.
52. Hall, T.A. BioEdit: A user-friendly biological sequence alignment editor and analysis program for Windows 95/98/NT. *Nucleic Acids Symp. Ser.* 41, 95, 1999.

35 Heterophyidae

Ron Dzikowski and Michael G. Levy

CONTENTS

35.1 INTRODUCTION

Millions of people worldwide are exposed to the risk of getting infected with zoonotic diseases. Among those, food-borne infections of various pathogens are a major global health concern since infected food products can be exported and can spread those pathogens around the world. Therefore, the requirement for reliable diagnostic tools both for humans and for contaminated food products is a global public health interest. The advances in molecular technologies provide an outstanding opportunity to develop very accurate diagnostic tools that will enable the identification of pathogens by their DNA traces even at very low quantities at a reasonable price. We have recently published a manuscript providing a comprehensive updated description of the current status of molecular diagnostics heterophyiasis, a food-borne infection caused by small trematodes of the heterophyidae family that affects humans around the globe. While looking through the recent literature, we came to realize that since our last publication a very few additional contributions were made to the "molecular diagnostic toolbox" of heterophyidae. We therefore made the required update for this chapter hoping that the development of new diagnostic tools for these pathogens will be accelerated and expanded for the betterment of humanity.

35.1.1 Classification and Biology

Digenea are heteroxenous platyhelminth parasites that require two or more hosts to complete their life cycle. Sexually mature adult stages are parasites of vertebrates. Infections by digeneans are the most common among fish-borne zoonoses, particularly by species that are less fastidious of their definitive vertebrate hosts. Until fairly recently, fish-borne zoonotic infections caused by digenea were mainly a problem of either developing countries or populations that traditionally consumed raw fish. Changes in global demographic and socio-economics coupled with improved transportation systems have opened global markets for trade and consumption of raw fish products, exposing additional populations to the risk of infection. Therefore, an increasing number of individuals have developed these zoonoses even in places where the source of infection does not naturally exist [1,2]. The current status of fish-borne trematode zoonoses has been recently reviewed by several authors [1,3,4].

Agents of human trematodiasis include the liver flukes such as *Fasciola hepatica*, *Clonorchis sinesis*, *Ophistorchis* spp., and *Metorchis conjunctus* as well as intestinal flukes such as echinostomes and heterophyids. Adult heterophyidae are one example of trematodes that are not fastidious in their definitive or adult stage hosts. These parasites may infect a large variety of vertebrate hosts, including piscivorous birds such as pelicans, herons, and cormorants, as well as a variety of wild and domesticated mammalian species, including cats, dogs, and humans. The heterophyidae are very small trematodes: adult worms found in the intestine of their definitive hosts range in length between 0.3 and 3 mm. The life cycle begins with embryonated eggs, each with a fully developed miracidium that are passed in the feces of the definitive host into aquatic habitats where they are ingested by particular aquatic snails. The miracidia develop in the host snail into sporocysts after which they develop into rediae. Each of the mature rediae may shed hundreds of free-swimming cercariae every day which are able to penetrate a fish second intermediate host, thus increasing their probability of finding and infecting fish that serve as the second intermediate hosts. The cercaria then shed their tails, migrate to a specific tissue (muscles, kidney, heart, gills, etc.) encapsulate, and develop into metacercariae. Upon ingestion of the infected fish by the final host, the metacercariae excyst from their capsule and develop into adult worms that migrate through the digestive tracts to the intestine.

Unlike their adult forms which are promiscuous and infect a variety of hosts, larval stages of heterophyids are more specific with regard to their snail and fish hosts. Additionally, most heterophyids exhibit high specificity to the fish tissue in which they will develop to metacercariae. For example, *Heterophyes heterophyes*, *H. dispar*, and *H. equalis* in the Middle East develop in the marine snails *Pirrenela conica* and *Cerithidae cingulata*, and the cercariae infect mainly fish of the families Mugilidae and Cichlidae (tilapia) and migrate to the muscles, where they develop into metacercariae. In endemic regions such as the Nile delta in Egypt, where these snails are abundant, each fish may harbor hundreds and even thousands of metacercariae. These minute metacercariae cannot be visually detected without a stereoscope; thus, people consuming the muscles of infected fish will not notice that they are heavily infected. The above reasons may explain why infection with *Heterophyes* spp. is one of the most common fish-borne zoonoses.

Different heterophyids develop into metacercariae in different organs. For example, *Centrocestus* spp. are less fastidious in their choice of fish hosts and infect several fish species, including Cichlidae, Cyprinidae, Poecelidae, and others. They are transmitted by the freshwater snail *Melanoides tuberculata* and develop into metacercariae in the gills. Other examples for larval migration of heterophyids to specific tissue are *Pygidiopsis genata*, which develops into metacercariae in the kidney, and *Phagicola longa*, which develops into metacercariae in the heart of freshwater fish such as cichlids (Figure 35.1).

35.1.2 Epidemiology

Food-borne trematode infections drew the attention of the World Health Organization (WHO) as a serious and growing public health concern estimated to affect 40 million people worldwide, whereas the number of people at risk of infection was estimated at over half a billion [5–7]. In Southeast Asia and the Western Pacific region, food-borne trematodiasis is an emerging public health problem [5]. The emergence of infections in these regions appears to be correlated with the exponential growth in aquaculture in these endemic regions. Heterophyiasis is not an exception to this trend and is important for emerging fish-borne zoonoses. Recently, heterophyiasis was documented in regions where it has not been known to be endemic in the past and in some cases the prevalence of infections in those regions reached remarkably high levels. A recent study in Vietnam has estimated that the prevalence of human infection with heterophyiasis in endemic regions was 100%. The most abundant species found in human were *Haplorchis pumilio* and *Haplorchis taichui*, in which the prevalence of infection was 70% and 100%, respectively [8]. The source of infections in those regions is consumption of freshwater fish harboring heterophyids metacercariae. A recent survey in northern Thailand also found heavy infection of *Haplorchis* metacercariae in fish from two freshwater reservoirs. The prevalence of infection of *H. taichui* and *Haplorchoides* sp. ranged between 74% and 84% [9]. In another study, cultured fish from the Mekong Delta, Vietnam, were found to be infected with *H. pumilio*, *H. taichui*, *Centrocestus formasanus*, and *Stellantchasmus falcatua* with infection rates ranging between 1.7% and 6.6%. Culture conditions as well as seasonal influences were determined to have affected these rates. A clinical survey in the southern Philippines found that 36% of patients that were complaining about abdominal discomfort were suffering from heterophyiasis [10]. In a later study, these authors reported an exceptionally high prevalence of infection (36%) with *H. taichui* in human population in another focus in Southern Philippines [11]. In Vietnam, in an agricultural province that is populated mainly with farmers with fish ponds integrated into their farming system, a random stool examination demonstrated that the prevalence of infection with small trematode eggs was 65%. Examination of adult worms recovered following antihelminthic treatment with praziquantel showed that the prevalence of infection with at least one heterophyid species was 100% [8]. The dominant species found in these examinations were *H. pumilio* and *H. taichui*, and *H. yokogawai* and *Stellanchasmus falcatus* were less abundant. The awareness of the epidemiological and clinical significance of various species of intestinal trematodes in the Republic of Korea has recently increased and resulted in several epidemiological studies. In a recent review of those studies, heterophyids and echinostomes are recognized as the two major groups with significance to public health in terms of both the number of species involved and the number of people infected [12]. In this survey, 12 species of heterophyids were documented in humans. The geographical distribution of human infection was found mostly along the eastern and southern coasts with inland endemic foci along certain rivers and streams. The prevalence of infection was found to be between 20% and 70% among the villagers living along these rivers. Surveys conducted at a later time point have demonstrated that new foci of heterophyiasis are emerging [13], mainly by piscivorous birds that migrate between habitats where the snails are thriving.

Similarly, high infection rates have been documented in other regions around the globe where heterophyid species are endemic and people are accustomed to eating raw fish. In some regions in Nile delta in Egypt, about a third of the population is infected. The intensity of infection was found to be significantly high among fishermen. This finding is not surprising given the high level of fish consumption, notably gray mullets eaten raw and either dried or mildly salted [14]. However, in other places in Egypt where the amount of raw fish consumption is less, the rate of infection was found to be significantly lower. A survey among 3180 patients attending Mansoura University Hospitals' Clinics, which were subjected to stool examination, found that heterophyiasis

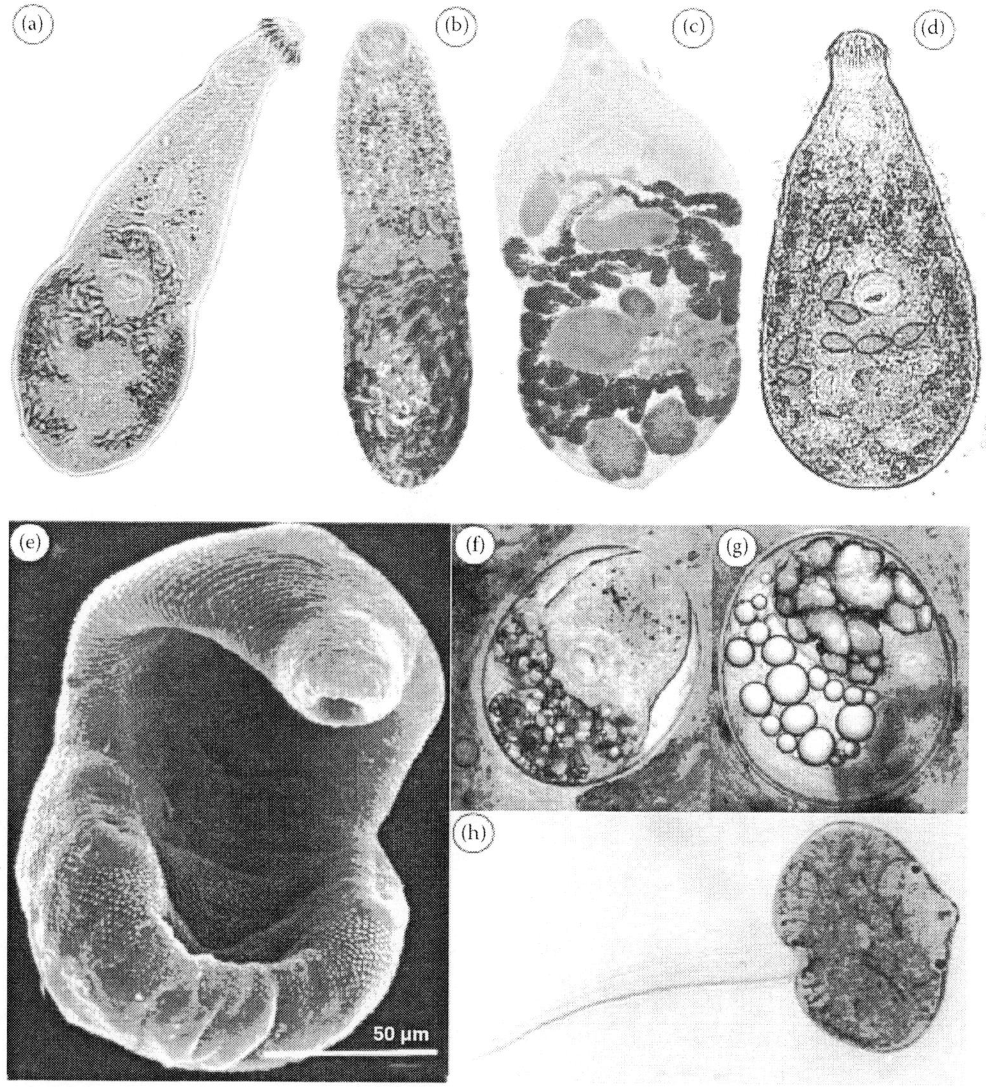

FIGURE 35.1 Heterophyid trematodes. (a–c) Adult specimens of *Phagicola longa*, *Haplorchis pumilio*, and *Dexiogonimus ciureanus* collected from cormorants (*Phalacrocorax carbo*). (d) Adult specimen of *Centrocestus* sp. collected from Little Egret (*Egretta garzetta*). (e) Scanning electron micrograph (SEM) of adult specimen of *Pygidiopsis genata* collected from cormorant (*P. carbo*). (f) Metacercaria of *P. genata* in the kidney of the cichlid fish *Tilapia zillii*. (g) Metacercaria of *P. longa* in the heart of *T. zillii*. (h) Pleurolophocercous cercaria of *Centocestus* sp. emerged from the snail *Melanoides tuberculata*.

infection rates were about 4.2% [15]. Interestingly, human infection with *H. heterophyes* and *H. dispar* was documented in Korea in people traveling to endemic regions in the Middle East and were reported as "imported" cases [12]. These cases emphasize the need for awareness to heterophyiasis as a global concern.

35.1.3 Clinical Manifestations

Intestinal heterophyiasis usually causes mild symptoms such as abdominal tenderness, chronic diarrhea, bloody diarrhea, abdominal pain, vomiting, and so on. [4] Symptoms of acid peptic and peptic ulcer disease are also common clinical manifestations of heterophyiasis. The degree of clinical symptoms seems to be related to the intensity of infection, that is, patients presenting with severe illness tend to have a higher intensity of infection. However, the severity of the symptoms may also be related to the susceptibility as well as the degree of acquired immunity of the individual patient [12]. Heavy infections may cause necrosis of the intestinal mucosa that allows penetration of heterophyid eggs into the mesenteric blood circulation. These eggs may reach different organs and have been reported to cause granulomatous lesions in the myocardium and the brain and spinal cord [4,16]. In some cases, patients with heterophyiasis also had cardiac symptoms that improved after antihelminthic treatment [12], suggesting a possible ectopic parasitism by heterophyids. Recently, experimental infections of *H. heterophyes* in rats have demonstrated that worms and/or eggs may cross the intestinal wall and spread to visceral organs such as the intestinal wall, lymph nodes, liver, and spleen [17]. These results suggest that under certain circumstances, humans—especially in immunocompromised patients—may be at risk of ectopic infection with *H. heterophyes*.

The intestinal histopathology caused by heterophyiasis has mainly been determined from studies using experimental infections in animal models such as mice, rats, cats, and dogs. The pathological features were characterized by different degrees of mucosal damage, villous atrophy, hyperplasia, and inflammatory reactions [12], with some evidence for differences in virulence between species. However, it is extremely difficult to distinguish between the clinical symptoms caused by each individual species in natural infection in humans. This is attributed to the fact that natural infections are usually composed of multiple species and the lack of species-specific diagnostic tools makes it difficult to associate each species with its corresponding symptom(s). In addition, once heterophyid eggs are found in a stool sample, humans are treated with praziquantel which is not species-specific but a general antitrematode treatment. However, the evidence for ectopic parasitism by some species and the speculation that these species may therefore be more virulent require a reliable diagnostic tool that would be able to detect heterophyiasis early enough to administer treatment before they cause damage to the central nervous system and without the need for tissue biopsy. In addition, useful diagnostic tools of early stages of heterophyids may have great potential as preventative measures. Early identification of heterophyids in snails and fish may be used by public health authorities to take measures that would prevent infected fish from reaching the markets and would minimize the prevalence of human infections.

35.1.4 Diagnosis

35.1.4.1 Phenotypic Techniques

The complex and variable life cycle of digenea is one of the challenging aspects in understanding the biology of these pathogens. Among digenea, heterophyids are among the most difficult to identify at the species level and require special expertise due to their minute size. In addition, precise species determination of early stages such as cercariae emerging from snails and metacercariae infecting fish is ambiguous because they lack the distinguishing morphological traits of the reproductive system that develop only in the adult worms. Moreover, our knowledge about the complete life cycles of many of these parasites is limited, as experimental demonstrations of life histories are laborious and often unachievable due to the unknown identity of the intermediate and/or definitive hosts. Therefore, currently available methodologies for accurate species diagnosis have serious limitations. The diagnosis of adult heterophyid worms in their definitive host such as dogs, cats, and human is performed by identifying the morphological characteristics of their eggs in stool samples. Staining with potassium permanganate may help us to differentiate between eggs of different species, as has been demonstrated for *O. viverrini*, *H. taichui*, and *Phaneropsolus bonnie* [18]. However, distinguishing between different species of heterophyid eggs is extremely difficult because of the high similarity in shape and size. Detection of heterophyid eggs in stool samples may be confirmed by recovering the adult flukes after antihelminthic treatment. Alternatively, serological tests have been used to screen humans for antibodies against parasitic infection [4,12]. Several immunodiagnostic tests were evaluated in dogs that were experimentally infected with *H. heterophyes* [19]. These authors concluded that the use of the skin ID test, which is independent of circulating antibodies, may be used for scientific and laboratory diagnoses in naturally infected animals and thus have a potential in the routine diagnosis of *Heterophyes* infection in humans. However, since the skin ID test did not offer optimal sensitivity and specificity, it is necessary to standardize procedures for antigen preparation and serum testing before firm recommendations can be made. A reported case of human infection with *Heterophyes nocens* was diagnosed by sectional morphology in a biopsy specimen of the small intestine [20]; however, this rare case obviously does not offer a procedure for routine diagnostic tests. These diagnosis difficulties not only limit our knowledge about the basic biology of these pathogens, but also inhibit disease prevention and treatment.

In view of the disease burden of heterophyiasis, the complex life histories of these worms, and the difficulties and limitations of the current diagnosis methodologies, new approaches to develop effective diagnostic tools for these pathogens are essential. Precise identification of the immature stages—eggs in human stool samples, cercariae from snails, and metacercariae in fish—is an even greater challenge considering the changes in their morphological traits. Although the morphology changes dramatically during the life cycle of digenea their genetic composition remains the same. This enables targeting of the specific nucleotide sequence of their DNA in order to develop genetic markers for precise identification of each species at any point of their life cycle. An additional advantage of this molecular approach is the ability to amplify DNA using polymerase chain reaction (PCR); therefore, only a very small amount of DNA template is required for these reactions. PCR-based diagnostics are therefore very sensitive and allow particular identification of a single cercaria, metacercaria, and adult worm [21], as well as detection of early sporocyst and rediae infections in snails even before they start shedding cercariae [22]. Several PCR-based methodologies have been developed and successfully applied for specific diagnostics and molecular taxonomy of fish pathogens (see the comprehensive up-to-date review by Ted Clark [23]). As a first step, PCR is used to amplify the gene of interest using gene-specific oligonucleotide primers. The PCR products are called amplicons and they may be subjected to direct sequencing. An alignment of the DNA sequence of a particular gene can be used to detect species-specific sequence identity or polymorphism.

As an alternative to direct sequencing, the amplicons may be used for a PCR-RFLP (restriction fragment length polymorphism) assay. In this assay, the amplicons are subjected to restriction digest that produce sequence-specific cutting patterns based on the affinity of different restriction enzymes for their specific binding sites. RFLP is a faster and more cost-effective way to detect polymorphisms, as there is no need for sequence information. However, it may not be as accurate as comparing the sequence itself. Using RFLP, sequence polymorphism can be detected only if it is at the site where the restriction enzyme binds to and cuts the DNA; therefore, lack of differences in the restriction patterns may be false negative if the polymorphisms are found away from the restriction site. It is necessary to perform several restriction digests with different enzymes. However, there is always a possibility that the same digest patterns may still be of different species. Therefore, to increase the accuracy of diagnoses, a specific RFLP assay should be designed to differentiate between known sequences that would result in different cutting patterns.

Another PCR-based methodology to detect genetic diversity is to use arbitrary primers in order to randomly amplify polymorphic DNA (RAPD). Sequence polymorphism in the genome of different organisms will result in different amplification patterns that can be visualized when separating the amplicons on an agarose gel, creating a typical genetic fingerprint for each organism. RAPD-PCR does not require prior knowledge of the nucleotide sequence and is a useful tool for studying population genetics. However, there is evidence that the sensitivity of RAPD-PCR may differentiate between populations of the same species, and therefore its suitability for species-specific diagnosis is not clear [24,25].

Unveiling species-specific sequence polymorphisms at a certain locus could be used for the development of species-specific molecular probes that will specifically anneal to a certain sequence. These probes can be used as species-specific primer sets for PCR amplification that will only amplify certain species or as multiplexes of primers that after PCR amplification will result in amplicons of different sizes, each specific for particular species. This method was developed for differentiation of several trematode species, including species of the heterophyidae family [21,22,25–27].

During the last 15 years, a considerable amount of data on trematodes' DNA sequences has accumulated and has been used to study their interrelationships and molecular phylogeny on the basis of sequence comparisons [28]. The accumulation of the available sequence data has also been used to elucidate life histories traits. This became possible by unveiling the genetic linkage through sequence identity between the adult worms that can be identified morphologically and sequenced to the sequence of the same locus in the immature stages. The life histories of a few marine and freshwater species have been elucidated using molecular techniques, demonstrating their advantages for specific diagnosis. An early example of the successful use of molecular techniques is the demonstration of the three-host life cycle of *Bivesicula claviformis* [29]. These authors were able to show in an elegant experiment that the mesocercaria found in the gut of the herbivorous fish *Thalassoma lunare* (Labridae) is genetically identical to the adult worm found in the gut of the carnivore *Epinephelus fasciatus* (Serranidae). Similar approaches were further used to study life histories traits of many other marine species. Anderson was able to unveil the link between the metacercaria and the adult of *Indodidymozoon pearsoni* (Digenea: Didymozoidae) by comparing the DNA sequence of fragments of their rDNA [30]. Similarly, Jousson and colleagues have elucidated the life cycles of some Mesometridae (Trematoda, Digenea) [31] and Monorchiidae species [32]. Sequence analysis was also used to link some undescribed larval stages of the digenean family Opecoelidae and the adult worms [33,34], and to distinguish between closely related species of the Opecoelidae family [35].

The most common gene targets for PCR amplification in diagnostic tests and phylogenetic studies of trematodes are probably the ribosomal RNA (rDNA) genes. The rDNA genes contain three coding regions that encode the three subunits: the 18S small subunit (SSU), the 5.8S, and the 28S large subunits (LSU). The noncoding sequence between the 18S and the 5.8S subunits is internal transcribed spacer 1 (ITS1) and that between the 5.8S and the 28S subunits is ITS2. The regions of the rDNA that have been mostly used for species differentiation are the 18S small subunit and the ITS regions. The slow evolution rate of these genes in eukaryotes is reflected in the high degree of conservation of their nucleotide sequence. This conservation among species makes them ideal for molecular taxonomy as well as for the development of molecular probes for species-specific diagnostics. Although the ITS regions have been extensively and successfully used to elucidate trematodes life histories [29–31], it seems that the ITS regions are more rapidly evolving than the 18S, and intraspecific variation has been found in these loci in some trematodes [36,37]. Therefore, while it is possible that the use of probes to the ITS regions may be able to distinguish between closely related species, it may also differentiate between strains of the same species. Therefore, it is possible that species-specific probes designed to hybridize to the ITS regions may achieve false-negative results and to effectively discriminate between different strains of the same species. Consequently, for species-specific diagnosis, it is recommended to use the 18S sequence as the main target for comparison, and the ITS regions can be used to confirm the 18S diagnosis or to detect variation among strains of poorly differentiated species.

35.1.4.2 Molecular Techniques

In spite of their importance as food-borne pathogens that affect millions of people worldwide, the number of species of heterophyid trematode that have been sequenced is still limited. However, in recent years, these data have been increasing continuously (Table 35.1), enabling further development of molecular diagnosis methodologies. Molecular genetics have also been used to demonstrate ecological phenomena. In recent studies, the sequences of ITS1 together with the mitochondrial cytochrome oxidase subunit 1 (mDNA COI) genes were used to differentiate cercariae at the species level. Cryptic species of the digeneans, *Cercaria batillariae* (Heterophyidae) and an undescribed philophthalmid, were identified using PCR-RFLP and sequence analyses of these genes [38]. Subsequent sequence comparisons of these cercariae were further used to demonstrate a marked reduction in the genetic variation in one of the species that were introduced to North America, suggesting that cryptic species

TABLE 35.1
List of Heterophyidae Sequences from GeneBank

Heterophyid Species	Genbank Accession Number	Sequence Description
Centrocestus sp.	AY245699	rDNA (18S-ITS1-5.8S-ITS2)
Centrocestus formosanus	AY245759	rDNA 18S
Haplorchis pumilio	AY245706	rDNA (18S-ITS1-5.8S-ITS2)
Haplorchis taichui	AY245705	rDNA (18S-ITS1-5.8S-ITS2)
Haplorchis taichui	EF055885	mRNA (cytochrome c oxidase subunit I)
Phagicola longa	AY245703	rDNA (18S-ITS1-5.8S-ITS2)
Galactosomum lacteum	AY222120	rDNA (18S)
Galactosomum lacteum	AY222227	rDNA (28S)
Pygidiopsis genata	AY245710	rDNA (18S-ITS1-5.8S-ITS2)
Dexiogonimus ciureanus	AY245702	rDNA (18S-ITS1-5.8S-ITS2)
Phagicolinae sp.	AY245766	rDNA (18S)
Cryptocotyle lingua	AJ287492	rDNA (18S)
Cryptocotyle lingua	AJ222228/AF023111	rDNA (28S)
Haplorchoides sp.	AJ287521	rDNA (18S)
Haplorchoides sp.	AY222226	rDNA (28S)
Heterophyes nocens	AF188119	mRNA (cytochrome c oxidase subunit I)
Heterophyes nocens	AF181890	rDNA (28S)
Metagonimus sp.	AF095333	rDNA (28S)
Metagonimus sp.	AF096232	mRNA (cytochrome c oxidase subunit I)
Metagonimus takahashi	AF095332	rDNA (28S)
Metagonimus takahashi	AF096231	mRNA (cytochrome c oxidase subunit I)
Metagonimus yokogawai	AF095331	rDNA (28S)
Metagonimus yokogawai	AF096230	mRNA (cytochrome c oxidase subunit I)
Metagonimus yokogawai	AF125905/6/7	Cysteine proteinase mRNA
Metagonimus yokogawai	AF121840/1/2/3/4	Cysteine proteinase mRNA
Pygidiopsis summa	AF181885	rDNA (28S)
Pygidiopsis summa	AF181884	mRNA (cytochrome c oxidase subunit I)
Stellanchasmus falcatus	AF181886	rDNA (28S)
Stellanchasmus falcatus	AF181887	mRNA (cytochrome c oxidase subunit I)
Stictodora lari	AY452519/20/21	Actin gene

exhibit different invasion pathways [39]. The 28S rDNA and mitochondrial cytochrome oxidase subunit I sequences were used to confirm that *Metagonimus yakogawai*, *M. takahashii*, and *M. myati* are separate species. The ability to amplify and identify even small numbers of initial DNA templates using PCR-based techniques enabled Traub et al. to develop a new PCR-based approach to detect *H. taichui* eggs from only 6 pg of DNA template. These authors used primers designated to the polymorphic ITS2 region of the rDNA to amplify DNA extracted from human stool samples containing small numbers of trematodes eggs [40]. Recent studies have shown that primers designated to polymorphic regions in the ITS1 and ITS2 were able to achieve PCR-based discrimination between *H. taichui* and *H. pumilio*. This molecular reaction was performed on adult trematodes, metacercariae as well as on the eggs where they showed 76.2% sensitivity for ITS1 and 95.2% sensitivity for ITS2 [41].

Alignments of heterophyids' rDNA showed a high level of sequence similarity. In fact, a comparison of the 18S sequence of six species of adult worms collected from the guts of piscivorous birds has shown that their base-pair content was 96–98% identical [21]. However, the alignment also revealed a polymorphic region in the sequence of the 18S subunit of the gene, about 1400 bp downstream of the start codon, in which all six species differ. This polymorphic region was the target sequence for designating species-specific nucleotide primers for specific PCR amplification of *Centrocestus* sp., *H. taichui*, *H. pumilio*, *Dexiogonimus ciureanus*, *P. longa*, and *P. genata*. In order to differentiate between *Haplorchis* spp. and *D. ciureanus*, it was necessary to perform PCR-RFLP on the entire 18S subunit due the high sequence similarities. Using this diagnosis assay, it was possible to genetically link single cercaria, metacercariae, and adult worms of *Centrocestus* sp., *H. pumilio*, *P. longa*, and *P. genata*. This study was the first demonstration of the complete life cycle of *P. genata*. Using this molecular technique, the adult worm collected from cormorants (*Phalacrocorax carbo*) was genetically linked to metacercaria from the fish *Tilapia zillii*. It also revealed that the snail *Melanopsis costata* is the first intermediate host by demonstrating that the cercaria it shed is genetically identical to the adult worm. The successful amplification and sequencing of the rDNA genes of these heterophyids was a prerequisite for the development of the species-specific molecular probes as well as the differentiating PCR-RFLP assay. This procedure has proved itself as a useful, cost-effective and, above all, accurate molecular tool for diagnosis of heterophyids that are pathogenic in both fish and human. The high sensitivity of these techniques, as well as the fact that the primers amplify only parasites' DNA and does not cross-react either with the snails or with the fish DNA, makes them a great diagnostic tool for early detection of possible risk of infection.

Recently, a RAPD-PCR reaction was designed to specifically identify *H. taichui* by HAT-RAPD (high annealing temperature random amplification polymorphic DNA) [42]. This group identified a 265-bp HAT-RAPD marker to which they designated an *H. taichui*-specific primer set. They have also demonstrated the great sensitivity of this reaction that was able to give correct diagnosis with only 0.1 pg of DNA template.

Applying these molecular tools in aquaculture systems and fisheries in endemic regions may have an important practical contribution to global public health interests. In most of these systems, fish are cultured in open ponds where the snail hosts are abundant and piscivorous bird may transmit infection through contaminating the ponds with heterophyid eggs in their feces. In addition to the economical risk that results from fish loss, there is a great risk for initiating heterophyiasis in humans as demonstrated by epidemiological surveys [8]. In order to be able to take preventative measures, the infection must be detected at early stages of the life cycle. In this regard, the complex life cycle of heterophyids is advantageous. DNA extracts from snail samples can be pooled and used as the template for PCR amplification. Using this approach, it is possible to sample snails routinely in order to detect early infections with heterophyid sporocysts and rediae even before they are infective to the fish, that is, before the snails start shedding ceracariae. Early detection of heterophyids in the snails in a certain pond could consequently result in preventative measures, such as transferring the fish and treating the pond against the snails, which will lead to reducing the risk of infection. Similarly, these diagnostic tests may be used by public health agencies to prevent consumption of raw fish infested with heterophyids metacercariae. DNA samples from the muscles and other organs of fish that are known to harbor heterophyids metacercariae, such as gray mullets (Mugilidae) and tilapia (Cichlidae), can be PCR amplified and tested for the presence of heterophyid metacercariae. Early detection of such infection would contribute to the reduction of the risk and the prevalence of infection by classifying these fish products as "risky" so that they are not consumed raw but only if properly cooked or deep frozen prior to consumption.

In addition to the PCR-based methods mentioned above that have been successfully applied, other approaches that have been used to diagnose other pathogens can be modified and be used for heterophyiasis diagnostics. For example, solid-phase hybridization techniques such as *in situ* hybridization (ISH) and fluorescence *in situ* hybridization (FISH) [23] may be considered as a useful tool for the identification of heterophyids in histological samples of snail and fish tissues and in some rare cases from human biopsies. However, at this point it seems that the PCR-based methodologies mentioned above are more convenient for routine diagnosis purposes.

Molecular techniques for the diagnosis of a number of zoonotic protozoan parasites such as *Cryptosporidium*, *Toxoplasma*, *Cyclospora*, *Giardia*, and Microsporidia have been developed and are utilized in developed countries. With the exception of *Toxoplasma* for which methods are available for detection of tissue stages as well as infectious oocysts, these methodologies have primarily focused on detection of infectious cysts contaminating food or water. Infectious cysts are generally present as surface contaminants, and culture methodologies, even when available, are not suitable for the biological amplification of parasite numbers. Therefore, most efforts at detection of protozoan parasites have focused on direct detection of these organisms either directly from food samples or following concentration and purification of infectious cysts. In contrast to various techniques developed for concentrating cystic forms of protozoan parasites that may be utilized in conjunction with molecular assays, manual removal of these minute heterophyid trematodes from tissue samples is laborious and therefore unlikely to be of value for the screening of fresh fish.

35.2 METHODS

In this chapter, we will focus on the methodologies that have already been proved to be successful for the diagnostics of species of heterophyid species. We will give detailed explanations of the techniques one can use to collect the genetic material, to extract DNA from different organisms and tissues, and to amplify the gene of interest by PCR in order to be able to sequence new species. We will also outline the basic rules for designing species-specific primer sets and multiplexes for specific diagnosis. We hope that this chapter will provide knowledge that would benefit anyone who wants to use a similar approach to expand the available molecular tool box for heterophyid diagnosis.

35.2.1 Sample Preparation

Development of specific PCR assays has been developed for a large number of trematode species, including heterophyids [21,22,28]. Adult trematodes that can be identified to species based on internal morphology are collected by sedimentation of the intestine content of the final host (piscivorous birds, cats, dogs, mice or rats) and then removed to clean physiological saline and preserved in 70% ethanol. Metacercariae of heterophyids are collected directly from the tissue of the infected fish (gills, subcutaneous, heart—conus arteriosus, kidney, liver). They should be removed gently from the surrounding tissue by needles under a stereoscope and preserved in 70% ethanol. Cercariae are collected from aquatic snails by exposing the infected snails to strong light. And then individual specimens can be either frozen at −70°C or fixed in 70% ethanol.

35.2.2 Detection Procedures

It is preferable to use freshly isolated or specimens preserved by freezing or in 70% alcohol rather than in solutions such as formalin that make recovery of DNA difficult. Ribosomal DNA genes are often the target of such efforts as the presence of multiple copies of these genes in the genome generally increases the sensitivity of the assays as well as a large number of published gene sequences useful for the development of oligonucleotide primers for initial sequencing. Once these genes are sequenced, they are aligned with similar genes from related species and unique sequences identified. Oligonucleotide primer sets are designed that complement these species-specific regions and are generally designed to bracket approximately a 150–1000 base-pair segment of the target gene.

Specific PCR assays for fish-infesting trematodes have generally utilized conventional PCR assays in which the final products are manually transferred to agarose or a similar gel matrix, electrophoresed, and the resulting DNA amplicons, if present, visualized following staining with a fluorochrome such as ethidium bromide whose fluorescence when excited with ultraviolet light increases approximately 20-fold when intercolated into double-stranded DNA. The use of real-time PCR (RT-PCR) that incorporates a fluorochrome directly into the PCR reaction mixture which is detected as the reaction progresses eliminates the need for post-PCR processing and offers significant saving of time as well as reduces the possibility of laboratory contamination with target DNA as reaction tubes need not be opened while increasing sample throughput. Although it is not yet possible to connect all stages of these parasites based on morphology, as previously stated, the DNA is constant between life cycle stages. Thus, once a PCR assay with the desired degree of specificity has been developed, it may be applied to the detection and identification of any stage of the parasite [21,22,26,29–32]. Molecular detection of trematodes in fish flesh will require extraction of DNA from representative samples of fish tissue. Following release of DNA, which requires one to several hours of proteinase K digestion, DNA is purified utilizing one of several available methods and the presence of heterophyid-specific DNA is determined. This is a time-consuming process, generally requiring greater than 24 h from the time samples are received in the laboratory, and cost constraints will likely mean that limited numbers of samples can be processed.

35.2.2.1 Procedure

35.2.2.1.1 PCR Amplification

This can be done on DNA extracted from a pool of trematodes or from individual worms, metacercariae, and cercariae. Specimens should be washed overnight in 10 mM Tris-EDTA buffer. Following two more washes (1 h each), a single metacercariae or cercaria is placed directly into a 0.2 mL PCR tube and gently crushed, prior to DNA extractions or direct amplification. DNA is extracted from single adult worms using a DNeasy tissue kit (Qiagen, Valencia, California) according to the manufacturer's instructions. PCRs are carried out in 20–50 µL volumes using 10× buffer, 2.5 U *Taq* polymerase, 10 mM of each deoxynucleotide triphosphate, and 100 ng of each primer.

The PCR conditions should be optimized for each of the primers that are used and according to the size of the expected amplicons. PCR amplicons are subjected to electrophoresis in 1.5% agarose gels, stained with ethidium bromide, and visualized under ultraviolet light. PCR products can be purified using a commercial PCR purification kit and sequenced directly. As an alternative the amplicons of interest can be cut out of the gels, purified, and cloned into the pGEM-T Easy Vector system, utilizing blue/white selection. Plasmids from white colonies should be screened for inserts using PCR amplification. Positive colonies are incubated overnight and then the plasmids are purified and sequenced.

35.2.2.1.2 Endonuclease Restriction Digest

A 5 µL aliquot of the rDNA 18S PCR product (1.5 µg) of each species can be used to perform the PCR-RFLP procedure and it can be digested with restriction enzymes according to the manufacturer's specifications, at 37°C for 4 h. Electrophoresis of the resulting fragments is performed on a 1.5% agarose gel and visualized with ethidium bromide.

35.2.2.2 Result Interpretation

Positive reactions are generally considered to be indicative of the presence of the suspected organism, but this is not always the case. Besides indicating a true positive, such an outcome may also result from the presence of contamination with DNA from another sample, laboratory contamination, and/or contamination with cross-reacting but "nontarget species" DNA should the PCR assay not be sufficiently specific. Careful sample handling from collection through final processing is required because parasite DNA is readily transferred due to the highly stable nature of DNA and the extreme sensitivity of PCR assays may enable detecting minute amounts of such contaminants. Similarly, negative reactions may indicate a true negative, but may also result from insufficient target material in the sample tested, DNA degradation, or the presence of inhibitors of the PCR assay.

35.3 CONCLUSION AND FURTHER PERSPECTIVE

Advances in molecular techniques in the last two decades as well as the increase in the amount of the genetic data available in open accessed computerized databases such as GeneBank has set the stage for the development of new diagnostic tools for pathogens. The field of molecular diagnosis of pathogen is evolving rapidly and new techniques are being developed for precise, cost–effective, and sometimes high-throughput diagnostics. These advances have been used to promote the development of molecular diagnostic tools for the precise identification of many causative agents of zoonoses, including fish-borne pathogens. The worldwide emergence of fish-borne zoonoses emphasizes the need for accurate diagnostic tools to determine the best course of clinical treatment for patients. In addition, the complexities of the life histories of trematodes can indeed be advantageous by providing an opportunity for the detection of infection foci of early-stages heterophyid even before they become infective. Therefore, the great potential of molecular diagnosis is that it would give health authorities a great monitoring tool, which is sensitive, specific, inexpensive, and with the right training, simple to use relying on automated equipment as much as possible. Upon early detection of infection, these tools will allow us to execute preventative measures to be executed in order to limit the scope of infection. In spite of its great potential, the development of molecular diagnostic tools for heterophyidae is still at its very early stage. It is surprising that although heterophyiasis is a disease that affects millions worldwide, the volume of molecular research of this particular trematode family is still limited. Consequently, the available sequence data, which are the basis for the development of molecular diagnostic tools, need to be expanded. Nevertheless, the methodologies that have been developed and applied successfully to sequence and specifically diagnose some of the species described in this chapter can be and hopefully will be applied for other heterophyid species. In order to develop these tools, researchers will first have to overcome the lack of sequence data of heterophyidae by sequencing the species of interest. Previously, the amplification of the rDNA from species lacking prior sequence data required the use of oligonucleotide primers that have been designated to anneal to the conserved regions flanking the gene of interest—in most cases, the 18S subunit of the rDNA gene. These conserved regions have been found by alignment of a known sequence of many other trematode species. However, some of the conserved primers that successfully amplified most trematode rDNA genes failed to amplify properly the heterophyids rDNA (Dzikowski, unpublished data). Thus, sequencing was possible only after purification and subcloning of the PCR fragments into pGEM-T vector. The accumulation of rDNA sequence data of heterophyids should help generate new alignments and allow the design of better primers for more efficient amplification and sequencing of additional heterophyid species. The foundations for the development of species-specific molecular probes set up in the work reviewed here will hopefully contribute to the expansion of these tools and to develop diagnostic tools for other heterophyid species that are important disease-causing agents in humans, such as *H. heterophyes*, *H. dispar*, *H. nocens*, *Pygidiopsis stanma*, and others. It will be possible in the near future to have molecular diagnostic tools for each of the heterophyid species. However, in order to reach that goal, a global effort in research and development will have to take place to collect specimen, sequence them, and to develop molecular diagnostic assays.

Expansion of the available database of heterophyid sequences will allow the development of more complex PCR-based methodologies that are sensitive enough to differentiate between species through the detection of single nucleotide polymorphism (SNPs) such as PCR/LDR [43] and high-resolution melt (HRM) [44]. Because of the high sequence similarity among the heterophyidae, these techniques may prove themselves in the future as the ultimate diagnostic tool for infections. For example, using PCR/LDR methodology, a number of species-specific primer sets can be designed that can differentiate between sequences with only one base-pair difference. In the future, one can envision a "heterophyidae multiplex" of primers that will give precise identification for each individual species in a single reaction.

In a recent report of the WHO, it was recommended that governments in countries in which fish-borne trematodiasis is endemic would implement control programs into the existing health care system, continue surveillance measures, and practice to eliminate the risk of infection from aquaculture systems and products [7]. These control procedures would greatly benefit from having good molecular diagnostic tools for the identification of larval stages of heterophyids in snails and fish in fisheries and aquaculture systems. Health care authorities will be able to accurately monitor the prevalence of infection among humans, snails, and fish. Early detection of infection foci will allow the implementation of preventative steps to protect the human population from heterophyiasis.

REFERENCES

1. Chai, J.Y., Murrell, K.D., and Lymbery, A.J. Fish-borne parasitic zoonoses: Status and issues. *Int. J. Parasitol.* 35, 1233–1254, 2005.
2. Macpherson, C.N. Human behaviour and the epidemiology of parasitic zoonoses. *Int. J. Parasitol.* 35, 1319–1331, 2005.
3. Fried, B., Graczyk, T.K., and Tamang, L. Food-borne intestinal trematodiases in humans. *Parasitol. Res.* 93, 159–170, 2004.
4. Ko, R.C. Fish-borne parasitic zoonoses. In: Woo, P.T.K. (ed), *Fish Diseases and Disorders*, Vol. 1, CAB International, Cambridge, MA, pp. 592–628, 2006.
5. Keiser, J., and Utzinger, J. Emerging foodborne trematodiasis. *Emerg. Infect. Dis.* 11, 1507–1514, 2005.
6. WHO. *Control of Foodborne Trematode Infections*, Report of a WHO Study Group in Technical Report Series 165, World Health Organization, 1995.
7. WHO. *Food-Borne Trematode Infections in Asia.* 55, WHO, Manila, Phillipines, 2004.
8. Trung Dung, D. et al. Fishborne zoonotic intestinal trematodes, Vietnam. *Emerg. Infect. Dis.* 13, 1828–1833, 2007.
9. Nithikathkul, C., and Wongsawad, C. Prevalence of *Haplorchis taichui* and *Haplorchoides* sp. metacercariae in freshwater fish from water reservoirs, Chiang Mai, Thailand. *Korean J. Parasitol.* 46, 109–112, 2008.
10. Belizario, V.Y., Jr. et al. Intestinal heterophyidiasis: An emerging food-borne parasitic zoonosis in southern Philippines. *Southeast Asian J. Trop. Med. Public Health* 32 Suppl 2, 36–42, 2001.
11. Belizario, V.Y., Jr. et al. A focus of human infection by *Haplorchis taichui* (Trematoda: Heterophyidae) in the southern Philippines. *J. Parasitol.* 90, 1165–1169, 2004.
12. Chai, J.-Y., and Lee, S.-H. Food-borne intestinal trematode infections in the Republic of Korea. *Parasitol. Int.* 51, 129–154, 2002.

13. Park, J.H. et al. A new endemic focus of *Heterophyes nocens* and other heterophyid infections in a coastal area of Gangjin-gun, Jeollanam-do. *Korean J. Parasitol.* 45, 33–38, 2007.
14. Abou-Basha, L.M., Abdel-Fattah, M., Orecchia, P., Di Cave, D., and Zaki, A. Epidemiological study of heterophyiasis among humans in an area of Egypt. *East Mediterr. Health J.* 6, 932–938, 2000.
15. El Shazly, A.M. et al. Intestinal parasites in Dakahlia governorate, with different techniques in diagnosing protozoa. *J. Egypt Soc. Parasitol.* 36, 1023–1034, 2006.
16. Africa, C.M., de Leon, W., and Garcia, E.Y. Visceral complications in intestinal heterophyidiasis of man. *Acta Medica Philippina, Monographic Series* 1, 1–132, 1940.
17. Elsheikha, H.M. Heterophyosis: Risk of ectopic infection. *Vet. Parasitol.* 147, 341–342, 2007.
18. Sukontason, K., Piangjai, S., and Chaithong, U. Potassium permanganate staining for differentiation the surface morphology of *Opisthorchis viverrini*, *Haplorchis taichui* and *Phaneropsolus bonnei* eggs. *Southeast Asian J. Trop. Med. Public Health* 30, 371–374, 1999.
19. Elshazly, A.M., Elsheikha, H.M., Rahbar, M.H., and Awad, S.I. Comparison of three immunodiagnostic tests for experimental *Heterophyes heterophyes* infection in dogs. *Vet. Parasitol.* 151, 196–202, 2008.
20. Ryang, Y.S., Lee, C.Y., Lee, K.J., Lee, S.H., and Chai, J.Y. An incidental case of human *Heterophyes nocens* infection diagnosed by sectional morphology in a biopsy specimen of the small intestine. *Korean J. Parasitol.* 37, 189–194, 1999.
21. Dzikowski, R., Levy, M.G., Poore, M.F., Flowers, J.R., and Paperna, I. Use of rDNA polymorphism for identification of Heterophyidae infecting freshwater fishes. *Dis. Aquat. Organ.* 59, 35–41, 2004.
22. Levy, M.G. et al. Morphologic, pathologic, and genetic investigations of *Bolbophorus* species affecting cultured channel catfish in the Mississippi delta. *J. Aquat. Animal Health* 14, 235–246, 2002.
23. Clark, T.K. Molecular approaches and techniques. In: Woo, P.T.K. (ed), *Fish Diseases and Disorders, Vol. 1 Protozoan and Metazoan Infections*, CAB International, Cambridge, MA, pp. 725–752, 2006.
24. Cunningham, C.O., and Mo, T.A. Random amplified polymorphic DNA (RAPD) analysis of three Norwegian *Gyrodactylus salaris* populations (Monogenea: Gyrodactylidae). *J. Parasitol.* 83, 311–314, 1997.
25. Martin-Sanchez, J., Diaz, M., Artacho, M.E., and Valero, A. Molecular arguments for considering *Hysterothylacium fabri* (Nematoda: Anisakidae) a complex of sibling species. *Parasitol. Res.* 89, 214–220, 2003.
26. Dzikowski, R., Levy, M.G., Poore, M.F., Flowers, J.R., and Paperna, I. *Clinostomum complanatum* and *Clinostomum marginatum* (Rudolphi, 1819) (Digenea: Clinostomidae) are separate species based on differences in ribosomal DNA. *J. Parasitol.* 90, 413–414, 2004.
27. Dzikowski, R., Levy, M.G., Poore, M.F., Flowers, J.R., and Paperna, I. Genetic and morphologic differentiation of *Bolbophorus confusus* and B-levantinus (Digenea: Diplostomatidae), based on rDNA SSU polymorphism and SEM. *Diseases of Aquatic Organisms* 57, 231–235, 2003.
28. Cribb, T.H.B., Bray, R.A., Littlewood, D.T.J., Pichelin, S.P., Herniou, E.A. The Digenea. In: Littlewood, D.T.J., and Bray, R.A. (eds), *Interrelationship of the Platyhelminthes*. Taylor & Francis Inc., London, pp. 168–185, 2001.
29. Cribb, T.H., Anderson, G.R., Adlard, R.D., and Bray, R.A. A DNA-based demonstration of a three-host life-cycle for the Bivesiculidae (Platyhelminthes: Digenea). *Int. J. Parasitol.* 28, 1791–1795, 1998.
30. Anderson, G.R. Identification and maturation of the metacercaria of *Indodidymozoon pearsoni*. *J. Helminthol.* 73, 21–26, 1999.
31. Jousson, O., Bartoli, P., Zaninetti, L., and Pawlowski, J. Use of the ITS rDNA for elucidation of some life-cycles of Mesometridae (Trematoda, Digenea). *Int. J. Parasitol.* 28, 1403–1411, 1998.
32. Bartoli, P., Jousson, O., and Russell-Pinto, F. The life cycle of *Monorchis parvus* (Digenea: Monorchiidae) demonstrated by developmental and molecular data. *J. Parasitol.* 86, 479–489, 2000.
33. Jousson, O., Bartoli, P., and Pawlowski, J. Molecular identification of developmental stages in Opecoelidae (Digenea). *Int. J. Parasitol.* 29, 1853–1858, 1999.
34. Jousson, O., and Bartoli, P. The life cycle of *Opecoeloides columbellae* (Pagenstecher, 1863) n. comb. (Digenea, opecoelidae): Evidence from molecules and morphology. *Int. J. Parasitol.* 30, 747–760, 2000.
35. Jousson, O., and Bartoli, P. Molecules, morphology and morphometrics of *Cainocreadium labracis* and *Cainocreadium dentecis* n. sp. (Digenea: Opecoelidae) parasitic in marine fishes. *Int. J. Parasitol.* 31, 706–714, 2001.
36. Sorensen, R.E., Curtis, J., and Minchella, D.J. Intraspecific variation in the rDNA its loci of 37-collar-spined echinostomes from North America: Implications for sequence-based diagnoses and phylogenetics. *J. Parasitol.* 84, 992–997, 1998.
37. van Herwerden, L., Blair, D., and Agatsuma, T. Intra- and interindividual variation in ITS1 of paragonimus westermani (Trematoda: Digenea) and related species: Implications for phylogenetic studies. *Mol. Phylogenet. Evol.* 12, 67–73, 1999.
38. Miura, O. et al. Molecular-genetic analyses reveal cryptic species of trematodes in the intertidal gastropod, *Batillaria cumingi* (Crosse). *Int. J. Parasitol.* 35, 793–801, 2005.
39. Miura, O., Torchin, M.E., Kuris, A.M., Hechinger, R.F., and Chiba, S. Introduced cryptic species of parasites exhibit different invasion pathways. *Proc. Natl. Acad. Sci. USA* 103, 19,818–19,823, 2006.
40. Traub, R.J. et al. A new PCR-based approach indicates the range of *Clonorchis sinensis* now extends to central Thailand. *PLoS Negl. Trop. Dis.* 3, e367, 2009.
41. Sato, M., Thaenkham, U., Dekumyoy, P., and Waikagul, J. Discrimination of *O. viverrini*, *C. sinensis*, *H. pumilio* and *H. taichui* using nuclear DNA-based PCR targeting ribosomal DNA ITS regions. *Acta Tropica.* 109, 81–83, 2009.
42. Wongsawad, C., Wongsawad, P., Chai, J.Y., and Anuntalabhochai, S. *Haplorchis taichui*, Witenberg, 1930: Development of a HAT-RAPD marker for the detection of minute intestinal fluke infection. *Exp. Parasitol.* 123, 158, 2009.
43. Favis, R., and Barany, F. Mutation detection in K-ras, BRCA1, BRCA2, and p53 using PCR/LDR and a universal DNA microarray. *Ann. N Y Acad. Sci.* 906, 39–43, 2000.
44. Liew, M. et al. Genotyping of single-nucleotide polymorphisms by high-resolution melting of small amplicons. *Clin. Chem.* 50, 1156–1164, 2004.

36 Metagonimus

Jae-Ran Yu and Jong-Yil Chai

CONTENTS

36.1 INTRODUCTION

Among the minute intestinal trematodes of the family Heterophyidae (= heterophyids), three genera, namely *Metagonimus*, *Heterophyes*, and *Haplorchis*, are highly important from medical and public health points of view [1,2]. The genus *Metagonimus* was created by F. Katsurada in 1912 based on adult flukes recovered from experimentally infected dogs in Taiwan with the flesh of sweetfish, *Plecoglossus altivelis*, and also from humans who consumed raw sweetfish [3]. Subsequently, several other species of *Metagonimus* have been reported; *Metagonimus takahashii* Suzuki, 1930 [4], *M. minutus* Katsuta, 1932 [5], *M. katsuradai* Izumi, 1935 [6], *M. miyatai* Saito et al., 1997 [7,8], and *M. otsurui* Saito and Shimizu, 1968 [9]. Among them, three species, namely *M. yokogawai*, *M. takahashii*, and *M. miyatai* are the major species known to infect humans [2,10]. *M. minutus* is listed among the human-infecting trematode species [11]; however, no literature background is available [2]. *M. katsuradai* was experimentally proven to be able to infect humans [6]; however, there has been no further report regarding human infections (Table 36.1).

Metagonimiasis in humans, an infection caused by one or more species belonging to the genus *Metagonimus*, is prevalent in the Far East, including the Republic of Korea (= Korea), China, Japan, and far eastern Russia [2,10]. The diagnosis of metagonimiasis is usually based on recovery of eggs in the feces of patients. However, in areas of mixed infections with different species of *Metagonimus*, for example, *M. takahashii* and *M. miyatai*, discrimination of each species in fecal examination is problematic [12]. It is also necessary at times, in light infection cases, to use immunological and molecular techniques [13,14].

36.1.1 Classification, Morphology, and Biology

The morphological characters of the genus *Metagonimus* include a small body (0.5–1.5 mm in length), a laterally deviated ventral sucker without a ventrogenital apparatus or a genital sucker, and two testes nearby the posterior extremity (Figure 36.1) [3]. They are morphologically different from other heterophyid flukes, such as *Heterophyes*, *Heterophyopsis*, *Pygidiopsis*, *Centrocestus*, *Stellantchasmus*, *Haplorchis*, *Procerovum*, and *Stictodora*, in several points.

TABLE 36.1
Species of *Metagonimus* Reported in the Literature

Species	Human Infection	Source of Infection
Metagonimus yokogawai	Natural	Sweetfish, *Plecoglossus altivelis* Dace, *Tribolodon* spp. Perch, *Lateolabrax japonicus*
Metagonimus takahashii	Natural	Crucian carp, *Carrassius carassius* Carp, *Cyprinus carpio* Dace, *Tribolodon taczanowskii* Perch, *Lateolabrax japonicus*
Metagonimus miyatai	Natural	Sweetfish, *Plecoglossus altivelis* Dace, *Tribolodon* spp. Common fat-minnow, *Morocco steindachneri* Pale chub, *Zacco platypus* Dark chub, *Zacco temmincki*
Metagonimus minutus	Uncertain[a]	Mullet, *Mugil cephalus*
Metagonimus katsuradai	Experimental	Freshwater fish, *Pseudorasbora parva* Pale chub, *Zacco platypus* Freshwater fish, *Acheilognathus lanceolata*
Metagonimus otsurui	No	Freshwater fish, *Tridentiger obscurus* Freshwater fish, *Chaenogobius castanea* Freshwater fish, *Chaenogobius urotaenia*

[a] Enlisted among the species infecting humans [11], but without literature background.

Heterophyes and *Heterophyopsis* differ from *Metagonimus* in having a bigger, median-located ventral sucker and a prominent genital sucker [2,15]. *Pygidiopsis* has a median-located ventral sucker connected anterolaterally with a ventrogenital apparatus [2,15]. *Centrocestus* has a median-located ventral sucker and small spines on the oral sucker [2,15]. *Stellantchasmus* has a small, laterally deviated ventral sucker, as in *Metagonimus*, but differs in the presence of an elongated sac-like seminal vesicle with a muscular expulsor at the opposite side of the ventral sucker [2,15]. *Haplorchis*, *Procerovum*, and *Stictodora* have an armed gonotyl superimposed on the submedially located ventral sucker. *Haplorchis* and *Procerovum* have only one testis, whereas *Metagonimus* and other heterophyids have two testes [2,5].

36.1.1.1 *Metagonimus yokogawai* (Katsurada, 1912)

Syn. *Loxotrema ovatum* Kobayashi, 1912; *Metagonimus ovatus* Yokogawa, 1913; *Loossia romanica* Ciurea, 1915; *Loossia parva* Ciurea 1915; *Loossia dobrogensis* Ciurea, 1915.

This species was originally described in Taiwan [16] and Japan [17], and is currently the most prevalent species of *Metagonimus*. This species has been reported to be distributed in Japan, Korea, China, Taiwan, Russia, Rumania, Israel, the Balkan states, and Spain [2,3,11]. Some of the old literatures on *M. yokogawai* actually dealt with *M. takahashii* or *M. miyatai*; therefore, caution is required to make a review on *M. yokogawai sensu stricto*.

The characteristic morphologies of *M. yokogawai* as a specific level include the presence of two testes closely adjacent to each other (more or less separated in *M. takahashii* and *M. miyatai*) near the posterior end of the body [2,12,18]. Another character of *M. yokogawai* is the distribution of vitelline follicles in the lateral fields between the level of the ovary and the posterior portion of the posterior testis (abundant distribution of vitellaria beyond the posterior testis in *M. takahashii*, and no distribution beyond the posterior testis in *M. miyatai*) [2,12,18]. The uterine tubule does not overlap or cross over the middle portion of the anterior testis in *M. yokogawai* (it crosses over and overlaps the whole anterior testis in *M. takahashii* and *M. miyatai*). The eggs of *M. yokogawai* are smaller (0.0260–0.0300 mm long and 0.0148–0.0178 mm wide) than those of *M. takahashii* (0.0300–0.0360 mm long and 0.0165–0.0195 mm wide) and *M. miyatai* (0.0285–0.0315 mm long and 0.0165–0.0195 mm wide), though there are some overlaps in the range [12]. The size of adult worms is slightly smaller in *M. yokogawai* (0.800–1.320 mm long and 0.419–0.536 mm wide) than those of *M. takahashii* (0.863–1.193 mm long and 0.444–0.571 mm wide) and *M. miyatai* (0.998–1.300 mm long and 0.462–0.627 mm wide) [8]. The adults of *M. yokogawai*, *M. takahashii*, *M. miyatai*, and *M. minutus* differ from those of *M. katsuradai* and *M. otsurui* in that they have a larger ventral sucker than the oral sucker [12]. *M. minutus* differs from *M. yokogawai*, *M. takahashii*, and *M. miyatai* in its smaller body size (0.457 mm long and 0.283 mm wide) and smaller egg size (0.023 mm long and 0.013 mm wide) [5,12].

The molluscan first intermediate host of *M. yokogawai* is a kind of freshwater snails, *Semisulcospira coreana* or *S. libertina* [19]. Ophthalmo-pleuro-lophocercous cercariae are shed from these snails, and they swim in water and infect freshwater fish, the second intermediate host. The most important fish host is the sweetfish *Plecoglossus altivelis*, but the dace *Tribolodon hokonensis* or *T. taczanowskii* and the perch *Lateolabrax japonicus* can also serve as the second intermediate hosts [2,8,20]. The metacercariae usually encyst in the muscle and rarely under the scale of these fish hosts [8,21]. They are alive in the fish hosts for at least 2.5 years or through the life span of the fish [3]. Mammals, including dogs [22], rats [23], cats [24], and foxes [25], and birds, including kites [25], were reported to be natural definitive hosts, although the significance of each animal host as the source of human infection (i.e., as reservoir hosts) has not been verified [2]. Mice, rats, cats, and dogs are experimental definitive hosts [2].

Humans become infected by eating raw flesh of the fish intermediate host, most notably the sweetfish and the dace

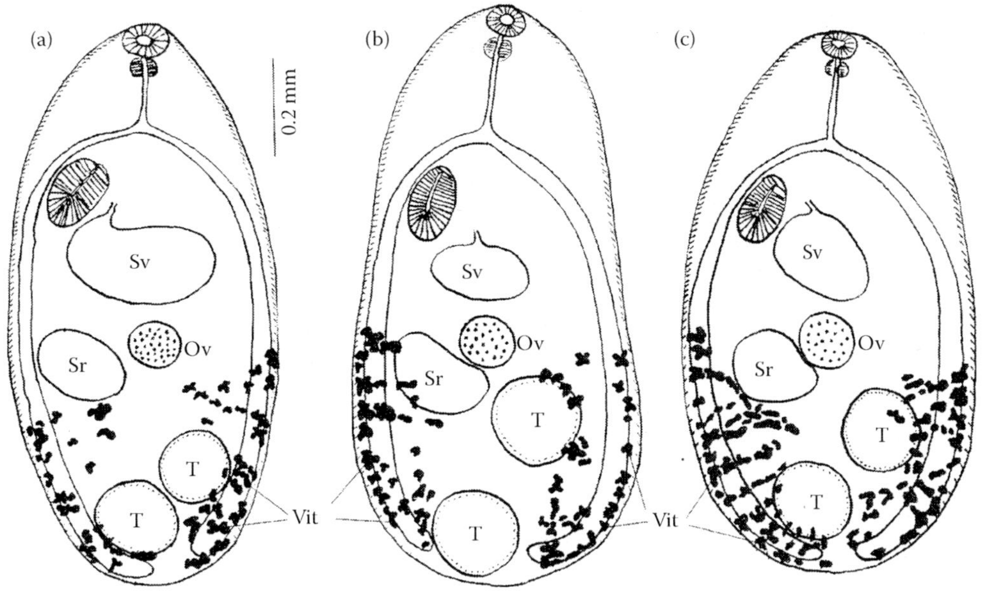

FIGURE 36.1 Three *Metagonimus* species infecting humans in Korea and Japan. A. *Metagonimus yokogawai*; B. *Metagonimus miyatai*; C. *Metagonimus takahashii*. Ov, ovary; T, testis; Sv, seminal vesicle; Sr, seminal receptable; Vit, vitelline follicles. (Reproduced with permission from Chai et al., *Korean J. Parasitol*. 31, 99, 1993.)

in the Far Eastern countries. Human infections are widely scattered along riverside areas where people traditionally eat raw fish dishes. In Korea, almost all large and small rivers and streams in eastern and southern coastal areas are endemic foci [1,2,26,27]. The Sumjin, Tamjin, and Boseong Rivers, Geoje Island, and Osip Stream (Gangwon-do) are the highest endemic areas with 20–70% egg positive rates among the villagers [2]. In some foci, this high prevalence is persisting, although the intensity of infection is slightly reduced [28]. Human infections have also been found in China (Guangdong, Anhui, Hubei, and Zhejiang provinces) and Taiwan [11]. In Japan, a survey around the Hamamatsu Lake, Shizuoka prefecture, showed 13.2% egg positive rate among 4524 lakeside people examined [29]. In sweetfish caught from 18 small rivers of Shizuoka prefecture, variable numbers of *M. yokogawai* metacercariae were detected [21]. In Russia, the prevalence in ethnic minority groups in endemic areas, including the Amur and Ussuri valleys of Khabarovsk territory, was reported to be between 20% and 70% [11]. In the north of Sakhalin Island, the infection rate was 1.5% among Russians and 10% among ethnic minorities, and sporadic cases were also reported in the Amur district and the Primorye territory [11].

36.1.1.2 *Metagonimus takahashii* (Suzuki, 1930)

Syn. *Metagonimus yokogawai* var. *takahashii* Asada, 1934; *Metagonimus yokogawai* Koga type Koga, 1938.

This species was first reported in Japan, by Takahashi [30], in the small intestine of mice and dogs fed *Metagonimus* metacercariae encysted in several species of freshwater fish, excluding the sweetfish. It was just called as a large egg-type *Metagonimus*, but 1 year later, reported to be a new species [4]. The validity of *M. takahashii* was questioned, and the name was compromised as *M. yokogawai* var. *takahashi* [31]. Consequently, however, its taxonomic validity became strongly supported not only by observing detailed differential morphological characters of larval and adult stages but also by observing different host specificities of the two species to their fist hosts by experimental infection with cercariae [32,33].

Morphologically, this species differs from *M. yokogawai* and *M. miyatai* in the position of the two testes, the distribution of vitelline follicles, and the egg size [12]. The Koga type of *Metagonimus* encysting in the dace *Tribolodon hakonensis* [34] is regarded as a synonym of *M. takahashii* [35]. The adults of *M. takahashii* differ from those of *M. katsuradai* and *M. otsurui* in that the latter two have a smaller ventral sucker than the oral sucker [12]. *M. takahashii* differs from *M. minutus* in its larger body and larger eggs [5,12].

Freshwater snails, *Semisulcospira* spp., such as *S. libertina*, *S. coreana*, and *Koreanomelania nodifila*, serve as the first intermediate host [19,32]. Freshwater fish, including the crucian carp *C. carassius*, carp *C. carpio*, dace *Tribolodon taczanowskii*, and perch *Lateolabrax japonicus*, harbor the metacercariae [2]. Natural infections have been reported in pelicans, kites, and other species of birds [36]. Dogs and cats were reported as natural and experimental definitive hosts [24,30,37]. Rats [38] and various strains of mice [39] were used as experimental definitive hosts.

After this species was first documented in Japan, its presence was reported also in Korea by recovery of adult flukes from experimental rabbits fed the metacercariae [40]. The presence of human infections in Korea was first demonstrated in 1988 from the riverside people around the Hongcheon River, Gangwon-do, by recovery of the adult worm [41]. Adult flukes were also collected in 1993 from the inhabitants of Umsong-gun, Chunchungnam-do, along the upper reaches of the Namhan River [12]. The inhabitants were found mixed-infected with *M. miyatai*, with an egg positive rate of 9.7% for both species [12]. It is presumed that *M. takahashii* is prevalent along many small rivers and streams in inland areas of Korea. The major source of human infection seems to be the crucian carp *Carassius carassius*.

36.1.1.3 *Metagonimus miyatai* (Saito et al., 1997)

This fluke was first reported in Japan in 1941 [7], but its taxonomic validity had long been questioned; and thus, it was designated as the "*Metagonimus yokogawai* Miyata type" [8]. A numerical taxonomy study on *Metagonimus* reported that *M. miyatai* (under the name of the Miyata type) was suggested to be a subspecies of *M. takahashii* [42]. However, its taxonomic significance as a specific level was acknowledged and described as a new species based on adult flukes collected from dogs and hamsters experimentally fed the metacercariae encysted in the sweetfish, dace, common fat-minnow *Morocco steindachneri*, pale chub *Zacco platypus*, and dark chub *Zacco temmincki* [8]. The new species description was also based on specimens collected from naturally infected humans in Japan and Korea [8].

M. miyatai is morphologically different from *M. yokogawai* and *M. takahashii* in the position of the posterior testis (markedly separated from the anterior testis), the distribution of vitelline follicles (never crossing over the posterior testis), and the intermediate size of eggs between *M. yokogawai* and *M. takahashi* [2,8]. *M. miyatai* is also genetically distinct from *M. yokogawai* and *M. takahashii* [14,43,44]. The adults of *M. miyatai* differ from those of *M. katsuradai* and *M. otsurui* in that the latter two have a smaller ventral sucker than the oral sucker [12]. *M. miyatai* differs from *M. minutus* in its larger body and egg size [5,12].

Freshwater snails, including *Semisulcospira globus* [45], *S. libertina*, and *S. dolorosa* [46], were reported to be the first intermediate host. Freshwater fish, *Zacco platypus* and *Zacco temminckii*, are important second intermediate hosts [47]. Other species of fish, namely *P. altivelis*, *T. hakonensis*, *T. taczanowskii*, *Opsariichthys bidens*, and *M. steindachneri* (natural), and *Phoxinus logowskii steindachneri* (experimental), were also reported to be the second intermediate host [8,46]. Mice, rats, hamsters, and dogs are experimental definitive hosts [8,35,45,46,48,49]. Natural definitive hosts include the dog, red fox, raccoon dog, and black-eared kite [8].

Human infections were reported in Korea under the name of *Metagonimus* sp. by detecting eggs slightly larger in size than those of *M. yokogawai* [48]. Later, adult worms (under the name *Metagonimus* Miyata type) were recovered from 32 riverside people living along the Namhan river in Umsong-gun (mixed- infected with *M. takahashii*) and Yongwol-gun (infected only with *M. miyatai*) [12]. In Yongwol-gun, the egg positive rate among 77 riverside people was as high as 48.1% [12]. Adult worms were also collected from inhabitants along the Hantan River, Chorwon-gun [50], and Talchongang (River), Chungwon-gun [51]. In Japan, the Hiroi River basin, Nagano Prefecture [46], and many small rivers of Shizuoka Prefecture [21] were reported to be endemic areas of *M. miyatai* infection.

36.1.1.4 *Metagonimus minutus* (Katsuta, 1932)

This fluke was first described in Taiwan in 1932 as a new species based on adult flukes recovered from the intestine of cats and mice experimentally infected with metacercariae from the brackish water fish, that is, mullet, *Mugil cephalus* [5]. This species has a smaller body and egg sizes than *M. yokogawai*, *M. takahashii*, or *M. miyatai* [2,3]. It is similar to *M. katsuradai* in the body size but differs in the egg size (bigger in *M. katsuradai*) and the relative size of the oral and ventral suckers (oral sucker > ventral sucker in *M. katsuradai*, and oral sucker < ventral sucker in *M. minutus*) [3,5,6]. *M. minutus* is enlisted among the human-infecting *Metagonimus* species without proper literature background [11].

36.1.1.5 *Metagonimus katsuradai* (Izumi, 1935)

This fluke was first described in Japan as a new species based on adult flukes recovered from the intestine of rats, mice, rabbits, dogs, and cats experimentally infected with metacercariae from freshwater fish, including *Pseudorasbora parva*, *Zacco platypus*, and *Acheilognathus lanceolata* [6]. The possibility of human infection was proven by recovery of eggs in the feces after experimental feeding with metacercariae to the author and family [6]. The first intermediate host was determined to be *S. libertina* in Japan [52] and *Juga tegulata* in Russia [53]. *M. katsuradai* differs from *M. yokogawai*, *M. takahashii*, *M. miyatai*, and *M. minutus* in having a smaller ventral sucker than the oral sucker [2,6]. *M. otsurui* differs from *M. katsuradai* in the position of the seminal receptacle; it is located in the right side of the ovary in *M. otsurui* and in the left side of the ovary in *M. katsuradai* [9].

36.1.2 Pathogenesis and Clinical Features

In *M. yokogawai* infection, histopathological studies of the intestine have been performed using animal models, including mice, rats, cats, and dogs [54–57]. In these animal hosts, adult flukes of *M. yokogawai* were seen to parasitize the middle part of the small intestine, and invade the crypt of Lieberkühn by day 2–3 postinfection, and localize between villi in later stages [54–57]. Villous atrophy and crypt hyperplasia were the two major histopathological features, and variable degrees of inflammatory reactions were accompanied [54]. The infected mucosa showed blunting and fusion of the villi, edema of the villus tips, congestion and inflammatory cell infiltrations in the villous stroma, and decreased villus/crypt height ratios [54]. Almost similar intestinal histopathology is seen in naturally infected humans with *M. yokogawai* [2]. The location of worms is confined to the intestinal mucosa in immunocompetent hosts [54]; however, immunosuppression may allow a deeper invasion of the worms into the submucosa [57]. Immunosuppression also enhances survival of the worms and prolongs the life spans in the same mouse strain [57,58]. Production of watery content seems to be a result of poor absorption of intestinal secretions from the secretory crypt cells [2]. In *M. miyatai* infection, the intestinal histopathology of experimentally infected mice was reported to be similar to, although the degree of mucosal damage was less severe than in *M. yokogawai* infection [59]. In other species of *Metagonimus*, pathogenicity and intestinal histopathology have not been studied in detail.

It has long been a question whether *Metagonimus* worms parasitize only the intestine of the hosts. In this respect, it is worth noting that several species of heterophyid flukes, namely *Stellantchamus falcatus*, *Haplorchis* spp., and *Procerovum* spp., were reported to cause erratic parasitism in man, which is often fatal, based on reports of Africa C.M. and coworkers in 1940 [2,47]. The heart valve, brain, and spinal cord were the three most frequently affected sites, where eggs and adult flukes originating from the intestinal mucosa embolized in the blood vessels. It seems that such erratic parasitism may have occurred in immunocompromised patients. With regard to *Metagonimus* spp., no direct evidence of erratic parasitism in humans has been reported. However, an interesting clinical case of *M. yokogawai* infection complicated with multiple intracerebral hemorrhages and diabetes mellitus was reported [60]. It was suggested that intracerebral hemorrhage might be an acute complication due to metagonimiasis, and diabetes mellitus a chronic sign [60]. In experimental mice, *M. yokogawai* adults were found to invade the submucosa of the small intestine when they were immunosuppressed [57].

The most frequent clinical symptoms reported in metagonimiasis patients include, unless complicated with other diseases, mild-to-moderate degrees of abdominal pain, diarrhea, lethargy, anorexia, and weight loss [2]. However, the severity of symptoms may vary and depend on several host-side factors. One is the intensity of infection, that is, the number of infected worms, in each patient; heavier infection cases tend to suffer from more severe illnesses [2]. Another factor includes immune status of each patient, that is, immunocompetent or immunodeficient, and susceptibility to the infection. A third is the history of previous exposures to infection that could have conferred acquired immunity of individual patients. A new visitor to an endemic area, for example, tends to suffer from a severe illness after a primary infection, whereas long-term residents in endemic areas generally complain of milder symptoms than those expected [2]. Inhabitants who have been living in an endemic area for a long time, and repeatedly infected, may protect against severe pathologies and symptoms caused by worms. However, in immunocompromised patients, more severe clinical manifestations, including erratic parasitisms in vital organs, may occur. A severe clinical case of *M. yokogawai* infection who revealed multiple intracerebral hemorrhages and diabetes mellitus is worth noting though it is unclear as to whether the patients were immunocompromised or not [60].

The drug of choice for all infections by *Metagonimus* spp. is praziquantel. The efficacy of a single oral dose of 10–20 mg/kg praziquantel is satisfactory, with up to a 95–100% cure rate for *M. yokogawai* infection [61]. The infectivity of metacercariae in fish can be controlled by gamma-irradiation at 200 Gy [62].

36.1.3 Diagnosis

36.1.3.1 Conventional Techniques

The diagnosis of *Metagonimus* infection is usually done by detecting eggs in fecal samples. Direct smear, cellophane thick smear, and Kato-Katz thick smear techniques can be applied chiefly in field studies, and concentration techniques, for example, formalin-ether sedimentation technique, can be performed in laboratories equipped with centrifuges.

Metagonimus eggs should be differentiated from *Clonorchis*, *Opisthorchis*, or heterophyid fluke eggs. However, it is not always easy. Eggs of *M. yokogawai* can be different from other heterophyid eggs in terms of their length of 26.9–31.6 mm, elliptical shape with length/width ratio of 1.5–2.1, clean shell surface, less prominent operculum, no shoulder rims, and dark yellow or brown color [63]. The eggs of *M. takahashii* and *M. miyatai* have similar morphology with those of *M. yokogawai*, with the exception of larger egg sizes in the two former species [12].

There could be false egg negative cases among the light infection cases with *M. yokogawai*, for example, with less than 100 worms in an infected person [15]. The number of eggs produced per day per worm for *M. yokogawai* is known to be only 14–64 eggs in the human host [38], so the probability to detect eggs in feces from such a case is almost negligible. In such cases, serological tests, including ELISA, are helpful [13].

36.1.3.2 Molecular Techniques

There are several studies supporting strong evidences of three *Metagonimus* species being genetically different [15]. The three species of *Metagonimus* are genetically differentiated by the polymerase chain reaction-based restriction fragment length polymorphism (PCR-RFLP) patterns [43]. The target genes were internal transcribed spacer 1 (ITS1) site of ribosomal RNA and mitochondrial cytochrome c oxidase I (mCOI). The restriction patterns of ITS1 gene with Rsa I, Alu I, and Msp I showed multiple fragmented bands of different sizes between three kinds of *Metagonimus* [43]. Especially in mCOI gene, restriction with Rsa I and Alu I produced differentially fragmented banding patterns [43]. PCR analysis based on random amplification of polymorphic DNA (RAPD) technique using random 10-mer oligonucleotide primers also showed distinguishable banding patterns between *M. yokogawai* and *M. miyatai* [64].

Analysis of the number of chromosomes and karyotypes of the genus *Metagonimus* supported strong evidence of genetic differences between *M. yokogawai* and *M. miyatai* [44]. They showed that the number of bivalents in the first meiotic division of *M. takahashii* was nine ($n = 9$) and the diploid number of *M. miyatai* and *M. yokogawai* were 18 ($2n = 18$) and 32 ($2n = 32$) [44]. In addition, chromosomes of *M. miyatai* were consisted of one pair of metacentric, seven pairs of submetacentric, and one pair of telocentric chromosomes and those of *M. yokogawai* were two pairs of metacentric, 11 pairs of submetacentric, and three pairs of subtelocentric chromosomes [44].

Simple sequence repeat anchored PCR using primers of 3′ or 5′ termini of the $(CA)_n$ repeats showed distinguishable banding patterns among the three species of *Metagonimus*, suggesting that they have different genotypes [14]. The study on sequence comparison of 28S ribosomal DNA and mitochondrial cytochrome c oxidase subunit I of three *Metagonimus* species also supported *M. miyatai* as a separate species [65]. The length of the 28S D1 region rDNA sequence of the three *Metagonimus* species is 248 bp and its G + C content is 52% [65]. The length of the mtCOI sequence averaged 400 bp with a G + C content ranging from 44% (*M. miyatai*), 46% (*M. takahashii*), to 47% (*M. yokogawai*). Nucleotide sequence differences between species were 23.0% (92/400 bp) between *M. miyatai* and *M. takahashii*, 16.2% (65/400 bp) between *M. miyatai* and *M. yokogawai*, and 13.2% (53/400 bp) between *M. takahashii* and *M. yokogawai* [65]. Nucleotide gaps sequence differences were 2.5% (10/400 bp) between *M. miyatai* and *M. takahashii*, 2.7% (11/400 bp) between *M. miyatai* and *M. yokogawai*, and 2.4% (11/400 bp) between *M. takahashii* and *M. yokogawai* [65]. The aligned sequences of three *Metagonimus* species showed high similarities with other comparative human intestinal trematodes (*P. summa* and *S. falcatus*) for 28S D1 rDNA (95.0% and 94%) and the mtCOI gene (68.5% and 78.0%) [65]. *M. takahashii* and *M. yokogawai* are placed in the same clade supported by DNA sequence and phylogenetic tree analysis for 28S D1 rDNA and mtCOI gene by neighbor-joining and parsimony method. *M. miyatai* is placed in different clade from two other *Metagonimus* species [65].

36.2 METHODS

36.2.1 Sample Preparation

36.2.1.1 Sample Collection

Metacercariae of *M. yokogawai* are collected by artificial digestion of *P. altivelis* caught at endemic areas, including Oshipcheon (stream), Samchok-gun, and Kangwon-do. Metacercariae of *M. miyatai* (formerly *Metagonimus* Miyata type) and *M. takahashii* are collected from *Z. platypus* and *C. carassius* caught at endemic areas, including Talchongang (River), Chungju, and Chungchongbuk-do. The adult flukes can be obtained from the intestine of rats or mice after oral infection. Worms recovered are stored at −20°C until use for DNA extraction.

36.2.1.2 DNA Extraction

1. Pulverize the adult worms of three kinds using an LN-cooled mortar and pestle.
2. Place a 1/10 volume of pulverized tissue in a 1.5 mL tube containing DNA lysis buffer (10 mM Tris-Cl pH 8.0, 100 mM EDTA, 0.5% SDS), and incubated overnight at 37°C.
3. Add proteinase K (100 μg/mL) and RNase A (20 μg/mL) and incubate for 2 h at 37°C.
4. Add an equal volume of phenol:chloroform:isoamyl alcohol and mix by inverting.
5. Precipitate total genomic DNA overnight with addition of 1/20 volume 5 M NaCl and 2 volume 99% ethanol at −70°C.
6. Centrifuge at 10,000 × *g* for 10 min.
7. Resuspend in 50 μL of TE buffer (10 mM Tris-Cl pH 8.0, 1 mM EDTA pH 8.0).
8. Add 0.1 volume of RNase A (10 mg/mL) and incubate for 1 h at 37°C.
9. Remove residual phenol by chloroform extraction. Resuspend the genomic DNA in 25 μL of TE buffer.
10. Store at 4°C.

36.2.2 Detection Procedures

36.2.2.1 PCR-RFLP

36.2.2.1.1 Principle

We present a PCR-based reverse fragment length polymorphism (PCR-RFLP) analysis for discrimination among the three species of *Metagonimus*, *M. yakogawai*, *M. takahashii*, and *M. miyatai* [43]. The conserved ITS1 sequence between 18S rRNA and 5.8S rRNA genes is amplified by PCR with primers BD1 (5-GTCGTAACAAGGTTCCGTA-3) and 4S (5-TCTAGATGCGTTCGAAGTGTCGATG-3). The ITS1 probe (including 5.8S rRNA) is prepared with primers 4S and 3S (5-GGTACCGGTGGATCACTCGGCTCG-3) and labeled with a DIG labeling kit. The mitochondrial COI is produced with primers JB3 (5-TTTTTTGGGCATCCTGAGTTTAT-3) and JB4.5 (5-TAAAGAAAGAACATAATGAAAATG-3). The ITS1 and mitochondrial COI fragments are digested with restriction enzymes *Rsa*I, *Alu*I, *Msp*I, *Cfo*I, and *Taq*I, separated on 3% agarose gel, transferred to nylon membrane, and probed with DIG-labeled ITS1 probe.

Southern blotting is done as follows [43]. PCR products of BD1 and 4S are electrophoresed through 1% agarose LE gels (Tris-acetate EDTA). DNA fragments are transferred onto a nylon membrane which is probed by digoxigenin-labeled 5.8S rRNA gene. DIG nucleic acid detection kit is used to detect the bands.

36.2.2.1.2 Procedure

1. Prepare PCR mixture (20 μL) containing
   - 1 × PCR buffer, 2.5 mM $MgCl_2$, 100 μM dNTP, 2.5 U of Taq DNA polymerase
   - 25 pmol of each primer pair (BD1 and 4S for ITS1, 4S and 3S for ITS1 probe, or JB3 and JB4.5 for mitochondrial COI)
   - 1 μL (90 ng) of template DNA, and distilled water to 20 μL
2. Perform PCR amplification in a thermocycler with the following cycling program:
   For ITS1: 45 cycles of 95°C for 45 s, 51°C for 1 min, and 72°C for 90 s
   For mCOI: 45 cycles of 94°C for 1 min, 50°C for 1 min, and 72°C for 2 min
3. Use DIG probe synthesis kit for making ITS1 probe (5.8S rRNA gene detecting ITS1).

Note: The results of ITS1 and mitochondrial COI digestion are expected to reveal unique banding patterns among the three species of *Metagonimus*. In fact, based on PCR-RFLP patterns of ITS1 and mitochondrial COI genes, it has been suggested that *M. miyatai* evolved at older times than the other two species of *Metagonimus*.

36.2.2.2 SSR-PCR

36.2.2.2.1 Principle

A simple sequence repeat-PCR (SSR-PCR) is also useful for comparison and differentiation among three *Metagonimus* species [14]. Metacercariae of *M. yokogawai*, *M. miyatai*, and *M. takahashii* are collected from freshwater fish, *P. altivelis*, *Z. platypus*, and *C. carassius*, respectively, caught at different localities. Metacercariae are isolated using an artificial digestion technique. Three kinds of adult worms are obtained from SD rats through oral infection and sacrifice on day 14 post-infection (PI). Collected worms are stored at –70°C until genomic DNA extraction using the method outlined in Section 36.2.1.2. SSR-PCR is performed with genomic DNA from the three *Metagonimus* species using 3′ or 5′ termini of $(CA)_n$ repeated primers (see below).

36.2.2.2.2 Procedure

1. Prepare PCR mixture (20 μL) containing
   - 1 × PCR buffer, 2.5 mM $MgCl_2$, 100 μM dNTP, 2.5 U of Taq DNA polymerase
   - 0.5 μM of each of the following primers with 3′ or 5′ termini of $(CA)_n$ repeats:
     Three kinds of 3′ anchored primers: $(CA)_8RG$, $(CA)_4RG$, and $(CA)_8RY$ (all at 0.5 μM/μL)
     Two kinds of 5′ anchored primers: $BDB(CA)_7C$, $DBDA(CA)_7$ (all at 0.5 μM/μL)
   - 1 μL (20 ng) of template DNA, and distilled water to 20 μL
2. Perform PCR amplification in a thermocycler with the following cycling program: 30 cycles of 94°C for 30 s, 52°C for 45 s, and 72°C for 2 min
3. Separate the amplified products on 1% agarose gel and detect by ethidium bromide staining

Note: Although SSR-PCR fingerprinting for genomic DNA generally shows a higher level of polymorphism than PCR-RFLP or RAPD analyses, it can be applied to analyze the phylogenic relations of three species of *Metagonimus*. The differences of SSR-PCR patterns among the three species might be related to their morphological differences and biological differences, for example, in taking different fish intermediate hosts [14].

36.2.2.3 RAPD

RAPD analysis can be applied to distinguish adult flukes of *M. yokogawai* and *M. miyatai* [63].

36.2.2.3.1 Procedure

1. Prepare PCR mixture (20 μL) containing
   - 1× PCR buffer, 2.5 mM $MgCl_2$, 100 μM dNTP, 2.5 U of Taq DNA polymerase
   - 25 pmol of each of the following random primer (in separate tube): OPA-02 (5-TGCCGAGCTG-3), OPA-09 (5-GGGTAACGCC-3), OPA-10 (5-GTGATCGCAG-3), OPA-11 (5-CAATC GCCGT-3), OPA-13 (5-CAGCACCCAC-3), OPA-17 (5-GACCGCTTGT-3), OPA-19 (5-CAAA CGTCGG-3), or OPA-20 (5-GTTGCGATCC-3)
   - 1 μL (90 ng) of template DNA, and distilled water to 20 μL
2. Perform PCR amplification in a thermocycler with the following cycling program: 45 cycles of 94°C for 1 min, 35°C for 1 min and 72°C for 2 min
3. Separate the amplified products on 8% polyacrylamide gel and detect by silver staining

Note: The above eight primers were proved to be useful for differentiating *M. yokogawai* from *M. miyatai* which have differential morphology and biology including the kinds of fish hosts involved [64].

36.2.2.4 Sequencing of 28S D1 rDNA and mtCOI PCR

36.2.2.4.1 Principle

Comparison of sequences of 28S D1 ribosomal DNA (28S D1 rDNA) and mitochondrial cytochrome c oxidase subunit I (mtCOI) fragments can be utilized to differentiate among the three species of *Metagonimus*: *M. yokogawai*, *M. takahashii*, and *M. miyatai* [65]. Adult specimens of *M. takahashii*, *M. miyatai*, and *M. yokogawai* are obtained from SD rats experimentally infected with metacercariae at day 7 PI. Metacercariae of each species are collected from freshwater fish, *C. carassius*, *Z. platypus*, and *P. altivelis*, respectively. Two species of the family Heterophyidae, *Pygidiopsis summa* and *Stellantchasmus falcatus*, are obtained from experimentally infected animals, and *S. falcatus* is used as an outgroup to infer phylogeny.

36.2.2.4.2 Procedure

1. Prepare PCR mixture (20 μL) for D1 rDNA and mtCOI PCR containing
   - 1× PCR buffer, 2.5 mM $MgCl_2$, 100 μM dNTP, 2.5 U of ExTaq enzyme
   - 25 pmol of each of the following random primer: 28S rDNA–JB10 (forward), 5′-GATTACCCGCTGAACTTAAGCATA-3′
     –JB9 (reverse), 5′-GCTGCATTCACAAACA CCCCCAC TC-3′
     mtCOI – JB3 (forward), 5′-TTTTTTGGGCATC CTGAGGT TTAT-3′ (2575)
     –JB4.5 (reverse), 5′-TAAAGAAAGAACATAA TGAAAA TG-3′ (3021)
     (primers from the nucleotide sequences of *Fasciola hepatica*)
   - 1 μL (10 ng) of template DNA, and distilled water to 20 μL.
2. Perform PCR amplification in a thermocycler with the following cycling program: 40 cycles of

denaturation at 95°C for 20 s, annealing at 55°C for 30 s (for 28S rDNA) or 48°C 30 s (for mtCOI), extension at 72°C for 30 s; and 1 cycle of final extension at 72°C for 6 min
3. Sequence the PCR products

Note: The sequences of 28S rDNA and mitochondrial COI genes are expected to be useful for discriminating the three species of *Metagonimus* [65]. The phylogenetic trees established by 28S rDNA and mitochondrial COI nucleotide sequences revealed that *M. yokogawai* is more closely related with *M. takahashii* than with *M. miyatai*, which supports the results of PCR-RFLP [43].

36.2.2.5 Karyotyping

36.2.2.5.1 Principle

Karyotyping of *M. yokogawai*, *M. takahashii*, and *M. miyatai* is performed to analyze the chromosome numbers and karyotypes [44]. Adult specimens of *M. takahashii*, *M. miyatai*, and *M. yokogawai* are obtained from SD rats at day 7 PI. Metacercariae of each species are collected from *C. carassius*, *Z. platypus*, and *P. altivelis*, respectively.

36.2.2.5.2 Procedure

1. Cultivate worms in 1× liquid MEM-199 medium containing 0.001% colchicines, at 37°C for 4 h.
2. Fix with modified Carnoy's fluid (1 part of glacial acetic acid, 3 parts of 95% ethanol).
3. Remove the testes of each specimen with a dissecting pin and transfer to the fixative using a fine forceps.
4. Remove pieces of the other tissues.
5. Mince the testes gently in 45% acetic acid to prepare cell suspensions.
6. Stain with aceticorcein solution for 10–30 min.
7. Rinse briefly overstained cell with 45% acetic acid and cover with cover glass, and then squash.
8. Heat the slide briefly with alcohol lamp to remove air bubbles.
9. Mount in Canada balsam.
10. Observe under a light microscope; measure the relative length of chromosomes, total chromosome length of the haploid set, and the positions of centromeres.

Note: The diploid chromosome number of *M. yokogawai* was shown to be 32 ($2n = 32$), whereas that of *M. miyatai* was 18 ($2n = 18$) [44]. Karyotyping of *M. takahashii* may be difficult, as it failed in a study [44].

36.3 CONCLUSION AND FUTURE PERSPECTIVES

Metagonimiasis is one of the important food-borne parasitic infections in the Far Eastern countries, including Korea, China, Taiwan, Japan, and Russia. The major etiological agents are *M. yokogawai*, *M. takahashii*, and *M. miyatai*. Two other species, namely *M. minutus* and *M. katsuradai*, may potentially infect humans. In Korea, an estimated about 240,000 infected cases with *M. yokogawai*, *M. takahashii*, or *M. miyatai*, are present along the rivers and streams running through the eastern and southern coastal areas. The snail host is *S. libertina* or *S. coreana*, and the fish host includes the sweetfish, *P. altivelis*, the dace, *Tribolodon* sp., and the perch, *L. japonicus*. Human infections can occur by consuming raw or improperly cooked flesh of these fishes. The adult flukes parasitize the small intestine and cause villous atrophy and crypt hyperplasia, with mucosal inflammation. Symptoms in infected patients include diarrhea, abdominal pain, anorexia, and weight loss. In immunocompromised patients, complications, for example, erratic parasitism in the heart, brain, and spinal cord, or intracerebral hemorrhage, may occur. The diagnosis is based on recovery of eggs in the feces. However, differential diagnosis of different *Metagonimus* species as well as differential diagnosis of *Metagonimus* from other trematodes, including *C. sinensis*, *O viverrini*, *Heterophyes nocens*, and *Pygidiopsis summa*, is difficult. Metagonimiasis can be treated with 10 mg/kg a single dose of praziquantel.

For taxonomic studies of *Metagonimus* species and for a diagnostic aid, molecular techniques have been applied. The techniques include PCR-RFLP, SSR-PCR, RAPD, sequencing of 28S rDNA and mitochondrial COI, and karyotyping. The results have shown that *M. yokogawai* and *M. takahashii* may be phylogenetically close to each other. *M. miyatai* is considered to have evolved at an earlier time than *M. yokogawai* or *M. takahashii*. Further studies are required to draw firm conclusions regarding the phylogenetic relationships among *Metagonimus* species.

REFERENCES

1. Chai, J.Y. Intestinal flukes. In: Murrell, K.D., and Fried, B. (eds), *Food-Borne Parasitic Zoonoses: Fish and Plant-Borne Parasites*, Chapter 2. Springer, New York, 2007.
2. Chai, J.Y., Shin, E.H., Lee, S.H., and Rim, H.J. Foodborne intestinal flukes in Southeast Asia. *Korean J. Parasitol.* 47, S103, 2009.
3. Ito, J. *Metagonimus* and other human hererophyid trematodes. *Progr. Med. Parasitol. Jpn.* 1, 314, 1964.
4. Suzuki, S. Yokogawa's *Metagonimus*. List of publications on special animals in Okayama prefecture, Okayama Prefectural Report, 146, 1930.
5. Katsuta, I. Studies on the metacercariae of Formosan brackish water fishes, 2.On a new species, *Metagonimus minutus* n. sp. parasitic in *Mugil cephalus*. *Taiwan Iggakai Zasshi* 31, 26, 1932. (in Japanese).
6. Izumi, M. Studies on a new species of *Metagonimus* and its life cycle. *Tokyo Iji Shinshi* 2929, 1224, 1935. (in Japanese).
7. Miyata, I. Comments on the classification of trematodes of the genus *Metagonimus*. *Dobutsugaku Zasshi* 56, 16, 1944. (in Japanese).
8. Saito, S. et al. *Metagonimus miyatai* sp. nov. (Digenea: Heterophyidae), a new intestinal trematode transmitted by freshwater fishes in Japan and Korea. *Korean J. Parasitol.* 35, 223, 1997.

9. Saito, S., and Shimizu, T. A new trematode, *Metagonimus otsurui* sp. nov. from the fresh-water fishes (Trematoda: Heterophyidae). *Jpn. J. Parasitol.* 17, 167, 1968.
10. Chai, J.Y., Murrell, K.D., and Lymbery, A.J. Fish-borne parasitic zoonoses: Status and issues. *Int. J. Parasitol.* 35, 1233, 2005.
11. Yu, S.H., and Mott, K.E. Epidemiology and morbidity of food-borne intestinal trematode infections. *Trop. Dis. Bull.* 91, R125, 1994.
12. Chai, J.Y. et al. An epidemiological study of metagonimiasis along the upper reaches of the Namhan River. *Korean J. Parasitol.* 31, 99, 1993.
13. Chai, J.Y. et al. An egg-negative patient of acute metagonimiasis diagnosed serologically by ELISA. *Seoul. J. Med.* 30, 139, 1989.
14. Yang, H.J. et al. Molecular differentiation of three species of *Metagonimus* by simple sequence repeat anchored polymerase chain reaction (SSR-PCR) amplification. *J. Parasitol.* 86, 1170, 2000.
15. Yu, J.R., and Chai, J.Y. *Metagonimus*. In: Liu, D. (ed), *Molecular Detection of Foodborne Pathogens*. CRC Press (Taylor & Francis Group), Boca Raton, FL, 2010.
16. Katsurada, F. On a new trematode. *Metagonimus yokogawai*, *Tokyo Iji Shinshi* 1796, 3483, 1912. (in Japanese).
17. Kobayashi, H. On a new genus of the trematode (a preliminary report). *Saikingaku Zasshi* 204, 780, 1912. (in Japanese).
18. Chai, J.Y. et al. Chronological observation on intestinal histopathology and intraepithelial lymphocytes in the intestine of rats infected with *Metagonimus yokogawai*. *Korean J. Parasitol.* 32, 215, 1994. (in Korean).
19. Cho, S.Y., Kang, S.Y., and Lee, J.B. Metagonimiasis in Korea. *Arzneimittelforschung* 34(9B), 1211, 1984.
20. Ahn, Y.K. *Lateolaborax japonicus*, a role of the second intermediate host of *Metagonimus* yokogawai. *New Med. J.* 26, 135, 1983. (in Korean).
21. Kino, H. et al. Geographical distribution of *Metagonimus yokogawai* and *M. miyatai* in Shizuoka Prefecture, Japan, and their site preferences in the sweetfish, *Plecoglossus altivelis*, and hamsters. *Parasitol. Int.* 55, 201, 2006.
22. Cho, S.Y., Kang, S.Y., and Ryang, Y.S. Helminthes infections in the small intestine of stray dogs in Ejungbu City, Kyunggi Do, Korea. *Korean J. Parasitol.* 19, 55, 1981.
23. Seo, B.S. et al. Studies on parasitic helminths of Korea 5. Survey on intestinal trematodes of house rats. *Korean J. Parasitol.* 19, 131, 1981.
24. Huh, S., Sohn, W.M., and Chai, J.Y. Intestinal parasites of cats purchased in Seoul. *Korean J. Parasitol*, 31, 371, 1993.
25. Miyamoto, K. Studies on zoonoses in Hokkaido 7. Survey of natural definitive hosts of *Metagonimus yokogawai*. *Jpn. J. Parasitol.* 34, 371, 1985. (in Japanese).
26. Ahn, Y.K. Epidemiological studies on *Metagonimus yokogawai* infection in Samcheok-gun, Kangwon-do, Korea. *Korean J. Parasitol.* 22, 161, 1984. (in Korean).
27. Chai, J.Y. et al. High endemicity of *Metagonimus yokogawai* infection along residents of Samcheok-shi, Kangwon-do. *Korean J. Parasitol.* 38, 33, 2000.
28. Lee, J.J. et al. Decrease of *Metagonimus yokogawai* endemicity along the Tamjin River basin. *Korean J. Parasitol.* 46, 289, 2008.
29. Ito, J. et al. On the prevalence of *Metagonimus* sp. among the inhabitants at Hamamatsu basin in Shizuoka prefecture, Japan. *Jpn. J. Parasitol.* 40, 274, 1991.
30. Takahashi, S. On the life history of *Metagonimus yokogawai*, a new species of *Metagonimus*, and *Exorchis major*, *Okayama Igakkai Zasshi* 41, 2687, 1929. (in Japanese).
31. Asada, S. On *Metagonimus yokogawai* and its related species. *Rinsho Igaku* (Clinical Medicine) 22, 43, 1934. (in Japanese).
32. Saito, S. On the differences between *Metagonimus yokogawai and Metagonimus takahashii* II. The morphological comparisons. *Jpn. J. Parasitol.* 21, 449, 1972. (in Japanese).
33. Saito, S. On the differences between *Metagonimus yokogawai and Metagonimus takahashii* II. The experimental infections to the second intermediate hosts. *Jpn. J. Parasitol.* 22, 39, 1973. (in Japanese).
34. Koga, G. Studies on the genus *Metagonimus*. *Igaku Kenkyu* 12, 3471, 1938. (in Japanese).
35. Chai, J.Y. et al. Three morphological types of the genus *Metagonimus* encysted in the dace, *Tribolodon taczanowskii*, caught from the Sumjin River. *Korean J. Parasitol.* 29, 217, 1991.
36. Yamaguti, S. Digenetic trematodes of vertebrates. *Systema Helminthum* 1, 716, 1958.
37. Kamiya et al. Helminths of stray dogs in Sapporo. II. *Jpn. J. Parasitol.* 24, 41, 1975. (in Japanese).
38. Ahn, Y.K. Intestinal flukes of genus *Metagonimus* and their second intermediate hosts in Kangwon-do, *Korean J. Parasitol.* 31, 331, 1993. (in Korean).
39. Guk, S.M. et al. Susceptibility of inbred mouse strains to infection with three species of *Metagonimus* prevalent in the Republic of Korea. *J. Parasitol.* 91, 12, 2005.
40. Chun, S.K. A study on the metacercaria of *Metagonimus takahashii* and *Exorchis oviformis* from *Carassius carassius*. *Bull. Pusan Fish. Coll.* 3, 31, 1960. (in Korean).
41. Ahn, Y.K., and Ryang, Y.S. Epidemiological studies on *Metagonimus* infection along the Hongcheon river, Kangwon Province. *Korean J. Parasitol.* 26, 207, 1988. (in Korean).
42. Kim, K.H. et al. A numerical taxonomic study on heterophyid trematodes. *Korean J. Parasitol.* 29, 55, 1991. (in Korean).
43. Yu, J.R. et al. PCR-RFLP pattern of three kinds of *Metagonimus* in Korea. *Korean J. Parasitol.* 35, 271, 1997.
44. Lee, S.U. et al. A cytogenetic study on human intestinal trematodes of the genus *Metagonimus* (Digenea: Heterophyidae) in Korea. *Korean J. Parasitol.* 37, 237, 1999.
45. Kim, C.H. et al. Studies on the *Metagonimus* fluke in the Daecheong reservoir and the upper stream of Geum River, Korea. *Korean J. Parasitol.* 25, 69, 1987. (in Korean).
46. Shimazu, T. Life cycle and morphology of *Metagonimus miyatai* (Digenea: Heterohpyidae) from Nagano, Japan. *Parasitol. Int.* 51, 271, 2002.
47. Chai, J.Y., and Lee, S.H. Food-borne intestinal trematode infections in the Republic of Korea. *Parasitol. Int.* 51, 129, 2002.
48. Kim, C.H. Study on the *Metagonimus* sp. in Gum River basin, Chungchung-nam Do, Korea. *Korean J. Parasitol.* 18, 215, 1980. (in Korean).
49. Ahn, Y.K., and Ryang, Y.S. Growth, egg-laying and life span of Miyata type of genus *Metagonimus* (Trematoda: Heterophyidae) in the final hosts. *J. Wonju Med. Coll.* 8, 1, 1995. (in Korean).
50. Park, M.S. et al. Intestinal parasite infections in the inhabitants along the Hantan River, Chorwon-gun. *Korean J. Parasitol.* 31, 375, 1993.
51. Yu, J.R., Kwon, S.O., and Lee, S.H. Clonorchiasis and metagonimiasis in the inhabitants along Talchongang (River), Chungwon-gun. *Korean J. Parasitol.* 32, 267, 1994.
52. Kurokawa, T. Studies on the genus *Metagonimus*, especially on the life history of *M. katsuradai* Izumi, 1935 and determination of its first intermediate host. *Tokyo Iji Shinshi* 3161, 2877, 1939. (in Japanese).

53. Besprozvannykh, V.V. et al. Occurrence of *Metagonimus katsuradai* Izumi, 1935 (Trematoda: Heterophyidae) in the southern Primor'e region. In: Yu, L. (ed), *Gel'minty I vyzyvaemye imi zabollevaniya, Mamaev*, Vladivostock, Russia, 1987.
54. Chai, J.Y. Study on *Metagonimus yokogawai* (Katsurada, 1912) in Korea V. Intestinal pathology in experimentally infected albino rats. *Seoul J. Med.* 20, 104, 1979.
55. Lee, J.B. et al. Study on the pathology of metagonimiasis in experimentally infected cat intestine. *Korean J. Parasitol.* 19, 109, 1981.
56. Kang, S.Y. et al. A study on intestinal lesions of experimentally reinfected dogs with *Metagonimus yokogawai. Korean J. Parasitol.* 21, 58, 1983.
57. Chai, J.Y., Kim, J., and Lee, S.H. Invasion of *Metagonimus yokogawai* into the submucosal layer of the small intestine of immunosuppressed mice. *Korean J. Parasitol.* 33, 313, 1995.
58. Chai, J.Y., Seo, B.S., and Lee, S.H. Study on *Metagonimus yokogawai* (Katsurada, 1912) in Korea VII. Susceptibility of various strains of mice to *Metagonimus* infection and effect of prednisolone. *Korean J. Parasitol.* 22, 153, 1984.
59. Yu, J.R., Myong, N., and Chai, J.Y. Expression patterns of proliferating cell nuclear antigen in the small intestine of mice infected with *Metagonimus yokogawai* and *Metagonimus* Miyata type. *Korean J. Parasitol.* 35, 239, 1997.
60. Yamada et al. A case of metagonimiasis complicated with multiple intracerebral hemorrhages and diabetes mellitus. *J. Nippon Med. Sch.* 75, 32–35, 2008.
61. Rim, H.J. et al. Anthelmintic effects of various drugs against metagonimiasis. *Korean J. Parasitol.* 16, 117, 1978. (in Korean).
62. Chai, J.Y. et al. Effects of gamma-irradiation on the survival and development of *Metagonimus yokogawai* metacercariae in rats. *Korean J. Parasitol.* 33, 297, 1995.
63. Lee, S.H. et al. Comparative morphology of eggs of heterophyids and *Clonorchis sinensis* causing human infections in Korea. *Korean J. Parasitol.* 22, 171, 1984.
64. Yu, J.R., Chung, J.S., and Chai, J.Y. Different RAPD patterns between *Metagonimus yokogawai* and *Metagonimus* Miyata type. *Korean J. Parasitol.* 35, 295, 1997.
65. Lee, S.U. et al. Sequence comparisons of 28S ribosomal DNA and mitochondrial cytochrome c oxidase subunit I of *Metagonimus yokogawai, M. takahashii* and *M. miyatai. Korean J. Parasitol.* 42, 129, 2004.

37 Nanophyetus

Dongyou Liu

CONTENTS

37.1 INTRODUCTION

Nanophyetus salmincola (fish flu fluke) in the genus *Nanophyetus* is a zoonotic parasite that is responsible for causing human nanophyetiasis (fish flu) in the coast of the US Pacific Northwest, Canada, and East Siberia and is also implicated in the transmission of salmon poisoning disease (SPD) in dogs.

This parasite has a complex life cycle involving two intermediate hosts (snail and fish) and a final host (fish-eating vertebrate). Although *N. salmincola* does not seem to have obvious adverse effects on most animal hosts (apart from dogs), its infection in humans can result in chronic diarrhea, nausea, abdominal discomfort, and a peripheral blood eosinophilia.

37.1.1 Classification, Morphology, Life Cycle, and Epidemiology

37.1.1.1 Classification

The genus *Nanophyetus* is classified in the family Nanophyetidae, suborder Troglotremata, order Plagiorchiida, subclass Digenea, class Trematoda, phylum Platyhelminthes. The family Nanophyetidae is separated into two genera: *Nanophyetus* and *Skrjabinophyetus*; and the genus *Nanophyetus* consists of a single species *N. salmincola* (synonyms: *Troglotrema salmincola* and *Nanophyetus schikhobalowi*; also known as fish flu fluke). Despite their seemingly identical morphological characteristics, *N. salmincola* and *N. schikhobalowi* have been considered by some authors as two distinct subspecies, that is, *N. salmincola salmincola* and *N. salmincola schikhobalowi*, in order to reflect their biological and geographic differences.

N. salmincola was first described in North America in 1926 by Chapin, who initially named the trematode as *Nanophyes salmincola* in family Heterophyidae. After further morphological examination, Chapin renamed the parasite as *N. salmincola* (synonym *T. salmincola*) in the family Troglotrematidae, since *Nanophyes* was already taken. Although *N. salmincola* has been implicated in the transmission of SPD (due to the infection of Gram-negative bacterium *Neorickettsia helminthoeca*) in canids, the parasite was not considered as a human pathogen until 1958 [1]. On the other hand, *N. schikhobalowi* was recognized as a cause of human nanophyetiasis (also called fish flu) in Eastern Siberia since 1931 [2].

37.1.1.2 Morphology

As a digenic trematode, *N. salmincola* is an unsegmented, dorsoventrally flattened ovioid fluke with a spiny tegument. The adult worm measures 0.8–2.5 mm in length and 0.3–0.8 mm in width. Being hermaphroditic, the worm possesses both male and female reproductive organs in the same organism. The worm has two large oval testes (0.2–0.3 mm in length) and one round ovary (0.07–0.11 mm in diameter). The uterus typically contains 5–16 eggs, the weight of which allows the fluke to sink rapidly in water. The male copulatory organ, but not seminal vesicle, is surrounded by a prominent cirrus pouch (or hollow organ). *N. salmincola* has an oral sucker (0.15–0.18 mm in diameter) and a ventral sucker (0.12–0.13

mm in diameter), which are used to grasp and crawl actively about the intestinal tissue of its host.

N. salmincola eggs (0.087–0.097 mm × 0.038–0.055 mm) are light brown, ovoid, and operculate at one end, with a small blunt projection at the other end.

37.1.1.3 Life Cycle

N. salmincola undergoes several stages of development (egg, larva, and adult) and requires two intermediate and one definitive (final) hosts during its life cycle. The first intermediate host is stream snail (*Oxytrema silicula*), and the second intermediate host is a salmonid fish (or some non-salmonid fish). The definitive (final) host is a fish-eating mammal, typically a canid, or one of the 30 other species (including felids, mustelids, bear, and humans as well as some birds) [3].

N. salmincola adult worms reside in the crypts of the small intestine of the definitive vertebrate hosts. The adult female lays unembryonated eggs within the vertebrate host, and the eggs are shed in host's feces. After hatching (embryonating) for 75–200 days (average three months) in water, the miracidia (the first larval stage) emerge from the eggs. The miracidia then penetrate the first intermediate host (stream snail *O. silicula*). Inside the stream snail, the miracidia develop into the rediae (the second larval stage, 0.45–3 mm in size), which contain germ cells that give rise to the cercariae (numbering up to 76). The rediae and cercariae are located mainly in the gonads, the digestive gland, and other tissues of the snail. Coming out of the snail, the cercariae (measuring 0.31–0.47 mm × 0.03–0.15 mm) penetrate the second intermediate host (i.e., the salmonid fish or some non-salmonid fish) within 30 s to 5 min, in which the cercariae develop into metacercaria (after losing their tails) and encyst within the kidneys and muscles of the salmonid fish (or the gills and fins of the non-salmonid fish). There may be as many as 1000–2000 metacercariae encysted in the fish tissues. The new cyst wall is thin, transparent, and easily ruptured. Upon ingestion by the final (definitive) host (e.g., canids and humans), the metacercaria encysted in the fish tissues are liberated. The metacercaria mature into adult worms in the host's small intestine, which lay eggs to be passed in the host's feces.

37.1.1.4 Epidemiology

Three hosts (including two intermediate and one definitive or final hosts) are needed for the completion of *N. salmincola* life cycle.

The first intermediate host for *N. salmincola* trematode is stream snail *O. silicula* (synonym: *Goniobasis plicifera* var. *silicula*), which is commonly present in the US Pacific Northwest and Canada.

The second intermediate host for *N. salmincola* is a salmonid fish (or some non-salmonid fish) [4]. Although 34 species of fresh and ocean water fish (including members of the families Salmonidae, Cottidae, and Cyprinidae such coastal cutthroat trout, rainbow trout, coho salmon, chum salmon, kokanee salmon, steelhead trout, and anadromous fish as well as lamprey, sculpin, redside shiner, tadpoles, and the Giant Pacific salamander) may harbor infective metacercariae, salmonid fish of the genera Salmo, Oncorhynchus, and Salvelinus play a significant role in the *N. salmincola* life cycle. Heavy metacercarial infections may kill some young fish within 24 h after contact with the cercariae and increase fish morbidity and mortality to viral infections. The liberation of cercariae-produced proteolytic enzymes, which in high concentrations are toxic, is thought to cause fish death. The cercariae encysted in the kidneys, muscles, fins, and other viscera of fish can remain viable for at least five years.

The definitive (final) host for *N. salmincola* is a fish-eating vertebrate, belonging to one of the 30 species of felids, mustelids, and bear as well as some birds and humans. While primary hosts include raccoons, spotted skunks, and minks, secondary hosts are dogs, cat, red fox, hamster, wood rat fish-eating birds, and humans. In general, *N. salmincola* infection causes little damage to definitive hosts except in very heavy infections, which may result in marked mucosal thickenings of the duodenum and jejunum, enteritis, and malnutrition [5]. However, because *N. salmincola* fluke is a vector for obligate intra-cytoplasmic bacterium *N. helminthoeca*, dogs and other canids may develop a severe disease called SPD with manifestations of fever, anorexia, vomiting, profuse diarrhea, necrosis of lymph follicles, ulceration, severe hemorrhage, and death (in 10–14 days) [6,7]. The mortality of SPD in dogs can be as high as 90% if left untreated [8].

Humans are also a definitive host for *N. salmincola*, which is transmitted via ingestion of raw, under-processed, and smoked salmon and steelhead trout. In addition, food-processing workers may become infected with the parasite by simple handling of fresh-killed salmon followed by hand-to-mouth contact. Fortunately, as humans are not susceptible for *N. helminthoeca* infection, human nanophyetiasis is generally non-fatal. Heavy infections in people may result in nausea, diarrhea, and abdominal pain, and *N. salmincola* eggs may be detected in the stool about one week after ingestion of contaminated fish, and *N. salmincola* commonly occurs in the US Pacific Northwest and East Siberia of Russia, due primarily to the availability (and geographic distribution) of intermediate hosts. Stream snails are found west of the Cascade Mountains in Oregon, north of the Olympic Peninsula in Washington, and in part of northern California. Thus, *N. salmincola* is endemic in Washington, Oregon, Northern California, the Pacific Northwest around the Columbia River Basin and tributaries, British Columbia, and Alaska. In Eastern Siberia, reported infection rates in some villages surpass 95%.

37.1.2 Clinical Features and Treatment

37.1.2.1 Clinical Features

Humans with *N. salmincola* infection may be asymptomatic, or show nonspecific symptoms that may be mistaken for indication of other gastrointestinal problems. Typical symptoms include diarrhea, abdominal pain, nausea, vomiting, weight loss, fatigue, and unexplained peripheral blood eosinophilia. Most patients may recall eating fish prior to onset of symptoms. *N. salmincola* eggs may appear in stools approximately one week after ingestion of infected fish [9,10].

The early incidences of naturally acquired *N. salmincola* infection in the the coastal US Pacific Northwest were noted between September 1974 and October 1985. The infected patients ($n = 10$) had recalled ingestion of undercooked or raw fish, and most of them had *N. salmincola* eggs in the stools, and displayed peripheral blood eosinophilia. In 1987, another 10 cases of human nanophyetiasis were reported. Five of these patients complained of gastrointestinal discomfort, and the remainder displayed peripheral eosinophilia. Nine of the 10 patients recalled eating inadequately cooked fish [9].

In 1990, an unusual case of human nanophyetiasis was described. The patient was a 19-year-old man in Washington state, whose job duties during a three-month study of coho salmon (*Oncorhynchus kisutch*) included removing the posterior one-third of the kidney of *N. salmincola*-infected coho salmon, preparing macerated wet mounts for microscopic examination, and enumerating metacercariae. Apparently, he had necropsied >1100 fresh-killed juvenile coho salmon without protective gloves. On admission, the patient complained of flu-like malaise, intermittent diarrhea, general abdominal discomfort, nausea, and vomiting. Examination of his blood sample revealed 43% eosinophilia, and a stool specimen was shown to contain the bipolar eggs of *N. salmincola*. The patient did not recall eating fish or other seafood products for several months before infection, and the infection was likely initiated by accidental hand-to-mouth ingestion of infectious metacercariae. After treatment with praziquantel at 20 mg/kg body weight, three times per day for one day for two months, the patient shed no fluke eggs in the stool, and eosinophils were not evident in peripheral blood [10].

Nanophysiasis in dogs tends to be more serious than in humans, with symptoms ranging from sharp fever (or pyrexia, 40–42°C, which may decrease gradually to normal or below normal over the next 4–8 days), anorexia, vomiting, dysentery, a serous to purulent ocular discharge, peri-orbital edema, weight loss, depression, dehydration, and polydipsia, with death occurring about 10–14 days after signs first appear. The main reason for the heightened severity of nanophysiasis in dogs is attributed to the fact that *N. salmincola* acts as a vector for *N. helminthoeca*, which is a rickettsial bacterium causing SPD in dogs, with clinical presentations of hemorrhagic enteritis and other pathological changes [11]. First reported in the Northwest region of Oregon, USA in 1814, SPD was once a deadly illness of dogs associated with the ingestion of uncooked salmon, killing 90% of infected dogs within a week or two.

37.1.2.2 Treatment

Human nanophysiasis can be treated successfully with anthelmintics such as bithionol (50 mg/kg orally on alternate days; two doses total), niclosamide (2 g orally on alternate days; three doses total), or praziquantel (20 mg/kg three times/day for 1 day) [12,13]. Praziquantel causes immobilized contraction of the worm, so that it can no longer grasp the intestinal walls, and thus can be eliminated from the body [14]. General supplements may be provided in order to maintain electrolyte balance and meet nutritional requirements if diarrhea recurs.

N. salmincola infection is easily preventable by cooking fish thoroughly, freezing fish for at least 24 h, fishing away from known snail endemic places, taking precautions when handling fish (to prevent hand-to-mouth transmission of the metacercariae), checking fish for cysts or sites of irritation from penetration of the cercarariae (relevant to fish handlers) and disposing off fish viscera safely, reporting sources of infected fish, eating raw fish only from trusted sources, using molluscicides in smaller bodies of water if practical, keeping dogs away from streams.

37.1.3 Laboratory Diagnosis

Presumptive diagnosis of nanophyetiasis may be made on the basis of the following clinical findings: chronic diarrhea, nausea, abdominal discomfort, a peripheral blood eosinophilia of 43%, a history of consuming raw or under-cooked fish, or cold smoked fish (particularly salmon or trout originating from the enzootic area of the Pacific Northwest), or traveling to East Siberia. Microscopic observation of characteristic bipolar eggs ($70 \times 40\ \mu m^2$, oval, yellowish brown, undeveloped with indistinct operculum) of *N. salmincola* in the stool samples is highly indicative of the infection. *N. salmincola* eggs need to be distinguished from those of *Diphyllobothrium latum*. The use of trichrome (Gomori's or Wheatly modified) staining rather than formalin–ethyl acetate concentrate provides a more sensitive way to identify *N. salmincola* eggs [15]. Nevertheless, patients with light infections are likely to have negative stool tests because only a few eggs are contained within each adult worm.

N. salmincola worm (1.5–2.0 mm, hemorrhagic) can be differentiated from other flatworms by morphological characteristics (e.g., the uterus with only a few loops and fewer than half a dozen eggs).

37.2 METHODS

37.2.1 Sample Preparation

Total DNA may be extracted from the worm or eggs using a QIAamp DNA kit (Qiagen) or a Wizard Genomic DNA Purification kit (Promega). The eluted DNA is stored at −20°C until testing. Alternatively, DNA is prepared from individual worms by placing the worm in 100 μL of 5% Chelex containing 0.2 mg/mL of proteinase K, incubated for 2 h at 56°C, and boiled at 100°C for 8 min.

37.2.2 Detection Procedures

37.2.2.1 Analysis of Cytochrome c Oxidase I (COI) Gene and ITS Sequences

Amor et al. [16] targeted the first (ITS-1), the 5.8S, and second (ITS-2) internal transcribed spacers (ITS) of nuclear ribosomal RNA (rRNA) genes and the mitochondrial COI gene for the identification and phylogenetic examination of trematodes.

The DNA region comprising ITS-1, 5.8S rRNA, and ITS-2 (ITS) is amplified by polymerase chain reaction (PCR) using primers BD1 (forward: 5′-GTCGTAACAAGGTTTCCGTA-3′) and BD2 (reverse: 5′-TATGCTTAAATTCAGCGGGT-3′). The partial COI gene is amplified with two conserved primers, Ita8 (forward; 5′-ACGTTGGATCATAAGCGTGT-3′) and Ita9 (reverse: 5′-CCTCATCCAACATAACCTCT-3′).

37.2.2.1.1 Procedure

1. PCR mixture (25 μL) is composed of 2.5 U AmpliTaq Gold, 10 mM Tris–HCl (pH 8.3), 50 mM KCl, 3 mM $MgCl_2$, 1 mM of dNTPs, 0.25 μM of each primer (BD1 and BD2 or Ita8 and Ita9), and 20–40 ng of purified DNA.
2. Amplification is performed in an Amplitron® PCR System II (Thermolyne) for one cycle at 94°C for 3 min; 45 cycles of 94°C for 40 s, 55°C (ITS) or 53°C (COI) for 45 s, 72°C for 1 min 40 s, and a final 72°C for 5 min at 72°C.
3. A volume of 5 μL of the amplification products are visualized on 1% ethidium-bromide-stained agarose gels to check the quality of amplification.
4. The PCR products are purified using the commercial kit NucleoSpin Extract (Macherey-Nagel), and sequenced with a capillary automated ABI PRISM 3700 DNA sequencer.
5. The GenBank Blast program is used for ITS and COI sequences comparisons. Sequences are corrected manually using Chromas 2.13 software and pair-wise aligned with available sequence data (ITS or COI) for select trematodes using ClustalW with the default gap and extension penalties used by this program.
6. ITS and COI sequences are entered in the MEGA for the construction of the phylogenetic trees using maximum parsimony and distance methods, namely, neighbor-joining, an unweighted pair-group method with arithmetic averages (UPGMA), and minimum evolution. Branch support is given using 1000 bootstrap replicates in MEGA.

Note: The ITS fragment amplified with primers BD1 and BD2 is expected to be approximately 1000 bp in length (comprising the complete ITS-1 and ITS-2 sequences).

37.2.2.2 Analysis of NADH-Dehydrogenase Subunit 1 (ND1) Gene Sequence

Criscione and Blouin [17] utilized primers MB352 (59-CGT AAG GGK CCT AAY AAG-39) and MB399 (59-CTT ACA AAA TAG TCA TAG CG-39) to amplify a 639 bp region of the NADH-dehydrogenase subunit 1 (ND1) mitochondrial gene for the identification and phylogenetic analysis of *N. salmincola* isolates.

37.2.2.2.1 Procedure

1. PCR (50 μL) is made up of 1× PCR buffer, 2.5 mM $MgCl_2$, 0.2 mM each of dNTP, 0.4 mM of each primer, and 1 U Taq DNA polymerase (Promega), and 5 μL of extraction supernatant (Chelex protocol).
2. Amplification is conducted with an initial 95°C for 3 min; 40 cycles of 94°C for 45 s, 54°C for 30 s, 72°C for 45 s; and a final 72°C for 7 min.
3. PCR products are purified with the UltraClean PCR clean-up kit (MO BIO Laboratories) and sequenced using PCR primers.

37.3 CONCLUSION

Human nanophyetiasis is a zoonotic disease caused by the digenetic trematode *N. salmincola*. Three hosts are required for the complete life cycle of *N. salmincola*: a stream snail, a salmonid fish (or non-salmonid fish), and a canid (or other vertebrates). The distribution of the parasite appears to be limited by the availability of stream snail host, which is commonly present in the US Pacific Northwest, Canada, and East Siberia. Transmission of *N. salmincola* to the definitive host such as humans and dogs is dependent on the ingestion of parasite-infected fish. In general, *N. salmincola* infection in humans induces non-specific (fish flu) symptoms ranging from abdominal pain, diarrhea, vomiting, and peripheral eosinophilia. It is important that human nanophyetiasis is distinguished from other parasitic diseases producing similar clinical symptoms. This will not only facilitate more effective treatment of the disease, but also help implement appropriate prevention strategies against the parasite. Given the challenges in using conventional procedures for the diagnosis of nanophyetiasis, molecular techniques will undoubtedly play a more prominent role in the identification and epidemiological tracking of *N. salmincola* parasite.

REFERENCES

1. Bennington, E., and Pratt, I., The life history of the salmon-poisoning fluke, *Nanophyetus salmincola* (Chapin). *J. Parasitol.*, 46, 91, 1960.
2. Skrjabin, K.J., and Podjapolskaja, W.P., *Nanophyetus schikhobalowi*, a new trematode from the human intestine. *Zentralbl. Bakteriol. Parasitkde Orig.*, 119, 294, 1931.
3. Millemann, R.E., and Knapp, S.E., Biology of *Nanophyetus salmincola* and salmon poisoning disease. *Adv. Parasitol.*, 8, 1, 1970.
4. Farrell, R.K., Soave, O.A., and Johnston, S.D., *Nanophyetus salmincola* infections in kippered salmon. *Am. J. Public Health*, 64, 808, 1974.
5. Kistner, T.P., Wyse, D., and Schmitz, J.A., Pathogenicity attributed to massive infection of *Nanophyteus salmincola* in a cougar. *J. Wildl. Dis.*, 14, 419, 1979.
6. Headley, S.A. et al., *Neorickettsia helminthoeca* in dogs in Brazil. *Emerg. Infect. Dis.*, 12, 1303, 2006.
7. Headley, S.A. et al., *Neorickettsia helminthoeca* and salmon poisoning disease: A review. *Vet. J.*, 187, 165, 2011.
8. Booth, A.J., Stogdale, L., and Grigor, J.A., Salmon poisoning disease in dogs in southern Vancouver Island. *Can. Vet. J.*, 25, 2, 1984.
9. Eastburn, R.L., Fritsche, T.R., and Terhune, C.A. Jr., Human intestinal infection with Nanophyetus salmincola from salmonid fishes. *Am. J. Trop. Med. Hyg.*, 36, 586, 1987.
10. Harrell, L.W., and Deardorff, T.L., Human nanophyetiasis: Transmission by handling naturally infected coho salmon (*Oncorhynchus kisutch*). *J. Infect. Dis.*, 161, 146, 1990.

11. Dumler, J.S. et al., Reorganization of genera in the families *Rickettsiaceae* and *Anaplasmataceae* in the order Rickettsiales: Unification of some species of *Ehrlichia* with *Anaplasma*, *Cowdria* with *Ehrlichia* and *Ehrlichia* with *Neorickettsia*, descriptions of six new species combinations and designation of *Ehrlichia equi* and 'HE agent' as subjective synonyms of *Ehrlichia phagocytophila*. *Int. J. Syst. Evol. Microbiol.*, 51, 2145, 2001.
12. Eastburn, R.L. et al., Praziquantel for treatment of human *Nanophyetus salmincola* (*Troglotrema salmincola*) infection. *J. Infect. Dis.*, 160, 896, 1989.
13. Fritsche, T.R. et al., Praziquantel for treatment of human *Nanophyetus salmincola* (*Troglotrema salmincola*) infection. *J. Infect. Dis.*, 160, 896, 1989.
14. Foreyt, W.J., and Gorham, J.R., Evaluation of praziquantel against induced *Nanophyetus salmincola* infections in coyotes and dogs. *Am. J. Vet. Res.*, 49, 563, 1988.
15. Johansen, M.V. et al., Towards improved diagnosis of zoonotic trematode infections in Southeast Asia. *Adv. Parasitol.*, 73, 171, 2010.
16. Amor, N. et al., Molecular characterization of *Fasciola gigantica* from Mauritania based on mitochondrial and nuclear ribosomal DNA sequences. *Exp. Parasitol.*, 129, 127, 2011.
17. Criscione, C.D., and Blouin, M.S., Life cycles shape parasite evolution: Comparative population genetics of salmon trematodes. *Evolution*, 58, 198, 2004.

38 Opisthorchis

Paiboon Sithithaworn, Thewarach Laha, and Ross H. Andrews

CONTENTS

38.1 INTRODUCTION

The liver fluke, *Opisthorchis viverrini* is a member of the fish-borne trematodes (FBT) that are endemic in Southeast Asia. The disease it causes, opisthorchiasis, is a neglected and overlooked disease because it is perceived as local and of limited occurrence. Comprehensive epidemiological studies and the extent of morbidity and disease burden as measured in Disability Adjusted Life Years [1], particularly in Southeast Asia, are lacking and greatly underestimated. Recent estimates of human populations at risk have indicated that ~70 million people in Southeast Asia alone, and ~700 millions in the rest of the world, are also at risk of infection by consumption of infected fish and aquaculture imports. *O. viverrini* infection causes asymptomatic hepatobiliary diseases (opisthorchiasis) which can lead to the bile duct cancer (cholangiocarcinoma, CCA). CCA is a devastating malignancy with high mortality, lacking early diagnosis and hence it has been significant public health disease in Southeast Asia for decades. Unlike other cancers, CCA is an infection-inflammation-induced cancer which in theory is preventable if exposure to the causative agent, *O. viverrini* is interrupted. Therefore, sensitive and diagnostically reliable detection methods for opisthorchiasis suitable to the current situation are required for human diagnosis as well as for other life stages of the parasite throughout its life cycle. Development of diagnostically accurate markers facilitates comprehensive epidemiological studies, development, and introduction of surveillance and control programs of the parasite in humans and reservoir hosts and will ensure food safety and security. Achieving these goals will eventually lead to lowering the risk of CCA which predominantly is an insidious disease of the poor in Asia.

38.1.1 Classification

Generally, liver flukes belong to the phylum platyhelminthes (flat worms), class trematoda (flukes), subclass digenea, and order opisthorchiformes. Human liver flukes comprise three families, namely; Opisthorchiidae, Fasiolidae, and Dicrocoeliidae. Causative agents for human infection are predominant members in the family Opisthorchiidae, consisting of *O. viverrini*, *Opisthorchis felineus*, and *Clonorchis sinensis* (oriental or Chinese liver fluke). Less common members in this family are *Opisthorchis conjunctus*, *Opisthorchis guayaquilensis*, *Opisthorchis noverca*, *Metorchis albidus*, *Metorchis comjunctus*, *Pseudamphistomum aethiopicum*, and *Pseudamphistomum truncatum*. Apart from *O. conjunctus* which has been reported to infect humans the rest are commonly found in animal hosts [2]. Four species of liver flukes in the families Fasciolidae and Dicrocoeliidae are commonly found in herbivores but they also cause liver diseases in humans, namely *Fasciola hepatica*, *Fasciola gigantica*, *Dicrocoelium dendriticum*, and *Dicrocoelium hospes*.

The generic name *Opisthorchis* was described in 1875 and *Clonorchis* in 1907. A recent study by Liu et al. [3] has shown that an ancient trematode sample from the gall bladder to be over 2000 years old and they identified the trematode as *C. sinensis*. Typically, classification is based on the morphology of adult worms, where the major differences separating them were the shape and position of testes and appearance of vitelline glands [4,5]. With respect to *C. sinensis*, adult worms have two highly branched testes situated tandemly, and vitelline glands are distributed continuously along both sides of the body. *O. viverrini* and *O. felineus* share similar morphological features by having lobular testes and vitelline glands arranged in clusters. However, *O. viverrini* has distinctive and deeper lobulation of testes as well as a lack of a transverse arrangement of vitelline follicles in comparison with *O. felineus*.

38.1.2 Biology and Life Cycle

O. viverrini is a hermaphroditic, dorso-ventrally flattened and leaf-like trematode. The body is equipped with two muscular suckers; the oral sucker is anteriorly situated while the ventral sucker is at the mid-body. The size of adult flukes measures 5.5–10 × 0.77–1.65 mm. *C. sinensis* is larger measuring 10–25 × 3–5 mm while *O. felineus* is the smallest measuring 7–12 × 2–3 mm [5]. The size of adult worms varies depending on the intensity of infection and the diameters of the bile ducts they inhabit. Adult worms occupy the biliary tract specifically intrahepatic bile ducts, the gall bladder and to a lesser extent the pancreatic duct [6]. They become hermaphroditic adult worms when they gain sexual maturity and they mate by out crossing as well as selfing fertilization as has been shown by population genetic studies using microsatellite DNA and multilocus enzyme electrophoresis analyses [7,8]. Studies of the biology of *O. viverrini* have shown that fertilized worms produce yellow-brown eggs which are discharged with bile fluid into the intestine and subsequently passed out with feces into the environment (Figure 38.1). The eggs measure 25–35 μm in length and 15–17 μm in width; and possess a distinct opercular shoulder surrounding the operculum at one end and a small knob or comma shape appendage at the abopercular end. They have a rough and irregular egg shell surface similar to musk-melon patterns when observed by electron microscopy [9,10]. Eggs of the three species of liver flukes have similar morphologies and are very difficult to differentiate. The number of eggs produced per worm depends on the worm burden, that is, a density-dependent effect and also pathological condition of the hepatobiliary system which may block egg passage. In humans each worm on average produces ~50–200 egg/g feces [6,11].

Once eggs reach a body of freshwater like small ponds, streams, and rivers, flooded rice fields they can be ingested

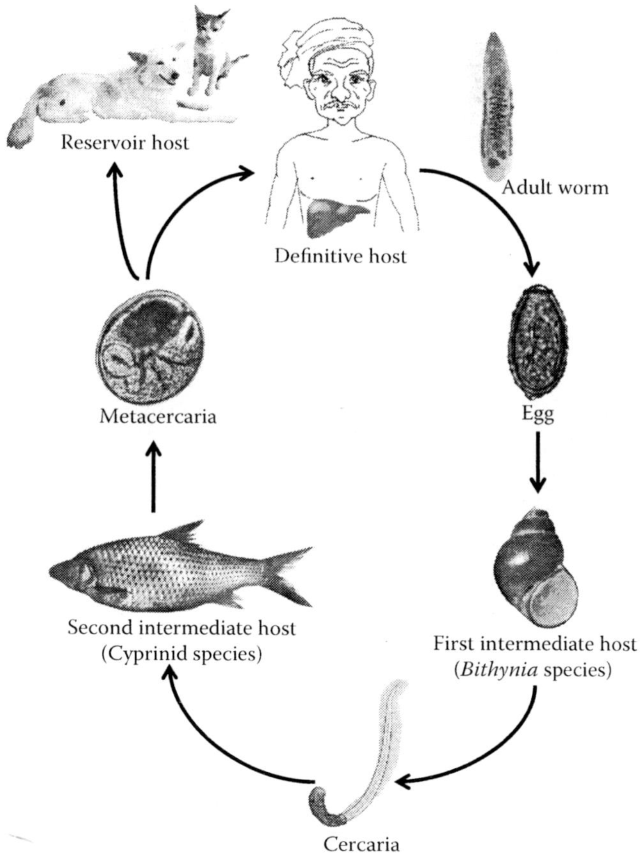

FIGURE 38.1 Life cycle of *O. viverrini*.

by snail intermediate hosts where the miracidium hatch and develop into sporocysts and rediae. The snails in the genus of *Bithynia* serve as the intermediate hosts, namely, *B. goniomphalos, B. siamensis siamensis*, and *B. funiculata* [4]. The rediae gave rise to cercariae which are released daily ~2 months after snail infection. The emergence of cercariae have diurnal periodicity with a peak between 8.00 and 10.00 a.m. and as high as 1728 cercariae are released per snail per day [12]. Our unpublished data supported this pattern of cercarial release under constant light conditions. Based on the assumption that a single egg is consumed by a snail this will result in a single miracidium, which will eventually produce more than a thousand cercariae. The infected snails may shed cercariae daily during the patent period which may last up to 2 months and hence capable of producing an approximate total number of 50,000–80,000 cercariae/miracidium.

Once exposed to appropriate species of fish intermediate host, the free-swimming cercariae shed their tails, penetrate into the tissues or the skin and become fully infective metacercariae in 3 weeks. Humans become infected by ingesting the metacercariae in raw or partially cooked fish. The ingested metacercariae excyst in the duodenum, enter the common bile duct, and migrate to the distal bile duct. Development of the juvenile worms in humans takes ~2 months in transforming to the sexually mature stage. Although the survivorship of the worms is not accurately known, it is thought that they may survive for several years. In the case of the closely related liver fluke, *C. sinensis*, it has been reported that they can survive for up to 26 years in a human host [13], but this observation may represent an extreme survival condition in an individual host. Based on the pattern of the age-intensity profile, it is anticipated that the average lifespan of *O. viverrini* in humans could be approximately up to 10 years. This estimate is based on the drop in intensity after 60 years of age and less consumption of raw or under-cooked fish.

Recent analyses of the systematics of natural populations *O. viverrini* in Southeast Asia have shown that it is a species complex in which adult worms show identical and indistinguishable morphologies but they are genetically very different (i.e., cryptic species). Within natural populations of *O. viverrini* sensu lato, current results indicate that there are at least two cryptic species from Northeast Thailand and Lao PDR [8,14]. Genetic evidence supporting the presence of cryptic species in *O. viverrini* arises from applications of allozyme electrophoresis and random amplified polymorphic DNA. Evidence from applications of these techniques has revealed the association of distinct genetic groups of *O. viverrini* within wetland river systems as well as correlations with specific genetically distinct geographical isolates of *Bithynia* snails [8]. For instance, recent studies have shown that there were at least six genetically distinct groups (genotypes) that are associated with different major river wetlands which included the rivers Wang, Mun, Chi, and Songkram in Thailand and the Nam Ngum River in Laos. These genetic groups have been shown to have different independent characteristics such as infectivity of metacercariae, growth, and fecundity of adult worms in experimental animals [15]. This report provides biological data supporting the significance of the genetic variation that has been detected in *O. viverrini sensu lato*, similar to that observed in other trematodes such as *Schistosoma japonicum* [16,17].

Preliminary evidence of genetic variation in a closely related liver fluke, *C. sinensis* has also been documented in which isolates from China and Korea have shown relatively low degrees of genetic difference. This is similar to the genetic differences detected between *C. sinensis* and *O. viverrini* based on nuclear ribosomal DNA and mitochondrial DNA sequences [18,19]. In addition to ribosomal and mitochondrial DNA, another genetic marker (paramyosin gene, Pm-int9) was recently applied to assess interspecific variation among liver flukes [20]. This study indicated that *O. viverrini* and *C. sinensis* were closely related evolutionarily to each other but appear to be separate from *O. felineus*. Until recently, limited information has been available on the systematics and phylogenetic relationships of *O. viverrini*, *C. sinensis* as well as *O. felineus* primarily due to the lack of suitable independent genetic markers systematic and phylogenetic relationships. Compared to the limited number of genes available currently for genetic and molecular characterization of *C. sinensis* and *O. felineus*, results from the recent application of the technique of allozyme electrophoresis have established up to 32 independent genetic markers that can now be used to examine the systematics and population structure of *O. viverrini* [8]. In addition, other DNA-based markers, for instance, microsatellite DNA markers, have been advocated to be particularly useful to examine the population genetics of parasitic trematodes. Application of these markers in *O. viverrini* sensu lato originating from localities in Thailand and Lao PDR is producing exciting data that are beginning to reveal essential population genetic indices regarding genetic diversity, mating polymorphism, and genetic differentiation [7].

Similar to other trematodes, the prevalence of *O. viverrini* infection in snail intermediate hosts (*Bithynia* sp.) is typically low [21]. Our recent study has shown variable prevalence in *Bithynia siamensis goniomphalos* which ranged from 0.22% to 6.93% per locality with a mean of 2.95% in Thailand and prevalence in Lao PDR was between 0.37% and 8.37% per locality with a mean of 2.44% (unpublished). These prevalence levels are dramatically higher than reported previously and may be due to different methods used to enable cercarial shedding, for example, potential differences in duration and exposure to light. In addition to *O. viverrini*, we have found that snails are infected with seven other zoonotic trematode groups, with some trematode groups showing 100% infection rate (unpublished). The density of snail populations displays strong seasonality and is dependent on rainfall. For example, snails are highly abundant in the rainy season and distributed extensively in shallow water and paddy fields; but disappearing rapidly by hibernation in the mud in the dry season. Because of such high fluctuations in number, parasite control via control of snail populations in nature does not appear to be a feasible approach. However, parasite control of snail populations in aquaculture conditions is more feasible and should be implemented as has been the case in Vietnam (Olsen, A., pers. comm.).

In contrast to the infection in snails, the prevalence of infection in fish intermediate hosts is much higher in some species of fish. As high as 90–95% of several species of cyprinid fish had been reported to harbor *O. viverrini* metacercariae [22]. The most common species of cyprinid fish (see Figure 38.2) is in the genera *Cyclocheilichthys* and *Hampala*. Differential susceptibility of fish species may be due to host-specific responses to cercarial challenge, differences in host genetics, and differences in ecological conditions. The host finding mechanism of cercariae is a complex process [23] and free swimming cercariae are very efficient in locating the appropriate species of fish in a large volume of water. This is important, because our preliminary evidence suggests that cercarial survival is limited to a maximum of 48 h (unpublished). The intensity of liver fluke infections in fish varies by season, species, individuals, and types of water bodies [22,24]. However, we have preliminary evidence that this is not always the case (unpublished). Most metacercariae are distributed throughout the body of fish but some metacercariae are found in the head of the fish. The metacercarial burden peaks in winter (around October–February) and becomes low in the rainy season and summer, and thus transmission of the parasite from fish to human is probably seasonal [25].

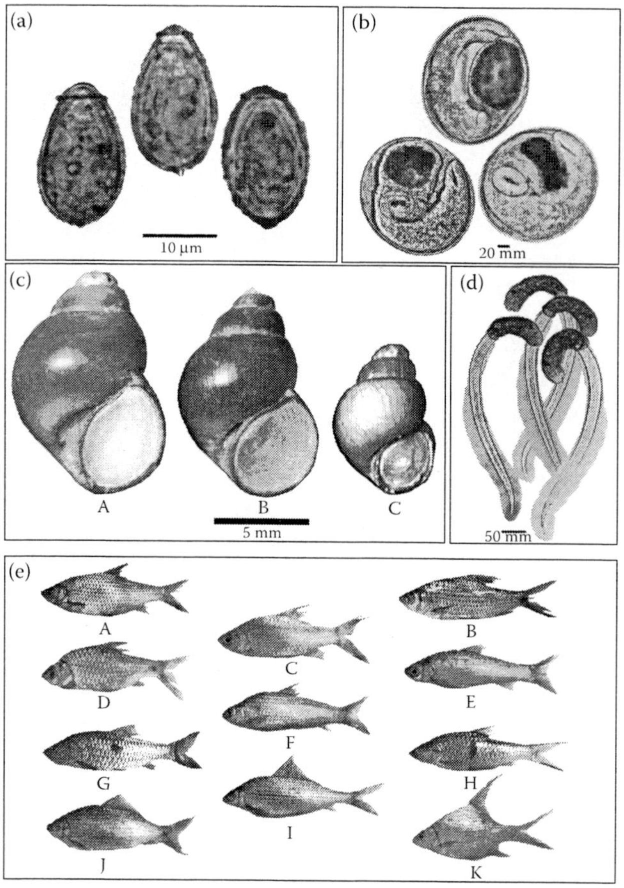

FIGURE 38.2 (a) *O. viverrini* eggs (iodine staining); (b) *O. viverrini* metacercariae; (c) *Bithynia* snails as the first intermediate host of *O. viverrini* (A; *Bithynia funiculata*, B; *B. s. goniomphalos*, C; *Bithynia siamensis siamensis*); (d) *O. viverrini* cercariae; (e) Fish intermediate host of *O. viverrini*. A: *Cyclocheilichthys armatus* (Saitan Takhao), B: *Barbonymus gonionotus* (Taphian Khao, Puntius carp), C: *Cyclocheilichthys apogon* (Saitan Tadaeng), D: *Puntius brevis* (Taphian Chut), E: *Henicorhynchus ornatipinnis* (Soi Pik Daeng), F: *Henicorhynchus siamensis* (Soi Khao), G: *Hampala dispar* (Hampala dispar), H: *Labiobarbus leptocheilus* (Sa), I: *Hampala macrolepidota* (Krasup Khit), J: *Osteochilus hasselti* (Soinokkhao), K: *Puntioplites falcifer* (Ka Mang).

The number of metacercariae reported in fish varies within a range of between one to hundreds. Frequency distribution patterns of metacercariae of *O. viverrini* in fish are highly dispersed; most fish have few metacercariae while a few fish harbor a heavy metacercarial load [26]. Recent findings of trematode metacercariae other than *O. viverrini* in cyprinid fish in Thailand [27,28] have indicated that the occurrence of mixed species of trematodes in a given fish species is common. Careful morphological identification of metacercariae from fish requires experience and trained personnel.

Dogs and cats serve as reservoir hosts in the life cycle of the liver fluke but the prevalence of infection in these animals is relatively low, variable, and dependent on local environmental condition [29]. In Lao PDR, study of feline reservoir hosts showed the prevalence of *O. viverrini* to be from 20% to 36% [30]. A recent survey in an endemic focus of opisthorchiasis in Khon Kaen province indicated that cats are the predominant reservoir host since 36.4% ($n = 22$) of them had *O. viverrini* infection while dogs in the same area had an infection rate of 3.8% ($n = 78$) [31]. Fecal contamination from infected animals undoubtedly contributes to parasite transmission to snails in the environment but its contribution relative to that of humans is unclear. In a situation where fecal contamination from humans is reduced or eliminated according to mass treatment and with improved sanitation, infection among reservoir hosts may play pivotal roles to maintain the source of parasite in transmission and maintenance of the life cycle. The contribution that reservoir hosts domestic and native play in the transmission of *O. viverrini* requires comprehensive examination throughout its endemic range.

38.1.3 Geographical Distribution

Among the three major human liver flukes, the current estimated total number of infections is 45 millions in Asia and Europe, of which 9 million people are infected with *O. viverrini*, 1.2 million with *O. felineus*, and 35 million with *C. sinensis* [32,33]. As much as 680 million people worldwide are at risk of infection by the three liver flukes. The hot spot of *O. viverrini* is in Northeast Thailand, Laos, and Cambodia and to a lesser extent, Vietnam. As many as 67 million people in Southeast Asia are estimated to be at risk of infection with *O. viverrini* [34], *C. sinensis* is found in Korea, China, Taiwan, Vietnam, and was previously found in Japan and in Russia [20], while *O. felineus* is endemic in Poland, Germany, Russia, Federation of Kazakhstan, and western Siberia. Being similar to other snail-borne trematodes, the presence of suitable snail intermediate hosts dictates the geographical distributions of these liver flukes. However, because of increasing population migration and extensive traveling plus growing importation of fresh water aquaculture products from Asia, it is not now increasingly common to detect infected people in nonendemic areas. For example, cases of opisthorchiasis in nonendemic areas, such as the United States and Europe, have been reported in immigrants from Asia. The cause of infection was due to the consumption of imported raw or undercooked fresh water fish containing metacercariae [35,36]. Recently, two outbreaks of *O. felineus* have been reported in Italy highlighting the increasing presence of liver fluke in Europe and direct infection from one species of fish (*Tinca tinca*) from one water source [37].

Similar to most parasitic heminths, high geographical variability has been observed in the distribution of *O. viverini* in a given endemic area. For example, *O. viverini* in Thailand is highly prevalent in the northeast in contrast to other regions of the country with lower prevalence [38]. In Lao PDR, *O. viverrini* has been found mainly in southern region such as Saravan, Suvannakhet, and Khammuan [39]. Very little information on the distribution of *O. viverrini* is available in Cambodia [40,41].

38.1.4 Epidemiology, Prevention, and Control

In 1915, detection of *O. viverrini* in Chiang Mai, Thailand was reported and subsequent studies in 1955 by Sadun [42] confirmed the identity of the liver fluke that was found to be endemic in Thailand was in fact *O. viverrini*. The pattern of infection in populations in endemic communities often shows that the prevalence and intensity increase with age of the population and thus middle-aged people tend to have high prevalence and intensity of infection. Likewise, associated morbidity in terms of hepatobiliary diseases frequently occur in this age group. The distribution of worm burden in humans is not uniform, since the majority of parasites are harbored by a minority group of people. Currently light infection group (egg per gram (EPG) <1000) show the most infection [26,43]. In addition, there is evidence of predisposition to heavy infection in which heavily infected individuals have a tendency to return to their pretreatment infection levels after treatment [44]. These epidemiological characteristics have an important bearing on the subsequent design and planning of control programs.

While snail control to disrupt the life cycle is not an attractive option for control of opisthorchiasis, chemotherapy by using praziquantel is the most common means for large-scale parasite control as well as individual treatment. The treatment program varies according to the local situation and constraint in each country. In Thailand, for example, a control program was implemented via the selective treatment approach (treatment given to infected individuals as opposed to mass drug treatment where treatments are given to all individuals in an endemic community). Although the efficacy of praziquantel is relatively high (90–95%) and to date no evidence of drug resistance has been reported, reinfections are a common phenomenon in many endemic areas [44,45]. Therefore, we recommend an integrated approach to parasite control where health education plus sanitary improvement is recommended [46]. A self-reliance control program was adopted at the beginning of the campaign during 1980–1987 in Thailand. This approach has contributed to significant reductions in the prevalence and intensity of infection [38]. In addition to chemotherapy, health education to avoid eating raw or under-cooked fish preparation is essential to reduce and prevent infection. This eating behavior, which has been deep rooted and a part of culture and tradition for decades in endemic areas of *O. viverrini* (Northeast Thailand and Lao PDR), is not easy to change. Thus, the health education message should be culturally sensitive and with suitable methods for implementation for a given community. For the current population of adults who have been exposed to the infection and are at risk of having hepatobiliary disease and CCA, health education may need to be focused on the importance of reinfection which can further increase the risk of CCA. Another approach for control is targeting the young generation, such as school-aged children, and hence realizing the future goal of a parasite-free generation that should have low or no risk of CCA.

Because of the growing fishery export trade from endemic areas of food-borne trematodes to nonendemic areas, both WHO and FAO are aware of the importance of an integrated, multidisciplinary approach to food safety/security and quality, which considers the entire food chain. In the case of aquaculture, the food chain approach required is a responsibility that the fish supplied is safe for human consumption by all involved with the production, processing, trade, and consumption of fish. Development and implementation of Good Aquaculture Practice, Good Hygienic Practice, and Hazard Analysis Critical Control Point are recommended for fish farming in order to reduce metacercarial contamination of fish and clean fishery products [32]. These measures are possible in intensive, industrial aquaculture practices but for small-scale subsistence fish farming a balance between the operation cost and potential profit needs to be considered. Evidence of FBT in cultured fish has clearly shown that fish inspection should include the process of screening for FBT [32]. In addition, choice of appropriate preventive measures should be made to reduce the risk of *O. viverrini* and other FBT in fish culture ponds [47]. Another important safety measure is to adequately freeze the fishery product (e.g., at –20°C for 7 days) or heat treatment (e.g., cooking).

38.1.5 Clinical Manifestation

Following infection with *O. viverrini*, the clinical manifestations in subjects are divided into two groups. Group one contains the majority of infected people and they are asymptomatic individuals. A continued and repeated infection with no treatment eventually leads to a symptomatic group. This group has been divided into three levels of symptoms, namely mild, moderate, and severe manifestation. But these symptomatic subjects usually occur in a minority of infected people. For example, a community-based morbidity study by standard physical examination in a village in Khon Kaen, Thailand, has reported that hepatomegaly at the midclavicular line occurred in 14% of the subjects sampled, mainly in the more heavily infected groups [48].

Abdominal-related symptoms may start to develop particularly in middle-aged individuals (30–40 years onward) as a consequence of long-term chronic infection. The symptoms are considered as mild infection when presenting with abdominal pain, flatulence, and/or right quadrant abdominal pain associated with a burning sensation. In the moderate group, the symptoms are more pronounced with liver and gall bladder enlargements but liver function tests have been shown to be normal. In the severe manifestation group, occasional jaundice with fever as a result of biliary obstruction and ascending cholangitis are frequently listed as common clinical manifestations. Liver function tests have shown signs of abnormalities such as elevated bilirubin in jaundice cases and in some cases liver enzymes may also be elevated.

The severity of clinical manifestations in opisthorchiasis is correlated with the duration and intensity of infection and complication of the disease [42] has classified the severity of infections by intensity of infection (fecal egg count) consisting of light or mild (egg/gm feces, EPG <1000), moderate or medium (EPG <10,000), heavy (EPG <30,000), and very heavy infection (EPG >30,000). From autopsy studies in Northeast Thailand, equivalent worm loads at each intensity classes of <62, <248, <479, and >479 worms, respectively, have been detected [6]. Fecal egg output has been used as a parameter to measure worm burden in the biliary system which is associated with severity of hepatobiliary abnormality. In general, observed morbidity in opisthorchiais is usually low, for example, Pungpak et al. [49] reported that only 88 cases (<1%) showed clinical manifestations among 15,243 infected people who presented at a hospital in Bangkok. Two large community-based studies by Upatham et al. [48,50] have shown significant increased frequencies of abdominal pain in the upper right quadrant, flatulence or dyspepsia, and weakness associated with increasing intensity of infection.

In a subsequent period, application of radiology imaging has aided in the diagnosis of sunclinical opisthorchiasis as well as associated complications. Ultrasonography (US) appears to be the most practical choice for investigation as it is a noninvasive examination and can operate under field conditions [51,52]. Previous community-based ultrasound studies in *O. viverrini* endemic areas of Northeastern Thailand suggest that hepatobiliary abnormalities such as enlargement of the left hepatic lobe and the gall bladder, loss of gall bladder contractility, presence of sludge, and increased periportal fibrosis are common [52,53]. Occasionally, biliary stones including gall bladder stones may be present. Additional characteristic sonographic findings showed active *C. sinensis* infection as floating echogenic foci in gall bladder or gall bladder sludge were detected [54].

Studies using US have demonstrated strong relationships between gall bladder enlargement, wall irregularities, sludge, enhanced echogenicity of portal vein radicles, and the intensity of infection particularly in moderate-to-heavy infection [53,55]. In another a large-scale study (sample size of 12,311 individuals) the results confirmed the above findings and showed the relationships in more quantitative terms [56]. With the intensity of infection of >3000 EPG feces, the adjusted odds ratio for distended gall bladder and portal vein radicles was between 14–20 and 10–28, respectively [57]. Reversibility of these abnormalities as seen by gall bladder size and its regain contractility was observed 11 months after praziquantel treatment [53,58]. In the current situation when a large proportion of infected people were treated with praziquanel, the pattern of hepatobiliary diseases as determined by US has changed considerably. Currently, the heavy infection group (EPG >3000) was rare, while most of the infected people had light infection with EPG <1000 and the disease frequency was no longer correlated with intensity of infection as in the past (unpublished). The ultimate complication in opisthorchiasis is CCA which may occur within symptomatic subjects but confirmation by radiological investigation is required. Conceivably, reduction in prevalence and intensity of *O. viverrini* as a result of the past and ongoing control program should contribute to lesser morbidity and risk of CCA.

38.1.6 Pathogenesis and Pathology

Pathological changes in liver fluke infection are confined to the bile ducts, liver, and gall bladder in humans as well as in animal models. Hypotheses proposed by Bhamarapravati et al. [59] regarding pathological mechanisms include (i) mechanical irritation due to migration and feeding of adult worms in the bile ducts, when oral and ventral suckers hook onto the biliary epithelium and damage tissue; (ii) chemical irritation through the action of metabolic products released from the worm which may cause direct damage to the biliary epithelium; and (iii) immunological responses, humoral, and cellular, to liver fluke antigen and subsequent inflammation. The extent of the pathology depends on the intensity, duration, and host susceptibility. In light infections, the liver appears grossly normal, whereas, in heavy infections, a localized dilation of the thickened peripheral bile ducts can be seen on the surface underneath the fibrotic capsule of the liver.

Hou [60] and Bhamarapravati et al. [59] have shown that histopathological changes in opisthorchiasis are generally similar to clonorchiasis. At early stages of infection, biliary epithelium frequently becomes edematous and desquamation and may be observed in areas of tissue in close proximity to the flukes. Periductal infiltrates of mononuclear cells are frequently detected; however, inflammation of bile duct walls is generally only slight in uncomplicated cases. Metaplasia of the biliary epithelial cells transforms into mucin-producing cells (goblet cells) and occurs during early infection, which may proliferate to produce many small gland-like structures in the mucosa (adenomatous hyperplasia). This results in persistent and excessively high mucus content in the bile. Persistent and chronic infections result in a gradual increase in the amount of fibrous tissues, giving the appearance of cholangiofibrosis (Figure 38.3). As fibrosis progresses, epithelial proliferation becomes milder. These histopathological changes are distinctive features particularly in clonorchiasis. Hence, when patients are observed in an area of endemic infection with proliferation of the ductal epithelium with metaplastic cells and periductal fibrosis, this is highly indicative of liver fluke infection on histological grounds [60].

No epithelial hyperplasia or fibrous proliferation has been reported during the acute or initial stage of infection (<4 weeks postinfection). In chronic infections, however, proliferation of epithelial cells is apparent with the formation of glandular acini, similar to the adenomatous changes in clonorchiasis, with varying degrees of periductal fibrosis. Major microscopic changes are confined to the large- and medium-sized bile ducts. The gross and microscopic characteristics of human opisthorchiasis are well-developed within 7–15 years after *O. viverrini* infection. In chronic and heavy infections, various degrees of cellular infiltration are caused by superimposed bacterial infection, which may result in suppurative cholangitis. Infection may also extend into the parenchyma of liver tissue, causing cholangiohepatitis and abscess formation. In heavy infections, adult worms are distributed within the gall bladder, the common bile duct, and

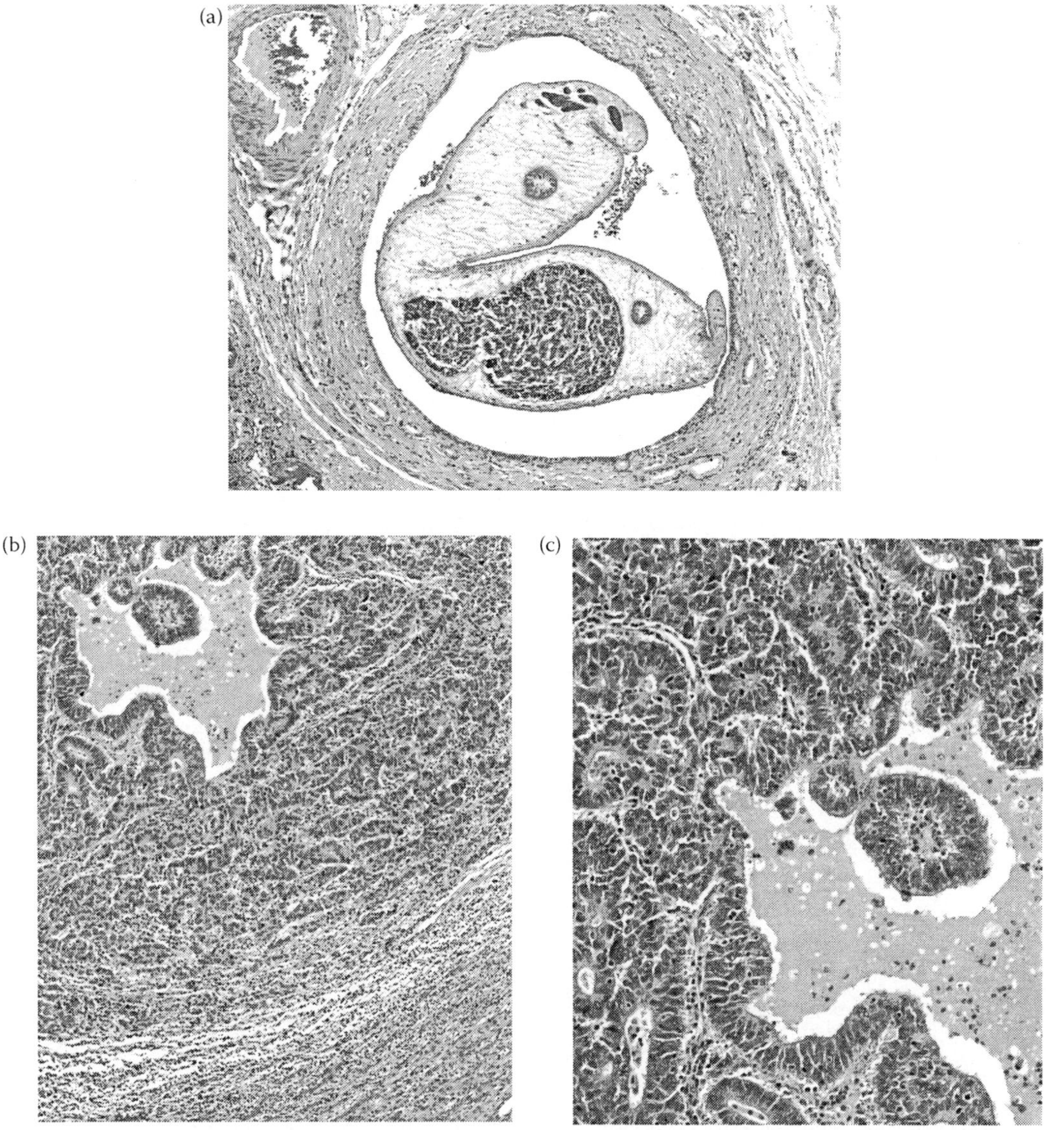

FIGURE 38.3 Histopathology in human opisthorchiasis and bile duct carcinoma (CCA). (a) Showing cholangiofibrosis with worm in the enlarged bile duct; (b) CCA; (c) Adenocarcinoma. (Courtesy Dr. C. Pairojkul, Khon Kaen University.)

the pancreatic duct. In large- and medium-sized bile ducts the worms induce chronic cholecystitis and with additional bacterial infection, empyema of the gall bladder may result. Unlike clonorchiaisis, stone formation in the bile duct or gall bladder has been rarely observed in postmortem opiosthorchiasis studies [61–63]. Parasite-induced mucin-secreting cells produce bile with a high mucin content and when combined with adult worms and eggs, serves as a nidus for bacterial superinfection and intrahepatic stone formation [64–66]. The mucin-rich bile together with worms and eggs in the bile duct cause cholestasis and provide a favorable environment for secondary bacterial infection. These are of enteric origin, with *Escherichia coli* identified most frequently as a pathogen. Enlargement of the gall bladder is commonly observed in opisthorchiasis at autopsy [62] and during ultrasonographic studies in patients with *O. viverrini* infection [53,64,65,67].

38.1.7 Host Immune Response

Even though *O. viverrini* infects the biliary system and does not invade into host tissues, *O. viverrini* infection elicits both local and systemic immune responses. To date, little information about specific immune responses to infection is available, especially cell-mediated immunity, the roles that T cells and cytokines play in protective immunity, and the pathogenesis of opisthorchiasis [68]. Marked humoral immune response (IgG, IgA, IgE) to parasite antigens in the serum and bile of humans and animals infected by the liver fluke has been observed [69–72]. The level of IgG against crude somatic antigen has been shown to correlate with hepatobiliary abnormalities diagnosed by US, but little correlation with the intensity of infection has been found [57]. In infected individuals, the level of serum antibodies (IgG, IgA, IgM) has been shown to change slowly and remain elevated up to

12 months after praziquantel treatment [73]. It is not clear as to whether this reflects long-lasting immunological memory or the presence of residual parasite antigen derived from treatment or cross-antigenic stimulations. Liver fluke DNA occurs in liver tissue of CCA patients suggesting that residual parasite material may remain within the liver [74]. In the hamster model, antibody responses have been first detected as early as 14 days after infection. These increased rapidly to a plateau at around 2 months postinfection and were relatively stable thereafter [75]. Reinfection with *O. viverrini* metacercariae elicits higher levels of IgG and more severe liver cell damage which correlates with the degree of periductal fibrosis [76]. In the case of cellular responses, *in vitro* proliferative responses of mononuclear cells to mitogen stimulation were unchanged after praziquantel treatment [68]. The role of cellular response in biliary pathology in *O. viverrini* infection is evident since the severity of bile duct inflammation diminished in T-cell-deprived hamsters [77]. A reduction of an *in vitro* lymphoproliferative response has also been observed in hamsters which are infected with *O. viverrini* suggesting a degree of immunosuppression [78].

In the acute phase of infection in an animal model, evidence of cellular immune response during acute phase was seen as abundant inflammatory cells at 3 weeks postinfection with *O. viverrini* due to the increase in infiltration of cells which included eosinophils, mononuclear cells, and neutrophils [75,79]. In the chronic phase (1–6 months) of infection, an increase in mononuclear cells and a decline of eosinophils occurs. Upon reinfection, a smaller number of infiltrated inflammatory cells have been detected but mainly in the lymphoid follicles in association with the progression of periductal fibrosis [76].

Although evidence presented above clearly showed that infection by *O. viverrini* stimulates systemic humoral and cell-mediated immune responses during the course of infection, the significance of these immune responses on protective immunity is not convincing. The fact that some individuals from the endemic area may harbor several thousand worms [56,80] indicates that reinfection does occur and the failure of immune responses to prevent reinfection by the same parasite. This conclusion was confirmed when no significant reduction in worm burden was detected in hamsters receiving immune spleen cells and serum compared to the control groups even though there was a substantial reduction in the fecal egg count [81]. It has been shown that primary infection of hamsters with *O. viverrini*, whether or not eliminated with praziquantel before challenge, conferred no protective immunity to reinfection [82]. However, both groups of investigators have demonstrated that previous exposure to *O. viverrini* depressed egg output in subsequent infections. Biliary obstruction could result in egg reduction, but egg reduction may not be due to immunological damage of the reproductive system of the flukes.

Lack of protective immunity is possibly due to immunosuppression of both cellular and humoral immune responses by *O. viverrini* infection to unrelated antigens, as evidenced by reduced phytohemagglutinin-induced lymphoproliferation and response to sheep red blood cell stimulation [83]. When specific antibody has been measured in infected animals, the titer was found to be depressed during the late stage of infection, particularly in the heavily infected group [84]. Furthermore, the parasites may be resistant to immune damage or have mechanisms to evade the host defense system. Not being able to kill worms by exposure to serum from infected animals or from patients with opisthorchiasis and the resistance to immune damage may be related to tegumental shedding and repair of the worms [85]. Cellular response studies in an animal model have indicated that IL-12 (Th1 type) dominates during acute inflammation of bile ducts but it is suppressed in chronic infection [86]. Results from this study also suggested that transforming growth factor (TGF)-b may play a role in switching from a Th1 to Th2—type response (IL-4) in the chronic infection stage that is a suppression of cellular immune response.

A number of inflammation-induced hepatobiliary pathologies are associated with liver fluke infection with *O. viverrini*, the most common of which is periductal fibrosis. A series of studies in endemic communities in Khon Kaen, Northeast Thailand, peripheral blood mononuclear cells (PBMC) proliferation assay using excretory secretory (ES) antigen of *O. viverrini* have indicated an relationship between an inflammatory cytokine and periductal fibrosis in liver fluke infection. IL-6 levels were found to be significantly higher (8 times) in individuals with advanced periductal fibrosis compared to *O. viverrini*-infected individuals without periductal fibrosis [87]. These results confirmed the roles of host response in pathogenesis of opisthorchiasis and CCA.

38.1.8 Liver Fluke and CCA

Infection and chronic inflammation contribute to carcinogenesis through an inflammation-related mechanism. Infection with hepatitis C virus, *Helicobacter pylori* and the liver flukes *O. viverrini* and *C. sinensis*, are important risk factors for hepatocellular carcinoma, gastric cancer, and CCA, respectively [88]. *O. viverrini* as well as *C. sinensis* have been classified as class 1 carcinogens to humans. Currently the evidence for *O. felineus*, however, is insufficient to assess its role in carcinogenesis.

CCA, generally referred to as peripheral CCA, is a malignant tumor stemming from the biliary epithelium of the intrahepatic biliary tree. It accounts for 10–15% of primary liver cancers and its geographical distribution overlaps with known endemic areas of *O. viverrini* [89]. Recent hypotheses on its etiology focus on chronic inflammation of the bile ducts as well as conditions associated with bile stasis that are predisposing factors to the development of CCA, including primary sclerosing cholangitis, liver fluke infection, and recurrent cholangitis with hepatolithiasis. In Thailand, CCA is 12 times more common in the northeast, the endemic area of *O. viverrini*, than in regions with low prevalence of *O. viverrini* [90]. The age-standardized incidence of CCA in Khon Kaen of 84.6 and 36.8 per 100,000 for male and female, respectively, is the highest in the world [91]. Cancer epidemiological studies revealed that evidence of current or past experience of liver fluke infections, in terms of intensity of infection and high-serum antibody against the liver fluke, are risk factors of CCA [91,92]. The frequency of suspected CCA is positively associated with fecal egg count where the odds ratio 14.1 for CCA was found in a group with EPG >6000. This is equivalent to the presence of >120 worms in the biliary tract [56]. Additional risk factors reported in CCA were the location of residency, alcohol consumption, age, and sex. More recently, host genetic polymorphism of glutathione *S*-transferase enzyme (GSTM1) in association with seropositivity for opisthorchiasis, has suggested that host genetic factors may also contribute as risk factors of CCA. These findings suggest a classic gene–environment interaction which could play a crucial role in an individual's susceptibility to cancer.

In endemic areas of liver fluke infection, histological features of CCA have been found to be similar to that observed in nonendemic regions [64]. Tumor cells produced from surface epithelial cell which line the peripheral bile duct invade the sinusoids of adjacent liver parenchyma and form a tumor nodule. In most clinical cases, the disease is in advanced stage and usually appears as a large, single, white firm tumor with a distinct margin in the noncirrhotic liver [93,94]. The most common histopathology of CCA (>95%) is adenocarcinoma that exhibits a glandular and/or papillary structure with a variable fibrous stroma. There is no dominant histological type of CCA in either endemic or nonendemic areas [95]. Atypical hyperplasia and dysplasia that lie adjacent to CCA are likely to be precancerous lesions [96].

Most authors of earlier studies suggest that liver flukes mediate tissue damage directly by mechanical and chemical irritation, but some studies have postulated that parasite-specific immune responses may play a major role [59,77,81,97]. At present, although the underlying enhancement of neoplasia by liver flukes is not fully understood, it is generally believed that carcinogenesis is a multistage process in which many factors are likely to be involved. However, the mechanistic explanation is gradually becoming clearer, particularly the hypothesis that the chronic infection and inflammation are risk factors of carcinogenesis due to the possible role that nitric oxide (NO) plays [98].

During liver fluke infection, macrophages and other cell types (e.g., mast cells, eosinophils, epithelial cells), activated by parasite-specific T-cell and cytokines synthesize NO from L-arginine via the induction of NO synthase (iNOS) in order to kill the parasite. NO is not only cytotoxic, but also genotoxic at physiological concentrations. Excess NO production plays a crucial role in a number of pathological processes, including cancer [98]. NO reacts with superoxide to form peroxynitrite, a highly reactive species causing nitrative and oxidative DNA damage. This can cause the formation of 8-oxodG and 8-nitroguanine, a marker of nitrative DNA damage [99–101].

Endogenous nitrosation caused by liver fluke infection has been studied in both animal models and humans. In *O. viverrini*-infected people, studies have shown significant increases in plasma and urinary nitrate and *N*-nitrosoproline (NPRO) in urine after proline loading, nitrate and NPRO, the indicators of endogenous NO synthesis and nitrosation reactions, respectively [102,103]. The site of the endogenous reaction has been shown to be extragastric, that is, within inflamed bile ducts [104]. Direct measurement of *N*-nitrosodimethylamine (NDMA), the carcinogenic product of nitrosation reaction, excreted in the urine appears to be associated with *in vitro* lymphoproliferative responses to liver fluke antigens. Following removal of flukes by praziquantel treatment, the association disappeared [105]. In multivariate analyses, NDMA levels were found to be related to urinary nitrates, stimulation indices for two T-cell responses to two parasite antigens (MW 37 and 110 kDa) and gall bladder dimensions. As shown in human and animal models, in active infection with *O. viverrini*, NDMA is further activated by isoforms of cytochrome P450 (CYP) enzymes, specifically CYP2E1 and CYP2A6, ultimately to become carcinogenic [106,107]. Increase in CYP 2A6-related enzyme activity may represent an important mechanistic link between inflammatory products of chronic liver fluke infection and the high risk of CCA faced by infected individuals.

There is evidence that chronic inflammation in opisthorchiasis induces oxidative and nitrative stress of the biliary epithelium which subsequently causes DNA damage. The DNA damage known as DNA adducts occurring in the nuclear and mitochondrial DNA are excreted in urine following DNA repair in the inflamed tissue. A recent study in an endemic community in Khon Kaen, Thailand, showed that evidence of *O. viverrini*-induced oxidative stress could be detected as 8-oxodG which accumulated in leukocyte DNA and is excreted in the urine of infected patients. This DNA adduct significantly decreased two months postpraziquantel treatment and the level was comparable with the levels detected in healthy subjects one year after treatment [108]. Also from this study, CCA patients had a significantly higher level of 8-oxodG than the *O. viverrini*-infected and healthy control subjects, the authors concluded that the 8-oxodG levels, particularly in urine, may serve as a useful biomarker to assess the risk of *O. viverrini*-associated CCA.

Therefore, NO and reactive oxygen species (ROS) may cause 8-oxodG and 8-nitroguanine formation through chronic inflammation reactions as a result of *O. viverrini* infections. Persistent cellular oxidative and nitrative stress also enhances lipid peroxidation (LPO) which has been implicated in inflammation-induced carcinogenesis (IARC, 1994). This LPO-derived DNA damage is excreted in urine as etheno DNA adducts (epsilondA, epsilondC) and could be another noninvasive biomarkers for CCA [109,110].

Formation of oxidative and nitrative DNA adducts in opisthorchiasis occurs through the Toll-like receptor 2 (TLR2)-mediated pathway. This leads to the activation of nuclear factor κB (NF-κB), a key player in tumor promotion in inflammation-associated cancer [111]. Additionally, a cofactor for DNA polymerase, proliferating cell nuclear antigen, is associated with DNA replication and long-patch base excision repair, and accumulates in the epithelium of bile ducts following repeated *O. viverrini* infection. This supports the hypothesis that cell proliferation is promoted by inflammation-mediated DNA damage [112]. It has been shown that the responses of a fibroblast cell line, National Institute of Health (NIH)-3T3, to ES product(s) released from *O. viverrini* in noncontact coculture conditions, increased markedly in cell proliferation [113]. The flukes have also been found to induce overexpression of mRNA encoding growth-promoting protein such as transforming growth factor [114].

During 1–3 weeks posttreatment with praziquantel reduces bile duct inflammation via TLR-2-mediated pathway as well as oxidative and nitrative DNA damage which may help prevent carcinogenesis [115]. However, subsequent studies suggest that short-term praziquantel treatment (6 h) induces inflammation and increases expression of apoptosis-related genes [116]. Expression levels return to normal after 24 h posttreatment. Praziquantel treatment clearly helps to inhibit iNOS-dependent DNA damage via elimination of parasites, but it also acts as a potential anti-inflammatory effect.

Currently it has been proposed that liver fluke-associated CCA development consists of four stages similar to those in colon cancer [107,117,118]. First, exposure to risk factors (opisthorchiasis) results in chronic inflammation and/or cholestasis through biochemical and/or mechanical processes. Second, genotoxic events may be the result of NO overproduction by inflammatory cells. Oxygen radicals and NO can inactivate and injure lipids, proteins, DNA, and carbohydrates within the cells, through direct oxidative damage. Free radicals caused by inflammatory cytokines can induce many subsequent events, for instance, gene expression activation, alteration of detoxification gene expression, and activation of carcinogen metabolism ultimately to form carcinogens, which increase DNA damage. Moreover, overproduction of NO has been reported to down-regulated lymphocyte proliferation. Third, apoptosis and dysregulation of DNA repair as well as genotoxic events with DNA damage usually result in either DNA mismatch repair or if the damage is not repaired, cell death will follow through apoptosis. Finally, histological changes to CCA may occur in a similar manner to that in the "hyperplasia–dysplasia–carcinoma" model, proposed for colon and gastric cancers [118].

38.1.9 Diagnosis of Opisthorchiasis

38.1.9.1 Parasitological Method

Demonstration of eggs in human feces, bile, or duodenal fluid and/or the recovery of adult flukes during transhepatic stent implantation or directly from the liver in postmortem autopsies are considered the gold standard diagnosis of *O. viverrini* infection [119]. However, fecal examination is the routine method for diagnosing *O. viverrini* infection due to their ease, noninvasiveness, and low cost. Fecal examination techniques that are used commonly include the formalin-ether concentration technique (FECT), the modified quick Kato–Katz thick smear, and Stoll's dilution egg count technique [67,120,121]. However, the sensitivity and specificity of fecal exam methods vary by the method of examination and the experience of the microscopist. The diagnostic value of these methods lies in the ability to detect light infections, for example, from individuals treated recently with praziquantel. Repeated stool examinations are needed, however, to improve diagnostic sensitivity. Even with repeated stool examination using a highly standardized method like FECT or Stoll's egg count, there can remain discrepancy between egg count and worm detection, that is, sufficient probability of a false-negative diagnosis. In an autopsy study [6], adult *O. viverrini* were recovered from 139 human livers (the gold standard). Subsequent fecal examinations of these same individuals showed that only 66.91% were positive for *O. viverrini* infection. It is commonly considered that conventional stool examination methods underestimate the actual prevalence by around 20% [6]. Moreover, diagnostic specificity of fecal examinations can also be problematic. Additionally, under light microscopy *O. viverrini* eggs are characterized by rough and thick egg shells, which are similar to several species of food-borne trematodes within the families Opisthorchiidae, Heterohyidae, and Lecithodendriidae that are endemic in the same areas. In Southeast Asia, the latter two families of trematodes are collectively referred to as minute intestinal flukes (MIFs), because of their small size compared to *O. viverrini* [9]. The Lecithodendriidae include *Phaneropsolus bonnei* and *Prostodendrium molenkampi* while the Heterohyidae include *Haplorchis taichui*, *Haplorchis pumilio*, and *Stellanchasmus falcatus* [122,123]. Within the heterophyids, *Metagonimus* spp and *Heterophyes* spp, *Haplorchis* spp, and *Pygidiopsis* spp occur in Laos and Korea [2]. The distinct resemblance of the morphology of MIF eggs with *O. viverrini* eggs raises the probability of a false-positive diagnosis by fecal exam method, with a concomitant lowering of diagnostic specificity. Indeed, in areas where *O. viverrini* coexists with heterophyid flukes including *Haplorchis* spp, the term "*O. viverrini* like egg" is sometimes used [124]. This dilemma clearly reflects the limitations of fecal examination methods particularly where diagnosis is based solely on the morphological of eggs as well as the urgent need for alternative diagnostic methods for *O. viverrini*.

38.1.9.2 Immunological Methods

Several types of serological tests have been developed over the last decade, with a view to provide a diagnostic assay with greater sensitivity and specificity than the fecal examination methods discussed above, and include the intradermal test, immunoelectrophoresis (IEP), indirect hemagglutination assay, indirect fluorescent antibody test, and indirect enzyme-linked immunosorbent assay (indirect ELISA) [68]. Of these, the indirect ELISA is favored for its ease and the proposed increase in diagnostic sensitivity and specificity gained through using an antibody-based system. However, serological tests have failed to achieve a higher level of diagnosis, mainly because of the complex composition of the crude antigen extracts used as detection methods. For instance, crude adult somatic extracts from *O. viverrini* have been used for IEP and ELISA, both of which are estimated to have a sensitivity of 76–100% for IgG, and markedly lower sensitivities for IgA (63%) and IgE (74%) [125,126]. Additionally, crude forms of ES antigen extracts from adult *O. viverrini* have also been used for diagnosis. An 89 kDa protein, among the most prominent bands in ES antigen of *O. viverrini*, has been shown to have immunodiagnostic potential, despite cross-reacting with *C. sinensis* [127]. Another serological-based method has been developed to act as a more diagnostically sensitive and specific test for *O. viverrini* infection. In this case, monoclonal antibody-based affinity-purified oval egg antigen showed high diagnostic specificity and sensitivity in the detection of serum antibody against *O. viverrini* by MAb-based dot-ELISA [126]. Attempts to utilize recombinant antigen for an IgG antibody detection kit in an ELISA, such as recombinant *O. viverrini* egg shall protein (rOVESP), have also shown increased diagnostic sensitivity and specificity over fecal examination methods [128].

A major limitation of antibody-based detection methods is the inability to differentiate past and present infections with *O. viverrini*, due to the long half-life of the antibody response to *O. viverrini*. This can persist in infected and non-infected hosts for months or years even after curative treatment [128,129]. With increasing availability of recombinant antigens the possibility increases that the serological tests may be improved to differentiate past and present infections with *O. viverrini*.

Recent molecular genetic studies of *O. viverrini* have identified a range of genes encoding different types of proteins that are stage specific. Some of these may be potential candidates for recombinant antigens for serodiagnosis, as well as to study their biological functions [130–133], Ruangsittichai et al. [128]. The *O. viverrini* asparaginyl endopeptidase (*Ov*-AEP-1) has been characterized by an immunoscreening method using infected human sera. The *Ov*-AEP-1 is characteristic of the C13 cysteine protease family. Recombinant *Ov*-AEP-1 has been produced in bacterial host cells and has been tested with infected human serum by Western blot analysis. Recombinant *Ov*-AEP-1 has been demonstrated to be recognized by serum antibodies from patients with opisthorchiasis but not to other helminth infections, yielding sensitivity and specificity of 85% and 100%, respectively. Positive and negative predictive values are 100% and 67%, respectively [132]. The availability of these newly described and more recombinant antigens will clearly improve the serodiagnostic performance in the near future.

Antigen-based detection systems have been shown to be informative with regard to infection status that is current. Several monoclonal antibodies to *O. viverrini* antigens have been produced against different liver fluke proteins, such as a 16 kDa tegumental protein, 89 kDa glycoprotein (excretory–secretory antigen), and a 90 kDa somatic protein. Monoclonal antibodies raised against these proteins have been used for coproantigen detection with varying success. The diagnostic sensitivity ranges from between 31% and 57% and diagnostic specificity has been shown to range between 70% and 100% [134,135]. Employing monoclonal antibody specific to 89 kDa antigen of *O. viverrini* in a monoclonal antibody based-ELISA (Mab-ELISA) for antigen detection in stool specimen diagnosis by the Mab-ELISA was found to be more sensitive than the conventional fecal examination particularly for the—lightly (EPG <500) infected cases [135]. This finding is in concordance with results from an autopsy study which demonstrated that fecal eggs were not detectable in most people harboring <20 worms in the liver [6]. It may be possible that Mab-based systems can offer an increase in diagnostic sensitivity, enough to detect secretory product and/or tegumental components from a few adult worms, even when eggs are not found by stool examination. Alternatively, detached tegumentary component of the worm present in fecal specimen may serve as source of DNA. Although this is a promising approach, there is need to sustain the Mab clone(s) by using conventional methods is expensive and may not be reliable. Production of recombinant Mab by using a genetic engineering approach is desirable and should help to resolve this problem in the future. Recently, new clones of Mab against ES product of *O. viverrini* which recognize 45–110 kDa have been established and developed for the Mab-ELISA for fecal antigen (Sithithaworn and coworkers, unpublished). This latest Mab-ELISA gave a sensitivity of 69.1% and specificity of 39.1%. Currently, tests are being conducted to determine whether antigen detection is in concordance with polymerase chain reaction (PCR)-based diagnosis. This analysis may shed insight into the development of alternative diagnostic methods in addition to the classical fecal examination. These alternative methods should be included as components of the gold standard method, in particular where cases are negative by conventional diagnosis.

38.1.9.3 Molecular Methods

In the case of DNA-based diagnosis, a probe designed from repetitive DNA elements of 334 base pairs has been used for the detection of egg DNA in fecal specimen [127]. Subsequently, a PCR-based detection of *O. viverrini* in human stools (based on a pair of primers complementary to the same target DNA) has been used to detect egg DNA [136,137]. From these studies, specificity of 97.8% and sensitivity of 100% were achieved in moderate-to-severe infections (stool EPG levels of >1000). However, the test was less sensitive (68% in light infections (<200 EPG)). Subsequent application of this PCR-based method for the detection of *O. viverrini* DNA in stool samples from Lao PDR yielded sensitivity of ~50% in samples with egg count >1000 EPG [138]. It was proposed that the presence of a PCR inhibitor in the fecal specimen could affect the sensitivity of diagnosis. Recent improvement of fecal DNA preparations by cetyltrimethylammonium bromide (CTAB) which removes PCR inhibitors was used. When compared with a commercial stool kit method, the sensitivity of the new test was 79.3% compared with 44.8% in the previous method [139]. In this study, there were several cases where the PCR test was positive whereas microscope-based stool examination was negative, which indicates superior performance in cases of light infections. A new *O. viverrini*-specific primer pair has recently been established to use in PCR tests. Results showed that the sensitivity of the test was 10^{-12} ng of adult worm DNA and the detection of three metacercariae in fish samples [140]. Additionally, species-specific PCR tests to identify different species of liver fluke are now available for *O. viverrini* [136,141] and *O. viverrini*, *C. sinensis* [142], and *O. felineus* [143]. If these tests are simplified and are reliable diagnostically to detect these infections, they will play significant roles in the assessment of cure, reinfection, and characterization of the endemic range of these food-borne trematodes in East Asia [144,145]. Furthermore, molecular diagnosis is applicable for food safety inspection in detecting other life stages of parasites in aquaculture products [140,146]. Moreover, it has been shown that based on molecular diagnostic approach, identification of coexisting FBT for accurate diagnosis in multiple infections is feasible [147,148].

38.2 METHODS

Here, we describe the methods of detection of DNA in stool samples using PCR technology with details on preparation of DNA template from stool samples and PCR amplification conditions.

38.2.1 Sample Preparation

38.2.1.1 Preparation of *O. viverrini* DNA Using Commercial Kit

DNA is prepared from stool sample using QIAamp® DNA Stool Mini Kit (Qiagen, Germany). Briefly, 100–500 mg of fecal samples is mixed with 1 mLl normal sterile saline (NSS) and 200 μL of ethyl acetate, and the mixture centrifuged at 8000 rpm for 5 min. The pellet was washed with 1.5 mL of NSS, mixed by vortex vigorously and incubated in 500 μL of 0.5 N NaOH for 60 min at room temperature. To break the eggs, the mixture was then autoclaved for 60 min (121°C), mixed and centrifuged for 5 min at 8000 rpm. Samples were evaluated for the presence of broken *O. viverrini* eggs after autoclaving. The supernatant was transferred to a 1.5 mL microcentrifuge tube. A volume of 200 μL of 3 M sodium acetate was added to the supernatant and pH was adjusted to 7–8. Half of one InhibitEX tablet (QIAamp DNA Stool Mini Kit, Qiagen) was added to reduce potential inhibition and the mixture was centrifuged for 5 min at 14,000 rpm. The supernatant was transferred into a new tube, and the pellet was discarded. Ten microliters of Proteinase K was added to the supernatant. Following the addition of 200 μL Buffer AL and mixing, the suspension was incubated at 70°C for 10 min. Finally, 200 μL of absolute ethanol was added, and the DNA solution cleaned and eluded by using a QIAamp DNAStool Mini Kit (Qiagen) according to the instructions of the manufacturers. Five microliters was used in the PCR [138].

38.2.1.2 Preparation of *O. viverrini* DNA Using CTAB Method

DNA from *O. viverrini* can also be prepared from stool samples using CTAB. The purpose of this method is to isolate DNA from *O. viverrini* eggs in human stool samples. The following manual protocol is modified from the DNA extraction method described by Ausubel (2002). Briefly, ~500 mg of fecal samples are mixed with 4 mL of sterilized normal saline and 400 μL of ethyl acetate. The mixture is shaken vigorously and then centrifuging at 4000 rpm for 10 min. The pellet is incubated in 1 mL of 0.5 N NaOH for 60 min at room temperature, autoclaved for 60 min at 121°C, mixed, and then centrifuged for 10 min at 4000 rpm. The supernatant is transferred to a 1.5 mL microcentrifuge tube, in which 56 μL of 10× Tris-ethylenediamine tetraacetic acid (EDTA) (TE; pH 8.0) was added and the pH adjusted to 7–8. Thirty microliters of 10% sodium dodecyl sulfate and 3 μL of 20 mg/mL proteinase K (60 μg) were added to the supernatant and the mixture was incubated at 65°C for 1 h. One hundred microliters of 5 M NaCl and 80 μL of CTAB—NaCl (0.7 M NaCl and 0.1 vol of CTAB) were added, and the mixture was incubated at 65°C for 30 min. The mixture was extracted twice with an equal volume of chloroform–isoamyl alcohol (24:1), followed by phenol–chloroform–isoamyl alcohol (25:24:1). DNA was precipitated by adding 0.6 vol of isopropanol and pelleted by centrifugation, washed once with 70% ethanol, dried, and dissolved in 50 μL of 1 × TE buffer. DNA concentrations were measured by a spectronanometer at OD260 [139].

38.2.2 Detection Procedures

Here we describe successful PCR detection procedures for human opisthorchiasis detection using two sets of specific primers, pOV-6F, and -R [136], and Trem25-F, and OV25-4R primers [140].

38.2.2.1 pOV-6 Primers

The pOV-6 primers were originally designed for the detection of *O. viverrini* in experimental infected hamsters. The assay has a specificity of 97.8%, a sensitivity of 100% in moderate–severe infections (more than 1000 EPG feces) and a maximum sensitivity of 68.2% in the detection of light infections (EPG <1000) with a detection limit of 200 eggs. With reference to FECT ($n = 58$) we have used these primers to detect human opisthorchiasis resulting in overall sensitivity of the modified PCR of 79.3% and specificity of 71.4%. The positive and negative predictive values were 88.5% and 55.6%, respectively.

The PCR procedures are as follows; 2 μL of the final DNA solution in TE buffer (see Section 38.2.1.2) was used in the PCR. *O. viverrini*-specific primers OV-6 (5′-CTGAATCTC TCGTTTGTTCA) and 5′-GTTCCAGGTGAGTCTCTCTA) were used for PCR. The primers were designed to amplify a repetitive DNA fragment specific for *O. viverrini* which showed no significant homology to other parasites. The amplification yielded a specific product with a size of 330 bp.

PCR assays were carried out in a final volume of 25 μL containing PCR buffer (20 mM Tris-HCl pH 8.4, 50 mM KCl, 1.5 mM $MgCl_2$), 200 μM each of deoxynucleoside triphosphate and 1 μM *Taq* DNA polymerase. PCR assays were performed in a DNA thermal cycler (MJ PTC-200) and adjusted to the following cycling conditions: an initial denaturation step at 94°C for 5 min, denaturation at 94°C for 30 s, annealing at 52°C for 30 s, and extension at 72°C for 45 s (30 cycles). Amplified products were analyzed by electrophoresis in a 2% agarose gel with 1× Tris-borate-EDTA buffer pH 8.0.

38.2.2.2 Trem25-F and OV25–4R Primers

Trem25-F and OV25-4R primers were developed by Parvathi et al. [140] based on a DNA fragment amplified with random amplification of polymorphic DNA (RAPD)-PCR. Two microliters of the final DNA solution in TE buffer was used in the PCR. Ten picomoles each of primers (Trem25-F: 5′-ACAGGAAGTGGAACCGTGTC-3′ and OV25–4R: 5′-AATGAACGGAAATCGTGACC-3′) mixed with 1 U Taq DNA polymerase, dNTP, 10× PCR buffer (20 mM Tris-HCl pH 8.4, 50 mM KCl, and 1.5 mM $MgCl_2$), and dH_2O were added. PCR amplification reaction was performed with the following cycling program: 1 cycle of 95°C for 5 min; 35 cycles of 94°C for 1 min, 60°C for 1 min, and 72°C for 1 min; 1 cycle of 72°C for 5 min. PCR product was subjected to agarose electrophoresis using 1.6% agarose gel containing 0.5 μg/mL ethidium bromide in 1× tris base-boric acid-ethylenediamine tetraacetic acid (TBE) buffer and a specific 229 bp product was visualized using a UV transluminator.

38.3 CONCLUSIONS AND FUTURE PERSPECTIVES

O. viverrini has been classified as type I carcinogen. *O. viverrini* is a zoonotic FBT that causes public health problem because of the massive numbers of people infected and induces serious morbidities such as hepatobiliary diseases and CCA. Although infections are identified throughout Southeast Asia, the hot spot is Northeast Thailand, where high prevalence coexists with a high incidence of CCA. It is likely that a similar situation exists in the endemic region of the liver fluke in nearby countries; data on incidence of CCA in Lao PDR and Cambodia are not available.

O. viverrini has a complex life cycle and requires three hosts to complete its life cycle. Human infection occurs through consumption of the infective metacerariae contained in raw and undercooked fresh water fish. Once eggs reach water, they are ingested by snail intermediate hosts in the genus *Bithynia*. The parasites multiply asexually in the snail and release the free swimming cercariae into water which subsequently enter the fish intermediate hosts. Common fish intermediate hosts are several species of cyprinid fish. Naturally bred fish species as well as cultured freshwater fish, all can act as the source of human and animal infection. Currently, there is little information on the role that fish and snail species play in transmission and genetic diversity in natural populations of *O. viverrini* within and between geographical regions and river wetlands. Comprehensive studies should be undertaken to test the hypothesis that *O. viverrini* is species complex and in particular to accurately define the number of species that occur within the complex, their geographical ranges, and species of snail and fish hosts in each cryptic *O. viverrini* species range. Furthermore, comprehensive genetic/molecular population structure studies need to be undertaken within each cryptic species. These studies are necessary and critical to detect and define the existence of unique pathogenic and nonpathogenic genotypes of *O. viverrini sensu lato* and whether these genotypes correlate with clinical manifestations and epidemiology of opisthorchiasis.

Although pathogenesis of opisthorchiasis is complex, recent studies have revealed that the pathogenesis feature in opisthorchiasis is largely a consequence of immunomodulation during the acute and chronic phase of infection. Long-term liver fluke infection gives rise to chronic inflammation of bile duct epithelium which initiates tumorigenesis by oxidative and nitrative DNA damage. To date no reliable biomarker for screening is available thus the best screening method is by US and other imaging radiology.

Diagnosis of opisthorchiasis is essential for treatment, parasite surveillance, and control. The standard conventional diagnosis is the parasitological method of demonstrating eggs in feces or bile specimens. Several immunological diagnostic methods by antibody or antigen detection are available but careful interpretation of the test result is required. Molecular diagnostic methods have been introduced as an alternative method for diagnosis with the advantage of high specificity and sensitivity. Diagnosis by detection of parasite DNA in fecal specimens is a promising method since it identifies positive tests even in a considerable number of egg-negative specimens and equally sensitive in egg-positive cases. The ability to discover positive cases in apparently egg negative cases as determined by the conventional method, has demonstrated that this PCR-based diagnostic approach is useful for both individual and mass screening diagnosis. It also provides an additional tool for evaluation of chemotherapy programs. With fish aquaculture playing a vital role in international aquaculture trade, this PCR-based method serves as a tool for fish inspection for metacercariae, which is a significant issue in food safety. The disadvantage of molecular diagnostic methods for the detection of parasite DNA is the requirement of specific molecular instrumentation, and thus the development of the detection system into a simplified kit format for routine field and laboratory use is required.

The target for long-term opisthorchis and CCA control should be directed toward the unexposed young generation through specifically designed and up-to-date programs in public health education. It is anticipated that such approaches will reduce the risk of *O. vivierrini* infection and CCA development in this generation hence going a long way to eliminate these diseases. With respect to a short-term control, the older generation who are already exposed to *O. viveriini* infection and potentially at risk of CCA development, early diagnostic markers are needed to identify the high-risk individuals for appropriate treatments.

Reducing the level of infection and reinfection will reduce morbidity and the risk of CCA development in the long term. Therefore, CCA can be prevented by appropriate control strategies, based on accurate information from integrated multidisciplinary studies (e.g., molecular/genetic/biochemical, etc.) in endemic areas in conjunction with up-to-date health education, especially focusing on the young generation. We hope that the ultimate outcome will be eliminating the type I carcinogen, *O. viverrini* sensu lato and the diseases it causes, opisthorchiasis and CCA. In a worldwide context, a similar multidisciplinary approach can be used to eliminate diseases caused by other fish-borne liver flukes, for example, *O. felineus* and *C. sinensis*.

REFERENCES

1. Hotez, P.J. et al. Rescuing the bottom billion through control of neglected tropical diseases. *Lancet* 373, 1570, 2009.
2. Chai, J.Y. et al. Fish-borne parasitic zoonoses: Status and issues. *Int. J. Parasitol.* 35, 1233, 2005.
3. Liu, W.Q. et al. Comparison of ancient and modern *Clonorchis sinensis* based on ITS1 and ITS2 sequences. *Acta Trop.* 101, 91, 2007.
4. Kaewkes, S. Taxonomy and biology of liver flukes. *Acta Trop.* 88, 177, 2003.
5. Beaver, P.C., Jung, R.C., and Cupp, E.W. *Clinical Parasitology*, 9th edn. Lea & Febiger, Philadelphia, 406pp., 1984.
6. Sithithaworn, P. et al. Relationship between faecal egg count and worm burden of *Opisthorchis viverrini* in human autopsy cases. *Parasitology* 102(Part 2), 277, 1991.
7. Laoprom, N. et al. Microsatellite loci in the carcinogenic liver fluke, *Opisthorchis viverrini* and their application as population genetic markers. *Infect. Genet. Evol.* 10, 146, 2010.
8. Saijuntha, W. et al. Evidence of a species complex within the food-borne trematode *Opisthorchis viverrini* and possible coevolution with their first intermediate hosts. *Int. J. Parasitol.* 37, 695, 2007.
9. Kaewkes, S. et al. Comparative studies on the morphology of the eggs of *Opisthorchis viverrini* and lecithodendriid trematodes. *Southeast Asian J. Trop. Med. Public Health* 22, 623, 1991.
10. Ditrich, O., Giboda, M., Scholz, T., and Beer, S.A. Comparative morphology of eggs of the Haplorchiinae (Trematoda: Heterophyidae) and some other medically important heterophyid and opisthorchid flukes. *Folia Parasitol.* 39, 123, 1992.
11. Elkins, D.B. et al. *Opisthorchis viverrini*: Relationships between egg counts, worms recovered and antibody levels within an endemic community in Northeast Thailand. *Parasitology* 102(Part 2), 283, 1991.
12. Phongsasakulchoti, P. et al. Emergence of *Opisthorchis viverrini* cercariae from naturally infected *Bithynia (Digoniostoma) siamensis goniomphalos*. *Southeast Asian J. Trop. Med. Public Health* 36(Suppl 4), 189, 2005.
13. Attwood, H.D., and Chou, S.T. The longevity of *Clonorchis sinensis*. *Pathology* 10, 153, 1978.
14. Sithithaworn, P. et al. Genetic variation in *Opisthorchis viverrini* (Trematoda: Opisthorchiidae) from Northeast Thailand and Laos PDR based on random amplified polymorphic DNA analyses. *Parasitol. Res.* 100, 613, 2007.
15. Laoprom, N. et al. Biological variation within *Opisthorchis viverrini* sensu lato in Thailand and Lao PDR. *J. Parasitol.* 95, 1307, 2009.
16. He, Y.X. et al. Strain complex of *Schistosoma japonicum* in the mainland of China. *Southeast Asian J. Trop. Med. Public Health* 25, 232, 1994.
17. He, Y.X. et al. Characteristics of different isolates of *Schistosoma japonicum* from China in the final hosts. *Southeast Asian J. Trop. Med. Public Health* 22, 240, 1991.
18. Park, G.M. Genetic comparison of liver flukes, *Clonorchis sinensis* and *Opisthorchis viverrini*, based on rDNA and mtDNA gene sequences. *Parasitol. Res.* 100, 351, 2007.
19. Lee, S.U. and Huh, S. Variation of nuclear and mitochondrial DNAs in Korean and Chinese isolates of *Clonorchis sinensis*. *Korean J. Parasitol.* 42, 145, 2004.
20. Shekhovtsov, S.V. et al. A novel nuclear marker, Pm-int9, for phylogenetic studies of *Opisthorchis felineus*, *Opisthorchis viverrini*, and *Clonorchis sinensis* (Opisthorchiidae, Trematoda). *Parasitol. Res.*, 106, 293, 2009.
21. Brockelman, W.Y. et al. Field studies on the transmission of the human liver fluke, *Opisthorchis viverrini*, in northeast Thailand: Population changes of the snail intermediate host. *Int. J. Parasitol.* 16, 545, 1986.
22. Vichasri, S. et al. *Opisthorchis viverrini*: Intensity and rates of infection in cyprinoid fish from an endemic focus in Northeast Thailand. *Southeast Asian J. Trop. Med. Public Health* 13, 138, 1982.
23. Haas, W. et al. *Opisthorchis viverrini*: Finding and recognition of the fish host by the cercariae. *Exp. Parasitol.* 71, 422, 1990.
24. Sithithaworn, P. et al. Seasonal variation of *Opisthorchis viverrini* infection in cyprinoid fish in north-east Thailand: Implications for parasite control and food safety. *Bull. World Health Organ* 75, 125, 1997.
25. Hinz, E. et al. Opisthorchiasis control in Northeast Thailand: Proposal for a new approach. *Appl. Parasitol.* 35, 118, 1994.
26. Sithithaworn, P., and Haswell-Elkins, M. Epidemiology of *Opisthorchis viverrini*. *Acta Trop.* 88, 187, 2003.
27. Srisawangwong, T. et al. Metacercariae isolated from cyprinoid fishes in Khon Kaen District by digestion technic. *Southeast Asian J. Trop. Med. Public Health* 28(Suppl 1), 224, 1997.
28. Waikagul, J. *Opisthorchis viverrini* metacercaria in Thai freshwater fish. *Southeast Asian J. Trop. Med. Public Health* 29, 324, 1998.
29. Hinz, E. Intestinal helminths in Bangkok stray dogs and their role in public health (author's transl). *Zentralbl Bakteriol. Mikrobiol. Hyg. [B]* 171, 79, 1980.
30. Scholz, T. et al. Helminth parasites of cats from the Vientiane province, Laos, as indicators of the occurrence of causative agents of human parasitoses. *Parasite* 10, 343, 2003.
31. Enes, J.E., Wages, A.J., Malone, J.B., and Teasana, S. Prevalence of *Opisthorchis viverrini* infection in the canine and feline in three villages, Khon kaen Province, Northeastern Thailand. *Southeast Asian J. Trop. Med. Public Health* 41, 36, 2010.
32. WHO. *Report of Joint WHO/FAO Workshop on Food-Borne Trematode Infections in Asia*, Hanoi, p. 1, 2004.
33. Rim, H.J. Clonorchiasis: An update. *J. Helminthol.* 79, 269, 2005.
34. Keiser, J., and Utzinger, J. Emerging food-borne trematodiasis. *Emerg. Infect. Dis.* 11, 1507, 2005.
35. Yossepowitch, O. et al. Opisthorchiasis from imported raw fish. *Emerg. Infect. Dis.* 10, 2122, 2004.

36. Stauffer, W.M. et al. Biliary liver flukes (Opisthorchiasis and Clonorchiasis) in immigrants in the United States: Often subtle and diagnosed years after arrival. *J. Travel. Med.* 11, 157, 2004.
37. Armignacco, O. et al. Human illnesses caused by *Opisthorchis felineus* flukes, Italy. *Emerg. Infect. Dis.* 14, 1902, 2008.
38. Jongsuksuntigul, P., and Imsomboon, T. Opisthorchiasis control in Thailand. *Acta Trop.* 88, 229, 2003.
39. Rim, H.J. et al. Prevalence of intestinal parasite infections on a national scale among primary schoolchildren in Laos. *Parasitol. Res.* 91, 267, 2003.
40. WHO. Control of food-borne trematode infections. Report of a WHO Study Group. *World Health Organ. Tech. Rep. Ser.* 849, 1, 1995.
41. Zhu, Y.C. et al. Application of dipstick dye immunoassay (DDIA) kit for the diagnosis of schistosomiasis mekongi. *Acta Trop.* 96, 137, 2005.
42. Sadun CMEH. Studies on the epidemiology of the human intestinal fluke, *Fasciolopsis buski* (Lankester) in central Thailand. *Am. J. Trop. Med. Hyg.* 2, 1070, 1953.
43. Sithithaworn, P. et al. Epidemiology of food-borne trematodes and other parasite infections in a fishing community on the Nam Ngum reservoir, Lao PDR. *Southeast Asian J. Trop. Med. Public Health* 37, 1083, 2006.
44. Upatham, E.S. et al. Rate of re-infection by *Opisthorchis viverrini* in an endemic Northeast Thai community after chemotherapy. *Int. J. Parasitol.* 18, 643, 1988.
45. Sornmani, S. et al. A pilot project for controlling *O. viverrini* infection in Nong Wai, Northeast Thailand, by applying praziquantel and other measures. *Arzneimittelforschung* 34, 1231, 1984.
46. Saowakontha, S. et al. Field trials in the control of *Opisthorchis viverrini* with an integrated programme in endemic areas of Northeast Thailand. *Parasitology* 106(Pt 3), 283, 1993.
47. Lan-Anh, N.T. et al. Animal reservoir hosts and fish-borne zoonotic trematode infections on fish farms, Vietnam. *Emerg. Infect. Dis.* 15, 540, 2009.
48. Upatham, E.S. et al. Morbidity in relation to intensity of infection in *Opisthorchiasis viverrini*: Study of a community in Khon Kaen, Thailand. *Am. J. Trop. Med. Hyg.* 31, 1156, 1982.
49. Pungpak, S. et al. Ultrasonographic study of the biliary system in opisthorchiasis patients after treatment with praziquantel. *Southeast Asian J. Trop. Med. Public Health* 20, 157, 1989.
50. Upatham, E.S. et al. Relationship between prevalence and intensity of *Opisthorchis viverrini* infection, and clinical symptoms and signs in a rural community in North-east Thailand. *Bull. World Health Organ* 62, 451, 1984.
51. Mairiang, E. et al. Ultrasound screening for *Opisthorchis viverrini*-associated cholangiocarcinomas: Experience in an endemic area. *Asian Pac. J. Cancer Prev.* 7, 431, 2006.
52. Mairiang, E., and Mairiang, P. Clinical manifestation of opisthorchiasis and treatment. *Acta Trop.* 88, 221, 2003.
53. Mairiang, E. et al. Relationship between intensity of *Opisthorchis viverrini* infection and hepatobiliary disease detected by ultrasonography. *J. Gastroenterol. Hepatol.* 7, 17, 1992.
54. Choi, D. et al. Sonographic findings of active *Clonorchis sinensis* infection. *J. Clin. Ultrasound* 32,17, 2004.
55. Dhiensiri, T. et al. Roentgenographically controlled healing of gall bladder lesions in opisthorchiasis after praziquantel treatment. *Arzneimittelforschung* 34, 1175, 1984.
56. Haswell-Elkins, M.R. et al. Cross-sectional study of *Opisthorchis viverrini* infection and cholangiocarcinoma in communities within a high-risk area in Northeast Thailand. *Int. J. Cancer* 59, 505, 1994.
57. Elkins, D.B. et al. Cross-sectional patterns of hepatobiliary abnormalities and possible precursor conditions of cholangiocarcinoma associated with *Opisthorchis viverrini* infection in humans. *Am. J. Trop. Med. Hyg.* 55, 295, 1996.
58. Mairiang, E. et al. Reversal of biliary tract abnormalities associated with *Opisthorchis viverrini* infection following praziquantel treatment. *Trans. R Soc. Trop. Med. Hyg.* 87, 194, 1993.
59. Bhamarapravati, N. et al. Liver changes in hamsters infected with a liver fluke of man, *Opisthorchis viverrini. Am. J. Trop. Med. Hyg.* 27, 787, 1978.
60. Hou, P.C. The pathology of *Clonorchis sinensis* infestation of the liver. *J. Pathol. Bacteriol.* 70, 53, 1955.
61. Koompirochana, C. et al. Opisthorchiasis: A clinicopathologic study of 154 autopsy cases. *Sotheast Asian J. Trop. Med. Public Health* 9, 60, 1978.
62. Riganti, M. et al. Human pathology of *Opisthorchis viverrini* infection: A comparison of adults and children. *Southeast Asian J. Trop. Med. Public Health* 20, 95, 1989.
63. Tansurat, P. Opisthorchiasis. In: Marcial-Rojas, R.A. (ed.), *Pathology of Protozoal and Helminthic Diseases*. Williams & Wilkins, Baltimore, 536pp., 1971.
64. Sripa, B. et al. Liver fluke induces cholangiocarcinoma. *PLoS Med.* 4, e201, 2007.
65. Riganti, M. et al. *Opisthorchis viverrini* eggs and adult flukes as nidus and composition of gallstones. *Southeast Asian J. Trop. Med. Public Health* 19, 633, 1988.
66. Sripa, B. et al. Opisthorchiasis-associated biliary stones: Light and scanning electron microscopic study. *World J. Gastroenterol.* 10, 3318, 2004.
67. Elkins, D.B. et al. A high frequency of hepatobiliary disease and suspected cholangiocarcinoma associated with heavy *Opisthorchis viverrini* infection in a small community in North-east Thailand. *Trans. R. Soc. Trop. Med. Hyg.* 84, 715, 1990.
68. Wongratanacheewin, S. et al. Immunology and molecular biology of *Opisthorchis viverrini* infection. *Acta Trop.* 88, 195, 2003.
69. Akai, P.S. et al. Serum antibody responses in opisthorchiasis. *Int. J. Parasitol.* 25, 971, 1995.
70. Akai, P.S. et al. Serum antibody response to *Opisthorchis viverrini* antigen as a marker for opisthorchiasis-associated cholangiocarcinoma. *Trans. R Soc. Trop. Med. Hyg.* 88, 471, 1994.
71. Itoh, M. et al. Association of antibodies to *Opisthorchis viverrini* with hepatobiliary disease in Northeastern Thailand. *Am. J. Trop. Med. Hyg.* 51, 424, 1994.
72. Wongratanacheewin, S., and Sirisinha, S. Analysis of *Opisthorchis viverrini* antigens: Physicochemical characterization and antigen localization. *Southeast Asian J. Trop. Med. Public Health* 18, 511, 1987.
73. Ruangkunaporn, Y. et al. Changes in serum antibodies to *Opisthorchis viverrini* in humans and hamsters following treatment of opisthorchiasis. *Asian Pac. J. Allergy Immunol.* 12, 83, 1994.
74. Suksumek, N. et al. TaqMan real-time PCR assay for specific detection of *Opisthorchis viverrini* DNA in Thai patients with hepatocellular carcinoma and cholangiocarcinoma. *Exp. Parasitol.* 119, 217, 2008.
75. Sripa, B., and Kaewkes, S. Localisation of parasite antigens and inflammatory responses in experimental opisthorchiasis. *Int. J. Parasitol.* 30, 735, 2000.

76. Pinlaor, S. et al. Hepatobiliary changes, antibody response, and alteration of liver enzymes in hamsters re-infected with *Opisthorchis viverrini*. *Exp. Parasitol.* 108, 32, 2004.
77. Flavell, D.J., and Flavell, S.U. *Opisthorchis viverrini*: Pathogenesis of infection in immunodeprived hamsters. *Parasite Immunol.* 8, 455, 1986.
78. Wongratanacheewin, S. et al. Immunodepression in hamsters experimentally infected with *Opisthorchis viverrini*. *J. Helminthol.* 61, 151, 1987.
79. Thamavit, W. et al. Effects of dimethylnitrosamine on induction of cholangiocarcinoma in *Opisthorchis viverrini*-infected Syrian golden hamsters. *Cancer Res.* 38, 4634, 1978.
80. Haswell-Elkins, M.R. et al. Distribution patterns of *Opisthorchis viverrini* within a human community. *Parasitology* 103(Part 1), 97, 1991.
81. Flavell, D.J. et al. *Opisthorchis viverrini*: Partial success in adoptively transferring immunity with spleen cells and serum in the hamster. *J. Helminthol.* 54, 191, 1980.
82. Sirisinha, S. et al. Attempts to induce protective immunity in hamsters against infection by a liver fluke of man (*Opisthorchis viverrini*). *Parasitology* 86(Pt 1), 127, 1983.
83. Wongratanacheewin, S. et al. Effect of praziquantel treatment on antibody levels and lymphoproliferative responses in patients with opisthorchiasis. *Southeast Asian J. Trop. Med. Public Health* 19, 109, 1988.
84. Sirisinha, S. et al. Humoral immune responses in hamsters infected with *Opisthorchis viverrini*. *Southeast Asian J. Trop. Med. Public Health* 14, 243, 1983.
85. Sirisinha, S., and Wongratanacheewin, S. Immunization of hamsters against *Opisthorchis viverrini* infection. *Southeast Asian J. Trop. Med. Public Health* 17, 567, 1986.
86. Jittimanee, J. et al. Cytokine expression in hamsters experimentally infected with *Opisthorchis viverrini*. *Parasite Immunol.* 29, 159, 2007.
87. Sripa, B. et al. Advanced periductal fibrosis from infection with the carcinogenic human liver fluke *Opisthorchis viverrini* correlates with elevated levels of interleukin-6. *Hepatology* 50, 1273, 2009.
88. IARC. *IARC Monographs on the Evaluation of Carcinogenic Risks to Humans.* International Agency for Research on Cancer, World Health Organization, Lyon, France, vol. 121, p. 61, 1994.
89. Parkin, D.M. et al. Cholangiocarcinoma: Epidemiology, mechanisms of carcinogenesis and prevention. *Cancer Epidemiol. Biomarkers Prev.* 2, 537, 1993.
90. Srivatanakul, P. et al. Epidemiology of liver cancer: An overview. *Asian Pac. J. Cancer. Prev.* 5, 118, 2004.
91. Lim, M.K. et al. *Clonorchis sinensis* infection and increasing risk of cholangiocarcinoma in the Republic of Korea. *Am. J. Trop. Med. Hyg.* 75, 93, 2006.
92. Honjo, S. et al. Genetic and environmental determinants of risk for cholangiocarcinoma via *Opisthorchis viverrini* in a densely infested area in Nakhon Phanom, Northeast Thailand. *Int. J. Cancer* 117, 854, 2005.
93. Uttaravichien, T. et al. Intrahepatic cholangiocarcinoma in Thailand. *J. Hepatobiliary Pancreat. Surg.* 6, 128, 1999.
94. Uttaravichien, T., and Buddhisawasdi, V. Experience of non-jaundiced cholangiocarcinoma. *Hepatogastroenterology* 37, 608, 1990.
95. Pairojkul, C. et al. Multistage carcinogenesis of liver-fluke-associated cholangiocarcinoma in Thailand. *Princess Takamatsu Symp.* 22, 77, 1991.
96. Hou, P.C. The relationship between primary carinoma of the liver and infestation with *Clonorchis sinesis*. *J. Pathol. Bact.* 72, 239, 1956.
97. Haswell-Elkins, M.R. et al. Immune responsiveness and parasite-specific antibody levels in human hepatobiliary disease associated with *Opisthorchis viverrini* infection. *Clin. Exp. Immunol.* 84, 213, 1991.
98. Ohshima, H., and Bartsch, H. Chronic infections and inflammatory processes as cancer risk factors: Possible role of nitric oxide in carcinogenesis. *Mutat. Res.* 305, 253, 1994.
99. Kawanishi, S. et al. Oxidative and nitrative DNA damage in animals and patients with inflammatory diseases in relation to inflammation-related carcinogenesis. *Biol. Chem.* 387, 365, 2006.
100. Ma, N. et al. Accumulation of 8-nitroguanine in human gastric epithelium induced by *Helicobacter pylori* infection. *Biochem. Biophys. Res. Commun.* 319, 506, 2004.
101. Pinlaor, S. et al. Nitrative and oxidative DNA damage in intrahepatic cholangiocarcinoma patients in relation to tumor invasion. *World J. Gastroenterol.* 11, 4644, 2005.
102. Srivatanakul, P. et al. *Opisthorchis viverrini* infestation and endogenous nitrosamines as risk factors for cholangiocarcinoma in Thailand. *Int. J. Cancer* 48,821, 1991.
103. Haswell-Elkins, M.R. et al. Liver fluke infection and cholangiocarcinoma: Model of endogenous nitric oxide and extragastric nitrosation in human carcinogenesis. *Mutat. Res.* 305, 241, 1994.
104. Satarug, S. et al. Thiocyanate-independent nitrosation in humans with carcinogenic parasite infection. *Carcinogenesis* 17,1075, 1996.
105. Satarug, S. et al. Relationships between the synthesis of *N*-nitrosodimethylamine and immune responses to chronic infection with the carcinogenic parasite, *Opisthorchis viverrini*, in men. *Carcinogenesis* 19, 485, 1998.
106. Satarug, S. et al. Induction of cytochrome P450 2A6 expression in humans by the carcinogenic parasite infection, *Opisthorchiasis viverrini*. *Cancer Epidemiol. Biomarkers. Prev.* 5, 795, 1996.
107. Ohshima, H. et al. Increased nitrosamine and nitrate biosynthesis mediated by nitric oxide synthase induced in hamsters infected with liver fluke (*Opisthorchis viverrini*). *Carcinogenesis* 15, 271, 1994.
108. Thanan, R. et al. Urinary 8-oxo-7,8-dihydro-2′-deoxyguanosine in patients with parasite infection and effect of antiparasitic drug in relation to cholangiocarcinogenesis. *Cancer Epidemiol. Biomarkers Prev.* 17, 518, 2008.
109. Dechakhamphu, S. et al. Lipid peroxidation and etheno DNA adducts in white blood cells of liver fluke-infected patients: Protection by plasma alpha-tocopherol and praziquantel. *Cancer Epidemiol. Biomarkers Prev.* 19, 310, 2009.
110. Dechakhamphu, S. et al. High excretion of etheno adducts in liver fluke-infected patients: Protection by praziquantel against DNA damage. *Cancer Epidemiol. Biomarker. Prev.* 17, 1658, 2008.
111. Pinlaor, S. et al. *Opisthorchis viverrini* antigen induces the expression of Toll-like receptor 2 in macrophage RAW cell line. *Int. J. Parasitol.* 35, 591, 2005.
112. Pinlaor, S. et al. Repeated infection with *Opisthorchis viverrini* induces accumulation of 8-nitroguanine and 8-oxo-7, 8-dihydro-2′-deoxyguanine in the bile duct of hamsters via inducible nitric oxide synthase. *Carcinogenesis* 25, 1535, 2004.
113. Thuwajit, C. et al. Increased cell proliferation of mouse fibroblast NIH-3T3 *in vitro* induced by excretory/secretory product(s) from *Opisthorchis viverrini*. *Parasitology* 129, 455, 2004.

114. Thuwajit, C. et al. Gene expression profiling defined pathways correlated with fibroblast cell proliferation induced by *Opisthorchis viverrini* excretory/secretory product. *World J. Gastroenterol.* 12, 3585, 2006.
115. Pinlaor, S. et al. iNOS-dependent DNA damage via NF-kappaB expression in hamsters infected with *Opisthorchis viverrini* and its suppression by the antihelminthic drug praziquantel. *Int. J. Cancer* 119, 1067, 2006.
116. Boonmars, T. et al. Apoptosis-related gene expression in hamster opisthorchiasis post praziquantel treatment. *Parasitol. Res.* 102, 447, 2008.
117. Maeda, H. Carcinogenesis via microbial infection. *Gan To Kagaku Ryoho* 25, 1474, 1998.
118. Holzinger, F. et al. Mechanisms of biliary carcinogenesis: A pathogenetic multi-stage cascade towards cholangiocarcinoma. *Ann. Oncol.* 10(Suppl 4), 122, 1999.
119. Sithithaworn, P., Yongvanit, P., Tesana, S., and Pairojkul, C. Liver flukes. In: Murrell, D., and Fried, B. (eds), *Parasitic Foodborne Zoonoses*. Springer-Verlag, Berlin, Germany, p. 1, 2008.
120. Hong, S.T. et al. The Kato–Katz method is reliable for diagnosis of *Clonorchis sinensis* infection. *Diagn. Microbiol. Infect. Dis.* 47, 345, 2003.
121. Viyanant, V. et al. A comparison of a modified quick-Kato technique and the Stoll dilution method for field examination for *Opisthorchis viverrini* eggs. *J. Helminthol.* 57, 191, 1983.
122. Radomyos, P., Bunnag, D., and Harinasuta, T. Worms recovered in stools following praziquantel treatment. *Drug Res.* 34, 1215, 1984.
123. Radomyos, P. et al. Multi-infection with helminths in adults from Northeast Thailand as determined by post-treatment fecal examination of adult worms. *Trop. Med. Parasitol.* 45, 133, 1994.
124. Chai, J.Y. et al. High prevalence of liver and intestinal fluke infections among residents of Savannakhet Province in Laos. *Korean J. Parasitol.* 45, 213, 2007.
125. Poopyruchpong, N. et al. Diagnosis of opisthorchiasis by enzyme-linked immunosorbent assay using partially purified antigens. *Asian Pac. J. Allergy. Immunol.* 8, 27, 1990.
126. Wongsaroj, T. et al. Affinity purified oval antigen for diagnosis of *Opisthorchiasis viverrini*. *Asian Pac. J. Allergy Immunol.* 19, 245, 2001.
127. Sirisinha, S. et al. Detection of *Opisthorchis viverrini* by monoclonal antibody-based ELISA and DNA hybridization. *Am. J. Trop. Med. Hyg.* 44, 140, 1991.
128. Ruangsittichai, J. et al. *Opisthorchis viverrini*: Identification of a glycine-tyrosine rich eggshell protein and its potential as a diagnostic tool for human opisthorchiasis. *Int. J. Parasitol.* 36, 1329, 2006.
129. Thammapalerd, N. et al. Detection of antibodies against *Opisthorchis viverrini* in patients before and after treatment with praziquantel. *Southeast Asian J. Trop. Med. Public Health* 19, 101, 1988.
130. Laha, T. et al. The bandit, a new DNA transposon from a hookworm—Possible horizontal genetic transfer between host and parasite. *PLoS Negl. Trop. Dis.* 1, e35, 2007.
131. Laha, T. et al. Gene discovery for the carcinogenic human liver fluke, *Opisthorchis viverrini*. *BMC Genomics* 8, 189, 2007.
132. Laha, T. et al. Asparaginyl endopeptidase from the carcinogenic liver fluke, *Opisthorchis viverrini*, and its potential for serodiagnosis. *Int. J. Infect. Dis.* 12, e49, 2008.
133. Suttiprapa, S. et al. Ov-APR-1, an aspartic protease from the carcinogenic liver fluke, *Opisthorchis viverrini*: Functional expression, immunolocalization and subsite specificity. *Int. J. Biochem. Cell Biol.*, 41, 1148, 2009.
134. Chaicumpa, W. et al. Specific monoclonal antibodies to *Opisthorchis viverrini*. *Int. J. Parasitol.* 21, 969, 1991.
135. Sirisinha, S. et al. Evaluation of a monoclonal antibody-based enzyme linked immunosorbent assay for the diagnosis of *Opisthorchis viverrini* infection in an endemic area. *Am. J. Trop. Med. Hyg.* 52, 521, 1995.
136. Wongratanacheewin, S. et al. Development of a PCR-based method for the detection of *Opisthorchis viverrini* in experimentally infected hamsters. *Parasitology* 122, 175, 2001.
137. Wongratanacheewin, S. et al. Detection of *Opisthorchis viverrini* in human stool specimens by PCR. *J. Clin. Microbiol.* 40, 3879, 2002.
138. Stensvold, C.R. et al. Evaluation of PCR based coprodiagnosis of human opisthorchiasis. *Acta Trop.* 97, 26, 2006.
139. Duenngai, K. et al. Improvement of PCR for detection of *Opisthorchis viverrini* DNA in human stool samples. *J. Clin. Microbiol.* 46, 366, 2008.
140. Parvathi, A. et al. Development and evaluation of a polymerase chain reaction (PCR) assay for the detection of *Opisthorchis viverrini* in fish. *Acta Trop.* 107, 13, 2008.
141. Ando, K. et al. Nucleotide sequence of mitochondrial CO I and ribosomal ITS II genes of *Opisthorchis viverrini* in Northeast Thailand. *Southeast Asian J. Trop. Med. Public Health* 32(Suppl 2), 17, 2001.
142. Le, T.H. et al. *Clonorchis sinensis* and *Opisthorchis viverrini*: Development of a mitochondrial-based multiplex PCR for their identification and discrimination. *Exp. Parasitol.* 112, 109, 2006.
143. Pauly, A. et al. Molecular characterization and differentiation of opisthorchiid trematodes of the species *Opisthorchis felineus* (Rivolta, 1884) and *Metorchis bilis* (Braun, 1790) using polymerase chain reaction. *Parasitol. Res.* 90, 409, 2003.
144. Touch, S. et al. Discovery of *Opisthorchis viverrini* metacercariae in freshwater fish in southern Cambodia. *Acta Trop.* 111, 108, 2009.
145. Traub, R.J. et al. A new PCR-based approach indicates the range of *Clonorchis sinensis* now extends to Central Thailand. *PLoS Negl. Trop. Dis.* 3, e367, 2009.
146. Parvathi, A. et al. *Clonorchis sinensis*: Development and evaluation of a nested polymerase chain reaction (PCR) assay. *Exp. Parasitol.* 115,291, 2007.
147. Sato, M. et al. Discrimination of *O. viverrini*, *C. sinensis*, *H. pumilio* and *H. taichui* using nuclear DNA-based PCR targeting ribosomal DNA ITS regions. *Acta Trop.* 109, 81, 2009.
148. Thaenkham, U. et al. Discrimination of *Opisthorchis viverrini* from *Haplorchis taichui* using COI sequence marker. *Acta Trop.* 103, 26, 2007.

39 Paragonimus

Hiromu Sugiyama, Takhellambam Shantikumar Singh, and Achariya Rangsiruji

CONTENTS

39.1 INTRODUCTION

The lung flukes of the genus *Paragonimus* cause infection primarily in the lungs and pleural cavity. This disease entity is called pulmonary paragonimiasis, which is difficult to differentiate from other respiratory diseases, especially pulmonary tuberculosis, due to its similar clinical and radiological features [1]. Except in highly endemic areas, paragonimiasis is never considered a public health problem, and misdiagnosis and mismanagement of pulmonary paragonimiasis are common. In addition, extrapulmonary paragonimiasis has also been reported as an important clinical manifestation [2].

About 50 species of *Paragonimus* trematodes (Distoma), including synonyms, are distributed in tropical, temperate, and sub-Arctic zones worldwide. Human cases of paragonimiasis occur mainly in countries with traditions of eating raw or undercooked second intermediate hosts, freshwater crabs, or crayfish. Consumption of raw or improperly cooked meat of paratenic hosts, such as wild boar, is another possible source of infection [3,4]. Through these types of food practices, it is estimated that about 293 million people are at risk for infection with *Paragonimus* spp. and several million people are infected worldwide [5,6].

The identification of *Paragonimus* spp. has relied on morphological characters of adult flukes and metacercariae, and sometimes eggs [7]. Recently, extensive studies of various *Paragonimus* spp. have been carried out using molecular techniques based on PCR and nucleotide sequencing. The generated DNA sequences have allowed the identification to the species level for specimens at any developmental stage. These sequence data are also used in taxonomic studies [8], improving our understanding of epidemiology of paragonimiasis. In this chapter, we provide an overview of the morphology and taxonomy of the important *Paragonimus* spp. that are involved in human disease. In addition to clinical manifestations in patients, diagnostics involving immunological and molecular techniques are also reviewed.

39.1.1 Classification and Morphology of *Paragonimus* spp.

Traditionally, morphological characterization has been regarded as an important criterion for identifying and establishing the taxonomy of the *Paragonimus* spp. Much emphasis was placed on the morphological features of adult worms at the macroscopic and light microscopic levels, including body shape and size, relative size of the oral sucker to the ventral sucker, patterns of lobulation of the ovary and testes, and arrangement of cuticular spines. At the metacercarial stage, morphological characteristics such as the presence of an inner and/or outer cyst wall(s), thickness of each cyst wall (if present), relative diameter of suckers, anterior extent of the excretory bladder, presence and length of a stylet on the oral sucker, presence of colored granules in the body, number of flame cells, body spination, and arrangement of papillae around suckers were employed as the key for species identification. At the egg stage, the size and eggshell sculpturing were the important keys for species identification and discrimination [6,7]. However, the extent of variation in morphology of adult worms, metacercariae, and eggs even among a single species [6] makes taxonomic classification difficult.

During the last decade, extensive studies of *Paragonimus* have been carried out using advanced molecular techniques based on PCR and DNA sequencing. Accumulated nuclear and mitochondrial sequence data have been employed to discriminate and identify the *Paragonimus* spp. Subsequent additions of sequence data for *Paragonimus* spp. have permitted reevaluation of the phylogenetic relationships of *Paragonimus* spp. [8–10]. Blair et al. used molecular phylogenetic analyses to divide *Paragonimus* spp. into a number of major groups, namely *P. westermani, P. skrjabini P. heterotremus, P. mexicanus* and *P. ohirai* [10]. Even though the number of valid species in the genus *Paragonimus* remains controversial, it is clear that the following seven species (or species-complexes) are causative agents in human infection [4,6].

39.1.1.1 *Paragonimus westermani* (Kerbert, 1878; Braun, 1899)

P. westermani is the type species of the genus that was first described by Kerbert in 1878 based on adult worm specimens obtained from the lungs of a Bengal tiger that died at the Amsterdam Zoo in 1877 [4,9]. The adult fluke is oval in shape. Its surface is covered with singly spaced cuticular spines. The oral sucker is usually slightly larger than the ventral sucker. The testes are divided into five or six lobes, while the ovary is simply branched into six lobes. The metacercariae are spherical, 280–450 μm in diameter depending on region of origin [4]. It has thick inner and thin outer cyst walls. Larval body is expanded to the entire cyst.

P. westermani is widely distributed from East Asia (China, Taiwan, Korea, Japan, and Southeast Siberia), Southeast Asia (the Philippines, Malaysia, Thailand, and Vietnam) to South Asia (Sri Lanka and India). Different populations isolated from various geographical regions show differences in animal and/or human susceptibilities [4,6]. Confirmed cases of human infection have been reported only from East Asian countries and the Philippines. Among *P. westermani* isolates, different ploidy levels, including diploid, triploid, and tetraploid, have been observed [9]. The triploid form has been reported only from East Asian countries and is regarded as being more pathogenic in humans [12]. The eggs of triploids are much bigger than those of diploids (92×50 μm on average for the triploid and 80×44 μm for the diploid from Japan), and the eggshell of the triploid form is often thickened at the nonoperculated end. The tetraploid form was found to be restricted to Northeast China [13] and their pathogenicity to mammals, including humans, has remained unclear.

From the isozyme studies carried out by Agatsuma et al. [14,15] of *P. westermani* in East and Southeast Asia, it was observed that genetic distances of diploid populations in East Asia (Japan and Taiwan) were different from those in Southeast Asia (Malaysia and the Philippines). The diploid populations from East Asia show relatively high similarity to one another and to the triploid populations, but in Southeast Asia, the diploid populations were genetically distant.

Molecular phylogenies based on the second internal transcribed spacer 2 (ITS2) of the nuclear ribosomal DNA (nrDNA) and cytochrome *c* oxidase subunit I (*cox1*) gene of mitochondrial DNA (mtDNA) have shown that *P. westermani* is partitioned into at least two groups [12,16]. One group comprises members from East Asia (China, Japan, Korea, and Taiwan) and includes both diploid and triploid forms. It was reported that this group exhibits relatively small molecular variation and, therefore, it is postulated that the triploid form arose somewhere in the East Asia group [17]. The second group includes isolates from Southeast Asia (Malaysia, Thailand, and the Philippines) that were genetically distant from one another. These results were in accordance with previous isozyme studies by Agatsuma et al. [14,15]. In 2008, additional *P. westermani* isolates, including isolates from Vietnam, India, and Sri Lanka, were analyzed using sequences of the ITS2 region and *cox1* gene [18], and the resulting phylogenetic trees revealed three groupings of *P. westermani* isolates originating in each of East, Southeast, and South Asia. This grouping seems to be in good agreement with the geographical distributions of *P. westermani* examined to date. In addition, according to Blair et al. [19], phylogenetic analyses of the ITS2 region and *cox1* gene also showed a close relationship between *P. westermani* and *P. siamensis*. All evidence from cytological as well as molecular studies have, therefore, indicated that *P. westermani* consists of a complex of cryptic/sibling species.

Sugiyama et al. [20] found two forms of *P. westermani* metacercariae in a single crab species during intensive field surveys in southern Thailand. The metacercariae were similar in morphology but were quite different in size from each other. The large metacercarial form was referred to as *P. westermani*, and the small form as *P. westermani*-like. In an attempt to obtain adult *P. westermani* and *P. westermani*-like, experimental infections to cats were carried out. Results indicated that the morphological features of *P. westermani*-like were fundamentally identical to those of *P. westermani* (diploid type) at the adult stage. However, the susceptibility of feline hosts to *P. westermani*-like was found to be different from that of *P. westermani*. *P. westermani*-like in the juvenile form were lodged predominantly in the liver, while some matured into adults in the lungs or pleural cavity. In contrast, adult or at least preadult *P. westermani* were detected only from the pleural cavity or lungs. Phylogenetic relationships based on combined sequence data of the ITS2 region and *cox1* gene showed that *P. westermani*-like should be placed as a sister group of *P. westermani* in Southeast and East Asia groups [21]. Taken together, data on the susceptibility and molecular analyses suggest that *P. westermani*-like could be regarded as another form within *P. westermani* species complex.

39.1.1.2 *Paragonimus skrjabini* (Chen, 1959) and *Paragonimus miyazakii* (Kamo et al., 1961)

The first reported specimens of *P. skrjabini* were isolated from the lungs of a viverrid in China. The adult fluke of *P. skrjabini* is long, slender, and spindle-shaped, and its surface is covered with singly spaced cuticular spines. The ventral sucker is slightly larger than the oral sucker and is located anteriorly at about one-third of the length of the body. The ovary and testes are delicately branched and rather small relative to the size of the body. The metacercariae are spherical, with a diameter of 400–450 μm and are enveloped with outer and inner cyst walls. This species occurs in various parts of China as well as in Thailand and the northeastern part of India [4,9,22].

The first reported adult specimens of *P. miyazakii* were isolated from a cat experimentally inoculated with metacercariae. After its discovery, it was confirmed that adults of *P. miyazakii* had been isolated from weasels and martens in western Japan [23]. *P. miyazakii* occurs in Japan.

Blair et al. [12] performed a phylogenetic study and determined that *P. skrjabini* from China was very closely related to *P. miyazakii*; for example, their ITS2 sequences were found to be identical. Blair et al. [11] then compared morphometric data with molecular phylogenetic findings to establish the identities of *P. skrjabini* and its related species, including *P. miyazakii*. They found that *P. skrjabini* from eastern China (Fujian Province) alone was phylogenetically very close to *P. miyazakii* from Japan, proposing that both populations should be referred to as the same subspecies as *P. skrjabini miyazakii*. It was also proposed that *P. skrjabini* populations from other parts of China should be named *P. skrjabini skrjabini*. They also regarded other related species as synonyms (e.g., *Paragonimus szechuanensis*) or cryptic/sibling species (e.g., *Paragonimus hokuoensis*) of *P. skrjabini* and concluded that *P. skrjabini* represents a separate species complex [11]. In this chapter, however, *P. miyazakii* from Japan is not referred to as *P. skrjabini miyazakii* but as *P. miyazakii* hereafter.

P. skrjabini was sometimes designated as *Pagumogonimus skrjabini* in old Chinese literature. In addition, the genus *Euparagonimus* was established for the lung fluke species *Euparagonimus cenocopiosus*. These two additional genera (*Pagumogonimus* and *Euparagonimus*) were created on the basis of adult characters. Recent molecular studies, however, indicated that these two additional genera were artificial and thus untenable [10].

39.1.1.3 *Paragonimus heterotremus* (Chen and Hsia, 1964)

This species occurs from south China to northeast India through Southeast Asian countries such as Vietnam, Laos, and Thailand [9]. The distinctive morphological feature of *P. heterotremus* at the adult stage is the large oral sucker that is more than twice as wide as the ventral one; the specific epithet of this lung fluke comes from that morphological characteristic. In most *Paragonimus* spp., that is, species other than *P. heterotremus* and *Paragonimus africanus* (see below), the oral and ventral suckers are similar in size. The ovary and testes are delicately branched and the latter is larger than the former. The cuticular spines are singly spaced. The metacercariae are readily identifiable due to small size, elliptical shape, and thickening of the inner cyst wall at both polar ends [4].

Recently, Waikagul [24] reported a newly described lung fluke species, *Paragonimus pseudoheterotremus*. This species is morphologically very similar to but genetically distinct from *P. heterotremus*, based on nucleotide sequence differences in *cox1* genes [25]. This finding suggests that *P. pseudoheterotremus* is a cryptic/sibling species to *P. heterotremus*.

39.1.1.4 *Paragonimus africanus* (Voelker and Vogel, 1965)

This species is distributed in sub-Saharan western and central African countries such as Cameroon, Nigeria, Equatorial Guinea, and Ivory Coast [9,26]. The adult fluke has an oral sucker that is considerably larger than the ventral one. The ovary and testes are both complexly branched; the testes are relatively larger than the ovary. The egg is large, with an average size of 91 × 49 μm. The metacercariae are spherical in shape with inner and outer cyst walls [4].

39.1.1.5 *Paragonimus uterobilateralis* (Voelker and Vogel, 1965)

This species is also distributed in sub-Saharan western and central African countries such as Cameroon, Nigeria, Liberia, and Gabon [9,26]. In the adult fluke, the oral and ventral suckers are almost the same size. The ovary is delicately branched while the testes are moderately branched and larger than the ovary. The cuticular spines are singly spaced. The eggs are much smaller than those of *P. africanus*, measuring 69 × 42 μm on average. The metacercariae are enveloped with a thin single cyst wall [4].

39.1.1.6 *Paragonimus mexicanus* (Miyazaki and Ishii, 1968)

This species is widespread in Central and South American countries from Mexico through Peru and is the only known species causing human paragonimiasis in those areas [9]. The adult fluke possesses profusely divided ovary and testes. The oral sucker is moderately larger than the ventral one, and egg size averages 79 × 48 μm with the surface of the shell showing marked undulation; eggshells of other species listed here have a smooth surface. The metacercariae have no cyst wall and are found free in the tissues of intermediate crab hosts [4].

39.1.1.7 *Paragonimus kellicotti* (Ward, 1908)

This species is found in the United States (east of the Rocky Mountains) and southern Canada [9]. The adult fluke has moderately branched ovary and testes; the former usually smaller than the latter. The oral sucker is slightly smaller than the ventral sucker. Cuticular spines are singly spaced. The egg is large, with an average size of 91 × 57 μm. The metacercariae are spherical in shape with inner and outer cyst walls and are exclusively parasitic in the heart of freshwater crayfish [7].

39.1.2 Biology and Pathogenesis

To complete the life cycle, *Paragonimus* spp. require three hosts [4]. Carnivorous or omnivorous animals and humans are the definitive hosts, freshwater snails are the first intermediate hosts, and freshwater crabs, and occasionally crayfish, are the second intermediate hosts. Adult worms inhabit worm cysts formed in the lungs of definitive hosts. Eggs laid by the adult worms become mixed with sputum, and if sputum is ingested, eggs may be excreted from the hosts to the external environment. In water, the eggs develop further, and miracidia hatch and enter into the first intermediate snail hosts, where they replicate asexually through stages of sporocysts and two generations of rediae and finally produce large numbers of cercariae. These cercariae infect the second intermediate crab or crayfish hosts, and become encysted as metacercariae in the gills, heart, midgut gland, genital organs, or muscle tissues. When the infected second intermediate hosts are eaten by an appropriate definitive host, the larvae excyst from metacercariae in the small intestine and enter through the intestinal wall into the peritoneal cavity and tissue. The growing larvae migrate through the diaphragm into the pleural cavity. Generally, the larvae invade the lung parenchyma after pairing with potential partners despite being hermaphroditic, grow to adult stage in a worm cyst and start laying eggs.

When an infected second intermediate host is ingested by an inappropriate host, larvae may not migrate into the thoracic cavity and inhabit the lung parenchyma, but instead remain in an immature stage, as in the case of *P. skrjabini* infection in humans [27]. Similarly, *P. westermani* larvae in the wild boar migrate to body muscles from the peritoneal cavity and remain there in a larval stage for a long period; these paratenic hosts are now becoming the major source of human infection in Japan [28].

In paragonimiasis, pathological lesions are due to migration of the worms through the tissue, toxic metabolites produced by the growing larvae, the presence of adult worms and eggs and the host's immune response leading to the formation of inflammatory granuloma and encapsulation in the lungs [2].

39.1.3 Clinical Features

Most *Paragonimus* infections are asymptomatic due to the mildness of infection, and spontaneous resolution is common, as demonstrated by high rates of immunologically positive cases in endemic areas [29]. The majority of symptomatic patients develop respiratory symptoms, while others develop symptoms related to the affected organ or tissue. Clinically, paragonimiasis can be classified into three main forms [30,31]: pulmonary, extrapulmonary, and pleuroplumonary.

39.1.3.1 Pulmonary Paragonimiasis

Pulmonary paragonimiasis is the most common form of paragonimiasis [31]. The major presenting symptoms include cough, chest pain, and expectoration of blood-stained or rusty brown sputum, or recurrent hemoptysis, which appear after an incubation period of 2–4 months. Hemoptysis is often increased on physical exertion, and may be severe enough to require hospitalization with blood transfusion.

The common associated symptoms are difficulty in breathing, fever, headache, and weakness, and the more uncommon symptoms are weight loss and anemia. The clinical and radiological features of pulmonary paragonimiasis are indistinguishable from pulmonary tuberculosis, so pulmonary paragonimiasis cases are typically first diagnosed as smear negative pulmonary tuberculosis [1,30,32–34]. Similarly, it is difficult to differentiate pulmonary paragonimiasis from lung cancer [35,36].

39.1.3.2 Extrapulmonary Paragonimiasis

Extrapulmonary paragonimiasis is usually associated with past or present pulmonary infection and is further classified by infestation site.

39.1.3.2.1 Pleural Paragonimiasis

This form, which presents as pleuritis or pleural effusion with or without pneumothorax, occurs in 17% of extrapulmonary paragonimiasis cases [31]. Pleural effusion is usually bilateral and is more common in children. The associated symptoms are chest pain, breathlessness, a cough, and sometimes fever. *Paragonimus* eggs are rarely detected in the pleural fluid aspirate; however, high eosinophil count, low pH, high protein level, low glucose level (<10 mg) and high lactate dehydrogenase (more than or equal to 1000 IU/L) are consistent with pleural paragonimiasis [37,38].

39.1.3.2.2 Cutaneous Paragonimiasis

This form accounts for about 16% of extrapulmonary paragonimiasis cases [31]. Aberrant parasites lodge in the subcutaneous tissue and form migratory subcutaneous nodules. The nodules initially appear on the chest or abdominal wall and then gradually migrate to the inguinal region, genitalia, and lower limbs. The nodules are characteristically mobile, nontender, firm in consistency, and measure 2–5 cm in diameter. Rarely, nodules remain at the location without further migration [39].

39.1.3.2.3 Cardiovascular Paragonimiasis

This is one of the rare but serious conditions in which pulmonary paragonimiasis may be associated with pericarditis, myocarditis, and congestive heart failure, especially in young children. The patient may die of congestive heart failure in spite of praziquantel therapy [40].

39.1.3.2.4 Abdominal Paragonimiasis

Gastrointestinal symptoms develop in most of the patients during the early period of the infection due to the growing larval flukes entering through the intestinal wall into the peritoneal cavity. Abdominal distension, multiple intra-abdominal nodules, hepatomegaly, and peritonitis may also occur due to migration of larval flukes, which probably mature into gravid adults, release eggs and then die [41].

39.1.3.2.5 Neuroparagonimiasis

Cerebral paragonimiasis is the most serious form of paragonimiasis commonly seen in younger age groups [42]. Clinical symptoms include epileptic fits, headache, nausea, vomiting, visual disturbances, blindness, aphasia, and hemiplegia or generalized muscular paresis [43]. Cerebral paragonimiasis can be difficult to differentiate from cerebral tuberculoma, and brain diseases caused by other parasites, such as *Schistosoma japonicum*, *Gnathostoma spinigerum*, and *Angiostrongylus cantonensis* [44,45]. Diagnosis can be made by serological tests and/or imaging techniques such as x-ray, computerized tomography (CT) scan and magnetic resonance imaging (MRI) [46,47].

39.1.3.2.6 Urinary and Ocular Types

Rarely, renal infection causing hematuria, painful scrotal swelling and ocular infection have been reported [27].

39.1.3.3 Pleuroplumonary Paragonimiasis

In about 6% of cases, infection occurs in both the pleural cavity and the lungs [31]. In this form, symptoms of pleural effusion such as pleuritic chest pain and dyspnea are associated with cough, expectoration of blood streaked sputum or recurrent hemoptysis. *Paragonimus* eggs are frequently present in the sputum smears.

39.1.3.4 Differential Diagnosis Based on Clinical Symptoms

The most important clinical symptom of pulmonary paragonimiasis is the expectoration of blood-stained sputum or recurrent hemoptysis. Differential diagnosis of infectious causes of hemoptysis should include pulmonary paragonimiasis, tuberculosis, bronchitis, bronchiectasis, pneumonia, hydatid disease, amebiasis, schistosomiasis, aspergillosis, and histoplasmosis. However, most standard text books hardly mention pulmonary paragonimiasis, and this may be the reason why clinicians usually fail to diagnose cases presenting with hemoptysis as pulmonary paragonimiasis. Careful history taking and thorough laboratory investigations are the preliminary steps in elucidating the cause of hemoptysis, although the cause may not be determined in about 50% of the cases. In areas with high incidences of paragonimiasis, pulmonary paragonimiasis should always be considered for patients presenting with hemoptysis who have a history of consuming crabs or crayfish [31,48].

39.1.4 Diagnosis

Conventional methods for the diagnosis of *Paragonimus* infections usually require detection of eggs discharged in sputum or stool samples. However, eggs are not always observed in clinical specimens, especially during the prepatent period of infection and in extrapulmonary paragonimiasis. To overcome the limitations of conventional methods in detecting eggs, immunological tests such as enzyme-linked immunosorbent assay (ELISA) have been applied to achieve a quick but adjunct diagnosis.

39.1.4.1 Conventional Techniques

39.1.4.1.1 Microscopic Examination of Clinical Samples for Paragonimus Eggs

A definitive diagnosis of paragonimiasis is usually made by finding characteristic golden brown, ellipsoidal or oval, and operculated *Paragonimus* eggs in the sputum or feces but rarely in the pleural fluid. Direct microscopic examination of at least three early-morning sputum samples is recommended for the detection of *Paragonimus* eggs in (pleuro)pulmonary paragonimiasis.

Similarly, three to four fresh or formalin-preserved fecal samples should be examined for the detection of *Paragonimus* eggs either by direct wet mount cover slip preparation or after AMS III concentration techniques [49] or formol-ether sedimentation technique (the modified method is recommended) [50] for a definitive diagnosis. Examination of stool samples is especially important in suspected, but apparently sputum-negative cases and in young children who usually do not expectorate but swallow sputum [30].

Charcot Leyden crystals (CLC) are usually present in the sputum of paragonimiasis cases, even when *Paragonimus* eggs are not detectable. Therefore, the presence of CLC in the sputum in a clinically and epidemiologically suspected case is a strong indication of pulmonary paragonimiasis [30].

39.1.4.1.2 Hematology Investigation

Paragonimiasis, especially in children, is characterized by marked eosinophilia with absolute eosinophil counts ranging from 650 to 4000/mm^3 and increased erythrocyte sedimentation rate. Therefore, elevated eosinophil count may strongly indicate diagnosis of paragonimiasis in a clinically and epidemiologically suspected case. Many cases of paragonimiasis present with leucocytosis characterized by a total white blood cell count varying from 10,200 to 16,350/mm^3 and mild lymphocytosis [30,38]. In contrast, hemoglobin values remain within normal limits despite frequent expectoration of blood-stained sputum or recurrent hemoptysis in most patients.

39.1.4.1.3 X-Ray and Other Imaging Techniques

The common abnormal findings in the chest roentgenograms are ill-defined patchy consolidations 5–60 mm in size (63%), pleural thickening or blunting of costophrenic angle (53%), nodular densities 1–4 cm in diameter (13%), cavities (11%), and pleural effusion that is usually bilateral (60%) [32]. Chest roentgenograms may be normal (15%), even in symptomatic patients with or without detectable *Paragonimus* eggs in sputum specimens [32]. CT and MRI are better techniques for visualization of the lesions in the lungs and lungs/brain, respectively [46,47,51].

39.1.4.1.4 Immunological Techniques

Immunological techniques play an important role in diagnosis during the prepatent period of infection and in extrapulmonary paragonimiasis in which eggs are not detectable in typical clinical specimens. Tests based on various immunological techniques have been used: intradermal (ID) test, complement fixation test (CFT), immunodiffusion test, indirect hemagglutination (IHA) test, ELISA, and immunoblotting.

39.1.4.1.5 ID Test

The ID test is simple, easy to perform, inexpensive, rapid, and highly sensitive; it was widely and effectively used over the past several decades in Japan [2]. The test has been used in countries such as India [29] and Laos [52] for mass screening in epidemiological surveillance of paragonimiasis as well as laboratory diagnosis. It is valuable in distinguishing pulmonary paragonimiasis from pulmonary tuberculosis, especially in areas where these are coendemic. The test is performed by intradermally inoculating of the test antigen on the volar aspect of the forearm and observing wheal formation diameter 15 min after inoculation. The major disadvantage of this test is the inability to distinguish between past and current infection as a positive result may remain as long as 10–20 years after successful treatment or spontaneous recovery [2]. It is also possible that cross reaction with other trematodiasis like schistosomiasis and clonorchiasis will produce a positive result if crude antigen is used, but this can be avoided by using purified and fractionated antigen; the sensitivity and specificity of the ID test can be up to 100% with the purified and fractionated antigen [53].

39.1.4.1.6 Complement Fixation Test

The CFT is a highly specific test used to diagnose active infection and to confirm ID positive cases. Because the test result becomes negative within 3–9 months after successful treatment, it can be used for therapeutic evaluation [54]. This test is not suitable for routine use owing to its lower sensitivity, long sampling-to-result time, labor intensiveness, and that it requires experienced technicians.

39.1.4.1.7 Immunodiffusion Test

Biguet et al. [55] first developed the immunodiffusion method for the diagnosis of paragonimiasis. The double-diffusion technique (Ouchterlony's method), immunoelectrophoresis, and counter-current electrophoresis are highly sensitive and specific tests and can be used for species differentiation by demonstration of specific precipitation bands [56–58].

39.1.4.1.8 IHA Test

The IHA test is another simple, rapid, and highly sensitive test used in the diagnosis of human paragonimiasis [59].

39.1.4.1.9 Enzyme-Linked Immunosorbent Assay

The first report of the application of ELISA in the diagnosis of paragonimiasis in Thailand was made in 1981 [60]. ELISA is now most widely used as serological tests of choice for their high sensitivity and specificity [61]. The tests are also applicable to mass screening [62]. ELISA tests based on different techniques and with different antigen preparations have been developed; other ELISA techniques include sandwich ELISA, which uses a monoclonal antibodies-based antigen detection assay [63] and multiple dot-ELISA, which is now commonly used in Japan for diagnostic purposes [64]. Different antigen preparations include F1 antigen, which attains 100% sensitivity and specificity in *P. heterotremus* infection [65]. ELISA test using excretory and secretory (ES) products of *Paragonimus* as the antigen also affords high sensitivity and specificity in the diagnosis of paragonimiasis [66]. Generally, although the ELISA has been used to detect parasite-specific IgG antibodies, the IgM ELISA has proven to be useful for the diagnosis of infection in its early stages [67] and IgE-ELISA was suggested to be useful in evaluating chemotherapy for paragonimiasis due to the short half-life of IgE antibodies [68]. An ELISA inhibition test has been established to differentiate infection-causing lung fluke species [36]. The test serum was incubated with either serially diluted *P. westermani* or *P. miyazakii* antigen, and the set of serum samples was used to measure the antibody titers against the respective antigens by ELISA. The reactivity of the test serum was inhibited more effectively after incubation with the homologous (causative species) antigen. The disadvantages of ELISA tests are that they are time consuming and require expensive equipment and experienced technicians.

39.1.4.1.10 Immunoblotting

Researchers have attempted to use sodium dodecyl sulfate polyacrylamide gel electrophoresis (SDS-PAGE) and immunoblotting techniques to characterize the antigenic substances that react with sera of infected patients and test animals. Using this technique, Sugiyama et al. [69] determined that the 27-kDa protein(s) in the adult fluke antigen is applicable for diagnosis of paragonimiasis westermani in not only definitive hosts but also in paratenic hosts. They also determined that the antigenic material is located in the gut of both adult and immature flukes by the immunoperoxidase staining technique [69], suggesting that the antigen(s) are probably ES products. The reactive band of 31.5 kDa in the immunoblot assay was described as a specific antigenic marker for *P. heterotremus* infection, which can be used for the serodiagnosis of infected patients [70] and also to evaluate the effect of praziquantel therapy [71]. The differential diagnosis between *P. heterotremus* and *P. westermani* infections was made using adult *P. heterotremus* extracts [72]; a 35-kDa molecule was found to be *P. heterotremus*-specific. Generally, though immunoblotting analyses detect parasite-specific IgG antibodies, IgE immunoblot was carried out and proved to reduce cross-reactions with other trematode infections [73].

39.1.4.1.11 Rapid Test

Recently, the dot-immunogold filtration assay (DIGFA) kit was developed in China for the detection of anti-*P. westermani* antibodies. The kit is based on the principle of a membrane-based flow-through immunoassay technique. Anti-human IgG conjugated with gold-colloids are applied after the test sera to the kit and the results are visualized within 10 min. The kit was prepared using *P. westermani* antigen and the sensitivity and specificity of the kit were reported to be up to 99% and 92%, respectively [74]. The kit was also found to be applicable in Japan to detect serum antibodies in patients infected with either *P. westermani* or *P. miyazakii* [75].

39.1.4.2 Molecular Techniques

The morphological characteristics of adults and metacercariae specimens, and sometimes eggs, are used to identify and discriminate the *Paragonimus* spp. (see Section 39.1.1). However, it is not always possible to make an identification of the causative species based only on the morphological characteristics of specimens available in clinical samples. In that case, molecular-based diagnostic techniques including PCR and DNA sequencing have been applied [76,77]. Other potential techniques, such as restriction fragment-length polymorphism (RFLP) analysis, PCR-RFLP, multiplex PCR, random amplified polymorphic DNA (RAPD) and DNA hybridization, have also been explored for the identification and discrimination of *Paragonimus* spp. using specimens at various life cycle stages as well as having damaged, incomplete, or indistinguishable morphological characteristics [6]. All these techniques have also been utilized for the genetic characterization and/or phylogenetic analysis of paragonimiasis causative agents. The data obtained from these investigations can be further applied to paragonimiasis control programs.

39.1.4.2.1 PCR and DNA Sequencing

In 2000, Chang et al. [77] amplified and sequenced the ITS2 region from eggs recovered from sputum of a Chinese patient with paragonimiasis. Sequence alignment showed 100% similarity with a sequence from *P. westermani* eggs obtained from an experimentally infected dog, indicating the causative agent to be *P. westermani*. A similar method was applied by Singh et al. [48,76] to determine the etiological agent of human paragonimiasis in India as *P. heterotremus*. In Vietnam, Le et al. [78] used *Paragonimus* DNA from different life cycle stages, including eggs, miracidia and adults recovered from a human host. Their phylogenetic trees, based on ITS2 region and *cox1* gene sequences, revealed a distinct group of *P. heterotremus* from Vietnam, Thailand, and China. These findings also demonstrated that, regardless of life cycle stage of the specimens used, molecular techniques were accurate and rapid in identifying this parasite as *P. heterotremus*. In addition, PCR methods to detect copro-DNA (genomic DNA derived from *Paragonimus* eggs in feces) have been developed. Using fecal samples, Intapan et al. [79] and Nkouawa et al. [80] demonstrated that the approaches were effective in diagnosing *P. heterotremus* and *P. africanus* infections in cats and humans, respectively.

39.1.4.2.2 Restriction Fragment-Length Polymorphism

Agatsuma et al. [81] examined genetic differences between the Japanese diploid and triploid *P. westermani*. RFLP analysis of the entire mitochondrial genome using *Pst*I endonuclease showed that all diploids produced two bands while only one band was generated by triploids. Blair et al. [12] mentioned that the parthenogenetic triploid form was more pathogenic than the diploid form, making it possible to use molecular techniques to distinguish these two medically important forms.

39.1.4.2.3 PCR-RFLP

The prevalence of lung fluke metacercariae, which are pathogens that cause human paragonimiasis, can be investigated using PCR-based techniques and DNA sequencing. The metacercariae of *P. westermani* and *P. miyazakii*, which are morphologically quite similar, were discriminated based on the ITS2 region (520 bp) using PCR-RFLP and PCR amplification with species-specific primers. For PCR-RFLP, restriction enzymes *Sna*BI and *Bss*SI were used; the former cleaved only the ITS2-amplified product from *P. westermani*, producing two fragments of about 420 and 100 bp, and the latter only digested *P. miyazakii* PCR products and generated two fragments of about 300 and 220 bp. Another technique, using either a *P. westermani*-specific primer (PwR1) or a *P. miyazakii*-specific primer (PmR1), yielded PCR products of about 420 bp only from the respective species when applied with a consensus forward primer (3S) [82].

39.1.4.2.4 Multiplex PCR

Molecular methods established to differentiate among species have relied mainly on nucleotide differences in the ITS2 region. Sugiyama et al. [83] developed a multiplex PCR method which used interspecies conserved and species-specific primers designed from the sequences of the ITS2 region, and which allowed species discrimination of individual metacercariae in a single-tube reaction. The newly designed forward primers, PwF1 and PmF1, were used in combination with the consensus reverse primer, A28, to obtain specific PCR products from *P. westermani* and *P.*

miyazakii, respectively. Although about 520 bp PCR products were obtained from both species, each product had a distinctive band, that is, about 140 bp for *P. westermani* and about 300 bp for *P. miyazakii*.

39.1.4.2.5 Random Amplified Polymorphic DNA

Intapan et al. [84] reported the use of another PCR-based technique, known as RAPD, to discriminate among five species of *Paragonimus* in Thailand. RAPD patterns obtained from amplification with an arbitrary primer, P2, showed that *P. heterotremus*, *P. harinasutai*, *P. siamensis*, *P. bangkokensis* and *P. westermani* were genetically distinct. A dendrogram constructed from the observed genetic polymorphisms revealed a close relationship between *P. heterotremus* and *P. harinasutai*.

39.1.4.2.6 DNA Hybridization

In 1997, Maleewong et al. [85] established a DNA hybridization technique for the detection of *P. heterotremus* DNA in fecal samples of experimentally infected cats. A radioactively labeled probe utilizing a highly repetitive DNA fragment (ca. 1500 bp; probably a portion of a retrotransposon [8]) derived from a *P. heterotremus* genomic DNA library could detect as few as two *P. heterotremus* eggs. This method has 100% specificity and sensitivity.

39.2 METHODS

Sample collection and preparation for molecular methods for identifying and discriminating *Paragonimus* spp. are described below. Detection procedures such as PCR, PCR-RFLP analysis, and cycle sequencing are also described.

39.2.1 Sample Preparation

39.2.1.1 Eggs

Detection and collection procedures of eggs from the sputa and/or feces of patients and definitive mammalian hosts are described above (see Section 39.1.4.1.1). If the sputum sample contains much mucus, the sample should be sedimented before isolating eggs. Briefly, digest the sputum with five volumes of 5% sodium hydroxide for 10 min and centrifuge for 5 min at 2000 × *g*. This procedure, as well as AMS III concentration or formol-ether sedimentation techniques, does not affect the integrity of eggs for further molecular examination.

A method has been developed to extract DNA from a small number of *P. westermani* eggs using proteinase K (Qiagen) and NP-40 (e.g., Tergitol-type NP-40®, Sigma-Aldrich Co.) for amplification by PCR [77]. More effective methods have also been applied in order to extract DNA from individual eggs [48,76]; a method using proteinase K and SDS was originally developed for extracting DNA from metacercariae [86]. Prior to the start of the extraction by these methods, the additional step of puncturing the shell of the egg with a dissecting needle is recommended [87]. This step gives the extraction solution access to the vitelline cells and oocytes inside the egg.

39.2.1.1.1 Reagents: Proteinase K-SDS solution (1 mL) [86]

1. 30 μL of 5 M NaCl.
2. 10 μL of 1 M Tris-Cl pH 8.0.
3. 20 μL of 0.5 M EDTA.
4. 5 μL of 20% SDS.
5. 10 μL of proteinase K (20 mg/mL).
6. 925 μL of distilled water to adjust the total volume of the working solution to 1 mL.

39.2.1.1.2 Procedure

1. Place several eggs on a concave slide (i.e., a glass microscope slide having a concave cell in the center of the slide) filled with distilled water.
2. Wash the eggs with several changes of distilled water using a micropipette under a dissecting microscope.
3. Puncture the shell of each egg with a dissecting needle.
4. Transfer each punctured egg in a small amount of water, for example, 1 μL, into each 1.5 mL microcentrifuge tube or 0.6 mL thin-walled microcentrifuge tube.
5. Add 7 μL of proteinase K-SDS solution to each tube and incubate the tubes at 60°C for 1 h to overnight (until the tissue in each tube becomes solubilized).
6. Add the following reagents to each tube and then hold the tube at −80°C for 10 min for ethanol precipitation.
   a. 0.8 μL of 3 M sodium acetate (pH 5.2).
   b. 20 μL of absolute ethanol.
7. Centrifuge the tubes at 4°C at 8000 × *g* for 10 min to pellet the extracted DNA; discard the supernatant.
8. Rinse the pellets with 0.1 mL of 70% ethanol.
9. Centrifuge for 10 min with the same conditions; discard the supernatant.
10. Place the tube at 60°C for 1 min to evaporate the remaining ethanol.
11. Add an appropriate amount of Tris-EDTA buffer, for example, 50 μL, to each tube and heat the tubes at 95°C for 10 min to inhibit proteinase K activity.
12. Determine the DNA concentration spectrophotometerically and proceed to PCR amplification.

39.2.1.2 Adult and Immature Flukes

Occasionally, adult worms are expectorated in the sputum [88] or recovered from the cyst in tissue removed by surgery [89] or at necropsy [90]. Immature (without eggs) or adult (with eggs) worms are isolated or found in the biopsy material of subcutaneous nodules in cases of cutaneous paragonimiasis [91]. Adult flukes are also isolated from worm cysts formed in the lungs of definitive mammalian hosts, naturally or experimentally infected with *Paragonimus* spp.; they are used for positive identification of species by molecular methods.

The above-mentioned proteinase K-SDS method is also applicable to the extraction of DNA from immature flukes and small fragments of adult flukes. DNA is also extracted from adult and immature flukes using commercially available kits, for example, DNeasy Blood and Tissue Kit® (Qiagen) or NucleoSpin Tissue Kit® (Macherey-Nagel). The E.Z.N.A. Mollusk DNA Isolation Kit® (Omega Biotek) is also recommended for extracting genomic DNA from lung flukes at different life cycle stages, even for specimens frozen or preserved in alcohol [6]. This kit is efficient for recovering genomic DNA from other invertebrates, including flatworms which have tissues rich in mucopolysaccharide, but the kit is disadvantageous in that it is necessary to prepare organic solvents other than ethanol, for example, chloroform and isoamyl alcohol.

39.2.1.3 Metacercariae

To gather epidemiological data from the source regions, *Paragonimus* metacercariae are collected by microscopic examination of crabs or crayfish, the second intermediate hosts, which also act as the infection sources to humans and other mammals. The preferred lodging sites of metacercariae in these crustacean hosts have been demonstrated in some *Paragonimus* spp. to be as follows: the body (cephalothorax) muscles for *P. westermani* (the diploid form in Japan) [92], the heart for *P. miyazakii* [93], and the leg (appendage) muscles for *P. heterotremus* (in Thailand) [94]. However, this information is limited to only a few species, and all the internal organs (gills, hearts, midgut gland, and genital organs) and muscle tissues (body and legs) are usually examined for each specimen. As the crustacean hosts are poikilothermic animals, they are immobilized by being placed on ice or immersed in chilled water. This facilitates the removal of the carapace and legs from the body by hand without the aid of any instruments. After removing the carapace, the internal organs and body muscles from each crustacean host are separately excised with tweezers, compressed between two glass plates (typically, 10 × 6 × 0.2 (thickness) cm), and examined for metacercariae under a stereoscopic microscope. The leg muscles are also removed with tweezers after breaking the shell with a nipper, and examined in a same manner. When *Paragonimus* metacercariae are identified, the glass plate covering the tissues is carefully removed and metacercariae in the compressed tissues are recovered using teasing needles or tweezers.

To gather general epidemiological information, the crustacean hosts are collected together and crushed in bulk using a kitchen blender or manual meat-mincing machine. Minced tissues are then washed with large volumes of tap water and repeatedly filtered through wire sieves, for example, 1-mm mesh. The washed sediment of the filtrate is examined under a stereoscopic microscope to detect *Paragonimus* metacercariae.

We use artificial gastric juice (acid-pepsin solution) consisting of 1 g of pepsin (e.g., Nacalai Tesque, Inc.; 1:10,000) in 1 L of 0.7% hydrochloric acid to facilitate the digestion of the host tissue [95]. Minced crab tissues are incubated with artificial gastric juice at 37°C for 1–2 h with shaking or stirring. The subsequent digest is filtered through a wire sieve, for example, 1-mm mesh, allowed to sediment, and washed with several changes of tap water followed by observation under a microscope to isolate metacercariae, as described above. Isolated metacercariae should be kept frozen (−20°C or lower) or in a tube with 70% ethanol at room temperature to extract DNA by the above-mentioned method [86].

39.2.2 Detection Procedures

DNA techniques utilizing genetic markers in nrDNA and mtDNA have been employed to resolve taxonomic issues related to several species of the genus *Paragonimus* at various life cycle stages. Blair et al. [11,12] suggested that the ITS2 region of nrDNA is a good interspecies marker while that of the partial *cox1* gene of mtDNA is a good intraspecies marker. The fragments for sequence analyses were generated by PCR of genomic DNA using universal primers reported previously or designed originally by researchers themselves. After successful PCR amplification, the amplified products were visualized as a band with a particular size on an agarose gel. The PCR products were further sequenced and analyzed by the procedures including a PCR-RFLP method.

For PCR amplification of the ITS2 region of *Paragonimus* spp., the following primer pair was used [12,96]:

1. An interspecies-conserved forward primer (3S): 5′-GGTACCGGTGGATCACTCGGCTCGTG-3′.
2. An interspecies-conserved reverse primer (A28): 5′-GGGATCCTGGTTAGTTTCTTTTCCTC CGC-3′.

   For PCR amplification of the partial *cox1* gene of *Paragonimus* spp., the following primer pair was used [96]:
3. An interspecies-conserved forward primer (JB3): 5′-TTTTTTGGGCATCCTGAGGTTTAT-3′.
4. An interspecies-conserved reverse primer (JB4.5): 5′-TAAAGAAAGAACATAATGAAAATG-3′.

39.2.2.1 Reagents for PCR

Reagents for PCR mixtures and the volume of each reagent required for a single reaction (50 μL) are prepared according to the manufacturer's instruction for TaKaRa EX *Taq*® (Takara Bio Inc.) as follows:

1. 0.5 μL of DNA polymerase, for example, TaKaRa EX *Taq* (1.25 U).
2. 1 μL of dNTPs (supplied by the manufacturer in 10 mM stocks).
3. 5 μL of 10× buffer (containing 100 mM Tris-HCl pH 8.3, 500 mM KCl, and 15 mM $MgCl_2$).
4. 1.25 μL of forward primer (20 μM stock).
5. 1.25 μL of a reverse primer (20 μM stock).
6. 1–10 μL of template DNA (prepared at 10–100 ng/μL).
7. 31–40 μL of distilled water to make each mixture volume up to 50 μL.

39.2.2.2 PCR Amplification and Visualization of the Amplified Products

PCR is carried out on a thermal cycler (e.g., TaKaRa PCR Thermal Cycler Dice Standard®) with 35 cycles of 98°C for 5 s, 55°C for 30 s, and 72°C for 60 s. A final extension is carried out at 72°C for 10 min.

The PCR products are separated by electrophoresis on 1.0% agarose gels in Tris-borate-EDTA buffer (TBE buffer) and visualized by illumination with short-wave ultraviolet light after ethidium bromide staining.

39.2.2.3 PCR-RFLP Analysis

For species identification and discrimination by PCR-RFLP, the restriction enzymes are selected on the basis of the theoretical restriction maps generated from the sequences of the target lung fluke species. The PCR products are digested with the selected restriction enzyme(s) according to the manufacturer's recommendations. The digested samples are then separated by electrophoresis on agarose gels and visualized as described above.

39.2.2.4 Cycle Sequencing

The PCR products are purified by removing primers from the PCR reactions using a commercially available kit, for example, NucleoSpin Extract II® (Macherey-Nagel). The purified products are sequenced using a BigDye Terminator Cycle Sequencing Kit® (Life Technologies) on an automated sequencer (e.g., ABI3100®, Life Technologies).

39.3 CONCLUSIONS AND PERSPECTIVES

Paragonimiasis caused by the lung fluke belonging to the genus *Paragonimus* is an important foodborne helminthic zoonosis; almost 300 million people around the world are estimated to be at risk for infection. The common mode of infection has been the consumption of raw or undercooked second intermediate hosts (freshwater crabs or crayfish) or raw or improperly cooked meat of paratenic hosts (e.g., wild boar). Pulmonary paragonimiasis is difficult to differentiate from pulmonary tuberculosis due to similarities in their clinical and radiological manifestations. Therefore, patients presenting with respiratory symptoms (e.g., hemoptysis and chronic cough) should also be investigated for paragonimiasis before prescribing antitubercular drugs for smear-negative pulmonary tuberculosis or multidrug-resistant tuberculosis. This will have a significant impact on the tuberculosis control program, especially in the areas where both diseases are sympatrically prevalent. Once diagnosed, paragonimiasis can be effectively treated with praziquantel [97].

The diagnosis of paragonimiasis is primarily based on the detection of eggs in sputa or feces. However, eggs are not always observed in clinical specimens, especially during the prepatent period of infection and in extrapulmonary paragonimiasis. To overcome these limitations, various serological methods including highly sensitive and specific tests have been developed for laboratory diagnosis of paragonimiasis. Molecular techniques have been explored for the identification of *Paragonimus* spp. at various life cycle stages, not limited to eggs but also adult and immature flukes, and metacercariae. Moreover, molecular approaches also have the potential to be used for species identification of specimens having damaged, incomplete, or indistinguishable morphological characteristics. The techniques used for these aims include PCR and DNA sequencing, PCR-RFLP, multiplex PCR, RAPD, and DNA hybridization.

Molecular techniques have also been applied for taxonomic studies. In this context, *P. westermani* has been extensively examined because this lung fluke species shows great diversity in geographical distribution, host susceptibility, pathogenicity, and even morphology, especially in metacercarial and egg stages. Molecular phylogenetic analyses consequently have revealed that the phylogenetic tree of *P. westermani* has three single clades, each comprising isolates from three distinct regions. This grouping is in good agreement with not only the geographical distribution, but also host susceptibility and cytological findings. Phylogenetic analyses also showed that *P. siamensis*, a species discriminated from *P. westermani* by morphology, has a close genetic relationship with *P. westermani*, indicating that *P. westermani* consists of species complex. In addition, *P. skrjabini* represents a separate species complex. Based on recent advances in molecular studies of *Paragonimus* spp., the concept of the species complex is now and will become more widely accepted [8,98]. To resolve the discrepancy between systematics in morphology and by phylogenetic analysis, it is suggested that the morphological and morphometric parameters traditionally applied for the speciations of *Paragonimus* spp. should be reevaluated.

The strategy for the prevention and control of paragonimiasis should include: (i) Information, Education and Communication (IEC) for the public with emphasis on safe habits of eating and the cooking of freshwater crabs, crayfish, and paratenic hosts, which transmit the infection, and the effects of the paragonimiasis on human health. (ii) IEC and adequate training for health care personnel about the epidemiology, life cycle, clinical features, prevention at all three levels (primary, secondary, and tertiary) and treatment of cases.

However experiences have shown little success as the eating traditions and cultural practices of people in endemic areas are difficult to change.

REFERENCES

1. Toscano, C. et al. Paragonimiasis and tuberculosis, diagnostic confusion: A review of the literature. *Trop. Dis. Bull.* 92, R1, 1995.
2. Yokogawa, M. Paragonimiasis. In: Hillyer, G.V., and Hopla, C.E. (eds.), *CRC Handbook Series in Zoonoses*, Vol. III. CRC Press, Inc., Boca Raton, p. 123, 1982.
3. Miyazaki, I. Paragonimiasis. In: Hillyer, G.V., and Hopla, C.E. (eds.), *CRC Handbook Series in Zoonoses*, Vol. III. CRC Press, Inc., Boca Raton, p. 143, 1982.
4. Miyazaki, I. Paragonimiasis. In: Miyazaki, I. (ed.), *An illustrated Book of Helminthic Zoonoses*. International Medical Foundation of Japan, Tokyo, p. 76, 1991.

5. Keiser, J., and Utzinger, J. Emerging foodborne trematodiasis. *Emerg. Infect. Dis.* 11, 1507, 2005.
6. Narain, K., Agatsuma, T., and Blair, D. *Paragonimus*. In: Liu, D.Y. (ed.), *Molecular Detection of Foodborne Pathogens*. Chapter 61. CRC Press, Inc., Boca Raton, p. 827, 2009.
7. Miyazaki, I. Lung flukes in the world–Morphology and life history. In: Sasa, M. (ed.), *A Symposium on Epidemiology of Parasitic Diseases*, International Medical Foundation of Japan, Tokyo, p. 101, 1974.
8. Blair, D., Agatsuma, T., and Wang, W. Paragonimiasis. In: Murrell, K.D., and Fried, B. (eds.), *World Class Parasites: Food-borne Parasitic Zoonoses: Fish and Plant-Borne Parasites*, Vol. 11. Springer, New York, p. 117, 2007.
9. Blair, D., Xu, Z.B., and Agatsuma, T. Paragonimiasis and the genus *Paragonimus*. *Adv. Parasitol.* 42, 113, 1999.
10. Blair, D. et al. A molecular perspective on the genera *Paragonimus* Braun, *Euparagonimus* Chen and *Pagumogonimus* Chen. *J. Helminthol.* 73, 295, 1999.
11. Blair, D. et al. *Paragonimus skrjabini* Chen, 1959 (Digenea: Paragonimidae) and related species in eastern Asia: A combined molecular and morphological approach to identification and taxonomy. *Syst. Parasitol.* 60, 1, 2005.
12. Blair, D. et al. Geographical genetic structure within the human lung fluke, *Paragonimus westermani*, detected from DNA sequences. *Parasitology* 115, 411, 1997.
13. Terasaki, K. et al. Tetraploids of the lung fluke *Paragonimus westermani* found in China. *Parasitol. Res.* 81, 627, 1995.
14. Agatsuma, T., and Habe, S. Electrophoretic studies on enzymes of diploid and triploid *Paragonimus westermani*. *Parasitology* 91, 489, 1985.
15. Agatsuma, T. et al. Genetic differentiation between Malaysian and other Asian *Paragonimus westermani*. *Trop. Biomed.* 10, 45, 1993.
16. Iwagami, M. et al. Molecular phylogeographic studies on *Paragonimus westermani* in Asia. *J. Helminthol.* 74, 315, 2000.
17. Park, G.M., Im, K.I., and Yong, T.S. Phylogenetic relationship of ribosomal ITS2 and mitochondrial COI among diploid and triploid *Paragonimus westermani* isolates. *Korean J. Parasitol.* 41, 47, 2003.
18. Doanh, P.N. et al. Discovery of *Paragonimus westermani* in Vietnam and its molecular phylogenetic status in *P. westermani* complex. *Parasitol. Res.* 104, 1149, 2009.
19. Blair, D. et al. Phylogenetic relationships among the Thai species of *Paragonimus* inferred from DNA sequences. In: Tada, I., Kojima, S., and Tsuji, M. (eds.), *IX International Congress of Parasitology (ICOPA IX)*, Monduzzi Editore, Bologna, p. 643, 1998.
20. Sugiyama, H. et al. New form of *Paragonimus westermani* discovered in Thailand: Morphological characteristics and host susceptibility. *Southeast Asian J. Trop. Med. Public Health* 38(Suppl. 1), 87, 2007.
21. Binchai, S. et al. Molecular systematics of a new form of *Paragonimus westermani* discovered in Thailand, *Southeast Asian J. Trop. Med. Public Health* 38(Suppl. 1), 92, 2007.
22. Singh, S.T., Singh, D.L., and Sugiyama, H. Possible discovery of Chinese lung fluke, *Paragonimus skrjabini* in Manipur, India. *Southeast Asian J. Trop. Med. Public Health* 37(Suppl. 3), 53, 2006.
23. Kamo, H. et al. On the occurrence of a new lung fluke, *Paragonimus miyazakii* n. sp. in Japan (Trematoda: Troglotrematidae). *Yonago Acta Med.* 5, 43, 1961.
24. Waikagul, J. A new species of *Paragonimus* (Trematoda: Troglotrematidae) from a cat infected with metacercariae from mountain crabs *Larnaudia larnaudii*. *J. Parasitol.* 93, 1496, 2007.
25. Thaenkham, U., and Waikagul, J. Molecular phylogenetic relationship of *Paragonimus pseudoheterotremus*. *Southeast Asian J. Trop. Med. Public Health* 39, 217, 2008.
26. Aka, N.A. et al. Human paragonimiasis in Africa. *Ann. Afr. Med.* 7, 153, 2008.
27. Zhong, H.L. et al. Recent progress in studies of *Paragonimus* and paragonimiasis control in China. *Chin. Med. J.* 94, 483, 1981.
28. Uchiyama, F., Morimoto, Y., and Nawa, Y. Re-emergence of paragonimiasis in Kyushu, Japan. *Southeast Asian J. Trop. Med. Public Health* 30, 686, 1999.
29. Singh, T.S. et al. Paragonimiasis in Manipur. *Indian J. Med. Res.* 97, 247, 1993.
30. Singh, T.S., Mutum, S.S., and Razaque, M.A. Pulmonary paragonimiasis: Clinical features, diagnosis and treatment of 39 cases in Manipur. *Trans. R. Soc. Trop. Med. Hyg.* 80, 967, 1986.
31. Singh, T.S., and Sugiyama, H. Paragonimiasis in India: A newly emerging food borne parasitic disease. *Clin. Parasitol.* 19, 95, 2008.
32. Razaque, M.A., Mutum, S.S., and Singh, T.S. Recurrent haemoptysis? Think of paragonimiasis. *Trop. Doct.* 21, 153, 1991.
33. Ogakwu, M., and Nwokolo, C. Radiological findings in pulmonary paragonimiasis as seen in Nigeria: A review based on one hundred cases. *Br. J. Radiol.* 46, 699, 1973.
34. Romeo, D.P., and Pollock, J.J. Pulmonary paragonimiasis: Diagnostic value of pleural fluid analysis. *South. Med. J.* 79, 241, 1986.
35. Watanabe, S. et al. Pulmonary paragonimiasis mimicking lung cancer on FDG-PET imaging. *Anticancer Res.* 23, 3437, 2003.
36. Yoshino, I. et al. Paragonimiasis westermani presenting as an asymptomatic nodular lesion in the lung: Report of a case. *Surg. Today* 28, 108, 1998.
37. Johnson, R.J., and Johnson, J.R. Paragonimiasis in Indochinese refugees. Roentgenographic findings with clinical correlations. *Am. Rev. Respir. Dis.* 128, 534, 1983.
38. Jeon, K. et al. Clinical features of recently diagnosed pulmonary paragonimiasis in Korea. *Chest* 128, 1423, 2005.
39. Chung, H.L., and Tsao, W.C. *Paragonimus westermani* (Szechuan variety) and a new species of lung fluke–*paragonimiasis szechuanensis*, part II. Studies on clinical aspects of *Paragonimiasis szechuanensis*: A new clinical entity. *Chin. Med. J.* 81, 419, 1962.
40. Singh, P.I., Singh, I.P., and Singh, L.B.M. Paragonimiasis: Review of 45 cases. *Indian J. Med. Microbiol.* 10, 243, 1992.
41. Jeong, M.G. et al. Retroperitoneal paragonimiasis: A case of ectopic paragonimiasis presenting as periureteral masses. *J. Comput. Assist. Tomogr.* 23, 696, 1999.
42. Singh, T.S., Khamo, V., and Sugiyama, H. Cerebral paragonimiasis mimicking tuberculoma: First case report in India. *Trop. Parasitol.* 1, 39, 2011.
43. Miyazaki, I., and Nishimura, K. Cerebral paragonimiasis. In: Hornabrook, R.W. (ed.), *Topics on Tropical Neurology*, Contemporary Neurology Series Vol. 12, Chapter 7. F.A. Davis Company, Philadelphia, p. 109, 1975.
44. Hughes, A.J., and Biggs, B.A. Parasitic worms of the central nervous system: An Australian perspective. *Intern. Med. J.* 32, 541, 2002.
45. Jaroonvesama, N. Differential diagnosis of eosinophilic meningitis. *Parasitol. Today* 4, 262, 1988.

46. Chang, K.H. et al. An imaging diagnosis of cerebral paragonimiasis: CT and MR findings and correlation with ELISA antibody test. *J. Korean Radiol. Soc.* 29, 345, 1993.
47. Choo, J.D. et al. Chronic cerebral paragonimiasis combined with aneurismal subarachnoid haemorrhage. *Am. J. Trop. Med. Hyg.* 69, 466, 2003.
48. Singh, T.S. et al. *Paragonimus heterotremus* infection in Nagaland: A new focus of paragonimiasis in India. *Indian J. Med. Microbiol.* 27, 123, 2009.
49. Komiya, Y., and Yokogawa, M. The recovering of *Paragonimus* eggs from stools of paragonimiasis patients by AMS III centrifuging technic. *Jpn. J. Med. Sci. Biol.* 6, 207, 1953.
50. Young, K.H. et al. Ethyl acetate as a substitute for diethyl ether in the formalin-ether sedimentation technique. *J. Clin. Microbiol.* 10, 852, 1979.
51. Kuroki, M. et al. High-resolution computed tomography findings of *P. westermani*. *J. Thorac. Imaging* 20, 210, 2005.
52. Song, H.O. et al. Skin test for paragonimiasis among schoolchildren and villagers in Namback District, Luangprabang province, Lao PDR. *Korean J. Parasitol.* 46, 179, 2008.
53. Sawada, T., Takei, K., and Yoneyama, K. Studies on the immunodiagnosis of paragonimiasis. II. Intradermal tests with fractionated antigens. *J. Infect. Dis.* 114, 315, 1964.
54. Yokogawa, M., and Tsuji, M. Immunological diagnosis as the screening method of paragonimiasis in the endemic area of paragonimiasis. In: *First Regional Symposium on Scientific Knowledge of Tropical Parasites, University of Singapore, 5 to 9 November, 1962*, UNESCO, Singapore, p. 194, 1962.
55. Biguet, J. et al. Contribution of immunoelectrophoretic analysis to the knowledge of worm antigens: Practical effects on their standardization, their purification and diagnosis of helminthiasis by immunoelectrophoresis. *Rev. Immunol. Ther. Antimicrob.* 29, 5, 1965.
56. Katamine, D., Imai, J., and Iwamoto, I. Immunological study on paragonimiasis: Agar-gel precipitin reactions in paragonimiasis. *Trop. Med.* 10, 28, 1968.
57. Tsuji, M., Capron, A., and Yokogawa, M. Immunodiffusion test in the diagnosis of paragonimiasis. In: *Second International Congress of Parasitology (ICOPA II)*, Colloquium on Immunodiagnosis, Washington, D.C., p. 6, 1970.
58. Tsuji, M. On the immunoelectrophoresis for helminthological researches. *Jpn. J. Parasitol.* 23, 335, 1974.
59. Pariyanonda, S. et al. Serodiagnosis of human paragonimiasis caused by *Paragonimus heterotremus. Southeast Asian J. Trop. Med. Public Health* 21, 103, 1990.
60. Quicho, L.P. et al. Humoral immune response of cats to *Paragonimus* infection. *Southeast Asian J. Trop. Med. Public Health* 12, 364, 1981.
61. Waikagul, J. Serodiagnosis of paragonimiasis by enzyme-linked immunosorbent assay and immunoelectrophoresis. *Southeast Asian J. Trop. Med. Public Health* 20, 243, 1989.
62. Imai, J. Evaluation of ELISA for the diagnosis of *paragonimiasis westermani Trans. R. Soc. Trop. Med. Hyg.* 81, 3, 1987.
63. Zhang, Z. et al. Diagnosis of active *Paragonimus westermani* infections with monoclonal antibody-based antigen detection assay. *Am. J. Trop. Med. Hyg.* 49, 329, 1993.
64. Itoh, M., and Sato, S. Multi-dot enzyme-linked immunosorbent assay for serodiagnosis of trematodiasis. *Southeast Asian J. Trop. Med. Public Health* 21, 471, 1990.
65. Indrawati, I. et al. Studies on immunodiagnosis of human paragonimiasis and specific antigen of *Paragonimus heterotremus. Int. J. Parasitol.* 21, 395, 1991.
66. Narain, K., Devi, K.R., and Mahanta, J. Development of enzyme-linked immunosorbent assay for serodiagnosis of human paragonimiasis. *Indian J. Med. Res.* 121, 739, 2005.
67. Nakamura-Uchiyama, F., Onah, D.N., and Nawa, Y. Clinical features of paragonimiasis cases recently found in Japan: Parasite-specific immunoglobulin M and G antibody classes. *Clin. Infect. Dis.* 32, e171, 2001.
68. Ikeda, T. et al. Parasite-specific IgE and IgG levels in the serum and pleural effusion of *Paragonimiasis westermani* patients. *Am. J. Trop. Med. Hyg.* 47, 104, 1992.
69. Sugiyama, H. et al. Characterization and localization of *Paragonimus westermani* antigen stimulating antibody formation in both the infected cat and rat. *J. Parasitol.* 73, 363, 1987.
70. Maleewong, W. et al. Antigenic components of *Paragonimus heterotremus* recognized by infected human serum. *Parasite Immunol.* 13, 89, 1991.
71. Maleewong, W. et al. Analysis of antibody levels before and after praziquantel treatment in human paragonimiasis heterotremus. *Asian Pac. J. Allergy Immunol.* 10, 69, 1992.
72. Dekumyoy, P., Waikagul, J., and Eom, K.S. Human lung fluke *Paragonimus heterotremus*: Differential diagnosis between *Paragonimus heterotremus* and *Paragonimus westermani* infections by EITB. *Trop. Med. Int. Health* 3, 52, 1998.
73. Kong, Y. et al. Immunoglobulin G (IgG) subclass and IgE responses in human paragonimiasis caused by three different species. *Clin. Diagn. Lab. Immunol.* 5, 474, 1998.
74. Gan, X.X. et al. Development of rapid diagnostic kit (Dot Immunogold Filtration Assay) for detection of antibodies against *Paragonimus westermani. Chin. J. Zoo.* 21, 988, 2005.
75. Qian, B.Z., and Sugiyama, H. Development of an immunological diagnostic kit for paragonimiasis and its use for an epidemiological survey of zoonotic paragonimiasis in Mainland China, 1. Multi-antigen dot immunogold filtration assay (Multi-DIGFA) for serodiagnosis of paragonimiasis. Final report of the JHSF project, The Japan Health Sciences Foundation, Tokyo, 2007.
76. Singh, T.S. et al. Morphological and molecular characterizations of *Paragonimus heterotremus*, the causative agent of human paragonimiasis in India. *Southeast Asian J. Trop. Med. Public Health* 38(Suppl. 1), 82, 2007.
77. Chang, Z.S. et al. Gene sequencing for identification of *Paragonimus* eggs from a human case. *Chin. J. Parasitol. Parasitic Dis.* 18, 213, 2000.
78. Le, T.H. et al. *Paragonimus heterotremus* Chen and Hsia, (1964), in Vietnam: A molecular identification and relationships of isolates from different hosts and geographical origins. *Acta Trop.* 98, 25, 2006.
79. Intapan, P.M. et al. Detection of *Paragonimus heterotremus* eggs in experimentally infected cats by a polymerase chain reaction-based method. *J. Parasitol.* 91, 195, 2005.
80. Nkouawa, A. et al. Paragonimiasis in Cameroon: Molecular identification, serodiagnosis and clinical manifestations. *Trans. R. Soc. Trop. Med. Hyg.* 103, 255, 2009.
81. Agatsuma, T. et al. Mitochondrial DNA differentiation of Japanese diploid and triploid *Paragonimus westermani. J. Helminthol.* 68, 7, 1994.
82. Sugiyama, H. et al. Molecular discrimination between individual metacercariae of *Paragonimus heterotremus* and *P. westermani* occurring in Thailand. *Southeast Asian J. Trop. Med. Public Health* 36(Suppl. 4), 102, 2005.
83. Sugiyama, H. et al. A multiplex PCR for discrimination between *Paragonimus westermani* and *P. miyazakii* at the

metacercarial stage. *Southeast Asian J. Trop. Med. Public Health* 35(Suppl. 1), 327, 2004.

84. Intapan, P.M. et al. Genomic characterization of lung flukes, *Paragonimus heterotremus*, *P. siamensis*, *P. harinasutai*, *P. westermani* and *P. bangkokensis* by RAPD markers. *Vet. Parasitol.* 124, 55, 2004.
85. Maleewong, W. et al. Detection of *Paragonimus heterotremus* in experimentally infected cat feces by antigen capture-ELISA and by DNA hybridization. *J. Parasitol.* 83, 1075, 1997.
86. Sugiyama, H. et al. Polymerase chain reaction (PCR)-based molecular discrimination between *Paragonimus westermani* and *P. miyazakii* at the metacercarial stage. *Mol. Cell. Probes* 16, 231, 2002.
87. Rowan, W.B. The mode of hatching of the egg of *Fasciola hepatica*. *Exp. Parasitol.* 5, 118, 1956.
88. Vanijanonta, S. et al. Pulmonary paragonimiasis with expectoration of worms: A case report. *Southeast Asian J. Trop. Med. Public Health* 12, 104, 1981.
89. Miyazaki, I., Arellano, R.C., and Grados, O. The first demonstration of the fluke, *Paragonimus* from man in Peru. *Jpn. J. Parasitol.* 21, 168, 1972.
90. Miyazaki, I., and Fontan, R. Mature *Paragonimus heterotremus* found from a man in Laos. *Jpn. J. Parasitol.* 19, 109, 1970.
91. Miyazaki, I., and Harinasuta, T. The first case of human paragonimiasis caused by *P. heterotremus* Chen et Hsia, 1964. *Ann. Trop. Med. Parasitol.* 60, 509, 1966.
92. Sugiyama, H. et al. Studies on the geographical distribution of the lung fluke, *Paragonimus westermani* (Kerbert, 1878) in the southern prefectures of the Kinki District, Japan: Observations on the incidence of encysted larvae of *P. westermani* in *Geothelphusa dehaani* in Higashi-Yoshino area, Nara Prefecture. *J. Vet. Med. Sci.* 45, 227, 1983.
93. Maejima, J., Kamo, H., and Hatsushika, R. Studies on *Paragonimus miyazakii* Kamo, Nishida, Hatsushika et Tomimura, 1961. 2. Second intermediate host and intracrab stage. *Yonago Acta Med.* 15, 35, 1971.
94. Komalamisra, C., Asavisanu, R., and Setasuban, P. Distribution of *Paragonimus heterotremus* metacercariae in fresh water crab, *Tiwaripotamon beusekomae* Bott 1970. *Southeast Asian J. Trop. Med. Public Health* 19, 337, 1988.
95. Rangsiruji, A. et al. A new record of *Paragonimus* other than *P. westermani* in southern Thailand. *Southeast Asian J. Trop. Med. Public Health* 37(Suppl. 3), 57, 2006.
96. Bowles, J., Blair, D., and McManus, D.P. A molecular phylogeny of the human schistosomes. *Mol. Phylogenet. Evol.* 4, 103, 1995.
97. Rim, H.J. Paragonimiasis: Experimental and clinical experience with praziquantel in Korea. *Arzneimittelforschung* 34(9B), 1197, 1984.
98. Nawa, Y., and Doanh, P.N. *Paragonimus bangkokensis* and *P. harinasutai*: Reappraisal of phylogenetic status. *J. Trop. Med. Parasitol.* 32, 87, 2009.

40 Plagiorchis

Sun Huh

CONTENTS

40.1 INTRODUCTION

Plagiorchiid species is one of the minute intestinal trematode of reptiles, birds, and mammals. Human infection with genus *Plagiorchis* is one of rare intestinal trematode infection occurred mainly in Southeast or East Asia. Twelve natural cases of human infection by the genus *Plagiorchis* have been reported [1].

Since plagiorchiid species does not cause clinical symptoms or signs, the diagnosis of this minute worm has not been considered at the first concern in the clinics or in the epidemiological fields. The detection of eggs or worms under microscope is the convenient diagnostic method. However, if only part of the worm is found or a histological section is only available, the molecular technique may be a valuable tool for the confirmation of species.

40.1.1 Classification, Morphology, and Biology

Plagiorchis is the genus of family Plagiorchiidae. Five plagiorchiid species were found from human.

Taxonomic position is as follows:

Class Trematoda
  Family Plagiorchiidae
    Genus Plagiorchis
      Species *Plagiorchis harinasutai* Radomyois, Bunnag & Harinasuta, 1989
      Plagiorchis javensis Sandground, 1940
      Plagiorchis muris (Tanabe, 1922) Shults & Skvortsov, 1931
      Plagiorchis philipinensis Sandground, 1940
      Plagiorchis vespertilionis (Muller, 1780) Braun, 1900.

Morphology of *P. muris* (Figure 40.1) and *P. vespertilionis* were described precisely in the literature [2–4]. These two species also have been studied extensibly in Korea, Japan, United States, and Europe. Table 40.1 shows the characteristics of these two species.

40.1.1.1 Life Cycle of *Plagiorchis vespertilionis parorchis*

P. vespertilionis parorchis is the subspecies of *P. vespertilionis* [5]. During the life cycle, the cercaria develops in *Lymnaea stagnalis*, the first intermediate host. The second intermediate hosts are mosquito larvae, caddis-fly larvae, mayfly larvae, and dragonfly nymphs. Newly formed metacercaria's size is 0.126×0.170 mm. The anterior end of the metacercaria is composed of fine spines. Seven days after the penetration to the second intermediate host, metacercariae can be infective when fed to mice. Seven days postinfection in mice, the worms had well-formed organs and sperm, but eggs had not yet appeared; in 16 days maximum size of mature worms was attained. However, worms were expelled soon from the host intestine.

HOST: *Plecotus auritus* (brown long-eared bat), *Meriones persicus* (Rodentia) [6].
LOCATION: Small intestine

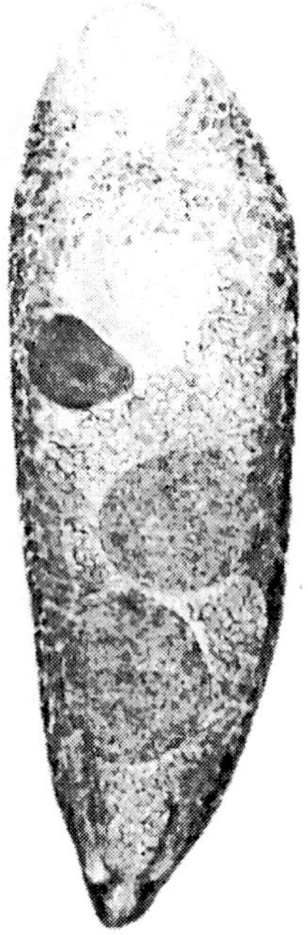

FIGURE 40.1 *Plagiorchis muris* adult specimen collected from experimental rat intestine 28 days postinfection. Acetocarmine stain. Size 2.13 × 0.69 mm. (Courtesy of Prof. Sung-Jong Hong, Chung-Ang University College of Medicine, Korea.)

40.1.1.2 Life Cycle of *P. muris*

Metacercariae of *P. muris*, obtained from dragonflies, *Sympetrum eroticum*, can be infective when fed to 4-week-old rats. The peak of egg production occurs at 14 days postinfection. At 28 days postinfection, the recovery rate dropped significantly. Mature flukes were found at the lower part of the small intestine 5 days postinfection [7]. Metacercariae of *P. muris*: 0.169 (0.165–0.174) mm × 0.158 (0.148–0.170) mm, was elliptical to spherical in shape with a hyaline thin wall. Adult worms can also be recovered from infected mouse intestines [7].

HOST: *Apodemus agrarius* [8], *Apodemus sylvaticus*, *Rattus r. alexandrinus*. *Procyon lotor* (raccoons) [9], *Felis silvestris catus* (cat)[10], *Neocaridina denticulata* (green neon shrimp).
LOCATION: Small intestine

In Thailand, four human cases of *P. harinasutai* infection were detected incidentally during the worm collection for the treatment of *Opisthorchis viverinii* in 1989 [11]. Three cases of *P. javensis* in Indonesia [12], two cases of *P. muris* in Japan and Korea [13,14], two cases of *P. philippinensis* in Philippines [15], and one case of *P. vespertilonis* in Korea [2] had been reported.

Humans are infected with plagiorchiid species probably by ingestion of insect larva and raw fresh water fishes, specifically snakehead mullet and gobies.

TABLE 40.1
Morphological Characteristics of *P. muris* and *P. vespertilionis*

Organs	*P. vespertilionis* [2]	*P. muris* [3]
Body length × width	2.6 × 0.7 mm	1.46–1.77 × 0.364–0.448 mm
Sucker ratio	1.0:1.0	1.0:1.0
Pharynx	0.14 × 0.15 mm	0.066–0.084 mm long × 0.076–0.084 mm
Esophagus	Nearly absent	Short
Ceca	Extend to near the posterior extremity	Extending to near the posterior end of the body
Cirrus sac	Extend from the ventral sucker to the anterior part of the ovary	Extends from genital aperture posteriad
Testis	Postovarian, oval	Oval, almost contiguous, slightly oblique in intracecal space
Ovary	Postacetabulum	Oval, almost contiguous with the ventral sucker
Vitellaria	Extend to the posterior end of the worm	Extending from the posterior margin of the oral sucker to the posterior tip of the body
Uterus	With descending and ascending loop	Forms a large "S" shape with descending and ascending loops
Eggs	32.5 + 1.3 × 17.5 + 1 0.5 μm	0.030–0.033 mm × 0.020–0.023 mm

40.1.2 Clinical Features and Pathogenesis

Plagiorchiid species are usually detected incidentally [14]. There are no specific or clinically noticeable symptoms or signs in the human infection. There was an experimental human infection by researcher [16]. He ingested 150 metaceracariae. However, he complains of no symptoms nor feels any gastrointestinal discomfort unlike other intestinal trematode infection during entire observational period. From 38th day postinfection, egg production was terminated. It is believed that adult worms were naturally expelled or they lost the egg productivity. If the number of worms is >1000 or 10,000, it may cause several gastrointestinal discomfort, specifically abdominal pain, diarrhea, or abdominal distension. Reported human cases did not mention such a large amount of worm burden. No symptom or sign of plagiorchiid species infection may reflect the low worm burden or suggest the low or negligible pathogenesis. In the animal examination with natural heavy infection of *P. muris*, there was no pathologic finding in the intestine. This phenomenon may support the later assumption.

40.1.3 Diagnosis

40.1.3.1 Conventional Techniques

Diagnosis of plagiorchiid species is possible by stool examination. Eggs of Plagiorchis can be detected by the microscopic examination. However, in the field or in the clinics, its human infection is usually not suspected due to no clinical symptom. Therefore, it usually can be detected after the worm collection incidentally or during autopsy. This minute worm recovered after expulsion by praziquantel treatment can be observed and identified under a microscope. For diagnosis of other intestinal trematode infection, serological technique such enzyme-linked immunosorbent assay may be applied. During the early phase of infection of *Metagonimus yokogwai*, although infected human complains of abdominal pain and diarrhea, eggs cannot be found in the stool since the worm is still in the larval stage [17]. Serological test can be powerful diagnosis tool during this early stage of infection. For plagiorchiid species, serologic test was not developed since its human infection is rare. Past history of eating raw freshwater fishes also cannot help diagnosis of plagiorchiid species since those are also the second intermediate host of many other intestinal trematodes.

40.1.3.2 Molecular Techniques

Since it is possible to confirm plagiorchiid species infection by conventional methods, molecular technique is not commonly applied in the field or in the clinics. However, if a part of the worm is only detected or only a section of the worm is available, the molecular identification may be possible. There are still no data on restriction-fragment-length polymorphism (RFLP) for plagiorchiid species. At present, for the confirmation of the plagiorchiid species, the sequence of the specific gene is required. Although there is no report on the identification of plagiorchiid species by molecular technique, the possible method of molecular detection is introduced. Those methods of gene sequencing of 28S D1 rDNA and mtCOI region of *P. muris* are from Lee et al. [18].

40.2 METHODS

40.2.1 Sample Preparation

40.2.1.1 DNA Preparation

Worms are kept in ethanol at –70 until assayed. Frozen worms are lyophilized and lysed with lysis buffer. DNA is extracted using the phenol/chloroform method and precipitated in ethanol.

40.2.2 Detection Procedures

40.2.2.1 Polymerase Chain Reaction for Specific Genes

The 28S D1 rDNA and mtCOI regions are amplified by polymerase chain reaction (PCR) using the primer sets described by Qu et al. [19]. PCR is done using extracted DNA as a template, and primers. PCR amplification is conducted over 40 cycles denaturing at 95°C for 20 s, annealing at 55°C (28S D1 rDNA) or 48°C (mtCOI) for 30 s, extending at 72°C for 30 s, followed by a final extension of 7 min.

40.2.2.2 Cloning

Amplified PCR products are extracted and purified and are cloned into a cloning vector using ligase and transformed into competent cells. Positive recombinant clones were picked, and grown in 2 mL of Luria–Bertani broth overnight at 37°C. The recombinant plasmid is screened. Positive plasmid DNA was purified.

40.2.2.3 DNA Sequencing

DNA sequencing is performed by an automated DNA sequencer. Each gene was sequenced in both orientations using the universal sequencing primers. At least two clones were sequenced per gene; additional clones were sequenced as necessary to resolve ambiguous sites.

The sequence data of 28S D1 rDNA region obtained can be compared for the identification of species. Those of the mtCOI region may not be suitable for the identification of species since its variability in each worm. The analyses of 28S D1 rDNA and mtCOI region sequences are determined by comparison with those of a range of other related trematodes. NCBI databases are used for sequence homology analysis using BLAST2. EMBL/GeneBank Data Libraries under Accession numbers for 28S rDNA sequence and mtCOI sequence of *P. muris* are AF096222 and AF096236, respectively.

40.3 CONCLUSIONS AND FUTURE PERSPECTIVES

Plagiorchid species is the trematode incidentally found in human intestine. It is not regarded as an important pathogen to human since it provokes no clinical symptoms or signs. Same phenomena were observed in the heavy infection of other zoonotic animals. It means that plagiorchiid species are well adapted to their natural host evolutionally. Only *P. muris* is much studied for the experimental and natural human infection. Detection of eggs or worms under a microscope is the common diagnosis method at present. In Korea where *P. muris* is found in human, the molecular technique specifically sequencing of the specific gene can be applied to confirm the species out of many intestinal trematodes. Molecular identification of intestinal trematodes in human is not a common method used in the field or laboratory. If only part of the worm is present for the identification, a common gene marker easily detectable by RFLP will be necessary to differentiate the intestinal trematode from other trematodes or cestodes. Thereafter, the specific diagnosis marker for each family will be necessary. For the specific identification of each species in each family, a single gene marker may be developed in the future.

REFERENCES

1. Chai, J.Y., Shin, E.H., Lee, S.H., and Rim, H.J. Foodborne intestinal flukes in Southeast Asia. *Korean J. Parasitol.* 47 (Suppl), S69–1S02, 2009.
2. Guk, S.M., Kim, J.L., Park, J.H., and Chai, J.Y. A human case of *Plagiorchis vespertilionis* (Digenea: Plagiorchiidae) infection in the Republic of Korea. *J. Parasitol.* 93(5), 1225–1227, 2007.
3. Seo, B.S., Rim, H.J., and Lee, C.W. Studies on the parasitic helmiths of Korea: I. Trematodes of rodents. *Korean J. Parasitol.* 2(1), 20–26, 1964.
4. Hong, S.J. Surface ultrastructure of *Plagiorchis muris* growth and developmental stages in rats, the final host. *Parasitol. Res.* 105(4), 1077–1083, 2009.
5. MACY, R.W. The life cycle of *Plagiorchis vespertilionis Parorchis*. N. SSP. (Trematoda: Plagiorchidae), and observations on the effects of light on the emergence of the cercaria. *J. Parasitol.* 46:337–345, 1960.
6. Groschaft, J., and Tenora, F. *Plagiorchis vespertilionis* (Müller, 1780) (Trematoda)—A parasite of *Meriones persicus* (Rodentia) in Afghanistan. *Folia Parasitol (Praha).* 18(1), 40, 1971.
7. Hong, S.J., Woo, H.C., Lee, S.U., and Huh, S. Infection status of dragonflies with *Plagiorchis muris* metacercariae in Korea. *Korean J. Parasitol.* 37(2), 65–70, 1999.
8. Rogan, M.T., Craig, P.S., Hide, G., Heath, S., Pickles, A., and Storey, D.M. The occurrence of the trematode *Plagiorchis muris* in the wood mouse *Apodemus sylvaticus* in North Yorkshire, UK. *J. Helminthol.* 81(1), 57–62, 2007.
9. Sato, H., and Suzuki, K. Gastrointestinal helminths of feral raccoons (*Procyon lotor*) in Wakayama Prefecture, Japan. *J. Vet. Med. Sci.* 68(4), 311–318, 2006.
10. Sohn, W.M., and Chai, J.Y. Infection status with helminthes in feral cats purchased from a market in Busan, Republic of Korea. *Korean J. Parasitol.* 43(3), 93–100, 2005.
11. Radomyos, P., Bunnag, D., and Harinasuta, T. A new intestinal fluke, *Plagiorchis harinasutai* n.sp. *Southeast Asian J. Trop. Med. Public Health* 20(1), 101–107, 1989.
12. Sandground, J.H. *Plagiorchis javensis* n. sp. a new trematode parasitic in man. *Rev. Med. Trop. Parasitol. (Habana)* 6, 207–211, 1940.
13. Asada, J.I., Otagaki, H., and Morita, M. et al. A case report on the human infection with *Plagiorchis muris* Tanabe, 1922 in Japan. *Jpn. J. Parasitol.* 11, 512–516, 1962.
14. Hong, S.J., Woo, H.C., and Chai, J.Y. A human case of *Plagiorchis muris* (Tanabe, 1922: Digenea) infection in the Republic of Korea: Freshwater fish as a possible source of infection. *J. Parasitol.* 82(4), 647–649, 1996.
15. Yu, S.H., and Mott, K.E. Epidemiology and morbidity of food-borne intestinal trematode infections. *Trop. Dis. Bull.* 91, R125–R152, 1994.
16. McMullen, D.B. An Experimental infection of *Plagiorchis muris* in man. *J. Parasitol.* 23, 113–115, 1937.
17. Chai, J.Y., Yu, J.R., Lee, S.H. et al. An egg-negative patient of acute metagonimiasis diagnosed serologically by ELISA. *Seoul. J. Med.* 30, 139–142, 1989.
18. Lee, S.U., Huh, S., and Sohn, W.M. Molecular phylogenic location of the *Plagiorchis muris* (Digenea, Plagiorchiidae) based on sequences of partial 28S D1 rDNA and mitochondrial cytochrome *C* oxidase subunit I. *Korean J. Parasitol.* 42(2), 71–75, 2004.
19. Qu, L.H., Nicoloso, M., and Bachellerie, J.P. Phylogenetic calibration of the 5′ terminal domain of large rRNA achieved by determining twenty eucaryotic sequences. *J. Mol. Evol.*, 28(1–2), 113–124, 1988.

41 Schistosoma and Schistosomiasis

Antonio Muro and Luís Pérez del Villar

CONTENTS

41.1 INTRODUCTION

Human schistosomiasis is a severe debilitating illness that represents one of the most important parasitic diseases in tropical areas. It affects over 200 million people across 74 countries [1]. A digenetic trematode worm ("fluke") of the genus *Schistosoma* that infects both humans and animals is the parasite responsible for schistosomiasis. This illness was first reported in 1798; Renault describes episodes of hematuria suffered by Napoleon's army soldiers in Egypt. However, the first report of one *Schistosoma* species was made in 1851 by Theodor von Siebold and Theodor Bilharz, who describe *Schistosoma haematobium* (this is the reason why schistosomiasis is also called "bilharziosis"). A year later, the relationship between *S. haematobium* and urinary pathology triggered by this *Schistosoma* species was established. *Schistosoma* is a neglected ancient pathogen whose evolution is associated with vertebrate host. In fact in 1910, Marc Arman Ruffer found *S. haematobium* eggs in two Egyptian mummies dated 1250–1000 years [2] B.C. A few years earlier, in 1902, Manson announced that intestinal schistosomiasis was caused by *Schistosoma mansoni*. Almost simultaneously, in 1904, Fujiro Katsurada discovered *Schistosoma japonicum* and Dario Fujii correlated this parasite with schistosomiasis found in people and animals of the Valley of Katayama. This fact determined the name of Katayama, attributable to the acute phase of the disease. In 1915, Robert Leipert described the intermediate hosts (in different snail species) completing the biological cycle. Other schistosome species like *Schistosoma intercalatum* or *Schistosoma mekongi* [3] were described in humans in the second part of the 20th century.

Recently, a new species denominated *Schistosoma guineensis* [4] has been described.

41.1.1 Structure and Biology of Schistosomes

41.1.1.1 Taxonomy

The genus *Schistosoma* belongs to the family Schistosomatidae, order Strigeiformes, subclass Digenea, class Trematoda, phylum Platyhelminthes, and kingdom Animalia [5]. The genus *Schistosoma* contains 21 species which are classified into four groups according to the geographic distribution, morphology of the parasite's eggs, and the intermediate host [6]. However, the taxonomy and phylogeny of *Schistosoma* sp. are in continuous review due to the large number of molecular studies being carried out [7]. Five major schistosome species are able to infect humans: *S. mansoni*, *S. haematobium*, *S. intercalatum*, *S. japonicum*, and *S. mekongi*, although human infections by *S. malayensis*, *S. mattheei*, and *S. guineensis* [4] have been described.

41.1.1.2 Morphology and Life Cycle

Schistosomiasis is acquired by contact with freshwater contaminated with cercariae (free-swimming larval forms of parasite). Cercarial larvae are disseminated by different species of snails. *S. mansoni* is transmitted by *Biomphalaria*; *S. haematobium* and *S. intercalatum* by *Bullinus*; *S. japonicum* by *Oncomelania*; and *S. mekongi* by *Neotricula* [3]. The cercariae actively penetrate mammal skin, discarding their bifurcated tail, and transform into the schistosomula phase. Schistosomula migrate toward the lungs and return through the general circulation to their final sites of parasitism; *S. mansoni*, *S. japonicum*, *S. mekongi*, and *S. intercalatum* migrate to mesenteric veins, whereas *S. haematobium* reaches the vesical veins of the bladder. In this location, male and female parasites complete their maturation; sexual reproduction occurs and oviposition begins. The females are able to produce hundreds to thousands of eggs per day. The majority of eggs migrate through the lumen of the bladder or the intestine and are excreted in the urine or feces in a period of time of 35–70 days after infection. The description of the more representative morphological features of eggs and adult worms are shown in Table 41.1. Eggs reach the water and release the miracidia that penetrate the snails to transform and multiply asexually until they become cercariae. The life cycle of *S. mansoni* kept at our laboratory is illustrated in Figure 41.1.

41.1.1.3 Genome, Transcriptome, and Proteome

In 1994, the World Health Organization (WHO) along with Schistosoma Genome Network began the project of sequencing the *genome* of *S. mansoni* in order to understand the processes involved in the transduction/translation and in the quantitative expression of potential genes as targets for designing vaccines and drugs. Finally, in 2009, the complete sequence and genome mapping of *S. mansoni* [8] was described. The genome size of *S. mansoni* is 363 Mb, larger than most pathogens sequenced to date. 11,809 genes have been identified with an average size of 4.7 kb encoding 13,197 transcripts. The genes described have a particular structure with large introns (average of 1692 bp) and shorter exons (217 bp). In addition, 5′ introns are much smaller than the introns of the 3′ end. 43% of the genome (distributed in 153 scaffolds) has been mapped using the technique of *in situ* hybridization (FISH) in eight pairs of chromosomes, seven autosomes, and one sex chromosome. Another important feature of the genome of *S. mansoni* is the high content of repetitive sequences (at least 40%), an important fact from the standpoint of evolutionary and phylogenetic relationship with the host. In the same year, the Chinese Human Genome Sequencing Center published and mapped the *S. japonicum* genome [9]. The genome of *S. japonicum* is 397 Mb, in which 13,469 genes encoding proteins have been identified. The

TABLE 41.1
Morphological Characteristics of *Schistosoma* spp.

	Morphology	*S. haematobium*	*S. intercalatum*	*S. mansoni*	*S. japonicum*	*S. mekongi*
Adults						
	Large (μm)	10–15	10–14	10–12	10–20	10–20
	Diameter (μm)	0.75–1	0.3–0.4	1.10	0.55	0.41
Male	Number of testes	4–5	2–7 (+4–5)	4–13 (+7–8)	6–7	6–7
	Tubercles	Fine	Fine	Prominent	Absent	Absent
Female	Ovary (position)	Hindquarters	Posterior half	Anterior third	Half	Posterior half
	Utero (position)	Anterior two-thirds	Anterior two-thirds	Anterior half	Anterior half	Anterior half
	Number of eggs *in utero*	100	10–50	5–60	5–200	10+
Mature eggs	Shape	Ovoid	Ovoid	Ovoid	Round	Round
	Size (μm)	62 × 150	61 × 176	61 × 140	50 × 100	57 × 66
	Spine	Terminal (prominent)	Terminal (prominent)	Lateral (prominent)	Lateral (reduced)	Lateral (reduced)
	Eggs/females/day	20–300	150–400	100–300	3500	—

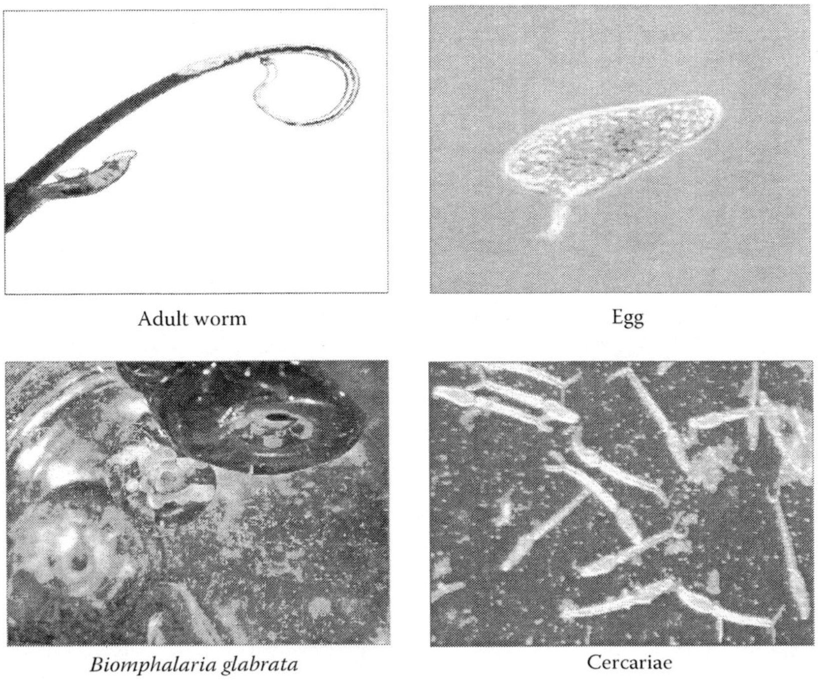

FIGURE 41.1 Biology of *Schistosoma mansoni*.

genome of *S. japonicum* conserves a large number of homologous sequences with the genome of *S. mansoni*, mostly those of protein coding regions [10]. Moreover, the *S. japonicum* genome has a high content of repetitive sequences (40.1% of the genome).

The first step in sequencing the genome of any organism is to generate *express sequencer tag* (EST), short sequences of randomly selected cDNA reflecting the transcriptional activity of organisms and tissues [11]. The *transcriptome* of *S. mansoni* [12] and *S. japonicum* [13] were completed in Brazil and China, respectively. In both projects, the ESTs were generated from a cDNA library obtained by RT-PCR which amplified the coding region of mRNA. The transcriptome obtained from six stages of the life cycle of *S. mansoni* presented 163,586 ESTs, grouped in 30.988 "or assembled EST clusters," whereas the transcriptome of *S. japonicum* consists of 43.707 ESTs, grouped into 13.131 clusters that cover most of the coding regions of proteins; 20% of these clusters show similarity to sequences described in mammals [13]. This fact provides information on the relationship established by schistosome parasites with the host and how these parasites exploit the endocrine and immune signals.

The term *proteome* was assigned in the mid-1990s to describe the total proteins expressed by the genome [14]. Proteome characterization of schistosome and other parasites are being currently developed using two powerful technologies: two-dimensional electrophoresis (2DE) and mass spectrometry. Many proteomic studies have compared the protein composition between different stages of the life cycle of *S. mansoni* [15] and between male and female adults of *S. japonicum* [16]. However, the numerous molecules of schistosomes in different organs and systems of interest are the molecules that are related to host–parasite interaction. These molecules are accessible to effector mechanisms of the host's immune system and may therefore be of great interest as the main targets for developing drugs and effective vaccines against these organisms. Three different large groups of proteins are the following: (i) Parasite surface proteins: nutrient transport proteins, enzymatic activity proteins, or receptors for host ligands. (ii) Host surface proteins: schistosomes acquire host's molecules in order to evade immune response. Major histocompatibility complex (MHC), low-density lipoproteins (LDL), C3 complement, and antibodies as molecules adhering to the parasite surface have been identified [17,18]. (iii) Secretion excretion proteins: proteins with a strong immunomodulatory character like proteases involved in the penetration of the parasite into the skin [19,20] or molecules involved in the granulomatous response produced by schistosomes eggs in tissues have been described [21].

41.1.2 Epidemiology of Schistosomiasis

41.1.2.1 Geographical Distribution and Current Status of Infection

The WHO announced in October 2001 that the epidemiology of schistosomiasis estimate should be recalculated. The data regarding this disease show 779 million people in 76 countries endemic for schistosomiasis. The geographical areas affected by the different species of human schistosomes are shown Figure 41.2. Furthermore, 250 million people are infected (80% in sub-Saharan Africa), 120 million have symptoms, and 20 million have severe disease and there is an annual mortality of 280,000. Some authors report the significant impact of the morbidity caused by schistosomiasis

FIGURE 41.2 Geographical distribution of schistosomiasis.

reflected in the loss of 1.53 million disability-adjusted life years [22].

There are a number of conditions necessary for an adequate transmission. The first is the water contamination with viable eggs deposited by the definitive host (humans or pets in the case of *S. japonicum*). The second is the conditions for the development of the intermediate hosts (snails). The snails are resistant to drought, climate change, and pollution. The construction of dams like Diama in Senegal, Akosombo in Ghana, and Three Gorges of the Yangtze River in China have allowed the dispersion of the snails and the risk of infection. Finally, the exposure of humans to cercariae-infected water bodies facilitates infection. Thus, children who are in contact with infected waters are most susceptible to contract the disease. In adults, people performing daily activities like washing clothes and utensils, farm workers, and fishermen are the populations most vulnerable to acquiring the infection. We must be aware that the transmission of schistosomiasis is local. Different levels of infection in various areas of the lake or between neighboring towns have been described [1].

41.1.2.2 Immigrants and Travelers

The absence of autochthonous cases of schistosomiasis in nontropical regions has caused it to be denominated as an "exotic disease." The rise of international tourism, increased migration, and international cooperation programs have led to an increase in schistosomiasis. The groups of patients diagnosed in developed countries belong to the following: (i) Travelers to rural areas and perirural endemic areas, especially those tourists practising "off track" adventure [23]. The database GeoSentinel considers schistosomiasis as a cause of disease in 4% of travelers returning from sub-Saharan Africa [24]. The highest prevalences in travelers from sub-Saharan Africa were detected by serology [25]. A study in Lake Malawi determined that travelers who have bathed for 10 days had an estimated 90% possibility of acquiring the disease [25]. Recently, an outbreak of schistosomiasis has been detected in travelers who were on a safari in Tanzania and had bathed in a natural lake near a luxury hotel [26]. (ii) Expatriates and immigrants from endemic countries [27]. In a study conducted in 788 sub-Saharan immigrants who had absolute eosinophilia, schistosomiasis was diagnosed in 17% of cases [25] and it was the first cause described in sub-Saharan immigrants who had relative eosinophilia [28].

41.1.3 Pathogenesis of Human Schistosomiasis: Immune and Genetic Aspects

The host's immune mechanisms involved in schistosomiasis are essential to the knowledge of this complex illness. The pathogenesis of schistosomiasis is developed in the context of host–parasite relationship where the development of severe disease is limited by the transmission of the parasite. The severe form of schistosomiasis is normally caused by excessive immune and inflammatory responses, although other factors, such as the intensity of the infection, nutritional status, or genetic background of the patients, may influence the severity of this complex illness. Two distinct phases can be differentiated in schistosomiasis infection: acute phase and chronic phase.

41.1.3.1 Acute Phase

The penetration of cercariae percutaneously is a complex process that requires the presence of chemical stimuli from the host and the release of proteases secreted by the cercariae that allow the digestion of structural proteins. The major component of these proteases is a serine protease of 30 kDa whose elastase activity degrades elastine in the skin, leading to the penetration of cercariae. Other excretory/secretory molecules (ES) are able to degrade other macromolecules present in the skin of the host, such as keratin, fibronectin, laminin, or collagen. Several studies of animal models show that cercariae ES products induce edema and neutrophil infiltration in the skin. However, this inflammatory reaction does not induce protective immunity in the host and favors the survival of the parasite. The initial interaction between cercariae ES products and *pattern recognition receptors* (PRRS) of the innate immune system is crucial in determining the type of adaptive immune response [29]. Toll-like receptors (TLRs) are the best characterized PRRs since they play a central role in the induction of inflammatory response against microbial infections [30]. In humans, 10 functional members have been identified (TLR1–TLR10). Depending on which TLRs are stimulated and the interactions that occur between them, different cellular responses [31] can be generated. There is considerable evidence that TLRs are involved in the recognition of *Schistosoma* antigens. Few *in vitro* studies have been carried out with thioglycollate-elicited peritoneal macrophages stimulated with cercarial ES products suggesting that the cytokine production of IL-6 and IL-12p40 is TLR-mediated signaling. However, it has also been demonstrated that other receptors are involved in the expression of these cytokines [32]. Furthermore, other ligands, particularly glycans, may interact simultaneously with other PRRs such as C-type lectins (C-TLs) or mannose receptors (MR) modulating the TLR activation [33]. During 4–6 weeks postinfection, human peripheral blood mononuclear cells (PBMCs) induce the generation of high levels of proinflammatory cytokines TNFα, IL-1, and IL-6 [34]. Moreover, there is an increased production of γIFN with a peak at 16 weeks postinfection [35]. This cytokine profile reflects a dominant *T helper 1* response that is characteristic of the acute phase of schistosomiasis.

41.1.3.2 Chronic Phase

The chronic phase begins with the oviposition at 8 weeks after infection. Although the majority of eggs are shed in urine or feces, a significant number are retained in the tissues causing a granulomatous response. The innate and adaptive immune system deals with antigens secreted by *Schistosoma* eggs. Molecules of eggs have been extensively studied by proteomic tools; so far, 188 proteins have been described, some of them have been implicated in cell signaling, redox balance, and energy production [21]. Deposition of schistosome eggs on the tissues is a stimulus to the influx of immune cells that leads to the development of a granulomatous reaction. It is clear that this immunological reaction protects the host and neutralizes the schistosome egg antigens and consequently it destroys the *Schistosoma* eggs. However, the granulomas are the most important pathogenic event in schistosomiasis since the deposition of collagen and the development of fibrosis cause the fibro-obstructive disease. The innate immune system again plays a key role in the initiation and development of immune response against the chronic phase of schistosomiasis. Several studies have examined the interaction between schistosome-soluble egg antigens (SEA) and TLR. Specifically, TLR-2 appears to regulate the activation of dendritic cells (DC) stimulating IL-10 production [36]. The progressive increase of IL-10 induces the suppression of *Th1* cytokines and consequently it facilitates the *switch* from *Th1* to *Th2* profile, an immune response characteristic of the chronic phase [37]. It is widely accepted that the development of granulomatous response is correlated with delayed-type hypersensitivity. $CD4^+$ T lymphocytes regulate the host's adaptive immune response in an environment for fibrogenesis and gene expression patterns associated with healing [38]. Studies performed in IL-$4^{-/-}$ mice infected with *S. mansoni* demonstrate that IL-4 has a direct implication in Th2 response regulation. Also, IL-4 has a central role in the size of granuloma lesion [39,40]. Moreover, IL-4 stimulates IL-5 and IL-13, whose role in the development of severe liver fibrosis has been demonstrated in animal models and humans [40–42]. The *Th2* response induces alternative activation of macrophages with collagen synthesis by the L-arginine pathway [43]. Moreover, IL-2 and IL-5 stimulate the eosinophil recruitment, a great source of IL-13 whose fibrogenic activity has already been discussed above [44].

Severe schistosomiasis was classically associated with a *Th1*-based immune response characterized by an increase in the levels of proinflammatory cytokines. It has been demonstrated that both IL-12 and γINF play a central role in this kind of immune response [45]. However, the recent description of *T-helper-17* response has brought a novel insight to the investigation of the schistosomiasis pathogenesis. *In vivo* studies have demonstrated that IL-23 and IL-17 (majority cytokines of *T-helper-17* immune response) are involved in developing severe schistosomiasis [46,47]. It is accepted that a sustained *Th2* response appears to protect against severe schistosomiasis. However, prolonged polarization of the immune response toward a *Th2* response contributes to the maintenance of schistosomiasis morbidity [48]. Consequently, the regulation of *Th1/Th2* and *Th17* response is essential to prevent the schistosomiasis malignancies. Finally, schistosomiasis occurs in endemic areas with a high prevalence of malaria, hepatitis C or B, or HIV infection modifying the immune response against *Schistosoma* spp.

41.1.3.3 Susceptibility and Resistance to Schistosomiasis

Schistosomiasis is a multifactorial illness with a strong genetic component influenced by socioeconomic, environmental, and ecological factors. Nevertheless, epidemiological genetic studies have identified genetic regions involved in the degree of susceptibility/resistance in individual infection as well as in the degree of hepatic fibrosis. The methodology undertaken to carry out these studies was segregation analysis.

Using 20 Brazilians, Marquet et al. obtained results consistent with the existence of a codominant major gene referred to as *SM1* that controls human susceptibility/resistance to infection by *S. mansoni*. This locus (*SM1*) was mapped on the chromosome region *5q31–q33* and included several interesting genes that encode IL-4, IL-5, and IL-13 [49]. Specifically, two single-nucleotide polymorphisms in the encoding region of IL-5 have been associated with the development of symptomatic infections of *S. japonicum* in the Chinese population [50]. Another study performed by Dessein et al. suggested that liver fibrosis schistosomiasis level is controlled by *SM2 (6q22–q23)* [51]. This *locus* contains genes encoding the 2α chain receptor of γIFN (*IFNGR-1*). Therefore, the polymorphism on the *IFNGR-1* may determine the severity of liver disease produced by *S. mansoni*. Subsequent studies conducted in an Egyptian population confirmed the association between severe schistosomiasis and *IFNGR-1* locus [52]. Furthermore, γIFN production by PBMCs stimulated with egg antigen is associated with protection from severe schistosomiasis [53]. Two polymorphisms in *IFNG* linked to advanced hepatic disease [54] have been described. This is consistent with the antifibrogenic role of γIFN. Both γIFN and IL-10 protect against fibrosis development; however, the expression of these molecules is controlled by different gene regions [55].

41.1.4 Clinical Disease

Clinical schistosomiasis can be divided into three different stages: cercarial dermatitis or swimmer's itch, acute Katayama syndrome, and chronic schistosomiasis. Moreover, we will review two specific clinical situations: the association between schistosomiasis and cancer and the interaction between schistosomiasis and other bacterial and viral infections.

41.1.4.1 Cercarial Dermatitis

Cercarial dermatitis is characterized by the development of pruritus in the first 24 h after the penetration of cercariae through the skin. Skin lesions are very common in people infected by avian schistosomiasis. This is due to the immunological reaction provoked by the death of the cercariae in the skin. It occurs between 7% and 36% of infected patients [56]. It is more observed in travelers and is unnoticed in persons resident in endemic areas.

41.1.4.2 Katayama Fever

"Katayama syndrome" or "acute schistosomiasis" occurs between 2 and 8 weeks after exposure and is due to the immune response triggered against the migration phase of the schistosomules and the start of egg lying by adult worms. The most chararacterisitic signs or symptoms of the disease are fever, cutaneous lesions (rash, urticaria), pulmonary symptoms (cough, dyspnea), and eosinophilia [57]. In general, it is autolimited, although on some occasions the symptoms can persist for more than 10 weeks, presenting diarrhea, weight loss, abdominal pain, hepatosplenomegaly, etc. Katayama fever is a clinical syndrome often seen in nonimmune individuals (tourists, immigrants).

41.1.4.3 Chronic Schistosomiasis

It appears months or years after the infection and is due to the granulomatous reaction formed around eggs trapped in the liver, spleen, intestine, and other locations, such as the lungs and nervous system. It appears with greater frequency in persons resident in areas of endemic schistosomiasis. The principal clinical symptoms can be ordered into three groups: (i) *Habitual symptoms*: The principal data that indicate the possibility of a chronic schistosomiasis are the presence of eosinophilia and microhematuria (in the infections caused by *S. haematobium*). (ii) *Frequent classical symptoms*: The most typical are urinary, hepatosplenic, and intestinal. The *urinary* symptoms are mainly produced by the localization of *S. haematobium* in the venous plexus of the urinary tract. They occur clinically with an irritative mictional syndrome with dysuria, urinary frequency, proteinuria, and especially terminal hematuria. Later stages of the disease can provoke thickening of the wall of the bladder, polyps, or vesical masses. Bacterial infections can occasionally be associated. If the disease progresses, it can provoke fibrosis and calcification of the ureters and bladder (Figure 41.3), producing hydroureter and hydronephrosis. If the parenchyma is affected, it can produce renal failure. The *hepatosplenic* symptoms are produced when the eggs of *Schistosoma* (fundamentally *S. mansoni*, *S. japonicum*, and *S. mekongi*) localized in the mesenteric veins reach the portal venous circulation, causing a gradual occlusion of the intrahepatic veins, which can cause Symers' fibrosis. This gives rise to a portal hypertension with ascites, splenomegaly (Figure 41.4), development of esophageal varicose veins with high digestive bleeding, and encephalopathy. Transaminases and bilirubin are at normal values except for the existence of coinfection with HVB or HVC [58]. The *intestinal* symptoms are produced when the eggs reach the intestine producing hyperplasia, ulceration, microabscesses, and polyposis. The infections by *S. intercalatum* are frequently associated with

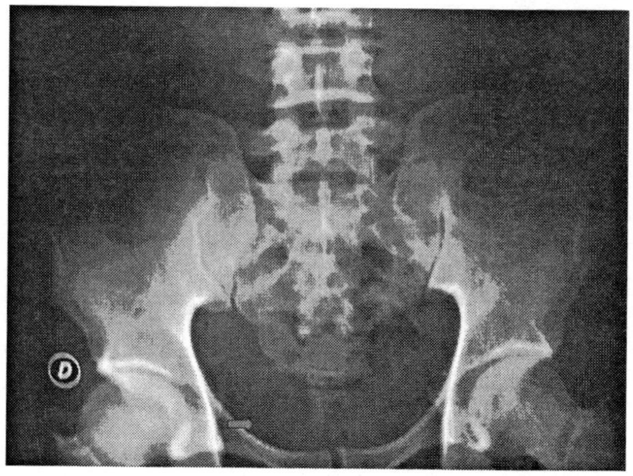

FIGURE 41.3 Bladder calcification in urinary schistosomiasis (Plain abdominal x-ray). (Courtesy of Dr. Perez-Arellano, Hospital Insular las Palmas de Gran Canaria, Spain.)

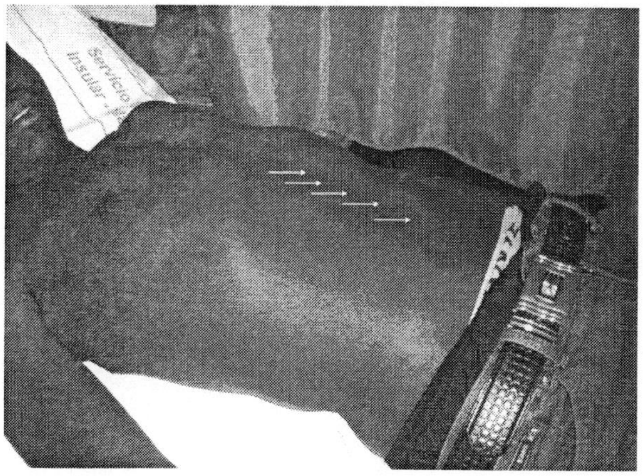

FIGURE 41.4 Splenomegaly in *Schistosoma mansoni* infection. (Courtesy of Dr. Perez-Arellano, Hospital Insular las Palmas de Gran Canaria, Spain.)

intestinal lesions. Clinically, it presents abdominal pain and inflammatory-type diarrhea with or without blood [59]. It can be complicated, producing an enteropathy losing proteins, an occlusive or subocclusive symptomology, anal prolapse, or even anorectal fistulas. All this requires a differential diagnosis with the intestinal inflammatory disease. (iii) *Infrequent classical symptoms*: These are also denominated ectopic schistosomiasis. The most frequent are cardiopulmonary, renal, genital, and those affecting the central nervous system. The *cardiopulmonary* symptoms are produced as a consequence of the eggs leaving the vesical veins (in the case of *S. haematobium*) or from the portal circulation through the porto-cava *shunt*. As a consequence of the localization of the eggs in the pulmonary circulation, fibrosis of the pulmonary arteries is produced, pulmonary hypertension being presented with right cardiac insufficiency [60]. Schistosomiasis can provoke lesions at the parenchymatous *renal* level, specifically affecting the glomerulus, causing glomerulonephritis [61]. In general, the glomerulonephritis is associated with the existence of hepatosplenic schistosomiasis with porto-systemic *shunt*. The most frequent type is the mesangial glomerulonephritis which is usually presented with a discreet proteinuria and microhematuria. The lesions at the *genital* level are of two types: inflammatory in relation to viable eggs and fibrous and hypertrophic in relation to nonviable or calcified eggs. In women, it affects the vulva, vagina, and neck of the uterus. They are not painful but in the case of ulcerating, fistulizing, or overinfection, they can produce dyspareunia and leucorrea [62]. In men, the most frequent are chronic prostatitis and the infection of the seminal vesicles. The most habitual symptomology is the alteration in ejaculation and hematospermia [23]. The two most typical symptoms of *neuroschistosomiasis* are medullar and cerebral syndromes. The localization of the eggs of *Schistosoma* spp. in cerebral territory can make their debut as a comicial crisis and usually progress to a cerebral focal encephalitis [63]. The medullar localization of the eggs of schistosomes is more frequent than the encephalic and can produce various forms of myelitis, the most characteristic being granulomatous myelitis, myeloradiculitis, and ischemic or vascular myelitis, the most frequent being transverse myelitis [64]. The patients sometimes present a neurogenic bladder associated with a sensitive motor level with weakness in the legs, incapacity for walking, lumbar pain, and paresthesias.

41.1.4.4 Association with Carcinogenesis

It has been demonstrated how chronic cystitis produced by *S. haematobium* behaves like precancerous lesions that in time degenerate into urinary bladder tumor [65]. This association has not been demonstrated with other species of schistosomes in relation with colon tumors or hepatocarcinomas.

41.1.4.5 Interaction with Other Infections

Patients with schistosomiasis present greater indices of chronic carriers of the Australia antigen of the virus of hepatitis B and antibodies against the virus of hepatitis C. It is thought that the campaigns of parenteral massive treatment in some countries, such as Egypt, contributed to the transmission of the infection of the virus of hepatitis C [66]. The coinfection of schistosomiasis and chronic infection by virus of hepatitis B and C is presented with an earlier and more intense fibrosis than in individuals who are not coinfected [67].

The HIV infection, on inducing an important deterioration of the cellular immunity, appears to condition an alteration in the migration of the eggs through the wall of the intestine. This leads to the decrease in the number of eggs in feces. This fact has not been demonstrated to imply a greater seriousness of the infection or a worse response to the treatment. It has been demonstrated that the HIV infection increases the susceptibility of reinfection in exposed patients previously treated with praziquantel [68]. In addition, chronic hyperstimulation of the immune response produced by the schistosomiasis can favor the pathogenesis of the infection by HIV. It is believed that the lesions of genital schistosomiasis can alter the mucocutaneous barrier favoring the transmission of sexually transmitted diseases, among which HIV infection is outstanding [62].

Finally, the association between piogenic hepatic abscesses of *Staphylococcus aureus* and the infection by *Schistosoma* [69] has been described, as have septicemias caused by enterobacteria and chronic intestinal schistosomiasis [70].

41.1.5 Diagnosis

The diagnosis of human schistosomiasis is based on clinical and epidemiological data. Nonetheless, these data are normally insufficient for an unequivocal diagnosis of this parasitosis. Therefore, distinct diagnostic techniques are employed, included as methods of conventional diagnosis and techniques of molecular diagnosis.

41.1.5.1 Conventional Techniques

This includes techniques of direct parasitological detection and techniques of detection of antigens or antibodies (immunological methods).

41.1.5.1.1 Parasitological Techniques

The parasitological diagnosis by the examination of samples of feces or urine is the most utilized method for the diagnosis of schistosomiasis, since it allows the identification of the different species of schistosomes, visualizing their characteristic morphology (Table 41.1). Moreover, this diagnosis allows the evaluation of the intensity of the infection. The technique utilized for the detection of *S. mansoni*, *S. japonicum*, *S. intercalatum*, and *S. mekongi* is the examination of feces by the Kato-Katz [71] technique, microscopically examining a fresh sample of 50 mg of feces, allowing the detection as positive of samples with a minimum of 20 eggs/g of feces. Furthermore, techniques of concentration (formol/ethyl acetate) can be used if the observation of the fresh sample gives negative results. If it is suspected that the patient presents a low number of worms, the eclosion test can be used, incubating the feces until the miracidia hatch from the eggs. The miracidia migrate toward a light source which must be applied to the sample. The infection by *S. haematobium* is not usually diagnosed in unconcentrated samples; therefore, methods of filtration or sedimentation of urine are employed [72].

The search for eggs of schistosomes represents a method of specific diagnosis with low cost and simplicity of performance, able to be used in laboratories with precarious structures and less trained personnel. These characteristics make this method applicable in field studies in endemic areas. Nonetheless, it presents serious inconveniences with regard to its sensitivity, especially when the intensity of the infection is low, as occurs in areas of low prevalence or in individuals with recent infections [73]. Moreover, it can only be carried out after the production and elimination of eggs, which begins at 2 months of infection. In areas of low transmission, where the sensitivity obtained by classical methods is low, the COPT (circumoval precipitation test) technique, described by Oliver-Gónzalez in 1954 [74], is used. This consists of incubating eggs of *S. mansoni* with sera of patients, and considered positive if there is more than 9% of precipitation around the mature eggs [5].

41.1.5.1.2 Immunological Techniques

The detection of the anodic and cathodic *circulating antigens* (CAA and CCA) is the most utilized method in direct immunological diagnosis [75]. These antigens are glycoconjugated and derive from adult worms. Both can be detected in blood and urine and the CCA in maternal milk, allowing the diagnosis of *S. mansoni*, *S. haematobium*, and *S. japonicum*. Nevertheless, they are shared by the named species and thus only permit diagnosis at the level of genus, although they present few cross-reactions with other parasites. Recently, a test to detect CCA in urine using strips of nitrocellulose (*dipstick*) has been commercialized. The sensitivity of this test has been low [76].

The infection by schistosomes is highly immunogenic and it is not difficult to demonstrate the *presence of antibodies* antischistosome in the infected subjects [77]. Numerous methods have been described, the most utilized being ELISA. Total extracts of adult worms and eggs (better than larval antigens), purified fractions of these extracts, or recombinant proteins have been assayed. The use of antigens of adult worms (the easiest to obtain), even from other schistosomes, such as *S. bovis*, have permitted the diagnosis of human infections with *S. mansoni*, *S. haematobium*, and *S. intercalatum* in sub-Saharan immigrants [25]. In addition, this test has been used to confirm the diagnosis of patients with Katayama fever [78]. Recently, a test denominated SEA-ELISA, in which soluble antigens of egg are used, has been commercialized. A study carried in 150 children in Zanzibar showed a sensitivity of 89% and a specificity of 70% [76]. The alternative for reducing cross-reactivity is, on the one hand, to eliminate the carbohydrates responsible [5] and, on the other, to isolate and purify more specific molecules. Thus, different preparations have been used from adult worms and microsomal extracts, polysaccharides associated with intestine, alkaline phosphatase, heat shock proteins, and Sm31 and Sm32 adult antigens, obtaining different results. Purified antigens of eggs, such as the antigens ω-1, α-1, and κ-5, have also been used. Recombinant antigens from adult worms, such as RP26 [79], or from eggs, such as CEF6, have still not solved the problem.

In conclusion, the methods of immunodiagnosis, both direct and indirect, present greater sensitivity than the techniques used for direct parasitological diagnosis. Nonetheless, immunodiagnosis continues to present a series of problems related to the obtaining of antigens—often complex—and the appearance of false positives and false negatives. Moreover, immunological detection of schistosomiasis is habitually delayed until the appearance of the corresponding antigens or antibodies, constituting a limited option for diagnosis in the acute phase of the disease. Finally, the persistence of antigens and especially of antibodies after efficient therapy frequently causes the detection of "false positives" corresponding to patients who have already eliminated the parasite.

41.1.5.2 Molecular Techniques

The technique of PCR has been little used in the diagnosis of human schistosomiasis, having been used principally for phylogenetic studies and for the detection of the parasite in contaminated waters or in its intermediary host [80]. The first PCR described for human diagnosis is due to the amplification of *S. mansoni* in human samples—serum and feces—of patients infected with this parasite [81,82]. These authors obtain an amplicon of 121 base pair (bp) using DNA of *S. mansoni*. This reaction is capable of detecting up to 1 fg of parasite DNA. This product is not detected from *Ascaris lumbricoides*, *Ancylostoma duodenale*, *Taenia solium*, and *Trichuris trichura* DNA extracts. However, these authors do not test the specificity of this reaction with DNA of other species of *Schistosoma* or of other parasites potentially present in the same ecological niche as *S. mansoni* (e.g., *Plasmodium* sp., *Leishmania* sp., *Trypanosoma cruzi*, etc.). In the application of the technique to feces of donors from an endemic area for *S. mansoni*, the authors detect a number of samples positive in PCR and negative in the direct parasitological diagnosis—Kato-Katz—and the inverse [83]. The first are attributed to patients with slight infection, yielding a negative diagnosis on examining the feces, but positive on employing a more sensitive technique, such as PCR. The false negatives are attributed either to the presence of inhibitors in particular samples or, more probably, to the fact that a portion of feces in which the eggs of the parasite can be absent is utilized.

Later, Sandoval et al. [84], using samples of urine from patients infected with different species of schistosomes, amplified a fragment of 877 pb genus-specific and one of 350 pb specific to *S. mansoni*. This reaction is capable of detecting up to 0.98 pg of parasite DNA, presenting a sensitivity of 94% with the genus-specific amplification and 100% when the fragment specific to *S. mansoni* were amplified. The specificity was also evaluated using sera diagnosed with different protozooses and helminthoses, obtaining a specificity of 99% genus-specific and 98% species-specific. Furthermore, in a controlled experiment carried out on mice infected with *S. mansoni* and comparing it with the Kato-Katz technique and detection of specific antibodies, these authors observe that parasite DNA can be amplified from the first week postinfection [85].

In recent years, rapid, simple tests are being developed based on oligochromatographics and real-time PCR techniques. *Oligochromatographic dipstick test* can be carried out in 10 min and can detect 10 fg of genomic DNA [86]. This method can be successfully applied in endemic areas in the subsequent years. Real-time PCR [87] has been developed from large volumes of plasma (10 mL) and has advantages of its validity in chronic schistosomiasis and in Katayama fever, in spite of the fact that the human samples utilized, up to the present, are scarce. It is difficult to apply in endemic areas due to its high cost. Nevertheless, it can be useful for the diagnosis of schistosomiasis imported into industrialized countries.

41.1.6 Treatment, Chemoprophylaxis, and Prevention

Praziquantel is the choice drug for the treatment of schistosomiasis, since oxamniquin (active against *S. mansoni*) and metriphonate (active against *S. haematobium*) are not commercially available at present [88]. Praziquantel is administered orally and is little toxic, and has very low cost. However, it does not prevent reinfection and for its administration, it is necessary to take into account the phase of the disease and the species responsible [89]. In the acute phase, praziquantel is highly efficient for adult worms, being less so in schistosomules. Praziquantel 20 mg/kg every 12 h together with dexametasone 20 mg/day for 3 days is the treatment of choice in this phase. It is necessary to repeat the dose of praziquantel at 3–4 weeks. In the chronic phase, the dose of praziquantel depends on the species involved. For infections by *S. mansoni* and *S. haematobium*, the recommended dose is 20 mg/kg every 12 h for 1 day. In the case of *S. japonicum*, 20 mg/kg every 8 h for 1 day is recommended.

The fact that the schistosomiasis treatment is limited to one single pharmaceutical presents the risk associated with the appearance of resistances. The WHO has recognized the necessity of identifying new compounds as alternatives to praziquantel. In the last decade, the derivatives of artemisine have appeared as a complement to the therapy against schistosomiasis. The use of these derivatives in combination with praziquantel can be a good strategy of control, since the derivatives of artemisine are effective against the juvenile phases of the parasite [88]. In addition, the activity of the derivatives of artemisine as chemotherapy has been demonstrated in numerous clinical assays [90]. Mefloquine has also been used against schistosomiasis, with good results in experimental models [91]. However, the use of these antimalarial drugs in the treatment and/or control of this disease could lead to the appearance of malaria resistant to these compounds.

Schistosomiasis can be eliminated by behavioral changes, sanitation, and safe water supply. Health educational programs can improve knowledge about schistosomiasis, avoiding contact with contaminated water. In schistosomiasis caused by *S. japonicum*, interventions on the animal reservoir are necessary. Molluscicides are toxic chemicals and are expensive. *Preventive treatment* has permitted the decrease in prevalence in certain zones. However, it has not been possible to eradicate the disease and, furthermore, they have generated resistances to praziquantel. Up to the present no effective human vaccine has been developed. Some authors indicate that the proteins of membrane expressed in the surface of the schistosomules or adult worms as key targets for the development of vaccines [92]. The sequencing of the genome of *S. mansoni* [8] and *S. japonicum* [9] will help to identify new candidates for vaccines, determined important advances toward the control and eradication of schistosomiasis.

41.2 METHODS

41.2.1 Sample Preparation

41.2.1.1 DNA Extraction from *Schistosoma* sp.

Adult schistosomes are collected by intracardiac perfusion of mice infected with cercariae. Adult worms are obtained by flowing out the portal vein. The parasites are washed in distilled water three times and they are cut into small pieces and homogenized.

DNA extraction is performed using NucleoSpin Tissue kit (MacheryNagel, Germany), using the manufacturer's instructions as follows:

1. Add 180 μL buffer T1 and 25 μL proteinase K solution. Vortex to mix for 20 s. Make sure that the samples are completely covered with the lysis solution.
2. Incubate at 56°C until complete lysis is obtained (from 1 h to overnight). Vortex occasionally during incubation.
3. Add 200 μL buffer B3, vortex vigorously, and incubate at 70°C for 10 min. Vortex briefly. If insoluble particles are visible, centrifuge for 5 min at 11,000× *g* and transfer the supernatant to a new microcentrifuge tube.

4. Add 210 μL ethanol 96% to the sample and vortex vigorously.
5. For each sample, place one column into a collection tube. Apply the sample to the column. Centrifuge for 1 min at 11,000× *g*. Discard the flow-through and place the column back into the collection tube. If the sample is not drawn completely through the matrix, repeat the centrifugation step at 11,000× *g*. Discard flow-through.
6. Add 500 μL buffer BW. Centrifuge for 1 min at 11,000× *g*. Discard flow-through and place the column back into the collection tube.
7. Add 600 μL buffer B5 to the column and centrifuge for 1 min at 11,000× *g*. Discard flow-through and place the column back into the collection tube.
8. Centrifuge the column for 1 min at 11,000× *g*.
9. Place the column into a new microcentrifuge tube and add 35 μL prewarmed buffer BE (70°C). Incubate at room temperature for 3 min.
10. Centrifuge for 1 min at 11,000× *g*.
11. Repeat steps (8) and (9). DNA is obtained at 70 μL final volume.

41.2.1.2 DNA Extraction from Urine Samples

DNA extraction from human urine (3 mL each) was performed with the NucleoSpin Trace kit (MacheryNagel, Germany), using the manufacturer's instructions:

1. Boil the urine samples for 5 min. Centrifuge for 5 min at 11,000× *g*.
2. Place the sample in a 15 mL centrifuge tube and add 4 mL lysis buffer FLB.
3. Add 50 μL proteinase K stock solution, mix by vortexing, and incubate at 56°C in a shaking water bath until complete lysis is obtained (1–3 h or overnight). Afterward, any insoluble particles remaining in the sample have to be removed by centrifugation for 10 min at 5000× *g*.
4. Add 3.5 mL ethanol (96%) to 4 mL cleared FLB-lysate and vortex the mixture.
5. Pipette the mixture onto the column. Centrifuge for 3 min at 3,000× *g*. Discard flow-through.
6. Add 2.5 mL buffer BW to the column. Centrifuge for 3 min at 3000× *g*. Discard flow-through.
7. Add 5 ml buffer B5 to the column. Centrifuge for 3 min at 3000× *g*. Discard flow-through.
8. Repeat step (6) once more.
9. Centrifuge additional 10 min at 3000× *g* in order to remove buffer B5 completely.
10. Attach the supplied elution tube with adaptor to the column. Add 35 μL elution buffer BE preheated to 70°C and incubate for 2 min at room temperature.
11. Centrifuge for 3 min at 3000× *g* to collect the nucleic acid-containing fraction.
12. Repeat steps (9) and (10) once more.
13. Remove the elution tube containing the nucleic acids and keep it for further use.

41.2.2 Detection Procedures

41.2.2.1 Principle

We present a sensitive and specific PCR-based approach for the amplification of defined regions from 28S ribosomal DNA of schistosome species. We use two primer pairs to amplify 28S rDNA: CF1/CR2 and CF2/CR2 giving genus-specific PCR products of 877 and 1032 bp, respectively; CF1 (5′-GAGTTGAACTGCAAGCTCTGG-3′), CR2 (5′-TTCGCCCCTATACTCACGTTAT-3′), CF2 (5′-GTTTGTGAATGCAGCCCAAAGT-3′). Specific primers for the amplification of the 28S rDNA region from *S. mansoni*: SmF/SmR giving species-specific PCR products of 350 bp; SmF (5′-GAGATCAAGTGTGACAGTTTTGC-3′), SmR (5′- ACAGTGCGCGCGTCGTAAGC-3′).

41.2.2.2 Procedure

1. PCR reactions were carried out in a final volume of 20 μL with 2 μL of 10× reaction buffer, 3 mM $MgCl_2$, 2.5 U *Taq* DNA polymerase, 2 mM of each primer, 0.5 mM dNTPs, and 1 μL of template DNA, either from schistosome DNA or from DNA extracted from human urine samples.
2. PCR was performed in 35 cycles, each consisting of 94 °C for 20 s, 65°C for 20 s, and 72°C for 30 s.
3. Corresponding PCR products were electrophoresed, together with the molecular weight markers in 1% agarose in Tris-borate EDTA gels with 0.5 μg/mL ethidium bromide, and visualized in a UV transilluminator.

41.3 CONCLUSION AND FUTURE PERSPECTIVE

Schistosomiasis is a tropical helminth disease due to a different schistosome species. 779 million people in 76 countries that are endemic could be infected. Furthermore, 250 million people are infected (80% in sub-Saharan Africa), 120 million have symptoms, and 20 million have severe disease and there is annual mortality of 280,000 with 1.53 million disability-adjusted life years. Moreover, schistosomiasis have been detected in *Schistosoma*-free areas, mainly due to the increase in off-track tourism, the spread of schistosomiasis in previously nonendemic areas, and the recent waves of immigrants from endemic countries. Cecariae penetration and migration of the schistomulae occurs generally 2–6 weeks after the exposure and combines fever, urticarial rash, pulmonary symptoms, and eosinophilia. This clinical outcome should be considered as suggestive for acute schistosomiasis or Katayama fever. In chronic stages of the infection, fibrosis and calcification of tissue trapped ova in the lower urinary tract, and liver and intestine lead several signs and symptoms, depending on the parasite load. Diagnosis of schistosomiasis is performed through the direct detection of parasite eggs in stool and urine. Immunological methods are usually applied to patients with schistosomiasis clinical signs when parasites cannot be directly detected. Nevertheless, both the above-mentioned diagnostic techniques present false-negative and false-positive results, especially in low parasite burden and acute phase of schistosomiasis. In the last few years, PCR-based diagnosis has been used. PCR amplifying an *S. mansoni*-specific DNA fragment has been applied to stool and serum from South American schistosomiasis patients. Recently, we have developed a new PCR approach allowing the amplification of both genus- and species-specific schistosome DNA in easy-to-handle human urine sample. Moreover, simple tests are being developed based on oligochromatographics and real-time PCR techniques. Praziquantel is the choice drug for the treatment of schistosomiasis, since oxamniquin and metriphonate are not commercially available. Praziquantel is administered orally, is little toxic, and has very low cost. The fact that schistosomiasis treatment is limited to one single pharmaceutical presents the risk associated with the appearance of resistances. In the last decade, only the derivatives of artemisine have appeared as a complement to the therapy against schistosomiasis. Actually, no effective human vaccine is available but the sequencing of the genome of *S. mansoni* and *S. japonicum* will help to identify new candidates for vaccines in the future.

REFERENCES

1. Vester, U. et al. Urinary tract morbidity due to *Schistosoma haematobium* infection in Mali. *Kidney Int.* 52, 478–481, 1997.
2. Cox, F.E. History of human parasitic diseases. *Infect. Dis. Clin. North Am.* 18, 171–188, table of contents, 2004.
3. Ohmae, H. et al. Schistosomiasis mekongi: From discovery to control. *Parasitol. Int.* 53, 135–142, 2004.
4. Webster, B.L., Southgate, V.R., and Littlewood, D.T. A revision of the interrelationships of *Schistosoma* including the recently described *Schistosoma guineensis. Int. J. Parasitol.* 36, 947–955, 2006.
5. Alarcón de Noya, B., and Noya, O. Schistosomosis. In: *Parasitología Médica*. México, 2008.
6. Rollinson, D., and Southgate, V.R. The genus *Schistosoma*: A taxonomic appraisal. In: Rollinson, D., and Simpson, A.J.G. (eds), *The Biology of Schistosomes. From Genes to Latrines*. Academic Press, London, 1987.
7. Morgan, J.A. et al. A newly-identified lineage of *Schistosoma. Int. J. Parasitol.* 33, 977–985, 2003.
8. Berriman, M. et al. The genome of the blood fluke *Schistosoma mansoni. Nature* 460, 352–358, 2009.
9. Liu, F. et al. The *Schistosoma japonicum* genome reveals features of host-parasite interplay. *Nature* 460, 345–351, 2009.
10. LoVerde, P.T. et al. *Schistosoma mansoni* genome project: An update. *Parasitol. Int.* 53, 183–92, 2004.
11. Adams, M.D. et al. Complementary DNA sequencing: Expressed sequence tags and human genome project. *Science* 252, 1651–1656, 1991.
12. Verjovski-Almeida, S. et al. Transcriptome analysis of the acoelomate human parasite *Schistosoma mansoni. Nat. Genet.* 35, 148–157, 2003.
13. Hu, W. et al. Evolutionary and biomedical implications of a *Schistosoma japonicum* complementary DNA resource. *Nat. Genet.* 35, 139–147, 2003.
14. Wilkins, M.R. et al. Progress with proteome projects: Why all proteins expressed by a genome should be identified and how to do it. *Biotechnol. Genet. Eng. Rev.* 13, 19–50, 1996.
15. Curwen, R.S. et al. The *Schistosoma mansoni* soluble proteome: A comparison across four life-cycle stages. *Mol. Biochem. Parasitol.* 138, 57–66, 2004.
16. Cheng, G.F. et al., Proteomic analysis of differentially expressed proteins between the male and female worm of *Schistosoma japonicum* after pairing. *Proteomics* 5, 511–521, 2005.
17. Braschi, S., and Wilson, R.A. Proteins exposed at the adult schistosome surface revealed by biotinylation. *Mol. Cell Proteomics* 5, 347–356, 2006.
18. Liu, F. et al. Insight into the host-parasite interplay by proteomic study of host proteins copurified with the human parasite, *Schistosoma japonicum. Proteomics* 7, 450–462, 2007.
19. Guillou, F. et al. Excretory-secretory proteome of larval *Schistosoma mansoni* and *Echinostoma caproni*, two parasites of *Biomphalaria glabrata. Mol. Biochem. Parasitol.* 155, 45–56, 2007.
20. Dvorak, J. et al. Differential use of protease families for invasion by schistosome cercariae. *Biochimie* 90, 345–358, 2008.
21. Cass, C.L. et al. Proteomic analysis of *Schistosoma mansoni* egg secretions. *Mol. Biochem. Parasitol.* 155, 84–93, 2007.
22. King, C.H., Dickman, K., and Tisch, D.J. Reassessment of the cost of chronic helmintic infection: A meta-analysis of disability-related outcomes in endemic schistosomiasis. *Lancet* 365, 1561–1569, 2005.
23. Corachan, M. et al. Hematospermia: A new etiology of clinical interest. *Am. J. Trop. Med. Hyg.* 50, 580–584, 1994.
24. Freedman, D.O. et al. Spectrum of disease and relation to place of exposure among ill returned travelers. *N. Engl. J. Med.* 354, 119–130, 2006.
25. Pardo, J. et al. Utility of *Schistosoma bovis* adult worm antigens for diagnosis of human schistosomiasis by enzyme-linked immunosorbent assay and electroimmunotransfer blot techniques. *Clin. Diagn. Lab. Immunol.* 11, 1165–1170, 2004.
26. Leshem, E. et al. Acute schistosomiasis outbreak: Clinical features and economic impact. *Clin. Infect. Dis.* 47, 1499–1506, 2008.
27. Jelinek, T., Nothdurft, H.D., and Loscher, T. Schistosomiasis in travelers and expatriates. *J. Travel. Med.* 3, 160–164, 1996.
28. Carranza-Rodriguez, C. et al. Cryptic parasite infection in recent West African immigrants with relative eosinophilia. *Clin. Infect. Dis.* 46, e48–e50, 2008.
29. Janeway, C.A., Jr., and Medzhitov, R. Innate immune recognition. *Annu. Rev. Immunol.* 20, 197–216, 2002.
30. Akira, S. Toll-like receptor signaling. *J. Biol. Chem.* 278, 38105–38108, 2003.
31. Beutler, B. Inferences, questions and possibilities in Toll-like receptor signalling. *Nature* 430, 257–263, 2004.
32. Jenkins, S.J. et al. Schistosome larvae stimulate macrophage cytokine production through TLR4-dependent and -independent pathways. *Int. Immunol.* 17, 1409–1418, 2005.
33. van Kooyk, Y., Engering, A., Lekkerkerker, A.N., Ludwig, I.S., and Geijtenbeek, T.B. Pathogens use carbohydrates to escape immunity induced by dendritic cells. *Curr. Opin. Immunol.* 16, 488–493, 2004.
34. Araujo, M.I. et al. Evidence of a T helper type 2 activation in human schistosomiasis. *Eur. J. Immunol.* 26, 1399–1403, 1996.
35. Caldas, I.R. et al. Human schistosomiasis mansoni: Immune responses during acute and chronic phases of the infection. *Acta Trop.* 108, 109–117, 2008.
36. van der Kleij, D. et al. A novel host-parasite lipid cross-talk. Schistosomal lyso-phosphatidylserine activates toll-like receptor 2 and affects immune polarization. *J. Biol. Chem.* 277, 48122–48189, 2002.

37. Montenegro, S.M. et al. Cytokine production in acute versus chronic human *Schistosomiasis mansoni*: The cross-regulatory role of interferon-gamma and interleukin-10 in the responses of peripheral blood mononuclear cells and splenocytes to parasite antigens. *J. Infect. Dis.* 179, 1502–1514, 1999.
38. Davies, S.J. et al. Modulation of blood fluke development in the liver by hepatic CD4+ lymphocytes. *Science* 294, 1358–1361, 2001.
39. Brunet, L.R. et al. IL-4 protects against TNF-alpha-mediated cachexia and death during acute schistosomiasis. *J. Immunol.* 159, 777–785, 1997.
40. Fallon, P.G. et al. Schistosome infection of transgenic mice defines distinct and contrasting pathogenic roles for IL-4 and IL-13: IL-13 is a profibrotic agent. *J. Immunol.* 164, 2585–2591, 2000.
41. Cheever, A.W. et al. Analysis of egg granuloma formation in *Schistosoma japonicum*-infected mice treated with antibodies to interleukin-5 and gamma interferon. *Infect. Immun.* 59, 4071–4074, 1991.
42. de Jesus, A.R. et al. Association of type 2 cytokines with hepatic fibrosis in human *Schistosoma mansoni* infection. *Infect. Immun.* 72, 3391–3397, 2004.
43. Hesse, M. et al. Differential regulation of nitric oxide synthase-2 and arginase-1 by type 1/type 2 cytokines in vivo: Granulomatous pathology is shaped by the pattern of L-arginine metabolism. *J. Immunol.* 167, 6533–6544, 2001.
44. Rumbley, C.A. et al. Activated eosinophils are the major source of Th2-associated cytokines in the schistosome granuloma. *J. Immunol.* 162, 1003–1009, 1999.
45. Pardo, J. et al. Helminth-related eosinophilia in African immigrants, Gran Canaria. *Emerg. Infect. Dis.* 12, 1587–1589, 2006.
46. Rutitzky, L.I., Lopes da Rosa, J.R., and Stadecker, M.J. Severe CD4 T cell-mediated immunopathology in murine schistosomiasis is dependent on IL-12p40 and correlates with high levels of IL-17. *J. Immunol.* 175, 3920–3926, 2005.
47. Rutitzky, L.I. et al. IL-23 is required for the development of severe egg-induced immunopathology in schistosomiasis and for lesional expression of IL-17. *J. Immunol.* 180, 2486–2495, 2008.
48. Hoffmann, K.F., Cheever, A.W., and Wynn, T.A. IL-10 and the dangers of immune polarization: Excessive type 1 and type 2 cytokine responses induce distinct forms of lethal immunopathology in murine schistosomiasis. *J. Immunol.* 164, 6406–6416, 2000.
49. Marquet, S. et al. Full results of the genome-wide scan which localises a locus controlling the intensity of infection by *Schistosoma mansoni* on chromosome 5q31-q33. *Eur. J. Hum. Genet.* 7, 88–97, 1999.
50. Burke, M.L. et al. Immunopathogenesis of human schistosomiasis. *Parasite Immunol.* 31, 163–176, 2009.
51. Dessein, A.J. et al. Severe hepatic fibrosis in *Schistosoma mansoni* infection is controlled by a major locus that is closely linked to the interferon-gamma receptor gene. *Am. J. Hum. Genet.* 65, 709–721, 1999.
52. Blanton, R.E. et al. Schistosomal hepatic fibrosis and the interferon gamma receptor: A linkage analysis using single-nucleotide polymorphic markers. *Eur. J. Hum. Genet.* 13, 660–668, 2005.
53. Henri, S. et al. Cytokine regulation of periportal fibrosis in humans infected with *Schistosoma mansoni*: IFN-gamma is associated with protection against fibrosis and TNF-alpha with aggravation of disease. *J. Immunol.* 169, 929–936, 2002.
54. Chevillard, C. et al. IFN-gamma polymorphisms (IFN-gamma +2109 and IFN-gamma +3810) are associated with severe hepatic fibrosis in human hepatic schistosomiasis (*Schistosoma mansoni*). *J. Immunol.* 171, 5596–5601, 2003.
55. Arnaud, V. et al. Regulatory role of interleukin-10 and interferon-gamma in severe hepatic central and peripheral fibrosis in humans infected with *Schistosoma japonicum*. *J. Infect. Dis.* 198, 418–426, 2008.
56. Meltzer, E. et al. Schistosomiasis among travelers: New aspects of an old disease. *Emerg. Infect. Dis.* 12, 1696–1700, 2006.
57. Ross, A.G., Vickers, D., Olds, G.R., Shah, S.M., and McManus, D.P. Katayama syndrome. *Lancet Infect. Dis.* 7, 218–224, 2007.
58. Bica, I., Hamer, D.H., and Stadecker, M.J. Hepatic schistosomiasis. *Infect. Dis. Clin. North Am.* 14, 583–604, viii, 2000.
59. Gryseels, B. et al. Human schistosomiasis. *Lancet* 368, 1106–1118, 2006.
60. Bethlem, E.P., Schettino Gde, P., and Carvalho, C.R. Pulmonary schistosomiasis. *Curr. Opin. Pulm. Med.* 3, 361–365, 1997.
61. Barsoum, R. The changing face of schistosomal glomerulopathy. *Kidney Int.* 66, 2472–2484, 2004.
62. Poggensee, G. et al. Female genital schistosomiasis of the lower genital tract: Prevalence and disease-associated morbidity in northern Tanzania. *J. Infect. Dis.* 181, 1210–1213, 2000.
63. Ferrari, T.C. Involvement of central nervous system in the schistosomiasis. *Mem. Inst. Oswaldo Cruz.* 99, 59–62, 2004.
64. Nobre, V. et al. Schistosomal myeloradiculopathy due to *Schistosoma mansoni*: Report on 23 cases. *Mem. Inst. Oswaldo Cruz.* 96(Suppl), 137–141, 2001.
65. Yosry, A. Schistosomiasis and neoplasia. *Contrib. Microbiol.* 13, 81–100, 2006.
66. Rao, M.R. et al. Further evidence for association of hepatitis C infection with parenteral schistosomiasis treatment in Egypt. *BMC Infect. Dis.* 2, 29, 2002.
67. Kamal, S.M. et al. Kinetics of intrahepatic hepatitis C virus (HCV)-specific CD4+ T cell responses in HCV and *Schistosoma mansoni* coinfection: Relation to progression of liver fibrosis. *J. Infect. Dis.* 189, 1140–1150, 2004.
68. Karanja, D. M. et al. Resistance to reinfection with *Schistosoma mansoni* in occupationally exposed adults and effect of HIV-1 co-infection on susceptibility to schistosomiasis: A longitudinal study. *Lancet* 360, 592–596, 2002.
69. Teixeira, R. et al. Schistosomiasis mansoni is associated with pyogenic liver abscesses in the state of Minas Gerais, Brazil. *Mem. Inst. Oswaldo Cruz.* 96(Suppl), 143–146, 2001.
70. Gendrel, D. et al. Nontyphoidal salmonellal septicemia in Gabonese children infected with *Schistosoma intercalatum*. *Clin. Infect. Dis.* 18, 103–105, 1994.
71. Katz, N., Chaves, A., and Pellegrino, J. A simple device for quantitative stool thick-smear technique in Schistosomiasis mansoni. *Rev. Inst. Med. Trop. Sao Paulo.* 14, 397–400, 1972.
72. Gyorkos, T.W. et al. Efficacy of new low-cost filtration device for recovering *Schistosoma haematobium* eggs from urine. *J. Clin. Microbiol.* 39, 2681–2682, 2001.
73. Berhe, N. et al. Variations in helminth faecal egg counts in Kato-Katz thick smears and their implications in assessing infection status with *Schistosoma mansoni*. *Acta Trop.* 92, 205–212, 2004.
74. Oliver-Gonzalez, J. Anti-egg precipitins in the serum of humans infected with *Schistosoma mansoni*. *J. Infect. Dis.* 95, 86–91, 1954.
75. van Lieshout, L., Polderman, A.M., and Deelder, A.M. Immunodiagnosis of schistosomiasis by determination of the

circulating antigens CAA and CCA, in particular in individuals with recent or light infections. *Acta Trop.* 77, 69–80, 2000.
76. Stothard, J.R. et al. An evaluation of urine-CCA strip test and fingerprick blood SEA-ELISA for detection of urinary schistosomiasis in schoolchildren in Zanzibar. *Acta Trop.* 111, 64–70, 2009.
77. Doenhoff, M.J., Chiodini, P.L., and Hamilton, J.V. Specific and sensitive diagnosis of schistosome infection: Can it be done with antibodies? *Trends Parasitol.* 20, 35–39, 2004.
78. Pardo, J. et al. Application of an ELISA test using *Schistosoma bovis* adult worm antigens in travellers and immigrants from a schistosomiasis endemic area and its correlation with clinical findings. *Scand. J. Infect. Dis.* 39, 435–440, 2007.
79. Makarova, E. et al. Detection of IgG binding to *Schistosoma mansoni* recombinant protein RP26 is a sensitive and specific method for acute schistosomiasis diagnosis. *Parasitol. Int.* 54, 69–74, 2005.
80. Hamburger, J. et al. Polymerase chain reaction assay based on a highly repeated sequence of *Schistosoma haematobium*: A potential tool for monitoring schistosome-infested water. *Am. J. Trop. Med. Hyg.* 65, 907–911, 2001.
81. Rabello, A., Pontes, L.A., and Dias-Neto, E. Recent advances in the diagnosis of *Schistosoma infection*: The detection of parasite DNA. *Mem. Inst. Oswaldo Cruz* .97(Suppl 1), 171–172, 2002.
82. Pontes, L.A., Dias-Neto, E., and Rabello, A. Detection by polymerase chain reaction of *Schistosoma mansoni* DNA in human serum and feces. *Am. J. Trop. Med. Hyg.* 66, 157–162, 2002.
83. Pontes, L.A. et al. Comparison of a polymerase chain reaction and the Kato-Katz technique for diagnosing infection with *Schistosoma mansoni. Am. J. Trop. Med. Hyg.* 68, 652–656, 2003.
84. Sandoval, N. et al. A new PCR-based approach for the specific amplification of DNA from different *Schistosoma* species applicable to human urine samples. *Parasitology* 133, 581–587, 2006.
85. Sandoval, N. et al. *Schistosoma mansoni*: A diagnostic approach to detect acute schistosomiasis infection in a murine model by PCR. *Exp. Parasitol.* 114, 84–88, 2006.
86. Akinwale, O.P. et al. Detection of schistosomes polymerase chain reaction amplified DNA by oligochromatographic dipstick. *Mol. Biochem. Parasitol.* 160, 167–170, 2008.
87. Wichmann, D. et al. Diagnosing schistosomiasis by detection of cell-free parasite DNA in human plasma. *PLoS Negl. Trop. Dis.* 3, e422, 2009.
88. Fenwick, A. et al. Drugs for the control of parasitic diseases: Current status and development in schistosomiasis. *Trends Parasitol.* 19, 509–515, 2003.
89. Pérez Arellano, J.L. et al. Tratamiento de las enfermedades parasitarias (II): Helmintosis y ectoparasitosis. *Información Terapéutica del Sistema Nacional de Salud* 31, 55, 2007.
90. Xiao, S.H. Development of antischistosomal drugs in China, with particular consideration to praziquantel and the artemisinins. *Acta Trop.* 96, 153–167, 2005.
91. Keiser, J. et al. Mefloquine-an aminoalcohol with promising antischistosomal properties in mice. *PLoS Negl. Trop. Dis.* 3, e350, 2009.
92. McManus, D.P., and Loukas, A. Current status of vaccines for schistosomiasis. *Clin. Microbiol. Rev.* 21, 225–242, 2008.

42 Trichobilharzia

P. Horák, F.M. Schets, L. Kolářová, and S.V. Brant

CONTENTS

42.1 INTRODUCTION

42.1.1 Classification and Biology

Schistosomes, members of the family Schistosomatidae (Platyhelminthes, Neodermata, Trematoda), represent well-known pathogens of birds and mammals, including humans. These mostly blood-inhabiting parasites belong to 13 genera. As far as human health is concerned, two genera are the most studied model organisms: *Schistosoma* sp. as a true human parasite causing schistosomiasis, a disease afflicting about 200 million people in tropical and subtropical areas, and *Trichobilharzia* sp. which is a bird parasite that can penetrate the skin of humans (accidental hosts) and cause cercarial dermatitis (swimmer's itch).

Avian schistosomes of the genus *Trichobilharzia* have a similar life cycle as their relatives parasitizing humans (genus *Schistosoma*). Eggs produced by the females embryonate within the host body, resulting in larvae of the first stage—miracidia. Fully developed eggs or even miracidia without egg envelops are expelled via feces or nasal secretions to the outer aquatic environment (visceral and nasal species, respectively). In the water body, the free-swimming miracidia search for a suitable snail host, penetrate its body surface, and transform to sporocysts. Sporocysts of the next (second) generation asexually produce cercariae, the last larval stage which leaves snails and attempts to reach a bird (definitive host). This free-swimming cercarial stage exhibits a high affinity for human skin (unsaturated fatty acids of the skin are highly attractive stimuli) [1] and therefore penetrates humans resulting in cercarial dermatitis.

It has been supposed that cercariae/schistosomula of bird schistosomes die soon after penetration into the mammalian (human) skin. Experimental infections of mammals by bird schistosomes of the genera *Trichobilharzia*, *Gigantobilharzia*, and *Ornithobilharzia* showed, however, that some parasites may escape from the skin and undergo a partial migration similar to that shown in bird hosts [2]. Schistosomula of visceral schistosomes were found in the lungs and liver, and juveniles of nasal *Trichobilharzia* in the central nervous system of various rodents. Bird schistosomes do not mature in mammals and die at various intervals post-infection (p.i.), for example, in immunocompetent and immunodeficient mice the migrating immature worms of *Trichobilharzia regenti* were recorded 21 days p.i. and 2 months p.i., respectively [3].

Members of the genus *Trichobilharzia* (the largest genus within schistosomes, with about 40 known species and some others awaiting characterization) use a few families of fresh water snails and several orders of birds for their development and reproduction. Unfortunately, using traditional methods of morphological identification, it is not always clear which adults/eggs from birds, and cercariae from snails belong to the same species. Generally, species-specific morphological characteristics of these stages are either missing or minute, and their recognition depends on personal experience.

Only some of these characteristics can be used for rough species identification, for example, position of internal organs and tegument spination of adults [2], morphology of eggs [4], or distribution of sensory papillae on cercariae [5]. In addition to the morphological features, there are some nonmorphological criteria that can be used to differentiate species, for example, behavior of cercariae in water column, compatibility of larvae with snail intermediate hosts and organ/tissue specificity of adults in birds.

Not only is there a high diversity of *Trichobilharzia* species yet to be characterized, but some of these species differ markedly in their pathogenicity toward birds and experimental mammals [2]. Therefore, from the epidemiological/medical viewpoint, there is a need for species delineation, and molecular techniques seem to be a suitable tool [6–8].

The ease with which we can now genetically match larval stages from snails to adult stages from birds has propelled us forward in understanding the epizootology and epidemiology of avian schistosomes [9–11]. Sequencing of gene regions such as nuclear internal transcribed spacer (ITS) region and the mitochondrial *cox*1 has not only linked life cycle stages, but has suggested that the diversity of avian schistosomes continues to grow [6–8,10,12]. Molecular phylogenetics has had a major impact on the taxonomy and discovery of new lineages of avian schistosomes, particularly the major etiological agents of cercarial dermatitis, species of *Trichobilharzia* [6–16].

To date, the nuclear ribosomal DNA D1-D2 regions of 28S, ITS, and the mitochondrial *cox*1 have been tested as molecular markers for systematic [6–10,13,17], epidemiology [18], and diagnostics [19,20]. While the molecular identification of avian schistosomes is still in the early stages, these markers have so far proved successful in attributing most samples to a known species. It is not just that few markers are available for systematics, epidemiology, and diagnostics, but also the availability of comparative DNA sequences in sequence databases such as GenBank. At this time microsatellite markers have not been developed that would allow detection of specific populations or host strains like there are for the human schistosomes. However, most mitochondrial genes with sufficient variation will give a desired level of diagnosis. Microsatellite markers might be more appropriate for epidemiological or population studies rather than detection and diagnosis.

The use of DNA sequences as a tool to identify species has been widely discussed. Yet, given the number and significance of pathogens, development of these tools has been relatively recent for many species [13,21–27]. While these methods provide a powerful scaffold for circumscribing species and, certainly, rapid detection, caution must be exercised as DNA identification alone is not sufficient in the absence of data on disease dynamics and morphology [28,29]. Molecular identification of a particular avian schistosome to the species level is only a partial characterization of the species, but is still a very useful piece of information. Identifying the species allows us access to the body of knowledge that exists about that species, or at least related species, especially if it has been framed within a phylogenetic context. For example, if the sample comes from a bird, what is the likely snail host? If the sample comes from a snail, what is the likely bird host? This can be addressed by knowing the life cycles of close relatives. An identification of the species provides clues as to the host source and biology and possible target for control. Nonetheless, there has been steady effort and testing of markers that have revealed some consistency and validity for species identification [9,22,23].

42.1.2 Epidemiology

42.1.2.1 Transmission via Intermediate and Definitive Hosts

Cercariae of species of *Trichobilharzia* were the first to be implicated in causing cercarial dermatitis [30] and since then it has been shown repeatedly that species of *Trichobilharzia* are the major group responsible for most dermatitis outbreaks in fresh water [7,16,18,31–38], although some other genera of avian schistosomes (e.g., *Bilharziella*) can also be a cause of swimmer's itch [11,39–44]. Although the frequency of occurrence of marine cercarial dermatitis is underreported, cases have been documented globally, for example, *Austrobilharzia* spp. are known as dermatitis agent in South Africa, Australia, and North America and *Gigantobilharzia* sp. is probably responsible for dermatitis of humans collecting marine molluscs in southern Europe—Venice lagoon, Italy [45].

Generally, *Trichobilharzia* infection rates in snails are low and do not exceed 5%, as known for infections in lymnaeids [2], physids [11], or planorbids [11]. For example, in Germany prevalence of 0.17% and 0.24% has been recorded in *Lymnaea stagnalis* and *Radix* spp., respectively [46], and in the United States 0.30–1.32% of snails of *Stagnicola emarginata* were infected by *Trichobilharzia stagnicolae* [38]. On the other hand, higher prevalence has occasionally been recorded, for example, 5–12% in *L. stagnalis* from the Netherlands [47], 24.5% in *Radix peregra* from Iceland [48], and 52.4% in *Chilina dombeyana* from Chile [49]. It is thought that in some cases these high values can result from disturbances of ecological balance of the lakes, including accelerated eutrophication leading to a higher abundance of susceptible snails [2].

Prevalence in birds is usually higher than that in snails, as seen from selected data for visceral *Trichobilharzia* spp.: 60.5% in spurwing goose in South Africa, 69.6% in domestic ducks in China, 83.9% in common merganser, 2.6% in Canada goose, and 4.9% in gulls from the United States [31], 94% for *T. querquedulae* in three anatid species from the United States [7], and 66.7% and 77% in mallards from Iceland and France, respectively [48,50]. Also the infections by nasal *Trichobilharzia* show high prevalence levels: 24.0% in wildfowl from Central Europe [51], 73.3% and 75% in mallards from France and Iceland, respectively [48,50]. Recent behavioral changes of definitive hosts might contribute to an intensified transmission of *Trichobilharzia* in some localities; some populations of European waterfowl stopped migrating between nesting and wintering areas and became resident [52], allowing an extended contact of birds, snails, and schistosomes at a locality.

42.1.2.2 Occurrence of Cercarial Dermatitis

Cercarial dermatitis is an increasing problem in many parts of the world [16,37,46,53–55], as noted in recent reports from Austria [56], Iceland [48], the United Kingdom [57], the Netherlands [58], Iran [33], China [59], Chile [49], and the United States [11]. Although the disease is recently considered as reemerging [52], no detailed information about the total number of afflicted persons is available. Locally, the incidence seems to be high: Dubois et al. [60], estimate that the disease can affect 15% of swimmers in the Savoy district, in Sweden approximately one-half of municipalities (125 out of 248) with lakes reported problems with dermatitis [61], and in the Netherlands more than 50% of recreational water-related outbreaks of skin incidents were associated with swimmer's itch in 2005 [58]. In Chile, the observation of 1500 swimmers in 2004–2005 revealed that 3% of them acquired cercarial dermatitis [49]. In Europe, the United States, Australia, and other areas, outbreaks of cercarial dermatitis are mostly associated with recreational contact with surface waters where parasite transmission via snail and bird hosts is enabled. These waters generally have a very good microbiological quality [58,62] according to bathing water legislation which is based on fecal contamination.

In a temperate climate the disease occurs seasonally with the highest prevalence during the warm summer months that corresponds to both the peak release of cercariae from snail hosts and increase in the number of people engaged in water activities. However, cercarial dermatitis has also been reported from cold lake areas at higher latitudes [63]; this might be a consequence of recent global climate changes [64], Under certain circumstances, the disease can be acquired during winter, for example, after bathing in geothermally warmed lakes [65].

42.1.2.3 Epidemiology and Outbreak Studies of Cercarial Dermatitis

Only a few epidemiological studies on cercarial dermatitis have been conducted. In 1993, a case–control study was performed among residents of Walloon Lake, Michigan [66]. Individuals who reported symptoms that met the case definition were included, and controls were selected from households residing on the lake. Participants completed either a case or a control questionnaire, providing information on their lake exposure. In 1996, a prospective cohort study conducted in Switzerland enrolled visitors from selected Geneva beaches [67]. A self-administered baseline questionnaire, which provided information on previous bathing and history of skin lesions, was followed (2–7 days later) by a questionnaire administered by the research team, inquiring about swimming behavior on the enrolment day and possible subsequent skin disorders. In another prospective study, performed in 2000 [68], 40 households of riparian Douglas Lake, Michigan, kept a daily diary in which they recorded their time spent in (shallow) water and the occurrence of symptoms. Cercarial dermatitis was self-evaluated based on explicit criteria that were provided by the research team. Lévesque et al. [62] and Schets et al. [58] studied outbreaks of cercarial dermatitis in Canada and the Netherlands, respectively. In both studies, questionnaires were distributed among potentially exposed groups and environmental samples were investigated.

In the above studies, Lindblade [66] found cases in 14% of the lake residences, Chamot et al. [67] reported dermatitis in 27.6% of the participants in the Swiss study, and Verbrugge et al. [68] reported dermatitis in 17.3% of the people that were in contact with Douglas Lake. A total of 53 people from 450 families likely to use Lac Beauport in Canada had symptoms [62], whereas 53% of the children that were exposed to dune lake Het Wed in the Netherlands contracted cercarial dermatitis [58]. In outbreaks, attack rates, that is, the number of exposed persons with symptoms compared to the total number of exposed persons, tended to be variable, but generally high: 55–82% [35].

In most studies, gender had no effect on the incidence of cercarial dermatitis [58,66–68]; however, Lévesque et al. [62] found that more males than females had symptoms. Age did not have any effect in two studies [67,68], whereas two other studies found that the mean age of cases was lower than that of controls [66] or that more than half of the cases was younger than 10 years of age [62]. Different effects observed for different age groups may reflect differences in susceptibility, but they may also be the result of different exposure: children tend to stay in the water for longer periods than adults and more likely swim or play in shallow water.

Both epidemiological and outbreak studies demonstrated an increased incidence of cercarial dermatitis with increased water contact, which was reflected by involvement in activities with extensive water contact, such as wading and swimming [66] and the number of days with water contact and/or the total time spent in the water [58,67,68]. Symptoms of cercarial dermatitis were more severe when more time was spent in the water, particularly when it was shallow [68]. It is obvious that the probability of contact of cercariae with the skin increases with increased time spent in the water. In addition, factors related to the biology of the parasite also had an effect on the incidence of cercarial dermatitis. For example, there were more cases in participants that had been swimming in the morning, when the release of cercariae from snails tends to be highest [66,67], and spent more time in shallow water, where the snail population is generally denser and the water temperature is higher, enhancing development of the parasite [62,66,68]. Moreover, an increased risk was observed for those who had a history of skin lesions [67].

42.1.2.4 General Approach to Investigating an Outbreak

Investigations into presumptive outbreaks of cercarial dermatitis generally start with the formulation of a case definition, in which public health professionals define who is included as a case in the outbreak investigation, followed by setting up a surveillance system that monitors the progress of symptoms in the exposed community, and designing an environmental investigation. Data from epidemiological studies can link cases to a specific body of water and can also provide insight into risk factors for cercarial dermatitis (see Sections 42.1.2.2 and 42.1.2.3). Epidemiological data may be obtained through surveillance via physicians. However, due to the generally benign pathology of cercarial dermatitis most patients will not visit a physician and will thus be missed in an epidemiological investigation. Therefore, a suitable approach is an on-site distribution of questionnaires among persons that are likely to be or have been exposed to the water suspected of schistosome-infected snails. The questionnaires may be self-administered or administered by a research team. When respondents evaluate their own symptoms, it is important to provide a clear case definition and evaluation criteria, otherwise similar symptoms from unrelated causes may be included in the study. In instances of self-reported and self-evaluated symptoms a visible inspection of each case or a request to submit a picture of affected body parts may help identify non-dermatitis pathology. It should be kept in mind that the overall incidence may be overestimated. For example, people with symptoms, or those that previously experienced episodes of cercarial dermatitis, may be more inclined to respond and those that were exposed, but did not have symptoms, may not be available for a follow-up.

42.1.2.5 Environmental Investigation

Environmental investigations generally involve collection and screening for the avian schistosomes in the snail hosts from the affected water. Snail collecting and determining the presence of schistosome cercaria by microscopy is an intensive laboratory procedure which is prone to errors, particularly in laboratories that do not have specific parasitological expertise. Erroneous identification of cercariae may result in ongoing exposure of bathers, or unnecessarily warning against swimming or even closing a swim area. In addition to the above difficulties, cercarial dermatitis may be reported in water where no snails are found [47,58] or collected snails do not shed cercariae perhaps due to low prevalence [37,46]. Hertel et al. [69] reported a method for polymerase chain reaction (PCR) detection of *Trichobilharzia* in lakes. Although their method detected *Trichobilharzia* cercariae that were added experimentally to snail and plankton samples, it was not tested on water samples taken from a lake. Schets et al. [47] suggested a novel approach for direct confirmation of *Trichobilharzia* in bathing water, including concentration of water samples by filtration and PCR detection of the parasite. The examination of snails and analysis of water samples appeared to be complementary procedures that enhanced the chance of parasite detection in water suspected of containing cercariae, and PCR of the ToSau3A (*Trichobilharzia* tandem repeated sequence) directly confirmed the parasite in the water [47]. PCR of the nuclear gene regions ITS [13] and D2 [16] not only amplified *Trichobilharzia*, but also *Echinostoma* and *Diplostomum*. Therefore, while these gene regions are useful in the identification of isolated samples of avian schistosomes (e.g., adults from a bird), they are not specific enough for samples that may contain other trematodes [16]. Both examination of snails and analysis of water samples have their limitations, but since examination of snails more often demonstrated the presence of *Trichobilharzia* than analysis of water samples, Schets et al. [47] concluded that environmental investigations of an outbreak of cercarial dermatitis should start with the examination of collected snails, followed by analysis of water samples in the case of negative results. In the absence of snails, analysis of water samples would also be a valuable alternative.

42.1.3 Diagnosis

42.1.3.1 Detection of Parasites in Birds, Snails, and Environment

There are a couple of methods for detecting schistosome infections in birds. Identification of visceral avian schistosomes is focused on examination of bird feces for eggs using various sedimentation methods (e.g., formol-ether technique) or Kato-Katz fecal smear technique. For nasal schistosomes, nasal cavities of the bird can be flushed out with saline or water to check for the presence of eggs or miracidia [2]. There is also evidence that schistosome infections can be detected by serological techniques, for example, indirect immunofluorescence test (IFAT) [70]. One caveat is that these techniques are not species-specific. Postmortem examination of birds is by examination of the arterial/venous system and the tissues of target organs, because bird schistosomes show preference for particular organs/tissues [2]. Final determination of *Trichobilharzia* species should reflect all data available, that is, host species, location, and morphological characters of adult worms and eggs. In addition, comparative molecular analyses must be performed to obtain reliable and convincing results of species identification (see Sections 42.1.1 and 42.2.2 for details).

Determination of the schistosome cercariae emerging from infected snails is the first step to identifying a schistosome infection, and identifying the species of snail may narrow pool of potential schistosome species. If no cercariae are shed at the time of examination, dissection of snails (snail hepatopancreases) is required. Cercariae may also be detected directly in water samples (without examination of snails), but their abundance in lakes is usually low. Therefore, water samples can be concentrated (e.g., filtration [47]; see also Section 42.1.2.5). Cercariae can also be "collected" by a trap—a glass microscope slide covered by a matrix containing linoleic acid as the substrate stimulating attachment and penetration of cercariae; then the worms can be visualized for counting [71]. However, it should be kept in mind that since the cercariae of *Trichobilharzia* are morphologically similar, final species determination should rely on molecular techniques (see Sections 42.1.1 and 42.2.2 for details).

Ideally, a voucher sample of *Trichobilharzia* should be preserved and deposited in a museum [72–74]. This is important for several reasons. First, in the event that the parasite sample does not match known species or has odd features, further work may be necessary. Access to the images and measurements, plus more of the sample(s) will be used in additional sequencing. Second, the samples might represent a new species, and thus its characterization will be imperative for the ongoing epidemiology of cercarial dermatitis. The samples can also be sent to an expert on that group of schistosomes for more detailed determinations, and any images and measurements will aid in resolving the diagnosis. Third, documentation of any species of schistosome coming through a clinic is an important record that can contribute to epidemiological studies, especially if it is associated with people affected by dermatitis.

42.1.3.2 Cercarial Dermatitis: Clinical Features and Pathogenesis

If a skin reaction occurs soon after swimming, then cercarial dermatitis should be considered. Severity and time course of cercarial dermatitis depend on the sensitivity of the afflicted persons and on the number of penetrating cercariae [2]. A person with a short exposure to the water containing cercariae will have a weak reaction. Initial infections (first contact) have been shown to be asymptomatic or a mild skin reaction characterized by development of maculae, or colored spots occurs. If papulae develop at this point they are small, innocuous, and inconspicuous, reaching 1–2 mm in diameter. Typical symptoms (Figure 42.1) develop after repeated infections (sensitization phenomenon [75]) when the disease manifests as an itchy maculopapular eruption. Shortly after cercarial penetration into the skin (within 4–20 min) a primary itching (like fine prickling) occurs persistently for up to 1 h. Later, transitory maculae (up to 10 mm in diameter) develop at the site of penetration for each cercaria. In a few hours (10–15 h p.i.), maculae usually disappear and are replaced by distinct and indurated papulae (about 3–8 mm in diameter) with a small central spot. Severe infections may result in the formation of papulae even during the first hour p.i. Papulae can be encompassed by an erythematous zone that can persist for three days. The size of the erythema usually reaches 10–100 mm in diameter, but can be markedly larger due to rubbing or scratching. The itching accompanying the eruption of papulae can range from negligible to unbearable, sometimes with insomnia. When papulae become confluent, the whole exposed area is elevated and edematous. On day 2–3 p.i. vesicles of 3–8 mm in diameter are formed on papulae of sensitized individuals and due to scratching they often rupture. Also excoriation may occur, and secondary infection may lead to the formation of pustules. Although regression of papulae is usually recorded after four days p.i., the symptoms may persist for about 20 days p.i. Around 10 days p.i. the papulae usually disappear and leave pigmented spots that can persist for weeks. Severe infections can be accompanied by fever, limb and lymph node swelling, nausea, and diarrhea. Rarely, the infection can be manifested by a generalized skin reaction which occurs on day 3 p.i. when small (about 2 mm) hard and pink papulae can be noted; the rash is accompanied by mild and/or intermittent itching and persists usually for two days. Sensitization can last for years, and there are no confirmed data on possible desensitization. Persons exposed repeatedly to cercariae (e.g., rice planters, pickers) can suffer from a chronic type of dermatitis ("koganbyo" in Japan) [76,77] which is accompanied by an extensive skin sloughing and injury recorded mainly on hands and legs.

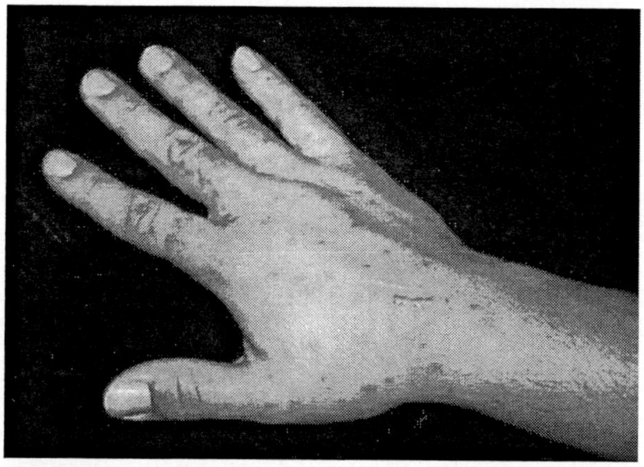

FIGURE 42.1 Cercarial dermatitis: maculopapular skin eruption caused by cercariae of *Trichobilharzia regenti*; six days p.i.

Histopathological studies [78] on skin biopsies of persons with well-developed cercarial dermatitis showed leukocytic (3–9 h p.i.), lymphocytic (24 h p.i.), and histiocytic (36–52 h p.i.) phases of the immune cell response against schistosomula of *Trichobilhazia*. Degradation of the parasite tegument was initiated 24 h p.i., and at 36 h p.i. the schistosomula were completely destroyed. In the area surrounding parasites, perivasculitis and spongiosis occurred. Subsequently, tissue repair started 72 h p.i., accompanied by parakeratosis, and reduced perivascular infiltration. Cell melanotic pigmentation has been noted in the surrounding tissue 100 h p.i.

Recent immunohistochemical studies revealed that cercarial dermatitis caused by bird schistosomes in mice comprises both immediate and late-phase cutaneous hypersensitivity reactions associated with a polarized Th2-type immune response [79]. Such reactions are probably stimulated by antigens of cercariae, for example, 34-kDa molecule recognized by a specific antibody of repeatedly infected mice and humans [80].

Studies on mice infected by *Trichobilharzia regenti or T. szidatii* showed that the parasites can escape from the skin and cause damage to internal organs. Neurotropic schistosomula of *T. regenti* found in the mouse spinal cord were surrounded by infiltrates of various immune cells, and dystrophic/necrotic changes of neurons were observed; perivascular eosinophilic inflammation developed in the matter [81]. Visceral migration of *T. szidati* in mice includes the lungs where the parasites are localized extravascularly in the alveolar walls; alveolar wall congestion, edema, and lymphocyte infiltrates were recorded [82].

Fortunately, repeated exposures to cercariae seem to trigger immune reactions preventing migration of schistosomula from the dermis [3,80]. It remains questionable, whether the above-mentioned migration and tissue/organ damage happen also in humans. Nevertheless, Bayssade-Dufour et al. [83] referred to an increase in pulmonary disorders of both the previously healthy volunteers and the asthmatic children after skin infection caused by bird schistosomes.

42.1.3.3 Diagnosis of Cercarial Dermatitis

Differential diagnosis of cercarial dermatitis is difficult. Grossly, the skin lesions resemble those resulting from dermatitis of bacterial etiology, contact dermatitis or insect bites from chiggers, fleas, and mosquitoes [84]. Therefore, to identify bird schistosome cercariae as the etiological agent, clinical, and epidemiological data, together with parasitological and immunological examinations, need to be evaluated. Among others, three criteria are significant: (a) history of contact with a natural water body within previous 96 h, (b) development of papular dermatitis with severe itching during 24 h after water contact, (c) distribution of lesions exclusively on body parts that were immersed in water [85].

Laboratory diagnosis includes histological and immunological techniques for direct and indirect identification of the disease agent. Histological examination of the skin biopsy taken from sites where papulae have developed provides direct proof. However, the biopsies should be taken shortly after exposure (until 48 h p.i.); on day 3 p.i. only destroyed parasites are detected, and later only an inflammatory reaction is recorded. Skin tests comprising injection of antigen from homogenized cercariae and evaluation of erythema around the site of puncture are not sufficiently specific and sensitive. Lack of specificity and sensitivity also restricts introduction of serological tests monitoring specific antibodies (e.g., "Cercarienhüllenreaktion," complement fixation test, IFAT, and ELISA). For example, the low specificity of serological tests does not enable differential diagnosis of cercarial dermatitis versus human schistosomiasis [2]; therefore, these tests are not routinely used in the field. It becomes apparent, however, that some cercarial antigens of *T. regenti* can provide sufficient sensitivity and specificity, for example, sera of patients with cercarial dermatitis contained specific IgG and IgE against 34 kDa antigen [80]. Antigens from other developmental stages of *Trichobilharzia* might also be (re)considered as a tool discriminating between skin lesions (and other symptoms) caused by human and bird schistosomes [70].

As cercarial dermatitis is an allergic reaction, laboratory tests may reveal increased eosinophil counts and elevated levels of total and specific IgE. It has been demonstrated that cercariae of *T. regenti* are a source of antigens that trigger IL-4 release from basophils of healthy volunteers, that is, cells associated with high levels of IgE during helminth infections. Unfortunately, data on total IgE and antigen-specific IgE from patients with cercarial dermatitis are not fully consistent and partly differ from those on mouse model; therefore, further experiments need to be performed [80].

42.2 METHODS

The current molecular techniques employed to identify species of avian schistosomes are straightforward and relatively fast. Currently, diagnosis of avian schistosome species is by sequencing one of the commonly used gene regions (28S, ITS, *cox*1) after amplification by PCR, and then comparing the results to sequences stored on a related database (i.e., GenBank). These methods apply to adult worms and eggs/miracidia from the bird host, larvae (cercariae and sporocysts) from the snail host, and possibly if found recently enough, small tissue samples taken from an entry point on the human skin. The nuclear gene region, 28S, is amplified consistently by use of universal primers that are widely used in digenetic trematode systematics (Table 42.1). Although 28S may not seem a likely marker for species differentiation, its usefulness should not be overlooked [86]. Even between closely related species, there is sufficient variation to delimit species (Table 42.2).

The nuclear ITS regions have proved effective in delineating species of avian schistosomes, particularly within *Trichobilharzia* that are responsible for outbreaks of dermatitis [7–9,13,16,20]. Moreover, there are three species of *Trichobilharzia* that are difficult to distinguish morphologically and are responsible for a major proportion of dermatitis outbreaks, *Trichobilharzia franki*, *Trichobilharzia physellae*, and *Trichobilharzia*

TABLE 42.1
List of Primers for PCR and Sequencing of Avian Schistosomes

Gene Region	Forward Primer	Reverse Primer	PCR Fragment	Sequencing Primers	Reference
28S	U178	L1642	1500 bp	U178	[91]
				L1642	
				ECD2	
				Dig12	[92]
28S D1-D2	C2'B	D2	560 bp	C2'B	[93]
				D2	
ITS	ITS5Trem	ITS4Trem	1400 bp	ITS5Trem	[13]
				ITS4Trem	
				4S	
				3S	
ITS-2	ITS3Trem	ITS4Trem	370 bp	ITS3Trem	[13]
				ITS4Trem	
*cox*1	CO1F15	CO1R15	1000 bp	CO1F15	[7]
				CO1R15	

TABLE 42.2
Average Intra- and Interspecific Pairwise Genetic Differences

	28S (D1-D2)	ITS1–5.8S-ITS2	*cox*1
Among avian schistosome genera	3.6–9.2%	4.3–10%	14.1–20.4%
Among species of *Trichobilharzia*	0.7–3.0%	0.7–3.1%	6.9–13.4%
Within a species of *Trichobilharzia*	0–0.5%	0–0.27%	0–0.4%

querquedulae. Using ITS genes, ITS1 in particular, average genetic differences among these three species range between 0.8 and 4.0% and within species 0–0.7%. If it is unclear that the species may belong to *Trichobilharzia*, then a more conserved section consisting of partial ITS1–5.8S-ITS2-partial 28S [13] easily discriminates species. The advantages of 28S are the availability of universal primers and ease of amplification. One disadvantage is that for closely related species, 28S may not be able to discriminate to a population level, if that is the target. While ITS may be more variable, often times multiple bands are obtained and cloning is required. However, for avian schistosomes, there are available one set of primers that work for most genera and species [6,7,12,13] (Table 42.1) and there exists a broad spectrum of digenetic trematodes for which there are ITS data [22,23]. One caveat to using universal primers is that if the sample must be amplified from snail tissue, then these primers will not only amplify the schistosome, but any other trematodes present in the snail.

Mitochondrial genes are ideal for the reason that they contain suitable variation for most levels of species discrimination. However, there is still difficulty in obtaining universal primers that will work for a wide range of taxa, or even species within a genus. Since the sequencing of the mitochondrial genome of *T. regenti*, there are promising genes that include *nad*6, *nad*4, *nad*2, and *cox*3 [87]. Currently, the 5′ end of the *cox*1 gene is most frequently sequenced and is used commonly for "barcoding" [7,10,11,88], but amplification and sequencing is not always straightforward. As with all mitochondrial genes, care must be taken to verify that the gene is from the mitochondrial genome and not a pseudogene [89]. There are several steps of precaution: (i) make sure the length of the fragment is the expected length (and a single band) using appropriate DNA ladders in the gel visualization of the PCR product, (ii) check for stop codons using the flatworm mitochondrial genetic code, (iii) sequence both forward and reverse primers to verify that the sequences are identical.

42.2.1 Sample Preparation

Samples for the detection of avian schistosomes will be collected from birds or snails in areas with outbreaks or suspected outbreaks of cercarial dermatitis. In humans, biopsy of the penetration site should be taken shortly after exposure (see Section 42.1.3.3), otherwise there remains insufficient parasite matter that would be detectable by PCR methods. It is usually several hours or the next day that a patient might contact a clinic regarding the dermatitis. Fresh material or molecular grade alcohol preserved material can be used and samples as small as a single egg or cercaria will yield sufficient DNA for diagnosis, especially with 28S or ITS gene regions [8]. While sequence data for diagnosis are the most preferred method for detection, morphological features should not be disregarded. Prior to preparing the sample for molecular analysis, a subsample of the worms, whether fresh or preserved, should be wet mounted on a microscope slide, images, and general measurements recorded (see also Section 42.1.3.1).

Adults or cercariae should be isolated and rinsed free of as much of the host tissue as possible (i.e., water or with a Tris-EDTA buffer). While most of the primers are specific for schistosomes, host tissue can lead to DNA contamination and problems in downstream DNA amplification and sequencing. If the samples are stored in alcohol, care should be taken that all of the alcohol has been evaporated. This can be done by centrifuging the sample to a pellet in the tube, followed by aspiration of most of the alcohol then leaving the tube open, or placing the tubes in a vacuum centrifuge. Traces of alcohol will inhibit DNA extraction. The sample procedure will be the same for fresh samples; centrifuge the sample to a pellet, then aspirate as much of the water/buffer as possible. The vacuum centrifuge with applied heat will accelerate the evaporation of trace fluid. In the case of water, trace amounts will not affect most standard DNA extraction methods.

A quick and inexpensive method for DNA extraction is the Hot Shot technique [90]. For parts of adult worms or 5–20 cercariae/miracidia, 25 μL or the lysis buffer and then 25 μL of the neutralizing buffer are sufficient. For smaller sample volumes such as an egg, miracidium, or single cercaria, it is recommended to use 5 μL of each reagent. Alternatively, for higher-quality DNA, several kits have been used successfully, for example, QIAamp DNA mini kit (Qiagen) or BioRad DNA isolation kits. Extracted DNA should optimally be stored in a −70–80°C ultracold freezer. Any samples not used for extraction should be kept minimally in a 5°C refrigerator.

42.2.2 Detection Procedures

After DNA extraction, the primers of choice are used in a PCR reaction. For most diagnostic tests, a fragment of about 300–600 base pairs (bp) is desirable, and decreases the sequencing costs since only two primers are required to obtain overlapping sequence. However, if a more thorough analysis, such as phylogenetics is desired, then primers that amplify a longer fragment are necessary. See Table 42.1 for a list of primers and the average sequence length. PCR should be completed and samples purified with any variety of clean-up protocols or kits. Takara Ex *Taq* kit (Takara Biomedicals) is a relatively inexpensive and reliable *Taq* polymerase. The PCR reactions and cycling conditions have been optimized using the suggested protocol that comes with the Takara *Taq*. The PCR reactions have been

TABLE 42.3
PCR Conditions for the Amplification of Schistosome DNA

Cycle	Step	Temp. (°C)	Time (s)	Number of Cycles
1	1	95	240	1
2	1	94	30	3
	2	55	30	
	3	72	120	
3	1	94	30	3
	2	54	30	
	3	72	120	
4	1	94	30	3
	2	53	30	
	3	72	120	
5	1	94	30	3
	2	52	30	
	3	72	120	
6	1	94	30	3
	2	51	30	
	3	72	120	
7	1	94	30	20
	2	50	30	
	3	72	150	
8	1	72	300	1

optimized for a reaction volume of 25 μL. Using thin-walled 0.2 mL reaction tubes the following is added: 15–100 ng DNA template, 0.2 μM of each primer, 2.5 μL of 10× buffer, 2.0 μL of 25 mM $MgCl_2$ (2 mM), 2.0 μL of 2.5 mM each of dNTP (0.8 mM), and 0.13 μL of 5 U/μL of Takara *Taq* DNA polymerase (0.65 U). The PCR cycling conditions for 28S and ITS primers can be found in Table 42.3. If amplifying *cox*1, decrease the annealing temperatures (°C) to 51°C, 50°C, 49°C, 48°C, 47°C, 46°C, and the rest of the program should remain the same. These are recommendations for Takara Ex *Taq* kit using an Eppendorf Mastercycler epgradient thermocycler as a result of different PCR reagents and thermocyclers may require slight adjustments. The sequencing will follow protocols standard for the institution and sequence editing can be done with a number of free programs, as well as ones for purchase, such as Sequencher. Copied nucleotide sequences should be entered into a sequencing database archive like GenBank and "Blasted" or compared to deposited sequences. Results show the coverage quality and percent sequence similarity, all useful for species diagnosis. However, as stated before, in the event that there is not a match to determine the specimen to species, morphological features or consulting of current experts are recommended.

42.3 CONCLUSIONS AND PERSPECTIVES

It is obvious that bird schistosomes in general and the genus *Trichobilharzia* in particular represent a rich group of parasites. In particular species of *Trichobilharzia* and their hosts should further be characterized. This seems to be essential not only due to their cosmopolitan distribution, exciting life cycles with sometimes cryptic/hidden phases of development within their bird hosts, or even due to phylogenetic study showing relations among schistosomes (including the human parasites of the genus *Schistosoma*), but also due to their role as pathogens of birds. Last but not least, members of the genus *Trichobilharzia* may represent an epidemiological problem and health risk to humans. Cercariae of *Trichobilharzia* are the causative agent of human cercarial dermatitis and, with regard to experiments with small laboratory mammals, the worms might leave the human skin (mainly in primary infection) and migrate to different parts of the body. In experimental animals this happens not only with visceral schistosomes, but also with neurotropic *T. regenti* occasionally causing paralysis of the animals. Unfortunately, consequences of such migration in humans are not understood.

It seems, therefore, that there are at least four main aims in the ongoing study of *Trichobilharzia*: (i) life cycles of *Trichobilharzia* spp. need to be recognized, including migration and pathogenicity in bird hosts and specificity toward snails. All this can contribute to our knowledge of parasite transmission and risks associated with particular species. (ii) A link between parasite species and transmission mode/pathogenicity can be made only if techniques allowing precise parasite identification/determination are employed; molecular tools are imperative. (iii) *Trichobilharzia* interactions with mammals should further be studied; this kind of research can allow predictions of what might happen in humans infected by cercariae (immune response, pathogenesis, etc.). (iv) In order to make any observation of people having contact with cercariae of bird schistosomes, and suffering from skin or other symptoms, specific and sensitive diagnostic tools (e.g., based on DNA detection) need to be developed and introduced. In this way, we could either make a reliable differential diagnosis of cercarial dermatitis, or control whether some (neurological, lung, etc.) disorders of yet unknown etiology are correlated with infections caused by *Trichobilharzia* spp.

ACKNOWLEDGMENTS

The work of coauthors was supported by the following grants/institutions: Czech Ministry of Education MSM 0021620828 (PH), MSM LC06009 (PH), MSM 0021620806 (LK), MSM 0021620812 (LK), Czech Science Foundation 206/09/0926 (PH), Charles University in Prague (UNCE and PRVOUK) (PH, LK), Ministry of Housing, Spatial Planning and the Environment—Directorate General of Environmental Protection in the Netherlands (FMS), College of Arts and Sciences at University of New Mexico and NIH grant RO1 AI44913 and P20 RR018754 to Dr. E.S. Loker (SVB).

REFERENCES

1. Haas, W., and van de Roemer, A. Invasion of the vertebrate skin by cercariae of *Trichobilharzia ocellata*: Penetration processes and stimulating host signals. *Parasitol. Res.* 84, 787, 1998.
2. Horák, P., Kolářová, L., and Adema, C. Biology of the schistosome genus *Trichobilharzia*. *Adv. Parasitol.* 52, 155, 2002.

3. Kouřilová, P., Syrůček, M., and Kolářová, L. The severity of mouse pathologies caused by the bird schistosome *Trichobilharzia regenti* in relation to host immune status. *Parasitol. Res.* 93, 534, 2004.
4. Skírnisson, K., and Kolářová, L. Diversity of bird schistosomes in anseriform birds in Iceland based on egg measurements and egg morphology. *Parasitol. Res.* 103, 43, 2008.
5. Podhorský, M. et al. Cercarial dimensions and surface structures as a tool for species determination of *Trichobilharzia* spp. *Acta Parasitol.* 54, 28, 2009.
6. Aldhoun, J.A. et al. Bird schistosome diversity in Iceland: Molecular evidence. *J. Helminthol.* 83, 173, 2009.
7. Brant, S.V., and Loker, E. S. Molecular systematics of the avian schistosome genus *Trichobilharzia* (Trematoda: Schistosomatidae) in North America. *J. Parasitol.* 95, 941, 2009.
8. Jouet, D. et al. Avian schistosomes in French aquatic birds: A molecular approach. *J. Helminthol.* 83, 181, 2009.
9. Rudolfová, J. et al. Validity reassessment of *Trichobilharzia* species using *Lymnaea stagnalis* as the intermediate host. *Parasitol. Res.* 95, 79, 2005.
10. Brant, S.V. et al. An approach to revealing blood fluke life cycles, taxonomy, and diversity: Provision of key reference data including DNA sequence from single life cycle stages. *J. Parasitol.* 92, 77, 2006.
11. Brant, S.V., and Loker, E.S. Schistosomes in the southwest United States and their potential for causing cercarial dermatitis or swimmer's itch. *J. Helminthol.* 83, 191, 2009.
12. Aldhoun J.A. et al. Schistosomes in the North: A unique finding from a prosobranch snail using molecular tools. *Parasitol. Int.* 58, 314, 2009.
13. Dvořák, J. et al. Comparison of European *Trichobilharzia* species based on ITS1 and ITS2 sequences. *Parasitology* 124, 307, 2002.
14. Bayssade-Dufour, C. et al. Seasonal morphological variations in bird schistosomes. *Parasite* 13, 205, 2006.
15. Kolářová, L. et al. *Allobilharzia visceralis* gen. nov. sp. nov. (Schistosomatidae-Trematoda) from *Cygnus cygnus* (L.) (Anatidae). *Parasitol. Int.* 55, 179, 2006.
16. Jouet, D. et al. *Trichobilharzia* spp. in natural conditions in Annecy Lake, France. *Parasitol. Res.* 103, 51, 2008.
17. Chrisanfova, G.G. et al. Genetic variability of bird schistosomes (class Trematoda, family Schistosomatidae) of Naroch lake: Identification of a new species in the *Trichobilharzia ocellata* group. *Doklady Biochem. Biophys.* 428, 268, 2009.
18. Ferte, H. et al. Presence of *Trichobilharzia szidati* in *Lymnaea stagnalis* and *T. franki* in *Radix auricularia* in northeastern France: Molecular evidence. *Parasitol. Res.* 95, 150, 2005.
19. Picard, D., and Jousson, O. Genetic variability among cercariae of the Schistosomatidae (Trematoda: Digenea) causing swimmer's itch in Europe. *Parasite* 8, 237, 2001.
20. Semyenova, S.K. et al. Individual and population variation in cercariae of bird schistosomes of the *Trichobilharzia ocellata* species group as revealed with the polymerase chain reaction. *Russ. J. Genet.* 41, 12, 2005.
21. Arnot, D.E., Roper, C., and Bayoumi, R.A.L. Digital codes from hypervariable tandemly repeated DNA sequences in the *Plasmodium falciparum* circumsporozoite gene can genetically barcode isolates. *Mol. Biochem. Parasitol.* 61, 15, 1993.
22. Nolan, M.J., and Cribb, T.J. The use and implications of ribosomal DNA sequencing for the discrimination of digenean species. *Adv. Parasitol.* 60, 101, 2005.
23. Vilas, R., Criscione, C., and Blouin, M. A comparison between mitochondrial DNA and the ribosomal internal transcribed regions in prospecting for cryptic species of platyhelminth parasites. *Parasitology* 131, 839, 2005.
24. Zarowiecki, M.Z., Huyse, T., and Littlewood, D.T.J. Making the most of mitochondrial genomes—Markers for phylogeny, molecular ecology and barcodes in *Schistosoma* (Platyhelminthes: Digenea). *Int. J. Parasitol.* 37, 1401, 2007.
25. Hamilton, P.B. et al. A novel, high-throughput technique for species identification reveals a new species of tsetse-transmitted trypanosome related to the *Trypanosoma brucei* subgenus *Trypanozoon. Infect. Genet. Evol.* 8, 26, 2008.
26. Ferri, E. et al. Integrated taxonomy: Traditional approach and DNA barcoding for the identification of filarioid worms and related parasites. *Front. Zool.* 6, 1, 2009.
27. Moszczynska, A. et al. Development of primers for the mitochondrial cytochrome c oxidase I gene in digenetic trematodes (Platyhelminthes) illustrates the challenge of barcoding parasitic helminths. *Mol. Ecol. Resources* 9, 75, 2009.
28. Besansky, N.J., Severson, D.W., and Ferdig, M.T. DNA barcoding of parasites and invertebrate disease vectors: What you don't know can hurt you. *Trends Parasitol.* 19, 545, 2003.
29. Hansen, H., Bakke, T.A., and Bachmann, L. DNA taxonomy and barcoding of monogenean parasites: Lessons from *Gyrodactylus. Trends Parasitol.* 23, 363, 2007.
30. Cort, W.W. Schistosome dermatitis in the United States (Michigan). *J. Am. Med. Assoc.* 90, 1027, 1928.
31. Loken, B.R., Spencer, C.N., and Granath, W.R. Jr., Prevalence and transmission of cercariae causing schistosome dermatitis in Flathead Lake, Montana. *J. Parasitol.* 81, 646, 1995.
32. Kolářová, L., Horák, P., and Sitko, J. Cercarial dermatitis in focus: Schistosomes in the Czech Republic. *Helminthologia* 34, 127, 1997.
33. Farahnak, A., and Essalat, M. A study on cercarial dermatitis in Khuzestan province, south western Iran. *BMC Public Health* 3, 35, 2003.
34. Voronin, M.V., and Beer, S.A. Morphological peculiarities of Schistosomatidae cercariae of *Trichobilharzia* cf. *ocellata* group occurring in Moscow and Saint-Petersburg populations. *Parazitologiia* 36, 60, 2002.
35. Bourée, P., and Caumes, E. La dermatitie cercarienne. *Presse Med.* 33, 490, 2004.
36. Sheng, S.C. et al. Preliminary study on the ecology of *Trichobilharzia* cercariae in the Huaihe river system. *Chin. J. Parasitol. Parasitol. Dis.* 22, 349, 2004.
37. Żbikowska, E. Infection of snails with bird schistosomes and the threat of swimmer's itch in selected Polish lakes. *Parasitol. Res.* 92, 30, 2004.
38. Coady, N.R. et al. Ubiquitous variability in the prevalence of *Trichobilharzia stagnicolae* (Schistosomatidae) infecting *Stagnicola emarginata* in three northern Michigan lakes. *J. Parasitol.* 92, 10, 2006.
39. Cort, W.W. Studies on schistosome dermatitis XI. Status of knowledge after more than twenty years. *Am. J. Hyg.* 52, 251, 1950.
40. Penner, L.R. *Cercaria littorinalinae* sp. nov., a dermatitis-producing schistosome larva from the marine snail, *Littorina planaxis* Philippi. *J. Parasitol.* 36, 466, 1950.
41. Stunkard, H.W., and Hinchliffe, M.C. The morphology and life history of *Microbilharzia variglandis* (Miller and Northup, 1926) Stunkard and Hinchliffe, 1951, avian blood flukes whose larvae cause "swimmer's itch" of ocean beaches. *J. Parasitol.* 38, 248, 1952.
42. Leigh, W.H. *Cercaria huttoni* sp. nov.: A dermatitis-producing schistosome larva from the marine snail, *Haminoea antillarum guadalupensis* Sowerby. *J. Parasitol.* 39, 625, 1953.

43. Chu, G.W.T.C. Pacific area distribution of freshwater and marine cercarial dermatitis. *Pac. Sci.* 12, 299, 1958.
44. Leedom, W.S., and Short, R.B. *Cercaria pomaceae* sp. n., a dermatitis-producing schistosome cercaria from *Pomacea paludosa*, the Florida apple snail. *J. Parasitol.* 67, 257, 1981.
45. Kolářová, L. Schistosomes causing cercarial dermatitis: A mini-review of current trends in systematics and host specificity and pathogenicity. *Folia Parasitol.* 54, 81, 2007.
46. Loy, C., and Haas, W. Prevalence of cercariae from *Lymnaea stagnalis* snails in a pond system in Southern Germany. *Parasitol. Res.* 87, 878, 2001.
47. Schets, F.M., Lodder, W.J., and de Roda Husman, A.M. Confirmation of the presence of *Trichobilharzia* by examination of water samples and snails following reports of cases of cercarial dermatitis. *Parasitology* 137, 77, 2010.
48. Skírnisson, K., Aldhoun, J.A., and Kolářová, L. A review on swimmers itch and the occurrence of bird schistosomes in Iceland. *J. Helminthol.* 83, 165, 2009.
49. Valdovinos, C., and Balboa, C. Cercarial dermatitis and lake eutrophication in South-Central Chile. *Epidemiol. Infect.* 136, 391, 2008.
50. Kolářová, L. et al. Avian schistosomes of the genus *Trichobilharzia* in Europe. *Bull. Scandinavian-Baltic Soc. Parasitol.* 14, 85, 2005.
51. Rudolfová, J., Sitko, J., and Horák, P. Nasal schistosomes in the Czech Republic. *Parasitol. Res.* 88, 1093, 2002.
52. Caumes, E. et al. Failure of an ointment based on IR3535 (ethylbutylacetylaminopropionate) to prevent an outbreak of cercarial dermatitis during swimming races across Lake Annecy, France. *Ann. Trop. Med. Parasitol.* 97, 157, 2003.
53. De Gentile, L. et al. La dermatite cercarienne en Europe: Un problème de santé publique nouveau? (Cercarial dermatitis in Europe: A new public health problem?). *WHO Bull.* 74, 159, 1996.
54. Kolářová, L., Skírnisson, K., and Horák, P. Schistosome cercariae as the causative agent of swimmer's itch in Iceland. *J. Helminthol.* 73, 215, 1999.
55. Rao, V.G. et al. Cercarial dermatitis in Central India: An emerging health problem among tribal communities. *Ann. Trop. Med. Parasitol.* 101, 409, 2007.
56. Hörweg, C., Sattmann, H., and Auer, H. Cercarial dermatitis in Austria: Questionnaires as useful tools to estimate risk factors? *Wien. Klin. Wochenschr.* 118, 177, 2006.
57. Fraser, S.J. et al. Cercarial dermatitis in the UK. *Clin. Exp. Dermatol.* 34, 344, 2009.
58. Schets, F.M. et al. Cercarial dermatitis in the Netherlands caused by *Trichobilharzia* spp. *J. Water Health* 06, 187, 2008.
59. Wang, C.R. et al. *Orientobilharzia* species: Neglected zoonotic agents. *Acta Trop.* 109, 171, 2008.
60. Dubois, J.-P. et al. Epidemiological studies related to cercarial dermatitis in lakes of the Savoy District (France). *Helminthologia* 38, 244, 2001.
61. Thors, C., and Linder, E. Swimmer's itch in Sweden. *Helminthologia* 38, 244, 2001.
62. Lévesque, B. et al. Investigation of an outbreak of cercarial dermatitis. *Epidemiol. Infect.* 129, 379, 2002.
63. Larsen, A.H., Bresciani, J., and Buchmann, K. Increasing frequency of cercarial dermatitis in higher latitudes. *Acta Parasitol.* 49, 217, 2004.
64. Mas-Coma, S., Valero, M.A., and Bargues, M.D. Effect of climate change on animal and zoonotic helminthiases. *Rev. Sci. Tech.* 27, 443, 2008.
65. Skírnisson, K., and Kolářová, L. Are nasal *Trichobilharzia* cercariae potential threat to human health? *Iceland Med. J.* 88, 739, 2002.
66. Lindblade, K.A. The epidemiology of cercarial dermatitis and its association with limnological characteristics of a northern Michigan lake. *J. Parasitol.* 84, 19, 1998.
67. Chamot, E., Toscani, L., and Rougemont, A. Public health importance and risk factors for cercarial dermatitis associated with swimming in Lake Leman at Geneva, Switzerland. *Epidemiol. Infect.* 120, 305, 1998.
68. Verbrugge, L.M. et al. Prospective study of swimmer's itch incidence and severity. *J. Parasitol.* 90, 697, 2004.
69. Hertel, J. et al. Detection of bird schistosomes in lakes by PCR and filter-hybridisation. *Exp. Parasitol.* 101, 57, 2002.
70. Kouřilová, P., and Kolářová, L. Variations in immunofluorescent antibody response against *Trichobilharzia* and *Schistosoma* antigens in compatible and noncompatible hosts. *Parasitol. Res.* 88, 513, 2002.
71. Graczyk, T.K., and Shiff, C.J. Recovery of avian schistosome cercariae from water using penetration stimulant matrix with an unsaturated fatty acid. *Am. J. Trop. Med. Hyg.* 63, 174, 2000.
72. Peterson, A.T. et al. The need for proper vouchering in phylogenetic studies of birds. *Mol. Phylogen. Evol.* 45, 1042, 2007.
73. Olsson, U. et al. What is the proper vouchering in phylogenetic studies of birds?—A reply to Peterson et al., 2007. *Mol. Phylogen. Evol.* 48, 383, 2008.
74. Hoberg, E.P., Pilitt, P.A., and Galbreath, K.E. Why museums matter: A tale of pinworms (Oxyuroidea: Heteroxynematidae) among pikas (*Ochotona princeps* and *O. collaris*) in the American West. *J. Parasitol.* 95, 790, 2009.
75. Olivier, L. Schistosome dermatitis, a sensitization phenomenon. *Am. J. Hyg.* 49, 290, 1949.
76. Oda, T. Schistosome dermatitis in Japan. In: Morishita, K., Komiya, Y., and Matsubayashi, H. (eds), *Progress in Medical Parasitology in Japan*, Vol. 2, Meguro Parasitological Museum, Tokyo, pp. 5–53, 1973.
77. Hunter, G.W., Ritchie, L.S., and Tanabe, H. The epidemiology of schistosome dermatitis ("Koganbyo") in Japan. *Trans. R. Soc. Trop. Med. Hyg.* 45, 103, 1951.
78. Haemmerli, U. Schistosomen-Dermatitis am Zürichsee. *Dermatologica* 107, 302, 1953.
79. Kouřilová, P. et al. Cercarial dermatitis caused by bird schistosomes comprises both immediate and late phase cutaneous hypersensitivity reactions. *J. Immunol.* 172, 3766, 2004.
80. Lichtenbergová, L. et al. Antibody responses induced by *Trichobilharzia regenti* antigens in murine and human hosts exhibiting cercarial dermatitis. *Parasite Immunol.* 30, 585, 2008.
81. Kolářová, L., Horák, P., and Čada, F. Histopathology of CNS and nasal infections caused by *Trichobilharzia regenti* in vertebrates *Parasitol. Res.* 87, 644, 2001.
82. Chanová, M., Vuong, S., and Horák, P. *Trichobilharzia szidati*: The lung phase of migration within avian and mammalian hosts. *Parasitol. Res.* 100, 1243, 2007.
83. Bayssade-Dufour, C., Martins, C., and Vuong, P.N. Histopathologie d'un modèle mammifère et dermatite cercarienne humaine. *Méd. Mal. Infect.* 31, 713, 2001.
84. Hoeffler, D.F. Swimmers' itch (cercarial dermatitis). *Cutis* 19, 461, 1977.
85. Appleton, C.C. Schistosome dermatitis—An unrecognized problem in South Africa. *South Afr. Med. J.* 65, 467, 1984.

86. Sonnenberg, R., Nolte, A.W., and Tautz, D. An evaluation of LSU rDNA D1-D2 sequences for their use in species identification. *Front. Zool.* 4, 6, 2007.
87. Webster, B.L. et al. The complete mitochondrial genome of the bird schistosome *Trichobilharzia regenti* (Platyhelminthes: Digenea), causative agent of cercarial dermatitis. *J. Parasitol.* 93, 553, 2007.
88. Brant, S.V. The occurrence of the avian schistosome *Allobilharzia visceralis* Kolářová, Rudolfová, Hampl et Skirnisson, 2006 (Schistosomatidae) in the tundra swan, *Cygnus columbianus* (Anatidae), from North America. *Folia Parasitol.* 54, 99, 2007.
89. Buhay, J.E. "CO1-LIKE" sequences are becoming problematic in molecular systematics and DNA barcoding studies. *J. Crust. Biol.* 29, 96, 2009.
90. Truett, G.E. et al. Preparation of PCR-quality mouse genomic DNA with hot sodium hydroxide and Tris (Hotshot). *Biotechniques* 29, 52, 2000.
91. Lockyer, A.E. et al. The phylogeny of the Schistosomatidae based on three genes with emphasis on the interrelationships of *Schistosoma* Weinland, 1858. *Parasitology* 126, 203, 2003.
92. Tkach, V.V., Pawlowski, J., and Mariaux, J. Phylogenetic analysis of the suborder Plagiorchiata (Platyhelminthes, Digenea) based on partial lsrDNA sequences. *Int. J. Parasitol.* 30, 83,, 2000.
93. Mollaret, I. et al. Phylogenetic analysis of the Monogenea and their relationships with Digenea and Eucestoda inferred from 28S rDNA sequences. *Mol. Biochem. Parasitol.* 90, 433, 1997.

Section V

Nematoda

43 Ancylostoma

Jaco J. Verweij

CONTENTS

43.1 INTRODUCTION

43.1.1 Classification and Morphology

Ancylostoma duodenale and *Necator americanus* are soil-transmitted helminths (STHs) responsible for an estimated 740 million human hookworm infections worldwide. *N. americanus* is considered the most common infection whereas *A. duodenale* infections appear to be restricted to certain scattered geographical areas. Moderate and heavy hookworms infections are a leading cause of anemia, which is directly correlated with the number of adult hookworms in the human intestine. [1,2]. Although the severity of anemia has also been shown to correlate with the hookworm species involved [3,4], accurate data on the precise distribution of the two species are lacking because species differentiation is usually not performed.

A. duodenale are helminths within the phylum Nematoda, the order Rhabditida and the family Ancylostomidae. Adult males measure 7–11 mm × 0.4–0.5 mm wide and females are slightly larger. These adult worms are rarely seen as they remain attached to the intestinal mucosa by large teeth on two ventral plates at the anterior margin of the buccal capsule. The *A. duodenale* female adult worms lay 10,000–30,000 eggs per day as compared with 3000–6000 eggs per day produced by an *N. americanus* female worm [5]. The eggs are not infectious but have to hatch and develop in moist, shady, and warm soil via several larval stages into the infective third-stage larvae (L3s) (Figure 43.1). These infective larvae are able to penetrate the intact skin and are transported with the bloodstream via the heart to the lungs. Subsequently, they migrate into the alveoli, up the airways and into the pharynx and after swallowing they reach the duodenum, where they develop into adult female and male worms. Infection by *A. duodenale* is also described through oral and transmammary routes.

A. duodenale eggs cannot be differentiated from eggs of other strongylid nematodes such as *N. americanus* and *Oesophagostomum bifurcum*. Therefore, coprocultures are required to allow eggs to hatch and develop into the characteristic L3s for the differential diagnosis. However, differentiation of the L3s, especially those of *A. duodenale* and *N. americanus*, requires highly specialist expertise and skills of the microscopist.

43.1.2 Clinical Features

The penetration of the *A. duodenale* larvae into the skin can cause a short-lasting itching dermatitis. The lung passage generally causes much less effect than *Ascaris* or *Strongyloides* infections. Symptoms in the intestinal phase are caused by necrosis of the intestinal tissue and blood loss through direct ingestion by the worms but more importantly continued blood loss from the attachment site, caused by anti-coagulants that are produced by the worm [6]. The amount of blood loss depends on several factors including the number of worms and the species involved of which *A. duodenale* causes more blood loss than *N. americanus* [3]. Hookworm infections are one of the major causes of anemia, especially in the developing world, with young children and pregnant women being the most affected [5–7].

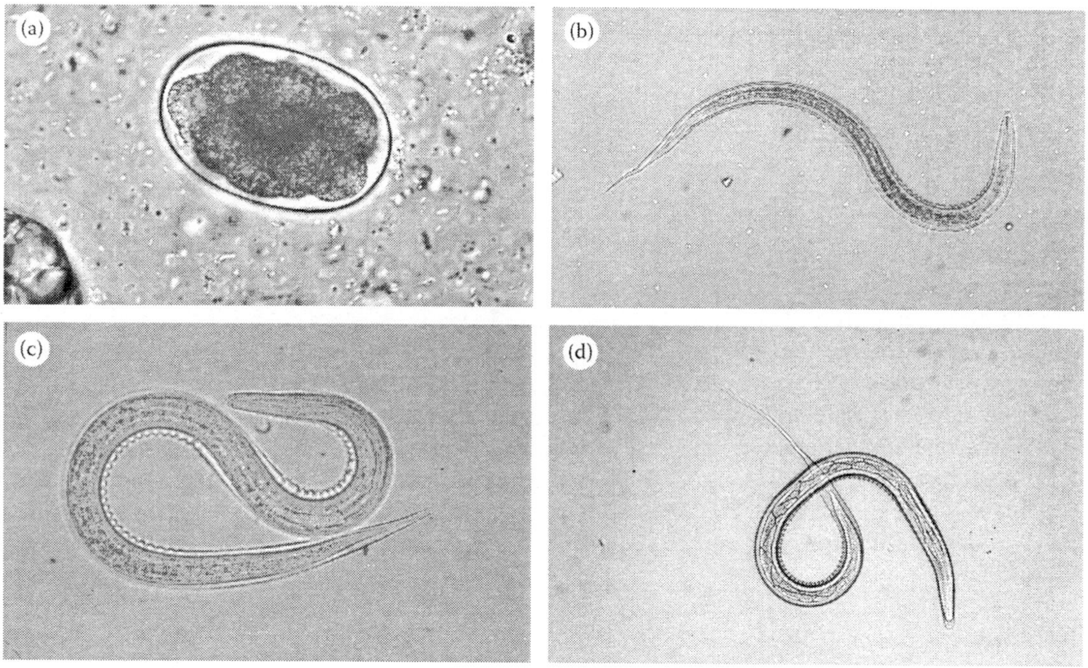

FIGURE 43.1 (a) Hookworm egg; (b) *Necator americanus* third-stage larvae; (c) *Ancylostoma duodenale;* (d) *Oesophagostomum bifurcum* third-stage larvae. (Courtesy of Eric A.T. Brienen, LUMC, Leiden.)

43.1.3 Laboratory Diagnosis

In patient diagnosis, direct smears from fecal samples and wet mount preparations of a formalin–ether concentrate are examined by microscopy for ova. The most common method used for describing all STH infections including hookworm (*N. americanus* and *A. duodenale*) within a human population involves counting the number of eggs per gram of feces by the Kato–Katz thick-smear technique [8]. This technique was originally described and widely used for the detection and quantification of *Schistosoma mansoni* eggs in feces. Microscopic examination of Kato–Katz slides for the presence of hookworm-like eggs is hampered by the fact that the slides must be read within 30 min or many eggs will no longer be visible. This is in contrast with *S. mansoni*, *Ascaris lumbricoides*, and *Trichuris trichiura* eggs, which can be detected best after 24–48 h. Moreover, species identification in hookworm infections is not possible due to the fact that the eggs of *N. americanus* and *A. duodenale* are morphologically indistinguishable. Therefore, coproculture is required to allow eggs to develop and hatch, producing the characteristic L3s needed for differential diagnosis [9,10]. However, coproculture-based diagnostics are completely dependent on the availability of fresh stool specimens, and the differentiation of the L3s of *A. duodenale* and *N. americanus* requires a highly skilled microscopist. Large-scale application of coproculture in epidemiological surveys is also hampered by the fact that these cultures have to be kept in closed Petri dishes for 1 week. At tropical temperatures, this frequently leads to the development of maggots and overgrowth of fungi, and coprocultures are performed in duplicate to obtain more reliable results [11]. As an alternative, Polymerase chain reaction (PCR)-based methods for specific detection of hookworm DNA in fecal samples have been used successfully for *A. duodenale* and *N. americanus* [12–14].

43.1.3.1 Conventional Techniques

43.1.3.1.1 Microscopy

The typical thin-shelled eggs with a clear space between the developing embryo and the shell measure ≈65 × 40 μm and can be detected in direct smears from fresh fecal samples and wet mount preparations of a formalin–ether concentrate without additional staining. The eggs of the different hookworm species are morphologically identical or very similar to those of other species of strongylids. The eggs may develop and hatch into first-stage larvae if the stools remain unpreserved for over 24 h. These larvae have to be differentiated from those of *Schistosoma stercoralis* and free-living ground nematodes that can appear quite frequently if stools are collected from contaminated soil. Kato–Katz slides used for microscopic detection and quantification of helminth eggs must be done within 30 min or many hookworm-like eggs will no longer be visible. This is in contrast with *S. mansoni*, *A. lumbricoides*, and *T. trichiura* eggs, which can be detected best after 24–48 h.

43.1.3.1.2 Coproculture

The eggs of *A. duodenale* and *N. americanus* cannot be differentiated morphologically and are identical or very similar to the eggs of other nematodes such as *Oesophagostomum* species and *Trichostrongylus*. Therefore, coproculture is needed to allow eggs to hatch and develop into L3s for differentiation at the species level [9]. Briefly, approximately 2 g of fresh stool is homogenized and mixed with an equal amount of vermiculite or ground charcoal. Filter paper is placed on an elevation in a Petri dish and tap water is added on the bottom. The stool–vermiculite mixture is brought onto the filter paper. The closed Petri dish is incubated at room temperature in a dark place. The culture has to be checked daily and stirred in case of fungal growth and fly larvae have to be removed. Water is added when necessary to avoid desiccation. After 5–7 days, the filter paper with the stool–vermiculite mixture is removed and the water is poured into a 50-mL tube. After a sedimentation time of 2 h the sediment is pipetted from the bottom of the tube and examined with 4 × 10 magnifications. Larvae that are found can be killed by adding a drop of iodine and examined with a higher magnification for a detailed examination of the morphological characteristics to achieve an accurate species determination. Another coproculture method that is frequently used for hookworm detection and species differentiation is the Harada–Mori filter paper method, in which an amount of fresh stool sample is incubated using filter paper soaked by distilled water inside a sealed plastic tube [15].

43.1.3.1.3 Serological Tests

Enzyme-linked immunosorbent assay and Western blot analysis have been used to measure human immune responses against crude larval and adult soluble extracts of adult excretory/secretory products. The antibody response consists predominantly of IgG1, IgG4, and IgE. Although IgE antibodies were found to be more specifically directed against *N. americanus* epitopes than the other Ig isotypes, also high levels of nonspecific IgE are found in hookworm-infected individuals. IgG4 against crude larval antigens has been suggested to be a marker of active infection [5]. However, serology cannot differentiate between the species of hookworm involved or between current and past infections. Moreover, cross reactions between hookworm antibody and other nematode antigens appear to be very frequent [16].

43.1.3.1.4 In Vivo Assays

In vivo assays are not used for patient diagnosis; however, animal models are used to study hookworm infections to achieve data that are not obtainable with *in vitro* culture and other related techniques. *A. duodenale* is highly host-specific (humans) and will develop in dogs only on the administration of exogenous steroids. Therefore it is not very reliable to extrapolate findings from these models to the human situation.

43.1.3.2 Molecular Techniques

Initially, PCR-based method for genetic differentiation have been employed on various target regions, such as ITS-1 and ITS-2 of ribosomal DNA (rDNA) and mitochondrial DNA (mtDNA) for the genetic identification of DNA isolated from adult worms [17,18].

43.1.3.2.1 Species-Specific Detection

For species-specific detection, the target DNA should show very limited variation within the species and be different enough to allow differentiation between closely related species. The cAMP-dependent protein kinase gene, the internal transcribed spaces, and the cytochrome *c* oxidase subunit 1 gene have been used with PCR–restriction-fragment-length polymorphism, PCR–single-stranded conformational polymorphism (SSCP), direct cycle sequencing and specific conventional and real-time PCR for species-specific detection and differentiation between different hookworm species [12–14,18–22]. Real-time PCR using fluorescent detection probes reduces the risk of contamination and substantially reduces labor time and reagent costs, through the possibility of combining assays for the detection of different targets into one reaction. Real-time PCR also offers the possibility of quantification using the threshold cycle; this is the amplification cycle in which the level of fluorescent signal exceeds background fluorescence, and this correlates with the initial copy number at the beginning of the reaction. A multiplex real-time PCR assay was established for the simultaneous detection and quantification of *A. duodenale*, *N. americanus*, and *O. bifurcum* DNA in fecal samples [14]. This PCR proved to be specific (100%) and sensitive (100%, 100%, and 86.7%) for the detection of *A. duodenale*, *N. americanus*, and *O. bifurcum*, respectively, using a set of well-defined control samples ($n = 150$) and known positive control samples ($n = 50$). Employing fecal samples ($n = 399$) from a region in Ghana where all three parasites are endemic, the assay again showed excellent sensitivity for the detection of *A. duodenale*, *N. americanus*, and *O. bifurcum* as compared to microscopy and coproculture. Moreover, a highly significant correlation was demonstrated between egg counts or the number of larvae found in coproculture and the C_t-values obtained in the species-specific real-time PCRs reflecting the amount of target DNA in the fecal sample.

43.1.3.2.2 Strains

Using a PCR–SSCP approach detecting variability in the cox1 gene, genetically distinct groups were detected within *A. duodenale* from China [23]. Sequence analysis using different methods of the ITS-1 and ITS of rDSNA or cAMP-dependent protein kinase usually does not provide sufficient levels of variability to identify genetic variants or "strains" within hookworm populations [18]. Fingerprinting methods such as random amplification of polymorphic DNA and amplified fragment-length polymorphism analysis have been successfully used to explore genetic diversity within *Necator* and *Oesophagostomum* species but have not been used in *A. duodenale* [24–26].

43.1.3.2.3 Susceptibility Testing

Resistance to benzimidazoles (BZ) anthelmintics in several strongylid nematodes of lifestock has been associated with single-nucleotide polymorphisms in the β-tubulin isotype-1 gene at codons 157 and 200 [27]. A real-time PCR based on the sequence analysis of the β-tubulin genes of *A. canium*, *A. duodenale*, and *N. americanus* for detecting polymorphisms in codons 137 and 200 in all three species was developed and used in hookworm specimens that were obtained from school children who had demonstrated a reduced response to treatment with mebendazole. In none of the samples, polymorphisms at these loci were detected and therefore changes in codons 167 and 200 of the β-tubulin genes could not be associated with resistance to BZ in human hookworms [28].

43.2 METHODS

43.2.1 Sample Preparation

43.2.1.1 Sample Collection

Unpreserved stool samples are preferred for molecular studies but have to be stored and transported frozen. Mixing stool samples with ethanol allows transportation and storage at room temperature for at least 3 months. If formalin-fixed fecal samples are used, the formalin has to be removed by washing/centrifuging steps with saline direct or after formalin–ether concentration. However, the sensitivity of amplification is known to decrease with time of formalin fixation.

43.2.1.2 DNA Isolation

Both in-house and commercial reagents have been used for the preparation of DNA from hookworm eggs. We present a method below that is used in our laboratory.

43.2.1.2.1 DNA Isolation from Feces Using QIAamp Spin Columns [13]*

1. Suspend 100 mg of feces in 200 μL of phosphate-buffered saline (PBS) with 2% PVPP† (polyvinylpolypyrolidone, Sigma P6755).‡
2. Freeze sample overnight at −20°C.
3. Heat feces suspension for 10 min in heatblock at 100°C.
4. Add an equal volume of tissue lysis buffer (ATL) containing 20 μL proteinase K (i.e., 180 μL of ATL and 20 μL of proteinase K from the kit) solution to the sample (vortex). Incubate for 2 h or overnight at 55°C in heatblock (after 1 h vortex briefly).
5. Add 400 μL AL buffer (with spike) to 400 μL sample, mix thoroughly by vortexing, and incubate at 70°C for 10 min.
6. Centrifuge for 30 s at full speed and transfer supernatant to an Eppendorf containing 400 μL of ethanol (96–100%), and mix thoroughly by vortexing (short spin).
7. Place a numbered spin column in a 2 mL collection tube. Carefully apply 600 μL of the mixture from step 5 to the spin column without moistening the rim, close the cap and centrifuge at 10,000 rpm for 1 min.
8. Place spin column in a clean 2 mL collection tube and repeat with the rest of the mixture from step 5.
9. Place the spin column in a clean 2 mL collection tube.
10. Carefully open the spin column and add 500 μL of AW1 buffer. Centrifuge at 10,000 rpm for 1 min.
11. Empty the collection tube and add 500 μL of AW2 buffer. Centrifuge for 3 min, for 1 min at 10,000 rpm and for 2 min at full speed.
12. Place the spin column in a numbered, clean 1.5 mL microfuge tube (with lid).
13. Carefully open the spin column, elute the DNA with 200 μL AE buffer, leave the buffer on the filter for 1 min and centrifuge at 8000 rpm for 1 min.

43.2.1.2.2 DNA Isolation after Ethanol Fixation [29]

1. Apply ~250 μL of an ethanol–feces suspension to a clean Eppendorf tube and centrifuge for 1 min (10,000 rpm).
2. Remove the ethanol with a micro-pipette and wash the pellet with 1000 μL of PBS and centrifuge for 1 min (10,000 rpm).
3. Remove the PBS with a micro-pipette and add 200 μL of 2% PVPP (shake the bottle before use).
4. Continue with the DNA isolation protocol above from step 2.

43.2.2 Detection Procedures

43.2.2.1 Semi-Nested PCR

43.2.2.1.1 Principle

A semi-nested PCR protocol has been reported for the detection and differentiation of *A. duodenale* from *N. americanus* infections. This PCR generates a 310-bp and a 450-bp fragment, respectively, of the ITS-2 region in the first-round amplification with primers NC1 and NC2 [30]. The second-round amplification is performed with primers AD, NA and NC2, resulting in a 130-bp and a 250-bp amplicons for *A. duodenale* and *N. americanus*, respectively (Table 43.1).

TABLE 43.1
Primers for Semi-Nested PCR

Oligo Name	Sequence
NC1	5′-ACG TCT GGT TCA GGG TTG TT-3′
NC2	5′-TTA GTT TCT TTT CCT CCG CT-3′
AD	5′-CGA CTT TAG AAC GTT TCG GC-3′
NA	5′-ATG TGC ACG CTG TTA TTC ACT-3′

* This adapted protocol using a QIAamp DNA minikit (cat. no. 51306) gives similar or even better results than the QIAamp stool kit.

† Or suspend 0.5 g of feces in 2 mL of PBS with 2% PVPP and sieve (filter). For the isolation of the DNA aliquot, separate an amount of 200 μL in an Eppendorf tube and store the rest at −70°C.

‡ PVPP does not dissolve; it stays like a suspension, so you have to stir regularly while using it.

43.2.2.1.2 Procedure

1. Prepare the first-round PCR mixture (20 µL) as follows, and add 5 µL DNA in the test tube (or distilled water in the negative control tube):

First-Round PCR		1× mix µL	20× mix µL
Hotstart mastermix		12.5	250
Primer NC1	25 µM	0.5	10
Primer NC2	25 µM	0.5	10
BSA	5 mg/mL	0.5	10
H_2O		6.5	120
Total		20.00	400

2. Run PCR amplification in a thermocycler for one cycle of 95°C for 15 min; 25 cycles of 95°C for 15 s, 55°C for 30 s, and 72°C for 30 s; one cycle of 72°C for 7 min.
3. Prepare the second-round PCR mixtures (20 µL) as follows, and add 5 µL of PCR product from the first-round PCR in each tube (or distilled water in the negative control tube).

Second-Round PCR *Ancylostoma duodenale*		1× mix µL	20× mix µL
Hotstart mastermix		12.5	250
Primer NC2	25 µM	0.5	10
Primer AD	25 µM	0.5	10
BSA	5 mg/mL	0.5	10
H_2O		6.5	120
Total		20.00	400

Second-Round PCR *Necator americanus*		1× mix µL	20× mix µL
Hotstart mastermix		12.5	250
Primer NC2	25 µM	0.5	10
Primer NA	25 µM	0.5	10
BSA	5 mg/mL	0.5	10
H_2O		6.5	120
Total		20.00	400

4. Perform PCR amplification as above (step 2).
5. Separate PCR products (10 µL) on 2% agarose gel containing ethidium bromide and visualize with UV transilluminator.

43.2.2.2 Real-Time PCR

43.2.2.2.1 Principle

A multiplex real-time PCR method was developed for the simultaneous detection and quantification of *Ancylostoma duodenale*, *Necator americanus*, and *Oesophagostomum bifurcum* in stool samples [14]. The *A. duodenale*-specific real-time PCR amplifies and detects a 71-bp fragment within the ITS2 sequence of *A. duodenale*. The *N. americanus* PCR primers and detection probe were designed on the ITS-2 sequence for *N. americanus*, amplifying and detecting a 103-bp fragment inside the ITS-2 sequence. The *O. bifurcum*-specific primers amplify a 105-bp fragment of the ITS2 sequence and the XS-probe detects the *O. bifurcum*-specific product.

This multiplex PCR included an internal control to detect inhibition of the amplification by fecal constituents in the sample (Table 43.2).

43.2.2.2.2 Procedure

1. Prepare PCR mixture (25 µL) containing PCR buffer (Hotstart Taq master mix, QIAgen), 5 mM $MgCl_2$, 2.5 µg bovine serum albumin (Roche Diagnostics), 5 pmol of each *A. duodenale*-specific primer and of each *N. americanus*-specific primer, 2.5 pmol of each *O. bifurcum*-specific primer and 3.75 pmol of each PhHV-1-specific primer, 2.5 pmol of *A. duodenale*-specific XS-probe and 1.25 pmol of each *N. americanus*-specific XS-probe, *O. bifurcum*-specific XS-probe and PhHV-1-specific double-labeled probe, and 5 µL of the DNA sample. Amplification consists of 15 min at 95°C followed by 50 cycles of 15 s at 95°C, 30 s at 60°C, and 30 s at 72°C. Amplification, detection, and analysis were performed with the CFX real-time detection system (Bio-Rad Laboratories). Negative and positive control samples are included in each amplification run and 5 µL of the DNA sample as follows:

			1× mix µL	100× mix µL
H_2O			1.58	158
$MgCl_2$ 25 mM	25 mM		3.50	350
BSA 5 mg/mL	5 mg/mL		0.50	50
Ad-125F	25 µm		0.20	20
Ad-195R 10 µm	25 µm		0.20	20
Ad-155 Texas Red	10 µm	Texas Red	0.25	25
Na-58F	25 µm		0.20	20
Na-158R	25 µm		0.20	20
Na-81 FAM	10 µm	FAM	0.125	13
Oeso-31F	25 µm		0.10	10
Oeso-136R	25 µm		0.10	10
Obif-63 VIC	10 µm	Yakima Yellow	0.125	13
PHHV-F	25 µm		0.15	15
PHHV-R	25 µm		0.15	15
PHHV-CY5	10 µm	Cy5	0.125	13
Hotstart mastermix			12.50	1250
Total			20	2000

2. Perform PCR in the CFX real-time detection system (Bio-Rad) for one cycle of 95°C for 15 min; 50 cycles of 95°C for 15 s, 60°C for 30 s, and 72°C for 30 s. (For amplification in ABI 7500, the cycling

TABLE 43.2
Oligonucleotide Primers and Detection Probes for Real-Time PCR for the Simultaneous Detection of *A. duodenale, N. americanus, O. bifurcum,* and Phocin Herpes Virus 1 as an Internal Control

Oligonucleotide Name	Oligonucleotide Sequence	GenBank Accession No. or the Literature Reference of Target Sequence
	Target Organism	
Ancylostoma duodenale		
Ad125F	5′-GAATGACAGCAAACTCGTTGTTG-3′	AJ001594
Ad195R	5′-ATACTAGCCACTGCCGAAACGT-3′	AJ001594
Ad155-XS	Texas Red-5′-ATCGTTTACCGACTTTAG-3′-XS	AJ001594
Necator americanus		
Na58F	5′-CTGTTTGTCGAACGGTACTTGC-3′	AJ001599
Na158R	5′-ATAACAGCGTGCACATGTTGC-3′	AJ001599
Na81Tmgb	FAM-5′-CTGTACTACGCATTGTATAC-3′-XS	AJ001599
Oesophagostomum bifurcum		
oeso31F	5′-GTTTGTCGAACGATGCTTACATTT-3′	Y11733
oeso136R	5′-CGAGGTCACGACATCCGTG-3′	Y11733
Obif63Tmgb	Yakima Yellow 5′-CCTCGTTCTAGATAAGAAAT-3′-XS	Y11733
Phocin Herpes Virus 1		
PhHV-267s	5′-GGGCGAATCACAGATTGAATC -3′	[31]
PhHV-337as	5′-GCGGTTCCAAACGTACCAA -3′	[31]
PhHV-305tq	Cy5 5′-TTTTTATGTGTCCGCCACCATCTGGATC-3′-BHQ2	[31]

program consists of 1 cycle of 95°C for 15 min; 50 cycles of 95°C for 15 s, and 60°C for 60 s).

3. Detect fluorescence of Texas Red, FAM, Yakima Yellow, and CY5 at their respective wavelengths during the annealing step of each cycle.

Note: The *A. duodenale* primers and probe were designed on the known sequence of *A. duodenale* (AJ001600). According to NCBI BLAST analysis, however, *Ancylostoma ceylanicum* will be amplified and detected also with this primer/probe combination. The *N. americanus* probe shows a two-base mismatch with the ITS-2 sequence of an *N. americanus* isolate from Sawarak (GenBank accession no. AJ001600). The detection of the PCR product of such a sequence will probably fail or at least be less efficient. The primers of the *O. bifurcum* PCR show 100% homology with *Oesophagostomum stephanostomum* and the probe only shows one mismatch with the ITS-2 sequence of *O. stephanostomum*. Therefore, *O. stephanostomum* will be amplified and also probably detected with this primer/probe combination.

43.3 CONCLUSION

In most epidemiological studies and in control programs of STHs, the species involved in hookworm infections are not differentiated to the species level. The paucity of such data makes it unclear whether the long-term effects of mass-treatment on the rate of infection, morbidity, and the emergence of resistance could be different between the diverse species involved. Real-time PCR offers highly sensitive and species-specific detection and quantification as a reflection of the intensity of infection, which makes it possible to obtain accurate and species-specific data on the epidemiology, biology, and the effect of repeated anthelmintic treatment in hookworm infections.

REFERENCES

1. Hotez, P.J. et al. Hookworm: "The great infection of mankind". *PLoS Med.* 2, e67, 2005.
2. Bethony, J. et al. Soil-transmitted helminth infections: Ascariasis, trichuriasis, and hookworm. *Lancet* 367, 1521, 2006.
3. Albonico, M. et al. Epidemiological evidence for a differential effect of hookworm species, *Ancylostoma duodenale* or *Necator americanus*, on iron status of children. *Int. J. Epidemiol.* 27, 530, 1998.
4. Nkhoma, E. et al. Severe anaemia in an 11-month-old girl. *Lancet* 365, 1202, 2005.
5. Brooker, S. et al. Human hookworm infection in the 21st century. *Adv. Parasitol.* 58, 197, 2004.
6. Hotez, P.J. et al. Hookworm infection. *N. Engl. J. Med.* 351, 799, 2004.
7. Guyatt, H.L. et al. Hookworm and anaemia prevalence. *Lancet* 356, 2101, 2000.
8. dKatz, N. et al. A simple device for quantitative stool thick-smear technique in *Schistosomiasis mansoni*. *Rev. Inst. Med. Trop. Sao Paulo* 14, 397, 1972.
9. Polderman, A.M. et al. Oesophagostomiasis, a common infection of man in northern Togo and Ghana. *Am. J. Trop. Med. Hyg.* 44, 336, 1991.

10. Blotkamp, J. et al. Observations on the morphology of adults and larval stages of *Oesophagostomum* sp. isolated from man in northern Togo and Ghana. *J. Helminthol.* 67, 49, 1993.
11. Krepel, H.P. et al. Quantitative interpretation of coprocultures in a population infected with *Oesophagostomum bifurcum. Trop. Geogr. Med.* 47, 157, 1995.
12. de Gruijter, J.M. et al. Polymerase chain reaction-based differential diagnosis of *Ancylostoma duodenale* and *Necator americanus* infections in humans in northern Ghana. *Trop. Med. Int. Health* 10, 574, 2005.
13. Verweij, J.J. et al. Determining the prevalence of *Oesophagostomum bifurcum* and *Necator americanus* infections using specific PCR amplification of DNA from faecal samples. *Trop. Med. Int. Health* 6, 726, 2001.
14. Verweij, J.J. et al. Simultaneous detection and quantification of *Ancylostoma duodenale, Necator americanus*, and *Oesophagostomum bifurcum* in fecal samples using multiplex real-time PCR. *Am. J. Trop. Med. Hyg.* 77, 685, 2007.
15. Harada, Y., and Mori, O. A new method for culturing hookworm. *Yonago Acta Med.* 1, 177, 1955.
16. Lindo, J.F. et al. Prospective evaluation of enzyme-linked immunosorbent assay and immunoblot methods for the diagnosis of endemic *Strongyloides stercoralis* infection. *Am. J. Trop. Med. Hyg.* 51, 175, 1994.
17. Gasser, R.B. Molecular tools—Advances, opportunities and prospects. *Vet. Parasitol.* 136, 69, 2006.
18. Gasser, R.B. et al. DNA technological progress toward advanced diagnostic tools to support human hookworm control. *Biotechnol. Adv.* 26, 35, 2008.
19. Hawdon, J.M. Differentiation between the human hookworms *Ancylostoma duodenale* and *Necator americanus* using PCR-RFLP. *J. Parasitol.* 82, 642, 1996.
20. Zhan, B. et al. Species-specific identification of human hookworms by PCR of the mitochondrial cytochrome oxidase I gene. *J. Parasitol.* 87, 1227, 2001.
21. Monti, J.R. et al. Specific amplification of *Necator americanus* or *Ancylostoma duodenale* DNA by PCR using markers in ITS-1 rDNA, and its implications. *Mol. Cell Probes* 12, 71, 1998.
22. Gasser, R.B. et al. Improved molecular diagnostic tools for human hookworms. *Expert Rev. Mol. Diagn.* 9, 17, 2009.
23. Hu, M. et al. Single-strand conformation polymorphism-based analysis of mitochondrial cytochrome c oxidase subunit 1 reveals significant substructuring in hookworm populations. *Electrophoresis* 23, 27, 2002.
24. de Gruijter, J.M. et al. AFLP fingerprinting for the analysis of genetic diversity within *Necator americanus. Mol. Cell Probes* 20, 317, 2006.
25. de Gruijter, J.M. et al. Genetic substructuring within *Oesophagostomum bifurcum* (Nematoda) from human and non-human primates from Ghana based on random amplified polymorphic DNA analysis. *Am. J. Trop. Med. Hyg.* 71, 227, 2004.
26. de Gruijter, J.M. et al. High resolution DNA fingerprinting by AFLP to study the genetic variation among *Oesophagostomum bifurcum* (Nematoda) from human and non-human primates from Ghana. *Parasitology* 130, 229, 2005.
27. Kaplan, R.M. Drug resistance in nematodes of veterinary importance: A status report. *Trends Parasitol.* 20, 477, 2004.
28. Schwenkenbecher, J.M. et al. Characterization of beta-tubulin genes in hookworms and investigation of resistance-associated mutations using real-time PCR. *Mol. Biochem. Parasitol.* 156, 167, 2007.
29. Verweij, J.J. et al. Prevalence of *Entamoeba histolytica* and *Entamoeba dispar* in northern Ghana. *Trop. Med. Int. Health* 8, 1153, 2003.
30. Gasser, R.B. et al. Rapid sequencing of rDNA from single worms and eggs of parasitic helminths. *Nucleic Acids Res.* 21, 2525, 1993.
31. Niesters, H.G. Clinical virology in real time. *J. Clin. Virol.* 25(Suppl 3), S3, 2002.

44 Angiostrongylus

Dongyou Liu

CONTENTS

44.1 INTRODUCTION

Angiostrongylus is a metastrongylid nematode genus that encompasses species found in insectivores, rodents, and carnivores. As a reemerging zoonotic pathogen, *Angiostrongylus cantonensis* (rat lungworm) is an important cause of eosinophilic meningitis in humans. While rats and mollusks act as the definitive and intermediate hosts, respectively, for *A. cantonensis*, humans acquire the infection through consumption of raw or undercooked infected molluscs (snails and slugs), or contaminated water and vegetables containing infective third-stage larvae, infected mollusks, or infected paratenic hosts. Although angiostrongyliasis is mostly self-limited, with full recovery within a month, it can lead to permanent brain and nerve damage as well as death in severe cases.

First identified in the lungs of rats in 1935, *A. cantonensis* was reported as a cause of human cerebral angiostrongyliasis in Southeast Asia. This parasite has since spread to other parts of the world via human cargo ships that carried infected rats and/or snails. To date, *A. cantonensis* infection is endemic around the Pacific Basin, the Hawaiian Islands, parts of Africa, Cuba, Puerto Rico, the Bahamas, and the Dominican Republic, with >2800 human cases being reported to date.

Another human pathogenic *Angiostrongylus* species is *Angiostrongylus costaricensis*. This parasite lives inside the mesenteric arterial system, causing abdominal disease (eosinophilic enteritis) in humans. Unlike *A. cantonensis* whose juvenile worms die in the human brain, *A. costaricensis* remains alive in the intestines. *A. costaricensis* infection occurs throughout the Americas, from the southern United States to northern Argentina.

It is likely that the true incidence of angiostrongyliasis is much higher, due to the limited sensitivity and specificity of traditional laboratory tests that have been applied to its identification. This is another area where use of molecular techniques will contribute to the improved detection and control of pathogens of increasing public health importance.

44.1.1 Classification, Morphology, Life Cycle, and Epidemiology

44.1.1.1 Classification

The genus *Angiostrongylus* is classified taxonomically in the family Angiostrongylidae, superfamily Metastrongyloidea, suborder Strongylida, order Rhabditida, class Chromadorea, phylum Nematoda.

The phylum Nematoda covers a large, diverse group of roundworms (nematodes) that possess tough outer cuticle, unsegmented bodies, and a fully-developed gastrointestinal (GI) tract. The suborder Strongylida includes hookworms and lungworms. The superfamily Metastrongyloidea consists of long, cylindrical, slender, threadlike worms that live inside arterial vessels and cardiac cavities in a vertebrate host (the definitive host), and develop infective larvae in a mollusk host (the intermediate host).

The family Angiostrongylidae is separated into three genera: *Aelurostrongylus*, *Angiostrongylus*, and *Didelphostrongylus*. In turn, the genus *Angiostrongylus* (synonym: *Parastrongylus*) is subdivided into at least 15 species, the most notable of which include: *A. cantonensis*, *Angiostrongylus chabaudi*, *A. costaricensis*, *Angiostrongylus daskalovi*, *Angiostrongylus dujardini*, *Angiostrongylus gubernaculatus*, *Angiostrongylus mackerrase*, *Angiostrongylus malaysiensis*, and *Angiostrongylus vasorum*.

Angiostrongylus species reside in the pulmonary arteries, the right ventricle of the heart, and the mesenteric veins of insectivores, rodents, and carnivores (including dogs, foxes, wolves, coyotes, red pandas, badgers, striped skunk, wild cat, rodents, pine marten, and beech marten) in various parts of the world. *A. cantonensis*, the rat lungworm commonly found in most tropical zones, is a major cause of human eosinophilic meningitis with central nervous system (CNS) symptoms. *A. costaricensis* is the causal agent of abdominal, or intestinal, angiostrongyliasis mainly in Central and South America. *A. malaysiensis* may be potentially pathogenic to humans in Southeast Asia [1].

44.1.1.2 Morphology

A. cantonensis adult worms are filiform, with three outer protective collagen layers, and a simple stomal opening with no lips or buccal cavity leading to a fully developed GI tract. The male measures 20–25 mm × 0.32–0.42 mm, and has a small copulatory bursa at the posterior. The female measures 22–34 mm × 0.34–0.56 mm, shows a "barber pole" shape down the middle of the body (which is created by the twisting together of the intestine and uterine tubules), and possesses a dark red digestive organ, two white reproductive organs, and a transparent cuticle. *A. cantonensis* juvenile worms are about 12–13 mm in length [2].

A. costaricensis adult worms are filiform in both sexes, tapering toward both ends. The caudal end of the male curves ventrally, with a well-developed copulatory bursa, while the caudal end of the female is roughly conical and slightly curved, with the tip of the tail bearing a small refringent spine. Cuticle is transversely striated and thicker toward both ends. Excretory pore is located immediately posterior to the base of the club-shaped esophagus. The male is of 15.0–17.9 mm in length and 0.07–0.09 mm in width (at the esophageal–intestinal junction). The female is of 26.9 mm in length and 0.10 mm in width (at base of esophagus).

A. vasorum (also known as French heartworm) is a parasite of dogs. Its adult worms are pinkish in color, and measure 14.0–20.5 mm in length and 0.17–0.30 mm in width [3,4].

Angiostrongylus adult worms from foxes and badgers are differentiable by the total worm length and the esophagus length. *A. daskalovi* is larger than *A. vasorum*, except for the male spicules.

The third-stage larvae of *A. cantonensis*, *A. vasorum*, *Aelurostrongylus abstrusus*, and *Anafilaroides rostratus* are distinguished from one another by a combination of the head features, variations in sheaths, distribution of refractive granules, and the morphology of the tip of the tail [5].

44.1.1.3 Life Cycle

While rodents are the definitive host for *A. cantonensis* and *A. costaricensis*, fish and crustaceans (mollusks) are their intermediate hosts. Various species of fish, frogs, crustaceans,. and flatworms can act as paratenic or transport hosts after they ingest infected mollusks. Humans represent an accidental intermediate host after consumption of infected fish and crustaceans.

A. cantonensis adult worms reside in the pulmonary arteries of rats, where the females lay eggs. The eggs hatch to become the first-stage (L1) larvae, which migrate via the trachea to the pharynx, and are swallowed. Excreted in the feces, the L1 larvae are taken up by a molluscan intermediate host (snail or slug). After two molts, the L1 larvae become the third-stage (L3) larvae, which are infective to vertebrate hosts. Rats and other vertebrate hosts acquire the infection after ingestion of mollusks containing the L3 larvae. The L3 penetrate the vertebrate intestinal wall, migrate through the circulatory system to the brain over the course of 2–3 days. After molting twice there, the L3 larvae develop into juvenile worms inside meningeal vessels, then move to the CNS, and finally arrive at the pulmonary arteries and right heart cavities to become sexually mature adult worms. Overall, it takes ~4 weeks from the initial intestinal penetration to the settlement in the final habitat.

Some vertebrate animals can act as paratenic (transport) hosts for *A. cantonensis* after ingestion of the infected snails containing L3 larvae. The L3 larvae remain alive in the paratenic hosts, and resume their development when the paratenic host is ingested by a definitive host [6].

As a nonpermissive, accidental host, humans acquire the infection by eating raw or undercooked snails or slugs containing L3 larvae (or infected paratenic animals such as crabs, and freshwater shrimps). The L3 pass through the vertebrate intestinal wall, and move via the circulatory system to the brain, where L3 larvae molt twice to become juvenile worms. The juvenile worms migrate hematogenously to the CNS, and burrow into neural tissue, without reaching the pulmonary arteries and right heart cavities. Thus, the juvenile worms do not complete their life cycle in humans, and their death leads to intense inflammatory lesions.

Intermediate hosts for *A. cantonensis* include: African giant land snail (*Achatina fulica*), golden apple snail (*Ampullarium canaliculatus*), and other aquatic or terrestrial snails (*Lymnaea catascopium*, *Thelidomus aspera*, *Satsuma mercatoria*, *Acusta despecta*, *Bradybaena circulus*, *Parmarion martensi*, *Pomacea canaliculata*, and *Pila* spp.);

slugs (*Limax maximus*, *Deroceras laeve*, *Deroceras reticulatum*, *Laevicaulis alte*, *Sarasinula plebeia*, and *Lehmannia valentiana*, etc) [7–10].

Paratenic hosts of *A cantonensis* include: crabs, shrimp, fish, frogs (*Hyla aurea*, *Rana* spp., and *Rhacophorus leucomystax*), and toads (*Bufo asiaticus*), planaria such as New Guinea flatworm (*Platydemus manokwar*), which preys on land mollusks.

The life cycle of *A. costaricensis* is similar, apart from the fact that the adult worms live in the arterioles of the ileocecal area of the definitive host (e.g., wild rodents *Sigmodon hispidus*, and *Rattus rattus*). In humans, *A. costaricensis* often reaches sexual maturity and release eggs into the intestinal tissues. However, the eggs and larvae degenerate and cause intense local inflammatory reactions and do not appear to be shed in the feces.

44.1.1.4 Epidemiology

A. cantonensis was first described from *Rattus norvegicus* (brown rats) and *Rattus rattus* (black rats) in Canton, China in 1935. *A. cantonensis* worms were identified in Taiwan in 1944 from the cerebrospinal fluid (CSF) of a patient with eosinophilic meningitis after eating raw food that may have been contaminated by rats. The life cycle of *A. cantonensis* was elucidated in 1955, involving rats as definitive host, snails and slugs as the intermediate host.

A. cantonensis spread throughout Southeast Asia, Western Pacific Islands (Micronesia, Melanesia, Polynesia), and Australia after following World War II, with cases of human angiostrongyliasis reported in Taiwan, Sumatra, the Philippines, New Caledonia, and Tahiti [11,12]. In the 1960s, cases of eosinophilic meningitis due to *A. cantonensis* were also reported from Vietnam, Cambodia, Thailand, Indonesia, Vanuatu, Guam, and Hawaii [13–15]. Since then, cases of angiostrongyliasis have appeared in Japan, New Britain, American Samoa, Western Samoa, Australia, Hong Kong, India, Fiji, and mainland China [16–26]. In addition, sporadic occurrences of *A. cantonensis* in rats have been noted in Madagascar, Cuba, Egypt, Puerto Rico, New Orleans, Louisiana, Nigeria, and South Africa [27–31].

Human infection with *A. cantonensis* occurs after ingestion of raw or undercooked infected mollusks (snail or slug) such as *L. catascopium*, slime produced by infected snails or slugs, infected paratenic hosts (e.g., frogs, freshwater prawns, crabs, fish, and planaria), or contaminated vegetables, water, or fruit juice [32–37]. Young children may become infected via hands carrying the L3 larvae directly to the mouth after playing with mollusks [38] Anonymous 2010]. To date, over 2800 cases of human angiostrongyliasis had been reported worldwide. *A. cantonensis* usually infects men in the third and fourth decades of life, and occasionally children.

Besides mammalian hosts, *A. cantonensis* has been shown to infect birds such as yellow-tailed black cockatoo (*Calyptorhynchus funereus*) and tawny frogmouths (*Podargus strigoides*) and cause neurological symptoms [39].

A. costaricensis lives inside the mesenteric arterial system, and causing abdominal disease (eosinophilic enteritis) in humans. Unlike *A. cantonensis* whose juvenile worms die in the human brain, *A. costaricensis* remains alive in the intestines, producing eggs and larvae that disintegrate afterward, and are not discharged in the feces [40,41]. *A. costaricensis* infection occurs throughout the Americas, from the southern United States to northern Argentina [42,43].

44.1.2 Clinical Features and Treatment

44.1.2.1 Clinical Features

A. cantonensis is an important cause of eosinophilic meningitis, which is marked by the inflammation of the meninges as the result of dying *A. cantonensis* larvae and the general presence of the young adult worm in the meninges and in the parenchyma of the medulla, pons, and cerebellum [44–48]. This is characterized by a predominance of infiltrating eosinophils (with a 10% or more increase in the volume of eosinophils in the CSF) along with localized suppurative necrosis and granulomatous reaction [49]. Clinical manifestations include headaches (a bitemporal character in the frontal or occipital lobe, resulting from increased intracranial pressure produced by the widespread inflammatory reaction in the meninges), nausea, vomiting, paresthesias (tingling, prickling, or numbing of skin), vertigo, low-grade fever or no fever, meningismus (neck stiffness and photophobia), hyperesthesia, paralysis, and blindness, leading to mental retardation, nerve damage, permanent brain damage, or death [50–53]. Symptoms typically develop after an incubation period of up to 2 weeks, and the disease often resolves spontaneously within in a few weeks [54,55].

Besides cerebral angiostrongyliasis, *A. cantonensis* may be involved occasionally in eye infection (ocular angiostrongyliasis), with permanent visual impairment and blindness as a consequence [56,57].

A. costaricensis infection tends to cause abdominal disease (eosinophilic enteritis) with manifestations ranging from acute abdominal pain, eosinophilia, eosinophilic vasculitis, leukocytosis, intestinal obstruction, perforation, to hemorrhage. The disease is limited to intestinal (ileocecal) location, affecting the mesenteric arterial branches and intestinal walls, and does not involve the CNS [58–61]. In a case of *A. costaricensis* infection involving a 57-year-old female in Venezuela, a palpable mass was noted at the lower right quadrant of the abdomen. WBC count was 16,600 cells/mm^3, with 46% eosinophils. Given the tumoral aspect of ileocolic area and peritoneal lymph nodes, a large area (covering the terminal ileum, cecum, part of the ascending colon, and a small part of the jejunum, where a small lesion was found) was resected. The pathological examination revealed thickened areas of the intestinal wall with areas of hemorrhage and a perforation of the cecum. Histological assessment uncovered intense eosinophil infiltration of the whole intestinal wall, granulomas with giant cells, and eosinophils. Some of the granuloma surrounded round or oval eggs with content characterized by a large empty area, cells, or embryo in the center, and occasionally nematode larvae. A cross section of an adult nematode worm was found inside a branch of mesenteric artery [62].

In a recent report, a 13-year-old male patient from Puriscal, Costa Rica complained of 15 days of abdominal pain in the right iliac fossa, abdominal cramps, diarrhea, vomits of ingested content, and fever. Laboratory tests revealed an increased number of eosinophils (39%). An exploratory laparotomy found abundant peritoneal fluid and necrotic patches without perforation 30 cm distal from the ileocecal valve, and hardened yellow-colored areas of 1–2 cm of length in large intestine, ileocecal region, and liver. Anti-*A. costaricensis* antibodies were detected on the serum by a latex agglutination test. Cecal appendix showed mild serositis, due to eosinophilic inflammatory infiltration. The small intestine was partially affected by ulcerated zones which showed inflammatory eosinophilic infiltration and extended from the submucosa to the subserosa. Some submucous arterioles, containing adult nematode structures in their lumen, presented wall fibrinoid necrosis and fibrin thrombi. Transversally sectioned nematode(s) showed a longitudinal muscle layer under the hypodermis, interrupted by lateral cords, and internal structures corresponding to the uterus and intestine of a female worm. Based upon the location, the morphologic features and the epidemiology, it was identified as *A. costaricensis* [63].

44.1.2.2 Treatment

Although human cerebral angiostrongyliasis is largely a self-limiting disease, with complete recovery in a few months without treatment, relief of symptoms (e.g., intense headaches) may be provided by administration of corticosteroid to control inflammation and reduce intracranial pressure [64]. For example, Liu et al. [65] used dexamethasone (a synthetic glucocorticoid hormone, 10 mg/d) and fluid infusion for the treatment of angiostrongyliasis with pleasing outcome. Application of prednisone or albendazole also helps alleviate persistent headaches [66]. Tsai et al. [67] prescribed prednisolone at a dose of 1 mg/kg/day for 14 days for successful treatment of angiostrongyliasis. A combination of adrenal cortical hormone, dehydration, and neurotrophic therapy may also be adopted. Combined therapy with albendazole and dexamethasone is effective. The most effective prevention strategy is to educate people not to eat raw or undercooked intermediate and paratenic hosts.

44.1.3 Laboratory Diagnosis

A. cantonensis represents one of several parasites that cause eosinophilic meningitis in humans, with the number of eosinophilia increasing 2–4 weeks after symptoms first appearing, then waning, returning again between weeks 6 and 8, and declining toward the end of the third month. Other common causes of eosinophilic meningitis (meningoencephalitis) in humans are *Gnathostoma* spp. (e.g., *Gnathostoma spinigerum*), *Paragonimus* spp., and *Taenia solium* [68]. In general, history of ingestion of uncooked fish, migratory swelling, subarachnoid hemorrhage, myeloencephalitis, and unusual intracerebral hemorrhage are suggestive of gnathostomiasis [69].

Presumptive diagnosis of human cerebral angiostrongyliasis is based on: (i) clinical presentation with meningitis showing severe headache, (ii) a history of consumption of raw mollusks, and (iii) a prominent (>10%) increase of eosinophilia in the CSF (and in the blood).

Magnetic resonance imaging (MRI) examination often reveals multiple micronodular enhancements in the brain and the spine. Computed tomography (CT) scan of the brain may show cerebral edema, or ventricular dilatation [67]. Complete resolution of abnormal MRI findings typically occur after 4–8 weeks.

Definitive diagnosis of human angiostrongyliasis is achieved by detecting *A. cantonensis* juvenile adults or L5 larvae in biopsy tissue, or in the CSF under microscope [70]. However, *A. cantonensis* is virtually undetectable in the CSF of 50% of the infected individuals [71].

Serological techniques such as enzyme-linked immunosorbent assay (ELISA) and Western blot (WB) have been applied to detect anti-*A. cantonensis* antibodies in CSF as well as in serum, and confirm the diagnosis of angiostrongyliasis [72–75]. These assays often relied on the use of crude somatic antigens or partially purified antigens of *A. cantonensis* adult worms, brain-stage larvae, or excretory–secretory products, which tended to show cross-reaction with other related parasites [73–75]. To enhance the specificity of serological assays, a number of stage-specific antigens have been identified. These include a 29-(31-, or 32-) kDa antigen from the adult worm, a 33-kDa antigen specific for the female adult worm, a 104-kDa antigen with specificity for larvae, a 204-kDa young adult worm antigen. Use of purified 31-kDa antigen in a dot-blot ELISA facilitated detection of *A. cantonensis* antibody in a highly sensitive (100%) and specific (100%) manner [76]. An AC32-ELISA kit for specific detection of immunoglobulin (Ig) G in serum samples of patients also offered a sensitive tool for immunodiagnosis of angiostrongyliasis. By employing crude extracts from adult *A. cantonensis* as the antigen and protein A conjugated with colloid gold as the detection marker, a simple and rapid dot-immunogold filtration assay enabled sensitive (90.5%) and specific (98.0%) detection of anti-*A. cantonensis* IgG antibodies in human serum samples. Furthermore, serological detection of *A. cantonensis* antigens with specific monoclonal antibodies allows a more rapid confirmation of acute or active angiostrongyliasis, although further improvement is needed in this type of assay [57].

More recently, molecular procedures have been developed for identification and differentiation of *Angiostrongylus* species [77–88]. By polymerase chain reaction (PCR) amplifying a 1134 base pair (bp) fragment in the *Angiostrongylus* 18S rRNA gene followed by immunological detection of the amplicons, an immuno-PCR assay offered a sensitive and specific means for confirming the presence of *A. cantonensis* [89]. A loop-mediated isothermal amplification (LAMP) method targeting the first internal transcribed spacer (ITS-1) of nuclear ribosomal RNA (rRNA) permitted specific detection of *A. cantonensis* in snail [90]. Real-time PCR assays further streamlined the species-specific identification of *A. cantonensis* in mollusk tissue and rat feces [88,91].

In addition, molecular methods have been applied for phylogenetic assessment of *Angiostrongylus* species originated from different geographical areas [92]. Through PCR amplification and sequencing analyses of the mitochondrial gene, cytochrome *c* oxidase subunit I (COI) gene, and a 66-kDa protein gene, it was clear that *A. cantonensis* and *A. malaysiensis* form one clade while *A. costaricensis* and *A. vasorum* form another. In the first major clade, *A. cantonensis* isolates of China and Hawaii appear to be different phylogenetically from those of Thailand. In the second major clade, *A. vasorum* isolates of Europe are distinct from those of Brazil isolates. Furthermore, *A. costaricensis* isolates of Costa Rica differ from those of Brazil [86,87].

44.2 METHODS

44.2.1 Sample Preparation

44.2.1.1 Sample Collection

A. cantonensis L1 larvae are collected from rat feces by using a sugar flotation technique and examined under a microscope.

Tissue pieces (~25 mg) are cut from the posterior end of the mollusk's foot and placed in 1 mL of 0.01% pepsin-0.7% HCl in individual wells of a 24-well culture dish (Corning) for digestion of the tissue to recover third-stage larvae for morphological identification or are used for DNA extraction.

To collect mucus secretions, slugs are kept individually in polystyrene petri dishes (60 × 15 mm; Becton Dickinson) overnight, with occasional prodding to stimulate secretion. The mucus samples are treated with pepsin-HCl to facilitate collection of larvae for identification by microscopy.

Hearts and lungs are dissected from infected rats for the extraction of Angiostrongylus adult worms. The worms are preserved in 70% ethanol, cleared in lactophenol, and identified using light microscopy.

44.2.1.2 DNA Extraction

DNA is extracted from purified nematode larvae and mucus samples by digestion with 0.1 μg proteinase K/mL in a buffer consisting of 50 mM Tris-HCl (pH 8.5), 1% laureth-12, and 1 mM EDTA overnight at 56°C, followed by incubation at 95°C for 10 min to inactivate the proteinase K. The mucus secretions require an additional purification step with QIAquick columns (QIAGEN) to remove PCR inhibitors.

DNA in intact fresh or frozen mollusk tissue is extracted using the FastDNA method (Q-Biogene). Disruption of samples in an FP120 cell disruptor is performed at speed 5.5 for 30 s. PCR inhibitors are removed by further purification with a QIAquick PCR purification kit (QIAGEN). Purified DNA is stored at 4°C until use.

For DNA extraction from preserved adult worms, worms are individually chopped with a sterile scalpel blade and resuspended in 100 μL of TE buffer (10 mM Tris and 1 mM EDTA, pH 8). The whole homogenate is digested with proteinase K (20 mg/mL) and DNA extraction is performed with Qiamp®DNA Blood mini kit (QIAGEN). Alternatively, the FTA card method (Whatman BioScience) may be used for DNA preparation from *Angiostrongylus* adult worms. Briefly, an individual worm is homogenized in 100–150 μL of sterilized phosphate-buffered saline (PBS) pH 7.4. The homogenate is applied and dried onto the FTA card according to the Whatman FTA tissue protocol. A sample disc is taken using a 2.0 mm diameter Harris micropunch, washed with FTA purification reagent, and used in PCR.

44.2.2 Detection Procedures

44.2.2.1 Real-Time PCR for *A. Cantonensis* Targeting ITS1

Qvarnstrom et al. [91] developed a real-time TaqMan PCR assay for specific detection of *A. cantonensis*, with primers and probe derived from the ITS1 region: AcanITS1F1 (5′-TTCATGGATGGCGAACTGATAG-3′), AcanITS1R1 (5′-GCGCCCATTGAAACATTATACTT-3′), and a TaqMan probe AcanITS1P1 (5′-6-carboxyfluorescein-ATCGCATATCT ACTATACGCATGTGACACCTG-BHQ-3′).

44.2.2.1.1 Procedure

1. The real-time PCR mixture (20 μL) consists of Platinum qPCR Supermix (Invitrogen,), 0.2 μM (each) primers, 0.05 μM the TaqMan probe.
2. The standard cycling conditions for TaqMan assays (i.e., 40 cycles of 95°C for 15 s and 60°C for 1 min) are used.

Note: The real-time PCR detects only DNA from *A. cantonensis* and not with DNA from other nematode species, with a sensitivity of down to 10 plasmid copies per reaction.

44.2.2.2 Real-Time PCR for *A. Cantonensis* Targeting ITS2

Fang et al. [88] reported a real-time TaqMan PCR assay for species-specific identification of *A. cantonensis*, with primers and probe from the ITS2 region: FP (5′-CCAGTTTTGGTGAAGAATAA-3′), RP (5′-ACACG ACGGTAACAATGACA-3′), and a dual-labeled probe AC (5′-FAM-GCTACATGTAATAATTCGACGATATGTG-TAMRA-3′). The FP/RP primer pair yields a 130 bp product.

44.2.2.2.1 Procedure

1. Real-time PCR mixture (20 μL) is composed of 2.5× RealMasterMix (Tianzen, China) 8 μL; 20× Probe Enhancer solution 1 μL; 0.5 μL of 10 μM forward and reverse primers, 0.25 μL of 10 μM dual-labeled probe, and 2 μL of template DNA.
2. Cycling conditions include an initial activation step of 95°C for 2 min followed by 40 cycles of 95°C for 15 s and 60°C for 40 s using a Rotor Gene 6000 quantitative PCR system (Corbett Research, AUS).

Note: This real-time PCR provides specific detection for *A. cantonensis*, and is negative with other common parasite species.

44.2.2.3 PCR-RFLP Identification of *Angiostrongylus* spp.

Caldeira et al. [77] described a PCR-restriction fragment length polymorphism (RFLP) assay for differentiation of *A. cantonensis*, *A. costaricensis*, and *A. vasorum*. The primers LCO (forward; 5′- ggtcaacaaatcataaagatattgg-3′) and HCO (reverse; 5′-taaacttcagggtgaccaaaaaatca-3′) came from the mtDNA COI region. After amplification, the PCR products (700 bp) were digested with restriction enzyme *Rsa*I, yielding distinct band patterns for *A. cantonensis*, *A. costaricensis*, and *A. vasorum*

44.2.2.3.1 Procedure

1. PCR mixture (10 μL) is composed of 10 mM Tris-HCl pH 8.5, 200 μM each dNTP, 1.5 mM $MgCl_2$, 0.8 U of Taq DNA polymerase, 50 mM KCl, 0.3 pmol of each primer, and 1–10 ng of template DNA.
2. The mixture is subjected to an initial denaturation at 95°C for 5 min; 40°C or 50°C for 1 min, 72°C for 2 min, 95°C for 45 s; and a final extension at 72°C for 5 min.
3. The PCR products (5 μL) are digested with 10 U restriction endonuclease *Rsa*I at 37°C for 3.5 h.
4. The digested products are separated on 6% silver-stained polyacrylamide gels, and the band patterns are recorded with a digital photodocumentation system.

44.2.2.4 Analysis of *Angiostrongylus* COI Gene

Eamsobhana et al. [86] utilized primers COI_F (5′ TAAAGAAAGAACATAATGAAAATG 3′) and COI_R (5′ TTTTTTGGGCATCCTGAGGTTTAT 3′) for amplification and analysis of a partial region of *Angiostrongylus* COI gene.

44.2.2.4.1 Procedure

1. The PCR mixture (25 μL) is made up of Tris-HCl, KCl, $(NH_4)_2SO_4$, 15 mM $MgCl_2$, 0.5 μL of dNTP mix (10 mM each), 0.5 μL of each primer (12.5 ng/μL), 0.1 μL of Taq DNA polymerase (5 U/μL), 21.4 μL of dH_2O, and a 2.0 mm DNA punched disc (FTA card) that contained extracted DNA from individual worm samples.
2. The thermocycling parameters consist of an initial denaturation at 94°C for 5 min, followed by 40 cycles, each at 94°C for 30 s, at 55°C for 30 s, at 72°C for 1 min, and a final extension at 72°C for 5 min.
3. The PCR products are separated by electrophoresis on 1.5% (w/v) agarose gel, stained with ethidium bromide, and visualized under ultraviolet light.
4. The amplified products are purified using a QIAquick PCR Purification Kit (QIAGEN), and sequenced using an ABI PrismDyeTerminator Cycle Sequencing Core kit (Applied Biosystems).
5. The resulting sequences are preliminarily aligned using the CLUSTAL X program and subsequently manually aligned. The aligned sequences are subjected to maximum-likelihood (ML), maximum-parsimony (MP), and neighbor-joining (NJ) analyses using phylogenetic analysis using parsimony (PAUP)* 4.0b10). ML analyses of the COI data are performed using PAUP* and a best-fitting evolution model. Tree likelihoods are estimated using a heuristic search with 100 random-addition-sequence replicates, tree bisection reconnection (TBR) branch swapping, and unordered and unweighted characters. Bootstrap percentage (BP) is computed with 1000 replications. NJ bootstrap values are estimated using 1000 replicates with Kimura's two-parameter model of substitution (K2P distance) evolution model. Bayesian analysis is performed using MrBayes 3.1.1, using GTR model with gamma distribution. The program was to start with a random starting tree, analysis using four chains of Markov chain Monte Carlo iterations simultaneously for 2,000,000 generations and sampling the data every 100 generations. The likelihood scores stabilized after 200,000 generations; hence, for these analyses a "burn-in" of 200,000 generations is used.

44.2.2.5 Analysis of *Angiostrongylus* ITS 2

Gerrikagoitia et al. [92] employed primers NC1 (5′-ACG TCTGGTTCAGGGTTGTT-3′) and NC2 (5′-TTAGTTTCT TTTCCTCCGCT-3′) to amplify and sequence a 600–650 bp region covering the ITS2 of rRNA.

44.2.2.5.1 Procedure

1. PCR mixture (25 μL) consists of 10 mM Tris–HCl pH 8.4, 50 mM KCl, 1.5 mM $MgCl_2$, 0.2 mM of each dNTP, 0.3 of each primer, and 0.8 Taq DNA Polymerase (Invitrogen), and 1–10 ng of template DNA.
2. Amplification is conducted with an initial 94°C for 90 s; 39 cycles of 94°C for 50 s, 58°C for 1 min, 72°C for 90 s; and a final extension step at 72°C for 10 min.
3. PCR products (10 μL) are visualized on a 3% agarose gel, purified using an Exo SAP-IT kit (Amersham Biosciences) and subjected to BigDye® Terminator v3.1 Cycle Sequencing Kit (Applied Biosystems) from both strands.
4. Sequences are compared with those available in GenBank database using BLAST. Multiple sequence alignments are performed using the program AlignX (Vector NTI 8.0 suite) with an engine based on the CLUSTALW algorithm. Similarity matrices

are then constructed from aligned sequence data by the Kimura's two-parameter model. Phylogenetic trees are constructed by the NJ method as implemented in the Mega 4 package. Stability or accuracy of inferred topologies is assessed via bootstrap analysis of 1000 replicates.

44.3 CONCLUSION

The genus *Angiostrongylus* consists of 15 roundworm species whose adult worms reside in the pulmonary or mesenteric arteries of rodents and carnivores, and whose larvae are found in mollusks (e.g., snail and slug) as well as paratenic (transport) hosts (e.g., fish, frog, crab, and shrimp) [93]. Two *Angiostrongylus* species are involved in human diseases, with *A. cantonensis* causing eosinophilic meningitis (or cerebral angiostrongyliasis), and *A. costaricensis* producing abdominal angiostrongyliasis.

Being an accidental or nonpermissive host for *Angiostrongylus*, humans acquire the infection via ingestion of raw or undercooked fish or frog meat. The presence of *A. cantonensis* larvae in the subarachnoid spaces and meninges in humans leads to nonfatal headaches and brain tissue damage [94–97]. On the other hand, the occurrence of *A. costaricensis* in the intestines contributes to acute abdominal pain, intestinal obstruction, perforation, or hemorrhage in addition to eosinophilia. While cerebral angiostrongyliasis is prevalent in the Pacific islands and Southeast Asia, abdominal angiostrongyliasis is common in South and Central America. With more frequent global trade and travel, there is a notable increase in the number of human angiostrongyliasis reported in other parts of the world.

Presumptive diagnosis of angiostrongyliasis may be made using the criteria of epidemiology, clinical symptoms, complete blood count, CSF investigation, immunologic examination, imaging examination, and etiologic examination. Detection of *Angiostrongylus* young adults and larvae from the biopsy specimens or CSF of infected patients provides a clear indication for human angiostrongyliasis. However, since nearly half of the patients are negative by this method, use of serological assays is necessary to confirm the diagnosis [98,99]. In recent years, molecular techniques have been applied to the identification and phylogenetic assessment of *Angiostrongylus* species. In particular, the development of real-time PCR assays has enabled rapid, sensitive, and precise detection of *Angiostrongylus* organisms in fecal samples from rats and tissues from mollusks, contributing to the more efficient monitoring and epidemiological surveillance of *Angiostrongylus* outbreaks in the future [91].

REFERENCES

1. Graeff-Teixeira, C., da Silva, A.C.A., and Yoshimura, K., Update on eosinophilic meningoencephalitis and its clinical relevance. *Clin. Microbiol. Rev.*, 22, 322, 2009.
2. Lv, S. et al., *Angiostrongylus cantonensis*: Morphological and behavioral investigation within the freshwater snail *Pomacea canaliculata*. *Parasitol. Res.*, 104, 1351, 2009.
3. Guilhon, J., and Cens, B., *Angiostrongilus vasorum* (Baillet, 1866). Biological and morphological study. *Ann. Parasitol. Hum. Comp.*, 48, 567, 1973.
4. Bolt, G. et al., Canine angiostrongylosis: A review. *Vet. Rec.*, 135, 447, 1994.
5. Ash, L.R., Diagnostic morphology of the third-stage larvae of *Angiostrongylus cantonensis*, *Angiostrongylus vasorum*, *Aelurostrongylus abstrusus*, and *Anafilaroides rostratus* (Nematoda: Metastrongyloidea). *J. Parasitol.*, 56, 249, 1970.
6. Bonetti, V.C., and Graeff-Teixeira, C., *Angiostrongylus costaricensis* and the intermediate hosts: Observations on elimination of L3 in the mucus and inoculation of L1 through the tegument of mollucs. *Rev. Soc. Bras. Med. Trop.*, 31, 289, 1998.
7. Ubelaker, J.E., Bullick, G.R., and Caruso, J., Emergence of third-stage larvae of *Angiostrongylus costaricensis* Morera and Cespedes 1971 from *Biomphalaria glabrata* (Say). *J. Parasitol.*, 66, 856, 1980.
8. Tesana, S. et al., Prevalence and intensity of infection with third stage larvae of *Angiostrongylus cantonensis* in mollusks from Northeast Thailand. *Am. J. Trop. Med. Hyg.*, 80, 983, 2009.
9. Thiengo, S.C. et al., The giant African snail *Achatina fulica* as natural intermediate host of *Angiostrongylus cantonensis* in Pernambuco, northeast Brazil. *Acta Trop.*, 115, 194, 2010.
10. Vitta, A. et al., Survey of *Angiostrongylus cantonensis* in rats and giant African land snails in Phitsanulok province, Thailand. *Asian Pac. J. Trop. Med.*, 4, 597, 2011.
11. Ash, L.R., Observations on the role of mollusks and planarians in the transmission of *Angiostrongylus cantonensis* infection to man in New Caledonia. *Rev. Biol. Trop.*, 24, 163, 1976.
12. Kliks, M.M., and Palumbo, N.E., Eosinophilic meningitis beyond the Pacific Basin: The global dispersal of a peridomestic zoonosis caused by *Angiostrongylus cantonensis*, the nematode lungworm of rats. *Soc. Sci. Med.*, 34, 199, 1992.
13. Chung, H., A letter regarding a case of eosinophilic meningitis. *Hawaii Med. J.*, 64, 60, 2005.
14. Hochberg, N.S. et al., Distribution of eosinophilic meningitis cases attributable to *Angiostrongylus cantonensis*, Hawaii. *Emerg. Infect. Dis.*, 13, 1675, 2007.
15. Hochberg, N.S. et al., Eosinophilic meningitis attributable to *Angiostrongylus cantonensis* infection in Hawaii: Clinical characteristics and potential exposures. *Am. J. Trop. Med. Hyg.*, 85, 685, 2011.
16. Lindo, J.F. et al., Enzootic *Angiostrongylus cantonensis* in rats and snails after an outbreak of human eosinophilic meningitis, Jamaica. *Emerg. Infect. Dis.*, 8, 324, 2002.
17. Slom, T.J. et al., An outbreak of eosinophilic meningitis caused by *Angiostrongylus cantonensis* in travelers returning from the Caribbean. *N. Engl. J. Med.*, 346, 668, 2002.
18. Raccurt, C.P., Blaise, J., and Durette-Desset, M.C., Presence of *Angiostrongylus cantonensis* in Haiti. *Trop. Med. Int. Health*, 8, 423, 2003.
19. Chen, X.G., Li, H., and Lun, Z.R., Angiostrongyliasis, mainland China. *Emerg. Infect. Dis.*, 11, 1645, 2005.
20. Lv, S. et al., Emerging angiostrongyliasis in Mainland China. *Emerg. Infect. Dis.*, 14, 161, 2008.
21. Lv, S. et al., Human angiostrongyliasis outbreak in Dali, China. *PLoS Negl. Trop. Dis.*, 3, e520, 2009.
22. Lv, S. et al., Invasive snails and an emerging infectious disease: Results from the first national survey on *Angiostrongylus cantonensis* in China. *PLoS Negl. Trop. Dis.*, 3, e368, 2009.

23. Wang, J. et al., An outbreak of angiostrongyliasis cantonensis in Beijing. *J. Parasitol.*, 96, 377, 2010.
24. Chen, D. et al., Epidemiological survey of *Angiostrongylus cantonensis* in the west-central region of Guangdong Province, China. *Parasitol. Res.*, 109, 305, 2011.
25. Deng, Z.H. et al., First provincial survey of *Angiostrongylus cantonensis* in Guangdong Province, China. *Trop. Med. Int. Health*, 2011 Sep 9.
26. Hu, X. et al., Epidemic status of *Angiostrongylus cantonensis* in Hainan Island, China. *Asian Pac. J. Trop. Med.*, 4, 275, 2011.
27. Campbell, B.G., and Little, M.D., The finding of *Angiostrongylus cantonensis* in rats in New Orleans. *Am. J. Trop. Med. Hyg.*, 38, 568, 1988.
28. Costa, L. et al., Verminous meningoencephalomyelitis by *Angiostrongylus* (= *Parastrongylus*) *cantonensis* in an American miniature horse. *Equine Vet. Educ.*, 12, 2, 2000.
29. Duffy, M.S. et al., *Parastrongylus cantonensis* in a nonhuman primate, Florida. *Emerg. Infect. Dis.*, 10, 2207, 2004.
30. Maldonado, Jr., A. et al., First report of *Angiostrongylus cantonensis* (Nematoda: Metastrongylidae) in *Achatina fulica* (Mollusca: Gastropoda) from Southeast and South Brazil. *Mem. Inst. Oswaldo Cruz*, 105, 938, 2010.
31. Archer, C.E. et al., The rat lung-worm *Angiostrongylus cantonensis*: A first report in South Africa. *S. Afr. Med. J.*, 101, 174, 2011.
32. New, D., Little, M.D., and Cross, J., *Angiostrongylus cantonensis* infection from eating raw snails. *N. Engl. J. Med.*, 332, 1105, 1995.
33. Marsh, C.M., Eosinophilic meningitis/angiostrongyliasis from eating aquaculture-raised snails: A case report. *Hawaii Med. J.*, 57, 652, 1998.
34. Lai, C.H. et al., Eosinophilic meningitis caused by *Angiostrongylus cantonensis* after ingestion of raw frogs. *Am. J. Trop. Med. Hyg.*, 76, 399, 2007.
35. Sawanyawisuth, K. et al., Clinical factors predictive of encephalitis caused by *Angiostrongylus cantonensis. Am. J. Trop. Med. Hyg.*, 81, 698, 2009.
36. Tsai, H.C. et al., Encephalitis caused by *Angiostrongylus cantonensis* after eating raw frogs mixed with wine as a health supplement. *Intern. Med.*, 50, 771, 2011.
37. Wang, J.J. et al., Eosinophilic meningitis risk associated with raw *Ampullarium canaliculatus* snails consumption. *Kaohsiung J. Med. Sci.*, 27, 184, 2011.
38. Anonimous, Predictive factors for encephalitis caused by *Angiostrongylus cantonensis* infection. *Am. J. Trop. Med. Hyg.*, 82, 516, 2010; author reply, 82, 517, 2010.
39. Gelis, S., Spratt, D.M., and Raidal, S.R., Neuroangiostrongyliasis and other parasites in tawny frogmouths (*Podargus strigoides*) in south-eastern Queensland. *Aust. Vet. J.*, 89, 47, 2011.
40. Mota, E.M., and Lenzi, H.L., *Angiostrongylus costaricensis* life cycle: A new proposal. *Mem. Inst. Oswaldo Cruz*, 90, 707, 1995.
41. Mota, E.M., and Lenzi, H.L., *Angiostrongylus costaricensis*: Complete redescription of the migratory pathways based on experimental *Sigmodon hispidus* infection. *Mem. Inst. Oswaldo Cruz*, 100, 407, 2005.
42. Hulbert, T.V., Larsen, R.A., and Chandrasoma, P.T., Abdominal Angiostrongyliasis mimicking acute appendicitis and Meckel's diverticulum: Report of a case in the United States and review. *Clin. Infect. Dis.*, 14, 836, 1992.
43. Graeff-Teixeira, C. et al., Longitudinal clinical and serological survey of abdominal angiostrongyliasis in Guaporé, southern Brazil, from 1995 to 1999. *Rev. Soc. Bras. Med. Trop.*, 38, 310, 2005.
44. Alicata, J.E., *Angiostrongylus cantonensis* and eosinophilic meningitis. *Lancet*, 2, 1367, 1988.
45. Panackel, C. et al., Eosinophilic meningitis due to *Angiostrongylus cantonensis. Indian J. Med. Microbiol.*, 24, 220, 2006.
46. Li, H. et al., A severe eosinophilic meningoencephalitis caused by infection of *Angiostrongylus cantonensis. Am. J. Trop. Med. Hyg.*, 79, 568, 2008.
47. Tsai, H.C. et al., Dynamic changes of hepatocyte growth factor in eosinophilic meningitis caused by *Angiostrongylus cantonensis* infection. *Am. J. Trop. Med. Hyg.*, 80, 980, 2009.
48. Lee, I.C., *Angiostrongylus cantonensis* meningitis in two developmentally delayed children: Findings in brain images. *Pediatr. Infect. Dis. J.*, 29, 90, 2010.
49. Cui, Y., Shen, M., and Meng, S., Lung CT findings of angiostrongyliasis cantonensis caused by *Angiostrongylus cantonensis. Clin. Imaging*, 35, 180, 2011.
50. Prociv, P., Spratt, D.M., and Carlisle, M.S., Neuroangiostrongyliasis: Unresolved issues. *Int. J. Parasitol.*, 30, 1295, 2000.
51. Maretić, T. et al., Meningitis and radiculomyelitis caused by *Angiostrongylus cantonensis. Emerg. Infect. Dis.*, 15, 996, 2009.
52. Dorta-Contreras, A.J. et al., Neuroimmunological findings of *Angiostrongylus cantonensis* meningitis in Ecuadorian patients. *Arq. Neuropsiquiatr.*, 69, 466, 2011.
53. Wang, Q.P. et al., Human *Angiostrongylus cantonensis*: An update. *Eur. J. Clin. Microbiol. Infect. Dis.*, 2011 Jul 5 [Epub ahead of print].
54. Alto, W., Human infections with *Angiostrongylus cantonensis. Pac. Health Dialog.*, 8, 176, 2001.
55. Diao, Z. et al., Human ocular angiostrongyliasis: A literature review. *Trop. Doct.*, 41, 76, 2011.
56. Liu, I.H. et al., Necrotizing retinitis induced by *Angiostrongylus cantonensis. Am. J. Ophthalmol.*, 141, 577, 2006.
57. Diao, Z. et al., International symposium on *Angiostrongylus* and *Angiostrongyliasis*, 2010. *Emerg. Infect. Dis.*, 17, e1, 2011.
58. Loría-Cortés, R., and Lobo-Sanahuja, J.F., Clinical abdominal angiostrongyliasis. A study of 116 children with intestinal eosinophilic granuloma caused by *Angiostrongylus costaricensis. Am. J. Trop. Med. Hyg.*, 29, 538, 1980.
59. Morera, P., Abdominal angiostrongyliasis: A problem of public health. *Parasitol. Today*, 1, 173, 1985.
60. Kramer, M.H. et al., First reported outbreak of abdominal angiostrongyliasis. *Clin. Infect. Dis.*, 26, 365, 1998.
61. Rodriguez, R. et al., Abdominal angiostrongyliasis: Report of two cases with different clinical presentations. *Rev. Inst. Med. Trop. Sao Paulo*, 50, 339, 2008.
62. Incani, R.N. et al., Human infection by *Angiostrongylus costaricensis* in Venezuela: First report of a confirmed case. *Rev. Inst. Med. Trop. Sao Paulo*, 49, 197, 2007.
63. Quirós, J.L. et al., Abdominal angiostrongyliasis with involvement of liver histopathologically confirmed: A case report. *Rev. Inst. Med. Trop. Sao Paulo*, 53, 219, 2011.
64. Sawanyawisuth, K., and Sawanyawisuth, K., Drug target in eosinophilic meningitis caused by *Angiostrongylus cantonensis. Infect. Disord. Drug Targets*, 10, 322, 2010.
65. Liu, J., Gao, J., and Zhou, C., A case report on eosinophilic meningitis caused by *Angiostrongylus cantonensis. Int. J. Med. Sci.*, 8, 510, 2011.
66. He, H.J. et al., Efficacy of combined treatment with albendazole and baicalein against eosinophilic meningitis induced by *Angiostrongylus cantonensis* in mice. *J. Helminthol.*, 85, 92, 2011.

67. Tsai, H.C. et al., Brain magnetic resonance imaging abnormalities in eosinophilic meningitis caused by *Angiostrongylus cantonensis* infection. *Vector Borne Zoonotic Dis.*, 12, 161, 2012.
68. Walker, M.D., and Zunt, J.R., Neuroparasitic infections: Nematodes. *Semin. Neurol.*, 25, 252, 2005.
69. Ramirez-Avila, L. et al., Eosinophilic meningitis due to *Angiostrongylus* and *Gnathostoma* species. *Clin. Infect. Dis.*, 48, 322, 2009.
70. Kuberski, T. et al., Recovery of *Angiostrongylus cantonensis* from cerebrospinal fluid of a child with eosinophilic meningitis. *J. Clin. Microbiol.*, 9, 629, 1979.
71. Graeff-Texeira, C., Camillo-Coura, L., and Lenzi, H.L., Histopathological criteria for the diagnosis of Angiostrongyliasis. *Parasitol. Res.*, 77, 606, 1991.
72. Vitta, A. et al., Application of recombinant SMR-domain containing protein of angiostrongylus cantonensis in immunoblot diagnosis of human angiostrongyliasis. *Southeast Asian J. Trop. Med. Public Health*, 41, 785, 2010.
73. Ben, R. et al., Use of heterologous antigens for the immunodiagnosis of abdominal angiostrongyliasis by an enzyme-linked immunosorbent assay. *Mem. Inst. Oswaldo Cruz*, 105, 914, 2010.
74. Chen, M.X. et al., Seroprevalence of *Angiostrongylus cantonensis* infection in humans in China. *J. Parasitol.*, 97, 144, 2011.
75. Chen, M.X. et al., Development of a double antibody sandwich ELISA assay for the diagnosis of angiostrongyliasis. *J. Parasitol.*, 97, 721, 2011.
76. Eamsobhana, P., and Hong, H.S., Immunological diagnosis of human angiostrongyliasis due to *Angiostrongylus cantonensis* (Nematoda: Angiostrongylidae). *Int. J. Infect. Dis.*, 13, 425, 2009.
77. Caldeira, R.L., Molecular differentiation of *Angiostrongylus costaricensis*, *A. cantonensis*, and *A. vasorum* by polymerase chain reaction-restriction fragment length polymorphism. *Mem. Inst. Oswaldo Cruz*, 98, 1039, 2003.
78. da Silva, A.C., Graeff-Teixeira, C., and Zaha, A., Diagnosis of abdominal angiostrongyliasis by PCR from sera of patients. *Rev. Inst. Med. Trop. Sao Paulo*, 45, 295, 2003.
79. Qvarnstrom, Y. et al., PCR-based detection of *Angiostrongylus cantonensis* in tissue and mucus secretions from molluscan hosts. *Appl. Environ. Microbiol.*, 73, 1415, 2007.
80. Fontanilla, I.K., and Wade, C.M., The small subunit (SSU) ribosomal (r) RNA gene as a genetic marker for identifying infective 3rd juvenile stage *Angiostrongylus cantonensis*. *Acta Trop.*, 105, 181, 2008.
81. He, H. et al., Preliminary molecular characterization of the human pathogen *Angiostrongylus cantonensis*. *BMC Mol. Biol.*, 10, 97, 2009.
82. Helm, J. et al., A case of canine *Angiostrongylus vasorum* in Scotland confirmed by PCR and sequence analysis. *J. Small Anim. Pract.*, 50, 255, 2009.
83. Jefferies, R., Morgan, E.R., and Shaw, S.E., A SYBR green real-time PCR assay for the detection of the nematode *Angiostrongylus vasorum* in definitive and intermediate hosts. *Vet. Parasitol.*, 166, 112, 2009.
84. Jefferies, R. et al., Improved detection of canine *Angiostrongylus vasorum* infection using real-time PCR and indirect ELISA. *Parasitol. Res.*, 109, 1577, 2011.
85. Al-Sabi, M.N. et al., PCR detection of *Angiostrongylus vasorum* in faecal samples of dogs and foxes. *Parasitol. Res.*, 107, 135, 2010.
86. Eamsobhana, P. et al., Molecular differentiation of *Angiostrongylus* taxa (Nematoda: Angiostrongylidae) by cytochrome c oxidase subunit I (COI) gene sequences. *Acta Trop.*, 116, 152, 2010.
87. Eamsobhana, P. et al., Molecular differentiation and phylogenetic relationships of three *Angiostrongylus* species and *Angiostrongylus cantonensis* geographical isolates based on a 66-kDa protein gene of *A. cantonensis* (Nematoda: Angiostrongylidae). *Exp. Parasitol.*, 126, 564, 2010.
88. Fang, W. et al., PCR assay for the cell-free copro-DNA detection of *Angiostrongylus cantonensis* in rat faeces. *Vet. Parasitol.*, 183, 299, 2012.
89. Chye, S.M. et al., Immuno-PCR for detection of antigen to *Angiostrongylus cantonensis* circulating fifth-stage worms. *Clin. Chem.*, 50, 51, 2004.
90. Liu, C.Y. et al., Specific detection of *Angiostrongylus cantonensis* in the snail *Achatina fulica* using a loop-mediated isothermal amplification (LAMP) assay. *Mol. Cell. Probes*, 25, 164, 2011.
91. Qvarnstrom, Y. et al., Improved molecular detection of *Angiostrongylus cantonensis* in mollusks and other environmental samples with a species-specific internal transcribed spacer 1-based TaqMan assay. *Appl. Environ. Microbiol.*, 76, 5287, 2010.
92. Gerrikagoitia, X., Barral, M., and Juste, R.A., *Angiostrongylus* species in wild carnivores in the Iberian Peninsula. *Vet. Parasitol.*, 174, 175, 2010.
93. Ubelaker, J.E., Systematics of species referred to the genus *Angiostrongylus*. *J. Parasitol.*, 72, 237, 1986.
94. Diao, Z., Jin, E., and Yin, C., *Angiostrongylus cantonensis*: Lesions in brain and spinal cord. *Am. J. Trop. Med. Hyg.*, 82, 519, 2010.
95. Diao, Z., Yin, C., and Jin, E., Images in clinical tropical medicine. Myelitis caused by infection of *Angiostrongylus cantonensis*. *Am. J. Trop. Med. Hyg.*, 83, 1176, 2010.
96. Tseng, Y.T. et al., Clinical manifestations of eosinophilic meningitis caused by *Angiostrongylus cantonensis*: 18 years' experience in a medical center in southern Taiwan. *J. Microbiol. Immunol. Infect.*, 44, 382, 2011.
97. Sawanyawisuth, K. et al., How can clinicians ensure the diagnosis of meningitic angiostrongyliasis? *Vector Borne Zoonotic Dis.*, 12, 73, 2012.
98. Abrahams-Sandi, E. et al., An indirect immunofluorescence antibody test employing whole eggs as the antigen for the diagnosis of abdominal angiostrongyliasis. *Mem. Inst. Oswaldo Cruz*, 106, 390, 2011.
99. Sawanyawisuth, K. et al., Specificity of immunoblotting analyses in eosinophilic meningitis. *Mem. Inst. Oswaldo Cruz*, 106, 570, 2011.

45 Anisakis

Stefano D'Amelio, Serena Cavallero, Marina Busi, Sofia Ingrosso, Elisabetta Giuffra, Sarra Farjallah, and Graça Costa

CONTENTS

45.1 INTRODUCTION

Anisakid nematodes are a homogeneous group constituted of species whose adult stages are recovered in fish, fish-eating birds, and marine mammals, while third-stage larvae are present in the body cavity and muscle of numerous fish species. Anisakids are of medical and economic importance, due to public health implications and for the subsequent effects on the marketability of fish products. Larval forms of anisakid nematodes, in particular those belonging to the genera *Anisakis* and *Pseudoterranova*, are in fact the main causative agents of human anisakidosis, a fish-borne zoonosis due to the ingestion of raw or undercooked fish or cephalopods, which are infected by these larvae.

Since the 1980s, pioneer studies on the genetic structure of anisakid nematodes has been carried out using allozyme markers, on a large number of taxa, sampled from many geographic regions from the boreal and austral hemispheres. Important achievements of these studies concern the detection of sibling species, the estimation of genetic divergence and of the relationships among taxa, the analysis of patterns of genetic variation in populations and species, and measurement of gene flow. In anisakid nematodes, cladogenetic events were accompanied by poor morphological differentiation, as ecological factors have led to a convergence to similar and well-fitting morpho-functional solutions. This gave origin to a large number of morphologically identical species (sibling species). Therefore, morphological traits do not provide definitive evidences to be used as landmarks for their identification. This scarcity of diagnostic characters in adult individuals is even more remarkable in larval forms, where the number of structural traits usable for diagnostic purposes is more limited. Moreover, it should be stressed that the larvae are responsible for human infection, and that their identification is essential for understanding the epidemiology of human anisakidosis. Also at higher taxonomic level, several morphological characters, even if strikingly differentiating one group from another, appear to be homoplastic and not always related to the phylogeny of species or genera. The poor taxonomic significance of some morphological characters and the occurrence of speciation processes without morphological differentiation undoubtedly advocate for the use of genetic and molecular approaches as efficient tools for the inference of the systematics and evolution of anisakid nematodes.

45.1.1 Classification of *Anisakis*

The taxonomy of anisakids is mainly based on the morphology of male adult specimens. The most significant structural characters for species identification are the distribution and pattern of caudal papillae, the spicule, and the morphology of cephalic end [1]. Anisakid larvae can be identified at genus level by light microscopy, mainly on the basis of the morphology of the digestive tract and of the excretory system.

By using multilocus allozyme electrophoresis, the analysis of the genetic structure of both adults and larvae of *Anisakis* species has provided interesting results at the taxonomic level [2–4]. At present, allozyme and polymerase chain reaction–restriction-fragment-length polymorphism (PCR–RFLP) studies have demonstrated that the genus *Anisakis* comprises at least nine distinct species, six belonging to Clade 1, corresponding to the larval Type I (sensu [5]), namely, *Anisakis simplex* sensu stricto, *Anisakis pegreffii*, *Anisakis simplex* C, *Anisakis typica*, *Anisakis ziphidarum*, and *Anisakis nascettii*, and three belonging to Clade 2, showing the larval Type II [5] that is, *Anisakis physeteris*, *Anisakis brevispiculata*, and *Anisakis paggiae*, which have different genetic structures, life cycles, and geographical distributions [6–15].

Recently, Palm et al. [16] have hypothesized the existence of one additional taxa, closely related to *A. typica*, recovered at larval stage in nonmigratory fish species in Balinese and Javanese waters, Indonesia.

Finally, *Anisakis schupakovi*, a parasite of the relict phocid species of the Caspian Sea, *Phoca caspica*, was considered as *inquirendae* by Davey [17] in his revision of the genus *Anisakis*, but retained as a valid species by Kurochkin [18].

45.1.2 Biology and Epidemiology

45.1.2.1 The Life Cycle

Adult forms of the genera *Anisakis* and *Pseudoterranova* live in the stomach of marine mammals. Specifically members of the genus *Anisakis* are parasites of cetaceans, with the exception of *A. schupakovi*, parasites of a relict seal species in the Caspian sea, *P. caspica*. As for the genus *Pseudoterranova*, the members of the *Pseudoterranova decipiens* complex are parasitic on seals.

The life cycle begins when eggs are passed through the feces of the definitive host. Once the eggs are passed, they molt into second-stage larvae. These must be consumed by an intermediate host, usually a crustacean. Fish or squid become infected after eating an infected crustacean. Since anisakids do not undergo any development or molt inside the fish or squid, these hosts should be regarded as paratenic hosts of the nematode. The life cycle is completed after the paratenic host is ingested by a definitive host. It is possible that small fishes and squids are predated by larger fish species, thus determining an additional passage in a new paratenic host. Inside the stomach or the intestine of its final mammalian host, the worm develops into a sexually mature adult.

While the above-mentioned life cycle is accepted by many scientists, there is considerable evidence that two molts actually occur during development in the egg of *A. simplex* and therefore the third larval stage hatches from the egg.

45.1.2.2 Host Preference

Anisakid larvae have been detected worldwide in a large variety of fish species. Among teleosts, they have been found in Gadiformes, Perciformes, Clupeiformes, Pleuronectiformes, Scorpaeniformes, Zeiformes, Bericiformes, Lophiiformes, Anguilliformes, and Atheriniformes. Anisakid larvae have also been detected in elasmobranchs and in a variety of cephalopods. Among these, they have been found in Octopodidae, Sepiidae, Loliginidae, and Ommastrephidae.

In the stomach of cetaceans, anisakid adults are often found in clusters of individuals embedded in the mucosa and submucosa. Ulcers of 5 cm × 3 cm sized associated with anisakids are found in the fundic portion of the stomach.

Six species have been grouped on the basis of the genetic relationships inferred from allozyme ad DNA sequence data, thus forming the Clade 1.

A. simplex s.s. has been frequently found at its adult stage in *Balaenoptera acutorostrata*, *Delphinapterus leucas*, *Delphinus delphis*, *Globicephala melaena*, *Lagenorhynchus albirostris*, *Orcinus orca*, *Pseudorca crassidens*, *Phocoena phocoena*, and more rarely in *Stenella coeruleoalba*.

Adults of *A. pegreffii* are frequent parasites of *D. delphis*, *Ziphius cavirostris*, and *Tursiops truncatus*, and have been recorded also in *Caperea marginata*.

The species *A. simplex* C has been identified as adult in the stomach of *L. albirostris*, *Lissodelphis borealis*, and *P. crassidens*.

Two related species, *A. ziphidarum* and *A. nascettii*, are parasitic specifically of cetaceans belonging to the family Ziphiidae. In particular, adult forms of *A. ziphidarum* have been found in *Z. cavirostris*, *Mesoplodon layardii*, *Mesoplodon europaeus*, and *Mesoplodon densirostris*. *A. nascettii* (previously described as *Anisakis* sp. A) has been found as adult or preadult in *M. layardii*, *M. densirostris*, *Mesoplodon mirus*, and *Mesoplodon grayi*.

The species *A. typica* has been recorded in several species of marine mammals. Mattiucci and Nascetti [19,20] and D'Amelio et al, [21] have reported adult forms of this species *in Globicephala macrorhynchus*, *Sotalia fluviatilis*, *Steno bredanensis*, *Stenella attenuata*, *Stenella longirostris*, *S. coeruleoalba*, *T. truncatus*, *Lagenodelphis hosei*, and *Mesoplodon bidens*.

Three species of the genus *Anisakis*, strictly related one to each other, form the so-called Clade 2. These species are *A. physeteris*, *A. brevispiculata*, and *A. paggiae*. The main definitive host for *A. physeteris* is the sperm whale, *Physeter macrocephalus*, but there are records of this species in the pigmy sperm whale, *Kogia breviceps*, *A. brevispiculata* and *A. paggiae* have been recorded in *K. breviceps* and in the dwarf sperm whale, *Kogia sima*. They are often found in several fish species although showing a particular predilection for cephalopods, which represent important prey items of *Physeter* and *Kogia* species.

45.1.2.3 Geographic Distribution

A. simplex s.s. has been recorded mainly in the North Atlantic and North Pacific Oceans, ranging from 35 °N to the Arctic Circle. This species has been rarely found in the Mediterranean Sea, probably as a result of migration of pelagic fishes through the Gibraltar Strait.

A. pegreffii is mainly distributed in the Mediterranean Sea and the South Atlantic Ocean. It is by far the most frequent species in the Mediterranean and the larvae of this species have been found in both pelagic and demersal fishes. Some records of this species have been reported in Japanese waters [22] and New Zealand waters [19].

The third species of the complex, *A. simplex* C, shows a "leopard skin" distribution, as it has been found in the Canadian and Californian Pacific waters, Chile, New Zealand, and the Eastern South Atlantic Ocean [19].

The species *A. typica* is generally found in warm temperate and tropical waters. Adults and larvae of this species have been found in several cetacean and fish species of the Gulf of Mexico, the Brazilian coasts of South America, coasts of Somalia, Mauritania, Morocco and in the central-eastern part of the Mediterranean Sea, including Cyprus, Tunisia, and Lybia [23]. The occurrence of *A. typica* in the eastern part of the Mediterranean has been related to migration of fishes from the Red Sea through the Suez Canal [19].

A. ziphidarum has been found in the coasts of South Africa, the Mediterranean Sea, and Central Atlantic waters up to the Carribean Sea, while its closely related species, *A. nascettii* has been recorded in the Central Atlantic Ocean, in the coasts of Madeira and north-western Spain, as well as in the waters of South Africa and New Zealand [10,14,24].

A. physeteris is distributed in the Mediterranean Sea, including the coasts of North Africa [23], the Atlantic coasts of Spain, Central Atlantic Ocean, and the Gulf of Mexico.

A. brevispiculata and *A. paggiae* show a geographic distribution along the warm-temperate waters of the Atlantic Ocean, from the Gulf of Mexico to the coasts of the Iberian Peninsula.

45.1.3 Human Anisakiasis

45.1.3.1 Pathogenesis and Medical Importance

Human anisakidosis, firstly reported in the Netherlands, has gained an increasing health and economic relevance in particular in countries such as Japan, where the consumption of raw fish and squids is frequent, although human cases are increasingly reported in the United States and many European countries (United Kingdom, France, Spain, and Italy).

The anisakidosis of the digestive tract is classified from the clinical point of view, as acute or moderate. On the basis of the localization, the disease is subdivided into gastric anisakidosis and intestinal anisakidosis. The acute variety is prevalently gastric, and is characterized by nausea, vomiting, and epigastric pain. These symptoms appear 4–6 h after the ingestion of the infected fish. In intestinal anisakidosis, acute signs appear after about 7 days after infection with abdominal pain, nausea, vomiting, fever, diarrhea, fecal occult blood, leukocytosis, and, rarely, eosinophilia. The moderate forms of gastric anisakiasis (GA) are characterized by appetite loss, epigastralgia, and the occurrence of gastric pseudotumor; the evolution of moderate intestinal anisakiasis (IA) is more malicious. Several rare extragastrointestinal localizations have been documented (abdominal cavity, mesenteries, and omentum).

In the past, reports of anisakidosis (human infection with anisakids due to the genera *Anisakis* and *Pseudoterranova*) were mainly from Asia [24–27] but in recent years, due to the popularity and increased consumption of Japanese dishes, as well as the consumption of pickled small fish, more and more cases of anisakiasis (disease due to infection with *Anisakis* species) or anisakidosis have been reported worldwide [28,29]. Thus, the consumption of pickled whole anchovies and sardines (*Engraulis encrasicholus*, *Sardina pilchardus*) and of marinated herring (*Clupea harengus*) have been the main sources of infection in Western Europe, including Spain [30,31]. Other fish species as chub mackerel (*Scomber japonicus*) and hake (*Merluccius merluccius*) are also sources of *Anisakis* infections [32,33].

45.1.3.2 Clinical Manifestation of GA

Abdominal pain is a frequent sign in acute GA and has been observed practically in 100% of cases during a large survey in Japan [34]. Other symptoms are nausea (68.3%), vomiting (39.0%), abdominal fullness or distension (29.3%), diarrhea (19.5%), anorexia (9.8%), and chest pain (4.9%). High fever is not a classical sign of GA, although it can be quite frequently slightly higher than normal (37.5°C). Other objective findings are leukocytosis, from slight to severe, and, sometimes, eosinophilia. As for the localization in the gastric mucosa, Shibata et al. [35], divided the stomach into four main parts, namely anterior wall, lesser curvature, posterior wall, and greater curvature. The majority of larvae (49%) were found in the greater curvature, followed by the posterior wall (22%). These authors also described several features of endoscopic findings, such as edematous hypertrophic gastric folds, increase of gastric secretion and peristalsis, and mucosal lesions, including edema, redness, coagulation, hemorrhage, and ulceration.

45.1.3.3 Clinical Manifestation of IA

In a parasitological survey of anisakiasis conducted by Ishikura [36] (1967), 27% were found at the intestinal level. Most of the cases of intestinal infection with *Anisakis* led to intestinal colic and sometimes peritonitis. Symptoms such as nausea, vomiting, and abdominal bulging and induration can occur, although numerous cases of asymptomatic IA have been observed. The mild form is characterized by eosinophilic granulomas forming tumors in the intestinal walls whereas the fulminant form has the symptoms of acute ileus. Cases of IA reported from Spain [37] and from Italy [38] were characterized by a clinical picture of acute abdominal pain, diarrhea, meteorism, and selective pain to palpitation in the right ileum with symptoms as nausea, fever, and vomiting. Although difficult to evaluate, some biochemical tests based on leukocyte counts, percentage of eosinophils, as well as of the enzymes GOT and GPT can be of help in the diagnosis of IA [39]. The condition can also be diagnosed by radiography or ultrasonography.

45.1.3.4 Pathophysiology of IA

On the basis of the histological picture, anisakiasis has been classified into four sequential stages: the first, or phlegmon, is typical of IA rather than GA; the second, or abscess, is rather frequent in GA and is characterized by abundant necrotic tissue around the larvae and by a rich population of eosinophils; the third, or granuloma-abscess, corresponds to the flogistic evolution of the disease, at least 6 months after the ingestion of the larva. At this stage the larva appears as few remnants invaded by the eosinophils, surrounded by giant cells and abundant inflammatory parvicellular infiltrate; the fourth or granuloma stage is the most advanced, not frequently observed, characterized by a further decrease in the presence of eosinophils, with abundance of lymphocytes, giant cells, and significant collagenization.

Although one of the main features of the local inflammatory lesions produced by *Anisakis* larvae is the presence of relevant eosinophilic infiltration in the tissues surrounding the parasite, systemic eosinophilia is associated with anisakidosis only in less than 30% of cases.

45.1.3.5 Immunodiagnosis of Anisakiasis

Live or dead anisakid larvae present in fish can lead to the onset of allergic reactions which are frequently associated with high levels of the immunoglobulin, IgE [28,40]. Additionally, cases of occupational *A. simplex* allergy were described in fishmongers, or related with exposure (either by contact or inhalation) to fish meals in chicken feed [41,42]. Clinical symptoms of allergic anisakiasis range from urticaria to anaphylactic shock. The diagnosis of anisakiasis can be complicated as infections with helminths, in general, are associated with high levels of IgE and other immunoglobulins. Additionally, cross reactivity occurs with other parasite antigens. This problem demanded the development of more sensitive immunological techniques for the diagnosis of allergy due to *A. simplex*, as the number of suspected cases of *Anisakis*-allergy showed an increasing tendency. In Spain, more than 50% of *A. simplex*-allergic patients required emergency treatment, with five out of 64 hospitalized due to respiratory failure [32]. Since 1995, more than 150 cases of allergy due to *Anisakis* were reported in Spain (see Audicana et al. [28]). Although the usual signs of anisakiasis are characterized by urticaria, anaphylactic shock and respiratory failure due to edema, Kikuchi et al. [43] referred to a possible association between *A. simplex* and rheumatic pathology.

Gastroallergic anisakiasis (GAA), the type of anisakiasis characterized by urticaria, angioedema and anaphylaxis, which consists of acute IgE-mediated generalized reaction, after the intake of raw or undercooked fish infected with *A. simplex*, is a frequent condition in Japan [44,45]. In this type of anisakiasis, the allergic reactions take place 24 h after the ingestion of fish. In Spain, gastric symptoms are mild, possibly because a different isoform of the allergen (namely *Ani s 1*) is involved in the allergic condition [46,47]. Other immunoglobulins such as IgA, IgG1, IgG2b and IgG2c can also be detected [45]. Allergic reactions occurred not only after consumption of undercooked fresh fish, but also of cooked or frozen fish. Therefore, Moneo et al. [48] investigated if intolerance was due to a particular antigen, testing the patient reactions against crude parasite extracts, excretory/secretory allergens, heated extract, and finally two gradient ethanol precipitates of the crude extract. Previously, Moneo et al. [49] and Shikamura et al. [47] have found that the *Anisakis* allergens, *Ani s 1* and *Ani s 4*, are heat-resistant, therefore withstanding cooking preparation.

To date nine *A. simplex* allergens were characterized: *Ani s 1* to *Ani s 9*. The *Ani s 2* and *Ani s 3* allergens have somatic location and all the others are located in excretory/secretory glands (ES) [29]. Anadón et al. [50] considered that allergen *Ani s 7* is the most important, being recognized by 100% of the infected patients. The results of this experimental work revealed that the antigenicity of the *nAni s 7* molecule was destroyed in the dead L3 larvae of *A. simplex*. They suggested that this antigen and other *Anisakis* ES allergens, stimulate the immune system only during a short time, which corresponds to the acute phase of infection, whereas other *Anisakis* antigens may retain their antigenic capacity during chronic stage of anisakiasis.

Healthy individuals can have high levels of anti-*Anisakis* IgE in their serum, without the development of allergic symptoms. On the other hand individuals with low level of specific IgE antibodies may show clinical manifestations. Del Pozo et al. [33] recommended four criteria for the diagnosis of *Anisakis* allergy: (1) urticaria/AE or anaphylaxis within 6 h after fish ingestion; (2) specific IgE antibodies to *A. simplex*; (3) positive skin prick test with *A. simplex* extract; (4) exclusion of fish allergy or other suspected causes.

45.1.3.6 Treatment

The endoscopic removal of the larvae in GA and the surgical removal of intestinal granulomas seem to be resolutive of the disease. The effective use of antiheminthic drugs is not supported by large surveys, although recent studies have evidenced a significant action of tiabendazole and albendazole against these nematodes.

45.1.3.7 Prevention

The most common preventive measures to avoid human infections with *Anisakis* larvae are summarized in Table 45.1.

45.1.4 Identification of *Anisakis* spp.

45.1.4.1 Limits of Conventional Identification of Anisakid Nematodes

The strong limit of the specific diagnosis of the etiological agents of anisakiasis on the basis of morphological examination relies mainly in several factors. The first is that larval forms of nematodes usually lack morphological characters useful for taxonomic purposes. Anisakid nematodes do not represent an exception to this assumption. The larvae belonging to the genus *Anisakis* can be easily discriminated from the other genera of the family Anisakidae, and can be subdivided, on the basis of the presence/absence of the mucron, in two distinct larval types, but cannot be identified at the species level, since each larval type is represented by more than one morphologically undistinguishable species.

TABLE 45.1
Conservative Technologies in Fish Products for the Prevention of Anisakidosis

Methods	Time of Exposition or Result
High Temperature	*Time*
70°C	1″
50°C	15″
45°C	78″
Low Temperature	*Time*
−35°C	15 h
−20°C	24–48 h
−10°C	Over 7 days
Smoking	*Result*
28°C	87.5% Survivors
40°C	14.4% Survivors
60°C	Complete devitalization
Marinade	*Result*
4% Acetic acid (vinegar) + 6% NaCl (salt)	Surviving of larvae after 26 days
Salting	*Result*
20° Baume' (saline density)	Complete devitalization after 14 days
Vinegar	*Result*
	Complete devitalization after 14 days
Soy Sauce	*Result*
	Complete devitalization after 21 days

Moreover, the widespread phenomenon of the existence of sibling species in anisakids makes the identification at species level rather problematic even in adult male specimens, where the number of morphological characters usable for taxonomic purposes is definitively much higher.

Finally, under diagnostic conditions, it happens quite often to recover only a portion of larva, which may lack the part of the body containing diagnostic features, or to receive, after intestinal surgery, a histological section from which it is sometimes very hard even to recognize a nematode.

These factors make the use of genetic and molecular methods absolutely necessary to identify the larvae of anisakid nematodes at the species level. The genetic, biochemical, and molecular methods so far applied in the taxonomy and molecular systematics of the genus *Anisakis* are summarized in the next paragraph.

45.1.4.2 Biochemical and Molecular Methods

45.1.4.2.1 Allozymes

Multilocus allozyme electrophoresis have been extensively used in anisakid nematodes since the beginning of the 1980s to address questions concerning the population genetics, evolutionary biology, host–parasite co-evolution, and the relationship between genetic variability and habitat disturbance. The most important achievements of these studies are at the population and species levels, while above these taxonomic levels, the methodology shows significant limitations. Among these results it is worth mentioning the detection of sibling species, that is, species morphologically very similar, at times identical, but genetically and ecologically differentiated; the estimation of the amount of genetic divergence and of the relationships among closely related taxa; the analysis of patterns of genetic variation in populations and species; and the measurement of gene flow within species [2–4,6,11,51–62].

45.1.4.2.2 PCR–RFLP

Notwithstanding the huge bulk of data obtained so far from the application of multilocus allozyme electrophoresis, the development of molecular markers for the accurate identification of related species using PCR-based approaches may be in some cases preferable, considering that this approach has shown a remarkable sensitivity in the detection of genetic variation requiring only small amounts of fresh or ethanol-fixed parasite material for analysis. For example, PCR–RFLP and sequence analyses of the ribosomal DNA (rDNA) internal transcribed spacers (ITS-1 and ITS-2) provides a useful approach for the specific identification of both distantly and closely related ascaridoid species, as these spacers showed high levels of interspecific sequence differences in the presence of low-level intraspecific variation.

PCR-based methods were therefore exploited to establish genetic markers in rDNA for the identification of anisakid nematodes belonging to the genera *Anisakis* (irrespective of developmental stages). In particular, RFLP markers were obtained for the three sibling species of the *A. simplex* complex and another four morphologically differentiated species of *Anisakis*, that is, *A. physeteris*, parasite of sperm whales, *A. typica*, parasite of cetaceans in warm waters, *A. ziphidarum*, a parasite of cetaceans of the family Ziphiidae and *A. schupakovi*, parasite of the Caspian Seal, *P. caspica* [12]. These markers have been successfully used in the identification of species of the genus *Anisakis* from fishes from Spain [13], from Madeira [14], from flatfish species from the coasts of Portugal [15], from the coasts of Japan [22], from the coasts of Tunisia [23], from migratory and nonmigratory fishes from Indonesia [16], fishes from marine mammals from Galician coasts of Spain [24], and from the Atlantic coasts of Morocco and Mauritania [63].

45.1.4.2.3 Multiplex and Species-Specific PCR

Umehara et al. [64] have developed a method based on multiplex PCR able to selectively amplify six different species of anisakids including two species of the *A. simplex* complex (*A. simplex* s.s. and *A. pegreffii*). The specificity of these primers to discriminate between the two sibling species was increased due to the introduction of artificial mismatched bases.

Recently, PCR-sequence-specific primers were developed by Abe [65] in order to establish a quick method for the discrimination between *A. simplex* s.s. and *A. pegreffii*.

45.1.4.2.4 DNA Sequencing

Direct DNA sequencing proved a fruitful tool for the identification of the different sibling and morphospecies of *Anisakis*. Actually, sequence data are available for almost all the species recognized within the genus with the only exception of *A. schupakovi*. In particular, the sequences available concern the region of the nuclear rDNA, spanning the final part of the 18S subunit, the first internal transcribed spacer (ITS-1), the 5.8S subunit, the second internal transcribed spacer (ITS-2), and the very beginning of the 28S subunit. The two ITS regions are highly variable, in particular among different morphospecies, and show a significant degree of variation also among closely related species, as the three members of the *A. simplex* complex (i.e., *A. simplex* s.s., *A. pegreffii*, and *A. simplex* C), and are useful for species discrimination. On the other hand, concerted evolution tends to minimize the intraspecific variation in this genomic region, thus allowing an unambiguous attribution of one specified sequence to one corresponding species.

As for the mitochondrial DNA (mtDNA), two regions have been so far studied. The mitochondrial gene *cox2* (cytochrome oxidase 2) was analyzed with the aim to clarify the phylogenetic relationships within the genus *Anisakis* [10,66]. Although a certain degree of intraspecific variation could interfere with the use of this marker for diagnostic (not phylogenetic) purposes, in our opinion this region is highly reliable for species discrimination.

The above markers have all been confirmed also by another mtDNA gene, the *rrnS* (the small subunit of the rDNA in the mitochondrial genome). Abollo et al. [13] reported fixed differences able to distinguish larvae of *A. simplex* s.s. and *A. pegreffii*.

45.1.4.3 Molecular Diagnosis of Human Anisakidosis

Only few instances of diagnosis of human anisakiasis performed by the application of molecular methods have been so far reported. The first molecular identification of a larva recovered following a gastric endoscopy was reported by D'Amelio et al. [67] in Italy. The paper describes PCR amplification of the entire ITS and subsequent RFLP analysis. The larva was identified as *A. pegreffii*, as expected since this is the most frequent species in fishes from Italian seas. The same approach was applied by Farjallah et al. [68] recognizing a larva recovered from the esophagus of a patient in Rome as belonging to *A .pegreffii*. Two additional cases from Italy were diagnosed as belonging to the same species by PCR amplification and sequencing of the mitochondrial *cox2* gene by Mattiucci et al. [69]. A widest survey was conducted by Umehara et al. [70] in Japan, where 99 larvae from human patients were identified as *A. simplex* s.s. and one as *A. pegreffii* using the same method described by D'Amelio et al. [12,67].

45.2 METHODS

The entire protocol, from the isolation of larvae to their specific identification is here described, using two genetic markers based on the different nucleotide sequences in the species of the genus *Anisakis*. The most reliable markers are the internal transcribed spacers of the nuclear rDNA (ITS-1 and ITS-2) as developed by D'Amelio et al. [12] and implemented by Pontes et al. [14] and the mitochondrial gene coding for the *rrnS* (the small subunit of the rDNA in the mitochondrial genome) developed by Abollo et al. [13] and here implemented.

Based on nucleotide differences, a taxonomic key based on two diagnostic restriction enzymes (*Hha*I and *Hinf*I) after amplification of the ITS genomic region has been developed for the identification of species of the genus *Anisakis*.

Additionally, species identification can be achieved, or confirmed, by direct sequencing of the PCR products after amplification using specific primers for the *rrnS* mitochondrial gene.

45.2.1 Sample Preparation

45.2.1.1 Isolation of Larvae from Human Cases

Anisakid larvae, due to their relatively large size, are easily recognizable during endoscopy from the esophagus or the stomach. They are usually removed using forceps jointly with the endoscopic device. However, it is possible to recover a sufficient quantity of DNA to perform PCR from intestinal histological sections, after treatment of slides to remove paraffin.

45.2.1.2 DNA Extraction

We report three different protocols for DNA purification aimed to obtain suitable templates to be used for PCR amplification. The first procedure has been developed in the laboratory, the second is based on the use of the *Wizard*® *SV*

Genomic DNA Purification System (Promega), while the third one has been developed for DNA extraction from formaline-fixed tissue.

45.2.1.2.1 Lab-Based Method

1. Introduce the larva in liquid nitrogen to be repeatedly frozen and thawed.
2. Pulverize the larva directly in a 1.5 mL plastic tube using a sterile pestle.
3. Add 100 µL of Holmes–Bonner solution (7 M urea, 100 mM Tris-HCl pH 8.0, 10 mM EDTA pH 8.0, 350 mM NaCl, 2% SDS); and pestle to complete homogenization.
4. Add 100 µL of phenol–chloroform–isoamyl alcohol (50:49:1) and agitate for 10 min.
5. Centrifuge for 10 min at 13,000 rpm.
6. Transfer the supernatant to a new sterile 1.5 mL tube.
7. Add 100 µL of chloroform and centrifuge for 5 min at 13,000 rpm, then transfer the supernatant to a new sterile 1.5 mL tube.
8. Repeat Step 7.
9. Add 200 µL of absolute ethanol, store at −20°C for 1 h.
10. Centrifuge for 20 min at 13000 rpm, rinse with 200 µL of ethanol 70%, and dry.
11. Resuspend pellet in 100 µL of TE containing RNase and store at room temperature for 30 min.
12. Run 5 µL on agarose gel to test.

45.2.1.2.2 Wizard® SV Genomic DNA Purification System

1. Cut a 0.5–1.2 cm length of tissue or weigh up to 20 mg of tissue sample and put it in a 1.5 mL microcentrifuge tube.
2. Add 275 µL of Digestion Solution Master Mix to each tube

Digestion Solution Master Mix	Volume per Sample
Nuclei lysis solution	500 µL
0.5 M EDTA pH 8.0	120 µL
Proteinase K 20 mg/mL	17.5 µL
RNase A solution 4 mg/mL	3 µL
Total volume	275 µL

3. Incubate the sample tubes overnight (16–18 h) in a 55°C heat block.
4. Add 250 µL of Wizard SV lysis buffer to each sample. Vortex.
5. Process the lysate as soon as possible after adding the lysis buffer. If frozen at −70°C, lysates should be thawed and heated at 55°C for 1 h prior to processing. Lysates must be warm for processing.
6. Transfer each sample lysate from the 1.5 mL tube to a separate Wizard SV Minicolumn Assembly.
7. Spin the Assembly at 13,000× *g* for 3 min.
8. Remove minicolumn from the Assembly and discard the liquid in the collection tube.
9. Replace the minicolumn into the collection tube.
10. Add 650 µL Wizard SV Wash Solution (with 95% ethanol added) to each assembly. Centrifuge at 13,000× *g* for 1 min. Discard the liquid from the collection tube. Repeat this step for a total of four washes.
11. Discard the liquid from the collection tube and reassemble the minicolumn assembly. Centrifuge for 2 min at 13,000× *g* to dry the banding matrix.
12. Transfer the Wizard SV Minicolumn to a new 1.5 mL tube. Add 250 µL of room temperature nuclease-free water. Incubate for 2 min at room temperature.
13. Centrifuge the minicolumn/elution tube assembly at 13.000× *g* for 1 min. Do not discard the liquid in the elution tube.
14. Add an additional 250 µL of nuclease-free water and incubate at room temperature for 2 min. Centrifuge the minicolumn/elution tube assembly at 13.000× *g* for 2 min.
15. Remove the minicolumn and store the purified DNA at −20°C to −70°C

45.2.1.2.3 DNA Extraction from Histological Sections

Since the most frequent finding related to IA is represented by histological section bearing a larva embedded in the paraffin block, the recovery of usable DNA from fixed tissues is a necessary step to carry out a molecular characterization of the specimen. The added value of such application is much more evident in the presence of highly spoiled samples, in which the structure itself of a nematode is hardly recognizable.

1. Cut paraffine block at 10 µm and collected in an autoclaved plastic microtube.
2. Add 1 mL of xylene to the microtube containing tissue sections for 30 min for two changes.
3. Add 100% and 75% ethanol for 30 min with two changes.
4. Wash with PBS for 15 min with two changes.
5. Add 500 µL of lysis buffer (50 µL of proteinase K 20 mg/mL, 10 µL of 1 M Tris-HCl, 2 µL of 0.5 M EDTA, 100 µL of 10% SDS, and distilled water up to 1 mL).
6. Add 500 µL of phenol–chlorophorm–isopropanol alcohol at 25:24:1 to the tissue and mix by vortex.
7. Centrifuge at 12.000× *g* for 10 min.
8. Transfer the supernatant fluid to an autoclaved microtube.
9. Add one volume of chlorophorm to the supernatant and mix by vortex.
10. Centrifuge at 12.000× *g* for 5 min.

11. Carefully remove the fluid supernatant to another fresh microtube.
12. Add 0.1 volume of sodium acetate 3 M to the new tube and mix by vortex.
13. Add one volume of isopropanol and incubate at −20°C overnight.
14. Centrifuge the precipitated DNA at 12.000 × *g* for 15 min at 4°C.
15. Discard the supernatant fluid and wash once with ethanol at 75%.
16. Centrifuge at 12.000 × *g* for 5 min.
17. Dissolve the final yield of DNA in 50 μL of distilled water after drying ethanol completely.

45.2.2 Detection Procedures

45.2.2.1 Primer Set for PCR Amplification

The entire ITS (ITS1, 5.8S rDNA gene, and ITS2) can be amplified in species of the genus *Anisakis* using the primers NC5 (forward: 5′-GTAGGTGAACCTGCGGAAG GATCATT-3′) and NC2 (reverse: 5′- TTAGTTTCTTCCT CCGCT-3′).

PCR amplification of the entire ITS in *Anisakis* species produces a fragment of about 1 kb.

It is possible, for sequencing analysis, to separately amplify the two internal spacers (ITS-1 and ITS-2) using the following primers:

The ITS-1 can be amplified in anisakid species using the primers NC5 (forward: 5′-GTAGGTGAACCTG CGGAAGGATCATT-3′) and NC13R (reverse: 5′-GCTGCGTTCTTCATCGAT-3′). PCR amplification of the ITS-1 in anisakid species produces a fragment of about 300 bp.

The ITS-2 can be amplified in anisakid species using the primers XZ1 (forward: 5′-ATTGCGCCAT CGGGTTCATTCC-3′) and NC2 (reverse: 5′-TTAG TTTCTTTTCCTCCGCT-3′). PCR amplification of the ITS-2 in anisakid species produces a fragment of about 400 bp.

The *rrnS* mitochondrial gene can be amplified using the primers MH3 (forward: 5′-TTGTTCCAGAA TAATCGGCTAGACTT-3′) and MH4 (reverse: 5′-TCTACTTTACTACAACTTACTCC-3′). PCR amplification of the *rrnS* in anisakid species produces a fragment of about 550 bp.

Each primer must be diluted at 50 pmol/μL in sterile H_2O.

45.2.2.2 PCR Protocol

Prepare a PCR mix multiplying each reagent for the number of the individuals to be examined plus two sets for a positive and a negative control (e.g., for five specimens, prepare a complete set for the amplification of seven samples) as follows:

1. Add sequentially, for each sample, 5 μL 10 mM PCR buffer (pH 8.3), 4 μL $MgCl_2$ 1.5 mM, 1 μL dNTPs 40 mM, 0.5 μL of each primer, 0.2 μL Taq DNA polymerase (1.0 U) and sterile H_2O up to 45 μL in a sterile 1.5 mL tube.
2. Add 5 μL of genomic DNA to each 0.2 mL thin-walled tube.
3. Add 45 μL of PCR mix to each 0.2 mL thin-walled tube. Place tubes on ice.
4. PCR cycle for ITS amplification: preamplification cycle at 95°C for 10 min, followed by 30 cycles of 30 s at 95°C, 40 s at 52°C, and 75 s at 72°C, followed by a final elongation of 7 min at 72°C.
5. PCR cycle for *rrnS* amplification: preamplification cycle at 95°C for 10 min, followed by 30 cycles of 30 s at 95°C, 30 sec at 55°C, and 30 s at 72°C, followed by a final elongation of 7 min at 72°C. Place tubes on ice.
6. Electrophoresis: load 5 μL of each amplification reaction to check the quality of amplification products and the fragment size on a 1% ethidiumbromide-stained agarose gel in TBE buffer at 10 V/cm.

45.2.2.3 RFLP Protocol

1. Add sequentially, up to a final volume of 15,2 μL, 10 μL of PCR-amplified DNA, 3 μL of distilled water, 0.5 μL of restriction enzyme (*Hinf*I and *Hha*I), 1.5 μL of enzyme buffer, and 0.2 μL of bovine serum albumin.
2. Incubate for 90 min at 37°C (with the exception of the endonuclease *Taq*I, to be incubated at 65°C).
3. Electrophoresis: load all the reaction volume on a 2% ethidiumbromide-stained agarose gel in TBE buffer at 10 V/cm.

Following the visualization on the gel of the PCR products after digestion with two restriction enzymes (*Hinf*I, and *Hha*I), useful as markers, the identification of larvae or adult at species level is made by comparison with the expected restriction patterns reported in Table 45.2.

TABLE 45.2
Expected Restriction Patterns after Digestion of the ITS Amplified Fragments Using the *Hinf*I and *Hha*I Endonucleases

	Fragment Sizes (bp)	
Species	***Hinf*I**	***Hha*I**
A. simplex s.s.	620–250–80	550–430
A. pegreffii	370–300–250	550–430
A. simplex C	620–250–80	550–300–130
A. typica	610–350	320–240–180–160
A. ziphidarum	370–320–290	550–430
A. nascettii	620–320	550–300–80
A. physeteris	380–290–270	540–420
A. brevispiculata	1000	400–320–200
A. paggiae	1000	520–400

45.2.2.4 Nucleotide Sequencing

The *rrnS* amplified fragments are purified and subjected to direct sequencing. The purification protocol is as follows:

The "NucleoSpin® Extract II (MACHEREY-NAGEL) is suitable for this purpose. The purification protocol is as follows:

1. Adjust DNA binding conditions. Mix two volumes of buffer NT with one volume of sample (e.g., 200 and 100 μL PCR reaction mix). For sample volumes <50 μL adjust the volume of the reaction mix to 50 μL using TE buffer pH 7.5.
2. Bind DNA. Place a NucleoSpin® Extract II column into a 2 mL collecting tube and load the sample. Centrifuge for 1 min at 11,000× *g*. Discard flow-through and place the NucleoSpin® Extract II column back into the collecting tube.
3. Wash silica membrane. Add 600 μL buffer NT3. Centrifuge for 1 min at 11,000× *g*. Discard flow-through and place the NucleoSpin® Extract II column back into the collecting tube.
4. Dry silica membrane. Centrifuge for 2 min at 11,000 × *g* to remove buffer NT3 quantitatively. Make sure the spin column does not come in contact with the flow-through while removing it from the centrifuge and the collecting tube. Residual ethanol from buffer NT3 would inhibit subsequent reactions and has to be removed in this step. In addition to centrifugation, total removal can be achieved by incubation of NucleoSpin® Extract II columns for 2–5 min at 70°C prior to elution.
5. Elute DNA. Place the NucleoSpin® Extract II column into a clean 1.5 mL microcentrifuge tube. Add 15–50 μL elution buffer NE, incubate at room temperature for 1 min to increase the yield of eluted DNA. Centrifuge for 1 min at 11,000× *g*.

Several services for DNA sequencing are available. It can be, in some conditions, more cost-effective to use these services rather than perform sequencing in your own lab. Therefore, a protocol for DNA sequencing is not included. Most sequencing services provide sequence results in different modalities including text files, pdf files, and electropherograms.

The DNA sequences of the *rrnS* mitochondrial gene can be used for species identification. The sequence alignment for species of genus *Anisakis* is reported in Figure 45.1. This method is built up on the basis of the paper published by Abollo et al. [14], where three variable sites were found to discriminate between the two sibling species *A. simplex* s.s. and *A. pegreffii*. Other fixed polymorphic positions are reported in the comparison of the sequences of the species belonging to the genus *Anisakis*, making the *rrnS* mitochondrial gene a new valid molecular marker to identify anisakid nematodes at species level.

45.3 CONCLUSIONS AND FUTURE PERSPECTIVES

The systematics and classification of ascaridoid nematodes and the relationships among genera and species belonging to this superfamily have been the subject of many speculations. The proposed classifications, based mostly on the analysis of morphological traits considered of high systematic significance, are far from unambiguous and are to some extent inconsistent one another. Since the 1980s, studies on the genetic structure of anisakid nematodes have been carried out on a large number of taxa, sampled from many geographic regions from the Boreal and Austral hemispheres. Important achievements of these studies concern the detection of sibling species, the estimation of genetic divergence and of the relationships among taxa, the analysis of patterns of genetic variation in populations and species, and measurement of gene flow. In anisakid nematodes, cladogenetic events were accompanied by poor morphological differentiation, as ecological factors have led to a convergence to similar and well-fitting morpho-functional solutions. This gave origin to a large number of morphologically identical species (sibling species). Also at higher taxonomic level, several morphological characters, even if strikingly differentiating one group from another, appear to be homoplastic and not related to the phylogeny of species or genera. Therefore, morphological traits do not provide definitive evidences to be used as landmarks for the inference of anisakid evolutionary histories. The poor phylogenetic significance of some morphological characters and the occurrence of speciation processes without morphological differentiation undoubtedly advocate for the use of genetic and molecular approaches as efficient tools for the inference of the systematics and evolution of anisakid nematodes [71], and consequently on the diagnosis of human cases at species level.

The discovery of new species in anisakid nematodes seems to be a continuous process, as the number of new sibling species detected by molecular methods increases as new geographical regions or new hosts are studied. This makes the systematics of the genus *Anisakis* and of related anisakid genera in a state of frequent rearrangement.

The main objective of the development of molecular methods for the characterization of anisakid nematodes, and in particular of the species of the genus *Anisakis*, has been so far to provide a quick and reliable tool for the identification of specimens at species level, irrespective of their sex or life history stage.

This has had a strong impact on the systematics of the genus and to the ecology of the different species, but, more importantly, has provided to physicians an effective method for the identification of larvae recovered after endoscopy. There are only few cases in which the nematodes obtained from human infections have been identified at species level, using nuclear [67,68,70] or mitochondrial [69] markers.

The identification of the etiological agent can aid physicians in diagnosis and treatment, in tracing the source of the infestation, in determining the area of origin of the

```
A.pegreffii01          GATATTGATTTTTAATTTTAGGGTGAAATCAGAAATTATTTTTATTTAAACTCTAAATTT
A.pegreffii02          ............................................................
A.simplexs.s.01        .G..........................................................
A.simplexs.s.02        .G..........................................................
A.ziphidarum01         .....................................T.........T............
A.ziphidarum02         .....................................T.........T............
A.nascettii01          .GCG.........................................C.T............
A.typica01             .T..G..........................................T..T........G
A.typica02             .T..G..........................................T..T........G
A.brevispiculata01     TT.T........................................C...............
A.brevispiculata02     TT.T........................................C...............
A.physeteris01         TT.T..........................................A.............
A.physeteris02         TT.T..........................................A.............
A.paggiae01            AC.T..........................................AC............
A.paggiae02            AC.T..........................................AC............

A.pegreffii01          TAAGATTTAGTAAACAAACCAGATTAGTACCTGGTTAGATAAAATTTAAAAGAGCAGGAG
A.pegreffii02          ............................................................
A.simplexs.s.01        ............................................................
A.simplexs.s.02        ............................................................
A.ziphidarum01         ............................................................
A.ziphidarum02         ............................................................
A.nascettii01          ..........C.................................................
A.typica01             .................G..........................................
A.typica02             .................G..........................................
A.brevispiculata01     ............................................................
A.brevispiculata02     ............................................................
A.physeteris01         ..........G.......................C.........................
A.physeteris02         ..........G.......................C.........................
A.paggiae01            ...............G............................................
A.paggiae02            ...............G............................................

A.pegreffii01          TAAAGTTATATTTAAACTGTAAGAATATTGGCAGGTTTTTAAATTATCTTTGGAGGCTGA
A.pegreffii02          ............................................................
A.simplexs.s.01        ............................................................
A.simplexs.s.02        ............................................................
A.ziphidarum01         .......G.G..................................................
A.ziphidarum02         .......G.G..................................................
A.nascettii01          .......G....................................................
A.typica01             .......G................................................T...
A.typica02             .......G................................................T...
A.brevispiculata01     .........G..........................C.......................
A.brevispiculata02     .........G..........................C.......................
A.physeteris01         .........G..................................................
A.physeteris02         .........G..................................................
A.paggiae01            .........G..................................................
A.paggiae02            .........G..................................................

A.pegreffii01          GTAGTGATTGAGAGCCCTCATTTTCAACTTTTTCTTTAGGCGCATGTATGATCGTTTATT
A.pegreffii02          ............................................................
A.simplexs.s.01        ............................................................
A.simplexs.s.02        ............................................................
A.ziphidarum01         .....A.......A..............................................
A.ziphidarum02         .....A.......A..............................................
A.nascettii01          .....A......................................................
A.typica01             .....A.......A........AA.........T.A.T......................
A.typica02             .....A.......A........AA.........T.A.T......................
A.brevispiculata01     .....A......................................................
A.brevispiculata02     .....A.......A..............................................
A.physeteris01         .....A......................................................
A.physeteris02         .....A......................................................
A.paggiae01            .....A.......A............................T.................
A.paggiae02            .....A.......A..............................................

A.pegreffii01          TTATTCTTAAGGATTGTAATTTTAGTCTTTG-CTATAAAAATAGATAGATATCTGGCTTA
A.pegreffii02          ...............................-............................
A.simplexs.s.01        ...............................-............................
A.simplexs.s.02        ...............................-.........C..................
A.ziphidarum01         ...............................-............................
A.ziphidarum02         ...............................-.....................C......
A.nascettii01          .....................C.........-..G.........................
A.typica01             ........................A.T....GT.G.........................
A.typica02             ........................A.T....GT.G.........................
A.brevispiculata01     ...............................-..G.........................
A.brevispiculata02     ...............................-..G.........................
A.physeteris01         ....................C.....T....-..G.....................T...
A.physeteris02         ....................C.....T....-..G......................C..
```

FIGURE 45.1 Alignment of *rrnS* sequences from species of the genus *Anisakis*.

```
A.paggiae01         ..............................-..G..........................
A.paggiae02         ..............................-..G..........................

A.pegreffii01       TGAGAAAGATTTAATTTGGCCTACAATAATTTAATTTTTGGATTCTAGTTTTAGAG-GCT
A.pegreffii02       ........................................................-...
A.simplexs.s.01     ........................................................-...
A.simplexs.s.02     ........................................................-...
A.ziphidarum01      ...AG....................C....G...............G...G.....-AT.
A.ziphidarum02      ...AG....................C....G...............G...G.....-AT.
A.nascettii01       ..G.G...........................GG........CC......G.....-A..
A.typica01          ...A..............A..........A.A.T...G.....G.CT.GC....T.GC.C
A.typica02          ...A..............A..........A.A.T...G.....G.CC.GC....T.GC.C
A.brevispiculata01  ..TA..........................G..T.........CTAG..A.G....GA..
A.brevispiculata02  ..TA..........................G..T.........CTAG..A.G....GA..
A.physeteris01      ..TA........................G...........A..CTAG.CA.G....G..C
A.physeteris02      ..TA........................G...........A..CTGG.CACG....G..C
A.paggiae01         ..TA........................G.G..T.C.......CTAG.GG.....AAA.C
A.paggiae02         ..TA........................G.G..T.C.......CTAG.GG.....AAA.C

A.pegreffii01       AGATGAAATTGTAAAAGACAGTAAGTTGCTTTTAATGGGTGGCTGAAGTTTATTTAAAAA
A.pegreffii02       ............................................................
A.simplexs.s.01     ........................................A...................
A.simplexs.s.02     ........................................A...................
A.ziphidarum01      .................................T......AA..................
A.ziphidarum02      .................................T......AA..................
A.nascettii01       ..G.A.........................CC............C....................
A.typica01          G.C.....................ACCT.....T.....G..T.................
A.typica02          G.C.....................ACCT.....T.....G..T.................
A.brevispiculata01  TTT........................A.....T..A...AA................G.
A.brevispiculata02  TTT........................A.....C..A...AA................G.
A.physeteris01      TAG........................A.....T..A...AA..................
A.physeteris02      TAG........................A.....T..A...AA..................
A.paggiae01         TAG........................A.....T..A...AA..................
A.paggiae02         TAG........................A.....T..A...AA..................

A.pegreffii01       CGGTACAAATCATCCATCAATTGCCTATAGGGGAGTGAGTT
A.pegreffii02       .........................................
A.simplexs.s.01     ...................T.....................
A.simplexs.s.02     .........................................
A.ziphidarum01      ....................................A....
A.ziphidarum02      ....................................A....
A.nascettii01       ....................................A....
A.typica01          ....................................A....
A.typica02          ....................................A....
A.brevispiculata01  ....................................A....
A.brevispiculata02  ....................................A....
A.physeteris01      ....................................A....
A.physeteris02      ....................................A....
A.paggiae01         ....................................A....
A.paggiae02         ....................................A....
```

FIGURE 45.1 (Continued) Alignment of *rrnS* sequences from species of the genus *Anisakis*.

contaminated food, and, especially, to provide a definitive diagnosis in the case of recovery of a small portion of a larva or of highly spoiled specimens in histological sections. It should be underlined that the occurrence of small portion of parasite material in fishes, even if adequate freezing and/or cooking can kill or inactivate anisakid larvae, can still cause sensitization and IgE-dependent hypersensitivity in humans.

However, these investigations on the identification of the larvae from human infections are based on recovered larvae after endoscopy (gastric) or histology (intestinal) and represent only partially a significant step forward as usable tools for the primary diagnosis of anisakiasis, as they are posterior to endoscopical or tissue-based observations.

In this context it appears needed to promote efforts in the direction of establishing direct methods for the detection of the parasite (or part of it) in human infections. A real-time PCR method for the detection and quantification of anisakid nematodes in fish products has been recently developed [72]. Lopez and Pardo [73] have also described a method based on real-time PCR applied to *A. simplex* allergens found in seafood products. Such methods may be, with some modifications, adapted to improve the diagnosis of human anisakidosis or for the detection of specific immune response to *Anisakis*.

It would also be interesting to investigate the reliability and effectiveness of methods that couple molecular and immunological approaches in a single diagnostic tool (e.g., immuno PCR).

Significant advancements in molecular diagnosis of infectious diseases are expected by the development of new miniaturized platforms for the analysis of nucleic acids and proteins. Moreover, DNA-chips may represent also an efficient technology for the rapid detection of genetic variations within pathogenic species.

REFERENCES

1. Fagerholm, H.P. Systematic implications of male caudal morphology in ascaridoid nematode parasites. *Syst. Parasitol.* 19, 215, 1991.
2. Mattiucci, S. et al. Genetic structure of *Anisakis physeteris*, and its differentiation from the *Anisakis simplex* complex (Ascaridida: Anisakidae). *Parasitology* 93, 83, 1986.
3. Mattiucci, S. et al. Genetic and ecological data on the *Anisakis simplex* complex, with evidence for a new species (Nematoda, Ascaridoidea, Anisakidae). *J. Parasitol.* 86, 401, 1997.
4. Nascetti, G. et al. Electrophoretic studies on the *Anisakis simplex* complex (Ascaridida: Anisakidae) from the Mediterranean and North-East Atlantic. *Int. J. Parasitol.* 16, 633, 1986.
5. Berland, B. Nematodes from some Norwegian marine fishes. *Sarsia* 2, 1, 1961.
6. Mattiucci, S. et al. Allozyme and morphological identification of *Anisakis*, *Contracaecum* and *Pseudoterranova* from Japanese waters (Nematoda, Ascaridoidea). *Syst. Parasitol.* 40, 81, 1998.
7. Mattiucci, S. et al. Genetic divergence and reproductive isolation between *Anisakis brevispiculata* and *Anisakis physeteris* (Nematoda: Anisakidae). *Int. J. Parasitol.* 31, 9, 2001.
8. Mattiucci, S. et al. Genetic markers in the study of *Anisakis typica* (Diesing, 1860): Larval identification and genetic relationships with other species of *Anisakis* Dujardin, 1845 (Nematoda: Anisakidae). *Syst. Parasitol.* 51, 159, 2002.
9. Mattiucci, S. et al. Evidence for a new species of *Anisakis* Dujardin, 1845: Morphological description and genetic relationships between congeners (Nematoda: Anisakidae). *Syst. Parasitol.* 61, 157, 2005.
10. Mattiucci, S. et al. *Anisakis nascettii* n. sp. (Nematoda: Anisakidae) from beaked whales of the southern hemisphere: Morphological description, genetic relationships between congeners and ecological data. *Syst. Parasitol.* 74, 199, 2009.
11. Paggi, L. et al. A new species of *Anisakis* Dujardin, 1845 (Nematoda: Anisakidae) from beaked whales (Ziphiidae): Allozyme and morphological evidence. *Syst. Parasitol.* 40, 161, 1998.
12. D'Amelio, S. et al. Genetic markers in ribosomal DNA for the identification of members of the genus *Anisakis* (Nematoda: Ascaridoidea) defined by polymerase chain reaction based restriction fragment length polymorphism. *Int. J. Parasitol.* 30, 223, 2000.
13. Abollo, E. et al. Occurrence of recombinant genotypes of *Anisakis simplex* s.s. and *Anisakis pegreffii* (Nematoda: Anisakidae) in an area of sympatry. *Infect. Genet. Evol.* 3, 175, 2003.
14. Pontes, T. et al. Molecular characterization of larval anisakid nematodes from marine fishes of Madeira by a PCR-based approach, with evidence for a new species. *J. Parasitol.* 91, 1430, 2005.
15. Marques, J.F. et al. Molecular identification of *Anisakis* species from Pleuronectiformes off the Portuguese coast. *J. Helminthol.* 80, 47, 2006.
16. Palm, H.W. et al. Molecular genotyping of *Anisakis* Dujardin, 1845 (Nematoda: Ascaridoidea: Anisakidae) larvae from marine fish of Balinese and Javanese waters, Indonesia. *Helminthologia* 45, 3, 2008.
17. Davey, J.T. A revision of the genus *Anisakis* Dujardin, 1845 (Nematoda: Ascaridida). *J. Helminthol.* 45, 51, 1971.
18. Kurochkin, Y.V. Parasites of the Caspian seal *Pusa caspica*. Rapport et proces-verbaux des reunion. *Cons. Int. Expl. Mer.* 169, 363, 1975.
19. Mattiucci, S., and Nascetti, G. Molecular systematics, phylogeny and ecology of anisakid nematodes of the genus *Anisakis* Dujardin, 1845: An update. *Parasite* 13, 99, 2006.
20. Mattiucci, S., and Nascetti, G. Genetic diversity and infection levels of anisakid nematodes parasitic in fish and marine mammals from Boreal and Austral hemispheres. *Vet. Parasitol.* 148, 43, 2007.
21. D'Amelio, S. et al. Molecular genotyping of anisakid nematodes from cetaceans of Florida waters. *European Cetacean Society Annual Conference*, Cetacean Social Organization. La Rochelle (France) 2–7 Aprile 32, 2005.
22. Umehara, A. et al. Molecular identification of *Anisakis simplex* sensu stricto and *Anisakis pegreffii* (Nematoda: Anisakidae) from fish and cetacean in Japanese waters. *Parasitol. Int.* 55, 267, 2006.
23. Farjallah, S. et al. Occurrence and molecular identification of *Anisakis* spp. from the North African coasts of Mediterranean Sea. *Parasitol. Res.* 102, 371, 2008.
24. Iglesias, R. et al. Molecular and morphological evidence of a new taxon of *Anisakis* Dujardin, 1845 (Nematoda, Anisakidae) from the Blainville's beaked whale (*Mesoplodon densirostris*). *J. Helminthol.* 82, 305, 2008.
25. Oshima, T. *Anisakis*-Is the sushi bar guilty? *Parasitol. Today* 3, 44, 1987.
26. Kasuya, S. et al. Mackerel-induced urticaria and *Anisakis*. *Lancet* 335, 665, 1990.
27. Kagei, N., and Isogaki, H. A case of abdominal syndrome caused by the presence of a large number of *Anisakis* larvae. *Int. J. Parasitol.* 22, 251, 1992.
28. Audicana, M.T. et al. *Anisakis simplex*: Dangerous dead and alive? *Trends Parasitol.* 18, 20, 2002.
29. Audicana, M.T., and Kennedy, M.W. *Anisakis simplex*: From obscure infectious worm to inducer of immune hypersensitivity. *Clin. Microbiol. Rev.* 21, 360, 2008.
30. Verhamme, M.A., and Ramboer, C.H. Anisakiasis caused by herring in vinegar: A little known medical problem. *Gut.* 29, 843, 1988.
31. Daschner, A. et al. Gastric anisakiasis: An underestimated cause of acute urticaria and angioedema. *Br. J. Dermatol.* 139, 822, 1998.
32. Fernández de Corres, L. et al. *Anisakis simplex* induces not only anisakiasis: Report n 28 cases of allergy caused by this nematode. *J. Investig. Allergol. Clin. Immunol.* 6, 315,, 1996.
33. Del Pozo, M.D. et al. *Anisakis simplex*, a relevant etiologic factor in acute urticaria. *Allergy* 52, 576, 1997.
34. Ohtaki, H., and Ohtaki, R. Clinical manifestation of gastric anisakisis. In: Ishikura, H., and Namiki, M. (eds), *Gastric Anisakiasis in Japan*, Springer-Verlag, Tokyo, p. 37, 1989.
35. Shibata, O., Uchida, Y., and Furusawa, T. Acute gastric anisakiasis with special analysis of the location of the worms penetrating the gastric mucosa. In: Ishikura, H., and Namiki, M. (eds) *Gastric Anisakiasis in Japan* Springer-Verlag, Tokyo, p. 53, 1989.

36. Ishikura, H., Kikuchi, Y., and Hayasaka, H. Pathological and clinical observation on intestinal anisakiasis. *Arch. Jpn. Chir.* 36, 663, 1967.
37. Rosales, M. et al. Acute intestinal anisakiasis in Spain: A fourth stage *Anisakis simplex* larva. *Mem. Inst. Oswaldo Cruz* 94, 823, 1999.
38. Moschella, C.M. et al. Intestinal anisakiasis in Italy: A case treated by emergency surgery. *G. Chir.* 26, 201, 2005.
39. Ishikura, H., and Kikuchi K. *Intestinal Anisakiasis in Japan*, Springer-Verlag, Tokyo, 1990.
40. Rodriguéz, E. et al. Novel sequences and epitopes of diagnostic value derived from the *Anisakis simplex Ani s 7* major allergen. *Allergy* 63, 219, 2008.
41. Anibarro, B., and Seoane, F.J. Occupational conjunctivitis caused by sensitizazion to *Anisakis simplex*. *J. All. Clin. Immunol.* 102, 331, 1998.
42. Armentia, A. et al. *Anisakis simplex* allergy after eating chicken meat. *J. Investig. Allergol. Clin. Immunol.* 16, 258, 2006.
43. Kikuchi, Y., Ishikura, H., and Kikuchi, K. Pathology of intestinal anisakiasis. Ishikura, H., and Kikuchi K. (eds), In *Intestinal Anisakiasis in Japan*. Springer-Verlag, Tokyo, p. 129, 1990.
44. Daschner, A. Usefulness of early serial measurement of specific and total IgE in the diagnostico of gastro-allergic anisakiasis. *Clin. Exp. All.* 29.1260, 1999.
45. Daschner, A. What does *Anisakis simplex* parasitism teach us about interpretating specific and total IgE values? *Allergol. Immunopathol.* 28, 67, 2000.
46. Moneo, I. et al. Isolation and characterization of a major allergen from the fish parasite *Anisakis simplex*. *J. All. Clin. Immunol.* 106, 177, 2000.
47. Shikamura, K. et al. Purification and molecular cloning of a major allergen from *Anisakis simplex*. *Mol. Biochem. Parasitol.* 135, 69, 2004.
48. Moneo, I. et al. Sensitization to the fish parasite *Anisakis simplex*: Clinical and laboratory aspects. *Parasitol. Res.* 101, 1051, 2007.
49. Moneo, I. et al. Isolation of a heat-resistant allergen from the fish parasite *Anisakis simplex*. *Parasitol. Res.* 96, 285, 2005.
50. Anádon, A.M. et al. The *Anisakis simplex Ani s 7* major allergen as an indicator of true *Anisakis* infections. *Clin. Exp. Immunol.* 156, 471, 2009.
51. Paggi, L. et al. Biochemical taxonomy of ascaridoid nematodes. *Parassitologia* 27, 105, 1985.
52. Paggi, L. et al. Genetic evidence for three species within *Pseudoterranova decipiens* (Nematoda, Ascaridida, Ascaridoidea) in the North Atlantic and Norwegian and Barents Seas. *Int. J. Parasitol.* 21, 195, 1991.
53. Paggi, L. et al. Nematodi del genere *Anisakis* in pesci, cefalopodi e cetacei del Mar Mediterraneo e dell'Oceano Atlantico e Pacifico. *Biol. Mar. Medit.* 5, 1585, 1998.
54. Paggi, L. et al. *Pseudoterranova decipiens* species A and B (Nematoda, Ascaridoidea): Nomenclatural designation, morphological diagnostic characters and genetic markers. *Syst. Parasitol.* 45, 185, 2000.
55. Bullini, L. et al. Genetic variation of ascaridoid worms with different life cycles. *Evolution* 40, 437, 1986.
56. Bullini, L. et al. Genetic and ecological studies on nematode endoparasites of the genera *Contracaecum* and *Pseudoterranova* in the Antarctic and Arctic-Boreal regions. In: *Proceedings of the 2nd Meeting on "Biology in Antarctica." Scienza e Cultura*, Numero Speciale. Edizioni Universitarie Patavine, p. 131, 1994.
57. Nascetti, G. et al. Three sibling species within *Contracaecum osculatum* (Nematoda, Ascaridida, Ascaridoidea) from the Atlantic Arctic-Boreal region: Reproductive isolation and host preferences. *Int. J. Parasitol.* 23, 105, 1993.
58. Orecchia, P. et al. Electrophoretic identification of larvae and adults of *Anisakis* (Ascaridida: Anisakidae). *J. Helminthol.* 60, 331, 1986.
59. Orecchia, P. et al. Two new members in the *Contracaecum osculatum* complex (Nematoda, Ascaridoidea) from the Antarctic. *Int. J. Parasitol.* 24, 367, 1994.
60. D'Amelio, S. et al. Genetic variation of a population of the *Pseudoterranova decipiens* complex from Japan. In: *Abstracts of the VIth European Multicolloquium of Parasitology*, The Hague, The Netherlands, 7, 45, 1992.
61. Paggi, L., and Bullini, L. Molecular taxonomy in anisakids. Symposium on "Parasites of biological and economical significance in the aquatic environment". *Bull. Scand. Soc. Parasitol.* 4, 25, 1994.
62. Arduino, P. et al. Isozyme variation and taxonomic rank of *Contracaecum radiatum* from the Antarctic Ocean (Nematoda, Ascaridoidea). *Syst. Parasitol.* 30, 1, 1995.
63. Farjallah, S. et al. Molecular characterization of larval Anisakid nematodes from marine fishes off the Moroccan and Mauritanian coasts. *Parasitol. Int.* 57, 430, 2008.
64. Umehara, A. et al. Multiplex PCR for the identification of *Anisakis simplex* sensu stricto, *Anisakis pegreffii* and the other anisakid nematodes. *Parasitol. Int.* 57, 49, 2008.
65. Abe, N. Application of the PCR-sequence-specific primers for the discrimination among larval *Anisakis simplex* complex. *Parasitol. Res.* 102, 1073, 2008.
66. Valentini, A. et al. Genetic relationships among *Anisakis* species (Nematoda: Anisakidae) inferred from mitochondrial cox2 sequences, and comparison with allozyme data. *J. Parasitol.* 92, 156, 2006.
67. D'Amelio, S. et al. Diagnosis of a case of gastric anisakidosis by PCR-based restriction fragment length polymorphism analysis. *Parassitologia* 41, 591, 1999.
68. Farjallah, S. et al. Diagnosis of a case of oesophageal anisakidosis by PCR-based restriction fragment length polymorphism analysis. *Parassitologia* 49, 217, 2007.
69. Mattiucci, S. et al. Human Anisakidosis in Italy: Molecular and histological identification of two cases. *Parassitologia*, 49, 226, 2007.
70. Umehara, A. et al. Molecular identification of the etiological agents of the human anisakiasis in Japan, *Parasitol. Int.* 56, 211, 2007.
71. D'Amelio, S. Phylogeny of anisakid nematodes: A review. *Helminthologia* 40, 87, 2003.
72. Mossali, C. et al. Sensitive detection and quantification of anisakid parasite residues in food products. *Foodborne Pathog. Dis.* 7, 391, 2010.
73. Lopez, I., and Pardo, M.A. Evaluation of a real time polymerase chain reaction (PCR) assay for detection of anisakis simplex parasite as a food-borne allergen source in seafood products. *J. Agric. Food Chem.* 58, 1469, 2010.

46 Ascaris

Dwight D. Bowman and Janice L. Liotta

CONTENTS

46.1 INTRODUCTION

46.1.1 Classification, Morphology, and Biology

Ascaris is a genus of nematode parasites that live in the small intestine of human and porcine hosts. Besides the two well-known species, *Ascaris lumbricoides* and *Ascaris suum* in people and pigs, respectively, other species of *Ascaris* have been described upon occasion. It appears, however, that the only two well-accepted species in the genus are the species in people and pigs. These forms have been observed many times, and are readily recognized as representative of the genus *Ascaris*. Some of the other described species, such as *Ascaris ovis*, are probably based on descriptions of worms from people or pigs that have undergone spurious development in atypical hosts. Some descriptions were based on misidentifications, such as *Ascaris castoris* of the beaver [1]. A majority of the described species represent forms that were either assigned to the genus *Ascaris* incorrectly or at a time when the distinctions between ascaridoid genera were not clearly drawn by many authors; this is what occurred with *Baylisascaris columnaris*, *Baylisascaris devosi*, *Baylisascaris procyonis*, *Baylisascaris schroederi*, *Bairdascaris dasypodina*, *Toxocara hippopotami*, and *Toxocara suricattae* [2–5]. Other species in rodents, such as *Ascaris joffi*, *Ascaris spalacis*, and *Ascaris tarbagan* (re-examined morphologically recently by Tenora and Shalayeva) [6], await confirmation with additional descriptions including molecular comparisons with other ascaridoid nematodes to verify their specific status. The species from the capuchin monkey, *Ascaris cebi*, needs to be redescribed. Thus, as of now, it appears that the genus *Ascaris* consists of only two valid species, *A. lumbricoides* and *A. suum*. The identification of a valid species in a rodent would be of significant interest in terms of the biology and evolution of the group.

The genus *Ascaris* is a member of the superfamily Ascaridoidea. According to Anderson [7], this is a superfamily within the order Ascaridida (along with the other related superfamilies Cosmocercoidea, Seuratoidea, Heterakoidea, and Subuluroidea). The Ascaridoidea has two families, the Anisakidae that are associated with aquatic environs and the Ascarididae that includes forms associated with terrestrial environs. The subfamily Ascaridinae contains *Ascaris*, *Bairdascaris*, *Baylisascaris*, *Lagochilascaris*, *Parascaris*, and *Toxascaris*. There is some thought that the Angusticaecinae that occur mainly in terrestrial reptiles should be placed together with the Ascarididae, but this remains unresolved [3]. The overview of the Nematoda based on small subunit (SSU) phylogenetic analysis places *A. suum* within the monophyletic Ascaridomorpha within the Spirurina of Clade A [8].

The species *A. lumbricoides* was described by Linnaeus in 1758, and *A. suum* was described by Goeze in 1782. The worms were considered virtually indistinguishable until 1952 when Sprent described detectable differences in the labial denticles. Others have since verified these results [9–13], but others have been unable to reproduce the morphological separation based on these same criteria [14]. Little [15] found that the denticles of worms from pigs were markedly larger than those of similar-sized worms from people, but that individual worms of similar size could not be distinguished as to species because the size and shape of the denticles varied with the age of the worms; the denticles wear as the worms age, especially in the center of the lips. However, Little also examined the shapes of the copulatory spicules from equal-sized worms and found that the spicules of *A. suum* worms from pigs were distinctly thinner and sharper at the tip than *A. lumbricoides* worms recovered from people [15]. Thus, it does appear that with careful and detailed comparative observations on the morphology of the worm, one can distinguish between these two species and with some certainty categorize a recovered worm as being either from pigs or from people. As will become evident from the text below, it appears that modern molecular methods have added significant support to the species being distinct biological entities that are for the most part restricted to their respective hosts with *A. lumbricoides* in people and *A. suum* in pigs. At the same time, the development of modern methods that would allow for the worms (or their eggs) to be rapidly assigned to either a pig or a human source would be welcome. The usefulness may not be great in human medicine where the specific diagnosis will not be likely to change the treatment of an individual patient, but in the realm of environmental sampling being able to determine whether an egg is from a pig or a person could means a great deal in terms of assessing potential sources of environmental contamination.

A single morphological description suffices for worms from pigs and people because of the morphological similarity of the two species. The adult worms are cream colored to pinkish and live within the small intestine of the host. Fully mature adult males are 15–31 cm long and 0.2–0.4 cm wide; fully mature adult females are 20–35 cm long (sometimes, adult females can reach lengths up to 49 cm) and 0.3–0.6 cm wide. The adults have three fleshy lips on the anterior end: a single dorsal lip with symmetrical and large lateral papillae, and a pair of subventral lips with large ventral papillae and small lateral papillae next to a fairly obscure amphidial opening. The anterior inner margin of the lips bears the small denticles that were initially found to differ between the two species. There are no cervical alae, that is, no cuticular wings along the lateral lines of the anterior end; such cervical alae are very prominent in the related genera *Toxocara* and *Toxascaris*. On fresh worms, light-colored lines can be observed along both the sides of the body that represent the syncytial lateral cords of the hypodermis. The shorter male has a tail that curls ventrad with the cloacal opening being near the posterior end of the body. The male tail has numerous preanal and postanal papillae with each typically having a respective pair on the other side of the ventral line of the body. Some of the caudal papillae are paired (double papillae) and some are singular; toward the anterior end of the worms the papillae may not remain symmetrical in number or location on the two sides of the worm. The tail does not have any rugose areas anterior to or posterior to the cloacal opening. A relaxed male *Ascaris* has the appearance of a shepherd's crook, with the posterior end of the body being curved around toward the anterior end via a hook-like bend in the ventral direction. The curved, posterior end of the male will curl around the female at the level of the vulva during copulation. The tail of the female tends to be straight with only a slight prolongation posteriad to the anus and with no significant ornamentation.

The reproductive system of these nematodes is massive to support the huge numbers of eggs that are created by the female on a daily basis. The reproductive tract of the male is a single tubular structure; the testis that coils through the anterior portion of the body empties into the cloaca through the glandular ejaculatory duct. In the cloaca are paired copulatory spicules that are solid and stout. These are inserted into the female during copulation; the size and shape of the spicules match the length and shape of the vagina vera (the cuticle-lined portion of the vagina) of the female. The sperm of *Ascaris* that is transferred during copulation, like other nematodes, is amoeboid rather than flagellate and moves with amoeba-like locomotion [16]. The vulva of the female opens on the ventral line about one-third of the body length toward the posterior end; the vulva can usually be observed without the need for additional magnification. There may or may not be a constriction around the body of the female at the level of the vulval opening that is believed to represent a guide for mating that is used by the male [17]; this is not a constriction caused by the embrace of the male as had been suggested by others. The female reproductive tract is opisthodelphic; the reproductive tract, both branches, run posteriad initially from the vulval opening. The female reproductive tract consists of the vagina (the vagina vera and vagina uterina) that runs posteriad into the uterus that branches into two genital tubules that each consists of a uterus, seminal receptacle, oviduct, and ovary. These tubules coil around through the middle and poster thirds of the body and are both several times the length of the worm. The combined uteri contain as many as 27 million eggs at any one time, and it has been shown that a single female can produce 200,000 eggs/day [18]. In a study in Brazil, it was found that in children the average output per female worm was about 3000 eggs/g or 3 eggs/mg [19].

Diagnosis of *A. lumbricoides* infections in people is usually relatively simple and straightforward. Whether adult worms are passed spontaneously or after treatment, their large size and light color make them apparent and unmistakable in the feces. Also, the eggs are produced by the female worms in such large numbers that diagnosis with typical fecal examination methods should almost never overlook an infection where only a few worms are present as long as one of the worms is a mature female. Female worms that have not been fertilized or which have utilized all the sperm from the last copulation will continue to produce eggs that can be recognized in feces by the trained observer who should be able to distinguish infertile from fertilized eggs. There are certainly cases where a few males are the only worms present in a host, but typically, this is an infection that can readily be diagnosed using simple fecal sedimentation methods or fecal flotation methods.

A. lumbricoides is a monoxenous parasite; that is, there is a single host involved in the life cycle. This parasite does not utilize any known paratenic hosts, including earthworms [20]. The infection of a host is initiated with the ingestion of an egg that contains a third-stage larva. There was considerable debate at one time as to the number of molts that occurred within the egg of *Ascaris* and the other ascaridoid nematodes. This has been discussed by a number of authors at significant length, and they are referred to in the text *Nematode Parasites of Vertebrates, Their Development and Transmission* by Roy C. Anderson who gave the issue a great deal of consideration as he prepared the second edition of this text [7]. Basically, this is an issue that should probably be considered at rest at this time.

The third-stage larva hatches from the eggshell in the intestine. It was believed that the majority of larvae hatched and penetrated the host through the small intestine, but it was recently shown in pigs that *A. suum* larvae actually enter the body through the mucosa of the cecum and colon [21]. After entering the mucosa, the larvae move to the liver via the hepatic portal system. From the liver, the larvae make their way to the lungs by 5–6 days after infection, break into the alveoli and migrate up the trachea to be swallowed, and are found in the intestine by 8 or 9 days after infection. The molt from the third to fourth stage occurs as the larvae migrate out of the lungs or after the larvae have entered the intestine and the intestinal mucosa around 10 or so days after infection. The molt to the adult stage occurs about 3–4 weeks after the ingestion of the eggs when the worms are about a 1 cm long in the intestinal lumen. The prepatent period is about 8 or 9 weeks, and at the beginning of patency, the females are about 15–20 cm long [22]. The adult *A. lumbricoides* can live as long as 20 months [23], but the usual life span is a year or less.

Humans tend to be the only host routinely infected naturally with *A. lumbricoides*. Experimentally, it has been possible to infect pigs with the eggs of *A. lumbricoides* from humans [24]. Also, the occurrence of sporadic cases of ascariasis, mainly in children, in areas where *A. lumbricoides* was not endemic, led people to presume that the worms when recovered were *A. suum* from pigs in areas where hogs were raised [25]. There have been instances where worms indistinguishable from *A. lumbricoides* have been recovered from nonhuman primates including the gorilla, orangutan, gibbon, macaques, and chimpanzees that have been infected sometimes with fatal heavy infections [26]. It has also been reported that some 40% of chimpanzees that were given 50–750 eggs of *A. lumbricoides* that had been recovered from human feces and cultured to the infective stage developed patent infections about 9 weeks later and that fertile eggs were found in the feces for several months [25]. Overall, it appears that the only natural reservoir of *A. lumbricoides* maintained in nature is in people.

Adult female *Ascaris* worms produce two types of eggs that can be found in the feces: those eggs that are recognized as being from fertile females and infertile eggs from young females that have not yet mated or females that have depleted their stored sperm. The eggs produced by fertilized female tend to be spherical to ovoid with thick, transparent shells. The shell is composed of a relatively impermeable lipoidal inner layer, a thick transparent middle layer, and an external mammillated albuminous layer that is applied by the female worm and which in feces is often stained golden brown by bile. These ovoid eggs measure about 50 μm × 75 μm and, when spherical, have a diameter of about 65 μm. The fertilized egg contains a single large nucleated zygote and is passed before any nuclear division has occurred. Mating occurs periodically within the host at undetermined intervals, and females can produce infertile eggs early in patency and periodically throughout the patent period. The infertile eggs tend to be elongate, up to 90 μm in length, and have a thin middle layer to the shell and often only a minimal albuminous layer. The infertile egg contains a mass of disorganized granules of various sizes. Typically, even a single infertile female can produce sufficient eggs for diagnosis by routine egg recovery methods.

The eggs that are passed in the feces require oxygen in order for development to continue. Eggs can be stored in the absence of oxygen for at least 4 months, and they will commence cell division and development when exposed to air [27]. Development to the infective stage occurs most efficiently at temperatures between 22°C and 30°C in moist shady locations. At 28°C, it takes about 18 days for the eggs to have developed to the fully infective third-stage larva. The eggs develop and survive best in soils with high clay content as opposed to highly sandy soils. The eggs in soil can persist for years when the soil protects them from desiccation, freezing, and direct sunlight. Persistence in soil has been found to be as long as 10–15 years [28,29]. It appears that many eggs in the soil are destroyed over time by soil-inhabiting fungi and other organisms.

46.1.2 Clinical Features and Pathogenesis

Ascariasis is not without effect on its human host. Sufficient evidence now exists to demonstrate that ascariasis is a significant public health problem causing debilitating morbidity which impairs the growth and development of several hundred million children in developing countries [30]. The major signs of infection are of three major types, those due to larval migration, mature worms in the intestine, and the unfortunate complications that may arise from atypical migration of adult worms or the presence of very large numbers of worms [31]. The migration of the larvae through the lungs on the way to the intestinal tract can cause pneumonitis, asthma, dyspnea, cough, and substernal pain. The worms living within the intestinal tract cause abdominal pain and distension, colic, nausea, vomiting, intermittent diarrhea, anorexia, restlessness, anal itching, enterocolitis, disordered small bowel patter, jejunal mucosal abnormalities, and vitamin A malabsorption. There can be complications and medical emergencies from the atypical aggregation or migration of adult worms in the intestine, resulting in intestinal obstruction, intussusception, volvulus, invasion of the bile duct, or acute pancreatitis, appendicitis, intestinal perforation, or peritonitis. Also, there can be hypersensitivity reactions to allergic antigens produced by the worms. In the nonlife-threatening cases, it is fairly well accepted that the infection causes decreased food intake and increased nutrient loss [32,33]. The intestinal pathology in infected children includes villous atrophy and cellular infiltration of the lamina propria of the intestine [34]. Treatment leads to improved appetite and weight gain, and anthelmintic treatment can improve growth rates when given to malnourished children with ascariasis. Increases in appetite and nutrient intake following anthelmintic treatment are the nutritional benefits most likely associated with worm removal [31]. In the field of veterinary medicine relative to nutritional deficits associated with *A. suum*, it has long been considered that infections with this parasite decrease the rate of gain of swine, causing farmers financial loss if the infection is not prevented or controlled [35].

Helminths, like most parasites within a community of hosts, tend to be clumped in a few hosts; that is, certain hosts have a number of worms, while a majority of hosts have a few to no worms [36,37]. This is also called overdispersion, and this prevalence of parasites within host populations is represented mathematically by a negative binomial distribution. It has also been found that those hosts that are infected with a large number of worms are often predisposed to becoming reinfected with a large number of worms after they have been cleared of their infections. This condition is known as predisposition. It has been shown in the field to occur in people infected with *A. lumbricoides* [38]. It has also been shown using pigs that the same condition occurs in pigs with *A. suum*; that is, there is a positive correlation between the number of worms present in individual pigs prior to treatment and after treatment when they are returned to pasture [39]. This has important treatment ramifications because it may be necessary to concentrate on certain "wormy persons" to successfully implement mass chemotherapy programs within a community [37].

Massive fatal infections in individual children represent the worst possible scenario in ascariasis, a disease that can be prevented by improved sanitation and chemotherapy with numerous products that are safe and efficacious. The statement by Little [40] about the general clinical condition associated with ascariasis is worth repeating. "When present in small numbers, adult worms generally cause no symptoms, although even a single ectopic worm may occasionally produce serious disease. Average infections in children cause intermittent colic, loss of appetite, fretfulness, and at times nervous symptoms. The abdomen is characteristically protuberant. The nutritional demands and space requirements of massive infections may be great, as in the case of a 2-year-old girl seen at autopsy in New Orleans, who weighed only 8 kg after having passed 80 worms in the hospital and an unrecorded number earlier and who still harbored 578 worms which weighed more than 800 g. The estimated volume displacement of the total mass of worms was ~1 L. Heavy *Ascaris* infection can lead to significant nutritional impairment, especially in children whose protein intake is low."

Ascariasis remains one of the most common infections of people throughout the world, mainly in people living under impoverished conditions or in situations where there is minimal sewage collection and treatment. The most recent estimate is that there are some 1.4 billion people infected or about 20% of the earth population [41]. A 1990 estimate was that some 29% of the world preschool age children (158 million) and 35% of the world's school-age children (320 million) are infected with *A. lumbricoides* [31]. It can be expected that the number of infections will decrease as the fertility rate associated with modernization continues to decrease and with increasing numbers of people moving out of subsistence life styles into areas with adequate sewage systems. In countries such as Japan, the decrease in infection rate may have been hastened by mandatory treatment of school children [42]; but Korea has seen a remarkable decline in cases without mandatory intervention [43]. It can be expected that marked declines in prevalence are also likely to occur in China, Brazil, and parts of India if their current rates of economic growth and prosperity continue on their current positive trajectories. Thus, to a great extent, infections with *A. lumbricoides* serve as a general measure of economic well-being among the world communities. Those nations with improved economic status and increased community infrastructures are going to have a lower prevalence of infection with this common infection.

46.1.3 Diagnosis

46.1.3.1 Conventional Techniques

As stated above, the detection of infections with *A. lumbricoides* is straightforward for the most part. If the adult worms are passed posttreatment, they are easily identifiable as to genus; a specific diagnosis would require fastidious study of the details of the denticles and spicules (if a male is present) or molecular methodologies.

For the routine diagnosis of infections, most laboratories today are performing sedimentation procedures using formalin-fixed specimens (for an in-depth discussion of all fecal examination methods, see Garcia) [44]. Other tests that are used are designed for use with fresh feces. These include the direct fecal smear, the Kato Thick Smear, and the zinc sulfate flotation. It should be noted that the required use of fresh and unfixed samples is often not the choice of many technical staff these days in our risk-adverse societies. There is an additional myriad of different methods for the microscopical diagnosis of parasites in fecal samples, and each has its proponents and detractors, and parasitologists can argue for hours about the details of any given test and the nuances of how an individual step in a procedure should be performed. Overall, this becomes almost as much an art as a science, but in the hands of an experienced and trained person, there are techniques that should be able to diagnose an *A. lumbricoides* infection in almost all cases. Currently, during the prepatent period, a specific diagnosis is not possible unless immature worms are expelled from the feces spontaneously or following treatment.

The routinely used formalin–ethyl acetate sedimentation procedures will prove successful for both fertile and infertile eggs. In this procedure, the feces that is fixed in 10% formalin is comminuted and sieved and placed in a conical centrifuge tube, typically a 15 mL tube. The tube is filled partially with formalin and then several milliliters of ethyl acetate are added. The tube is then shaken and centrifuged at 800 × *g*. A plug of detritus will form at the interface during centrifugation, and after the tube is removed from the centrifuge, the plug is dislodged from the wall using an applicator and the supernatant is discarded. The pellet is then examined microscopically for the presence of eggs.

A single female produces about 200,000 eggs per day, and this is sufficient to have several characteristic eggs present in an average direct fecal smear that usually utilizes 2 mg of feces. Basically, the 2 mg or so of feces is mixed with saline on a slide using a wooden applicator stick or equivalent, and then a coverslip is placed on the mixture. The entire field should be examined, and this method will detect both fertile and infertile *Ascaris* eggs.

The Kato Thick Smear technique has been shown to produce meaningful correlations between egg counts and patient worm burdens if performed in a quantitative manner [45]. This method was used to verify success in the mass treatment campaign in Japan directed at school children [46]. This is a method whereby a 50 mg sample of feces is screened through a 105-mesh stainless-steel cloth and then placed on a slide. To the slide is added a cellophane coverslip that has been soaked in glycerol and malachite green. The slide is then turned upside down on an absorbent surface and pressed until the feces covers a 20–25 mm diameter area. The slide is then allowed to stand face up for about 1 h at room temperature with moderate relative humidity or for 20–30 min in a dry incubator or under an electric lamp. The glycerin clears the fecal matter but not the *Ascaris* eggs. The malachite green reduces eye strain, but is not essential for performing the assay. The test has been evaluated numerous times and has also been formulated such that it can be utilized in a quantitative fashion [47]. Katz et al. [48] developed the so-called Kato–Katz modification of the method for use in schistosomiasis. As with the direct smear, this method will detect both fertile and infertile *Ascaris* eggs.

Zinc sulfate flotation is a method that was developed to allow the identification of both protozoa and many helminth eggs in fecal samples. For *Ascaris*, this procedure works well for fertilized eggs, but the infertile eggs will typically not float at the usual specific gravity of 1.18 that is used; this is perhaps a combination of both their specific gravity and the lack of a lipoidal membrane that prevents the equilibration of the interior of the fertile egg with the external flotation media. These methods work by the fact that the buoyant density of the eggs is such that they will float in the elevated specific gravity of the zinc sulfate, while a majority of other materials will sink to the bottom of the centrifuge tube. The process involves stirring the feces into water, straining through gauze, centrifuging at 800 × *g* to remove excess fine particulates and bacteria that do not sediment. The pellet is then resuspended in the zinc sulfate solution, and recentrifugation occurs. The material of interest is at the top of the tube, and is removed typically with a wire loop and transferred to a slide. This can be a quick and useful method, often faster to scan than sedimentation, and will reveal the presence of most cyst-forming protozoa in a fecal sample. However, it again needs to be reiterated that it is not useful for the detection of infertile *Ascaris* eggs.

46.1.3.2 Molecular Techniques

Molecular methodologies are not likely to replace routine diagnostic methods any time in the near future unless they become parts of larger panels that function as routine screening devices. However, if a fecal methodology were developed that utilized something such as real-time multiplex polymerase chain reaction (PCR), there is the chance that the molecular panel could supersede the microscopic fecal analysis for routine pathogens, for example, *Giardia*, *Cryptosporidium*, *Entamoeba histolytica*, *Ascaris*, and *Trichuris*. An unrelated but real complication, however, is that most laboratories have got used to having all fecal samples submitted in formalin for routine pathogen inactivation, and formalin has so far been found to interfere significantly with the extraction of useful DNA for molecular methods. Similarly, many countries require that samples entering from other nations be treated with formalin to inactivate any contained pathogens. Thus, it seems that it may be difficult to reinstate the shipment of unfixed samples to different diagnostic centers. Also, it would seem that the development of alternative pathogen inactivation methods would lead to significant improvement in the development of molecular diagnostics for fecal samples by providing a means of protecting the DNA for future analysis.

Ascaris detection methods have not been applied for the most part for the simple sake of a diagnosis, but they have been used to ask more complicated questions about the biology of the organism. Molecular methods used for *Ascaris* infections have been for the study of the population structure of *Ascaris* infections in people and pigs around the world, to identify adult worms collected from occasional autochthonous humans infections around the world, for the determination of the maternal source of individual larvae of *Ascaris* recovered from a host, to verify the development of eggs following the testing of disinfection processes on egg inactivation, and to identify the eggs of *Ascaris* in human coprolites from different parts of the world. As has been shown with the utilization of the methodology for the identification of *Ascaris* DNA in coprolites, the assays are easily transferred to feces for the purpose of using as a diagnostic. The first such report has occurred with the development of assays for nuclear and mitochondrial DNA that are targeted toward detection of infection and specific diagnosis using human fecal samples.

The molecular methods that have been used in the diagnosis of *Ascaris* infections have been of several different general types: mitochondrial DNA (mtDNA) sequence variation with restriction mapping, restriction-fragment-length polymorphism (PCR–RFLP) using the internal transcribed spacer region of the ribosomal gene (ITS), RFLP of a number of introns of nuclear genes and mtDNA variation, ribosomal ITS amplification with single-strand conformation polymorphism as analyzed by gel electrophoresis (PCR–SSCP), random amplified polymorphic DNA (RAPD), DNA fingerprinting (AFLP), microsatellite markers, and a real-time PCR for quantifying viability.

The work on the molecular detection of *Ascaris* has been driven forward for reasons secondary to improved diagnostics. Excellent techniques have been developed for the detection on *A. lumbricoides* in fecal samples, but they have been developed for other applications. Thus, the methods employed can distinguish between the species in pigs and people, has answered questions as to the pre-Columbian distribution of *Ascaris* in the world, can trace a single egg (larva) back to its mother, and can be used for the purpose of mechanizing the testing of the viability of *Ascaris* eggs following various disinfection protocols. A method for the routine detection of *Ascaris* DNA in a fecal sample simply awaits the need for the application; a method is in place that is ready for this application [49].

The methods that have been developed and used for population genetic studies have included PCR–RFLP, PCR–SSCP, RAPD, AFLP, and microsatellite markers. The work represents significant effort from the Americas, Asia, and Europe [50–63]. The work overall supports the fact that the population of worms in people is likely separate from that of pigs; thus, it seems fairly safe to say that people are infected most of the time with *A. lumbricoides* and pigs with *A. suum* even when the two worms cycle within the same environment.

Work on the molecular biology of *Ascaris* as it is related to detection began initially as a study of the *Ascaris* populations present in people and pigs living in close proximity in Guatemala. The work showed using PCR–RFLP on mtDNA genes that the people were infected with the human form and the pigs with a separate swine form [50]. Further analysis revealed distinct clustering of specific mtDNA genotypes within hosts [51]; this was not unlike that found by Nadler using RAPD in *A. suum* of pigs in the Midwestern United States [56]. Further analysis using additional nuclear genes and worms from different locations around the world [52]. This work led to the statements that "programmes aiming to control *Ascaris* infections in the human population can safely ignore zoonotic infection from pigs" and "human and pig-associated *Ascaris* populations should continue to be regarded as sibling species (*A. suum* in pigs and *A. lumbricoides* in humans), both for convenience, and because there are evidently strong reproductive isolating mechanisms limiting gene flow between them" [52]. It was believed that since some of the mtDNA sequences were shared by pigs and people as opposed to the nuclear loci, the different divergent lineages of mtDNA were best explained as retained ancestral polymorphisms. Zhu et al. and Peng et al. have reported similar work from China using mtDNA and ribosomal DNA in PCR–SSCP analyses and ultimately concluded that "pigs are not a significant source of *Ascaris* infection in humans in endemic regions" [58–63]. More recent work with microsatellite markers has indicated that there may be rare occurrences of hybrid worms produced (4% from Guatemala and 7% from China) which suggest that there is contemporary cross-transmission between populations of *A. lumbricoides* and *A. suum* [54,55]. In the case of the analysis of *Ascaris* in pigs from Denmark using DNA fingerprinting (AFLP), there was substantial single generation inbreeding, a marked founder effect, and probably significant sharing of worms between farms throughout the country [57].

While it appears that for the most part that *A. suum* is in pigs and *A. lumbricoides* is in people, in areas where human infections are rare, people sometimes do become infected with *A. suum*. Studies have been used to identify individual worms from people as being from swine sources [64,65]. In both cases PCR–RFLP was used showing that the human forms had the same RFLP banding patterns as seen in pigs. In Denmark using AFLP, the *A. lumbricoides* from people in Bangladesh and Nepal, and DNA from worms from people in Guatemala segregated into distinct groups, while the *A. suum* from pigs and people in Denmark segregated together. Thus, it appears that if people are naïve relative to an infection with *Ascaris* and are living in areas where *A. suum* is common or in the surrounding environ that they can become infected with *A. suum* from swine.

Using PCR–RFLP it has been possible to identify maternal haplotypes of mtDNA and to show in larvae from experimentally infected pigs that this mtDNA can be used to identify the maternal source of the larvae collected [66]. The identification of individual worms as to their maternal mtDNA haplotype was done by harvesting larvae from the small intestine of four pigs infected with eggs from one of four female *A. suum* that has had their mtDNA characterized. Basically, 100 larvae were analyzed from each pig, and they were all found to be of the same haplotype as their maternal female. Thus, it is now possible to define the individual source of an infection if potential female worms can be identified. This also verified the maternal inheritance of mitochondria in *Ascaris* [67].

Pecson et al. [68] applied molecular diagnostics to assess the viability of *A. suum* eggs. These authors used real-time quantitative PCR to determine whether *A. suum* eggs were viable after treatment with inactivation processes (high heat, ammonia, and ultraviolet). *A. suum* and *A. lumbricoides* eggs are model organisms that are used to look at various parasite inactivation methods for the purpose of killing helminths in sewage sludges and other pathogens in waste inactivation processes. Typically, the assay for egg viability posttreatment requires the examination of eggs that have undergone studies on their potential inactivation following culture at 28°C for 4 weeks, and the eggs are then scored as to whether or not they have undergone successful embryonation. The method described by for this purpose uses real-time PCR to determine viability by examining the copy number of the genes that increase in number as the cells within the viable eggs multiply during embryonation. This method makes it possible to ultimately have the samples read via automation and allows testing to occur before the full developmental period of 28 days has passed.

There have been a number of papers where *Ascaris* eggs have been recovered from ancient coprolites; this is often done on the basis of finding the morphologically recognizable eggs in rehydrated fecal matter [69]. Methods now for rDNA and cytochrome *b* (*cytb*) gene targets have been used to isolate DNA from coprolite extractions and from eggs that have been isolated from coprolites [70,71]. These papers have been of significance because they indicate the presence of *Ascaris* in the Americas before the arrival of Columbus and the post-1492 colonization and even in specimens that are believed to have been from remains deposited before the pig was domesticated in the Americas.

46.2 METHODS

46.2.1 Sample Preparation

Leles et al. [49] developed a method for the detection of *Ascaris* eggs in the feces of people. They defined assays using two different target genes, the mtDNA cytochrome *b* gene and the nuclear ITS rDNA region. The assay has been verified using extracted eggs in water (1, 5, 10, 30, 50, and 80 eggs) and human fecal samples containing *A. lumbricoides* eggs varying in number from a theoretical number of 4 to 2,760 eggs as determined by a parallel Kato–Katz technique run on the same samples.

There are numerous published DNA extraction methods. Three successful methods of DNA extraction for *Ascaris* eggs include physical disruption of the eggs (five boil/freeze cycles) in water. Another method for eggs or eggs in feces (150 μL fecal sediment), boil/freeze cycles were followed by chemical treatment, including the addition of 400 μL digestion buffer (NaCl 100 mM, Tris–HCl 50 mM, SDS 1%, and EDTA 50 mM, pH 8.0) to which was added 20 μL DTT 1 M, 60 μL proteinase K 10 mg/mL, and 100 μL SDS 10%. The reaction was incubated at 55–60°C for 3–24 h with occasional homogenization. The DNA was extracted using phenol/chloroform and precipitated in 95% ethanol. The pellet was air dried and diluted in 50 μL ddH_2O or TE buffer (Tris–HCl 10 mM and EDTA 1 mM, pH 8.0). A third method followed the boil/freeze cycles with the use of a QIAmp DNA Stool Mini kit (Qiagen) and minor modifications (Proteinase K digestion at 55–60°C for 4 h) [49].

46.2.2 Detection Procedures

Both the mtDNA and the ITS1 regions amplified well in almost all samples; the commercial kit seems to have been less successful than routine samples extraction. The ITS1 region used was that described by Zhu et al. [63]. The authors also employed a nested PCR in those cases where the ITS1 product did not apparently amplify using the larger 480 bp primers. Following the nested PCR using primers for the 407 bp segment, restriction was undertaken using HaeIII (Invitrogen). The RFLP produced the patterns representative of the G1–G5 of Peng et al. (2003). In the paper the products were also sequenced, and the PCR products were also cloned in the pGEM-T Easy Vector System (Promega) with three clones from each sample being sequenced.

The PCR–RFLP analysis revealed the identical pattern of G1, the most prevalent human isolate, in all human samples (or the less probable G4 isolate which bands the same way), and it was confirmed to be G1 by sequencing. It was also revealed that three of the 12 human samples that were tested appeared to have a new genotype that is being called G6 with a deletion of nucleotide T at position 127. The sequence analysis of the *cytb* gene (142 bp) revealed that all Brazilian isolates were identical and had the nucleotide T from *A. lumbricoides* at position 5522 instead of the C as in *A. suum*. It would seem that this and other similar systems have great applicability to the routine examination of fecal samples by diagnostic laboratories around the world and will rapidly provide a great deal of new information about the biology of *Ascaris*. Table 46.1 provides basic information to extract, amplify and sequence DNA from *Ascaris*. In addition, several protocols have been briefly outlined in some detail below.

46.2.2.1 PCR and RFLP Method (ITS1 407 bp)

A PCR amplification of the ITS1 (480 bp) rDNA region described by a semi-nested ITS1 (407 bp) is conducted in a 50 μL reaction solution composed of 10 mM Tris–HCl, pH 8.4, 50 mM KCl, 3 mM $MgCl_2$, 250 μM dNTP, 2 U Taq DNA polymerase, 20 or 100 pmol of each primer (XZ5F and XZ1R, or XZ5F and NC13R), and 1 μL (5–10 ng) genomic DNA template with the following thermocycling conditions: initial denaturation at 94°C for 5 min, followed by 30 cycles of 94°C for 30 s (denaturation), 60°C for 30 s (annealing) and 72°C for 30 s (extension) and then a final extension of 72°C for 5 min [49,63]. Amplification products are then subjected to electrophoresis, stained with ethidium bromide and viewed under ultraviolet light. Positive amplification products are used in RFLP analysis, where the amplicon is digested with HaeIII endonuclease to determine *Ascaris* ITS1 genotypes G1–G5.

46.2.2.2 Cytochrome *b* mtDNA Amplification

To amplify cytochrome *b* (142 bp) mtDNA, DNA template is added to a 50 μL reaction volume containing 50 mM KCl, 10 mM Tris–HCl, 2 mM $MgCl_2$, 200 mM each of dNTPs, 1 mg/mL bovine serum albumin, 20 pmol of each primer (Asc1 and Asc2), and 5 U Taq with the following thermocycling conditions: one initial denaturing step of 94°C for 10 min, followed by 50 cycle of 94°C for 1 min, 52–55°C for 45 s and 72°C for 45 s. PCR products are then subjected to electrophoresis, stained with ethidium bromide and viewed under ultraviolet light [70].

46.2.2.3 Microsatellite Marker Amplification

Microsatellite marker amplification is conducted in a 5 μL reaction volume containing a final concentration 1× PCR buffer, 2.5 mM $MgCl_2$, 0.4 mM of each primer (see the reference for a complete list of 35 microsatellite loci), 0.12 U Taq and 2 μL template in a PCR ramp-down thermocycling method beginning with an initial denaturing step (96°C for 5 min), followed by 5 cycles of (96°C for 45 s, 55°C for 30 s, 72°C for 1 min), then 35 cycles of low-temperature PCR (96°C for 45 s, 47°C for 30 s, 72°C for 1 min), and a final step of 72°C for 7 min [54]. PCR products are subjected to electrophoresis, stained with ethidium bromide, and viewed under ultraviolet light.

TABLE 46.1
Primers and Probes for PCR Identification of *Ascaris*

Site	Name	Sequence (5′–3′)	Reference(s)	Notes
cytb (cytochrome *b*) (142 bp)	Asc1	GTTAGGTTACCGTCTAGTAAGG	49,70	mtDNA, fecal and coprolite egg extraction
	Asc2	CACTCAAAAAGGCCAAAGCACC		
ITS1 (480 bp)	XZ5F	TGATGTAATAGCAGTCGGCG	49,63,72	rDNA, fecal extraction
	XZ1R	GGAATGAACCCGATGGCGCAAT		
ITS1 (407 bp) semi-nested uses ITS1 480 bp amplicon as template	XZ5F	TGATGTAATAGCAGTCGGCG	49,63	rDNA fecal extraction
	NC13R	GGCTGCGTTCTTCATCGAT		
mtDNA (2299 bp)		CCGTTGGCTTTAACGGTGCGTTT	50	mtDNA RFLP
		GGAATAGCACGCAAAATAGCATAAGCAA		
mtDNA (2738 bp)		TGAGGCCTGTTCATATTGTGCCTGAGTG		
		GACCAGGTTTAGCCAATTCCAAACGAAT		
mtDNA (2912 bp)		GATTCGTTTGGAATTGGCTAAACCTGG		
		CTTAATGTCCTCACGCTAAGACTGCCA		
633 bp (includes HADH ND4 and COI)		GGGAATTTTATCATGTTTCTGGTAGGCG	52	mtDNA, worms fresh a preserved
		GACCAGGTTTAGCCAATTCCAAACGAAT		
MYO (Myo D)		GGGATGCCGATCATTAAG	52	rDNA
		TGCTGCCTTCCGCCTGTCC		
HEM (hemoglobin)		GAGTCGGTTCCTCCTGGCG	52	rDNA
		TTCTTGCGGCGAACCATGC		
MSP (major sperm protein)		GAGGCGAGACAGGATGGCA	52	rDNA
		TCCCGGTCATCGTAGGTGG		
TUB (α-tubulin)		GGCGCGTAAGTGGCCAGAG	52	rDNA
		GCGTTCTGCGAATCTCTGC		
COL (collagen)		TCAACTGGCGCCAAAGGCA	52	rDNA
		GGCAAACCAGGCACACCGG		
G4		GCGCCAAAGCAGCCGTCG	52	rDNA
		TGCCTTTGCCACTTTCTGCTTC		
G9		TTCATTCGCCACGCAAACG	52	rDNA
		TGTGCCGATGATTGGCGCT		
G12		GGCCAGCAGCCAGGTGGG	52	rDNA
		TGCGTACGGCACTGCGAAA		
ITS1 and ITS2 (950 bp)		TTGAACCGGGTAAAAGTCGT	52,58,64,67	rDNA PCR RFLP
		TTAGTTTCTTTTCCTCCGCT		
		See reference for complete list	54	Microsatellite markers

RAPD	#302	CGGCCCACGT	56	RAPD
	#306	GTCCTCGTAG		
	#308	AGCGGCTAGG		
	#314	ACTTCCTCCA		
	#316	CCTCACCTGT		
	#322	GCCGCTACTA		
	#337	TCCCGAACCG		
	#345	GCGTGACCCG		
	#348	CACGGCTGCG		
AFLP		EcoRI, GACTGCGTACCAATTCC (E_C)	57,65	gDNA fingerprinting
PCR Primers used after digestion and ligation (see reference for details)		MseI, GATGAGTCCTGAGTAAC (M_C); EcoRI, GACTGCGTACCAATTCA (E_A), MseI, GATGAGTCCTGAGTAAC (M_C)		AFLP
mtDNA (633 bp)		GGGAATTTTATCATGTTTCTGGTAGGCG	58,67,67	mtDNA, PCR–RFLP
		GACCAGGTTTAGCCAATTCCAAACGAAT		
pITS1 (300 bp)	NC5 XZ6R	GTAGGTGAACCTGCGGAAGGATCATT	59	rDNA, PCR–SSCP
		ATAGCTCGGTGAAGCGTAGA		
pnad1 (400 bp of the dehydrogenase	MH5F	TATGAGCGTCATTTATTGGG	60	mtDNA, PCR–SSCP
subunit 1 (*nad*1) genes)	As-NDR	GCATCACAATAGCCAACAAATAC		
Cytochrome *c* oxidase	As-Co1F	TTTTTTGGTCATCCTGAGGTTTAT	60	mtDNA, PCR–SSCP
subunit 1 (*cox*1) genes	As-Co1R	ACATAATGAAAATGACTAACAAC		
ITS1		ITS-1 forward primer TGCACATAAGTACTATTTGCGCGTAT, ITS-1 reverse primer TGATGTAATAGCAGTCGGCGG, and ITS-1 probe 6-carboxyfluorescein–CGTGAGCCACATAGTAAATTGCACACAAATG–6-carboxytetramethylrhodamine)	68	rDNA, qPCR
18S rRNA gene (123 bp)	Asc6	CGAACGGCTCATTACAACAG	70	gDNA extracted coprolite
	Asc7	TCTAATAGATGCGCTCGTC		
18S rRNA gene (99 bp)	Asc8	ATACATGCACCAAAGCTCCG	70	gDNA extracted coprolite
	Asc9	GCTATAGTTATTCAGAGTCACC		
18S rRNA gene (147 bp)	Asc10	CCATGCATGTCTAAGTTCAA	70	gDNA extracted coprolite
	Asc11	CARAAAWTCGGAGCTTTGGT		

46.3 CONCLUSIONS AND FUTURE PERSPECTIVES

In the case of *A. lumbricoides*, the methodology is in place for the routine examination of fecal samples using methods that will be able to differentiate *A. lumbricoides* and *A. suum* eggs in the feces of people (and pigs). The only thing currently holding back the regular use of these assays is the current ease of diagnosing *Ascaris* infections using routine microscopical methods and perhaps the fixation of fecal samples with formalin for the purpose of submission. It seems very likely that this will all change shortly and that methods will be routinely utilized that will be able to distinguish *Ascaris* eggs readily and easily in fecal samples. With the development of automation and the combination of probes to allow for the detection of multiple species simultaneously, it is expected that shortly there is a good chance that the microscopical diagnosis of the routine parasites in feces will be supplanted in many laboratories by molecular methods.

There is little doubt that *A. lumbricoides* infections in people can be readily diagnosed with molecular methods. The work on the experimental infection of pigs with *A. lumbricoides* (G1) of people and *A. suum* (G3) of swine supported the host-associated development of these parasites within the constraints bound by their host-adapted genotypes [61]. Although work on *Ascaris* population biology has suggested that there is a potential for rare hybridization occurring between the human and porcine populations of *A. lumbricoides* and *A. suum* [55], there is no reason to think that this will interfere with routine diagnosis. Similarly, although Leles et al. [72] recently suggested that the ITS region may not be sufficient for specific identification of the G1–G6 genotypes because of variability in individual *Ascaris* eggs. This should not interfere with the ability of diagnostic laboratories to make routine diagnoses of an *Ascaris* infection. It would be of interest in terms of control programs to examine whether the defined genotypes may explain some of the differences in fecundity that have been observed with different *A. lumbricoides* infections around the world [73]. However, in the end, it is expected that the development of easily automated fecal analysis of samples for the routine soil-transmitted helminths will greatly aid international control projects, and their use along with the increased sanitation that development brings to the world will aid in even more rapid infections with these parasites than has occurred even in the recent past.

REFERENCES

1. Tenora, F., and Borkovcová, M. Corrections in the review of helminthofauna of *Ondatra zibethica* (L., 1785) (Rodentia) in the Czech Republic. *Acta Univ. Agric. Silvic. Mendelianae Brunensis* 51, 65, 2003.
2. Maung, M. An ascaridoid nematode from a hippopotamus. *Int. J. Parasitol.* 5, 431, 1975.
3. Sprent, J.F.A. Ascaridoid nematodes of South American mammals, with a definition of a new genus. *J. Helminthol.* 56, 275, 1982.
4. Sprent, J.F.A. Notes on *Ascaris* and *Toxascaris*, with a definition of *Baylisascaris* gen. nov. *Parasitology* 58, 185, 1968.
5. Warren, G. Studies on the morphology and taxonomy of the genera *Toxocara stiles*, 1905 and *Neoascaris travassos*, 1927. *Zool. Anz.* 185, 393, 1970.
6. Tenora, F., and Shalayeva, N. Scanning electron microscopy of *Ascaris tarbagan* (Nematoda). *Acta Soc. Zool. Bohemoslov.* 54, 130, 1990.
7. Anderson, R.C. *Nematode Parasites of Vertebrates, Their Development and Transmission*, 2nd edn. CABI, Wallingford, UK, 650pp., 2000.
8. Meldal, B.H. et al. An improved molecular phylogeny of the Nematoda with special emphasis on marine taxa. *Mol. Phylogen. Evol.* 42, 622, 2007.
9. Sprent, J.F.A. Anatomical distinction between human and pig strains of *Ascaris. Nature* 170, 627, 1952.
10. Abdulrachman, S., and Joe, L.K. Morphological differences between *Ascaris* from man and pigs. *Doc. Med. Geo. Tropica* 6, 342, 1954.
11. Ubelaker, J.E., and Allison, V.F. Scanning electron microscopy of the denticles and eggs of *Ascaris lumbricoides* and *Ascaris suum. Thirtieth Annual Proceedings Electron Microscopy Society America*, 1972.
12. Maung, M. *Ascaris lumbricoides* Linne, 1758 and *Ascaris suum* Goeze, 1782: Morphological differences between specimens obtained from man and pig. *Southeast Asian J. Trop. Med. Public Health* 4, 41, 1973.
13. Weise, R.W. *Ascaris suum*: A scanning electron microscope study. *J. Parasitol.* 59, 141, 1973.
14. Kurimoto, H. Morphological, biochemical and immunological studies on the differences between *Ascaris lumbricoides* Linnaeus, 1758 and *Ascaris suum* Goeze, 1782. *Jpn. J. Parasitol.* 23, 251, 1974.
15. Little, M.D. Comparison of human and pig *Ascaris. Eighth International Congress Tropical Medicine Malaria.* Abstracts & Reviews, pp. 181, 1968.
16. Yi, K. et al. Dephosphorylation of major sperm protein (MSP) fiber protein 3 by protein phosphatase 2A during cell body retraction in the MSP-based amoeboid motility of *Ascaris* sperm. *Mol. Biol. Cell* 20, 3200, 2009.
17. Beaver, P.C., and Little, M.D. The genital girdle in relation to estrus and mating in *Ascaris* lumbricoides. *J. Parasitol.* 50, 128, 1964.
18. Brown, H.W., and Cort, W.W. The egg production of *Ascaris lumbricoides. J. Parasitol.* 14, 88, 1927.
19. Mello, D.A. A note on egg production of *Ascaris lumbricoides. J. Parasitol.* 60, 380, 1974.
20. Roepstorff, A. The earthworm *Lumbricus terrestris* as a possible paratenic or intermediate host of the pig parasite *Ascaris suum. Comp. Parasitol.* 69, 206, 2002.
21. Murrell, K.D. et al. *Ascaris suum*: A revision of its early migratory path and implications for human ascariasis. *J. Parasitol.* 83, 255, 1997.
22. Seo, B.S. Ascariasis and its control problems in Korea. *Seoul J. Med.* 22, 323, 1981.
23. Hobo, B. Epidemiological studies on *Ascaris* infection among prisoners and the length of life of *Ascaris lumbricoides* in human host. *Jpn. J. Natl Health* 25, 1, 1956.
24. Galvin, T.J. Development of human and pig ascaris in the pig and rabbit. *J. Parasitol.* 54, 1085, 1968.
25. Beaver, P.C., Jung, R.C., and Cupp, E.W. *Clinical Parasitology.* Lea and Febiger, Philadelphia, 825pp., 1986.

26. Orihel, T.C. Primates as models for parasitological research. In: Goldsmith, E.I. and Moor-Jankowski, J. (eds), *Medical Primatology*, S. Karger, Basel, 772pp., 1971.
27. Kmetec, E., Beaver, P.C., and Bueding, E. Succinoxidase activities and cytochrome oxidase of extracts of *Ascaris* eggs. *Comp. Biochem. Physiol.* 9, 115, 1963.
28. Brudastov, A.N. et al. Infectivity of *Ascaris lumbricoides* eggs to man and guinea-pigs after 10 years in the soil. *Med. Parasitol. Parasit. Dis.* 39, 447, 1970.
29. Krasnonos, L.L. Many-year viability of ascarid eggs (*Ascaris lumbricoides* L. 1758) in soil of Sarkand. *Med. Parasitol. Parasit. Dis.* 47, 103, 1978.
30. Crompton, D.W. *Ascaris* and ascariasis. *Adv. Parasitol.* 48, 285, 2001.
31. Stephenson, L.S. Pathophysiology of intestinal nematodes, In Holland, C.V. and Kennedy, M.W. (eds), *World Class Parasites: Volume 2. The Geohelminths: Ascaris, Trichuris, and Hookworm*, Springer, Heidelberg, Germany, pp. 39–61, 2002.
32. Hadju, V. et al. Improvements of growth, appetite, and physical activity in helminth-infected schoolboys 6 months after single dose of albendazole. *Asia Pac. J. Clin. Nutr.* 7, 170, 1998.
33. Hadju, V. et al. Improvements in appetite and growth in helminth-infected schoolboys three and seven weeks after a single dose of *pyrantel pamoate. Parasitology* 113, 497, 1996.
34. Tripathy, K. et al. Malabsorption syndrome in ascariasis. *Am. J. Clin. Nutr.* 25, 1276, 1972.
35. McOrist, S., Blunt, R., and Wiseman, J. *Ascaris suum* parasitism in finisher pigs: Welfare and economic impacts and control options. *Pig J.* 61, 74, 2008.
36. Crofton, H.D. A quantitative approach to parasitism. *Parasitology* 62, 179, 1971.
37. Holland, C., and Boes, J. Distribution and predisposition: People and pigs, in Holland, C.V. and Kennedy, M.W. (eds), *World Class Parasites: Volume 2. The Geohelminths: Ascaris, Trichuris, and Hookworm*, Springer, Heidelberg, Germany, pp. 1–24, 2002.
38. Elkins, D.B., Haswell-Elkins, M., and Anderson, R.M. The epidemiology and control of intestinal nematodes in the Pulicat Lake region of Southern India. I. Study design and pre- and post-treatment observations of *Ascaris lumbricoides* infection. *Trans. R. Soc. Trop. Med. Hyg.* 80, 774, 1986.
39. Boes, J. et al. Distribution of *Ascaris suum* in experimentally and naturally infected pigs and comparison with *Ascaris lumbricoides* infections in humans. *Parasitology* 117, 589, 1998.
40. Little, M.D. Nematodes of the digestive tract and related species. In: Beaver, P.C. and Jung, R.C. (eds), *Animal Agents and Vectors of Human Disease*, Lea & Febiger, Philadelphia, pp. 127–170, 1985.
41. Crompton, D.W.T. How much human helminthiasis is there in the world? *J. Parasitol.* 85, 397, 1999.
42. Kurosawa, C.M. et al. The half-life of *Ascaris lumbricoides* prevalence in Japanese school children. *Acta Med. Okayama* 62, 303, 2008.
43. Chai, J.Y. et al. Reduction in the incidence of biliary and other surgical complications of ascariasis according to the decrease of its national egg prevalence in Korea. *Korean J. Parasitol.* 29, 101, 1991.
44. Garcia, L.S. *Diagnostic Medical Parasitology*, 5th edn. ASM Press, Washington, DC, pp. 1092, 2006.
45. WHO. *Bench Aids for the Diagnosis of Intestinal Parasites*. Prepared by LR Ash, TC Orihel, L Savioli. World Health Organization, Geneva.
46. Komiya, Y., and Kobayashi, A. Evaluation of Kato's thick smear technic with a cellophane cover for helminth eggs in feces. *Jpn. J. Med. Sci. Biol.* 19, 59, 1966.
47. Martin, L.K., and Beaver, P.C. Evaluation of Kato thick-smear technique for quantitative diagnosis of helminth infections. *Am. J. Trop. Med. Hyg.* 17, 382, 1968.
48. Katz, N., Coelho, P.M.Z., and Pellegrino, J. Evaluation of Kato's quantitative method through the recovery of *Schistosoma mansoni* eggs added to human feces. *J. Parasitol.* 56, 1032, 1970.
49. Leles, D. et al. Molecular diagnosis of ascariasis from human feces and description of a new *Ascaris* sp. genotype in Brazil. *Vet. Parasitol.* 163, 167, 2009.
50. Anderson, T.J.C., Romero-Abal, M.E., and Jaenike, J. Genetic structure and epidemiology of *Ascaris* populations: Patterns of host affiliation in Guatemala. *Parasitology* 107, 319, 1993.
51. Anderson, T.J.C., Romero-Abal, M., and Jaenike, J. Mitochondrial DNA and *Ascaris* microepidemiology: The composition of parasite populations from individual hosts, families and villages. *Parasitology* 110, 221, 1995.
52. Anderson, T.J.C., and Jaenike, J. Host specificity, evolutionary relationships and macrogeographic differentiation among *Ascaris* populations from humans and pigs. *Parasitology* 115, 325, 1997.
53. Anderson, T.J.C. The dangers of using single locus markers in parasite epidemiology: *Ascaris* as a case study. *Trends Parasitol.* 17, 183, 2001.
54. Criscione, C.D. et al. Microsatellite markers for the human nematode parasite *Ascaris lumbricoides*: Development and assessment of utility. *J. Parasitol.* 93, 704, 2007.
55. Criscione, C.D. et al. Disentangling hybridization and host colonization in parasitic roundworms of humans and pigs. *Proc. R. Soc. Lond. B.* 274, 2669, 2007.
56. Nadler, S.A., Lindquist, R.L., and Near, T.J. Genetic structure of midwestern *Ascaris suum* populations: A comparison of isoenzyme and RAPD markers. *J. Parasitol.* 81, 385, 1995.
57. Nejsum, P. et al. Population structure in *Ascaris suum* (Nematoda) among domestic swine in Denmark as measured by whole genome DNA fingerprinting. *Hereditas* 142, 7, 2005.
58. Peng, W. et al. Genetic variation in sympatric *Ascaris* populations from humans and pigs in China. *Parasitology* 117, 355, 1998.
59. Peng, W. et al. Molecular epidemiological investigation of *Ascaris* genotypes in China based on single-strand conformation polymorphism analysis of ribosomal DNA. *Electrophoresis* 24, 2308, 2003.
60. Peng, W. et al. Mutation scanning-coupled analysis of haplotypic variability in mitochondrial DNA regions reveals low gene flow between human and porcine *Ascaris* in endemic regions of China. *Electrophoresis* 26, 4317, 2005.
61. Peng, W. et al. Experimental infections of pigs and mice with selected genotypes of *Ascaris. Parasitology* 133, 651, 2006.
62. Peng, W. et al. Recent insights into the epidemiology and genetics of *Ascaris* in China using molecular tools. *Parasitology* 134, 325, 2007.
63. Zhu, X. et al. Characterisation of *Ascaris* from human and pig hosts by nuclear ribosomal DNA sequences. *Int. J. Parasitol.* 29, 469, 1999.
64. Anderson T.J.C. *Ascaris* infections in humans from North America: Molecular evidence for cross-infection. *Parasitology* 110, 215, 1995.
65. Nejsum, P. et al. Ascariasis is a zoonosis in Denmark. *J. Clin. Microbiol.* 43, 1142, 2005.

66. Nejsum, P. et al. A novel technique for identification of *Ascaris suum* cohorts in pigs. *Vet. Parasitol.* 154, 171, 2008.
67. Anderson, T.J.C. et al. Are mitochondria inherited paternally in *Ascaris*? *Int. J. Parasitol.* 25, 1001, 1995.
68. Pecson, B.M. et al. A real-time PCR method for quantifying viable *Ascaris* eggs using the first internally transcribed spacer region of ribosomal DNA. *Appl. Environ. Microbiol.* 72, 7864, 2006.
69. Gonçalves, M.L.C., Araujo, A., and Ferreira, L.F. Human intestinal parasites in the past: New findings and a review. *Mem. Inst. Oswaldo Cruz* 98, 103, 2003.
70. Loreille, O. et al. Ancient DNA from *Ascaris*: Extraction amplification and sequences from eggs collected in coprolites. *Int. J. Parasitol.* 31, 1101, 2001.
71. Leles, D. et al. Molecular paleoparasitological diagnosis of *Ascaris* sp. from coprolites: New scenery of ascariasis in pre-Colombian South America times. *Mem. Inst. Oswaldo Cruz* 103, 106, 2008.
72. Leles, D. et al. ITS1 intra-individual variability of *Ascaris* isolates from Brazil. *Parasitol. Int.* 59, 93, 2010.
73. Hall, A., and Holland, C. Geographical variation in *Ascaris lumbricoides* fecundity and its implications for helminth control. *Parasitol. Today* 16, 540, 2000.

47 Baylisascaris

Dongyou Liu

CONTENTS

47.1 INTRODUCTION

Ascariases are common, important parasitic infections of humans and animals that are caused by roundworms (Ascaridida, Ascarididae) belonging to the genera of *Ascaris* (e.g., *Ascaris lumbricoides*), *Toxocara* (e.g., *Toxocara canis* and *Toxocara cati*), and *Baylisascaris* (e.g., *Baylisascaris procyonis*). While *A. lumbricoides* affects about one billion people worldwide, *Toxocara canis* and *Toxocara cati* chiefly infect dogs and cats, respectively, and *B. procyonis* occurs naturally in raccoons.

B. procyonis is a remarkable parasite in that it has the ability to produce a large quantity of highly resistant eggs, that its larval stage induces severe, neurologically damaging, and potential fatal diseases (neural larva migrans (NLM), ocular larva migrans (OLM), and visceral larva migrans (VLM)) in intermediate hosts including humans (especially infants and young children). Given the shared clinical features (e.g., eosinophilic meningoencephalitis) between *B. procyonis* and other infections, there is an urgent need to develop noninvasive laboratory techniques for improved diagnosis of human baylisascariasis. In addition, further understanding of *B. procyonis* pathogenesis and development of effective treatment measures are integral to the control of this highly dangerous pathogen.

47.1.1 Classification, Morphology, and Biology

47.1.1.1 Classification

The genus *Baylisascaris* is classified in the family Ascarididae, superfamily Ascaridoidea, order Ascaridida, class Chromadorea, phylum Nematoda. The family Ascarididae contains eight species: *Angusticaecum, Ascaris, Baylisascaris, Krefftascaris, Parascaris, Porrocaecum, Rostellascaris*, and *Toxascaris*. In turn, the genus *Baylisascaris* is divided into at least seven species: *Baylisascaris columnaris, Baylisascaris devosi, Baylisascaris laevis, Baylisascaris melis, B. procyonis, Baylisascaris schroederi*, and *Baylisascaris transfuga* [1].

First isolated as *Ascaris columnaris* from raccoons (*Procyon lotor*) in 1931, *B. procyonis* was later redesignated as a distinct species (*Ascaris procyonis*). The genus *Baylisascaris* was created (in honor of British parasitologist H.A. Baylis) in 1968 to cover a number of recognized as well as provisional species previously classified as members of the *Ascaris* or *Toxascaris* genus [2]. While *B. procyonis* causes subclinical infection in raccoons, it is extremely harmful to animals and humans in extraintestinal infection, in which migrating larvae induce visceral, neural, or OLM syndromes.

B. columnaris is a parasite of skunks and American badgers, and an important cause of mortality in pet skunks; *Baylisascaris d.* parasitizes martens; *B. laevis* infects marmots; *B. melis* is a parasite of European badgers; *B. transfuga* is a roundworm of bears, and its larvae are responsible for causing visceral, neural, or OLM syndromes in mice, gerbils, guinea pigs, and chickens.

47.1.1.2 Morphology

B. procyonis adult worms are large and tan-colored, with the female being of 20–22 cm in length, and the male being of 9–11 cm in length.

B. procyonis eggs measure from 63 to 88 μm × 50 to 70 μm in size, and appear ellipsoidal and dark brown. The egg contains a large single-celled embryo surrounded by a thick shell with a finely granular surface. In contrast, the eggs of the closely related *Toxocara* species are slightly larger in size (85 × 75 μm), appear lighter in color, and have a coarsely pitted shell.

B. procyonis second-stage larvae contained within the hatched eggs are 300 μm in length, and the third-stage larvae grown in intermediate hosts are of 1500–1900 μm in length by 60–80 μm in maximal diameter.

47.1.1.3 Biology

Raccoons are the natural host of *B. procyonis*, whose adult worms reside in the small intestine. With the female adult worms producing between 115,000 and 179,000 eggs/worm/day, the infected raccoons shed an average of 20,000–26,000 (with a maximum of >250,000) eggs per gram of feces. The eggs are noninfective immediately after shedding, and develop into infective second-stage larvae in 2–4 weeks under suitable temperature and humidity. Being resistant to decontamination and environmental degradation (heat and cold), *B. procyonis* eggs remain viable in moist soil for years. In addition, *B. procyonis* eggs are capable of sticking and adhering to animal fur, surfaces, objects (e.g., toy), and the fingers of infants. Together, these factors contribute to the success of *B. procyonis* as a parasite [3].

Following ingestion by intermediate hosts (which include over 100 species of animals and birds, such as rodents, rabbits, primates, and humans) during forage for food at latrines (preferred sites of raccoon defecation), *B. procyonis* infective eggs hatch in the small intestine to become larvae, which gain entry into the intestinal mucosa, migrate via the portal circulation through the liver to the lungs, reach to the left side of the heart via the pulmonary veins, and move to the tissues by the systemic circulation, with 5–7% of larvae entering the central nervous system (CNS) after penetrating cerebral blood vessels. Larvae encapsulate within eosinophilic granulomas, develop onto the third-stage, and remain viable there until they are ingested by raccoons.

Adult raccoons acquire the infection by preying intermediate host infested with the third-stage larvae encysted in tissues. The larvae hatch in the small intestine, stay within the lumen, and mature into adult worms (about 35 days, with 12–22 worms each). Young raccoons (of the first few months in life) acquire the infection through ingestion of the infective eggs sticking to their mother's fur or the den. The eggs hatch in the small intestine to become second-stage larvae. Larvae penetrate the intestinal mucosa, undergo further development, reemerge into the lumen, and mature into adult worms (in about 63 days, with 48–62 worms each). Occasionally, dog becomes infected by eating intermediate host containing the third-stage larvae encysted in tissues or ingestion of the infective eggs, and develops intestinal infection. Given their close proximity to humans and their indiscriminate defecation habit, the dogs with intestinal infection of *B. procyonis* lay eggs, which upon ingestion may infect humans.

Native to the Americas from Canada to Panama, raccoons are common free-ranging mammals, and thrive in areas with a reliable water supply, a ready availability of food, and suitable sites for dens. Being indigenous in North American raccoons, *B. procyonis* is prevalent (up to 82% of raccoon populations) in the Midwest, Northeast, and West Coast of the United States [4–6]. Raccoons were introduced into France, Germany, the Netherlands, Russia, and Japan for the commercial fur trade and as pets [7,8]. The prevalence of *B. procyonis* infection is high in wild raccoons in Germany and zoo pets in Japan.

Raccoons tend to defecate in preferred communal sites (termed as latrines), such as flat surfaces in the forest areas (e.g., stumps and limbs of trees, large logs, downed timber and rocks, and at the bases or in the crotches of trees), or lofts, attics and chimneys, and on flat roofs, wooden decks, woodpiles, and patios in the suburb/urban settings [9]. A single latrine may hold between 370 and 750 g of feces, which contain eggs with infective second-stage larvae [10]. Soil near latrines is also heavily contaminated with infective eggs. Raccoon latrines are routinely foraged by small mammals and birds for undigested seeds and grain, and are favorite place for children with geophagia or pica to play. In addition, egg-laden raccoon feces in roof latrines are washed down gutters by rainfall and contaminate the surrounding ground below, providing another source of infection in humans. Other potential source of infection include exposure to baby or pet raccoons; indoor storage of contaminated downed timber, wood chips, or bark for firewood; indoor contamination by raccoon dens in chimneys and fireplaces; outdoor contamination of children's play areas by raccoon feces.

47.1.2 Clinical Features and Treatment

47.1.2.1 Clinical Features

As accidental intermediate hosts for *B. procyonis*, humans acquire the infection after ingestion of *B. procyonis* eggs containing the infective second-stage larvae. *B. procyonis* larvae migrate to viscera, the CNS, and eyes. Depending on the number of eggs ingested (which has a bearing on the number of larvae entering the CNS) and the size of the brain, the clinical spectrum of human baylisascariasis ranges from asymptomatic infection, VLM, NLM, to OLM syndromes [11–14]. The most common manifestations include: (i) skin irritations from larvae migrating within the skin; (ii) respiratory discomfort, liver enlargement, and fever due to reaction to larvae migration; (iii) eye and brain tissue damage due to the random migration of the larvae; (iv) nausea, lethargy, incoordination, and loss of eyesight; (v) severe neurological signs (e.g., imbalance, circling and abnormal behavior, seizures, and coma) and death, as a result of extensive tissue damage caused by larval migration through the brain [1,9].

47.1.2.1.1 Visceral Larva Migrans

B. procyonis larvae demonstrate somatic tropism for the cranial rather than caudal regions of intermediate hosts. Therefore, *B. procyonis* VLM is characterized by preferential concentration of larvae in the head, neck, and thoracic regions, as a result of larval migration from the left side of the heart via the first major arterial branches of the systemic circulation arising from the aortic arch. In infants and young children with in severe infections, *B. procyonis* larvae migrate to various visceral organs (e.g., the heart, mediastinal soft tissues, pleura and lungs, small and large bowel walls, and mesentery and mesenteric lymph nodes) and induce the formation of numerous granulomata. Accompanied with this migration are nonspecific clinical manifestations such as macular rash (predominantly on the face and trunk), pneumonitis, and hepatomegaly. Macroscopically, VLM presents multiple small whitish nodules (of 1–1.5 mm in diameter) on pleural surfaces, soft tissues surrounding the laryngo-tracheobronchial tree, lung hila, epicardium and myocardium, the wall of the colon and ileocecal regions, small bowel mesentery, and mesenteric lymph nodes. Microscopically, the nodules consist of granulomata harboring coiled *B. procyonis* larvae surrounded by macrophages, eosinophils, plasma cells, and dense fibrous collagenous tissue.

In a case of *B. procyonis* infection involving a mildly mentally handicapped 10-year-old boy from Massachusetts, an inflammatory polypoid mass in the left ventricular myocardium and marked peripheral eosinophilia were noted. The intracardiac mass consisted of several degenerating ascarid larvae (of 60–70 μm in diameter) surrounded by eosinophilic granules and granulomata.

47.1.2.1.2 Neural Larva Migrans

If sufficient numbers of *B. procyonis* eggs (containing the infective second-stage larvae) are ingested, the larvae spread rapidly and undergo continued growth in host tissues, resulting in extensive damage. Fulminant NLM (characterized by acute eosinophilic meningoencephalitis and accompanied by widespread visceral involvement) develops within 2–4 weeks after a proportion (amounting to several thousands) of the ingested larvae (5–7%) enter the CNS [15,16]. Somatic larval migration causes mechanical tissue damage and necrosis and provokes an intense eosinophilic and granulomatous inflammatory reaction. The initial clinical features of NLM include low-grade fever, ataxia, increasing lethargy, somnolence, and periods of increased irritability. This is followed by progression to extensor posturing, increasing spasticity with hemi- or quadriparesis, ocular or cranial nerve involvement, and seizures. Rapid deterioration of neurologic status leads to stupor, coma, and death. Survivors of NLM are left in a persistent vegetative state or with severe residual deficits, including blindness and visual impairment and requiring total nursing care.

NLM appears to affect predominantly infants and young children, especially males, due to their relatively small brain size and their tendency to play outdoors, which increases the risk of *B. procyonis* infection. The first confirmed cases of human *B. procyonis* NLM involving 10- and 18-month-old boys from Pennsylvania and central Illinois, respectively, were described in 1984 and 1985 [17,18]. Both patients succumbed to rapidly progressive and ultimately fatal eosinophilic meningoencephalitis. In older patients with developmental delay and geophagia, whose brains are larger than those of infants and young children, a more subacute, indolent encephalopathy (well-defined granulomata) may occur.

47.1.2.1.3 Ocular Larva Migrans

Ocular involvement in baylisascariasis is frequently associated with severe NLM and VLM. It may also form a standalone feature. The patients show choroidioretinitis, optic neuritis, or atrophy with the presence of occasionally motile larvae within the retina as examined by ophthalmoscopy. This leads to the destruction of the visual cortex, with visual impairment or blindness as a consequence. Compared to *Toxocara* larvae (measuring 350–445 μm × 20 μm), which represent a more common cause of isolated OLM, *B. procyonis* larvae (measuring 1500–2000 μm × 60–70 μm) are larger. While NLM is almost exclusively restricted to infants and young children, isolated OLM may occur in adults after ingestion of relatively few *B. procyonis* eggs with a single larva entering into the eye [19].

B. procyonis is also responsible for causing diffuse unilateral subacute neuroretinitis (DUSN), another form of OLM, which is associated with progressive monocular visual loss and changes in retinal pigmentation and optic nerve anatomy. Histological features of DUSN include retinitis, retinal and optic nerve perivasculitis and atrophy, peripheral retinal degeneration and depigmentation, and low-grade choroiditis. This localized ocular damage is attributable to host reactions to the invading *B. procyonis* larvae [1].

As the most common cause of clinical larva migrans in animals, *B. procyonis* is associated with fatal or severe neurological disease. Compared to other organisms that also cause larval migrans in animals, *B. procyonis* demonstrates the propensity for aggressive somatic migration, larval invasion of the CNS, and the capability for continued larval growth within intermediate hosts.

47.1.2.2 Treatment

Baylisascaris adult worms in the intestines can be treated with anthelmintics (e.g., albendazole, diethylcarbazine, thiabendazole, mebendazole, fenbendazole, tetramisole, or ivermectin) [20]. Ivermectin can reduce CNS eosinophilia but does not alter unfavorable disease course, and corticosteroids may be used to decrease the potentially deleterious eosinophilic inflammatory response. A combination of laser photocoagulation and systemic corticosteroids has been used successfully to kill intraretinal larvae and decrease any resulting intraocular inflammatory response in *B. procyonis* OLM and DUSN.

Considering the limited availability and effectiveness of treatment options for baylisascariasis, prevention represents the best approach to reducing *B. procyonis* infection [21]. This is achieved through: (i) education of the public regarding the potential dangers of contact with raccoons or their feces; (ii) keeping infants or toddlers with geophagia or pica away from raccoon latrines or environment contaminated by infected raccoon feces; (iii) careful handwashing after contact with animal feces; (iv) decontamination of large or small areas by flaming (with a propane flame-gun), boiling water, steam-cleaning, or fire; (vi) cleaning contaminated surfaces with a xylene–ethanol mixture; (v) wearing disposable overalls, gloves, and mask and eye protection during cleaning.

47.1.3 Laboratory Diagnosis

B. procyonis infections in humans typically induce VLM, NLM, to OLM syndromes [22]. Although it is considered as the hallmarks of *B. procyonis* NLM, peripheral and CSF eosinophilia may also occur under other infectious and noninfectious conditions [23,24]. For example, roundworms *Angiostrongylus cantonensis* and *Gnathostoma spinigerum* can also cause eosinophilic meningoencephalitis. However, angiostrongyliasis is generally associated with meningitis (which has a relatively benign course and usually a good prognosis) rather than meningoencephalitis (which is highly vicious and has a poor prognosis) as in the case of baylisascariasis. Gnathostomiasis is associated with severe neurologic sequelae and a poor prognosis but is characterized by myeloencephalitis, focal cerebral hemorrhage (with xanthochromia), painful radiculopathy, and migrating cutaneous swellings. Other less common causes of eosinophilic meningoencephalitis include neurocysticercosis, cerebral paragonimiasis, cerebral toxocariasis, neurotrichinosis, and CNS schistosomiasis. It is notable that *Toxocara* spp. are a more common cause of VLM and OLM in humans, but are a very uncommon cause of NLM. In addition, disseminated *Coccidioides immitis* infection and acute disseminated encephalomyelitis may also cause eosinophilic meningitis and related illnesses.

To differentiate human baylisascariasis from other infections and medical conditions, a variety of laboratory procedures have been utilized. These include microscopy, neuroimaging, electroencephalography, biochemical tests, serology, and molecular techniques.

47.1.3.1 Microscopy

Morphologic identification of characteristic *B. procyonis* larvae (of 60–80 μm in diameter) in tissue (e.g., transverse section of the midbody/midintestinal region) represents a useful approach to diagnose human baylisascariasis. Other histological features include a large, centrally located, and laterally compressed intestine; paired triangular lateral excretory columns flanking the intestine, and prominent single lateral cuticular alae on either side of the body. Antemortem diagnosis of *B. procyonis* NLM can also be made by demonstration of characteristic *B. procyonis* larvae in brain biopsies (particularly of deep white matter site) [25]. Importantly, because *B. procyonis* does not complete its life cycle in humans, eggs or larvae are not shed in the feces of infected patients. The detection of eosinophilia, particularly eosinophilic meningitis (which is defined as the presence of 10 or more eosinophils/μL or eosinophilia of at least 10% in CSF), in cytocentrifuged CSF specimens using Wright's or Giemsa stain, suggests the possibility of a parasitic etiology.

47.1.3.2 Neuroimaging

Neuroimaging can be used to detect nonspecific diffuse white matter abnormalities, particularly of cerebral periventricular and deep cerebellar regions, which may be associated with changes of hydrocephalus, cortical edema, and loss of gray–white matter differentiation, with little or no meningeal or parenchymal enhancement. The magnetic resonance imaging finding of diffuse white matter disease on neuroimaging together with encephalopathy with CSF and peripheral eosinophilia strongly suggests the diagnosis of *Baylisascaris* NLM. A history of exposure to raccoons or their feces is also helpful for confirmation of the disease.

47.1.3.3 Electroencephalography

Electroencephalography may help demonstrate nonspecific diffuse slow-wave activity consistent with generalized encephalitis associated with NLM.

47.1.3.4 Biochemical Tests

The finding of elevated serum isohemagglutinins provides an additional clue to the diagnosis of baylisascariasis.

47.1.3.5 Serology

Anti-*Baylisascaris* antibodies is detectable in CSF and serum by indirect immunofluorescence, enzyme-linked immunosorbent assay, and Western blotting, using excretory/secretary antigens and recombinant proteins [26–30].

47.1.3.6 Molecular Techniques

Polymerase chain reaction (PCR) amplification and sequencing analysis of large (28S) ribosomal RNA, internal transcribed spacer region 1 and 2 (ITS 1 and ITS 2), and mitochondrial (cytochrome *c* oxidase subunit 1 (*cox*1) and cytochrome c oxidase subunit 2 (*cox*2)) target regions offers a new approach for reliable detection and phylogenetic assessment of *Baylisascaris* species and related parasites [31–38]. A SYBR Green dye-based real-time PCR targeting mitochondrial cox-2 gene allowed highly sensitive detection of *B. procyonis* eggs in fecal specimens [39].

47.2 METHODS

47.2.1 Sample Preparation

DNA is extracted from 0.5-g soil (or 500-μL volumes of concentrated water) using an UltraClean Soil DNA Isolation kit (MoBio Laboratories). The extracted DNA is stored at −20°C. Alternatively, genomic DNA is extracted from tissues previously treated with liquid nitrogen, using QIAamp Tissue Kit (Qiagen), eluted in 200 μL of sterile distilled water, and stored at −20°C until molecular testing.

47.2.2 Detection Procedures

47.2.2.1 Analysis of ITS Sequence

Testini et al. [38] utilized primers 18sF (forward: 5′-ACTGCTGTTTCGAGACCTTTCGAG-3′) and 28sR (reverse: 5′-TAGCACCTTCTTTGGACTATAGCC-3′) from the 18S and 28S regions of *B. transfuga* sequences for PCR amplification and comparative analysis of *Baylisascaris* ITS sequences.

47.2.2.1.1 Procedure

1. PCR mixture (50 μL) is made up of 2.5 mM $MgCl_2$, 10 mM Tris-HCl pH 8.3, 50 mM KCl, 250 μM of each dNTP, 50 pmol of each primer, 1.25 U of Ampli *Taq* Gold (Applied Biosystems), and 4 μL of genomic DNA.
2. The PCR amplification is carried out using the following cycling protocol: an initial 94°C for 12 min; 30 cycles of 94°C for 30 s, 58°C for 45 s, and 72°C for 45 s; and a final 72°C for 7 min.
3. Amplicons are resolved in 2% ethidium bromide-stained agarose gel and sized by comparison with a DNA marker (100 bp DNA Ladder, Sharp Mass).
4. Amplicons produced are purified using Ultrafree-DNA columns (Jetquick Genomed) and then sequenced directly in both directions (with the same PCR primers individually) using Taq DyeDeoxy Terminator Cycle Sequencing Kit (v.2, Applied Biosystems) in an automated sequencer (ABI-PRISM 377).
5. The electropherograms are verified by eye. Sequences are aligned using ClustalX and compared to sequences available in GenBank using Basic Local Alignment Search Tool (BLAST 2.2.22). Pairwise comparisons of nucleotide sequence differences (*D*) between the consensus sequences of each gene and those of *Baylisascaris* spp. available in GenBank database are made. Variable, informative and singleton sites occurring in the amino acid sequences and in all DNA regions herein examined are also evaluated with MEGA.

47.2.2.2 Real-Time PCR Identification

Gatcombe et al. [39] described a molecular beacon probe-based real-time PCR assay targeting the cytochrome oxidase subunit 2 (cox-2) gene for rapid and specific detection of eggs of *Baylisascaris* spp. The forward primer (SrF1: 5′-GAGTTATGAGTTTAGTGATATTCCTGGA-3′) has a $T_m = 60$°C. The reverse primer (JVBPR: 5′-GCAAA GCCCAAGAATGAATCAC-3′) has a $T_m = 66$°C. The molecular beacon probe (JVBPP: 5′-cgcgatcTATTAACATCA-CAAGGTACAACACAACGgatcgcg-3′) has a $T_m = 75$°C and a stem $\Delta G = -4.96$. The molecular beacon probe contains an FAM fluorophore label at the 5′ end and a Black Hole Quencher molecule as a quencher at the 3′ end.

Primers SrF1 and JVBPR amplify a 146 bp fragment, and possess no secondary structures, self-annealing, or significant identity to non-*Baylisascaris* sequences as assessed BLAST searching. The probe JVBPP is designed to form a stem–loop structure to bring into close proximity the 5′ and 3′ ends of the probe, resulting in minimal fluorescence for the nonhybridized probe. In the presence of a complementary target sequence, the probe hybridizes to the target, separating the reporter dye from the quencher and resulting in a measurable increase in fluorescence.

47.2.2.2.1 Procedure

1. Real-time PCR is performed in a volume of 20 μL, containing 10 μL of 2× QuantiTect Probe PCR kit (Qiagen), 250 nM of each specific primer, 100 nM probe, and 2 μL of the genomic DNA (from the water or soil sample extract).
2. Amplification is conducted with an initial 95°C for 15 min followed by 45 cycles of 95°C for 10 s, 55°C for 30 s (with fluorescence data collection), and 72°C fir 20 s using the ABI 7500 Real-Time PCR System (Applied Biosystems).

Note: Although the probe does not differentiate *B. procyonis* from the related parasites *B. columnaris* and *B. transfuga*, the primers recognize only *B. procyonis* and *B. columnaris*. While *B. columnaris* has been involved in a limited number of clinical cases in animals, *B. procyonis* accounts for an estimated 96% of recognized cases and outbreaks in humans and animals.

The real-time PCR assay is capable of detecting 100 fg of the parasite genomic DNA, a single *in vitro* hatched *B. procyonis* larva, and as few as 20 unembryonated *B. procyonis* eggs per gram of feces. No DNA amplification is obtained when the genomic DNA of related ascarids (including B. transfuga) and a hookworm is used as template in the PCR; however, amplification is noted with the very closely related *B. columnaris*.

Thus, this real-time PCR assay allows sensitive detection of *B. procyonis* in a wide range of soil types, and represents a valuable tool for investigations of soil or water potentially contaminated with eggs of this parasite.

47.3 CONCLUSION

The genus *Baylisascaris* covers seven roundworm species that parasitize a range of animals including mammals, birds, and humans. Of these, *B. procyonis* commonly occurs in raccoons (*P. lotor*), which are native to North America, but have been transported to Europe, Japan, and other parts of the world for fur production and as pet. Within its natural host raccoons, *B. procyonis* adult worms inhabit the small intestines and cause little harm to the definitive host. However, *B. procyonis* eggs discharged in raccoon feces are infective to over 90 animal species including humans. Upon ingestion by intermediate hosts, *B. procyonis* eggs hatch to become larvae, which pass through intestinal mucosa and migrate to visceral organs, brain, and eyes. The migration and continuing growth of *B. procyonis* larvae induce clinical neural, ocular, and VLM (NLM, OLM, and VLM) syndromes in humans, which are difficult-to-treat, severe, potentially fatal, neurologically devastating diseases, particularly in infants and young children [40].

Although the incidence of human baylisascariasis is currently low, the disease is poised to increase in the coming years. This may be attributed to the fact that *B. procyonis* adult worms have the ability to produce enormous number of eggs (with a single *B. procyonis* adult female worm producing 115,000–877,000 eggs per day, and an infected raccoon shedding as many as 45,000,000 eggs daily); that *B. procyonis* eggs are highly resistant to decontamination procedures and can survive in the environment for years; that *B. procyonis* infection requires a relatively low infectious dose (estimated to be <5000 eggs; and that *B. procyonis* infected raccoon populations (accounting for up to 70% adult raccoons and exceeding 90% juvenile raccoons) living in close proximity to human residential populations have shown a notable rise [41]. In addition, being capable of acting as final host for *B. procyonis*, the infected pet dogs pose an additional health risk to human populations.

In the absence of effective treatment, prevention of *B. procyonis* infection is critical. Besides educating the public about the potential dangers of *B. procyonis* infection, and adopting appropriate measures to avoid exposure to *B. procyonis* eggs, it is important to develop highly sensitive and rapid techniques for the detection and identification of *B. procyonis* eggs, larvae, and adult worms in feces, soil, water, and tissue specimens. Traditionally, a presumptive diagnosis of human baylisascariasis is based on clinical (meningoencephalitis, DUSN, pseudotumor), epidemiologic (raccoon exposure), radiologic (white matter disease), and laboratory results (blood and CNS eosinophilia). Further confirmation is dependent on microscopic identification of *B. procyonis* larvae in tissue sections. However, due to the morphological resemblances of *B. procyonis* larvae to other larval nematode (e.g., *Toxocara canis*, *T. oxocara cati*, *A. lumbricoides*, *Gnathastoma*, *Angiostrongylus*, and *Ancylostoma*) and cestode agents (e.g., cysticercosis and echinococcosis), accurate diagnosis demands specialized skills. In addition, microscopic identification requires a biopsy specimen containing an adequate cross-section of a larva. A biopsy without the presence of *B. procyonis* larvae will give a negative finding. While serologic procedures may be utilized to support the clinical diagnosis, cross-reactions between *B. procyonis* and related parasites remain an issue. Therefore, the development and application of molecular techniques offer a new alternative approach for identification and diagnosis of *B. procyonis* organism and human baylisascariasis. This is even more so in an era of increasing concern about bioterrorism, as certain characteristics of *B. procyonis* (e.g., low infection dose and hardiness) make it a potential bioterrorist agent.

REFERENCES

1. Gavin, P.J., Kazacos, K.R., and Shulman, S.T., Baylisascariasis. *Clin. Microbiol. Rev.*, 18, 703, 2005.
2. Averbeck, G. et al., Differentiation of *Baylisascaris* species, *Toxocara canis*, and *Toxascaris leonina* infections in dogs. *Compendium*, 17, 475, 1995.
3. Shafir, S.C. et al., Viability of *Baylisascaris procyonis* eggs. *Emerg. Infect. Dis.*, 17, 1293, 2011.
4. Snyder, D.E., and Fitzgerald, P.R., The relationship of *Baylisascaris procyonis* to Illinois raccoons (*Procyon lotor*). *J. Parasitol.*, 71, 596, 1985.
5. Eberhard, M.L. et al., *Baylisascaris procyonis* in the metropolitan Atlanta area. *Emerg. Infect. Dis.*, 9, 1636, 2003.
6. Moore, L. et al., *Baylisascaris procyonis* in California. *Emerg. Infect. Dis.*, 10, 1693, 2004.
7. Blizzard, E.L. et al., Distribution, prevalence, and genetic characterization of *Baylisascaris procyonis* in selected areas of Georgia. *J. Parasitol.*, 96, 1128, 2010.
8. Blizzard, E.L. et al., Geographic expansion of *Baylisascaris procyonis* roundworms, Florida, USA. *Emerg. Infect. Dis.*, 16, 1803, 2010.
9. Park, S.Y. et al., Raccoon roundworm (*Baylisascaeis procyonis*) encephalitis: Case report and field investigation. *Pediatrics*, 106, 56, 2000.
10. Evans, R., *Baylisascaris procyonis* (Nematoda; Ascaridoidea) eggs in raccoon (*Procyon lotor*) latrine scats in Orange County, California. *J. Parasitol.*, 88, 189, 2002.
11. Kazacos, K., Raccoon ascarids as a cause of larval migrans. *Parasitol. Today*, 2, 253, 1986.
12. Kazacos, K.R., and Boyce, W.M., *Baylisascaris* larva migrans. *J. Am. Vet. Med. Assoc.*, 195, 894, 1989.
13. Goldberg, M.A. et al., Diffuse unilateral subacute neuritis. Morphometric, serologic and epidemiologic support for *Baylisascaris* as a causative agent. *Ophthalmology*, 100, 1695, 1993.
14. Murray, W.J., and Kazacos, K.R., Raccoon roundworm encephalitis. *Clin. Infect. Dis.*, 39, 1484, 2004.
15. Gavin, P.J. et al., Neural larval migrans caused by the raccoon roundworm *Baylisascaris procyonis*. *Pediatr. Infect. Dis. J.*, 21, 971, 2002.
16. Kelly, T.G. et al., Spinal cord involvement in a child with raccoon roundworm (*Baylisascaris procyonis*) meningoencephalitis. *Pediatr. Radiol.*, 2011 Jun 1. [Epub ahead of print]
17. Huff, D.S. et al., The first fatal *Baylisascaris* infection in humans. *Pediatr. Pathol.*, 2, 345, 1984.
18. Fox, A.S. et al., Fatal eosinophilic meningoencephalitis and visceral larva migrans caused by the raccoon ascarid *Baylisascaris procyonis*. *N. Engl. J. Med.*, 312, 1619, 1985.
19. Mets, M.B. et al., Eye findings of diffuse unilateral subacute neuroretinitis and multiple choroidal infiltrates associated with neural larva migrans due to *Baylisascaris procyonis*. *Am. J. Ophthalmol.*, 135, 888, 2003.
20. Baeur, C., and Gey, A., Efficacy of six anthelmintics against luminal stages of *Baylisascaris procyonis* in naturally infected raccoons (*Procyon lotor*). *Vet. Parasitol.*, 60, 155, 2002.

21. Page, L.K., Swihart, R.K., and Kazacos, K.R., Implications of raccoon latrines in the epizootiology of baylisascariasis. *J. Wildl. Dis.*, 35, 474, 1999.
22. Sorvillo, F. et al., *Baylisascaris procyonis*: An emerging helminthic zoonosis. *Emerg. Infect. Dis.*, 8, 355, 2002.
23. Weller, P.F., Eosinophilic meningitis. *Am. J. Med.*, 95, 250, 1993.
24. Moertel, C.L. et al., Eosinophil-associated inflammation and elaboration of eosinophil-derived proteins in two children with raccoon roundworm (*Baylisascaris procyonis*) encephalitis. *Pediatrics*, 108, e93, 2001.
25. Cunningham, C.K. et al., Diagnosis and management of *Baylisascaris procyonis* infection in an infant with nonfatal meningoencephalitis. *Clin. Infect. Dis.*, 18, 868, 1994.
26. Boyce, W.M., Branstetter, B.A., and Kazacos, K.R., *In vitro* culture of *Baylisascaris procyonis* and initial analysis of larval excretory-secretory antigens. *Proc. Helminthol. Soc. Washington*, 55, 15, 1988.
27. Boyce, W.M. et al., Physiochemical characterization and monoclonal and polyclonal antibody recognition of *Baylisascaris procyonis* larval excretory-secretory antigens. *J. Parasitol.*, 75, 540, 1989.
28. Hamann, K.J. et al., Immunofluorescent localization of eosinophil granule major basic protein in fatal human cases of *Baylisascaris procyonis* infection. *Am. J. Trop. Med. Hyg.*, 40, 291, 1989.
29. Dangoudoubiyam, S. et al., Molecular cloning of an immunogenic protein of *Baylisascaris procyonis* and expression in *Escherichia coli* for use in developing improved serodiagnostic assays. *Clin. Vaccine Immunol.*, 17, 1933, 2010.
30. Dangoudoubiyam, S. et al., Recombinant antigen-based enzyme-linked immunosorbent assay for diagnosis of *Baylisascaris procyonis* larva migrans. *Clin. Vaccine Immunol.*, 18, 1650, 2011.
31. Pontes, T. et al., Molecular characterization of larval anisakid nematodes from marine fishes of Madeira by a PCR-based approach, with evidence for a new species. *J. Parasitol.*, 91, 1430, 2005.
32. Abe, N., Tominaga, K., and Kimata, I., Usefulness of PCR-restriction fragment length polymorphism analysis of the internal transcribed spacer region of rDNA for identification of *Anisakis simplex* complex. *Jpn. J. Infect. Dis.*, 59, 60, 2006.
33. Chen, Q. et al., Specific PCR assays for the identification of common anisakid nematodes with zoonotic potential. *Parasitol. Res.*, 104, 79, 2008.
34. Umehara, A. et al., Multiplex PCR for the identification of *Anisakis simplex* sensu stricto, *Anisakis pegreffii* and the other anisakid nematodes. *Parasitol. Int.*, 57, 49, 2008.
35. Dangoudoubiyam, S., Vemulapalli, R., and Kazacos, K.R., PCR assays for detection of *Baylisascaris procyonis* eggs and larvae. *J. Parasitol.*, 95, 571, 2009.
36. Fang, W. et al., Multiple primer PCR for the identification of anisakid nematodes from Taiwan Strait. *Exp. Parasitol.*, 124, 197, 2010.
37. Leles, D. et al., ITS1 intra-individual variability of *Ascaris* isolates from Brazil. *Parasitol. Int.*, 59, 93, 2010.
38. Testini, G. et al., New insights into the morphology, molecular characterization and identification of *Baylisascaris transfuga* (Ascaridida, Ascarididae). *Vet. Parasitol.*, 175, 97, 2011.
39. Gatcombe, R.R. et al., Evaluation of a molecular beacon real-time PCR assay for detection of *Baylisascaris procyonis* in different soil types and water samples. *Parasitol. Res.*, 106, 499, 2010.
40. Walker, M.D., and Zunt, J.R., Neuroparasitic infections: Nematodes. *Semin. Neurol.*, 25, 252, 2005.
41. Page, K. et al., Reducing *Baylisascaris procyonis* roundworm larvae in raccoon latrines. *Emerg. Infect. Dis.*, 17, 90, 2011.

48 Brugia

Kosum Chansiri and Supatra Areekit

CONTENTS

48.1 INTRODUCTION

Lymphatic filariasis, more commonly known as elephantiasis, is a significant human health problem in many developing tropical and subtropical countries around the world, infecting ~120 million people. It is mainly caused by three species of filaria nematodes, *Wuchereria bancrofti*, *Brugia malayi*, and *Brugia timori*, which are localized in human lymphatics and are associated with the development of pathological lesions of lymphatic filariasis. All three species are transmitted by mosquitoes and the most number of cases of lymphatic filariasis occur in Southeast Asia, the Indian subcontinent, and Africa. Of the three parasite species, *W. bancrofti* causes 90% and is widely distributed in endemic regions such as Africa, India, the Pacific islands, South America, and Southeast Asia. Ten percent are caused by *B. malayi*, which is widely distributed in Asian countries such as China, Korea, India, Indonesia, Malaysia, the Philippines, and Sri Lanka. *B. timori* infection occurs in Indonesia (islands of Alor, Flores, and Timor) [1,2]. The adult filarial parasites live in the lymphatic vessels near the lymph nodes where they distort the vessels and cause local inflammation. As the parasites accumulate in the blood vessels, they can restrict circulation and cause the surrounding tissue to become enlarged in the excessively infected area. The Malayan filariasis caused by *B. malayi* affects the legs, below the knees. Repeated episodes of inflammation lead to blockade of the lymphatic system, especially in the genitals and legs. The worms in the lymphatic system cause tissue changes which restrict normal flow of lymph and result in swelling, fibrosis, and eventually secondary infection in the affected tissue; the lower extremities and groin are the parts most likely to be affected. This causes the most common, visible signs of infection, with the excessively affected area becoming grossly enlarged with thickened and coarse skin, a clinical syndrome that is conveniently known as elephantiasis [3,4].

The *B. malayi* that infect human have been classified on the basis of their microfilarial periodicity into "periodic" and "subperiodic" forms. These variants have minor morphological differences but can be distinguished through their vectors and mammalian host range and distribution. There are some differences in mosquito vectors of *B. malayi* infection. The nocturnally periodic microfilaria type is transmitted by the mosquitoes which include *Mansonia uniformis* and *Mansonia indiana Indiana* group or some *Anopheles* species, while the nocturnally subperiodic microfilaria Malayan type is transmitted by *Mansonia dives* and *Mansonia bonneae*. The prevalence of the latter is facilitated by many animal reservoirs including domestic cats, forest-leaf monkeys, and some wild animals that contribute to the difficulty and ineffectiveness in the control program of filariasis in the country.

48.1.1 Classification, Morphology, and Biology

48.1.1.1 Classification

Human filariasis is classified as shown below [2].

Phylum	Nematoda
Class	Secernentea
Subclass	Spiruria
Order	Spirurida
Suborder	Spirurina
Superfamily	Filarioidea
Family	Filariidae
Genus	Brugia spp.
Species	*B. malayi, Brugia pahangi, and B. timori*

48.1.1.2 Morphology

The adult worms of *B. malayi* are smaller than those of *W. bancrofti*. There are also some differences in the pattern of papillae and spicules. The morphology and size of the copulatory spicules are one of the most important diagnostic characters among the filariae. All Malayan genera have dissimilar and usually unequal spicules. Female worms are 43–55 mm in length by 130–170 μm in width, and males are 13–23 mm in length by 70–80 μm in width. Adults produce microfilariae, measuring 177–230 μm in length and 5–7 μm in width. The microfilariae of *B. malayi* have two terminal nuclei that are distinctly separated from the other nuclei in the tail [5,6]. The last terminal nucleus is quite small and located at the tip of the tail. The sheath stains deep purple with hematoxylin stain (Figures 48.1 and 48.2).

48.1.1.3 Morphology of *B. malayi* and *B. pahangi*

The basic structure of microfilaria for brugian filariae has been investigated by a number of workers. However, Fulleborn proposed a standard method for the description and measurement of various specialized structures of microfilaria in 1913. Certain "fixed points" were designated, and their relative positions in percentages to the total length were found to be stable figures for each species, regardless of the variations in size. The following are the main characteristics useful in the identification of species. The microfilaria of *Brugia* can be differentiated by the measurement of the body length, body weight, mouth diameter, hook length, annulations width, tail length, and morphology of sheath as shown in Table 48.1 [8].

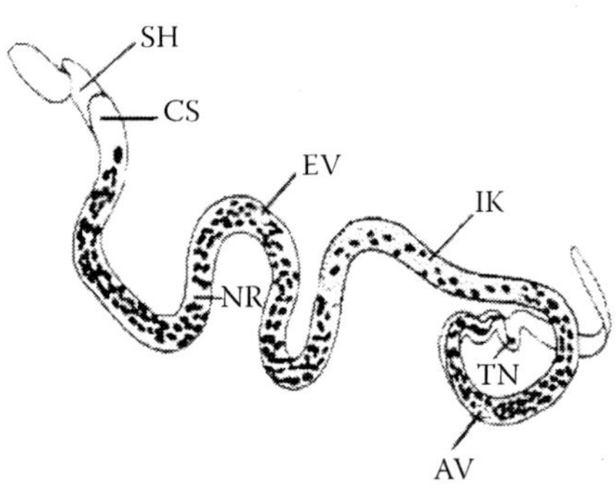

FIGURE 48.1 Microfilaria of *B. malayi*. AV = anal vessicle, CS = cephalic space, EV = excretory vessicle, IK = innenkoper, NR = nerve ring, SH = sheath, and TN = terminal nuclei.

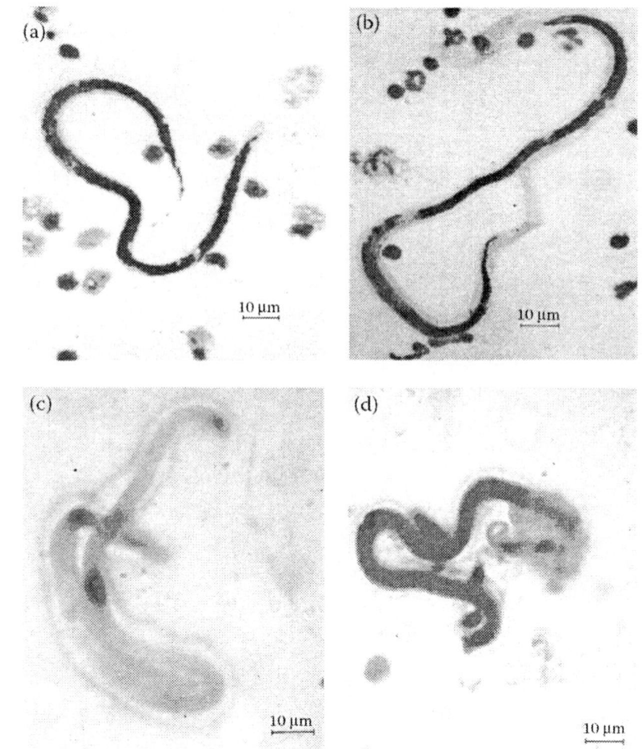

FIGURE 48.2 Geimsa stain of blood smear. Microfilaria of *B. malayi* in human (a) and of *B. pahangi* in cat (b). Acid phosphatase stain of blood smear. Microfilaria of *B. malayi* in human (c) and of *B. pahangi* in cat (d). (Adapted from Chansiri, K. et al. *Mol. Cell. Probes* 16, 129, 2002.)

The apparent difference in the length of the "Innenkorper" of *B. pahangi* and subperiodic *B. malayi* was also been considered. The "Innenkorper" of *B. pahangi* being 21.7% (1/4.6) of the whole length of the microfilaria as against only 13.8% (1/7.3) of subperiodic *B. malayi*. Comparison of the absolute length of the "Innenkorper," 53.1 μm (range 44–63 μm) in

TABLE 48.1
Comparison of the Size and the Dimension (in μm) of Surface Structures of Microfilaria of *B. malayi* and *B. pahangi*

	B. malayi	*B. pahangi*
1. Body length (mm)	210 ± 15	218 ± 32
2. Body width (μm)	3.88 ± 0.68	3.59 ± 0.46
3. Mouth diameter (μm)	0.26	0.26
4. Hook length (μm)	0.76	0.46
5. Annulations width (μm)	0.34 ± 0.04	0.36 ± 0.02
6. Tail length (μm)	5.41 ± 1.34	3.06 ± 0.51
7. Morphology of sheath (μm)	Translucent, smooth	Granular sheath

Source: Adapted from Saitongdee, P. In: *Anatomy*, Mahidol University, pp. 148, 1989.

B. pahangi and 30.7 μm (range 24–34 μm) in subperiodic *B. malayi*, also indicates a marked difference; there is no overlap between the longest measurement in the latter and the shortest in the former [10].

Identification of *B. malayi* and *B. pahangi* by the use of their adult worms is normally performed by detailed examination of the male spicules. The adults of *B. malayi* are delicate, whitish, thread-like round worms and coiled up in pairs in dilated lymphatics. Mature females vary in length from 43.5 to 55 mm and in breadth from 130 to 170 μm. The mature males vary in length from 13.5 to 23.3 mm and in breadth from 70 to 80 μm.

The adults of *B. pahangi* are whitish and thread-like round worms. The body cuticle is unstained throughout (except in the ventral surface of the male tail) and the bulbous head and papillae are of *Wuchereria* type. The esophagus is more clearly visible into two parts in both sexes than in *B. malayi*. The anterior muscular part is thinner than the glandular part and the junction of the two can easily be seen. The female measures 40 mm in length and 100 μm in breadth. The males vary in length from 12.6 to 17.3 mm and in breadth from 70 to 80 μm. The spicule of *B. pahangi*, a very marked structural difference, is seen in the left spicule whose distal part is much shorter and somewhat thinner than that of the *B. malayi* with the lack of spatulate termination characteristic [11].

The microfilarial stage of *B. malayi* and *B. pahangi* is engulfed with a sheath. The sheath could be used to differentiate species since the sheath of *B. pahangi* was granular [12], whereas that of *B. malayi* is translucent and rather smooth. The unsheathed microfilaria of both species shared a common pattern of the cuticular surface structure. The anterior end carries the cephalic cap with the hook, mouth, amphidial openings, and spines. The cuticle shows annulations.

At the third-stage larvae, the structure at the anterior extremity of *B. pahangi* infective larva is very similar to that of *B. malayi* [13]. The anterior ends of *B. malayi* and *B. pahangi* show two circles of four cephalic papillae each; each circle consists of two subdosal and two subventral papillae [14] around the mouth and two amphidial opening lined at the inner circle. The anal opening is situated at posterior region. The cuticle is transverse annulations. The posterior end has two ventrolateral papillae and one dorsoterminal papillae. The ventrolateral papillae in *Brugia* have an indentation around its base which is wider in *B. malayi* than in *B. pahangi*. Futhermore, the ventrolateral papillae of *Brugia* is round-shaped and those of *B. malayi* are more protruded and bigger than that of *B. pahangi*.

48.1.1.4 Life Cycle

The life cycle of the nematodes' parasitic filarial worms *W. bancrofti*, *B. malayi*, and *B. timori* that cause lymphatic filariasis is similar. They are dwelling in the lymphatic system, the network of nodes and vessels that maintain the delicate fluid balance in the tissues, blood and an essential component of the body's immune defense system. They live for years and produced millions of immature microfilaria that circulate in the blood. When they are picked up by mosquitoes, the microfilariae develop to infective larva stage (L3) and can be transmitted to the other.

Adult male and female of *B. malayi* are located in the lumen of lymphatic. They may live for 7–10 years. Cases of 40 year longevity have been reported [15]. Adult female filariae are viviparous, producing thousands of microfilariae which are circular in the blood, after fertilization by adult male parasite. Microfilariae of *B. malayi* retain the egg membrane as a sheath and migrate from the lymphatic into the bloodstream. The microfilaria are ingested by the mosquito where they lose the sheath and within an hour they escape from the gut contents by penetrating the surrounding epithelium. A hook-like device may serve to breach the midgut wall [16] and eventually reach the thoracic musculature or fatty tissues of the mosquitoes. In the thoracic muscles, the parasite is shortened by metamorphosis into first stage larvae (L1). A sausage-shaped L1 larva of *B. malayi* at 2.5 days has a mean length of 143.7 ± 3.8 μm and a maximum width of 16.5 ± 0.7 μm; at about 5 days the L1 moults become the second stage (L2) larva with a mean length of 441 ± 27 μm and a maximum width of 29.4 ± 1.8 μm. It is slightly more active than the L1. By the 9th to 10th day, the L2 moults become the third or infective stage (L3) with a mean length of 1503 ± 40.6 μm and a maximum width 30.5 ± 0.7 mm. The L3, which is a very active stage at maturity, migrates mainly to the proboscis but can also be seen in other parts of the insect body. When the infective mosquito bites a susceptibility host, the L3 larvae are deposited on the skin surface during the process of probing by mosquito prior to puncture with the proboscis. After withdrawal of the proboscis, the L3 actively migrates into the wound and travels to afferent lymphatic and subcapsular sinus. Approximately 9–10 days after entry into the definitive host, the L3 mounts to the fourth stage (L4). In 2-week-old experimental infections in the leaf monkey *Presbytis melalophos*, *B. malayi* female and male L4 larvae have a mean length of 4710 ± 214.3 and 4236 ± 120.8 μm with a maximum width of 38.6 ± 2.7 and 37.9 ± 1.9 μm, respectively. The final moult (L5) occurs ~35–40 days after initial infection; by the 5th week, juvenile adults of *P. melalophos* have mean lengths of 13.7 ± 0.3 μm and 12.6 ± 0.5 mm in female and male [17] with the maximum width of 76.4 ± 2.8 and 64.7 ± 2.2 μm, respectively. Sexual maturity occurs 2–3 months postinfection in *B. malayi* with mean prepatent periods of 67 days in *P. melalophos* and 80 days in cats and jirds [18] and 3.5 months in human [19] (Figure 48.3).

48.1.2 Clinical Features and Pathogenesis

The pathology associated with lymphatic filariasis results from a complex interplay of the pathogenic potential of the parasite, the immune response of the host. The adult worms cause lymphatic inflammation and obstruction in the scrotum, spermatic cords, epididymis, testis, arms, legs, abdomen, and retroperitoneal area. Lymphangitis and lymphadenitis are thought to be allergic reactions to worm secretions or antigens liberated from dead filariae periodically so that the attacks of chill and fever

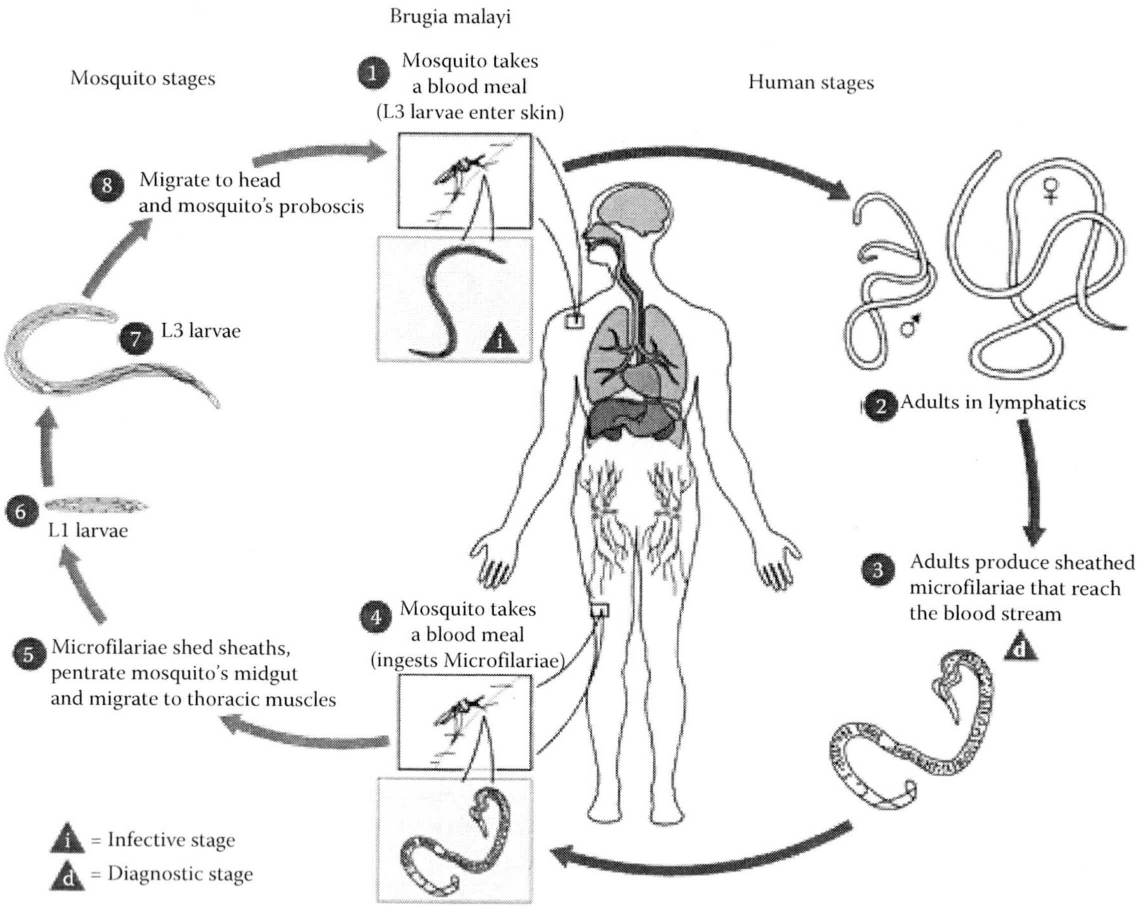

FIGURE 48.3 Life cycle of *B. malayi*. (Adapted from Life cycle of *B. malayi*, from URL:http://www.dpd.cdc.gov/dpdx/HTML/ImageLibrary/Filariasis_il.asp?body = AF/Filariasis/body_Filariasis_il17.htm. Filariasis, ed. Image Library, 2005.)

are recurrent. Obstruction almost certainly results from simple mechanical blockade by the worm. Lymph varicose is among the early signs of lymphatic obstruction, and if they rupture, lymph (chyle) is released into the peritoneum, kidneys, bladder, scrotum, or bowel. Another pathological condition associated with lymphatic filariasis is chyluria. Chyluria and chylous as cited are manifestations of this phenomenon. Impediment of the smaller subcutaneous lymphatic vessels produces lymphedema of the scrotum, vulva, legs, arms, and breast. As lymphedema leads to subcutaneous fibrosis and hyperkeratosis, the skin takes on the character of an elephant hide and becomes thick, hard, and dry from loss of sweat glands.

The position of elephantiasis relies on which lymph nodes are obstructed. Obstruction of the para-aortic nodes blocks drainage of lymph from the tunica virginals, epididymis, and spermatic cord in order that hydrocele develops [21] exclusive of edema of the scrotal skin, which drains into the superficial inguinal glands. In patients with elephantiasis of the leg, the obstruction is in the inguinal or iliac group of nodes.

Contrary to the adult worms, the microfilaria contributes little to the pathology of this disease. Often, there is a negative correlation between lymphatic obstruction and positive blood smears or microfilaria. Sometimes, perhaps 20,000 microfilaria per millilitere of blood are observed from people showing no clinical symptoms, while patients with elephantiasis may have negative blood smears. It is probably because the adults have died before lymphatic injury and become severe enough to produce clinical disturbances. A positive correlation exists between clinical severity and the development of rapid infection with many worms.

48.1.3 Diagnosis

48.1.3.1 Conventional Techniques

The diagnosis of lymphatic filariasis has traditionally depended on the laboratory examination of blood extracted between 10 p.m. and 2 p.m. when the worm is most active. Parasitological techniques involve the demonstration and identification of the presence of circulating microfilariae in various body fluids, for example, hydrocele fluid and chylous urine. Normally, 60 μL thick blood smear films are prepared from peripheral blood taken at night and are then allowed to dry preferably overnight and stained with diluted Giemsa of Field's stain.

48.1.3.1.1 Thick Blood Film Technique (Giemsa Stain Technique)

The conventional thick blood film is still the best method in the field work, but it is limited, which depends on the presence of microfilariae in blood at night time. Despite its simplicity and low cost, the sensitivity is dependent on the volume of blood sample (typically 20–60 μL), the time of blood collection, and skill and dedication of the microscopy. In addition, the method cannot morphologically distinguish between *B. malayi* and *B. pahangi.*

48.1.3.1.2 Filtration Technique

One of the diagnostic tools for the identification of filaria parasites is the membrane filtration. It is more sensitive than thick blood smear but is also labor-intensive and time consuming. The capillary tube technique is a rather rapid assessment to work efficiently in the field when an immediate result is required.

48.1.3.1.3 Acid Phosphatase Activity Staining

Blood samples of infected cats containing microfilaria of *Brugia* spp were smeared on a glass slide and air dried. Then, the slides were fixed in acetone for 1 min at 4°C prior to incubation in Naphthol AS-TR phosphate and pararosaniline hydrochloride solution, and the slides were washed with distilled water for 2–3 times, transferred to 95% alcohol solution and observed under a light microscope. The technique has been employed for the differentiation of these two species, but it is not reproducible and is complicated.

48.1.3.1.4 Scanning Electron Microscopy

The scanning electron microscopic study of the structural differentiation among the three stages: microfilaria, third-stage larva, and adults of *B. malayi* and *B. pahangi* give more valuable information. Therefore, this method might be utilized for precise characterization of species. The life cycles of human and animal filariae are known to be potentially zoonotic, especially in areas where more than one host species for the parasite co-exist.

48.1.3.1.5 Antibody Test

Antibody test seems to be an alternative method for detection of infected people. New antibody tests based on clone antigens may overcome the cross-reactivity difficulty. These may be able to replace night blood sampling. In addition, they may also be useful tools in epidemiological surveys [4] but do not distinguish between active and past infections [22,23]. Antibody titers are not correlated with infection intensity (or worm load) in the patient.

48.1.3.2 Molecular Techniques

48.1.3.2.1 PCR-Based Methods Technique

Alternatively, filarial DNA can be detected by PCR which is very sensitive and species specific but is imperfectly performed with contamination in the field condition or differentiation of the genetically closely related species such as *B. malayi* and *B. pahangi.* The PCR-based methods such as PCR-RFLP have been introduced to use as tools for discrimination of *B. malayi* and *B. pahangi* [7,24–27]. Despite the sensitivity and specificity of PCR, the analysis based on this method may not be effective for the identification of mixed infection between closely related species. Moreover, the method is expensive and impractical for routine use in endemic area.

48.1.3.2.2 Molecular Phylogenetic Studies

The phylogenetic analysis of these specific and informative genes or regions can explain the relationship and diversity of the interested organisms. Among those genes, the 18S rDNA has been widely used for the classification and identification of inter-species. Nevertheless, the previous examination of 18S DNA phylogenetic trees using Neighbor-Joining and DNA maximum likelihood analysis could not differentiate *B. malayi* from *B. pahangi.* The nucleotide comparison of 18S rDNA between these two species revealed only some of single-nucleotide polymorphism (SNP), which is not sufficient for using them as the specific markers of each species. The use of the internal transcribed spacer (ITS) region has been demonstrated to be informative for its utility in the analysis of the intra-species due to its variability of some specific nucleotide sequences.

In 2009, Areekit et al. reported the genetic diversity of *Brugian* filaria in cat reservoirs based on ITS information [28]. The techniques can be used for diagnosis of a mixed infection of *B. malayi* and *B. pahangi* in a single domestic cat using the ITS1 region [29].

48.1.3.2.3 Real-Time PCR and High-Resolution Melting

High-resolution melting (HRM) real-time PCR was first demonstrated by Ririe and coworkers with the double-stranded DNA-specific dye SYBR Green I and has since seen widespread adoption in real-time PCR applications [30]. Melting-curve analysis provides immediate practical benefits in real-time PCR, obviating the need for gel electrophoresis by providing a reproducible signature of the amplified DNA sequence that may be used for typing PCR products [31]. Typing is typically achieved by examining the first derivative of the melting curve and identifying the characteristic "melt peak" (T_m), which is the temperature at which the rate of fluorescence change (DNA denaturation) is highest and is observed in the raw data as a sudden decrease in fluorescence [31]. The melting-curve assays can be applied for targeting SNPs. The technique is probe-free and dramatically reduces the amount of sequencing. This assay also needs no multiplexing and a decreased risk of PCR carryover contamination providing a new approach for genetic screening of closely related species in a clinical laboratory. HRM is a simple and rapid scanning method for the identification of two closely related *B. malayi* and *B. pahangi* [32].

48.2 METHODS

48.2.1 Sample Preparation

48.2.1.1 Isolation of Parasites

Five milliliters of microfilaria-infected blood are taken from host and transferred to a test tube containing 7 mg/mL of EDTA as an anticoagulant agent. The blood is diluted with an equal volume of phosphate-buffered saline, pH 7.0 (PBS; 0.137 M NaCl, 10 mM Na_2HPO_4, 3.2 mM KH_2PO_4), and filtered through a 5-μm polycarbonate membrane (Minipore). Microfilariae is then resuspended in PBS and centrifuged at 5000 rpm for 10 min at 4°C. The pellet of microfilariae is washed with PBS for three times prior to storage at −70°C until use.

48.2.1.2 Isolation of Genomic DNA

The microfilaria pellet is digested in 300 μL of cell lysis solution (0.01 M Tris, 2% SDS, 0.01 M EDTA, pH 8) and 1.5 μL of proteinase K solution (200 mg/mL) at 55°C for 4–6 h. After that, the lysate is further digested by the addition of 0.12 unit of RNase A solution (1 mg/mL in 5 mM Tris-HCl, pH 8.0) and incubated at 37°C for 15–60 min. A 100 μL portion of protein precipitation solution (3 M ammonium acetate, pH 7.4) is added to the cell lysate, quickly spinned at high speed for 20 s and centrifuged at 13,000 rpm for 3 min. Supernatant containing DNA is transferred to a new tube, precipitated by the addition of 300 μL of 100% isopropanol, and centrifuged at 13,000 rpm for 3 min. The supernatant is removed and DNA pellet was washed once with 300 μL of 70% ethanol. The sample is mixed and centrifuged at 13,000 rpm for 3 min. DNA pellet is dried (air dried) for 10 min, dissolved in 20 μL of DNA hydration solution (TE; 0.01 M Tris, 0.001 M EDTA, pH 7.4) and incubated at 65°C for 60 min. The genomic DNA of parasites is stored at 4°C until use. DNA concentration is determined by using an ultraviolet (UV) spectrophotometer and the absorbance is measured at 260 nm. Molar extinction of DNA is calculated as 1 O.D. at 260 nm = DNA concentration of 50 ng/mL.

48.2.2 Detection Procedures

48.2.2.1 PCR Detection

Primers that are derived from the genes encoding internal transcribed sequences (ITS) are useful for investigating the intraspecies variation among *Brugia* spp. These primers are designed on the basis of the nucleotide sequence comparison among ITS genes of *Brugia* spp. The ITS primers for PCR amplification consist of ITS1F (5′-GGT GAA CCT GCG GAA GGA TC-3′) and ITS2R (5′-AGC GGG TAA TCA CGA CTG -3′).

All reactions were manipulated in 25 μL volume containing 50 ng of genomic DNA in 10× PCR buffer (200 mM Tris-HCl, pH 8.4, 500 mM KCl), 1 μM each of primers, 100 μM each of dATP, dTTP, dCTP, and dGTP, 1.5 mM $MgCl_2$ and 1.5 U of *Taq* DNA polymerase. Sterile distilled water is added to adjust volume to 25 μL. PCR is performed by using DNA thermal cycler for 30 cycles. Each cycle consists of denaturation at 94°C for 1 min, annealing at 65°C for 1 min, and extension at 72°C for 1 min. The PCR products are analyzed by electrophoresis on 1.5% agarose gel at 110 V for ~30 min prior to staining with ethidium bromide and observation under UV light.

48.2.2.2 Analysis of the Nucleotide Sequence of the ITS Regions

The partial and complete nucleotide sequences of ITS were analyzed by using BLAST program from http://www.ncbi.nlm.gov/BLAST. The nucleotide sequence of interested gene fragments was compared to those of various organisms that have been previously submitted in GenBank. The percentage of nucleotide and amino acid sequence homology in comparison to various organisms was reported. All sequence data were analyzed by using Clustal W sequence alignment program for the European Molecular Biology Laboratory (EMBL, U.K.) at http://www.ebi.ac.uk/. All alignment data were inspected with sequence data in GenBank by using PAUP program version 4.0b10 [33] for the construction of evolutionary trees of ITS.

48.2.2.3 Real-Time PCR and HRM

The two closely related sequences used in this study are amplified by normal PCR using specific primers. Real-time PCR mixture is prepared using 50 ng of DNA, 10 μL of 2× PCR mixture (10× PCR buffer, 1 μM each of primers, 100 μM each of dATP, dTTP, dCTP, and dGTP, 1.5 mM $MgCl_2$ and 1.5 U of *Taq* DNA polymerase), 2 μM of dye (e.g., SYTO 9) and primers were mixed to the final concentration of 20 μM. Real-time PCR reactions and HRM analysis were performed using HRM apparatus. The real-time PCR profile comprises one initial cycle of 95°C for 2 min, followed by 40 cycles of denaturation at 95°C for 10 s, annealing at a temperature dependent on the primer set for 15 s, and extention at 72°C for 20 s. After real-time PCR amplification is completed, HRM is performed using melting profile from 72°C to 95°C rising at 0.2°C per second. The melting curves are normalized by software provided by the company.

The specificity of primers used in the experiment is confirmed by a single amplicon obtained using melting analysis. Data from HRM analysis were determined using specific software.

48.3 CONCLUSIONS AND FUTURE PERSPECTIVES

It is known that *B. malayi* and *B. pahangi* are very similar in their morphologies. Species identification by Giemsa staining may not be sufficient, although it is cheap to perform. In addition, most of the coding and noncoding genes of these two species are highly homologous. Previously, molecular techniques such as PCR-based techniques have been used in filarial parasite detection and differentiation. The PCR-RFLP of the ITS1 region could distinguish human *B. malayi* from feline *B. pahangi* [25,34]. Alternatively, the use of gene amplification in real-time PCR analysis and HRM of five key genes/regions, H*ha*I repetitive region (HR), small heat-shock protein (SHP), small subunit ribosomal DNA (18S rDNA), ITS region, and trans-spliced leading Exon I gene

(SLX1) is able to discriminate *B. malayi and B. pahangi* [32]. Nevertheless, PCR-based analysis of the genes mentioned above could not illustrate the genetic diversity of *Brugia* spp. within a specific host. To study the intraspecies variation of *Brugia* spp, the ITS region has been investigated due to its information regarding discrimination of very closely related species. Since the ITS of organisms have ~100–200 copies throughout the genome, differences are likely to be genetic variation rather than mixed infection. In 2009, Areekit and coworkers indicated that the tree inferred from ITS1, ITS2, and completed ITS data can discriminate *B. malayi* and *B. pahangi*. The technique can be applied for the investigation of genetic diversity of *Brugia* spp in cat reservoirs which are beneficial not only for the classification of these nematodes but also for studying the dynamics of these parasites, epidemiological study, and control and eradication of lymphatic filarial in endemic area. To achieve rapid identification of brugian filaria, especially in field work, specific, sensitive, convenient, and low-cost techniques should be explored. Currently, there has been increasing interest in biosensor technology for rapid and sensitive detection. Among them, the piezoelectric biosensor has its own advantages that the detection method is label-free from radioactive or fluorescent tags [35]. In addition, this technique could tenuously exhibit the qualitative and quantitative analysis. It shows high sensitivity to mass on the surface of the quartz crystal with the high specificity of a bioreaction [36]. The thin oscillating gold quartz crystal surface generates the intrinsic resonance frequency by mass attached, adhered or deposited onto the piezoelectric active surface. DNA-based biosensor technology that uses the oligonucleotide hybridization detection method, does not require labeling and is easy to use is desirable. The piezoelectric quartz crystal is one of the candidate devices of biosensor technology for the detection of DNA hybridization. The piezoelectric DNA-based biosensor is used for the determination of change between the frequency of the oligonucleotides probe immobilized on quartz crystal and the frequency after the hybridization of DNA target. The advantage of this technique is in using a nonamplified genomic bacterial DNA target for reducing the risk of PCR carryover. This method will reduce time-consumption, costs, the tedious step of amplification and can be applied for field sample diagnosis.

REFERENCES

1. WHO Lymphatic filariasis. Fourth report of the WHO Expert Committee on Filariasis. *World Health Organ. Tech. Rep. Ser.* 702, 3–112, 1984.
2. Baird, J.K. et al. North American brugian filariasis: Report of nine infections of humans. *Am. J. Trop. Med. Hyg.* 35, 1205, 1986.
3. Cox, F.E. Elimination of lymphatic filariasis as a public health problem. *Parasitol. Today* 16, 135, 2000.
4. Despommier, D. et al. Lymphatic filariae: *Wuchereria bancrofti*, Cobbold 1877) and *Brugia malayi* (Brug 1927). In: Despommier, D.D. and Gwadz, R.W. (eds.), *Parasitic Diseases*, 3 edn. Springer-Verlag, New York, pp. 40–47, 1995.
5. Markell, E. et al. The blood and tissue dwelling nematodes. In: Markell, E.K., John, D.T., and Williams, A. (eds.), *Medical Parasitology*.W.B. Saunders Company, London, p. 304, 1999.
6. Arcari, M., Baxendine, A., and Bennett, C.E. The blood nematodes. http://www.southampton.ac.uk/~ceb/Diagnosis/Vol10.htm
7. Chansiri, K. et al. PCR based method for identification of zoonostic *Brugia malayi* microfilariae in domestic cats. *Mol. Cell. Probes* 16, 129, 2002.
8. Sucharit, S. et al. The differentiation of *brugia malayi, B. pahangi, B. tupaiae* and *Wuchereria bancrofti. Southeast Asian J. Trop. Med. Public Health* 6, 549, 1975.
9. Saitongdee, P. Scanning electron microscopic study of microfilaria, third stage larva and adult of *Brugia malayi, B. pahangi* and third stage larva of *Wuchereria bancrofti*. In: *Anatomy*, Mahidol University, pp. 148, 1989.
10. Sivandam, S., and Fredericks, H. The Innenkorper in differentation between the microfilariae of *Brugia pahangi* and *B. malayi* (superiodic form). *Med. J. Malaya* 20, 337, 1966.
11. Buckley, J.J., and Edeson J.F. On the adult morphology of *Wuchereria* sp. (malayi?) from a monkey (*Macaca irus*) and from cats in Malaya, and on *Wuchereria pahangi* n. sp. from a dog and a cat. *J. Helminthol.* 30, 1, 1956.
12. Suguri, S. Ultrastructure of the microfilaria of *Brugia pahangi* (Buckley and Edeson, 1956) Buckley, 1958. *Acta Med. Okayama* 31, 295, 1977.
13. Lim P. et al. Scanning electron microscope study of subperiodic *Brugia malayi* and *Brugia pahangi* infective larvae (Filarioidae: Onchocercidae). *Trop. Bio Med.* 1, 29, 1984.
14. Aoki Y. et al., Scanning electron microscopy of third- and fourth-stage larvae and adults of *Brugia pahangi* (Nematoda: Filarioidea). *J. Parasitol.* 66, 449, 1980.
15. Leeuwin, R.S. Microfilaraemia in Surinamese living in Amsterdam. *Trop. Geogr. Med.* 14, 355, 1962.
16. Esslinger, J.H. Development of *Porocephalus crotali* (Humboldt, 1808) (Pentastomida) in experimental intermediate hosts. *J. Parasitol.* 48, 452, 1962.
17. The World Health Report 1995—Bridging the gaps. *World Health Forum* 16, 377, 1995.
18. Schacher, J.F. Morphology of the microfilaria of *Brugia pahangi* and of the larval stages in the mosquito. *J. Parasitol.* 48, 679, 1962.
19. Smyth, J. *Introduction to Animal Parasitology: Nematoda: General Account*, 3rd edn. Cambridge University Press, Cambridge, p. 369, 1994.
20. Life cycle of *B. malayi*. In: from URL:http://www.dpd.cdc.gov/ dpdx/HTML/ImageLibrary/Filariasis_il.asp?body = AF/Filariasis/body_Filariasis_il17.htm. Filariasis, ed. Image Library, 2005.
21. Jordan, P. Notes on elephantiasis and hydrocoele due to *Wuchereria bancrofti. J. Trop. Med. Hyg.* 58, 113, 1955.
22. Chandrashekar, R. et al. Molecular cloning of *Brugia malayi* antigens for diagnosis of lymphatic filariasis. *Mol. Biochem. Parasitol.* 64, 261, 1994.
23. Dissanayake, S. et al. Evaluation of a recombinant parasite antigen for the diagnosis of lymphatic filariasis. *Am. J. Trop. Med. Hyg.* 50, 727, 1994.
24. Fischer, P. et al. PCR-based detection and identification of the filarial parasite *Brugia timori* from Alor Island, Indonesia. *Ann. Trop. Med. Parasitol.* 96, 809, 2002.
25. Nuchprayoon, S. et al. Detection and differentiation of filarial parasites by universal primers and polymerase chain reaction-restriction fragment length polymorphism analysis. *Am. J. Trop. Med. Hyg.* 73, 895, 2005.

26. Rishniw, M. et al. Discrimination between six species of canine microfilariae by a single polymerase chain reaction. *Vet. Parasitol.* 135, 303, 2006.
27. Thanomsub, B. et al. Differential diagnosis of human lymphatic filariasis using PCR-RFLP. *Mol. Cell. Probes* 14, 41, 2000.
28. Areekit, S. et al. Intraspecies variation of *Brugia* spp. in cat reservoirs using complete ITS sequences. *Parasitol. Res.* 104, 1465, 2009.
29. Areekit, S. et al. Molecular genetics analysis for co-infection of *Brugia malayi* and *Brugia pahangi* in cat reservoirs based on internal transcribed spacer region 1. *Southeast Asian J. Trop. Med. Public Health* 40, 30, 2009.
30. Wilhelm, J., and Pingoud, A. Real-time polymerase chain reaction. *Chembiochem* 4, 1120, 2003.
31. Ririe K.M. et al. Product differentiation by analysis of DNA melting curves during the polymerase chain reaction. *Anal. Biochem.* 245, 154, 1997.
32. Areekit, S. et al. High resolution melting real-time PCR for rapid discrimination between *Brugia malayi* and *Brugia pahangi*. *J. Med. Assoc. Thai.* 92, 24, 2009.
33. Swofford, D.L. *PAUP (version 4.0b10)*. Sinauer Associates, Sunderland, MA, 2002.
34. Nuchprayoon, S. et al. Detection of filarial parasites in domestic cats by PCR-RFLP of ITS1. *Vet. Parasitol.* 140, 366, 2006.
35. Yao, C. et al. Hybridization assay of hepatitis B virus by QCM peptide nucleic acid biosensor. *Biosens. Bioelectron.* 23, 879, 2008
36. Zhou, X. et al. Microgravimetric DNA sensor based on quartz crystal microbalance: Comparison of oligonucleotide immobilization methods and the application in genetic diagnosis. *Biosens. Bioelectron.* 16, 85, 2001.

49 Capillaria

Dongyou Liu

CONTENTS

49.1 INTRODUCTION

The genus *Capillaria* comprises about 300 roundworm species that commonly occur in nature and are capable of parasitizing mammals, birds, fish, invertebrate animals, and plants. Of the three well-known human pathogenic *Capillaria* species, *Capillaria aerophila* resides in the bronchial tissues and the lungs, causing pulmonary capillariasis; *Capillaria hepatica* inhabits the liver, and is responsible for hepatic capillariasis; *Capillaria philippinensis* lives in the intestines and causes intestinal capillariasis.

In general, the human-infecting *Capillaria* species have a relatively complex life cycle, and often induce nonspecific clinical symptoms in humans. Without correct diagnosis and prompt treatment, human capillariasis can have serious consequences, sometimes death. Given their structural similarity, identification of *Capillaria* species based on morphological criteria is often challenging. For this reason, the development and application of immunological and molecular diagnostic procedures have not only helped overcome the difficulty associated with morphological cauterization, but also opened a new avenue for epidemiological tracking and phylogenetic analysis of *Capillaria* isolates recovered from various hosts.

49.1.1 Classification, Morphology, and Biology

49.1.1.1 Classification

The genus *Capillaria* encompasses a large number of round worm species that are classified in the family Trichuridae, superfamily Trichinelloidae, order Trichocephalida, subclass Enoplia, class Enoplea, phylum Nematoda. Being one of the four genera (i.e., *Anatrichosoma*, *Capillaria*, *Cystoopsis*, and *Trichuris*) in the family Trichuridae, the genus *Capillaria* is separated into >300 species [1–3]. Of these, *C. aerophila* (obsolete synonyms: *Trichosoma aerophila*, *Eucoleus aerophilus*, and *Thominx aerophilum*), *C. hepatica* (synonym: *Calodium hepaticum*), and *C. philippinensis* (obsolete synonym: *Aonchotheca philippinensis*; synonym: *Paracapillaria philippinensis*) are involved in human pulmonary, hepatic, and intestinal capillariasis, respectively.

It is notable that while *C. philippinensis* was transferred to the genus *Paracapillaria* (known as *P. philippinensis*), the name of *C. philippinensis* has continued to be widely used in the medical literature [4]. The other important human pathogenic genera in the superfamily Trichinelloidae are *Trichuris* and *Trichinella*.

49.1.1.2 Morphology

Members of the family Trichuridae are characterized by having an esophagus that is surrounded by glandular cells or stichocytes. This esophageal pattern is called stichosomal esophagus.

C. aerophila adult males measure up to 25 mm in length and 62 μm in width; adult females up to 32 mm in length and 105 μm in width. *C. aerophila* eggs (of 64 μm × 29 μm in size) appear lemon-shaped and have a double-layered shell, with radial striations on the outer shell and two polar plugs.

Adult *C. hepatica* has a slender body, whose anterior part is narrow, and whose posterior part swells gradually. The females are of 53–78 mm × 0.11–0.20 mm, while the males are of 24–37 mm × 0.07–0.10 mm. The long esophagus occupies half of the female body, but a third of the male body. The cauda of *C. hepatica* male bears a copulatory spicule and sheath. *C. hepatica* eggs (of 48–66 μm × 28–36 μm in size) appear peanut-shaped with flattened bipolar plugs and a striated shell with numerous minipores in the outer shell, resembling *Trichuris trichiura* eggs, but differing in size.

Adult *C. philippinensis* is very small, with males being 1.5–3.9 mm × 23–28 μm, and females being 2.3–5.3 mm × 29–47 μm. *C. philippinensis* eggs (of 36–45 μm × 20 μm) appear peanut-shaped with a striated shell.

49.1.1.3 Biology

C. aerophila is a common nematode that parasitizes the trachea and bronchi of carnivorous mammals, and causes pulmonary infection in humans. Infestation by *C. aerophila* is referred to as pulmonary capillariasis (occasionally spelled capillariosis), bronchial capillariasis, or thominxosis, as its predilection site is the respiratory tract. *C. aerophila* has a direct life cycle, and can complete its life cycle in a single host [5].

C. aerophila adult worms reside in the mucosa of the trachea and bronchi. Females deposit nonembryonated ova in the lungs, which are coughed up, swallowed, and passed in the feces. The eggs develop (or embryonate) in the soil under favorable conditions. Following ingestion by earthworms, the embryonated eggs hatch to become larvae. Subsequent ingestion of earthworms by animals results in infection. Infection may also be transmitted directly by the ingestion of embryonated eggs (e.g., consumption of insufficiently washed vegetables). Inside the host, the larvae emerging from the embryonated eggs invade the mucosa and migrate to the lungs and the respiratory tissue through the blood and lymphatic vessels.

Originally described in 1839 from specimens found in the trachea of a wolf, *C. aerophila* shows a cosmopolitan distribution in both humans and wild mammals [6]. The following mammals have been reported as hosts for *C. aerophila*: cat (*Felis catus*) and dog (*Canis lupus familiaris*) in Asia, Europe, North America, and South America; American marten (*Martes americana*), American black bear (*Ursus americanus*), badger (*Taxidea taxus*), bobcat (*Lynx rufus*), coyote (*Canis latrans*), gray fox (*Urocyon cinereoargenteus*), kit fox (*Vulpes macrotis*), opossum (*Didelphis virginiana*), raccoon (*Procyon lotor*), wildcat (*Felis silvestris*), and Florida panther (*Puma concolor coryi*) in North America; crab-eating fox (*Cerdocyon thous*) in South America; Egyptian mongoose (*Herpestes ichneumon*), beech marten (or Stone Marten; *Martes foina*); Eurasian Lynx (*Lynx lynx*), European pine marten (*Martes martes*), European polecat (*Mustela putorius*), Iberian lynx (*Lynx pardinus*), and hedgehog (*Erinaceus* sp.) in Europe; golden jackal (*Canis aureus*) in Azerbaijan; arctic fox (*Alopex lagopus*) in Iceland; brown bear (*Ursus arctos caucasicus*) in Russia; Iriomote cat (*Prionailurus iriomotensis*), Japanese marten (*Martes melampus*), and leopard cat (*Prionailurus bengalensis euptailurus*) in Japan; raccoon dog (*Nyctereutes procyonoides*) in Europe and Japan; red fox (*Vulpes vulpes*) in Europe and North America; sable (*Martes zibellina*) in Asia; and human (*Homo sapiens*) in Europe, Iran, and Morocco.

C. hepatica has a direct life cycle with a marked affinity for the liver. The embryonated eggs are ingested by a mammalian host, and hatch in the cecum to become larvae. The larvae invade the intestinal mucosa and migrate via the mesenteric vein and portal vein to the liver. Within the parenchyma of the liver, the larvae undertake three stages of development in 16 days and turn into the fourth larval stage, when sexual differentiation starts. Sexually mature males appear 15–18 days postinfection (p.i.), and sexually mature females appear 15–20 days p.i. After mating, the females begin to lay eggs in 28 days. Overall, the life span of adult worms ranges from 18 to 60 days p.i. Eggs deposited in the liver parenchyma develop to the eight-cell stage and further cleavage of eggs takes place after their release from the liver tissue. Eggs are excreted in the feces either of carnivores or of scavengers feeding on infected mammals, or are released through the decay of an infected carcass. Eggs embryonate and become infective in the soil within 5–8 weeks under appropriate temperature (25–30°C) and moisture. Embryonated eggs remain viable for 25 months at room temperature. The life cycle is completed when the embryonated eggs are ingested by a new host [7]. Interestingly, the ingestion of immature (nonembryonated) eggs does not lead to hepatic capillariasis, nor clinical signs (the so-called the spurious infections), as eggs (ova) simply pass through the gastrointestinal tract of the host and are excreted in the feces [8].

The principal hosts of *C. hepatica* are rodents (*Mus musculus* and *Rattus novergicus*), lagomorphs, *Tamias striatus*, squirrel, mole, shrew, opossum, weasel, skunk, and so on [9–19]. To date, 37 cases of hepatica capillariasis have been reported predominantly in Japan, India, America, Canada, Brazil, Germany, Italy, Korea, China, Thailand, and Czechoslovakia [20–25].

The life cycle of *C. philippinensis* involves freshwater fish as intermediate hosts and fish-eating birds as definitive hosts. Adult *C. philippinensis* inhabits in the intestines of fish-eating birds, and embryonated eggs are shed in bird feces. The eggs embryonate in water. After ingestion by freshwater fish (e.g., *Amaurornis phoenicurus* and *Ardeola bacchus*), the eggs hatch as larvae in the fish intestines. The larvae develop into males and egg-producing female worms in 10–11 days, and the first-generation females produce eggs within 22–24 days. Being part of the life cycle, *C. philippinensis* may autoinfect (i.e., the offspring produced by adults reinfest the same host) leading to hyperinfection. Humans acquire the infection by eating raw or undercooked freshwater fish. Inside the human digestive tract, *C. philippinensis* adult worms are usually situated in the mucosa of the small intestine (between the lower third of the duodenum and the proximal third of the ileum).

C. philippinensis was first identified as a cause of human intestinal capillariasis in the Philippines in 1963 and then in Thailand in 1973. Although *C. philippinensis* is also found in Japan, Laos, Indonesia, Korea, Taiwan, Iran, and Egypt, the major outbreaks have remained in the Philippines and Thailand, especially the northeastern Thailand (including Buri Ram, Khon Kaen, Nakhon Ratchasima, Roi Et, Sakon Nakhon,

Surin, and Udon Thani) [26–36]. Of the 82 accumulated cases of intestinal capillariasis reported in Thailand from 1994 to 2006, about half have consumed raw or undercooked fish such as Koi pla and Taab Kaab bagsit (*Hypseleotris bipartita*). In Thailand, children ages less than 1–9 years had a higher infection rate with *C. philippinensis* than other population groups, whereas in the Philippines and Egypt, most intestinal capillariasis cases were found in middle-aged or the active age group.

Under experimental conditions, fish such as *Cyprinus carpio*, *Puntius gonionotus*, *Rasbora boraperensis* (Pla Sew), *Elotris melanosoma* (birut), *Ambassis commersoni* (bagsang), and *Apagon* sp. (bagsit) yields infective larvae after ingestion of *C. philippinensis* eggs. Similarly, *C. philippinensis* larvae eaten by birds such as *A. phoenicurus*, *A. bacchus*, *Nycticorax nycticorax*, *Bubulcus ibis*, *Ixobrychus sinensis*, *Gallinula chloropus*, and *Rostratula benghalensis* become mature adults.

49.1.2 Clinical Features and Treatment

49.1.2.1 Clinical Features

C. aerophila is responsible for causing pulmonary capillariasis in humans. In the 11 reported cases of human pulmonary capillariasis reported to date, clinical symptoms range from irritation of the respiratory tract, bronchitis, coughing, mucoid or blood-tinged sputum, fever, dyspnea, pneumonia, to eosinophilia [37].

In general, *C. aerophila* infestations do not induce serious clinical symptoms in dogs and cats. However, heavy infestations may result in inflammation of the respiratory tract (rhinitis, tracheitis, or bronchitis), or secondary bacterial infections of the respiratory tract (e.g., bronchopneumonia), with the animals making a whistling sound when breathing, and showing frequent sneezing, or a deep, wheezing cough [8].

In a case of pulmonary capillariasis reported in Serbia, a 68-year-old woman with a 1-month history of mild fever (38°C) and mild weight loss presented with a productive cough with purulent expectorates, fatigue, and anal and skin pruritus. Previous antibiotics (garamycin, azitromycine, cyprofloxacine, and cephachlor) for bronchopneumonia had not relieved her coughing. An x-ray image revealed bilateral bronchopneumonic shadows, especially intense in the right basal region. Computed tomography scan confirmed the shadows as spotted, partially confluent, nodular, and banding infiltrative lesions of the pulmonary parenchyma, and detected lymphadenopathy in the retrocaval space near the tracheal carina with lymph nodes of up to 20 mm in diameter. Bronchoscopy uncovered macro- and microtumors on the right side as well as many lemon-shaped parasite eggs with numerous eosinophils and Charcot–Leyden crystals in the necrotic detritus. Histological examination detected well-preserved nonembryonated eggs, egg shells, and immature eggs with incomplete shells. No larval stage or adult parasites were present in the sputum, biopsy specimens, or stool; no egg was detected in the sputum or stool. On the basis of the egg morphology and size and the localization of the nematode in the bronchus, the parasite was identified as a *C. aerophila*, and confirmed by immunfluorescent antibody test (IFAT).

C. hepatica (syn. *C. hepaticum*) is the cause of hepatic capillariasis that has been reported in less than 40 cases around the world [23,24,38–45]. *C. hepatica* invades primarily the sinus hepaticus, where it matures and lays egg. Both the worms and their eggs induce focal chronic inflammation (attracting macrophages, eosinophils, and some multinucleate giant cells) in the liver, which may persist until the final formation of encapsulation or calcification of dead worms [46]. The encapsulated parasitic lesions continue to release disintegrated products, which activate the Kupffer cells, and promote the development of fibrosis in the liver [47–49]. Classically, hepatic capillariasis displays severe symptoms that mimic acute hepatitis [50,51].

C. philippinensis causes intestinal capillariasis that manifests intermittent or continuous diarrhea, borborygmus, weight loss, abdominal pain, anorexia, vomiting, and general weakness [52–57]. Without treatment, the disease can result in severe muscle wasting, cachexia, and bipedal edema. Chronic intestinal capillariasis leads to malabsorption, protein and electrolyte loss, and death results from irreversible effects of septicemia and heart failure [58–62].

The first case of *C. philippinensis* in Thailand involved an 18-month-old girl, who had diarrhea 2 or 3 times per day for at least 6 months and edema for 3 days. Laboratory examination revealed low albumin and potassium in blood, and *C. philippinensis* eggs, larvae and adult worms in stool. The second case of *C. philippinensis* infection in Thailand concerned a 46-year-old man with watery diarrhea 4–5 times a day. Similar to the first case, the patient's blood contained low level of albumin and potassium, and his feces showed a large number of *C. philippinensis* eggs and adult worms [63].

49.1.2.2 Treatment

The drug of choice for treating capillariasis is mebendazole (200 mg orally twice a day for 20 days) or albendazole (400 mg 1×/day for 10 days). Albendazole is especially useful for treating human intestinal capillariasis as it is effective against *C. philippinensis* eggs, larvae, and adult worms. Other anthelmintics (e.g., levamisole or fenbendazole) are sufficient to cure *C. aerophila* infestations [64]. In addition to anthelmintic drugs, electrolyte replacement and administration of an antidiarrheal agent should be considered.

Unsanitary practices, poor hygienic conditions, and the presence of dense rodent populations are predisposing factors for human capillariasis. To prevent *C. philippinensis* and related infections, it is important to promote the consumption of cooked fish and to avoid defecation into a water resource, and to educate children or adults about the danger of eating soil (geophagia, pica) or soil-to-hand-to-mouth contact in the development of human capillariasis.

49.1.3 Laboratory Diagnosis

As human capillariasis tends to produce nonspecific clinical symptoms, it is difficult to achieve a correct diagnosis without relying on laboratory techniques that detect the eggs and adult worms of *Capillaria* species in specimens [65].

Identification of *Capillaria* eggs and adult worms in clinical specimens (the nasal or tracheal lavage, sputum, or the feces of infested individuals) represents a valid approach for diagnosis of human capillariasis. However, because of their small size and the limited range of characteristic morphological features, species-specific determination of *Capillaria* on the basis of morphological criteria is technically challenging, given the pronounced variation in *Capillaria* morphometrics, host specificity, and habitat (tissue) specificity. Additional differentiations need to be made for *T. trichiura*, *Toxocara canis*, or *Toxocara cati* (the dog and cat roundworms, respectively; symptom, larva migrans visceralis), *Fasciola hepatica*, and *Echinococcus granulosus* or *Echinococcus multilocularis*, *Schistosoma japonicum*. As in the case of *C. hepatica* infection, apart from spurious infection, both adult worms and eggs are absent in the peripheral blood and stools of infected hosts. This may lead to potential confusion with hepatitis B virus and hepatitis C virus infections. Although hepatic capillariasis is verifiable through liver biopsy, the process can be traumatic.

Therefore, the use of immunological and molecular techniques for diagnosis of human capillariasis has considerable appeals in terms of improved convenience and efficiency. Toward this end, indirect immunofluorescence test and enzyme-linked immunosorbent assay have been successfully developed for sensitive and specific diagnose of *C. hepatica* infection [66–70]. Through an analysis of the nucleotide variations in mitochondrial DNA (mtDNA) (e.g., the cytochrome *c* oxidase subunit I (COI)), ribosomal RNA (rRNA), and other gene regions in *Capillaria*, the phylogenetic relationship among *Capillaria* species and isolates can be established [71].

49.2 METHODS

49.2.1 Sample Preparation

49.2.1.1 Sample Collection

Capillaria specimens are dissected from host epithelial tissues, washed several times in physiological saline, and separated into aliquots for morphological and molecular studies. Material for morphological study is fixed in hot (10%), neutral-buffered formalin and stored in small McCartney bottles. Individual nematodes are cleared in lactophenol and allocated to morphospecies based on a suite of morphological characteristics from male and female worms and the ornamentation of the egg shell [71]. Specimens for molecular analysis are placed in individual Eppendorf tubes and frozen immediately in liquid nitrogen. On return to the laboratory, all frozen specimens are transferred to a biofreezer at –70°C [71].

49.2.1.2 DNA Extraction

Genomic DNA is extracted from individual nematodes by sodium dodecyl-sulfate/proteinase K treatment and direct purification over spin columns (Wizard™ DNA CleanUp, Promega) [71].

49.2.2 Detection Procedures

Zhu et al. [71] employed primers JB3 (forward: 5′-TTTT TTGGGCATCCTGAGGTTTAT-3′) and JB4.5 (reverse: 5′-TAAAGAAAGAACATAATGAAAATG-3′) for amplification and sequencing analysis of a portion (~450 bp) of the mitochondrial COI gene.

49.2.2.1 Procedure

1. Polymerase chain reaction (PCR) mixture (50 μL) is composed of 10 mM Tris–HCl pH 8.4, 50 mM KCl, 4 mM $MgCl_2$, 250 μM of each dNTP, 100 pmol of each primer (JB3 and JB4.5), 2 U *Taq* polymerase (Promega), and 10–20 ng of parasite DNA template.
2. Amplification is conducted with an initial 94°C for 5 min, then 30 cycles of 94°C for 30 s, 55°C for 30 s, and 72°C for 30 s, followed by a final 72°C for 5 min.
3. Each PCR product of 5 μL (~450 bp) are examined by agarose gel electrophoresis, and remainder PCR products are purified and subjected to automated DNA sequencing using the same primers as for primary amplification.
4. A 374 bp sequence of COI is aligned manually. Pairwise comparisons are made of the level of sequence differences (*D*) using the formula $D = 1 - (M/L)$, where M is the number of alignment positions at which the two OTUs have a base in common, and L is the total number of alignment positions over which the two OTUs are compared. A phenogram is constructed from the sequence data using the unweighted pair group method using arithmetic averages.

49.3 CONCLUSION

Capillariasis due to infection with *C. aerophila*, *C. hepatica*, and/or *C. philippinensis* is a zoonotic disease that tends to induce nonspecific clinical symptoms and have serious consequences if left untreated. *C. aerophila* adult worms colonize the respiratory system (causing pulmonary capillariasis), and lay eggs in the lungs that are coughed up and swallowed by the host, and are then passed in the feces. The eggs embryonate in the soil, and the embryonated eggs can be ingested by another mammalian host or the same host (direct life cycle), in which the larvae emerge from the embryonated eggs, pass through the intestine tissues, and migrate to the lungs. Alternatively, the embryonated eggs are picked up by earthworm (intermediate hosts) and develop into the infective larvae, and mammalian hosts become infected after consuming the earthworm containing the infective larvae. *C. hepatica* also has a direct lifecycle and establishes in the mammalian hosts after its embryonated eggs are consumed by the hosts. Within the host, the eggs hatch to become larvae. The larvae migrate to the liver and mature into adults, producing eggs, and causing hepatic capillariasis (with manifestations of persistent fever, hepatomegaly, eosinophilia, and death). Of about 40 cases of human infections documented to date, half have been diagnosed by liver biopsy, and the other half have been diagnosed at autopsy. *C.* (*Paracapillaria*) *philippinensis* is a parasitic nematode involving fish as intermediate host and fish-eating birds as definitive host. Humans acquire *C. philippinensis* by eating raw or undercooked fish meat. Because of their relatively complex life cycles, and their lack of morphological distinctions, identification of *Capillaria* species on the basis of morphological characteristics is difficult and technically demanding. Use of molecular procedures offers the opportunity to streamline the diagnosis of capillariasis, and to facilitate the implementation of effective, tailor-made treatments against the diseases.

REFERENCES

1. Butterworth, E.W., and Beverley-Burton, M., The taxonomy of *Capillaria* spp. (Nematoda: Trichuroidea) in carnivorous mammals from Ontario, Canada. *Syst. Parasitol.*, 1, 211, 1980.
2. Moravec, F., Proposal of a new systematic arrangement of nematodes of the family Capillariidae. *Folia Parasitol. (Praha)*, 29, 119, 1982.
3. Moravec, F., Nagasawa, K., and Madinabeitia, I., A new species of *Capillaria* (Nematoda: Capillariidae) from the intestine of the marine fish *Acanthopagrus schlegelii schlegelii* (Sparidae) from Japan. *J. Parasitol.*, 96, 771, 2010.
4. Moravec, F., Redescription and systematic status of *Capillaria philippinensis*, an intestinal parasite of human beings. *J. Parasitol.*, 87, 161, 2001.
5. Traversa, D. et al., New insights into morphological and biological features of *Capillaria aerophila* (Trichocephalida, Trichuridae). *Parasitol. Res.*, 109, S97, 2011.
6. Camargo, L.M. et al., Capillariaisis (Trichurida, Trichinellidae, *Capillaria hepatica*) in the Brazilian Amazon: Low pathogenicity, low infectivity and a novel mode of transmission. *Parasit. Vectors*, 3, 11, 2010.
7. Jeong, W.I. et al., Macrophages, myofibroblasts and mast cells in a rat liver infected with *Capillaria hepatica. J. Vet. Sci.*, 9, 211, 2008.
8. Fan, P.C., Chung, W.C., and Chen, E.R., *Capillaria hepatica*: A spurious case with a brief review. *Kaohsiung J. Med. Sci.*, 16, 360, 2000.
9. Gomes, A.T. et al., *Capillaria hepatica* in rats: Focal parasitic hepatic lesions and septal fibrosis run independent courses. *Mem. Inst. Oswaldo Cruz*, 101, 895, 2006.
10. Hamir, A.N., and Rupprecht, C.E., Hepatic capillariasis (*Capillaria hepatica*) in porcupines (*Erethizon dorsatum*) in Pennsylvania. *J. Vet. Diagn. Invest.*, 12, 463, 2000.
11. Oliveira, L., de Souza, M.M., and Andrade, Z.A., *Capillaria hepatica*-induced hepatic fibrosis in rats: Paradoxical effect of repeated infections. *Rev. Soc. Bras. Med. Trop.*, 37, 123, 2004.
12. Redrobe, S.P., and Patterson-Kane, J.C., *Calodium hepaticum* (syn. *Capillaria hepatica*) in captive rodents in a zoological garden. *J. Comp. Pathol.*, 133, 73, 2005.
13. Reperant, L.A., and Deplazes, P., Cluster of *Capillaria hepatica* infections in non-commensal rodents from the canton of Geneva, Switzerland. *Parasitol. Res.*, 96, 340, 2005.
14. Santos, C.C., Onofre-Nunes, Z., and Andrade, Z.A., Role of partial hepatectomy on Capillaria hepatica-induced hepatic fibrosis in rats. *Rev. Soc. Bras. Med. Trop.*, 40, 495, 2007.
15. Pizzi, R. et al., *Capillaria hepatica* (syn. *Calodium hepaticum*) in primates in a zoological collection in the UK. *Vet. Rec.*, 163, 690, 2008.
16. Mowat, V. et al. Histopathological features of *Capillaria hepatica* infection in laboratory rabbits. *Toxicol. Pathol.*, 37, 661, 2009.
17. Quadros, R.M. et al., *Capillaria hepatica* in Puma concolor: First report in Brazil. *J. Zoo Wildl. Med.*, 40, 586, 2009.
18. Resendes, A.R. et al., Prevalence of *Calodium hepaticum* (syn. *Capillaria hepatica*) in house mice (*Mus musculus*) in the Azores archipelago. *Vet. Parasitol.*, 160, 340, 2009.
19. Stidworthy, M.F. et al., *Capillaria hepatica* in primates in zoological collections in the British Isles. *Vet. Rec.*, 164, 66, 2009.
20. Sekikawa, H. et al., First human case of *Capillaria hepatica* infection in Japan. *Jpn. J. Parasitol.*, 40, 528, 1991.
21. Choe, G. et al., Hepatic capillariasis: First case report in the Republic of Korea. *Am. J. Trop. Med. Hyg.*, 48, 610, 1993.
22. Bhattacharya, D. et al., *Capillaria hepatica*, a parasite of zoonotic importance–a brief overview. *J. Commun. Dis.*, 31, 267, 1999.
23. Tesana, S., Puapairoj, A., and Saeseow, O., Granulomatous, hepatolithiasis and hepatomegaly caused by *Capillaria hepatica* infection: First case report of Thailand. *Southeast Asian J. Trop. Med. Public Health*, 38, 636, 2007.
24. Li, C.D., Yang, H.L., and Wang, Y., *Capillaria hepatica* in China. *World J. Gastroenterol.*, 16, 698, 2010.
25. Carvalho-Costa, F.A. et al., Pseudoparasitism *by Calodium hepaticum (syn. Capillaria hepatica*; *Hepaticola hepatica*) in the Negro River, Brazilian Amazon. *Trans. R. Soc. Trop. Med. Hyg.*, 103, 1071, 2009.
26. Mansour, N.S., Anis, M.H., and Mikhail, E.M., Human intestinal capillariasis in Egypt. *Trans. R. Soc. Trop. Med. Hyg.*, 84, 114, 1990.
27. Chichino, G., Bernuzzi, A.M., and Bruno, A., Intestinal capillariasis (*Capillaria philippinensis*) acquired in Indonesia: A case report. *Am. J. Trop. Med. Hyg.*, 47, 10, 1992.
28. Dronda, F. et al., Human intestinal capillariasis in an area of nonendemicity: Case report and review. *Clin. Infect. Dis.*, 17, 909, 1993.
29. Kang, G. et al., Human intestinal capillariasis: First report from India. *Trans. R. Soc. Trop. Med. Hyg.*, 88, 204, 1994.
30. Hoghooghi-Rad, N., Maraghi, S., and Narenj-Zadeh, A., *Capillaria philippinensis* infection in Khoozestan Province, Iran: Case report. *Am. J. Trop. Med. Hyg.*, 37, 135, 1987.
31. Ahmed, L. et al., *Capillaria philippinensis*: An emerging parasite causing severe diarrhoea in Egypt. *J. Egypt. Soc. Parasitol.*, 29, 483, 1999.
32. Austin, D.N. et al., Intestinal capillariasis acquired in Egypt. *Eur. J. Gastroenterol. Hepatol.*, 11, 935, 1999.
33. Belizario, V.Y. Jr. et al., Intestinal capillariasis, western Mindanao, the Philippines. *Emerg. Infect. Dis.*, 16, 736, 2010.
34. Wongsawasdi, L., Ukarapol, N., and Lertprasertsuk, N., The endoscopic diagnosis of intestinal capillariasis in a child: A case report. *Southeast Asian J. Trop. Med. Public Health*, 33, 730, 2002.
35. Lu, L.H. et al., Human intestinal capillariasis (*Capillaria philippinensis*) in Taiwan. *Am. J. Trop. Med. Hyg.*, 74, 810, 2006.
36. Soukhathammavong, P. et al., Three cases of intestinal capillariasis in Lao People's Democratic Republic. *Am. J. Trop. Med. Hyg.*, 79, 735, 2008.
37. Lalosević, D. et al., Pulmonary capillariasis miming bronchial carcinoma. *Am. J. Trop. Med. Hyg.*, 78, 14, 2008.

38. Berger, T. et al., Hepatic capillariasis in a 1-year-old child. *Eur. J. Pediatr.*, 149, 333, 1990.
39. Kokai, G.K. et al., *Capillaria hepatica* infestation in a 2-year old girl. *Histopathology*, 17, 275, 1990.
40. Sawamura, R. et al., Hepatic capillariasis in children: Report of 3 cases in Brazil. *Am. J. Trop. Med. Hyg.*, 61, 642, 1999.
41. Gonzalez-Barranco, D. et al., Third case of human hepatic capillariasis in Mexico. *Rev. Investig. Clin.*, 48, 301, 1996.
42. Govil, H., and Desai, M., *Capillaria hepatica* parasitism. *Indian J. Pediatr.*, 63, 698, 1996.
43. Pampiglione, S., and Gustinelli, A., Human hepatic capillariasis: A second case occurred in Korea. *J. Korean Med. Sci.*, 23, 560, 2008.
44. Nabi, F. et al., *Capillaria hepatica* infestation. *Indian Pediatr.*, 44, 781, 2007.
45. Fuehrer, H.P., Igel, P., and Auer, H., *Capillaria hepatica* in man-an overview of hepatic capillariosis and spurious infections. *Parasitol. Res.*, 109, 969, 2011.
46. Bustamente-Sarabia, J., Suswillo, R., and Contreras-Moreno, E., *Capillaria hepatica* in a 19-month old boy, third case in Mexico. *Bol. Med. Hosp. Infant Mex.*, 53, 395, 1996.
47. Ferreira, L.A., and Andrade, Z.A., *Capillaria hepatica*: A cause of septal fibrosis of the liver. *Mem. Inst. Oswaldo Cruz*, 88, 441, 1993.
48. Lemos, Q.T., Magalhães-Santos, I.F., and Andrade, Z.A., Immunological basis of septal fibrosis of the liver in *Capillaria hepatica*-infected rats. *Braz. J. Med. Biol. Res.*, 36, 1201, 2003.
49. Andrade, S.B., and Andrade, Z.A., Experimental hepatic fibrosis due to *Capillaria hepatica* infection (differential features presented by rats and mice). *Mem. Inst. Oswaldo Cruz*, 99, 399, 2004.
50. Kohatsu, H. et al., A space-occupying lesion in the liver due to *Capillaria* infection. *Am. J. Trop. Med. Hyg.*, 52, 414, 1995.
51. Attah, E.B. et al., Hepatic capillariasis. *Am. J. Clin. Pathol.*, 79, 127, 1983.
52. Chunlertrith, K., Mairiang, P., and Sukeepaisarnjaroen, W., Intestinal capillariasis: A cause of chronic diarrhea and hypoalbuminemia. *Southeast Asian J. Trop. Med. Public Health*, 23, 433, 1992.
53. Lee, S.H. et al., A case of intestinal capillariasis in the Republic of Korea. *Am. J. Trop. Med. Hyg.*, 48, 542, 1993.
54. Hong, S.T. et al., Two cases of intestinal capillariasis in Korea. *Korean J. Parasitol.*, 32, 43, 1994.
55. Wilairatana, P., Capillariasis with chronic intestinal pseudo-obstruction. *Southeast Asian J. Trop. Med. Public Health*, 25, 600, 1994.
56. Sangchan, A. et al., The endoscopic-pathologic findings in intestinal capillariais: A case report. *J. Med. Assoc. Thai.*, 90, 175, 2007.
57. Kusolsuk, T. et al., Capillariasis: Chronic watery diarrhea–not only from microorganisms. *Southeast Asian J. Trop. Med. Public Health*, 39, 1000, 2008.
58. Cross, J.H., Intestinal capillariasis. *Parasitol. Today*, 6, 26, 1990.
59. Cross, J.H., and Basaca-Sevilla, V., Capillariasis philippinensis: A fish-borne parasitic zoonosis. *Southeast Asian J. Trop. Med. Public Health*, 22, S153, 1991.
60. Cross, J.H., Intestinal capillariasis. *Clin. Microbiol. Rev.*, 5, 120, 1992.
61. Bair, M.J. et al., Clinical features of human intestinal capillariasis in Taiwan. *World J. Gastroenterol.*, 10, 2391, 2004.
62. el-Karaksy, H. et al., *Capillaria philippinensis*: A cause of fatal diarrhea in one of two infected Egyptian sisters. *J. Trop. Pediatr.*, 50, 57, 2004.
63. Saichua, P., Nithikathkul, C., and Kaewpitoon, N., Human intestinal capillariasis in Thailand. *World J. Gastroenterol.*, 14, 506, 2008.
64. Pereira, V.G., and Mattosinho Franca, L.C., Successful treatment of *Capillaria hepatica* infection in an acutely ill adult. *Am. J. Trop. Med. Hyg.*, 32, 1272, 1983.
65. Rana, S.S. et al., Human intestinal capillariasis: Diagnosis by jejunal fluid analysis obtained at enteroscopy and reversal of subtotal villous atrophy after treatment. *Endoscopy*, 41 (suppl 2), E102, 2009.
66. Juncker-Voss, M. et al., Serological detection of *Capillaria hepatica* by indirect immunofluorescence assay. *J. Clin. Microbiol.*, 38, 431, 2000.
67. Assis, B.C. et al., A contribution to the diagnosis of *Capillaria hepatica* infection by indirect immunofluorescence test. *Mem. Inst. Oswaldo Cruz*, 99, 173, 2004.
68. El Dib, N.A. et al., Evaluation of *Capillaria philippinensis* coproantigen in the diagnosis of infection. *J. Egypt. Soc. Parasitol.*, 34, 97, 2004.
69. Intapan, P.M. et al., Potential use of *Trichinella spiralis* antigen for serodiagnosis of human capillariasis philippinensis by immunoblot analysis. *Parasitol. Res.*, 98, 227, 2006.
70. Intapan, P.M. et al., An enzyme-linked immunosorbent assay as screening tool for human intestinal capillariasis. *Southeast Asian J. Trop. Med. Public Health*, 41, 298, 2010.
71. Zhu, X. et al., Mitochondrial DNA polymorphism within and among species of Capillaria sensu lato from Australian marsupials and rodents. *Int. J. Parasitol.*, 30, 933, 2000.

50 Dioctophyme

Dongyou Liu

CONTENTS

50.1 INTRODUCTION

Dioctophyme renale (giant kidney worm) is the only species in the genus *Dioctophyme* that represents one of the largest nematodes to parasitize humans, with its adult males measuring up to 35 cm by 3–6 mm, and adult females measuring up to 103 cm by 5–12 mm. The life cycle of *D. renale* is complex, involving an aquatic oligochete (or annelid) as an intermediate host, fish and frogs as paratenic hosts, and fish-eating carnivores as the definitive hosts. As an accidental host, humans acquire the infection through consumption of raw or undercooked fish or frog meat containing *D. renale* infective larvae. As its name suggests, the predilection site of the giant kidney worm is the kidneys, causing obstruction, hydronephrosis, and destruction of the renal parenchyma. Occasionally, the parasite may sometimes find its way to other parts of the body. Despite its widespread distribution and frequent occurrence in animal hosts, the incidence of dioctophymatosis in humans is relatively low, with <20 confirmed cases being reported to date. Nevertheless, with an increasing reliance on fish as a source of food supply, there is a likelihood that prevalence of human dioctophymatosis may trend higher in the future. Thus, our awareness of *D. renale* and preparedness in the diagnosis and prevention of dioctophymatosis will go a long way toward reducing its negative impact on human populations.

50.1.1 Classification, Morphology, and Biology

50.1.1.1 Classification

D. renale is the only member in the genus *Dioctophyme* (synonym: *Dioctophyma*), family Dioctophymatidae, superfamily Dioctophymatoidea, subclass Enoplia, class Enoplea, phylum Nematoda. Besides *Dioctophyme*, the family Dioctophymatidae includes another genus *Eustrongylides*.

D. renale (giant kidney worm) is a parasitic roundworm whose larval stage is found in aquatic oligochetes (or annelids) as well as fishes and frogs, and whose adult worms reside in fish-eating carnivores and occasionally humans. First discovered in 1583, *D. renale* was fully described by Johann Goeze in 1782 from the specimens collected in the kidney of an infected dog. Its association with human hosts is undoubtedly long, as *D. renale* eggs were identified in human coprolites (fossilized feces) dated to 3384–3370 BC [1].

50.1.1.2 Morphology

As the largest nematode to parasitize humans, *D. renale* adult worms have a blood-red color, and are cylindrical, tapering at both the anterior and posterior ends (with the tail being blunt). The males measure up to 35 cm by 0.3–0.6 cm and the females up to 103 cm by 0.5–1.2 cm.

The mouth of *D. renale* adult worms is simple, without lips, and is surrounded by two circles of six papillae, consisting of an inner and an outer circle. The esophagus is long, narrow, and lacteous in color. The male's bursa (whose main function is to facilitate mating activity) is bell-shaped, muscular, without supporting rays, and has a single bristle-like spicule. The female's vulva is located toward the posterior end of the esophagus [2].

D. renale worms are covered by a cuticle (which protects itself during invasion of the digestive tracts of hosts) with three or more main outer layers made of collagen and other compounds. The outer layers (secreted by the epidermis) are noncellular. Longitudinal muscles (obliquely arranged in bands) line the body wall. Dorsal, ventral, and longitudinal nerve cords are connected to the main body of the muscle. Similar to other nematodes, *D. renale* worms employ papillae, setae (mechanoreceptors), and amphids (chemoreceptors) as the main sense organs.

D. renale larvae are yellow to rusty in color, threadlike, and 6–10 mm in length and 0.1–0.2 mm in width.

The eggs (each containing an embryo) measure 60–84 μm by 38–52 μm, are unicellular, elliptical (lemon- or barrel-shaped), yellowish, and have a thick shell (whose surface appears to be pitted except at the poles). The shell's outer layer is uneven and has a transparent zone at both ends [3].

50.1.1.3 Biology

The life cycle of *D. renale* is complex, requiring intermediate (aquatic oligochete annelids or earthworms), paratenic (fishes or frogs), and definite (wild fish-eating carnivores) hosts [4]. Humans represent an accidental/terminal host for the parasite. The entire life cycle may take up to 2 years to complete, and the prepatent period is approximately 155 days [5].

When discharged by the definitive hosts (wild fish-eating carnivores, e.g., mink) in the urine, *D. renale* eggs embryonate in 2 weeks to 3 months (depending on the temperature). Upon ingestion by the intermediate host (aquatic oligochete *Lumbriculus variegatus* or earthworm), the eggs hatch to become first-stage larvae, which penetrate the ventral blood vessel of the annelid host and molt twice to become the third-stage larvae over a period of 50 days, which are directly infective to mammal hosts. The annelids harboring the third-stage larvae may also be ingested by a fish (e.g., pike and bullhead) or a frog, and the larvae encyst in abdominal muscle or wall of the digestive tube and remain there until being taken up by mammal hosts. Additionally, the infected annelids parasitic on crayfish (*Cambarus* spp.) may be ingested together with the crayfish by fish or frogs, which act as transporters rather than true intermediate hosts. Definite hosts (wild fish-eating carnivores) acquire the infection by ingestion of infected oligochetes (annelids) or paratenic/transporter hosts. In the definitive hosts, the larvae penetrate the duodenal wall and stay in the stomach muscle for about 2 weeks. Afterward, the larvae move to the liver and feed on the liver tissue for about a month; then they enter the abdominal cavity and migrate through the bloodstream to the kidney, where they mature into adult worms, and remain alive for up to 5 years (causing damages to the kidney parenchyma, chronic peritonitis, adhesions, and damage on the surface of the liver). Females may produce pheromones to attract males, and a male coils around a female with its curved area over the female genital pore. The male uses its spicule to hold the female during copulation and ejects amoeboid-like, flagella-less sperm. After mating, the male may live for some time and die, and the oviparous female lays egg in the kidneys (often the right kidney due to its proximity to the stomach and liver) of the definitive host and the eggs are passed to the urinary bladder.

D. renale is distributed worldwide, especially in areas of temperate climate, with regions around the Caspian Sea (e.g., Iran) having the highest number of cases, and Africa and Oceana showing relatively few cases. *D. renale* infections are also commonly found in areas where freshwater fish is a dietary mainstay.

D. renale is infective to a wide range of mammalian host species, especially fish-eating mammals, including dog (*Canis lupus familiaris*), maned wolf (*Chrysocyon brachyurus*), coati (*Nasua nasua*), fox (e.g., crab-eating fox *Cerdocyon thous*), grison (*Galictis cuja* and *G. vittata*), marten (*Martes americana*), rat, cat, cheetah, bear, oxen, raccoon, weasel (*Mustela sibirica coreana*), mink (*Mustela vison*), otter, seals minnow, horse, swine, and human [6–14]. Mammals (e.g., horses) drinking water containing infected annelid intermediate hosts can also become infected.

While dogs represent a common definitive host for *D. renale*, mink (*Mustela vison*) is regarded as the principal definite host in North America, and mustelids (*Galictis vittata* and *G. cuja*) may act as a definitive host in South America [15]. As an accidental host, humans acquire the infection by eating raw or undercooked fish or frog meat containing *D. renale* larvae.

D. renale generally feeds on blood and tissue cells, and its predilection site is the kidneys of mammalian hosts. Occasionally, *D. renale* is located in the abdominal cavity, scrotum, urethra, uterus, ovary, urinary bladder, urinary canal, subcutaneous tissues, stomach, mammary gland, and mesenteric lymph nodes.

50.1.2 Clinical Features and Treatment

50.1.2.1 Clinical Features

D. renale is a common parasite of fish-eating mammals, and humans may become infected occasionally through consumption of raw or undercooked fish or frog meat. Since the early 1900s, <20 confirmed cases of human dioctophymatosis have been reported worldwide. While several cases involved *D. renale* adult worms in kidneys, others had the larvae ending up in subcutaneous nodules and did not undergo further development [16–25].

Individuals with *D. renale* infection often show unspecific symptoms such as hematuria (blood in urine), nephritis, loin pain (around the kidney area), renal enlargement, and/or renal colic (intermittent pain in the kidney area, as a result of the migration of worms through ureters). The presence of the parasite in the kidney may lead to the development of excess fibrous connective tissue (fibrosis). In extreme circumstances, parenchymal inflammation (i.e., inflammation of the functional parts of an organ) may result in death. Many infected hosts suffer from pressure necrosis caused by growing *D. renale* worms and their feeding activities, which contribute to macroscopic alterations in the kidneys as well as in the abdominal cavities and reduce the infected kidney to a thin-walled, ineffective organ. Loss of kidney function and uremic poisoning are some of the most severe side effects.

Li et al. [26] documented a case of dioctophymatosis in a 51-year-old woman, who presented with unspecific clinical signs. Giant kidney worms were observed in the urine and histological examination confirmed the *D. renale* infection. In another report, a *D. renale* larva was found in the subcutaneous tissue of the abdomen of a 23-year-old woman. The patient may have acquired the infection through ingestion of raw fish [27]. Additionally, a 67-year-old Javanese man complained of having expelled a large worm from his urethra during urination and feeling discomfort on the upper-left and back side of his abdomen for 2 months. On cystoscopic examination, a worm was seen emerging from the left ureter, and many worms were found in the bladder. MRI revealed enlargement of the left kidney, and many worms were found in the cortex and pelvio-calices of the left kidney and bladder. The worms were of 2–3 mm in diameter and 35–40 cm in length, and were identified as *D. renale*. Despite medical intervention including nephrectomy, the patient's health deteriorated and he died at day 8 of admission to the hospital [28].

In animals, clinical signs of dioctophymatosis may include hematuria (blood in urine), recurrent urinary tract infections, painful urination, and frequent urination. Other symptoms include weakness, reluctance to walk, anorexia, convulsions, anemia, ascites, renal and abdominal colic, irritability, polydipsia, depression, micturition, and proteinuria. Renal parasitism by *D. renale* can cause progressive destruction of the renal parenchyma (leaving only a thin capsule containing the worm and hemorrhagic exudates inside), atrophy and fibrosis of renal tubules, dilation of the renal pelvis, urethral obstruction, bone formation, and renal calculi. Unaffected kidney has compensatory hypertrophy. The presence of *D. renale* worms in the abdominal cavity can lead to adhesions and peritonitis. Severe abdominal pain (intermittent pain in the kidney area), paresis of the hindquarters, sudden weight loss, and vomiting may also be present [29].

50.1.2.2 Treatment

Currently, there is no effective drug against *D. renale* parasite. Treatment is limited to surgical removal of the parasite and the affected kidney. Nephrectomy should be performed quickly before other complications occur.

Human infection with *D. renale* is easy to avoid via thorough cooking of fish or frogs and boiling of water, as a majority of *D. renale* infections have resulted from the consumption of undercooked or raw freshwater fish. Proper sanitation is the key to preventing *D. renale* infection.

Reduction of exposure to the intermediate hosts (earthworms, fish, and frogs) and avoiding raw or improperly cooked fish are critical for preventing dioctophymatosis in animals.

50.1.3 Laboratory Diagnosis

Imaging techniques (e.g., radiology and sonography) may be employed to search for enlarged or calcified kidneys and to confirm the presence of the parasite in the renal parenchyma. A definitive diagnosis of dioctophymatosis is made through the observation of *D. renale* eggs (ova) in the urine. However, the eggs will only appear in the urine sediment, if a gravid parasite female is in the kidney causing the infection. Urinalysis will likely show hematuria, and blood tests may reveal eosinophilia.

In domestic animals, postmortem finding of *D. renale* parasitism along with peritonitis and adhesions is still the most frequent way for verifying the infection. If the parasite does not reach the urinary system, its eggs may be seen in ascitic liquid.

Molecular approaches offer a valuable means of confirming the identity of *D. renale* parasites, and assessing the phylogenetic relationships of isolates from different hosts.

50.2 METHODS

50.2.1 Sample Preparation

50.2.1.1 Sample Collection

D. renale worms are placed in saline solution and then fixed in alcohol-formalin-acetic acid solution. Eggs are collected from the uterus of the female worm. Identification of adult worms and eggs is based on the key of Anderson and Bain [30].

50.2.1.2 DNA Isolation

Genomic DNA from *D. renale* adult worms and eggs may be extracted according to a NaOH lysis protocol [2]. Namely, samples are placed in 1.5 mL Eppendorf tubes containing 50 μL of 50 mM NaOH and heated at 95°C for 30 min. After centrifugation at 1300 × *g* for 20 min, the supernatant of each mixture is transferred to a new tube containing 50 μL of 1 M Tris HCl acid solution (pH 8.0). Alternatively, DNA may be extracted from the parasite using QIAamp DNA tissue kit (Qiagen).

50.2.2 Detection Procedures

Tokiwa et al. [2] utilized primer sets Sobo18S-F/R (5′-TTTGG TTTTCGGATCTGAGG-3′/5′-GTACAAAGGGCAGGGAC GTA-3′) and SSU18A/26R (5′-AAAGTTAAGCCATGCATG-3′/5′-CATTCTTGGCAAATGCTTTCG-3′), for PCR amplification and sequencing analysis of partial small subunit (18S) rRNA gene of *D. renale*.

50.2.2.1 Procedure

1. PCR mixture (20 μL) is made up of 2 μL of 10× PCR buffer, 1.6 μL dNTPs, 0.2 μL of each primer, 0.1 μL of Taq DNA polymerase, 14.8 μL of water, and 1 μL of sample DNA.
2. For primer set Sobo18S-F/R, the amplification conditions consist of initial denaturing at 94°C for 2 min; 25 cycles of 94°C for 30 s, at 60°C for 30 s, and at 72°C for 2 min; and a final extension at 72°C for 10 min. For primer set SSU18A/26R, the amplification conditions include initial denaturing at 94°C for 1 min;

34 cycles of 94°C for 30 s, at 53°C for 15 s, and at 72°C for 30 s; and a final extension at 72°C for 10 min.
3. PCR products are analyzed by electrophoresis on a 1.5% agarose gel and visualized by UV illumination with ethidium bromide staining.
4. The PCR products are purified by Exosap-IT® (GE Healthcare), and sequenced by using the BigDye® Terminator v3.1 Cycle Sequencing kit (Applied Biosystems) and the ABI3100 system (Applied Biosystems). Sequences are visualized by electropherograms and manually corrected using Finch TV software (Geospiza).

Note: The SSU rRNA sequences generated from *D. renale* with the primers above are of 1526 bp in length. Upon BLAST search, the identity of the sequences is confirmed as belonging to *D. renale*.

50.3 CONCLUSION

D. renale (giant kidney worm) is a large nematode that parasitizes a wide range of primarily fish-eating carnivorous mammals (including dog, maned wolf, coati, fox, grison, marten, rat, cat, cheetah, bear, oxen, raccoon, weasel, mink, otter, seals minnow, horse, and swine). Occasionally, humans become infected with the parasite after eating raw or undercooked fish or frog meat containing *D. renale* infective larvae. Laboratory diagnosis of dioctophymatosis relies on the detection of *D. renale* eggs in urine. Application of molecular techniques provides a valuable approach for confirmation of parasite identity and phylogenetic assessment of *D. renale* isolate from different hosts.

REFERENCES

1. Bailly, M.L., Leuzinger, U., and Bouchet, F., Dioctophymidae eggs in coprolites from neolithic site of Arbon-Bleiche 3 (Switzerland). *J. Parasitol.*, 89, 1073, 2003.
2. Tokiwa, T. et al., *Dioctophyme renale* (Nematoda: Dioctophymatoidea) in the abdominal cavity of *Rattus norvegicus* in Japan. *Parasitol. Int.*, 60, 324, 2011.
3. Pedrassani, D. et al., Morphology of eggs of *Dioctophyme renale* Goeze, 1782 (Nematoda: Dioctophymatidae) and influences of temperature on development of first-stage larvae in the eggs. *Rev. Bras. Parasitol. Vet.*, 18, 15, 2009.
4. Measures, L.N., and Anderson, R.C., Centrarchid fish as paratenic hosts of the giant kidney worm, *Dioctophyma renale* (Goeze, 1782). *J. Wildl. Dis.*, 21, 11, 1985.
5. Mace, T.F., and Anderson, R.C., Development of the giant kidney worm, *Dioctophyma renale* (Goeze, 1782) (Nematoda: Dioctophymatoidea). *Can. J. Zool.*, 53, 1552, 1975.
6. Hodesson, D.L., Daniels, R.B., and Richards, B., *Dioctophyma renale* in a dog (a case report). *Vet. Med. Small Anim. Clin.*, 75, 825, 1980.
7. Alvarenga, J. et al., *Dioctophyma renale* in a dog. *Mod. Vet. Pract.*, 65, 125, 1984.
8. Barros, D.M., Lorini, M.L., and Persson, V.G., Dioctophymosis in the Little Grison (Galictis cuja). *J. Wildl. Dis.*, 26, 538, 1990.
9. Dyer, N.W., *Dioctophyma renale* in ranch mink. *J. Vet. Diagn. Invest.*, 10, 111, 1998.
10. Mech, L.D., and Tracy, S.P., Prevalence of giant kidney worm (*Dioctophyma renale*) in wild mink (Mustela vison) in Minnesota. *Am. Midland Natural.*, 145, 206, 2001.
11. Nakagawa, T.L. et al., Giant kidney worm (*Dioctophyma renale*) infections in dogs from Northern Paraná, Brazil. *Vet. Parasitol.*, 145, 366, 2007.
12. Ribeiro, C.T., Verocai, G.G., and Tavares, L.E., *Dioctophyme renale* (Nematoda, Dioctophymatidae) infection in the crab-eating fox (Cerdocyon thous) from Brazil. *J. Wildl. Dis.*, 45, 248, 2009.
13. Verocai, G.G. et al., *Dioctophyme renale* (Goeze, 1782) in the abdominal cavity of a domestic cat from Brazil. *Vet. Parasitol.*, 161, 342, 2009.
14. Ishizaki, M.N. et al., *Dioctophyma renale* (Goeze, 1782) in the abdominal cavity of a capuchin monkey (Cebus apella), Brazil. *Vet. Parasitol.*, 173, 340, 2010.
15. Jorde, D.G., Occurrence of *Dioctophyma renale* (Goeze, 1782) in mink from North Dakota. *J. Wildl. Dis.*, 16, 381, 1980.
16. Hanjani, A.A. et al., The first report of human infection with *Dioctophyma renale* in Iran. *Trans. R. Soc. Trop. Med. Hyg.*, 62, 647, 1968.
17. Myers, B.J., Nematodes transmitted to man by fish and aquatic mammals. *J. Wildl. Dis.*, 6, 266, 1970.
18. Beaver, P.C., and Khamboonruang, C., *Dioctophyma*-like nematode in a subcutaneous nodule from man in northern Thailand. *Am. J. Trop. Med. Hyg.*, 33, 1032, 1984.
19. Beaver, P.C. and Theis, J.H., Dioctophymatid larval nematode in a subcutaneous nodule from man in California. *Am. J. Trop. Med. Hyg.*, 28, 206, 1979.
20. Fernando, S.S., The giant kidney worm (*Dioctophyma renale*) infection in man in Australia. *Am. J. Surg. Pathol.*, 7, 281, 1983.
21. Sun, T. et al., Giant kidney worm (*Dioctophyma renale*) infection mimicking retroperitoneal neoplasm. *Am. J. Surg. Pathol.*, 10, 508, 1986.
22. Narváez, J.A. et al., Hyperdense renal cystic lesions caused by *Dioctophyma renale*. *Am. J. Roentgenol.*, 163, 997, 1994.
23. Urano, Z. et al., Dioctophymatid nematode larva found from human skin with creeping eruption. *J. Parasitol.*, 87, 462, 2001.
24. Ignjatovic, I. et al., Infestation of the human kidney with *Dioctophyma renale*. *Urol. Int.*, 70, 70, 2003.
25. Pilsczek, F.H., Helminthic infections mimicking malignancy: A review of published case reports. *J. Infect. Dev. Ctries.*, 4, 425, 2010.
26. Li, G. et al., Fatal bilateral dioctophymatosis. *J. Parasitol.*, 96, 1152, 2010.
27. Gutierrez, Y., Cohen, M., and Machicao, C.N., *Dioctophyme* larva in the subcutaneous tissues of a woman in Ohio. *Am. J. Surg. Pathol.*, 13, 800, 1989.
28. Sardjono, T.W. et al., *Dioctophymatosis renalis* in humans: First case report from Indonesia. *Proc. ASEAN Congr. Trop. Med. Parasitol.*, 3, 90, 2008.
29. Ferreira, V.L. et al., *Dioctophyma renale* in a dog: Clinical diagnosis and surgical treatment. *Vet. Parasitol.*, 168, 151, 2010.
30. Anderson, R.C., and Bain, O., Keys to genera of the superfamilies Rhabditoidea, Dioctophymatoidea, Trichinelloidea and Muspiceoidea. In: Anderson, R.C., Chabaud, A.G., and Willmott, S. (eds.). *CIH Keys to the Nematode Parasites of Vertebrates. No. 9*, Commonwealth Agricultural Bureaux, Farnham Royal, Bucking-hamshire, U.K. pp. 1–26, 1982.

51 Dirofilaria

Andrea Gustinelli, Maria Letizia Fioravanti, Monica Caffara, Lorella Garagnani, and Francesco Rivasi

CONTENTS

51.1 INTRODUCTION

Filarial nematodes of the genus *Dirofilaria*, natural parasites of wild and domestic animals, can occasionally infect humans and could experience aberrant or partial development, rarely reaching sexual maturity and sometimes causing the formation of subcutaneous nodules or pulmonary lesions.

Among the zoonotic filariae, *Dirofilaria immitis* and *D.* (*Nochtiella*) *repens* have, above all, aroused in recent decades a growing interest in the medical and parasitological field since they are responsible for hundreds of human cases. Sporadic cases are also ascribed to *D.* (*Nochtiella*) *tenuis*, more rarely to *D.* (*Nochtiella*) *ursi* (Table 51.1), and isolated cases to *D.* (*Nochtiella*) *subdermata* (*ursi*-like) and *D.* (*Nochtiella*) *striata*.

The parasite is transmitted to the vertebrate host by hematophagous arthropods, represented by several species of culicid mosquitoes and some species of black flies.

The epidemiology of human dirofilariasis depends on three main factors: the prevalence of the parasite in a suitable definitive host (reservoir), density of the hematophagous arthropod (intermediate host/biological vector), and human activities at high risk of exposure to the vector.

Until 50 years ago, human dirofilariasis was considered a rather sporadic event but, in recent years, the increase in the number of cases (almost 400 cases between 1995 and 2000) has led scientist to reconsider dirofilariasis as an emerging zoonosis.

Dirofilaria localization in man depends on the species involved and is prevalently at the subcutaneous level for *D.* (*Nochtiella*) *repens*, reported for the first time in Italy at the end of the nineteenth century [1] and, subsequently, also reported in atypical locations, while *D. immitis* would seem to demonstrate greater tropism for the lungs, a site described for the first time in the United States [2].

Since the first report, hundreds of cases have been reported in the literature, for the most part ascribed to *D.* (*Nochtiella*) *repens* in Europe and Asia (with the exception of Japan) and *D. immitis* on the American continent.

The greatest problems correlated to human dirofilariasis are to be attributed to the frequent diagnostic errors which often require costly investigative procedures to reach a correct diagnosis.

TABLE 51.1
Hosts, Localization, and Geographic Distribution of the Most Important Zoonotic *Dirofilaria* Species

	D. (Dirofilaria) immitis (Leidy, 1856) Railliet & Henry, 1911	*D. (Nochtiella) repens* Railliet & Henry, 1911	*D. (Nochtiella) tenuis* Chandler, 1942	*D. (Nochtiella) ursi* Yamaguti, 1941
Main vertebrate host	*Canis* spp.	*Canis* spp.	*Procyon lotor*	*Ursus* spp.
Localization	Right heart, pulmonary circle	Subcutaneous tissue	Subcutaneous tissue	Subcutaneous tissue
Invertebrate hosts	Mosquitoes	Mosquitoes	Mosquitoes	Black flies
Geographic distribution	Cosmopolitan	Europe, Asia, Africa	USA	Japan, Russia North America

Furthermore, the necessity of a diagnostic tool capable of distinguishing between the different *Dirofilaria* species and, in particular, the zoonotic ones, as *D. immitis* and *D.* (*Nochtiella*) *repens*, has always been a problem of great importance. The lack of a reliable serological test and the necessity of working on bioptic samples, often with parasites in regression, and/or fixed in formalin and/or embedded in paraffin, together with the specialization required to read the histological sections, have prompted research toward finding a molecularly based method capable of overcoming these difficulties.

51.1.1 Classification and Morphology

51.1.1.1 Classification of *Dirofilaria*

Order: Spirurida
Suborder: Spirurina
Superfamily: Filarioidea
Family: Onchocercidae
Subfamily: Dirofilariinae
Genus: *Dirofilaria*

According to Anderson [3], Filariodea nematodes include vertebrate parasites (with the exception of fish) transmitted through hematophagous arthropods. The Onchocercidae family includes 80 genera which are fundamentally characterized by the production of larvae called microfilariae (even if some species of the Filariidae family produce a microfilaria-like stage too) which represents the strategic factor for the transmission of the parasitosis.

Numerous genera of medical interest in tropical countries belong to the Onchocercidae family, for example, *Wuchereria* spp., *Brugia* spp., *Onchocerca* spp., and *Loa* spp., and some genera of animals parasites with high zoonotic potential, above all *Dirofilaria* spp.

The genus *Dirofilaria* Railliet & Henry, 1911 includes two subgenera *Dirofilaria* and *Nochtiella* [4]. The *Dirofilaria* subgenus is currently composed of five species, considered valid at this time (*D. ailure*, *D. freitasi*, *D. immitis*, *D. lutrae*, and *D. spectans*), and another seven *species inquirendae*. The subgenus *Nochtiella* consists of 22 valid species [5,6].

51.1.1.2 Morphology

Macroscopically, the adults appear to be whitish round worms, rather long and thin, with a rounded cephalic portion, buccal cavity without lips, and a short tail. The nematodes belonging to the *Dirofilaria* subgenus usually have greater dimensions with respect to those of the *Nochtiella* subgenus (e.g., the females of *D.* (*Dirofilaria*) *immitis* can reach lengths of 30 cm versus 17 cm of the females of *D.* (*Nochtiella*) *repens*).

Anatomically, they are characterized by simple cephalic structures, without pseudolips and usually with developed cephalic papillae. In particular, they have eight small cephalic papillae arranged in groups of four in two circles and caudal papillae, the number and position of which are of taxonomic importance with respect to the anus, a digestive system extending for almost the entire length of the body with the esophagus divided into a muscular part and a glandular part which are not distinguishable from each other and the anus in a subterminal position, numerous polymyarian-type muscular cells, cuticle with four longitudinal thickenings of which two are lateral containing well-developed excretory tubules, one dorsal and one ventral, usually less conspicuous and containing the nerve cords.

They are helminths with sexual dimorphism; the males are usually smaller in size and have a pointed spiral posterior end, perianal papillae, developed caudal wings, and asymmetric spicules structurally different from each other of variable length according to the species with a subterminal dilation and a pointed end. The females are characterized by their larger size, a more rounded posterior extremity, didelphic uterus with seminal vesicles in the terminal portions of each branch, and the vulva positioned near the buccal opening [3].

The specimens belonging to the *Dirofilaria* subgenus are characterized, besides their greater dimensions, by the absence of external cuticular ridges (ECR) except on the ventral surface of the caudal extremity of the male where evident striations are present, spaced 13–19 μm from each other [7].

Nematodes of the *Nochtiella* species have a cuticle with thin transversal ridges and marked longitudinal ridges. In particular, in *D.* (*Nochtiella*) *repens*, the cuticle in both sexes is finely striated transversally and has raised longitudinal ridges spaced 4–14 μm from each other, with the exception of the extremities which have a smooth surface except for the ventro-caudal region of the male which has fine ridges.

D. (*Nochtiella*) *tenuis* has a transversally striated cuticle, with longitudinal ridges spaced 10 μm from each other. *D.* (*Nochtiella*) *ursi* has a finely striated cuticle longitudinally along the entire body [8]. The principal morphometric

TABLE 51.2
Main Morphometric Characteristics of the Most Important Zoonotic Species of *Dirofilaria*

		D. (Dirofilaria) immitis	*D. (Nochtiella) repens*	*D. (Nochtiella) tenuis*	*D. (Nochtiella) ursi*
			Adults		
Body length	Female	250–300 mm	100–170 mm	80–130 mm	188–231 mm
	Male	120–200 mm	50–70 mm	40–48 mm	63–93 mm
Body width	Female	1.0–1.3 mm	0.45–0.65 mm	0.26–0.36 mm	0.65–0.75 mm
	Male	0.7–0.9 mm	0.37–0.45 mm	0.19–0.26 mm	0.33–0.63 mm
Oesophagus length	Female	1.1–1.6 mm	0.91–1.53 mm	1.12–1.4 mm	1.5–1.6 mm
	Male	1.46 mm	0.75 mm	—	1.1–1.5 mm
Cephalic papillae + amphids	8 + 2	8 + 2	8 + 2	8 + 2	
Caudal papillae (pairs)[a]	5 + 0 + 6	2–6 + 0 + 4–5	6–9 + 1 + 4–5	7–8 + 2 + 6	
Right spicule length	175–229 μm	180–210 μm	100–130 μm	100–200 μm	
Left spicule length	300–375 μm	460–590 μm	210–270 μm	300–610 μm	
Distance cephalic end-vulva	2.7 mm	1.15–1.62 mm	0.98–1.6 mm	1.7–3.3 mm	
			Microfilariae		
Body length	290–330 μm	320–370 μm	220–250 μm	185–292 μm	
Body width	4–6.2 μm	6–8.3 μm	5–7 μm	3–6.5 μm	
Cephalic end	Roundish	Cone-shaped	Roundish	Roundish	
Tail	Straight	Hooked	Short and rounded	Filamentous	

[a] Preanal, perianal, and postanal respectively.

differential characteristics for *D.* (*Dirofilaria*) *immitis*, *D.* (*Nochtiella*) *repens*, *D.* (*Nochtiella*) *tenuis*, and *D.* (*Nochtiella*) *ursi*, the species most frequently reported in humans, are presented in Table 51.2.

51.1.2 Biology and Epidemiology

51.1.2.1 Life Cycle

The life cycle of the *Dirofilaria* parasites follows the typical model of Filarioidea nematodes in which the hematophagous arthropods are intermediate/vector hosts/biological vectors and different species of homeothermal vertebrates are definitive hosts (Figure 51.1). Humans take part of the life cycle as accidental hosts, being infected by parasitized arthropods but without representing a suitable host for the development of sexually competent adults.

The females of *Dirofilaria* are ovoviviparous and produce microfilariae (L1) without cuticle which develop inside the uterus and have morphological differences according to the species (Table 51.2). Once they enter the circulatory system of a suitable definitive host, the microfilariae must be ingested by blood-feeding arthropods in order continue their biological development.

Inside the vector, the microfilariae migrate in the malpighian tubules where they develop to L2 and then to L3 in about 2 weeks at ≥26°C for *D. immitis* and in 10–21 days at 24–27°C for *D.* (*Nochtiella*) *repens* [9]. Once they have reached the L3 stage, the parasite migrates toward the buccal apparatus (*labium*) of the mosquito, perforating the malpighian tubules and passing through the hemocoel. The length of this phase depends on environmental parameters, such as temperature and humidity, as well as on the species of insect involved.

When the parasitized mosquito feeds on the blood of a suitable vertebrate host, the L3 come out of the labial sheath of the mouth and enter the bite wound, penetrate the connective tissue, and migrate to the subcutaneous and muscular tissue where they molt to L4 and then to the preadult stage L5 after 60–90 days. Through the bloodstream, *D. immitis* will reach the cardiac and pulmonary arteries while the other species reach the subcutaneous tissues, for the final molt into mature adults.

The time of prepatency, from the moment of infection to the appearance of the microfilaremia, is 174–191 days for *D. immitis*, 175–238 days for *D. repens*, and about 210 days for *D. tenuis*.

In humans, as a nonsuitable accidental host, after inoculation by arthropods, the parasite is usually isolated, inactivated, and destroyed by the immune system, or its development continues until the preadult or adult stage, localizing at the subcutaneous and pulmonary levels or in ectopic sites. Except for a unique case reported by Nozais et al. [10] there is no production of microfilariae.

51.1.2.2 Epidemiology

The presence of *Dirofilaria* spp. is usually closely linked to the presence/abundance of microfilaremic vertebrate hosts and presence/density of the suitable arthropod

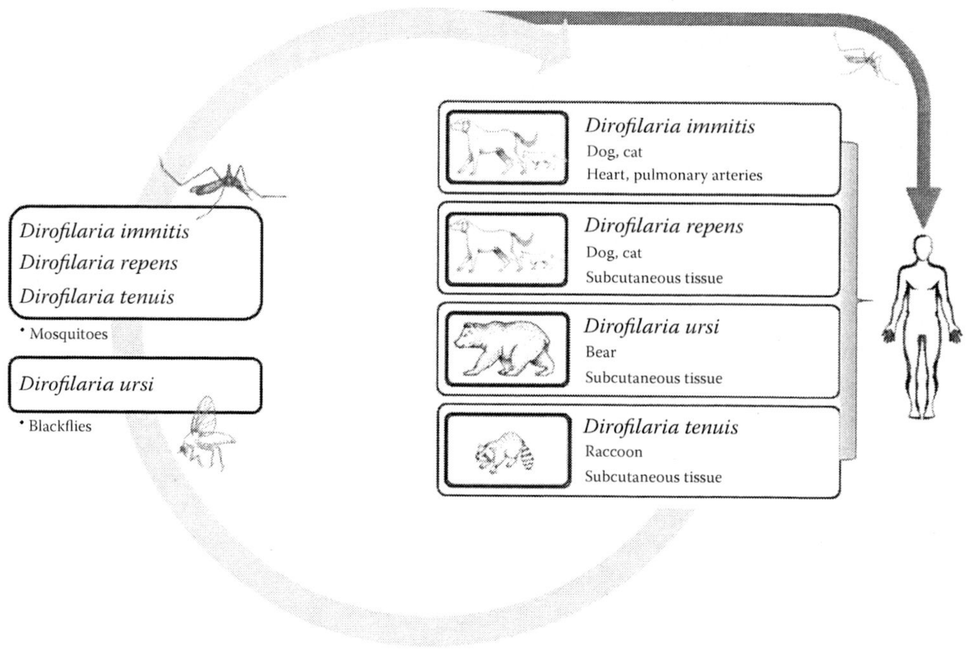

FIGURE 51.1 Life of zoonotic *Dirofilaria* spp.

vectors. Among these, the genera *Aedes*, *Anopheles*, *Culex*, and *Mansonia* can be considered the most important for *D.* (*Nochtiella*) *repens* according to Coluzzi [11] while for *D. immitis* more than 60 species of potential vectors have been identified [12], even if the most reported genera are *Aedes* and *Culex*. Human activities entailing a frequent exposure to the vector represent an important risk factor for zoonotic dirofilariasis. Another important factor is currently represented by "global warming" which has led to the wider geographic distribution of some vectors, that is, *Aedes albopictus* in Europe, originally closely linked to the equatorial areas and, in recent years, in rapid expansion into temperate areas [13].

51.1.2.3 Host Preference and Geographic Distribution

D. (*Nochtiella*) *repens* is a common parasite of the subcutaneous tissue in dogs and other carnivores in Eurasia and Africa. Cats, and felines in general, are less susceptible, with infection intensity very low (one or two nematodes), and transitory and low microfilaremia. Natural infections are usually amicrofilaremic [14], although it has been proven experimentally that mosquitoes which feed on microfilaremic cats are capable of transmitting infective L3 to dogs [15].

D. immitis occurs in cardiopulmonary system of more than 30 species of mammals, including mustelids, primates, marine mammals, rodents, and ungulates [16] with cosmopolitan distribution. Among definitive host, dogs represent the most suitable one, with prevalence and infection intensity influenced by size, age, habits (outdoor dogs) [17,18]. In enzootic areas, the prevalence for *D. immitis* in cats is 5–20% of that reported in dogs [19].

D. (*Nochtiella*) *tenuis* is a parasite of the subcutaneous tissue of raccoon (*Procyon lotor*) in southeastern United States, where it is considered enzootic [3]. The parasite is transmitted by different species of culicid mosquitoes.

D. (*Nochtiella*) *ursi* Yamaguti, 1941 is a parasite of the subcutaneous tissue, in the peritoneal cavity, and in numerous other body locations, and is commonly reported in *Ursus torquatus japonicus* and *U. americanus* in Japan, Russia, and North America. The parasite is transmitted by black flies (Simuliids) [3].

51.1.3 Clinical Features of Human Dirofilariasis

Currently, approximately 300 cases of human pulmonary dirofilariasis, almost exclusively caused by *D. immitis*, have been reported worldwide; the majority have been described in the United States [20], Brazil [21,22], and Japan [23]. Instead, approximately 800 cases have a subcutaneous/subconjunctival location [24] of which almost all have been attributed to *D.* (*Nochtiella*) *repens*, the majority being found in Italy, Russia, France, and Turkey [25,26].

Concerning *D.* (*Nochtiella*) *repens* infection, the age group most involved is that between 40 and 50 years with no significant difference regarding sex, even if women are infected to a slightly greater extent, while people living in rural areas are notably more susceptible with respect to those living in metropolitan areas, probably because of the greater possibility of the former coming into contact with the arthropod vectors [26]. Dissanaike et al. [27] in Sri Lanka found elevated prevalence of *D.* (*Nochtiella*) *repens* infection in children under 9 years, in consequence of often being bare bodies.

D. (*Nochtiella*) *tenuis* is responsible for numerous human subcutaneous infections in the United States, especially in Texas and Florida [28].

Several human cases from *D.* (*Nochtiella*) *ursi* have been reported in Canada and the United States [28,29] while, in the other zones where it has been found in bears, that is, Japan and Russia, no human cases have been reported so far [30].

Sporadic cases of human subcutaneous infection have been reported in North America due to *D.* (*Nochtiella*) *subdermata*, parasite of the porcupine belonging to the group *D. ursi*-like, and to *D. striata*, parasite of the bobcat, puma, and other carnivores in North and South America [30] and responsible for a case of orbital localization in a child in North Carolina [31].

In pulmonary cases, prevalently due to *D. immitis*, half of the patients have no respiratory symptoms while, in the other half, cough, fever, dyspnea, hemoptysis, and chest pain have been reported [32].

In human dirofilariasis by *D. immitis*, the infective stages of the parasite incidentally penetrate humans during the blood feeding of an infected vector. Many of them die at the site of infection and the few which survive are capable of migrating into the pulmonary circulation to then stop at this level and die, generating an area of pulmonary infarct and developing a granulomatous reaction which, at the end, assumes the radiological aspect of the so-called coin lesion, having a diameter of 1–3 cm [28]. The confirmation of such a lesion is usually completely incidental and consequent to check-ups based on imaging diagnostics.

Besides the typical pulmonary location, other unusual sites for *D. immitis* have been reported, such as the abdominal cavity, the spermatic cord, and subcutaneous tissue [33].

From a clinical point of view, the parasitosis has almost always been interpreted as a benign neoplasia (rarely malignant) or, when in extrapulmonary sites as a sebaceous cyst, hernia, foreign body, abscess, suspected appendicitis, other filariasis or another lesion of undetermined nature.

As regards dirofilariasis by *D.* (*Nochtiella*) *repens*, the parasite can be localized in all body regions at the level of the subcutaneous tissue, more frequently in the upper half of the body. The parasite can also migrate from the initial point of infection to a secondary site, as in the case of subconjunctival localization and deep organs, such as the lungs, mesentery, omentum, male sexual organs, and the female breast.

The nematode is usually blocked by the host defensive reaction, forming a nodule of 1–3 cm in diameter. Depending on host reactivity, the symptomatology ranges from an almost or completely asymptomatic form in which the lesion is discovered incidentally to forms in which inflammatory-allergic manifestations are present; they are usually local (recurrent edema, such as *Calabar swelling*, erythema, urticaria-like patches, and itching) but are seldom diffuse.

In many cases, there is only a solitary subcutaneous nodule, more or less movable, with a consistency of varying elasticity, sometimes spontaneously painful or sometimes painful under pressure, in some cases in suppurative phase. A slight persistent fever lasting a couple of months or satellite lymphadenopathy can be present together with the abovementioned symptoms. In some cases, episodes of intense pain in the arm have been observed in the case of mammary localization and in the scrotum in patients parasitized at the level of the epididymis [34]. In subconjunctival and periorbital localizations, the symptomatology can include hyperemia, lachrymation, the sensation of a foreign body, burning, and palpebral edema. The nematode is sometimes visible due to the transparency of these tissues, to a lesser or greater extent rolled up and movable, in some cases with a spontaneous emission of the parasite. In the literature, secondary complications have been reported, represented by uveitis, detachment of the retina, crystalline lesions, and permanent decrease of the *visus* [26].

In cases of localization in the deep organs, that is, the omentum, the finding is completely incidental, with the discovery of the nodule during an exploratory laparotomy following manifestations of pain which had aroused suspicion of acute abdomen.

Concerning to human cases by *D.* (*Nochtiella*) *tenuis* and *D.* (*Nochtiella*) *ursi*, the parasites are encapsulated at the level of the subcutaneous tissue in nodules having 1–3 cm of diameter. The lesion can sometimes be painful or cause itching, manifesting various degrees of redness.

In *D.* (*Nochtiella*) *tenuis*, the colonization of the oral mucosa [35] or the involvement of the palpebral, periorbital, or even subconjunctival tissue has sometimes been observed [36].

Human cases from *D.* (*Nochtiella*) *ursi* in North America have been summarized by Beaver et al. [29], who reported 10 cases, all in women and having a subcutaneous localization, one in particular at the palpebral level. In all cases except for one, the lesions were not itchy or painful. Currently, although *D.* (*Nochtiella*) *ursi* is also present in bears in Japan and Siberia, human cases have not been reported in these regions so far [30].

51.1.4 Diagnoses of Human Dirofilariasis

51.1.4.1 Parasitological Diagnosis

While a parasitological diagnosis is rather simple in dogs, based on research and identification of the microfilariae in the blood, in humans, apart from one particular case [10], microfilaremia has never been observed, even in cases in which gravid females have been discovered.

In human pulmonary dirofilariasis from *D. immitis*, only lesions sometimes containing residue portions of the devitalized parasite are usually found; therefore, the diagnosis is essentially based on histology.

In the case of subcutaneous and subconjunctival dirofilariasis, a diagnosis based on morphology is also possible in humans (see Table 51.2) if surgical excision of a still-alive or well-conserved parasite is carried out.

In humans, *D.* (*Nochtiella*) *repens* is often immature and always solitary inside the lesions, with a females:males ratio of 5:1. Gravid females have rarely been observed. The parasite sometimes shows more or less marked signs of regression, due to its destruction by the inflammatory reaction of the host some weeks or months before the ablation of the nodule.

In cases in which the morphology of the nematode is perfectly preserved, it can be hypothesized that the specimens were alive at the moment of their removal.

51.1.4.2 Histopathological Diagnosis

In cases of dirofilariasis due to *D.* (*Nochtiella*) *repens* with a subcutaneous localization, the histological picture is characterized by a chronic granulomatous abscess-forming inflammatory process. The lesion is usually well defined, nodular, and has dimensions varying from 0.5 to 3 cm in diameter; only in some cases it shows a pseudolobate aspect or undefined margins, with a central area usually abscessed. In the central area of the lesion, one or more sections of the nematode with an indented cuticle, or necrotic or calcified fragments can be found, with an inflammatory infiltrate composed of neutrophils, eosinophils, hystiocytes, and giant Langhans-type multinucleate cells. In the majority of cases, reactive tissue can be found all around the central abscess; this tissue is made up of fibroblasts, epithelioid cells, plasma cells, lymphocytes, and eosinophils, forming an enclosing wall, sometimes thin, sometimes thick. More peripherally, inflammatory cells, prevalently acute or chronic, irradiating and infiltrating the surrounding soft tissue, can be found. Instead, in a few cases, reactive tissue consisting almost exclusively of lymphoid-type elements can be observed all around the central area, giving the lesions the aspect of pseudo-lymphomatous lymphoid hyperplasia.

In pulmonary human infections, the nematode (*D. immitis* or *D.* (*Nochtiella*) *repens*), partially necrotic and calcified, is found inside a thrombotic lesion in pulmonary artery, giving origin to a small rounded infarct area surrounded by fibrous tissue ("coin lesion"). The adjacent pulmonary parenchyma can show prevalently acute inflammatory infiltrates with a marked presence of eosinophils.

In cases with an omental localization, the inflammatory reaction around the nematode appears to be of a chronic abscess-forming nature with chronic-type demarcation tissue, a rich lymphocytic and plasmacytic component and with the presence of a fibroblastic reaction.

In histological sections, the two subgenera *Dirofilaria* and *Nochtiella* can be distinguished on the basis of the presence or absence of ECR; in fact, the subgenus *Dirofilaria* has a smooth cuticle, except for the posterior ventral part of the male, while the subgenus *Nochtiella* is characterized by the presence of ECR on the entire cuticular surface with the exception of the anterior and posterior extremities in both sexes (Figure 51.2).

The main histological characteristics of the most important zoonotic *Dirofilaria* species are reported in Table 51.3.

The parasite can remain viable during the granulomatous reaction for very long periods of time (up to some years) before dying and being attacked by the reparative processes of the host. This often permits us to have excellent histological samples and facilitates the diagnosis on the basis of the anatomical characteristics of the species. Moreover, the inflammatory processes are often developed in a centripetal way starting from the natural orifices of the parasite, allowing the maintenance of the cuticular structure for a long time.

Unfortunately, during the routinary diagnoses based on histopathological findings, the conditions of the parasite can sometimes be in such a state of regression that the main anatomical features useful to differentiate the species are difficult to find or are completely lacking [37].

In these cases, it is necessary to have a very high number of sections in order to search for eventual well-conserved segments of the parasite.

Furthermore, as regards pulmonary dirofilariasis, frequently due to male parasites, the presence of ECR in a single section can be deceptive and carry to a misidentification, since *D. immitis* males has ECR at the level of the posterior portion [37,38].

In the differential diagnosis between the various species belonging to the *Nochtiella* subgenus, all characterized by an indented cuticle, although with a different appearance, the anamnesis plays an important role in excluding the eventual permanence of the patient in geographic areas where *Nochtiella* species other than *D. repens* are enzootic.

51.1.4.3 Serologic Diagnosis

Infection with *D.* (*Nochtiella*) *repens* induces high levels of specific IgG in humans [39], while high levels of specific IgM, IgG, and IgE have been detected in patients with active *D. immitis* disease and also in individuals exposed to *Dirofilaria* species [39,40].

Further studies using ELISA with somatic antigens from adult worms confirmed that people living in endemic areas for canine dirofilariasis develop specific antibody such as IgM, IgG, and IgE [41].

Currently, serology represents a useful instrument in epidemiology and risk assessment of dirofilariasis by

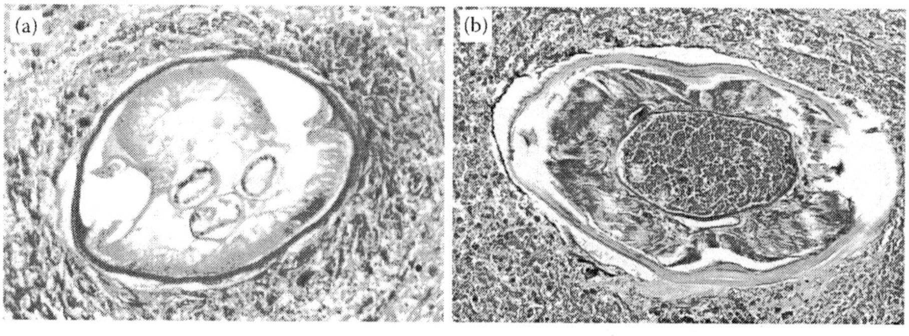

FIGURE 51.2 Cross-sections of pulmonary dirofilariasis by *Dirofilaria immitis* (a, Grocott stain, 10×) and *D.* (*Nochtiella*) *repens* (b, Ematorylin Eosin stain, 10×).

TABLE 51.3
Morphological Characteristics Used to Differentiate the Various *Dirofilaria* Species Found in Histological Sections (Eosin Stain, 10×)

	Sex	Maximum Diameter	Cuticle Thickness	Ridge Distance	Other Features of the Cuticle	Lateral Chords	Somatic Muscle
D. repens	Female	450–660 μm	8–40 μm	7–12 μm	Prominent external longitudinal ridges	2–5 nuclei per histology section	Well developed
	Male	370–450 μm	9–12 μm				
D. immitis	Female	125–325 μm	5–25 μm	—	Smooth, ECR present in the posterior part of the male	Large and prominent	Well developed
	Male	140–200 μm	5–15 μm				
D. tenuis	Female	280–330 μm	8–10 μm	10 μm	Low and rounded, spaces narrower than the ridges	Prominent	Large
	Male	180 μm	8–10 μm				
D. ursi	Female	80–260 μm	3–6 μm	10 μm	Tall, sharply crested, narrow ridges	Not prominent	Well developed
	Male	200 μm	3–6 μm				

D. immitis in humans but it cannot be considered a routinely diagnostic tool [20]. As an example, in Japan, only four out of the seven cases of pulmonary dirofilariasis described by Miyoshi et al. [23] showed antibody response against *Dirofilaria* antigens.

Moreover, the antigens of *D. immitis* can cross-react with antigens of other nematodes structurally similar but belonging to different species, that is, *Toxocara canis* [42]. Regarding *D. (Nochtiella) repens*, there is still a lack of knowledge about the antibody response against this species. Despite the fact that serological diagnosis has been improved so far, reliable, quick, and simple test are not yet available [26].

Furthermore, the description of an anti-Wolbachia surface protein antibody toward the surface antigens of *Wolbachia* endosymbiont in patients affected by pulmonary dirofilariasis due to *D. immitis* [43] and subcutaneous dirofilariasis due to *D.* (*Nochtiella*) *repens* [44] have suggested the use of this antibody response in the differential diagnosis of the disease [43] and could be a future application in serological diagnosis in human dirofilariasis.

51.1.4.4 Molecular Diagnosis

The requirement of a diagnostic tool capable of distinguishing between the various species of *Dirofilaria* and, in particular, that of considerable zoonotic importance, namely *D. immitis* and *D.* (*Nochtiella*) *repens*, has always been a problem of fundamental importance.

In fact, the lack of a reliable serological test and the necessity of working on bioptic samples, often containing parasites in regression, fixed in formalin and/or embedded in paraffin, together with the extreme specialization necessary to read the histological preparations, have prompted research toward finding a molecular-based method useful to overcome these difficulties.

In 1991, Cancrini et al. [45] carried out a multilocus analysis of the gene–enzyme system, using diagnostic isozymes capable of distinguishing *D. immitis* from *D.* (*Nochtiella*) *repens*. However, this technique can be applied only to parasite frozen immediately after removal from the host and for this reason is scarcely suitable for routine diagnosis of human dirofilariasis.

To overcome these diagnostic limit, DNA technology has provided an alternative approach for identification of the filarial parasites [46].

Several PCR-based techniques capable of identifying any developmental stages of the parasite in both vector and suitable and/or accidental host have been developed.

In Italy, a country enzootic for both species of *Dirofilaria* in dogs and with the highest number of human cases by *D.* (*Nochtiella*) *repens* in the world, a diagnostic PCR was developed by Favia et al. [47]. The authors exploited recombinant DNA technologies for detecting and identifying the two *Dirofilaria* species, and designing two sets of specific primers. For *D. immitis*, they used nucleotide sequences of the cuticular antigen-containing multiple tandem repeats [48,49], and for *D. repens* the highly repetitive DNA sequences characterized by 176-bp units arranged as direct tandem repeats clustered in the parasite genome, constituting approximately 3% of the entire genome [50].

The frequent preservation in formalin of the diagnostic samples collected from *Dirofilaria*-infected patients arouse the necessity of developing protocols for extracting the parasite DNA from different sources and preservation conditions. In 1999, Vakalis et al. [51] developed and improved a PCR protocol for the amplification of *D.* (*Nochtiella*) *repens*-specific repetitive sequences from a formalin-fixed cutaneous biopsy. Successively, Rivasi et al. [52] carried out DNA extraction from paraffin-embedded pulmonary tissue and identified *D.* (*Nochtiella*) *repens* by using PCR techniques with primers that amplify DNA fragments of 246 bp for *D.* (*Nochtiella*) *repens* [51] and 378 bp for *D. immitis* [49].

Currently, the molecular identification techniques based on the study of different genes (e.g., ITS1 and ITS2 rRNA, COI and 12S mtDNA) are being implemented; the amplifications of these genes may be also useful for conducting research of phylogeny aimed to study the epidemiology of the different species of *Dirofilaria*, including the zoonotic ones [53–55].

51.2 METHODS

51.2.1 Sample Preparation

We report a protocol for DNA purification aimed to obtain a suitable template to be used for PCR amplification. The protocol is routinely applied in the Department of Pathological Anatomy of the Hospital of Modena (Italy) for the molecular diagnosis of human dirofilariasis.

We also suggest straightforward procedures reported by Wright and Manos [56] and the QIAamp DNA mini kit (QIAGEN, UK) for the isolation of genomic DNA from formalin-fixed PET.

51.2.1.1 DNA Extraction from Paraffin-Embedded Tissues

1. Cut with a microtome three 5 μm sections from paraffin-embedded tissues (to avoid cross-contamination of samples, clean the microtome blade with xylene between each block).
2. Deparaffinize sections by washing twice with xylene and twice with 100% ethanol.
3. Let the ethanol evaporate.
4. Add 200 μL of lysis buffer (50 mM Tris (pH 8.5), 1 mM EDTA, 0.5% Tween 20 and 200 μg/mL of proteinase K).
5. Incubate the mixture for 3 h at 56°C.
6. Incubate the samples at 95°C for 10 min to inactivate proteinase K.
7. Add 10 μL of extracted DNA to the PCR mixtures.
8. Store the DNA at −20°C.

51.2.1.2 DNA Extraction from Fresh or Frozen Tissues

For the DNA purification from fresh or frozen materials, the protocols reported by Strauss [57] and Sambrook [58] are used, based on a traditional organic extraction protocols using phenol–chloroform. The obtained DNA is subsequently concentrated and purified by precipitation in a cold mixture of salt and ethanol. Finally, DNA is solubilized in Tris-EDTA buffer or in water.

51.2.2 Detection Procedures

PCR does not require high-molecular-weight DNA and therefore allows the analysis of deteriorated specimens. The use of PCR to examine the DNA in fixed, paraffin-embedded tissues provides a relatively simple and extremely sensitive method for examining large numbers of samples. We have noticed a length limit on the amplification product from extracted, digested tissue from formalin-fixed, paraffin-embedded sections. PCR products in the range of 450 to 650 bp can be amplified from most sections of this type (unless the blocks are very old). Larger products can be amplified only from some of these samples. Our targets for human dirofilariasis detection are 246 bp for *D.* (*Nochtiella*) *repens* and 378 for *D. immitis*.

51.2.2.1 PCR Assay for *D. (Nochtiella) repens*

The PCR primer sets used are: DIR3 (position: 54–73): 5′-CCGGTAGACCATGGCATTAT-3′ (forward) and DIR4 (position: 299–280): 5′-CGGTCTTGGACGTTTGGTTA-3′ (reverse) specific amplifying a 246-bp fragment of the IpS insert of *D. repens* [51].

1. Total PCR volume 25 μL containing: 50 mM KCl, 10 mM Tris (pH 8.3), 200 μM dNTPs (GE Heathcare), 1.5 mM $MgCl_2$, 25 pmol/μL of each primer, 1 U of AmpliTaq Gold DNA Polymerase (Applera, Italia).
2. Add 10 μL of template DNA.
3. PCR cycle: initial denaturation step at 94°C for 10 min coupled with a 48 cycle of 30 s at 94°C, 30 s at 51°C and 1 min at 72°C, followed by an extended elongation step at 72°C for 5 min [52].

51.2.2.2 PCR Assay for *D. immitis*

PCR is performed by the *D. immitis*-specific primers: 5′-ACGTATCTGAGCTGGCTCAC-3′ (forward, position: 13–32) and 5′-ATGATCATTCCGCTTACGCC-3′ (reverse, position: 390–371) [49], amplifying 378-bp-specific fragment of the cuticular surface antigen gene.

1. Total PCR volume 25 μL containing 50 mM KCl, 10 mM Tris (pH 8.3), 200 μM dNTPs, 1.5 mM $MgCl_2$ (GE Heathcare), 25 pmol/μL of each primer, 1 U of AmpliTaq Gold DNA Polymerase (Applera).
2. Add 10 μL of template DNA.
3. The PCR consists of a denaturation step at 94°C for 10 min and 40 cycles at 94°C for 30 s, 55°C for 40 s, 72°C for 40 s, and a final extension step at 72°C for 7 min.
4. DNAase-free water was used as a negative control; the positive controls were plasmid DNA containing genomic sequences of *D. immitis* or *D. repens*.

The PCR products for both species are detected after electrophoresis in 2% agarose gel (120 V/45 min), revealed by staining with ethidium bromide and visualized under UV light.

51.3 CONCLUSIONS AND FUTURE PERSPECTIVE

Up to now, the diagnosis of dirofilariasis in humans has prevalently been linked to the occasional discovery of the parasite/lesion during diagnostic investigation carried out to clarify the causes of nodular lesions located at the pulmonary or subcutaneous level or in ectopic sites. Instead, the subconjunctival localization has represented the only case of diagnosis carried out following direct observation of the parasite.

The increase in human cases of dirofilariasis recorded worldwide, due to both increased diagnostic attention and a real increase of incidence caused by climatic changes, requires, however, the formulation of methods which can direct the diagnosis in preventive and/or resolutive terms.

In this regard, the most serious lack is, in the field of serology which has not yet devised techniques specific enough to permit a reliable diagnosis to be applied in a preoperative diagnosis or in epidemiological investigation on the human population [20].

Standardized methods of species-specific PCR from both fresh and formalin-fixed material [47,51,52] are currently available, representing a useful tool especially in the species identification of samples in regression and in epidemiological studies, but still ancillary in routine diagnosis of human dirofilariasis.

In the future, molecular studies on different genes may help in clarifying the existence of *Dirofilaria* strains of different virulence for human as suggested by Pampiglione et al. [38] with reference to the different geographic distribution of human dirofilariasis due to *D. immitis*, almost exclusively present in the United States and Japan although it shows a very high prevalence in dogs also in Europe and others geographic areas.

An additional challenge for research in this direction could be represented by the development of proteomics which could finally clarify many doubts which, even today, render the epidemiology and pathogenesis of human dirofilariasis fascinating but still unclear.

ACKNOWLEDGMENT

The present work is dedicated to Professor Silvio Pampiglione, colleague, great friend, and great teacher who passed away recently, who had been working on human dirofilariasis for 26 years leaving a fundamental contribution to the knowledge of this zoonosis.

REFERENCES

1. Addario, C. Su un nematode dell'occhio umano. *Ann. Oftalm.* 14, 134, 1885.
2. Dashiell, G.F. A case of dirofilariasis involving the lung. *Am. J. Trop. Med. Hyg.* 10, 37, 1961.
3. Anderson, R.C. Description and relationship of *Dirofilaria ursi* Yamaguti, 1941, and a review of the genus *Dirofilaria* Railliet and Henry, 1911. *Trans. Royal Can. Inst.* 29, 35, 1952.
4. Faust, E.C. Mammalian heart worms of the genus *Dirofilaria*, In: *Festschrift Berhanrd Nocht zum 80.* Geburtstag, Hamburg, p. 131, 1937.
5. Canestri Trotti, G. et al. The species of genus *Dirofilaria* Railliet & Henry, 1911. *Parassitologia*, 39, 369, 1997.
6. Manfredi, M.T. et al. Phylogeny, systematics and structural aspects. In: Simon, F., and Genchi, C. (eds), *Heartworm Infection in Humans and Animals*, Ediciones Universidad de Salamanca, Salamanca, España, p. 19, 2001.
7. Wong, M.M., and Brummer, M.E.G. Cuticular morphology of five species of *Dirofilaria*: A scanning electron microscope study. *J. Parasitol.* 64, 108, 1978.
8. Anderson, R.C. *Nematode Parasites of Vertebrates. Their Development and Transmission.* CAB International Publishing, Wallingford, Oxon, UK, 467pp, 2000.
9. Manfredi, M.T. et al. Biology of filarial worms parasitizing dogs and cats. *Proceedings First European Dirofilaria Days*, Zagreb, February 22–25, 2007.
10. Nozais, J.P. et al. A case of subcutaneous *Dirofilaria* (*Nochtiella*) *repens* with microfilaremia originating in Corsica. *Bull. Soc. Pathol. Exot.* 87, 183, 1994.
11. Coluzzi, M. Osservazioni sperimentali sul comportamento di *Dirofilaria repens* in diversi gruppi di artropodi ematofagi. *Parassitologia* 6, 57, 1964.
12. Ludlam, K.W. et al. Potential vectors of *Dirofilaria immitis*. *JAVMA* 157, 1354, 1970.
13. Genchi, C. et al. Climate and *Dirofilaria* infection in Europe. *Vet. Parasitol.* 163, 286, 2009.
14. Prieto, G. et al. Seroepidemiology of *Dirofilaria immitis* and *Dirofilaria repens* in humans from three areas of southern Europe. *Res. Rev. Parasitol.* 60, 95, 2000.
15. Donahoe, J.M.R. Clinical aspects of feline dirofilariasis. In: *Proceedings of Heartworm Symposium*. VM Publishing, Bonner Spring, p. 59, 1975.
16. Otto, G.F. Changing geographic distribution of heartworm disease in the United States. In: Morgan, H.C. *Proceedings Heartworm Symposium '74.* Veterinary Medicine, Bonner Springs, Kansas, USA, 1975.
17. Genchi, C. et al. Epidemiology of canine and feline dirofilariasis: A global view. In: Simon, F., and Genchi, C. (eds), *Heartworm Infection in Humans and Animals.* Ediciones Universidad de Salamanca, Salamanca, España, p. 121, 2001.
18. Bolio-Gonzalez, M.E. et al. Prevalence of *Dirofilaria immitis* infection in dogs from Meridia, Yucatan, Mexico. *Vet. Parasitol.* 148, 166, 2007.
19. Ryan, W.G., and Newcomb, K.M. Prevalence of feline heartworm disease—A global review. In: Soll, M.D. (ed.), *Proceedings of the Heartworm Symposium* '95, Auburn, AL, The American Heartworm Society, Batavia, IL, p. 79, 1995.
20. McCall, J.W. et al. Heartworm disease in animals and humans: A review. *Adv. Parasitol.* 66, 193, 2008.
21. Milanez de Campos, J.R. et al. Human pulmonary dirofilariasis: Analysis of 24 cases from Sao Paulo, Brazil. *Chest*, 112, 729, 1997.
22. Vezzani, D. et al. Dirofilariasis in Argentina: Historical review and first report of *Dirofilaria immitis* in a natural mosquito population. *Vet. Parasitol.* 136, 259, 2006.
23. Miyoshi, T. et al. Human pulmonary dirofilariasis: A case report and review of the recent Japanese literature. *Respirology* 11, 343, 2006.
24. Simón, F. et al. What is new about animal and human dirofilariosis? *Trends Parasitol.* 25, 404, 2009.
25. Pampiglione, S. et al. Human dirofilariasis due to *Dirofilaria* (*Nochtiella*) *repens*: A review of world literature. *Parassitologia* 37, 149, 1995.
26. Pampiglione, S., and Rivasi, F. Human dirofilariasis due to *Dirofilaria* (*Nochtiella*) *repens*: An update of world literature from 1995 to 2000. *Parassitologia* 42, 231, 2000.
27. Dissanaike, A.S. et al. Human infection caused by *Dirofilaria* (*Nochtiella*) *repens* in Sri Lanka. *Parassitologia* 39, 375, 1997.
28. Orihel, T.C., and Ash, L. *Parasites in Human Tissues.* ASCPL [American Society of Clinical Pathologists] Press, Chicago, p. 386, 1995.
29. Beaver, P.C. et al. *Dirofilaria ursi*-like parasites acquired by humans in the northern United States and Canada: Report of two cases and brief review. *Am. J. Trop. Med. Hyg.* 37, 357, 1987.

30. Orihel, T.C., and Eberhard, M.L. Zoonotic filariasis. *Clin. Microbiol. Rev.* 11, 366, 1998.
31. Orihel, T.C., and Isbey, E.K. *Dirofilaria striata* infection in a North Carolina child. *Am. J. Trop. Med. Hyg.* 42, 124, 1990.
32. Beaver, P.C. et al. In: Faust, E.C., Campbell, H.E., and Kellogg, C.R. (eds), *Clinical Parasitology*. 9th edn. Lea & Febiger, Philadelphia, PA, 1984.
33. Theis, J.H. et al. Distribution, prevalence, and relative risk of filariasis in dogs from the state of Washington (1997–1999). *J. Am. An. Hosp. Ass.* 37, 339, 2001.
34. Pampiglione, S. et al. Human *Dirofilaria repens* infection in Hungary: A case in the spermatic cord and a review of the literature. *Acta Vet. Acad. Sci. Hung.* 47, 77, 1999.
35. Collins, B.M. et al. *Dirofilaria tenuis* infection of the oral mucosa and cheek. *J. Oral Maxillofac. Surg.* 51, 1037, 1993.
36. Marty, A.M., and Neafie, R.C. *Pathology of Infectious Diseases, Vol. I, Helminthiases*. Washington, DC, Armed Forces Institute of Pathology, 2000.
37. Pampiglione, S. et al. Pitfalls and difficulties in histological diagnosis of human dirofilariasis due to *Dirofilaria* (*Nochtiella*) *repens*. *Diag. Microbiol. Infect. Dis.* 34, 57, 1999.
38. Pampiglione, S. et al. Dirofilarial human cases in the Old World, attributed to *Dirofilaria immitis*: A critical analysis. *Histopathology* 54, 192, 2009.
39. Simon, F. et al. Human humoral immune response to *Dirofilaria* species. *Parassitologia* 39, 397, 1997.
40. Santamaria, B. et al. Evaluation of *Dirofilaria immitis* excretory/secretory products for seroepidemiological studies on human dirofilariasis. *Parasite* 2, 269, 1995.
41. Simón, F. et al. Immunity in the vertebrate hosts. Simón, F., and Genchi, C. (eds), *Ediciones Universidad de Salamanca*. Salamanca, España, 83, 2001.
42. Song, K.H. et al. Cross-reactivity between sera from dogs experimentally infected with *Dirofilaria immitis* and crude extract of *Toxocara canis*. *Korean J. Parasitol.* 40, 195, 2002.
43. Simón, F. et al. Immunoglobulin G antibodies against the endosymbionts of filarial nematodos (*Wolbachia*) in patients with pulmonary dirofilariasis. *Clin. Diagn. Lab. Immunol.* 10, 180, 2003.
44. Grandi, G. *Wolbachia* in *Dirofilaria repens* causing human subcutaneous dirofilariosis. *J. Parasitol.* 94, 1421, 2008.
45. Cancrini, G. et al. Identification of *Dirofilaria* in man by multi locus electrophoretic analysis. *Ann. Trop. Med. Pad.* 85, 529, 1991.
46. Thanomsub, B.W. et al. Differential diagnosis of human lymphatic filariasis using PCR-RFLP. *Mol. Cell. Probes* 14, 41, 2000.
47. Favia, G. et al. Polymerase chain reaction-identification of *Dirofilaria repens* and *Dirofilaria immitis*. *Parasitology*, 113, 567, 1996.
48. Poole, C.B. et al. Cloning of a cuticular antigen that contains multiple tandem repeats from the filarial parasite *Dirofilaria immitis*. *Proc. Natl. Acad. Sci. USA* 89, 5986, 1992.
49. Scoles, A.G., and Kambhamptati, S. Polymerase chain reaction-based method for detection of canine heartworm (Filaria: Onchocercidae) in mosquitoes (Diptera: Culicidae) and vertebrate hosts. *J. Med. Entomol.* 32, 864, 1995.
50. Chandrasekharan, N.V. et al. *Dirofilaria repens*: Cloning and characterization of a repeated DNA sequence for the diagnosis of dirofilariasis in dogs, *Canis familiaris*. *Exp. Parasitol.* 78, 279, 1994.
51. Vakalis, N. et al. Improved detection of *Dirofilaria repens* DNA by direct polymerase chain reaction. *Parasitol. Int.* 48, 145, 1999.
52. Rivasi, F. et al. Detection of *Dirofilaria* (*Nochtiella*) *repens* DNA by polymerase chain reaction in embedded paraffin tissues from two human pulmonary locations. *APMIS*, 114, 566, 2006.
53. Mar, P.H. et al. Specific polymerase chain reaction for differential diagnosis of *Dirofilaria immitis* and *Dirofilaria reconditum* using primers derived from internal transcribed spacer region 2 (ITS2). *Vet. Parasitol.* 106, 243, 2002.
54. Nuchprayoon, S. et al. Detection of filarial parasites in domestic cats by PCR-RFLP of ITS1. *Vet. Parasitol.* 140, 366, 2006.
55. Huang, H. et al. Molecular characterization and phylogenetic analysis of *Dirofilaria immitis* of China based on COI and 12S rDNA genes. *Vet. Parasitol.* 160, 175, 2009.
56. Wright, D.K., and Manos, M. Sample preparation from paraffin-embedded tissue. In: Innis, M.A., Glefand, D.H., and Sninsky, J.J. (eds), *PCR Protocol: A Guide to Methods and Application*. Academic, San Diego, pp. 153–158, 1990.
57. Strauss, W.M. Preparation of genomic DNA from mammalian tissue, in Current Protocols. In: Ausubel, F.M., Brent, R., Kingston, R.E., Moore, D.D., Seidnan, J.G., Smith, J.A., and Struhl, K. (eds), *Molecular Biology*. John Wiley & Sons, New York, pp. 2.2.1–2.2.3, 1995.
58. Sambrook, J., Fritsch, E.F., and Maniatis, T. *Molecular Cloning: A Laboratory Manual*, 2nd edn. Cold Spring Harbor Laboratory Press, Cold Spring Harbor, NY, pp. 9.14–9.23, 1989.

52 Dracunculus

Dongyou Liu

CONTENTS

52.1 INTRODUCTION

Being the only species within the genus *Dracunculus* that affects humans, *Dracunculus medinensis* is a waterborne filarial nematode that causes subcutaneous filariasis (also known as Guinea worm disease, dracunculiasis or dracunculosis). Other filarial nematodes also responsible for subcutaneous filariasis in humans are *Loa loa* (the African eye worm), *Mansonella streptocerca*, and *Onchocerca volvulus* (river blindness).

Dracunculiasis has been known as a human parasitic disease since ancient Egypt. The Guinea worm *D. medinensis* is supposed to represent the snake on the Aeskulaps stick, an ancient symbol of medicine. The disease once plagued a large number of tropical and subtropical countries in Africa, Asia, and South America (e.g., Algeria, Egypt, Gambia, Guinea, Iraq, Iran, Brazil, Uzbekistan, and the West Indies), but the disease died out spontaneously or was eradicated in many of these countries. Currently, dracunculiasis (Guinea worm disease) remains endemic in Ethiopia, Ghana, Mali, and Sudan. In addition, a recent study indicates that there is a renewed transmission of dracunculiasis in Chad. Therefore, the availability of rapid and specific diagnostic techniques is essential for the improved detection, epidemiological tracking, and control of dracunculiasis in human populations.

52.1.1 Classification, Morphology, and Biology

52.1.1.1 Classification

The genus *Dracunculus* is a filarial nematode classified in the family Dracunculidae, superfamily Dracunculoidea, order Spirurida, class Chromadorea, phylum Nematoda. The family Dracunculidae comprises the genera *Dracunculus* and *Micropleura*. In turn, the genus *Dracunculus* can be separated into at least 12 species: *Dracunculus brasiliensis*, *Dracunculus dahomensis*, *Dracunculus doi*, *Dracunculus fuelleborni*, *Dracunculus globocephalus*, *Dracunculus houdemeri*, *Dracunculus insignis*, *Dracunculus lutrae*, *Dracunculus medinensis*, *Dracunculus mulbus*, *Dracunculus oesophageus*, and *Dracunculus ophidensis* [1–4].

Among the *Dracunculus* species, only *D. medinensis* (little dragon from Medina, referring to its one-time high incidence in the city of Medina) infects humans, causing Guinea worm disease (due to a similar past high incidence along the Guinea coast of West Africa; also known as dracunculiasis or dracunculosis, Latin for affliction with little dragons), although both Medina and Guinea are now free of Guinea worm [5–10]. Other *Dracunculus* species mainly reside in the internal tissues and body cavities of non-human mammals and reptiles (snakes and turtles) [11]. While *Dracunculus* worms recovered from mammals in the Old World and South America are regarded as *D. medinensis*, those from mammals in North America are considered as *D. insignis*. Additionally, specimens recovered from otter in North America are assumed to be *D. lutrae*. Recently, a new *Dracunculus* species, *D. brasiliensis*, was identified from the body cavity of the anaconda *Eunectes murinus* in Brazil [4].

As a member of spirurids (which are tissue parasites producing eggs containing larvae or free larvae requiring arthropod intermediate hosts), *Dracunculus* (superfamily Dracunculoidea) has been previously included among the filariae within the superfamily Filarioidea, which covers important human filarial parasites *Wuchereria bancrofti*, *Brugia malayi*, *M. streptocerca*, *O. volvulus*, and *L. loa*. *Dracunculus* differs from these filarial parasites in the disparate sizes of the sexes and life history. Molecular evidence based on the analyses of small-subunit (18S) ribosomal RNA (rRNA) sequences also lends support for the distinct relationship between *Dracunculus* and members of the superfamily Filarioidea [12].

52.1.1.2 Morphology

The *D. medinensis* adult female is one of the longest nematodes known to exist, measuring 70–120 cm in length and 1–2 mm in width. The body cavity of the *D. medinensis* adult female is filled with the uterus, which extends anteriorly and posteriorly and contains 1–3 million first-stage larvae. The vulva is located halfway down the body and is closed with a plug. The gut is flattened and is nonfunctional. The first-stage larvae measure 643 (490–737) μm × 23 (18–24) μm with a pointed tail and a fully formed gut, although they do not feed.

The *D. medinensis* adult male is 15–40 mm in length and 0.4 mm in width, and its tail has 3–6 pairs of preanal papillae and 4–6 pairs of postanal papillae, with subequal spicules (490–750 μm long) and an accessory organ, the gubernaculum (about 117 μm).

By contrast, *D. brasiliensis* female is 130 mm in length and 1.77 mm in width, and appears whitish, filiform, without any anterior constriction. The female also possesses unusually large, anteriorly protruding dorsal and ventral double cephalic papillae and much smaller lateral papillae of the internal circle, a widely rounded caudal end, the excretory pore situated just posterior to the nerve ring, and a distinctly transversely striated cuticle. *D. brasiliensis* larvae are 396–429 μm in length [4].

D. mulbus, a parasite residing in the body cavity of the water python *Liasis fuscus* in northern Australia, differs from other *Dracunculus* species in the position of the papillae at the posterior end in males and in the possession of thick, narrow caudal alae. Submedian cephalic papillae are single in both sexes [3].

Similar to *D. medinensis*, the tail in other *Dracunculus* spp (with the exception of *D. lutrae* and *D. brasiliensis*) is conical, with a pointed tip. The female excretory pore in *D. globocephalus*, *D. insignis*, *D. lutrae*, and *D. oesophageus* is situated some distance posterior to the level of the nerve ring. Additionally, the larvae of *D. fuelleborni*, *D. insignis*, *D. lutrae*, *D. medinensis*, and *D. globocephalus* are 581–857 μm in length, whereas the larvae of *D. fuelleborni* are 300–430 μm in length; the larvae of *D. oesophageus* are of 475–570 μm as compared with those of *D. brasiliensis* (396–429 μm) and *D. medinensis* (490–737 μm).

52.1.1.3 Biology

D. medinensis requires two hosts (copepods as intermediate host and humans as final host) for its complete life cycle. Humans (the final host) become infected by drinking water from stagnant ponds, pools, cisterns, or open wells that contains fresh water copepods (Cyclops or water fleas) harboring *D. medinensis* infective larvae. The stomach fluid breaks down the copepods to release the larvae. The larvae then move to the small intestine, penetrate the intestinal wall, across the peritoneal cavity, and migrate into the wall of the abdomen and thorax by 15 days. After two molts the sexually mature *D. medinensis* males and females (both of comparable size) meet and mate in 60–90 days after infection. The males remain in the tissues and become encapsulated and die in a few months. The females migrate in the subcutaneous tissues toward the surface of the skin. About 10–14 months later, the gravid adult female worm(s) (70–120 cm long) forms a painful ulcerating blister on the skin (typically on the distal lower extremity such as foot or leg, and occasionally on other parts of the body such as the head, torso, upper extremities, buttocks, and genitalia) [13]. The female worm extrudes the skin and becomes exposed to the environment. When the patient seeks to relieve the local discomfort by placing their foot in fresh water, the gravid adult female contracts, causing a loop of the worm's uterus (containing a million or more larvae) to burst and expel hundreds of thousands of motile first-stage larvae. The first-stage larvae do not feed and can survive in water for a few days. These larvae are then ingested by water copepods (1–2 mm in size), belonging to the genera *Mesocyclops* (*M. aequatorialis* and *M. kieferi*), *Metacyclops* (*M. margaretae*), and *Thermocyclops* (*T. crassus*, *T. incisus*, *T. inopinus*, and *T. oblongatus*). Inside the intermediate host copepods, the first-stage larvae develop into infective third-stage larvae after 14 days and two molts. A new infection cycle begins when humans acquire the parasite by drinking water containing copepods infected with *D. medinensis* third-stage larvae. On rare occasions, the infection can be acquired by eating a fish paratenic host (as in the presumably zoonotic cases of dracunculiasis in human patients reported in Japan, Korea, and Indonesia) [14]. The process from ingestion of a larva to emergence of an adult worm typically takes about a year.

Dracunculiasis is a disabling disease of poor people who do not have access to safe drinking water. The disease was common in much of southern Asia, North, West, and East Africa during the nineteenth and twentieth centuries. After a series of eradication campaigns, endemic dracunculiasis was confined to 17 African and three Asian countries (i.e., Benin, Burkina Faso, Cameroon, Chad, Côte d'Ivoire, Ethiopia, Ghana, India, Kenya, Mali, Mauritania, Niger, Nigeria, Pakistan, Senegal, Sudan, Togo, Uganda, Yemen, and Central African Republic) during the 1980s and 1990s [15–23]. By the year 2009, dracunculiasis remains endemic in only four African countries (Ethiopia, Ghana, Mali, and Sudan) [24,25]. Asia is now free of dracunculiasis [26,27].

Cases of dracunculiasis most commonly occur among young adults (aged 15–45), and are distributed equally among males and females. Farmers tend to be most frequently infected. The occurrence of cases of dracunculiasis often coincides with harvest or planting seasons [28,29]. In arid areas, peak transmission often coincides with the rainy season, when surface water is available. In wet areas, transmission is most intense in the dry season, when sources of drinking water are limited [1,2].

52.1.2 Clinical Features and Treatment

52.1.2.1 Clinical Features

Clinical symptoms of dracunculiasis appear 10–14 months after infection with the formation of a blister (preceded by fever and allergic symptoms and accompanied by redness and induration) and the subsequent skin lesion. Other symptoms associated with dracunculiasis include local erythema, an urticarial rash with intense itching, nausea, vomiting, diarrhea, and dizziness. In 80–90% of cases, the parasite emerges from the lower extremities, and in the rest cases, the parasite comes out of other parts of the body, such as the head, torso, upper extremities, buttocks, and genitalia [30,31].

The irritation and a burning sensation induced by the formation and subsequent burst of the blister cause the patient to seek relief by immersing the affected limb or by pouring water over the lesion. This expedites the sloughing-off of the skin over the blister and exposes the anterior end of the Guinea worm to the environment outside its human host, allowing the parasite to discharge larvae into water. With the Guinea worm extruding through the skin, patients may seek to pull it out, often by winding a few centimeters of the worm each day on a small stick. After the worm is completely extruded, the lesion then resolves quickly. However, in about 50% of cases, secondary bacterial infections may occur along the track of the worm, leading to severe local inflammation, pain, abscesses, septic arthritis, ankylosis of the joint, or tetanus, severely incapacitating the patients [32–34]. In case the Guinea worm breaks during manual extraction, an intense inflammatory reaction occurs along with more pain, swelling, and cellulitis along the worm tract [35]. Death of adult worms in joints may lead to arthritis and paralysis in the spinal cord. Infertile females or males stimulate a weak inflammatory reaction and sometimes calcify. On average, dracunculiasis results in about 8.5 weeks (ranging from 2 to 16 weeks) of incapacitation in the patients. As dracunculiasis does not induce acquired immunity, patients may become reinfected again.

52.1.2.2 Treatment

Application of wet compresses to the lesion may relieve pain during the worm's emergence. Covering the wound with an occlusive bandage keeps it clean. Oral administration of aspirin may alleviate the associated pain, and topical antiseptics or antibiotic ointment may help minimize the risk of secondary bacterial infections and reduce inflammation [36]. Removal of the worm by rolling it up on a stick a few centimeters each day remains the most useful treatment, especially when combined with a clean dressing and antibiotic ointment to prevent secondary bacterial infection. The whole process may take approximately four weeks or more. When the entire worm is near the surface of the skin, surgical removal may be possible.

Prevention of dracunculiasis should be focused on: (i) education of persons with emerged worms to avoid bathing affected body parts in sources of drinking water; (ii) boiling or filtration of potentially contaminated drinking water through a fine-mesh cloth filter such as nylon (to remove the diseased worm-containing crustacean); (iii) treatment of potentially contaminated surface water with a larvicide such as temephos (Abate) (to kill copepods); (iv) provision of safe drinking water from borehole or hand-dug wells [1,2,26].

52.1.3 Diagnosis

Diagnosis of dracunculiasis is achieved by the observation of *D. medinensis* worms emerging from the lesions (following burst of the blister accompanied by sharp pain) on the legs of infected individuals and by microscopic examination of the larvae that are released after exposure of lesions to cold water.

Specific antibodies to *D. medinensis* may be detected in infected patients up to six months before emergence by enzyme-linked immunosorbent assay (ELISA) or dot-enzyme-linked immunosorbent assay (dot-ELISA) using whole-worm antigens [37–40]. As measured by ELISA, individuals with prepatent or postpatent infections generated a significantly lower level of specific IgG_1 and IgG_4 than those with a patent infection, but had a significantly higher level of specific IgG_1 and IgG_4 than endemic normal individuals. Furthermore, individuals with a patent infection showed a significantly lower serum level of total IgE than prepatent, postpatent, and endemic normal individuals [41,42]. The detection of parasite-specific IgG_1 and IgG_4 subclasses may help distinguish newly exposed, patent, and post-patent *D. medinensis* infections [43].

Molecular procedures have been proven useful in distinguishing *D. medinensis* from other tissue-dwelling nematodes (including other *Dracunculus* species such as *D. insignis* that infect humans and other animals) [44,45]. While the small-subunit ribosomal RNA (18S rRNA) gene of *D. medinensis* contains 1819 bp, that of *D. insignis* is 1821 bp [12].

52.2 METHODS

52.2.1 Sample Preparation

D. medinensis female worms are collected and preserved in 70% ethanol. After the careful decanting off of any ethanol or saline, each *D. medinensis* worm specimen is washed three times with 0.01 M phosphate-buffered saline (pH 7.2) containing 1 mM EDTA. The sample is then soaked overnight (12–18 h), at 37 C, in 1 mL of CLS-TC buffer from the FastDNA kit (Q-BIOgene) to facilitate the rehydration of any DNA. Approximately 20 mg of larvae or 2–3 cm of adult worm is used for each DNA extraction. Then, each 2–3-cm-long piece of adult worm is placed in a Petri dish with 1 mL CLS-TC buffer and cut into small pieces with a disposable scalpel blade. Each sample of larvae or chopped adult worm is then homogenized, with an electric homogenizer, in a 2.0-mL microcentrifuge tube. The FastDNA kit and the FastPrep cell disruptor (Q-BIOgene) are used to extract the DNA. The cell disruptor is run, for 20 s at setting 5.0, either once on each sample or, if intact tissue is still visible, twice. The residual PCR inhibitors in the extracted DNA may be removed using the QIAquick PCR purification kit (Qiagen) [12]. Alternatively, small pieces (2–4 mm) of *D. medinensis* are cut with sterile scissors, mechanically homogenized, centrifuged, and resuspended in lysis buffer. Genomic DNA is then isolated using the Jetquick DNA isolation kit (Genomed) [44].

52.2.2 Detection Procedures

52.2.2.1 Analysis of 18S rRNA Gene Sequence (Protocol of Bimi et al. 2005)

Bimi et al. [12] utilized primers NEMFG1 (5′-TCT CCG ATT GAT TCT GTC GGC GAT TAT ATG-3′) and CRYPTOR (5′-GCT TGA TCC TTC TGC AGG TTC ACC TAC-3′) for PCR amplification of *Dracunculus* 18S rRNA gene. The resulting PCR amplicons were sequenced using primers using primers NEMFG1 and CRYPTOR as well as the internal primers NEM1F (5′-CTG CCT TAT CAA CTT TCG ATG-3′), NEM1R (5′-CAT CGA AAG TTG ATA AGG CAG-3′), NEM2F (5′-GCG GTT AAA AAG CTC GTA GTT GG-3′), NEM2R (5′-CCA ACT ACG AGC TTT TTA ACC GC-3′), NEM3F (5′-GCG GCT TAA TTT GAC TCA ACA C-3′), NEM3R (5′-GTG TTG AGT CAA ATT AAG CCG C-3′), NEM4F (5′-CCG GGA CTG AGC CGT TTC GAG-39), and NEM4R (59-CTC GAA ACG GCT CAG TCC CGG-3′).

52.2.2.1.1 Procedure

1. PCR mixture (50-μL) is composed of 1× PCR buffer (10 mM Tris-HCl, pH 9.0, 1.5 mM $MgCl_2$, 50 mM KCl), 200 μM each of dATP, dCTP, dGTP, and dTTP, 2.5 U of AmpliTaq Gold DNA polymerase (Applied Biosystems), 25 pmol of each primer (NEMFG1 and CRYPTOR), and 10–20 ng of parasite DNA template.
2. Amplification is conducted with an initial denaturation at 94°C for 15 min; 25 cycles of 94°C for 30 s, 60°C for 30 s, and 72°C for 90 s; and a final extension at 72°C for 10 min, using an automated thermocycler (GeneAmp PCR System 9700; Applied Biosystems).
3. The amplicons are resolved by electrophoresis in 1% agarose gel, stained with ethidium bromide, and visualized on an ultraviolet trans-illuminator (Fotodyne, Hartland, WI). A 100-bp ladder (Invitrogen, Carlsbad, CA) is used as the size standard.
4. The products of expected size are purified using the StrataPrep PCR purification kit (Stratagene), eluted in 50 mL ultraviolet-irradiated water, and sequenced from both strands using PCR primers as well as internal primers in a Model 3100 genetic analyser (Applied Biosystems).
5. The resulting sequences are assembled and analyzed using SeqManII program from the LasergeneH software package (DNAStar). Phylogenetic analysis is performed with version 5.2 of TREE-PUZZLE program, version 4.0b10 of PAUP program and programs from version 3.63 of the PHYLIP software package after aligning the sequences using version 1.83 of CLUSTAL W program. Phylogenetic trees are then created using version 1.6.6 of TreeView program.

52.2.2.2 Analysis of 18S rRNA Gene Sequence (Protocol of Wijová et al. 2005)

Wijová et al. [44] targeted the 18S rRNA gene for phylogenetic characterization of *D. medinensis*. The primers consist of D-1F (5′-GCCTATAATGGTGAAACCGCGAAC-3′) and D-1R (5′-CCGGTTCAAGCCA CTGCGATTA-3′).

52.2.2.2.1 Procedure

1. About 20 ng of genomic DNA is used for PCR amplification of the 18S rRNA gene using Taq Purple polymerase (Top-Bio) under the following conditions (6 cycles of 95°C for 1 min, 44°C for 1 min, and 72°C for 2 min; 24 cycles with the annealing temperature increased to 48°C).
2. The amplicons of expected size are gel purified using a Jetquick gel extraction kit (Genomed) and cloned into the pDrive Cloning Vector (Qiagen).
3. Both strands are sequenced using a Beckman Coulter automatic sequencer.
4. Sequence fragments are completed with Editseq and Seqman (DNASTAR) and the sequence alignment is created and further edited using ClustalX 1.81 and BioEdit software, respectively.
5. Alignments are analyzed by maximum parsimony (MP) and maximum likelihood (ML), performed using PHYLIP (SEQBOOT), PAUP* and PHYML software. The tree is rooted with a monhysterid (*D. procerus*) and a strongylid (*S. chitwoodorum*).

Note: Phylogenetic analysis of 18S rRNA gene sequences revealed a distinct relationship between *Dracunculus* spp and other spirurids (*B. malayi* and *W. bancrofti*), and showed that *D. oesophageus* represents a different species in spite of its morphological similarity to the human Guinea worm *D. medinensis*.

52.3 CONCLUSION

The genus *Dracunculus* comprises at least 12 species of filarial nematodes that live in the tissues of mammals and reptiles. Of these, *Dracunculus medinensis* is an important pathogen, whose female adult worm causes a painful and debilitating disease (dracunculiasis or Guinea worm disease) in humans. Transmitted to humans through drinking water containing copepods (water fleas) that are infected with the larvae of *D. medinensis*, dracunculiasis was once widespread in tropical and subtropical countries in Africa, Asia, and South America. With improved education and water supply along with extensive control efforts, dracunculiasis has been eradicated in most countries. To date, the only countries where dracunculiasis remains endemic are Ethiopia, Ghana, Mali, and Sudan, although a more recent report suggests a renewed transmission of the disease in Chad [46]. Diagnosis of dracunculiasis relies on the observation of female adult worm(s) emerging from lesions on the legs or other body parts of infected individuals. The use of molecular techniques provides a reliable and precise means of identification and phylogenetic assessment of various *Dracunculus* species.

REFERENCES

1. Muller, R., *Dracunculus* and dracunculiasis. *Adv. Parasitol.*, 9, 73, 1971.
2. Muller, R., Guinea worm disease; epidemiology, control and treatment. *Bull. WHO*, 57, 683, 1979.
3. Jones, H.I., and Mulder, E., *Dracunculus mulbus* n. sp. (Nematoda: Spirurida) from the water python *Liasis fuscus* (Serpentes: Boidae) in northern Australia. *Syst. Parasitol.*, 66, 195, 2007.
4. Moravec, F., and Santos, C.P., *Dracunculus brasiliensis* sp. n. (Nematoda: Dracunculidae) from the anaconda, *Eunectes murinus* (Ophidia: Boidae). *Parasitol. Res.*, 104, 589, 2009.
5. Watts, S.J., Dracunculiasis in Africa: Its geographic extent, incidence, and at-risk population. *Am. J. Trop. Med. Hyg.*, 37, 119, 1987.
6. Adamson, P., Dracontiasis in antiquity. *Med. Hist.*, 32, 204, 1988.
7. Eberhard, M.L., Rab, M.A., and Dilshad, M.N., Red *Dracunculus medinensis*. *Am. J. Trop. Med. Hyg.*, 41, 479, 1989.
8. Greenaway, C., Dracunculiasis (guinea worm disease). *CMAJ*, 170, 495, 2004.
9. Moravec, F., Some aspects of the taxonomy and biology of dracunculoid nematodes parasitic in fishes: A review. *Folia Parasitol.*, 51, 1, 2004.
10. Moravec, F., and Little, M.D., Redescription of *Dracunculus globocephalus* Mackin, 1927 (Nematoda: Dracunculidae), a parasite of the snapping turtle, *Chelydra serpentina*. *Folia Parasitol.*, 51, 339, 2004.
11. Bimi, L., Potential vector species of Guinea worm (*Dracunculus medinensis*) in Northern Ghana. *Vector Borne Zoonotic Dis.*, 7, 324, 2007.
12. Bimi, L. et al., Differentiating *Dracunculus medinensis* from *D. insignis*, by the sequence analysis of the 18S rRNA gene. *Ann. Trop. Med. Parasitol.*, 99, 511, 2005.
13. Birare, S.D. et al., Guinea worm infection of urinary bladder manifesting as obstructive uropathy in rural Maharashtra. *Trop. Doct.*, 35, 242, 2005.
14. Eberhard, M.L., and Brandt, F.H., The role of tadpoles and frogs as paratenic hosts in the life cycle of *Dracunculus insignis* (Nematoda: Dracunculoidea). *J. Parasitol.*, 81, 792, 1995.
15. Edungbola, L.D. et al., A method of rapid assessment of the distribution and endemicity of dracunculiasis in Nigeria. *Soc. Sci. Med.*, 23, 550, 1986.
16. Cairncross, S., Anemana, S.D., and Olsen, A., Towards the eradication of guinea worm; a Danish-Ghanaian collaboration. *Parasitol. Today*, 15, 127, 1999.
17. Cairncross, S., Muller, R., and Zagaria, N., Dracunculiasis (guinea worm disease) and the eradication initiative. *Clin. Microbiol. Rev.*, 15, 223, 2002.
18. Watts, S.J., Dracunculiasis in the Caribbean and South America: A contribution to the history of dracunculiasis eradication. *Med. Hist.*, 45, 227, 2000.
19. Menon, T., Incidental finding of *Dracunculus medinensis* in Australia. *Med. J. Aust.*, 183, 51, 2005.
20. Ruiz-Tiben, E., and Hopkins, D.R., Dracunculiasis (Guinea worm disease) eradication. *Adv. Parasitol.*, 61, 275, 2006.
21. Rwakimari, J.B., Hopkins, D.R., and Ruiz-Tiben, E., Uganda's successful Guinea Worm Eradication Program. *Am. J. Trop. Med. Hyg.*, 75, 3, 2006.
22. Mathur, S., and Gahlot, R.S., Calcified guinea worms in breast. *J. Assoc. Physicians India*, 54, 201, 2006.
23. Miri, E. et al., Nigeria's triumph: Dracunculiasis eradicated. *Am. J. Trop. Med. Hyg.*, 83, 215, 2010.
24. Centers for Disease Control and Prevention (CDC), Progress toward global eradication of dracunculiasis, January 2009–June 2010. *Morb. Mortal. Wkly. Rep.*, 59, 1239, 2010.
25. Fabiansen, C., Harboe, Z.B., and Christensen, V., Dracunculiasis in South Sudan. *Am. J. Trop. Med. Hyg.*, 82, 757, 2010.
26. Barry, M., The tail end of guinea worm—global eradication without a drug or a vaccine. *N. Engl. J. Med.*, 356, 2561, 2007.
27. Iriemenam, N.C. et al., Dracunculiasis—the saddle is virtually ended. *Parasitol. Res.*, 102, 343, 2008.
28. Brieger, W.R., Watts, S., and Yacoob, M., Guinea worm, maternal morbidity and child health. *J. Trop. Paediatr.*, 35, 285, 1989.
29. Adeyeba, O.A., and Kale, O.O., Epidemiology of dracunculiasis and its socio-economic impact in a village in south-west Nigeria. *West Afr. J. Med.*, 10, 208, 1991.
30. Hakim, F.A., and Khan, N.N., Dracunculiasis. An incidental diagnosis in a Saudi female. *Saudi Med. J.*, 28, 1438, 2007.
31. Hopkins, D.R. et al., Dracunculiasis, onchocerciasis, schistosomiasis, and trachoma. *Ann. N.Y. Acad. Sci.*, 1136, 45, 2007.
32. Adeyeba, O.A., Secondary infection in dracunculiasis: Bacteria and morbidity. *Int. J. Zoonoses*, 12, 147, 1985.
33. Smith, G.S. et al., Disability from dracunculiasis: Effect on mobility. *Ann. Trop. Med. Parasitol.*, 83, 151, 1989.
34. Imtiaz, R., Hopkins, D.R., and Ruiz-Tiben, E., Permanent disability from dracunculiasis. *Lancet*, 336, 630, 1990.
35. Glenshaw, M.T. et al., Guinea worm disease outcomes in Ghana: Determinants of broken worms. *Am. J. Trop. Med. Hyg.*, 81, 305, 2009.
36. Chippaux, J.P., Mebendazole treatment of dracunculiasis. *Trans. R. Soc. Trop. Med. Hyg.*, 85, 280, 1991.
37. Bloch, P., Simonsen, P.E., and Vennervald, B.J., The antibody response to *Dracunculus medinensis* in an endemic human population of northern Ghana. *J. Helminthol.*, 67, 37, 1993.
38. Bloch, P., and Simonsen, P.E., Studies on immunodiagnosis of dracunculiasis. I. Detection of specific serum antibodies. *Acta Trop.*, 70, 73, 1998.
39. Bloch, P., Vennervald, B.J., and Simonsen, P.E., Studies on immunodiagnosis of dracunculiasis. II. Search for circulating antigens. *Acta Trop.*, 70, 303, 1998.
40. Bapna, S., and Renapurkar, D.M., Immunodiagnosis of early dracunculiasis. *J. Commun. Dis.*, 28, 33, 1996.
41. Bloch, P., and Simonsen, P.E., Immunoepidemiology of *Dracunculus medinensis* infections I. Antibody responses in relation to infection status. *Am. J. Trop. Med. Hyg.*, 59, 978, 1998.

42. Bloch, P., and Simonsen, P.E., Immunoepidemiology of *Dracunculus medinensis* infections II. Variation in antibody responses in relation to transmission season and patency. *Am. J. Trop. Med. Hyg.*, 59, 985, 1998.
43. Knopp, S. et al., Antibody and cytokine responses in *Dracunculus medinensis* patients at distinct states of infection. *Trans. R. Soc. Trop. Med. Hyg.*, 102, 277, 2008.
44. Wijová, M. et al., Phylogenetic position of *Dracunculus medinensis* and some related nematodes inferred from 18S rRNA. *Parasitol. Res.*, 96, 133, 2005.
45. Elsasser, S.C. et al., Species identification of North American guinea worms (Nematoda: Dracunculus) with DNA barcoding. *Mol. Ecol. Resour.*, 9, 707, 2009.
46. Centers for Disease Control and Prevention (CDC), Renewed transmission of dracunculiasis—Chad, 2010. *Morb. Mortal. Wkly. Rep.*, 60, 744, 2011.

53 Enterobius

Dongyou Liu

CONTENTS

53.1 INTRODUCTION

The genus *Enterobius* consists of two small nematode (roundworm) species (*Enterobius vermicularis* and *Enterobius anthropopitheci*) that are parasitic to humans and/or chimpanzees. Commonly known as human pinworm (in the United States), threadworm (in the United Kingdom, where pinworm refers to *Strongyloides*), or seatworm, *E. vermicularis* is one of the oldest parasites known to infect humans, especially children, institutional populations, and homosexuals. The clinical manifestations of the resulting disease (enterobiasis or oxyuriasis) range from itching of the anus (pruritus ani), disturbed sleep, chronic or acute appendicitis, genitourinary infections, abdominal pain, and abscesses in the rectum (due to ectopic pinworm migrations). Affecting about 1000 million people worldwide, *E. vermicularis* (human pinworm/threadworm) represents a common, yet probably underestimated and highly neglected human intestinal nematode, due to the relative insensitivity of current parasitological diagnostic methods and the mildness of its symptoms that cause medical resources to be directed to other relevant human parasites instead.

53.1.1 Classification, Morphology, Life Cycle, and Epidemiology

53.1.1.1 Classification

The genus *Enterobius* is classified in the family Oxyuridae, superfamily Oxyuroidea, order Oxyurida, class Chromadorea, phylum Nematoda. The family Oxyuridae encompasses 10 genera: *Enterobius*, *Lemuricola*, *Oxyuris*, *Passalurus*, *Pongobius*, *Protozoophaga*, *Skrjabinema*, *Syphacia*, *Trypanoxyuris*, and *Wellcomia*. Within the genus *Enterobius*, two species are recognized: *E. vermicularis* (formerly *Oxyuris vermicularis*) and *E. anthropopitheci*. While *E. vermicularis* (human pinworm) is a pinworm affecting humans and captive chimpanzees (e.g., *Pan troglodytes*), *E. anthropopitheci* (chimpanzee pinworm) is limited to chimpanzees [1,2]. Because *E. vermicularis* from captive chimpanzees is somewhat smaller than that from humans, a subspecies, *E. vermicularis microbulbus*, was suggested, but its validity has been questioned. In addition, a proposal was put forward to erect a third *Enterobius* species (*Enterobius gregorii*) within the genus, as this human infective parasite shows a slightly smaller spicule (i.e., sexual organ) than *E. vermicularis* [3]. However, recent reports indicate that *E. gregorii* possibly represents a younger stage of *E. vermicularis* [4]. Indeed, considering their identical internal transcribed spacer 2 (ITS2) sequence and similar cytochrome *c* oxidase subunit 1 (CO1) gene, *E. gregorii* and *E. vermicularis* are two different names for the same pinworm [5].

53.1.1.2 Morphology

E. vermicularis (human pinworm) is a white, small, and spindle-shaped nematode (roundworm). The adult female (8–13 mm in length and 0.5 mm in width) has a sharply pointed posterior end and a T-shaped reproductive system containing characteristic eggs within the uteri. The adult male (2–5 mm in length and 0.2 mm in width) has a ventrally curved tail, with caudal alae, and a single large copulatory spicule. Cross sections of adult worms often show lateral spines and ova inside the gravid female. The acid mucopolysaccharide-containing lateral spines of the worms are readily visualized using a Movat pentachrome stain, giving them a distinct green color.

E. vermicularis eggs (50–60 μm by 20–30 μm) are translucent, with a smooth, clear outer capsule (shell), which is flattened on one side, giving an asymmetric appearance. The eggs have three distinct membranes. The outermost, thick, albuminous layer (which adheres to environmental objects or surfaces) readily dissolves in gastric or pancreatic juices; the next layer, a thin, hyaline, chitin layer, gives mechanical support; and the third layer is a lipoidal embryonic membrane. At oviposition, *E. vermicularis* larvae within eggs are immature and noninfective. They become infective within 6 h at body temperature and with atmospheric oxygen exposure, and reach 140–150 μm in length.

E. anthropopitheci (chimpanzee pinworm) has a small body (females 3.33–4.73 mm long and males 1.13–1.83 mm long), a straight spicule with a ventral membranous formation in males, double-crested lateral alae in females, small eggs (53–58 μm by 24–28 μm), and a smooth eggshell with three longitudinal thickenings [6]. In contrast, *E. vermicularis* male has a spicule with a hooked distal end, and the female has a single-crested lateral ala [7].

53.1.1.3 Life Cycle

The entire life cycle of *Enterobius*, from egg, larva, to adult, takes place in the gastrointestinal tract of a single human host in about 2–8 weeks. After ingestion via nail biting, poor hygiene, inadequate hand washing, or inhalation of dust containing eggs, *Enterobius* eggs hatch in the duodenum (the first part of the small intestine) and emerge as larvae, which grow rapidly to 140–150 μm in size, molt twice, and become adults by the time of their arrival at the ileocecal junction. Developed from fertilized eggs, females are diploid and survive for 5–13 weeks; developed parthenogenetically, males are haploid, and survive for about 7 weeks. Congregating in the ileocecal region, adult worms sometimes wander throughout the gastrointestinal tract. After copulation with female pinworms in the ileum (the last part of the small intestine), the male pinworms usually die, and are passed out with stool. When females become gravid, they migrate within the lumen and settle in the ileum, cecum (beginning of the large intestine), appendix, and ascending colon, where they attach their heads onto the mucosa and feed on colonic contents (including epithelial cells, bacteria, etc.). The number of eggs in a gravid female pinworm ranges from about 11,000 to 16,000. Approximately 5 weeks after initial ingestion of pinworm eggs by the human host, the egg-laying process begins. The gravid female pinworms migrate through the colon toward the rectum at a rate of 12–14 cm/h. Emerging from the anus (to obtain the oxygen necessary for the maturation of the eggs), the female pinworms move onto the perianal skin and deposit eggs through (i) contracting and expelling the eggs, (ii) dying and then disintegrating, or (iii) bodily rupture due to the host scratching the worm. After oviposition, females become opaque and die. Within 6 h at body temperature, eggs containing a partially developed juvenile are infective and remain so for up to 20 days. Some pinworm larvae that hatch on the anus mucosa can then migrate up the rectum and colon to mature and mate. This is a process called retroinfection, which can lead to an extremely heavy parasite load within the host. Thus, autoinfection (i.e., infection from the original host to itself) either through the anus-to-mouth route or through retroinfection enables the pinworms to inhabit the same host indefinitely, despite their relatively short life span (up to 13 weeks) [8].

53.1.1.4 Epidemiology

E. vermicularis (human pinworm) is an obligate human parasite that has been identified only in humans and captive chimpanzees (e.g., *Pan troglodytes*), but not in wild chimpanzees [9]. Humans acquire the parasite via four principal routes: (i) direct infection from the anal and perianal regions by fingernail contamination (autoinfection) and soiled night clothes; (ii) exposure to viable eggs on soiled bed linen, water, food, or fomites; (iii) inhalation (and subsequent swallowing of pharyngeal or phlegm-containing eggs) of airborne eggs that are dislodged easily from bed linens and clothes or contained in the dust; and (iv) retroinfection following retrograde migration of hatched juveniles from the anal mucosa to the sigmoid colon, and cecum.

E. vermicularis eggs are sticky and can remain viable (i.e., infectious) in a moist environment for up to 3 weeks. While tolerating chlorination in swimming pools, hydrocyanic gas, paradichlorobenzene, and naphthalene well, the eggs are sensitive to ultraviolet lamps, sunlight, and dry heat as well as low temperatures. Clothing, bedding, curtains, walls, carpets, bathroom fixtures, furniture, and toys represent some common sources of reinfection. Being refractory to human pinworm infection, dogs and cats are not directly involved in its transmission. However, their fur can pick up eggs from the environment and become a source of infection. In addition, flies and cockroaches are potential carriers of human parasites including *E. vermicularis*.

E. vermicularis is contagious with transmission occurring from person to person via the fecal–oral route, and affects all socioeconomic groups, particularly school- and preschool-aged children worldwide. It is particularly common in the temperate and cool climates, where reduced sun exposure, heavy clothing, and fewer baths are thought to contribute to its high prevalence. Risk factors for *E. vermicularis* infection include finger sucking and exposure in group settings (day care or acute hospitalization, orphanages, and mental health institutions) [10–12]. In adolescents and adults, both male-to-male sexual behavior and heterosexual coitus may transmit infection [13–15]. The estimated prevalence of *E. vermicularis* in children has been reported to be as high as 11% in the United States, 29% in Denmark, 31% in Korea, 37% in

Sweden, 39% in Thailand, 50% in England, and 61% in India [16–28]. The infection rates of 80–90% have been recorded at summer camps and various institutions.

53.1.2 Clinical Features and Treatment

53.1.2.1 Clinical Features

E. vermicularis is one of the oldest parasites known to infect humans, and its DNA has been detected in pre-Columbian human coprolites [29,30]. The first human infection of the vermiform appendix was noted in 1898, and inflammation in the ileocolic region due to *E. vermicularis* was documented in 1919.

Human infection with *E. vermicularis* is often asymptomatic (accounting for about one-third of the cases). When present, the most common symptoms are either pruritus ani or perineal pruritus (itchiness) [31]. Caused by pinworms migrating to the area around the rectum to lay their eggs, this itching is usually worse at night. The perianal pruritus may cause local "tickling," acute pain, restlessness, loss of appetite (anorexia), nose picking, whining, tooth grinding, insomnia, tiredness, impaired concentration, irritability, enuresis, incontinence, and weight loss. When the itchy area is scratched, local skin trauma, excoriation, eczema, or bacterial dermatitis and folliculitis around the rectum may appear [32]. In girls, pinworm infection can spread to the vagina and cause vulval psoriasis and vaginal discharge [33].

Occasionally, adult worms migrate from the genital tract to the peritoneum, enter the peritoneal cavity through a perforated bowel wall, cause serious gastrointestinal complications (e.g., abdominal pain, intestinal obstruction, intestinal perforation, appendicitis, Meckel diverticulitis, intussusceptions, eosinophilic colitis, mesenteric abscess, enteric ulceration, or eosinophilic granulomas) and ectopic infections (e.g., keratoconjunctivitis, gingivitis, epididymitis, hepatic granuloma, pulmonary embolus, and nodules) [34–69].

In the case of chronic *E. vermicularis* infection in genitourinary tract, burning on micturition and crawling sensations in vagina and urethra with profuse pyuria have been reported. Other clinical symptoms associated with pinworm infections in female patients include dysuria, nocturia, enuresis, vulvovaginitis, salpingitis, and pelvic inflammatory disease (both acute and chronic), abnormal vaginal discharge, vaginal bleeding during pregnancy, postmenopausal bleeding, endometritis, pelvic peritonitis, paraovarian abscess, granulomas of female reproductive organs (vulva, cervix, uterus, fallopian tubes), and infection of human embryo [70–93].

Young et al. [94] presented a case of an *E. vermicularis* infection of the fallopian tube in a 30-year-old woman that resulted in inflammation, tubal obstruction, and infertility. The patient sought treatment for infertility. Laser laparoscopy with hydrotubation with methylene blue dye showed a left fallopian tube obstruction and extensive pelvic adhesions. Microscopic examination of the fallopian tube obtained from salpingectomy revealed numerous calcified and noncalcified ova associated with a moderate infiltrate of lymphocytes, plasma cells, and eosinophils. The ova measured about 50 μm × 20 μm with one elongated side and no terminal or lateral spines.

Pampiglione and Rivasi [95] documented two ectopic cases of enterobiasis producing tumor-like lesions. The first case involved a 57-year-old woman with abdominal pain. Upon surgery, a bilateral tubal-ovarian abscess, associated with diffuse peritonitis, was found in abdomen. Histological examination of the nodular lesion revealed lymphocytic and plasmacellular chronic infiltration and eosinophils at the periphery and amorphous material in the center containing numerous eggs (50 μm in length and 20 μm in width), oval in shape with one flattened side, sometimes calcified or altered by regressive process, suggestive of *E. vermicularis*. The second case related to a 59-year-old man with perianal swelling and pain. On physical examination, a rounded mass of 2 cm in diameter located in the perianal subcutaneous tissue with a hard-elastic consistency was externally palpable in the posterior quadrants of the perineum, and highly painful at palpation. Histological examination of the excised nodular tumor-like lesion showed a plurinucleate giant cell granuloma, foreign body type, fibrous tissue, lymphocytes, and numerous eosinophilic granulocytes mingled with necrotic material. In the center of the suppurative tissue, egg-shaped, asymmetric structures 52–55 μm × 20–25 μm in size, identifiable by the dimensions and form as *E. vermicularis* eggs, were observed.

Interestingly, while *E. vermicularis* generally causes no symptoms or mild symptoms such as perianal pruritus in humans, it tends to induce severe, sometimes fatal, symptoms in captive chimpanzees [96].

53.1.2.2 Treatment

Pinworms are readily treated by mebendazole and other benzimidazole compounds (which inhibit microtubule function in the adult worms and cause glycogen depletion). Alternative chemotherapeutic agents are piperazine, pyrantel embonate, viprynium (pyrvinium) embonate, ivermectin, and *Artemisia herba-alba* extract [97–99]. An asymptomatic infection in another family member(s) can serve as a reservoir of infection, and should be treated simultaneously with a single dose of mebendazole (100 mg for all ages), albendazole (400 mg for all ages), or pyrantel pamoate (5 mg/kg, maximum 350 mg) and repeated in 2 weeks.

It is important to prevent reinfection during treatment. Careful hand washing and finger (nails must be kept short) scrubbing after defecation and before meals are essential. Bed covers, sleeping garments, and hand towels should be changed daily to disinfect infective eggs and the bedroom floor should be kept clean. Children should wear gloves while asleep. Food should be covered to limit contamination with dust-borne eggs.

53.1.3 Laboratory Diagnosis

Laboratory diagnosis of enterobiasis has relied on the identification of adult worms, eggs, or both. Sometimes, *E. vermicularis* pinworms with the appearance of light-colored, moving threads can be seen in the area around a child's rectum (particularly in the early morning when a host's body temperature is at its diurnal nadir) or in the stool (occurring in about 5% of infected individuals); *E. vermicularis* eggs and larvae may be present in urine and vaginal discharge [100].

By pressing the adhesive side of a strip of cellophane tape (Scotch tape or Sellotape) several times against anal or perianal skin and then onto (sticky side downward) to a microscope slide (followed by a drop of toluene to clean debris), *E. vermicularis* worms and eggs are easily visualized under low-power magnification (10× to 45× objective) [101]. Alternatively, after cellophane tape is applied to a clean glass slide, a drop of cedar oil is added for improved detection of eggs and worms [102]. The swabbing is best carried out in the early morning (or even better at 1–2 h before the usual waking time), shortly after waking, before defecation or bathing. Repeat investigation (on separate days) gives a higher yield of positive results. While a single cellophane tape test will allow detection in 50% of infected children, three swabs detect 90%, and five swabs detect 99% of infections. Six consecutive negative results on separate days help rule out the infection. Eggs may also be detected from beneath the fingernails. Stool analysis for ova and parasites is usually inadequate, but endoscopy can be helpful. Compared to *Enterobius* eggs, *Schistosoma* eggs are typically larger (measuring 70–175 μm × 40–70 μm), and possess the lateral or terminal spines and a thin shell, in contrast to the absence of spines and the existence of a thick double shell in *E. vermicularis* eggs. At laparotomy, peritoneal involvement can be confused with tuberculosis, schistosomiasis, or metastatic neoplasia (carcinomatosis).

In recent years, molecular techniques have been applied for the accurate identification and phylogenetic analysis of the human pinworm *E. vermicularis*. Examination of the spacer region of 5S ribosomal RNA (rRNA) genes allowed the molecular identification of *E. vermicularis* implicated in occult enterobiasis [103,104]. The mitochondrial DNA genome of *E. vermicularis* is a 14,010-bp circular DNA molecule that encodes 36 genes (12 proteins, 22 tRNAs, and 2 rRNAs). Phylogenetic analyses (maximum parsimony, maximum likelihood, neighbor joining, and Bayesian inference) of nucleotide sequences for the 12 protein-coding genes of 25 nematode species placed *E. vermicularis*, a representative of the order Oxyurida, as sister to the main Ascaridida + Rhabditida group [105,106]. Combined analyses of the mitochondrial CO1 gene, the ITS2 region of nuclear rRNA, and the nuclear 5S rRNA facilitated elucidation of the phylogenetic relationship between pinworms from captive chimpanzees and those from humans [107].

53.2 METHODS

53.2.1 Sample Preparation

53.2.1.1 Sample Collection

E. vermicularis eggs and pinworms are recovered from adhesive tape samples and feces. To isolate the eggs from pinworm-positive film, the film is soaked in 95% ethanol in a petri dish and shaken lightly to detach the eggs from the adhesive surface of the film. The eggs are collected under a stereomicroscope using a fine capillary pipette, transferred to 100% ethanol, and stored at −20°C until use [5].

The feces are fixed in 100% ethanol, and washed with running tap water through a fine strainer with an aperture size of 100 μm in diameter. Residue left on the strainer is transferred to a petri dish and examined under a stereomicroscope; worms are collected using a fine curved needle attached to the end of a glass rod, and stored at −20°C until DNA extraction. The tail of each male pinworm is cut before DNA extraction and mounted on a glass slide with 50% glycerol to distinguish *E. vermicularis* from *E. gregorii* [5].

53.2.1.2 DNA Isolation

The individual adult pinworm, 50–200 eggs or 150 μL of fecal sediment is added in a tube containing 400 μL digestion buffer (100 mM NaCl, 50 mM Tris-HCl, 1% SDS, and 50 mM EDTA, pH 8.0), and 20 μL of 1 M DTT, 60 μL of 10 mg/mL proteinase K, and 100 μL of 10% SDS. The tube is incubated at 55–60°C for 3–24 h with occasional inversion. The DNA is extracted using the phenol/chloroform method and precipitated in 95% ethanol. The pellet is air dried and diluted in 50 μL sterile H_2O or TE buffer (10 mM Tris-HCl and 1 mM EDTA, pH 8.0). The DNA is purified using a silica resin column (Glass Max DNA Isolation Spring Cartridge System, Gibco-BRL) and quantified on a spectrophotometer by absorbance at 260 nm [5].

Alternatively, DNA is extracted from the ethanol-fixed adults and eggs using the QIAamp DNA Mini kit (Qiagen), according to the manufacturer's instructions, with the following modifications: a random piece of tape of 1–2 cm^2 is excised from each slide, immersed in ATL buffer containing 20 μL of proteinase K, and then incubated at 56°C for 24 h. Following the addition of AL buffer, the sample is incubated for an additional 10 min at 70°C. DNA is eluted in 100 μL of AE buffer [5].

53.2.2 Detection Procedures

53.2.2.1 PCR Identification Targeting 5S rRNA Spacer Region

Iñiguez et al. [104] designed primers from the 5S rRNA spacer region of *E. vermicularis*: Entf (5′-CACTTGCTATACCAACAACAC-3′) and Entr (5′-GCGCTACTAAACCATAGACG-3′) for its specific identification in fecal samples. PCR is performed using 2.5 μL of 10× buffer (200 mM Tris-HCl and 500 mM KCl, pH 8.0, Gibco-BRL), 3 mM $MgCl_2$ (Gibco-BRL), 500 ng of each oligonucleotide, 0.2 mM of each dNTPs, 1.5 U of *Taq* polymerase (Gibco-BRL), and 50–100 ng DNA extract in a final volume of 25 μL. The mixture is subjected to an initial cycle of 5 min at 98°C, followed by 35 cycles of 98°C for 30 s, 50–55°C for 30 s, and 72°C for 30 s. Products are electrophoresed through 1.5% agarose gels and visualized by ethidium

TABLE 53.1
Primers Targeting *Enterobius vermicularis* CO1, ITS2, and 5S rRNA Genes

Gene	Primer	Sequence (5′-3′)	PCR Conditions
CO1	pr-a	TGGTTTTTTGTGCATCCTGAGGTTTA	1 × 94°C for 1 min;
	pr-b	AGAAAGAACGTAATGAAAATGAGCAAC	30 × 94°C for 1 min, 40°C for 1 min, 72°C for 2 min; 1 × 72°C for 7 min
ITS2	LC1	CGAGTATCGATGAAGAACGCAGC	1 × 94°C for 1 min;
	HC2	ATATGCTTAAGTTCAGCGGG	30 × 94°C for 1 min, 52°C for 3 min, 72°C for 2 min; 1 × 72°C for 7 min
5S rRNA	Ent 5SF	GCGAATTCTTGGATCGGAGACGGCCTG	1 × 94°C for 1 min;
	Ent 5SR	GCTCTAGACGAGATGTCGTGCTTTCAACG	30 × 94°C for 1 min, 40°C for 1 min, 72°C for 2 min; 1 × 72°C for 7 min

bromide staining. A unique PCR band of 420 bp is obtained from *E. vermicularis* DNA.

53.2.2.2 Sequence Analysis of CO1, ITS2, and 5S rRNA Genes

Nakano et al. [5] utilized primers from the mitochondrial CO1 gene, nuclear ITS2 region of rRNA, and 5S rRNA for identification and phylogenetic investigation of *E. vermicularis* (Table 53.1). As the nucleotide sequences of CO1 and ITS2 regions are variable, they provide valuable markers for species of many organisms, including nematodes.

After PCR, the amplicons are electrophoresed in a 1.5% agarose gel and visualized by ethidium bromide staining. Subsequently, the PCR products are recovered from the gel using SUPREC-01 (Takara Bio), precipitated with ethanol, and vacuum-dried, or purified using a MiniElute PCR purification kit (Qiagen). Then they are amplified with the BigDye Terminator Cycle Sequencing kit (Applied Biosystems) using the PCR primers. The products are purified using Centri-Sep spin columns (Princeton Separations) and sequenced by an ABI-PRISM310 Genetic Analyzer (Applied Biosystems). The resulting nucleotide sequences are aligned using the CLUSTAL W computer program. The evolutionary distances are computed by Kimura's two-parameter method. The phylogenetic tree is constructed by the neighbor-joining method and the parsimony method using MEGA version 3.1, and evaluated using the bootstrap test based on 1000 resamplings. Phylogenetic trees are outgroup-rooted using the nucleotide sequences from *Ascaris suum* Goeze 1782 (GenBank X54253).

53.3 CONCLUSION

E. vermicularis is an obligate human parasite (commonly known as pinworm, threadworm, or seatworm) whose life cycle takes place entirely within the lumen of the human gastrointestinal tract, with the fecal–oral route being the main mode of transmission. Affecting over 30% of children worldwide, especially those aged 5–12 years and girls 12 years of age, enterobiasis represents one of the most common parasitic infection in humans. Overcrowding, poor sanitation, and travel to underdeveloped countries increase the risk of infection.

While *E. vermicularis* infection is asymptomatic in one-third of patients, reported symptoms of enterobiasis include perianal, perineal, or vulvar irritation due to ova deposition, abdominal pain, nausea, vomiting, diarrhea, cervicitis, endometritis, salpingitis, oophoritis, and generalized peritonitis and cystitis, with the formation of caseating and noncaseating granulomata.

Clinical diagnosis of *E. vermicularis* gastrointestinal infection depends on direct visualization of the adult worm or microscopic detection of the eggs collected from the perianal skin using transparent adhesive tape. The use of molecular techniques permits accurate and specific identification and phylogenetic analysis of *E. vermicularis*, facilitating the monitoring of treatment efficacy and potential relapses of infection.

REFERENCES

1. Russell, L.J., The pinworm, *Enterobius vermicularis*. *Prim. Care*, 18, 13, 1991.
2. Hugot, J.P., Redescription of *Enterobius anthropopitheci* (Gedoelst, 1916) (Nematoda, Oxyurida), a parasite of chimpanzees. *System. Parasitol.*, 26, 201, 1993.
3. Hugot, J.P. et al., Human enterobiasis in evolution: Origin, specificity and transmission. *Parasite*, 6, 201, 1999.
4. Hasegawa, H. et al., Is *Enterobius gregorii* Hugot, 1983 (Nemotoda: Oxyuridae) a distinct species? *J. Parasitol.*, 84, 131, 1998.
5. Nakano, T. et al., Mitochondrial cytochrome c oxidase subunit 1 gene and nuclear rDNA regions of *Enterobius vermicularis* parasitic in captive chimpanzees with special reference to its relationship with pinworms in humans. *Parasitol. Res.*, 100, 51, 2006.
6. Hasegawa, H. et al., Morphology of chimpanzee pinworms, *Enterobius* (*Enterobius*) *anthropopitheci* (Gedoelst, 1916) (Nematoda: Oxyuridae), collected from chimpanzees, *Pan troglodytes*, on Rubondo Island, Tanzania. *J. Parasitol.*, 91, 1314, 2005.

7. Hasegawa, H., and Udono, T., Chimpanzee pinworm, *Enterobius anthropopitheci* (Nematoda: Oxyuridae), maintained for more than twenty years in captive chimpanzees in Japan. *J. Parasitol.*, 93, 850, 2007.
8. Vermund, S.H., and Wilson, C.M., Pinworm (*Enterobius vermicularis*). *Sem. Pediat. Infect. Dis.*, 11, 252, 2000.
9. Grencis, R.K., and Cooper, E.S., *Enterobius, Trichuris, Capillaria*, and hookworm including *Ancylostoma caninum*. *Gastroenterol. Clin. North. Am.*, 25, 579, 1996.
10. Ashford, R.W., Hart, C.A., and Williams, R.G., *Enterobius vermicularis* infection in a children's ward. *J. Hosp. Infect.*, 12, 221, 1988.
11. Herrstrom, P. et al., *Enterobius vermicularis* and finger sucking in young Swedish children. *Scand. J. Prim. Health Care*, 15, 146, 1997.
12. Lohiya, G.S. et al., Epidemiology and control of enterobiasis in a developmental center. *West J. Med.*, 172, 305, 2000.
13. Ogunji, F.O., Post coital pinworm infection. *J. Hyg. Epidemiol. Microbiol. Immunol.*, 27, 103, 1983.
14. Williams, D.J., and Dixon, M.F., Sex, *Enterobius vermicularis* and the appendix. *Br. J. Surg.*, 75, 1225, 1988.
15. Abdolrasouli, A., and Hart, J., Oral-anal intercourse and sexual transmission of *Enterobius vermicularis*; do we need to screen for other intestinal parasites? *Int. J. STD AIDS*, 20, 739, 2009.
16. Mahdi, N.K., and al Khfaji, A.A., Prevalence and seasonal variation of enterobiasis in children of Iraq. *Southeast Asian J. Trop. Med. Public Health*, 21, 135, 1990.
17. Fan, P.C., Review of enterobiasis in Taiwan and offshore islands. *J. Microbiol. Immunol. Infect.*, 31, 203, 1998.
18. Lee, K.J., Lee, I.Y., and Im, K.I., *Enterobius vermicularis* egg positive rate in a primary school in Chungchongnam-do (Province) in Korea. *Korean J. Parasitol.*, 38, 177, 2000.
19. Sirivichayakul, C. et al., Prevalence of enterobiasis and its incidence after blanket chemotherapy in a male orphanage. *Southeast Asian J. Trop. Med. Public Health*, 31, 144, 2000.
20. Sung, J.F. et al., Pinworm control and risk factors of pinworm infection among primary-school children in Taiwan. *Am. J. Trop. Med. Hyg.*, 65, 558, 2001.
21. Kim, B.J. et al., Egg positive rate of *Enterobius vermicularis* of primary school children in Geoje island. *Korean J. Parasitol.*, 41, 75, 2003.
22. Song, H.J. et al., Prevalence and risk factors for enterobiasis among preschool children in a metropolitan city in Korea. *Parasitol. Res.*, 91, 46, 2003.
23. Park, J.H. et al., A survey of *Enterobius vermicularis* infection among children on western and southern coastal islands of the Republic of Korea. *Korean J. Parasitol.*, 43, 129, 2005.
24. Remm, M., Distribution of enterobiasis among nursery school children in SE Estonia and of other helminthiases in Estonia. *Parasitol. Res.*, 99, 729, 2006.
25. Jang, M.H. et al., Prevalence of *Enterobius vermicularis* infection among children attending preschools in Taipei City. *Taipei City Med. J.*, 4, 565, 2007.
26. Chang, T.K. et al., Prevalence of *Enterobius vermicularis* infection among preschool children in kindergartens of Taipei City, Taiwan in 2008. *Korean J. Parasitol.*, 47, 185, 2009.
27. Wang, C.C. et al., Current status of *Enterobius vermicularis* infection in primary schoolchildren in Miaoli County and Taichung County, Taiwan. *J. Microbiol. Immunol. Infect.*, 42, 420, 2009.
28. Wang, L.C., Hwang, K.P., and Chen, E.R., *Enterobius vermicularis* infection in schoolchildren: A large-scale survey 6 years after a population-based control. *Epidemiol. Infect.*, 138, 28, 2010.
29. Iñiguez, A.M. et al., *Enterobius vermicularis*: Ancient DNA from North and South American human coprolites. *Mem. Inst. Oswaldo Cruz*, 98, S67, 2003.
30. Iñiguez, A.M. et al., SL1 RNA gene recovery from *Enterobius vermicularis* ancient DNA in pre-Columbian human coprolites. *Int. J. Parasitol.*, 36, 1419, 2006.
31. Ganor, S., In whom does pinworm infection itch? *Int. J. Dermnatol.*, 26, 667, 1987.
32. Raghallaigh, S.N., and Powell, F.C., *Enterobius vermicularis* dermatitis. *Clin. Exp. Dermatol.*, 35, e32, 2010.
33. Cook, G.C., *Enterobius vermicularis* infection. *Gut*, 35, 1159, 1994.
34. Vafai, M., and Mohit, P., Granuloma of the anal canal due to *Enterobius vermicularis*. Report of a case. *Dis. Colon Rectum*, 26, 349, 1983.
35. Mayayo, E. et al., Pelvic oxyuriasis. *Acta Obstet. Gynecol. Scand.*, 65, 805, 1986.
36. Gupta, S.C. et al., Pathology of tropical appendicitis. *J. Clin. Pathol.*, 42, 1169, 1989.
37. Mondou, E.N., and Gnepp, D.R., Hepatic granuloma resulting from *Enterobius vermicularis*. *Am. J. Clin. Pathol.*, 91, 97, 1989.
38. Goldenberg, S.P., and Marignani, P., The endoscopic diagnosis of colonic enterobiasis. *Gastrointest. Endosc.*, 36, 309, 1990.
39. Sinniah, B. et al., Enterobiasis: A histopathological study of 259 patients. *Ann. Trop. Med. Parasitol.*, 85, 625, 1991.
40. Wiebe, B.M., Appendicitis and *Enterobius vermicularis*. *Scand. J. Gastroenterol.*, 26, 336, 1991.
41. Kollias, G., Kyriakopoulos, M., and Tiniakos, G., Epididymitis from *Enterobius vermicularis*: Case report. *J. Urol.*, 147, 1114, 1992.
42. Mattia, A.R., Perianal mass and recurrent cellulitis due to *Enterobius vermicularis*. *Am. J. Trop. Med. Hyg.*, 47, 811, 1992.
43. Tornieporth, N.G. et al., Ectopic enterobiasis: A case report and review. *J. Infect.*, 24, 87, 1992.
44. Dalimi, A., and Khoshzaban, F., Comparative study of two methods for the diagnosis of *Enterobius vermicularis* in the appendix. *J. Helminthol.*, 67, 85, 1993.
45. Patterson, L.A. et al., Perforation of the ileum secondary to *Enterobius vermicularis:* Report of a rare case. *Mod. Pathol.*, 6, 781, 1993.
46. Dahlstrom, J.E., and Macarthur, E.B., *Enterobius vermicularis*: A possible cause symptoms resembling appendicitis. *Aust. N. Z. J. Surg.*, 64, 692, 1994.
47. Sterba, J., and Vicek, M., Appendiceal enterobiasis—its incidence and relationships to appendicitis. *Folia Parasitol.*, 31, 311, 1984.
48. Liu, L.X. et al., Eosinophilic colitis associated with larvae of the pinworm *Enterobius vermicularis*. *Lancet*, 346, 410, 1995.
49. Surmont, I., and Liu, L.X., Enteritis, eosinophilia, and *Enterobius vermicularis*. *Lancet*, 346, 1167, 1995.
50. Beattie, R., Walker-Smith, A., and Domizio, P., Ileal and colonic ulceration due to *Enterobius*. *J. Paediatr.*, 132, 1055, 1995.
51. Gazvani, M.R., and Emery, S.J., Intraperitoneal *Enterobius vermicularis* infection: A case report. *Infect. Dis. Obstet. Gynecol.*, 4, 28, 1996.
52. Zoorob, R.J., Appendiceal colic caused by *Enterobius vermicularis*. *J. Am. Board Fam. Pract.*, 9, 57, 1996.
53. Ajao, O.G. et al., *Enterobius vermicularis* (pin worm) causing symptoms of appendicitis. *Trop. Doct.*, 27, 182, 1997.
54. Cacopardo, B. et al., Eosinophilic ileocolitis by *Enterobius vermicularis:* A description of two rare cases. *Ital. J. Gastroenterol. Hepatol.*, 29, 51, 1997.

55. Arora, V.K. et al., Fine needle aspiration diagnosis of a subcutaneous abscess from *Enterobius vermicularis* infestation. A case report. *Acta Cytol.*, 41, 1845, 1997.
56. Gatti, S. et al., Intestinal parasitic infections in an institution for the mentally retarded. *Ann. Trop. Med. Parasitol.*, 94, 453, 2000.
57. Macedo, T., and MacCarty, R., Eosinophilic ileocolitis secondary to *Enterobius vermicularis* case report, *Abdom. Imaging*, 25, 530, 2000.
58. Saxena, A.K. et al., Laparoscopic appendectomy in children with *Enterobius vermicularis. Surg. Laparosc. Endosc. Percutan. Tech.*, 11, 284, 2001.
59. Kumar, N. et al., Perianal nodule due to enterobiasis: An aspiration cytologic diagnosis. *Diagn. Cytopathol.*, 28, 58, 2003.
60. Mahomed, A.A. et al., *Enterobius vermicularis* and perianal sepsis in children. *Pediat. Surg. Int.*, 19, 740, 2003.
61. Arca, M.J. et al., Clinical manifestations of appendiceal pinworms in children: An institutional experience and a review of literature. *Pediatr. Surg. Int.*, 20, 372, 2004.
62. Nackley, A.C. et al., Appendiceal *Enterobius vermicularis* infestation associated with right-sided chronic pelvic pain. *JSLS*, 8, 171, 2004.
63. Nordstrand, A.J., and Jayasekera, L.K., *Enterobius vermicularis* and clinical appendicitis: Worms in the vermiform appendix. *Aust. N. Z. J. Surg.*, 74, 1024, 2004.
64. Petro, M., Iavu, K., and Minocha, A., Unusual endoscopic and microscopic view of *Enterobius vermicularis*: A case report with a review of the literature. *South Med. J.*, 98, 927, 2005.
65. Jardine, M., Kokai, G., and Dalzell, M., *Enterobius vermicularis* and colitis in children. *J. Pediatr. Gastroenterol. Nutr.*, 43, 610, 2006.
66. Quasem, A., and Salam, A., Ectopic enterobiasis: A case report and review of literature. *Pak. J. Med. Sci.*, 23, 785, 2007.
67. Ariyarathenam, A.V. et al., *Enterobius vermicularis* infestation of the appendix and management at the time of laparoscopic appendicectomy: Case series and literature review. *Int. J. Surg.*, 8, 466, 2010.
68. Cateau, E. et al., *Enterobius vermicularis* in the kidney: An unusual location. *J. Med. Microbiol.*, 59, 860, 2010.
69. Hirai, Y. et al., Unusual colonoscopic view of *Enterobius vermicularis. Intern. Med.*, 50, 657, 2011.
70. Beckman, E.N., and Holland, J.B., Ovarian enterobiasis—a proposed pathogenesis. *Am. J. Trop. Med. Hyg.*, 30, 74, 1981.
71. Neri, A. et al., *Enterobius (Oxyuris) vermicularis* of the pelvic peritoneum—a cause of infertility. *Eur. J. Obstet. Gyn. RB*, 23, 239, 1986.
72. Mendoza, E. et al., Invasion of human embryo by *Enterobius vermicularis. Arch. Pathol. Lab. Med.*, 111, 761, 1987.
73. Knuth, K.R. et al., Pinworm infestation of the genital tract. *Am. Fam. Phys.*, 38, 127, 1988.
74. Gokalp, A. et al., Relation between *Enterobius vermicularis* infestation and dysuria, nocturia, enuresis nocturna and bacteriuria in primary school girls. *Indian Pediatr.*, 28, 948, 1991.
75. Sun, T. et al., *Enterobius* egg granuloma of the vulva and peritoneum: Review of the literature. *Am. J. Trop. Med. Hyg.*, 45, 249, 1991.
76. Schnell, V.L. et al., *Enterobius vermicularis* salpingitis: A distant episode from precipitating appendicitis. *Obstet. Gynecol.*, 80, 553, 1992.
77. Hamblin, J., and Connor, P.D., Pinworms in pregnancy. *J. Am. Board Fam. Pract.*, 8, 321, 1995.
78. McCabe, K. et al., Enterobiasis of the ovary in a patient with cervical carcinoma in situ. *Infect. Dis. Obstet. Gynecol.*, 2, 231, 1995.
79. Chung, D.I. et al., Live female *Enterobius vermicularis* in the posterior fornix of the vagina of a Korean woman. *Korean J. Parasitol.*, 35, 67, 1997.
80. al Rufaie, H.K. et al., Pinworms and postmenopausal bleeding. *J. Clin. Pathol.*, 51, 401, 1998.
81. Avolio, L. et al., Perianal granuloma caused by *Enterobius vermicularis*: Report of a new observation and review of the literature. *J. Paediatr.*, 132, 1055, 1998.
82. Dundas, K.C., Calder, A.A., and Alyusuf, R., *Enterobius vermicularis* threadworm infestation of paraovarian tissue in a woman who has had a hysterectomy. *Brit. J. Obstet. Gynaecol.*, 106, 605, 1999.
83. Ok, U.Z. et al., Relationship between pinworm and urinary tract infections in young girls. *APMIS*, 107, 474, 1999.
84. Erhan, V. et al., Unilateral salpingitis due to *Enterobius vermicularis. Int. J. Gynecol. Pathol.*, 19, 188, 2000.
85. Das, D.K. et al., Pelvic abscess from *Enterobius vermicularis*: Report of a case with cytologic detection of eggs and worms. *Acta Cytol.*, 45, 425, 2001.
86. Tandan, T. et al., Pelvic inflammatory disease associated with *Enterobius vermicularis. Arch. Dis. Child.*, 86, 439, 2002.
87. Smolyakov, R. et al., *Enterobius vermicularis* infection of female genital tract: A report of three cases and review of the literature. *Eur. J. Obstet. Gynecol. Reprod. Biol.*, 107, 220, 2003.
88. Sizer, A.R. et al., A pelvic mass due to infestation of the fallopian tube with *Enterobius vermicularis. J. Obstet. Gynaecol.*, 24, 462, 2004.
89. Thomson, J.C., Pelvic pain caused by intraperitoneal *Enterobius vermicularis* (threadworm) ova with an associated systemic autoimmune reaction. *J. Obstet. Gynaecol. Res.*, 30, 90, 2004.
90. Burkhart, C.N., and Burkhart, C.G., Assessment of frequency, transmission, and genitourinary complications of enterobiasis (pinworms). *Int. J. Dermatol.*, 44, 837, 2005.
91. Siochou, A., Birtsou, H., and Papazahariadou, M., *Enterobius vermicularis* infection of female genital tract. *Int. J. Immunopathol. Pharmacol.*, 21, 1031, 2008.
92. Craggs, B. et al., *Enterobius vermicularis* infection with tubo-ovarian abscess and peritonitis occurring during pregnancy. *Surg. Infect.* (Larchmt), 10, 545, 2009.
93. Konanahalli, P. et al., *Enterobious vermicularis* (pinworm) infestation of the vulva: Report of 2 cases of a pseudoneoplastic lesion mimicking squamous carcinoma. *Int. J. Gynecol. Pathol.*, 29, 490, 2010.
94. Young, C. et al., *Enterobius vermicularis* infection of the fallopian tube in an infertile female. *Pathol. Res. Pract.*, 206, 405, 2010.
95. Pampiglione, S., and Rivasi, F., Enterobiasis in ectopic locations mimicking tumor-like lesions. *Int. J. Microbiol.*, 2009, 642481, 2009.
96. Murata, K. et al., Fatal infection with human pinworm, *Enterobius vermicularis*, in a captive chimpanzee. *J. Med. Primatol.*, 31, 104, 2002.
97. Singh, S., and Samantaray, J.C., Topical anthelmintic treatment of recurrent genitourinary enterobiasis. *Genitourin. Med.*, 65, 284, 1989.
98. Yang, Y.S. et al., Chemotherapeutic trial to control enterobiasis in schoolchildren. *Korean J. Parasitol.*, 35, 265, 1997.
99. St Georgiev, V., Chemotherapy of enterobiasis (oxyuriasis). *Expert Opin. Pharmacother.*, 2, 267, 2001.
100. Adungo, N.I., Ondijo, S.O., and Pamba, H.O., Observation of *Enterobius vermicularis* ova in urine: 3 case reports. *East Afr. Med. J.*, 63, 676, 1986.

101. Cho, S.Y., A female *Enterobius vermicularis* picked up in a cellotape anal swab. *Korean J. Parasitol.*, 35, 299, 1997.
102. Khalil, H.M. et al., Modifications in diagnosis and treatment of *Enterobius vermicularis. J. Egypt. Soc. Parasitol.*, 17, 643, 1987.
103. Liu, L.X., Blaxter, M.L., and Shi, A., The 5S ribosomal RNA intergenic region of parasitic nematodes: Variation in size and presence of SL1 RNA. *Mol. Biochem. Parasitol.*, 83, 235, 1996.
104. Iñiguez, A.M. et al., *Enterobius vermicularis*: Specific detection by amplification of an internal region of 5S ribosomal RNA intergenic spacer and trans-splicing leader RNA analysis. *E. vermicularis*: Specific detection by PCR and SL1 RNA analysis. *Exp. Parasitol.*, 102, 218, 2002.
105. Kang, S. et al., The mitochondrial genome sequence of *Enterobius vermicularis* (Nematoda: Oxyurida)–an idiosyncratic gene order and phylogenetic information for chromadorean nematodes. *Gene*, 429, 87, 2009.
106. Zelck, U.E., Bialek, R., and Weiß, M., Molecular phylogenetic analysis of *Enterobius vermicularis* and development of an 18S ribosomal DNA-rargeted diagnostic PCR. *J. Clin. Microbiol.*, 49, 1602, 2011.
107. Piperaki, E.T. et al., Characterization of *Enterobius vermicularis* in a human population, employing a molecular-based method from adhesive tape samples. *Mol. Cell. Probes*, 25, 121, 2011.

54 Gnathostoma

Paron Dekumyoy, Tippayarat Yoonuan, and Jitra Waikagul

CONTENTS

54.1 INTRODUCTION

Parasitic nematodes in the genus *Gnathostoma* cause a nuisance disease among humans, normally manifested by a migratory swelling in the cutaneous tissues, when parasite larvae or immature stages migrate through the human face, limb, or trunk. Nematode larvae/immatures may migrate through the internal organs, causing varying degrees of disease severity depending on the site of infection, such as the genitourinary tract, pulmonary tissues, throat, ears, and eyes; infection may sometimes be fatal if a vital organ is affected, such as the brain. Gnathostomiasis is a food-borne parasitic zoonosis, and the life cycle involves many species of animal hosts, including fish/frogs as second intermediate hosts, and snakes, birds, and other paratenic hosts, harboring infective larvae. The disease is distributed throughout areas where raw fish dishes are favored by humans. The disease is currently very prevalent in Ecuador and Mexico (in Latin America), and Laos, Thailand, and Vietnam (in the Mekong region of Southeast Asia). Several *Gnathostoma* species have been reported as causative agents of disease, but only two species are frequently found, *G. binucleatum* in Latin America and *G. spinigerum* in Southeast Asia. International travel has spread the disease globally into nonendemic areas. Rapid and accurate diagnosis is important for prompt treatment and appropriate case management. A reliable laboratory is useful in providing diagnostic services and advice for physicians in nonendemic countries, so that imported cases can be dealt with efficiently [1,2].

54.1.1 Classification, Morphology, and Biology

The taxonomy of the genus *Gnathostoma* is as follows: Phylum Nematoda, Class Secernentea, Order Spirurida, Family Gnathostomatidae. Of the 23 proposed species, only 13 are now recognized as valid. Five species have been reported as causative agents of disease among humans—*G. binucleatum* Almeyda-Artigas, 1991, *G. doloresi* Tubangui, 1925, *G. hispidum* Fedschenko, 1872, *G. nipponicum* Yamaguti, 1941, and *G. spinigerum* Owen, 1836 [1,2]. Human infection with *G. binucleatum* has been found to be distributed throughout many districts of Mexico [3], in Guayaqil State, Ecuador [4], and in central Peru [5], and Argentina [6]. Human infections with *G. doloresi* and *G. nipponicum* have been reported from Japan [7,8]. Human gnathostomiasis has been reported in Myanmar more than once. In 2000, two Japanese were found infected after returning from Myanmar; the suspected causative agent was *G. malaysiae* [9]. In 2003, an outbreak occurred among Korean emigrants who had eaten snakehead fish in a Yangon restaurant; the causative agent was reportedly *G. spinigerum* [10]. Human infection with *G. hispidum* has been found in Japan and Spain [11,12]. Infection with *G. spinigerum* is widely distributed throughout Asia, except for Korea, down to Australia. Recently, transmission to humans has been reported in Africa, among travelers to Botswana, Tanzania, and Zambia [13,14]. Adult *G. spinigerum* worms were reported in the stomach of a lion from North Rhodesia in 1958 [15], indicating the existence of this parasite in Africa long before human cases were first recorded, in 1991.

Gnathostome adults are usually about 2–3 cm long and milky white to pinkish-cream in color. The mouth has two lips on a cephalic bulb. The body is covered with rows of spines, from the cephalic bulb to the posterior end. The shape and number of spines are marked features of each species. Adult males and females emerge as a colony in a tumor formed on the stomach wall (esophagus, liver, or kidney, depending on the species) of the carnivorous definitive host [1]. Ovoid eggs with knob(s) and pitted shells are expelled from the host in the feces. After 12 days in fresh water, the larvae inside the eggs hatch, and are ingested by a copepod crustacean (the first intermediate host), and develop into third-stage larvae. Fish or amphibians eat the contaminated copepods, *Cyclops* sp., and the gnathostome larvae develop into advanced third-stage larvae (AL3). Cats/dogs, or other carnivorous animals, eat the infected fish, and the AL3 develop into adults. Humans are not the definitive host of gnathostomes, and the AL3 do not develop into fully grown adults in humans, but only to the immature stage. Gnathostoma have a complicated life cycle, in which many species of fish-eating animals can act as paratenic hosts, where the AL3 do not develop further; they can survive for some time by forming a cyst in the host muscle. Encysted larvae become infective when ingested by the next host. Gnathostome larvae can enter the human body by ingestion or skin penetration. When humans consume the exotic raw-fish dish, "Koi Pla" or "Som Phac" in Thailand, or "ceviche" in Latin America, AL3 can enter the stomach/intestine, penetrate the tract wall, and migrate freely through the human body. Experiments have confirmed the risk of infection by consuming infected copepods. AL3 can also enter the skin by direct penetration when handling infected fish with bare hands, such as while preparing fish for cooking or preservation. A human fetus or embryo can be infected by trans-placental migration of a worm from an infected pregnant mother [1,2].

AL3 are normally about 0.4 mm long, and are found as free or encysted larvae in the host muscles or liver. About four rows of cuticular hooklets present on the cephalic head bulb (Figure 54.1). The shapes and numbers of cephalic hooklets and body spines are used to identify AL3 to species. A simplified key to the identification of AL3 species infecting humans follows:

A. 3 rows of cephalic hooklets.................... *G. nipponicum*
   4 rows of cephalic hooklets .... (B)
B. <40 cephalic hooklets in each row................ *G. doloresi*
   >40 cephalic hooklets in each row .... (C)
C. Cephalic hooklets; rectangular shaped .... *G. hispidum*
   Cephalic hooklets; oblong shaped .... (D)
D. Nuclei of intestinal cell, 3–7 .... *G. spinigerum*
   Nuclei of intestinal cell, 2–5 .... *G. binucleatum*

The AL3 of *G. spinigerum* and *G. binucleatum* are similar, and may be differentiated only after cross-section through the intestine, and counting the nuclei of the intestinal cells. However, the two species differ in geographical distribution, with *G. spinigerum* in Asia and *G. binucleatum* in America [16,17].

Humans usually acquire *Gnathostoma* infection by ingesting raw or undercooked infected freshwater fish or other infected paratenic hosts. In Thailand, 48 species of animals have been reported as possible hosts for the transmission of infective gnathostome larvae to humans. Also in Thailand, the highest recorded infection rate was for the swamp eel (*Monopterus albus*) at 30.1%, while another seven fish species had infection rates <10%—*Anabas testudineus, Channa lucius, Channa micropeltes, Channa striata,*

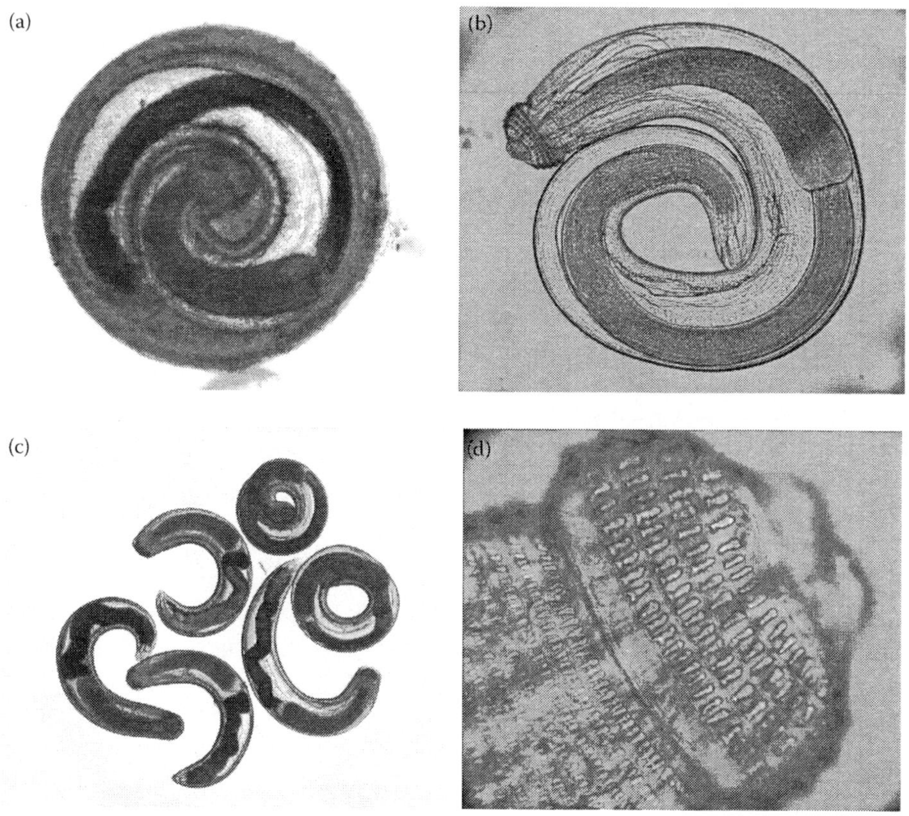

FIGURE 54.1 *Gnathostoma spinigerum* AL3 larvae, showing 4 rows of cephalic hooklets: (a) encysted larva; (b) excysted larva; (c) excysted larvae; (d) cephalic bulb with four rows of hooklets.

Clarias batrachus, *Clarias macrocephalus*, and *Ompok krattensis* [18]. The number of larvae per fish ranged between 1 and 9, but in some eels, record highs reached 698–2582 larvae [19]. Fortunately, eating raw swamp eel is not popular; if it were, the incidence of human cases would be several times the present annual incidence of 100–400. In Mexico, catfish have the highest infection rate, with *Eleotris pictus* (31.8%), and *Dormitator latifrons*, *Gobiomorus maculates*, *Cichlasoma trimaculatum*, and *Cathorops caerulescens* with <7.2% [20]. Recently, the aquatic apple snails, *Pila ampullacea* and *Pomacea canaliculata*, were found in experiments to be suitable paratenic hosts for the AL3 of *G. spinigerum* [21]. Several investigations tested larval survival in different extreme conditions. In the freezer (a household refrigerator, −2 to −4°C), at 48 hours' being frozen, the free encysted larvae lost their infectivity completely, while encysted larvae in meat retained 20% infectivity in experimental mice. AL3 are killed after soaking for 6 h in vinegar or 12 h in soy sauce, but survive for a long time in lime juice [22].

54.1.2 Clinical Features and Pathogenesis

Symptoms normally occur within 24–48 h of ingesting the larva, and may include nausea, vomiting, diarrhea, fever, arthralgia, myalgia, malaise, anorexia, and epigastric pain. The symptoms of later gnathostomiasis manifest as two types: (1) cutaneous: the cutaneous type occurs when larvae migrate into the muscles and subcutaneous tissues, causing intermittent migratory swelling that does not retract when pressed. The swelling is red, painful, and itchy, and may last for 2–7 days, after which it subsides and reappears elsewhere, near the same site. When a larva moves closer to the human skin, the moving track looks similar to the creeping eruption or cutaneous larva migrans caused by animal hookworms. The visceral type, which is more dangerous than the cutaneous type, occurs when a larva penetrates deep into the internal organs. Severity and symptoms vary according to the organ(s) the larva may invade, particularly organs above the neck. AL3 invasion of the eye may result in cellulites and visual impairment. AL3 entering the brain or nervous system may cause paralysis, seizure, and death, due to severe brain damage. Invasion of the lungs may causes pneumothorax or hemoptysis, or hematuria when AL3 invade the kidneys [1,23]. Gnathostomiasis is difficult to treat, and formerly, the only cure was surgical. Since 1990, two drugs have had reportedly high cure rates: albendazole, at a dose of 400 mg once or twice daily for 21 days, gave a cure rate of 93.9–94.1%; ivermectin, 0.2 mg/kg for one or two consecutive days, also gave a 90% cure rate [24,25]. Praziquantel (20 mg/kg single dose) was administered after multiple doses of albendazole, for better results [10]. Frequently, after drug treatment, the AL3 moves to the skin surface, where it can be seen easily, and removed by picking with a needle. Sometimes, repeated treatments become necessary, when symptoms recur [24].

Prevention is very important in any infectious disease, and the most effective preventive measure for gnathostomiasis is to avoid eating raw or improperly cooked animal flesh, especially freshwater fish, frog, snake, monitor lizard, chicken, duck, bird, and rodent. Meat of the second intermediate host or paratenic host, frozen for more than three days, may be considered safe to eat raw. However, the consumption of well-cooked meat is recommended to completely prevent food-borne diseases, including gnathostomiasis.

54.1.3 Diagnosis

54.1.3.1 Conventional Techniques

54.1.3.1.1 Eosinophil Detection

Eosinophilia has been described as one of the diagnostic characteristics of gnathostomiasis, in combination with specific symptoms of migratory swelling [1]. However, in some reports, it has been observed that eosinophil levels remained normal; they did not increase even in worm-proven *Gnathostoma* patients or suspected gnathostomiasis cases [26,27].

54.1.3.1.2 Worm Removal

During migration in the subcutaneous tissues, gnathostome larvae may sometimes be seen clearly, even with the naked eye. The patient or physician can remove a larva using a sharp needle to prick the skin where the larva is visible. This is an excellent way to eliminate a *Gnathostoma* worm and to confirm a diagnosis of gnathostomiasis. The removed larva can be identified to species morphologically, or by DNA extraction and confirmation by polymerase chain reaction (PCR). Anthelmintics frequently stimulate the parasite to move up toward the skin surface. After treatment with albendazole (and probably ivermectin), gnathostome larvae were found in patients' skin, more often than not after they had taken an anthelmintic drug and could easily be extracted by a physician and processed for specific diagnosis [24].

54.1.3.1.3 Immunodiagnosis

Human gnathostomiasis cases have been confirmed by antibody detection using native antigens, such as crude antigens, excretory–secretory (ES) products, or partially purified antigens from adult worms or larvae. Serodiagnosis of gnathostomiasis has been conducted in Asian countries, especially Thailand. In Central America and Mexico, researchers are working on the antibody detection of gnathostomiasis binucleatum. In Europe, a few countries have developed sero-tests for *Gnathostoma* infection. Globally, human gnathostomiasis is caused by five species, that is, *Gnathostoma spinigerum*, *G. hispidum*, *G. doloresi*, *G. nipponicum*, and *G. binucleatum*. *Gnathostoma spinigerum* is reportedly the main causative agent of human gnathostomiasis. Thailand is a gnathostomiasis-endemic area, so that *Gnathostoma* worms are not difficult to obtain and prepare as a source of antigen for antibody detection; therefore, sero-tests are available to detect gnathostomiasis antibodies among patients, and diagnostic services are widely available.

Diagnostic testing has used crude somatic antigen widely, but the specificity of the test varies, depending on the number of heterologous serum samples and the technique chosen.

To date, enzyme-linked immunosorbent assay (ELISA) and immunoblot tests have reportedly been used for the detection of gnathostomiasis. Early diagnoses of human gnathostomiasis used IgG-ELISA with a water-extracted antigen of *G. spinigerum* larvae obtained from experimental mice, to detect antibody in the sera of gnathostomiasis patients with cutaneous migratory swelling, sera from healthy controls, and other parasitic infections. The test demonstrated 56% sensitivity and 84% specificity [28]. The test showed ~100% sensitivity and specificity with marginal cross-reaction using antigen prepared from a normal saline extraction of larvae obtained from naturally infected eels, to react with sera of patients with intermittent cutaneous migratory swelling and a small number of nematode-infected sera [29]. Total IgG antibody has been the focus of gnathostomiasis detection for several decades. Recently, a study on detection of IgG subclasses was conducted, seeking to reduce cross-reactivity, and improve detection specificity. Analysis of crude antigen with IgG subclass showed that among the subclasses, IgG1 antibody provides the highest sensitivity (98%) to the antigen of infective *G. spinigerum* larvae, but the highest specificity (88%) was obtained from IgG2. A screening test for gnathostomiasis could thus use IgG1 antibody, and be confirmed using IgG2 antibody [30].

To improve sensitivity and specificity, purified *G. spinigerum* larvae antigens were prepared using more sophisticated techniques. Purified *G. spinigerum* larvae antigen was prepared by DE52 column chromatography, determined at pI 8.5 by the isoelectro-focusing technique, and showed 100% sensitivity, specificity, and positive and negative predictive values by IgG-ELISA; excellent specificity was produced with 17 heterologous diseases [31]. The monoclonal antibody affinity chromatography-prepared antigen, GN6/24-monoclone, can bind with a 24 kDa specific antigen to an antibody against *Gnathostoma spinigerum* infection [32], and SK-6C4 produced a monoclone, but the eluted antigens showed similar results to the crude antigen by ELISA [33].

Other *Gnathostoma* species have been used to detect the gnathostomiases spinigerum and binucleatum. A crude adult *G. doloresi* worm extract was used to detect antibody against sera of Mexican patients in Culiacan, Mexico, and controls. Indirect ELISA showed 279/300 of patients positive and 147/152 of healthy control sera negative [34]. Korean patients were diagnosed by ELISA using *G. doloresi*, the patients were infected in Yangon, Myanmar, where *G. spinigerum* was detected from a catfish in a local market [10]. By competitive-inhibition ELISA, individual extracts from *G. spinigerum*, *G. doloresi*, and *G. hispidum* were used to detect antibodies from gnathostomiasis cases in Mexico, where *G. binucleatum* is endemic. OD-ELISA values were nearly identical [35]. The useful characteristics of the serodiagnostic tests for detecting human gnathostomiasis are summarized in Table 54.1.

As mentioned above regarding the use of crude antigen, cross-reactions are common in most techniques, but the use of immunoblot can eliminate this problem, since one or more specific band(s) in the immunoblot test may be used to detect

TABLE 54.1
Sensitivity and Specificity of Sero-Tests in the IgG Detection of Gnathostomiasis Patients Using Antigens Prepared from *Gnathostoma spinigerum* Advanced Third-Stage Larvae from Naturally Infected Eels

Test	Antigen Preparation	Positive Control	Sensitivity (%)	Specificity (%)	Remarks	Reference
ELISA	Crude Ag/extraction buffer not known	Ocular visceral gnathostomiasis	7/8	96.7%	Specificity determined with 12 diseases	[42]
	ES/RPMI 1640	Ocular visceral gnathostomiasis	7/8	97.4%	Specificity determined with 12 diseases	[42]
	Partially purified Ag (24 kDa) and nonimmunogenic proteins	*G.*-confirmed and clinically diagnosed gnathostomiasis cases	100%	100%	Specificity determined with 17 diseases	[31]
	Fractionated somatic Ag, 27 kDa	*G.*-confirmed and suspected cases	100%	100%	Specificity determined with 7 diseases	[43]
	Fractionated ES antigen, <29 kDa	*G.*-confirmed and suspected cases	100%	92%	Specificity determined with 7 diseases	[43]
	Crude Adult Ag of *G. doloresi*/PBS extraction	Cutaneous larva migrants	93%	98.7%	Compared between cases and healthy volunteers	[34]
	Crude Ag/NSS extraction	*G*-confirmed and suspected cases	98% (IgG1) screening test	88% (IgG2) confirmed test	Specificity determined with 10 diseases	[30]
Immunoblot	Crude Ag/DW extraction	*G.*-confirmed and suspected cases	100%	99.8%	Specificity determined with 16 diseases	[41]
	Crude Ag/DW extraction	*G.*-confirmed and suspected cases	100% IgG4	100%	Specificity determined with 15 diseases	[36]
	Recombinant antigen	*G.*-confirmed and suspected cases	100%	100%	Specificity determined with 7 diseases	[38]

a specific infection. For gnathostomiasis, the 24 kDa band is reacted with total IgG. In a differential study with subclasses of IgG, the crude extract of advanced third-stage larvae was identified by three interesting bands at 20, 21, and 24 kDa in the IgG4 diagnosis of gnathostomiasis. The sensitivity of the immunoblot was 100%, with a strong reaction at 24 kDa, but weak at 20 and 21 kDa. However, the 20- and 21-kDa diagnostic bands can help interpret *Gnathostoma spinigerum* infection. Other subclasses (IgG1, IgG2, and IgG3) were unable to identify any specific band from the reactions [36]. Using the same infective-stage *Gnathostoma*, total IgG was proven to be an immune detector to 24 kDa antigen, with the highest immunoblot sensitivity (91.6%), while IgG 1–4 showed lower sensitivity. However, IgG4 detection was more specific (93.9%) than total IgG [37]. In conclusion, using the immunoblot method, total IgG antibody is best for detecting gnathostomiasis.

The use of molecular technology in antibody detection for *Gnathostoma* infection is not well established, but recently, a cDNA clone was detected by monoclonal antibody against 24 kDa of infective *G. spinigerum* larvae (mAb, mentioned in Tapchaisri et al. [38], and the insert gene encoded a protein similar to 33–39% of the matrix metalloproteinases (MMPs) of *Caenorhabditis elegans* and other vertebrates. This was found to be a secretory protein involved in the ES products of infective *G. spinigerum* larvae [39]. Several researchers have studied the MMPs of this parasite. One presented a MMP-like protein containing a protein of approximately MW 23–24 kDa, at pI 8.1–9.3. This fusion protein was produced by *G. spinigerum* advanced-third-stage larvae obtained from experimental mice; it was focused and characterized by two-dimensional gel electrophoresis (2DE), and its amino acid sequence was identified by liquid chromatography/tandem mass spectrometry (LC-MS/MS). This peptide sequence matched the *G. spinigerum* MMP-like protein sequence, and may prove useful in the production of recombinant antigenic protein [40]. Based on the original finding of the specific 24 kDa diagnostic band for gnathostomiasis spinigerum using native infective larvae extract, and following the study by Tapchaisri et al. [41], a recombinant *G. spinigerum* antigen MMP was produced and characterized by gel based-proteomics and LC-MS/MS. This synthesized protein has a range of molecular weights, from 27.9 to 28.6 kDa, and carries different pIs, ranging from 5.6 to 7.7. The specific bands of both the recombinant MMP and native-advanced third-stage larvae antigens were compared immunogenically using the same serum groups. Both antigens showed 100% sensitivity, specificity, and accuracy by immunoblot [38].

In addition, IgM and IgE antibodies have been predetermined by both indirect ELISA and immunoblot using crude extract of *G. spinigerum* infective larvae. It was found that tests cannot differentiate between sera from confirmed gnathostomiasis cases accompanied by clinically immunologically confirmed cases and healthy controls (Dekumyoy P, unpublished data). However, elevated levels of specific IgE antibody and total IgE have been reported in serum samples from gnathostomiasis cases [44]. Specific IgE antibody to somatic extract, and ES antigens from *G. spinigerum* infective larvae, showed 71.87 and 87.50% sensitivity, and 57.53 and 69.86% specificity, by ELISA, respectively. However, a few heterologous diseases were used in analyzing the specificity of the tests [45].

54.1.3.1.4 Image Diagnosis

Computerized tomography (CT) may help in locating a pathogenic feature with a nodular lesion, or hemorrhage with tracks in the brain, and possibly in an organ [46,47].

Magnetic resonance imaging (MRI) scanning was used in a suspected case of medullary gnathostomiasis, but MRI must be accompanied by a serological test. The images showed lesions of the thoracic spinal cord [48], cervical cords, and lumbosacral spine [49,50]. This technique was used to exclude intracranial involvement in *Gnathostoma* infection [51].

54.1.3.1.5 Ultrasonography

Ultrasonic waves, applied to surface locations such as the cheek, can reveal a diffuse extragladular space-occupying lesion with nonhomogenous echoes [51].

54.1.3.2 Molecular Techniques

Because human *Gnathostoma* infection cannot be detected by the presence of eggs or larvae in the feces, PCR cannot be used directly to diagnose a stool sample. Moreover, the gold-standard detection method still needs improvement. Recently, the molecular technique has only been used to address gnathostome identification problems, via DNA sequencing. The cytochrome oxidase I (COI), internal transcribed spacer region 1 (ITS1), ITS2, 5.8S, 18S, and 28S regions were determined and analyzed for their usefulness in identifying gnathostome species.

Almeyda-Artigas et al. (2000) selected the ITS1, ITS2, and 5.8S genes for a study of the validity of *G. binucleatum* and identified three undecided species of *Gnathostoma* in Mexico. The study found that the 5.8S sequences of all the studied species were identical and could thus not be used to differentiate gnathostome species. However, the sizes of the ITS1 and ITS2 regions differed consistently between the species, so that this may be used to differentiate gnathostome species [16].

Ando et al. studied the sequences of six nucleotide regions—COI, ITS1, ITS2, 5.8S, 18S, and 28S—from five gnathostome species reported to infect humans in Asia and the Americas. They found that, since the sequence lengths, 5.8S (158 bp), 18S (613 bp), and 28S (598 bp), among the five species were almost identical, they could be used to differentiate species. However, the ITS1, ITS2, and COI sequence sizes did differ. They concluded that ITS2 and COI regions were useful genetic markers for the identification of gnathostome species [52].

Several researchers have employed ITS2 as a genetic tool to identify gnathostome larvae found in fish hosts. Kifune et al. used the ITS2 sequence to identify gnathostome larvae collected from four species of freshwater fish from Pantanos de Centla, Tabasco, Mexico. It was found that all of the collected larvae were the same species, *G. binucleatum*; *Petenia splendida* and *Gobiomorus dormitor* were first recorded as intermediate hosts of *G. binucleatum* [53].

The ITS2 sequence was used to analyze the species of six morphologically different types of cephalic hooklets on *G. binucleatum* larvae. It was concluded that the morphological variations observed were intraspecific [54].

Bertoni-Ruiz et al. demonstrated the ITS2 sequence of a new species of gnathostome, *G. lamothei*, and found that the sequence was >50% different from *G. binucleatum* and *G. turgidum* sequences [55].

Hernández-Gómez et al. used ITS2 sequences to discriminate the advanced third-stage gnathostome larvae collected from fish. They recovered *G. lamothei* from a commercial freshwater fish, *Gobiomorus dormitor*. That was the first report of the first-intermediate host of this species, and showed that this fish species is the host of >1 species of gnathostome, since *G. binucleatum* has also been found in this fish species. The authors also suggested that *G. lamothei* might be able to infect humans [56].

Ngarmamonpirat et al. studied the partial COI genes of morphologically typical and atypical *Gnathostoma spinigerum* advanced third-stage larvae, using single-strand conformation polymorphism (SSCP) and sequence variation analysis. High sequence variations were detected in 25.6% (8/32) of AL3, which was higher than 11.5% (72/565) for morphological variations, and they were not equally distributed. However, enzyme activity was unaffected by these sequence variations. A distinct SSCP pattern was also detected on the COI region. These results showed that genetic variations are present in the studied *Gnathostoma* population, and further study is needed to investigate complexity within *G. spinigerum* [17].

The COI gene was used to confirm gnathostome species in larval infections in both wild and cultured swamp eels in southern Vietnam. Eels were purchased from Long An Province and Hoc Mon market, Ho Chi Minh City, from September 2006 to August 2007. A high prevalence was found in August–October in Long An Province, and June–November in Ho Chi Minh City. The collected gnathostome larvae were morphologically similar to *G. spinigerum*, and the COI sequences of the sampled larvae were all identical to the *G. spinigerum* sequence deposited in Genbank [57].

In conclusion, although the COI and ITS2 regions are both considered genetic markers for identifying gnathostome species, ITS2 is more often utilized to differentiate gnathostome species. The neighborhood joining phylogenic tree of the genus *Gnathostoma* shows that the common human species—*G. spinigerum* and *G. binucleatum*—which are similar morphologically, are in the same clade, showing that they are genetically related.

54.2 METHODS

54.2.1 Sample Preparation

Advanced third-stage larvae or adult *Gnathostoma* sp. fixed in 50–100% were used for DNA extraction, by standard phenol/chloroform extraction and ethanol precipitation techniques, [58] or using a DNA mini kit, according to the manufacturer's instructions.

54.2.1.1 Phenol/Chloroform Extraction

1. The larva was broken open with a surgical blade and then placed in a 1.5 ml microcentrifuge tube, suspended in 300 μL of lysis buffer; 3 μL of proteinase K (10 mg/ml) was added and incubated for 3 h at 50°C in a water bath.
2. DNA was extracted by standard phenol/chloroform extraction and ethanol precipitation, according to Sambrook et al. [57].
3. The DNA was resuspended in 100 μL of TE buffer, pH 8.0.

54.2.2 Detection Procedures

Detection involved standard procedures of amplification, purification, and quantification of PCR products, sequencing, and alignment. The primers for the ITS2 region were News2 (forward) 5′-tgt gtc gat gaa gaa cgc ag-3′ and ITS2-RIXO (reverse) 5′-ttc tat gct taa att cag gg-3′ and the primers for the COI region were JB3 (forward) 5′-ttt ttt cgg cat cct gag gtt tat-3′ and JB4.5 (reverse) 5′-taa aga aag aac ata atg aaa atg-3′.

54.3 CONCLUSIONS AND FUTURE PERSPECTIVES

Gnathostomiasis has recently been recognized worldwide for trans-regional transmission. It has increasingly been found among populations of nonendemic countries, where the disease became evident after visits to endemic countries. It is important that information about endemicity and mode of transmission be made available to visitors to endemic areas, for example by providing information about risky foods; this would raise travelers' awareness to reduce the risk of acquiring gnathostomiasis by simply avoiding dishes made from raw fish and the meat of paratenic hosts. Case consultancy and diagnostic services are now available in Mexico and Thailand, the endemic countries (www.tm.mahidol.ac.th). Gnathostomiasis is a tissue infection; therefore, diagnosis is mainly made by clinical symptoms and serological tests. The diagnostic tests currently used focus on antibody detection in serum and cerebrospinal fluid (CSF) samples of clinically suspected cases. Both specific symptoms and antibody detection are required to confirm suspected cases, before treatment is administered. The methodology for detecting circulating antigen is not yet completely developed, and it is rather complicated, requiring specific or monoclonal antibody. Antigen detection required better specificity for current gnathostomiasis infection. At present, most laboratories continue to focus on the detection of antibodies to *Gnathostoma* infection, and cannot differentiate between past and current infections.

The species' identification of *Gnathostoma* larvae is possible, mainly by external morphology—the shape and number of cephalic hooklets, the number of rows of cephalic hooklets, the shape of body spines and body size. The two common human species *G. spinigerum* and *G. binucleatum*, are similar in morphology and are only differentiated by the number of nuclei in the intestinal cells. In these circumstances, molecular differentiation is very useful, and laboratories with the necessary equipment can do it easily. DNA can be extracted from a worm removed from a patient, and the sequences obtained can be compared with those in GenBank, to determine species. Broad discrimination based on morphology and molecular differentiation is useful in identifying parasite species when larvae are found and removed from the patient. Where no worm is recovered from a patient, and the gnathostome species needs to be identified specifically, infections among the definitive hosts (domestic cats and dogs), and the second intermediate hosts (fish, frogs) in the patient's hometown should be examined. However, the infection rates and worm burdens among these hosts may be low, and the number of samples for examination should be high. The detection of gnathostomes in hosts is more feasible in hypo-endemic areas. Infection in these hosts is seasonal, and the best season is between the end of the rainy season and the beginning of the dry season. It would be appropriate to develop molecular testing for *Gnathostoma* eggs in definitive hosts, to support epidemiological studies in gnathostomiasis-endemic areas.

REFERENCES

1. Daengsvang, S. A monograph on the genus *Gnathostoma* & gnathostomiasis in Thailand, p. 1, SEAMIC/IMFJ, Tokyo, 1980.
2. Waikagul, J., and Diaz Chamacho, S.P. Gnathostomiasis. In: Darwin, K.D., and Fried, B. (eds), *Food-Borne Parasitic Zoonoses, Fish and Plant-Borne Parasites, World Class Parasites*, Chapter 11, p. 235, Springer, 2007.
3. Magana, M. et al. Gnathostomiasis: Clinicopathologic study. *Am. J. Dermatopathol.* 26, 91, 2004.
4. Ollaque, W., Ollaque, J., and Guevara de Veliz, A. Eosinophilic migratory nodular paniculitis in (human gnathostomiasis in Ecuador). 1st finding of parasites in South America. *Med. Cutan. Ibero. Lat. Am.* 10, 73, 1982.
5. Villar de Ciprini, E. Paniculitis migratoria eosinofilica en el Peru, *Gnathostoma* como agente causal. *Rev. Peru Med. Exp. Salud Publica* 20, 220, 2003.
6. Kaminsky, C.A. et al. Eosinophilic migratory nodular panniculitis (human gnathostomiasis). *Med. Cutan. Ibero. Lat. Am.* 17, 158, 1989.
7. Nawa, Y., Ogata, K., and Imai, J. Gnathostomiasis by *Gnathostoma doloresi. Saishin Igaku* 44, 807, 1989. (In Japanese).
8. Ando, H. et al. Two human cases of gnathostomiasis and discovery of a second intermediate host of *Gnathostoma nipponicum* in Japan. *J. Parasitol.* 74, 626, 1988.
9. Nomura, Y. et al. Gnathostomiasis possibly caused by *Gnathostoma malaysiae. Tokai J. Exp. Clin. Med.* 25, 1, 2000.
10. Chai, J.Y. et al. An outbreak of gnathostomiasis among Korean emigrants in Myanmar. *Am. J. Trop. Med. Hyg.* 69, 67, 2003.
11. Akahane, H., Iwata, K., and Miyazaki, I. Studies on *Gnathostoma hispidum* Fedtchenko, 1872 parasitic loaches imported from China. *Jpn. J. Parasitol.* 31, 507, 1982. (In Japanese with English summary).
12. Montero, E. et al. Human gnathostomiasis in Spain: First report in humans. *Acta Trop.* 78, 59, 2001.
13. Herman, J.S. et al. Gnathostomiasis acquired by British tourists in Botswana. *Emerg. Infect. Dis.* 15, 594, 2009.
14. Hale, D.C., Blumberg, L., and Frean, J. Case report: Gnathostomiasis in two travelers to Zambia. *Am. J. Trop. Med. Hyg.* 68, 707, 2003.
15. LeRoux, P.L. *Gnathostoma spinigerum* in a lion in Northern Rhodesia. *Trans. R. Soc. Trop. Med. Hyg.* 52, 14, 1958.
16. Almeyda-Artigas, R.J., Bargues, M.D., and Mas-Coma, S. ITS-2 rDNA sequencing of *Gnathostoma* species (Nematoda) and elucidating of the species causing gnathostomiasis in the Americas. *J. Parasitol.* 86, 537, 2000.
17. Ngarmamonpirat, C. et al. Analysis of sequence variation in *Gnathostoma spinigerum* mitochondrial DNA by single-strand conformation polymorphism analysis and DNA sequence. *Parasitol. Int.* 54, 65, 2005.
18. Rojekittikhun, W. On the biology of *Gnathostoma spinigerum. J. Trop. Med. Parasitol.* 25, 91, 2002.
19. Rojekittikhun, W. et al. Record numbers of *Gnathostoma* larvae in swamp eels. *J. Trop. Med. Parasitol.* 24, 79, 2001.
20. Leon-Regagnon, V. et al. Gnathostomiasis in fish from Tres Palos Lagoon, Guerrero, Mexico. *Emerg. Infect. Dis.* Letter to the editor, 2000.
21. Komalamisra, C., Nuamtanong, S., and Dekumyoy, P. *Pila ampullacea* and *Pomacea canaliculata*, as new paratenic hosts of *Gnathostoma spinigerum. Southeast Asian J. Trop. Med. Public Health* 40, 243, 2009.
22. Rojekittikhun, R., and Buchachart, K. The infectivitity of frozen *Gnathostoma spinigerum* encysted larvae in mice. *J. Trop. Med. Parasitol.* 25, 79, 2002.
23. Miyazaki, I. On the genus *Gnathostoma* and human gnathostomiasis, with special reference to Japan. *Exp. Parasitol.* 9, 338, 1960.
24. Nontasut, P. et al. Comparison of ivermectin and albendazole treatment for gnathostomiasis. *Southeast Asian J. Trop. Med. Public Health* 31, 374, 2000.
25. Chappuis, F., Farineli, T., and Loutan, L. Ivermectin treatment of a traveler who returned from Peru with cutaneous gnathostomiasis. *Clin. Infect. Dis.* 33, 17, 2001.
26. Parola, P. et al. Clinical picture: Gnathostomiasis. *Lancet* 357, 1011, 2001.
27. Dekumyoy, P. et al. A seven-year retrospective evaluation of gnathostomiasis and diagnostic specificity by immunoblot. *Proceedings of the first International Meeting on Gnathostomiasis*, p. 45, Culiacan, Sinaloa, Mexico, March 2002.
28. Suntharasamai, P. et al. ELISA for immunodiagnosis of human gnathostomiasis. *Southeast Asian J. Trop. Med. Public Health* 16, 274, 1985.
29. Dhamkrong-at, A. et al. Enzyme-linked immunosorbent assay for detection of antibody to *Gnathostoma* antigen in patients with intermittent cutaneous migratory swelling. *J. Clin. Microbiol.* 23, 847, 1986.
30. Nuchprayoon, S. et al. Study of specific IgG subclass antibodies for diagnosis of *Gnathostoma spinigerum. Parasitol. Res.* 91,137, 2003.
31. Nopparatana, C. et al. Purification of *Gnathostoma spinigerum* specific antigen and immunodiagnosis of human gnathostomiasis. *Int. J. Parasitol.* 21, 677, 1991.
32. Chaicumpa, W. et al. Monoclonal antibody to a diagnostic Mr 24,000 antigen of *Gnathostoma spinigerum. Int. J. Parasitol.* 21, 677, 1991.

33. Rojekittikhun, W., Saito, S., and Sendo, F. Crude and monoclonal antibody affinity-purified larval *Gnathostoma spinigerum* larval antigens for immunodiagnosis of human gnathostomiasis. *Southeast Asian J. Trop. Med. Public Health* 26, 439, 1995.
34. Diaz-Camacho, S.P. et al. Clinical manifestations and immunodiagnosis of gnathostomiasis in Culiacan, Mexico. *Am. J. Trop. Med. Hyg.* 59, 908, 1998.
35. Ishiwata, K. et al. Evaluation of the antigenic similarities of adult-worm extracts from three *Gnathostoma* species, using sera from Mexican and Japanese patients with *Gnathostoma* infections. *Ann. Trop. Med. Parasitol.* 97, 629, 2003.
36. Anantaphuti, M.T., Nuamtanong, S., and Dekumyoy, P. Diagnostic value of IgG4 in gnathostomiasis. *Trop. Med. Int. Health* 10, 1013, 2005.
37. Laummaunwai, P. et al. Evaluation of human IgG class and subclass antibodies to a 24 kDa antigenic component of *Gnathostoma spinigerum* for the serodiagnosis of gnathostomiasis. *Parasitol. Res.* 101, 703, 2007.
38. Tapchaisri, P. et al. *Gnathostoma spinigerum* specific recombinant antigen for immunodiagnosis of human gnathostomiasis. *6th Seminar on Food-and Water-borne Parasitic Zoonoses (FBPZ6) incorporating the 3rd International Meeting on Gnathostomiasis (abstract).* Bangkok, Thailand p. 52, 2–4 December 2009.
39. Uparananukraw, P. et al. Molecular cloning of a gene encoding matrix metalloproteinase-like protein from *Gnathostoma spinigerum. Parasitol. Res.* 87, 751, 2001.
40. Laummaunwai, P., Intapan, P.M., Wongkham, C., Lulitanond, V., and Maleewong, W. Identification of antigenic components of *Gnathostoma spinigerum* advanced-third stage larvae by two-dimensional gel electrophoresis and mass spectrometry. *Southeast Asian J. Trop. Med. Public Health*, 39 (suppl. 1), 19, 2008.
41. Tapchaisri, P. et al. Specific antigen of *Gnathostoma spinigerum* for immunodiagnosis of human gnathostomiasis. *Int. J. Parasitol.* 21, 315, 1991.
42. Maleewong, W. et al. Serodiagnosis of human gnathostomiasis. *Southeast Asian J. Trop. Med. Public Health* 19, 201, 1988.
43. Tuntipopipat, S., Chawengkirttikul, R., and Sirisinha, S. A simplified method for the fractionation of *Gnathostoma*-specific antigens for serodiagnosis of human gnathostomosis. *J. Helminthol.* 67, 297, 1993.
44. Soesatyo, M.H. et al. IgE responses in human gnathostomiasis. *Trans. R. Soc. Trop. Med. Hyg.* 81, 799, 1987.
45. Saksirisampant, W. et al. Specific IgE antibody responses to somatic and excretory–secretory antigens of third stage *G. spinigerum* larvae in human gnathostomiasis. *J. Med. Assoc. Thai.* 84, S173, 2001.
46. Boongird, P. et al. Neurological manifestation of gnathostomiasis. *J. Neurol. Sci.* 31, 279, 1977.
47. Rusnak, R.M., and Lucey, D.R. Clinical Gnathostomiasis. Case report and review of the English language literature. *Clin. Infect. Dis.* 16, 33, 1993.
48. Chandenier, J. et al. Medullary gnathostomiasis in a white patient: Use of immunodiagnosis and magnetic resonance imaging. *Clin. Infect. Dis.* 32, 154, 2001.
49. Sawanyawisuth, K. et al. MR imaging findings in cerebrospinal gnathostomiasis. *Am. J. Neuroradiol.* 25, 446, 2004.
50. Sawanyawisuth, K. et al. MR imaging findings in cauda equina gnathostomiasis. *AJNR Am. J. Neuroradiol.* 26, 39, 2005.
51. Hennies, F. et al. Gnathostomiasis: Import from Laos. *JDDG* 4, 414, 2006.
52. Ando, K. et al. Comparative study on DNA sequences of ribosomal DNA and cytochrome c oxidase subunit 1 of mitochondrial DNA among five species of gnathostomes. *J. Helminthol.* 80, 7, 2006.
53. Kifune, T. et al. *Gnathostoma binucleatum* (Spirurida: Gnathostomatidae) from the freshwater fish in Tabasco, Mexico. *Rev. Biol. Trop.* 52, 371, 2004.
54. Martı´nez-Salazar, E.A., and León-Règagnon, V. Confirmation of *Gnathostoma binucleatum* Almeyda-Artigas, 1991, advanced third-stage larvae in Tres Palos Lagoon, Mexico, by morphological and molecular data. *J. Parasitol.* 91, 962, 2005.
55. Bertoni-Ruiz, F. et al. A new species of *Gnathostoma* (Nematoda: Gnathostomatidae) in *Procyon lotor hernandezii* from Mexico. *J. Parasitol.* 91, 1143, 2005.
56. Hernández-Gómez, R.E. et al. Molecular identification of the advanced third-stage larvae (ADV L3) of *Gnathostoma lamothei* in Tabasco, Mexico. *Parasitol. Int.* 59, 97, 2010.
57. Sieu, T.P.M. et al. Prevalence of *Gnathostoma spinigerum* infection in wild and cultured swamp eels in Vietnam. *J. Parasitol.* 98, 246, 2009.
58. Sambrook, J., Fritsch, E.F., and Maniatis, T. *Molecular Cloning: A Laboratory Manual*, 2nd ed., Cold Spring Harbor Laboratory, Cold Spring Harbor, New York 1989.

55 Haemonchus

Michael B. Hildreth and Aaron Harmon

CONTENTS

55.1 INTRODUCTION

55.1.1 Classification, Morphology, and Biology

Haemonchus contortus is one of the 12 known *Haemonchus* species, as recently reviewed by Hoberg et al. [1]. Three species in the genus *Haemonchus* (i.e., *H. contortus*, *H. placei*, and *H. similis*) have global distributions and are among the most economically important helminths in sheep, goats, and cattle. The fourth species infecting domesticated ungulates, *Haemonchus longistipes*, is a common parasite of camels (i.e., Camelidae), and its expanding geographic distribution is resulting from the anthropogenic spread of its hosts. Distributions among the remaining eight *Haemonchus* species are much more limited, being found primarily in wild Antilopinae (e.g., true antelopes) and Bovinae (e.g., water buffalo and yaks) ungulates in sub-Saharan Africa [1].

The genus *Haemonchus* is one of 22 genera grouped into the family Trichostrongylidae. Hoberg and Lichtenfels [2] group these 22 genera into six subfamilies, arising from two clades, with the genus *Haemonchus* being in the Hemonchinae subfamily and part of the Graphidiinae clade. Other members of this clade include the subfamilies: Graphidiinae and Ostertagiinae. The remaining subfamilies (i.e., Cooperiinae, Libyostrongylinae, and Trichostrongylinae) are listed as part of the Cooperiinae clade. For nematode taxonomic classifications above the family level, several competing systems have been proposed. A recent system proposed by Blaxter [3] and by De Ley and Blaxter [4] is based upon sequence analysis of the ribosomal DNA (rDNA) small subunit genes from a large number of nematode species. Their molecular-based system is similar to other morphology-based systems except that several traditional order-level taxa were adjusted to a lower level. Regardless of the system used, the family Trichostrongylidae share linage with numerous other bursate nematodes such as hookworms (family Ancylostomidae), horse strongyles, nodular worms (family Strongylidae), poultry gapeworms (family Syngamidae), and lungworms (e.g., families Dictyocaulidae and Angiostrongylidae). The common feature of all these groups is the presence of a copulatory bursa on the caudal end of males. Each bursa consists of dorsal and ventral expansions of the body (lobes) along with two lateral lobes. Each lobe is supported by one or more muscular processes (rays). The appearance of each lobe and its rays are very important taxonomic characters. The morphology and size of the copulatory spicules are also valuable characters used to differentiate these worms at the species level. Blaxter [3] groups these bursate nematodes into the superfamily Strongyloidea, infraorder Rhabditomorpha, suborder Rhabditina, order Rhabditida (=Phasmidea; =Secernentea), and class Chromadorea.

Among the various gastrointestinal, bursate nematode families (i.e., families Trichostrongylidae, Ancylostomidae, and Strongylidae), there are significant differences in host specificity, routes of transmission, and adult morphology that are characteristic of each family and are of diagnostic importance. For example, members of Ancylostomidae are generally intestinal parasites of carnivores or omnivores (including humans), and adult hookworms have a large, sclerotized buccal cavity that is generally armed with teeth, cutting plates, lancets, or a dorsal cone. The infective third-stage (J_3) juveniles from this group actively burrow through the skin of their hosts, gaining entry into blood vessels below the skin and eventually entering the respiratory tree via the lungs. Passing up the trachea to the epiglottis, these juveniles quickly pass through the stomach and finally into the intestinal tract. The two main human hookworms, *Ancylostoma duodenale* (Old World Hookworm) and *Necator americanus* (New World Hookworm), infect more than a billion people globally [5]. *Ancylostoma ceylanicum* normally parasitizes carnivores, but has been occasionally found in humans.

In contrast to hookworms, trichostrongyle nematodes (family Trichostrongylidae) are parasites generally of herbivores, primarily ruminants, and reside either in the host's abomasum or intestine. Gibbons and Khalil [6] composed a taxonomic key to the six subfamilies and more than 35 genera of trichostrongyles based upon morphological characteristics of the adult stage. The buccal cavities of trichostrongyles are relatively small, and teeth are found in only a few members from this group. Four of the six genera in this family possess a notable system of cuticular ridges distributed longitudinally on the nematode's surface; collectively these ridges are termed a synlophe. The synlophe's appearance, particularly along the anterior half of the nematode, is one of the most useful characters for classifying genera and higher taxa in the family Trichostrongylidae [7]. Infective trichostrongyle juveniles lack the ability to penetrate through skin and instead must be swallowed to initiate an infection. Ruminants become infected with these trichostrongyles as they graze on infected pastures. On rare occasions, trichostrongyles have been diagnosed in humans, the result of accidental ingestion of juveniles. Most of these cases have involved members of the genus *Trichostrongylus*, and *T. colubriformis* and *T. orientalis* seem to be the two most common species.

Human infections with *H. contortus* have been specifically reported in a very few primary publications. In 1973, Ghadirian and Arfaa [8] described finding one *H. contortus* male and two females in a 60-year-old woman from the rural area of Isfahan, Iran. These adult worms were recovered from the patient's fecal samples following treatment with bephenium hydroxynaphthoate. This description is the only human case of haemonchosis referenced in the MEDLINE/Pubmed Database (http://www.ncbi.nlm.nih.gov/pubmed/), and according to Ghadirian and Arfaa [8], their reported case from Iran was only the fifth published report of *H. contortus* in humans. The authors stated that three of the other cases came from Australia and one from Brazil, referencing *Craig and Faust's Clinical Parasitology* [9] book for the other four cases. For the Brazilian case, Faust et al. [9] referenced "Magalhães" but included no citation. They almost certainly were referring to Magalhães' 1908 report of an adult *H. contortus* that was submitted to him for identification from a Dr. Von Basseviz, of Vitória Santa do Palmar, Rio Grande Do Sul (southernmost tip of Brazil) [10]. The patient had been suffering from hookworm-like symptoms, and the *Haemonchus* adults were recovered following treatment with thymol. Symptoms ended soon after treatment. Fleury et al. [11] state that there is another reference to a human case of haemonchosis from Brazil: a vague mention of some human-derived *H. contortus* worms that were examined by Travassos [12]. These worms were given to him by Dr. Alolfo Lutz, a Brazilian physician/parasitologist. No clinical information (e.g., method of recovery and the date worms were recovered, location of the patient) is provided regarding this second possible case, and it even seems possible that the worms sent to Travassos by Lutz were the same worms reported by Magalhães. For the three Australian cases, Faust et al. [9] referenced "Sweet, 1924" but again included no citation. In this case, they are almost certainly referring to Sweet's 1924 report of finding *H. contortus* eggs in three aboriginal persons from western Australia [13]. Heydon and Green [14] question the legitimacy of using Sweet's report as unqualified human cases of haemonchosis, noting that Sweet based his identification only on the egg stage, and pointing out that the fecal samples had been collected from the ground and possibly contaminated by sheep feces. Another suspected case of human haemonchosis was reported by Heydon and Green [14]. This report was based upon finding *H. contortus* adults in a goat kid fed third-stage juveniles cultured from pooled fecal samples containing hookworm-like eggs collected primarily from one family living in northern Queensland. Extreme care had been taken to minimize the chances of this goat kid having its own *H. contortus* natural infection, but Heydon and Green do concede that they cannot exclude this possibility, especially since no *Haemonchus* juveniles were seen in observations of the juveniles prior to the infection. In summary, the 1908 case in Brazil and the 1973 case in Iran appear to be the only two definitive human haemonchosis cases reported in the literature so far. The four cases in Australia and the 1921 report from Brazil should be listed as possible cases.

Members of the family Strongylidae are parasites of the large intestine, primarily of equine. Similar to the trichostrongyles, all Strongylidae genera infect their host through the ingestion of infective juveniles. Adult strongyles have large- to moderate-sized buccal cavities that are often surrounded by internal and/or external leaf-crowns. Buccal teeth or cutting plates can also be present. Not all strongyles infect equine; a few species of *Oesophagostomum* infect the large intestines of primates, including humans. The species *Oesophagostomum bifurcum* is highly prevalent in primates, including humans in a small area of west-central Africa, primarily the countries of Tonga and Ghana.

The egg stage for virtually all gastrointestinal, bursate nematodes are morphologically very similar, and are typically referred to either as strongyle-type or as hookworm-type eggs. These eggs are generally ovoid or ellipsoid shaped, consisting of a multiple blastomeric embryo (morula) surrounded by a smooth, thin egg shell. Their existence in fecal samples is used diagnostically to reveal the presence of these worms in their hosts. Hookworm eggs tend to appear in feces at the morula stage composed of 4–8 blastomeres, whereas trichostrongyle and strongyle eggs tend to be released while at the 16-cell stage. Freshly excreted eggs of *H. contortus* typically have 16–32 blastomeres [15]. Embryonation of the morula continues as eggs are excreted from the host, and therefore, blastomere numbers are of no diagnostic value unless great care is given to observe the eggs immediately after being excreted or stopping development with fixation. The size and shape of eggs from hookworms, trichostrongyles, and strongyles vary somewhat, even within the same species, and attempts have been made to use morphometry to identify these eggs to the genus level. Georgi and McCulloch [16] measured the lengths of 100 *H. contortus* eggs, and found that they averaged 77 μm and ranged between 69 and 88 μm [16]. Based on length alone, *H. contortus* could be separated from *T. colubriformis* 86% of the time if eggs of 81 μm and less were assumed to be *H. contortus*. They also found that among 12 sheep trichostrongyle species, *H. contortus* eggs could only be correctly identified 67% of the time when eight geometric parameters of total egg shape were subjected to a stepwise discrimination analysis.

H. contortus adults exhibit the typical trichostrongyle characteristics described above, but there are also distinctive features that are characteristic of the subfamily Haemonchinae and the genus *Haemonchus* [6]. Notable of the subfamily Haemonchinae is the presence of a large pair of cervical papillae projecting from the surface along the anterior region of nematode body. The most diagnostically useful characteristic of the genus is the presence of a single, slender dorsal tooth or lancet inside its relatively small buccal cavity. This lancet is present in both males and females, and is easily visible with a conventional light microscope. These adults feed on blood, and so the lancet is used to pierce capillaries beneath the host's epithelium to release blood into the lumen. The white, egg-filled uterus of the female worm wraps around the red, blood-filled intestine of these worms and can be seen through the body wall. This is reminiscent of the historical barber-pole composed of a pole with a helix of red and white stripes. For this reason, one of the common names frequently used for this genus is the "barber-pole worm." The twisting nature of the worm's uterus and intestine, and its location in the stomachs of ruminant, has also caused people to commonly refer to *Haemonchus* as the "twisted stomach worm" or "wire worm." The vulva of this genus is located towards the anterior end of the posterior-most quarter of the female body length and is generally covered by prominent vulva flap. These female characteristics, and the relatively large size of both the females (18–30 mm) and the males (10–20 mm), makes it possible to more easily recognize *Haemonchus*-infected animals at necropsy than infections with most other trichostrongyles. The male bursa for this genus has large lateral lobes, but the dorsal lobe is small and asymmetrical. Spiracles from this genus are short, thick, unbranched, containing one or two projections at the distal end. The *Haemonchus* synlophe is well developed, but restricted to the anterior end of the body. The phylogeny work of Hoberg et al. [1] provides evidence that the 12 *Haemonchus* species can be individually recognized based upon a few morphological features of the adult, such as the total length and the appearances of the synlophe, and the male bursa and spicules. Differentiation of the four species commonly found in domesticated ungulates (i.e., *H. contortus*, *H. placei*, *H. similis*, and *H. longistipes*) can be accomplished based upon the total length of the male, the number of ridges composing the synlophe at the level of the esophagus, how far the synlophe extends down the body length, and the length of the male spicules. *H. contortus* is a medium-sized species (males about 15 mm), with 34 ridges that end about three-fourth of the body length. Its spiracles are 400–450 μm in length [1].

Even though the eggs of hookworms and trichostrongyles are morphologically indistinguishable, it is possible to differentiate members of each genera based upon morphological characteristics of the third stage (J_3) juvenile. Under natural conditions, the first juvenile stage (J_1) develops within the eggshell of each excreted egg, and eventually hatches and molts to the J_2 stage. The J_2 typically remains in the fecal sample feeding on bacteria until it acquires sufficient nutrients to molt to the infective juvenile stage (J_3). The remnant J_2 cuticle remains intact over the J_3 in the form of a cuticular sheath, protecting this infective stage as it migrates out of the feces and onto vegetation for consumption by the next host. Standard laboratory methods are available for culturing excreted eggs to the J_3 stage, and these recovered juveniles can then be identified with conventional bright-field microscopy. The digestive tract of this juvenile stage consists of three regions: the esophagus, the intestine, and the rectum. Variations in the shapes of these regions have diagnostic importance. For example, the triangular shape and alternating arrangement of the intestinal cells from trichostrongyle juveniles give the intestinal lumen of this group a zigzag arrangement [17] (Figure 55.1). In contrast, the intestinal lumen of hookworm J_3 juveniles is straight. The total length of the juvenile, the shape of the head and tail, the number of intestinal cells, and the presence and appearance of the cuticular sheath all aid in identifying genera within the trichostrongyle juveniles. *Haemonchus* has a medium-sized, sheathed J_3 with bullet-shaped head and medium-sized sheath extension that is generally kinked at the tip of the tail (Figure 55.1); 16 cells comprise the intestine [18,19].

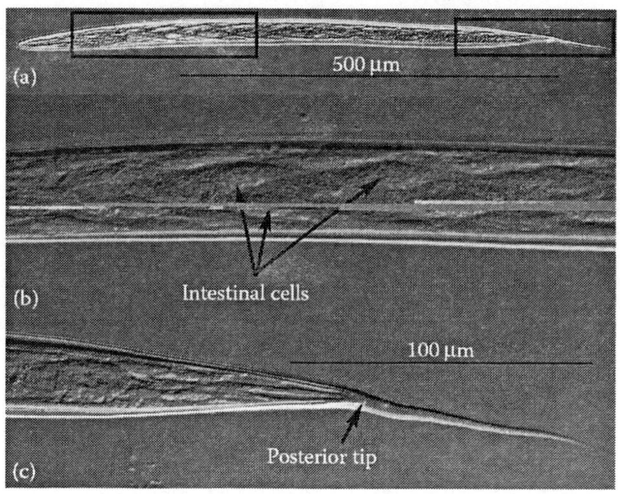

FIGURE 55.1 *H. contortus* third-stage juvenile photographed with differential interference microscopy. (a) Entire juvenile with anterior end oriented to the left side. Left box shows region represented by (b); right box shows region represented by (c). Even at this low magnification, the pattern of intestinal cells is apparent. (b) Higher magnification of the juvenile toward the anterior end. Alternating arrangement of the triangular intestinal cells is more obvious at this magnification. (c) Higher magnification of the posterior end. The posterior tip of the tail is indicated by an arrow, and a medium-sized sheath extends beyond the tail which is exhibits a slight kink. Magnification is the same as for (b).

Although found infrequently in humans, *H. contortus* is an extremely common parasite of sheep and goats throughout the world. The global distribution of *H. contortus* has resulted from its broad host specificity among a wide variety of wild and domesticated ruminants (particularly members of the subfamily Caprinae), and also from the widespread translocation of the domesticated hosts throughout the world. The intensity of *H. contortus* infections tend to be higher in warmer and wetter tropical and subtropical climates than in the cooler and dryer temperate climates largely because the free-living, infectious, juvenile stage of this species survives poorly under cold and dry conditions. A form of arrested development, commonly known at hypobiosis, seems to assist the perpetuation of this life cycle into geographic areas that are not optimum for juvenile survival. With this approach, juveniles ingested just prior to the onset of harsh environmental conditions (e.g., cold winters) can delay the development of their short-lived, egg-producing adult stage until weather conditions become acceptable for the survival of its free-living juvenile stages [20]. With these survival mechanisms, low numbers of *H. contortus* have even been found as far north as the arctic circle [21]. Because there have been so few cases of *Haemonchus* infections in humans, it has been impossible to directly document factors affecting human transmission, and yet, it is safe to assume that the consumption of plant materials contaminated with *Haemonchus* juveniles would be the primary factor. For human cases of trichostrongylosis, the consumption of garden products contaminated with ruminant feces has been shown as a primary route of transmission [22]. Human cases of haemonchosis should have the greatest probability of occurring in tropical and subtropical areas where sheep and/or goats are heavily infected with *H. contortus* and where fresh manure is frequently used as a garden fertilizer.

55.1.2 Clinical Features and Pathogenesis

In addition to being one of the most common parasites in sheep and goats throughout most areas of the world, it can become one of the most lethal diseases of sheep and goats. The primary pathogenesis occurring in *H. contortus* infections results from anemia caused by the blood-feeding adult worms. The clinical effects of haemonchosis are generally proportionate to the number of worms present and the rate of infection. A sudden infusion of numerous infective juveniles into young susceptible lambs quickly leads to the acute form of the disease, consisting of severe anemia and hypoprotenaemia (i.e., bottle-jaw). These infected individuals can die suddenly depending on the number of worms and the immunological and nutritional health of the animals. Blood loss not only results from the ruptured submucosal capillaries pierced by the single buccal-tooth from each adult worm, but also from the reduced blood clotting caused by an anti-haemostatic cocktail which is secreted into the wound by each worm [23]. Part of the escaping blood is ingested by the worms, but much of it flows into the lumen and contributes to the dark-colored feces, often characteristic of haemonchosis. Until the haemopoietic system of the host begins compensating for this blood loss, packed cell volume falls dramatically and the symptoms of anemia begin. This can even occur before worm egg production begins. Mobilization of the hemopoietic system soon begins limiting the progress of anemia until the iron reserves become depleted. Once depleted, the progression of anemic symptoms continues until the immune system begins expelling adults or until the hosts are treated or until the animal dies. The hypoproteinemia associated with loss of serum proteins from haemonchosis causes a condition known as bottle-jaw. In chronic infections, the animals lose weight, and look malnourished; old sheep lose wool, and young animals show stunted growth. These latter problems resemble those caused by other trichostrongyles in ruminants resulting from a rise in abomasal pH and increased plasma pepsinogen.

The anemia-associated symptoms of haemonchosis in sheep resemble those of hookworm disease in humans, and the only mention of symptoms caused by a human *H. contortus* infection indicated that they mimicked a hookworm infection [10]. Eosinophilia is commonly associated with most helminth infections, and the combination of eosinophilia and anemia can be a primary clinical indicator of hookworm disease. The common symptoms between hookworm disease and haemonchosis and the fact that *Haemonchus* eggs are difficult to differentiate from hookworm eggs provide a possibility that some human *H. contortus* are being misdiagnosed as hookworm infections.

55.1.3 Diagnosis

55.1.3.1 Conventional Techniques

So far, the two definitively diagnosed human haemonchosis cases, from Brazil and Iran, were diagnosed based upon recovery and identification of the adult worms following anthelmintic treatment. Until the arrival of molecular diagnostic techniques, this posttreatment recovery of adult worms was the most common way to document this type of nematode infection to the species level. When adult bursate nematodes are recovered from human fecal samples, they are normally assumed to be human hookworms, either *A. dueodenale* or *N. americanus*. Hopefully, an experienced observer would likely notice any *H. contortus* adults present in the sample because they are twice as large as hookworms and the females have the characteristic barber-pole appearance. Confirmed identification of the genus *Haemonchus* could be based upon the presence of the dorsal tooth inside a relatively small buccal cavity. Species identification to *H. contortus* would then be based upon the appearance of the male bursa and the number of ridges composing the synlophe at the level of the esophagus.

Gastrointestinal helminth infections are most typically diagnosed based upon finding their characteristic eggs and/or juveniles when examining a fecal sample, either by means of a direct fecal smear or after the application of a concentration technique. In many cases, the eggs from the major helminth taxa look very similar, and species-level identification is frequently not reliable with conventional microscopy, particularly eggs from bursate nematodes (i.e., hookworms and trichostrongyles). In developing countries where convenience is critically important, the Kato-Katz technique is most commonly used for detecting schistosome, *Ascaris*, and whipworm eggs, but this approach is only usable for diagnosing hookworm eggs when slides are examined immediately after preparation. In human diagnostic laboratories from developed countries, the formol/ethyl acetate sedimentation technique is the more commonly used concentration method for diagnosing parasites because of its thoroughness with a wide variety of helminth eggs and protozoan cysts. Although thorough, sedimentation techniques are more expensive because they require more time and training than other fecal examinations, and they use hazardous materials and specialty equipment. Concentration methods based upon the floatation of eggs are more commonly used in veterinary medicine, because they are more time efficient than sedimentation methods. Young sheep typically shed large numbers of trichostrongyle eggs and so the quick but less sensitive (upto 50 eggs/g) McMaster technique is most frequently used for this host–parasite combination [24]. To detect the lower number of trichostrongyle eggs found in cows or recently treated sheep, a more sensitive centrifugation technique, such as the Wisconsin sugar floatation method, is a more popular diagnostic approach [25]. The Wisconsin method is a centrifuge-dependent method that utilizes a sucrose solution (specific gravity 1.25 or greater) to float the eggs onto a cover-slip [26]. Goodman et al. [27] compared several traditional human and veterinary fecal egg concentration techniques in an attempt to find the most effective method for identifying infants infected with low numbers of ascarids, whipworms, and hookworms. He concluded that the Wisconsin floatation technique was more sensitive than both the Kato–Katz or formol/ethyl acetate fecal exams. The Wisconsin technique might also be more useful in situations where eggs need to be later used for molecular analysis. Eggs are more difficult to recover from the Kato–Katz and McMaster slides, and the formalin used in formol/ethyl acetate techniques inhibits DNA recovery from these eggs [28].

As a practical matter, a case of human haemonchosis will first appear as a suspected human hookworm infection based upon the presence of strongyle-like eggs in a routine fecal exam. Oyerinde (1978) found that *N. americanus* hookworm eggs measured 60–70 µm in length, whereas the smaller *A. duodenale* hookworm eggs measured 47–56 µm. These hookworm eggs are significantly smaller than the 69–88 µm eggs of *H. contortus* eggs as reported by Georgi and McCulloch [16]. Because fecal eggs from trichostrongyles tend to be larger and contain 16 or more blastomeres instead of the 4–8 blastomeres seen in the smaller hookworm eggs, a trichostrongyle infection might be suspected simply based upon these differences. As described earlier in this review, *Haemonchus* eggs might be suspected over *Trichostrongylus* eggs if sufficient *H. contortus* eggs are available in the sample to correctly reflect the size difference.

Although not used so far for the previously described cases of human haemonchosis, it is possible to identify hookworms and trichostrongyle genera by observing the J_3 stage cultured from suspicious-looking eggs isolated from a patient. This approach has been used effectively to document human cases of trichostrongylosis [22], and methods used for culturing hookworm eggs to the J_3 stage are also effective for the various trichostrongyles. Advantages exist among the various culturing methods, and the Harada–Mori method tends to be used more for human patients, while Baermann's method tends to be used more for ruminants [19,29,30]. The disadvantage of this approach is the time needed (both in delay for results and in the labor needed for microscopic identification of larvae), and the poor efficiency of the hatching and development step. One study involving sheep trichostrongyles found that only 62% of cultured eggs hatched and molted into J_3s from larval cultures and only 87% of the subsequent larvae could be collected with a Baermann's funnel [31]. Harvested juveniles can be fixed and stored for later identification if necessary. van Wyk et al. [18] suggest fixing them with formalin at a final concentration of 1–2%, and heating them to 55–57°C for about 1 min. They also suggest staining the juveniles with diluted Lugol's iodine (LI) solution prior to the addition of a coverslip.

Pawlowski et al. [29] includes a taxonomic key designed to identify J_3s from the two most common human hookworm species (i.e., *A. duodenale* and *N. americanus*) and three trichostrongyle genera commonly isolated from human fecal samples (i.e., *Trichostrongylus*, *Ternidens*, and *Oesophagostomum*). The zigzag appearance of the intestinal lumen of trichostrongyle juveniles is used to separate these juveniles from those of hookworms (Figure 55.1). Even though *Haemonchus* is not included in this or other human-related J_3 keys, Pawlowski's method for separating trichostrongyles from hookworm juveniles is also valid for *Haemonchus* [17]. A ruminant trichostrongyle J_3 key is available that can be used to differentiate *Haemonchus* from the other genera [18], and Bowman [19] provides comparative photographs and tables that support this key. Characteristics used for differentiating these genera include: overall length, tail sheath extension, cephalic shape, the presence or absence of refractory bodies in the cephalic region, and the number of intestinal cells [18]. Significant skill is required to properly identify the different juvenile genera.

55.1.3.2 Molecular Techniques

Molecular biology has revolutionized diagnostic medicine, including the diagnosis of helminth infections in both humans and animals [32,33]. In most cases, these emerging molecular methods have simply involved DNA analysis of the transmittal stages traditionally used for diagnosis. In the case of intestinal helminthes, this has generally involved polymerase chain reaction (PCR) amplification of DNA isolated from fecal eggs. So far, particular attention has been given to helminth groups, such as hookworms and trichostrongyles, whose eggs are difficult to differentiate morphologically. Mitreva et al. [34] list 14 species of nematodes with genome sequencing projects under way or planned (as of December 2006), and five of those are intestinal trichostrongyles of ruminants. *Ancyclostoma caninum* was the only hookworm listed.

In 1989, the first DNA-based assay for identifying trichostrongyles simply used restriction digestion patterns of DNA isolated from adult worms to differentiate *H. contortus*, *O. circumcincta*, and *T. colubriformis* [35]. Improved sensitivity was achieved in 1994 through the use of genus-specific, radioactively labeled DNA probes in a Southern hybridization format to differentiate *Ostertagia*, *Cooperia*, *Haemonchus*, and *Oesophagostomum* eggs [36]. This assay was able to detect as few as 25 eggs. At the same time, PCR amplification approaches were being developed based upon sequence analysis of markers chosen because of their expected copy number and their potential for species-specific sequence variations. Based upon these criteria, a few mitochondrial genes have been evaluated for their diagnostic potential, but most researchers have focused on the fragments of genomic DNA involved in ribosome biogenesis [32,33].

In eukaryotes, rDNA is composed of hundreds of tandem repeats, with each repeat consisting of an intergenic, nontranscribed spacer region (IGS) followed by a region that is transcribed initially as a precursor molecule to be later processed into the three rRNA molecules (18S, 5.8S, and 28S rRNA) used in the production of ribosomes. The hundreds of rDNA repeats in nematodes tend to be distributed on different chromosomes, but the intra-genomic variability among these various repeats is significantly reduced (homogenized) through a process known as concerted evolution [37]. In addition to the three rRNA genes, each rDNA repeat also contains three transcribed spacer regions. The external transcribed spacer (ETS) separates IGS from the 18S gene, the internal transcribed spacer 1 (ITS-1) separates the 18S gene from the 5.8 gene, and the ITS-2 separates the 5.8 gene from the 28S gene (Figure 55.2). Because they are located between three very conserved genes, the ITS regions tend to be more conserved than most intergenic DNA, but do contain sufficient sequence variability, at the genus and species level, to make them valuable diagnostically. The sequence conservation among the three rRNA genes facilitates the selection of "universal" PCR primers at the ends of these genes which can be used to amplify the internal spacers regions from any nematode species [38]. Therefore, it is easy to generate enough PCR product from these ITS regions to sequence them either directly or after being cloned into a plasmid vector [39–43]. For this reason, sequences from the ITS regions of numerous bursate nematodes are now available on GenBank™. Oligonucleotide primers can then be specifically designed from these sequences to selectively amplify target DNA from appropriate nematode species in a PCR format. For example, Zarlenga et al. [44] designed five sets of forward and reverse primers for simultaneous use in a standard multiplex PCR assay for differentiating sample DNA from five different genera of cattle trichostrongyles. In this assay, ITS and ETS primers were selected such that they produced PCR products of differing molecular weights, which could then be identified in a standard agarose gel. Primer sets and fluorescent oligonucleotide probes have also been developed for use in real-time PCR formats. Based on differences in the ITS-2 region, a genus-level real-time PCR simplex assay was developed by von Samson-Himmelstjerna et al. [45] to detect first-stage juveniles (hatched from egg) of *H. contortus* and three other sheep trichostrongyles. They also found that their real-time assay could be used in duplex to simultaneously detect *H. contortus* and *T. colubriformis*.

The preparation of human fecal sample for the molecular detection of *Haemonchus* eggs simply involves the isolation of genomic DNA from the egg stage. Studies have shown that hookworm and trichostrongyle DNA can be isolated directly from either frozen or ethanol preserved human feces without first harvesting the eggs from the samples [42,46,47]. As a practical matter, however, these eggs are often already concentrated from unfrozen fecal samples as part of the microscopic portion of the routine diagnostic process. An effective method for recovering trichostrongyle DNA from eggs isolated from ruminant feces using the Wisconsin sugar-floatation method was developed by Harmon et al. [48]. Because the Wisconsin method works well for isolating hookworm eggs from human fecal samples [27], Harmon's DNA isolation method [48] should also be effective for hookworm and trichostrongyle eggs in human feces. At times, it might be useful to store fecal samples until analysis is possible, and it has been shown that ruminant fecal samples added to LI, neutral buffered formalin (NBF), or sodium azide (SA) preserved *H. contortus* egg morphology over time, and did not interfere with egg recovery from these samples [28]. PCR detection of these eggs was also not affected by storage in LI or SA, but was inhibited by NBF. This suggests that the formol/ethyl acetate sedimentation technique should not be

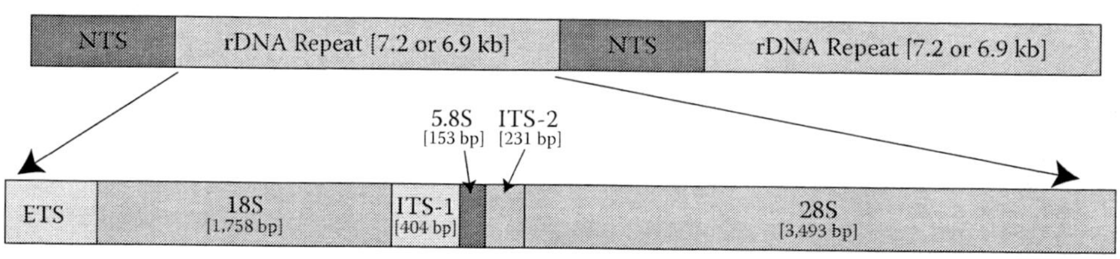

FIGURE 55.2 Diagram of the *H. contortus* rDNA repeat array showing the tandem repeats along with a "blow-up" of one repeat array. Base-pair (bp) sizes for each region are indicated in brackets, and except for ETS, are proportional to the length of the bar. Data came from GenBank and other sources [39,44,69,70].

used with hookworm and trichostrongyle diagnostic cases if a PCR assay might also be used. One challenge in using PCR to detect trichostrongyle eggs isolated from ruminant feces is the potential presence of PCR inhibitors also isolated from fecal samples. The most likely source of these inhibitors is the partially digested plant material that serves as the major component of ruminant feces [49–51]. Depending on a patient's diet, these plant materials may also be present in sufficient quantities to affect PCR efficiencies, and the use of a fecal–egg concentration method might help limit the quantity of these inhibitors.

Recent efforts have focused on the use of PCR assays to detect species of trichostrongyle eggs occasionally found in human fecal samples. One study used a multiplex real-time PCR assay to detect *O. bifurcum* eggs, differentiating them from *A. duodenale* and *N. americanus* eggs based upon their unique ITS-2 sequences [46]. For this study, separate PCR primers were used for each species, and the nematode DNA was isolated directly from fecal samples without first harvesting eggs. A second study sequenced rDNA from *T. colubriformi* adult worms obtained from several infected people living in Laos [40]. A unique ITS-1 primer sequence was identified and used in a gel-based PCR assay to differentiate this species from human hookworms (i.e., *A. duodenale* and *N. americanus*). This assay used a universal reverse primer located in the 28S gene (NC2 primer) and three different forward primers from the ITS-1 region to generate *T. colubriformi*, *A. duodenale*, and *N. americanus* PCR products of differing molecular weights (482, 690, and 870 bp, respectively). These amplicons can then be differentiated with agarose gel electrophoresis. This latter assay is one of the least-expensive PCR approaches available, requiring only a standard thermocycler, and a horizontal get box with appropriate power supply. Because it used a universal reverse primer for all the species tested, the only new reagent needed to add another nematode species to their assay (e.g., *H. contortus*) is a new forward primer specific to the trichostrongyle species added. A similar PCR approach involving a specific forward primer and universal reverse primer (NC2) has just been reported by Bott et al. [52] for sheep trichostrongyles. Because their reverse primer is the same as that used in the Laos study, and because their ITS-2 forward primer produces an amplicon with a molecular weight (165 bp) significantly different from those produced for *T. colubriformis*, *A. duodenale*, and *N. americanus*, it should be possible to simply add the *Haemonchus* primer to the human trichostrongyle assay to identify *H. contortus* infections in humans. This is the approach that we describe below.

55.2 METHODS

55.2.1 Sample Preparation

To detect *Haemonchus* DNA using PCR on human fecal samples, first the condition of the sample and the availability of fresh samples must be considered. If fresh fecal material is available, then it would be better to immediately harvest eggs from this sample and isolate DNA from these eggs. If the sample has been frozen, then egg recovery is not possible, and egg DNA should be isolated as described by Verweij et al. [46]. This method has also been used for samples fixed in ethanol. Because formalin inhibits PCR amplification, methods such as the formol/ethyl acetate sedimentation technique are not recommended for assays involving PCR [53]. For samples that are already fixed in formalin but need to be analyzed with PCR, the eggs should be isolated and washed well before isolating the DNA, but even with egg isolation, some inhibition will still occur, as shown with *Haemonchus* eggs fixed in formalin and floated from the fecal sample in sucrose [28]. Preservatives such as LI (5% iodine and 10% potassium iodide) and SA (0.15%, w/v, in distilled water) do not inhibit PCR amplification when added to fecal samples containing *Haemonchus* eggs, especially when added prior to egg isolation [28].

As explained previously in this review, a modified Wisconsin sucrose-flotation method [26] should be used to isolate eggs from the fecal samples. For this method, fecal samples are diluted into water, centrifuged, and a saturated sucrose solution (Sp.Grav. 1.27), is added to the resultant pellet. Once the pellet has been vortexed into the sucrose solution, it should be centrifuged again at roughly 1500*g* for 10 min. Enough sucrose solution is then be added to each tube to form a meniscus so that a microscope cover-slip can be floated on top for about 30 min. During this time, eggs will float up through the solution and remain attached as each coverslip is removed and placed on microscope slides for egg quantification. These eggs can easily be reisolated by adding each slide to a 50 mL centrifuge tube containing 45 mL distilled water, vortexing and centrifuging the tube at 2000*g* [48]. The pelleted eggs are then suspended in 1.5 mL of distilled water and can be analyzed or frozen for later analysis.

A variety of different approaches have been used to disrupt trichostrongyle and hookworm egg shells to optimize DNA recovery from the morula. Using *Ostertagia ostertagi* eggs, Harmon et al. [48]. compared several egg disruption procedures including: probe sonication, bath sonication, bead-beating, boiling, microwaving, proteinase K/sodium dodecyl sulfate (SDS) digestion, freezing, and various combinations of the above with the incorporation of SDS. Results showed that a Mini-BeadBeater-8 (BioSpec, Bartlesville, OK) operated at the homogenized setting with a 6.5 mm ceramic sphere and a 0.25 g garnet matrix (Lysis Matrix A, QBiogene, Irvine, CA, USA) resulted in 100% egg disruption within 1 min; the remaining techniques were less effective. These disruption procedures were also evaluated in conjunction with four commercial DNA extraction kits: a DNA Stool mini kit, a DNeasy Plant kit (Qiagen, Valencia, CA, USA), a Fast DNA kit (QBiogene), and the MAP extraction system (Tetracore). When DNA extraction was preceded by egg isolation from feces, all procedures except the Fast DNA kit produced PCR-ready DNA from two or more eggs. The DNeasy Plant kit allowed consistent detection of DNA released from one egg. A quantitative PCR study involving

H. contortus eggs found that DNA recovery using the Mini-BeadBeater-8/Lysis Matrix A/DNeasy Plant kit protocol was proportional to egg number up to at least 75 eggs [48]. This protocol is therefore recommended for recovering DNA from trichostrongyle and hookworm eggs.

55.2.2 Detection Procedures

Once sample DNA is recovered from isolated eggs, it can be detected through the PCR amplification of a specific portion of template DNA and the visualization of this appropriately sized amplicon with gel electrophoresis. The total volume of PCR reaction mixture varies from 25 to 100 μL, and sample DNA volume is dependent on the total volume and DNA concentration. Sample volumes ranging from 2 to 10 μL are most common. The sample and primers are added to a buffered aqueous mixture containing the deoxynucleotides (dNTPs), KCl, $MgCl_2$, and DNA Taq polymerase. Various commercial formulations of this mixture are available and add convenience and uniformity to the assay.

The specificity of this assay is created by the PCR forward primers that are diagnostic for each species. The reverse primer used for this assay is a universal primer (NC2) located toward the 5′ end of the 28S gene, and therefore will amplify all four species included in this assay. The NC2 primer consists of the following sequence: 5′-TTAGTTTCTTTTCCTCCGCT-3′ [54].

The forward primers are listed in Table 55.1 along with the amplicon size produced with the NC2 reverse primer. The primer sequence used for *H. contortus* in this assay is identical to the comparable ITS-2 region in *H. placei* [55]; therefore, this primer will amplify DNA from either species, and the size of the amplicon will not differ significantly. Optimum primer concentrations for most PCR systems tend to range between 0.1 μM and 1 μM. Additional nematode species can be added to this assay as long as the amplicon size can be differentiated in agarose gels. The hookworm/trichostrongyle assay described by Yong et al. [40] used a reaction volume of 50 μL, with 5–50 ng of sample DNA. The reaction mixture also contained 12.5 pmol of each primer, 1 U of Taq polymerase, 250 μM dNTPs, 10% glycerol, 0.01% gelatin, and 1 mM $MgCl_2$. The HotstarTaq master mix (Qiagen) and Amplitaq Gold PCR Mastermix (0.25 U/μL Taq polymerase, 200 μM each of dNTPs, and 2.5 mM $MgCl_2$, Applied Biosystems) have been used in similar assays by Verweij et al. [46] and Harmon et al. [48], respectively, and would add convenience to the proposed *Haemonchus* assay.

The amplification conditions chosen for this protocol are based upon the previously described human *Trichostrongylus*/hookworm assay [40], and consisted of a 5 min initial denaturation step at 94°C, followed by 30 cycles of 94°C for 1 min, 52°C for 40 s, 72°C for 90 s, and then a final 72°C extension step for 15 min. These conditions differed only slightly from those used with the *Haemonchus* primers in the Bott et al. [52] assay; denaturation and extension temperatures were identical, and the annealing in the recommended assay was only slightly lower. Therefore, these conditions should be equally effective in amplifying *H. contortus*.

Visual detection of the oligonucleotides produced through PCR is typically accomplished by electrophoretically separating them on an agarose gel, staining the gel with ethidium bromide, and then observing the gel with ultraviolet light. Commercial oligonucleotides markers of known base-pair (bp) size (and molecular weights) are electrophoresed alongside the sample amplicons to enable size verifications to be made for the samples. A 2% agarose gel is typically used for separating amplicons ranging from 200 bp to 1 kb, making this concentration appropriate for the amplicons generated in the present assay. Powdered agarose is diluted into the same solution used as the running buffer; the running buffer typically contains 1–2 mM ethylenediaminetetraacetic acid (EDTA) and 45–90 mM Tris-borate (TBE) at a pH of 8.0. The ionic strength and pH can vary slightly depending on the specific applications. In most mini-gel systems, 10–20 μL of PCR product is added into the wells of the gel along with the loading buffer containing a tracking dye. The voltage used for band-separation depends on numerous factors including the size of the gel. The lower the voltage, the slower the separation, but overheating can occur if a voltage above 5 V/cm is used. Gels can be stained with aqueous (0.5 μg/mL) ethidium bromide after electrophoresis, or ethidium bromide can be added directly to the melted agarose (0.5 μg/mL) just prior to pouring the gel. The amplicons in the gel can then be visualized and photographed on an ultraviolet (UV) (~254 nm) transilluminator.

55.3 CONCLUSIONS AND FUTURE PERSPECTIVES

At present, no molecular assays have yet been developed specifically for diagnosing haemonchosis in humans. This absence of a tested assay is because *Haemonchus* is rarely diagnosed in humans. It is possible that some human cases of haemonchosis are not receiving any medical attention or are being misdiagnosed as hookworm infections. If used on

TABLE 55.1
Forward Primers for Nematodes Included in Proposed *Haemonchus* Assay

Species	Oligo Name	Sequence	Amplicon Length (bp)	GenBank Access. No.
Ancyclostoma duodenale	jmAD	5′-TGCGAAGTTCGCGTTTCGCTGAGCT-3′	690	AJ001679
Necator americanus	jmNA	5′-CGTTAACATTGTATACCTGTACATAC-3′	870	AJ001680
Trichostrongylus spp.	jhTSP	5′-TTATGTGCCACAAATGAAGA-3′	482	S69220
Haemonchus spp.	HAE	5′-CAAATGGCATTTGTCTTTTAG-3′	165	X78803

a regular basis, the proposed PCR assay for *H. contortus* may help identify the true prevalence of this problem in humans. By combining components of an existing PCR assay for *H. contortus* in ungulates with a PCR assay for hookworm and *Trichostronglus* in humans, we have attempted to maximize the chances that this assay correctly identifies *Haemonchus* eggs in human samples. An additional step toward validating the proposed assay would be to "spike" human fecal samples with varying numbers of *H. contortus* eggs and then testing these samples with the proposed assay. If problems with assay specificity are observed at any point, unique reverse primers could be developed for each genus or species to increase specificity. Due to the bp differences of their amplicon, it should be possible to perform the proposed hookworm-*Trichostrongylus*-*Haemonchus* assay as a multiplex PCR similar to that developed by Zarlenga et al. [44] for cattle trichostrongyles. Other potential hookworm and trichostrongyle species could also be added to this human assay depending on species present in the various geographic regions. By using fluorescently labeled oligonucleotide probes that are unique to the amplicon produced for each genus/species, a multiplex real-time PCR could be developed for this human assay. Studies by von Samson-Himmelstjerna et al. [45], Harmon et al. [56], and Learmount et al. [57] have demonstrated the advantages of real-time PCR assays for diagnosing trichostrongyle eggs. These advantages include the ability to quantify relative numbers of eggs from each trichostrongyle species and to facilitate high-throughput analysis in a "closed-tube" format for the large number of samples required for survey studies [33].

The *H. contortus* real-time primer and probes from the sheep studies could easily be incorporated into the multiplex real-time PCR assay that already exists for detecting trichostrongyle eggs from *O. bifurcum* and hookworm eggs from *A. duodenale* and *N. americanus* [46].

Anthelmintic resistance is already a serious global problem among trichostrongyles of small ruminants, particularly in regions where sheep and/or goat production is high and multiple drug resistance is present [58]. This is especially true for *H. contortus*, where resistance to the benzimidazoles and macrocylic lactones is common. As a result, research to describe the gene mutations responsible for this resistance has increased greatly, and significant progress has been made, at least with benzimidazole resistance. In this class of anthelmintics, molecular methods are currently available for measuring resistant-gene frequency in *H. contortus* [59,60]. Less progress has been made with identifying gene mutations responsible for resistance to the macrocyclic lactones. Although frequently suspected, anthelmintic resistance has not yet been clearly demonstrated for human hookworms [61,62], and less progress has been made in developing diagnostic molecular tools for detecting resistance in hookworms than with the trichostrongyles. Given the high prevalence of drug resistance in *Haemonchus*, it is highly possible that human cases of haemonchosis would include resistant worms. If necessary, existing molecular assays could be used to measure the level of resistance in these human infections.

Apart from molecular approaches to detecting *Haemonchus* eggs in human fecal samples, a few studies suggest that the differential binding of certain fluorescently labeled lectins might support routine microscopy in the identification of different species of trichostrongyle eggs. Palmer and McCombe [63] found that peanut agglutinin lectin (PNA) bound uniquely to eggs of three different isolates of *H. contortus* but not to those of *T. circumcincta* or *T. colubriformis*. A second study involving mixed trichostrongyle infections in sheep confirmed the strong binding of PNA to *Haemonchus* spp. and found that *Trichostrongylus* spp. bound strongly to *Maclura pomifera* agglutinin (MPA). Recognition of *T. circumcincta* was based upon the lack of PNA and MPA binding to those eggs. Using field samples from naturally infected sheep and goats, Jurasek et al. [64] found that the relative number of PNA-positive eggs correlated well with the percentage of *Haemonchus* juveniles harvested from coprocultures prepared from the same fecal samples. Unfortunately, there have been no published lectin studies involving human hookworm eggs, but if PNA does not bind to *A. duodenale* or *N. americanus* eggs, then PNA or some other differentially binding lectin might be a very fast, easy, and inexpensive approach to identify *Haemonchus* eggs in human fecal samples.

Although not useful in diagnosing active cases of human haemonchosis, serological assays eventually might be useful in identifying present and past human cases and thereby assessing populations at risk for human haemonchosis. So far, serodiagnosis has not become an effective diagnostic tool for either trichostrongyles or human hookworm infections. Both somatic and excretory–secretory antigens have been tested for their specificity against infected sheep and goat sera, and comparisons with crude and purified antigens of *H. contortus* have demonstrated that appropriate purifications can improve the effectiveness of serodiagnosis, but cross-reactivity with other nematodes is still a problem. A purified somatic antigen (Hc26) has shown some usefulness in several sheep breeds when used in a Western blot analysis [65,66], but considerable improvements are needed before this approach will become a practical diagnostic tool in ruminants. Similarly, several studies have shown that serum antibodies from infected humans will bind to various antigens from either *A. duodenale* or *N. americanus*, but that these antibodies also frequently cross-react with other nematodes [67,68]. An improved relationship between infection intensity and antibody levels has been noted with specific isotypes within human immunoglobulin subclasses [67,68]; however, further improvement in specificity is needed before serology might serve as an effective approach to any of these intestinal-dwelling nematodes.

REFERENCES

1. Hoberg, E.R., Lichtenfels, J.R., and Gibbons, L. Phylogeny for species of *Haemonchus* (Nematoda: Trichostrongyloidea): Considerations of their evolutionary history and global biogeography among Camelidae and Pecora (Artiodactyla). *J. Parasitol.* 90, 1085, 2004.

2. Hoberg, E.P., and Lichtenfels, J.R. Phylogenetic systematic analysis of the Trichostrongylidae (Nematoda), with an initial assessment of coevolution and biogeography. *J. Parasitol.* 80, 976, 1994.
3. Blaxter, M.L. Nematoda: Genes, genomes and the evolution of parasitism. *Adv. Parasitol.* 54, 101, 2003.
4. De Ley, P.B. Systematic position and phylogeny. In: Lee, D. (ed.), *The Biology of Nematodes*, Taylor & Francis, London, pp. 1–30, 2002.
5. Chan, M.S. The global burden of intestinal nematode infections—Fifty years on. *Parasitol. Today* 13, 438, 1997.
6. Gibbons, L.M.K. A key for the identification of genera of the nematode family Trichostrongylidae Leiper, 1912. *J. Helminthol.* 56, 185, 1982.
7. Lichtenfels, J.R., and Pilitt, P.A. Synlophe patterns of the Haemonchinae of ruminants (Nematoda: Trichostrongyloidea). *J. Parasitol.* 86, 1093, 2000.
8. Ghadirian, E., and Arfaa, F. First report of human infection with *Haemonchus contortus*, *Ostertagia ostertagi*, and *Marshallagia marshalli* (family Trichostrongylidae) in Iran. *J. Parasitol.* 59, 1144, 1973.
9. Faust E.C., Russell P.F., and Rodney, C. J., *Craig and Faust's Clinical Parasitology*, Lea & Febiger, Philadelphia, pp. 890, 1970.
10. Magalhães, P.S. Notes d'helminthologie bresilienne—L'*Haemonchus contortus* comme parasite accidentel de l'homme. *Arch. Parasitol.* 12, 283, 1908.
11. Fleury, G.C., Correa, M.O.A., and Neto, V.A. Identificação do *Trichostrongylus colubriformis* como parasita do homem. *Rev. Inst. Meã. trop. São Paulo* 12, 288, 1970.
12. Travassos, L., Ensaio monografico da familia Trichostrongilidae. *Mem. Inst. Oswaldo Cruz* 13, 1, 1921.
13. Sweet, W.C. The intestinal parasites of man in Australia and its dependencies as found by the Australian hookworm campaign. *Med. J. Aust.* 1, 353, 1924.
14. Heydon, G.M., and Green, A.K. Some worm infectations of man in Australia. *Med. J. Aust.* 1, 619, 1931.
15. Soulsby, E.J.L. *Helminths, Arthropods and Protozoa of domesticated animals.* Bailliere Tindall, London, 809pp., 1982.
16. Georgi, J.R., and McCulloch, C.E., Diagnositic morphometry: Identification of helminth eggs by discriminant analysis of morphometric data. *Proc. Helminthol. Soc. Washington* 56, 44, 1989.
17. Veglia, F. The anatomy and life-history of *Haemonchus contortus* (Rud.). *Third and Fourth Reports of the Director of Veterinary Research* 1915, 347, 1916.
18. van Wyk, J.A., Cabaret, J., and Michael, L.M. Morphological identification of nematode larvae of small ruminants and cattle simplified. *Vet. Parasitol.* 119, 277, 2004.
19. Bowman, D.D. (ed.), In: *Georgis' Parasitology for Veterinarians*, Saunders Elsevier, St. Louis, 464pp., 2009.
20. Gibbs, H.C. Hypobiosis in parasitic nematodes—An update. *Adv. Parasitol.* 25, 129, 1986.
21. Lindqvist, A. et al. The dynamics, prevalence and impact of nematode infections in organically raised sheep in Sweden. *Acta Vet. Scand.* 42, 377, 2001.
22. Boreham, R.E. et al. Human trichostrongyliasis in Queensland. *Pathology* 27, 182, 1995.
23. Crab, A. et al. The parasitic hematophagous worm *Haemonchus contortus* inhibits human platelet aggregation and adhesion: Partial purification of a platelet inhibitor. *Thromb. Haemost.* 87, 899, 2002.
24. Koňáková, I. Use of modified McMaster method for the diagnosis of intestinal helminth infections and estimating parasitic egg load in human faecal samples. *Helminthologia* 46, 62, 2009.
25. Egwang, T.G., and Slocombe, J.O.D. Efficiency and sensitivity of techniques for recovering nematode eggs from bovine feces. *Can. J. Comp. Med.* 45, 243, 1981.
26. Cox, D.D., and Todd, A.C. Survey of gastrointestinal parasitism in Wisconsin dairy cattle. *JAVMA* 141, 706, 1962.
27. Goodman, D. et al. A comparison of methods for detecting the eggs of *Ascaris*, *Trichuris*, and hookworm in infant stool, and the epidemiology of infection in Zanzibari infants. *Am. J. Trop. Med. Hyg.* 76, 725, 2007.
28. Harmon, A.F. et al. Comparison of three different preservatives for morphological and real-time PCR analyses of *Haemonchus contortus* eggs. *Vet. Parasitol.* 145, 361, 2007.
29. Pawlowski, Z.S., and Stott, G.J. *Hookworm Infection and Anemia: Approaches to Prevention and Control*, Vol. 96. World Health Organization, Geneva, 1991.
30. Garcia, L.S. *Practical Guide to Diagnostic Parasitology*, ASM Press, Washington, DC, 448pp., 2009.
31. Hubert, J., and Kerboeuf, D. A new method for culture of larvae used in diagnosis of ruminant gastrointestinal strongylosis: Comparison with fecal cultures. *Can. J. Comp. Med.* 48, 63, 1984.
32. Gasser, R.B., Cantacessi, C., and Loukas, A. DNA technological progress toward advanced diagnostic tools to support human hookworm control. *Biotechnol. Adv.* 26, 35, 2008.
33. Gasser, R.B. et al. Toward practical, DNA-based diagnostic methods for parasitic nematodes of livestock—Bionomic and biotechnological implications. *Biotechnol. Adv.* 26, 325, 2008.
34. Mitreva, M. et al. Parasitic nematodes—From genomes to control. *Vet. Parasitol.* 148, 31, 2007.
35. Beh, K.J., Foley, R.C., and Goodwin, E.J. Restriction fragment length patterns of DNA from parasitic nematodes of sheep, *Res. Vet. Sci.* 46, 127, 1989.
36. Christensen, C.M., Zarlenga, D.S., and Gasbarre, L.C. *Ostertagia*, *Haemonchus*, *Cooperia*, and *Oesophagostomum*: Construction and characterization of genus-specific DNA probes to differentiate important parasites of cattle, *Exp. Parasitol.* 78, 93, 1994.
37. Mes, T.H., and Cornelissen, A.W. Non-homogenized ITS regions in the parasitic nematode *Cooperia oncophora*, *Parasitology* 129, 213, 2004.
38. Hillis, D.M., and Dixon, M.T. Ribosomal DNA: Molecular evolution and phylogenetic inference. *Q. Rev. Biol.* 66, 411, 1991.
39. Heise, M., Epe, C., and Schnieder, T. Differences in the second internal transcribed spacer (ITS-2) of eight species of gastrointestinal nematodes of ruminants. *J. Parasitol.* 85, 431, 1999.
40. Yong, T.S. et al. Differential diagnosis of *Trichostrongylus* and hookworm eggs via PCR using ITS-1 sequence, *Korean J. Parasitol.* 45, 69, 2007.
41. Garretson, P.D. et al., Anthelmintic resistant *Haemonchus contortus* in a giraffe (*Giraffa camelopardalis*) in Florida, *J. Zoo Wildl. Med.* 40, 131, 2009.
42. Traub, R.J. et al. PCR-based coprodiagnostic tools reveal dogs as reservoirs of zoonotic ancylostomiasis caused by *Ancylostoma ceylanicum* in temple communities in Bangkok. *Vet. Parasitol.* 155, 67, 2008.

43. Traub, R.J., Monis, P.T., and Robertson, I.D. Molecular epidemiology: A multidisciplinary approach to understanding parasitic zoonoses. *Int. J. Parasitol.* 35, 1295, 2005.
44. Zarlenga, D.S. et al. A multiplex PCR assay for differentiating economically important gastrointestinal nematodes of cattle. *Vet. Parasitol.* 97, 199, 2001.
45. von Samson-Himmelstjerna, G., Harder, A., and Schnieder, T. Quantitative analysis of ITS2 sequences in trichostrongyle parasites. *Int. J. Parasitol.* 32, 1529, 2002.
46. Verweij, J.J. et al. Simultaneous detection and quantification of *Ancylostoma duodenale*, *Necator americanus*, and *Oesophagostomum bifurcum* in fecal samples using multiplex real-time PCR. *Am. J. Trop. Med. Hyg.* 77, 685, 2007.
47. de Gruijter, J.M. et al. Polymerase chain reaction-based differential diagnosis of *Ancyclostoma duodenale* and *Nacator americanus* infections in humans in northern Ghana. *Trop. Med. Int. Health* 10, 574, 2005.
48. Harmon, A.F., Zarlenga, D.S., and Hildreth, M.B., Improved methods for isolating DNA from *Ostertagia ostertagi* eggs in cattle feces. *Vet. Parasitol.* 135, 297, 2006.
49. Monteiro, L. et al. Complex polysaccharides as PCR inhibitors in feces: *Helicobacter pylori* model. *J. Clin. Microbiol.* 35, 995, 1997.
50. Thornton, C.G., and Passen, S. Inhibition of PCR amplification by phytic acid, and treatment of bovine fecal specimens with phytase to reduce inhibition. *J. Microbiol. Methods* 59, 43, 2004.
51. Kreader, C.A. Relief of amplification inhibition in PCR with bovine serum albumin or T4 gene 32 protein. *Appl. Environ. Microbiol.* 62, 1102, 1996.
52. Bott, N.J. et al. A combined microscopic-molecular method for the diagnosis of strongylid infections in sheep. *Int. J. Parasitol.* 39, 1277, 2009.
53. Quach, N., Goodman, M.F., and Shibata, D. *In vitro* mutation artifacts after formalin fixation and error prone translesion synthesis during PCR. *BMC Clin. Pathol.* 4, 1, 2004.
54. Gasser, R.B., Stewart, L.E., and Speare, R. Genetic markers in ribosomal DNA for hookworm identification. *Acta Trop.* 62, 15, 1996.
55. Stevenson, L.A., Chilton, N.B., and Gasser, R.B. Differentiation of *Haemonchus placei* from *H. contortus* (Nematoda: Trichostrongylidae) by the ribosomal DNA second internal transcribed spacer. *Int. J. Parasitol.* 25, 483, 1995.
56. Harmon, A.F. et al. Real-time PCR for quantifying *Haemonchus contortus* eggs and potential limiting factors. *Parasitol. Res.* 101, 71, 2007.
57. Learmount, J. et al. Development and validation of real-time PCR methods for diagnosis of *Teladorsagia circumcincta* and *Haemonchus contortus* in sheep. *Vet. Parasitol.* 166, 268, 2009.
58. Wolstenholme, A.J. et al. Drug resistance in veterinary helminths. *Trends Parasitol.* 20, 469, 2004.
59. Coles, G.C. et al. The detection of anthelmintic resistance in nematodes of veterinary importance. *Vet. Parasitol.* 136, 167, 2006.
60. von Samson-Himmelstjerna, G. Molecular diagnosis of anthelmintic resistance. *Vet. Parasitol.* 136, 99, 2006.
61. Zheng, Q. et al. The control of hookworm infection in China. *Parasit. Vectors* 2, 44, 2009.
62. Keiser, J., and Utzinger, J. Efficacy of current drugs against soil-transmitted helminth infections: Systematic review and meta-analysis. *JAMA.* 299, 1937, 2008.
63. Palmer, D.G., and McCombe, I.L. Lectin staining of trichostrongylid nematode eggs of sheep: Rapid identification of *Haemonchus contortus* eggs with peanut agglutinin. *Int. J. Parasitol.* 26, 447, 1996.
64. Jurasek, M.E. et al. Modification and further evaluation of a fluorescein-labeled peanut agglutinin test for identification of *Haemonchus contortus* eggs. *Vet. Parasitol.* 169, 209, 2010.
65. Gomez-Munoz, M.T., Cuquerella, M., and Alunda, J.M. Identification and partial purification of a 26 kilodalton antigen of adult *Haemonchus contortus*. *Int. J. Parasitol.* 26, 311, 1996.
66. Gomez-Munoz, M.T. et al. Serodiagnosis of haemonchosis with a somatic antigen (Hc26) in several breeds of sheep. *J. Vet. Diagn. Invest.* 12, 354, 2000.
67. Mahmoud, M.S., Abou Gamra, M.M., and Elkhayat, M.M. *Ancylostoma duodenale* infection: A study of serum immunoglobulin G4 response to the excretory secretory antigen of adult worm. *J. Egypt Soc. Parasitol.* 35, 1, 2005.
68. Palmer, D.R., Bradley, M., and Bundy, D.A. IgG4 responses to antigens of adult *Necator americanus*: Potential for use in large-scale epidemiological studies. *Bull. World Health Organ.* 74, 381, 1996.
69. Zarlenga, D.S. et al. Cloning and characterization of ribosomal RNA genes from three species of *Haemonchus* (Nematoda: Trichostrongyloidea) and identification of PCR primers for rapid differentiation. *Exp. Parasitol.* 78, 28, 1994.
70. Chilton, N.B. et al. The evolutionary origins of nematodes within the order Strongylida are related to predilection sites within hosts. *Mol. Phylogenet. Evol.* 40, 118, 2006.

56 Loa

Fousseyni S. Touré Ndouo

CONTENTS

56.1 INTRODUCTION

56.1.1 Classification, Morphology, and Biology

Filariasis is a chronic tropical disease caused by nematodes belonging to the Filariidae family. Eight species are currently known to infect humans, namely *Wuchereria bancrofti*, *Brugia malayi*, *Brugia timoni*, *Onchocerca volvulus*, *Loa loa*, *Mansonella perstans*, *Mansonella ozzardi*, and *Mansonella streptocerca*. Three groups of filariasis have been distinguished on the basis of their human target tissues: lymphatic filariasis (wuchereriasis and brugiasis); cutaneous dermal filariasis (loaiosis, onchocerciasis, and *streptocerca* mansonelliasis); and serous filariasis (*perstans* and *ozzardi* mansonelliasis) [1]. The vectors are blood-sucking flies and female mosquitoes. The microfilariae or embryos from peripheral blood are ingested by the vector when it bites an infected human. These microfilariae become infective stage L3 larvae after two successive mutes within the vector, and are transmitted to a new human host through a new blood meal or bite (Figure 56.1).

More than 3.3 billion people are exposed to filariasis, and an estimated 300 million people are infected [2]. Loaiosis occurs in Africa, brugiasis in South Asia, wuchereriasis in Africa and Asia, onchocerciasis in Africa, Central and South America, and Asia (Yemen), *perstans* mansonelliasis in Africa and Central and South America, *ozzardi* mansonelliasis in Central and South America, and *streptocerca* mansonelliasis in Africa [3] (Figure 56.2).

Filariasis due to *L. loa* (loaiosis) was described for the first time by Mongin in a female slave originating from West Africa and living in Saint Domingue Island in 1770 [5]. In 1864, Guyon found the same worm in Gabon [6]. Loaiosis has been described in detail by Brumpt [7] and Connor [8]. Although initially discovered in the French West Indies, loaiosis now only occurs in Africa [1].

Loaiosis is a dermal helminthiasis. The adult worms live under the skin for up to 15 years or even longer [1]. The adult male worms of *L. loa* are considerably smaller than their female counterparts, measuring, respectively, about 30–34 mm × 0.35–0.42 mm compared to 40–70 mm × 0.5 mm. The adults mate in the host's subcutaneous tissue and produce worm-like eggs or embryos called microfilariae, which are 250–300 μm long and 6–8 μm wide. These eggs can be distinguished morphologically from those of other filariae, as they are sheathed and contain body nuclei that extend to the tip of the tail [9].

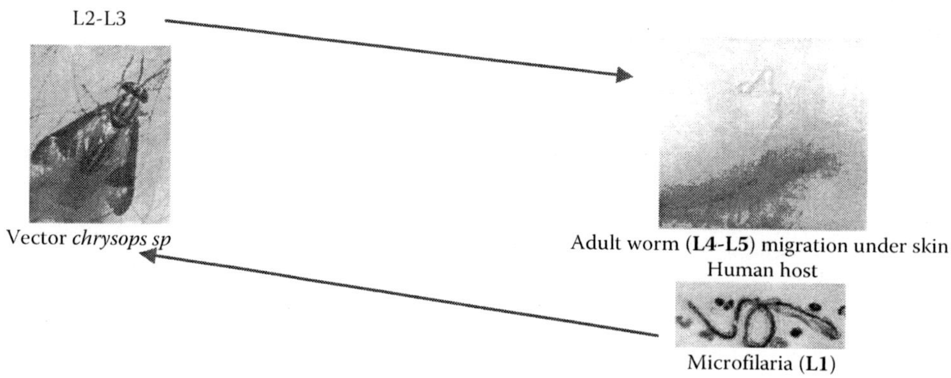

FIGURE 56.1 *L. loa* life cycle.

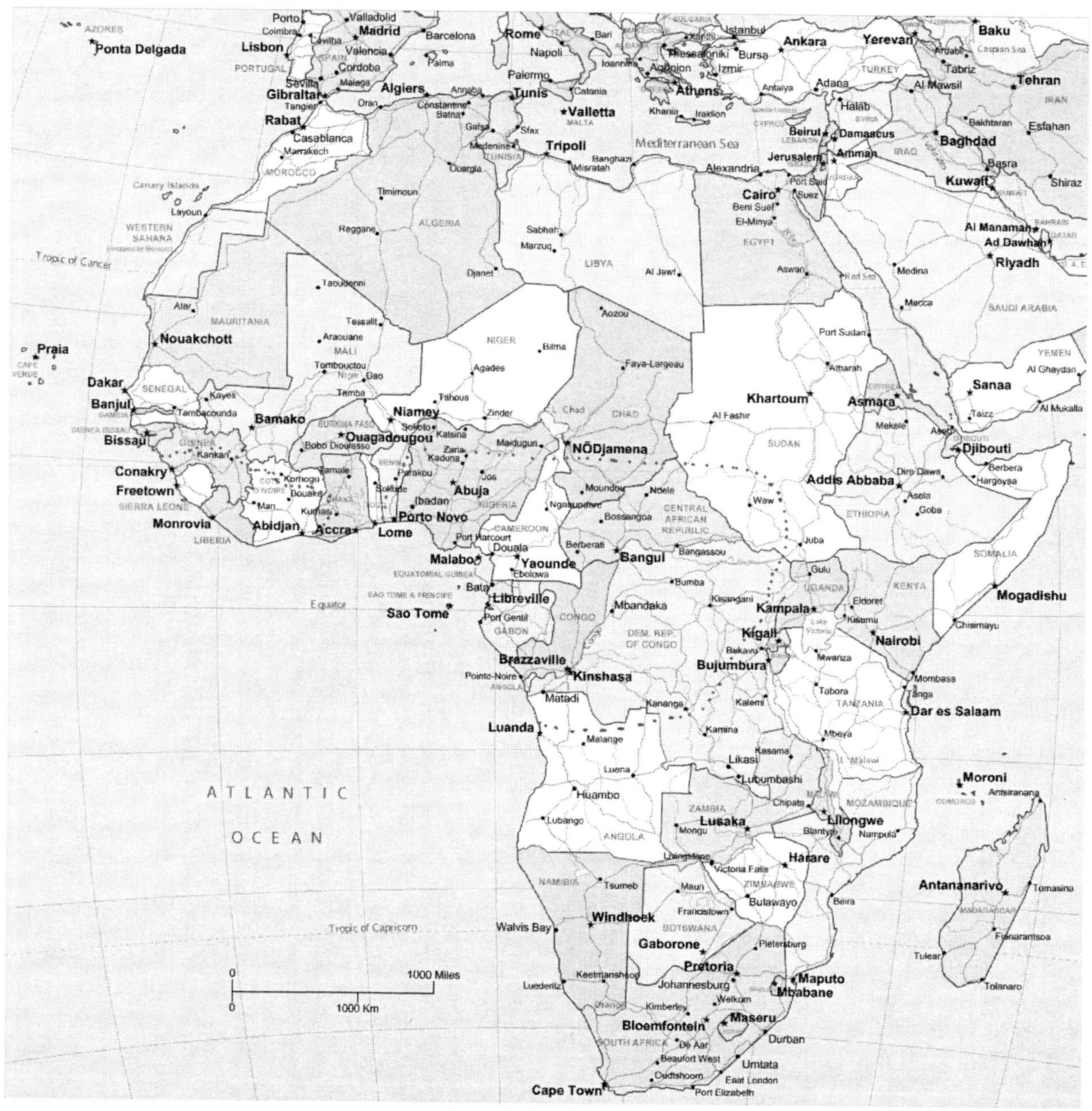

FIGURE 56.2 Geographical distribution of *L. loa*, Loaiosis endemic areas. Map from www.geographicguide.com/. . ./maps/map-africa.gif. (Adapted from Muller, R. *Worms and Disease*. Heinemann Medical Books, London, UK, 1975.)

L. loa is well adapted to its tabanide vectors and host, and has no free-living stage. The vectors are only found in the hot and wet forests of western and central Africa. The main vector species are *Chrysops dimidiata* and *Chrysops silacea*, two forest species often present in the same hearth. Only the females are hematophagous, biting during the day mainly between 9 a.m. and 4 p.m. Once inside the host, the adult female worms release sheathed microfilariae which circulate in peripheral blood during daytime. Microfilariae ingested by *Chrysops* vectors continue their normal development in the vector's digestive tract, becoming infective larvae (L3) after 12 days. When the vector bites a new host, the larvae escape from the horn of the fly and actively cross the host's skin. The development of such infective larvae into adult worms able to emit embryos in their human host takes about three months. The adult worms live in subcutaneous and deep connective tissues, but can also migrate to other tissues, as witnessed by subconjunctival passage. The microfilariae remain in the peripheral bloodstream for between 5 a.m. and 19 p.m., with a peak between 9 a.m. and 17 p.m [10,11]. There they can be ingested by *Chrysops* flies, which only bite during the day.

In 1920, Treagold discovered a new species of filaria (*Loa papionis*) in baboons (*Papio cynocephalus*) in Equatorial Guinea [12]. It differed from *L. loa* by the larger size of the adult worms and by the night periodicity of the microfilairia. In 1936, Sandground, on examining filarial parasites isolated from a monkey (*Cercocebus albigena johnstoni*) captured in Zaire (Democratic Republic of Congo), concluded that they were closely related to *L. loa* [13]. Sandground considered that "*Loa papionis*" described by Treagold was in fact *L. loa*. Some other filarial parasites attributed to the *L. loa* species were announced by Gordon in 1950 and then by Duke in 1958, in three very widespread monkey species (*Mandrillus leukophaeus*, *Cercopithecus nictitans martinii*, and *Cercopithecus mona mona*) living in the rain forest in the area of Kumba, southwestern Cameroun [14,15]. These findings pointed to a monkey reservoir for *L. loa*, but experimental infection showed that simian *L. loa* was transmitted exclusively by *Chrysops langi* and *Chrysops centurionis*, two species with twilight activity that do not bite humans [15–17]. Also, studies of *C. dimidiata* and *C. silecea* showed that 89–90% of blood meals were of human origin, and no traces of simian blood were found [18]. This worked showed, however, that the mandrill is a good experimental model of human *L. loa* infection. In natural conditions the simian and human parasites each have specific "host–parasite–vector" complexes, making exchanges between them rather unlikely [19]. However, further observations are necessary before concluding that humans are the sole reservoir of *L. loa* [19].

Loaiosis is only found in Africa [20]. Although some cases have been described in intertropical Africa, the disease seems to be limited to the forest block of Central Africa and West Africa (Benin, Nigeria, Cameroun, Centrafrican Republic, Democratic Republic of Congo, Congo, Rwanda, Equatorial Guinea, Gabon, and Angola). In these countries more than 13 million people are infected, with prevalence rates of 3–35%, even reaching 100% in certain forest villages [19]. Some cases have been reported among expatriates living in endemic areas [21,22].

56.1.2 Clinical Features and Pathogenesis

Loaiosis is characterized clinically by calabar edema (swelling) and conjunctivitis due to ocular passage of adult worms. Calabar edema is transient and located on the face, limbs, and back of the hands and fingers. It may last for 1–3 days and is sometimes accompanied by local urticaria and pruritus. *L. loa* is also called the "African eye worm." Passage beneath the conjunctiva usually takes less than 15 min. The frequency of this complication increases with the victim's age. Dead worms can cause chronic abscesses, which may lead to granulomatous reactions and fibrosis [9]. Meningoencephalic complications are an adverse effect of diethylcarbamazine treatment for hypermicrofilaremia [23–25]. Other complications include nephropathies [26], endocarditis [27], retinopathies [28], neuropathies [29], and pneumonitis [30]. Symptoms are more frequent in expatriates [31]. Immunologically, loaiosis is characterized by hypergammaglobulinemia, hypereosinophilia, and high IgE levels responsible for allergic symptoms [32]. In endemic areas loaiosis is the third reason for medical consultations in rural settings, although many microfilaremic subjects are asymptomatic [33].

Occult loaiosis (amicrofilaremic (AMF) infection) is defined as infection by the adult worm, without peripheral microfilaremia (MF) observable by standard microscopy. AMF status is common among autochthonous residents and may be due to sequestration of microfilaria or to their massive destruction by the immune system. This form of infection is most common in endemic areas. Other AMF subjects are thought to be resistant. There is currently no way of discriminating between these two AMF subgroups in the absence of (transient) ocular passage of adult worms.

Three immunodominant antigens have been characterized in *L. loa*: the 29–31 kDa antigen expressed on the surface of adult worms, the 23 kDa antigen on the surface of microfilaria and the 15 kDa antigen expressed by all developmental stages of the parasite [34–36]. The presence of antibodies to the 29–31 and 23 kDa antigens in individuals living in endemic zones is associated with AMF status.

Recently, Western blot studies showed that specific IgG1 from AMF subjects selectively recognizes an adult worm antigen of 28–31 kDa and a doublet of 21–22 kDa expressed on microfilaria [37]. Sera from AMF subjects, contrary to microfilaremic subjects, can lyse *L. loa* microfilaria *in vitro* by antibody-dependent cellular adherence [38]. Preliminary studies show that the control of MF involves CD4 T helper cell responses (proliferation and production of Th1 cytokines) as well as strong production of Th2 cytokines [39].

By contrast, microfilaremic status is accompanied by suppression of the cellular response, possibly due to immune tolerance induced by microfilaria. The presence of circulating microfilaria is also associated with a lack of T cells responses (proliferation and cytokine production) in the mandrill model of human loaiosis [40]. Residents of endemic areas develop a specific immune response, directed particularly against microfilaria, thus partly explaining the predominance of AMF status. While the immune mechanisms related to the suppression of microfilaria are starting to be elucidated, those involved in the destruction of the adult worms and infective larvae remain obscure.

56.1.3 Diagnosis

56.1.3.1 Conventional Techniques

56.1.3.1.1 Clinical Diagnosis

Clinically, *L. loa* infection is diagnosed when migration of adult worms under the conjunctiva and/or skin is observed or when a patient presents with classical symptoms.

56.1.3.1.2 Microscopic Examination

Diagnosis is classically based on standard microscopy. Microfilariae are the blood stage of *L. loa*.

Thick smears are prepared with venous blood and stained with Giemsa or hematoxylin–eosin.

Microfilariae can also be detected in fresh blood drops between a slide and coverslip, but sensitivity can be enhanced by leukoconcentration, notably after centrifugation of the blood sample lysed in 2% formalin or 2% saponin (Knott's technique), or filtration through a Nucleopore membrane.

The quantitative coating buffer method is also used to detect *L. loa* microfilariae.

For microscopic detection of *L. loa* microfilariae, blood samples must be collected during the day, given the diurnal periodicity of human loaiosis.

56.1.3.1.3 Antibody Assay

The first serological studies based on the detection of total IgM and IgG directed against *L. loa* crude antigens raised hopes that occult forms could be detected. Seroepidemiological investigations showed that 80% of AMF subjects living in endemic areas had total IgG directed against *L. loa* antigens [21,41,42]. Very quickly, it was noted that tests based on IgG detection were not very specific, however, because of cross-reactions between the different filarial species.

IgG4 predominates in lymphatic filariasis, onchocerciasis, and loaiosis [36,43–48]. In loaiosis, IgG4 subclass antibodies recognize low-molecular-weight antigens of 15–30 kDa [36]. IgG4 antibody production requires chronic antigen stimulation [21].

In loaiosis endemic zones, permanent residents with microfilaremic or occult infection can produce IgG4 antibodies if they are not genetically restricted. It is reasonable to assume that nonpatently infected permanent residents and nonrestricted can produce the same IgG4; those are being usually exposed to the bites of *Chrysops* and thus to reinfection by larvae L3. In the former case these antibodies should be considered as indirect markers of active infection and, in the latter, as indirect markers of exposure.

Using adult worm crude antigens, 12/13 microfilairemic subjects and 10/13 subjects with occult infection were found to have specific IgG4, with a specificity of 94% relative to sympatric *M. perstans* [36].

The average optical density of IgG4 in the two groups was similar, suggesting that IgG4 production can be triggered by adult worms and not only by blood microfilaria. IgG4 serum levels in healthy individuals are normally very low (like IgE levels), and IgG4 is a marker of Th2 type responses, as the switch from IgM to IgG4 is induced by the Th2 cytokines IL-4 and IL-13 [49]. However, in several studies of lymphatic filariasis [50] and more recently loaiosis [39], a Th1 cytokine profile was found to be associated with AMF status or undetectable antigenemia.

56.1.3.1.4 Diagnostic Challenges

The frequency of symptomatic AMF loaiosis among residents in endemic areas implies the need for a specific and sensitive detection method. Indeed, in endemic areas microfilaremic subjects represent only a small percentage of infected individuals [51,52]. The prevalence of loaiosis is still based on microscopic examination alone, which underestimates the true prevalence of the disease.

Because of cross-reactions, serological tests, and especially those based on total IgG detection, tend to overestimate the prevalence. We studied specific IgG4 against microfilaria antigens, which are more accessible than those of adult worms, in serum samples collected from 222 inhabitants of three villages (Djoutou, Ndjokaye, and Okoumbi). We found that 55% of AMF individuals (97/178) produced specific IgG4. However, only 80% of microfilaremic subjects (35/44) were positive, with an estimated specificity of 78% relative to *M. perstans* [53].

The difference in specificity and sensitivity between this study and that carried out by Dr. Akué 1994 may be due to the sample size rather than to the antigens used (microfilaria crude extract versus adult worms). The results also show that 20% of MF subjects do not produce IgG4, suggesting possible genetic restriction of this antibody subclass.

It is unclear whether IgG4 antibodies are markers of active *L. loa* infection. Circulating *L. loa* antigens and/or DNA are the only direct markers, and should soon form the basis for diagnostic tests. The prevalence of loaiosis is also high in Gabon [54] and this disease may have had a negative effect on the demography of this country [55]. Loaiosis has been a research priority for CIRMF (International Centre for Medical Research in Franceville) since its creation. CIRMF teams have studied loaiosis in humans as well as in mandrills, the simian model of human loaiosis. Certain specific *L. loa* antigens have been characterized, and the development of a specific and sensitive diagnostic method for occult loaiosis has been considered a major priority.

Antigen detection based on *immunoassay*. A recent rapid luciferase immunoprecipitation assay (QLIPS) detects *L. loa* infection in 15 min, with 97% and 100% sensitivity and specificity with respect to sympatric filariae [56]. This test remains to be evaluated in the field, however.

56.1.3.2 Molecular Techniques

56.1.3.2.1 DNA Hybridization

In 1990, considerable progress was made in filarial genome research, with the main objective of identifying specific molecular markers of each filarial parasite. Certain species-specific sequences have been identified and characterized, permitting the development of oligonucleotide probes for molecular hybridization methods designed to detect filarial parasites [57,58]. Other probes have since been developed for *W. bancrofti*, *O. volvulus*, and *L. loa* [59–62]. These specific probes have facilitated the identification of infected vectors and parasites in biological samples such as blood and skin biopsies.

56.1.3.2.2 Polymerase Chain Reaction (PCR)

In 1993, the use of highly sensitive PCR methods to diagnose filarial infections, including occult infection, was discussed by WHO experts.

PCR target DNA sequence: The *L. loa* 15 kDa protein is a member of the nematode polyprotein allergen family [63]. The polyprotein is composed of five tandem repeats of 15 kDa monomers that are cleaved at arginine-rich sites to generate a ladder-like array on Western blot. Nematode polyproteins are antigenic or allergenic. The gene encoding the *L. loa* 15 kDa polyprotein counts 1741 base pairs (bp) (Table 56.1) and is organized into five repeat regions (r1, r2, r3, r4, and r5) [64].

TABLE 56.1
Nucleotide Sequences of *L. loa* Gene Encoding 15 kDa Polyprotein and 15r3 Primers

***L. loa* Gene Encoding 15 kDa Polyprotein:**	
1	GATCATCATG AACACAATTT GGATGAGTAC TTCCGAACGC ATCTAAGCTG GCTCACGGAT
61	ATCCAGAAAG ATGAAATTAG GAAAATGAAG GAAGAAGGAA AACCGAAAGC GGATATGCAA
121	AAGAAAATTT TTGATTATTA CGAAAGTCTA ACCGGTGATG AAAAGAAGGA AGCTAGTGAG
181	AAGCTTCGAG AAGGCTGCCG CGCATTGCTG AAGCAAATTG TTGGTGATGA GAAGATGGCT
241	GAACTAAAAC AGATGAAAGA TTCGGGCGAT GGATATGAGG AACTGATAGC CAAAGTGGAT
301	CATATGTTGG AACATGTTAC TGATGAGCCA AAAAAGGAAA AAATTACGGA ATATGGTCCT
361	GCATGCCGTA AAATCTATGG GGATCGACAT AAGCGGGATC ATCATGAACA CAATTTGGAT
421	GAGTACTTCC GAACGCATCT AAGCTGGCTC ACGGATATCC AGAAAGATGA AATTAGGAAA
481	ATGAAGGAAG AAGGAAAACC GAAAGCGGAT ATGCAAAAGA AAATTTTTGA TTATTACGAA
541	AGTCTAACCG GTGATGAAAA GAAGGAAGCT AGTGAGAAGC TTCGAGAAGG CTGCCGCGCA
601	TTGCTGAAGC AAATTGTTGG TGATGAGAAG ATGGCTGAAC TAAAACAGAT GAAAGATTCG
661	GGCGTAGGTA TAGAGGAACT GATAGCCAAA GTGGATCATA TGTTGGAACA TGTTACTGAT
721	GAGCCAAAAA AGGAAAAAAT TCAGGAATAT GGTCCTGCAT GCCATAAAAT CTACGGAACA
781	CCTGTGATGC GGAATAAGCG GAATCAGGCA AATAATGGCA CAAAACACTG CAGCAGTCCT
841	TACTTGCAAT GGCTTACTGA CGAAGAAAAA GGCGAAATTC GAAAAATTAT AGGGGGAAAC
901	AAATCAAGAG CCGATATACT GAAAGCTATC TTTTTTTACT ATCAAATTCT TCCTGGAAAG
961	GAGAAGAAAA ATGCGGGAGA GCGCTTGAGG TTAAGTTGCG AAGAAATCGT TCGCAGTTTG
1021	GTCTACGAAA ATCGATTAAG TGAATTAGAA GCTTTGGAAA ATGATAGTTC TACAATGAGC
1081	GAGATGATGA AAATGTTGTC AGCTGCAACG GATAGATCAA AGTTTGCACA AATTAAGGCA
1141	TATCAAACCG CTTGTAACAG AATCTTCGAT TTGAGACAGC TGGTGAGAAG AAAACGCGAA
1201	CACAAAGGTAATTCAATCGATAAATTATCTT GAGAAGAAATT TAAAGTGGCT TTCGGTGGAG
1261	CAGAGGGAGG AATTGAGGGA AATGAAGAAG AATGGTAAAT CGAGGGCTGA TATGATTGCT
1321	AAAATGTTTC ATTATTACGA AGAATTATTT GGAGAAGCAA AGCAGCATGT CACTGAACGG
1381	TTATACGACG GTTGTCGACA AATTTTAAAA GATGTTGTTG GTGGAGATCG CTACAATGAA
1441	TTGGCAAAAA TGAAGGATTC GGGTGCAAAT ATGAATGATT TAAAAGCGAA AGCTGATGCA
1501	ATGTTGAATG AAATAATAGA CGAAGAAAAG AAGGAAAAAA TTAAAATTTA CGGATCTGGA
1561	TGCATGAGAA TTTTTACAAG GATGGGTCAT AAACATTCAT TGGAGGAACA TTTCAAGACA
1621	GATCTGAAAT GGCTCACAAA GGAACAAAAG GATGAGCTAC TAAAGATGAA GGAGGAAAAC
1681	AAATCGGAAG CAGATATTCG AGAAAAAGTA TTGCACTTCT ACGAAAGTCT AAATGAAGAA
1741	GCTAAAAAAG AAACGGCTGA ATTC
15r3 Primers:	
15r3-1:	5′-AAT CAG GCA AAT AAT GGC ACA AAA-3′
15r3-2:	5′-GCG TTT TCT TCT CAC CAG CTG TCT-3′
15r3-3:	5′-GGC ACA AAA CAC TGC AGC AGT CCT-3′
15r3-4:	5′-CAG CTG TCT CAA ATC GAA GAT TCT-3′
15r3-probe:	5′-TCA AGA GCC GAT ATA CTG AAA GCT ATC-3′

In *Caenorhabditis elegans* and *Dirofilaria immitis* this gene is composed of a core of conserved repeats, the last three having divergent sequences (McReynolds, personal communication). It is likely that repeat 3 in *L. loa* is one of the C-terminal divergent repeats. The third repeat was chosen as a target sequence because it was most dissimilar from other sequences, showing only 34% amino acid identity with other nematode polyproteins, compared to 52–83% for the other repeats [64].

In 1997, a PCR method (15r3-PCR) was developed to detect the repeat 3 region of the gene encoding the *L. loa* polyprotein in blood samples [65,66]. The 15r3-PCR assay can be followed by hybridization, or takes the form of a nested PCR method, for detecting repeat region 3 of the gene coding for the *L. loa* 15 kDa polyprotein. It was developed using blood lysates from patently infected subjects.

Microfilaremia. All microfilaremic samples (from human and mandrills) tested by 15r3-PCR were positive. Template DNA for PCR was extracted from 100 μL of whole blood, theoretically containing at least one microfilairia (mf) in microfilaremic subjects with >10 mf/mL of blood. One *L. loa* microfilairia contains approximately 100 pg of genomic DNA. This implies that parasite DNA detected in microfilaremic subjects can be derived from *L. loa* microfilariae. To test this possibility we added 16, 8, 4, 2, and 1 pg of *L. loa* microfilarial DNA in tubes containing 100 μL of uninfected whole blood and immediately extracted a DNA template. All the samples, including that spiked with 1 pg of DNA were positive by 15r3-PCR in ethidium bromide staining gels. Identical results were obtained with *L. loa* adult worm genomic DNA. However, most patient samples containing between 1 and 8 mf/mL of blood were negative by single-round PCR with ethidium bromide staining and only became positive after hybridization or nested PCR on the first-round PCR products. It is conceivable that the 100 μL of whole blood used to prepare the template DNA contained no microfilariae, implying that this approach can also detect circulating *L. loa* DNA.

Occult infection (*L. loa* adult worms but no circulating microfilariae). AMF status, commonly observed in endemic areas, can be due to the destruction of microfilariae by the host immune system, or to sequestration, or even to the absence of egg production because of adult worm sterility. One or more ocular passages of adult worms occasionally occur in AMF individuals. In most AMF subjects, massive destruction of microfilariae occurs, thereby releasing parasite DNA into the bloodstream. This DNA can be present in free form in plasma, or associated with cell surface proteins, or contained in phagocytic cells. PCR can detect a specific target sequence in the *L. loa* genome if: (a) microfilarial DNA is not completely degraded or removed by the immune system; (b) DNA contained in whole blood is preserved during template preparation; and (c) the concentration of the DNA target in the template is at least equal to the detection limit of the assay.

Adult worms can also release DNA when they lay nonviable eggs or are killed by host defenses, although they appear to be far less vulnerable than microfilariae. At all events, the quantity of DNA released, whether from nonviable eggs, microfilariae and/or adults, is related to the adult worm load.

The 15r3-PCR assay was applied to blood lysates from 20 AMF subjects with proven ocular passage of adult worms and 19 of them were positive. This shows that PCR positivity does not require the presence of blood microfilariae and confirms that the assay detects circulating *L. loa* DNA [67,68]. Consequently, all AMF individuals positive by 15r3-PCR assay must be regarded as infected.

The pre-patent phase of experimental mandrill infection by human *L. loa* L3 stages cannot be detected by 15r 3-PCR. This period, estimated to last at least three months, precedes the appearance of microfilariae in blood. We attempted to detect this period by amplifying DNA from developing larvae and/or early adult worms. In this experimental situation only 100 L3 larvae were inoculated to each mandrill (this species has approximately 2 L of blood).

The 15r3-PCR assay had a sensitivity of 95% with respect to detection of ocular passage of *L. loa* adult worms and 100% compared to detection of MF. Specificity, determined on blood samples from subjects living in areas nonendemic for *L. loa* and infected by other filarial species, was 100% with respect to *W. bancrofti*, *O. volvulus*, and *M. perstans*.

56.2 METHODS

56.2.1 Sample Preparation

56.2.1.1 Blood Sampling

Blood is collected by venipuncture into Vacutainer tubes containing ethylenediaminetetraacetic acid (EDTA).

56.2.1.2 Leukoconcentration

A volume of 1 mL of each sample is added to a 15-mL tube containing 9 mL of phosphate buffered saline, in duplicate. The mixture is treated with 600 μL of 2% saponin at room temperature for 15 min to lyse red blood cells, followed by centrifugation at 1000 × *g* for 15 min at 4°C. The supernatant is discarded and the pellet is examined microscopically for microfilariae. The distinction between *L. loa* and *M. perstans* microfilariae is based on size, motility, and by the presence of a sheath (*L. loa*).

56.2.1.3 Whole Blood Lysate Processing (Template DNA)

For each sample, 100 μL of whole blood is mixed with 500 μL of TE buffer (10 mM Tris, pH 8, and 0.1 mM EDTA, pH 8) and spun at 10,000 × *g* twice for 2 min. The supernatant is discarded at each step. The pellet is resuspended in 500 μL of sucrose buffer (10 mM Tris, pH 7.6, 5 mM $MgCl_2$, 1 M sucrose, and 1% Triton × 100) and spun at 10,000 × *g* twice for 2 min. After the final wash the supernatant is discarded and the pellet is resuspended in 200 μL of prewarmed (56°C) proteinase K buffer containing 20 mM Tris, pH 8, 50 mM KCl, 2.5 mM $MgCl_2$, 100 μg/mL proteinase K, and 0.5% Tween20, incubated at 56°C for two hours, then heated at 90°C for 10 min. The DNA can be stored at 4°C for several days or at −20°C until required.

56.2.2 Detection Procedures

56.2.2.1 PCR

Primers corresponding to the 5′ and 3′ ends of the repeat 3 sequence of the gene coding for the *L. loa* 15 kDa antigen are used (Table 56.1). The sequences of these primers are as follows: forward (24-mer): 5′-AAT-CAG-GCA-AAT-AAT-GGC-ACA-AAA-3′, reverse (24-mer): 5′-GCG-TTT-TCT-TCT-CAC-CAG-CTG-TCT-3′. A 27-mer oligonucleotide (5′-TCA-AGA-GCC-GAT-ATA-CTG-AAA-GCT-ATC-3′) homologous to an internal sequence of the repeat 3 region is used as hybridization probe.

PCR mixture (50-μL) is composed of 1× PCR buffer (200 mM Tris-HCl, pH 8.7, 100 mM KCl, 100 mM $(NH_4)_2SO_4$, 20 mM $MgSO_4$, 1% Triton × 100, and 1 mg/mL bovine serum albumin) supplied by the manufacturer (Amersham), 200 μM each of dATP, dCTP, dGTP, and dTTP, 1 μM of each primer, 1 U of DNA polymerase, and 2 μL of each blood lysate. Amplification is carried out in a Perkin Elmer thermal cycler for 40 cycles at 94°C for 1 min (denaturation), 65°C for 1 min (annealing), and 72°C for 2 min (extension), preceded by a "hot start" cycle at 96°C for 10 min, 80°C for 5 min, and 94°C for 30 s.

56.2.2.2 Nested PCR

Two primers (5′-GGC ACA AAA CAC TGC AGC AGT CCT-3′ sense and 5′-CAG CTG TCT CAA ATC GAA GAT TCT-3′ antisense) are used for nested PCR. 1 μL of product from the first amplification is amplified in the above conditions for 30 cycles.

56.2.2.3 Analysis of PCR and Nested PCR Products

After amplification, 10 μL of each PCR product is mixed with 1 μL of loading dye (0.25% bromophenol blue, 0.25% xylene cyanol, and 40% *w/v* sucrose in water) and analyzed by electrophoresis on 1.5% agarose gel. The gel is then stained with ethidium bromide or FluoProbes Gel Red (Interchim Montlucon, France) and DNA is visualized and photographed under ultraviolet (UV) light. Standard and nested PCRs yield bands of 396 and 366 bp, respectively (Figures 56.3a, 56.4a, and 56.5).

56.2.2.4 Southern Blotting

DNA fragments on the gels are transferred to Zeta Probe membranes (Biorad Laboratories, Richmond, CA). Prior to prehybridization and hybridization (50°C for 1 h and overnight, respectively), the oligonucleotide probe is labeled with $\alpha^{32}P$ dCTP by using Terminal transferase (Boehringer). Prehybridization and hybridization are performed using a buffer containing 75 mM sodium citrate

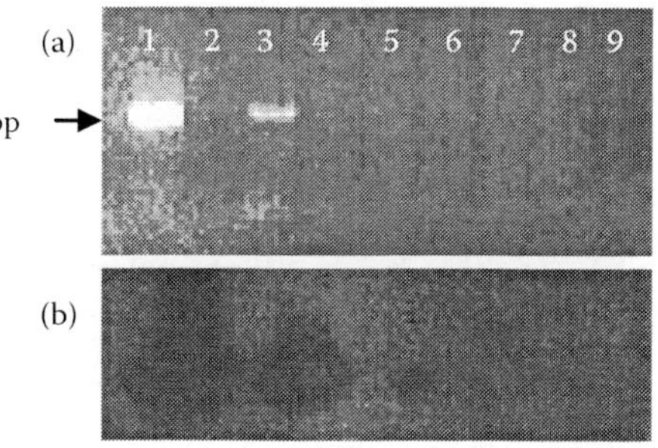

FIGURE 56.3 (a) Representative 1.5% agarose gel electrophoresis patterns generated with 15r3-PCR products from two *L. loa* microfilaremic individuals (lanes 1 and 3) and six AMF individuals (lanes 4–9). Lane 2 shows the negative PCR control (without template). Only the two microfilaremic samples are PCR-positive with ethidium bromide staining. (b) Southern hybridization of the samples in Figure 56.1a. All *L. loa* Mf samples (lanes 1 and 3) are positive. Four AMF individuals (lanes 5–8) are also positive by hybridization. Two AMF individuals are negative both on the gels and by hybridization.

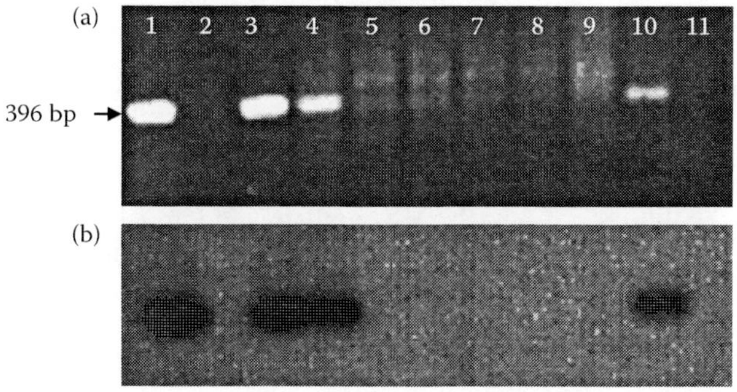

FIGURE 56.4 (a) 15r3-PCR products from: (1) three *L. loa* microfilaremics (Mf) (lane 1, 70 mf/mL blood; lane 3, 50 mf/mL, and lane 4, 30 mf/mL); (2) two 3 pg of *O. volvulus* genomic DNA (lanes 5 and 6); (3) two *W. bancrofti* microfilaremics (lane 7, 4352 mf/mL blood, and lane 8, 1020 mf/mL); (4) one *M. perstans* microfilaremics (lane 9, 20,000 mf/mL); and (5) one *L. loa* and *M. perstans* microfilaremics (lane 10, with, respectively, 25 and 1500 mf/mL). Lanes 2 and 11 show the negative PCR control (without template). (b) Southern hybridization on samples from Figure 56.3a. Only *L. loa* samples give specific band of 396 bp DNA and therefore positive.

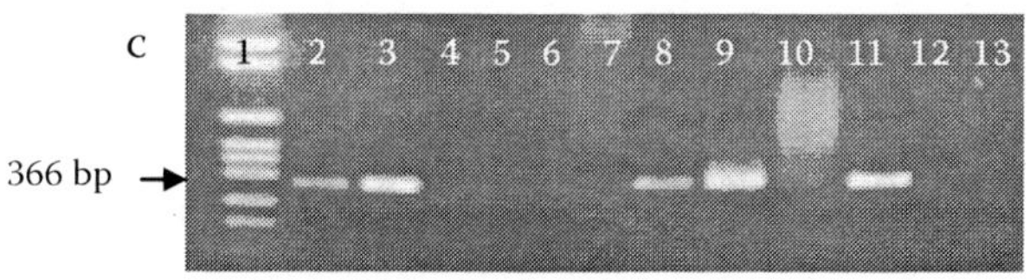

FIGURE 56.5 Typical and representative 1.5% agarose gel electrophoresis pattern of the 15r3 nested PCR products of 10 *L. loa* AMF individuals. Lanes 5 and 13 represent PCR-negative controls (without template); 5 μL of each nested PCR product are applied to each lane, and the result is revealed using UV transillumination after ethidium bromide staining. A fragment of 366 bp DNA is observed with positive samples (lanes 2, 3, 8, 9, and 11). By contrast the samples 4, 6, 7, 10, and 12 are negative. Lane 1 represents DNA molecular weight marker VI (Boehringer).

(SSC), 20 mM NaH_2PO_4, 7% (sodium dodecyl sulfate (SDS), 10× Denhardt's solution, and 100 μg of denatured herring sperm DNA. After hybridization the membranes are washed at 50°C in wash 1 solution (45 mM SSC, 10× Denhardt's, 5% SDS, and 25 mM NaH_2PO_4) for 30 min and in wash 2 solution (15 mM SSC and 1% SDS) for 30 min. Hybridizing bands are revealed by autoradiography on x-ray film (Kodak). The radiographs are developed after at least two days of exposure.

Alternatively, it is also possible to perform Southern hybridization with a nonradioactive method such as ECL Gold Hybridization (Amersham Pharmacia) (Figures 56.3b and 56.4b).

56.3 CONCLUSIONS AND FUTURE PERSPECTIVES

56.3.1 Prevalence Studies

Based on microscopic examination, the estimated prevalence of loaiosis in Gabon is approximately 20% [54]. The prevalence of MF is low, regardless of the intensity of transmission. For example, in villages studied in southeastern Gabon, with different transmission rates, the prevalence was 20%, 21%, 18%, and 19%, respectively, in Okoumbi, Ndjokaye, Moyabi, and Djoutou [53]. Note that these figures do not take into account individuals with occult infection. The 15r3-PCR assay was evaluated on samples from 157 AMF subjects living in these *L. loa* endemic villages located within 50 km of Franceville (southeastern Gabon). The test detected 106 cases of occult infection, for a prevalence of 68% [69]. This shows that the PCR test is a powerful tool for detecting occult infection and therefore for estimating the overall prevalence of loaiosis. We also found a correlation between the intensity of transmission and the prevalence of occult infection. Indeed, the highest prevalence rates were observed in Okoumbi (80%) and Djokaye (79%), where each inhabitant receives, respectively, 100 and 30 *Chrysops* bites per day on average. In Djoutou and Moyabi, where only three bites per day are received, the respective prevalence rates are 52% and 38% [70]. This test must now be evaluated in other parts of Gabon, and elsewhere, to determine its precise performance.

56.3.2 Treatment

The absence of peripheral microfilaria in subjects with occult infection may be due to larval destruction or sequestration. Sequestration is difficult to detect with traditional methods and the patients consequently go untreated. These patients' circulating parasite DNA, detected by PCR, probably comes from the destruction of microfilaria and/or adult worms. If detection of *L. loa* DNA is a marker of active infection, all subjects positive by PCR should be treated. This would not have a major impact on health at the individual level but could reduce the parasite burden in the community, in turn reducing the intensity of transmission and resulting in public health benefits.

In addition, this PCR method can be used to evaluate the effectiveness of filaricides after mass treatment and that of new drugs in animal models. Pinder in 1994 showed that experimental mandrill infection (*Mandrillus sphinx*) by human *L. loa* isolates led to the same parasitological characteristics as the natural human infection [71]. Thus, mandrills with occult infection, absence of microfilarae but presence of adult worms, as shown by 15r3-PCR positivity, can be used to evaluate macrofilaricide drugs [72].

56.3.3 Characterization of Endemic "Resistance"

Human loaiosis differs from other filariasis by the fact that most infected individuals do not have blood microfilariae detectable by standard microscopy. Since the first description of this filariasis, many epidemiologists have found a low prevalence of microfilaria despite local vector abundance. The notion that most patients clear their MF but continue to have (occult) infection rests primarily on the observation of adult worms during eye passage. The assumption that endemic resistant subjects also exist (subjects able to completely eliminate *L. loa* infection) rests on the same observations. Only a sensitive diagnostic test can confirm these assumptions. Our results suggest that 15r3-PCR will be suitable for discriminating among endemic groups (microfilaremics, occult infected individuals (occults) and resistant subjects), as the results should be positive in the first two groups and negative in the last. This needs to be shown in a longitudinal study, however, as *L. loa* infection is characterized by its relative stability in humans and mandrills, the adult worm having a lifespan of about 15 years [1,71].

Studies of resistant individuals may provide interesting immunological information. Marked differences in humoral and cellular immune responses have already been noted between microfilaremic and AMF patients [35,37,39], as well as in the mandrill model [40]. However, lacking a reliable method for diagnosing occult infection, it is not known if this difference is due to immunity directed against adult worms or against microfilaria. The identification of endemic groups ("microfilaremics," "occults," and "resistants") by the 15r3-PCR method should allow immunological studies to be carried out with sera and cells from each endemic group, using antigens of each developmental stage of *L. loa*, and particularly infective larvae and adult worms. Such studies could help us to identify possible cellular or humoral markers involved in resistance to infection, as well as the underlying mechanisms, including host genetic factors. These studies would open the way to investigations of the underlying molecular mechanisms.

56.3.4 Evaluation of Serological Tests

Serological tests using purified recombinant antigens or peptides offer much better specificity than those using crude antigens. When these antigens become available for loaiosis, immunoenzymatic methods such as IgG4 enzyme linked immunosorbent assay (ELISA) will reach acceptable specificity. Comparison of ELISA and PCR results could show whether or not specific IgG4 antibodies are markers of active *L. loa* infection.

56.3.5 Conclusion

PCR can now be used to detect the r3 repeat region of the gene encoding the *L. loa* allergenic polyprotein in blood from subjects with microfilaremic and occult infection. The use of nested PCR has largely simplified this approach, as it obviates the need to transfer amplification products to nitrocellulose membranes and also removes the hybridization steps.

The results obtained in the four villages that we have studied show that the 15r3-PCR assay is a powerful tool for detecting occult loaiosis and that it is well adapted to epidemiological studies.

In conclusion, 15r3-PCR offers many exciting perspectives, including (i) use in prevalence studies; (ii) immunological characterization of endemic resistance; (iii) evaluation of serological diagnostic methods; (iv) evaluation of filaricide treatment efficacy; and (v) *in vivo* evaluation of new macrofilaricides in animal models of occult loaiosis.

REFERENCES

1. Gentillini, M., and Duflo, B. Filarioses. *Médecine Tropicale*, Flammarion, Paris, pp. 173–195, 1982.
2. World Health Organization. *WHO Expert Commitee on Filariasis*. Fourth Report, WHO Technical Series, 702, 1984.
3. World Health Organization. *Lymphatic Filariasis: The Disease and Its Control. Fifth Report of the WHO Expert Commitee on Filariasis*. WHO Technical Report Series, 821, 1992.
4. Muller, R. *Worms and Disease*. Heinemann Medical Books, London, UK, 1975.
5. Mongin. Observations sur un ver trouvé dans la conjonctive à maribou. île Saint Domingue. *J. Med. Chir. Pharm. etc (Paris)* 32, 338–339, 1770.
6. Guyon, J. Sur un nouveau cas de filaire, sous conjonctive, ou Filaria occuli des auteurs, observe au Gabon (Côte occidentale d''frique). *C. R. Séances Acad. Sci.* 59, 743–748, 1964.
7. Brumpt, E. Les filarioses humaines en Afrique. *C. R. Soc. Biol.* 56, 758–760, 1904.
8. Connor, D.H., Neafie, R.C., and Meyers, W.M. *Loiasis: Pathology of Tropical and Extra-Ordinary Diseases*. Armed Forces Institute of Pathology, Washington, pp. 356–359, 1976.
9. John, D.T., Petri, W.A., Markell, E.K., Voge, M. *Markell and Voge's Medical Parasitology*. 9th edition, Elsevier Health Sciences, USA, 2006.
10. Kershaw, W.E. Studies on the epidemiology of filariasis in West Africa, with special reference to the British Cameroons and the Niger delta 1. Methods of survey for infections with *Loa loa* and *Acanthocheilonema pesrtans. Ann. Trop. Med. Parasitol.* 44, 361–378, 1950.
11. Kershaw, W.E. Studies on the epidemiology of filariasis in West Africa, with special reference to the British Cameroons and the Niger delta. The intensity of of infections with *Loa loa* and with and *Acanthocheilonema pesrtans. Ann. Trop. Med. Parasitol.* 24, 110–120, 1954.
12. Treadgold, C.H. On a filarial *Loa papionis* n. sp. parasitic in *Papio cynocephalus. Parasitology* 12, 113, 1920.
13. Sandround, J.H. On the occurrence of a species of *Loa* in monkeys in the *Begian Congo. Ann. Soc. Belge Med. Trop.* 16, 273–278, 1936.
14. Gordon, R.M. The problem of loiasis in West Africa with special reference to recent investigations at Kumba in the British Cameroons and at Sapele in Southern Nigeria. *Trans R. Soc. Trop. Med. Hyg.* 44, 11–47, 1950.
15. Duke, B.O.L., and Wijres, D.J.B. Studies on loiasis in monkeys. I. The relationship between human and simian *Loa* in the rain-forest zone of the British Cameroons. *Ann. Trop. Med. Parasitol.* 52, 158–175, 1958.
16. Duke, B.O.L. The development of *Loa* in flies of genus *Chrysops* and the probable significance of the different species in the transmission of loiasis. *Trans. R. Soc. Trop. Med. Hyg.* 49, 115–121, 1955.
17. Duke, B.O.L. Behavioural aspects of the life cycle of *loa. Zoolog. J. Linnean Soc.* 51, 97–107, 1972.
18. Gouteux, J.P., Noireau, F., and Staak, C. The host preferences of *Chrysops silacea* and *Chrysops dimidiata* (*Diptera*: Tabanidae) in an endemic area of *Loa loa* in the Congo. *Ann. Trop. Med. Parasitol.* 83, 167–172, 1989.
19. Fain, A. Les problèmes actuels de la loase. *Bull. Organ. Mondiale Santé* 56, 155–167, 1978.
20. Rhodain, F. A propos de la distribution géographique de la loase. *Med. Mal. Infect.* 3, 429–436, 1973.
21. Fain, A. Epidémiologie et pathologie de la loase. *Ann. Soc. Belge Med. Trop.* 61, 277–285, 1981.
22. Nutman, T.B., Miller, M., and Ottesen, E.A. *Loa loa* infection in temporary residents of endemic regions; recognition of hyperresponsive syndrome with characteristic clinical manifestations. *J. Infect. Dis.* 154,10–18, 1986.
23. Van-Bogaert, L. et al. Encephalitis in *Loa loa* filariasis. *J. Neurol. Neurosurg. Psychiatry* 18, 103–119, 1955.
24. Cauchie, R. et al. Encéphalite à *Loa loa* traitée par l'association de cortisone et de Carbamazine. *Rev. Belge Pathol. Méd. Exp.* 31, 232–244, 1965.
25. Carme, B. et al. Essai thérapeutique de l'ivermectine au cours de la loase à moyenne et à forte microfilarémie. *Ann. Soc. Belge Med. Trop.* 71, 47–50, 1991.
26. Pillay, V.K.G., Kirch, E., and Kurtzman, N.A., 1973. Glomerulopathy associeted with filarial loiasis. *JAMA*. 225, 179, 1973.
27. Andy, J.J. et al. Loiasis as a possible trigger of African endomiocardial fibrosis: A case report from Nigeria. *Acta Trop.* 38, 179–186, 1981.
28. Toussaint, D., and Danis, P. Retinopathy in generalized *Loa loa* filariasis. A clinicopathological study. *Arch. Ophthalmol.* 74, 470–476, 1965.
29. Schofield, F.D. Two cases of loiasis with peripheral nerve involvement. *Trans. R. Soc. Trop. Med. Hyg.* 49, 588–589, 1955.

30. Hulin, C. et al. Atteinte pulmonaire d'évolution favorable au cours d'une filariose de type *Loa loa*. *Bull. Soc. Path. Exp.* 87, 248–250, 1994.
31. Klion, A.D. et al. Loiasis in endemic and non endemic populations: Immunologically mediated differences in clinical presentation. *J. Infect. Dis.* 163, 1318–1325, 1991b.
32. Nutman, T.B. et al. Immunologic correlates of the hyperresponsive syndrome of loiasis. *J. Infect. Dis.* 157, 544–550, 1988.
33. Boulestiex, G., and Carme, B. Encephalite au cours du traitement de la filariose à loase par diethylcarbamazine. A propos de 6 observations. *Bull. Soc. Pathol. Exot.* 79, 649–654, 1986.
34. Egwang, T.G. et al. The indentification and partial characterization of an immunodominant 29–31 kilodalton surface antigen expressed by adult worms of the human filaria *Loa loa*. *Mol. Biochem. Parasitol.* 31, 263–277, 1988.
35. Pinder, M., Dupont, A., and Egwang, T.G. Identification of a surface antigen on *Loa loa* microfilariae the recognition of which correlates with the amicrofilaremic state in man. *J. Immunol.* 141, 2480–2486, 1988.
36. Akué, J.P., Egwang, T.G., and Devaney, E. High levels of parasite- specific IgG4 in the absence of microfilaremia in *Loa loa* infection. *Trop. Med. Parasitol.* 45, 246–248, 1994.
37. Akué, J.P., Hommel, M., and Devaney, E. High levels of parasite- specific IgG1 correlate with the amicrofilaremic state in *Loa loa* infection. *J. Infect. Dis.* 175, 158–163, 1997.
38. Pinder, M., Leclaire, A., and Everaere, S. Antibody-dependent cell-mediated immune reactions to *Loa loa* microfilariae in amicrofilaremic subjects. *Parasite Immunol.* 14, 541–556, 1992.
39. Baize, S. et al. T helper responsiveness in human *Loa loa* infection: Defective specific proliferation and cytokine production by CD4+ T cells from microfilaremic subjects compared with amicrofilaremics. *Clin. Exp. Immunol.* 108, 272–278, 1997.
40. Leroy, E. et al. Experimental infection of a nonhuman primate with Loa loa induces transient strong immune activation followed by peripheral unresponsiveness of helper T cells. *Infect. Immun.* 65, 1876–1882, 1997.
41. Goussard, B. et al. Age of appearance of IgG, IgGM and IgE antibodies specific for *Loa loa* in Gabonese children. *Microbiol. Immunol.* 28, 787–792, 1984.
42. Van Hoegaerden, M. et al. Filariasis due to *Loa loa* and *Mansonella perstans*: Distribution in the region of Okondja, Haut Ogooué Province, Gabon, with parasitological and serological follow-up over one year. *Trans. R. Soc. Trop. Med. Hyg.* 81, 441–446, 1987.
43. Lal, R.B., and Ottesen, E.A. Enhanced diagnostic specificity in human filariasis by IgG4 antibody assessment. *J. Infect. Dis.* 158,1034, 1983.
44. Lal, R.B. et al. Lack of IgG4 antibody response to carbohydrate antigens in patients with lymphatic filariasis. *Immunology* 74, 333–337, 1991.
45. Ottesen, E.A. et al. Prominence of IgG4 in the IgG antibody response to human filariasis. *J. Immunol.* 134, 2707–2712, 1985.
46. Lucius, R. et al. Specific and sensitive IgG4 immunodiagnosis of onchocerciasis with a recombinant 33Kd *Onchocerca volvulus* protein (OV33). *Trop. Med. Parasitol.* 43, 139–145, 1992.
47. Weil, G.J. et al. IgG4 subclass antibody serology for onchocerciasis. *J. Infect. Dis.* 161, 549–554, 1990.
48. Akué, J.P. et al. Analysis of specific IgG subclasses in a population naturally exposed to *Loa loa*. *Am. J. Trop. Med. Hyg.* 49(suppl), 209, 1993.
49. Garraud, O., and Nutman, T.B. The role of cytokines in human B-cell differentiation into immunoglobulin-secreting cells. *Bull. Inst. Pasteur* 94, 285–309, 1996.
50. Dimock, K.A., Eberhard, M.L., and Lamma, P.J. Th1-Like antifilarial immune responses predominate in antigen-negative persons. *Infect. Immun.* 64, 2962–2967, 1996.
51. Eveland, L.K., Yermakov, V., and Kennedy, M. *Loa loa* infection without microfilariae. *Trans. R. Soc. Trop. Med. Hyg.* 69, 354–355, 1975.
52. Dupont, A., Zué-Ndong, J., and Pinder, M. Common occurence of amicrofilaremic *Loa loa* filariasis within the endemic region. *Trans. R. Soc. Trop. Med. Hyg.* 82, 730, 1988.
53. Touré, F.S. et al. IgG4 serology of loiasis in three endemic villages in south-eastern Gabon. *Trop. Med. Int. Health* 3, 313–317, 1998.
54. Richard-Lenoble, D. et al. Prévalence des filarioses humaines sanguicoles au Gabon. *Bull. Soc. Pathol. Exot.* 73, 192–199, 1980.
55. Languillat, G. et al. Enquête sur l'étiologie de l'hypofécondité au Gabon Oriental: Endémie filarienne (*Loa loa* et *D. perstans*). Fréquences des microfilaires dans les hydrocèles. *Rev. Epidemiol. Santé Publique* 26, 273–283, 1978.
56. Burbelo, P.D. et al. Rapid antibody quantification and generation of whole proteome antibody response profiles using LIPS (luciferase immunoprecipitation systems). *Biochem. Biophys. Res. Commun.* 352, 889–895, 2007.
57. McReynolds, L.A., DeSimone, S.M., and Williams, S.A. Cloning and comparison of repeated DNA sequences from the human filarial parasite *Brugia malayi* and the animal parasite *Brugia pahangi*. *Proc. Natl Acad. Sci. USA* 83, 797–801, 1986.
58. Williams, S.A., DeSimone, S.M., and McReynolds, L.A. Species-specific oligonucleotide probes for the identification of human filarial parasites. *Mol. Biochem. Parasitol.* 28, 163–169, 1988.
59. Williams, S.A. et al. A polymerase chain reaction assay for the detection of *Wuchereria bancrofti* in blood samples from French Polynesia. *Trans. R. Soc. Trop. Med. Hyg.* 90, 384–387, 1996.
60. Erttmann, K.D. et al. A DNA sequence specific for forest form *Onchocerca volvulus*. *Nature* 327, 415–417, 1987.
61. Klion, A.D. et al. Cloning and characterization of a species-specific repetitive DNA sequence from *Loa loa*. *Mol. Biochem. Parasitol.* 45, 297–305, 1991. Erratum in: *Mol. Biochem. Parasitol.* 47, 265.
62. Egwang, T.G., Ajuh, P.M., and Akue, J.P. Cloning and characterization of a *Loa loa*-specific repetitive DNA. *Mol. Biochem. Parasitol.* 56, 189–196, 1992.
63. McReynolds, L.A., Kennedy, M.W., and Selkirk, M.E. The polyprotein allergens of nematodes. *Parasitol. Today* 9, 403–406, 1993.
64. Ajuh, P.M. et al. *Loa loa*: Structural diversity of a 15-kDa repetitive antigen. *Exp. Parasitol.* 81, 145–153, 1995.
65. Touré, F.S. et al. Species-specific sequence in the repeat 3 region of the gene encoding a putative *Loa loa* allergen: A diagnostic tool for occult loiasis. *Am. J. Trop. Med. Hyg.* 56, 57–60, 1997a.

66. Touré, F.S. et al. Detection of *Loa loa*-specific DNA in blood from occult-infected individuals. *Exp. Parasitol.* 86, 163–170, 1997b.
67. Lizotte, M.R. et al. A polymerase chain reaction assay for the detection of *Brugia malayi* in blood. *Am. J. Trop. Med. Hyg.* 51, 314–321, 1994.
68. Zimmermann, P.A. et al. Polymerase chain reaction-based diagnosis of *Onchocerca volvulus* infection: Improved detection of patients with Onchocerciasis. *J. Infect. Dis.* 169, 686–689, 1994.
69. Touré, F.S. et al. Human occult loiasis: Field evaluation of a nested polymerase chain reaction assay for the detection of occult infection. *Trop. Med. Int. Health* 3, 505–511, 1998.
70. Touré, F.S. et al. Relationship between the intensity of *Loa loa* filariasis transmission and prevalence of infections. *Med. Trop.* 59, 249–252, 1998. [In French].
71. Pinder, M., Everaere, S., and Roelants, G.E. *Loa loa*: Immunological responses during experimental infections in mandrills (*Mandrillus sphinx*). *Exp. Parasitol.* 79, 126–136, 1994.
72. Touré, F.S. et al. Use of polymerase chain reaction for accurate follow-up of *Loa loa* experimental infection in *Mandrillus sphinx*. *Am. J. Trop. Med. Hyg.* 61, 956–959, 1999.

57 Mansonella

Ramiro Morales-Hojas

CONTENTS

57.1 INTRODUCTION

The genus *Mansonella* (Nemata: Onchocercidae) contains three species that parasitize humans: *M. ozzardi*, *M. perstans*, and *M. streptocerca*. Infection with these species is generally considered asymptomatic and thus the epidemiology and pathogenesis of mansonelliasis have not been comprehensively studied. The geographic distribution of these *Mansonella* human parasites is sympatric with that of *Onchocerca volvulus*, the causative agent of river blindness, in some regions. Therefore, molecular methods for the detection and differentiation of *Mansonella* spp. and *O. volvulus* have been developed. This chapter will serve as an introduction to this genus and will describe the DNA-based methods developed for the detection of *M. ozzardi* and *M. streptocerca*.

57.1.1 Classification and Morphology

The genus *Mansonella*, named *Tetrapetalonema* before the re-description of *M. ozzardi* by Orihel and Eberhard [1], is taxonomically classified within the family Onchocercidae, order Spirurida, class Secernentea and phylum Nemata. It comprises 29 species characterized by the absence of the cuticular pre-esophageal ring, the presence of four caudal lappets in the adult females, a simple, undivided thread-like esophagus, the regression of the buccal capsule in all members of the genus with one exception, *Sandnema sunci*, and unsheathed microfilariae [2,3].

Mansonella is further subdivided into six subgenera (*Sandnema*, *Tupainema*, *Tetrapetalonema*, *Mansonella*, *Esslingeria*, and *Cutifilaria*) according to different morphological characteristics [4,5]. The subgenus *Sandnema* shows the most primitive morphological characters. It includes two species parasitic of an insectivore and a cercopithecid monkey in Asia. It is characterized by a rounded cephalic extremity not oriented along the lateral or median axis. Males have the tip of the tail not flattened, their caudal papillae are not located around the cloaca and show a spoon-shaped right spicule, which is similar to that of *Acanthocheilonema*. Females have the vulva near the base of the esophagus and the microfilariae have nuclei to the tip of the tail. Furthermore, *S. sunci* is the only species of the genus that presents a tubular buccal capsule. The subgenus *Tupainema* comprises a single species, *Mansonella (T.) dunni*, a parasite of tupaids. It shows a cephalic extremity with a lateral orientation, the body is constricted to form the cephalic extremity in the shape of the hemisphere and narrower than the body behind it, the tip of the tail of males is not flattened and the right spicule is spoon-shaped, and females have the vulva at the posterior to the base of the esophagus. The microfilariae have no nuclei at the tip of the tail. The subgenus *Tetrapetalonema* includes species that have the cephalic extremity either with a dorso-ventral orientation or the body is constricted to form the cephalic extremity in the shape of the hemisphere and narrower than the body behind it. Males have no flattened tip of the tail and the blade of the right spicule can be either flattened and spatulated or filiform. Females

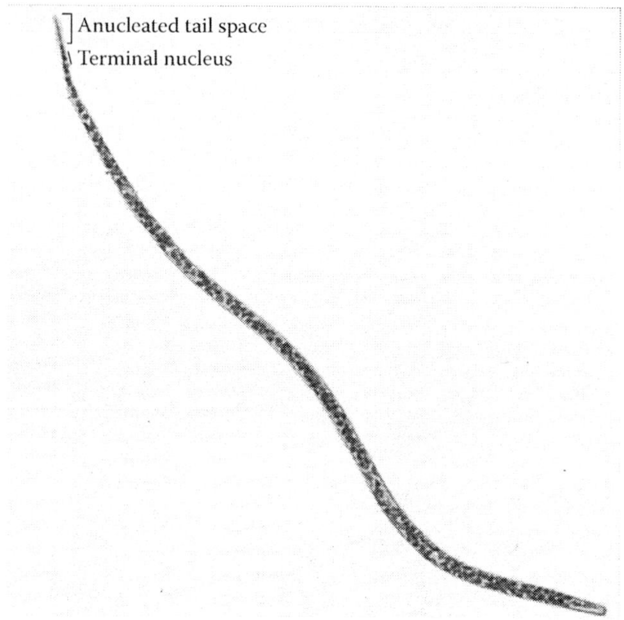

FIGURE 57.1 Microfilaria of *M. ozzardi* stained with orcein. The head of the microfilaria is on the lower right-hand corner.

have the vulva in the mid-esophageal region. The microfilariae have nuclei distributed along the body to the tip of the tail. The subgenus *Mansonella* comprises parasites of carnivores in the Holarctic region and parasites of sciurids and humans (*M. (M.) ozzardi*) in South America. Adults show a cephalic extremity with a dorso-ventral orientation, the tip of the tail in males is flattened dorso-ventrally and the right spicule has a subterminal heel, the females have the vulva in the mid-oesophageal region, and nuclei in the microfilariae do not extend to the tip of the tail (Figure 57.1). The subgenus *Esslingeria* comprises parasites of anthropoids and humans (*M. (E.) perstans* and *M. (E.) streptocerca*) in Africa and caviomorph rodents in South America. Adults show a cephalic extremity with lateral orientation, the tip of the tail of males can be flattened or not and their right spicule is flattened and spatulated or filiform, the vulva is found in the mid-esophageal region and microfilariae show nuclei to the tip of the tail. The *Cutifilaria* have recently been reassigned as a new subgenus of *Mansonella* based on the description of *Mansonella (C.) perforata*, a dermal parasite of *Cervus nippon nippon* in Japan [6]. This new species showed morphological affinities between the *Cutifilaria* and *Mansonella* such as the presence of an area rugosa composed of transverse bands with points, the presence of 14–16 papillae around the cloaca, two rhomboidal subterminal papillae, males with a spoon-shaped right spicule and females with cuticular pores near the vulva, the absence of buccal capsule, a thin, thread-like esophagus, and four lappets in the tip of the tail of females.

57.1.2 Biology and Epidemiology

The host range of the genus *Mansonella* is very broad, comprising parasites of insectivora, Tupaiidae, carnivora, caviomorpha, Sciuridae, ungulates, and primates including humans. Also the genus shows a very broad geographical distribution being present in the Oriental, Palearctic, Afrotropical, and Neotropical regions. The evolutionary history of these parasites is poorly known. Being parasites of insectivores, ruminant ungulates, carnivores and primates, it is suggested to have a remote Asiatic origin, where the most primitive species exist, with a subsequent expansion into Africa and America [3,5]. African anthropoid primates harbor six species of *Mansonella (E.)*, including the human parasites *M. (E.) perstans* and *M. (E.) streptocerca*, and the radiation of these species has been dated to the Oligocene in concurrence with the radiation of their anthropoid hosts [3,5]. The South American form of the human parasite, *M. (M.) ozzardi*, is morphologically similar to species parasites of carnivores and sciurids, and it has been suggested that the human parasite originated from a capture [3,5]. However, comprehensive phylogenetic analyses of this genus, preferentially using DNA markers, should be done in order to corroborate the proposed evolutionary history and taxonomy.

The biology of the filarial worms of the family Onchocercidae is highly specialized. They are tissue dwellers, their life cycle involves two hosts, one intermediate that serves as vector, usually a hematophagous species of Diptera, and the vertebrate definitive host, but probably the most distinctive feature is that adult females lay mobile microfilariae (Figure 57.1). This characteristic is likely an adaptation to the constraint of adults to living within the host's tissues with no direct contact with the exterior, and according to some authors it has probably evolved several times in evolution [3,5]. The adults of the genus *Mansonella* can be found embedded in different tissues of the definitive host (see Anderson [7] and references therein). For example, adults of the human parasite *M. perstans* are found in the body cavities such as the peritoneal or pleural cavities, and less frequently in the pericardium; the adults of *M. ozzardi* commonly live in the subcutaneous tissues of the human host, although they have also been found in the peritoneal cavity; and those of *M. streptocerca*, another parasite of humans, dwell in the dermis and subcutaneous tissues. The female adult worms lay mobile embryos, known as microfilariae, which migrate to more external tissues to facilitate their intake by the blood feeding vectors. Microfilariae of different species can also be found in different tissues. Thus, the microfilariae of *M. ozzardi* and *M. perstans* circulate in the periferal blood while those of *M. streptocerca* are found in the skin (see Anderson [7] and references therein). Once the vector ingests the microfilariae with the blood meal, they penetrate the vector's midgut and migrate within 24 h to the thoracic muscles where they develop into infective L3 larvae between 6 and 7 days postinfection [8–11]. The L3 larvae migrate to the arthropod's proboscis from where they enter a new definitive vertebrate host when the vector takes a subsequent blood meal. The species that serve as vectors of *Mansonella* can vary within the genus and even within species. Thus, *M. streptocerca* and *M. perstans* are transmitted by members of the genus *Culicoides* (Diptera: Ceratopogonidae), *M. streptocerca* by *Culicoides grahami*, while *M. perstans* can be transmitted by both *C. grahami* and *Culicoides austeni*, although there are other minor vectors [10–13]. *M. ozzardi* is transmitted by different species of *Culicoides* in the Caribbean region and some parts of South America (e.g., North Argentina) and by several members of the genus *Simulium* (Diptera: Simuliidae) in the Amazon region and North Argentina [8,9,14–17].

Three species of the genus are parasites of humans: *M. (E.) streptocerca* in Africa, *M. (E.) perstans* in Africa, and recently introduced to South America by human migration, and *M. (M.) ozzardi* in South and Central America. The prevalence of infection with these *Mansonella* spp can vary in different regions, but it can be as high as 100% in some endemic regions of Africa for *M. streptocerca* and *M. perstans*, such as the rain forest of Cameroon or the middle part of the river Zaire [18]. The highest infection rate recorded for *M. ozzardi* was 96% among Indians of the Comisaría del Vaupes in Colombia and 94% in a particular area of the state of Bolívar in Venezuela [19,20]. Prevalence increases with age, especially after the second decade of life, and the number of microfilariae appears to increase little with age. Indeed, in a study of *M. streptocerca* infection in Uganda it was observed that the microfilaria density did not increase with age [21]. Nevertheless, in another study the microfilarial density of *M. ozzardi* increased with age, probably due to an increase in the frequency of exposure [22]. In fact, experimental studies of the patency of *M. ozzardi* infection suggested that the high microfilaremia levels in adults are likely the result of successive infections, probably due to continued exposure [23].

57.1.3 Clinical Features and Pathogenesis

Infections with *Mansonella* spp. are generally considered to have little pathogenic effects in humans or to be entirely asymptomatic [24]. However, several authors have associated different pathologies to infection with *Mansonella* spp. Different skin lesions, such as pruritus, papular eruptions or changes in the skin pigmentation as well as chronic lymphadenitis, have been associated with infection with *M. streptocerca* [21,25,26]. In the case of *M. ozzardi* and *M. perstans* several pathologies such as eosinophilia, arthralgias, headaches, moderate fever, and coldness of the legs have been described in association with infection with these species [20,27]. Furthermore, corneal lesions and visual impairment have also been associated with infection with *M. ozzardi* in the Amazon region [28–30] and it has been suggested in one case of infection with *M. perstans* [31]. Nevertheless, mansonelliasis is considered a minor form of human filariasis because of the lack of a clear pathology; thus, the specific pathogenesis has been little studied and there is a lack of information in the literature.

57.1.4 Diagnosis

57.1.4.1 Conventional Techniques

The most commonly used techniques for the detection of mansonelliasis rely on the microscopic visualization of microfilariae (Figure 57.1). This can be done either in blood films as in the case of *M. ozzardi* and *M. perstans* that have blood-dwelling microfilariae or in skin biopsies in the case of *M. streptocerca* whose microfilariae are skin-dwellers. The microscopy methodologies have two differentiated parts: (i) the collection and preparation of samples and (ii) the identification of the microfilariae to the species level. There are different methods for the preparation of blood samples prior to the microscopic visualization. In some cases unmeasured blood drops are collected for posterior examination. Analysis of unmeasured blood samples can be practical in surveys interested only in knowing the number of positives and negatives in populations. However, for epidemiological studies and control programs it is more useful to record the number of microfilariae in a given volume of blood. Collection of measured blood samples from finger or ear-lobe pricks can be easily done with a pipette, and the amount collected can vary between 20 (20 μL) and 60 mm^3 (60 μL). Blood samples can be examined as one single thick smear or they can be subdivided into smears of similar sizes. The method described by Sasa [32] is commonly used in the case of mansonelliasis. This author suggests the collection of 30 mm^3 (30 μL) and the analysis of three smears of 10 mm^3 each (~50 mm long × 3 mm wide). Alternatively, thick blood smears can be done in a circular or rectangular form with the aid of the corner of another slide [33]. The smears should be left to air-dry at room temperature before examination. Microscopic visualization of the blood films can be done without staining, but staining the blood films prior to examination produces higher microfilariae counts and is necessary for the posterior species identification. Giemsa (1–5%) or Orcein are commonly used stains [22,34–36] although Sasa [32] recommends the use of azure II. Dried films should be dehemoglobinized and fixed prior to staining. Hemoglobin is removed from the blood films by placing the slide under water for about 3–5 min and once it is air-dried the film can be fixed in methanol for 30–60 s. For the azure II staining method, the dehemoglobinized slide has to be incubated for 1 h in a water solution containing 0.02% azure II and 0.01% eosin; the excess of stain is removed by washing the slide first in water and then in 0.1% acetic acid for 30 s; finally the slide is washed in water and left to air-dry vertically. In the case of Giemsa stain, dehemoglobinized, fixed slides are stained for 45 min in a 1:50 dilution of Giemsa and then washed with water for 3–5 min before it is left to dry in a vertical position.

Detection of microfilariae can be improved by using concentration procedures that allow the examination of larger amounts of blood (typically 1 mL). One of the most commonly used methods of concentration is the Knott method. Red cells are lysed by the addition of 10 mL of formalin to the 1 mL blood sample and shaking the tube. The sample is then centrifuged for 2 min at 300× *g* to precipitate the microfilariae (which sediment together with the white blood cells). Alternatively, the tube might be left standing in the vertical position for 12 h until the cells and microfilariae are deposited. The supernatant can be removed and a portion of the sediment is spread into the slide as a thick film and left to dry. This sample can be then stained as described above.

In the case of the species whose microfilariae dwell in the dermis, *M. streptocerca*, skin biopsies have to be examined. Given that there is a higher density of microfilariae in the shoulder than in the buttocks [21,26], it is recommended that samples be taken preferentially from the shoulder [21]. Following the World Health Organization [33] recommendation, two skin snips should be taken from the shoulder and buttocks. Skin snips are taken with the aid of Holth sclerocorneal punches, which take a skin biopsy of ~2.5 mm. These samples are then incubated overnight in saline solution in the wells of microtiter plates. The solution is then pipetted onto a slide for staining and microscopic examination. Staining procedures are the same as described above with the exception that the microfilariae extracted from the skin biopsies do not need to be dehemoglobinized. The percentage of microfilariae that remain inside the skin snips with this method is large (68% in the case of *M. streptocerca*) and the probability of false negatives increases as the microfilaremia decreases [21]. Thus, a more sensitive method involves the incubation of skin snips with 0.5% collagenase in saline solution at 30°C or 37°C for 24–72 h.

Once the samples are prepared the examination of the slides can be done under a microscope with low-magnification objectives (5× or 10×), which allow for wider visual fields. Results should be annotated including the number of microfilariae counted per sample, as this information is important for epidemiological purposes and control programs.

The major problem of conventional techniques is the morphological similarity between microfilariae. Most standard texts refer to the length of the microfilariae and the pattern of nucleation of the tail as characters for the identification of species. However, in the case of the *Mansonella* spp the lengths of the microfilariae are overlapping, which renders this character useless for identification purposes. Thus, *M. ozzardi*, *M. perstans*, and *M. streptocerca* microfilariae have lengths in the ranges of 200–225, 190–200, and 180–260 μm, respectively [21,37]. Furthermore, the length of the microfilariae can vary depending on the method used for sample preparation [33]. Nevertheless, both the morphology of the tail and the presence/absence of nuclei to the tip are reliable characters for differentiating between the three *Mansonella* species that infect humans. In *M. ozzardi* the tail is long and slender and lacks nuclei; *M. perstans*, however, has a blunt-rounded tail with nuclei to its tip; microfilariae of *M. streptocerca* occur in the skin and are readily identified by the slender, hooked tail with nuclei to the tip.

57.1.4.2 Molecular Techniques

Identification of filarial parasites based on morphological characters can be difficult because of the impoverished morphology and high levels of homoplasy that they show. Immunodiagnostic tests provide useful methods for the differential diagnosis of infections with different filarial parasites, although in some cases the specificity of the assays is not very high due to cross-reactivity. Furthermore, there are no specific antibody detection tests for any of the three species of *Mansonella* parasites that infect humans. On the other hand, polymerase chain reaction (PCR)-based techniques provide powerful tools for the detection of parasites, as the appropriate DNA markers are species-specific and allow identification of any parasitic stage. Also, different DNA-based methods have been described in the literature for the identification of two of the species of the genus *Mansonella* that infect humans. Infection with species members of this genus are generally considered to be mostly asymptomatic, with the exception of *M. streptocerca* that can produce some skin lesions and pruritus. Nevertheless, as these species are found in different regions in sympatry with *O. volvulus* (the etiological agent of river blindness), DNA-based methods for the identification of *M. streptocerca* and *M. ozzardi* have been developed with the purpose of differential diagnosis with respect to *O. volvulus*. The primers used in these methods are listed in Table 57.1.

Fischer et al. [38] developed a nested PCR assay to specifically detect *M. streptocerca* in areas where *O. volvulus*

TABLE 57.1
Primers Used for the Identification of *Mansonella* spp

Primer Pair	Sequence (5′–3′)	Gene	Species	Reference
S2	GTTAAGCAACGTTGGGCCTGG	5S rDNA	Filariae universal	[39]
S16	TTGACAGATCGGACGAGATG			
MsU	TTTTTGTTTTCAACAGCATT	5S rDNA spacer region	*M. streptocerca*	[38]
MsL	AGTTCCCATCAATTGTCC			
MSPEx1	ATGGCGCAATCGGTTCCACC	Msp	*M. ozzardi*; *O. volvulus*[a]	[40]
MSPEx2	CTTAAGATTTTTGCGACGAACCAT			
NC2[b]	TTAGTTTCTTTTCCTCCGCT			
MoITS2	CTTATCATCAGGTGATATTAAT	ITS2 rDNA	*M. ozzardi*; *O.volvulus*	[43,44]
OvITS2	TTCATACATATATAAATGTAGC			

Note: Primer pair refers to the set of primers used jointly in the same PCR; the gene column indicates the amplification target of the primers; the species column gives the species that the primers amplify and identify.

[a] These primers were designed based on the *O. volvulus* gene sequence, and they are likely to work with other species as well.

[b] Reverse nematode primer from Reference 44.

and *M. perstans* were also present. The assay consists of a first amplification of the 5S rDNA spacer region using the conserved primers S2 and S16 described by Xie et al. [39], and a second PCR using the species-specific internal primers MsU and MsL with 1 μL of the previous PCR product as template. The optimal annealing temperature for the second, species-specific PCR was 55°C and the product size was of ~200 bp. The authors showed that the PCR-based assay was specific for *M. streptocerca* and that it was more sensitive than the microscopic examination of skin biopsies.

There are two different PCR-based methods described in the literature for the detection and identification of infection with *M. ozzardi* in the Brazilian Amazonia, where it is sympatric with *O. volvulus*. Hojas and Post [40] developed a PCR-based method for the differential identification of *M. ozzardi* and *O. volvulus* in this region using the major sperm protein (MSP) gene (*msp*). The MSP is a fine fiber protein found exclusively in the nematode sperm cells. It is coded by a short gene that contains two conserved exons separated by a variable intron. There are several copies of the *msp* in the genome, which makes it easy to amplify. All these characteristics make this gene a good marker for species identification [41,42]. Using a set of two conserved primers (MSPEx1 and MSPEx2), Hojas and Post [40] showed that the *msp* genes of *M. ozzardi* and *O. volvulus* had different intron lengths (233 bp and 153–156 bp in *M. ozzardi* and *O. volvulus*, respectively), and that this difference in length could be observed in 1% agarose gels. This allows the simple diagnosis of infections with either or both of these parasites using a simple PCR of the *msp* gene.

A second method for the differential diagnosis of infections with *M. ozzardi* and/or *O. volvulus* in the Brazilian region of the Amazonia uses a multiplex PCR assay. The method involves a single PCR reaction containing two species-specific primers located in the rDNA internal transcribed spacer 2 (ITS2) and a conserved primer located in the 28S rDNA gene [43,44]. The amplification products were 295 and 195 bp long for *M. ozzardi* and *O. volvulus*, respectively, and could be distinguished in 1% agarose gels. This method was shown to be specific for the two species in samples from the Brazilian Amazonia and to be more sensitive than the microscopic examination of skin biopsies and blood samples.

Recently, a new nested PCR-based method has been described that attempts to detect and identify three sympatric filarial parasites in the Brazilian Amazonia, namely *M. perstans*, *M. ozzardi*, and *O. volvulus* [45]. The method involves a first PCR of the complete internal transcribed spacer 1 (ITS1) using a universal forward primer in the 18S gene and a reverse primer located in the 5.8S gene. A second, nested PCR uses the amplification product from the first as template to amplify a fragment of the ITS1 with a forward primer on the 18S and a reverse primer in the ITS1. Identification of the species relies on the different lengths of the PCR products. Nevertheless, the reported lengths of the products of the two Mansonella species are very similar (305 bp for *M. ozzardi* and 312 bp for *M. perstans*) and identification in an agarose gel will certainly prove difficult. Alternatively, the 5S rDNA spacer region of *M. perstans* is already sequenced and it should be simple to develop a nested PCR assay specific for this species similar to that described for *M. streptocerca*. This would require designing specific primers within this region and a posterior validation of the specificity ad sensitivity of the method

57.2 METHODS

57.2.1 Sample Preparation

Preparation of samples for the DNA-based detection of infection with different species of *Mansonella* can vary slightly in relation to the tissue from where the DNA is being isolated. Furthermore, there is no standard method for DNA extraction from the same tissue in the literature; therefore, different methods from the one described here can be used. Also, methods described in the chapter for *Onchocerca* are also appropriate for *Mansonella* and vice versa.

57.2.1.1 Preparation of DNA from Skin Biopsies

1. Shred the skin biopsy with a sterile scalpel to facilitate the exit of microfilariae.
2. Incubate the skin snip 1 h at 56°C in 25 μL of nucleolysis buffer (0.05 M NaCl, 0.01 M EDTA, 0.05 M Tris base), 2.5 μL of 10% sodium dodecyl sulfate (SDS), and 0.25 μL of 1 M dithiothreitol (DTT).
3. After the incubation add 100 μL of nucleolysis buffer and 1 μL proteinase K (20 mg/mL); incubate at 56°C for 2 h.
4. Add an equal volume of phenol:chloroform (1:1) and mix thoroughly by inverting the tube for 5 min.
5. Centrifuge at 5000 rpm for 5 min and transfer the upper, aqueous layer to a clean microcentrifuge tube.
6. Add an equal volume of chloroform and mix by inverting the tube during 2 min.
7. Centrifuge at 5000 rpm for 5 min and transfer the upper, aqueous layer to a clean microcentrifuge tube.
8. Add 1/10 of the total volume of 3 M Na acetate (pH 5.6) and 2× the total volume of 95–100% EtOH. Incubate at −20°C for 1 h (or overnight) to precipitate the DNA.
9. Centrifuge the sample at 14,000 rpm for 30 min to pellet the DNA. Remove the supernatant.
10. Add 500 μL of 75% EtOH and centrifuge at 14,000 rpm for 2 min. Remove supernatant.
11. Dry the pellet and resuspend in 50 μL of autoclaved H_2O (DNA can be alternatively resuspended in Tris-EDTA (TE) buffer).

57.2.1.2 Preparation of DNA from Blood Samples

1. Add 500 μL of TE buffer to a sample of 100 μL of whole blood and mix. Centrifuge for 2 min at 10,000× *g* and discard supernatant.
2. Repeat step 1.
3. Resuspend the pellet in 500 μL of red cell lysis buffer (10 mM Tris, pH 7.6, 5 mM $MgCl_2$, 1 M sucrose,

1% Triton X-100). Centrifuge for 2 min at 10,000× *g* and discard supernatant.
4. Repeat step 3.
5. Resuspend the pellet in 200 μL of prewarmed (56°C) proteinase K buffer (20 mM Tris, pH 8, 50 mM KCl, 2.5 mM $MgCl_2$, 100 μg/mL proteinase K, 0.5% Tween 20).
6. Incubate at 56°C for 2 h.
7. Incubate for 10 min at 90°C.

57.2.2 Detection Procedures

Three detection procedures follow in this subsection, those described by Fischer et al. [38], Hojas and Post [40], and Tang et al. [45] for the detection and differentiation of *M. streptocerca*, *M. ozzardi*, and *M. perstans*, respectively. It should be noted that another method is described in Chapter 61 of the book that can differentially detect infections with *M. ozzardi* and/or *O. volvulus*.

57.2.2.1 PCR Detection of *M. streptocerca*

1. Prepare a PCR reaction containing in a total volume of 50 μL: 1× PCR buffer, 1.5 mM $MgCl_2$, 500 nmol of each dNTP, 20 pmol of primer S2 (5′-GTTA AGCAACGTTGGGCCTGG-3′), 20 pmol of primer S16 (5′-TTGACAGATCGGACGAGATG-3′), 2 U of *Taq* polymerase and 2 μL of the extracted DNA.
2. The reactions are then performed in a thermocycler with the following program: 1 cycle of 10 min at 94°C; 25–35 cycles of 1 min at 90°C, 1 min at 55°C and 1 min 72°C; 1 cycle of 5 min at 72°C.
3. Perform a second, nested PCR using the same reaction as in step 1 using the primers MsU (5′-TTTTTGTTTTCAACAGCATT-3′) and MsL (5′-AGTTCCCATCAATTGTCC-3′) instead of S2 and S16 and with 1 μL of the first PCR product instead of 2 μL of the extracted DNA. The cycling program is the same as in step 2.
4. Run PCR products in a 1.5% agarose gel and stain with ethidium bromide to visualize on a UV light transilluminator.

Note: An amplification product of 200 bp should be observed if the sample is positive for the presence of *M. streptocerca*.

57.2.2.2 PCR Detection of *M. ozzardi*

1. Prepare a PCR reaction containing a total volume of 25 μL: 1× PCR buffer, 2 mM $MgCl_2$, 60 μM of each dNTP, 0.2 μM of primer MSPE × 1 (5′-ATGGCGCAATCGGTTCCACC-3′), 0.2 μM of primer MSPE × 2 (5′-CTTAAGATT TTTGCGACGAACCAT-3′), 0.5 U of *Taq* polymerase, and 1 μL of the extracted DNA.
2. The reactions are then run in a thermocycler with the following program: 1 cycle of 3 min at 94°C; 35 cycles of 30 s at 94°C, 45 s at 53°C and 30 s at 72°C; 1 cycle of 5 min at 72°C.
3. Run PCR products in a 1% agarose gel and stain with ethidium bromide to visualize on a UV light transilluminator.

Note: The amplification product for *M. ozzardi* is 555 bp long. The primers MSPEx1 and MSPEx2 are not specific for *M. ozzardi*; therefore, in case of sympatry with other filarial species an amplification product could correspond to other species than *M. ozzardi*. The amplification product of *O. volvulus*, found in sympatry with *M. ozzardi* in the Brazilian Amazonia, shows a different length (480 bp), which allows for the differentiation of the two species by running the amplification products in a single gel [40].

57.2.2.3 PCR Detection of *M. perstans*

1. Prepare a PCR reaction containing a total volume of 50 μL: 75 mM Tris HCl (pH 9.0), 2 mM $MgCl_2$, 50 mM KCl, 20 mM $(NH_4)_2SO_4$, 200 mM dNTPs, primer FIL-1F (5′-GTGCTGT AACCATTACCGAAAGG-3′), primer UNI-1R (5′-CGCAGCTAGCTGCGTTCTTCATCG-3′), 1.25 U of Taq polymerase, and 5 μL of extracted DNA template.
2. Run the reactions in a thermocycler with the following program: 1 cycle of 7 min at 94°C; 40 cycles of 20 s at 94°C, 20 s at 60°C and 30 s at 72°C; 1 cycle of 10 min at 72°C.
3. Prepare a second, nested PCR using the same reaction mixture using instead primers FIL-2F (5′-GGTGAACCTGCGGAAGGATC-3′) and FIL-2R (5′-TGCTTATTAAGTCTACTTAA-3′), and 2 μL of the first PCR product as template.
4. Run the reactions in a thermocycler with the following program: 1 cycle of 7 min at 94°C; 35 cycles of 20 s at 94°C, 20 s at 50°C and 20 s at 72°C; 1 cycle of 10 min at 72°C.
5. Run amplification products in 2.5% agarose gel and stain with ethidium bromide to visualize on a UV light transilluminator.

Note: It should be noted that the concentrations of the primers according to the authors should be established empirically. The expected size of the amplification product of *M. perstans* is 312 bp, and that of *M. ozzardi* is 305 bp. These lengths are very similar and could be difficult to differentiate in a simple agarose gel. The expected amplification product of *O. volvulus* is 344 bp in length, which allows for the differentiation of infection with this parasite from those with *Mansonella* spp.

57.3 CONCLUSIONS AND FUTURE PERSPECTIVES

Reliable detection and identification of human parasites is of great importance for any control program of parasitic disease. The advancement of the molecular methodologies has allowed the development of faster and more reliable techniques for the detection and differentiation of the etiological agents of infection than microscopic examination. Among these the PCR-based methods show a high sensitivity, as they are capable of detecting minute amounts of parasite material. But there are other advantages to DNA-based methods such as the identification of any parasitic stage even when the morphology of all life stages of the parasite is not characterized. Species of the genus *Mansonella* are generally considered to be not very pathogenic and have therefore been less studied than other filarial species. However, the *Mansonella* human parasites are found in sympatry in some geographical regions with *O. volvulus*, the etiological agent of river blindness. This renders the diagnosis of onchocerciasis difficult and thus the development of specific methods for the detection of *O. volvulus* has been of major relevance. It was the sympatric distribution of *Mansonella* spp and *O. volvulus* that generated interest in the detection and identification of *Mansonella* spp [38,40,43]. The DNA-based methods described in the literature were shown to be specific and sensitive. Furthermore, the methods described for *M. ozzardi* were also able to differentially detect infection with both species. Nevertheless, none of the described methods have been extensively validated in epidemiological studies, which continue to use preferentially the microscopic examination and morphological identification of microfilariae as the standard method of detection of mansonelliasis. Microfilariae of *M. ozzardi* and *M. perstans* are blood dwelling while those of *M. streptocerca* and *O. volvulus* are skin dwelling. Nevertheless, *M. ozzardi* microfilariae can be found in the skin [46–48]. Furthermore, *O. volvulus* microfilariae have also be observed in the blood [49–51]. Thus, developing a DNA-based method for the specific identification of *M. perstans* would likely be of interest.

The described PCR-based methods have not been tested for the detection of *Mansonella* spp in the vectors. However, the PCR-based methods described in this chapter are most likely to work in DNA extracted from vector flies if the DNA extraction procedure is adequate (see Chapter 61 for DNA extraction methods from vectors). Detection of filarial species in the vectors is important for some aspects of disease epidemiology, such as the estimation of the annual transmission potential. As some vectors are capable of transmitting different filarial nematodes, it would be desirable that the molecular methods developed for the detection of infection with *Mansonella* spp and described in this chapter were also validated for their use in the vectors.

Apart from the epidemiological-related interest of *Mansonella* spp, molecular phylogenetic studies should be carried out to confirm the proposed evolutionary history of the genus. It is interesting to note that while the African human parasites may have evolved through cospeciation with their hosts, as there are related species present in other anthropoid primates, the South American species parasite of humans is the result of a host switch. Dating the molecular phylogeny of the group would help in understanding when and why these events occurred. Knowledge of the evolutionary history of human parasites is of relevance for disease control [52].

ACKNOWLEDGMENT

Figure 57.1 was kindly provided by Dr. Rory J. Post.

REFERENCES

1. Orihel, T.C., and Eberhard, M.L. *Mansonella ozzardi*: A redescription with comments on its taxonomic relationships. *Am. J. Trop. Med. Hyg.* 31, 1142, 1982.
2. Bain, O. et al. The nematoda Filarioidea: Critical analysis linking molecular and traditional approaches. *Parasite* 15, 342, 2008.
3. Chabaud, A.G., and Bain, O. The evolutionary expansion of the Spirurida. *Int. J. Parasitol.* 24, 1179, 1994.
4. Eberhard, M.L., and Orihel, T.C. The genus *Mansonella* (syn. *Tetrapetalonema*): A new classification. *Ann. Parasitol. Hum. Comp.* 59, 483, 1984.
5. Bain, O. Evolutionary relationships among filarial nematodes. In: Klei, T.R., and Rajan, T.V. (eds), *World Class Parasites. The Filaria*, Vol. 5. Kluwer Academic Publishers, Boston, p. 21, 2002.
6. Uni, S., Bain, O., and Takaoka, H. Affinities between *Cutifilaria* (Nematoda: Filarioidea), parasites of deer, and *Mansonella* as seen in a new onchocercid, *M. (C.) perforata* n. sp., from Japan. *Parasite* 11, 131, 2004.
7. Anderson, R.C. *Nematode Parasites of Vertebrates: Their Development and Transmission*, 2nd edn. CABI Publishing, Cambridge, MA, 2000.
8. Buckley, J.J. A note on the development of *Filaria ozzardi* in *Culicoides furens* Poey. *J. Helminthol.* 11, 257, 1933.
9. Buckley, J.J. On the development in *Culicoides furens* Poey, of *Filaria (=Mansonella) ozzardi* Manson, 1897. *J. Helminthol.* 12, 99, 1934.
10. Sharp, N.A.D. *Filaria perstans*: Its development in *Culicoides austeni*. *Trans. R. Soc. Trop. Med. Hyg.* 21, 371, 1928.
11. Chardome, M., and Peel, E. La répartition des filaires dans la région de Coquilhatville et la transmission de *Dipetalonema streptocerca* par *Culicoides grahamii*. *Ann. Soc. Belge Méd. Trop.* 29, 21, 1949.
12. Duke, B.O. The intake of the microfilariae of *Acanthocheilonema perstans* by *Culicoides grahamii* and *C. inornatipennis*, and their subsequent development. *Ann. Trop. Med. Parasitol.* 50, 32, 1956.
13. Linley, J.R., Hoch, A.L., and Pinheiro, F.P. Biting midges (Diptera: Ceratopogonidae) and human health. *J. Med. Entomol.* 20, 347, 1983.
14. Shelley, A.J., and Coscaron, S. Simuliid blackflies (Diptera: Simuliidae) and ceratopogonid midges (Diptera: Ceratopogonidae) as vectors of *Mansonella ozzardi* (Nematoda: Onchocercidae) in northern Argentina. *Mem. Inst. Oswaldo Cruz* 96, 451, 2001.
15. Shelley, A.J., and Shelley, A. Further evidence for the transmission of *Mansonella ozzardi* by *Simulium amazonicum* in Brazil. *Ann. Trop. Med. Parasitol.* 70, 213, 1976.
16. Tidwell, M.A., and Tidwell, M.A. Development of *Mansonella ozzardi* in *Simulium amazonicum*, *S. argentiscutum*, and *Culicoides insinuatus* from Amazonas, Colombia. *Am. J. Trop. Med. Hyg.* 31, 1137, 1982.
17. Tidwell, M.A., Tidwell, M.A., and Munoz de Hoyos, P. Development of *Mansonella ozzardi* in a black fly species of the *Simulium sanguineum* group from Eastern Vaupes, Colombia. *Am. J. Trop. Med. Hyg.* 29, 1209, 1980.
18. Berghe, L.V.D., and Chardome, M. The geographical distribution of *Acanthocheilonema streptocerca* in the Belgian Congo. *Trans. R. Soc. Trop. Med. Hyg.* 46, 99, 1952.

19. Godoy, G.A. et al. *Mansonella ozzardi* infections in Indians of the Southwestern part of the state of Bolivar, Venezuela. *Am. J. Trop. Med. Hyg.* 29, 373, 1980.
20. Marinkelle, C.J., and German, E. Mansonelliasis in the Comisaria del Vaupes of Colombia. *Trop. Geogr. Med.* 22, 101, 1970.
21. Fischer, P., Bamuhiiga, J., and Buttner, D.W. Occurrence and diagnosis of *Mansonella streptocerca* in Uganda. *Acta Trop.* 63, 43, 1997.
22. Bartoloni, A. et al. *Mansonella ozzardi* infection in Bolivia: Prevalence and clinical associations in the Chaco region. *Am. J. Trop. Med. Hyg.* 61, 830, 1999.
23. Orihel, T.C., Eberhard, M.L., and Lowrie, R.C. Jr. *Mansonella ozzardi*: The course of patency in experimentally-infected patas monkeys. *Trop. Med. Parasitol.* 44, 49, 1993.
24. Sasa, M. *Human filariasis: A Global Survey of Epidemiology and Control*. University Park Press, Baltimore, 1976.
25. Duke, B.O. A case of streptocerciasis in a European. *Ann. Trop. Med. Parasitol.* 51, 364, 1957.
26. Meyers, W.M. et al. Human streptocerciasis. A clinico-pathologic study of 40 Africans (Zairians) including identification of the adult filaria. *Am. J. Trop. Med. Hyg.* 21, 528, 1972.
27. Nutman, T.B., Nash, T.E., and Ottesen, E.A. Ivermectin in the successful treatment of a patient with *Mansonella ozzardi* infection. *J. Infect. Dis.* 156, 662, 1987.
28. Branco, B.C. et al. Achados oculares entre habitantes do municipio de Pauini e possível associação entre lesões corneanas e mansonelose na Amazônia. *Arq. Bras. Oftalmol.* 61, 674, 1998.
29. Cohen, J.M., Ribeiro, J.A., and Martins, M. Ocular manifestations in mansonelliasis. *Arq. Bras. Oftalmol.* 71, 167, 2008.
30. Garrido, C., and Campos, M. First report of presumed parasitic keratitis in Indians from the Brazilian Amazon. *Cornea* 19, 817, 2000.
31. Bregani, E.R. et al. Case report: Intraocular localization of *Mansonella perstans* in a patient from south Chad. *Trans. R. Soc. Trop. Med. Hyg.* 96, 654, 2002.
32. Sasa, M. Microfilaria survey methods and analysis of survey data in filariasis control programmes. *Bull. World Health Organ.* 37, 629, 1967.
33. WHO. *Bench Aids for the Diagnosis of Filarial Infections*. WHO, Geneva, 1997.
34. Medeiros, J.F. et al. Current profile of *Mansonella ozzardi* (Nematoda: Onchocercidae) in communities along the Ituxi river, Labrea municipality, Amazonas, Brazil. *Mem. Inst. Oswaldo Cruz* 103, 409, 2008.
35. Post, R.J. et al. The morphological discrimination of microfilariae of *Onchocerca volvulus* from *Mansonella ozzardi*. *Parasitology* 127, 21, 2003.
36. Wanji, S. et al. Epidemiology of concomitant infections due to *Loa loa*, *Mansonella perstans*, and *Onchocerca volvulus* in rain forest villages of Cameroon. *Med. Microbiol. Immunol.* 192, 15, 2003.
37. Crewe, W. Mansonelliasis. In: Service, M.W. (ed), *The Encyclopedia of Arthropod-Transmitted Infections*. CABI Publishing, Wallingford, p. 327, 2001.
38. Fischer, P. et al. Detection of the filarial parasite *Mansonella streptocerca* in skin biopsies by a nested polymerase chain reaction-based assay. *Am. J. Trop. Med. Hyg.* 58, 816, 1998.
39. Xie, H., Bain, O., and Williams, S.A. Molecular phylogenetic studies on filarial parasites based on 5S ribosomal spacer sequences. *Parasite* 1, 141, 1994.
40. Hojas, R.M., and Post, R.J. Regional genetic variation in the major sperm protein genes of *Onchocerca volvulus* and *Mansonella ozzardi* (Nematoda: Filarioidea). *Int. J. Parasitol.* 30, 1459, 2000.
41. Scott, A.L. et al. Major sperm protein and actin genes in free-living and parasitic nematodes. *Parasitology* 98, 471, 1989.
42. Setterquist, R.A. et al. Diagnostic probes targeting the major sperm protein gene that may be useful in the molecular identification of nematodes. *J. Nematol.* 28, 414, 1996.
43. Morales-Hojas, R. et al. Characterisation of nuclear ribosomal DNA sequences from *Onchocerca volvulus* and *Mansonella ozzardi* (Nematoda: Filarioidea) and development of a PCR-based method for their detection in skin biopsies. *Int. J. Parasitol.* 31, 169, 2001.
44. Newton, L.A. et al. Systematic relationships of some members of the genera *Oesophagostomum* and *Chabertia* (Nematoda: Chabertiidae) based on ribosomal DNA sequence data. *Int. J. Parasitol.* 28, 1781, 1998.
45. Tang, T.H. et al. Nested PCR to detect and distinguish the sympatric filarial species *Onchocerca volvulus, Mansonella ozzardi and Mansonella perstans* in the Amazon region. *Mem. Inst. Oswaldo Cruz* 105, 823, 2010.
46. Moraes, M.A. et al. The concentration of *Mansonella ozzardi* microfilariae in skin capillaries. *Trans. R. Soc. Trop. Med. Hyg.* 77, 463, 1983.
47. Moraes, M.A.P. *Mansonella ozzardi* microfilariae in skin snips. *Trans. R. Soc. Trop. Med. Hyg.* 70, 16, 1976.
48. Ewert, A., Smith, J.H., and Corredor, A. Microfilariae of *Mansonella ozzardi* in human skin biopsies. *Am. J. Trop. Med. Hyg.* 30, 988, 1981.
49. Fuglsang, H., and Anderson, J. Microfilariae of *Onchocerca volvulus* in blood and urine before, during, and after treatment with diethylcarbamazine. *J. Helminthol.* 48, 93, 1974.
50. Anderson, R.I., Fazen, L.E., and Buck, A.A. Onchocerciasis in Guatemala. II. Microfilariae in urine, blood, and sputum after diethylcarbamazine. *Am. J. Trop. Med. Hyg.* 24, 58, 1975.
51. Duke, B.O., Moore, P.J., and Vincelette, J. Factors influencing the passage of *Onchocerca volvulus* microfilariae into the urine. *Tropenmed. Parasitol.* 26, 449, 1975.
52. Morales-Hojas, R. Molecular systematics of filarial parasites, with an emphasis on groups of medical and veterinary importance, and its relevance for epidemiology. *Infect. Genet. Evol.* 9, 748, 2009.

58 Marshallagia

Jesús M. Pérez and Luca Rossi

CONTENTS

58.1 INTRODUCTION

Despite their relatively less variable structure, nematodes constitute one of the most diverse groups within invertebrates. Also impressive are the proportion of described nematode taxa known to occur as parasites (over 30% of extant species) and the ways in which they achieve it [1–4]. In the last few decades, significant progress was made on knowledge about the impact of gastrointestinal strongyles in both domestic [5–7] and wild ruminants [8]. Within strongylate nematodes, trichostrongylids constitute a "dominant" part in terms of diversity and their potential pathogenicity [8,9].

Trichostrongylids (family Trichostrongylidae) are small bursate nematodes that parasitize the digestive tract of all groups of vertebrates except fishes [10]. An exception is the genus *Dictyocaulus*, which is a parasite of the respiratory tract. Ruminants (both domestic and wild) are the mammalian group more frequently infected and harbor the great majority of these nematodes. Their life cycle is usually direct: eggs pass to the exterior with host feces, and the infective phase, L3 larvae, are ingested by new hosts while grazing. Trichostrongylids may produce high morbidity and mortality rates and a zooeconomic impact in domesticated host species.

From a veterinary viewpoint, the most important genera are: *Ostergagia*, *Teladorsagia*, and *Marshallagia*, parasites of the abomasum of domestic and wild ungulates; *Haemonchus*, including large-sized hematophagous species located in the abomassum of these hosts; *Mecistocirrus*, another hematophagous abomasal parasite of cattle and buffaloes from Asia; *Trichostrongylus*, depending on the species involved, can be found in the abomassum or the small intestine of both domestic and wild ungulates; *Cooperia*, a parasite of the small intestine of ruminants; *Hyostrongylus*, a parasite of the stomach of swine hosts; and *Nematodirus*, which is found in the small intestine of domestic and wild ruminants from temperate regions [11].

These parasites, especially in cases of acute infections, may cause not only anemia, weight loss, or death, but also biochemical alterations of pathological concerns, such as changes in the pH of the gastric juice up to values close to 7, thus lacking its bacteriostatic properties. Therefore, these gastrointestinal nematodes may cause production losses in farmed ruminants as they have a significant impact on their growth. They also can produce negative effects on the condition and fecundity of wild hosts [12].

58.1.1 Classification, Morphology, and Biology

Nematodes constitute a highly diverse group of invertebrates and therefore their systematics is complex. The same occurs within trichostrongylids [13–15], particularly nowadays, when molecular methods are increasingly being used. The classification herein used (see below) is that proposed by De Ley and Blaxter [16].

Phylum: Nematoda Potts, 1932
Class: Chromadorea Inglis, 1983
Subclass: Chromadoria Pearse, 1942
Order: Rhabditida Chitwood, 1933
Suborder: Rhabditina Chitwood, 1933

Superfamily: Strongyloidea Baird, 1853
Family: Trichostrongylidae Witenberg, 1925
Subfamily: Ostertagiinae López-Neyra, 1947
Genus: *Marshallagia* Orloff, 1933

Synonyms of *M. marshalli* (reviewed by Hoberg et al. [8]): *Ostertagia marshalli* (Ransom, 1907); *O. bignatiaca* Blanchard, 1909; *O. tricuspis* Marotel, 1912; *O. orientalis* Bhalero, 1932; *Haemonchus* sp. Marshall, 1904.

Synonyms of *M. occidentalis* (reviewed by Hoberg et al. [8]): *Ostertagiella occidentalis* (Ransom, 1907); *Grosspiculagia occidentalis* Jansen, 1958; *G. trifida* (Cuillé, Marotel, and Panisset, 1912); *G. skrjabini* (Kamenski, 1929).

Polymorphic species are relatively common within the trichostrongylid subfamily Ostertagiinae. This polymorphism consists of the cooccurrence of pairs of morphs, one of them more abundant (major morph) and the other scarcer (minor morph). While the females of both forms are identical males may differ in certain features such as the shape of spicules, gubernaculum, and genital cone [17]. *M. marshalli* (major morph) and *M. occidentalis* (minor morph) are examples of these polymorphic species, which parasitize a variety of host species, including man. It is usually found in the abomasum of a wide range of domestic and wild ruminants, mainly in the northern hemisphere (see Table 58.1); hence, it is considered as a "generalist" species as opposed to "specialists," which infect only one or a relatively few host species [18,19]. In fact, within the broad distribution range of *M. marshalli*, the majority of the parasitic population is found in one or a few "primary" host species but the capability is retained by this nematode to "spill over" occasionally into other sympatric hosts [20]. As an example, *M. marshalli* is perceived in the Alps as the dominant abomasal parasite of wild Caprinae (*Rupicapra rupicapra* and *Capra ibex*), but it is also found with much lower intensity and prevalence in sympatric domestic sheep and goats, mouflons (*Ovis aries*), and Cervids (*Capreolus capreolus* and *Cervus elaphus*) [21]. Likewise, other wild mountain Caprinae (*Ovis Canadensis*, *O. dalli*, and *Oreamnus americanus*) are "primary" hosts of *M. marshalli* in North America [22–26] and the musk ox (*Ovibos moschatus*) and caribou (*Rangifer tarandus*) play this role in the Arctic [8,19]. However, domestic sheep and goats are true "primary" hosts of *M. marshalli* in several countries from Morocco eastward to Mongolia, and it seems that no regular input form infected wildlife—albeit a likely event on a local basis—is necessary to sustain the transmission of this nematode in dry habitats [27,28].

Within Ostertagiinae, the genus *Marshallagia* is characterized by the males showing a wide elongate copulatory bursa and a particularly long and thin dorsal ray, bifurcating in its very distal portion, with posterior rays being longer than the ventral ones [45,46]. *M. marshalli* grossly resembles *Ostertagia* and *Teladorsagia* spp., but adults of the former reach greater length (up to 20.0 mm) [11] and width. They can be identified on the basis of genital features of males (namely, copulatory bursa, and spicules) and by the synlophe in both sexes. Males belonging to *M. marshalli* have bifurcated

TABLE 58.1
Geographical Distribution of *Marshallagia marshalli* Reported from Both Domestic and Wild Hosts Other than Man (Selected References Only)

Host Species	Family	Geographic Origin	Reference
Giraffa camelopardalis	Giraffidae	Europe (zoo)	[29]
Camelus dromedarius	Camelidae	Asia, Africa, Australia	[30]
Camelus bactrianus	Camelidae	Asia	[30,31]
Llama spp.	Camelidae	South America, North America	[30,32]
Vicugna spp.	Camelidae	South America, North America	[30,32]
Antilocapra americana	Antilocapridae	NW North America	[25,33]
Rangifer tarandus	Cervidae	North America; Svalbard (Norway)	[8,19]
Odocoileus hemionus	Cervidae	NW North America	[25]
Cervus elaphus	Cervidae	North America, NE Italy	[22,34]
Capreolus capreolus	Cervidae	Europe	[35]
Saiga tatarica	Bovidae	Asia	[36]
Ovibos moschatus	Bovidae	NW North America	[25,37]
Bos taurus	Bovidae	NW North America	[25]
Ovis aries	Bovidae	North America, Africa, Asia, Europe, Australia	[33,30]
Ovis aries[a]	Bovidae	Europe, Antarctica (Kerguelen Archipelago)	[38,39]
Ovis canadensis	Bovidae	North America	[22,23]
Ovis dalli	Bovidae	North America	[24]
Ovis orientalis	Bovidae	Asia	[40]
Oreamnus americanus[b]	Bovidae	North America	[25,26]
Rupicapra rupicapra	Bovidae	Europe	[41]
Rupicapra pyrenaica	Bovidae	Europe	[42]
Capra hircus	Bovidae	Europe	[28]
Capra ibex	Bovidae	Europe	[43]
Capra pyrenaica	Bovidae	Europe	[44]
Capra aegagrus	Bovidae	Asia	[31]

[a] European mouflon.
[b] May represent a distinct species [25].

spicules with a main stem. The width of the stem is similar to that of branches, which show rounded distal tips. The genital cone is characterized by a pair of ventral raylets surrounded by a bilobulate bursal membrane. Spicules of the major morph, *M. marsalli*, are slender and curved distally toward the middle and the genital cone is not prominent; males of the minor morph, *M. occidentalis*, have massive and straight spicules (Figure 58.1) and a prominent genital cone. Females of this polymorphic species bear a lateral synlophe near the cervical papillae and their oesophagus has three or five lateral ridges; at midbody, the cuticle of *M. marshalli* females has 50–56 ridges. Eggs are very large (172×82 μm) (Figure 58.1e) and, although resembling those of *Nematodirus battus*, harbor a more developed morula, thus being easily distinguished from other trichostrongyle eggs [11,14,17,25,45,47].

M. marshalli has a direct cycle (Figure 58.2), consisting of several stages: egg, five larval or juvenile instars (L1–L5), and adult. Remarkably, this genus differs from the remaining Trichostrongylidae in that the first molt (from L1 to L2) occurs within the egg. Thus, the first-instar larvae enjoy a protected environment and the same occurs for the second-instar larvae until hatched. The large egg and hatching of the second- rather than the first-instar larval stage confer an obvious advantage with respect to survival and transmission under unfavorable conditions, especially dryness and/or extreme cold [5]. The infectious stages of *Marshallagia* (free-living "ensheathed" L3s) are themselves resistant to harsh environmental stress. Development of the egg to the infective stage may take 2–3 weeks or more in favorable conditions. After being ingested by a foraging competent host, the L3s reach the forestomachs and are stimulated to exsheath, and then migrate to the gland layer of the gastric wall, where they molt twice before reaching the adult stage [48]. The whole cycle, from larval ingestion to adulthood, may be completed in 3–4 weeks, but periods of arrested larval development may occur, as suggested by the bimodal size distribution of these parasites in the Arctic [19].

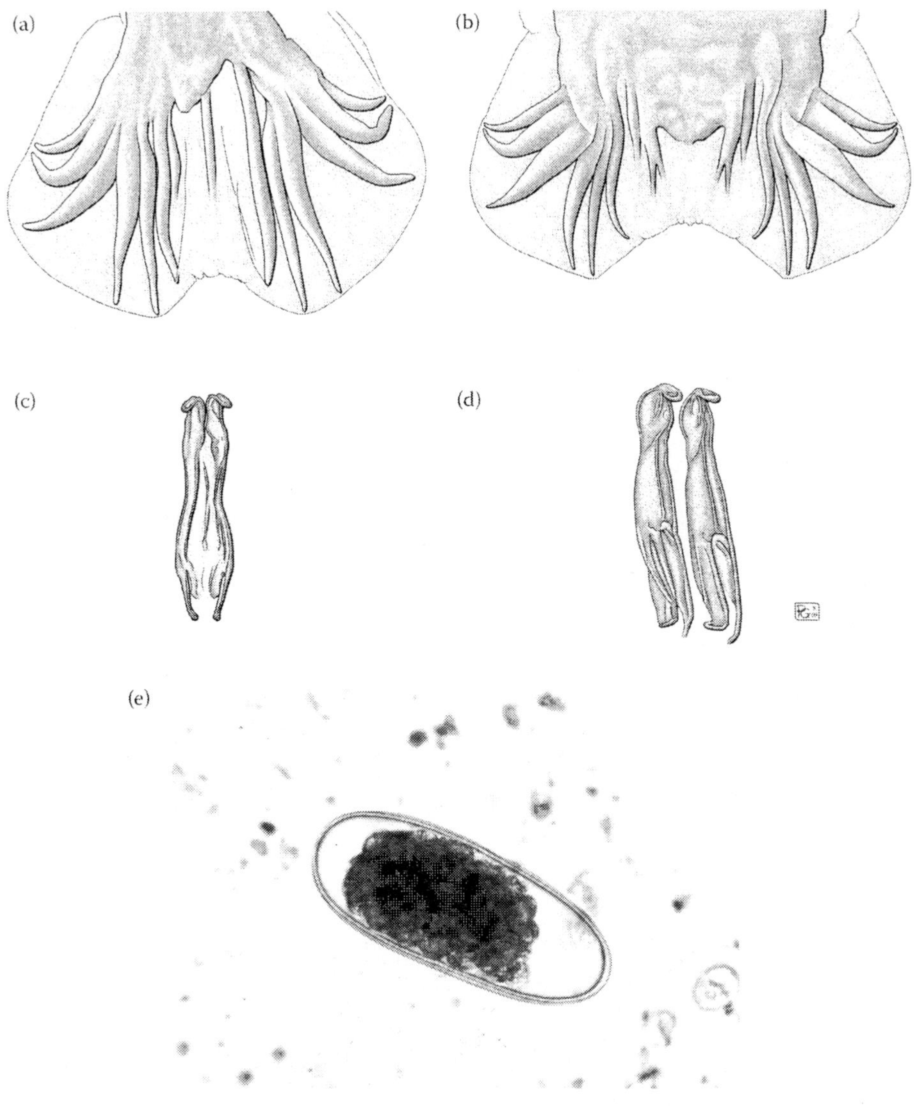

FIGURE 58.1 (a) Copulatory bursa of *Marshallagia marshalli*. (b) Copulatory bursa of *M. occidentalis*. (c) Spicules of *M. marshalli*. (d) Spicules of *M. occidentalis*. (e) *Marshallagia* egg. (Photo by Luca Rossi.)

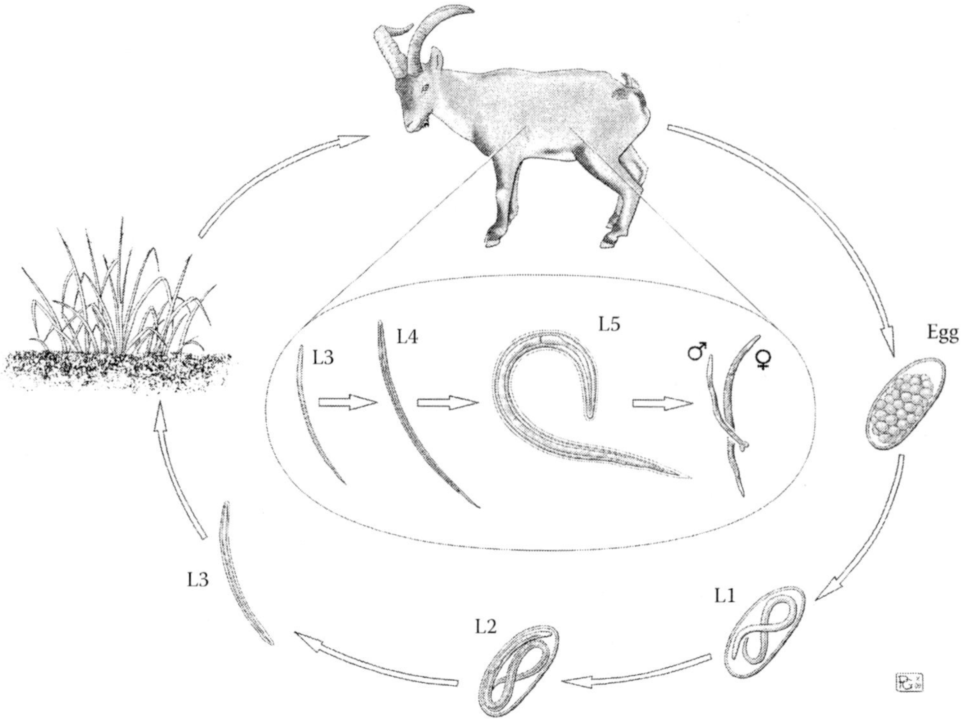

FIGURE 58.2 Life cycle of *Marshallagia marshalli*.

Within their ruminant hosts, many ostertagiinae species show epidemiological patterns characterized by low intensity of parasitation coupling with high prevalence values and strong seasonal dynamics [19,21]. *M. marshalli* is no exception to this, although the peculiar timing of the transmission peaks in seasons other than summer is different from other important and well-studied related taxa, namely *Ostertagia* and *Teladorsagia* spp. [49]. In the Alps [43] and the Arctic [47], efficient transmission of *M. marshalli* in the autumn and winter, likely due to the development and persistence of cold-resistant L3s on pastures [19], mirrored on peak intensities being recorded in winter and early spring. Stien et al. [12] suggested a short life expectancy (<3 months) for adults of this species. Population dynamics of *M. marshalli* does not reveal density-dependence regulation through fecundity [50]. Irvine et al. [50] found a positive relation between fecal egg output and nematode burden, but no relation between the first and the mean number of eggs *in utero*; moreover, host body condition did not influence worm fecundity. In reindeer, abundance of *M. marshalli* increases with host age, suggesting that an efficient immune response is not mounted or may be worth mounting under selected circumstances [47]. In general, it has been suggested that abundance may limit competitive interactions with other nematode species, at the community level, by means of contrasted life histories which involve contrasted epidemiological patterns [47,51]. Testosterone proved to have an immunosuppressive effect on hosts. Decristophoris et al. [52] evidenced a strong positive effect of this hormone on the fecal parasite egg counts in male Alpine ibex, *Capra ibex*. On the other hand, host diet quality, particularly crude protein intake, also plays an important role in developing resistance to gastrointestinal nematode infections [53].

There is a single citation of human infection by *M. marshalli*. Ghadirian and Arfaa [54] carried out an extensive epidemiological survey of intestinal helminthiases affecting people in a rural area of Isfahan (Iran). A 45-year-old patient who had been treated with bephenium hydroxynaphthoate (Alcopar) expelled one *M. marshalli* male, 12 worms belonging to the genus *Ascaris* and 693 *Trichostrongylus* specimens belonging to the species *orientalis* and *colubriformis*. These authors considered the close contact with domestic animals (mainly sheep, goats, and cattle) and the local habit of handling and using the manure of these animals as combustible for heating as the main factors predisposing human infection with these parasites. These factors should explain high prevalence values (up to 69%) and intensity of parasitation by several *Trichostrongylus* species in humans in many parts of Iran. *T. orientalis* and *T. probolurus* have been cited as parasites of man in Egypt [10]. Human infections by *Haemonchus contortus* have also been reported by Ghadirian and Arfaa [54] in Iran and by Faust et al. [55] in patients from Australia and Brazil. Other trichostrongylid species, such as *Ostertagia ostertagi*, have also been found once parasitizing a person from Azerbaijan (Kasimov, *Med. Parasit. Parasit. Dis.*, 12, 81, 1943, cited by Ghadirian and Arfaa [54]).

58.1.2 Clinical Features and Pathogenesis

High burdens of gastrointestinal nematodes cause clinical disease characterized by diarrhea and weight loss [11]. Usually, infected hosts harbor a variety of helminth species and each of them differs significantly in its pathogenicity and susceptibility to antihelminthic drugs. Usually, mixed infections produce a severer impact on hosts than infections caused by a single parasite species [56], and species composition is considered to play an important role in the severity of infection [57].

M. marshalli L3s penetrate the gastric glands of the hosts, eventually resulting in nodule formation. These nodules, with a diameter ranging from 2.0 to 4.0 mm, contain 3–4 developing larvae. Nevertheless, the pathogenicity of this parasitic species in its ungulate hosts is still rather unknown [11]. It might be low, since several studies did not evidence negative effects of this parasite on both domestic and wild hosts [12,58]. Regarding infections in man by this trichostrongylid, since it has been reported only once (by a single adult male worm), pathogenicity is also presumably low, although practically no information about this subject is currently available.

58.1.3 Diagnosis

The increasing use of molecular markers has improved sensitivity and specificity when diagnosing parasitic infections. The "epizootiological probes" derived from molecular biology and systematics have become a powerful tool for monitoring such conditions [8].

58.1.3.1 Conventional Techniques

Direct diagnosis of *Marshallagia* infection in naturally infested hosts is carried out through necropsy, removal and opening of the abomasum, observation of diluted abomasal content, washings of abomasal mucosa, and identification of nematodes found, mainly adult males [59–61]. The use of nonstandardized methods (e.g., analysis of digesta or not, or the way in which the organ is washed and its mucosa treated) may lead to significant differences in the estimation of nematode burden or intensity of parasitation [61].

A less invasive technique consists of collecting feces from the rectum of the host or directly from soil, provided they are freshly deposited. Regarding human samples, we can use a gloved finger or ask patients for collecting fecal specimens into a clean wide-mouthed container. Usually samples weighing 10–20 g are enough for routine examination. These can be refrigerated (3–5°C) for some days as eggs of Ostertagiinae usually retain their characteristic morphology in feces during this period, but the samples should not be frozen because of the risk of shell fragilization and altered flotation in diagnostic solutions, and the impossibility of using them for a successive coproculture, when indicated. For prolonged storage, glass containers are recommended. If examination must be delayed, a preferable alternative is to preserve feces with a fixative, such as Schaudinn's fixative, formalin solution, PVA (polyvinyl alcohol), or MIF (merthiolate–iodine–formalin), among others [62].

Different concentration procedures, such as sedimentation (by centrifugation or by the simple gravity), flotation, or a combination of both, may be used to recover even a small number of eggs from fresh and fixed samples [62–64]. Several techniques have been described, and commonly used among them are the following: the formalin–ethyl acetate or formalin–ether sedimentation (which may be modified for small amounts of feces); the acid–ether sedimentation; the simple gravity sedimentation in tap water; the zinc sulfate flotation or Sheather's sugar flotation, among others. A modified formalin–ethyl acetate or formalin–ether sedimentation technique may be used for material fixed in formalin, PVA, or other fixatives. An adapted zinc sulfate flotation technique can be employed for concentrating helminth eggs from formalin-preserved fecal samples.

Once Ostertagiine eggs are recovered, their identification to the genus level can be attempted based on their size, shape, and content. However, eggs of the genus *Marshallagia* are the only ones that are readily and reliably identified based on the above features. For the remaining taxa, the use of coproculture techniques is recommended [65]. For this purpose, nematode eggs (without fecal debris) must be placed in a solution of 0.5% formalin or 0.1 N sulfuric acid (1–2 cm deep) into a cotton-tapered Erlenmeyer flask of a volume ranging from 250 to 1000 mL, and then incubated at 22°C for 3 weeks (when infective larvae have been developed). It is recommended to keep the culture in the dark and check that the solution level remains constant by adding solution as needed [62].

58.1.3.2 Molecular Techniques

Molecular methods allow us to improve diagnosis and species identification and have other important applications, such as the study of gene flow between parasite populations or parasite resistance against treatments, among others [66–71].

To illustrate this, we will give an example. Benzimidazoles, as broad-spectrum antihelminthic drugs, are frequently used for the treatment of gastrointestinal nematodosis in small domestic ruminants. But, because of their continued use, trichostrongylids have developed resistance to these molecules. This kind of resistance has been profusely studied, even at the molecular level. Álvarez-Sánchez et al. [71] developed a real-time PCR protocol for detecting resistance of several trichostrongylid species to these drugs, based on detecting mutations of position 200 on isotype 1 of β-tubulin. Such mutations produce a replacement of Phe (TTC) by Tyr (TAC). The resistance level of a parasite population is determined by the proportion of β-tubulin codon 200 TAC. In resistant strains or populations, the frequency of this allele reaches values between 71% and 86%, while parasites susceptible to benzimidazoles show values ranging from 24% to 32%. This procedure allows estimating the proportion of each allele by kinetically combining real-time quantitative PCR with specific amplifications of each allele and, thus, no post-PCR protocol is required. Therefore, this protocol proved to be faster and cheaper than conventional *in vivo* and *in vitro* tests.

On the other hand, molecular markers have been successfully used to assess diversity of parasitic nematode species by applying PCR-based assays to eggs and cultured larvae from host feces [72] or adult worms. Apart from their adequate specificity and sensitivity, these methods have the advantage that they are noninvasive. As an example, PCR analyses using mixtures of single species-purified nematode eggs reached a sensitivity of less than 0.5 egg-DNA equivalent per species [64]. Real-time PCR from ITS-2 allows the identification of 0.01 single larval genomic DNA equivalent [73].

Regarding taxonomy of trichostrongylids, sequences of particular markers, such as internal transcribed spacers (ITS-1 and ITS-2), have demonstrated that different morphs of polymorphic species belong, in fact, to the same taxon [74–76]. ITS-1 of different *Trichostrongylus* species, with length ranging from 387 to 390 pb, shows little or no intraspecific variation even when isolates from different geographic areas are compared. On the contrary, interspecific variation in sequence of this marker reached 1.3–5.7% [77]. Hoste et al. [78] compared sequences of ITS-2 (237–238 pb) from five *Trichostrongylus* species and found interspecific differences between 1.3% and 7.6% positions. These authors also found variations at five positions (2.1%) in individual sequences of *Trichostrongylus vitrinus* and at one position (0.4%) within *T. axei* isolates.

Moreover, a multiplex PCR assay has been developed to simultaneously diagnose mixed infections by trichostrongylid nematodes of cattle [64].

58.2 METHODS

58.2.1 Sample Preparation

58.2.1.1 Preparation of Genomic DNA from Single Nematodes

Worms are collected after necropsy as described before. They can be stored in 70% ethanol at 4°C. For DNA isolation, individual nematodes are picked into single tubes and incubated in a lysis buffer containing 100 μL of 10 mM Tris-HCl, 0.1 mM EDTA, pH 8.0, and 20 μg/mL proteinase K for 18 h at 55°C, followed by heating at 95°C for 10 min [75,76,79]. A modified lysis buffer can also be used as described by Grillo et al. [55]: 50 mM KCl, 10 mM Tris, pH 8.3, 2.5 mM $MgCl_2$, 0.45% Nonidet P-40, 0.45% Tween 20, 0.01% (w/v) gelatine, and proteinase K at a concentration of 200 μg/mL. Lysis products are then maintained at −80°C for at least 10 min before being incubated at 60°C for 98 min and posteriorly at 94°C for 20 min. DNA samples can be stored at −20°C.

DNA isolation from single adult nematodes can be achieved by proteinase K (2 mg/mL), sodium dodecyl sulfate (SDS) (2%) digestion at 65°C for 30 min, followed by organic extraction with phenol, phenol/chloroform, and chloroform, and posterior precipitation in 0.3 M sodium acetate and 2.2 vol. 95% ethanol, according to Zarlenga et al. [49].

58.2.1.2 Preparation of Genomic DNA from Fecal Samples

Genomic DNA from *Ostertagia ostertagi* eggs may be extracted by heat-treating frozen eggs, dilution to a final volume of 500 μL with DNA extraction buffer, and posterior sonication for 20 s [67]. Regarding the same nematode species, Harmon et al. [65] were able to disrupt 100% of eggs by combining a bead beater system (Mini BeadBeater-8 (BioSpec, Bartlesville, OK)) with 0.9 g zirconia/silica beads (Tetracore, Gaithersburg, MD) for 3 min or 0.25 g ceramic beads (Qbiogene Lysing Matrix A) for just 1 min. They obtained amplifiable DNA from purified samples containing one egg by using the DNeasy Plant Mini kit (Qiagen).

Genomic DNA from disrupted trychostrongylid eggs and cultured larvae obtained from fecal samples may also be extracted by using a standard phenol–chloroform extraction procedure [72,80]. For this purpose, an extraction buffer is prepared with 288 mL ddH_2O, 3 mL 1 M Tris (pH 8.0), and 6 mL 0.5 M NaCl. This mixture is autoclaved and then 2 g of solid SDS is added. Ten microliters of proteinase K (20 mg/mL), 10 μL DDT (1 M in 0.01 M sodium acetate, pH 5.2), and 2 μL RNAase A (10 mg/mL) are finally added to the extraction buffer. Samples (an amount of 10–30 μL) are digested by incubating them into this mixture at 50–60°C for at least 3 h or overnight. After digestion, an equal volume of PCI (phenol/chloroform/isoamyl alcohol) is added to each DNA solution for purification in a 1.5 mL microcentrifuge tube. After mixing for 5 min in a rocking platform, the mixture is microcentrifuged at 10,000 rpm for 10 min at room temperature. The aqueous phase (at the top), containing the DNA, is then carefully removed using a pipette and transferred to a new tube. These purification steps must be repeated once or twice (if after first repetition a white precipitate is present in the aqueous–organic interface). Then, an equal volume of CI (chloroform/isoamyl alcohol 24:1) is added and gently mixed for 2 min and centrifuged at 10,000 rpm for 1 min. After this step, the top (aqueous) phase, containing the DNA, is carefully removed with a 1000 μL pipette and transferred to a new tube. Posteriorly, 2–2.5 vol. of ice-cold 100% ethanol is added to this solution and the product is gently mixed and maintained at −20°C overnight or at −70°C during 1 h. Then, the mixture is centrifuged for 20 min at maximum speed and the supernatant is removed and the pellet is dried at room temperature. Finally, it is dissolved in 20–25 μL Tris-EDTA buffer, pH 8.0.

58.2.2 Detection Procedures

58.2.2.1 Amplification

ITS-1 and ITS-2 from individual nematodes can be amplified as described by Dallas et al. [79]. They amplified ITS-1 of individual worms using the following primers: NC5 (5′-GTAGGTGAACCTGGGGAAGGATCATT-3′) and NC1R (5′-AACAACCCTGAACCAGACGT-3′). Specific PCR primers for detecting a 110-bp ITS-2 fragment of *M. marshalli* were also designed by these authors: Mm-ITS2-F (5′-TCCTGAATGATATGAATGTATTACC-3′), Mm-ITS-R (5′-CAATACAAATGATATATTGAACATACAG-3′).

Reagents for PCR probes (50 μL) include 1× NH_4 buffer, 0.5 U *Taq* DNA polymerase (BioLine), 2.5 mM $MgCl_2$, 200 μM dATP, dCTP, dGTP, and dTTP, 0.5 μM of each primer, and 10 μL of DNA solution. General PCR conditions included: 90°C for 2 min 15 s; 35 or 40 cycles of 90°C for 15 s, 55°C for 15 s, and 72°C for 1 min.

TABLE 58.2
Molecular Markers for *M. marshalli/M. occidentalis*

Marker	Product Length (bp)	GenBank™ Accession Number	Author/s and/or Reference
Marshallagia marshalli			
ITS-1	384	AY013242	[76]
ITS-2	274	AJ577469	Leignel and Cabaret[a]
26S rRNA	50	AJ400715	[79]
5.8S rRNA	133	AY013242	[76]
Marshallagia occidentalis			
ITS-1	384	AY013243	[76]
ITS-2	235	AY013244	[76]
26S rRNA	50	AY013244	[76]
5.8S rRNA	133	AY013243	[76]

[a] Unpublished data.

58.2.2.2 Electrophoresis and Sequencing

PCR products are separated by agarose gel electrophoresis, purified using Quiaquick spin columns (Qiagen), and finally sequenced using an automated DNA sequencer [79].

Information about molecular markers characterized for *M. marshalli/M. occidentalis* is summarized in Table 58.2.

58.3 CONCLUSION AND FUTURE PERSPECTIVES

Trichostrongylids are parasitic nematodes with a low host specificity, which, under particular circumstances, may cause zoonotic infections in humans. Educational programs oriented to prevent them by improving hygiene habits could contribute to solving this problem.

Despite the above-mentioned advance in knowledge about these parasites and the diseases they produce, a significant number of problems remain unsolved, including the standardization of sampling and preservation methods and hosts' necropsy protocols as well, or consistency in nematode taxonomy, among others [81]. The identification of parasitic nematode eggs, free-living infesting and parasitic larvae, and adult females to the level of genus and species is of concern [6], and difficult and time-consuming when conventional (morphobiometrical) features are used. Molecular markers, therefore, are a versatile and powerful tool for achieving these diagnostic goals.

Future research perspectives would focus on characterizing the genetic structure of these parasitic populations and communities in order to better understand host–parasite relationships. This is basic for improving the understanding of the biology and ecology of these nematodes, as well as for developing effective control of the diseases they cause. For this goal, characterization and development of "new" markers for this parasitic nematode group, such as mitochondrial markers and microsatellites, would be useful [70].

ACKNOWLEDGMENTS

The authors would like to thank Dr. Antonio Sánchez from the Área of Genetics of the Department of Experimental Biology, Jaén University for his constructive comments that helped in improving an original draft of the manuscript. They are also indebted to Dr. Pablo Guerrero, Andalusian Nematologists Group, for original artwork.

REFERENCES

1. Anderson, R.C. The origins of zooparasitic nematodes. *Can. J. Zool.* 62, 317, 1984.
2. Anderson, R.C. *Nematode Parasites of Vertebrates. Their Development and Transmission.* CAB International, Wallingford, 1992.
3. Malakhov, V.V. *Nematodes. Structure, Development, Classification, and Phylogeny.* Smithsonian Institution Press, Washington, 1994.
4. Chen, Z.X., Chen, S.Y., and Dickson, D.W. (eds), *Nematology. Advances and Perspectives. Volume I. Nematode Morphology, Physiology and Ecology.* CABI Publishing, Cambridge, MA, 2004.
5. Levine, N. *Nematode Parasites of Domestic Animals and Man.* 2nd ed. Burguess Publishing Co., Minneapolis, 1980.
6. Lichtenfels, J.R., Hoberg, E.P., and Zarlenga, D.S., Systematics of gastrointestinal nematodes of domestic ruminants: Advances 1992–1995, and proposals for future research. *Vet. Parasitol.* 72, 225, 1997.
7. Hoberg, E.P., and Lichtenfels, J.R. Phylogenetic systematic analysis of the Trichostrongylidae (Nematoda), with an initial assessment of coevolution and biogeography. *J. Parasitol.* 80, 976, 1994.
8. Hoberg, E.P., Kocan, A.A., and Rickard, L.G. Gastrointestinal strongyles in wild ruminants. In: Samuel, W.M., Pybus, M.J., and Kocan, A.A. (eds), *Parasitic Diseases of Wild Mammals.* 2nd ed., Chapter 8. Manson Publishing/The Veterinary Press, London, 2001.
9. Durette-Desset, M.C. Trichostrongyloid nematodes and their vertebrate hosts: Reconstruction of the phylogeny of a parasitic group. *Adv. Parasitol.* 24, 239, 1985.
10. Cheng, T.C. *General Parasitology.* 2nd ed. Academic Press, Orlando, 1986.
11. Urquhart, G.M. et al. *Veterinary Parasitology.* 2nd ed. Blackwell Science, Oxford, 1996.
12. Stien, A. et al. The impact of gastrointestinal nematodes on wild reindeer: Experimental and cross-sectional studies. *J. Anim. Ecol.* 71, 937, 2002.
13. Dróżdz, J. Studies on helminth and helminthiasis in Cervidae I. Revision of the subfamily Ostertagiinae Sarwar, and an attempt to explain the phylogenesis of its representatives, *Acta Parasitol. Pol.* 13, 445, 1965.
14. Lichtenfels, J.R., and Hoberg, E.P. The systematics of nematodes that cause ostertagiasis in domestic and wild ruminants in North America: An update and a key to species. *Vet. Parasitol.* 46, 33, 1993.
15. Durette-Desset, M.C. et al. A cladistic analysis of the Trichostrongyloidea (Nematoda). *Int. J. Parasitol.* 29, 1065, 1999.

16. De Ley, P., and Blaxter, M. Systematic position and phylogeny. In: Lee, D.L. (ed.), *The Biology of Nematodes*. Chapter 1. Taylor & Francis, London and New York, 2002.
17. Dróżdż, J. Polymorphism in the Ostertagiinae Lopez-Neyra, 1947 and comments on the systematics of these nematodes. *Syst. Parasitol.* 32, 91, 1995.
18. Suarez, V.H., and Cabaret, J. Similarities between species of the Ostertagiinae (Nematoda: Trichostrongyloidea) in relation to host-specificity and climatic environment. *Syst. Parasitol.* 20, 179, 1991.
19. Halvorsen, O., and Bye, K. Parasites, biodiversity, and population dynamics in an ecosystem in the high arctic, *Vet. Parasitol.* 84, 205, 1999.
20. Poulin, R. *Evolutionary Ecology of Parasites: From Individuals to Communities.* Chapman and Hall, London, 1998.
21. Zaffaroni, E. et al. Host specificity of abomasal nematodes in free ranging alpine ruminants. *Vet. Parasitol.* 90, 221, 2000.
22. Bergstrom, R.C. Prevalence of *Marshallagia marshalli* Orloff, 1933 in wild ruminants in Wyoming. *Proc. Oklahoma Acad. Sci.* 55, 101, 1975.
23. Kistner, T.P. et al. Helminth parasites of bighorn sheep in Oregon, *J. Wildl. Dis.* 13, 125, 1977.
24. Nielsen, C.A., and Neiland, K.A. *Sheep Disease Report*, vol. 14, Alaska Department of Fish and Game, Juneau, 1974.
25. Lichtenfels, J.R., and Pilitt, P.A. Cuticular ridge patterns of *Marshallagia marshalli* and *Ostertagia occidentalis* (Nematoda: Trichostrongyloidea) parasitic in ruminants of North America. *Proc. Helminthol. Soc. Wash.* 56, 173, 1989.
26. Kerr, G.R., and Holmes, J.C. Parasites of mountain goats in west-central Alberta. *J. Wildl. Manag.* 30, 786, 1966.
27. Jacquiet, P. et al. Dry areas: An example of seasonal evolution of helminth infection on sheep and goats in southern Mauritania. *Vet. Parasitol.* 56, 137, 1995.
28. Valcárcel, F., and García-Romero, C. Prevalence and seasonal pattern of caprinae trichostrongyles in a dry area of central Spain. *J. Vet. Med. B* 46, 673, 1999.
29. Garijo, M.M., Ortiz, J.M., and De Ibáñez, M.R.R. Helminths in a giraffe (*Giraffa camelopardalis giraffa*) from a zoo in Spain. *Onderst. J. Vet. Res.* 71, 153, 2004.
30. Taylor, M.A. et al. *Veterinary Parasitology.* 3rd ed. Blackwell Publishing, Oxford, 2007.
31. Skrjabin, K.I., Shikhobalova, N.P., and Shultz, R.S. *Essentials of Nematology. Vol. IV. Trichostrongylids of Animals and Man.* Academy of Sciences of the URSS, Moscow, 1954. [Translated by the Israel Program for Scientific Translations, 1960].
32. Rickard-Ballweber, L. Ecto- and endoparasites of new world Camelids. *Vet. Clin. North Am. Food Anim. Prac.* 25, 2959, 2009.
33. Bergstrom, R.C. Incidence of *Marshallagia marshalli* Orloff, 1933 in Wyoming sheep, *Ovis aries*, and pronghorn antelope. *Antilocapra Americana*, *Proc. Helminthol. Soc. Wash.* 42, 61, 1975.
34. Manfredi, M.T. et al. Abomasal nematodes of the red deer *Cervus elaphus* in north-eastern Italy. *J. Helminthol.* 81, 247, 2007.
35. Rossi, L., Eckel, B., and Ferroglio, E. A survey of the gastrointestinal nematodes of roe deer (*Capreolus capreolus*) in a mountain habitat. *Parasitologia* 39, 303, 1997.
36. Morgan, E.R. et al. Helminths of saiga antelope in Kazakhstan: Implications for conservation and livestock production. *J. Wildl. Dis.* 41, 149, 2005.
37. Webster, W.A., and Rowell, J. Some helminths parasites from the small intestine of free-ranging muskoxen *Ovibos moschatus* (Zimmermann) of Devon and Ellsemere Islands, Northwest Territories. *Can. J. Zool.* 58, 304, 1980.
38. Pisanu, B. et al. Inventaire préliminaire des helminthes gastrointestinaux du mouflon de Corse (*Ovis amon musimon*) de Lile Haute, Archipel de Kerguelen. *Gibier Faun. Sauv.* 11, 327, 1994.
39. Rossi, L. et al. La simpatria muflone/camoscio modifica sensibilmente le rispettive comunità elmintiche? *Suppl. Ric. Biol. Selvaggina* 24, 79, 1996.
40. Eslami, A. et al. Gastrointestinal nematodes of wild sheep (*Ovis orientalis*) from Iran. *J. Wildl. Dis.* 15, 263, 1979.
41. Rossi, L. et al. Elmintofauna del camoscio (*Rupicapra rupicapra*) nel Parco Naturale Argentera. *Ann. Fac. Med. Vet. Torino.* 33, 1 (1988–1989).
42. Cancrini, G. et al. Occurrence of pulmonary and gastrointestinal nematodes in the Abruzzo chamois. In: Lovari, S. (ed.), *The Biology and Management of Mountain Ungulates.* Chapter 29. Croom Helm, London, 1985.
43. Lanfranchi, P. et al. A three-year evaluation of abomasal helminths of Piz Albris ibex colony. *Z. Jagdwiss.* 41, 24 (1995) [in German].
44. Pérez, J.M. et al. A survey of the gastrointestinal nematodes of Spanish ibex (*Capra pyrenaica*) in a high mountain habitat. *J. Parasitol.* 89, 315, 2003.
45. Gibbons, L.M., and Khalil, L.F. A key for the identification of genera of the nematode family Trichostrongylidae Leiper, 1912. *J. Helminthol.* 56, 185, 1982.
46. Durette-Desset, M.C. Keys to genera of the superfamily Trichostrongyloidea. In:. Anderson, R.C., and Chabaud, A.G. (eds), *CIH Keys to the Nematode Parasites of Vertebrates.* CAB, Farnham Royal, 1983.
47. Irvine, R.J. et al. Life-history strategies and population dynamics of abomasal nematodes in Svalbard reindeer (*Rangifer tarandus platyrhynchus*). *Parasitology* 120, 297, 2000.
48. Lee, D.L. Life cycles. In: Lee, D.L. (ed.), *The Biology of Nematodes.* Chapter 3. Taylor & Francis, London, 2002.
49. Smith, G., and Grenfell, B.T. The population biology of *Ostertagia ostertagi. Parasitol. Today* 1, 76, 1985.
50. Irvine, R.J. et al. Contrasting regulation of fecundity in two abomasal nematodes of Svalbard reindeer (*Rangifer tarandus platyrhynchus*). *Parasitology* 122, 673, 2001.
51. Côté, S.D. et al. Resistance to abomasal nematodes and individual genetic variability in reindeer. *Mol. Ecol.* 14, 4159, 2005.
52. Decristophoris, P.M.A., von Hardenberg, A., and McElligott, A.G., Testosterone is positively related to the output of nematode eggs in male Alpine ibex (*Capra ibex*) faeces. *Evol. Ecol. Res.* 9, 1, 2007.
53. Ezenwa, V.O. Interactions among host diet, nutritional status and gastrointestinal parasite infection in wild bovids. *Int. J. Parasitol.* 34, 535, 2004.
54. Ghadirian, E., and Arfaa, F. First report of human infection with *Haemonchus contortus*, *Ostertagia ostertagi*, and *Marshallagia marshalli* (Family Trichostrongylidae) in Iran. *J. Parasitol.* 59, 1144, 1973.
55. Faust, E.C., Russell, P.F., and Jung, R.C. *Craig and Faust's Clinical Parasitology*. Lea & Febiger, Philadelphia, 1970.
56. Stear, M.J. et al. The processes influencing the distribution of parasitic nematodes among naturally infected lambs. *Parasitology* 117, 165, 1998.
57. Sykes, A.R. Parasitism and production in farm-animals. *Anim. Prod.* 59, 155, 1994.

58. Al-Khalidi, N.W. et al. Pathogenicity of *Marshallagia marshalli* (Ransom) in sheep, *Ind. J. Anim. Sci.* 59, 917, 1989.
59. MAFF. *Manual of Veterinary Parasitology Laboratory Techniques*. HMSO, London, 1986.
60. Eysker, M., and Kooyman, F.N.J. Notes on necropsy and herbage processing techniques for gastrointestinal nematodes of ruminants. *Vet. Parasitol.* 46, 205–213, 1993.
61. Gaba, S. et al. Estimation of abomasum strongyle nematode infections in sheep at necropsy: Tentative proposals for a simplified technique. *Vet. Parasitol.* 140, 105, 2006.
62. Ash, L.R., and Orihel, T.C. *Parasites: A Guide to Laboratory Procedures and Identification*. ASCP Press, Chicago, 1987.
63. Thienpont, D., Rochette, F., and Vanparijs, O.F.J. *Diagnosing Helminthiasis by Coprological Examination*. Jansen Research Foundation, Beerse, 1986.
64. Zarlenga, D.S. et al. A multiplex PCR assay for differentiating economically important gastrointestinal nematodes of cattle. *Vet. Parasitol.* 97, 199, 2001.
65. Harmon, A.F., Zarlenga, D.S., and Hildreth, M.B. Improved methods for isolating DNA from *Ostertagia ostertagi* eggs in cattle feces. *Vet. Parasitol.* 135, 297, 2006.
66. Callaghan, M.J., and Beh, K.J. A tandemly repetitive DNA sequence is present at diverse locations in the genome of *Ostertagia circumcincta*. *Gene* 174, 273, 1996.
67. Zarlenga, D.S. et al. Identification and semi-quantification of *Ostertagia ostertagi* eggs by enzymatic amplification of ITS 1 sequences. *Vet. Parasitol.* 77, 245, 1998.
68. Otsen, M. et al. Microsatellite diversity of isolates of the parasitic nematode *Haemonchus contortus*. *Mol. Biochem. Parasitol.* 110, 69, 2000.
69. Prichard, R. Genetic variability following selection of *Haemonchus contortus* with antihelmintics. *Trends Parasitol.* 17, 445, 2001.
70. Grillo, V., Jackson, F., and Gilleard, J.S. Characterisation of *Teladorsagia circumcincta* microsatellites and their development as population genetic markers. *Mol. Biochem. Parasitol.* 148, 181, 2006.
71. Álvarez-Sánchez, M.A. et al. Real time PCR for the diagnosis of benzimidazole resistance in trichostrongylids of sheep. *Vet. Parasitol.* 129, 291, 2005.
72. Wimmer, B. et al. Non-invasive assessment of parasitic nematode species diversity in wild Soay sheep using molecular markers. *Int. J. Parasitol.* 34, 625, 2004.
73. von Samson-Himmelstjerna, G., Harder, A., and Schnieder, T. Quantitative analysis of ITS2 sequences in trichostrongyle parasites. *Int. J. Parasitol.* 32, 1529, 2002.
74. Stevenson, L.A., Gasser, R.B., and Chilton, N.B. The ITS-2 rDNA of *Teladorsagia circumcincta, T. trifurcata* and *T. davtiani* (Nematoda: Trichostrongylidae) indicates that these taxa are one species. *Int. J. Parasitol.* 26, 1123, 1996.
75. Dallas, J.F., Irvine, R.J., and Halvorsen, O. DNA evidence that *Ostertagia gruehneri* and *Ostertagia arctica* (Nematoda: Ostertagiinae) in reindeer from Norway and Svalbard are conspecific. *Int. J. Parasitol.* 30, 655, 2000.
76. Dallas, J.F., Irvine, R.J., and Halvorsen, O. DNA evidence that *Marshallagia marshalli* Ransom, 1907 and *M. occidentalis* Ransom, 1907 (Nematoda: Ostertagiinae) from Svalbard reindeer are cospecific. *Syst. Parasitol.* 50, 101, 2001.
77. Hoste, H. et al. A comparison of the first internal transcribed spacer of ribosomal DNA in seven species of *Trichostrongylus* (Nematoda: Trichostrongylidae). *Int. J. Parasitol.* 28, 1251, 1998.
78. Hoste, H. et al. Differences in the second internal transcribed spacer (ribosomal DNA) between five species of *Trichostrongylus* (Nematoda: Trichostrongylidae). *Int. J. Parasitol.* 25, 75, 1995.
79. Dallas, J.F. et al. Identification by polymerase chain reaction (PCR) of *Marshallagia marshalli* and *Ostertagia guehneri* from Svalbard reindeer. *Int. J. Parasitol.* 30, 863, 2000.
80. Sambrook, J., MacCallum, P., and Russell, D. *Molecular Cloning: A Laboratory Manual*. 3rd ed. Cold Spring Harbor Laboratory Press, New York, 2001.
81. Brooks, D.R., and Hoberg, E.P. Triage for the biosphere: The need and rationale for taxonomic inventories and phylogenetic studies of parasites. *Comp. Parasitol.* 67, 1, 2000.

59 Necator

Jan M. Schwenkenbecher

CONTENTS

59.1 INTRODUCTION

Nematodes are the most numerous multicellular organisms on earth. About 20,000 species are described, but too little attention has been paid to these organisms and their true number may be closer to 500,000. About 15,000 known nematode parasites have adapted to parasitism, infecting all known vertebrate species; phylogenetic analysis suggests that they developed a parasitic lifestyle several times in evolution. Among the parasites that have adapted to a parasitic lifestyle in the human intestines is the hookworm *Necator americanus*, causing one of the most important tropical diseases in humans. *N. americanus* is a plexing species because some features of its infection suggest that parasitic lifestyles and symbiotic relationships are sometimes not easy to distinguish. Yet, heavy hookworm infections are a major public health issue; morbidity caused by these parasites is estimated at 22 million disability-adjusted life years.

59.1.1 Classification and Morphology

59.1.1.1 Classification

Human hookworms are members of the family Ancylostomatidae, superfamily Ancylostomatoidae, order Strongylida, subclass Rhabditida, class Secernentia, phylum Nematoda, subkingdom Eumetazoa, kingdom Animalia. *N. americanus* is the single species in the genus *Necator*, being phylogenetically distinct from the other human hookworm *Ancylostoma duodenale*, which is one of several species in the genus *Ancylostoma* [1]. Along with the roundworm, *Ascaris lumbricoides*, and the whipworm, *Trichuris trichiura*, hookworms are often referred to as soil-transmitted helminths (STHs) [2].

59.1.1.2 Morphology

Adult *N. americanus* have a cylindrical body. The anterior end is often found to be formed like a hook, hence the name hookworm. The worms are up to 11 mm long. The body has a cuticle with three main outer layers of collagen and other compounds, which protects the hookworm so it can persist in the digestive tracts of animals [3].

The interior of hookworms consists basically of the digestive tract and the reproductive organs. Female hookworms, with their uterus containing a number of eggs, are usually bigger than males. A typical morphological feature of male hookworms is the copulatory bursa at the posterior end, which they use to attach themselves to the female for fertilization. In the

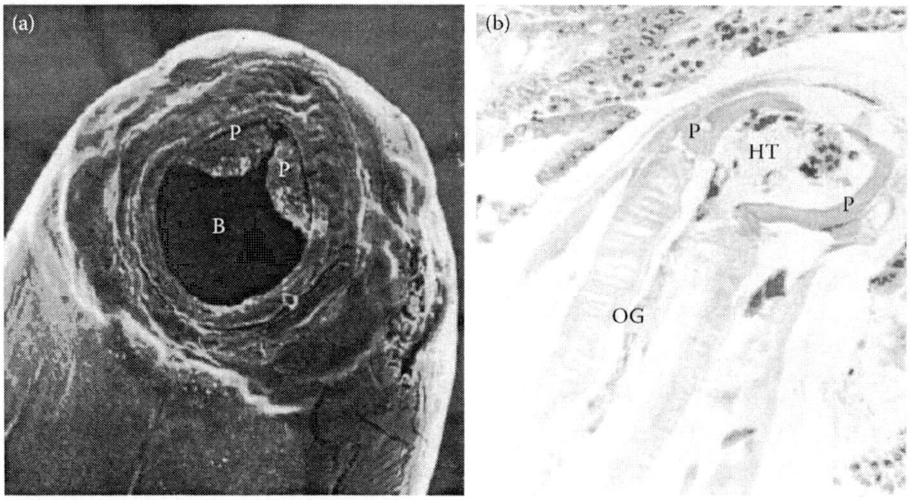

FIGURE 59.1 (a) Anterior end of the *Necator americanus* electron microscopy image, showing the buccal cavity (b) and the cutting plates (P). (b) *N. americanus* attached to the intestinal mucosa, stained with hematoxylin and eosin. Host tissue (HT) is drawn into the buccal cavity prior to digestion in the esophagus (OE). In addition, parasite-protective molecules are secreted at the site of attachment. (Reprinted from *Trends Parasitol.*,17, Pritchard, D.I., and Brown, A. Is *Necator americanus* approaching a mutualistic relationship with humans? 169, Copyright 2001, with permission from Elsevier.)

opening of their buccal cavity, *N. americanus* have two dorsal and two ventral curved cutting plates (Figure 59.1a); this feature can be used to distinguish them from *Ancylostoma* spp., which have two pairs of teeth in their mouth.

Eggs of hookworms are oval, about 60 μm long and 40 μm wide, with a thin shell. They may be unsegmented or show typical embryonic cleavages of up to eight stages. Eggs that have been in the air for several hours often contain developing larvae which are about to hatch. The morphology of *N. americanus* larvae changes through several stages as described in the next chapter.

59.1.2 Life Cycle and Epidemiology

59.1.2.1 Life Cycle

N. americanus is an obligate intestinal parasite. Up to 30,000 fertilized eggs can be shed by a female per day, which exit the body in the feces [4]. In the air, embryos develop and rhabditiform stage L1 larvae hatch from the eggs under favorable conditions (moisture, warmth, and shade) within 24–48 h. The L1 larvae grow and molt to reach the L2 stage and then again develop into filariform infective L3 larvae 10 days after hatching. At this stage, larvae move through the superficial soil layer and can survive in moist, temperate soil for 3–4 weeks. However, larvae die in freezing or hot, dry conditions. Upon contact with their human host, the larvae penetrate the exposed skin. They move passively through the blood vessels to the heart and then to the lungs. There they move through the tissue into the pulmonary alveoli, ascend the bronchial tree to the pharynx, where they are coughed up and swallowed so they reach the small intestine, where they molt twice more, living through stages L4 and L5. They attach to the mucosal wall of the small intestine and feed on blood (Figure 59.1b). Unless removed by medical treatment, they may live for several years [4,5].

59.1.2.2 Epidemiology

Hookworm disease is the second most common human intestinal helminth infection (after ascariasis). Hookworms are spread worldwide in areas with tropical and subtropical climate. They prefer warm, moist, and shadowy conditions. Its name implicates that *N. americanus* predominates in the Americas, where it was first described ("American killer"). However, the species is found in Africa, Asia, and Australia as well and is the predominant etiology of human hookworm infection [6]. Its areas of occurrence overlap with *A. duodenale*, for instance in sub-Saharan Africa and eastern Asia, where the highest hookworm prevalence has been found.

Generally, hookworm disease can be considered a disease of the poor, as a high rate of infection occurs in areas of rural poverty in the tropics. A striking relationship between hookworm prevalence and low socioeconomic status has been found [7]. Hookworm infection is spread in areas with poor sanitation, where people have no latrines, so that the soil is contaminated with infective larvae. As they cannot afford shoes, people walk with bare feet and become infected, as will be children playing in the soil with their bare hands. It has been found that the practice of using the content of latrines for fertilizing crops in Southern China and Vietnam has also been the cause of infection with *N. americanus* [8,9].

The intensity of hookworm infections is estimated by using the fecal egg count as a marker for worm burden. High-intensity infections are those with a count of 4000 or more eggs per gram of feces, medium intensity between 2000 and 4000, and mild infections below 2000 [4]. In contrast to other STH infections, the intensities of which peak in childhood and adolescence, the intensity of hookworm infections rises steadily with human age or plateaus in adulthood [10,11].

An interesting epidemiological issue is the extensive geographic overlap of malaria and STH infections. Human coinfection with *Plasmodium falciparum* and hookworm is extremely common across the African continent, because the environmental conditions favor a wider distribution of hookworms compared to *Trichuris* and *Ascaris* [12]. School-age children are most at-risk of *Plasmodium*–helminth coinfection and thereby at greatest risk of its impact on hemoglobin levels. Malaria contributes to the development of anemia through a number of mechanisms, principally by destruction and removal of parasitized cells and the shortening of the life span of nonparasitized red cells, and decreasing the rate of erythrocyte production in the bone marrow [13]. The contribution of hookworms to anemia by intestinal blood loss suggests that coinfection of *P. falciparum* and hookworm has an additive impact on hemoglobin, exacerbating anemia-related malarial disease burden. While children and pregnant women are at highest risk of anemia, in areas of high malaria transmission the disease burden due to coinfection is most significant among school-aged children, which showed lower hemoglobin levels than children with single infections [12].

59.1.3 Clinical Features, Pathogenesis, Prevention, and Treatment

59.1.3.1 Clinical Features and Pathogenesis

Hookworms are voracious blood feeders. Attached to the mucosal tissue, their feeding results in significant gastrointestinal hemorrhage, loss of serum proteins, and intestinal inflammation [6,14]. The blood loss causes iron deficiency anemia and protein malnutrition, which is the most common symptom of hookworm infection. The blood loss can be accompanied by cardiac problems. Gastrointestinal and metabolic symptoms can also occur. Local dermatological manifestations can occur during penetration by the infectious larvae (ground itch). Larval migration through the lungs may cause respiratory problems.

Malnutrition and anemia caused by heavy infections put children and women in child-bearing age at risk, as the parasite intensity and reductions in host hemoglobin correlate; both groups are at an increased risk of anemia in areas where the prevalence of hookworm and malaria overlap [4,15]. In pregnant women, blood loss caused by heavy infections can cause a variety of adverse maternal-fetal outcome [16]. In children, a heavy worm burden can lead to developmental disorders such as stunted growth and mental retardation, leading to decreased future economic productivity [17,18]. Chronic infection with hookworm promotes long-term disability and increases the likelihood that an afflicted population will remain in poverty [4]. Measured in disability-adjusted life years, global disease burden from hookworm exceeds all other major tropical diseases with the exception of malaria, leishmaniasis, and lymphatic filariasis [4].

59.1.3.2 Prevention and Treatment

Hookworms have had a historical presence in Western society. So, infection with *N. americanus* contributed to the slow economic development of the southern United States, which was the motivation for founding the Rockefeller Sanitary Commission for Eradicating Hookworm Disease [6]. This demonstrates that a control program can indeed reduce parasite burden. Hookworms do not replicate in humans, so the easiest way should be to disrupt the transmission by improving the sanitation, footwear, and health education. Because hookworm impact is worst in areas of poverty and slow economic development, this has been difficult to accomplish [19]. Therefore, control strategies have been shifted toward using anthelmintic drugs.

Since the 1990s, a major approach to hookworm control has been treatment of school-aged children with benzimidazoles. Mebendazole and albendazole are the most commonly used drugs for treatment; the latter is the most effective drug against *N. americanus*. Other effective drugs used in STH control campaigns are levamisole and pyrantel. They are cheap, can be given as a single-dose tablet, the dosage is not weight-dependent and pills are easily administered in mass drug administration (MDA). The drugs are quite effective; hookworm prevalence has been significantly reduced in treatment programs directed directly against hookworm infection. Long-term programs have shown efficient deworming effects, and regular treatment produces immediate as well as long-term benefits, significantly contributing to the development of affected individuals, particularly children [20]. These treatments also significantly reduce the morbidity of other STHs [20,21]. Likewise, hookworm prevalence has been reduced significantly as a by-product of the global program for elimination of lymphatic filariasis [22,23].

The heavy drug pressure placed on hookworm populations by MDA bear the threat of selecting for resistance to the drugs. This has been seen in veterinary medicine, where the widespread and intense use of anthelmintic drugs has resulted in the appearance of drug resistance in parasites of every livestock host and to every anthelmintic class [24]. Drug resistance is a heritable trait and is defined as an increase in the frequency of individuals able to tolerate elevated drug doses that are typically lethal in a normal population [25]. Benzimidazole drugs have been shown to exert the nematodical effect by binding to beta-tubulin, thereby inhibiting polymerization of the alpha/beta-tubulin heterodimer subunits into microtubules. It is generally accepted that mutations in codon positions 167 [26,27] or 200 [28], characterized by a phenylalanine (TTT/TTC) to tyrosine (TAT/TAC) substitution or a glutamate (GAG) to alanine (GCG) substitution in codon position 198 [29] in beta-tubulin isotype-1, are the major determinant of BZ resistance by preventing binding of the drugs to the protein. The usual gold standard for monitoring drug efficacy and resistance is based on the reduction in fecal egg counts in drug-exposed communities as compared to the egg count before drug exposure began. Progress has been made in the development of egg hatch assays and larval development assays to investigate drug resistance to BZ in hookworms under field conditions [30,31], and a larval motility assay appears to detect resistance to ivermectin [31,32]. However, these classical methods can only detect drug resistance once it already exists. With the knowledge of the genetic basis it is possible to monitor developing resistance in its early stages before whole nematode populations are rendered irresponsive to drug treatment. Cases of reduced levels of treatment efficacy of *N. americanus* infection have been reported of mebendazole in Mali [33]. Although these cases fall short of conclusive evidence due to methodological weakness [34], they highlight the possible emergence of anthelmintic drug resistance in human hookworms. Albonico et al. [35] reported reduced sensitivity of hookworm eggs recovered from children on Pemba Island who have been exposed to treatment with benzimidazoles over a period of 5 years. Interestingly, neither sequencing of cDNAs [35] nor real-time PCR assays [36] could detect elevated levels of BZ-resistance alleles in this population of *N. americanus*. Frequent treatment was expected to select for resistance alleles, but a big reservoir (eggs and larvae in the soil, people not subjected to treatment) may contribute to an ease of the drug pressure compared to MDA. One can also assume that mutations in other beta-tubulin isotypes or microtubule-associated proteins provide alternative mechanisms contributing to resistance to BZ. Thus, the subject of developing resistance of hookworms to BZ and other drugs remains an important part of public health surveys and is a highly challenging field of research [37].

59.1.3.3 Vaccination

Hookworms do not replicate in humans, so if an efficient treatment removes all adult worms from the intestines, the infections should be cleared. However, because school-based treatment programs would have little impact in areas with heavy infection in both children and adults, reinfection can occur with 4–12 months [38]. An additional tool for the control of hookworm, the Human Hookworm Vaccine Initiative (HHVI) is developing a vaccine targeting the hookworm larval antigen ASP-2 (ancylostoma secreted protein 2) [5], which is directed against infective L3 larvae, hence reinfection after elimination of adult hookworms. The protein has been successfully isolated from the secreted product of *N. americanus* (Na-Asp-2), which is currently the leading larval-stage hookworm vaccine candidate. Process development, good manufacturing practice and control, and preclinical testing of Na-ASP-2 were completed in 2004 [5], and a randomized, double-blind, placebo-controlled study has been performed in which 36 healthy adults without a history of hookworm infection were given three intramuscular injections of three different concentrations of Na-ASP-2, which induced a significant immune response [39]. The development is carried further to phase one and two trials to assess the feasibility and success of the vaccine in human patients [40].

Another vaccination strategy, which is directed against adult hookworms in the gut, is to inhibit the aspartic protease of the hookworm, APR-1, which starts the proteolytic cascade to digest hemoglobin from erythrocytes. An enzymatically inactivated hemoglobinase from *N. americanus* induces a neutralizing antibody response against multiple hookworm species; the human vaccine, Na-APR1, significantly reduced egg burden and blood loss in dogs that were challenged with the *A. caninum* [41]. Other antigens of adult hookworms which are suited as targets for vaccination are the cysteine protease-hemoglobinase CP2 [42] and a glutathione-*S*-transferase [43].

59.1.3.4 Immunological Aspects of *N. americanus* Infections

The immunology of hookworm infection has been extensively investigated for *N. americanus*. There is increasing evidence for protective immunity and interactions with other diseases.

Important antigens that induce immunological activity are proteinases [44], acetylcholinesterase [45], and cuticular proteins [46]. Serum antibody responses of all five isotypes were detected in naturally infected populations [47,48], which showed cross-reactivity with other hookworm species and other helminths [49]. In aspects of lymphocyte proliferation and cytokine profile, hookworms are typical gastrointestinal nematodes, inducing a T helper type (Th2) immune response, characterized by significant production of IL-3, IL-4, IL-5, and IL-13 [50]. Eosinophilia is commonly observed and in concert with IgE antibodies it is linked to the control of parasite size and fecundity [45]. It is remarkable, however, that these immune responses fail to expulse the parasites. As mentioned in Section 59.1.4, burden of *N. americanus* is accumulating with the age of the host and reinfection is quickly reaching pretreatment levels after complete clearance, indicating that the worms must have an efficient strategy to evade immune effector mechanisms. Moderate hookworm infections have been shown to be tolerated by the organism. Moreover, they appear to have beneficial effects on hosts suffering from diseases linked to overactive immune systems. It is now a commonly accepted hypothesis that hookworms of the species *N. americanus* have developed strategies to evade host reactions induced by the worm's migration and feeding. Strategies to evade host immunity include: (i) Induction of apoptosis in activated T-cells by releasing a small proapoptotic factor [51]. T-cells can also be modulated by releasing a kaliseptine-like molecule, which has the potential to modulate the function of T-cells by blocking potassium channels [52]. (ii) Controlling eosinophilia by neutralizing effector cells and molecules. This involves the secretion of superoxide dismutase (SOD), which generates hydrogen peroxide (H_2O_2) from superoxide anions released by eosinophils [53,54]. H_2O_2 will damage host tissue but spare the parasite, which protects itself against lipid degradation by secretion of neutralizing gluthathione-*S*-transferase. The secreted SOD might also act as an antihemostatic agent by neutralizing O_2, which in turn reduces the expression of tissue factor and hence coagulation. The neutralization of O_2 might also reduce the destruction of endothelium-derived relaxing factor, inhibiting blood coagulation and thereby increasing blood flow at the site of attachment [55]. Parasite defense also includes the secretion of proteinases which digest eosinophil agonist such as eotaxin [56], and of calreticulin, which binds to complement components and thereby inhibits hemolysis [57]. (iii) Inhibition of signaling in mast cells via the induction of excess polyclonal IgE and IgG autoantibodies to IgE to co-cross-link antibodies on the cell membrane [58].

By secretion of proteinases and other proteins, *N. americanus* generates an allergic-like response. Yet, it survives in an environment that is highly hostile by neutralizing the inflammatory response, which is on the other hand sufficient to control worm growth and fecundity [59]. This could explain the silent character of the infection, characterized by longevity of the hookworm in the immunologically hostile environment of its human host, yet being under control by Th2 cell activity which influences worm weight and reproduction. It has been hypothesized that *N. americanus* and humans are approaching a mutualistic symbiotic relationship, where parasite and host defense molecules effectively neutralize each other [58]. It may indeed be true that moderate infections have beneficial effects on the host's health by modulation of the immune system. By down-regulation of Th1 response, namely cytokines such as IFN-gamma, hookworms are beneficial in the treatment of autoimmune diseases such as Morbus Crohn or diabetes [60]. Because of their anti-inflammatory activity and their influence on the balance of the immune system, chronic hookworm infection is also thought to help host suffering from diseases linked to overactive immune systems, such as asthma or hay fever [61,62]; by the way, the allergenic microenvironment created by *N. americanus* has biochemical attributes in common with environmental antigens. It has also been argued that gut nematodes are beneficial in malarious areas by promoting neuroprotective cytokine profiles, thereby preventing cerebral malaria, and by manipulating iron availability [63,64].

In fact, not all of these effects are specific for *N. americanus*, but have been observed for other gastrointestinal parasites as well. Contacts with infectious agents are part of the immune system's maturation, that is, to learn which parasites are dangerous or not. Studies in Africa have shown that it may be beneficial to have some parasites and that the effects of worm eradication in indigenous populations should be carefully considered [65]. Nevertheless, a high hookworm burden has a strong impact on the health and development of children and pregnant women, and treatment strategies should be carefully balanced.

59.1.4 Diagnosis

59.1.4.1 Conventional Techniques

Infection with *N. americanus* is diagnosed by microscopic identification of eggs in the feces. Hookworm eggs are well distinguishable from the eggs of other STHs, namely *Trichuris* and *Ascaris*. Feces are prepared as described in Section 59.2.1. The morphology of hookworm eggs is described under Section 59.1.1. How to prepare stool samples and identify hookworm eggs is described in Section 59.2.1.

Eggs of the two human hookworms *N. americanus* and *A. duodenale*, as well as various other hookworm species, are indistinguishable from each other. In terms of treatment purposes, they are considered to be identical; however, there are significant life history differences between the species that must be considered for the rational design of therapeutic and control strategies. A way of identifying different parasites is coproculture, in which feces are cultured for the purpose of hatching parasite eggs and identification is based on the differences between the filariform larvae of the two hookworms [66]. Fecal cultivation is also used to diagnose infection with human parasites as *Oesophagum bifurcum*, eggs of which have similar morphological features as hookworm eggs and which are found in some areas of hookworm prevalence [67,68]. The major drawback of coprocultures is that diagnosis depends on the availability of fresh stool specimens, and they are easily contaminated with other microorganisms [69]; the use of molecular tools to identify parasites is an alternative to conventional techniques.

59.1.4.2 Molecular Techniques

Advances in DNA technology facilitated the identification of many closely related species which are difficult to differentiate by morphological features on the basis of their DNA sequences (barcoding), in conjunction with the identification of taxonomic groupings and clarification of phylogenetic relationships [70–72]. A direct approach is to amplify predefined DNA segments by PCR using oligonucleotide primers derived from conserved sequences. While sequencing single specimens can become costly when investigating many isolates, characteristic DNA sequences can be used to design species-specific primer sets which amplify PCR products of a certain size that corresponds to a species identity. This has been achieved using polymorphisms at the mitochondrial cytochrome oxidase I (COI), facilitating the identification of many infectious isolates [73]. Zhan et al. [74] developed a species-specific PCR assay to differentiate between *N. americanus* and *A. duodenale*. Polymorphisms in COI have also been used to investigate the genetic structure of populations of *N. americanus* [75]. PCR assays which included the identification of other nematode parasites, eggs of which are difficult to distinguish from those of hookworms, were designed for *Trichostrongylus colubriformis* [76], using the nuclear internal transcribed spacer 1 region (ITS1), and for *O. bifurcum* [77], using ITS2.

A technique that relies on the principle that the electrophoretic mobility of a single-stranded DNA molecule in a nondenaturating gel is dependent on its structure and size is the PCR-based single-strand conformation polymorphism (SSCP). As the conformation of a DNA molecule depends on its size and sequence, differences of a single nucleotide can change the mobility [78]. Using appropriate sequences such as ITS-1 and ITS-2, this method has been used to differentiate several nematode species, including hookworm species of several life stages [79,80]. Genetic variability within *N. americanus* populations has been investigated by SSCP of COI [81].

Another PCR-based technique for species identification is restriction fragment length polymorphism (RFLP), which is based on the detection of DNA sequence variation by digesting with restriction enzymes and analyzing the size of the resulting fragments by gel electrophoresis [82]. This technique has been applied for the differentiation of hookworm species by PCR amplification of a fragment of the 3′ untranslated region of the cAMP-dependent protein kinase catalytic subunit gene and digestion of the amplified DNA with different restriction enzymes, which generated specific restriction fragment patterns unique to each species [83]. This method is explained in detail in Section 59.2.2.

59.2 METHODS

59.2.1 Sample Preparation

59.2.1.1 Sample Collection and Microscopic Diagnosis

Adult human hookworms can be retrieved from stools of people who have received anthelmintic treatment the day before. The stools should be diluted with tap water so that the worms can be picked by eyesight. As an alternative, although not very common in practice, hookworms can be removed from the duodenum by surgery or post mortem.

Most often, the material to work with are hookworm eggs, which are found in fresh stools and need to be spotted using a microscope. A quick method, although not very sensitive, is to investigate simple stool smear. Because the material is not transparent, smears need to be very thin and only little material can be investigated. That way, light and medium infections are possibly not identified, and little material for molecular investigations can be retrieved that way. Therefore, it is better to enrich the stools for helminth eggs prior to microscopic investigation. A variant of thick smear techniques is the Kato-Katz method, which is commonly used for the diagnosis of intestinal helminth infections [84,85]. However, it has been found that it is not suitable for the identification of hookworm eggs [86,87] and will not be discussed in detail in this chapter.

A simple method to perform even under field conditions in tropical countries is the sugar gradient. In a beaker, 5 g of stools is mixed with 15 mL of tap water using a spatula. Upon stirring, the mix turns into a suspension (slurry) which is filtered through a coarse gauze mesh into a 15 mL plastic Falcon tube. After centrifugation of the filtrate at 1200 rpm for 10 min, the supernatant is discarded and the pellet is resuspended with a saturated sugar solution (specific gravity 1.26). A microscope coverslip is placed on top of the tube, which is again centrifuged at 1200 rpm for 10 min. If working under field conditions with poor technical equipment or if no centrifuge is available, it is sufficient to keep the tubes with the coverslips standing on the bench for at least 15 min or, if a centrifuge is used, to place the coverslip on the tubes after centrifuging. Helminth eggs float atop of the solution and can be transferred to a microscopic slide within a drop of liquid sticking to the coverslips which should be taken off carefully. A second coverslip should be placed on the tube to collect all eggs swimming at the surface. The slides are examined for characteristic hookworm eggs under the microscope. If found, eggs are counted for estimation of the intensity of infection; it is always worth keeping a protocol of other helminth eggs. The number of eggs is divided by the grams of stool used for the investigation. The number of eggs per gram is a measure of the intensity of the infection. For genetic species verification and other molecular analyses, hookworm eggs are recovered by rinsing them off the coverslips after microscopic analysis. The egg samples are washed twice in dH_2O to remove residual sugar, and can be stored in 70% ethanol for later analysis.

Another concentration technique is the sedimentation. The stool specimen (5 g) is fixed in 6 mL 10% formalin in a 15 mL plastic Falcon tube. For sedimentation, 2 mL of ethyl acetate is added and the tube is centrifuged at 500 rpm for 5 min. A spatula is used to transfer a part of the sediment on a microscopic slide, which is then investigated for hookworm eggs under the microscope. Formalin preserves specimens by preventing degradation by microorganisms, but has a major shortcoming. Formalin degrades the DNA by double-strand breaks so that it cannot be used for subsequent molecular analyses. PCR amplification of short products using DNA extracted from formalin-treated samples may work if the samples have been stored only shortly [88], but storage over a longer period of time damages the DNA beyond repair, so it is actually common sense not to store hookworm eggs in formalin if later molecular analysis is planned.

59.2.1.2 DNA and RNA Preparation from Adult *N. americanus*

For the preparation of genomic DNA from adult *N. americanus*, commercial kits can be used (QIAGEN). An alternative is the isolation of DNA with phenol and chloroform. One adult worm is transferred to a tube containing 200 μL extraction buffer (100 mM NaCl, 50 mM Tris, pH 8, 50 mM ethylenediamine tetraacetic acid, 1% sodium dodecyl sulfate (SDS), 1% beta-mercaptoethanol, 250 mg/mL proteinase K), crushed with a micropestle and incubated at 65°C for 2 h. After incubation, the lysate is extracted once with 200 μL phenol/chloroform/isoamyl alcohol, followed by extraction with 200 μL chloroform. The DNA is precipitated with 10 μL 3 M sodium acetate and 500 μL 96% ethanol. After centrifugation at 15,000 rpm for 10 min, the DNA is washed with 70% ethanol, centrifuged again and air-dried. The pellet is resuspended in 25 μL TE, pH 8 (Tris-EDTA), or water. DNA is more stable in TE than in water, because EDTA inhibits enzymes which may degrade the DNA. However, it is important to keep in mind that EDTA may also inhibit enzymes as the Taq polymerase.

For the isolation of RNA, it is important to use fresh specimens because RNA degrades more quickly than DNA. It is sufficient to keep the samples in 70% ethanol, although there are commercially available reagents such as RNAlater (QIAGEN). The procedure is similar to DNA preparation, but rather than using phenol the use of Trizol Reagent

(Invitrogen) is recommended. The samples are crushed with a pestle in 200 μL of Trizol, followed by 10 min incubation at room temperature. The samples are extracted with 200 μL chloroform. The RNA is precipitated with 200 μL of isopropanol. After centrifugation at 15,000 rpm for 10 min, the DNA is washed with 70% ethanol, centrifuged again, and air-dried. The RNA is resuspended in 10 μL water and used for analysis subsequently. RNA can be stored at −80°C.

59.2.1.3 DNA Preparation from *N. americanus* Eggs

For isolation of DNA from one or more eggs, the specimens are transferred to 50 μL extraction buffer as described in Section 59.2.1.2. After sonication in an ultrasonic bath for 15 min, SDS is added to a final concentration of 1%. The eggs will be boiled at 95°C for 10 min in a thermocycler, followed by freezing at −80°C. The procedure is repeated. Proteinase K is added to a final concentration of 250 μg/mL and the lysate is incubated at 65°C for 2 h. After the incubation period, the preparation is continued as described for adult specimens.

59.2.2 Detection Procedures

There are a variety of techniques available for molecular species diagnostics mentioned under Section 59.1.4.2. What they all have in common is the use of PCR, which is nowadays a widely used method for which a number of commercial kits and various kinds of thermocyclers are available. Since all the methods are described within publications these will not be described in this chapter except for one example. A nice application for differentiation of *N. americanus* from other hookworm species is the RFLP method described by John Hawdon [83]. As described under Section 59.1.3, the 3′ untranslated region of the cAMP-dependent protein kinase catalytic subunit gene is amplified.

The RFLP is carried out as follows:

1. About 20–50 ng of the prepared DNA is used as a template for PCR. For PCR, preferably a kit such as the HotStarTaq Mastermix kit (QIAGEN) is used, whereby in a reaction volume of 50 μL, 25 μL of the Mastermix are set up with 0.4 μM each of primer PKA5′4 (5′CTCGCCAGCATCTCTGCG) and primer PKA3′1 (5′GCCTACAAACTTGTCAGCAGG), the genomic DNA, as well as ddH_2O to the final volume. Alternatively, if one prepares a PCR mixture which consists of the single components, a reaction volume of 50 μL should contain 20–50 ng of genomic DNA, 20 μL of each primer, 50 nM of each dNTP, 1.5 mM $MgCl_2$, 2 U Taq DNA polymerase, and ddH_2O ad the final volume. The PCR is performed in the following conditions: initial denaturion at 94°C for 2 min (15 min with HotStarTaq) and then 32 cycles of 94°C for 30 s, annealing at 55°C for 30 s, and elongation at 72°C for 1 min, which is followed by a final elongation at 72°C for 6 min.

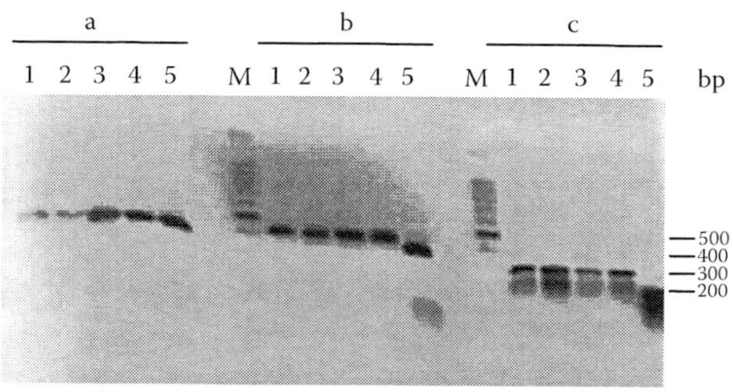

FIGURE 59.2 Molecular differentiation of hookworm species by RFLP of a PCR-amplified part of the 3′UTR of PKA (a) digested with *Hpa*II, (b) digested with *Mbo*I, (c) digested with *Taq*I. 1–4: different samples of *Necator americanus*, 5: *Ancylostoma duodenale*.

2. To test the sample quality, 5 μL of the reaction can be tested by a run on an agarose gel that is stained with ethidium bromide to visualize the DNA in UV light.
3. Digestions of the amplified DNA with the restriction enzymes *Hpa*II, *Mbo*I, *Taq*I, and *Tha*I generate specific restriction fragment patterns unique to each species. For digestion, 5 μL of the PCR product is set up with 2 μL of 10× reaction buffer, 1 μL of restriction enzyme, and 12 μL ddH_2O. Reactions with HpaII, MboI, and ThaI are incubated at 37°C for 1 h; TaqI is incubated at 65°C for 1 h. After incubation, the whole reaction volume is run on an agarose gel, which should have a concentration of 2% to separate the small fragments well enough for analysis. The patterns of MboI and TaqI are particularly suited to differentiate *N. americanus* from *A. duodenale* (Figure 59.2).

59.3 CONCLUSIONS AND FUTURE PERSPECTIVES

In the twenty-first century, infection with hookworm in general, and with *N. americanus* in particular as the predominant etiology, remains one of the great plagues of mankind. Although this chapter implicates that a few worms do no harm and that they may indeed be good for us, heavy infections have a major impact on public health, particularly on the development of children, the nutrition of the unborn in the uterus, and the well-being of women in child-bearing age. It is a disease of the poor; in all regions with high transmission levels, there is a striking relationship between hookworm prevalence and low socioeconomic status. As the eradication of *N. americanus* in the southern United States shows, the parasite can be eradicated. Because hookworms have a direct life cycle without a vector or intermediate host and do not replicate in humans, interruption of the transmission by wearing shoes, using latrines, and educating people would prevent new infection. However, this is difficult to accomplish in areas of comprehensive economic development, so that hookworm has proven to be extremely difficult to eliminate or eradicate in areas of poverty and poor sanitation. The major efforts to reduce morbidity caused by hookworm infection are therefore the mass treatment of affected populations with anthelmintic drugs.

There is now clear evidence that regular treatment of helminth infections produces immediate as well as long-term benefits, contributing to the development of affected individuals, particularly children. As discussed in this chapter, the most common drugs administered are of the class of benzimidazoles. As effective and useful as they are, it has to be remembered that MDA bear the danger of developing resistance toward the drugs, which can render whole parasite populations irresponsive to the drug. Assays have been developed to investigate drug resistance in hookworms under field conditions. Having learned from experience in veterinary health, measures have been taken to monitor the molecular basis responsible for BZ resistance. To date, human hookworms remain one of the few parasitic nematode species in which no elevated levels of alleles corresponding to BZ resistance have been found. To maintain this situation, further monitoring of treated populations will implicate change of treatment regimes as soon as elevated levels of resistance alleles are found.

The high rates of reinfection after treatment with drugs and the limited effect that school-based eradication programs may have on the whole population lead to the idea of vaccine development. As discussed in this chapter, the point is the creation of a vaccine that reduces severity of infection after treatment. The world's leading experts in vaccine development are located at the George Washington University, London School of Hygiene and Tropical Medicine, and the Oswaldo Cruz Foundation, which are organized in the HHVI and sponsored by the Sabin Vaccine Institute and the Gates Foundation, which dedicated 34 million US dollars to the global network for neglected tropical diseases for the reduction of the burden of most neglected diseases. As discussed, larval and adult hookworm antigens have been identified as potential candidates for vaccines. It will be interesting to see how efficient and practical vaccination will be under conditions of poverty-stricken tropical and subtropical countries.

Human infection with *N. americanus* goes back a long way in history and will be an issue for a long time. It is an interesting parasite, which has adapted to life in the human guts so well that light infections are basically without symptoms. As nice as it would be to use these hookworms for therapeutic aspects, heavy infections in developing countries remain a major public health concern and will be subject to further research in immunology, drug resistance, epidemiology, and population genetics.

REFERENCES

1. Lichtenfels, J.R. *Keys to the Genera of the Superfamilies Ancylostomatoidea and Diaphanocephaliodea*. Farnham Royal, Commonwealth Agricultural Bureaux, 1980.
2. Bethony, J. et al. Soil-transmitted helminth infections: Ascariasis, trichuriasis, and hookworm. *Lancet* 367, 1521, 2006.
3. John, D.T., and Petri, W.A., Jr. *Markell and Voge's Medical Parasitology*, 9th edn. Saunders Elsevier, St. Louis, 2006.
4. Hotez, P.J. et al. Hookworm: The great infection of mankind. *PLoS Med.* 2, e67, 2005.
5. CDC PDx-Lab Assistance: Hookworm Infection. www.dpd.cdc.gov/dpdx/HTML/Hookworm.htm
6. Hotez, P.J. et al. Hookworm infection. *N. Engl. J. Med.* 351, 799, 2004.
7. de Silva, N.R. Impact of mass chemotherapy for the control of filariasis on geohelminth infections in Sri Lanka. *Ann. Trop. Med. Parasitol.* 97, 421, 2003.
8. Liu, C.H. et al. Epidemiology of human hookworm infections among adult villagers in Hejiang and Santai Counties, Sichuan Province, China. *Acta Trop.* 73, 243, 1999.
9. Olsen, A. et al. Cross-sectional parasitological survey for helminth infections among fish farmers in Nghe An province, Vietnam. *Acta Trop.* 100, 199, 2006.
10. Bethony, J. et al. Emerging patterns of hookworm infections: Influence of aging on the intensity of *Necator* infection in Hainan Province, People's Republic of China. *Clin. Inf. Dis.* 35, 1336, 2002.
11. Hotez, P.J. et al. Current concepts: Hookworm infection. *N. Engl. J. Med.* 351, 799, 2004.
12. Brooker, S. et al. Epidemiology of *Plasmodium*-helminth co-infection in Africa: Populations at risk, potential impact on anemia and prospects for combining control. *Am. J. Trop. Med. Hyg.* 77(6 suppl.), 88, 2007.
13. McDevitt, M.A. et al. The anemia of malaria infection: Role of inflammatory cytokines. *Curr. Hematol. Rep.* 3, 97, 2004.
14. Williamson, A.L., Brindley, P.J., and Loukas, A. Hookworm cathepsin D aspartic proteases: Contributing roles in the host-specific degradation of serum proteins and skin macromolecules. *Parasitology* 126, 179, 2003.
15. Stoltzfus, R.J. et al. Malaria, hookworms and recent fever are related to anemia and iron status indicators in 0- to 5-y old Zanzibari children and these relationships change with age. *J. Nutr.* 130, 1724, 2000.
16. Bundy, D.A., Chan, M.S., and Savioli, L. Hookworm infection in pregnancy. *Trans. R. Soc. Trop. Med. Hyg.* 89, 521, 1995.
17. Sakti, H. et al. Evidence for an association between hookworm infection and cognitive function in Indonesian school children. *Trop. Med. Int. Health* 4, 322, 1999.
18. Christian, P., Khatry, S.K., and West, J.P., Jr. Antenatal anthelminthic treatment, birthweight, and infant survival in rural Nepal. *Lancet* 364, 981, 2004.
19. Brooker, S., Bethony, J., and Hotez, P.J. Hookworm infection in the 21st century. *Adv. Parasitol.* 58, 197, 2004.
20. Albonico, M. et al. Controlling soil-transmitted helminthiasis in pre-school-age children through preventive chemotherapy. *PLoS Negl. Trop. Dis.* 2, e126, 2008.
21. Albonico, M. et al. Intervention for the control of soil-transmitted helminthiasis in the community. *Adv. Parasitol.* 61, 311, 2006.
22. De Rochars, M.B. et al. Community-wide reduction in prevalence and intensity of intestinal helminths as a collateral benefit of lymphatic filariasis elimination programs. *Am. J. Trop. Med. Hyg.* 71, 466, 2004.
23. Fox, L.M. et al. Ultrasonographic examination of Haitian children with lymphatic filariasis: A longitudinal assessment in the context of antifilarial drug treatment. *Am. J. Trop. Med. Hyg.* 72, 642, 2005.
24. Kaplan, R.M. Drug resistance in nematodes of veterinary importance: A status report. *Trends Parasitol.* 20, 477, 2004.

25. Prichard, R.K. et al. The problem of anthelmintic resistance in nematodes. *Aust. Vet. J.* 56, 239, 1980.
26. Silvestre, A., and Cabaret, J. Mutation in position 167 of isotype 1 β-tubulin gene of Trichostrongylid nematodes: Role in benzimidazole resistance? *Mol. Biochem. Parasitol.* 120, 297, 2002.
27. Hodgkinson, J.E. et al. The role of polymorphisms at β-tubulin isotype 1 codons 167 and 200 in benzimidazole resistance in cyathostomins. *Int. J. Parasitol.* 38, 1149, 2008.
28. Kwa, M.S., Veenstra, J.G., and Roos, M.H. Benzimidazole resistance in *Haemonchus contoruts* is correlated with a conserved mutation at amino acid 200 in β-tubulin isotype 1. *Mol. Biochem. Parasitol.* 63, 299, 1994.
29. Ghisi, M., Kaminsky, R., and Maser, P. Phenotyping and genotyping of *Haemonchus contortus* isolates reveals a new putative candidate mutation for benzimidazole resistance in nematodes. *Vet. Parasitol.* 144, 313, 2007.
30. Albonico, M. et al. Development of the egg hatch assay for detection of anthelminthic resistance in human hookworms. *Int. J. Parasitol.* 35, 803, 2005.
31. Kotze, A.C. et al. Field evaluation of anthelmintic drug sensitivity using *in vitro* egg hatch and larval motility assays with *Necator americanus* recovered from human clinical isolates. *Int. J. Parasitol.* 35, 445, 2005.
32. Kotze, A.C. An *in vitro* larval motility assay to determine anthelmintic sensitivity for human hookworm and *Strongyloides* species. *Am. J. Trop. Med. Hyg.* 71, 608, 2004.
33 De Clercq, D. et al. Failure of mebendazole in treatment of human hookworm infections in the southern region of Mali. *Am. J. Trop. Med. Hyg.* 57, 25, 1997.
34. Geerts S., and Gryseels B. Drug resistance in human helminths: Current situation and lessons from livestock. *Clin. Microbiol. Rev.* 13, 207, 2000.
35. Albonico, M., Wright, V., and Bickle, Q. Molecular analysis of the β-tubulin gene of human hookworms as a basis for possible benzimidazole resistance on Pemba Island. *Mol. Biochem. Parasitol.* 134, 281, 2004.
36. Schwenkenbecher, J.M. et al. Characterization of beta-tubulin genes in hookworms and investigation of resistance-associated mutations using real-time PCR. *Mol. Biochem. Parasitol.* 156, 167–174, 2007.
37. Churcher, T.S. et al. Mass treatment of parasitic diseases: Implications for the spread of anthelmintic resistance. In: Weber, J.T. (ed.), *Antimicrobial Resistance—Beyond the Breakpoint. Issues Infect. Dis.*, vol. 6, pp. 120–137, Karger, Basel, 2010.
38. Albonico, M. et al. Rate of re-infection with intestinal nematodes after treatment of children with mebendazole or albendazole in a highly endemic area. *Trans. R. Soc. Trop. Med. Hyg.* 89, 538, 1995.
39. Bethony, J.M. et al. Randomized, placebo-controlled, double-blind trial of the Na-ASP-2 hookworm vaccine in unexposed adults. *Vaccine* 26, 2408, 2008.
40. Diemert, D.J., Bethony, J.M., and Hotez, P.J. Hookworm vaccines. *Vaccines* 46, 282, 2008.
41. Pearson, M.S. An enzymatically inactivated hemoglobinase from *Necator americanus* induces neutralizing antibodies against multiple hookworm species and protects dogs against heterologous hookworm infection. *FASEB J.* 23, 3007, 2009.
42. Loukas, A. et al. Vaccination with recombinant aspartic hemoglobinase reduces parasite load and blood loss after hookworm infection in dogs. *PLoS Med.* 2, e295, 2005.
43. Zhan, B. et al. Biochemical characterization and vaccine potential of a heme-binding glutathione transferase from the adult hookworm *Ancylostoma caninum. Infect. Immun.* 73, 6903, 2005.
44. Brown, A. et al. An initial characterization of the proteolytic enzymes secreted by the adult stage of the human hookworm *Necator americanus. Parasitology* 110, 555, 1995.
45. Brown, A., and Pritchard, D.I. The imunogenicity of hookworm (*Necator americanus*) acetylcholinesterase (AChE) in man. *Parasite Immunol.* 15, 195, 1993.
46 Pritchard, D.I. et al. Epidemiology and immunology of *Necator americanus* infection in a community in Papua New Guinea: Humoral responses to excretory-secretory and cuticular collagen antigens. *Parasitology* 100, 317, 1990.
47. Pritchard, D.I. et al. Isotypic variation in antibody responses in a community in Papua New Guinea to larval and adult antigens during infection, and following reinfection, with the hookworm *Necator americanus. Parasite Immunol.* 14, 617, 1992.
48. Palmer, D.R., Bradley, M., and Bundy, D.A. IgG4 responses to antigens of adult *Necator americanus*: Potential for use in large-scale epidemiological studies. *Bull. World Health Org.* 74, 381, 1996.
49. Quinnell, R.J., Bethony, J., and Pritchard, D.I. The immunoepidemiology of human hookworm infection. *Parasite Immunol.* 26, 443, 2004.
50. Loukas, A., Constant, S.L., and Bethony, J.M. Immunobiology of hookworm infection. *FEMS Immunol. Med. Microbiol.* 43, 115, 2005.
51. Chow, S.C. The human hookworm pathogen *Necator americanus* induces apoptosis in T lymphocytes. *Parasitology* 22, 21, 2000.
52. Daub, J. et al. A survey of genes expressed in adults of the human hookworm, *Necator americanus. Parasitology* 120, 171, 2000.
53. Brophy, P.M. et al. Secretory nematode SOD-offensive or defensive? *Int. J. Parasitol.* 25, 865, 1995.
54. Taiwo, F.A. et al. Cu/Zn superoxide dismutase in excretory-secretory products of the human hookworm *Necator americanus*: An electron paramagnetic study. *Eur. J. Biochem.* 264, 434, 1999.
55. Pritchard, D.I. Do haematophagous parasites secrete SOD and promote blood flow? *Int. J. Parasitol.* 26, 1339, 1996.
56. Culley, F.J. et al. Eotaxin is specifically cleaved by hookworm metalloproteases preventing its action *in vitro* and *in vivo*. *J. Immunol.* 165, 6447, 2000.
57. Pritchard, D.I. et al. A hookworm allergen which strongly resembles calreticulin. *Parasite Immunol.* 21, 439, 1999.
58. Pritchard, D.I., and Brown, A. Is *Necator americanus* approaching a mutualistic relationship with humans? *Trends Parasitol.* 17, 169, 2001.
59. Pritchard, D.I. et al. Immunity in humans to *N. americanus*: IgE, parasite weight and fecundity. *Parasite Immunol.* 71, 71, 1995.
60. Croese, J. et al. A proof of concept study establishing *Necator americanus* in Crohn's patients and reservoir donors. *Gut* 55, 136, 2006.
61. Lynch, N.R. et al. Clinical improvement of asthma after anthelminthic treatment in a tropical situation. *Am. J. Respir. Crit. Care Med.* 156, 50, 1997.
62. Pritchard, D.I. The pro-allergic influences of helminth parasites. *Mem. Inst. Oswaldo Cruz* 92, 15, 1997.
63. Oppenheimer, S.J. Iron and infection: A clinical review. *J. Singap. Paediatr. Soc.* 36, 1, 1994.

64 Nacher, M. et al. Short report: Hookworm infection is associated with decreased body temperature during mild *Plasmodium falciparum* malaria. *Am. J. Trop. Med. Hyg.* 65, 136, 2001.
65 Cooper, P.J. et al. Human infection with *Ascaris lumbricoides* is associated with a polarized cytokine response. *J. Infect. Dis.* 182, 1207, 2000.
66. Harada, Y., and Mori, O. A new method for culturing hookworm. *Yonago Acta Med.* 1, 177, 1955.
67. Polderman, A.M. et al. Oesophagostomiasis, a common infection of man in northern Togo and Ghana. *Am. J. Trop. Med. Hyg.* 44, 336, 1991.
68. Blotkamp, J. et al. Observations on the morphology of adults and larval stages of *Oesophagostomum* sp. isolated from man in northern Togo and Ghana. *J. Helminthol.* 67, 49, 1993.
69. Krepel, H.P. et al. Quantitative interpretation of coprocultures in a population infected with *Oesophagostomum bifurcum. Trop. Geogr. Med.* 47, 157, 1995.
70. Hebert, P.D. et al. Biological identification though DNA barcodes. *Proc. Biol. Sci.* 270, 313, 2003.
71. Blaxter, M.L. The promise of a DNA taxonomy. *Philos. Trans. R. Soc. Lond. B. Biol. Sci.* 359, 669, 2004.
72. Schwenkenbecher, J.M., Mordue, A.J., and Piertney, S.B. Phylogenetic analysis indicates that *Culicoides dewulfi* should not be considered part of the *Culicoides obsoletus* complex. *Bull. Entomol. Res.* 99, 371, 2009.
73. Schwenkenbecher, J.M. et al. Discrimination of *Culicoides* midge larvae using multiplex polymerase chain reaction assays based on DNA sequence variation at the mitochondrial cytochrome C oxidase I gene. *J. Med. Entomol.*, 46, 610, 2009.
74. Zhan, B. et al. Species-specific identification of human hookworms by PCR of the mitochondrial cytochrome oxidase I gene. *J. Parasitol.* 87, 1227, 2001.
75. Hawdon, J.M. et al. Genetic structure of populations of the human hookworm, *Necator americanus*, in China. *Mol. Ecol.* 10, 1433, 2001.
76. Yong, T.S. et al. Differential diagnosis of *Trichostrongylus* and hookworm eggs via PCR using ITS-1 sequences. *Korean J. Parasitol.* 45, 69, 2007.
77. Verweij, J.J. et al. Simultaneous detection and quantification of *Ancylostoma duodenale*, *Necator americanus*, and *Oesophagostomum bifurcum* in fecal samples using multiplex real-time PCR. *Am. J. Trop. Med. Hyg.* 77, 685, 2007.
78. Orita, M. et al. Detection of polymorphisms of human DNA by gel electrophoresis as single-strand conformation polymorphisms. *Proc. Natl. Acad. Sci. USA* 86, 2766, 1989.
79. Gasser, R.B., and Monti, J.R. Identification of parasitic nematodes by PCR-SSCP of ITS-2 rDNA. *Mol. Cell. Probes* 11, 201, 1997.
80. Gasser, R.B. et al. A mutation scanning approach for the identification of hookworm species and analysis of population variation, *Mol. Biochem. Parasitol.* 92, 303, 1998.
81. Hu, M. et al. Single-strand conformation polymorphism-based analysis of mitochondrial cytochrome c oxidase subunit 1 reveals significant substructuring in hookworm populations. *Electrophoresis* 23, 27, 2002.
82. Saiki, R.K. et al. Enzymatic amplification of beta-globin genomic sequences and restriction site analysis for diagnosis of sickle cell anemia. *Science* 230, 1350, 1985.
83. Hawdon, J.M. Differentiation between the human hookworms *Ancylostoma duodenale* and *Necator americanus* using PCR-RFLP. *J. Parasitol.* 82, 642, 1996.
84. Kato, K., and Miura, M. Comparative examinations. *Jpn. J. Parasitol.* 3, 35, 1954.
85. Katz, N., Chavez, A., and Pellegrino, J. A simple device for quantitative stool thick-smear technique in *Schistosomiasis mansoni. Rev. Inst. Med. Trop. Sao Paulo* 14, 397, 1972.
86. Engels, D., Nahimana, S., and Gryseels, B. Comparison of the direct faecal smear and two thick smear techniques for the diagnosis of intestinal parasitic infections, *Trans. R. Soc. Trop. Med. Hyg.* 90, 523, 1996.
87. Santos, F.L., Cerqueira, E.J., and Soares, N.M., Comparison of the thick smear and Kato-Katz techniques for diagnosis of intestinal helminth infections. *Rev. Soc. Bras. Med. Trop.* 38, 196, 2005.
88. Rivera, W.L., Tachibana, H., and Kanbara, H. Application of the polymerase chain reaction (PCR) in the epidemiology of *Entamoeba histolytica* and *Entamoeba dispar* infections, *Tokaj J. Exp. Clin. Med.* 23, 413, 1998.

60 Oesophagostomum

Jaco J. Verweij

CONTENTS

60.1 INTRODUCTION

60.1.1 Classification, Morphology, and Biology

Although several reports on human infections with *Oesophagostoma* species had been published [1] starting as early as 1905 [2], human oesophagostomiasis has been considered for a long time as a rare zoononis. However, in 1964 Haaf and Van Soest reported on nine cases of human *Oesophagostomum* infections in Bawku in northern Ghana [3] and in 1987 Gigase and others reported on 54 cases in the northern region of Togo [4]. These publications suggested that human infections with *Oesophagostoma* species could be more than a rare zoonotic infection in this region of West Africa. To answer the questions raised on the epidemiology and biology of human *Oesophagostomum* infections several large studies on the taxonomy, prevalence, and clinical presentation of these infections have been performed. However, study on the prevalence of this infection cannot be carried out by simple microscopic examination of stool samples with Kato–Katz for the presence of eggs, which is a normal approach for other intestinal helminth infections. *Oesophagostomum bifurcum* eggs cannot be differentiated from eggs of other strongylid nematodes such as *Ancylostoma duodenale* and *Necator americanus*. So, coprocultures are required to allow eggs to hatch and develop for the characteristic third-stage larvae (L3s) for the differential diagnosis. In the first survey, stool samples from more than 3000 individuals in 25 villages in northern Togo and 16 villages in northern Ghana were examined using coproculture for the presence of the characteristic L3 of *Oesophagostomum* species and *N. americanus*. Prevalence of *Oesophagostomum* infections reached 50% or more in some villages and up to 100% for hookworm infections [5,6]. In later years, more data on the distribution and biology of *O. bifurcum* infections were collected using coproculture as the diagnostic method [7–10].

In an elaborated study on the taxonomy of adult worms, L1, and L3 of the *Oesophagostomum* species found, it was shown that *O. bifurcum* was the species involved in these infections [1]. These findings are later confirmed by genetic analysis of the internal transcribed spacer (ITS) region and the cytochrome *c* oxidase I gene of worms obtained from humans and nonhuman primates in which only little sequence variation was found [11,12]. *O. bifurcum* are helminths within the phylum of the Nematoda, order of the Rhabditida, and family of the Oesophagostominae. Adult males measure 8.6–15.1 mm × 0.2–0.5 mm wide and females are slightly larger [1]. The *O. bifurcum* adult female worms produce ≈5000 eggs per day [7]. The eggs are excreted with the feces and will hatch and develop in moist, shady, and warm soil via several larval stages into the infective L3s. It is known that these larvae are resistant to environmental stresses like desiccation or freezing [13,14]. In contrast to hookworm infection, it is not likely that these larvae can penetrate the skin. Infection is assumed to take place through the ingestion of L3s in contaminated water, food, or soil. After ingestion, they enter the colonic mucosa and form small nodules, in which they develop into fourth-stage larvae. These larvae can remain in the nodules (i.e., arrested state) or migrate to the intestinal lumen where they develop into adult male and female worms [14].

It has been suggested that nonhuman primates could act as the reservoir for human infections with *O. bifurcum*. In the last decade, however, accumulating studies suggests strongly that cross-infection does not occur between nonhuman primates and humans [15]. Population diversity within *O. bifurcum* was already suggested based on the variation in the size of adult worms retrieved from humans and different species of monkeys [1,16]. Furthermore, experimental animal studies showed that the parasite obtained from humans is poorly adapted to nonhuman primates [17]. However, in two early studies assessing the genetic variability in nuclear ribosomal and mitochondrial DNA regions within *O. bifurcum* from human and primate hosts no clear sequence differences were detected in the ITS-2 and in a part of the cytochrome *c* oxidase subunit 1 gene (p*cox*1), respectively [11,12]. Finally, random amplification of polymorphic DNA (RAPD) and amplified fragment-length polymorphism (AFLP) methods (Figure 60.1) showed distinct clusters when applied on DNA originated

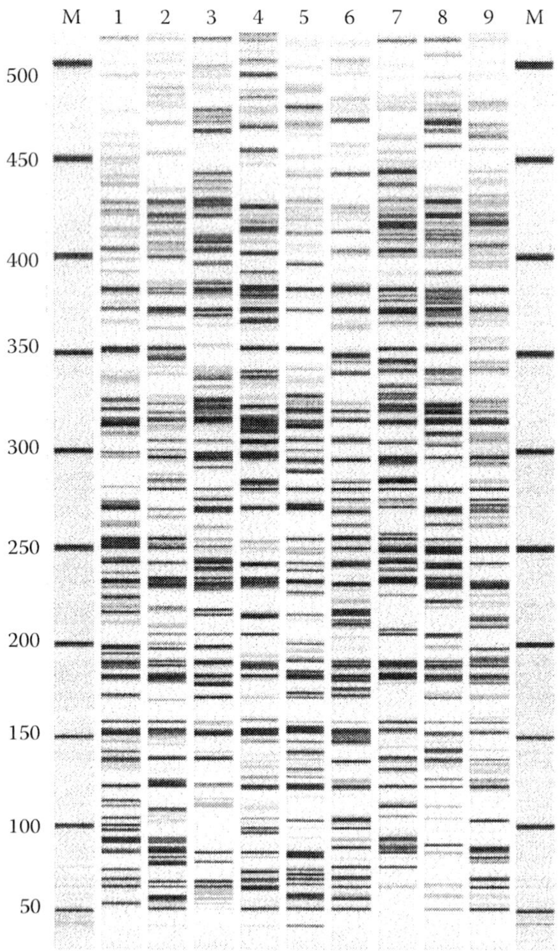

FIGURE 60.1 Example of the AFLP banding patterns obtained for *O. bifurcum* from humans from Ghana (lanes 1–9) using the restriction enzyme and primer combination HindIII + AG/BglII + AC. M represents the ALFexpress sizer 50–500 molecular weight marker. The numbers on the left indicate the fragment size in base pairs.

from humans and nonhuman primates. Moreover, these techniques even revealed genetically distinct groups of *O. bifurcum* isolates from Patas monkey, from the Mona monkey, and from the Olive baboon [18,19] (Figure 60.2). This shows that *O. bifurcum* infections are caused by host species-affiliated variants of *O. bifurcum*. Therefore, it is not likely that cross-infection with these *O. bifurcum* variants occurs between nonhuman primates and humans.

60.1.2 Clinical Features

In veterinary medicine, nodular disease or oesophagostomiasis is well known in herbivores and primates. The disease is characterized by nodules in the intestinal wall; the nodules are filled with caseous material and result from host response to encystment of the larvae of *Oesophagostomum* species. In human oesophagostomiasis, two distinct clinical presentations of the disease exist, a uni-nodular presentation which is named "Dapaong tumour" and multinodular disease [20]. Surgical intervention may be needed in cases of luminal narrowing of the colon, or bowel obstruction secondary to inflammatory adhesions. Ultrasound detection of *O. bifurcum*-induced colonic pathology proved to be a diagnostic as well as an epidemiological tool in the description of the disease [21–24]. Anthelmintic treatment with albendazole has been shown to be effective against the lumen dwelling adult worms and for the nonsurgical treatment of nonacute cases of clinical oesophagostomiasis [8,25].

60.1.3 Laboratory Diagnosis

In patient diagnosis, direct smears from fecal samples and wet mount preparations of a formalin–ether concentrate are examined by microscopic examination for ova. The most common method used to describe all soil transmitted helminth (STH) infections, including hookworm (*N. americanus* and *A. duodenale*) within a human population, involves counting the number of eggs per gram of feces by the Kato–Katz thick-smear technique [26]. This technique was originally described and widely used for the detection and quantification of *Schistosoma mansoni* eggs in feces. Microscopic examination of Kato–Katz slides for the presence of hookworm-like eggs must be read within 30 min or many eggs will no longer be visible. This is in contrast with *S. mansoni*, *Ascaris lumbricoides*, and *Trichuris trichiura* eggs which can be detected best after 24–48 h. Specific detection of *Oesophagostomum* infections is hampered by the fact that the eggs of *Oesophagostomum* species are morphologically indistinguishable from those of hookworm species. Therefore, coprocultures are needed to allow eggs to hatch and develop for the characteristic L3s for the differential diagnosis [1,5].

However, coproculture-based diagnostics are completely dependent on the availability of fresh stool specimens and differentiation of the L3s, of *O. bifurcum*, *A. duodenale*, and *N. americanus*, requires a highly skilled microscopist. Large-scale application of coproculture in epidemiological surveys is also hampered by the fact that these cultures

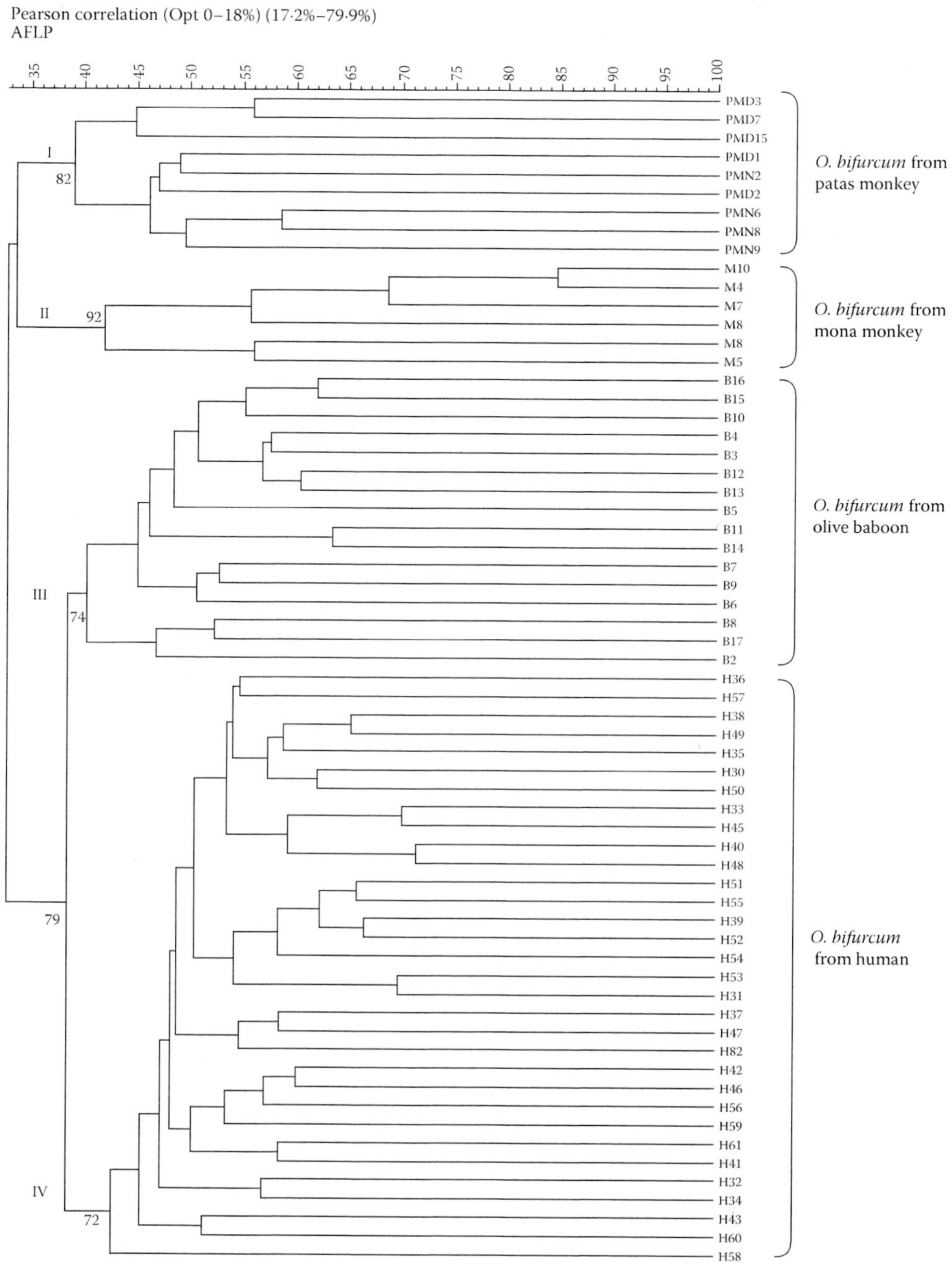

FIGURE 60.2 Dendrogram based on AFLP analysis of 63 *O bifurcum* from human ($n = 32$), Patas monkey ($n = 9$), Mona monkey ($n = 6$), and Olive baboon ($n = 16$) using the HindIII + AG/BglII + AC. The scale represents the Pearson's product-moment correlation coefficient (r) in percentage. The numbers represent cophenetic values. (Figures reprinted with permission (Cambridge University Press) from de Gruijter, J.M. et al. High resolution DNA fingerprinting by AFLP to study the genetic variation among *Oesophagostomum bifurcum* (Nematoda) from human and non-human primates from Ghana. *Parasitology* 130, 229, 2005.)

have to be kept in closed Petri dishes for 1 week. At tropical temperatures, this frequently leads to the development of maggots and overgrowth of fungi, and coprocultures are performed in duplicate to obtain more reliable results [10]. As an alternative, polymerase chain reaction (PCR)-based methods for specific detection of *O. bifurcum* DNA in fecal samples have been used successfully for species-specific detection and differentiation of *O. bifurcum*, *A. duodenale*, and *N. americanus* [27–30].

60.1.3.1 Conventional Techniques

60.1.3.1.1 Microscopy

The typical thin-shelled eggs with a clear space between the developing embryo and the shell measure ≈65 × 40 μm and can be detected in direct smears from fresh fecal samples and wet mount preparations of a formalin–ether concentrate without additional staining. The eggs of *O. bifurcum* and those of the different hookworm species are morphological identical and

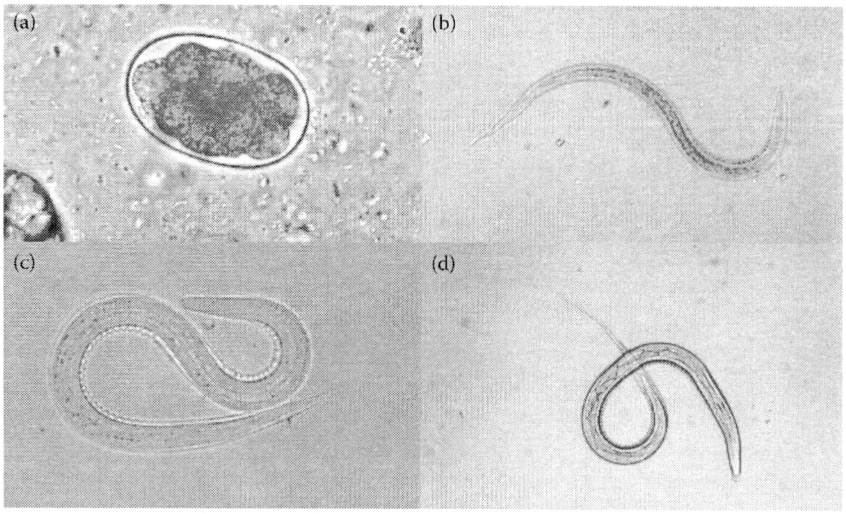

FIGURE 60.3 (a) Hookworm egg; (b) *N. americanus* L3s; (c) *A. duodenale*; (d) *O, bifurcum* L3s. (Courtesy of Eric A.T. Brienen, LUMC, Leiden.)

similar to those of *Trichostrongylis* species. The eggs may develop and hatch into first-stage larvae if the stools remain unpreserved for over 24 h. These larvae have to be differentiated from those of the different hookworm species, *S. stercoralis*, and free-living ground nematodes that can appear quit frequently if stools are collected from contaminated soil (Figure 60.3). Kato–Katz slides used for microscopic detection and quantification of helminth eggs must be read within 30 min or many hookworm-like eggs will not longer be visible. This in contrast with *S. mansoni*, *Ascaris lumbricoides*, and *Trichuris trichiura* eggs which can be detected best after 24–48 h.

60.1.3.1.2 Coproculture

The eggs of *O. bifurcum* cannot be differentiated morphologically of other nematodes such as *A. duodenale*, and *N. americanus*. Therefore, coproculture is needed to allow eggs to hatch and develop into L3s for differentiation at the species level [5]. Briefly, ~2 g of fresh stool is homogenized and mixed with an equal amount of vermiculite or ground charcoal. Filter paper is placed on an elevation in a Petri dish and tap water is added on the bottom. The stool–vermiculite mixture is brought onto the filter paper. The closed Petri dish is incubated at room temperature in a dark place. The culture has to be checked daily and stirred in case of fungal growth and fly larvae have to be removed. Water has to be added when necessary to avoid desiccation. After 5–7 days, the filter paper with the stool–vermiculite mixture is removed and the water is poured into a 50 mL tube. After sedimentation time of 2 h the sediment is pipetted from the bottom of the tube and examined with 4×10 magnifications. Larvae that are found can be killed by adding a drop of iodine and examined with a higher magnification for detailed examination of the morphological characteristics to achieve an accurate species determination.

60.1.3.1.3 Serological Tests

Alternative diagnostic methods using serological tests based on the detection of specific IgG_4 and IgE antibodies gave promising results [31,32]. Preabsorption of sera with *N. americanus* antigens showed much lower optical densities in the detection of IgG4 antibodies against *O. bifurcum* antigens confirming the nonspecific binding of *N. americanus*-specific antibodies to *O. bifurcum* antigens. IgE antibodies against *O. bifurcum* and *N. americanus* antigens appeared more specific [31]. However, serology has not become a major diagnostic tool because in the *O. bifurcum* endemic areas, drawing blood is little accepted, and serology cannot differentiate between current and past infections.

60.1.3.1.4 In Vivo Assays

They are not being used for diagnosis but have contributed in elucidating gaps in the knowledge of the transmission and epidemiology of *O. bifurcum* infections [17].

60.1.3.2 Molecular Techniques

Genetic markers were found for *O. bifurcum* and *N. americanus* in the ITS-2 of ribosomal DNA [33]. Nuclear ribosomal and mitochondrial DNA regions were used to assess the genetic variability by mutation scanning within *O. bifurcum* from human and monkey from Ghana [11,12]. Finally, DNA fingerprinting using RAPD and AFLP revealed a distinct genetic substructuring within *O. bifurcum* from human and nonhuman primates [18,19,34].

60.1.3.2.1 Species-Specific Detection

For species-specific detection, the target DNA should show very limited variation within the species and are different enough to allow differentiation between closely related species. Species-specific primers were designed from the ITS-2 of ribosomal DNA for the specific amplification of *O. bifurcum* and *N. americanus* DNA in a PCR. Initially, the efficiency and specificity of this method were tested using small amounts of DNA derived from adult worms of *O. bifurcum* and *N. americanus* and a range of control samples [33]. It was shown that this PCR had an excellent performance to specifically detect minute amounts of parasitic DNA. However, it did not achieve effective amplification using DNA isolated from fecal samples. Subsequently, a two-step semi-nested PCR proved to be highly sensitive and specific for the detection of *O. bifurcum* and *N. americanus* DNA isolated from stool samples [27,28].

Real-time PCR using fluorescent detection probes reduces the risk of contamination and substantially reduces labor time and reagent costs, through the possibility of combining assays for the detection of different targets into one reaction. Real-time PCR also offers the possibility of quantification using the threshold cycle; this is the amplification cycle in which the level of fluorescent signal exceeds background fluorescence, and this correlates with the initial copy number at the beginning of the reaction. Recently, a real-time PCR for the simultaneous detection and quantification of *A. duodenale*, *N. americanus*, and *O. bifurcum* DNA showed an excellent performance using fecal samples from an area in northern Ghana where human infections with all three species are endemic [29]. Thus PCR proved to be specific (100%) using a set of well-defined control samples ($n = 150$) and sensitive (100, 100, and 86.7%) for the detection of *A. duodenale*, *N. americanus*, and *O. bifurcum*, respectively, as compared with microscopy and coproculture using fecal samples ($n = 399$) from a region in Ghana where all three parasites are endemic.

60.1.3.2.2 Strains

PCR-coupled single-strand conformation polymorphism (SSCP) in combinations with DNA sequencing analysis of the ITS-2 and the p*cox*1 gene did not provide sufficient levels of variability to indentify genetic variants or "strains" within *O. bifurcum* populations [11,12,35]. Fingerprinting methods such as RAPD and AFLP analysis have been successfully used to explore genetic diversity within *Oesophagostomum* species. Cluster analysis of the RAPD profile showed that *O. bifurcum* from human and nonhuman primates represent genetically distinct groups but there was no association between the *O. bifurcum* genotype and geographical origin of the host species. These results were confirmed and even more distinct using AFLP showing that *O. bifurcum* from human and different species of nonhuman primates represent genetically distinct groups [16,18,19,34].

60.1.3.2.3 Susceptibility Testing

Resistance to benzimidazoles (BZ) anthelmintics in several *Oesophagostomum* species of livestock has been reported. Until now, however, no genetic markers are known that are associated with anthelmintics resistance in *Oesophagostomum*.

60.2 METHODS

60.2.1 Sample Preparation

60.2.1.1 Sample Collection

Unpreserved stool samples are preferred for molecular studies but have to be stored and transported frozen, mixing stool samples with ethanol allows transportation and storage at room temperature for at least three months. If formalin-fixed fecal samples are used the formalin has to be removed by washing/centrifuging steps with saline direct or after formalin–ether concentration. However, sensitivity of amplification is known to decrease with type of fixation.

60.2.1.2 DNA Isolation

Both in-house and commercial reagents have been used for the preparation of DNA from hookworm eggs. We present a method below that is used in our laboratory.

60.2.1.2.1 DNA Isolation from Feces Using QIAamp Spin Columns [28]*

1. Suspend 100 mg of feces in 200 μL of phosphate-buffered saline (PBS) with 2% PVPP† (polyvinylpolypyrolidone, Sigma P6755).‡
2. Freeze sample overnight at −20°C.
3. Heat feces suspension for 10 min in heatblock at 100°C.
4. Add an equal volume of Tissue Lysis Buffer (ATL) containing 20 μL proteinase K (i.e., 180 μL of ATL and 20 μL of proteinase K from the kit) solution to the sample (vortex). Incubate for 2 h or overnight at 55°C in heatblock (after 1 h vortex briefly).
5. Add 400 μL AL Buffer (with spike) to 400 μL sample, mix thoroughly by vortexing, and incubate at 70°C for 10 min.
6. Centrifuge at 14000 rpm for 30 s full speed and transfer supernatant to an Eppendorf containing 400 μL of ethanol (96–100%), and mix thoroughly by vortexing (short spin).
7. Place a numbered spin column in a 2 mL collection tube. Carefully apply 600 μL of the mixture from step 5 to the spin column without moistening the rim, close the cap, and centrifuge at 10,000 rpm for 1 min.
8. Place spin column in a clean 2 mL collection tube and repeat with the rest of the mixture from step 5.
9. Place the spin column in a clean 2 mL collection tube.
10. Carefully open the spin column and add 500 μL of AW1 buffer. Centrifuge at 10,000 rpm for 1 min.
11. Empty the collection tube and add 500 μL of AW2 buffer. Centrifuge for 3 min–1 min at 10,000 rpm and 2 min at 14,000 rpm full speed.
12. Place the spin column in a numbered, clean 1.5 mL microfuge tube (with lit).
13. Carefully open the spin column, elute the DNA with 200 μL AE buffer, leave the buffer on the filter for 1 min and centrifuge at 8000 rpm for 1 min.

* QIAamp DNA minikit cat. No. 51306 gives better results than QIAamp stool kit.

† Or suspend 0.5 g of feces in 2 mL of PBS with 2% PVPP and sieve (filter). For isolation of the DNA aliquot a separate amount of 200 μL in an Eppendorf tube, and store the rest at −70°C.

‡ PVPP does not dissolve; it stays like a suspension so you have to stir regularly while using it.

60.2.1.2.2 DNA Isolation after Ethanol Fixation [36]

1. Apply ~250 µL of an ethanol–feces suspension to a clean Eppendorf tube and centrifuge for 1 min (10,000 rpm).
2. Remove the ethanol with a micropipette and wash the pellet with 1000 µL of PBS and centrifuge for 1 min (10,000 rpm).
3. Remove the PBS with micropipette and add 200 µL of 2% PVPP (shake bottle before use).
4. Continue with DNA isolation protocol above from step 2.

60.2.2 Detection Procedures

60.2.2.1 Semi-Nested PCR [27,28]

60.2.2.1.1 Principle

A semi-nested PCR protocol has been reported for the detection and differentiation of *O. bifurcum* from *N. americanus* infections. This PCR generates a 310- and 450-bp fragment, respectively, of the ITS-2 region in the first-round amplification with primers NC1 and NC2 [37]. The second-round amplification is performed with primers OB, NA, and NC2 resulting in a 220- and 250-bp amplicons for *O. bifurcum* and *N. americanus*, respectively (Table 60.1).

60.2.2.1.2 Procedure

1. Prepare the first-round PCR mixture (20 µL) as follows, and add 5 µL DNA in the test tube (or distilled water in the negative control tube):

First-Round PCR		1 × mix µL	20 × mix µL
Hotstart mastermix		12.5	250
Primer NC1	25 µM	0.5	10
Primer NC2	25 µM	0.5	10
BSA	5 mg/mL	0.5	10
H_2O		6.0	120
Total		20.0	400

2. Run PCR amplification in a thermocycler for 1 cycle of 95°C for 15 min; 25 cycles of 95°C for 15 s, 55°C for 30 s, and 72°C for 30 s; 1 cycle of 72°C for 7 min.
3. Prepare the second-round PCR mixture (20 µL) as follows, and add 5 µL of PCR product from the first-round PCR in each tube (or distilled water in the negative control tube).

Second-Round PCR		1 × mix µL	20 × mix µL
Hotstart mastermix		12.5	250
Primer NC2	25 µM	0.5	10
Primer OB	25 µM	0.5	10
BSA	5 mg/mL	0.5	10
H_2O		6.0	120
Total		20.0	400
Hotstart mastermix		12.5	250
Primer NC2	25 µM	0.5	10
Primer NA	25 µM	0.5	10
BSA	5 mg/mL	0.5	10
H_2O		6.0	120
Total		20.0	400

4. Perform PCR amplification as above (step 2).
5. Separate PCR products (10 µL) on 2% agarose gel containing ethidium bromide and visualize with a UV transilluminator.

TABLE 60.1
Primers for Semi-Nested PCR

Oligo Name	Sequence
NC1	5′-ACG TCT GGT TCA GGG TTG TT-3′
NC2	5′-TTA GTT TCT TTT CCT CCG CT-3′
OB	5′-TAT ATT GCA ACA GGT ATT TTG GTA C-3′
NA	5′-ATG TGC ACG CTG TTA TTC ACT-3′

60.2.2.2 Real-Time PCR [29]

60.2.2.2.1 Principle

A multiplex real-time PCR method was developed for the simultaneous detection and quantification of *A. duodenale*, *N. americanus*, and *O. bifurcum* in stool samples [29]. The *A. duodenale*-specific real-time PCR amplifies and detects a 71-bp fragment within the ITS-2 sequence of *A. duodenale*. The *N. americanus* PCR primers and detection probe were designed on the ITS-2 sequence for *N. americanus*, amplifying and detecting a 103-bp fragment inside the ITS-2 sequence. The *O. bifurucm*-specific primers amplify a 105-bp fragment of the ITS2 sequence and the XS-probe detects the *O. bifurcum*-specific product.

This multiplex PCR included an internal control to detect inhibition of the amplification by fecal constituents in the sample (Table 60.2).

60.2.2.2.2 Procedure

1. Prepare a PCR mixture (25 µL) containing PCR buffer (Hotstart Taq master mix, QIAgen), 5 mM $MgCl_2$, 2.5 µg Bovine Serum Albumin (Roche Diagnostics Nederland B.V., Almere, the Netherlands), 5 pmol of each *A. duodenale*-specific primer, and of each *N. americanus*-specific primer, 2.5 pmol of each *O. bifurcum*-specific primer, and 3.75 pmol of each PhHV-1-specific primer, 2.5 pmol of A. duodenale-specific XS-probe, and 1.25 pmol of each, *N. americanus*-specific XS-probe, *O. bifurcum*-specific XS-probe, PhHV-1-specific double-labeled probe, and 5 µL of the DNA sample. Amplification consists of 15 min at 95°C followed by 50 cycles of 15 s at 95°C, 30 s at 60°C, and 30 s at 72°C. Amplification, detection, and analysis were performed with the CFX real-time detection system (Bio-Rad laboratories). Negative

TABLE 60.2
Oligonucleotide Primers and Detection Probes for Real-Time PCR for the Simultaneous Detection of *Enterocytozoon Bieneusi, Encephalitozoon Spec,* and Phocin Herpes Virus 1 as an Internal Control

Oligonucleotide Name	Oligonucleotide Sequence	Genbank Accession no. or Literature Reference of Target Sequence
	Target Organism	
A. duodenale		
Ad125F	5′-GAATGACAGCAAACTCGTTGTTG-3′	AJ001594
Ad195R	5′-ATACTAGCCACTGCCGAAACGT-3′	AJ001594
Ad155-XS	Texas Red-5′-ATCGTTTACCGACTTTAG-3′-XS	AJ001594
N. americanus		
Na58F	5′-CTGTTTGTCGAACGGTACTTGC-3′	AJ001599
Na158R	5′-ATAACAGCGTGCACATGTTGC-3′	AJ001599
Na81Tmgb	FAM-5′-CTGTACTACGCATTGTATAC-3′-XS	AJ001599
O. bifurcum		
oeso31F	5′-GTTTGTCGAACGATGCTTACATTT-3′	Y11733
oeso136R	5′-CGAGGTCACGACATCCGTG-3′	Y11733
Obif63Tmgb	Yakima Yellow 5′-CCTCGTTCTAGATAAGAAAT-3′-XS	Y11733
Phocin Herpes Virus 1		
PhHV-267s	5′-GGGCGAATCACAGATTGAATC -3′	[38]
PhHV-337as	5′-GCGGTTCCAAACGTACCAA -3′	[38]
PhHV-305tq	Cy5 5′-TTTTTATGTGTCCGCCACCATCTGGATC-3′-BHQ2	[38]

and positive control samples are included in each run, and 5 μL of the DNA sample as follows:

			1 × mix μL	100 × mix μL
H_2O			1.53	153
$MgCl_2$ 25 mM	25 mM		3.50	350
BSA 5 mg/mL	5 mg/mL		0.50	50
Ad-125F	25 μm		0.20	20
Ad-195R 10 μm	25 μm		0.20	20
Ad-155 Texas Red	10 μm	Texas Red	0.25	25
Na-58F	25 μm		0.20	20
Na-158R	25 μm		0.20	20
Na-81 FAM	10 μm	FAM	0.125	13
Oeso-31F	25 μm		0.10	10
Oeso-136R	25 μm		0.10	10
Obif-63 VIC	10 μm	Yakima Yellow	0.125	13
PHHV-F	25 μm		0.15	15
PHHV-R	25 μm		0.15	15
PHHV-CY5	10 μm	Cy5	0.125	13
Hotstart mastermix			12.50	1250
Total			20.00	2000

2. Perform PCR in CFX real-time detection system (BioRad, Hercules, CA) for 1 cycle of 95°C for 15 min; 50 cycles of 95°C for 15 s, 60°C for 30 s, and 72°C for 30 s. (For amplification in ABI 7500, the cycling program consists of 1 cycle of 95°C for 15 min; 50 cycles of 95°C for 15 s, and 60°C for 60 s.)
3. Detect fluorescence of Texas Red, FAM, Yakima Yellow, and CY5 at their respective wavelengths during the annealing step of each cycle.

Note: The primers of the *O. bifurcum* PCR show 100% homology with *Oesophagostomum stephanostomum* and the probe only shows one mismatch with the ITS-2 sequence of *O. stephanostomum*. Therefore, *O. stephanostomum* will be amplified and also probably detected with this primer/probe combination. The *A. duodenale* primers and probe were designed on the known sequence of *A. duodenale* (AJ001600). According to NCBI BLAST analysis, however, *Ancylostoma ceylanicum* will be amplified and detected also with this primer/probe combination. The *N. americanus* probe shows a two-base mismatch with the ITS-2 sequence of a *N. americanus* isolate from Sawarak (Genbank accession AJ001600). The detection of the PCR product of such an isolate will probably fail or at least be less efficient.

60.3 CONCLUSION

Reports of isolated human cases with oesophagostomiasis have been reported worldwide. Until now, only in northern Togo and Ghana infections of humans with *O. bifurcum* have been shown to be extremely common and prevalence reached

50% or more in some villages [14]. In future, real-time PCR would enable a rapid screening of individuals in the immediate surrounding of a clinical case of human oesophagostomiasis and discover or dismiss new endemic foci of this disease. Recently, it has been shown that the infection rates of *O. bifurcum* have decreased dramatically one and two years after mass treatment [39–41]. It has been predicted that human oesophagostomiasis can be eliminated as a public health problem and even can be eradicated completely as a human infection in northern Ghana through a series of mass treatments with albendazole as part of the lymphatic filariasis eradication program (LFEP) [40]. Real-time PCR provides a simple method for monitoring the effect of treatment in intervention programs and it would be interesting to use it after such programs have been stopped to detect if *O. bifurcum* infections will reemerge.

REFERENCES

1. Blotkamp, J. et al. Observations on the morphology of adults and larval stages of *Oesophagostomum* sp. isolated from man in northern Togo and Ghana. *J. Helminthol.* 67, 49, 1993.
2. Railliet, A., and Henry, A. Encore un nouveau sclerostomien (*Oesophagostomum brumpti* nov. sp.) parasite de l'homme. *Comtes Rendus De Seances De La Societe De Biologie* 58, 643, 1905.
3. Haaf, E., and van Soest, A.H. Oesophagostomiasis in man in North Ghana. *Trop. Geogr. Med.* 16, 49, 1964.
4. Gigase, P. et al. Frequency of symptomatic human oesophagostomiasis (helminthoma) in Northern Togo. In: Geerts, S., Kumar, V., and Brandt, J. (eds), *Helminth Zoonosis*, p. 223, Martinus Nijhoff Publ., Dordrecht, 1987.
5. Polderman, A.M. et al. Oesophagostomiasis, a common infection of man in northern Togo and Ghana. *Am. J. Trop. Med. Hyg.* 44, 336, 1991.
6. Krepel, H.P. et al. Human *Oesophagostomum* infection in northern Togo and Ghana: Epidemiological aspects. *Ann. Trop. Med. Parasitol.* 86, 289, 1992.
7. Krepel, H.P., and Polderman, A.M. Egg production of *Oesophagostomum bifurcum*, a locally common parasite of humans in Togo. *Am. J. Trop. Med. Hyg.* 46, 469, 1992.
8. Krepel, H.P. et al. Treatment of mixed *Oesophagostomum* and hookworm infection: Effect of albendazole, pyrantel pamoate, levamisole and thiabendazole. *Trans. R. Soc. Trop. Med. Hyg.* 87, 87, 1993.
9. Krepel, H.P. et al. Reinfection patterns of *Oesophagostomum bifurcum* after anthelmintic treatment. *Trop. Geogr. Med.* 47, 160, 1995.
10. Krepel, H.P. et al. Quantitative interpretation of coprocultures in a population infected with *Oesophagostomum bifurcum. Trop. Geogr. Med.* 47, 157, 1995.
11. de Gruijter, J.M. et al. Screening for haplotypic variability within *Oesophagostomum bifurcum* (Nematoda) employing a single-strand conformation polymorphism approach. *Mol. Cell Probes* 16, 185, 2002.
12. Gasser, R.B. et al. Screening for nucleotide variations in ribosomal DNA arrays of *Oesophagostomum bifurcum* by polymerase chain reaction-coupled single-strand conformation polymorphism. *Electrophoresis* 20, 1486, 1999.
13. Pit, D.S. et al. The capacity of the third-stage larvae of *Oesophagostomum bifurcum* to survive adverse conditions. *Ann. Trop. Med. Parasitol.* 94, 165, 2000.
14. Polderman, A.M., and Blotkamp, J. *Oesophagostomum* infections in humans. *Parasitol. Today* 11, 451, 1995.
15. Gasser, R.B. et al. Insights into the epidemiology and genetic make-up of *Oesophagostomum bifurcum* from human and non-human primates using molecular tools. *Parasitology* 132, 453, 2006.
16. de Gruijter, J.M. et al., Morphological variability within *Oesophagostomum bifurcum* among different primate species from Ghana. *J. Helminthol.*, 80, 357, 2006.
17. Eberhard, M.L. et al. Experimental *Oesophagostomum bifurcum* in monkeys. *J. Helminthol.* 75, 51, 2001.
18. de Gruijter, J.M. et al. High resolution DNA fingerprinting by AFLP to study the genetic variation among *Oesophagostomum bifurcum* (Nematoda) from human and non-human primates from Ghana. *Parasitology* 130, 229, 2005.
19. de Gruijter, J.M. et al. Genetic substructuring within *Oesophagostomum bifurcum* (Nematoda) from human and non-human primates from Ghana based on random amplified polymorphic DNA analysis. *Am. J. Trop. Med. Hyg.* 71, 227, 2004.
20. Storey, P.A. et al. Clinical epidemiology and classification of human oesophagostomiasis. *Trans. Roy. Soc. Trop. Med. Hyg.* 94, 177, 2000.
21. Storey, P.A. et al. Ultrasound diagnosis of oesophagostomiasis. *Br. J. Radiol.* 73, 328, 2000.
22. Storey, P.A. et al. Ultrasound appearance of preclinical *Oesophagostomum bifurcum* induced colonic pathology. *GUT* 48, 565, 2001.
23. Storey, P.A. et al. Ultrasonographic detection and assessment of preclinical *Oesophagostomum bifurcum*—Induced colonic pathology. *Clin. Inf. Dis.* 33, 166, 2001.
24. Storey, P.A. et al. Intraobserver and interobserver variation of ultrasound diagnosis of *Oesophagostomum bifurcum* colon lesions. *Am. J. Trop. Med. Hyg.* 67, 680, 2002.
25. Storey, P.A. et al. The effect of albendazole on *Oesophagostomum bifurcum* infection and pathology in children from rural northern Ghana. *Ann. Trop. Med. Parasitol.* 95, 87, 2001.
26. Katz, N. et al. A simple device for quantitative stool thick-smear technique in *Schistosomiasis mansoni. Rev. Inst. Med. Trop. Sao Paulo* 14, 397, 1972.
27. Verweij, J.J. et al. PCR assay for the specific amplification of *Oesophagostomum bifurcum* DNA from human faeces. *Int. J. Parasitol.* 30, 137, 2000.
28. Verweij, J.J. et al. Determining the prevalence of *Oesophagostomum bifurcum* and *Necator americanus* infections using specific PCR amplification of DNA from faecal samples. *Trop. Med. Int. Health* 6, 726, 2001.
29. Verweij, J.J. et al. Simultaneous detection and quantification of *Ancylostoma duodenale*, *Necator americanus*, and *Oesophagostomum bifurcum* in fecal samples using multiplex real-time PCR. *Am. J. Trop. Med. Hyg.* 77, 685, 2007.
30. de Gruijter, J.M. et al. Polymerase chain reaction-based differential diagnosis of *Ancylostoma duodenale* and *Necator americanus* infections in humans in northern Ghana. *Trop. Med. Int. Health* 10, 574, 2005.
31. Pit, D.S.S. et al. Parasite-specific antibody and cellular immune responses in human infected with *Necator americanus* and *Oesophagostomum bifurcum. Parasitol. Res.* 87, 722, 2001.
32. Polderman, A.M. et al. Serological diagnosis of *Oesophagostomum* infections. *Trans. R. Soc. Trop. Med. Hyg.* 87, 433, 1993.
33. Romstad, A. et al. Differentiation of *Oesophagostomum bifurcum* from *Necator americanus* by PCR using genetic markers in spacer ribosomal DNA. *Mol. Cell Probes* 11, 169, 1997.

34. de Gruijter, J.M. et al. AFLP fingerprinting for the analysis of genetic diversity within *Necator americanus*. *Mol. Cell Probes* 20, 317, 2006.
35. Gasser, R.B. et al. DNA technological progress toward advanced diagnostic tools to support human hookworm control. *Biotechnol. Adv.* 26, 35, 2008.
36. Verweij, J.J. et al. Prevalence of *Entamoeba histolytica* and *Entamoeba dispar* in northern Ghana. *Trop. Med. Int. Health* 8, 1153, 2003.
37. Gasser, R.B. et al. Rapid sequencing of rDNA from single worms and eggs of parasitic helminths. *Nucleic Acids Res.* 21, 2525, 1993.
38. Niesters, H.G. Clinical virology in real time. *J. Clin. Virol.* 25(Suppl 3), S3, 2002.
39. Ziem, J.B. et al. Annual mass treatment with albendazole might eliminate human oesophagostomiasis from the endemic focus in northern Ghana. *Trop. Med. Int. Health* 11, 1759, 2006.
40. Ziem, J.B. et al. Impact of repeated mass treatment on human *Oesophagostomum* and hookworm infections in northern Ghana. *Trop. Med. Int. Health* 11, 1764, 2006.
41. Ziem, J.B. et al. Mass treatment with albendazole reduces the prevalence and severity of *Oesophagostomum*-induced nodular pathology in northern Ghana. *Trans. R. Soc. Trop. Med. Hyg.* 100, 760, 2006.

61 Onchocerca

Ramiro Morales-Hojas

CONTENTS

61.1 INTRODUCTION

The genus *Onchocerca* (Nemata: Onchocercidae) is comprised of species that are parasites of ungulates with the exception of *O. volvulus*, the etiological agent of human river blindness. Onchocerciasis is distributed mainly in sub-Saharan Africa with smaller foci in Yemen and the Americas. It has three different pathologies, dermal, lymphatic, and ocular, but there are differences in the clinical manifestations depending on the geographical region. Thus, in West Africa, levels of blindness are higher in the savannah than in the forest. This has been explained by the existence of two differentiated strains. It is estimated that approximately 17 million people are affected, of which 500,000 are visually impaired and 270,000 blind. Given the relevance of onchocerciasis, the World Health Organization has put in place different control programs to eradicate the disease. This has led to a wealth of methods for the detection of infection and even for the differentiation between the two West African strains. This chapter introduces these parasites to the reader and describes some of the DNA-based methods used for the detection of onchocerciasis.

61.1.1 Classification and Morphology

The genus *Onchocerca* comprises approximately 30 species of parasites of ungulates [1]. The most relevant exception is *O. volvulus*, a human parasite and etiological agent of river blindness in large areas of sub-Saharan Africa and some foci in Yemen and South and Central America. A second exception found in dogs has been proposed, *O. lupi* [2–5a], although some authors consider canine onchocerciasis as a result of aberrant infections with *O. lienalis* [6,7]. The genus is distributed worldwide and the distribution of each species is restricted to that of its host [8]. The genus is phylogenetically close to the *Dirofilaria*, although the phylogeny of the family Onchocercidae is not well resolved [9]. The phylogenetic relationships between the species of the genus *Onchocerca* are not well resolved either, and the proposed origin and evolutionary history of the group have been based mainly on the geographical distribution and evolutionary history of the hosts. The origin of the group has been placed in Africa because the morphologically most primitive species (*O. raillieti*) is a parasite of an African equid [1,10,11]. This is further supported by the fact that the greatest species diversity is found in that continent [11]. However, molecular phylogenetic analyses do not place the African species in the basal branches of the genus, although this is likely due to incomplete species sampling as *O. raillieti* has not been included in any analysis yet [9,12,13]. It has also been suggested that the geographical distribution of the species is more relevant to the evolution of the genus than the evolutionary relationships of the host [11]. However, there is no clear grouping in the species phylogenetic tree reflecting this geographical distribution [9,12,13]. Nevertheless, host switching appears to have played a major role in the evolution of the genus [9,13], which suggests that geographical distribution is relevant in the evolutionary relationships of the genus. The hosts harboring the greatest number of species are the cervids and bovids [1]. Within the African bovid there are two lineages of *Onchocerca* spp: one more primitive found in the forested regions and a second derived line found in the savannah whose adult females are confined within nodules. The human parasite *O. volvulus* is the sister species of the nodule-inhabiting dermal bovid parasite *O. ochengi*, as confirmed by molecular phylogenetic studies [12,13].

Morphologically, these worms are long, thin nematodes that live intricately embedded in the tissues of the host. Given this lifestyle of the adult worms they have developed a flexible, but resistant, cuticle, which is differentiated in two layers, the external one which is regularly thickened with transversal ridges and a mid-layer divided by striae. Some authors consider the distance between the ridges and the number of striations between each ridge the most important character for the taxonomy of the genus *Onchocerca* [14]. These characters can separate the species at least in two groups, one that shows a distance of ~25 μm between ridges and 2–3 striations between each, and a second group in which the ridges are separated by ~50 μm with 4–5 striations between each. The first group includes species such as *O. ochengi*, *O. lienalis*, *O. gibsoni*, and *O. reticulata* and the human parasite *O. volvulus*, and the second group includes species such as *O. gutturosa* and *O. cervicalis*. There is a third group that shows a smooth cuticle that includes species such as *O. jakutensis*, *O. cervipedis*, and *O. raillieti*. On the contrary, some other authors recognize the caudal papillae as the most relevant character for the taxonomy of the genus [10,11]. *Onchocerca raillieti* would represent the most primitive papillae distribution: four precloacal pairs plus one impair papilla, two postcloacal pairs, and four terminal pairs. In other species the distribution, number, and morphology of the papillae are modified from this primitive plan. In the human parasite *O. volvulus* the precloacal papillae are grouped toward the cloaca. In its sister species, *O. ochengi*, the terminal papillae are modified into cuticular points, a modification which is also observed in *O. dukei*, *O. tarsicola*, and *O. gutturosa*. The terminal papillae in some species have migrated toward the cloaca (*O. reticulata* and *O. armillata*) and in others there is a reduction in their number (*O. jakutensis*). The vulva in some of the species is located at approximately the end of the muscular part of the esophagus, which is divided in a muscular and a glandular part, not very distant from the nervous ring such as in the *Onchocerca* from the equids as well as in *O. flexuosa* and *O. armillata*. In other species the vulva moves toward the end of the esophagus as observed in *O. jakutensis*, *O. garmsi*, or *O. tarsicola*. Beneath the cuticle, the epitheliomuscular sheath shows lateral chords that are narrow and muscles well developed in most species, but in those species that are encapsulated within nodules (e.g., *O. gibsoni*, *O. ochengi*, and *O. volvulus*) the muscles become atrophied and the hypodermis, mainly the lateral chords, hypertrophied.

Fain et al. [15] and Muller [14] classified the species of *Onchocerca* within five groups depending on the location of the adult worms in the host. The first group included those species that induce the encapsulation of adults within fibrous nodules, such as *O. volvulus*, *O. fasciata*, *O. dukei*, *O. gibsoni*, *O. cervipedis*, and *O. flexuosa*. There is a second group comprised of *O. armillata*, whose adults inhabit the wall of the aorta. A third group includes those species whose adults dwell in the connective tissues of the gastrosplenic and cervical ligaments as *O. gutturosa*, *O. lienalis*, *O. beaveri*, and *O. cervicalis*. The fourth group includes those species that live in the subcutaneous connective tissues, and comprises *O. reticulata*, *O. raillieti*, *O. tarsicola*, and *O. dewittei*, for example. The last group includes species that are intradermal such as *O. ochengi* and *O. cebei*. As already mentioned, adults of the human parasite *O. volvulus* inhabit fibrous nodules. The nodules are generally subcutaneous although there is evidence that many adults lie deep within the body [16]. The location of the subcutaneous nodules in the body can vary depending on the geographical region, but they are mainly observed around the pelvic region, in the head, ribs, or shoulders.

61.1.2 Biology and Epidemiology

Adult females of the genus *Onchocerca* lay unsheathed mobile embryos called microfilariae, which have pointed tails (Figure 61.1). These are always found in the skin of the hosts, although there is some contradictory information regarding the microfilariae of *O. armillata*, which some authors describe as blood-dwellers [8,17] and others as skin-dwellers [18,19]. The microfilariae tend to concentrate in specific regions of the skin. The localization of the adults in the host may influence the distribution of the microfilariae, although this is not always the case. It has been suggested that

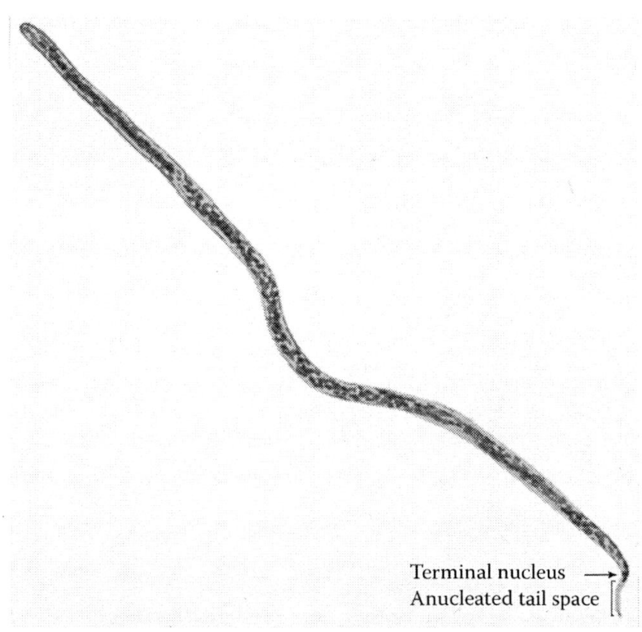

FIGURE 61.1 Microfilaria of *Onchocerca volvulus* stained with orcein. The head is in the upper left-hand corner.

the microfilariae also tend to concentrate on the preferred biting sites of the vectors [14]. Both *Simulium* spp. (Diptera: Simuliidae) and *Culicoides* spp. (Diptera: Ceratopogonidae) have been incriminated in the transmission of different species of *Onchocerca*. The human parasite *O. volvulus* is transmitted in Africa by several species of the *Simulium damnosum* complex, and by *S. neavei* s.s. and *S. woodi* [20]. In the Americas the vector species are different in the different foci [20]. Thus, in Mexico and Guatemala the primary vectors are members of the *S. ochraceum* group, with secondary vectors being *S. metallicum* s.l., *S. callidum*, *S. haematopotum*, *S. gonzalezi* s.l., and *S. veracruzanum* s.l.. In Northern Venezuela the primary vector is *S. metallicum* s.l., and *S. exiguum* s.l. act as the secondary vector. However, in Ecuador and Colombia members of the *S. exiguum* group are the main vectors. In the Amazonia focus of Southern Venezuela and Brazil, the primary vectors are *S. guianense* s.l., *S. oyapockense* s.l., and *S. roraimense*.

Once taken up by the vector, the *O. volvulus* microfilariae migrate from the midgut to the thoracic muscles where they develop into infective L3s. Within 28 h microfilariae differentiate into L1 and in 96 h they become L2. By day 7 the microfilariae have become infective L3s and migrate to the proboscis, from where they enter the host when the vector blood-feeds again, although the moulting time can take up to 12 days depending on the temperature [20]. Once inside the definitive human host the L3s moult to L4 within one week and into adults in 1–3 months [20,21]. The immature adults are attracted to already established nodules and may settle on their surface to become satellite nodules. The mechanism by which juveniles locate other nodules is not known. Alternatively, females can induce nodule formation themselves. The reproductive lifespan of adult females varies between 9 and 14 years and while they remain within the nodule their entire life, males seem to move from nodule to nodule, presumably to mate with the different females [22,23]. Females independently of the presence of males are continuously producing oocytes, and those not fertilized are reabsorbed [23,24]. The development of the microfilariae takes place within the females' uteri and takes around three to four weeks before they leave actively the female, rather than being expelled. The average number of released microfilariae per female is 700–1500 per day [23]. Microfilariae leave the nodules and migrate to the skin and eyes where their lifespan ranges from 6 to 24 months.

Human onchocerciasis is mostly distributed across sub-Saharan Africa, although other smaller foci are also found in Yemen and in Latin America, where the presence of the disease is likely the result of a recent introduction [25–29]. It has been estimated that around 17 million people are affected, of which 500,000 are visually impaired and 270,000 blind [30]. There are two differentiated strains in Africa according to the clinical manifestations [31,32]. In the savannah regions the levels of blindness can reach up to 15% of the population while in the forested areas the main manifestations are in the form of skin lesions and severe itching, with less than 2% of blindness [33]. In America, levels of blindness are low, with less than 1% of the infected people becoming blind [20]. The clinical manifestations appear with the accumulation of microfilariae in the body, and blindness peaks at the age range of 40–50 years. There are no apparent sex differences in the acquisition of the parasite, although in some regions the prevalence appears higher among the male population probably due to a higher exposure to vector biting [34].

The different pathologies observed in West Africa have been shown to be the result, at least partly, of the existence of two parasite strains, one in the savannah and another in the forest. Of these, the savannah form was shown to be more pathogenic [31]. It is also known that each strain of the parasite is more efficiently transmitted by the vectors from the same ecosystem than by vectors from allopatric regions, which led to the concept of different vector–parasite complexes [35–39]. This suggests strong local adaptations between the parasite and vector populations [40]. There are also other vector factors that affect the transmission of the parasite. Vector abundance and feeding habits (preference of biting human versus other animals) determine the rates of transmission and the location of the nodules [33]. Thus, nodules in Africa are mainly located in the lower regions of the body, as the primary vector *S. damnosum* s.l. tends to bite the lower parts of the body. On the other hand, nodules of the Guatemalan, Mexican, and Amazonian foci are distributed on the head and upper parts of the body where the main vectors *S. ochraceum* s.l. and *S. oyapockense* s.l. have a tendency to bite. The frequency of infective bites per annum, known as the annual transmission potential (ATP), is used as an estimate of the number of L3s deposited in a person if exposed for 11 h/day during 365 days. An ATP value of 100 was chosen as the lower limit for the risk of blindness in a community being less than 1% [41].

Great efforts have been made to control onchocerciasis and several programs have been put in place by the World Health Organization (WHO): the Onchocerciasis Control Programme (OCP), African Programme for Onchocerciasis Control (APOC), and the Onchocerciasis Elimination Programme in the Americas (OEPA). Control of onchocerciasis has been implemented through vector control and/or mass chemotherapy, and parasite transmission has been reduced in many regions of Africa and Latin America [42].

61.1.3 Clinical Features and Pathogenesis

Human onchocerciasis has three different clinical manifestations: dermal, lymphatic, and ocular [20,26,33]. The presence of subcutaneous, fibrous nodules generally over bony prominences and easily palpable is characteristic of onchocerciasis. The general distribution of the nodules varies from region to region. Thus, in Africa nodules are often located in the iliac crests, ribs, greater trochanters, and ischial tuberosities. Additionally, some nodules may be observed in the knees, elbows, patella, and scalp. In Guatemala and Mexico, fewer nodules are observed and generally located in the scalp.

The cutaneous onchocerciasis (onchodermatitis) is usually the first symptom of onchocerciasis apart from the presence of visible nodules. It begins with an intense pruritus (itching) and progresses to the development of papular rashes known as *craw-craw* in some parts of Africa. This condition can evolve into chronic papular dermatitis, which is characterized by large papules and development of hyper-pigmented and thickening of the skin. This is normally followed by an asymmetric lichenization known as *sowda* in some regions of Africa and the advanced stage is characterized by a spotty depigmentation known as leopard skin. The skin, with time, can become atrophied and lose elasticity. In Guatemala and Mexico, onchodermatitis are characterized by a depigmentation of the skin, which can become purplish (a symptom called *mal morado*). Here follows a brief description of the classification scheme for recording dermal onchocerciasis recommended by the WHO [20]:

i. Acute papular onchodermatitis (APOD): it is characterized by small (~1 mm in diameter), scattered pruritic papules. Vesicles or pustules in more severe cases and edema of the skin may be seen on a single limb or area of the trunk.
ii. Chronic papular onchocermatitis (CPOD): it is characterized by scattered, pruritic papules of 3–9 mm in diameter, which are hyperpigmented and flat-topped.
iii. Lichenified onchodermatitis (LOD): it is characterized by pruritic, hyperpigmented, and hyperkeratotic papular plaques, which become more confluent as the severity increases. In later stages the skin becomes lichenified. Its distribution is generally asymmetric involving one limb (a condition known as *sowda*). It may be associated to enlarged draining lymph nodes (lymphadenopathy).
iv. Atrophy (ATR): it is characterized by excessively wrinkled and dry skin, which has lost the elasticity. When the edge of a finger is firmly pushed along the skin, many additional fine wrinkles are revealed. As this condition reflects many of the characteristics of aging, ATR is only scored as a significant abnormality in individuals less than 50 years old.
v. Depigmentation (DPM): it is also known as leopard skin. It is characterized by patches of pigment loss associated with spots of normal pigmentation generally around hair follicles.

The lymphatic symptomatology of onchocerciasis is characterized by an inflammation of the lymph nodes due to the passing of the microfilariae. It is characterized by enlarged, fibrotic lymphatic nodes, the inguinal and femoral nodes being the most commonly affected. These may be localized in hanging pockets of the skin, a condition known as hanging groin, and could eventually lead to elephantiasis.

Ocular lesions are caused by the invasion of the eye by microfilariae. An early infection is characterized by inflammatory infiltrates surrounding dead microfilariae in the cornea, resulting in punctate opacities (keratitis). As the infection progresses different types of ocular damages are observed. The cornea can be damaged with sclerosing keratitis as a result of the early opacity covering the entire cornea. If microfilariae are present in the anterior chamber and cornea, the iris and ciliary body can also be damaged. The intraocular pressure can increase due to the presence of the microfilariae and secondary cataracts and glaucoma can occur. The retina and choroids can also be affected by the presence of microfilariae. Disturbances of the retinal pigment epithelium lead to sclerosis of the choroid and retinal vessels, with their subsequent degeneration and atrophy of the optic nerve. The optic nerve can also be atrophied by the invasion of microfilariae, which leads to the reduction of the visual field, key-hole vision, and total blindness. There are geographical differences in the type of ocular onchocerciasis. In the savannah of West Africa, blindness is usually attributable to corneal sclerosing keratitis. Nevertheless, there are some localities where the atrophy of the optic nerve plays a significant role in blindness. In the forest regions of West Africa as well as in Central and East Africa, the most common cause of blindness is by lesions of the posterior segments of the eye. Nevertheless, the prevalence of blindness in East and Central Africa is higher than in the forest of West Africa. In the Americas few serious ocular onchocerciasis occurs nowadays.

The pathogenesis of onchocerciasis is related to the significant inflammation observed around dead parasites, mainly the microfilariae. In fact, living parasites, whether microfilariae or adults encapsulated in the nodules, result in little immune response by the host [43]. It is well documented that the presence of filarial antigens induces the Th2 anti-inflammatory response [43–46]. Nevertheless, it has recently been shown that the immune response that causes onchocerciasis is, at least partially, a reaction to the *Wolbachia* endosymbiotic bacteria found in the parasites. Thus, there is increasing evidence for the involvement of the *Wolbachia* endobacteria in inducing the innate immune response as well as the neutrophil and Th1 pro-inflammatory responses [47–52]. The *Wolbachia* endosymbionts have also been shown to be essential in the biology of filarial nematodes, which makes them ideal targets for filarial disease control [53,54].

61.1.4 Diagnosis

61.1.4.1 Conventional Techniques

Traditionally, onchocerciasis is first diagnosed by evaluating the clinical manifestations of the disease such as the presence of subcutaneous nodules, leopard skin, hanging groin, and skin atrophy [20]. Nevertheless, some of the skin lesions can be due to other parasites such as *Mansonella streptocerca* or due to insect bites, and cannot be used as definitive diagnostic characters. The presence of pruritus can also be indicative of onchocercal infection although it can also have other causes. Ocular onchocerciasis can be ophthalmologically evaluated with slit-lamp examination of the eye after the patients have been with the head down for at least 2 min. At a magnification of ×16 the microfilariae of the anterior chamber and the dead microfilariae in the cornea can be visualized. With a magnification of ×25 and the use of retroillumination, live corneal microfilariae can be visualized. Although, slit-lamp examination allows the evaluation of the onchocercal anterior ocular lesions (punctate keratitis, sclerosing keratitis, limbitis, and anterior uveitis), these have to be differentiated from other causes before definitive diagnosis of onchocerciasis is given. On the other hand, the presence of microfilariae in the eye allows for the definitive diagnosis of onchocerciasis.

Onchocerciasis is most commonly diagnosed by the detection and specific identification of microfilariae by microscopic examination of skin biopsies [55]. The size and weight of the skin biopsies have been standardized in order to be able to compare microfilariae counts between samples. Thus, skin snips are obtained using either sclerocorneal Holth or Walser punches that take samples of a uniform size, ~2.3–2.5 mm in diameter. In order to maximize the efficacy of the sampling, in each geographical region skin biopsies should be taken preferentially from those body sites where the microfilariae concentrate. Thus, in Africa the iliac crest and buttocks should be sampled preferably; in Latin America, the iliac crest and scapular body part; and in Yemen the lower calves. Nevertheless, it is recommended by the WHO [55] that two samples from each of these three body regions should be taken. Once the snips have been obtained, they are placed in flat-bottom 96-well plates and incubated for 30 min to 3 h with a drop of normal saline, distilled water, or tissue culture medium as described by Prost and Prod'hon [56]. It has been reported that 66–78% of the *O. volvulus* microfilariae leave the skin snip using this methodology [57,58]. In those cases where the biopsy is negative, incubation should be continued overnight either at 37°C or room temperature [55]. Alternatively, skin snips could be incubated in a saline solution with 0.5% collagenase for 24–72 h [59]. In order to make permanent preparations of the microfilariae, the fluid is transferred to a slide and let to dry. After the sample is dried the mcirofilariae are fixed in methanol and stained. Some staining methods have already been described in Chapter 57. Given here is Delafield's hematoxylin stain procedure. Once dried, the slides are stained for 10–15 min in Delafield's hematoxylin solution. The excess stain is removed by rinsing the slides with distilled water. Slides are then introduced for 1 min in 0.1% (w/v) aqueous hydrochloric acid. Rinse the slide for 1 min with distilled water. Introduce the slide in tap water with several drops of ammonia and leave for 3–5 min until the films become dark blue. Rinse in tap water for 2–5 min and dry.

Once the preparations are ready, morphological identification of microfilariae is done under the microscope with low magnification (5× or 10×). The microfilariae of *O. volvulus* are unsheathed and have pointed, anucleated tails (Figure 61.1). The length range measured in different studies varies from 186 to 313 μm [60,61], although the length can vary depending on the preparation method [55]. Morphological identification of *O. volvulus* microfilariae can be difficult in regions where they are sympatric with other filarial nematodes. In particular, specific identification can be compromised by the presence of other skin dwelling microfilariae such as those of *M. streptocerca*. In this case, discrimination between the two species can be easily done because of the hooked end, nucleated tails of *M. streptocerca* microfilariae. In South America, however, misidentification of *O. volvulus* for *M. ozzardi* can be problematic. Despite *M. ozzardi* microfilariae being blood-dwelling, they have also been reported from the skin, for example [62–64]. The microfilariae of *O. volvulus* are typically larger than those of *M. ozzardi* in several body measurements (body length, length of the tail space, length of the terminal nucleus, cephalic space length, and width at the nerve ring), although the ranges have been shown to overlap, for example [61], However, the ratio of the length of the tail space to the length of the terminal nucleus was found to be a specific character for *O. volvulus* (range 2.00–6.33) and *M. ozzardi* (range 0.53–1.64). Thus, this ratio can be used as a diagnostic for the two species [61].

Individual adult worms free from tissue can be needed for different reasons such as morphological identification or DNA isolation. These are only available from excised nodules from where adult worms have to be isolated and separated [59]. For that, nodules are incubated either in medium 199 or phosphate-buffered saline (PBS) with 1–5 mg/mL collagenase depending on the need to obtain live adults. Low collagenase concentrations (1 mg/mL) require longer incubations, and therefore adult worms are less likely to survive. Gentomicin is added to a final concentration of 0.2 mg/mL to prevent bacterial growth. The nodule is incubated at 37°C for a period of time that depends on the size of the nodules and the concentration of collagenase used. It can range from about 8 h up to three days. With low collagenase concentrations the incubation normally takes more than 24 h. Every 4–8 h nodules are examined to remove any loose tissue and any loose worm before continuing the incubation in a fresh solution.

61.1.4.2 Molecular Techniques

The creation of programs to control and eliminate onchocerciasis by the WHO has generated a wealth of studies dedicated to the development of molecular techniques to specifically detect *O. volvulus* infections and differentiate between the different strains in West Africa. Those methods developed include immunological diagnosis techniques as well as DNA-based methods. The first type of techniques aims at detecting antigens of onchocercal origin or *Onchocerca*-specific antibodies circulating in the blood. However, this chapter will only be concerned with the DNA-based methods for the detection of *O. volvulus*.

Polymerase chain reaction (PCR)-based techniques for the detection of parasites provide a powerful tool for epidemiologists as they can be highly specific. Furthermore, DNA-based methods allow for the detection of any parasitic stage. However, while DNA-based methods have been shown to be more sensitive than microscopic examination of biopsies, they do not correlate well with the intensity of the infection. Therefore, these techniques should not be used to determine the intensity of individual infections with *O. volvulus*.

Several studies have identified and developed DNA probes for the detection of *O. volvulus* infection since the WHO [33] recognized the potential that DNA methods had for its identification in the human host and vectors. With this purpose, DNA probes were characterized from tandemly repeated DNA family, with a repetition unit of 148–149 base pairs (bp), denominated the O-150 family [65]. The probes derived from the O-150 family showed different levels of specificities, and some detected DNA from several *Onchocerca* spp. tested, indicating that the O-150 family is not exclusive of the human parasite. However, the individual repeats group in different clusters, some of which were exclusive of *O. volvulus* and others were described to be strain-specific [65–69]. Thus, probes were developed with different specificities. The number of repeat units per genome was estimated to range from 6000 to 9000 [65,70]. The most commonly used method involves the amplification of the O-150 family using either the primers S3 (5′-ATCAATTTTGCAAAATGCG-3′) and S4 (5′-AATAACTGATGACCTATGACC-3′) designed from the consensus sequence of the *O. volvulus*-specific DNA sequence pOvs134, or the O-150-specific primers upstream (5′-GATTYTTCCGRCGAANARCGC-3′) and downstream (5′-GCNRTRTAAATNTGNAAATTC-3′) [65,69,71]. The annealing temperatures for these reactions are 60°C and 37°C for the former and latter primer sets, respectively. The amplification products are then run in a 2% (v/w) agarose gel and then transferred to a membrane for a posterior hybridization with O-150 probes. Different probes can be used depending on the specificity required. Thus, OVS2 is specific for *O. volvulus* while the probe Och is used for the detection of *O. ochengi*, the sympatric sister species of the human parasite [72]. For the differentiation of the two West African strains, pFS-1 and pSS-1BT can be used for the specific detection of the forest and savannah strains, respectively. Table 61.1 shows the primer sets and probes used for onchocerciasis detection and their specificities. Some modifications to the method have been introduced so that the detection of the PCR products is not done by Southern blot, but instead they are detected by a PCR enzyme-linked immunosorbent assay (ELISA) [73]. This method has been successfully tested in skin biopsies, blood samples, and pools of vectors. Nevertheless, the blinding and nonblinding strain differentiation has shown some problems with samples from Uganda [60].

Other studies have attempted to develop species-specific primers for the identification of *O. volvulus* and *M. ozzardi* infections in regions where they are sympatric such as South America [28,74]. These methods have not been as fully validated as those described above but they are easier to implement, as they require a single PCR. Hojas and Post [74] showed that the *msp* genes of *M. ozzardi* and *O. volvulus* had different intron lengths (233 bp and 153–156 bp

TABLE 61.1
Primers and Probes Described for the Detection of *O. volvulus* and Other Related Species

Type	DNA Region	Name	Sequence	Specificity	Reference
Primer	O-150	S3	5′-ATCAATTTTGCAAAATGCG-3′	*Onchocerca* spp.	—
		S4	5′-AATAACTGATGACCTATGACC-3′		
Primer	O-150	Upstream	5′-GATTYTTCCGRCGAANARCGC-3′[a]	*Onchocerca* spp.	[69]
		–downstream	5′-GCNRTRTAAATNTGNAAATTC-3′[a]		
Probe	O-150	OVS2	5′-AATCTCAAAAAACGGGTACATA-3′	*O. volvulus*	[72]
Probe	O-150	S9	5′-AAATTGATTATTAACAGATGACCT ATGACATATAA-3′	*O. volvulus*	[69]
Probe	O-150	pFS-1 107	5′-CTGTTCGTCGGAAAAATCGCGGC ATAAATCTGCAAATTCACCAAAAT ATAGTCGAATATTTTTTTTAGCACT CAATTTGAAGGTATGTACCCGTTT CTTGAAATTAG-3′	*O. volvulus* West African forest strain	[67]
Probe	O-150	pSS-1BT	5′-GTGACCTATGACCCCTAATTCCAA GAAACGGGAACGTGCCTCCAAA CTGAGTCCTAAAGAAAATATTC-3′	*O. volvulus* West African savannah strain	[66]
Probe	O-150	Och	5′-CAAAAAACGTTCAAAAAACGGG AAAATAC-3′	*O. ochengi*	[72]
Primer	*msp*	MSPEx1	5′-ATGGCGCAATCGGTTCCACC-3′	*M. ozzardi*;	[74]
		MSPEx2	5′-CTTAAGATTTTTGCGACGAACCAT-3′	*O. volvulus*	
Primer	ITS2 rDNA	NC2[b]	5′-TTAGTTTCTTTTCCTCCGCT-3′	*M. ozzardi*;	[28]
		MoITS2	5′-CTTATCATCAGGTGATATTAAT-3′	*O. volvulus*	
		OvITS2	5′-TTCATACATATATAAATGTAGC-3′		

[a] R = A or G; Y = C or T; N = A, G, C or T.
[b] Reverse nematode primer from Newton, L.A. et al. *Int. J. Parasitol.* 28, 1781, 1998.
Type indicates whether the oligo is a primer for PCR amplification of the DNA region indicated, or a probe for hybridization; the specificity indicates the species and/or strain that the primers or probes identify.

in *M. ozzardi* and *O. volvulus*, respectively), and that this difference in length could be observed in 1% agarose gels. This allows the simple diagnosis of infections with either or both of these parasites using a simple PCR of the *msp* gene. Morales-Hojas et al. [28] describe a multiplex PCR assay that uses two species-specific primers located in the rDNA ITS2 and a conserved primer located in the 28S rDNA gene. The amplification products obtained were 295 and 195 bp long for *M. ozzardi* and *O. volvulus*, respectively, and could be distinguished by running the PCR products in a 1% agarose gel. Primers and their specificities are shown in Table 61.1.

Recently, a new nested PCR-based method has been described that attempts to detect and identify three sympatric filarial parasites in the Brazilian Amazonia, namely *O. volvulus*, *M. perstans*, and *M. ozzardi* [75]. The method involves a first PCR of the complete Internal Transcribed Spacer 1 (ITS1) using a universal forward primer located in the 18S gene and a reverse primer in the 5.8S gene. A second, nested PCR uses the amplification product from the first as template to amplify a fragment of the ITS1 with a forward primer on the 18S and a reverse primer in the ITS1. Identification of the species relies on the different lengths of the PCR products. The expected amplification product of *O. volvulus* is 344 bp in length, and can be easily discerned from the PCR products of Mansonella spp. (305–312 bp) in an agarose gel. This method is described in Chapter 57 (Mansonella).

61.2 METHODS

61.2.1 Sample Preparation

Detection of infection with *O. volvulus* is normally done analyzing skin biopsies. Thus, as this is the material regularly collected, DNA extraction from skin snips is described below. Furthermore, as the ATP, which is a measure of the number of infective larvae to which a person living in an onchocerciasis endemic area is exposed, it is of importance to estimate the prevalence of infected flies in the total vector population. Thus, a protocol for *O. volvulus* DNA extraction from blackfly vectors is also given below. It should be noted that the methods described for *Mansonella* spp. in Chapter 57 are also valid for *O. volvulus*. DNA has also been successfully extracted from an adult worm of *Onchocerca* sp. using QIAamp DNA Mini Kit from QIAGEN (Izasa Portugal, Lda.).

61.2.1.1 Preparation of DNA from Skin Biopsies

1. Depending on the preextraction treatment of the skin biopsy, proceed as follows:
   a. If the skin snip has been incubated in saline, place the solution containing the microfilariae in a microcentrifuge tube and centrifuge at 14,000 rpm for 5 min to sediment the microfilariae. Remove the supernatant.
   b. If the skin snip has not been incubated, shred with a sterile scalpel.
2. Add to the sample 500 μL of 50 mM or 100 mM EDTA, 5 μL of 10 mg/mL proteinase K, and 5 μL of 10% sodium dodecyl sulfate (SDS). Incubate for 1 h at 56°C.
3. Add 10 μL of 1 M dithiothreitol (DTT) and incubate at 100°C for 30 min.
4. Freeze and thaw the sample three times.
5. Add 1 mL of 100% EtOH and 100 mM sodium acetate and incubate at −20°C from 1 h to overnight.
6. Centrifuge the sample at 14,000 rpm for 20 min, remove the supernatant, and let the pellet dry.
7. Resuspend pellet in 25 μL of autoclaved water.

61.2.1.2 Preparation of DNA from Pools of Blackfly Vectors

Extraction of DNA from *O. volvulus* infective larvae can be done from either individual flies or pools of flies. Different methods have been proposed, but here a modification of that of Oskam et al. is presented [76].

1. Homogenize individual or pooled blackflies (1–80) in 15 μL/fly of lysis buffer (50 mM NaCl, 10 mM EDTA pH 8.0, 50 mM Tris-HCl pH 7.4) containing 1% triton X-100 and 10 mM DTT.
2. Freeze and thaw the sample four times using liquid nitrogen for freezing and a water bath at 60°C for thawing.
3. Incubate for 1 h at 60°C.
4. Add 80 μL/fly of lysis buffer containing 200 μg/mL proteinase K and 1% triton X-100; incubate for 3 h at 60°C.
5. Centrifuge the samples at 12000 × *g* for 10 min and transfer supernatant to a new tube.
6. The DNA solution can be diluted 10× and heated at 100°C for 5 min to prevent inhibition of the PCR by the fly's proteins.
   (Alternatively, the following optional steps can be done before step 6 to remove the proteins).
7. Add 400 μL of phenol:chloroform (1:1) and mix thoroughly.
8. Centrifuge at 5000 rpm for 5 min and transfer the upper, aqueous layer to a clean microcentrifuge tube.
9. Repeat steps 4 and 5.
10. Add an equal volume of chloroform and mix by inverting the tube for 2 min.
11. Centrifuge at 5000 rpm for 5 min and transfer the upper, aqueous layer to a clean microcentrifuge tube.
12. Add 1/10th of the total volume of 3 M Na acetate (pH 5.6) and 2× the total volume of 95–100% EtOH. Incubate at −20°C for 1 h (or overnight) to precipitate the DNA.
13. Centrifuge the sample at 14,000 rpm for 30 min to pellet the DNA. Remove the supernatant.
14. Add 500 μL of 75% EtOH and centrifuge at 14,000 rpm for 2 min. Remove supernatant.
15. Dry the pellet and resuspend in 50 μL of autoclaved H_2O.

61.2.2 Detection Procedures

Two different PCR-based methods for the detection and identification of *O. volvulus* infections will be described in the following subsections. The first one is based on the amplification of the O-150 repeat family followed by the detection of the PCR products by an ELISA assay [73,77]. The second method here described is that of Morales-Hojas et al. [28], who described a multiplex PCR of the rDNA ITS2 to detect and differentiate infections with *O. volvulus* and *M. ozzardi*.

61.2.2.1 The O-150 DNA Family-Based Method

1. Prepare a PCR in a total volume of 50 μL containing: 1× PCR buffer, 2 mM $MgCl_2$, 0.2 mM of each dNTP, 0.5 μM of primer O-150 upstream (5′-GATTYTTCCGRCGAANARCGC-3′), 0.5 μM of primer O-150 downstream (5′-GCNRTRTAAAT NTGNAAATTC-3′), 2.5 U of *Taq* polymerase, and 2.5 μL of extracted DNA
2. Run the cycling reactions in a thermocycler with the following program: 5 cycles of 1 min at 94°C, 2 min at 37°C, and 30 s at 72°C; 35 cycles of 30 s at 94°C, 30 s at 37°C, and 30 s at 72°C.
3. To detect the PCR products by ELISA follow the next steps:
   a. Coat overnight a 96-well microtiter plate covering each well with 50 μL of coating buffer (22.65 mL of 1 M $NaHCO_3$ and 9.1 mL of 1M Na_2CO_3) containing 300 μg/mL streptavidin.
   b. Rinse the wells six times with tris-buffered saline-tween (TBST) solution (0.5 M Tris-HCl pH 7.5, 1.5 M NaCl, 0.05% Tween 20).
   c. Add 40 μL of hybridization buffer (6× saline sodium phosphate EDTA (SSPE), 0.1% SDS, 5× Denhardt's solution, and 0.1% sodium sarcosine) to each well and mix well.
   d. Add 10 μL of each PCR product to each well and incubate for 30 min at room temperature.
   e. Wash the wells six times with TBST.
   f. Add 100 μL of 1 M NaOH per well and incubate for 1–2 min at room temperature. Wash the wells six times with TBST.
   g. Add 50 μL of hybridization buffer containing 50 ng/mL of the *O. volvulus*-specific oligonucleotide probe OVS2 (5′-AATCTCAAAAAA CGGGTACATA-fluorescein-3′) marked with fluorescein and incubate at 37°C for 30 min. Wash the wells six times with TBST (room temperature) to remove the unbound probe. (For the discrimination of the forest and savannah strains the forest-specific probe OVF (5′-AATTCCTAATTTCAAGAAGC-fluorescein-3′) can be used, at a concentration of 400 ng/mL.)
   h. Add 100 μL of prewarmed (56°C) 1× SSPE, 0.1% SDS solution to each well and incubate at 56°C for 5 min. Wash the wells six times with prewarmed (56°C) TBST. (The temperature appropriate for OVF is 40°C.)
   i. Dilute 1/10,000 alkaline phosphatase-conjugated antifluorescein FAB fragment in TBS/BSA solution (0.5 M Tris-HCl pH 7.5, 1.5 M NaCl, 0.1% bovine serum albumin). Add 50 μL of this solution to each well and incubate at 37°C for 30 min.
   j. Wash the wells six times with TBST at room temperature.
   k. Detect the alkaline phosphatase bound to the probe using any commercial alkaline phosphatase detection kit.

61.2.2.2 The Multiplex PCR Method

This multiplex PCR-based method was developed to detect and identify infections with *O. volvulus* and *M. ozzardi* in a region where these parasites are sympatric.

1. Prepare a PCR mix (25 μL) containing 1× PCR buffer, 1.5 mM $MgCl_2$, 60 μM of each dNTP, 5 pmol primer NC2 (5′-TTAGTTTCTTTTCCTCCGCT-3′), 10 pmol primer MoITS2 (5′-CTTATCATCAGG TGATATTAAT-3′), 2.5 pmol primer OvITS2 (5′-TTCATACATATATAAATGTAGC-3′), 0.5 U of *Taq* polymerase, and 1 μL of the extracted DNA
2. The reactions are then performed in a thermocycler with the following program: 1 cycle of 2 min at 94°C; 35 cycles of 15 s at 94°C, 30 s at 52°C, and 30 s at 72°C; 1 cycle of 10 min at 72°C.
3. Run PCR products in a 1% agarose gel and stain with ethidium bromide to visualize on a UV-light transilluminator.

Note: The amplification products for each species show different lengths, 295 bp for *M. ozzardi* and 195 bp for *O. volvulus*. These sizes can be distinguished in a 1% agarose gel.

61.3 CONCLUSIONS AND FUTURE PERSPECTIVES

Onchocerciasis is a major disease affecting a large population primarily in sub-Saharan Africa, but also in Yemen and South and Central America. The different control programs put in place for its eradication have resulted in a profusion of studies. These have been mainly concerned with different aspects of the epidemiology of the disease. Methods for the detection of infection with *O. volvulus* require specificity so that the nature of the infection is not misidentified for other filariasis. But they also need to be sensitive enough to be able to detect low-intensity infections, as during the early stages of infection or after Ivermectin (Mectizan®, Merck & Co.) treatment, when microfilariae are not present any more in large numbers. While microscopic examination of skin biopsies is still widely used, it is a time-consuming method. Furthermore, the difficulty of morphological identification of microfilariae could diminish the specificity of the method in areas where different filarial species are sympatric. Also, the sensitivity of the method is not very high and false negatives are always a possibility. For these reasons, the development of molecular detection methods has been central in onchocerciasis research. These can be highly specific and detect minute amounts of parasite material or the products of the host response to infection. Additionally, DNA based methods have been developed that detect the parasite in the vector population, allowing the estimation of other epidemiological parameters of relevance such as the ATP. Furthermore, different DNA probes have been developed for the differentiation of the blinding and nonblinding West African strains of the parasite.

Despite all the research carried out in *O. volvulus*, there is still a paucity of evolutionary research. Thus, the population structure of the parasite or its genetic diversity is not known. These types of studies are of relevance to control reinfection of treated areas or to determine the genetic differentiation level between the different strains, for example. Other aspects of the epidemiology of the disease that can be better recognized from an evolutionary point of view are the appearance and spread of resistance to antihelmintics. This will be influenced by the genetic diversity and patterns of gene flow between populations. Also, there is ample evidence for the coevolution of vector and parasite populations, the *Simulium–Onchocerca* transmission complexes [36,37,39,40]. Understanding these adaptations could provide useful information toward a definitive control of transmission. In conclusion, future research on onchocerciasis should have an evolutionary focus. This type of knowledge will be of great assistance toward the definitive control and eradication of the disease [9].

ACKNOWLEDGMENT

Figure 61.1 was kindly provided by Dr. Rory J. Post.

REFERENCES

1. Bain, O. Evolutionary relationships among filarial nematodes. In: Klei, T.R., and Rajan, T.V. (eds) *World Class Parasites. The Filaria*, Vol. 5, Kluwer Academic Publishers, Boston, p. 21, 2002.
2. Egyed, Z. et al. Morphologic and genetic characterization of *Onchocerca lupi* infecting dogs. *Vet. Parasitol.* 102, 309, 2001.
3. Egyed, Z. et al. Molecular phylogenetic analysis of *Onchocerca lupi* and its *Wolbachia* endosymbiont. *Vet. Parasitol.* 108, 153, 2002.
4. Sreter-Lancz, Z., Szell, Z., and Sreter, T. Molecular genetic comparison of *Onchocerca* sp. infecting dogs in Europe with other spirurid nematodes including *Onchocerca lienalis. Vet. Parasitol.* 148, 365, 2007.
5. Sreter, T., and Szell, Z. Onchocercosis: A newly recognized disease in dogs. *Vet. Parasitol.* 151, 1, 2008.

5a. Faísca, P. et al. A case of canine ocular onchocercosis in Portugal. *Vet. Ophthalmol.* 13, 117, 2010.

6. Eberhard, M.L. et al. Ocular *Onchocerca* infections in two dogs in western United States. *Vet. Parasitol.* 90, 333, 2000.
7. Orihel, T.C. et al. Onchocerciasis in a California dog. *Am. J. Trop. Med. Hyg.* 44, 513, 1991.
8. Anderson, R.C. *Nematode Parasites of Vertebrates: Their Development and Transmission*, 2nd. CABI Publishing, Cambridge, MA, 2000.
9. Morales-Hojas, R. Molecular systematics of filarial parasites, with an emphasis on groups of medical and veterinary importance, and its relevance for epidemiology. *Infect. Genet. Evol.* 9, 748, 2009.
10. Bain, O. Le genre *Onchocerca*: Hypothèses sur son évolution et clé dichotomique des espèces. *Ann. Parasitol. Hum. Comp.* 56, 503, 1981.
11. Chabaud, A.G., and Bain, O. The evolutionary expansion of the Spirurida. *Int. J. Parasitol.* 24, 1179, 1994.
12. Krueger, A., Fischer, P., and Morales-Hojas, R. Molecular phylogeny of the filaria genus *Onchocerca* with special emphasis on Afrotropical human and bovine parasites. *Acta Trop.* 101, 1, 2007.
13. Morales-Hojas, R., Cheke, R.A., and Post, R.J. Molecular systematics of five *Onchocerca* species (Nematoda: Filarioidea) including the human parasite, *O. volvulus*, suggest sympatric speciation. *J. Helminthol.* 80, 281, 2006.
14. Muller, R. Identification of *Onchocerca*. In: *Proceedings 17th Symposium of the British Society for Parasitology, Problems in the Identification ofParasites and their Vectors*, Blackwell, Oxford, p. 175, 1979.
15. Fain, A., Herin, V., and Thienpont, D. Filariases of cattle in Ruanda-Urundi. III. Parasitological study. B. Filariae of the genera *Setaria* and *Onchocerca*, and microfilariae in blood and skin. *Ann. Soc. Belg. Med. Trop.* 35, 555, 1955.
16. Duke, B.O. The population dynamics of *Onchocerca volvulus* in the human host. *Trop. Med. Parasitol.* 44, 61, 1993.
17. Patnaik, B. Onchocerciasis due to *O. armillata* in cattle in Orissa. *J. Helminthol.* 36, 313, 1962.
18. Bain, O. Redescription of five species of *Onchocerca* (author's translation). *Ann. Parasitol. Hum. Comp.* 50, 763, 1975.
19. ElBihari, S., and Hussein, H.S. Location of the microfilariae of *Onchocerca armillata. J. Parasitol.* 61, 656, 1975.
20. WHO. Onchocerciasis and its control. *World Health Organ. Tech. Rep. Ser.* 852, 1, 1995.
21. Burnham, G. Onchocerciasis. *Lancet* 351, 1341, 1998.
22. Plaisier, A.P. et al. The reproductive lifespan of *Onchocerca volvulus* in West African savanna. *Acta Trop.* 48, 271, 1991.
23. Schulz-Key, H., and Soboslay, P.T. Reproductive biology and population dynamics of *Onchocerca volvulus* in the vertebrate host. *Parasite*, 1, S53, 1994.
24. Schulz-Key, H. Observations on the reproductive biology of *Onchocerca volvulus. Acta Leiden* 59, 27, 1990.
25. Dalmat, H.T. The blackflies (Diptera, Simuliidae) of Guatemala and their role as vectors of onchocerciasis. *Smithson. Misc. Collect.* 125, 1425, 1955.
26. Duke, B.O. Human onchocerciasis—An overview of the disease. *Acta Leiden* 59, 9, 1990.
27. Zimmerman, P.A. et al. Recent evolutionary history of American *Onchocerca volvulus*, based on analysis of a tandemly repeated DNA sequence family. *Mol. Biol. Evol.* 11, 384, 1994.
28. Morales-Hojas, R. et al. Characterisation of nuclear ribosomal DNA sequences from *Onchocerca volvulus* and *Mansonella ozzardi* (Nematoda: Filarioidea) and development of a PCR-based method for their detection in skin biopsies. *Int. J. Parasitol.* 31, 169, 2001.

29. Morales-Hojas, R., Cheke, R.A., and Post, R.J. A preliminary analysis of the population genetics and molecular phylogenetics of *Onchocerca volvulus* (Nematoda: Filarioidea) using nuclear ribosomal second internal transcribed spacer sequences. *Mem. Inst. Oswaldo Cruz* 102, 879, 2007.
30. Hoerauf, A. et al. Onchocerciasis. *BMJ* 326, 207, 2003.
31. Duke, B.O., and Anderson, J. A comparison of the lesions produced in the cornea of the rabbit eye by microfilariae of the forest and Sudan-savanna strains of *Onchocerca volvulus* from Cameroon. I. The clinical picture. *Z. Tropenmed. Parasitol.* 23, 354, 1972.
32. Anderson, J. et al. Studies on onchocerciasis in the United Cameroon Republic. II. Comparison of onchocerciasis in rain-forest and Sudan-savanna. *Trans. R. Soc. Trop. Med. Hyg.* 68, 209, 1974.
33. WHO. WHO expert committee on onchocerciasis. *World Health Organ. Tech. Rep. Ser.* 752, 1, 1987.
34. Anosike, J.C. et al. The prevalence, intensity and clinical manifestations of *Onchocerca volvulus* infection in Toro local government area of Bauchi State, Nigeria. *Int. J. Hyg. Environ. Health* 203, 459, 2001.
35. De Leon, J.R., and Duke, B.O. Experimental studies on the transmission of Guatemalan and West African strains of *Onchocerca volvulus* by *Simulium ochraceum*, *S. metallicum* and *S. callidum*. *Trans. R. Soc. Trop. Med. Hyg.* 60, 735, 1966.
36. Duke, B.O., Moore, P.J., and De Leon, J.R. *Onchocerca–Simulium* complexes. V. The intake and subsequent fate of microfilariae of a Guatemalan strain of *Onchocerca volvulus* in forest and Sudan-savanna forms of West African *Simulium damnosum*. *Ann. Trop. Med. Parasitol.* 61, 332, 1967.
37. Duke, B.O., Lewis, D.J., and Moore, P.J. *Onchocerca–Simulium* complexes. I. Transmission of forest and Sudan-savanna strains of *Onchocerca volvulus*, from Cameroon, by *Simulium damnosum* from various West African bioclimatic zones. *Ann. Trop. Med. Parasitol.* 60, 318, 1966.
38. Duke, B.O. *Onchocerca-Simulium* complexes. VI. Experimental studies on the transmission of Venezuelan and West African strains of *Onchocerca volvulus* by *Simulium metallicum* and *S. exiguum* in Venezuela. *Ann. Trop. Med. Parasitol.* 64, 421, 1970.
39. Basáñez, M.G. et al. *Onchocerca-Simulium* complexes in Venezuela: Can human onchocerciasis spread outside its present endemic areas? *Parasitology* 120, 143, 2000.
40. Basáñez, M.G., Churcher, T.S., and Grillet, M.E. *Onchocerca–Simulium* interactions and the population and evolutionary biology of *Onchocerca volvulus*. *Adv. Parasitol.* 68, 263, 2009.
41. Walsh, J.F. et al., Standardization of criteria for assessing the effect of *Simulium* control in onchocerciasis control programmes. *Trans. R. Soc. Trop. Med. Hyg.* 72, 675, 1978.
42. Molyneux, D.H. Onchocerciasis control and elimination: Coming of age in resource-constrained health systems. *Trends Parasitol.* 21, 525, 2005.
43. Ottesen, E.A. Immune responsiveness and the pathogenesis of human onchocerciasis. *J. Infect. Dis.* 171, 659, 1995.
44. Turaga, P.S. et al. Immunity to onchocerciasis: Cells from putatively immune individuals produce enhanced levels of interleukin-5, gamma interferon, and granulocyte–macrophage colony-stimulating factor in response to *Onchocerca volvulus* larval and male worm antigens. *Infect. Immun.* 68, 1905, 2000.
45. Abraham, D., Lucius, R., and Trees, A.J. Immunity to *Onchocerca* spp. in animal hosts. *Trends Parasitol.* 18, 164, 2002.
46. Brattig, N.W. et al. *Onchocerca volvulus*-exposed persons fail to produce interferon-gamma in response to *O. volvulus* antigen but mount proliferative responses with interleukin-5 and IL-13 production that decrease with increasing microfilarial density. *J. Infect. Dis.* 185, 1148, 2002.
47. Saint Andre, A. et al. The role of endosymbiotic *Wolbachia* bacteria in the pathogenesis of river blindness. *Science* 295, 1892, 2002.
48. Brattig, N.W., Buttner, D.W., and Hoerauf, A. Neutrophil accumulation around *Onchocerca* worms and chemotaxis of neutrophils are dependent on *Wolbachia* endobacteria. *Microbes Infect.* 3, 439, 2001.
49. Brattig, N.W. et al. Lipopolysaccharide-like molecules derived from *Wolbachia* endobacteria of the filaria *Onchocerca volvulus* are candidate mediators in the sequence of inflammatory and anti-inflammatory responses of human monocytes. *Microbes Infect.* 2, 1147, 2000.
50. Cross, H.F. et al. Severe reactions to filarial chemotherapy and release of *Wolbachia* endosymbionts into blood. *Lancet* 358, 1873, 2001.
51. Taylor, M.J. et al. *Wolbachia* bacteria in filarial immunity and disease. *Parasite Immunol.* 23, 401, 2001.
52. Fischer, P. et al. An aspartate aminotransferase of *Wolbachia* endobacteria from *Onchocerca volvulus* is recognized by IgG1 antibodies from residents of endemic areas. *Parasitol. Res.* 90, 38, 2003.
53. Taylor, M.J., Bandi, C., and Hoerauf, A. *Wolbachia* bacterial endosymbionts of filarial nematodes. *Adv. Parasitol.* 60, 245, 2005.
54. Johnston, K.L., and Taylor, M.J. *Wolbachia* in filarial parasites: Targets for filarial infection and disease control. *Curr. Infect. Dis. Rep.* 9, 55, 2007.
55. WHO. *Bench Aids for the Diagnosis of Filarial Infections*. WHO, Geneva, 1997.
56. Prost, A., and Prod'hon, J. Parasitological diagnosis of onchocerciasis. A critical review of present methods (author's transl). *Med. Trop. (Mars)* 38, 519, 1978.
57. Schulz-Key, H., and Karam, M. Quantitative assessment of microfilariae and adults of *Onchocerca volvulus* in ethanol-fixed biopsies and nodules. *Trans. R. Soc. Trop. Med. Hyg.* 78, 157, 1984.
58. Fischer, P. et al. Parasitological and clinical characterization of *Simulium neavei*-transmitted onchocerciasis in western Uganda. *Trop. Med. Parasitol.* 44, 311, 1993.
59. Schulz-Key, H., Albiez, E.J., and Buttner, D.W. Isolation of living adult *Onchocerca volvulus* from nodules. *Tropenmed. Parasitol.* 28, 428, 1977.
60. Fischer, P. et al. Strain differentiation of *Onchocerca volvulus* from Uganda using DNA probes. *Parasitology* 112, 401, 1996.
61. Post, R.J. et al. The morphological discrimination of microfilariae of *Onchocerca volvulus* from *Mansonella ozzardi*. *Parasitology* 127, 21, 2003.
62. Ewert, A., Smith, J.H., and Corredor, A. Microfilariae of *Mansonella ozzardi* in human skin biopsies. *Am. J. Trop. Med. Hyg.* 30, 988, 1981.
63. Moraes, M.A. et al. The concentration of *Mansonella ozzardi* microfilariae in skin capillaries. *Trans. R. Soc. Trop. Med. Hyg.* 77, 463, 1983.
64. Moraes, M.A.P. *Mansonella ozzardi* microfilariae in skin snips. *Trans. R. Soc. Trop. Med. Hyg.* 70, 16, 1976.
65. Meredith, S.E. et al. Cloning and characterization of an *Onchocerca volvulus* specific DNA sequence. *Mol. Biochem. Parasitol.* 36, 1, 1989.

66. Erttmann, K.D. et al. Isolation and characterization of form specific DNA sequences of *O. volvulus*. *Acta Leiden* 59, 253, 1990.
67. Erttmann, K.D. et al. A DNA sequence specific for forest form *Onchocerca volvulus*. *Nature* 327, 415, 1987.
68. Harnett, W. et al. An oligonucleotide probe specific for *Onchocerca volvulus*. *Mol. Biochem. Parasitol.* 35, 119, 1989.
69. Meredith, S.E. et al. *Onchocerca volvulus*: Application of the polymerase chain reaction to identification and strain differentiation of the parasite. *Exp. Parasitol.* 73, 335, 1991.
70. Shah, J.S. et al. Characterization of an *Onchocerca*-specific DNA clone from *Onchocerca volvulus*. *Am. J. Trop. Med. Hyg.* 37, 376, 1987.
71. Zimmerman, P.A. et al. Polymerase chain reaction-based diagnosis of *Onchocerca volvulus* infection: Improved detection of patients with onchocerciasis. *J. Infect. Dis.* 169, 686, 1994.
72. Zimmerman, P.A., Toe, L., and Unnasch, T.R. Design of *Onchocerca* DNA probes based upon analysis of a repeated sequence family. *Mol. Biochem. Parasitol.* 58, 259, 1993.
73. Unnasch, T.R., and Meredith, S.E. The use of degenerate primers in conjunction with strain and species oligonucleotides to classify *Onchocerca volvulus*. *Methods Mol. Biol.* 50, 293, 1996.
74. Hojas, R.M., and Post, R.J. Regional genetic variation in the major sperm protein genes of *Onchocerca volvulus* and *Mansonella ozzardi* (Nematoda: Filarioidea). *Int. J. Parasitol.* 30, 1459, 2000.
75. Tang, T.H. et al. Nested PCR to detect and distinguish the sympatric filarial species *Onchocerca volvulus, Mansonella ozzardi* and *Mansonella perstans* in the Amazon Region. *Mem. Inst. Oswaldo Cruz* 105, 823, 2010.
76. Oskam, L. et al. Polymerase chain reaction for detecting *Onchocerca volvulus* in pools of blackflies. *Trop. Med. Int. Health* 1, 522, 1996.
77. Nutman, T.B. et al. A universally applicable diagnostic approach to filarial and other infections. *Parasitol. Today* 10, 239, 1994.
78. Newton, L.A. et al. Systematic relationships of some members of the *Oesophagostomum* and *Chabertia* (Nematoda: Chabertiidae) based on ribosomal DNA sequence data. *Int. J. Parasitol.* 28, 1781, 1998.

62 *Pseudoterranova* and *Contracaecum*

Simonetta Mattiucci, Michela Paoletti, Stephen C. Webb, and Giuseppe Nascetti

CONTENTS

62.1 INTRODUCTION

The fish-borne parasitic zoonoses previously known as Terranovasis and Contracaeciasis, being related to the larval nematodes of the genera *Pseudoterranova* and *Contracaecum*, respectively, as causative agents of the disease in humans, are currently better recognized in the general term of "Anisakidosis." The last includes, indeed, the disease provoked by the nematodes belonging to the Family Anisakidae, and thus, also those zoonotic aspects produced by *Anisakis* spp larvae. Historically, the "codworm" or "sealworm" *Pseudoterranova decipiens s.l.* (= *Phocanema decipiens*) has been considered to have the potential to cause the fish-borne zoonosis Anisakidosis. However, far fewer cases of anisakidosis ascribed to *Pseudoterranova* are known in the literature compared with those attributed to *Anisakis* spp larvae.

The first reports of human *P. decipiens s.l.* infections were from North America, Alaska, and California [1–3], followed by cases described as transient infections in California [4]. In Korea, human anisakidosis caused by *Pseudoterranova* was first described by Lee et al. [5], other reports followed [6,7] and more recently, several cases have been reported [8]. Other human cases of this zoonosis have been reported from Chile and Northern Europe.

Cases of anisakidosis caused by members of the genus *Pseudoterranova* are sparse and as yet, there are no records of human infections with larvae of the genus *Contracaecum*. This is despite experimental demonstration that larvae of *Contracaecum* could have a potential zoonotic role in vertebrates [9].

So far, human cases of anisakidosis due to *Pseudoterranova* have been restricted to those populations where raw and/or undercooked fish species infected by the larval forms of *P. decipiens s.l.* are habitually eaten. However, there are prospects for the expansion of these fish-borne diseases as a result of growing of international fish markets, improved transportation system for fresh fish, and changes of human eating habits.

This chapter will deal mainly with molecular diagnosis of species of *Pseudoterranova* and *Contracaecum* (here considering only the species of these genera maturing in seals), parasites, at adult stage, of pinnipeds and, at larval stages, of fish. Such methods could be useful since in some geographic areas members of both these genera might produce coinfections in the same fish host.

In particular, the chapter deals with the species of the *P. decipiens* and *Contracaecum osculatum* complexes which have been genetically characterized by genetic/molecular methodologies, and whose larvae have zoonotic potential. Before giving a detailed description of the diagnostic procedures to be applied in the larval recognition of these anisakids, their current taxonomy ecology, epidemiology and diagnosis to species level in humans are reviewed.

62.1.1 Taxonomy, Host Preference, and Geographical Distribution

Adults of most species of the genera *Pseudoterranova* and *Contracaecum* are parasites of the alimentary tract in pinnipeds (here considering the species of genus *Contracaecum* parasites of seals). They display indirect life cycles in aquatic ecosystems and exploit various hosts at different levels in food webs. Marine mammals (seals, sea lions and fur seals) are definitive hosts; while fish are intermediate or paratenic hosts. Crustaceans serve as first intermediate hosts. The study of anisakids has been greatly facilitated in the last 20 years by the use of genetic molecular methodologies for their recognition to the species level. As a result, their taxonomy, hosts and geographical distribution recently have been reviewed [10].

62.1.1.1 The *P. decipiens* Species Group

The species *decipiens* was first described as *Ascaris decipiens*. This taxon was later linked with *Terranova* Leiper and Atkinson, 1914, a genus erected for parasites of elasmobranchs by Baylis [11], because of the presence of an intestinal caecum. Subsequently, Baylis linked this species within *Porrocaecum* Railliet and Henry, 1912 (a group of ascarids now including only parasites of terrestrial birds) because he considered it a senior synonym of *Terranova*. *Terranova* was resurrected at full generic level by Johnston and Mawson [12] and as a subgenus by Karokhin [13], the latter being the first to use the combination *T. decipiens*. However, because *Terranova* included species maturing in different definitive hosts (elasmobranchs, reptiles, marine mammals), Myers [14] erected *Phocanema* (with *decipiens* as the type and only species) for those species included in *Terranova* which were parasites of marine mammals on the base of some morphological features at the cephalic end (structure of the labia) and the male caudal end. Accordingly, the species *decipiens* and *Terranova kogiae* Johnston and Mawson [12], (described from the pygmy sperm whale, *Kogia breviceps)* were included in *Phocanema*. Later, *Phocanema* was considered [15] as a synonym of *Pseudoterranova*, previously erected by Mozgovoi [16] (listed in Mozgovoi, without indication) for parasites from sperm whales, and *decipiens* was transferred to this genus (combination first used in Gibson and Colin [17]). According to the same authors, the group of species of *Pseudoterranova* from marine mammals included: *Pseudoterranova kogiae* [12] *Pseudoterranova ceticola* [18] and *P. decipiens*.

A summary of the sibling species so far recognized in the *P. decipiens (s.l.)* complex are reported below with data on their hosts and geographical range.

P. decipiens (*sensu stricto*) (= *P. decipiens* B of [19]): The species first genetically detected on the basis of allozymes, was formally described by morphological features considered to be of diagnostic value with respect to the other members included in the *P. decipiens* complex [20]. Its geographical range appears to be wide and mainly in the Arctic and sub-Arctic regions, including the North East Atlantic (comprising Scottish, Faroes, southern Icelandic, and Norwegian waters), the Canadian Atlantic (including Newfoundland waters and the Gulf of St. Lawrence [19,21] and Canadian Pacific waters [22–24]. Areas of sympatry between *P. decipiens* (*s.s.*) and other species of the *P. decipiens* complex have been detected: in the North Eastern Atlantic waters in mixed infections with both *Pseudoterranova krabbei* and *Pseudoterranova bulbosa* and in Canadian Atlantic waters with *P. bulbosa* [19]. *P. decipiens* (*s.s.*) was recognized, using genetic markers, as a parasite at the adult stage in the common seal, *Phoca vitulina*, gray seal, *Halichoerus grypus*, in both North and West Atlantic waters, and in *P. vitulina richardsii* and *Zalophus californianus* in northern Pacific waters [25]. Its larval forms have been identified in some gadoid fish species of the North Atlantic Ocean [19,22].

P. krabbei Paggi, Mattiucci, Gibson, Berland, Nascetti, Cianchi, and Bullini [20]: This species was previously indicated on the basis of allozymes as *P. decipiens* A [19]. *P. krabbei* exhibits, at the adult stage, in comparison with *P. decipiens* (*s.s.*), the following diagnostic morphological features in male specimens: shorter spicules; a proximal papilla (nomenclature according to Fagerholm [26]) smaller than d1 *versus* the same papilla size in *P. decipiens (s.s.);* distal papillae 1, 2, and 4 closer to each other; and caudal plates of a similar width and narrower than in *P. decipiens (s.s.)* where plates (wp) 1 (wp1) and 2 (wp2) are of a similar width but plate 3 (wp3) is narrower [20]. To date it has been found only in the North-East Atlantic (including off Scotland, the Faroes, southern Iceland, and Norway). It is mainly an adult parasite of the gray seal, *H. grypus*, and the common seal, *P. vitulina*, often found in mixed infections in the same individual host with *P. decipiens (s.s.)*. However, in the eastern Atlantic waters, when detected in sympatry with *P. decipiens* (*s.s.*), *P. krabbei* has shown higher relative proportion in the gray seal rather than in the common seal. *P. krabbei* has been recognized genetically at the larval stage as a parasite of *Gadus morhua*, *Melanogrammus aeglefinus*, and *Pollachius virens* [19].

P. bulbosa (Cobb, 1888) (= *P. decipiens* C of [19]): First detected genetically on the basis of allozymes the species was later morphologically described and the name *P. bulbosa* was resurrected from synonymy with *P. decipiens* [14,22,27]. It differs morphologically, at the adult stage, from the other members of the *P. decipiens* complex in having the following features in adult male specimens: longer spicules; caudal plates (wp) of unequal width, with plate 2 (wp2) narrower than 1(wp1) and 3 (wp3); and a different pattern of distal caudal papillae, with d2 far apart from d4 [22]. *P. bulbosa* has been recorded from the Barents and Norwegian Seas, the Canadian Atlantic and the Sea of Japan, between 40 and 80°N [19,21,22]. Its main definitive host so far detected is the bearded seal, *Erignathus barbatus*. In this phocid host from the Otaru Sea (Sea of Japan), *P. bulbosa* has been identified as occurring in mixed infections with *Pseudoterranova azarasi* [22]. The benthic flatfishes *Hippoglossoides platessoides*, and *Reinhardtius hippoglossoides* are reported as its intermediate hosts [19].

P. azarasi (Yamaguti and Arima, 1942) (= *P. decipiens* D of [22]): This species was first genetically disclosed and named *P. decipiens* D in hosts from Japanese waters and later morphologically described and the name *P. azarasi* was resurrected from synonymy with *P. decipiens* [14,27]. This species differs morphologically from the other members of the *P. decipiens* complex in having: shorter spicules; caudal plates (wp) of increasing width from 1 to 3; and a different pattern of distal papillae (*sensu* [26]), with d2 closer to d4 than in *P. bulbosa* but more distant than in *P. krabbei* and *P. decipiens (s.s.)* [22]. Its main definitive host has been identified as the Steller's sea lion, *Eumetopias jubatus*. Its geographical distribution appears limited to Japanese and Sakhalinese waters of the North Pacific Ocean [22].

P. decipiens E of Bullini, Arduino, Cianchi, Nascetti, D'Amelio, Mattiucci, Paggi, Orecchia, Plötz, Smith, and Brattey [28]. This taxon was genetically detected based on allozymes in the Antarctic Weddell seal, *Leptonychotes weddellii* [28]. Scanty data are so far available in relation to the nature of its intermediate host, although larvae corresponding to this sibling species have been genetically identified as parasites of some benthic fish hosts in sub-Antarctic waters and the genetic relationships of *P. decipiens* E with respect to the other siblings has been genetically established based on the mitochondrial DNA cytochrome oxidase 2 (mtDNA *cox-2*) sequences analysis (unpublished data).

Pseudoterranova cattani George-Nascimento and Urrutia [29]: This species was detected by those authors based on allozymes and some morphological diagnostic features with respect to the other siblings of the *P. decipiens* complex, as an adult in *Otaria byronia* on the Chilean coast. Using molecular markers in the internal transcribed spacers of ribosomal DNA (ITS-rDNA), this species was recently shown to cluster with the *P. decipiens* complex [30].

Host and geographical ranges so far reported for the species of the *P. decipiens* complex are reviewed in detail in Mattiucci and Nascetti [10].

62.1.1.2 The *C. osculatum* Species Group

The genus *Contracaecum* was originally defined by Railliet and Henry (1912) based on the morphology of the esophago-intestinal region. *Contracaecum spiculigerum* (Rudolphi, 1809) (= *Ascaris spiculigerum*, Rudolphi, 1809) was designated as the type-species. The original type-material was later identified as *Contracaecum microcephalum* (Rudolphi, 1809) by Hartwich [31] and this species was accordingly defined as the new type-species of the genus. At present, the genus comprises some 50 nominal species, most of which are adults in pinnipeds and fish-eating birds. Hartwich [32] defined the genus based on morphological details; important features are the presence of interlabia and the position of the excretory pore at the base of the ventral interlabium. However, several later authors suggested that the genus should be divided or amended [16]. The genus *Phocascaris* Höst, 1932 included the following definition "*interlabia present, reduced or absent parasites of the digestive tract of seals*," *Phocascaris phocae* being the type-species of the genus. According to that definition, Berland [33], when describing *Phocascaris cystophorae* from the seal *Cystophora cristata*, suggested that the species of *Contracaecum* "with opposite caeca from seals" should be transferred to the genus *Phocascaris*. His proposal was to retain within *Contracaecum* species from fish-eating birds. Berland's [33] hypothesis was mainly based on the life-cycle patterns of these anisakid nematodes and their definitive hosts, irrespective of their morphological phenotype. However, a taxonomic proposal was not generally accepted by taxonomists, and Hartwich [32] did not take into account those features, still considering a morphological character (relating to the presence/absence of interlabia) as a valid criterion for distinguishing the species belonging to *Contracaecum* and *Phocascaris*.

Several years later the same morphological character was shown to be inconsistent and not of useful systematic value for differentiating the two groups of species included in *Contracaecum*. Indeed, based on allozymes (22 enzyme loci), Orecchia et al. [34] and Nascetti et al. [35] used species *Contracaecum* from seals and fish-eating birds to show that all of the species of the *C. osculatum* complex are genetically much more closely related to species of *Phocascaris* (i.e., *P. phocae* and *P. cystophorae*) from seals than to the congeneric species from fish-eating birds, demonstrating high levels of differentiation (*Contracaecum* species from pinnipeds share no alleles in common with those from fish-eating birds). Clustering methods based on allozyme markers, showed that *P. phocae* and *P. cystophorae* form a clade with the species of *Contracaecum* from seals [35] suggesting an evolutionary hypothesis for the systematic status of these species. A phylogenetic hypothesis, based on different clustering analyses of the ITS-rDNA [36] and the mtDNA *cox-2* [37] sequence data of several taxa of the genera *Contracaecum* and *Phocascaris*, strongly supported that based on allozymes, according to which species of *Phocascaris* are nested within the clade formed by the *Contracaecum* species hosted by phocid seals and are thus closely related to species of the *C. osculatum* complex [36]. All of the phylogenetic analyses also support the hypothesis advanced by Berland [33], based on the biology of these anisakid nematodes, in terms of the monophyly of the species of *Contracaecum* and *Phocascaris* from phocids.

To date, five members of the *C. osculatum* complex have been recognized genetically by the use of allozyme markers. They are the three Arctic sibling species referred to as *C. osculatum* A, *C. osculatum* B, and *C. osculatum* (*s.s.*) [38], and the two Antarctic members, *C. osculatum* D and *C. osculatum* E [39].

C. osculatum sp. A of Nascetti, Cianchi, Mattiucci, D'Amelio, Orecchia, Paggi, Battey, Berland, Smith, and Bullini [38]. This species occurs in the Norwegian and Barents Seas, Canadian Atlantic, Icelandic, Canadian Pacific waters, and the Sea of Japan, between 40 and 80°N. Its main definitive host is the bearded seal, *E. barbatus*, in both the eastern and western part of the North Atlantic Ocean, but the gray seal, *H. grypus*, has also been recorded as a host [21,38]. Other hosts have been detected genetically in Steller'sea lion, *E. jubatus*, in the Western Pacific Ocean (Sea of Japan) [22,25]. Its larvae have been identified in the gadoid fish *Theragra chalcogramma* from the Sea of Japan [22].

C. osculatum sp. B of Nascetti Cianchi, Mattiucci, D'Amelio, Orecchia, Paggi, Battey, Berland, Smith and Bullini [38]. This species was first detected genetically as an adult in the phocid seals *Pagophilus groenlandicus, P. vitulina, and H. grypus* from the North-eastern and north-western Atlantic Ocean [21,38]; reproductive isolation was demonstrated by allozyme markers between this species and *C. osculatum* A and *C. osculatum s.s.* [38]. Later, *C. osculatum* B it was identified, based on diagnostic allozyme markers, as a parasite in the adult stage of the *P. vitulina* and *Z. californianus* in northern Pacific waters [22].

C. osculatum (sensu stricto) (= *C. osculatum* C by Nascetti et al. [38]) was genetically characterized from the gray seal, *H. grypus*; its reproductive isolation from *C. osculatum* B was shown in the sympatric situation of individual seal hosts (i.e., the gray seal) from Iceland waters [38]. It is the only species of the *C. osculatum* complex present in the Baltic Sea. So far, this species has not been reported from the western part of the Atlantic or from Pacific waters.

C. osculatum sp. D of Orecchia, Mattiucci, D'Amelio, Paggi, Plotz, Cianchi, Nascetti, Arduino and Bullini [39]. This represents one of the two Antarctic members of the *C. osculatum* complex. Its reproductive isolation from *C. osculatum* E was demonstrated by allozyme markers: they occur sympatrically in the same definitive host, the Weddell seal, *L. weddellii*, and have so far been reported from both the Weddell and the Ross Seas (Antarctica). The larval stages of *C. osculatum* D have been identified by allozyme markers as a main parasite of pelagic and meso-pelagic fish hosts belonging to the families Channicthydae and Bathydraconidae [40].

C. osculatum sp. E of Orecchia, Mattiucci, D'Amelio, Paggi, Plotz, Cianchi, Nascetti, Arduino, and Bullini [39]. While at the adult stage the species *C. osculatum* E is a parasite of the Antarctic Weddell seal, at the larval stages it has been identified by allozyme markers as the main parasite of Antarctic benthic fish hosts belonging to the family Nothotenidae [40].

A detailed host list and geographical ranges of the species included in the *C. osculatum* complex is reviewed in Mattiucci and Nascetti [10].

62.1.2 Epidemiology, Clinical Features, and Pathogenesis

As stated above, most human infections by *Pseudoterranova* spp have been so far described in Japan, the United States, and Chile. In Japan, where seafood is a major component of the diet and traditionally served raw, 335 cases of gastric anisakidosis due to *Pseudoterranova* were reported among the 12,000 cases of clinical Anisakidosis [41]. It was found more frequently in the northern region rather than in the southern region, and mainly related with the fish halibut and flat-fish from the northern area, infected by *P. decipiens s.l.* larvae in [42].

Human infections involving *Pseudoterranova* worms were also found in some cases of anisakidosis reported from North America, Alaska, and California [1–3]; in all those cases the infections were transient, presumably luminal, with no evidence for a significant tissue penetration. In addition to these cases, other transient infections were reported also in California [4].

In Korea, after the first case of anisakidosis by *Pseudoterranova* described in humans [5], other cases followed [6,7] and more recently, 12 human cases of the zoonosis caused by *Pseudoterranova* were reported [8]. In Chile, after a first case documented [43], more recently seven new cases were reported [44].

The popularity of ethnic dishes based on raw fish such as "ceviche" "sashimi," "sushi," "oka," "carpaccio," "poison cru," and the import-export of fish from all the world have increased the potential for infection with *P. decipiens s.l.* larvae in those areas where its life-cycle is currently absent. Thus, in Europe, up to now the disease has been reported mainly from the north. Now it is also being encountered in southern Europe—terranovasis was reported in France [45] and in Italy a case of human infection by *Pseudoterranova* sp. was documented in a woman infected by *P. decipiens s.l.* larva acquired by eating raw imported flat-fish [46].

Pseudoterranova spp. larvae have been implicated in human infections [47,48], and a few species of *Contracaecum* were found experimentally to be potentially infective [9]. It is thus opportune to review and comment on reports concerning human infections attributable to these worms.

Third-stage *Pseudoterranova* larvae have been recovered from human infections or shown in a few cases to be capable of invading gastrointestinal tissue of experimentally infected animals [49]. However, according to the literature, the clinical symptoms of infection caused by *Pseudoterranova* sp. larvae in humans are weaker than those observed for *Anisakis* spp. larvae, and may even be asymptomatic, being the infection accidentally discovered in patients affected by pain [42] from other causes. Nevertheless, clinical symptoms of anisakidosis related to *Pseudoterranova* can appear from 2 to 3 h up to some days (7 days) after the ingestion of the infected fish.

It seems that terranovasis can manifest as "mild" or "acute or fulminant." There are various degrees of acute symptomatology and with different organs involved. The "fulminant" form is so far described, for instance, for those anisakidoses due to *P. decipiens s.l.* larvae reported both in Chile, California, and North America cases. In the "fulminant" form, *P. decipiens s.l.* larvae have been often found attached to the oral, pharynx, and digestive mucosa or, at least, free in the lumen of the stomach. Most patients of those Chilean cases showed nausea, irritation of the pharynx, and in some cases the larvae were coughed up between 36 h and 7 days after eating fish. In several described cases, the larval stages were expelled per *os*, or vomited, or coughed up, or found dead in the feces as a "transient luminal infection." [50] Nausea and persistent pain of the pharyngeal zone have been described in these cases. The latter is generally asymptomatic except for a tingling sensation in the throat or mild abdominal discomfort, and the nematodes are expelled by coughing, vomiting, and with the feces. In some patients, irritation of the pharynx and/or some allergic symptoms have been described [3,44,51]. Reports of a "mild form" related to the cases so far described from those geographical areas, are very few and not fully documented. In one case, a third-stage larva was reported emerging from the neck region after trans-esophageal penetration [52].

In Japan, documented cases of anisakidosis due to *P. decipiens s.l.* have involved only the stomach. Indeed, there are no reports of the occurrence of the nematode in the human intestine, and thus the larvae of *P. decipiens s.l.* are implicated mostly in causing "gastric anisakidosis." In gastric disease a "fulminant form" presenting a severe and acute disease characterized by gastric symptoms is present, while a "mild form" which mimics the presence of a gastric granuloma, is extremely rare and so far not adequately documented [42]. Larvae of *P. decipiens s.l.* have been detected by endoscopy in Japan. They have been found as possibly penetrating gastric and intestinal walls but they are less invasive than *Anisakis simplex s.l.* However, in most cases the infection remains luminal, with no evidence of significant tissue penetration. Some acute gastric mucosal lesions caused by the penetration of the larvae of *P. decipiens s.l.* present in the stomach have been reported [42]. In these infections, it has been observed that *P. decipiens s.l.* larvae usually penetrate the body of the stomach especially at the level of the major curvature side of the cardia [43]. Endoscopic images reveal that the mucosa is edematous: biopsy of the penetration site of the larvae showed epithelial edema, fibrination, cellular infiltration but not eosinophilic penetration and, thus, there is not the formation of an eosinophilic granuloma typical of *Anisakis* spp larvae. In Europe, gastric terranovasis has been reported, where the worms were found free in the lumen. Therefore, there is no report of larval penetration into the intestinal mucosa or submucosa, and thus no occurrence of eosinophilic granuloma in the intestine associated with the infection by *P. decipiens s.l.* larvae. For this reason, the infection by *P. decipiens s.l.* seems to be milder than that caused by *Anisakis* spp. larvae.

A series of experiments carried out by Asaishi et al. [53] revealed that *A. simplex s.l.* larvae and *P. decipiens s.l.* share common antigens. Based on this finding, those authors suggested that the severe acute gastric symptoms due to *P. decipiens s.l.* larvae could have arisen as cross reaction to a secondary infection after previous exposure to *A. simplex s.l.* larvae. The allergic hypersensitivity that has been seen in infection by *Anisakis*, is not apparent in infections with *P. decipiens s.l.* larvae. This finding is probably associated with the absence of the eosinophilic granuloma formed around *P. decipiens s.l.* larvae.

62.1.3 Diagnosis

62.1.3.1 Conventional Techniques

Because the main pathological significance described so far for anisakidosis due to *P. decipiens* larvae is an "acute/fulminant" and/or "chronic/mild" pathology, rapid diagnosis and accurate identification of the larva are important. Larvae may be obtained by endoscopy or from vomit or feces.

If an intact larva is obtained from a human infection, it may be possible by macroscopic inspection to distinguish larval *P. decipiens* from *Anisakis* Type I [54] larva because the first is longer, thicker, creamy white-yellow, reddish, or brown. However, *Anisakis* Type II larvae [54] could confound those findings, because they are as whitish-red and as long as *P. decipiens* larvae. Therefore, a microscopic diagnosis is required. This procedure involves the optical observation of the larvae after their clarification by lactophenol. The main morphological feature on which a differential diagnosis is possible is, in *P. decipiens* larvae, the presence of the intestine which, immediately behind the ventriculus is produced forward as an intestinal caecum. The intestinal caecum is lacking in *Anisakis* spp. larvae. *Contracaecum* spp. larvae possess an intestinal caecum and a ventriculus appendix expanding immediately after the ventriculus toward the intestine. Therefore, in case of given intact specimens and in good condition, the larval stages of *Pseudoterranova* and *Contracaecum* could be at least distinguished at the genus level from those of *Anisakis* larvae, based on those gastrointestinal tract morphological features. Nevertheless, based on those microscopic morphological features, the differential diagnosis between *Pseudoterranova* and *Contracaecum* larvae with respect to *Anisakis* larvae, is possible only at the genus level. Indeed, it must be underlined that there are no morphological characters that identify to species level any member of the genera *Pseudoterranova*, *Contracaecum*, or *Anisakis*.

62.1.3.2 Molecular Techniques

Because in most human infections, the anisakid worms removed are incomplete the morphological features mentioned above, are not available. Since there are no additional characters to be used for their identification, molecular approaches must be advocated.

The varied epidemiology of anisakidosis (due to *Pseudoterranova* spp) seems to indicate the existence of different clinical aspects of this zoonotic disease in humans. Further, the existence of different sibling species of the *P. decipiens* complex, as a source of infection to humans, often occurring in sympatry in some geographical areas and in mixed infections in the same fish host is documented. This makes impossible the identification of the pathogens causing the anisakidosis, and could mask the epidemiological role of the anisakid species involved in a geographical area. Different species of anisakid worms could be associated with different clinical forms of anisakidosis in humans. Thus, the identification to species level of an anisakid larval stage is a fundamental prerequisite for any epidemiological study to be carried out in a particular area.

62.1.3.2.1 Allozyme Data

Allozyme typically consisting of allele frequencies obtained from starch gel electrophoresis of proteins—have been considered an important source of characters for understanding the genetic structure of a population of anisakid nematodes. One such analysis is Multilocus Allozyme Electrophoresis (MAE). The data consist of multiple nuclear loci, with each locus providing independent estimate of the species differentiation. It is relatively cheap and easy to survey allozyme variation for a large number of individual anisakid nematodes; further, it is possible to test for reproductive isolation between populations of anisakid nematodes. Although species of anisakid nematodes still continue to be detected and described based on allozyme markers [37,55,56], one of the limits of this methodology lies in the fact that even if it was standardized on the different anisakids, some differences could still be found in different laboratories.

62.1.3.2.2 PCR-DNA Analysis

Thus, in recent years, the results achieved by this methodology in the detection of anisakid taxa belonging to the genera *Pseudoterranova* and *Contracaecum* and to infer their phylogenetic relationships have been supplemented by other molecular/genetic approaches using DNA [10,24,37,56–58]. For instance, ITS-rDNA markers[30] were produced for individual specimens previously characterized by allozymes, belonging to the taxa: *P. decipiens (s.s.)*, *P. krabbei*, *P. bulbosa*, *P. azarasi*, *P. cattani*, and a taxon indicated as *P. decipiens* Ca1; the last likely corresponding to *P. decipiens* E, but this last correspondence has not been verified [30]. Single-strand conformation polymorphism-based identification of members of the *C. osculatum* complex using genetic markers in the ITS-rDNA [57] and in three mitochondrial DNA regions (mitochondrial cytochrome oxidase 1 or mtDNA cox1 (COI), small subunit of the mitochondrial ribosomal RNA gene (ssrRNA), large-subunit of the mitochondrial ribosomal RNA gene (lsrRNA)) [58] allowed the unequivocal differentiation of all the taxa previously disclosed by allozyme markers, except that of the two Antarctic species, *C. osculatum* D and *C. osculatum* E [57,58].

Despite the large informative methodology based on allozymes to acquire genetic characterization and genetic variation of anisakid species, however, as stated above, MAE is not suitable for employment in all laboratories. On the other hand, the sequences of the mtDNA *cox-2* gene have been used, in conjunction with nuclear markers from allozymes, in both the species groups here considered, to recognize all the species included in the *P. decipiens* and *C. osculatum* complexes and to support their taxonomic status [10,37].

Fortunately, all the species belonging to *Pseudoterranova* and *Contracaecum* have been genetically characterized on the basis of the direct DNA sequencing of the mtDNA *cox-2* gene and its genetic variation has been also studied by the direct sequencing of several specimens sequenced and deposited in GenBank [10,37,55], it is possible to use that mitochondrial gene as a molecular "barcode" for the species identification of the *P. decipiens* and *C. osculatum* groups of species genetically characterized so far. Therefore, we will here report, in detail, the methodological procedures and DNA sequencing analysis at the mtDNA *cox-2* gene for those anisakids, which have been previously also identified at species level, by means of allozyme markers.

62.2 METHODS

The entire protocol, from the isolation of the larvae from fish and/or from humans to their specific identification is here described in detail based on the nucleotide differences in the mitochondrial gene coding for the *cox-2* as developed by Mattiucci et al. [37] and unpublished. The species identification can be then performed by direct DNA sequencing of the PCR products after amplification using specific primers designed for the mtDNA *cox-2* gene.

62.2.1 Preparation of Samples

62.2.1.1 Isolation of Larvae from Fish and/or Human Cases

Because of their size, larvae of the genera *Pseudoterranova* and *Contracaecum* (here considering only those maturing in seals), could be readily seen "by eye" as elongated or coiled, in the body cavity of their fish hosts, generally on the surface of internal organs, such as liver, stomach, gonads and, in the muscles (mostly *Pseudoterranova*) [59]. The larvae could be easily removed after opening the fish from the anus to the gills. The presence of larvae in muscle could be easily detected by a transilluminator or by light-candling or by ultraviolet-frozen transilluminator [60]. After their removal from natural hosts, the larvae can be stored as frozen, or, alternatively, fixed in absolute alcohol.

Larvae of *Pseudoterranova* from humans could be removed by endoscopy and colonscopy. Alternatively, they could be collected after being coughed up, or recovered in patients' feces. In most cases, the larvae are fragmentary or deteriorated. After removal the larvae must be washed in a saline solution and quickly frozen or fixed in absolute alcohol. Formalin is not recommended as this fixative reduces the quality of DNA and could render identification impossible. For those stored in ethanol (absolute or 70%), it is advisable to leave the samples in phosphate-buffered saline for 2–3 h and then wash them with distilled water before DNA purification.

62.2.1.2 DNA Purification

Two protocols to isolate the genomic DNA for PCR amplification are here described.

62.2.1.2.1 Cetyltrimethylammonium Bromide (CTAB)/Phenol-Chloroform Modified Method [61]

62.2.1.2.1.1 Procedure

1. Add 500 μL of CTAB solution to a 1.5 mL microcentrifuge tube in addition to 5–10 μL of proteinase K.
2. Add 0.5–1 cm of length of sample worm or 15–20 mg of tissue to the mixture (CTAB/proteinase K) and pestle the sample to homogenize the tissue.
3. Incubate the sample tube at 56°C for 5 h or at 50°C overnight, in the heating bath.
4. Add 500 μL of phenol:chloroform:isoamyl alcohol 25:24:1 (saturated with 10 mM Tris, pH 8.0, 1 mM EDTA).

5. Mix for 5 min on a shaker and then centrifuge for 10–15 min at 13,000× *g*.
6. Transfer the supernatant to a clean 1.5 mL microcentrifuge tube and repeat the procedure described at step 4 and 5.
7. Transfer the supernatant to a clean 1.5 mL microcentrifuge tube and add 500 μL of chloroform.
8. Mix for 5 min on a shaker and then centrifuge for 10–15 min at 13,000× *g*.
9. Transfer the supernatant to a clean 1.5 mL microcentrifuge tube and add 1000 μL absolute ethanol (stored at −20°C) and 20 μL of NaCl 5 M.
10. Store the sample for 30–60 min at −20°C.
11. Centrifuge for 10–15 min at 13,000× *g*.
12. Remove all the ethanol with care, leaving the pellet on the bottom of the tube.
13. Add 500 μL of 70% ethanol (stored at −20°C) and centrifuge 10–15 min at 13,000× *g*.
14. Invert the tube to discard the ethanol and air-dry overnight or leave the sample 10–15 min in the thermalblock at 60–65°C.
15. Resuspend the DNA in 50–100 μL TE buffer (pH = 8) or ultrapure water.
16. Store the resuspended DNA at −20°C.

62.2.1.2.2 Wizard® Genomic DNA Purification Kit (Promega)

62.2.1.2.2.1 Procedure

1. Add 600 μL of nuclei lysis solution to a 1.5 mL centrifuge tube and chill on ice.
2. Add 10–20 mg of fresh or thawed tissue to the chilled nuclei lysis solution and homogenize for 10 s using a small homogenizer. Transfer the lysate to a 1.5 mL microcentrifuge tube. Alternatively, grind tissue in liquid nitrogen using a mortar and pestle that has been prechilled in liquid nitrogen. After grinding, allow the liquid nitrogen to evaporate and transfer ~10–20 mg of the ground tissue to 600 μL of nuclei lysis solution in a 1.5 mL microcentrifuge tube.
3. Incubate the lysate at 65°C for 15–30 min.
4. Add 3 μL of RNAase solution to the nuclear lysate and mix the sample by inverting the tube 2–5 times. Incubate the mixture for 15–30 min at 37°C. Allow the sample to cool to room temperature for 5 min before proceeding.
5. To the room temperature sample, add 200 μL of protein precipitation solution and vortex vigorously at high speed for 20 s. Chill sample on ice for 5 min.
6. Centrifuge for 4 min at 13,000–16,000 × *g*. The precipitated proteins will form a tight white pellet.
7. Carefully remove the supernatant containing the DNA (leaving the protein pellet behind) and transfer it to a clean 1.5 mL microcentrifuge tube containing 600 μL of room-temperature isopropanol.
8. Gently mix the solution by inversion until the white thread-like strands of DNA form a visible mass.
9. Centrifuge for 1 min at 13,000–16,000× *g* at room temperature. The DNA will be visible as a small white pellet. Carefully decant the supernatant.
10. Add 600 μL of room temperature 70% ethanol and gently invert the tube several times to wash the DNA. Centrifuge for 1 min at 13,000–16,000× *g* at room temperature.
11. Carefully aspirate the ethanol using either a drawn Pasteur pipette or a sequencing pipette tip. The DNA pellet is very loose at this point and care must be used to avoid aspirating the pellet into the pipette.
12. Invert the tube on clean absorbent paper and air-dry the pellet for 10–15 min.
13. Add 100 μL of DNA rehydration solution and rehydrate the DNA by incubating at 65°C for 1 h. Periodically mix the solution by gently tapping the tube. Alternatively, rehydrate the DNA incubating the solution overnight at room temperature or at 4°C.

62.2.2 Detection Procedures

62.2.2.1 PCR for *Contracaecum* spp. and *Pseudoterranova* spp.

The mtDNA *cox-2* gene can be amplified for detecting species of the genus *Contracaecum* and *Pseudoterranova* using primers 211F (forward: 5′-TTT TCT AGT TAT ATA GAT TGR TTY AT-3′) and 210R (reverse: 5′-CAC CAA CTC TTA AAA TTA TC-3′) [55,62]; PCR amplification produces a fragment of about 520 bp.

62.2.2.1.1 Procedure

1. Prepare the PCR master-mix (final volume of 45 μL) multiplying each reagent for the samples number to amplify, plus two additional sets for the positive and negative control (e.g., for six specimens, prepare the master-mix for eight samples) as follows: (i) Add in order, for each sample, 5 μL $MgCl_2$, 5 μL 10× $MgCl_2$ buffer, 1 μL dNTPs, 1 μL for each primer, 1 μL dimethyl-sulfoxide (99.9%), 0.3 μL Taq DNA polymerase and then ultrapure water up to 45 μL, in a 1.5 mL sterile tube. (ii) Prepare 0.2 mL PCR tubes for each sample plus two additional 0.2 mL tubes (as specified above) and add 45 μL of PCR master-mix to each. (iii) Add 5 μL of genomic DNA to the same 0.2 mL PCR tubes, up to 50 μL of final volume.
2. PCR cycle: 3 min at 94°C; 34 cycles of 94°C for 30 s, 46°C for 1 min, 72°C for 1.5 min; and final extension at 72°C for 10 min.
3. Electrophoresis: to test the PCR amplification quality, add 5 μL of each PCR product to 2 μL of blue/orange loading dye and load the entire volume on a 1% agarose gel containing 1.5–3 μL of biotium gelred, in TBE buffer (0.5×) at 10–15 V/cm.
4. The *cox-2* amplification products are purified with Wizard® SV Gel and PCR Clean-Up System (Promega) and sequenced. Apply the purification protocol as follows: (i) Place on SV Minicolumn in

a collection tube for each dissolved PCR reaction. (ii) Transfer the prepared PCR product to the SV Minicolumn assembly and incubate for 1 min at room temperature. (iii) Centrifuge the SV Minicolumn assembly in a microcentrifuge at 10,000× *g* for 1 min. Remove the SV Minicolumn from the Spin Minicolumn assembly and discard the liquid in the collection tube. Return the SV minicolumn to the collection tube. (iv) Wash the column by adding 700 μL of membrane wash solution, previously diluted with 95% ethanol (add the indicated volume of 95% ethanol to the membrane wash solution prior to beginning the procedure). Tightly close the bottle cap after each use to prevent evaporation), to the SV Minicolumn. Centrifuge the SV Minicolumn assembly for 1 min at 10,000× *g*. Empty the collection tube as before and place the SV Minicolumn back in the collection tube. Repeat the wash with 500 μL of membrane wash solution and centrifuge the SV Minicolumn assembly for 5 min at 10,000× *g*. (v)

```
                            10        20        30        40        50        60        70        80
                    ....|....|....|....|....|....|....|....|....|....|....|....|....|....|....|....|
P. decipiens (s.s.) GTTTTTACTTTAAGAGGAAAAAAATTGAGTATCAATTGGGTGAACTTTTGTGTAGTATTTTTCCTACTTTGATTTTGGTA
P. krabbei          .A.....T........T..G.......................G...................................G
P. bulbosa          ......................G...........G..A..........................................G
P. azarasi          ................................................................................
P. decipiens E      .......T...........G..G.....A........A..A...................................A..G

                            90       100       110       120       130       140       150       160
                    ....|....|....|....|....|....|....|....|....|....|....|....|....|....|....|....|
P. decipiens (s.s.) GCTCAGATGGTGCCGTCTTTAAGTCTTCTTTATTATTATGGCTTGATAAATTTGGACAGTAATTTAACTGTTAAAGTTAC
P. krabbei          ...........A..T.....G....................T....................C..............C..
P. bulbosa          ...........A..T..........................T.................G..C...........G.....
P. azarasi          .........................................T...................................C..
P. decipiens E      ...........A..T.....G....................T..A...........T..G...........G..G..A..

                           170       180       190       200       210       220       230       240
                    ....|....|....|....|....|....|....|....|....|....|....|....|....|....|....|....|
P. decipiens (s.s.) TGGTCATCAATGATATTGGAGTTATGAGTTTAGGGATATTCCTGGTTTGGAGTTTGATTCTTATATAAAATCTGTTGATC
P. krabbei          ...C.....G..G..........................C..G...........C.........................
P. bulbosa          ............G.....A.......................G.....................................
P. azarasi          ..........................................C.....................................
P. decipiens E      C...........G.....A..............A..C.............................G..G..........

                           250       260       270       280       290       300       310       320
                    ....|....|....|....|....|....|....|....|....|....|....|....|....|....|....|....|
P. decipiens (s.s.) AATTAGAGTTGGGGGAGCCACGTTTGTTAGAGGTTGATAATCGTTGTGTTGTTCCTTGTGATACTAATATTCGTTTTTGT
P. krabbei          .G..G..............G..............G..........................C..................
P. bulbosa          ...................G..............G..C..C.......................................
P. azarasi          ...................G.........................................C..................
P. decipiens E      .G..G..A.....A....................C.............................................

                           330       340       350       360       370       380       390       400
                    ....|....|....|....|....|....|....|....|....|....|....|....|....|....|....|....|
P. decipiens (s.s.) ATTACTTCTGGCGATGTAATTCATTCTTGAGCTTTGCCCAGAATATCTATTAAGTTGGATGCTATGAGAGGTATTTTATC
P. krabbei          ...........T.................G........T..A..........................G........G..
P. bulbosa          ...........T..........................A..A..G................................G..
P. azarasi          ........C..................................................C....................
P. decipiens E      ...........T.................G........T.....G....................A..G........G..

                           410       420       430       440       450       460       470       480
                    ....|....|....|....|....|....|....|....|....|....|....|....|....|....|....|....|
P. decipiens (s.s.) CACTTTATCTTATAGCTTTCCTACTGTGGGTGTTTTTTATGGTCAGTGTTCTGAGATTTGTGGGGCTAATCATAGTTTTA
P. krabbei          T..............T................................................................
P. bulbosa          T.....G........T.........................................C...........C.....G....
P. azarasi          T.................C.............................................................
P. decipiens E      T..............T................................C.....A.........T.T.............

                           490       500       510
                    ....|....|....|....|....|....|....|....
P. decipiens (s.s.) TACCTATTGCTTTAGAAGTAACTTTACTTGATAATTTTA
P. krabbei          .......C..C............................
P. bulbosa          .............G.........C.G.............
P. azarasi          .......................................
P. decipiens E      ................G..G...................
```

FIGURE 62.1 Alignment of mtDNA *cox-2* (519 bp) sequences for the five species of the *P. decipiens* complex so far genetically characterized, with their corresponding accession number in GenBank. They are the following: *P. decipiens s.s.*-HM147278; *P. krabbei*-HM147279; *P. bulbosa*-HM147280; *P. azarasi*-HM147281; *P. decipiens* E-HM147282.

Remove the SV Minicolumn assembly from the centrifuge, being careful not to wet the bottom of the column with the flowthrough. Empty the collection tube and recentrifuge the column assembly for 1 min. (vi) Carefully transfer the SV Minicolumn to a clean 1.5 mL microcentrifuge tube. Apply 50 μL of nuclease-free water directly to the center of the column without touching the membrane with the pipette tip. Incubate at room temperature for 1 min. Centrifuge for 1 min at 10,000 × *g*. (vii) Discard the SV Minicolumn and store the microcentrifuge tube containing the eluted DNA at 4°C or −20°C.

5. PCR product is sequenced using an external service. Commercial services can provide fast accurate sequencing at reasonable cost. Results may be available in formats such as phenograms, txt files, and pdf files.

62.2.2.2 Species Identification Based on mtDNA *cox-2* Sequences

Larvae of the *P. decipiens* and *C. osculatum* complexes can be identified by performing a direct comparison of the mtDNA *cox-2* sequences obtained on that specimen with those sequences available in GenBank by using Bioedit software [63]. A maximum parsimony phylogenetic analysis, by using the software PAUP* [64] should be applied, including the specimen to be identified, and sequences of related taxa of the genera *Pseudoterranova* and *Contracaecum* available in GenBank. Further, because

```
                                        10        20        30        40        50        60        70        80
                               ....|....|....|....|....|....|....|....|....|....|....|....|....|....|....|....|
C. osculatum (s.s.) (EU477206) GTTTTTATTTTAAAAGTAAGAAGATCGAGTATCAATTTGGGGAGCTTTTGTGTAGTGTGTTTCCTACTTTAATTTTGGTT
C. osculatum A (EU477203)      ..........C..G...........T..................T.G.............................A...
C. osculatum B (EU477204)      ................G........T..................T.G.....C.....T...........G.....A..C
C. osculatum D (EU477205)      .............G...........T.....C.....C......T.GC.A.................C..G.........
C. osculatum E (EU477207)      A............G...........T.....C............T.G..A....................G.........

                                        90       100       110       120       130       140       150       160
                               ....|....|....|....|....|....|....|....|....|....|....|....|....|....|....|....|
C. osculatum (s.s.) (EU477206) ATGCAGATGGTACCTTCTTTAAGTTTGTTGTATTATTATGGTCTTATAAATCTTGACAGTAGCTTAACTGTTAAGGTAAC
C. osculatum A (EU477203)      ..A........G........G...................................T.....T........G..A..T..
C. osculatum B (EU477204)      ...........G..........................C............T.G..T.....T........A........
C. osculatum D (EU477205)      G..........G.......................C....................T......C.......A.....T..
C. osculatum E (EU477207)      G..........G.......................C..............C.....T.....T...........A..G..

                                       170       180       190       200       210       220       230       240
                               ....|....|....|....|....|....|....|....|....|....|....|....|....|....|....|....|
C. osculatum (s.s.) (EU477206) TGGTCACCAATGGTATTGAAGTTATGAGTTTAGGGACATCCCTGGACTTGAGTTTGATTCTTATATGAAATCTCTGGATC
C. osculatum A (EU477203)      .................................A..T..T.....G.......................G...T......
C. osculatum B (EU477204)      G..G..T.................C........T..T........G.......................G...T....C.
C. osculatum D (EU477205)      A.....T.............................T..T.....T...........................T......
C. osculatum E (EU477207)      ......T.............................T..T.....T.C.....................G...T......

                                       250       260       270       280       290       300       310       320
                               ....|....|....|....|....|....|....|....|....|....|....|....|....|....|....|....|
C. osculatum (s.s.) (EU477206) AGTTGGAGTTGGGGGAGCCTCGTTTATTGGAAGTAGATAACCGTTGTGTTATTCCTTGTGATACTAATATCCGGTTTTGT
C. osculatum A (EU477203)      .A........................C.T..G..C.............................................
C. osculatum B (EU477204)      ....A.....................C.T..G..T......T......................................
C. osculatum D (EU477205)      ....A.....A.................A..G..T......T........G.............................
C. osculatum E (EU477207)      ....A.......................A..G..C......T........G.............................

                                       330       340       350       360       370       380       390       400
                               ....|....|....|....|....|....|....|....|....|....|....|....|....|....|....|....|
C. osculatum (s.s.) (EU477206) ATTACTTCTGGGGATGTGATCCACTCTTGGGCTTTGCCCAGGCTATCTATTAAACTTGATGCTATGAGGGGCATTTTATC
C. osculatum A (EU477203)      ...........T........T..T.....A.....A......T.G........GT.A..............T........
C. osculatum B (EU477204)      .......................T...............T....G...........A.......................
C. osculatum D (EU477205)      ...........T........T..T.........C.A...T..T.G.........T.A..............T........
C. osculatum E (EU477207)      ..C........T........T..T...........A........G.........T.A...........A..T........

                                       410       420       430       440       450       460       470       480
                               ....|....|....|....|....|....|....|....|....|....|....|....|....|....|....|....|
C. osculatum (s.s.) (EU477206) TACTTTGTCTTATAGGTTTCCTGTGGTGGGCGTGTTTTATGGCCAGTGTTCAGAAATTTGTGGTGCTAACCATAGTTTTA
C. osculatum A (EU477203)      ...............C..............G..A..C.....T..A.....G.............A..............
C. osculatum B (EU477204)      ...............T..............A.....C.....G........G............................
C. osculatum D (EU477205)      ...............T...........A..A..A.....C..G..A..C..T..G..............T..........
C. osculatum E (EU477207)      ............C..T...........A..A..A........G..A.....T..G..............T..........

                                       490       500       510
                               ....|....|....|....|....|....|....|....
C. osculatum (s.s.) (EU477206) TGCCTATTGCTTTGGAGGTAACTTTGTTGGATAATTTTA
C. osculatum A (EU477203)      .......................................
C. osculatum B (EU477204)      .......C...............................
C. osculatum D (EU477205)      ...................G...................
C. osculatum E (EU477207)      .A.....................................
```

FIGURE 62.2 Alignment of mtDNA *cox-2* (519 bp) sequences for the five species of *C. osculatum* complex so far genetically characterized, with their corresponding accession number in GenBank. They are the following: *C. osculatum s.s.*-EU477206; *C. osculatum* A-EU477203; *C. osculatum* B-EU477204; *C. osculatum* D-EU477205; *C. osculatum* E-EU477207.

in human infection, the larval stage could belong to the species of the genus *Anisakis*, it is advisable to compare the mtDNA *cox-2* sequence obtained on the specimen to be identified, with those of all the species of *Anisakis* so far genetically characterized based on the same mtDNA *cox-2* gene and deposited in GenBank [55].

An explicative sequences alignment based on mtDNA *cox-2* genes, both for *P. decipiens* and *C. osculatum* complexes here considered, is reported in Figures 62.1 and 62.2, respectively.

62.3 CONCLUSIONS AND FUTURE PROSPECTS

Reports of human infections with anisakids are accumulating at an increasing rate and this increase appears set to continue. Several factors may be contributing to this: (i) elucidation of the biology of these parasites, including taxonomy and identification by molecular markers, allowing greater insight into the zoonotic potential of the worms; (ii) the introduction of the fiberscope in the routine endoscopy as a direct diagnostic technique which has allowed visual detection of larvae *in situ* followed by removal for identification, thus avoiding the misdiagnosis which has accompanied and confounded clinical work on anisakidosis; (iii) the enlarged demand for seafood all over the world which has extended fish species to markets far beyond the original geographical area of seafood capture; (iv) the growing habits and preference by humans to consume dishes based on raw or lightly cooked fish, thus increasing the risk of exposure.

In the main, reported occurrence of Anisakidosis in humans is related, at the patients' anamnesthic diagnosis, to the consumption of traditional dishes, such as Scandinavian "gravlax," Hawaiian "lomi-lomi," Latin American "ceviche," and others. In areas where anisakidosis is rare, it more likely to go undiagnosed. Therefore, knowledge of the epidemiological aspects of these parasites should also be extended to other geographical areas so far not deeply investigated by recently available molecular methodologies. This would include areas such as Mexico, the central Pacific, and New Zealand, where scanty data are so far existing [65–67]. This may contribute to the correct diagnosis of anisakidosis should it occur in these less anisakid-aware regions.

Since human infection with anisakids starts by ingestion of larvae carried in the fish host, greater knowledge about the distribution of these parasites in the host tissue would be desirable. Some studies suggest that larval migration can occur after fish capture when they move from the visceral cavity to the musculature. However, this has been suggested only for some fish species; in other cases this migration has not been demonstrated [68]. The issue of migration remains unresolved but it is of central interest to human health especially since better knowledge could inform appropriate prophylaxis measures, such as freezing and cooking before consumption, or the immediate evisceration of the fish directly on board, to prevent *postmortem* migration of the larvae.

In recent years, anisakid nematodes have been associated with allergic reactions after fish consumption, even after the larvae were removed from fish host. Those cases have been so far documented mostly due to *A. simplex s.l.* in Spain and have involved a specific antigen (An4), identified by immunoblotting analysis, reacting with human sera from patients affected by general allergy [69]. The An4 antigen seems also to have a high thermostability, being highly resistant to heat [70]. This would mean that standard precautions of freezing and cooking fish may not provide protection.

However, such hypersensitivity reaction to other anisakid nematodes such as *Pseudoterranova* and *Contracaecum* is not yet known. Recent work has demonstrated that *P. decipiens* larvae do possess cold tolerance capacity [71]. According to these authors, even if the *P. decipiens* L3 have the ability to survive to temperatures at low as −15°C, they are not able to withstand inoculative freezing at higher subzero temperatures; the authors further emphasize in their work the importance of adequate freezing conditions for fish commonly infected by these nematodes.

Species identification in the anisakids has traditionally been complicated by the lack of distinguishing morphological characters; this is particularly evident in larval stages. Molecular genetic analysis of these parasites has provided essential tools for their basic species recognition and their ecology. Even so, we need more information about the full extent of their geographical distribution, host range, and epidemiology of the species of *Pseudoterranova* and *Contracaecum* in both definitive and intermediate hosts, in order to further clarify the dynamics of transmission to humans and the epidemiology of the disease. As to diagnosis of human Pseudoterranoviasis, there is one documented case in which the etiological agent has been identified by genetic molecular markers [72]. Furthermore, molecular markers are required to ascertain if different pathogenicity and/or differential sites of infection are associated with different anisakid larvae. *Anisakis*, *Pseudoterranova*, and *Contracaecum* can have coinfections in the intermediate/paratenic host leading to potential coinfections in the accidental (human) host.

REFERENCES

1. Margolis, L. Public health aspects of "codworm" infection: A review. *J. Fish. Res. Board Can.* 34, 887, 1977.
2. Smith, J.W., and Wootten, R. *Anisakis* and anisakiasis. *Adv. Parasitol.* 16, 93, 1978.
3. Kliks, M.M. Anisakiasis in the Western United States: Four new case reports from California. *Am. J. Trop. Med. Hyg.* 32, 526, 1983.
4. Myers, B.J. Anisakine nematodes in fresh commercial fish from waters along the Washington, Oregon and California coasts. *J. Food Prot.* 42, 380, 1979.
5. Lee, A. et al. A case of human infection with the larva of Terranova type A. *Korean J. Pathol.*, 19, 463, 1985. [In Korean].
6. Lee, I.H. et al. A case of human infection of the larvae from *Pseudoterranova decipiens*. *Korean J. Gastrointest. Endosc.* 18, 732, 1998.

7. Koh, M.S. et al. A case of gastric pseudoterranoviasis in a 43-year old man in Korea. *Korean J. Parasitol.* 37, 47, 1999. [In Korean].
8. Yu, J.R. et al. A human case of gastric infection by *Pseudoterranova decipiens* larva. *Korean J. Parasitol.* 39, 193, 2001.
9. Vidal-Martinez, V.M. et al. Experimental infection of *Contracaecum multipapillatum* (Nematoda, Anisakidae) from México in the domestic cat. *J. Parasitol.* 80, 576, 1994.
10. Mattiucci, S., and Nascetti, G. Advances and trends in the molecular systematics of Anisakid nematodes, with implications for their evolutionary ecology and host-parasite coevolutionary processes. *Adv. Parasitol.* 66, 47, 2008.
11. Baylis, H.A. Some ascarids in the British Museum (Nat. Hist.). *Parasitology* 8, 360, 1916.
12. Johnston, T.H., and Mawson, P.M. Parasitic nematodes. *Rep. Br. N. Z. Aust. Antartic Res. Exp.* 5, 74, 1945.
13. Karokhin, V.I. Two new species of *Porrocacum* from Siberian birds of prey. In: Pod'yanolskaya, V.P. (ed.), *Helminthological Collection*. Nauka, Moscow, pp. 132–141, 1946. [In Russian].
14. Myers, B.J. *Phocanema*, a new genus for the anisakid nematode of seals. *Can. J. Zool.* 37, 459, 1959.
15. Gibson, D.I. The systematics of ascaridoid nematodes-a current assessment. In: Stone, A.R., Platt, H.M., and Khalil, L.F. (eds), *Concepts in Nematode Systematics*. Academic Press, New York, pp. 321–338, 1983.
16. Mozgovoi, A.A., Ascaridata. In: Skrjabin, K.I., Shikhobalova, N.P., and Mozgovoi, A.A., (eds), *Descriptive Catalogue of Parasitic Nematodes*, Vol. II. Press of Academy of Sciences of USSR, Moscow, Russia, pp. 537–542, 1951. [In Russian]
17. Gibson, D.I., and Colin, J.A. The *Terranova* enigma. *Parasitology* 85, 36, 1982.
18. Deardorff, T.L., and Overstreet, R.M. *Terranova ceticola* n. sp. (Nematoda: Anisakidae) from the dwarf sperm whale; *Kogia simus* (Owen), in the Gulf of Mexico. *Syst. Parasitol.* 3, 25, 1981.
19. Paggi, L. et al. Genetic evidence for three species within *Pseudoterranova decipiens* (Nematoda, Ascaridida, Ascaridoidea) in the North Atlantic and Norwegian and Barents Seas. *Int. J. Parasitol.* 21, 195, 1991.
20. Paggi, L. et al. *Pseudoterranova decipiens* species A and B (Nematoda: Ascaridoidea): Nomenclatural designation, morphological diagnostic characters and genetic markers. *Syst. Parasitol.* 45, 185, 2000.
21. Brattey, J., and Stenson, G.B. Host specificity and abundance of parasitic nematodes (Ascaridoidea) from the stomachs of five phocid species from Newfoundland and Labrador. *Can. J. Zool.* 71, 2156, 1993.
22. Mattiucci, S. et al. Allozyme and morphological identification of *Anisakis*, *Contracaecum* and *Pseudoterranova* from Japanese waters (Nematoda: Ascaridoidea). *Syst. Parasitol.* 40, 81, 1998.
23. Paggi, L. et al. A new species of *Anisakis* Dujardin, 1845 (Nematoda: Anisakidae) from beaked whale (Ziphiidae): Allozyme and morphological evidence. *Syst. Parasitol.* 40, 161, 1998.
24. Nadler, S.A. et al. Molecular phylogenetics and diagnosis of *Anisakis*, *Pseudoterranova*, and *Contracaecum* from northern pacific marine mammals. *J. Parasitol.* 91, 1413, 2005.
25. Paggi, L. et al. Molecular genetics in anisakid nematodes from the Pacific Boreal Region. In: Ishikura, H., Aikawa, M., Itakura, H., and Kikuchi, K., (eds), *Host Response to International Parasitic Zoonoses.* Springer-Verlag., Tokyo, 83, 1998.
26. Fagerholm, H.P. Intra-specific variability of the morphology in a single population of the zeal parasite *Contracaecum osculatum* (Rudolphi) (Nematoda, Ascaridoidea), with a redescription of the species. *Zool. Scripta.* 18, 33, 1989.
27. Margolis, L. Report on parasite studies of sockeye and pink salmon collected in 1955, with special reference to the utilization of parasites as a means of distinguishing between Asiatic and American stocks of salmon on the high seas—A progress report on work being carried out as part of F.R.B.'s commitments to INPFC. *Fish. Res. Board Can. Rep.* 24, 36, 1956.
28. Bullini, L. et al. Genetic and ecological research on Anisakid endoparasites of fish and marine mammals in the Antarctic and Arctic-Boreal regions. In Battaglia, B., Valencia, J., and Walton, D.W.H. (eds), *Antarctic Communities: Species, Structure and Survival.*, University Press, Cambridge, pp. 39–44, 1997.
29. George-Nascimento, M., and Urrutia, X. *Pseudoterranova cattani* sp. n. (Nematoda, Ascaridoidea, Anisakidae), a gastric parasite of South American sea lion *Otaria byronia* De Blainville in Chile. *Rev. Chil. Hist. Nat.* 73, 93, 2000.
30. Zhu, X.Q. et al. SSCP-based identification of members within the *Pseudoterranova decipiens* complex (Nematoda: Ascaridoidea: Anisakidae) using genetic markers in the internal transcribed spacers of ribosomal DNA. *Parasitology* 124, 615, 2002.
31. Hartwich, G. Revision der vogelparasitischen Nematoden Mitteleuropas II.—Die Gattung *Contracaecum* Railliet and Henry, 1912 (Ascaridoidea). *Mitt. Zool. Mus. Berl.* 40, 15, 1964.
32. Hartwich, G. Rhabditida und Ascaridida. *Tierwelt Dtl.* 62, 1, 1975.
33. Berland, B. *Phocascaris cystophorae* sp. nov. (Nematoda) from the hooded seal, with an emendation of the genus. *Årbok Univ. Bergen Mat.-naturv., Serie 1963* 17, 1, 1964.
34. Orecchia, P. et al. Divergenza genetica di specie dei generi *Contracaecum* e *Phocascaris* (Ascaridida, Anisakidae) con diverso ciclo biologico. *Ann. Ist. Super. Sanita* 22, 345, 1986.
35. Nascetti, G. et al. Genetic relationships among anisakid species belonging to the genera *Contracaecum* and *Phocascaris. Bull. Soc. Fran. Parasitol.* 8, 261, 1990.
36. Nadler, S.A. et al. Phylogenetic relationships among species of *Contracaecum* Railliet & Henry, 1912 and *Phocascaris* Host, 1932 (Nematoda: Ascaridoidea) based on nuclear rDNA sequence data. *Parasitology* 121, 455, 2000.
37. Mattiucci, S. et al. Genetic relationships among species of *Contracaecum* Railliet & Henry, 1912 and *Phocascaris* Höst, 1932 (Nematoda: Anisakidae) from pinnipeds inferred from mitochondrial *cox2* sequences, and congruence with allozyme data. *Parasite* 15, 408, 2008.
38. Nascetti, G. et al. Three sibling species within *Contracaecum osculatum* (Nematoda, Ascaridida, Ascaridoidea) from the Atlantic Arctic-boreal region: Reproductive isolation and host preferences. *Int. J. Parasitol.* 23, 105, 1993.
39. Orecchia, P. et al. Two new members in the *Contracaecum osculatum* complex (Nematoda, Ascaridoidea) from the Antarctic. *Int. J. Parasitol.* 24, 367, 1994.
40. Mattiucci, S., and Nascetti, G. Genetic diversity and infection levels of anisakid nematodes parasitic in fish and marine mammals from Boreal and Austral hemispheres. *Vet. Parasitol.* 148, 43, 2007.
41. Ishikura, H. et al. Anisakidae and anisakidosis. In Sun, T., (ed), *Progress in Clinical Parasitology* Springer-Verlag, New York, pp. 43–102, 1993.

42. Nagano, K. Gastric terranovasis. In: Ishikura, H.M., and Namiki, M., (eds), *Gastric anisakiasis in Japan*, Springer-Verlag, Tokyo, pp. 133–140, 1989.
43. Mercado, R. et al. Human case of gastric infection by a fourth larval stage of *Pseudoterranova decipiens* (Nematoda, Anisakidae). *Rev. Saude Publica* 31, 178, 1997.
44. Mercado, R. et al. Human infection by *Pseudoterranova decipiens* (Nematoda, Anisakidae) in Chile: Report of seven cases. *Mem. Inst. Oswaldo Cruz* 96, 653, 2001.
45. Pinel, C. et al. Gastric anisakidosis due to *Pseudoterranova decipiens* larva. *Lancet* 347, 1829, 1996.
46. Fioravanti, M.L. et al. *Pseudoterranova decipiens* (Nematoda: Anisakidae), human infection in Italy, *Parassitologia* 46, S112, 2004.
47. Oshima, T. Anisakiasis—Is the sushi bar guilty? *Parasitology* 3, 44, 1987.
48. Adams, A.M. et al. Parasites of risks to public health. *Rev. Sci. Tech. Off. Int. Epiz.* 16, 652, 1997.
49. Deardorff, T.L. et al. Histopathology induced by larval *Terranova* (Type HA) (Nematoda: Anisakinae) in experimentally infected rats. *J. Parasitol.* 69, 191, 1983.
50. Smith, J.W. Ascaridoid nematodes and pathology of the alimentary tract and its associated organs in vertebrates, including man: A literature review. *Helminthol. Abstr.* 68, 49, 1999.
51. Little, M.D., and Most, H. Anisakid larva from the throat of a woman in New York. *Am. J. Trop. Med. Hyg.* 22, 609, 1973.
52. Amin, O.M. et al. An unusual case of anisakiasis in California, U.S.A. *Comp. Parasitol.* 67, 71, 2000.
53. Asaishi, K. et al. Studies on the pathogenic mechanisms of anisakiasis: II Histological changes of stomach wall due to the injection of worm extracts. *Jpn. J. Parasitol.* 23, 34, 1974. (in Japanese)
54. Berland, B. Nematodes from some Norwegian marine fishes. *Sarsia* 2, 1, 1961.
55. Mattiucci, S. et al. *Anisakis nascettii* n. sp. (Nematoda: Anisakidae) from beaked whales of the southern hemisphere: Morphological description, genetic relationships between congeners and ecological data. *Syst. Parasitol.* 74, 199, 2009.
56. Mattiucci, S. et al. *Contracaecum gibsoni* n. sp. and *C. overstreeti* n. sp. (Nematoda: Anisakidae) from Dalmatian pelican *Pelecanus crispus* (L.) in Greek waters: Genetic and morphological evidence. *Syst. Parasitol.* 75, 207, 2010.
57. Zhu, X.Q. et al. Assessing sequence variation in the internal transcribed spacers of ribosomal DNA within and among members of the *Contracaecum osculatum* complex (Nematoda: Ascaridoidea: Anisakidae). *Parasitol. Res.* 86, 677, 2000.
58. Hu, M. et al. Mutation scanning for sequence variation in three mitochondrial DNA regions for members of the *Contracaecum osculatum* (Nematoda:Ascaridoidea) complex. *Electrophoresis* 22, 1069, 2001.
59. McClelland, G. The trouble with sealworms (*Pseudoterranova decipiens* species complex, Nematoda): A Review. *Parasitology* 124, 183, 2002.
60. Levsen, A. et al. Low detection efficiency of candling as a commonly recommended inspection method for nematode larvae in the flesh of pelagic fish. *J. Food Prot.* 68, 828, 2005.
61. Murray, M.G., and Thompson, W.F. Rapid isolation of high-molecular-weight plant DNA. *Nucleic Acids Res.* 8, 4321, 1980.
62. Valentini, A. et al. Genetic relationships among *Anisakis* species (Nematoda: Anisakidae) inferred from mitochondrial *cox-2* sequences, and comparison with allozyme data. *J. Parasitol.* 92, 156, 2006.
63. Hall, T.A. BioEdit: A user-friendly biological sequence alignment and analysis program for Windows 95/98/NT. *Nucleic Acids Symp. Ser.* 41, 95, 1999.
64. Swofford, D.L. *PAUP*. Phylogenetic Analysis using Parsimony (*and Other Methods).* Sinauer Associates, Sunderland, Massachusetts, 2003.
65. Paltridge, G.P. et al. Intestinal anisakiasis : A New Zealand disease. *N. Z. Med. J.* 97, 558, 1984.
66. Palm, H.W. et al. Molecular genotyping of *Anisakis* Dujardin, 1845 (Nematoda: Ascaridoidea: Anisakidae) larvae from marine fish of Balinese and Javanese waters, Indonesia. *Helminthologia* 45, 3, 2008.
67. Mattiucci, S. et al. Human Anisakidosis in Italy: Molecular and histological identification of two new cases. *Parassitologia* 49, S226, 2007.
68. Cattan, P.E., and Carvajal, J. A study of the migration of larval *Anisakis simplex* (Nematoda: Ascaridida) in the Chilean hake, *Merluccius gayi* (Guichenot). *J. Fish Biol.* 24, 649, 1984.
69. Rodriguez-Mahillo, A.I. et al. Cloning and characterization of the *Anisakis simplex* allergen Ani s4 as a cysteine-protease inhibitor. *Int. J. Parasitol.* 37, 907, 2007.
70. Moneo, I. et al. Isolation of a heat-resistant allergen from fish parasite *Anisakis simplex. Parasitol. Res.* 96, 285, 2005.
71. Stormo, S.K. et al. Cold tolerance in sealworm (*Pseudoterranova decipiens*) due to heat-shock adaptations. *Parasitology* 136, 1317, 2009.
72. Arizono, N. et al., Human infection with *Pseudoterranova azarasi roundworm. Emerg. Infect. Dis.*, 17, 3, March 2011.

63 Strongyloides

Dongyou Liu

CONTENTS

63.1 INTRODUCTION

The genus *Strongyloides* (commonly known as threadworm in the United States, or pinworm in the United Kingdom, where threadworm refers to *Enterobius*) consists of >50 species of obligate gastrointestinal roundworms of vertebrates (e.g., mammals, birds, reptiles, and amphibians). Of the two human-infecting *Strongyloides* species, *S. stercoralis* parasitizes humans in tropical and subtropical regions, and *S. fuelleborni* mainly occurs in primates as well as humans in Africa. In addition, an *S. fuelleborni* subspecies (i.e., *S. fuelleborni kellyi*) has been shown to infect humans in Papua New Guinea.

While *Strongyloides* infection (strongyloidiasis) exerts a negative impact mainly on the nutritional status of immunocompetent individuals, the disease (in the form of fulminant hyperinfection syndrome or disseminated strongyloidiasis) can have serious and sometimes fatal consequences in immunocompromised patients. *S. fuelleborni kellyi* is responsible for causing baby (or belly) syndrome in Papua New Guinea, which is also fatal in the absence of anti-*Strongyloides* therapy.

Strongyloidiasis due to *S. stercoralis* is endemic in Southeast Asia, Latin America, sub-Saharan Africa, and parts of the southeastern United States, affecting >100 million people worldwide. Given its ability to replicate in the human host and to undergo cycles of autoinfection, the detrimental effects of *S. stercoralis* on human health and well-being can be both acute and long-lasting.

63.1.1 Classification, Morphology, and Biology

63.1.1.1 Classification

The genus *Strongyloides* is classified in the family Strongyloides, superfamily Panagrolaimoidea, order Rhabditida, class Chromadorea, phylum Nematoda. The family is divided into two genera *Parastrongyloides* and *Strongyloides* [1]. In turn, the genus *Strongyloides* is separated into >50 species, the most notable of which include *Strongyloides akbari*, *Strongyloides callosciureus*, *Strongyloides canis*, *Strongyloides cebus*, *Strongyloides fuelleborni* (consisting of two subspecies *S. fuelleborni fuelleborni* and *S. fuelleborni kellyi*), *Strongyloides mirzai*, *Strongyloides myopotami*, *Strongyloides ophidiae*, *Strongyloides papillosus*, *Strongyloides planiceps*, *Strongyloides procyonis*, *Strongyloides ransomi*, *Strongyloides ratti*, *Strongyloides robustus*, *Strongyloides stercoralis*, *Strongyloides suis*, *Strongyloides venezuelensis*, *Strongyloides vituli*, and *Strongyloides westeri* [2,3].

Within the genus *Strongyloides*, two species (i.e., *Strongyloides stercoralis* and *Strongyloides fuelleborni*) are known to infect humans. First reported in 1876 in the stools of French soldiers returning home from expeditions in IndoChina who suffered from severe diarrhea (known as Cochin-China diarrhea at the time), *Strongyloides stercoralis* is the type of the genus whose parasitic adult stage inhabits the tunnels (crypts) in the mucosa of the small intestine. In addition, the parasite shows a free-living (dioecious) adult stage, which was once described as distinct species of *Anguillula* from the parasitic adult stage. When the parasitic and free-living stages of this parasite were later discovered to be separate stages of one life cycle, they were unified under the genus *Strongyloides*.

Strongyloides fuelleborni fuelleborni (or *Strongyloides fuelleborni* for short) is a zoonosis that primarily infects various primates and humans in Africa [4], and *Strongyloides fuelleborni kellyi* infects humans in Papua New Guinea [5]. *S. fuelleborni kelleyi* is distinguished from *S. fuelleborni fuelleborni* by the appearance of the perivulval cuticle of the parasitic females, by the position of the phasmidial pore of free-living males, and by the geographical isolation of *S. fuelleborni kelleyi* [5,6].

Being the sister genus to *Strongyloides*, *Parastrongyloides* is present mainly in moles and marsupials. The life cycle of *Parastrongyloides* is similar to that of *Strongyloides* except that both the male and female adults of *Parastrongyloides* are parasitic (while only the female adult of *Strongyloides* is parasitic), and the free-living forms of *Parastrongyloides* are able to undergo multiple, sequential generations outside of the host indefinitely (while *Strongyloides* generally undergoes one generation outside of the host).

63.1.1.2 Morphology

Strongyloides spp. undergo parasitic and free-living phases in their life cycle which demonstrate notable morphological differences [7]. The parasitic females of *Strongyloides* measure 2–3 mm in length by 30–50 μm, with blunt-ended tails, an elongated, straight-sided (filariform) esophagus (which covers 1/3 of the body length), a didelphic ovary, and a vulva (which is located at approximately 2/3 of the body length). The free-living adult stages of *Strongyloides* measure about 1 mm in length, with the female (1 mm in length and 50–75 μm in width) being slightly larger than the male (0.9 mm in length). Both sexes possess a tiny buccal capsule and rhabditiform (cylindrical) esophagus without a posterior bulb; the free-living female has a didelphic ovary and a vulva at the mid-point of the body; the free-living male has spicules and gubernaculum.

The key diagnostic features of the parasitic females include the disposition of the ovaries (parallel to or spiraled around the intestine); the en face view of the stoma; the stage passed in host feces (i.e., eggs, larvae, or eggs and larvae); the morphology of the perivulval lateral cuticle; and the morphometrics (e.g., body size). For the free-living adult males and females, morphometrics as well as the arrangement of the pericloacal papillae of the male are informative.

Strongyloides eggs (50–60 μm by 30–35 μm) are ellipsoid, with a thin wall containing a morula- to tadpole-stage embryo or larva (characteristic of *Strongyloides*). Rhabditoid larvae (L1) (180–380 μm by 14–20 μm) emerge from eggs in the lumen, and are excreted with feces in the external environment, where they transform into either filariform larvae (L3, the infective stage, 300–600 μm by 10–20 μm) via an L2 stage (the direct development cycle) or into free-living adult males and females in the soil (the indirect development cycle). Filariform (infective) larvae (L3) are radially constricted, with a filariform esophagus about 2/5 of the body length, and a tripartite pointed tail (forked tail), which is diagnostic for *Strongyloides*.

63.1.1.3 Biology

Strongyloides is a soil-dwelling nematode with a complex life cycle, which may range from direct development cycle (a homogenic, parasitic life cycle), an indirect development cycle (a heterogonic, free-living life cycle), to an autoinfection (or hyperinfection) cycle, depending on the prevalent conditions.

During the direct development (parasitic) cycle, *S. stercoralis* parthenogenetic females (with the capability of asexual reproduction where development of embryos occurs without fertilization by a male) embed in the intestinal mucosa (often close to the crypts of Lieberkühn in the proximal part of the small intestine) and lay embryonated eggs that hatch internally. The resultant first-stage larvae (L1; rhabditiform free-living larvae) pass out in the feces and molt twice to become second- (L2) and third (L3; filariform)-stage larvae. Filariform infective larvae (L3) are attracted by urocanic acid on the skin, penetrate the skin when there is contact with the soils, enter dermal vessels, and migrate through the bloodstream or lymphatic channels to the lungs and heart. In the lungs, the filariform larvae break out of the capillaries into alveolar spaces to become air borne, ascend the respiratory tract into the pharynx, are swallowed, and reach the small intestine (the duodenum and proximal jejunum) where they mature to adult parthenogenetic females. The females then shed eggs which hatch to become rhabditiform larvae. The larvae are either passed out in the stools to continue the soil-based cycle or remain inside the host to cause autoinfection. It takes about 2 weeks from the initial skin penetration of filariform L3 larvae to egg development.

During the indirect development (free-living) cycle, rhabditiform L1 emerging from eggs are passed in stool, and undergo 1–9 reproductive cycles of the free-living forms (e.g., one for *S. stercoralis*; nine for *S. planiceps*, with decreasing fecundity) to become free-living adult males and females in the soil [8]. The free-living adults reproduce sexually and die after one generation. The eggs produced by free-living adult females develop over several weeks into rhabditiform larvae (L1) and into infectious third-stage larvae (L3). L3 penetrate the skin of human host, pass through the circulation to the lungs, enter the airways, be swallowed, and finally reach the intestine, where they mature into adult egg-laying females. Alternatively, rhabditiform larvae may develop into free-living adult forms and continue the free-living cycle. The heterogonic life cycle allows reproduction for one or more generations in the absence of a host.

During the autoinfection cycle, rhabditiform L1 emerge from eggs, molt twice in the intestinal lumen of the host (instead of being shed in the stool) to become filariform larvae that penetrate the intestinal mucosa or perianal skin (internal and external autoinfection, respectively), and then migrate to the definitive site in the small intestine or to parenteral sites (e.g., lungs), causing autoinfection (hyperinfection). Whereas hyperinfection syndrome due to *S. stercoralis* is limited to respiratory and gastrointestinal tracts, disseminated infection tends to affect other organs (e.g., brain). The unique ability of *S. stercoralis* to establish a cycle of repeated endogenous reinfection within the host leads to chronic infection that can last for 75 years, and the possible multiplication within the host (hyperinfection), with a tremendous increase in worm burdens [9,10]. Such hyperchronic infections often go undiagnosed due to the absence of larvae in the feces, the presence of dormant larvae in the tissues, or senescent or nonreproductive female worms in the intestine, with immunosuppression triggering either reactivation of dormant larvae or resumption of egg laying by barren parasitic females. Hypobiosis or dormancy of *Strongyloides* L3 is key to transmammary transmission, as infective L3 presumably arrest their development and migration in the mammary glands, and then reactivate at lactation. Only one other species in the *Strongyloides* genus, *S. felis*, has the trait of autoinfection.

As a parasite of humans and other primates, *S. stercoralis* affects an estimated 100 million people in >70 countries with temperate, tropical, and subtropical climates [11,12]. The infection is endemic in Southern, Eastern, and Central Europe, Islands of the Caribbean, Latin America, sub-Saharan Africa, and Southeast Asia, and is also reported in Japan, Papua New Guinea, and the southeastern United States [13,14].

In endemic areas, up to 10% of humans may be infected with *S. stercoralis*, especially in younger individuals (whose behavior favors the transmission of soil-based helminths). The risk factors for hyperinfection and disseminated disease are (i) immunosuppressive therapy (particularly steroids), (i) human T-lymphotropic virus-1 (HTLV-1) or human immunodeficiency virus (HIV) infection, (iii) solid organ transplantation, (iv) hematologic malignant disease, especially lymphoma, (v) hypogammaglobulinemia, (vi) chronic alcohol consumption, (vii) uremia (or renal failure), (viii) severe malnutrition, (ix) diabetes mellitus, and (x) travel to areas of endemicity [15–17]. In particular, patients with chronic obstructive pulmonary disease receiving corticosteroids often develop pulmonary strongyloidiasis that mimics an exacerbation of the underlying disease.

Strongyloides fuelleborni is distributed widely among Old World primates, including humans. While *S. fuelleborni* mainly occurs in monkeys in Asian regions, a human parasitic strain is maintained among humans in Africa.

63.1.2 Clinical Features and Treatment

63.1.2.1 Clinical Features

Strongyloidiasis was first noted by Fulleborn in 1926, and infection due to *Strongyloides stercoralis* was first documented in 1876. The disease is usually associated with anemia, hypocholesterolemia, and hypoalbuminemia, and sometimes eosinophilia, with clinical presentations ranging from asymptomatic, acute, chronic, hyperinfection, to disseminated strongyloidiasis [18–24].

63.1.2.1.1 Acute Strongyloidiasis

The entry of the infective filariform larvae into the human host often induces serpiginous urticarial rash, petechiae, purpure, pruritic, erythematous, morbiliform eruption, or intensely itching prurigo (at the site of entry, especially in the legs) [25]. Subsequent extraintestinal migration may lead to pulmonary symptoms a few days later: cough, dyspnea, wheezing (Löffler's syndrome), and tracheal irritation (mimicking bronchitis from migration of the larvae through the lungs) with peripheral eosinophilia (which is seen in patients with intact immune systems but absent in immunosuppressed patients) [26–28]. Gastrointestinal (GI) symptoms develop several weeks later (just before the appearance of the larvae in the stool): abdominal cramping with bloating, watery diarrhea, and constipation (due to lodging of the larvae and maturation into adult females in the small intestine, and peripheral eosinophilia. Eosinophilia is present in up to 35% during the acute phase, and is less frequent in patients with chronic or disseminated disease [29–32]. Without correct diagnosis and effective treatment, the host harboring the parasite may develop chronic strongyloidiasis.

63.1.2.1.2 Chronic Strongyloidiasis

Although >50% of patients with the chronic form of strongyloidiasis are asymptomatic (apart from eosinophilia), especially those with light infections, some may complain about mild gastrointestinal symptoms (light intermittent abdominal pain and diarrhea), or pruritus of anal skin or larva currens (which is a rapidly advancing maculopapular or urticarial serpiginous rash appearing as a red line that moves >5 cm/day and then quickly disappears, due to the subcutaneous migration of larvae from the anal area down the upper thighs or the pelvic area) [33]. Larva currens is pathognomonic for *Strongyloides* infection. Patients with heavy infections tend to show more severe intestinal symptoms (nausea, vomiting, diarrhea), malabsorption, epigastric pain with tenderness, constipation, weight loss, gastrointestinal bleeding, jejunal perforation, paralytic ileus, granulomatous hepatitis, and eosinophilic ascites with marked eosinophilia [34–40]. Patients with duodenal obstruction often present low eosinophil count, indicating a chronic infection [41]. Intestinal obstruction may result from severe mucosal edema and inflammation with significant narrowing of duodenal lumen, as well as from an extrinsic compression of the duodenum by the superior mesenteric neurovascular bundle [42–47]. However, duodenal obstruction may also be caused by tuberculosis, primary intestinal lymphoma, Crohn's disease, eosinophilic gastroenteritis, and gastrointestinal stromal tumor.

63.1.2.1.3 Hyperinfection Syndrome

The hallmark of hyperinfection syndrome is an increase in the number of larvae in the stool and/or sputum along with manifestations confined to respiratory and gastrointestinal systems along with peritoneum [48]. Hyperinfection involving the intestine may induce cramping abdominal pain, watery diarrhea, weight loss, nausea, vomiting, gastrointestinal bleeding, paralytic ileus, and ulcerative enteritis, while hyperinfection in the extraintestinal sites include cough, wheezing, pneumonia, and pulmonary hemorrhage with diffuse bilateral infiltrates on the chest x ray, as well as eosinophilic pleural effusions and eosinophilic granulomatous enterocolitis [49–55].

Hyperinfection syndrome occurs in 1.5–2.5% of the patients with strongyloidiasis. These patients often have suppressed immune functions due to use of systemic corticosteroids (for chronic obstructive pulmonary disease or asthma, systemic lupus erythematosus, rheumatoid arthritis, autoimmune hemolytic anemia, chronic active hepatitis, etc.) or due to intercurrent illness (such as HTLV-1 and HIV infection, organ transplantation, and other infectious diseases like kala-azar) [56–95].

63.1.2.1.4 Disseminated Strongyloidiasis

Fatal disseminated infection involving multiple organ systems other than the respiratory and gastrointestinal systems often occurs in patients with immunosuppression, chronic infection, and malnutrition [96–103]. Glomerulonephritis and alveolar hemorrhage are rare complications of disseminated disease [104]. Involvement of the central nervous system may lead to headache, altered mental state, seizures, and coma. When larvae travel through the venous system to other locations in the body, they facilitate translocation of intestinal bacteria (e.g., *Streptococcus bovis*, *Escherichia coli*, *Streptococcus fecalis*, *Klebsiella pneumoniae*, or *Enterobacter* sp.), leading to concomitant Gram-negative bacteremia (meningitis, cholecystitis, liver abscess, and pancreatitis) [105–110]. The mortality rate of disseminated infection can be as high as 50–90%.

63.1.2.2 Treatment

Albendazole, mebendazole, piperazine, thiabendazole, and ivermectin are effective against adult worms (but not autoinfective migrating larvae) of *Strongyloides stercoralis* [111–120]. The recommended treatment for strongyloidiasis is either ivermectin (200 μg/kg body weight in a single dose) or albendazole (400 mg daily for 3 days) [121–124]. Ivermectin has a higher cure rate than thiabendazole, which also tends to cause more side effects (dizziness, vomiting, nausea, malaise) [125–131]. Mebendazole and piperazine are commonly used in the United Kingdom. To ensure infection eradication, follow-up stool examinations should be performed over a period of 3 months after treatment.

Because *Strongyloides stercoralis* usually enters human host when an individual walks barefoot in contaminated soil, proper protection of skin to avoid direct contact with contaminated soil represents a most effective way to prevent the infection. At the community level, proper disposal of human excreta, community education about protective and hygienic measures, and prompt treatment of diagnosed cases are helpful in the prevention of the disease [132].

63.1.3 Laboratory Diagnosis

Accurate diagnosis of strongyloidiasis represents a vital step to control and prevent this infection. A careful history of residence in or travel to areas of endemicity often provides a clue of potential infection. A characteristic radiographic feature of strongyloidiasis is the reflux of oral contrast into the biliary tree, possibly due to an incompetent sphincter of Oddi caused by severe inflammation of the duodenal wall. Strongyloidiasis (particularly the hyperinfection syndrome and more systemic disseminated infection) needs to be differentiated from pneumonia, polyarteritis nodosa, malignant mediastinal neoplasia, eosinophilic folliculitis, relapse of lymphoma, primary intestinal lymphoma, systemic lupus erythematosus, peptic ulcer disease, and ulcerative colitis or Crohn's disease. Laboratory techniques such as direct fecal smears, agar plate culture, serology, and molecular testing as well as duodenal fumigation are helpful for the confirmation of strongyloidiasis [133–136].

63.1.3.1 Parasitological Techniques

Identification of *S. stercoralis* eggs and first-stage rhabditoid (L1) larvae in feces and other samples represents a useful way to confirm strongyloidiasis [137,138]. *S. stercoralis* eggs are ellipsoid, 40–85 μm in length, with a thin wall containing a larva; and L1 larvae are 180–380 by 14–20 μm in size, with a short buccal cavity, an attenuated tail and a prominent genital primordium. These morphologic features are useful for differentiation from hookworm rhabditoid larvae. As *S. stercoralis* is the only helminth that secretes larvae in the stools, *S. stercoralis* and *S. fuelleborni* infections can be differentiated by the presence of larvae or eggs, respectively.

Common parasitological techniques for the detection of *S. stercoralis* larvae in stool samples include direct smear of feces in saline–Lugol iodine stain, Baermann concentration, formalin-ethyl acetate concentration, Harada–Mori filter paper culture, and nutrient agar plate cultures [139,140]. In the Baermann procedure, fecal sample is placed on mesh screen and a coarse fabric in a funnel that is filled with warm water and connected to a clamped tubing. After an hour of incubation, larvae crawl out of the fecal suspension, migrate into the warm water, and are collected by centrifugation. In the Harada–Mori technique, filter paper containing fresh fecal material is placed in a test tube with water that continuously soaks the filter paper by capillary action. Incubation at 30°C facilitates the development of larvae, which migrate to either side of the filter paper. In the agar culture method, the fecal sample is placed on a nutrient agar plate and incubated for at least 2 days. As the larvae crawl over the agar, they carry bacteria with them, creating visible tracks.

In general, direct fecal smear has a sensitivity of 30%, which increases to 50% with three stool examinations and 100% with seven serial stool samples. Baermann concentration technique is the method of choice for the detection of the parasite, although it may miss light (chronic) infections. Formalin-ether concentration method and Harada–Mori filter paper culture method demonstrate a sensitivity of up to 55%, and agar plate culture technique has a sensitivity of 78–100% [141,142]. On the other hand, the Kato-Katz technique detects helminth eggs in feces, but not larvae. Microscopic examination of duodenal fluid appears to be more sensitive than wet-mount analysis of stools samples for the detection of larvae.

In hyperinfection and disseminated infection, filariform larvae may be seen in wet-mount preparations from sputum, bronchoalveolar lavage fluid, bronchial washings, cerebrospinal fluid, liver and lung biopsy specimens, urine and fecal samples or examination of pleural fluid with the help of Gram, Papanicolaou, or acid-fast (auramine O and Kinyoun) stains [143]. Biopsy with duodenal intubation and Enterotest™ is a frequent method for demonstration of rhabditiform larvae in transplant recipients. Larvae may also be demonstrated in skin biopsy specimens collected from the site of a larva currens rash.

Strongyloides is relatively easy to maintain in the laboratory. To grow the free-living generation, feces (which may be mixed with wetted activated charcoal) are maintained damp and at a slightly elevated temperature. Larvae recovered from fecal cultures can also be maintained on modified NGM agar plates.

63.1.3.2 Biochemical and Serological Techniques

Eosinophilia is an important laboratory finding in patients with strongyloidiasis. However, the eosinophil count alone is not sufficiently sensitive to screen for strongyloidiasis, as patients with chronic infection, hyperinfection syndrome, and disseminated infection tend to have low or normal eosinophil counts [144].

Several immunological tests [e.g., enzyme-linked immuno-sorbent assay (ELISA), gelatin particle indirect agglutination (GPIA), indirect fluorescent antibody test (IFAT), and Western blot] are useful for the diagnosis of strongyloidiasis in fecal, sputum, and serum specimens [135,145,146]. However, serological techniques lack sufficient sensitivity to diagnose recently acquired infections, and sometimes display cross-reactivity with other helminths such as filaria, schistosoma, and *Ascaris lumbricoides* [147].

63.1.3.3 Molecular Techniques

Nucleic acid-based procedures offer a promising alternative approach for diagnosis, epidemiological investigation, and drug resistance of *Strongyloides*. The common gene targets consist of the hypervariable region IV (HVR-IV) of 18S ribosomal DNA, mitochondrial cytochrome *c*-oxidase subunit 1 gene (*cox1*), and beta-tubulin gene [148–152]. Use of real-time PCR permits sensitive detection of *Strongyloides* larva [153,154].

63.2 METHODS

63.2.1 Sample Preparation

63.2.1.1 Sample Collection

Feces from the patient are examined by direct wet smear and Kato-Katz preparation within 4 h of collection. Alternatively, routine formalin-ether concentration and filter paper culture are used for initial assessment of fecal samples. *S. stercoralis* larvae are collected from fecal material by a Baermann funnel or by nematode egg concentration techniques on formalin-preserved stool samples. Fecal samples are also grown in fecal cultures to obtain infective L3 larvae. Larvae are fixed in 100% ethanol and stored at −20°C until use [151].

63.2.1.2 DNA Isolation

S. stercoralis DNA can be extracted from the ethanol-fixed worms using a NaOH method. Worms are incubated in 20 μL of 0.02 N NaOH solution at 95°C for 10 min, and then 2 μL of this solution is used as a template for the 50 μL PCR solution [151].

For preparation of DNA from fecal specimens, about 100 mg unpreserved feces are suspended in 200 μL PBS containing 2% polyvinylpolypyrolidone (PVPP; Sigma). After heating at 100°C for 10 min, suspensions are treated with sodium dodecyl sulfate-proteinase K at 55°C for 2 h. DNA is then isolated using QIAamp Tissue Kit spin columns (QIAgen). Within the isolation lysis buffer, 103 PFU/mL phocine herpes virus 1 (PhHV-1) is added to serve as an internal control. The amplification is considered to be hampered by fecal inhibitory factors if the expected cycle threshold (Ct) value of 33 in the PhHV-specific PCR is increased by more than 3.3 cycles [155].

63.2.2 Detection Procedures

63.2.2.1 Real-Time PCR Identification

63.2.2.1.1 *Protocol of Verweij et al.*

Verweij et al. [155] described a real-time PCR method targeting the 18S rRNA gene for the detection of *Strongyloides stercoralis* DNA in fecal samples. The forward primer (5′-GAATTCCAAGTAAACGTAAGTCATTAGC-3′) and reverse primer (5′-TGCCTCTGGATATTGCTCAGTTC-3′) amplify a 101-bp fragment, which is detected with the probe FAM-5′-ACACACCGGCCGTCGCTGC-3′-BHQ1.

63.2.2.1.2 *Procedure*

1. PCR mixture (25 μL) consists of PCR buffer (Hotstar Taq master mix, QIAgen), 5 mM $MgCl_2$, 2.5 μg bovine serum albumin (Roche Diagnostics), 1.5 pmol of each *S. stercoralis*-specific primer, 2.5 pmol of *S. stercoralis*-specific double-labeled probe, and 5 μL of DNA extracted from stool specimens.
2. Amplification is carried out with 1 cycle of 15 min at 95°C followed by 50 cycles of 15 s at 95°C and 1 min at 60°C. Amplification, detection, and data analysis are performed with the Applied Biosystems 7500 Real-Time PCR system and sequence detection software version 1.2.2.

Note: With a twofold increase in the detection rate as compared with the Baermann sedimentation method, this assay provides a rapid, sensitive, and specific means for diagnosis and monitoring of *S. stercoralis* infections.

63.2.2.1.3 Protocol of Basuni et al.

Basuni et al. [153] utilized primers from 18S rRNA gene for specific identification of *S. stercoralis*. The forward primer Stro18S-1530F (5′-GAA TTC CAA GTA AAC GTA AGT CAT TAG C-3′) and reverse primer Stro18S-1630R (5′-TGC CTC TGG ATA TTG CTC AGT TC-3′) amplify a 101-bp fragment, which is detected with the probe Stro18S-1586T (Alexa 680-5′-ACA CAC CGG CCG TCG CTG C-3′-black hole quencher 3).

63.2.2.1.4 Procedure

1. PCR mixture (25 μL) is composed of PCR buffer (HotStar Taq master mix, Qiagen), 5 mM $MgCl_2$, 0.1 mg/mL bovine serum albumin (Sigma Aldrich), 300 nM of each species-specific primer, 250 nM of probe, and 5 μL of template DNA.
2. Amplification is performed for 15 min at 95°C followed by 50 cycles of 9 s at 95°C, and 60 s at 60°C. Amplification, detection, and data analysis are performed with the Rotor Gene 6000 (Rotorgene-Q) real-time analyzer. Fluorescence emitted at 710 nm is measured during the annealing step of each cycle. The cycle threshold (Ct) value is defined as the number of PCR cycles required for the detection of fluorescence signal of the amplified products to exceed the set threshold value. As a consequence, higher quantities of DNA will result in lower Ct values and vice versa.

63.2.2.2 Analysis of *Cox1* Gene Sequences

Hasegawa et al. [151] utilized primers from *cox1* genes for identification and phylogenetic examination of *Strongyloides* spp. The primers StrCoxAfrF (5'-GTGGTTTTGGTAATTGAATGGTT-3') and StrCoxAfrR (5'- TTACTATAGGTGGTATAACTGGT-3') amplify a 722-bp fragment from *S. stercoralis cox1* gene. Subsequent analysis of the nucleotide sequence allows precise speciation of *Strongyloides* under investigation.

63.2.2.2.1 Procedure

1. PCR cycling conditions consist of an initial denaturation at 94°C for 2 min, followed by 10 cycles at 94°C for 1 min, 40°C for 1 min, 68°C for 2 min; 30 cycles of 94°C for 1 min, 45°C for 2 min, 68°C for 2 min; and a postamplification extension at 68°C for 7 min.
2. The PCR products are mixed with EZ-Vision™ three DNA dye (Amresco), electrophoresed in a 1.5% agarose gel and visualized by a UV illuminator.
3. The PCR products are then recovered from the gel using SUPREC™-01 (Takara Bio), precipitated with ethanol, dissolved in distilled water, and amplified using one of the primer pairs used in the initial amplification with the BigDye™ Terminator Cycle Sequencing Kit Version 3.1 (Applied Biosystems). The products are purified using Centri-Sep™ spin columns (Princeton Separations) and sequenced by an ABI-PRISM 3130 Genetic Analyzer (Applied Biosystems).
4. For phylogenetic analysis, the DNA sequences are aligned using the CLUSTAL W computer program. Maximum likelihood (ML) analysis is applied with the program PhyML provided on the "phylogeny.fr" website (http://www.phylogeny.fr/). The probability of inferred branch is assessed by the approximate likelihood-ratio test (aLRT) with SH-like algorithm, an alternative to the nonparametric bootstrap estimation of branch support. *N. americanus* (GenBank AJ417719) is used as an outgroup for the construction of phylogenetic trees.

63.3 CONCLUSION

The genus *Strongyloides* encompasses >50 species of obligate gastrointestinal roundworms of vertebrates, of which at least two are infective to humans. While *S. stercoralis* affects humans in tropical and subtropical regions, *S. fuelleborni* mainly infects primates as well as humans in Africa. In addition, *S. fuelleborni kellyi* has been detected from humans in Papua New Guinea.

Although strongyloidiasis often produces mild, nonspecific symptoms in immunocompetent individuals, it has the potential to fulminant hyperinfection syndrome or disseminated infection in immunocompromised patients. Considering their ability to replicate in the human host and to undergo cycles of autoinfection, the impact of *S. stercoralis* and related species on human health and well-being can be both acute and long-lasting.

Therefore, it is critical to identify and treat patients who have strongyloides, especially if they are likely to undergo immunosuppression by chemotherapy or high-dose steroids. Traditionally, laboratory identification of *Strongyloides* spp. relies on stool microscopy, charcoal culture, agar plate, hairy string test, and serology. Recent application of molecular techniques such as PCR and sequencing has opened a new avenue for improved diagnosis of strongyloidiasis and for differentiation between species and strains from different hosts and geographical origin.

REFERENCES

1. Blaxter, M.L. et al., A molecular evolutionary framework for the phylum Nematoda. *Nature*, 392, 71, 1998.
2. Dorris, M., Viney, M.E., and Blaxter, M.L., Molecular phylogenetic analysis of the genus *Strongyloides* and related nematodes. *Int. J. Parasitol.*, 32, 1507, 2002.
3. dos Santos, K.R. et al., Morphological and molecular characterization of *Strongyloides ophidiae* (Nematoda, Strongyloididae). *J. Helminthol.*, 84, 136, 2010.
4. Hira, P.R., and Patel, B.G., Human strongyloidiasis due to the primate species *Strongyloides fuelleborni*. *Trop. Geogr. Med.*, 32, 23, 1980.
5. Ashford, R.W., Barnish, G., and Viney, M.E., *Strongyloides fuelleborni kellyi*: Infection and disease in Papua New Guinea. *Trends Parasitol.*, 8, 314, 1992.

6. King, S.E., and Mascie-Taylor, C.G., *Strongyloides fuelleborni kellyi* and other intestinal helminths in children from Papua New Guinea: Associations with nutritional status and socioeconomic factors. *P. N. G. Med. J.*, 47, 181, 2004.
7. Little, M.D., Comparative morphology of six species of *Strongyloides* (Nematoda) and redefinition of the genus. *J. Parasitol.*, 52, 69, 1966.
8. Yamada, M. et al., Species-specific differences in heterogonic development of serially transferred free-living generations of *Strongyloides planiceps* and *Strongyloides stercoralis*. *J. Parasitol.*, 77, 592, 1991.
9. Gill, G.V. et al., A British Second World War veteran with disseminated strongyloidiasis. *Trans. R. Soc. Trop. Med. Hyg.*, 98, 382, 2004.
10. Prendki, V. et al., Strongyloidiasis in man 75 years after initial exposure. *Emerg. Infect. Dis.*, 17, 5, 2011.
11. Olsen, A. et al., Strongyloidiasis—the most neglected of the neglected tropical diseases? *Trans. R. Soc. Trop. Med. Hyg.*, 103, 967, 2009.
12. Montes, M., Sawhney, C., and Barros, N., *Strongyloides stercoralis*: There but not seen. *Curr. Opin. Infect. Dis.*, 23, 500, 2010.
13. Safdar, A. et al., Strongyloidiasis in patients at a comprehensive cancer center in the United States: A retrospective study covering the years 1971–2003. *Cancer*, 100, 1531, 2004.
14. Steinmann, P. et al., Occurrence of *Strongyloides stercoralis* in Yunnan Province, China, and comparison of diagnostic methods. *PLoS Negl. Trop. Dis.*, 1, e75, 2007.
15. DeVault, G.A. et al., Opportunistic infections with *Strongyloides stercoralis* in renal transplantation. *Rev. Infect. Dis.*, 12, 653, 1990.
16. Roman-Sanchez, P. et al., High prevalence of *Strongyloides stercoralis* among farm workers on the Mediterranean coast of Spain: Analysis of the predictive factors of infection in developed countries. *Am. J. Trop. Med. Hyg.*, 69, 336, 2003.
17. Czachor, J.S., and Jonas, A.P., Transmission of *Strongyloides steracolis* person to person. *J. Travel Med.*, 7, 211, 2000.
18. Liu, L.X., and Weller, P.F., Strongyloidiasis and other intestinal nematode infections. *Infect. Dis. Clin. North. Am.*, 7, 655, 1993.
19. Mahmoud, A.A., Strongyloidiasis. *Clin. Infect. Dis.*, 23, 949, 1996.
20. Sudarshi, S. et al., Clinical presentation and diagnostic sensitivity of laboratory tests for *Strongyloides stercoralis* in travellers compared with immigrants in a non-endemic country. *Trop. Med. Int. Health*, 8, 728, 2003.
21. Keiser, P.B., and Nutman, T.B., *Strongyloides stercoralis* in the immunocompromised population. *Clin. Microbiol. Rev.*, 17, 208, 2004.
22. Lim, S. et al., Complicated and fatal *Strongyloides* infection in Canadians: Risk factors, diagnosis and management. *CMAJ*, 171, 479, 2004.
23. Segarra-Newnham, M., Manifestations, diagnosis, and treatment of Strongyloides stercoralis infection. *Ann. Pharmacother.*, 41, 1992, 2007.
24. Greiner, K., Bettencourt, J., and Semolic, C., Strongyloidiasis: A review and update by case example. *Clin. Lab. Sci.*, 21, 82, 2008.
25. von Kuster, L.C., and Genta, R.M., Cutaneous manifestations of strongyloidiasis. *Arch. Dermatol.*, 124, 1826, 1988.
26. Livneh, A. et al., *Strongyloides stercoralis* hyperinfection mimicking systemic lupus erythematosus flare. *Arthritis Rheum.*, 31, 930, 1988.
27. Chu, E., Whitlock, W.L., and Dietrich, R.A., Pulmonary hyperinfection syndrome with Strongyloides stercoralis. *Chest*, 97, 1475, 1990.
28. Ly, M.N. et al., Cutaneous *Strongyloides stercoralis* infection: An unusual presentation. *J. Am. Acad. Dermatol.*, 49, S157, 2003.
29. Gutierrez, Y. et al., *Strongyloides stercoralis* eosinophilic granulomatous enterocolitis. *Am. J. Surg. Pathol.*, 20, 603, 1996.
30. Cannavo, S.P., Guarneri, F., and Guarneri, C., Strongyloidiasis histologically mimicking eosinophilic folliculitis. *Eur. J. Dermatol.*, 14, 264, 2004.
31. Currie, B.J., and McCarthy, J.S., *Strongyloides stercoralis* infection as a manifestation of immune restoration syndrome. *Clin. Infect. Dis.*, 40, 635, 2005.
32. Vithayasai, N., and Jennuvat, S., Gastrointestinal manifestations in severe strongyloidiasis: Report of 3 cases and literature review. *Southeast Asian J. Trop. Med. Public Health*, 41, 22, 2010.
33. Mansfield, L.S. et al., *Strongyloides stercoralis*: Maintenance of exceedingly chronic infections. *Am. J. Trop. Med. Hyg.*, 55, 617, 1996.
34. Sandlund, J.T., Kauffman, W., and Flynn, P.M., *Strongyloides stercoralis* infection mimicking relapse in a child with small noncleaved cell lymphoma. *Am. J. Clin. Oncol.*, 20, 215, 1997.
35. Thompson, B.F. et al., The spectrum of GI strongyloidiasis: An endoscopic-pathologic study. *Gastrointest. Endosc.*, 59, 906, 2004.
36. Concha, R., Harrington, W.J., and Rogers, A.I., Intestinal strongyloidiasis: Recognition, management, and determinants of outcome. *J. Clin. Gastroenterol.*, 39, 203, 2005.
37. Csermely, L. et al., *Strongyloides* hyper-infection causing life-threatening gastrointestinal bleeding. *World J. Gastroenterol.*, 12, 6401, 2006.
38. Ghoshal, U.C. et al., *Strongyloides stercoralis* infestation in a patient with severe ulcerative colitis. *Indian J. Med. Sci.*, 60, 106, 2006.
39. Yoshida, H. et al., Recurrent paralytic ileus associated with strongyloidiasis in a patient with systemic lupus erythematosus. *Mod. Rheumatol.*, 16, 44, 2006.
40. Qu, Z. et al., *Strongyloides colitis* is a lethal mimic of ulcerative colitis: The key morphologic differential diagnosis. *Hum. Pathol.*, 40, 572, 2009.
41. Bannon, J.P., Fater, M., and Solit, R., Intestinal ileus secondary to *Strongyloides stercoralis* infection: Case report and review of the literature. *Am. Surg.*, 61, 377, 1995.
42. Al-Bahrani, Z.R., Al-Saleem, T., and Al-Gailani, M.A., Subacute intestinal obstruction by *Strongyloides stercolaris*. *J. Infect.*, 30, 47, 1995.
43. Friedenberg, F. et al., Duodenal obstruction caused by *Strongyloides stercoralis* enteritis in an HTLV-1-infected host. *Dig. Dis. Sci.*, 44, 1184, 1999.
44. Suvarna, D. et al., Infiltrating *Strongyloides stercoralis* presenting as duodenal obstruction. *Indian J. Gastroenterol.*, 24, 173, 2005.
45. Harish, K. et al., Strongyloidiasis presenting as duodenal obstruction. *Trop. Gastroenterol.*, 26, 201, 2005.
46. Juchems, M.S. et al., *Strongyloides stercoralis*: A rare cause of obstructive duodenal stenosis. *Digestion*, 77, 141, 2008.
47. Cruz, R.J. Jr., Vincenzi, R., and Ketzer, B.M., Duodenal obstruction—an unusual presentation of *Strongyloides stercoralis* enteritis: A case report. *World J. Emerg. Surg.*, 5, 23, 2010.
48. Marcos, L.A. et al., *Strongyloides* hyperinfection syndrome: An emerging global infectious disease. *Trans. R. Soc. Trop. Med. Hyg.*, 102, 314, 2008.

49. Igra-Siegman, Y. et al., Syndrome of hyperinfection with *Strongyloides stercoralis*. *Rev. Infect. Dis.*, 3, 397, 1981.
50. Nonaka, D. et al., Paralytic ileus due to strongyloidiasis: Case report and review of the literature. *Am. J. Trop. Med. Hyg.*, 59, 535, 1998.
51. Lemos, L.B. et al., Hyperinfection syndrome in strongyloidiasis: Report of two cases. *Ann. Diagn. Pathol.*, 7, 87, 2003.
52. Potter, A., Stephens, D., and De Keulenaer, B., *Strongyloides* hyper-infection: A case for awareness. *Ann. Trop. Med. Parasitol.*, 97, 855, 2003.
53. Namisato, S. et al., Pulmonary strongyloidiasis in a patient receiving prednisolone therapy. *Intern. Med.*, 43, 731, 2004.
54. Schaeffer, M.W. et al., *Strongyloides* hyperinfection syndrome after heart transplantation: Case report and review of the literature. *J. Heart Lung Transplant.*, 23, 905, 2004.
55. Mayayo, E. et al., *Strongyloides stercolaris* infection mimicking a malignant tumour in a non-immunocompromised patient. Diagnosis by bronchoalveolar cytology. *J. Clin. Pathol.*, 58, 420, 2005.
56. Kaslow, J.E. et al., Disseminated strongyloidiasis: An unheralded risk of corticosteroid therapy. *J. Allergy Clin. Immunol.*, 86, 138, 1990.
57. Newton, R.C. et al., *Strongyloides stercoralis* hyperinfection in a carrier of HTLV-I virus with evidence of selective immunosuppression. *Am. J. Med.*, 92, 202, 1992.
58. Morgello, S. et al., Central nervous system *Strongyloides stercoralis* in acquired immunodeficiency syndrome: A report of two cases and review of the literature. *Acta Neuropathol. (Berl)*, 86, 285, 1993.
59. Nandy, A. et al., Fulminating strongyloidiasis complicating Indian kala-azar. *Trop. Geogr. Med.*, 47, 139, 1995.
60. Nucci, M. et al., Strongyloidiasis in patients with hematologic malignancies. *Clin. Infect. Dis.*, 21, 675, 1995.
61. Vadlamudi, R.S., Chi, D.S., and Krishnaswamy, G., Intestinal strongyloidiasis and hyperinfection syndrome. *Clin. Mol. Allergy*, 4, 8, 2006.
62. Palau, L.A., and Pankey, G.A., *Strongyloides* hyperinfection in a renal transplant recipient receiving cyclosporine: Possible *Strongyloides stercoralis* transmission by kidney transplant. *Am. J. Trop. Med. Hyg.*, 57, 413, 1997.
63. Gotuzzo, E. et al., *Strongyloides stercoralis* hyperinfection associated with human T cell lymphotropic virus type-1 infection in Peru. *Am. J. Trop. Med. Hyg.*, 60, 146, 1999.
64. Adedayo, C.O., Grell, G.A., and Bellot, P., Case study: Fatal strongyloides associated with human T-cell lymphotropic virus type 1 infection. *Am. J. Trop. Med. Hyg.*, 65, 650, 2001.
65. Adedayo, O., Grell, G., and Bellot, P., Hyperinfective strongyloidiasis in the medical ward: A review of 27 cases in 5 years. *South Med. J.*, 95, 711, 2002.
66. Feitosa, G. et al., High prevalence of giardiasis and stronglyloidiasis among HIV-infected patients in Bahia, Brazil. *Braz. J. Infect. Dis.*, 5, 339, 2001.
67. Rodrigues, M.A. et al., Invasive enteritis by *Strongyloides stercoralis* presenting as acute abdominal distress under corticosteroid therapy. *Rev. Hosp. Clin. Fac. Med. Sao Paulo*, 56, 103, 2001.
68. Singh, S., Human strongyloidiasis in AIDS era: Its zoonotic importance. *J. Assoc. Physicians India*, 50, 415, 2002.
69. Orlent, H. et al., Strongyloidiasis pre and post autologous peripheral blood stem cell transplantation. *Bone Marrow Transplant.*, 32, 115, 2003.
70. Carvalho, E.M., and Da Fonseca Porto, A., Epidemiological and clinical interaction between HTLV-1 and *Strongyloides stercoralis*. *Parasite Immunol.*, 26, 487, 2004.
71. Ohnishi, K. et al., Strongyloidiasis in a patient with acquired immunodeficiency syndrome. *J. Infect. Chemother.*, 10, 178, 2004.
72. Viney, M.E. et al., Why does HIV infection not lead to disseminated strongyloidiasis? *J. Infect. Dis.*, 190, 2175, 2004.
73. Arsić-Arsenijević, V. et al., Fatal *Strongyloides sterocoralis* infection in a young woman with lupus glomerulonephritis. *J. Nephrol.*, 18, 787, 2005.
74. Ben-Youssef, R. et al., *Strongyloides stercoralis* infection from pancreas allograft: Case report. *Transplantation*, 80, 997, 2005.
75. El Masry, H.Z., and O'Donnell, J., Fatal *Strongyloides* hyperinfection in heart transplantation. *J. Heart Lung Transplant*, 24, 1980, 2005.
76. Hauber, H.P. et al., Fatal outcome of a hyperinfection syndrome despite successful eradication of *Strongyloides* with subcutaneous ivermectin. Infection, 33, 383, 2005.
77. Lanzafame, M. et al., Strongyloidiasis in an HIV-1–infected patient after highly active antiretroviral therapy–induced immune restoration (letter). *J. Infect. Dis.*, 191, 1027, 2005.
78. Reddy, I.S., and Swarnalata, G., Fatal disseminated strongyloidiasis in patients on immunosuppressive therapy. *Indian J. Dermatol. Venerol. Leprol.*, 71, 38, 2005.
79. Satyanarayana, S. et al., Disseminated *Strongyloides stercoralis* in AIDS: A report from India. *Indian J. Pathol. Microbiol.*, 48, 472, 2005.
80. Seet, R.C.S., Lau, L.G., and Tambyah, P.A., *Strongyloides* hyperinfection and hypogammaglobulinemia. *Clin. Diagn. Lab. Immunol.*, 12, 680, 2005.
81. Brown, M., Cartledge, J.D., and Miller, R.F., Dissemination of *Strongyloides stercoralis* as an immune restoration phenomenon in an HIV-1–infected man on antiretroviral therapy. *Int. J. STD AIDS*, 17, 560, 2006.
82. Mora, C.S., Segami, M.I., and Hidalgo, J.A., *Strongyloides stercoralis* hyperinfection in systemic lupus erythematosus and the antiphospholipid syndrome. *Semin. Arthritis Rheum.*, 36, 135, 2006.
83. Qazilbash, M.H. et al., Strongyloidiasis after unrelated nonmyeloablative allogeneic stem cell transplantation. *Bone Marrow Transplant.*, 38, 393, 2006.
84. Ramdial, P.K., Hlatshwayo, N.H., and Singh, B., *Strongyloides stercoralis* mesenteric lymphadenopathy: Clue to the etiopathogenesis of intestinal pseudo-obstruction in HIV-infected patients. *Ann. Diagn. Pathol.*, 10, 209, 2006.
85. Fardet, L. et al., Severe strongyloidiasis in corticosteroid-treated patients: Case series and literature review. *J. Infect.*, 54, 18, 2007.
86. Said, T. et al., Hyperinfection strongyloidiasis: An anticipated outbreak in kidney transplant recipients in Kuwait. *Transplant. Proc.*, 39, 1014, 2007.
87. Taylor, C.L. et al., Immune reconstitution syndrome to *Strongyloides stercoralis* infection (letter). *AIDS*, 21, 649, 2007.
88. Kia, E.B. et al., A case of fatal strongyloidiasis in a patient with chronic lymphocytic leukemia and molecular characterization of the isolate. *Korean J. Parasitol.*, 46, 261, 2008.
89. Patel, G. et al., *Strongyloides* hyperinfection syndrome after intestinal transplantation. *Transpl. Infect. Dis.*, 10, 137, 2008.
90. Dulley, F.L. et al., *Strongyloides stercoralis* hyperinfection after allogeneic stem cell transplantation. *Bone Marrow Transplant.*, 43, 741, 2009.
91. Mizuno, S. et al., *Strongyloides* hyperinfection syndrome following simultaneous heart and kidney transplantation. *Transpl. Int.*, 22, 251, 2009.

92. Roxby, A.C., Gottlieb, G.S., and Limaye, A.P., Strongyloidiasis in transplant patients. *Clin. Infect. Dis.*, 49, 1411, 2009.
93. Vilela, E.G. et al., *Strongyloides stercoralis* hyperinfection syndrome after liver transplantation: Case report and literature review. *Transpl. Infect. Dis.*, 11, 132, 2009.
94. Wirk, B., and Wingard, J.R., *Strongyloides stercoralis* hyperinfection in hematopoietic stem cell transplantation. *Transpl. Infect. Dis.*, 11, 143, 2009.
95. Iriemenam, N.C. et al., *Strongyloides stercoralis* and the immune response. *Parasitol. Int.*, 59, 9, 2010.
96. Thomas, M.C., and Costello, S.A., Disseminated strongyloidiasis arising from a single dose of dexamethasone before stereotactic radiosurgery. *Int. J. Clin. Pract.*, 52, 520, 1998.
97. Wong, T.Y.H. et al., Nephrotic syndrome in strongyloidiasis: Remission after eradication with antihelminthic agents. *Nephron*, 79, 333, 1998.
98. Hughes, R., and McGuire, G., Delayed diagnosis of disseminated strongyloidiasis. *Intensive Care Med.*, 27, 310, 2001.
99. Kim, A.C., and Lupatkin, H.C., *Strongyloides stercoralis* infection as a manifestation of immune restoration syndrome (letter). *Clin. Infect. Dis.*, 39, 439, 2004.
100. Giavina-Bianchi, P. et al., A rare intestinal manifestation in a patient with common variable immunodeficiency and strongyloidiasis. *Int. Arch. Allergy Immunol.*, 140, 199, 2006.
101. Lam, C.S. et al., Disseminated strongyloidiasis: A retrospective study of clinical course and outcome. *Eur. J. Clin. Microbiol. Infect. Dis.*, 25, 14, 2006.
102. Galimberti, R. et al., Disseminated strongyloidiasis in immunocompromised patients–report of three cases. *Int. J. Dermatol.*, 48, 975, 2009.
103. Basile, A. et al., Disseminated *Strongyloides stercoralis*: hyperinfection during medical immunosuppression. *J. Am. Acad. Dermatol.*, 63, 896, 2010.
104. Hsieh, Y.P., Won, Y.K., and Chen, M.L., Minimal change nephrotic syndrome in association with strongyloidiasis. *Clin. Nephrol.*, 66, 459, 2006.
105. Jain, A.K., Agarwal, S.K., and el Sadr, W., *Streptococcus bovis* bacteremia and meningitis associated with *Strongyloides stercoralis* colitis in a patient infected with human immunodeficiency virus. *Clin. Infect. Dis.*, 18, 253, 1994.
106. Link, K., and Orenstein, R., Bacterial complications of strongyloidiasis: *Streptococcus bovis* meningitis. *South Med. J.*, 92, 728, 1999.
107. Linder, J.D. et al., *Streptococcus bovis* bacteremia associated with *Strongyloides stercoralis* colitis. *Gastrointest. Endosc.*, 52, 796, 2000.
108. de Silva, T., Raychaudhuri, M., and Poulton, M., HIV infection associated with *Strongyloides stercoralis* colitis resulting in *Streptococcus bovis* bacteraemia and meningitis. *Sex. Transm. Infect.*, 81, 276, 2005.
109. Newberry, A.M. et al., *Strongyloides* hyperinfection presented as acute respiratory failure and Gram-negative sepsis. *Chest*, 128, 3681, 2005.
110. Shorman, M., and Al-Tawfiq, J.A., *Strongyloides stercoralis* hyperinfection presenting as acute respiratory failure and Gram-negative sepsis in a patient with astrocytoma. *Int. J. Infect. Dis.*, 13, e288, 2009.
111. Pitisuttithum, P., Supanaranond, W., and Chindanond, D., A randomized comparative study of albendazole and thiabendazole in chronic strongyloidiasis. *Southeast Asian J. Trop. Med. Public Health*, 26, 735, 1995.
112. Ashraf, M., Gue, C.L., and Baddour, L.M., Case report: Strongyloidiasis refractory to treatment with ivermectin. *Am. J. Med. Sci.*, 311, 178, 1996.
113. Toma, H. et al., Comparative studies on the efficacy of three antihelminthics on treatment of human strongyloidiasis in Okinawa, Japan. *Southeast Asian J. Trop. Med. Public Health*, 31, 147, 2000.
114. Zaha, O. et al., Strongyloidiasis—progress in diagnosis and treatment. *Intern. Med.*, 39, 695, 2000.
115. Zaha, O. et al., Efficacy of ivermectin for chronic strongyloidiasis: Two single doses given 2 weeks apart. *J. Infect. Chemother.*, 8, 94, 2002.
116. Marty, F.M. et al., Treatment of human disseminated strongyloidiasis with a parenteral veterinary formulation of ivermectin. *Clin. Infect. Dis.*, 41, e5, 2005.
117. Pacanowski, J. et al., Subcutaneous ivermectin as a safe salvage therapy in *Strongyloides stercoralis* hyperinfection syndrome: A case report. *Am. J. Trop. Med. Hyg.*, 73, 122, 2005.
118. Boulware, D.R. et al., Maltreatment of *Strongyloides* infection: Case series and worldwide physicians-in-training survey. *Am. J. Med.*, 120, e1, 2007.
119. Balagopal, A. et al., Detection and treatment of *Strongyloides* hyperinfection syndrome following lung transplantation. *Transpl. Infect. Dis.*, 11, 149, 2009.
120. Huston, J.M. et al., Treatment of *Strongyloides stercoralis* hyperinfection-associated septic shock and acute respiratory distress syndrome with drotrecogin alfa (activated) in a renal transplant recipient. *Transpl. Infect. Dis.*, 11, 277, 2009.
121. Datry, A. et al., Treatment of *Strongyloides stercoralis* infection with ivermectin compared with albendazole: Results of an open study of 60 cases. *Trans. R. Soc. Trop. Med. Hyg.*, 88, 344, 1994.
122. Chiodini, P.L. et al., Parenteral ivermectin in *Strongyloides* hyperinfection. *Lancet*, 355, 43, 2000.
123. Orem, J. et al., *Strongyloides stercoralis* hyperinfection in a patient with AIDS in Uganda successfully treated with ivermectin. *Clin. Infect. Dis.*, 37, 152, 2003.
124. Gulbas, Z. et al., Successful ivermectin treatment of hepatic strongyloidiasis presenting with severe eosinophilia. *South Med. J.*, 97, 907, 2004.
125. Gann, P.H., Neva, F.A., and Gam, A.A., A randomized trial of single- and two-dose ivermectin versus thiabendazole for treatment of strongyloidiasis. *J. Infect. Dis.*, 169, 1076, 1994.
126. Tarr, P. et al., Case report: Rectal administration of ivermectin to a patient with *Strongyloides* hyperinfection syndrome. *Am. J. Trop. Med. Hyg.*, 68, 453, 2003.
127. Muennig, P. et al., The cost-effectiveness of ivermectin vs. albendazole in the presumptive treatment of strongyloidiasis in immigrants to the United States. *Epidemiol. Infect.*, 132, 1055, 2004.
128. Pornsuriyasak, P., Niticharoenpong, K., and Sakapibunnan, A., Disseminated strongyloidiasis successfully treated with extended duration ivermectin combined with albendazole: A case report of intractable strongyloidiasis. *Southeast Asian J. Trop. Med. Public Health*, 35, 531, 2004.
129. Turner, S.A. et al., Parenteral administration of ivermectin in a patient with disseminated strongyloidiasis. *Am. J. Trop. Med. Hyg.*, 73, 911, 2005.
130. Lichtenberger, P. et al., Hyperinfection strongyloidiasis in a liver transplant recipient treated with parenteral ivermectin. *Transpl. Infect. Dis.*, 11, 137, 2009.
131. Grein, J.D. et al., Serum ivermectin levels after enteral and subcutaneous administration for *Strongyloides* hyperinfection: A case report. *Scand. J. Infect. Dis.*, 42, 234, 2010.
132. Conway, D.J. et al., Towards effective control of *Strongyloides stercoralis*. *Trends Parasitol.*, 11, 420, 1995.

133. Takayanagui, O.M. et al., Detection of *Strongyloides stercoralis* in the cerebrospinal fluid of a patient with acquired immunodeficiency syndrome. *Neurology*, 45, 193, 1995.
134. Marchi, B.J., and Cantos, G.A., Evaluation of techniques for the diagnosis of *Strongyloides stercoralis* in human immunodeficiency virus (HIV) positive and HIV negative individuals in the city of Itajai, Brazil. *Braz. J. Infect. Dis.*, 7, 402, 2003.
135. Siddiqui, A.A., and Berk, S.L., Diagnosis of *Strongyloides stercoralis* infection. *Clin. Infect. Dis.*, 33, 1040, 2001.
136. Kishimoto, K. et al., Endoscopic and histopathological study on the duodenum of *Strongyloides stercoralis* hyperinfection. *World J. Gastroenterol.*, 14, 1768, 2008.
137. Uparanukraw, P., Phongsri, S., and Morakote, N., Fluctuations of larval excretion in *Strongyloides stercoralis* infection. *Am. J. Trop. Med. Hyg.*, 60, 967, 1999.
138. Steiner, B. et al., *Strongyloides stercoralis* eggs in a urethral smear after bone marrow transplantation (letter). *Clin. Infect. Dis.*, 34, 1280, 2002.
139. Sato, Y. et al., Efficacy of stool examination for detection of *Strongyloides* infection. *Am. J. Trop. Med. Hyg.*, 53, 248, 1995.
140. de Paula, F.M. et al., Parasitological and immunological diagnoses of strongyloidiasis in immunocompromised and non-immunocompromised children at Uberlandia, State of Minas Gerais, Brazil. *Rev. Inst. Med. Trop. Sao Paulo*, 42, 51, 2000.
141. Koga, K. et al., An evaluation of the agar plate method for the detection of *Strongyloides stercoralis* in northern Thailand. *J. Trop. Med. Hyg.*, 93, 183, 1990.
142. Hirata, T. et al., Increased detection rate of *Strongyloides stercoralis* by repeated stool examinations using the agar plate culture method. *Am. J. Trop. Med. Hyg.*, 77, 683, 2007.
143. Grapsa, D. et al., *Strongyloides stercoralis* in a bronchial washing specimen processed as conventional and Thin-Prep smears: report of a case and a review of the literature. *Diagn. Cytopathol.*, 37, 903, 2009.
144. Loutfy, M.R. et al., Serology and eosinophil count in the diagnosis and management of strongyloidiasis in a non-endemic area. *Am. J. Trop. Med. Hyg.*, 66, 749, 2002.
145. Genta, R.M., Predictive value of an enzyme-linked immunosorbent assay (ELISA) for the serodiagnosis of strongyloidiasis. *Am. J. Clin. Pathol.*, 89, 391, 1988.
146. Huaman, M.C. et al., Gelatin particle indirect agglutination and enzyme-linked immunosorbent assay for diagnosis of strongyloidiasis using *Strongyloides venezuelensis* antigen. *Trans. R. Soc. Trop. Med. Hyg.*, 97, 535, 2003.
147. Lindo, J.F. et al., Prospective evaluation of enzyme-linked immunosorbent assay and immunoblot methods for the diagnosis of endemic *Strongyloides stercoralis* infection. *Am. J. Trop. Med. Hyg.*, 51, 175, 1994.
148. Melville, L.A., Sykes, A.M., and McCarthy, J.S., The beta-tubulin genes of two *Strongyloides* species. *Exp. Parasitol.*, 112, 144, 2006.
149. ten Hove, R.J. et al., Molecular diagnostics of intestinal parasites in returning travellers. *Eur. J. Clin. Microbiol. Infect. Dis.*, 28, 1045, 2009.
150. Hasegawa, H. et al., Hyper-variable regions in 18S rDNA of *Strongyloides* spp. as markers for species-specific diagnosis. *Parasitol. Res.*, 104, 869, 2009.
151. Hasegawa, H. et al., Molecular identification of the causative agent of human strongyloidiasis acquired in Tanzania: Dispersal and diversity of *Strongyloides* spp. and their hosts. *Parasitol. Int.*, 59, 407, 2010.
152. Taniuchi, M. et al., High throughput multiplex PCR and probe-based detection with Luminex beads for seven intestinal parasites. *Am. J. Trop. Med. Hyg.*, 84, 332, 2011.
153. Basuni, M. et al., A pentaplex real-time polymerase chain reaction assay for detection of four species of soil-transmitted helminths. *Am. J. Trop. Med. Hyg.*, 84, 338, 2011.
154. Kramme, S. et al., Novel real-time PCR for the universal detection of *Strongyloides* spp. *J. Med. Microbiol.*, 60, 454, 2011.
155. Verweij, J.J. et al., Molecular diagnosis of *Strongyloides stercoralis* in faecal samples using real-time PCR. *Trans. R. Soc. Trop. Med. Hyg.*, 103, 342, 2009; Erratum: 104, 378, 2010.

64 Syphacia

Sanford H. Feldman and John H. Porter

CONTENTS

64.1 INTRODUCTION

Syphacia obvelata (SO) is a mouse pinworm in the family Oxyuridae, superfamily Oxyuroidea. This superfamily contains a single known human parasite *Enterobius vermicularis*, which resides in the vermiform appendage and is not highly pathogenic. In a case report published in 1919 by Riley [1], SO was noted in the stool of a young girl from the Philippines Islands. That author was convinced that "the food of the child and others in the family were grossly contaminated by mice or rats"; however, it is also possible that these rodents were a component of the diet of the infected family. In addition, *Syphacia muris* eggs were reportedly found in the feces of two children, as detailed in a personal letter from Dr. E. C. Faust at Tulane University, dated January 6, 1965 [2].

SO [3, identified by Rudophi in 1802] is an obligate parasite of rodents of the genus *Mus* with a direct life cycle. It is a common parasite encountered in mice raised in pet shops and a nuisance in laboratory mouse colonies maintained at biomedical research facilities. Rats, hamsters, gerbils, and wild rodents have also been reported to harbor this parasite [4].

In this chapter, molecular detection of SO is described as well as molecular detection of *Aspiculuris tetraptera* (AT), a related pinworm equally detrimental in laboratory mouse research colonies. The development of molecular methods for detection of *E. vermicularis* should take advantage of SO in mice as a model system for the development of methods to detect members of the family Oxyuridae. There are significant differences between SO and *E. vermicularis* regarding the sensitivity of embryonation to environmental parameters, the fecundity of gravid females, and the length of the life cycle in the primary host [5]. Our molecular methods for the detection of SO (and AT) are detailed below.

64.1.1 Classification, Morphology, and Biology

Nematode parasites in the superfamily Oxyuroidea are commonly known as pinworms. Based upon morphologic characters, these parasites of vertebrates can be grouped into three families: the Pharyngodonidae (parasites of fish and reptiles), Oxyuridae (parasites of primates and rodents), and Heteroxynematidae (parasites of rodents and birds) [6]. SO belongs to the family Oxyuridae. Members of the family Oxyuridae are distinguished from other Oxyuroidea based upon genital characters such as reduction in the number of genital papillae and differentiation of cuticularized genital pieces, and cephalic characters such as reduction of pharyngeal teeth and the development of labial and interlabial structures [6]. Morphologically, Oxyuridae have nonpedunculate amphids, a genital cone without sclerotized supporting structure, the tails of males have irregular shape being bluntly truncate with or without dorsal points, sometimes males have a large digitiform papillae extending into the caudal alae.

The gross appearance of SO is a very small, pointed, and white threadworm [7]. The male is 1.1–1.5 mm by 120–140 μm and the female is 3.4–5.8 mm by 240–400 μm. The mouth has three distinct lips without a buccal capsule; the esophagus cylindrical anterior has a prebulbar swelling and a posterior, globular bulb. Cervical alae are present. In the male, the tail is long, sharply pointed, curved ventrally, equal in length to the width of the body, and ends abruptly after the cloaca in a fine terminal point. The vulva of the female is in the anterior 6th of the body. The egg is flat on one side, measures 118–153 μm by 33–55 μm with a small opercular cap near one end on the convex side. Additional morphologic characters of SO include: identical cephalic structures in both male and female; cephalic papillae are not arranged in a square, and are located more or less lateral in close proximity to amphids; there are 2–4 ventral cuticular swellings (mammalons) present in the males.

The life cycle of SO is direct and requires only 11–15 days for completion [8]. Gravid females migrate from the cecum to deposit their eggs in the perianal area and die. The eggs become infective after 6 h. Transmission occurs by ingestion of the embryonated egg. Twenty-four hours after ingestion by a mouse, the eggs hatch in the small intestine and the larvae reach the cecum. The larvae mature for 10–11 days in the cecum where they mature and mate. The eggs become disseminated in the environment and can survive for weeks in moderate temperature and humidity conditions.

Nucleotide sequences of the gene encoding the small ribosomal subunit (SSU) have been analyzed from a diverse set of nematodes to reexamine the phylogeny of the phylum Nematoda [9]. Accumulations of single-nucleotide mutations in the hypervariable regions of the SSU gene occur at a predictable rate over time. Therefore, phylogenetic analysis using the variable SSU gene regions can be viewed as examination of a molecular clock. This method of nematode molecular phylogenetic analysis showed similarities in gene sequence that clustered parasitic nematodes of warm-blooded vertebrates into clade III [9]. These authors concluded that their analysis did not agree with classical taxonomy based on morphologic characters, that three major clades exist in the phylum Secernentea rather than the nine orders classically described. Moreover, clade III represented a novel grouping of vertebrate and arthropod parasitic taxa with diverse life histories and life cycle patterns from the orders Ascaridida, Spirurida, Oxyurida, and Rhignonematida. In this same analysis of a large number of nematodes in clade III, early evolutionary divergence appears to have given rise to the two closely related sister orders Oxyurida and Ascardida, while the third evolutionary branch gave rise to the remaining clade III orders Ascaridoidea, Spirurina, Dracunculoidea, and Camallanoidea [10]. Bayesian analysis of SSU sequence alignments examining additional members of the superfamily Oxyuroidae of laboratory rodents clearly established that SO and *Syphacia muris* belong to the family Oxyuridae whereas *Passaluris ambiguus* and *Aspiculuris tetraptera* belong to the family Heteroxynematidae [11]. Molecular phylogenetic analysis of the mitochondrial cytochrome oxidase type I gene from additional nematodes of the genus *Syphacia* found in wild rodents of Japan [12] revealed that host switching likely commonly occurs between Murinae and Microtinae rodents. This host switching has resulted in relatively distantly related *Syphacia* species infecting the same or closely related rodents within a class, rather than closely related *Syphacia* species exclusively infecting closely related rodent genera.

64.1.2 Clinical Features and Pathogenesis

Experimental infections in mice with SO are clinically silent [7]. The prevalence of pinworms in an infected rodent population is determined in part by age, sex, host immune status, host strain, and SO egg burden in the environment. Within an enzootically infected laboratory mouse colony, weanling animals develop the greatest parasite loads with male mice having a greater parasite burden than female mice, and individual parasitic burden diminishing with age [13]. Infection with SO failed to induce goblet cell hyperplasia [14]. Infection of CBA mice with SO increased mouse bone marrow erythropoiesis and myelopoieisis in part due to increased sensitivity to IL-17 [15]. Enzootic infection with SO of a transgenic mouse colony lacking the major histocompatibility complex II (MHC-II) revealed a prevalence of 67–69% in all ages and significant, lifelong retention of the parasite [16]. In this same investigation, C3H/He mice (which have a spontaneous mutation in the toll-like receptor 4 gene) had an SO prevalence of 18–23% and were more heavily burdened retaining adult parasites longer than C57BL/6 (prevalence 1–2%). In C57BL/6 mice, infection was generally found in mice <2 months of age. Mice lacking both MHC-II and TLR-4 had the greatest prevalence of infection (73%). Transgenic mice lacking IL-13 and/or the IL-4α receptor subunit (found in both the IL-4 and IL-13 receptor complexes) had a 10-fold increase in adult worm burden and were unable expel the parasites leading to a chronic infection [17]. Infection with SO induces a Th2 immune response (increased IL-4, IL-5, and IL-13) that diminishes the response of infected mice to other exogenous antigens that normally induce Th2 response, such as ovalbumin [17]. This phenomenon is the basis of the speculation about the clinical suppression of Th2 allergic reactions of humans infected with helminth parasites, the so-called hygiene hypothesis [18]. Adult B6AF1 mice normally develop strong Th1 (monocytic, increase interferon-γ) autoimmune ovarian disease after inoculation with recombinant mouse zona pellucida peptide 3 (ZP3) in complete Freund's adjuvant [19]. The presence of enzootic SO infection in the colony induced a strong Th2 heralded by eosinophilic autoimmune ovarian disease after neonatal administration of ZP3 in water. This Th2 response was further exacerbated upon rechallenge of SO-infected mice at 8 weeks of age. Induction of Th2 response was associated with increased levels of IL-4 and IL-5 as well as fivefold increase in IgG1 antigen-specific antibodies when compared with IgG2 levels. The Th2 ovarian pathology was specific to the priming antigen as neonatal priming with moth cytochrome *c* peptide followed by ZP3 administration to the adult mouse induced the anticipated Th1 ovarian response. Athymic nude mice (lacking T-cells) are unable to expel SO and clear the infection [20].

64.1.3 Diagnosis

Conventional methods to diagnose SO infection use light microscopic examination of samples obtained by perianal tape impression, anal swab, fecal flotation, necropsy with microscopic examination of cecal and colonic contents, and histological examination. A recent evaluation of perianal tape testing (touching clear adhesive tape to the perianal area for light microscopic examination) calculated a sensitivity of 85.5% comparing the results with isolation of adults and larvae by sieving as the gold standard [21]. Goncalves et al. [22], who developed the anal swab method, calculated its sensitivity to be 93% for the detection of nematode parasitic infection. However, that study was performed on animals coinfected with SO and AT where detection of either parasite was considered a true positive. A study in our laboratory compared the following methods for detection of SO and AT infections in 3-week-old mice: perianal tape testing, fecal flotation, direct cecal and colon examination, and a molecular beacon real-time polymerase chain reaction (PCR) assays [11]. The real-time PCR assay for the detection of SO targeted a portion of the intervening sequence 2 (ITS2) located between the 5.8S rRNA and 28S rRNA genes. The PCR assay for the detection of AT targeted the mitochondrial cytochrome oxidase type 1 (MCO-1) gene. In our study, the sensitivity of perianal tape testing was 0% likely due to lack of maturity of the nematodes in the young mice. The molecular assay detected 63.6% of infections when compared with direct cecal examination for nematodes as the definitive test [11]. Others have similarly proposed using SSU sequences [23] or PCR-induced fingerprints [24] derived from randomly amplified polymorphic DNA for PCR-based detection of SO in mice.

64.2 METHODS

64.2.1 Sample Preparation

We found that proteinase-K treatment is ineffective in digesting these parasitic nematodes for subsequent DNA recovery. Consequently, we process pinworm and fecal specimens using a glass-bead pulverization method. Q-biogene's (MP Biomedical, Irvine, CA) FASTDNA for soil and Q-biogene's FASTPrep FP120 orbital shaker coupled with Q-biogene's FASTDNA for soil has proven to be our method of choice for the isolation of nematode parasitic DNA. The kit is used according to the manufacturer's recommendations with an orbital shaker speed setting of 5.5 applied for 30 s. Total fecal DNA is eluted from the FASTDNA silica matrix using 50 μL of water.

64.2.2 Detection Procedures

The PCR assay for the detection of SO targets a portion of the ITS2 located between the 5.8S rRNA and 28S rRNA genes [18]. The primer sequences are SO ITS2 sense F1728: 5′-GTTGTCGACTGACTGCTAATATATACATGCA-3′ and SO ITS2 antisense R1977: 5′-AAATCATGTCTGTCAA CCACGTTATGATA-3′. This SO-specific reaction amplifies a 249 base pair (bp) fragment of the ITS2 region. The molecular beacon nucleotide sequence that hybridizes to this fragment is 5′-FAM-CGCGATCGCACAACGCAGCACGA CTATACTGATCGCG-BHQ1–3′ (black hole quencher).

The molecular beacon assay that detects AT targets a region of the MCO-1 gene. The primer sequences for this beacon assay are AT MCO1 sense F121: 5′-ATTATTTCTATTGCTTTGATTGGAAGG-3′ and AT MCO1 antisense R306 5′-AAGCAAAGGCTGAACACCA TATT-3′. PCR using these AT primers amplify a 185-bp fragment of the MCO-1 gene. The AT molecular beacon sequence is 5′-CY3-CGCGATCCTATAATTATTGCTGTT CCTACTGGTGATCGCG-BHQ2-3′.

The reaction condition for the two molecular beacon assays are 0.5 μM of each primer, 0.3 μM of the beacon, 2 mM $MgCl_2$, 1.25 U of Taq polymerase (HotStar, Qiagen, Valencia, CA), and 1 μL sample (~100 ng DNA) in a 25 μL volume. The cycling conditions are 95°C for 14.5 min (hot-start) followed by 40 cycles of 95°C for 30 s and 55°C for 30 s using an IQ5 optical system thermal cycler and optical analysis software (BioRad, Hercules, CA).

The limit of detection of the SO and AT molecular beacon assays is determined by cloning the respective amplicons into plasmid pCR-4 TOPO (Invitrogen Corp, Carlsbad, CA), the plasmids made linear by Bgl-1 restriction, and the linear targets serial 10-fold diluted to below one target copy per reaction. These control serial dilutions of cloned, linear amplicon are subjected to the molecular beacon assays. The results of this sensitivity testing are shown in Figure 64.1. Relative fluorescence intensity versus copy number at cycle 35 is graphically represented. To be certain that each assay will uniquely detect the species for which it is designed, DNA obtained from other pinworm species (*Syphacia muris* and *Passalurus ambiguus*) as well as total fecal DNA from pinworm negative animals are tested as negative controls for each assay.

The molecular assays produce amplicons of the predicted sizes, and each molecular assay detects DNA of the species for which it is designed and not the other pinworms studied (data not shown). The limit of detection for each of the assays is 6–7 copies of the PCR target in the PCR assay (Figure 64.1). The limit of detection of the assays diminishes 100-fold when the reactions are combined and run as a multiplex (data not shown). The sensitivity of the SO molecular beacon assay is 63.6% and its specificity is 100%. The sensitivity of the AT molecular beacon assay is 35.7% and its specificity is 100%. The molecular beacon assays are superior to the perianal tape method for the detection of SO and the fecal flotation method for detection of AT (Fisher's exact test, $p < 0.01$). Microscopic examination of the cecum and colon for the detection of the pinworms is superior to the molecular beacon assays (Fisher exact test, $p < 0.01$).

64.3 CONCLUSION AND FUTURE PERSPECTIVES

PCR detection of gastrointestinal parasitic nematodes has been reported for *Ancyclostoma caninum* and *Ancyclostoma braziliense* [25], *Ancyclostoma duodenale* and *Necator americanus* [26], and *Oesophagostomum bifurcum* [27]. Only in the case of *O. bifurcum* were assay sensitivity and specificity reported; the seminested PCR assay sensitivity was 94.6% with a specificity of 100%. In our investigation of molecular detection of pinworms in mice, we report a specificity of 100% for both assays; however, the sensitivities were low, 63.6% for SO and 35.7% for AT, respectively. We made several assumptions regarding the presence of parasite DNA in the feces of mice that may account for this. These assumptions were: (i) parasitic ova would be detected if they were present, (ii) parasites being expulsed would be detected, and (iii) endoparasites slough cells from their cuticle and/or gastrointestinal tract could be detected. Since no ova were observed on perianal tape and fecal floatation tests, we conclude that the most likely sources of parasite DNA in the stool

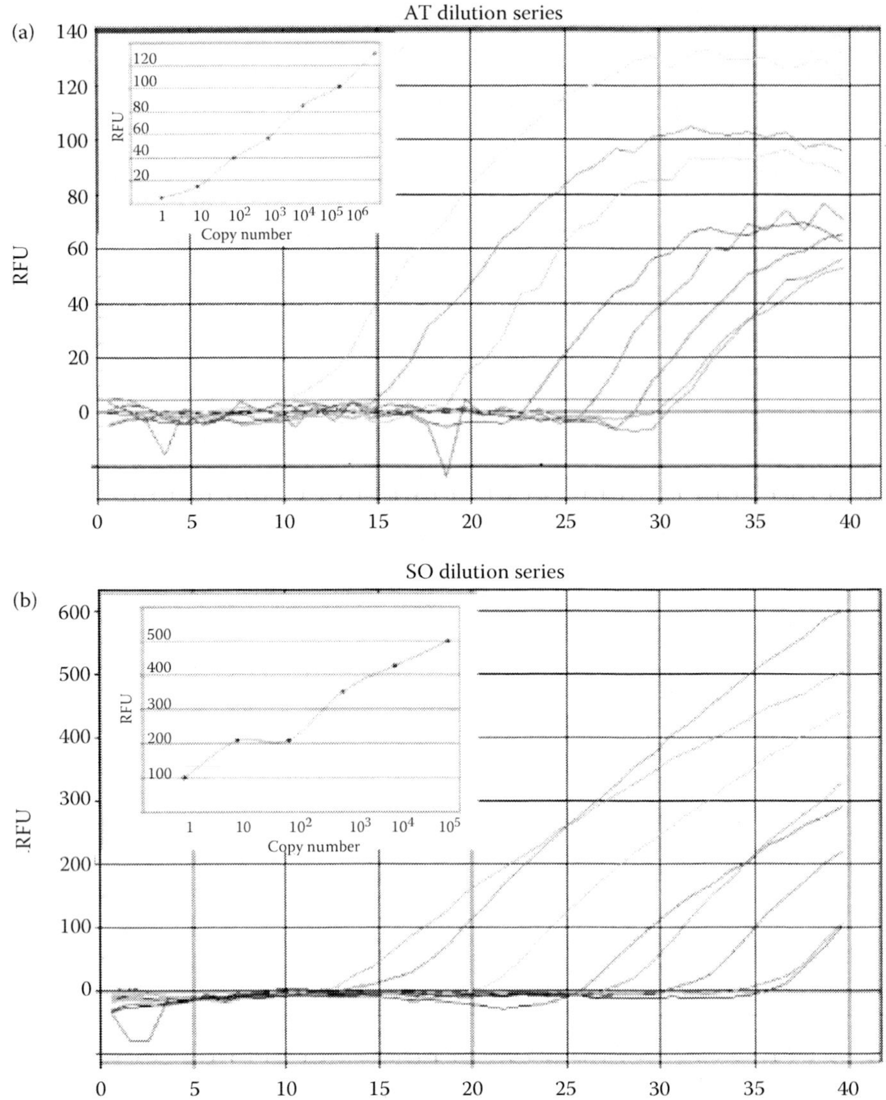

FIGURE 64.1 Graphical depiction of the limit of detection for the molecular beacon assays for *A. tetraptera* (a) and *S. obvelata* (b). The relative fluorescence intensity with each cycle in a series of reactions using serial 10-fold dilutions of plasmid containing the target sequence. The smaller inserted semilog graphs relate the relative fluorescence intensity at cycle 35 resulting from the log of the copy number in each reaction. The limit of detection at cycle 35 for each assay is between 1 and 7 copies per reaction.

were expulsed worms or sloughed cells. One possible explanation for the low sensitivities reported is that the sloughing of cells and expulsion of worms is episodic and somewhat rare. Circadian rhythm of egg excretion has been characterized in *Passalurus ambiguus* infection of laboratory rabbits with highest egg excretion occurring during the dark phase of the light cycle [28]. In our study using molecular beacon assays, we examined weanling mice few of which contained adult gravid females and the experiment was performed between 10 AM and noon.

There are approximately 55 tandem repeats of the ribosomal gene complex in *C. elegans* [29]. If we assume a similar number of copies of the ribosomal genes in SO and assuming 100% recovery of parasite DNA from a fecal pellet, while knowing that we eluted the DNA from each fecal pellet with 50 μL of water we should have at least one copy of the gene in each reaction if a single pinworm cell was present in the fecal sample.

A highly sensitive method of detection of pinworm infestation in laboratory rats and mice that does not require euthanasia of the animal and examination of the content of the cecum and large intestine would facilitate accurate, high-throughput screening of transgenic mice. Our data demonstrate a high specificity for molecular detection of SO and AT; however, sensitivity was low. Traditional methods of perianal tape and fecal flotation tests also have sensitivity less than 100%. The anal swab method, although superior to tape and flotation method, is not definitive [22]. In a study of SO infestation of Swiss Webster and Balb/c athymic nude mice, the prepatent period between exposure and detection by perianal tape detection of ova was 2.5–3 weeks [20]. Furthermore, data from this same study showed that the length of SO infestation was up to 19 weeks in both strains with intermittent periods when no ova were detected by perianal tape testing. In experiments using SO detection by molecular beacon assay, the majority of the mice were young juveniles whose parasites were likely in the prepatent period (period before mature parasite produce ova). During this prepatent period, our molecular detection assays were vastly more sensitive for the detection of rodent pinworms when compared with perianal tape and fecal flotation methods; however, only necropsy and microscopic examination of the cecum and ascending colon were definitive. We attribute the diminished sensitivity to our error in assuming that there is continual and substantial sloughing of parasite cells into the intestinal lumen providing DNA that can be detected in the stool.

REFERENCES

1. Riley, W.A. A mouse oxyurid, *Syphacia obvelata*, as a parasite of man. *J. Parasitol.* 6, 89, 1919.
2. Stone, W.B., and Manwell, R.D. Potential helminth infections in humans from pet or laboratory mice and hamsters. *Public Health Rep.* 81, 647, 1966.
3. Seurat, L.G. Sur les Oxyures de Mammifères, *Comptes Rendus de la Société Biologique (Paris)* 79, 64, 1916.
4. Committee on Infectious Diseases of Rats and Mice, National Research Council, *Infectious Disease of Mice and Rats*, National Academy Press, Washington, DC, 156p, 1991.
5. Grice, R.L., and Prociv, P. *In vitro* embryonation of *Syphacia obvelata* eggs. *Int. J. Parasitol.* 23, 257, 1993.
6. Petter, A.J., and Quentin, J-D. Keys to genera of the Oxyuroidea. In: Anderson, R.C., Chabaud, A.G., and Willmott, S. (eds), *CIH Keys to the Nematode Parasites of Vertebrates*. Commonwealth Agricultural Bureaux, No. 4, 1976.
7. Flynn, R.J. *Parasites of Laboratory Animals*. Iowa State Press, Ames, IA, p. 236, 1973.
8. Taffs, L.F. Pinworm infections in laboratory rodents: A review. *Lab. Anim.* 10, 1, 1976.
9. Blaxter, M.L. et al. A molecular evolutionary framework for the phylum Nematoda. *Nature* 392, 71, 1998.
10. Nadler, S.A. et al. Molecular phylogeny of clade III nematodes reveals multiple origins of tissue parasitism. *Parasitology* 134, 1421, 2007.
11. Feldman, S.H., and Bowman, S.G. Molecular phylogeny of the pinworms of mice, rats and rabbits, and its use to develop molecular beacon assays for the detection of pinworms in mice. *Lab. Anim.* 36, 1, 2007.
12. Okamakoto, M. et al. Phylogenetic relationships of rodent pinworms (genus *Syphacia*) in Japan inferred from mitochondrial CO1 gene sequences. *J. Vet. Med. Sci.* 69, 545, 2007.
13. Wescott, R.B., Malczewski, A., and Van Hoosier, G.L. The influence of filter top caging on transmission of pinworm infections in mice. *Lab. Anim. Sci.* 26, 742, 1976.
14. Marillier, R.G. et al. IL-4/IL-13 independent goblet cell hyperplasia in experimental helminth infections. *BMC Immunol.* 9, 11, 2008.
15. Bugarski, D. et al. Hematopoietic changes and altered reactivity to IL-17 in *Syphacia obvelata*-infected mice. *Parasitol. Int.* 55, 91, 2006.
16. Stewart, P.W., and Chapes, S.K. Role of major histocompatibility complex class II in resistance of mice to naturally acquired infection with *Syphacia obvelata*. *Comp. Med.* 53, 70, 2003.
17. Michels, C. et al. Infection with *Syphacia obvelata* (pinworm) induces protective Th2 immune responses and influences ovalbumin-induced allergic reactions. *Infect. Immun.* 74, 5926, 2006.
18. Fallon, P.G., and Mangan, N.E. Suppression of the T_h2-type allergic reactions by helminth infection. *Nat. Rev. Microbiol.* 7, 220, 2007.
19. Tung, K.S.K. et al. Regulatory T-cell, endogenous antigen and neonatal environment in the prevent and induction of autoimmune disease. *Immunol. Rev.* 182, 135, 2001.
20. Clarke, C.L., and Perdue, K.A. Detection and clearance of *Syphacia obvelata* infection in Swiss Webster and athymic nude mice. *Contemp. Top. Lab. Anim. Sci.* 34, 9, 2004.
21. Hill, W.A., Randolph, M.M., and Mandrell, T.D. Sensitivity of perianal tape impressions to diagnose pinworm (*Syphacia* spp.) infections in rats (*Rattus norvegicus*) and mice (*Mus musculus*). *J. Am Assoc. Lab. Anim. Sci.* 48, 378, 2009.
22. Goncalves, L. et al. Helminth parasites of conventionally maintained laboratory mice—II. Inbred strains with an adaptation of the anal swab technique. *Mem. Inst. Oswaldo Cruz, Rio De Janieiro*, 93, 121, 1998.
23. Parel, J.D.C., Galula, J.U., and Ooi, H.-K. Characterization of rDNA sequences from *Syphacia obvelata*, *Syphacia muris*, and *Aspiculuris tetraptera* and development of a PCR-based method for identification. *Vet. Parasitol.* 153, 379, 2008.
24. Jobet, E. et al. Use of random amplified polymorphic DNA (RAPD) for generating specific DNA probes for oxyuroid species (Nematoda). *Parasite* 5, 47, 1998.
25. Traub, T.K.J. et al. Application of a species-specific PCR-RFLP to identify *Ancylostoma* eggs directly from canine faeces. *Vet. Parasitol.* 123, 245, 2004.
26. de Gruijter, J.M. et al. Polymerase chain reaction-based differential diagnosis of *Ancylostoma duodenale* and *Necator americanus* infections in humans in northern Ghana. *Trop. Med. Int. Health* 10, 574, 2005.
27. Verweij, J.J. et al. PCR assay for the specific amplification of *Oesophagostomum bifurcum* DNA from human faeces. *Int. J. Parasitol.* 30, 137, 2000.
28. Rinaldi, L., Russo, T., and Schioppi, M. *Passalurus ambiguus*: New insights into the copromicroscopic diagnosis and circadian rhythm of egg excretion. *Parasitol. Res.* 101, 557, 2007.
29. Ellis, R.E., Sulston, J.E., and Coulson, A.R. The rDNA of *E. elegans*: Sequence and structure. *Nucleic Acids Res.* 14, 2345, 1986.

65 Thelazia

Domenico Otranto and Donato Traversa

CONTENTS

65.1 INTRODUCTION

The genus *Thelazia* (Spirurida, Thelaziidae) includes 16 species of parasitic nematodes commonly known as "eyeworms," which affect the eyes and associated tissues (e.g., orbital cavities, areas under the eyelid, and nictitating membrane) of several animals, including canids, felids, ruminants, equids, and human beings. *Thelazia* species are transmitted by invertebrate vectors (i.e., secretophagous, nonbiting flies) which feed on the lachrymal secretions from the definitive hosts [1].

Among the *Thelazia* species causing ocular infections in animals, *Thelazia californiensis* Price 1930 and *Thelazia callipaeda* Railliet and Henry 1910 are parasites of carnivores and, importantly, may be zoonotic agents. While *T. californiensis* plays a minor role in human infections, being confined to western United States [2,3], *T. callipaeda* has been reported to infect humans in several geographic areas. Also known as "the oriental eyeworm," the latter species can indeed parasitize several hosts (e.g., dogs, cats, foxes, wolves, rabbits, hares, and humans) and has been for a long time thought to be confined only in Asia (i.e., Indonesia, Thailand, China, Korea, Birmania, India, and Japan) [4–6]. Nonetheless, recent reports of thelaziosis by *T. callipaeda* in Europe (i.e., Italy, France, Portugal, Spain, Switzerland, and Germany), mainly in dogs, foxes, cats, but also in humans, have raised interest in this nematode infection within the field of medical and veterinary parasitology. Hence, given the veterinary and medical importance of such infection, the present chapter focuses only on this eyeworm species while readers interested in *Thelazia* spp. eyeworms are suggested to refer to Reference 1.

65.1.1 Classification and Morphology

65.1.1.1 Classification

The classification and taxonomy of *Thelazia* nematodes (order Spirurida, superfamily Thelazioidea, family Thelaziidae) have been debated for a long time. Anderson [7] claimed that eyeworms were ancestors of certain spirurid species, including those currently linked to onchocercids [8,9]. This hypothesis seems to be supported by the position of *Thelazia* spp. in an external branch of the phylogenetic trees recently generated for spirurid nematodes [8–10]. Nonetheless, in a recent phylogenetic analysis, *Thelazia* spp. clustered with the spirurid *Spirocerca lupi* resulting in a clade well separated from Filarioidea and representing the superfamily Thelazioidea [11]. Therefore, the hypothesis that eyeworms are ancestors of filarial nematodes [7] is not supported anymore. Conversely, Thelazioidea and Filarioidea appear as separate clades originating from *Filaria* spp., whose specimens live in the subcutaneous tissues of wild animals [7,12]. Therefore, it has been suggested that *Filaria* spp. could be the ancestor of *Thelazia* nematodes [11,13].

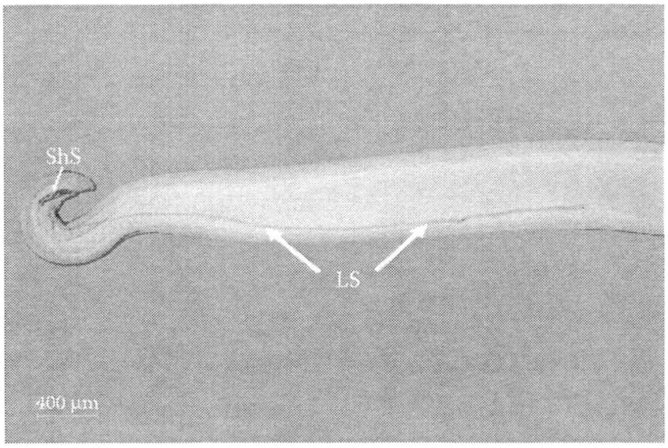

FIGURE 65.1 Posterior end of male *Thelazia callipaeda*: two spicules differing in shape and size: longer right (LS) and shorter left spicule (ShS). (From Otranto, D. et al. *Parassitologia* 45, 125–133, 2003. With permission.)

65.1.1.2 Morphology

T. callipaeda measures 7.5–13 and 12–18.5 mm in length and 340–430 and 370–510 μm in width, in males and females, respectively [14]. Adult nematodes are characterized by the presence of a buccal capsule and their mouth opening has a hexagonal profile. The internal margins of the buccal capsule are everted and subdivided by excavations into six festoons [14]. The males have a ventrally curved caudal end, without caudal alae and with 15 pairs of papillae on the ventral surface, 10 of which are cloacal and 5 postcloacal [14]. The testes are found from the mid-portion of the tail. The two spicules (Figure 65.1) differ from each other in terms of shape and size, with the left spicule being longer and characterized by a well-defined shape of the anterior extremity whereas the right spicule has a typical crescent shape [14]. The vulva of the females has a short flap located in the anterior region of the body. The vagina opens at about 62–160 μm anteriorly to the esophagus–intestinal junction. In the posterior half of the body, immature eggs or germ cells fill the uterine tubules and two sensory phasmids are present on the tip of the tail. The first-stage larvae (L1s) are 382–400 μm long (Figure 65.2); they present a shell membrane (i.e., embryonated egg) and are usually arranged in a row in the distal uterus of the adult females [14].

65.1.2 Biology and Epidemiology

65.1.2.1 Biology

The adult stages of *T. callipaeda* live under the nictitating membrane of the eye of the definitive hosts (Figure 65.3), where the mature females release the L1s into the lachrymal secretions. The biological cycle requires an arthropod intermediate host, represented by a nonbiting dipteran, which feeds on the lachrymal secretions of animals [1]. In the intermediate host, the L1s undergo development until the infective third-stage larvae (L3s). After being ingested by the dipteran, the larvae remain encapsulated in the testes of the vector and then migrate through the arthropod coeloma to the *labella*. The infectious stages emerge from the *labella* of infected

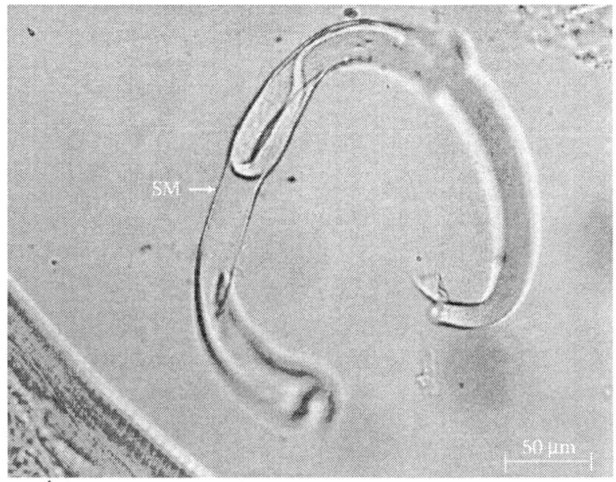

FIGURE 65.2 First-stage larvae of *Thelazia callipaeda* presenting a shell membrane (SM). (From Otranto, D. et al. *Parassitologia* 45, 125–133, 2003. With permission.)

flies as soon as these insects feed on the lachrymal secretions of a receptive animal and then develop into the adult stage in the ocular cavity within ~35 days [15].

Until not long time ago, little information was available about the intermediate hosts of *T. callipaeda*. One single report described *Musca domestica* (Diptera, Muscidae) as vector of *T. callipaeda* in the Asian Pacific regions [16]. However, a recent study in which *M. domestica* flies were allowed to feed directly on the eyes of an infected dog or were directly fed with L1s of *T. callipaeda* demonstrated that this species of fly is unlikely to act as a vector of *T. callipaeda* [17].

Following the first report by Kozlov [18], *Phortica variegata* Fallén 1823 (Drosophilidae, Steganinae) and *Phortica okadai* Okada 1956 have been shown, under experimental conditions, to act as vectors of *T. callipaeda* in Europe and China, respectively [15,19]. This finding is of relevance considering that, unlike the vectors of other *Thelazia* species (i.e., *Musca* spp.), *Phortica* flies are drosophilids (commonly known as "fruitflies") which display an

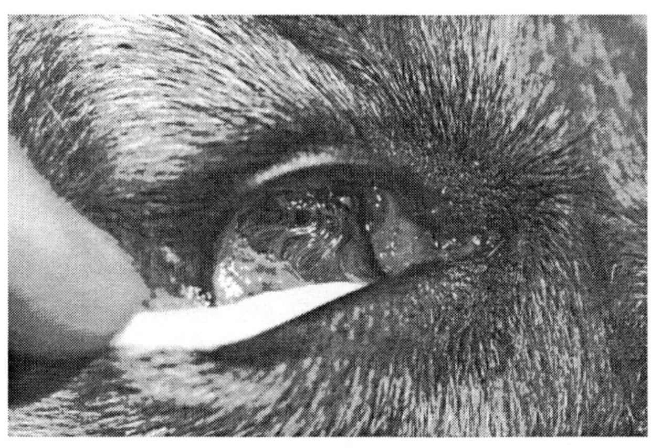

FIGURE 65.3 Eye of a dog with a natural infection by *Thelazia callipaeda*. (From Otranto, D. et al. *Vet. Parasitol.* 116, 315–325, 2003. With permission.)

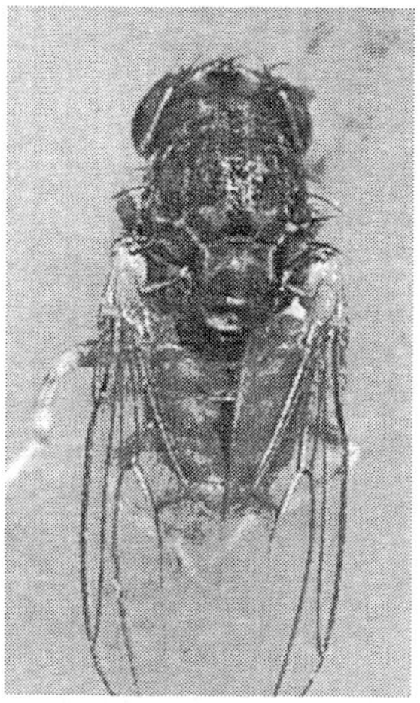

FIGURE 65.4 *Phortica variegata* fly. (From Otranto, D. et al. *Medical and Veterinary Entomology* 20, 358–364, 2006. With permission.)

"unusual" behavior since they feed on lachrymal secretions of humans and carnivores. Furthermore, it has been recently demonstrated that only males of *P. variegata* (Drosophilidae, Steganinae) (Figure 65.4) are the main biological vectors of *T. callipaeda* under natural conditions [20]. Indeed, the absence of *T. callipaeda*-infected drosophilid females supports the proposal that the zoophilic behavior of *P. variegata* is linked specifically to the male gender as well as to environmental and biological factors and to dietary needs (e.g., high-protein supplementation). Interestingly, a study that analyzed the life cycle of *T. callipaeda* in naturally infected dogs in Italy demonstrated the existence of seasonality in the reproductive cycle of female nematodes, according to the presence/absence of the vector/s, thus reinforcing the theory of a strong coevolution between this parasite and its vector [21].

65.1.2.2 Epidemiology

With regard to the infection in animals, the first epidemiological data collected in Europe are from Italy, where *T. callipaeda* has been reported initially in dogs and cats from the Italian Northern regions of Piedmont and Lombardy [22,23] and later, in dogs from the southern region of Basilicata [24]. Thereafter, an epidemiological study in Northern and Southern Italy investigated the occurrence of *T. callipaeda* in dogs, cats, and foxes [25]. The rate of infected dogs from Southern Italy resulted almost double (41.76%) than the rate detected in the North (23.07%), with a peak of prevalence (68.59%) in a single municipality of the Basilicata region. The presence of the nematode in cats and foxes showed that *T. callipaeda* is widespread throughout the Italian peninsula and may infect different host species [25]. Additionally, in Southern Italy, wolves were also found to be affected by *T. callipaeda* and the existence of a sylvatic life cycle of the nematode was also confirmed by its detection in foxes from areas with a high prevalence of canine thelaziosis, thus indicating that wild carnivores may act as reservoir of *T. callipaeda* for pets [25–27]. In the past few years, new cases of canine thelaziosis have been registered in France [28] and Germany [29] in animals which had spent some time, mostly during summer, in Northern Italy. Nonetheless, autochthonous cases of canine and feline thelaziosis have also been described in South-Western France (Dordogne area) [30] and Switzerland [31]. Recently, *T. callipaeda* has been found for the first time from other wild animal species (i.e., beech martens, brown hares, and wild cats), thus supporting the low species specificity of this parasite [32]. The molecular analysis of *T. callipaeda* collected from different species of wild hosts confirmed that the only haplotype (see Section 1.3.2) circulating in Europe is present in wild animals living in that area [32]. Such a finding indicates that *T. callipaeda* displays the broadest spectrum of definitive hosts among all eyeworms and, interestingly, suggests a high degree of specificity of the nematode for its vector together with a low affinity for its definitive hosts [25,32]. The wide range of hosts could also contribute to the spreading of *T. callipaeda* in endemic areas and accounts for the increasing number of reports of canine infection from different parts of Europe in the past few years [25,29,32]. In fact, wild animals like beech martens or foxes that maintain the sylvatic life cycle of the nematode have been known to colonize anthropogenic landscapes [32,33] and, analogously, the capability of drosophilids to colonize human habitats might concur in spreading the infection in human beings apart from forest ecosystems [32].

So far, human infections by *T. callipaeda* have received only little attention, despite the high prevalence recorded in some Asiatic countries [34]. Since its first report in Beijing (China) [35], human thelaziosis by *T. callipaeda* was described in several areas of the former Soviet Union [36] and Asian continent, including China [37], Korea [38], Japan [39], Indonesia [6], and Thailand [4,40]. The infection occurs predominantly in poor, rural communities with low health and socioeconomic standards particularly where domestic dogs and other animals (e.g., cats and foxes) are heavily infected [19,41]. Despite its clinical importance, human thelaziosis remains a neglected disease for the majority of ophthalmologists and physicians, probably being underestimated due to the misleading clinical symptoms and the fact that most clinical reports have been published in Chinese, Russian, or Korean and thus are not accessible to a good proportion of the international scientific community [42]. The prevalence of human thelaziosis has increased in China in the past two decades [42]. From 1990 to 2005, a total of 371 cases of thelaziosis by *T. callipaeda* in men were reported in China [42]. An epidemiological investigation of almost 200 human cases in China (the vast majority coming from the rural areas in poor socioeconomic settings and whose complete histories were available) revealed that most individuals were children of less than 6 years of age (64.2%) [42–44]. Recently, *T. callipaeda* infections in humans (Figure 65.5) were diagnosed for the first time in Europe, specifically in Italy and France [45]. In particular, four cases of human thelaziosis have been diagnosed in Cuneo (Piedmont region, Italy) in patients coming from an area of North-Western Italy and South-Eastern France [45], where the infection had been previously reported in dogs, cats, and foxes [25,30]. All these cases of human thelaziosis

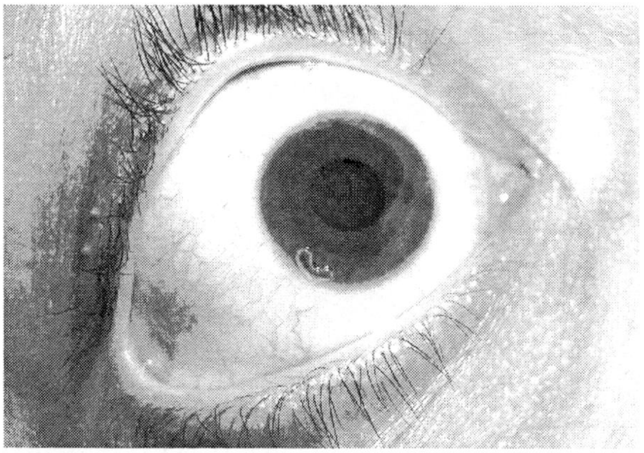

FIGURE 65.5 Eye of a man with a natural infection by *Thelazia callipaeda*. (From Shen, J.L. et al. J. *Parasitol*. 92, 186–190, 2006. With permission.)

were reported during summer (from June to August) [45] when *P. variegata* is most active [20].

65.1.3 Clinical Features and Pathogenesis

The pathogenesis of *T. callipaeda* infection has been experimentally investigated in rabbits by Wang and Yang [46], who conducted studies aiming to establish its clinical signs. Adult parasites transplanted into the eyes of rabbits induced an inflammatory reaction within 3–5 h, followed by swelling of the eyelids, congestion of the conjunctiva, and yellow secretions from affected eyes. In a subsequent study Wang et al. [19] showed that heavy experimental infection in rabbits (with 40–50 worms) may lead to blindness as a consequence of leukoma.

Symptoms of thelaziosis are mainly due to the lateral serration of *T. callipaeda* cuticle, which is responsible for mechanical damage of the conjunctival and corneal epithelium. In general, the most common clinical manifestations showed by human patients are exudative conjunctivitis, follicular hypertrophy of the conjunctiva, foreign body sensation, excessive lachrymation and ocular secretions, itchiness, congestion, swelling, hypersensitivity to light and keratitis, lasting from a few days to weeks before being referred for medical examination [42]. In most of the cases, patients complain of floating "filaments" on the eye surface [45]. The intensity of the symptoms depends on the number of nematodes that are present in the eye, their location, and the host response [43,46,47]. Sometimes, patients show redness of the eye, discharge and they might have decreased vision and "floaters" within the eye chamber [48]. Follicles, ulcer, nubecula, and occasionally even paralysis of the eye muscles, ectropion, and papilloma have also been reported [49]. The localization to anterior chamber, the vitreous body, or the retina may be associated with the development of fibrous tissue, inducing more serious clinical manifestations, like black spots in the visual field, ocular congestion, and sometimes purulent exudation under the anterior chamber [19,37]. The ocular lesions may even worsen if bacterial secondary infections occur, especially when *Pasteurella*, *Chlamydia*, and/or *Staphylococcus* spp. are involved [50]. Secondary infections are more frequent in infants, who rub their eyes complicating the primary problem as well as the diagnosis.

With regard to treatment and control of thelaziosis, adults and larvae may be removed mechanically by rinsing the conjunctival sac with sterilized saline and fluids or picking them out with fine forceps or cotton swabs. Prevention of human thelaziosis includes strategies aiming to control the fly populations. Use of bed nets to prevent the contact with the fly intermediate host could also be useful, as well as treatment of all infected domestic animals. The topic instillation of organophosphates [51] or moxidectin 1% [52] is highly efficacious against *T. callipaeda* in dogs. Imidacloprid 10% and moxidectin 2.5% in spot-on formulation showed to be effective for the control of dog thelaziosis within 5 (90.47%) to 9 (95.23%) days after treatment and it has been demonstrated to overcome problems linked to the mechanical removal of parasites or to the restraining of the animals for the local instillation of drugs in the eyes [53]. Recently, the administration of an injectable sustained-release formulation of moxidectin provided one season protection against *T. callipaeda* infection in dogs from an endemic area of Northern Italy [54]. The above-mentioned chemoprophylactic approach would more likely reduce the prevalence of dog thelaziosis, and therefore the risk for human infections in endemic areas.

65.1.4 Diagnosis

The diagnosis of thelaziosis relies on the detection of the nematodes in the eye and associated tissues of the infected host, and on their morphological and/or molecular identification.

65.1.4.1 Conventional Techniques

A thorough clinical and ophthalmological examination of infected patients reveals the presence of adult and/or larval nematodes located mostly in the conjunctival sac or in the medial or lateral canthus of the eye. Knowing the seasonality of human thelaziosis may help the correct etiological diagnosis of the disease, also considering that spring and summer are the seasons when allergic conjunctivitis (often caused by pollens) occurs most frequently, thus causing troubles for the differential diagnosis. This is particularly important in the early phases of the infection, when even though located in the conjuctival sac of the infected host, the larvae are difficult to be detected and identified (by the naked eye) due to their small size. In general, larval or adult specimens appear as thin, creamy-white threads in the conjunctival sac and they can be collected by flushing saline from the medial canthus. In the latter case, local anesthesia with 1% dicaine is recommended for detection and recovery of *T. callipaeda* [34].

The specific morphological identification of the parasite is central to the diagnosis and is based on key morphological features (see Section 65.1.1). Alternatively, the conventional diagnosis may be achieved microscopically by detecting the typical newborn L1s in eye secretions, which may be taken from medial canthus and diluted with a drop of saline.

Usually, the clinical diagnosis of thelaziosis is easy and may present some difficulties only when caused by a small number of nematodes, since clinical signs relate to an inflammatory response linked predominantly to the presence of developing L3s and/or fourth-stage larvae (L4s), similar to an allergic conjunctivitis [34].

65.1.4.2 Molecular Techniques

The relatively easy detection and morphological identification of *T. callipaeda* make molecular studies mainly useful for researches on the epidemiology and ecology of this nematode species, rather than for the specific diagnosis of the infection. The first molecular study on *T. callipaeda* was carried out to evaluate the interspecific and intraspecific differences in the sequence of the first internal transcribed spacer (ITS-1) of the ribosomal DNA (rDNA) among this species and other four common eyeworms [55]. This research showed that molecular data on the ITS-1 of *Thelazia* spp. (i.e., high AT content, variation in length within a species, presence of nucleotide repeats) are similar to the findings reported for other Spirurida and appear to be typical of this taxon. Since the intraspecific differences showed to be markedly lower than the interspecific variations, the ITS-1 was demonstrated to be a powerful genetic marker for the molecular identification of eyeworms and a useful tool for developing molecular epidemiological techniques to study *Thelazia* transmission patterns and prevalence in definitive and intermediate hosts [55]. This ribosomal target has indeed been used to investigate the vector of *T. callipaeda* under natural conditions [20].

A recent molecular investigation based on the analysis of DNA markers within the mitochondrial gene encoding for the cytochrome *c* oxidase subunit 1 (*cox1*), showed that significant genetic variability exists within *T. callipaeda* [56]. The study was carried out by both PCR-coupled sequencing and PCR-based single-strand conformation polymorphism (SSCP) of the most variable portion (v-*cox1*) of the *cox1* gene. Eight different haplotypes were determined for individual adult specimens of *T. callipaeda* from dogs, foxes, or cats from Italy, Germany, and the Netherlands, and from dogs from China and Korea. In particular, the results demonstrated the absence of genetic variation among individual nematodes from Europe, and the existence of seven different haplotypes within *T. callipaeda* isolates from Asia, with distinct genetic differences between the European and Asian specimens. The results of the molecular characterization of the nematodes collected in Europe were consistent with a structured population with low within-population diversity and with a genetic homogeneity within *T. callipaeda* from dogs, foxes, and cats from three countries in Europe (i.e., no affiliation to host species), while the small number of worms from disparate areas in Asia prevented detailed interpretations from the data. Nonetheless, the results for the SSCP analysis and sequencing were concordant indicating that the mutation scanning approach provides a useful tool for investigating the population genetics and molecular ecology of *T. callipaeda* [56].

65.2 Methods

In the following sections, a number of protocols that have been successfully used on *T. callipaeda* are described.

65.2.1 Sample Preparation

Specimens of *T. callipaeda* are washed extensively in physiological saline and then stored in 70% or absolute ethanol for both morphological and molecular investigations. For light microscopy observation, the nematodes are then fixed in Amman's solution.

65.2.2 Detection Procedures

65.2.2.1 Morphological Identification

Diagnostic morphological characters for adult specimens of *T. callipaeda* have been described and discussed in Otranto et al. [14]. On the whole, according to the morphological description of Yamaguti [57], eyeworms are characterized by (i) male: blunt tail, curved without caudal alae, pre- and postanal papillae, and dissimilar spicules (longer and shorter spicules); (ii) female: bluntly rounded tail with a pair of lateral papillae near its extremity and vulva in the esophageal region. All species are considered viviparous.

When found in human beings, *T. callipaeda* needs to be differentiated from *T. californiensis*. The differentiation of *T. callipaeda* and *T. californiensis* males is based on the presence of five pairs of postcloacal papillae: three ventral, one subventral, and one lateral assimilated to the phasmids [14] in *T. callipaeda* and only three pairs of postcloacal papillae in *T. californiensis* [58,59]. In *T. callipaeda* females, the vulva is located anteriorly to the esophagus–intestinal junction, thus being a diagnostic character with respect to *T. californiensis* females in which the vulva is located posteriorly to the esophagus–intestinal junction [58].

65.2.2.2 PCR Amplification

Genomic DNA from single nematodes is isolated using commercial kits and eluted in distilled water. The ribosomal ITS-1 (plus 18S and 5.8S flanking regions) is amplified using conserved primers, namely rDNA-A and rDNA-C in a total volume of 25 μL of a PCR mixture containing 25 mM $MgCl_2$, 1× buffer (100 mM Tris-HCl, pH 8.3, and 500 mM KCl), 1.25 mM of each dNTP, 50 pmol of each primer, and 1.25 U of polymerase.

Reactions are performed as follows: 94°C for 12 min, 30 cycles at 94°C for 30 s (denaturation)–58°C for 45 s (annealing)–72°C for 45 s (extension), with 7 min of extension at 72 °C in the final cycle [55].

The informative 689-bp-long region internal to the *cox1* gene is amplified using primers NTF and NTR. For SSCP analysis, the 5′ part of the *cox1* (v-*cox1*; first 307 bp) is amplified using NTF and the primer rv-*cox1*. For the PCRs, genomic DNA (4 μL) is added to the reaction mix (total of 50 μL), containing 2.5 mM $MgCl_2$, 10 mM Tris-HCl, pH 8.3, and 50 mM KCl, 250 μM of each dNTP, 50 pmol of each primer, and 1.25 U polymerase.

TABLE 65.1
List of PCR, PCR-SSCP, and Sequencing Primers for *Thelazia callipaeda*

Site	Direction	Name	Sequence (5′–3′)	Reference
18S	Forward	rDNA-A	AGGTGAACCTGCGGAAGGA	[55]
5.8S	Reverse	rDNA-C	CACATTAATTCTCGCAGCTAGC	[55]
cox1	Forward	NTF	TGATTGGTGGTTTTGGTAA	[56]
cox1	Reverse	NTR	ATAAGTACGAGTATCAATATC	[56]
cox1	Reverse	rv-*cox1*	ACATGCGTTTAACTGCTGTTAC	[56]

PCRs are performed using the following cycling protocol: 94°C for 12 min (polymerase activation), followed by 40 cycles of 95°C for 1 min (denaturation); 48°C for 1 min (annealing); 72°C for 1 min (extension), followed by 7 min at 72°C (final extension) [56].

The sequences of the primers used for the characterization of ribosomal and mitochondrial sequences of *T. callipaeda* are given in Table 65.1 and corresponding references [55,56].

65.2.2.3 Single-Strand Conformation Polymorphism Analysis

For the SSCP analysis of v-*cox1* amplicons individual products (in general 10 μL) are mixed with an equal volume of loading buffer (10 mM NaOH, 95% formamide, 0.05% of both bromophenol blue and xylene cyanole). After denaturation at 94°C for 15 min and snap cooling on a freeze block (−20°C), individual samples (12 μL) are loaded into the wells of a 96 × 261 mm precast GMA™ S-50 gel and subjected to electrophoresis for 17 h at 74 V and 7.4°C (constant) in a horizontal SEA2000™ apparatus connected to a cooling system. After electrophoresis, gels are stained for 20 min with ethidium bromide (10 mg/mL), destained in water for the same time, and then visualized upon ultraviolet transillumination [56,60].

65.3 Conclusions and Future Perspectives

Recent reports of thelaziosis in several carnivorous species in Europe (e.g., Italy, France, and Germany) [14,28,29,32] and in humans in Italy and France as well [45] have raised interest in this infection in both the veterinary and medical fields, suggesting that humans living in endemic areas may be at risk of this disease. In fact, the close relationship between infected animals and humans is considered to be central to the transmission of eyeworm infection in rural environments [61–63]. Human thelaziosis by *T. callipaeda* is still a neglected and underestimated disease, especially in European countries. Undoubtedly, the disease is better known under epidemiological and clinical standpoints in Asiatic areas, where the number of reported cases has increased in the last two decades reaching 371 reports by the 2005, in contrast to the small number of infections diagnosed in the last century [34,37,43,64]. However, recent cases of human thelaziosis described in Europe indicate that *T. callipaeda* has also to be considered among the eye parasitic infections of patients from the Old Continent [45].

Therefore, the eyeworm infection seems to be one of the several vector-borne parasitosis, emerging in the past few years as a more likely effect of the increase in the prevalence of its intermediate hosts (fly vectors) [67] and/or the increased awareness of the disease due to enhanced knowledge (through education campaigns and publications) and the improved diagnostic skills of physicians. Therefore, an integration of medical and veterinary expertise is necessary to improve the scientific knowledge to control this parasitosis and to increase the public health standards mainly in rural areas of Asia.

REFERENCES

1. Otranto, D., and Traversa, D. *Thelazia* eyeworm: An original "endo-ecto" parasitic nematode. *Trends Parasitol.* 21, 1, 2005.
2. Skrjabin, K.I., Sobolev, A.A., and Ivashkin, V.M. *Principles of Nematology Vol. IX. Spirurata of Animals and Man and the Disease Caused by Them. Part. 4: Thelazioidea*, Israel Program for Scientific Translation, pp. 1–54, 1967.
3. Soulsby, E.J.L. *Helminths, Arthropods and Protozoa of Domesticated Animals*, 7th edn. Bailliere Tindall, Oxford, UK, pp. 289–290, 1982.
4. Bhaibulaya, M., Prasertsilpa, S., and Vajrasthira, S. *Thelazia callipaeda* Railliet and Henry, 1910, in man and dog in Thailand. *Am. J. Trop. Med. Hyg.* 19, 476–479, 1970.
5. Hong, S.T., Lee, S.H., and Kim, S.I. A human case of *Thelazia callipaeda* infection with references to its internal structures, *Kor. J. Parasitol.* 2, 137–139, 1988.
6. Kosin, E., Kosman, M.L., and Depary, A.A. First case of human thelaziasis in Indonesia. *South Asian J. Trop. Med. Pub. Health* 20, 233–236, 1989.
7. Anderson, R.C. The life cycles of dipetalonematid Nematodes (Filarioidea, Dipetalonematidae): The problem of their evolution. *J. Helminthol.* 31, 203–224, 1957.
8. Casiraghi, M. et al. A phylogenetic analysis of filarial nematodes: Comparison with the phylogeny of *Wolbachia* endosymbionts. *Parasitology* 122, 93–103, 2001.
9. Casiraghi, M. et al. Mapping the presence of *Wolbachia pipientis* on the phylogeny of filarial nematodes: Evidence for symbiont loss during evolution. *Int. J. Parasitol.* 34, 191–203, 2004.
10. Traversa, D. et al. Mitochondrial cytochrome *c* oxidase subunit 1 (*cox1*) gene sequence of *Spirocerca lupi* (Nematoda, Spirurida) from different geographical areas: Avenues for potential implications. *Vet. Parasitol.* 146, 263–270, 2007.
11. Iorio, R. et al. Phylogenetic relationships of *Habronema microstoma* and *Habronema muscae* (Nematoda, Habronematoidea) within the order Spirurida inferred using mitochondrial

cytochrome *c* oxidase subunit 1 (*cox1*) gene analysis. *Parasitol. Res.* 104(5), 979–984, 2009.
12. Anderson, R.C. A study of *Filaria martis* Gmelin 1790, from *Martes foina* and *Pedetes caffer. Can. J. Zool.* 38, 157–167, 1960.
13. Chabaud, A.G., and Bain, O. The evolutionary expansion of the Spirurida. *Int. J. Parasitol.* 24, 1179–1201, 1994.
14. Otranto, D. et al. *Thelazia callipaeda* (Spirurida, Thelaziidae) of carnivores and humans: Morphological study by light and scanning electron microscopy. *Parassitologia* 45, 125–133, 2003.
15. Otranto, D. et al. Nematode biology and larval development of *Thelazia callipaeda* (Spirurida, Thelaziidae) in the drosophilid intermediate host in Europe and China. *Parasitology* 131, 847–855, 2005.
16. Shi, Y.E. et al. *Thelazia callipaeda* (Nematoda: Spirurida): Transmission by flies from dogs to children in Hubei, China. *Trans. R. Soc. Trop. Med. Hyg.* 82, 627, 1988.
17. Otranto, D. et al. *Musca domestica* is not a vector of *Thelazia callipaeda* in experimental or natural conditions. *Med. Vet. Entomol.* 19, 1–5, 2005.
18. Kozlov, D.P. Study on the biology of *Thelazia callipaeda*. Papers of Helminthologic Laboratory. *Academy of Sciences USSR* 142, 732–733, 1963.
19. Wang, Z.X., Shen, J.L., and Wang, K.C. Further studies on the life cycle if *Thelazia callipaeda* in China. *Chin. J. Parasitol. Paras. Dis.* 20, 328–331, 2002.
20. Otranto, D. et al. *Phortica variegata* is an intermediate host of *Thelazia callipaeda* under natural conditions: Evidence for pathogen transmission by a male arthropod vector. *Int. J. Parasitol.* 36, 1167–1173, 2006.
21. Otranto, D. et al. Biology of *Thelazia callipaeda* (Spirurida, Thelaziidae) eyeworms in naturally infected definitive hosts. *Parasitology* 129, 627–633, 2004.
22. Rossi, L. and Bertaglia, P. Presence of *Thelazia callipaeda* Raillet & Henry, 1910 in Piedmont, Italy. *Parassitologia* 31, 167–172, 1989.
23. Di Sacco, B., Ciocca, A., and Sirtori, G. *Thelazia callipaeda* (Railliet and Henry, 1910) nel sacco congiuntivale di un gatto di Milano. *Veterinaria* 4, 81–84, 1995.
24. Lia, R. et al. First report of *Thelazia callipaeda* in southern Italy- Basilicata region. *Proceedings of the Eight Conference of the European Federation of Parasitologist. Acta Parasitol.* 45, 178, 2000.
25. Otranto, D. et al. Current status and epidemiological observations of *Thelazia callipaeda* (Spirurida, Thelaziidae) in dogs, cats and foxes in Italy: A "coincidence" or a parasitic disease of the Old Continent? *Vet. Parasitol.* 116, 315–325, 2003.
26. Rossi, L. et al. *Thelazia callipaeda* in foxes from North-West Italy. *Parassitologia* 44, 159, 2002.
27. Otranto, D., Cantacessi, C., and Mallia, E. First report of *Thelazia callipaeda* (Spirurida, Thelaziidae) in wolves (*Canis lupus*) in Italy. *J. Wildl. Dis.* 43, 508–511, 2007.
28. Chermette, R., Guillot, J., and Bussiéras, J. Canine ocular thelaziosis in Europe. *Vet. Rec.* 154, 248, 2004.
29. Hermosilla, C., Herrmann, B., and Bauer, C. First case of *Thelazia callipaeda* infection in a dog in Germany. *Vet. Rec.* 154, 568–569, 2004.
30. Dorchies, Ph. et al. First reports of autochthonous eyeworm infection by *Thelazia callipaeda* (Spirurida, Thelaziidae) in dogs and cat from France. *Vet. Parasitol.* 149, 294–297, 2007.
31. Malacrida, F. et al. Emergence of canine ocular Thelaziosis caused by *Thelazia callipaeda* in southern Switzerland. *Vet. Parasitol.* 157(3–4), 321–327, 2008.
32. Otranto, D. et al. *Thelazia callipaeda* (Spirurida, Thelaziidae) in wild animals: Report of new host species and ecological implications. *Vet. Parasitol.* 166, 262–267, 2009.
33. Herr, J., Schley, L., and Roper, T.J. Socio-spatial organization of urban stone martens. *J. Zool.* 277, 54–62, 2009.
34. Shen, J.L. et al. Human thelaziosis—A neglected parasitic disease of the eye. *J. Parasitol.* 92, 186–190, 2006.
35. Stuckey, E.J. Circumocular filariasis. *Chin. Med. J.* 31, 24, 1917.
36. Miroshnichenko, V.A. et al. A case of ocular thelaziasis in a 3-year-old child. *Vestnik Oftalmologii* 104, 64, 1988 [In Russian].
37. Chen, C.C., and Zhang, Y.M. A case report of human thelaziosis in the eye anterior chamber. *Chin. J. Ophthalm.* 4, 466, 1954.
38. Min, S., Jae, R.Y., and Hyun Y.P. Enzooticity of the dogs, the reservoir host of *Thelazia callipaeda*, in Korea. *Kor. J. Parasitol.* 40, 101–103, 2002.
39. Koyama, Y. et al. Five cases of thelaziasis. *Br. J. Ophthalmol.* 84, 441, 2000.
40. Yospaiboon, Y. et al. Ocular thelaziasis in Thailand: A case report. *J. Med. Assoc. Thail.* 72, 469–473, 1989.
41. Wang, Z.X., Shen, J.L., and Du, J.S. Studies on the relationship between alternative generation of house dogs and control of human thelaziosis. *J. Trop. Dis. Parasitol.* 1, 204–207, 2003.
42. Shen, Y.P. et al. Human infections with *Thelazia callipaeda* in China. *Jiangsu Med.* 4, 41–42, 1979 [In Chinese].
43. Zhonghua, Y.Z. Analysis of 83 cases of human thelaziosis in China. *Zhonghua Yan Ke Za Zhi* (*Chin. J. Ophthalmol.*) 16, 64–67, 1980 [In Chinese].
44. Jiang, Z.X., Xu, L.Q., and Yu, L.Q. Human thelaziasis in China. *Zhongguo Ji Sheng Chong Bing Fang Zhi Za Zhi* (*Chin. J. Parasitic Dis. Control*) 4, 48–51, 1991 [In Chinese].
45. Otranto, D., and Dutto, M. Human thelaziasis, Europe. *Emerg. Infect. Dis.* 4, 647–649, 2008.
46. Wang, Z.X., and Yang, Z.X. Studies on the biology of *Thelazia callipaeda* and its experimental pathogenic observations on rabbits. *Zhongguo Ji Sheng Chong Xue Yu Ji Sheng Chong Bing Za Zhi* (*Chinese J. Parasitol. Parasitic Dis.*) 3, 128–130, 1985 [In Chinese].
47. Wang, K.C., Wang, Z.X., and Shen, J.L. Canine infection with *Thelazia callipaeda* and human thelaziosis. *J. Trop. Dis. Parasitol.* 28, 216–218, 1999 [In Chinese].
48. Zakir, R. et al. Intraocular infestation with the worm, *Thelazia callipaeda. Br. J. Ophthalmol.* 83, 1994, 1999.
49. Ohira, A. Five cases of thelaziasis. *Br. J. Ophthalmol.* 84, 439, 2000.
50. Smith, J. Bacterial conjunctivitis. *Clin. Evid.* 12, 926–932, 2004.
51. Rossi, L., and Peruccio, C. Thelaziosi oculare nel cane: Aspetti clinici e terapeutici. *Veterinaria*, 2, 47–50, 1989 [In Italian, with abstract in English].
52. Lia, R.P. et al. Field efficacy of moxidectin 1 per cent against *Thelazia callipaeda* in naturally infected dogs. *Vet. Rec.* 154(5), 143–145, 2004.
53. Bianciardi, P., and Otranto, D. Treatment of dog thelaziosis caused by *Thelazia callipaeda* (Spirurida, Thelaziidae) using a topical formulation of imidacloprid 10% and moxidectin 2.5%. *Vet. Parasitol.* 129(1–2), 89–93, 2005.
54. Rossi, L. et al. Use of sustained-release moxidectin to prevent eyeworm (*Thelazia callipaeda*) infection in dogs. *Vet. Rec.* 161, 820–821, 2007.
55. Otranto, D., and Traversa, D. Molecular characterization of the first internal transcribed spacer of ribosomal DNA of the most common species of eyeworms (Thelazioidea: *Thelazia*). *J. Parasitol.* 90, 185–188, 2004.

56. Otranto, D. et al. Analysis of genetic variability within *Thelazia callipaeda* (Nematoda: Thelazioidea) from Europe and Asia by sequencing and mutation scanning of mitochondrial cytochrome *c* oxidase subunit 1. *Mol. Cell. Probes* 19, 306–313, 2005.
57. Yamaguti, S. *The Nematodes of Vertebrates Part I. Systema Helminthum. Vol. III.* Interscience Publisher, Inc., New York, USA, pp. 634–636, 1961.
58. Price, E.W. A new nematode parasitic in the eyes of dogs in the United States. *J. Parasitol.* 17, 112–113, 1930.
59. Price, E.W. A note on the occurrences of eye worms in dogs in the United States. *North. Am. Vet.* 41, 49–58, 1931.
60. Gasser, R.B. et al. Single-strand conformation polymorphism (SSCP) for the analysis of genetic variation. *Nat. Prot.* 1, 3121–3128, 2006.
61. Okada, T. Bleeding sap preference of the Drosophilid flies. *Jpn. J. Appl. Entomol. Zool.* 3, 216–229, 1962.
62. Kimura, M.T. et al. Breeding sites of Drosophilid flies in and near Sapporo, Northern Japan, with supplementary notes on adult feeding habits. *Kontyû* 45, 571–582, 1977.
63. Máca, J. Revision of Palearctic species of *Amiota* subg. *Phortica* (Diptera, Drosophilidae). *Acta Entomol. Bohemoslovaca* 74, 115–130, 1977.
64. Jin, D. Parasites of human eyes in China. *Zhonghua Yan Ke Za Zhi* (*Chin. J. Ophthalmol.*) 7, 227, 1957 [In Chinese].
65. Otranto, D., Brianti, E., Cantacessi, C., Lia, R. P., and Máca, J. The zoophilic fruitfly Phortica variegata: Morphology, ecology and biological niche. *Medical and Veterinary Entomology* 20, 358–364, 2006.

66 Toxocara

Anna Borecka

CONTENTS

66.1 INTRODUCTION

66.1.1 Classification, Morphology, and Epidemiology

Nematodes from *Toxocara* genus belong to the Ascarididae family. Twenty-eight species rank to *Toxocara* genus, four of them: *T. canis* (from canids), *T. cati* (from felids), *T. vitulorum* (from ruminants), and *T. pteropodis* (from bats) are the etiological agents of human toxocariasis [1–4]. Other species are infective to canidae (*T. marginata*), felidae (*T. canarisi, T. genettae, T. lyncis, T. malaysiensis, T. mystax*), rodents (*T. apodemi, T. mackerrasae*), and another carnivora (*T. alienata, T. anakumae, T. herpestum, T. paradoxura, T. sprenti, T. suricattae, T. tanuki, T. vajrasthirae, T. vincenti*).

Toxocara adult worms are large parasites, and the female is bigger than the male. The adult female of *T. canis* is 12–18 cm, and the adult male is 10–12 cm in length; while the adult female of *T. cati* is up to 12 cm and the adult male is 3–7 cm in length. The infection is typical for young animals (up to six months old), and almost 100% of puppies are born infected.

The life cycle of *T. canis* is simple and involves direct or via paratenic host transmission. Adult worms live in the small intestine of their hosts, producing up to 200,000 eggs/day [5], which are passed unembryonated with feces. Excreted eggs require a period of time in the environment to develop into infectivity. At a temperature of 25–30°C eggs become infective after 2 weeks, but at a temperature of 12–18°C they need 8 weeks for development. Larvae molt twice inside the eggs to become the infective third-stage larvae (L_3) [6]. In young animals, infective ova hatch in the host's intestine, penetrate through the intestinal wall and are transported via bloodstream to the liver and lungs, where they molt into L_4. The L_4 finally reach, via the trachea, the intestine, and matures. For this reason, a large majority of puppies and kittens are born infected with *Toxocara* spp. In adult dogs and cats, paratenic hosts (rodents) or accidental hosts (humans), larval development progresses no further than the infective stage (L_3). During pregnancy the dormant larvae are activated by bitch hormones, reenter the circulatory system, are carried to the placenta, and penetrate into the fetus, which thus becomes prenatally infected. Infection with the mother's milk is common for *T. canis* and *T. cati*.

Red foxes (*Vulpes vulpes*) are also *T. canis'* main hosts. They visit domiciles in rural and urban areas and spread *Toxocara* ova. The highest reported prevalence rate of *T. canis* in foxes in European countries was detected in Great Britain (61.6%) and in the Netherlands (73.3%) [7,8].

The main source of infection with *Toxocara* spp for specific, paratenic, or accidental (human) host is infective eggs compound in natural environment. Hosts become infected following the ingestion of infective eggs-contaminated food (fruits, vegetables) or water. The occurrence of *Toxocara* eggs in urban and rural areas were investigated by many authors. The highest soil contamination rates were observed in urban settlements where backyards (45% Czech Republic, 26.4% Poland), family gardens (25% Poland, 38% Ireland), parks (67% Spain), playgrounds (54.5% Nigeria), and sandpits (92.3% Japan, 87.1% German) were found to be heavily contaminated [9–15]. The accumulation of population of stray and household dogs and cats is the underlying reason for this situation. Fillaux et al. [16] mentioned three exposure parameters interacting with transmission of human toxocariasis, namely rural residence (common dog ownership and poor hygiene), socio-economic status (poor sanitation level in disadvantage areas, e.g., lack of sewage system), and favorable climatic condition for embryonation and survival of *Toxocara* eggs (humid and warm climate).

Additionally geophagia (soil eating) or onychophagia (nail biting) increase the risk of toxocariasis, especially in children living in homes with puppies that have not been dewormed. Also consumption of raw fruits or vegetables grown in contaminated kitchen gardens may result in chronic low-dose infections [17]. Studies on helminthological quality of vegetables (lettuce, parsley, green onion, cucumber) in Ankara (Turkey) performed by Kozan et al. [18] showed presence of *Toxocara* spp eggs on 1.5% unwashed samples. Nematode eggs detected on the vegetable surface probably originated from soil contaminated with animal feces, utilizing wastewater or contaminated compost for plants treatment. Gaspard et al. [19] detected ascarid eggs in compost samples, of which *Toxocara* eggs accounted for 43%.

It was also proved that contaminated mango fruits were the source of *T. pteropodis* eggs responsible for an outbreak of acute hepatitis/gastroenteritis among the children of Palm Island [20].

Direct contact between humans and infected animals is also hazardous because *T. canis* eggs were found in the hair of 25% of 60 dogs examined. The maximum densities of the embryonating and embryonated eggs were 180 and 20 eggs/g of hair, respectively, which were much higher than the densities reported for soil samples [21].

Zoonotic toxocariasis is associated with consumption of raw liver from potential paratenic hosts, such chickens, ducks, pigs, cattles, rabbits [22–24]. Transmission following dispersion of ova by filth flies should also be included. *Toxocara* ova have been found in nearly 2% of wild-caught naturally infected flies in Nigeria [25].

Toxocariasis is a worldwide infection. Seroprevalence surveys in Western countries found that 2–5% of adults from urban areas had a positive result compared with 14.2–37% of adults in rural areas [26]. In tropical countries the seroprevalence of *Toxocara* infection was higher, namely 63.2% in Bali [27], and 92.8% in La Reunion (French Overseas Territories, Indian Ocean) [28].

66.1.2 Clinical Features and Pathogenesis

Toxocariasis was recognized as a zoonotic disease about 60 years ago and described as clinical condition of visceral larva migrans (VLM) in children [1]. The spectrum of clinical manifestation of toxocariasis varies widely from asymptomatic cases to clinical forms. The manifestation and clinical course are determined by the size of the inoculum, frequency of reinfections, localization of *Toxocara* larvae, and the host immune status. In 2001, Pawłowski [29] proposed a new scheme of clarifying clinical forms of human toxocariasis. The proposed general classification was a compromise between an observed clinical status and the location of *Toxocara* larvae. The classification divides human toxocariasis into four major forms namely: systematic (classical VLM, incomplete visceral larva migrans—iVLM); compartmentalized (ocular larva migrans—OLM, neurological toxocariasis—NLM); covert toxocariasis (CT), and asymptomatic. The classic VLM is a systemic form of toxocariasis, characterized by high eosinophilia, hepatosplenomegaly, fever, hypergammaglobulinemia, and pulmonary involvement. Incidents of VLM are uncommon and occur almost always in young children. More common is VLM, in which only some symptoms of classic iVLM syndrome may appear.

OLM and NLM forms are categorized separately from other syndromes on the basis that a specific organ (eye or brain) is the final site of larval migration. The story of OLM began from Wilder [30], who found nematode larvae in 24 of the 46 eyes, removed because of a diagnosis of retinoblastoma. Later Nichols [31] identified the larvae as *T. canis* in section from four out of five of the eyes examined by Wilder [30]. OLM typically occurs unilaterally in children and young adults. The common symptom is visual loss, uveitis, endophthalmitis, papillitis, retinal granulomatous lesions, and inflammatory masses in the peripheral vitreous. Ocular infection may also be subclinical and is only detected during a routine eye examination.

Neurotoxocariasis is a very rare form of toxocariasis. Magnaval et al. [32], after reviewing the English literature from the early 1950s to 1997, reported only 12 published cases of neurological toxocariasis as determined by finding *Toxocara* larvae in the cerebrospinal fluid (CSF) or in the brain and/or by immunodiagnosis of the CSF. In any case, migration of parasites in the human brain does not frequently induce recognizable neurological symptoms or signs.

CT is the most common form of toxocariasis. It is characterized by nonspecific symptoms and signs, which do not fall into the categories of VLM, iVLM, OLM, or NLM. CT is an organ-oriented immunopathological host response to stimulation of the host immune system by parasite antigens. The symptoms of CT form include pulmonary involvement, such as asthma, acute bronchitis, pulmonitis, with or without a Loeffler syndrome, dermatological disorders such as chronic urticaria or eczema, lymphadenopathy, myositis, and pseudorheumatic syndrome such as arthralgia [29].

Asymptomatic toxocariasis is diagnosed by positive serology, which is characteristic for light or old infections, may be accompanied by eosinophilia but most frequently is not [33].

66.1.3 Diagnosis

66.1.3.1 Conventional Techniques

66.1.3.1.1 Radiology

Various imaging studies are applied to confirm a diagnosis of toxocariasis. Imaging appearance depends on the stage or activity of the underlying disease. Abdominal ultrasound revealed multiple hypoechoic areas in the livers of 15 patients from 18 analyzed (83.3%) who presented hepatomegaly (72.7%) and splenomegaly (50%) [34]. Vidal et al. [35] presented a case of a healthy two-year-old boy who was admitted after 24 h of severe neurological symptoms with marked eosinophilic pleocytosis in the CSF and a single subcortical lesion detected by cerebral magnetic resonance imaging. Ultrasound was useful in examination of a 10-year-old girl with OLM and revealed an elevated mass. Enzyme-linked immunosorbent assay (ELISA) result for *T. canis* was positive. Further history revealed contact with puppies. The ELISA ruled out tumor [36].

66.1.3.1.2 Biopsy and Autopsy

A direct diagnosis of human toxocariasis can be made by identification of *Toxocara* larvae by biopsy. The method is rarely performed to confirm the presence of *Toxocara* larvae, but more often is used as an indirect method suggesting that a *Toxocara* infection may be responsible for disease present in a particular case. Hossack et al. [37] revealed the presence of eosinophilic granulomatous lesions with central necrosis in case of adult hepatic toxocariasis, but a diagnosis of VLM form of toxocariasis was confirmed by the detection of elevated levels of antibodies against L_2 *Toxocara* stage in the serum using an ELISA. In OLM cases ELISA on aqueous or vitreous biopsy specimens may facilitate the diagnosis. The presence of eosinophils in aqueous or vitreous biopsy specimens also suggests the diagnosis of toxocariasis.

66.1.3.1.3 Peripheral Blood Eosinophilia

The term eosinophilia is defined as an increase in peripheral blood eosinophilic leukocytes to >600 cells/μL of blood. Emphasis is placed on the number of eosinophils circulating in the peripheral blood, although an increase in eosinophils can be observed in other body fluids and many body tissues. In toxocariasis eosinophilia is connected with VLM [1]. In patients with OLM, this sign is often absent, probably due to low larval burden.

66.1.3.1.4 Eosinophil Cationic Protein

Eosinophil cationic protein (ECP) is one of the mediators released during eosinophil activation. In an attempt to distinguish between active and past toxocara infection, the subject's level of ECP was assessed by Magnaval et al. [38]. An elevated ECP level was significantly associated with both cough and rhinitis in patients with diagnosed toxocariasis.

66.1.3.1.5 Enzyme-Linked Immunosorbent Assay

ELISA is a biochemical technique used mainly in immunology to detect the presence of an antibody or an antigen in a sample. In ELISA an unknown amount of antigen is affixed to a surface, and then a specific antibody is added over the surface so that it can bind to the antigen. This antibody is linked to an enzyme, and in the final step a substance is added so that the enzyme can convert to some detectable signal.

Toxocariasis is one of the few human parasitic diseases whose serodiagnosis uses a standardized antigen (i.e., TES). This standardization assists in determining the global occurrence of disease, and infection or exposure, in the absence of relevant clinical data. The indirect TES–ELISA for antibody detection is currently the most common diagnostic method and has increased specificity over previously used tests [39].

De Savigny [40] was the first to describe the *in vitro* production of TES antigen (TES-Ag) for use in serodiagnostic tests. These antigens were analyzed and physico-chemical properties described. Fifteen bands between 29 and 94 kDa were detected by silver stain, all of them were recognized by antibodies in serum of a patient with VLM [41]. TES-Ag was divided into two groups: low-molecular-weight (LMW) (>35 kDa) and higher-molecular-weight (HMW). Cross reactivity between the HMW components of TES-Ag with antibodies from other nematode infection (ascaridiasis, filariasis, strongyloidiasis) occurs with native, unfractionated TES-Ag, whereas LMW TES-Ag displays high specificity for *Toxocara* infection [42]. The generation of two recombinant TES-Ag (TES-30 and TES-120) further aids the serodiagnosis of human toxocariasis [43,44].

While diagnostic ELISA based on the detection of IgE or IgM have been described, IgG remains the antibody isotype of choice, because of its abundance in serum and its increased avidity and affinity.

Obwaller et al. [45] analyzed the expression of specific IgE an IgG subclasses (IgG1-4) in patients with symptomatic (VLM, OLM) and asymptomatic course of infection. TES-Ag-specific IgG subclass reactivity was similar in symptomatic and asymptomatic VLM cases (IgG1 > IgG2 > IgG4 > IgG3). Subclass IgG1, 2, 4 showed significant differences between patients with VLM-associated symptoms and asymptomatic persons ($P < 0.001$), but not between patients with OLM-associated symptoms and asymptomatic. Significantly elevated levels of IgE/anti-IgE immune complexes were detected in sera of patients with symptomatic course of the disease, both VLM and OLM ($P < 0.001$).

Serodiagnosis of OLM, using the IgG TES–ELISA is more challenging than serodiagnosis of VLM and CT since serum antibody levels are usually low or undetectable. When OLM is suspected, aqueous or vitreous fluids offer the best diagnostic options.

66.1.3.1.6 IgG Antibody Avidity

The IgG avidity test was developed to help discriminate between past and recently acquired infection. Following antigenic challenge the IgG antibodies produced initially bind weakly to the antigen (low avidity). The avidity increases with time after antibody challenge (high avidity). Hübner et al. [46] developed method to distinguish acute and chronic *Toxocara* infection. This method detected the antibodies for months or even years after infection. The index of avidity was calculated as the ratio of IgG values in sera treated with urea and the value of IgG in nontreated sera, multiplied by 100. An index up to 40 is considered as low avidity, which means freshly acquired infection (36–40 borderline), and >40 is high avidity. In the group of 1376 patients only 5.09% low avidities were found. It means that predominantly patients in the chronic stage of infection attend examination.

66.1.3.1.7 Western Blotting

Western blotting (WB) is a technique used to identify and localize proteins based on their ability to bind to specific antibodies. WB analysis can detect protein of interest from a mixture of a great number of proteins. Besides providing information about the size of protein (in comparison with a size marker or ladder in kDa), WB also shows the relative level of protein expressed (in comparison with a control such as untreated sample or another cell type or tissue). Magnaval et al. [42] demonstrated that a positive TES-ELISA can be confirmed by WB which was sensitive as TES-ELISA, and specific. The WB assay performed on positive sera (VLM cases) revealed a typical pattern of seven bands divided into two groups, the first of which included the LMW fractions (24, 28, 30, and 35 kDa) and other, the HMW fractions (132, 147, and 20 kDa). The WB analysis using aqueous fluid from OLM patient showed the same seven-band pattern.

66.1.3.1.8 Radioimmunoassay

Radioimmunoassay (RIA) is a very sensitive technique used to measure concentrations of antigens without the need to use a bioassay. The RIA technique requires special precautions, since radioactive substances are involved. Therefore, it has been largely supplanted by the ELISA method, where the antigen–antibody reaction is measured using colorimetric signals instead of a radioactive signal. Park et al. [47] reported five cases of ocular toxocariasis in Korean adults. The total IgE antibodies levels in the patient sera were measured by IgE RIA. The serum levels of total IgE antibodies were elevated in three of five cases.

66.1.3.1.9 Radioallergosorbent Test

Radioallergosorbent test (RAST) is a blood test used to determine to what substances a person is allergic. This is different from a skin allergy test, and is more sensitive, specific, and simpler to use. A RAST test, using a person's extracted blood, detects the amount of IgE that reacts specifically with suspected or known allergens. RAST has been evaluated for serological diagnosis of toxocariasis in Italy and resulted positive in 4/5 cases of visceral form of toxocariasis [48].

66.1.3.1.10 Paper Radioimmunosorbent Test

Paper radioimmunosorbent test (PRIST) is most commonly used for measuring total IgE. This test is a double-antibody sandwich-type RIA in which the IgE is the antigen. Serum IgE levels may be increased in persons with parasitic infections. Smith et al. [49] reported that PRIST is sensitive and reproducible when used with ES antigen of *T. canis* second-stage larvae. Sera from patients with toxocariasis produced counts of up to 4–6-great as controls.

66.1.3.1.11 Counter-Immuno-Electrophoresis

Counter-immuno-electrophoresis (CIEP) is immunoelectrophoresis in which the antigen and antibody migrate in opposite directions. Twenty-five suspected cases of visceral toxocariasis were tested in the CIEP test, where adult *Toxocara* worm antigen was used. Twenty-four percent of analyzed patients were positive [50].

66.1.3.1.12 Bentonite Flocculation Test

Bentonite flocculation test (BF) is an agglutination technique where antigen adsorbed on particles of bentonite, when the antigen is added to serum containing antibodies flocculation occurs. BF was evaluated using sera sent to the Center for Disease Control from patients with a presumptive diagnosis of VLM. Patients having 5–6 of the clinical or laboratory criteria for VLM were designated as cases while those with 0–2 criteria served as controls. The sensitivity of the BF was 25.8%, compared to 78.3% for ELISA. Based on these findings, the BF appears not to be the serodiagnostic method of choice for VLM [51].

66.1.3.1.13 Ouchterlony Double Immuno Diffusion

In this technique a gel plate is cut to form a series of holes in the gel. A parasite's antigen is placed in the center well, whereas the patient's serum is placed in one (or more) of the outer wells and the plate left for 48 h to develop. During this time the antigens and antibodies migrate in a radial manner out from its wells. When the two antigen fronts meet, if there are antibodies in the patient serum to any of the antigens, they will create an immune complex. This immune complex precipitates in the gel to give a thin white line. Sera from 12 of the 17 patients with suspected toxocariasis were judge positive by the Ouchterlony test employing somatic antigens from adult *T. canis* and *T. cati* worms. The frequency of the two *Toxocara* species in the 12 patients was comparable [52].

66.1.3.1.14 Hemagglutination Test

Hemagglutination (HA) is a specific form of agglutination that involves red blood cells. The rationale for this test is that *T. canis* larvae contain components of human blood group substances that may produce a response in some patients infected with these larvae. Huntley et al. [53] studied a group of 51 patients thought to have VLM and compared them with a group of controls. In the infected patients group there was a hyperglobulinemia with significant elevation of the gamma globulins relative to the controls, while the HA test was positive to *Toxocara* antigens in only 30%. In another study the same author [53] demonstrated a significant level of anti-gamma globulin factor in 26 of 59 children during the acute phase of their disease. They also found anti-A and anti-B titers to be higher in a group of patients with VLM as compared to controls.

66.1.3.1.15 Indirect Hemagglutination Test

In Indirect hemagglutination test (IHA) (passive agglutination), red blood cells are used to adsorb soluble antigen onto their surfaces. The red blood cells then agglutinate in the presence of antiserum specific for the adsorbed antigen. In a study of 237 patients by Krupp [54] using the IHA with antigens from larvae and adult worms, antibodies to *T. canis* were found in one-third of patients believed to have VLM.

66.1.3.1.16 Indirect Fluorescent Antibody Test

Indirect fluorescent antibody test (IFAT) enables the detection of antibodies to specific antigenic material using fluorescent microscopy. Bisseru and Woodruff [55] used IFAT to detect

Toxocara antibodies quantitatively in human sera. Larvae exposed to immune serum, followed by fluorescein-labeled antiglobulin gave a bright greenish-yellow fluorescence of the cuticle. The intensity of fluorescence was graded. Sera from patients with toxocariasis affect motility of the larvae after a few hours and the larvae survive about 24 h in *Toxocara* serum. The results obtained correlate well with those obtained with the toxocara skin test.

66.1.3.1.17 Larval Precipitation Test

Matthes and Buchwalder [56] performed larval precipitation test on different laboratory animals experimentally infected with *T. canis* for the serological evidence of specific antibodies. By using living *T. canis* larvae from infected animals as antigen prepatent *T. canis* infections in laboratory animals were serologically proved because of precipitates on oral and excretion porus. In this method, antibodies conjugated with fluorescein isothiocyanate can be used. Fluorescent antibodies simplify observation worms under microscope.

66.1.3.1.18 Skin Test

Woodruff [57] performed the skin test using antigen made from both larval and adult forms of *T. canis*. In a British survey of 485 healthy persons ~2% gave a positive reaction to a *Toxocara* skin test indicating that they probably had a *Toxocara* infection sometime in the past. Additionally, this author found that antigens made from adult were more sensitive indicators of *Toxocara* infection than antigens made from larvae.

66.1.3.2 Molecular Techniques

Conventional parasitological techniques have significant limitation with regard to identification of developmental stages (e.g., eggs, larvae) of *Toxocara* spp. Molecular methods provide a useful approach to the genetic characterization (irrespective of their developmental stage), and have implications for diagnosis, detailed studies of their life cycle, systematics, and epidemiology.

Various target regions, including repetitive DNA, mitochondrial DNA (mtDNA) and nuclear ribosomal DNA (rDNA), have been employed to achieve the identification of parasites to species level. In the 1990s, studies developed by many authors demonstrated that the first (ITS-1) and second (ITS-2) internal transcribed spacers (ITS) of rDNA provide reliable genetic markers for species of *Toxocara* and range of other nematodes. Molecular techniques are especially useful for phylogenetic studies of nematodes from *Toxocara* genus, for the investigation of environmental condition with *Toxocara* spp eggs, and also for the detection of *Toxocara* DNA in natural- or experimental-infected animals.

66.1.3.2.1 Molecular Characterization and Identification of Toxocara *spp*

Recent studies of *Toxocara* spp at the molecular level were dealing with polymorphism, genetic distance, intra- and interspecific difference and variability and molecular phylogenetics. Nadler [58] estimated of genetic variability from population samples of the *T. canis*, *T. cati*, and *Parascaris equorum*. He found a much higher-level (8 ± 5 ± 13 ± 7%) mean heterozygosity, and reported that 22 ± 39% of all loci tested showed some degree of allelic polymorphism across the population as a whole. These data led Nadler [58] to conclude that ascarids do display considerable genetic diversity. In 1990, Nadler [59] realized phylogenetic studies; based on full-length sequences 18S and 28S of rRNA for ascaridoid (including *T. canis* and *T. cati*) and created gene tree. In the same time Turčeková and Dubinský [60] found high interspecific differences between *T. canis* and *T. cati* nematodes after digestion their total DNA with *Bgl*I, *Pst*I, and *Hind*III restriction endonucleases.

When polymerase chain reaction (PCR) became popular many authors began constructing primers for *Toxocara* spp. Wu et al. [61] designed specific primers for PCR detection of genomic DNA from *T. canis* and *T. cati*. However, Jacobs and coworkers [62] created primers to regions of the ITS-2 sequences of the rDNA of *T. canis*, *T. cati*, and *Toxascaris leonina*. Representative ITS-2 products for each nematode species were identified as genetic markers in PCR-linked restriction fragment-length polymorphism (RFLP) assay. Using these primers, three species can be distinguished from each other and from other ascaridoids by the use of two endonucleases, *Hinf*I and *Rsa*I. An alternative approach to detection and identification of dog and cat nematodes was a direct PCR method with three specific primer sets constructed by the same authors [62].

Next research was devoted molecular characterization of *Toxocara* variant from cats in Malaysia [63]. This parasite was previously identified morphologically as *Toxocara* sp. cf. *canis*. The regions of the ITS-1 and ITS-2 of the rDNA, and 5.8S gene were amplified and sequenced. Obtained sequences from nematode from Malaysian cats were compared with those for *T. cati* and *T. canis*. The sequence data showed that this taxon was genetically more similar to *T. cati* than to *T. canis*, which is in contrast to the original reports suggesting that, morphologically *Toxocara* sp. cf. *canis* was the same species as *T. canis*. The PCR-linked RFLP and single-strand conformation polymorphism (SSCP) methods were employed for the unequivocal differentiation of *Toxocara* sp. cf. *canis*, *T. canis* and *T. cati*. Results confirmed the presence of a separate species to *T. canis* and *T. cati*- *T. malaysiensis*, and demonstrating the existence of a cryptic (genetic distinct but morphologically similar) species of *Toxocara* in Malaysian cats.

PCR could be also used to identify of parasite isolates from different hosts. Epe et al. [64] showed similarity of *T. canis* isolates from red foxes and dogs. The random amplification of polymorphic DNA-PCR showed a similarity coefficient of 0.95, similar to the range of intraspecific variation in *T. cati* and *T. leonina* specimens as outgroups. Therefore, the study demonstrated that the red fox and the dog isolate of *T. canis* were identical in molecular structure of the conserved ITS-2 region. The sequence comparison of both isolates showed an identity of 100%.

Also Li et al. [65] utilized the PCR-SSCP method and sequence analyses of the ITS-2 of rDNA to show that *T. canis*,

T. cati, *T. malaysiensis*, and *T. leonina* from China are genetically the same as those from other geographical origins.

In 2008, Jex et al. [66] amplified mitochondrial genome of *T. canis* by long-range PCR (long PCR) and sequenced using a primer-walking strategy. In their opinion the mitochondrial genome of *T. canis* provides genetic markers due to their high mutation rates and will be useful for systematic, population, genetic, and epidemiology studies. This circular mitochondrial genome can be used in prospecting for cryptic species and for exploring host specificity and zoonotic potential, prevention and control of toxocariasis in humans and other hosts.

At the same time mt genome sequences for three *Toxocara* species—*T. canis*, *T. cati*, and *T. malaysiensis*, have been described [67]. Additionally, phylogenetic analyses based on concatenated amino acid sequences of 12 protein-coding genes revealed that the newly described species *T. malaysiensis* was more closely related to *T. cati* than to *T. canis*, consistent with results of a previous study using sequences of ITS-2 as genetic markers.

Morphological observation is the traditional approach for the identification of dogs and cats nematodes, but this approach can have limitation in the differentiation of *T. malaysiensis* from the other *Toxocara* spp and determination of specific identity of ascaridoid larvae recovered from animal or human tissues. Therefore, the established specific PCR assays should provide useful tools for the diagnosis and epidemiological investigations of nematodes from *Toxocara* genus.

66.1.3.2.2 *PCR Detection and Differentiation of* Toxocara *spp Eggs in Environmental Samples*

Many studies have been undertaken worldwide to determine the prevalence of *Toxocara* spp eggs in the soil of public places, because of their medical importance. High level of the soil contamination with *Toxocara* eggs in pairs with their considerable vivacity cause significant risk of infection for animals and humans.

Identification of closely related taxa such as *T. canis* and *T. cati* is possible only by applying molecular techniques, because they are similar both in size and morphology. Until 2004, differentiation of *Toxocara* eggs isolated from the soil was conducted using morphometric methods [68], but the obtained results were not credible. Adapting molecular methods, for example, PCR or PCR-RFLP, to the delineation of *Toxocara* eggs has increased the robustness of results.

For the first time Borecka [69] used PCR-linked RFLP method for differentiation of *Toxocara* spp eggs isolated from the soil samples to a species level (*T. canis* or *T. cati*). Applied in that investigation primers for ITS-2 of the rDNA of *T. canis* and *T. cati* were designed by Jacobs et al. [62]. However, the greatest problem encountered was connected with the extraction of embryos from eggs. *Toxocara* spp, like other geohelminths, have eggs with thick and complex layers, which protect them from the influence of environmental factors. The application of low temperatures (–70°C), boiling and incubation with proteinase K did not lead to disintegration of egg shells. For this reason, embryonic material was removed from eggs by needle [69].

Other authors [70] crushed eggs by applying pressure to coverslip on a microscope slide with the fingers. The both methods were time consuming and need experienced technician, for that reason they were not suitable for routine investigations. Modification was required. In 2008, Borecka and Gawor [71] proposed simplification and direct extraction of DNA from the soil samples, without necessity of *Toxocara* eggs isolation by floatation. In described study the modified protocol of a commercial kit was used. The sand samples were digested overnight with proteinase K until complete lysis was achieved. DNA extraction was confirmed by specific products of amplification of ITS-2 region of *Toxocara*. The application of DNA extraction directly from soil enables the examination of many samples in a short time and is useful for routine analysis of soil, sand, or sediments samples. Additionally elaborated method of gDNA isolation makes possible to examine small samples (2.5 g) with low level of contamination with *Toxocara* eggs.

66.1.3.2.3 *Molecular Detection and Identification of* Toxocara *Larvae Present in Animal Tissues*

The diagnosis of human toxocariasis often relies on the used of serological tests, but these tests only provide evidence of infection; however, they do not provide a means of differentiating closely related taxa, such as *T. canis* and *T. cati*.

The molecular techniques offer an excellent tool for the detection and identification of nematodes in human and animal tissues. The first report of practical application of PCR technique in the diagnosis of VLM in a mouse model came from 1997 [72]. Liver samples obtained from mice experimentally infected with *T. canis*, *T. cati*, and *Ascaris suum* were examined by double PCR using the primer SB12 and after electrophoresis clear-cut bands were observed. Similarly, Borecka et al. [73] adapted the PCR method for the detection of *T. canis* larvae in liver biopsy materials from laboratory-infected Mongolian gerbils. The gDNA of nematode was found in the liver of all infected animals at 3, 5, 7, and 14 days postinfection. Whereas Ishiwata et al. [74] performed study on naturally infected turkey. Utilization of the PCR amplification and sequencing of ITS-2 regions of rDNA has demonstrated the presence of *T. canis* larvae embedded in the livers. The results of molecular techniques affirmed the results of microscopy examination, when nematode larvae were detected in the lesions of four out of five turkey livers. The results of preliminary studies indicate that the PCR method has a potential as a supporting technique for diagnosis of human toxocariasis.

66.2 METHODS

66.2.1 SAMPLE PREPARATION

Sample preparation constitutes an essential stage in the molecular detection and identification of *Toxocara* spp in environmental samples or in animal tissues base on PCR technique.

66.2.1.1 *Toxocara* spp Eggs in Environmental Samples

Molecular method for routine examination of soil, sand, or sediments for the presence of *Toxocara* spp eggs were described by Borecka and Gawor [71]. The samples are collected from sandpits supposed to be contaminated with *Toxocara* eggs. The *Toxocara* DNA isolation is performed using a commercial tissue kit (Eurx, Poland). A measure of 2.5 g of sand from the sample is taken for molecular testing. The sand samples are incubated with proteinase K overnight (16 h) until complete lysis is achieved. Next steps of DNA isolation are identical to those described in kit protocol. The extracted DNA can be used directly for PCR or can be stored at −20°C.

A previous study noted that extraction of embryos from eggs is most problematic [69]. Freezing in liquid nitrogen, boiling for 10 min, and incubation with proteinase K at 65°C for 2 h did not lead to the disintegration of eggshells. For this reason, embryonic material was removed from the eggs using a needle [69]. Other authors [70] crushed eggs under a coverslip. Another difficulty is the presence of organic and inorganic compounds in soil, which inhibit the amplification of nucleic acids by PCR. They can interfere with the reaction at several levels, leading to different degrees of attenuation and even to complete inhibition [75]. The use of enzymatic lysis (proteinase K), bead beating, or sonication can inactivate these inhibitors [76].

66.2.1.2 *Toxocara* Larvae Present in Animal Tissues

PCR has been used to detect *Toxocara* DNA in the liver of experimental-infected mice or gerbils, but not yet for detecting *Toxocara* DNA in human biopsy material. The study of Borecka and Gawor [71] was undertaken to adapt the PCR technique for the detection of *T. canis* larvae in the liver of experimentally infected Mongolian gerbils (*Meriones unguiculatus*), used as an animal model for human toxocariasis.

The animals were inoculated with 1000 eggs of *T. canis* by a stomach tube under light anesthesia. At 3, 5, 7, and 14 days postinfection, the gerbils were killed and the livers were removed for molecular analysis. The isolation was performed using commercial tissue kit (Eurx, Poland). A measure of 25 mg of the liver were homogenized and mixed with buffers and proteinase K solution was added. The samples were incubated at 56°C until complete lysis was obtained (3 h). In the next steps authors proceed in accordance with the kit protocol. The extracted DNA was used for PCR. The obtained results indicate that the PCR method has potential to be used for the detection of *T. canis* and *T. cati* larvae in humans, in particular in ocular cases difficult for diagnosis.

66.2.2 Detection Procedures

DNA extracted from environmental samples or from biopsy material is used for PCR amplification of ITS-2 as described by Jacobs et al. [62]. The region spanning ITS-2 is amplified from the gDNA (10–15 ng) by PCR using oligonucleotide species-specific primers: Tcan1 (5′-AGTATGATGGGCGCGCCAAT-3′) and NC2 (5′-TAGTTTCTTTTCCTCCGCT-3′) for *T. canis*; and primers Tcat1 (5′-GGAGAAGTAAACTC-3′) and NC2 for *T. cati*. The Tcan1–NC2 primer set amplifies a PCR product of c. 380 bp from *T. canis*, and the Tcat1–NC2 set amplifies the product of 370 bp from *T. cati*.

PCR mixture (50 μL) is composed of 10 mM Tris–HCl pH 8.4, 50 mM KCl, 3.0 mM $MgCl_2$, 250 mM of each of dATP, dCTP, dGTP, dTTP; 100 pmol of each primer with 2 U Taq polymerase, and 1 μL (usually equal to 10–15 ng) of sample DNA. PCR mixture without DNA is included in each run to exclude "carry-over" contamination.

The cycling conditions comprise 30 cycles of 94°C for 30 s, 55°C for 30 s, and 72°C for 30 s. The obtained fragments are separated on 2.5% agarose gels containing ethidium bromide. The pUC 1444-80 marker (BTL, Poland) is used to estimate the size of the obtained fragments. Gels are transilluminated in UV light and photographed.

66.3 CONCLUSION AND FUTURE PERSPECTIVE

T. canis and *T. cati* are cosmopolitan gastrointestinal parasites of canids and felids, respectively. Females of these parasites shed resistant ova into environment, which embryonate and can remain infective, under suitable conditions, for many years. This environmental reservoir is a source of infection for final hosts and for paratenic hosts, including man. Human become infected after the ingestion of infective ova or, less frequently, larvae. Human toxocariasis can be divided into four clinical forms: VLM, OLM, CT, and asymptomatic toxocariasis. At present, the most effective laboratory diagnostic approach for confirming *Toxocara* as the etiological agent of disease in clinically suspected cases is by serology. The best serodiagnostic options are using the IgG TES–ELISA as a screening test (with confirmation by IgE TES–ELISA), but these tests only provide evidence of infection but do not provide a means of differentiating closely related taxa, such as *T. canis* and *T. cati*. PCR has been used to detect *Toxocara* DNA in the liver of experimental-infected mice and gerbils, but not yet for detection of parasite DNA in human biopsy material, because of the difficulty in acquiring samples. From an epidemiological and ecological perspective, it would be interesting to confirm, employing a DNA method, the involvement of *Toxocara* in human cases. It would also be crucial to establish whether specific genotypes of *T. canis* and *T. cati* have an affinity to the human host and/or predilection sites in tissues to cause different types of toxocariasis. PCR also offers a valuable means to surveying environmental contamination with *Toxocara* eggs, especially when there is a clear connection between the degree of soil contamination with *Toxocara* eggs and the prevalence of toxocariasis among people.

REFERENCES

1. Baever, P.C. et al. Chronic eosinophilia due to visceral larva migrans. *Pediatrics* 9, 7, 1952.
2. Fordham von Reyn et al. Infection of an infant with an adult. *Toxocara cati* (Nematoda). *J. Pediatr.* 93, 247, 1978.

3. Iddawela, D.R. et al. A seroepidemiological study of toxocariasis and risk factors for infection in children in Sri Lanka. *Southeast Asian J. Trop. Med. Public Health* 34, 7, 2003.
4. Prociv, P. Observations on the transmission and development of *Toxocara pteropodis* (Ascaridoidea: Nematoda) in the Australian grey-headed flying-fox, *Pteropus poliocephalus* (Pteropodidae: Megachiroptera). *Z. Parasitenkd.* 69, 773, 1983.
5. Glickman, L.T., Schantz, P.M., and Cypess, R.H. Canine and human toxocariasis: Review of transmission, pathogenesis and clinical disease. *J. Am. Vet. Med. Ass.* 175, 1265, 1979.
6. Brunanska, M., Dubinsky, P., and Reiterova, K. *Toxocara canis*: Ultrastructural aspects of larval moulting in the maturing eggs. *Int. J. Parasitol.* 25, 683, 1995.
7. Borgsteede, F.H.M. Heminth parasites of wild foxes (*Vulpes vulpes*) in the Netherlands. *Z. Parasitenkd.* 70, 527, 1984.
8. Smith, G.C. et al. Prevalence of zoonotic important parasites in the red fox (*Vulpes vulpes*) in Great Britain. *Vet. Parasitol.* 118, 133, 2003.
9. Dubná, S. et al. Contamination of soil with *Toxocara* eggs in urban (Praque) and rural areas in the Czech Republic. *Vet. Parasitol.* 144, 81, 2007.
10. Düwel, D. The prevalence of Toxocara eggs in the sand in children's playgrounds in Frankfurt/M. *Ann. Trop. Med. Parasitol.* 78, 633, 1984.
11. Emehelu, C.O., and Fakae, B.B. Prevalence of *Toxocara canis* ova on playgrounds of nursery schools in Nsukka, Nigeria. *Int. J. Zoonoses* 13, 158, 1986.
12. Gawor, J. et al. Environmental and personal risk factors for toxocariasis in children diagnosed disease in urban and rural areas of central Poland. *Vet. Parasitol.* 155, 217, 2008.
13. Holland, C. et al. Families, parks, gardens and toxocariasis. *Scand. J. Infect. Dis.* 23, 225, 1991.
14. Ruiz de Ybáñez, M.R., Garijo, M.M., and Alonso, F.D. Prevalence and viability of eggs of *Toxocara* spp. and *Toxascaris leonina* in public parks in eastern Spain. *J. Helminthol.* 75, 169, 2001.
15. Uga, S. Prevalence of *Toxocara* eggs and number of faecal deposits from dogs and cats in sandpits of public parks in Japan. *J. Helminthol.* 67, 78, 1993.
16. Fillaux, J. et al. Epidemiology of toxocariasis in a steppe environment: The Patagonia study. *Am. J. Trop. Med. Hyg.* 76, 1144, 2007.
17. Magnaval, J.-F. et al. Highlights of human toxocariasis. *Kor. J. Parasitol.* 39, 1, 2001.
18. Kozan, E. et al. Prevalence of helminth eggs on raw vegetables used for salads. *Food Contr.* 16, 239, 2005.
19. Gaspard, P., Wiart, J., and Schwartzbrod, J. Parasitological contamination of urban sludge used for agricultural purposes. *Waste Manag. Res.* 15, 429, 1997.
20. Moorhouse, D.E. Toxocariasis—A possible cause of the Palm Island mystery disease. *Med. J. Aust.* 1, 172, 1982.
21. Wolfe, A., and Wright, I.P. Human toxocariasis and direct contact with dogs. *Vet. Rec.* 152, 419, 2003.
22. Hoffmeister, B. et al. Cerebral toxocariasis after consumption of raw duck liver. *Am. J. Trop. Med. Hyg.* 76, 600, 2007.
23. Lee, K.-T. et al. Studies on the including possibility of human visceral larva migrans associated with eating habit of raw liver of domestic animals. *Kor. J. Parasitol.* 14, 51, 1976.
24. Stürchler, D. Weiss, N., and Gassner, M. Transmission of toxocariasis. *J. Infect. Dis.* 162, 571, 1990.
25. Umeche, N., and Mandah, L.E. Musca domestica as carrier of intestinal helminths in Calabor, Nigeria. *East Afr. Med. J.* 66, 349, 1989.
26. Magnaval, J.-F., Glickman, L.T., and Dorchies, Ph. La toxocarose, une zoonose helminthique majeure. *Rev. Med. Vet.* 145, 611, 1994).
27. Chomel, B.B. et al. Serosurvey of some major zoonotic infections in children and teenagers in Bali, Indonesia. *Southeast Asian J. Trop. Med. Public Health* 24, 321, 1993.
28. Magnaval, J.F. et al. Epidemiology of human toxocariasis in La Réunion. *Trans. R. Soc. Trop. Med. Hyg.* 88, 531, 1994.
29. Pawłowski, Z. Toxocariasis in humans: Clinical expression and treatment dilemma. *J. Helminthol.* 75, 299, 2001.
30. Wilder, H.C. Nematode endophthalmitis. *Trans. Am. Acad. Ophthalmol. Otolaryngol.* 55, 99, 1950.
31. Nichols, R.L. The etiology of visceral larva migrans. *J. Parasitol.* 42, 349, 1956.
32. Magnaval, J.F. et al. Human *Toxocara* infection of the central nervous system and neurological disorders: A case–control study. *Parasitology* 115, 537, 1997.
33. Bass, J.L. et al. Asymptomatic toxocariasis in children. *Clin. Pediatr.* 26, 441, 1987.
34. Baldisserotto, M. et al. Ultrasound findings in children with toxocariasis: Report on 18 cases. *Pediatr. Radiol.* 29, 316, 1999.
35. Vidal, J., Sztajnbok, J., and Seguro, A.C. Meningoencephalitis due to *Toxocara canis*: Case report and review of the literature. *Am. J. Trop. Med. Hyg.* 69, 341, 2003.
36. Kennedy, J.J., and Defeo, E. Ocular toxocariasis demonstrated by ultrasound. *Ann. Ophthalmol.* 13, 1357, 1981.
37. Hossack, J. et al. A case of adult hepatic toxocariasis. *Natl. Clin. Pract. Gastroenterol. Hepatol.* 5, 344, 2008.
38. Magnaval, J.-F. et al. Eosinophil cationic protein as a possible marker of active human *Toxocara* infection. *Allergy* 56, 1096, 2001.
39. Smith, H. et al. How common is human toxocariasis? Towards standardizing our knowledge. *Trends Parasitol.* 25, 182, 2009.
40. De Savigny, D.H. *In vitro* maintenance of *Toxocara canis* larvae and a simple method for the production of *Toxocara* ES antigen for use in serodiagnostic tests for visceral larva migrans. *J. Parasitol.* 61, 781, 1979.
41. Badley, J.E. et al. Analysis of *Toxocara canis* larval excretory–secretory antigens: Physicochemical characterization ant antibody recognition. *J. Parasitol.* 73, 593, 1987.
42. Magnaval, J.F. et al. Application of the western blotting procedure for immunodiagnosis of human toxocariasis. *Parasitol. Res.* 77, 697, 1991.
43. Fong, M.-Y., and Lau, Y.L. Recombinant expression of the larval excretory-secretory antigen TES-120 of *Toxocara canis* in the methylotrophic yeast *Pichia pastoris*. *Parasitol. Res.* 92, 173, 2004.
44. Yamasaki, H. et al. Development of a highly specific recombinant *Toxocara canis* second-stage larva excretory–secretory antigen for immunodiagnosis of human toxocariasis. *J. Clin. Microbiol.* 38, 1409, 2000.
45. Obwaller, A. et al. *Toxocara* infestations in humans: Symptomatic course of toxocarosis correlates significantly with levels of IgE/anti-IgE immune complexes. *Parasite Immunol.* 20, 311, 1998.
46. Hübner, J., Uhlikova, M., and Leissova, M. Diagnosis of the early phase of larval toxocariasis using IgG avidity. *Epidemiol. Mikrobiol. Immunol.* 50, 67, 2001.
47. Park, S.-P. et al. Five cases of ocular toxocariasis confirmed by serology. *Kor. J. Parasitol.* 38, 267, 2000.
48. Tinneli, M. et al. RAST and ELISA in three autochthonous parasitic infections in Italy: Toxocariasis, hydatidosis and fascioliasis. *Boll. Ist. Sieroter Milan* 66, 38, 1987.

49. Smith, H.V. et al. A paper radioimmunosorbent test (PRIST) for the detection of larva-specific antibodies to *Toxocara canis* in human sera. *J. Immunol. Methods* 37, 47, 1980.
50. Khalil, H.M. et al. Immunodiagnostic of visceral toxocarias. *J. Egypt. Soc. Parasitol.* 19, 381, 1989.
51. Glickaman, L.T. et al. Evaluation of serodiagnostic tests for visceral larva migrans. *Am. J. Trop. Med. Hyg.* 27, 492, 1978.
52. Nagakura, K. et al. Serologic differentiation between *Toxocara canis* and *Toxocara cati*. *J. Infect. Dis.* 162, 1227, 1980.
53. Huntley, C.C., Costas, M.C., and Lyerly, A. Visceral larva migrans syndrome: Clinical characteristics and immunologic studies in 51 patients. *Pediatrics* 36, 523, 1965.
54. Krupp, I.M. Hemagglutination test for the detection of antibodies specific for *Ascaris* and *Toxocara* antigens in patients with suspected visceral larva migrans. *Am. J. Trop. Med. Hyg.* 23, 378, 1974.
55. Bisseru, B., and Woodruff, A.W. The detection of circulating antibody in human toxocara infections using the indirect fluorescent antibody test. *J. Clin. Pathol.* 21, 449, 1968.
56. Matthes, H.F., and Buchwalder, R. The larva precipitation test with *Ascaris suum* and *Toxocara canis* larvae. *Angew. Parasitol.* 23, 1, 1982.
57. Woodruff, A.W. Toxocariasis. *Br. Med. J.* 3, 663, 1970.
58. Nadler, S.A. Advantages and disadvantages of molecular phylogenetics: A case study of ascaridoid nematodes. *J. Nematol.* 27, 423, 1995.
59. Nadler, S.A. Biochemical polymorphism in *Parascaris equorum*, *Toxocara canis* and *Toxocara cati*. *Mol. Biochem. Parasitol.* 18, 45, 1986.
60. Turčeková, Ľ., and Dubinský, P. Differentiation between *Toxocara canis* and *T. cati* using restriction profiles and ribosomal gene probe. *Helminthologia* 33, 223, 1996.
61. Wu, Z. et al. Primers for polymerase chain reaction to detect genomic DNA of *Toxocara canis* and *T. cati*. *J. Helminthol.* 71, 77, 1997.
62. Jacobs, D.E. et al. PCR-based methods for identification of potentially zoonotic ascaridoid parasites of the dog, fox and cat. *Acta Trop.* 68, 191, 1997.
64. Epe, C. et al. Transmission trials, ITS2-PCR and PAPR-PCR show identity of *Toxocara canis* isolates from red fox and dog. *Vet. Parasitol.* 84, 101, 1999.
65. Li, M.-W. et al. The occurence of *Toxocara malaysiensis* in cats in China, confirmed by sequence-based analyses of ribosomal DNA. *Parasitol. Res.* 99, 554, 2006.
65. Zhu, X.Q. et al. Molecular characterization of a *Toxocara* variant from cats in Kuala Lumpur, Malaysia. *Parasitology* 117, 155, 1998.
66. Jex, A.R. et al. The mitochondrial genome of *Toxocara canis*. *Plos Negl. Trop. Dis.* 2, e273, 2008.
67. Li, M.-W. et al. The complete mitochodrial genomes for three *Toxocara* species of human and animal health significance. *BMC Genomics* 9, 224, 2005.
68. Mizgajska, H., and Rejmenciak, A. Differentiation of eggs of *Toxocara canis* and *Toxocara cati*—Parasites of dog and cat. *Wiad. Parazyt.* 43, 435, 1997.
69. Borecka, A. Differentiation of *Toxocara* spp eggs isolated from the soil by the PCR-linked RFLP method. *Helminthologia* 41, 185, 2004.
70. Fogt-Wyrwas, R., Jarosz, W., and Mizgajska-Wiktor, H. Utilizing a polymerase reaction method for the detection of *Toxocara canis* and *T. cati* eggs in soil. *J. Helminthol.* 81, 75, 2007.
71. Borecka, A., and Gawor, J. Modification of gDNA extraction from soil for PCR designed for the routine examination of soil contaminated with *Toxocara* spp. eggs. *J. Helminthol.* 82, 119, 2008.
72. Rai, S.K. et al. Use of polymerase chain reaction in the diagnosis of toxocariasis: An experimental study. *Southeast Asian J. Trop. Med. Public Health* 28, 541, 1997.
73. Borecka, A. et al. Detection of *Toxocara canis* larvae by PCR in the liver of experimentally infected Mongolian gerbils (*Meriones unguiculatus*). *Helminthologia* 45, 147, 2008.
74. Ishiwata, K. et al. Identification of tissue-embedded ascarid larvae by ribosomal DNA sequencing. *Parasitol. Res.* 92, 50, 2004.
75. Moreira, D. Efficient removal of PCR inhibitors using agarose-embedded DNA preparations. *Nucleic Acids Res.* 26, 3309, 1998.
76. Yeates, C. et al. Methods for microbial DNA extraction from soil for PCR amplification. *Biol. Proced. Online* 1, 40, 1998.

67 Trichinella

Edoardo Pozio and Giuseppe La Rosa

CONTENTS

67.1 INTRODUCTION

Nematode worms belonging to the genus *Trichinella* are the etiological agents of the zoonosis trichinellosis, which is the appropriate term for the human disease also known as "trichinosis" or "trichiniasis." On all continents but for Antarctica, these parasites are widespread among wildlife, and in many countries they are widespread among domestic pigs [1]. Human infection occurs following the consumption of raw or undercooked meat and meat products from different animals (e.g., pork, horse, game). The main distinguishing feature of the life cycle of these parasites is that two generations occur in the same host. The first generation (from the infective larva to the adult) is present in the gut, and the second generation (from the newborn larva to the infective larva) is present in the cell of striated muscles, which is modified by the larva (referred to as a "nurse cell") [2].

67.1.1 Taxonomy

Until 1972, *Trichinella spiralis* was the only known species in the *Trichinella* genus, and for this reason it was the only species considered in most of the studies conducted through the 1990s. At present, there are eight known species and four additional genotypes that have not been characterized at the species level.

Morphologically, the individual species and genotypes cannot be distinguished, yet they can be categorized into two clades based on whether or not the nurse cells in the host's muscles are surrounded by a collagen capsule, which is important for the long-term survival of the larva in decaying muscle tissue (i.e., when the larva is not protected by the host). If the collagen capsule is present, then the species/genotype belongs to the encapsulated clade, which consists of five species and the four genotypes; otherwise it belongs to the nonencapsulated clade. However, the nonencapsulated larvae induce the same modifications as the encapsulated larvae to the muscle cell, though the amount of collagen produced is very low. A brief description of the encapsulated and nonencapsulated species and genotypes is provided subsequently.

67.1.1.1 The Encapsulated Clade

T. spiralis is the species that has been characterized the most, not only because it was the first to have been discovered but also because of its importance both as a cause of human disease and as a model for basic biological research, particularly in the field of immunology, due to the relatively high prevalence of infection in domestic and wild animals and the marked infectivity for laboratory animals. The species originated in East Asia [3], and its worldwide dissemination was facilitated by the European colonization of the Americas, New Zealand, Hawaii, and Egypt from the sixteenth century to the twentieth century (Table 67.1) [4]. The predominant hosts are domestic and wild swine (*Sus scrofa*), synanthropic animals, other domestic animals, and diverse species of wild carnivores. *T. spiralis* is the etiological agent of most human infections worldwide. It is more pathogenic than the other species and genotypes because the female produces a higher number of newborn larvae [5] and because it induces a greater immune response in humans [6,7].

Trichinella nativa is the etiological agent of infection in sylvatic carnivores in frigid zones of Asia (China, Kazakhstan, Mongolia, Russia), North America (Canada, Greenland, and the United States), and Europe (Estonia, Finland, Lithuania, Norway, Russia, and Sweden) (Table 67.1). The southern boundary of distribution has been tentatively identified to lie between the isotherms −5°C and −4°C in January. The main hosts are polar, brown and black bears; wolverines; raccoons; lynxes; wolves; foxes; and walruses [8,9]. One of this species' biological characteristics is the larva's ability to survive in frozen muscles of carnivores for up to 5 years [9]. Human infection has been reported in frigid zones of Canada, Greenland, Siberia, and Kamchatka [10–12] and among hunters from Europe and the United States who consume raw or undercooked meat from bears hunted in arctic or subarctic regions [13].

The genotype *Trichinella* T6 is closely related to *T. nativa* and has been detected in several regions of Canada (British

TABLE 67.1
Principal Features of *Trichinella* Species and Genotypes and Amplicon Sizes by PCR-RFLP

Trichinella Species or Genotype	Distribution	Cycle	Hosts	Amplicon Size by PCR-RFLP (*Mse* I)
		Encapsulated		
T. spiralis	Cosmopolitan[a]	Domestic and sylvatic	Swine, rats, seldom carnivores	22, 70, 126, 201 bp
T. nativa	Arctic and subarctic areas of Holoarctic region[b]	Sylvatic	Terrestrial and marine carnivores	22, 70, 327 bp
Trichinella T6	Canada and the United States	Sylvatic	Carnivores	22, 70, 327 bp
T. britovi	Temperate areas of Palearctic region,[c] northern and western Africa	Sylvatic, seldom domestic	Carnivores, seldom swine	22, 62, 64, 70, 201 bp
Trichinella T8	South Africa and Namibia	Sylvatic	Carnivores	22, 62, 64, 70, 201 bp
T. murrelli	Temperate areas of Nearctic region	Sylvatic	Carnivores	92, 126, 201 bp
Trichinella T9	Japan	Sylvatic	Carnivores	92, 327 bp
T. nelsoni	Ethiopic region	Sylvatic	Carnivores, seldom swine	62, 64, 70, 223 bp
Trichinella T12	Argentina	Sylvatic	Carnivores	Unknown
		Nonencapsulated		
T. pseudospiralis	Cosmopolitan	Sylvatic, seldom domestic	Mammals and birds	419 bp[d] unknown[e]
T. papuae	PNG Thailand	Sylvatic, seldom domestic	Swine, Saltwater crocodiles	unknown
T. zimbabwensis	Ethiopia, Mozambique, South Africa, Zimbabwe	Sylvatic and domestic	Nile crocodiles, Nile monitor lizards	unknown

Source: Nagano, I. et al. *Int. J. Parasitol.* 29, 1113, 1999.

[a] This species has not been detected in arctic regions.
[b] The isotherm −4°C in January is the southern limit of distribution.
[c] The isotherm −6°C in January is the northern limit of distribution.
[d] Palearctic region.
[e] Nearctic and Australian (Tasmania) regions.

Columbia, Ontario, Manitoba, and Nunavut) and in the United States (Alaska, Montana, Idaho, and Pennsylvania) [14]. The larvae of this genotype, like those of *T. nativa*, can survive in frozen muscles of carnivores for long periods of time [9].

Trichinella britovi is the etiological agent of infection in sylvatic carnivores (e.g., foxes, wolves, badgers, stone martens, jackals, bears, lynxes, cats, and dogs) and omnivores (in most cases the wild boar) in temperate areas of Europe and Asia (from the Iberian Peninsula to Kazakhstan) and in North and West Africa (Table 67.1). The northern boundary of distribution appears to be between the isotherms −6°C and −5°C in January. The larvae of *T. britovi* can survive in frozen muscles of carnivores for up to 11 months and in frozen muscles of swine for up to 3 weeks, which, however, is a shorter period of time compared to *T. nativa* larvae [9]. Human infection caused by the consumption of free-ranging pigs, game, and horse meat has been documented in Europe, Asia, and Africa north of the Sahara [15–19]. The clinical course is relatively benign, and no deaths have been documented [20].

The genotype *Trichinella* T8 is related to *T. britovi*. It has been detected in sylvatic carnivores of South Africa and Namibia (Table 67.1) [21]. No human infections have been documented to date.

Trichinella murrelli is the etiological agent of infection in wild carnivores in temperate areas of North America. The isotherm −6°C in January seems to be the northern boundary of distribution (Table 67.1). Because this species shows very low infectivity to swine and rats, it does not pose a risk for domestic pigs [22]. Human infection has occurred for the consumption of bear meat in the United States and of horse meat exported from the United States to France [1]. The genotype *Trichinella* T9, which is phylogenetically related to *T. murrelli*, has been detected in wild carnivores in Japan [23]. No human infections have been documented.

Trichinella nelsoni is the etiological agent of infection of wild carnivores (e.g., spotted and striped hyenas, lions, leopards, servals, bat eared foxes, and cheetahs) in East Africa, from Kenya to South Africa (Table 67.1) [24]. This species occasionally infects wild swine (e.g., warthogs and bush pigs), whose meat can be the source of human infection. Infection has never been detected in domestic pigs. To date, fewer than 100 human infections have been documented. This species shows low pathogenicity in humans, and deaths have only been reported in persons with >4000 larvae/g of muscle [25,26].

Trichinella T12 is a newly discovered genotype, which was detected in a mountain lion (*Puma concolor*) from Trapalco (66°59′ W, 39°34′ S) (Patagonia, Río Negro, Argentina) (Table 67.1). This taxon is considered as a separate genotype based on the molecular structure of two noncoding sequences (expansion segment V (ESV), 5S ribosomal DNA intergenic spacer region) and one coding sequence (cytochrome *c*-oxidase subunit I), which are different from those of the other 11 species/genotypes [27].

67.1.1.2 The Nonencapsulated Clade

Trichinella pseudospiralis is a cosmopolitan species which infects both domestic and wild mammals; it is the only species that infects birds, both wild carnivorous and omnivorous (Table 67.1). Three populations, which can be distinguished on a molecular basis, have been detected: one population in Europe and Asia, one in North America, and one in Tasmania [28]. *T. pseudospiralis* has been found in 14 mammal species and seven bird species [29]; however, relatively few bird species have been examined, and the number of examined birds has also been low. In humans, the clinical severity of infection can range from moderate to severe; one death has been reported [30–33].

Trichinella papuae has been detected in Papua New Guinea (PNG) and Thailand (Table 67.1) [34–36]. It can infect both mammals (i.e., wild pigs) and reptiles (i.e., saltwater crocodiles and monitor lizards). Wild pigs have been implicated in human infection [37]. In PNG, it was reported that a high percentage of persons who consume wild pig were serologically positive for *Trichinella* antigens yet had not clinical signs of the disease, suggesting that they have frequent contact with this parasite [37].

Trichinella zimbabwensis has been detected in farmed Nile crocodiles of Zimbabwe [38] and Ethiopia, in wild crocodiles of Mozambique and South Africa, and in wild monitor lizards of Zimbabwe (Table 67.1) [39]. In laboratory conditions, this species can infect mammals, although no naturally infected mammals have ever been detected.

67.1.2 Biology

A peculiarity of the biology of *Trichinella* nematodes is that two generations (i.e., adult worms and newborn larvae) are present in the host. Female worms embedded in the intestinal mucosa give birth to larvae, which migrate directly into the lymphatic vessels; they then enter the blood vessels and are spread throughout the entire body. Once the larvae reach the striated muscle cells, they penetrate the cell using a stiletto apparatus and lytic enzymes; this occurs in all striated muscles except for those of the heart. After about 15 days, the larvae develop to the L1 infective stage (at which point they are referred to as "muscle larvae"). In the muscle cells, larvae can survive for years (e.g., for over 20 years in polar bears and for up to 40 years in humans). When a new host ingests infected muscle tissue, the cells are digested in the stomach, releasing the larvae; when the larvae reach the duodenum, they penetrate the villa and within two days undergo four moults, developing to the adult stage. Males and females copulate, and from six to seven days postinfection (d.p.i.), the females begin to produce newborn larvae (i.e., the new generation), which continue to be produced for at least one or two weeks, or more, depending on the host's immune response at the gut level, which results in the expulsion of adult worms.

All *Trichinella* species and genotypes have a sylvatic cycle (i.e., natural transmission occurs among wildlife), though *T. spiralis* is primarily transmitted by a domestic cycle involving domestic pigs, synanthropic rats and, on rare occasions, horses [1]. A domestic cycle involving pigs and rats has also been reported for *T. britovi* and *T. pseudospiralis*, though rarely [1]. The prevalence of infection among carnivorous mammals (e.g., foxes, wolves, racoon dogs, lynxes, bears, and mustelids) ranges from 0.01% to 90%, depending on the animal's habitat and the extent of human influence [40,41]. Given that *Trichinella* infection in sylvatic and domestic animals is mainly asymptomatic, even when thousands of larvae per gram of muscle tissue are present, there are no clinical signs that can help to identify infected animals.

67.1.3 Clinical Features and Pathogenesis

Humans acquire trichinellosis by consuming raw or undercooked meat from pigs, horses, dogs, and game animals (e.g., wild boar, fox, jackal, badger, bear, and walrus); the presence of *Trichinella* in domestic or wild animals is not in itself sufficient for transmission to humans. The clinical picture and prognosis of human infection depend on the number of infective larvae ingested, the *Trichinella* species, and the person's immune response [42]. About 0.2% of cases result in death, though the percentage has been reported to range from 0.1% to 1.0%. Death is usually due to respiratory or cardiovascular complications or neurological disorders [43].

The pathological picture of infection is complex because, as mentioned, there exist two generations in the host: adults, which induce pathological alterations in the intestine; and larvae, which, during their development, induce alterations in the muscle. The extent of the immunological response is proportional to the number of larvae ingested, and it greatly depends on individual host type I hypersensitivity, which plays a major role in the early intestinal stage of the disease. The adult worms induce a severe intestinal pathology, including villus atrophy and crypt and goblet cell hyperplasia. This intestinal pathology is directly correlated with the inflammatory response inducing the spontaneous expulsion of adult worms (referred to as "self-expulsion"). In this process, the antibody production and cytokines produced by Th-2 cells play a major role. The activation of the Th-2 cytokines induces the stem cell factor (SCF) production, followed by the SCF differentiation, preferentially along the goblet cell lineage [44].

Unlike other muscle parasites, *Trichinella* larvae transform the muscle cell, altering its metabolism and transforming the cell into a nurse cell, which allows the larva to survive and mature. The parasite elicits angiogenesis directly or indirectly by inducing changes in the nurse cells, which, in turn, stimulate the expansion of the network of blood vessels [44]. This expansion could be explained by increased levels of vascular endothelial growth factor [45]. In the heart muscle, the larvae and high levels of eosinophils induce transient myocarditis, whereas damage to the central nervous system (CNS) may be far more severe. The pathogenesis of CNS lesions could be linked to the toxicity of eosinophiles for endothelial cells, leading to vasculitis and the infiltration of the pia-arachnoid with lymphocytes, macrophages, fibroblasts, and gitter cells. Nodules of glial cells and small hemorrhages may be present in the periventricular area and other areas of the white matter.

Worldwide, an average of 10,000 cases of clinical disease in humans has been estimated to occur each year, with a mortality rate of about 0.2% [2]. However, in many countries the number of infections is underreported because appropriate serological tests are not available and because physicians are often not familiar with the disease.

The acute stage of the disease corresponds to the phase in which the newborn larvae migrate from the lymphatic vessels and invade the muscle cells. In varying percentages of infected individuals, the acute stage is preceded by loose stools or diarrhea. These symptoms may be accompanied by flatulence, moderately intense abdominal pain, loss of appetite and, at times, vomiting [43].

The signs and symptoms of the acute stage of trichinellosis develop and disappear at varying intervals and are of varying intensity. In most persons, the onset of the acute stage is sudden, with general weakness, chills, headache, fever (up to 40°C), excessive sweating, and tachycardia. In the majority of cases, symmetrical eyelid and periocular edema occur, and edema frequently affects the entire face [43]. The blood vessels of the conjunctivae become inflamed and in some persons petechiae, intraconjunctival hemorrhage, and hemorrhage of nail beds occur. These symptoms are accompanied by eosinophilia and usually by leukocytosis. This symptomatology is followed by pain in various muscle groups, which may restrict motility. Of these signs and symptoms, those considered to be the principal manifestations are: fever, eyelid, and periocular edema, and muscle pain [43]. In severe cases, fever may be continuous and persist for up to 3 weeks; in mild cases, very mild fever is common. The intensity of periocular, eyelid and facial edema depends on the host's allergic reaction. The intensity of muscle pain reflects the severity of the disease. Pain occurs upon movement; spontaneous pain is less common, and according to some authors it occurs more frequently in persons with severe disease or as a result of complications in the form of thrombophlebitis. The symptoms gradually disappear during convalescence, and complete regression may be achieved with physical therapy [42].

It has not been firmly established whether a chronic form of trichinellosis actually exists. Nonetheless, there have been reports of persons who, months or even years after the acute stage, continue to suffer from chronic pain, general discomfort, tingling, numbness, and excessive sweating, and who show signs of paranoia and a syndrome of persecution. The persistence of these symptoms has been more frequently observed among persons who had suffered from severe trichinellosis. There have been reports of impaired muscle strength, conjunctivitis, impaired coordination and the presence of IgG antibodies up to 10 years after infection, and live larvae have been detected in muscles up to 39 years after infection, yet without clinical signs or symptoms. All cases in which trichinellosis has been defined as "chronic" have been reported among persons who had not been treated in a timely manner [43].

67.1.4 Diagnosis

In humans, which are the only host in which *Trichinella* can induce clinical disease, diagnosis is based on the following: (i) epidemiological data (i.e., the source of infection, the amount of infected meat ingested, the number of larvae present in the infected meat, and the number of cases in the epidemic focus); (ii) clinical evaluation (i.e., signs and symptoms and their clinical severity, which significantly affects the choice of treatment); and (iii) laboratory tests (i.e., immunodiagnosis and/or detection of larvae in a muscle biopsy using HCl-pepsin digestion). In animals, which show no clinical signs, the diagnosis is based exclusively on the detection of *Trichinella* larvae in muscle tissues using HCl-pepsin digestion.

Given that morphological characters cannot be used to distinguish the taxon and that the biological and biochemical characters (e.g., allozymes) do not allow the species to be easily or rapidly identified, methods based on polymerase chain reaction (PCR) must be applied to the species or genotype determination. Identifying the species is extremely important in tracing the source of infection, in determining and predicting the clinical course of infection, in interpreting the clinical and laboratory data, in estimating the potential risk for pigs, in establishing appropriate strategies for control and eradication, and in better understanding the epidemiology of the infection.

PCR has proven to be important in simplifying the identification of *Trichinella* isolates from different host species and geographical regions (Table 67.1) [46–61]. The high sensitivity of PCR allows the species to be identified using a single larva, which is important because often only one larva is detected in samples from humans and animals. That the species can be identified with a single larva is also important because it allows different *Trichinella* species to be identified in the same host, a finding that has contributed to understanding the potential for gene flow between sympatric species or genotypes [1]. PCR also allows the species to be identified based on a fraction of the larva (as small as 1/15th), which is important because in histological sections only a fraction of a larva is present.

67.1.4.1 Molecular Probes

The first molecular approaches for distinguishing among *Trichinella* species and genotypes were based on restriction enzyme analyses of genomic DNA and DNA probes. However, this method, in addition to being time consuming, requires a large number of parasites. This entails performing isolation and propagation in laboratory animals, which, however, can result in the loss of genetic variability and failure to identify mixed infections.

67.1.4.2 Random Amplified Polymorphic DNA

A plethora of PCR-derived methods have been developed to identify *Trichinella* larvae at the species or genotype level. These methods can be divided into those using arbitrary primers (random amplified polymorphic DNA (RAPD)) and those using DNA sequence-derived genotype-specific primers. The fingerprint pattern obtained with RAPD is a simple and effective tool for rapidly generating and detecting genetic markers; however, this method can be imprecise, producing bands that are not always reproducible and thus creating confusion in interpreting the results [49]. Furthermore, problems associated with PCR signal reproducibility due to variations in DNA quality and contamination, reaction conditions, and instrumentation, can generate substantial sample-to-sample deviation [62]. This, in turn, requires extensive controls to be performed with each analysis. An RAPD-derived approach has been developed to distinguish between *T. spiralis* and *T. pseudospiralis* [53]. In this work, RAPD has been produced from the two species using purified genomic DNA and an arbitrary 10-mer primer. Two species-specific fragments have been isolated, sequenced, and used to design two primer sets that identify the two species using a standard PCR. In an expanded yet similar set of experiments, primer pairs with a high degree of specificity for *T. spiralis*, *T. nativa*, *T. britovi*, *T. nelsoni*, and *T. pseudospiralis* have been generated [54]. Problems and inconsistencies in using RAPD and associated technologies can lead to a more deliberate approach toward designing genotype-specific PCR primers.

67.1.4.3 Single-Strand Conformational Polymorphism

Single-strand conformational polymorphism (SSCP) can be used to characterize and differentiate eight distinct taxa of *Trichinella* [52]. In SSCP, heat denaturation is used to induce strand separation in PCR-amplified products, which are then subjected to nondenaturing gel electrophoresis. During electrophoresis, even minor conformational changes induced by innate point mutations in the DNA strands can alter gel migration patterns of individual DNA strands and present a means by which genotypic markers can be generated and evaluated. The results indicate that a high level of sequence variation exist in the ESV region of the large subunit ribosomal DNA (lsrDNA) [63] and in the D3 domain of nuclear ribosomal DNA. Based on these two markers, all the recognized species and genotypes can be distinguished, with the exception of the genotype T12 which has not been tested. The results also show substantial sequence variation among isolates of the same taxon. Thus, if used to analyze a variable region of DNA such as the *Trichinella* ESV sequences, SSCP shows great promise for defining both inter- and intraspecific variation.

67.1.4.4 PCR-Restriction-Fragment-Length Polymorphism

For reasons described above [62], researchers continue to evaluate sequence variation within distinct genes to develop species-specific PCR primers or to generate unique restriction enzyme banding patterns from PCR-amplified gene fragments using genus-specific primers [55,57]. Degenerate primers based on the cytochrome oxidase I (COX I) gene of mitochondrial DNA (mt DNA) have been used to amplify and sequence DNA from nine *Trichinella* taxa and subsequently differentiated them by restriction fragment length polymorphism (RFLP) analysis [55].

67.1.4.5 Multiplex PCR

To unequivocally distinguish all currently recognized *Trichinella* species and the T6 genotype, a multiplex PCR test that uses five primer sets based on the ESV sequences and the internal transcribed spacer (ITS) regions 1 and 2 has been developed [59]. The multiplex PCR eliminates the need to perform multiple PCR reactions. When the five primer sets are combined into a single mix and PCR is performed on individual parasites, a distinct and simple agarose gel profile is generated for each of the eight species and the T6 genotype. Each taxon is characterized by no more than two PCR fragments. The ability to perform this test on individual larvae makes it ideal for samples where often only a few larvae are available. The technology, as originally described, used nested PCR for the individual larval analysis [59]; however, the protocol has been improved and a single-round PCR has been generated to work on single larvae [60].

The molecular identification of single larvae is carried out with a multiplex-PCR analysis, which allows for the identification of eight species (*T. spiralis*, *T. nativa*, *T. britovi*, *T. pseudospiralis*, *T. murrelli*, *T. nelsoni*, *T. papuae*, and *T. zimbabwensis*), the T6 genotype, and three populations of *T. pseudospiralis* (from the Australian, Nearctic, and Palearctic regions).

Since the genotypes T8 and T9 show the same band pattern as *T. britovi* by multiplex PCR, a PCR-RFLP analysis of the gene encoding for the CO I protein can be used to distinguish between *T. britovi* and genotype T9 [55]. To distinguish between *T. britovi* and genotype T8, a standard PCR based on the ITS2 sequence can be used [24].

67.1.4.6 Reverse Line Blot Hybridization

A reverse line blot (RLB) assay has been developed to identify different *Trichinella* genotypes [61]. The RLB assay can be used to detect and identify the most of *Trichinella* taxa (*T. spiralis*, *T. britovi*, *T. nativa*, *T. murrelli*, *T. nelsoni*, and *T. pseudospiralis*, and *Trichinella* T6 and T8); it is based on the hybridization of the amplified 5S ribosomal DNA intergenic spacer regions to specific, membrane-bound oligonucleotide probes.

67.2 METHODS

67.2.1 Sample Preparation

67.2.1.1 Reference Material

A major resource of vital parasite material and information is the International *Trichinella* Reference Centre (ITRC) in Rome, Italy (www.iss.it/site/Trichinella/index.asp), which was established in 1990 and whose biobank currently contains >4000 isolates from around the world. For each isolate, the ITRC database includes information on the results of molecular typing, the host species, and the geographical origin, as well as other epidemiological information [64].

67.2.1.2 Hosts and Preferential Muscles

The best candidates for detecting *Trichinella* are large carnivorous and omnivorous animals (mammals, birds, and reptiles) that engage in scavenger and cannibalistic behavior and are at the top of the food chain. The most common candidates in Nearctic and Palearctic regions are the polar bear, brown bear, black bear, wolf, red fox, lynx, bob cat, and racoon dog; in other regions, some examples of the most common candidates are viverrides, spotted hyena, and lion in the Ethiopian region and carnivorous marsupials in Tasmania. At the farm level (i.e., in the domestic habitat), the best candidate is the domestic pig; other candidates include synanthropic animals such as rats, foxes, mustelids, armadillos, domestic cats and dogs, and farmed crocodiles. *Trichinella* infection has also been detected, though rarely, in horses and sheep, which were the source of human infection, yet infected herbivorous animals represent an offshoot of the infection occurring in pigs [1].

As mentioned, *Trichinella* is most commonly identified based on the analysis of muscle larvae, which are very easy to collect from both animals and humans. Although PCR-derived methods have been used to perform molecular diagnosis of *Trichinella* infection by detecting DNA directly in the muscle (i.e., without performing artificial digestion to collect the larvae), these methods have been shown to have a low sensitivity because of the huge amount of muscle DNA, which dilutes the larva target sequence DNA. These methods are also unfeasible because of the great amount of Proteinase K that would be needed to digest the muscle sample.

Muscle samples should be collected from the preferential muscles (i.e., those with the highest density of larvae), which vary by host species. However, as a general rule, the base of the tongue can be considered as the preferential muscle, though the superficial layer should be removed before sample preparation, since it cannot be digested. Other important muscles are: the pillar of the diaphragm and the masseter in swine and rodents; the anterior tibial in carnivorous mammals; and the masseter in carnivores, horses, crocodiles, monitor lizards. Human biopsies are generally taken from the deltoid muscle. Other important considerations when choosing the specific muscle to be used for digestion, are the digestibility of the muscle, which can vary greatly [65], and the amount of muscle, free of fat and fascia, which should be tested. Because the average number of larvae per gram of preferential muscle tissue generally ranges from 0.1 to 1.0, the chance of detecting larvae increases with the amount of tissue sampled. For meat products (e.g., sausage), before digestion the muscle tissue should be separated from the fat and other tissues and then cut into small pieces and rehydrated (if necessary) at +4°C in a saline solution for 1–2 days.

If digestion is not performed immediately, yet within 2 weeks of the animal's death, then the samples should be stored at +4°C. If digestion is performed more than 2 weeks after the animal's death, then the samples should either be frozen at −20°C, without interrupting the cold chain before digestion, or stored at room temperature for 5–6 months in a 0.1% trimerosal (merthiolate) solution.

67.2.1.3 Parasite Isolation and Preservation

The magnetic stirrer method is considered the gold standard for isolating larvae because it is the only method that has been subjected to validation studies [66].

A maximum of 100 g of muscle tissue from the preferential muscle should be tested. The sample should be taken near the muscle insertion, where the greatest concentration of larvae is located. The ratio of muscle (weight) to digestion fluid (volume) should be 1:20. The quantities reported below are for digesting 100 g of muscle tissue. If <100 g of meat is used, the amount of digestion fluid should be proportionally reduced.

1. In a beaker, add 16 mL of hydrochloric acid to 2.0 L of tap water that has been preheated to 44–45°C.
2. Place the beaker on the preheated plate of the magnetic stirrer and stir with the Teflon stirring rod.
3. Add 10 g of pepsin.
4. Cut the muscle sample into small pieces (0.5–2 g each), removing all nonmuscle tissue (tendons, fat, etc.).
5. Place the pieces of muscle in a professional blender and mince.
6. Transfer the minced meat to the beaker containing the water, pepsin, and hydrochloric acid, leaving the beaker on the heated plate; to remove any remaining meat from the blender, remove the blade and immerse it in the beaker containing the digestion fluid; rinse the blender glass pitcher with a small quantity of digestion fluid and pour into the beaker.
7. Cover the beaker with aluminum foil. The magnetic stirrer must be adjusted so that the digestion fluid maintains a constant temperature of 38–40°C throughout the procedure. The digestion fluid must be stirred at a sufficiently high speed to create a deep whirl without splashing (during this process, the muscles are digested and the larvae remain in the fluid). The digestion time is generally around 30 min, though more time may be needed (not to exceed 60 min) for certain types of meat (e.g., tongue, game meat). The process is complete when undigested muscle is no longer present in the fluid.
8. Place the sieve in the glass funnel, which has been placed in the separatory funnel.
9. Pour the digestion fluid through the sieve (which retains any undigested tissues but allows the passage of the larvae). The digestion process can be considered as successful if no >5% of the initial sample (in terms of weight) remains on the sieve; allow the larvae to sediment for 30 min.
10. Open the safety plug of the separatory funnel and release 40 mL of the sediment into a 50 mL tube.
11. Allow the larvae in the separatory funnel to sediment for another 10 min, to clarify the suspension.
12. Discard 30 mL of supernatant and pour the remaining 10 mL of sediment into a gridded Petri dish.
13. Rinse the 50 mL tube with 10 mL of tap water, shake, and add to the sediment in the gridded Petri dish.
14. Examine the sample in the Petri dish by a stereomicroscope (15–40× magnification) for the presence of *Trichinella* larvae, as soon as possible after sedimentation. Larvae should then be stored in absolute ethanol, which should be done immediately if the larvae are dead (C-shaped or comma shaped), to avoid DNA destruction.

67.2.1.4 DNA Purification from Single Larvae

This procedure is for larvae that have been isolated according to the method described in Section 67.2.1.3. The procedure recommended for the kit is slightly modified for use with single *Trichinella* larvae.

1. Transfer the ethanol containing the larvae into a Petri dish.
2. Under the microscope, collect single larvae using a 2 μL tip and place one larva each in a 1.5 mL conical tube.
3. Spin the tubes at maximum speed for a few seconds and then make sure that the larva is on the bottom of the tube, using the microscope.
4. Allow the ethanol to evaporate completely, leaving the tube open for a few minutes.
5. Prepare a quantity of IB+ and LB+ that is sufficient for the number of samples to be analyzed.
6. Add 20 μL of IB+.
7. Incubate at 55°C for 30 min in the thermoblock, shaking at 1400 vibrations/min.
8. Spin the tubes.
9. Add 40 μL of LB+ and mix by pipetting a few times.
10. Add 4 μL of the paramagnetic resin (Tissue and Hair Extraction Kit) and resuspend it by slow vortexing, avoiding the creation of bubbles.
11. Incubate for 5 min at 25°C in the thermoblock, shaking at 1400 vibrations/min.
12. Place the tubes in the magnetic separation stand and wait for 60 s.
13. Discard the liquid phase by aspiration, avoiding the dislodging of the resin particles.
14. Add 100 μL of LB+ and resuspend the resin particles by vortexing.
15. Repeat steps 12 and 13.
16. Add 100 μL of 1× Wash Buffer (Tissue and Hair Extraction Kit) and resuspend the resin particles by vortexing.
17. Repeat steps 12 and 13.
18. Repeat two times steps 16, 12, and 13.
19. Leave the tubes open for 15–20 min to allow the resin particles to dry.
20. Add 50 μL of the eluting buffer (Tissue and Hair Extraction Kit) and gently resuspend the resin particles, without vortexing.
21. Incubate at 65°C for 5 min and shake at 1400 vibrations/min.
22. Repeat step 12.
23. Collect the liquid phase (about 40–50 μL) and transfer it to a 1.5 mL conical tube. The DNA cannot be checked using a gel because the total amount is very low.
24. Store the DNA samples frozen (–20°C).

67.2.2 Detection Procedures

67.2.2.1 Multiplex PCR

Reagents: (i) 2× PCR Master Mix commercial solution (standard composition: 400 µM dATP, 400 µM dCTP, 400 µM dGTP, 400 µM dTTP, 3 mM $MgCl_2$, 50 U/mL Taq DNA polymerase; (ii) oligonucleotides: (a) ESV target locus, 5′-GTTCCATGTGAACAGCAGT-3′, 5′-CGAAAACATACGACAACTGC-3′; (b) ITS1 target locus, 5′-GCTACATCCTTTTGATCTGTT-3′, 5′-AGACACAATATCAACCACAGTACA-3′, and 5′-GCGGAAGGATCATTATCGTGTA-3′, 5′-TGGATTACAAAGAAAACCATCACT-3′; and (c) ITS2 target locus, 5′-GTGAGCGTAATAAAGGTGCAG-3′, 5′-TTCATCACACATCTTCCACTA-3′, 5′-CAATTGAAAACCGCTTAGCGTGTTT-3′, and 5′-TGATCTGAGGTCGACATTTCC-3′. These lyophilized products are reconstituted with 0.1× TE at a concentration of 100 pmol/µL; (iii) SetB, the oligonucleotide mixture used for the multiplex-PCR. The mixture is obtained combining an equal volume of the 10 oligonucleotides (final concentration of 10 pmol/µL); (iv) genomic DNA (10 ng/µL) purified from a pool of larvae of a *Trichinella* reference strain (ITRC, Rome, Italy), considered as positive control DNA; and (v) purified DNA of the larvae to be tested (see Section 67.2.1.4).

Equipment: a thermocycler.

Procedure: The total volume of the amplification reaction is 30 µL. Before the procedure, thaw the following on ice: DNA samples purified from single larvae, 2× PCR Master Mix, SetB, and positive control DNA.

1. Prepare the amplification solution by adding sequentially: 15 µL 2× PCR Master Mix, 4 µL water, 1 µL SetB, and 10 µL of DNA purified from single larvae to be tested (see Section 67.2.1.3) or control DNA. When working with multiple samples, a quantity of amplification solution that is sufficient for all of the samples can be prepared all at once; aliquot the solution in the PCR tubes containing the DNA samples to be tested.
2. Close the tubes, mix by vortexing, and centrifuge at maximum speed for a few seconds.
3. Perform amplification: 1 cycle of 95°C for 4 min; 35 cycles of 95°C for 20 s, 55°C for 30 s and 72°C for 30 s; 1 cycle of 72°C for 3 min. It is recommended that a hot start Taq DNA polymerase be used, instead of standard enzymes.
4. At the end of the amplification, centrifuge the tubes at maximum speed for a few seconds.
5. Add the DNA loading buffer and load the samples onto a 2% agarose gel to view the result.
6. Evaluate the size of the amplification bands by comparison with the molecular weight markers. The species is identified by comparing the size of the band(s) shown by the sample with those in Table 67.2 and Figure 67.1.

67.2.2.2 PCR-Restriction-Fragment-Length Polymorphism to Distinguish between *T. britovi* and *Trichinella* T9

When using multiplex-PCR, larvae belonging to the T9 genotype display the same band pattern as *T. britovi* larvae. To distinguish between these two taxa, a PCR-RFLP method based on the CO I mitochondrial gene sequence has been developed [55]. According to the available information, the T9 genotype is circulating only in Japan; for this reason, it is suggested that this method be used only if the larvae with a *T. britovi* band pattern by multiplex-PCR originated from this country.

Reagents: (i) 2× PCR Master Mix; (ii) oligonucleotides for CO I, as target locus: L6625, F 5′-TTYTGGTTYTTCGGKCACCC-3′;H7005,R5′-ACGACGTAGTA

TABLE 67.2
Dimension of the Expected Amplification Products (in Base Pairs) for Each Taxon by Multiplex PCR

	T. spiralis	*T. nativa*[a]	*T. britovi*[b]	*T. pseudospiralis* Palearctic isolates	*T. pseudospiralis* Nearctic isolates	*T. pseudospiralis* Tasmanian isolates	*T. murrelli*	*Trichinella* T6	*T. nelsoni*	*T. papuae*	*T. zimbabwensis*
Genotype	T1	T2, T12	T3, T8, T9	T4A	T4B	T4C	T5	T6	T7	T10	T11
ESV	173	127	127	310	340	360	127	127	127	240	264
ITS1			253					210			
ITS2							316		404		

[a] Not only *T. nativa* but also *Trichinella* T12 display a single band of 127 bp.
[b] Not only *T. britovi*, but also *Trichinella* T8 and *Trichinella* T9 display two bands of 127 and 253 bp.

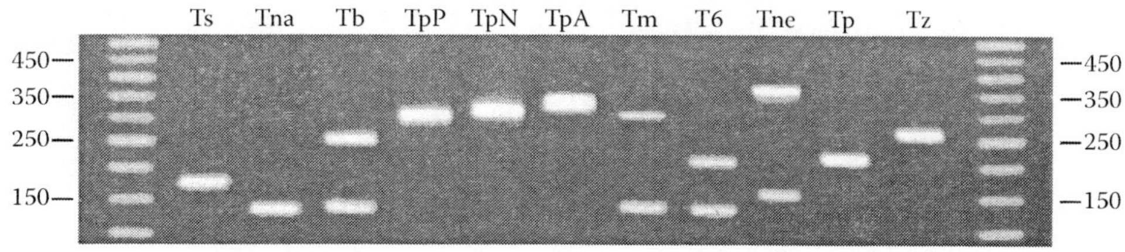

FIGURE 67.1 Photograph of an ethidium bromide-stained 2.5% agarose gel under ultraviolet light illumination showing the multiplex PCR amplification of single larvae of 11 taxa of *Trichinella*. The samples are as follows: Marker (ladder 50), sizes are in base pairs; Ts, *T. spiralis*; Tna, *T. nativa*; Tb, *T. britovi*; TpP, *T. pseudospiralis* (Palearctic isolate); TpN, *T. pseudospiralis* (Nearctic isolate); TpA, *T. pseudospiralis* (Tasmanian isolate); Tm, *T. murrelli*; T6, *Trichinella* T6; Tne, *T. nelsoni*; Tp, *T. papuae*; and Tz, *T. zimbabwensis*. The genotypes *Trichinella* T8 and *Trichinella* T9 show the same band pattern of *T. britovi*. The genotype *Trichinella* T12 shows the same band pattern of *T. nativa*.

GGTRTCRTG-3′; (iii) reference DNA from *T. britovi* and *Trichinella* T9 (ITRC, Rome, Italy); and (iv) purified DNA of the larvae to be tested (see Section 67.2.1.4).

Equipment: a thermocycler.

Procedures: The procedure consists of two steps: (i) PCR amplification of samples; and (ii) restriction analysis of the PCR products.

67.2.2.2.1 PCR Amplification

The total volume of the amplification reaction is of 50 μL.

1. Prepare the amplification solution by adding sequentially: 25 μL 2× PCR Master Mix, 13 μL water, 1 μL of each primer, and 10 μL of DNA extract from the single larvae to be tested (see Section 67.2.1.3).
2. Apply the following cycling program: 1 cycle of 94°C for 4 min; 35 cycles of 94°C for 1 min, 48°C for 1 min, and 72°C for 1 min; 1 cycle of 72°C for 3 min. It is recommended that a hot start Taq DNA polymerase be used instead of standard enzymes.
3. At the end of the amplification, centrifuge the tubes at maximum speed for a few seconds.
4. Add the DNA loading buffer and load the samples onto a 2% agarose gel to view the result (both *T. britovi* and *Trichinella* T9 display a 419 bp band). The amount of DNA product present in each sample should be evaluated by eye.
5. Store the amplified products at −20°C until use.

67.2.2.2.2 Restriction Analysis

Set the reactions keeping the tube on ice.

1. For each sample, transfer at least 100 ng of the amplified product into a 1.5 mL reaction tube.
2. Set the volume to 10 μL with water.
3. Add to the reaction tube: 5 μL of 10× *Mse*I reaction buffer, 1 μL of *Mse*I (10 U/μL) restriction enzyme, and water up to 50 μL. When working with multiple samples, a quantity of amplification solution that is sufficient for all of the samples can be prepared all at once; aliquot the solution in 1.5 mL tubes containing the PCR products to be tested.
4. Mix and incubate at 37°C for 60 min.
5. Stop the reaction by heating at 65°C for 20 min and then spinning for a few seconds.
6. Visualize the results by 2% agarose gel electrophoresis.
7. Evaluate the results according to the expected size pattern: for *T. britovi* five bands of 22, 62, 64, 70, and 201 bp; for *Trichinella* T9, two bands of 92 and 327 bp (Table 67.1).

67.2.2.3 PCR to Distinguish between *T. britovi* and *Trichinella* T8

By multiplex-PCR, the larvae of *Trichinella* T8 display the same band pattern as *T. britovi* larvae. To distinguish between these two taxa, a PCR method based on a 21 bp deletion in the ITS2 sequence of *Trichinella* T8 has been developed [24]. According to the available information, the T8 genotype is circulating only in sub-Saharan Africa; it is thus recommended that this method only be used if the larvae with a *T. britovi* band pattern by multiplex-PCR originate from this region.

Reagents: (i) oligonucleotides: ITS2, as target: ITS2G.F 5′-CCGGTGAGCGTAATAAAG-3′, ITS2G.R, 5′-TACACACAACGCAACGAT-3′ [24]; (ii) reference DNA from *T. britovi* and from *Trichinella* T8 (ITRC, Rome, Italy); and (iii) purified DNA of *Trichinella* larvae to be tested (see Section 67.2.1.4).

Equipment: a thermocycler.

Procedure: The total volume of the amplification reaction should be of 30 μL.

1. Prepare the amplification solution by adding sequentially: 15 μL 2× PCR Master Mix, 3 μL water, 1 μL of each primer, and 10 μL of DNA extract from single larvae or control DNA. When working with multiple samples, a quantity of amplification solution that is sufficient for all of the samples can be prepared all at once; aliquot the solution in the PCR tubes containing the DNA samples to be tested.
2. Close the tubes, mix by a vortex, and centrifuge at maximum speed for a few seconds.
3. Use the following cycling program: 1 cycle of 94°C for 4 min; 35 cycles of 94°C for 30 sec, 51°C for 30 sec, and 72°C for 1 min; 1 cycle of 72°C for

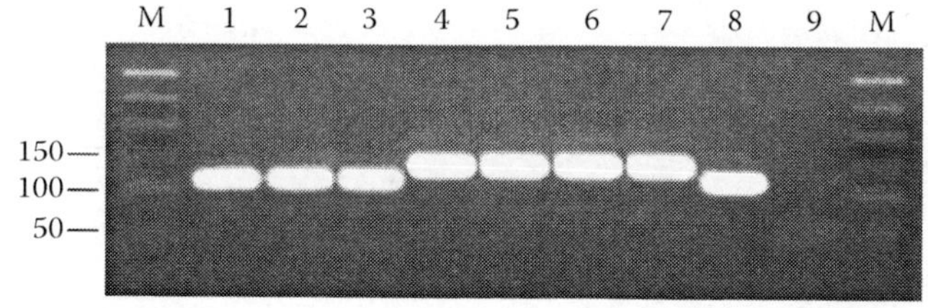

FIGURE 67.2 Photograph of an ethidium bromide-stained 2.5% agarose gel under ultraviolet light illumination showing the PCR amplification of single larvae of *Trichinella britovi* and *Trichinella* T8. The samples are as follows: M, (ladder 50), sizes are in base pairs; lines 1-3 and 8, *Trichinella* T8; lines 4-7, *T. britovi;* line 9, negative control.

3 min. It is recommended that a hot start Taq DNA polymerase be used instead of standard enzymes.

4. At the end of the amplification, centrifuge the tubes at maximum speed for a few seconds.
5. Add the DNA loading buffer, and visualize the results by 2% agarose gel electrophoresis.
6. Evaluate the result according to the expected size pattern: *T. britovi*, band of 125 bp; *Trichinella* T8, band of 104 bp (Figure 67.2).

67.2.2.4 Single-Strand Conformation Polymorphism

A simple, nonisotopic ("cold") SSCP approach for the analysis of genetic variation within and among the currently recognized taxa of the *Trichinella* genus using the variable ES5 as the target region has been developed [52].

Reagents: (i) 2× PCR master mix solution (400 μM dATP, 400 μM dCTP, 400 μM dGTP, 400 μM dTTP, 3 mM $MgCl_2$, 50 U/mL Taq DNA polymerase, Promega code M7501); (ii) denaturing DNA loading buffer (10 mM NaOH, 95% formamide, 0.05% of both bromophenol blue and xylene cyanole); (iii) GMA™ S-50 precast gels; (iv) oligonucleotides: ESV target locus: oTsr4 5′-GTTCCATGTGAACAGCAGT-3′ and oTsr3 5′-GCGATTGAGTTGAACGC-3′ [63]; D3 domain target locus: D328–26 5′-ACCCGTCTTGAAACACG GA-3′ and D328–25R 5′-GATTAGTCTTTCGCCCCTA-3′; (v) genomic purified DNA (10 ng/μL) from a pool of *Trichinella* larvae of each taxon (ITRC, Rome, Italy).

Equipment: (i) a thermocycler; and (ii) horizontal electrophoresis apparatus connected to a cooling system.

Procedure: The total volume of the amplification reaction should be of 50 μL.

1. Prepare the amplification solution by adding sequentially: 25 μL 2× PCR Master Mix, 13 μL water, 1.0 μL of each primer (25 pmol), and 10 μL of DNA extract from single larvae or control DNA. When working with multiple samples, a quantity of amplification solution that is sufficient for all the samples can be prepared all at once; aliquot the solution in the PCR tubes containing the DNA samples to be tested.
2. Close the tubes, mix by vortexing, and centrifuge at maximum speed for a few seconds.
3. Apply the following cycling program: 1 cycle of 95°C for 5 min; 35 cycles of 94°C for 30 s, 55°C for 30 s, and 72°C for 30 s; 1 cycle of 72°C for 5 min. It is recommended that hot start Taq DNA polymerase be used instead of standard enzymes.
4. At the end of the amplification, centrifuge the tubes at maximum speed for a few seconds.
5. Add 1 volume of denaturing DNA loading buffer into each sample tube.
6. Load the samples onto a 2% agarose gel to check the amplification products. Evaluate by eye the amount of DNA product in each sample, and evaluate the volume to submit to the SSCP run to have signals of similar intensity.
7. Denaturate the samples at 94°C for 15 min.
8. Perform snap-cooling on individual samples (12 mL) on a freezer block (–20°C).
9. Load samples onto precast GMA™ S-50 gels (96 × 261 mm).
10. Run electrophoresis for 14 h at 72 V and constant temperature of 7.2°C in a horizontal apparatus connected to a cooling system.
11. Stain the gel with ethidium bromide (0.5 μg/mL) for 15 min.
12. Destain the gel in water for 15 min and take a photograph of the results.
13. For the identification of the taxa, compare the band patterns of the tested samples with those of the reference DNAs.

67.3 CONCLUSIONS AND FUTURE PERSPECTIVES

The molecular methods used to identify parasites of the genus *Trichinella* have been developed with muscle larvae, which can be easily collected from infected animals, meat samples, and human biopsies. Attempts to detect migrating newborn larvae in blood or buffy coat samples, using a PCR-derived method, have failed because the newborn larva are produced intermittently, making it difficult to identify the period of larval production by females parasitizing the gut. This problem could potentially be resolved by improving methods for extracting DNA from blood, which would allow the infection to be diagnosed earlier.

With regard to the possibility of using a PCR-derived method directly on muscle samples (i.e., without first performing artificial digestion), this is not feasible, given that, on average, there are only 1–10 larvae (or less)/g of muscle, so that the amount of larva DNA would be greatly exceeded by the amount of host DNA and thus not detectable. Only if the quantity of muscle larvae were exceptionally high (e.g., 100 larvae/g or more) would the primers succeed in recognizing the target DNA. Thus, PCR-derived methods cannot be used to diagnose *Trichinella* infection but only to identify the parasite at the species or genotype level. A possible goal for the future would be to develop new PCR-derived methods that would allow a diagnosis to be made on muscle samples with a low worm burden.

One of the advantages of PCR-derived methods is that they are sensitive enough to identify the species or genotype based on the analysis of a single larva or fragment of larva. This is extremely important for detecting mixed infections, which can provide important information on host–parasite contacts and on the role of immunity in preventing the development of a secondary infection [1].

Since the discovery that the genus *Trichinella* is characterized by more than one species, scientists have developed new and simple methods for identifying the various taxa. At present, PCR-derived methods have proven to have both the highest sensitivity and the highest specificity. However, when using these methods, it should be kept in mind that the ability to differentiate isolates should not be equated with their classification as a unique taxon.

As mentioned, some PCR-derived methods are so sensitive that the species or genotype of the parasite can be identified based on only 1/15 of larval DNA. This amount of DNA could probably be further reduced using new technologies such as real-time PCR, although the cost–benefit ratio of applying such methods would obviously be quite high.

Anomalies can occur when analyzing individual larvae by multiplex PCR. Genotype-specific bands generated from the ITS primer pairs are occasionally absent from the band pattern. This is most likely the result of a poor amplification of the ITS region, together with poor-quality genomic DNA extracted from individual larvae. Alternatively, absolute sequence conservation may not exist among individual parasites within a given population at the genotype-specific primer binding sites in the multiplex PCR. Less than 100% sequence similarity at the 3′-end could also cause periodic loss of bands.

Since *Trichinella* parasites reach humans through the consumption of meat, it would be very useful to identify microsatellite loci which could be used as isolate markers to trace the source of infection. This would contribute to the control and eradication of foci in the domestic habitat and to identifying the region of origin of infected wild animals, so as to inform local hunters of the risk of zoonotic transmission constituted by certain practices, such as that of leaving carcasses in the field after skinning, or discarding entrails [1]. Tracing the source of infection has become increasingly important with the expansion of the global market of food products. For example, in a large outbreak of trichinellosis for the consumption of horse meat in France, the identification of the etiological agent as *T. murrelli* allowed the origin of infection to be traced back to the United States, given that this species only exists in North America.

REFERENCES

1. Pozio, E., and Murrell, K.D. Systematics and epidemiology of *Trichinella*. *Adv. Parasitol.* 63, 367, 2006.
2. Pozio, E. Taxonomy, biology and epidemiology of *Trichinella* parasites. In: Dupouy-Camet, J., and Murrell, K.D. (eds.), *FAO/WHO/OIE Guidelines for the Surveillance, Management, Prevention and Control of Trichinellosis*, Chapter 1. World Organisation for Animal Health, Paris, 2007.
3. Zarlenga, D.S. et al. Post-Miocene expansion, colonization, and host switching drove speciation among extant nematodes of the archaic genus *Trichinella*. *Proc. Natl. Acad. Sci. USA*. 103, 7354, 2006.
4. Pozio, E., and Zarlenga, D.S. Recent advances on the taxonomy, systematics and epidemiology of *Trichinella*. *Int. J. Parasitol.* 35, 1191, 2005.
5. Pozio, E. et al. Taxonomic revision of the genus *Trichinella*. *J. Parasitol.* 78, 654, 1992.
6. Bruschi, F. et al. Anaphylactic response to parasite antigens: IgE and IgG1 independently induce death in *Trichinella*-infected mice. *Int. Arch. Allergy. Immunol.* 119, 291, 1999.
7. Gomez Morales, M.A. et al. Increased CD8+-T-cell expression and a type 2 cytokine pattern during the muscular phase of *Trichinella* infection in humans. *Inf. Immun.* 70, 233, 2002.
8. Forbes, L.B. Zoonotic aspects of trichinellosis. In: Rhode, K. (ed.), *Marine Parasitology*, Chapter 11. CSIRO, Melbourne, 2005.
9. Dick, T.A., and Pozio, E. *Trichinella* spp. and trichinellosis. In: Samuel, W.M., Pybus, M.J., and Kocan, A.A. (eds), *Parasitic Diseases of Wild Mammals*, Chapter 15. Iowa State University Press, Ames, 2001.
10. Serhir, B. et al. Outbreak of trichinellosis associated with arctic walruses in northern Canada. *Can. Com. Dis. Rep.* 27, 31, 2001.
11. Schellenberg, R.S. et al. An outbreak of trichinellosis due to consumption of bear meat infected with *Trichinella nativa*, in 2 northern Saskatchewan communities. *J. Infec. Dis.* 188, 835, 2003.
12. Nelson, M. et al. A common-source outbreak of trichinosis from consumption of bear meat. *J. Environ. Health* 65, 16, 2003.
13. Ancelle, T. et al. Outbreak of trichinellosis due to consumption of bear meat from Canada, France, September 2005. *Eurosurveillance* 10, 10, 2005.
14. La Rosa, G. et al. Molecular identification of natural hybrids between *Trichinella nativa* and *Trichinella* T6 provides evidence of gene flow and ongoing genetic divergence. *Int. J. Parasitol.* 33, 209, 2003.
15. Gari-Toussaint, M. et al. Human trichinellosis due to *Trichinella britovi* in southern France after consumption of frozen wild boar meat. *Eur. Surveill.* 10, 117, 2005.
16. Rodriguez de las Parras, E. et al. Trichinellosis outbreaks in Spain (1990–2001). *Enferm. Infec. Microbiol. Clin.* 22, 70, 2004.
17. Ozdemir, D. et al. Acute trichinellosis in children compared with adults. *Pediatric. Infect. Dis. J.* 24, 897, 2005.
18. Nezri, M. et al. Premiere observation d'un cas de humain de trichinellose a *Trichinella britovi* en Algerie après consummation de viande de chacal (*Canis aureus*). *Bull. Soc. Pathol. Exotique* 99, 94, 2006.
19. Pozio, E. et al. Human outbreak of trichinellosis in the Mediterranean island of Sardinia, Italy. *Vet. Parasitol.* 140, 177, 2006.
20. Pozio, E., Gomez Morales, M.A., and Dupouy Camet, J. Clinical aspects, diagnosis and treatment of trichinellosis. *Exp. Rev. Anti-infect. Ther.* 1, 471, 2003.
21. Marucci, G. et al. *Trichinella nelsoni* and *Trichinella* T8 mixed infection in a lion (*Panthera leo*) of the Kruger National Park (South Africa). *Vet. Parasitol.* 159, 225, 2009.
22. Kapel, C.M.O., and Gamble, H.R. Infectivity, persistence, and antibody response to domestic and sylvatic *Trichinella* spp. in experimentally infected pigs. *Int. J. Parasitol.* 30, 215, 2000.

23. Kanai, Y. et al. Epizootiological survey of *Trichinella* spp. infection in carnivores, rodents and insectivores in Hokkaido, Japan. *Jpn. J. Vet. Res.* 54, 175, 2007.
24. Pozio, E. et al. *Trichinella britovi* etiological agent of sylvatic trichinellosis in the Republic of Guinea (West Africa) and a re-evaluation of geographical distribution for encapsulated species in Africa. *Int. J. Parasitol.* 35, 955, 2005.
25. Forrester, A.T. et al. The first record of an outbreak of trichinosis in Africa south of the Sahara. *Trans. Roy. Soc. Trop. Med. Hyg.* 55, 503, 1961.
26. Bura, M.W., and Willett, W.C. An outbreak of trichinosis in Tanzania. *East African Med. J.* 54, 185, 1977.
27. Krivokapich, S. et al. Molecular evidence for a novel encapsulated genotype of *Trichinella* from Patagonia, Argentina, Abstract book, p. 13, *12th International Conference on Trichinellosis*, National Park Plitvice Lakes, Croatia, pp. 25–30, September, 2007.
28. La Rosa, G. et al. *Trichinella pseudospiralis* populations of the Palearctic region and their relationship with populations of the Nearctic and Australian regions. *Int. J. Parasitol.* 31, 297, 2001.
29. Pozio, E. The broad spectrum of *Trichinella* hosts: From cold- to warm-blooded animals. *Vet. Parasitol.* 132, 3, 2005.
30. Andrews J.R.H. et al. Identification of *Trichinella pseudospiralis* from a human case using random amplified polymorphic DNA. *Am. J. Trop. Med. Hyg.* 53, 185, 1995.
31. Britov, V.A. Trichinellosis in Kamchatka. *Wiad. Parazytolog* 43, 287, 1997.
32. Jongwutiwes, S. et al. First outbreak of human trichinellosis caused by *Trichinella pseudospiralis*. *Clin. Infect. Dis.* 26, 111, 1998.
33. Ranque, S. et al. *Trichinella pseudospiralis* outbreak in France. *Em. Infect. Dis.* 6, 543, 2000.
34. Pozio, E. et al. *Trichinella papuae* n. sp. (Nematoda), a new non-encapsulated species from domestic and sylvatic swine of Papua New Guinea. *Int. J. Parasitol.* 29, 1825, 1999.
35. Pozio, E. et al. Inappropriate feeding practice favors the transmission of *Trichinella papuae* from wild pigs to saltwater crocodiles in Papua New Guinea. *Vet. Parasitol.* 127, 245, 2005.
36. Kusolsuk, T. et al. *Trichinella papuae* in Ban-Rai district, Uthaithani province, Thailand, Abstract book, p. 47, *5th Food and Waterborne Parasitic Zoonoses*, Bangkok, Thailand, pp. 28–30, November, 2006.
37. Owen, I.L. et al. *Trichinella* infection in a hunting population of Papua New Guinea suggests an ancient relationship of *Trichinella* with human beings. *Trans. Roy. Soc. Trop. Med. Hyg.* 99, 618, 2005.
38. Pozio, E. et al. *Trichinella zimbabwensis* n.sp. (Nematoda), a new non-encapsulated species from crocodiles (*Crocodylus niloticus*) in Zimbabwe also infecting mammals. *Int. J. Parasitol.* 32, 1787, 2002.
39. Pozio, E. et al. *Trichinella zimbabwensis* in wild reptiles of Zimbabwe and Mozambique and in farmed reptiles of Ethiopia. *Vet. Parasitol.* 143, 305, 2007.
40. Pozio, E. et al. Trichinellosis in the European Union: Epidemiology, ecology and economic impact. *Parasitol. Today* 14, 35, 1998.
41. Pozio, E. et al. Hunting practices increase the prevalence of *Trichinella* infection in wolves from European Russia. *J. Parasitol.* 87, 1498, 2001.
42. Dupouy-Camet, J. et al. Opinion on the diagnosis and treatment of human trichinellosis. *Exp. Opinion Pharmacother.* 3, 1117, 2002.
43. Dupouy-Camet, J., and Bruschi, F. Management and diagnosis of human trichinellosis. In: Dupouy-Camet, J., and Murrell, K.D. (eds), *FAO/WHO/OIE Guidelines for the Surveillance, Management, Prevention and Control of Trichinellosis*, Chapter 2. World Organisation for Animal Health, Paris, 2007.
44. Kociecka, W., Trichinellosis: Human disease, diagnosis and treatment. *Vet. Parasitol.* 93, 365, 2000.
45. Capó, V.A., Despommier, D.D., and Polvere, R.I. *Trichinella spiralis*: Vascular endothelial growth factor is up-regulated within the nurse cell during the early phase of its formation. *J. Parasitol.* 84, 209, 1998.
46. Dick, T.A. et al. The use of the polymerase chain reaction to identify porcine isolates of *Trichinella*. *J. Parasitol.* 78, 145, 1992.
47. Soulé, C. et al. Differentiation of *Trichinella* isolates by polymerase chain reaction. *Parasitol. Res.* 79, 461, 1993.
48. Bandi, C. et al. Random amplified polymorphic DNA technique for the identification of *Trichinella* species. *Parasitology* 107, 419, 1993.
49. Bandi, C. et al. Random amplified polymorphic DNA fingerprints of the eight taxa of *Trichinella* and their comparison with allozyme analysis. *Parasitology* 110, 401, 1995.
50. Wang, H. et al. Restriction fragment length polymorphism (RFLP) analysis of genomic DNA of 5 strains of *Trichinella spiralis* in China. *Chin. Med. Sci. J.* 10, 131, 1995.
51. Gasser, R.B. et al. PCR-SSCP of rDNA for the identification of *Trichinella* isolates from mainland China. *Mol. Cell. Probes* 12, 27, 1998.
52. Gasser, R.B. et al. Nonisotopic single-strand conformation polymorphism analysis of sequence variability in ribosomal DNA expansion segments within the genus *Trichinella* (Nematoda: Adenophorea). *Electrophoresis* 25, 3357, 2004.
53. Wu, Z. et al. Polymerase chain reaction primers to identify *Trichinella spiralis* or *T. pseudospiralis*. *Parasitol. Int.* 46, 149, 1997.
54. Wu, Z., Nagano, I., and Takahashi, Y. The detection of *Trichinella* with polymerase chain reaction (PCR) primers constructed using sequences of random amplified polymorphic DNA (RAPD) or sequences of complementary DNA encoding excretory-secretory (E–S) glycoproteins. *Parasitology* 117, 173, 1998.
55. Nagano, I. et al. Identification of *Trichinella* isolates by polymerase chain reaction—Restriction fragment length polymorphism of the mitochondrial cytochrome *c*-oxidase subunit I gene. *Int. J. Parasitol.* 29, 1113, 1999.
56. Appleyard, G.D. et al. Differentiation of *Trichinella* genotypes by polymerase chain reaction using sequence-specific primers. *J. Parasitol.* 85, 556, 1999.
57. Wu, Z. et al. Polymerase chain reaction-restriction fragment length polymorphism (PCR-RFLP) for the identification of *Trichinella* isolates. *Parasitology*. 118, 211, 1999.
58. Wu, Z. et al. DNA fingerprints of *Trichinella* as revealed by restriction fragment length polymorphism and single-strand conformational polymorphism (RFLP-SSCP). *Mol. Cell. Probes* 14, 291, 2000.
59. Zarlenga, D.S. et al. A single multiplex PCR for unequivocal differentiation of six distinct genotypes of *Trichinella* and three geographical genotypes of *Trichinella pseudospiralis*. *Int. J. Parasitol.* 29, 1859, 1999.
60. Pozio, E., and La Rosa, G. PCR-derived methods for the identification of *Trichinella* parasites from animal and human samples. *Methods Mol. Biol.* 216, 299, 2003.

61. Rombout, Y.B., Bosch, S., and Van Der Giessen, J.W. Detection and identification of eight *Trichinella* genotypes by reverse line blot hybridization. *J. Clin. Microbiol.* 39, 642, 2001.
62. MacPherson, J.M. et al. Variability of the random amplified polymorphic DNA assay among thermal cyclers, and effects of primer and DNA concentration. *Mol. Cell. Probes* 7, 293, 1993.
63. Zarlenga, D.S., and Dame, J.B. The identification and characterization of a break within the large subunit ribosomal RNA of *Trichinella spiralis*: Comparison of gap sequences within the genus, *Mol. Biochem. Parasitol.* 51, 281, 1992.
64. Pozio, E. et al. Twelve years of activity of the International *Trichinella* Reference Centre. *Parasite*, 8, S44, 2001.
65. Kapel, C.M., Webster, P., and Gamble, H.R. Muscle distribution of sylvatic and domestic *Trichinella* larvae in production animals and wildlife. *Vet. Parasitol.* 132, 101, 2005.
66. Nöckler, K., and Kapel, C.M.O. Detection and surveillance for *Trichinella* meat inspection and hygiene, and legislation. In: Dupouy-Camet, J., and Murrell, K.D. (eds), *FAO/WHO/OIE Guidelines for the Surveillance, Management, Prevention and Control of Trichinellosis*, Chapter 3. World Organisation for Animal Health, Paris, 2007.

68 Trichostrongylus

Tai-Soon Yong

CONTENTS

68.1 INTRODUCTION

The genus *Trichostrongylus* is a primarily parasitic nematode in the digestive tract of herbivores throughout the world. Therefore, *Trichostrongylus* infections are important in veterinary parasitology. Human infections associated with several *Trichostrongylus* spp. have been occasionally reported from many regions, which are mostly incidental [1]. However, it is believed that there are many more human cases of *Trichostrongylus* infection. Human *Trichostrongylus* infections are presumably underestimated in certain endemic areas where hookworm infections are prevalent. One of the reasons of overlooking is a technical difficulty in differential diagnosis between hookworm and *Trichostrongylus* infections by morphological examination of eggs contained in the stool specimens. Hookworm infections would cause more significant clinical symptoms. Hookworms are believed to be more prevalent and lay more eggs than *Trichostrongylus* spp. in humans. Consequently, *Trichostrongylus* infections are easily neglected than hookworm infections.

68.1.1 Morphology and Classification

Trichostrongylus worms are small and slender (male: $3.8 \sim 4.8 \times 0.07 \sim 0.08$ mm; female: $4.9 \sim 6.7 \times 0.075 \sim 0.087$ mm) with pale reddish color. No specially developed head part was found. The worms lack a buccal capsule, and have a conspicuous ventral notch in the cervical region where the excretory pore is usually situated. The males have a copulatory bursa with well-developed long lateral lobes and poorly developed dorsal lobe. The two spicules are brownish, stout, and developed differently in each species. There is one gubernaculum. These features of male reproductive organs constitute the most useful morphological characters for species identification. In the female, the uteri are opposed, and paired genital tubules open through a vulva, which is situated a short distance behind the middle of the body. Both uteri contain only a small number of eggs, generally 10 or less in *Trichostrongylus orientalis* [2].

The typical eggs are elliptical shaped and thin-shelled, which resemble those of hookworms, except for the longer length ($73 \sim 95 \times 40 \sim 50$ μm) and pointed ends. The eggs are usually laid segmenting in the morula stage of embryonation. Nevertheless, differential diagnosis between hookworm and *Trichostrongylus* spp. by egg morphology is difficult. Some eggs are almost practically indistinguishable on cellophane-thick smears (Kato–Katz stool preparation), one of the most commonly used stool examination methods. The larvae have a characteristic small caudal knob at the posterior end.

The genus *Trichostrongylus* comprises more than 30 species that are primarily parasites of herbivores. Species identification of the genus *Trichostrongylus* is almost entirely dependent on the morphological difference of their two spicules and bursa in males. It is impossible to differentiate individual species of *Trichostrongylus* female worms according to their morphology. According to the literature, eight (or nine) species of *Trichostrongylus* have been recorded from humans, which are *T. orientalis*, *T. colubriformis* (synonym *T. instabilis*), *T. vitrinus*, *T. axei*, *T. capricola*,

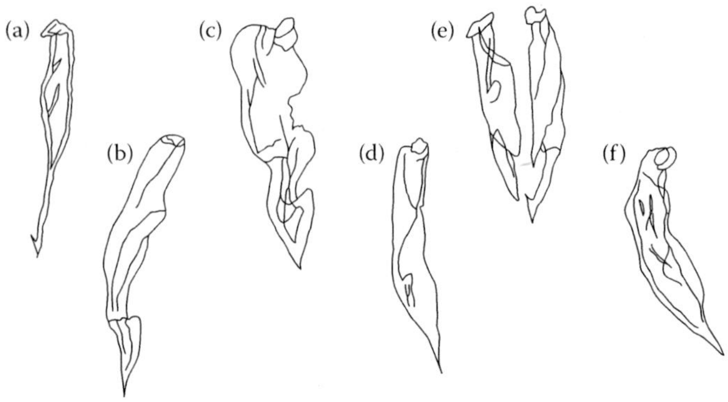

FIGURE 68.1 Spicules of *Trichostrongylus* spp. found in humans. (a) *T. orientalis*, (b) *T. colubriformis*, (c) *T. probolurus*, (d) *T. vitrinus*, (e) *T. axei*, (f) *T. capricola*. Both spicules of *T. axei* are shown where those are not similar. (Adapted from Yong, T.S. et al. *Korean J. Parasitol.* 45, 69, 2007.)

T. probolurus, *T. skrjabini*, and *T. brevis*. The spicules of some species of *Trichostrongylus*, which occur in humans, are shown in Figure 68.1. Some species have two equal spicules while others have unequal ones.

68.1.2 Biology and Epidemiology

The eggs, being in the morula stage, are passed in the feces of herbivores or humans. They mature within 24 h at the earliest, and the first-stage rhabditoid larvae develop. The larvae grow further into infective third-stage larvae having sheath usually in about 4–6 days under optimal conditions with adequate humidity and temperature (27°C). Lower temperatures delay development and almost no development occurs below 9°C [2]. *Trichostrongylus* worms are principally zoonotic parasites of herbivores. For example, *T. colubriformis* occurs in the small intestine and sometimes in the abomasum of sheep, goat, cattle, camel, and various antelopes. It has also been recorded in rabbit, pig, dog, and man. However, where the use of human feces for fertilizer is a common practice, as in East and Southeast Asia, man-to-man transmission is highly possible. Humans are reported to be the most important host for *T. orientalis* parasitizing in the small intestine. The infective larvae migrate on blades of grass to wait for being eaten by herbivores, especially during early morning or early evening when temperature, humidity, and light intensity are favorable for their survival. Oral infection is the major route to final hosts, including humans. However, in some rare occasions, they are believed to infect humans through the skin, and reach the small intestine via lung migration [3]. There are no intermediate hosts or vectors for *Trichostrongylus* transmission.

T. orientalis has been frequently reported from humans in Korea, Japan, China, and Armenia [1]. Human infections with *T. orientalis*, *T. colubriformis*, *T. vitrinus*, *T. axei*, *T. capricola*, *T. probolurus*, and *T. skrjabini* have been reported previously in Iran, of which *T. orientalis* and *T. colubriformis* were more frequently detected than others [4]. *Trichostrongylus* adult worms were collected after anthelmintics treatment such as pyrantel pamoate, and each parasite species was identified morphologically.

The reports on *Trichostrongylus* in humans are relatively few and therefore the actual situation of worldwide human *Trichostrongylus* infections is difficult to estimate. *Trichostrongylus* are primarily parasites of herbivores, so human infections might be a trivial problem comparatively. However, there are other causes to becloud the situation. This lack of information is due in part to the difficulty inherent in the differentiation of *Trichostrongylus* from the two most important human hookworm species, *Ancylostoma duodenale* and *Necator americanus*, at certain life cycle stages (eggs and larvae) via the comparison of morphological features. The differentiation of eggs from those of hookworms can be accomplished when eggs are clearly observed via microscopy after the application of the formalin–ether or formalin–ethyl acetate concentration technique. It is difficult to distinguish the eggs from those of the hookworm, especially on cellophane-thick smears. It is impossible to identify eggs from various species of the genus *Trichostrongylus*. Although the diagnosis of *Trichostrongylus* infection is essential for the control of parasitic infections, current methods for the detection of this parasite from feces are normally predicated only on the observations of light microscopy. For example, stool examinations were conducted nationwide in Laos for the purposes of intestinal parasite control in schoolchildren between 2000 and 2005 [5] in which the overall hookworm infection rate was shown to be 19.1% without a specific description on *Trichostrongylus* infections. In the report, *Trichostrongylus* infections were neglected, so the infection rate of hookworm would have included *Trichostrongylus*-infected cases. The authors realized later that human *Trichostrongylus* infections in Laos nationwide were not a small number, especially in those regions where hookworm infections are prevalent [6]. So far, most human *Trichostrongylus* infections have been identified only morphologically via microscopy of the eggs contained in the fecal specimens or adult worms obtained by anthelmintic treatment and purgation in rare cases.

68.1.3 Clinical Features and Pathogenesis

The larvae do not undergo lung migration following oral infection, but develop to adults in the small intestine. The worms penetrate into the intestinal mucosa, producing desquamation with hemorrhage at the sites of infection. Abdominal pain, diarrhea, anorexia, headache, fatigue, anemia, and eosinophilia (<10% up to 30% approximately) may appear. *Trichostrongylus* may or may not ingest blood. However, clinical signs of blood loss, such as an anemia or emaciation, appear in heavy infections by several hundred worms, even though the worms do not suck blood. Anemia may occur due to shortening of the life span of the red blood cells, impaired erythropoiesis, or the reduction of the amino acid pool. In light infections with *Trichostrongylus*, no clinical manifestations are provoked [7].

68.1.4 Diagnosis

68.1.4.1 Conventional Techniques

A diagnosis is usually made by observing characteristic eggs in the stool specimen by microscopic examination. When eggs are sparse in the stool specimens, diagnosis is made by culturing feces and recovering the larvae. Since the number of eggs contained in the stool is usually small, it is necessary to concentrate them by the floating method using saturated saline or zinc sulfate, or by precipitation method using formalin–ether or formalin–ethyl acetate concentration techniques. Differentiation between hookworm and *Trichostrongylus* eggs rely on the size, shape, and number of the egg cells described above. The eggs do not always have a typical shape, so sometimes it is difficult to differentiate *Trichostrongylus* eggs from those of hookworms, especially by Kato–Katz technique for stool examination. Cultured larvae require differentiation from those of hookworms and *Strongyloides stercoralis*. *Trichostrongylus* larvae are sheathed, and their tail ends are nonpointed. Species identification of individual *Trichostrongylus* spp. with eggs and/or cultured larvae is impossible by using conventional techniques based on the morphological characters.

68.1.4.2 Molecular Techniques

It is necessary to develop and apply molecular biological techniques for the diagnosis of *Trichostrongylus* infections since accurate morphological detection and differentiation are not feasible. Moreover, female worms of *Trichostrongylus* are impossible to identify since morphological characters are not distinct among different species of the genus *Trichostrongylus*. Their egg output is relatively small in comparison with that of hookworms. *Trichostrongylus* eggs are morphologically similar with those of hookworms. So far there is no available or practically applicable molecular technique for the field survey.

In the veterinary field, species identification or classification of *Trichostrongylus* spp. from a livestock has been attempted recently by several research groups using molecular biological techniques. Molecular detection of *Trichostrongylus* infection of humans, however, has seldom been applied thus far.

68.2 METHODS

68.2.1 Sample Preparation

68.2.1.1 Collection of Adult Worms/Eggs and Preparation of Genomic DNA

Trichostrongylus spp. adult worms are collected from human or animal hosts after anthelmintic treatment such as pyrantel pamoate and purgation with magnesium sulfate. Alternatively, the worms can be recovered from the intestine of the slaughtered animals. Stereomicroscopic observation is required to harvest worms after several rounds of rinsing of fresh stools. In many cases, the worms are alive and movable for several hours at room temperature after being expelled from their habitat of the host by pyrantel pamoate treatment and purgation. The eggs inside the uterus of gravid females can be removed under a stereomicroscope.

Genomic DNA can be extracted from worms according to Gasser et al. [8] as follows. Worms are suspended in 20 mM Tris-HCl, pH 8.0, 100 mM EDTA, 1% sodium dodecyl sulfate containing 500 μg/mL proteinase K, and homogenized with a polytron homogenizer (Omni 1000, FSE) for 1 min and incubated for 10 min at 37°C. After homogenization and incubation, the suspension is centrifuged (10,000 × *g*) for 3 min. The supernatant is transferred to a fresh tube and extraction is performed with phenol/chloroform/isoamyl alcohol (25/24/1). The aqueous phase is precipitated with absolute alcohol, and the precipitate of DNA is suspended in the distilled water.

68.2.1.2 Preparation of Eggs and DNA from Eggs/Larvae in Fecal Worms

The isolation of minute amounts of parasite DNA from different biological samples or the removal of substances inhibitory to the optimal enzymatic amplification process is critical for molecular diagnosis. There has been no report which describes the preparation of *Trichostrongylus* DNA directly from the stool specimens without preparation of eggs. Usually, eggs or larvae are harvested, and fecal contaminants are removed. Then, the DNA for molecular diagnosis can be easily prepared from the specimens by any standard procedures.

It would be possible to recover the concentrated eggs by using a saturated zinc sulfate or brine flotation technique, and prepare the DNA from the eggs. For example, the preparation of fecal eggs was described by Schnieder et al. [9] as follows. Ten grams of feces is suspended in the water followed by a 2 h sedimentation and subsequent flotation of the sediment in saturated sodium chloride solution. Five milliliters is taken from the surface, transferred to a 50 mL tube and diluted in 200 mL of distilled water. The procedure is repeated once. The surface (5 mL) is transferred into a 1.5 mL Eppendorf tube followed by washings with distilled water. The supernatant is removed, and a sediment volume of 10 μL containing prepared eggs is used in the polymerase chain reaction (PCR).

L3 larvae can be harvested using the Baermann apparatus following fecal cultures [10] or Harada and Mori's filter paper strip method [1]. However, molecular diagnosis has not been tried using the larvae obtained by cumbersome *in vitro* cultivation.

68.2.2 Detection Procedures

68.2.2.1 PCR Detection of *Trichostrongylus* spp. in Animal Hosts

Trichostrongyle nematodes of ruminants are of major animal health and economic importance. Diagnosis of infections with these parasites is primarily based on the identification of eggs in fecal specimens. However, eggs of most species are morphologically indistinguishable at the generic level [11]. The development of molecular techniques for the identification of nematode species was required and significant, and there were some reports showing genus- or species-specific genes for the accurate identification or delineating these parasites. The second internal transcribed spacer (ITS-2) and the first internal transcribed spacer (ITS-1) have been studied for the purpose as follows:

68.2.2.1.1 Second Internal Transcribed Spacer

Restriction fragment length polymorphisms (RFLP) of the ITS-2 of ribosomal DNA, in which no intraspecific variation was found previously [12], were examined for six species of ruminant nematodes suggesting that this technique has considerable potential for the identification of eggs of nematode species in fecal samples [11]. The simplified procedure is as follows:

1. Target gene: ITS-2 gene region plus the first 68 bases of the 5′ end of the 28s rDNA
2. Primers: NC1: 5′-ACGTCTGGTTCAGGGTTG TT-3′ (forward) and NC2: 5′-TTAGTTTCTTTT CCTCCGCT-3′ (reverse)
3. PCR conditions: 1 cycle at 94°C for 5 min, 35 cycles at 94°C for 1 min (denaturation), 55°C for 1 min (annealing) and 72°C for 1 min (extension)
4. RFLP analysis: digestion for 12 h with *DraI*, *Hinfl*, *RsaI*, or *VspI* and agarose gel electrophoresis

68.2.2.1.2 First Internal Transcribed Spacer

The ITS-1 of seven species of *Trichostrongylus* was sequenced, and little variation was detected in *Trichostrongylus axei*, *T. colubriformis*, and *T. vitrinus* indicating this region of rDNA may also be useful as ITS-2 gene for systematic studies in trichostrongylid nematodes [13]. Adult worms removed from the abomasum and the small intestine were stored frozen at –70°C. Genomic DNA was extracted from 15–20 pooled worms of each species according to Gasser et al. [8].

1. Target gene: center of the ITS-1
2. Primers: NC5:5′-GTAGGTGAACCTGCGGAA GGATCATT-3′ (forward), NC2: 5′-TTAGTTTC TTTTCCTCCGCT-3′ (reverse)
3. Cycle sequencing with primers NC5 (forward) and NC1r: 5′-AACAACCCTGAAGCAGACGT-3′ (reverse) and in a second step with primers HH1: 5′-GTTGTAT GTACGGTACCTGG-3′ (forward) and HH1r: 5′-CC AGGTACCGTACA TA CAAC-3′ (reverse)

Alternatively, genus-specific PCR using ITS-2 with prepared eggs from 10 g of fecal specimens was performed. Single eggs or larvae from feces could be differentiated without previous DNA extraction. However, it was suggested that quantification of the PCR result is difficult because the DNA content between eggs from fresh or 24-h-old feces varied considerably [9]. No species-specific PCR has been successfully developed for identification of single worms or eggs in the fecal specimens of the genus *Trichostrongylus*.

A combined microscopic–molecular approach was reported using specific PCR primers located in the ITS-2 for a real-time PCR of key strongylid infections, including *Trichostrongylus* spp. in sheep [14]. For the combined approach, both ends of each parasite were cut off and preserved in polyvinyl lactophenol as a permanent mount for reference. Nematode specimens for DNA preparation were frozen. Isolation of genomic DNA was performed from adult nematodes, feces, diced tissue, or packed cells. PCR primers previously designed for ITS-2 were used, and applied for a real-time PCR. The result showed the technique to be useful for diagnosis and epidemiological surveys. However, it was not suitable for a field application practically.

68.2.2.2 PCR Detection of *Trichostrongylus* spp. in Humans

There are very few scientific reports showing the possibility and necessity of the molecular technique application for the detection and identification of *Trichostrongylus* in humans. It would have been thought that it is not very necessary to detect or differentiate *Trichostrongylus* spp. in humans. The reason would be that the parasite is not frequently found in humans, and it is treatable with anthelmintics easily without identification of individual species. Another reason would be that *Trichostrongylus* is amiable in the pathogenicity compared with hookworms.

Yong et al. [6] reported that they applied the morphological examination and molecular techniques together to identify *Trichostrongylus* adult worms obtained from Laotians. The experiment showed the possible usefulness of molecular biological techniques in the detection and differentiation of human parasitic nematodes, including hookworms and *Trichostrongylus*. The DNA sequencing data matched relatively well with the morphological feature. However, the result was not very definitely conclusive since a comparison with *T. orientalis* had not been possible. No *T. orientalis* sequence was available in the database so far, which is believed to be a main human parasite species among *Trichostrongylus* spp. in East Asia. In fact, molecular biological or genetic studies on *T. orientalis* are necessary for more understanding of human trichostrongyliasis. PCR using species-specific primers based on the ITS-1 sequence evidenced specifically amplified PCR products of *Trichostrongylus* sp., *A. duodenale*, and *N. americanus* from the eggs of each female parasite (520 bp, 690 bp, and 870 bp, respectively) (Figure 68.2a and b). So far, no concrete attempt has been reported to use the eggs isolated from fecal specimens to identify *Trichostrongylus* species. However, molecular techniques may prove useful in the detection of *Trichostrongylus* infection in humans, and detailed studies of the epidemiology and population biology of *Trichostrongylus* spp. in endemic areas in the future.

68.3 CONCLUSIONS AND FUTURE PERSPECTIVES

Classical methods of diagnosis of parasitic infections are important; however, molecular diagnostic techniques for the reliable detection and identification of parasites are also of great importance. Molecular diagnostic tools can be used

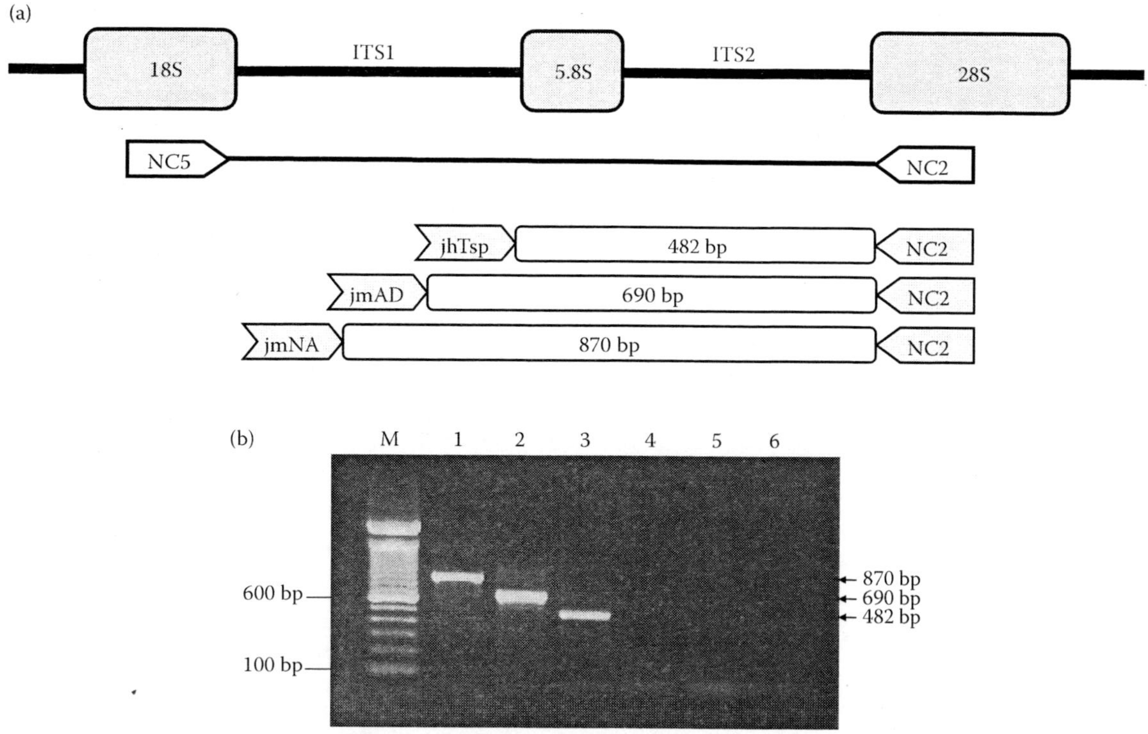

FIGURE 68.2 (a) Schematic representation of part of the rRNA transcriptional unit and relative locations of primers (NC5, NC2, jmAD, jmNA, and jhTsp) used in this study. The region harboring ITS-1, ITS-2, and 5.8S rDNA is amplified with NC5 and NC2 primers. The species-specific primers, jmAD, jmNA, and jhTsp target the region within the ITS-1 sequence. (b) Specificity of ITS-PCR in the recognition of the genomic DNA of other bacteria and protozoa. Lane M, 100 bp ladder, 600 bp and 100 bp bands are shown. Lane 1, *N. americanus*; lane 2, *A. duodenale*; lane 3, *Trichostrongylus colubriformis*; lane 4, *Escherichia coli*; lane 5, *Entamoeba histolytica*; lane 6, *Giardia lamblia*. Only specifically amplified PCR products are detected on lanes 1–3. (From Yong, T.S. et al. *Korean J. Parasitol.* 45, 69, 2007. With permission.)

successfully and effectively for epidemiological studies and the control program of parasitic diseases. The development of the molecular techniques would allow the diagnosis to be faster, easier, and more sensitive to detect minute amount of parasite materials than the microscopic or morphological examination.

By morphological examination, identification of individual species of *Trichostrongylus* is possible only for adult males based on different characters of their reproductive organs. It is, however, impossible to identify individual species of *Trichostrongylus* female worms. Eggs of *Trichostrongylus* spp. passed in the feces of the infected hosts are indistinguishable morphologically, even with those of hookworms. Development of molecular techniques for the identification of these nematode species is required and is significant.

There have been some significant reports which describe the accurate identification or delineation of these parasites using PCR amplification and nucleotide sequencing of genus- or species-specific gene sequences, such as ITS-1 and ITS-2 genes in the veterinary field. Compared with trichostrongyliais in herbivores, human infections associated with several *Trichostrongylus* spp. have been only occasionally reported by the morphological examination mostly. However, the usefulness of molecular biological techniques in the detection and differentiation of human parasitic nematodes, including *Trichostrongylus* and hookworms, was reported.

In the near future, more technical development in the molecular diagnosis is very necessary for the use in epidemiological and/or phylogenetic studies, especially on *Trichostrongylus* infections in humans, which would have been neglected among the neglected tropical diseases.

ACKNOWLEDGMENT

I thank Professor Jong-Yil Chai, Department of Parasitology and Tropical Medicine, Seoul National University College of Medicine, Korea, who provided several informative references and valuable comments.

REFERENCES

1. John, D.T., and Petri, W.A. *Markell and Voge's Medical Parasitology*. 9th ed. Elsevier, St. Louis, MO, p. 266, 2006.
2. Soulsby, E.J.L. *Helminths, Arthropods and Protozoa of Domesticated Animals*. 6th ed. The Williams and Wilkins Co., Baltimore, MD, p. 221, 1968.
3. Miyazaki, I. *An Illustrated Book of Helminthic Zoonoses*. International Medical Foundation of Japan, Tokyo, Japan, p. 344, 1991.
4. Ghadirian, E., and Arfaa, F. Present status of trichostrongyliasis in Iran. *Am. J. Trop. Med. Hyg.* 24, 935, 1975.

5. Rim, H.J. et al. Prevalence of intestinal parasite infections on a national scale among primary schoolchildren in Laos. *Parasitol. Res.* 91, 267, 2003.
6. Yong, T.S. et al. Differential diagnosis of *Trichostrongylus* and hookworm eggs via PCR using ITS-1 sequence. *Korean J. Parasitol.* 45, 69, 2007.
7. Faust, E.C., Russell, P.F., and Jung, R.C. *Craig and Faust's Clinical Parasitology*. 8th ed. Lea & Febiger, Philadelphia, p. 289, 1970.
8. Gasser, R.B. et al. Rapid sequencing of rDNA from single worms and eggs of parasitic helminthes. *Nucleic Acids Res.* 21, 2525, 1993.
9. Schnieder, T., Heise, M., and Epe, C. Genus-specific PCR for the differentiation of eggs or larvae from gastrointestinal nematodes of ruminants. *Parasitol. Res.* 85, 895, 1999.
10. McMurtry, L.W. et al. Distinguishing morphological features of the third larval stage of ovine *Trichostrongylus* spp. *Vet. Parasitol.* 90, 73, 2000.
11. Gasser, R.B. et al. Species identification of trichostrongyle nematodes by PCR-linked RFLP. *Int. J. Parasitol.* 24, 291, 1994.
12. Hoste, H. et al. Lack of intraspecific variation in the second internal transcribed spacer (ITS-2) of *Trichostrongylus colubriformis* ribosomal DNA. *Int. J. Parasitol.* 23, 1069, 1993.
13. Hoste, H. et al. A comparison of the first internal transcribed spacer of ribosomal DNA in seven species of *Trichostrongylus* (Nematoda: Trichostrongylidae). *Int. J. Parasitol.* 28, 1251, 1998.
14. Bott, N.J. et al. A combined microscopic-molecular method for the diagnosis of strongylid infections in sheep. *Int. J. Parasitol.* 39, 1277, 2009.

69 Trichuris

C. Cutillas, M. de Rojas, and R. Callejón

CONTENTS

69.1 INTRODUCTION

Trichuris trichiura is a nematode parasite causing a parasitism known as "tricuriasis" that affects millions of people in the world. First described by Roederer in 1761, *T. trichiura* is the third most common nematode worldwide after *Ascaris* and *Enterobius*. Since prehistoric age *T. trichiura* was present in humans, adapted from the original primate ancestor [1], and coevolved with humans in the Old World tropical and subtropical regions; therefore, eggs of *T. trichiura* have been found in different Old World archaeological sites [2,3]. Since humans migrated out from Africa to other tropical regions, the parasite was carried inside its human host to different worldwide tropical regions where the environmental conditions enabled transmission. Thus, different archaeological records appoint to the existence of this species in North and South America 800–6000 years ago [4,5].

69.1.1 Classification

T. trichiura is a parasite located preferentially in the cecum, although in heavy infections, whipworm can be seen throughout the colon and rectum in humans and primates. Taxonomically, according to Chabaud [6] this species is classified within the class Nematoda, subclass Adenophorea (formerly called Aphasmidia), order Enoplida, superfamily Trichuroidea, and family Trichuridae. Nevertheless, de Ley and Blaxter [7] updated the classification of the Phylum Nematoda using molecular data available by sequencing the small subunit (SSU) rRNA. The molecular data confirmed the existence of two main classes: Chromadorea and Enoplea divided through three nematode lineages or subclasses: Chromadoria and, Dorylaimia and Enoplia, respectively [8–10]. Thus, *T. trichiura* is included on the Enoplea, Dorylaimia, and Trichinellida [10].

Specific morphological differentiation of species of the genus *Trichuris* (Roederer, 1761) is very difficult since some of them can hardly be differentiated from one another because the ranges of most or all the characters in these species compared are overlapped [11]. Different studies have been carried out in order to clarify the systematics of this genus. The first review for *Trichuris* was carried out by Dujardin [12]. Chandler [13] stated that the spicule length was the most reliable character for species differentiation, but that other morphological characters were needed for confirmation. Several morphological studies have been carried out on *Trichuris suis* and *T. trichiura* and the early morphological studies on the adult parasites considered that the two trichurid nematodes belonged to the same species [14–16]. Further studies stated that these two parasites were distinct species [17–19]. Other authors considered that the eggs and the infective larvae of *T. trichiura* were significantly smaller than those of *T. suis* [20]. Beer [21] found no visible morphological differences between the ova and larvae of the two species. It was assumed that the main morphological criterion to differentiate *T. suis* and *T. trichiura* is the existence of typical caudal papillae in *T. trichiura* and the absence of these in *T. suis*. In order to clarify the taxonomic status of both species, Cutillas et al. [22] carried out a morphological, biometrical, and molecular study of them. They concluded that *T. trichiura* and *T. suis* are morphologically identical but genetically different species and they can be differentiated by digestion with specific restriction endonucleases.

69.1.2 Morphology

Adults of *T. trichiura* are commonly known as whipworm due to the typical whip-like shape with a long, narrow anterior end containing the esophagus and the shorter, more thickly posterior end that contains the reproductive organs and intestine. The generic name *Trichuris* was assumed under the impression that the thin region of the body was the caudal one. Latterly, when it was realized that the "hair" was the anterior end rather than the tail, the name *Trichocephalus* (*Trichos* = a hair + *kephale* = head), was given to the genus but it was not taxonomically accepted since *Trichuris* has priority [23].

T. trichiura is 30–50 mm long, with males being somewhat smaller than females (Figure 69.1). The moniliform esophagus is very long and thin, occupying about two-thirds of the body length. The esophagus contains a reduced anterior muscular portion while the rest of the esophagus consists of a thin-walled tube surrounded by large, unicellular glands, or stichocytes [24]. The mouth is a simple opening lacking lips. The entire structure is often referred to as a stichosome. Both sexes have a single gonad, and the anus is near the tip of the tail. Sexual dimorphism is easily indicated by the posterior region of the body (Figure 69.1), since the male shows the posterior end coiled (Figure 69.1a). In both sexes, an excretory system is absent. The ventral surface of the esophageal region bears a wide band of minute pores, leading to underlying glandular and nonglandular cells. This bacillary band is typical of the order. Although the function of the cells in the bacillary band is unknown, their ultrastructure suggests that the gland cells may have a role for osmotic or ion regulation, and nongland cells may function in cuticle formation and food storage [25,26].

The posterior end of males is usually bent spiral-like dorsally (Figure 69.2a). The genital apparatus is a long tube whose different sections differ from each other in structure and bear different functions. Only one, elongated with a pointed tip, spicule is present. It is surrounded by a retractile spicular sheath which is covered by densely located spines (Figure 69.2b). One pair caudal papillae are visible when the spicule is invaginated. On both lateral sides closed to the pericloacal region, a cluster of small papillae is present [22, 27–30] (Figure 69.2c).

In the case of the females, the posterior region of females is occupied by the intestine, whose anus is terminal, and with a single reproductive organ presenting from the distal region the ovary–oviduct–uterus and vagina (Figure 69.3a); the vagina ends in a ventral vulva, located at the beginning of the thickened region of the body (Figure 69.3b). The vulva is a rigid ridge protruding above the surface of the body and surrounding the aperture of the vagina. The surface of the ridge is densely beset with spines, which resemble those on the spicular sheath of the males [25] (Figure 69.3c). The uterus contains many nonembryonated eggs (Figure 69.3a).

The eggs of *T. trichiura* have the typical barrel-shape form and in addition to the vitelline membrane, have a triple deck at which the outermost layer is usually found impregnated with bile. Also, they have a prominent opercular plug in both poles, which has the appearance of mucoidales plugs. The measurements of the eggs are about 50–54 × 22–23 μm^2 and contain an uncleaved zygote at oviposition [31] (Figure 69.4a).

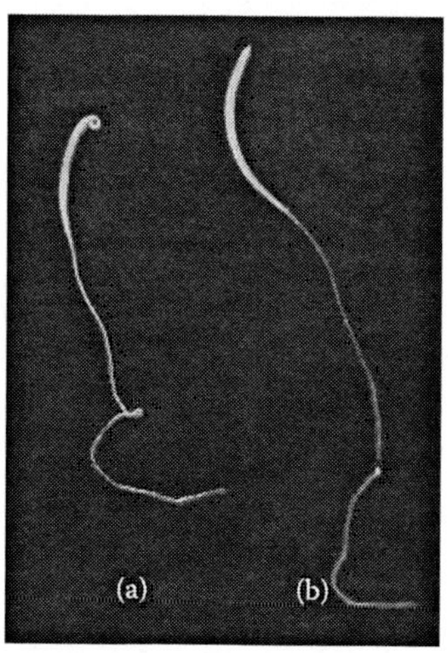

FIGURE 69.1 Male (a) and female (b) of *T. trichiura*.

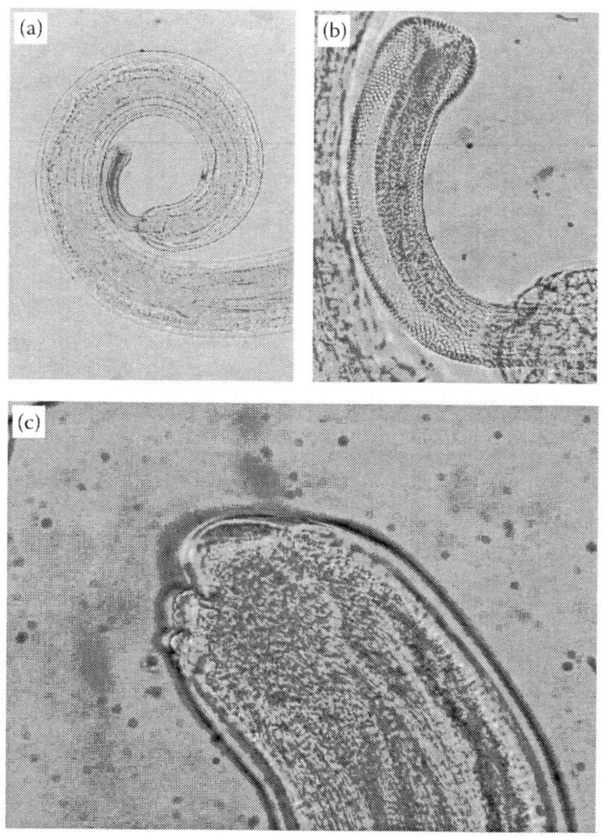

FIGURE 69.2 Male of *T. trichiura*. (a) Coiled posterior end showing the spicular sheath. (b) Spicular sheath showing typical spines. (c) Cloacal papillae.

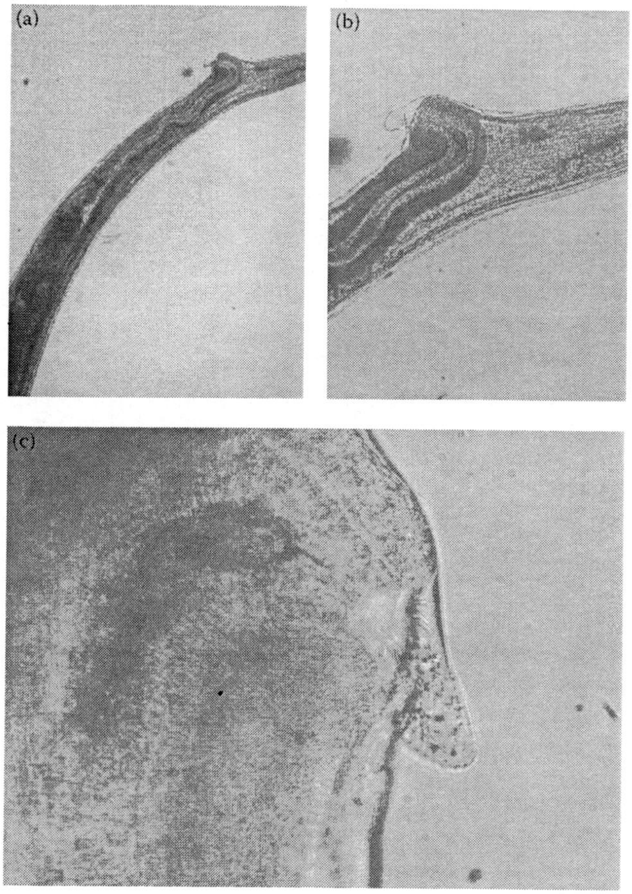

FIGURE 69.3 Female of *T. trichiura* (a). (b) Vulva and vagina. (c) Vulva.

The invasive larva of *T. trichiura* lies in the eggs. The body of the larva is elongated, and the cephalic end is rounded and it is provided with a sharp stylet capable of protrusions from the buccal aperture; the caudal end of the larva bears a small protruding tubercle. The average length of larva, which is artificially liberated from the eggs shell by a slight pressure on the microscopic slide, is 0.115–0.155 mm. The larvae are 0.009–0.115 mm wide. The cephalic end of the larvae is provided with a small but distinct stylet [25].

69.1.3 Biology

T. trichiura has a direct cycle. Adults are located in the cecum and colon of humans where they live embedded in epithelial tunnels within the intestinal mucosa and the posterior end located in the lumen. The female carries out the ovoposition and each female produces thousands of eggs per day. Typical unembryonated eggs are passed out in the feces to the external environmental. Adequate moisture and warm temperatures (about 20–25°C) favor the embryonation (Figure 69.4b) and the formation of the first larvae (Figure 69.4c) inside the egg. Under optimal conditions *T. trichiura* develop to the infective second larval stage (Figure 69.4d). In the soil, the eggs mature in about 3 weeks. The fully developed eggs containing the infective larvae are ingested and hatch in the small intestine of humans. The released larvae enter the intestinal crypts [32]; after a brief period of maturation, the larvae may migrate to the proximal colon where they reside and over 1–3 months mature into adult whipworms [32–35]. The time period from infection to egg production (the prepatent period) has not been determined for *T. trichiura* in humans, although a period of 60 days is widely quoted [36].

T. trichiura is responsible of a disease known as trichuriasis or "whipworm disease," one of the most important geohelminthiasis or soil-transmitted helminthiasis (STHs) widespread in the tropical regions of South of Asia, Africa, and South of America; nevertheless, *Trichuris* infections reach their highest prevalence in Central Africa, Southern India, and Southeast Asia [37]. Trichuriasis is associated with a poor hygiene which favors the direct transmission affecting millions of persons in the tropical regions; especially children are vulnerable because of their high exposure risk. The distribution of the worms for host is such that, under the appropriate conditions for geohelminth transmission about 10% of a population of young children would be expected to have sufficiently intense infections to lead to chronic colitis and growth retardation [38]. Recent estimates suggest that *T. trichiura* infects 1.049 billion persons, including 114 million preschool-age children and 233 million school-age children [39]. However, even within the developing world, wide differences in prevalence rates are also apparent [37]. In developed countries, tricuriasis is only present when the

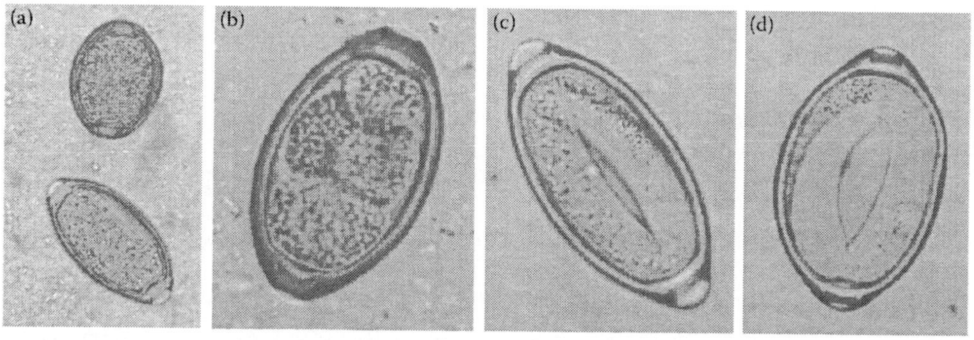

FIGURE 69.4 Eggs of *T. trichiura*. (a) Different aspects of unembryonated eggs. (b) Eggs with cells in division. (c) Egg containing the first larva. (d) Infecting egg.

sanitation is compromised affecting the most deprived communities. Intensity of infection is measured by the number of eggs per gram of feces, generally by the Kato–Katz fecal thick-smear technique [40]. The most intense infections are in children aged 5–15 years, with a decline in intensity and frequency in adulthood. Whether such age dependency indicates changes in exposure, acquired immunity, or a combination of both remain controversial [40,41]. *T. trichiura* is the etiologic agent of tricuriasis, disease considered as one of the STHs together with ascariasis and hookworms. The infection is caused by the ingestion of mature eggs contaminating food, water, or soil [42]. In the developing world, inadequate water and sanitation and crowded living conditions, combined with lack of access to health care and low levels of education, make the poor particularly susceptible to this disease. In areas where there are no latrine systems, the soil (and water) around the village or community becomes contaminated with feces containing worm eggs. Persons infected with STH have parasite eggs in their feces. Thus, *T. trichiura* can infect humans in different ways: by ingestion of the eggs present in vegetables (not washed, cooked, or peeled) which have been irrigated with fecal waters or fertilized with human feces, by ingestion of contaminated water or directly with dirty hands by a poor hygiene. Furthermore, the practice of geophagia may be a major contributor to the transmission in some tropical communities [42]. Therefore, human is the main source of infection to other humans. Nevertheless, zoonotic infections have been reported [43–46]. Thus, *Trichuris vulpis*, the whipworm of dogs, have been recovered from humans [43–46] from different localities attending to the presence of eggs of this species in human stools. Nevertheless, Yoshikawa et al. [47] reported the presence of eggs of *T. trichiura* with two different sizes obtained from the uteri of the adult worms. Furthermore, it is noteworthy that some anthelminthics have an effect on the reproductive system of *T. trichiura* female, resulting in the production of abnormal eggs, which could lead to misdiagnosis of the infection, since they can be mistaken as eggs of other parasites or artifacts [48]. Furthermore, Kenney and Yermakov [49] identified one gravid female and one male of *T. vulpis* in a histopathologic section of an appendix of a man in a postmortem examination. The zoonotic infection from nonhuman primates (monkeys) has been experimentally studied by Horii and Usui [50] in human volunteers. They reported a prepatent period of 127 days of this *T. trichiura* infection in man and eggs were slightly more slender in shape than those from monkeys. On the other hand, Beer [21] achieved for the first time the cross infections of man with *T. suis* ova, the whipworm of pigs, and of pigs with *T. trichiura* ova.

69.1.4 Clinical Features and Pathogenesis

Trichuriasis, or "whipworm disease," becomes symptomatic only in patients with heavy parasite burden. Since there is no larval migration in the life cycle of this nematode, the clinical features of trichuriasis are associated with the acute and chronic manifestations resulting from parasitism of the gastrointestinal tract by adult worms [40].

Adult whipworms occur primarily in the human host's colon but may inhabit the appendix and rectum as well. The anterior end of the worms is embedded in the colonic mucosa and, apparently, producing little damage to the host since parasite burden, frequently, is not heavy. Thus, light infections are confined to the cecum and ascending colon. Involvement of the extreme end of the small bowel, the terminal ileum, may also prove to be important in heavy infections [40].

A parasitic load of fewer than 100 worms causes hardly any symptoms and most infections are asymptomatic. Small children are particularly susceptible to heavy infections that may involve more than 1000 worms [38].

Inflammation at the site of attachment from large numbers of whipworms results in colitis. Longstanding colitis produces a clinical disorder that resembles inflammatory bowel disease, including chronic abdominal pain and diarrhea, as well as the sequelae of impaired growth, anemia of chronic disease and finger-clubbing. *Trichuris* dysentery syndrome is an even more serious manifestation of heavy whipworm infection, resulting in chronic dysentery and rectal prolapse. However, extraintestinal effects such as loss of appetite, nausea, vomiting, retarded growth, and malnutrition can also be detected.

Chronic processes such as the inflammatory bowel diseases and Crohn's disease may be differentiated from trichuriasis where all the pathological manifestations of the parasitic disease seem to be reversible after a short course of therapy [36].

Colitis (an inflammation of the mucosa of the colon) is produced by the traumatic effect on the intestinal epithelium and the submucosa produced by the slender anterior ends of the worms that remains embedded in the gut mucosa. Furthermore, Th2 immune responses to helminth infections may also cause host pathology and organ damage. Thus, Th2 immunity can mediate pathological fibrotic responses in chronic parasitic infections [51].

There is an increase in macrophages in the colonic lamina propria and increased tumor necrosis factor (TNF) concentration. There is also an increase in degranulating mast cells. These facts suggest that inflammation in *Trichuris* colitis may be considered a local tissue anaphylactic response [52].

Another common symptom in strong infection is the rectal prolapse. The mechanism of the prolapse amounts to that of a recto-rectal intussusception, that is an invagination of the bowel into itself. This mechanism differs from the rectal prolapse associated with a weakening of the pelvis musculature and explains why the rectal prolapse in trichuriasis enjoys a good prognosis after the expulsion of the worms. Finally, appendicitis brought about by blockage of the lumen of that organ by worms has been frequently reported.

An odd thickening of the ends of the digits, which is found in several diseases, can also be detected in trichuriasis, and is called finger-clubbing. According to Kitis et al. [53], the origin of this condition is due to the vagus nerve forming the afferent limb of a reflex arc, and suggests that the involvement of the ileum and proximal colon, innervated by the vagus, is responsible for this symptom. Furthermore, moderate-to-heavy infections adversely affect cognitive function in children [54].

Finally, it is worth emphasizing that, in the majority of cases, the immune responses of the hosts-to-worm infections are dominated by Th2-like T-helper cell responses with a significant production of interleukins (IL): IL-4, IL-5, IL-10, IL-13, IL-25, and IL-31 [55,56,57]. Th2 cells activate B cell and antibody production. Th2 immune responses to helminth infections can prevent the survival of invading parasites during a secondary infection [58] and expel adult parasites from the gut. However, Th2 responses can also attenuate damaging Th1-driven inflammatory responses in the host [59,60]. However, suppression of Th1 response may also lead to the impairment of host protective Th1 immunity against concurrent infections caused by bacterial, viral, and parasite pathogens, exacerbating the diseases they induce [61].

69.1.5 Diagnosis

69.1.5.1 Conventional Techniques

Whipworm infection has been considered as one of the STH affecting the poor and developing countries. One of the most important control measurements of this parasitism is by the administration of single-dose anthelminthic treatment to high-risk groups, usually without prior diagnosis [62]. This "preventive chemotherapy" lies in an uncompleted cure and the existence of light infections with a decreasing number of worms and, therefore, in a decline of egg excretion, reducing the environmental contamination and transmission [63]. The success of repeated drug administration can be assessed by measuring the decline in the prevalence and intensity of worm infections [63]; Knopp et al. [63] reported that for accurate diagnosis of *T. trichiura* up to 20 stool samples were needed to be examined in this low-intensity setting. Therefore, diagnostic methods with higher sensitivity and low technical demands are urgently needed.

Examination of human feces for helminth eggs is the most widely used diagnostic technique for the detection of STH infection. Several stool examination techniques are available to detect the presence of STH eggs in fecal samples, including direct fecal smear, the formalin-ether concentration technique [64,65], the Kato–Katz technique [66,67], McMaster technique [68], and the most recently described FLOTAC® technique [69].

Among the diagnostic methods that are used to detect the presence of STH ova in stool samples, the quick Kato–Katz [67] is the most commonly utilized technique [63,70,71]. Due to the ease of its use in the field and relatively low cost, the Kato–Katz technique is recommended by the World Health Organization (WHO) for surveillance and epidemiological field survey of STH infections [72,73].

FLOTAC technique has been described recently both for human and veterinary medicine [69]. This method is facilitated by the FLOTAC apparatus, which has been designed to carry out flotation in a centrifuge, followed by a transversal cut (i.e., translation) of the apical portion of the floating suspension. This technique allows quantification of eggs and/or larvae of nematodes and trematodes as well as cysts and oocysts of intestinal protozoa in up to 1 g of feces [71].

Different authors have reported that the presence of eggs in fecal samples varies day-to-day [74–77] and, therefore, the sensitivity of the diagnostic technique used to determine the whipworm infection is limited to this variation and, hence estimating the presence of infection. Thus, several authors [71,77] recommended analyzing fecal samples during consecutive days to improve the sensitivity of the test, but the higher sensitivity the lower specificity of this technique. As FLOTAC uses a much larger amount of stool for the detection of helminth eggs than does the Kato–Katz method (up to 1 g vs. 41.7 mg) there is a greater chance of detecting helminth eggs, especially if contained in the sample at low density [78]. Thus, Knopp [78] demonstrated that in a random sample of human stools, the examination of a single stool sample with the FLOTAC method more accurately diagnosed infections with the three main STHs than the examination of three consecutive stool samples with the Kato–Katz technique. Furthermore, the advantage of examining several-fold larger amounts of stool with the FLOTAC apparatus than with the widely used Kato–Katz method was most evident for the diagnosis of light-intensity infections with *T. trichiura*, since this nematode produces fewer eggs than *Ascaris*, the probability of detecting their eggs in a larger amount of stool is higher than in a smaller one [78].

Tarafder et al. [79] reviewed six studies that reported the sensitivity and/or specificity of the Kato–Katz technique for detecting STH infection [34,63,70,71,77,80]. Four of these studies estimated the sensitivity and/or specificity of the Kato–Katz against other tests or a combination of tests, none of which is 100% accurate ("pseudo gold standard") [34,70,71,80]. Moreover, although two of these studies [70,80] collected stool samples on more than two consecutive days, they did not account for the day-to-day variation in egg output in their estimation of sensitivity and specificity of the test, and they assumed 100% specificity (constrained) of the Kato–Katz technique. This is unrealistic since false positives can arise if STH eggs are confused with eggs of other helminth species, for example, *Capillaria philippinensis* eggs [81].

Goodman et al. [70] reported that the combination of the Kato–Katz and formol ethyl acetate methods showed the highest prevalence of *Trichuris*, which was equivalent to the prevalence found by all methods. However, this combination did not result in a prevalence that was significantly greater than the duplicate Kato–Katz method alone.

Knopp et al. [63] used a mathematical model to calculate the sensitivity of the Kato–Katz technique for STH infections assuming a perfect specificity for the test [63,76,82]. They concluded that in epidemiologic settings characterized by low-infection intensities of STH, it is important to examine multiple stool samples in order to avoid underestimating the "true" prevalence of STH infections, and hence their transmission potential. Their results indicated that for rigorous epidemiologic surveillance, a combination of methods is required to more accurately assess the situation. Booth et al. [77] used a Bayesian latent class model to estimate the sensitivity of the Kato–Katz technique in detecting hookworm infections [77].

Tarafder et al. [79] reported the first study of sensitivity and specificity of the Kato–Katz stool examination technique for all three STH human infections (hookworms, *Ascaris*, and *T. trichiura*) using a Bayesian statistical method. They demonstrated that this approach could be used to estimate the sensitivity and specificity of the Kato–Katz technique when not all participants provide stool sample over several consecutive days. They concluded that the Kato–Katz technique is currently the method of choice for the detection of whipworm infection in studies conducted in rural environments with poor infrastructure. Thus, their results show that this technique permits diagnosing STH infections with reasonably low proportions of error, particularly with 2 days of stool collection. Correct diagnosis and quantification of STH infections are of considerable importance for successful control of these helminths.

Levecke et al. [83] studied, comparatively, four techniques for the qualitative and quantitative detection of *Trichuris* in simians stool samples as well as their feasibility for mass diagnosis under field conditions. Nonhuman primates are an appropriate model since these animals share the same STHs and have a comparable fecal composition [84,85]. Thus, the ether-based concentration technique, the Parasep SF, the FLOTAC, and the McMaster were compared for the detection of *Trichuris* eggs in stool of simians, focusing on validity, feasibility under field conditions, and ability to estimate the "true" drug efficacy using a statistical simulation. Although FLOTAC is the most sensitive technique, followed by the Parasep SF and the ether-based concentration technique, McMaster and the FLOTAC detected significantly more eggs and the time taken for preparing samples and cleaning between samples was the lowest for McMaster, the FLOTAC technique being the most laborious one. Thus, these authors concluded that McMaster can be considered as the method of choice for monitoring drug efficacy, since sensitivity appeared to be a less important criterion. It is a quantitative technique that can be easily performed under field conditions [75].

Furthermore, routinely, the detection of parasites in fecal specimens is enhanced by the use of concentration procedures, of which formalin–ether sedimentation [86] is generally used in most clinical parasitology or investigation laboratories [87,31]. A variety of reagents, such as ether, ethylacetate, detergent, and acetone, have been employed in stool concentration techniques [86,88–91]. Of these, ether is one of the more commonly useful fat solvents in the concentration of the stool for parasites [92]. This technique also provides certain advantages, including less alteration to organisms and increased recovery of helminth eggs and protozoan cysts [92]. Unfortunately, diethyl ether, an essential component in the technique, is extremely flammable, is highly volatile, produces anesthetic vapor, and forms explosive peroxides when exposed to light [93].

Recently, Ahmadi and Pakdad [87] improved a new concentration technique based in a formalin–Tween method that proved to be as good as or better than formalin–ether technique in concentrating parasite eggs and cysts, as well as in maintaining characteristic morphology, the amount of fine precipitate in the sediment was more than in formalin–ether and concealed parasites from detection. According to the results of this study, Ahmadi and Damraj [94] found the formalin–gasoline method to be as sensitive as the formalin–ether technique for concentration of parasite ova and cysts in stool specimens.

In addition to these techniques, serodiagnostic tools for the detection of STHs, although rarely used, are also available. The serodiagnostic approach has some disadvantages including its invasive nature (collection of blood sample), persistence of antibodies after cure (long half-life of antibodies to helminths) which makes it inefficient for the detection of current infection status, and cross-reactivity with other helminths (nematodes) [63]. In particular, serological diagnosis (indirect diagnosis) is useful especially in the cases of toxocarosis, trichinellosis, echinococcosis, cysticercosis, some filariasis, and schistosomiasis. When possible, for infections caused by helminths, the "gold standard" is represented by direct diagnosis performed by microscopic and/or macroscopic observation of the parasite [95].

Although none of these tests (coprological or serological) is considered as a "gold standard," the Kato–Katz technique is recommended by the WHO for field diagnosis of STHs due to its low cost, noninvasive nature and the relatively low level of technical skill required.

The McMaster technique, which has the advantage of not needing any apparatus other than the McMaster counting chamber, is the most common technique used in veterinary parasitology [68]. Two recent studies claimed higher sensitivity of the FLOTAC technique compared with the Kato–Katz technique, but the authors used a combination of tests as the "gold standard" instead of a true "gold standard" test [71,78]. Also, the feasibility of its use in resource-poor settings is yet to be addressed [71]. A "gold standard" test is not available for the detection of STH infections and the development of such a test may not be feasible [96,97].

The diagnosis of parasite infections by stool examination may be difficult in cases of infection only with a few male parasites. Twelve cases of colonoscopic diagnosis of *T. trichiura* have been reported in different countries [98]. In developed countries, detection of whipworm during colonoscopic examination is unusual. *T. trichiura* developed sporadically in western countries and mostly immigrants from endemic areas. In recent years, direct inspection of whipworms at colonoscopy is increasing, presumably because of an increase in both the prevalence of STH and the use of colonoscopy [98]. Tokmak et al. [99] concluded that colonoscopy with biopsy is the gold standard for disease such as trichuriasis.

69.1.5.2 Molecular Techniques

To date, the microscopic analysis of samples is the only direct method that allows all identification in short times, at a reduced cost, independently from geographical origin and peculiar status of the patient. It has to be regarded as the first step in diagnostic procedures for all laboratories. Some molecular techniques have greater sensitivity than traditional methods, but at least at the present time, their costs may well preclude their routine use [100].

The utilization of Molecular Biology techniques brought new to different biomedical areas. Methods to extract DNA of different evolutive forms of parasites in fecal samples of human beings [101], or other animals [102] have allowed the use of molecular tools for the diagnosis of several organisms. With the application of the polymerase chain reaction (PCR) it was possible to extend these studies [103]. Thus, Machado et al. [103] carried out a random amplified polymorphic DNA analysis to DNAs extracted from *T. trichiura* eggs recovered from human fecal samples. Furthermore, the PCR technique has been used to detect DNA of diverse organisms, facilitating the biological diagnosis of the parasites in tissues and secretions [104]. Another important application of the PCR is the differentiation of helminth species that are morphologically indistinct [22,105,106] or synonymous [107].

The detection and specific identification of *Trichuris* eggs in fecal samples by molecular techniques have been carried out by Cutillas et al. [22]. These authors amplified and sequenced the internal transcribed spacer (ITS) 1 and 2 (ITS1–5.8S-ITS2 fragment) of the ribosomal DNA from eggs of *T. suis* from swine and wild boar, and eggs of *T. trichiura* isolated from feces of primates. They concluded that *T. suis* and *T. trichiura* can be differentially detected by specific enzymatic digestion of DNA products amplified from ITS1, 5.8S and ITS2 sequences.

69.2 METHODS

69.2.1 Sample Preparation

The fecal flotation technique is used routinely for the diagnosis of most parasites that reside within the gastrointestinal tract. The specific gravity and type of solution used will influence the results obtained. A sugar flotation solution with a specific gravity of 1.27 is one that will float a majority of parasite stages with minimal distortion [108].

69.2.1.1 Concentration of Eggs from Feces

The detection of eggs of *T. trichiura* in fecal samples is, firstly, performed by direct smear method and consecutively, different concentration techniques [102,103,109] have been improved (Callejón et al., in preparation).

Eggs of *T. trichiura* are isolated from feces of primates and concentrated with saturated sodium chloride protocol ([102], modified):

- Dilution of a portion of feces (about 10 g) in 600 mL of a saturated salt solution.
- After homogenizing during 15 min, the homogenized solution is filtered through a folded gauze into a beaker and let to float.
- Flotation time varies depending on the type of parasite eggs. Flotation is complete in 2 h. If it is more convenient, flotation can be continued overnight, since this will not adversely affect them.
- The supernatant (~200 mL) is transferred to a new beaker and diluted in an equal volume of distilled water.
- Sedimentation is performed for 1–2 h. The supernatant is discarded and the sediment (~100 mL) centrifuged for 5 min at 10,000 × *g*.
- The resultant pellet is passed to a clean 1.5 mL or 2 mL microcentrifuge tube, washed with distilled water, and centrifuged for 5 min at 10,000 × *g* several times.
- The residue is transferred to a new clean microcentrifuge tube and water was added until a final volume of 1 mL.
- For quantification, 20 μL of the sediment are observed under a light microscope with 40× objective lens.

69.2.1.2 Culture Conditions

After quantification, the material is cultured at 32°C for 3–4 weeks in tubes with 0.2% potassium dichromate solution [50]. About 75–100 eggs containing fully formed infective larvae are, manually, disrupted and subsequently used for molecular purposes [22].

69.2.2 Detection Procedures

Genomic DNA from eggs is extracted using a Dneasy Blood & Tissue Kit (Qiagen). Genomic DNA is detected after 0.8% agarose gel electrophoresis using ethidium bromide. The rDNA region is amplified by PCR using a thermocycler and the following PCR mix: 10 μL 10× PCR buffer, 2 μL 10 mM dNTP mixture (0.2 mM each), 3 μL 50 mM $MgCl_2$, 5 μL primer mix (0.5 mM each), 1.5 μL template DNA, 0.5 μL *Taq* DNA polymerase (2.5 U), and autoclaved distilled water to 100 μL. The following conditions are used: 94°C 3 min (denaturing), 40 cycles at 94°C for 1 min, 57°C for 1 min, 72°C for 1 min, followed by 10 min at 72°C. DNA sequences for *T. suis* of the forward primer NC5 (5′-GTAGGTGAACCTGCGGAAGGATCATT-3′) and reverse primer NC2 (5′-TTAGTTTCTTTTCCTCCGCT-3′) correspond to the conserved 3′-5′ ends of the ITS1–5.8S-ITS2 flanking the 18S and 28S regions [110], while for the amplification of the ITS1–5.8S-ITS2 fragment of *T. trichiura*, we designed a forward TTF (5′-CTACGTCCCTGCCCTTTGTA-3′) attending to a conserved sequence of the 18S gene of *T. trichiura* and reverse TTR (5′-TCTTTTCCTCCGCTTAACGAT-3′) attending to a conserved sequence of the 28S gene of *T. suis*. For each set of PCR reactions and extraction of the DNA, samples without DNA (negative) and a known (positive) control DNA sample must also be included.

The PCR products are checked on ethidium bromide-stained 2% Tris-Borate-EDTA (TBE) agarose gels. Bands are eluted from the agarose by using the Wizard SV Gel and PCR Clean-Up System Kit (Promega). The isolated DNA is cloned into *E. coli* (DH5α) using pGEM-T Easy vector system (Promega). Transformed cells are selected by overnight incubation at 37°C on LBB/Amp/X-gal/IPTG plates. In order to check for successful cloning and to study the intra-individual variation, at least ten single recombinants (clones) are screened for the DNA insert and sequenced. The 10 clones containing the correct insert are used to inoculate 5 mL of LBB/Amp broth and incubated, shaking, at 37°C overnight. Plasmid is purified using a Wizard Plus SV Miniprep (Promega) and sequenced.

69.3 CONCLUSIONS AND FUTURE PERSPECTIVES

The update presented by de Silva et al. [37] clearly indicates that STH infections must be considered as the most prevalent infections of humankind. Although, in some regions there has been a precipitous decline in STH prevalence primarily because of economic development and specific control, in many cases the prevalence rates are equivalent to those first estimated by Norman Stoll more than 50 years ago [111]. The extraordinary numbers of STH infections, which approach two billion, are a reflection of a remarkably successful adaptation to parasitism by *Ascaris*, *Trichuris*, *Necator*, and *Ancylostoma* nematodes [37]. Short of dramatic improvements in the global economy, it appears unlikely that the prevalence of infection will decrease any time soon. However, despite this situation, the application on a large-scale of available simple and low-cost interventions can significantly control the morbidity due to these infections in the vulnerable groups [37].

Vaccination remains the method of choice to control STH infection, because it offers the possibility of a simple, single step for the interruption of infection, disease, and transmission. Several substantial obstacles impede vaccine development against STHs [112], including the lack of good animal models and a poor understanding of the events that permit STHs to endure for years in their human host in the face of a potent immune response [40].

STH infections in people will remain a worldwide public-health threat for as long as poverty persists in the developing world. The UN agencies have appropriately recognized the health and educational effect of these infections in children, and have taken steps to distribute anthelmintic drugs in schools and to undertake chemotherapy programs on an unprecedented scale [40]. Large-scale deworming is necessary to reduce the worldwide morbidity of these infections, but without improved water supplies and sanitation this approach cannot be relied on for sustainable reductions in parasite frequency or intensity of infection [40]. The infrastructure that has been established for deworming of children in schools is expected, however, to facilitate introduction of new anthelmintic vaccines and other control tools [113], and some of the proposed interventions for the integrated control of endemic neglected tropical coinfections such as lymphatic filariasis, onchocerciasis, schistosomiasis, and trachoma [114]. Such strategies could result in substantial reductions in the worldwide disease burden in the years to come [40].

Despite progress made in recent years, there is still no vaccine against STH [115]. In May 2001, preventive chemotherapy was endorsed by World Health Assembly resolution WHA54.19, urging member states to control morbidity due to STH through regular administration of anthelminthic drugs. The declared aim is to regularly target at least 75% of school-aged children and other high-risk groups by the year 2010 [62,116]. Four anthelminthics are currently on the WHO model list of essential medicines for the treatment and control of STH: albendazole, mebendazole, levamisole, and pyrantel pamoate [116]. The former two are benzimidazoles, which are widely used against STH, often in combination with other drugs to form an integrated approach targeting the so-called neglected tropical diseases [117–119]. However, there is considerable concern that large-scale administration of anthelminthics might result in the development and spread of drug-resistant nematodes, which is already a significant problem in veterinary medicine [120]. Olsen et al. [121] found a low efficacy of the mebendazole and albendazole used as the treatment of trichuriasis. They concluded that the low cure and egg reduction rates of albendazole, and the fact that the egg reduction is only temporary, suggest that alternative drugs and/or alternative regimens in current control programs against *T. trichiura* need renewed attention.

The diagnosis of intestinal parasitic infections in returned travelers traditionally relies on time-consuming analyses by experienced microscopists. Nowadays, the increasing number of travelers to a variety of exotic countries calls for new diagnostic approaches for the efficient processing of samples. Thus, ten Hove et al. [122] carried out a multiplex real-time PCR comparatively with the routine approach of microscopy and antigen-based methods, focusing on four target parasites, *E. histolytica*, *G. lamblia*, *Cryptosporidium* and *S. stercoralis*, respectively. These authors showed PCR to be more sensitive for the specific detection of *E. histolytica*, *G. lamblia*, *C. hominis*/*C. parvum*, and *S. stercoralis*. Furthermore, they discussed the use of multiplex real-time PCR as the diagnostic technique of the most important intestinal parasitic infections and concluded that fully automated procedures and combination with additional targets might replace microscopy in the future to a large extent.

Nevertheless, since trichuriasis is related with the poverty and therefore with the lowest economic (developing society) social class, many years will pass before molecular techniques substitute the conventional diagnostic methods.

ACKNOWLEDGMENT

Research carried out at the laboratory of the authors was supported by a grant from the Ministry of Science and Technology (CGL2008–01459/BOS).

REFERENCES

1. Glen, D.R., and Brooks, D.R. Parasitological evidence pertaining to the phylogeny of the hominoid primates. *Biol. J. Linn. Soc.* 27, 331, 1986.

2. Gonçalves, M.L. et al. Human intestinal parasites in the past: New findings and a review. *Mem. Ints. Oswaldo Cruz* 98, 103, 2003.
3. Harter-Lailheugue, S., and Bouchet, F. Palaeoparasitological study of a typical elements of the low and high Nile Valley. *Bull. Soc. Pathol. Exot.* 99, 53, 2006.
4. Araujo, A. et al. Parasites as probes for prehistoric human migrations? *Trends Parasitol.* 24, 112, 2008.
5. Reinhard, K.J., Hevly, R.H., and Anderson, G.A. Helminth remains from prehistoric Indian coprolites on the Colorado Plateau. *J. Parasitol.* 73, 630, 1987.
6. Chabaud, A.G. CIH keys to the nematode parasites of vertebrates. In: Anderson, R.C., Chabaud, A.G., and Willmott, S. (eds), *Class Nematoda*. Commonwealth Agricultural Bureaux, Slough, Inglaterra, pp. 7–17, 1974.
7. de Ley, P., and Blaxter, M.L. Systematic position and phylogeny. In: Lee, D.L. (ed.), *The Biology of Nematode*. Taylor & Francis, London, pp. 1–30, 2002.
8. Lorenzen, S. Entwurf eines phylogenetischen systems der freilebenden nematoden. *Veröffentlichungen des Institut für Meeresforshungen Bremerhaven* 7, 472, 1981.
9. Inglis, W.G. An outline classification of the phylum Nematoda. *Aust. J. Zool.* 31, 243, 1983.
10. de Ley, P.A. *A Quick Tour of Nematode Diversity and Backbone of Nematode Phylogeny*. WormBook, pp. 1–8, 2006. Website: http://www.wormbook.org. USA.
11. Gagarin, V.G. Critical evaluation of measured characters in the differentiation of helminth species (on the model of trichocephalids) (in Russian). In: Tokobaev, M.M. (ed), *Fauna gelmintov zivotnych y rastenij kirgizii*. ILIM, Frunze, pp. 3–9, 1974.
12. Dujardin, F. *Histoire naturalle des helminthes ou vers intestinaux*. Paris, 1841.
13. Chandler, A.C. Specific characters in the genus *Trichuris* with a description of the new species *Trichuris tenuis* from a camel. *J. Parasitol.*, 16, 198, 1930.
14. Creplin, F.C.H. Cited by Schwartz (1926), 1825.
15. Leuckart, R. Cited by Schwartz (1926), 1876.
16. Schwartz, B. *J. Agric. Res.*, Washington DC, 33, 311, 1926.
17. Shneider, A. Cited by Schwartz (1926), 1866.
18. Sondak, V.A. On the question of species independence of whipworms of man and swine. *Parazitol. Sbornik. Zool. Inst. AN. SSSR*, 10:195–204.
19. Pavlovski, E.N., and Sondak, V.A. On the specific differences between the whipworm of man and swine. *Parazitol. Sbornik. Zool. Inst. AN. SSSR*, 13:35–44.
20. Dinnik, N.N. Differences between *Trichocephalus trichiura* and *Trichocephalus suis* (Schrank, 1938). *Med. Parazit.* 1, 907, 1938.
21. Beer, R.J. The relationship between *Trichuris trichiura* (Linnaeus 1758) of man and *Trichuris suis* (Schrank 1889 of the pig). *Res. Vet. Sci.* 20, 47, 1976.
22. Cutillas, C. et al. *Trichuris trichiura* and *Trichuris suis* are different nematode species. *Acta Trop.* 111, 299, 2009.
23. Gállego Berenguer, J. *Manual de parasitología. Morfología y biología de los parásitos de interés sanitario*, Ed. Universitat de Barcelona, Spain, 2nd ed., pp. 342–346, 2003.
24. Bogitsh, B.J, Carter, C.E., and Oeltmann, T.N. *Human Parasitology*, 3rd ed., Elsevier Academic Press, USA/UK, pp. 324–327, 2005.
25. Skrjabin, K.I, Shikhobalova, N.P., and Orlow, I.V. *Tricocephalidae and Capillariidae of Animals and the Man and the Diseases Caused by them. Izd.* Akademii Nauk SSRR, Moskva, 587, 1957.
26. Schmidt, G.D., and Robert´s, L.S. *Foundations of Parasitology*, 7th ed., McGraw-Hill, New York, USA, pp. 397–400, 2005.
27. Tenora, F., Hovorka, I., and Hejlková, D. A supplement to the scanning electron microscopy of some *Trichocephalus* spp (Nematoda). *Helminthologia* 25, 227, 1988.
28. Tenora, F. et al. Some novel features of male posterior end of *Trichuris discolor* as revealed by scanning electron microscopy. *Jpn. J. Parasitol.* 41, 487, 1992.
29. Ooi, H.K. et al. Comparative study of *Trichuris trichiura* from non-human primates and from man and their difference with *T. suis*. *J. Vet. Med. Sci.* 55, 363, 1993.
30. Mahmoud, L.H. Scanning electron microscopy of *Trichuris trichiura*. *J. Egypt. Soc. Parasitol.* 32, 469, 2002.
31. Beaver, P.C., Jung, R.C., and Cupp, E.W. *Parasitología Clínica*, Ed. Salvat, Barcelona, 2nd ed., pp. 241–245, 1994.
32. Monroe, L.S. Gastrointestinal parasites. In: Haubrich, W.S., Schaffner, F., and Berk, J.E. (eds), *Bockus Gastroenterology*, 5th ed., Phil. Saunders, Philadelphia, USA, 3113, 1995.
33. Fishman, J.A., and Perrone, T.L. Colonic obstruction and perforation due to *Trichuris trichiura*. *Am. J. Med.* 77, 154, 1984.
34. Santos, F.L, Cerqueira, E.J., and Soares, N.M. Comparison of the thick smear and Kato–Katz techniques for diagnosis of intestinal helminth infections. *Rev. Soc. Bras. Med. Trop.*, 38, 196, 2005.
35. Chandra, B., and Long, J.D. Diagnosis of *Trichuris trichiura* (whipworm) by colonoscopic extraction. *J. Clin. Gastroenterol.* 27, 152, 1998.
36. Bundy, D.A.P., and Cooper, E.S. *Trichuris* and trichuriasis in humans. *Adv. Parasitol.* 28, 107, 1989.
37. de Silva, N.R. et al. Soil-transmitted helminth infections: Updating the global picture. *Trends Parasitol.* 19, 547, 2003.
38. Cooper, E.S., and Bundy, D.A.P. *Trichuris* is not trivial. *Parasitol. Today* 4, 301, 1988.
39. Khuroo, M.S. et al. *Trichuris* dysentery syndrome: A common cause of chronic iron deficiency anemia in adults in an endemic area (with videos). *Gastrointest. Endoscopy* 71, 200, 2010.
40. Bethony, J. et al. Soil-transmitted helminth infections: Ascariasis, trichuriasis, and hookworm. *Seminar* 367, 1521, 2006.
41. Galvani, A.P. Age-dependent epidemiological patterns and strain diversity in helminth parasites. *J. Parasitol.* 91, 24, 2005.
42. Cooper, E.S., and Bundy, D.A.P. Trichuriasis. In: Pawloswki, Z.-S. (ed.), *Intestinal Helminthic Infections, Baillière's Clin.*, Bailliere Tindall, London and Philadelphia, *Trop. Med. Com. Dis.* 2, 629, 1987.
43. Dunn, J.J. et al. *Trichuris vulpis* recovered from a patient with chronic diarrhea and five dogs. *J. Clin. Microbiol.* 40, 2703, 2002.
44. Mirdha, B.R. et al. *Trichuris vulpis* infection in slum children. *Indian J. Gastroenterol.* 17, 154, 1998.
45. Vásquez, O. et al. Infección Mixta por *Trichuris trichiura* y por *Trichuris vulpis*. *Rev. Gastroenterol. Perú* 17, 255–258, 1997.
46. Singh, S. et al. *Trichuris vulpis* infection in an Indian tribal population. *J. Parasitol.* 79, 457, 1993.
47. Yoshikawa, H. et al., Variations in egg size of *Trichuris trichiura*. *Parasitol. Res.* 75, 649, 1989.
48. Ferrer-Rodríguez, I., and Kozek, W.J. Abnormal *Trichuris trichiura* eggs detected during an epidemiological survey. *P. R. Health Sci. J.* 26, 237, 2007.
49. Kenney M., and Yermakov V. Infection of man with *Trichuris vulpis*, the whipworm of dogs. *Am. J. Trop. Med. Hyg.* 29, 1205, 1980.
50. Horii, Y., and Usui, M. Experimental transmission of *Trichuris* ova from monkeys to man. *Trans. Roy. Soc. Trop. Med. Hyg.* 79, 423, 1984.

51. Zaccone, P. et al. *Schistosoma mansoni* antigens modulate the activity of the innate immune response and prevent onset of type 1 diabetes. *Eur. J. Inmunol.* 33, 1439, 2003.
52. Cooper, E.S. et al. Intestinal nematode infections in children: The pathophysiological price paid. *Parasitology* 104, S91, 1992.
53. Kitis, G., Thompson, H., and Allan, R.N. Finger clubbing in inflammatory bowel disease: Its prevalence and pathogenesis. *Br. Med. J.* 2(6194), 825–828, 1979.
54. Nokes, C. et al., Moderate to heavy infections of *Trichuris trichiura* affect cognitive function in Jamaican school children. *Parasitology* 104, 539, 1992.
55. Fort, M.M. et al. IL-25 induces IL-4, IL-5, and IL-13 and Th2-associated pathologies in vivo. *Immunity* 15, 985, 2001.
56. Harnett, W., and Harnett, M.M. Molecular basis of worm-induced immunomodulation. *Parasite Immunol.* 28, 535, 2006.
57. Finkelman, F.D. Interleukin-4 and interleukin-13 mediated host protection against intestinal nematode parasites. *Inmunol. Rev.* 201, 139, 2004.
58. Voehringer, D. et al. Type 2 immunity is controlled by IL-4/IL-13 expression in hematopoietic non-eosinophil cells of the innate immune system. *J. Exp. Med.* 203, 1435, 2006.
59. Fox, J.G. et al. Concurrent enteric helminth infection modulates inflammation and gastric immune responses and reduces helicobacter-induced gastric atrophy. *Nat. Med.* 6, 536, 2000.
60. Summers, R.W. et al. *Trichuris suis* therapy for active ulcerative colitis: A randomized controlled trial. *Gastroenterology* 128, 825, 2005.
61. Wang, L.J., Cao, Y., and Shi, H.N. Helminth infections and intestinal inflammation. *World J. Gastroenterol.* 14, 5125, 2008.
62. WHO. *Preventive Chemotheraphy in Human Helminthiasis: Coordinated use of Anthelminthic Drugs in Control Interventions: A Manual for Health Profesionals and Programme Managers*. World Health Organization, Geneva, pp. 1–62, 2006.
63. Knopp, S. Diagnosis of soil-transmitted helminths in the era of preventive chemotherapy: Effect of multiple stool sampling and use of different diagnostic techniques. *PloS Negl. Trop. Dis.* 2, 331, 2008.
64. Ridley, D.S., and Hawgood, B.C. The value of formol-ether concentration of faecal cysts and ova. *J. Clin. Pathol.* 9, 74, 1956.
65. Allen, A.V.H., and Ridley, D.S. Further observations on the formol-ether concentration technique for faecal parasites. *J. Clin. Pathol.* 23, 545, 1970.
66. Katz, N., Chaves, A., and Pellegrino, J. A simple device for quantitative stool thick-smear technique in *schistosomiasis mansoni. Rev. Inst. Med. Trop. São Paulo* 14, 397, 1972.
67. Peters, P.A. et al. Quick Kato smear for field quantification of *Schistosoma mansoni* eggs. *Am. J. Trop. Med. Hyg.* 29, 217, 1980.
68. WHO. *Monitoring Anthelmitic Efficacy for Soil Transmitted Helminths (STH)*. World Health Organization, Geneva, 2008.
69. Cringoli, G. FLOTAC, a novel apparatus for a multivalent faecal egg count technique. *Parassitologia* 48, 381, 2006.
70. Goodman, D. et al. A comparison of methods for detecting the eggs of *Ascaris, Trichuris*, and hookworm in infant stool, and the epidemiology of infection in Zanzibari infants. *Am. J. Trop. Med. Hyg.* 74, 725, 2007.
71. Utzinger, J. et al. FLOTAC: A new sensitive technique for the diagnosis of hookworm infections in humans. *Trans. Roy. Soc. Trop. Med. Hyg.* 102, 84, 2008.
72. WHO. *Bench Aids for the Diagnosis of Intestinal Parasites*. World Heath Organization, Geneva, 1994.
73. Montresor, A. et al. *Guidelines for the Evaluation of Soil-Transmitted Helthminthiasis and Schistosomiasis at Community Level. A Guide for Managers of Control Programmes*. World Heath Organization, Geneva, pp. 1–45, 1998.
74. Hall, A. Quantitative variability of nematode egg counts in faeces: A study among rural Kenyans. *Trans. R. Soc. Trop. Med. Hyg.* 75, 682, 1981.
75. Anderson, R.M., and Schad G.A. Hookworm burdens and faecal egg counts: An analysis of the biological basis variation. *Trans. R. Soc. Trop. Med. Hyg.* 79, 812, 1985.
76. Marti, H.P., and Koella, J.C. Multiple stool examinations for ova and parasites and rate of false-negative results. *J. Clin. Microbiol.* 31, 3044, 1993.
77. Booth, M. et al. The influence of sampling effort and performance of the Kato–Katz technique in diagnosing *Schistosoma mansoni* and hookworm co-infections in rural Cộte dÍvore. *Parasitology* 127, 525, 2003.
78. Knopp, S. et al. A single FLOTAC is more sensitive than triplicate Kato–Katz for the diagnosis of low-intensity soil-transmitted helminth infections. *Trans. Roy. Soc. Trop. Med. Hyg.* 103, 347, 2009.
79. Tarafder, M.R. et al. Estimating the sensitivity and specificity of Kato–Katz stool examination technique for detection of hookworms, *Ascaris lumbricoides* and *Trichuris trichiura* infections in humans in the absence of a `gold standard`. *Int. J. Parasitol.* 40, 399, 2010.
80. Steinmann, P. et al. Extensive multiparasitism in a village of Yunnan province, Peoplés Republic of China, revealed by a suit of diagnostic methods. *Am. J. Top. Med. Hyg.* 78, 760, 2008.
81. Walden, J. Parasitic diseases. Other roundworms. *Trichuris*, hookworm and *Strongyloides. Prim. Care* 18, 53, 1991.
82. Mullen, K., and Prost, A. Decreased microfilarial load and its effect on the calculation of prevalence and the rate of false negatives in the detection of onchocerciasis. *Int. J. Epidemiol.* 12, 102, 1983.
83. Levecke, B. et al. Field validity and feasibility of four techniques for the detection of *Trichuris* in simians: A model for monitoring drug efficacy in public health? *Negl. Trop. Dis.* 3, 366, 2009.
84. Munene, E. et al. Helminth and protozoan gastrointestinal tract parasites in captive and wild-trapped African non-human primates. *Vet. Parasitol.* 78, 195, 1998.
85. Muriuki, S.M.K. et al. Some gastro-intestinal parasites of zoonotic (public health) importance commonly observed in old world non-human primates in Kenya. *Acta Trop.* 71, 73, 1998.
86. Ritchie, L.S. An ether sedimentation technique for routine stool examinations. *Bull. U.S. Army Med. Dept.* 8, 326, 1948.
87. Ahmadi, N.A., and Pakdad, K. Tween as a substitute for diethyl ether in the formalin-ether sedimentation technique. *Iranian J. Publ. Health* 36, 91, 2007.
88. Young, K.H. et al. Ethyl acetate, as a substitute for diethyl ether in the formalin-ether sedimentation technique. *J. Clin. Microbiol.* 10, 852, 1979.
89. Kightlinger, L., and Kightlinger, M.B. Examination of faecal specimens by the formalin-detergent technique. *Trans. R. Soc. Trop. Med. Hyg.* 84, 417, 1990.
90. Parija, S.C. et al. Evaluation of formalin-acetone sedimentation in the concentration of stool for intestinal parasites. *Trop. Doct.* 33, 163, 2003.
91. Ahmadi, N.A. et al. Potency of wet mount, formalin-acetone and formalin-ether methods in detection of intestinal parasitic infections. *Iranian J. Dis. Trop. Med.* 12, 43, 2007.

92. Cheesbrough, M. *District Laboratory Practice in Tropical Countries, Part 1*. Cambridge University Press, Cambridge, 1998.
93. Boswell, M.V., and Collins, V.J. Diethyl ether and chloroform. In: Collins, V.J. (ed), *Physiologic and Pharmacologic bases of Anesthesia*. Williams & Wilkins, Pennsylvania, pp. 650–662, 1996.
94. Ahmadi, N.A., and Damraj, F-A. A field evaluation of formalin-gasoline technique in the concentration of stool for detection of intestinal parasites. *Parasitol. Res.* 104, 553, 2009.
95. Bruschi, F., and Castagna, B. The serodiagnosis of parasitic infections. *Parassitologia* 46, 141, 2004.
96. Joseph, L. et al. Bayesian estimation of disease prevalence and the parameters of diagnostic tests in the absence of a gold standard. *Am. J. Epidemiol.*, 141, 263, 1995.
97. Dendukuri, N. et al. Bayesian sample size determination for prevalence and diagnostic test studies in the absence of a gold standard test. *Biometrics* 60, 388, 2004.
98. Kyung-Sun, O. et al. *Trichuris trichiura* infection diagnosed by colonoscopy: Case reports and review of literature. *Korean J. Parasitol.*, 47, 273, 2009.
99. Tokmak, N. et al. Computed tomographic findings of trichuriasis. *World J. Gastroenterol.* 12, 4270, 2006.
100. Cancrini, G., and Iori, A. Traditional and innovative diagnostic tools: When and why they should be applied. *Parassitologia* 46, 173, 2004.
101. Silva, A.J. et al. Fast reliable extraction of protozoan parasite DNA from fecal specimens. *Mol. Diag.* 4, 57, 1999.
102. Shnieder, T., Heise, M., and Epe, C. Genus-specific PCR for differentiation of eggs or larvae from gastrointestinal nematodes of ruminants. *Parasitol. Res.* 85, 895, 1999.
103. Machado, E. et al. Random amplified polymorphic DNA analysis of DNA extracted from *Trichuris trichiura* (Linnaeus, 1771) eggs and its prospective application to paleoparasitological studies. *Mem. Inst. Oswaldo Cruz* 98, 59, 2003.
104. Nagano, I. et al. Identification of *Trichinella* isolates by polymerase chain reaction-restriction fragment length polymorphism of the mitochondrial cytochome c oxidase subunit I gene. *Int. J. Parasitol.* 29, 1113, 1999.
105. Zarlenga, D.S. et al. Characterization and detection of a newly described Asian Taeniid using closed ribosomal DNA fragments and sequence amplification by polimerase chain reaction. *Exp. Parasitol.* 72, 174, 1991.
106. Zarlenga, D.S. Cloning and characterization of ribosomal RNA genes from three species of *Haemonchus* (Nematoda: Trichostrongyloidea) and identification of PCR primers for rapid differentiation. *Exp. Parasitol.* 8, 28, 1994.
107. Oliveros, R. et al. Characterization of four species of *Trichuris* (Nematoda: Enoplida) by their second internal transcribed spacer ribosomal DNA sequence. *Parasitol. Res.* 86, 1008, 2000.
108. Foreyt, W.J. Diagnostic parasitology. *Vet. Clin. Am. Small Anim. Pract.* 19, 979, 1989.
109. Mochizuki, R. et al. PCR-based species-specific amplification of ITS of *Mecistocirrus digitatus* and its application in identification of G.I. nematode eggs in bovine faeces. *J. Vet. Med. Sci.* 68, 345, 2006.
110. Gasser, R.B. et al. Species markers for equine strongyles detected in intergenic spacer rDNA by PCR-RFLP. *Mol. Cell. Prob.* 10, 371, 1996.
111. Stoll, N.R. This wormy world. *J. Parasitol.* 33, 1–18, 1947.
112. Maizels, R.M. et al. Vaccination against helminth parasites: The ultimate challenge for vaccinologists? *Inmunol.* 171, 125, 1999.
113. Hotez, P.J. et al. Helminth infections: Soil-transmitted helminth infections and schistosomiasis. In: Jamison, D.T., Breman, J.G., Measham, A.R., Alleyne, G., Claeson, M., Evans, D.B., Jha, P., Mills, A., and Musgrove, P. (eds). *Disease Control Priorities in Developing Countries*, 2nd edition. Oxford University Press, New York, Chapter 24, pp. 467–482, 2006.
114. Molyneux, D.H., Hotez, P.J., and Fenwick, A. Rapid-impact interventions: How a policy of integrated control for Africás neglected tropical diseases could benefit the poor. *PloS Med.* 2, 336, 2005.
115. Loukas, A. et al. Hookworm vaccines: Past, present, and future. *Lancet Infect. Dis.* 6, 733, 2006.
116. Utzinger, J., and Keiser, J., Schistosomiasis and soil-transmitted helminthiasis: Common drugs for treatment and control. *Expert Opin. Pharmacother.* 5, 263, 2004.
117. Lammie, P.J., Fenwick, A., and Utzinger, J. A blueprint for success: Integration of neglected tropical disease control programmes. *Trends Parasitol.* 22, 313, 2006.
118. Hotez, P.J. et al. Control of neglected tropical diseases. *N. Engl. J. Med.* 357, 1018, 2007.
119. Kabaterine, N.B. The control of schistosomiasis and soil-transmitted helminths in East Africa. *Trends Parasitol.* 22, 332, 2006.
120. Keiser, J., and Utzinger, J. Efficacy of current drugs against soil-transmitted helminth infections. *J. Am. Med. Assoc.* 299, 23, 2008.
121. Olsen, A. et al. Albendazole and mebendazole have low efficacy against *Trichuris trichiura* in school-age children in Kabale District, Uganda. *Trans. Roy. Soc. Trop. Med.* 103, 443, 2009.
122. ten Hove, R.J. et al. Molecular diagnostics of intestinal parasites in returning travellers. *Eur. J. Clin. Microbiol. Infect. Dis.* 28, 1045, 2009.

70 Wuchereria

Dongyou Liu and Frank W. Austin

CONTENTS

70.1 INTRODUCTION

Wuchereria bancrofti, *Brugia malayi*, and *Brugia timori* are three filarial (thread-like) nematodes (roundworms) that cause lymphatic filariasis (LF). In particular, *W. bancrofti* is responsible for 90% of LF (i.e., Bancroftian filariasis), while *B. malayi* and *B. timori* account for the remaining 10%. The clinical symptoms of LF comprise psychiatric illness, and chronic, disabling conditions such as lymphedema (LE), elephantiasis (i.e., thickening and hardening of the skin as a result of blockage of lymph fluid due to worms inside lymph vessels), and genital deformities [1].

As a mosquito-borne disease, LF infects an estimated 120 million people, and puts 1.2 billion people over 80 countries at risk of infection. The disease is endemic across the broad equatorial belt, including Africa, the Nile Delta, Turkey, India, the East Indies, Southeast Asia, the Philippines, Oceanic Islands, Australia, and parts of South America (e.g., Guiana, Haiti, Dominican Republic, and Brazil) [1].

70.1.1 Classification, Morphology, Life Cycle, and Epidemiology

70.1.1.1 Classification

Named in honor of Otto Wucherer and Joseph Bancroft, both of whom extensively studied filarial infections, *W. bancrofti* is a thread-like nematode classified in the genus *Wuchereria*, family Onchocercidae, suborder Filarioidea, order Spirurida, class Chromadorea, phylum Nematoda. Besides *Wuchereria*, the family Onchocercidae includes the genera of *Acanthocheilonema*, *Brugia*, *Cercopithifilaria*, *Dipetalonema*, *Dirofilaria*, *Filaria*, *Foleyella*, *Litomosa*, *Litomosoides*, *Loa*, *Loxodontofilaria*, *Mansonella*, *Ochoterenella*, *Onchocerca*, *Piratuba*, *Sarconema*, and *Waltonella*, of which several (e.g., *Brugia*, *Dirofilaria*, *Loa*, *Mansonella*, *Onchocerca*, *Waltonella*, and *Wuchereria*) contain well-known human pathogens. For example, *W. bancrofti*, *B. malayi*, and *B. timori* reside in regional lymphatic vessels, causing LF (lymphangitis, elephantiasis); *Onchocerca volvulus* and *Loa loa* inhabit subcutaneous nodules, resulting in loss of vision or dermatitis; *Mansonella* spp. parasitize skin, body cavities, subcutaneous, and connective tissues, leading to dermatitis and other malaises. Within the genus *Wuchereria*, two species are recognized: *W. bancrofti* and *Wuchereria kalimantani*, with *W. bancrofti* being the causative agent of Bancroftian filariasis in humans.

70.1.1.2 Morphology

The *W. bancrofti* adult worm is elongate, slender, smooth, and has a round ended. It has a short cephalic region, with nuclei dispersed throughout its body cavity, but not extended to the tail tip. Exhibiting sexual dimorphism, the male worm measures 40 × 0.1 mm with a curved tail, while the female measures 80–100 mm × 0.24–0.30 mm, and is two times longer and nearly three times larger in diameter than the male. Being ovoviviparous, *W. bancrofti* females can produce thousands of juveniles known as microfilariae, which measure 240–300 µm in length and 7.5–10 µm in width (equaling in diameter to a red blood cell). Often considered advanced embryos, *W. bancrofti* microfilariae have a gently curved body with a tapered or pointed tail, and retain the egg membrane as a sheath. Similar to adult worms, the nuclear columns (cells) in microfilariae are loosely packed along the body, and are absent at the tip of the tail.

70.1.1.3 Life Cycle

W. bancrofti relies on two hosts for completion of its life cycle, with humans as the definitive host and mosquitoes as the intermediate host. Within the definitive host, *W. bancrofti* adult male and female worms inhabit regional lymphatic vessels, where after mating, the female produces a large number of microfilariae (first-stage larvae or prelarval eggs). Microfilariae circulate in the blood, and exhibit circadian behavior, namely, when the host is awake, they stay in deep blood vessels; during sleep, they migrate near the surface in peripheral blood vessels (especially between 10 p.m. and 2 a.m.), facilitating their ingestion by a night-biting mosquito (intermediate host). Interestingly, if the host sleeps by day and is awake at night, the periodic appearance of microfilariae is reversed. It is possible that physiological changes associated with sleeping (e.g., lowered body temperature, oxygen tension and adrenal activity, and an increased carbon dioxide tension) may function as signals for the rhythmic behavior of microfilarial parasites. Within the mosquito, microfilariae pass through the wall of the proventriculus and cardiac portion of the midgut, and arrive at the thoracic muscles. Within 1–2 weeks, microfilariae mature into first-stage larvae and then infective third-stage larvae (L3), which move through the hemocoel to the mosquito's proboscis (mouthparts). During mosquito feeding, the infective larvae (L3) leave the proboscis and enter the puncture wound. Via the lymphatic system, the L3 larvae reach regional lymph nodes (especially in the legs and genital area), where they develop into adult worms and achieve sexual maturity in the afferent lymphatic vessels within 6–8 months. The time from infection until microfilariae are detectable in the blood varies from 3 to many months. The average life span of adult worms that release microfilariae may be up to 10 years.

70.1.1.4 Epidemiology

W. bancrofti is prevalent in many parts of the tropics and subtropics. Humans are considered as the only hosts for *W. bancrofti*, since no naturally infected reservoirs other than humans have been identified to date. However, several species of monkeys can be infected experimentally with infective stage *W. bancrofti* larvae recovered from mosquitoes that have fed on infected human volunteers.

Among the mosquito vectors of *W. bancrofti* are: *Culex* species (*C. annulirostris*, *C. bitaeniorhynchus*, *C. pipiens*, and *C. quinquefasciatus*), *Anopheles* species (*A. arabinensis*, *A. bancroftii*, *A. farauti*, *A. funestus*, *A. gambiae*, *A. koliensis*, *A. melas*, *A. merus*, *A. punctulatus*, and *A. wellcomei*), *Aedes* species (*A. aegypti*, *A. aquasalis*, *A. bellator*, *A. cooki*, *A. darlingi*, *A. kochi*, *A. polynesiensis*, *A. pseudoscutellaris*, *A. rotumae*, *A. scapularis*, and *A. vigilax*), *Mansonia* species (*M. pseudotitillans* and *M. uniformis*), and *Coquillettidia juxtamansonia*.

Members of the common house mosquito genera *Anopheles* (in isolated rural areas of Africa), *Culex* (in urban centers of the Americas, particularly *Culex quinquefasciatus*), *Aedes*, and *Mansonia* (in the Pacific and in Asia) play a crucial role in the development of the larvae into the human infective stage, and for transmission to the human hosts. In the South Pacific, *W. bancrofti* infection showing diurnal periodicity is known as periodic.

70.1.2 Clinical Features, Pathogenesis, and Treatment

70.1.2.1 Clinical Features

The key clinical features of *W. bancrofti* infection (i.e., the enlarged limbs and cracked skin of an infected individual resembling that of an elephant) have been documented early in Greek and Roman times. However, the common term "elephantiasis" for the disease is a misnomer, as elephantiasis literally translates to "a disease caused by elephants." LF remains a major public health problem in the tropics and subtropics, and the World Health Organization (WHO) ranks the disease as the second leading cause of permanent and long-term disability, with a goal for its elimination by 2020.

The following clinical signs are generally attributable to Bancroftian filariasis: fever, headache, myalgia, lymphadenitis, lymphadenopathy, inflammation of lymphatics, adenolymphangitis, lymphatic dysfunction, progressive swelling of limbs, progressive swelling of genitalia, chyluria, and elephantiasis. An asymptomatic phase of the disease without clinical signs can occur.

Although *W. bancrofti* infection is usually asymptomatic (especially when it involves small numbers of worms), it can evolve into LE and swelling (in the legs, and sometimes also in the arms, genitalia, and breasts) in some patients. In the endemic areas, 12.5% of infected individuals suffer from LE and 21% of the male patients suffer from hydrocele (i.e., swelling of genitalia).

LF typically displays three distinct phases: asymptomatic phase, inflammatory (acute) phase, and obstructive (chronic) phase.

The asymptomatic phase is seen in individuals born in the endemic areas who demonstrate high microfilaremia but no clinical symptoms, possibly as a result of the cytokine IL-4 suppressing the activity of TH1 cells in the human immune system. This status may remain for years until the inflammatory reaction rises again.

The inflammatory (acute) phase is characterized by fever, chills, skin infections, painful lymph nodes, and tender skin of the lymphedematous extremity. Early symptoms of LF consist of intermittent fever and enlarged, tender lymph nodes (particularly in the inguinal lymph nodes). The afferent lymphatic vessels (draining into the lymph nodes) with the developing and adult worms may also become inflamed and painful. In more chronic infections, orchitis inflammation of the testes is accompanied by painful immediate enlargement and epididymitis, as a consequence of the inflammation of the spermatic cord. Interestingly, adults who have not been previously infected with LF tend to show clinical signs, but not circulating microfilariae, following infection with large numbers of infective larvae [2,3]. In addition, patients with elephantiasis do not usually have circulating microfilariae, but they do have high levels of antibody to microfilariae and increased lymphocyte proliferative responses to adult worm antigens.

The obstructive (chronic) phase is marked by lymph varices, lymph scrotum, hydrocele, chyluria (lymph in urine), and elephantiasis, with scar formation from affected areas. Swollen lymphatics may burst and drain into the genitourinary system, leading to chyluria; the LE, such as thickening of the skin and elephantiasis (in the legs, arms, and scrotum in men, the legs and arms in women), develops after disruption of the flow of the lymph channels by the parasites.

Histologically, inflammation occurs around adult worms when the worms die, either spontaneously or drug-induced. Granulomas arise around the dead worms, with macrophages developing into giant cells, as well as infiltration of plasma cells, eosinophils, and neutrophils. Starting from the location of worm death, filarial fever leads to lymphangitis (painful, with swelling) and lymphadenitis. After 1 week, the fever abruptly ends, and the skin begins to scale off.

The swelling and decreased flow of the lymph fluid render the body susceptible to skin and lymph system infections, and pulmonary tropical eosinophilia syndrome (e.g., cough, shortness of breath, and wheezing). The thickening and hardening of the skin (i.e., elephantiasis) may have fatal consequences.

70.1.2.2 Pathogenesis

W. bancrofti infective larvae from a feeding mosquito migrate to the regional lymphatic vessels and induce a host inflammatory reaction at the same time. The host reaction to the parasite intensifies when the parasites molt, then the developing females first begin to produce microfilariae, and the parasites die and degenerate. Besides antigens from *W. bancrofti* worms, those from its endosymbiont *Wolbachia* bacteria, a member of the family *Rickettsiaceae*, may also play a role in pathogenesis, since their depletion by doxycycline contributes to a reduction of vascular endothelial growth factors (VEGF) in plasma, which precedes a reduction of lymph vessel dilation. Consequently, lymphatic vessels are partially or completely blocked by lymphatic thrombi, masses of dead worms, endothelial proliferation, fibrin deposition, or granulomatous reactions. The initial inflammation of regional lymph nodes and major lymphatic vessels is followed by a prolonged asymptomatic period and then by recurring attacks of lymphangitis and "filarial fever" over a period of years. Elephantiasis (a grotesque enlargement of the infected area) develops when recurring attacks of lymphangitis lead to production of abnormally large amounts of collagenous material and to fibrosis of the tissue around the affected lymphatics, resulting in permanent lymphatic blockage and LE.

The enlarged lymph vessels are inefficient at transporting lymph from the periphery, leading to fluid extravasation, particularly in the lower limbs and rendering the host vulnerable to exogenous bacterial (e.g., streptococcal) and fungal infections following minor injuries. The secondary infections trigger more inflammation/lymphangiogenesis, presenting as acute dermatolymphangioadenitis (ADLA), resulting eventually in LE.

On the other hand, hydrocele (i.e., accumulation of fluid between the two folia of the tunica vaginalis in the scrotum) results from a chronic inflammatory stimulus for the tunica vaginalis to secrete fluid between the two folia, leading to lymph vessel enlargement and later obliteration, independent of exogenous bacteria.

In essence, the pathogenesis of *W. bancrofti* infection encompasses physical blockage of lymph vessels by a growing number of parasites and the host's immune responses to developing and adult worms and their antigens. These responses result in the activation of VEGF, which control angiogenesis and lymphangiogenesis in humans, thus promoting endothelial cell proliferation, lymphatic inflammation and dysfunction, thrombus formation within the lymphatic vessels, LE, fibrosis, and eventually elephantiasis (which is a buildup of excess tissue in the affected area). It is noteworthy that patients with LE or hydrocele (swelling of the scrotal/groin area) typically have few or no parasites but vigorous specific immune reactions. However, there are microfilaremic patients with high parasite numbers and downregulated cell-mediated immune responses. For example, neonates born from microfilaremic mothers tend to develop asymptomatic infection with high parasite loads and microfilaremia. At the molecular level, a lack of filarial-specific T-cell proliferation and production of IL-10 or TGF-β is associated with the permissive status, while higher levels of IFN-γ, IL-5, and IL-13, as well as IL-17, correlate with a lower likelihood to become infected in an endemic area.

70.1.2.3 Treatment

Diethylcarbamazine (DEC) eliminates the microfilariae from the blood and also kills the adult worms in the lymph vessels with a dosage of 6 mg/kg semiannually or annually. The side effects of DEC include dizziness, fever, headache, nausea, lymphatic inflammation, and muscle and joint pain. DEC should only be used for *W. bancrofti* infection as it can worsen onchocerciasis (an eye disease due to *Onchocerca volvulus*) and can cause encephalopathy and death in people with *Loa loa* infection.

Ivermectin (IVM) and albendazole also kill microfilariae, but have limited, long-term effects on the adult worms. In areas where the filarial nematode *Onchocerca volvulus* coexists with *W. bancrofti*, a combination of IVM and albendazole may be administered. In areas where only *W. bancrofti* or *Brugia* spp. is present, DEC and albendazole are administered. A polytherapy treatment comprising IVM with DEC or albendazole tends to be more effective than each drug alone.

Because *Wolbachia* bacterium lives inside *W. bancrofti* in a symbiotic relationship, elimination of this bacterium with antibiotics contributes to the death of the parasite. This may present another useful approach for the control of *W. bancrofti* parasite.

Exercising the swollen leg or arm helps improve the lymph flow and relieve LE. Retention of the swollen skin by washing it with soap and water daily not only reduces secondary bacterial infections but also enhances flow of fluids and thus immune defenses.

Avoiding infective mosquitoes between dusk and dawn (the time when they mostly feed) provides an easy way to prevent new infections. Use of mosquito nets, long trousers and sleeves, and application of mosquito repellent on skin also help prevent the mosquito transmission.

70.1.3 Laboratory Diagnosis

W. bancrofti adult worms inhabit the lymphatic ducts and are present in the lymph glands of the afferent lymphatic channels in the lower part of the body. Patients in an endemic area with enlarged and tender lymph nodes, especially in the inguinal region, or inflammation of lymphatic vessels in the extremities along with irregular fever are indicative of filariasis, and microscopic examination of thick blood smears and skin snips for microfilariae are employed for a definitive diagnosis [4]. Other laboratory techniques may also be used. For example, ultrasonography can detect the movements and noises caused by the movement of adult worms [5]; radiographic examinations can detect dead, calcified worms; serological methods can detect *W. bancrofti* antigens and antibodies; molecular procedures such as polymerase chain reaction (PCR) can detect the nucleic acids of *W. bancrofti* parasites [6].

Diagnosis for LF is traditionally performed with Giemsa-stained thick blood smear by microscopic examination. The sample is taken during the night to ensure the microfilariae are present in the bloodstream. Light infections are detected by using one of several concentration methods such as membrane filtration (MF) technique (i.e., filtration of 1–20 mL of venous blood through a polycarbonate membrane) [7]. Species identification is based on the presence of a characteristic sheath and the position of terminal nuclei and the size of the cephalic space. However, microscopic differentiation of closely related filarial parasites can be subjective and challenging. In addition, microfilariae may not be present in the blood during the early and late stages of the disease.

Serologic tests are useful for detection of *W. bancrofti* antigens or antibodies against antigens excreted by adult female *W. bancrofti* worms. While detection of specific circulating filarial antigens is considered by the WHO as the "gold standard" for diagnosis of LF, this approach has been limited by the cost of the antigen tests. The recent development of a highly sensitive immunochromatographic test (ICT) or "card test" facilitates rapid detection of circulating filarial antigens from finger-prick blood droplets without laboratory equipment [8]. Although antifilarial IgG4 antibody is associated with active LF infections, cross-reactivity with other filarial parasites is commonly observed [9,10].

Molecular techniques targeting *W. bancrofti* nucleic acids are increasingly utilized for rapid, sensitive, and specific identification and differentiation of filarial parasites in humans, animal reservoir hosts, and mosquito vectors [11–22]. Over the past two decades, molecular technologies for *W. bancrofti* detection have evolved from early nonamplified protocols (e.g., DNA hybridization) to nucleic acid amplification procedure such as PCR. Several specific primer sets [e.g., Ssp I repeat, pWb12 repeat, pWb-35 repeat, and LDR repeat, abundant larval transcript-2 (*alt*-2) gene] as well as universal primers [targeting ribosomal RNA and internal transcribed spacer (ITS) regions] have been reported for PCR amplification and identification of *W. bancrofti* [23,24]. A range of amplification and detection formats [e.g., single round PCR, nested PCR, multiplex PCR, PCR-restriction fragment length polymorphism (PCR-RFLP), PCR-enzyme-linked immunosorbent assay (PCR-ELISA), reverse transcriptase PCR, and real-time PCR] have been developed [25–36]. Use of these techniques has allowed accurate detection, phylogenetic analysis, epidemiological investigation, and antidrug resistance of *W. bancrofti* parasites [37–42].

70.2 METHODS

70.2.1 Sample Preparation

70.2.1.1 Sample Collection

Blood is collected by finger prick or by venipuncture between 9:00 p.m. and 1:00 a.m. due to the nocturnal periodicity of microfilariae in peripheral blood of infected individuals [43]. Mosquitoes are captured using a suction apparatus (Castro device), between 8 a.m. and 9 a.m., at the house of the microfilaremic patients. Alternatively, mosquitoes are collected by using centers for disease control (CDC) light traps without CO_2 placed inside houses. Female mosquitoes are sorted into two separate pools (engorged or gravid versus host-seeking) from each collection site.

70.2.1.2 Microscopy

W. bancrofti microfilariae are detected by microscopic examination of Giemsa-stained 50-μL-thick smears (TS). Briefly, one drop (50 μL) of blood is added onto a glass slide, dried at room temperature, and fixed with absolute methanol at 37°C for 10 min. Slides are then immersed in a 2.5% Giemsa stain in phosphate-buffered saline (PBS, pH 7.2) for 25 min, washed with water, dried and cover-slipped before microscopy.

In addition to TS, the MF technique (i.e., filtration of 1–20 mL of venous blood through polycarbonate membrane for microscopy) and rapid immunochromatographic tests (ICT "card test," ICT filariasis; AMRAD ICT, French's Forest, NSW, Australia, which detect the presence of *W. bancrofti* circulating antigens) are employed to increase the sensitivity in the diagnosis of microfilaremic patients [30].

70.2.1.3 DNA Isolation

W. bancrofti DNA is recovered from dried nucleopore membranes (5 μm pore size; Nucleopore, Pleasanton, CA) that had been used to filter venous blood from humans with microfilaremia. DNA isolated from these filters is assumed to be largely parasite DNA with some human DNA from cells trapped on the filters. Genomic DNA (gDNA) is extracted using Wizard Genomic DNA kits (Promega) into 200 μL of water as per the manufacturer's instructions. The quality and quantity of gDNA is assessed by spectrophotometry [30].

DNA may be also extracted from dried blood in sample application pads from used filarial antigen card tests. Sample application pads contain cells and microfilariae (when present) from 100 μL blood samples. Total gDNA is extracted from these pads using QIAamp DNA kits (Qiagen) [30].

70.2.2 Detection Procedures

70.2.2.1 Conventional PCR-RFLP Identification of *Wuchereria bancrofti*

Zhong et al. [23] developed a conventional PCR for specific amplification of a 188-bp product from the *Ssp*I repeat DNA (with 300 copies per haploid genome) of *W. bancrofti* using primers NV1 (5′-CGTGATGGCATCAAAGTAGCG-3′) and NV2 (5′-CCCTCACTTACCATAAGACAAC-3′). The cycling parameters consist of 35 cycles of 92°C, 55°C, and 72°C for 1 min each. The resulting 188-bp product is detected through 1.5% agarose gel electrophoresis.

70.2.2.2 PCR-RFLP Identification of Filarial Species

Nuchprayoon et al. [39] reported a PCR-RFLP-based method for detection and differentiation of a broad range of filarial species, including *W. bancrofti*, *B. malayi*, and *B. pahangi*. The primers (ITS1-F: 5′-GGTGAACCTGCGGAAGGATC-3′ and ITS1-R: 5′-GCGAATTGCAGACGCATTGAG-3′) are designed from the conserved sites at 18S and 5.8S rRNA sequence (covering ITS1 region). After a single round of PCR, the amplified ITS1 fragment is digested with the restriction endonuclease *Ase* I, generating a fragment profile that allows species-specific differentiation of *W. bancrofti*, *B. malayi*, *B. pahangi*, *Dirofilaria immitis*, and *D. repens*.

70.2.2.2.1 Procedure

1. The PCR mixture (50 μL) is composed of 1× PCR buffer (10 mM Tris-HCl, pH 9.0, 1.5 mM $MgCl_2$, 50 mM KCl), 200 μM each of dATP, dCTP, dGTP, and dTTP, 0.625 U of *Taq* DNA polymerase (Amersham Phamacia), 5 pmol of each primer (ITS1-F and ITS1-R), and 10–20 ng of parasite DNA template.
2. After incubation at 94°C for 5 min, amplification is conducted with 35 cycles of denaturation at 94°C for 30 s, annealing at 58°C for 30 s, and extension at 72°C for 45 s, followed by a final extension at 72°C for 10 min.
3. The PCR products are digested with 5 U of *Ase* I according to the manufacturer's protocols (New England Biolabs).
4. The resulting DNA fragments are separated by agarose gel electrophoresis, stained with ethidium bromide, and visualized under ultraviolet light.

Note: The ITS1-F/ITS1-R primer set yields PCR products of 482, 504, 510, 595, and 602 bp from *W. bancrofti*, *B. malayi*, *B. pahangi*, *D. immitis*, and *D. repens*, respectively. Subsequent digestion of the ITS1 PCR products with the restriction enzyme *Ase* I, followed by gel electrophoresis, permits the unequivocal differentiation of the five species tested.

70.2.2.3 Real-Time PCR Identification of *Wuchereria bancrofti*

Rao et al. [30] described a real-time PCR assay based on the *W. bancrofti* "LDR" repeat DNA sequence for detection in human blood and in mosquitoes. The primers consist of LDR1 (5′-ATTTTGATCATCTGGGAACGTTAATA-3′) and LDR2 (5′-CGACTGTCTAATCCATTCAGAGTGA-3′), and TaqMan probe (5′-FAM-ATCTGCCCATAGAAATAA CTACGGTGGATCTCTG-TAMRA-3′) are labeled with the reporter dye FAM (6-carboxyfluorescein) at the 5′ end and the quencher dye TAMRA (6-carboxytetramethylrhodamine) at the 3′ end.

70.2.2.3.1 Procedure

1. The real-time PCR mixture (25 μL) is made up of 12.5 μL of TaqMan master mix (Applied Biosystems) along with 450 nmol/L of each primer and 125 nmol/L probe and allocated into in 96-well MicroAmp optical plates (Applied Biosystems).
2. Two microliters of genomic DNA isolated from mosquitoes, from used nucleopore membranes, or from used filariasis card test sample application pads is added to the PCR mix.
3. Water is used as a negative control, and DNA from *W. bancrofti* microfilariae serves as a positive control sample in all real-time PCR runs.
4. Real-time PCR is carried out in duplicate with an ABI Prism 7000 instrument (Applied Biosystems), and cycle threshold (Ct) values for each sample are determined using SDS software according to the manufacturer's instructions.

70.3 CONCLUSION

W. bancrofti is one of the most important filarial nematodes (roundworms) that cause LF (so called elephantiasis) in the tropic and subtropic areas. Transmitted by mosquitoes, *W. bancrofti* adult worms inhabit lymph vessels where they mate and produce microfilariae. Physical blockage of lymph vessels by a growing number of parasites and host's immune responses to developing and adult worms and their antigens promote endothelial cell proliferation, lymphatic inflammation and dysfunction, LE, fibrosis, and eventually elephantiasis (a buildup of excess tissue in the affected area), sometimes with fatal consequences.

Correct diagnosis of LF is vital for effective treatment of infected individuals, for epidemiological assessment, and for monitoring of the control program. Traditionally, laboratory identification of *W. bancrofti* relies on microscopic observation of microfilariae in blood specimens. However, this technique suffers from low sensitivity and specificity. Serological assays have been developed for detection of specific circulating filarial antigens and antifilarial IgG4 antibody levels. Again, widespread application of these procedures is hindered by their high costs and their cross-reactivity with other common filarial parasites.

Nucleic acid-based techniques offer the opportunity for enhancing laboratory identification and differentiation of *W. bancrofti* and other filarial parasites. In particular, the application of PCR and its derivatives permits rapid, sensitive, and specific diagnosis of LF. In addition, these state-of-the-art procedures provide a powerful tool for epidemiological tracking of *W. bancrofti* parasites and for monitoring the efficacy of drug therapy against the disease [44–46].

REFERENCES

1. Pfarr, K.M. et al., Filariasis and lymphoedema. *Parasite Immunol.*, 31, 664, 2009.
2. Southgate, B., The significance of low density microfilaremia in the transmission of lymphatic filarial parasites. *J. Trop. Med. Hyg.*, 95, 79, 1992.
3. Leite, A.B. et al., Assessment of family and neighbors of an individual infected with *Wuchereria bancrofti* from a non-endemic area in the city of Maceió, Brazil. *Braz. J. Infect. Dis.*, 14, 125, 2010.
4. Rosenblatt, J.E., Laboratory diagnosis of infections due to blood and tissue parasites. *Clin. Infect. Dis.*, 49, 1103, 2009.
5. Amaral, F. et al., Live adult worms detected by ultrasonography in human Bancroftian filariasis. *Am. J. Trop. Med. Hyg.*, 50, 753, 1994.
6. Weil, G.J., and Ramzy, R.M.R., Diagnostic tools for filariasis elimination programmes. *Trends Parasitol.*, 23, 78, 2007.
7. Chularerk, P., and Desowitz, R.S., A simplified membrane filtration technique for the diagnosis of microfilaremia. *J. Parasitol.*, 56, 623, 1970.
8. Hoti, S.L. et al., A method for detecting microfilaraemia, filarial specific antigens and antibodies and typing of parasites for drug resistance and genotypes using finger prick blood sample. *Acta Trop.*, 107, 268, 2008.
9. Nuchprayoon, S. et al., Endemic bancroftian filariasis in Thailand: Detection by Og4C3 antigen capture ELISA and the polymerase chain reaction. *J. Med. Assoc. Thai.*, 84, 1300, 2001.
10. Lammie, P.J. et al., Recombinant antigen-based antibody assays for the diagnosis and surveillance of lymphatic filariasis—a multicenter trial. *Filaria J.*, 3, 9, 2004.
11. Ramzy, R.M., Field application of PCR-based assays for monitoring *Wuchereria bancrofti* infection in Africa. *Ann. Trop. Med. Parasitol.*, 96, S55, 2002.
12. Ramzy, R.M.R. et al., A polymerase chain reaction-based assay for detection of *Wuchereria bancrofti* in human blood and *Culex pipiens*. *Trans. R. Soc. Trop. Med. Hyg.*, 91, 156, 1997.
13. Williams, S.A. et al., Development and standardization of a rapid, PCR-based method for the detection of *Wuchereria bancrofti* in mosquitoes, for xenomonitoring the human prevalence of bancroftian filariasis. *Ann. Trop. Med. Parasitol.*, 96, S41, 2002.
14. Fischer, P., Boakye, D., and Hamburger, J., Polymerase chain reaction-based detection of lymphatic filariasis. *Med. Microbiol. Immunol.*, 192, 3, 2003.
15. Vasuki, V. et al., A simple and rapid DNA extraction method for the detection of *Wuchereria bancrofti* infection in the vector mosquito, *Culex quinquefasciatus* by Ssp I PCR assay. *Acta Trop.*, 86, 109, 2003.
16. Vasuki, V., Hoti, S.L., and Patra, K.P., RT-PCR assay for the detection of infective (L3) larvae of lymphatic filarial parasite, *Wuchereria bancrofti*, in vector mosquito *Culex quinquefasciatus*. *J. Vector Borne Dis.*, 45, 207, 2008.
17. Helmy, H. et al., Test strip detection of *Wuchereria bancrofti* amplified DNA in wild-caught *Culex pipiens* and estimation of infection rate by a PoolScreen algorithm. *Trop. Med. Int. Health*, 9, 158, 2004.
18. Supali, T. et al., Estimation of the prevalence of lymphatic filariasis by a pool screen PCR assay using blood spots collected on filter paper. *Trans. R. Soc. Trop. Med. Hyg.*, 100, 753, 2006.
19. Boakye, D.A. et al., Monitoring lymphatic filariasis interventions: Adult mosquito sampling, and improved PCR-based pool screening method for *Wuchereria bancrofti* infection in Anopheles mosquitoes. *Filaria J.*, 6, 13, 2007.
20. Kagai, J.M. et al., Molecular technique utilising sputum for detecting *Wuchereria bancrofti* infections in Malindi, Kenya. *East Afr. Med. J.*, 85, 118, 2008.
21. Chambers, E.W. et al., Xenomonitoring of *Wuchereria bancrofti* and *Dirofilaria immitis* infections in mosquitoes from American Samoa: Trapping considerations and a comparison of polymerase chain reaction assays with dissection. *Am. J. Trop. Med. Hyg.*, 80, 774, 2009.
22. Intapan, P.M. et al., Rapid detection of *Wuchereria bancrofti* and *Brugia malayi* in mosquito vectors (Diptera: Culicidae) using a real-time fluorescence resonance energy transfer multiplex PCR and melting curve analysis. *J. Med. Entomol.*, 46, 158, 2009.
23. Zhong, M. et al., A PCR assay for detection of the parasite *Wuchereria bancrofti* in human blood samples. *Am. J. Trop. Med. Hyg.*, 54, 357, 1996.
24. Dassanayake, R.S., Chandrasekharan, N.V., and Karunanayake, E.H., Trans-spliced leader RNA, 5S-rRNA genes and novel variant orphan spliced-leader of the lymphatic filarial nematode *Wuchereria bancrofti*, and a sensitive polymerase chain reaction based detection assay. *Gene*, 269, 185, 2001.
25. Chansiri, K. et al., Detection of *Plasmodium falciparum* and *Wuchereria bancrofti* infected blood samples using multiplex PCR. *Mol. Cell. Probes*, 15, 201, 2001.
26. Chansiri, K., and Phantana, S., A polymerase chain reaction assay for the survey of bancroftian filariasis. *Southeast Asian J. Trop. Med. Public Health*, 33, 504, 2002.
27. Lulitanond, V. et al., Rapid detection of *Wuchereria bancrofti* in mosquitoes by LightCycler polymerase chain reaction and melting curve analysis. *Parasitol. Res.*, 94, 337, 2004.
28. Kanjanavas, P. et al., Detection of lymphatic *Wuchereria bancrofti* in carriers and long-term storage blood samples using semi-nested PCR. *Mol. Cell. Probes*, 19, 169, 2005.

29. Mishra, K. et al., Combined detection of *Brugia malayi* and *Wuchereria bancrofti* using single PCR. *Acta Trop.*, 93, 233, 2005.
30. Rao, R.U. et al., A real-time PCR-based assay for detection of *Wuchereria bancrofti* DNA in blood and mosquitoes. *Am. J. Trop. Med. Hyg.*, 74, 826, 2006.
31. Rao, R.U. et al., A qPCR-based multiplex assay for the detection of *Wuchereria bancrofti*, *Plasmodium falciparum* and *Plasmodium vivax* DNA. *Trans. R. Soc. Trop. Med. Hyg.*, 103, 365, 2009.
32. Nuchprayoon, S., Junpee, A., and Poovorawan, Y., Random amplified polymorphic DNA (RAPD) for differentiation between Thai and Myanmar strains of *Wuchereria bancrofti*. *Filaria J.*, 6, 6, 2007.
33. Kamal, I.H. et al., Evaluation of a PCR-ELISA to detect *Wuchereria bancrofti* in *Culex pipiens* from an Egyptian village with a low prevalence of filariasis. *Ann. Trop. Med. Parasitol.*, 95, 833, 2001.
34. Jiménez, M. et al., Detection and discrimination of *Loa loa, Mansonella perstans* and *Wuchereria bancrofti* by PCR-RFLP and nested-PCR of ribosomal DNA ITS1 region. *Exp. Parasitol.*, 127, 282, 2011.
35. Laney, S.J. et al., Detection of *Wuchereria bancrofti* L3 larvae in mosquitoes: A reverse transcriptase PCR assay evaluating infection and infectivity. *PLoS Negl. Trop. Dis.*, 4, e602, 2010.
36. Mehlotra, R.K. et al., Molecular-based assay for simultaneous detection of four *Plasmodium* spp. and *Wuchereria bancrofti* infections. *Am. J. Trop. Med. Hyg.*, 82, 1030, 2010.
37. Torres, E.P. et al., Detection of bancroftian filariasis in human blood samples from Sorsogon province, the Philippines by polymerase chain reaction. *Parasitol. Res.*, 87, 677, 2001.
38. Bhandari, Y. et al., Analysis of polymorphism of 18S rRNA gene in *Wuchereria bancrofti* microfilariae. *Microbiol. Immunol.*, 49, 909, 2005.
39. Nuchprayoon, S. et al., Detection and differentiation of filarial parasites by universal primers and polymerase chain reaction-restriction fragment length polymorphism analysis. *Am. J. Trop. Med. Hyg.*, 73, 895, 2005.
40. Farid, H.A. et al., A critical appraisal of molecular xenomonitoring as a tool for assessing progress toward elimination of lymphatic filariasis. *Am. J. Trop. Med. Hyg.*, 77, 593, 2007.
41. Hoti, S.L. et al., Laboratory evaluation of Ssp I PCR assay for the detection of *Wuchereria bancrofti* infection in *Culex quinquefasciatus*. *Indian J. Med. Res.*, 114, 59, 2001.
42. Sakthidevi, M. et al., Lymphatic filarial species differentiation using evolutionarily modified tandem repeats: Generation of new genetic markers. *Infect. Genet. Evol.*, 10, 591, 2010.
43. Van Hoegaerden, M., and Ivanoff, B., A rapid, simple method for isolation of viable microfilariae. *Am. J. Trop. Med. Hyg.*, 35, 148, 1986.
44. Nuchprayoon, S., DNA-based diagnosis of lymphatic filariasis. *Southeast Asian J. Trop. Med. Public Health*, 40, 904, 2009.
45. Hoti, S.L. et al., An allele specific PCR assay for screening for drug resistance among *Wuchereria bancrofti* populations in India. *Indian J. Med. Res.*, 130, 193, 2009.
46. Palumbo, E., Filariasis: Diagnosis, treatment and prevention. *Acta Biomed.*, 79, 106, 2008.

Section VI

Arthropoda: Arachnida

71 Argasidae (Soft Ticks)

Dongyou Liu

CONTENTS

71.1 INTRODUCTION

Ticks are relatively large-bodied, bloodsucking, ectoparasitic arthropods that feed on mammals, birds, and reptiles in most parts of the world. The capacity of ticks to cause damage to vertebrate hosts has been long recognized, and the first case of tick bite-associated disease was documented in 1893 involving a blood-borne parasite *Babesia bigemina* that was transmitted by a tick *Rhipicephalus (Boophilus) annulatus*, resulting in Texas cattle fever. Since then many tick species have been shown to play a role in the transmission of pathogenic bacteria, viruses, and parasites in animals and humans. In addition, tick bites may cause tick paralysis, toxicoses, and allergic reactions.

In general, ticks are of two main types: hard-bodied ticks (classified in the family Ixodidae) and soft-bodied ticks (classified in the family Argasidae). Furthermore, there exists a distinct tick species classified in the family Nuttalliellidae [1]. The hard tick family Ixodidae consists of 702 species, which are characterized by having a sclerotized dorsal shield in the shape of a fingernail (also known as the scutum), by feeding for hours to days, with disease transmission usually occurring near the end of a meal, after the tick becoming full of blood, and by having a relatively short life span (2–6 years). The soft tick family Argasidae comprises 193 species, which are characterized by having a flexible and leathery exoskeleton (absence of the scutum), by feeding for less than 1 h, with disease transmission occurring in less than a minute of a meal, and by having a relatively short life span (up to 20 years) [2–4].

71.1.1 Classification, Morphology, and Biology

71.1.1.1 Classification

The soft tick family Argasidae is classified in the order Ixodida, superorder Parasitiformes, subclass Acari, class Arachnida, phylum Arthropoda. The family is separated into two subfamilies: Argasinae (consisting of a single genus *Argas*) and Ornithodorinae (consisting of four genera: *Antricola*, *Nothoaspis*, *Ornithodoros*, and *Otobius*) [5]. The total number of species in the family Argasidae stands at 193, with the genera *Argas* and *Ornithodoros* (incorporating the former genus *Carios*) accounting for a majority of these species (Table 71.1) [1].

The most notable members of the Argasidae family are *Argas brevipes*, *Argas hermanni*, *Argas keiransi*, *Argas lagenoplastis*, *Argas lahorensis*, *Argas monachus*, *Argas monolakensis*, *Argas neghmei*, *Argas persicus*, *Argas polonicus*, *Argas reflexus*, *Argas robertsi*, *Argas vulgaris*, and *Argas vespertilionis* in the genus *Argas*; *Antricola delacruzi*, *Antricola guglielmonei*, *Antricola marginatus*, and *Antricola mexicanus* in the genus *Antricola*; *Ornithodoros brasiliensis*, *Ornithodoros coriaceus*, *Ornithodoros decoloratus*,

TABLE 71.1
Classification of the Soft Tick Family Argasidae

Family	Subfamily	Genus	Number of Species
Argasidae	Argasinae	Argas	61
	Ornithodorinae	Antricola	17
		Nothoaspis	1
		Ornithodoros	112
		Otobius	2
			193

Ornithodoros erraticus, *Ornithodoros gurneyi*, *Ornithodoros hermsi*, *Ornithodoros macmillani*, *Ornithodoros marinkellei*, *Ornithodorus maritimus*, *Ornithodoros moubata*, *Ornithodoros papillipes*, *Ornithodoros parkeri*, *Ornithodoros peruvianus*, *Ornithodoros porcinus*, *Ornithodoros rioplatensis*, *Ornithodoros rostratus*, *Ornithodoros savignyi*, *Ornithodoros sonrai*, *Ornithodoros tartakovskyi*, and *Ornithodoros turicata* in the genus *Ornithodoros*; and *Otobius lagophilus* and *Otobius megnini* (formerly *Argas megnini*) in the genus *Otobius* [5]. Many medically relevant soft ticks belong to the genera *Argas*, *Ornithodoros*, and *Otobius*.

71.1.1.2 Morphology

Ticks undergo four developmental stages in their life cycle: egg, larva, nymph (with 2–7 instars or growth stages), and adult, with each demonstrating distinct morphological features. Soft ticks appear oval (pear-shaped), and are made up of two parts, the capitulum (gnathosoma) and the body (idiosoma), to which six legs (larvae) or eight legs (nymphs and adults) are attached. Unfed adult ticks range from 2 to 20 mm in length, while blood-engorged females may measure 25–30 mm in length [2,3].

The capitulum bearing the mouthparts is located beneath the anterior margin of the body and is thus invisible from a dorsal view (although in larvae it is located in an anterior position). The mouthparts consist of two four-segmented palps, two sclerotized, two-segmented tubular chelicerae (cutting mandibles) extending from the basis capituli (an integumental ring encircling the mouthparts) and located medially to the palps, and two highly moveable and sharp cutting digits located at the extremities of the cheliceral shafts. The medially positioned hypostome with ventral, backward-pointing denticles (teeth) on its external surface has its internal surface covered dorsally by the chelicerae and is used as a holdfast organ and feeding tube. Blood passes from the host through this food canal, formed by the hypostome ventrally, and by the chelicerae dorsally.

The body (idiosoma) lacks the ixodid scutum and is covered with a leathery cuticle of rough texture with tiny protuberances (called mammillae). The anterior part of the body bears the six-segmented legs and genital pore, with Haller's organ on the tarsus of the first leg detecting temperature, air currents, odors, and chemicals. The coxal pores (or the openings of the coxal glands) are located between the coxae of the first and second pairs of legs. The genital aperture is closed in larvae and nymphs and open in adults. The posterior part of the body possesses the spiracles (or respiratory openings) and anus. The spiracles lie on the sides of the body above the third and fourth pairs of legs. When present, eyes occur on the lateral surface of the body, and the ventral organ appears as a broad transverse ridge posterior to the anus. Males and females are distinguished by the shape of the genital opening which in males is circular or crescent-shaped and in females is a transverse split, wider than long.

While the genus *Argas* shows a definite sutural line along the lateral margin of the body, the genus *Ornithodoros* and other members of the family Argasidae display a rounded abdominal margin, and thus no definite sutural line.

71.1.1.3 Biology

The life stages of soft ticks are not easily distinguished. Larvae with six legs emerge from eggs, take a blood meal from a host, and molt to the first nymphal stage. After going through multiple nymphal stages (unlike hard ticks which have a single nymphal stage), and gradually increasing in size, nymphs finally molt to adults. Females lay multiple small batches of eggs between blood meals during their lives [6].

Most soft ticks are nest-dwelling or nidicolous parasites, residing in the nests (and dens) of the hosts such as burrows, caves, rock ledges, or crevices, and feeding rapidly when the host returns and disturbs the contents. Humans are usually bitten by entering a cave or a large burrow, crawling under a building or sleeping in primitive cabins, where host rodents nests are located.

However, some soft tick species are field-dwelling (or nonnidicolous) and seek hosts by questing on low-lying vegetation. Carbon dioxide, heat, vibration, and shadowing serve as stimuli for host-seeking behavior. Soft ticks feed for short periods of time (from several minutes to days) on hosts, and often are long-lived (up to 20 years), with the ability to fast for months or longer and survive years until a suitable host is encountered.

The sequence of events leading to successful feeding by a tick includes nine major steps: (i) appetence (questing or hunting for a host); (ii) engagement (adherence to the skin or fur of the host); (iii) exploration (searching on the skin for a suitable attachment site); (iv) penetration (insertion of the mouthparts or more specifically the hypostome into the host's epidermis and dermis); (v) attachment (establishing feeding site); (vi) ingestion (uptake of blood and other fluids); (vii) engorgement (partial or complete meals of blood); (viii) detachment (withdrawal of the mouthparts); and (ix) disengagement (dropping off of the host).

Ticks produce pheromones that aid their aggregation, attachment, and sexual reproduction. These chemicals facilitate tick assembly in microhabitats conducive for survival (e.g., caves, ledges, cracks, crevices, and sites frequented by hosts), host finding, attachment to specific feeding locations, and mating [7].

Soft ticks are most common in the tropics and subtropics, with fewer species inhabiting temperate climes, and even fewer being present in arctic environments [8–12].

Argas reflexus, *Argas plonicus*, and *Argas vulgaris* are found in bird nests in Europe [13,14]. The life cycle of *Argas reflexus* (the pigeon tick) includes the egg, larva, 2–4 nymphal instars, and adult stages. Nymphs II, nymphs I, and larvae feed later than in mid-July, August, or September, respectively, entered a state of diapause, and overwinter in the engorged state [15].

Argas monolakensis inhabits the nest of California gulls and may transmit *Coxiella* bacteria [16]; *Argas moreli* infects chickens or wild birds in Peru; and *Argas brumpti* is found in caves where bats live [17].

Ornithodoros moubata (the African or eyeless tampan) inhabits cracks of mud or clay beds and feeds on humans living in huts with a central cooking fire in arid sub-Saharan Africa. This species is responsible for transmitting *Borrelia duttoni* (relapsing fever) [18]. Ticks concealed in sleeping mats, spare clothing, or baskets are carried by mobile humans into new areas [19].

Ornithodoros hermsi lives in coniferous forests at elevations of 1500–8000 feet in Idaho, Oregon, Washington, California, Nevada, northern Arizona, northwestern New Mexico, Utah, Montana, and Colorado, where it is associated with nests of small mammals such as the western chipmunk (*Eutamias*), ground squirrels, tree squirrels (*Tamiasciurus*), and rats. Humans (especially tourists) may be bitten by *O. hermsi* when sleeping in infested cabins, and acquire *Borrelia hermsii* infection.

Ornithodoros turicata is found in caves, holes made by burrowing animals, including rodents, gopher tortoises, snakes, terrapins, cattle, pigs, ground squirrels, and prairie dogs, and at campsites in California, Utah, Colorado, New Mexico, Arizona, Oklahoma, Kansas, Florida, and Texas as well as Florida [20]. This species is an efficient vector of *Borrelia turicatae*, causing relapsing fever, and people usually become infected while visiting limestone caves in Texas. *O. turicata* is larger than *O. hermsi*, has an integument with conspicuous conical mammillae, and has dorsal projections on the distal segment (tarsus) of the front legs.

Ornithodoros parkeri is found in burrows, rodent nests, and caves in arid regions, grasslands, or treeless grassy plains throughout much of the western United States. This species attacks rodents and humans, and transmits relapsing fever (*Borrelia parkeri* infection) and Rocky Mountain spotted fever.

Ornithodoros sonrai feeds on rodents and insectivores and transmits the spirochete *Borrelia crocidurae* to humans in West Africa. This species has an elongated body with parallel lateral margins and broadly rounded posterior margin; distinct mammillae on the dorsal integument; hypostome with denticles and palpi slightly exceeding hypostome in length; eyes absent; genital area absent for nymphs and forming a semicircle apron with smooth surface for males and a transverse slit for females; dorsoventral, preanal, and transverse postanal grooves present; subapical dorsal protuberances on tarsi absent; and pulvilli absent.

Ornithodoros savignyi is implicated in the transmission of Alkhurma hemorrhagic fever virus (AHFV) in Saudi Arabia [21,22].

Ornithodoros erraticus is found in pig pens in the Middle East and Mediterranean such as the provinces of Salamanca, Badajoz, and Huelva in Spain. It is one of the more common soft ticks to bite humans.

Ornithodoros talaje occurs in southern United States. *Ornithodoros denmarki* inhabits sea bird nests in Florida.

Soft ticks are usually collected via dry ice traps, as carbon dioxide emitted by dry ice represents a host-seeking stimulant. Traps are set in and around nesting areas of animal hosts. Soft ticks running along the surface of the ground toward the trap are collected by hand, or inside a collection chamber in the trap. Soft ticks can be artificially fed in the laboratory, with sausage casings filled with mammal blood heated up to 37°C (e.g., *Ornithodorus coriaceus*) or various types of membranes in the presence of biochemical and environmental feeding stimulants such as ATP, carbon dioxide, heat, and animal fur.

71.1.2 Clinical Features

Soft ticks pose a considerable health risk to the human population. In addition to cause local redness, itching, burning, and other discomfort [23], soft ticks are vectors of several human pathogens that may result in a variety of clinical diseases (Table 71.2) [2,3,24,25].

The pigeon tick *Argas reflexus* is an ectoparasite of urban pigeons whose bites are responsible for inducing polymorphous skin manifestations, including wheals with a central hemorrhagic point, crusted-ulcerative lesions, pustular lesions, generalized urticaria, angioedema, dyspnea, cardiovascular dysregulation, unconsciousness, and atopy in human patients [14,26–35].

Ornithodoros erraticus feeds on pigs and is one of the more common soft ticks to bite humans, particularly in the Middle East and Mediterranean. *Ornithodoros hermsi*, *Ornithodoros parkeri*, and *Ornithodoros turicata* are associated with tick-borne relapsing fever (TBRF) due to *Borrelia hermsii*, *Borrelia parkeri*, and *Borrelia turicatae* in the western United States [36–38]. Being unaware of the tick bite, humans usually become very ill, with symptoms ranging from headache, myalgia, chills, arthralgia, vomiting, abdominal pain, confusion, dry cough, eye pain, diarrhea, dizziness, photophobia, neck pain, rash, dysuria, jaundice, hepatomegaly, splenomegaly, conjunctival injection, eschar, meningitis, and nuchal rigidity [37]. In Iran, *Ornithodoros tholozani* transmits *Borrelia persica* [39].

71.1.3 Diagnosis and Treatment

Soft ticks often feed on human hosts for a short period, and because of the secretion of neurotoxin that lessens the pain of their bites, many people do not feel or recall the tick bites. Careful examination of the entire body for ticks still attached, rashes, and other signs is important for the

TABLE 71.2
Geographic Distribution of Soft Ticks and Their Reported Roles in the Transmission of Human Disease-Causing Agents

Genus	Species	Geographic Distribution	Transmission of
Carios	*C. erraticus*	North and East Africa, Near East, southeastern Europe	*Borrelia crocidurae*, *Borrelia hispanica*
	C. kelleyi	Midwest United States	*Rickettsia*
	C. rudis	Panama, South America	*Borrelia venezuelensis*
Ornithodoros	*O. asperus*	Caucasus, Iraq	*Borrelia caucasica*
	O. erraticus	Middle East, Mediterranean, Northern Africa	*Borrelia crocidurae*, *Borrelia hispanica*, African swine fever virus
	O. hermsi	Western North America	*Borrelia hermsii*
	O. moubata	Eastern, central, southern Africa	*Borrelia duttoni*
	O. parkeri	Western North America	Tick-borne relapsing fever
	O. porcinus	East and South Africa	*Borrelia duttoni*
	O. savignyi	Saudi Arabia	Alkhurma hemorrhagic fever virus
	O. sonrai	West Africa, Northern Africa	*Borrelia crocidurae*
	O. tartakovskyi	Central Asia	*Borrelia latyschewii*
	O. tholozani	Central Asia	*Borrelia persica*
	O. turicata	Mexico, Central America, southwest United States	*Borrelia turicatae*

Source: Anderson, J.F., *Infect. Dis. Clin. North Am.* 22, 195, 2008.

diagnosis of tick bites and tick-caused diseases (e.g., Lyme disease, Rocky Mountain spotted fever, babesiosis, ehrlichiosis, tularemia, and TBRF). If still attached on skin, soft ticks are sprayed twice within a minute with aerosol insect repellent containing pyrethrin or a pyrethroid, and left for 24 h before removal with a pair of bent-nose tweezers for further identification based on morphological criteria using a series of dichotomous keys [40]. Take care not to puncture the body of the tick, as this can release more harmful bacteria and toxins.

Identification of soft ticks is aided with a magnifying lens or microscope on low power. Six morphological features including shape, size, mouthparts, color, dorsal shield, and festoons in combination with knowing endemic areas generally allow determination of tick types affecting humans.

For more accurate and objective determination of soft tick identity, molecular techniques based on amplification and analysis of the ribosomal rRNA (18S and 28S, ITS1 and ITS2), mitochondrial rRNA (12S and 16S), nuclear ribosomal RNA (18S and 28S), and other gene sequences can be utilized [18,41–52].

71.1.3.1 Treatment

If a tick is found still attached to the skin, do not apply methylated spirits or other chemicals onto the tick. Instead, an aerosol insect repellent preferably containing pyrethrin or a pyrethroid (or Lyclear, a scabies cream containing permethrin) is sprayed (or dabbed). After 1 min, the tick is sprayed with aerosol insect repellent (or dabbed with the Lyclear) again and left for 24 h. Hydrocarbons and pyrethrin present in aerosol insect repellent function as a narcotic and a toxicant, and prevent the tick from injecting its saliva. The tick should drop off naturally or be gently removed with fine-tipped forceps or tweezers. A tick bite may remain slightly itchy for several weeks. The irritation and other allergic conditions associated with bites can be alleviated with an antipruritic or antihistamines.

If aerosol insect repellent is not available, attempts may be made to remove the tick (and portions of its head and mouthparts embedded in the skin) with small forceps or tweezers. By carefully flipping the tick over onto its back, and grasping the tick firmly with the tweezers as close to the skin as possible, the tick is gently pulled free. Twisting or turning the tick should be avoided as this may break off its head and mouthparts, thus increasing the chances for infection. The removed tick can be kept in a tightly closed jar or taped to a piece of paper for further examination. If portions of the head or mouthparts remain, they should be removed surgically. Wearing gloves helps prevent the spread of pathogens from the tick to hands. Instruments that touch a tick are cleaned and disinfected. Local cleansing and antibiotic cream are applied. Preparations containing diphenhydramine (Benadryl) can be applied directly to the skin for itching or administered orally by tablets. Antibiotics may be prescribed for oral or IV administration.

The best method of avoiding ticks is to stay away from known tick-infested areas. If visiting such an area, light-colored clothing should be worn, as ticks will be much easier to detect. An insect repellent containing *N,N*-Diethyl-meta-Toluamide (DEET) or Picaridin should be applied, with a cream repellent applied to the skin and a spray repellent to footwear and clothing.

71.2 METHODS

71.2.1 Sample Preparation

71.2.1.1 Sample Collection

Adult ticks are collected by introducing a flexible tube inside burrows and aspirating their contents using a portable petrol-powered aspirator. The contents are exposed to sunlight on sorting trays, and selected soft tick specimens are stored in absolute ethanol at 4°C until determination and DNA extraction. Use of dry ice traps is also useful. Ticks are identified by the shape of the spiracular plates, the relative size and number of the goblets within the spiracular plates, and other morphological features. Ticks are maintained at 22°C, under a 14-h light/10-h dark photoperiod. Ticks are held in 16-mL glass vials (Wheaton Glass, Millville, NJ) with a mesh top over a supersaturated solution of potassium nitrate. Ticks are frozen at −70°C until required.

71.2.1.2 DNA Extraction

Living adult ticks are killed by freezing at −70°C or by immersion in isopropanol. Genomic DNA is extracted from individual ticks. Briefly, the ticks are homogenized in 400 µL of TET buffer (50 mM Tris-HCl (pH 7.5), 25 mM EDTA, 100 mM NaCl, and 0.4% Triton X-100), and 10 µL (20 mg/mL) of proteinase K. After incubation at 56°C for 2 h, the DNA is extracted by shaking in 200 µL of phenol/chloroform/isoamylic alcohol (25:24:1) and precipitated by adding two volumes of ice-cold ethanol overnight. After centrifugation (12,000 × *g* for 5 min), the pellet is resuspended in 35 µL of TE (10 mM Tris (pH 8.0) and 1 mM EDTA), and the DNA is used as a template for PCR.

DNA from ticks can also be isolated using QIAamp DNA Mini Kit (Qiagen). Individual ticks are placed in 2 mL screw-top centrifuge tubes and placed in liquid nitrogen for 60 s. The frozen tick is then transferred to an identical tube containing three sizes of glass beads (2, 0.5, 0.1 mm) and crushed by vigorous shaking with a Mini-BeadBeater-8 (BioSpec Products, Inc., Bartlesville, OK) for two cycles of 90 s. The tubes are then centrifuged at 10,000× *g* for 10 s at room temperature, and the samples are resuspended in 180 µL Buffer ATL (QIAamp DNA Mini Kit). DNA is eluted in 200 µL of elution buffer. The concentration of isolated DNA is determined using a SmartSpec Plus spectrophotometer (Bio-Rad Laboratories, Hercules, CA), and the samples are stored at −20°C [53,54].

Alternatively, individual ticks are cleaned by sonication for 3–5 min in ethanol solution and then washed twice in sterile distilled water. Then, individual ticks are dissected in pieces and placed in a microcentrifuge tubes filled with 180 µL lysing buffer solution and then homogenized with a sterile tissue grinder (catalogue no. 358103, Wheaton Scientific Products, Millville, NJ). The homogenate is centrifuged at room temperature and the supernatant fluid is further processed by a DNeasy Tissue Kit™, as instructed by the manufacturer (Qiagen) [55].

In addition, genomic DNA is extracted from either one or two legs, or the complete body of adult ticks using the DNeasy Tissue Kit (Qiagen). The use of only one or two legs permits the remainder of the specimen to be preserved for morphological examination or to be tested for the presence of pathogenic organisms. The leg or complete body of an individual tick is placed into a 1.5 mL micropestle tube with 180 µL of ATL Buffer (Qiagen) and homogenized using a micropestle attached to a cordless drill. Proteinase K (20 µL of 15 µg/µL) was then added to the micropestle tube and the sample incubated overnight at 55°C. 200 µL of AL Buffer (Qiagen) is added to the sample, vortexed, and incubated for 10 min at 70°C. Then, 200 µL of 100% ethanol is added, and the solution is applied to a spin column. After rinsing the columns with the wash buffers AW1 and AW2 (Qiagen), DNA is eluted with 100 µL AE buffer (Qiagen) and stored at −70°C [56].

71.2.2 Detection Procedures

71.2.2.1 Sequence Analysis of Tick 16S rRNA Gene

Mangold et al. [44] employed primers (forward, 5′-CCG GTC TGA ACT CAG ATC AAG T-3′ and reverse, 5′-GCT CAA TGA TTT TTT AAA TTG CTG T-3′) to amplify a 460-bp fragment of the 16S rRNA gene. This fragment corresponds to the *Drosophila yakuba* 16S rRNA gene sequence between positions 12,887 and 13,431. Subsequent analysis of the amplicon sequence enables identification of the organisms under investigation.

71.2.2.1.1 Procedure

1. PCR mixture (100 µL) is made up of 1× PCR buffer (10 mM Tris-HCl (pH 9.0), 1.5 mM $MgCl_2$, 50 mM KCl), 200 µM each of dATP, dCTP, dGTP, and dTTP, 2 U of *Taq* DNA polymerase (Promega), 50 pmol of each primers, and 10–20 ng of parasite DNA template.
2. Amplification is conducted with an initial denaturation step at 94°C for 2 min followed by 35 cycles for 30 s at 94°C, 30 s for primer annealing, and 45 s for primer extension at 72°C. The annealing temperature of the first seven cycles is increased by 0.3°C every second cycle from 47°C to 48.8°C, followed by 28 cycles using an annealing temperature of 50°C. A final extension step is carried out for 7 min at 72°C.
3. The PCR products (10 µL) are examined by 1% agarose gel electrophoresis, stained with ethidium bromide, and visualized under ultraviolet light.
4. The remaining amplicon is purified using the Wizard PCR Preps DNA Purification System (Promega) and resuspended in 50 µL of 10 mM TE (pH 7.6). The DNA sequence is analyzed using the same PCR primers.
5. The resulting sequences are aligned using CLUSTAL V, and phylogenetic relationships are analyzed using distance and maximal parsimony methods.

71.2.2.2 Sequence Analysis of Tick 16S and 18S rRNA Genes

Vial et al. [48] utilized primers from the mitochondrial 16S rRNA and nuclear 18S rRNA genes for phylogenetic analyses of *Ornithodoros sonrai* group as well as other closely related groups. To amplify the mitochondrial 16S rRNA gene sequence, primers Tm16S + 1 (5′-CTGCTCAATGATTTTTTAAATTGC-3′) and Tm16S-1

(5′-CCGGTCTGAACTCAGATCATGTA-3′) are used, which result in a product of approximately 475 bp long. To amplify the nuclear 18S rRNA gene sequence, primers NS3 (5′-GCAAGTCTGGTGCCAGCAGCC-3′) and NS4 (5′-CTTCCGTCAATTCCTTTAAG-3′) are used, which result in a product of approximately 600 bp.

71.2.2.2.1 Procedure

1. PCR mixture (25 μL) is prepared with the Multiplex PCR Kit (Qiagen) containing 12.5 μL of 2× Multiplex PCR Master Mix (HotStarTaq DNA Polymerase/Multiplex PCR buffer/dNTP Mix), 2.5 μL of 10× Primer Mix (2 μM of each primer/AE buffer), 2.5 μL of 5× Q-solution, and 4 μL (40–100 ng/μL) of DNA template.
2. For 16S rRNA PCR, the mixture is subjected to an initial denaturation at 95°C for 15 min; 10 cycles of 1 min at 92°C, 1.5 min at 48°C, and 1.5 min at 72°C; 32 cycles of 1 min at 92°C, 1.5 min of 54°C, and 1.5 min of 72°C; and a final extension at 72°C for 10 min. For 18S rRNA PCR, the mixture is subjected to an initial denaturation at 95°C for 15 min; 30 cycles of 1 min at 92°C, 1.5 min at 54°C, and 1.5 min at 72°C; and a final extension at 72°C for 10 min.
3. The amplified products (3 μL) are separated by electrophoresis on 2% agarose gel in TAE 0.5× buffer (40 mM Tris-acetate, 2 mM EDTA, pH 8.3) and stained with ethidium bromide. The remaining reaction mixtures are stored at –20°C for direct sequencing.
4. If extra bands or a smear appears around the band, PCR products are separated in a 1.2% agarose gel, the desired band is excised, and the DNA is extracted from the gel (Qiaquick Gel Extraction Kit, Qiagen). Amplified DNA is sequenced using an ABI PRISM 3730xl DNA analyzer.
5. Sequences are edited and manually aligned with Seaview. A heuristic search was conducted using PAUP 4 to generate the most parsimonious trees (MPTs) with random stepwise addition (10 replicates), deletion treated as fifth base, TBR (tree bisection-reconnection) branch swapping and branches with maximum length zero collapsed to yield polytomies. Tree robustness is determined in terms of bootstrap (BS) proportions with 10,000 replicates. Two tick species are used as the outgroup: the hard tick *Ixodes ricinus* from the Ixodidae family and the soft tick *Ornithodoros moubata* from the Ornithodorinae subfamily, Argasidae family.

71.2.2.3 Sequence Analysis of Tick ITS2 Region

Marrelli et al. [50] utilized primers 5.8S (5′-CCATCGATGTGAAYTGCAGGACA-3′) and 28S (5′-GTGAATTCTATGCTTAAATTCAGGGGGT-3′) to amplify and analyze the ribosomal nucleotide sequences of the second internal transcribed spacer (ITS2) of ticks.

71.2.2.3.1 Procedure

1. PCR mixture (50 μL) is composed of 1× PCR buffer (10 mM Tris-HCl (pH 9.0), 1.5 mM $MgCl_2$, 50 mM KCl), 200 μM each of dNTPs, 1 U of *Taq* DNA polymerase (Promega), 50 pmol each of primers, and 10–20 ng of tick DNA.
2. Amplification is conducted with 35 cycles of 94°C for 45 s, 55°C for 1 min, and 72°C for 1.5 min.
3. The PCR products (10 μL) are examined by 0.8% agarose gel electrophoresis, stained with ethidium bromide, and visualized under ultraviolet light.
4. The amplicons containing the ITS2 and part of the flanking 5.8S and 28S genes are cloned into pGEM-T™ easy vector plasmid (Promega) and used to transform *Escherichia coli* Dh5α cells. The cloned fragments are sequenced in both forward and reverse direction by using ABI Prism dGTP BigDye Terminator Ready Reaction kit (Applied Biosystems).
5. The sequences are aligned with the CLUSTAL W software, version 1.60. The boundaries of the ITS2 are deduced by comparisons with rRNA gene sequences from other ticks species. Only the ITS2 sequences are used for CG content determination and phylogenetic analyses.
6. Aligned sequences are examined with the MEGA software (Molecular Evolutionary Genetics Analysis, version 3.01), and a divergence matrix is constructed. Relationships between individuals are assessed by three methods: neighbor-joining (NJ) method with Kimura two-parameter distances with 1000 replications in the BS test, maximum parsimony (MP) with 1000 BS replicates with the Min-mini branch-and-bound search option, and maximum likelihood, also with the Kimura two-parameter model (ML) (PHYLIP: Phylogeny Inference Package, version 3.6). An additional analysis by the ML method is also performed with PAUP*, and the model is used to reconstruct the ML tree (General Time Reversible model with gamma-distributed rates).

71.2.2.4 Sequence Analysis of Tick *cox1* and *nad5* Genes

Chitimia et al. [52] targeted portions of mitochondrial cytochrome *c* oxidase subunit 1 gene (*pcox1*) and nicotinamide adenine dinucleotide dehydrogenase subunit 5 gene (*pnad5*) for identification and phylogenetic analysis of tick species. A 732-bp fragment of *pcox1* gene is amplified with primers cox1F (5′-GGAACAATATATTTAATTTTTGG-3′) and cox1R (5′-ATCTATCCCTACTGTAAATATATG-3′); and a 519-bp fragment of *pnad5* gene is amplified with primers nd5F (5′-TCTAAAATTAAATCCTTTGAAT-3′) and nd5R (5′-AAGAGCCCAAATTCCATTTTC-3′) separately from individual ticks by PCR. The amplicons are sequenced and analyzed.

71.2.2.4.1 Procedure

1. PCR mixture (25 μL) is composed of 0.25 μL of each primer (50 μM), 1.25 U of r*Taq* DNA polymerase (Takara), 2 mM $MgCl_2$, 200 μM each of dNTPs, and 1 μL of DNA.
2. Amplification is carried out with an initial 95°C for 5 min, followed by 40 cycles of 95°C for 30 s, 55°C for 1 min, 72°C for 1 min, and a final extension of 72°C for 5 min.
3. The PCR products (2 μL) are examined by 1% agarose gel electrophoresis, stained with ethidium bromide, and visualized under ultraviolet light.
4. PCR products are purified and cloned into pGEM-T Easy vector (Promega,) and transformed into *Escherichia coli* competent JM109 cells. The recombinant bacteria are screened, confirmed by PCR amplification, and then sequenced.
5. Amino acid sequences of *pcox1* and *pnad5* are separately aligned using Clustal X, which is subsequently utilized as a scaffold to align the corresponding nucleotide sequences with the program CodonAlign 2.0 (http://homepage.mac.com/barryghall/CodonAlign.html). The aligned nucleotide sequences are then concatenated into single alignments for phylogenetic analyses. Dataset congruence between gene partitions is tested in a MP framework using the partition homogeneity test as implemented in PAUP* 4.0b10.
6. Data are analyzed using three approaches, namely MP, ML, and NJ, and *Varroa destructor* is used as outgroup in all analyses. Standard unweighted MP are performed in PAUP* using heuristic searches with TBR branch swapping and 1000 random-addition sequence replicates with 10 trees held at each step. Prior to ML analyses, best-fit models of nucleotide substitution are selected by the Akaike Information Criterion as implemented by Modeltest 3.5. ML topologies are then inferred with PAUP* and PhyML using TBR branch swapping under best-fit model). The NJ approach is performed by MEGA 4.0 using Jukes–Cantor as a substitution model. Branch supports in ML tree are estimated by BS analysis of 500 replicates, while this is done with 1000 replicates in both MP and NJ analyses.

71.3 CONCLUSION

Ticks are a group of obligate, bloodsucking, ectoparasitic arthropods that are classified under two major families, Ixodidae (hard-bodied ticks) and Argasidae (soft-bodied ticks). Whereas the hard tick family Ixodidae contains 702 species and is characterized by having a sclerotized dorsal shield (the scutum), the soft tick family Argasidae covers 193 species and is characterized by having a flexible and leathery exoskeleton (without the scutum). Ticks lead a complex life and undergo four developmental stages (i.e., egg, larva, nymph, and adult), with hard ticks having a single nymphal stage and soft ticks having 2–8 nymphal instars (stages).

Although ticks are mostly associated with vertebrate animal hosts, some may attack humans when animal hosts are unavailable. Besides causing local redness, itching, burning, and other discomfort through their bites, ticks play an important role in the transmission of pathogenic bacteria, viruses, and parasites to humans, resulting in a range of clinical disease. Within the soft tick family Argasidae, *Argas* species are responsible for inducing irritations and skin manifestations through bites, and *Ornithodoros* species transmit a number of bacterial and viral pathogens to humans.

Given the close morphological similarities between soft and hard ticks as well as among soft tick species, it has been a challenge to accurately determine the species identity of ticks implicated in the transmission of disease-causing agents. The availability of this information is critical to the effective treatment and control of infections. The recent development and application of molecular techniques help streamline the identification and phylogenetic assessment of ticks, making it possible to speciate ticks in a much fast, precise, and sensitive manner.

REFERENCES

1. Guglielmone, A.A. et al., The Argasidae, Ixodidae and Nuttalliellidae (Acari: Ixodida): A world list of valid tick names. *Zootaxa*, 2528, 1, 2010.
2. Anderson, J.F., The natural history of ticks. *Med. Clin. North. Am.*, 86, 205, 2002.
3. Anderson, J.F., Biology of ticks. *Infect. Dis. Clin. North. Am.*, 22, 195, 2008.
4. Nava, S., Guglielmone, A.A., and Mangold, A.J., An overview of systematics and evolution of ticks. *Front. Biosci.*, 14, 2857, 2009.
5. http://www.uniprot.org/taxonomy/; last accessed on 10 October 2011.
6. Estrada-Peña, A., and Jongejan, F., Ticks feeding on humans: A review of records on human-biting Ixodoidea with special reference to pathogen transmission. *Exp. Appl. Acarol.*, 23, 688, 1999.
7. Klompen, J.S.H. et al., Evolution of ticks. *Annu. Rev. Entomol.*, 41, 141, 1996.
8. Yamaguchi, N., Geographic variation of ixodid ticks in human cases. *Med. Entomol. Zool.*, 45, 198, 1994.
9. Evans, D.E., Martins, J.R., and Guglielmone, A.A., A review of the ticks (Acari, ixodida) of Brazil, their hosts and geographic distribution-1. The state of Rio Grande do Sul, southern Brazil. *Mem. Inst. Oswaldo Cruz*, 95, 453, 2000.
10. Barker, R.W. et al., Occurrence of the gulf coast tick (Acari: Ixodidae) on wild and domestic mammals in north-central Oklahoma. *J. Med. Entomol.*, 41, 170, 2004.
11. González-Acuña, D., and Guglielmone, A.A., Ticks (Acari: Ixodoidea: Argasidae, Ixodidae) of Chile. *Exp. Appl. Acarol.*, 35, 147, 2005.
12. Aydin, L., and Bakirci, S., Geographical distribution of ticks in Turkey. *Parasitol. Res.*, 101, S163, 2007.

13. Thewes, M. et al., Argas reflexus (the pigeon tick)–a household pest. *Acta Derm. Venereol.*, 77, 173-4, 1997.
14. Haag-Wackernagel, D., and Bircher, A.J., Ectoparasites from feral pigeons affecting humans. *Dermatology*, 220, 82, 2010.
15. Dautel, H., Scheurer, S., and Kahl, O., The pigeon tick (*Argas reflexus*): Its biology, ecology, and epidemiological aspects. *Zentralbl. Bakteriol.*, 289, 745, 1999.
16. Reeves, W.K., Molecular evidence for a novel *Coxiella* from *Argas monolakensis* (Acari: Argasidae) from Mono Lake, California, USA. *Exp. Appl. Acarol.*, 44, 57, 2008.
17. Condy, J.B. et al., The effects of the bites of *Argas brumpti* (Acarina: Argasidae) on humans. *Cent. Afr. J. Med.*, 26, 212, 1980.
18. Fukunaga, M. et al. Molecular phylogenetic analysis of Ixodid ticks based on the ribosomal DNA spacer, internal transcribed spacer 2, sequences. *J. Parasitol.*, 86, 38, 2000.
19. Cutler, S.J., Browning, P., and Scott, J.C., *Ornithodoros moubata*, a soft tick vector for Rickettsia in east Africa? *Ann. N. Y. Acad. Sci.*, 1078, 373, 2006.
20. Beck, A.F., Holscher, K.H., and Butler, J.F., Life cycle of *Ornithodoros turicata americanus* (Acari: Argasidae) in the laboratory. *J. Med. Entomol.*, 23, 313, 1986.
21. Helmy, N., Seasonal abundance of *Ornithodoros (O.) savignyi* and prevalence of infection with *Borrelia* spirochetes in Egypt. *J. Egypt. Soc. Parasitol.*, 30, 607, 2000.
22. Charrel, R.N. et al., Alkhurma hemorrhagic fever virus in *Ornithodoros savignyi* ticks. *Emerg. Infect. Dis.*, 13, 153, 2007.
23. Krinsky, W.L., Dermatoses associated with the bites of mites and ticks (Arthropoda: Acari). *Int. J. Dermatol.*, 22, 75, 1983.
24. Loftis, A.D. et al., Detection of *Rickettsia, Borrelia*, and *Bartonella* in *Carios kelleyi* (Acari: Argasidae). *J. Med. Entomol.*, 42, 473, 2005.
25. Mumcuoglu, K.Y. et al., Argasid ticks as possible vectors of West Nile virus in Israel. *Vector Borne Zoonotic Dis.*, 5, 65, 2005.
26. Tosti, A., Peluso, A.M., and Spedicato, S., Urticaria–angioedema syndrome caused by an *Argas reflexus* sting. *Contact Dermatitis*, 19, 315, 1988.
27. Veraldi, S., Scarabelli, G., and Grimalt, R., Acute urticaria caused by pigeon ticks (*Argas reflexus*). *Int. J. Dermatol.*, 35, 34, 1996.
28. Veraldi, S. et al., Skin manifestations caused by pigeon ticks (*Argas reflexus*). *Cutis*, 61, 38, 1998.
29. Sirianni, M.C. et al., Anaphylaxis after *Argas reflexus* bite. *Allergy*, 55, 303, 2000.
30. Rolla, G. et al., Allergy to pigeon tick (*Argas reflexus*): Demonstration of specific IgE-binding components. *Int. Arch. Allergy Immunol.*, 135, 293, 2004.
31. Hilger, C. et al., IgE-mediated anaphylaxis caused by bites of the pigeon tick Argas reflexus: Cloning and expression of the major allergen Arg r 1. *J. Allergy Clin. Immunol.*, 115, 617, 2005.
32. Quercia, O. et al., Anaphylactic shock to *Argas reflexus* bite. *Eur. Ann. Allergy Clin. Immunol.*, 37, 66, 2005.
33. Kleine-Tebbe, J. et al., Bites of the European pigeon tick (*Argas reflexus*): Risk of IgE-mediated sensitizations and anaphylactic reactions. *J. Allergy Clin. Immunol.*, 117, 190, 2006.
34. Spiewak, R. et al., Allergy to pigeon tick (*Argas reflexus*) in Upper Silesia, Poland. *Ann. Agric. Environ. Med.*, 13, 107, 2006.
35. Weckesser, S. et al., Anaphylactic reactions to bites of the pigeon tick *Argas reflexus*. *Eur. J. Dermatol.*, 20, 244, 2010.
36. Dworkin, M.S., Schwan, T.G., and Anderson, D.E. Jr., Tick-borne relapsing fever in North America. *Med. Clin. North. Am.*, 86, 413, 2002.
37. Dworkin, M.S., Tick-borne relapsing fever. *Infect. Dis. Clin. North. Am.*, 22, 449, 2008.
38. Aher, A.R. et al., A case report of relapsing fever. *Indian J. Pathol. Microbiol.*, 51, 292, 2008.
39. Moemenbellah-Fard, M.D. et al., Tick-borne relapsing fever in a new highland endemic focus of western Iran. *Ann. Trop. Med. Parasitol.*, 103, 529, 2009.
40. Hoskins, J.D., Ixodid and argasid ticks. Keys to their identification. *Vet. Clin. North. Am. Small. Anim. Pract.*, 21, 185, 1991.
41. Black, W.C., and Piesman, J., Phylogeny of hard- and soft-tick taxa (Acari: Ixodida) based on mitochondrial 16S rDNA sequences. *Proc. Natl. Acad. Sci. USA*, 91, 10034, 1994.
42. Black, W.C., Klompen, J.S.H., and Keirans, J.E., Phylogenetic relationships among tick subfamilies (Ixodida: Ixodidae: Argasidae) based on the 18S nuclear rDNA gene. *Mol. Phylog. Evol.*, 7, 129, 1997.
43. Crampton, A., McKay, Y., and Barker, S.C., Phylogeny of ticks (Ixodida) inferred from nuclear ribosomal DNA. *Int. J. Parasitol.*, 26, 511, 1996.
44. Mangold, A.J., Bargues, M.D., and Mas-Coma, S., Mitochondrial 16S rRNA sequences and phylogenetic relationships of *Rhipicephalus* and other tick genera among Metastriata (Acari: Ixodidae). *Parasitol. Res.*, 84, 478, 1998.
45. Hwang, U.W., and Kim, W., General properties and phylogenetic utilities of nuclear ribosomal DNA and mitochondrial DNA commonly used in molecular systematics, *Korean J. Parasitol.*, 37, 215, 1999.
46. Cruickshank, R.H., Molecular marker for the phylogenetics of mites and ticks. *Syst. Appl. Acarol.*, 3, 1, 2002.
47. Anderson, J.M., Ammerman, N.C., and Norris, D.E., Molecular differentiation of metastriate tick immatures. *Vector Borne Zoonotic Dis.*, 4, 334, 2004.
48. Vial, L. et al., Molecular divergences of the *Ornithodoros sonrai* soft tick species, a vector of human relapsing fever in West Africa. *Microbes Infect.*, 8, 2605, 2006.
49. Geraci, N.S. et al., Variation in genome size of argasid and ixodid ticks. *Insect. Biochem. Mol. Biol.*, 37, 399, 2007.
50. Marrelli, M.T. et al., Taxonomic and phylogenetic relationships between neotropical species of ticks from genus *Amblyomma* (Acari: Ixodidae) inferred from second internal transcribed spacer sequences of rDNA. *J. Med. Entomol.*, 44, 222, 2007.
51. Chitimia, L. et al., Molecular characterization of hard and soft ticks from Romania by sequences of the internal transcribed spacers of ribosomal DNA. *Parasitol. Res.*, 105, 907, 2009.
52. Chitimia, L. et al., Genetic characterization of ticks from southwestern Romania by sequences of mitochondrial cox1 and nad5 genes. *Exp. Appl. Acarol.*, 52, 305, 2010.
53. Halos, L. et al., Determination of an efficient and reliable method for DNA extraction from ticks. *Vet. Res.* 5, 1, 2004.
54. Shone, S.M. et al., A novel real-time PCR assay for the speciation of medically important ticks. *Vector Borne Zoonotic Dis.*, 6, 152, 2006.
55. Chao, L.L., Wu, W.J., and Shih, C.M., Species identification of *Ixodes granulatus* (Acari: Ixodidae) based on internal transcribed spacer 2 (ITS2) sequences. *Exp. Appl. Acarol.*, 54, 51, 2011.
56. Dergousoff, S.J., and Chilton, N.B., Differentiation of three species of ixodid tick, *Dermacentor andersoni, D. variabilis* and *D. albipictus*, by PCR-based approaches using markers in ribosomal DNA. *Mol. Cell. Probes*, 21, 343, 2007.

72 Demodex (Hair Follicle Mite)

Dongyou Liu

CONTENTS

72.1 INTRODUCTION

The genus *Demodex* covers >100 host-specific species of mites of mammals, including cats, dogs, sheep, cattle, pigs, goats, deer, bats, hamsters, rats, mice, primates, and humans. Owing to their tendency to live in or near hair follicles, *Demodex* spp are commonly known as hair follicle mites. To date, two *Demodex* species (i.e., *Demodex folliculorum* and *Demodex brevis*) have been shown to inhabit the human hair follicles and eyelashes (thus sometimes called eyelash mites). Although infestation with *Demodex* mites is common and usually asymptomatic, some skin diseases (demodicidosis and other disorders) may emerge, particularly in the immunosuppressed individuals. In addition, hair follicle mites may be involved in the transmission of bacterial and other infectious diseases to humans.

72.1.1 Classification, Morphology, and Biology

72.1.1.1 Classification

The genus *Demodex* (in Greek, "demo" = lard; "dex" = boring worm) is classified in the family Demodicidae, order Trombidiformes, subclass Acari (Acarina), class Arachnida, phylum Arthropoda. Being the only member in the family Demodicidae, the genus *Demodex* consists of more than 100 recognized species, the most notable of which include *Demodex antechini*, *D. brevis*, *Demodex bovis*, *Demodex canis*, *Demodex caprae*, *Demodex cati*, *Demodex equi*, *D. folliculorum*, *Demodex macroglossi*, *Demodex ovis*, *Demodex phyloides*, and *Demodex zalophi* [1].

First discovered in humans in 1841, members of the genus *Demodex* are host-specific mites that inhabit the hair follicles of mammals (the so-called hair follicle mites), with one or two *Demodex* species found in each mammal species. *D. folliculorum* (obsolete synonyms: *Acarus folliculorum*, *Entozoon folliculorum*, and *D. folliculorum longus*) and *D. brevis* (obsolete synonym: *D. folliculorum brevis*) are the two human infecting species that are commonly identified from the skin and eyelashes of human host (thus sometimes referred to as eyelash mites) [2,3]. On rare occasions, *D. canis*, an ectoparasite of domestic dog, may be found in humans [4].

72.1.1.2 Morphology

Demodex adults have a semi-transparent elongated body that consists of a head, neck, body, and tail, which are fused to form two noticeable segments: cephalothorax and abdomens. Eight (four pairs) short segmented legs are attached to the cephalothorax. The body is covered with scales for anchoring itself in the hair follicle, and the mite has pin-like mouth parts for eating skin cells, hormones, and oils (sebum) which accumulate in the hair follicles. Both male and female *Demodex* mites have a genital opening, and fertilization is internal. *Demodex* larva and nymph are similar to adult, but larva and/or nymph possesses six (three pairs) legs instead of eight legs.

D. folliculorum adults measure about 0.3–0.4 mm in length, with the females being somewhat shorter and rounder than the males, and they often gather as a group in the follicle area of the hair or lash. The mouthparts of *D. folliculorum* are sharp, with the capability of piercing the adipose tissue. The body and tail region of *D. folliculorum* is striated [5].

D. brevis adults are about 0.2–0.3 mm in length, have a spindle shape and stubby legs, and reside solitarily in the sebaceous gland.

72.1.1.3 Biology

The life cycle of a *Demodex* mite was first elucidated by Frank French in the 1960s during the study of *D. canis*, a hair follicle mite in the domestic dog, with the observation of the following developmental stages: ovum, hexapod larva, hexapod protonymph, octopod nymph, and adult [6]. Overall, the development of *Demodex* from an egg to the larval stage of protonymph to deutonymph and finally to the adult stage takes ~14–24 days.

Demodex mites can walk slowly (at a speed of between 8 and 16 cm/h) from their usual habitat, the hair follicles, to the skin, especially at night as they are photosensitive. After *Demodex* adult male and female mites mate in the opening of the hair follicle (near the skin surface), the gravid female migrates to the sebaceous gland or inside the hair follicle to deposit 20–24 eggs. Each egg gives rise to a larva and then a protonymph in the sebaceous canal. Moving to the opening of the hair follicle, a protonymph matures into a deutonymph, which migrates to the skin surface, then re-enters a hair follicle to become an adult. Generally, it takes 3–4 days for an egg to turn into larva, and about 7 days for the larva to develop into adult. Because their relatively short life span, the ability of *Demodex* mites to maintain and increase their population depends on their successful copulation and production of the next generation.

The main food source of *Demodex* sp. is sebum (oil), although follicular and glandular epithelial cells may serve as an alternative source of energy. Not surprisingly, *D. folliculorum* and *D. brevis* often gather in the face, cheeks, forehead, nose, nasolabial fold, ear canal, eyelashes, eyelids, alar region, nipple, knee, ectopic sebaceous glands on the tongue, scalp, chest, buttock, mons veneris, and penis (as these regions are rich in predisposing pilosebaceous units). The active excretion of sebum in these areas creates favorable habitats and breeding conditions for *Demodex* species.

Within the pilosebaceous unit, *D. folliculorum* lives in the hair follicles above the level of the sebaceous glands, whereas *D. brevis* resides in the sebaceous glands of the hair follicles. Owing to their susceptibility to desiccation, *D. folliculorum* and *D. brevis* do not survive long outside the living body (about one week). This feature can be exploited in forensic entomology (acarology) to determine the time of death of a crime victim [7,8].

Human hair follicle mites *D. folliculorum* and *D. brevis* are cosmopolitan, obligatory parasites of the pilosebaceous complex in humans [9]. They occur in all races and age groups, except the new-born, with over 50% of populations being carriers [10]. Men tend to be more heavily infested than women, probably due to the fact that men have more abundant sebaceous glands producing more food for the mites. It is estimated that two-third of elderly people, half of adults, and about one-third of children, and young adults, harbor these mites. The lower prevalence of hair follicle mites in children may be attributable to their reduced production of sebum. In one study, *Demodex* mites were identified in 13% of human patients with blepharitis (an inflammation of eyelid) aged between 3 and 15 years, 69% aged between 31 and 50 years, and 95% by age 71.

D. folliculorum and *D. brevis* species are most common on the face, and are detected in about 10% of biopsy specimens of the facial skin, and in about 12% of all follicles. *Demodex* mites can be found by carefully removing an eyelash or eyebrow hair and observing under a microscope. While *D. folliculorum* is a predominant mite, *D. brevis* has a wider distribution on the body. In a recent survey, 31% of Caucasians and African-Americans were found to harbor *D. brevis* only, 11% with *D. folliculorum* only, and 14% with both *D. brevis* and *D. folliculorum*.

D. folliculorum and *D. brevis* species are transmitted between human hosts through contact of hair, eyebrows, and of the sebaceous glands on the nose. As different species of mammals host different species of *Demodex*, human hair follicle mites *D. folliculorum* and *D. brevis* are not transferable to other animals. However, there are rare incidences of dog hair follicle mite *D. canis* being found in humans.

Human hair follicle mites show varied rates of survival under different temperatures: at 16–20°C (*D. folliculorum* 68.0 h, *D. brevis* 88.0 h), at 5°C (*D. folliculorum* 110 h, *D. brevis* 145 h), at 0°C (*D. folliculorum* 23 h, *D. brevis* 35 h), and at −15°C (*D. folliculorum* 5.5 h, *D. brevis* 5 h). At 25–26°C, *D. folliculorum* and *D. brevis* mites become active, and have a shorter survival time than at 16–20°C. At 37°C or above, *D. folliculorum* and *D. brevis* mites do not survive as long and appear extraordinarily restless. At 54°C, *D. folliculorum* and *D. brevis* mites can only survive for 3–5 min, and at 58°C, *D. folliculorum* and *D. brevis* mites die in 1 min [11,12].

D. folliculorum and *D. brevis* mites live inside the sebaceous glands and hair follicles of human host, sucking nutrients from the hair roots and damaging the cell walls. These activities generally have limited adverse effects on healthy individuals [13]. However, in patients with a suppressed immune system, human hair follicle mite populations can increase dramatically, resulting in a condition known as demodicosis or *Demodex* mite bite, with clinical presentations of itching, inflammation of the eyelids (blepharitis), and other skin disorders (e.g., pityriasis and rosacea, which may be further exacerbated by bacteria *Bacillus oleronius* and *Wolbachia* found in the mites). Indeed, rosacea predisposes patients to blepharitis by congesting the oil-producing glands on the skin. Other factors contributing to mite's proliferation include sunlight exposure, alcohol intake, smoking, stress, hot beverages, spicy food, and abrupt temperature changes.

72.1.2 Clinical Features, Pathogenesis, and Treatment

72.1.2.1 Clinical Features

As common, obligate ectoparasites of mammals, *Demodex* mites reside in the pilosebasceous unit of hair follicles in various parts of the body [14–16]. Although infestation of *Demodex* mites often causes negligible damages to the host, there is a growing body of evidence linking *Demodex* mites with demodicidosis (characterized by the presence of an erythematous papule–pustular rash mainly on the face) and other skin disorders, especially in immunocompromised individuals after topical or systemic administration of steroids or other immunosuppressive agents, or with underlying diseases, such as leukemia and HIV [17–36].

In the human skin, *D. folliculorum* and *D. brevis* have been associated with facial itching or pruritus (with or without erythema), tightness and burning sensation in the skin, dry skin, acne (inflammatory papule), perioral dermatitis (a facial rash occurring around the mouth), follicular distention, mild perifolliculitis, rosacea (a chronic inflammatory dermatosis of the convexities of the central face characterized by the presence of multiple small dome-shaped erythematous papules and papulopustules arising on a background of fixed inflammatory erythema, leading to papullo–pustular rosecea and granulomatous rosecea), pityriasis folliculorum (typically found in older women who do not use soap but apply large amounts of makeup), folliculitis (characterized by the posterior portions of numerous *D. folliculorum* mites protruding from the follicles), external otitis, scabies-like eruptions, facial pigmentation, comedons (blackheads), eruptions of the bald scalp, demodicosis gravis (dermal granulomas resulting from mite remnants phagocytosed by foreign-body giant cells), basal cell carcinoma, spongiosis, and lymphoid inflammation [37–57]. Interestingly, *D. folliculorum* is more closely related to erythema and scale in the T zone, and *D. brevis* is more closely related to a symmetrical malar and papulopustular eruption superimposed on existing skin disease.

D. folliculorum and *D. brevis* mite infestation in the face easily spreads and flourishes in the eyelid since the eyelid is not as accessible as other protruding body parts (e.g., the nose, the brow, and the cheek) of the face to daily cleansing hygiene. Frequent habitants of the sebaceous and apocrine glands of human eyelids, *D. folliculorum* is found in the lash follicle, and *D. brevis* burrows deep into the lash sebaceous gland and the meibomian gland. The meibomian gland is a producer of meibum that forms the superficial layer of the preocular tear film. Apocrine glands secrete an oily fluid, which is rich in odoriferous substances (cholesterol, triglycerides, fatty acids, cholesterol esters, and squelene) and contains androgens, carbohydrates, ammonia, and ferric iron.

In the human eye, *D. folliculorum* and *D. brevis* have been linked to blepharitis (inflammation of eyelid), chronic eczematous blepharitis (blepharitis acarica), madarosis (loss of eyelashes), hyperkeratosis, and localized keratitis [58–66]. In addition to itching, burning, foreign body sensation, crusting and redness of the lid margin, and blurry vision, other common clinical signs often appear. These include cylindrical dandruff, disorders of eyelashes (malalignment, trichiasis, madarosis, punctate epithelial erosions, corneal ulceration, and pannus formation), lid margin inflammation (blepharitis), meibomian gland dysfunction (blocking, swelling, hordeolum, or chalazion), blepharoconjunctivitis, and blepharokeratitis [67–69].

Cylindrical dandruff (CD) on the skin surface where lashes emit is a common clinical sign in some blepharitis patients, indicative of *Demodex* infestation of the eyelash. The presence of *Demodex* mites close to the follicles disorganizes the orientation of lashes, which can turn into touch the ocular surface (trichiasis), or result in the loss of lashes (madarosis). The inflammation of meibomian glands can lead to plugging of the orifices. The combined involvement of lashes and meibomian glands of the lid margin can be spill over to the conjunctiva (conjunctivitis) and cause blinding corneal lesions such as superficial corneal vascularization, marginal infiltration, phlyctenule-like lesion, superficial opacity, and nodular scar.

In fact, about 58% of individuals with hair follicle mites show evidence of chronic blepharitis. Furthermore, papulovesicular rosacea-like lesions and spiny blepharitis often respond to agents that reduce *Demodex* numbers [45,70–72].

Benessahraoui et al. [73] reported a case of demodicidosis in a 22-month-old girl undergoing six-month chemotherapy (mercaptopurine, 50 mg daily) for chronic myelomonocytic leukemia associated with xanthoma and type-1 neurofibromatosis. The patient presented with a facial eruption. Physical examination revealed erythematous papules, perioral and subpalpebral pustula with a few lesions on the trunk, along with splenomegaly and hepatomegaly. A 10% potassium hydroxide preparation of skin material uncovered the presence of *Demodex follicularum* mites. The eruption cleared within 2 weeks after oral and topical metronidazole therapy (oral metronidazole, 250 mg/day for one month associated with topical formulation containing metronidazole 0.75% gel).

Türk et al. [74] demonstrated that *D. folliculorum* was present in 11 of 37 (29.72%) patients with blepharitis, in one of 11 (9.09%) patients with blepharoconjunctivitis, and in 2 of 48 (4.16%) persons in healthy control group. *Staphylococcus aureus* was cultivated in five patients and *D. folliculorum* was found two of them.

In animals, *Demodex* mites represent a well-established cause of mange, producing two types of lesions: squamous (with dry alopecia and thickening of the skin) and pustular (due to secondary *Staphylococcus* infection, resulting in the characteristic red, numerous pustules, and wrinkling of the skin).

72.1.2.2 Pathogenesis

Demodex mites commonly occur in the skin and eyelashes of humans, and do not inflict significant damage to the habitat that they live in. Demodicosis only appears with the multiplication of *Demodex* spp, as a result of hypervascular tissue, poor hygiene, and immune deficiency.

Anatomically, *Demodex* blepharitis can be classified into anterior and posterior blepharitis. While anterior blepharitis is an infestation of eyelashes and follicles by *D. folliculorum*, clustering to the root of the lashes, posterior blepharitis is an infestation of the meibomian gland preferentially by *D. brevis*. The pathogenesis of *Demodex* blepharitis may be attributed to direct damage, inflammation/hypersensitivity reaction, and transmission of bacteria [75].

1. Direct damage. *D. folliculorum* consume epithelial cells at the hair follicle leading to follicular distension, and formation of loose or misdirected lashes. The mite's claws cause microabrasions inducing epithelial hyperplasia and reactive hyperkeratinization around the base of the lashes, and forming cylindrical dandruff. With an increasing number, *D. brevis* may block the orifices of meibomian glands mechanically, causing meibomian gland dysfunction with lipid tear deficiency.
2. Inflammation/hypersensitivity reaction. Located deeply in the meibomian glands, *D. brevis* chitinous exoskeleton debris and wastes may function as a foreign substance and elicit granulomatous reaction and inflammatory responses via a delayed hypersensitivity reaction or an innate immune response, resulting in recurrent and refractory chalazia, epithelial hyperplasia and hyperkeratinization.
3. Transmission of infectious agents. *Demodex* mites carry several bacteria on its surface and inside its body, including *Staphylococcus albus*, *Streptococci*, and *B. oleronius* [76]. *Staphylococcus albus* present on the mite skin surface is transported by *Demodex* mites from follicle to follicle. Staphylococcal superantigens may play a role in the induction of rosacea. The 62- and 83-kDa proteins of *B. oleronius* are capable of stimulating proliferation of peripheral blood mononuclear cells in patients with rosacea [77]. A strong correlation has also been observed among positive serum immunoreactivity to these two proteins, ocular *Demodex* infestation, facial rosacea, and blepharitis [78]. Microbial involvement in the pathogenesis of mite-infested diseases is further supported by the observation that the skin inflammation in rosacea can be improved markedly by topical metronidazole or oral antibiotics (e.g., tetracycline), which do not kill mites.

Other mechanisms of *Demodex* pathogenecity include genotype susceptibility and immune dysfunction. Individuals with the human leukocyte antigen (HLA) phenotype Cw2 phenotype show increased vulnerability to demodicosis, with deep papular and papulopustular lesions [79,80]. In patients with HIV infection and myelocytic leukemia, *Demodex* mites may cause a pruritic eruption in the face, as well as the presternal and interscapular areas. In addition, *Demodex* spp have the ability to suppress the normal T-lymphocyte response of human host, further enhancing their pathogenic potency [81].

72.1.2.3 Treatment

Demodex adult worms are resistant to common antiseptic solutions including 75% alcohol and 10% povidone–iodine, but are susceptible to tea tree oil (TTO) at a dose-dependent manner [82]. As a natural essential oil steam-distilled from the leaf of the tea tree *Melaleuca alternifolia*, TTO demonstrates antibacterial, antifungal, antiparasitic, and antiinflammatory activities. While 5% TTO interrupts *Demodex* life cycle by preventing mating, 50% TTO has direct killing effect on the mites. Besides cleansing CD from the lash root, lid scrub with TTO stimulates embedded mites to migrate out to the skin at the same time. Thus, weekly lid scrub with 50% TTO and daily lid scrub with tea tree shampoo is effective in eradicating ocular *Demodex* infestation and bringing the *Demodex* count down to zero in 4 weeks [82].

For treatment of *Demodex* skin infections, application of topical sulfur ointment, sulfacetamide, camphor oil, crotamiton, choline esterase inhibitors, metronidazole, permethrin, ivermectin, lindane, mercury pamoad, pilocarpin gel, benzoyl benzoate, and Amitraz is valuable [83–91]. Facial *Demodex* infestations may be treated with topical dilute camphor oil with oral metronidazole (an agent known to be effective for rosacea). Use of antibiotics (e.g., tetracycline) is valuable to control secondary bacterial infection associated with pustular demodicosis. Mechanical cleaning, general hygiene, nutritional improvement, and attention to any possible underlying immune dysfunction also help mitigate the disease.

72.1.3 Laboratory Diagnosis

Diagnosis of *Demodex* infections has traditionally reliant on assessment of clinical history followed by microscopic confirmation. Despite their ubiquitous presence in humans, *Demodex* spp do not usually cause disease, unless the host immune functions have been dramatically altered by chemotherapy or underlying diseases. For instance, individuals with leukemia undergoing chemotherapy often develop unusual facial eruptions, for which any of the following may be responsible: demodicosis/folliculitis, dermatophytosis, candidiasis, impetigo, papulopustular or erythematotelangiectatic rosacea, seborrhoeic dermatitis, perioral dermatitis, contact dermatitis, acneiform eruptions, facial flushing/blushing, erythema, telangiectasia, scaling, facial skin roughness on palpation, centrofacial inflammatory lesions, and zinc deficiency. Therefore, microscopic identification of *Demodex* eggs, larvae, and adult mites in the skin and epilated lashes is critical to verify the disease.

For qualitative detection of *Demodex* in human skin, several scrapings of the skin are taken (using the cellophane tape method or sterile scalpel blade), transferred onto a slide, clarified with a drop of glycerine, covered with a coverslip, and viewed under a light microscope [92]. For skin sampling using the cellophane tape method, the subjects' face is washed with warm water, and 2 × 7 cm pieces of cellophane tape are applied to the forehead, cheeks, nose, and chin just prior to night sleep. Next morning, the tape is removed and pressed onto slides for observation and counting by light microscopy. For hair sampling (epilation), hair is picked with a forceps one by one. Collected hair follicles are covered with glycerin over a slide and examined by light microscopy (×40 and ×100). *D. folliculorum* has a cigar-shaped long body with an abdomen forming two-thirds of its length, short and obtuse legs and cone-shaped termination of the body. For eyelash sampling, two lashes, one from each half of each lid, are removed by fine forceps under a slit-lamp microscope at a magnification of ×25 and placed separately on each end of glass slides. A coverslip is mounted onto each lash before slowly pipetting 20 μL of saline to the edge of the coverslip to surround the lash. This results in preservation of the *Demodex* that has a loose contact with the lash at the tip. Under the microscope (magnification of ×40 and ×100), the number of *Demodex* is recorded. If a compacted CD is preserved, 20 μL of 100% alcohol or 0.25% fluorescein drops is pipetted into the edge of the coverslip, and the counting time is prolonged up to 20 min to allow the embedded *Demodex* to migrate from the cylindrical dandruff.

For quantitative measurement of the density of *Demodex*, two methods are commonly used: standardized skin surface biopsy (SSSB) and direct microscopic examination of fresh secretions from sebaceous glands (DME). Before sampling, patients are instructed not to wash their face for 12 h.

In the SSSB method, an area of 1 cm^2 is drawn on a microscope slide with a waterproof pen; a drop of cyanoacrylic adhesive (superglue) is placed on the slide in the 1 cm^2 area; the adhesive-bearing surface of the slide is applied to the lesional skin (on cheek or forehead); the slide is removed gently after it has been allowed to dry (about 1 min); the sample (containing the surface keratin layer, top of the pilosebaceous follicle, and its content) is clarified with 2–3 drops of immersion oil (or glycerine) and covered with a coverslip; *Demodex* is located in follicles under a light microscope (magnification ×40 and ×100) [93]. Because hairs together with the superficial horny layer are sampled with minimal sebaceous glands, *D. folliculorum* is primarily found.

In the direct microscopic examination of fresh secretions from sebaceous glands (DME), an area of 1 cm^2 of affected skin is squeezed between the thumb and index finger and then scraped using the blunt end of a sterile scalpel (to obtain fresh secretions from sebaceous glands, follicular plugs and pustules, and skin scrapings); the material is transferred to a microscope slide, mixed with a drop of glycerine and covered with a coverslip; *Demodex* is observed under a light microscope (no follicles) (magnification ×40 and ×100) [79].

A diagnosis of demodicosis is confirmed when a high number of *Demodex* mites (>5 mites/cm^2) are measured using SSSB and/or DME along with clinical manifestations. The optimal threshold of >5 mites/cm^2 has a specificity of 98%, and a sensitivity of 55% [79,93]. A recent comparison of SSSB and DME using the threshold of 5 mites/cm^2 indicated that SSSB demonstrates a higher sensitivity than DME, and therefore represents a better diagnostic tool for demodicosis [94]. If live mites are present, treatment is undertaken, until two subsequent samples reveal no mites, or no live mites.

In recent years, molecular techniques have been increasingly applied for improved identification and phylogenetic analysis of pathogenic mites [95,96]. Nonetheless, to date relatively few molecular studies have been reported concerning human hair follicle mites *D. folliculorum* and *D. brevis*, apart from three reports dealing with *D. canis* [97–99]. A search of GenBank database revealed that the nucleotide sequences of partial mitochondrial cytochrome-c oxidase subunit I (COI) gene, mitochondrial 16S ribosomal RNA gene, 18S ribosomal RNA gene, 5.8S rRNA gene, 28S rRNA gene, internal transcribed spacer 1 (ITS 1) and ITS2, and chitin synthase gene from *D. folliculorum* and *D. brevis* are available. This provides a foundation for future development of PCR and sequence based approaches for determination of these mites.

72.2 METHODS

72.2.1 Sample Preparation

72.2.1.1 Sample Collection

Demodex mite samples are collected from human skin and hair using the cellophane tape method, the SSSB method, or a sterile scalpel (see Section 72.1.3 for details). The samples are mounted on slides and observed under light microscope (magnification ×40 or ×100).

72.2.1.2 DNA Isolation

DNA extraction from human skin scrapings is carried out using the Qiagen DNeasy® Tissue Kit (Qiagen) following the manufacturer's recommendations. Briefly, human specimens are resuspended in 180 μL ATL buffer, transferred into a 1.5 mL tube and incubated at 56°C overnight with 20 μL proteinase K. After proteinase K digestion, undigested material is removed by centrifugation and the supernatant is used for subsequent DNA extraction [97].

For DNA preparation from mites, individual mites are crushed with a sterile plastic pestle in a 1.5 mL tube containing 200 μL of extraction buffer (2% CTAB, 1.4 M NaCl, 20 mM EDTA, 100 mM Tris–HCl, pH 8.0, and 0.2% 2-mercaptoethanol), preheated to 65°C. This homogenate is incubated at 65°C for 1 h. Proteins are removed with one volume of chloroform/isoamyl alcohol. DNA is precipitated from the aqueous layer with one volume of isopropanol for 2 h at –20°C and spun at 18,000 × *g* for 30 min. The DNA pellet is then washed with 300 μL of 70% ethanol, dried for 10–15 min at 50°C, resuspended in 20 μL of ultrapure water and stored at –20°C [100].

72.2.2 Detection Procedures

72.2.2.1 Sequencing Analysis of Partial 18S rRNA Gene

The partial sequence (877 bp) of *D. folliculorum* 18S ribosomal RNA gene is available at GenBank (Accession No. EU861211.1). Because the 5′ and 3′ ends of this sequence is highly conserved among various eukaryotic organisms, a pair of primers (Demodex_18SF: 5′-ATCC AAGGAAGGCAGCCAGGCA-3′ and Demodex_18SR: 5′-CCATGCACCACCACCCACTGAA-3′) are designed to amplify a 873-bp fragment from *Demodex* DNA. Subsequent analysis of the nucleotide sequence permits precise determination of *Demodex* identity.

72.2.2.1.1 Procedure

1. PCR mixture (50 μL) is made up of 1× PCR buffer (10 mM Tris–HCl pH 9.0, 1.5 mM $MgCl_2$, and 50 mM KCl), 200 μM each of dATP, dCTP, dGTP, and dTTP, 2 U of *Taq* DNA polymerase (Promega), 50 pmol of each primers, and 10–20 ng of parasite DNA template.
2. The mixture is subjected to an initial step at 94°C for 2 min; 35 cycles for 30 s at 94°C, 30 s at 55°C, and 60 s at 72°C; and a final step for 7 min at 72°C.
3. The resulting PCR product is electrophoresed in a 1.5% agarose gel, stained with ethidium bromide, and visualized under UV light.
4. The PCR product is recovered from the gel and sequenced using one of the PCR primers with the BigDye™ Terminator Cycle Sequencing Kit Version 3.1 (Applied Biosystems) on an ABI-PRISM 3130 Genetic Analyzer (Applied Biosystems).

72.2.2.2 Sequencing Analysis of ITS Regions

A gene region (654 bp) covering *D. folliculorum* partial 18S ribosomal RNA gene, ITS1, 5.8S rRNA gene, ITS2, and partial 28S rRNA gene is available at GenBank (Accession No. AM904564.1). Use of primers (Demodex_ITSF: 5′-AGA GGAAGTAAAAGTCGTAACA-3′ and Demodex_ITSR: 5′-GGGTTGTCTGACTTCATTTGAG-3′) allows amplification of a 637-bp fragment from *Demodex* DNA. Subsequent comparison of the ITS1 and ITS2 sequences enables specific identification of *Demodex* spp.

72.2.2.2.1 Procedure

1. PCR mixture (50 μL) is made up of 1× PCR buffer (10 mM Tris–HCl pH 9.0, 1.5 mM $MgCl_2$, 50 mM KCl), 200 μM each of dNTPs, 2 U of *Taq* DNA polymerase (Promega), 50 pmol of each primers, and 10–20 ng of parasite DNA template.
2. The mixture is subjected to an initial step at 94°C for 2 min; 35 cycles for 30 s at 94°C, 30 s at 55°C, and 45 s at 72°C; and a final step for 7 min at 72°C.
3. The resulting PCR product is electrophoresed in a 1.5% agarose gel, stained with ethidium bromide, and visualized under UV light.
4. The PCR product is recovered from the gel and sequenced using one of the PCR primers with the BigDye™ Terminator Cycle Sequencing Kit Version 3.1 (Applied Biosystems) on an ABI-PRISM 3130 Genetic Analyzer (Applied Biosystems).

72.3 CONCLUSION

The genus *Demodex* encompasses >100 host-specific, tiny mites that feed on sebum (oil), follicular, and glandular epithelial cells of mammals. Two *Demodex* species, *D. folliculorum* and *D. brevis*, have been shown to occur exclusively in humans, forming part of normal fauna in the skin (especially the face, cheeks, forehead, nose, and external ear tract, etc.), where numerous hair follicles sebaceous glands provide these mites an ideal hiding place and a steady food supply. Usually the total number of *Demodex* mites is kept in balance by host's immune systems, and the presence of these mites causes minimal harm to the host. However, when the immune function of the host becomes compromised due to underlying diseases, chemotherapy or poor hygiene, *Demodex* mites often multiply out of control, and cause significant clinical and pathological changes (collectively referred to as demodicosis).

In the human skin, *D. folliculorum* and *D. brevis* have been implicated in acne, follicular pityriasis, pustular folliculitis, and rosacea, and so on. In the human eye, *D. folliculorum* and *D. brevis* have been associated with blepharitis (inflammation of eyelid), hyperkeratosis, blurry vision, blepharoconjunctivitis, and blepharokeratitis, and so on [101–103]. Because clinical manifestations of demodicosis are similar to a range of skill disorders due to other infectious agents and disease processes, it is critical to correctly identify the responsible *Demodex* organisms before effective treatment and prevention measures are implemented. Considering the specialized skills required for the identification of *Demodex* species on the basis of phenotypic criteria [104], molecular techniques such as PCR amplification and DNA sequencing offer a valuable alternative for improved determination and phylogenetic analysis of *Demodex* mites.

REFERENCES

1. http://www.uniprot.org/taxonomy/; last accessed on 10 October 2011.
2. Rufli, T., and Mumcuoglu, Y., The hair follicle mites *Demodex folliculorum* and *Demodex brevis*: Biology and medical importance. *Dermatologica*, 162, 1, 1981.
3. Elston, D.M., *Demodex* mites: Facts and controversies. *Clin. Dermatol.*, 28, 502, 2010.
4. Morsy, T.A. et al., *Demodex* (follicular mite) infesting a boy and his pet dog. *J. Egypt. Soc. Parasitol.*, 25, 509, 1995.
5. Desch, C.E., and Nutting, W.B., *Demodex folliculorum* (Simon) and *D. brevis* Akbulatova of man: Redescription and reevaluation. *J. Parasitol.*, 58, 169, 1972.
6. French, F.E., *Demodex canis* in cainine tissues. *Cornell Vet.*, 54, 270, 1964.
7. Desch, C.E., Human hair follicle mites and forensic acarology. *Exp. Appl. Acarol.*, 49, 143, 2009.
8. O'Connor, B.M., Astigmatid mites (Acari: Sarcoptiformes) of forensic interest. *Exp. Appl. Acarol.*, 49, 125, 2009.
9. Uğraş M. et al., The prevalence of *Demodex folliculorum* on the scrotum and male perineal skin. *Turkiye Parazitol. Derg.*, 33, 28, 2009.

10. Ozdemir, M.H. et al., Hilal, Prevalence of *Demodex* in health personnel working in the autopsy room. *Am. J. Forensic Med. Pathol.*, 26, 18, 2005.
11. Yao, Y.E., Guo, N., and Wu, L.P., The effect of temperature on the viability of *Demodex folliculorum* and *Demodex brevis*. *Parasitol. Res.*, 105, 1623, 2009.
12. Zhao, Y.E., Guo, N., and Wu, L.P., Influence of temperature and medium on viability of *Demodex folliculorum* and *Demodex brevis* (Acari: Demodicidae). *Exp. Appl. Acarol.*, 54, 421, 2011.
13. Kemal, M. et al., The prevalence of *Demodex folliculorum* in blepharitis patients and the normal population. *Ophthalmic Epidemiol.*, 12, 287, 2005.
14. Aylesworth, R., and Vance, J.C., *Demodex folliculorum* and *Demodex brevis* in cutaneous biopsies. *J. Am. Acad. Dermatol.*, 7, 583, 1982.
15. Bonnar, E., Eustace, P., and Powell, F.C., *Demodex* mite in normal skin. *Lancet*, 337, 1168, 1991.
16. Pena, G.P., and Andrade Filho, J.S., Is *Demodex* really nonpathogenic? *Rev. Inst. Med. Trop Sao Paulo*, 42, 171, 2000.
17. Ayres, S. Jr., *Demodex folliculorum* as a pathogen. *Cutis*, 37, 441, 1986.
18. Heacock, C.E., Clinical manifestations of demodicosis, *J. Am. Optom. Assoc.*, 57, 914, 1986.
19. Sahn, E.E., and Sheridan, D.M., Demodicidosis in a child with leukaemia. *J. Am. Acad. Dermatol.*, 27, 799, 1992.
20. Redondo Mateo, J. et al., *Demodex*-attributed rosacea-like lesions in AIDS. *Acta Derm. Venereol.*, 73, 437, 1993.
21. Barrio, J. et al., Rosacea-like demodicidosis in an HIV-positive child. *Dermatology*, 192, 143, 1996.
22. Castanet, J. et al., Demodicidosis in an immunodeficient child. *Pediatr. Dermatol.*, 14, 219, 1997.
23. Patrizi, A. et al., Demodicidosis in immunocompetent young children: Report of eight cases. *Dermatology*, 195, 239, 1997.
24. Pallotta, S. et al., Unilateral demodicidosis. *Eur. J. Dermatol.*, 8, 191, 1998.
25. Aydıngoz, I.E., Dervent, B., and Guney, O., *Demodex folliculorum* in pregnancy. *Int. J. Dermatol.*, 39, 743, 2000.
26. Jansen, T. et al., Rosacea-like demodicidosis associated with acquired immunodeficiency syndrome. *Br. J. Dermatol.*, 144, 139, 2001.
27. Kastner, U., Kreuter, A., and Altmeyer, P., Rosacea-like demodicidosis associated with acquired immunodeficiency syndrome. *Br. J. Dermatol.*, 144, 139, 2001.
28. Akdeniz, S. et al., Is *Demodex folliculorum* larger in diabetic patients? *J. Eur. Acad. Dermatol. Venereol.*, 16, 539, 2002.
29. Baima, B., and Sticherling, M., Demodicidosis revisited. *Acta Derm. Venereol.*, 82, 3, 2002.
30. Seyhan, M.E. et al., Density of *Demodex folliculorum* in haematological malignancies. *J. Int. Med. Res.*, 32, 411, 2004.
31. Akilov, O.E., Butov, Y.S., and Mumcuoglu, K.Y., A clinicopathological approach to the classification of human demodicosis. *J. Dtsch. Dermatol. Ges.*, 3, 607, 2005.
32. Ciftci, I.H. et al., *Demodex folliculorum* in patients with rheumatoid arthritis. *Acta Parasitol.*, 2007, 52, 70–73.
33. Ozcelik, S. et al., The incidence of *Demodex folliculorum* in patients with chronic kidney deficiency. *Turkiye Parazitol. Derg.*, 31, 66, 2007.
34. Kulac, M. et al., Clinical importance of *Demodex folliculorum* in patients receiving phototherapy. *Int. J. Dermatol.*, 47, 72, 2008.
35. Hsu, C.K., Hsu, M.M., and Lee, J.Y., Demodicosis: A clinicopathological study. *J. Am. Acad. Dermatol.*, 60, 453, 2009.
36. Kligman, A.M., and Christensen, M.S., *Demodex folliculorum*: Requirements for understanding its role in human skin disease. *J. Invest. Dermatol.*, 131, 8, 2011.
37. Purcell, S.M., Hayes, T.J., and Dixon, S.L., Pustular folliculitis associated with *Demodex folliculorum*. *J. Am. Acad. Dermatol.*, 15, 1159, 1986.
38. Shelley, W.B., Shelley, E.D., and Burmeister, V., Unilateral demodectic rosacea. *J. Am. Acad. Dermatol.*, 20, 915, 1989.
39. Sibenge, S., and Gawkrodger, D.J., Rosacea: A study of clinical patterns, blood flow, and the role of *Demodex folliculorum*. *J. Am. Acad. Dermatol.*, 26, 590, 1992.
40. Bonnar, E., Eustace, P., and Powell, F.C., The *Demodex* mite population in rosacea. *J. Am. Acad. Dermatol.*, 28, 443, 1993.
41. Vollmer, R.T., Demodex-associated folliculitis. *Am. J. Dermatopathol.*, 18, 589, 1996.
42. Hwang, S.M. et al., Demodecidosis manifested on the external genitalia. *Int. J. Dermatol.*, 37, 634, 1998.
43. Georgala, S. et al., Increased density of *Demodex folliculorum* and evidence of delayed hypersensitivity reaction in subjects with papulopustular rosacea. *J. Eur. Acad. Dermatol. Venereol.*, 15, 441, 2001.
44. Cribier, B., Pathophysiology of rosacea: Redness, telangiectasia, and rosacea. *Ann. Dermatol. Venereol.*, 138, S184, 2011.
45. Erbagci, Z., and Ozgoztasi, O., The significance of *Demodex folliculorum* density in rosacea, *Int. J. Dermatol.*, 37, 421, 1998.
46. Forton, F. et al., Demodicosis and rosacea: Epidemiology and significance in daily dermatologic practice. *J. Am. Acad. Dermatol.*, 52, 74, 2005.
47. Morras, P.G. et al., Rosacea-like demodicidosis in an immunocompromised child. *Pediatr. Dermatol.*, 20, 8, 2003.
48. Schaller, M., Sander, C.A., and Plewig, G., *Demodex* abscesses: Clinical and therapeutic challenges. *J. Am. Acad. Dermatol.*, 49, S272, 2003.
49. Karincaoglu, Y. et al., The clinical importance of *Demodex folliculorum* presenting with nonspecific facial signs and symptoms. *J. Dermatol.*, 31, 618, 2004.
50. Dolenc-Voljc, M., Pohar, M., and Lunder, T., Density of *Demodex folliculorum* in perioral dermatitis, *Acta Derm. Venereol.*, 85, 211, 2005.
51. el-Bassiouni, S.O. et al., A study on *Demodex folliculorum* mite density and immune response in patients with facial dermatoses. *J. Egypt. Soc. Parasitol.*, 35, 899, 2005.
52. Okyay, P. et al., Prevalence of *Demodex folliculorum* in young adults: Relation with sociodemographic/hygienic factors and acne vulgaris. *J. Eur. Acad. Dermatol. Venereol.*, 20, 474, 2006.
53. García-Vargas, A., Mayorga-Rodríguez, J.A., and Sandoval-Tress, C., Scalp demodicidosis mimicking favus in a 6-year-old boy. *J. Am. Acad. Dermatol.*, 57, S19, 2007.
54. Lee, J.Y., and Hsu, C.K., Granulomatous rosacea-like demodicidosis. *Dermatol. Online J.*, 13, 9, 2007.
55. Moravvej, H. et al., Association of rosacea with demodicosis. *Arch. Iran. Med.*, 10, 199, 2007.
56. Fernandez-Flores, A., and Alija, A., Scalp folliculitis with *Demodex*: Innocent observer or pathogen? *Braz. J. Infect. Dis.*, 13, 81, 2009.
57. Zhao, Y.E. et al., Retrospective analysis of the association between *Demodex* infestation and rosacea. *Arch. Dermatol.*, 146, 896, 2010; Erratum: 146, 1412, 2010.
58. Humiczewska, M., *Demodex folliculorum* and *Demodex brevis* (Acarida) as the factors of chronic marginal blepharitis. *Wiad. Parazytol.*, 37, 127, 1991.

59. Kamoun, B. et al., Blepharitis due to Demodex: myth or reality? *J. Fr. Ophtalmol.*, 22, 525, 1999.
60. Koksal, M. et al., A rare agent of chalazion: Demodectic mites. *Can. J. Ophthalmol.*, 38, 605, 2003.
61. Arıcı, M.K. et al., The Prevalence of *Demodex folliculorum* in blepharitis patients and the normal population. *Ophthalmic Epidemiol.*, 12, 287, 2005.
62. Rodriguez, A.E., Ferrer, C., and Alio, J.L., Chronic blepharitis and *Demodex. Arch. Soc. Esp. Ophthalmol.*, 80, 635, 2005.
63. Czepita, D. et al., *Demodex folliculorum* and *Demodex brevis* as a cause of chronic marginal blepharitis. *Ann. Acad. Med. Stetin.*, 53, 63, 2007.
64. Lee, S.H. et al., The relationship between demodex and ocular discomfort. *Invest. Ophthalmol. Vis. Sci.*, 51, 2906, 2010.
65. Lacey, N., Kavanagh, K., and Tseng, S.C., Under the lash: *Demodex* mites in human diseases. *Biochemistry* (Lond), 31, 2, 2009.
66. Liang, L. et al., Ocular demodicosis as a potential cause of pediatric blepharoconjunctivitis. *Cornea*, 29, 1386, 2010.
67. English, F.P. et al., The presence of the parasite Demodex folliculorum on the skin surface of the eyelid. *Aust. N. Z. J. Ophthalmol.*, 19, 229, 1991.
68. Morfin Maciel, B.M., Demodicidosis in a female patient treated as allergic blepharoconjunctivitis. A case report. *Rev. Alerg. Mex.*, 50, 232, 2003.
69. Kheirkhah, A. et al., Corneal manifestations of ocular *Demodex* infestation *Am. J. Ophthalmol.*, 143, 743, 2007.
70. Hoekzema, R., Hulsebosch, H.J., and Bos, J.D., Demodicidosis or rosacea: What did we treat? *Br. J. Dermatol.*, 133, 294, 1995.
71. Crawford, G.H., Pelle, M.T., and James, W.D., Rosacea: I Etiology, pathogenesis, and subtype classification. *J. Am. Acad. Dermatol.*, 51, 327, 2004.
72. Buechner, S.A., Rosacea: An update, *Dermatology*, 210, 100, 2005.
73. Benessahraoui, M. et al., Demodicidosis in a child with xantholeukaemia associated with type 1 neurofibromatosis. *Eur. J. Dermatol.*, 13, 311, 2003.
74. Türk, M. et al., Comparison of incidence of *Demodex folliculorum* on the eyelash follicule in normal people and blepharitis patients. *Turkiye Parazitol. Derg.*, 31, 296, 2007.
75. Liu, J., Sheha, H., and Tseng, S.C., Pathogenic role of *Demodex* mites in blepharitis. *Curr. Opin. Allergy Clin. Immunol.*, 10, 505, 2010.
76. Clifford, C.W., and Fulk, G.W., Association of diabetes, lash loss, and Staphylococcus aureus with infestation of eyelids by *Demodex folliculorum* (Acari: Demodicidae). *J. Med. Entomol.*, 27, 467, 1990.
77. Li, J. et al., Correlation between ocular *Demodex* infestation and serum immunoreactivity to bacillus proteins in patients with facial rosacea. *Ophthalmology*, 117, 870, 2010.
78. Lacey, N. et al., Mite-related bacterial antigens stimulate inflammatory cells in rosacea. *Br. J. Dermatol.*, 157, 474, 2007.
79. Akilov, O.E., and Mumcuoglu K.Y., Association between human demodicosis and HLA class I. *Clin. Exp. Dermatol.*, 28, 70, 2003.
80. Mumcuoglu, K.Y., and Akilov, O.E., The role of HLA A2 and Cw2 in the pathogenesis of human demodicosis. *Dermatology*, 210, 109, 2005.
81. Akilov, O.E., and Mumcuoglu, K.Y., Immune response in demodicosis. *J. Eur. Acad. Dermatol. Venereol.*, 18, 440, 2004.
82. Gao, Y.-Y. et al., Clinical treatment of ocular demodicosis by lid scrub with tea tree oil, *Cornea*, 26, 136, 2007.
83. Demmler, M. et al., Blepharitis. *Demodex folliculorum*, associated pathogen spectrum and specific therapy. *Ophthalmologe*, 94, 191, 1997.
84. Damian, D., and Rogers, M., *Demodex* infestation in a child with leukaemia: Treatment with ivermectin and permethrin. *Int. J. Dermatol.*, 42, 724, 2003.
85. Antille, C., Saurat, J.H., and Lubbe, J., Induction of rosaceiform dermatitis during treatment of facial inflammatory dermatoses with tacrolimus ointment. *Arch. Dermatol.*, 140, 457, 2004.
86. El-Shazly, A.M. et al., Treatment of human *Demodex folliculorum* by camphor oil and metronidazole. *J. Egypt. Soc. Parasitol.*, 34, 107, 2004.
87. Clyti, E., Sayavong, K., and Chanthavisouk, K., Demodecidosis in a patient infected by HIV: Successful treatment with ivermectin. *Ann. Dermatol. Venereol.*, 132, 459, 2005.
88. Yagdiran Düzgün, O., and Aytekin, S., Comparison of *Demodex folliculorum* density in haemodialysis patients with a control group. *J. Eur. Acad. Dermatol. Venereol.*, 21, 480, 2007.
89. Bikowski, J.B., and Del Rosso, J.Q., *Demodex* dermatitis: A retrospective analysis of clinical diagnosis and successful treatment with topical crotamiton. *J. Clin. Aesthet. Dermatol.*, 2, 20, 2009.
90. Inceboz, T. et al., Diagnosis and treatment of demodectic blepharitis. *Turkiye Parazitol. Derg.*, 33, 32, 2009.
91. Nara, T. et al., Eosinophilic folliculitis with a *Demodex folliculorum* infestation successfully treated with ivermectin in a man infected with human immunodeficiency virus. *Clin. Exp. Dermatol.*, 34, e981, 2009.
92. Kheirkhah, A. et al., Fluorescein dye improves microscopic evaluation and counting of *Demodex* in blepharitis with cylindrical dandruff. *Cornea*, 26, 697, 2007.
93. Forton, F., and Seys, B., Density of *Demodex folliculorum* in rosacea: A case-control study using standardized skin-surface biopsy. *Br. J. Dermatol.*, 128, 650, 1993.
94. Aşkin, U., and Seçkin, D., Comparison of the two techniques for measurement of the density of *Demodex folliculorum*: Standardized skin surface biopsy and direct microscopic examination. *Br. J. Dermatol.*, 162, 1124, 2010.
95. Noge, K. et al., Identification of astigmatid mites using the second internal transcribed spacer (ITS2) region and its application for phylogenetic study. *Exp. Appl. Acarol.*, 35, 29, 2005.
96. Wong, S.F. et al., Molecular identification of house dust mites and storage mites. *Exp. Appl. Acarol.*, 55, 123, 2011.
97. Borgo, S.N. et al., PCR analysis for *Wolbachia* in human and canine *Demodex* mites. *Arch. Dermatol. Res.*, 301, 747, 2009.
98. Toops, E. et al., Extraction and characterization of DNA from *Demodex canis. Int. J. Appl. Res. Vet. Med.*, 8, 31, 2010.
99. Ravera, I. et al., Development of a real-time PCR to detect *Demodex canis* DNA in different tissue samples. *Parasitol. Res.*, 108, 305, 2011.
100. Desloire, S. et al., Comparison of four methods of extracting DNA from *D. gallinae* (Acari: Dermanyssidae). *Vet. Res.*, 37, 725, 2006.

101. Basta-Juzbasic, A., Subic, J.S., and Ljubojevic, S., *Demodex folliculorum* in development of dermatitis rosaceiformis steroidica and rosacea-related diseases. *Clin. Dermatol.*, 20, 135, 2002.
102. Erbagci, Z., Erbagci, I., and Erkilic, S., High incidence of demodicidosis in eyelid basal cell carcinomas. *Int. J. Dermatol.*, 42, 567, 2003.
103. Gao, Y.-Y. et al., High prevalence of ocular *Demodex* in eyelashes with cylindrical dandruffs. *Invest. Ophthalmol. Vis. Sci.*, 46, 3089, 2005.
104. Forton, F., and Song, M., Limitation of standardized skin surface biopsy in measurement of the density of *Demodex folliculorum*. A case report. *Br. J. Dermatol.*, 139, 697, 1998.

73 Dermanyssus, Ornithonyssus, and Trombicula (Red, Fowl, and Harvest Mites)

Dongyou Liu

CONTENTS

73.1 INTRODUCTION

Mites in the subclass Acari (Acarina) are abundant in nature, and obtain food from animals or plants. In the absence of their natural animal hosts, these mites may attack humans for blood meals (e.g., *Dermanyssus* and *Ornithonyssus* species) or host tissues/cells (e.g., *Trombicula* species), causing pruritic dermatitis and other skin disorders, and contributing to secondary infections in human populations. As these mites show distinct biological and ecological features, and induce the clinical diseases that are indistinguishable from those caused by pathogenic organisms, it is important to accurately identify these mites for treatment and control purposes.

73.1.1 Classification, Morphology, and Biology

73.1.1.1 The Genus *Dermanyssus*

73.1.1.1.1 Classification

The genus *Dermanyssus* is classified in the family Dermanyssidae, superfamily Dermanyssoidea, suborder Dermanyssina, order Monogynaspida, superorder Parasitiformes, subclass Acari, class Arachnida, pylum Arthropoda. The superfamily Dermanyssoidea encompasses the families of Dermanyssidae, Haemogamasidae, Laelapidae, Macronyssidae, Raillietiidae, Rhinonyssidae, Spelaeorhynchidae, Spinturnicidae, and Varroidae. The family Dermanyssidae consists of the genera *Acanthonyssus*, *Dermanyssus*, *Draconyssus*, *Laelaspisella*, and *Liponyssoides*.

Currently, the genus *Dermanyssus* comprises 25 recognized hematophagous mite species: *Dermanyssus americanus*, *Dermanyssus antillarum*, *Dermanyssus apodis*, *Dermanyssus brevirivulus*, *Dermanyssus brevis*, *Dermanyssus carpathicus*, *Dermanyssus chelidonis*, *Dermanyssus faralloni*, *Dermanyssus gallinae*, *Dermanyssus gallinoides*, *Dermanyssus grochovskae*, *Dermanyssus hirsutus*, *Dermanyssus hirundinis*, *Dermanyssus lacertarum*, *Dermanyssus longipes*, *Dermanyssus nipponensis*, *Dermanyssus passerinus*, *Dermanyssus prognephilus*, *Dermanyssus quintus*, *Dermanyssus richiardii*, *Dermanyssus rwandae*, *Dermanyssus transvaalensis*, *Dermanyssus triscutatus*, *Dermanyssus trochilinis*, and *Dermanyssus wutaiensis*, in addition to a few unassigned species [1–4]. Of these, *D. gallinae*, also known as the poultry red mite, red fowl mite, chicken mite, or roost mite, is a blood parasite of wild birds and a serious pest in poultry industry. Further, *D. americanus* may be occasionally implicated in human parasitism [5].

73.1.1.1.2 Morphology

Based on their morphological and ecological features, *Dermanyssus* species fall into two major groups: the gallinae-group and hirsutus-group. Consisting of 14 species and causing major problems in poultry industry, the gallinae-group (e.g., *D. gallinae* and *D. hirundinis*) demonstrate high levels of morphological variability, are nidicolous in nature (laying eggs and spending much time off the host), are nearly impossible to distinguish, and have a very large host spectrum (involving a variety of bird families). On the other hand, the hirsutus-group (e.g., *D. quintus* and *D. hirsutus*) show consistent characters, are typical ectoparasites (living and developing on the host), are easily distinguishable, and have a narrow host spectrum [6].

First identified by De Geer in 1778, *D. gallinae* adults measure about 0.5–0.7 mm and show oval, flattened dorsoventral forms. *D. gallinae* turns yellow-brown when hungry and turns redblack when full. The anal plate of *D. gallinae* is a broad shield.

73.1.1.1.3 Biology

D. gallinae undergoes five stages of development in its life cycle: egg, larva, protonymph, deutonymph, and adult. In the field, *D. gallinae* lays its eggs on the host or in the nest. In the laboratory, *D. gallinae* lays most of its eggs in the litter away from its hosts. *D. gallinae* eggs usually hatch within 3 days; the larva molts in about 17 h; the protonymph molts in 1–2 days, and the deutonymph molts about a day to become adult. Under optimal conditions (at 20–25°C temperature and 70% humidity), *D. gallinae* can complete a biological cycle in about 1 week, and live up to 8 months without feeding. This mite is resistant to dry weather and but intolerant to high humidity [7].

As blood feeder, *D. gallinae* mite feeds the resting birds in the nests at night since only a few mites are found on birds that are flying about. The mite only visits its host for bloodmeals about every third day. After feeding, the mite (a nidicolous mite that resides in the nest) hides in cracks and crevices away from daylight, where female mates with male and lays eggs. The mite prefers the fluffy downy feathers and is numerous about the vent, accumulating on a few feathers. The mites may also attack mammals (e.g., dogs, cats, rodents, horses, and humans) for blood sucking (and causing dermatitis and intense pruritus) in case birds are not available [8]. In addition, humans handling the infested chicken will acquire the infection. However, the mite needs an avian host to reproduce. While *D. gallinae* protonymph, deutonymph, and adult stages are capable of blood-feeding, its larva is nonfeeding.

The blood-feeding mite *D. gallinae* has worldwide distribution and its preferred hosts include pigeons, hens, starlings, and lovebirds. With a short life cycle (7 days), *D. gallinae* can establish a dense population of mites rapidly and represent an economically important parasitic pest of domestic chickens. Affected birds show decreased production, irritation and, in severe infection, anemia leading to death. *D. gallinae* may be found throughout all seasons; however, human infestation cases are generally reported in late spring and summer months [9,10].

D. gallinae has been suggested as a potential vector of pathogens causing erysipelas (*Erysipelothrix rhusiopathiae*), fowl spirochaetosis, chicken pox, salmonellosis, encephalitis, fowl cholera, and Newcastle disease [11,12].

73.1.1.2 The Genus *Ornithonyssus*

The genus *Ornithonyssus* is classified in the family Macronyssidae, superfamily Dermanyssoidea, suborder Dermanyssina, order Monogynaspida, superorder Parasitiformes, subclass Acari, class Arachnida, pylum Arthropoda. The family Macronyssidae comprises seven genera (*Chiroptonyssus, Ophionyssus, Ornithonyssus, Parichoronyssus, Pellonyssus, Radfordiella, and Steatonyssus*) in addition to unclassified Macronyssidae. In turn, the genus *Ornithonyssus* is separated into five species, *Ornithonyssus bacoti*, *Ornithonyssus bursa* (obsolete synonyms: *Leiognathus bursa* and *Liponyssus bursa*), *Ornithonyssus dasyure*, *Ornithonyssus sylviarum*, and *Ornithonyssus wernecki*. While *O. bursa* (also known as the tropical fowl mite, bird mite, or starling mite) and *O. sylviarum* (the northern fowl mite) mainly affect avian species, *O. bacoti* (the tropical rat mite) often occurs in rats.

Morphologically, *Ornithonyssus* spp are gray-brown to red (when blood engorged) or black with tapering holodorsal shields. Sternal shield is entire or eroded, bearing 1–3 pairs of setae; genital shield is tongue- or knife-shaped with 0–1 pairs of setae; anal shield is small, with three circumanal setae. Tarsus I has claws; trochanter I has six setae; tibia and genu I each has 6/3 dorsal/ventral setae; genu IV has 5/2 dorsal/ventral setae. Chelicerae are elongated and edentate, containing a movable digit without excrescences. Palp genu contains six setae; palp apotele is 2-tined; corniculi are membranous [6].

The tropical fowl mite *O. bursa* is a small (0.8 mm in size), oval-shaped mite, with eight legs (except the larva that has six) and a sparse covering of short hairs. Nonfed bird mite is semitransparent in color, and after ingestion of blood, it turns reddish to blackish. The northern fowl mite *O. sylviarum* appears similar to *O. bursa*; however, the posterior end of the dorsal plate tapers acutely in *O. sylviarum* but more evenly in *O. bursa*. In addition, there are two pairs of setae on the sternal plate in *O. sylviarum* and three pairs in *O. bursa*.

O. bursa and *O. sylviarum* mites have a life cycle of ~7 days, and normally remain on birds or in bird nests. Eggs laid in nests or on feathers hatch in 2–3 days and adults appear about 5 days later if birds are present.

Being widespread in warmer regions of the world, and most active during spring and early summer, these mites are blood-feeding parasites of common birds including pigeons, starlings, sparrows, Indian mynahs, poultry, and some wild birds, but not wild mammals. Besides the unfeathered nestlings, the mites also feed on the adult birds. The nesting material provides the mites with an ideal environment for hiding and reproduction. Most mites will die within 3–6 weeks without a blood meal from a bird host.

O. bursa and *O. sylviarum* mites gain access to humans when birds enter homes, factories, barns, and other dwellings and construct their nests in early spring or summer. Birds roosting on the outside of dwellings such as window ledges or awnings represent another source of human infestations. After the birds abandon their nests, the mites seek other hosts and move into the building through windows, doors, and vents and bite humans. Being unable to reproduce on human blood, these mites require birds to complete their life cycle, and rarely transmit diseases.

O. bacoti (the tropical rat mite) builds up large populations on feral mice and also infests native mammals. The mite can live for up to 10 days off its rodent hosts. On leaving the dead rodents, the mite often congregates around heat sources (e.g., hot pipes and stoves) and seeks alternative food sources, including humans. *O. bacoti* bite often causes tiny, clear blisters and rash, but does not appear to transmit any human diseases.

73.1.1.3 The Genus *Trombicula*

73.1.1.3.1 Classification

Trombicula is a genus of harvest mites (also known as red mite, red bug, scrub-itch mite, berry bug or, in their larval stage, as chigger or chigoe) belonging to family Trombiculidae, superfamily Trombidioidea, order Trombidiformes, Subclass Acari, Class Arachnida, Phylum Arthropoda.

The family Trombiculidae (Greek "tromb," to tremble; Latin "culex," gnat or midge), first described as an independent family by H.E. Ewing in 1944, comprises at least 26 genera: *Acomatacarus*, *Anahuacia*, *Ascoschoengastia*, *Axiogastia*, *Blankaartia*, *Brunehaldia*, *Chatia*, *Cheladonta*, *Doloisia*, *Euschoengastia*, *Eutrombicula*, *Gahrliepia*, *Guntherana*, *Guntheria*, *Hannemania*, *Heaslipia*, *Hirsutiella*, *Kayella*, *Leptotrombidium*, *Microtrombicula*, *Miyatrombicula*, *Neoschoengastia*, *Neotrombicula*, *Novotrombicula*, *Ornithogastia*, *Parasecia*, *Pseudoschoengastia*, *Schoengastiella*, *Schoutedenichia*, *Speleocola*, *Trombicula*, and *Whartonia*.

Within the genus *Trombicula*, 16 species are recorded from North America: *Trombicula acuitlapanensis*, *Trombicula alfreddugesi*, *Trombicula autumnalis*, *Trombicula boneti*, *Trombicula formicarum*, *Trombicula geniticula*, *Trombicula halidasys*, *Trombicula imperfecta*, *Trombicula jamesoni*, *Trombicula jonesae*, *Trombicula nagayoi*, *Trombicula patrizii*, *Trombicula pumilis*, *Trombicula spinosa*, *Trombicula tecta*, and *Trombicula zeta* [13]. Other *Trombicula* species of note include *Trombicula batatas* (obsolete synonym; *Acarus batatas*) and *Trombicula hirsti*. It is notable that *T. alfreddugesi* (referred to as scrub-itch mite) commonly occurs in Northern America, *T. autumnalis* (called harvest mite) is prevalent in the United Kingdom, and *T. (Eutrombicula) hirsti* (also referred to as scrub-itch mite) is found in Australia.

73.1.1.3.2 Morphology

Trombicula undergoes four stages of development during its life cycle: eggs, larva, nymph, and adult. *Trombicula* eggs are round in shape. *Trombicula* larvae (which can be separated into three phases: the deutovum, unfed larva, and engorged larva), commonly called chiggers, are of 0.17–0.21 mm in diameter, white to orange red in color, covered in hairs, and motile with six legs. *Trombicula* nymph (which can be also separated into three phases: protonymph, deutonymph, and tritonymph) is unusual in species of Trombiculidae. The protonymph is an inactive transitional stage that combines larval and protonymph characteristics with deutonymph and tritonymph morphology. The active deutonymph develops an additional pair of legs (for a total of eight). The tritonymph is also inactive before growing into an adult (about 0.4 mm in size). The adult and nymph show a marked constriction in the front part of the body [6].

It is noteworthy that chigger (a corruption of chigoe) is also an alternate term for the chigoe flea *Tunga penetrans*, a sand flea found in tropical and subtropical climates in the Americas and Africa [14].

73.1.1.3.3 Biology

The life cycle of trombiculid mites (including egg, larva, nymph, and adult stages) lasts for 2–12 months, with multiple generations occurring in tropical climates, and only 2–3 generation each season in subtropical/temperate climates. Being the only parasitic stage, the six-legged larva feeds on a variety of mammals (including humans, who are accidental hosts), birds, amphibians, reptiles, and even some insects. After crawling onto its host, the larva thrusts its small hooked chelicerae (fangs) into the surface layers of the skin, injects enzymes into the skin to break down skin cells and to cause surrounding tissues to harden forming a straw-like feeding tube of hardened flesh (stylostome), and chews up tiny parts of the inner skin, instead of biting, burrowing into skin, or sucking blood [15]. This may cause severe irritation (itching) and swelling (red pimple-like bumps, papules, or hives and skin rash or lesions on a sun-exposed area) in 3–6 h, which may linger for over a week without treatment. After feeding for 4 days, the larva drops to the ground (leaving a red welt with a white, hard central area on the skin that itches severely and may later develop into dermatitis as a result of scratching and secondary infections), becomes eight-legged nymph after a period of 5 or 6 weeks, and then matures into an adult (which also has eight legs). Both nymph and adult are nonparasitic to humans and animals, and feed on plant juices and small insects. Following mating, the female lays 3–15 eggs per day in a clutch, usually on a leaf or under the roots of a plant, and lives up to 1 year (usually dies by autumn). The eggs are dormant for about 6 days and then hatch into the six-legged larvae (called deutovum), which are enclosed in a membrane in addition to the eggshell. After hatching, the unfed larvae congregate in groups on small clods of earth, in matted vegetation and even on low bushes and plants, and wait for a host.

Trombicula mites are distributed throughout the world, especially in the tropical and subtropical areas [16–18]. They are found in the forests, grasslands, woodlands, berry bushes, orchards, along lakes and streams, lawns, golf courses, and parks. Interestingly, *Trombicula* mites tend to stay in tall grass and other vegetation during wet seasons, and remain underneath brush and shady areas during dry seasons. The adults overwinter in protected places such as slightly below the soil. When the soil temperature reaches above 16°C in the spring, the female lays up to 15 eggs per day in vegetation. Emerging from the eggs, the larvae are most abundant in early summer. During April through early autumn up until the first frost, humans are prone to chigger bites while walking in mite-infested areas, sitting, or lying on infested ground. The preferred feeding locations on people are the belt line, waistline, navel area, the groin, back of the knees, the ankles, under girdles, under socks, the armpits, the forearms, the nape, and the neck, where the flesh is thin, tender, or wrinkled.

73.1.2 Clinical Features and Treatment

73.1.2.1 Clinical Features

The poultry red mite *D. gallinae* is a blood-feeding parasite of birds (including pigeons and sparrows). In the absence of natural host population, it may attack humans, causing redness, itching, papules, vesicles, and pruritic dermatitis [19–30]. The latter is sometimes mistaken for scabies, pediculosis, erythematous maculopapular, or papulovesicular lesions, bites forming urticaria, and otitis externa [31]. The mite may function as a vector for bacteria (e.g., *Borrelia* species, *Pasteurella multocida*, *Salmonella gallinarum*, and *Rickettsia* species) and viruses (Hantaviruses, St. Louis, Eastern equine, and Western equine encephalitis viruses, and West Nile virus) [32,33].

A 40-year-old resident on the top floor of a building where a few pigeons used the roof of air ventilation hole as their nest, presented with tense itching on her body during evening hours. Physical examination revealed pruritic dermatitis lesions on her shoulders, neck, and arms. During the examination of her residence uncovered the existence of white-, yellow-, brown-, and black-colored arthropods of <1 mm, which were subsequently identified by microscopy as *D. gallinae*. After blocking the entry of pigeons to the premises and disinfecting the area with pulverized dichlorvos (10 mL/L), no *D. gallinae* mites were detected in subsequent inspections.

Bellanger et al. [34] reported a case of submammary and auricular pruritus due to *D. gallinae* in a 66-year-old woman with fracture and luxation of the elbow. Many small insects were found on the patient and in her bed, which were identified microscopically as *D. gallinae*. After destruction of a pigeon nest located 5 m from the patient's bed, a spray of the patient's room and the main corridor of the ward with permethrin, and washing of her sheets and clothes, the patient's itching subsided spontaneously within 2 days, without any specific treatment.

The tropical fowl mite *O. bursa* and the northern fowl mite (*O. sylviarum*) are also blood sucking mites of birds that may cause pruritic dermatoses in humans [35]. Watson [36] reported of a case of human infestation of bird mite belonging to *Ornithonyssus* sp. The individual suffered hundreds of bites, with itchy red papules 3–4 mm in diameter. Tiny mobile parasites (<1 mm) collected from the skin and adjacent bedroom wall were identified as *Ornithonyssus* sp. The source of the infestation was a starling nest under the eaves adjacent to the bedroom.

Orton et al. [37] described a case of pruritic dermatosis due to infestation with the northern fowl mite (*O. sylviarum*) from an abandoned birds nest in the roof of a patient's home. The patient presented with a pruritic erythematous papular eruption on exposed sites, which was treated with topical steroids.

There are also reports of dermatitis caused by the tropical rat mite *O. bacoti*. Baumstark et al. [38] documented five cases of hematophagous mite *O. bacoti* infection in individuals living in a home for disabled persons in Germany. The patients developed pruritic, red papules on the upper extremities, neck, upper trunk, and face. Inspection of their home and the sheds nearby disclosed infestation with rat populations and mites. The mites were subsequently identified as *O. bacoti*. After elimination of the rats, disinfection of the rooms, and treatment of the patients with topical corticosteroids and antihistamines, new cases of tropical rat mite dermatitis occurred. Beck [39] described tropical rat mite dermatitis due to *O. bacoti* in three children (4, 11, and 15 years old) of a family and a 23-year-old medical. An arthropod obtained from the flats of the patients (husbandry of gerbils) was identified as *O. bacoti*.

The harvest mites belonging to *Trombicula* species occasionally attack humans by chewing up tiny parts of the inner skin, and causing itching that may not develop until 24–48 h after the attack [40]. In a case of harvest mite infestation, a 45-year-old woman presented with pruritus after traveling to South America. A small erythematous area with a well-circumscribed papule in the sulcus of the second digit of the right foot was incised and drained, from which harvest (chigger) mite was identified.

73.1.2.2 Treatment

As mites belonging to the genera *Dermanyssus*, *Ornithonyssus*, and *Trombicula* do not live and reproduce underneath the skin of human host, the treatment of their infestations in humans is centered on the relief of dermatitis symptoms with topical corticosteroids (e.g., benzocaine, hydrocortisone, calamine lotion), antipruritics (e.g. crotamiton), and antihistamines. Antibiotics may be administered if secondary bacterial infections are apparent.

However, unless the sources of infestation are identified and handled appropriately, the problem will persist. This includes the destruction of bird nests nearby, fixing up of potential bird entry points, physical removal of mites with a vacuum cleaner (and then freezing the bag to prevent the mites from crawling back out) or wipe them up with a moist cloth, spray of insecticides (e.g., bifenthrin, permethrin, and amitraz) in rooms and buildings, thorough cleaning of bed linens and clothes (e.g., using insect repellent *N*,*N*-diethyl-meta-toluamide on clothing and skin) [41,42].

For prevention, mowing of briars, weeds, and thick vegetation and close clipping of lawns help eliminate shade and moisture. This will permit sunlight and air to circulate freely, and thus reduce mite larvae populations. Treat grass around picnic tables, lawn chairs, or recreational equipment with insecticides. As harvest mite larvae tend to concentrate in "mite islands" while nearby spots are free of them, black cardboard may be used to determine which spots (e.g., grass, dead leaves, briars, weeds, etc.) are more heavily infested. Namely, place six-inch squares of black cardboard (which attracts heat to make it warmer than the surrounding grass) on edge in the grass and observe for a few minutes. Any small, yellowish or pinkish mite larvae present will climb rapidly to the top of the square and congregate there.

Before going into an area where these mite larvae may be present, it is important to apply repellent or permethrin on uncovered hands, arms, or legs, and clothing openings at cuffs, neck, waistband, and upper edges of socks. Do not sit or lie down in a sunny spot at midday with temperatures above 16°C, and never sit on a freshly used brush hog. If possible, stick to roads and trails. After returning from a chigger/harvest mite-infested area, take a good hot bath or shower and soap repeatedly, and launder the field clothes in soapy, hot water for about half an hour.

73.1.3 Laboratory Diagnosis

Dermanyssus, *Ornithonyssus*, and *Trombicula* mites are opportunistic pests and pathogens that attack humans when their natural hosts (e.g., birds and mites) are not available. The common clinical symptoms of mite infestations include dermatitis, which can sometimes be confused with dermatological conditions due to bedbug bites, flea bites, scabies, pediculosis, and urticaria, and so on.

As these mites have distinct biological and ecological feature, it is critical to correctly identify them for effective control and prevention purposes. Microscopic examination of mites collected from patients is valuable for identification of these ectoparasites. Morphologically, *Ornithonyssus* and *Dermanyssus* are distinguishable by their size, pilosity, shape of the anal and the genital shield, as well as the position of the anal orifice on the anal shield. In addition, *Ornithonyssus* species (the tropical fowl mite *O. bursa*, the northern fowl mite *O. sylviarum*, and the tropical rat mite *O. bacoti*), whose whole life cycles occur in the host, induce more irritating lesions than *D. gallinae* (poultry red mite) does.

In recent years, molecular procedures have been increasingly utilized for identification of the poultry red, bird, and harvest mites [43]. The commonly targeted gene regions include 18S ribosomal RNA (rRNA) (small subunit rRNA, SSU rRNA), 28S rRNA (large subunit rRNA, LSU rRNA), 5.8S rRNA, ribosomal internal transcribed spacers 1 and 2 (ITS1 and ITS2), mitochondrial cytochrome oxidase subunit 1 (mt-COI), elongation factor 1-α, and tropomyosin intron *n* gene regions, and so on [43–52].

For example, Roy et al. [4] employed primers to amplify a 700–800 bp fragment from the mt-COI gene of various *Dermanyssus* species (Table 73.1). Following sequencing, the 558 nucleotides of mt-COI gene were analyzed. Comparison of the 216 bp parsimony-informative sequences facilitated the division of two main clades in the species of *Dermanyssus* tested. One clade consists of *D. carpathicus*, *D. hirundinis*, *D. longipes*, and *D. Apodis*, while the second clade includes the various lineages of *D. gallinae*.

TABLE 73.1
PCR Primers for Amplification of mt-COI Gene

Primer Orientation	Primer	Sequence 5′–3′
Forward	CO1RQF1	GAAAGAGGAACAGGAACAGG
	CO1LCF	GAAAGAGGAGCAGGCACTGG
	COF1bis	CTGCACCTGACATGGCTTTCCCAC
	CO1F4	CACCTGACATGGCTTTCCCACGAT
	RhipiCOIF	CGAATAAATAATATAAGATTTTGA
	SKPOFa2diagF1	CTTTTTAGATCTTTAATTGAAA
Reverse	COIGOR	GTTGGAATtGCAATAAT
	RQ-COI-R	CCAGTAATACCTCCAATTGTAAAT
	ObCOIF-rev	GTGGGAATHGCAATAAT
	TyphloCOIR	GCTAATCAAGAAAAAATTTTAAT

Note: Primer pair CO1RQF1 + RQ-COI-R is used for *D. carpathicus*, *D. hirundinis*, and *D. Longipes*; primer pairs SKPOFa2diag F1 + RQCOIR, CO1F4 + RQCOIR, COF1bis + RQ-COIR, COF1bis + ObCOIF-rev, and CO1LCF + RQ-COI-R for *D. Gallinae*; primer pair RhipiCOIF + TyphloCOIR for outgroups; and primer pair CO1LCF + COIGOR for *D. apodis*.

73.2 METHODS

73.2.1 Sample Preparation

73.2.1.1 Sample Collection

Mite isolates are sampled from wild bird nests or from farms using 7 × 10-cm semitransparent corrugated plastic traps. Traps are placed in the nests for 7 days, and afterward are put into plastic bags. Engorged female mites are selected and placed in wells of a round-bottomed 96-well plate, and stored at room temperature for 8–10 days to allow larvae to hatch. Both larvae and protonymphs are collected [50]. The mites are identified on the basis of microscopic features, and the verified specimens are stored in 70% ethanol at −20°C until used for DNA extraction.

Nests from wild birds are collected from experimental nesting boxes, and mites were recovered by placing the nest in double plastic bags. The inner bag is opened and the outer bag closed, allowing those live mites emerging into the outer bag to be collected. Dead and live mites still in the nest are recovered by flotation. The nests are pressed to the bottom of a bucket with a wire net, and water is added. After 2 h, floating mites are collected with a small strainer [50].

73.2.1.2 DNA Isolation

Individual mites are crushed with a sterile plastic pestle in a 1.5 mL microcentrifuge tube bottom containing 200 μL of extraction buffer (2% CTAB, 1.4 M NaCl, 20 mM EDTA, 100 mM Tris-HCl, pH 8.0, 0.2% 2-mercaptoethanol), preheated to 65°C. This homogenate is incubated at 65°C for 1 h. Proteins are removed with one volume of chloroform/isoamyl alcohol. DNA is then precipitated from the aqueous layer by adding one volume of isopropanol, left standing for

2 or 3 h at −20°C and spun at 18,000 × *g* for 30 min. The DNA pellet is washed with 300 μL of ethanol 70%, dried for 10–15 min at 50°C, and resuspended in 20 μL of ultrapure water. Tubes are placed for 2 h at 4°C and DNA were resuspended again before being stored at −20°C [53].

Alternatively, DNA is extracted with the QIAamp DNA extraction kit (Qiagen) with slight modifications. Briefly, mites are placed in clean Eppendorf tubes containing 20 μL of phosphate-buffered saline (8 g/L NaCl, 1.44 g/L Na_2HPO_4, 0.2 g/L KH_2PO_4, 0.2 g/L KCl), plunged into liquid nitrogen, and thawed. This freeze–thawing step is repeated three times. The volume is then adjusted to 100 μL with ATL buffer and incubated with proteinase K at 56°C for at least 5 h or overnight. Other reactive volumes are also modified (10 μL proteinase K, 100 μL ethanol, 100 μL AL buffer, 300 μL AW1 and AW2 buffer). Genetic material was finally eluted from the column with 50 μL of AE buffer [49].

In addition, DNA may be prepared from individual mite using PrepMan™ Ultra Sample Preparation Reagent (Applied Biosystems). The mite is placed in an Eppendorf tube and 30 μL PrepMan™ is added. The preparation is homogenized for 2 × 30 s directly in the tube with a mini-pestle, and thereafter boiled for 10 min and cooled on ice for 2 min. After cooling, the tube is centrifuged for 3 min at 17,000 × *g*, and the supernatant containing the DNA template is removed and stored at 4°C [50].

73.2.2 Detection Procedures

73.2.2.1 PCR-Restriction Fragment-Length Polymorphism Identification of *D. gallinae*

Potenza et al. [49] developed a PCR-RFLP procedure for rapid of determination of *D. gallinae* identity. The PCR primers ITS2-F (5′-GCGTGTCTATGCTGCATTTG-3′) and ITS2-R (5′-GGGGTCGTCACACTTGATTT-3′) amplify a 154 bp fragment from the ITS2 region. Subsequent digestion of the PCR product with restriction enzyme *Cla* I and *Eco* RV gives rise to a two-band fingerprint of 48 and 104 bp with *Cla* I, and of 50 and 102 bp with *Eco* RV.

73.2.2.1.1 Procedure

1. PCR mixture (25 μL) is made up of 0.5 μM of each of the primers ITS2-F and ITS2-R, 200 μM dNTP's, 2.5 μL of 10× buffer, 1.5 U *Taq* DNA polymerase (GE Healthcare), and 5 μL of the genomic DNA.
2. The mixture is subjected to an initial 95°C for 10 min followed by 35 cycles of 95°C for 30 s, 60°C for 30 s, and 72°C for 30 s, with final extension at 72°C for 10 min.
3. A portion (5 μL) of the amplification product (154 bp) is electrophoresed on 2% agarose gel in 0.5× TBE, stained with ethidium bromide (0.3 μg/mL), and detected under UV light.
4. Then, 10 μL PCR product is each digested with *Cla* I and *Eco* RV in separate tubes, and separated on 2% agarose gel.

Note: Two-band fingerprint are obtained after restriction digestion: 48 and 104 bp with *Cla* I, 50 and 102 bp with *Eco* RV.

73.2.2.2 Analysis of ITS 1 and ITS 2 Sequences

Potenza et al. [49] targeted the complete ITS1, 5.8S ribosomal DNA, and ITS2 region of the ribosomal DNA for phylogenetic analysis of *D. gallinae* and other mites. The PCR primers consist of forward ITS1 (5′-TCCGTAGGTGAACCTGCGG-3′) and reverse ITS-R (5′-AGAGGAAGTAAAAGTCGTAACA-3′), which are located on the 3′end of 18S and the 3′start of 28S ribosomal DNA, respectively.

73.2.2.2.1 Procedure

1. PCR mixture (25 μL) is made up of 0.5 μM of each of the primers ITS1 and ITS-R, 200 μM dNTP's, 2.5 μL of 10× buffer, 1.5 U *Taq* DNA polymerase (GE Healthcare), and 5 μL of the genomic DNA.
2. The mixture is subjected to an initial 95°C for 10 min followed by 35 cycles of 95°C for 30 s, 55°C for 30 s, and 72°C for 30 s, with final extension at 72°C for 10 min.
3. The amplification product (~550 bp) is electrophoresed on 0.8% agarose gel in 0.5× TBE, stained with ethidium bromide (0.3 μg/mL), and detected under UV light. The specific amplicon is then sequenced using one of the PCR primers and the ABI PRISM BigDye Terminator Sequence Kit on an ABI 3700 automated sequencer. The resulting nucleotide sequences are aligned using ClustalW2 and nBLAST (basic local alignment search tool).

Note: The 5.8S and ITS2 region of *D. gallinae* are relatively conserved, while the ITS1 region shows slight intraspecies variations involving substitutions as well as deletions.

73.2.2.3 Analysis of 18S rRNA Gene Sequences

Brännström et al. [50] employed universal primers targeting the start, the central part and the end of the rRNA SSU gene to amplify the SSU region of the mite DNA, with one forward primer (NSF4:5′-CTG GTT GAT [C,T]CT GCC AGT-3′) and two reverse primers (NSR951:5′-TTG G[C,T][A,G] AAT GCT TTC GC-3′, and NSR1787:5′-C[C,T]G CAG GTT CAC CTA C[A,G]G-3′).

73.2.2.3.1 Procedure

1. PCR mixture (50 μL) is composed of 5 μL GeneAmp® 10× PCR buffer (100 mM Tris-HCl pH 8.3), 500 mM KCl, 15 mM $MgCl_2$, 0.01% [w/v] gelatin; Applied Biosystems), 1 U AmpliTaq DNA polymerase (Applied Biosystems), 200 μM of each deoxynucleotide (GE Healthcare, Uppsala, Sweden), dH_2O, 20 pmol of each primer and 1 μL DNA template.
2. Amplification is conducted with an initial 95°C for 2 min; 40 cycles of 95°C for 15 s, 55°C for 39 s, 72°C for 2 min; and a final 72°C for 4 min.

3. The PCR product is purified from gel using GENECLEAN® II KIT (Qbiogene), or from solution using JetQuick (SaveenWerner) eluted with 10 μL dH_2O.
4. DNA is sequenced using ABI PRISM BigDye® Terminator v3.1 Cycle Sequencing Kit (Applied Biosystems). The sequence reaction mixture of 20 μL contains 2 μL BigDye solution, 3 μL BigDye 5× sequencing buffer, dH_2O, 5 pmol primer (shown above), and 1 μL purified PCR product. After the sequence reaction, unincorporated terminators are removed by ethanol precipitation. The sequencing reactions are separated on an ABI PRISM 3100 Genetic Analyzer (Applied Biosystems). Sequences are analyzed using the Vector NTI program, Suite 9 (Informax).

73.2.2.4 Analysis of COI Gene Sequences

Marangi et al. [51] employed primers COIF (5′-TGATTTTT TGGTCACCCAGAAG-3′) and COIR (5′-TACAGCTCCTA TAGATAAAAC-3′) tő amplify a 400 bp fragment from the mitochondrial COI gene for phylogenetic analysis of *D. gallinae* mites.

73.2.2.4.1 Procedure

1. PCR mixture (25 μL) is made up of 10 μm of each oligonucleotide primer, 12.5 μL of REDTaq ReadyMix PCR Reaction Mix (Sigma-Aldrich), 1.25 μL of BSA, and 5 μL of DNA extract.
2. The cycling program consists of an initial 94°C for 4 min; 44 cycles of 94°C for 1 min, 53°C for 1 min, 72°C for 1 min, and a final 72°C for 4 min.
3. PCR products are separated by a 1.2% agarose gel electrophoresis containing ethidium bromide, and DNA fragments are visualized with UV illumination.
4. PCR products are purified using a PCR Montage KIT (Millipore) and are directly sequenced using the ABI PRIMS BygDye Terminator sequences KIT v. 3.1 (Applied Biosystems) according to the manufacturer's instructions. Sequences are analyzed on an ABI 3130 automated DNA sequencer. Double-stranded sequences are obtained for all samples for accuracy.
5. Nucleotide sequences are aligned using the sequence alignment program Clustal X\and compared to the sequences obtained from the GenBank database. Sequences that differ by one or more nucleotides are considered as different haplotypes, while sequences exhibiting identical SNPs at the same nucleotide positions are considered as similar haplotypes. Mitochondrial DNA COI partial amino acid sequence alignment is also carried out using the sequence alignment program Clustal X.
6. Phylogenetic analyses are conducted using PHYLIP (Phylogeny Inference Package) version 3.5c. This software is used to calculate numbers of variable sites, uncorrected genetic distance, nucleotides composition, and transition and transversion ratios. The genetic distance, uncorrected *P*, is calculated for each pair of sequences, and then transitional and transversional substitution are compared with these genetic distances. The phylogenetic tree is constructed based on neighbor-joining (NJ) algorithms (*P*-distance) using PHYLIP. For the NJ analyses robustness of nodes is assessed with 1000 NJ-bootstrap replicates.

Note: The COI gene is highly polymorphic within species, and is therefore suitable for analyses of intraspecific variations of *D. gallinae* and other mites.

73.3 CONCLUSION

Members of the genera *Dermanyssus*, *Ornithonyssus*, and *Trombicula* are blood-sucking or tissue-eating mites whose natural hosts consist of birds, mammals, and other animals as well as plants. When their natural animal hosts are not available, these mites may attack humans, causing various skin irritation and disorders (e.g., itching, papules, vesicles, or pruritic dermatitis) and contributing to secondary infections. Given their morphological similarity and their tendency to induce nonspecific clinical symptoms, it is necessary to apply laboratory means for their diagnosis, in order to implement appropriate control and prevention measures. Currently, laboratory identification of Acari mites including *Dermanyssus*, *Ornithonyssus*, and *Trombicula* species relies heavily on microscopic examination of morphological characteristics. However, this testing scheme is not only labor intensive, but also demands highly specialized skills. The recent development and application of molecular techniques offer a rapid, sensitive, and specific alternative for identification and phylogenetic analysis of these mites.

REFERENCES

1. Moss, W.W., An illustrated key to the species of the Acarine genus *Dermanyssus* (Mesostigmata: Laelapoidea: Dermanyssidae). *J. Med. Entomol.*, 1, 67, 1968.
2. Moss, W.W., The mite genus *Dermanyssus*: A survey, with description of *Dermanyssus trochilinis*, n. sp., and a revised key to the species (Acari: Mesostigmata: Dermanyssidae). *J. Med. Entomol.*, 14, 627, 1978.
3. Roy, L., and Chauve, C.M., Historical review of the genus *Dermanyssus* Dugès, 1834 (Acari: Mesostigmata: Dermanyssidae). *Parasite*, 14, 87, 2007.
4. Roy, L. et al., Delimiting species boundaries within *Dermanyssus* Dugès, 1834 (Acari:Dermanyssidae) using a total evidence approach. *Mol. Phylogenet. Evol.*, 50, 446, 2009.
5. Naltsas, S. et al., Eczematous dermatitis caused by *Dermanyssus americanus*. *Cutis*, 25, 429, 1980.

6. Baker, A.S., Mites and ticks of domestic animals: An identification guide and information source. London: The Stationary Office, pp. 134–136, 1999.
7. Hutcheson, H.J., and Oliver, J.H. Jr., Spermiogenesis and reproductive biology of *Dermanyssus gallinae* (DeGeer) (Parasitiformes: Dermanyssidae). *J. Med. Entomol.*, 25, 321, 1988.
8. Lucky, A.W. et al., Avian mite bites acquired from a new source-pet gerbils. *Arch. Dermatol.*, 137, 167, 2001.
9. Oliver, J.H., Cytogenetics of mites and ticks. *Annu. Rev. Entomol.*, 22, 407, 1977.
10. Sparagano, O. et al., Prevalence and key figures for the poultry red mite Dermanyssus gallinae infections in poultry farm systems. *Exp. Appl. Acarol.*, 48, 3, 2009.
11. Valiente-Moro, C., Chauve, C., and Zenner, L., Vectorial role of some Dermanyssoid mites, Parasite, 12, 99, 2005.
12. Valiente Moro, C. et al., The poultry red mite (*Dermanyssus gallinae*): A potential vector of pathogenic agents. *Exp. Appl. Acarol.*, 48, 93, 2009.
13. Daniel, M., and Stekol'nikov, A.A., Chigger mites (Acari: Trombiculidae) from Makalu region in Nepal Himalaya, with a description of three new species. *J. Med. Entomol.*, 46, 753, 2009.
14. Brown, W.A., Description of six new species of the genus *Leptotrombidium* from Mindanao Island and notes on the medically important genera of chiggers (Acari: Trombiculidae) of the Philippine Islands. *J. Med. Entomol.*, 29, 284, 1992.
15. Shatrov, A.B., Stylostome formation in trombiculid mites (Acariformes: Trombiculidae). *Exp. Appl. Acarol.*, 49, 261, 2009.
16. Iwasa, M. et al., Trombiculid mites (Acari: Trombiculidae) and *Rickettsia tsutsugamushi* isolated from wild rodents in a new endemic area of Japan. *J. Med. Entomol.*, 27, 501, 1990.
17. Takahashi, M. et al., A new member of the trombiculid mite family Neotrombicula nagayoi (Acari: Trombiculidae) induces human dermatitis. *Southeast Asian J. Trop. Med. Public Health*, 35, 113, 2004.
18. Daniela, M. et al., Chigger mites (Acari, Trombiculidae) parasitizing small mammals in the Eastern Hindu Kush and some other Afghan areas. *Parasitol. Res.*, 107, 1221, 2010.
19. Vargo, J., Ginsberg, M., and Mizrahi, R., Human infestation by the pigeon mite: A case report. *Am. J. Infect. Control*, 11, 24, 1983.
20. Regan, A.M., Metersky, M.L., and Craven, D.E., Nosocomial dermatitis and pruritus caused by pigeon mite infestation. *Arch. Intern. Med.*, 174, 2185, 1987.
21. Yassien, N.A. et al., Immunohistopathological status of the skin in cases infested with three species of mites. *J. Egypt. Soc. Parasitol.*, 26, 567, 1996.
22. Rossiter, A., Occipational otitis externa in chicken catchers. *J. Laryngol. Otol.*, 111, 366, 1997.
23. Chauve, C., The poultry red mite *Dermanyssus gallinae*: Current situation and future prospects. *Vet. Parasitol.*, 8, 364, 1998.
24. Rosen, S., Yeruham, I., and Braverman, Y., Dermatitis in humans associated with the mites *Pyemotes tritici, Dermanyssus gallinae, Ornithonyssus bacoti* and *Androlaelaps casalis* in Israel. *Med. Vet. Entomol.*, 16, 442, 2002.
25. Nisbet, A.J. et al., A house dust mite allergen homologue from poultry red mite *Dermanyssus gallinae* (De Geer). *Parasite Immunol.*, 28, 401, 2006.
26. Green, B., Sparling, J., and Sparling, M.B., What's eating you? Pigeon mite (*Dermanyssus gallinae*). *Cutis*, 80, 461, 2007.
27. Akdemir, C., Gülcan, E., and Tanritanir, P., Case report: *Dermanyssus gallinae* in a patient with pruritus and skin lesions. *Turkiye Parazitol. Derg.*, 33, 242, 2009.
28. McClain, D., Dana, A.N., and Goldenberg, G., Mite infestations. *Dermatol. Therapy*, 22, 327, 2009.
29. Dogramaci, A.C., Culha, G., and Ozçelik, S., *Dermanyssus gallinae* infestation: An unusual cause of scalp pruritus treated with permethrin shampoo. *J. Dermatol. Treat.*, 21, 319, 2010.
30. Haag-Wackernagel, D., and Bircher, A.J., Ectoparasites from feral pigeons affecting humans. *Dermatology*, 220, 82, 2010.
31. Cafiero, M.A. et al., Pseudoscabies caused by *Dermanyssus gallinae* in Italian city dwellers: A new setting for an old dermatitis. *J. Eur. Acad. Dermatol. Venereol.*, 22, 1382, 2008.
32. Chirico, J. et al., The poultry red mite, *Dermanyssus gallinae*, a potential vector of *Erysipelothrix rhusiopathiae* causing erysipelas in hens. *Med. Vet. Entomol.*, 17, 232, 2003.
33. De Luna, C.J. et al., The poultry red mite *Dermanyssus gallinae* as a potential carrier of vector-borne diseases. *Ann. N. Y. Acad. Sci.*, 1149, 255, 2008.
34. Bellanger, A.P. et al., Nosocomial dermatitis caused by *Dermanyssus gallinae. Infect. Control Hosp. Epidemiol.*, 29, 282, 2008.
35. Gupta, A.K., Billings, J.K., and Ellis, C.N., Chronic pruritus: An uncommon cause. Avian mite dermatitis caused by *Ornithonyssus sylviarum* (Northern fowl mite). *Arch. Dermatol.*, 124, 1102, 1988.
36. Watson, C.R., Human infestation with bird mites in Wollongong. *Commun. Dis. Intel.*, 27, 259, 2003.
37. Orton, D.I., Warren, L.J., and Wilkinson, J.D., Avian mite dermatitis. *Clin. Exp. Dermatol.*, 25, 129, 2000.
38. Baumstark, J., Beck, W., and Hofmann, H., Outbreak of tropical rat mite (*Ornithonyssus bacoti*) dermatitis in a home for disabled persons. *Dermatology*, 215, 66, 2007.
39. Beck, W., Occurrence of a house-infesting tropical rat mite (*Ornithonyssus bacoti*) on murides and human beings. *Travel Med. Infect. Dis.*, 6, 245, 2008.
40. Axman, W.R., and Brummer, J.J., Chigger mite infestation. *J. Am. Podiatr. Med. Assoc.*, 93, 399, 2003.
41. Thind, B.B., and Ford, H.L., Assessment of susceptibility of the poultry red mite *Dermanyssus gallinae* (Acari: Dermanyssidae) to some acaricides using an adapted filter paper based bioassay. *Vet. Parasitol.*, 144, 344, 2007.
42. Marangi, M. et al., Evaluation of the poultry red mite, *Dermanyssus gallinae* (Acari: Dermanyssidae) susceptibility to some acaricides in field populations from Italy. *Exp. Appl. Acarol.*, 48, 11, 2009.
43. Roy, L. et al., Diversity of phylogenetic information according to the locus and the taxonomic level: An example from a parasitic mesostigmatid mite genus. *Int. J. Mol. Sci.*, 11, 1704, 2010.
44. Söller, R. et al., Phylogenetic relationships within terrestrial mites (Acari: Prostigmata, Parasitengona) inferred from comparative DNA sequence analysis of the mitochondrial cytochrome oxidase subunit I gene. *Mol. Phylogenet. Evol.*, 18, 47, 2001.
45. Cruickshank, R.H., Molecular markers for the phylogenetics of mites and ticks. *Syst. Appl. Acarol.*, 7, 3, 2002.
46. Jeyaprakash, A., and Hoy M.A., Mitochondrial 12S rRNA sequences used to design a molecular ladder assay to identify six commercially available phytoseiids (Acari: Phytoseiidae). *Biol. Control*, 25, 136, 2002.
47. Wuyts, J. et al., The European ribosomal RNA database. *Nucleic Acids Res.*, 32, D101, 2004.

48. Klompen, H., Lekveishvili, M., and Black, W.C., Phylogeny of parasitiform mites (Acari) based on rRNA. *Mol. Phylogenet. Evol.*, 43, 936, 2007.
49. Potenza, L. et al., Characterization of *Dermanyssus gallinae* (Acarina: Dermanissydae) by sequence analysis of the ribosomal internal transcribed spacer regions. *Vet. Res. Commun.*, 33, 611, 2009.
50. Brännström, S. et al., Genetic differences in internal transcribed spacer 1 between *Dermanyssus gallinae* from wild birds and domestic chickens. *Med. Vet. Entomol.*, 22, 152, 2008.
51. Marangi, M. et al., Phylogenetic relationship between *Dermanyssus gallinae* populations in European countries based on mitochondrial COI gene sequences. *Exp. Appl. Acarol.*, 48, 143, 2009.
52. Stekolnikov, A.A., and Klimov, P.B., A revision of chiggers of the minuta species-group (Acari: Trombiculidae: Neotrombicula Hirst, 1925) using multivariate morphometrics. *Syst. Parasitol.*, 77, 55, 2010.
53. Desloire, S. et al., Comparison of four methods of extracting DNA from *D. gallinae* (Acari: Dermanyssidae). *Vet. Res.*, 37, 725, 2006.

74 Ixodidae (Hard Ticks)

Dongyou Liu and Chong Yin

CONTENTS

74.1 INTRODUCTION

Ticks are ectoparasites (external parasites) that survive by hematophagy on the blood of mammals, birds, and occasionally reptiles and amphibians. Besides inducing discomfort and dermatological conditions, ticks are responsible for transmitting a number of viral, bacterial, and protozoan diseases among birds and mammals, including humans. Indeed, ticks are the leading carriers (vectors) of human diseases in the United States, second only to mosquitoes worldwide.

To date, 896 tick species have been described, and a vast majority of them are assigned in the families of Ixodidae (hard ticks, covering 702 species) and Argasidae (soft ticks, covering 193 species), with a single species (*Nuttalliella namaqua*) being included in a separate family Nuttalliellidae [1]. Hard ticks are characterized by the possession of a tough back plate or scutum that covers the anterior part of the tick's body and by the tendency to attach and feed for hours to days, with disease transmission usually occurring near the end of a meal and after the tick becoming fully engorged (with blood). Soft ticks typically have more rounded bodies without the hard scutum as seen in hard ticks, and usually feed for less than 1 h, with disease transmission occurring within a minute of a meal. In addition, while hard ticks can live 2–6 years, soft-bodied ticks can live as long as 20 years [2,3].

Members of the hard tick family Ixodidae are further organized into 14 genera, of which the genera *Amblyomma*, *Dermacentor*, *Haemaphysalis*, *Hyalomma*, *Ixodes*, and *Rhipicephalus* harbor many medically important species. On the other hand, members of the soft tick family Argasidae are arranged in four genera *Antricola*, *Argas*, *Ornithodoros*, and *Otobius*, of which *Argas* and *Ornithodoros* are of medical significance.

74.1.1 Classification, Morphology, Biology, and Epidemiology

Taxonomically, the hard tick family Ixodidae is assigned in the order Ixodida, superorder Parasitiformes, subclass Acari, class Arachnida, phylum Arthropoda. The family Ixodidae is separated into 14 genera which together contain 702 species (Table 74.1) [1,4,5]. In addition, a subclassification of the family has been used by some authors, which includes six subfamilies Amblyomminae, Bothriocrotoninae, Haemaphysalinae, Hyalomminae, Ixodinae, and Rhipicephalinae [6]. Many medically important species are found in the genera *Amblyomma*, *Dermacentor*, *Haemaphysalis*, *Hyalomma*, *Ixodes*, and *Rhipicephalus* [2,3].

Hard ticks typically undergo four distinct developmental stages in their life cycle: egg, larva, nymph, and adult. Structurally, adults possess capitulum (mouth parts) in the anterior portion, scutum in the dorsal portion, and spiracular plate, genital aperture, anus, and four pairs of legs in the ventral portion. Male and female adult ticks are distinctly different in that the scutum in the male hard tick is large, and completely covers the dorsal surface, and the scutum in the female hard tick covers only a part of the dorsal surface and is almost obscured after engorgement. The capitulum of hard ticks extends forward from the anterior end of the body, resembling a true head, in contrast to that of soft ticks, which is found on the underside and is not usually visible from a dorsal view. The spiracles of hard ticks lie behind the fourth pair of coxae, or basal segments of the leg. Larvae and nymphs may possess similar structures to those of adults, except that larvae have six legs; nymphs have eight legs. Larvae are sometimes referred to as "seed ticks" or "grass ticks" because of their resemblance to small plant seeds [2,3].

The mouthparts (or capitulum) of hard ticks consists of the highly mobile palps, which are two jointed parts outside, and paired chelicerae in between, which protect the central rod-shaped structure, the hypostome, which is a single feeding tube with many beak-like projections. During feeding, the palps move laterally and do not enter the skin of the host, while the chelicerae (cutting mandibles) cut a channel in skin with their horizontal cutting action. These rip and tear at the epidermal layers, resulting in the formation of a small pool of blood. The *hypostome* (feeding tube) is then inserted. The recurved teeth covering the hypostome serve as anchors. The process by which Ixodid and Argasid ticks feed is referred to as telmophagy (pool feeding), in contrast to the process of solenophagy by mosquitoes, in which feeding is direct from a small venule. Many ticks secrete a cement-like substance in the salivary glands which helps attach their mouthparts and head more firmly to the host; the substance dissolves after completion of feeding. Ticks may secrete or regurgitate small amounts of saliva that contain neurotoxins, preventing the host from feeling the pain and irritation of the bite. The saliva may contain a blood thinner, facilitating the flow of blood during blood meal [2,3].

Ticks are found in most wooded or forested regions of the world [7–14]. Ticks are especially abundant near water, where warm-blooded animals come to drink, in areas where deer trails or human tracks exist, and in meadows wherever shrubs and brush provide woody surfaces and cover. To seek feeding hosts, hard ticks often climb onto vegetation and wave forelegs slowly until a host comes within reach (a behavior known as "questing"). While female tick quests for a blood meal, male tick searches for a host for a female tick in order to mate and feed from her (by piercing the cuticle of female with mouthparts to feed on the hemolymph). Changes in temperature and day length are the factors signaling a tick to seek a host. Carbon dioxide (hence the use of "dry ice" baits), heat, and movement of passing hosts serve as stimuli for ticks to climb on to a potential host. Leaning against a tree or sitting on an old log represents the quickest way to acquire ticks (about 30 s) in tick-infested areas. Hard ticks feed for extended periods of time on their hosts, varying from several days to weeks. The outside surface, or cuticle, of hard ticks actually grows to accommodate the large volume of blood ingested. The common flagging technique used to collect hard ticks for research exploits this property. The "flags" or "drags" are made from 1 m square pieces of roughly textured fabric (e.g., fleece or flannel) attached to a rod handle. The flags are slowly dragged across the surface of vegetation to allow tick attachment and collection.

During the life cycle of hard ticks, larvae with six legs emerge from eggs. After obtaining a blood meal from a vertebrate host, larvae molt to nymphs with eight legs. Nymphs feed and molt to the next and final stage—adults, which also have eight legs. After feeding once more, the adult female hard ticks lay one batch of eggs (1000–6000) and die. Only one blood meal is taken during each of larval, nymphal, and adult stages, and the time to complete the entire life cycle may vary from less than a year in tropical regions to over 3 years in cold climates, depending on the availability of hosts. Some ticks (called one-host ticks) feed on only one host throughout larval, nymphal, and adult stages, with the

TABLE 74.1
Classification of the Hard Tick Family Ixodidae

Family	Genus	Number of Species
Ixodidae	*Amblyomma*	130
	Anomalohimalaya	3
	Bothriocroton	7
	Compluriscutula	1
	Cornupalpatum	1
	Cosmiomma	1
	Dermacentor	34
	Haemaphysalis	166
	Hyalomma	27
	Ixodes	243
	Margaropus	3
	Nosomma	2
	Rhipicentor	2
	Rhipicephalus (Boophilus)	82
		702

larva and nymph remaining in one host until becoming adult, when the female drops off the host after feeding to lay its batch of eggs. Other ticks (called two-host ticks) feed on two hosts, with the larva and nymph remaining on one host until becoming adult, which then drops off and attaches to a different host for its final blood meal, and the adult female then drops off after feeding to lay eggs. Many ticks feed on three hosts (called three-host ticks), with the tick dropping off and reattaching to a new host during each life stage, until finally the adult female lays its batch of eggs. The female dies after laying one batch of eggs, and the male also dies after reproduction [2,3]. Many medically important ticks feed on three different animals.

74.1.1.1 The Genus *Amblyomma*

74.1.1.1.1 Classification

The genus *Amblyomma* is classified in the subfamily Amblyomminae, family Ixodidae. The subfamily Amblyomminae currently includes a single genus *Amblyomma*. A previous genus *Aponomma* (consisting of several species such as *Aponomma auruginans*, *Aponomma concolor*, *Aponomma exornatum*, *Aponomma fimbriatum*, *Aponomma latum*, and *Aponomma varanensis*) in this subfamily has been absorbed into the genera *Bothriocroton* and *Amblyomma* [1], although the genus name *Aponomma* has been continued to be used by some authors.

The genus *Amblyomma* covers 130 species that are organized into 11 subgenera: *Amblyomma*, *Amerindia*, *Anastosiella*, *Brasiliana*, *Cernyonuna*, *Fifippovania*, *Keiransiella*, *Koloninum*, *Macintosheilla*, *Theileriana*, and *Walkeriana* [6]. Nearly half of all *Amblyomma* species occur in the American continent, and the most notable species in the genus *Amblyomma* include *Amblyomma americanum*, *Amblyomma brasiliense*, *Amblyomma darwini*, *Amblyomma dubitatum* (synonyms: *Amblyomma cooperi*, *Amblyomma lutzi*, *Amblyomma ypsilophorum*), *Amblyomma extraoculatum*, *Amblyomma hebraeum*, *Amblyomma incisum*, *Amblyomma maculatum*, *Amblyomma romitii*, *Amblyomma testudinarium*, and *Amblyomma variegatum* [15–26].

Amblyomma americanum (the lone star tick), the first tick to be described in the United States in 1754, is widespread in wooded areas, particularly in forests with thick underbrush of the United States [27–31]. The parasite can be a vector of diseases, including human granulocytic ehrlichiosis (*Ehrlichia chaffeensis*) [32,33], canine and human granulocytic ehrlichiosis (*Ehrlichia ewingii*) [34], tularemia (*Francisella tularensis*), Q fever (*Coxiella burnetii*) [35], and Southern tick-associated rash illness (STARI, possibly caused by spirochete *Borrelia lonestari*), the latter of which induces less severe rash than that caused by *Borrelia burgdorferi* (Lyme disease) (Table 74.2) [36].

Amblyomma brasiliense is reported as one of the most aggressive ticks attacking humans in Brazil.

Amblyomma dubitatum is a parasite of large mammals such as the South American tapir (*Tapirus terrestris*) and cattle. Although *Rickettsia belli* and *Rickettsia parkeri* have been detected in *A. dubitatum* ticks in Brazil, the role of this parasite in human disease has not been clearly established [21].

Amblyomma incisum is a major tick species in the Atlantic Forest of Brazil, and is also present in non-Amazonian areas of Argentina, Paraguay, and southern Brazil. Its adults are infective to the South American tapir (*Tapirus terrestris*) and horse, and its larvae and nymphs infect small red-brocket deer (*Mazama bororo*) and rabbit. *A. incisum* nymphs show aggressiveness to humans [37].

Amblyomma maculatum (the Gulf Coast tick) is associated with a deformity called "gotch ear" in cattle wherein the ears become thickened and curled with damage to the muscles and cartilage. The smell of fresh blood in these wounds attracts screwworm flies to feed and lay eggs in them, cause festering sores that can kill cattle. Common hosts of *A. maculatum* include livestock, deer, dogs, and man, and this tick is present along the Gulf and South Atlantic coastlines of the United States.

74.1.1.1.2 Morphology

The genus *Amblyomma* is one of the largest ticks among the hard ticks, and shows four developmental stages in its life cycle: egg, larva, nymph, and adult. *Amblyomma* adults measure 1–9 mm in length before engorgement, and reach >20 mm in length after full engorgement (so called replete). Most *Amblyomma* species display whitish markings, long palpi, a long hypostome, and festoons on the posterior margin of the abdomen.

Amblyomma americanum adult females are distinguished by a conspicuous whitish spot or "lone star" on the back near the posterior end of the scutum (which is the hard shield extending over roughly the anterior third of a female hard tick's dorsal surface and may be obscured in an engorged tick). Males lack this conspicuous white spot (although some may show dots and white streaks on the edge of their bodies) and the scutum extends over most of the dorsal surface. *A. americanum* larva and nymph are about the size of a pinhead.

Amblyomma maculatum shows spurs on the second, third, and fourth pairs of legs and more diffuse pale markings on the female than does the lone star tick.

Amblyomma testudinarium adult females measure 5–20 mm depending on the stages of engorgement, and may reach >20 mm in length after feeding for 30 days or longer. The pedipalp of the capitulum is slender with the second segment much longer than the third. The scutum is ornate with nonprotruding eyes at the edge. The comma-shaped spiracular plate is located posterior to the coxa IV. Genital opening is located anterior to the level of the coxa II. The anus is surrounded posteriorly by the anal groove. The coxa of the first leg has subequal two spurs (the external one is slightly larger than internal). The spine of coxa on the fourth leg is longer than those on the second and third legs. The tarsi of the third and fourth legs have two distinct subapical ventral spurs [38,39].

Amblyomma brasiliense larvae display rectangular basis capituli, short palpi, rounded idiosoma, coxa I with two spurs with the external one being longer than the internal one, and

TABLE 74.2
Geographic Distribution of Hard Ticks and Their Reported Roles in the Transmission of Human Disease-Causing Agents

Genus	Species	Geographic Distribution	Transmission of
Amblyomma	*A. americanum*	Eastern United States	*Ehrlichia chaffeensis*, *Ehrlichia ewingii*, *Coxiella burnetii*, *Francisella tularensis*, *Borrelia lonestari*
	A. cajennense	Texas, Central and South America	*Rickettsia rickettsii*
	A. hebraeum	Southern Africa	*Rickettsia africae*
	A. variegatum	Africa, Caribbean Islands	Crimean Congo hemorrhagic fever virus
Dermacentor	*D. andersoni*	Western North America	*Rickettsia rickettsii*, *Francisella tularensis*, *Coxiella burnetii*, Colorado tick fever virus
	D. marginatus	Eurasia	*Rickettsia slovaca*, *Rickettsia sibirica*, *Coxiella burnetti*, tick-borne encephalitis virus, Omsk hemorrhagic fever virus
	D. nuttalli	Russia, Mongolia, China	*Rickettsia sibirica*, *Francisella tularensis*
	D. occidentalis	California, Oregon, Baja Mexico	*Rickettsia rickettsii*, *Coxiella burnetii*, *Francisella tularensis*, Colorado tick fever virus
	D. reticulatus	Eurasia	*Rickettsia conorii*, *Rickettsia sibirica*, *Francisella tularensis*, Omsk hemorrhagic fever virus
	D. silvarum	Russia, Eastern Europe, Mongolia	Tick-borne encephalitis virus
	D. variabilis	Eastern and western North America	*Rickettsia rickettsii*, *Francisella tularensis*
Haemaphysalis	*H. concinna*	Central Europe, Russia, Far East	Tick-borne encephalitis virus
Hyalomma	*H. anatolicum*	Russia, southern Europe, Near East	Crimean Congo hemorrhagic fever virus
	H. marginatum	Turkey, Israel, Yemen, Russia	Crimean Congo hemorrhagic fever virus
Ixodes	*I. cookei*	North America	Powassan virus
	I. holocyclus	Eastern Australia	*Rickettsia australis*, *Borrelia* sp.
	I. pacificus	Western North America	*Borrelia burgdorferi*, *Anaplasma phagocytophilum*, *Babesia* spp.
	I. persulcatus	Eurasia	*Borrelia* spp., tick-borne encephalitis virus
	I. ricinus	Europe	*Borrelia* spp., *Rickettsia conori*, *Anaplasma phagocytophilum*, *Babesia divergens*, tick-borne encephalitis virus, Crimean Congo hemorrhagic fever virus
	I. scapularis	Eastern, midwestern North America	*Borrelia burgdorferi*, *Anaplasma phagocytophilum*, *Babesia microti*
Rhipicephalus	*R. sanguineus*	Cosmopolitan	*Rickettsia conori*, *Rickettsia rickettsii*, *Coxiella burnetii*

Source: Anderson, J.F., Biology of ticks. *Infect. Dis. Clin. North Am.* 22, 195, 2008.

coxae II and III each with one short spur. Nymphs shows rectangular basis capituli with a sharp pointed cornua, oval idiosoma with scutum reaching coxa III, coxa I with two evident spurs with the external one being longer than the internal one, coxae II–III each with one short spur, and coxa IV with a very short spur and chitinous tubercles on internal surface of posterior border of idiosoma [40].

74.1.1.1.3 Biology and Epidemiology

Amblyomma americanum (the lone star tick) is primarily identified in Texas and other parts of the southeastern United States (including Missouri, New Jersey, and New York). *A. americanum* adults are parasites of medium- and large-sized mammals (including cattle and white-tailed deer), and its larvae and nymphs feed on small to large mammals and ground-feeding birds. All three motile life stages of *A. americanum* bite people (although the nymph is the most common stage found on people) and play a role in the transmission of bacterial, viral, and protozoan diseases to humans. In general, *A. americanum* adults appear in February, peak in May, or June, and diminish by the end of July. Nymphs may appear in March after overwintering as flat nymphs and remain active until August or September. Larvae are active from June and peak in August. Because of its aggressiveness, its nonspecific feeding habits and its high population densities, the lone star tick is one of the most medically and economically important ticks in the United States.

Amblyomma dubitatum infects capybara (*Hydrochaeris hydrochaeris*), which has the capacity to act as an amplifier host for *Rickettsia rickettsii*. Other large mammals (e.g., cattle and horse) and small- and medium-sized mammals such as opossums (*D. albiventris* and *D. aurita*), Cricetidae rodents (*L. molitor*, *S. tumidus*), and Chinchillidae rodent *L. maximus* represent alternative hosts for the immature stages of *A. dubitatum*.

Amblyomma incisum is a tick species with a long life cycle lasting a minimum of 262 days at 27°C, 85% relative humidity, and a photoperiod regimen 00:24 (light:darkness), although its egg hatching rate is improved under natural conditions (with higher humidity, lower temperature, natural mold control, and some daylight). *A. incisum* adults are mostly active during spring–summer months and *A. incisum* larvae and nymphs feed well on deer, humans, and rabbits.

Amblyomma testudinarium is a tropical tick and found mainly in India, South East Asia (including Myanmar, Thailand, Malaysia, Indonesia, and the Philippines), Taiwan, Korea, and Japan. In Japan, the parasite is restricted to the southwestern region where the warm temperature with heavy rainfall and the subtropical vegetation fauna provide a proper environmental condition for this tick. The frequent location of *A. testudinarium* infestations in humans is the axillary, anogenital, or inguinal region with rich apocrine glands [41].

74.1.1.2 The Genus *Dermacentor*

The genus *Dermacentor* is classified in the subfamily Rhipicephalinae, family Ixodidae. The subfamily Rhipicephalinae is further divided into three genera: *Dermacentor*, *Rhipicentor*, and *Rhipicephalus*, and the genus *Dermacentor* currently includes 34 species, with the following species being most notable: *Dermacentor albipictus*, *Dermacentor andersoni*, *Dermacentor auratus*, *Dermacentor halli*, *Dermacentor hunter*, *Dermacentor imitans*, *Dermacentor marginatus*, *Dermacentor nitens*, *Dermacentor niveus*, *Dermacentor occidentalis*, *Dermacentor parumapertus*, *Dermacentor reticulatus*, *Dermacentor rhinocerinus*, *Dermacentor steini*, and *Dermacentor variabilis* [6,42,43]. Recent surveys showed that *Dermacentor marginatus* is one of the most common ixodid tick species infesting humans in Turkey [44,45], whereas *Dermacentor reticulatus* commonly occurs in humans in Bosnia and Herzegovina [46].

Morphologically, *Dermacentor* ticks are ornate, with pale markings on the scutum, eleven festoons, eyes, and short palpi. The basis capituli is quadrangular dorsally. *Dermacentor variabilis* (the American dog tick) females may increase in size from about 5 mm long and 2.5 mm wide to 13 mm long and 10 mm wide. *D. variabilis* larvae are about 0.6 mm long, without spiracular plates, and appear yellow with red markings near the eyes, while engorged larvae are about 1.5 mm long and are slate gray to black. *D. variabilis* nymphs have four pairs of legs and are light yellowish brown with red markings near the eyes, while engorged nymphs are slate gray and about 4 mm in length. *Dermacentor andersoni* (the Rocky Mountain wood tick) adults in general have paler coloring and larger goblets on the spiracular plates than the American dog tick. Compared to *D. variabilis*, the spiracular plates of *D. andersoni* have a more pronounced dorsal prolongation and contain fewer but larger goblets. *Dermacentor albipictus* (the winter tick) shows oval spiracular plates with very large goblets. *Dermacentor occidentalis* (the Pacific Coast tick) has conspicuous tooth-like dorsal projections on the posterior margin of its basis capituli (known as cornua). Compared to *D. andersoni* or *D. variabilis*, adult *D. albipictus* lacks a dorsal prolongation on the spiracular plates and has fewer but larger goblets within the spiracular plates.

Dermacentor ticks are found mainly in open landscapes: steppes, forest-steppes, savannas, and sparsely forested areas throughout North and Central America, Africa, Eurasia, and New Guinea. Almost all *Dermacentor* species have a three-host life cycle although *D. albipictus*, *D. nitens*, and possibly *D. dissimilis* have a one-host life cycle. *Dermacentor* adults feed on large mammals, including livestock, and immature ticks parasitize small mammals, but not birds.

The Rocky Mountain wood tick *Dermacentor andersoni* is a three-host tick that is known to spread Rocky Mountain spotted fever (*Rickettsia rickettsii* infection), Q fever (*Coxiella burnetii* infection), Colorado tick fever (an RNA virus *Coltivirus* infection), and Powassan encephalitis (an RNA arbovirus infection) (Table 74.2).

The American dog tick *Dermacentor variabilis* commonly occurs in meadows, old field-forest ecotones, second growth forest, along edges of roads, trails, power lines, and clearings around homes through the Atlantic coast to the Plains states in North America. In the south parts, larvae emerge from the soil in early spring, feed on small mammals (i.e., voles and mice) during April and May, and then August and September, when larvae hatch from eggs, which do not feed until the following early spring. Nymphs also feed on small mammals usually in May and June, and adults feed on medium-sized mammals (i.e., dogs and raccoons), including humans and are most abundant from May to July. The generation time is 1 year in the south, and 2 years in the north. This tick is a vector of several rickettsial species, including *R. rickettsii*, the causative agent of Rocky Mountain spotted fever (RMSF), and *Francisella tularensis*, the bacterium causing tularemia. *D. variabilis* bites have been associated with occurrences of tick paralysis through the production of a neurotoxin that induces rapidly progressive flaccid quadriparesis. This neurotoxin prevents presynaptic release of acetylcholine from neuromuscular junctions (Table 74.2) [47].

The winter tick *Dermacentor albipictus* is a one-host species that parasitizes cattle, horses, moose, elk, deer, and sometimes man. Because of their sheer number, *Dermacentor albipictus* ticks may cause blood loss leading to anemia and even death. This species is widely distributed throughout North America.

The Pacific Coast tick *Dermacentor occidentalis* is a three-host tick parasitizing small rodents and large mammals (cattle as well as man). It is a vector of tularemia and a suspected vector of RMSF.

74.1.1.3 The Genus *Haemaphysalis*

The genus *Haemaphysalis* is assigned in the subfamily Haemaphysalinae, family Ixodidae. The subfamily Haemaphysalinae consists of a single genus *Haemaphysalis*, which is further separated into 166 species, with the following species being most notable: *Haemaphysalis bispinosa*, *Haemaphysalis calcarata*, *Haemaphysalis concinna*, *Haemaphysalis cornigera*, *Haemaphysalis cretica*, *Haemaphysalis elliptica*, *Haemaphysalis flava*, *H. houyi*, *Haemaphysalis humerosa*, *Haemaphysalis inermis*, *Haemaphysalis juxtakochi*, *Haemaphysalis leachi*, *Haemaphysalis leporispalustris*, *Haemaphysalis longicornis*, *Haemaphysalis parva*, *Haemaphysalis petrogalis*, *Haemaphysalis punctata*, and *Haemaphysalis qinghaiensis* [48,49]. According to recent surveys, *Haemaphysalis punctata* and *Haemaphysalis parva* are among the most common ixodid tick species infesting humans in Turkey [44,45]. *Haemaphysalis punctata* has been involved in the transmission of tick-borne encephalitis virus in Central Europe, Russia, and the Far East.

74.1.1.4 The Genus *Hyalomma*

The genus *Hyalomma* is classified in the subfamily Hyalomminae, family Ixodidae. The subfamily Hyalomminae contains two genera *Hyalomma* and *Nosomma*, with the genus *Hyalomma* covering 27 species. The most notable *Hyalomma* species include *Hyalomma aegyptium*, *Hyalomma anatolicum*, *Hyalomma asiaticum*, *Hyalomma detritum*, *Hyalomma dromedarii*, *Hyalomma franchinii*, *Hyalomma hussaini*, *Hyalomma impeltatum*, *Hyalomma isacii*, *Hyalomma lusitanicum*, *Hyalomma marginatum*, and *Hyalomma truncatum*. The camel ticks *Hyalomma dromedarii* is known for being an active seeker of host and has been observed emerging from barren sand and running across hot ground to attack humans and livestock. Recent surveys showed that *Hyalomma marginatum*, *Hyalomma detritum*, *Hyalomma isacii*, *Hyalomma aegyptium*, and *Hyalomma franchinii* are among the most common human infecting ixodid tick species in Turkey [44,45]. *Hyalomma anatolicum* and *Hyalomma marginatum* have been associated with Crimean Congo hemorrhagic fever virus infection in Russia, southern Europe, Turkey, Israel, and Yemen.

74.1.1.5 The Genus *Ixodes*

The genus *Ixodes*, the only member within the subfamily Ixodinae, family Ixodidae, consists of 243 species of three-host ticks [6]. Ticks in the genus *Ixodes* are recognized by the position of the anal groove, which lies in front of the anus and extends from one side of the body to the other. Males show a complicated arrangement of plates on the ventral side of the abdomen, and most species possess enlarged club-like palps. Unlike other members of the family Ixodidae, which are field-dwelling (nonnidicolous) parasites, *Ixodes* species are nidicolous (nest-dwelling) parasites that infest caves, nests, rock ledges, crevices, or burrows, or hide in soil or cracks or crevices of tree bark or wood nearby host-occupied sheltered sites.

The most notable species in the genus *Ixodes* include *Ixodes aulacodi*, *Ixodes caledonicus*, *Ixodes hexagonus*, *Ixodes holocyclus*, *Ixodes laguri*, *Ixodes pacificus*, *Ixodes persulcatus*, *Ixodes ricinus*, *Ixodes scapularis*, and *Ixodes uriae* [50–52]. Recent surveys showed that *Ixodes laguri* and *Ixodes ricinus* are among the most common ixodid tick species infesting humans in Turkey [44,45], whereas *Ixodes hexagonus* commonly occurs in human population in Bosnia and Herzegovina [46].

Ixodes scapularis (formerly *Ixodes dammini*), the black-legged tick or deer tick, parasitizes a broad range (at least 126 species) of hosts, including mammals, birds, and reptiles in the eastern central parts of the United States, and shows a 2-year life cycle. Larvae hatch from eggs in July and August, feed on host for 2.5–5 days, detach, drop to the ground, and remain in soil until they molt into nymphs in May and June of the following calendar year. Nymphs peak in late May till July, feed on animals for 3–5 days, then detach, drop off to the ground, and crawl to the duff, until they molt into adults in October. Adult females are abundant from mid-October through November, and are actively seeking hosts on relatively warm days during winter, and in late March through mid-May of the following year to feed (7–10 days until engorged). Males and females are often found mating on deer, dogs, and other large mammals in the fall and winter. Females lay 1000–3000 eggs in a single cluster during June, and then die. Adult male ticks do not feed. Larvae, nymphs, and adult females also feed on humans (who appear in the yards) and serve as the vector for *Borrelia burgdorferi*, the bacterium that causes Lyme disease; *Babesia microti*, the parasite that causes human babesiosis; *Ehrlichia*, the etiologic agent for human ehrlichiosis; and Powassan encephalitis virus.

Ixodes pacificus (the western black-legged tick) inhabits the western part of North America and parasitizes a wide range of animals, including birds, lizards, and mammals, with a preference for livestock such as cows as its adult host. This tick is the primary vector of *B. burgdorferi* (Lyme disease) in the western United States, and has also been implicated in the transmission of *Anaplasma phagocytophilum* and *Ehrlichia chaffeensis* (the etiologic agent of human monocytic ehrlichiosis).

Ixodes ricinus (the sheep tick) occurs in Eurasia, where it overlaps with *Ixodes persulcatus*. Larvae and nymphs feed on small and large mammals, lizards, and ground-inhabiting birds, while adults feed only on large mammals (e.g., deer and livestock). To date, this tick has been reported from at least 237 different species of birds, mammals, and lizards. The generation time can vary from 2 to 6 years depending upon the habitat (with an average of 3 years). This species is commonly found in leaf litter and low-growing vegetation in deciduous and mixed woodlands, as it requires a relative humidity of 80% or greater for survival.

Ixodes persulcatus inhabits boreal coniferous, deciduous, and mixed forests in northeastern Europe, northern Asia, and Japan. Larvae and nymphs climb to the tips of low-growing vegetation at night to seek small mammals (e.g., voles and shrews) and ground-inhabiting birds for feeding, while adults feed on large wild and domestic mammals. At least 212 species of mammals, birds, and lizards have been involved in the life cycle. Hares are suitable hosts for all three feeding stages. The generation time ranges from 2 to 6 years (averaging 3 years).

Ixodes holocyclus (the Australian paralysis tick) is distributed in a 20-km band along the eastern coastline of Australia, including New South Wales and South-East Queensland. The natural hosts of *Ixodes holocyclus* include bandicoots, wallabies, kangaroos, and other marsupials, which are not affected by the tick's toxin. However, the neurotoxin can cause progressive motor paralysis, respiratory depression, and death in animals (e.g., dogs, cats, cattle, and horses) and may cause paralysis, tick typhus, and severe allergic reactions in humans. In addition, this tick transmits Australian spotted fever (*Rickettsia australis*) and Lyme disease (*Borrelia*) to humans.

Morphologically, capitulum of female *Ixodes holocyclus* measures 0.2 mm in length, basis triangular, about 0.16 mm wide, palpi elongate, and slender. Its hypostome is apically rounded, 0.14 mm in length, dentition 2/2 of 10–12 teeth, the teeth of the inner file blunt and small, and some minute dentils apically. Its body is broadly oval, 0.5 × 0.4 mm (unfed) to 1.15 × 1.0 mm (engorged). Its scutum is about as long as broad, 0.31 by 0.32 mm and widest a little anterior to mid-length, lateral carina present but faint; anterolateral margins usually convex and posterolateral margins concave; and cervical grooves short but well defined. Its anal grooves are ill defined anteriorly and do not converge behind the legs. All coxae contain small external spurs, with tarsus I of 0.14 mm in length. *Ixodes holocyclus* has no eyes.

Ixodes holocyclus adult females lay 2000–6000 eggs, usually in leaf and branch litter, under the scaly or fibrous bark of certain trees and shrubs, or in dense fine foliage near the tips of branches. They wax the eggs to form a cohesive mass, and a small fraction of the eggs survive and hatch to larvae after 40–110 days with suitable warmth and high humidity. Larvae (merely visible to the naked eye) emerge from the eggs and move toward lateral branches. After undergoing 7–44 days of hardening, larvae then climb the tips of leaves, and attach to a passing host. Following a 4–6 day feed, larvae drop to ground, and molt to become nymphs over a further 19–41 day period, which is temperature dependent. Nymphs undergo molting and attach to another host 5–6 days (up to 31 days) later. After feeding for 4–7 days, nymphs drop to the ground. After 3–11 weeks, nymphs molt to become adult males or females. The adult female attaches to the final host after 7–9 (up to 77) days. After insemination by a male tick (sometimes before) a female feeds on blood for a period of 6–30 days in order to gain the nutrients for the egg mass to mature. At the same time, the engorging adult female tick injects a holocyclotoxin into the animal host within 3–6 days of attachment. The fully engorged (replete) adult female drops off the host to the ground and after 11–20 days the gravid female lays a batch of eggs (20–200 eggs per day) over 16–34 days. The female tick dies 1–2 days after egg laying is complete. The adult male is yellowish-brown, flat, oval, and smaller than the female and spends its time wandering around on the host looking for a female with which it can mate.

74.1.1.6 The Genus *Rhipicephalus*

The genus *Rhipicephalus* is assigned in the subfamily Rhipicephalinae, family Ixodidae. The subfamily Rhipicephalinae is further divided into three genera: *Dermacentor*, *Rhipicentor*, and *Rhipicephalus*, with the genus *Rhipicephalus* comprising 82 species [6,53]. Members of the genus *Rhipicephalus* have a hexagonal basis capituli, eyes on the scutum and festoons on the posterior margin of the abdomen. Males are distinguished from most other common ticks by the adanal plates and accessory plates on either side of the anus.

The most notable *Rhipicephalus* species include *Rhipicephalus appendiculatus*, *Rhipicephalus (Boophilus) annulatus*, *Rhipicephalus bursa*, *Rhipicephalus (Boophilus) decoloratus*, *Rhipicephalus (Boophilus) kohlsi*, *Rhipicephalus maculatum*, *Rhipicephalus (Boophilus) microplus*, *Rhipicephalus simpsoni*, *Rhipicephalus sanguineus*, and *Rhipicephalus turanicum*. Many *Rhipicephalus* species occur in Africa and Asia where they are important vectors of disease to man and animals. Recent surveys indicated that *Rhipicephalus bursa*, *Rhipicephalus sanguineus*, and *Rhipicephalus (Boophilus) annulatus* are among the most common ixodid tick species infesting humans in Turkey [44,45].

Rhipicephalus sanguineus (the brown dog tick) is a reddish-brown species that attacks dogs (especially the ears and between the toes) and other mammals including man. The engorged females are about a 13 mm long, and often crawl on walls or around baseboards and cracks, looking for protected areas in which to deposit 1000–3000 eggs. This species is a vector of ehrlichiosis (*Ehrlichia canis*) and babesiosis (*Babesia canis*) in dogs and RMSF (*Rickettsia rickettsii*) and Q fever (*Coxiella burnetii*) in humans [54,55]. This tick also transmits *Rickettsia conorii* (known as Mediterranean spotted fever, boutenneuse fever, or tick typhus) in parts of Europe, Asia, and Africa [56,57]. Morphologically, *Rhipicephalus sanguineus* isolates from Argentina and Brazil demonstrate considerable differences in terms of body size (3.8 × 1.7 mm versus 2.7 × 1.1 mm), shape of genital aperture (U-shaped groove versus V-shaped groove), and size of chelicerae (0.5 mm versus 0.25 mm) [58].

Rhipicephalus (Boophilus) microplus (the southern cattle tick) attacks a relatively narrow range of hosts, mainly cattle, and to a lesser extent, sheep, horses, and goats. This species is responsible for causing huge loss to cattle producers worldwide each year.

74.1.2 Clinical Features

Tick bites are generally painless, and many people do not notice the bite and never find the tick (especially hard tick larva and nymph, which are pinhead-sized) if it falls off [59,60]. However, some ticks may secrete a neurotoxin at the time of attachment, causing local redness (erythema), fluid swelling (edema), itching (maddening rash or scrub itch), burning, throat (tracheopharyngeal) compression, muscle weakness, or paralysis before or after the tick drops off. Given that ticks are vectors for a number of human pathogens, individuals may develop symptoms that are associated with specific pathogen(s) after tick bites: flu-like symptoms, fever, numbness, rash, confusion, weakness, pain and swelling in joints, palpitations, shortness of breath, and nausea and vomiting.

74.1.2.1 Tick Bites

Yamada et al. [41] documented a case of *Amblyomma testudinarium* bite on the glans penis of a 40-year-old Japanese man. Nakamura-Uchiyama et al. [61] reported two cases of tick bite in Japan due to *A. testudinarium*. The first case involved an 86-year-old farmer infested with a fully engorged adult tick attached on his inguinal region. The second case was a 57-year-old male infested with a large number of larval ticks (>100 larvae). Kim et al. [39] documented a first case of *A. testudinarium* tick bite in Korea. The patient was a 74-year-old Korean woman who was attacked by the tick while working in her vegetable garden. On admission, she presented with mild discomfort, itching, and a large mass on her left inguinal region. A dark green, round tick with head partially burrowed under the skin was removed from the inguinal region using a pair of blunt forceps. The causative tick measured 23 mm in length, 19 mm in width, and 12 mm in height, and possessed slender pedipalps, ornate scutum with eyes at the edge and four pairs of legs. The distinct location and shape of its genital aperture, spiracular plate (comma-shaped), anus, coxae, and tarsi permitted identification of the parasite as the fully engorged adult female of *A. testudinarium*.

74.1.2.2 Tick Paralysis

When an engorged and gravid (egg-laden) female tick injects a neurotoxin from its salivary glands into the host's bloodstream during feeding, a clinical syndrome known as tick paralysis may eventuate. The major neurotoxin secreted by *Ixodes holocyclus* is a protein of 40–60 kD with ability to inhibit production of neurotransmitter acetylcholine at neuromuscular junctions. Tick paralysis kills thousands of animals such as cows, sheep, and pets in various parts of the world, and human cases of tick paralysis tends to occur in children under the age of 10. The tick species most commonly associated with tick paralysis in North America are the Rocky Mountain wood tick (*Dermacentor andersoni*) and the American dog tick (*Dermacentor variabilis*). In Australia, tick paralysis is caused by *Ixodes holocyclus*.

74.1.2.3 Tick-Transmitted Diseases

Members of Ixodidae are vectors of a number of pathogenic bacteria (e.g., *Borrelia*, *Rickettsia*, *Anaplasma*, *Ehrlichia*, *Francisella*, and *Coxiella*), viruses (e.g., West Nile virus), and parasites (e.g., *Babesia*) [62–70]. For instance, *Ixodes holocyclus* transmits *Rickettsia australis* and *Rickettsia honei*, which cause rickettsial spotted fever (Queensland tick typhus) and Flinders Island spotted fever, respectively. As *R. australis* proliferates within the endothelial cells of small blood vessels, its infection may lead to vasculitis, along with high fever, rash, black spot (of 2–5 mm in diameter at the site of the tickbite, also known as an eschar), and enlarged and painful lymph glands. Other symptoms include headache, stiff neck, nausea, vomiting, mental confusion, and aching muscles and joints. The disease often runs its course in about 2 weeks, although it can be cured more quickly with tetracyclines.

Lyme disease is an illness caused by spirochetes *Borrelia burgdorferi* and other *Borrelia* species that are transmitted by Ixodidae hard ticks. Early symptoms include a rash or red patch (up to 50 cm or more in diameter) at the site of tick bite. This rash (also called erythema migrans or EM) occurs in 20% of patients with Lyme disease and often has a "target" or "bullseye" appearance, clearing in the middle with a more obvious red edge. More common symptoms of Lyme disease occur after spirochete invasion of the nervous system, heart, muscles, and joints, resulting in flu-like illness, fever, headache, sore throat, fatigue, aching muscles and joints, meningitis, Bell's palsy (weakness of the face muscles), swelling of joints, and heart problems with palpitations and breathlessness.

74.1.3 Diagnosis, Treatment, and Prevention

Ticks are blood-sucking arthropods that parasitize various species of vertebrates throughout the world. Because tick bites are often painless, and diseases transmitted by ticks may take days and weeks to develop after the tick bites, a tick-related illness is not suspected and diagnosed until much later. Keeping physicians informed of a tick bite and prior outdoor activity (camping, hiking, etc.) is important for diagnosis of tick-transmitted diseases.

A first step in the diagnosis of a tick-caused disease includes searching entire body for ticks still attached, rashes, and other pathophysiological changes (e.g., peripheral nerve dysfunction, esophageal dysfunction, respiratory difficulty, hypertension, and body temperature). Further blood tests may be undertaken for hematologic and biochemical changes associated with diseases such as Lyme disease, RMSF, ehrlichiosis, and tularemia, and examination of blood under a microscope is necessary to diagnose babesiosis.

Identification and differentiation of hard ticks belonging to the family Ixodidae have been traditionally based on morphological criteria using a series of dichotomous keys. However, because members of the Ixodidae family show limited variations and even overlap among different species, and the difficulty to differentiate specimens that are small or damaged, considerable efforts have been made on devising other methods of identification.

Protein electrophoresis and cuticular hydrocarbon analysis may be used for taxonomy studies of several tick species and populations [71]. Molecular techniques based on DNA sequence analysis offer a most effective method for the identification and phylogenetic characterization of ticks [72–79]. The common molecular targets range from nuclear ribosomal RNA (18S and 28S, ITS1 and ITS2), mitochondrial rRNA (12S and 16S), to other gene sequences [47,80–99].

Use of PCR-RFLP-based technique allowed identification and differentiation of various tick species. Application of TaqMan® chemistry and primers from the second internal transcribed spacer region (ITS2) of the conserved ribosomal RNA (rRNA) gene enabled the development of a novel, real-time PCR assay for the differentiation of *Ixodes scapularis*, *I. pacificus*, *Dermacentor variabilis*, and *Amblyomma americanum* ticks of medical importance in the United States [100]. A PCR assay based on part of the second internal transcribed spacer ribosomal RNA (pITS-2 rRNA) was reported for discrimination of *D. andersoni* from *D. variabilis* [101]. In addition, single-strand conformation polymorphism (SSCP) analysis of the pITS-2 rDNA provided a reliable method of distinguishing specimens of the three ixodid ticks (*D. andersoni*, *D. variabilis*, and *D. albipictus*) [101].

74.1.3.1 Treatment and Prevention

Physical removal of ticks provides a useful approach in the treatment of the illness and prevention of disease transmission. Because the tick is attached firmly to hosts, use of a small forceps or tweezers is essential to remove portions of its head and mouthparts embedded in the skin. Application of an aerosol insect repellent containing pyrethrin or a pyrethroid is helpful before physical removal (see Chapter 71). By carefully flipping the tick over onto its back, and grasping the tick firmly with the tweezers as close to the skin as possible, the tick is gently pulled free. Twisting or turning the tick should be avoided as this may break off its head and mouthparts, thus increasing the chances for infection. The removed tick can be kept in a tightly closed jar or taped to a piece of paper for further examination. If portions of the head or mouthparts remain, they should be removed surgically. Wearing gloves helps prevent the spread of pathogens from the tick to hands. Instruments that touch a tick are cleaned and disinfected. Local cleansing and antibiotic cream are applied. Preparations containing diphenhydramine (Benadryl) can be applied directly to the skin for itching or administered orally by tablets. Antibiotics may be prescribed for oral or IV administration.

Prevention of tick bite is best achieved by avoiding grassy areas and shrubs where ticks are lying in wait to tag a ride on a potential meal. During the active tick season (the months of April through September in the United States), it is important to stay away from outdoor areas where ticks thrive. Wear light-colored clothing so ticks can be easily seen, and brushed off. Tucking pants into boots or socks, and applying insect repellent, specifically the brands also help reduce tick bite. Repellents containing permethrins is applied to clothing but not to skin, and DEET-containing repellents may need to be reapplied more frequently in high tick areas, than for repelling mosquitoes. Pets are also sprayed with flea and tick repellents. People from a tick-infested area with a fever in the previous 2 months should not donate blood. Application of acaricides to animal hosts by using baited tubes, boxes, and feeding stations is also helpful in areas where infected ticks are endemic.

74.2 METHODS

74.2.1 Sample Preparation

74.2.1.1 Sample Collection

Adult ticks are collected by flagging grassy and shrubby vegetation. Ticks are identified by the shape of the spiracular plates, the relative size and number of the goblets within the spiracular plates, and other morphological features. Ticks are maintained at 22°C, under a 14-h light/10-h dark photoperiod. Ticks are held in 16-mL glass vials (Wheaton Glass, Millville, NJ) with a mesh top over a supersaturated solution of potassium nitrate. Ticks are frozen at −70°C until required.

74.2.1.2 DNA Extraction

After taxonomic identification, living adult ticks were killed by freezing at −70°C or by immersion in isopropanol. Genomic DNA is extracted from individual ticks using a modified protocol of Zahler et al. [42]. The ticks are homogenized in 400 μL of TET buffer (50 mM Tris-HCl, pH 7.5, 25 mM EDTA, 100 mM NaCl, and 0.4% Triton X-100) and 10 μL (20 mg/mL) of proteinase K. After incubation at 56°C for 2 h, the DNA is extracted by shaking in 200 μL of phenol/chloroform/isoamylic alcohol (25:24:1) and precipitated by adding two volumes of ice-cold ethanol overnight. After centrifugation (12,000 × *g* for 5 min), the pellet is resuspended in 35 μL of TE (10 mM Tris, pH 8.0, and 1 mM EDTA), and the DNA is used as a template for PCR.

DNA from ticks can also be isolated according to "Method 3" in Halos et al. [100]. Briefly, individual ticks are placed in 2 mL screw-top centrifuge tubes and placed in liquid nitrogen for 60 s. The frozen tick is then transferred to an identical tube containing three sizes of glass beads (2 mm, 0.5 mm, 0.1 mm) and crushed by vigorous shaking with a Mini-BeadBeater-8 (BioSpec Products, Inc., Bartlesville, OK) for two cycles of 90 s. The tubes are then centrifuged at 10,000 × *g* for 10 s at room temperature, and the samples are resuspended in 180 μL Buffer ATL (QIAamp DNA Mini Kit). DNA is eluted in 200 μL of elution buffer. The concentration of isolated DNA is determined using a SmartSpec Plus spectrophotometer (Bio-Rad Laboratories, Hercules, CA), and the samples are stored at −20°C [101].

Alternatively, genomic DNA is extracted from either one or two legs, or the complete body of adult ticks using the DNeasy Tissue Kit™ (Qiagen). The use of only one or two legs permits the remainder of the specimen to be preserved for morphological examination or to be tested for the presence of pathogenic organisms. The leg or complete body of an individual tick is placed into a 1.5-mL micropestle tube with 180 μL of ATL Buffer (Qiagen) and homogenized using a micropestle attached to a cordless drill. Proteinase K (20 μL of 15 μg/μL) was then added to the micropestle tube and the sample incubated overnight at 55°C. 200 μL of AL Buffer (Qiagen) is added to the sample, vortexed and incubated for 10 min at 70°C. Then, 200 μL of 100% ethanol is added, and the solution is applied to a spin column. After rinsing the columns with the wash buffers AW1 and AW2 (Qiagen), DNA is eluted with 100 μL AE buffer (Qiagen) and stored at −70°C [102].

Total genomic DNA was extracted from individual tick specimen used in this study. Briefly, tick specimens were cleaned by sonication for 3–5 min in ethanol solution and then washed twice in sterile distilled water. Afterward, individual tick specimen dissected in pieces was placed in a microcentrifuge tube filled with 180 μL lysing buffer solution and then homogenized with a sterile tissue grinder (catalogue no. 358103, Wheaton Scientific Products, Millville, NJ). The homogenate was centrifuged at room temperature and the supernatant fluid was further processed by a DNeasy Tissue Kit, as instructed by the manufacturer (Qiagen) [97].

74.2.2 Detection Procedures

74.2.2.1 Real-Time PCR Identification of *Amblyomma americanum, Dermacentor variabilis, Ixodes pacificus,* and *Ixodes scapularis*

Shone et al. [101] developed a real-time PCR assay for the identification of *Amblyomma americanum, Dermacentor variabilis, Ixodes pacificus,* and *Ixodes scapularis,* with primers and probes from the ITS2 of the conserved rRNA gene (Table 74.3). In addition, primers and probe from tick 16S rRNA gene are included as an assay control.

74.2.2.1.1 Procedure

1. Real-time PCR mixture (25 μL) is composed of 5 mM $MgCl_2$, 0.4 μM each of forward and reverse primers and the dual-labeled probe (Table 74.1), and 12.5 μL of LightCycler® FastStart DNA Master Hybridization Probes (Roche Diagnostics Corp).
2. Amplification is carried out with Standard TaqMan conditions (95°C for 10 min followed by 40 cycles of 95°C for 15 s, 60°C for 1 min) using an ABI 7000 Sequence Detection System (Applied Biosystems).
3. Samples demonstrating fluorescence above the background prior to cycle 40 are considered to be positive.

74.2.2.2 PCR Identification of *Dermacentor andersoni* and *Dermacentor variabilis*

Dergousoff and Chilton [102] described the use of primers DAVF (5′-TCACATATCAAGAGAGCCTT-3′) and DAVR (5′-ACGTACTTCGAAGGCAAACA-3′) for specific amplification of partial ITS-2 (pITS-2) rRNA sequence of *Dermacentor andersoni* and *Dermacentor variabilis.*

74.2.2.2.1 Procedure

1. PCR mixture (25 μL) is composed of 200 μM of each dNTP (Promega), 3 mM $MgCl_2$, 50 pmol of each primer, and 0.5 U of *Taq* polymerase (Promega).
2. Amplification is performed with an initial 95°C for 5 min, 30 cycles of 95°C for 30 s, 52°C for 30 s, and 74°C for 30 s, and a final 74°C for 5 min using a thermal cycler (Bio-Rad iCycler).
3. The amplicons are electrophoresed on SYBR-safe™ (Molecular Probes) stained 2% agarose-TBE (89 mM Tris, 89 mM boric acid, 2 mM EDTA, pH 8.3) gels, along with a 100-bp TrackIt™ DNA ladder (Invitrogen) as a size standard.

Note: Using primers DAVF and DAVR, a 430-bp fragment is generated from *D. andersoni* DNA while a 360-bp fragment is produced from *D. variabilis* DNA. The PCR assay does not discriminate between *D. variabilis* and *D. albipictus* because their respective pITS-2 amplicons are of an equivalent size on agarose gel. If desired, SSCP on the pITS-2 amplicons may be conducted to discriminate *D. andersoni, D. variabilis,* and *D. albipictus* [102].

74.2.2.3 PCR Identification of *Ixodes hexagonus, Ixodes persulcatus, Ixodes ricinus,* and *Dermacentor reticulatus*

Rumer et al. [98] developed PCR-based assays for the discrimination of Euro-Asian tick species of high medical

TABLE 74.3
Primers and Probes for Real-Time PCR Identification of Hard Ticks

Targeting Organism	Primer/Probe	Sequence (5′–3′)
Amblyomma americanum	AamerITS2b-F	AAGCCCGCGCTCCAAGC
	AamerITS2c-R	GCAGCAGTTCGGCTACACGTA
	AamerITS2-P	FAM-ACGACGTAACGCGGGACGGC-BHQ1
Dermacentor variabilis	DvalTS2b-F	CTGAAGATTCTTTGCGAGGAGCGG
	DvalTS2b-R	GCGTCAGCTCGGCCAAC
	DvarITS2-P	FAM-AGAAGGGCGTGCCCCGAAAGCGG-BHQ1
Ixodes pacificus	IpacITS2-F	CTTCCTACCCGGTTTGTGTCGTC
	IpacITS2-R	CGTACTCGTTCTCTCCCCAGC
	IpacITS2-P	FAM-AAGACCTTTGGGGACGGATGATGGA-BHQ1
Ixodes scapularis	IscapITS2-F	TGCGTTTTCTTTGAGCAAATGCACGAG
	IscapITS2-R	GTACGGGATTTTCCACAAACGGTATCCA
	IscapITS2-P	FAM-TGCGCTTAACCAGTCCTCCTCCTCCTACGA-BHQ1
Tick	Tick16S-F	CTGCTCAATGATTTTTTAAATTGCTGTGGT
	Tick16S-R	CCGGTCTGAACTCAGATCAAGTAGGA
	Tick16S-P	FAM-AAATAGTTTGCGACCTCGATGTTGGATTAGGAT-BHQ1

Note: FAM, carboxyfluorescein.

importance: *Ixodes ricinus*, *Ixodes persulcatus*, *Ixodes hexagonus*, and *Dermacentor reticulatus*, with primers from the ITS2 of the conserved rRNA gene (Table 74.4). While primers dITS29 and rITS800 recognize all tick species (generating a 778-bp fragment for *Ixodes* sp. and a 1045-bp fragment for *Dermacentor reticulatus*), primers dITS678 and rITS814 as well as primers dITS555 and rITS754 are specific for *Ixodes ricinus* (producing amplicons of 137 bp and 200 bp, respectively). In addition, primers dITS118 and rITS436 amplify a 319-bp fragment from *Ixodes persulcatus*, and primers dITSh277 and rITSh422 amplify a 146-bp fragment from *Ixodes hexagonus*. Use of 16S rRNA gene primers 16Sa and 16Sb provides a useful control for these assays.

74.2.2.3.1 Procedure

1. PCR mixture (10 μL) is made up of 1× PCR buffer, 1.5 mM $MgCl_2$, 200 μM of each dNTPs, 0.05 U/μL of AmpliTaq DNA polymerase (Invitrogen), 0.4 pm/μL of each primers (with individual primer pairs in separate tubes), and tick DNA.
2. Amplification is carried out with an initial 94°C for 3 min; 35 cycles of 94°C for 40 s, 50–55°C for 1 min (depending on primer pair, see Table 74.2), and 72°C for 1 min; and a final 72°C for 10 min.
3. PCR products are analyzed by electrophoresis in 2% agarose-TBE gel, containing 0.5 mg/mL ethidium bromide. A 100-bp DNA ladder (Invitrogen GmbH) is used as the size standard.

Note: While *Ixodes ricinus*, *Ixodes persulcatus*, and *Ixodes hexagonus* are identified by species-specific PCR, *Dermacentor reticulatus* is differentiated from *Ixodes* spp. by the fact that *D. reticulatus* forms a 1045-bp fragment and *Ixodes* spp. show a 778-bp product using primers dITS29 and rITS800. The identity of PCR amplicons can be verified by direct sequencing after gel extraction and purification with QIAquick Gel Extraction kit (Qiagen).

74.2.2.4 Sequence Analysis of Tick 16S rRNA Gene

Mangold et al. [88] employed primers (forward, 5′-CCG GTC TGA ACT CAG ATC AAG T-3′ and reverse, 5′-GCT CAA TGA TTT TTT AAA TTG CTG T-3′) to amplify a 460-bp fragment of the 16S rRNA gene. This fragment corresponds to the *Drosophila yakuba* 16S rRNA gene sequence between positions 12,887 and 13,431. Subsequent analysis of the amplicon sequence enables identification of the organisms under investigation.

74.2.2.4.1 Procedure

1. PCR mixture (100 μL) is made up of 1× PCR buffer (10 mM Tris-HCl, pH 9.0, 1.5 mM $MgCl_2$, 50 mM KCl), 200 μM each of dATP, dCTP, dGTP, and dTTP, 2 U of *Taq* DNA polymerase (Promega), 50 pmol of each primers, and tick DNA.
2. Amplification is conducted with an initial denaturation step at 94°C for 2 min followed by 35 cycles

TABLE 74.4
PCR Primers for Identification of Euro-Asian Tick Species

Targeting Organism/Gene	Primer/Probe	Sequence (5′–3′)	Expected Product (bp)	Annealing Temperature (°C)
Ticks	dITS29	CCTTCCCGTGGCTTCGTCTGT	778 for *Ixodes* sp.; 1045 bp for *Dermacentor reticulatus*	55
	rITS800	GGGGGTTGTCTCGCCTGATGT		
Ixodes hexagonus	dITSh277	GGCCGCCGTTGGGATTTAC	146	56
	rITSh422	ACAACATCGCCCCAACAAACAGAC		
Ixodes persulcatus	dITS118	TGGGCAGCACGCATTTTT	319	55
	rITS436	TGTACATCCGTCCATTTAGGC		
Ixodes ricinus	dITS678	CGTGTGTACAACAACACTTTTCGA	137	52
	rITS814	TATGCTTAAATTCAGGGGGTTGTC		
Ixodes ricinus	dITS555	TGCGTCGTAGCCTTCCGTCAGTCT	200	55
	rITS754	GCGCTTTCCAAACAAAAACTTTGA		
16S rRNA	16Sa	CGCCTGTTTATCAAAAACAT	509	50
	16Sb	CTCCGGTTTGAACTCAGATC		

for 30 s at 94°C, 30 s for primer annealing, and 45 s for primer extension at 72°C. The annealing temperature of the first seven cycles is increased by 0.3°C every second cycle from 47°C to 48.8°C, followed by 28 cycles using an annealing temperature of 50°C. A final extension step is carried out for 7 min at 72°C.

3. The PCR products (10 μL) are examined by 1% agarose gel electrophoresis, stained with ethidium bromide, and visualized under ultraviolet light.
4. The remaining amplicon is purified using the Wizard PCR Preps DNA Purification System (Promega) and resuspended in 50 μL of 10 mM TE (pH 7.6). The DNA sequence is determined by analysis using the same PCR primers.
5. The resulting sequences are aligned using CLUSTAL V, and phylogenetic relationships are analyzed using distance and maximal parsimony methods.

74.2.2.5 Sequence Analysis of Tick ITS2 Region

Marrelli et al. [94] utilized primers 5.8S (5′-CCATCGATG TGAAYTGCAGGACA-3′) and 28S (5′-GTGAATTCTATG CTTAAATTCAGGGGGT-3′) to amplify and analyze the ribosomal nucleotide sequences of the ITS2 of hard tick species of the genus *Amblyomma*.

74.2.2.5.1 Procedure

1. PCR mixture (50 μL) is composed of 1× PCR buffer (10 mM Tris-HCl, pH 9.0, 1.5 mM $MgCl_2$, 50 mM KCl), 200 μM each of dNTPs, 1 U of *Taq* DNA polymerase (Promega), 50 pmol each of primers, and 10–20 ng of tick DNA.
2. Amplification is conducted with 35 cycles of 94°C for 45 s, 55°C for 1 min, and 72°C for 1.5 min.
3. The PCR products (10 μL) are examined by 0.8% agarose gel electrophoresis, stained with ethidium bromide, and visualized under ultraviolet light.
4. The amplicons containing the ITS2 and part of the flanking 5.8S and 28S genes are cloned into pGEM-T™ easy vector plasmid (Promega) and used to transform *Escherichia coli* Dh5α cells. The cloned fragments are sequenced in both forward and reverse direction by using ABI Prism dGTP BigDye Terminator Ready Reaction kit (Applied Biosystems).
5. The sequences are aligned with the CLUSTAL W software, version 1.60. The boundaries of the ITS2 are deduced by comparisons with rRNA gene sequences from other ticks species. Only the ITS2 sequences are used for CG content determination and phylogenetic analyses.
6. Aligned sequences are examined with the MEGA software (Molecular Evolutionary Genetics Analysis, version 3.01), and a divergence matrix is constructed. Relationships between individuals are assessed by three methods: neighbor-joining (NJ) method with Kimura two-parameter distances with 1000 replications in the bootstrap test, maximum parsimony (MP) with 1000 bootstrap replicates with the Min-mini branch-and-bound search option, and maximum likelihood, also with the Kimura two-parameter model (ML) (PHYLIP: Phylogeny Inference Package, version 3.6). An additional analysis by the ML method is also performed with PAUP*, and the model is used to reconstruct the ML tree (General Time Reversible model with gamma distributed rates). The trees are rooted using the ITS2 sequences from *Amblyomma vikirri* (GenBank accession no. AF199112). ITS2 sequences and from *Amblyomma americanum* (GenBank accession no. AF548539).

Note: The lengths of the ITS2 sequences from *Amblyomma* species under investigation range from 956 to 1207 bp, with GC content of 62.4% to 66.9%. The ITS2 region proves useful for *Amblyomma*-specific identification.

74.2.2.6 Sequence Analysis of Tick *cox1* and *nad5* Genes

Chitimia et al. [96] targeted portions of mitochondrial cytochrome *c* oxidase subunit 1 gene (*pcox1*) and nicotinamide adenine dinucleotide dehydrogenase subunit 5 gene (*pnad5*) for identification and phylogenetic analysis of tick species. A 732-bp fragment of *pcox1* gene is amplified with primers cox1F (5′-GGAACAATATATTTAATTTTTGG-3′) and cox1R (5′-ATCTATCCCTACTGTAAATATATG-3′); and a 519-bp fragment of *pnad5* gene is amplified with primers nd5F (5′-TCTAAAATTAAATCCTTTGAAT-3′) and nd5R (5′-AAGAGCCCAAATTCCATTTTC-3′) separately from individual ticks by PCR. The amplicons are sequenced and analyzed.

74.2.2.6.1 Procedure

1. PCR mixture (25 μL) is composed of 0.25 μL of each 50 μM PCR primer, 1.25 U of r*Taq* DNA polymerase (Takara), 2 mM $MgCl_2$, 200 μM each of dNTPs, and 1 μL of DNA.
2. Amplification is carried out with an initial 95°C for 5 min, followed by 40 cycles of 95°C for 30 s, 55°C for 1 min, 72°C for 1 min, and a final extension of 72°C for 5 min.
3. The PCR products (2 μL) are examined by 1% agarose gel electrophoresis, stained with ethidium bromide, and visualized under ultraviolet light.
4. PCR products are purified and cloned into pGEM-T Easy vector (Promega,) and transformed into *Escherichia coli* competent JM109 cells. The recombinant bacteria are screened, confirmed by PCR amplification, and then sequenced.

5. Amino acid sequences of p*cox1* and p*nad5* are separately aligned using Clustal X, which is subsequently utilized as a scaffold to align the corresponding nucleotide sequences with the program CodonAlign 2.0 (http://homepage.mac.com/barryghall/CodonAlign.html). The aligned nucleotide sequences are then concatenated into single alignments for phylogenetic analyses. Dataset congruence between gene partitions is tested in an MP framework using the partition homogeneity test as implemented in PAUP* 4.0b10.
6. Data are analyzed using three approaches, namely, MP, maximum likelihood (ML), and neighbor joining (NJ), and *Varroa destructor* is used as outgroup in all analyses. Standard unweighted MP are performed in PAUP* using heuristic searches with tree-bisection-reconnection (TBR) branch swapping and 1000 random-addition sequence replicates with 10 trees held at each step. Prior to ML analyses, best-fit models of nucleotide substitution are selected by the Akaike Information Criterion as implemented by Modeltest 3.5. ML topologies are then inferred with PAUP* and PhyML using TBR branch swapping under best-fit model). The NJ approach is performed by MEGA 4.0 using Jukes–Cantor as a substitution model. Branch supports in ML tree is estimated by bootstrap analysis of 500 replicates, while 1000 replicates in both MP and NJ analyses.

74.3 CONCLUSION

The hard tick family Ixodidae consists of >702 species that are grouped under six subfamilies and 14 genera, with many medically important species found in the genera of *Amblyomma*, *Dermacentor*, *Haemaphysalis*, *Hyalomma*, *Ixodes*, and *Rhipicephalus*. Hard ticks are characterized by having a tough back plate or scutum and by feeding for a lengthy period (hours to days), with disease transmission usually occurring near the end of a meal. This is in clear contrast to soft ticks belonging to the family Argasidae that typically have rounded bodies without the hard scutum, and feed for a short time (>1 h), with disease transmission occurring within a minute of a meal. Apart from being a source of considerable irritation and discomfort through their bites, hard ticks are vectors of a number of bacterial, viral, and parasitic pathogens of humans and animals, transmitting Lyme disease, ehrlichiosis, tularemia, tick fever, Q fever, babesiosis, and other serious illnesses.

Effective treatment and control of tick-borne diseases rest on the accurate identification of the vector ticks. Since morphological characterization of ticks is difficult and prone to misidentification, molecular techniques provide a useful alternative for tick identification and phylogenetic assessment. Through amplification and sequence analyses of nuclear ribosomal RNA (18S and 28S, ITS1 and ITS2), mitochondrial rRNA (12S and 16S), and other gene sequences, the species identity of ticks implicated in disease transmission can be reliably and rapidly determined.

REFERENCES

1. Guglielmone, A.A. et al., The Argasidae, Ixodidae and Nuttalliellidae (Acari: Ixodida): A world list of valid tick names. *Zootaxa*, 2528, 1, 2010.
2. Anderson, J.F., The natural history of ticks. *Med. Clin. North Am.*, 86, 205, 2002
3. Anderson, J.F., Biology of ticks. *Infect. Dis. Clin. North Am.*, 22, 195, 2008.
4. Klompen, J.S.H. et al., Evolution of ticks. *Annu. Rev. Entomol.*, 41, 141, 1996.
5. Klompen, H., Dobson, S.J., and Barker, S.C., A new subfamily, Bothriocrotoninae n. subfam., for the genus *Bothriocroton* Keirans, King & Sharrad, 1994 status amend. (Ixodida: Ixodidae), and the synonymy of *Aponomma* Neumann, 1899 with *Amblyomma* Koch, 1844. *Syst. Parasitol.*, 53, 101, 2002.
6. http://www.uniprot.org/taxonomy/; last accessed on 10 October 2011.
7. Yamaguchi, N., Geographic variation of ixodid ticks in human cases. *Med. Entomol. Zool.*, 45, 198, 1994.
8. Keirans, J.E., and Lancombe, E.H., First records of *Amblyomma americanum, Ixodes dentatus*, and *Ixodes uriae* from Maine. *J. Parasitol.* 84, 629, 1998.
9. Barker, R.W. et al., Occurrence of the gulf coast tick (Acari: Ixodidae) on wild and domestic mammals in north-central Oklahoma. *J. Med. Entomol.*, 41, 170, 2004.
10. Guglielmone, A.A. et al., Ticks (Ixodidae) on humans in South America. *Exp. Appl. Acarol.*, 40, 83, 2006.
11. Lynen, G. et al., Cattle ticks of the genera *Rhipicephalus* and *Amblyomma* of economic importance in Tanzania: Distribution assessed with GIS based on an extensive field survey. *Exp. Appl. Acarol.*, 43, 303, 2007.
12. Briciu, V.T. et al., First survey on hard ticks (Ixodidae) collected from humans in Romania: Possible risks for tick-borne diseases. *Exp. Appl. Acarol.*, 54, 199, 2011.
13. Chen, Z. et al., Ticks (Acari: Ixodoidea: Argasidae, Ixodidae) of China. *Exp. Appl. Acarol.*, 51, 393, 2010.
14. Karaer, Z. et al., Ticks on humans in Ankara, Turkey. *Exp. Appl. Acarol.*, 54, 85, 2011.
15. Lemos, E.R.S. et al., Primary isolation of spotted fever group rickettsiae from *Amblyomma cooperi* collected from *Hydrochaeris hydrochaeris* in Brazil. *Mem. Inst. Oswaldo Cruz*, 91, 273, 1997.
16. Arzua, M. et al., *Amblyoma aureolatum* and *Ixodes auritulus* (Acari: Ixodidae) on birds in southern Brazil, with notes on their ecology. *Exp. Appl. Acarol.*, 31, 283, 2002.
17. Estrada Peña, A., Venzal, J.M., and Guglielmone, A.A., *Amblyomma dubitatum* Neumann: Description of nymph and redescription of adults, together with the description of the immature stages of *A. triste* Koch. *Acarologia*, 42, 323, 2002.
18. Estrada-Peña, A., Guglielmone, A.A., and Mangold, A.J., The distribution and ecological "preferences" of the tick *Amblyomma cajennense* (Acari: Ixodidae), an ectoparasite of humans and other mammals in the Americas. *Ann. Trop. Med. Parasitol.*, 98, 283, 2004.
19. Guglielmone, A.A. et al., *Amblyomma aureolatum* (Pallas, 1772) and *Amblyomma ovale* Koch, 1844 (Acari: Ixodidae): DNA sequences, hosts and distribution. *Vet. Parasitol.*, 113, 273, 2003.

20. Labruna, M.B. et al., *Amblyomma latepunctatum*, a valid tick species (Acari: Ixodidae) long misidentified with both *Amblyomma incisum* and *Amblyomma scalpturatum*. *J. Parasitol.*, 91, 527, 2005.
21. Labruna, M.B. et al., Human parasitism by the capibara tick, *Amblyomma dubitatum* (Acari: Ixodida). *Entomol. News*, 118, 77, 2007.
22. Labruna, M.B. et al., Redescription of the female, description of the male, and several new records of *Amblyomma parkeri* (Acari: Ixodidae), a South America tick species. *Exp. Appl. Acarol.*, 49, 243, 2009.
23. Aguirre, D.H. et al., *Amblyomma tigrinum* (Acari: Ixodidae): New data on hosts and biology of immature stages and on DNA composition. *Vet. Parasitol.*, 134, 267, 2005.
24. Nava, S., Mangold, A.J., and Guglielmone, A.A., The natural hosts of larvae and nymphs of *Amblyomma tigrinum* Koch, 1844 (Acari: Ixodidae). *Vet. Parasitol.*, 140, 124, 2006.
25. Nava, S. et al., Distribution, hosts, 16S rDNA sequences and phylogenetic position of the Neotropical tick *Amblyomma parvum* (Acari: Ixodidae). *Ann. Trop. Med. Parasitol.*, 102, 409, 2008.
26. Nava, S. et al., Ecology of *Amblyomma neumanni* (Acari: Ixodidae). *Acta Trop.*, 111, 226, 2009.
27. Goddard, J., A review of the disease agents harbored and transmitted by the lone star tick, *Amblyomma americanum* (L.). *Southwest. Entomol.*, 12, 158, 1987.
28. Goddard, J., and Varela-Stokes, A.S., Role of the lone star tick, *Amblyomma americanum* (L.), in human and animal diseases. *Vet. Parasitol.*, 160, 1, 2009.
29. Childs, J.E., and Paddock, C.D., The ascendancy of *Amblyomma americanum* as a vector of pathogens affecting humans in the United States. *Ann. Rev. Entomol.*, 48, 307, 2003.
30. Paddock, C.D., and Yabsley, M.J., Ecological havoc, the rise of white-tailed deer, and the emergence of *Amblyomma americanum*-associated zoonoses in the United States. *Curr. Top. Microbiol. Immunol.*, 315, 289, 2007.
31. Trout, R.T., Steelman, C.D., and Szalanski, A.L., Population genetics of *Amblyomma americanum* (Acari; Ixodidae) collected from Arkansas. *J. Med. Entomol.*, 47, 152, 2010.
32. Lockhart, J.M. et al., Temporal association of *Amblyomma americanum* with the presence of *Ehrlichia chaffeensis* reactive antibodies in white-tailed deer. *J. Wildl. Dis.* 31, 119, 1995.
33. Loftis, A.D. et al., Geographic distribution and genetic diversity of the Ehrlichia sp. from Panola Mountain in *Amblyomma americanum*. *BMC Infect. Dis.* 8, 54, 2008.
34. Anderson, B.E. et al. *Amblyomma americanum*: A potential vector of human ehrlichiosis. *Am. J. Trop. Med. Hyg.* 49, 239, 1993.
35. Jasinskas, A., Zhong, J., and Barbour, A.G., Highly prevalent *Coxiella* sp. bacterium in the tick vector *Amblyomma americanum*. *Appl. Environ. Microbiol.*, 73, 334, 2007.
36. Mixson, T.R. et al., Prevalence of *Ehrlichia*, *Borrelia*, and Rickettsial agents in *Amblyomma americanum* (Acari: Ixodidae) collected from nine states. *J. Med. Entomol.*, 43, 1261, 2006.
37. Szabo, M.P.J. et al., Biology and life cycle of *Amblyomma incisum* (Acari: Ixodidae). *Exp. Appl. Acarol.*, 48, 263, 2009.
38. Kang, Y.B. et al. *Amblyomma testudinarium* Koch, 1844: Discovery and record in Korea, and indentification and redescription of male tick. *Korean J. Vet. Res.*, 21, 65, 1981.
39. Kim, J. et al., A case of *Amblyomma testudinarium* tick bite in a Korean woman. *Korean J. Parasitol.*, 48, 313, 2010.
40. Sanches, G.S., Bechara, G.H., and Camargo-Mathias, M.I., Morphological description of *Amblyomma brasiliense* Aragão, 1908 (Acari: Ixodidae) larvae and nymphs. *Rev. Bras. Parasitol. Vet.*, 18, 15, 2009.
41. Yamada, Y. et al., A case of tick bite from *Amblyomma testudinarium* on the glans penis. *J. Dermatol.*, 23, 136, 1996.
42. Zahler, M., Gothe, R., and Rinder, H., Genetic evidence against a morphologically suggestive conspecificity of *Dermacentor reticulatus* and *D. marginatus* (Acari: Ixodidae). *Int. J. Parasitol.*, 25, 1413, 1995.
43. Zahler, M., and Gothe, R., Evidence for the reproductive isolation of *Dermacentor marginatus* and *Dermacentor reticulatus* (Acari: Ixodidae) ticks based on cross-breeding, morphology and molecular studies. *Exp. Appl. Acarol.*, 21, 685, 1997.
44. Bursali, A. et al., Ixodid ticks (Acari: Ixodidae) infesting humans in Tokat Province of Turkey: species diversity and seasonal activity. *J. Vector Ecol.*, 35, 180, 2010.
45. Bursali, A. et al., Species diversity of ixodid ticks feeding on humans in Amasya, Turkey: Seasonal abundance and presence of Crimean-Congo hemorrhagic fever virus. *J. Med. Entomol.*, 48, 85, 2011.
46. Omeragic, J., Ixodid ticks in Bosnia and Herzegovina. *Exp. Appl. Acarol.*, 53, 301, 2011.
47. Patterson, E.I., Dergousoff, S.J., and Chilton, N.B., Genetic variation in the 16S mitochondrial DNA gene of two Canadian populations of *Dermacentor andersoni* (Acari: Ixodidae). *J. Med. Entomol.*, 46, 475, 2009.
48. Ahmed, B.M., Taha, K.M., and Hussein, A.M., Life cycle of *Hyalomma anatolicum* Koch, 1844 (Acari: Ixodidae) fed on rabbits, sheep and goats. *Vet. Parasitol.*, 177, 353, 2011.
49. Zheng, H. et al., Development and biological characteristics of *Haemaphysalis longicornis* (Acari: Ixodidae) under field conditions. *Exp. Appl. Acarol.*, 53, 377, 2011.
50. Poucher, K.L. et al. Molecular genetic key for the identification of 17 *Ixodes* species of the United States (Acari: Ixodidae): A methods model. *J. Parasitol.*, 85, 623, 1999.
51. Labruna, M.B. et al. Study of the taxonomic status of *Ixodes didelphidis* (Acari: Ixodidae). *J. Med. Entomol.*, 39, 242, 2002.
52. Xu, G. et al., Molecular phylogenetic analysis indicate that the *Ixodes ricinus* complex is a paraphyletic group. *J. Parasitol.*, 89, 452, 2003.
53. Eremeeva, M.E., and Stromdahl, E.Y., Short report: New spotted fever group *Rickettsia* in a *Rhipicephalus turanicus* tick removed from a child in eastern Sicily, Italy. *Am. J. Trop. Med. Hyg.*, 84, 99, 2011.
54. Wikswo, M.E. et al., Detection of *Rickettsia rickettsii* and *Bartonella henselae* in *Rhipicephalus sanguineus* ticks from California. *J. Med. Entomol.*, 44, 158, 2007.
55. Dantas-Torres, F., The brown dog tick, *Rhipicephalus sanguineus* (Latreille, 1806) (Acari: Ixodidae): From taxonomy to control. *Vet. Parasitol.*, 152, 173, 2008.
56. Demma, L.J. et al., An outbreak of Rocky Mountain spotted fever associated with a novel tick vector, *Rhipicephalus sanguineus*, in Arizona, 2004: Preliminary report. *Ann. N. Y. Acad. Sci.*, 1078, 342, 2006.
57. Parola, P., Socolovschi, C., and Raoult, D., Deciphering the relationships between *Rickettsia conorii conorii* and *Rhipicephalus sanguineus* in the ecology and epidemiology of Mediterranean spotted fever. *Ann. N. Y. Acad. Sci.*, 1166, 49, 2009.
58. de Oliveira, P.R. et al., Comparison of the external morphology of *Rhipicephalus sanguineus* (Latreille, 1806) (Acari: Ixodidae) ticks from Brazil and Argentina. *Vet. Parasitol.*, 129, 139, 2005.

59. Yamasaki, T., Oda, T., and Kitaoka, S., A case of human infestation with a female of *Amblyomma testudinarium* Koch, 1844 (Acarina; Ixodidae). *Trop. Med.*, 21, 167, 1979.
60. Krinsky, W.L., Dermatoses associated with the bites of mites and ticks (Arthropoda: Acari). *Int. J. Dermatol.*, 22, 75, 1983.
61. Nakamura-Uchiyama, F. et al., *Amblyomma testudinarium* tick bite: One case of engorged adult and a case of extraordinary number of larval tick infestation. *J. Dermatol.*, 27, 774, 2000.
62. Kahl, O. et al., Tick infection rates with *Borrelia*: *Ixodes ricinus* versus *Haemaphysalis concinna* and *Dermacentor reticulatus* in two locations in eastern Germany. *Med. Vet. Entomol.*, 6, 363, 1992.
63. Chao, L.L., Wu, W.J., and Shih, C.M., Molecular analysis of *Ixodes granulatus*, a possible vector tick for *Borrelia burgdorferi* sensu lato in Taiwan. *Exp. Appl. Acarol.*, 48, 329, 2009.
64. Dumler, J.S., and Bakken, J.S., Human ehrlichioses: Newly recognized infections transmitted by ticks. *Annu. Rev. Med.* 49, 201, 1998.
65. Felek, S. et al., Sensitive detection of *Ehrlichia chaffeensis* in cell culture, blood, and tick specimens by reverse transcription-PCR. *J. Clin. Microbiol.,* 39, 460, 2001.
66. Steiert, J.G., and Gilfoy, F., Infection rates of *Amblyomma americanum* and *Dermacentor variabilis* by *Ehrlichia chaffeensis* and *Ehrlichia ewingii* in southwest Missouri. *Vector Borne Zoonotic Dis.*, 2, 53, 2002.
67. Anderson, J.F. et al., Transstadial transfer of West Nile virus by three species of ixodid ticks (Acari: Ixodidae). *J. Med. Entomol.*, 40, 528, 2003.
68. Apperson, C.S. et al., Tick-borne diseases in North Carolina: Is "*Rickettsia amblyommii*" a possible cause of rickettsiosis reported as Rocky Mountain spotted fever? *Vector Borne Zoonotic Dis.,* 8, 597, 2008.
69. Venzal, J.M. et al., Guglielmone AA *Amblyomma triste* Koch, 1844 (Acari: Ixodidae): Hosts and seasonality of the vector of *Rickettsia parkeri* in Uruguay. *Vet. Parasitol.*, 155, 104, 2008.
70. Oliveira, K.A. et al., *Amblyomma imitator* ticks as vectors of *Rickettsia rickettsii*, Mexico. *Emerg. Infect. Dis.*, 16, 1282, 2010.
71. Hilburn, L.R., and Sattler, P.W., Electrophoretically detectable protein variation in natural population of the lone star tick, *Amblyomma americanum* (Acari: Ixodidae). *Heredity*, 56, 67, 1986.
72. Wesson, D.M. et al., Investigation of the validity of species status of *Ixodes dammini* (Acari: Ixodidae) using rDNA. *Proc. Natl. Acad. Sci. USA*, 90, 10221, 1993.
73. Barker, S.C., Distinguishing species and populations of *Rhipicephalini* ticks with ITS 2 ribosomal RNA. *J. Parasitol.*, 84, 887, 1998.
74. Fournier, P.E. et al., Genetic identification of rickettsiae isolated from ticks in Japan. *J. Clin. Microbiol.*, 40, 2176, 2002.
75. Anderson, J.M., Ammerman, N.C., and Norris, D.E., Molecular differentiation of metastriate tick immatures. *Vector Borne Zoonotic Dis.*, 4, 334, 2004.
76. Geraci, N.S. et al., Variation in genome size of argasid and ixodid ticks. *Insect Biochem. Mol. Biol.*, 37, 399, 2007.
77. Pagel Van Zee, J. et al., Tick genomics: The *Ixodes* genome project and beyond. *Int. J. Parasitol.*, 37, 1297, 2007.
78. Nava, S., Guglielmone, A.A., and Mangold, A.J., An overview of systematics and evolution of ticks. *Front. Biosci.*, 14, 2857, 2009.
79. Dharmarajan, G., and Rhodes, O.E., Evaluating levels of PCR efficiency and genotyping error in DNA extracted from engorged and non-engorged female *Dermacentor variabilis* ticks. *Med. Vet. Entomol.*, 25, 109, 2011.
80. Black, W.C., and Piesman, J., Phylogeny of hard- and soft-tick taxa (Acari: Ixodida) based on mitochondrial 16S rDNA sequences. *Proc. Natl. Acad. Sci. USA*, 91, 10034, 1994.
81. Black, W.C., Klompen, J.S.H., and Keirans, J.E., Phylogenetic relationships among tick subfamilies (Ixodida: Ixodidae: Argasidae) based on the 18S nuclear rDNA gene. *Mol. Phylogenet. Evol.*, 7, 129, 1997.
82. Caporale, D.A. et al., Discriminating between *Ixodes* ticks by means of mitochondrial DNA sequences. *Mol. Phylogenet. Evol.*, 4, 361, 1995.
83. McLain, D.K. et al., Variation in ribosomal DNA internal transcribed spaces 1 among eastern populations of *Ixodes scapularis* (Acari: Ixodidae). *J. Med. Entomol.*, 32, 353, 1995.
84. Rich, S.M. et al., Distribution of the Ixodes ricinus-like ticks of eastern North America. *Proc. Natl. Acad. Sci. USA*, 92, 6284, 1995.
85. Crampton, A., McKay, Y., and Barker, S.C., Phylogeny of ticks (Ixodida) inferred from nuclear ribosomal DNA. *Int. J. Parasitol.*, 26, 511, 1996.
86. Norris, D.E. et al., Population genetics of *Ixodes scapularis* (Acari: Ixodidae) based on mitochondrial 16S and 12S genes. *J. Med. Entomol.*, 33, 78, 1996.
87. Norris, D.E. et al., Taxonomic status of *Ixodes neotomae* and *I. spinipalpis* (Acari: Ixodidae) based on mitochondrial DNA evidence. *J. Med. Entomol.*, 34, 696, 1997.
88. Mangold, A.J., Bargues, M.D., and Mas-Coma, S., Mitochondrial 16S rRNA sequences and phylogenetic relationships of *Rhipicephalus* and other tick genera among Metastriata (Acari: Ixodidae). *Parasitol. Res.*, 84, 478, 1998.
89. Fukunaga, M. et al., Molecular phylogenetic analysis of Ixodid ticks based on the ribosomal DNA spacer, internal transcribed spacer 2, sequences. *J. Parasitol.*, 86, 38, 2000.
90. Cruickshank, R.H., Molecular marker for the phylogenetics of mites and ticks. *Syst. Appl. Acarol.*, 3, 1, 2002.
91. Hlinka, O., Murrell, A., and Barker, S.C., Evolution of the secondary structure of the rRNA internal transcribed spacer 2 (ITS2) in hard tick (Ixodidae, Arthropoda). *Heredity*, 88, 275, 2002.
92. Estrada-Peña, A. et al., The *Amblyomma maculatum* Koch, 1844 (Acari: Ixodidae: Amblyomminae) tick group: Diagnostic characters, description of the larva of *A. parvitarsum* Neumann, 1901, 16S rDNA sequences, distribution and hosts. *Syst. Parasitol.*, 60, 99, 2005.
93. Reichard, M.V. et al., Sequence variation of the ribosomal DNA second internal transcribed spacer region in two spatially distinct population of *Amblyomma americanum* (L.) (Acari: Ixodidae). *J. Parasitol.*, 91, 260, 2005.
94. Marrelli, M.T. et al., Taxonomic and phylogenetic relationships between neotropical species of ticks from genus *Amblyomma* (Acari: Ixodidae) inferred from second internal transcribed spacer sequences of rDNA. *J. Med. Entomol.*, 44, 222, 2007.
95. Chitimia, L. et al., Molecular characterization of hard and soft ticks from Romania by sequences of the internal transcribed spacers of ribosomal DNA. *Parasitol. Res.,* 105, 907, 2009.
96. Chitimia, L. et al., Genetic characterization of ticks from southwestern Romania by sequences of mitochondrial cox1 and nad5 genes. *Exp. Appl. Acarol.,* 52, 305, 2010.

97. Chao, L.L., Wu, W.J., and Shih, C.M., Species identification of *Ixodes granulatus* (Acari: Ixodidae) based on internal transcribed spacer 2 (ITS2) sequences. *Exp. Appl. Acarol.*, 54, 51, 2010.
98. Rumer, L. et al., Differentiation of medically important Euro-Asian tick species *Ixodes ricinus, Ixodes persulcatus, Ixodes hexagonus*, and *Dermacentor reticulatus* by polymerase chain reaction. *Vector Borne Zoonotic Dis.*, 11, 899, 2010.
99. Tian, Z. et al., Discrimination between *Haemaphysalis longicornis* and *H. qinghaiensis* based on the partial 16S rDNA and the second internal transcribed spacer (ITS-2). *Exp. Appl. Acarol.*, 54, 165, 2011.
100. Halos, L. et al., Determination of an efficient and reliable method for DNA extraction from ticks. *Vet. Res.*, 5, 1, 2004.
101. Shone, S.M. et al., A novel real-time PCR assay for the speciation of medically important ticks. *Vector Borne Zoonotic Dis.*, 6, 152, 2006.
102. Dergousoff, S.J., and Chilton, N.B., Differentiation of three species of ixodid tick, *Dermacentor andersoni*, *D. variabilis* and *D. albipictus*, by PCR-based approaches using markers in ribosomal DNA. *Mol. Cell. Probes*, 21, 343, 2007.

75 Pyroglyphidae and Glycyphagidae (House Dust and Storage Mites)

Dongyou Liu

CONTENTS

75.1 INTRODUCTION

The order Astigmata within the subclasss Acari (Acarina), class Arachnida covers 65 mite families. While members of the Astigmata families are largely decomposers of organic residues and are relatively harmless, some may cause direct or indirect damage to human health and animal welfare.

House dust mites and storage mites (collectively known as domestic mites or dust mites) represent some examples of the Astigmata taxa that are detrimental to humans by provoking the host's immune system and inducing various immunological disorders. These mites are widespread in nature, and abundant in soil, stored products, live plants, and dwellings. They survive by scavenging on dander, shed skin particles (containing collagen and keratin), other human excretions and secretions, as well as grains, fungi, and some bacteria. Repeated exposure to allergens (e.g., fatty acid-binding proteins, tropomysin and paramyosin homologues, apoliphorine-like proteins, and alpha-tubulines) produced by these mites can lead to allergy and allergic diseases (e.g., asthma, rhinitis, and atopic dermatitis). Farm workers and bakers who handle grains are particularly prone to mite-related allergy and occupational respiratory diseases.

House dust mites (or bed mites) belong to the family Pyroglyphidae, and frequently occur in warm, damp indoor environments (e.g., mattresses, sofas, carpets, and blankets). The medically important house dust mite species include *Dermatophagoides pteronyssinus*, *Dermatophagoides farinae*, and *Euroglyphus maynei*. Other less known house dust mite species are *Dermatophagoides siboney*, *Dermatophagoides microceras*, and *Gymnoglyphus longior*.

Storage mites (or stored-products mites) belong to several families of the order Astigmata (e.g., Glycyphagidae, Echimyopodidae, Chortoglyphidae, Ebertiidae, Acaridae, and Cheyletidae), and are found on poorly stored food and grains (e.g., wheat, corn, oats, barley, and hay), in kitchen floor dust, cupboards, and pantries. Of >150 species in this group, about 20 are considered economically and medically important. These include *Blomia tropicalis* (abundant in tropical and subtropical regions), *Lepidoglyphus destructor* (commonly found in barns), *Acarus siro*, *Glycyphagus domesticus*, *Tyrophagus putrescentiae*, *Aleuroglyphus ovatus*, *Suidasia medanensis*, and *Thyreophagus entomophagus*. Other less-known storage mites of medical significance comprise *Acarus farris*, *Austroglycyphagus malaysiensis*,

Blomia kulagini, *Blomia tjibodas*, *Cheyletus eruditus*, *Chortoglyphus arcuatus*, *Gohieria fusca*, *Tyrophagus longior*, and *T. putrescentiae*.

75.1.1 Classification, Morphology, and Biology

75.1.1.1 Classification

The house dust mite family Pyroglyphidae is classified in the superfamily Analgoidea, suborder Psoroptidia (Sarcoptidia), order Astigmata, superorder Acariformes, subclass Acari (or Acarina), class Arachnida, phylum Arthropoda.

Within the order Astigmata, seven suborders are recognized: Acaroidea, Canestrinioidea, Glycyphagoidea, Hemisarcoptoidea, Histiostomatoidea, Hypoderatoidea, and Psoroptidia (Sarcoptidia).

The suborder Psoroptidia (Sarcoptidia) consists of the superfamilies Analgoidea, Freyanoidea, Pterolichoidea, and Sarcoptoidea. The superfamily Analgoidea is further separated into the families of Alloptidae, Analgidae, Avenzoariidae, Dermationidae, Dermoglyphidae, Epidermoptidae, Knemidokoptidae, Proctophyllodidae, Psoroptoididae, Pteronyssidae, Pyroglyphidae, Trouessartiidae, Turbinoptidae, and Xolalgidae.

The house dust mite family Pyroglyphidae is divided into two subfamilies, Pyroglyphinae (containing the genera of *Pyroglyphus* and *Euroglyphus*) and Dermatophagoidinae (containing the genus *Dermatophagoides*). The genus *Dermatophagoides* comprises six recognized species (i.e., *Dermatophagoides evansi*, *Dermatophagoides farinae*, *Dermatophagoides microceras*, *Dermatophagoides neotropicalis*, *Dermatophagoides pteronyssinus*, and *Dermatophagoides siboney*), whereas the genus *Euroglyphus* has three recognized species (i.e., *Euroglyphus longior*, *Euroglyphus maynei*, and *Euroglyphus osu*) [1].

Dermatophagoides pteronyssinus (the European house dust mite), *Dermatophagoides farina* (the American house dust mite), and *Euroglyphus maynei* are distributed widely throughout the world. *Dermatophagoides siboney* occurs mainly in Cuba, and *Dermatophagoides microcerus*, often misidentified as *Dermatophagoides farinae*, is predominantly found in Europe.

The storage mite families Glycyphagidae (suborder Glycyphagoidea), Echimyopodidae (suborder Glycyphagoidea), Chortoglyphidae (suborder Glycyphagoidea), Suidasiidae (suborder Acaroidea), and Acaridae (suborder Acaroidea) are classified in the suborders Acaroidea and Glycyphagoidea, order Astigmata [2].

The medically important storage mite species include *Glycyphagus domesticus*, *G. privates*, *Gohieria fusca*, *Lepidoglyphus destructor*, and *Aeroglyphus robustus* (of the family Glycyphagidae); *Blomia tropicalis*, *B. kulagini*, *B. tjibodas*, and *B. thori* (of the family Echimyopodidae); *Chortoglyphus arcuatus* (of the family Chortoglyphidae); *Suidasia medanensis* (of the family Suidasiidae), *Tyrophagus putrescentiae*, *T. longior*, *Acarus siro*, *A. farris*, *Thyreophagus entomophagus*, and *Aleuroglyphus ovatus* (of the family Acaridae) [3–5].

All in all, the house dust mites *Dermatophagoides farinae*, *D. pteronyssinus*, *Euroglyphus maynei*, and the storage mite *Blomia tropicalis* represent the most common and important allergy-causing mites found in homes worldwide [6].

75.1.1.2 Morphology

Dust mites undergo five developmental stages in their life: egg, six-legged larva, two eight-legged nymph stages, and a breeding adult. The adults possess a sucker and hooks, but have no eyes or organized respiratory structures. Within the house dust mite family Pyroglyphidae, the subfamily Pyroglyphinae has short scapular setae and an acute anterior margin of the prodorsum, whereas the subfamily Dermatophagoidinae has long external scapular setae and a rounded, intact anterior margin of the prodorsum [7,8]. Interestingly, although *Dermatophagoides pteronyssinus* and *D. farinae* belong to the same genus, *D. pteronyssinus* appears evolutionarily more similar to *Euroglyphus maynei* than to *D. farinae* [9].

Both male and female adults of house dust mite *D. pteronyssinus* are globular in shape, creamy white/blue in color, and possess a striated cuticle. The female is about 420 μm in length and 320 μm in width, and the male is 420 μm in length and 245 μm in width. The male is more sclerotized with enlarged legs I and III, and the male aedeagus is located between the apodemes of leg IV. Additionally, the male has a pair of suckers on the ventral posterior idiosoma which is used to grasp the female during copulation. The female shows a genital opening anterior to the bursa copulatrix located near the anus [7,8].

Under the microscope, storage mite eggs appear ovoid, globular, or elliptical, and are of the centrolecithal type, with the yolk in the center and cytoplasm in the periphery. Cytoplasm is bounded in the vitelline envelope which underlies the one- or two-layered shell (endo- and exochorion). Storage mites eggs are small (<165 μm long), with *Glycyphagus domesticus* and *Caloglyphus redikorzevi* eggs being >150 μm long; *Acarus siro*, *Acarus gracilis*, *Aeroglyphus robustus*, *Carpoglyphus lactis*, *Tyrophagus putrescentiae*, *Tyrophagus brevicrinatus*, *Tyroborus lini*, *Aleuroglyphus ovatus*, and *Caloglyphus oudemansi* eggs being <150 μm long; *Lepidoglyphus destructor* being >128 μm long; and eggs of *Chortoglyphus arcuatus* being eggs being <128 μm long [10].

75.1.1.3 Biology

House dust mite *Dermatophagoides* shows the following stages in its life cycle: egg, active larva, resting larva (pharate tritonymph), active tritonymph, resting tritonymph (pharate adult), and breeding adult. Depending upon the temperature and humidity, the complete life cycle of *Dermatophagoides* takes between 19 and 30 days [11].

Dermatophagoides male attaches to a tritonymph female and mates when female reaches the adult stage. *Dermatophagoides* mated female lives about 50–70 days, and male survives for 15 days. *Dermatophagoides pteronyssinus* female lays about 80 eggs over a 45-day period, while *Dermatophagoides farinae* lays eggs over a 30-day period. The hatched egg produces a larva with six legs, which then transforms into a nymph with eight legs. After the nymph stages, an adult with eight legs emerges.

Storage mite *Acarus siro* lays on average 435 eggs with a maximum of 858 eggs at 15°C and 90% relatively humidity (RH). Storage mite *Tyrophagus putrescentiae* lays about 450–500 eggs/female and completes the generation in 2–3 weeks at 20–25°C and near-saturation humidity.

House dust mites live in colonies in damp, warm, still environments such as old bedding, fabric, furniture, or carpets with long fibers [12–14]. A typical used mattress may hold up to 10 million living dust mites [15]. They are scavengers that feed on organic debris such as dander (flakes of skin), fungal spores, pollen grains, plant fibers, and insect scales. In fact, an average human individual sheds 0.5 to 1.0 g of skin daily. Without sight and unable to drink, dust mites live by absorbing moisture and oxygen from the atmosphere. Adult dust mites have no "stomach" but a chambered gut (or a sac or pouch-like diverticulae) that secretes digestive enzymes to break down scraps of food.

House dust mites favor high humidity levels (65–80%) and constant warm temperatures (c. 25°C), and often stop growing and die out at 60% or lower of relative humidity. Their numbers are reduced at high altitudes and dry climates [16].

Although house dust mites do not bite or sting, they harbor at least 22 allergens in their carcases, body parts, secretions, and excreta, which are capable of eliciting IgE-mediated reactions and triggering respiratory and dermatological complaints in some humans upon repeated contact. Adult mites use sucker and hooks on each of their eight legs to travel about and take in oxygen through their waxy shell-like cover. Without sight, dust mites navigate using their sense of smell and signaling vibrations from other mites. Because of their light weight, dust mites can become airborne during bed making [17].

House dust mites *Dermatophagoides farinae*, *D. pteronyssinus*, *Euroglyphus maynei*, and storage mites *Blomia tropicalis* and *Tyrophagus putrescentiae* occur with high frequency and at high levels of infestation in houses throughout tropical and subtropical areas of the world. In contrast to *Dermatophagoides* spp., *Blomia tropicalis* is more commonly found in tropical compared with temperate climates [18]. In Cuba, *D. pteronyssinus*, *D. siboney*, *D. farinae*, *Acarus siro*, and *B. tropicalis* are abundant mite species in house dust.

Storage mites are present in storage facilities for grains such as wheat, corn, oats, barley, and hay. They are also found in kitchen floor dust, cupboards, and pantries [3,4,19]. When processed foods made from the grains (e.g., flour, cereals, and baking mixes) become moist or are stored in humid environments, they provide ideal media for storage mites to invade and prosper [20].

75.1.2 Clinical Features, Pathogenesis, and Treatment

75.1.2.1 Clinical Features

Dust mites contain at least 22 allergens (triggers) in their body parts, saliva, or droppings [21–24]. Mite droppings comprise active enzymes, scraps of food, and living fungi. Disturbed by movement or cleaning, dust mite droppings and allergens are pushed into the air and drift on air currents. Human exposure to dust mites and their droppings may develop asthma, rhinitis (hayfever), conjunctivitis, sinusitis, atopic dermatitis (eczema), or any combination of these disorders [25–34].

Dust mite exposures often affect the skin, lungs, nose, and eyes, with symptoms ranging from itchy, watery eyes, runny nose, itching and clogging of the ear canals, and asthma. Asthmatic symptoms include respiratory congestion, wheezing, and severe shortness of breath. Many allergic symptoms appear at night, when millions of dust mites inhabiting the mattresses and pillows become active.

Dust mites *Dermatophagoides pteronyssinus*, *Dermatophagoides farinae*, or *Blomia tropicalis* have been linked with allergic disorders such as Pancake syndrome [20,35–38].

Müsken et al. [20] investigated the frequency of sensitization and potential importance of mite species in 86 German farmers with rhinitis and/or asthma using skin prick testing and/or enzyme allergosorbent test (EAST) for *Blomia tjibodas*, *Blomia tropicalis*, *Blomia kulagini*, *Glycyphagus domesticus*, *Thyreophagus entomophagus*, *Euroglyphus maynei*, *Chortoglyphus arcuatus*, *Dermatophagoides pteronyssinus*, *Dermatophagoides farinae*, *Acarus siro*, *Lepidoglyphus destructor*, *Tyrophagus putrescentiae*, *Acarus farris*, and *Cheyletus eruditus*. The most frequent sensitizations determined by skin tests were due to three *Blomia* species, *E. maynei*, and *G. domesticus*. Twelve patients (14%) gave a positive EAST with the predator mite *C. eruditus*. A total of 22 patients gave positive EAST results with the *Dermatophagoides* species.

Edston and van Hage-Hamsten [39] described a case of anaphylaxis leading to sudden death of a 47-year-old farmer. The patient had previously been diagnosed with an allergy to house dust mites, and experienced infrequent episodes of airway symptoms, nausea, hypotension, and diarrhea usually after going to bed. Serum tryptase in postmortem blood was substantially elevated in two samples (170 and >200 μg/L). Analysis of allergen-specific IgE revealed high values for *Dermatophagoides pteronyssinus* and *Dermatophagoides farinae*. High mite allergen levels were detected in dust obtained from the patient's mattress. The results of the immunological tests support the assumption that he died of anaphylactic shock due to exposure to mite-contaminated food and/or exposure to mite allergens in bed.

Tay et al. [38] reported two cases of *Dermatophagoides farinae*-related anaphylaxis after eating mite-contaminated food. Patient A showed urticaria, throat discomfort, acute dyspnea, and severe wheezing 20 min after the ingestion of home-made fried fish and prawns coated with Japanese tempura flour, while Patient B developed bilateral periorbital angioedema, wheezing, and a feeling of impending doom 1 h after the ingestion of home-baked scones. Both patients had a history of allergic rhinitis and showed a strong positive result for dust mites and for extracts prepared from the ingested flour in skin prick tests. Microscopic examination of flour samples revealed large numbers of live *Dermatophagoides farinae*.

Iglesias-Souto et al. [40] documented a case of storage mite *Thyreophagus entomophagus*-related allergy in a 13-year-old boy. The patient presented with wheals, itching and diffuse erythema, cough, and wheeze immediately after consuming a home-made crêpe, made with wheat flour, which was stored in the kitchen for weeks. He had a previous history of mild persistent allergic rhinoconjunctivitis and sensitization to house dust mite and facial angioedema, urticaria, and bronchospasm after Ibuprofen, but not any history about food allergy. Microscopic examination of the culprit flour revealed mite contamination by *Thyreophagus entomophagus* (104 mites/g). Specific IgE against the extract of *Thyreophagus entomophagus* was also demonstrated *in vitro* by specific immunological assays. His symptoms disappeared in 2 h after treatment with intravenous fluids, diphenhydramine, epinephrine, and methylprednisolone.

75.1.2.2 Pathogenesis

Dust mites generate at least 22 allergens that are responsible for causing a variety of allergic symptoms in humans [41–49]. These allergens share the biochemical properties and functions with cysteine protease, trypsin, amylase, chymotrypsin, glutathione-*S*-transferase, collagenolytic serine protease, tropomyosin, fatty acid-binding protein, vitellogenin/apolipophorin-like protein, chitinase, gelsolin, Ca-binding EF protein, and antimicrobial peptide [50–55].

In *Dermatophagoides pteronyssinus* extracts, 14 groups of allergens (Der p1–14) have been identified. Der p1 is a major dust mite allergen that is produced in the guts and appears in the droppings of *Dermatophagoides pteronyssinus*, *Dermatophagoides farinae*, *Dermatophagoides microceaus*, *Dermatophagoides siboney*, *Euroglyphus maynei*, and *Blomia tropicals*. Being a glycoprotein with sequence homology and thiol protease function similar to the enzymes papain (a cysteine protease), actinidin (also a cysteine protease), bromelain (a proteolytic enzyme), and cathepsins B and H, Der p1 has the ability to break down the adhesion molecules binding the epithelial cells together, increase the permeability of the lung, and breach its defense system. This leads to apoptosis (cell death) in the lungs, which in turn gives Der p1 access to the other parts of body, including bloodstream and the placenta, and facilitates further invasion by inhaled allergens or other microorganisms. Der p1 is considered a major allergen since many people are sensitive or "specifically" allergic to its harmful effects [23].

Besides the lungs, the eye represents another common site for the development of allergic inflammation as allergens and other irritants are easily deposited directly onto its surface. Without a mechanical barrier to prevent the impact that allergens create, the eye is relatively defenseless except for the release of repellent mediators (e.g., histamine, which makes the eye red, sore, or itchy) and tears (which may even carry allergens absorbed from other parts of the body onto the eye).

The nose is also a vulnerable site for allergic response and malaise. The nose is connected to the ear by the Eustachian tube, whose main functions are to help balance the air pressure on the inside of the eardrum, and to drain away any fluids produced by the ear itself. In response to various allergens, the nose secretes an increasing amount of mucus, which disrupts the normal function of the nose and blocks the Eustachian tube as well as traps mucus and the air in the middle ear. In addition to allergic rhinitis and otitis media, this may lead to a certain degree of deafness (referred to as "glue ear"), with very young children showing a tendency to scratch or pull at their ears.

Another common site of allergic reaction is the skin, in which constant scratching may contribute to secondary microbial infections (e.g., *Staphylococcus aureus*), leading to eczema and the weeping or crusting of diseased skin.

75.1.2.3 Treatment

There are three ways to control and prevent the harmful effects of dust mites: reduction of the mite population, avoidance of dust mites and allergens, and immunotherapy using mite extracts [56–58].

1. Reduction of the mite population. Dust mites in patient's home and surroundings can be reduced or eliminated through boiling/washing, freezing, spraying, exclusion, or drying. Boiling/washing helps remove dust mites and droppings from linen and clothes, as temperatures over 60°C are fatal for dust mites. Freezing (at <0°C) also kills dust mites. Spraying with benzyl benzoate or using disodium octaborate tetrahydrate powder also helps eradicate dust mites at home and surroundings [59]. Exclusion of mite larvae and droppings (allergens) is possible with the use of bedding/cloth pore size woven to <10 μm to stop them from passing through fabric. Drying by increasing air circulation and ventilation reduces the humidity (<51%) also kills dust mites.
2. Avoidance of dust mites and allergens. Use of bedding acts as a barrier to the dust mite and its allergens represents a good first step. The bedding should also be woven to 10 μm or less, breathable, and resistant to frequent washing. Wall-to-wall carpet should be reduced, or replaced with easy-to-clean, smooth hard surfaces. Mattresses, carpets, sofas, and chairs should be vacuum cleaned on a regular basis. Increased air circulation and ventilation throughout the house will reduce humidity levels and thus the number of dust mites. Mites take in oxygen through tiny holes located on their shell-like covering.
3. Immunotherapy using mite extracts. This is another useful option to increase resistance to dust mite allergens. Recent reports suggested that high-dose *D. farinae* sublingual immunotherapy (SLIT) increases serum D farinae-specific IgG4 levels, and improves the bronchial threshold to allergen challenge [60], and that a codon-optimized DNA vaccine encoding mature Der p1 in a lipoplex formulation represents a promising hypoallergenic vaccine candidate for safer immunotherapy against house dust mite allergy [61].

75.1.3 Laboratory Diagnosis

The presence of dust mites in a sample can be easily observed by putting a small amount of dust on the surface of water and examining under 20× magnification. Live and dead mites will remain on the surface of the water. Alternatively, 0.05 g sample of dust is placed in 30 mL of a saturated NaCl solution, and five drops of detergent is added. The dust is teased apart and subjected to ultrasonic treatment for 20 min. The suspension is rinsed through a 45 μm mesh sieve, stained with crystal violet, and examined under a stereomicroscope. Another way to detect the presence of dust mites is to leave A4 sheets of white paper on places where the dust mites are suspected. The paper is examined in the mornings and evenings by holding the paper over a lamp, as the dust mites will cast a shadow [62].

For species-specific identification of astigmatid mites, microscopic examination of their morphological characteristics of mature and immature stages along with their behavioral, ecological, and biochemical properties is helpful [10,63,64]. However, due to the close morphological similarity among various dust mite species (e.g., *D. farinae* and *D. microceras*, or *Acarus* and *Tyrophagus*), it is difficult even for a highly trained acorologist to accurately determine the species identity of these organisms. In addition, identification of mites based on morphology is even more challenging when only one gender is obtained, or when the specimen is not intact with its legs missing.

To test for allergy triggers in patients, skin prick testing or blood test is often used [65,66]. Skin prick testing measures the sensitivity of IgE-driven histamine reactivity in the host to foreign proteins (allergens) that may have found their way into the body. It involves a safe, slight pin-prick delivery of an allergen or allergens just under the skin (usually on the arm). Patients undertaking this type of allergy test need to avoid using antihistamine medication for several days prior to the test. A positive reaction (usually a red itchy wheal or bump) indicates the presence of IgE-released histamine, but not necessarily the confirmation of an allergic disease. People can show positive skin prick test reactions without allergy symptoms. Blood test (often called RAST, CAP-RAST, or specific-IgE) measures the amount of circulating levels of IgE antibody in the blood to specific allergens (e.g., house dust mites) [67–69]. Unlike skin prick testing, there is no need to stop taking antihistamines for blood test [9,64].

Recent developments in molecular biology offer unprecedented opportunity to improve the speciation and phylogenetic analysis of dust mites [70–72]. Among the gene regions targeted by molecular identification procedures, the ribosomal RNA (rRNA) region has proven very useful. The rRNA region in eukaryotes comprises three highly conserved regions (18S rRNA, 5.8S rRNA, and 28S rRNA) and two rapidly evolving regions [the internal transcribed spacer regions (ITS1 and ITS2)]. While the conserved regions are valuable for the genus- and family-level verification, the latter rapidly evolving regions are useful for species-specific determination and low-level phylogenetic relationships [73–76]. The other gene regions of interest include mitochondrial genes (cytochrome oxidase, 12S and 16S rRNA) [77]. Being highly conserved, the central part of the mitochondrial cytochrome oxidase subunit I (mtCOI) region is frequently focused in the investigation of intra- and interspecific variations [78]. The mitochondrial 12S rRNA and 16S rRNA genes are also relatively conserved and informative for discrimination between species belonging to different families, although the 12S rRNA sequence analysis is not sufficiently revealing in the differentiation between the families Pyroglyphidae and Acaridae [79].

75.2 METHODS

75.2.1 Sample Preparation

75.2.1.1 Sample Collection

Samples (sample area 1 m^2) are collected with portable vacuum cleaners, on a specially constructed dust trap filter attached to the end of the hose. A new filter is used for each sample, and samples are kept separately. After vacuuming, dust samples are weighed in a 150 mL beaker, soaked in 75% ethanol for 4 h, suspended in a saturated NaCl solution and a few drops of soap, stirred with a magnetic stirrer, and held in flotation for 24 h. Then supernatants are poured over a filter paper. Sediments are again suspended in NaCl solution—the procedure was repeated at least three times. Filters with the material retained are placed in Petri dishes onto which a saturated NaCl solution is poured. The surface of the liquid and of the filter paper is carefully examined for mites under a binocular stereomicroscope, starting 1–2 h after pouring. Mites are mounted on microscope slides in Hoyer's medium. Damaged mites are assumed dead at the time of sampling, whereas intact mites are assumed alive at this time. Mite density of each taxon is calculated as number of specimens per g of dust or as number of specimens per sample [33].

For maintenance of house dust and storage mites, *Dermatophagoides farinae*, *Dermatophagoides pteronyssinus*, and *Blomia tropicalis* are reared at 70% relative humidity on a mixture of dry yeast and dried fish meat (1:1). *Carpoglyphus lactis* is maintained on a mixture of dry yeast and sugar (1:1). *Suidasia medanensis*, *Aleuroglyphus ovatus*, *Tyrophagus* sp., and *Histiostoma laboratorium* are kept submerged and reared on dry yeast at 70% relative humidity. Other dust mites are reared on an agar medium composed of dry yeast and corn powder in a Petri dish (85 mm in diameter, 20 mm in height).

Alternatively, laboratory cultures of the storage mite species *A. siro*, *Acarus gracilis*, *T. putrescentiae*, *Tyrophagus brevicrinatus*, *Tyroborus lini*, *Aleuroglyphus ovatus*, *Caloglyphus redikorzevi*, *C. oudemansi*, *Lepidoglyphus destructor*, *Glycyphagus domesticus*, *Aeroglyphus robustus*, *Chortoglyphus arcuatus*, and *Carpoglyphus lactis* are reared in darkness at 20°C and 85% relative humidity. A mixture of powdered oat flakes, wheat germ, and yeast in the ratio 5:5:1 is used as breeding diet. Small breeding cells (1 cm in diameter) are established for egg collection in which five females of each species are allowed to oviposit for 1–3 days and are then removed. The eggs are collected gently with a brush, cleaned in a drop of distilled water, and subjected to examination with a light or electron microscope [80].

To isolate individual mites, 8.6 g of live ready cultures is added to 30 g fish food flakes (Tetra) previously weighted and treated with UV (three rounds of 90 s) in a crystal 9″ round plate. Dust mites are grown for 20 days and then processed in a 300 mm sieve with exposure to the heat generated by a 40 W bulb. The live mites migrate from the sieve located on the top of a funnel to a 50 mL sterile tube located at the bottom (heat-escaped bodies). The collected bodies are immediately frozen in liquid nitrogen and stored at 20°C prior to DNA extraction procedures.

75.2.1.2 DNA Isolation

Mite DNA is isolated from single individuals preserved in 100% ethanol after field collection or *in vitro* culture. After removing ethanol by evaporation, single individuals are coarsely crushed with a plastic pestle in a 1.5 mL microcentrifuge tube containing 200 μL of extraction buffer (2% CTAB, 1.4 NaCl, 0.2% 2-β-mercaptoethanol, 20 mM EDTA, 100 mM Tris-HCl, pH 8). The tube is incubated at 60°C for 45 min. Proteins are removed with one volume of chloroform:isoamyl alcohol (24:1). DNA is precipitated with one volume of isopropanol, and the pellet is washed with cold 70% ethanol. The pellet is resuspended in 20 μL of double-distilled water [70].

Alternatively, the mite cultures are harvested with a heat escape method. The mites are resuspended in phosphate-buffered saline (PBS) and then homogenized with Microsmash (Tomy) in the presence of glass beads (1.0 mm diameter; Tomy) at 5000 rpm for 15 s (or ground into a fine particulate matter using a previously cooled mortar and pestle). The homogenate is cooled at −80°C for 3 min before being subjected to further homogenization for nine cycles. Genomic DNA is extracted with a Qiagen DNeasy tissue kit (Qiagen) according to the manufacturer's instruction. Purity and concentration of the DNA are determined using a spectrophotometer at 260 and 280 nm wavelength. Pure DNA has an A_{260}/A_{280} ratio of 1.8–2.0. The extracted DNA is either used directly or stored at −20°C [75].

75.2.2 Detection Procedures

75.2.2.1 PCR–RFLP Identification of Dust Mites

Wong et al. [75] described a polymerase chain reaction–restriction fragment length polymorphism (PCR–RFLP) targeting the ITS2 gene region for the species-specific identification of dust mites *Dermatophagoides pteronyssinus*, *D. farinae*, *Blomia tropicalis*, *Tyrophagus putrescentiae*, *Aleuroglyphus ovatus*, and *Glycycometus malaysiensis*. Use of the ITS2 forward (5′-CGACTTTCGAACGCATATTGC-3′) and reverse (5′- GCTTAAATTCAGGGGGTAATCTCG-3′) primers allows amplification of 260–490-bp products from these species. Subsequent digestion with restriction enzymes facilitates species-specific identification of dust mites.

75.2.2.1.1 Procedure

1. PCR mixture (50 μL) is made up of 25 μL of Taq Master Mix (Qiagen), 0.25 μL of each ITS2 forward and reverse primer (100 μM), 23.5 μL of ultrapure water, and 2 ng DNA.
2. The reaction mixture is subjected to an initial 94°C for 2 min, followed by 25 cycles of 94°C for 30 s, 56°C for 30 s, and 72°C for 50 s.
3. PCR products (5 μL) are analyzed by size fractionation using 1% agarose gels stained with 1 μL of ethidium bromide (10 mg/mL) in TBE buffer (89 mM Tris-borate, 89 mM boric acid, and 2 mM EDTA) with an electrophoresis apparatus (Mini Sub-Cell GT, Bio-Rad, USA) at 80 V for 45 min. The gel is visualized by a gel documentation system (Biovision) and photographed with Biovision 1000 superbright transilluminator software.
4. PCR products (8 μL each) are digested individually with restriction enzymes *Hae*III (10 U), *Bfa*I (5 U), *Dde*I (10 U), *Alu*I (10 U), *Hin*fI (10 U), and *Ple*I (5 U) (New England Biolabs). The mixtures are in a total volume of 20 μL each and incubated overnight at 37°C.
5. The resulting restriction fragments were separated with sodium dodecyl sulfate–polyacrylamide gel electrophoresis (SDS–PAGE) using Mini PROTEAN Tetra (Bio-Rad) at a constant voltage of 80 V. The resolving gel contained final concentrations of 12% acrylamide, *N,N*′-methylene-bis-acrylamide, 0.375 M Tris-HCl (pH 8.8), 0.1% SDS, 0.05% TEMED, and 0.05% ammonium persulfate (APS). Electrophoresis is performed in the electrophoresis running buffer (pH 8.3). The gel is stained with ethidium bromide, visualized, and photographed using a gel documentation system (Biovision).

Note: The ITS2 region for house dust mites (*D. pteronyssinus*, *D. farinae*, and *B. tropicalis*) and storage mites (*T. putrescentiae*, *A. ovatus*, and *G. malaysiensis*) ranges from 260 to 490 bp, that is, *Dermatophagoides pteronyssinus* (324 bp), *D. farinae* (335 bp), *Blomia tropicalis* (326 bp), *Glycycometus malaysiensis* (256/261 bp), *Tyrophagus putrescentiae* (488 bp), and *Aleuroglyphus ovatus* (432 bp). The amplified products from *T. putrescentiae* and *A. ovatus* are larger (>400 bp) than those from the rest of the species and can be easily distinguished from the rest with agarose gel electrophoresis.

Based on characteristic restriction enzyme digestion patterns, AluI, HinfI, and PleI differentiate *D. pteronyssinus* from *B. tropicalis*. BfaI, HinfI, and PleI digestions differentiate *D. pteronyssinus* from *G. malaysiensis*. HaeIII, HinfI, and PleI digestions differentiate *D. pteronyssinus* from *T. putrescentiae*. HaeIII and HinfI digestions differentiate *D. pteronyssinus* from *A. ovatus*. AluI digestion differentiates *D. farinae* from *B. tropicalis*. BfaI digestion differentiates *D. farinae* from *G. malaysiensis*. HaeIII digestion distinguishes *D. farinae* from *T. putrescentiae*. HaeIII and PleI digestions differentiate *D. farinae* from *A. ovatus*. HaeIII and AluI digestions differentiate *B. tropicalis* from *T. putrescentiae*. HaeIII, AluI, and PleI digestions differentiate *B. tropicalis* from *A. ovatus*. HaeIII and BfaI digestions differentiate *G. malaysiensis* for *T. putrescentiae*. HaeIII, BfaI, and PleI digestions differentiate *G. malaysiensis* from *A. ovatus*. HinfI and PleI digestions differentiate *D. pteronyssinus* from *D. farinae*. BfaI and AluI digestions differentiate *B. tropicalis* and *G. malaysiensis*. PleI digestion differentiates *T. putrescentiae* from *A. ovatus*. BfaI digestion differentiates *G. malaysiensis* from the rest of the species.

75.2.2.2 Sequence Analysis of ITS2 Region

Yang et al. [76] utilized primers (forward, 5′-CGACTTTCG AACGCATATTGC-3′; reverse, 5′-GCTTAAATTCAGGGG GTAATCTCG-3′) spanning a portion of the 5.8S recombinant RNA (rRNA), the full-length ITS2 region, and a part of the 28S rRNA for PCR amplification and sequence analysis of dust mites.

75.2.2.2.1 Procedure

1. PCR mixture (50 µL) contains 1.25 U *Pyrobest* DNA Polymerase (TaKaRa Bio), 1× *Pyrobest* buffer, 0.1 mM dNTP mixture, 0.4 µM each of the two oligonucleotide primers, and 5 ng template DNA.
2. Amplification is conducted with an initial 94°C for 3 min, followed by 35 cycles at 94°C for 30 s, 55°C for 30 s, and 72°C for 1 min, and a final 72°C for 7 min.
3. The PCR products are sequenced using primers mentioned above.
4. The nucleotide sequences are aligned using the multiple alignment program ClustalX (version 1.83), with the gap-opening penalty of 15 and the gap-extension penalty of 6.66. The fast pairwise alignment parameters are as follows: gap penalty = 5, k-tuple size = 2, top diagonals = 4, and window size = 4. The aligned sequences are also checked manually. Phylogenetic analyses are carried out using the PAUP* version 4.0b10 for maximum parsimony (MP) analysis and PHYLIP version 3.68 for neighbor-joining (NJ) analysis. In the MP analysis, each base is treated as an unordered character with equal weights, whereas gaps are treated as missing data. Statistical support for produced phylogeny is determined by bootstrap resampling of 1000 replicated datasets. A phylogenetic tree is produced by NJ analysis, based on the calculated values obtained from Kimura's two-parameter distance method, with 1000 bootstrap replications. Furthermore, the genetic distance is calculated using DNADIST from PHYLIP version 3.68 with Kimura's two-parameter distance method. For the MP and NJ analyses, two sequences from each species are chosen randomly as representative sequences. The sequences of *Tyrophagus longior* (AB104997) and *Acarus immobilis* (AY525562) are used as out-groups.

Note: The boundaries of the ITS2 sequences are estimated by comparison with those of *A. ovatus* (GenBank accession numbers AB104984 and EU078966) and *D. pteronyssinus* (GenBank accession numbers AB105008 and GU365873). The lengths of the ITS2 sequences vary between 316 and 488 bp. Analyses of the sequences allows identification of astigmatid mites with similar morphological characteristics.

75.2.2.3 Sequence Analysis of Mitochondrial 16S rRNA Gene

de Rojas et al. [77] targeted the mitochondrial 16S rRNA gene for differentiation and phylogenetic analysis of dust mites. The sequences of primers are as follows: forward, 5′-CCG GTC TGA ACT CAG ATC ATG-3′ and reverse, 5′-GCT CAA TGA TTT TTT AAA TTG CTG-3′.

75.2.2.3.1 Procedure

1. PCR mixture (50 µL) consists of 10 µL 10× PCR buffer, 1 µL 10 mM dNTP, 1.5 µL 50 mM Mg_2Cl, 1.5 µL each of forward and reverse primer 10 mM, 5 µL template DNA, 0.5 µL Taq DNA polymerase (2.5 U), and double-distilled water.
2. PCR conditions include an initial step at 94°C for 2 min; 35 cycles of 94°C for 20 s, 56°C for 30 s, and at 72°C for 45 s; and a final step at 72°C for 7 min.
3. PCR products are separated and excised from the agarose gel and DNA purified using the QIAEX II Gel Extraction Kit. The isolated DNA is cloned into pGEM T® easy vector (Promega). The recombinant clone containing the PCR product is purified using a Plasmid Midi Kit (Qiagen) and sequenced using automated DNA cycle sequencing (LiCor). Both strands of a single copy of 16S rRNA from three mites per species and origin are sequenced.
4. Sequence alignments are obtained using the PileUp program of the Wisconsin Genetics Computer Group (GCG) package. Alignments are adjusted according to the model for secondary structure of the mitochondrial 16S rRNA in *Haemaphysalis cretica*. The *Dermanyssus gallinae* sequence (Gen Bank access number L34326) is included as an outgroup. A distance matrix is generated using Kimura's two-parameter model (DNADIST) and the neighbor-joining method (NEIGHBOR) is applied to construct a phylogenetic tree. The solidity of nodes of the tree is tested by bootstrap analysis (SEQBOOT) using 1000 replicates. MP analysis is performed using DNAPARS. All the programs used for these analyses are part of the PHYLIP 3.5c package.

75.2.2.4 Sequence Analysis of Mitochondrial 12S rRNA Gene

Suarez-Martinez et al. [79] examined the partial mtDNA 12SrRNA sequences (391–396 bp) for elucidation of evolutionary relationships among the four dust mite families Pyroglyphidae (*Dermatophagoides pteronyssinus*), Glycyphagoidea (*Glycyphagus privatus*), Acaridae (*Aleuroglyphus ovatus*), and Echimyopodidae (*Blomia tropicalis*). The primer set consists of 12S forward primer 5′-TACTATGTTAGCACTTA-3′ and 12S reverse primer 5′-AAACTAGGATTAGATACCA-3′. The resulting sequence data permit discrimination between species belonging to different families of dust mites.

75.2.2.4.1 Procedure

1. PCR mixture (100 μL) contains 200 ng mite DNA, 22.5 pmol of each of the primers, 8 μL of 10 mM dNTPs, 10 μL of 10× *Taq* buffer, and 2.5 U *Taq*. Positive and negative control reactions are *Bombyx mori* silk gland DNA and sterile water, respectively.
2. PCR parameters include an initial 95°C for 10 min; 40 cycles of 95°C for 1 min, 38°C for 1 min, and 72°C for 1 min; and a final 72°C for 10 min. If not immediately electrophoresed, samples are stored at 22°C.
3. The resulting PCR products (391–396 bp) are separated by agarose gel electrophoresis, excised, and cloned into pCR2.1 linearized plasmid vector (TA Cloning kit, Invitrogen). Recombinant plasmids containing PCR products are sequenced using the ABI Prism Big Dye Terminator Cycle Sequencing Kit, M13 forward or reverse primers (Promega). The sequencing reactions are performed in a 9600 Perkin-Elmer thermocycler for 30 cycles as specified by the ABI Prism Protocol. The cycle sequencing products are ethanol precipitated, vacuum-dried, and stored at 22°C until needed. Sequences are obtained using the ABI Prism 377 Sequencer (Perkin-Elmer). The resultant sequences are edited and analyzed using the ABI Autoassembler Program. A BLAST data bank search confirms the obtained sequencing product.

75.3 CONCLUSION

House dust mites (e.g., *Dermatophagoides pteronyssinus*, *Dermatophagoides farinae*, and *Euroglyphus maynei*) and storage mites (e.g., *Blomia tropicalis*) are ubiquitously present in soil, stored products, live plants, and dwellings. As scavengers, these organisms live on dander, shed skin particles (containing collagen and keratin), excretions and secretions of humans and animals as well as grains, fungi, and some bacteria. Although they are not directly involved in human disease, the carcasses, body parts, and droppings of these mites contain a variety of allergens that are capable of provoking host's immune systems and inducing various immunological disorders in humans [81–84].

In order to implement effective control and prevention measures, it is critical to correctly identify house dust mites and storage mites. Due to their close resemblances, identification of mites on the basis of morphological characteristics is challenging and requires considerable specialized skills. Application of molecular techniques such as PCR and sequencing provides a reliable means for their differentiation with minimal acarological expertise. In addition, sequence analysis of relatively conserved ribosomal and mitochondrial RNA genes open a valuable approach for the elucidation of phylogenetic relationship among various mite species.

REFERENCES

1. Wharton, G.W., House dust mites. *J. Med. Entomol.*, 12, 577, 1976.
2. van Hage-Hamsten, M., and Johansson, S.G., Storage mites. *Exp. Appl. Acarol.*, 16, 117, 1992.
3. van Hage-Hamsten, M. et al., Storage mite allergy is common in a farming population. *Clin. Allergy*, 15, 555, 1985.
4. van Hage-Hamsten, M., *Dermatophagoides siboney* and *Blomia tropicalis*—dust mites of subtropical and tropical areas. *Clin. Exp. Allergy*, 25, 905, 1995.
5. Thomas, W.R., Hales, B.J., and Smith, W., *Blomia tropicalis*: More than just another source of mite allergens. *Clin. Exp. Allergy*, 33, 416, 2003.
6. van Bronswijk, J.E., de Cock, A.W., and Oshima, S., The genus *Blomia* Oudemans (Acari: Glycyphagidae) I. Description of *Blomia tropicalis* sp. n. from house dust in tropical and subtropical regions. *Acarologia*, 15, 477, 1974.
7. Colloff, M.J., and Spieksma, F.T., Pictorial keys of the identification of domestic mites. *Clin. Exp. Allergy*, 22, 823, 1992.
8. Colloff, M.J., Taxonomy and identification of dust mites. *Allergy*, 53 (48 suppl.), 7, 1998.
9. Cui, Y.B. et al., Phylogenetic analysis of house dust mites. *Central Eur. J. Med.*, 5, 69, 2009.
10. Kucerova, Z., and Stejskal, V., Morphological diagnosis of the eggs of stored-products mites. *Exp. Appl. Acarol.*, 49, 173, 2009.
11. Suggars, A.L., House dust mites: A review. *J. Entomol. Sci.*, supplement 1, 3, 1987.
12. Suto C. et al., Distribution and abundance of house dust mites, *Dermatophagoides* spp. (Acari: Pyroglyphidae), influenced by environmental conditions in wooden and apartment houses in Nagoya. *Jpn. J. Sanit. Zool.*, 42, 255, 1991.
13. Babe, K.S. Jr. et al., House dust mite (*Dermatophagoides farinae* and *Dermatophagoides pteronyssinus*) prevalence in the rooms and hallways of a tertiary care hospital. *J. Allergy Clin. Immunol.*, 95, 801, 1995.
14. Morsy, T.A. et al., Four species of house dust mites recovered from houses of patients with allergic respiratory diseases. *J. Egypt. Soc. Parasitol.*, 25, 195, 1995.
15. de Oliveira, C.H. et al., Comparison of house dust mites found on different mattress surfaces. *Ann. Allergy Asthma Immunol.*, 91, 559, 2003.
16. Arlian, L.G., Neal, J.S., and Bacon S.W., Survival, fecundity and development of *Dermatophagoides farinae* (Acari: Pyroglyphidae) at fluctuating relative humidity. *J. Med. Entomol.*, 35, 962, 1998.
17. O'Connor, B.M., Evolutionary ecology of astigmatid mites. *Annu. Rev. Entomol.*, 27, 385, 1982.

18. Fernández-Caldas, E., and Lockey, R.F., *Blomia tropicalis*, a mite whose time has come. *Allergy*, 59, 1161, 2004.
19. Thind, B.B., and Clarke, P.G., The occurrence of mites in cereal-based foods destined for human consumption and possible consequences of infestation. *Exp. Appl. Acarol.*, 25, 203, 2001.
20. Müsken, H. et al., Sensitization to different mite species in German farmers: Clinical aspects. *J. Investig. Allergol. Clin. Immunol.*, 10, 346, 2000.
21. Arlian, L.G. et al., Cross-antigenicity between the scabies mite, *Sarcoptes scabiei*, and the house dust mite, *Dermatophagoides pteronyssinus*. *J. Invest. Dermatol.*, 96, 349, 1991.
22. Arlian, L.G. et al., Allergenicity of the mite, *Blomia tropicalis*. *J. Allergy Clin. Immunol.*, 91, 1042, 1993.
23. Arlian, L.G., Morgan, M.S., and Neal, J.S., Dust mite allergens: Ecology and distribution. *Curr. Allergy Asthma Rep.*, 2, 401, 2002.
24. Arlian, L.G. et al., Cross-reactivity between storage and dust mites and between mites and shrimp. *Exp. Appl. Acarol.*, 47, 159, 2009.
25. Erben, A.M. et al., Anaphylaxis after ingestion of beignets contaminated with Dermatophagoides farinae. *J. Allergy Clin. Immunol.*, 92, 846, 1993.
26. Matsumoto, T. et al., Systemic anaphylaxis after eating storage-mite-contaminated food. *Int. Arch. Allergy Immunol.*, 109, 197, 1996.
27. Stanaland, B.E. et al., Positive nasal challenge response to *Blomia tropicalis*. *J. Allergy Clin. Immunol.*, 97, 1045, 1996.
28. Li, C.P. et al., Acaroid mite, intestinal and urinary acariasis. *World J. Gastroenterol.*, 9, 874, 2003.
29. Milián, E., and Díaz, A.M., Allergy to house dust mites and asthma. *P.R. Health Sci. J.*, 23, 47, 2004.
30. Wen, D.C. et al., Systemic anaphylaxis after the ingestion of pancake contaminated with the storage mite *Blomia freemani*. *Ann. Allergy Asthma Immunol.*, 95, 612, 2005.
31. Kuljanac, I., The role of *Dermatophagoides pteronyssinus* in atopic dermatitis. *Acta Dermatovenerol. Croat.*, 14, 86, 2006.
32. Nuttall, T.J. et al., House dust and forage mite allergens and their role in human and canine atopic dermatitis. *Vet. Dermatol.*, 17, 223, 2006.
33. Solarz, K., Indoor mites and forensic acarology. *Exp. Appl. Acarol.*, 49, 135, 2009.
34. Sade, K., Roitman, D., and Kivity, S., Sensitization to *Dermatophagoides*, *Blomia tropicalis*, and other mites in atopic patients. *J. Asthma*, 47, 849, 2010.
35. Spiegel, W.A. et al., Anaphylaxis associated with dust mite ingestion. *Ann. Allergy*, 72, 56, 1994.
36. Matsumoto, T., and Satoh, A., The occurrence of mite-containing wheat flour. *Pediatr. Allergy Immunol.*, 15, 469, 2004.
37. Sanchez-Borges, M., Capriles-Hulett, A., and Caballero-Fonesca, F., Oral mite anaphylaxis (pancake syndrome) also observed in children. *Ann. Allergy Asthma Immunol.*, 96, 755, 2006.
38. Tay, S.Y. et al., Anaphylaxis following the ingestion of flour contaminated by house dust mites—a report of two cases from Singapore. *Asian Pac. J. Allergy Immunol.*, 26, 165, 2008.
39. van Hage-Hamsten, M., and Johansson, E., Clinical and immunologic aspects of storage mite allergy. *Allergy*, 53, 49, 1998.
40. Iglesias-Souto, J. et al., Oral mite anaphylaxis by *Thyreophagus entomophagus* in a child: A case report. *Clin. Mol. Allergy*, 7, 1, 2009.
41. Griffin, P. et al., Allergenic and antigenic relationship between three species of storage mite and the house dust mite, *Dermatophagoides pteronyssinus*. *J. Allergy Clin. Immunol.*, 84, 108, 1989.
42. Puerta, L. et al., Cross-reactivity among *Blomia tropicalis, Lepidoglyphus destructor* and *Dermatophagoides farinae. J. Allergy Clin. Immunol.*, 87, 190, 1991.
43. Arruda, L.K. et al., Sensitisation to *Blomia tropicalis* in patients with asthma: Molecular cloning and expression of a major allergen (Blo t 5) with homology to *Dermatophagoides pteronyssinus* Der p 5. *Am. J. Respir. Crit. Care Med.*, 155, 343, 1997.
44. Simpson, A., Arruda, K., and Chapman, M., Antigenic interrelationships among mite allergens (*Blomia* and *Dermatophagoides* spp). *Clin. Rev. Allergy Immunol.*, 15, 461, 1997.
45. Flores, I. et al., Cloning and molecular characterization of a cDNA from *Blomia tropicalis* homologous to dust mite group 3 allergens (trypsin-like proteases). *Int. Arch. Allergy Immunol.*, 130, 12, 2003.
46. Mora, C. et al., Cloning and expression of Blo t 1, a novel allergen from the dust mite *Blomia tropicalis*, homologous to cysteine proteases. *Clin. Exp. Allergy*, 33, 28, 2003.
47. Boquete, M. et al., *In vivo* and *in vitro* allergenicity of the domestic mite *Chortoglyphus arcuatus*. *Ann. Allergy Asthma Immunol.*, 97, 203, 2006.
48. Chua, K.Y. et al., The *Blomia tropicalis* allergens. *Protein Pept. Lett.*, 14, 325, 2007.
49. Liao, E.C. et al., *Dermatophagoides pteronyssinus* and *Tyrophagus putrescentiae* allergy in allergic rhinitis caused by cross-reactivity not dual-sensitization. *J. Clin. Immunol.*, 30, 830, 2010.
50. O'Brien, R. et al., Immunologic responses following respiratory sensitization to house dust mite allergens in mice. *Immunol. Cell Biol.*, 74, 174, 1996.
51. Morgan, M.S., Arlian, L.G., and Fernández-Caldas, E., Cross-allergenicity of the house dust mites *Euroglyphus maynei* and *Blomia tropicalis*. *Ann. Allergy Asthma Immunol.*, 77, 386, 1996.
52. Caraballo, L. et al., Cloning and IgE binding of a recombinant allergen from the mite *Blomia tropicalis*, homologous with fatty acid-binding proteins. *Int. Arch. Allergy Immunol.*, 112, 341, 1997.
53. Chew, F.T. et al., Allergenic differences between the domestic mites *Blomia tropicalis* and *Dermatophagoides pteronyssinus*. *Clin. Exp. Allergy*, 29, 982, 1999.
54. Yi, F.C. et al., Identification of shared and unique immunoglobulin E epitopes of the highly conserved tropomyosins in *Blomia tropicalis* and *Dermatophagoides pteronyssinus*. *Clin. Exp. Allergy*, 32, 1203, 2002.
55. Yi, F.C. et al., Molecular cloning of *Blomia tropicalis* allergens—a major source of dust mite allergens in the tropics and subtropics. *Inflamm. Allergy Drug Targets*, 5, 261, 2006.
56. Boner, A.L. et al., Skin and serum reactivity to some storage mites in children sensitive to Dermatophagoides pteronyssinus. *Ann. Allergy*, 63, 82, 1989.
57. Boner, A., Pescollderungg, L., and Silverman, M., The role of house dust mite elimination in the management of childhood asthma: An unresolved issue. *Allergy*, 57 Suppl 74, 23, 2002.
58. Sheikh, A., and Hurwitz, B., House dust mite avoidance measures for perennial allergic rhinitis: A systematic review of efficacy. *Br. J. Gen. Pract.*, 53, 318, 2003.

59. Mitchell, E.B. et al., Reduction of house dust mite allergen levels in the home: Uses of the acaracide, pirimiphos methyl. *Clin. Allergy*, 15, 235, 1985.
60. Bush, R.K. et al., House dust mite sublingual immunotherapy: Results of a US trial. *J. Allergy Clin. Immunol.*, 127, 974, 2011.
61. Pulsawat, P. et al., Production and immunogenicity of hypoallergenic codon-optimized DNA vaccine encoding mature Der p 1 allergen. *J. Investig. Allergol Clin. Immunol.*, 20, 582, 2010.
62. Hirschmann, W., Simple method for identifying live house dust mites. *Pract. Pest Controller*, 38, 1, 1986.
63. Fernández-Caldas, E., Iraola, V., and Carnés, J., Molecular and biochemical properties of storage mites (except *Blomia* species). *Protein Pept. Lett.*, 14, 954, 2007.
64. Dunn, J.A. et al., Rapid method for the detection of storage mites in cereals: Feasibility of an ELISA based approach. *Bull. Entomol. Res.*, 98, 207, 2008.
65. Arruda, L.K., and Chapman, M.D., A review of recent immunochemical studies of *Blomia tropicalis* and *Euroglyphus maynei* allergens. *Exp. Appl. Acarol.*, 16, 129, 1992.
66. Jung, Y.G. et al., Comparison of the skin-prick test and Phadia ImmunoCAP as tools to diagnose house-dust mite allergy. *Am. J. Rhinol. Allergy*, 24, 226, 2010.
67. Simpson, A. et al., Skin test reactivity to natural and recombinant *Blomia* and *Dermatophagoides* spp. allergens among mite allergic patients in the UK. *Allergy*, 58, 53, 2003.
68. Puccio, F.A. et al., Importance of including *Blomia tropicalis* in the routine diagnosis of Venezuelan patients with persistent allergic symptoms. *Allergy*, 59, 753, 2004.
69. Arias-Irigoyen, J. et al., Limited IgE cross-reactivity between *Dermatophagoides pteronyssinus* and *Glycyphagus domesticus* in patients naturally exposed to both mite species. *J. Allergy Clin. Immunol.*, 120, 98, 2007.
70. Navajas, M., and Fenton, B., The application of molecular markers in the study of diversity in acarology: A review. *Exp. Appl. Acarol.*, 24, 751, 2000.
71. Angus, A.C., Ong, S.T., and Chew, F.T., Sequence tag catalogs of dust mite-expressed genomes: Utility in allergen and acarologic studies. *Am. J. Pharmacogenomics*, 4, 357, 2004.
72. Jeyaprakash, A., and Hoy, M.A., A DNA extraction procedure that allows mite specimens to be slide mounted: Phytoseiid species evaluated as a model. *Exp. Appl. Acarol.*, 52, 131, 2010.
73. Noge, K. et al., Identification of astigmatid mites using the second internal transcribed spacer (ITS2) region and its application for phylogenetic study. *Exp. Appl. Acarol.*, 35, 29, 2005.
74. de Rojas, M. et al., Utility of ITS1–5.8S–ITS2 and 16S mitochondrial DNA sequences for species identification and phylogenetic inference within the *Rhinonyssus coniventris* species complex (Acari: Rhinonyssidae). *Parasitol. Res.*, 100, 1041, 2006.
75. Wong, S.F. et al., Molecular identification of house dust mites and storage mites. *Exp. Appl. Acarol.*, 55, 123, 2011.
76. Yang, B., Cai, J., and Cheng, X., Identification of astigmatid mites using ITS2 and COI regions. *Parasitol. Res.*, 108, 497, 2011.
77. de Rojas, M. et al., Phylogenetic relationships in rhinonyssid mites (Acari: Rhinonyssidae) based on mitochondrial 16S rDNA sequences. *Exp. Appl. Acarol.*, 25, 957, 2001.
78. Ros, V.I.D., and Breeuwer, J.A.J., Spider mite (Acari: Tetranychidae) mitochondrial COI phylogeny reviewed: Host plant relationships, phylogeography, reproductive parasites and barcoding. *Exp. Appl. Acarol.*, 42, 239, 2007.
79. Suarez-Martinez, E.B. et al., Molecular identification of pathogenic house dust mites using 12S rRNA sequences. *Electrophoresis*, 26, 2927, 2005.
80. Miyamoto, J., and Ishii, A., and Sasa, M., A successful method for mass culture of the house dust mite, *Dermatophagoides pteronyssinus* (Troussart, 1897). *Jpn. J. Exp. Med.*, 45, 133, 1975.
81. Ferrándiz, R., Casas, R., and Dreborg, S., Sensitization to *Dermatophagoides siboney*, *Blomia tropicalis*, and other domestic mites in asthmatic patients. *Allergy*, 51, 501, 1996.
82. Puerta, L. et al., Nucleotide sequence analysis of a complementary DNA coding for a Blomia tropicalis allergen. *J. Allergy Clin. Immunol.*, 98, 932, 1996.
83. Wayne, R. et al., Characterization and immunobiology of house dust mite allergens. *Int. Arch. Allergy Immunol.*, 129, 1, 2002.
84. Edston, E., and van Hage-Hamsten, M., Death in anaphylaxis in a man with house dust mite allergy. *Int. J. Legal Med.*, 117, 299, 2003.

76 Sarcoptes (Itch Mite)

Dongyou Liu

CONTENTS

76.1 INTRODUCTION

The genus *Sarcoptes* consists of a single mite species *Sarcoptes scabiei* that can be separated into over 15 varieties. *S. scabiei* varieties are responsible for causing acariasis, a skin condition often known as scabies (from Latin word scabere, meaning scratch) in humans and sarcoptic mange in animals.

Scabies has been known to humankind since ancient times. Aristotle (384–322 BC) utilized the term "akari" to describe scabies mite, and the term "scabies" was used to depict the disease and its characteristic features during Roman period. *Sarcoptes scabiei*, the etiologic agent for scabies, was discovered in mid-1800s.

Being an obligate parasite, *S. scabiei* (scab mite) burrows into the skin of humans and other hosts and lives on host epidermis and sera. During the process, scab mite and its metabolic products elicit strong inflammatory and allergy-like reactions in the host, leading to intense itching and secondary skin infections followed by weight loss and unthriftiness.

Given its widespread distribution and its propensity to cause severe irritation in humans and economically important animals, it is essential that *S. scabiei* is correctly identified and promptly treated. While microscopic examination of scabies specimens remains an indispensable tool for the diagnosis of the disease, molecular techniques offer a potentially more sensitive and accurate approach for the identification and phylogenetic analysis of *S. scabiei*.

76.1.1 Classification, Morphology, and Biology

76.1.1.1 Classification

The genus *Sarcoptes* is classified in the family Sarcoptidae, superfamily Sarcoptoidea (Psoroptoidea or Listrophoroidea), order Astigmata, subclass Acari (Acarina), class Arachnida, phylum Arthropoda.

The order Astigmata encompasses seven suborders: Acaroidea, Canestrinioidea, Glycyphagoidea, Hemisarcoptoidea, Histiostomatoidea, Hypoderatoidea, and Sarcoptidia (Psoroptidia); the suborder Sarcoptidia (Psoroptidia) contains four superfamilies: Analgoidea, Freyanoidea, Pterolichoidea, and Sarcoptoidea; and the superfamily Sarcoptoidea (Psoroptidia) comprises nine families: Atopomelidae, Chirodiscidae, Gastronyssidae, Listrophoridae, Lobalgidae, Myocoptidae, Psoroptidae, Rhyncoptidae, and Sarcoptidae.

Besides *Sarcoptes*, the family Sarcoptidae includes another genus *Notoedres* (with a single species *Notoedres cati*). The genus *Sarcoptes* consists of one recognized species *Sarcoptes scabiei*, which is subdivided into >15 varieties, the most notable of which are *S. scabiei* var. *bovis*, *S. scabiei* var. *canis* (synonym: *Sarcoptes canis*), *S. scabiei* var. *caprae*, *S. scabiei* var. *equi* (synonym: *Sarcoptes equi*), *S. scabiei* var. *hominis*, *S. scabiei* var. *ovis*, and *S. scabiei* var. *suis*. *Sarcoptes* is related to *Psoroptes*, a scab mite belonging to the family Psoroptidae, superfamily Sarcoptoidea that infests the skin of domestic animals [1–4].

S. scabiei varieties are morphologically similar; however, they demonstrate clear physiological and genetic differences, as well as host specificity. The variety specific to man (*S. scabiei* var. *hominis*) is generally referred to as the "itch mite" or "scab mite," with its infection being commonly called "scabies." Humans may also be susceptible to mite carried by small dogs (*S. scabiei* var. *canis*) [5].

76.1.1.2 Morphology

The order Astigmata covers relatively slow-moving mites with thinly sclerotized integuments and without detectable spiracles or tracheal systems.

Adult *S. scabiei* appears creamy white (or translucent, dirty-white), and has a spherical, ventrally flattened and dorsally convex tortoise-like body, with six or seven pairs of cuticular spines on dorsal surfaces. The body consists of two regions separated by a fold in the integument; the anterior portion possesses mouthparts, is eyeless, and bears two pairs of legs; the posterior portion also bears two pairs of very short legs, which do not extend as far as the margins of the body. *S. scabiei* females are 0.33–0.45 mm long and 0.25–0.35 mm wide, and males are about 0.20–0.24 mm long by 0.2 mm wide. The adult male is distinguished from the female by its smaller size, darker color, and the presence of stalked pulvilli (suckers) on leg 4 (as leg 4 in the adult female ends in long setae or trailing bristle).

S. scabiei larvae have six legs, whereas nymphs and adults have eight legs, with stalked pulvilli (suckers) present on legs 1 and 2 of both the male and female adult mites. *S. scabiei* eggs are oval, measuring 0.1–0.15 mm in size.

76.1.1.3 Biology

The scabies mite, *S. scabiei*, undergoes five stages of development in its life cycle: egg, larva, protonymph, tritonymph, and adult, with the duration of each stage being: egg, 3–4 days; larva, 3 days; protonymph, 3–4 days; tritonymph, 3–4 days; the adult female, up to 2 months; the adult male, 9–11 days (often die after mating). Thus it takes 14–17 days for *S. scabiei* eggs to turn into adults [6].

As an obligate parasite, the adult female burrows into the stratum corneum of the skin, where it deposits 2–3 eggs per day for up to 6 weeks (with a total of 40–50 eggs), resulting in raised papules on the skin's surface. Eggs hatch in 3–4 days to become six-legged larvae, which migrate to the skin surface and then burrow into molting pouches (e.g., hair follicles, which are shorter and smaller than the adult burrows). After 3–4 days, the larvae molt (discard old cuticles to grow), and turn into eight-legged protonymphs, which molt a second time into slightly larger tritonymphs (or adult males), before a final molt into adult females (in hair follicles). It is estimated that only 10% of the laid eggs develop into adult mites. Because the male requires one less molt than the female, it takes shorter 9–11 days instead of 17 days for a male to reach adulthood. After adult mites move out hair follicles and mate, the male penetrates the molting pouch of the female. The female remains fertile for the rest of her life (1–2 months). The impregnated female leaves the molting pouch (crawling at the rate of 2.5 cm/min on warm skin) in search of a permanent burrow (e.g., the interdigital spaces, the ulnar margins of the wrists, the elbows, the anterior folds of the axillae, the penis, scrotum, buttocks, the back of the knee, the ankles and toes, and the undersides of the breasts). Once found, the female creates her characteristic S-shaped burrow, laying eggs in the process. The female continues to lengthen her burrow (reaching a length of 5–15 mm), lays eggs and stays there for the duration of her life (about 30 days). The period between the beginning of burrow formation and the appearance of the first larva varies from 71 to 78 h, and adult mites emerge on the surface of the skin after ~2 weeks.

S. scabiei causes scabies in humans, and mange (sarcoptic mange) in dogs, cats, ungulates, boars, cattle, sheep, wombats, koalas, and great apes, as well as scaly leg in fowls [7–9]. *S. scabiei* varieties are host specific, and cross-infection does not usually occur between mites from different host species. Occasional cases of human scabies following exposure to animal scabies (e.g., *S. scabiei* var. *canis*) are generally self-limiting, with no evidence of long-term reproduction occurring on the aberrant host [10–14].

S. scabiei digests and consumes the horny layer of the epidermis and the sera that seeps into the burrow from the dermis. The average mite in a normal patient is 11, and in case of crusted scabies, the number of mites is huge because of uncontrolled proliferation. Even a relatively small number of scab mites can cause agonizing itching, especially at night, and severe damage to the skin as a result of scratching and secondary bacterial infection, leading to impetigo or eczema.

Scabies is highly contagious and can be spread by scratching, picking up the mites under the fingernails or by touching someone with scabies for a prolonged period of time. Transmission is aided by close, prolonged skin-to-skin contact with an infected person (e.g., bed partners, schoolmates, daycare, dancing, flirtation, and ordinary intimate contact between members of a family). It may be spread by clothing, bed sheets, towels, or other fomites particularly when mite burden is high as in cases of crusted (severe) scabies. Transfer from undergarments and bedclothes only occurs if they have been contaminated by infested persons immediately beforehand. It has been shown on several occasions that the staffs in nursing homes acquire the infection by changing residents' bedding and mattress.

Owing to its lack of sight, scab mite uses odor and thermal stimuli for active host taxis. After dislodgement from its host, scab mite can survive for 24–36 h at room temperature with normal humidity (21°C and 40–80% relative humidity) and even longer (2–7 days) at lower temperatures with high humidity. As scab mite desiccates very easily, putting clothing away for 2 or 3 days at ordinary room temperature is sufficient to rid off it.

The incubation period before the onset of symptoms is 3–6 weeks for primary infestation, and 1–3 days in reinfestation. The onset of symptoms is associated with the burrowing of adult mite. The burrowing activity is assisted by scab mite's mouthparts and special cutting surfaces on the front legs, and it takes about 2–5 min for a scab mite to burrow into the skin. The intense itch caused by scabies is due mainly to the action of the mites moving within the skin and on the skin. The presence of feces (scybala) and eggs of the scab mites may also contribute to the allergic reactions.

S. scabiei is distributed throughout the world, affecting hundreds of millions of people of all ages, races, and socioeconomic classes in all climates. Scabies is endemic in Africa, Egypt, Central and South America, Northern and Central Australia, the Caribbean Islands, India, and Southeast Asia. In other regions, scabies is observed primarily in sporadic individual cases and institutional outbreaks, where crowded conditions facilitate the spread of infection. Scabies is commonly seen in the very young, followed by older children and young adults. Mothers of young children and the elderly in nursing homes with compromised immune systems are also prone to infection.

76.1.2 Clinical Features and Treatment

76.1.2.1 Clinical Features

Scabies is a contagious disease caused by *S. scabiei* that burrows into the stratum corneum of the skin, and induces a series of allergic reactions in the host [15]. Depending on the intactness of immune functions of the hosts, the clinical manifestations of scabies can be differentiated into two categories: classic scabies and atypical scabies [16–18].

i. Classic scabies. Classic scabies accounts for bulk of the clinical cases and often occurs in individuals with fully functional immune systems. Clinical symptoms with a primary infestation of scabies emerge about 4–6 weeks after infection, but with reinfestation, the associated lesions and pruritus are evident within 24–48 h. Symptoms include generalized and intense itching, especially at night. The lesions containing pruritic papules or follicular papules (small blister at the base of a hair) are located in the webs of the fingers, the flexor aspects of the wrists, the extensor aspects of the elbows, the periumbilical skin, the buttocks, the ankles, the penis in males, the periareolar region in females, and the breasts of women. However, the lesions are generally absent on the face and head. Scabies-specific burrows and nodules may be present in the genital regions and axillae. The number of mites per patient amounts to 10–12 in primary infection, and it can be much reduced in repeated infection (reinfection). Nonspecific secondary lesions (e.g., excoriation, eczematization, and impetiginization) may also occur. Even though only a few mites may be present in restricted locations between the fingers, the rash may occur over much of the body. Clinical sign for the diagnosis of classic scabies is the burrow, which presents as serpiginous, grayish lines of 5 mm or longer. Prolonged skin-to-skin contact with the infested person plays a central role in the transmission of classic scabies.

Complications of classic scabies include cellulitis (localized swelling, redness, and fever), and pyoderma due to secondary infection with group A streptococcus and *S. aureus*. Early scabies infestation may resemble other skin diseases, including insect bites (e.g., midges, fleas, and bedbugs), microbial infections (e.g., folliculitis, impetigo, tinea, syphilis, and viral exanthema), eczema, contact dermatitis, and allergic disorders (e.g., papular urticaria, bullous pemphigoid, and pityriasis rosea).

ii. Atypical scabies. Atypical scabies often occurs in infants, with lesions extending to the face, scalp, palms, and soles, including blisters and pustules on the palms and soles of the feet (acropustulosis). Other atypical scabies include atypical papular scabies in the elderly, crusted scabies in immunosuppressed patients, and impetigo in patients with scabies superinfection [19].

Crusted scabies (or Norwegian scabies) is a severe, debilitating disease due to the uncontrolled proliferation in the skin of mites belonging to the same variety as that causes classic scabies [20]. It is characterized by the formation of a scaly rash, slight itching and thickened crusts of skin containing large numbers of mites and eggs (totaling over a million in severe cases), and the absence of itching. The hyperkeratotic skin crusts may be either loose, scaly, and flaky or thick and adherent. The distribution of crusts over the body may be localized or extensive (including the neck, scalp, face, eyelids, and the area under the nails). The lesions often present as horny plaques with a distinct erythematous appearance. The plaque is composed of a parakeratotic crust of 3–15 mm in thickness. With a creamy, gray, yellow brown, or yellow green in color, the crusts are adherent and firm. After the crusts are removed, the lesions show a porous appearance akin to a pumice stone. With very large and thick lesions, the surface may be broken up into slabs by fissures. Upon removal of the crust, the undersurface appears smooth, red, moist, and velvety. Such crusts tend to occur over the palms, on the extensor aspect of the elbows, on the scalp and ears and on the soles of the feet and the toes. In addition, crusted scabies may present as psoriasiform dermatitis, and warty dermatosis of the hands and feet with hyperkeratosis of the nails [21–28].

Because of the involvement of large number of mites, crusted scabies is more infectious and more difficult to treat than classic scabies. For this reason, patient care staff and support staff (e.g., cleaning staff and laundry employees) in institutional settings should be considered to have been exposed to infested persons [29].

Predisposing conditions for crusted scabies include leprosy, infection with human T-cell leukemia virus type 1 and human immunodeficiency virus, and immunosuppression by medication, as well as paralyzed limb or a limb with sensory neuropathy. Patients with cognitive deficiency and institutionalized patients may also develop crusted scabies, due presumably to the absence of itch or the inability to scratch [30].

Complications of crusted scabies include secondary bacterial infection by *Staphylococcus aureus*, impetigo, ecthyma, cellulitis, lymphangitis, generalized lymphadenopathy, and septicemia. Crusted scabies can be confused with psoriasis, eczema, seborrheic dermatitis, Darier's disease, pityriasis rubra pilaris, lichen planus, and cutaneous lymphoma.

The clinical signs of sarcoptic mange in animals are slightly raised red papules on the sparsely haired regions of the body, and intense pruritus, with consequent scratching, excoriation, and skin inflammation. Left untreated, sarcoptic mange can lead to loss of hair, scaling, and crusting of the skin with dried exudate of serum. Secondary pyoderma may occur [31].

76.1.2.2 Treatment

Scabies can be treated by topical and oral administration of scabicidal drugs.

1. Permethrin. As a synthetic pyrethrin derivative, permethrin acts to disrupt the neuronal sodium channel repolarization and cause paralysis and eventual death of the mite. Being very effective against the mites with a low mammalian toxicity, permethrin 5% cream is currently the preferred topical scabicidal agent, and is applied for 8–14 h or overnight to the entire body surface (including the finger and toe creases, nails, navel, and cleft of the buttocks) and showered off in the morning. One application of permethrin is normally sufficient to cure an infection, although for crusted scabies, repeated applications in combination with oral ivermectin are necessary. On the whole, permethrin cream 5% is effective and well-tolerated, but expensive [32].
2. Benzyl benzoate. Synthetically derived from the esterification of benzoic acid with benzyl alcohol, benzyl benzoate is neurotoxic to the mites. Benzyl benzoate 25% emulsion is diluted with three parts water for children under 2 years and with equal parts of water for Permethrin sensitive adults and children 2–10 years; with a contact time of 24–36 h. Benzyl benzoate is effective in the management of permethrin-resistant crusted scabies and may be used in combination with ivermectin in patients with relapses after a single treatment with ivermectin. Benzyl benzoate lotion (10–25%) represents a cheap and effective therapy for scabies, but requires application on more than 1 day (two to three applications).
3. Sulfur 6–33% ointment in petroleum jelly. This is an oldest antiscabietic in use and still represents an effective, very cheap and safe topical treatment for scabies. However, it stains clothing and requires application on three successive days. For children <2 months of age, apply topically sulfur 5% cream daily for 2–3 days [33].
4. Lindane (gammabenzene hexachloride) lotion. This is a second-line treatment where first-line medications, such as permethrin have either failed to kill the scabies mites or ruled out in special situation. Due to increased percutaneous absorption and the risk of neurotoxicity following repeated application, lindane should be avoided in crusted scabies [34,35]. In fact, European Medicines agency withdrew lindane from its list of recommended drugs in 2008 [36].
5. Crotamiton 10% cream (Crotamitex® Gel). This is applied daily for 3–5 days to kill the scabies mites and also to relieve itch (pruritus).
6. Malathion. This is applied topically for 24 h for effective killing of both *S. scabiei* adults and eggs.
7. Neem (*Azadirachta indica*) oil may be used for patients who are sensitive to permethrin, or for mites that are resistant to other drugs. Other topical scabicidals include 5% tea tree oil (from the tree *Melaleuca alternifolia*, which is only partially effective), 20% lippia oil (*Lippia multiflora*), turmeric (*Curcuma longa*), camphor oil (*Eucalyptus globulus*), and a repellent containing coconut and jojoba oil [37].
8. Ivermectin. Ivermectin is a semisynthetic antihelminthic agent derived from *Streptomyces avirmitilis* that inhibits the gamma amino benzoic acid (GABA) transmission and binds the glutamate-gated chloride ion channels in the nerve and muscle cells, leading to the paralysis and death of parasites. As it does not cross the intact blood–brain barrier, ivermectin is safe with minimal side effects in humans (e.g., mild abdominal pain, nausea, vomiting, myalgia, and/or arthralgia) [38,39]. Ivermectin is given orally at 200 μg per kg in patients weighing more than 15 kg and repeated in 2 weeks. Alternatively, it is given orally with a single dose of 400 μg per kg [40]. For treatment of crusted scabies, Ivermectin may be used as monotherapy or in combination with other scabicidals (e.g., permethrin) [41]. Ivermectin is contraindicated in a patient with an allergy to ivermectin and central nervous system disorders [42,43].

In addition to topical scabicidals, treatment of crusted scabies should also include keratolytic agents (e.g., 5–10% salicylic acid in petrolatum, 40% urea, or by soaking in a hot bath) to remove the thick crusts, which prevent the adequate penetration of topical scabicidals. Concomitant topical and oral treatments (e.g., topical permethrin 5% cream with oral ivermectin, or topical benzyl benzoate emulsion with oral ivermectin) are beneficial for patients with crusted scabies [44].

Symptomatic treatment includes antihistamines for pruritus and an emollient for dry skin as well as broad-spectrum antibiotics for secondary bacterial infection or septicemia.

Along with topical/oral scabicidal treatment, the nails of the patients should be clipped and scabicidals applied. The environment should be cleaned (e.g., treating furniture and bedding; vacuuming floors, carpets, and rugs; mopping floor and bathroom surfaces; cleaning shower/bath tub after each use, etc.). All clothes and bed linen worn/used within the previous 48 h should be washed in hot water (at 60°C). Any items that cannot be washed in hot water (e.g., blankets, jumpers, or lounge chairs) can be placed in a sealed plastic bag for one week. Itching persisting for 1–2 weeks after treatment should not be interpreted as a sign of treatment failure or re-infestation due to the fact that the dead mites and eggs are still under the skin [45–47].

Moreover, patients should be advised to avoid close contact with their partners until completion of treatment. Persons (including hospital staff) with protracted or frequent skin contact (e.g., via the hands) should be also treated using permethrin 5% cream as first line followed by oral ivermectin, even if asymptomatic. Persons with indirect contact (e.g., via bedding) only require treatment in cases of crusted scabies [48–50].

76.1.3 Laboratory Diagnosis

Clinically, scabies causes severe itch (pruritus), especially at night, which usually develops 2–6 weeks after infestation. In a reinfected person the itch recurs within 1–4 days. However, crusted scabies (Norwegian scabies) shows no or reduced itching. Skin lesions induced by scabies include papules, pustules, burrows, nodules, and occasionally plaques, which are prominently around finger webs, anterior surfaces of wrists and elbows, anterior axillary folds, belt line. Other commonly affected areas include thighs and external genitalia of men; nipples, abdomen, and the lower portion of the buttocks in women; and the head, neck, palms, and soles in children. Eczema and impetigo are consequences of itch. Clinical signs of crusted scabies typically consist of hyperkeratotic/crusted plaques, papules, and nodules (particularly on the palms of the hands and the soles of the feet as well as the axillae, scalp, buttocks, and genitalia). Occasionally, psoriasiform or eczematous lesions with fine, powder-like scaling and redness (generally on a dry skin) are seen [51,52].

As scabies does not cease until the mites is eliminated, it is important to diagnose the disease correctly for efficacious treatment. Generally, diagnosis is made by finding burrows of *S. scabiei* females, which often appear as characteristic silvery lines (about 0.5 cm in length) in the skin (typically between the fingers and on the sides of hands and feet, on wrists, extensor surface of the elbow, female nipples, anterior axillary folds, penis, and scrotum). The suspicious area may be rubbed with ink from a fountain pen (or a topical tetracycline solution that will glow under a special light), and the surface is then wiped off with an alcohol pad. The characteristic zigzag or "S" pattern of the burrow across the skin will appear if the person is infected with scabies. The mite can usually be identified as a dull-white spot toward the end of the burrow, and can be removed with a needle. In case that burrows are obscured by scratch marks, eczema or impetigo, it becomes difficult to diagnose [53].

When a suspected burrow is found, diagnosis may be confirmed by microscopy of surface scrapings taken with a scalpel or curette, or a burrow may be probed with a needle to obtain material [54]. The material is then placed on a glass slide in glycerol or mineral oil and contained with a coverslip. Potassium hydroxide is better avoided because it may dissolve fecal pellets. Positive diagnosis is made when the mite, eggs, eggshell fragments, or mite fecal pellets from skin scrapings (e.g., from scabietic papules or from under the fingernails) or by the detection of the mite at the end of its burrow. Actual detection of scabies often requires the scraping of dozens of suspicious lesions down to the superficial dermis. This will result in minor bleeding in spots. Even a negative (unsuccessful) scraping will not completely rule out scabies. For crusted scabies, a scraping under the fingernail is often diagnostic. Histopathology can be diagnostic if mites or parts of the mite are identified in the section.

For morphological assessment of *S. scabiei* female mites, the arrangement of scales in the posterior dorsal field, the shape of the caudal margin of the propodosomal shield, and the presence or absence of ventrolateral scales are determined. The structure of the posterior dorsal field of scales with respect to a bare area is defined as either absent, rudimentary (1–5 missing scales), small (6–10 missing scales), or large (more than 10 missing scales). The caudal margin of the propodosomal shield is characterized either as convex, straight, or concave [55].

Epiluminescence microscopy and high-resolution videodermatoscopy are noninvasive techniques that allow detailed inspection of the patient's skin, from the surface to the superficial papillary dermis [56,57]. Diagnosis is by observations of the "jet-with-contrail" pattern in the skin representing a mite and its burrow. However, use of dermatoscopy is limited by the high cost of the equipment [18,58–60].

Enzyme-linked immunosorbent assays have been developed for the detection of antibodies to *S. scabiei* in pigs and dogs and are commercially available in Europe. Recombinant *S. scabiei* antigens have been experimented for diagnosis of human scabies. These include mature forms of both active and inactive homologs of the cysteine protease group 1 allergens, mature forms of active and inactive homologs of the serine protease group 3 allergens, a mu class and a delta class glutathione *S*-transferase group 8 allergen, and a homolog to the C terminus of an apolipoprotein group 14 allergen.

In recent years, molecular techniques have been developed for identification and phylogenetic analysis of *S. scabiei* [61–68]. Using the second internal transcribed spacer (ITS-2) of ribosomal DNA (rDNA) as genetic marker, Zahler et al. [55] showed that *Sarcoptes* consists of a single, heterogeneous species. By targeting rDNA and mitochondrial DNA sequences, Berrilli et al. [69] detected genetic differences between mites derived from red foxes in North Italy and Northwest Spain. In addition, using microsatellite markers, genetic differences between geographically distinct populations were observed [70–74]. The ability to amplify and detect *S. scabiei* DNA from clinical specimens offers a valuable approach for diagnosis of scabies and for clarification of clinically atypical scabies infection [75].

76.2 METHODS

76.2.1 Sample Preparation

76.2.1.1 Sample Collection

Cutaneous scales are collected into sterile Petri dishes. Migrated mites, firmly attached to the dish by their ambulacral suckers, are manually collected by means of fine needles or small blades. Skin biopsies are taken prior to therapy and divided, one part is macerated in 10% KOH at room temperature for 1–2 h and examined microscopically, and the other part for molecular analysis [55,76].

For extraction of living mites from scabietic skin samples (crusts or deep skin scrapings), a 50-mL plastic syringe (with the syringe cone being cut off ~5 mm from the base) is connected with a 1.5-mL Eppendorf tube. The source material (2 g of crusts cut into 1-cm^2 pieces) is introduced in the syringe. The syringe/Eppendorf tube device is kept horizontally (supported with paper or other packing material) in a board box (shoe box) containing a hole (about the size of the cylindrical part of the tube) on the short side of the box), with the conical part of the tube protruding from the hole. The box is kept overnight at 30 cm from the lamp (a 40 W reading table lamp) at room temperature and mites migrate toward a heat source during a minimum of 36 h, and the tube (dotted with mites) is detached and the mites collected for examination or DNA extraction. To maximize harvesting, a new may be inserted to the syringe cone. Another technique uses a small strip of PVC adhesive tape to fix mites with the assistance of tweezers. Then, mite and strip are plunged in the lyses buffer and heated twice to disrupt the chitinous exoskeleton [76].

76.2.1.2 DNA Extraction

DNA from mites can be extracted by using the Qiagen Tissue Kit (Qiagen). Briefly, cutaneous scales and biopsies each are suspended in 540 μL of buffer ATL, incubated overnight with 60 μL proteinase K at 55°C, mixed with 600 μL of buffer ATL and incubated at 70°C for 10 min. Then 630 μL of absolute ethanol are added to this reaction and mixed. Of this mixture 600 μL are loaded on to a QiaAmp column containing a silicate fleece, and centrifuged at room temperature for 60 s. This centrifugation step is repeated twice using all the extraction solution. The columns are washed twice with washing buffer AW and eluted with 100 μL elution buffer. The DNA is stored at −20°C [76].

Alternatively, individual *S. scabei* mites are collected from crusty skin lesions and stored in 70% alcohol. Fixed mites are instilled on a glass slide, and each mite is transferred from a glass slide into a 1,5-mL Eppendorf tube added with T1 buffer (180 μL, from NucleoSpin Tissue DNA extraction kit, Macherey–Negel) with the help of a small strip of polyvinyl choride (PVC) adhesive tape (0.5 × 0.2 mm) operated with a pair of tweezers. Then, the mite is plunged in the buffer with the strip. The tube is subjected to a heat shock, consisting of a −80°C step for 2 min followed by a 70°C step for 2 min, repeated three times. After addition of proteinase K (25 μL, from NucleoSpin Tissue DNA extraction kit, Macherey–Negel), the tube is incubated at 56°C overnight. DNA is then extracted using the NucleoSpin Tissue DNA extraction kit (Macherey–Negel), and the DNA is eluted with 50 μL of BE buffer [64].

For DNA extraction from skin scraping, up to 100 mg of specimens (or isolated mites) are frozen in liquid nitrogen and ground to a fine powder in a mortar. The powder is resuspended in 400 μL TET buffer (50 mM Tris-HCl pH 8.0, 25 mM EDTA, 100 mM NaCl, and 0.4% Triton X-100) and 5 μL (20 mg/mL) proteinase K. After incubation at 50°C for 2 h, the DNA is extracted by shaking with 400 μL phenol: chloroform:isoamyl alcohol (25:24:1) saturated with 10 mM Tris-HCl (pH 8.0) containing 1 mM EDTA. After centrifugation, the DNA is purified from the aqueous phase by silica gel adsorption (Geneclean; Bio 101) and eluted in 120 μL water [55].

76.2.2 Detection Procedures

76.2.2.1 Analysis of ITS 2 Sequence

Zahler et al. [55] utilized primers RIB-18 (5′-GGG CTG CAG TAT CCG ATG GCT TCG T-3′) and RIB-3 (5′-CGG GAT CCT TC(A,G) CTC GCC G(C,T)T ACT-3′) to amplify ITS-2 fragment flanked by 5.8S and 28S sequences. Subsequent analysis of the PCR amplicon sequence allowed precise identification of *S. scabei* mites.

76.2.2.1.1 Procedure

1. PCR mixture (50 μL) is made up of 5 μL buffer [100 mM Tris–HCl pH 8.3, 500 mM KCl], 0.5 μL dNTP-mix (25 mM each), 2.2 μL 50 mM $MgCl_2$, 1 μL each of both primers (50 μM each), 0.5 μL (5 U/μL) *Taq* DNA polymerase (Amersham) and extracted DNA.
2. Amplification is conducted with 45 cycles of 92°C for 60 s, 64°C for 60 s and 72°C for 90 s.
3. PCR products are visualized in a 1.3% agarose gel containing 0.2 μg/mL ethidium bromide. PCR products (0.45 kb) are excised from the agarose gel, purified by silica gel adsorption, and sequenced.
4. The resulting DNA sequences are compared to those available at GenBank via BLAST search.

Note: ITS-2 rDNA is a suitable genetic marker for distinguishing closely related species and examining phylogenetic relationships within genera but not for genetic characterization within a species for mites. For characterization of subpopulations or strain typing within *S. scabiei*, genetic markers other than ITS-2 rDNA, such as microsatellite DNA and/or mitochondrial DNA should be used [67].

76.2.2.2 Analysis of Mitochondrial 16S RNA Sequence

Berrilli et al. [69] targeted mitochondrial 16S rRNA gene for phylogenetic assessment of *S. scabiei*. Use of PCR primers 16S-F (5′-CCG GTC TGA ACT CAG ATC AAG T-3′) and

16S-R (5′-GCT CAA TGA TTT TTT AAA TTG CTG T-3′) facilitates amplification of a ~460 bp fragment. Sequence analysis of the resulting products permits identification of the parasite.

76.2.2.2.1 Procedure

1. PCR mixture (50 μL) is made up of 1 × PCR buffer, 0.2 mM dNTPs, 5 U of *Taq* DNA polymerase (Promega), 2 μL of each primer (10 mM), and 0.2 μL of extracted DNA.
2. Amplification is performed with an initial denaturation of 10 min at 95°C, followed by 40 cycles of 45 s at 95°C, 90 s for the annealing step, 2 min at 72°C and a final elongation of 7 min at 72°C. The annealing temperature is increased by 0.3°C for the first seven cycles starting at 40°C, whereas for the other 33 cycles the annealing temperature is 45°C.
3. PCR products are assessed on a 1% agarose gel, stained with ethidium bromide and visualized under UV light.
4. PCR products are then purified for direct sequencing by enzymatic treatment using exonuclease I and shrimp alkaline phosphatase (PCR Product Presequencing Kit, Amersham).
5. Sequences are aligned using ClustalW and adjusted manually. Phenograms showing the genetic relationships among individual mites are constructed using the unweighted pair group method using arithmetic means (UPGMA) in MEGA version 2.0, starting from a distance matrix based on the Kimura 2-parameter index. The genetic differentiation between the populations studied is evaluated by analyzing the average number of nucleotide substitutions using DnaSP version 3. The same program is used to calculate the amount of gene flow among populations.

76.2.2.3 PCR–Enzyme-Linked Immunosorbent Assay (ELISA)

Bezold et al. [75] described a PCR-ELISA for detection of *S. scabiei* DNA from cutaneous scales. PCR primers from highly conserved regions of the *S. scabiei* microsatellite 15 (Sarms15) consist of 5′-ATTAAA TCATTGCACAATAGAGCG-3′ (upstream) and 5′-CTCA TTAATTTTTTCCACCCTC-3′ (downstream) for scabies; 5′-GAAGAGC-CAAGGACAGGTAC-3′ (upstream) and 5′-CAACTTCATCCACGTTCACC-3′ (downstream) for β-globin. The downstream primers are 5′-labeled with digoxigenin. The 5′-biotinylated probes consist of 5′-TCTCTCTCTCTCTCTCTCTCTCTC-3′ (scabies), and 5′-ACACAACTGTGTTCACTAGC-3′ (β-globin) and 5′-CAAGTTTCGTGAGCTGATTG-3′ (both for β-globin mimic and scabies mimic).

76.2.2.3.1 Procedure

1. PCR mixture (50 μL) is composed of 2.5 mM $MgCl_2$, 10 mM Tris, 50 mM KCl, 0.5 μM of each primer, 200 μM each of dNTPs and 2.5 U *Taq* polymerase (Boehringer Mannheim).
2. Amplification is conducted with an initial heating step (95°C for 5 min), followed by 42 amplification cycles (95°C for 15 s, 60°C for 30 s, and 72°C for 30 s) and one final extension step (72°C for 7 min).
3. PCR fragments are separated on a 3% agarose gel containing 0.5 μg/mL ethidium bromide and visualized under ultraviolet radiation. Fragment lengths are 268 bp for β-globin, 546 bp for the globin mimic, and 348 bp for the scabies mimic. Fragment lengths for scabies are between 138 and 178 bp.
4. PCR–ELISA is performed on the basis of a commercially available system (PCR ELISA, Boehringer). Briefly, two parts of 10 μL each of the respective PCR product are denatured with 10 μL alkaline solution (denaturation solution) for 10 min at room temperature and hybridized with 0.75 pmol of a 5′-biotinylated oligonucleotide specific for the target (scabies or β-globin) or the mimic, respectively, in a total volume of 120 μL. Then 100 μL of this mixture is incubated in a streptavidin-coated microtiter plate at 55°C for 1 h, washed five times (washing buffer included in the kit), incubated with 100 μL of a peroxidase-conjugated antidigoxigenin antibody at 37°C for 30 min, washed five times and incubated with 100 μL of the substrate for peroxidase, 3,3′–5,5′-tetramethylbenzidine, at room temperature for 15 min. The reaction is stopped by adding 100 μL of 2 mol L^{-1} hydrochloric acid, and the optical density is measured at a wavelength of 450 nm.

76.3 CONCLUSION

S. scabiei is an obligate ectoparasite that burrows into the host epidermis and causes scabies in humans and mange in wild, domestic, and farmed animals worldwide. Clinical symptoms of *S. scabiei* infestations include intensely itchy lesions that often lead to secondary bacterial septicemia and post-streptococcal glomerulonephritis, as well as other complications. Early and accurate diagnosis of scabies plays a crucial role in the control and prevention of this treatable disease. Because conventional methods based on morphological characteristics of *S. scabiei* is time-consuming and technically demanding, molecular procedures have been developed and applied for improved identification and phylogenetic analysis of *S. scabiei* isolates. Use of the molecular techniques has contributed to significant reduction in the prevalence of scabies across various communities.

REFERENCES

1. Arlian, L.G. et al., Survival and infestivity of *Sarcoptes scabiei* var. *canis* and var. *hominis*. *J. Am. Acad. Dermatol.*, 11, 210, 1984.
2. Arlian, L.G., and Vyszenski-Moher, D.L., Life cycle of *Sarcoptes scabiei* var *canis*. *J. Parasitol.*, 74, 427, 1988.

3. Arlian, L., Biology, host relations and epidemiology of *Sarcoptes scabiei*. *Ann. Rev. Entomol.*, 34: 139, 1989.
4. Walton, S.F. et al., Scabies: New future for a neglected disease. *Adv. Parasitol.*, 57, 309, 2004.
5. Kemp, D. et al., The scourge of scabies. *Biologist*, 49, 19, 2002.
6. Burgess, I., Biology and epidemiology of scabies. *Curr. Opin. Infect. Dis.*, 12, 177, 1999.
7. Montesu, M.A. et al., Discoverers of the parasitic origin of scabies. *Am. J. Dermatopathol.*, 13, 425, 1991.
8. Bates, P., Sarcoptic mange (*Sarcoptes scabiei* var. *vulpes*) in a red fox (*Vulpes vulpes*) population in north-west Surrey. *Vet. Rec.*, 152, 112, 2003.
9. Dagleish, M.P. et al., Fatal *Sarcoptes scabiei* infection of blue sheep (*Pseudois nayaur*) in Pakistan. *J. Wildl. Dis.*, 43, 512, 2007.
10. Roncalli, R.A., The history of scabies in veterinary and human medicine from biblical to modern times. *Vet. Parasitol.*, 25, 193, 1987.
11. Green, M.S., Epidemiology of scabies. *Epidemiol. Rev.*, 11, 126, 1989.
12. Pence, D.B., and Ueckermann, E., Sarcoptic mange in wildlife. *Rev. Sci. Tech. Off. Int. Epizoot.*, 21, 385, 2002.
13. Chosidow, O., Scabies and pediculosis. *Lancet*, 355, 819, 2000.
14. Chosidow, O., Scabies. *N. Engl. J. Med.*, 354, 1718, 2006.
15. Menzano, A., Rambozi, L., and Rossi, L., Outbreak of scabies in human beings, acquired from chamois (*Rupicapra rupicapra*). *Vet. Rec.*, 155, 568, 2004.
16. Cabrera, R., and Dahl, M.V., The immunology of scabies. *Semin. Dermatol.*, 12, 15, 1993.
17. Holt, D.C. et al., Mechanisms for a novel immune evasion strategy in the scabies mite *Sarcoptes scabiei*: A multigene family of inactivated serine proteases. *J. Investig. Dermatol.*, 121, 1419, 2003.
18. Walton, S.F., The immunology of susceptibility and resistance to scabies. *Parasite Immunol.*, 32, 532, 2010.
19. Hengge, U.R. et al., Scabies: A ubiquitous neglected skin disease. *Lancet Infect. Dis.*, 6, 769, 2006.
20. Anbar, T.S. et al., Scaly scalp associated with crusted scabies: Case series. *Dermatol. Online J.*, 13, 18, 2007.
21. Parish, L.C., and Lomholt, G., Crusted scabies alias Norwegian scabies. *Int. J. Dermatol.*, 15, 747, 1976.
22. Guggisberg, D. et al., Norwegian scabies in a patient with acquired immunodeficiency syndrome. *Dermatology*, 197, 306, 1998.
23. Van Der Wal, V.B. et al., Crusted (Norwegian) scabies in a patient with dystrophic epidermolysis bullosa. *Br. J. Dermatol.*, 141, 918, 1999.
24. Walton, S.F. et al., New insights into disease pathogenesis in crusted (Norwegian) scabies: The skin immune response in crusted scabies. *Br. J. Dermatol.*, 158, 1247, 2008.
25. Roberts, L.J. et al., Crusted scabies: Clinical and immunological findings in seventy-eight patients and a review of the literature. *J. Infect.*, 50, 375, 2005.
26. Karthikeyan, K., Crusted scabies. *Indian J. Dermatol. Venereol. Leprol.*, 75, 340, 2009.
27. Sampathkumar, K. et al., Norwegian scabies in a renal transplant patient. *Indian J. Nephrol.*, 20, 89, 2010.
28. Subramaniam, G. et al., Norwegian scabies in a malnourished young adult: A case report. *J. Infect. Dev. Ctries.*, 4, 349, 2010.
29. Heukelbach, J., and Feldmeier, H., Scabies. *Lancet*, 367, 1767, 2006.
30. Guldbakke, K.K., and Khachemoune, A., Crusted scabies: A clinical review. *J. Drugs Dermatol.*, 5, 221, 2006.
31. Smets, K., and Vercruysse, J., Evaluation of different methods for the diagnosis of scabies in swine. *Vet. Parasitol.*, 90, 137, 2000.
32. Taplin, D. et al., Comparison of crotamiton 10% cream (Eurax) and permethrin 5% cream (Elimite) for the treatment of scabies in children. *Pediatr. Dermatol.*, 7, 67, 1990.
33. Pruksachatkunakorn, C., Damrongsak, M., and Sinthupuan, S., Sulfur for scabies outbreaks in orphanages. *Pediatric Dermatol.*, 19, 448, 2002.
34. Chouela, E.N. et al., Equivalent therapeutic efficacy and safety of ivermectin and lindane in the treatment of human scabies. *Arch. Dermatol.*, 135, 651, 1999.
35. Jin, S.P. et al., Scabies in a 2-month-old infant successfully treated with lindane. *Ann. Dermatol.*, 21, 200, 2009.
36. Scott, G.R., and Chosidow, O., European guideline for the management of scabies. *Int. J. STD AIDS*, 22, 301, 2010.
37. Leone, P.A., Scabies and pediculosis pubis: An update of treatment regimens and general review. *Clin. Infect. Dis.*, 44, S153, 2007.
38. Huffam, S.E., and Currie, B.J., Ivermectin for *Sarcoptes scabiei* hyperinfestation. *Int. J. Infect. Dis.*, 2, 152, 1998.
39. Santoro, A.F., Rezac, M.A., and Lee, J.B., Current trend in ivermectin usage for scabies. *J. Drugs Dermatol.*, 2, 397, 2003.
40. Alberici, F. et al., Ivermectin alone or in combination with benzyl benzoate in the treatment of human immunodeficiency virus associated scabies. *Br. J. Dermatol.*, 142, 969, 2000.
41. Aubin, F., and Humbart, P., Ivermectin for crusted (Norwegian) scabies. *N. Eng. J. Med.*, 333, 26, 1995.
42. Currie, B.J. et al., First documentation of *in vivo* and *in vitro* ivermectin resistance in *Sarcoptes scabiei*. *Clin. Infect. Dis.*, 39, e8, 2004.
43. Currie, B.J., and McCarthy, J.S., Permethrin and ivermectin for scabies. *N. Eng. J. Med.*, 362, 717, 2010.
44. Walton, S.F., Myerscough, M.R., and Currie, B.J., Studies *in vitro* on the relative efficacy of current acaricides for *Sarcoptes scabiei* var. *hominis*. *Trans. R. Soc. Trop. Med. Hyg.*, 94, 92, 2000.
45. Meinking, T.L., and Elgart, W.E., Scabies therapy for the millenium. *Pediatric Dermatol.*, 2, 154, 2000.
46. Scheinfeld, N., Controlling scabies in institutional settings: A review of medications, treatment models, and implementation. *Am. J. Clin. Dermatol.*, 5, 31, 2004.
47. Karthikeyan, K., Treatment of scabies: Newer perspectives. *Postgrad. Med. J.*, 81, 7, 2005.
48. Mumcuoglu, K.Y., and Gilead, L., Treatment of scabies infestations. *Parasite*, 15, 248, 2008.
49. Gould, D., Prevention, control and treatment of scabies. *Nurs. Stand.*, 25, 42, 2010.
50. Gilmore, S.J., Control strategies for endemic childhood scabies. *PLoS One*, 6, e15990, 2011.
51. McCarthy, J.S. et al., Scabies: More than just an irritation. *Postgrad. Med. J.*, 80, 382, 2004.
52. Walton, S.F., and Currie, B.J., Problems in diagnosing scabies, a global disease in human and animal populations. *Clin. Microbiol. Rev.*, 20, 268, 2007.
53. Johnston, G., and Sladden, M., Scabies: Diagnosis and treatment. *BMJ*, 331, 619, 2005.
54. Leibowitz, M., Shave biopsy an underutilised technique in the diagnosis of scabies. *N. Z. Med. J.*, 107, 111, 1994.
55. Zahler, M. et al., Molecular analyses suggest monospecificity of the genus *Sarcoptes* (Acari: Sarcoptidae). *Int. J. Parasitol.*, 29, 759, 1999.

56. Katsumata, K., and Katsumata, K., Simple method of detecting *Sarcoptes Scabiei* var *hominis* mites among bedridden elderly patients suffering from severe scabies infestation using an adhesive-tape. *Intern. Med.*, 45, 857, 2006.
57. Turan, E. et al., The detection of *Sarcoptes scabiei* in human skin by *in vivo* confocal microscopy. *Eur. J. Dermatol.*, 21, 1004, 2011.
58. Dupuy, A. et al., Accuracy of standard dermoscopy for diagnosing scabies. *J. Am. Acad. Dermatol.*, 59, 530, 2008.
59. Griffin, J.R., and Newman, C.C., Clinical, dermatoscopic, and microscopic findings of iInfestation with *Sarcoptes scabiei* var. *hominis*. *Mayo Clin. Proc.*, 86, e47, 2011.
60. Walter, B. et al., Comparison of dermoscopy, skin scraping, and the adhesive tape test for the diagnosis of scabies in a resource-poor setting. *Arch. Dermatol.*, 147, 468, 2011.
61. Skerratt, L.F. et al., The mitochondrial 12S gene is a suitable marker of populations of *Sarcoptes scabiei* from wombats, dogs and humans in Australia. *Parasitol. Res.*, 88, 376, 2002.
62. Cruickshank, R.H., Molecular markers for the phylogenetics of mites and ticks. *Syst. Appl. Acarol.*, 7, 3, 2002.
63. Alasaad, S. et al., *Sarcoptes* mite from collection to DNA extraction: The lost realm of the neglected parasite. *Parasitol. Res.*, 104, 723, 2008.
64. Alasaad, S. et al., HotSHOT Plus ThermalSHOCK, a new and efficient technique for preparation of PCR-quality *Sarcoptes* mite genomic DNA. *Parasitol. Res.*, 103, 1455, 2008.
65. Alasaad, S. et al., Skin-scale genetic structure of *Sarcoptes scabiei* populations from individual hosts: Empirical evidence from Iberian ibex-derived mites. *Parasitol. Res.*, 104, 101, 2008.
66. Alasaad, S. et al., Effectiveness of postponed isolation (post-frozen isolation) method for PCR-quality *Sarcoptes* mite gDNA. *Exp. Appl. Acarol.*, 47, 173, 2009.
67. Alasaad, S. et al., Is ITS-2 rDNA suitable marker for genetic characterization of *Sarcoptes* mites from different wild animals in different geographic areas? *Vet. Parasitol.*, 159, 181, 2009.
68. Alasaad, S. et al., *Sarcoptes*-World Molecular Network (*Sarcoptes*-WMN): Integrating research on scabies. *Int. J. Infect. Dis.*, 15, e294, 2011.
69. Berrilli, F., D'Amelio, S., and Rossi, L., Ribosomal and mitochondrial DNA sequence variation in *Sarcoptes* mites from different hosts and geographical regions. *Parasitol. Res.*, 88, 772, 2002.
70. Walton, S.F., Currie, B.J., and Kemp, D.J., A DNA fingerprinting system for the ectoparasite *Sarcoptes scabiei*. *Mol. Biochem. Parasitol.*, 85, 187, 1997.
71. Walton, S.F. et al., Genetically distinct dog-derived and human-derived *Sarcoptes scabiei* in scabies-endemic communities in nörthern Australia. *Am. J. Trop. Med. Hyg.*, 61, 542, 1999.
72. Walton, S.F. et al., Crusted scabies: A molecular analysis of *Sarcoptes scabiei* variety *hominis* population from patients with repeated infestations. *Clin. Infect. Dis.*, 29, 1226, 1999.
73. Walton, S.F. et al., Genetic epidemiology of *Sarcoptes scabiei* (Acari: Sarcoptidae) in northern Australia. *Int. J. Parasitol.*, 34, 839, 2004.
74. Soglia, D. et al., Microsatellites as markers for comparison among different populations of *Sarcoptes scabiei*. *It. J. Anim. Sci.*, 7, 214, 2007.
75. Bezold, G. et al., Hidden scabies: Diagnosis by polymerase chain reaction. *Br. J. Dermatol.*, 144, 614, 2001.
76. Soglia, D. et al., Two simple techniques for the safe *Sarcoptes* collection and individual mite DNA extraction. *Parasitol. Res.*, 105, 1465, 2009.

Section VII

Arthropoda: Insecta

77 Calliphoridae, Oestridae, and Sarcophagidae (Myiasis-Causing Flies)

Dongyou Liu

CONTENTS

77.1 INTRODUCTION

Myiasis (myia, meaning fly in Greek) is an infestation of live vertebrates (humans and/or animals) with dipterous larvae (commonly referred to as maggots), leading to a range of clinical diseases (e.g., dermal and subdermal, facial cavity, wound or traumatic, gastrointestinal, vaginal, and generalized myiasis). Among the myiasis causing Diptera flies (the two-winged true flies), members of the families Calliphoridae (blowflies), Sarcophagidae (flesh flies), and Oestridae (botflies and warble flies) are most important [1,2]. Anisopodidae, Piophilidae, Stratiomyidae, and Syrphidae are other Diptera families that are occasionally involved in myiasis. However, these families will not be covered in this chapter.

Being recognized since ancient times, myiases are prevalent in the tropics and subtropics, predominately Central and South America, and Africa. As an ephemoral, self-limiting infection, the majority of myiasis cases are nonfatal. Whereas oestrid larvae are obligate parasites that feed on live hosts for weeks/months in the nasopharyngeal tract, gut system, and internal organs, as well as in subcutaneous tissues, calliphorid, and sarcophagid larvae are obligate and/or facultative parasites on living and dead organic matter, inducing myiases within 4–7 days of infection [3,4].

77.1.1 CLASSIFICATION, MORPHOLOGY, AND BIOLOGY

77.1.1.1 The Family Calliphoridae

77.1.1.1.1 Classification

The family Calliphoridae (commonly known as blowflies, carrion flies, bluebottles, greenbottles, or cluster flies) is classified in the superfamily Oestroidea, order Diptera, class Insecta, phylum Arthropoda. The name blow-fly comes from an older English term for meat that had eggs laid on it (i.e., fly blown or flystrike). The family is nonmonophyletic, and consists of >1000 species worldwide that are organized in around 150 genera and 2–12 subfamilies (e.g., Auchmeromyiinae, Calliphorinae, Chrysomyinae, Helicoboscinae, Luciliinae, Mesembrinellinae, Polleniinae, Rhiniinae, and Toxotarsinae) [5–7].

The blowfly genera of medically and veterinary importance mostly belong to the subfamilies Auchmeromyiinae (e.g., *Auchmeromyia* and *Cordylobia*), Calliphorinae (e.g., *Calliphora*, *Lynomyia*, *Eucalliphora*, and *Onesia*), Chrysomyinae (e.g., *Chrysomyia*, *Cochliomyia*, *Compsomyiops*, *Phormia*, *Protophormia*, and *Protocalliphora*), and Luciliinae (e.g., *Lucia*) [8].

Auchmeromyia sengalensis (or *Auchmeromyia luteola*, the Congo floor maggot) in the genus *Auchmeromyia* (which consists of five species) does not live on or in the living host, but sucks the blood of burrow-dwelling wild pigs, warthogs, aardvark, hyena, and occasionally sleeping humans (sanguinivorous myiasis). The larvae inhabit soils in tropical Africa, and feed on persons sleeping on dirty floors.

The genus *Cordylobia* consists of three species: *Cordylobia anthropophaga* (also known as Tumbu fly, putzi fly, mango fly, or skin maggot fly; with *anthropophaga* derived from the Greek word *anthropophagus*, meaning "eater of men"), *Cordylobia (Stassisia) rodhaini* (Lund's fly), and *Cordylobia ruandae*. *Cordylobia* larvae are parasites of various mammals, particularly rodents, although *C. anthropophaga* and occasionally *C. rodhaini* parasitize humans.

The genus *Calliphora* comprises of >24 recognized species (*Calliphora albifrontalis*, *Calliphora augur*, *Calliphora centralis*, *Calliphora dubia*, *Calliphora fallax*, *Calliphora fulvicoxa*, *Calliphora fuscofemorata*, *Calliphora gilesi*, *Calliphora hilli*, *Calliphora lata*, *Calliphora livida*, *Calliphora loewi*, *Calliphora macleayi*, *Calliphora maritime*, *Calliphora nigribarbis*, *Calliphora nigrithorax*, *Calliphora ochracea*, *Calliphora pattoni*, *Calliphora quadrimaculata*, *Calliphora sternalis*, *Calliphora stygia*, *Calliphora varifrons*, *Calliphora vicina*, and *Calliphora vomitoria*). Of these, about 10 species are associated with human and animal myiasis, including *C. livida*, *C. vicina*, and *C. vomitoria* (the common bluebottle fly).

The genus *Chrysomyia* contains 26 valid species (i.e., *Chrysomya albiceps*, *Chrysomya bezziana*, *Chrysomya cabrerai*, *Chrysomya chain*, *Chrysomya chloropyga*, *Chrysomya defixa*, *Chrysomya flavifrons*, *Chrysomya greenbergi*, *Chrysomya incisuralis*, *Chrysomya inclinata*, *Chrysomya latifrons*, *Chrysomya marginalis*, *Chrysomya megacephala*, *Chrysomya nigripes*, *Chrysomya norrisi*, *Chrysomya pacifica*, *Chrysomya phaonis*, *Chrysomya pinguis*, *Chrysomya putoria*, *Chrysomya rufifacies*, *Chrysomya saffranea*, *Chrysomya semimetallica*, *Chrysomya thanomthini*, *Chrysomya varipes*, *Chrysomya villeneuvi*, and *Chrysomya yayukae*), of which at least eight are implicated in human and animal myiasis including *Chrysomyia bezziana* (Old World screwworm), *C. rufifacies*, and *C. megacephala*.

The genus *Cochliomyia* (commonly referred to as the New World screwworm fly) contains four species: *Cochliomyia macellaria*, *Cochliomyia hominivorax*, *Cochliomyia aldrichi*, and *Cochliomyia minima*, of which *C. hominivorax* and *C. macellaria* are important. *C. hominivorax* is a primary screwworm whose larvae (of 15–17 mm in length, with bands of spines encircling the anterior margin of each body segment) feed on living tissue, resulting in deep, pocket-like lesions in the skin of animal host. *C. macellaria* is a secondary screwworm whose larvae feed only on necrotic tissue. Because of its association with dead bodies and carcasses, *C. macellaria* is of forensic importance [9].

The genus *Lucia* covers at least 19 valid species (i.e., *Lucilia adisoemartoi*, *Lucilia ampullaceal*, *Lucilia bazini*, *Lucilia bufonivora*, *Lucilia Caesar*, *Lucilia cluvia*, *Lucilia coeruleiviridis*, *Lucilia cuprina*, *Lucilia eximia*, *Lucilia hainanensis*, *Lucilia illustris*, *Lucilia Mexicana*, *Lucilia papuensis*, *Lucilia porphyrina*, *Lucilia richardsi*, *Lucilia sericata*, *Lucilia silvarum*, *Lucilia taiyuanensis*, and *Lucilia thatuna*), of which six are involved in animal and human myiasis (including *L. cuprina*, *L. sericata*, and *L. illustris* that are primarily responsible for sheep strike and are important to forensic entomology) [10–12]. *L. sericata* is a primary facultative ectoparasite of sheep in the United Kingdom and other parts of Europe, but functions mostly as a secondary species in Australia. *L. cuprina* is the most aggressive agent of myiasis in Australia [13].

Other notable members of the blowfly family Calliphoridae include *Cynomya mortuorum*, *Phormia regina*, and *Protophormia terraenovae*.

77.1.1.1.2 Morphology

The life cycle of the blowfly family Calliphoridae encompasses four developmental stages: egg, larva, or maggot (three instars), puparium, and adult (fly). Like all other dipterous flies, the body of Calliphoridae adult consists of three sections: head (including compound eyes, arista, and antenna), mesothorax (including prescutm, scutum, scutellum, balancer, and wings), and abdomen.

Calliphoridae adults are shiny with metallic coloring, often with blue, green, or black thoraxes and abdomen. Antennae are 3-segmented, aristate. The terminal antennal segment (arista) is plumose (feathery), and the second antennal segment is distinctly grooved. In addition, Calliphoridae adults possess branched vein, frontal suture, and well-developed calypters. The thorax has the continuous dorsal suture across the middle, along with well-defined posterior calli. The postscutellum is absent or weakly developed. The costa is unbroken and the subcosta is apparent. The characteristics and arrangement of hairs are useful for differentiation between members of the Calliphoridae family. All blowflies have bristles located on the meron, with two notopleural bristles and a hindmost posthumeral bristle located lateral to presutural bristle.

The Tumbu fly (*C. anthropophaga*) and Lund's fly (*C. rodhaini*) in the subfamily Auchmeromyiinae can be differentiated morphologically based on the shape of the posterior spiracle and the morphology of the slit. *C. anthropophaga* larvae are oval, of 11–15 mm in size, with three curved slits in spiracles, and numerous small black spines. *C. rodhaini* adult flies are large and robust, with a nonmetallic red–brown to black abdomen and yellow coloration on the thorax and head. *C. rodhaini* larvae are up to 23 mm in length, with scattered spines, and three sinuous slits in each posterior spiracle.

The subfamily Calliphorinae shows bare stem vein, bristly lower calypter and proepisternal depression, and essentially bare suprasquamal. The thorax is dull and bears fine hairs, and the abdomen is shining blue in color. *Calliphora* adult flies are metallic blue, green, copper, or black. The subfamily Luciliinae is characterized by its shining green, blue or bronze thorax and abdomen, suprasquamal ridge with a setae cluster, and a bare lower calypter. The subfamilies

Melanomyinae and Polleniinae are distinguished by their dull color.

77.1.1.1.3 Biology

Calliphoridae female flies lay 150–200 eggs (of 1.5×0.4 mm^2 in size) per batch and around 2000 eggs during the course of her life on the body's natural orifices (ears, eyes, nose, and genitals) or open wounds. Hatching from eggs in 6–48 h, larva (maggot) undergoes three stages of development (called instars, involving one molt per stage). The first instar appears in 2 days and measures 5 mm in length; the second instar emerges after 2.5 days and measures 10 mm in length; and the third instar appears after 4–5 days and measures 17 mm in length. After the third stage of development, the larva (maggot) leaves the food source (carrion) to pupate in soil. With its body parts becoming obscured after 2–7 days, a prepupa (about 12 mm long 8–12 days after egg laying) emerges. The prepupa turns to a pupa (9 mm long) 18–24 days after egg laying. Adult fly appears 10–17 days after the formation of the puparium. In all, the development from egg to adult takes 16–35 days.

Members of the family Calliphoridae are distributed in all continents except for Antarctica, and are of major economic and medical importance as pests of humans and livestock. Calliphoridae larvae demonstrate the ability to develop in dung or animal carcasses (saprophagous), and parasitize facultatively or obligately on both vertebrates and invertebrates (sanguinivorous).

C. anthropophaga (the tumbu fly) is a parasite of large mammals (including humans) during its larval stage, and is the most common cause of myiasis in humans in the sub-Saharan tropical Africa. *C. anthropophaga* adult females deposit eggs on sand, soil, bedding, or clothing (when soiled with feces or urine); their larvae (known as Cayor worms) hatch in 7–20 days, use oral hooks to infect the unbroken skin of a human host, and lie in the subcutaneous tissue, causing the formation of tumors. Afterward, the larvae leave, fall to the ground, bury themselves, and pupate there.

C. (*S.*) *rodhaini* (Lund's fly) is found in tropical Africa, especially areas of rainforest from Senegal, through Central Africa, to Angola, and Zimbabwe. Adult flies feed on rotting fruits, vegetables and feces, and are more abundant throughout the wet season and most active in the early morning and evening. Adult female deposits its eggs in soil, after hatching, the larvae may penetrate the skin of barefoot persons.

Chrysomyia flies are distributed across Africa, Australia, and Asia. Infestation with larvae of the Old World screwworm, *C. bezziana*, can be disfiguring. The term screwworm is derived from the fact that maggots embed themselves into the flesh of their host in a screw-like fashion. Like the New World screw worm, *C. bezziana* can invade bone. Infestation of the eye is particularly dangerous. Morphologically, *C. bezziana* adult (of 8 and 12 mm in length) is typically metallic green or blue, with yellow faces and soft fine yellow hairs. Its abdominal tergites (the segments of the dorsal portion) have narrow darker bands and the legs are black or dark brown. Only the bases of the wings are infuscated; the wings have a glassy appearance. The anterior spiracles (small openings on the surface for the respiratory system) range from dark brown to orange. *C. bezziana* third-instar larvae are smooth, lacking obvious body processes except on the last segment. Its posterior spiracles are not hidden in the body cavity, and the peritreme of these spiracles is open [14]. *C. megacephala* occurs in many of the same regions as *C. bezziana*. While *C. bezziana* male does not have a boundary dividing the upper- and lower-eye facets, *C. megacephala* male has a boundary distinguishing the upper and lower portions. In addition, while *C. bezziana* female has a more parallel frons, *C. megacephala* female has a frons that is distinctly narrow in the center.

The New World screwworm, *C. hominivorax*, is found throughout the southern United States, Central America, and tropical regions of South America [15,16]. Adult females deposit eggs in open wounds or discharging orifices (e.g., the nose). As a primary screwworm, *C. hominivorax* larvae invade adjacent living tissue, including cartilage and bone. The larvae feed on the necrotic or damaged tissue, and then bore into adjoining normal tissue and cause ulceration. Because infestation of ears and nose provides the larvae with access to brain tissue, myiasis caused by *C. hominivorax* is more dangerous than infestation with other agents. *C. macellaria* is a secondary screwworm feeding only on necrotic tissue.

C. hominivorax is distinguished from *C. bezziana* by the dorsal tracheal trunks in the larvae. The dorsal tracheal trunks of *C. hominivorax* larvae are darkly pigmented from the 12th segment to the 10th or sometimes 9th segment, whereas the dorsal tracheal trunks of *C. bezziana* larvae are only pigmented for the last half of the 12th segment, and its anterior spiracle has 4–6 lobes, with seven on occasion [9].

77.1.1.2 The Family Oestridae

The family Oestridae (commonly known as botflies and warble flies) is classified in the superfamily Oestroidea, order Diptera, class Insecta, pyhlum Arthropoda. The family is separated into five subfamilies: Cephenemyiinae (with a single genus *Cephenemyia*), Cuterebrinae (with two genera *Cuterebra* and *Dermatobia*), Gasterophilinae (with a single genus *Gasterophilus*), Hypodermatinae (with three genera *Hypoderma*, *Oestromyia*, and *Przhevalskiana*), and Oestrinae (with two genera *Oestrus* and *Rhinoestrus*), which together cover 151 species [17].

The genus *Dermatobia* consists of a single species *Dermatobia hominis* (the human botfly). *D. hominis* is found in warm, humid lowland forests ranging from northern Argentina to southern Mexico, and causes cutaneous myiasis in humans and other mammals (especially cattle). *D. hominis* lifecycle begins with the female botfly capturing a day-biting mosquito or other blood-sucking arthropod, laying 10–50 eggs on its abdomen, cementing them with a glue-like secretion all in midair (a process called phoresis). The blood-sucking arthropods then transmit the infectious larvae to human or animal host. The eggs hatch for 5–12 weeks in the host tissues, and the newly hatched larvae migrate through the bite wound, or along a hair follicle, into the cutaneous tissue of the human host. The larvae undertake three stages of development over a period of 5–10 weeks inside the host tissue cavity. The mature third instar larva (of 18–25 mm in length, with pair of flower like anterior spiracles, and rows of spines) utilizes its spines as a burr to widen the central punctum, emerges from the lesion, and falls to the ground. The larva burrows into the soil, pupates, and molts after 4–11 weeks, producing an adult fly. Facultative myiasis is preventable by properly covering open wounds.

The genus *Gasterophilus* consists of several species that affect horses and their relatives (i.e., *Gasterophilus haemorrhoidalis*, *Gasterophilus inermis*, *Gasterophilus intestinalis*, *Gasterophilus nasalis*, and *Gasterophilus pecorum*), as well as *Gyrostigma pavesii*, which parasitizes rhinoceroses. *Gasterophilus intestinalis* (the horse botfly) causes a migratory form of cutaneous myiasis (also known as creeping eruption), which may resemble migratory helminthiasis.

The genus *Oestrus* (botflies) includes the sheep botfly (*Oestrus ovis*) that parasitizes sheep and goats and is a major pest in the Australian sheep industry.

The genus *Rhinoestrus* contains *Rhinoestrus purpureus*, which infests horses and occasionally causes ocular myiasis in human hosts.

The genus *Hypoderma* (warble flies, heel flies, bomb flies, and gad flies; whose larvae are often called cattle grubs or wolves) consists of seven species: *Hypoderma actaeon*, *Hypoderma bovis*, *Hypoderma diana*, *Hypoderma lineatum*, *Hypoderma pantholopsum*, *Hypoderma sinense*, and *Hypoderma tarandi*. Of these *H. lineatum* (horse), *H. bovis* (cattle) and *H. tarandi* (reindeer) are of economic importance. *Hypoderma* infections in human hosts produce migratory subcutaneous swellings or harmful invasion of the eye.

Hypoderma spp are distributed widely in the northern hemisphere and North Africa. The warble fly adults are large, hairy, and beelike, and appear brown, orange, or yellow in color. With vestigial mouthparts, the adults do not feed during their short lifespan (as short as five days). The adults lay eggs on the foreleg of the affected cattle, which are subsequently ingested by licking. The eggs pass through esophagus muscles and spinal cord before reemerging subcutaneously as larvae, which cause many swellings ("warbles") under the skin. The larvae do not burrow into the flesh, but stay under the skin (thus called *Hypoderma*). When accidentally destroyed by pressure, the larvae can cause large purulent swellings. The warbles puncture the skin when coming out of the subcutaneous swellings, producing numerous holes in the hide that reduce its value. The migrating larvae can also damage meat as the tunnels they make fill with a substance known as butcher's jelly. The affected animals have a lower daily weight gain, and reduced milk yields. The pour-on application of an organophosphorus compound has helped eradication of warble fly in Denmark, Germany, and the United Kingdom. Preventive treatment involves the subcutaneous use of ivermectin.

77.1.1.3 The Family Sarcophagidae

The family Sarcophagidae (flesh flies, from the Greek sarco = flesh, phage = eating) is classified in the superfamily Oestroidea, section Schizophora, order Diptera, class Insecta, and phylum Arthropoda. The family is further differentiated into three subfamilies: Miltogramminae, Paramacronychiinae, and Sarcophaginae, with 108 genera between them.

The subfamily Sarcophaginae consists of 27 genera (*Alisarcophaga*, *Australopierretia*, *Bellieriomima*, *Bercaea*, *Bercaeopsis*, *Boettcherisca*, *Fengia*, *Harpagophalla*, *Helicophagella*, *Heteronychia*, *Hosarcophaga*, *Iranihindia*, *Lioproctia*, *Liopygia*, *Liosarcophaga*, *Neobellieria*, *Pandelleana*, *Pandelleisca*, *Parasarcophaga*, *Pseudothyrsocnema*, *Robineauella*, *Sarcophaga*, *Sarcorohdendorfia*, *Sarcorohdendrofia*, *Seniorwhitea*, *Thyrsocnema*, and *Wohlfahrtiopsis*), of which the genus *Sarcophaga* (with three species *Sarcophaga carnaria*, *Sarcophaga subvicina*, and *Sarcophaga variegate*) is of medical importance.

The subfamily Paramacronychiinae comprises five genera (*Brachicoma*, *Goniophyto*, *Nyctia*, *Sarcophila*, and *Wohlfahrtia*), of which the genus *Wohlfahrtia* (containing *Wohlfahrtia aschersoni*, *Wohlfahrtia bella*, *Wohlfahrtia magnifica*, *Wohlfahrtia monegrosensis*, *Wohlfahrtia nuba*, *Wohlfahrtia trina*, and *Wohlfahrtia vigil*) is of medical importance. *W. magnifica* is found in southeastern Europe, southern and Asiatic Russia, the Middle East, and North Africa, and its larvae infest the ear, eye, and nose, causing damages to living tissues. The larvae of the North American species, *W. vigil* and *Wohlfahrtia opaca*, infect only infants, as they are incapable of penetrating adult skin.

Sarcophagidae adults are medium-sized flies (about 12 mm from head to tail) with black and gray longitudinal stripes on the thorax and checkerboard-like pattern on the abdomen. The antennae are 3-segmented, with an arista, branched vein, frontal suture, well-developed calypters, and four notopleural bristles (setae) (short, long, short, long, from front to rear). The males of the Sarcophagidae family can be identified by examination of dissected genitalia.

Flesh flies differ from the tachinid flies by their lack of the postscutellum, and their large swelling underneath the scutellum on the thorax. Flesh flies differ from the Muscid flies by their possession of two sets of bristles (setae) on each side of the thorax and under the base of the wing, as Muscid flies rarely have both sets. Flesh flies differ from blowflies by having highly contrasting black and gray stripes on the thorax and a checkerboard-like pattern on the abdomen, and never being metallic colored.

W. magnifica (the spotted flesh fly, or sometimes called the screwworm fly) adults measure 6–10 mm in length, three instar larvae measure 5–7 mm in length. *W. magnifica* larvae infest genitalia or open wounds in sheep, cattle, goat, and horses, and the ear, eye, mouth, nose, and open wounds (including post surgery) in human, causing damages to living tissues, and blood poisoning. *W. magnifica* is found in southeastern Europe, southern and Asiatic Russia, China, the Middle East, and North Africa.

The larvae of the North American species, *W. vigil* and *W. opaca*, affects only infants, as they are incapable of penetrating adult skin.

Flesh flies often breed in carrion, dung, or decaying material, but a few species lay eggs in the open wounds of mammals. Being viviparous, flesh flies frequently give birth to live young on corpses of human and other animals. Some flesh fly larvae (maggots) are internal parasites of other insects. Occasionally, flesh fly maggots eat other smaller larvae (of grasshoppers, beetles, snails, and caterpillars, especially the forest tent caterpillar) that get in the way. Flesh flies and their larvae also eat decaying vegetable matter and excrement and thus are found around compost piles and pit latrines.

Flesh-flies may carry asymptomatic leprosy infections and transmit intestinal pseudomyiasis to people who eat the flesh-fly larvae.

77.1.2 Clinical Features, Pathogenesis, and Treatment

77.1.2.1 Clinical Features

Over 50 fly species have been reported to cause human myiasis in various parts of the world [18–24]. Depending on their preference of host living or dead tissues and decaying plants, myiasis-causing flies are separated into obligatory (specific), facultative (semispecific), or accidental (nonspecific) parasites. In general, blowflies and screwworm flies (Calliphoridae) and flesh flies (Sarcophagidae) are obligate and/or facultative parasites that live in host wounds or body orifices for a relatively short period (4–7 days), whereas botflies (Oestridae) are obligate parasites that persist in host nasopharyngeal tract, gut system, and/or internal organs and subcutaneous tissue, for several weeks to months [25].

77.1.2.1.1 Obligatory (Specific) Parasites

The human botfly (*D. hominis*) and the tumbu fly (*C. anthropophaga*) in the blowfly family Calliphoridae cause obligatory myiasis, and their larval stages occur only in the living tissue of animal or human hosts. Other obligatory flies include *Oestrus ovis* (sheep botfly), *Hypoderma* spp (cattle botflies or ox warbles), *Gasterophilus* spp (horse botfly), *C. hominivorax* (New World screwworm fly), *C. bezziana* (Old World screwworm fly), *Cuterebra* spp (rodent and rabbit botfly), *Rhinoestrus* spp, and *Wohlfahrtia* spp.

77.1.2.1.2 Facultative (Semispecific) Parasites

Members of the family Sarcophagidae are facultative parasites that normally develop in decomposing tissue, and their larvae parasitize wounds and other damaged tissues. However, some species in the genera *Calliphora* (blowfly) and *Lucilia* (green-bottle fly) may invade living tissues adjacent to the wound. Other semispecific flies include *Cochliomyia* spp (blue-bottle fly), *Phormia* spp (black-bottle fly), and *Sarcophaga* spp (flesh fly or sarcophagids).

77.1.2.1.3 Accidental (Nonspecific) Parasites

Some Sarcophagidae species cause accidental myiasis when egg-stage flies are ingested on contaminated food or come in contact with the genitourinary tract (as flies are attracted to feces and urine). Ingestion of food contaminated with fly larvae may produce vomiting, diarrhea, abdominal pain, and anal pruritus.

Based on the site (location, or area) of infection in the host, myiasis cause by flies may be of the following categories:

i. Cutaneous myiasis. Cutaneous myiasis can be divided into furuncular (furuncle- or boil-like), creeping (migratory), and wound (traumatic) myiasis [26]. The flies responsible for furuncular myiasis include *D. hominis*, *C. anthropophaga*, *W. vigil*, and *Cuterebra* species [27–40]. The flies causing creeping myiasis are *Gasterophilus* and *Hypoderma*. *H. bovis* and *H. lineatum* can cause creeping or migratory myiasis when persons come in close contact with infected cattle. The flies associated with wound myiasis include *C. hominivorax* and *C. bezziana*, and *W. magnifica* [41–43].

Myiasis due to the tumbu fly, *C. anthropophaga*, commonly appears on the trunk, buttocks, and thighs, causing erythematous nodules on the skin, and crusted, odoriferous, purulent, or serosanguinous discharge, whereas myiasis due to the human botfly, *D. hominis*, occurs on exposed sites (e.g., the scalp, face, forearms, and legs). Abrasions and wounds may be inhabited by several larvae (sometimes of different species), and furuncular and migratory myiatic lesions may contain only one or few larvae. While some species of larvae remain within the decaying tissue, other species may feed on living tissue nearby, producing subcutaneous nodules.

Larval infestation of cutaneous tissue may lead to the development of a pruritic papule (of 2–3 mm in diameter) within 24 h of initial contact with the larvae. The papule extends gradually to form a lesion with a diameter of 1.0–3.5 cm, and a height of 0.5 and 1.0 cm. The respiratory sinuses and occasionally the posterior end of the larva may be observed through a central punctum of 2–3 mm in diameter. The patient may show a sensation of pain caused by the tearing of host tissues as the larva feeds or from the outer spines irritating the surrounding skin as the larva moves.

Lane et al. [44] presented a case of furuncular myiasis caused by *D. hominis* in a 54-year-old patient who has a travel history to an endemic region in Central and South America.

Fujisaki et al. [45] described a case of exotic myiasis due to the tumbu fly (*C. anthropophaga*) in Japan. A 40-year-old man acquired the infection after traveling in Namibia, and presented with discharge in the left anterior pectoral region and left dorsal region. A hard red bump of 5–8 cm in diameter was noted around each site of infestation, and part of the white body of a larva (the spiracle) was seen in the center of the lesion. Eighteen fly larvae (with the largest being 12 mm in length) were removed from the lesions with forceps. The larvae preserved in 10% formalin showed a barrel-shaped body, with many mucrones on the segments under light microscopy. No fleshy appendages were detected around the posterior spiracle, and the slit on the posterior spiracle was very linear. These features suggested that they were the third stage (final stage) of *C. anthropophaga*. The sites of infestation subsequently healed. In a separate study, Geary et al. [46] documented a case of furuncular myiasis in a 57-year-old woman due to a third-instar larva of the Lund's fly *C. rodhaini*. The patient returned to Australia from a four-week holiday in East Africa and presented with a painful, discharging lesion on the outer aspect of her right thigh. On examination, something moving was noted at the base of the lesion. After occlusion with petroleum jelly, a fly larva protruded from the lesion and was removed with forceps. The larva was identified as belonging to *C. rodhaini* (Lund's fly). The lesion healed over the following two weeks.

Royce et al. [47] reported a case in an infant myiasis due to the horse botfly (*Gasterophilus haemorroidalis*), which occurred in a rural area of the Pacific Northwest (Washington) in late summer. The infant developed a red papular lesion (0.75 × 1 cm) on chest wall, and a larva was later recovered from the lesion, which was identified as the second instar of horse botfly *G. haemorroidalis*.

Karaman et al. [48] described a case of otomyiasis in a 57-year-old woman caused by *W. magnifica*. A combination of suctioning and alligator forceps was used to remove larvae. In addition, Tuygun et al. [49] reported a case of cutaneous myiasis due to a second-stage larva of *W. magnifica* in an 8-year-old boy. Following removal of the larva, the clinical and hematological manifestations returned to normal.

Romero-Cabello et al. [50] documented a case of cutaneous myiasis in a 62-year-old woman due to *C. bezziana* larvae. The patient had a complex vascular cutaneous anomaly in her lower right extremity for 8 years, and had a nonlimping walk with pink and painful feet and an ulcerative lesion on the internal surface of the right leg above the internal malleolus. The ulcer was large, clean, without evidence of infection, and had tissue in the process of granulation. Adjacent to the upper edge of this lesion was a second, crater-like ulcer of 2.5 cm in diameter from which drained an abundant, highly purulent, serohematic material containing worms. The worms were subsequently identified as *C. bezziana* larvae.

Starr et al. [51] described a protracted illness caused by infestation with *H. lineatum* in a 35-year-old man who had a 9-month multisystemic illness with pronounced eosinophilia, pleuritis, pericarditis, and myositis. Myiasis was recognized when a worm emerged from the patient's skin; after a second worm emerged, the patient's symptoms disappeared rapidly.

ii. Myiasis of body cavities. The human botfly (*D. hominis*), the Old World screwworm (*C. bezziana*) and the New World screwworm (*C. hominivorax*) may invade the eye, orbit, nose, ear canal, or nasopharynx, in addition to wounds and open sores, causing invasive myiasis [52–56]. When larvae gain access to the brain, meningitis and even death may result. For example, *C. bezziana* eggs are often laid on open sores, ulcers, scratches externally, tick bite wound, or mucus membranes throughout the body. After the egg hatching, the larvae burrow as deep as 15 cm into the host's living tissue to feed, causing tissue damage, and a characteristic odor.

Ng et al. [57] described a case of human infestation of *C. bezziana* in Hong Kong involving an 89-year-old woman who had previously had a stroke. The patient presented fever, a small fissure at the labial gingiva of the upper incisors, and several ulcerative lesions at the hard palate. A live maggot was seen protruding from the small fissure, and a total of seven maggots were removed by forceps. Computed tomography and magnetic resonance imaging of the oral cavity revealed an ulcerative soft-tissue lesion over the anterior palate, with a fistula communicating to the labial gingiva. Manual removal of maggots and surgical debridement contributed the patient's recovery. Sharma and Hedge [58] also reported a case of oral myiasis caused by *C. bezziana* in a 12-year-old boy with neurodegenerative disease with seizures. Maggots were noted in a soft-tissue pocket in the left buccal mucosa and a pocket under the palatal mucosa. After manual removal of maggots followed by irrigation and application of turpentine oil along with a single-dose Ivermectin 3 mg, the patient showed complete healing of the lesion within 10 days without any recurrence in a follow-up period of 1 year. In addition, Gealh et al. [59] presented two cases of oral myiasis due to *C. hominivorax*, which was successfully treated with ivermectin.

iii. Intestinal myiasis. When fly eggs or larvae previously deposited in food are ingested and survive in the gastrointestinal tract, it is called intestinal myiasis. While some infested patients are asymptomatic, others may display abdominal pain, vomiting, and diarrhea. It appears that the internal myiasis is one of the most common forms of diseases due to flies. This is followed by cutaneous, aural, nasopharyngeal, ocular, wound, urinary myiasis. Most myiasis cases occur during the warmer months (April through September), when fly populations are abundant.

Das et al. [60] reported a human case of intestinal myiasis caused by Sarcophaga larvae in a 25-year-old male. The patient presented with the symptom of passage of live worms in stool. The saline-washed fresh worms (about 4–5 mm long and 0.5–1 mm wide) were small, white, with segmented structures. Under microscope, the larvae revealed 12 segments, each with short spines at its posterior margin. It had a broader posterior end and a tapering anterior end with two oral hooks and mouth brushes. The posterior end of the larvae was cut and immersed in 10% NaOH (clearing agent) for 6 h, and posterior spiracles showed two parallel slits indicative of a second instar larva. The posterior spiracles were located inside a fossa (hidden spiracles), characteristic of genus *Sarcophaga*. The patient was treated with osmotic purgative polyethylene glycol (PEG) consisting of 137.15 g of PEG in 2 L of water during a period of 2–3 h, resulting in 3–5 loose motions in the next 4–5 h. The patient recovered completely with cessation of maggots in stool after treatment.

77.1.2.2 Pathogenesis

As obligate, host-specific gastrointestinal or subdermal larval parasites, members of the Oestridae family usually show high host-specificity and low levels of pathogenicity. Besides showing the capacity to survive the immune challenge, the high body temperatures of vertebrates and the differences in diet presented by a living vertebrate host, the Oestridae family adapts to a parasitic lifestyle by undergoing changes in larval morphology (with the formation of large mouth-hooks for attaching and rows of well-developed spines for migrating within the host) and enzymic repertoire (through the secretion of proteases for survival in the host immune system).

On the other hand, species of the Calliphoridae and Sarcophagidae families tend to display lower host-specificity, shorter periods of larval feeding, a superficial or cutaneous location in the host, and relatively high levels of pathogenicity. Despite the fact that the saprophagous obligate or facultatively parasitic sister species of the Calliphoridae and Sarcophagidae families are all morphologically similar and are distinguished by only minor character differences, they behave quite differently in their temperature tolerance, response to immune challenge, production of enzymes (particularly those associated with feeding on the tissues of a living host), and possible responses to semichemical cues that enable the parasitic species to locate and oviposit on a living host.

77.1.2.3 Treatment

Although myiasis is a self-limiting disease that often poses no danger to the host, the psychological distress associated with maggot infestation necessitates the treatment. Methods for treating myiasis include occlusive salves and dressings followed by manual removal for cutaneous myiasis; manual removal of larvae in aural, genitourinary, and nasopharyngeal myiasis; application of a 15% chloroform in light vegetable oil solution followed by manual removal in wound myiasis; administration of a mild cathartic agent in intestinal myiasis; use of steroids, photocoagulation, and surgery in ocular myiasis; use of machine oil to increase the transparency of the skin followed by removal of the larva with a needle in creeping or migratory myiasis; and washing of larvae from the wound, or surgical removal in wound myiasis [61].

Specifically, surgical debridgement and biomedical treatment, occlusion/suffocation, and other approaches can be utilized to treat furuncular cutaneous myiasis [62].

77.1.2.3.1 Surgical Debridgement

In case the larva has died inside the lesion, surgical incision and extraction of the larva are often performed under local anesthesia. Precaution to prevent laceration of the larva is important, which may produce inflammatory response, bacterial infection, or the formation of a granuloma. Following surgical removal of larvae, administration of antimicrobials to control secondary infections and tetanus prophylaxis are valuable.

77.1.2.3.2 Occlusion/Suffocation

These approaches utilize petroleum jelly, heavy oil, liquid paraffin, beeswax, raw meat, nail polish, adhesive tape, butter, chewing gum, and mineral oil to block the larva's respiratory sinuses in the central punctum, forcing the larva to the surface in search of air and allowing removal with the aid of forceps or tweezers.

77.1.2.3.3 Other Strategies

The injection of lidocaine at the base of the tissue cavity which the larva inhabits causes local swelling which forces the larva to the surface for easy grasp and removal. Alternatively, by exerting manual pressure on the burrow with two wooden spatulas may also force the larva out.

Prevention of myiasis involves controlling the source of the larvae and the ovipositing female fly through the application of insecticides (organophosphorus or organochlorine) and insect repellents containing diethyltoluamide (DEET) and the use of mosquito netting. Other important contributors to the elimination of myiasis in human populations include improved access to primary healthcare, improvements in housing, improved sanitation, and reduction in the proximity of human habitations to livestock.

The fact that maggots of facultative species infest decomposing wound tissue but do not feed on adjacent living tissue provides the theoretical foundation for maggot debridement therapy for cleaning nonhealing wounds, such as pressure ulcers, diabetic foot wounds, venous stasis ulcers, and postsurgical wounds. Medicinal maggots eat away the dead, infected tissue, kill off the bacteria, and stimulate wound healing. The common green bottle fly (*Lucilia sericata* or *Phaenicia sericata*) is the preferred species for the maggot therapy because it does not eat healthy tissue. Medicinal maggots are placed on the wound and covered with a sterile dressing of gauze and nylon mesh. Too many larvae placed on the wound may result in healthy tissue being devoured [63].

77.1.3 Diagnosis

Clinical diagnosis of myiasis is based on several key features, including (i) recent travel to an endemic area; (ii) nonhealing lesions on exposed areas of skin; (iii) serous, serosanguineous, or seropurulent drainage from a central punctum; (iv) visible small, white, threadlike structure(s) (larva) protruding from the lesion; (v) local symptoms of pruritis, pain, movement, or tenderness [64–67]. Differential diagnosis includes early-stage leishmaniasis, onchocerciasis, or tungiasis, pyogenic infection, tropical ulcer, adenopathy, cellulitis or furunculosis, skin abscess, insect bites, and subcutaneous cysts. Use of ultrasound may further assist the discrimination of myiasis from other diseases [68].

In general, the species identity of Calliphoridae and Sarcophagidae responsible for the infestation cannot accurately be determined on the basis of the fly's larval stage because of their close morphological resemblance and variations within specimens of larvae collected from different animals and regions. Therefore, larvae obtained from a patient are reared on meat or a synthetic medium to pupate and develop into an adult fly, which can then be identified by their distinct morphological characteristics [69].

Given that the culture techniques for myiasis-causing flies are time consuming, and that in case of improperly preserved adult fly specimens, the morphological differentiation is difficult if not impossible, several nonculture-based methods have been described as alternative for diagnosis of myiases. For example, immunodiagnostic methods using excretory–secretory (ES) products and other antigenic preparations provide a useful approach for clinical and postmortem examinations of myiasis-causing flies [70–73]. More recently, molecular procedures targeting various gene regions have been increasingly applied for their identification and phylogenetic analysis [74–78].

Among the commonly targeted genes for molecular studies of myiasis-causing Oestridae, Calliphoridae, and Sarcophagidae species are nuclear 16S rRNA and 18S rRNA, 12S mitochondrial DNA (mtDNA), cytochrome oxidase I and II (COI and COII), transfer RNA (tRNA), NADH dehydrogenase, and other genes [45,79–88].

Nuclear and mitochondrial genes and intergene regions demonstrate notable sequence variations, with introns and noncoding regions, especially the internal transcribed spacer (ITS) of rRNA, exhibiting a higher mutational rate than the coding genes of mtDNA. For species-specific differentiation, analysis of a target gene or region with a low intraspecific variability (e.g., 18S, 5.8S, and 28S rRNA genes) is desirable. On the other hand, for the discrimination of two larval populations of the same species originating from two well-distinguished geographic areas, examination of a target gene with a high level of intraspecific variation (e.g., rRNA ITS-1 and ITS-2) is preferable [87].

Because of its high copy number and its inclusion of conserved sequences (two rRNA genes, 13 protein coding genes, and 22 tRNA genes), mtDNA has been used extensively for molecular identification of insects [89–93]. The mitochondrial gene encoding COI is particularly suitable as a molecular marker, due to its large size, with both highly conserved and variable regions. Indeed, the COI amino acid sequence comprises 12 transmembrane helices (M1–M12), six external loops (E1–E6), five internal loops (I1–I5), one carboxyl (COOH), and one amino (NH_2) terminal. While the COOH terminal region is the most variable part, the E1, M3, E2, I2, I4, M9, and M12 regions are not far behind. However, the fact that some mitochondrial pseudogenes are integrated into a nuclear genome may sometimes impact on the reliability of mitochondria-based studies [87,94].

By targeting a large fragment of the COI gene (1300 bp spanning from M3 to the terminal carboxylic region), Otranto et al. [86] showed that the Oestridae family contained two branches: a branch for Gasterophilinae (*G. intestinalis*) and a branch including Hypodermatinae (*H. bovis*, *H. lineatum*, and *P. silenus*) and Oestrinae (*O. ovis*). Subsequent examination of a smaller target region spanning from E4 to the COOH sequence (688 bp) permitted the molecular differentiation of larvae of five species of genus *Hypoderma* (i.e., *H. bovis*, *H. lineatum*, *H. tarandi*, *H. diana*, and *H. actaeon*), and confirmed the monophylic characteristics of Calliphoridae and Oestridae families.

77.2 METHODS

77.2.1 Sample Preparation

77.2.1.1 Sample Collection

Myiasis-causing larvae are collected from patients by forceps, by gentle pressure around the site of infestation, by occlusion of the opening (to prevent larval respiration), or by injection of anesthetic (e.g., lidocaine hydrochloride) into the cavity below the larva to force it out [95–97].

For light microscopic examination, mouthparts, spiracles, and other material are first cleared by macerating the specimen in a 10% aqueous solution of potassium hydroxide (KOH) at room temperature for at least 15 min. Specimens that have been in alcohol for six months or more need a longer period (up to 12 h) in KOH at room temperature or a shorter time in warmed KOH. Small larvae are put into the solution whole, with punctures to allow its penetration; larger larvae are dissected and only the required parts macerated (taking care to keep all the parts of a single specimen together). The muscle and fat-body are cleared away with fine forceps or sharp needles. Care is taken to avoid destroying the sclerites. The specimens are then placed in glacial acetic acid for at least 15 min to neutralize any residual KOH, and are rinsed well with 80% ethanol. Thorough rinsing with ethanol is sufficient if acetic acid is not available. The specimens are now ready for examination, mounting, or storage. A common way to mount the specimen on slides is to dehydrate the samples in absolute alcohol, transfer to clove oil to clear, and mount in Canada balsam. Alternatively, the specimens can be mounted directly from absolute alcohol into Euparal [97].

Because larvae are more difficult to identify than adults, it is valuable to keep some larvae alive and to rear to adulthood for confirmation. Thus, larvae may be placed in a jar with sawdust or paper tissue and a perforated lid (to enable gaseous exchange) and maintained at room temperature. If the larvae are very young and of a facultative myiasis-causing species meat (e.g., liver, minced meat, or canned dog food) is added to enable larvae to feed and complete their development (with full color and rigidity). Adult flies are killed by placing them in a deep freezer for 1 h. Good-quality specimen are obtained by first killing the larvae by immersion in near boiling water (>80°C) for 15–30 s. After killing, the larvae are transferred to 80% ethanol for long-term storage. The preservative medium is changed after 24 h as it will be diluted by the water on and in the larvae. Dead or wet insects will quickly decompose if not dry out, while dry insects will keep well without preservative [97].

77.2.1.2 DNA Extraction

Individual larvae (or flies) are washed in phosphate-buffered saline (PBS) and longitudinally cut to retrieve the internal organs. DNA is prepared from individual larvae using the QIAamp Tissue kit (Qiagen) and kept at 4°C until analysis [86].

TABLE 77.1
PCR Primers from COI Gene

Primer	Sequence (5′–3′)	Expected Product
UEA1	GAATAATGAATAATTCCCATAAATAGATTTACA	1350 bp (UEA8); 1630 bp (UEA10)
UEA3	TATAGCATTCCCACGACGAATAAATAA	1020 bp (UEA8); 1300 bp (UEA10)
UEA5	AGTTTTAGCAGGAGCAATTACTAT	690 bp (UEA8); 970 bp (UEA10)
UEA8	AAAAATGTTGAGGGAAAAATGTTA	
UEA10	TCCAATGCACTAATCTGCCATATTA	

77.2.2 Detection Procedures

77.2.2.1 Polymerase Chain Reaction–Restriction Fragment-Length Polymorphism Identification of Oestridae Larvae

Otranto et al. [86] described a PCR-restriction fragment length polymorphism (PCR–RFLP assay for the identification of Oestridae larvae causing myiasis, that is, *Gasterophilus* spp (equine gasterophilosis), *H. bovis* and *H. lineatum* (bovine hypodermosis), *Oestrus ovis* (sheep and goat oestrosis), and *Przhevalskiana silenus* (goat warble fly infestation). The primers were designed from the COI gene (Table 77.1). After PCR amplification, the products were digested with restriction enzymes *Taq*$^{\alpha}$I and *Rsa*I.

77.2.2.1.1 Procedure

1. PCR mixture (25 μL) is made up of 2.5 mM $MgCl_2$, 200 μM of each dNTP, 0.5 U of Taq Gold polymerase (Perkin-Elmer), 1× PCR reaction buffer, 50 pmol of each primer, and 10–20 ng of DNA template.
2. Amplification is conducted with 40 cycles of 1 min at 95°C, 1 min at 52°C, and 1 min at 72°C followed by a final 7 min at 72°C in a thermal cycler.
3. PCR products (5 μL) are resolved on 1.4% tris base/boric acid/ethylenediaminetetraacetic acid (TBE) agarose gel and visualized under ultra violet (UV) light by ethidium bromide staining. The six primer combinations yield amplicons of 690–1630 bp.
4. Restriction analysis is set up with amplicons (1020–1630 bp) from primer combinations UEA1–UEA8, UEA1–UEA10, UEA3–UEA8, and UEA3–UEA10. The digestion mixture (20 μL) consists of 15 μL of PCR product, 3 U of restriction enzyme (*Taq*$^{\alpha}$I, *Rsa*I) and 2 μL of 10× enzyme buffer. The mixture is incubated for 2 h at 37°C (*Rsa*I) or at 65°C (*Taq*$^{\alpha}$I). Digestion patterns are visualized on a 2% Ambion agarose gel after ethidium bromide staining, along with 100 bp DNA ladder (New England Biolabs), and photographed.

Note: PCR–RFLP assay based on *Taq*$^{\alpha}$I and *Rsa*I digestion of the COI gene products allows the differentiation of larvae from four Oestridae genera (*Gasterophilus*, *Hypoderma*, *Oestrus*, and *Przhevalskiana*). For example, *Taq*$^{\alpha}$I digestion of PCR amplicons from the primer combination of UEA3–UEA10 results in the production of fragments of 420 and 670 bp for *Gasterophilus* spp, 280, 400, and 700 bp for *Hypoderma* spp, 280, 420, and 670 bp for *Oestrus*, and 480 and 500 bp for *Przhevalskiana*. In a separate study, *Bfa*I enzyme digestion of the PCR amplicons from the COI terminal region of five larvae belonging to the *Hypoderma* genus facilitates their differentiation (producing fragments of ~700 bp in *H. bovis*, 250 and 450 bp in *H. lineatum*, 100 and 600 bp in *H. diana*, 100, 220, 480 bp in *H. actaeon*, and 130 and 570 bp in *H. tarandi*) [87].

77.2.2.2 Sequence Analysis of 28S rRNA D1–D2 Region

Fujisaki et al. [45] utilized primers (forward: 5′-CCC CCT GAA TTT AAG CAT AT-3′ and reverse: 5′ GTT AGA CTC CTT GGT CCG TG-3′) to amplify and sequence the 28S rRNA D1–D2 region for identification of myiasis-causing flies such as *C. anthropophaga*.

77.2.2.2.1 Procedure

1. PCR mixture (25 μL) is made up of 2.5 mM $MgCl_2$, 200 μM of each dNTP, 1 U of Taq DNA polymerase, 1× PCR reaction buffer, 50 pmol of each primer, and 10–20 ng of DNA template.
2. The mixture is subjected to 30 cycles of 94°C for 1 min, 55°C for 2 min, 72°C for 1.5 min. This is followed by a final extension at 72°C for 10 min.
3. A small aliquot of the amplified product (620 bp) is separated on 1.2% agarose gel electrophoresis, stained with ethidium bromide and visualized under UV light.
4. The remaining PCR product is purified, sequenced with a DNA sequencing kit (Applied Biosystems), and analyzed on an automated sequencer (ABI Genetic Analyzer 310; Applied Biosystems).
5. The resulting sequence (620 bp) is compared with those at DDBJ/EMBL/GenBank databases using BLAST (Basic Local Alignment Search Tool) analysis.

77.3 CONCLUSION

Myiasis is an infestation of living or dead tissues (or liquid body substances) of living vertebrates by dipterous (two-winged) larvae (or maggots) that commonly occurs in tropical and subtropical regions of the world. Although this infection is largely self-limiting and seldom results in death, its economical impact on animal production, its damage to hides, and its treatment are significant. In addition, myiasis has also attracted the attention of forensic entomologists because some human parasitic flies provide potential clues to crime investigations.

The causative agents for myiasis belong to three dipteran families: Calliphoridae (blowflies), Sarcophagidae (flesh flies), and Oestridae (botflies and warble flies). The Calliphoridae and Sarcophagidae families are characterized by having musciform larvae that are segmented and pointed at the anterior end with no head capsule, and broad and usually truncate at the posterior end. Their head and first thoracic segment contain the cephaloskeleton bearing the mouth hooks, and the last segment bears the posterior spiracles and the anus.

While the Oestridae family contains obligate host-specific endoparasites with relatively high host specificity and relatively low pathogenicity, the Calliphoridae and Sarcophagidae families are often less host specific, with shorter periods of larval feeding and higher pathogenicity, together with a broad range of interaction with their hosts, from saprophagous, facultative, to obligate parasitism.

Considering the difficulty in the morphological identification of myiasis-causing larvae and the time it takes to cultivate larvae to adult flies for accurate determination, immunological and molecular techniques have been developed and increasingly applied for the diagnosis of myiasis and for phylogenetic assessment of myiasis-causing flies. In particular, by targeting nuclear 16S rRNA and 18S rRNA, 12S mtDNA, COI and COII, tRNA, NADH dehydrogenase, and other genes, it is possible to determine the species identity and phylogenetic relationship of various myiasis-causing Oestridae, Calliphoridae, and Sarcophagidae parasites rapidly, sensitively, and accurately.

REFERENCES

1. Noutsis, C., and Millikan, L.E., Myiasis. *Dermatol. Clin.*, 12, 729, 1994.
2. Millikan, L.E., Myiasis. *Clin. Dermatol.*, 17, 191, 1999.
3. Stevens, J.R., and Wallman, J.F., The evolution of myiasis in humans and other animals in the Old and New Worlds (part I): Phylogenetic analyses. *Trends Parasitol.*, 22, 129, 2006.
4. Stevens, J.R. et al., The evolution of myiasis in humans and other animals in the Old and New Worlds (part II): Biological and life-history studies. *Trends Parasitol.*, 22, 181, 2006.
5. Erzinçlioglu, Y.Z., The origin of parasitism in blowflies. *Brit. J. Entomol. Nat. Hist.*, 2, 125, 1989.
6. Rognes, K., The Calliphoridae (blowflies) (Diptera: Oestroidea) are not a monophyletic group. *Cladistic*, 13, 27, 1997.
7. Stevens, J.R., The evolution of myiasis in blowflies (Calliphoridae). *Int. J. Parasitol.*, 33, 1105, 2003.
8. Wallman, J.F., Third-instar larvae of common carrion-breeding blowflies of the genus *Calliphora* in South Australia. *Invertebr. Taxon*, 15, 37, 2001.
9. Alexander, J.L., Screwworms. *J. Am. Vet. Med. Assoc.*, 228, 357, 2006.
10. Stevens, J., and Wall, R., Species, subspecies and hybrid populations of the blowflies *Lucilia cuprina* and *Lucilia sericata* (Diptera: Calliphoridae). *Proc. Biol. Sci.*, 263, 1335, 1996.
11. Stevens, J., and Wall, R., The evolution of ectoparasitism in the genus *Lucilia* (Diptera: Calliphoridae). *Int. J. Parasitol.*, 27, 51, 1997.
12. Stevens, J., and Wall, R., Genetic relationships between blowflies (Calliphoridae) of forensic importance. *Forensic Sci. Int.*, 120, 116, 2001.
13. Norris, K.R., Evidence for the multiple exotic origin of Australian populations of the sheep blowfly, *Lucilia cuprina* (Wiedemann) (Diptera: Calliphoridae). *Aust. J. Zool.*, 38, 635, 1990.
14. Hall, M.J.R. et al., Old World screwworm fly, *Chrysomya bezziana*, occurs as two geographical races. *Med. Vet. Entomol.*, 15, 393, 2001.
15. Wyss, J., Screwworm eradication in the Americas. *Ann. N.Y. Acad. Sci.*, 916, 186, 2000.
16. Coronado, A., and Kowalski, A., Current status of the New World screwworm Cochliomyia hominivorax in Venezuela. *Med. Vet. Entomol.*, 23 (Suppl 1), 106, 2009.
17. Pape, T., Phylogeny of Oestridae (Insecta: Diptera). *Syst. Entomol.*, 26, 133, 2001.
18. Hall, M.J.R., and Wall, R., Myiasis of humans and domestic animals. *Adv. Parasitol.*, 35, 257, 1995.
19. Powers, N., and Yorgensen, M., Myiasis in humans: An overview and a report of two cases in the Republic of Panama. *Mil. Med.*, 161, 495, 1996.
20. MacNamara, A., and Durham, S., *Dermatobia hominis* in the accident and emergency department: "I've got you under my skin." *J. Accid. Emerg. Med.*, 14, 179, 1997.
21. Bapat, S., Neonatal myiasis. *Pediatrics*, 106, e6, 2000.
22. Jiang, C., A collective analysis on 54 cases of human myiasis in China from 1995-2001. *Chin. Med. J.* (Engl), 115, 1445, 2002.
23. Cestari, T.F., Pessato, S., and Ramos-e-Silva, M., Tungiasis and myiasis. *Clin, Dermatol.*, 25, 158, 2007.
24. Paris, L.A. et al., Pin-site myiasis: A rare complication of a treated open fracture of tibia. *Surg. Infect.* (Larchmt), 9, 403, 2008.
25. Otranto, D., The immunology of myiasis: Parasite survival and host defense strategies. *Trends Parasitol.*, 17, 176, 2001.
26. Robbins, K., and Khachemoune, A., Cutaneous myiasis: A review of the common types of myiasis. *Int. J. Dermatol.*, 49, 1092, 2010.
27. Gordon, P. et al., Cutaneous myiasis due to *Dematobia hominis*: A report of six cases. *Brit. J. Dermatol.*, 132, 811, 1995.
28. Kpea, N., and Zywocinski, C., Flies in the flesh: A case report and review of cutaneous myiasis. *Cutis*, 55, 47, 1995.
29. Swetter, S., Stewart, M., and Smoller, B., Cutaneous myiasis following travel to Belize. *Int. J. Dermatol.*, 35, 118, 1996.
30. Tsuda, S. et al., Furuncular cutaneous myiasis caused by *Dermatobia hominis* larvae following travel to Brazil. *Int. J. Dermatol.*, 35, 121, 1996.
31. Shorter, N. et al., Furuncular cuterebrid myiasis. *J. Pediatr. Surg.*, 22, 1511, 1997.
32. Hasegawa, M. et al., An imported case of furuncular myiasis due to *Cordylobia anthropophaga* which emerged in Japan. *Br. J. Dermatol.*, 143, 912, 2000.
33. Jelinek, T. et al., Cutaneous myiasis: Review of 13 cases of travelers returning from tropical countries. *Int. J. Dermatol.*, 39, 689, 2000.
34. Robert, L., and Yelton, J., Imported furuncular myiasis in Germany. *Mil. Med.*, 167, 990, 2002.
35. Haruki, K. et al., Myiasis with Dermatobia hominis in a traveler returning from Costa Rica: Review of 33 cases imported from South America to Japan. *J. Travel Med.*, 12, 285, 2005.

36. Kokcam, I., and Saki, C.E., A case of cutaneous myiasis caused by *Wohlfahrtia magnifica*. *J. Dermatol.*, 32, 459, 2005.
37. Curtis, S.J. et al., Case of the month: Cutaneous myiasis in a returning traveler from the Algarve: First report of tumbu maggots, *Cordylobia anthropophaga*, acquired in Portugal. *Emerg. Med. J.*, 23, 236, 2006.
38. Delshad, E. et al., Cuterebra cutaneous myiasis: Case report and world literature review. *Int. J. Dermatol.*, 47, 363, 2008.
39. McGraw, T.A., and Turiansky, G.W., Cutaneous myiasis. *J. Am. Acad. Dermatol.*, 58, 907, 2008; quiz 58, 927, 2008.
40. Krajewski, A. et al., Cutaneous myiasis. *J. Plast Reconstr. Aesthet. Surg.*, 62, e383, 2009.
41. Sherman, R.A., Wound myiasis in urban and suburban United States. *Arch. Intern. Med.*, 160, 2004, 2000.
42. Hemmings, S.C., Matthews, K.J., and Alexander, J., Human myiasis in western Jamaica: Five years after the implementation of a screwworm eradication programme. *West Indian Med. J.*, 56, 341, 2007.
43. Sesterhenn, A.M. et al., Cutaneous manifestation of myiasis in malignant wounds of the head and neck. *Eur. J. Dermatol.*, 19, 64, 2009.
44. Lane, J.E. et al., Furuncular myiasis secondary to Dermatobia hominis. *J. Drugs Dermatol.*, 4, 365, 2005.
45. Fujisaki, R. et al., Exotic myiasis caused by 19 larvae of *Cordylobia anthropophaga* in Namibia and identified using molecular methods in Japan. *Trans. Royal Soc. Trop. Med. Hyg.*, 102, 599, 2008.
46. Geary, M.J. et al., Exotic myiasis with Lund's fly (*Cordylobia rodhaini*). *Med. J. Aust.*, 171, 654, 1999.
47. Royce, L.A. et al., Recovery of a second instar Gasterophilus larva in a human infant: A case report. *Am. J. Trop. Med. Hyg.*, 60, 403, 1999.
48. Karaman, E. et al., Otomyiasis by *Wohlfahrtia magnifica*. *J. Craniofac. Surg.*, 20, 2123, 2009.
49. Tuygun, N. et al., Furuncular myiasis in a child caused by *Wohlfahrtia magnifica* (Diptera: Sarcophagidae) associated with eosinophilia. *Turk. J. Pediatr.*, 51, 279, 2009.
50. Romero-Cabello, R. et al., Cutaneous myiasis caused by *Chrysomya bezziana* larvae, Mexico. *Emerg. Infect. Dis.*, 16, 2014, 2010.
51. Starr, J. et al., Myiasis due to *Hypoderma lineatum* infection mimicking the hypereosinophilic syndrome. *Mayo Clin. Proc.*, 75, 755, 2000.
52. Syrdalen, P., Nitter, T., and Mehl, R., Ophthalmomyiasis interna posterior: Report of case caused by the reindeer warble fly larva and review of previous reported cases. *Br. J. Ophthalmol.*, 66, 589, 1982.
53. Passos, M.R. et al., Vulvar myiasis. *Infect. Dis. Obstet. Gynecol.*, 6, 69, 1998.
54. Droma, E.B. et al., Oral myiasis: A case report and literature review. *Oral Surg. Oral. Med. Oral Pathol. Oral Radiol. Endod.*, 103, 92, 2007.
55. Lagacé-Wiens, P.R. et al., Human ophthalmomyiasis interna caused by *Hypoderma tarandi*, Northern Canada. *Emerg. Infect. Dis.*, 14, 64, 2008.
56. Terterov, S. et al., Posttraumatic human cerebral myiasis. *World Neurosurg.*, 73, 557, 2010.
57. Ng, K.H. et al., A case of oral myiasis due to *Chrysomya bezziana*. *Hong Kong Med. J.*, 9, 454, 2003.
58. Sharma, A., and Hedge, A., Primary oral myiasis due to *Chrysomya bezziana* treated with Ivermectin. A case report. *J. Clin. Pediatr. Dent.*, 34, 259, 2010.
59. Gealh, W.C. et al., Treatment of oral myiasis caused by *Cochliomyia hominivorax*: Two cases treated with ivermectin. *Br. J. Oral Maxillofac. Surg.*, 47, 23, 2009.
60. Das, A. et al., Accidental intestinal myiasis caused by genus *Sarcophaga*. *Indian J. Med. Microbiol.*, 28, 176, 2010.
61. Brewer, T. et al., Bacon therapy and furuncular myiasis. *JAMA*, 270, 2087, 1993.
62. Caissie, R. et al., Cutaneous myiasis: Diagnosis, treatment, and prevention. *J. Oral Maxillofac. Surg.*, 66, 560, 2008.
63. Sherman, R.A. et al., Medicinal maggots: An ancient remedy for some contemporary afflictions. *Annu. Rev. Entomol.*, 45, 55, 2000.
64. Kitching, J., Tropical myiasis: An unwanted holiday souvenir. *J. Accid. Emerg. Med.*, 14, 178, 1997.
65. Lucchina, L., Wilson, M., and Drake, L., Dermatology and the recently returned traveler: Infectious diseases with dermatologic manifestations. *Int. J. Dermatol.*, 36, 167, 1997.
66. Diaz, J.H., The epidemiology, diagnosis, management, and prevention of ectoparasitic diseases in travelers. *J. Travel Med.*, 13, 100, 2006.
67. Hakeem, M.J., and Bhattacharyya, D.N., Exotic human myiasis. *Travel Med. Infect. Dis.*, 7, 198, 2009.
68. Bowry, R., and Cottingham, R., Use of ultrasound to aid management of late presentation of *Dermatobia hominis* larva infestation. *J. Accid. Emerg. Med.*, 14, 177, 1997.
69. Greenberg, B., and Singh, D., Species identification of calliphorid (Diptera) eggs. *Med. Entomol.*, 32, 21, 1995.
70. Moiré, N. et al., Sequencing and gene expression of hypodermins A, B, C in larval stages of *Hypoderma lineatum*. *Mol. Biochem. Parasitol.*, 66, 233, 1994.
71. Boulard, C., Villejoubert, C., and Moiré, N., Cross-reactive, stage-specific antigens in the Oestridae family. *Vet. Res.*, 27, 535, 1996.
72. Goddard, P., Bates, P., and Webster, K.A., Evaluation of a direct ELISA for the serodiagnosis of *Oestrus ovis* infections in sheep. *Vet. Rec.*, 144, 497, 1999.
73. Figarola, J.L. et al., Identification of screwworm, *Cochliomyia hominivorax* (Coquerel) (Diptera: Calliphoridae), with a monoclonal antibody-based enzyme-linked immunosorbent assay (MAb-ELISA). *Vet. Parasitol.*, 102, 341, 2001.
74. Kocher T.D., and Xiong, B., Comparison of mitochondrial DNA sequences of seven morphospecies of black flies (Diptera: Simulidae). *Genome*, 34, 306, 1991.
75. Sperling, F.A.H., Anderson, G.S., and Hickey, D.A., A DNA-based approach to the identification of insect species used for post-mortem interval estimation. *J. Forensic Sci.*, 39, 418, 1994.
76. Infante-Malachias, M.E. et al., Random amplified polymorphic DNA of screwworm fly populations (Diptera: Calliphoridae) from Southeastern Brazil and Northern Argentina. *Genome*, 42, 772, 1999.
77. Caterino, M.S., Cho, S., and Sperling, F.A.H., The current state of insect molecular systematics. *Annu. Rev. Entomol.*, 45, 1, 2000.
78. Nirmala, X. et al., Molecular phylogeny of Calyptratae (Diptera: Brachycera): The evolution of 18S and 16S ribosomal rDNAs in higher dipterans and their use in phylogenetic inference. *Insect Mol. Biol.*, 10, 475, 2001.
79. Hillis, D.M., and Dixon, M.T., Ribosomal DNA: Molecular evolution and phylogenetic inference. *Q. Rev. Biol.*, 66, 411, 1991.
80. Lunt, D.H. et al., The insect cytochrome oxidase I gene: Evolutionary patterns and conserved primers for phylogenetic studies. *Ins. Mol. Biol.*, 5, 153, 1996.

81. Wells, J.D., and Sperling, F.A., Molecular phylogeny of *Chrysomya albiceps* and *C. rufifacies*. *J. Med. Entomol.*, 36, 222, 1999.
82. Wells, J.D. et al., DNA-based identification and molecular systematics of forensically important Sarcophagidae (Diptera). *J. Forensic Sci.*, 46, 1098, 2001.
83. Wells, J.D. et al., Human and insect mitochondrial DNA analysis from maggots. *J. Forensic Sci.*, 46, 685, 2001.
84. Lessinger, A.C., and Azeredo-Espin, A.M.L., Evolution and structural organisation of mitochondrial DNA control region of myiasis-causing flies. *Med. Vet. Entomol.*, 14, 71, 2000.
85. Lessinger, A.C. et al., The mitochondrial genome of the primary screwworm fly *Cochliomyia hominivorax* (Diptera: Calliphoridae). *Insect Mol. Biol.*, 9, 521, 2000.
86. Otranto, D. et al., Differentiation by polymerase chain reaction-restriction fragment length polymorphism of some Oestridae larvae causing myiasis. *Vet. Parasitol.*, 90, 305, 2000.
87. Otranto D., and Stevens, J.R., Molecular approaches to the study of myiasis-causing larvae. *Int. J. Parasitol.*, 32, 1345, 2002.
88. Otranto, D. et al., Utility of mitochondrial and ribosomal genes for the differentiation and phylogenesis of species of gastrointestinal botflies. *J. Econ. Entomol.*, 98, 2235, 2005.
89. Simon, C. et al., Evolution, weighting and phylogenetic utility of mitochondrial gene sequences and a compilation of conserved polymerase chain reaction primers. *Ann. Entomol. Soc. Am.*, 87, 651, 1994.
90. Taylor, D.B., Szalanski, A.L., and Peterson, R.D., Mitochondrial DNA variation in screwworm. *Med. Vet. Entomol.*, 10, 161, 1996.
91. Litjens, P., Lessinger, A.C., and Azeredo-Espin, A.M.L., Characterization of the screwworm flies *Cochliomyia hominivorax* and *Cochliomyia macellaria* by PCR-RFLP of mitochondrial DNA. *Med. Vet. Entomol.*, 15, 183, 2001.
92. Wallman, J.F., and Donnellan, S.C., The utility of mitochondrial DNA sequences for the identification of forensically important blowflies (Diptera: Calliphoridae) in southeastern Australia. *Forensic Sci. Int.*, 120, 60, 2001.
93. Wallman, J.F. et al., Molecular systematics of Australian carrion-breeding blowflies (Diptera: Calliphoridae) based on mitochondrial DNA. *Invert. Syst.*, 19, 1, 2005.
94. Zhang, D.X., and Hewitt, G.M., Assessment of the universality and utility of a set of conserved mitochondrial COI primers in insects. *Insect Mol. Biol.*, 6, 143, 1996.
95. Li Loong, P.T., Lui, H., and Buck, H.W., Cutaneous myiasis: A simple and effective technique for extraction of *Dermatoba hominis* larvae. *Int. J. Dermatol.*, 31, 657, 1992.
96. Olumide, Y., Cutaneous miasis: A simple and effective technique for extraction of *Dermatobia hominis* larvae. *Int. J. Dermatol.*, 33, 148, 1994.
97. Adams, Z.J.O., and Hall, M.J.R., Methods used for killing and preservation of blowfly larvae, and their effect on postmortem larval length. *Forensic Sci. Int.*, 138, 50, 2003.

78 Cimex (Bedbug)

Dongyou Liu

CONTENTS

78.1 INTRODUCTION

The family Cimicidae encompasses a large group of small, oval, flat, wingless insects (true bugs) that are obligate hematophagous ectoparasites of warm-blooded animals, including bats, birds, and mammals. Within the family Cimicidae, *Cimex lectularius* (also known as common bedbug) is one of the widely recognized pest insects, whose bites induce various cutaneous and systematic reactions in humans. Having originated possibly in caves inhabited by bats and humans in the Middle East, *C. lectularius* spread to northern regions in cold climates following the development of housing and clothing. Prior to the mid-twentieth century, bedbugs were very common. However, extensive use of chlorinated hydrocarbon and other synthetic insecticides such as dichlorodiphenyl-trichloroethane (DDT) during the 1940s–1950s has contributed to a dramatic decline of bedbugs throughout the world. In recent decades, along with increased international travel and immigration, changes in pest control practices and insecticide resistance, bedbug infestations have shown resurgence, especially in hotel rooms, hospitals, and dormitories. This highlights the need to have rapid and sensitive diagnostic methods available for identification and surveillance of bedbugs to facilitate the control and prevention of this pathogen.

78.1.1 Classification, Morphology, and Biology

78.1.1.1 Classification

The genus *Cimex* is classified taxonomically in the subfamily Cimicinae, family Cimicidae, superfamily Cimicoidea, infraorder Cimicomorpha, suborder Heteroptera, order Hemiptera, class Insecta, phylum Arthropoda. The family Cimicidae comprises over 90 species of true bugs that are organized in six subfamilies (Afrociminae, Cimicinae, Cacodminae, Haematosiphoninae, Latrocimicinae, and Primicimicinae). The subfamily Cimicinae consists of five genera: *Bertilia*, *Cimex*, *Oeciacus*, *Paracimex*, and *Propicimex*, and the genus *Cimex* is further divided into 10 species: *Cimex adjunctus*, *Cimex antennatus*, *Cimex brevis*, *Cimex columbarius*, *Cimex hemipterus* (synonym: *Cimex rotundatus*), *Cimex incrassatus*, *Cimex latipennis*, *Cimex lectularius*, *Cimex pilosellus*, and *Cimex pipistrella*.

Among the *Cimex* species, *Cimex lectularius* (*Cimex* is the Roman designation for bug and *lectularius* is the Latin name for couch or bed) and *Cimex hemipterus* are parasitic for humans [1]. While *Cimex lectularius* (common bedbug) is present in temperate climates throughout the world, *Cimex hemipterus* (tropical bedbug) occurs mainly in tropical regions and also affects poultry and bats [2]. *Cimex pilosellus* and *Cimex pipistrella* primarily infest bats. Other names for *C. lectularius* (common bedbug) include wall louse, mahogany flat, crimson rambler, heavy dragoon, chinche, and redcoat [3].

78.1.1.2 Morphology

C. lectularius adults possess an oval, flattened (paper thin), wingless body. The head of bedbug is pyramid shaped, with prominent compound eyes, slender antennae, and a long proboscis tucked underneath the head and thorax. Due to the presence of hairs, bedbug has a banded appearance. The unfed bedbug measures 4–5 mm in length and 1.5–3 mm in width and is reddish-brown (chestnut) in color. After a blood meal, bedbug may increase its length by 30–50% and its weight by 150–200%, and its color changes to dull red. Bedbug eggs (1/32 inch length) are elongated and white colored, with a distinct cap at the end. The eggs are laid singly or in clusters and stick to wood, fabrics, or other surfaces in places where bedbug hides. Newly hatched bedbug nymphs are smaller than the adults, and appear translucent, light-colored (pale yellow) and become browned as they mature.

In comparison with the oval body of *C. lectularius*, *Cimex hemipterus* has a diamond-shaped body.

78.1.1.3 Biology

C. lectularius (common bedbug) adults have a life span of 6–12 months and are able to survive up to a year without feeding. After mating, female bedbug lays three or four eggs per day (with a life time total of 200–500 eggs) in the hiding place. The eggs attach on the substrate, and hatch in 6–17 days to become nymphs. The newly hatched nymphs begin to feed immediately and undergo five molts (with one feed between each molt, and growing larger after each molt), and mature into adults in 1 or 2 months [4].

Cimex hemipterus (tropical bedbug) has a fecundity of up to 50 eggs per lifetime, over 11–14 oviposition cycles. Increased feeding improves fecundity. After feeding and mating, adult females take 2–3 days to produce a first batch of eggs. The oviposition period lasts 2–7 days.. The eggs hatch for 5–7 days before first instars emerge. The nymphs undergo five stadia (the first four take 3–4 days each, and the last takes 4–5 days) before becoming adults at a sex ratio of 1:1. More than five blood meals are required by the nymphs for successful molt [5].

Bedbugs often hide during the day to avoid light, and feed at night. Their favorite hiding places include seams in mattresses, crevices in box springs, joints in the bed, behind the headboard, beneath loose carpeting, behind pictures, behind wallpaper, cracks in plaster, light fittings, and plug sockets, and so on. Bedbugs only come out of hiding places to seek warm-blooded hosts when hungry. They can sense the body warmth of the host at distances of 5–10 cm. After seeing a human host, some bedbugs may even climb to the ceiling and drop on the bed where host sleeps. An adult bedbug can suck up to seven times its own weight in blood in the course of 3–5 min. Although both male and female bedbugs are blood feeders, female bedbug is capable of taking up more blood per meal than the male counterpart. After feeding, bedbug retreats to its hiding place to digest the blood meal, mate, and lay eggs until it becomes hungry again (at intervals of a few days to a week) [6]. It is notable that *C. lectularius* secretes calcium- or magnesium-activated plasma membrane-bound enzymes (apyrases), which prevent blood clotting during feeding, and facilitate its blood-sucking activity on a wide range of hosts. The fact that common bedbug *C. lectularius* is more efficient than tropical bedbug *C. hemipterus* in obtaining blood from animal hosts may contribute to the former's higher levels of infestation and prevalence than the latter [6].

Bedbugs are widely distributed in the world, and are very common. Being wingless, bedbugs spread mainly by passive dispersal, including hitchhike in furniture, mattresses, clothing, suitcases, and other personal possessions, pets, and wild animals (e.g., bats, birds). They may also move from one apartment or hotel room to another through holes in walls, water pipes, or gutters [7–9]. Bedbugs from infested poultry farms are transported by human shoes, clothes, egg boxes, production equipment, and vehicles.

With the extensive use of insecticides such as DDT and organophosphates and invention of vacuum cleaner, bedbug infestations have decreased between 1930s and 1980s. However, since the 1980s, bedbugs have shown a remarkable comeback, possibly due to the more frequent international travel, migration, resistance to insecticides, use of broad-spectrum insecticides that do not harm bedbugs, and movement of bedbugs from birds and bats to humans [10–20].

78.1.2 Clinical Features and Treatment

78.1.2.1 Clinical Features

In many cases, a bedbug bite only induces a barely visible punctum at the location of the bite, without other noticeable reactions. In about 30% of the cases, cutaneous and systematic reactions occur as a result of bedbug biting [21–25].

i. Cutaneous reactions. Pruritic maculopapular, erythematous lesions (2–5 mm in size) appear at bedbug feeding sites. These lesions often resolve within a week if not abraded. Individuals experiencing repeated bites may display lesions of increasing size and pruritus. Some individuals develop pruritic wheals (local urticaria) around a central punctum, papular urticaria, and diffuse urticaria at bite sites, with bullous rashes occurring days later. Pruritic papules (nodules or macules) may become superinfected (impetiginous) after scratching, and secondary infection may result in impetigo, ecthyma, folliculitis, cellulitis, and lymphanigites [26–28].

A number of bedbug salivary proteins responsible for cutaneous reactions in mammalian hosts have been identified. These include a nitric oxide-liberating heme protein (nitrophorin), a 17-kDa anticoagulant (Factor X), and a 40-kDa apyrase-like nucleotide-binding enzyme [29]. Human patients with bedbug exposure developed specific IgE antibodies to nitrophorin, and IgG antibodies to other *C. lectularius* antigens [30,31].

ii. Systemic reactions. Asthma, generalized urticaria, and anaphylaxis are common systemic reactions in humans suffering from bedbug bites. Some patients may develop angioedema and hypotension. Patients with bedbug bites may also suffer anxiety, insomnia, worsening of an existing mental health condition, and/or anemia [32].

Additionally, *C. lectularius* may be involved in mechanical transmission of infectious agents, including typhus, plague, smallpox, anthrax, Q-fever, tuberculosis, leprosy, yellow fever, relapsing fever, hepatitis B virus, human immunodeficiency virus, filariasis, kala azar (leishmaniasis), and Chagas disease (*Trypanasoma cruzi*) as well as cancer.

78.1.2.2 Treatment

Topical application of antipruritic agents (paroxime, doxepin) or corticosteroids (triamcinolone) may be helpful for relieving pruritic lesions from bedbug bites. Topical mupirocin or systemic antibiotics may be used for superinfected lesions. Intramuscular epinephrine, antihistamines, and corticosteroids may be needed to treat systemic reactions to bedbug bites.

Bedbug eradication is dependent on proper identification of the bedbug species, education of individuals concerned, thorough inspection of infested areas, implementation of chemical (pyrethroids, dichlorvos, malathion, chlorfenapyr, and hydroprene) and nonchemical (vacuuming, heat or steam treatments) control strategies, and follow-up to evaluation [33,34]. Bedbug adults, nymphs, and eggs in clothes and linen can be eliminated with 60°C washing cycle, >40°C tumble drying cycle for at least 30 min, dry cleaning with perchloroethylene, or freezing at −17°C for at least 2 h [35,36].

Avoidance represents a best approach for the prevention of bedbug bites. As bedbugs and their fecal matters are easily visible, it is prudent to check the premises (hotels or other unfamiliar environments) for bed bugs or their excreta before sleeping. Attention should be paid to potential bedbug hideouts such as mattress cords, cracks, and crevices in box springs, and the back of headboards. Furthermore, used mattresses, box springs, and bedding should be carefully inspected for bedbugs before they are brought into homes [37–39].

Natural enemies of bedbugs consist of the masked hunter (also known as masked bedbug hunter), cockroaches, ants, spiders, mites, and centipedes. The Pharaoh ant's (*Monomorium pharaonis*) venom is lethal to bedbugs. However, it is impractical to use these agents for eradication and prevention of bedbugs in home environments.

78.1.3 Laboratory Diagnosis

Diagnosis of bedbug infestations involves the finding/identification of bedbugs and the observation of compatible symptoms (bites, fecal spots, blood smears on sheets, and molts). Attractant devices such as heat and/or carbon dioxide may be used to attract bedbugs from their hiding places.

Considering technical difficulty in the identification of bedbugs on the basis of morphological criteria, molecular techniques have been developed in recent years to streamline the diagnostic process [40]. Analysis of the mitochondrial DNA (mtDNA) 16S rRNA gene and the nuclear rRNA internal transcribed spacer (ITS) revealed a high level of genetic variation in *C. lectularius* mtDNA sequence and a lack of genetic variation in its nuclear ITS1 sequence, providing support for a recent spread and interbreeding of bed bug populations [41].

78.2 METHODS

78.2.1 Sample Preparation

78.2.1.1 Sample Collection

Adult and nymphal bedbugs are collected and preserved in alcohol, dried, or frozen at −20°C.

78.2.1.2 DNA Isolation

DNA may be extracted from individual bedbugs using the Puregene DNA isolation kit D-5000A (Gentra Systems). Extracted DNA is resuspended in 50 μL of Tris-EDTA and stored at −20°C [41,42].

78.2.2 Detection Procedures

Szalanski et al. [41] utilized two sets of primers from mitochondrial and nuclear DNA genes for phylogenetic analysis of *C. lectularius* populations. Primers LR-J-13007 (5′-TTA CGC TGT TAT CCC TAA-3′) and LR-N-13398 (5′-CGC CTG TTT ATC AAA AAC AT-3′) amplify a 428-bp fragment of the insect mtDNA 16S rRNA gene. Primers rDNA2 (5′-TTGATTACGTCCCTGCCCTTT-3′) and rDNA 1.58S (5′-GCCACCTAGTGAGCCGAGCA-3′) amplify a 1500-bp fragment covering the 3′ portion of 18S rRNA, all of the ITS1, and the 5′ portion of 5.8S. Furthermore, another primer ITS1-R (5′-TCTCTTTCCGTGTGTTTAAAATGT-3′) located at the 3′ prime end of the ITS1 region is used with primer rDNA2 to amplify a 589-bp fragment from the ITS1 region.

78.2.2.1 Procedure

1. PCR is conducted with 2 μL of the extracted DNA. For amplification of mtDNA 16S rRNA, the thermocycler profile consists of 94°C for 45 s, 46°C for 45 s, and 72°C for 60 s for a total of 35 cycles. For amplification of ITS1 marker, the thermal cycler profile consists of 94°C for 45 s, 53°C for 1 min, and 72°C for 1 min for a total of 40 cycles.
2. The resulting PCR amplicons are resolved on 1% agarose gels, purified, concentrated using Microcon-PCR Filter Units (Millipore Corporation), and sequenced in both directions.
3. Phylogenetic analysis is conducted using maximum parsimony (MP) analysis with the best-fitting evolutionary model as implemented in PAUP*. Bootstrapping is performed using neighbor joining or MP (1000 replicates) to determine the reliability of the obtained topologies. Phylogenetic trees are obtained using Bayesian inference with the GTR + G model by using Bayesian Evolutionary Analysis Sampling Trees (BEAST) version 1.4.2 software. For Bayesian inference, four Markov chains run for 106 generations with a burn-in of 2 × 104 are used to reconstruct the consensus tree.

Note: A 397-bp sequence from the 428-bp fragment of the insect mtDNA 16S rRNA gene and a 449-bp sequence from the 589-bp ITS1 fragment (excluding 140 bp from the 18S rRNA gene) are used for the phylogenetic studies.

78.3 CONCLUSION

The family Cimicidae covers a large number of flightless, obligate hematophagous ectoparasites of mammals, of which *Cimex lectularius* (common bedbug) and *Cimex hemipterus* (tropical bedbug) are involved in human illnesses [43]. Whereas *C. lectularius* is widespread in temperate regions, *C. hemipterus* is predominant in tropical regions. Overall, *C. lectularius* is more frequently reported than *C. hemipterus*, possibly due to the fact that the former is a more efficient blood feeder than the latter. Although bedbug infestations are generally not life-threatening, they pose grave quality-of-life issues and economic concerns for households. Indeed, the cutaneous and systematic reactions elicited by bedbug bites are undoubtedly unpleasant for the sufferers to say the least. Along with increasing international travel and migration as well as insecticide resistance buildup in recent decades, bedbug populations have shown a remarkable resurgence. This has necessitated the development of improved diagnostic methods for identification and tracking of bedbug outbreaks. Application of molecular techniques offers a rapid, sensitive, and specific means for determination and phylogenetic assessment of bedbugs and other pest insects, and contributes to the effective control and prevention measures against bedbug infestations in humans and animals.

REFERENCES

1. Thomas, I., Kihiczak, G.G., and Schwartz, R.A., Bed bug bites: A review. *Int. J. Dermal.*, 43, 430, 2004.
2. Harlan, H.J., Bed bugs 101: The basics of *Cimex lectularius*. *Am. Entomol.*, 52, 99, 2006.
3. Kolb, A. et al., Bedbugs. *Dermatol. Ther.*, 22, 347, 2009.
4. Reinhardt, K., and Siva-Jothy, M.T., The biology of bedbugs (Cimicidae). *Ann. Rev. Entomol.*, 52, 351, 2007.
5. How, Y.F., and Lee, C.Y., Fecundity, nymphal development and longevity of field-collected tropical bedbugs, *Cimex hemipterus*. *Med. Vet. Entomol.*, 24, 108, 2010.
6. Araujo, R.N. et al., The feeding process of *Cimex lectularius* (Linnaeus 1758) and *Cimex hemipterus* (Fabricius 1803) on different bloodmeal sources. *J. Insect Physiol.*, 55, 1151, 2009.
7. Ryan, N., Peters, B., and Miller, P., A survey of bed bugs in short-stay lodges. *NSW Public Health Bull.*, 15, 215, 2004.
8. Hwang, S.W. et al., Bed bug infestations in an urban environment. *Emerg. Infect. Dis.*, 11, 533, 2005.
9. Wang, C. et al., Characteristics of *Cimex lectularius* (Hemiptera: Cimicidae), infestation and dispersal in a high-rise apartment building. *J. Econ. Entomol.*, 103, 172, 2010.
10. Boase, C.J., Bedbugs: Back from the brink. *Pest Outl.*, 12, 159, 2001.
11. Myamba, J. et al., Pyrethroid resistance in tropical bedbugs, *Cimex hemipterus*, associated with use of treated bed nets. *Med. Vet. Entomol.*, 16, 448, 2002.
12. Doggett, S.L., Geary, M.J., and Russell, R.C., The resurgence of bed bugs in Australia, with notes on their ecology and control. *Environ. Health*, 4, 30, 2004.
13. Ter Poorten, M.C., and Prose, N.S., The return of the common bedbug. *Pediatr. Dermatol.*, 22, 183, 2005.
14. Karunaratne, S.H.P.P. et al., Insecticide resistance in the tropical bedbug *Cimex hemipterus*. *Pest Biochem. Physiol.*, 88, 102, 2007.
15. Romero, A. et al., Insecticide resistance in the bed bug: A factor in the pest's sudden resurgence? *J. Med. Entomol.*, 44, 175, 2007.
16. Lee, I.Y. et al., Reemergence of the bedbug *Cimex lectularius* in Seoul, Korea. *Korean J. Parasitol.*, 46, 269, 2008.
17. Anderson, A.L., and Leffler, K., Bedbug infestations in the news: A picture of an emerging public health problem in the United States. *J. Environ. Health*, 70, 24, 2008.
18. Levy Bencheton, A. et al., Resurgence of bedbugs in southern France: A local problem or the tip of the iceberg? *J. Eur. Acad. Dermatol. Venereol.*, 25, 599, 2011.
19. Manuel, J., Invasion of the bedbugs. *Environ. Health Perspect.*, 118, A429, 2010.
20. Zhu, F. et al., Widespread distribution of knockdown resistance mutations in the bed bug, *Cimex lectularius* (Hemiptera: Cimicidae), populations in the United States. *Arch. Insect Biochem. Physiol.*, 73, 245, 2010.
21. Stibich, A.S., Carbonaro, P.A., and Schwartz, R.A., Insect bite reactions: An update. *Dermatology*, 202, 193, 2001.
22. Doggett, S.L., and Russell, R., Bed bugs—What the GP needs to know. *Aust. Fam. Physician*, 38, 880, 2009.
23. Goddard, J., and deShazo, R., Bed bugs (*Cimex lectularius*) and clinical consequences of their bites. *J. Am. Med. Assoc.*, 301, 1358, 2009.
24. Lorenzen, R., Bedbugs and health. *CMAJ*, 181, 617, 2009.
25. Reinhardt, D. et al., Sensitivity to bites by the bedbug, *Cimex lectularius*. *Med. Vet. Entomol.*, 23, 163, 2009.
26. Scarupa, M.D., and Economides, A., Bedbug bites masquerading as urticaria. *J. Allergy Clin. Immunol.*, 117, 1508, 2006.
27. Stucki, A., and Ludwig, R., Images in clinical medicine: Bedbug bites. *N. Engl. J. Med.*, 359, 1047, 2008.
28. Delaunay, P. et al., Bedbugs and infectious diseases. *Clin. Infect. Dis.*, 52, 200, 2011.
29. Reinhardt, K., Wong, C.H., and Georgiou, A.S., Detection of seminal fluid proteins in the bed bug, *Cimex lectularius*, using two-dimensional gel electrophoresis and mass spectrometry. *Parasitology*, 136, 283, 2009.
30. Abdel-Naser, M.B. et al., Patients with papular urticaria have IgG antibodies to bedbug (*Cimex lectularius*) antigens. *Parasitol. Res.*, 98, 550, 2006.
31. Leverkus, M. et al., Bullous allergic hypersensitivity to bed bug bites mediated by IgE against salivary nitrophorin. *J. Invest. Dermatol.*, 126, 91, 2006.
32. Pritchard, M.J., and Hwang, S.W., Cases: Severe anemia from bedbugs. *CMAJ*, 181, 287, 2009.
33. Moore, D.J., and Miller, D.M., Laboratory evaluations of insecticide product efficacy for control of *Cimex lectularius*. *J. Econ. Entomol.*, 99, 2080, 2006.
34. Kweka, E.J. et al., Insecticides susceptibility status of the bedbugs (*Cimex lectularius*) in a rural area of Magugu, Northern Tanzania. *J. Glob. Infect. Dis.*, 1, 102, 2009.
35. Gangloff-Kaufmann, J. et al., Bed bugs in America: A pest management industry survey. *Am. Entomol.*, 52, 105, 2006.
36. Naylor, R.A., and Boase, C.J., Practical solutions for treating laundry infested with *Cimex lectularius* (Hemiptera: Cimicidae). *J. Econ. Entomol.*, 103, 136, 2010.

37. Potter, M.F., A bed bug state of mind: Emerging issues in bed bug management. *Pest Control Technol.*, 33, 82, 2005.
38. Delaunay, P. et al., Bedbugs and healthcare-associated dermatitis, France. *Emerg. Infect. Dis.*, 15, 989, 2009.
39. Rossi, L., and Jennings, S., Bed bugs: A public health problem in need of a collaborative solution. *J. Environ. Health*, 72, 34, 2010.
40. Szalanski, A.L. et al., Time course analysis of bed bug, *Cimex lectularius* L. (Hemiptera: Cimicidae) blood meals using PCR. *J. Agr. Urban Entomol.*, 23, 237, 2006.
41. Szalanski, A.L. et al. Mitochondrial and ribosomal internal transcribed spacer (ITS1) diversity of the bed bug *Cimex lectularius* L. (Heteroptera: Cimicidae). *J. Med. Entomol.*, 45, 229, 2008.
42. Szalanski, A.L. et al. Isolation and characterization of human DNA from the bed bug, *Cimex lectularius* L. (Heteroptera: Cimicidae). *J. Agr. Urban Entomol.*, 23, 189, 2006.
43. Panagiotakopulu, E., and Buckland, P.C., *Cimex lectularius* L., the common bed bug from Pharaonic Egypt. *Antiquity*, 73, 908, 1999.

79 Pediculus (Body and Head Lice)

Dongyou Liu and Sacha Stelzer-Braid

CONTENTS

79.1 INTRODUCTION

Lice are small insects whose infestations in humans and animals have been recognized for thousands of years. With over 3000 described species, lice are classified in the order Phthiraptera (lice), which is further separated into four suborders (groups): Anoplura (sucking lice), Rhyncophthirina, Amblycera, and Ischnocera. Within the suborder Anoplura, 540 wingless species of obligate hematophagous ectoparasites on nearly all groups of mammals have been identified. Of these, the most important sucking lice in humans are head louse (*Pediculus humanus capitis*) and body louse (*Pediculus humanus humanus*) in the family Pediculidae, and pubic louse (or crab louse, *Phthirus pubis*) in the family Phthiridae.

The condition associated with louse infestation such as head louse, body louse, or pubic louse is commonly referred to as pediculosis. While pediculosis capitis (also called head louse infestation or cooties) results from the colonization of the hair and skin in the head or scalp by *Pediculus humanus capitis* and occurs in all countries, pediculosis corporis (also called pediculosis vestimenti or vagabond's disease) is a cutaneous infection of *Pediculus humanus humanus* that prospers in conditions such as cold weather, lack of hygiene, or war. Besides causing irritation, the body louse plays a role in the transmission of several bacterial pathogens. The pubic (crab) louse is largely a sexually transmitted organism, with occasional appearance in eyebrows and eyelashes, which is covered in Chapter 80.

79.1.1 Classification, Morphology, Life Cycle, and Epidemiology

79.1.1.1 Classification

The genus *Pediculus* is classified in the family Pediculidae, suborder Anoplura (sucking lice with no wings and incomplete metamorphosis), order Phthiraptera (lice), superorder Psocodea (lice and fleas), class Insecta (insects with three body parts and six legs), phylum Arthropoda (arthropods with segmented body, exoskeleton, and jointed appendages). The suborder Anoplura encompasses the families of Echinophthiriidae, Haematopinidae, Hoplopleuridae, Linognathidae, Pedicinidae, Pediculidae, Polyplacidae, and Pthiridae, with >540 species of sucking lice parasitizing mammals. The family Pediculidae consists of a single genus *Pediculus*, which is subdivided into the species/varieties *Pediculus humanus capitis* (human head louse; synonym: *Pediculus capitis*), *Pediculus humanus humanus* (human body louse; synonyms: *Pediculu humanus corporis*, *Pediculus corporis*, and *Pediculus vestimenti*), and *Pediculus schaeffi* (chimpanzee louse) [1].

Phylogenetic studies suggest that the ancestor of all human lice may have emerged about 770,000 years ago, and the human body louse (*Pediculus humanus humanus*) may have originated from the head louse (*Pediculus humanus capitis*) about 107,000 years ago when humans adopted frequent use of clothing [2–11].

79.1.1.2 Morphology

Sucking lice (Anoplura) are small wingless dorsoventrally flattened insects with three body parts (head, thorax, and abdomen) and leathery, grayish skin (Figure 79.1). The head contains mouthparts and two antennae, each with five segments, protruding from the head. Adapted for piercing skin and sucking blood, the mouthparts are introduced directly into the host blood vessel (solenophage mode of feeding) and are retracted in the narrow head when not feeding. Two eyes are present in the *Pediculidae* family, but are reduced or absent in most other members of the suborder Anoplura. The thorax has six legs arranged in three bilateral pairs, which are short and terminate with a single claw and opposing thumb. These adaptations facilitate tight grasping of the hair of a host between claw and thumb, but somehow render the louse incapable of jumping, or walking efficiently on flat surfaces. The abdomen has seven visible segments: the first six segments each have a pair of spiracles through which the insect breathes, and the last segment contains the anus and the genitalia (separately). The abdomen becomes distended after feeding.

Males are slightly smaller than females and are characterized by a pointed end of the abdomen and a well-developed genital apparatus visible inside the abdomen. In males, the two front legs are slightly larger than the other four, and are used for efficient holding of a female during copulation. Females are characterized by two gonopods in the shape of a W at the end of the abdomen.

The head louse *Pediculus humanus capitis* (*Pediculus capitus*), also known colloquially as cooties, greybacks, or mechanized dandruff, forms three developmental stages: egg, nymph, and adult. The egg (commonly called nit; some parasitologists refer the hatched and empty egg shell as a "nit" and the developing embryonated egg as an "egg") harbors a single embryo and has an ellipsoidal operculate body (0.8 × 0.3 mm), which glues near the base of a host hair shaft. The egg is coffee-colored when containing an embryo, but transparent after hatching. The nymph is similar in appearance to an adult, but is smaller (1–2 mm in length). The adult has an elongated, dorsoventrally flattened body (male: 2 mm long; female: 3 mm long) which appears opaque although darker internal organs can be seen in the abdomen (Figure 79.1a and b).

The body louse *Pediculus humanus humanus* is of 2–4 mm in size (about the size of a sesame seed; being slightly larger than head louse, with longer body and antennae) and is tan to grayish-white in color (Figure 79.1c). Its morphology is similar to that of the head louse *Pediculus humanus capitis*, and can only be distinguished with certainty by the habits. Namely, the body louse differs from the head louse in preferring to attach eggs to clothing rather than to scalp hair.

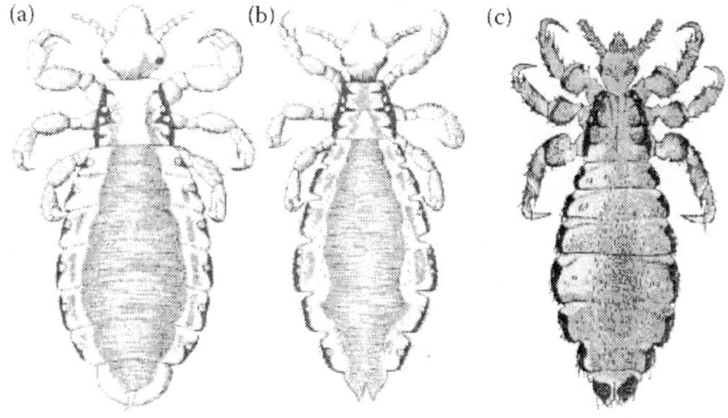

FIGURE 79.1 Head and body lice. (a) Adult male head louse (*Pediculus humanus capititis*). (b) Adult female head louse (*Pediculus humanus capititis*). (c) Adult body louse (*Pediculus humanus humanus*). (Adapted from www.ces.csiro.au/aicn/ (copyright free images).)

79.1.1.3 Life Cycle and Epidemiology

All lice undergo gradual metamorphosis evolving from egg, nymph, and adult, with no free-living stages present. Eggs are cemented to host hairs whereas nymphs and adults cling to hairs using enlarged tarsal claws.

The life span of sucking lice is about 2–4 weeks, and females lay 10 large, yellowish-white eggs per day (with a life time total of 50–150 eggs). The egg hatches in a week, and the nymph (young louse) inflates itself with air, pushes off the top of the egg, and crawls out. The newly hatched louse (nymph) starts to feed immediately. After undergoing gradual metamorphosis for 8 days whereby the nymph molts several times to adult form, the louse is ready to mate and lay eggs. Sucking lice feed exclusively on blood twice a day, do not burrow into the skin, and only survive for a few days in the absence of a host. Sucking lice are sensitive to temperature changes, and will move off quickly from a host with fever, and also from a cold, dead person.

Head louse *Pediculus humanus capitus* is highly host-specific for humans and does not infest other animals. Being highly site-specific, all developmental stages (egg, three nymphal instars, and adult or imago) of head louse can be found attached to, or grasping, hairs on the head, especially at the back of the neck and behind the ears. The head louse is repelled by light and often moves toward shadows or dark-colored objects in its vicinity. The only visible differences between different instars and the adult, other than size, is the relative length of the abdomen, which increases with each molt. In cool climates, eggs are laid within 1 cm of the scalp surface, whereas in warm climates, eggs are laid 15 cm or more down the hair shaft. The adult female secretes a glue (made of proteins similar to hair keratin) from her reproductive organ. This glue quickly hardens into a "nit sheath" that covers the hair shaft and the entire egg except for the operculum, a cap through which the embryo breathes. After hatching, the nymph leaves behind its egg shell. The empty egg shell remains in place until physically removed by abrasion or the host, or until it slowly disintegrates in months or years. The head louse does not have free-living stages and cannot survive for long away from its hosts. Therefore, head louse infestations are transmitted between hosts by direct physical contact (head to head), although transmissions via contaminated combs, brushes, towels, hats, clothing, beds, or closets are possible. Head louse populations often exhibit pronounced increase during winter months (due to crowding), particularly in temperate regions. Head lice infestation is endemic throughout the world. Children between 4 and 13 years of age are the most frequently infested group. Girls (with long hair) are more frequently infested than boys. Head louse causes irritation, but is not known to be a carrier of other infectious diseases.

Body louse *Pediculus humanus humanus* is less host-specific, and can be bred in the laboratory after adaptation to feeding on rabbits. Body louse spends most of the time in host clothing, particularly around the waistline and under armpits, and lays eggs (of oval shape and usually yellow to white color) on or near the seams of clothing. The eggs hatch in 7 days, and the nymph feeds on blood and molts three times to become an adult. The adult feeds on blood, and displays a reddish color after feeding. Nymphs and adults of both sexes feed by piercing the skin and sucking blood about every 2–3 h, causing irritation. Body louse occurs worldwide and infests people of all races. Body louse is transmitted through prolonged direct physical contact with an infested person or through contact with articles such as clothing, beds, bed linens, or towels that have been in contact with a person already carrying lice. Thus, body louse infestations can spread rapidly under crowded, primitive living conditions where hygiene is poor (homeless, refugees, victims of war, or natural disasters). Body louse is hardier than head louse and at 23°C it can live for 4 days without access to blood. Body louse is known to transmit human bacterial pathogens *Borrelia recurrentis* (the agent of relapsing fever), *Bartonella quintana* (an agent of bacillary angiomatosis bacteremia, trench fever, endocarditis, and chronic lymphadenopathy among the homeless), and *Rickettsia prowazekii* (the agent of epidemic typhus) [12].

79.1.2 Clinical Features and Treatment

79.1.2.1 Clinical Features

Commonly known as pediculosis, louse infections in humans usually cause non-life-threatening pruritus (itching) and irritations [13].

The most common symptom of head louse infestation (pediculosis capitis or pediculosis) is pruritus (itching) on the head, which intensifies 3–4 weeks after the initial infestation, with bites observed on the neck of long-haired individuals when the hair is pushed aside. In rare cases, itching may induce scratching that can lead to secondary infection with impetigo and pyoderma, and occasional swelling of the local lymph nodes and fever.

Body louse infestation also causes annoying irritations, resulting from the feeding activity (skin piercing and blood sucking) of nymphs and adults of both sexes that occurs every 2–3 h. Light infestations may only lead to moderate itching, while heavy infestations (often associated with crowded conditions and poor sanitation) may cause considerable discomfort as the bites produce red papules, fever, aches, and intense pruritus. Subsequent scratching can lead to dermatitis and secondary infections. However, because of its role in the transmission of typhus fever, trench fever, and louse-borne relapsing fever in human populations, body louse can be a significant cause of public health problems [14].

79.1.2.2 Treatment

Treatment of human louse infestations in a community setting involves the use of insecticides (pediculicides) and removal or cleaning of infected material (e.g., hair combing or shaving, and clothes washing, etc). Efforts to treat head and body lice should focus on the hair or body (or clothes), and not on the home environment [15–17].

Recommended pediculocide treatments for pediculosis include (i) 0.33% pyrethrins + PBO shampoo (pet basic original shampoo, apply to hair, and wash off after 10 min), (ii) 1–5% permethrin cream rinse (1% creme rinse, wash off after 10 min), (iii) 0.5% malathion lotion shampoo (flammable and organophosphate poisoning risk), (iv) 1% lindane lotion and shampoo (potential CNS toxicity from organochlorine poisoning; only use as last resort; wash off after 4 min), and (v) ivermectin 0.8% shampoo (apply to hair, wash off after 5 min).

For head louse infestation, daily grooming with nit combs is recommended to remove eggs and lice. As most insecticides have limited activity against eggs, repeat washing is required within 10 days, and treatment may be repeated several times at intervals of 4–5 days. A fine-toothed comb may be used a day or two after the final application of insecticide to verify the efficacy of the treatment. Shaving off the hair in the affected region and an intensive course of hygiene are also beneficial [18].

For body louse infestation, clothing and bedding items are laundered appropriately (using hot water at least 54°C) and machine dried (using the hot cycle) at least once a week or boiled. A temperature of 54°C for 5 min is sufficient to kill most of the adults and prevent eggs from hatching. Powder dusting of all clothes and blankets with 10% DDT (dichlorodiphenyltrichloroethane), 1% malathion, or 1% permethrin is also effective. Clothes, towels, bedding, combs, and brushes can also be disinfected by leaving them outside for at least 3 days. An insecticidal treatment of the house and furniture is unnecessary due to the risk of toxicity to humans if inhaled or absorbed.

79.1.3 Laboratory Diagnosis

The head louse attaches its eggs firmly to hairs close to the scalp, which can only be removed by using a fine tooth comb. The empty egg shells (which are more transparent unhatched eggs, with the operculum missing) remain on the hair and move outward as it grows (at a rate of 1 cm per month). Diagnosis of head louse infestation is by finding live lice or empty egg shells in the hair either by direct visual examination or by using a fine-toothed nit comb (after applying hair conditioner to untangle hairs and trap lice). Direct visual examination is performed by parting the hair at 2 cm intervals to look for moving lice near the scalp, especially near the ears and the nape of the neck, for at least 5 min, as most infections have fewer than 10 lice, mostly small nymphs 1–2 mm long [18]. The use of a nit comb (with the teeth of the comb being 0.2–0.3 mm apart to trap lice) is the most effective way to detect living lice. The hair is first combed or brushed to remove tangles, and the fine-toothed louse comb is then inserted near the crown until it touches the scalp, and drawn firmly down. The entire head of hair should be combed systematically at least twice; the comb should be examined for lice after each stroke [18]. Examination of the material collected between the teeth of the comb using a magnifying glass helps prevent misdiagnosis. If lice are detected, the entire family (especially children up to the age of 13 years) must be checked with a louse comb, and only those with living (crawling) lice should be treated. Treatment of children with louse eggs/nits on the hair or itchy scalp may be unnecessary [19].

Body louse is similar to head louse in appearance, but it is found in the body, not on the head. Another hair-clinging pubic louse (or crab louse, *Pthirus pubis*) is quite distinctive, with shorter bodies and pincer-like claws (colloquially known as "crabs").

Recently, molecular techniques have been applied for identification and phylogenetic analysis of head and body lice [20–24]. By sequencing portions of the mitochondrial DNA (mtDNA) *ND4* (579 bp), *CYTB* (440 bp), and other genes from head and body lice as well as chimpanzee louse *Pediculus schaeffi*, it was shown that three distinct clade phylotypes of *Pediculus humanus* are found among modern humans. The most common mtDNA phylotype includes both head and body lice (type A) and is worldwide in distribution. The second mtDNA group (type B) occurs only in head lice and is present in South and North America, Europe, and Australia. The third type (type C) exists only among head lice from Nepal and Ethiopia [2,25,26]. It is apparent that the mtDNA sequences associate the origin of body lice (and, by inference, clothing) with modern humans, and that body lice may have originated from head lice. Additionally, the greater diversity in African lice implies an African origin for lice [2,25].

79.2 METHODS

79.2.1 Sample Preparation

79.2.1.1 Sample Collection

Head and body lice are collected and preserved in 95% ethanol and stored at −80°C, or air-dried.

79.2.1.2 DNA Isolation

A modified silica-based extraction method may be used for the extraction of louse DNA. Two to five lice are ground together in liquid nitrogen and incubated for 48 h with agitation at 55°C in 600 µL of extraction buffer (7.5 mol/L guanidinium thiocyanate, 0.1 mol/L Tris hydrochloride pH 6.4, 0.02 mol/L EDTA pH 8.0, and 1.3% Triton X-100). After centrifugation, 500 µL of supernatant is removed to a second tube containing an additional 500 µL of extraction buffer and 40 µL of saturated silica suspension (SiO_2 in water). DNA is bound to the silica for 10 min at 27°C and then pelleted by centrifugation, washed twice with extraction buffer and twice with 70% ethanol supplemented with 10 mmol/L sodium chloride, and eluted in two 75 µL volumes of Tris-EDTA buffer at 60°C [23].

Alternatively, lice are rinsed with distilled water, dried on sterile filter paper, and crushed individually in sterile Eppendorf tube with a sterile pipette tip. DNA is extracted using the QIAamp Tissue Kit (Qiagen) according to the manufacturer's instructions, except that incubation in lysis buffer was performed overnight at 37°C [9,26].

79.2.2 Detection Procedures

79.2.2.1 Sequence Analysis of 18S rRNA Gene

Leo and Barker [6] utilized forward primers NS1 (5′-GTAGTCATATGCTTGTCTC-3′) and NS10 (5′-AGGCT CTGCAATCGGAATG-3′) together with reverse primers NS2a (5′-CGCGGCTGCTGGCACCAGACTTGC-3′) and NS11 (5′-GTCAAATTAAGCCGCAAGC-3′), respectively, to amplify and analyze the 18S rRNA gene from both head and body lice.

79.2.2.1.1 Procedure

1. PCR amplification is carried out with one cycle of 95°C for 3 min; 10 cycles of 93°C for 30 s, 40°C for 30 s, and 72°C for 1 min 30 s; 25 cycles of 93°C for 30 s, 50°C for 30 s, and 72°C for 1 min 30 s; and one cycle of 72°C for 5 min.
2. Nucleotide sequences are aligned with Clustal X, and the alignments are checked by eye. The program Gblocks v. 0.91b is used to remove ambiguous sites from the 1554-bp alignment and the final alignment of 1195 bp is put into MODELTEST. Phylogenetic trees are inferred by using parsimony (PAUP) v. 4.0b10 with neighbor-joining (NJ) and maximum parsimony (MP) methods.

Note: The alternative 18S rRNA primers 18sai (5′-CCTGAGAAACGGCTACCACATC-3′) and 18 sbi5.0 (5′-TAACCGCAACAACTTTAAT-3′) may also be applied for molecular identification of insects, including lice [27].

79.2.2.2 Sequence Analysis of *ND4*, *CYTB*, *COX1*, *EF-1α*, and *RPII* genes

79.2.2.2.1 Protocol of Kittler et al.

Kittler et al. [2] targeted mtDNA *ND4*, cytochrome *b* (*CYTB*), cytochrome oxidase subunit 1 (*COX1*), elongation factor-1 (*EF-1α*), and RNA polymerase II (*RPII*) genes for identification and phylogenetic assessment of human louse species (Table 79.1).

79.2.2.2.1.1 Procedure

1. PCR mixture (50 µL) is composed of 1 µL DNA extract, 0.2 µM of each primer, 0.2 mM dNTPs, and 2.5 U of AmpliTaq Gold (Perkin Elmer) with the supplied buffer.
2. PCR cycling conditions include an initial denaturation at 95°C for 11 min, then 40 cycles of denaturation at 94°C for 30 s, annealing at 52°C (*ND4* and *CYTB*), 40°C (*COX1*), or 57°C (*EF-1α*) or 59°C (*RPII*) for 30 s, and elongation at 72°C for 45 s, followed by a final extension step for 10 min at 72°C.

TABLE 79.1
Primers for Amplification and Sequencing Analysis of Human Louse Species

Gene	Primer Sequence (5′–3′)	Product (bp)	Note
ND4	F: GAGCTTTAGGTAATCAAGAATG R: AACAAATCTCAACTCGTTTCA	741	Additional sequencing primer (reverse): 5′-TAGCGCAAGAGGAAACTTAG-3′
CYTB	F: GGTTATGTTCTTCCTTGAGGT R: CCCTCCTAATTTGTTAGGTACA	483	
COX1	F: GGAGGATTTGGAAATTGATTAGTTCC R: ACTGTAAATATATGATGAGCTCA	659	
EF-1α	F: AGCCTCTTCGACTGCCATTA R: GCGGAAATTCGGAGAAAGAT	610	Additional sequencing primer (forward): 5′-CGGTACTGTGCCTGTTGGTA-3′
RPII	F: TGTCAGGATCCAAAGGATCAA R: CTTTGCAAGTTACAGGGCAAT	643	

3. PCR products are sequenced directly from both strands using the BigDye cycle sequencing kit (Applied Biosystems) and an ABI 3700 automated DNA sequencer (Applied Biosystems).
4. Sequence trace files are analyzed with the program Sequence Analysis (DNAStar) and aligned with ClustalW; all polymorphic positions are confirmed by manual inspection of the trace files. Neighbor-joining trees are constructed, using Kimura two parameter distances, with MEGA [28]. Tajima's D and the nucleotide diversity based on the number of polymorphic sites, θ_W are calculated with DnaSP 3.51.

79.2.2.2.2 Protocol of Reed et al.

Reed et al. [26] employed degenerate primers for PCR amplification and sequencing of a portion of mtDNA cytochrome *b* (*CYTB*), cytochrome *c* oxidase subunit I gene (Cox1), and the nuclear elongation factor 1 alpha gene (EF-1α) for identification of house lice (Table 79.2).

79.2.2.2.2.1 Procedure

1. For amplification of mtDNA genes COXI (854 bp) and Cytb (671 bp) using degenerate primers C1-J-1718 and H7005 (COXI) and L11122 and H11823 (Cytb), PCR parameters consist of 94°C for 30 s, 48°C for 30 s, and 72°C for 1.5 min (five cycles), then 30 cycles of 94°C for 30 s, 52°C for 30 s, and 72°C for 1.5 min. To ensure that nuclear copies of mitochondrial genes are amplified, additional PCR amplifications using nested sets of overlapping primers may be carried out.
2. For amplification of a 379-bp region of the mitochondrial COI gene (using degenerate primers L6625 and H7005; with the 3′ position of each primer relating to the human mitochondrial genome) and 392-bp EF1α gene (using primer EF1-For3 with the 5′ end corresponding to position 1496 in *Apis mellifera*, and primer Cho10 with the 5′ end corresponding to position 1887 in *Apis mellifera*), PCR parameters include four cycles of 1 min of denaturation (95°C), 1 min of annealing (45°C), and 1 min of extension (72°C), followed by 30 cycles at reduced denaturation temperatures (93°C) and increased annealing temperatures (60°C).
3. Amplified fragments are purified using ExoSAP-IT (USB Corporation) and sequenced in both directions. Sequences are edited using Sequencher v. 4.2.2 (Gene Codes Corporation, Ann Arbor, Michigan) and aligned by eye using Se-Al v2.0a11 http://evolve.zps.ox.ac.uk/Se-Al/Se-Al.html. Primer sequences are removed and sequences are trimmed in reference to the translated protein sequence using Se-AL v2.01a11 and MacClade 4.0.

TABLE 79.2
Degenerate Primers for Amplification and Sequencing Analysis of Human Louse Species

Gene	Primer Sequence (5′–3′)	Product (bp)
CYTB	L11122: GAAATTTTGGGTCWTTRCTNGG H11823: GGCATATGCGAATARGAARTATCA	671
COX1	C1-J-1718: GGAGGTTTTGCTAATTGATTAG H7005: CCGGATCCACNACRTARTANGTRTCRTG	854
COX1	L6625: COGGATCCTTYTGRTTYTTYGGNCAYCC H7005: CCGGATCCACNACRTARTANGTRTCRTG	379
EF-1α	For3: GGN GAC AA[C/T] GTT GG[T/C] TTC AAC G Cho10: AC [A/G]GC [A/G/C]AC [G/T]GT [T/C]TG [A/T/C]C[T/G] CAT GTC	345

79.2.2.2.3 Protocol of Raoult et al.

Raoult et al. [9] used primers (L6625 and H7005) to amplify the Cox1 (cytochrome oxidase subunit 1) gene (Table 79.2) and primers CytbF1 (5′-GAGCGACTGTAATTACTA ATC-3′), CytbR1 (5′-CAA CAA AAT TAT CCG GGT CC-3′), CytbF2 (5'-GAG GAG GGT TTT CAG TTA-3′), and CytbR2 (5′-ACT TTA TCA CTA TCC AAA TC-3′) to amplify of the Cytb (cytochrome *b*) gene for phylogenetic study of human louse species.

79.2.2.2.3.1 Procedure

1. PCR mixture (25 μL) is composed of 12.5 pmol of each primer, 200 μM each of dATP, dCTP, dGTP, and dTTP, 1 U of *Taq* DNA polymerase in 1× PCR buffer, 0.8 μL of 25 mM $MgCl_2$ (Gibco BRL), and 2.5 μL of the extracted DNA.
2. Amplification is conducted with an initial 3-min denaturation step at 95°C, 40 cycles of denaturation at 95°C (30 s), annealing at 50°C (30 s), and extension at 72°C (1 min); and a final extension at 72°C for 7 min.
3. PCR products are purified with Qiagen columns (QIAquick Spin PCR purification kit; QIAGEN), and sequenced.
4. The sequence data are examined using NJ with a best-fit model of nucleotide evolution in Modeltest. Levels of topological support are calculated from 100 bootstrap replicates.

79.3 CONCLUSION

The genus *Pediculus* in the family Pediculidae consists of two obligate hematophagous ectoparasites of humans, that is, head louse (*Pediculus humanus capitis*) and body louse (*Pediculus humanus humanus*). Together with pubic louse (or crab louse) *Phthirus pubis*) in the family Phthiridae, these louse species/varieties are responsible for causing intense pruritus and irritations in human hosts. In addition, the body louse also acts as vector for several human bacterial pathogens.

Correct identification of louse species is crucial for the effective treatment and control of their infestations in humans. Traditionally, human-infecting louse species are diagnosed through detection of the adult parasites and eggs, in association with their habits. As the diagnostic approach requires considerable technical expertise, molecular procedures have been developed in recent years to aid their determination. Indeed, PCR amplification and sequencing analysis of mitochondrial genes [e.g., cytochrome *b* (*CYTB*) and cytochrome oxidase subunit 1 (*COX1*)] and genetic regions (e.g., elongation factor-1 (*EF-1α*), RNA polymerase II (*RPII*), and ribosomal RNA genes) have provided an unequivocal way for speciation and phylogenetic discrimination of human-infecting lice.

REFERENCES

1. Light, J.E., Toups, M.A., and Reed, D.L., What's in a name: The taxonomic status of human head and body lice. *Mol. Phylogenet. Evol.*, 47, 1203, 2008.
2. Kittler, R., Kayser, M., and Stoneking, M., Molecular evolution of *Pediculus humanus* and the origin of clothing. *Curr. Biol.*, 13, 1414, 2003.
3. Yong, Z. et al., The geographical segregation of human lice preceded that of *Pediculus humanus capitis* and *Pediculus humanus humanus*. *C. R. Biol.*, 326, 565, 2003.
4. Reed, D.L. et al., Genetic analysis of lice supports direct contact between modern and archaic humans. *PloS Biol.*, 2, e340, 2004.
5. Reed, D.L. et al., Pair of lice lost or parasites regained: The evolutionary history of anthropoid primate lice. *BMC Biol.*, 5, 7, 2007.
6. Leo, N.P., and Barker, S.C., Unravelling the evolution of the head lice and body lice of humans. *Parasitol. Res.*, 98, 44, 2005.
7. Leo, N.P. et al., The head and body lice of humans are genetically distinct (Insecta: Phthiraptera, Pediculidae): Evidence from double infestations. *Heredity*, 95, 34, 2005.
8. Light, J.E. et al., Geographic distributions and origins of human head lice (*Pediculus humanus capitis*) based on mitochondrial data. *J. Parasitol.*, 94, 1275, 2008.
9. Raoult, D. et al., Molecular identification of lice from pre–Columbian mummies. *J. Infect. Dis.*, 197, 535, 2008.
10. Kirkness, E.F. et al., Genome sequences of the human body louse and its primary endosymbiont provide insights into the permanent parasitic lifestyle. *Proc. Natl. Acad. Sci.*, 107(27), 12168, 2010.
11. Li, W. et al., Genotyping of human lice suggests multiple emergencies of body lice from local head louse populations. *PLoS Negl. Trop. Dis.*, 4, e641, 2010.
12. Roux, V., and Raoult, D., Body lice as tools for diagnosis and surveillance of reemerging diseases. *J. Clin. Microbiol.*, 37, 596, 1999.
13. Elston, D.M., What's eating you? Pediculus humanus (head louse and body louse). *Cutis*, 63, 259, 1999.
14. Raoult, D., and Roux, V., The body louse as a vector of reemerging human diseases. *Clin. Infect. Dis.*, 29, 888, 1999.
15. Burgess, I.F., Human lice and their management. *Adv. Parasitol.*, 36, 271, 1995.
16. Namazi, M.R., Levamisole: A safe and economical weapon against pediculosis. *Int. J. Dermatol.*, 40, 292, 2001. Erratum in: *Int. J. Dermatol.*, 40, 794, 2001.
17. Foucault, C. et al., Oral Ivermectin in the treatment of body lice. *J. Infect. Dis.*, 193, 474, 2006.
18. Roberts, R.J., Clinical practice. Head lice. *N. Engl. J. Med.*, 346, 1645, 2002.
19. Ibarra, J. et al., The impact of well-developed preventative strategies on the eradication of head lice. *Perspect. Public Health*, 129, 165, 2009.
20. Leo, N.P., and Barker, S.C., Intragenomic variation in ITS2 rDNA in the louse of humans, *Pediculus humanus*: ITS2 is not a suitable marker for population studies in this species. *Insect Mol. Biol.*, 11, 651, 2002.
21. Leo, N.P. et al., Evidence from mitochondrial DNA that head lice and body lice of humans (Phthiraptera: Pediculidae) are conspecific. *J. Med. Entomol.*, 39, 662, 2002.

22. Yong, Z. et al., The geographical separation of human lice preceded that of *Pediculus humanus capitis* and *Pediculus humanus humanus*. *C. R. Biol.*, 326, 565, 2003.
23. Raoult, D. et al., Evidence for louse-transmitted diseases in soldiers of Napoleon's grand army in Vilnius. *J. Infect. Dis.*, 193, 112, 2006.
24. Odronitz, F., Becker, S., and Kollmar, M., Reconstructing the phylogeny of 21 completely sequenced arthropod species based on their motor proteins. *BMC Genomics*, 10, 173, 2009.
27. Desalle, R. et al., DNA sequences from a fossil termite in Oligo-Miocene amber and their phylogenetic implications. *Science*, 257, 1933, 1992.
28. Tamura, K. et al., MEGA5: Molecular evolutionary genetics analysis using maximum likelihood, evolutionary distance, and maximum parsimony methods. *Mol. Biol. Evol.*, 28, 2731, 2011.

80 Phthirus (Crab Louse)

Dongyou Liu

CONTENTS

80.1 INTRODUCTION

Pthirus is one of the two sucking louse genera (i.e., *Pthirus* and *Pediculus*) that contain human pathogenic species/varieties. While *Pthirus pubis* (pubic louse or crab louse) in the genus *Pthirus* is notorious for infesting human genitals, *Pediculus humanus captitis* (head louse) and *Pediculus humanus humanus* (or *Pediculus humanus corpotitis*, body louse) in the genus *Pediculus* occur in human head and body, respectively.

Considered largely as a sexually transmitted organism, pubic (crab) louse (*Pthirus pubis*) is often present in pubic hair (a condition known as phthiriasis pubis). However, this parasite occasionally invades other areas with hair such as armpits, beard, moustache, eyebrows, and eyelashes (with pubic louse infestation of eyelashes being known as phthiriasis palpebrarum). By contrast, head louse (*Pediculus humanus capitus*) is transmitted through close head-to-head contact and commonly affects schoolchildren, whereas body louse (*Pediculus humanus humanus*) is transmitted via close contact with infected individual or clothes carrying the parasite and associated typically with poor hygiene, socioeconomic status, disasters, and war. In addition, body louse has the capacity to act as a vector for human bacterial diseases such as trench fever, epidemic typhus, and louse-borne relapsing fever (see Chapter 79 for details).

80.1.1 Classification, Morphology, and Biology

80.1.1.1 Classification

The genus *Pthirus* is classified in the family Pthiridae, suborder Anoplura (sucking lice), order Phthiraptera (lice), subclass Neoptera, class Insecta, phylum Arthropoda. The suborder Anoplura (sucking lice) contains at least eight families (i.e., Echinophthiriidae, Haematopinidae, Hoplopleuridae, Linognathidae, Pedicinidae, Pediculidae, Polyplacidae, and Pthiridae). As the only member recognized in the family Pthiridae, the genus *Pthirus* comprises the species of *Pthirus pubis* (synonym: *Pediculus pubis*; commonly known as pubic louse, crab louse, or crabs) and *Pthirus gorillae* (gorilla louse).

80.1.1.2 Morphology

The pubic (crab) louse has a greenish color, and is almost as broad as long (1.5–2 mm), with stout middle and hind legs, giving a crab-like appearance (Figure 80.1). The third pair of legs is equipped with a claw that helps it to hang on to coarse pubic hair.

80.1.1.3 Biology

The life cycle of pubic louse includes three distinct stages, from egg, larva (nymph) to adult, taking about 3 weeks in all. The relative proportion of female to male pubic lice on the host is about 3:2. Female pubic louse lays up to three eggs per day

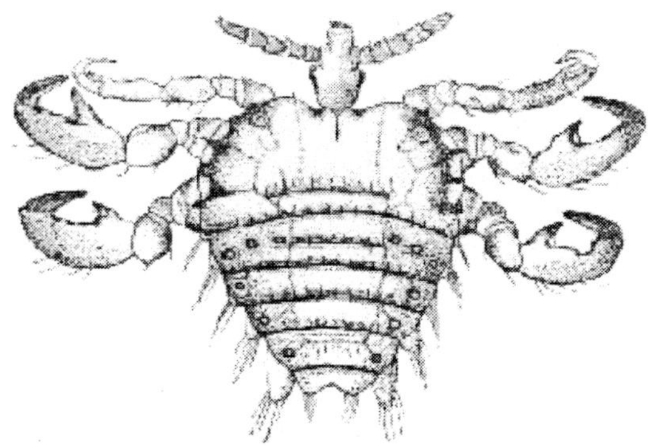

FIGURE 80.1 Adult pubic louse (*Pthirus pubis*). (Adapted from www.ces.csiro.au/aicn/ (copyright free image).)

(with a life-time total of 26 eggs), each of which firmly attaches to a hair [1]. The egg hatches in 7–8 days. On emerging, the young larva clasps a single hair and feeds at its base, and after one day or more it may clasp two hairs. The larva stops feeding to molt on the host; after molting it proceeds to feed again, shifting slightly to a fresh feeding ground. After undergoing three molts (with three larval or nymphal stages), pubic louse becomes sexually mature (adult). Pubic louse requires continuous feeding from the host. Young unfed larvae often die within 10 h of emergence, and adult louse may survive for only a short time, away from the warmth and humidity of the human body. A male survives 22 days and a female 17 days on human [2].

Utilizing its large claws, pubic louse seizes the host's body hair (particularly in the pubic region, and occasionally in the armpits, in beards, and sometimes on eyelids and eyebrows), bores into the skin with its mouthparts, and sucks blood several times in succession (with short intervals). The louse has adapted to a sedentary lifestyle, and tends to remain confined throughout its life to a limited area on the host's skin, with mouthparts inserted in skin and the bites causing intense pruritus unless disturbed [3,4].

Transmission of public louse from one human to another mostly takes place during intimate relationships (involving body contact or sexual contact), but the louse can be also transferred in crowded situations (e.g., barracks or billets) [5–10]. A detached louse promptly clings to any hair with which it comes in contact. *Phthirus pubis* often infests the strong hairs of the pubic and perianal areas, and occasionally the legs, forearms, chest, axillary hair, beard, eyebrows, and eyelashes (phthiriasis palpebrarum). There are records of small children carrying crab lice among the fine hair on the heads due to parent-to-child transmission through routes of shared towels, clothing, beds, or closets. The insect may pass from one infant to another through sharing a cradle [11]. Because the scalp hairs are crowded close together and finer, the head is probably a less suitable habitat for pubic louse. On the other hand, the hairs on the pubis and in the axillae are sparser (2 mm or more apart), and are within reach between the extended two hind leg-pairs of the adult insect (about 2 mm) [12].

Pubic louse affects approximately 2% of the adult population worldwide [13,14]. The incidence of pubic louse infestation in patients with STD or travellers is often higher than that in the general population [15]. The available evidence suggests that the louse infestation has existed in humans since ancient times. The oldest known pubic lice are from the Roman period in Britain and from a 2000-year-old South American mummy [16,17]. Examination of the head hair and pubic hair of the artificial mummy of Ferdinand II of Aragon (1467–1496), King of Naples, uncovered infestations with *Pediculus capitis* (head louse) and *Pthirus pubis* (pubic louse) [18].

80.1.2 Clinical Features and Treatment

80.1.2.1 Clinical Features

Infestation with pubic louse (crab louse, also known as "crabs" due to its grasping tarsi reminiscent of crab pincers) commonly occurs in the pubic and perianal area (the so-called phthiriasis pubis). Occasionally, it may spread to the abdomen, breast, armpits, beard, moustache, eyebrows, eyelashes, thighs, and legs. Pubic louse infestation of eyelashes is called phthiriasis palpebrarum, which is likely transferred by hand contact from the genital area to the eye [19–22]. In rare, generalized infestation, only the head and neck, the hands and feet are spared [23]. In infants, the infestation may occur in the eyelids, with the nits (eggs) being found on the eyelashes, or upon the eyebrows and hairy portions of the scalp [24–29].

The main symptom of public louse infestation is itching (and the formation of itching red papules), usually in the pubic hair area, due to hypersensitive reaction to louse saliva, which can become more intense a few weeks after initial infestation. A characteristic gray-blue or slate coloration (maculae caeruleae) may appear at the feeding sites for days, particularly on the lower abdomen and thighs. Nits and/or lice may be found attached to hairs.

De Fazio et al. [30] described a generalized case of public louse infestation in a 50-year-old person. The patient presented with a 1-month history of mild, generalized pruritus. Examination of the patient's skin revealed small, yellowish-brown flecks in the hair of the upper arm, chest, and axilla on the left side and in the pubic area. Dermoscopy showed pubic lice (*Phthirus pubis*) which have the typical broad body of the crab and the large middle and hind legs with thick claws for grasping. Treatment of the patient and household members with two total-body applications of topical permethrin 5% cream, 1 week apart, resulted in complete resolution without recurrence.

Turgut et al. [31] reported a case of phthiriasis palpebrarum mimicking lid eczema and blepharitis in a 68-year-old woman with diabetes mellitus. The patient displayed moderate itching in both eyelids. On external inspection of the periorbital region, the condition resembled lid eczema or blepharitis due to the findings similar to exfoliative lesions and color changes in eyelids and to excretions over eyelashes. Slit-lamp examination revealed two lice on the right eyelashes and one louse around the base of the left one. Many translucent oval eggs and clipping eyelashes were also observed in both eyelids. Further investigation uncovered the exfoliation, laceration, maceration, and thinning in the lid skin and protuberances due to the buried lice into the base of the eyelashes or follicles. The patient was treated with pilocarpine hydrochloride 4% drops for 7 days, applied four times daily. The patient and her family were advised to avoid close body contact until the completion of treatment and follow-up. At the end of the first week, no louse or nit was present.

Ma et al. [32] documented a case of *Pthirus pubis* infestation of the eyelashes (phthiriasis palpebrum) in a 36-year-old woman with a 3-week history of persistent itching of the eyelids of the right eye. Her husband had suffered from pubic pruritus for about 1 month. Examination revealed a few lice and nits at the roots of the woman's eyelashes, and the right conjunctiva appeared reddish. Numerous lice and nits were also noted in the pubic area. The louse specimen was verified microscopically as *Phthirus pubis*. After removal of the lice from the eyelashes with fine forceps, petrolatum was applied twice daily. The pubic area was shaved and treated with 10% sulfur ointment once daily for 3 days. Her husband was treated simultaneously. The patient was advised to wash all clothing and fomites. The itching resolved 2 weeks later, and the eyelashes and pubic area remained free of parasites.

Additionally, Keklikci et al. [33] described a case of phthiriasis palpebrarum caused by *Phthirus pubis* in the right upper eyelid of a 9-month-old male infant. The patient recovered after mechanical removal of all lice and nits from the eyelashes.

80.1.2.2 Treatment

Common treatment options for pubic louse infestation include mechanical removal (e.g., shaving off or grooming hair in the affected areas with a fine-toothed comb, trimming or plucking of eyelashes, and removing lice and eggs on eyelashes with fine forceps), chemotherapy (e.g., fluorescein eye drops 20%, physostigmine 0.25%, lindane 1%, petroleum gel, yellow mercuric oxide ointment 1%, malathion drops 1%, malathion shampoo 1%, pilocarpine gel 4%, permethrin 1%, pyrethrins, phenothrin, and oral ivermectin), cryotherapy, and argon laser photocoagulation, etc. [34–41].

For pubic louse infestation of the eyelashes, after removal of lice with tweezers or forceps, 1% permethrin lotion (petrolatum) may be used twice daily for 8–10 days. Alternatively, permethrin 5% cream may be applied with a cotton swab to the eyelashes and washed off after 10 min. Permethrin does not irritate the eyes (but the eyelids should be kept closed throughout the treatment), and is effective in eradicating lice (but not the eggs). A pilocarpine 4% drop is capable of destroying pubic lice and eggs, possibly due to its direct cholinergic action causing paralysis of the lice or to its direct pediculicidal action. Other treatments for phthiriasis palpebrarum include local application of yellow mercuric oxide N.F. 1% ophthalmic ointment twice daily for 1 week or 0.25% physostigmine (Eserine) ointment twice daily for a minimum of 10 days. In addition, vaseline eye patch may be applied twice daily for 8–10 days [42].

Body hair infestation with lice can be treated with gamma benzene hexachloride shampoo, which should be used with caution in infants, children, and pregnant women. The pubic area may be shaved and treated with 10% sulfur ointment once daily for 3 days. Malathion 0.5% lotion may be applied on dry hair and washed out 12 h after application. Permethrin 5% cream may be applied on wet hair and washed out after 10 min. Lindane 1% should not be used immediately after a bath or shower and by persons with extensive dermatitis, pregnant or lactating women, or children under 2 years of age.

Permethrin 1% cream rinse and pyrethrins are the drugs of choice for treating pubic lice in pregnant or lactating women. These agents should be applied to the affected areas and washed off after 10 min. A second treatment after 10 days is recommended.

Patients with *Pthirus pubis* infestation should be checked 7 days after treatment has begun. If the infestation has not been eradicated, another therapy should be started. Family members, sexual contacts, and close companions should be examined and treated appropriately. Patients should be advised to avoid close body contact until they and their partner(s) have completed treatment and follow-up.

Clothing, towels, and bedding used by the patient within 2–3 days before treatment began should be machine washed (with water of at least 55°C) and dried on the hot cycle for 5–10 min. Items that cannot be washed or dry-cleaned can be stored in a sealed plastic bag for 2 weeks [43].

80.1.3 Laboratory Diagnosis

Pubic louse infestation is usually diagnosed by careful examination of boy hair such as pubic hair for nits, nymphs, and adult lice. Lice and nits can be removed either with forceps or by cutting the infested hair with scissors. A magnifying glass or a microscope can be used for louse identification if there is diagnostic uncertainty. Empty egg cases attached to hair shafts are not diagnostic of an active infection [24].

Although *Pthirus pubis* can affect any hair-bearing area, infestation of the eyelashes (phthiriasis palpebrum) is uncommon in healthy adults. All hair-bearing areas should be examined carefully if infestation is found in one area. Because *Pthirus pubis* in adults is frequently transmitted sexually, a screen for other sexually transmitted infections should be carried out. In children, eyelashes are the most common site of infestation, usually through direct passage of the lice from their parents or other infected contacts [44].

Although lice are difficult to detect due to its semitransparency and deep burrowing in the lid margin, their slow movement may be noted by careful and prolonged observation. Accumulation of translucent oval nits and feces as the red dish-brown granular material on the base of the lashes may be suggestive of phthiriasis palpebrarum. As a rare cause of blepharoconjunctivitis, phthiriasis palpebrarum may result in blepharitis with marked conjunctival inflammation, preauricular lymphadenopathy, and secondary infection at the site of louse bite [45–48]. To avoid misdiagnosis, patients with the symptom of pruritus of the eyelids and with clinical findings resembling exfoliation on the surface of lid skin and seborrhea accumulation on eyelashes need careful examination using slit lamp [49]. Buried lice into the lid skin and into the base of follicle may be overlooked during biomicroscopic examination. Furthermore, the translucent oval nits located in the bases of the eyelashes and on the cilia may be confused with the crusty excretions of seborrheic blepharitis.

In recent years, molecular techniques have been employed for the identification and phylogenetic analysis of lice and other insect parasites [50–52].

80.2 METHODS

80.2.1 Sample Preparation

80.2.1.1 Sample Collection

Pubic lice are collected and preserved in 95% ethanol and stored at −80°C, or air-dried.

80.2.1.2 DNA Isolation

One modified silica-based extraction method is useful for the extraction of louse DNA. Two to five lice are ground together in liquid nitrogen and incubated for 48 h with agitation at 55°C in 600 μL of extraction buffer (7.5 mol/L guanidinium thiocyanate, 0.1 mol/L Tris hydrochloride, pH 6.4, 0.02 mol/L EDTA, pH 8.0, and 1.3% Triton X-100). After centrifugation, 500 μL of supernatant is removed to a second tube containing an additional 500 μL of extraction buffer and 40 μL of saturated silica suspension (SiO_2 in water). DNA is bound to the silica for 10 min at 27°C and then pelleted by centrifugation, washed twice with extraction buffer and twice with 70% ethanol supplemented with 10 mmol/L sodium chloride, and eluted in two 75 μL volumes of Tris-EDTA buffer at 60°C.

Alternatively, lice are rinsed with distilled water, dried on sterile filter paper, and crushed individually in a sterile Eppendorf tube with a sterile pipette tip. DNA is extracted using the QIAamp Tissue kit (QIAGEN) according to the manufacturer's instructions, except that incubation in lysis buffer was performed overnight at 37°C [53].

80.2.2 Detection Procedures

80.2.2.1 Sequence Analysis of 18S rRNA Gene

Leo and Barker [52] utilized forward primers NS1 (5′-GTAGTCATATGCTTGTCTC-3′) and NS10 (5′-AGGCTC TGCAATCGGAATG-3′) together with reverse primers NS2a (5′-CGCGGCTGCTGGCACCAGACTTGC-3′) and NS11 (5′-GTCAAATTAAGCCGCAAGC-3′), respectively, to amplify and analyze 18S rRNA gene from human-infecting lice.

80.2.2.1.1 Procedure

1. PCR amplification is carried out with one cycle of 95°C for 3 min; 10 cycles of 93°C for 30 s, 40°C for 30 s, and 72°C for 1 min 30 s; 25 cycles of 93°C for 30 s, 50°C for 30 s, and 72°C for 1 min 30 s; and one cycle of 72°C for 5 min.
2. Nucleotide sequences are aligned with Clustal X, and the alignments are checked by eye. The program Gblocks v. 0.91b is used to remove ambiguous sites from the 1554-bp alignment and the final alignment of 1195 bp is put into MODELTEST. Phylogenetic trees are inferred by using parsimony (PAUP) v.4.0b10 with neighbor-joining (NJ) and maximum parsimony (MP) methods as for COI.

Note: The alternative 18S rRNA primers 18sai (5′-CCTGAGAAACGGCTACCACATC-3′) and 18 sbi5.0 (5′- TAACCGCAACAACTTTAAT-3′) may also be applied for the molecular identification of insects including lice [50].

80.2.2.2 Sequence Analysis of *COX1* and *CYTB* Genes

Raoult et al. [51] used primers L6625 (5′-COGGATCCTTY TGRTTYTTYGGNCAYCC-3′) and H7005 (5′-CCGGAT CCACNACRTARTANGTRTCRTG-3′) to amplify the *cox*1 (cytochrome oxidase subunit 1) gene (Table 79.2) and primers CytbF1 (5′-GAGCGACTGTAATTACTA ATC-3′), CytbR1 (5′-CAA CAA AAT TAT CCG GGT CC-3′), CytbF2 (5′-GAG GAG GGT TTT CAG TTA-3′), and CytbR2 (5′-ACT TTA TCA CTA TCC AAA TC-3′) to amplify the *cytb* (cytochrome b) gene for phylogenetic study of human louse species.

80.2.2.2.1 Procedure

1. PCR mixture (25 μL) is made up of 12.5 pmol of each primer, 200 μM each of dATP, dCTP, dGTP, and dTTP, 1 U of *Taq* DNA polymerase in 1× PCR buffer, 0.8 μL of 25 mM $MgCl_2$ (Gibco BRL), and 2.5 μL of the extracted DNA.
2. Amplification is conducted with an initial 3-min denaturation step at 95°C, 40 cycles of denaturation at 95°C (30 s), annealing at 50°C (30 s), and extension at 72°C (1 min), and a final extension at 72°C for 7 min.
3. PCR products are purified with Qiagen columns (QIAquick Spin PCR purification kit; QIAGEN), and sequenced.
4. The sequence data are examined using NJ with a best-fit model of nucleotide evolution in Modeltest. Levels of topological support are calculated from 100 bootstrap replicates.

80.3 CONCLUSION

Pthirus pubis (pubic louse, crab louse, or crabs) in the genus *Pthirus* commonly affects human pubic and perianal area (a condition referred to as phthiriasis pubis). Occasionally, this parasite may appear in other areas with hair such as armpits, beard, moustache, eyebrows, and eyelashes. Pubic louse infestation of eyelashes is known as phthiriasis palpebrarum. Itching and the formation of itching red papules, along with a characteristic gray-blue or slate coloration (maculae caeruleae) at the feeding sites, are the main symptoms of public louse infestation, as a result of host hypersensitive reaction to louse saliva. Clinical diagnosis of public louse infestation is dependent on the finding and identification of nits, nymphs, and adult lice in body hair in pubic and other areas. In recent years, molecular techniques have been developed for phylogenetic assessment of human-infecting lice, including pubic louse, head louse, and body louse. This has facilitated the improved tracking and control of pubic louse.

REFERENCES

1. Burkhart, C.N., Gunning, W., and Burkhart, C.G., Scanning electron microscopic examination of the egg of the pubic louse (Anoplura: *Pthirus pubis*). *Int. J. Dermatol.*, 39, 201, 2000.
2. Anderson, A.L., and Chaney, E., Pubic Lice (*Pthirus pubis*): History, biology and treatment vs. knowledge and beliefs of US College students. *Int. J. Env. Res. Public Health*, 6, 592, 2009.
3. Ko, C.J., and Elston, D.M., Pediculosis. *J. Am. Acad. Dermatol.*, 50, 1, 2004.
4. Puri, P.K. et al., The histopathology of *Phthirus pubis*. *J. Cutan. Pathol.*, 36, 80, 2009.
5. Routh, H.B. et al.; Ectoparasites as sexually transmitted diseases. *Semin. Dermatol.*, 13, 243, 1994.
6. Skinner, C.J., Viswalingam, N.D., and Goh, B.T., *Phthirus pubis* infestation of the eyelids: A marker for sexually transmitted diseases. *Int. J. STD AIDS*, 6, 451, 1995.
7. Mimouni, D. et al., Secular trends in epidemiology of pediculosis capitis and pubis among Israeli soldiers: A 27-year follow-up. *Int. J. Derm.*, 40, 637, 2001.
8. Pierzchalski, J.L., Bretl, D.A., and Matson, S.C., *Phthirus pubis* as a predictor for chlamydia infections in adolescents. *Sex. Transm. Dis.*, 29, 331, 2002.
9. Varela, J.A. et al., *Pthirus pubis* in a sexually transmitted diseases unit: A study of 14 years. *Sex. Transm. Dis.*, 30, 292, 2003.
10. Orion, E., Hagit, M., and Wolf, R., Ectoparasitic sexually transmitted diseases: Scabies and Pediculosis. *Clin. Dermatol.*, 22, 513, 2004.
11. Mimouni, D. et al., Seasonality trends of *Pediculosis capitis* and *Phthirus pubis* in a young adult population: Follow-up of 20 years. *J. Eur. Acad. Dermatol. Venereol.*, 16, 257, 2002.
12. Flinders, D.C., and DeSchweinitz, P., Pediculosis and scabies. *Am. Fam. Phys.*, 69, 341, 2004.
13. Poudel, S.K., and Barker, S.C., Infestation of people with lice in Kathmandu and Pokhara, Nepal. *Med. Vet. Entomol.*, 18, 212, 2004.
14. Pakeer, O. et al., Four cases of pediculosis caused by *Pthirus pubis* Linnaeus, 1758 (Diptera: Anoplura) from peninsular Malaysia. *Trop. Biomed.*, 24, 101, 2007.
15. Lee, H.L. et al., Review of forensically important entomological specimens in the period of 1972–2002. *Trop. Biomed.*, 21, 69, 2004.
16. Kenward, H., Pubic lice in Roman and Medieval Britain. *Trends Parasitol.*, 17, 167, 2001.
17. Rick, F.M. et al., Crab louse infestation in pre-Columbian America. *J. Parasitol.*, 88, 1266, 2002.
18. Fornaciari, G. et al., 'Royal' pediculosis in Renaissance Italy: Lice in the mummy of the King of Naples Ferdinand II of Aragon (1467–1496). *Mem. Inst. Oswaldo Cruz*, 104, 671, 2009.
19. Manjunatha, N.P. et al., Pediculosis pubis: Presentation to ophthalmologist as phthriasis palpebrarum associated with corneal epithelial keratitis. *Int. J. STD AIDS*, 17, 424, 2006.
20. Ngai, J.W., Yuen, H.K., and Li, F.C., An unusual case of eye itchiness. *Hong Kong Med. J.*, 14, 414, 2008.
21. Pinckney, J. 2nd. et al., Phthiriasis palpebrarum: A common culprit with uncommon presentation. *Dermatol. Online J.*, 14, 7, 2008.
22. Niazi, M.K., and Arain, M.A., Phthiriasis palpebrarum. *J. Coll. Physicians Surg. Pak.*, 19, 589, 2009.
23. Klaus, S., Shvil, Y., and Mumcuoglu, K.Y., Generalized infestation of a 31/2-year-old girl with the pubic louse. *Pediatr. Dermatol.*, 11, 26, 1994.
24. Couch, J.M. et al., Diagnosing and treating *Phthirus pubis* palpebrarum. *Surv. Ophthalmol.*, 26, 219, 1982.
24. Silburt, B.S., and Parsons, W.L., Scalp infestation by *Phthirus pubis* in a 6-week-old infant. *Pediatr. Dermatol.*, 7, 205, 1990.
25. Singh, S. et al, *Phthirus pubis* infestation of the scalp: Report of three cases. *Rev. Infect. Dis.*, 12, 560, 1990.
26. Mansour, A.M., Photo essay: Phthiriasis palpebrarum. *Arch. Ophthalmol.*, 118, 1458, 2000.
27. Bessis, D. et al., Erythema annulare centrifugum induced by generalized *Phthirus pubis* infestation. *Br. J. Dermatol.*, 149, 1291, 2003.
28. Ikeda, N. et al., *Phthirus pubis* infestation of the eyelashes and scalp hairs in a girl. *Pediatr. Dermatol.*, 20, 356, 2003.
29. DeFazio, J.L., and Spencer, P., Images in clinical medicine. Dermoscopy of phthiriasis. *N. Engl. J. Med.*, 362, e33, 2010.
30. Turgut, B. et al., Phthriasis palpebrarum mimicking lid eczema and blepharitis. *J. Ophthalmol.*, 2009, 803951, 2009.
31. Ma, D.L., and Vano-Galvan, S., Infestation of the eyelashes with *Phthirus pubis*. *CMAJ*, 182, E187, 2010.
32. Keklikci, U. et al., Phthiriasis palpebrarum in an infant. *J. Pediatr. Ophthalmol. Strabismus*, 46, 173, 2009.
33. Newsome, J.H., Fiore, J.L., Jr., and Hackett, E., Treatment of infestation with *Phthirus pubis*: Comparative efficacies of synergized pyrethrins and gamma-benzene hexachloride. *Sex. Transm. Dis.*, 6, 203, 1979.
35. Ashkenazi, I., Desatnik, H.R., and Abraham, F.A., Yellow mercuric oxide: A treatment of choice for Phthriasis palpebrarum. *Brit. J. Ophthalmol.*, 75, 356, 1991.
36. Burkhart, C.N., and Burkhart, C.G., Oral ivermectin therapy for phthiriasis palpebrum. *Arch. Ophthalmol.*, 118, 134, 2000.
37. Kumar, N., Dong, B., and Jenkins, C., Pubic lice effectively treated with Pilogel. *Eye*, 17, 538, 2003.
38. Yoon, K.C. et al. Mechanical treatment of phthiriasis palpebrarum. *Korean J. Ophthalmol.*, 17, 71, 2003.
39. Elston, D.M., Drugs used in the treatment of pediculosis. *J. Drugs Dermatol.*, 4, 207, 2005.
40. Leone, P.A., Scabies and pediculosis pubis: An update of treatment regimens and general review. *Clin. Infect. Dis.*, 44, S153, 2007.
41. Scott, G.R., and Chosidow, O., European guideline for the management of pediculosis pubis, 2010. *Int. J. STD AIDS*, 22, 304, 2011.
42. Das, J.C., Sharma, P., and Singla, R., A new treatment modality for phthiriasis palpebrarum. *J. Pediatr. Ophthalmol. Strabismus*, 40, 304, 2003.
43. Burgess, I.F., Human lice and their management. *Adv. Parasitol.*, 36, 271, 1995.
44. Galiczynski, E.M., Jr., and Elston, D.M., What's eating you? Pubic lice (*Pthirus pubis*). *Cutis*, 81, 109, 2008.

45. Turow, V.D., Phthiriasis palpebrarum: An unusual cause of blepharitis. *Arch. Pediatr. Adolesc. Med.*, 149, 704, 1995.
46. Lin, Y.-C. et al., Phthiriasis palpebrarum: An unusual blepharoconjunctivitis. *Chin. Med. J.*, 65, 498, 2003.
47. Vandeweghe, K., and Zeyen, P., Phthiriasis palpebrarum: 2 case reports. *Bull. Soc. Belge Ophtalmol.*, 300, 27, 2006.
48. Zvandasara, T., and Lam, F.C., Conjunctivitis due to *Phthirus pubis*: A trap for the unwary. *Int. J. STD AIDS*, 21, 519, 2010.
49. Rundle, P.A., and Hughes, D.S., Phthirus pubis infestation of the eyelids. *Brit. J. Ophthalmol.*, 77, 815, 1993.
50. Desalle, R. et al. DNA sequences from a fossil termite in Oligo-Miocene amber and their phylogenetic implications. *Science*, 257, 1933, 1992.
51. Raoult, D. et al., Molecular identification of lice from pre-Columbian mummies. *J. Infect. Dis.*, 197, 535, 2008.
52. Leo, N.P., and Barker, S.C., Unravelling the evolution of the head lice and body lice of humans. *Parasitol. Res.*, 98, 44, 2005.
53. Reed, D.L. et al. Genetic analysis of lice supports direct contact between modern and archaic humans. *PloS Biol.*, 2, e340, 2004.

81 Pulex, Xenopsylla, and Ctenocephalides (Human, Rat, and Dog/Cat Fleas)

Dongyou Liu

CONTENTS

81.1 INTRODUCTION

Fleas are small, wingless insects (1–10 mm in length) whose 2500 or so recognized species/subspecies are placed in the order Siphonoptera (consisting of 16 families). Fleas are characterized by having a laterally compressed body (allowing rapid movement among hair or feathers), well-developed mouthparts (for piercing host skin and sucking blood), enlarged hind legs (adapted for jumping), strong tarsal claws (adapted for holding onto their hosts), backward pointing hairs and bristles (for ease of movement through the hair of a host), and small antennae (tucking away into special groves in the head). Being hematophagous ectoparasites, fleas do not cause significant damage to their hosts, and are generally considered as a nuisance. However, due to their potential roles in the transmission of other important microbial pathogens, flea infestations demand adequate attention, including correct identification and prompt treatment, to protect the well-being of humans and animals. With medically important fleas being mainly in the genera *Pulex*, *Xenopsylla*, *Ctenocephalides*, and *Tunga*, this chapter is devoted to the genera of *Pulex*, *Xenopsylla*, and *Ctenocephalides*, while Chapter 82 of this book focuses on the genus *Tunga*.

81.1.1 Classification, Morphology, and Biology

81.1.1.1 The Genus *Pulex*

81.1.1.1.1 Classification

The genus *Pulex* is classified taxonomically in the subfamily Pulicinae, family Pulicidae, order Siphonaptera, class Insecta, phylum Arthropoda. The family Pulicidae consists of six subfamilies: Archaeopsyllinae, Parapulex, Pulicinae, Spilopsyllinae, Tunginae, and Xenopsyllinae; and the subfamily Pulicinae is separated into two genera: *Echidnophaga* and *Pulex*. In turn, the genus *Pulex* includes about 10 species: *Pulex alvarezi*, *Pulex conepati*, *Pulex dugesii*, *Pulex echidnophagoides*, *Pulex irritans*, *Pulex larimerius*, *Pulex porcinus*, *Pulex simulans*, *Pulex sinoculus*, and *Pulex vulgaris*, whereas the genus *Echidnophaga* contains the species of *Echidnophaga gallinacea* (poultry sticktight flea) and *Echidnophaga iberica*, which are seldom involved in human illness.

P. irritans (synonyms: *P. irritans bahiensis*, *P. irritans fulvus*, and *Pulex orientalis*; commonly called human flea) is a cosmopolitan flea species with a wide host spectrum. The species is thought to have originated in South America, where its original host may have been the guinea pig or peccary [1,2].

81.1.1.1.2 Morphology

P. irritans demonstrates varied morphological characteristics during its four distinct stages of development (egg, larva, pupa, and adult) [3]. *P. irritans* adults (often accounting for 5% of living fleas) measure 1.5–4 mm in length and are laterally flattened, black to brownish-black in color, and wingless. The body of *P. irritans* contains three regions: head, thorax, and abdomen. The adults have a curved or rounded head bearing short, 3-jointed, clubbed antennae, and piercing-sucking mouthparts (which consist of the grooved labrum and a pair of sharp, sword-shaped mandibles). The adults also possess stout and spiny legs, which consist of four parts (coxa, femur, tibia, and tarsus). With the help of its strong legs, the flea can jump up to 200 times its own body length (about 50 cm, which is equivalent to a man jumping 350 m) and 130 times its own height (about 33 cm).

P. irritans eggs are about 0.5 mm in length, smooth, oval or rounded, and light-colored. *P. irritans* larvae are hairy, maggot-like, and creamy/yellow in color. The larvae have a distinct, brownish head without eyes or legs, and have three thoracic and 10 abdominal segments, each surrounded by a single row of bristles. The larvae are about 1.5 mm long initially and grow through three instars to about 5 mm long. *P. irritans* pupae, contained within loosely woven debris-collecting cocoons, are about 4 × 2 mm, appear white initially and change to brownish before the adults emerge. The pupae display the general shape and characteristics of the adult. The adults emerge from the pupae in a week or two (under favorable conditions) or a year (under adverse conditions).

81.1.1.1.3 Biology

Fleas are holometabolous and undergo four developmental stages: egg (embryo), larva, pupa, and adult (imago). *P. irritans* adult female lays four to eight eggs after each blood meal (with a life time total of >400 eggs). The eggs are deposited on the body, the bedding, or the nest of the host. As the eggs are nonsticky, they are easily brushed off, and are found in crevices in the floor, under the edges of carpeting, in sofas, or in cat or dog boxes, kennels. The fertilized eggs usually hatch in 10 days or less to become larvae. The larvae move forward by using their backward-projecting bristles and a pair of hooked, chitinous processes located at the end of the abdomen. They begin to feed immediately on a variety of debris, including dried excrement, dried blood, dead mites, dried bits of skin, and other organic remains (including their own cast skins) and may remain in this stage for up to 200 days. When disturbed, they may flip in circles to escape. The larvae molt three times (going through three instars) within a week to several months, and then spin pupal cocoons covered with grains of sand, dust, or organic debris. The cocoons protect the pupae from external stresses, and the pupae can last in this form from 10 days to 50 weeks. The emergence of adults from the pupae is stimulated by heat, carbon dioxide, or movement (both human and/or pet activity). The life span of adult fleas ranges from a few weeks to a year.

P. irritans (human flea) persists as a house flea in many parts of the world, and is commonly found in refugee and homeless populations as well as other disadvantaged communities with poor hygiene [1,4]. Besides humans, *P. irritans* bites other mammals and birds, including pigs, canids, felids, goats, monkeys in captivity, chickens, burrowing owls, deer, skunks, badgers, ground squirrels, opossums, black and Norwegian rats, wild rodents, and free-tailed bats [5–8]. As *P. irritans* breeds in profusion in pigsties, piggery workers or visitors can readily pick up the flea and acquire the infection. Apart from bites, *P. irritans* (human flea) can serve as the intermediate host of the dog tapeworm (*Dipylidium caninum*), a vector for bubonic plague, tularemia, and erysipeloid [9–12].

81.1.1.2 The Genus *Xenopsylla*

81.1.1.2.1 Classification

The genus *Xenopsylla* is classified in the subfamily Xenopsyllinae, family Pulicidae, order Siphonaptera, class Insecta, phylum Arthropoda. Being one of the six subfamilies (i.e., Archaeopsyllinae, Parapulex, Pulicinae, Spilopsyllinae, Tunginae, and Xenopsyllinae) within the family Pulicidae, the subfamily Xenopsyllinae contains a single genus *Xenopsylla*, which is subdivided into >80 species and subspecies. The most notable *Xenopsylla* species include *Xenopsylla cheopis* (named after the Cheops pyramids; also known as oriental rat flea, or tropical rat flea), *Xenopsylla cunicularis*, and *Xenopsylla gratiosa*.

81.1.1.2.2 Morphology

X. cheopis adults measure about 2.5 mm in length, have laterally compressed body, are wingless and appear light brown to dark brown in color. The absence of both genal and pronotal ctendium (combs of bristles in the front and back) in *X. cheopis* adults allows differentiation of oriental rat flea from the cat flea, dog flea, and other fleas. *X. cheopis* females have dark-colored spermatheca (resembling small sacs), which represents a distinguishing characteristic of this species. *X. cheopis* males have complex genitalia that are easily distinguishable from the females'. *X. cheopis* adults possess very large legs for jumping from one nest to another, and have a sensory organ (pygidium) on the dorsal side for detecting vibrations and air currents. Pupae also use these signals to time their emergence from cocoons [13].

X. cheopis eggs are small and white-colored. *X. cheopis* larvae (2 mm long) are slender, eyeless, legless, and white-colored. The larvae have mandibles (for feeding on dead skin cells, flea droppings, and other smaller parasites, but not drinking blood) and 14 bristled segments. During the last larval instar, the larva molts and makes a silken cocoon around itself and pupates. The flea remains a pupa from 1 week to 6 months, and emerges as an adult, which then sucks blood from the host and mates with other fleas.

81.1.1.2.3 Biology

X. cheopis female can mate once and lay eggs every day after blood meal (up to 50 eggs per day). Since fleas are nidiculous parasites and live on the host's nests, the eggs usually hatch on their nests rather than on their hosts. *X. cheopis* eggs normally hatch in 2–12 days, and the larvae go through three molts in 9–15 days (up to 200 days under unfavorable conditions). The larva spins a silk cocoon where it remains until it is finished pupating. The flea's development rate during the pupal stage is affected by temperature and moisture. Emerging from the pupa, the adult can survive up to 100 days. The maximum life span for *X. cheopis* is 376 days.

X. cheopis can sense exhaled carbon dioxide from humans and animals and jumps rapidly (with the assistance of its small powerful legs) to the source to feed on the newly found host (up to 0.5 mL^3 blood per meal). The flea uses an elastic protein called resilin to store up energy and then release 97% of it in a single jump. Both *X. cheopis* male and female adults feed on blood of *Rattus rattus* (black rat) and other mammals, including humans. *X. cheopis* relies on its external mouthparts, that is, maxillary lacunae to pierce the host skin and an epipharynx to enter the wound and to inject salvia. Saliva contains special chemicals (e.g., histamine-like compounds, enzymes, polypeptides, and amino acids) to prevent the host's blood from coagulating and to provoke the host's allergic response. A canal formed by the maxillary laciniae and the epipharynx then sucks up blood from a pool (telmophagy). *X. cheopis* then uses its proventriclus down the gut to break down blood cells and digest the blood meal. *X. cheopis* dislikes temperatures close to human body temperatures, and tends to leave the host when not feeding.

X. cheopis (oriental rat flea) is a parasite of *Rattus* spp and other mammals including humans. The species is distributed in tropical and subtropical as well as temperate regions of the world, but is rarely found in cold areas since it requires a tropical/subtropical climate to pupate.

X. cheopis is a vector for pathogens such as *Yersina pestis* (plague), *Rickettsia typhi* (murine typhus), and *Bartonella*, and also acts as a host for *Hymenolepis diminuta* (rat tapeworm) and *Hymenolepis nana* (mouse tapeworm) [14–17]. The transmission of bacterial pathogens occurs when the flea feeds on an infected rodent, and bacteria enter into the flea's gut. The bacteria multiply rapidly and block the flea's proventriculus. Without its stomach unfilled, the flea searches for a new host for blood meal. After biting the host, the flea regurgitates and mixes the blood of an uninfected host with the bacteria in the flea's stomach, and then expels infected blood back through the proboscis into the wound, infecting a new host. Hosts may also become infected by consuming fecal material or dead remnants of an infected *X. cheopis*. Diseases can be transmitted from one generation of fleas to the next through the eggs.

81.1.1.3 The Genus *Ctenocephalides*

81.1.1.3.1 Classification

The genus *Ctenocephalides* is classified in the subfamily Archaeopsyllinae, family Pulicidae, order Siphonaptera, class Insecta, phylum Arthropoda. Forming one of the six subfamilies (i.e., Archaeopsyllinae, Parapulex, Pulicinae, Spilopsyllinae, Tunginae, and Xenopsyllinae) in the family Pulicidae, the subfamily Archaeopsyllinae comprises two genera: *Archaeopsylla* and *Ctenocephalides*. The genus *Ctenocephalides* is further separated into two species: *Ctenocephalides canis* (dog flea) and *Ctenocephalides felis* (cat flea), with the latter species being most common and important [18,19].

81.1.1.3.2 Morphology

C. felis (cat flea) is a small, wingless insect of a few millimeters in length. *C. felis* eggs (about 1/4 mm long and visible to the naked eye) are pearly white, oval with rounded ends, and 0.5 mm long. The larvae (1–5 mm in length) are white, blind, and worm-like with a row of bristles on each segment [20].

81.1.1.3.3 Biology

C. felis mates on its host animal after feeding, and the female begins egg production within 24–48 h after its first blood meal. *C. felis* female lays 4–8 eggs in groups (with a lifetime total of several hundreds) either on the pelage of the host (from where the eggs readily drop onto bedding, carpet, or soil) or directly in the nest. The eggs hatch in about 10 days at room temperature. The larvae are negatively phototaxic and tend to hide from light in the substrate (grass, branches, leaves, or soil). They consume the dead skin and other dirt and dust from the host animal as well as the feces of the adult fleas (which contain undigested blood). After going through three molts (instars) over 8–150 days depending on availability of food and climatic conditions, the fully grown larvae pupate in a cocoon (being ovoid, ~0.5 cm long, and whitish) spun from a salivary secretion and covered with dust. The pupal stage may last from 1 week to several months, and the adult flea may remain in the cocoon for a long time (years) until an animal comes nearby. Stimulated by vibration or physical pressure, carbon dioxide, or heat, the adult emerges from the cocoon, jumps to a new host within seconds, and begins feeding on host blood within minutes (consuming 13.6 μL of blood daily). A newly emerged cat flea can survive 24–72 h without a blood meal, and may move to the top of the carpet pile or vegetation to wait for a passing host. Under most household conditions, *C. felis* completes its life cycle in 3–6 weeks.

C. felis (cat flea) is one of the most abundant and widespread flea species. Its primary host is the domestic cat and it also feeds dogs, poultry, goats, cattle, sheep, rats, opossums, raccoons, foxes, and humans. Humans represent an aberrant host for *C. felis*, which can bite, but not infect human host. Thus the cat flea cannot be sustained on humans, unlike human flea, which is able to breed on a diet of human blood only [21].

Besides causing flea allergy dermatitis or flea bite hypersensitivity, *C. felis* can transmit plague (*Y. pestis*), murine typhus (*R. typhi* and *Rickettsia felis*), *Bartonella* and ape dermatitis as well as *Borrelia burgdorferi* to dogs and cats and also to humans [22–34]. The tapeworm *D. caninum* is transmitted when an infected cat flea (harboring cysticercoids) is swallowed by pets or humans. Cat flea can also act as the intermediate host of the nonpathogenic subcutaneous filarid nematode of dogs, *Dipetalonema reconditum* [35].

81.1.2 Clinical Features and Treatment

81.1.2.1 Clinical Features

In general, bites and blood sucking by *P. irritans* (human flea), *X. cheopis* (oriental rat flea), and *C. felis* (cat flea) cause insignificant harm and represent a nuisance to humans. While the bites of the cat flea often concentrate on the lower parts of the legs (the legs and ankles, usually two or three bites in a row), those of the human flea may be distributed over the body. The bites are felt immediately, but may become increasingly irritating and sore for up to a week. A small, red spot surrounded by a red halo without much swelling appears near the site where the flea's mandibles have penetrated the skin. Flea bite may be a transient wheal or cause long-lasting symptoms. Children <10 years old are more sensitive to the bites than older persons.

The enzymes and histamine-like substances contained in the flea's saliva may cause allergic reactions leading to intense itching, restlessness, irritation, and scratching of the skin, rash, and urticarial papules on the hands, forearms, and lower legs. Histologically, a wedge-shaped dermal infiltrate surrounds both venules and arterioles. Epidermal necrosis, spongiosis, and intraepidermal or subepidermal bullae may be present. However, the importance of these fleas lies in their ability to act as vectors for a variety of microbial pathogens including those causing plague (*Y. pestis*), murine typhus (*R. typhi*), tularemia, and tapeworm, and induce clinical disease associated with these pathogens.

81.1.2.2 Treatment

Treatment measures for fleas consist of flea removal, anti-inflammatory/cooling agents, and antibiotics for secondary pyoderma.

All animals (cats, dogs, and pigs) should be treated with insecticides such as fipronil (with or without the addition of *S*-methoprene), imidacloprid, nitenpyram, selamectin, pyrethroids, lufenuron (with pyrethroid spray or nitenpyram tablets), selamectin, pyriproxyfen, or organophosphates [36]. Orally administered nitenpyram eliminates fleas within 3–4 h, while topical application of spot-on formulations containing fipronil, imidacloprid, or selamectin takes 12–42 h. The combination of fipronil/(*S*)-methoprene or other adulticidal/ovicidal products is effective against adult fleas and eggs. Topically applied methoprene or pyriproxyfen and orally administered or injectable (cats only) lufenuron also have ovicidal activity.

Carpets should be shampooed and vacuumed; room foggers may be used in the house; the bedding, cracks, grooves in hardwood floors, behind baseboards, under the edge of rugs, beneath furniture (beds, tables, and sofas), and within closets should be washed in hot water, dusted, or sprayed. Outdoor areas (e.g., dog houses, within garages, under porches, and in animal lounging areas beneath shrubs or other shaded areas) may be treated with cyfluthrin or fenvalerate. A second treatment may be necessary 7–10 days later to eliminate adult fleas emerging from cocoons [18].

Topical application of cooling preparations (e.g., carbolated vaseline, propylene glycol, menthol, camphor, pramoxine, or calamine solution) or a piece of ice on the affected area may produce some relief from the itching of flea bites. Topical steroids (e.g., glucocorticoids and prednisolone) under occlusion may be useful for short-term treatment. Systemic antibiotics may be used to control the pyoderma and reduce the associated inflammation and pruritus. Lufenuron, an orally administered growth regulator, does not kill adult fleas, but prevents flea reproduction. Topical fipronil is also highly effective.

81.1.3 Laboratory Diagnosis

Diagnosis of flea infestations relies on the finding and identification of the flea species from the host. All fleas are characterized morphologically by laterally compressed bodies and large hind legs. Additionally, *P. irritans* (human flea) lacks the ctenidia (combs) and has the appearance of a mustache (genal comb) and mane of hair (pronotal comb) in relation to *C. felis* (cat flea). *P. irritans* also has a rounded frons (forehead), permitting differentiation from the anteriorly flattened head of the sticktight flea. The head of *P. irritans* has only a single pair of setae (hairs) behind the antennae, and shows no pleural rod, a useful feature for differentiating it from *X. cheopis* (oriental rat flea). On the other hand, *X. cheopis* adults lack both genal and pronotal ctendium (combs of bristles in the front and back), permitting their differentiation from the cat flea, dog flea, and other fleas. Furthermore, *X. cheopis* females show dark-colored spermatheca (resembling small sacs), which is a diagnostic feature of this species.

Serological testing of IgE directed against flea-specific salivary antigens may be used to aid in the diagnosis of flea allergic disease. In recent years, molecular techniques have increasingly been utilized for the reliable identification and phylogenetic assessment of flea species. Through amplification and sequence analysis of 18S ribosomal RNA, 28S ribosomal RNA, cytochrome oxidase II, elongation factor 1-α and other gene regions, the correct identity of flea species/subspecies can be rapidly and precisely ascertained [2,37–39].

81.2 METHODS

81.2.1 Sample Preparation

Genomic DNA from fleas can be isolated using the DNeasy Tissue Kit (Qiagen). Two flea individuals per population are frozen in individual preparations in liquid nitrogen and pulverized in a steel mortar. The powder is then mixed with lysis buffer, and genomic DNA preparation is performed according to the protocol for animal tissues [38].

81.2.2 Detection Procedures

Vobis et al. [38] employed primers from the internal transcribed spacer 1 (ITS1), internal transcribed spacer 2 (ITS2) and mitochondrial 16S ribosomal RNA gene (mt16S-rRNA) for the identification and phylogenetic analysis of 31 different flea isolates (including *P. irritans*, *C. felis*, *C. canis*, *X. cheopis*, and *Tunga penetrans*) (Table 81.1).

81.2.2.1 Procedure

1. PCR mixture (50 μL) is composed of 1.4 mM of each primer, 2.5 U of *Taq* polymerase (Taq PCR Core Kit, Qiagen), 200 μM of each dNTP and 2 mM $MgCl_2$, and 10–20 ng of parasite DNA template.

TABLE 81.1
Primers Used for Phylogenetic Analysis of Flea Species

Primer	Sequence (5′–3′)
Sen-ITS1	GTA CAC ACC GCC CGT GCG TAC T
Rev-ITS1	GCT GCG TTC TTC ATC GAC CC
Sen-ITS2	GGG TCG ATG AAG AAC GCA GC
Rev-ITS2	GCG CAC ATG CTA GAC TCC GTG GTT CAA G
Sen-mt16S	TAC ATA ACA CGA GAA GAC C
Rev-mt16S	GTG ATT GCG CTG TTA TCC

2. Amplification is conducted with 30 cycles of 94°C, 54°C, and 72°C for 1 min each.
3. The resulting amplicons are purified from agarose gels using the QIAquick Gel Extraction Kit (Qiagen) and sequenced.

Note: The ITS1 region defined by primers Sen-ITS1 and Rev-ITS1 is ~1000 bp in *X. cheopis*, 950 bp in *P. irritans*, 800 bp in *C. felis*, and 700 or 800 bp in *T. penetrans*. The ITS2 region defined by primers Sen-ITS2 and Rev-ITS2 is ~540 bp in *X. cheopis* and *P. Irritans*, 500 bp in *C. canis* and *C. felis*, and 530 bp in *T. penetrans*. Hypervariable section of mt16S-rRNA defined by primers Sen-mt16S and Rev-mt16S is ~200 bp in *X. cheopis*, *C. felis*, *C. canis*, and *P. irritans*.

81.3 CONCLUSION

Fleas (Siphonoptera) are small, wingless, hematophagous ectoparasites of mammals including humans, whose biting and blood-sucking do not usually cause significant harm to the hosts themselves. However, due to their ability to transmit a number of human pathogenic organisms, including those causing plague (*Y. pestis*), murine typhus (*R. typhi*), and other pathogens, infestations by fleas such as *P. irritans* (human flea), *X. cheopis* (oriental rat flea), and *C. felis* (cat flea) cannot be ignored. Traditionally, diagnosis of flea infestations is based on the assessment of the history of flea exposure, clinical signs, and presence of fleas or flea excrement. Given the generalized clinical symptoms induced by various flea species/subspecies, correct diagnosis of flea infestations relies heavily on accurate identification of fleas species involved. While it is possible to determine the species identity of fleas using morphological criteria, the process is time-consuming and demands considerable technical expertise. The application of PCR and sequencing techniques offers a rapid, sensitive, and specific approach for the identification of fleas and other insects. It is envisaged that future development of species-specific primers and probes will further streamline the molecular detection and diagnosis of flea infestations in humans and animals.

REFERENCES

1. Buckland, P.C., and Sadler, J.P., A biogeography of the human flea, *Pulex irritans* L. (Siphonaptera: Pulicidae). *J. Biogeography*, 16, 115, 1989.
2. Whiting, M.F. et al., A molecular phylogeny of fleas (Insecta: Siphonaptera): Origins and host associations. *Cladistics* 24, 1, 2008.
3. Elston, D.M., What's eating you? Pulex irritans. *Cutis*, 66, 175, 2000.
4. Brouqui, P., and Raoult, D., Arthropod-borne diseases in homeless. *Ann. N. Y. Acad. Sci.*, 1078, 223, 2006.
5. Gracia, M.J. et al., *Pulex irritans* infestation in dogs. *Vet. Rec.*, 147, 748, 2000.
6. Ménier, K., Infestation of dairy goats with *Pulex irritans*. *Vet. Rec.*, 153, 128, 2003.
7. Christodoulopoulos, G. et al., Biological, seasonal and environmental factors associated with *Pulex irritans* infestation of dairy goats in Greece. *Vet. Parasitol.*, 137, 137, 2006.
8. Yin, J.X. et al., Predictors for abundance of host flea and floor flea in households of villages with endemic commensal rodent plague, Yunnan Province, China. *PLoS Negl. Trop. Dis.*, 5, e997, 2011.
9. Engelthaler, D.M., and Gage, K.L., Quantities of *Yersinia pestis* in fleas (Siphonaptera: Pulicidae, Ceratophyllidae, and Hystrichopsyllidae) collected from areas of known or suspected plague activity. *J. Med. Entomol.*, 37, 422, 2000.
10. Loftis, A.D. et al., Surveillance of Egyptian fleas for agents of public health significance: *Anaplasma, Bartonella, Coxiella, Ehrlichia, Rickettsia*, and *Yersinia pestis*. *Am. J. Trop. Med. Hyg.*, 75, 41, 2006.
11. Laudisoit, A. et al., Plague and the human flea, Tanzania. *Emerg. Infect. Dis.*, 13, 687, 2007.
12. Hornok, S. et al., Molecular investigation of hard ticks (Acari: Ixodidae) and fleas (Siphonaptera: Pulicidae) as potential vectors of rickettsial and mycoplasmal agents. *Vet. Microbiol.*, 140, 98, 2010.
13. Krug, L.E., and Elston, D.M., What's eating you? Oriental rat flea (*Xenopsylla cheopis*). *Cutis*, 86, 282, 2010.
14. Eremeeva, M.E. et al., *Rickettsia typhi* and *R. felis* in rat fleas (*Xenopsylla cheopis*), Oahu, Hawaii. *Emerg. Infect. Dis.*, 14, 1613, 2008.
15. Billeter, S.A. et al., Molecular detection and identification of *Bartonella* species in *Xenopsylla cheopis* (Siphonaptera: Pulicidae) collected from *Rattus norvegicus* in Los Angeles, California. *Appl. Env. Microbiol.*, 77, 7850, 2011.
16. Morick, D. et al., Investigation of *Bartonella* acquisition and transmission in *Xenopsylla ramesis* fleas (Siphonaptera: Pulicidae). *Mol. Ecol.*, 20, 2864, 2011.
17. Schotthoefer, A.M. et al., Effects of temperature on early-phase transmission of *Yersina pestis* by the flea, *Xenopsylla cheopis*. *J. Med. Entomol.*, 48, 411, 2011.
18. Dryden, M.W., and Rust, M.K., The cat flea: Biology, ecology and control. *Vet. Parasitol.*, 52, 1, 1994.
19. Cañón-Franco, W.A., and Pérez-Bedoya, J.L., Siphonaptera (Pulicidae) in dogs and cats of Colombia: Clinical and epidemiological aspects. *Vet. Parasitol.*, 173, 353, 2010.
20. Elston, D.M., and Do, H., What's eating you? Cat flea (*Ctenocephalides felis*), Part 1: Clinical features and role as a disease vector. *Cutis*, 85, 231, 2010.
21. Rust, M.K., and Dryden, M.W., The biology, ecology, and management of the cat flea. *Annu. Rev. Entomol.*, 42, 451, 1997.
22. Kelly, P.J., A review of bacterial pathogens in *Ctenocephalides felis* in New Zealand. *N. Z. Vet. J.*, 52, 352, 2004.
23. Shaw, S.E. et al., Pathogen carriage by the cat flea *Ctenocephalides felis* (Bouché) in the United Kingdom. *Vet. Microbiol.*, 102, 183, 2004.

24. Hawley, J.R., Shaw, S.E., and Lappin, M.R., Prevalence of *Rickettsia felis* DNA in the blood of cats and their fleas in the United States. *J. Feline Med. Surg.*, 9, 258, 2007.
25. Reeves, W.K. et al., Association of *Bartonella* with the fleas (Siphonaptera) of rodents and bats using molecular techniques. *J. Vector Ecol.*, 32, 118, 2007.
26. Eisen, R.J. et al., Early-phase transmission of *Yersinia pestis* by cat fleas (*Ctenocephalides felis*) and their potential role as vectors in a plague-endemic region of Uganda. *Am. J. Trop. Med. Hyg.*, 78, 949, 2008.
27. Eisen, R.J., Eisen, L., and Gage, K.L., Studies of vector competency and efficiency of North American fleas for *Yersinia pestis*: State of the field and future research needs. *J. Med. Entomol.*, 46, 737, 2009.
28. Oliveira, K.A. et al., Molecular identification of *Rickettsia felis* in ticks and fleas from an endemic area for Brazilian Spotted Fever. *Mem. Inst. Oswaldo Cruz*, 103, 191, 2008.
29. Reif, K.E., and Macaluso, K.R., Ecology of *Rickettsia felis*: A review. *J. Med. Entomol.*, 46, 723, 2009.
30. Kaewmongkol, G. et al., Zoonotic *Bartonella* species in fleas and blood from red foxes in Australia. *Vector Borne Zoonotic Dis.*, 11, 1549, 2011.
31. Kernif, T. et al., *Bartonella clarridgeiae* in fleas, Tahiti, French Polynesia. *Emerg. Infect. Dis.*, 17, 1773, 2011.
32. Parola, P., *Rickettsia felis*: From a rare disease in the USA to a common cause of fever in sub-Saharan Africa. *Clin. Microbiol. Infect.*, 17, 996, 2011.
33. Tsai, K.H. et al., Prevalence of *Rickettsia felis* and the first identification of *Bartonella henselae* Fizz/CAL-1 in cat fleas (Siphonaptera: Pulicidae) from Taiwan. *J. Med. Entomol.*, 48, 445, 2011.
34. Williams, M. et al., First probable Australian cases of human infection with *Rickettsia felis* (cat-flea typhus). *Med. J. Aust.*, 194, 41, 2011.
35. Ugbomoiko, U.S., Ariza, L., and Heukelbach, J., Parasites of importance for human health in Nigerian dogs: High prevalence and limited knowledge of pet owners. *BMC Vet. Res.*, 4, 49, 2008.
36. Rust, M.K., Advances in the control of *Ctenocephalides felis* (cat flea) on cats and dogs. *Trends Parasitol.*, 21, 232, 2005.
37. Dittmar, K. et al., Techniques of DNA-studies on prehispanic ectoparasites (*Pulex* sp., Pulicidae, Siphonaptera) from animal mummies of the Chiribaya culture, southern Peru. *Mem. Inst. Oswaldo Cruz*, 98, S53, 2003.
38. Vobis, M. et al., Molecular phylogeny of isolates of *Ctenocephalides felis* and related species based on analysis of ITS1, ITS2 and mitochondrial 16S rDNA sequences and random binding primers. *Parasitol. Res.*, 94, 219, 2006.
39. Gamerschlag, S. et al., Repetitive sequences in the ITS1 region of the ribosomal DNA of *Tunga penetrans* and other flea species (Insecta, Siphonaptera). *Parasitol. Res.*, 102, 193, 2008.

82 Tunga (Jigger Flea)

Dongyou Liu

CONTENTS

82.1 INTRODUCTION

The genus *Tunga* encompasses 10 sand flea species (Siphonaptera) that are ectoparasites of animals and humans. While both the male and female fleas are hematophagous, only the breeding female penetrates the exposed skin of mammals and causes inflammation, irritation, itching, pain, and ulceration.

In domestic animals, infestation of *Tunga* species induces lesions on the legs, leading to walking difficulties and discomfort, and affecting weight gain and milk production. Painful lesions on the teats of pigs can result in the death of offspring due to the mother's inability to nurse. In humans, *Tunga* species (i.e., *T. penetrans* and *T. trimamillata*) infect the feet, resulting in notable ambulatory disturbances and loss of working capacity as well as serious local and general complications from secondary infections that necessitate medical and surgical intervention.

Native to the West Indies, *T. penetrans* is prevalent in South and Central America and the Caribbean islands, as well as in sub-Saharan Africa and India. *T. trimamillata* is a newly described species that has been reported only in Ecuador and Peru. *Tunga* fleas put hundreds of millions of people at risk of infection and represent a significant cause of morbidity in economically disadvantaged communities.

82.1.1 Classification, Morphology, Life Cycle, and Epidemiology

82.1.1.1 Classification

The genus *Tunga* comprises a group of sand fleas belonging to the subfamily Tunginae, family Pulicidae, order Siphonaptera, class Insecta, phylum Arthropoda. Currently, the genus consists of 10 recognized species: *Tunga bondari*, *Tunga caecata*, *Tunga caecigena*, *Tunga callida*, *Tunga libis*, *Tunga monositus*, *Tunga penetrans*, *Tunga terasma*, *Tunga travassosi*, and *Tunga trimamillata* (Table 82.1). Of these, *T. penetrans* and *T. trimamillata* cause tungiasis in humans and domestic animals (e.g., pigs and dogs), and the remainder are parasitic to a number of wild mammal species (e.g., armadillos, giant anteaters, and rodents) [1].

First described by Linnaeus in the eighteenth century, *T. penetrans* (synonyms: *Pulex minimus cutem penetrans*, *Pulex minutissimus nigricans*, *Acarus fuscus sub-cutem nidulans*, *Pulex penetrans*, *Pulex reptans*, *Rhynchoprion penetrans*, *Dermatophilus penetrans*, and *Sarcopsylla penetrans*) is commonly referred to as chiggers, although true chiggers are harvest mites that are classified in the genus *Trombicula*, family Trombiculidae, superfamily Trombidioidea, order Trombidiformes, subclass Acari, class Arachnida, phylum Arthropoda (see Chapter 73). In contrast to harvest mites that

TABLE 82.1
Characteristics of *Tunga* Species

Species	Dimensions (Width, Height, Length)	Head and Thorax (Profile)	Geographical Distribution	Host
Tunga bondari	Not described	Not visible, embedded in abdomen	Brazil	Anteater (*Tamandua tetradactyla*)
Tunga caecata	5 × 5 × 5 mm	Not visible, embedded in abdomen	Brazil	*Rattus rattus*, *Rattus norvegicus*, *Mus musculus*
Tunga caecigena	5 – 6 × 7 – 10 mm	Not visible, among four humps	China, Japan	*Rattus rattus*, *Rattus norvegicus*
Tunga callida	4.5 × 4.5 × 4.5 mm	Not visible, among four humps	China	Rodents
Tunga libis	Not described	Not described	Ecuador, Chile	Soft grass mouse (*Akodon mollis*)
Tunga monositus	5.4 × 6 × 5 mm	Not visible, among eight humps	Mexico	Rodents
Tunga penetrans	4 × 5 × 6 mm	Visible	Latin America, Africa	Human, swine, dog, bovine, horse, donkey, etc.
Tunga terasma	8 × 14 × 10 mm	Visible	Brazil	Armadillo (*Cabassous unicinctus*)
Tunga travassosi	9 × 9 – 10 × 13 mm	Visible	Brazil	Armadillo (*Dasypus novemcinctus*)
Tunga trimamillata	5 × 5 × 12 mm	Not visible, among three humps	Ecuador, Peru	Bovine, goat, sheep, swine, human

Source: Pampiglione, S. et al. *Med. Vet. Entomol.* 23, 172–186, 2009.

penetrate the skin to suck blood but do not lay eggs, the breeding females of *T. penetrans* penetrate the skin to take blood meal and lay eggs. Additionally, only *T. penetrans* adults feed on mammals and only the females stay attached to the host, whereas both adult and larval forms of harvest mites feed on animals [2]. The second human-infecting *Tunga* species, *T. trimamillata*, was only identified in 2002 [3,4].

82.1.1.2 Morphology

T. penetrans is a small flea of <1 mm in length. The flea has a relatively flattened head, a well-developed lacinia and epipharynx, which facilitate penetrating and burrowing through the epidermal and dermal layers. Additionally, the flea possesses a hypertrophic region between tergites 2 and 3, which along with the abdominal sternites can stretch considerably and are bent apart. Chitinous clasps located in the abdominal enlargement hold onto the hypertrophic zone, giving the appearance of a three-leafed clover. Females are generally smaller than males and show a depression or groove at the abdominal end; males have protrusive copulatory organs in the same region [5].

T. penetrans eggs are of 604 μm in length. Young larvae (1500 μm in length) emerge from eggs after hatching and molt to become first instar larvae (1150 μm in length), which molt again to be second instar larvae (2900 μm in length). Unlike *T. penetrans*, other *Tunga* spp. undergo a third instar larval stage.

The key structural differences between *T. trimamillata* and *T. penetrans* include: (i) The presence of three rounded protuberances on the anterior end of the gravid females in *T. trimamillata*, which appear in the form of a clover leaf in *T. penetrans* gravid females. (ii) In *T. trimamillata*, the first segment of the maxillary palps is longer, while in *T. penetrans*, the second segment of the maxillary palps is longer. (iii) *T. trimamillata* has a greater number of spines on both the first tarsal segment and the medial border of the tibia of the third pair of legs than *T. penetrans*. (iv) The gena in *T. penetrans* shows a deep groove, which is absent in *T. trimamillata*. (v) The last abdominal tergite in *T. trimamillata* nongravid females protrudes outward, forming a sort of small "tail," while that in *T. penetrans* nongravid females does not [6].

82.1.1.3 Life Cycle

Fertilized eggs released by a *T. penetrans* gravid female hatch within 1–6 days to produce young larvae, which molt twice to become first and then second instar larvae. Other *Tunga* spp. go through a third instar larval stage. Between days 10 and 18, *Tunga* larvae burrow into the soil or sand to pupate and build cocoons around themselves, feeding on organic residues. Sand helps stabilize the cocoon and promote its development [7]. Adult *Tunga* fleas emerge from the puparium in 9–15 days. Therefore, after oviposition, the time required for *T. penetrans* to complete its life cycle is about 17–19 days [1].

As outlined by Eisele et al. [8], the development of *T. penetrans* infection (tungiasis) can be separated into five stages. (i) Stage 1 (within a few minutes or a few hours): the female flea utilizes its posterior legs to push its body upward by an angle between 45° and 90°, and inserts its proboscis through the epidermis to the basal layer of the chorion. (ii) Stage 2 (days 1–2): the female flea burrows most of its body into the skin, with only the anus, the copulatory organs, and four rear air holes (called stigmatas) remaining on the outside of the epidermis through an orifice. This orifice allows the flea to breathe and defecate while feeding on blood vessels in the cutaneous and subcutaneous dermal layer. During this time, the expansion of the hypertrophic zone between tergites 2 and 3 in the abdominal region gives the appearance of a life belt, and the flea begins to feed on the host's blood. (iii) Stage 3: the female flea achieves maximum hypertrophy (around 2000 times its original size in 6 days) in its midsection (to the size of a pea), with the outer layer of the skin stretching thin, giving the appearance of a globular white mass (5–10 mm in diameter) with a black dot (rear end of the flea) at the center of the lesion. Subsequently, the chitin exoskeleton of tergites 2 and 3 increases in thickness and gives the appearance of a mini caldera. The burrowing and excretion of fecal coils through the anus may attract the male fleas for mating (which takes place only once in a lifetime), as the male and female do not show interest in each other in the wild. After copulation, the male flea dies, and the female flea undergoes *in vivo* ectodevelopment. The female may release up to 200 eggs one after another within a week, which often stick to the skin. (iv) Stage 4 (at about the 3rd week after penetration): the female flea loses its signs of vitality and appears near death, with the lesion shrinking in size, appearing brown in color and wrinkled. After the death of the flea (around day 25 post-penetration), the host begins to eliminate the parasite through skin repair mechanisms (e.g., shedding and subsequent skin repair), a process that may last up to 6 weeks, with the lesion appearing brown or black. (v) Stage 5: the carcass of *T. penetrans* flea is expelled, and circular skin residues of the infection remain. If embedded fleas are not removed, lesions often become superinfected. Overall, *T. penetrans* eggs hatch and mature into adult fleas within 3–4 weeks. The life cycle of *T. trimamillata* is similar to that of *T. penetrans*.

82.1.1.4 Epidemiology

Commonly known as chigoe flea, sand flea, chigger flea, or jigger flea in English, chique in French, Sandfloh in German, and bicho do pè in Portuguese, *T. penetrans* is native to the West Indies. *Tunga* infection (tungiasis) is also referred to as nigua (the Caribbean, Ecuador, and Mexico), kuti, suthipique (Bolivia), ogri eye (Surinam), bicho de pé, bicho de porco (Brazil), chica (Colombia and Venezuela), sikka (Guyana), pique (Argentina, Chile, Uruguay, and Paraguay), and huacos (Peru).

The South American ancestors of the Incas depicted tungiasis on clay jars, pottery, and ceramics long before the European arrival. Gonzalez Fernandez de Oviedo y Valdes is thought to be the first European to document tungiasis on crew members from Columbus's Santa Maria after they were shipwrecked on Haiti in 1525. Through ship routes and further expeditions, *T. penetrans* flea was spread to the rest of Latin America subsequently and sub-Saharan Africa (including Zanzibar and Madagascar to the east and the Cape Verde Islands and São Tomé to the west) as well as India throughout the seventeenth and nineteenth centuries [9,10].

The second human-infecting *Tunga* species, *T. trimamillata*, occurring in humans, bovines, goats, sheep, and pigs, was identified in 2002 [11–14]. To date, *Tunga* fleas put hundreds of millions of people at risk of infection in at least 88 countries, particularly in highly endemic areas (e.g., Nigeria, Trinidad, Tobago, and Brazil), and represent a significant cause of morbidity in economically disadvantaged communities [10,15–26]. Cases of tungiasis infections have also been reported among travelers/migrants returning/coming from disease-endemic areas [27–41].

T. penetrans has been described in domestic animals (e.g., pigs, bovines, goats, sheep, horses, donkeys, mules, llamas, dogs, and cats) and some wild animals such as agouti (*Dasyprocta punctata*), paca (*Agouti paca*), acouchi (*Myoprocta acouchy*), armadillo (*Euphractes sexcinctus*), tapir (*Tapirus terrestris*, *Tapirus americanus*), chiropterans, and various rats and mice in Latin America [42–45]. In Africa, *T. penetrans* has been reported in elephants, gorillas, and many species of monkey. While pigs are considered as the main reservoir, humans may act as both biological vectors and definitive hosts for *T. penetrans* and *T. trimamillata*.

T. penetrans fleas tend to inhabit 2–5 cm below sand and soil, feeding on organic detritus, as the surface temperature is generally too hot for the larvae to develop and the deeper level has inadequate oxygen. Adult females can live without food for a month or more. Although 100–200 eggs are expelled by the gravid female, only 15% of the eggs are found to develop into larvae, and of those, only 14% form a cocoon. Moreover, only half of the pupae reach the adult phase, resulting in gender disequilibrium (with a female bias), perhaps as a result of *Wolbachia* infection [46].

Intermittently, the flea feeds on warm-blooded hosts, with the female burrowing into any mammal (including humans) and causing inflammatory skin disease (called tungiasis). With the ability to jump only 20 cm and to move rapidly by crawling, *Tunga* fleas and tungiasis lesions are mostly found in the periungual region of the toes. Human infections are linked to the practice of walking barefoot or wearing only sandals, to the lack of personal and soil hygiene, and to the free movement of animals (pigs, dogs, rats) between and into houses. Tungiasis was more common in males than in females, and children 5–9 years of age had the highest prevalence rates [22,47,48].

Rate of incidence is significantly higher during the peak of the dry season, and noticeably lower in the rainy season. Rain may simply wash away all stages (eggs, larvae, pupae, nymphs, and adults) of *T. penetrans* due its small size of 1 mm, and the high humidity in the soil also impairs the development of free-living stages of *T. penetrans* during the rainy season [49].

82.1.2 Clinical Features and Treatment

82.1.2.1 Clinical Features

During the first day or two of human tungiasis, a mild local itching or irritation may occur, which often passes quickly (and even goes unnoticed) as the area around the flea calluses becomes insensitive. Around the 3rd to 6th day after penetration, the females embedded in the skin begin to increase in size (due to the production of eggs), and press neighboring nerves or blood vessels, causing skin irritation and tenderness, erythema and insomnia (due to intense itching or pruritus). At the site of infestation, the lesion often appears as a blister or furuncular nodule (5–10 mm in size) characterized by a black dot (representing the female flea's exposed hind legs, respiratory spiracles, and reproductive organs) at the center of a swollen red lesion (erythema), surrounded by a white halo (formed by thinned epidermis). Desquamation (with crusty or wart-like appearance) of the skin often occurs, especially after the flea expands during hypertrophy. This is accompanied by soreness or pain (ranging from very slight to intense), depending on the sensitivity of the site (e.g., around the nail bed of the toes, other parts of the foot, hands, gluteus, scrotum, penis, groin, face, the subconjunctival tissue near the cornea, etc.), and especially at night or, if the nodule is on the foot, while walking [50]. Along with egg release, a watery secretion can be observed. At the 3rd week after penetration, the flea is near death, and the lesion becomes smaller and more wrinkled, with fecal and water secretion and egg release all coming to an end. Around the 25th day after penetration, the lesion looks like a black crust and the flea's carcass is removed by host repair mechanisms and the skin begins to heal [51,52].

Manual extraction or expulsion of the parasite may leave a sore which, if not protected, is liable to secondary bacterial or mycotic infections (e.g., tetanus and gangrene) [47]. The lesions with secondary infections may become enlarged (due to severe local inflammation), leading to the formation of abscesses, phlegmons, thrombophlebitis, lymphangitis and phagedenic ulcers, fissures, loss of nails, spontaneous mutilation of the toes (as a result of gangrene), and walking difficulties. Patients lacking immunization may die from tetanus (*Clostridium tetani*) [53].

In the Andean regions, human tungiasis can be differentiated into two types: a less annoying one is caused by a small sand flea *T. penetrans* ("niguas de cerro" or "swine sand fleas") and a more painful one caused by a larger sand flea *T. trimamillata* ("niguas de vaca" or "cattle sand fleas").

In pigs, the site preferences of *T. penetrans* are the feet and legs, udder, scrotum, snout, and tail. The lesions are often multiple and agglomerate, and are covered with thick crusts. *T. trimamillata* infestation has been described in swine and occasionally in ruminants (often as a coinfection with *T. penetrans*). In bovines, goats, and sheep, *T. trimamillata* tends to embed along the coronary band, in the coronary and digital cushions, and in the sole of the hoof. This may lead to the formation of conspicuous cavities in both the soft and hard tissues of the hoof, and mechanical compression and secondary infections may further exacerbate the condition, resulting in abscesses and hoof or foot rot. In heavily infested goats and sheep, a striking limp is often observed. In addition, *T. trimamillata* may also infect the calcaneal region and scrotum in males and the udder in females [9].

82.1.2.2 Treatment

A self-limiting disease, tungiasis may heal even without treatment after the burrowed fleas die within 5 weeks and are naturally sloughed off as the skin sheds. To relieve the symptoms of uncomplicated cases of human tungiasis, the simplest therapy involves extraction of the parasite by breaking the skin around the caudal end with a sterile needle and pulling the entire parasite out with tweezers. In fact, extraction with a needle (auguring) or a thorn (a sharpened stick) was practiced for centuries among the Indian populations of Latin America. Care should be taken to avoid tearing the flea during the extraction procedure, as part of the flea left behind often stimulates the hyperkeratotic reaction and inflammation. Disinfection of the small residual ulceration, protection with sterile gauze, and application of a topical antibiotic is important to avoid secondary infections (e.g., tetanus). Other topical treatments include cryotherapy and electrodesiccation of the lesion [54].

T. penetrans flea may be also suffocated using occlusive petrolatum or vaseline. Poultices showing varied efficacy include a base of cassava, a decoction of *Tournefortia hirsutissima*, palm oil with the juice of *Bixa orleana*, tincture (called rocou) of *Carapa guineensis*, tobacco powder or infusion, a mush of the fruit of *Anona muricata*, salted butter, castor oil, turpentine, mercury ointment, and so on. Other topical treatment methods consist of the application of an aqueous solution of 1% picric acid, a footbath in 2% Neguvon solution or petroleum, a spray with ethyl chloride, a compound of salicylic acid (2.50 g) with ichthyol (10 g) and yellow vaseline (10 g), metrifonate ointment, a liniment (made up of citronella oil, petroleum, coconut oil, phenic acid, and menthol), and Zanzarin® (a derivative of coconut oil, jojoba oil, and aloe vera). Thiabendazole (25–50 mg/kg) or ivermectin (200 μg/kg) or niridazole may be administered orally. For superinfections, trimethoprim, sulfamethoxazole, metronidazole, and amoxicillin (with/without clavulanate) may be used to control secondary infections [55–57].

Public health and prevention strategies against tungiasis should target the elimination and avoidance of the flea. These include reducing exposure by wearing closed shoes (as opposed to sandals or slippers), watering the floors within houses regularly, building a cemented rather than a sand floor in shacks and some favelas and washing it often, control of rodent populations, disinfection of soil in houses and stables (with insecticides such as naphthalene and pyrethrum), and the prohibition of free movement by pigs, goats, sheep, or bovines in the vicinity of houses [58].

82.1.3 Diagnosis

Clinical diagnosis of uncomplicated human tungiasis is based on characteristic skin lesions and a history of travel to an endemic area. The typical hyperkeratotic appearance (located in the finger pad or nail border of the toes or interdigital spaces, plantar zone or heel, glutei, hands, elbows, or other sites) with a black pore in the middle of the skin lesion surrounded by a whitish halo and a slight inflammatory reddening is diagnostic. While one or a very few lesions are often noted, occasionally, multiple lesions located close together in parts of the surface may resemble a honeycomb [54,59].

Under a dermoscope, the central brown spot visible to the naked eye can appear as a typical brown pigmented ring with a central pore (corresponding to the front portion of the parasite's exoskeleton) [60,61]. A dermoscopic "gray-blue blotch" relates to developing eggs [62–65]. Cabrera and Daza [66] utilized dermoscopy to show eggs of *T. penetrans* in a 19-year-old woman with a 6-mm-wide whitish papule, with a central brown spot, in the periungual area of the third toe of her left foot after a trip to Brazil. Upon sequential and careful shaving of the epidermis and gently compressing the edges of the wound, a jelly-like bag full of eggs emerged from the cavity and a dark brown blotch (feces of the parasite) were observed through dermoscopy.

For lesions complicated by secondary infections, differential diagnosis may include warts, mycotic lesions, granulomas of foreign bodies, subungual exostosis, early melanoma, tick bite, ingrown toenail, fire ant bite, dracontiasis, some furuncular myiasis, or any other pustular condition.

In animals, the differential diagnosis for tungiasis consists of infectious diseases affecting the skin in the foot or hoof, perineal and genital areas, udder, and tail. These include infectious vesicular diseases (mainly in swine and ruminants), mastitis, footrot, interdigital and pododermatitis, mange, pustular dermatitis, and abscesses. In addition, foreign body granulomas may also be considered.

Microscopic examination of specimens permanently mounted on slides or preserved in 70% ethanol provides a useful means of identification of the parasite [67]. This is largely based on morphological criteria such as the shape of abdomen, the presence/absence of eyes, or the size of the flea. Eggs of 650 μm in length may also be visualized under the microscope at ×20 magnification with the section stained with hematoxylin and eosin. Nonetheless, while the morphological differentiation of gravid females is very easy, the identification of nongravid females, males, or body fragments is technically demanding.

The use of mitochondrial, nuclear, and other DNA markers permits precise determination of the taxonomic status of *T. penetrans* and *T. trimamillata* [68–71]. Sequencing analyses of the mitochondrial genes, cytochrome oxidase II (COII), the large ribosomal subunit (16S), and the nuclear ribosomal internal transcribed spacer (ITS1 and ITS2) loci enable the clarification of genetic distances and phylogenetic relationships of *Tunga* species [71–73]. In addition, the repeat units (of 10–99 bp) in the ITS1 of various flea species have also been shown to serve as a valuable tool for phylogeographic studies of *Tunga* fleas [72].

The application of a polymerase chain reaction-restricted fragment length polymorphism (PCR-RFLP) method targeting the ribosomal ITS2 region followed by the digestion of the PCR product with two restriction enzymes (MspI and RsaI) offers a reliable and rapid approach for unambiguous discrimination between *T. penetrans* and *T. trimamillata* [73].

82.2 METHODS

82.2.1 Sample Preparation

82.2.1.1 DNA Isolation

Genomic DNA from fleas can be isolated using the DNeasy Tissue Kit (Qiagen). One individual flea is frozen in liquid nitrogen and pulverized in a steel mortar (mechanically shredded). The powder is then mixed with lysis buffer containing proteinase K and incubated for 4 h. Purification of genomic DNA from the resulting lysate is then performed according to the protocol for animal tissue provided by the manufacturer [69].

82.2.2 Detection Procedures

82.2.2.1 Sequencing Analysis of Mitochondrial 16S rDNA

Vobis et al. [69] targeted a hypervariable section of mitochondrial 16S rDNA for the identification and phylogenetic analysis of *Tunga* fleas. The primers consist of Sen-mt16S (5′-TAC ATA ACA CGA GAA GAC C-3′) and Rev-mt16S (5′-GTG ATT GCG CTG TTA TCC-3′). The resulting PCR product is then sequenced.

82.2.2.1.1 Procedure

1. PCR mixture (50 μL) is prepared with the HotStar Master Mix Kit (Qiagen), containing 0.4 mM of each primer, 2.5 U of *Taq* polymerase, 200 μM of each dNTP, and 2 mM $MgCl_2$.
2. Amplification is conducted with 30 cycles of denaturation for 45 s at 94°C, primer annealing at 40°C for 45 s, and elongation at 72°C for 45 s.
3. PCR product (of approximately 200 bp) is examined on 1.5% agarose gel, excised, purified using the QIAquick Gel Extraction Kit (Qiagen), eluted in 30 μL H_2O, and sequenced.
4. The multiple sequence alignment program Clustal X version 1.8 is used to obtain a nucleotide sequence alignment file. The phylogenetic tree is constructed by the neighbor joining method using Kimura 2 parameter distances in MEGA 2.1 software. Bootstrap analysis is employed to test the reliability of the phylogenetic tree (1000 replications).

Note: Avelar and Linardi [73] further improved the technique by employing the multiple displacement amplification (MDA) technique that uses the bacteriophage φ29 DNA polymerase and exonuclease-resistant, thiophosphate-modified, degenerate hexamers to produce a large quantity of high-quality genomic DNA from low amounts of the original DNA template (e.g., 70% ethanol at room temperature).

TABLE 82.2
Identity and Sequence of ITS1 Primers

Primer	Sequence (5′–3′)	Position
ITS1 sen (1)	GTACACACCGCCCGTGCGTACT	End of the 18S rDNA from *C. felis*
ITS1 rev (1)	GCTGCGTTCTTCATCGACCC	Beginning of the 5.8S rDNA from *Anopheles pseudopunctipennis*
ITS1 sen (2)	CTTTGTACACACCGCCCGTCGCTAC	End of the 18S rDNA from *T. monositus*
ITS1 rev (2)	CCAGGTGTGGTCCGGGAACAGTATC	End of the 5.8S rDNA from *T. penetrans* and *T. trimamillata*
ITS1 sen (3)	GTTGCTGGGAAGATGCCCAAACTTG	End of the 18S rDNA from *T. monositus*, *Echidnophaga*, and *Pulex*
ITS1 sen (4)	CAAGGTTTCCGTAGGTGAACCTGC	End of the 18S rDNA from *T. monositus*, *Echidnophaga*, and *Pulex*

82.2.2.2 Sequencing Analysis of ITS1 Gene

Gamerschlag et al. [72] utilized primers matching the 5′ end of the 5.8S rDNA and the 3′ end of the 18S rDNA coding regions to amplify the ITS1 region (about 878 bp) for comparative analysis of *Tunga* flea species (Table 82.2).

82.2.2.2.1 Procedure

1. PCR (50 μL) is composed of 0.4 mM of each primer, 2.5 U of *Taq* polymerase, 200 μM deoxynucleotide triphosphates, and 2 mM $MgCl_2$ (the Hot Star *Taq* Master Mix Kit, Qiagen).
2. The mixture is subjected to a denaturation time of 5 min at 94°C, 35 cycles of denaturation for 30 s at 94°C, annealing at 58°C for 30 s, and elongation at 72°C for 90 s, followed by a final elongation for 5 min at 72°C.
3. PCR products (about 878–1076 bp) are examined on 1.5% agarose gels, purified by the QIAquick Gel Extraction Kit (Qiagen), and sequenced.
4. The multiple sequence alignment program Clustal X and the program MAFFT are used to obtain a nucleotide sequence alignment file. The program SEAVIEW is used to optimize the alignments; alignments are also edited by eyes. The program DOTMATCHER from the European Molecular Biology Open Software Suite package is used to find repeats within the sequences. Various known sequences are obtained from GenBank. The Basic Local Alignment Search Tool search is accomplished via NCBI.

Note: The ITS1 rDNA of *T. penetrans* encompasses 99 bp repeat units, which are present twice within the sequence of the South American *T. penetrans* (positions 114–313) and four times within the ITS1 rDNA sequence of the African *T. penetrans* (positions 114–513).

82.3 CONCLUSION

The genus *Tunga* encompasses 10 small sand flea species that are ectoparasites of mammals, including humans, dogs, cats, pigs, and rodents [74]. Whereas *Tunga* males and nonfertilized females are hematophagous without penetrating permanently, pregnant (breeding) females burrow into the exposed skin of mammals, and remain there to feed and lay eggs. Besides inducing inflammation, irritation, itching, pain, and ulceration, *Tunga* infestation (tungiasis) is associated with the risk of tetanus and other secondary infections [75]. Among the 10 recognized species, two (*T. penetrans* and *T. trimamillata*) are involved in human diseases. While *T. penetrans* is found in the Caribbean islands, South and Central America as well as sub-Saharan Africa and India, *T. trimamillata* is limited to Ecuador and Peru. Together, these fleas represent a significant cause of morbidity in economically disadvantaged communities [76,77]. Correct identification of *Tunga* fleas is vital for their effective control and prevention. Although microscopic examination of morphological features provides a valuable approach for the differentiation of these fleas, its requirement for technical expertise limits its application. Therefore, the use of molecular techniques offers a convenient alternative for the determination of the species identity of *Tunga* fleas.

REFERENCES

1. Pampiglione, S. et al., Sand flea (*Tunga* spp.) infections in humans and domestic animals: State of the art. *Med. Vet. Entomol.*, 23, 172, 2009.
2. Witt, L.H. et al., Blood-feeding of *Tunga penetrans* males. *Med. Vet. Entomol.*, 18, 439, 2004.
3. Pampiglione, S. et al., Additional description of a new species of *Tunga* (Siphonaptera) from Ecuador. *Parasite*, 10, 9, 2003.
4. Pampiglione, S. et al., Anatomy of *Tunga trimamillata* Pampiglione et al., 2002 (Insecta, Siphonaptera, Tungidae) and developmental phases of the gravid female. *Parasite*, 12, 241, 2005.
5. Nagy, N. et al., Investigations on the life cycle and morphology of *Tunga penetrans* in Brazil. *Parasitol. Res.*, 101, S233, 2007.
6. Pampiglione, S. et al., Differential diagnosis between *Tunga penetrans* (L., 1758) and *T. trimamillata* Pampiglione et al., 2002 (Insecta, Siphonaptera), the two species of the genus *Tunga* parasitic in man. *Parasite*, 11, 51, 2004.
7. Linardi, P.M. et al., Occurrence of the off-host life stages of *Tunga penetrans* (Siphonaptera) in various environments in Brazil. *Ann. Trop. Med. Parasitol.*, 104, 337, 2010.
8. Eisele, M. et al., Investigations on the biology, epidemiology, pathology and control of *Tunga penetrans* in Brazil: I. Natural history of tungiasis in man. *Parasitol. Res.*, 90, 87, 2003.

9. Collins, G. et al., Tungiasis: A neglected health problem in rural Cameroon. *Int. J. Collaborat. Res. Intern. Med. Public Health*, 1, 2, 2009.
10. Maco, V., Tantaleán, M., and Gotuzzo, E., Evidence of tungiasis in pre-Hispanic America. *Emerg. Infect. Dis.*, 17, 5, 2011.
11. Fioravanti, M.L., Pampiglione, S., and Trentini, M., A second species of *Tunga* (Insecta, Siphonaptera) infecting man: *Tunga trimamillata. Parasite*, 10, 282, 2003.
12. Fioravanti, M.I. et al., Presence of *Tunga trimamillata* (Insecta, Siphonaptera) in Peru. *Parasite*, 13, 85, 2006.
13. Fioravanti, M.L. et al., Sand flea (*Tunga* spp.) infections in humans and domestic animals: State of the art. *Med. Vet. Entomol.*, 23, 172, 2009.
14. Maco, V., Maco, V.P., and Gotuzzo, E., An ectopic case of *Tunga* spp. infection in Peru. *Am. J. Trop. Med. Hyg.*, 82, 1076, 2010.
15. Heukelbach, J. et al., Tungiasis: A neglected health problem of poor communities. *Trop. Med. Int. Health*, 6, 267, 2001.
16. Heukelbach, J. et al., The animal reservoir of *Tunga penetrans* in severely affected communities of north-east Brazil. *Med. Vet. Entomol.*, 18, 329, 2004.
17. Heukelbach, J., Franck, S., and Feldmeier, H., High attack rate of *Tunga penetrans* (Linnaeus 1758) infestation in an impoverished Brazilian community. *Trans. R. Soc. Trop. Med. Hyg.*, 98, 431, 2004.
18. Wilcke, T. et al., High prevalence of tungiasis in a poor neighbourhood in Fortaleza, Northeast Brazil. *Acta Trop.*, 83, 255, 2002.
19. de Carvalho, R.W. et al., The patterns of tungiasis in Araruama township, state of Rio de Janeiro, Brazil. *Mem. Inst. Oswaldo Cruz*, 98, 31, 2003.
20. Feldmeier, H. et al., Investigations on the biology, epidemiology, pathology and control of *Tunga penetrans* in Brazil: III. Cytokine levels in peripheral blood of infected humans. *Parasitol. Res.*, 91, 298, 2003.
21. Feldmeier, H. et al., Investigations on the biology, epidemiology, pathology and control of *Tunga penetrans* in Brazil: IV. Clinical and histopathology. *Parasitol. Res.*, 94, 275, 2004.
22. Heemskerk, J., van Empel, I., and Jakimowicz, J.J., *Tunga penetrans*. A case report and review of the literature. *Acta Chir. Belg.*, 105, 548, 2005.
23. Ugbomoiko, U.S., Ofoezie, I.E., and Heukelbach, J., Tungiasis: High prevalence, parasite load, and morbidity in a rural community in Lagos State, Nigeria. *Int. J. Dermatol.*, 46, 475, 2007.
24. Valentín, S.M. et al., Tungiasis: A case report. *P.R. Health Sci. J.*, 26, 423, 2007.
25. Damazio, O.R., and Silva, M.V., Tungiasis in schoolchildren in Criciúma, Santa Catarina State, South Brazil. *Rev. Inst. Med. Trop. Sao Paulo*, 51, 103, 2009.
26. Mazigo, H.D. et al., Tungiasis infestation in Tanzania. *J. Infect. Dev. Ctries.*, 4, 187, 2010.
27. Fein, H. et al., Tungiasis in North America: A report of 2 cases in internationally adopted children. *J. Pediatr.*, 139, 744, 2001.
28. Pospísilová, A., and Pirochtová, K., Tungiasis (tungosis) comes to the Czech Republic. *J. Cosmet. Dermatol.*, 1, 216, 2002.
29. Franck, S., Feldmeier, H., and Heukelbach, J., Tungiasis: More than an exotic nuisance. *Travel Med. Infect. Dis.*, 1, 159, 2003.
30. Caputo, V. et al., Tungiasis: A report of four cases and a review of imported cases in Italy in the last 30 years. *G. Ital. Med. Trop.*, 10, 33, 2005.
31. Sachse, M.M., Guldbakke, K.K., and Khachemoune, A., *Tunga penetrans*: A stowaway from around the world. *J. Eur. Acad. Dermatol. Venereol.*, 21, 11, 2007.
32. Veraldi, S., and Valsecchi, M., Imported tungiasis: A report of 19 cases and review of the literature. *Int. J. Dermatol.*, 46, 1061, 2007.
33. Escamilla-Martinez, E. et al., Tungiasis–traveler's ectoparasitosis of the foot: A case report. *Foot Ankle Int.*, 29, 354, 2008.
34. Hager, J. et al., Tungiasis in the United States: A travel souvenir. *Dermatol. Online J.*, 14, 3, 2008.
35. Dialynas, M. et al., Imported human tungiasis in Greece. *Travel Med. Infect. Dis.*, 7, 375, 2009.
36. Ferreira, L.A. et al., *Tunga penetrans* as a traveler's disease. *Travel Med. Infect. Dis.*, 7, 381, 2009.
37. Scalvenzi, M. et al. Tungiasis: Case report of a traveller to Kenya. *Case Rep. Dermatol.*, 1, 29, 2009.
38. Hakeem, M.J. et al., Tungiasis—a cause of painful feet in a tropical traveller. *Travel Med. Infect. Dis.*, 8, 29, 2010.
39. Rosmaninho, A. et al., *Tunga penetrans*: Painful lesions on the feet-the first imported case from Guinea-bissau. *Case Report Med.*, 2010, 681302, 2010.
40. Wiwanitkit, V., Tungiasis among traveller. *Travel Med. Infect. Dis.*, 8, 273, 2010.
41. Yotsu, R.R. et al., Imported tungiasis in a Japanese student returning from East Africa. *J. Dermatol.*, 38, 185, 2011.
42. Feldmeier, H. et al., Investigations on the biology, epidemiology, pathology and control of *Tunga penetrans* in Brazil: VI. Natural history of the infestation in laboratory-raised Wistar rats. *Parasitol. Res.*, 102, 1, 2007.
43. Witt, L. et al., Infestation of Wistar rats with *Tunga penetrans* in different microenvironments. *Am. J. Trop. Med. Hyg.*, 76, 666, 2007.
44. Pilger, D. et al., Investigations on the biology, epidemiology, pathology, and control of *Tunga penetrans* in Brazil: VII. The importance of animal reservoirs for human infestation. *Parasitol. Res.*, 102, 875, 2008.
45. Ugbomoiko, U.S., Ariza, L., and Heukelbach, J., Pigs are the most important animal reservoir for *Tunga penetrans* (jigger flea) in rural Nigeria. *Trop. Doct.*, 38, 226, 2008.
46. Luchetti, A. et al., Wolbachia infection in the newly described Ecuadorian sand flea, *Tunga trimamillata. Exp. Parasitol.*, 108, 18, 2004.
47. Heukelbach, J. et al., Ectopic localization of tungiasis. *Am. J. Trop. Med. Hyg.*, 67, 214, 2002.
48. Feldmeier, H. et al., High exposure to *Tunga penetrans* (Linnaeus, 1758) correlates with intensity of infestation. *Mem. Inst. Oswaldo Cruz*, 101, 65, 2006.
49. Winter, B. et al., Tungiasis-related knowledge and treatment practices in two endemic communities in northeast Brazil. *J. Infect. Dev. Ctries.*, 3, 458, 2009.
50. Kaimbo, D.K. et al., Upper eyelid localisation of *Tunga penetrans. Ophthalmologica*, 221, 439, 2007.
51. Macías, P.C., and Sashida, P.M., Cutaneous infestation by *Tunga penetrans. Int. J. Dermatol.*, 39, 296, 2000.
52. Kehr, J.D. et al., Morbidity assessment in sand flea disease (tungiasis). *Parasitol. Res.*, 100, 413, 2007.
53. Cestari, T.F., Pessato, S., and Ramos-e-Silva, M., Tungiasis and myiasis. *Clin. Dermatol.* 25, 158, 2007.
54. Gibbs, S.S., The diagnosis and treatment of tungiasis. *Br. J. Dermatol.*, 159, 981, 2008.
55. Ade-Serrano, M.A., Olomolehin, O.G., and Adenwunmi, A., Treatment of human tungiasis with niridazole (Ambilhar) in a double-blind placebo-controlled trial. *Ann. Trop. Med. Parasitol.*, 76, 89, 1982.

56. Heukelbach, J. et al., Topical treatment of tungiasis: A randomized, controlled trial. *Ann. Trop. Med. Parasitol.*, 97, 743, 2003.
57. Klimpel, S. et al., Field trial of the efficacy of a combination of imidacloprid and permethrin against *Tunga penetrans* (sand flea, jigger flea) in dogs in Brazil. *Parasitol. Res.*, 97, S113, 2005.
58. Heukelbach, J., Revision on tungiasis: Treatment options and prevention. *Expert Rev. Anti Infect. Ther.*, 4, 151, 2006.
59. Ariza, L. et al., A simple method for rapid community assessment of tungiasis. *Trop. Med. Int. Health*, 15, 856, 2010.
60. Di Stefani, A. et al., An additional dermoscopic feature of tungiasis. *Arch. Dermatol.*, 141, 1045, 2005.
61. Marazza, G. et al., *Tunga penetrans*: Description of a new dermoscopic sign—the radial crown. *Arch. Dermatol.*, 145, 348, 2009.
62. Bauer, J. et al., Dermoscopy of tungiasis. *Arch. Dermatol.*, 40, 761, 2004.
63. Bakos, R.M., and Bakos, L., 'Whitish chains': A remarkable in vivo dermoscopic finding of tungiasis. *Br. J. Dermatol.*, 159, 991, 2008.
64. Cabrera, R., and Daza, F., Dermoscopy in the diagnosis of tungiasis. *Br. J. Dermatol.*, 160, 1136, 2009.
65. Dunn, R., Asher, R., and Bowling, J., *Tunga penetrans*—egg head? *Dermatol. Online J.*, 16, 14, 2010.
66. Cabrera, R., and Daza, F., Eggs seen with dermoscopy. *Br. J. Dermatol.*, 158, 635, 2008.
67. Smith, M.D., and Procop, G.W., Typical histologic features of *Tunga penetrans* in skin biopsies. *Arch. Pathol. Lab. Med.*, 126, 714, 2002.
68. Vobis, M. et al., Molecular biological investigations of Brazilian *Tunga* sp. isolates from man, dogs, cats, pigs and rats. *Parasitol. Res.*, 96, 107, 2005.
69. Vobis, M. et al., Molecular phylogeny of isolates of *Ctenocephalides felis* and related species based on analysis of ITS1, ITS2 and mitochondrial 16S rDNA sequences and random binding primers. *Parasitol. Res.*, 94, 219, 2004.
70. Luchetti A. et al., Molecular characterization of *Tunga trimamillata* and *T. penetrans* (Insecta, Siphonaptera, Tungidae): Taxonomy and genetic variability. *Parasite*, 12, 123, 2005.
71. Luchetti, A. et al., Genetic variability of *Tunga penetrans* (Siphonaptera, Tungidae) sand fleas across South America and Africa. *Parasitol. Res.*, 100, 593, 2007.
72. Gamerschlag, S. et al., Repetitive sequences in the ITS1 region of the ribosomal DNA of Tunga penetrans and other flea species (Insecta, Siphonaptera). *Parasitol. Res.*, 102, 193, 2008.
73. Avelar, D.M., and Linardi, P.M., Use of multiple displacement amplification as pre-polymerase chain reaction (pre-PCR) to amplify genomic DNA of siphonapterids preserved for long periods in scientific collections. *Parasites Vectors*, 3, 86, 2010.
74. Heukelbach, J., Tungiasis. *Rev. Inst. Med. Trop. Sao Paulo*, 47, 307, 2005.
75. Feldmeier, H. et al., Bacterial superinfection in human tungiasis. *Trop. Med. Int. Health*, 7, 559, 2002.
76. Heukelbach, J. et al., Seasonal variation of tungiasis in an endemic community. *Am. J. Trop. Med. Hyg.*, 72, 145, 2005.
77. Muehlstaedt, M., Images in clinical medicine. Periungual tungiasis. *N. Engl. J. Med.*, 359, e30, 2008.

Section VIII

Arthropoda: Pentastomida

83 Armillifer, Linguatula, and Porocephalus (Tongue Worms)

Dongyou Liu

CONTENTS

83.1 INTRODUCTION

Pentastomids are worm-like organisms with characteristics of both arthropods and annelids, whose closest known relatives are branchiuran fish lice (e.g., *Argulus*). Being dioecious obligate parasites, adult pentastomids (commonly known as "tongue worms") usually reside in the respiratory tract of the reptile (e.g., snake, frog, lizard, and crocodile), mammal (e.g., dog and other carnivores), or bird final hosts, whereas pentastomid larvae inhabit the internal organs of vertebrate or arthropod intermediate hosts.

Currently, 10 pentastomid species belonging to the genera of *Armillifer*, *Linguatula*, *Porocephalus*, *Raillietiella*, and *Sebekia* are known to infect humans, causing armilliferosis, linguatulosis, porocephalosis, raillietiellosis, and sebekiosis (which are collectively referred to as pentastomiasis) [1]. Among them, *Linguatula* and *Armillifer* are most commonly implicated in zoonosis, with >90% of the human cases of pentastomiasis caused by the larvae (nymphs) of two species: *L. serrata* and *A. armillatus*. *Porocephalus* has also been associated with a small number of human clinical cases of pentastomiasis.

As a food- or water-borne parasitic zoonosis, human pentastomiasis may present as either visceral pentastomiasis or nasopharyngeal pentastomiasis (linguatuliasis), depending on whether pentastomid eggs or nymphs are involved. In visceral pentastomiasis, humans act as an intermediate host and acquire the infection through ingestion of pentastomid eggs or ova from the respiratory secretions or feces from the final hosts (dogs and other carnivores for *Linguatula*, and several species of large snakes for *Armillifer* and *Porocephalus*). In the digestive tract of the human host, the eggs hatch, and the larvae emerge and migrate to liver and other parts of the body, forming cysts or granulomas which contain viable or calcified pentastomid larvae. In nasopharyngeal pentastomiasis, humans function as an accidental final host after consumption of raw or poorly cooked meat containing pentastomid larvae or nymphs. The larvae penetrate the intestinal wall, migrate to nasopharyngeal tissues, and develop into adult worms.

83.1.1 Classification, Morphology, and Biology

83.1.1.1 Classification

Until relatively recently, the taxonomy of pentastomids has been in a state of confusion. Over the years, this group of organisms has been associated with a variety of metazoan phyla, including Arthropoda, Tardigrada, Annelida, Platyhelminthes, and Nematoda. It is only after molecular analysis that their relationship with Arthropods/Crustacea is generally accepted [2–8].

Currently, pentastomids are classified in subclass Pentastomida, class Maxillopoda, subphylum Crustacea, phylum Arthropoda. The subclass Pentastomida is divided into two orders: Cephalobaenida and Porocephalida. Most human-infecting pentastomids belong to the families Armilliferidae, Linguatulidae, and Porocephalidae, which have all been grouped under the order Porocephalida previously. However, a recent analysis of 18S rRNA gene sequence suggests that the family Linguatulidae may be placed in the order Cephalobaenida [9]. The fact that typical porocephalids (e.g., the genus *Armillifer*) have hooks horizontally lined up whereas cephalobaenids (e.g., the genus *Linguatula*) have hooks pairwise obliquely arranged provides additional support for this placement.

The family Armilliferidae contains a single genus *Armillifer*, of which at least four species (*Armillifer agkistrodontis*, *Armillifer armillatus*, *Armillifer grandis*, and *Armillifer moniliformis*) are known to infect humans [10,11]. Whereas *A. armillatus* is distributed in Africa and the Arabian Peninsula, *A. grandis* is present in Africa, *A. agkistrodontis* in China, and *A. moniliformis* in Southeast Asia.

The family Linguatulidae also covers a single genus *Linguatula*, which includes a sole member *Linguatula serrata*. Along with *A. armillatus*, *L. serrata* accounts for >90% of all reported cases of human pentastomiasis.

The family Porocephalidae consists of a single genus *Porocephalus*, which encompasses a recognized species *Porocephalus crotali* and an unassigned species *Porocephalus* sp. RS-2007.

Other pentastomids that are occasionally involved in human infections are found in the genera *Raillietiella* and *Sebekia* [12].

83.1.1.2 Morphology

Adult pentastomids are dorsoventrally flattened, worm-like parasites tapering backward resembling a vertebrate tongue (the so-called "tongue worm"). The males are 1.8–2.0 cm in length and the females are 8.0–13.0 cm in length. The body is segmented and covered in a chitinous cuticle. The anterior portion has five protuberances, four of which are clawed legs, and the fifth one bears the mouth (with the appearance of muscular pump). Being a blood-feeding parasite, the digestive tract of pentastomids is tubular. Similar to that of other arthropods, the pentastomid nervous system includes a ventral nerve cord with ganglia in each segment. Despite the presence of a hemocoel, the body of pentastomids contains no circulatory, respiratory or excretory organs.

The infective nymphs of pentastomids possess a ventrally located mouth, typical large acidophilic glands in the anterior region, a digestive system, and primordial genital organs. Two pairs of jointed chitinous hooks are observable circumorally. The combination of a centrally located mouth surrounded by four hooks gives the impression of an organism with five mouths (hence the name "pentastomes"). Indeed, detection of these four oral hooks provides a typical criterion for the diagnosis of pentastomes in tissue sections.

Among the human-infecting pentastomes, *A. agkistrodontis* adult males measure 23–25 mm × 3.2–4.8 mm and adult females measure 33–48 mm × 4.1–6.3 mm. The body appears translucent and is covered with small mastoid process-like spines (giving a screw-like appearance). Located on the abdominal side of the cephalothorax, the oval mouth has two horizontal pairs of hamuli on each side. The chest cavity is opaque and yellow-whitish, whereas the bilateral body is transparent and filled with eggs. Both females and males have seven to nine abdominal annuli with a cutin structure in bangle shape on the abdominal side. The anus is located terminally, behind the gonopore [13].

A. agkistrodontis larvae (at 22 weeks) measure 15–19 mm × 2.5–3.0 mm and have a cylinder-shaped body widened at the forehead and narrowed toward the posterior end. They possess seven to nine abdominal annuli, which gradually increase in length from the head toward the tail. Wave-like folds are observed on the body surface between the abdominal annuli. The oral aperture is located in the forehead; hamuli and the excretory pore are located on the abdominal side of the rear body. A pair of curved hamuli is observed on each side of the oral aperture. Larvae appear transparent when alive, but turn into a creamy color after death. Living larvae in the mouse are C-shaped and covered by a thin pouch with a fibrous structure [13]. Among the *Armillifer* species, *A. grandis* nymphs are the smallest.

A. agkistrodontis eggs are of oval shape and show a transparent outer layer with a whitish color and a yellow inner layer under a light microscope and three layers (i.e., an outer membrane, a transparent layer, and a medullary layer) under a transmission electron microscope. While immature eggs (of 126.74 × 119.73 μm) contain yolk cells, mature eggs (of 131.28 × 119.01 μm) harbor a typical arthropod larva. A pair of curved hamuli is observed on both sides of the oral aperture of larvae. Several sucker-like sensory organs of different sizes are distributed around the mouth, outside the hooks, and on the abdominal annuli. Waved folds are present on the body surface between the abdominal annuli. Granulation-like excretory pores appear on the abdominal side of the front part. Seven cortical layers (i.e., the outer cortex, lucidity layer, inner cortex, homogenous layer, medulla, striated layer, and base plate) are distinguishable under a transmission electron microscope.

L. serrata (tongue worm) adults show hooks in the ventral anterior end; primordial uterus, genital porus, and intestine in the posterior end; and cuticular structures of apical papillae, chitinoid oral clasp, and insertions of oral muscles in the dorsal anterior end. *L. serrata* nymphs are flat, slightly annulated, and have 72–92 body segments, with prominent rows of spines on the chitinous cuticle, ranging from 32 μm in the anterior part to 16 μm in the posterior part.

P. crotali nymphs are cylindrical and annulated and have 38–40 body segments. Under the microscope, *P. crotali* nymphs show a pseudosegmented body, a chitinous cuticle with pit-like openings to skin glands, chitinous hooked mouth parts, a prominent body cavity, numerous acidophilic glands, metamerically arranged striated muscle, and a digestive tract with villi lined by low columnar epithelium. *P.*

crotali nymphs are characterized by the lack of an expanded head and a U-shaped segment of chitin associated with the pharynx, the possession of double nymphal hooks, and the presence of simple sclerotized ring morphology of the mouth parts [14–17].

Morphologically, the cephalic part of *A. agkistrodontis* larvae is columniform, whereas that of *L. serrata* and *Porocephalus taiwana* is conic or round, respectively. The oral aperture of *A. agkistrodontis* is oval and located on the abdominal side in the center of the forehead, whereas those of *L. serrata* and *P. taiwana* are square or annular, respectively. The number of abdominal annuli of *A. agkistrodontis* is 7–9, whereas *A. armillatus*, *L. serrata*, and *P. taiwana* have about 14–22, 80 (72–108), and 10–11, respectively. These features can be useful for their differentiation [13].

83.1.1.3 Biology

Pentastomids undergo three stages of development (i.e., egg, nymph, and adult) and typically require two different hosts (intermediate and final/definitive hosts) to complete their life cycle, with pentastomid adults residing in the final host and their larvae living in the intermediate host (Table 83.1) [18]. In a typical pentastomid life cycle, the adult female mates with the adult male and produces hundreds of thousands of eggs. The eggs are expelled from the trachea of a definitive (final) host and subsequently ingested by an intermediate host. The eggs hatch to become first instar larvae (with legs and claws), and develop into nymphs in multiple tissues. The adult worms may also be swallowed by the final host, with the eggs being shed in the feces. The definitive host ingests the intermediate host containing nymphs, and the nymphs molt several times to become the infective nymphs, which resemble the adult pentastomes in shape and finally mature into adults in the lungs [17,19–21].

Armillifer adults typically occur in snakes, which discharge *Armillifer* eggs in their secretion and feces, contaminating vegetation. Mice and other mammals including humans become infected by ingestion of *Armillifer* eggs, which develop into larva. When a snake feeds on the rodent, the life cycle is completed. Human represents a dead-end host for the *Armillifer* parasite.

A. agkistrodontis adults reside in the abdominal cavity 5–6 months postinfection but move to the lungs about 9–10 months after infection [13]. In experimental mice fed with *A. agkistrodontis* eggs, one to two nodules appear in the liver and spleen by week 6, and the size of the nodules increases significantly in subsequent weeks. The first larvae appear 11 weeks postinfection, and larvae reach the infectious stage after 16 weeks. Twenty-two weeks postinfection, larvae are found in the mesentery (66.0% of the total number of larvae), liver (15.0%), spleen (13.9%), thoracic cavity (4.4%), and in the lung (0.9%), with a recovery rate of 79.2%. In experimental rats, the larval recovery rate is 51.2%; 70.0% of the larvae resided in the mesentery 11 weeks post-infection and another 18.6% in the spleen, 4.0% in the liver, and 3.2% in the kidney and thoracic cavity. No larvae are found in the rats' lungs. Thus, *A. agkistrodontis* requires a minimum period of 14 months to complete the life cycle, as mature eggs appear in the feces of the final host snakes (i.e., *A. acutus* and *P. molurus*) after a prepatent period of 10 months, and an infectious larva emerges from an egg in about 4 months [13].

Drinking water contaminated with infectious eggs and eating uncooked snake blood and gallbladders are the likely sources of *Armillifer* infections in humans. While *A. armillatus* is a dominant human-infecting species in West Africa, Central Africa, and Malaysia, where snakes are locally prepared for food, a number of human infections with *A. agkistrodontis* have been reported in China in recent years [22].

L. serrata (commonly called the tongue worm) is a worm-like, blood-sucking parasite that inhabits the upper respiratory tract of terrestrial, carnivorous vertebrates (e.g., cats, dogs, and foxes), reptiles, and birds (resulting in nasopharyngeal linguatuliasis). With dog or jackal as its most common final host, the species is sometimes known as the dog tongue worm. The intermediate hosts of *L. serrata* are usually sheep, cattle, or rodents (resulting in visceral linguatuliasis), although any mammal (including human) may act as a potential intermediate host.

TABLE 83.1
Characteristics of Pentastomid Species Infecting Humans

Characteristics	*A. agkistrodontis*	*A. armillatus*	*A. grandis*	*A. moniliformis*	*L. serrata*	*P. crotali*
Definitive host	*Agkistrodon acutus* snakes	Pythons	Pythons	Cobras	Dogs, wolves	Rattlesnakes
Intermediate host	Rodents	Rodents, monkeys	Rodents, monkeys	Rodents, monkeys	Ruminants	Rodents, monkeys
Nymphs	Cylindrical, spiral rings; 15–19 mm body length	Cylindrical, spiral rings; 13–23 mm body length	Cylindrical, spiral rings; 9–15 mm body length	Cylindrical, spiral rings; 12–20 mm body length	Flat, annulated; row of spines on each annulus; 4–6 mm body length	Cylindrical, annulated; 12 mm body length
Geographic distribution	China	West Africa	Central Africa	Southeast Asia	Cosmopolitan	America

Sources: Adapted from Tappe, D., and Büttner, D.W. *PLoS Negl. Trop. Dis.*, 3, e320, 2009; Chen, S.H. et al. *PLoS Negl. Trop. Dis.*, 4, e647, 2010.

In visceral linguatuliasis, a herbivorous mammal (the intermediate host) including humans ingests *Linguatula* eggs that are sneezed or defecated out by the mammalian final host. The eggs hatch in the intestines and the resulting four-legged, first instar larvae (resembling a mite) penetrate the intestinal wall, and lodge in tissues, particularly in lungs, liver, and lymph nodes, where they form cysts or granulomas (visceral larva migrans) and molt to the third larval stage. Occasionally, the larvae of both *Lingulatula* and *Armillifer* may enter the eye accidentally.

In nasopharyngeal linguatuliasis, a mammalian final host consumes raw or poorly cooked viscera (i.e., liver, lungs, and trachea) containing the encysted *L. serrata* larvae (nymphs). Inside the stomach, the larvae liberate from cysts within a couple of hours, then crawl up the esophagus, establish in the nose, pharynx, or lungs with the help of the hooks flanking the mouth, and mature into adults. The adults embed in the nasopharyngeal mucosa, feeding on blood and fluids. Females begin egg production in about 6 months and produce millions of eggs during their life span (up to 2 years). Eggs exit the host in nasal secretion or, if swallowed, with feces.

Humans can be an aberrant final host for *L. serrata*. Human nasopharyngeal linguatuliasis is known as the Halzoun syndrome in the Middle East or as the Marrara syndrome in Sudan where Marrara is a popular local dish prepared from raw offal. *L. serrata* has a worldwide distribution, and is especially prevalent in warm subtropical and temperate regions such as the Middle East, Africa, and Central and South America [23–30]. Nasopharyngeal linguatuliasis affects up to 20% of the population in some parts of Sudan. Liver autopsies conducted from the early part of the 20th century in Berlin indicate an *L. serrata* infection rate of nearly 12% [31].

First described by Humboldt in 1808, *P. crotali* has been reported in the Southeastern United States in snakes of the genera *Crotalus* and *Agkistrodon*, with mammals serving as intermediate hosts [17].

Interestingly, while dogs can serve as intermediate for *P. crotali* and as definitive host for *L. serrata*, humans can act as both intermediate and final host for *L. serrata*. Close contact with dogs and their secretions predisposes people to infection with *L. serrata*, whereas people who eat snake meat, workers at snake farms, snake keepers in zoos and pet shops, veterinarians, and owners of snakes (e.g., pythons, vipers, cobras, and rattlesnakes) may be exposed to ova of *Armillifer* and *Porocephalus*.

83.1.2 Clinical Features

Pentastomiasis is a food- or water-borne parasitic zoonosis that was first described in animal (snake) in 1847 [32]. After entry into the human host via food or water contaminated with the respiratory secretions or feces of the final hosts (dogs and other carnivores for *Linguatula*, several species of large snakes for *Armillifer* and *Porocephalus*), pentastomid eggs hatch in the intestine. The minute four-legged first instar larvae then penetrate through the intestinal mucosa, and by way of lymphatic spread, attach to the visceral peritoneum and/or pleura, over the surfaces of liver, intestine, spleen, lungs, and into the mesenteric and pulmonary lymph nodes (and other organs including esophagus, stomach, duodenum, jejunum, ileum, colon, appendix, rectum, gallbladder, broad ligament, pancreas, omentum, bladder, adrenal gland, heart, skin, epididymis, eye, brain, and inguinal hernia sac, etc.) [33,34]. The larvae then molt and develop into nymphs. This is followed by local encystment of the nymphs, nymphal migration due to high level of infection and reinfection, and finally nymphal death. The dead nymphs release a large number of foreign proteins, which stimulate either an acute inflammatory reaction or a chronic granulomatous response, leading to the development of central necrosis and formation of fibro-hyalinized and fibrocalcified nodules, cysts, or granulomas [35–37]. In a majority of cases of visceral larva migrans, nodules of 2–8 mm in diameter containing pentastomid larvae are located in the subperitoneal tissue around the liver, mesentery, spleen, and in the intestinal wall [38–42].

Although visceral pentastomiasis is largely asymptomatic in humans and the parasite may be incidentally found in the peritoneal cavity during surgery or at autopsy, its clinical significance is related to the organs involved and the degree of infection [43]. Patients infected with a few *L. serrata* nymphs may show symptoms resembling hepatic or pulmonary malignancy clinically and on radiological assessments, and develop chronic cough or night sweats. On the other hand, patients with a heavy *L. serrata* infestation may show chronic fever, ascites, abdominal pain, diarrhea, mild anemia, hepatosplenomegaly, eosinophilia in the bone marrow and blood, multiple polyps in the whole colon, and even death. In nasopharyngeal linguatuliasis, the presence of the parasite can induce headaches, itching of throat and ears, coughing, nasal discharge, hemoptysis, vomiting, sneezing, bleeding, dyspnea, and inflammation. Submaxillary and cervical lymph nodes sometimes enlarge and the neck is swollen. Other complications include abscesses in the auditory canals, facial paralysis, and enlarged tonsils producing asphyxiation. In heavy infections with *Armillifer* spp., death may result due to secondary septicemia, pneumonia, or severe enterocolitis.

Macroscopically, the lesions of visceral pentastomiasis are nodular or cyst-like, and are sometimes linked by a short stalk to the serosa or organ surface. In *L. serrata* infections, the lesions are called "*Linguatula* nodules," which measure 2–8 mm in size. Histologically, the nodules are granulomatous, with a target-like appearance (due to the presence of fibrous or hyalinized concentric rings of connective tissue surrounding the parasite). Inside the nodule, the primary larva loses its legs and claws, and next-developmental-stage nymph emerges as a flat, C-shaped spiral (which is of diagnostic significance).

Besides viable nymph, the nodules in human patients often show necrotic pentastomid granuloma containing a calcified, dead nymph with the presence of oral hooks or the chitinous cuticle, especially in long-standing infections. Occasionally, only acellular, partly hyalinized fibrous tissue surrounding a central mass of amorphous or calcified material is visible [44].

83.1.2.1 *Armillifer* Infections

Herzog et al. [39] described a case of visceral pentastomiasis in a 34-year-old native woman from Cameroon. The patient complained of colicky, epigastric abdominal pain, nausea, vomiting, constipation, and retention of flatus. At laparotomy, numerous small cystic nodules (3–8 mm in diameter) were found in the mesentery and under the intestinal and parietal peritoneum. Histological examination revealed slightly enlarged mesenteric lymph nodes containing several parasitic pseudocysts. The parasites were diagnosed as encysted, *A. armillatus* nymphs. Gardiner et al. [38] reported a case of visceral pentastomiasis due to *A. armillatus* in a 28-year-old man originally from Nigeria. The patient died as a result of a motor vehicle accident and numerous small cystic nodules (3–9 mm in diameter) were discovered in the liver, pleura, lungs, and mesentery and under the intestinal and parietal peritoneum during autopsy. The parasites were confirmed as encysted *A. armillatus* nymphs. Additionally, Dakubo et al. [29] documented a case of visceral pentastomiasis in a 55-year-old farmer who had a mechanical small intestinal obstruction. During laparotomy, encysted parasites and degenerative granulomas were found in the omentum and in the subserosal layer of the small intestine. These parasites were diagnosed as *A. armillatus* larvae. The patient showed an uneventful recovery after the operation. In a separate study, Chen et al. [13] reported a case of *A. agkistrodontis*-related visceral pentastomiasis in a patient with a history of drinking *A. acutus* gall and blood. The juvenile parasites recovered from the patient measured about 21 × 1.3 mm and had seven abdominal annuli consisting of a cutin structure in bangle shape on their abdominal side. The parasite was enlarged in its anterior section, and had ventral hooks. The observation of anterior hooks, a cylinder-shaped larval body, an oval oral aperture with one pair of equal-sized hamuli on either side, and seven to nine abdominal annuli with sucker-like sensory organs suggested that the parasite was *A. agkistrodontis*, not *A. armillatus*.

83.1.2.2 *Linguatula* Infections

Lang et al. [45] presented a case of visceral pentastomiasis due to *L. serrata* in an Israeli 12-year-old boy, who suffered from iritis, subluxation of the lens, and secondary glaucoma. A single secondary pentastomid larva was found in the anterior chamber of the right eye, attached loosely to the pupil's border by a fibrinous mass. Morsy et al. [46] described a case of nasopharyngeal linguatuliasis in a 20-year-old male in Egypt. The patient presented with fever, urticaria (face and neck), coughing, vomiting, and passage of small worm-like structures (<1 cm in length) in his nasal discharge and vomitus. The patient was treated with a single dose of praziquantel. Separately, Koehsler et al. [9] reported a case of *L. serrata*-related pentastomiasis in a 14-year-old girl in Austria, who had redness, pain, and progressive visual loss (to 0.1 Snellen) in the right eye. The patient had regular contact with domestic animals: two dogs, cats, and one turtle. Slit lamp examination revealed a mobile parasite swimming like a fish in the anterior chamber of the eye. The viable parasite was removed and examined. It measured 4.5 mm long, appeared whitish, and had a flattened (tongue-like) body shape with a rounded anterior (0.9 mm wide) and a pointed posterior end (0.35 mm). It showed curved hooks with sharp tips on the anterior ventral side, and a chitinoid oral clasp in the anterior. The cuticle had 94 rings with conspicuous spicules at the posterior margins of the cuticle rings. These morphological features enabled its identification as a third instar larva of *L. serrata* tongue worm. Further examination of 18S rRNA gene sequence (1834 bp) using six primers (P1fw, P1rev, P2fw, P2rev, P3fw, and P3rev) confirmed its *L. serrata* identity. After removal of the parasite, the eye was free of irritation within a month, and implantation of an artificial intraocular lens 3 months later restored the visual acuity to 1.0 Snellen.

83.1.2.3 *Porocephalus* Infections

Yao et al. [47] described a case of visceral pentastomiasis due to *P. taiwana* in a 13-year-old boy in China. The patient had persistent fever, abdominal pain, and diarrhea for 3 months. Abdomen CT revealed hepatomegaly and multiple nodular low-density pathological changes. At laparotomy, yellow and turbid ascites were seen in the abdominal cavity and miliary nodules were observed on the surface of the omentum majus, liver, and small intestine wall. Histological examination revealed parenchymal tubercles containing several worms. Pathological diagnosis was parasitic granuloma. The parasites were identified as *P. taiwana* sp. nov. The patient recovered well after therapy.

In asymptomatic patients no treatment is necessary, as the parasites degenerate after ~2 years. Although praziquantel or mebendazole may be used to treat pentastomiasis with some success, the surgical approach represents a most effective way of getting rid of the parasites, especially in symptomatic infections with numerous parasites. Visceral pentastomiasis can be prevented by proper hand washing after contact with dog saliva or feces and snake secretions or meat. Nasopharyngeal pentastomiasis (linguatuliasis) can be prevented by proper cooking of meat prior to consumption [46,48].

83.1.3 Laboratory Diagnosis

Although human pentastomiasis is often asymptomatic, it may induce some nonspecific clinical symptoms, including chronic fever, abdominal pain, diarrhea, mild anemia, and hepatosplenomegaly (as in the case of visceral pentastomiasis), or headaches, and itching of throat and ears (as in the case of nasopharyngeal pentastomiasis/linguatuliasis).

Differential diagnoses for visceral pentastomiasis with hepatic involvement (e.g., liver enlargement), high fever, and abdominal pain include diffuse liver metastasis, diffuse hepatocellular carcinoma, lymphoma, and diffuse granulomatosis [49]. Detection of the numerous pearl-like nodules over the peritoneal surface of the abdominal cavity together with the centrally located calcification forms the basis of the presumptive diagnosis of pentastomiasis by roentgenography, ultrasonography and CT scan [50]. In addition, calcified nymphs of *Armillifer* spp and less often *L. serrata* may show a horseshoe or C-shaped structure upon radiological examination.

Other differential diagnoses for visceral pentastomiasis consist of tuberculosis, the ova of *Schistosoma japonicum*, cysticerci, spargana, tissue-inhabiting arthropods, and the anisakine larvae. In contrast to tuberculosis lesions, those of visceral pentastomiasis are characterized by the protuberance of the lesions over the visceral peritoneum and pleura; the concentric or targetoid appearance of the hyalinization and calcium deposition in the fibronodular lesions; the foreign-body giant cell reaction, instead of the Langhans type, in the foci of chronic granulomatous inflammation. Although calcified circumoral hooks may be observed in infection with *Cysticercus cellulosae*, *Echinococcus granulosus*, and *Diptera* larvae (myiasis), the hooklets of both cysticerci and echinococci are more numerous (25–50 in number) and much smaller than those of pentastomes, and those of *Diptera* larvae (maggots) have only one pair of mouth hooks each.

Owing to the lack of specific serological tests, a definitive diagnosis of human pentastomiasis relies heavily on histopathological examination of biopsied lesions. The key criteria for diagnosing visceral pentastomiasis include the morphological characteristics of pentastome nymphs (e.g., the encysted nymph, the dislocated and calcified cuticular spines, sclerotized openings, and circumoral hooks), the anatomical distribution of the lesion, and the surrounding tissue reaction, as well as the geographical distribution of the different pentastome species and the patient's regional provenance. The use of Movat's pentachrome or Masson's trichrome stain is helpful for the morphological identification of pentastomids. While the presence or absence of cuticular spines differentiates between the nymphs of *L. serrata* and *A. armillatus*, the sclerotized openings are often found in the lesions of both linguatuliosis and armilliferiosis.

Given that often incomplete embedded pentastomid nymph or degenerated and/or calcified body structures of the embedded nymph (relics, such as the cuticular spines, sclerotized openings, and/or the circumoral hooks) may be observed in the lesion materials, the pathological identification of human pentastomiasis should not be limited to etio-pathological diagnosis by finding the entire nymph in the tissue section of the lesion, but should include both subetio-pathological diagnosis and presumptive pathological diagnosis of this disease when the overall structure of the etiological agent or even a fragment cannot be revealed [44].

In recent years, molecular techniques such as polymerase chain reaction (PCR) have been utilized for the accurate identification and phylogenetic analysis of pentastomid parasites. Several reports showed that 18S rRNA and Cox1 gene may be targeted for pentastomid identification. The examination of the full 18S rRNA sequence demonstrated the close genetic relationship between *A. agkistrodontis* and *Raillietiella* spp., whereas sequence analysis of the Cox1 gene indicated that *A. agkistrodontis* is related to *A. armillatus*. This line of phylogenetic analyses has provided evidence that the pentastomids belong to the phylum Arthropoda [13].

83.2 METHODS

83.2.1 Sample Preparation

83.2.1.1 Sample Collection

Pentastomid parasites are collected from snakes, and mature eggs are recovered after culture for 1–2 h in an incubator at 26 ± 0.5°C and 80% relative humidity.

83.2.1.2 DNA Isolation

DNA from pentastomid parasites is extracted by using a DNEasy kit (Qiagen) [17].

83.2.2 Detection Procedures

83.2.2.1 Analysis of Partial 18S rRNA Gene Sequence

Brookins et al. [17] designed primers Pent629F (5′-CGG TTAAAAAGCTCGTAGTTGG-3′) and Pent1011R (5′-GGC ATCGTTTATGGTTAGAACTAGGG-3′) from a conserved region of pentastomid small subunit ribosomal RNA gene (18S rRNA) for PCR amplification and sequence analysis of a 425-bp segment (corresponding to bases 638–1062 of 18S rRNA gene) from pentastomid/branchiuran organisms.

83.2.2.1.1 Procedure

1. PCR mixture (20 μL) is composed of 1 μM each of forward primer Pent629F and reverse primer Pent1011R, 200 μM each of dATP, dCTP, dGTP, and dTTP, 0.5 U of Platinum Taq DNA polymerase (Invitrogen), 5% DMSO, and 2 μL PCR buffer plus magnesium (Invitrogen), and 2 μL DNA (extracted by a DNEasy kit).
2. Amplification is conducted with an initial denaturation at 94°C for 5 min, followed by 40 cycles of 94°C for 30 s, 58°C for 60 s, and 72°C for 60 s, and a final extension step at 72°C for 7 min.
3. The PCR product is separated in a 1% agarose gel, excised and purified using the QIAquick gel extraction kit (Qiagen). The product is sequenced directly in both directions using the Big-Dye Terminator Kit (Perkin-Elmer) and analyzed on an ABI 377 automated DNA sequencer.
4. The sequence (383 bp after removal of primer sequences) is compared to known sequences in GenBank, EMBL, and Data Bank of Japan databases using BLASTN.

83.2.2.2 Analysis of 18S rRNA and Cox1 Gene Sequences

Chen et al. [13] utilized primers from the 18S rRNA and Cox1 gene of an *A. armillatus* mitochondrial sequence for molecular analysis of pentastomids. The primers consist of 18S rRNA-F: (5′-AAC CTG GTT GAT CCT GCC AGT AG -3′) and 18S rRNA-R: (5′-GAT CCT TCT GCA GGT TCA CCT AC-3′) for amplification of full-length 18S rRNA gene (about 1835 bp); Cox1-F: (5′-CTG CGA CAA TGA CTA TTT TCA AC-3′) and Cox1-R: (5′-ATA TGG GAA GTT CTG AGT AGG-3′) for amplification of Cox1 gene (about 1525 bp).

83.2.2.2.1 Procedure

1. PCR is performed on a C_{1000}™ Thermal Cycler (Bio-Rad) using 2.0 μL pentastomid genomic DNA as template.
2. The reaction conditions consist of 96°C for 3 min followed by 33 cycles of 94°C for 30 s, 58°C for 30 s, and 72°C for 2 min. The last step is at 72°C for 10 min.
3. The product is visualized on an agarose gel using ethidium bromide staining.
4. The target DNA fragments are cloned into pGEM-T easy vector (Promega). The plasmid is transformed into *E. coli* DH5α, and sequenced.
5. The sequence is compared to available sequences in the NCBI GenBank using the BLAST procedure.

83.3 CONCLUSION

Pentastomids (commonly known as "tongue worms") are an unusual group of organisms that display the characteristics of both arthropods and annelids. For many years, the taxonomy of pentastomids has remained unsolved. Following a recent application of molecular techniques, it became clear that pentastomids belong to the phylum Arthropoda.

As dioecious obligate parasites, pentastomids typically require two different hosts (intermediate and final/definitive hosts) to complete their life cycle. While pentastomid adults live in the respiratory tract of the reptiles (e.g., snake, frog, lizard, and crocodile), carnivores (e.g., dog and cats), or birds (as final/definitive hosts), pentastomid larvae/nymphs reside in the internal organs of vertebrates (e.g., rodents) or arthropods (as intermediate hosts).

Being an accidental, dead-end host for pentastomids, humans become infected through ingestion of pentastomid eggs that are discharged from the trachea (or feces) of final/definitive hosts (e.g., dogs for *Linguatula* and snakes for *Armillifer* and *Porocephalus*), and develop visceral pentastomiasis. Alternatively, human infection with pentastomids can occur after consumption of raw or poorly cooked meat/offals containing pentastomid larvae/nymphs, and develop nasopharyngeal pentastomiasis (linguatuliasis).

Although human pentastomiasis is often asymptomatic, it can induce fever, stomach pain, and diarrhea (as in the case of visceral pentastomiasis), or chronic cough and vomiting (as in the case of nasopharyngeal pentastomiasis/linguatuliasis). The infection is generally limited to the tropics and subtropics (including Middle East, Africa, Latin America, and Asia), but can occur in travelers from other parts of the world.

Diagnosis of human pentastomiasis relies on the identification of viable or calcified nymphs or other characteristic structures (e.g., oral hooks) of pentastomids in the lesion materials [51]. The recent application of molecular procedures has not only helped clarify the taxonomic status of pentastomids, but also provided a valuable approach for epidemiological investigation of pentastomid species involved in human pentastomiasis.

REFERENCES

1. Berenji, F., Fata, A., and Hosseininejad, Z., A case of *Moniliformis moniliformis* (Acanthocephala) infection in Iran. *Korean J. Parasitol.*, 45, 145, 2007.
2. Abele, L.G., Kim, W., and Felgenhauer, B.E., Molecular evidence for inclusion of the phylum *Pentastomida* in the *Crustacea*. *Mol. Biol. Evol.*, 6, 685, 1989.
3. Almeida, W.O. et al., Prevalence and intensity of pentastomid infection in two species of snakes from northeastern Brazil. *Braz. J. Biol.*, 67, 759, 2007.
4. Almeida, W.O. et al., Morphological support for the phylogenetic positioning of *Pentastomida* and related fossils. *Rev. Biotemas*, 21, 81, 2008.
5. Lavrov, D.V., Brown, W.M., and Boore, J.L., Phylogenetic position of the *Pentastomida* and (pan) crustacean relationships. *Proc. Biol. Sci.*, 271, 537, 2004.
6. Waloszek, D., and Maas, A., The evolutionary history of crustacean segmentation: A fossil-based perspective. *Evol. Dev.*, 7, 515, 2005.
7. Qiu, M.H., and Jiang, Y.Y., Advances in studies of human pentastomiasis. *Int. J. Med. Parasit. Dis.*, 33, 281, 2006.
8. Møller, O.S. et al., First maxillae suction discs in branchiura (Crustacea): Development and evolution in light of the first molecular phylogeny of branchiura, pentastomida, and other "maxillopoda." *Arthropod Struct. Dev.*, 37, 333, 2008.
9. Koehsler, M. et al., *Linguatula serrata* tongue worm in human eye, Austria. *Emerg. Infect. Dis.*, 17, 870, 2011.
10. Fain, A., The *Pentastomida* parasitic in man. *Ann. Soc. Belge Med. Trop.*, 55, 59, 1975.
11. Cagnard, V. et al., Generalized and fatal pentastomaisis due to *Armillifer grandis*. *Bull. Soc. Pathol. Exot. Filiales*, 72, 345, 1979.
12. Stender-Seidel, S., Thomas, G., and Bockeler, W., Investigation of different ontogenetic stages of *Raillietiella* sp. (Pentastomida: Cephalobaenida): Suboral gland and frontal gland. *Parasitol. Res.*, 86, 385, 2000.
13. Chen, S.H. et al., Multi-host model-based identification of *Armillifer agkistrodontis* (Pentastomida), a new zoonotic parasite from China. *PLoS Negl. Trop. Dis.*, 4, e647, 2010.
14. Hunter, W.S., and Higgins, R.P., An unusual case of human porocephaliasis. *J. Parasitol.*, 48, 68, 1960.
15. Esslinger, J.H., Development of *Porocephalus crotali* (Humboldt, 1808) (Pentastomida) in experimental intermediate hosts. *J. Parasitol.*, 48, 452, 1962.
16. Buckle, A.C., Riley, J., and Hill, G.F., The *in vitro* development of the pentastomid *Porocephalus crotali* from the infective instar to the adult stage. *Parasitology*, 115, 503, 1997.
17. Brookins, M.D. et al., Massive visceral pentastomiasis caused by *Porocephalus crotali* in a dog. *Vet. Pathol.*, 46, 460, 2009.
18. Ravindran, R. et al., Prevalance of *Linguatula serrata* in domestic ruminants in South India. *Southeast Asian J. Trop. Med. Public Health*, 39, 808, 2008.
19. Riley, J., The biology of pentastomids. *Adv. Parasitol.*, 25, 45, 1986.
20. Riley, J., and Henderson, R.J., Pentastomids and the tetrapod lung. *Parasitology*, 119, S89, 1999.
21. Almeida, W.O., and Christoffersson, M.L., A cladistic approach to relationships in Pentastomida. *J. Parasitol.*, 85, 695, 1999.
22. Zhang, Q., Wang, B., and Huang, M., *Armillifer agkistrodontis* disease, case report. *Chin. Med. J.*, 35, 747, 1996.

23. Khalil, G.M., *Linguatula serrata* (Pentastomida) parasitizing humans and animals in Egypt, neighbouring countries, and elsewhere: A review. *J. Egypt. Public Health Assoc.*, 47, 364, 1972.
24. Gardiner, S.N. et al., Pentastomiasis in Canada. *Arch. Pathol. Lab. Med.*, 115, 515, 1991.
25. Yagi, H. et al., The Marrara syndrome: A hypersensitivity reaction of the upper respiratory tract and buccopharyngeal mucosa to nymphs of *Linguatula serrata*. *Acta Trop.*, 62, 127, 1996.
26. De Meneghi, D., Pentastomes (Pentastomiasis, *Amillifer armillatus* Wyman, 1848) in snakes from Zambia. *Parasitogia*, 41, 573, 1999.
27. Lazo, R.F. et al., Ocular linguatuliasis in Ecuador: Case report and morphometric study of the larva of *Linguatula serrata*. *Am. J. Trop. Med. Hyg.*, 60, 405, 1999.
28. Siavashi, M.R., Assmar, M., and Vatankhah, A., Nasopharyngeal pentastomiasis (Halzoun): Report of 3 cases. *Iran J. Mol. Sci.*, 27, 191, 2002.
29. Dakubo, J.C. et al., Human pentastomiasis: A case report. *West Afr. J. Med.*, 25, 166, 2006.
30. Dakubo, J., Naaeder, S., and Kumodji, R., Totemism and the transmission of human pentastomiasis. *Ghana Med. J.*, 42, 165, 2008.
31. Tappe, D. et al., Linguatuliasis in Germany. *Emerg. Infect. Dis.*, 12, 1034, 2006.
32. Drabick, J.J., Pentastomiasis. *Rev. Infect. Dis.*, 9, 1087, 1987.
33. Deweese, M.W., Murrah, W.F., and Caruthers, S.B., Case report of a tongue worm (*Linguatula serrata*) in the anterior chamber. *Arch. Ophthalmol.*, 68, 587, 1962.
34. Sousaefaro, B., and Pinhao, R.C., An isolated case of ocular parasitosis caused by *Linguatula serrata*. *J. Soc. Cienc. Med. Lisb.*, 128, 401, 1964.
35. Smith, J.A., Oladiran, B., and Lagundoye, S.B., Pentastomiasis and malignancy. *Ann. Trop. Med. Parasitol.*, 69, 503, 1975.
36. Buslau, M., Kuhne, U., and Marsch, W.C., Dermatological signs of nasopharyngeal linguatulosis (Halzoun, Marrara syndrome): The possible role of a major basic protein. *Dermatologica*, 181, 327, 1990.
37. Pampiglione, S. et al., A nodular pulmonary lesion due to *Linguatula serrata* in an HIV-positive man. *Parassitologia*, 43, 105, 2001.
38. Gardiner, C.H., Dyke, J.W., and Shirley, S.F., Hepatic granuloma due to a nymph of *Linguatula serrata* in a woman from Michigan: A case report and review of the literature. *Am. J. Trop. Med. Hyg.*, 33, 187, 1984.
39. Herzog, U., Marty, P., and Zak, F., Pentastomiasis: Case report of an acute abdominal emergency. *Acta Trop.*, 42, 261, 1985.
40. Baird, J.K., Kassebaum, L.J., and Ludwig, G.K., Hepatic granuloma in a man from North America caused by a nymph of *Linguatula serrata*. *Pathology*, 20, 198, 1988.
41. du Plessis, V. et al. Pentastomiasis (*Armillifer armillatus* infestation). *S. Afr. Med. J.*, 97, 928, 2007.
42. Tappe, D., and Büttner, D.W., Diagnosis of human visceral pentastomiasis. *PLoS Negl. Trop. Dis.*, 3, E320, 2009.
43. Yapo Ette, H. et al., Human pentastomiasis discovered postmortem. *Forensic Sci. Int.*, 137, 52, 2003.
44. Ma, K.C., Qiu, M.H., and Rong, Y.L., Pathological differentiation of suspected cases of pentastomiasis in China. *Trop. Med. Int. Health*, 7, 166, 2002.
45. Lang, Y. et al., Intraocular pentastomiasis causing unilateral glaucoma. *Br. J. Ophthalmol.*, 71, 391, 1987.
46. Morsy, T.A., El-Sharkawy, I.M., and Lashin, A.H., Human nasopharyngeal linguatuliasis (Pentasomida) caused by *Linguatula serrata*. *J. Egypt. Soc. Parasitol.*, 29, 787, 1999.
47. Yao, M.H., Wu, F., and Tang, L.F., Human pentastomiasis in China: Case report and literature review. *J. Parasitol.*, 94, 1295, 2008.
48. Magnino, S. et al., Biological risks associated with consumption of reptile products. *Int. J. Food Microbiol.*, 134, 163, 2009.
49. Machado, M.A.C. et al., Unusual case of pentastomiasis mimicking liver tumour. *J. Gastroenterol. Hepatol.*, 21, 1218, 2006.
50. Lai, C. et al., Imaging features of pediatric pentastomiasis infection: A case report. *Korean J. Radiol.*, 11, 480, 2010.
51. Mourra, N., Huerre, M., and Tiret, E., Liver pentastomiasis associated with rectal adenocarcinoma. *Colorectal Dis.*, 12, 261, 2010.

Index

K

L

M

N

Q

R

S

X

Y

Z